QUANTITATIVE CHEMICAL ANALYSIS

"The Experiment" by Sempé © C. Charillon, Paris

QUANTITATIVE CHEMICAL ANALYSIS

Eighth Edition

Daniel C. Harris

Michelson Laboratory
China Lake, California

W. H. Freeman and Company
New York

Publisher: Clancy Marshall
Senior Acquisitions Editor: Jessica Fiorillo
Marketing Manager: John Britch
Media Editor: Dave Quinn
Editorial Assistant: Kristina Treadway
Photo Editor: Ted Szczepanski
Cover and Text Designer: Vicki Tomaselli
Senior Project Editor: Mary Louise Byrd
Illustrations: Network Graphics, Precision Graphics
Illustration Coordinators: Bill Page, Eleanor Jaekel
Production Coordinator: Julia DeRosa
Composition and Text Layout: Aptara, Inc.
Printing and Binding: RR Donnelley

Library of Congress Control Number: 2009943186
ISBN-13: 978-1-4292-1815-3
ISBN-10: 1-4292-1815-0

Printed in the United States of America
First Printing

W. H. Freeman and Company
41 Madison Avenue
New York, NY 10010

Houndmills, Basingstoke RG21 6XS, England
www.whfreeman.com

BRIEF CONTENTS

CONTENTS

Experiments

Experiments are found at the Web site
www.whfreeman.com/qca8e

Spreadsheet Topics

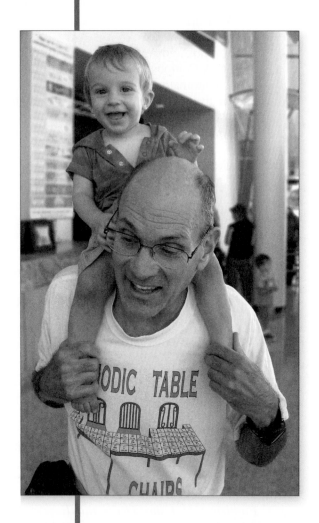

Dan's grandson Samuel discovers that the periodic table can take you to great places.

Goals of This Book

My goals are to provide a sound physical understanding of the principles of analytical chemistry and to show how these principles are applied in chemistry and related disciplines—especially in life sciences and environmental science. I have attempted to present the subject in a rigorous, readable, and interesting manner that will appeal to students whether or not their primary interest is chemistry. I intend the material to be lucid enough for nonchemistry majors, yet to contain the depth required by advanced undergraduates. This book grew out of an introductory analytical chemistry course that I taught mainly for nonmajors at the University of California at Davis and from a course for third-year chemistry students at Franklin and Marshall College in Lancaster, Pennsylvania.

What's New?

A significant change in this edition that instructors will discover is that the old Chapter 7 on titrations from earlier editions is missing, but its content is dispersed throughout this edition. My motive was to remove precipitation titrations from the critical learning path. Precipitation titrations have decreased in importance and they have not appeared in the last two versions of the American Chemical Society examination in quantitative analysis.* The introduction to titrations comes in Chapter 1. Kjeldahl analysis is grouped with acid-base titrations in Chapter 10. Spectrophotometric titrations appear in Chapter 17 with spectrophotometry. Efficiency in titrimetric experimental design is now with quality assurance in Chapter 5. Precipitation titrations appear with gravimetric analysis in Chapter 26. Gravimetric analysis and precipitation titrations remain self-contained topics that can be covered at any point in the course.

A new feature of this edition is a short "Test Yourself" question at the end of each worked example. If you understand the worked example, you should be able to answer the Test Yourself question. Compare your answer with mine to see if we agree.

Chapter 0 begins with a biographical account of Charles David Keeling's measurement of atmospheric carbon dioxide. His results have been described as "the single most important environmental data set taken in the 20th century." Boxes in Chapters 3 and 19 provide detail on Keeling's precise manometric and spectrometric techniques. Box 9-1 discusses ocean acidification by atmospheric carbon dioxide.

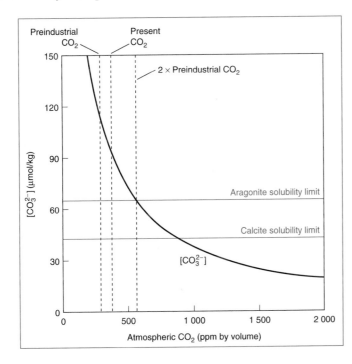

Effect of increasing atmospheric CO_2 on the ability of marine organisms to make calcium carbonate shells and skeletons (Box 9-1).

---

*P. R. Griffiths, "Whither 'Quant'? An Examination of the Curriculum and Testing Methods for Quantitative Analysis Courses Taught in Universities and Colleges in the Western USA," *Anal. Bioanal. Chem.* **2008,** *391,* 875.

Phoenix Mars Lander discovered perchlorate in Martian soil with ion-selective electrodes (Chapter 14).

New boxed applications include biochemical measurements with a nano-electrode (Chapter 1), the quartz crystal microbalance in medical diagnosis (Chapter 2), a case study of systematic error (Chapter 3), choosing the null hypothesis in epidemiology (Chapter 4), a lab-on-a-chip example of isoelectric focusing (Chapter 9), Kjeldahl nitrogen analysis in the headlines (Chapter 10), lithium-ion batteries (Chapter 13), measuring selectivity coefficients of ion-selective electrodes (Chapter 14), how perchlorate was discovered on Mars (Chapter 14), an updated description of the Clark oxygen electrode (Chapter 16), Rayleigh and Raman scattering (Chapter 17), spectroscopic upconversion (Chapter 18), trace elements in the ocean (Chapter 20), phase transfer agents (Chapter 22), gas chromatography on a chip (Chapter 23), paleothermometry (Chapter 24), structure of the solvent-bonded phase interface (Chapter 24), and measuring illicit drug use by analyzing river water (Chapter 27).

Spreadsheet instructions are updated to Excel 2007, but instructions for earlier versions of Excel are retained. A new section in Chapter 2 describes how electronic balances work. Rectangular and triangular uncertainty distributions for systematic error are introduced in Chapter 3. Chapter 4 includes discussion of standard deviation of the mean and "tails" in probability distributions. The Grubbs test replaces the Dixon Q test for outliers in Chapter 4. Reporting limits are illustrated with *trans* fat analysis in food in Chapter 5.

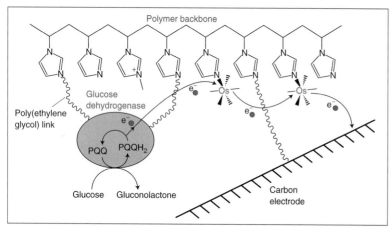

"Wired" enzymes described in Section 16-4 are at the heart of sensitive personal blood glucose monitors.

Elementary discussion of the systematic treatment of equilibrium in Chapter 7 is enhanced with a discussion of ammonia acid-base chemistry. Chapter 8 and the appendix now include pK_a for acids at an ionic strength of 0.1 M in addition to an ionic strength of 0. Discussion of selectivity coefficients was improved in Chapter 14 and the iridium oxide pH electrode is introduced. "Wired" enzymes and mediators for coulometric blood glucose monitoring are described in Chapter 16. Voltammetry in Chapter 16 now includes a microelectrode array for biological measurements. There is a completely new section on flow injection analysis and sequential injection in Chapter 18, and these techniques appear again in later examples. Chapter 19 on spectrophotometers is heavily updated. Laser-induced breakdown and dynamic reaction cells for atomic spectrometry are introduced in Chapter 20. Mass spectrometry in Chapter 21 now includes the linear ion trap and the orbitrap, electron-transfer dissociation for protein sequencing, and open-air sampling methods.

Numerous chromatography updates are found throughout Chapters 22–25. Stir-bar sorption was added to sample preparation in Chapter 23. Polar embedded group stationary phases, hydrophilic interaction chromatography, and the charged aerosol detector were added to Chapter 24. There is a discussion of the linear solvent strength model in liquid chromatography and a new section that teaches how to use a spreadsheet to predict the effect of solvent composition in isocratic elution. The supplement at **www.whfreeman.com/qca** gives a spreadsheet for simulating gradient elution. Chapter 25 describes hydrophilic interaction chromatography for ion exchange, hydrophobic interaction chromatography for protein purification, analyzing heparin contamination by electrophoresis, wall charge control in electrophoresis, an update on DNA sequencing by electrophoresis, and microdialysis/electrophoresis of neurotransmitters with a lab-on-a-chip. Data from a round-robin study of precision and accuracy of combustion analysis are included in Chapter 26. The 96-well plate for solid-phase extraction sample preparation was added to Chapter 27.

Interaction of (R)- and (S)-naproxen with (S,S) stationary phase

Naphthalene group

Dinitrophenyl group

(S)-Naproxen

(R)-Naproxen

(S,S) More stable adsorbate

(S,S) Less stable adsorbate

Chiral stationary phase separates enantiomers of the drug naproxen by high-performance liquid chromatography (Figure 24-10).

There is a new discussion of the operation of an electronic balance in Chapter 2, Tools of the Trade.

Applications

A basic tenet of this book is to introduce and illustrate topics with concrete, interesting examples. In addition to their pedagogic value, Chapter Openers, Boxes, Demonstrations, and Color Plates are intended to help lighten the load of a very dense subject. I hope you will find these features interesting and informative. **Chapter Openers** show the relevance of analytical chemistry to the real world and to other disciplines of science. I can't come to your classroom to present **Chemical Demonstrations,** but I can tell you about some of my favorites and show you color photos of how they look. **Color Plates** are located near the center of the book. **Boxes** discuss interesting topics related to what you are studying or amplify points in the text.

Problem Solving

Nobody can do your learning for you. The two most important ways to master this course are to work problems and to gain experience in the laboratory. **Worked Examples** are a principal pedagogic tool designed to teach problem solving and to illustrate how to apply what you have just read. Each worked example ends with a **Test Yourself** question that asks you to apply what you learned in the example. **Exercises** are the minimum set of problems that apply most major concepts of each chapter. Please struggle mightily with an Exercise before consulting the solution at the back of the book. **Problems** at the end of the chapter cover the entire content of the book. **Short answers** to numerical problems are at the back of the book and complete solutions appear in the **Solutions Manual** that can be made available for purchase if your instructor so chooses.

> **EXAMPLE** **How Many Tablets Should We Analyze?**
>
> In a gravimetric analysis, we need enough product to weigh accurately. Each tablet provides ~15 mg of iron. How many tablets should we analyze to provide 0.25 g of Fe_2O_3?
>
> ⋮
>
> *Test Yourself* If each tablet provides ~20 mg of iron, how many tablets should we analyze to provide ~0.50 g of Fe_2O_3? (*Answer:* 18)

Spreadsheets are indispensable tools for science and engineering. You can cover this book without using spreadsheets, but you will never regret taking the time to learn to use them. The text explains how to use spreadsheets and some problems ask you to apply them. If you are comfortable with spreadsheets, you will use them even when the problem does not ask you to. A few of the powerful built-in features of Microsoft Excel are described as they are needed. These features include graphing in Chapters 2 and 4, statistical functions and regression in Chapter 4,

Spreadsheets are introduced as an important problem-solving tool.

	A	B	C	D
1	Mg(OH)$_2$ Solubility			
2				
3	K_{sp} =		[OH$^-$]$_{guess}$ =	[OH$^-$]^3/(2 + K$_1$[OH$^-$]) =
4	7.1E-12		0.0002459	7.1000E-12
5	K$_1$ =			
6	3.8E+02		[Mg^{2+}] =	[MgOH$^+$] =
7			0.0001174	0.0000110
8				
9	D4 = C4^3/(2+A6*C4)			
10	C7 = A4/C4^2			
11	D8 = A6*C7*C4			

Goal Seek ? ☒

Set cell: D4

To value: 7.1E-12

By changing cell: C4

OK   Cancel

multiple regression for experimental design in Chapter 5, solving equations with Goal Seek in Chapters 7, 8, and 12, Solver in Chapters 12 and 18, and matrix operations in Chapter 18.

Other Features of This Book

Terms to Understand Essential vocabulary, highlighted in **bold** in the text, is collected at the end of the chapter. Other unfamiliar or new terms are *italic* in the text, but not listed at the end of the chapter.

Glossary All bold vocabulary terms and many of the italic terms are defined in the glossary.

Appendixes Tables of solubility products, acid dissociation constants, redox potentials, and formation constants appear at the back of the book. You will also find discussions of logarithms and exponents, equations of a straight line, propagation of error, balancing redox equations, normality, and analytical standards.

Notes and References Citations in the chapters appear at the end of the book.

Supplements

The *Solutions Manual for Quantitative Chemical Analysis* (ISBN 1-4292-3123-8) contains complete solutions to all problems.

The **student Web site, www.whfreeman.com/qca8e,** has **directions for experiments,** which may be reproduced for your use. "Green chemistry" is introduced in Chapter 2 of the textbook and "green profiles" of student experiments are included in the instructions for experiments at the Web site. There are instructions for two new experiments on fitting an acid-base titration curve with a spreadsheet and liquid carbon dioxide extraction of lemon peel oil. At the Web site, you will also find **lists of experiments** from the *Journal of Chemical Education.* **Supplementary topics** at the Web site include spreadsheets for precipitation titrations, microequilibrium constants, spreadsheets for redox titrations curves, analysis of variance, and spreadsheet simulation of gradient liquid chromatography. **Online quizzing** helps students reinforce their understanding of the chapter content.

The **instructors' Web site, www.whfreeman.com/qca8e,** has all **artwork** and **tables** from the book in preformatted PowerPoint slides and as JPG files, an online quizzing gradebook, and more.

For instructors interested in online homework management, W. H. Freeman and WebAssign have partnered to deliver **WebAssign Premium.** WebAssign Premium combines over 600 questions with a fully interactive DynamicBook at an affordable price. To learn more or sign up for a faculty demo account, visit **www.webassign.net.**

DynamicBook for *Quantitative Chemical Analysis, Eighth Edition,* is an electronic version of the text that gives you the flexibility to fully tailor content to your presentation of course material. It can be used in conjunction with the printed text, or it can be adopted on its own. Please go to **www.dynamicbooks.com** for more information, or speak with your W. H. Freeman sales representative.

WebAssign Premium logo.

The People

A book of this size and complexity is the work of many people. Jodi Simpson—the most thoughtful and meticulous copy editor—read every word with a critical eye and improved the exposition in innumerable ways. At W. H. Freeman and Company, Jessica Fiorillo provided overall guidance and was especially helpful in ferreting out opinions from instructors. Mary Louise Byrd shepherded the manuscript through production with her magic wand. Kristina Treadway managed the process of moving the book into production, and Anthony Petrites coordinated the reviewing of every chapter. Ted Sczcepanski located several hard-to-find photographs for the book. Dave Quinn made sure that the supplements were out on time and that the Web site was up and running with all its supporting resources active. Katalin Newman, at Aptara, did an outstanding job of proofreading.

At the Scripps Institution of Oceanography, Ralph Keeling, Peter Guenther, David Moss, Lynne Merchant, and Alane Bollenbacher shared their knowledge of atmospheric CO_2 measurements and graciously provided access to Keeling family photographs. I am especially delighted to have had feedback from Louise Keeling on my story of her husband, Charles David Keeling. This material opens the book in Chapter 0. Sam Kounaves of Tufts University devoted a day to telling me about the *Phoenix Mars Lander* Wet Chemistry Laboratory, which is featured in Chapter 14. Jarda Ruzika of the University of Washington brought the importance of flow injection and sequential injection to my attention, provided an excellent tutorial, and reviewed my description of these topics in Chapters 18 and 19. David Sparkman of the University of the Pacific had detailed comments and suggestions for Chapter 21 on mass spectrometry. Joerg Barankewitz of Sartorius AG provided information and graphics on balances that you will find in Chapter 2.

Solutions to problems and exercises were checked by two wonderfully careful students, Cassandra Churchill and Linda Lait of the University of Lethbridge in Canada. Eric Erickson and Greg Ostrom provided helpful information and discussions at Michelson Lab.

My wife, Sally, works on every aspect of every edition of this book and the Solutions Manual. She contributes mightily to whatever clarity and accuracy we have achieved.

In Closing

This book is dedicated to the students who use it, who occasionally smile when they read it, who gain new insight, and who feel satisfaction after struggling to solve a problem. I have been successful if this book helps you develop critical, independent reasoning that you can apply to new problems. I truly relish your comments, criticisms, suggestions, and corrections. Please address correspondence to me at the Chemistry Division (Mail Stop 6303), Research Department, Michelson Laboratory, China Lake CA 93555.

Acknowledgments

I am indebted to many people who asked questions and provided suggestions and new information for this edition. They include Robert Weinberger *(CE Technologies)*, Tom Betts *(Kutztown University)*, Paul Rosenberg *(Rochester Institute of Technology)*, Barbara Belmont *(California State University, Dominguez Hills)*, David Chen *(University of British Columbia)*, John Birks *(2B Technologies)*, Bob Kennedy *(University of Michigan)*, D. Brynn Hibbert *(University of New South Wales)*, Kris Varazo *(Francis Marion University)*, Chongmok Lee *(Ewha Womans University, Korea)*, Michael Blades *(University of British Columbia)*, D. J. Asa *(ESA, Inc.)*, F. N. Castellano and T. N. Singh-Rachford *(Bowling Green State University)*, J. M. Kelly and D. Ledwith *(Trinity College, University of Dublin)*, Justin Ries *(University of North Carolina)*, Gregory A. Cutter *(Old Dominion University)*, Masoud Agah *(Virginia Tech)*, Michael E. Rybak *(U.S. Centers for Disease Control and Prevention)*, James Harnly *(U.S. Department of Agriculture)*, Andrew Shalliker *(University of Western Sydney)*, R. Graham Cooks *(Purdue University)*, Alexander Makarov *(Thermo Fisher Scientific, Bremen)*, Richard Mathies *(University of California, Berkeley)*, A. J. Pezhathinal and R. Chan-Yu-King *(University of Science and Arts of Oklahoma)*, Peter Licence *(University of Nottingham)*, and Geert Van Biesen *(Memorial University of Newfoundland)*.

People who reviewed parts of the eighth edition manuscript or who reviewed the seventh edition to make suggestions for the eighth edition include Rosemari Chinni *(Alvernia College)*, Shelly Minteer *(St. Louis University)*, Charles Cornett *(University of Wisconsin–Platteville)*, Anthony Borgerding *(St. Thomas College)*, Jeremy Mitchell-Koch *(Emporia State University)*, Kenneth Metz *(Boston College)*, John K. Young *(Mississippi State University)*, Abdul Malik *(University of Southern Florida)*, Colin F. Poole *(Wayne State University)*, Marcin Majda *(University of California, Berkeley)*, Carlos Garcia *(University of Texas, San Antonio)*, Elizabeth Binamira-Soriaga *(Texas A&M University)*, Erin Gross *(Creighton University)*, Dale Wood *(Bishop's University)*, Xin Wen *(California State University, Los Angeles)*, Benny Chan *(The College of New Jersey)*, Pierre Herckes *(Arizona State University)*, Daniel Bombick *(Wright State University)*, Sidney Katz *(Rutgers University)*, Nelly Matteva *(Florida A&M University)*, Michael Johnson *(University of Kansas)*, Dmitri Pappas *(Texas Tech University)*, Jeremy Lessmann *(Washington State University)*, Alexa Serfis *(Saint Louis University)*, Stephen Wolf *(Indiana State University)*, Stuart Chalk *(University of North Florida)*, Barry Lavine *(Oklahoma State University)*, Katherine Pettigrew *(George Mason University)*, Blair Miller *(Grand Valley State University)*, Nathalie Wall *(Washington State University)*, Kris Varazo *(Francis Marion University)*, Carrie Brennan *(Austin Peay State University)*, Lisa Ponton *(Elon University)*, Feng Chen *(Rider University)*, Eric Ball *(Metropolitan State College of Denver)*, Russ Barrows *(Metropolitan State College of Denver)*, and Mary Sohn *(Florida Institute of Technology)*.

The Analytical Process

Atmospheric CO_2 has been measured since 1958 at Mauna Loa Observatory, 3 400 meters above sea level on a volcano in Hawaii. [Forrest M. Mims III, www.forrestmims.org/maunaloaobservatory.html, photo taken in 2006.]

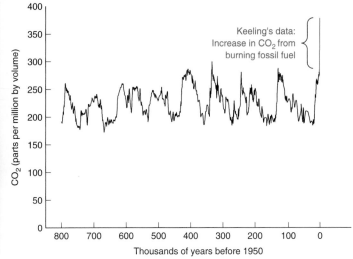

Historic atmospheric CO_2 data are derived from analyzing air bubbles trapped in ice drilled from Antarctica. Keeling's measurements of atmospheric CO_2 give the vertical line at the right side of the graph. [Ice core data from D. Lüthi et al., *Nature*, **2008**, *453*, 379. Mauna Loa data from http://scrippsco2.ucsd.edu/data/in_situ_co2/monthly_mlo.csv.]

In 1958, Charles David Keeling began a series of precise measurements of atmospheric carbon dioxide that have been called "the single most important environmental data set taken in the 20th century."* A half century of observations now shows that human beings have increased the amount of CO_2 in the atmosphere by more than 40% over the average value that existed for the last 800 000 years. On a geologic time scale, we are unlocking all of the carbon sequestered in coal and oil in one brief moment, an outpouring that is jarring the Earth away from its previous condition.

The vertical line at the upper right of the graph shows what we have done. This line will continue on its vertical trajectory until we have consumed all of the fossil fuel on Earth. The consequences will be discovered by future generations, beginning with yours.

---

*C. F. Kennel, Scripps Institution of Oceanography.

I n the last century, humans abruptly changed the composition of Earth's atmosphere. We begin our study of quantitative chemical analysis with a biographical account of how Charles David Keeling came to measure atmospheric CO_2. Then we proceed to discuss the general nature of the analytical process.

0-1 Charles David Keeling and the Measurement of Atmospheric CO_2

Notes and references appear after the last chapter of the book.

Charles David Keeling (1928–2005, Figure 0-1) grew up near Chicago during the Great Depression.[1] His investment banker father excited an interest in astronomy in 5-year-old Keeling. His mother gave him a lifelong love of music. Though "not predominantly interested in science," Keeling took all the science available in high school, including a wartime course in aeronautics that exposed him to aerodynamics and meteorology. In 1945, he enrolled in a

FIGURE 0-1 Charles David Keeling and his wife, Louise, circa 1970. [Courtesy Ralph Keeling, Scripps Institution of Oceanography, University of California, San Diego.]

FIGURE 0-2 A manometer made from a glass U-tube. The difference in height between the mercury on the left and the right gives the pressure of the gas in millimeters of mercury. Box 3-2 provides more detail.

summer session at the University of Illinois prior to his anticipated draft into the army. When World War II ended that summer, Keeling continued at Illinois, where he "drifted into chemistry."

Upon graduation in 1948, Professor Malcolm Dole of Northwestern University, who had known Keeling as a precocious child, offered him a graduate fellowship in chemistry. On Keeling's second day in the lab, Dole taught him how to make careful measurements with an analytical balance. Keeling went on to conduct research in polymer chemistry, though he had no special attraction to polymers or to chemistry.

A requirement for graduate study was a minor outside of chemistry. Keeling noticed the book *Glacial Geology and the Pleistocene Epoch* on a friend's bookshelf. It was so interesting that he bought a copy and read it between experiments in the lab. He imagined himself "climbing mountains while measuring the physical properties of glaciers." In graduate school, Keeling completed most of the undergraduate curriculum in geology and twice interrupted his research to hike and climb mountains.

In 1953, Ph.D. polymer chemists were in demand for the new plastics industry. Keeling had job offers from manufacturers in the eastern United States, but he "had trouble seeing the future this way." He had acquired a working knowledge of geology and loved the outdoors. Professor Dole considered it "foolhardy" to pass up high-paying jobs for a low-paying postdoctoral position. Nonetheless, Keeling wrote letters seeking a postdoctoral position as a chemist "exclusively to geology departments west of the North American continental divide." He became the first postdoctoral fellow in the new Department of Geochemistry in Harrison Brown's laboratory at Caltech in Pasadena, California.

One day, "Brown illustrated the power of applying chemical principles to geology. He suggested that the amount of carbonate in surface water . . . might be estimated by assuming the water to be in chemical equilibrium with both limestone [$CaCO_3$] and atmospheric carbon dioxide." Keeling decided to test this idea. He "could fashion chemical apparatus to function in the real environment" and "the work could take place outdoors."

Keeling built a vacuum system to isolate CO_2 from air or acidified water. The CO_2 in dried air was trapped as a solid in the vacuum system by using liquid nitrogen, "which had recently become available commercially." Keeling built a manometer to measure gaseous CO_2 by confining the gas in a known volume at a known pressure and temperature (Figure 0-2 and Box 3-2). The measurement was precise (reproducible) to 0.1%, which was as good or better than other procedures for measuring CO_2.

Keeling prepared for a field experiment at Big Sur. The area is rich in calcite ($CaCO_3$), which would, presumably, be in contact with groundwater. Keeling "began to worry . . . about assuming a specified concentration for CO_2 in air." This concentration had to be known for his experiments. Published values varied widely, so he decided to make his own

CHAPTER 0 The Analytical Process

measurements. He had a dozen 5-liter flasks built with stopcocks that would hold a vacuum. He weighed each flask empty and filled with water. From the mass of water it held, he could calculate the volume of each flask. To rehearse for field experiments, Keeling measured air samples in Pasadena. Concentrations of CO_2 varied significantly, apparently affected by urban emissions.

Not being certain that CO_2 in pristine air next to the Pacific Ocean at Big Sur would be constant, he collected air samples every few hours over a full day and night. He also collected water samples and brought everything back to the lab to measure CO_2. At the suggestion of Professor Sam Epstein, Keeling provided samples of CO_2 for Epstein's group to measure carbon and oxygen isotopes with their newly built isotope ratio mass spectrometer. "I did not anticipate that the procedures established in this first experiment would be the basis for much of the research that I would pursue over the next forty-odd years," recounted Keeling. Contrary to hypothesis, Keeling found that river water and groundwater contained more dissolved CO_2 than expected if the dissolved CO_2 were in equilibrium with the CO_2 in the air.

Keeling's attention was drawn to the diurnal pattern that he observed in atmospheric CO_2. Air in the afternoon had an almost constant CO_2 content of 310 parts per million (ppm) by volume of dry air. The concentration of CO_2 at night was higher and variable. Also, the higher the CO_2 content, the lower the $^{13}C/^{12}C$ ratio. It was thought that photosynthesis by plants would draw down atmospheric CO_2 near the ground during the day and respiration would restore CO_2 to the air at night. However, samples collected in daytime from many locations had nearly the same 310 ppm CO_2.

Diurnal means the pattern varies between night and day.

Keeling found an explanation in a book entitled *The Climate Near the Ground*. All of his samples were collected in fair weather, when solar heating induces afternoon turbulence that mixes air near the ground with air higher in the atmosphere. At night, air cools and forms a stable layer near the ground that becomes rich in CO_2 from respiration of plants. Keeling had discovered that CO_2 is near 310 ppm in the free atmosphere over large regions of the Northern Hemisphere. By 1956, his findings were firm enough to be told to others, including Dr. Oliver Wulf of the U.S. Weather Bureau, who was working at Caltech.

Wulf passed Keeling's results to Harry Wexler, Head of Meteorological Research at the Weather Bureau. Wexler invited Keeling to Washington, DC, where he explained that the International Geophysical Year commencing in July 1957 was intended to collect worldwide geophysical data for a period of 18 months. The Bureau had just built an observatory near the top of Mauna Loa volcano in Hawaii, and Wexler was anxious to put it to use. The Bureau wanted to measure atmospheric CO_2 at remote locations around the world.

Keeling explained that measurements in the scientific literature might be unreliable. He proposed to measure CO_2 with an infrared spectrometer that would be precisely calibrated with gas measured by a manometer. The manometer is the most reliable way to measure CO_2, but each measurement requires half a day of work. The spectrometer could measure several samples per hour but must be calibrated with reliable standards.

Wexler liked Keeling's proposal and declared that infrared measurements should be made on Mauna Loa and in Antarctica. The next day, Wexler offered Keeling a job. Keeling described what happened next: "I was escorted to where I might work . . . in the dim basement of the Naval Observatory where the only activity seemed to be a cloud-seeding study being conducted by a solitary scientist."

Fortunately, Keeling's CO_2 results had also been brought to the attention of Roger Revelle, Director of the Scripps Institution of Oceanography near San Diego, California. Revelle invited Keeling for a job interview. He was given lunch outdoors "in brilliant sunshine wafted by a gentle sea breeze." Keeling thought to himself, "dim basement or brilliant sunshine and sea breeze?" He chose Scripps, and Wexler graciously provided funding to support CO_2 measurements.

Keeling identified several continuous gas analyzers and tested one from "the only company in which [he] was able to get past a salesman and talk directly with an engineer." He went to great lengths to calibrate the infrared instrument with precisely measured gas standards. Keeling painstakingly constructed a manometer whose results were reproducible to 1 part in 4 000 (0.025%), thus enabling atmospheric CO_2 measurements to be reproducible to 0.1 ppm. Contemporary experts questioned the need for such precision because existing literature indicated that CO_2 in the air varied by a factor of 2. Furthermore, there was concern that measurements on Mauna Loa would be confounded by CO_2 emitted by the volcano.

Scripps pier, wafted by a gentle sea breeze.

Roger Revelle of Scripps believed that the main value of the measurements would be to establish a "snapshot" of CO_2 around the world in 1957, which could be compared with

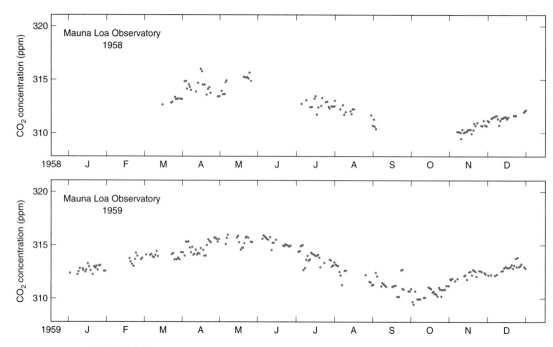

FIGURE 0-3 Atmospheric CO_2 measurements from Mauna Loa in 1958–1959. [J. D. Pales and C. D. Keeling, "The Concentration of Atmospheric Carbon Dioxide in Hawaii," *J. Geophys. Res.* **1965**, *70*, 6053.]

another snapshot taken 20 years later to see if CO_2 concentration was changing. People had considered that burning of fossil fuel could increase atmospheric CO_2, but it was thought that a good deal of this CO_2 would be absorbed by the ocean. No meaningful measurements existed to evaluate any hypothesis.

In March 1958, Ben Harlan of Scripps and Jack Pales of the Weather Bureau installed Keeling's infrared instrument on Mauna Loa. The first day's reading was within 1 ppm of the 313-ppm value expected by Keeling from his measurements made on the pier at Scripps. Concentrations in Figure 0-3 rose between March and May, when operation was interrupted by a power failure. Concentrations were falling in September when power failed again. Keeling was then allowed to make his first trip to Mauna Loa to restart the equipment. Concentrations steadily rose from November to May 1959, before gradually falling again. Data for the full year 1959 in Figure 0-3 reproduced the pattern from 1958. *These patterns could not have been detected if Keeling's measurements had not been made so carefully.*

Maximum CO_2 was observed just before plants in the temperate zone of the Northern Hemisphere put on new leaves in May. Minimum CO_2 was observed at the end of the growing season in October. Keeling concluded that "we were witnessing for the first time nature's withdrawing CO_2 from the air for plant growth during the summer and returning it each succeeding winter."

Figure 0-4, known as the *Keeling curve*, shows the results of half a century of CO_2 monitoring on Mauna Loa. Seasonal oscillations are superimposed on a steady rise.

FIGURE 0-4 Monthly average atmospheric CO_2 measured on Mauna Loa. This graph, known as the *Keeling curve*, shows seasonal oscillations superimposed on rising CO_2. [Data from http://scrippsco2.ucsd.edu/data/ in_situ_co2/monthly_mlo.csv.]

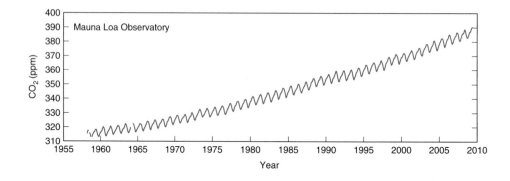

CHAPTER 0 The Analytical Process

Approximately half of the CO_2 produced by the burning of fossil fuel (principally coal, oil, and natural gas) in the last half century resides in the atmosphere. Most of the remainder was absorbed by the ocean.

In the atmosphere, CO_2 absorbs infrared radiation from the surface of the Earth and reradiates part of that energy back to the ground (Figure 0-5). This *greenhouse effect* warms the Earth's surface and might produce climate change. In the ocean, CO_2 forms carbonic acid, H_2CO_3, which makes the ocean more acidic. Fossil fuel burning has already lowered the pH of ocean surface waters by 0.1 unit from preindustrial values. Combustion during the twenty-first century is expected to acidify the ocean by another 0.3–0.4 pH units—threatening marine life whose calcium carbonate shells dissolve in acid (Box 9-1). The entire ocean food chain is jeopardized by ocean acidification.[2]

The significance of the Keeling curve is apparent by appending Keeling's data to the 800 000-year record of atmospheric CO_2 and temperature preserved in Antarctic ice. Figure 0-6 shows that temperature and CO_2 experienced peaks roughly every 100 000 years, as marked by arrows.

Cyclic changes in Earth's orbit and tilt cause cyclic temperature change. Small increases in temperature drive CO_2 from the ocean into the atmosphere. Increased atmospheric CO_2 further increases warming by the greenhouse effect. Cooling brought on by orbital changes redissolves CO_2 in the ocean, thereby causing further cooling. Temperature and CO_2 have followed each other for 800 000 years.

Burning fossil fuel in the last 150 years increased CO_2 from its historic cyclic peak of 280 ppm to today's 380 ppm. No conceivable action in the present century will prevent CO_2 from climbing to several times its historic high, which might significantly affect climate. The longer we take to reduce fossil fuel use, the longer this unintended global experiment will continue. Increasing population exacerbates this and many other problems.

Keeling's CO_2 measurement program was jeopardized many times by funding decisions at government agencies. His persistence ensured the continuity and quality of the measurements. Manometrically measured calibration standards are labor intensive and costly. Funding agencies tried to reduce the cost by finding substitutes for manometry, but no method provided the same precision. The analytical quality of Keeling's data has enabled subtle trends, such as the effect of El Niño ocean temperature patterns, to be teased out of the overriding pattern of increasing CO_2 and seasonal oscillations.

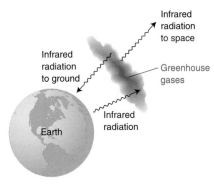

FIGURE 0-5 *Greenhouse effect.* The sun warms the Earth mainly with visible radiation. Earth emits infrared radiation, which would all go into space in the absence of the atmosphere. Greenhouse gases in the atmosphere absorb some of the infrared radiation and emit some of that radiation back to the Earth. Radiation directed back to Earth by greenhouse gases keeps the Earth warmer than it would be in the absence of greenhouse gases.

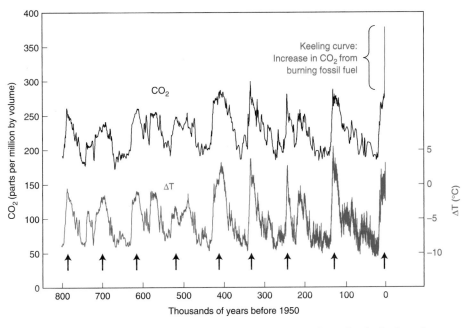

FIGURE 0-6 Significance of the Keeling curve (upper right, color) is shown by plotting it on the same graph with atmospheric CO_2 measured in air bubbles trapped in ice cores drilled from Antarctica. Atmospheric temperature at the level where precipitation forms is deduced from hydrogen and oxygen isotopic composition of the ice. [Vostok ice core data from J. M. Barnola, D. Raynaud, C. Lorius, and N. I. Barkov, http://cdiac.esd.ornl.gov/ftp/trends/co2/vostok.icecore.co2.]

Chocolate is great to eat, but not so easy to analyze. [W. H. Freeman photo by K. Bendo.]

A *diuretic* makes you urinate.
A *vasodilator* enlarges blood vessels.

Chemical Abstracts is the most comprehensive source for locating articles published in chemistry journals. *SciFinder* is software that accesses *Chemical Abstracts*.

Bold terms should be learned. They are listed at the end of the chapter and in the Glossary at the back of the book. *Italicized* words are less important, but many of their definitions are also found in the Glossary.

Homogeneous: same throughout
Heterogeneous: differs from region to region

0-2 The Analytical Chemist's Job

Chocolate[3] has been the savior of many a student on the long night before a major assignment was due. My favorite chocolate bar, jammed with 33% fat and 47% sugar, propels me over mountains in California's Sierra Nevada. In addition to its high energy content, chocolate packs an extra punch with the stimulant caffeine and its biochemical precursor, theobromine.

Theobromine (from Greek "food of the gods")
A diuretic, smooth muscle relaxant, cardiac stimulant, and vasodilator

Caffeine
A central nervous system stimulant

Too much caffeine is harmful for many people, and even small amounts cannot be tolerated by some unlucky individuals. How much caffeine is in a chocolate bar? How does that amount compare with the quantity in coffee or soft drinks? At Bates College in Maine, Professor Tom Wenzel teaches his students chemical problem solving through questions such as these.[4]

But, how *do* you measure the caffeine content of a chocolate bar? Two students, Denby and Scott, began their quest with a computer search for analytical methods. Searching with the key words "caffeine" and "chocolate," they uncovered numerous articles in chemistry journals. Two reports entitled "High-Pressure Liquid Chromatographic Determination of Theobromine and Caffeine in Cocoa and Chocolate Products"[5] described a procedure suitable for the equipment in their laboratory.[6]

Sampling

The first step in any chemical analysis is procuring a representative sample to measure—a process called **sampling.** Is all chocolate the same? Of course not. Denby and Scott bought one chocolate bar in the neighborhood store and analyzed pieces of it. If you wanted to make broad statements about "caffeine in chocolate," you would need to analyze a variety of chocolates from different manufacturers. You would also need to measure multiple samples of each type to determine the range of caffeine in each kind of chocolate.

A pure chocolate bar is fairly **homogeneous,** which means that its composition is the same everywhere. It might be safe to assume that a piece from one end has the same caffeine content as a piece from the other end. Chocolate with a macadamia nut in the middle is an example of a **heterogeneous** material—one whose composition differs from place to place. The nut is different from the chocolate. To sample a heterogeneous material, you need to use a strategy different from that used to sample a homogeneous material. You would need to know the average mass of chocolate and the average mass of nuts in many candies. You would need to know the average caffeine content of the chocolate and of the macadamia nut (if it has any caffeine). Only then could you make a statement about the average caffeine content of macadamia chocolate.

Sample Preparation

The first step in the procedure calls for weighing out some chocolate and extracting fat from it by dissolving the fat in a hydrocarbon solvent. Fat needs to be removed because it would interfere with chromatography later in the analysis. Unfortunately, if you just shake a chunk of chocolate with solvent, extraction is not very effective, because the solvent has no access to the inside of the chocolate. So, our resourceful students sliced the chocolate into small bits and placed the pieces into a mortar and pestle (Figure 0-7), thinking they would grind the solid into small particles.

Imagine trying to grind chocolate! The solid is too soft to be ground. So Denby and Scott froze the mortar and pestle with its load of sliced chocolate. Once the chocolate was cold, it was brittle enough to grind. Then small pieces were placed in a preweighed 15-milliliter (mL) centrifuge tube, and their mass was noted.

Pestle

Mortar

FIGURE 0-7 Ceramic mortar and pestle used to grind solids into fine powders.

Figure 0-8 shows the next part of the procedure. A 10-mL portion of the solvent, petroleum ether, was added to the tube, and the top was capped with a stopper. The tube was shaken vigorously to dissolve fat from the solid chocolate into the solvent. Caffeine and theobromine are insoluble in this solvent. The mixture of liquid and fine particles was then spun in a centrifuge to pack the chocolate at the bottom of the tube. The clear liquid, containing dissolved fat, could now be **decanted** (poured off) and discarded. Extraction with fresh portions of solvent was repeated twice more to ensure complete removal of fat from the chocolate. Residual solvent in the chocolate was finally removed by heating the centrifuge tube in a beaker of boiling water. The mass of chocolate residue could be calculated by weighing the centrifuge tube plus its content of defatted chocolate residue and subtracting the known mass of the empty tube.

Substances being measured—caffeine and theobromine in this case—are called **analytes.** The next step in the sample preparation procedure was to make a **quantitative transfer** (a complete transfer) of the fat-free chocolate residue to an Erlenmeyer flask and to dissolve the analytes in water for the chemical analysis. If any residue were not transferred from the tube to the flask, then the final analysis would be in error because not all of the analyte would be present. To perform the quantitative transfer, Denby and Scott added a few milliliters of pure water to the centrifuge tube and used stirring and heating to dissolve or suspend as much of the chocolate as possible. Then they poured the **slurry** (a suspension of solid in a liquid) into a 50-mL flask. They repeated the procedure several times with fresh portions of water to ensure that every bit of chocolate was transferred from the centrifuge tube to the flask.

To complete the dissolution of analytes, Denby and Scott added water to bring the volume up to about 30 mL. They heated the flask in a boiling water bath to extract all the caffeine and theobromine from the chocolate into the water. To compute the quantity of analyte later, the total mass of solvent (water) must be accurately known. Denby and Scott knew the mass of chocolate residue in the centrifuge tube and they knew the mass of the empty Erlenmeyer flask. So they put the flask on a balance and added water drop by drop until there were exactly 33.3 g of water in the flask. Later, they would compare known solutions of pure analyte in water with the unknown solution containing 33.3 g of water.

A solution of anything in water is called an **aqueous** solution.

Before Denby and Scott could inject the unknown solution into a chromatograph for the chemical analysis, they had to clean up the unknown even further (Figure 0-9). The slurry of chocolate residue in water contained tiny solid particles that would surely clog their expensive chromatography column and ruin it. So they transferred a portion of the slurry to a centrifuge tube and centrifuged the mixture to pack as much of the solid as possible at the bottom of the tube. The cloudy, tan, **supernatant liquid** (liquid above the packed solid) was then filtered in a further attempt to remove tiny particles of solid from the liquid.

It is critical to avoid injecting solids into a chromatography column, but the tan liquid still looked cloudy. So Denby and Scott took turns between classes to repeat the centrifugation and filtration five times. After each cycle in which the supernatant liquid was filtered and centrifuged, it became a little cleaner. But the liquid was never completely clear. Given enough time, more solid always seemed to precipitate from the filtered solution.

Real-life samples rarely cooperate with you!

The tedious procedure described so far is called **sample preparation**—transforming a sample into a state that is suitable for analysis. In this case, fat had to be removed from the chocolate, analytes had to be extracted into water, and residual solid had to be separated from the water.

FIGURE 0-9 Centrifugation and filtration are used to separate undesired solid residue from the aqueous solution of analytes.

The Chemical Analysis (At Last!)

Chromatography solvent is selected by a systematic trial-and-error process described in Chapter 24. Acetic acid reacts with negative oxygen atoms on the silica surface. When not neutralized, these oxygen atoms tightly bind a small fraction of caffeine and theobromine.

$$\text{silica-O}^- \xrightarrow{\text{acetic acid}} \text{silica-OH}$$

Binds analytes very tightly   Does not bind analytes strongly

Denby and Scott finally decided that the solution of analytes was as clean as they could make it in the time available. The next step was to inject solution into a *chromatography* column, which would separate the analytes and measure the quantity of each. The column in Figure 0-10a is packed with tiny particles of silica (SiO_2) to which are attached long hydrocarbon molecules. Twenty microliters (20.0×10^{-6} liters) of the chocolate extract were injected into the column and washed through with a solvent made by mixing 79 mL of pure water, 20 mL of methanol, and 1 mL of acetic acid. Caffeine is more soluble than theobromine in the hydrocarbon on the silica surface. Therefore, caffeine "sticks" to the coated silica particles in the column more strongly than theobromine does. When both

FIGURE 0-10 Principle of liquid chromatography. (*a*) Chromatography apparatus with an ultraviolet absorbance monitor to detect analytes at the column outlet. (*b*) Separation of caffeine and theobromine by chromatography. Caffeine is more soluble than theobromine in the hydrocarbon layer on the particles in the column. Therefore, caffeine is retained more strongly and moves through the column more slowly than theobromine.

analytes are flushed through the column by solvent, theobromine reaches the outlet before caffeine (Figure 0-10b).

Analytes are detected at the outlet by their ability to absorb ultraviolet radiation from the lamp in Figure 0-10a. The graph of detector response versus time in Figure 0-11 is called a *chromatogram*. Theobromine and caffeine are the major peaks in the chromatogram. Small peaks arise from other substances extracted from the chocolate.

The chromatogram alone does not tell us what compounds are present. One way to identify individual peaks is to measure spectral characteristics of each one as it emerges from the column. Another way is to add an authentic sample of either caffeine or theobromine to the unknown and see whether one of the peaks grows in magnitude.

Identifying *what* is in an unknown is called **qualitative analysis.** Identifying *how much* is present is called **quantitative analysis.** The vast majority of this book deals with quantitative analysis.

In Figure 0-11, the *area* under each peak is proportional to the quantity of compound passing through the detector. The best way to measure area is with a computer that receives output from the chromatography detector. Denby and Scott did not have a computer linked to their chromatograph, so they measured the *height* of each peak instead.

Calibration Curves

In general, analytes with equal concentrations give different detector responses. Therefore, the response must be measured for known concentrations of each analyte. A graph of detector response as a function of analyte concentration is called a **calibration curve** or a *standard curve*. To construct such a curve, **standard solutions** containing known concentrations of pure theobromine or caffeine were prepared and injected into the column, and the resulting peak heights were measured. Figure 0-12 is a chromatogram of one of the standard solutions, and Figure 0-13 shows calibration curves made by injecting solutions containing 10.0, 25.0, 50.0, or 100.0 micrograms of each analyte per gram of solution.

Straight lines drawn through the calibration points could then be used to find the concentrations of theobromine and caffeine in an unknown. From the equation of the theobromine line in Figure 0-13, we can say that, if the observed peak height of theobromine from an unknown solution is 15.0 cm, then the concentration is 76.9 micrograms per gram of solution.

Interpreting the Results

Knowing how much analyte is in the aqueous extract of the chocolate, Denby and Scott could calculate how much theobromine and caffeine were in the original chocolate. Results for dark and white chocolates are shown in Table 0-1. The quantities found in white chocolate are only about 2% as great as the quantities in dark chocolate.

Only substances that absorb ultraviolet radiation at a wavelength of 254 nanometers are observed in Figure 0-11. The major components in the aqueous extract are sugars, but they are not detected in this experiment.

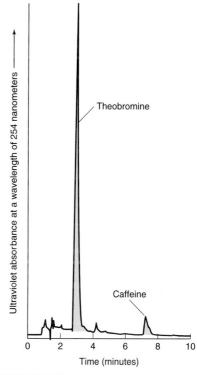

FIGURE 0-11 Chromatogram of 20.0 microliters of dark chocolate extract. A 4.6-mm-diameter × 150-mm-long column, packed with 5-micrometer particles of Hypersil ODS, was eluted (washed) with water:methanol:acetic acid (79:20:1 by volume) at a rate of 1.0 mL per minute.

FIGURE 0-12 Chromatogram of 20.0 microliters of a standard solution containing 50.0 micrograms of theobromine and 50.0 micrograms of caffeine per gram of solution.

FIGURE 0-13 Calibration curves, showing observed peak heights for known concentrations of pure compounds. One *part per million* is one microgram of analyte per gram of solution. Equations of the straight lines drawn through the experimental data points were determined by the *method of least squares* described in Chapter 4.

The table also reports the *standard deviation* of three replicate measurements for each sample. Standard deviation, discussed in Chapter 4, is a measure of the reproducibility of the results. If three samples were to give identical results, the standard deviation would be 0. If results are not very reproducible, then the standard deviation is large. For theobromine in dark chocolate, the standard deviation (0.002) is less than 1% of the average (0.392), so we say the measurement is reproducible. For theobromine in white chocolate, the standard deviation (0.007) is nearly as great as the average (0.010), so the measurement is poorly reproducible.

TABLE 0-1   Analyses of dark and white chocolate

| | Grams of analyte per 100 grams of chocolate | |
Analyte	Dark chocolate	White chocolate
Theobromine	0.392 ± 0.002	0.010 ± 0.007
Caffeine	0.050 ± 0.003	$0.000\ 9 \pm 0.001\ 4$

Average ± standard deviation of three replicate injections of each extract.

The purpose of an analysis is to reach some conclusion. The questions posed earlier were "How much caffeine is in a chocolate bar?" and "How does it compare with the quantity in coffee or soft drinks?" After all this work, Denby and Scott discovered how much caffeine is in the *one* particular chocolate bar that they analyzed. It would take a great deal more work to sample many chocolate bars of the same type and many different types of chocolate to gain a more universal view. Table 0-2 compares results from analyses of different sources of caffeine. A can of soft drink or a cup of tea contains less than one-half of the caffeine in a small cup

TABLE 0-2   Caffeine content of beverages and foods

Source	Caffeine (milligrams per serving)	Serving size[a] (ounces)
Regular coffee	106–164	5
Decaffeinated coffee	2–5	5
Tea	21–50	5
Cocoa beverage	2–8	6
Baking chocolate	35	1
Sweet chocolate	20	1
Milk chocolate	6	1
Caffeinated soft drinks	36–57	12
Red Bull	80	8.2

a. 1 ounce = 28.35 grams.
SOURCES: http://www.holymtn.com/tea/caffeine_content.htm. Red Bull from http://wilstar.com/caffeine.htm.

of coffee. Chocolate contains even less caffeine, but a hungry backpacker eating enough baking chocolate can get a pretty good jolt!

0-3 General Steps in a Chemical Analysis

The analytical process often begins with a question that is not phrased in terms of a chemical analysis. The question could be "Is this water safe to drink?" or "Does emission testing of automobiles reduce air pollution?" A scientist translates such questions into the need for particular measurements. An analytical chemist then chooses or invents a procedure to carry out those measurements.

When the analysis is complete, the analyst must translate the results into terms that can be understood by others—preferably by the general public. A most important feature of any result is its limitations. What is the statistical uncertainty in reported results? If you took samples in a different manner, would you obtain the same results? Is a tiny amount (a *trace*) of analyte found in a sample really there or is it contamination? Only after we understand the results and their limitations can we draw conclusions.

We can now summarize general steps in the analytical process:

Formulating the question	Translate general questions into specific questions to be answered through chemical measurements.
Selecting analytical procedures	Search the chemical literature to find appropriate procedures or, if necessary, devise new procedures to make the required measurements.
Sampling	*Sampling* is the process of selecting representative material to analyze. Box 0-1 provides some ideas on how to do so. If you begin with a poorly chosen sample or if the sample changes between the time it is collected and the time it is analyzed, the results are meaningless. "Garbage in, garbage out!"
Sample preparation	*Sample preparation* is the process of converting a representative sample into a form suitable for chemical analysis, which usually means dissolving the sample. Samples with a low concentration of analyte may need to be concentrated prior to analysis. It may be necessary to remove or *mask* species that interfere with the chemical analysis. For a chocolate bar, sample preparation consisted of removing fat and dissolving the desired analytes. Fat was removed because it would interfere with chromatography.
Analysis	Measure the concentration of analyte in several identical **aliquots** (portions). The purpose of *replicate measurements* (repeated measurements) is to assess the variability (uncertainty) in the analysis and to guard against a gross error in the analysis of a single aliquot. *The uncertainty of a measurement is as important as the measurement itself,* because it tells us how reliable the measurement is. If necessary, use different analytical methods on similar samples to make sure that all methods give the same result and that the choice of analytical method is not biasing the result. You may also wish to construct and analyze several different bulk samples to see what variations arise from your sampling procedure.
Reporting and interpretation	Deliver a clearly written, complete report of your results, highlighting any limitations that you attach to them. Your report might be written to be read only by a specialist (such as your instructor), or it might be written for a general audience (perhaps your mother). Be sure the report is appropriate for its intended audience.
Drawing conclusions	Once a report is written, the analyst might not be involved in what is done with the information, such as modifying the raw material supply for a factory or creating new laws to regulate food additives. The more clearly a report is written, the less likely it is to be misinterpreted by those who use it.

Chemists use the term **species** to refer to any chemical of interest. Species is both singular and plural. **Interference** occurs when a species other than analyte increases or decreases the response of the analytical method and makes it appear that there is more or less analyte than is actually present. **Masking** is the transformation of an interfering species into a form that is not detected. For example, Ca^{2+} in lake water can be measured with a reagent called EDTA. Al^{3+} interferes with this analysis because it also reacts with EDTA. Al^{3+} can be masked with excess F^- to form AlF_6^{3-}, which does not react with EDTA.

Most of this book deals with measuring chemical concentrations in homogeneous aliquots of an unknown. The analysis is meaningless unless you have collected the sample properly, you have taken measures to ensure the reliability of the analytical method, and you communicate your results clearly and completely. The chemical analysis is only the middle portion of a process that begins with a question and ends with a conclusion.

BOX 0-1 Constructing a Representative Sample

In a **random heterogeneous material,** differences in composition occur randomly and on a fine scale. When you collect a portion of the material for analysis, you obtain some of each of the different compositions. To construct a representative sample from a heterogeneous material, you can first visually divide the material into segments. A **random sample** is collected by taking portions from the desired number of segments chosen at random. If you want to measure the magnesium content of the grass in the 10-meter × 20-meter field in panel *a*, you could divide the field into 20 000 small patches that are 10 centimeters on a side. After assigning a number to each small patch, you could use a computer program to pick 100 numbers at random from 1 to 20 000. Then harvest and combine the grass from each of these 100 patches to construct a representative bulk sample for analysis.

For a **segregated heterogeneous material** (in which large regions have obviously different compositions), a representative **composite sample** must be constructed. For example, the field in panel *b* has three different types of grass segregated into regions A, B, and C. You could draw a map of the field on graph paper and measure the area in each region. In this case, 66% of the area lies in region A, 14% lies in region B, and 20% lies in region C. To construct a representative bulk sample from this segregated material, take 66 of the small patches from region A, 14 from region B, and 20 from region C. You could do so by drawing random numbers from 1 to 20 000 to select patches until you have the desired number from each region.

(*a*)

(*b*)

Terms to Understand

Terms are introduced in **bold** type in the chapter and are also defined in the Glossary.

aliquot	heterogeneous	quantitative transfer	segregated heterogeneous
analyte	homogeneous	random heterogeneous	material
aqueous	interference	material	slurry
calibration curve	masking	random sample	species
composite sample	qualitative analysis	sample preparation	standard solution
decant	quantitative analysis	sampling	supernatant liquid

Problems

Complete solutions to Problems can be found in the *Solutions Manual*. Short answers to numerical problems are at the back of the book.

0-1. What is the difference between *qualitative* and *quantitative* analysis?

0-2. List the steps in a chemical analysis.

0-3. What does it mean to *mask* an interfering species?

0-4. What is the purpose of a calibration curve?

0-5. (a) What is the difference between a homogeneous material and a heterogeneous material?

(b) After reading Box 0-1, state the difference between a segregated heterogeneous material and a random heterogeneous material.

(c) How would you construct a representative sample from each type of material?

0-6. The iodide (I^-) content of a commercial mineral water was measured by two methods that produced wildy different results.[7] Method A found 0.23 milligrams of I^- per liter (mg/L) and method B found 0.009 mg/L. When Mn^{2+} was added to the water, the I^- content found by method A increased each time more Mn^{2+} was added, but results from method B were unchanged. Which of the *Terms to Understand* describes what is occurring in these measurements?

Chemical Measurements

BIOCHEMICAL MEASUREMENTS WITH A NANOELECTRODE

(a) Carbon-fiber electrode with a 100-nanometer-diameter (100 × 10⁻⁹ meter) tip extending from glass capillary. The marker bar is 200 micrometers (200 × 10⁻⁶ meter). [From W.-H. Huang, D.-W. Pang, H. Tong, Z.-L. Wang, and J.-K. Cheng, *Anal. Chem.* **2001**, *73*, 1048.] (b) Electrode positioned adjacent to a cell detects release of the neurotransmitter, dopamine, from the cell. A nearby, larger counterelectrode is not shown. (c) Bursts of electric current detected when dopamine is released. Insets are enlargements. [From W.-Z. Wu, W.-H. Huang, W. Wang, Z.-L. Wang, J.-K. Cheng, T. Xu, R.-Y. Zhang, Y. Chen, and J. Liu, *J. Am. Chem. Soc.* **2005**, *127*, 8914.]

Cell   Electrode

(a)   (b)

(c)

An electrode with a tip smaller than a single cell allows us to measure neurotransmitter molecules released by a nerve cell in response to a chemical stimulus. We call the electrode a *nanoelectrode* because its active region has dimensions of nanometers (10^{-9} meters). Neurotransmitter molecules released from one *vesicle* (a small compartment) of a nerve cell diffuse to the electrode, where they donate or accept electrons, thereby generating an electric current measured in picoamperes (10^{-12} amperes) for a period of milliseconds (10^{-3} seconds). This chapter discusses units that describe chemical and physical measurements of objects ranging in size from atoms to galaxies.

Neurotransmitter measurements illustrate the need for units of measurement covering many *orders of magnitude* (powers of 10) in range. This chapter introduces those units and reviews chemical concentrations, solution preparation, stoichiometry, and fundamentals of titrations.

For readability, we insert a space after every third digit on either side of the decimal point. Commas are not used, because in some parts of the world a comma has the same meaning as a decimal point. Examples:

Speed of light: 299 792 458 m/s

Avogadro's number: 6.022 141 79 × 10²³ mol⁻¹

1-1   SI Units

SI units of measurement, used by scientists around the world, derive their name from the French *Système International d'Unités*. *Fundamental units* (base units) from which all others are derived are listed in Table 1-1. Standards of length, mass, and time are the *meter* (m), *kilogram* (kg), and *second* (s), respectively. Temperature is measured in *kelvins* (K), amount of substance in *moles* (mol), and electric current in *amperes* (A).

TABLE 1-1 Fundamental SI units

Quantity	Unit (symbol)	Definition
Length	meter (m)	One meter is the distance light travels in a vacuum during $\frac{1}{299\ 792\ 458}$ of a second.
Mass	kilogram (kg)	One kilogram is the mass of the Pt-Ir alloy prototype kilogram made in 1885 and kept under an inert atmosphere at Sèvres, France. This object has been removed from its protective enclosure only in 1890, 1948, and 1992 to weigh secondary standards kept in several countries. Unfortunately, the mass of the prototype kilogram can change slowly over time by chemical reaction with the atmosphere or from mechanical wear. Work in progress will replace the prototype kilogram with a standard based on unchanging properties of nature that can be measured within an uncertainty of 1 part in 10^8. See I. Robinson, "Weighty Matters," *Scientific American*, December 2006, p.102.
Time	second (s)	One second is the duration of 9 192 631 770 periods of the radiation corresponding to a certain atomic transition of ^{133}Cs.
Electric current	ampere (A)	One ampere of current produces a force of 2×10^{-7} newtons per meter of length when maintained in two straight, parallel conductors of infinite length and negligible cross section, separated by 1 meter in a vacuum.
Temperature	kelvin (K)	Temperature is defined such that the triple point of water (at which solid, liquid, and gaseous water are in equilibrium) is 273.16 K, and the temperature of absolute zero is 0 K.
Luminous intensity	candela (cd)	Candela is a measure of luminous intensity visible to the human eye.
Amount of substance	mole (mol)	One mole is the number of particles equal to the number of atoms in exactly 0.012 kg of ^{12}C (approximately 6.022×10^{23}).
Plane angle	radian (rad)	There are 2π radians in a circle.
Solid angle	steradian (sr)	There are 4π steradians in a sphere.

Pressure is force per unit area: 1 pascal (Pa) = 1 N/m^2. The pressure of the atmosphere is approximately 100 000 Pa.

Table 1-2 lists some quantities that are defined in terms of the fundamental quantities. For example, force is measured in *newtons* (N), pressure in *pascals* (Pa), and energy in *joules* (J), each of which can be expressed in terms of length, time, and mass.

Using Prefixes as Multipliers

Rather than using exponential notation, we often use prefixes from Table 1-3 to express large or small quantities. As an example, consider the pressure of ozone (O_3) in the upper atmosphere (Figure 1-1). Ozone is important because it absorbs ultraviolet radiation from the sun that damages many organisms and causes skin cancer. Each spring, a great deal of ozone disappears from the Antarctic stratosphere, thereby creating what is called an ozone "hole." The opening of Chapter 17 discusses the chemistry behind this process.

At an altitude of 1.7×10^4 meters above the Earth's surface, the pressure of ozone over Antarctica reaches a peak of 0.019 Pa. Let's express these numbers with prefixes from Table 1-3. We customarily use prefixes for every third power of ten (10^{-9}, 10^{-6}, 10^{-3}, 10^3,

TABLE 1-2 SI-derived units with special names

Quantity	Unit	Symbol	Expression in terms of other units	Expression in terms of SI base units
Frequency	hertz	Hz		1/s
Force	newton	N		$m \cdot kg/s^2$
Pressure	pascal	Pa	N/m^2	$kg/(m \cdot s^2)$
Energy, work, quantity of heat	joule	J	$N \cdot m$	$m^2 \cdot kg/s^2$
Power, radiant flux	watt	W	J/s	$m^2 \cdot kg/s^3$
Quantity of electricity, electric charge	coulomb	C		$s \cdot A$
Electric potential, potential difference, electromotive force	volt	V	W/A	$m^2 \cdot kg/(s^3 \cdot A)$
Electric resistance	ohm	Ω	V/A	$m^2 \cdot kg/(s^3 \cdot A^2)$
Electric capacitance	farad	F	C/V	$s^4 \cdot A^2/(m^2 \cdot kg)$

Frequency is the number of cycles per unit time for a repetitive event. Force is the product mass $\times$ acceleration. Pressure is force per unit area. Energy or work is force $\times$ distance = mass $\times$ acceleration $\times$ distance. Power is energy per unit time. The electric potential difference between two points is the work required to move a unit of positive charge between the two points. Electric resistance is the potential difference required to move one unit of charge per unit time between two points. The electric capacitance of two parallel surfaces is the quantity of electric charge on each surface when there is a unit of electric potential difference between the two surfaces.

TABLE 1-3   Prefixes

Prefix	Symbol	Factor	Prefix	Symbol	Factor
yotta	Y	10^{24}	deci	d	10^{-1}
zetta	Z	10^{21}	centi	c	10^{-2}
exa	E	10^{18}	milli	m	10^{-3}
peta	P	10^{15}	micro	μ	10^{-6}
tera	T	10^{12}	nano	n	10^{-9}
giga	G	10^{9}	pico	p	10^{-12}
mega	M	10^{6}	femto	f	10^{-15}
kilo	k	10^{3}	atto	a	10^{-18}
hecto	h	10^{2}	zepto	z	10^{-21}
deca	da	10^{1}	yocto	y	10^{-24}

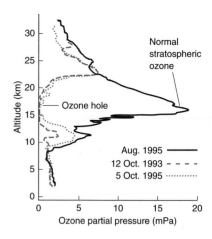

FIGURE 1-1 An ozone "hole" forms each year in the stratosphere over the South Pole at the beginning of spring in October. The graph compares ozone pressure in August, when there is no hole, with the pressure in October, when the hole is deepest. Less severe ozone loss is observed at the North Pole. [Data from National Oceanic and Atmospheric Administration.]

10^6, 10^9, and so on). The number 1.7×10^4 m is more than 10^3 m and less than 10^6 m, so we use a multiple of 10^3 m (= kilometers, km):

$$1.7 \times 10^4 \text{ m} \times \frac{1 \text{ km}}{10^3 \text{ m}} = 17 \text{ km}$$

The number 0.019 Pa is more than 10^{-3} Pa and less than 10^0 Pa, so we use a multiple of 10^{-3} Pa (= millipascals, mPa):

$$0.019 \text{ Pa} \times \frac{1 \text{ mPa}}{10^{-3} \text{ Pa}} = 19 \text{ mPa}$$

Figure 1-1 is labeled with km on the y-axis and mPa on the x-axis. The y-axis of a graph is called the **ordinate** and the x-axis is called the **abscissa.**

Of course you recall that $10^0 = 1$.

It is a fabulous idea to write units beside each number in a calculation and to cancel identical units in the numerator and denominator. This practice ensures that you know the units for your answer. If you intend to calculate pressure and your answer comes out with units other than pascals (N/m^2 or kg/[m · s^2] or other units of force/area), then you have made a mistake.

Converting Between Units

Although SI is the internationally accepted system of measurement in science, other units are encountered. Useful conversion factors are found in Table 1-4. For example, common non-SI

TABLE 1-4   Conversion factors

Quantity	Unit	Symbol	SI equivalent[a]
Volume	liter	L	$*10^{-3}$ m^3
	milliliter	mL	$*10^{-6}$ m^3
Length	angstrom	Å	$*10^{-10}$ m
	inch	in.	*0.025 4 m
Mass	pound	lb	*0.453 592 37 kg
	metric ton		*1 000 kg
Force	dyne	dyn	$*10^{-5}$ N
Pressure	bar	bar	$*10^5$ Pa
	atmosphere	atm	*101 325 Pa
	atmosphere	atm	*1.013 25 bar
	torr (= 1 mm Hg)	Torr	133.322 Pa
	pound/in.^{2}	psi	6 894.76 Pa
Energy	erg	erg	$*10^{-7}$ J
	electron volt	eV	1.602 176 487 × 10^{-19} J
	calorie, thermochemical	cal	*4.184 J
	Calorie (with a capital C)	Cal	*1 000 cal = 4.184 kJ
	British thermal unit	Btu	1 055.06 J
Power	horsepower		745.700 W
Temperature	centigrade (= Celsius)	°C	*K − 273.15
	Fahrenheit	°F	*1.8(K − 273.15) + 32

a. An asterisk (*) indicates that the conversion is exact (by definition).

Oops! In 1999, the $125 million Mars Climate Orbiter spacecraft was lost when it entered the Martian atmosphere 100 km lower than planned. *The navigation error would have been avoided if people had written their units of measurement.* Engineers who built the spacecraft calculated thrust in the English unit, pounds of force. Jet Propulsion Laboratory engineers thought they were receiving the information in the metric unit, newtons. Nobody caught the error.

One *calorie* is the energy required to heat 1 gram of water from 14.5° to 15.5°C.

One *joule* is the energy expended when a force of 1 newton acts over a distance of 1 meter. This much energy can raise 102 g (about $\frac{1}{4}$ pound) by 1 meter.

$$1 \text{ cal} = 4.184 \text{ J}$$

1 pound (mass) $\approx$ 0.453 6 kg
1 mile $\approx$ 1.609 km
The symbol $\approx$ is read "is approximately equal to."

Significant figures are discussed in Chapter 3. For multiplication and division, the number with the fewest digits determines how many digits should be in the answer. The number 91 kcal at the beginning of this problem limits the answer to 2 digits.

A *homogeneous* substance has a uniform composition. Sugar dissolved in water is homogeneous. A mixture that is not the same everywhere (such as orange juice, which has suspended solids) is *heterogeneous*.

Avogadro's number =
number of atoms in 12 g of ^{12}C

$$\text{Molarity (M)} = \frac{\text{moles of solute}}{\text{liters of solution}}$$

Atomic masses are shown in the periodic table inside the cover of this book. Physical constants such as Avogadro's number are also listed inside the cover.

Strong electrolyte: mostly dissociated into ions in solution

Weak electrolyte: partially dissociated into ions in solution

$MgCl^+$ is called an *ion pair*. See Box 7-1.

units for energy are the *calorie* (cal) and the *Calorie* (written with a capital C and standing for 1 000 calories, or 1 kcal). Table 1-4 states that 1 cal is exactly 4.184 J (joules).

Your *basal metabolism* uses approximately 46 Calories per hour (h) per 100 pounds (lb) of body mass just to carry out basic functions required for life, apart from doing any kind of exercise. A person walking at 2 miles per hour on a level path uses approximately 45 Calories per hour per 100 pounds of body mass beyond basal metabolism. The same person swimming at 2 miles per hour consumes 360 Calories per hour per 100 pounds beyond basal metabolism.

EXAMPLE **Unit Conversions**

Express the rate of energy used by a person walking 2 miles per hour (46 + 45 = 91 Calories per hour per 100 pounds of body mass) in kilojoules per hour per kilogram of body mass.

Solution We will convert each non-SI unit separately. First, note that 91 Calories = 91 kcal. Table 1-4 states that 1 cal = 4.184 J; so 1 kcal = 4.184 kJ, and

$$91 \text{ kcal} \times 4.184 \frac{\text{kJ}}{\text{kcal}} = 3.8 \times 10^2 \text{ kJ}$$

Table 1-4 also says that 1 lb is 0.453 6 kg; so 100 lb = 45.36 kg. The rate of energy consumption is therefore

$$\frac{91 \text{ kcal/h}}{100 \text{ lb}} = \frac{3.8 \times 10^2 \text{ kJ/h}}{45.36 \text{ kg}} = 8.4 \frac{\text{kJ/h}}{\text{kg}}$$

We could have written this as one long calculation:

$$\text{Rate} = \frac{91 \text{ kcal/h}}{100 \text{ lb}} \times 4.184 \frac{\text{kJ}}{\text{kcal}} \times \frac{1 \text{ lb}}{0.453 6 \text{ kg}} = 8.4 \frac{\text{kJ/h}}{\text{kg}}$$

Test Yourself A person swimming at 2 miles per hour requires 360 + 46 Calories per hour per 100 pounds of body mass. Express the energy use in kJ/h per kg of body mass. (*Answer:* 37 kJ/h per kg)

1-2 Chemical Concentrations

A *solution* is a *homogeneous* mixture of two or more substances. A minor species in a solution is called **solute**, and the major species is the **solvent**. In this book, most discussions concern *aqueous solutions*, in which the solvent is water. **Concentration** states how much solute is contained in a given volume or mass of solution or solvent.

Molarity and Molality

A **mole** (mol) is *Avogadro's number* of particles (atoms, molecules, ions, or anything else). **Molarity** (M) is the number of moles of a substance per liter of solution. A **liter** (L) is the volume of a cube that is 10 cm on each edge. Because 10 cm = 0.1 m, 1 L = $(0.1 \text{ m})^3$ = 10^{-3} m³. Chemical concentrations, denoted with square brackets, are usually expressed in moles per liter (M). Thus "[H^+]" means "the concentration of H^+."

The **atomic mass** of an element is the number of grams containing Avogadro's number of atoms.[1] The **molecular mass** of a compound is the sum of atomic masses of the atoms in the molecule. It is the number of grams containing Avogadro's number of molecules.

An **electrolyte** is a substance that dissociates into ions in solution. In general, electrolytes are more dissociated in water than in other solvents. We refer to a compound that is mostly dissociated into ions as a *strong electrolyte*. One that is partially dissociated is called a *weak electrolyte*.

Magnesium chloride is a strong electrolyte. In 0.44 M $MgCl_2$ solution, 70% of the magnesium is free Mg^{2+} and 30% is $MgCl^+$. The concentration of $MgCl_2$ molecules is close to 0. Sometimes the molarity of a strong electrolyte is called the **formal concentration** (F), to emphasize that the substance is really converted into other species in solution. When we say that the "concentration" of $MgCl_2$ is 0.054 M in seawater, we are really speaking of its formal concentration (0.054 F). The "molecular mass" of a strong electrolyte is called the **formula mass** (FM), because it is the sum of atomic masses of atoms in the formula, even though there are very few molecules with that formula. *We are going to use the abbreviation FM for both formula mass and molecular mass.*

EXAMPLE **Molarity of Salts in the Sea**

(a) Typical seawater contains 2.7 g of salt (sodium chloride, NaCl) per 100 mL (= 100×10^{-3} L). What is the molarity of NaCl in the ocean? **(b)** $MgCl_2$ has a concentration of 0.054 M in the ocean. How many grams of $MgCl_2$ are present in 25 mL of seawater?

Solution **(a)** The molecular mass of NaCl is 22.99 g/mol (Na) + 35.45 g/mol (Cl) = 58.44 g/mol. The moles of salt in 2.7 g are (2.7 g)/(58.44 g/mol) = 0.046 mol, so the molarity is

$$\text{Molarity of NaCl} = \frac{\text{mol NaCl}}{\text{L of seawater}} = \frac{0.046 \text{ mol}}{100 \times 10^{-3} \text{ L}} = 0.46 \text{ M}$$

(b) The molecular mass of $MgCl_2$ is 24.30 g/mol (Mg) + 2×35.45 g/mol (Cl) = 95.20 g/mol. The number of grams in 25 mL is

$$\text{Grams of } MgCl_2 = \left(0.054 \frac{\text{mol}}{\text{L}}\right)\left(95.20 \frac{\text{g}}{\text{mol}}\right)(25 \times 10^{-3} \text{ L}) = 0.13 \text{ g}$$

Test Yourself Calculate the formula mass of $CaSO_4$. What is the molarity of $CaSO_4$ in a solution containing 1.2 g of $CaSO_4$ in a volume of 50 mL? How many grams of $CaSO_4$ are in 50 mL of 0.086 M $CaSO_4$? (**Answer:** 136.14 g/mol, 0.18 M, 0.59 g)

For a *weak electrolyte* such as acetic acid, CH_3CO_2H, some of the molecules dissociate into ions in solution:

	Formal concentration	Percent dissociated
	0.10 F	1.3%
	0.010 F	4.1%
	0.001 0 F	12%

Molality (*m*) is concentration expressed as moles of substance per kilogram of solvent (not total solution). Molality is independent of temperature. Molarity changes with temperature because the volume of a solution usually increases when it is heated.

Percent Composition

The percentage of a component in a mixture or solution is usually expressed as a **weight percent** (wt%):

$$\text{Weight percent} = \frac{\text{mass of solute}}{\text{mass of total solution or mixture}} \times 100 \qquad (1\text{-}1)$$

A common form of ethanol (CH_3CH_2OH) is 95 wt%; this expression means 95 g of ethanol per 100 g of total solution. The remainder is water. **Volume percent** (vol%) is defined as

$$\text{Volume percent} = \frac{\text{volume of solute}}{\text{volume of total solution}} \times 100 \qquad (1\text{-}2)$$

Although units of mass or volume should always be expressed to avoid ambiguity, mass is usually implied when units are absent.

Confusing abbreviations:

mol = moles

$$M = \text{molarity} = \frac{\text{mol solute}}{\text{L solution}}$$

$$m = \text{molality} = \frac{\text{mol solute}}{\text{kg solvent}}$$

$$Density = \frac{mass}{volume} = \frac{g}{mL}$$

A closely related dimensionless quantity is

$$Specific\ gravity = \frac{density\ of\ a\ substance}{density\ of\ water\ at\ 4°C}$$

The density of water at 4°C is close to 1 g/mL, so specific gravity is nearly the same as density.

If you divide 1.01/0.063 0, you get 16.0. Dan got 16.1 because he kept all the digits in his calculator and did not round off until the end. The number 1.01 was really 1.014 8 and (1.014 8)/(0.063 0) = 16.1.

Converting Weight Percent into Molarity and Molality

Find the molarity and molality of 37.0 wt% HCl. The **density** of a substance is the mass per unit volume. The table inside the back cover of this book tells us that the density of the reagent is 1.19 g/mL.

Solution For *molarity*, we need to find the moles of HCl per liter of solution. The mass of a liter of solution is (1.19 g/mL)(1 000 mL) = 1.19×10^3 g. The mass of HCl in a liter is

$$\text{Mass of HCl per liter} = \left(1.19 \times 10^3 \frac{\text{g solution}}{\text{L}}\right)\left(\underbrace{0.370 \frac{\text{g HCl}}{\text{g solution}}}_{\substack{\text{This is what} \\ \text{37.0 wt\% means}}}\right) = 4.40 \times 10^2 \frac{\text{g HCl}}{\text{L}}$$

The molecular mass of HCl is 36.46 g/mol, so the molarity is

$$\text{Molarity} = \frac{\text{mol HCl}}{\text{L solution}} = \frac{4.40 \times 10^2 \text{ g HCl/L}}{36.46 \text{ g HCl/mol}} = 12.1 \frac{\text{mol}}{\text{L}} = 12.1 \text{ M}$$

For *molality*, we need to find the moles of HCl per kilogram of solvent (which is H_2O). The solution is 37.0 wt% HCl, so we know that 100.0 g of solution contains 37.0 g of HCl and $100.0 - 37.0 = 63.0$ g of H_2O (= 0.063 0 kg). But 37.0 g of HCl contains 37.0 g/(36.46 g/mol) = 1.01 mol. The molality is therefore

$$\text{Molality} = \frac{\text{mol HCl}}{\text{kg of solvent}} = \frac{1.01 \text{ mol HCl}}{0.063 \text{ 0 kg } H_2O} = 16.1 \text{ m}$$

Test Yourself Calculate the molarity and molality of 49.0 wt% HF, using the density given inside the back cover of this book. (***Answer:*** 31.8 M, 48.0 *m*)

Figure 1-2 illustrates a weight percent measurement in the application of analytical chemistry to archaeology. Gold and silver are found together in nature. Dots in Figure 1-2 show the weight percent of gold in more than 1 300 silver coins minted over a 500-year period. Prior to A.D. 500, it was rare for the gold content to be below 0.3 wt%. By A.D. 600, people developed techniques for removing more gold from the silver, so some coins had as little as 0.02 wt% gold. Colored squares in Figure 1-2 represent known, modern forgeries made from silver whose gold content is always less than the prevailing gold content in the years A.D. 200 to 500. Chemical analysis makes it easy to detect the forgeries.

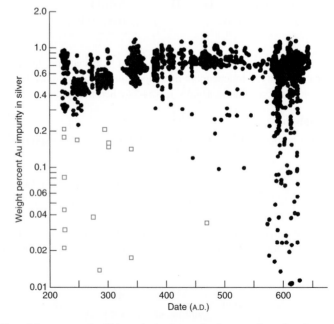

FIGURE 1-2 Weight percent of gold impurity in silver coins from Persia. Colored squares are known, modern forgeries. Note that the ordinate scale is logarithmic. [A. A. Gordus and J. P. Gordus, *Archaeological Chemistry, Adv. Chem.* No. 138, American Chemical Society, Washington, DC, 1974, pp. 124–147.]

Parts per Million and Parts per Billion

Sometimes composition is expressed as **parts per million** (ppm) or **parts per billion** (ppb), which mean grams of substance per million or billion grams of total solution or mixture. Because the density of a dilute aqueous solution is close to 1.00 g/mL, *we frequently equate 1 g of water with 1 mL of water*, although this equivalence is only approximate. Therefore, 1 ppm corresponds to 1 μg/mL (= 1 mg/L) and 1 ppb is 1 ng/mL (= 1 μg/L). For gases, ppm usually refers to volume rather than mass. Atmospheric CO_2 has a concentration near 380 ppm, which means 380 μL CO_2 per liter of air. It is best to label units to avoid confusion.

$$ppm = \frac{\text{mass of substance}}{\text{mass of sample}} \times 10^6$$

$$ppb = \frac{\text{mass of substance}}{\text{mass of sample}} \times 10^9$$

Question What does one part per thousand mean?

EXAMPLE Converting Parts per Billion into Molarity

Normal alkanes are hydrocarbons with the formula C_nH_{2n+2}. Plants selectively synthesize alkanes with an odd number of carbon atoms. The concentration of $C_{29}H_{60}$ in summer rainwater collected in Hannover, Germany is 34 ppb. Find the molarity of $C_{29}H_{60}$ and express the answer with a prefix from Table 1-3.

Solution A concentration of 34 ppb means there are 34 ng of $C_{29}H_{60}$ per gram of rainwater, a value that we equate to 34 ng/mL. Multiplying nanograms and milliliters by 1 000 gives 34 μg of $C_{29}H_{60}$ per liter of rainwater. The molecular mass of $C_{29}H_{60}$ is 408.8 g/mol, so the molarity is

$$\text{Molarity of } C_{29}H_{60} \text{ in rainwater} = \frac{34 \times 10^{-6}\,\text{g/L}}{408.8\,\text{g/mol}} = 8.3 \times 10^{-8}\,\text{M}$$

An appropriate prefix from Table 1-3 would be nano (n), which is a multiple of 10^{-9}:

$$8.3 \times 10^{-8}\,\text{M} \left(\frac{1\,\text{nM}}{10^{-9}\,\text{M}}\right) = 83\,\text{nM}$$

nM = nanomoles per liter

Test Yourself How many ppm of $C_{29}H_{60}$ are in 23 μM $C_{29}H_{60}$? (**Answer:** 9.4 ppm)

1-3 | Preparing Solutions

To prepare a solution with a desired molarity from a pure solid or liquid, we weigh out the correct mass of reagent and dissolve it in a *volumetric flask* (Figure 1-3).

EXAMPLE Preparing a Solution with a Desired Molarity

Copper(II) sulfate pentahydrate, $CuSO_4 \cdot 5H_2O$, has 5 moles of H_2O for each mole of $CuSO_4$ in the solid crystal. The formula mass of $CuSO_4 \cdot 5H_2O$ (= $CuSO_9H_{10}$) is 249.68 g/mol. (Copper(II) sulfate without water in the crystal has the formula $CuSO_4$ and is said to be **anhydrous.**) How many grams of $CuSO_4 \cdot 5H_2O$ should be dissolved in a volume of 500.0 mL to make 8.00 mM Cu^{2+}?

Solution An 8.00 mM solution contains 8.00×10^{-3} mol/L. We need

$$8.00 \times 10^{-3}\,\frac{\text{mol}}{\text{L}} \times 0.5000\,\text{L} = 4.00 \times 10^{-3}\,\text{mol } CuSO_4 \cdot 5H_2O$$

The mass of reagent is $(4.00 \times 10^{-3}\,\text{mol}) \times (249.68\,\text{g/mol}) = 0.999\,\text{g}$.

Using a volumetric flask: The procedure is to place 0.999 g of solid $CuSO_4 \cdot 5H_2O$ into a 500-mL volumetric flask, add about 400 mL of distilled water, and swirl to dissolve the reagent. Then dilute with distilled water up to the 500-mL mark and invert the flask several times to ensure complete mixing.

Test Yourself Find the formula mass of anhydrous $CuSO_4$. How many grams should be dissolved in 250.0 mL to make a 16.0 mM solution? (**Answer:** 159.61 g/mol, 0.638 g)

500-mL mark

TC 20°C   500 mL

FIGURE 1-3 A *volumetric flask* contains a specified volume when the liquid level is adjusted to the middle of the mark in the thin neck of the flask. Use of this flask is described in Section 2-5.

Dilution

Dilute solutions can be prepared from concentrated solutions. A volume of the concentrated solution is transferred to a fresh vessel and diluted to the desired final volume. The number of moles of reagent in V liters containing M moles per liter is the product $M \cdot V = \text{mol}/\cancel{L} \cdot \cancel{L}$, so we equate the number of moles in the concentrated (conc) and dilute (dil) solutions:

Dilution formula:

$$M_{conc} \cdot V_{conc} = M_{dil} \cdot V_{dil} \qquad (1\text{-}3)$$

Moles taken from concentrated solution   Moles placed in dilute solution

In Equation 1-3, you can use any units for concentration per unit volume (such as mmol/L or g/mL) and any units for volume (such as mL or μL), as long as you use the same units on both sides. We frequently use mL for volume.

EXAMPLE Preparing 0.100 M HCl

The molarity of "concentrated" HCl purchased for laboratory use is approximately 12.1 M. How many milliliters of this reagent should be diluted to 1.000 L to make 0.100 M HCl?

Solution The dilution formula handles this problem directly:

$$M_{conc} \cdot V_{conc} = M_{dil} \cdot V_{dil}$$
$$(12.1 \text{ M})(x \text{ mL}) = (0.100 \text{ M})(1\,000 \text{ mL}) \Rightarrow x = 8.26 \text{ mL}$$

To make 0.100 M HCl, we would dilute 8.26 mL of concentrated HCl up to 1.000 L. The concentration will not be exactly 0.100 M because the reagent is not exactly 12.1 M. A table inside the cover of this book gives volumes of common reagents required to make 1.0 M solutions.

The symbol ⇒ is read "implies that."

Test Yourself With information on the inside cover of the book, calculate how many mL of 70.4 wt% nitric acid should be diluted to 0.250 L to make 3.00 M HNO$_3$. (***Answer:*** 47.5 mL)

EXAMPLE A More Complicated Dilution Calculation

A solution of ammonia in water is called "ammonium hydroxide" because of the equilibrium

$$NH_3 \; + \; H_2O \; \rightleftharpoons \; NH_4^+ \; + \; OH^- \qquad (1\text{-}4)$$

Ammonia   Ammonium   Hydroxide

In a chemical reaction, species on the left side are called **reactants** and species on the right are called **products**. NH$_3$ is a reactant and NH$_4^+$ is a product in Reaction 1-4.

The density of concentrated ammonium hydroxide, which contains 28.0 wt% NH$_3$, is 0.899 g/mL. What volume of this reagent should be diluted to 500.0 mL to make 0.250 M NH$_3$?

Solution To use Equation 1-3, we need to know the molarity of the concentrated reagent. The solution contains 0.899 g of solution per milliliter and there is 0.280 g of NH$_3$ per gram of solution (28.0 wt%), so we can write

$$\text{Molarity of NH}_3 = \frac{899 \dfrac{\cancel{\text{g solution}}}{L} \times 0.280 \dfrac{\cancel{\text{g NH}_3}}{\cancel{\text{g solution}}}}{17.03 \dfrac{\cancel{\text{g NH}_3}}{\text{mol NH}_3}} = 14.8 \text{ M}$$

Now we find the volume of 14.8 M NH$_3$ required to prepare 500.0 mL of 0.250 M NH$_3$:

$$M_{conc} \cdot V_{conc} = M_{dil} \cdot V_{dil}$$
$$14.8 \text{ M} \times V_{conc} = 0.250 \text{ M} \times 500.0 \text{ mL} \Rightarrow V_{conc} = 8.46 \text{ mL}$$

The procedure is to place 8.46 mL of concentrated reagent in a 500-mL volumetric flask, add about 400 mL of water, and swirl to mix. Then dilute to exactly 500 mL with water and invert the flask many times to mix well.

Test Yourself From the density of 70.4 wt% HNO$_3$ on the inside cover, calculate the molarity of HNO$_3$. (***Answer:*** 15.8 M)

CHAPTER 1 Chemical Measurements

1-4 Stoichiometry Calculations for Gravimetric Analysis

Chemical analysis based on weighing a final product is called **gravimetric analysis.** Iron from a dietary supplement tablet can be measured by dissolving the tablet and then converting the iron into solid Fe_2O_3. From the mass of Fe_2O_3, we can calculate the mass of iron in the original tablet.

Here are the steps in the procedure:

Step 1 Tablets containing iron(II) fumarate ($Fe^{2+}C_4H_2O_4^{2-}$) and inert binder are mixed with 150 mL of 0.100 M HCl to dissolve the Fe^{2+}. The solution is filtered to remove insoluble binder.

Step 2 Iron(II) in the clear liquid is oxidized to iron(III) with excess hydrogen peroxide:

$$2Fe^{2+} \;+\; H_2O_2 \;+\; 2H^+ \;\rightarrow\; 2Fe^{3+} \;+\; 2H_2O \qquad (1\text{-}5)$$

Iron(II)   Hydrogen peroxide   Iron(III)
(ferrous ion)   FM 34.01   (ferric ion)

Step 3 Ammonium hydroxide is added to precipitate hydrous iron(III) oxide, which is a gel. The gel is filtered and heated in a furnace to convert it to pure solid Fe_2O_3.

$$Fe^{3+} + 3OH^- + (x-1)H_2O \longrightarrow FeOOH \cdot xH_2O(s) \xrightarrow{900°C} Fe_2O_3(s) \qquad (1\text{-}6)$$

Hydroxide   Hydrous iron(III) oxide   Iron(III) oxide
FM 159.69

We now work through some practical laboratory calculations for this analysis.

EXAMPLE   How Many Tablets Should We Analyze?

In a gravimetric analysis, we need enough product to weigh accurately. Each tablet provides ~15 mg of iron. How many tablets should we analyze to provide 0.25 g of Fe_2O_3 product?

Solution We can answer the question if we know how many grams of iron are in 0.25 g of Fe_2O_3. The formula mass of Fe_2O_3 is 159.69 g/mol, so 0.25 g is equal to

$$\text{mol } Fe_2O_3 = \frac{0.25 \text{ g}}{159.69 \text{ g/mol}} = 1.6 \times 10^{-3} \text{ mol}$$

Each mol of Fe_2O_3 has 2 mol of Fe, so 0.25 g of Fe_2O_3 contains

$$1.6 \times 10^{-3} \text{ mol } Fe_2O_3 \times \frac{2 \text{ mol Fe}}{1 \text{ mol } Fe_2O_3} = 3.2 \times 10^{-3} \text{ mol Fe}$$

The mass of Fe is

$$3.2 \times 10^{-3} \text{ mol Fe} \times \frac{55.845 \text{ g Fe}}{\text{mol Fe}} = 0.18 \text{ g Fe}$$

If each tablet contains 15 mg Fe, the number of tablets required is

$$\text{Number of tablets} = \frac{0.18 \text{ g Fe}}{0.015 \text{ g Fe/tablet}} = 12 \text{ tablets}$$

Test Yourself If each tablet provides ~20 mg of iron, how many tablets should we analyze to provide ~0.50 g of Fe_2O_3? (**Answer:** 18)

EXAMPLE   How Much H₂O₂ Is Required?

What mass of 3.0 wt% H_2O_2 solution is required to provide a 50% excess of reagent for Reaction 1-5 with 12 dietary iron tablets?

Solution Twelve tablets provide 12 tablets $\times$ (0.015 g Fe^{2+}/tablet) = 0.18 g Fe^{2+}, or (0.18 g Fe^{2+})/(55.845 g Fe^{2+}/mol Fe^{2+}) = 3.2×10^{-3} mol Fe^{2+}. Reaction 1-5 requires 1 mol of H_2O_2 for every 2 mol of Fe^{2+}. Therefore 3.2×10^{-3} mol Fe^{2+} requires (3.2×10^{-3} mol Fe^{2+})(1 mol H_2O_2/2 mol Fe^{2+}) = 1.6×10^{-3} mol H_2O_2. A 50% excess means

---

Stoichiometry is the calculation of quantities of substances involved in a chemical reaction. It is derived from the Greek *stoicheion* (simplest components) and *metrisi* (to measure).

Fumarate anion, $C_4H_2O_4^{2-}$

The units of formula mass (FM) are g/mol.

$Fe_2O_3(s)$ means that Fe_2O_3 is a *solid*. Other abbreviations for phases are (*l*) for *liquid*, (*g*) for *gas*, and (*aq*) for *aqueous* (meaning "dissolved in water").

The symbol ~ is read "approximately."

$$\text{mol} = \frac{\text{grams}}{\text{grams per mol}} = \frac{\text{grams}}{\text{formula mass}}$$

The atomic mass of Fe, 55.845 g/mol, is in the periodic table inside the cover.

$$\text{Moles} = \frac{\text{grams}}{\text{formula mass}} = \frac{g}{g/mol}$$

You should be able to use this relationship in your sleep.

that we want to use 1.50 times the stoichiometric quantity: $(1.50)(1.6 \times 10^{-3} \text{ mol } H_2O_2) = 2.4 \times 10^{-3} \text{ mol } H_2O_2$. The formula mass of H_2O_2 is 34.01 g/mol, so the required mass of pure H_2O_2 is $(2.4 \times 10^{-3} \text{ mol})(34.01 \text{ g/mol}) = 0.082$ g. But hydrogen peroxide is available as a 3.0 wt% solution, so the required mass of solution is

$$\text{Mass of } H_2O_2 \text{ solution} = \frac{0.082 \text{ g } H_2O_2}{0.030 \text{ g } H_2O_2/\text{g solution}} = 2.7 \text{ g solution}$$

Test Yourself What mass of 3.0 wt% H_2O_2 solution is required to provide a 25% excess of reagent for Reaction 1-5 with 12 dietary iron tablets? (**Answer:** 2.3 g)

EXAMPLE The Gravimetric Calculation

The mass of Fe_2O_3 isolated at the end of the experiment was 0.277 g. What is the average mass of iron per dietary tablet?

Solution The moles of isolated Fe_2O_3 are $(0.277 \text{ g})/(159.69 \text{ g/mol}) = 1.73 \times 10^{-3}$ mol. There are 2 mol Fe per formula unit, so the moles of Fe in the product are

$$(1.73 \times 10^{-3} \text{ mol Fe}_2O_3)\left(\frac{2 \text{ mol Fe}}{1 \text{ mol Fe}_2O_3}\right) = 3.47 \times 10^{-3} \text{ mol Fe}$$

The mass of Fe is $(3.47 \times 10^{-3} \text{ mol Fe})(55.845 \text{ g Fe/mol Fe}) = 0.194$ g Fe. Each of the 12 tablets therefore contains an average of $(0.194 \text{ g Fe})/12 = 0.016 \text{ 1 g} = 16.1$ mg.

Test Yourself If the mass of isolated Fe_2O_3 were 0.300 g, what is the average mass of iron per tablet? (**Answer:** 17.5 mg)

1-5 Introduction to Titrations

Procedures in which we measure the volume of reagent needed to react with *analyte* (the substance being measured) are called **volumetric analysis.** A **titration** is a form of volumetric analysis in which increments of reagent solution—the **titrant**—are added to analyte until their reaction is complete. From the quantity of titrant required, we calculate the quantity of analyte that must have been present. Titrant is usually delivered from a *buret*, which is a glass tube with marks indicating volume (Figure 1-4). Each increment of titrant is completely and quickly consumed by analyte until analyte is used up.

The **equivalence point** occurs when the quantity of added titrant is the exact amount necessary for stoichiometric reaction with the analyte. For example, 5 mol of oxalic acid react with 2 mol of permanganate in hot acidic solution:

$$5HO-\overset{\overset{\displaystyle O}{\|}}{C}-\overset{\overset{\displaystyle O}{\|}}{C}-OH + 2MnO_4^- + 6H^+ \longrightarrow 10CO_2 + 2Mn^{2+} + 8H_2O \qquad (1\text{-}7)$$

Analyte   **Titrant**
Oxalic acid   Permanganate
colorless   purple   colorless   colorless

If the unknown contains 5.000 mmol of oxalic acid, the equivalence point is reached when 2.000 mmol of MnO_4^- have been added.

The equivalence point is the ideal (theoretical) result we seek in a titration. What we actually measure is the **end point,** which is marked by a sudden change in a physical property of the solution. In Reaction 1-7, a convenient end point is the abrupt appearance of the purple color of permanganate in the flask. Prior to the equivalence point, permanganate is consumed by oxalic acid, and the titration solution remains colorless. After the equivalence point, unreacted MnO_4^- accumulates until there is enough to see. The *first trace* of purple color is the end point. The better your eyes, the closer will be your measured end point to the true equivalence point. Here, the end point cannot exactly equal the equivalence point, because extra MnO_4^-—some small amount beyond that needed to react with oxalic acid—is required to exhibit purple color.

Methods for determining when analyte has been consumed include (1) detecting a sudden change in the voltage or current between a pair of electrodes, (2) monitoring absorp-

Retain all digits in your calculator during a series of calculations. The product 1.73×2 is not 3.47; but, with extra digits in the calculator, the answer is 3.47.

Level of titrant

Buret clamp

Buret

Stopcock

Flask

Solution of analyte

Magnetic stirring bar

Magnetic stirrer

FIGURE 1-4 Typical setup for a titration. Analyte is contained in the flask, and titrant is in the buret. The stirring bar is a magnet coated with Teflon, which is inert to most solutions. The bar is spun by a rotating magnet inside the stirring motor. Burets are described in Section 2-4.

BOX 1-1   **Reagent Chemicals and Primary Standards**

Chemicals are sold in many grades of purity. For analytical chemistry, we usually use **reagent grade** chemicals with purity set by the American Chemical Society (ACS) Committee on Analytical Reagents.[2] Sometimes "reagent grade" simply meets purity standards set by the manufacturer. A lot analysis for specified impurities should appear on the reagent bottle. Here is a lot analysis from a bottle of zinc sulfate:

$ZnSO_4$	ACS Reagent	Lot Analysis:
Assay: 100.6%	Fe: 0.000 5%	Ca: 0.001%
Insoluble matter: 0.002%	Pb: 0.002 8%	Mg: 0.000 3%
pH of 5% solution at 25°C: 5.6	Mn: 0.6 ppm	K: 0.002%
Ammonium: 0.000 8%	Nitrate: 0.000 4%	Na: 0.003%
Chloride: 1.5 ppm		

The assay value of 100.6% means that a specified analysis for one of the major components produced 100.6% of the theoretical value. If $ZnSO_4$ is contaminated with the lower-molecular-mass $Zn(OH)_2$, the assay for Zn^{2+} will be higher than the value for pure $ZnSO_4$. Less pure chemicals, generally unsuitable for analytical chemistry, carry designations such as "chemically pure" (CP), "practical," "purified," or "technical."

A few chemicals have high enough purity to be *primary standard grade*. Whereas reagent grade potassium dichromate has a lot assay of $\geq 99.0\%$, primary standard grade $K_2Cr_2O_7$ is in the range 99.95–100.05%. Primary standards must be indefinitely stable.

For **trace analysis** (ppm and lower levels), impurities in reagent chemicals must be extremely low. For this purpose, we use very-high-purity, expensive grades of acids such as "trace metal grade" HNO_3 or HCl to dissolve samples. We must pay careful attention to reagents and vessels whose impurity levels could be greater than the quantity of analyte we seek to measure.

To protect the purity of chemical reagents, you should

- Avoid putting a spatula into a bottle. Instead, pour chemical out of the bottle into a clean container (or onto weighing paper) and dispense the chemical from the clean container.
- Never pour unused chemical back into the reagent bottle.
- Replace the cap on the bottle immediately to keep dust out.
- Never put a stopper from a liquid-reagent container down on the lab bench. Either hold the stopper or place it in a clean beaker or watchglass while you dispense reagent.
- Store chemicals in a cool, dark place away from sunlight.

tion of light by reactants or products, and (3) observing an indicator color change. An **indicator** is a compound with a physical property (usually color) that changes abruptly near the equivalence point. The change is caused by disappearance of analyte or appearance of excess titrant.

The difference between the end point and the equivalence point is an inescapable **titration error.** We estimate the titration error with a **blank titration,** in which we carry out the same procedure *without* analyte. For example, we can titrate a solution containing no oxalic acid to see how much MnO_4^- is needed to produce observable purple color. We then subtract this volume of MnO_4^- from the volume observed in the analytical titration.

The validity of an analytical result depends on knowing the amount of one of the reactants used. If titrant is prepared by dissolving a weighed amount of pure reagent in a known volume of solution, its concentration can be calculated. We call such a reagent a **primary standard,** because it is pure enough to be weighed and used directly. A primary standard should be 99.9% pure, or better. It should not decompose under ordinary storage, and it should be stable when dried by heat or vacuum, because drying is required to remove traces of water adsorbed from the atmosphere. Primary standards for many elements are given in Appendix K. Box 1-1 discusses reagent purity.

Many reagents used as titrants, such as HCl, are not available as primary standards. Instead, we prepare titrant with approximately the desired concentration and use it to titrate an analyte that is a primary standard. By this procedure, called **standardization,** we determine the concentration of titrant. We then say that the titrant is a **standard solution.** The validity of the analytical result ultimately depends on knowing the composition of a primary standard. Sodium oxalate ($Na_2C_2O_4$) is a commercially available primary standard for generating oxalic acid to standardize permanganate in Reaction 1-7.

In a **direct titration,** we add titrant to analyte until the reaction is complete. Occasionally, we perform a **back titration,** in which we add a known *excess* of one standard reagent to the analyte. (A "standard" reagent is one whose concentration is known.) Then we titrate the excess reagent with a second standard reagent. A back titration is useful when its end point is clearer than the end point of the direct titration, or when an excess of the first reagent is required for complete reaction with analyte. To appreciate the difference between direct and back titrations, consider first the addition of permanganate titrant to oxalic acid analyte in Reaction 1-7; this reaction is a direct titration. Alternatively, to perform a back titration, we could add a known *excess* of permanganate to consume oxalic acid. Then we could back titrate the excess permanganate with standard Fe^{2+} to measure how much permanganate was left after reaction with the oxalic acid.

In a **gravimetric titration,** titrant is measured by mass, not volume. Titrant can be delivered from a pipet. Titrant concentration is expressed as moles of reagent per kilogram of solution. Precision is improved from 0.3% attainable with a buret to 0.1% with a balance. Experiments by Guenther and by Butler and Swift at the Web site for this book (www.whfreeman.com/qca) provide examples. "Gravimetric titrations should become the gold standard, and volumetric glassware should be seen in museums only."[3]

1-6 Titration Calculations

Here is an example to illustrate stoichiometry calculations in volumetric analysis. The key step is to *relate moles of titrant to moles of analyte.*

EXAMPLE Standardization of Titrant Followed by Analysis of Unknown

The calcium content of urine can be determined by the following procedure:

Step 1 Precipitate Ca^{2+} with oxalate in basic solution:

$$Ca^{2+} + \underset{\text{Oxalate}}{C_2O_4^{2-}} \rightarrow \underset{\text{Calcium oxalate}}{Ca(C_2O_4) \cdot H_2O(s)}$$

Step 2 Wash the precipitate with ice-cold water to remove free oxalate and then dissolve the solid in acid to obtain Ca^{2+} and $H_2C_2O_4$ in solution.

Step 3 Heat the solution to 60°C and titrate the oxalate with standardized potassium permanganate until the purple end point of Reaction 1-7 is observed.

Standardization Suppose that 0.356 2 g of $Na_2C_2O_4$ is dissolved in a 250.0-mL volumetric flask. If 10.00 mL of this solution require 48.36 mL of $KMnO_4$ solution for titration by Reaction 1-7, what is the molarity of the permanganate solution?

> Retain all digits in your calculator during a calculation. Chapter 3 discusses rounding to the proper number of digits at the end of the calculation. In this example, extra digits beyond those justified by the measurements are subscripted.

Solution The concentration of oxalate solution is

$$\frac{0.356\ 2\ \text{g Na}_2\text{C}_2\text{O}_4/(134.00\ \text{g Na}_2\text{C}_2\text{O}_4/\text{mol})}{0.250\ 0\ \text{L}} = 0.010\ 63_3\ \text{M}$$

The moles of $C_2O_4^{2-}$ in 10.00 mL are $(0.010\ 63_3\ \text{mol/L})(0.010\ 00\ \text{L}) = 1.063_3 \times 10^{-4}\ \text{mol} = 0.106\ 3_3$ mmol. Reaction 1-7 requires 2 mol of permanganate for 5 mol of oxalate, so the MnO_4^- delivered must have been

> Reaction 1-7 requires 2 mol MnO_4^- for 5 mol $C_2O_4^{2-}$.

$$\text{Moles of MnO}_4^- = \left(\frac{2\ \text{mol MnO}_4^-}{5\ \text{mol C}_2\text{O}_4^{2-}}\right)(\text{mol C}_2\text{O}_4^{2-}) = 0.042\ 53_1\ \text{mmol}$$

The concentration of MnO_4^- in the titrant is therefore

> Note that $\dfrac{\text{mmol}}{\text{mL}}$ is the same as $\dfrac{\text{mol}}{\text{L}}$.

$$\text{Molarity of MnO}_4^- = \frac{0.042\ 53_1\ \text{mmol}}{48.36\ \text{mL}} = 8.794_7 \times 10^{-4}\ \text{M}$$

Analysis of Unknown Calcium in a 5.00-mL urine sample was precipitated with $C_2O_4^{2-}$ and redissolved. The redissolved $C_2O_4^{2-}$ required 16.17 mL of standard MnO_4^- solution to reach the end point. Find the concentration of Ca^{2+} in the urine.

Solution In 16.17 mL of MnO_4^-, there are $(0.016\ 17\ \text{L})(8.794_7 \times 10^{-4}\ \text{mol/L}) = 1.422_1 \times 10^{-5}$ mol MnO_4^-. This quantity will react with

> Reaction 1-7 requires 5 mol $C_2O_4^{2-}$ for 2 mol MnO_4^-.

$$\text{Moles of C}_2\text{O}_4^{2-} = \left(\frac{5\ \text{mol C}_2\text{O}_4^{2-}}{2\ \text{mol MnO}_4^-}\right)(\text{mol MnO}_4^-) = 0.035\ 55_3\ \text{mmol}$$

Because there is one oxalate ion for each calcium ion in $Ca(C_2O_4) \cdot H_2O$, there must have been 0.035 55$_3$ mmol of Ca^{2+} in 5.00 mL of urine:

$$[Ca^{2+}] = \frac{0.035\ 55_3\ \text{mmol}}{5.00\ \text{mL}} = 0.007\ 11_1\ \text{M}$$

Test Yourself In standardization, 10.00 mL of $Na_2C_2O_4$ solution required 39.17 mL of $KMnO_4$. Find the molarity of $KMnO_4$. The unknown required 14.44 mL of MnO_4^-. Find $[Ca^{2+}]$ in the urine. (***Answer:*** 1.086×10^{-3} M, 7.840×10^{-3} M)

Terms to Understand

Terms are introduced in **bold** type in the chapter and are also defined in the Glossary.

abscissa	formal concentration	ppb (parts per billion)	stoichiometry
anhydrous	formula mass	ppm (parts per million)	titrant
atomic mass	gravimetric analysis	primary standard	titration
back titration	gravimetric titration	product	titration error
blank titration	indicator	reactant	trace analysis
concentration	liter	reagent grade chemical	volume percent
density	molality	SI units	volumetric analysis
direct titration	molarity	solute	weight percent
electrolyte	mole	solvent	
end point	molecular mass	standardization	
equivalence point	ordinate	standard solution	

Summary

SI base units include the meter (m), kilogram (kg), second (s), ampere (A), kelvin (K), and mole (mol). Derived quantities such as force (newton, N), pressure (pascal, Pa), and energy (joule, J) can be expressed in terms of base units. In calculations, units should be carried along with the numbers. Prefixes such as kilo- and milli- are used to denote multiples of units. Common expressions of concentration are molarity (moles of solute per liter of solution), molality (moles of solute per kilogram of solvent), formal concentration (formula units per liter), percent composition, and parts per million. To calculate quantities of reagents needed to prepare solutions, the relation $M_{conc} \cdot V_{conc} = M_{dil} \cdot V_{dil}$ is useful because it equates the moles of reagent removed from a stock solution to the moles delivered into a new solution. You should be able to use stoichiometry relationships to calculate required masses or volumes of reagents for chemical reactions. From the mass of the product of a reaction, you should be able to compute how much reactant was consumed.

The volume of titrant required for stoichiometric reaction of analyte is measured in volumetric analysis. The stoichiometric point of the reaction is the equivalence point. What we measure by an abrupt change in a physical property (such as indicator color or electrode potential) is the end point. The difference between the end point and the equivalence point is a titration error. This error can be reduced by either subtracting results of a blank titration, in which the same procedure is carried out in the absence of analyte, or standardizing the titrant by using the same reaction and a similar volume as that used for analyte.

The validity of an analytical result depends on knowing the amount of a primary standard. A solution with an approximately desired concentration can be standardized by titrating a primary standard. In a direct titration, titrant is added to analyte until the reaction is complete. In a back titration, a known excess of reagent is added to analyte, and the excess is titrated with a second standard reagent. Calculations of volumetric analysis relate known moles of titrant to unknown moles of analyte.

Exercises

Complete solutions to *Exercises* are provided at the back of the book, whereas only numerical answers to *Problems* are provided. Complete solutions to *Problems* are in the *Solutions Manual*. *Exercises* cover many of the major ideas in each chapter.

1-A. A solution with a final volume of 500.0 mL was prepared by dissolving 25.00 mL of methanol (CH_3OH, density = 0.791 4 g/mL) in chloroform.

(a) Calculate the *molarity* of methanol in the solution.

(b) The solution has a density of 1.454 g/mL. Find the *molality* of methanol.

1-B. A 48.0 wt% solution of HBr in water has a density of 1.50 g/mL.

(a) Find the formal concentration of HBr.

(b) What mass of solution contains 36.0 g of HBr?

(c) What volume of solution contains 233 mmol of HBr?

(d) How much solution is required to prepare 0.250 L of 0.160 M HBr?

1-C. A solution contains 12.6 ppm of dissolved $Ca(NO_3)_2$ (which dissociates into $Ca^{2+} + 2NO_3^-$). Find the concentration of NO_3^- in parts per million.

1-D. Ascorbic acid (vitamin C, page 354) reacts with I_3^- according to the equation

$$\text{Ascorbic acid} + I_3^- + H_2O \rightarrow \text{dehydroascorbic acid} + 3I^- + 2H^+$$
$$\quad\quad C_6H_8O_6 \quad\quad\quad\quad\quad\quad\quad\quad\quad\quad\quad C_6H_8O_7$$

Starch is used as an indicator in the reaction. The end point is marked by the appearance of a deep blue starch-iodine complex when the first fraction of a drop of unreacted I_3^- remains in the solution.

(a) Use atomic masses from the periodic table on the inside cover of this book to find the formula mass of ascorbic acid.

(b) If 29.41 mL of I_3^- solution are required to react with 0.197 0 g of pure ascorbic acid, what is the molarity of the I_3^- solution?

(c) A vitamin C tablet containing ascorbic acid plus inert binder was ground to a powder, and 0.424 2 g was titrated by 31.63 mL of I_3^-. Find the weight percent of ascorbic acid in the tablet.

1-E. A solution of NaOH was standardized by titration of a known quantity of the primary standard, potassium hydrogen phthalate (page 223):

$$C_8H_5O_4K + NaOH \longrightarrow C_8H_4O_4NaK + H_2O$$

Potassium hydrogen
phthalate (FM 204.221)

The NaOH was then used to find the concentration of an unknown solution of H_2SO_4:

$$H_2SO_4 + 2NaOH \longrightarrow Na_2SO_4 + 2H_2O$$

(a) Titration of 0.824 g of potassium hydrogen phthalate required 38.314 g of NaOH solution to reach the end point detected by phenolphthalein indicator. Find the concentration of NaOH (mol NaOH/kg solution).

(b) A 10.00-mL aliquot of H_2SO_4 solution required 57.911 g of NaOH solution to reach the phenolphthalein end point. Find the molarity of H_2SO_4.

Problems

Units and Conversions

1-1. (a) List the SI units of length, mass, time, electric current, temperature, and amount of substance; write the abbreviation for each.

(b) Write the units and symbols for frequency, force, pressure, energy, and power.

1-2. Write the names and abbreviations for each of the prefixes from 10^{-24} to 10^{24}. Which abbreviations are capitalized?

1-3. Write the name and number represented by each symbol. For example, for kW you should write kW = kilowatt = 10^3 watts.

(a) mW	**(c)** kΩ	**(e)** TJ	**(g)** fg
(b) pm	**(d)** μF	**(f)** ns	**(h)** dPa

1-4. Express the following quantities with abbreviations for units and prefixes from Tables 1-1 through 1-3:

(a) 10^{-13} joules	**(d)** 10^{-10} meters
(b) 4.317 28 $\times$ 10^{-8} farads	**(e)** 2.1 $\times$ 10^{13} watts
(c) 2.997 9 $\times$ 10^{14} hertz	**(f)** 48.3 $\times$ 10^{-20} moles

1-5. During the 1980s, the average emission of carbon from burning fossil fuels on Earth was 5.4 petagrams (Pg) of carbon per year in the form of CO_2.[4]

(a) How many kg of C were placed in the atmosphere each year?

(b) How many kg of CO_2 were placed in the atmosphere each year?

(c) A metric ton is 1 000 kg. How many metric tons of CO_2 were placed in the atmosphere each year? If there were 5 billion people on Earth, how many tons of CO_2 were produced for each person?

1-6. How many joules per second and how many calories per hour are produced by a 100.0-horsepower engine?

1-7. A 120-pound woman working in an office consumes about 2.2 $\times$ 10^3 kcal/day, whereas the same woman climbing a mountain needs 3.4 $\times$ 10^3 kcal/day.

(a) Express these numbers in terms of joules per second per kilogram of body mass (= watts per kilogram).

(b) Which consumes more power (watts), the office worker or a 100-W light bulb?

1-8. How many joules per second (J/s) are used by a device that requires 5.00 $\times$ 10^3 British thermal units per hour (Btu/h)? How many watts (W) does this device use?

1-9. The table shows fuel efficiency for several automobiles.

Car model	Fuel consumption (L/100 km)	CO_2 emission (g CO_2/km)
Gasoline Engine		
Peugeot 107	4.6	109
Audi Cabriolet	11.1	266
Chevrolet Tahoe	14.6	346
Diesel Engine		
Peugeot 107	4.1	109
Audi Cabriolet	8.4	223

SOURCE: M. T. Oliver-Hoyo and G. Pinto, "Using the Relationship Between Vehicle Fuel Consumption and CO_2 Emissions to Illustrate Chemical Principles," J. Chem. Ed. **2008**, 85, 218.

(a) A mile is 5 280 feet and a foot is 12 inches. Use Table 1-4 to find how many miles are in 1 km.

(b) The gasoline-engine Peugeot 107 consumes 4.6 L of fuel per 100 km. Express the fuel efficiency in miles per gallon. A U.S. liquid gallon is 3.785 4 L.

(c) The diesel Cabriolet is more efficient than the gasoline Cabriolet. How many metric tons of CO_2 are produced by the diesel and gasoline Cabriolets in 15 000 miles of driving? A metric ton is 1 000 kg.

1-10. Newton's law states that force = mass $\times$ acceleration. Acceleration has the units m/s^2. You also know that energy = force $\times$ distance and pressure = force/area. From these relations, derive the dimensions of newtons, joules, and pascals in terms of the fundamental SI units in Table 1-1. Check your answers in Table 1-2.

1-11. Dust falls on Chicago at a rate of 65 mg m^{-2} day^{-1}. Major metallic elements in the dust include Al, Mg, Cu, Zn, Mn, and Pb.[5] Pb accumulates at a rate of 0.03 mg m^{-2} day^{-1}. How many metric tons (1 metric ton = 1 000 kg) of Pb fall on the 535 square kilometers of Chicago in 1 year?

Chemical Concentrations

1-12. Define the following terms:

(a) molarity	**(e)** volume percent
(b) molality	**(f)** parts per million
(c) density	**(g)** parts per billion
(d) weight percent	**(h)** formal concentration

1-13. Why is it more accurate to say that the concentration of a solution of acetic acid is 0.01 F rather than 0.01 M? (Despite this distinction, we will usually write 0.01 M.)

1-14. What is the formal concentration (expressed as mol/L = M) of NaCl when 32.0 g are dissolved in water and diluted to 0.500 L?

1-15. How many grams of methanol (CH_3OH, FM 32.04) are contained in 0.100 L of 1.71 M aqueous methanol (i.e., 1.71 mol CH_3OH/L solution)?

1-16. The concentration of a gas is related to its pressure by the *ideal gas law:*

$$\text{Concentration}\left(\frac{\text{mol}}{\text{L}}\right) = \frac{n}{V} = \frac{P}{RT},$$

$$R = \text{gas constant} = 0.083\ 14\frac{\text{L·bar}}{\text{mol·K}}$$

where n is the number of moles, V is volume (L), P is pressure (bar), and T is temperature (K).

(a) The maximum pressure of ozone in the Antarctic stratosphere in Figure 1-1 is 19 mPa. Convert this pressure into bars.

(b) Find the molar concentration of ozone in part **(a)** if the temperature is −70°C.

1-17. Any dilute aqueous solution has a density near 1.00 g/mL. Suppose the solution contains 1 ppm of solute; express the concentration of solute in g/L, μg/L, μg/mL, and mg/L.

1-18. The concentration of the alkane $C_{20}H_{42}$ (FM 282.55) in a particular sample of rainwater is 0.2 ppb. Assume that the density of rainwater is close to 1.00 g/mL and find the molar concentration of $C_{20}H_{42}$.

1-19. How many grams of perchloric acid, $HClO_4$, are contained in 37.6 g of 70.5 wt% aqueous perchloric acid? How many grams of water are in the same solution?

1-20. The density of 70.5 wt% aqueous perchloric acid is 1.67 g/mL. Recall that grams refers to grams of *solution* (= g $HClO_4$ + g H_2O).

(a) How many grams of solution are in 1.000 L?

(b) How many grams of $HClO_4$ are in 1.000 L?

(c) How many moles of $HClO_4$ are in 1.000 L?

1-21. An aqueous solution containing 20.0 wt% KI has a density of 1.168 g/mL. Find the molality (m, not M) of the KI solution.

1-22. A cell in your adrenal gland has about 2.5×10^4 tiny compartments called *vesicles* that contain the hormone epinephrine (also called adrenaline).

(a) An entire cell has about 150 fmol of epinephrine. How many attomoles (amol) of epinephrine are in each vesicle?

(b) How many molecules of epinephrine are in each vesicle?

(c) The volume of a sphere of radius r is $\frac{4}{3}\pi r^3$. Find the volume of a spherical vesicle of radius 200 nm. Express your answer in cubic meters (m^3) and liters, remembering that $1\ \text{L} = 10^{-3}\ m^3$.

(d) Find the molar concentration of epinephrine in the vesicle if it contains 10 amol of epinephrine.

1-23. The concentration of sugar (glucose, $C_6H_{12}O_6$) in human blood ranges from about 80 mg/100 mL before meals to 120 mg/100 mL after eating. Find the molarity of glucose in blood before and after eating.

1-24. An aqueous solution of antifreeze contains 6.067 M ethylene glycol ($HOCH_2CH_2OH$, FM 62.07) and has a density of 1.046 g/mL.

(a) Find the mass of 1.000 L of this solution and the number of grams of ethylene glycol per liter.

(b) Find the molality of ethylene glycol in this solution.

1-25. Protein and carbohydrates provide 4.0 Cal/g, whereas fat gives 9.0 Cal/g. (Remember that 1 Calorie, with a capital C, is really 1 kcal.) The weight percent of these components in some foods are

Food	wt% protein	wt% carbohydrate	wt% fat
Shredded wheat	9.9	79.9	—
Doughnut	4.6	51.4	18.6
Hamburger (cooked)	24.2	—	20.3
Apple	—	12.0	—

Calculate the number of calories per gram and calories per ounce in each of these foods. (Use Table 1-4 to convert grams into ounces, remembering that there are 16 ounces in 1 pound.)

1-26. It is recommended that drinking water contain 1.6 ppm fluoride (F^-) for prevention of tooth decay. Consider a reservoir with a diameter of 4.50×10^2 m and a depth of 10.0 m. (The volume is $\pi r^2 h$, where r is the radius and h is the height.) How many grams of F^- should be added to give 1.6 ppm? Fluoride is provided by hydrogen hexafluorosilicate, H_2SiF_6. How many grams of H_2SiF_6 contain this much F^-?

1-27. Noble gases (Group 18 in the periodic table) have the following volume concentrations in dry air: He, 5.24 ppm; Ne, 18.2 ppm; Ar, 0.934%; Kr, 1.14 ppm; Xe, 87 ppb.

(a) A concentration of 5.24 ppm He means 5.24 μL of He per liter of air. Using the ideal gas law in Problem 1-16, find how many moles of He are contained in 5.24 μL at 25.00°C (298.15 K) and 1.000 bar. This number is the molarity of He in the air.

(b) Find the molar concentrations of Ar, Kr, and Xe in air at 25°C and 1 bar.

Preparing Solutions

1-28. How many grams of boric acid, $B(OH)_3$ (FM 61.83), should be used to make 2.00 L of 0.050 0 M solution? What kind of flask is used to prepare this solution?

1-29. Describe how you would prepare approximately 2 L of 0.050 0 *m* boric acid, $B(OH)_3$.

1-30. What is the maximum volume of 0.25 M sodium hypochlorite solution (NaOCl, laundry bleach) that can be prepared by dilution of 1.00 L of 0.80 M NaOCl?

1-31. How many grams of 50 wt% NaOH (FM 40.00) should be diluted to 1.00 L to make 0.10 M NaOH? (Answer with two digits.)

1-32. A bottle of concentrated aqueous sulfuric acid, labeled 98.0 wt% H_2SO_4, has a concentration of 18.0 M.

(a) How many milliliters of reagent should be diluted to 1.000 L to give 1.00 M H_2SO_4?

(b) Calculate the density of 98.0 wt% H_2SO_4.

1-33. What is the density of 53.4 wt% aqueous NaOH (FM 40.00) if 16.7 mL of the solution diluted to 2.00 L gives 0.169 M NaOH?

Stoichiometry Calculations

1-34. How many milliliters of 3.00 M H_2SO_4 are required to react with 4.35 g of solid containing 23.2 wt% $Ba(NO_3)_2$ if the reaction is $Ba^{2+} + SO_4^2 \rightarrow BaSO_4(s)$?

1-35. How many grams of 0.491 wt% aqueous HF are required to provide a 50% excess to react with 25.0 mL of 0.023 6 M Th^{4+} by the reaction $Th^{4+} + 4F^- \rightarrow ThF_4(s)$?

Titrations

1-36. Explain the following statement: "The validity of an analytical result ultimately depends on knowing the composition of some primary standard."

1-37. Distinguish between the terms *end point* and *equivalence point*.

1-38. How does a blank titration reduce titration error?

1-39. What is the difference between a direct titration and a back titration?

1-40. What is the difference between a reagent grade chemical and a primary standard?

1-41. Why are ultrapure acid solvents required to dissolve samples for trace analysis?

1-42. How many milliliters of 0.100 M KI are needed to react with 40.0 mL of 0.040 0 M $Hg_2(NO_3)_2$ if the reaction is $Hg_2^{2+} + 2I^- \rightarrow Hg_2I_2(s)$?

1-43. For Reaction 1-7, how many milliliters of 0.165 0 M $KMnO_4$ are needed to react with 108.0 mL of 0.165 0 M oxalic acid? How many milliliters of 0.165 0 M oxalic acid are required to react with 108.0 mL of 0.165 0 M $KMnO_4$?

1-44. Ammonia reacts with hypobromite, OBr^-, by the reaction $2NH_3 + 3OBr^- \rightarrow N_2 + 3Br^- + 3H_2O$. What is the molarity of a hypobromite solution if 1.00 mL of the OBr^- solution reacts with 1.69 mg of NH_3?

1-45. Sulfamic acid is a primary standard that can be used to standardize NaOH.

$$^+H_3NSO_3^- + OH^- \longrightarrow H_2NSO_3^- + H_2O$$

Sulfamic acid
FM 97.094

What is the molarity of a sodium hydroxide solution if 34.26 mL react with 0.333 7 g of sulfamic acid?

1-46. Limestone consists mainly of the mineral calcite, $CaCO_3$. The carbonate content of 0.541 3 g of powdered limestone was measured by suspending the powder in water, adding 10.00 mL of 1.396 M HCl, and heating to dissolve the solid and expel CO_2:

$$CaCO_3(s) + 2H^+ \longrightarrow Ca^{2+} + CO_2\uparrow + H_2O$$

Calcium carbonate
FM 100.087

The excess acid required 39.96 mL of 0.100 4 M NaOH for complete titration. Find the weight percent of calcite in the limestone.

QUARTZ CRYSTAL MICROBALANCE IN MEDICAL DIAGNOSIS

Slice of quartz (a) used to make microbalance (b). (c) Change in quartz oscillator frequency when 1 μM folate binding protein binds to folate attached to Au surface. After washing, most protein remains irreversibly bound to folate [Photo courtesy Dave Garvey/LapTech. Graph from W. A. Henne, D. D. Doorneweerd, J. Lee, P. S. Low, and C. Savran, "Detection of Folate Binding Protein with Enhanced Sensitivity Using a Functionalized Quartz Crystal Microbalance Sensor," *Anal. Chem.* 2006, *78*, 4880.]

Measuring folate binding protein with quartz crystal microbalance. Folate is attached to Au surface in (d). Folate binding protein then binds to folate in (e). To increase mass, an antibody tethered to a gold nanoparticle binds to folate binding protein in (f).

Folic acid is a vitamin essential to many metabolic pathways. Folate binding protein expressed by cancer cells appears in blood at elevated levels not found in healthy people. Measurement of folate binding protein is a tool for cancer diagnosis, but "elevated" levels are only ~20 pM (pM = picomolar = 10^{-12} M) and are difficult to measure.

A quartz crystal vibrating at its resonant frequency keeps time in your wristwatch. A quartz crystal microbalance consists of a slice of quartz sandwiched between two thin Au electrodes.[1,2] Application of an oscillating electric field causes the quartz to oscillate. Binding of 10 ng (nanograms = 10^{-9} g) of matter to a 1 cm^2 area of a gold electrode lowers the 5-MHz resonant frequency of the quartz by an observable 1 Hz.[3,4] A substance whose dimensions change when an electric field is applied is said to be *piezoelectric*.

To measure folate binding protein, folate is attached to the Au surface in panel (d). When exposed to serum containing folate binding protein, the protein binds to the folate, increasing the mass on the Au surface and decreasing the oscillation frequency of the quartz. The mass of protein is too little to provide a sufficiently low detection level. So, clever chemists then add an antibody that binds specifically to folate binding protein (e). The antibody is covalently attached to a 20-nm-diameter Au particle (f). The combined mass of antibody and Au nanoparticle lowers the detection limit from 30 nM to 50 pM. Another factor-of-10 decrease in detection limit is still required to reach a clinically meaningful level.

FIGURE 2-1 Goggles or safety glasses should be worn at all times in the lab.

Why we wear lab coats: In 2008, 23-year-old University of California research assistant Sheharbano Sangji was withdrawing *t*-butyllithium with a syringe from a bottle. The plunger came out of the syringe and the pyrophoric liquid burst into flames, igniting her sweater and gloves. Burns on 40% of her body proved fatal. A flame-resistant coat might have protected her.

Limitations of gloves: In 1997, Dartmouth College professor Karen Wetterhahn, age 48, died from a drop of dimethylmercury absorbed through her latex gloves. Many organic compounds readily penetrate latex. Wetterhahn was an expert in the biochemistry of metals, the first female professor of chemistry at Dartmouth, and played a major role in bringing more women into science.

Analytical chemistry extends from simple "wet" chemical procedures to elaborate instrumental methods. This chapter describes basic laboratory apparatus and manipulations associated with chemical measurements.[5] We also introduce spreadsheets, which have become essential to everyone who manipulates quantitative data.

2-1 Safe, Ethical Handling of Chemicals and Waste

Chemical experimentation, like driving a car or operating a household, creates hazards. *The primary safety rule is to familiarize yourself with the hazards and then to do nothing that you (or your instructor or supervisor) consider to be dangerous.* If you believe that an operation is hazardous, discuss it first and do not proceed until sensible precautions are in place.

Before working, familiarize yourself with safety features of your laboratory. You should wear goggles (Figure 2-1) or safety glasses with side shields at all times in the lab to protect your eyes from liquids and glass, which fly around when least expected. Contact lenses are not recommended in the lab because vapors can be trapped between the lens and your eye. You can protect your skin from spills and flames by wearing a flame-resistant lab coat. Use rubber gloves when pouring concentrated acids. Do not eat or drink in the lab.

Organic solvents, concentrated acids, and concentrated ammonia should be handled in a fume hood. Air flowing into the hood keeps fumes out of the lab and dilutes the fumes before expelling them from the roof. Never generate large quantities of toxic fumes that are allowed to escape through the hood. Wear a respirator when handling fine powders, which could produce a cloud of dust that might be inhaled.

Clean up spills immediately to prevent accidental contact by the next person who comes along. Treat spills on your skin first by flooding with water. In anticipation of splashes on your body or in your eyes, know where to find and how to operate the emergency shower and eyewash. If the sink is closer than an eyewash, use the sink first for splashes in your eyes. Know how to operate the fire extinguisher and how to use an emergency blanket to extinguish burning clothing. A first aid kit should be available, and you should know how and where to seek emergency medical assistance.

Label all vessels to indicate what they contain. An unlabeled bottle left and forgotten in a refrigerator or cabinet presents an expensive disposal problem, because the contents must be analyzed before they can be legally discarded. National Fire Protection Association labels (Figure 2-2) identify hazards associated with reagents. A Material Safety Data Sheet provided

FIGURE 2-2 Chemical hazards label used by the National Fire Protection Association.

with each chemical sold in the United States lists hazards and precautions for that chemical. It gives first aid procedures and instructions for handling spills.

If we want our grandchildren to inherit a habitable planet, we need to minimize waste production and dispose of chemical waste in a responsible manner. When it is economically feasible, recycling of chemicals is preferable to waste disposal.[6] Carcinogenic dichromate $(Cr_2O_7^{2-})$ waste provides an example of an accepted disposal strategy. Cr(VI) from dichromate should be reduced to the less toxic Cr(III) with sodium hydrogen sulfite ($NaHSO_3$) and precipitated with hydroxide as insoluble $Cr(OH)_3$. The solution is evaporated to dryness and the solid is discarded in an approved landfill that is lined to prevent escape of the chemicals. Wastes such as silver and gold that can be economically recycled should be chemically treated to recover the metal.[7]

Green chemistry is a set of principles intended to change our behavior to help sustain a habitable planet.[8] Examples of unsustainable behavior are consumption of a limited resource and careless disposal of waste. Green chemistry seeks to design chemical products and processes to reduce the use of resources and energy and the generation of hazardous waste. It is better to design a process to prevent waste than to dispose of waste. For example, NH_3 can be measured with an ion-selective electrode instead of using the spectrophotometric Nessler procedure, which generates HgI_2 waste. "Microscale" classroom experiments are encouraged to reduce the cost of reagents and the generation of waste.

Compact fluorescent lights, which reduce energy use, must be treated as hazardous waste because they contain mercury. *Fluorescent bulbs must not be discarded as ordinary waste.* Light-emitting diodes (LEDs) are even more efficient than fluorescent lamps, contain no mercury, and will replace fluorescent lights soon.

2-2 The Lab Notebook

The critical functions of your lab notebook are to state *what you did* and *what you observed*, and it should be *understandable by a stranger*. The greatest error, made even by experienced scientists, is writing incomplete or unintelligible notebooks. Using *complete sentences* is an excellent way to prevent incomplete descriptions.

Beginning students often find it useful to write a complete description of an experiment, with sections dealing with purpose, methods, results, and conclusions. Arranging a notebook to accept numerical data prior to coming to the lab is an excellent way to prepare for an experiment. It is good practice to write a balanced chemical equation for every reaction you use. This practice helps you understand what you are doing and may point out what you do not understand about what you are doing.

The measure of scientific "truth" is the ability of different people to reproduce an experiment. A good lab notebook will state everything that was done and what you observed and will allow you or anyone else to repeat the experiment.

Record in your notebook the names of computer files where programs and data are stored. Paste hard copies of important data into your notebook. The lifetime of a printed page is 10 to 100 times longer than that of a computer file.

The lab notebook must
1. State what was done
2. State what was observed
3. Be understandable to someone else

One fine future day, you or one of your classmates will make an important discovery and will seek a patent. The lab notebook is your legal record of your discovery. For this purpose, each page in your notebook should be signed and dated. Anything of potential importance should also be signed and dated by a second person.

2-3 Analytical Balance

An *electronic balance* uses electromagnetic force compensation to balance the load on the pan. Figure 2-3 shows a typical analytical balance with a capacity of 100–200 g and a readability of 0.01–0.1 mg. *Readability* is the smallest increment of mass that can be indicated. A *microbalance* weighs milligram quantities with a readability of 1 μg.

To weigh a chemical, first place a clean receiving vessel on the balance pan. The mass of the empty vessel is called the **tare**. On most balances, you can press a button to reset the tare to 0. Add the chemical to the vessel and read its mass. If there is no automatic tare operation, subtract the tare mass from that of the filled vessel. To protect the balance from corrosion, *chemicals should never be placed directly on the weighing pan.* Also, be careful never to spill chemicals into the mechanism below the balance pan.

An alternate procedure, called *weighing by difference,* is necessary for **hygroscopic** reagents, which rapidly absorb moisture from the air. First weigh a capped bottle containing dry reagent. Then quickly pour some reagent from the weighing bottle into a receiver. Cap the weighing bottle and weigh it again. The difference is the mass of reagent delivered from the weighing bottle. With an electronic balance, set the initial mass of the weighing bottle to 0 with the tare button. Then deliver reagent from the bottle and reweigh the bottle. The negative reading on the balance is the mass of reagent delivered from the bottle.[9]

FIGURE 2-3 Electronic analytical balance measures mass down to 0.1 mg. [Courtesy Fisher Scientific, Pittsburgh, PA.]

FIGURE 2-4 Equal-long-arm nineteenth-century balance. [Reproduced from *Fresenius' Quantitative Chemical Analysis*, 2nd American ed., 1881.]

How a Mechanical Balance Works

The classic *mechanical balance* in Figure 2-4 has two pans suspended on opposite ends of an equal-arm lever balanced at its center on a knife edge. An unknown mass is placed on one pan and standard masses are placed on the other pan. When the balance is restored to its level position, the mass of the standards is equal to the mass of the unknown.

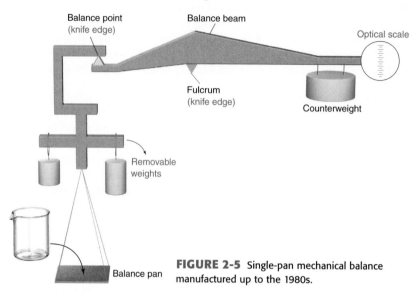

FIGURE 2-5 Single-pan mechanical balance manufactured up to the 1980s.

The single-pan mechanical balance in Figure 2-5 evolved from the two-pan balance and operates by the substitution principle. The mass of the pan at the left is balanced by a counterweight at the right. An object to be weighed is placed on the pan. We then rotate mechanical knobs to detach calibrated internal weights until the balance beam is restored as close as possible to its original position. The remaining small deflection is read on the optical scale and this reading is added to that of the removed weights.

A mechanical balance should be in its arrested position when you load or unload the pan and in the half-arrested position when you are dialing weights. This practice minimizes wear on the knife edges in Figure 2-5.

How an Electronic Balance Works

An object placed on the electronic balance in Figure 2-3 pushes the pan down with a force $m \times g$, where m is the mass of the object and g is the acceleration of gravity. Figure 2-6

FIGURE 2-6 Schematic diagram of electronic balance. [Adapted from C. Berg, *The Fundamentals of Weighing Technology* (Göttingen, Germany: Sartorius AG, 1996).]

Weighing pan

Joint flexure

Null position sensor

Upper guide

Joint flexure

Coil

Servo input

Force-transmitting lever

Servo output to coil

Load receptor (pan support)

Thin linkage

System support

Joint flexure   Lower guide

(a)

Internal, electropolished, nonmagnetic calibration weights

(b)

FIGURE 2-7 (*a*) Mechanical layout of an electronic balance. The lever ratio is such that the electromagnetic force must be only ~10% of the load on the pan. [Adapted from C. Berg, *The Fundamentals of Weighing Technology* (Göttingen, Germany: Sartorius AG, 1996).] (*b*) Internal components of Sartorius analytical balance with a capacity of 300 g and readability of 0.1 mg. The monolithic (one-piece) metal weighing system has nonmagnetic calibration weights, which are placed gently on the load receptor by a microprocessor-activated motor. Calibration is automatically activated by temperature changes. [Courtesy J. Barankewitz, Sartorius AG, Göttingen, Germany.]

shows how the balance works. The pan pushes down on a load receptor attached to parallel guides. The force of the sample pushes the left side of the force-transmitting lever down and moves the right side of the lever up. The null position sensor on the far right detects the smallest movement of the lever arm away from its equilibrium (null) position. When the null sensor detects displacement of the lever arm, the servo amplifier sends electric current through the force compensation wire coil in the field of a permanent magnet. The enlargement at the lower left shows part of the coil and magnet. Electric current in the coil interacts with the magnetic field to produce a downward force. The servo amplifier provides current that exactly compensates for the upward force on the lever arm to maintain a null position. Current flowing through the coil creates a voltage across the precision resistor, which is converted to a digital signal and ultimately to a readout in grams. The conversion between current and mass is accomplished by measuring the current required to balance an internal calibration mass. Figure 2-7 shows the layout of components inside a balance.

Weighing Errors

Samples must be at *ambient temperature* (the temperature of the surroundings) to prevent errors due to convective air currents. Warm samples might appear lighter and cool samples might appear heavier than their actual mass. The warmth of your hand and your fingerprints can affect the apparent mass of an object, so tweezers or a tissue are recommended for placing objects on a balance. A sample that has been dried in an oven takes about 30 min to cool to room temperature. Place the sample in a desiccator during cooling to prevent accumulation of moisture. Close the glass doors of the balance in Figure 2-3 to prevent drafts from affecting the reading. Many top-loading balances have a plastic fence around the pan to protect against drafts. Analytical balances should be located on a heavy table, such as a marble slab, to minimize vibrations. The balance has adjustable feet and a bubble meter that allow you to keep it level. If the balance is not level, force is not transmitted directly down the load receptor in Figure 2-6 and an error results. The balance must be recalibrated after the level is adjusted. Keep the object to be weighed close to the center of the pan.

Errors in weighing magnetic objects might be evident from a change of the indicated mass as the object is moved around on the weighing pan.[10] It is best to weigh magnetic objects on a spacer such as an upside-down beaker to minimize attraction to the stainless steel parts of the balance. Electrostatic charge on the object being weighed interferes with the measurement and could be evident from unidirectional drift of the indicated mass as the object slowly discharges.

TABLE 2-1 Tolerances for laboratory balance weights[a]

Denomination	Tolerance (mg)		Denomination	Tolerance (mg)	
Grams	Class 1	Class 2	Milligrams	Class 1	Class 2
500	1.2	2.5	500	0.010	0.025
200	0.50	1.0	200	0.010	0.025
100	0.25	0.50	100	0.010	0.025
50	0.12	0.25	50	0.010	0.014
20	0.074	0.10	20	0.010	0.014
10	0.050	0.074	10	0.010	0.014
5	0.034	0.054	5	0.010	0.014
2	0.034	0.054	2	0.010	0.014
1	0.034	0.054	1	0.010	0.014

a. Tolerances are defined in ASTM (American Society for Testing and Materials) Standard E 617. Classes 1 and 2 are the most accurate. Larger tolerances exist for Classes 3–6, which are not given in this table.

Analytical balances have built-in calibration. A motor gently places an internal mass on the load receptor beneath the balance pan (see Figure 2-7b). Electric current required to balance this mass is measured. For external calibration, you should periodically weigh standard masses and verify that the reading is within allowed limits. *Tolerances* (allowable deviations) for standard masses are listed in Table 2-1. Another test for a balance is to weigh a standard mass six times and calculate the standard deviation (Section 4-1). Variations are partly from the balance but also reflect factors such as drafts and vibrations.

Linearity error (or *linearity*) of a balance is the maximum error that can occur as a result of nonlinear response of the system to added mass after the balance has been calibrated (Figure 2-8). A balance with a capacity of 220 g and a readability of 0.1 mg might have a linearity of ± 0.2 mg. Even though the scale can be read to 0.1 mg, the error in mass can be as large as ± 0.2 mg in some part of the allowed range.

After a balance is calibrated, the reading might drift if room temperature changes. If a balance has a temperature coefficient of sensitivity of 2 ppm/°C, and temperature changes by 4°C, the apparent mass will change by (4°C)(2 ppm/°C) = 8 ppm. For a mass of 100 g, 8 ppm is $(100 \text{ g})(8 \times 10^{-6}) = 0.8$ mg. You can recalibrate the balance at its current temperature by pressing the calibrate button. For temperature stability, it is best to leave a balance in its standby mode when not in use.

Buoyancy

You can float in water because your weight when swimming is nearly zero. **Buoyancy** is the upward force exerted on an object in a liquid or gaseous fluid.[11] An object weighed in air appears lighter than its actual mass by an amount equal to the mass of air that it displaces. True mass is the mass measured in vacuum. A standard mass in a balance is also affected by buoyancy, so it weighs less in air than in vacuum. A buoyancy error occurs whenever the density of the object being weighed is not equal to the density of the standard mass.

FIGURE 2-8 *Linearity error.* Dashed line is ideal, linear response proportional to mass on the balance, which has been calibrated at 0 and 200 g. Real response deviates from the straight line. Linearity error is the maximum deviation, which is greatly exaggerated in this drawing.

If mass m' is read on a balance, the true mass m of the object weighed in vacuum is[12]

Buoyancy equation:

$$m = \frac{m'\left(1 - \dfrac{d_a}{d_w}\right)}{\left(1 - \dfrac{d_a}{d}\right)} \qquad (2\text{-}1)$$

where d_a is the density of air (0.001 2 g/mL near 1 bar and 25°C),[13] d_w is the density of the calibration weights (8.0 g/mL), and d is the density of the object being weighed.

EXAMPLE **Buoyancy Correction**

A pure compound called "tris" is used as a *primary standard* to measure concentrations of acids. The volume of acid that reacts with a known mass of tris tells us the concentration of acid. Find the true mass of tris (density = 1.33 g/mL) if the apparent mass weighed in air is 100.00 g.

Solution If the density of air is 0.001 2 g/mL, we find the true mass by using Equation 2-1:

$$m = \frac{100.00\ g\left(1 - \dfrac{0.001\ 2\ g/mL}{8.0\ g/mL}\right)}{1 - \dfrac{0.001\ 2\ g/mL}{1.33\ g/mL}} = 100.08\ g$$

Unless we correct for buoyancy, we would think that the mass of tris is 0.08% less than its actual mass and we would think that the molarity of acid reacting with the tris is 0.08% less than the actual molarity.

FIGURE 2-10 Glass buret with Teflon stopcock. Enlargement shows meniscus at 9.68 mL. Estimate the reading of any scale to the nearest tenth of a division. This buret has 0.1-mL divisions, so we estimate the reading to the nearest 0.01 mL.

FIGURE 2-9 Buoyancy correction, assuming d_a = 0.001 2 g/mL and d_w = 8.0 g/mL. The apparent mass measured in air (1.000 0 g) is multiplied by the buoyancy correction to find the true mass.

Figure 2-9 shows buoyancy corrections for several substances. When you weigh water with a density of 1.00 g/mL, the true mass is 1.001 1 g when the balance reads 1.000 0 g. The error is 0.11%. For NaCl with a density of 2.16 g/mL, the error is 0.04%; and for $AgNO_3$ with a density of 4.45 g/mL, the error is only 0.01%.

2-4 Burets

The **buret** in Figure 2-10 is a precisely manufactured glass tube with graduations enabling you to measure the volume of liquid delivered through the stopcock (the valve) at the bottom. The 0-mL mark is near the top. If the initial liquid level is 0.83 mL and the final level is 27.16 mL, then you have delivered 27.16 − 0.83 = 26.33 mL. Class A burets (the most accurate grade) are certified to meet tolerances in Table 2-2. If the reading of a 50-mL buret is 27.16 mL, the true volume can be anywhere in the range 27.21 to 27.11 mL and still be within the tolerance of ±0.05 mL.

When reading the liquid level in a buret, your eye should be at the same height as the top of the liquid. If your eye is too high, the liquid seems to be higher than it really is. If your eye

TABLE 2-2 Tolerances of Class A burets

Buret volume (mL)	Smallest graduation (mL)	Tolerance (mL)
5	0.01	±0.01
10	0.05 or 0.02	±0.02
25	0.1	±0.03
50	0.1	±0.05
100	0.2	±0.10

Operating a buret:

- Wash buret with new solution
- Eliminate air bubble before use
- Drain liquid slowly
- Deliver fraction of a drop near end point
- Read bottom of concave meniscus
- Estimate reading to 1/10 of a division
- Avoid parallax
- Account for graduation thickness in readings

In a *titration,* increments of reagent in the buret are added to analyte until reaction is complete. From the volume delivered, we calculate the quantity of analyte.

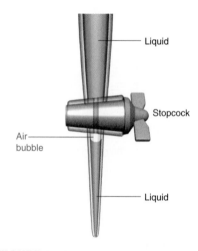

FIGURE 2-11 An air bubble trapped beneath the stopcock should be expelled before you use the buret.

FIGURE 2-12 Autotitrator delivers reagent from the bottle at the left to the beaker at the right. The electrode immersed in the beaker monitors pH or the concentrations of specific ions. Volume and pH readings can go directly to a spreadsheet. [Schott Instruments, Mainz, Germany, and Cole-Parmer Instruments, Vernon Hills, IL.]

is too low, the liquid appears too low. The error that occurs when your eye is not at the same height as the liquid is called **parallax.**

The surface of most liquids forms a concave **meniscus** like that shown at the right side of Figure 2-10.[14] It is helpful to use black tape on a white card as a background for locating the precise position of the meniscus. Move the black strip up the buret to approach the meniscus. The bottom of the meniscus turns dark as the black strip approaches, thus making the meniscus more easily readable. Highly colored solutions may appear to have two meniscuses; either one may be used. Because volumes are determined by subtracting one reading from another, the important point is to read the position of the meniscus reproducibly. Estimate the reading to the nearest tenth of a division between marks.

The thickness of the markings on a 50-mL buret corresponds to about 0.02 mL. For best accuracy, select one portion of the marking to be called zero. For example, you can say that the liquid level is *at* the mark when the bottom of the meniscus just touches the top of the mark. When the meniscus is at the *bottom* of the same mark, the reading is 0.02 mL greater.

For precise location of the end of a *titration,* we deliver less than one drop at a time from the buret near the end point. (A drop from a 50-mL buret is about 0.05 mL.) To deliver a fraction of a drop, carefully open the stopcock until part of a drop is hanging from the buret tip. (Some people prefer to rotate the stopcock rapidly through the open position to expel part of a drop.) Then touch the inside glass wall of the receiving flask to the buret tip to transfer the droplet to the wall of the flask. Carefully tip the flask so that the main body of liquid washes over the newly added droplet. Swirl the flask to mix the contents. Near the end of a titration, tip and rotate the flask often to ensure that droplets on the wall containing unreacted analyte contact the bulk solution.

Liquid should drain evenly down the wall of a buret. The tendency of liquid to stick to glass is reduced by draining the buret slowly (<20 mL/min). If many droplets stick to the wall, then clean the buret with detergent and a buret brush. If this cleaning is insufficient, soak the buret in peroxydisulfate–sulfuric acid cleaning solution,[15] which eats clothing and people as well as grease in the buret. Never soak volumetric glassware in alkaline solutions, which attack glass. A 5 wt% NaOH solution at 95°C dissolves Pyrex glass at a rate of 9 μm/h.

Error can be caused by failure to expel the bubble of air often found directly beneath the stopcock (Figure 2-11). If the bubble becomes filled with liquid during the titration, then some volume that drained from the graduated portion of the buret did not reach the titration vessel. The bubble can be dislodged by draining the buret for a second or two with the stopcock wide open. You can expel a tenacious bubble by abruptly shaking the buret while draining it into a sink.

Before you fill a buret with fresh solution, it is a wonderful idea to rinse the buret several times with small portions of the new solution, discarding each wash. It is not necessary to fill the buret with wash solution. Simply tilt the buret to allow all surfaces to contact the wash liquid. This same technique should be used with any vessel (such as a spectrophotometer cuvet or a pipet) that is reused without drying.

The labor of conducting a titration is greatly reduced by using an autotitrator (Figure 2-12) instead of a buret. This device delivers reagent from a reservoir and records the volume of reagent and the response of an electrode immersed in the solution being titrated. Output can go directly to a computer for manipulation in a spreadsheet.

Mass Titrations and Microscale Titrations

For best precision, measure the *mass* of reagent, instead of the volume, delivered from a buret or syringe.[16] Mass can be measured more precisely than can volume.

For procedures that can tolerate poor precision, "microscale" student experiments reduce consumption of reagents and generation of waste. An inexpensive student buret can be constructed from a 2-mL pipet graduated in 0.01-mL intervals.[17] Volume can be read to 0.001 mL, and titrations can be carried out with a precision of 1%.

Precision refers to reproducibility.

2-5 Volumetric Flasks

A **volumetric flask** is calibrated to contain a particular volume of solution at 20°C when the bottom of the meniscus is adjusted to the center of the mark on the neck of the flask (Figure 2-13, Table 2-3). Most flasks bear the label "TC 20°C," which means *to contain* at 20°C. (Pipets and burets are calibrated *to deliver*, "TD," their indicated volume.) The temperature of the container is relevant because both liquid and glass expand when heated.

To use a volumetric flask, dissolve the desired mass of reagent in the flask by swirling with *less* than the final volume of liquid. Then add more liquid and swirl the solution again. Adjust the final volume with as much well-mixed liquid in the flask as possible. (When two different liquids are mixed, there is generally a small volume change. The total volume is *not* the sum of the two volumes that were mixed. By swirling the liquid in a nearly full volumetric flask before liquid reaches the thin neck, you minimize the change in volume when the last liquid is added.) For good control, add the final drops of liquid with a pipet, *not a squirt bottle*. After adjusting the liquid to the correct level, hold the cap firmly in place and invert the flask several times to complete mixing. Before the liquid is homogeneous, we observe streaks (called *schlieren*) arising from regions that refract light differently. After the schlieren are gone, invert the flask a few more times to ensure complete mixing.

Figure 2-13 shows how liquid appears when it is at the *center* of the mark of a volumetric flask or a pipet. Adjust the liquid level while viewing the flask from above or below the level of the mark. The front and back of the mark describe an ellipse with the meniscus at the center.

Glass is notorious for *adsorbing* traces of chemicals—especially cations. **Adsorption** is the process in which a substance sticks to a surface. (In contrast, **absorption** is the process in which a substance is taken inside another, as water is taken into a sponge.) For critical work, you should **acid wash** glassware to replace low concentrations of cations on the surface with H^+. To do this, soak already thoroughly cleaned glassware in 3–6 M HCl or HNO_3 (in a fume hood) for >1 h. Then rinse it well with distilled water and, finally, soak it in distilled water. Acid can be

Volumetric glassware made of Pyrex, Kimax, or other low-expansion glass can be safely dried in an oven heated to at least 320°C without harm,[18] although there is rarely reason to go above 150°C.

Example of acid washing: High-purity HNO_3 delivered from an acid-washed glass pipet had no detectable level of Ti, Cr, Mn, Fe, Co, Ni, Cu, and Zn (<0.01 ppb). The same acid delivered from a clean—but not acid-washed—pipet contained each metal at a level of 0.5 to 9 ppb.[19]

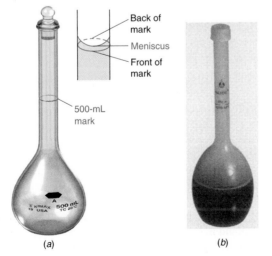

(a)   (b)

TABLE 2-3 Tolerances of Class A volumetric flasks

Flask capacity (mL)	Tolerance (mL)
1	±0.02
2	±0.02
5	±0.02
10	±0.02
25	±0.03
50	±0.05
100	±0.08
200	±0.10
250	±0.12
500	±0.20
1 000	±0.30
2 000	±0.50

FIGURE 2-13 (a) Class A glass volumetric flask showing proper position of the meniscus—at the center of the ellipse formed by the front and back of the calibration mark when viewed from above or below. Volumetric flasks and transfer pipets are calibrated to this position. (b) Class B polypropylene plastic volumetric flask for trace analysis. [Courtesy Fisher Scientific, Pittsburgh, PA.] Class A flasks meet tolerances of Table 2-3. Class B tolerances are twice as big as Class A tolerances.

reused many times, as long as it is only used for clean glassware. Acid washing is *especially* appropriate for new glassware, which you should assume is not clean. The polypropylene plastic volumetric flask in Figure 2-13b is designed for trace analysis (parts per billion concentrations) in which analyte might be lost by adsorption on the walls of a glass flask.

2-6 Pipets and Syringes

Pipets deliver known volumes of liquid. The *transfer pipet* in Figure 2-14a is calibrated to deliver one fixed volume. The last drop does not drain out of the pipet and *should not be blown out*. The *measuring pipet* in Figure 2-14b is calibrated like a buret. It is used to deliver a variable volume, such as 5.6 mL by starting delivery at the 1.0-mL mark and terminating at the 6.6-mL mark. The transfer pipet is more accurate, with tolerances listed in Table 2-4.

FIGURE 2-14 (*a*) Transfer pipet and (*b*) measuring (Mohr) pipet. [Courtesy A. H. Thomas Co., Philadelphia, PA.]

Using a Transfer Pipet

Using a rubber bulb or other pipet suction device, *not your mouth*, suck liquid up past the calibration mark. Discard one or two pipet volumes of liquid to rinse traces of previous reagents from the pipet. After taking up a third volume past the calibration mark, quickly replace the bulb with your index finger at the end of the pipet. Gently pressing the pipet against the bottom of the vessel while removing the rubber bulb helps prevent liquid from draining below the mark while you put your finger in place. (Alternatively, you can use an automatic suction device that remains attached to the pipet.) Wipe the excess liquid off the outside of the pipet with a clean tissue. *Touch the tip of the pipet to the side of a beaker* and drain the liquid until the bottom of the meniscus just reaches the center of the mark, as in Figure 2-13. Touching the beaker draws liquid from the pipet without leaving part of a drop hanging when the liquid reaches the calibration mark.

Transfer the pipet to a receiving vessel and drain it by gravity *while holding the tip against the wall of the vessel.* After liquid stops draining, hold the pipet to the wall for a few more seconds to complete draining. *Do not blow out the last drop.* The pipet should be nearly vertical at the end of delivery. When you finish with a pipet, you should rinse it with distilled water or soak it until you are ready to clean it. Solutions should never be allowed to dry inside a pipet because removing internal deposits is very difficult.

Micropipets

Micropipets (Figure 2-15) deliver volumes of 1 to 1 000 μL (1 μL = 10^{-6} L). Liquid is contained in the disposable polypropylene tip, which is stable to most aqueous solutions and many organic solvents except chloroform ($CHCl_3$). The tip also is not resistant to concentrated nitric or sulfuric acids. To prevent aerosols from entering the pipet shaft, tips are available with polyethylene filters. Aerosols can corrode mechanical parts of the pipet or cross-contaminate biological experiments.

To use a micropipet, place a fresh tip tightly on the barrel. Keep tips in their package or dispenser so that you do not contaminate the tips with your fingers. Set the desired volume with the knob at the top of the pipet. Depress the plunger to the first stop, which corresponds to the selected volume. Hold the pipet *vertically*, dip it 3–5 mm into the reagent solution, and *slowly* release the plunger to suck up liquid. Leave the tip in the liquid for a few seconds to allow the aspiration of liquid into the tip to go to completion. Withdraw the pipet vertically from the liquid without touching the tip to the side of the vessel. The volume of liquid taken into the tip depends on the angle at which the pipet is held and how far beneath the liquid surface the tip is held during uptake. To dispense liquid, touch the tip to the wall of the receiver and gently depress the plunger to the first stop. Wait a few seconds to allow liquid to drain down the tip, and then depress the plunger further to squirt out the last liquid. Prior to dispensing liquid, clean and wet a fresh tip by taking up and discarding three squirts of

---

Do not blow the last drop out of a transfer pipet.

TABLE 2-4 Tolerances of Class A transfer pipets

Volume (mL)	Tolerance (mL)
0.5	±0.006
1	±0.006
2	±0.006
3	±0.01
4	±0.01
5	±0.01
10	±0.02
15	±0.03
20	±0.03
25	±0.03
50	±0.05
100	±0.08

An *aerosol* is a suspension of fine liquid droplets or solid particles in the gas phase.

Sources of error for micropipets:[20]

- Use tip recommended by manufacturer. Other tips might make inadequate seal.
- Take up and expel liquid three times before delivery to wet pipet tip and equilibrate the inside with vapor.
- Unnecessary wiping of the tip can cause loss of sample.
- Liquid must be at same temperature as pipet. Less than the indicated volume of cold liquid is delivered and more than the indicated volume of warm liquid is delivered. Errors are greatest for smallest volumes.
- Micropipets are calibrated at sea level pressure. They are out of calibration at higher elevations. Errors are greatest for smallest volumes. Calibrate your pipet at your elevation.

reagent. The tip can be discarded or rinsed well with a squirt bottle and reused. A tip with a filter (Figure 2-15b) cannot be cleaned for reuse.

The procedure we just described for *aspirating* (sucking in) and delivering liquids is called "forward mode." The plunger is depressed to the first stop and liquid is then taken up. To expel liquid, the plunger is depressed beyond the first stop. In "reverse mode," the plunger is depressed beyond the first stop and *excess* liquid is taken in. To deliver the correct volume, depress the plunger to the first stop and *not beyond*. Reverse mode is good for foamy (protein or surfactant solutions) and viscous (syrupy) liquids.[21]

Table 2-5 lists tolerances for micropipets from one manufacturer. As internal parts wear out, both precision and accuracy can decline by an order of magnitude. In a study[22] of 54 micropipets in use at a biomedical lab, 12 were accurate and precise to ≤1%. Five of 54 had errors ≥10%. When 54 quality control technicians at four pharmaceutical companies used a properly functioning micropipet, 10 people were accurate and precise to ≤1%. Six were inaccurate by ≥10%. Micropipets require periodic calibration and maintenance (cleaning, seal replacement, and lubrication), and operators require certification. If the mean time between falling out of tolerance for micropipets is 2 years, calibration is required every *2 months* to be confident that 95% of the micropipets in a laboratory are operating within specifications.[23] You can calibrate a micropipet by measuring the mass of water it delivers, as described in Section 2-9, or with a commercial colorimetric kit.[24]

Piston

Shaft

Filter

Aerosol

Disposable
polypropylene
tip

(a)   (b)

TABLE 2-5   **Manufacturer's tolerances for micropipets**

Pipet volume (μL)	At 10% of pipet volume		At 100% of pipet volume	
	Accuracy (%)	Precision (%)	Accuracy (%)	Precision (%)
Adjustable Volume				
0.2–2	±8	±4	±1.2	±0.6
1–10	±2.5	±1.2	±0.8	±0.4
2.5–25	±4.5	±1.5	±0.8	±0.2
10–100	±1.8	±0.7	±0.6	±0.15
30–300	±1.2	±0.4	±0.4	±0.15
100–1 000	±1.6	±0.5	±0.3	±0.12
Fixed Volume				
10			±0.8	±0.4
25			±0.8	±0.3
100			±0.5	±0.2
500			±0.4	±0.18
1 000			±0.3	±0.12

SOURCE: *Data from Hamilton Co., Reno, NV.*

(c)

FIGURE 2-15 (*a*) Micropipet with disposable plastic tip. (*b*) Enlarged view of disposable tip containing polyethylene filter to prevent aerosol from contaminating the shaft of the pipet. (*c*) Volume selection dial set to 150 μL. [Courtesy Rainin Instrument Co., Emeryville, CA.]

Syringes

Microliter *syringes*, such as that in Figure 2-16, come in sizes from 1 to 500 μL and have an accuracy and precision near 1%. When using a syringe, take up and discard several volumes of liquid to wash the glass walls and to remove air bubbles from the barrel. The steel needle is attacked by strong acid and will contaminate strongly acidic solutions with iron. A syringe is more reliable than a micropipet, but the syringe requires more care in handling and cleaning.

Accuracy refers to nearness to the true value. *Precision* refers to reproducibility.

Needle   Barrel   Plunger

FIGURE 2-16 Hamilton syringe with a volume of 1 μL and divisions of 0.02 μL on the glass barrel. [Courtesy Hamilton Co., Reno, NV.]

2-7 Filtration

In *gravimetric analysis*, the mass of product from a reaction is measured to determine how much unknown was present. Precipitates from gravimetric analyses are collected by filtration, washed, and then dried. Most precipitates are collected in a *fritted-glass funnel* (also called a Gooch filter crucible), with suction applied to speed filtration (Figure 2-17). The porous glass plate in the funnel allows liquid to pass but retains solids. The empty funnel is first dried at 110°C and weighed. After collecting solid and drying again, the funnel and its contents are weighed a second time to determine the mass of collected solid. Liquid from which a substance precipitates or crystallizes is called the **mother liquor.** Liquid that passes through the filter is called **filtrate.**

Gooch filter crucible

Porous glass

Gooch filter crucible

Rubber adaptor

Glass funnel

To air

To house vacuum

Suction flask

Trap

FIGURE 2-17 Filtration with a Gooch crucible that has a porous (*fritted*) glass disk through which liquid can pass. The trap prevents liquid from being accidentally sucked into the vacuum system.

In some gravimetric procedures, **ignition** (heating at high temperature over a burner or in a furnace) is used to convert a precipitate into a known, constant composition. For example, Fe^{3+} precipitates as hydrous ferric oxide, $FeOOH \cdot xH_2O$, with variable composition. Ignition converts it into pure Fe_2O_3 prior to weighing. When a precipitate is to be ignited, it is collected in **ashless filter paper,** which leaves little residue when burned.

To use filter paper with a conical glass funnel, fold the paper into quarters, tear off one corner (to allow a firm fit into the funnel), and place the paper in the funnel (Figure 2-18). The filter paper should fit snugly and be seated with some distilled water. When liquid is poured in, an unbroken stream of liquid should fill the stem of the funnel (Figure 2-19). The weight of liquid in the stem helps speed filtration.

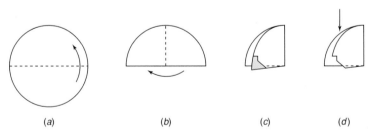

(a)   (b)   (c)   (d)

FIGURE 2-18 Folding filter paper for a conical funnel. (*a*) Fold the paper in half. (*b*) Then fold it in half again. (*c*) Tear off a corner to allow better seating of the paper in the funnel. (*d*) Open the side that was not torn when fitting the paper in the funnel.

For filtration, pour the slurry of precipitate down a glass rod to prevent splattering (Figure 2-19). (A **slurry** is a suspension of solid in liquid.) Particles adhering to the beaker or rod can be dislodged with a *rubber policeman*, which is a flattened piece of rubber at the end of a glass rod. Use a jet of appropriate wash liquid from a squirt bottle to transfer particles from the rubber and glassware to the filter. If the precipitate is going to be ignited, particles remaining in the beaker should be wiped onto a small piece of moist filter paper. Add that paper to the filter to be ignited.

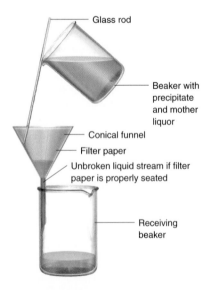

Glass rod

Beaker with precipitate and mother liquor

Conical funnel

Filter paper

Unbroken liquid stream if filter paper is properly seated

Receiving beaker

FIGURE 2-19 Filtering a precipitate. The conical funnel is supported by a metal ring attached to a ring stand, neither of which is shown.

Watchglass   Bent glass hook

Weighing bottle
with cap ajar

Beaker   Reagent

FIGURE 2-20 Use a watchglass as a dust cover while drying reagents or crucibles in the oven.

(a)   (b)

FIGURE 2-21 (a) Ordinary desiccator. (b) Vacuum desiccator that can be evacuated through the side arm at the top and then sealed by rotating the joint containing the side arm. Drying is more efficient at low pressure. [Courtesy A. H. Thomas Co., Philadelphia, PA.]

2-8 Drying

Reagents, precipitates, and glassware are conveniently dried in an oven at 110°C. (Some chemicals require other temperatures.) *Anything that you put in the oven should be labeled.* Use a beaker and watchglass (Figure 2-20) to minimize contamination by dust during drying. It is good practice to cover all vessels on the benchtop to prevent dust contamination.

The mass of a gravimetric precipitate is measured by weighing a dry, empty filter crucible before the procedure and reweighing the same crucible filled with dry product after the procedure. To weigh the empty crucible, first bring it to "constant mass" by drying it in the oven for 1 h or longer and then cooling it for 30 min in a desiccator. Weigh the crucible and then heat it again for about 30 min. Cool it and reweigh it. When successive weighings agree to ± 0.3 mg, the filter has reached "constant mass." If the crucible is warm when it is weighed, it creates convection currents that give a false weight. You can use a microwave oven instead of an electric oven for drying reagents and crucibles. Try an initial heating time of 4 min, with subsequent 2-min heatings. Use a 15-min cooldown before weighing.

A **desiccator** (Figure 2-21) is a closed chamber containing a drying agent called a **desiccant** (Table 2-6). The lid is greased to make an airtight seal, and desiccant is placed in the bottom beneath the perforated disk. Another useful desiccant that is not in the table is 98 wt% sulfuric acid. After placing a hot object in the desiccator, leave the lid cracked open for a minute until the object has cooled slightly. This practice prevents the lid from popping open when the air inside warms up. To open a desiccator, slide the lid sideways rather than trying to pull it straight up.

Dust is a source of contamination in all experiments, so . . .

> **Cover all vessels whenever possible.**

TABLE 2-6 Efficiencies of drying agents

Agent	Formula	Water left in atmosphere ($\mu g\ H_2O/L$)[a]
Magnesium perchlorate, anhydrous	$Mg(ClO_4)_2$	0.2
"Anhydrone"	$Mg(ClO_4)_2 \cdot 1-1.5H_2O$	1.5
Barium oxide	BaO	2.8
Alumina	Al_2O_3	2.9
Phosphorus pentoxide	P_4O_{10}	3.6
Calcium sulfate (Drierite)[b]	$CaSO_4$	67
Silica gel	SiO_2	70

a. Moist nitrogen was passed over each desiccant, and the water remaining in the gas was condensed and weighed. [A. I. Vogel, A Textbook of Quantitative Inorganic Analysis, 3rd ed. (New York: Wiley. 1961), p. 178.] For drying gases, the gas can be passed through a 60-cm-long Nafion tube. At 25°C, the residual moisture is 10 μg/L. If the drier is held at 0°C, the residual moisture is 0.8 μg/L. [K. J. Leckrone and J. M. Hayes, "Efficiency and Temperature Dependence of Water Removal by Membrane Dryers," Anal. Chem. **1997,** 69, 911.]

b. Used Drierite can be regenerated by irradiating 1.5-kg batches in a 100 × 190 mm Pyrex crystallizing dish in a microwave oven for 15 min. Stir the solid, heat a second time for 15 min. and place the hot, dry material back in its original container. Use small glass spacers between the crystallizing dish and the glass tray of the oven to protect the tray. [J. A. Green and R. W. Goetz, "Recycling Drierite," J. Chem. Ed. **1991,** 68, 429.]

2-9 Calibration of Volumetric Glassware

Each instrument that we use has a scale of some sort to measure a quantity such as mass, volume, force, or electric current. Manufacturers usually certify that the indicated quantity lies within a certain *tolerance* from the true quantity. For example, a Class A transfer pipet is certified to deliver 10.00 ± 0.02 mL when you use it properly. Your individual pipet might always deliver 10.016 ± 0.004 mL in a series of trials. That is, your pipet delivers an average of 0.016 mL more than the indicated volume in repeated trials. **Calibration** is the process of measuring the actual quantity that corresponds to an indicated quantity on the scale of an instrument.

For greatest accuracy, we calibrate volumetric glassware to measure the volume actually contained in or delivered by a particular piece of equipment. We do this by measuring the mass of water contained or delivered by the vessel and using the density of water to convert mass into volume.

In the most careful work, it is necessary to account for thermal expansion of solutions and glassware with changing temperature. For this purpose, you should know the lab temperature when a solution was prepared and when it is used. Table 2-7 shows that water expands 0.02% per degree near 20°C. Because the concentration of a solution is proportional to its density, we can write

Correction for thermal expansion: $$\frac{c'}{d'} = \frac{c}{d} \tag{2-2}$$

where c' and d' are the concentration and density at temperature T', and c and d apply at temperature T.

Page 49 gives a detailed procedure for calibrating a buret.

Concentration decreases when the temperature increases.

TABLE 2-7 Density of water

Temperature (°C)	Density (g/mL)	Volume of 1 g of water (mL)	
		At temperature shown[a]	Corrected to 20°C[b]
10	0.999 702 6	1.001 4	1.001 5
11	0.999 608 4	1.001 5	1.001 6
12	0.999 500 4	1.001 6	1.001 7
13	0.999 380 1	1.001 7	1.001 8
14	0.999 247 4	1.001 8	1.001 9
15	0.999 102 6	1.002 0	1.002 0
16	0.998 946 0	1.002 1	1.002 1
17	0.998 777 9	1.002 3	1.002 3
18	0.998 598 6	1.002 5	1.002 5
19	0.998 408 2	1.002 7	1.002 7
20	0.998 207 1	1.002 9	1.002 9
21	0.997 995 5	1.003 1	1.003 1
22	0.997 773 5	1.003 3	1.003 3
23	0.997 541 5	1.003 5	1.003 5
24	0.997 299 5	1.003 8	1.003 8
25	0.997 047 9	1.004 0	1.004 0
26	0.996 786 7	1.004 3	1.004 2
27	0.996 516 2	1.004 6	1.004 5
28	0.996 236 5	1.004 8	1.004 7
29	0.995 947 8	1.005 1	1.005 0
30	0.995 650 2	1.005 4	1.005 3

a. Corrected for buoyancy with Equation 2-1.

b. Corrected for buoyancy and expansion of borosilicate glass (0.001 0% K⁻¹).

b. Corrected for buoyancy and expansion of borosilicate glass ($0.001\ 0\%\ K^{-1}$).

EXAMPLE Effect of Temperature on Solution Concentration

A 0.031 46 M aqueous solution was prepared in winter when the lab temperature was 17°C. What is the molarity of the solution on a warm day when the temperature is 25°C?

Solution We assume that the thermal expansion of a dilute solution is equal to the thermal expansion of pure water. Then, using Equation 2-2 and densities from Table 2-7, we write

$$\frac{c' \text{ at } 25°C}{0.997\ 05 \text{ g/mL}} = \frac{0.031\ 46 \text{ M}}{0.998\ 78 \text{ g/mL}} \implies c' = 0.031\ 41 \text{ M}$$

The concentration has decreased by 0.16% on the warm day.

Pyrex and other borosilicate glasses expand by 0.001 0% per degree near room temperature. If the temperature increases by 10°C, the volume of a piece of glassware increases by (10°C)(0.001 0%/°C) = 0.010%. For most work, this expansion is insignificant.

To calibrate a 25-mL transfer pipet, first weigh an empty weighing bottle like the one in Figure 2-20. Then fill the pipet to the mark with distilled water, drain it into the weighing bottle, and cap the bottle to prevent evaporation. Weigh the bottle again to find the mass of water delivered from the pipet. Finally, use Equation 2-3 to convert mass into volume.

> Small or odd-shaped vessels can be calibrated with Hg, which is easier than water to pour out of glass and is 13.6 times denser than water. This procedure is for researchers, not students.

$$\text{True volume} = (\text{grams of water}) \times (\text{volume of 1 g of } H_2O \text{ in Table 2-7}) \qquad (2\text{-}3)$$

EXAMPLE Calibration of a Pipet

An empty weighing bottle had a mass of 10.313 g. After the addition of water from a 25-mL pipet, the mass was 35.225 g. If the lab temperature was 27°C, find the volume of water delivered by the pipet.

Solution The mass of water is 35.225 − 10.313 = 24.912 g. From Equation 2-3 and the next-to-last column of Table 2-7, the volume of water is (24.912 g)(1.004 6 mL/g) = 25.027 mL at 27°C. The last column in Table 2-7 tells us what the volume would be if the pipet were at 20°C. This pipet would deliver (24.912 g)(1.004 5 mL/g) = 25.024 mL at 20°C.

> The pipet delivers less volume at 20°C than at 27°C because glass contracts slightly as the temperature is lowered. Volumetric glassware is usually calibrated at 20°C.

2-10 Introduction to Microsoft Excel®

If you already use a spreadsheet, you can skip this section. The computer spreadsheet is an essential tool for manipulating quantitative information. In analytical chemistry, spreadsheets can help us with calibration curves, statistical analysis, titration curves, and equilibrium problems. Spreadsheets allow us to conduct "what if" experiments such as investigating the effect of a stronger acid or a different ionic strength on a titration curve. We use Microsoft Excel in this book as a tool for solving problems in analytical chemistry.[25] Although you can skip over spreadsheets with no loss of continuity, spreadsheets will enrich your understanding of chemistry and provide a valuable tool for use outside this course.

Getting Started: Calculating the Density of Water

Let's prepare a spreadsheet to compute the density of water from the equation

$$\text{Density (g/mL)} = a_0 + a_1*T + a_2*T^2 + a_3*T^3 \qquad (2\text{-}4)$$

where T is temperature (°C) and $a_0 = 0.999\ 89$, $a_1 = 5.332\ 2 \times 10^{-5}$, $a_2 = -7.589\ 9 \times 10^{-6}$, and $a_3 = 3.671\ 9 \times 10^{-8}$.

> This equation is accurate to five decimal places over the range 4° to 40°C.

The blank spreadsheet in Figure 2-22a has columns labeled A, B, C and rows numbered 1, 2, 3, . . . , 12. The box in column B, row 4 is called *cell* B4.

Begin each spreadsheet with a title to help make the spreadsheet more readable. In Figure 2-22b, we click in cell A1 and type "Calculating Density of H2O with Equation 2-4". Then we click in cell A2 and write "(from the delightful book by Dan Harris)" without quotation marks. The computer automatically spreads the text to adjoining cells. To save your worksheet, click on the Office button at the upper left and select Save As. Give your

Columns

(a)

	A	B	C
1			
2			
3			
4		cell B4	
5			
6			
7			
8			
9			
10			
11			
12			

Rows

(b)

	A	B	C
1	Calculating Density of H2O with Equation 2-4		
2	(from the delightful book by Dan Harris)		
3			
4	Constants:		
5	a0 =		
6	0.99989		
7	a1 =		
8	5.3322E-05		
9	a2 =		
10	-7.5899E-06		
11	a3 =		
12	3.6719E-08		

(c)

	A	B	C
1	Calculating Density of H2O with Equation 2-4		
2	(from the delightful book by Dan Harris)		
3			
4	Constants:	Temp (°C)	Density (g/mL)
5	a0 =	5	0.99997
6	0.99989	10	
7	a1 =	15	
8	5.3322E-05	20	
9	a2 =	25	
10	-7.5899E-06	30	
11	a3 =	35	
12	3.6719E-08	40	

(d)

	A	B	C
1	Calculating Density of H2O with Equation 2-4		
2	(from the delightful book by Dan Harris)		
3			
4	Constants:	Temp (°C)	Density (g/mL)
5	a0 =	5	0.99997
6	0.99989	10	0.99970
7	a1 =	15	0.99911
8	5.3322E-05	20	0.99821
9	a2 =	25	0.99705
10	-7.5899E-06	30	0.99565
11	a3 =	35	0.99403
12	3.6719E-08	40	0.99223
13			
14	Formula:		
15	C5 = \$A\$6+\$A\$8*B5+\$A\$10*B5^2+\$A\$12*B5^3		

FIGURE 2-22 Evolution of a spreadsheet for computing the density of water.

spreadsheet a descriptive name that will tell you what is in it long after you have forgotten about it. File it in a location that you can find in the future. Information in your computer is only as good as your ability to retrieve it. In earlier versions of Excel, select the File menu to find Save As.

We adopt a convention in this book in which constants are collected in column A. Type "Constants:" in cell A4. Then select cell A5 and type "a0 =". Now select cell A6 and type the number 0.99989 (without extra spaces). In cells A7 to A12, enter the remaining constants. Powers of 10 are written, for example, as E-5 for 10^{-5}.

After you type "5.3322E-5" in cell A8, the spreadsheet probably displays "5.33E-5" even though additional digits are kept in memory. To display a desired number of digits in scientific format, click in cell A8 and select the Home ribbon at the top of the spreadsheet in Excel 2007. Go to the part of the ribbon that says Number and click on the little arrow at the lower right. The Format Cells window should appear. Select Scientific and 4 decimal places. When you click OK, the entry in cell A8 will be "5.3322E-5". If you need more room for the number of digits, grab the vertical line at the top of the column with your mouse and resize the column. In earlier versions of Excel, go to Format and then select Cells to obtain the formatting window. You can format all numbers in column A by clicking on the top of the column before setting the format. Your spreadsheet should now look like Figure 2-22b.

In cell B4, write the heading "Temp (°C)". In Excel 2007, you can find the degree sign on the Insert ribbon by clicking Symbol. In earlier versions of Excel, go to the Insert menu and select Symbol. Now enter temperatures from 5 through 40 in cells B5 through B12. This is our *input* to the spreadsheet. The *output* will be computed values of density in column C.

In cell C4, enter the heading "Density (g/mL)". Cell C5 is the most important one in the table. In this one, you will write the *formula*

$$= \$A\$6 + \$A\$8*B5 + \$A\$10*B5^2 + \$A\$12*B5^3$$

It doesn't matter whether or not you use spaces around the arithmetic operators. When you hit Return, the number 0.99997 appears in cell C5. The formula above is the spreadsheet translation of Equation 2-4. \$A\$6 refers to the constant in cell A6. We will explain the dollar signs shortly. B5 refers to the temperature in cell B5. The times sign is * and the exponentiation sign is ^. For example, the term "\$A\$12*B5^3" means "(contents of cell A12) $\times$ (contents of cell B5)^3."

Now comes the most magical property of a spreadsheet. Highlight cell C5 and the empty cells below it from C6 to C12. In the Home ribbon, at the right side above Editing, click on the down arrow and select Down. In earlier versions of Excel, go to the Edit menu and select Fill Down. Excel copies the formula from C5 into the cells below it and evaluates the numbers in each of the selected cells. The density of water at each temperature now appears in column C in Figure 2-22d. You can make numbers appear as decimals, rather than scientific notation, by clicking on the arrow at the lower right in Number in the Home ribbon. In Format Cells, select Number and 5 decimal places.

In this example, we made three types of entries. *Labels* such as "a0 =" were typed in as text. An entry that does not begin with a digit or an equal sign is treated as text. *Numbers*, such as 25, were typed in some cells. The spreadsheet treats a number differently from text. In cell C5, we entered a *formula* that necessarily begins with an equal sign.

Formulas begin with an equal sign. Arithmetic operations in a spreadsheet are

$+$ addition

$-$ subtraction

$*$ multiplication

$/$ division

$\wedge$ exponentiation

Three kinds of entries:

label	a3 =
number	4.4E-05
formula	= \$A\$8*B5

Arithmetic Operations and Functions

Addition, subtraction, multiplication, division, and exponentiation have the symbols $+$, $-$, $*$, $/$, and $\wedge$. *Functions* such as Exp($\cdot$) can be typed or can be selected from the Formula ribbon in Excel 2007. In earlier versions of Excel, select the Insert menu and choose Function. Exp($\cdot$) raises e to the power in parentheses. Other functions such as Ln($\cdot$), Log($\cdot$), Sin($\cdot$), and Cos($\cdot$) are also available.

The order of arithmetic operations in formulas is negation first, followed by $\wedge$, followed by $*$ and $/$ (evaluated in order from left to right as they appear), finally followed by $+$ and $-$ (also evaluated from left to right). Make liberal use of parentheses to be sure that the computer does what you intend. The contents of parentheses are evaluated first, before carrying out operations outside the parentheses. Here are some examples:

$$9/5*100+32 = (9/5)*100+32 = (1.8)*100+32 = (1.8*100)+32 = (180)+32 = 212$$
$$9/5*(100+32) = 9/5*(132) = (1.8)*(132) = 237.6$$
$$9+5*100/32 = 9+(5*100)/32 = 9+(500)/32 = 9+(500/32) = 9+(15.625) = 24.625$$
$$9/5^2+32 = 9/(5^2)+32 = (9/25)+32 = (0.36)+32 = 32.36$$
$$-2^2 = 4 \quad \text{but} \quad -(2^2) = -4$$

When in doubt about how an expression will be evaluated, use parentheses to force what you intend.

Order of operations:

1. Negation (a minus sign before a term)
2. Exponentiation
3. Multiplication and division (in order from left to right)
4. Addition and subtraction (in order from left to right)

Operations within parentheses are evaluated first, from the innermost set.

Documentation and Readability

The first important *documentation* in the spreadsheet is the name of the file. A name such as "Expt 10 Gran Plot" is more meaningful than "Chem Lab". The next important feature is a title at the top of the spreadsheet, which tells its purpose. To tell what formulas were used in the spreadsheet, we added text (labels) at the bottom. In cell A14, write "Formula:" and in cell A15 write "C5 = \$A\$6+\$A\$8*B5+\$A\$10*B5^2+\$A\$12*B5^3". The surest way to document a formula is to copy the text from the formula bar for cell C5. Go to cell A15, type "C5" and then paste in the text you copied.

We improved the *readability* of data in the spreadsheet by selecting the number (decimal) or scientific format and specifying how many decimal places would be shown. The spreadsheet retains more digits in its memory, even though just 5 might be displayed.

Documentation means labeling. If your spreadsheet cannot be read by another person without your help, it needs better documentation. (The same is true of your lab notebook!)

Save your files frequently while you are working and make a backup file of anything that you don't want to lose.

Absolute and Relative References

The formula "= $A$8*B5" refers to cells A8 and B5 in different manners. $A$8 is an *absolute reference* to the contents of cell A8. No matter where cell $A$8 is called from in the spreadsheet, the computer goes to cell A8 to look for a number. "B5" is a *relative reference* in the formula in cell C5. When called from cell C5, the computer goes to cell B5 to find a number. When called from cell C6, the computer goes to cell B6 to look for a number. If called from cell C19, the computer would look in cell B19. This is why the cell written without dollar signs is called a relative reference. If you want the computer to always look only in cell B5, then you should write "$B$5".

2-11 Graphing with Microsoft Excel

Graphs are critical to understanding quantitative relations. To make a graph in Excel 2007 from the spreadsheet in Figure 2-22d, go to the Insert ribbon and select Chart. Click on Scatter and select the icon for Scatter with Smooth Lines and Markers. The other most common graph we will make is Scatter with only Markers. Grab the blank chart with your mouse and move it to the right of the data. In Chart Tools, select Design and click on Select Data. Click on Add. For Series name, write "Density" (without quotation marks). For X values, highlight cells B5:B12. For Y values, delete what was in the box and highlight cells C5:C12. Click OK twice. Click inside the plot area and select the Chart Tools Format ribbon. In Plot Area, Format Selection provides options for the border and fill color of the graph. For Fill, select Solid fill and Color white. For Border Color, select Solid line and Color black. We now have a white graph surrounded by a black border.

To add an X axis title, select Chart Tools Layout. Click on Axis Titles and Primary Horizontal Axis Title. Click on Title Below Axis. A generic axis title appears on the graph. Highlight it and type "Temperature (°C)" over the title. Get the degree sign from Insert Symbol. To put a title on the Y axis, select Chart Tools Layout again. Click on Axis Titles and Primary Vertical Axis Title. Click on Rotated Title. Then type "Density (g/mL)" for the title. Select the title that appears above the graph and remove it with the delete key. Your graph probably looks like the one in Figure 2-23 now.

Let's change the graph so that it looks like Figure 2-24. Click on the curve on the graph to highlight all data points. If only one point is highlighted, click elsewhere on the line. Select Chart Tools Format. In Current Selection, choose Format Selection. A Format Data Series window appears. For Marker Options, choose Built-in. Select the Type circle and Size 6. For Marker Fill, select Solid fill and a Color of your choice. Select Marker Line Color, then Solid line, then the same Color as the marker. To change the appearance of the curve on the graph, use Line Color and Line Style. Create a solid black line with a width of 1.5 points.

	A	B	C	D	E	F	G	H
1	Calculating Density of H2O with Equation 2-4							
2	(from the delightful book by Dan Harris)							
3								
4	Constants:	Temp (C)	Density (g/mL)					
5	a0 =	5	0.99997					
6	0.99989	10	0.99970					
7	a1 =	15	0.99911					
8	5.3322E-05	20	0.99821					
9	a2 =	25	0.99705					
10	−7.5899E-06	30	0.99565					
11	a3 =	35	0.99403					
12	3.6719E-08	40	0.99223					
13								
14	Formula:							
15	C5 = $A$6+$A$8*B5+$A$10*B5^2+$A$12*B5^3							
16								
17								

FIGURE 2-23 Initial density chart drawn by Excel.

To change the appearance of the Y axis, click any number on the Y axis and they will all be highlighted. Select Chart Tools and Format and Format Selection. The Format Axis box appears. For Axis Options, Minimum, click on Fixed and set the value to 0.992. For Axis Options, Maximum, click on Fixed and set the value to 1.000. For Major unit, click on Fixed and set the value to 0.002. For Minor unit, click on Fixed and set the value to 0.0004. Set Minor tick mark type to Outside. In the Format Axis window, select Number and set a display of 3 decimal places. Close the Format Axis window to finish with the vertical axis.

In a similar manner, select a number on the X axis and change the appearance so that it looks like Figure 2-24 with a Minimum of 0, Maximum of 40, Major unit of 10, and Minor unit of 5. Place Minor tick marks Outside. To add vertical grid lines, go to Chart Tools and select Layout and Grid Lines. Select Primary Vertical Gridlines and Major Gridlines.

Add a title back to the chart. In Chart Tools Layout, select Chart Title and highlight Above Chart. Type "Density of Water". In the Home ribbon, select a font size of 10 points. Your chart ought to look much like Figure 2-24 now. You can resize the chart from its lower right corner. To draw on an Excel worksheet, select Insert and then Shapes.

To write on the chart, go to the Insert ribbon and select Text Box. Click in the chart and begin typing. Drag the text box where you want it to be. To format the box, click on its border. Go to the Format ribbon and use Shape Fill and Shape Outline. To add arrows or lines, go to the Insert ribbon and select Shapes. To change the data point symbol, click on one point. On the Format Ribbon, click on Format Selection. The box that apppears lets you change the appearance of the points and the line.

Graphing in Earlier Versions of Excel

To make a graph from the spreadsheet in Figure 2-22d, go to the Insert menu and select Chart. A window appears with a variety of options. The one you will almost always want is XY (Scatter). Highlight XY (Scatter) and several options appear. Select the one that shows data points connected by a smooth curve. Click Next to move to a window called Chart Source Data. Click on the Series tab and click Add. Boxes appear for three inputs. For Name, type "Density" (without quotation marks). For X Values, enter B5:B12 or just highlight those cells on the spreadsheet and they will be automatically entered. For Y Values, highlight cells C5:C12. Click Next.

Now a small graph of your data appears. If it does not look as expected, make sure you selected the correct data, with x before y. The new window asks you for axis labels and an optional title for the graph. For the title, write "Density of Water". For the x-axis, enter "Temperature (°C)" and for the y-axis write "Density (g/mL)". Click Next.

Now you are given the option of drawing the graph on a new sheet or on the same sheet that is already open. Select "As object in Sheet 1". Click Finish and the chart will appear on your spreadsheet. Grab the chart with your mouse and move it to the right of the data.

Excel gives us many options for changing features of the graph. Here are a few, but you should experiment with the graph to discover other formatting options. Double click on the y-axis and a window appears. Select the Patterns tab. Change Minor tick mark type from None to Outside and click OK. You will see new tick marks appear on the y-axis. Double click on the y-axis again and select the Number tab. Change the decimal places to 3 and click OK. Double click on the y-axis again and select the Scale tab. Set the minimum to 0.992 and the maximum to 1.000 and click OK.

Double click on the x-axis and select Patterns. Change the Minor tick mark type from None to Outside. Select Scale and set the maximum to 40, the major unit to 10, the minor unit to 5, and click OK.

Double click on the gray area of the graph and a window called Patterns appears. Select Automatic for the Border and None for the Area. This removes the gray background and gives a solid line around the graph. To add vertical lines at the major tick marks, select the graph with the mouse. Then go to the Chart menu and select Chart Options. In the window that appears, select Gridlines. For the Value (X) axis, check Major gridlines. Then select the tab for Legend and remove the check mark from Show Legend. The legend will disappear. Click OK. You should be getting the idea that you can format virtually any part of the chart.

Click on the outer border of the chart and handles appear. Grab the one on the right and resize the chart so that it does not extend past column F of the spreadsheet. Grab the handle at the bottom and resize the chart so that it does not extend below row 15. When you resized the chart, letters and numbers shrank. Double click on each set of numbers and change the font to 8 points. Double click on the labels and change the letters to 9 points. Your chart should look much like the one in Figure 2-24.

FIGURE 2-24 Density chart after reformatting.

To write on the chart, go to View and select Toolbars and Drawing. Select Text Box from the Drawing toolbar, click inside your chart, and begin typing. Draw arrows with the Arrow tool. Double click on a data point to get options to change plotting symbols.

Terms to Understand

absorption	buret	green chemistry	parallax
acid wash	calibration	hygroscopic	pipet
adsorption	desiccant	ignition	slurry
ashless filter paper	desiccator	meniscus	tare
buoyancy	filtrate	mother liquor	volumetric flask

Summary

Safety requires you to think in advance about what you will do; never do anything that seems dangerous. Know how to use safety equipment such as goggles, fume hood, lab coat, gloves, emergency shower, eyewash, and fire extinguisher. Chemicals should be stored and used in a manner that minimizes contact of solids, liquids, and vapors with people. Environmentally acceptable disposal procedures should be established in advance for every chemical that you use. Your lab notebook tells what you did and what you observed; it should be understandable to other people. It also should allow you to repeat an experiment in the same manner in the future. You should understand the principles of operation of balances and treat them as delicate equipment. Buoyancy corrections are required in accurate work. Burets should be read in a reproducible manner and drained slowly for best results. Always interpolate between markings to obtain accuracy one decimal place

beyond the graduations. Volumetric flasks are used to prepare solutions with known volume. Transfer pipets deliver fixed volumes; less accurate measuring pipets deliver variable volumes. Do not be lulled into complacency by the nice digital reading on a micropipet. Unless your pipet has been calibrated recently and your personal technique tested, micropipets can have gross errors. Filtration and collection of precipitates require careful technique, as does the drying of reagents, precipitates, and glassware in ovens and desiccators. Volumetric glassware is calibrated by weighing water contained in or delivered by the vessel. In the most careful work, solution concentrations and volumes of vessels should be corrected for changes in temperature.

If you plan to use spreadsheets in this course, you should know how to enter formulas in a spreadsheet and how to draw a graph of data from a spreadsheet.

Exercises

2-A. What is the true mass of water if the measured mass in the atmosphere is 5.397 4 g? When you look up the density of water, assume that the lab temperature is **(a)** 15°C and **(b)** 25°C. Take the density of air to be 0.001 2 g/mL and the density of balance weights to be 8.0 g/mL.

2-B. A sample of ferric oxide (Fe_2O_3, density = 5.24 g/mL) obtained from ignition of a gravimetric precipitate weighed 0.296 1 g in the atmosphere. What is the true mass in vacuum?

2-C. A solution of potassium permanganate ($KMnO_4$) was found by titration to be 0.051 38 M at 24°C. What is the molarity when the lab temperature drops to 16°C?

2-D. Water was drained from a buret between the 0.12- and 15.78-mL marks. The apparent volume delivered was $15.78 - 0.12 = 15.66$ mL. Measured in the air at 22°C, the mass of water delivered was 15.569 g. What was the true volume?

2-E. 🖩 Reproduce the spreadsheet in Figure 2-23 and the graph in Figure 2-24.

Problems

Safety and Lab Notebook

2-1. What is the primary safety rule and what is your implied responsibility to make it work?

2-2. After safety features and safety procedures in your laboratory have been explained to you, make a list of them.

2-3. For chemical disposal, why is dichromate converted to $Cr(OH)_3(s)$?

2-4. Explain what each of the three numbered hazard ratings means for 37 wt% HCl in Figure 2-2.

2-5. State three essential attributes of a lab notebook.

Analytical Balance

2-6. Explain the principles of operation of electronic and mechanical balances.

2-7. Why is the buoyancy correction equal to 1 in Figure 2-9 when the density of the object being weighed is 8.0 g/mL?

2-8. Pentane (C_5H_{12}) is a liquid with a density of 0.626 g/mL near 25°C. Find the true mass of pentane when the mass in air is 14.82 g. Assume air density = 0.001 2 g/mL.

2-9. The densities (g/mL) of several substances are: acetic acid, 1.05; CCl_4, 1.59; S, 2.07; Li, 0.53; Hg, 13.5; PbO_2, 9.4; Pb, 11.4; Ir, 22.5. From Figure 2-9, predict which substances will have the smallest and largest buoyancy corrections.

2-10. Potassium hydrogen phthalate is a primary standard used to measure the concentration of NaOH solutions. Find the true mass of potassium hydrogen phthalate (density = 1.636 g/mL) if the mass weighed in air is 4.236 6 g. If you did not correct the mass for buoyancy, would the calculated molarity of NaOH be too high or too low? By what percentage?

2-11. (a) Use the ideal gas law (Problem 1-16) to calculate the density (g/mL) of helium at 20°C and 1.00 bar.

(b) Find the true mass of Na (density = 0.97 g/mL) weighed in a glove box with a He atmosphere, if the apparent mass is 0.823 g.

2-12. (a) The equilibrium vapor pressure of water at 20°C is 2 330 Pa. What is the vapor pressure of water in the air at 20°C if the relative humidity is 42%? (*Relative humidity* is the percentage of the equilibrium water vapor pressure in the air.)

(b) Use note 13 for Chapter 2 at the end of the book to find the air density (g/mL, not g/L) under the conditions of part **(a)** if the barometric pressure is 94.0 kPa.

(c) What is the true mass of water in part **(b)** if the mass in air is 1.000 0 g?

2-13. *Effect of altitude on electronic balance.* If an object weighs m_a grams at distance r_a from the center of the Earth, it will weigh $m_b = m_a(r_a^2/r_b^2)$ when raised to r_b. An object weighs 100.000 0 g on the first floor of a building at r_a = 6 370 km. How much will it weigh on the tenth floor, which is 30 m higher?

Glassware and Thermal Expansion

2-14. What do the symbols "TD" and "TC" mean on volumetric glassware?

2-15. Describe how to prepare 250.0 mL of 0.150 0 M K_2SO_4 with a volumetric flask.

2-16. When is it preferable to use a plastic volumetric flask instead of a more accurate glass flask?

2-17. (a) Describe how to deliver 5.00 mL of liquid by using a transfer pipet.

(b) Which is more accurate, a transfer pipet or a measuring pipet?

2-18. (a) Describe how to deliver 50.0 μL by using a 100-μL adjustable micropipet.

(b) What would you do differently in **(a)** if the liquid foams.

2-19. What is the purpose of the trap in Figure 2-17 and the watchglass in Figure 2-20?

2-20. Which drying agent is more efficient, Drierite or phosphorus pentoxide?

2-21. An empty 10-mL volumetric flask weighs 10.263 4 g. When the flask is filled to the mark with distilled water and weighed again in the air at 20°C, the mass is 20.214 4 g. What is the true volume of the flask at 20°C?

2-22. By what percentage does a dilute aqueous solution expand when heated from 15° to 25°C? If a 0.500 0 M solution is prepared at 15°C, what would its molarity be at 25°C?

2-23. The true volume of a 50-mL volumetric flask is 50.037 mL at 20°C. What mass of water measured **(a)** in vacuum and **(b)** in air at 20°C would be contained in the flask?

2-24. You want to prepare 500.0 mL of 1.000 M KNO_3 at 20°C, but the lab (and water) temperature is 24°C at the time of preparation. How many grams of solid KNO_3 (density = 2.109 g/mL) should be dissolved in a volume of 500.0 mL at 24°C to give a concentration of 1.000 M at 20°C? What apparent mass of KNO_3 weighed in air is required?

2-25. A simple model for the fraction of micropipets that operate within specifications after time t is

$$\text{Fraction within specifications} = e^{-t(\ln 2)/t_m}$$

where t_m is the mean time between failure (the time when the fraction within specifications is reduced to 50%). Suppose that t_m = 2.00 years.

(a) Show that the equation predicts that the time at which 50% remain within specifications is 2 yr if t_m = 2.00 yr.

(b) Find the time t at which pipets should be recalibrated (and repaired, if necessary) so that 95% of all pipets will operate within specifications.

2-26. Glass is a notorious source of metal ion contamination. Three glass bottles were crushed and sieved to collect 1-mm pieces.[26] To see how much Al^{3+} could be extracted, 200 mL of a 0.05 M solution of the metal-binding compound EDTA were stirred with 0.50 g of ~1-mm glass particles in a polyethylene flask. The Al content of the solution after 2 months was 5.2 μM. The total Al content of the glass, measured after completely dissolving some glass in 48 wt% HF with microwave heating, was 0.80 wt%. What fraction of the Al was extracted from glass by EDTA?

2-27. ▦ The efficiency of a gas chromatography column is measured by a parameter called plate height (H, mm) which is related to the gas flow rate (u, mL/min) by the van Deemter equation: $H = A + B/u + Cu$, where A, B, and C are constants. Prepare a spreadsheet with a graph showing values of H as a function of u for u = 4, 6, 8, 10, 20, 30, 40, 50, 60, 70, 80, 90, and 100 mL/min. Use the values A = 1.65 mm, B = 25.8 mm · mL/min, and C = 0.023 6 mm · min/mL.

Reference Procedure: Calibrating a 50-mL Buret

> This procedure tells how to construct a graph such as Figure 3-3 to convert the measured volume delivered by a buret to the true volume delivered at 20°C.

0. Measure the temperature in the laboratory. Distilled water for this experiment must be at laboratory temperature.

1. Fill the buret with distilled water and force any air bubbles out the tip. See whether the buret drains without leaving drops on the walls. If drops are left, clean the buret with soap and water or soak it with cleaning solution.[15] Adjust the meniscus to be at or slightly below 0.00 mL,

and touch the buret tip to a beaker to remove the suspended drop of water. Allow the buret to stand for 5 min while you weigh a 125-mL flask fitted with a rubber stopper. (Hold the flask with a tissue or paper towel, not with your hands, to prevent fingerprint residue from changing its mass.) If the level of the liquid in the buret has changed, tighten the stopcock and repeat the procedure. Record the level of the liquid.

2. Drain approximately 10 mL of water at a rate <20 mL/min into the weighed flask, and cap it tightly to prevent evaporation. Allow about 30 s for the film of liquid on the walls to descend before you read the buret. Estimate all readings to the nearest 0.01 mL. Weigh the flask again to determine the mass of water delivered.

3. Now drain the buret from 10 to 20 mL, and measure the mass of water delivered. Repeat the procedure for 30, 40, and 50 mL. Then do the entire procedure (10, 20, 30, 40, 50 mL) a second time.

4. Use Table 2-7 to convert the mass of water into the volume delivered. Repeat any set of duplicate buret corrections that do not agree to within 0.04 mL. Prepare a calibration graph like that in Figure 3-3, showing the correction factor at each 10-mL interval.

EXAMPLE   **Buret Calibration**

When draining the buret at 24°C, you observe the following values:

Final reading	10.01	10.08 mL
Initial reading	0.03	0.04
Difference	9.98	10.04 mL
Mass	9.984	10.056 g
Actual volume delivered	10.02	10.09 mL
Correction	+0.04	+0.05 mL
Average correction		+0.045 mL

To calculate the actual volume delivered when 9.984 g of water are delivered at 24°C, look at the column of Table 2-7 headed "Corrected to 20°C." In the row for 24°C, you find that 1.000 0 g

of water occupies 1.003 8 mL. Therefore, 9.984 g occupies (9.984 g)(1.003 8 mL/g) = 10.02 mL. The average correction for both sets of data is +0.045 mL.

To obtain the correction for a volume greater than 10 mL, add successive masses of water collected in the flask. Suppose that the following masses were measured:

Volume interval (mL)	Mass delivered (g)
0.03–10.01	9.984
10.01–19.90	9.835
19.90–30.06	10.071
Sum 30.03 mL	29.890 g

The total volume of water delivered is (29.890 g)(1.003 8 mL/g) = 30.00 mL. Because the indicated volume is 30.03 mL, the buret correction at 30 mL is −0.03 mL.

What does this mean? Suppose that Figure 3-3 applies to your buret. If you begin a titration at 0.04 mL and end at 29.00 mL, you would deliver 28.96 mL if the buret were perfect. Figure 3-3 tells you that the buret delivers 0.03 mL less than the indicated amount, so only 28.93 mL were actually delivered. To use the calibration curve, either begin all titrations near 0.00 mL or correct both the initial and the final readings. Use the calibration curve whenever you use your buret.

3 Experimental Error

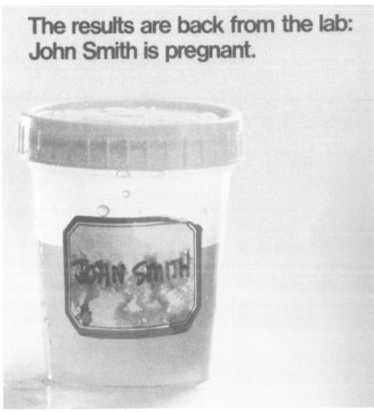

The results are back from the lab: John Smith is pregnant.

[Courtesy 3M Company, St. Paul, MN.]

Some laboratory errors are more obvious than others, but there is error associated with every measurement. There is no way to measure the "true" value of anything. The best we can do in a chemical analysis is to carefully apply a technique that experience tells us is reliable. Repetition of one method of measurement several times tells us the *precision* (reproducibility) of the measurement. If the results of measuring the same quantity by different methods agree with one another, then we become confident that the results are *accurate*, which means they are near the "true" value.

Suppose that you determine the density of a mineral by measuring its mass (4.635 ± 0.002 g) and volume (1.13 ± 0.05 mL). Density is mass per unit volume: 4.635 g/1.13 mL = 4.101 8 g/mL. The uncertainties in measured mass and volume are ±0.002 g and ±0.05 mL, but what is the uncertainty in the computed density? And how many significant figures should be used for the density? This chapter discusses the propagation of uncertainty in lab calculations.

3-1 Significant Figures

Significant figures: minimum number of digits required to express a value in scientific notation without loss of precision

The number of **significant figures** is the minimum number of digits needed to write a given value in scientific notation without loss of precision. The number 142.7 has four significant figures, because it can be written 1.427×10^2. If you write $1.427\ 0 \times 10^2$, you imply that you know the value of the digit after 7, which is not the case for the number 142.7. The number $1.427\ 0 \times 10^2$ has five significant figures.

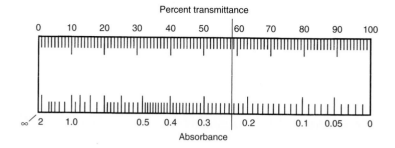

FIGURE 3-1 Analog scale of Bausch and Lomb Spectronic 20 spectrophotometer. Percent transmittance is a linear scale and absorbance is a logarithmic scale.

The number 6.302×10^{-6} has four significant figures, because all four digits are necessary. You could write the same number as 0.000 006 302, which also has just *four* significant figures. The zeros to the left of the 6 are merely holding decimal places. The number 92 500 is ambiguous. It could mean any of the following:

9.25×10^4	3 significant figures
9.250×10^4	4 significant figures
$9.250\,0 \times 10^4$	5 significant figures

You should write one of the three numbers above, instead of 92 500, to indicate how many figures are actually known.

Zeros are significant when they occur (1) in the middle of a number or (2) at the end of a number on the right-hand side of a decimal point.

Significant zeros below are **bold**:

106   0.010 **6**   0.106   0.106 **0**

The last significant digit (farthest to the right) in a measured quantity always has some associated uncertainty. The minimum uncertainty is ± 1 in the last digit. The scale of a Spectronic 20 spectrophotometer is drawn in Figure 3-1. The needle in the figure appears to be at an absorbance of 0.234. We say that this number has three significant figures because the numbers 2 and 3 are completely certain and the number 4 is an estimate. The value might be read 0.233 or 0.235 by other people. The percent transmittance is near 58.3. Because the transmittance scale is smaller than the absorbance scale at this point, there is more uncertainty in the last digit of transmittance. A reasonable estimate of uncertainty might be 58.3 ± 0.2. There are three significant figures in the number 58.3.

When reading the scale of any apparatus, try to estimate to the nearest tenth of a division. On a 50-mL buret, which is graduated to 0.1 mL, read the level to the nearest 0.01 mL. For a ruler calibrated in millimeters, estimate distances to the nearest 0.1 mm.

Interpolation: Estimate all readings to the nearest tenth of the distance between scale divisions.

There is uncertainty in any *measured* quantity, even if the measuring instrument has a digital readout that does not fluctuate. When a digital pH meter indicates a pH of 3.51, there is uncertainty in the digit 1 (and maybe even in the digit 5). By contrast, integers are exact. To calculate the average height of four people, you would divide the sum of heights (which is a measured quantity with some uncertainty) by the integer 4. There are exactly 4 people, not 4.000 ± 0.002 people!

3-2 | Significant Figures in Arithmetic

We now consider how many digits to retain in the answer after you have performed arithmetic operations with your data. Rounding should only be done on the *final answer* (not intermediate results), to avoid accumulating round-off errors.

Addition and Subtraction

If the numbers to be added or subtracted have equal numbers of digits, the answer goes to the *same decimal place* as in any of the individual numbers:

$$
\begin{array}{r}
1.362 \times 10^{-4} \\
+\ 3.111 \times 10^{-4} \\
\hline
4.473 \times 10^{-4}
\end{array}
$$

The number of significant figures in the answer may exceed or be less than that in the original data.

$$
\begin{array}{r}
5.345 \\
+\ 6.728 \\
\hline
12.073
\end{array}
\qquad
\begin{array}{r}
7.26 \times 10^{14} \\
-\ 6.69 \times 10^{14} \\
\hline
0.57 \times 10^{14}
\end{array}
$$

If the numbers being added do not have the same number of significant figures, we are limited by the least certain one. For example, the molecular mass of KrF_2 is known only to the third decimal place, because we know the atomic mass of Kr to only three decimal places:

$$
\begin{array}{rl}
18.998\ 403\ 2 & (F) \\
+\ 18.998\ 403\ 2 & (F) \\
+\ 83.798 & (Kr) \\
\hline
121.794\ 806\ 4 &
\end{array}
$$

⏝
Not significant

The periodic table inside the cover of this book gives uncertainty in the last digit of atomic mass:

F: 18.998 403 2 ± 0.000 000 5
Kr: 83.798 ± 0.002

The number 121.794 806 4 should be rounded to 121.795 as the final answer.

When rounding off, look at *all* the digits *beyond* the last place desired. In the preceding example, the digits 806 4 lie beyond the last significant decimal place. Because this number is more than halfway to the next higher digit, we round the 4 up to 5 (that is, we round up to 121.795 instead of down to 121.794). If the insignificant figures were less than halfway, we would round down. For example, 121.794 3 is rounded to 121.794.

Rules for rounding off numbers

In the special case where the number is exactly halfway, round to the nearest *even* digit. Thus, 43.55 is rounded to 43.6, if we can only have three significant figures. If we are retaining only three figures, 1.425×10^{-9} becomes 1.42×10^{-9}. The number $1.425\ 01 \times 10^{-9}$ would become 1.43×10^{-9}, because 501 is more than halfway to the next digit. The rationale for rounding to an even digit is to avoid systematically increasing or decreasing results through successive round-off errors. Half the round-offs will be up and half down.

In the addition or subtraction of numbers expressed in scientific notation, all numbers should first be expressed with the same exponent:

$$
\begin{array}{r}
1.632 \times 10^5 \\
+\ 4.107 \times 10^3 \\
+\ 0.984 \times 10^6 \\
\hline
\end{array}
\qquad \rightarrow \qquad
\begin{array}{r}
1.632\ \ \ \times 10^5 \\
+\ 0.041\ 07 \times 10^5 \\
+\ 9.84\ \ \ \times 10^5 \\
\hline
11.51\ \ \ \times 10^5
\end{array}
$$

Addition and subtraction: Express all numbers with the same exponent and align all numbers with respect to the decimal point. Round off the answer according to the number of decimal places in the number with the fewest decimal places.

The sum $11.513\ 07 \times 10^5$ is rounded to 11.51×10^5 because the number 9.84×10^5 limits us to two decimal places when all numbers are expressed as multiples of 10^5.

Multiplication and Division

In multiplication and division, we are normally limited to the number of digits contained in the number with the fewest significant figures:

$$
\begin{array}{r}
3.26 \times 10^{-5} \\
\times\ 1.78 \\
\hline
5.80 \times 10^{-5}
\end{array}
\qquad
\begin{array}{r}
4.317\ 9 \times 10^{12} \\
\times\ 3.6\ \ \ \times 10^{-19} \\
\hline
1.6\ \ \ \times 10^{-6}
\end{array}
\qquad
\begin{array}{r}
34.60 \\
\div\ 2.462\ 87 \\
\hline
14.05
\end{array}
$$

The power of 10 has no influence on the number of figures that should be retained. The section on the real rule for significant figures on page 59 explains why it is reasonable to keep an extra digit when the first digit of the answer is 1. The middle product above could be expressed as 1.55×10^{-6} instead of 1.6×10^{-6} to avoid throwing away some of the precision of the factor 3.6 in the multiplication.

Logarithms and Antilogarithms

If $n = 10^a$, then we say that a is the base 10 **logarithm** of n:

Logarithm of n:
$$n = 10^a \quad \text{means that} \quad \log n = a \tag{3-1}$$

For example, 2 is the logarithm of 100 because $100 = 10^2$. The logarithm of 0.001 is -3 because $0.001 = 10^{-3}$. To find the logarithm of a number with your calculator, enter the number and press the *log* function.

$10^{-3} = \dfrac{1}{10^3} = \dfrac{1}{1\ 000} = 0.001$

3-2 **Significant Figures in Arithmetic**

In Equation 3-1, the number n is said to be the **antilogarithm** of a. That is, the antilogarithm of 2 is 100 because $10^2 = 100$, and the antilogarithm of -3 is 0.001 because $10^{-3} = 0.001$. Your calculator has either a *10^x* key or an *antilog* key. To find the antilogarithm of a number, enter it in your calculator and press *10^x* (or *antilog*).

A logarithm is composed of a **characteristic** and a **mantissa.** The characteristic is the integer part and the mantissa is the decimal part:

$$\log 339 = 2.530 \qquad \log 3.39 \times 10^{-5} = -4.470$$

Characteristic Mantissa   Characteristic Mantissa
$= 2$ $= 0.530$ $= -4$ $= 0.470$

The number 339 can be written 3.39×10^2. *The number of digits in the mantissa of log 339 should equal the number of significant figures in 339.* The logarithm of 339 is properly expressed as 2.530. The *characteristic*, 2, corresponds to the exponent in 3.39×10^2.

To see that the third decimal place is the last significant place, consider the following results:

$$10^{2.531} = 340 \,(339.6)$$
$$10^{2.530} = 339 \,(338.8)$$
$$10^{2.529} = 338 \,(338.1)$$

The numbers in parentheses are the results prior to rounding to three figures. Changing the exponent in the third decimal place changes the answer in the third place of 339.

In the conversion of a logarithm into its antilogarithm, *the number of significant figures in the antilogarithm should equal the number of digits in the mantissa.* Thus,

$$\text{antilog} \,(-3.42) = 10^{-3.42} = 3.8 \times 10^{-4}$$

2 digits   2 digits   2 digits

Here are several examples showing the proper use of significant figures:

$$\log 0.001\,237 = -2.907\,6 \qquad \text{antilog } 4.37 = 2.3 \times 10^4$$
$$\log 1\,237 = 3.092\,4 \qquad\qquad 10^{4.37} = 2.3 \times 10^4$$
$$\log 3.2 = 0.51 \qquad\qquad\qquad 10^{-2.600} = 2.51 \times 10^{-3}$$

Significant Figures and Graphs

When drawing a graph on a computer, consider whether the graph is meant to display qualitative behavior of the data (Figure 3-2) or precise values that must be read with several significant figures. If someone will use the graph (such as Figure 3-3) to read points, it should at least have tick marks on both sides of the horizontal and vertical scales. Better still is a fine grid superimposed on the graph.

Number of digits in **mantissa** of $\log x$ = number of significant figures in x:

$$\log(5.403 \times 10^{-8}) = -7.267\,4$$

4 digits   4 digits

Number of digits in antilog x (= 10^x) = number of significant figures in **mantissa** of x:

$$10^{6.142} = 1.39 \times 10^6$$

3 digits   3 digits

FIGURE 3-2 Example of a graph intended to show the qualitative behavior of the function $y = e^{-x/6} \cos x$. You are not expected to be able to read coordinates accurately on this graph.

Problem 3-8 shows you how to control gridlines in an Excel graph.

FIGURE 3-3 Calibration curve for a 50-mL buret. The volume delivered can be read to the nearest 0.1 mL. If your buret reading is 29.43 mL, you can find the correction factor accurately enough by locating 29.4 mL on the graph. The correction factor on the ordinate (*y*-axis) for 29.4 mL on the abscissa (*x*-axis) is -0.03 mL (to the nearest 0.01 mL).

3-3 Types of Error

Every measurement has some uncertainty, which is called *experimental error*. Conclusions can be expressed with a high or a low degree of confidence, but never with complete certainty. Experimental error is classified as either *systematic* or *random*.

Systematic Error

Systematic error, also called **determinate error,** arises from a flaw in equipment or the design of an experiment. If you conduct the experiment again in exactly the same manner, the error is reproducible. In principle, systematic error can be discovered and corrected, although this may not be easy.

For example, a pH meter that has been standardized incorrectly produces a systematic error. Suppose you think that the pH of the buffer used to standardize the meter is 7.00, but it is really 7.08. Then all your pH readings will be 0.08 pH unit too low. When you read a pH of 5.60, the actual pH of the sample is 5.68. This systematic error could be discovered by using a second buffer of known pH to test the meter.

> *Systematic error* is a consistent error that can be detected and corrected. Box 3-1 provides an example from environmental analysis.

BOX 3-1   Case Study in Ethics: Systematic Error in Ozone Measurement

Ozone (O_3) is an oxidizing, corrosive gas that harms your lungs and all forms of life. It is formed near the surface of the Earth by the action of sunlight on air pollutants largely derived from automobile exhaust. The U.S. Environmental Protection Agency sets an 8-h average O_3 limit of 80 ppb (80 nL/L by volume) in air. Regions that fail to meet this standard can be required to reduce sources of pollution that contribute to O_3 formation. Error in ozone measurement can have serious consequences for the health and economy of a region.

To monitor compliance, a variety of instruments are used. The instrument in the diagram pumps air through a cell with a pathlength of 15 cm. Ultraviolet radiation from a mercury lamp is partially absorbed by O_3. The more O_3 in the air, the less radiation reaches the detector. From the measured absorbance, the instrument computes O_3 concentration. In routine use, the operator only adjusts the zero control, which sets the meter to read zero when O_3-free air is drawn through the instrument. Periodically, the instrument is recalibrated with a source of known O_3.

A study of commercial O_3 monitors found that controlled changes in humidity led to *systematic errors in the apparent O_3 concentration of tens to hundreds of ppb* (errors several times greater than the O_3 being measured). Increasing humidity produced systematic *positive* errors in some types of instruments and systematic *negative* errors in other instruments.

Water does not absorb the ultraviolet wavelength measured by the detector, so humidity is not interfering by absorbing radiation. A perceptive analysis of the problem led to the hypothesis that *adsorption* of moisture on the inside surface of the measurement cell changed the reflectivity of that surface. In one type of instrument, water adsorbed inside a quartz cell reflects less light than dry quartz and thus increases the amount of light lost by absorption in black paint on the outside of the cell. This instrument produces a false, *high* O_3 reading. Another instrument has a highly reflective aluminum cell coated on the inside with polyvinylidene fluoride. Adsorption of moisture on polyvinylidene fluoride reduces total internal reflection within the coating and increases the radiant energy reaching the detector, giving a false, *low* O_3 reading. These effects need not be large. A 0.03% change in light intensity reaching the detector corresponds to an O_3 change of 100 ppb. The solution to the problem was to install a length of water-permeable tubing just before the absorption cell to equalize the humidity in air being measured and that used to zero the instrument.

Optical path of 2B Technologies Model 202 Ozone Monitor. The solenoid alternately admits ambient air or air that has been scrubbed free of O_3. Absorbance of ultraviolet radiation from the Hg lamp is proportional to O_3 concentration. [Diagram from www.twobtech.com/manuals/model_202_new.pdf. Story from K. L. Wilson and J. W. Birks, "Mechanism and Elimination of a Water Vapor Interference in the Measurement of Ozone by UV Absorbance," *Environ. Sci. Technol.* 2006, *40*, 6361.]

Prior to understanding the effect of humidity on O_3 measurement, it was known that O_3 monitors often exhibit erratic behavior on hot, humid days. It was conjectured by some people that half of the regions deemed to be out of compliance with the O_3 standard might actually have been under the legal limit. This error could force expensive remediation measures when none were required. Conversely, there were rumors that some unscrupulous operators of O_3 monitors were aware that zeroing their instrument at night when the humidity is higher produced lower O_3 readings the next day, thereby reducing the number of days when a region is deemed out of compliance.

Ways to detect systematic error:

1. Analyze a known sample, such as a certified reference material. Your method should reproduce the known answer. (See Box 14-1 for an example.)

2. Analyze blank samples containing no analyte being sought. If you observe a nonzero result, your method responds to more than you intend. Section 5-1 discusses different kinds of blanks.

3. Use different analytical methods to measure the same quantity. If results do not agree, there is error in one (or more) of the methods.

4. *Round robin* experiment: Different people in several laboratories analyze identical samples by the same or different methods. Disagreement beyond the estimated random error is systematic error.

Random error cannot be eliminated, but it might be reduced by a better experiment.

Precision: reproducibility

Accuracy: nearness to the "truth"

Another systematic error arises from an uncalibrated buret. The manufacturer's tolerance for a Class A 50-mL buret is ±0.05 mL. When you think you have delivered 29.43 mL, the real volume could be anywhere from 29.38 to 29.48 mL and still be within tolerance. One way to correct for an error of this type is to construct a calibration curve, such as that in Figure 3-3, by the procedure on page 49. To do this, deliver distilled water from the buret into a flask and weigh it. Determine the volume of water from its mass by using Table 2-7. Figure 3-3 tells us to apply a correction factor of −0.03 mL to the measured value of 29.43 mL. The actual volume delivered is 29.43 − 0.03 = 29.40 mL.

A key feature of systematic error is that it is reproducible. For the buret just discussed, the error is always −0.03 mL when the buret reading is 29.43 mL. Systematic error may always be positive in some regions and always negative in others. With care and cleverness, you can detect and correct a systematic error.

Random Error

Random error, also called **indeterminate error,** arises from uncontrolled (and maybe uncontrollable) variables in the measurement. Random error has an equal chance of being positive or negative. It is always present and cannot be corrected. There is random error associated with reading a scale. Different people reading the scale in Figure 3-1 report a range of values representing their subjective interpolation between the markings. One person reading the same instrument several times might report several different readings. Another random error results from electrical noise in an instrument. Positive and negative fluctuations occur with approximately equal frequency and cannot be completely eliminated.

Precision and Accuracy

Precision describes the reproducibility of a result. If you measure a quantity several times and the values agree closely with one another, your measurement is precise. If the values vary widely, your measurement is not precise. **Accuracy** describes how close a measured value is to the "true" value. If a known standard is available, accuracy is how close your value is to the known value.

The U.S. National Institute of Standards and Technology and national standards laboratories in other countries sell **certified reference materials** (called *Standard Reference Materials* in the U.S.), such as clinical and environmental standards and engineering materials that you can use to test the accuracy of your analytical procedures.[1] The quantity of analyte in a reference material is certified—with painstaking care—to lie in a stated range.

A measurement might be reproducible, but wrong. If you made a mistake preparing a solution for a titration, you might do a series of reproducible titrations but report an incorrect result because the concentration of the titrating solution was not what you intended. In this case, precision is good but accuracy is poor. Conversely, it is possible to make poorly reproducible measurements clustered around the correct value. For this case, precision is poor but accuracy is good. An ideal procedure is both precise and accurate.

Accuracy is defined as nearness to the "true" value. *True* is in quotes because somebody must *measure* the "true" value, and there is error associated with *every* measurement. The "true" value is best obtained by an experienced person using a well-tested procedure. It is desirable to test the result by using different procedures, because systematic error could lead to poor agreement between methods. Good agreement among several methods affords us confidence, but never proof, that results are accurate.

Absolute and Relative Uncertainty

Absolute uncertainty expresses the margin of uncertainty associated with a measurement. If the estimated uncertainty in reading a calibrated buret is ±0.02 mL, we say that ±0.02 mL is the absolute uncertainty associated with the reading.

Relative uncertainty compares the size of the absolute uncertainty with the size of its associated measurement. The relative uncertainty of a buret reading of 12.35 ± 0.02 mL is a dimensionless quotient:

An uncertainty of ±0.02 mL means that, when the reading is 13.33 mL, the true value could be anywhere in the range 13.31 to 13.35 mL.

Relative uncertainty:

$$\text{Relative uncertainty} = \frac{\text{absolute uncertainty}}{\text{magnitude of measurement}} \qquad (3\text{-}2)$$

$$= \frac{0.02 \text{ mL}}{12.35 \text{ mL}} = 0.002$$

The percent relative uncertainty is simply

Percent relative uncertainty:

$$\text{Percent relative uncertainty} = 100 \times \text{relative uncertainty} \qquad (3\text{-}3)$$

$$= 100 \times 0.002 = 0.2\%$$

If the absolute uncertainty in reading a buret is constant at ± 0.02 mL, the percent relative uncertainty is 0.2% for a volume of 10 mL and 0.1% for a volume of 20 mL.

If you use a 50-mL buret, design your titration to require 20–40 mL of reagent to produce a small relative uncertainty of 0.1–0.05%.

3-4 Propagation of Uncertainty from Random Error[2]

We can usually estimate or measure the random error associated with a measurement, such as the length of an object or the temperature of a solution. Uncertainty might be based on how well we can read an instrument or on our experience with a particular method. If possible, uncertainty will be expressed as the *standard deviation of the mean* or a *confidence interval*, which we discuss in Chapter 4. This section applies only to random error. We assume that systematic error has been detected and corrected.

For most experiments, we need to perform arithmetic operations on several numbers, each of which has a random error. The most likely uncertainty in the result is not the sum of individual errors, because some of them are likely to be positive and some negative. We expect some cancellation of errors.

In a gravimetric analysis, plan to have enough precipitate for a low relative uncertainty. If weighing precision is ± 0.3 mg, a 100-mg precipitate has a relative weighing error of 0.3% and a 300-mg precipitate has an uncertainty of 0.1%.

Most propagation of uncertainty computations that you will encounter deal with random error, not systematic error. Our goal is always to eliminate systematic error.

Addition and Subtraction

Suppose you wish to perform the following arithmetic, in which the experimental uncertainties, designated e_1, e_2, and e_3, are given in parentheses.

$$
\begin{aligned}
1.76 \,(&\pm 0.03) \leftarrow e_1 \\
+\; 1.89 \,(&\pm 0.02) \leftarrow e_2 \\
-\; 0.59 \,(&\pm 0.02) \leftarrow e_3 \\
\hline
3.06 \,(&\pm e_4)
\end{aligned}
\qquad (3\text{-}4)
$$

The arithmetic answer is 3.06. But what is the uncertainty associated with this result?

For addition and subtraction, the uncertainty in the answer is obtained from the *absolute uncertainties* of the individual terms as follows:

Uncertainty in addition and subtraction:

$$e_4 = \sqrt{e_1^2 + e_2^2 + e_3^2} \qquad (3\text{-}5)$$

For addition and subtraction, use *absolute* uncertainty.

For the sum in Equation 3-4, we can write

$$e_4 = \sqrt{(0.03)^2 + (0.02)^2 + (0.02)^2} = 0.04_1$$

The absolute uncertainty e_4 is ± 0.04, and we express the answer as 3.06 ± 0.04. Although there is only one significant figure in the uncertainty, we wrote it initially as 0.04_1, with the first insignificant figure subscripted. We retain one or more insignificant figures to avoid introducing round-off errors into later calculations through the number 0.04_1. The insignificant figure was subscripted to remind us where the last significant figure should be when we conclude the calculations.

To find the percent relative uncertainty in the sum of Equation 3-4, we write

$$\text{Percent relative uncertainty} = \frac{0.04_1}{3.06} \times 100 = 1._3\%$$

The uncertainty, 0.04_1, is $1._3\%$ of the result, 3.06. The subscript 3 in $1._3\%$ is not significant. It is sensible to drop the insignificant figures now and express the final result as

$$
\begin{aligned}
3.06 \,(&\pm 0.04) \qquad \text{(absolute uncertainty)} \\
3.06 \,(&\pm 1\%) \qquad \text{(relative uncertainty)}
\end{aligned}
$$

For addition and subtraction, use absolute uncertainty. Relative uncertainty can be found at the end of the calculation.

EXAMPLE Uncertainty in a Buret Reading

The volume delivered by a buret is the difference between final and initial readings. If the uncertainty in each reading is ± 0.02 mL, what is the uncertainty in the volume delivered?

3-4 Propagation of Uncertainty from Random Error **57**

Solution Suppose that the initial reading is 0.05 (± 0.02) mL and the final reading is 17.88 (± 0.02) mL. The volume delivered is the difference:

$$\begin{array}{r} 17.88 \ (\pm 0.02) \\ - \ 0.05 \ (\pm 0.02) \\ \hline 17.83 \ (\pm e) \end{array} \qquad e = \sqrt{0.02^2 + 0.02^2} = 0.02_8 \approx 0.03$$

Regardless of the initial and final readings, if the uncertainty in each one is ± 0.02 mL, the uncertainty in volume delivered is ± 0.03 mL.

Test Yourself What would be the uncertainty in volume delivered if the uncertainty in each reading were 0.03 mL? (**Answer:** ± 0.04 mL)

Multiplication and Division

For multiplication and division, first convert all uncertainties into percent relative uncertainties. Then calculate the error of the product or quotient as follows:

For multiplication and division, use percent relative uncertainty.

Uncertainty in multiplication and division:
$$\% e_4 = \sqrt{(\% e_1)^2 + (\% e_2)^2 + (\% e_3)^2} \qquad (3\text{-}6)$$

For example, consider the following operations:

$$\frac{1.76 \ (\pm 0.03) \times 1.89 \ (\pm 0.02)}{0.59 \ (\pm 0.02)} = 5.64 \pm e_4$$

First convert absolute uncertainties into percent relative uncertainties.

$$\frac{1.76 \ (\pm 1._7\%) \times 1.89 \ (\pm 1._1\%)}{0.59 \ (\pm 3._4\%)} = 5.64 \pm e_4$$

Then find the percent relative uncertainty of the answer by using Equation 3-6.

Advice Retain one or more extra insignificant figures until you have finished your entire calculation. Then round to the correct number of digits. When storing intermediate results in a calculator, keep all digits without rounding.

$$\% e_4 = \sqrt{(1._7)^2 + (1._1)^2 + (3._4)^2} = 4._0\%$$

The answer is $5.6_4 \ (\pm 4._0\%)$.

To convert relative uncertainty into absolute uncertainty, find $4._0\%$ of the answer.

$$4._0\% \times 5.6_4 = 0.04_0 \times 5.6_4 = 0.2_3$$

The answer is $5.6_4 \ (\pm 0.2_3)$. Finally, drop the insignificant digits.

For multiplication and division, use percent relative uncertainty. Absolute uncertainty can be found at the end of the calculation.

$$5.6 \ (\pm 0.2) \qquad \text{(absolute uncertainty)}$$
$$5.6 \ (\pm 4\%) \qquad \text{(relative uncertainty)}$$

The denominator of the original problem, 0.59, limits the answer to two digits.

Mixed Operations

Now consider a computation containing subtraction and division:

$$\frac{[1.76 \ (\pm 0.03) - 0.59 \ (\pm 0.02)]}{1.89 \ (\pm 0.02)} = 0.619_0 \pm \ ?$$

First work out the difference in the numerator, using absolute uncertainties. Thus,

$$1.76 \ (\pm 0.03) - 0.59 \ (\pm 0.02) = 1.17 \ (\pm 0.03_6)$$

because $\sqrt{(0.03)^2 + (0.02)^2} = 0.03_6$.

Then convert into percent relative uncertainties. Thus,

$$\frac{1.17 \ (\pm 0.03_6)}{1.89 \ (\pm 0.02)} = \frac{1.17 \ (\pm 3._1\%)}{1.89 \ (\pm 1._1\%)} = 0.619_0 \ (\pm 3._3\%)$$

because $\sqrt{(3._1\%)^2 + (1._1\%)^2} = 3._3\%$.

The percent relative uncertainty is $3._3\%$, so the absolute uncertainty is $0.03_3 \times 0.619_0 = 0.02_0$. The final answer can be written as

$$0.619\ (\pm 0.02_0) \qquad \text{(absolute uncertainty)}$$
$$0.619\ (\pm 3._3\%) \qquad \text{(relative uncertainty)}$$

Because the uncertainty begins in the 0.01 decimal place, it is reasonable to round the result to the 0.01 decimal place:

The result of a calculation ought to be written in a manner consistent with its uncertainty.

$$0.62\ (\pm 0.02) \qquad \text{(absolute uncertainty)}$$
$$0.62\ (\pm 3\%) \qquad \text{(relative uncertainty)}$$

The *Real* Rule for Significant Figures

The first digit of the absolute uncertainty is the last significant digit in the answer. For example, in the quotient

The real rule: The first uncertain figure is the last significant figure.

$$\frac{0.002\ 364\ (\pm 0.000\ 003)}{0.025\ 00\ (\pm 0.000\ 05)} = 0.094\ 6\ (\pm 0.000\ 2)$$

the uncertainty ($\pm 0.000\ 2$) occurs in the fourth decimal place. Therefore, the answer 0.094 6 is properly expressed with *three* significant figures, even though the original data have four figures. The first uncertain figure of the answer is the last significant figure. The quotient

$$\frac{0.002\ 664\ (\pm 0.000\ 003)}{0.025\ 00\ (\pm 0.000\ 05)} = 0.106\ 6\ (\pm 0.000\ 2)$$

is expressed with *four* significant figures because the uncertainty occurs in the fourth place. The quotient

$$\frac{0.821\ (\pm 0.002)}{0.803\ (\pm 0.002)} = 1.022\ (\pm 0.004)$$

is expressed with *four* figures even though the dividend and divisor each have *three* figures.

Now you can appreciate why *it is alright to keep one extra digit when the first digit of an answer lies between 1 and 2.* The quotient 82/80 is better written as 1.02 than 1.0. If the uncertainties in 82 and 80 are in the ones place, the uncertainty is of the order of 1%, which is in the second decimal place of 1.02. If I write 1.0, you can surmise that the uncertainty is at least $1.0 \pm 0.1 = \pm 10\%$, which is much larger than the actual uncertainty.

In multiplication and division, keep an extra digit when the answer lies between 1 and 2.

EXAMPLE **Significant Figures in Laboratory Work**

You prepared a 0.250 M NH_3 solution by diluting 8.45 (± 0.04) mL of 28.0 (± 0.5) wt% NH_3 [density = 0.899 (± 0.003) g/mL] up to 500.0 (± 0.2) mL. Find the uncertainty in 0.250 M. The molecular mass of NH_3, 17.030 5 g/mol, has negligible uncertainty relative to other uncertainties in this problem.

Solution To find the uncertainty in molarity, we need to find the uncertainty in moles delivered to the 500-mL flask. The concentrated reagent contains 0.899 (± 0.003) g of solution per milliliter. Weight percent tells us that the reagent contains 0.280 (± 0.005) g of NH_3 per gram of solution. In our calculations, we retain extra insignificant digits and round off only at the end.

$$\begin{aligned} \text{Grams of } NH_3 \text{ per mL} & = 0.899\ (\pm 0.003)\ \frac{\text{g solution}}{\text{mL}} \times 0.280\ (\pm 0.005)\ \frac{\text{g } NH_3}{\text{g solution}} \\[4pt] & = 0.899\ (\pm 0.334\%)\ \frac{\text{g solution}}{\text{mL}} \times 0.280\ (\pm 1.79\%)\ \frac{\text{g } NH_3}{\text{g solution}} \\[4pt] & = 0.251\ 7\ (\pm 1.82\%)\ \frac{\text{g } NH_3}{\text{mL}} \end{aligned}$$

Convert absolute uncertainty into percent relative uncertainty for multiplication.

because $\sqrt{(0.334\%)^2 + (1.79\%)^2} = 1.82\%$.

3-4 Propagation of Uncertainty from Random Error

59

BOX 3-2 Keeling's Exquisitely Precise Measurement of CO_2

Atmospheric CO_2 measurements on Mauna Loa, introduced in Section 0-1, are made several times per hour by infrared absorption (Box 19-3). Accuracy depends on calibration gases consisting of CO_2 in dry N_2 or dry air. Calibration gas is measured at Scripps Institution of Oceanography by extremely careful manometry in which temperature, pressure, and volume of CO_2 in the gas are measured to find the number of moles of gas.

The diagram depicts the principle of the measurement. Calibration gas, passed through Dry Ice traps to freeze out traces of H_2O, goes to a chamber in panel *a* whose volume is 5.013 8_2 ± 0.000 5 L and whose temperature and pressure are measured. Pressure is measured within ±0.02 mm Hg by the difference in the level of Hg on either side of a U-tube. Mercury height is read with a *cathetometer*, which is a telescope mounted on a precision threaded shaft. The shaft is rotated to move the telescope up or down until the top of the Hg meniscus is centered in the cross hairs. Height is indicated by the number of turns of the shaft.

In panel *a*, CO_2 from the 5-L volume of calibration gas is *quantitatively* (completely) collected by freezing onto the inside wall of the manometer with liquid N_2 (77 K). Gas that does not freeze is pumped out through stopcocks 1 and 2. With CO_2 frozen in the evacuated chamber, N_2 is admitted through stopcock 3 so that Hg rises and seals off port 1.

In panel *b*, CO_2 has warmed to room temperature. Pressure in the Hg reservoir is adjusted to bring the level of Hg exactly to the tip of the glass pointer in the inset. The previously measured volume above the tip of the pointer is 3.793 0 ± 0.000 9 mL. Pressure inside this volume is the difference in height between the Hg columns, labeled Δ*P* in panel *b*. From pressure, temperature, and volume, moles of CO_2 can be calculated.

The *ideal gas law* does not provide enough accuracy, so the *virial equation* is used:

Virial equation of state: $$PV_m = RT\left(1 + \frac{B}{V_m} + \frac{C}{V_m^2} + \cdots\right)$$

where P is pressure, R is the gas constant, T is temperature (K), and V_m is the volume of 1 mol of gas. B is the *second virial coefficient* and C is the *third virial coefficient*. If the coefficients were zero, the virial equation would be identical to the ideal gas law. Virial coefficients are known as a function of temperature. The coefficient B alone provides sufficient accuracy for measuring CO_2 in air.

Air contains 0.3 ppm of N_2O, which condenses with CO_2 and interferes with the measurement. The ratio N_2O/CO_2 in air is measured by gas chromatography and the N_2O is subtracted from the sum of $CO_2 + N_2O$ measured in the manometer.

To measure the 5.013 8_2-L volume in panel *a*, a 4-L flask that can be closed with a vacuum stopcock is weighed empty and weighed again filled with water. Both weights are corrected for buoyancy. From the mass of water and the temperature, the flask's volume is computed. The calibrated 4-L flask is then filled with CO_2 at known pressure and temperature. The CO_2 is quantitatively transferred to the 5-L compartment and pressure and temperature are measured. From the pressure and temperature of this known quantity of gas, the volume of the 5-L compartment is calculated. By a similar procedure, the volume of the 3.793 0-mL compartment is measured by transferring CO_2 from an external 2-mL volume. The volume of the 2-mL vessel is measured by weighing it empty and weighing it filled with mercury. Measurements conducted over many years have a standard deviation of 0.010% for the 5-L volume and 0.025% for the 4-mL volume.

The 0.025% random error in measuring the 4-mL volume limits the precision of measuring CO_2 in air to 0.025%, or about 0.1 ppm for CO_2 levels of 400 ppm. To estimate systematic error, CO_2 in a calibration standard was measured by two different methods at Scripps and, independently, at the National Oceanographic and Atmospheric Administration. The three measurements agreed within 0.2 ppm CO_2.

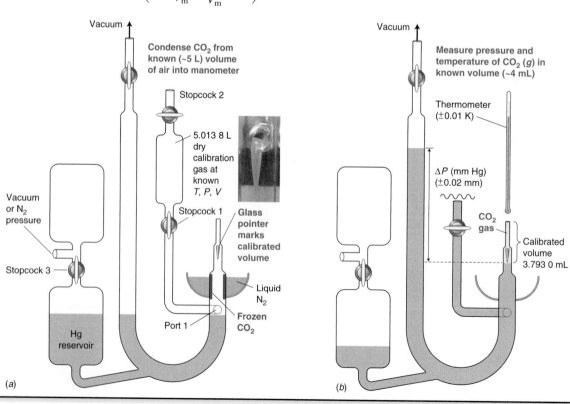

(a)

(b)

Next, we find the moles of ammonia contained in 8.45 (± 0.04) mL of concentrated reagent. The relative uncertainty in volume is 0.04/8.45 = 0.473%.

$$\text{mol NH}_3 = \frac{0.251\ 7\ (\pm 1.82\%)\ \dfrac{\text{g NH}_3}{\text{mL}} \times 8.45\ (\pm 0.473\%)\ \text{mL}}{17.030\ 5\ (\pm 0\%)\ \dfrac{\text{g NH}_3}{\text{mol}}}$$

$$= 0.124\ 9\ (\pm 1.88\%)\ \text{mol}$$

because $\sqrt{(1.82\%)^2 + (0.473\%)^2 + (0\%)^2} = 1.88\%$.

This much ammonia was diluted to 0.500 0 ($\pm 0.000\ 2$) L. The relative uncertainty in the final volume is 0.000 2/0.500 0 = 0.04%. The molarity is

$$\frac{\text{mol NH}_3}{\text{L}} = \frac{0.124\ 9\ (\pm 1.88\%)\ \text{mol}}{0.500\ 0\ (\pm 0.04\%)\ \text{L}}$$

$$= 0.249\ 8\ (\pm 1.88\%)\ \text{M}$$

because $\sqrt{(1.88\%)^2 + (0.04\%)^2} = 1.88\%$. The absolute uncertainty is 1.88% of 0.249 8 M = 0.004 7 M. The uncertainty in molarity is in the third decimal place, so our final, rounded answer is

$$[\text{NH}_3] = 0.250\ (\pm 0.005)\ \text{M}$$

Test Yourself Find the uncertainty in $[\text{NH}_3]$ if the starting reagent is 28.0 (± 0.7) wt% NH_3. (*Answer:* 0.250 (± 0.006) M)

Exponents and Logarithms

For the function $y = x^a$, the percent relative uncertainty in y ($\% e_y$) is equal to a times the percent relative uncertainty in x ($\% e_x$):

Uncertainty for powers and roots:
$$y = x^a \implies \% e_y = a(\% e_x) \tag{3-7}$$

For example, if $y = \sqrt{x} = x^{1/2}$, a 2% uncertainty in x will yield a $\left(\frac{1}{2}\right)(2\%) = 1\%$ uncertainty in y. If $y = x^2$, a 3% uncertainty in x leads to a $(2)(3\%) = 6\%$ uncertainty in y.

If y is the base 10 logarithm of x, then the absolute uncertainty in y (e_y) is proportional to the relative uncertainty in x, which is e_x/x:

Uncertainty for logarithm:
$$y = \log x \implies e_y = \frac{1}{\ln 10}\frac{e_x}{x} \approx 0.434\ 29\ \frac{e_x}{x} \tag{3-8}$$

You should not work with percent relative uncertainty [$100 \times (e_x/x)$] in calculations with logs and antilogs, because one side of Equation 3-8 has relative uncertainty and the other has absolute uncertainty.

The **natural logarithm** (ln) of x is the number y, whose value is such that $x = e^y$, where e (= 2.718 28 . . .) is called the base of the natural logarithm. The absolute uncertainty in y is equal to the relative uncertainty in x.

Uncertainty for natural logarithm:
$$y = \ln x \implies e_y = \frac{e_x}{x} \tag{3-9}$$

Now consider $y = \text{antilog } x$, which is the same as saying $y = 10^x$. In this case, the relative uncertainty in y is proportional to the absolute uncertainty in x.

Uncertainty for 10^x:
$$y = 10^x \implies \frac{e_y}{y} = (\ln 10)e_x \approx 2.302\ 6\ e_x \tag{3-10}$$

If $y = e^x$, the relative uncertainty in y equals the absolute uncertainty in x.

Uncertainty for e^x:
$$y = e^x \implies \frac{e_y}{y} = e_x \tag{3-11}$$

Table 3-1 summarizes rules for propagation of uncertainty. You need not memorize the rules for exponents, logs, and antilogs, but you should be able to use them.

Box 3-2 describes measurements of CO_2 standards by Charles David Keeling for atmospheric CO_2 measurements. You should appreciate how precisely pressure, volume, and temperature had to be measured for an overall precision of 1 part in 4 000 (0.025%).

To calculate a power or root on your calculator, use the y^x button. For example, to find a cube root ($y^{1/3}$), raise y to the 0.333 333 333 . . . power with the y^x button. In Excel, y^x is $y\^x$. The cube root is $y\^(1/3)$.

Use relative uncertainty (e_x/x), not percent relative uncertainty [$100 \times (e_x/x)$], in calculations involving $\log x$, $\ln x$, 10^x, and e^x.

In Excel, the base 10 logarithm is $\log(x)$. The natural logarithm is $\ln(x)$. The expression 10^x is $10\^x$ and the expression e^x is $\exp(x)$.

Appendix C gives a general rule for propagation of random uncertainty for any function.

TABLE 3-1 Summary of rules for propagation of uncertainty

Function	Uncertainty	Function[a]	Uncertainty[b]
$y = x_1 + x_2$	$e_y = \sqrt{e_{x_1}^2 + e_{x_2}^2}$	$y = x^a$	$\%e_y = a\%e_x$
$y = x_1 - x_2$	$e_y = \sqrt{e_{x_1}^2 + e_{x_2}^2}$	$y = \log x$	$e_y = \dfrac{1}{\ln 10}\dfrac{e_x}{x} \approx 0.434\,29\,\dfrac{e_x}{x}$
$y = x_1 \cdot x_2$	$\%e_y = \sqrt{\%e_{x_1}^2 + \%e_{x_2}^2}$	$y = \ln x$	$e_y = \dfrac{e_x}{x}$
$y = \dfrac{x_1}{x_2}$	$\%e_y = \sqrt{\%e_{x_1}^2 + \%e_{x_2}^2}$	$y = 10^x$	$\dfrac{e_y}{y} = (\ln 10)e_x \approx 2.302\,6\,e_x$
		$y = e^x$	$\dfrac{e_y}{y} = e_x$

a. x represents a variable and a represents a constant that has no uncertainty.
b. e_x/x is the relative error in x and $\%e_x$ is $100 \times e_x/x$.

$-\log[H^+] = pH$
$\log[H^+] = -pH$
$10^{\log[H^+]} = 10^{-pH}$
$[H^+] = 10^{-pH}$

EXAMPLE **Uncertainty in H$^+$ Concentration**

Consider the function $pH = -\log[H^+]$, where $[H^+]$ is the molarity of H^+. For $pH = 5.21 \pm 0.03$, find $[H^+]$ and its uncertainty.

Solution First solve the equation $pH = -\log[H^+]$ for $[H^+]$: Whenever $a = b$, then $10^a = 10^b$. If $pH = -\log[H^+]$, then $\log[H^+] = -pH$ and $10^{\log[H^+]} = 10^{-pH}$. But $10^{\log[H^+]} = [H^+]$. We therefore need to find the uncertainty in the equation

$$[H^+] = 10^{-pH} = 10^{-(5.21\pm0.03)}$$

In Table 3-1, the relevant function is $y = 10^x$, in which $y = [H^+]$ and $x = -(5.21 \pm 0.03)$. For $y = 10^x$, the table tells us that $e_y/y = 2.302\,6\,e_x$.

$$\frac{e_y}{y} = 2.302\,6\,e_x = (2.302\,6)(0.03) = 0.069\,1 \qquad (3\text{-}12)$$

The relative uncertainty in $y\,(= e_y/y)$ is 0.069 1. Inserting the value $y = 10^{-5.21} = 6.17 \times 10^{-6}$ into Equation 3-12 gives the answer:

$$\frac{e_y}{y} = \frac{e_y}{6.17 \times 10^{-6}} = 0.069\,1 \quad \Rightarrow \quad e_y = 4.26 \times 10^{-7}$$

The concentration of H^+ is $6.17\,(\pm0.426) \times 10^{-6} = 6.2\,(\pm0.4) \times 10^{-6}$ M. An uncertainty of 0.03 in pH gives an uncertainty of 7% in $[H^+]$. Notice that extra digits were retained in the intermediate results and were not rounded off until the final answer.

Test Yourself If uncertainty in pH is doubled to ±0.06, what is the relative uncertainty in $[H^+]$? (**Answer:** 14%)

3-5 Propagation of Uncertainty from Systematic Error

Systematic error occurs in some common situations and is treated differently from random error in arithmetic operations. We will look at examples of systematic error in molecular mass and volumetric glassware.[3]

Uncertainty in Atomic Mass: The Rectangular Distribution

The periodic table inside the cover of this book gives the atomic mass of oxygen as $15.999\,4 \pm 0.000\,3$ g/mol. The uncertainty is *not* mainly from random error in measuring the atomic mass. The uncertainty is predominantly from isotopic variation in samples of oxygen from different sources. That is, oxygen from one source could have a mean atomic mass of 15.999 1 and oxygen from another source could have an atomic mass of 15.999 7. The atomic mass of oxygen in a particular lot of reagent has a *systematic* uncertainty. It could be relatively constant at 15.999 7 or 15.999 1, or any value in between, with only a small random variation around the mean value.

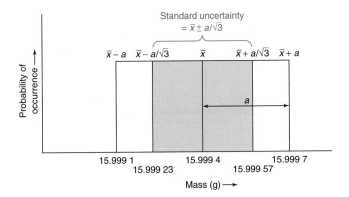

Oxygen atomic mass from different sources approximates a *rectangular distribution* (Figure 3-4). There is approximately equal probability of finding any atomic mass between 15.999 1 and 15.999 7 and negligible probability of finding an atomic mass outside of this range. We call the mean value $\bar{x}$ (= 15.999 4) and the range on either side of the mean a (= 0.000 3). The *standard deviation* (defined in Section 4-1) for this distribution, also called the *standard uncertainty*, is $\pm a/\sqrt{3} = \pm 0.000\ 3/\sqrt{3} = \pm 0.000\ 17$. The standard deviation is used as a measure of the uncertainty in atomic mass.

Uncertainty in Molecular Mass

What is the uncertainty in molecular mass of O_2? If the mass of each oxygen atom were at the upper limit of the standard uncertainty range in Figure 3-4 (15.999 57), the mass of O_2 is 2 × 15.999 57 = 31.999 14 g/mol. If the mass of each oxygen atom were at the lower limit of the standard uncertainty range (15.999 23), the mass of O_2 is 2 × 15.999 23 = 31.998 46 g/mol. The mass of O_2 is somewhere in the range 31.998 8 ± 0.000 34. The uncertainty of the mass of n atoms is n × (standard uncertainty of one atom) = 2 × (±0.000 17) = ±0.000 34. The uncertainty is not $\pm \sqrt{0.000\ 17^2 + 0.000\ 17^2} = \pm 0.000\ 24$. *For systematic uncertainty, we add the uncertainties of each term in a sum or difference.*

Now let's find the standard uncertainty in molecular mass of C_2H_4. Uncertainties in the periodic table are 0.000 8 g/mol for carbon and 0.000 07 g/mol for hydrogen. To find the standard uncertainties, divide these uncertainties by $\sqrt{3}$:

$$\text{Atomic mass of C} = 12.010\ 7 \pm 0.000\ 8/\sqrt{3} = 12.010\ 7 \pm 0.000\ 46$$

$$\text{Atomic mass of H} = 1.007\ 94 \pm 0.000\ 07/\sqrt{3} = 1.007\ 94 \pm 0.000\ 040$$

Uncertainties in the masses of the atoms in C_2H_4 are obtained by multiplying the standard uncertainties by the number of atoms of each type:

$$
\begin{array}{llll}
2C: & 2(12.010\ 7 \pm 0.000\ 46) & = 24.021\ 4 & \pm\ 0.000\ 92 \leftarrow 2 \times 0.000\ 46 \\
4H: & 4(1.007\ 94 \pm 0.000\ 040) & = \underline{4.031\ 76 \pm 0.000\ 16} \leftarrow 4 \times 0.000\ 040 \\
& & 28.053\ 16 \pm\ ?
\end{array}
$$

(3-13)

For the uncertainty in the sum of the masses of 2C + 4H, we use Equation 3-5, which applies to random error, because the uncertainties in the masses of C and H are independent of each other. One might be positive and one might be negative. So the molecular mass of C_2H_4 is

$$28.053\ 16 \pm \sqrt{0.000\ 92^2 + 0.000\ 16^2}$$

$$28.053\ 16 \pm 0.000\ 93$$

$$28.053\ 2 \pm 0.000\ 9 \text{ g/mol}$$

In the last step, we rounded off so that the last significant digit is the first uncertain digit.

Propagation of systematic uncertainty:
Uncertainty in mass of n identical atoms
$= n$ × (standard uncertainty in atomic mass)
$= n$ × (uncertainty listed in periodic table)/$\sqrt{3}$

Use the rule for propagation of random uncertainty for the sum of atomic masses of different elements because uncertainties for different elements are independent.

Multiple Deliveries from One Pipet: The Triangular Distribution

A 25-mL Class A volumetric pipet is certified by the manufacturer to deliver 25.00 ± 0.03 mL. The volume delivered by a given pipet is reproducible but can be in the range 24.97 to 25.03 mL. However, the pipet manufacturer works hard to make the volume close to 25.00 mL. In such case, we approximate the volumes of a large number of pipets by the *triangular distribution* in Figure 3-5. There is highest probability that the pipet will deliver

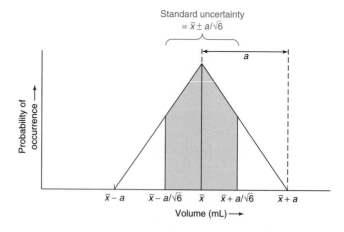

FIGURE 3-5 *Triangular distribution* for volumetric glassware including volumetric flasks and transfer pipets. The standard uncertainty interval (standard deviation) shown in color is $a/\sqrt{6}$.

Standard uncertainty
$= \bar{x} \pm a/\sqrt{6}$

Probability of occurrence →

$\bar{x} - a$ $\bar{x} - a/\sqrt{6}$ $\bar{x}$ $\bar{x} + a/\sqrt{6}$ $\bar{x} + a$

Volume (mL) →

25.00 mL. The probability falls off approximately in a linear manner as the volume deviates from 25.00 mL. There is negligible probability that a volume outside of 25.00 ± 0.03 mL will be delivered. The *standard uncertainty* (standard deviation) in the triangular distribution is $\pm a/\sqrt{6} = \pm 0.03/\sqrt{6} = \pm 0.012$ mL.

If you use an uncalibrated 25-mL Class A volumetric pipet four times to deliver a total of 100 mL, what is the uncertainty in 100 mL? The uncertainty is a systematic error, so the uncertainty in four pipet volumes is like the uncertainty in the mass of 4 mol of oxygen: The standard uncertainty is $\pm 4 \times 0.012 = \pm 0.048$ mL, not $\pm \sqrt{0.012^2 + 0.012^2 + 0.012^2 + 0.012^2} = \pm 0.024$ mL.

The difference between 25.00 mL and the actual volume delivered by a particular pipet is a *systematic* error. It is always the same, within a small random error. You could calibrate a pipet by weighing the water it delivers, as in Section 2-9. Calibration eliminates systematic error, because we would know that the pipet always delivers, say, 24.991 ± 0.006 mL. The uncertainty (± 0.006 mL) is random error.

Calibration improves certainty by removing systematic error. If a calibrated pipet delivers a mean volume of 24.991 mL with a standard uncertainty of ± 0.006 mL, and you deliver four aliquots, the volume delivered is $4 \times 24.991 = 99.964$ mL and the uncertainty is $\pm \sqrt{0.006^2 + 0.006^2 + 0.006^2 + 0.006^2} = \pm 0.012$ mL. For an uncalibrated pipet, the uncertainty is ± 0.048 mL.

Calibrated pipet volume $= 99.964 \pm 0.012$ mL

Uncalibrated pipet volume $= 100.00 \pm 0.05$ mL

0.006 mL is the standard deviation (defined in Chapter 4) measured for multiple deliveries of water.

Terms to Understand

absolute uncertainty	characteristic	mantissa	relative uncertainty
accuracy	determinate error	natural logarithm	significant figure
antilogarithm	indeterminate error	precision	systematic error
certified reference material	logarithm	random error	

Summary

The number of significant digits in a number is the minimum required to write the number in scientific notation. The first uncertain digit is the last significant figure. In addition and subtraction, the last significant figure is determined by the number with the fewest decimal places (when all exponents are equal). In multiplication and division, the number of figures is usually limited by the factor with the smallest number of digits. The number of figures in the mantissa of the logarithm of a quantity should equal the number of significant figures in the quantity. Random (indeterminate) error affects the precision (reproducibility) of a result, whereas systematic (determinate) error affects the accuracy (nearness to the "true" value). Systematic error can be discovered and eliminated by a clever person, but some random error is always present. We must strive to eliminate systematic errors in all measurements. For random errors, propagation of uncertainty in addition and subtraction requires absolute uncertainties ($e_3 = \sqrt{e_1^2 + e_2^2}$), whereas multiplication and division utilize relative uncertainties ($\%e_3 = \sqrt{\%e_1^2 + \%e_2^2}$). Other rules for propagation of random error are found in Table 3-1. Always retain more digits than necessary during a calculation and round off to the appropriate number of digits at the end. Systematic error in the mass on n atoms of one element is n times the standard uncertainty in mass of that element. For a rectangular distribution of atomic mass, standard uncertainty in atomic mass is the uncertainty listed in the periodic table divided by $\sqrt{3}$. Uncertainty in the mass of a molecule with several elements is computed from the sum of squares of the systematic uncertainty for each element.

Exercises

3-A. Write each answer with a reasonable number of figures. Find the absolute and percent relative uncertainty for each answer.

(a) $[12.41 (\pm 0.09) \div 4.16 (\pm 0.01)] \times 7.068\ 2 (\pm 0.000\ 4) = ?$

(b) $[3.26 (\pm 0.10) \times 8.47 (\pm 0.05)] - 0.18 (\pm 0.06) = ?$

(c) $6.843 (\pm 0.008) \times 10^4 \div [2.09 (\pm 0.04) - 1.63 (\pm 0.01)] = ?$

(d) $\sqrt{3.24 \pm 0.08} = ?$

(e) $(3.24 \pm 0.08)^4 = ?$

(f) $\log (3.24 \pm 0.08) = ?$

(g) $10^{3.24 \pm 0.08} = ?$

3-B. (a) How many milliliters of $53.4 (\pm 0.4)$ wt% NaOH with a density of $1.52 (\pm 0.01)$ g/mL will you need to prepare 2.000 L of 0.169 M NaOH?

(b) If the uncertainty in delivering NaOH is ± 0.01 mL, calculate the absolute uncertainty in the molarity (0.169 M). Assume there is negligible uncertainty in the formula mass of NaOH and in the final volume (2.000 L).

3-C. We have a $37.0 (\pm 0.5)$ wt% HCl solution with a density of $1.18 (\pm 0.01)$ g/mL. To deliver 0.050 0 mol of HCl requires 4.18 mL of solution. If the uncertainty that can be tolerated in 0.050 0 mol is $\pm 2\%$, how big can the absolute uncertainty in 4.18 mL be? (*Caution:* In this problem, you have to work backward. You would normally compute the uncertainty in mol HCl from the uncertainty in volume:

$$\text{mol HCl} = \frac{\text{mL solution} \times \dfrac{\text{g solution}}{\text{mL solution}} \times \dfrac{\text{g HCl}}{\text{g solution}}}{\dfrac{\text{g HCl}}{\text{mol HCl}}}$$

But, in this case, we know the uncertainty in mol HCl (2%) and we need to find what uncertainty in mL solution leads to that 2% uncertainty. The arithmetic has the form $a = b \times c \times d$, for which $\%e_a^2 = \%e_b^2 + \%e_c^2 + \%e_d^2$. If we know $\%e_a$, $\%e_c$, and $\%e_d$, we can find $\%e_b$ by subtraction: $\%e_b^2 = \%e_a^2 - \%e_c^2 - \%e_d^2$.)

3-D. Compute the molecular mass and its standard uncertainty for NH_3. What is the percent relative uncertainty in molecular mass?

Problems

Significant Figures

3-1. How many significant figures are there in the following numbers?

(a) 1.903 0   (b) 0.039 10   (c) 1.40×10^4

3-2. Round each number as indicated:

(a) 1.236 7 to 4 significant figures

(b) 1.238 4 to 4 significant figures

(c) 0.135 2 to 3 significant figures

(d) 2.051 to 2 significant figures

(e) 2.005 0 to 3 significant figures

3-3. Round each number to three significant figures:

(a) 0.216 74   (b) 0.216 5   (c) 0.216 500 3

3-4. *Vernier scale.* The figure below shows a scale found on instruments such as a micrometer caliper used for accurately measuring dimensions of objects. The lower scale slides along the upper scale and is used to interpolate between the markings on the upper scale. In (a), the reading (at the left-hand 0 of the lower scale) is between 1.4 and 1.5 on the upper scale. To find the exact reading, observe which mark on the lower scale is aligned with a mark on the upper scale. Because the 6 on the lower scale is aligned with the upper scale, the correct reading is 1.46. Write the correct readings in (b) and (c) and indicate how many significant figures are in each reading.

3-5. Write each answer with the correct number of digits.

(a) $1.021 + 2.69 = 3.711$

(b) $12.3 - 1.63 = 10.67$

(c) $4.34 \times 9.2 = 39.928$

(d) $0.060\ 2 \div (2.113 \times 10^4) = 2.849\ 03 \times 10^{-6}$

(e) $\log(4.218 \times 10^{12}) = ?$

(f) antilog $(-3.22) = ?$

(g) $10^{2.384} = ?$

3-6. Write the formula mass of (a) BaF_2 and (b) $C_6H_4O_4$ with a reasonable number of digits. Use the periodic table inside the cover of this book to find atomic masses.

Figure for Problem 3-4.

3-7. Write each answer with the correct number of significant figures.

(a) $1.0 + 2.1 + 3.4 + 5.8 = 12.300\ 0$

(b) $106.9 - 31.4 = 75.500\ 0$

(c) $107.868 - (2.113 \times 10^2) + (5.623 \times 10^3) = 5\ 519.568$

(d) $(26.14/37.62) \times 4.38 = 3.043\ 413$

(e) $(26.14/(37.62 \times 10^8)) \times (4.38 \times 10^{-2}) = 3.043\ 413 \times 10^{-10}$

(f) $(26.14/3.38) + 4.2 = 11.933\ 7$

(g) $\log(3.98 \times 10^4) = 4.599\ 9$

(h) $10^{-6.31} = 4.897\ 79 \times 10^{-7}$

3-8. 🖩 *Controlling the appearance of a graph.* Figure 3-3 requires gridlines to read buret corrections. In this exercise, you will format a graph so that it looks like Figure 3-3. Follow the procedure in Section 2-11 to graph the data in the following table. For Excel 2007, insert a Chart of the type Scatter with data points connected by straight lines. Delete the legend and title. With Chart Tools, Layout, Axis Titles, add labels for both axes. Click any number on the abscissa (*x*-axis) and go to Chart Tools, Format. In Format Selection, Axis Options, choose Minimum = 0, Maximum = 50, Major unit = 10, and Minor unit = 1. In Format Selection, Number, choose Number and set Decimal places = 0. In a similar manner, set the ordinate (*y*-axis) to run from -0.04 to $+0.05$ with a Major unit of 0.02 and a Minor unit of 0.01. To display gridlines, go to Chart Tools, Layout, and Gridlines. For Primary Horizontal Gridlines, select Major & Minor Gridlines. For Primary Vertical Gridlines, select Major & Minor Gridlines. To move *x*-axis labels from the middle of the chart to the bottom, click a number on the *y*-axis (not the *x*-axis) and select Chart Tools, Layout, Format Selection. In Axis Options, choose Horizontal axis crosses Axis value and type in -0.04. Close the Format Axis window and your graph should look like Figure 3-3.

For earlier versions of Excel, select Chart Type XY (Scatter) showing data points connected by straight lines. Double click the *x*-axis and select the Scale tab. Set Minimum = 0, Maximum = 50, Major unit = 10, and Minor unit = 1. Select the Number tab and highlight Number. Set Decimal places = 0. In a similar manner, set the ordinate to run from -0.04 to $+0.05$ with a Major unit of 0.02 and a Minor unit of 0.01, as in Figure 3-3. The spreadsheet might overrule you. Continue to reset the limits you want and click OK each time until the graph looks the way you intend. To add gridlines, click inside the graph, go to the Chart menu and select Chart Options. Select the Gridlines tab and check both sets of Major gridlines and Minor gridlines and click OK. In Chart Options, select the Legend tab and deselect Show legend. To move *x*-axis numbers to the bottom of the chart, double click the *y*-axis (not the *x*-axis) and

select the Scale tab. Set "Value (x) axis crosses at" to -0.04. Click OK and your graph should look like Figure 3-3.

Volume (mL)	Correction (mL)
0.03	0.00
10.04	0.04
20.03	0.02
29.98	-0.03
40.00	0.00
49.97	0.03

Types of Error

3-9. Why do we use quotation marks around the word *true* in the statement that accuracy refers to how close a measured value is to the "true" value?

3-10. Explain the difference between systematic and random error.

3-11. Suppose that in a gravimetric analysis, you forget to dry the filter crucibles before collecting precipitate. After filtering the product, you dry the product and crucible thoroughly before weighing them. Is the apparent mass of product always high or always low? Is the error in mass systematic or random?

3-12. State whether the errors in (a)–(d) are random or systematic:

(a) A 25-mL transfer pipet consistently delivers 25.031 ± 0.009 mL.

(b) A 10-mL buret consistently delivers 1.98 ± 0.01 mL when drained from exactly 0 to exactly 2 mL and consistently delivers 2.03 mL ± 0.02 mL when drained from 2 to 4 mL.

(c) A 10-mL buret delivered $1.983\ 9$ g of water when drained from exactly 0.00 to 2.00 mL. The next time I delivered water from the 0.00 to the 2.00 mL mark, the delivered mass was $1.990\ 0$ g.

(d) Four consecutive 20.0-μL injections of a solution into a chromatograph were made and the area of a particular peak was 4 383, 4 410, 4 401, and 4 390 units.

3-13. Cheryl, Cynthia, Carmen, and Chastity shot the targets below at Girl Scout camp. Match each target with the proper description.

(a) accurate and precise

(b) accurate but not precise

(c) precise but not accurate

(d) neither precise nor accurate

3-14. Rewrite the number 3.123 56 ($\pm 0.167\ 89\%$) in the forms (a) number ($\pm$ absolute uncertainty) and (b) number ($\pm$ percent relative uncertainty) with an appropriate number of digits.

Cheryl   Cynthia   Carmen   Chastity

Figure for Problem 3-13.

Propagation of Uncertainty

3-15. Find the absolute and percent relative uncertainty and express each answer with a reasonable number of significant figures.

(a) $6.2 (\pm 0.2) - 4.1 (\pm 0.1) = ?$

(b) $9.43 (\pm 0.05) \times 0.016 (\pm 0.001) = ?$

(c) $[6.2 (\pm 0.2) - 4.1 (\pm 0.1)] \div 9.43 (\pm 0.05) = ?$

(d) $9.43 (\pm 0.05) \times \{[6.2 (\pm 0.2) \times 10^{-3}] + [4.1 (\pm 0.1) \times 10^{-3}]\} = ?$

3-16. Find the absolute and percent relative uncertainty and express each answer with a reasonable number of significant figures.

(a) $9.23 (\pm 0.03) + 4.21 (\pm 0.02) - 3.26 (\pm 0.06) = ?$

(b) $91.3 (\pm 1.0) \times 40.3 (\pm 0.2)/21.1 (\pm 0.2) = ?$

(c) $[4.97 (\pm 0.05) - 1.86 (\pm 0.01)]/21.1 (\pm 0.2) = ?$

(d) $2.016 4 (\pm 0.000 8) + 1.233 (\pm 0.002) + 4.61 (\pm 0.01) = ?$

(e) $2.016 4 (\pm 0.000 8) \times 10^3 + 1.233 (\pm 0.002) \times 10^2 + 4.61 (\pm 0.01) \times 10^1 = ?$

(f) $[3.14 (\pm 0.05)]^{1/3} = ?$

(g) $\log [3.14 (\pm 0.05)] = ?$

3-17. Verify the following calculations:

(a) $\sqrt{3.141 5 (\pm 0.001 1)} = 1.772 4_3 (\pm 0.000 3_1)$

(b) $\log[3.141 5 (\pm 0.001 1)] = 0.497 1_4 (\pm 0.000 1_5)$

(c) $\text{antilog}[3.141 5 (\pm 0.001 1)] = 1.385_2 (\pm 0.003_5) \times 10^3$

(d) $\ln[3.141 5 (\pm 0.001 1)] = 1.144 7_0 (\pm 0.000 3_5)$

(e) $\log\left(\dfrac{\sqrt{0.104(\pm 0.006)}}{0.051 1(\pm 0.000 9)}\right) = 0.80_0 (\pm 0.01_5)$

3-18. (a) Show that the formula mass of NaCl is $58.443 (\pm 0.001_2)$ g/mol.

(b) To prepare a solution of NaCl, you weigh out $2.634 (\pm 0.002)$ g and dissolve it in a volumetric flask whose volume is $100.00 (\pm 0.08)$ mL. Express the molarity of the solution, along with its uncertainty, with an appropriate number of digits.

3-19. What is the true mass of water weighed at 24°C in the air if the apparent mass is $1.034 6 \pm 0.000 2$ g? The density of air is $0.001 2 \pm 0.000 1$ g/mL and the density of balance weights is 8.0 ± 0.5 g/mL. The uncertainty in the density of water in Table 2-7 is negligible in comparison to the uncertainty in the density of air.

3-20. Twelve dietary iron tablets were analyzed by the gravimetric procedure in Section 1-4 and the final mass of Fe_2O_3 (FM 159.688) was $0.277_4 \pm 0.001_8$ g. Find the average mass of Fe per tablet. (Relative uncertainties in atomic masses are small compared with relative uncertainty in the mass of Fe_2O_3. Neglect uncertainties in atomic masses in this problem.)

3-21. We can measure the concentration of HCl solution by reaction with pure sodium carbonate: $2H^+ + Na_2CO_3 \rightarrow 2Na^+ + H_2O + CO_2$. Complete reaction with $0.967 4 \pm 0.000 9$ g of Na_2CO_3 (FM $105.988 4 \pm 0.000 7$) required 27.35 ± 0.04 mL of HCl. Find the molarity of the HCl and its absolute uncertainty.

3-22. Avogadro's number can be computed from the following measured properties of pure crystalline silicon:[4] (1) atomic mass (obtained from the mass and abundance of each isotope), (2) density of the crystal, (3) size of the unit cell (the smallest repeating unit in the crystal), and (4) number of atoms in the unit cell. For silicon, the mass is $m_{Si} = 28.085 384 2 (35)$ g/mol, where 35 is the uncertainty (standard deviation) in the last two digits. The density is $\rho = 2.329 031 9 (18)$ g/cm^3, the size of the cubic unit cell is $c_0 = 5.431 020 36 (33) \times 10^{-8}$ cm, and there are 8 atoms per unit cell. Avogadro's number is computed from the equation

$$N_A = \frac{m_{Si}}{(\rho c_0^3)/8}$$

From the measured properties and their uncertainties, compute Avogadro's number and its uncertainty. To find the uncertainty of c_0^3, use the function $y = x^a$ in Table 3-1.

3-23. Express the molecular mass ($\pm$uncertainty) of $C_9H_9O_6N_3$ with the correct number of significant figures.

3-24. *Estimating uncertainty in Keeling's CO_2 measurements.* The manometer in Box 3-2 has the following key uncertainties:

Volume of large vessel: $5.013 8_2 \pm 0.000 5$ L
Volume of small vessel: $3.793 0 \pm 0.000 9$ mL
Pressure: ± 0.03 mm Hg in ~400 mm Hg
Temperature: ± 0.03 K in ~300 K

Which factor has the largest percent relative uncertainty? If one uncertainty is sufficiently larger than the others, then the uncertainty in measuring CO_2 is governed by that factor. Considering just the most uncertain factor, estimate the uncertainty in 400 ppm CO_2 in air.

4 | Statistics

Red blood cells, also called erythrocytes, transport oxygen bound to the protein hemoglobin. [Susumu Nishinaga/Photo Researchers.]

All measurements contain experimental error, so it is never possible to be completely certain of a result. Nevertheless, we often seek the answers to questions such as "Is my red blood cell count today higher than usual?" If today's count is twice as high as usual, it is probably truly higher than normal. But what if the "high" count is not excessively above "normal" counts?

Count on "normal" days	Today's count
5.1	
5.3	
4.8 $\left. \right\} \times 10^6$ cells/μL	5.6×10^6 cells/μL
5.4	
5.2	
Average = 5.16	

The number 5.6 is higher than the five normal values, but the random variation in normal values might lead us to expect that 5.6 will be observed on some "normal" days.

We will learn in Section 4-3 that there is only a 1.3% random chance of observing a value as far from the average as 5.6 on a "normal" day. It is still up to you to decide what to do with this information.

Experimental measurements always contain some variability, so no conclusion can be drawn with certainty. Statistics gives us tools to accept conclusions that have a high probability of being correct and to reject conclusions that do not.[1]

4-1 Gaussian Distribution

We say that the variation in experimental data is *normally distributed* when replicate measurements exhibit the bell-shaped distribution in Figure 4-1. It is equally probable that a measurement will be higher or lower than the mean. The probability of observing any value decreases as its distance from the mean increases.

If an experiment is repeated a great many times and if the errors are purely random, then the results tend to cluster symmetrically about the average value (Figure 4-1). The more times the experiment is repeated, the more closely the results approach an ideal smooth curve called the **Gaussian distribution.** In general, we cannot make so many measurements in a lab experiment. We are more likely to repeat an experiment 3 to 5 times than 2 000 times. However, from the small set of results, we can estimate parameters that describe the large set. We can then make estimates of statistical behavior from the small number of measurements.

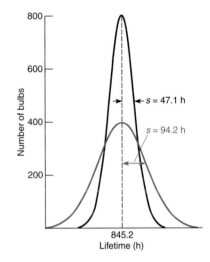

FIGURE 4-1 Bar graph and Gaussian curve describing the lifetimes of a hypothetical set of incandescent light bulbs. The smooth curve has the same mean, standard deviation, and area as the bar graph. Any finite set of data, however, will differ from the bell-shaped curve. The more measurements we make, the closer the results will come to the smooth curve.

FIGURE 4-2 Gaussian curves for two sets of light bulbs, one having a standard deviation half as great as the other. The number of bulbs described by each curve is the same.

Mean Value and Standard Deviation

In the hypothetical case in Figure 4-1, a manufacturer tested the lifetimes of 4 768 electric light bulbs. The bar graph shows the number of bulbs with a lifetime in each 20-h interval. Lifetimes approximate a Gaussian distribution because variations in the construction of light bulbs, such as filament thickness and quality of attachments, are random. The smooth curve is the Gaussian distribution that best fits the data. Any finite set of data will vary somewhat from the Gaussian curve.

Light bulb lifetimes, and the corresponding Gaussian curve, are characterized by two parameters. The arithmetic **mean,** $\bar{x}$—also called the **average**—is the sum of the measured values divided by n, the number of measurements:

Mean:
$$\bar{x} = \frac{\sum_i x_i}{n}$$
(4-1)

The mean gives the center of the distribution. The standard deviation measures the width of the distribution.

where x_i is the lifetime of an individual bulb. The Greek capital sigma, Σ, means summation: $\sum_i x_i = x_1 + x_2 + x_3 + \cdots + x_n$. In Figure 4-1, the mean value is 845.2 h.

The **standard deviation,** s, measures how closely the data are clustered about the mean. *The smaller the standard deviation, the more closely the data are clustered about the mean* (Figure 4-2).

Standard deviation:
$$s = \sqrt{\frac{\sum_i (x_i - \bar{x})^2}{n - 1}}$$
(4-2)

An experiment that produces a small standard deviation is more *precise* than one that produces a large standard deviation. Greater precision does not necessarily imply greater *accuracy*, which is nearness to the "truth."

In Figure 4-1, $s = 94.2$ h. A set of light bulbs having a small standard deviation in lifetime is more uniformly manufactured than a set with a large standard deviation.

For an *infinite* set of data, the mean is designated by the lowercase Greek letter mu, μ (the population mean), and the standard deviation is written as a lowercase Greek sigma, σ (the population standard deviation). We can never measure μ and σ, but the values of $\bar{x}$ and s approach μ and σ as the number of measurements increases.

As the number of measurements increases, $\bar{x}$ approaches μ, *if there is no systematic error.*

The quantity $n - 1$ in Equation 4-2 is called the **degrees of freedom.** The square of the standard deviation is called the **variance.** The standard deviation expressed as a percentage of the mean value ($= 100 \times s/\bar{x}$) is called the *relative standard deviation* or the *coefficient of variation.*

Coefficient of variation $= 100 \times \dfrac{s}{\bar{x}}$

EXAMPLE Mean and Standard Deviation

Find the average, standard deviation, and coefficient of variation for 821, 783, 834, and 855.

Solution The average is

$$\bar{x} = \frac{821 + 783 + 834 + 855}{4} = 823._2$$

To avoid accumulating round-off errors, retain one more digit than was present in the original data. The standard deviation is

$$s = \sqrt{\frac{(821 - 823.2)^2 + (783 - 823.2)^2 + (834 - 823.2)^2 + (855 - 823.2)^2}{(4 - 1)}} = 30._3$$

The average and the standard deviation should both end at the *same decimal place*. For $\bar{x} = 823._2$, we will write $s = 30._3$. The coefficient of variation is the percent relative uncertainty:

$$\text{Coefficient of variation} = 100 \times \frac{s}{\bar{x}} = 100 \times \frac{30._3}{823._2} = 3.7\%$$

Test Yourself If each of the four numbers 821, 783, 834, and 855 in the example is divided by 2, how will the mean, standard deviation, and coefficient of variation be affected? (***Answer:*** $\bar{x}$ and s will be divided by 2, but the coefficient of variation is unchanged)

Spreadsheets have built-in functions for the average and standard deviation. In the adjacent spreadsheet, data points are entered in cells B1 through B4. The average in cell B5 is computed with the statement "= AVERAGE(B1:B4)". B1:B4 means cells B1, B2, B3, and B4. The standard deviation in cell B6 is computed with "= STDEV(B1:B4)".

For ease of reading, cells B5 and B6 were set to display two decimal places. A heavy line was placed beneath cell B4 in Excel 2007 by highlighting the cell, going to Home, Font, and selecting the Border icon. In earlier versions of Excel, highlight cell B4, go to Format, Cells, and select Border.

Significant Figures in Mean and Standard Deviation

We commonly express experimental results in the form $\bar{x} \pm s(n = \_)$, where n is the number of data points. It is sensible to write the preceding result as 823 ± 30 ($n = 4$) or even $8.2\ (\pm 0.3) \times 10^2$ ($n = 4$) to indicate that the mean has just two significant figures. The expressions 823 ± 30 and $8.2\ (\pm 0.3) \times 10^2$ are not suitable for continued calculations in which $\bar{x}$ and s are intermediate results. Retain one or more insignificant digits to avoid introducing round-off errors into subsequent work. Try not to go into cardiac arrest over significant figures when you see $823._2 \pm 30._3$ as the answer to a problem in this book.

Standard Deviation and Probability

The formula for a Gaussian curve is

Gaussian curve:
$$y = \frac{1}{\sigma\sqrt{2\pi}}\, e^{-(x-\mu)^2/2\sigma^2} \qquad (4\text{-}3)$$

where e (= 2.718 28 . . .) is the base of the natural logarithm. For a finite set of data, we approximate μ by $\bar{x}$ and σ by s. A graph of Equation 4-3 is shown in Figure 4-3, in which the values $\sigma = 1$ and $\mu = 0$ are used for simplicity. The maximum value of y is at $x = \mu$ and the curve is symmetric about $x = \mu$.

It is useful to express deviations from the mean value in multiples, z, of the standard deviation. That is, we transform x into z, given by

$$z = \frac{x - \mu}{\sigma} \approx \frac{x - \bar{x}}{s} \qquad (4\text{-}4)$$

Learn to use the standard deviation function on your calculator and see that you get $s = 30.269\ 6 \ldots$.

	A	B
1		821
2		783
3		834
4		855
5	Average =	823.25
6	Std dev =	30.27
7	B5 = AVERAGE(B1:B4)	
8	B6 = STDEV(B1:B4)	

Do not round off during a calculation. Retain all the extra digits in your calculator.

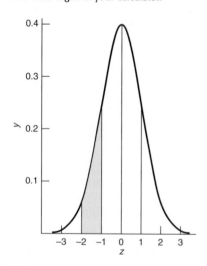

FIGURE 4-3 A Gaussian curve in which $\mu = 0$ and $\sigma = 1$. A Gaussian curve whose area is unity is called a normal error curve. The abscissa $z = (x - \mu)/\sigma$ is the distance away from the mean, measured in units of the standard deviation. When $z = 2$, we are two standard deviations away from the mean.

TABLE 4–1 Ordinate and area for the normal (Gaussian) error curve, $y = \dfrac{1}{\sqrt{2\pi}}e^{-z^2/2}$

$|z|^a$	y	Area^b	$|z|$	y	Area	$|z|$	y	Area
0.0	0.398 9	0.000 0	1.4	0.149 7	0.419 2	2.8	0.007 9	0.497 4
0.1	0.397 0	0.039 8	1.5	0.129 5	0.433 2	2.9	0.006 0	0.498 1
0.2	0.391 0	0.079 3	1.6	0.110 9	0.445 2	3.0	0.004 4	0.498 650
0.3	0.381 4	0.117 9	1.7	0.094 1	0.455 4	3.1	0.003 3	0.499 032
0.4	0.368 3	0.155 4	1.8	0.079 0	0.464 1	3.2	0.002 4	0.499 313
0.5	0.352 1	0.191 5	1.9	0.065 6	0.471 3	3.3	0.001 7	0.499 517
0.6	0.333 2	0.225 8	2.0	0.054 0	0.477 3	3.4	0.001 2	0.499 663
0.7	0.312 3	0.258 0	2.1	0.044 0	0.482 1	3.5	0.000 9	0.499 767
0.8	0.289 7	0.288 1	2.2	0.035 5	0.486 1	3.6	0.000 6	0.499 841
0.9	0.266 1	0.315 9	2.3	0.028 3	0.489 3	3.7	0.000 4	0.499 904
1.0	0.242 0	0.341 3	2.4	0.022 4	0.491 8	3.8	0.000 3	0.499 928
1.1	0.217 9	0.364 3	2.5	0.017 5	0.493 8	3.9	0.000 2	0.499 952
1.2	0.194 2	0.384 9	2.6	0.013 6	0.495 3	4.0	0.000 1	0.499 968
1.3	0.171 4	0.403 2	2.7	0.010 4	0.496 5	∞	0	0.5

a. $z = (x - \mu)/\sigma$.

b. The area refers to the area between $z = 0$ and $z =$ the value in the table. Thus the area from $z = 0$ to $z = 1.4$ is 0.419 2. The area from $z = -0.7$ to $z = 0$ is the same as from $z = 0$ to $z = 0.7$. The area from $z = -0.5$ to $z = +0.3$ is $(0.191\ 5 + 0.117\ 9) = 0.309\ 4$. The total area between $z = -\infty$ and $z = +\infty$ is unity.

The probability of measuring z in a certain *range* is equal to the *area* of that range. For example, the probability of observing z between –2 and –1 is 0.136. This probability corresponds to the shaded area in Figure 4-3. The area under each portion of the Gaussian curve is given in Table 4-1. Because the sum of the probabilities of all the measurements must be unity, the area under the whole curve from $z = -\infty$ to $z = +\infty$ must be unity. The number $1/(\sigma\sqrt{2\pi})$ in Equation 4-3 is called the *normalization factor*. It guarantees that the area under the entire curve is unity. A Gaussian curve with unit area is called a *normal error curve*.

When $z = +1$, x is one standard deviation above the mean. When $z = -2$, x is two standard deviations below the mean.

EXAMPLE Area Under a Gaussian Curve

Suppose the manufacturer of the bulbs used for Figure 4-1 offers to replace free of charge any bulb that burns out in less than 600 hours. If she plans to sell a million bulbs, how many extra bulbs should she keep available as replacements?

Solution We need to express the desired interval in multiples of the standard deviation and then find the area of the interval in Table 4-1. Because $\bar{x} = 845.2$ and $s = 94.2$, $z = (600 - 845.2)/94.2 = -2.60$. The area under the curve between the mean value and $z = -2.60$ is 0.495 3 in Table 4-1. The entire area from $-\infty$ to the mean value is 0.500 0, so the area from $-\infty$ to –2.60 must be 0.500 0 – 0.495 3 = 0.004 7. The area to the left of 600 hours in Figure 4-1 is only 0.47% of the entire area under the curve. Only 0.47% of the bulbs are expected to fail in fewer than 600 h. If the manufacturer sells 1 million bulbs a year, she should make 4 700 extra bulbs to meet the replacement demand.

Test Yourself If the manufacturer will replace bulbs that burn out in less than 620 hours, how many extras should she make? (*Answer:* $z \approx -2.4$, area $\approx 0.008\ 2 = 8\ 200$ bulbs)

EXAMPLE   Using a Spreadsheet to Find Area Under a Gaussian Curve

What fraction of bulbs is expected to have a lifetime between 900 and 1 000 h?

Solution *We need to find the fraction of the area of the Gaussian curve between $x = 900$ and $x = 1\ 000$ h.* The function NORMDIST in Excel gives the area of the curve from $-\infty$ up to a specified point, x. Here is the strategy: We will find the area from $-\infty$ to 900 h, which is the shaded area to the left of 900 h in Figure 4-4. Then we will find the area from $-\infty$ to 1 000 h, which is all the shaded area to the left of 1 000 h in Figure 4-4. The difference between the two is the area from 900 to 1 000 h:

Area from 900 to 1 000 = (area from $-\infty$ to 1 000) – (area from $-\infty$ to 900)

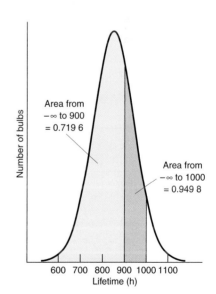

FIGURE 4-4 Use of the Gaussian curve to find the fraction of bulbs with a lifetime between 900 and 1 000 h. We find the area between $-\infty$ and 1 000 h and subtract the area between $-\infty$ and 900 h.

In a spreadsheet, enter the mean in cell A2 and the standard deviation in cell B2. To find the area under the Gaussian curve from $-\infty$ to 900 h in cell C4, select cell C4 and, in Excel 2007, go to Formulas, Insert Function. In earlier versions of Excel, go to the Insert menu and choose Function. In the window that appears, select Statistical functions and find NORMDIST from the list of possibilities. Double click on NORMDIST and another window appears asking for four values that will be used by NORMDIST. (If you click on Help, you will find a cryptic explanation of how to use NORMDIST.)

	A	**B**	**C**
1	Mean =	Std dev =	
2	845.2	94.2	
3			
4	Area from $-\infty$ to 900 =		0.7196
5	Area from $-\infty$ to 1000 =		0.9498
6	Area from 900 to 1000		0.2302
7			
8	C4 = NORMDIST(900,$A$2,$B$2,TRUE)		
9	C5 = NORMDIST(1000,$A$2,$B$2,TRUE)		
10	C6 = C5-C4		

Values provided to the function NORMDIST(x,mean,standard_dev,cumulative) are called *arguments* of the function. The first argument is x, which is 900. The second argument is the mean, which is 845.2. You can enter 845.2 for the mean or enter A2, which is the cell containing 845.2. We will enter $A$2 so that we can move the entry to other cells and still always refer to cell A2. The third argument is the standard deviation, for which we enter $B$2. The last argument is called "cumulative." When it has the value TRUE, NORMDIST gives the area under the Gaussian curve. When cumulative is FALSE, NORMDIST gives the ordinate (the y-value) of the Gaussian curve. We want area, so enter TRUE. The formula "= NORMDIST(900,$A$2,$B$2,TRUE)" in cell C4 returns 0.719 6. This is the area under the Gaussian curve from $-\infty$ to 900 h. To get the area from $-\infty$ to 1 000 h, write "= NORMDIST(1000,$A$2,$B$2,TRUE)" in cell C5. The value returned is 0.949 8. Then subtract the areas (C5 − C4) to obtain 0.230 2, which is the area from 900 to 1 000. That is, 23.02% of the area lies in the range 900 to 1 000 h. We expect 23% of the bulbs to have a lifetime of 900 to 1 000 h.

Test Yourself Find the area from 800 to 1 000 hours. (***Answer:*** 0.634 2)

The standard deviation measures the width of the Gaussian curve. The larger the value of σ, the broader the curve. In any Gaussian curve, 68.3% of the area is in the range from $\mu - 1\sigma$ to $\mu + 1\sigma$. That is, more than two-thirds of the measurements are expected to lie within one standard deviation of the mean. Also, 95.5% of the area lies within $\mu \pm 2\sigma$, and 99.7% of the area lies within $\mu \pm 3\sigma$. Suppose that you use two different techniques to measure sulfur in coal: Method A has a standard deviation of 0.4%, and method B has a standard deviation of 1.1%. You can expect that approximately two-thirds of measurements from method A will lie within 0.4% of the mean. For method B, two-thirds will lie within 1.1% of the mean.

Range	Percentage of measurements
$\mu \pm 1\sigma$	68.3
$\mu \pm 2\sigma$	95.5
$\mu \pm 3\sigma$	99.7

Standard Deviation of the Mean

To measure the mean lifetime of a large number of light bulbs, we could select one at a time and measure its lifetime. Alternatively, we could select, say, four at a time, measure the life of each, and compute the average of the four. We repeat this process of measuring four at a time many times and compute a mean μ and a standard deviation that we label σ_4 because it is based on sets of four bulbs. The mean of many sets of four bulbs is the same as the population mean. However, the standard deviation of the means of sets of four bulbs is smaller than the population standard deviation, σ. The relation is $\sigma_4 = \sigma/\sqrt{4}$. We call σ_4 the **standard deviation of the mean** of sets of four samples. In general, the standard deviation of the mean for sets of n samples is

Standard deviation of the mean of sets of n values:
$$\sigma_n = \frac{\sigma}{\sqrt{n}}$$
(4-5)

The more times you measure a quantity, the more confident you can be that the average is close to the population mean. Uncertainty decreases in proportion to $1/\sqrt{n}$, where n is the number of measurements. You can decrease uncertainty by a factor of $2(=\sqrt{4})$ by making 4 times as many measurements and by a factor of $10\ (=\sqrt{100})$ by making 100 times as many measurements.

Instruments with rapid data acquisition allow us to average many experiments in a short time to improve precision.

4-2 Confidence Intervals

Student's t is a statistical tool used most frequently to express confidence intervals and to compare results from different experiments. It is the tool you could use to evaluate the probability that your red blood cell count will be found in a certain range on "normal" days.

"Student" was the pseudonym of W. S. Gosset, whose employer, the Guinness breweries of Ireland, restricted publications for proprietary reasons. Because of the importance of Gosset's work, he was allowed to publish it (*Biometrika* **1908**, 6, 1), but under an assumed name.

Calculating Confidence Intervals

From a limited number of measurements (n), we cannot find the true population mean, μ, or the true standard deviation, σ. What we determine are $\bar{x}$ and s, the sample mean and the sample standard deviation. The **confidence interval** is computed from the equation

$$\text{Confidence interval} = \bar{x} \pm \frac{ts}{\sqrt{n}} \qquad (4\text{-}6)$$

where t is Student's t, taken from Table 4-2 for a desired level of confidence, such as 95%. The meaning of the confidence interval is this: If we were to repeat the n measurements many times to compute the mean and standard deviation, the 95% confidence interval would include the true population mean (whose value we do not know) in 95% of the sets of n measurements. We say (somewhat imprecisely) that "we are 95% confident that the true mean lies within the confidence interval."

EXAMPLE Calculating Confidence Intervals

The carbohydrate content of a glycoprotein (a protein with sugars attached to it) is found to be 12.6, 11.9, 13.0, 12.7, and 12.5 wt% (g carbohydrate/100 g glycoprotein) in replicate analyses. Find the 50% and 90% confidence intervals for the carbohydrate content.

TABLE 4-2 Values of Student's t

Degrees of freedom	Confidence level (%)						
	50	90	95	98	99	99.5	99.9
1	1.000	6.314	12.706	31.821	63.656	127.321	636.578
2	0.816	2.920	4.303	6.965	9.925	14.089	31.598
3	0.765	2.353	3.182	4.541	5.841	7.453	12.924
4	0.741	2.132	2.776	3.747	4.604	5.598	8.610
5	0.727	2.015	2.571	3.365	4.032	4.773	6.869
6	0.718	1.943	2.447	3.143	3.707	4.317	5.959
7	0.711	1.895	2.365	2.998	3.500	4.029	5.408
8	0.706	1.860	2.306	2.896	3.355	3.832	5.041
9	0.703	1.833	2.262	2.821	3.250	3.690	4.781
10	0.700	1.812	2.228	2.764	3.169	3.581	4.587
15	0.691	1.753	2.131	2.602	2.947	3.252	4.073
20	0.687	1.725	2.086	2.528	2.845	3.153	3.850
25	0.684	1.708	2.060	2.485	2.787	3.078	3.725
30	0.683	1.697	2.042	2.457	2.750	3.030	3.646
40	0.681	1.684	2.021	2.423	2.704	2.971	3.551
60	0.679	1.671	2.000	2.390	2.660	2.915	3.460
120	0.677	1.658	1.980	2.358	2.617	2.860	3.373
∞	0.674	1.645	1.960	2.326	2.576	2.807	3.291

In calculating confidence intervals, σ may be substituted for s in Equation 4-6 if you have a great deal of experience with a particular method and have therefore determined its "true" population standard deviation. If σ is used instead of s, the value of t to use in Equation 4-6 comes from the bottom row of Table 4-2.

Values of t in this table apply to two-tailed tests illustrated in Figure 4-9a. The 95% confidence level specifies the regions containing 2.5% of the area in each wing of the curve. For a one-tailed test, we use values of t listed for 90% confidence. Each wing outside of t for 90% confidence contains 5% of the area of the curve.

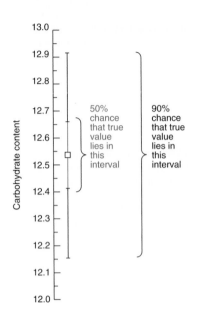

Solution First calculate $\bar{x}\ (=12.5_4)$ and $s\ (=0.4_0)$ for the five measurements. For the 50% confidence interval, look up t in Table 4-2 under 50 and across from *four* degrees of freedom (degrees of freedom $= n - 1$). The value of t is 0.741, so the 50% confidence interval is

$$50\%\ \text{confidence interval} = \bar{x} \pm \frac{ts}{\sqrt{n}} = 12.5_4 \pm \frac{(0.741)(0.4_0)}{\sqrt{5}} = 12.5_4 \pm 0.1_3\ \text{wt\%}$$

The 90% confidence interval is

$$90\%\ \text{confidence interval} = \bar{x} \pm \frac{ts}{\sqrt{n}} = 12.5_4 \pm \frac{(2.132)(0.4_0)}{\sqrt{5}} = 12.5_4 \pm 0.3_8\ \text{wt\%}$$

If we repeated sets of five measurements many times, half of 50% confidence intervals are expected to include the true mean, μ. Nine-tenths of the 90% confidence intervals are expected to include the true mean, μ.

Test Yourself Carbohydrate measured on one more sample was 12.3 wt%. Using six results, find the 90% confidence interval. (***Answer:*** $12.5_0 \pm (2.015)(0.3_7)/\sqrt{6} = 12.5_0 \pm 0.3_1\ \text{wt\%}$)

The Meaning of a Confidence Interval

Figure 4-5 illustrates the meaning of confidence intervals. A computer chose numbers at random from a Gaussian population with a population mean (μ) of 10 000 and a population standard deviation (σ) of 1 000 in Equation 4-3. In trial 1, four numbers were chosen, and their mean and standard deviation were calculated with Equations 4-1 and 4-2. The 50% confidence interval was then calculated with Equation 4-6, using $t = 0.765$ from Table 4-2 (50% confidence, 3 degrees of freedom). This trial is plotted as the first point at the left in Figure 4-5a; the square is centered at the mean value of 9 526, and the error bar extends from the lower limit to the upper limit of the 50% confidence interval (±290). The experiment was repeated 100 times to produce the points in Figure 4-5a.

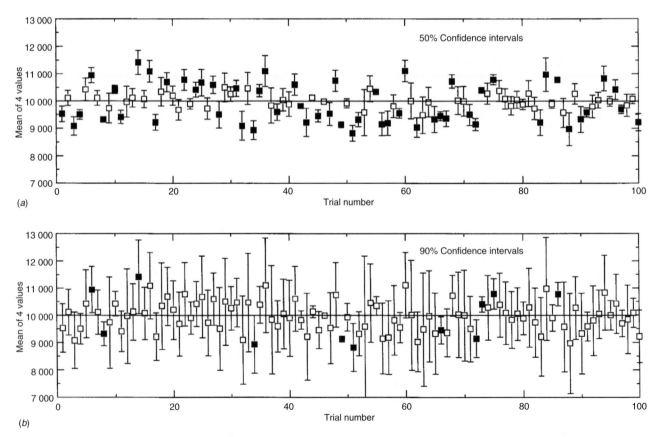

FIGURE 4-5 50% and 90% confidence intervals for the same set of random data. Filled squares are the data points whose confidence interval does not include the true population mean of 10 000.

The 50% confidence interval is defined such that, if we repeated this experiment an infinite number of times, 50% of the error bars in Figure 4-5a would include the true population mean of 10 000. In fact, I did the experiment 100 times, and 45 of the error bars in Figure 4-5a pass through the horizontal line at 10 000.

Figure 4-5b shows the same experiment with the same set of random numbers, but this time the 90% confidence interval was calculated. For an infinite number of experiments, we would expect 90% of the confidence intervals to include the population mean of 10 000. In Figure 4-5b, 89 of the 100 error bars cross the horizontal line at 10 000.

Standard Deviation and Confidence Interval as Estimates of Experimental Uncertainty

Chapter 3 gave rules for propagation of uncertainty in calculations. For example, if we were dividing a mass by a volume to find density, the uncertainty in density is derived from the uncertainties in mass and volume. Common estimates of uncertainty are the standard deviation and the confidence interval.

Suppose you measure the volume of a vessel five times and observe values of 6.375, 6.372, 6.374, 6.377, and 6.375 mL. The average is $\bar{x} = 6.374_6$ mL and the standard deviation is $s = 0.001_8$ mL. We could report a volume of $6.374_6 \pm 0.001_8$ mL ($n = 5$), where n is the number of measurements.

You could choose a confidence interval (such as 95%) for the estimate of uncertainty. Using Equation 4-6 with 4 degrees of freedom, you find that the 95% confidence interval is $\pm ts/\sqrt{n} = \pm(2.776)(0.001_8)/\sqrt{5} = \pm0.002_5$. By this criterion, the uncertainty in volume is $\pm0.002_5$ mL. *It is critical to specify what kind of uncertainty you are reporting*, such as the standard deviation for n measurements or the 95% confidence interval for n measurements.

We reduce uncertainty by making more measurements. If we make 21 measurements and have the same mean and standard deviation, the 95% confidence interval is reduced from $\pm0.002_5$ to $\pm(2.086)(0.001_8)/\sqrt{21} = \pm0.000\ 8$ mL.

Finding Confidence Intervals with Excel

Excel has a built-in function to compute Student's t. In Figure 4-6, we enter data in the block of cells A4:A13. We reserved 10 cells for data, but you could modify a spreadsheet to accept more data. For the five data points entered in Figure 4-6, the mean is computed in cell C3 with the statement "=AVERAGE(A4:A13)" even though some of the cells in the range A4:A13 are empty. Excel ignores empty cells and does not consider them to be 0, which would produce an incorrect average. Standard deviation is computed in cell C4. We find the number of data points, n, with the statement in cell C5 "=COUNT(A4:A13)". Degrees of freedom are computed as $n - 1$ in cell C7. Cell C9 has the desired confidence level (0.95), which is our only input other than the data.

The function for finding Student's t in cell C11 is "=TINV(probability,deg_freedom). Probability in this function is $1 -$ confidence level $= 1 - 0.95 = 0.05$. So, the statement in cell C11 is "=TINV(1-C9,C7)", which returns Student's t for 95% confidence and 4 degrees of freedom. Cell C13 gives the confidence interval computed with Equation 4-6.

"Analytical chemists must always emphasize to the public that *the single most important characteristic of any result obtained from one or more analytical measurements is an adequate statement of its uncertainty interval*."[2]

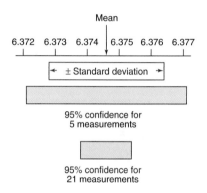

Verify that Student's t in cell C11 is the same as the value in Table 4-2.

	A	B	C	D	E	F
1	Confidence Inverval					
2						
3	Data	mean =	6.3746		= AVERAGE (A4:A13)	
4	6.375	stdev =	0.0018		= STDEV (A4:A13)	
5	6.372	n =	5		= COUNT (A4:A13)	
6	6.374	degrees of				
7	6.377	freedom =	4		= C5-1	
8	6.375	confidence				
9		level =	0.95			
10		Student's				
11		t =	2.776		= TINV (1-C9,C7)	
12		confidence				
13		interval =	0.0023		= C11*C4/SQRT(C5)	

FIGURE 4-6 Spreadsheet for finding confidence interval.

4-3 Comparison of Means with Student's t

Confidence limits and the t test (and, later in this chapter, the Grubbs test) assume that data follow a Gaussian distribution.

If you make two sets of measurements of the same quantity, the mean value from one set will generally not be equal to the mean value from the other set because of small, random variations in the measurements. We use a t **test** to compare one mean value with another to decide whether there is a statistically significant difference between the two. That is, do the two means agree "within experimental error"? The *null hypothesis* in statistics states that the mean values from two sets of measurements are *not* different. Statistics gives us a probability that the observed difference between two means arises from random measurement error. *We customarily reject the null hypothesis if there is less than a 5% chance that the observed difference arises from random variations.* With this criterion, we have a 95% chance that our conclusion is correct. One time out of 20 when we conclude that two means are not different we will be wrong.

Here are three cases that are handled in slightly different manners:

Case 1   We measure a quantity several times, obtaining an average value and standard deviation. We need to compare our answer with an accepted answer. The average is not exactly the same as the accepted answer. Does our measured answer agree with the accepted answer "within experimental error"?

Case 2   We measure a quantity multiple times by two different methods that give two different answers, each with its own standard deviation. Do the two results agree with each other "within experimental error"?

Case 3   Sample A is measured once by method 1 and once by method 2; the two measurements do not give exactly the same result. Then a different sample, designated B, is measured once by method 1 and once by method 2; and, again, the results are not exactly equal. The procedure is repeated for n different samples. Do the two methods agree with each other "within experimental error"?

Case 1. Comparing a Measured Result with a "Known" Value

You purchased a Standard Reference Material coal sample certified by the National Institute of Standards and Technology to contain 3.19 wt% sulfur. You are testing a new analytical method to see whether it can reproduce the known value. The measured values are 3.29, 3.22, 3.30, and 3.23 wt% sulfur, giving a mean of $\bar{x} = 3.26_0$ and a standard deviation of $s = 0.04_1$. Does your answer agree with the known answer? To find out, *compute the 95% confidence interval for your answer and see if that range includes the known answer*. If the known answer is not within your 95% confidence interval, then the results do not agree.

If the "known" answer does not lie within the 95% confidence interval, then the two methods give "different" results.

So here we go. For four measurements, there are 3 degrees of freedom and $t_{95\%} = 3.182$ in Table 4-2. The 95% confidence interval is

Retain many digits in this calculation.

$$\text{95\% confidence interval} = \bar{x} \pm \frac{ts}{\sqrt{n}} = 3.26_0 \pm \frac{(3.182)(0.04_1)}{\sqrt{4}} = 3.26_0 \pm 0.06_5 \quad (4\text{-}7)$$

$$\text{95\% confidence interval} = 3.19_5 \text{ to } 3.32_5 \text{ wt\%}$$

The known answer (3.19 wt%) is just outside the 95% confidence interval. Therefore we conclude that there is less than a 5% chance that our method agrees with the known answer.

We conclude that our method gives a "different" result from the known result. However, in this case, the 95% confidence interval is so close to including the known result that it would be prudent to make more measurements before concluding that our new method is not accurate.

Case 2. Comparing Replicate Measurements

Do the results of two different sets of measurements agree "within experimental error"?[3] An example comes from the work of Lord Rayleigh (John W. Strutt), who is remembered today for landmark investigations of light scattering, blackbody radiation, and elastic waves in solids. His Nobel Prize in 1904 was received for discovering the inert gas argon. This discovery occurred when he noticed a small discrepancy between two sets of measurements of the density of nitrogen gas.

In Rayleigh's time, it was known that dry air was composed of about one-fifth oxygen and four-fifths nitrogen. Rayleigh removed all O_2 from air by passing the air through red-hot copper (to make solid CuO). He then measured the density of the remaining gas by collecting it in a fixed volume at constant temperature and pressure. He also prepared the same volume

FIGURE 4-7 Lord Rayleigh's measurements of the mass of constant volumes of gas (at constant temperature and pressure) isolated by removing oxygen from air or generated by decomposition of nitrogen compounds. Rayleigh recognized that the difference between the two clusters was outside of his experimental error and deduced that a heavier component, which turned out to be argon, was present in gas isolated from air.

of pure N_2 by chemical decomposition of nitrous oxide (N_2O), nitric oxide (NO), or ammonium nitrite ($NH_4^+NO_2^-$). Table 4-3 and Figure 4-7 show the mass of gas collected in each experiment. The average mass collected from air (2.310 11 g) is 0.46% greater than the average mass of the same volume of gas from chemical sources (2.299 47 g).

If Rayleigh's measurements had not been performed with care, this difference might have been attributed to experimental error. Instead, Rayleigh understood that the discrepancy was outside his margin of error, and he postulated that gas collected from the air was a mixture of nitrogen with a small amount of a heavier gas, which turned out to be argon.

Let's see how to use a t test to decide whether gas isolated from air is "significantly" heavier than nitrogen isolated from chemical sources. In this case, we have two sets of measurements, each with its own uncertainty and no "known" value. We assume that the population standard deviation (σ) for each method is essentially the same.

For two sets of data consisting of n_1 and n_2 measurements (with averages $\bar{x}_1$ and $\bar{x}_2$), we calculate a value of t with the formula

$$t_{calculated} = \frac{|\bar{x}_1 - \bar{x}_2|}{s_{pooled}} \sqrt{\frac{n_1 n_2}{n_1 + n_2}} \qquad (4\text{-}8)$$

where $|\bar{x}_1 - \bar{x}_2|$ is the absolute value of the difference (a positive number) and s_{pooled} is a *pooled* standard deviation making use of both sets of data:

$$s_{pooled} = \sqrt{\frac{\sum_{set\ 1}(x_i - \bar{x}_1)^2 + \sum_{set\ 2}(x_j - \bar{x}_2)^2}{n_1 + n_2 - 2}} = \sqrt{\frac{s_1^2(n_1 - 1) + s_2^2(n_2 - 1)}{n_1 + n_2 - 2}} \qquad (4\text{-}9)$$

$t_{calculated}$ from Equation 4-8 is compared with t in Table 4-2 for $n_1 + n_2 - 2$ degrees of freedom. *If $t_{calculated}$ is greater than t_{table} at the 95% confidence level, the two results are considered to be different.* There is less than a 5% chance that the two sets of data were drawn from populations with the same population mean.

TABLE 4-3 Masses of gas isolated by Lord Rayleigh

From air (g)	From chemical decomposition (g)
2.310 17	2.301 43
2.309 86	2.298 90
2.310 10	2.298 16
2.310 01	2.301 82
2.310 24	2.298 69
2.310 10	2.299 40
2.310 28	2.298 49
—	2.298 89
Average	
2.310 11	2.299 47
Standard deviation	
0.000 14$_3$	0.001 38

SOURCE: R. D. Larsen, *J. Chem. Ed.* **1990**, 67, 925; see also C. J. Giunta, *J. Chem. Ed.* **1998**, 75, 1322.

If $t_{calculated} > t_{table}$ (95%), the difference is significant.

<table>
<tr><td colspan="2">EXAMPLE   Is Lord Rayleigh's Gas from Air Denser Than N₂ from Chemicals?</td></tr>
</table>

EXAMPLE **Is Lord Rayleigh's Gas from Air Denser Than N_2 from Chemicals?**

The average mass of gas from air in Table 4-3 is $\bar{x}_1 = 2.310\ 11$ g, with a standard deviation of $s_1 = 0.000\ 14_3$ (for $n_1 = 7$ measurements). The mass of gas from chemical sources is $\bar{x}_2 = 2.299\ 47$ g, with $s_2 = 0.001\ 38$ ($n_2 = 8$ measurements).

Solution To answer the question, we calculate s_{pooled} with Equation 4-9,

$$s_{pooled} = \sqrt{\frac{(0.000\ 14_3)^2(7 - 1) + (0.001\ 38)^2(8 - 1)}{7 + 8 - 2}} = 0.001\ 02$$

and $t_{calculated}$ with Equation 4-8:

$$t_{calculated} = \frac{|2.310\ 11 - 2.299\ 47|}{0.001\ 02} \sqrt{\frac{7 \cdot 8}{7 + 8}} = 20.2$$

For $7 + 8 - 2 = 13$ degrees of freedom in Table 4-2, t_{table} lies between 2.228 and 2.131 for 95% confidence. Because $t_{calculated} > t_{table}$, the difference is significant. In fact, t_{table} for

99.9% confidence is ~4.3. The difference is significant beyond the 99.9% confidence level. Our eyes were not lying to us in Figure 4-7: Gas from the air is undoubtedly denser than N_2 from chemical sources. This observation led Rayleigh to discover a heavy constituent of air.

Test Yourself If $|\bar{x}_1 - \bar{x}_2|$ is only half as great, but s_{pooled} is unchanged, are the two means significantly different? (*Answer:* $t_{\text{calculated}} = 10.1$ and the difference is still significant)

Section 4-4 uses for *F* test to see if two standard deviations are significantly different.

Equations 4-8 and 4-9 assume that the population standard deviation is the same for both sets of measurements. If this is not true, then we use the equations[4]

$$t_{\text{calculated}} = \frac{|\bar{x}_1 - \bar{x}_2|}{\sqrt{s_1^2/n_1 + s_2^2/n_2}} \tag{4-8a}$$

$$\text{Degrees of freedom} = \frac{(s_1^2/n_1 + s_2^2/n_2)^2}{\dfrac{(s_1^2/n_1)^2}{n_1 - 1} + \dfrac{(s_2^2/n_2)^2}{n_2 - 1}} \tag{4-9a}$$

Equation 4-9a is slightly different from its ancestor in previous editions of this book. The new and old equations are both approximations.

For Rayleigh's data in Figure 4-7, we suspect that the population standard deviation from air is smaller than that from chemical sources. Using Equations 4-8a and 4-9a, we find that $t_{\text{calculated}} = 21.7$ and degrees of freedom $= 7.17 \approx 7$. This value of $t_{\text{calculated}}$ still far exceeds values in Table 4-2 for 7 degrees of freedom at 95% or 99.9% confidence.

Case 3. Paired *t* Test for Comparing Individual Differences

In this case, we use two methods to make single measurements on several different samples. No measurement has been duplicated. Do the two methods give the same answer "within experimental error"? Figure 4-8 shows measurements of aluminum in 11 samples of drinking water. Results for Method 1 are in column B and results for Method 2 are in column C. For each sample, the two results are similar, but not identical.

To see if there is a significant difference between the methods, we use the paired *t* test. First, column D computes the difference (d_i) between the two results for each sample. The mean of the 11 differences ($\bar{d} = -2.4_{91}$) is computed in cell D16 and the standard deviation of the 11 differences (s_d) is computed in cell D17.

	A	B	C	D
1	Comparison of two methods for measuring Al			
2				
3	Sample	Method 1	Method 2	Difference
4	number	(μg/L)	(μg/L)	(d_i)
5	1	17.2	14.2	−3.0
6	2	23.1	27.9	4.8
7	3	28.5	21.2	−7.3
8	4	15.3	15.9	0.6
9	5	23.1	32.1	9.0
10	6	32.5	22.0	−10.5
11	7	39.5	37.0	−2.5
12	8	38.7	41.5	2.8
13	9	52.5	42.6	−9.9
14	10	42.6	42.8	0.2
15	11	52.7	41.1	−11.6
16			mean =	−2.491
17			std dev =	6.748
18			$t_{\text{calculated}} =$	1.224
19	D5 = C5-B5			
20	D16 = AVERAGE(D5:D15)			
21	D17 = STDEV(D5:D15)			
22	D18 = ABS(D16)*SQRT(A15)/D17			
23		ABS = absolute value		

FIGURE 4-8 Measurement of Al by two methods. [Data from P. T. Srinivasan, T. Viraraghavan, and K. S. Subramanian, "Method Development for Drinking Water Aluminum Measurement Using a Graphite Furnace Atomic Absorption Spectrophotometer," *Am. Lab.,* February 2000, p. 76.]

BOX 4-1   Choosing the Null Hypothesis in Epidemiology

One fine morning, I found myself seated on a cross-country flight next to Malcolm Pike, an epidemiologist at the University of Southern California. Epidemiologists employ methods of statistics to guide practices in medicine. Pike was studying the relation between menopausal oestrogen–progestin hormone therapy and breast cancer in women. His study concluded that there was a 7.6% increase in breast cancer risk per year of oestrogen–progestin hormone therapy.[5]

How could such therapy have been approved? Pike explained that tests required by the U.S. Food and Drug Administration are designed to test the null hypothesis that "the treatment does no harm." Instead, he said, the null hypothesis should be "the treatment increases the likelihood of causing breast cancer."

What did he mean? In the field of statistics, the null hypothesis is assumed to be true. Unless you find strong evidence that it is not true, you continue to believe that it is true. In the U.S. legal system, the null hypothesis is that the accused person is innocent. It is up to the prosecution to produce compelling evidence that the accused person

is not innocent; failing that, the jury must acquit the defendant. In the hormone therapy example, the null hypothesis is that the treatment does not cause cancer. The burden of the test is to provide compelling evidence that the treatment *does cause* cancer. Pike is saying that if you believe a treatment causes cancer, the null hypothesis should be that the treatment causes cancer. Then it is up to the proponent of the treatment to provide compelling evidence that the treatment *does not* cause cancer. In Pike's words, test the hypothesis that "the obvious is likely to be true!"

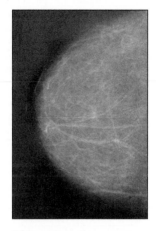

White regions of mammogram are denser tissue than dark regions. [allOver photography/Alamy.]

$$s_d = \sqrt{\frac{\sum(d_i - \bar{d})^2}{n-1}}$$

$$s_d = \sqrt{\frac{(-3.0 - \bar{d})^2 + (4.8 - \bar{d})^2 + \ldots + (0.2 - \bar{d})^2 + (-11.6 - \bar{d})^2}{11-1}} = 6.7_{48} \qquad (4\text{-}10)$$

Once you have the mean and standard deviation, compute $t_{\text{calculated}}$ with the formula

$$t_{\text{calculated}} = \frac{|\bar{d}|}{s_d}\sqrt{n} \qquad (4\text{-}11)$$

where $|\bar{d}|$ is the absolute value of the mean difference, so that $t_{\text{calculated}}$ is always positive. Inserting numbers into Equation 4-11 gives

$$t_{\text{calculated}} = \frac{2.4_{91}}{6.7_{48}}\sqrt{11} = 1.2_{24}$$

We find that $t_{\text{calculated}}$ (1.2_{24}) is less than t_{table} (2.228) listed in Table 4-2 for 95% confidence and 10 degrees of freedom. *There is more than a 5% chance that the two sets of results lie "within experimental error" of each other, so we conclude that the results are not significantly different.* You have now read enough to appreciate Box 4-1.

One-Tailed and Two-Tailed Significance Tests

In Equation 4-7, we sought to compare the mean from four replicate measurements with a certified value. The curve in Figure 4-9a is the t distribution for 3 degrees of freedom. If the certified value lies in the outer 5% of the area under the curve, we reject the null hypothesis and conclude with 95% confidence that the measured mean is not equivalent to the certified value. The critical value of t for rejecting the null hypothesis is 3.182 for 3 degrees of freedom in Table 4-2. In Figure 4-9a, *2.5% of the area* beneath the curve lies above $t = 3.182$ and *2.5% of the area* lies below $t = -3.182$. We call this a *two-tailed test* because we reject the null hypothesis if the certified value lies in the low-probability region on either side of the mean.

If we have a preconceived reason to believe that our method gives systematically low values, we could use the *one-tailed t test* in Figure 4-9b. In this case, we reject the null hypothesis (which states that there is no significant difference between the measured and certified values) if $t_{\text{calculated}}$ is greater than 2.353. Figure 4-9b shows that *5% of the area* beneath the curve lies above $t = 2.353$. We did not consider area at the left side of the curve because we had a reason to believe that our method produces low results, not high results.

FIGURE 4-9 Student's *t* distribution for 3 degrees of freedom. In panel *a*, each shaded tail contains 2.5% of the area under the curve. In panel *b*, the single shaded tail contains 5% of the area. The fewer the degrees of freedom, the broader the distribution. As the degrees of freedom increase, the curve approaches a Gaussian curve.

(a)

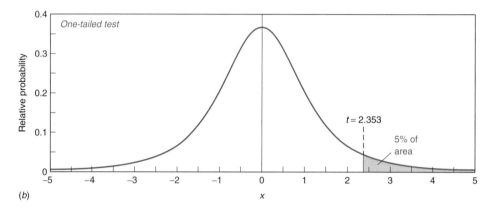

(b)

How can you find the value of *t* that bounds the upper 5% of the area of the curve? Because the *t* distribution is symmetric, the two-tailed value *t* = 2.353 for 90% confidence in Table 4-2 must be the value we seek, because 5% of the area lies above *t* = 2.353 and 5% of the area lies below *t* = −2.353.

The purpose of this discussion was to inform you of the distinction between one- and two-tailed tests. All *t* tests in this book will be two-tailed.

Is My Red Blood Cell Count High Today?

At the opening of this chapter, red cell counts on five "normal" days were 5.1, 5.3, 4.8, 5.4, and 5.2×10^6 cells/μL. The question was whether today's count of 5.6×10^6 cells/μL is "significantly" higher than normal? Disregarding the factor of 10^6, the mean of the normal values is $\bar{x} = 5.16$ and the standard deviation is $s = 0.23$. For today's value of 5.6,

$$t_{\text{calculated}} = \frac{|\text{today's count} - \bar{x}|}{s} \sqrt{n} = \frac{|5.16 - 5.6|}{0.23} \sqrt{5} = 4.28$$

What is the probability of finding *t* = 4.28 for 4 degrees of freedom?

In Table 4-2, looking across the row for 4 degrees of freedom, we see that 4.28 lies between the 98% (*t* = 3.747) and 99% (*t* = 4.604) confidence levels. Today's red cell count lies in the upper tail of the curve containing less than 2% of the area of the curve. There is less than a 2% probability of observing a count of 5.6×10^6 cells/μL on "normal" days. It is reasonable to conclude that today's count is elevated.

Table 4-2 places the probability of today's red cell count between 1 and 2%. Excel provides a probability with the function TDIST(*x*,deg_freedom,tails), where *x* is $t_{\text{calculated}}$, deg_freedom = 4, and tails = 2. The function TDIST(4.28,4,2) returns the value 0.013. Today's red cell count lies in the upper 1.3% of the area of the *t*-distribution.

Find probability with Excel:
TDIST(*x*,deg_freedom,tails)

4-4 Comparison of Standard Deviations with the *F* Test

To decide whether Rayleigh's two sets of nitrogen masses in Figure 4-7 are "significantly" different from each other, we used the *t* test. If the standard deviations of two data sets are not significantly different from each other, then we use Equation 4-8 for the *t* test. If the standard deviations are significantly different, then we use Equation 4-8a instead.

TABLE 4-4 **Critical values of $F = s_1^2/s_2^2$ at 95% confidence level**

Degrees of freedom for s_2	Degrees of freedom for s_1													
	2	3	4	5	6	7	8	9	10	12	15	20	30	∞
2	19.0	19.2	19.2	19.3	19.3	19.4	19.4	19.4	19.4	19.4	19.4	19.4	19.5	19.5
3	9.55	9.28	9.12	9.01	8.94	8.89	8.84	8.81	8.79	8.74	8.70	8.66	8.62	8.53
4	6.94	6.59	6.39	6.26	6.16	6.09	6.04	6.00	5.96	5.91	5.86	5.80	5.75	5.63
5	5.79	5.41	5.19	5.05	4.95	4.88	4.82	4.77	4.74	4.68	4.62	4.56	4.50	4.36
6	5.14	4.76	4.53	4.39	4.28	4.21	4.15	4.10	4.06	4.00	3.94	3.87	3.81	3.67
7	4.74	4.35	4.12	3.97	3.87	3.79	3.73	3.68	3.64	3.58	3.51	3.44	3.38	3.23
8	4.46	4.07	3.84	3.69	3.58	3.50	3.44	3.39	3.35	3.28	3.22	3.15	3.08	2.93
9	4.26	3.86	3.63	3.48	3.37	3.29	3.23	3.18	3.14	3.07	3.01	2.94	2.86	2.71
10	4.10	3.71	3.48	3.33	3.22	3.14	3.07	3.02	2.98	2.91	2.84	2.77	2.70	2.54
11	3.98	3.59	3.36	3.20	3.10	3.01	2.95	2.90	2.85	2.79	2.72	2.65	2.57	2.40
12	3.88	3.49	3.26	3.11	3.00	2.91	2.85	2.80	2.75	2.69	2.62	2.54	2.47	2.30
13	3.81	3.41	3.18	3.02	2.92	2.83	2.77	2.71	2.67	2.60	2.53	2.46	2.38	2.21
14	3.74	3.34	3.11	2.96	2.85	2.76	2.70	2.65	2.60	2.53	2.46	2.39	2.31	2.13
15	3.68	3.29	3.06	2.90	2.79	2.71	2.64	2.59	2.54	2.48	2.40	2.33	2.25	2.07
16	3.63	3.24	3.01	2.85	2.74	2.66	2.59	2.54	2.49	2.42	2.35	2.28	2.19	2.01
17	3.59	3.20	2.96	2.81	2.70	2.61	2.55	2.49	2.45	2.38	2.31	2.23	2.15	1.96
18	3.56	3.16	2.93	2.77	2.66	2.58	2.51	2.46	2.41	2.34	2.27	2.19	2.11	1.92
19	3.52	3.13	2.90	2.74	2.63	2.54	2.48	2.42	2.38	2.31	2.23	2.16	2.07	1.88
20	3.49	3.10	2.87	2.71	2.60	2.51	2.45	2.39	2.35	2.28	2.20	2.12	2.04	1.84
30	3.32	2.92	2.69	2.53	2.42	2.33	2.27	2.21	2.16	2.09	2.01	1.93	1.84	1.62
∞	3.00	2.60	2.37	2.21	2.10	2.01	1.94	1.88	1.83	1.75	1.67	1.57	1.46	1.00

Critical values of F for a one-tailed test of the hypothesis that $s_1 > s_2$. There is a 5% probability of observing F above the tabulated value.

You can compute F for a chosen level of confidence with the Excel function FINV(probability,deg_freedom1,deg_freedom2). The statement "=FINV(0.05,7,6)" reproduces the value F = 4.21 in this table. The statement "=FINV(0.1,7,6)" gives F = 3.01 for 90% confidence.

The **F test** tells us whether two standard deviations are "significantly" different from each other. F is the quotient of the squares of the standard deviations:

$$F_{\text{calculated}} = \frac{s_1^2}{s_2^2} \qquad (4\text{-}12)$$

We always put the larger standard deviation in the numerator so that $F \geq 1$. We test the hypothesis that $s_1 > s_2$ by using the one-tailed F test in Table 4-4. If $F_{\text{calculated}} > F_{\text{table}}$, then the difference is significant.

Use the F test for Case 2 in comparison of means in Section 4-3.

If $F_{\text{calculated}} < F_{\text{table}}$, use Equation 4-8.
If $F_{\text{calculated}} > F_{\text{table}}$, use Equation 4-8a.

The square of the standard deviation is called the *variance*.

EXAMPLE **Is the Standard Deviation from Chemical Decomposition Significantly Greater Than the Standard Deviation from Air in Rayleigh's Data?**

In Table 4-3, the standard deviation from chemical decomposition is $s_1 = 0.001\ 38$ ($n_1 = 8$ measurements) and the standard deviation from air is $s_2 = 0.000\ 14_3$ ($n_2 = 7$ measurements).

Solution To answer the question, find F with Equation 4-12:

$$F_{\text{calculated}} = \frac{s_1^2}{s_2^2} = \frac{(0.001\ 38)^2}{(0.000\ 14_3)^2} = 93._1$$

In Table 4-4, look for F_{table} in the column with 7 degrees of freedom for s_1 (because degrees of freedom $= n - 1$) and the row with 6 degrees of freedom for s_2. *Because $F_{\text{calculated}} (= 93.1) > F_{\text{table}} (= 4.21)$, we accept the hypothesis that $s_1 > s_2$ above the 95% confidence level.* The obvious difference in scatter of the two data sets in Figure 4-7 is highly significant.

Test Yourself If two variances were $s_1^2 = (0.002\ 00)^2$ (7 degrees of freedom) and $s_2^2 = (0.001\ 00)^2$ (6 degrees of freedom), is the difference significant? (*Answer:* No. $F_{\text{calculated}} = 4.00 < F_{\text{table}} = 4.21$)

To compare Rayleigh's two sets of results in Table 4-3, enter his data in columns B and C of a spreadsheet (Figure 4-10). In rows 13 and 14, we computed the averages and standard deviations, but we did not need to do this.

In Excel 2007, in the Data ribbon, you might find Data Analysis as an option. If not, click the Microsoft Office Button at the upper left of the Excel window. Click Excel Options and Add-Ins. Select Analysis ToolPak and click OK to load the Analysis ToolPak. For future use, follow the same steps to load Solver Add-In. In earlier versions of Excel, you might find Data Analysis in the Tools Menu. If not, go to Tools, Add-Ins, and check both Analysis ToolPak and Solver Add-In. Click OK to load the applications.

Returning to Figure 4-10, we want to test the null hypothesis that the two sets of data are not statistically different. In Excel 2007, in the Data ribbon, select Data Analysis. In earlier versions of Excel, go to the Tools Menu and select Data Analysis. In the window that appears, select t-Test: Two-Sample Assuming Equal Variances. Click OK. The next window asks you in which cells the data are located. Write B5:B12 for Variable 1 and C5:C12 for Variable 2. The routine ignores the blank space in cell B12. For the Hypothesized Mean Difference enter 0 and for Alpha enter 0.05. Alpha is the level of probability to which we are testing the difference in the means. With Alpha = 0.05, we are at the 95% confidence level. For Output Range, select cell E1 and click OK.

Excel goes to work and prints results in cells E1 to G13 of Figure 4-10. Mean values are in cells F3 and G3. Cells F4 and G4 give *variance*, which is the square of the standard deviation. Cell F6 gives *pooled variance* computed with Equation 4-9. That equation was painful to use by hand. Cell F8 shows degrees of freedom (df) = 13 and $t_{calculated}$ (t Stat) = 20.2 from Equation 4-8 appears in cell F9.

At this point in Section 4-3, we consulted Table 4-2 to find that t_{table} lies between 2.228 and 2.131 for 95% confidence and 13 degrees of freedom. Excel gives us the critical value of t (2.160) in cell F13 of Figure 4-10. Because $t_{calculated}$ (= 20.2) > t_{table} (= 2.160), we conclude that the two means are not the same. The difference is significant. Cell F12 states that the probability of observing these two mean values and standard deviations by random chance if the mean values were really the same is 3×10^{-11}. The difference is *highly* significant. For any value of $P \leq 0.05$ in cell F12, we reject the *null hypothesis* and conclude that the means *are different*.

	A	B	C	D	E	F	G
1	Analysis of Rayleigh's Data				t-Test: Two-Sample Assuming Equal Variances		
2						Variable 1	Variable 2
3		Mass of gas (g) collected from			Mean	2.310109	2.299473
4		air	chemical		Variance	2.03E-08	1.9E-06
5		2.31017	2.30143		Observations	7	8
6		2.30986	2.29890		Pooled Variance	1.03E-06	
7		2.31010	2.29816		Hypothesized Mean Diff	0	
8		2.31001	2.30182		df	13	
9		2.31024	2.29869		t Stat	20.21372	
10		2.31010	2.29940		P(T<=t) one-tail	1.66E-11	
11		2.31028	2.29849		t Critical one-tail	1.770932	
12			2.29889		P(T<=t) two-tail	3.32E-11	
13	Average	2.31011	2.29947		t Critical two-tail	2.160368	
14	Std Dev	0.00014	0.00138				
15					t-Test: Two-Sample Assuming Unequal Variances		
16	B13 = AVERAGE(B5:B12)					Variable 1	Variable 2
17	B14 = STDEV(B5:B12)				Mean	2.310109	2.299473
18					Variance	2.03E-08	1.9E-06
19					Observations	7	8
20					Hypothesized Mean Diff	0	
21					df	7	
22					t Stat	21.68022	
23					P(T<=t) one-tail	5.6E-08	
24					t Critical one-tail	1.894578	
25					P(T<=t) two-tail	1.12E-07	
26					t Critical two-tail	2.364623	

FIGURE 4-10 Spreadsheet for comparing mean values of Rayleigh's measurements in Table 4-3.

The F test in Equation 4-12 told us that the standard deviations of Rayleigh's two experiments are different. Therefore, we can select the other t test found in the Analysis ToolPak. Select t-Test: Two-Sample Assuming Unequal Variances and fill in the blanks as before. Results based on Equations 4-8a and 4-9a are displayed in cells E15 to G26 of Figure 4-10. Just as we found in Section 4-3, the degrees of freedom are 7 (cell F21) and $t_{calculated} = 21.7$ (cell F22). Because $t_{calculated}$ is greater than the critical value of t (2.36 in cell F26), we reject the null hypothesis and conclude that the two means *are* significantly different.

4-6 Grubbs Test for an Outlier

Students dissolved zinc from a galvanized nail and measured the mass lost by the nail to tell how much of the nail was zinc. Here are 12 results:

Mass loss (%): 10.2, 10.8, 11.6, 9.9, 9.4, 7.8, 10.0, 9.2, 11.3, 9.5, 10.6, 11.6

The value 7.8 appears out of line. A datum that is far from other points is called an **outlier.** Should 7.8 be discarded before averaging the rest of the data or should 7.8 be retained?

We answer this question with the **Grubbs test.** First compute the average ($\bar{x} = 10.16$) and the standard deviation ($s = 1.11$) of the complete data set (all 12 points in this example). Then compute the Grubbs statistic G, defined as

Grubbs test:
$$G_{calculated} = \frac{|\text{questionable value} - \bar{x}|}{s}$$
(4-13)

where the numerator is the absolute value of the difference between the suspected outlier and the mean value. *If $G_{calculated}$ is greater than G in Table 4-5, the questionable point should be discarded.*

In our example, $G_{calculated} = |7.8 - 10.16| / 1.11 = 2.13$. In Table 4-5, $G_{table} = 2.285$ for 12 observations. *Because $G_{calculated} < G_{table}$, the questionable point should be retained.* There is more than a 5% chance that the value 7.8 is a member of the same population as the other measurements.

Common sense must prevail. If you know that a result is low because you spilled part of a solution, then the probability that the result is wrong is 100% and the datum should be discarded. Any datum based on a faulty procedure should be discarded, no matter how well it fits the rest of the data.

TABLE 4-5 Critical values of G for rejection of outlier

Number of observations	G (95% confidence)
4	1.463
5	1.672
6	1.822
7	1.938
8	2.032
9	2.110
10	2.176
11	2.234
12	2.285
15	2.409
20	2.557

$G_{calculated} = |\text{questionable value} - \text{mean}|/s$. If $G_{calculated} > G_{table}$, the value in question can be rejected with 95% confidence. Values in this table are for a one-tailed test, as recommended by ASTM.

SOURCE: ASTM E 178-02 Standard Practice for Dealing with Outlying Observations, http://webstore.ansi.org; F. E. Grubbs and G. Beck, Technometrics **1972**, 14, 847.

4-7 The Method of Least Squares

For most chemical analyses, the response of the procedure must be evaluated for known quantities of analyte (called *standards*) so that the response to an unknown quantity can be interpreted. For this purpose, we commonly prepare a **calibration curve,** such as the one for caffeine in Figure 0-13. Most often, we work in a region where the calibration curve is a straight line.

We use the **method of least squares** to draw the "best" straight line through experimental data points that have some scatter and do not lie perfectly on a straight line.[6] The best line will be such that some of the points lie above and some lie below the line. We will learn to estimate the uncertainty in a chemical analysis from the uncertainties in the calibration curve and in the measured response to replicate samples of unknown.

Finding the Equation of the Line

The procedure we use assumes that the errors in the y values are substantially greater than the errors in the x values.[7] This condition is often true in a calibration curve in which the experimental response (y values) is less certain than the quantity of analyte (x values). A second assumption is that uncertainties (standard deviations) in all y values are similar.

Suppose we seek to draw the best straight line through the points in Figure 4-11 by minimizing the vertical deviations between the points and the line. We minimize only the vertical deviations because we assume that uncertainties in y values are much greater than uncertainties in x values.

Let the equation of the line be

Equation of straight line: $\qquad y = mx + b$ $\qquad$ (4-14)

FIGURE 4-11 Least-squares curve fitting. The Gaussian curve drawn over the point (3,3) is a schematic indication of the fact that each value of y_i is normally distributed about the straight line. That is, the most probable value of y will fall on the line, but there is a finite probability of measuring y some distance from the line.

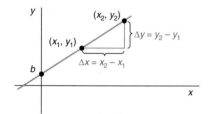

Equation for a straight line: $y = mx + b$

Slope $(m) = \dfrac{\Delta y}{\Delta x} = \dfrac{y_2 - y_1}{x_2 - x_1}$

y-Intercept (b) = crossing point on y-axis

To evaluate the determinant, multiply the diagonal elements $e \times h$ and then subtract the product of the other diagonal elements $f \times g$.

Translation of least-squares equations:

$$m = \frac{n\Sigma(x_i y_i) - \Sigma x_i \Sigma y_i}{n\Sigma(x_i^2) - (\Sigma x_i)^2}$$

$$b = \frac{\Sigma(x_i^2)\Sigma y_i - \Sigma(x_i y_i)\Sigma x_i}{n\Sigma(x_i^2) - (\Sigma x_i)^2}$$

in which m is the **slope** and b is the **y-intercept.** The vertical deviation for the point (x_i, y_i) in Figure 4-11 is $y_i - y$, where y is the ordinate of the straight line when $x = x_i$.

$$\text{Vertical deviation} = d_i = y_i - y = y_i - (mx_i + b) \qquad \textbf{(4-15)}$$

Some of the deviations are positive and some are negative. Because we wish to minimize the magnitude of the deviations irrespective of their signs, we square all the deviations so that we are dealing only with positive numbers:

$$d_i^2 = (y_i - y)^2 = (y_i - mx_i - b)^2$$

Because we minimize the squares of the deviations, this is called the *method of least squares.* It can be shown that minimizing the squares of the deviations (rather than simply their magnitudes) corresponds to assuming that the set of y values is the most probable set.

Finding values of m and b that minimize the sum of the squares of the vertical deviations involves some calculus, which we omit. We express the final solution for slope and intercept in terms of *determinants*, which summarize certain arithmetic operations. The **determinant** $\begin{vmatrix} e & f \\ g & h \end{vmatrix}$ represents the value $eh - fg$. So, for example,

$$\begin{vmatrix} 6 & 5 \\ 4 & 3 \end{vmatrix} = (6 \times 3) - (5 \times 4) = -2$$

The slope and the intercept of the "best" straight line are found to be

Least-squares "best" line

$$\text{Slope:} \qquad m = \begin{vmatrix} \Sigma(x_i y_i) & \Sigma x_i \\ \Sigma y_i & n \end{vmatrix} \div D \qquad \textbf{(4-16)}$$

$$\text{Intercept:} \qquad b = \begin{vmatrix} \Sigma(x_i^2) & \Sigma(x_i y_i) \\ \Sigma x_i & \Sigma y_i \end{vmatrix} \div D \qquad \textbf{(4-17)}$$

where D is

$$D = \begin{vmatrix} \Sigma(x_i^2) & \Sigma x_i \\ \Sigma x_i & n \end{vmatrix} \qquad \textbf{(4-18)}$$

and n is the number of points.

Let's use these equations to find the slope and intercept of the best straight line through the four points in Figure 4-11. The work is set out in Table 4-6. Noting that $n = 4$ and putting the various sums into the determinants in Equations 4-16, 4-17, and 4-18 gives

$$m = \begin{vmatrix} 57 & 14 \\ 14 & 4 \end{vmatrix} \div \begin{vmatrix} 62 & 14 \\ 14 & 4 \end{vmatrix} = \frac{(57 \times 4) - (14 \times 14)}{(62 \times 4) - (14 \times 14)} = \frac{32}{52} = 0.615\,38$$

$$b = \begin{vmatrix} 62 & 57 \\ 14 & 14 \end{vmatrix} \div \begin{vmatrix} 62 & 14 \\ 14 & 4 \end{vmatrix} = \frac{(62 \times 14) - (57 \times 14)}{(62 \times 4) - (14 \times 14)} = \frac{70}{52} = 1.346\,15$$

TABLE 4-6 Calculations for least-squares analysis

x_i	y_i	$x_i y_i$	x_i^2	$d_i (= y_i - m x_i - b)$	d_i^2
1	2	2	1	0.038 46	0.001 479 3
3	3	9	9	−0.192 31	0.036 982
4	4	16	16	0.192 31	0.036 982
6	5	30	36	−0.038 46	0.001 479 3
$\Sigma x_i = 14$	$\Sigma y_i = 14$	$\Sigma(x_i y_i) = 57$	$\Sigma(x_i^2) = 62$		$\Sigma(d_i^2) = 0.076\ 923$

The equation of the best straight line through the points in Figure 4-11 is therefore

$$y = 0.615\ 38x + 1.346\ 15$$

We tackle the question of significant figures for m and b in the next section.

EXAMPLE 📊 **Finding Slope and Intercept with a Spreadsheet**

Excel has functions called SLOPE and INTERCEPT whose use is illustrated here:

	A	B	C	D	E	F
1	x	y			Formulas:	
2	1	2		slope =		
3	3	3		0.61538	D3 = SLOPE(B2:B5,A2:A5)	
4	4	4		intercept =		
5	6	5		1.34615	D5 = INTERCEPT(B2:B5,A2:A5)	

The slope in cell D3 is computed with the formula "=SLOPE(B2:B5,A2:A5)", where B2:B5 is the range containing the y values and A2:A5 is the range containing x values.

Test Yourself   Change the second value of x from 3 to 3.5 and find the slope and intercept. (*Answer:* 0.610 84, 1.285 71)

How Reliable Are Least-Squares Parameters?

To estimate the uncertainties (expressed as standard deviations) in the slope and intercept, an uncertainty analysis must be performed on Equations 4-16 and 4-17. Because the uncertainties in m and b are related to the uncertainty in measuring each value of y, we first estimate the standard deviation that describes the population of y values. This standard deviation, σ_y, characterizes the little Gaussian curve inscribed in Figure 4-11.

We estimate σ_y, the population standard deviation of all y values, by calculating s_y, the standard deviation, for the four measured values of y. The deviation of each value of y_i from the center of its Gaussian curve is $d_i = y_i - y = y_i - (mx_i + b)$. The standard deviation of these vertical deviations is

$$\sigma_y \approx s_y = \sqrt{\frac{\Sigma(d_i - \bar{d})^2}{(\text{degrees of freedom})}} \qquad (4\text{-}19)$$

Equation 4-19 is analogous to Equation 4-2.

But the average deviation, $\bar{d}$, is 0 for the best straight line, so the numerator of Equation 4-19 reduces to $\Sigma(d_i^2)$.

The *degrees of freedom* is the number of independent pieces of information available. For n data points, there are n degrees of freedom. If you were calculating the standard deviation of n points, you would first find the average to use in Equation 4-2. This calculation leaves $n - 1$ degrees of freedom in Equation 4-2 because only $n - 1$ pieces of information are available in addition to the average. If you know $n - 1$ values and you also know their average, then the nth value is fixed and you can calculate it.

For Equation 4-19, we began with n points. Two degrees of freedom were lost in determining the slope and the intercept. Therefore, $n - 2$ degrees of freedom remain. Equation 4-19 becomes

$$s_y = \sqrt{\frac{\Sigma(d_i^2)}{n - 2}} \qquad (4\text{-}20)$$

where d_i is given by Equation 4-15.

Uncertainty analysis for Equations 4-16 and 4-17 leads to the following results:

$$\text{Standard deviation of slope and intercept} \quad \begin{cases} s_m^2 = \dfrac{s_y^2 n}{D} & \text{(4-21)} \\[2em] s_b^2 = \dfrac{s_y^2 \sum (x_i^2)}{D} & \text{(4-22)} \end{cases}$$

where s_m is an estimate of the standard deviation of the slope, s_b is an estimate of the standard deviation of the intercept, s_y is given by Equation 4-20, and D is given by Equation 4-18.

At last, we can assign significant figures to the slope and the intercept in Figure 4-11. In Table 4-6, we see that $\sum(d_i^2) = 0.076\ 923$. Putting this number into Equation 4-20 gives

$$s_y^2 = \frac{0.076\ 923}{4 - 2} = 0.038\ 462$$

Now, we can plug numbers into Equations 4-21 and 4-22 to find

$$s_m^2 = \frac{s_y^2 n}{D} = \frac{(0.038\ 462)(4)}{52} = 0.002\ 958\ 6 \Rightarrow s_m = 0.054\ 39$$

$$s_b^2 = \frac{s_y^2 \sum (x_i^2)}{D} = \frac{(0.038\ 462)(62)}{52} = 0.045\ 859 \Rightarrow s_b = 0.214\ 15$$

Combining the results for m, s_m, b, and s_b, we write

The first digit of the uncertainty is the last significant figure. We often retain extra, insignificant digits to prevent round-off errors in further calculations.

$$\text{Slope:} \quad \begin{matrix} 0.615\ 38 \\ \pm 0.054\ 39 \end{matrix} = 0.62 \pm 0.05 \quad \text{or} \quad 0.61_5 \pm 0.05_4 \qquad \text{(4-23)}$$

$$\text{Intercept:} \quad \begin{matrix} 1.346\ 15 \\ \pm 0.214\ 15 \end{matrix} = 1.3 \pm 0.2 \quad \text{or} \quad 1.3_5 \pm 0.2_1 \qquad \text{(4-24)}$$

where the uncertainties represent one standard deviation. *The first decimal place of the standard deviation is the last significant figure of the slope or intercept.* Many scientists write results such as 1.35 ± 0.21 to avoid excessive round-off.

The 95% confidence interval for the slope is

$$\pm t s_m = \pm (4.303)(0.054) = \pm 0.23$$

based on degrees of freedom $= n - 2 = 2$. The confidence interval is $\pm t s_m$, not $\pm t s_m/\sqrt{n}$, because $\sqrt{n}$ is already implicit in s_m.

If you want to express the uncertainty as a confidence interval instead of one standard deviation, multiply the uncertainties in Equations 4-23 and 4-24 by the appropriate value of Student's t from Table 4-2 for $n - 2$ degrees of freedom.

EXAMPLE   Finding s_y, s_m, and s_b with a Spreadsheet

The Excel function LINEST returns the slope and intercept and their uncertainties in a table (a *matrix*). As an example, enter x and y values in columns A and B. Then highlight the 3-row × 2-column region E3:F5 with your mouse. This block of cells is selected to contain the output of the LINEST function. In the Insert Function window, go to Statistical and double click on LINEST. The new window asks for four inputs to the function. For y values, enter B2:B5. Then enter A2:A5 for x values. The next two entries are both "TRUE". The first TRUE tells Excel that we want to compute the y-intercept of the least-squares line and not force the intercept to be 0. The second TRUE tells Excel to print out the standard deviations as well as the slope and intercept. The formula you have just entered is "=LINEST(B2:B5,A2:A5,TRUE,TRUE)". Click OK and the slope appears in cell E3.

	A	B	C	D	E	F	G
1	x	y			Output from LINEST		
2	1	2			Slope	Intercept	
3	3	3		Parameter	0.61538	1.34615	
4	4	4		Std Dev	0.05439	0.21414	
5	6	5		R^2	0.98462	0.19612	Std Dev (y)
6							
7	Highlight cells E3:F5						
8	Type "= LINEST(B2:B5,A2:A5,TRUE,TRUE)"						
9	Press CTRL+SHIFT+ENTER (on PC)						
10	Press COMMAND+RETURN (on Mac)						

The output of LINEST should be a matrix, not a single number. What went wrong? To tell the computer that you want a matrix, go back and highlight cells E3:F5. "=LINEST(B2:B5,A2:A5,TRUE,TRUE)" appears once again in the formula line. Now press CONTROL+SHIFT+ENTER on a PC or COMMAND(⌘)+RETURN on a Mac. Excel dutifully prints out a matrix in cells E3:F5. Write labels around the block to indicate what is in each cell. The slope and intercept are on the top line. The second line contains s_m and s_b. Cell F5 contains s_y and cell E5 contains a quantity called R^2, which is defined in Equation 5-2 and is a measure of the goodness of fit of the data to the line. The closer R^2 is to unity, the better the fit.

Test Yourself Change the second value of x from 3 to 3.5 and apply LINEST. What is the value of s_y from LINEST? (***Answer:*** 0.364 70)

4-8 Calibration Curves

A *calibration curve* shows the response of an analytical method to known quantities of analyte.[8] Table 4-7 gives real data from a protein analysis that produces a colored product. A *spectrophotometer* measures the absorbance of light, which is proportional to the quantity of protein analyzed. Solutions containing known concentrations of analyte are called **standard solutions.** Solutions containing all reagents and solvents used in the analysis, but no deliberately added analyte, are called **blank solutions.** Blanks measure the response of the analytical procedure to impurities or interfering species in the reagents.

When we scan across the three absorbance values in each row of Table 4-7, the number 0.392 seems out of line: It is inconsistent with the other values for 15.0 μg, and the range of values for the 15.0-μg samples is much bigger than the range for the other samples. The linear relation between the average values of absorbance up to the 20.0-μg sample also indicates that the value 0.392 is in error (Figure 4-12). We choose to omit 0.392 from subsequent calculations.

It is reasonable to ask whether all three absorbances for the 25.0-μg samples are low for some unknown reason, because this point falls below the straight line in Figure 4-12. Repetition of this analysis shows that the 25.0-μg point is consistently below the straight line and there is nothing "wrong" with the data in Table 4-7.

Constructing a Calibration Curve

We adopt the following procedure for constructing a calibration curve:

Step 1   Prepare known samples of analyte covering a range of concentrations expected for unknowns. Measure the response of the analytical procedure to these standards to generate data like the left half of Table 4-7.

Step 2   Subtract the average absorbance (0.099_3) of the *blank* samples from each measured absorbance to obtain *corrected absorbance*. The blank measures the response of the procedure when no protein is present.

Step 3   Make a graph of corrected absorbance versus quantity of protein analyzed (Figure 4-13). Use the least-squares procedure to find the best straight line through the linear portion of the data, up to and including 20.0 μg of protein (14 points, including the 3 corrected blanks, in the shaded portion of Table 4-7). Find the slope and intercept and uncertainties with Equations 4-16, 4-17, 4-20, 4-21, and 4-22. The results are

$$m = 0.016\,3_0 \qquad s_m = 0.000\,2_2 \qquad s_y = 0.005_9$$
$$b = 0.004_7 \qquad s_b = 0.002_6$$

Sections 17-1 and 17-2 discuss absorption of light and define the term *absorbance*. We will use concepts from these two sections throughout this book. You may want to read these sections for background.

FIGURE 4-12 Average absorbance values in Table 4-7 versus micrograms of protein analyzed. Averages for 0 to 20 μg of protein lie on a straight line if the questionable datum 0.392 at 15 μg is omitted.

Absorbance of the blank can arise from the color of starting reagents, reactions of impurities, and reactions of interfering species. Blank values can vary from one set of reagents to another, but corrected absorbance should not.

TABLE 4-7   Spectrophotometer data used to construct calibration curve

Amount of protein (μg)	Absorbance of independent samples			Range	Corrected absorbance		
0	0.099	0.099	0.100	0.001	-0.000_3	-0.000_3	0.000_7
5.0	0.185	0.187	0.188	0.003	0.085_7	0.087_7	0.088_7
10.0	0.282	0.272	0.272	0.010	0.182_7	0.172_7	0.172_7
15.0	0.345	0.347	(0.392)	0.047	0.245_7	0.247_7	—
20.0	0.425	0.425	0.430	0.005	0.325_7	0.325_7	0.330_7
25.0	0.483	0.488	0.496	0.013	0.383_7	0.388_7	0.396_7

FIGURE 4-13 Calibration curve for protein analysis in Table 4-7. The equation of the solid straight line fitting the 14 data points (open circles) from 0 to 20 μg, derived by the method of least squares, is $y = 0.016\,3_0\,(\pm0.000\,2_2)x + 0.004_7\,(\pm0.002_6)$ with $s_y = 0.005_9$. The equation of the dashed quadratic curve that fits all 17 data points from 0 to 25 μg, determined by a nonlinear least squares procedure[6] is $y = -1.1_7(\pm0.2_1) \times 10^{-4}\,x^2 + 0.018\,5_8\,(\pm0.000\,4_6)x - 0.000\,7\,(\pm0.001\,0)$ with $s_y = 0.004_6$.

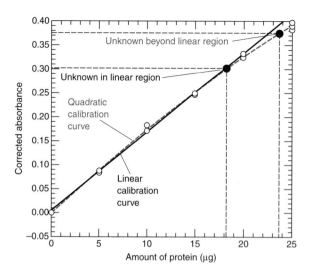

Equation of calibration line:

$$y(\pm s_y) = [m(\pm s_m)]x + [b(\pm s_b)]$$

The equation of the linear calibration line is

$$\underbrace{\text{absorbance}}_{y} = m \times \underbrace{(\mu\text{g of protein})}_{x} + b$$

$$= (0.016\,3_0)(\mu\text{g of protein}) + 0.004_7 \qquad (4\text{-}25)$$

where y is the corrected absorbance ($=$ observed absorbance $-$ blank absorbance).

Step 4 If you analyze an unknown at a future time, run a blank at the same time. Subtract the new blank absorbance from the unknown absorbance to obtain corrected absorbance.

EXAMPLE **Using a Linear Calibration Curve**

An unknown protein sample gave an absorbance of 0.406 and a blank had an absorbance of 0.104. How many micrograms of protein are in the unknown?

Solution The corrected absorbance is $0.406 - 0.104 = 0.302$, which lies on the linear portion of the calibration curve in Figure 4-13. Equation 4-25 therefore becomes

$$\mu\text{g of protein} = \frac{\text{absorbance} - 0.004_7}{0.016\,3_0} = \frac{0.302 - 0.004_7}{0.016\,3_0} = 18.2_4\,\mu\text{g} \qquad (4\text{-}26)$$

Test Yourself What mass of protein gives a corrected absorbance of 0.250? (**Answer:** $15.0_5\,\mu$g)

We prefer calibration procedures with a **linear response,** in which the corrected analytical signal ($=$ signal from sample $-$ signal from blank) is proportional to the quantity of analyte. Although we try to work in the linear range, you can obtain valid results beyond the linear region ($>20\,\mu$g) in Figure 4-13. The dashed curve that goes up to 25 μg of protein comes from a least-squares fit of the data to the equation $y = ax^2 + bx + c$ (Box 4-2).

BOX 4-2 Using a Nonlinear Calibration Curve

Consider an unknown whose corrected absorbance of 0.375 lies beyond the linear region in Figure 4-13. We can fit all the data points with the quadratic equation[6]

$$y = -1.17 \times 10^{-4}x^2 + 0.018\,58x - 0.000\,7 \qquad \text{(a)}$$

To find the quantity of protein, substitute the corrected absorbance into Equation a:

$$0.375 = -1.17 \times 10^{-4}x^2 + 0.018\,58x - 0.000\,7$$

This equation can be rearranged to

$$1.17 \times 10^{-4}x^2 - 0.018\,58x + 0.375\,7 = 0$$

which is a quadratic equation of the form

$$ax^2 + bx + c = 0$$

whose two possible solutions are

$$x = \frac{-b + \sqrt{b^2 - 4ac}}{2a} \qquad x = \frac{-b - \sqrt{b^2 - 4ac}}{2a}$$

Substituting $a = 1.17 \times 10^{-4}$, $b = -0.018\,58$, and $c = 0.375\,7$ into these equations gives

$$x = 135\,\mu\text{g} \qquad x = 23.8\,\mu\text{g}$$

Figure 4-13 tells us that the correct choice is 23.8 μg, not 135 μg.

The **linear range** of an analytical method is the analyte concentration range over which response is proportional to concentration. A related quantity in Figure 4-14 is **dynamic range**—the concentration range over which there is a measurable response to analyte, even if the response is not linear.

FIGURE 4-14 Calibration curve illustrating linear and dynamic ranges.

Good Practice

Always make a graph of your data. The graph gives you an opportunity to reject bad data or the stimulus to repeat a measurement or decide that a straight line is not an appropriate function.

It is not reliable to extrapolate any calibration curve, linear or nonlinear, beyond the measured range of standards. Measure standards in the entire concentration range of interest.

At least six calibration concentrations and two replicate measurements of unknown are recommended. The most rigorous procedure is to make each calibration solution independently from a certified material. Avoid serial dilution of a single stock solution. Serial dilution propagates any systematic error in the stock solution. Measure calibration solutions in random order, not in consecutive order by increasing concentration.

Propagation of Uncertainty with a Calibration Curve

In the preceding example, an unknown with a corrected absorbance of $y = 0.302$ had a protein content of $x = 18.2_4$ µg. What is the uncertainty in the number 18.2_4? Propagation of uncertainty for fitting the equation $y = mx + b$ (but not $y = mx$) gives the following result:[1,9]

$$\text{Uncertainty in } x \,(= s_x) = \frac{s_y}{|m|}\sqrt{\frac{1}{k} + \frac{1}{n} + \frac{(y - \bar{y})^2}{m^2 \sum (x_i - \bar{x})^2}} \qquad (4\text{-}27)$$

where s_y is the standard deviation of y (Equation 4-20), $|m|$ is the absolute value of the slope, k is the number of replicate measurements of the unknown, n is the number of data points for the calibration line (14 in Table 4-7), $\bar{y}$ is the mean value of y for the points on the calibration line, x_i are the individual values of x for the points on the calibration line, and $\bar{x}$ is the mean value of x for the points on the calibration line. For a single measurement of the unknown, $k = 1$ and Equation 4-27 gives $s_x = \pm 0.3_9$ µg. Four replicate unknowns ($k = 4$) with an average corrected absorbance of 0.302 reduce the uncertainty to $\pm 0.2_3$ µg.

The confidence interval for x is $\pm t s_x$, where t is Student's t (Table 4-2) for $n - 2$ degrees of freedom. If $s_x = 0.2_3$ µg and $n = 14$ points (12 degrees of freedom), the 95% confidence interval for x is $\pm t s_x = \pm(2.179)(0.2_3) = \pm 0.5_0$ µg.

y = corrected absorbance of unknown = 0.302
x_i = µg of protein in standards in Table 4-7
   = (0, 0, 0, 5.0, 5.0, 5.0, 10.0, 10.0, 10.0, 15.0, 15.0, 20.0, 20.0, 20.0)
$\bar{y}$ = average of 14 y values = 0.161_8
$\bar{x}$ = average of 14 x values = 9.64_3 µg

To find values of t that are not in Table 4-2, use the Excel function TINV. For 12 degrees of freedom and 95% confidence, the function TINV(0.05,12) returns $t = 2.179$.

4-9   📊   A Spreadsheet for Least Squares

Figure 4-15 implements least-squares analysis, including propagation of error with Equation 4-27. Enter values of x and y in columns B and C. Then select cells B10:C12. Enter the formula "=LINEST(C4:C7,B4:B7,TRUE,TRUE)" and press CONTROL+SHIFT+ENTER on a PC or COMMAND(⌘)+RETURN on a Mac. LINEST returns m, b, s_m, s_b, R^2, and s_y in cells B10:C12. Write labels in cells A10:A12 and D10:D12 so that you know what the numbers in cells B10:C12 mean.

Cell B14 gives the number of data points with the formula "=COUNT(B4:B7)". Cell B15 computes the mean value of y. Cell B16 computes the sum $\sum (x_i - \bar{x})^2$ that we need for Equation 4-27. This sum is common enough that Excel has a built-in function called DEVSQ that you can find in the Statistics menu of the Insert Function window.

Enter the measured mean value of y for replicate measurements of the unknown in cell B18. In cell B19, enter the number of replicate measurements of the unknown. Cell B20 computes the value of x corresponding to the measured mean value of y. Cell B21 uses Equation 4-27 to find the uncertainty (the standard deviation) in the value of x for the unknown. If you want a confidence interval for x, multiply s_x times Student's t from Table 4-2 for $n - 2$ degrees of freedom and the desired confidence level.

We always want a graph to see if the calibration points lie on a straight line. Follow the instructions in Section 2-11 to plot the calibration data as a scatter plot with only markers (no line yet). To add a straight line in Excel 2007, click on the chart to obtain the Chart Tools ribbon. Select Layout, then Trendline and Choose More Trendline Options. Select Linear and Display Equation on chart. The least-squares straight line and its equation appear on the

95% confidence interval for x in Figure 4-15:
$$x \pm t s_x = 2.232\,5 \pm (4.303)(0.373\,5)$$
$$= 2.2 \pm 1.6$$
(degrees of freedom $= n - 2 = 2$)

	A	B	C	D	E	F	G	H	I
1	Least-Squares Spreadsheet								
2									
3	Highlight cells B10:C12	x	y						
4	Type "= LINEST(C4:C7,	1	2						
5	B4:B7,TRUE,TRUE)	3	3						
6	For PC, press	4	4						
7	CTRL+SHIFT+ENTER	6	5						
8	For Mac, press								
9	COMMAND+RETURN	LINEST output:							
10	m	0.6154	1.3462	b					
11	s_m	0.0544	0.2141	s_b					
12	R^2	0.9846	0.1961	s_y					
13									
14	$n =$	4	B14 = COUNT(B4:B7)						
15	Mean y =	3.5	B15 = AVERAGE(C4:C7)						
16	$\Sigma(x_i - \text{mean } x)^2 =$	13	B16 = DEVSQ(B4:B7)						
17									
18	Measured y =	2.72	Input						
19	k = Number of replicate measurements of y =	1	Input						
20	Derived x =	2.2325	B20 = (B18-C10)/B10						
21	$s_x =$	0.3735	B21 = (C12/B10)*SQRT((1/B19)+(1/B14)+((B18-B15)^2)/(B10^2*B16))						

FIGURE 4-15 Spreadsheet for linear least-squares analysis.

graph. Use the Forecast section of the Format Trendline box to extend the line forward or backward beyond the range of data. The Format Trendline box also allows you to select the color and style of the line.

In earlier versions of Excel, click on one data point and they will all be highlighted. In the Chart menu, select Add Trendline. In the window that appears, select Linear. Go to Options in the Trendline box and select Display Equation on Chart. Double click on the line and you can adjust its thickness and appearance. Double clicking on the equation allows you to modify its format. Double click on the straight line and select Options. In the Forecast box, you can extend the trendline Forward and Backward as far as you like.

Adding Error Bars to a Graph

Error bars on a graph help us judge the quality of the data and the fit of a curve to the data. Consider the data in Table 4-7. Let's plot the mean absorbance of columns 2 to 4 versus sample mass in column 1. Then we will add error bars corresponding to the 95% confidence interval for each point. Figure 4-16 lists mass in column A and mean absorbance in column B. The standard deviation of absorbance is given in column C. The 95% confidence interval for absorbance is computed in column D with the formula in the margin. Student's $t = 4.303$ can be found for 95% confidence and $3 - 1 = 2$ degrees of freedom in Table 4-2. Alternatively, compute Student's t with the function TINV(0.05,2) in cell B11. The parameters for TINV are 0.05 for 95% confidence and 2 for degrees of freedom. The 95% confidence interval in cell D4 is computed with "=$B$11*C4/SQRT(3)". You should be able to plot mean absorbance (y) in column B versus protein mass (x) in column A.

To add error bars in Excel 2007, click on one of the points to highlight all points on the graph. In Chart Tools, Layout, select Error Bars and choose More Error Bars Options. For Error Amount, choose Custom and Specify Value. For both Positive Value and Negative Value, enter D4:D9. You just told the spreadsheet to use 95% confidence intervals for error bars. When you click OK, the graph has both x and y error bars. Click on any x error bar and press Delete to remove all x error bars.

To add error bars in earlier versions of Excel, click on one of the points to highlight all points on the graph. In the Format menu, choose Selected Data Series. Select Y Error Bars and a window appears. Click Custom. Click in the positive error bars box and select cells D4:D9. Click the negative error bars box and select cells D4:D9 again. You just told Excel to use values in cells D4:D9 for the lengths of the error bars. Click OK and error bars appear in the graph.

Confidence interval $= \pm ts/\sqrt{n}$

t = Student's t for 95% confidence and $n - 1 = 2$ degrees of freedom

s = standard deviation

n = number of values in average = 3

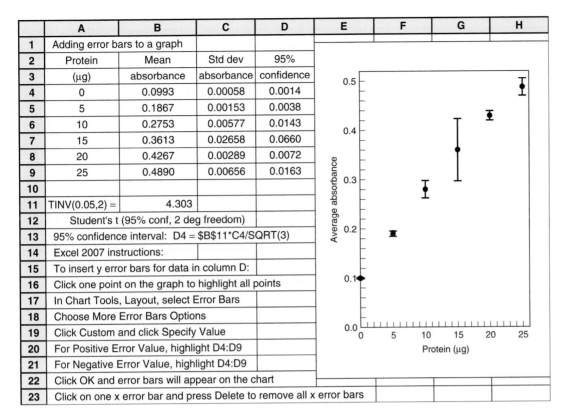

	A	B	C	D	E	F	G	H
1	Adding error bars to a graph							
2	Protein	Mean	Std dev	95%				
3	(µg)	absorbance	absorbance	confidence				
4	0	0.0993	0.00058	0.0014				
5	5	0.1867	0.00153	0.0038				
6	10	0.2753	0.00577	0.0143				
7	15	0.3613	0.02658	0.0660				
8	20	0.4267	0.00289	0.0072				
9	25	0.4890	0.00656	0.0163				
10								
11	TINV(0.05,2) =	4.303						
12	Student's t (95% conf, 2 deg freedom)							
13	95% confidence interval: D4 = $B$11*C4/SQRT(3)							
14	Excel 2007 instructions:							
15	To insert y error bars for data in column D:							
16	Click one point on the graph to highlight all points							
17	In Chart Tools, Layout, select Error Bars							
18	Choose More Error Bars Options							
19	Click Custom and click Specify Value							
20	For Positive Error Value, highlight D4:D9							
21	For Negative Error Value, highlight D4:D9							
22	Click OK and error bars will appear on the chart							
23	Click on one x error bar and press Delete to remove all x error bars							

FIGURE 4-16 Adding 95% confidence error bars to a graph.

Terms to Understand

average	dynamic range	linear response	standard deviation of
blank solution	F test	mean	the mean
calibration curve	Gaussian distribution	method of least squares	standard solution
confidence interval	Grubbs test	outlier	Student's t
degrees of freedom	intercept	slope	t test
determinant	linear range	standard deviation	variance

Summary

The results of many measurements of an experimental quantity follow a Gaussian distribution. The measured mean, $\bar{x}$, approaches the true mean, μ, as the number of measurements becomes very large. The broader the distribution, the greater is σ, the standard deviation. For n measurements, an estimate of the standard deviation is given by $s = \sqrt{[\Sigma(x_i + \bar{x})^2]/(n - 1)}$. About two-thirds of all measurements lie within $\pm 1\sigma$, and 95% lie within 2σ. The probability of observing a value within a certain interval is proportional to the area of that interval.

Student's t is used to find confidence intervals ($\mu = \bar{x} \pm ts/\sqrt{n}$) and to compare mean values measured by different methods. The F test is used to decide whether two standard deviations are significantly different from each other. The Grubbs test helps you to decide whether or not a questionable datum should be discarded. It is best to repeat the measurement several times to increase the probability that your decision to accept or reject a datum is correct.

A calibration curve shows the response of a chemical analysis to known quantities (standard solutions) of analyte. When there is a linear response, the corrected analytical signal (= signal from sample − signal from blank) is proportional to the quantity of analyte. Blank solutions are prepared from the same reagents and solvents used to prepare standards and unknowns, but blanks have no intentionally added analyte. The blank tells us the response of the procedure to impurities or interfering species in the reagents. The blank value is subtracted from measured values of standards prior to constructing the calibration curve. The blank value is subtracted from the response of an unknown prior to computing the quantity of analyte in the unknown.

The method of least squares is used to determine the equation of the "best" straight line through experimental data points. Equations 4-16 to 4-18 and 4-20 to 4-22 provide the least-squares slope and intercept and their standard deviations. Equation 4-27 estimates the uncertainty in x from a measured value of y with a calibration curve. A spreadsheet simplifies least-squares calculations and graphical display of the results.

Exercises

4-A. For the numbers 116.0, 97.9, 114.2, 106.8, and 108.3, find the mean, standard deviation, range, and 90% confidence interval for the mean. Using the Grubbs test, decide whether the number 97.9 should be discarded.

4-B. 🖳 *Spreadsheet for standard deviation*. Let's create a spreadsheet to compute the mean and standard deviation of a column of numbers in two different ways. The spreadsheet here is a template for this exercise.

	A	B	C	D
1	Computing standard deviation			
2				
3		Data = x	x-mean	(x-mean)^2
4		17.4		
5		18.1		
6		18.2		
7		17.9		
8		17.6		
9	sum =			
10	mean =			
11	std dev =			
12				
13	Formulas:	B9 =		
14		B10 =		
15		B11 =		
16		C4 =		
17		D4 =		
18		D9 =		
19				
20	Calculations using built-in functions:			
21	sum =			
22	mean =			
23	std dev =			

(a) Reproduce the template on your spreadsheet. Cells B4 to B8 contain the data (x values) whose mean and standard deviation we will compute.

(b) Write a formula in cell B9 to compute the sum of numbers in B4 to B8.

(c) Write a formula in cell B10 to compute the mean value.

(d) Write a formula in cell C4 to compute (x − mean), where x is in cell B4 and the mean is in cell B10. Use Fill Down to compute values in cells C5 to C8.

(e) Write a formula in cell D4 to compute the square of the value in cell C4. Use Fill Down to compute values in cells D5 to D8.

(f) Write a formula in cell D9 to compute the sum of the numbers in cells D4 to D8.

(g) Write a formula in cell B11 to compute the standard deviation.

(h) Use cells B13 to B18 to document your formulas.

(i) Now we are going to simplify life by using formulas built into the spreadsheet. In cell B21 type "=SUM(B4:B8)", which means find the sum of numbers in cells B4 to B8. Cell B21 should display the same number as cell B9. In general, you will not know what functions are available and how to write them. In Excel 2007, use the

Formulas ribbon and Insert Function to find SUM. In earlier versions of Excel, find the Function menu under Insert.

(j) Select cell B22. Go to Insert Function and find AVERAGE. When you type "=AVERAGE(B4:B8)" in cell B22, its value should be the same as B10.

(k) For cell B23, find the standard deviation function ("=STDEV(B4:B8)") and check that the value agrees with cell B11.

4-C. Use Table 4-1 for this exercise. Suppose that the mileage at which 10 000 sets of automobile brakes had been 80% worn through was recorded. The average was 62 700, and the standard deviation was 10 400 miles.

(a) What fraction of brakes is expected to be 80% worn in less than 40 860 miles?

(b) What fraction is expected to be 80% worn at a mileage between 57 500 and 71 020 miles?

4-D. 🖳 Use the NORMDIST spreadsheet function to answer these questions about the brakes described in Exercise 4-C:

(a) What fraction of brakes is expected to be 80% worn in less than 45 800 miles?

(b) What fraction is expected to be 80% worn at a mileage between 60 000 and 70 000 miles?

4-E. A reliable assay shows that the ATP (adenosine triphosphate) content of a certain cell type is 111 µmol/100 mL. You developed a new assay, which gave the following values for replicate analyses: 117, 119, 111, 115, 120 µmol/100 mL (average = $116._4$). Does your result agree with the known value at the 95% confidence level?

4-F. Traces of toxic, man-made hexachlorohexanes in North Sea sediments were extracted by a known process and by two new procedures, and measured by chromatography.

Method	Concentration found (pg/g)	Standard deviation (pg/g)	Number of replications
Conventional	34.4	3.6	6
Procedure A	42.9	1.2	6
Procedure B	51.1	4.6	6

SOURCE: *D. Sterzenbach, B. W. Wenclawiak, and V. Weigelt, Anal. Chem.* **1997**, *69, 831.*

(a) Are the concentrations parts per million, parts per billion, or something else?

(b) Is the standard deviation for procedure B significantly different from that of the conventional procedure?

(c) Is the mean concentration found by procedure B significantly different from that of the conventional procedure?

(d) Answer the same two questions as parts **(b)** and **(c)** to compare procedure A to the conventional procedure.

4-G. 🖳 *Calibration curve.* (You can do this exercise with your calculator, but it is more easily done by the spreadsheet in Figure 4-15.) In the Bradford protein determination, the color of a dye changes from brown to blue when it binds to protein. Absorbance of light is measured.

Protein (µg):	0.00	9.36	18.72	28.08	37.44
Absorbance at 595 nm:	0.466	0.676	0.883	1.086	1.280

(a) Find the equation of the least-squares straight line through these points in the form $y = [m(\pm s_m)]x + [b(\pm s_b)]$ with a reasonable number of significant figures.

(b) Make a graph showing the experimental data and the calculated straight line.

Problems

Gaussian Distribution

4-1. What is the relation between the standard deviation and the precision of a procedure? What is the relation between standard deviation and accuracy?

4-2. Use Table 4-1 to state what fraction of a Gaussian population lies within the following intervals:

(a) $\mu \pm \sigma$ **(c)** μ to $+\sigma$ **(e)** $-\sigma$ to -0.5σ
(b) $\mu \pm 2\sigma$ **(d)** μ to $+0.5\sigma$

4-3. The ratio of the number of atoms of the isotopes ^{69}Ga and ^{71}Ga in eight samples from different sources was measured in an effort to understand differences in reported values of the atomic mass of gallium:

Sample	^{69}Ga/^{71}Ga	Sample	^{69}Ga/^{71}Ga
1	1.526 60	5	1.528 94
2	1.529 74	6	1.528 04
3	1.525 92	7	1.526 85
4	1.527 31	8	1.527 93

SOURCE: *J. W. Gramlich and L. A. Machlan, Anal. Chem.* **1985**, *57*, 1788.

Find the **(a)** mean, **(b)** standard deviation, and **(c)** variance. **(d)** Write the mean and standard deviation together with an appropriate number of significant digits.

4-4. (a) Calculate the fraction of bulbs in Figure 4-1 expected to have a lifetime greater than 1 005.3 h.

(b) What fraction of bulbs is expected to have a lifetime between 798.1 and 901.7 h?

(c) 🖩 Use the Excel NORMDIST function to find the fraction of bulbs expected to have a lifetime between 800 and 900 h.

4-5. 🖩 Blood plasma proteins of patients with malignant breast tumors differ from proteins of healthy people in their solubility in the presence of various polymers. When the polymers dextran and poly(ethylene glycol) are mixed with water, a two-phase mixture is formed. When plasma proteins of tumor patients are added, the distribution of proteins between the two phases is different from that of plasma proteins of a healthy person. The distribution coefficient (K) for any substance is defined as K = [concentration of the substance in phase A] / [concentration of the substance in phase B]. Proteins of healthy people have a mean distribution coefficient of 0.75 with a standard deviation of 0.07. For the proteins of people with cancer, the mean is 0.92 with a standard deviation of 0.11.

(a) Suppose that K were used as a diagnostic tool and that a positive indication of cancer is taken as $K \geq 0.92$. What fraction of people with tumors would have a false negative indication of cancer because $K < 0.92$?

(b) What fraction of healthy people would have a false positive indication of cancer? This number is the fraction of healthy people with $K \geq 0.92$, shown by the shaded area in the adjoining graph.

Estimate an answer with Table 4-1 and obtain a more exact result with the NORMDIST function in Excel.

(c) Vary the first argument of the NORMDIST function to select a distribution coefficient that would identify 75% of people with tumors. That is, 75% of patients with tumors would have K above the selected distribution coefficient. With this value of K, what fraction of healthy people would have a false positive result indicating they have a tumor?

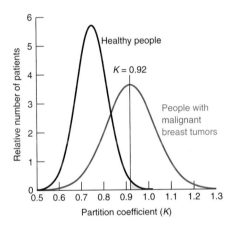

Distribution coefficients of plasma proteins from healthy people and from people with malignant breast tumors. [Data from B. Y. Zaslavsky, "Bioanalytical Applications of Partitioning in Aqueous Polymer Two-Phase Systems," *Anal. Chem.* **1992**, *64*, 765A.]

4-6. 🖩 The equation for the Gaussian curve in Figure 4-1 is

$$y = \frac{(\text{total bulbs})(\text{hours per bar})}{s\sqrt{2\pi}} e^{-(x-\bar{x})^2/2s^2}$$

where $\bar{x}$ is the mean value (845.2 h), s is the standard deviation (94.2 h), total bulbs = 4 768, and hours per bar (= 20) is the width of each bar in Figure 4-1. Set up a spreadsheet like the one on the next page to calculate the coordinates of the Gaussian curve in Figure 4-1 from 500 to 1 200 h in 25-h intervals. Note the heavy use of parentheses in the formula at the bottom of the spreadsheet to force the computer to do the arithmetic as intended. Use Excel to graph your results.

4-7. 🖩 Repeat Problem 4-6 but use the values 50, 100, and 150 for the standard deviation. Superimpose all three curves on a single graph.

Confidence Intervals, *t* Test, *F* Test, and Grubbs Test

4-8. What is the meaning of a confidence interval?

4-9. What fraction of vertical bars in Figure 4-5a is expected to include the population mean (10 000) if many experiments are carried out? Why are the 90% confidence interval bars longer than the 50% bars in Figure 4-5?

	A	B	C
1	Gaussian curve for light bulbs (Fig 4-1)		
2			
3	mean =	x (hours)	y (bulbs)
4	845.2	500	0.49
5	std dev =	525	1.25
6	94.2	550	2.98
7	total bulbs =	600	13.64
8	4768	700	123.11
9	hours per bar =	800	359.94
10	20	845.2	403.85
11	sqrt(2 pi) =	900	340.99
12	2.506628	1000	104.67
13		1100	10.41
14		1200	0.34
15	Formula for cell C4 =		
16	($A$8*$A$10/($A$6*$A$12))*		
17	EXP(-((B4-$A$4)^2)/(2*$A$6^2))		

Spreadsheet for Problem 4-6

4-10. List the three different cases that we studied for comparison of means, and write the equations used in each case.

4-11. The percentage of an additive in gasoline was measured six times with the following results: 0.13, 0.12, 0.16, 0.17, 0.20, 0.11%. Find the 90% and 99% confidence intervals for the percentage of the additive.

4-12. Sample 8 of Problem 4-3 was analyzed seven times, with $\bar{x} = 1.527\,93$ and $s = 0.000\,07$. Find the 99% confidence interval for sample 8.

4-13. A trainee in a medical lab will be released to work on her own when her results agree with those of an experienced worker at the 95% confidence level. Results for a blood urea nitrogen analysis are shown below.

Trainee: $\bar{x} = 14.5_7$ mg/dL $\quad s = 0.5_3$ mg/dL $\quad n = 6$ samples

Experienced worker: $\bar{x} = 13.9_5$ mg/dL $\quad s = 0.4_2$ mg/dL $\quad n = 5$ samples

(a) What does the abbreviation dL stand for?

(b) Should the trainee be released to work alone?

4-14. The CdSe content (g/L) of nanocrystals was measured by two methods for six different samples. Do the two methods differ significantly at the 95% confidence level?

Sample	Method 1 Anodic stripping	Method 2 Atomic absorption
A	0.88	0.83
B	1.15	1.04
C	1.22	1.39
D	0.93	0.91
E	1.17	1.08
F	1.51	1.31

SOURCE: E. Kuçur, F. M. Boldt, S. Cavaliere-Jaricont, J. Ziegler, and T. Nann, Anal. Chem. **2007**, 79, 8987.

4-15. Now we use a built-in routine in Excel for the paired t test to see if the two methods in Problem 4-14 produce significantly different results. Enter the data for Methods 1 and 2 into two columns of a spreadsheet. For Excel 2007, find Data Analysis in the Data ribbon. In earlier versions of Excel, find Data Analysis in the Tools menu. If Data Analysis does not appear, follow the instructions at the beginning of Section 4-5 to load this software. Select Data Analysis and then select t-Test: Paired Two Sample for Means. Follow the instructions of Section 4-5, and the routine will print out information including $t_{calculated}$ (labeled t Stat) and t_{table} (labeled t Critical two-tail). You should reproduce the results of Problem 4-14.

4-16. Two methods were used to measure fluorescence lifetime of a dye. Are the standard deviations significantly different? Are the means significantly different?

Quantity	Method 1	Method 2
Mean lifetime (ns)	1.382	1.346
Standard deviation (ns)	0.025	0.039
Number of measurements	4	4

SOURCE: N. Boens et al., Anal. Chem. **2007**, 79, 2137.

4-17. Do the following two sets of measurements of $^6Li/^7Li$ in a Standard Reference Material give statistically equivalent results?

Method 1	Method 2
0.082 601	0.081 83
0.082 621	0.081 86
0.082 589	0.082 05
0.082 617	0.082 06
0.082 598	0.082 15
	0.082 08

SOURCE: S. Ahmed, N. Jabeen, and E. ur Rehman, Anal. Chem. **2002**, 74, 4133; L. W. Green, J. J. Leppinen, and N. L. Elliot, Anal. Chem. **1988**, 60, 34.

4-18. If you measure a quantity four times and the standard deviation is 1.0% of the average, can you be 90% confident that the true value is within 1.2% of the measured average?

4-19. Students measured the concentration of HCl in a solution by titrating with different indicators to find the end point.

Indicator	Mean HCl concentration (M) (± standard deviation)	Number of measurements
1. Bromothymol blue	0.095 65 ± 0.002 25	28
2. Methyl red	0.086 86 ± 0.000 98	18
3. Bromocresol green	0.086 41 ± 0.001 13	29

SOURCE: D. T. Harvey, J. Chem. Ed. **1991**, 68, 329.

Is the difference between indicators 1 and 2 significant at the 95% confidence level? Answer the same question for indicators 2 and 3.

4-20. Hydrocarbons in the cab of an automobile were measured during trips on the New Jersey Turnpike and trips through the Lincoln Tunnel connecting New York and New Jersey.[10] The concentrations (± standard deviations) of m- and p-xylene were

Turnpike: 31.4 ± 30.0 µg/m³ (32 measurements)

Tunnel: 52.9 ± 29.8 µg/m³ (32 measurements)

Do these results differ at the 95% confidence level? At the 99% confidence level?

4-21. A Standard Reference Material is certified to contain 94.6 ppm of an organic contaminant in soil. Your analysis gives values of 98.6, 98.4, 97.2, 94.6, and 96.2 ppm. Do your results differ from the expected result at the 95% confidence level? If you made one more measurement and found 94.5, would your conclusion change?

4-22. Nitrite (NO_2^-) was measured by two methods in rainwater and unchlorinated drinking water. The results ± standard deviation (number of samples) are

Sample source	Gas chromatography	Spectrophotometry
Rainwater	0.069 ± 0.005 mg/L ($n = 7$)	0.063 ± 0.008 mg/L ($n = 5$)
Drinking water	0.078 ± 0.007 mg/L ($n = 5$)	0.087 ± 0.008 mg/L ($n = 5$)

SOURCE: I. Sarudi and I. Nagy, Talanta **1995**, 42, 1099.

(a) Do the two methods agree with each other at the 95% confidence level for both rainwater and drinking water?

(b) For each method, does the drinking water contain significantly more nitrite than the rainwater (at the 95% confidence level)?

4-23. Should the value 216 be rejected from the set of results 192, 216, 202, 195, and 204?

Linear Least Squares

4-24. A straight line is drawn through the points $(3.0, -3.87 \times 10^4)$, $(10.0, -12.99 \times 10^4)$, $(20.0, -25.93 \times 10^4)$, $(30.0, -38.89 \times 10^4)$, and $(40.0, -51.96 \times 10^4)$ to give $m = -1.298\ 72 \times 10^4$, $b = 256.695$, $s_m = 13.190$, $s_b = 323.57$, and $s_y = 392.9$. Express the slope and intercept and their uncertainties with reasonable significant figures.

4-25. Here is a least-squares problem that you can do by hand with a calculator. Find the slope and intercept and their standard deviations for the straight line drawn through the points $(x,y) = (0,1)$, $(2,2)$, and $(3,3)$. Make a graph showing the three points and the line. Place error bars ($\pm s_y$) on the points.

4-26. Set up a spreadsheet to reproduce Figure 4-15. *Add error bars:* Follow the procedure on page 90. Use s_y for the + and − error.

4-27. *Excel LINEST function.* Enter the following data in a spreadsheet and use LINEST to find slope, intercept, and standard deviations. Use Excel to draw a graph of the data and add a trendline. Draw error bars of $\pm s_y$ on the points.

x:	3.0	10.0	20.0	30.0	40.0
y:	−0.074	−1.411	−2.584	−3.750	−5.407

Calibration Curves

4-28. Explain the following statement: "The validity of a chemical analysis ultimately depends on measuring the response of the analytical procedure to known standards."

4-29. Suppose that you carry out an analytical procedure to generate a linear calibration curve like that shown in Figure 4-13. Then you analyze an unknown and find an absorbance that gives a negative concentration for the analyte. What does this mean?

4-30. Using the linear calibration curve in Figure 4-13, find the quantity of unknown protein that gives a measured absorbance of 0.264 when a blank has an absorbance of 0.095.

4-31. Consider the least-squares problem in Figure 4-11.

(a) Suppose that a single new measurement produces a y value of 2.58. Find the corresponding x value and its uncertainty.

(b) Suppose you measure y four times and the average is 2.58. Calculate the uncertainty based on four measurements, not one.

4-32. Consider the linear calibration curve in Figure 4-13, which is derived from the 14 corrected absorbances in the shaded region at the right side of Table 4-7. Create a least-squares spreadsheet like Figure 4-15 to compute the equation of the line and the standard deviations of the parameters. Suppose that you find absorbance values of 0.265, 0.269, 0.272, and 0.258 for four identical samples of unknown and absorbances of 0.099, 0.091, 0.101, and 0.097 for four blanks. Find the corrected absorbance by subtracting the average blank from the average absorbance of the unknown. Calculate the amount of protein and its uncertainty in the unknown.

4-33. Here are mass spectrometric signals for methane in H_2:

CH$_4$ (vol%):	0	0.062	0.122	0.245	0.486	0.971	1.921
Signal (mV):	9.1	47.5	95.6	193.8	387.5	812.5	1 671.9

(a) Subtract the blank value (9.1) from all other values. Then use the method of least squares to find the slope and intercept and their uncertainties. Construct a calibration curve.

(b) Replicate measurements of an unknown gave 152.1, 154.9, 153.9, and 155.1 mV, and a blank gave 8.2, 9.4, 10.6, and 7.8 mV. Subtract the average blank from the average unknown to find the average corrected signal for the unknown.

(c) Find the concentration of the unknown and its uncertainty.

4-34. *Nonlinear calibration curve.* Following the procedure in Box 4-2, find how many micrograms (µg) of protein are contained in a sample with a corrected absorbance of 0.350 in Figure 4-13.

4-35. *Logarithmic calibration curve.* Calibration data spanning five orders of magnitude for an electrochemical determination of p-nitrophenol are given in the table. (The blank has already been subtracted from the measured current.) If you try to plot these data on a linear graph extending from 0 to 310 µg/mL and from 0 to 5 260 nA, most of the points will be bunched up near the origin. To handle data with such a large range, a logarithmic plot is helpful.

p-Nitrophenol (µg/mL)	Current (nA)	p-Nitrophenol (µg/mL)	Current (nA)
0.010 0	0.215	3.00	66.7
0.029 9	0.846	10.4	224
0.117	2.65	31.2	621
0.311	7.41	107	2 020
1.02	20.8	310	5 260

SOURCE: Data from Figure 4 of L. R. Taylor, Am. Lab., February 1993, p. 44.

(a) Make a graph of log(current) versus log(concentration). Over what range is the log-log calibration linear?

(b) Find the equation of the line in the form log (current) $= m \times$ log (concentration) $+ b$.

(c) Find the concentration of p-nitrophenol corresponding to a signal of 99.9 nA.

4-36. *Confidence interval for calibration curve.* To use a calibration curve based on n points, we measure a new value of y and calculate the corresponding value of x. The one-standard-deviation uncertainty in x, s_x, is given by Equation 4-27. We express a *confidence interval* for x, using Student's t:

$$\text{Confidence interval} = x \pm ts_x$$

where t is taken from Table 4-2 for $n - 2$ degrees of freedom.

A calibration curve based on $n = 10$ known points was used to measure the protein in an unknown. The results were protein $= 15.2_2$ ($\pm 0.4_6$) µg, where $s_x = 0.4_6$ µg. Find the 90% and 99% confidence intervals for protein in the unknown.

5 Quality Assurance and Calibration Methods

(a)

(b)

(a) Scattered measurements of lead in river water by different laboratories, each of which employed a recognized quality management system. (b) Reproducible results from national measurement institutes. [From P. De Bièvre and P. D. P. Taylor, "'Demonstration' vs. 'Designation' of Measurement Competence: The Need to Link Accreditation to Metrology," *Fresenius J. Anal. Chem.* 2000, *368*, 567.]

The Institute for Reference Materials and Measurements in Belgium conducts an International Measurement Evaluation Program to allow laboratories to assess the reliability of their analyses. Panel *a* shows results for lead in river water. Of 181 labs, 18 reported results more than 50% above and 4 reported results more than 50% below the certified level of 62.3 ± 1.3 nM. Even though most labs in the study employed recognized quality management procedures, a large fraction of results did not include the certified range. Panel *b* shows that when this same river water was analyzed by nine different national measurement institutes, where the most care is taken, all results were close to the certified range.

This example illustrates that there is no guarantee that results are reliable, even if they are obtained by "accredited" laboratories using accepted procedures. A good way to assess the reliability of a lab working for you is to provide the lab with "blind" samples—similar to your unknowns—for which you know the "right" answer but the analyst does not. If the lab does not find the known result, there is a problem. Periodic blind check samples are required to demonstrate continuing reliability.

Data quality standards:
• Get the right data
• Get the data right
• Keep the data right

[Nancy W. Wentworth, U.S. Environmental Protection Agency.[1]]

Quality assurance is what we do to get the right answer for our purpose. The answer should have sufficient accuracy and precision to support subsequent decisions. There is no point in spending extra money to obtain a more accurate or more precise answer if it is not necessary. This chapter describes basic issues and procedures in quality assurance[2] and introduces two more calibration methods. In Chapter 4, we discussed how to make a calibration curve. In this chapter, we describe *standard addition* and *internal standards*.

5-1 Basics of Quality Assurance

"Suppose you are cooking for some friends. While making spaghetti sauce, you taste it, season it, taste it some more. Each tasting is a sampling event with a quality control test. You can taste the whole batch because there is only one batch. Now suppose you run a spaghetti sauce plant that makes 1 000 jars a day. You can't taste each one, so you decide to taste three a day, one each at 11 A.M., 2 P.M., and 5 P.M. If the three jars all taste OK, you conclude all 1 000 are OK. Unfortunately, that may not be true, but the relative risk—that a jar has too much or too little seasoning—is not very important because you agree to refund the money of any customer who is dissatisfied. If the number of refunds is small, say, 100 a year, there is no apparent benefit in tasting 4 jars a day." There would be 365 additional tests to avoid refunds on 100 jars, giving a net loss of 265 jars worth of profit.

Quotation from Ed Urbansky. Section 5-1 is adapted from a description written by Ed Urbansky.

In analytical chemistry, the product is not spaghetti sauce but, rather, raw data, treated data, and results. *Raw data* are individual values of a measured quantity, such as peak areas from a chromatogram or volumes from a buret. *Treated data* are concentrations or amounts found by applying a calibration procedure to the raw data. *Results* are what we ultimately report, such as the mean, standard deviation, and confidence interval, after applying statistics to treated data.

Raw data: individual measurements

Treated data: concentrations derived from raw data by use of calibration method

Results: quantities reported after statistical analysis of treated data

Use Objectives

An important goal of quality assurance is making sure that results meet the customer's needs. If you manufacture a drug whose therapeutic dose is just a little less than the lethal dose, you should be more careful than if you make spaghetti sauce. The kind of data that you collect and the way in which you collect them depend on how you plan to use those data. A bathroom scale does not have to measure mass to the nearest milligram, but a drug tablet required to contain 2 mg of active ingredient probably cannot contain 2 ± 1 mg. Writing clear, concise **use objectives** for data and results is a critical step in quality assurance and helps prevent misuse of data and results.

Use objective: states purpose for which results will be used

Here is an example of a use objective. Drinking water is usually disinfected with chlorine, which kills microorganisms. Unfortunately, chlorine also reacts with organic matter in water to produce "disinfection by-products"—compounds that might harm humans. A disinfection facility was planning to introduce a new chlorination process and wrote the following analytical use objective:

> Analytical data and results shall be used to determine whether the modified chlorination process results in at least a 10% reduction of formation of selected disinfection by-products.

The new process was expected to decrease the disinfection by-products. The use objective says that uncertainty in the analysis must be small enough that a 10% decrease in selected by-products is clearly distinguishable from experimental error. In other words, is an observed decrease of 10% real?

Specifications

Once you have use objectives, you are ready to write **specifications** stating how good the numbers need to be and what precautions are required in the analytical procedure. How shall samples be taken and how many are needed? Are special precautions required to protect samples and ensure that they are not degraded? Within practical restraints, such as cost, time, and limited amounts of material available for analysis, what level of accuracy and precision will satisfy the use objectives? What rate of false positives or false negatives is acceptable? These questions need to be answered in detailed specifications.

Quality assurance begins with sampling. We must collect representative samples, and analyte must be preserved after sample is collected. If our sample is not representative or if analyte is lost after collection, then even the most accurate analysis is meaningless.

What do we mean by *false positives* and *false negatives*? Suppose you must certify that a contaminant in drinking water is below a legal limit. A **false positive** says that the concentration exceeds the legal limit when, in fact, the concentration is below the limit. A **false negative** says that the concentration is below the limit when it is actually above the limit. Even well-executed procedures produce some false conclusions because of the statistical nature of sampling and measurement. For drinking water, it is more important to have a low rate of false negatives than a low rate of false positives. It would be worse to certify that contaminated water is safe than to certify that safe water is contaminated. Drug testing of

Specifications might include
- sampling requirements
- accuracy and precision
- rate of false results
- selectivity
- sensitivity
- acceptable blank values
- recovery of fortification
- calibration checks
- quality control samples

athletes is designed to minimize false positives so that an innocent athlete is not falsely accused of doping. In Section 5-2, we will see that there is a tradeoff between false positives, false negatives, and the *detection limit* of an analytical method.

In choosing a method, we also consider selectivity and sensitivity. **Selectivity** (also called *specificity*) means being able to distinguish analyte from other species in the sample (avoiding interference). **Sensitivity** is the capability of responding reliably and measurably to changes in analyte concentration. The *detection limit* of an analytical method must be lower than the concentrations to be measured.

Specifications could include required accuracy and precision, reagent purity, tolerances for apparatus, the use of certified reference materials, and acceptable values for blanks. *Certified reference materials* contain certified levels of analyte in realistic materials that you might be analyzing, such as blood or coal or metal alloys. Your analytical method should produce an answer acceptably close to the certified level or there is something wrong with the accuracy of your method.

Blanks account for interference by other species in the sample and for traces of analyte found in reagents used for sample preservation, preparation, and analysis. Frequent measurements of blanks detect whether analyte from previous samples is carried into subsequent analyses by adhering to vessels or instruments.

A **method blank** is a sample containing all components except analyte, and it is taken through all steps of the analytical procedure. We subtract the response of the method blank from the response of a real sample prior to calculating the quantity of analyte in the sample. A **reagent blank** is similar to a method blank, but it has not been subjected to all sample preparation procedures. The method blank is a more complete estimate of the blank contribution to the analytical response.

A **field blank** is similar to a method blank, but it has been exposed to the site of sampling. For example, to analyze particulates in air, a certain volume of air could be sucked through a filter, which is then dissolved and analyzed. A field blank would be a filter carried to the collection site in the same package with the collection filters. The filter for the blank would be taken out of its package in the field and placed in the same kind of sealed container used for collection filters. The difference between the blank and the collection filters is that air was not sucked through the blank filter. Volatile organic compounds encountered during transportation or in the field are conceivable contaminants of a field blank.

Another performance requirement often specified is *spike recovery*. Sometimes, response to analyte can be decreased or increased by something else in the sample. **Matrix** refers to everything in the sample other than analyte. A **spike,** also called a *fortification*, is a known quantity of analyte added to a sample to test whether the response to a sample is the same as that expected from a calibration curve. Spiked samples are analyzed in the same manner as unknowns. For example, if drinking water is found to contain 10.0 μg/L of nitrate, a spike of 5.0 μg/L could be added. Ideally, the concentration in the spiked portion found by analysis will be 15.0 μg/L. If a number other than 15.0 μg/L is found, then the matrix could be interfering with the analysis.

> *Sensitivity*
>
> = slope of calibration curve
>
> $= \dfrac{\text{change in signal}}{\text{change in analyte concentration}}$

> Add a small volume of concentrated standard to avoid changing the volume of the sample significantly. For example, add 50.5 μL of 500 μg/L standard to 5.00 mL of sample to increase analyte by 5.00 μg/L.
>
> Final concentration
>
> = initial concentration × dilution factor
>
> $= \left(500\dfrac{\mu g}{L}\right)\left(\dfrac{50.5\ \mu L}{5050.5\ \mu L}\right) = 5.00\dfrac{\mu g}{L}$

> Matrix effects can decrease (Figure 5-4 and Problem 5-26) or increase (Problem 5-33) response to analyte.

EXAMPLE   Spike Recovery

Let C stand for concentration. One definition of spike recovery is

$$\%\ \text{recovery} = \frac{C_{\text{spiked sample}} - C_{\text{unspiked sample}}}{C_{\text{added}}} \times 100 \tag{5-1}$$

An unknown was found to contain 10.0 μg of analyte per liter. A spike of 5.0 μg/L was added to a replicate portion of unknown. Analysis of the spiked sample gave a concentration of 14.6 μg/L. Find the percent recovery of the spike.

Solution The percent of the spike found by analysis is

$$\%\ \text{recovery} = \frac{14.6\ \mu g/L - 10.0\ \mu g/L}{5.0\ \mu g/L} \times 100 = 92\%$$

If the acceptable recovery is specified to be in the range from 96% to 104%, then 92% is unacceptable. Something in your method or techniques needs improvement.

Test Yourself   Find % recovery if the spiked sample gave a concentration of 15.3 μg/L. (***Answer:*** 106%)

BOX 5-1 | **Control Charts**

A **control chart** is a visual representation of confidence intervals for a Gaussian distribution. The chart warns us when a monitored property strays dangerously far from an intended *target value*.

Consider a manufacturer making vitamin C tablets intended to have a target value of μ milligrams of vitamin C per tablet. Many analyses over a long time tell us the population standard deviation, σ, associated with the manufacturing process.

For quality control, 25 tablets are removed at random from the manufacturing line each hour and analyzed. The mean value of vitamin C in the 25 tablets is shown by a data point on the control chart.

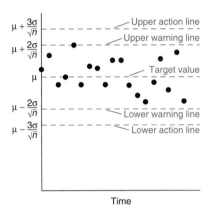

For a Gaussian distribution, 95.5% of all observations are within $\pm 2\sigma/\sqrt{n}$ from the mean and 99.7% are within $\pm 3\sigma/\sqrt{n}$, where n is the number of tablets ($= 25$) that are averaged each hour. Recall that $\sigma/\sqrt{n}$ is the standard deviation of the mean discussed in Section 4-1. The $\pm 2\sigma/\sqrt{n}$ limits are *warning lines* and the $\pm 3\sigma/\sqrt{n}$ limits are *action lines*. We expect ~4.5% of measurements to be outside the warning lines and ~0.3% to be outside the action lines. It is unlikely that we would observe two consecutive measurements at the warning line (probability $= 0.045 \times 0.045 = 0.002\,0$).

The following conditions are considered to be so unlikely that, if they occur, the process should be shut down for troubleshooting:

- 1 observation outside the action lines
- 2 out of 3 consecutive measurements between the warning and action lines
- 7 consecutive measurements all above or all below the center line
- 6 consecutive measurements all increasing or all decreasing, wherever they are located
- 14 consecutive points alternating up and down, regardless of where they are located
- an obvious nonrandom pattern

For quality assessment of an analytical process, a control chart could show the relative deviation of measured values of calibration check samples or quality control samples from their known values. Another control chart could display the precision of replicate analyses of unknowns or standards as a function of time.

When dealing with large numbers of samples and replicates, we perform periodic calibration checks to make sure that our instrument continues to work properly and the calibration remains valid. In a **calibration check,** we analyze solutions formulated to contain known concentrations of analyte. A specification might, for example, call for one calibration check for every 10 samples. Solutions for calibration checks should be different from the ones used to prepare the original calibration curve. This practice helps verify that the initial calibration standards were made properly.

Performance test samples (also called *quality control samples* or *blind samples*) are a quality control measure to help eliminate bias introduced by an analyst who knows the concentration of the calibration check sample. These samples of known composition are provided to the analyst as unknowns. Results are then compared with the known values, usually by a quality assurance manager. For example, the U.S. Department of Agriculture maintains a bank of quality control homogenized food samples for distribution as blind samples to test laboratories that measure nutrients in foods.[3]

Together, raw data and results from calibration checks, spike recoveries, quality control samples, and blanks are used to gauge accuracy. Analytical performance on replicate samples and replicate portions of the same sample measures precision. Fortification also helps ensure that qualitative identification of analyte is correct. If you spike the unknown in Figure 0-11 with extra caffeine and the area of a chromatographic peak not thought to be caffeine increases, then you have misidentified the caffeine peak.

Standard operating procedures stating what steps will be taken and how they will be carried out are the bulwark of quality assurance. For example, if a reagent has "gone bad" for some reason, control experiments built into your normal procedures should detect that something is wrong and your results should not be reported. It is implicit that everyone follows the standard operating procedures. Adhering to these procedures guards against the normal human desire to take shortcuts based on assumptions that could be false.

A meaningful analysis requires a meaningful sample that represents what is to be analyzed. It must be stored in containers and under conditions that do not allow relevant chemical characteristics to change. Protection might be needed to prevent oxidation, photodecomposition, or growth of organisms. The *chain of custody* is the trail followed by a sample from the time

To gauge accuracy:
- calibration checks
- fortification recoveries
- quality control samples
- blanks

To gauge precision:
- replicate samples
- replicate portions of same sample

In drug testing of athletes, the chain of custody necessarily uses different people to collect samples and to analyze samples. The identity of the athlete is known to the collector, but not to the analyst, so that the analyst will not deliberately falsify a result to favor or incriminate a particular person or team.

it is collected to the time it is analyzed and, possibly, archived. Documents are signed each time the material changes hands to indicate who is responsible for the sample. Each person in the chain of custody follows a written procedure telling how the sample is to be handled and stored. Each person receiving a sample should inspect it to see that it is in the expected condition in an appropriate container. If the original sample was a homogeneous liquid, but it contains a precipitate when you receive it, the standard operating procedure may dictate that you reject that sample.

Standard operating procedures specify how instruments are to be maintained and calibrated to ensure their reliability. Many labs have their own standard practices, such as recording temperatures of refrigerators, calibrating balances, conducting routine instrument maintenance, or replacing reagents. These practices are part of the overall quality management plan. The rationale behind standard practices is that some equipment is used by many people for different analyses. We save money by having one program to ensure that the most rigorous needs are met.

Assessment

Assessment is the process of (1) collecting data to show that analytical procedures are operating within specified limits and (2) verifying that final results meet use objectives.

Documentation is critical for assessment. Standard *protocols* provide directions for what must be documented and how the documentation is to be done, including how to record information in notebooks. For labs that rely on manuals of standard practices, it is imperative that tasks done to comply with the manuals be monitored and recorded. *Control charts* (Box 5-1) can be used to monitor performance on blanks, calibration checks, and spiked samples to see if results are stable over time or to compare the work of different employees. Control charts can also monitor sensitivity or selectivity, especially if a laboratory encounters a wide variety of matrixes.

Government agencies such as the U.S. Environmental Protection Agency set requirements for quality assurance for their own labs and for certification of other labs. Published standard methods specify precision, accuracy, numbers of blanks, replicates, and calibration checks. To monitor drinking water, regulations state how often and how many samples are to be taken. Documentation is necessary to demonstrate that all requirements have been met. Table 5-1 summarizes the quality assurance process.

TABLE 5-1   Quality assurance process

Question	Actions
Use Objectives Why do you want the data and results and how will you use the results?	• Write use objectives
Specifications How good do the numbers have to be?	• Write specifications • Pick methods to meet specifications • Consider sampling, precision, accuracy, selectivity, sensitivity, detection limit, robustness, rate of false results • Employ blanks, fortification, calibration checks, quality control samples, and control charts to monitor performance • Write and follow standard operating procedures
Assessment Were the specifications achieved?	• Compare data and results with specifications • Document procedures and keep records suitable to meet use objectives • Verify that use objectives were met

5-2 Method Validation

Method validation is the process of proving that an analytical method is acceptable for its intended purpose.[4] In pharmaceutical chemistry, method validation requirements for regulatory submission include studies of *method specificity, linearity, accuracy, precision, range, limit of detection, limit of quantitation*, and *robustness*.

FIGURE 5-1 Electropherogram of the drug cefotaxime (peak 4) spiked with known impurities (peaks 2, 3, 5–9) from synthesis of the drug. Peak 1 is a marker for electroosmotic flow. Smaller peaks from unknown impurities are also observed. Separation was performed by micellar electrophoretic capillary chromatography (Section 25-7). [From H. Fabre and K. D. Altria, "Key Points for Validating CE Methods, Particularly in Pharmaceutical Analysis," *LCGC* **2001**, *19*, 498.]

Specificity

Specificity is the ability of an analytical method to distinguish analyte from everything else that might be in the sample. *Electrophoresis* is an analytical method in which substances are separated from one another by their differing rates of migration in a strong electric field. An *electropherogram* is a graph of detector response versus time in an electrophoretic separation. Figure 5-1 shows an electropherogram of the drug cefotaxime (peak 4) spiked with 0.2 wt% of known impurities normally present from the synthesis. A reasonable requirement for specificity might be that there is baseline separation of analyte (cefotaxime) from all impurities that might be present. *Baseline separation* means that the detector signal returns to its baseline before the next compound reaches the detector.

In Figure 5-1, impurity peak 3 is not completely resolved from the cefotaxime. In this case, another reasonable criterion for specificity might be that unresolved impurities at their maximum expected concentration do not affect the assay of cefotaxime by more than 0.5%. If we were trying to measure impurities, as opposed to assaying cefotaxime, a reasonable criterion for specificity is that all impurities having >0.1% of the area in the electropherogram are baseline separated from cefotaxime. Figure 5-1 does not meet this criterion.

When developing an assay, we need to decide what impurities to deliberately add to test for specificity. For analysis of a drug formulation, we would want to compare the pure drug with one containing additions of all possible synthetic by-products and intermediates, degradation products, and *excipients* (substances added to give desirable form or consistency). Degradation products might be introduced by exposing pure material to heat, light, humidity, acid, base, and oxidants to decompose ~20% of the original material.

Linearity

Linearity measures how well a calibration curve follows a straight line, showing that response is proportional to the quantity of analyte. If you know the target concentration of analyte in a drug preparation, for example, test the calibration curve for linearity with five standard solutions spanning the range from 0.5 to 1.5 times the expected analyte concentration. Each standard should be prepared and analyzed three times. (This procedure requires $3 \times 5 = 15$ standards plus three blanks.) To prepare a calibration curve for an impurity that might be present at, say, 0.1 to 1 wt%, you might prepare a calibration curve with five standards spanning the range 0.05 to 2 wt%.

A superficial, but common, measure of linearity is the *square of the correlation coefficient, R^2*:

Square of correlation coefficient:
$$R^2 = \frac{[\sum(x_i - \bar{x})(y_i - \bar{y})]^2}{\sum(x_i - \bar{x})^2 \sum(y_i - \bar{y})^2}$$

(5-2)

where $\bar{x}$ is the mean of all the x values and $\bar{y}$ is the mean of all the y values. An easy way to find R^2 is with the LINEST function in Excel. In the example on page 86, values of x and y are entered in columns A and B. LINEST produces a table in cells E3:F5 that contains R^2 in cell E5. R^2 must be very close to 1 to represent a linear fit. For a major component of an unknown, a value of R^2 above 0.995 or, perhaps, 0.999, is deemed a good fit for many purposes.[5] For data in Figure 4-11, which do not lie very close to the straight line, $R^2 = 0.985$.

R^2 can be used as a diagnostic. If R^2 decreases after a method is established, something has gone wrong with the procedure.

Another criterion for linearity is that the y-intercept of the calibration curve (after the response of the blank has been subtracted from each standard) should be close to 0. An acceptable degree of "closeness to 0" might be 2% of the response for the target value of analyte. For the assay of impurities, which are present at concentrations lower than that of the major component, an acceptable value of R^2 might be ≥ 0.98 for the range 0.1 to 2 wt% and the y-intercept should be $\leq 10\%$ of the response for the 2 wt% standard.

Accuracy

Accuracy is "nearness to the truth." Ways to demonstrate accuracy include

1. Analyze a *certified reference material* in a matrix similar to that of your unknown. Your method should find the certified value for analyte in the reference material, within the precision (random uncertainty) of your method.
2. Compare results from two or more different analytical methods. They should agree within their expected precision.
3. Analyze a blank sample spiked with a known addition of analyte. The matrix must be the same as your unknown. When assaying a major component, three replicate samples at each of three levels ranging from 0.5 to 1.5 times the expected sample concentration are customary. For impurities, spikes could cover three levels spanning an expected range of concentrations, such as 0.1 to 2 wt%.
4. If you cannot prepare a blank with the same matrix as the unknown, then it is appropriate to make *standard additions* (Section 5-3) of analyte to the unknown. An accurate assay will find the known amount of analyte that was added.

Spiking is the most common method to evaluate accuracy because reference materials are not usually available and a second analytical method may not be readily available. Spiking ensures that the matrix remains nearly constant.

An example of a specification for accuracy is that the analysis will recover $100 \pm 2\%$ of the spike of a major constituent. For an impurity, the specification might be that recovery is within 0.1 wt% absolute or $\pm 10\%$ relative.

Precision

Precision is how well replicate measurements agree with one another, usually expressed as a standard deviation. Many types of precision are distinguished.

Instrument precision, also called *injection precision*, is the reproducibility observed when the same quantity of one sample is repeatedly introduced (≥ 10 times) into an instrument. Variability could arise from variation in the injected quantity and variation of instrument response.

Intra-assay precision is evaluated by analyzing aliquots of a homogeneous material several times by one person on one day with the same equipment. Each analysis is independent, so the intra-assay precision is telling us how reproducible the analytical method can be. Intra-assay variability is greater than instrument variability, because more steps are involved. Examples of specifications might be that instrument precision is $\leq 1\%$ and intra-assay precision is $\leq 2\%$.

Intermediate precision, formerly called *ruggedness*, is the variation observed when an assay is performed by different people on different instruments on different days in the same lab. Each analysis might incorporate fresh reagents and different chromatography columns.

Interlaboratory precision, also called *reproducibility*, is the most general measure of reproducibility observed when aliquots of the same sample are analyzed by different people in different laboratories. Interlaboratory precision becomes poorer as the level of analyte decreases (Box 5-2).

Range

Range is the concentration interval over which linearity, accuracy, and precision are all acceptable. An example of a specification for range for a major component of a mixture is the concentration interval providing a correlation coefficient of $R^2 \geq 0.995$ (a measure of linearity), spike recovery of $100 \pm 2\%$ (a measure of accuracy), and interlaboratory precision of $\pm 3\%$. For an impurity, an acceptable range might provide a correlation coefficient of $R^2 \geq 0.98$, spike recovery of $100 \pm 10\%$, and interlaboratory precision of $\pm 15\%$.

Autosamplers in chromatography and graphite furnace atomic spectroscopy, for example, have improved injection precision by a factor of 3–10 compared with that attained by humans.

Confusing terms:

Linear range: concentration range over which calibration curve is linear (Figure 4-14)

Dynamic range: concentration range over which there is measurable response

Range: concentration range over which linearity, accuracy, and precision meet specifications for analytical method

BOX 5-2   The Horwitz Trumpet: Variation In Interlaboratory Precision

Interlaboratory tests are routinely used to validate new analytical procedures—especially those intended for regulatory use. Typically, 5 to 10 laboratories are given identical samples and the same written procedure. If all results are "similar," and there is no serious systematic error, then the method is considered "reliable."

The **coefficient of variation** is the standard deviation divided by the mean, usually expressed as a percentage: $CV(\%) = 100 \times s/\bar{x}$. The smaller the coefficient of variation, the more precise is a set of measurements.

In reviewing more than 150 interlaboratory studies with different analytes measured by different techniques, it was observed that the coefficient of variation of mean values reported by different laboratories increased as analyte concentration decreased. At best, the coefficient of variation never seemed to be better than[6]

Horwitz curve: $CV(\%) \approx 2^{(1 - 0.5 \log C)}$

where C is the fraction of analyte in the sample (C = g analyte/ g sample). The coefficient of variation within a laboratory is about one-half to two-thirds of the between-laboratory variation. Experimental results varied from the idealized curve by about a factor of 2 in the vertical direction and a factor of 10 in the horizontal direction. About 5–15% of all interlaboratory results were "outliers"— clearly outside the cluster of other results. This incidence of outliers is above the statistical expectation.

The Horwitz curve predicts that when the concentration of analyte is 1 ppm, the coefficient of variation between laboratories is ~16%. When the concentration is 1 ppb, the coefficient of variation is ~45%. If, perchance, you become a regulation writer one day, acceptable analyte levels should allow for variation among laboratories. The Gaussian distribution tells us that ~5% of measurements lie above $\bar{x} + 1.65s$ (Section 4-1). If the target allowable level of analyte is 1.0 ppb, the allowed observed amount might be set at

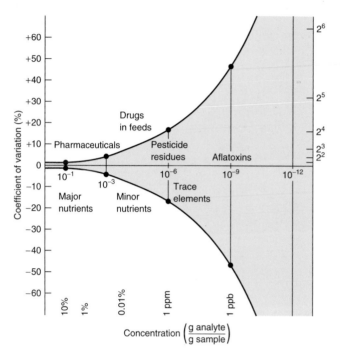

Coefficient of variation of interlaboratory results as a function of sample concentration (expressed as g analyte/g sample). The shaded region has been referred to as the "Horwitz trumpet" because of the way it flares open. [From W. Horwitz, "Evaluation of Analytical Methods Used for Regulation of Foods and Drugs," *Anal. Chem.* 1982, *54*, 67A.]

$1 + 1.65 \times 0.45$ ppb, or about 1.7 ppb. This level gives a 5% rate of false positives that exceed the allowed value even though the true value is below 1.0 ppb.

Limits of Detection and Quantitation

The **detection limit** (also called the *lower limit of detection*) is the smallest quantity of analyte that is "significantly different" from the blank.[7] Here is a procedure that produces a detection limit with ~99% chance of being greater than the blank. That is, only ~1% of samples containing no analyte will give a signal greater than the detection limit (Figure 5-2). We say that there is a ~1% rate of *false positives* in Figure 5-2. We assume that the standard deviation of the signal from samples near the detection limit is similar to the standard deviation from blanks.

1. After estimating the detection limit from previous experience with the method, prepare a sample whose concentration is ~1 to 5 times the detection limit.
2. Measure the signal from *n* replicate samples ($n \geq 7$).
3. Compute the standard deviation (*s*) of the *n* measurements.
4. Measure the signal from *n* blanks (containing no analyte) and find the mean value, y_{blank}.
5. The minimum detectable signal, y_{dl}, is defined as

   Signal detection limit: $y_{dl} = y_{blank} + 3s$ (5-3)

6. The corrected signal, $y_{sample} - y_{blank}$, is proportional to sample concentration:

   Calibration line: $y_{sample} - y_{blank} = m \times$ sample concentration   (5-4)

   where y_{sample} is the signal observed for the sample and *m* is the slope of the linear calibration curve. The *minimum detectable concentration*, also called the detection limit, is obtained by substituting y_{dl} from Equation 5-3 for y_{sample} in Equation 5-4:

   Detection limit: Minimum detectable concentration $= \dfrac{3s}{m}$ (5-5)

5-2 Method Validation   **103**

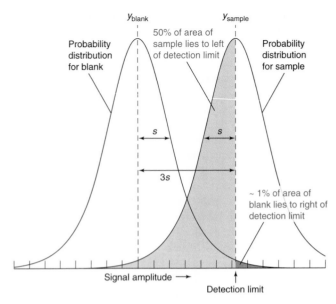

FIGURE 5-2 Detection limit. Distribution of measurements for a blank and a sample whose concentration is at the detection limit. The area of any region is proportional to the number of measurements in that region. Only ~1% of measurements for a blank are expected to exceed the detection limit. However, 50% of measurements for a sample containing analyte at the detection limit will be below the detection limit. There is a 1% chance of concluding that a blank has analyte above the detection limit (*false positive*). If a sample contains analyte at the detection limit, there is a 50% chance of concluding that analyte is *absent* because its signal is below the detection limit (*false negative*). Curves are Student's t distributions for 6 degrees of freedom and are broader than the corresponding Gaussian distributions.

Slope = 0.229 nA/μM

Concentration (μM) →

Detector signal (nA) →

EXAMPLE **Detection Limit**

From previous measurements of a low concentration of analyte, the signal detection limit was estimated to be in the low nanoampere range. Signals from seven replicate samples with a concentration about three times the detection limit were 5.0, 5.0, 5.2, 4.2, 4.6, 6.0, and 4.9 nA. Reagent blanks gave values of 1.4, 2.2, 1.7, 0.9, 0.4, 1.5, and 0.7 nA. The slope of the calibration curve for higher concentrations is $m = 0.229$ nA/μM. **(a)** Find the signal detection limit and the minimum detectable concentration. **(b)** What is the concentration of analyte in a sample that gave a signal of 7.0 nA?

Solution **(a)** First compute the mean for the blanks and the standard deviation of the samples. Retain extra, insignificant digits to reduce round-off errors.

Blank: Average = $y_{blank} = 1.2_6$ nA
Sample: Standard deviation = $s = 0.5_6$ nA

The signal detection limit from Equation 5-3 is

$$y_{dl} = y_{blank} + 3s = 1.2_6 \text{ nA} + (3)(0.5_6 \text{ nA}) = 2.9_4 \text{ nA}$$

The minimum detectable concentration is obtained from Equation 5-5:

$$\text{Detection limit} = \frac{3s}{m} = \frac{(3)(0.5_6 \text{ nA})}{0.229 \text{ nA/μM}} = 7._3 \text{ μM}$$

(b) To find the concentration of a sample whose signal is 7.0 nA, use Equation 5-4:

$$y_{sample} - y_{blank} = m \times \text{concentration}$$

$$\Rightarrow \quad \text{Concentration} = \frac{y_{sample} - y_{blank}}{m} = \frac{7.0 \text{ nA} - 1.2_6 \text{ nA}}{0.229 \text{ nA/μM}} = 25._1 \text{ μM}$$

Test Yourself Find the minimum detectable concentration if the average of the blanks is 1.0_5 nA and $s = 0.6_3$ nA. (**Answer:** $8._3$ μM)

Another common way to define detection limit is from the least-squares equation of a calibration curve: signal detection limit $= b + 3s_y$, where b is the y-intercept and s_y is given by Equation 4-20. A more rigorous procedure is described in the notes for this chapter.[8]

The lower limit of detection in Equation 5-5 is $3s/m$, where s is the standard deviation of a low-concentration sample and m is the slope of the calibration curve. The standard deviation is a measure of the *noise* (random variation) in a blank or a small signal. When the signal is 3 times greater than the noise, it is detectable, but still too small for accurate measurement. A signal that is 10 times greater than the noise is defined as the **lower limit of quantitation,** or the smallest amount that can be measured with reasonable accuracy.

$$\text{Lower limit of quantitation} \equiv \frac{10s}{m} \qquad (5\text{-}6)$$

The *instrument detection limit* is obtained by replicate measurements ($n \geq 7$) of aliquots from one sample. The *method detection limit*, which is greater than the instrument detection limit, is obtained by preparing $n \geq 7$ individual samples and analyzing each one once.

The **reporting limit** is the concentration below which regulations say that a given analyte is reported as "not detected," which does not mean that analyte is not observed. It means that analyte is below a prescribed level. Reporting limits are set at least 5 to 10 times higher than the detection limit, so that detecting analyte at the reporting limit is not ambiguous.

Labels on U.S. packaged foods must state how much *trans* fat is present. This type of fat is derived mainly from partial hydrogenation of vegetable oil and is a major component of margarine and shortening. *Trans* fat is thought to increase risk of heart disease, stroke, and some cancers. However, the *reporting limit* for *trans* fat is 0.5 g per serving. If the concentration is <0.5 g/serving, it is reported as 0, as in Figure 5-3. By reducing the serving size, a manufacturer can state that *trans* fat content is 0. If your favorite snack food is made with partially hydrogenated oil, it contains *trans* fat even if the label says otherwise.

$$\text{Detection limit} \equiv \frac{3s}{m}$$

$$\text{Quantitation limit} \equiv \frac{10s}{m}$$

The symbol $\equiv$ means "is defined as."

FIGURE 5-3 Nutritional label from a package of crackers. The *reporting limit* for *trans* fat is 0.5 g/serving. Any amount less than this is reported as 0. The end of Chapter 6 explains the shorthand used to draw these 18-carbon compounds.

Robustness

Robustness is the ability of an analytical method to be unaffected by small, deliberate changes in operating parameters. For example, a chromatographic method is robust if it gives acceptable results when small changes are made in solvent composition, pH, buffer concentration, temperature, injection volume, and detector wavelength. In tests for robustness, the organic solvent content in the mobile phase could be varied by, say, ±2%, the eluent pH varied by ±0.1, and column temperature varied by ±5°C. If acceptable results are obtained, the written procedure should state that these variations are tolerable. Capillary electrophoresis requires such small volumes that a given solution could conceivably be used for months before it is used up. Therefore, solution stability (shelf life) should be evaluated for robustness.

FIGURE 5-4 Calibration curves for perchlorate in pure water and in groundwater. [Data from C. J. Koester, H. R. Beller, and R. U. Halden, "Analysis of Perchlorate in Groundwater by Electrospray Ionization Mass Spectrometry/Mass Spectrometry," *Environ. Sci. Technol.* **2000**, *34*, 1862.]

The matrix affects the magnitude of the analytical signal. In standard addition, all samples are in the same matrix.

Derivation of Equation 5-7:

$I_X = k[X]_i$, where k is a constant of proportionality

$I_{S+X} = k([S]_f + [X]_f)$, where k is the same constant

Dividing one equation by the other gives

$$\frac{I_X}{I_{S+X}} = \frac{k[X]_i}{k([S]_f + [X]_f)} = \frac{[X]_i}{[S]_f + [X]_f}$$

5-3 Standard Addition[9]

In **standard addition**, known quantities of analyte are added to the unknown. From the increase in signal, we deduce how much analyte was in the original unknown. This method requires a linear response to analyte. As in titrations, higher precision can be achieved when standards are added by mass instead of volume.[10]

Standard addition is especially appropriate when the sample composition is unknown or complex and affects the analytical signal. The *matrix* is everything in the unknown, other than analyte. A **matrix effect** is a change in the analytical signal caused by anything in the sample other than analyte.

Figure 5-4 shows a strong matrix effect in the analysis of perchlorate (ClO_4^-) by mass spectrometry. Perchlorate at a level above 18 μg/L in drinking water is of concern because it can reduce thyroid hormone production. Standard solutions of ClO_4^- in pure water gave the upper calibration curve in Figure 5-4. The slope in the lower curve for standard solutions in groundwater was 15 times less. Reduction of the ClO_4^- signal is a *matrix effect* attributed to other anions present in the groundwater.

Different groundwaters have different concentrations of many anions, so there is no way to construct a calibration curve for this analysis that would apply to more than one specific groundwater. Hence, the method of standard addition is required. When we add a small volume of concentrated standard to an existing unknown, we do not change the concentration of the matrix very much.

Consider a standard addition in which a sample with unknown initial concentration of analyte $[X]_i$ gives a signal intensity I_X. Then a known concentration of standard, S, is added to an aliquot of the sample and a signal I_{S+X} is observed for this second solution. Addition of standard to the unknown changes the concentration of the original analyte because of dilution. Let's call the diluted concentration of analyte $[X]_f$, where "f" stands for "final." We designate the concentration of standard in the final solution as $[S]_f$. (Bear in mind that the chemical species X and S are the same.)

Signal is directly proportional to analyte concentration, so

$$\frac{\text{Concentration of analyte in initial solution}}{\text{Concentration of analyte plus standard in final solution}} = \frac{\text{signal from initial solution}}{\text{signal from final solution}}$$

Standard addition equation:
$$\frac{[X]_i}{[S]_f + [X]_f} = \frac{I_X}{I_{S+X}} \tag{5-7}$$

For an initial volume V_o of unknown and added volume V_S of standard with concentration $[S]_i$, the total volume is $V = V_o + V_S$ and the concentrations in Equation 5-7 are

$$[X]_f = [X]_i \left(\frac{V_o}{V}\right) \qquad [S]_f = [S]_i \left(\frac{V_S}{V}\right) \tag{5-8}$$

The quotient (initial volume/final volume), which relates final concentration to initial concentration, is called the **dilution factor.** It comes directly from Equation 1-3.

By expressing the diluted concentration of analyte, $[X]_f$, in terms of the initial concentration of analyte, $[X]_i$, we can solve for $[X]_i$, because everything else in Equation 5-7 is known.

EXAMPLE Standard Addition

Serum containing Na^+ gave a signal of 4.27 mV in an atomic emission analysis. Then 5.00 mL of 2.08 M NaCl were added to 95.0 mL of serum. This spiked serum gave a signal of 7.98 mV. Find the original concentration of Na^+ in the serum.

Solution From Equation 5-8, the final concentration of Na^+ after dilution with the standard is $[X]_f = [X]_i(V_o/V) = [X]_i(95.0 \text{ mL}/100.0 \text{ mL})$. The final concentration of added standard is $[S]_f = [S]_i(V_S/V) = (2.08 \text{ M})(5.00 \text{ mL}/100.0 \text{ mL}) = 0.104$ M. Equation 5-7 becomes

$$\frac{[Na^+]_i}{[0.104 \text{ M}] + 0.950[Na^+]_i} = \frac{4.27 \text{ mV}}{7.98 \text{ mV}} \Rightarrow [Na^+]_i = 0.113 \text{ M}$$

Test Yourself If spiked serum gave a signal of 6.50 mV, what was the original concentration of Na^+? (***Answer:*** 0.182 M)

CHAPTER 5 Quality Assurance and Calibration Methods

Graphical Procedure for Standard Addition to a Single Solution

There are two common methods to perform standard addition. If the analysis does not consume solution, we begin with an unknown solution and measure the analytical signal. Then we add a small volume of concentrated standard and measure the signal again. We add several more small volumes of standard and measure the signal after each addition. Standard should be concentrated so that only small volumes are added and the sample matrix is not appreciably altered. Added standards should increase the analytical signal by a factor of 1.5 to 3. The other common procedure is described in the next section.

	A	B	C	D	E
1	Vitamin C standard addition experiment				
2	Add 0.279 M ascorbic acid to 50.0 mL orange juice				
3					
4		Vs =			
5	Vo (mL) =	mL ascorbic	I(s+x) =	x-axis function	y-axis function
6	50	acid added	signal (μA)	Si*Vs/Vo	I(s+x)*V/Vo
7	[S]i (mM) =	0.000	1.78	0.000	1.780
8	279	0.050	2.00	0.279	2.002
9		0.250	2.81	1.395	2.824
10		0.400	3.35	2.232	3.377
11		0.550	3.88	3.069	3.923
12		0.700	4.37	3.906	4.431
13		0.850	4.86	4.743	4.943
14		1.000	5.33	5.580	5.437
15		1.150	5.82	6.417	5.954
16					
17	D7 = $A$8*B7/$A$6		E7 = C7*($A$6+B7)/$A$6		

FIGURE 5-5 Data for standard addition experiment with *variable total volume*.

Figure 5-5 shows data for an experiment in which ascorbic acid (vitamin C) was measured in orange juice by an electrochemical method. The current between a pair of electrodes immersed in the juice is proportional to the concentration of ascorbic acid. Eight standard additions increased current from 1.78 to 5.82 μA (column C), which is at the upper end of the desired range of 1.5- to 3-fold increase in analytical signal.

$y = 0.6463x + 1.8687$

The equation of a line is $y = mx + b$. The *x*-intercept is obtained by setting $y = 0$:

$$0 = mx + b$$
$$x = -b/m$$

FIGURE 5-6 Graphical treatment of standard additions to a single solution *with variable total volume*. Data from Figure 5-5. Standard additions should increase the analytical signal to between 1.5 and 3 times its original value (that is, B = 0.5A to 2A).

Figure 5-6 allows us to find the original concentration of unknown. The theoretical response is derived by substituting expressions for $[X]_f$ and $[S]_f$ from Equations 5-8 into Equation 5-7. Following a little rearrangement, we find

For successive standard additions to one solution:

$$\underbrace{I_{S+X}\left(\frac{V}{V_o}\right)}_{\substack{\text{Function to plot} \\ \text{on } y\text{-axis}}} = I_X + \underbrace{\frac{I_X}{[X]_i}[S]_i\left(\frac{V_s}{V_o}\right)}_{\substack{\text{Function to plot} \\ \text{on } x\text{-axis}}}$$

(5-9)

Successive standard additions to one solution:

Plot $I_{S+X}\left(\dfrac{V}{V_o}\right)$ versus $[S]_i\left(\dfrac{V_s}{V_o}\right)$

x-intercept is $[X]_i$

A graph of $I_{S+X}(V/V_o)$ (the *corrected response*) on the *y*-axis versus $[S]_i(V_S/V_o)$ on the *x*-axis should be a straight line. The data plotted in Figure 5-6 are computed in columns D and E of Figure 5-5. The right side of Equation 5-9 is 0 when $[S]_i(V_S/V_o) = -[X]_i$. The magnitude of the intercept on the *x*-axis is the *original* concentration of unknown, $[X]_i = 2.89$ mM in Figure 5-6.

The uncertainty in the *x*-intercept is[11]

$$\text{Standard deviation of } x\text{-intercept} = \frac{s_y}{|m|}\sqrt{\frac{1}{n} + \frac{\bar{y}^2}{m^2 \sum(x_i - \bar{x})^2}} \qquad (5\text{-}10)$$

where s_y is the standard deviation of *y* (Equation 4-20), $|m|$ is the absolute value of the slope of the least-squares line (Equation 4-16), *n* is the number of data points (nine in Figure 5-6), $\bar{y}$ is the mean value of *y* for the nine points, x_i are the individual values of *x* for the nine points, and $\bar{x}$ is the mean value of *x* for the nine points. For the points in Figure 5-6, the uncertainty in the *x*-intercept is 0.09_8 mM.

The confidence interval is $\pm t \times$ (standard deviation of *x*-intercept) where *t* is Student's *t* (Table 4-2) for $n - 2$ degrees of freedom. The 95% confidence interval for the intercept in Figure 5-6 is $\pm(2.365)(0.09_8 \text{ mM}) = \pm 0.23$ mM. The value $t = 2.365$ was taken from Table 4-2 for $9 - 2 = 7$ degrees of freedom.

Graphical Procedure for Multiple Solutions with Constant Volume

The second common standard addition procedure is shown in Figure 5-7. Equal volumes of unknown are pipetted into several volumetric flasks. Increasing volumes of standard are added to each flask and each is diluted to the *same final volume*. Every flask contains the same concentration of unknown and differing concentrations of standard. For each flask, a measurement of analytical signal, I_{S+X}, is then made. The method in Figure 5-7 is necessary when the analysis consumes some of the solution.

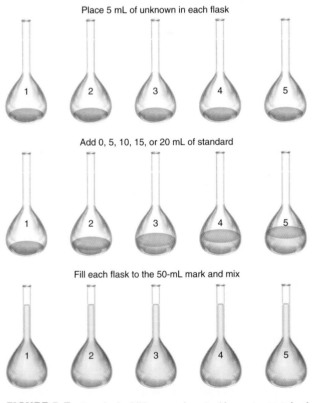

FIGURE 5-7 Standard addition experiment with *constant total volume*.

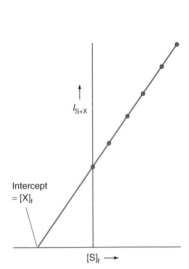

FIGURE 5-8 Graphical treatment of standard addition with *constant total volume*. Plot I_{S+X} versus $[S]_f$, and the *x*-intercept is $[X]_f$. The lines in Figures 5-6 and 5-8 are both derived from Equation 5-9.

If all standard additions are made to a constant final volume, plot the signal I_{S+X} versus the concentration of diluted standard, $[S]_f$ (Figure 5-8). In this case, the *x*-intercept is the *final* concentration of unknown, $[X]_f$, after dilution to the final sample volume. Equation 5-10 still applies to the uncertainty. The initial concentration of unknown, $[X]_i$, is calculated from the dilution that was applied to make the final sample.

5-4 Internal Standards

An **internal standard** is a known amount of a compound, different from analyte, that is added to the unknown. Signal from analyte is compared with signal from the internal standard to find out how much analyte is present.

Internal standards are especially useful for analyses in which the quantity of sample analyzed or the instrument response varies slightly from run to run. For example, gas or liquid flow rates that vary by a few percent in a chromatography experiment (Figure 0-10) could change the detector response. A calibration curve is accurate only for the one set of conditions under which it was obtained. However, the *relative* response of the detector to the analyte and standard is usually constant over a range of conditions. If signal from the standard increases by 8.4% because of a change in flow rate, signal from the analyte usually increases by 8.4% also. As long as the concentration of standard is known, the correct concentration of analyte can be derived. Internal standards are used in chromatography because the small quantity of sample injected into the chromatograph is not reproducible.

Internal standards are desirable when sample loss can occur during sample preparation steps prior to analysis. If a known quantity of standard is added to the unknown prior to any manipulations, the ratio of standard to analyte remains constant because the same fraction of each is lost in any operation.

To use an internal standard, we prepare a known mixture of standard and analyte to measure the relative response of the detector to the two species. In the chromatogram in Figure 5-9, the area, A, under each peak is proportional to the concentration of the species injected into the column. However, the detector generally has a different response to each component. For example, if both analyte (X) and internal standard (S) have concentrations of 10.0 mM, the area under the analyte peak might be 2.30 times greater than the area under the standard peak. We say that the **response factor,** F, is 2.30 times greater for X than for S.

Response factor:
$$\frac{\text{Area of analyte signal}}{\text{Concentration of analyte}} = F\left(\frac{\text{area of standard signal}}{\text{concentration of standard}}\right) \qquad (5\text{-}11)$$

$$\frac{A_X}{[X]} = F\left(\frac{A_S}{[S]}\right)$$

[X] and [S] are the concentrations of analyte and standard *after they have been mixed together*. Equation 5-11 is predicated on linear response to analyte and standard.

EXAMPLE Using an Internal Standard

In a preliminary experiment, a solution containing 0.083 7 M X and 0.066 6 M S gave peak areas of $A_X = 423$ and $A_S = 347$. (Areas are measured in arbitrary units by the instrument's computer.) To analyze the unknown, 10.0 mL of 0.146 M S were added to 10.0 mL of unknown, and the mixture was diluted to 25.0 mL in a volumetric flask. This mixture gave the chromatogram in Figure 5-9, for which $A_X = 553$ and $A_S = 582$. Find the concentration of X in the unknown.

Solution First use the standard mixture to find the response factor in Equation 5-11:

Standard mixture:
$$\frac{A_X}{[X]} = F\left(\frac{A_S}{[S]}\right)$$

$$\frac{423}{0.083\ 7} = F\left(\frac{347}{0.066\ 6}\right) \Rightarrow F = 0.970_0$$

In the mixture of unknown plus standard, the concentration of S is

$$[S] = \underbrace{(0.146\ \text{M})}_{\substack{\text{Initial}\\\text{concentration}}}\underbrace{\left(\frac{10.0}{25.0}\right)}_{\substack{\text{Dilution}\\\text{factor}}} = 0.058\ 4\ \text{M}$$

In *standard addition*, the standard is the same substance as the analyte. An *internal standard* is different from the analyte.

The assumption that relative response to analyte and standard remains constant over a range of concentrations must be verified.

If the detector responds equally to standard and analyte, $F = 1$. If the detector responds twice as much to analyte as to standard, $F = 2$. If the detector responds half as much to analyte as to standard, $F = 0.5$.

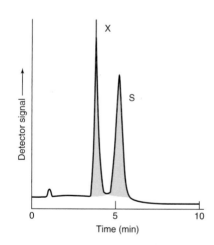

FIGURE 5-9 Chromatographic separation of unknown (X) and internal standard (S). A known amount of S was added to the unknown. The relative areas of the signals from X and S allow us to find out how much X is in the mixture. It is necessary first to measure the relative response of the detector to each compound.

The dilution factor $\dfrac{\text{initial volume}}{\text{final volume}}$ converts initial concentration into final concentration.

Using the known response factor, substitute back into Equation 5-11 to find the concentration of unknown in the mixture:

Unknown mixture:
$$\frac{A_X}{[X]} = F\left(\frac{A_S}{[S]}\right)$$

$$\frac{553}{[X]} = 0.970_0\left(\frac{582}{0.058\ 4}\right) \Rightarrow [X] = 0.057\ 2_1\ M$$

Because X was diluted from 10.0 to 25.0 mL when the mixture with S was prepared, the original concentration of X in the unknown was $(25.0\ mL/10.0\ mL)(0.057\ 2_1\ M)$ = 0.143 M.

Test Yourself Suppose that peak areas for the known mixture were $A_X = 423$ and $A_S = 447$. Find [X] in the unknown. (**Answer:** $F = 0.753_0$, [X] = 0.184 M)

5-5 ▦ Efficiency In Experimental Design

Operating parameters usually need to be optimized when we develop an analytical method. The least efficient way to do this is to vary one parameter at a time while keeping everything else constant. More efficient procedures are called *fractional factorial experimental design*[12] and *simplex optimization*.[13] We now discuss one example of experimental design intended to provide maximum information in the fewest number of trials.

Suppose we have three different unknown solutions of acid, designated A, B, and C. If we titrate each one once with base, we find its concentration, but have no estimate of uncertainty. If we titrate each solution three times, for a total of nine measurements, we would find each concentration and its standard deviation.

A more efficient experimental design provides concentrations and standard deviations in fewer than nine experiments.[14] One of many efficient designs is shown in Figure 5-10. Instead of titrating each acid by itself, we titrate mixtures of the acids. For example, in row 5 of the spreadsheet, a mixture containing 2 mL A, 2 mL B, and 2 mL C required 23.29 mL of 0.120 4 M NaOH, which amounts to 2.804 mmol of OH^-. In row 6, the acid mixture contained 2 mL A, 3 mL B, and 1 mL C. Other permutations are titrated in rows 7 and 8. Then row 5 is repeated independently in row 9. Column E gives mmol of base for each run.

Acids could be delivered by transfer pipets whose tolerances are given in Table 2-4. So 2 mL means 2.000 mL with uncertainty in the third decimal place.

	A	B	C	D	E
1	Experimental Design				
2					
3	Volumes of unknown acids (mL)			mL NaOH	mmol
4	A	B	C	(0.1204 M)	NaOH
5	2	2	2	23.29	2.804
6	2	3	1	20.01	2.409
7	3	1	2	21.72	2.615
8	1	2	3	28.51	3.433
9	2	2	2	23.26	2.801
10					
11			[C]	[B]	[A]
12		Molarity	0.8099	0.4001	0.1962
13		Std. dev.	0.0062	0.0062	0.0062
14			0.9994	0.0130	#N/A
15			R^2	S_y	
16	Highlight cells C12:E14				
17	Type "= LINEST(E5:E9,A5:C9,FALSE,TRUE)"				
18	Press CTRL +SHIFT+ENTER (on PC)				
19	Press COMMAND+RETURN (on Mac)				

FIGURE 5-10 Spreadsheet for efficient experimental design uses Excel LINEST routine to fit the function $y = m_A x_A + m_B x_B + m_C x_C$ to experimental data by a least-squares procedure.

For each experiment, mmol of base consumed equals mmol of acid in the mixture:

$$\underbrace{\text{mmol OH}^-}_{y} = \underbrace{[A]V_A}_{m_A x_A} + \underbrace{[B]V_B}_{m_B x_B} + \underbrace{[C]V_C}_{m_C x_C} \qquad (5\text{-}12)$$

$$\left(\frac{\text{mol}}{\text{L}}\right) \cdot \text{mL} = \text{mmol}$$

where [A] is the concentration of acid A (mol/L) and V_A is the volume of A in mL. Rows 5 through 9 of the spreadsheet are equivalent to the following equalities:

$$\left.\begin{aligned}
2.804 &= [A] \cdot 2 + [B] \cdot 2 + [C] \cdot 2 \\
2.409 &= [A] \cdot 2 + [B] \cdot 3 + [C] \cdot 1 \\
2.615 &= [A] \cdot 3 + [B] \cdot 1 + [C] \cdot 2 \\
3.433 &= [A] \cdot 1 + [B] \cdot 2 + [C] \cdot 3 \\
2.801 &= [A] \cdot 2 + [B] \cdot 2 + [C] \cdot 2
\end{aligned}\right\} \qquad (5\text{-}13)$$

Our problem is to find the best values for the molarities [A], [B], and [C].

As luck would have it, Excel will find these values for us by the least-squares procedure LINEST. On page 86, we used LINEST to find the slope and intercept in the equation $y = mx + b$. In Figure 5-10, we use LINEST to find the slopes for $y = m_A x_A + m_B x_B + m_C x_C + b$ (where the intercept, b, is 0). To execute LINEST, highlight cells C12:E14 and type "=LINEST(E5:E9,A5:C9,FALSE,TRUE)". Then press CTRL+SHIFT+ENTER on a PC or COMMAND(⌘)+RETURN on a Mac. The first argument of LINEST, E5:E9, contains values of y (= mmol OH$^-$). The second argument, A5:C9, contains values of x (= volumes of acid). The third argument (FALSE) tells the computer to set the intercept (b) to zero, and the fourth argument (TRUE) says that we want statistics to be computed.

Excel finds the least-squares slopes in row 12 and their uncertainties in row 13. These slopes are the molarites [C], [B], and [A], in reverse order. So, for example, cells C12 and C13 tell us that the molarity of acid C is $0.809_9 \pm 0.006_2$ M. Uncertainty is derived from the quality of the least-squares fit to Equations 5-13, which is related to uncertainties in volumes and estimation of the equivalence point of the titrations.

We require at least n equations to solve for n unknowns. In this example, we have five equations (5-13), but just three unknowns ([A], [B], [C]). The two extra equations allow us to estimate uncertainty in the unknowns. If you do more experiments, you will generally decrease the uncertainty in concentration. An efficient experimental design with five experiments leaves us with more uncertainty than if we had done nine experiments. However, we have cut our effort nearly in half with the efficient design.

For five equations and three unknowns, there are $5 - 3 = 2$ *degrees of freedom*. With no degrees of freedom, there is no information with which to estimate uncertainty.

Terms to Understand

assessment	field blank	performance test sample	sensitivity
calibration check	internal standard	quality assurance	specifications
coefficient of variation	linearity	range	specificity
control chart	lower limit of quantitation	reagent blank	spike
detection limit	matrix	reporting limit	standard addition
dilution factor	matrix effect	response factor	standard operating procedure
false negative	method blank	robustness	use objectives
false positive	method validation	selectivity	

Summary

Quality assurance is what we do to get the right answer for our purpose. We begin by writing use objectives, from which specifications for data quality can be derived. Specifications could include requirements for sampling, accuracy, precision, specificity, detection limit, standards, and blank values. For any meaningful analysis, we must first collect a representative sample. A method blank containing all components except analyte is taken through all steps of the analytical procedure. We subtract the response of the method blank from the response of a real sample prior to calculating the quantity of analyte in the sample. A field blank tells us if analyte is inadvertently picked up by exposure to field conditions. Accuracy can be assessed by analyzing certified standards, by calibration checks performed by the analyst, with spikes made by the analyst, and by analyzing blind quality control samples. Written standard operating procedures must be followed rigorously to avoid inadvertent changes in procedure that could affect the outcome. Assessment is the process of (1) collecting data to show that analytical procedures are operating within specified limits and (2) verifying that final results meet use objectives. Control charts can be used to monitor accuracy, precision, or instrument performance as a function of time.

Method validation is the process of proving that an analytical method is acceptable for its intended purpose. In validating a method, we typically demonstrate that requirements are met for specificity, linearity, accuracy, precision, range, limit of detection, limit of quantitation, and robustness. Specificity is the ability to distinguish analyte from anything else. Linearity is usually measured by the square of the correlation coefficient for the calibration curve. Types of precision include instrument precision, intra-assay precision, intermediate precision, and, most generally, interlaboratory precision. The "Horwitz trumpet" is an empirical statement that precision becomes poorer as analyte concentration decreases. Range is the concentration interval over which linearity, accuracy, and precision are acceptable. The detection limit is usually taken as 3 times the standard deviation of the blank. The lower limit of quantitation is 10 times the standard deviation of the blank. The reporting limit is the concentration below which regulations say that analyte is reported as "not detected," even when it is observed. Robustness is the ability of an analytical method to be unaffected by small changes in operating parameters.

A standard addition is a known quantity of analyte added to an unknown to increase the concentration of analyte. Standard additions are especially useful when matrix effects are important. A matrix effect is a change in the analytical signal caused by anything in the sample other than analyte. Use Equation 5-7 to compute the quantity of analyte after a single standard addition. For multiple standard additions to a single solution, use Equation 5-9 to construct the graph in Figure 5-6, in which the x-intercept gives us the concentration of analyte. For multiple solutions made up to the same final volume, the slightly different graph in Figure 5-8 is used. Equation 5-10 gives the x-intercept uncertainty in either graph.

An internal standard is a known amount of a compound, different from analyte, that is added to the unknown. Signal from analyte is compared with signal from the internal standard to find out how much analyte is present. Internal standards are useful when the quantity of sample analyzed is not reproducible, when instrument response varies from run to run, or when sample losses occur in sample preparation. The response factor in Equation 5-11 is the relative response to analyte and standard.

Efficient experimental design decreases the number of experiments needed to obtain required information and an estimate of uncertainty in that information. A tradeoff is that the fewer experiments we do, the greater the uncertainty in the results.

Exercises

5-A. *Detection limit.* In spectrophotometry, we measure the concentration of analyte by its absorbance of light. A low-concentration sample was prepared, and nine replicate measurements gave absorbances of 0.004 7, 0.005 4, 0.006 2, 0.006 0, 0.004 6, 0.005 6, 0.005 2, 0.004 4, and 0.005 8. Nine reagent blanks gave values of 0.000 6, 0.001 2, 0.002 2, 0.000 5, 0.001 6, 0.000 8, 0.001 7, 0.001 0, and 0.001 1.

(a) Find the absorbance detection limit with Equation 5-3.

(b) The calibration curve is a graph of absorbance versus concentration. Absorbance is a dimensionless quantity. The slope of the calibration curve is $m = 2.24 \times 10^4$ M^{-1}. Find the concentration detection limit with Equation 5-5.

(c) Find the lower limit of quantitation with Equation 5-6.

5-B. *Standard addition.* An unknown sample of Ni^{2+} gave a current of 2.36 μA in an electrochemical analysis. When 0.500 mL of solution containing 0.028 7 M Ni^{2+} was added to 25.0 mL of unknown, the current increased to 3.79 μA.

(a) Denoting the initial, unknown concentration as [Ni^{2+}]$_i$, write an expression for the final concentration, [Ni^{2+}]$_f$, after 25.0 mL of unknown were mixed with 0.500 mL of standard. Use the dilution factor for this calculation.

(b) In a similar manner, write the final concentration of added standard Ni^{2+}, designated as [S]$_f$.

(c) Find [Ni^{2+}]$_i$ in the unknown.

5-C. *Internal standard.* A solution was prepared by mixing 5.00 mL of unknown element X with 2.00 mL of solution containing 4.13 μg of standard element S per milliliter, and diluting to 10.0 mL. The signal ratio in atomic absorption spectrometry was (signal from X)/(signal from S) = 0.808. In a separate experiment, with equal concentrations of X and S, (signal from X)/(signal from S) = 1.31. Find the concentration of X in the unknown.

5-D. In Figure 5-6, the x-intercept is −2.89 mM and its standard deviation is 0.09$_8$ mM. Find the 90% and 99% confidence intervals for the intercept.

5-E. *Control chart.* Volatile compounds in human blood serum were measured by purge and trap gas chromatography/mass spectrometry. For quality control, serum was periodically spiked with a constant amount of 1,2-dichlorobenzene and the concentration (ng/g = ppb) was measured. Find the mean and standard deviation for the following spike data and prepare a control chart. State whether or not the observations (Obs.) meet each criterion for stability in a control chart.

Day	Obs. (ppb)	Day	Obs. (ppb)	Day	Obs. (ppb)	Day	Obs. (ppb)	Day	Obs. (ppb)
0	1.05	91	1.13	147	0.83	212	1.03	290	1.04
1	0.70	101	1.64	149	0.88	218	0.90	294	0.85
3	0.42	104	0.79	154	0.89	220	0.86	296	0.59
6	0.95	106	0.66	156	0.72	237	1.05	300	0.83
7	0.55	112	0.88	161	1.18	251	0.79	302	0.67
30	0.68	113	0.79	167	0.75	259	0.94	304	0.66
70	0.83	115	1.07	175	0.76	262	0.77	308	1.04
72	0.97	119	0.60	182	0.93	277	0.85	311	0.86
76	0.60	125	0.80	185	0.72	282	0.72	317	0.88
80	0.87	128	0.81	189	0.87	286	0.68	321	0.67
84	1.03	134	0.84	199	0.85	288	0.86	323	0.68

SOURCE: D. L. Ashley, M. A. Bonin, F. L. Cardinali, J. M. McCraw, J. S. Holler, L. L. Needham, and D. G. Patterson, Jr., "Determining Volatile Organic Compounds in Blood by Using Purge and Trap Gas Chromatography/Mass Spectrometry," Anal. Chem. **1992**, 64, 1021.

Problems

Quality Assurance and Method Validation

5-1. Explain the meaning of the quotation at the beginning of this chapter: "Get the right data. Get the data right. Keep the data right."

5-2. What are the three parts of quality assurance? What questions are asked in each part and what actions are taken in each part?

5-3. How can you validate precision and accuracy?

5-4. Distinguish *raw data*, *treated data*, and *results*.

5-5. What is the difference between a *calibration check* and a *performance test sample*?

5-6. What is the purpose of a blank? Distinguish *method blank*, *reagent blank*, and *field blank*.

5-7. Distinguish *linear range*, *dynamic range*, and *range*.

5-8. What is the difference between a *false positive* and a *false negative*?

5-9. Consider a sample that contains analyte at the detection limit defined in Figure 5-2. Explain the following statements: There is approximately a 1% chance of falsely concluding that a sample containing no analyte contains analyte above the detection limit. There is a 50% chance of concluding that a sample that really contains analyte at the detection limit does not contain analyte above the detection limit.

5-10. How is a control chart used? State six indications that a process is going out of control.

5-11. Here is a use objective for a chemical analysis to be performed at a drinking water purification plant: "Data and results collected quarterly shall be used to determine whether the concentrations of haloacetates in the treated water demonstrate compliance with the levels set by the Stage 1 Disinfection By-products Rule using Method 552.2" (a specification that sets precision, accuracy, and other requirements). Which one of the following questions best summarizes the meaning of the use objective?

(a) Are haloacetate concentrations known within specified precision and accuracy?

(b) Are any haloacetates detectable in the water?

(c) Do any haloacetate concentrations exceed the regulatory limit?

5-12. What is the difference between an instrument detection limit and a method detection limit? What is the difference between robustness and intermediate precision?

5-13. Define the following terms: instrument precision, injection precision, intra-assay precision, intermediate precision, and interlaboratory precision.

5-14. *Control chart.* A laboratory monitoring perchlorate (ClO_4^-) in urine measured quality control samples made from synthetic urine spiked with (ClO_4^-). The graph shows consecutive quality control measurements. Are any troubleshooting conditions from Box 5-1 observed in these data?

Control chart for ClO_4^- in urine. [Data from L. Valentin-Blasini, J. P. Mauldin, D. Maple, and B. C. Blount, "Analysis of Perchlorate in Human Urine Using Ion Chromatography and Electrospray Tandem Mass Spectrometry," *Anal. Chem.* **2005**, *77*, 2475.]

5-15. 🖩 *Correlation coefficient and Excel graphing.* Synthetic data are given below for a calibration curve in which random Gaussian noise with a magnitude of 80 was superimposed on y values that follow the equation $y = 26.4x + 1.37$. This exercise shows that a high value of R^2 does not guarantee that data quality is excellent.

(a) Enter concentration in column A and signal in column B of a spreadsheet. Prepare an XY (Scatter) chart of signal versus concentration without a line as described in Section 2-11. Use LINEST (Section 4-7) to find the least-squares parameters including R^2.

(b) Now insert the Trendline by following instructions in Section 4-9. In the Options window used to select the Trendline, select Display Equation and Display R-Squared. Verify that Trendline and LINEST give identical results.

(c) Add 95% confidence interval y error bars following the instructions at the end of Section 4-9. The 95% confidence interval is $\pm t s_y$, where s_y comes from LINEST and Student's t comes from Table 4-2 for 95% confidence and $11 - 2 = 9$ degrees of freedom. Also, compute t with the statement "=TINV(0.05,9)".

Concentration (x)	Signal (y)	Concentration (x)	Signal (y)
0	14	60	1 573
10	350	70	1 732
20	566	80	2 180
30	957	90	2 330
40	1 067	100	2 508
50	1 354		

5-16. In a murder trial in the 1990s, the defendant's blood was found at the crime scene. The prosecutor argued that blood was left by the defendant during the crime. The defense argued that police "planted" the defendant's blood from a sample collected later. Blood is normally collected in a vial containing the metal-binding compound EDTA as an anticoagulant with a concentration of ~4.5 mM after the vial has been filled with blood. At the time of the trial, procedures to measure EDTA in blood were not well established. Even though the amount of EDTA found in the crime-scene blood was orders of magnitude below 4.5 mM, the jury acquitted the defendant. This trial motivated the development of a new method to measure EDTA in blood.

(a) *Precision and accuracy.* To measure accuracy and precision of the method, blood was fortified with EDTA to known levels.

$$\text{Accuracy} = 100 \times \frac{\text{mean value found} - \text{known value}}{\text{known value}}$$

$$\text{Precision} = 100 \times \frac{\text{standard deviation}}{\text{mean}} \equiv \textit{coefficient of variation}$$

For each of the three spike levels in the table, find the precision and accuracy of the quality control samples.

EDTA measurements (ng/mL) at three fortification levels

Spike:	22.2 ng/mL	88.2 ng/mL	314 ng/mL
Found:	33.3	83.6	322
	19.5	69.0	305
	23.9	83.4	282
	20.8	100.0	329
	20.8	76.4	276

SOURCE: R. L. Sheppard and J. Henion, *Anal. Chem.* **1997**, *69*, 477A, 2901.

(b) *Detection and quantitation limits.* Low concentrations of EDTA near the detection limit gave the following dimensionless instrument readings: 175, 104, 164, 193, 131, 189, 155, 133, 151, and 176. Ten blanks had a mean reading of $45._0$. The slope of the calibration curve is $1.75 \times 10^9 \ M^{-1}$. Estimate the signal and concentration detection limits and the lower limit of quantitation for EDTA.

5-17. (a) From Box 5-2, estimate the minimum expected coefficient of variation, CV(%), for interlaboratory results when the analyte concentration is (i) 1 wt% or (ii) 1 part per trillion.

(b) The coefficient of variation within a laboratory is typically ~0.5–0.7 of the between-laboratory variation. If your class analyzes an unknown containing 10 wt% NH_3, what is the minimum expected coefficient of variation for the class?

5-18. *Spike recovery and detection limit.* Species of arsenic found in drinking water include AsO_3^{3-} (arsenite), AsO_4^{3-} (arsenate), $(CH_3)_2AsO_2^-$ (dimethylarsinate), and $(CH_3)AsO_3^{2-}$ (methylarsonate). Pure water containing no arsenic was spiked with 0.40 μg arsenate/L. Seven replicate determinations gave 0.39, 0.40, 0.38, 0.41, 0.36, 0.35, and 0.39 μg/L.[15] Find the mean percent recovery of the spike and the concentration detection limit (μg/L).

5-19. *Detection limit.* Low concentrations of Ni^{2+}-EDTA near the detection limit gave the following counts in a mass spectral measurement: 175, 104, 164, 193, 131, 189, 155, 133, 151, 176. Ten measurements of a blank had a mean of 45 counts. A sample containing 1.00 μM Ni^{2+}-EDTA gave 1 797 counts. Estimate the detection limit for Ni-EDTA.

5-20. *Detection limit.* A sensitive chromatographic method was developed to measure sub-part-per-billion levels of the disinfectant by-products iodate (IO_3^-), chlorite (ClO_2^-), and bromate (BrO_3^-) in drinking water. As the oxyhalides emerge from the column, they react with Br^- to make Br_3^-, which is measured by its strong absorption at 267 nm. For example, each mole of bromate makes 3 mol of Br_3^- by the reaction

$$BrO_3^- + 8Br^- + 6H^+ \rightarrow 3Br_3^- + 3H_2O$$

Bromate near its detection limit gave the following chromatographic peak heights and standard deviations (s). For each concentration, estimate the detection limit. Find the mean detection limit. The blank is 0 because chromatographic peak height is measured from the baseline adjacent to the peak. Because blank = 0, relative standard deviation applies to both peak height and concentration, which are proportional to each other. Detection limit is $3s$ for peak height or concentration.

Bromate concentration (μg/L)	Peak height (arbitrary units)	Relative standard deviation (%)	Number of measurements
0.2	17	14.4	8
0.5	31	6.8	7
1.0	56	3.2	7
2.0	111	1.9	7

SOURCE: *H. S. Weinberg and H. Yamada, "Post-Ion-Chromatography Derivatization for the Determination of Oxyhalides at Sub-PPB Levels in Drinking Water," Anal. Chem.* **1998,** *70, 1.*

5-21. Olympic athletes are tested to see if they are using illegal performance-enhancing drugs. Suppose that urine samples are taken and analyzed and the rate of false positive results is 1%. Suppose also that it is too expensive to refine the method to reduce the rate of false positive results. We do not want to accuse innocent people of using illegal drugs. What can you do to reduce the rate of false accusations even though the test always has a false positive rate of 1%?

5-22. *Blind samples.* The U.S. Department of Agriculture provided homogenized baby food samples to three labs for analysis.[3] Results agreed well for protein, fat, zinc, riboflavin, and palmitic acid. Results for iron were questionable: Lab A, 1.59 ± 0.14 (13); Lab B, 1.65 ± 0.56 (8); Lab C, 2.68 ± 0.78 (3) mg/100 g. Uncertainty is the standard deviation, with the number of replicate analyses in parentheses. Use two separate t tests to compare results from Lab C with those from Lab A and Lab B at the 95% confidence level. Comment on the sensibility of the t test results. Offer your own conclusions.

Standard Addition

5-23. Why is it desirable in the method of standard addition to add a small volume of concentrated standard rather than a large volume of dilute standard?

5-24. An unknown sample of Cu^{2+} gave an absorbance of 0.262 in an atomic absorption analysis. Then 1.00 mL of solution containing 100.0 ppm (= μg/mL) Cu^{2+} was mixed with 95.0 mL of unknown, and the mixture was diluted to 100.0 mL in a volumetric flask. The absorbance of the new solution was 0.500.

(a) Denoting the initial, unknown concentration as $[Cu^{2+}]_i$, write an expression for the final concentration, $[Cu^{2+}]_f$, after dilution. Units of concentration are ppm.

(b) In a similar manner, write the final concentration of added standard Cu^{2+}, designated as $[S]_f$.

(c) Find $[Cu^{2+}]_i$ in the unknown.

5-25. 🖩 *Standard addition graph.* Tooth enamel consists mainly of the mineral calcium hydroxyapatite, $Ca_{10}(PO_4)_6(OH)_2$. Trace elements in teeth of archeological specimens provide anthropologists with clues about diet and diseases of ancient people. Students at Hamline University measured strontium in enamel from extracted wisdom teeth by atomic absorption spectroscopy. Solutions were prepared with a constant total volume of 10.0 mL containing 0.750 mg of dissolved tooth enamel plus variable concentrations of added Sr.

Added Sr (ng/mL = ppb)	Signal (arbitrary units)
0	28.0
2.50	34.3
5.00	42.8
7.50	51.5
10.00	58.6

SOURCE: *V. J. Porter, P. M. Sanft, J. C. Dempich, D. D. Dettmer, A. E. Erickson, N. A. Dubauskie, S. T. Myster, E. H. Matts, and E. T. Smith, "Elemental Analysis of Wisdom Teeth by Atomic Spectroscopy Using Standard Addition," J. Chem. Ed.* **2002,** *79, 1114.*

(a) Find the concentration of Sr and its uncertainty in the 10-mL sample solution in parts per billion = ng/mL.

(b) Find the concentration of Sr in tooth enamel in parts per million = μg/g.

(c) If the standard addition intercept is the major source of uncertainty, find the uncertainty in the concentration of Sr in tooth enamel in parts per million.

(d) Find the 95% confidence interval for Sr in tooth enamel.

5-26. Europium is a lanthanide element found in parts per billion levels in natural waters. It can be measured from the intensity of orange light emitted when a solution is illuminated with ultraviolet radiation. Certain organic compounds that bind Eu(III) are required to enhance the emission. The figure shows standard addition experiments in which 10.00 mL of sample and 20.00 mL containing a large excess of organic additive were placed in 50-mL volumetric flasks. Then Eu(III) standards (0, 5.00, 10.00, or 15.00 mL) were added and the flasks were diluted to 50.0 mL with H_2O. Standards added to tap water contained 0.152 ng/mL (ppb) of Eu(III), but those added to pond water were 100 times more concentrated (15.2 ng/mL).

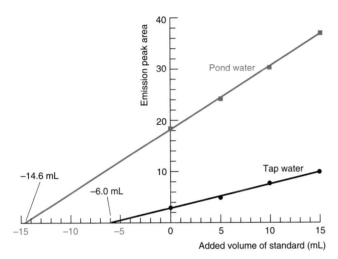

Standard addition of Eu(III) to pond water or tap water. [Data from A. L. Jenkins and G. M. Murray, "Enhanced Luminescence of Lanthanides," *J. Chem. Ed.* **1998**, *75*, 227.]

(a) Calculate the concentration of Eu(III) (ng/mL) in pond water and tap water.

(b) For tap water, emission peak area increases by 4.61 units when 10.00 mL of 0.152 ng/mL standard are added. This response is 4.61 units/1.52 ng = 3.03 units per ng of Eu(III). For pond water, the response is 12.5 units when 10.00 mL of 15.2 ng/mL standard are added, or 0.082 2 units per ng. How would you explain these observations? Why was standard addition necessary for this analysis?

5-27. ▦ *Standard addition graph.* Students performed an experiment like that in Figure 5-7 in which each flask contained 25.00 mL of serum, varying additions of 2.640 M NaCl standard, and a total volume of 50.00 mL.

Flask	Volume of standard (mL)	Na^+ atomic emission signal (mV)
1	0	3.13
2	1.000	5.40
3	2.000	7.89
4	3.000	10.30
5	4.000	12.48

(a) Prepare a standard addition graph and find $[Na^+]$ in the serum.

(b) Find the standard deviation and 95% confidence interval for $[Na^+]$.

5-28. ▦ *Standard addition graph.* Allicin is a major component in garlic (0.4 wt% in fresh garlic) with antimicrobial and possibly anticancer and antioxidant activity. It is unstable and therefore difficult to measure. An assay was developed in which the stable precursor alliin is added to freshly crushed garlic and rapidly converted to allicin by the enzyme alliinase found in garlic. Components of the crushed garlic are extracted and measured by liquid chromatogra-

phy. The graph shows standard additions reported as mg alliin added per gram of garlic. The chromatographic peak is allicin from the conversion of alliin.

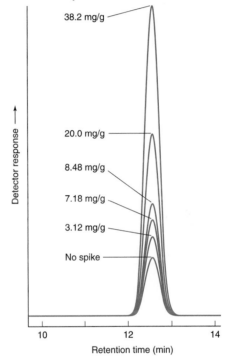

(a) The standard addition procedure has a constant total volume. Prepare a graph to find how much alliin equivalent was in the unspiked garlic. The units of your answer will be mg alliin/g garlic. Include the standard deviation in your answer.

Chromatographic measurement of alliin after standard addition to garlic. [From M. E. Rybak, E. M. Calvey, and J. M. Harnly, "Quantitative Determination of Allicin in Garlic: Supercritical Fluid Extraction and Standard Addition of Alliin," *J. Agric. Food Chem.* **2004**, *52*, 682.]

(b) Given that 2 mol of alliin are converted to 1 mol of allicin in the assay, find the allicin content of garlic expressed as mg allicin/g garlic, including the standard deviation.

Internal Standards

5-29. State when standard additions and internal standards, instead of a calibration curve, are desirable, and why.

5-30. A solution containing 3.47 mM X (analyte) and 1.72 mM S (standard) gave peak areas of 3 473 and 10 222, respectively, in a chromatographic analysis. Then 1.00 mL of 8.47 mM S was added to 5.00 mL of unknown X, and the mixture was diluted to 10.0 mL. This solution gave peak areas of 5 428 and 4 431 for X and S, respectively.

(a) Calculate the response factor for the analyte.

(b) Find the concentration of S (mM) in the 10.0 mL of mixed solution.

(c) Find the concentration of X (mM) in the 10.0 mL of mixed solution.

(d) Find the concentration of X in the original unknown.

5-31. Chloroform is an internal standard in the determination of the pesticide DDT in a polarographic analysis in which each compound is reduced at an electrode surface. A mixture containing 0.500 mM chloroform and 0.800 mM DDT gave signals of 15.3 μA for chloroform and 10.1 μA for DDT. An unknown solution (10.0 mL) containing DDT was placed in a 100-mL volumetric flask and 10.2 μL of chloroform (FM 119.39, density = 1.484 g/mL) were added. After dilution to the mark with solvent, polarographic signals of 29.4 and 8.7 μA were observed for the chloroform and DDT, respectively. Find the concentration of DDT in the unknown.

5-32. *Verifying constant response for an internal standard.* When we develop a method using an internal standard, it is important to verify that the response factor is constant over the calibration range. Data are shown below for a chromatographic analysis of naphthalene ($C_{10}H_8$), using deuterated naphthalene ($C_{10}D_8$ in which D is the isotope ^{2}H) as an internal standard. The two compounds emerge from the column at almost identical times and are measured by a mass spectrometer, which distinguishes them by molecular mass. From the definition of response factor in Equation 5-11, we can write

$$\frac{\text{Area of analyte signal}}{\text{Area of standard signal}} = F\left(\frac{\text{concentration of analyte}}{\text{concentration of standard}}\right)$$

Prepare a graph of peak area ratio ($C_{10}H_8/C_{10}D_8$) versus concentration ratio ($[C_{10}H_8]/[C_{10}D_8]$) and find the slope, which is the response factor. Evaluate F for each of the three samples and find the standard deviation of F to see how "constant" it is.

Sample	$C_{10}H_8$ (ppm)	$C_{10}D_8$ (ppm)	$C_{10}H_8$ peak area	$C_{10}D_8$ peak area
1	1.0	10.0	303	2 992
2	5.0	10.0	3 519	6 141
3	10.0	10.0	3 023	2 819

The volume of solution injected into the column was different in all three runs.

5-33. *Correcting for matrix effects with an internal standard.* The appearance of pharmaceuticals in municipal wastewater (sewage) is an increasing problem that is likely to have adverse effects on our drinking water supply. Sewage is a complex matrix. When the drug carbamazepine was spiked into sewage at a concentration of 5 ppb, chromatographic analysis gave an apparent spike recovery of 154%.[16] When deuterated carbamazepine was used as an internal standard for the analysis, the apparent recovery was 98%. Explain how the internal standard is used in this analysis and rationalize why it works so well to correct for matrix effects.

Carbamazepine   Carbamazepine-d_{10}

Experimental Design

5-34.   Acid-base titrations similar to those in Section 5-5 had volumes and results shown in the table.[14] Use Excel LINEST to find the concentrations of acids A, B, and C and estimate their uncertainty.

A	B	C	mmol of OH^- required
2	2	2	3.015
0	2	2	1.385
2	0	2	2.180
2	2	0	2.548
2	2	2	3.140

Acid volume (mL) spans columns A, B, C above.

6 Chemical Equilibrium

Paper mill on the Potomac River near Westernport, Maryland, neutralizes acid mine drainage in the water. Upstream of the mill, the river is acidic and lifeless; below the mill, the river teems with life. [Photo courtesy C. Dalpra, Potomac River Basin Commission.]

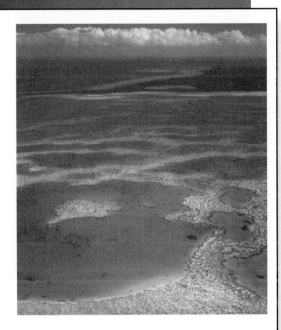

Great Barrier Reef and other coral reefs are threatened with extinction by rising atmospheric CO_2. [Copyright Jon Arnold Images/Almay.]

Part of the North Branch of the Potomac River runs crystal clear through the scenic Appalachian Mountains, but it is lifeless—a victim of acid drainage from abandoned coal mines. As the river passes a paper mill and a wastewater treatment plant near Westernport, Maryland, the pH rises from an acidic, lethal value of 4.5 to a neutral value of 7.2, at which fish and plants thrive. This happy "accident" comes about because calcium carbonate exiting the paper mill equilibrates with massive quantities of carbon dioxide from bacterial respiration at the sewage treatment plant. The resulting soluble bicarbonate neutralizes the acidic river and restores life downstream of the plant.[1] In the absence of CO_2, solid $CaCO_3$ would be trapped at the treatment plant and would never enter the river.

$$CaCO_3(s) + CO_2(aq) + H_2O(l) \rightleftharpoons Ca^{2+}(aq) + 2HCO_3^-(aq)$$

| Calcium carbonate trapped at treatment plant | Dissolved calcium bicarbonate enters river and neutralizes acid |

$$HCO_3^-(aq) + H^+(aq) \xrightarrow{\text{neutralization}} CO_2(g) + H_2O(l)$$

Bicarbonate   Acid in river

The chemistry that helps the Potomac River endangers coral reefs, which are largely $CaCO_3$. Burning of fossil fuel has increased CO_2 in the atmosphere from 280 ppm when Captain Cook first sighted the Great Barrier Reef in 1770 to 380 ppm today (Figure 0-6). CO_2 in the atmosphere adds CO_2 to the ocean, dissolving $CaCO_3$ from coral. Rising CO_2 and, perhaps, rising atmospheric temperature from the greenhouse effect threaten coral reefs with extinction.[2] CO_2 has lowered the average pH of the ocean from its preindustrial value of 8.16 to 8.04 today.[3] Without changes in man's activities, the pH could be 7.7 by 2100.

Chemical equilibrium provides a foundation not only for chemical analysis, but also for other subjects such as biochemistry, geology, and oceanography. This chapter introduces equilibria for the solubility of ionic compounds, complex formation, and acid-base reactions.

6-1 The Equilibrium Constant

For the reaction

$$a\text{A} + b\text{B} \rightleftharpoons c\text{C} + d\text{D} \tag{6-1}$$

we write the **equilibrium constant,** K, in the form

Equilibrium constant:
$$K = \frac{[\text{C}]^c[\text{D}]^d}{[\text{A}]^a[\text{B}]^b} \tag{6-2}$$

where the lowercase superscript letters denote stoichiometry coefficients and each capital letter stands for a chemical species. The symbol [A] stands for the concentration of A relative to its standard state (defined next). By definition, *a reaction is favored whenever* $K > 1$.

In the thermodynamic derivation of the equilibrium constant, each quantity in Equation 6-2 is expressed as the *ratio* of the concentration of a species to its concentration in its **standard state.** For solutes, the standard state is 1 M. For gases, the standard state is 1 bar ($\equiv 10^5$ Pa; 1 atm $\equiv$ 1.013 25 bar), and for solids and liquids, the standard states are the pure solid or liquid. It is understood that [A] in Equation 6-2 really means [A]/(1 M) if A is a solute. If D is a gas, [D] really means (pressure of D in bars)/(1 bar). To emphasize that [D] means pressure of D, we usually write P_D in place of [D]. The terms in Equation 6-2 are actually dimensionless; therefore, all equilibrium constants are dimensionless.

For the ratios [A]/(1 M) and [D]/(1 bar) to be dimensionless, [A] *must* be expressed in moles per liter (M), and [D] *must* be expressed in bars. If C were a pure liquid or solid, the ratio [C]/(concentration of C in its standard state) would be unity (1) because the standard state is the pure liquid or solid. If C is a solvent, the concentration is so close to that of pure liquid C that the value of [C] is still essentially 1.

The take-home lesson is this: When you evaluate an equilibrium constant,

1. Concentrations of solutes should be expressed as moles per liter.
2. Concentrations of gases should be expressed in bars.
3. Concentrations of pure solids, pure liquids, and solvents are omitted because they are unity.

These conventions are arbitrary, but you must use them if you wish to use tabulated values of equilibrium constants, standard reduction potentials, and free energies.

Manipulating Equilibrium Constants

Consider the reaction

$$\text{HA} \rightleftharpoons \text{H}^+ + \text{A}^- \qquad K_1 = \frac{[\text{H}^+][\text{A}^-]}{[\text{HA}]}$$

If the direction of a reaction is reversed, the new value of K is simply the reciprocal of the original value of K.

Equilibrium constant for reverse reaction:
$$\text{H}^+ + \text{A}^- \rightleftharpoons \text{HA} \qquad K_1' = \frac{[\text{HA}]}{[\text{H}^+][\text{A}^-]} = 1/K_1$$

If two reactions are added, the new K is the product of the two individual values:

$$\begin{array}{ll} \text{HA} \rightleftharpoons \cancel{\text{H}^+} + \text{A}^- & K_1 \\ \cancel{\text{H}^+} + \text{C} \rightleftharpoons \text{CH}^+ & K_2 \\ \hline \text{HA} + \text{C} \rightleftharpoons \text{A}^- + \text{CH}^+ & K_3 \end{array}$$

Equilibrium constant for sum of reactions:
$$K_3 = K_1 K_2 = \frac{\cancel{[\text{H}^+]}[\text{A}^-]}{[\text{HA}]} \cdot \frac{[\text{CH}^+]}{\cancel{[\text{H}^+]}[\text{C}]} = \frac{[\text{A}^-][\text{CH}^+]}{[\text{HA}][\text{C}]}$$

If n reactions are added, the overall equilibrium constant is the product of n individual equilibrium constants.

Equation 6-2, the *law of mass action*, was formulated by the Norwegians C. M. Guldenberg and P. Waage in 1864. Their derivation was based on the idea that the forward and reverse rates of a reaction at equilibrium must be equal.[4]

The equilibrium constant is more correctly expressed as a ratio of *activities* rather than of concentrations. We reserve the discussion of activity for Chapter 7.

Equilibrium constants are dimensionless.

Equilibrium constants are dimensionless but, when specifying concentrations, you must use units of molarity (M) for solutes and bars for gases.

Throughout this book, assume that all species in chemical equations are in aqueous solution, unless otherwise specified.

If a reaction is reversed, then $K' = 1/K$. If two reactions are added, then $K_3 = K_1 K_2$.

EXAMPLE Combining Equilibrium Constants

The equilibrium constant for the reaction $H_2O \rightleftharpoons H^+ + OH^-$ is called K_w ($= [H^+][OH^-]$) and has the value 1.0×10^{-14} at 25°C. Given that $K_{NH_3} = 1.8 \times 10^{-5}$ for the reaction $NH_3(aq) + H_2O \rightleftharpoons NH_4^+ + OH^-$, find K for the reaction $NH_4^+ \rightleftharpoons NH_3(aq) + H^+$.

Solution The third reaction can be obtained by reversing the second reaction and adding it to the first reaction:

$$H_2O \rightleftharpoons H^+ + OH^- \qquad\qquad K = K_w$$
$$\underline{NH_4^+ + OH^- \rightleftharpoons NH_3(aq) + H_2O} \qquad K = 1/K_{NH_3}$$
$$NH_4^+ \rightleftharpoons H^+ + NH_3(aq) \qquad\qquad K = K_w \cdot \frac{1}{K_{NH_3}} = 5.6 \times 10^{-10}$$

Test Yourself For the reaction $Li^+ + H_2O \rightleftharpoons Li(OH)(aq) + H^+$, $K_{Li}' = 2.3 \times 10^{-14}$. Combine this reaction with the K_w reaction to find the equilibrium constant for the reaction $Li^+ + OH^- \rightleftharpoons Li(OH)(aq)$. (**Answer:** 2.3)

6-2 Equilibrium and Thermodynamics

Equilibrium is controlled by the thermodynamics of a chemical reaction. The heat absorbed or released (*enthalpy*) and the degree of disorder of reactants and products (*entropy*) independently contribute to the degree to which the reaction is favored or disfavored.

Enthalpy

The **enthalpy change,** ΔH, for a reaction is the heat absorbed or released when the reaction takes place under constant applied pressure.[5] The *standard enthalpy change*, $\Delta H°$, refers to the heat absorbed when all reactants and products are in their standard states:[†]

$$HCl(g) \rightleftharpoons H^+(aq) + Cl^-(aq) \qquad \Delta H° = -74.85 \text{ kJ/mol at 25°C} \qquad (6\text{-}3)$$

The negative sign of $\Delta H°$ indicates that heat is released by Reaction 6-3—the solution becomes warmer. For other reactions, ΔH is positive, which means that heat is absorbed. Consequently, the solution gets colder during the reaction. A reaction for which ΔH is positive is said to be **endothermic.** Whenever ΔH is negative, the reaction is **exothermic.**

$\Delta H = (+)$
Heat is absorbed
Endothermic

$\Delta H = (-)$
Heat is liberated
Exothermic

Entropy

The **entropy,** S, of a substance is a measure of its "disorder," which we will not attempt to define in a quantitative way. The greater the disorder, the greater the entropy. In general, a gas is more disordered (has higher entropy) than a liquid, which is more disordered than a solid. Ions in aqueous solution are normally more disordered than in their solid salt:

$$KCl(s) \rightleftharpoons K^+(aq) + Cl^-(aq) \qquad \Delta S° = +76.4 \text{ J/(K·mol) at 25°C} \qquad (6\text{-}4)$$

$\Delta S°$ is the change in entropy (entropy of products minus entropy of reactants) when all species are in their standard states. The positive value of $\Delta S°$ indicates that a mole of $K^+(aq)$ plus a mole of $Cl^-(aq)$ is more disordered than a mole of $KCl(s)$. For Reaction 6-3, $\Delta S° = -130.4 \text{ J/(K·mol)}$ at 25°C. The aqueous ions are less disordered than gaseous HCl.

$\Delta S = (+)$
Products more disordered than reactants

$\Delta S = (-)$
Products less disordered than reactants

Free Energy

Systems at constant temperature and pressure, which are common laboratory conditions, have a tendency toward lower enthalpy and higher entropy. A chemical reaction is driven toward the formation of products by a *negative* value of ΔH (heat given off) or a *positive* value of ΔS (more disorder) or both. When ΔH is negative and ΔS is positive, the reaction is clearly favored. When ΔH is positive and ΔS is negative, the reaction is clearly disfavored.

[†]The definition of the standard state contains subtleties beyond the scope of this book. For Reaction 6-3, the standard state of H^+ or Cl^- is the hypothetical state in which each ion is present at a concentration of 1 M but behaves as if it were in an infinitely dilute solution. That is, the standard concentration is 1 M, but the standard behavior is what would be observed in a very dilute solution in which each ion is unaffected by surrounding ions.

When ΔH and ΔS are both positive or both negative, what decides whether a reaction will be favored? The change in **Gibbs free energy,** ΔG, is the arbiter between opposing tendencies of ΔH and ΔS. At constant temperature, T,

Free energy:
$$\Delta G = \Delta H - T\Delta S \qquad (6\text{-}5)$$

A *reaction is favored if* ΔG *is negative.*

For the dissociation of HCl (Reaction 6-3) when all species are in their standard states, $\Delta H°$ favors the reaction and $\Delta S°$ disfavors it. To find the net result, we evaluate $\Delta G°$:

$$\begin{aligned} \Delta G° &= \Delta H° - T\Delta S° \\ &= (-74.85 \times 10^3 \text{ J/mol}) - (298.15 \text{ K})(-130.4 \text{ J/K}\cdot\text{mol}) \\ &= -35.97 \text{ kJ/mol} \end{aligned}$$

$\Delta G°$ is negative, so the reaction is favored when all species are in their standard states. The favorable influence of $\Delta H°$ is greater than the unfavorable influence of $\Delta S°$ in this case.

The point of discussing free energy is to relate the equilibrium constant to the energetics ($\Delta H°$ and $\Delta S°$) of a reaction. The equilibrium constant depends on $\Delta G°$ in the following manner:

Free energy and equilibrium:
$$K = e^{-\Delta G°/RT} \qquad (6\text{-}6)$$

where R is the gas constant $[= 8.314\ 472 \text{ J/(K}\cdot\text{mol)}]$ and T is temperature (Kelvin). The more negative the value of $\Delta G°$, the larger is the equilibrium constant. For Reaction 6-3,

$$K = e^{-(-35.97 \times 10^3 \text{ J/mol})/[8.314\ 472 \text{ J/(K}\cdot\text{mol)}](298.15 \text{ K})} = 2.00 \times 10^6$$

Because the equilibrium constant is large, HCl(*g*) is very soluble in water and is nearly completely ionized to H^+ and Cl^- when it dissolves.

To summarize, a chemical reaction is favored by the liberation of heat (ΔH negative) and by an increase in disorder (ΔS positive). ΔG takes both effects into account to determine whether or not a reaction is favorable. We say that a reaction is *spontaneous* under standard conditions if $\Delta G°$ is negative or, equivalently, if $K > 1$. The reaction is not spontaneous if $\Delta G°$ is positive ($K < 1$). You should be able to calculate K from $\Delta G°$ and vice versa.

Le Châtelier's Principle

Suppose that a system at equilibrium is subjected to a change that disturbs the system. **Le Châtelier's principle** states that the direction in which the system proceeds back to equilibrium is such that the change is partially offset.

To see what this statement means, let's see what happens when we attempt to change the concentration of one species in the reaction

$$\text{BrO}_3^- + 2\text{Cr}^{3+} + 4\text{H}_2\text{O} \rightleftharpoons \text{Br}^- + \text{Cr}_2\text{O}_7^{2-} + 8\text{H}^+ \qquad (6\text{-}7)$$

$$\begin{array}{cc} \text{Bromate} & \text{Chromium(III)} & & \text{Dichromate} \end{array}$$

$$K = \frac{[\text{Br}^-][\text{Cr}_2\text{O}_7^{2-}][\text{H}^+]^8}{[\text{BrO}_3^-][\text{Cr}^{3+}]^2} = 1 \times 10^{11} \text{ at } 25°\text{C}$$

In one particular equilibrium state of this system, the following concentrations exist: $[\text{H}^+] = 5.0 \text{ M}$, $[\text{Cr}_2\text{O}_7^{2-}] = 0.10 \text{ M}$, $[\text{Cr}^{3+}] = 0.003\ 0 \text{ M}$, $[\text{Br}^-] = 1.0 \text{ M}$, and $[\text{BrO}_3^-] = 0.043 \text{ M}$. Suppose that the equilibrium is disturbed by adding dichromate to the solution to increase the concentration of $[\text{Cr}_2\text{O}_7^{2-}]$ from 0.10 to 0.20 M. In what direction will the reaction proceed to reach equilibrium?

According to the principle of Le Châtelier, the reaction should go back to the left to partially offset the increase in dichromate, which appears on the right side of Reaction 6-7. We can verify this algebraically by setting up the **reaction quotient,** Q, which has the same form as the equilibrium constant. The only difference is that Q is evaluated with whatever concentrations happen to exist, even though the solution is not at equilibrium. When the system reaches equilibrium, $Q = K$. For Reaction 6-7,

$$Q = \frac{(1.0)(0.20)(5.0)^8}{(0.043)(0.003\ 0)^2} = 2 \times 10^{11} > K$$

Note that 25.00°C = 298.15 K.

Challenge Satisfy yourself that $K > 1$ if $\Delta G°$ is negative.

$\Delta G = (+)$
Reaction is disfavored

$\Delta G = (-)$
Reaction is favored

Notice that H_2O is omitted from K because it is the solvent.

The reaction quotient has the same form as the equilibrium constant, but the concentrations are generally not the equilibrium concentrations.

Because $Q > K$, the reaction must go to the left to decrease the numerator and increase the denominator, until $Q = K$.

If $Q < K$, then the reaction must proceed to the right to reach equilibrium. If $Q > K$, then the reaction must proceed to the left to reach equilibrium.

1. If a reaction is at equilibrium and products are added (or reactants are removed), the reaction goes to the left.
2. If a reaction is at equilibrium and reactants are added (or products are removed), the reaction goes to the right.

When the temperature of a system is changed, so is the equilibrium constant. Equations 6-5 and 6-6 can be combined to predict the effect of temperature on K:

$$K = e^{-\Delta G°/RT} = e^{-(\Delta H° - T\Delta S°)/RT} = e^{(-\Delta H°/RT + \Delta S°/R)}$$
$$= e^{-\Delta H°/RT} \cdot e^{\Delta S°/R} \qquad (6\text{-}8)$$

$$e^{(a+b)} = e^a \cdot e^b$$

The term $e^{\Delta S°/R}$ is independent of T (at least over a limited temperature range in which $\Delta S°$ is constant). The term $e^{-\Delta H°/RT}$ increases with increasing temperature if $\Delta H°$ is positive and decreases if $\Delta H°$ is negative. Therefore,

1. The equilibrium constant of an endothermic reaction ($\Delta H° = +$) increases if the temperature is raised.
2. The equilibrium constant of an exothermic reaction ($\Delta H° = -$) decreases if the temperature is raised.

These statements can be understood in terms of Le Châtelier's principle as follows. Consider an endothermic reaction:

$$\text{Heat + reactants} \rightleftharpoons \text{products}$$

Heat can be treated as if it were a reactant in an endothermic reaction and a product in an exothermic reaction.

If the temperature is raised, then heat is added to the system. The reaction proceeds to the right to partially offset this change.[6]

In dealing with equilibrium problems, we are making *thermodynamic* predictions, not *kinetic* predictions. We are calculating what must happen for a system to reach equilibrium, but not how long it will take. Some reactions are over in an instant; others will not reach equilibrium in a million years. For example, dynamite remains unchanged indefinitely, until a spark sets off the spontaneous, explosive decomposition. The size of an equilibrium constant tells us nothing about the rate (the kinetics) of the reaction. A large equilibrium constant does not imply that a reaction is fast.

6-3 Solubility Product

In chemical analysis, we encounter solubility in precipitation titrations, electrochemical reference cells, and gravimetric analysis. The effect of acid on the solubility of minerals and the effect of atmospheric CO_2 on the solubility (and death) of coral reefs are important in environmental science.

The **solubility product** is the equilibrium constant for the reaction in which a solid salt dissolves to give its constituent ions in solution. Solid is omitted from the equilibrium constant because it is in its standard state. Appendix F lists solubility products.

As an example, consider the dissolution of mercury(I) chloride (Hg_2Cl_2, also called mercurous chloride) in water. The reaction is

$$Hg_2Cl_2(s) \rightleftharpoons Hg_2^{2+} + 2Cl^- \qquad (6\text{-}9)$$

for which the solubility product, K_{sp}, is

$$K_{sp} = [Hg_2^{2+}][Cl^-]^2 = 1.2 \times 10^{-18} \qquad (6\text{-}10)$$

A solution containing excess, undissolved solid is said to be **saturated** with that solid. The solution contains all the solid capable of being dissolved under the prevailing conditions.

The physical meaning of the solubility product is this: If an aqueous solution is left in contact with excess solid Hg_2Cl_2, the solid will dissolve until the condition $[Hg_2^{2+}][Cl^-]^2 = K_{sp}$ is satisfied. Thereafter, the amount of undissolved solid remains constant. Unless excess solid remains, there is no guarantee that $[Hg_2^{2+}][Cl^-]^2 = K_{sp}$. If Hg_2^{2+} and Cl^- are mixed together (with appropriate counterions) such that the product $[Hg_2^{2+}][Cl^-]^2$ exceeds K_{sp}, then Hg_2Cl_2 will precipitate.

Mercurous ion, Hg_2^{2+}, is a *dimer* (pronounced DIE mer), which means that it consists of two identical units bound together:

250 pm

$[Hg-Hg]^{2+}$
+1 oxidation state of mercury

OH^-, S^{2-}, and CN^- stabilize Hg(II), thereby converting Hg(I) into Hg(0) and Hg(II):

$$Hg_2^{2+} + 2CN^- \rightarrow Hg(CN)_2(aq) + Hg(l)$$

Hg(I)   Hg(II)   Hg(0)

Disproportionation is the process in which an element in an intermediate oxidation state gives products in both higher and lower oxidation states.

BOX 6-1 Solubility Is Governed by More Than the Solubility Product

If we want to know how much Hg_2^{2+} is dissolved in a saturated solution of Hg_2Cl_2, we are tempted to look at Reaction 6-9 and to note that two Cl^- are created for each Hg_2^{2+}. If we let x be the concentration of Hg_2^{2+}, we could say that the concentration of Cl^- is $2x$. Substituting these values into the solubility product expression 6-10, we could write $K_{sp} = [Hg_2^{2+}][Cl^-]^2 = (x)(2x)^2$, and we would find $[Hg_2^{2+}] = x = 6.7 \times 10^{-7}$ M.

However, this answer is incorrect because we have not accounted for other reactions, such as

Hydrolysis: $\quad Hg_2^{2+} + H_2O \rightleftharpoons Hg_2OH^+ + H^+ \quad K = 10^{-5.3}$

Disproportionation: $\quad 2Hg_2^{2+} \rightleftharpoons Hg^{2+} + Hg(l) \quad K = 10^{-2.1}$

Both reactions consume Hg_2^{2+}. By Le Chatelier's principle, if Hg_2^{2+} is consumed, more Hg_2Cl_2 will dissolve. We need to know all significant chemical reactions to compute the solubility of a compound.

We most commonly use the solubility product to find the concentration of one ion when the concentration of the other is known or fixed by some means. For example, what is the concentration of Hg_2^{2+} in equilibrium with 0.10 M Cl^- in a solution of KCl containing excess, undissolved $Hg_2Cl_2(s)$? To answer this question, we rearrange Equation 6-10 to find

$$[Hg_2^{2+}] = \frac{K_{sp}}{[Cl^-]^2} = \frac{1.2 \times 10^{-18}}{0.10^2} = 1.2 \times 10^{-16} \text{ M}$$

Because Hg_2Cl_2 is so slightly soluble, additional Cl^- obtained from Hg_2Cl_2 is negligible compared with 0.10 M Cl^-.

The solubility product does not tell the entire story of solubility. In addition to complications described in Box 6-1, most salts form soluble *ion pairs* to some extent. That is, $MX(s)$ can give $MX(aq)$ as well as $M^+(aq)$ and $X^-(aq)$. In a saturated solution of $CaSO_4$, for example, two-thirds of the dissolved calcium is Ca^{2+} and one third is $CaSO_4(aq)$.[7] The $CaSO_4(aq)$ **ion pair** is a closely associated pair of ions that behaves as one species in solution. Appendix J and Box 7-1 provide information on ion pairs.[8]

A *salt* is any ionic solid, such as Hg_2Cl_2 or $CaSO_4$.

Common Ion Effect

For the ionic solubility reaction

$$CaSO_4(s) \rightleftharpoons Ca^{2+} + SO_4^{2-} \qquad K_{sp} = 2.4 \times 10^{-5}$$

the product $[Ca^{2+}][SO_4^{2-}]$ is constant at equilibrium in the presence of excess solid $CaSO_4$. If the concentration of Ca^{2+} were increased by adding another source of Ca^{2+}, such as $CaCl_2$, then the concentration of SO_4^{2-} must decrease so that the product $[Ca^{2+}][SO_4^{2-}]$ remains constant. In other words, less $CaSO_4(s)$ will dissolve if Ca^{2+} or SO_4^{2-} is already present from some other source. Figure 6-1 shows how the solubility of $CaSO_4$ decreases in the presence of dissolved $CaCl_2$.

This application of Le Chatelier's principle is called the **common ion effect.** *A salt will be less soluble if one of its constituent ions is already present in the solution.*

Common ion effect: A salt is less soluble if one of its ions is already present in the solution. Demonstration 6-1 illustrates the common ion effect.

DEMONSTRATION 6-1 Common Ion Effect[9,10]

Fill two large test tubes about one-third full with saturated aqueous KCl containing no excess solid. The solubility of KCl is approximately 3.7 M, so the solubility product (ignoring activity effects introduced later) is

$$K_{sp} \approx [K^+][Cl^-] = (3.7)(3.7) = 13.7$$

Now add one-third of a test tube of 6 M HCl to one test tube and an equal volume of 12 M HCl to the other. Even though a common ion, Cl^-, is added in each case, KCl precipitates only in one tube.

To understand your observations, calculate the concentrations of K^+ and Cl^- in each tube after HCl addition. Then evaluate the reaction quotient, $Q = [K^+][Cl^-]$, for each tube. Explain your observations.

Add 1 volume of 6 M HCl(aq)
Add 1 volume of 12 M HCl(aq)

Saturated KCl(aq) (no excess solid)
Homogeneous solution
KCl precipitate

CHAPTER 6 Chemical Equilibrium

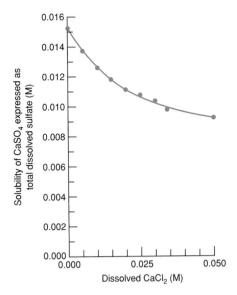

FIGURE 6-1 Solubility of CaSO₄ in solutions containing dissolved CaCl₂. Solubility is expressed as total dissolved sulfate, which is present as free SO_4^{2-} and as the ion pair, CaSO₄(aq). [Data from reference 8.]

FIGURE 6-2 A yellow solid, lead(II) iodide (PbI₂), precipitates when a colorless solution of lead nitrate (Pb(NO₃)₂) is added to a colorless solution of potassium iodide (KI). [Photo by Chip Clark.]

Separation by Precipitation

Precipitation can sometimes be used to separate ions from each other.[11] For example, consider a solution containing lead(II) (Pb^{2+}) and mercury(I) (Hg_2^{2+}) ions, each at a concentration of 0.010 M. Each forms an insoluble iodide (Figure 6-2), but the mercury(I) iodide is considerably less soluble, as indicated by the smaller value of K_{sp}:

$$PbI_2(s) \rightleftharpoons Pb^{2+} + 2I^- \qquad K_{sp} = 7.9 \times 10^{-9}$$

$$Hg_2I_2(s) \rightleftharpoons Hg_2^{2+} + 2I^- \qquad K_{sp} = 4.6 \times 10^{-29}$$

The smaller K_{sp} implies a lower solubility for Hg_2I_2 because the stoichiometries of the two reactions are the same. If the stoichiometries were different, it does not follow that the smaller K_{sp} would imply lower solubility.

Is it possible to lower the concentration of Hg_2^{2+} by 99.990% by selective precipitation with I^-, without precipitating Pb^{2+}?

We are asking whether we can lower $[Hg_2^{2+}]$ to 0.010% of 0.010 M = 1.0×10^{-6} M without precipitating Pb^{2+}. Here is the experiment: We add enough I^- to precipitate 99.990% of Hg_2^{2+} if all the I^- reacts with Hg_2^{2+} and none reacts with Pb^{2+}. To see if any Pb^{2+} should precipitate, we need to know the concentration of I^- in equilibrium with precipitated $Hg_2I_2(s)$ plus the remaining 1.0×10^{-6} M Hg_2^{2+}.

$$Hg_2I_2(s) \xrightleftharpoons{K_{sp}} Hg_2^{2+} + 2I^-$$

$$[Hg_2^{2+}][I^-]^2 = K_{sp}$$

$$(1.0 \times 10^{-6})[I^-]^2 = 4.6 \times 10^{-29}$$

$$[I^-] = \sqrt{\frac{4.6 \times 10^{-29}}{1.0 \times 10^{-6}}} = 6.8 \times 10^{-12} \text{ M}$$

Will this concentration of I^- cause 0.010 M Pb^{2+} to precipitate? That is, is the solubility product of PbI_2 exceeded?

$$Q = [Pb^{2+}][I^-]^2 = (0.010)(6.8 \times 10^{-12})^2$$

$$= 4.6 \times 10^{-25} < K_{sp} \quad (\text{for } PbI_2)$$

The reaction quotient, $Q = 4.6 \times 10^{-25}$ is less than K_{sp} for $PbI_2 = 7.9 \times 10^{-9}$. Therefore, Pb^{2+} will not precipitate and separation of Pb^{2+} and Hg_2^{2+} is feasible. We predict that adding I^- to a solution of Pb^{2+} and Hg_2^{2+} will precipitate virtually all the mercury(I) before any lead(II) precipitates.

Life should be so easy! We have just made a thermodynamic prediction. If the system comes to equilibrium, we can achieve the desired separation. However, occasionally one substance *coprecipitates* with the other. In **coprecipitation,** a substance whose solubility is not

Question If you want to know whether a small amount of Pb^{2+} coprecipitates with Hg_2I_2, should you measure the Pb^{2+} concentration in the mother liquor (the solution) or the Pb^{2+} concentration in the precipitate? Which measurement is more sensitive? By "sensitive," we mean responsive to a small amount of coprecipitation. (Answer: Measure Pb^{2+} in precipitate.)

exceeded precipitates along with another substance whose solubility is exceeded. For example, some Pb^{2+} might become adsorbed on the surface of the Hg_2I_2 crystal or might even occupy sites within the crystal. Our calculation says that the separation is worth trying. *Only an experiment can show whether the separation actually works.*

6-4 Complex Formation

If anion X^- precipitates metal M^+, it is often observed that a high concentration of X^- causes solid MX to redissolve. The increased solubility arises from formation of **complex ions,** such as MX_2^-, which consist of two or more simple ions bonded to one another.

Lewis Acids and Bases

In complex ions such as PbI^+, PbI_3^-, and PbI_4^{2-}, iodide is said to be the *ligand* of Pb^{2+}. A **ligand** is any atom or group of atoms attached to the species of interest. We say that Pb^{2+} acts as a *Lewis acid* and I^- acts as a *Lewis base* in these complexes. A **Lewis acid** accepts a pair of electrons from a **Lewis base** when the two form a bond:

Lewis acid + Lewis base ⇌ adduct
Electron pair   Electron pair
acceptor   donor

$$^{++}Pb\,\square + \boxed{:}\,\ddot{\underset{..}{I}}:^- \quad \rightarrow \quad [Pb - \ddot{\underset{..}{I}}:]^+$$

Room to   A pair of
accept   electrons to
electrons   be donated

The product of the reaction between a Lewis acid and a Lewis base is called an *adduct.* The bond between a Lewis acid and a Lewis base is called a *dative* or *coordinate covalent* bond.

Effect of Complex Ion Formation on Solubility[12]

If Pb^{2+} and I^- only reacted to form solid PbI_2, then the solubility of Pb^{2+} would always be very low in the presence of excess I^-:

$$PbI_2(s) \xrightleftharpoons{K_{sp}} Pb^{2+} + 2I^- \qquad K_{sp} = [Pb^{2+}][I^-]^2 = 7.9 \times 10^{-9} \qquad (6\text{-}11)$$

We observe, however, that high concentrations of I^- cause solid PbI_2 to dissolve. We explain this by the formation of a series of complex ions:

$$Pb^{2+} + I^- \xrightleftharpoons{K_1} PbI^+ \qquad\qquad K_1 = [PbI^+]/[Pb^{2+}][I^-] = 1.0 \times 10^2 \qquad (6\text{-}12)$$

$$Pb^{2+} + 2I^- \xrightleftharpoons{\beta_2} PbI_2(aq) \qquad \beta_2 = [PbI_2(aq)]/[Pb^{2+}][I^-]^2 = 1.4 \times 10^3 \quad (6\text{-}13)$$

$$Pb^{2+} + 3I^- \xrightleftharpoons{\beta_3} PbI_3^- \qquad\qquad \beta_3 = [PbI_3^-]/[Pb^{2+}][I^-]^3 = 8.3 \times 10^3 \qquad (6\text{-}14)$$

$$Pb^{2+} + 4I^- \xrightleftharpoons{\beta_4} PbI_4^{2-} \qquad\qquad \beta_4 = [PbI_4^{2-}]/[Pb^{2+}][I^-]^4 = 3.0 \times 10^4 \qquad (6\text{-}15)$$

The species $PbI_2(aq)$ in Reaction 6-13 is *dissolved* PbI_2, containing two iodine atoms bound to a lead atom. Reaction 6-13 is *not* the reverse of Reaction 6-11, in which the species is solid PbI_2.

Notation for these equilibrium constants is discussed in Box 6-2.

BOX 6-2 Notation For Formation Constants

Formation constants are the equilibrium constants for complex ion formation. The **stepwise formation constants,** designated K_i, are defined as follows:

$$M + X \xrightleftharpoons{K_1} MX \qquad K_1 = [MX]/[M][X]$$

$$MX + X \xrightleftharpoons{K_2} MX_2 \qquad K_2 = [MX_2]/[MX][X]$$

$$MX_{n-1} + X \xrightleftharpoons{K_n} MX_n \qquad K_n = [MX_n]/[MX_{n-1}][X]$$

The **overall,** or **cumulative, formation constants** are denoted β_i:

$$M + 2X \xrightleftharpoons{\beta_2} MX_2 \qquad \beta_2 = [MX_2]/[M][X]^2$$

$$M + nX \xrightleftharpoons{\beta_n} MX_n \qquad \beta_n = [MX_n]/[M][X]^n$$

A useful relation is that $\beta_n = K_1 K_2 \cdots K_n$. Some formation constants can be found in Appendix I.

At low I^- concentrations, the solubility of lead is governed by precipitation of $PbI_2(s)$. At high I^- concentrations, Reactions 6-12 through 6-15 are driven to the right (Le Châtelier's principle), and the total concentration of dissolved lead is considerably greater than that of Pb^{2+} alone (Figure 6-3).

A most useful characteristic of chemical equilibrium is that *all equilibria are satisfied simultaneously.* If we know $[I^-]$, we can calculate $[Pb^{2+}]$ by substituting the value of $[I^-]$ into

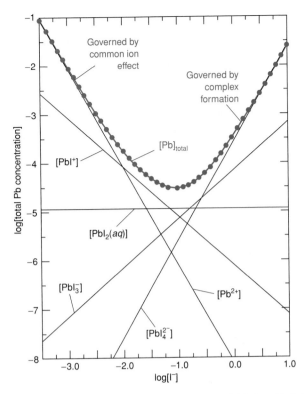

FIGURE 6-3 Total solubility of lead(II) (curve with circles) and solubilities of individual species (straight lines) as a function of the concentration of free iodide. To the left of the minimum, $[Pb]_{total}$ is governed by the solubility product for $PbI_2(s)$. As $[I^-]$ increases, $[Pb]_{total}$ decreases because of the common ion effect. At high values of $[I^-]$, $PbI_2(s)$ redissolves because it reacts with I^- to form soluble complex ions, such as PbI_4^{2-}. Note logarithmic scales. The solution is made slightly acidic so that $[PbOH^+]$ is negligible.

the equilibrium constant expression for Reaction 6-11, regardless of whether there are other reactions involving Pb^{2+}. *The concentration of Pb^{2+} that satisfies any one equilibrium must satisfy all equilibria. There can be only one concentration of Pb^{2+} in the solution.*

EXAMPLE **Effect of I^- on the Solubility of Pb^{2+}**

Find the concentrations of PbI^+, $PbI_2(aq)$, PbI_3^-, and PbI_4^{2-} in a solution saturated with $PbI_2(s)$ and containing dissolved I^- with a concentration of **(a)** 0.001 0 M and **(b)** 1.0 M.

Solution **(a)** From K_{sp} for Reaction 6-11, we calculate

$$[Pb^{2+}] = K_{sp}/[I^-]^2 = (7.9 \times 10^{-9})/(0.001\ 0)^2 = 7.9 \times 10^{-3}\ M$$

From Reactions 6-12 through 6-15, we then calculate the concentrations of the other lead-containing species:

$$[PbI^+] = K_1[Pb^{2+}][I^-] = (1.0 \times 10^2)(7.9 \times 10^{-3})(1.0 \times 10^{-3})$$
$$= 7.9 \times 10^{-4}\ M$$
$$[PbI_2(aq)] = \beta_2[Pb^{2+}][I^-]^2 = 1.1 \times 10^{-5}\ M$$
$$[PbI_3^-] = \beta_3[Pb^{2+}][I^-]^3 = 6.6 \times 10^{-8}\ M$$
$$[PbI_4^{2-}] = \beta_4[Pb^{2+}][I^-]^4 = 2.4 \times 10^{-10}\ M$$

(b) If, instead, we take $[I^-] = 1.0\ M$, then analogous computations show that

$$[Pb^{2+}] = 7.9 \times 10^{-9}\ M \qquad [PbI_3^-] = 6.6 \times 10^{-5}\ M$$
$$[PbI^+] = 7.9 \times 10^{-7}\ M \qquad [PbI_4^{2-}] = 2.4 \times 10^{-4}\ M$$
$$[PbI_2(aq)] = 1.1 \times 10^{-5}\ M$$

Test Yourself Find $[Pb^{2+}]$, $[PbI_2(aq)]$, and $[PbI_3^-]$, in a saturated solution of $PbI_2(s)$ with $[I^-] = 0.10\ M$. (*Answer:* 7.9×10^{-7}, 1.1×10^{-5}, $6.6 \times 10^{-6}\ M$)

The total concentration of dissolved lead in the preceding example is

$$[Pb]_{total} = [Pb^{2+}] + [PbI^+] + [PbI_2(aq)] + [PbI_3^-] + [PbI_4^{2-}]$$

When $[I^-] = 10^{-3}$ M, $[Pb]_{total} = 8.7 \times 10^{-3}$ M, of which 91% is Pb^{2+}. As $[I^-]$ increases, $[Pb]_{total}$ decreases by the common ion effect operating in Reaction 6-11. However, at sufficiently high $[I^-]$, complex formation takes over and $[Pb]_{total}$ increases in Figure 6-3. When $[I^-] = 1.0$ M, $[Pb]_{total} = 3.2 \times 10^{-4}$ M, of which 76% is PbI_4^{2-}.

Tabulated Equilibrium Constants Are Usually Not "Constant"

If you look up the equilibrium constant of a chemical reaction in two different books, there is an excellent chance that the values will be different (sometimes by a factor of 10 or more).[13] This discrepancy occurs because the constant may have been determined under different conditions and, perhaps, by using different techniques.

A common source of variation in the reported value of K is the ionic composition of the solution. Note whether K is reported for a particular ionic composition (for example, 1 M $NaClO_4$) or whether K has been extrapolated to zero ion concentration. If you need an equilibrium constant for your own work, choose a value of K measured under conditions as close as possible to those you will employ.

The effect of dissolved ions on chemical equilibria is the subject of Chapter 7.

6-5 Protic Acids and Bases

Understanding the behavior of acids and bases is essential to every branch of science having anything to do with chemistry. In analytical chemistry, we almost always need to account for the effect of pH on analytical reactions involving complex formation or oxidation-reduction. pH can affect molecular charge and shape—factors that help determine which molecules can be separated from others in chromatography and electrophoresis and which molecules will be detected in some types of mass spectrometry.

In aqueous chemistry, an **acid** is a substance that increases the concentration of H_3O^+ **(hydronium ion)** when added to water. Conversely, a **base** decreases the concentration of H_3O^+. We will see that a decrease in H_3O^+ concentration necessarily requires an increase in OH^- concentration. Therefore, a base increases the concentration of OH^- in aqueous solution.

The word *protic* refers to chemistry involving transfer of H^+ from one molecule to another. The species H^+ is also called a *proton* because it is what remains when a hydrogen atom loses its electron. Hydronium ion, H_3O^+, is a combination of H^+ with H_2O. Although H_3O^+ is a more accurate representation than H^+ for the hydrogen ion in aqueous solution, we will use H_3O^+ and H^+ interchangeably in this book.

Brønsted-Lowry Acids and Bases

Brønsted-Lowry acid: proton donor

Brønsted-Lowry base: proton acceptor

J. N. Brønsted (1879–1947) of the University of Copenhagen and T. M. Lowry (1874–1936) of Cambridge University independently published their definitions of acids and bases in 1923.

Brønsted and Lowry classified *acids* as *proton donors* and *bases* as *proton acceptors*. HCl is an acid (a proton donor), and it increases the concentration of H_3O^+ in water:

$$HCl + H_2O \rightleftharpoons H_3O^+ + Cl^-$$

The Brønsted-Lowry definition does not require that H_3O^+ be formed. This definition can therefore be extended to nonaqueous solvents and to the gas phase:

$$\underset{\substack{\text{Hydrochloric acid} \\ \text{(acid)}}}{HCl(g)} + \underset{\substack{\text{Ammonia} \\ \text{(base)}}}{NH_3(g)} \rightleftharpoons \underset{\substack{\text{Ammonium chloride} \\ \text{(salt)}}}{NH_4^+Cl^-(s)}$$

For the remainder of this book, when we speak of acids and bases, we are speaking of Brønsted-Lowry acids and bases.

Salts

Any ionic solid, such as ammonium chloride, is called a **salt.** In a formal sense, a salt can be thought of as the product of an acid-base reaction. When an acid and a base react, they are said to **neutralize** each other. Most salts containing cations and anions with single positive and negative charges are strong electrolytes—they dissociate nearly completely into ions in dilute aqueous solution. Thus, ammonium chloride gives NH_4^+ and Cl^- in water:

$$NH_4^+Cl^-(s) \rightarrow NH_4^+(aq) + Cl^-(aq)$$

Conjugate Acids and Bases

The products of a reaction between an acid and a base are also classified as acids and bases:

Acetic acid   Methylamine   Acetate ion   Methylammonium ion

Acid   Base   Base   Acid

Conjugate pair

Conjugate pair

Conjugate acids and bases are related by the gain or loss of one proton. In these structures, a solid wedge is a bond coming out of the plane of the page and a dashed wedge is a bond to an atom behind the page.

Acetate is a base because it can accept a proton to make acetic acid. Methylammonium ion is an acid because it can donate a proton and become methylamine. Acetic acid and the acetate ion are said to be a **conjugate acid-base pair.** Methylamine and methylammonium ion are likewise conjugate. *Conjugate acids and bases are related to each other by the gain or loss of one H^+.*

The Nature of H^+ and OH^-

The proton does not exist by itself in water. The simplest formula found in some crystalline salts is H_3O^+. For example, crystals of perchloric acid monohydrate contain pyramidal hydronium (also called *hydroxonium*) ions:

$$HClO_4 \cdot H_2O \text{ is really}$$

Hydronium   Perchlorate

The formula $HClO_4 \cdot H_2O$ is a way of specifying the composition of the substance when we are ignorant of its structure. A more accurate formula would be $H_3O^+ClO_4^-$.

Average dimensions of the H_3O^+ cation in many crystals are shown in Figure 6-4. In aqueous solution, H_3O^+ is tightly associated with three molecules of water through exceptionally strong hydrogen bonds (Figure 6-5). The $H_5O_2^+$ cation is another simple species in which a hydrogen ion is shared by two water molecules.[16,17]

Zundel structure of $(H_3O^+ \cdot H_2O)$

← 243 pm →

In the gas phase, H_3O^+ can be surrounded by 20 molecules of H_2O in a regular dodecahedron held together by 30 hydrogen bonds.[18] In a salt containing the discrete cation $(C_6H_6)_3H_3O^+$, and in benzene solution, hydrogen atoms of the pyramidal H_3O^+ ion are each attracted toward the center of the pi electron cloud of a benzene ring (Figure 6-6).

The ion $H_3O_2^-$ ($OH^- \cdot H_2O$) has been observed by X-ray crystallography.[19] The central $O\cdots H\cdots O$ linkage contains the shortest hydrogen bond involving H_2O that has ever been observed.

← 229 pm →

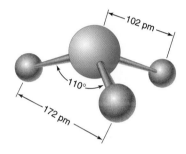

Eigen structure of H_3O^+

FIGURE 6-4 Structure of hydronium ion, H_3O^+, proposed by M. Eigen and found in many crystals.[14] The bond enthalpy (heat needed to break the OH bond) of H_3O^+ is 544 kJ/mol, about 84 kJ/mol greater than the bond enthalpy in H_2O.

FIGURE 6-5 Environment of aqueous H_3O^+.[14] Three H_2O molecules are bound to H_3O^+ by strong hydrogen bonds (dotted lines), and one H_2O (at the top) is held by weaker ion-dipole attraction (dashed line). The $O—H\cdots O$ hydrogen-bonded distance of 252 pm (picometers, 10^{-12} m) compares with an $O—H\cdots O$ distance of 283 pm between hydrogen-bonded water molecules. The discrete cation $(H_2O)_3H_3O^+$ found in some crystals is similar in structure to $(H_2O)_4H_3O^+$, with the weakly bonded H_2O at the top removed.[15]

FIGURE 6-6 $H_3O^+ \cdot 3C_6H_6$ cation found in the crystal structure of $[(C_6H_6)_3H_3O^+]$ $[CHB_{11}Cl_{11}^-]$. [From E. S. Stoyanov, K.-C. Kim, and C. A. Reed, "The Nature of the H_3O^+ Hydronium Ion in Benzene and Chlorinated Hydrocarbon Solvents," *J. Am. Chem. Soc.* **2006,** *128*, 1948.]

We will ordinarily write H^+ in most chemical equations, although we really mean H_3O^+. To emphasize the chemistry of water, we will write H_3O^+. For example, water can be either an acid or a base. Water is an acid with respect to methoxide:

$$H-\overset{..}{\underset{..}{O}}-H \; + \; CH_3-\overset{..}{\underset{..}{O}}^- \; \rightleftharpoons \; H-\overset{..}{\underset{..}{O}}^- \; + \; CH_3-\overset{..}{\underset{..}{O}}-H$$

Water　　　　Methoxide　　　　Hydroxide　　　　Methanol

But with respect to hydrogen bromide, water is a base:

$$H_2O \; + \; HBr \; \rightleftharpoons \; H_3O^+ \; + \; Br^-$$

Water　　Hydrogen　　Hydronium　　Bromide
　　　　bromide　　　　ion

Autoprotolysis

Water undergoes self-ionization, called **autoprotolysis,** in which it acts as both an acid and a base:

$$H_2O + H_2O \rightleftharpoons H_3O^+ + OH^- \qquad (6\text{-}16)$$

or

$$H_2O \rightleftharpoons H^+ + OH^- \qquad (6\text{-}17)$$

Reactions 6-16 and 6-17 mean the same thing.

Protic solvents have a reactive H^+, and all protic solvents undergo autoprotolysis. An example is acetic acid:

$$2CH_3\overset{O}{\overset{\|}{C}}OH \; \rightleftharpoons \; CH_3\overset{OH}{\underset{OH}{C^+}} \; + \; CH_3\overset{O}{\overset{\|}{C}}-O^- \quad \text{(in acetic acid)} \qquad (6\text{-}18)$$

The extent of these reactions is very small. The *autoprotolysis constants* (equilibrium constants) for Reactions 6-17 and 6-18 are 1.0×10^{-14} and 3.5×10^{-15}, respectively, at 25°C.

6-6 pH

The autoprotolysis constant for H_2O has the special symbol K_w, where "w" stands for water:

Autoprotolysis of water: $\qquad H_2O \overset{K_w}{\rightleftharpoons} H^+ + OH^- \qquad K_w = [H^+][OH^-] \qquad (6\text{-}19)$

Table 6-1 shows how K_w varies with temperature. Its value at 25.0°C is 1.01×10^{-14}.

We will write H^+ when we really mean H_3O^+.

Examples of protic solvents (acidic proton **bold**):

　H_2O　　　　CH_3CH_2OH
　Water　　　　Ethanol

Examples of **aprotic solvents** (no acidic protons):

　$CH_3CH_2OCH_2CH_3$　　　CH_3CN
　Diethyl ether　　　　Acetonitrile

Recall that H_2O (the solvent) is omitted from the equilibrium constant. The value $K_w = 1.0 \times 10^{-14}$ at 25°C is accurate enough for problems in this book.

TABLE 6-1   Temperature dependence of $K_w{}^a$

Temperature (°C)	K_w	$pK_w = -\log K_w$	Temperature (°C)	K_w	$pK_w = -\log K_w$
0	1.15×10^{-15}	14.938	40	2.88×10^{-14}	13.541
5	1.88×10^{-15}	14.726	45	3.94×10^{-14}	13.405
10	2.97×10^{-15}	14.527	50	5.31×10^{-14}	13.275
15	4.57×10^{-15}	14.340	100	5.43×10^{-13}	12.265
20	6.88×10^{-15}	14.163	150	2.30×10^{-12}	11.638
25	1.01×10^{-14}	13.995	200	5.14×10^{-12}	11.289
30	1.46×10^{-14}	13.836	250	6.44×10^{-12}	11.191
35	2.07×10^{-14}	13.685	300	3.93×10^{-12}	11.406

a. Concentrations in the product $[H^+][OH^-]$ in this table are expressed in molality rather than in molarity. Accuracy of $\log K_w$ is ±0.01. To convert molality (mol/kg) into molarity (mol/L), multiply by the density of H_2O at each temperature. At 25°C, $K_w = 10^{-13.995}$ $(mol/kg)^2(0.997\ 05\ kg/L)^2 = 10^{-13.998}(mol/L)^2$.

SOURCE: *W. L. Marshall and E. U. Franck, "Ion Product of Water Substance, 0–1 000°C, 1–10,000 Bars," J. Phys. Chem. Ref. Data **1981**, 10, 295. For values of K_w over a temperature range of 0°–800°C and a density range of 0–1.2 g/cm³, see A. V. Bandura and S. N. Lvov, "The Ionization Constant of Water over Wide Ranges of Temperature and Pressure," J. Phys. Chem. Ref. Data **2006**, 35, 15.*

EXAMPLE   Concentration of H^+ and OH^- in Pure Water at 25°C

Calculate the concentrations of H^+ and OH^- in pure water at 25°C.

Solution The stoichiometry of Reaction 6-19 tells us that H^+ and OH^- are produced in a 1:1 molar ratio. Their concentrations must be equal. Calling each concentration x, we can write

$$K_w = 1.0 \times 10^{-14} = [H^+][OH^-] = [x][x] \Rightarrow x = 1.0 \times 10^{-7}\ M$$

The concentrations of H^+ and OH^- are both 1.0×10^{-7} M in pure water.

Test Yourself Use Table 6-1 to find the $[H^+]$ in water at 100°C and at 0°C. (**Answer:** 7.4×10^{-7} and 3.4×10^{-8} M)

EXAMPLE   Concentration of OH^- When $[H^+]$ Is Known

What is the concentration of OH^- if $[H^+] = 1.0 \times 10^{-3}$ M? (From now on, assume that the temperature is 25°C unless otherwise stated.)

Solution Putting $[H^+] = 1.0 \times 10^{-3}$ M into the K_w expression gives

$$K_w = 1.0 \times 10^{-14} = (1.0 \times 10^{-3})[OH^-] \Rightarrow [OH^-] = 1.0 \times 10^{-11}\ M$$

A concentration of $[H^+] = 1.0 \times 10^{-3}$ M gives $[OH^-] = 1.0 \times 10^{-11}$ M. *As the concentration of H^+ increases, the concentration of OH^- necessarily decreases, and vice versa.* A concentration of $[OH^-] = 1.0 \times 10^{-3}$ M gives $[H^+] = 1.0 \times 10^{-11}$ M.

Test Yourself Find $[OH^-]$ if $[H^+] = 1.0 \times 10^{-4}$ M. (**Answer:** 1.0×10^{-10} M)

An approximate definition of **pH** is the negative logarithm of the H^+ concentration.

Approximate definition of pH: $pH \approx -\log[H^+]$ (6-20)

Chapter 7 defines pH more accurately in terms of *activities*, but, for many purposes, Equation 6-20 is a good definition. The measurement of pH with glass electrodes, and buffers used by the U.S. National Institute of Standards and Technology to define the pH scale, are described in Chapter 14.

In pure water at 25°C with $[H^+] = 1.0 \times 10^{-7}$ M, the pH is $-\log(1.0 \times 10^{-7}) = 7.00$. If $[OH^-] = 1.0 \times 10^{-3}$ M, then $[H^+] = 1.0 \times 10^{-11}$ M and the pH is 11.00. A useful relation between $[H^+]$ and $[OH^-]$ is

$$pH + pOH = -\log(K_w) = 14.00 \text{ at } 25°C \qquad (6\text{-}21)$$

where $pOH = -\log[OH^-]$, just as $pH = -\log[H^+]$. Equation 6-21 is a fancy way of saying that if pH = 3.58, then pOH = 14.00 – 3.58 = 10.42, or $[OH^-] = 10^{-10.42} = 3.8 \times 10^{-11}$ M.

$pH \approx -\log[H^+]$. The term "pH" was introduced in 1909 by the Danish biochemist S. P. L. Sørensen, who called it the "hydrogen ion exponent."[20]

Take the log of both sides of the K_w expression to derive Equation 6-21:

$$K_w = [H^+][OH^-]$$
$$\log K_w = \log[H^+] + \log[OH^-]$$
$$-\log K_w = -\log[H^+] - \log[OH^-]$$
$$14.00 = pH + pOH \quad \text{(at 25°C)}$$

FIGURE 6-7 pH of various substances. [From *Chem. Eng. News*, 14 September 1981.] **The most acidic rainfall (Box 14-1) is a stronger acid than lemon juice. The most acidic natural waters known are mine waters, with total dissolved metal concentrations of 200 g/L and sulfate concentrations of 760 g/L.[22] The pH of this water, -3.6, does not mean that $[H^+] = 10^{3.6}$ M $= 4\,000$ M! It means that the *activity* of H^+ (discussed in Chapter 7) is $10^{3.6}$.**

pH is measured with a *glass electrode* described in Chapter 14.

The surface of water or ice is ~2 pH units more acidic than the bulk because H_3O^+ is more stable on the surface. Surface acidity could be important to the chemistry of atmospheric clouds.[21]

A solution is **acidic** if $[H^+] > [OH^-]$. A solution is **basic** if $[H^+] < [OH^-]$. At 25°C, an acidic solution has a pH below 7 and a basic solution has a pH above 7.

$$
\begin{array}{c}
\text{pH} \quad \underset{\longleftarrow \text{Acidic} \longrightarrow \quad \uparrow \quad \longleftarrow \text{Basic} \longrightarrow}{\underline{-1 \quad 0 \quad 1 \quad 2 \quad 3 \quad 4 \quad 5 \quad 6 \quad 7 \quad 8 \quad 9 \quad 10 \quad 11 \quad 12 \quad 13 \quad 14 \quad 15}} \\
\text{Neutral}
\end{array}
$$

pH values for various common substances are shown in Figure 6-7.

Although pH generally falls in the range 0 to 14, these are not the limits of pH. A pH of -1, for example, means $-\log[H^+] = -1$; or $[H^+] = 10$ M. This concentration is attained in a concentrated solution of a strong acid such as HCl.

Is There Such a Thing as Pure Water?

In most labs, the answer is "No." Pure water at 25°C should have a pH of 7.00. Distilled water from the tap in most labs is acidic because it contains CO_2 from the atmosphere. CO_2 is an acid by virtue of the reaction

$$CO_2 + H_2O \rightleftharpoons \underset{\text{Bicarbonate}}{HCO_3^-} + H^+ \tag{6-22}$$

CO_2 can be largely removed by boiling water and then protecting it from the atmosphere.

More than a century ago, careful measurements of the conductivity of water were made by F. Kohlrausch and his students. To remove impurities, they found it necessary to distill water *42 consecutive times* under vacuum to reduce conductivity to a limiting value.

6-7 Strengths of Acids and Bases

Acids and bases are commonly classified as strong or weak, depending on whether they react nearly "completely" or only "partially" to produce H^+ or OH^-. Although there is no sharp distinction between weak and strong, some acids or bases react so completely that they are easily classified as strong—and, by convention, everything else is termed weak.

Strong Acids and Bases

Common strong acids and bases are listed in Table 6-2, which you need to memorize. By definition, a strong acid or base is completely dissociated in aqueous solution. That is, the equilibrium constants for the following reactions are large.

$$HCl(aq) \rightleftharpoons H^+ + Cl^-$$

$$KOH(aq) \rightleftharpoons K^+ + OH^-$$

Virtually no undissociated HCl or KOH exists in aqueous solution. Demonstration 6-2 shows a consequence of the strong-acid behavior of HCl.

TABLE 6-2 Common strong acids and bases

Formula	Name
Acids	
HCl	Hydrochloric acid (hydrogen chloride)
HBr	Hydrogen bromide
HI	Hydrogen iodide
$H_2SO_4{}^a$	Sulfuric acid
HNO_3	Nitric acid
$HClO_4$	Perchloric acid
Bases	
LiOH	Lithium hydroxide
NaOH	Sodium hydroxide
KOH	Potassium hydroxide
RbOH	Rubidium hydroxide
CsOH	Cesium hydroxide
R_4NOH^b	Quaternary ammonium hydroxide

a. For H_2SO_4, only the first proton ionization is complete. Dissociation of the second proton has an equilibrium constant of 1.0×10^{-2}.

b. This is a general formula for any hydroxide salt of an ammonium cation containing four organic groups. An example is tetrabutylammonium hydroxide: $(CH_3CH_2CH_2CH_2)_4N^+OH^-$.

The complete dissociation of HCl into H^+ and Cl^- makes $HCl(g)$ extremely soluble in water.

$$HCl(g) \rightleftharpoons HCl(aq) \qquad \text{(A)}$$

$$\underline{HCl(aq) \rightleftharpoons H^+(aq) + Cl^-(aq)} \qquad \text{(B)}$$

Net reaction: $\qquad HCl(g) \rightleftharpoons H^+(aq) + Cl^-(aq) \qquad \text{(C)}$

Because the equilibrium of Reaction B lies far to the right, it pulls Reaction A to the right as well.

Challenge The standard free energy change (ΔG°) for Reaction C is -36.0 kJ/mol. Show that the equilibrium constant is 2.0×10^6.

The extreme solubility of $HCl(g)$ in water is the basis for the HCl fountain,[23] assembled as shown below. In Figure *a*, an inverted 250-mL round-bottom flask containing air is set up with its inlet tube leading to a source of $HCl(g)$ and its outlet tube directed into an inverted bottle of water. As HCl is admitted to the flask, air is displaced. When the bottle is filled with air, the flask is filled mostly with $HCl(g)$.

The hoses are disconnected and replaced with a beaker of indicator and a rubber bulb (Figure *b*). For an indicator, we use slightly alkaline, commercial methyl purple solution, which is green above pH 5.4 and purple below pH 4.8. When ~1 mL of water is squirted from the rubber bulb into the flask, a vacuum is created and indicator solution is drawn up into the flask, making a fascinating fountain (Color Plate 1).

Question Why is a vacuum created when water is squirted into the flask, and why does the indicator change color when it enters the flask?

Constriction

250-mL round-bottom flask

Glass tubes

Rubber stopper

250-mL bottle

Hoses

HCl(g) in from tank

Beaker containing your favorite indicator

2-mL rubber bulb

(a) Water

(b)

Even though the hydrogen halides HCl, HBr, and HI are strong acids, HF is *not* a strong acid, as explained in Box 6-3. For most purposes, the hydroxides of the alkaline earth metals (Mg^{2+}, Ca^{2+}, Sr^{2+}, and Ba^{2+}) can be considered to be strong bases, although they are far less soluble than alkali metal hydroxides and have some tendency to form MOH^+ complexes (Table 6-3). The strongest known base is the gas-phase molecule LiO^-.[25]

Weak Acids and Bases

All weak acids, denoted HA, react with water by donating a proton to H_2O:

Dissociation of weak acid: $\qquad HA + H_2O \overset{K_a}{\rightleftharpoons} H_3O^+ + A^- \qquad$ (6-23)

which means exactly the same as

Dissociation of weak acid: $\qquad HA \overset{K_a}{\rightleftharpoons} H^+ + A^- \qquad K_a = \dfrac{[H^+][A^-]}{[HA]} \qquad$ (6-24)

The equilibrium constant is called K_a, the **acid dissociation constant.** By definition, a weak acid is one that is only partially dissociated in water, so K_a is "small."

TABLE 6-3 Equilibria of alkaline earth metal hydroxides

$M(OH)_2(s) \rightleftharpoons M^{2+} + 2OH^-$		
$K_{sp} = [M^{2+}][OH^-]^2$		
$M^{2+} + OH^- \rightleftharpoons MOH^+$		
$K_1 = [MOH^+]/[M^{2+}][OH^-]$		
Metal	**log K_{sp}**	**log K_1**
Mg^{2+}	-11.15	2.58
Ca^{2+}	-5.19	1.30
Sr^{2+}	—	0.82
Ba^{2+}	—	0.64

NOTE: 25°C and ionic strength = 0.

Acid dissociation constant: $K_a = \dfrac{[H^+][A^-]}{[HA]}$.

The hydrogen halides HCl, HBr, and HI are all strong acids, which means that the reactions

$$HX(aq) + H_2O \rightarrow H_3O^+ + X^-$$

(X = Cl, Br, I) all go to completion. Why, then, does HF behave as a weak acid?

The answer is odd. First, HF does completely give up its proton to H_2O:

$$HF(aq) \rightarrow \underset{\substack{\text{Hydronium} \\ \text{ion}}}{H_3O^+} + \underset{\substack{\text{Fluoride} \\ \text{ion}}}{F^-}$$

But fluoride forms the strongest hydrogen bond of any ion. The hydronium ion remains tightly associated with F^- through a hydrogen bond. We call such an association an **ion pair.**

$$\underset{\text{An ion pair}}{H_3O^+ + F^- \rightleftharpoons F^- \cdots H_3O^+}$$

Ion pairs are common in aqueous solutions of any ion with a charge greater than 1. Ion pairs are the rule in nonaqueous solvents, which cannot promote ion dissociation as well as water.

Thus, HF does not behave as a strong acid because F^- and H_3O^+ remain associated with each other. Dissolving one mole of the strong acid HCl in water creates one mole of free H_3O^+. Dissolving one mole of the "weak" acid HF in water creates little free H_3O^+.

HF is not unique in its propensity to form ion pairs. Many moderately strong acids, such as those below, are thought to exist predominantly as ion pairs in aqueous solution (HA + H_2O ⇌ $A^- \cdots H_3O^+$).[24]

CF_3CO_2H
Trifluoroacetic acid
$K_a = 0.31$

Squaric acid
$K_a = 0.29$

Weak bases, B, react with water by abstracting a proton from H_2O:

Base hydrolysis constant: $K_b = \dfrac{[BH^+][OH^-]}{[B]}$

Hydrolysis refers to any reaction with water.

Base hydrolysis:

$$B + H_2O \overset{K_b}{\rightleftharpoons} BH^+ + OH^- \qquad K_b = \frac{[BH^+][OH^-]}{B} \qquad (6\text{-}25)$$

The equilibrium constant K_b is the **base hydrolysis constant,** which is "small" for a weak base.

Common Classes of Weak Acids and Bases

Acetic acid is a typical weak acid.

$$\underset{\substack{\text{Acetic acid} \\ (\text{HA})}}{CH_3-C(=O)(O-H)} \rightleftharpoons \underset{\substack{\text{Acetate} \\ (A^-)}}{CH_3-C(=O)(O^-)} + H^+ \qquad K_a = 1.75 \times 10^{-5} \qquad (6\text{-}26)$$

Acetic acid is a representative carboxylic acid, which has the general formula RCO_2H, where R is an organic substituent. *Most* **carboxylic acids** *are weak acids, and most* **carboxylate anions** *are weak bases.*

A carboxylic acid
(weak acid, HA)

A carboxylate anion
(weak base, A^-)

Methylamine is a typical weak base.

Carboxylic acids (RCO_2H) and ammonium ions (R_3NH^+) are weak acids.

Carboxylate anions (RCO_2^-) and amines (R_3N) are weak bases.

$$\underset{\substack{\text{Methylamine} \\ B}}{} + H_2O \rightleftharpoons \underset{\substack{\text{Methylammonium ion} \\ BH^+}}{} + OH^- \qquad K_b = 4.47 \times 10^{-4} \qquad (6\text{-}27)$$

Amines are nitrogen-containing compounds:

$\overset{..}{R}NH_2$ a primary amine RNH_3^+

$R_2\overset{..}{N}H$ a secondary amine $R_2NH_2^+$ } ammonium ions

$R_3\overset{..}{N}$ a tertiary amine R_3NH^+

Li+ 13.64	Be													
Na+ 13.9	Mg2+ 11.4											Al3+ 5.00		
K	Ca2+ 12.70	Sc3+ 4.3	Ti3+ 1.3	VO2+ 5.7	Cr2+ 5.5^a / Cr3+ 3.66	Mn2+ 10.6	Fe2+ 9.4 / Fe3+ 2.19	Co2+ 9.7 / Co3+ 0.5^b	Ni2+ 9.9	Cu2+ 7.5	Zn2+ 9.0	Ga3+ 2.6	Ge	
Rb	Sr 13.18	Y3+ 7.7	Zr4+ −0.3	Nb	Mo	Tc	Ru	Rh3+ 3.33^c	Pd2+ 1.0	Ag+ 12.0	Cd2+ 10.1	In3+ 3.9	Sn2+ 3.4	Sb
Cs	Ba2+ 13.36	La3+ 8.5	Hf	Ta	W	Re	Os	Ir	Pt	Au	Hg$_2^{2+}$ 5.3^d / Hg2+ 3.40	Tl+ 13.21	Pb2+ 7.6	Bi3+ 1.1

Stronger acid →

Ce3+ 9.1^b	Pr3+ 9.4^b	Nd3+ 8.7^b	Pm	Sm3+ 8.6^b	Eu3+ 8.6^d	Gd3+ 9.1^b	Tb3+ 8.4^d	Dy3+ 8.4^d	Ho3+ 8.3	Er3+ 9.1^b	Tm3+ 8.2^d	Yb3+ 8.4^b	Lu3+ 8.2^d

Ionic strength = 0 unless noted by superscript
a. Ionic strength = 1 M b. Ionic strength = 3 M c. Ionic strength = 2.5 M d. Ionic strength = 0.5 M

FIGURE 6-8 Acid dissociation constants ($-\log K_a$) for aqueous metal ions: $M^{n+} + H_2O \overset{K_a}{\rightleftharpoons} MOH^{(n-1)+} + H^+$. For example, for Li+, $K_a = 10^{-13.64}$. In Chapter 8, we will learn that the numbers in this table are called pK_a. Darkest shades are strongest acids. [Data from R. M. Smith, A. E. Martell, and R. J. Motekaitis, *NIST Critical Stability Constants of Metal Complexes Database 46* (Gaithersburg, MD: National Institute of Standards and Technology, 2001).]

Amines *are weak bases, and* **ammonium ions** *are weak acids*. The "parent" of all amines is ammonia, NH_3. When a base such as methylamine reacts with water, the product is the conjugate acid. That is, methylammonium ion produced in Reaction 6-27 is a weak acid:

$$\underset{BH^+}{CH_3\overset{+}{N}H_3} \overset{K_a}{\rightleftharpoons} \underset{B}{CH_3\ddot{N}H_2} + H^+ \qquad K_a = 2.26 \times 10^{-11} \qquad (6\text{-}28)$$

The methylammonium ion is the conjugate acid of methylamine.

You should learn to recognize whether a compound is acidic or basic. The salt methylammonium chloride, for example, dissociates in aqueous solution to give methylammonium cation and chloride:

$$\underset{\substack{\text{Methylammonium} \\ \text{chloride}}}{CH_3\overset{+}{N}H_3Cl^-(s)} \rightarrow \underset{\substack{\text{Methylammonium} \\ \text{ion}}}{CH_3\overset{+}{N}H_3(aq)} + Cl^-(aq) \qquad (6\text{-}29)$$

Methylammonium ion, being the conjugate acid of methylamine, is a weak acid (Reaction 6-28). Chloride is the conjugate base of HCl, a strong acid. In other words, Cl^- *has virtually no tendency to associate with* H^+, or else HCl would not be a strong acid. Methylammonium chloride is acidic because methylammonium ion is an acid and Cl^- is not a base.

Metal cations, M^{n+} act as weak acids by *acid hydrolysis* to form $M(OH)^{(n-1)+}$.[26] Figure 6-8 shows acid dissociation constants for the reaction

$$M^{n+} + H_2O \overset{K_a}{\rightleftharpoons} MOH^{(n-1)+} + H^+$$

Monovalent metal ions are very weak acids (Na+, $K_a = 10^{-13.9}$). Divalent ions tend to be stronger (Fe^{2+}, $K_a = 10^{-9.4}$) and trivalent ions are stronger yet (Fe^{3+}, $K_a = 10^{-2.19}$).

Polyprotic Acids and Bases

Polyprotic acids and bases are compounds that can donate or accept more than one proton. For example, oxalic acid is diprotic and phosphate is tribasic:

$$\underset{\substack{\text{Oxalic} \\ \text{acid}}}{\overset{\overset{O\ O}{||\ ||}}{HOCCOH}} \rightleftharpoons H^+ + \underset{\substack{\text{Monohydrogen} \\ \text{oxalate}}}{\overset{\overset{O\ O}{||\ ||}}{{}^-OCCOH}} \qquad K_{a1} = 5.37 \times 10^{-2} \qquad (6\text{-}30)$$

Although we will usually write a **base** as **B** and an **acid** as **HA**, it is important to realize that BH^+ is also an **acid** and A^- is also a **base**.

Methylammonium chloride is a weak acid because

1. It dissociates into $CH_3NH_3^+$ and Cl^-.
2. $CH_3NH_3^+$ is a weak acid, being conjugate to CH_3NH_2, a weak base.
3. Cl^- has no basic properties. It is conjugate to HCl, a strong acid. That is, HCl dissociates completely.

Challenge Phenol (C_6H_5OH) is a weak acid. Explain why a solution of the ionic compound potassium phenolate ($C_6H_5O^-K^+$) is basic.

Aqueous metal ions are associated with (*hydrated* by) several H_2O molecules, so a more accurate way to write the acid dissociation reaction is

$$M(H_2O)_x^{n+} \rightleftharpoons M(H_2O)_{x-1}(OH)^{(n-1)} + H^+$$

$$\text{}^{-}OCCOH \rightleftharpoons H^+ + {}^{-}OCCO^{-} \qquad K_{a2} = 5.42 \times 10^{-5} \quad (6\text{-}31)$$
Oxalate

$$\text{Phosphate} + H_2O \rightleftharpoons \text{Monohydrogen phosphate} + OH^{-} \qquad K_{b1} = 2.3 \times 10^{-2} \quad (6\text{-}32)$$

$$\text{} + H_2O \rightleftharpoons \text{Dihydrogen phosphate} + OH^{-} \qquad K_{b2} = 1.60 \times 10^{-7} \quad (6\text{-}33)$$

$$\text{} + H_2O \rightleftharpoons \text{Phosphoric acid} + OH^{-} \qquad K_{b3} = 1.42 \times 10^{-12} \quad (6\text{-}34)$$

Standard notation for successive acid dissociation constants of a polyprotic acid is K_1, K_2, K_3, and so on, with the subscript "a" usually omitted. We retain or omit the subscript as dictated by clarity. For successive base hydrolysis constants, we retain the subscript "b." The preceding examples illustrate that K_{a1} (or K_1) *refers to the acidic species with the most protons, and K_{b1} refers to the basic species with the least number of protons.* Carbonic acid, a very important diprotic carboxylic acid derived from CO_2, is described in Box 6-4.

BOX 6-4 Carbonic Acid[27]

Carbonic acid is formed by the reaction of carbon dioxide with water:

$$CO_2(g) \rightleftharpoons CO_2(aq) \qquad K = \frac{[CO_2(aq)]}{P_{CO_2}} = 0.034\,4$$

$$CO_2(aq) + H_2O \rightleftharpoons \underset{\text{Carbonic acid}}{\text{}} \qquad K = \frac{[H_2CO_3]}{[CO_2(aq)]} \approx 0.002$$

$$H_2CO_3 \rightleftharpoons \underset{\text{Bicarbonate}}{HCO_3^{-}} + H^+ \qquad K_{a1} = 4.46 \times 10^{-7}$$

$$HCO_3^{-} \rightleftharpoons \underset{\text{Carbonate}}{CO_3^{2-}} + H^+ \qquad K_{a2} = 4.69 \times 10^{-11}$$

Its behavior as a diprotic acid appears anomalous at first, because the value of K_{a1} is about 10^2 to 10^4 times smaller than K_a for other carboxylic acids.

CH_3CO_2H
Acetic acid
$K_a = 1.75 \times 10^{-5}$

HCO_2H
Formic acid
$K_a = 1.80 \times 10^{-4}$

$N\equiv CCH_2CO_2H$
Cyanoacetic acid
$K_a = 3.37 \times 10^{-3}$

$HOCH_2CO_2H$
Glycolic acid
$K_a = 1.48 \times 10^{-4}$

The reason for this anomaly is not that H_2CO_3 is unusual but, rather, that the value commonly given for K_{a1} applies to the equation

$$\text{All dissolved } CO_2 \rightleftharpoons HCO_3^{-} + H^+$$
$$(= CO_2(aq) + H_2CO_3)$$

$$K_{a1} = \frac{[HCO_3^{-}][H^+]}{[CO_2(aq) + H_2CO_3]} = 4.46 \times 10^{-7}$$

Only about 0.2% of dissolved CO_2 is in the form H_2CO_3. When the true value of $[H_2CO_3]$ is used instead of the value $[H_2CO_3 + CO_2(aq)]$, the value of the equilibrium constant becomes

$$K_{a1} = \frac{[HCO_3^{-}][H^+]}{[H_2CO_3]} = 2 \times 10^{-4}$$

The hydration of CO_2 (reaction of CO_2 with H_2O) and dehydration of H_2CO_3 are slow reactions that can be demonstrated in a classroom.[27] Living cells utilize the enzyme *carbonic anhydrase* to speed the rate at which H_2CO_3 and CO_2 equilibrate in order to process this key metabolite. The enzyme provides an environment just right for the reaction of CO_2 with OH^{-}, lowering the *activation energy* (the energy barrier for the reaction) from 50 down to 26 kJ/mol and increasing the rate of reaction by more than a factor of 10^6.

Carbonic acid has limited stability in aqueous solution and is not well characterized. The dimer $(H_2CO_3)_2$ or oligomers $(H_2CO_3)_n$ appear to exist in the solid state.[28]

Relation Between K_a and K_b

A most important relation exists between K_a and K_b of a conjugate acid-base pair in aqueous solution. We can derive this result with the acid HA and its conjugate base A$^-$.

$$\text{HA} \rightleftharpoons \text{H}^+ + \text{A}^- \qquad K_a = \frac{[\text{H}^+][\text{A}^-]}{[\text{HA}]}$$

$$\underline{\text{A}^- + \text{H}_2\text{O} \rightleftharpoons \text{HA} + \text{OH}^- \qquad K_b = \frac{[\text{HA}][\text{OH}^-]}{[\text{A}^-]}}$$

$$\text{H}_2\text{O} \rightleftharpoons \text{H}^+ + \text{OH}^- \qquad K_w = K_a \cdot K_b$$

$$= \frac{[\text{H}^+][\text{A}^-]}{[\text{HA}]} \cdot \frac{[\text{HA}][\text{OH}^-]}{[\text{A}^-]}$$

When the reactions are added, their equilibrium constants are multiplied to give

Relation between K_a and K_b
for a conjugate pair:
$$K_a \cdot K_b = K_w \qquad (6\text{-}35)$$

$K_a \cdot K_b = K_w$ for a conjugate acid-base pair in aqueous solution.

Equation 6-35 applies to any acid and its conjugate base in aqueous solution.

EXAMPLE **Finding K_b for the Conjugate Base**

K_a for acetic acid is 1.75×10^{-5} (Reaction 6-26). Find K_b for acetate ion.

Solution This one is trivial:[†]

$$K_b = \frac{K_w}{K_a} = \frac{1.0 \times 10^{-14}}{1.75 \times 10^{-5}} = 5.7 \times 10^{-10}$$

Test Yourself K_a for chloroacetic acid is 1.36×10^{-3}. Find K_b for chloroacetate ion. (**Answer:** 7.4×10^{-12})

EXAMPLE **Finding K_a for the Conjugate Acid**

K_b for methylamine is 4.47×10^{-4} (Reaction 6-27). Find K_a for methylammonium ion.

Solution Once again,

$$K_a = \frac{K_w}{K_b} = 2.2 \times 10^{-11}$$

Test Yourself K_b for dimethylamine is 5.9×10^{-4}. Find K_a for dimethylammonium ion. (**Answer:** 1.7×10^{-11})

For a diprotic acid, we can derive relationships between each of two acids and their conjugate bases:

$$\text{H}_2\text{A} \rightleftharpoons \text{H}^+ + \text{HA}^- \quad K_{a1} \qquad\qquad \text{HA}^- \rightleftharpoons \text{H}^+ + \text{A}^{2-} \quad K_{a2}$$

$$\underline{\text{HA}^- + \text{H}_2\text{O} \rightleftharpoons \text{H}_2\text{A} + \text{OH}^- \quad K_{b2}} \qquad \underline{\text{A}^{2-} + \text{H}_2\text{O} \rightleftharpoons \text{HA}^- + \text{OH}^- \quad K_{b1}}$$

$$\text{H}_2\text{O} \rightleftharpoons \text{H}^+ + \text{OH}^- \quad K_w \qquad\qquad\quad \text{H}_2\text{O} \rightleftharpoons \text{H}^+ + \text{OH}^- \quad K_w$$

The final results are

General relation
between K_a and K_b:
$$K_{a1} \cdot K_{b2} = K_w \qquad (6\text{-}36)$$

$$K_{a2} \cdot K_{b1} = K_w \qquad (6\text{-}37)$$

[†] In this text, we use $K_w = 10^{-14.00} = 1.0 \times 10^{-14}$ at 25°C. The more accurate value from Table 6-1 is $K_w = 10^{-13.995}$. For acetic acid with $K_a = 10^{-4.756}$, the accurate value of K_b is $10^{-(13.995 - 4.756)} = 10^{-9.239} = 5.77 \times 10^{-10}$.

Challenge Derive the following results for a triprotic acid:

$$K_{a1} \cdot K_{b3} = K_w \tag{6-38}$$

$$K_{a2} \cdot K_{b2} = K_w \tag{6-39}$$

$$K_{a3} \cdot K_{b1} = K_w \tag{6-40}$$

Shorthand for Organic Structures

We are beginning to encounter organic (carbon-containing) compounds in this book. Chemists and biochemists use simple conventions for drawing molecules to avoid writing every atom. Each vertex of a structure is understood to be a carbon atom, unless otherwise labeled. In the shorthand, we usually omit bonds from carbon to hydrogen. Carbon forms four chemical bonds. If you see carbon forming fewer than four bonds, the remaining bonds are assumed to go to hydrogen atoms that are not written. Here is an example:

Epinephrine (also called adrenaline)

Shorthand drawing of epinephrine

Benzene, C_6H_6, has two equivalent resonance structures, so all C–C bonds are equivalent. We often draw benzene rings with a circle in place of three double bonds.

Benzene
C_6H_6

The shorthand shows that the carbon atom at the top right of the six-membered benzene ring forms three bonds to other carbon atoms (one single bond and one double bond), so there must be a hydrogen atom attached to this carbon atom. The carbon atom at the left side of the benzene ring forms three bonds to other carbon atoms and one bond to an oxygen atom. There is no hidden hydrogen atom attached to this carbon. In the CH_2 group adjacent to nitrogen, both hydrogen atoms are omitted in the shorthand structure.

Terms to Understand

acid	Brønsted-Lowry base	entropy	overall formation constant
acid dissociation constant (K_a)	carboxylate anion	equilibrium constant	pH
acidic solution	carboxylic acid	exothermic	polyprotic acids
amine	common ion effect	Gibbs free energy	polyprotic bases
ammonium ion	complex ion	hydronium ion	protic solvent
aprotic solvent	conjugate acid-base pair	ion pair	reaction quotient
autoprotolysis	coprecipitation	Le Châtelier's principle	salt
base	cumulative formation constant	Lewis acid	saturated solution
base hydrolysis constant (K_b)	disproportionation	Lewis base	solubility product
basic solution	endothermic	ligand	standard state
Brønsted-Lowry acid	enthalpy change	neutralization	stepwise formation constant

Summary

For the reaction $aA + bB \rightleftharpoons cC + dD$, the equilibrium constant is $K = [C]^c[D]^d/[A]^a[B]^b$. Solute concentrations should be expressed in moles per liter; gas concentrations should be in bars; and the concentrations of pure solids, liquids, and solvents are omitted. If the direction of a reaction is changed, $K' = 1/K$. If two reactions are added, $K_3 = K_1K_2$. The equilibrium constant can be calculated from the free-energy change for a chemical reaction: $K = e^{-\Delta G°/RT}$. The equation $\Delta G = \Delta H - T\Delta S$ summarizes the observations that a

reaction is favored if it liberates heat (exothermic, negative ΔH) or increases disorder (positive ΔS). Le Châtelier's principle predicts the effect on a chemical reaction when reactants or products are added or temperature is changed. The reaction quotient, Q, tells how a system must change to reach equilibrium.

The solubility product is the equilibrium constant for the dissolution of a solid salt into its constituent ions in aqueous solution. The common ion effect is the observation that, if one of the ions of that salt is already present in the solution, the solubility of a salt is decreased. Sometimes, we can selectively precipitate one ion from a solution containing other ions by adding a suitable counterion. At high concentration of ligand, a precipitated metal ion may redissolve by forming soluble complex ions. In a metal-ion complex, the metal is a Lewis acid (electron pair acceptor) and the ligand is a Lewis base (electron pair donor).

Brønsted-Lowry acids are proton donors, and Brønsted-Lowry bases are proton acceptors. An acid increases the concentration of H_3O^+ in aqueous solution, and a base increases the concentration of OH^-. An acid-base pair related through the gain or loss of a single proton is described as conjugate. When a proton is transferred from one molecule to another molecule of a protic solvent, the reaction is called autoprotolysis.

The definition $pH = -\log[H^+]$ will be modified to include activity later. K_a is the equilibrium constant for the dissociation of an acid: $HA + H_2O \rightleftharpoons H_3O^+ + A^-$. The base hydrolysis constant for the reaction $B + H_2O \rightleftharpoons BH^+ + OH^-$ is K_b. When either K_a

or K_b is large, the acid or base is said to be strong; otherwise, the acid or base is weak. Common strong acids and bases are listed in Table 6-2, which you should memorize. The most common weak acids are carboxylic acids (RCO_2H), and the most common weak bases are amines (R_3N:). Carboxylate anions (RCO_2^-) are weak bases, and ammonium ions (R_3NH^+) are weak acids. Metal cations are weak acids. For a conjugate acid-base pair in water, $K_a \cdot K_b = K_w$. For polyprotic acids, we denote successive acid dissociation constants as $K_{a1}, K_{a2}, K_{a3}, \cdots$, or just $K_1, K_2, K_3, \cdots$. For polybasic species, we denote successive hydrolysis constants $K_{b1}, K_{b2}, K_{b3}, \cdots$. For a diprotic system, the relations between successive acid and base equilibrium constants are $K_{a1} \cdot K_{b2} = K_w$ and $K_{a2} \cdot K_{b1} = K_w$. For a triprotic system, the relations are $K_{a1} \cdot K_{b3} = K_w$, $K_{a2} \cdot K_{b2} = K_w$, and $K_{a3} \cdot K_{b1} = K_w$.

In the chemists' shorthand for organic structures, each vertex is a carbon atom. If fewer than four bonds to that carbon are shown, it is understood that H atoms are attached to the carbon so that it makes four bonds.

Exercises

6-A. Consider the following equilibria in aqueous solution:

(1) $Ag^+ + Cl^- \rightleftharpoons AgCl(aq)$ $K = 2.0 \times 10^3$

(2) $AgCl(aq) + Cl^- \rightleftharpoons AgCl_2^-$ $K = 9.3 \times 10^1$

(3) $AgCl(s) \rightleftharpoons Ag^+ + Cl^-$ $K = 1.8 \times 10^{-10}$

(a) Calculate the numerical value of the equilibrium constant for the reaction $AgCl(s) \rightleftharpoons AgCl(aq)$.

(b) Calculate the concentration of $AgCl(aq)$ in equilibrium with excess undissolved solid AgCl.

(c) Find the numerical value of K for the reaction $AgCl_2^- \rightleftharpoons AgCl(s) + Cl^-$.

6-B. Reaction 6-7 is allowed to come to equilibrium in a solution initially containing 0.010 0 M BrO_3^-, 0.010 0 M Cr^{3+}, and 1.00 M H^+. To find the concentrations at equilibrium, we can construct a table showing initial and final concentrations (see below). We use the stoichiometry coefficients of the reaction to say that if x mol of Br^- are created, then we must also make x mol of $Cr_2O_7^{2-}$ and $8x$ mol of H^+. To produce x mol of Br^-, we must have consumed x mol of BrO_3^- and $2x$ mol of Cr^{3+}.

(a) Write the equilibrium constant expression that you would use to solve for x to find the concentrations at equilibrium. Do not try to solve the equation.

(b) Because $K = 1 \times 10^{11}$, we suppose that the reaction will go nearly "to completion." That is, we expect both the concentration of Br^- and $Cr_2O_7^{2-}$ to be close to 0.005 00 M at equilibrium. (Why?) That is, $x \approx 0.005\ 00$ M. With this value of x, $[H^+] = 1.00 + 8x = 1.04$ M and $[BrO_3^-] = 0.010\ 0 - x = 0.005\ 0$ M. However, we cannot say $[Cr^{3+}] = 0.010\ 0 - 2x = 0$, because there must be some small concentration of Cr^{3+} at equilibrium. Write $[Cr^{3+}]$ for the concentration of Cr^{3+} and solve for $[Cr^{3+}]$. The *limiting reagent* in this example is Cr^{3+}. The reaction uses up Cr^{3+} before consuming BrO_3^-.

6-C. Find $[La^{3+}]$ in the solution when excess solid lanthanum iodate, $La(IO_3)_3$, is stirred with 0.050 M $LiIO_3$ until the system reaches equilibrium. Assume that IO_3^- from $La(IO_3)_3$ is negligible compared with IO_3^- from $LiIO_3$.

6-D. Which will be more soluble (moles of metal dissolved per liter of solution), $Ba(IO_3)_2$ ($K_{sp} = 1.5 \times 10^{-9}$) or $Ca(IO_3)_2$ ($K_{sp} = 7.1 \times 10^{-7}$)? Give an example of a chemical reaction that might occur that would reverse the predicted solubilities.

6-E. Fe(III) precipitates from acidic solution by addition of OH^- to form $Fe(OH)_3(s)$. At what concentration of OH^- will [Fe(III)] be reduced to 1.0×10^{-10} M? If Fe(II) is used instead, what concentration of OH^- will reduce [Fe(II)] to 1.0×10^{-10} M?

6-F. Is it possible to precipitate 99.0% of 0.010 M Ce^{3+} by adding oxalate ($C_2O_4^{2-}$) without precipitating 0.010 M Ca^{2+}?

$$CaC_2O_4 \qquad K_{sp} = 1.3 \times 10^{-8}$$
$$Ce_2(C_2O_4)_3 \qquad K_{sp} = 5.9 \times 10^{-30}$$

6-G. For a solution of Ni^{2+} and ethylenediamine, the following equilibrium constants apply at 20°C:

$$Ni^{2+} + \underset{\substack{\text{Ethylenediamine}\\\text{(abbreviated en)}}}{H_2NCH_2CH_2NH_2} \rightleftharpoons Ni(en)^{2+} \quad \log K_1 = 7.52$$

$$Ni(en)^{2+} + en \rightleftharpoons Ni(en)_2^{2+} \qquad \log K_2 = 6.32$$

$$Ni(en)_2^{2+} + en \rightleftharpoons Ni(en)_3^{2+} \qquad \log K_3 = 4.49$$

Calculate the concentration of free Ni^{2+} in a solution prepared by mixing 0.100 mol of en plus 1.00 mL of 0.010 0 M Ni^{2+} and diluting to 1.00 L with dilute base (which keeps all the en in its unprotonated form. Assume that nearly all nickel is in the form $Ni(en)_3^{2+}$, so $[Ni(en)_3^{2+}] = 1.00 \times 10^{-5}$ M. Calculate the concentrations of $Ni(en)^{2+}$ and $Ni(en)_2^{2+}$ to verify that they are negligible in comparison with $Ni(en)_3^{2+}$.

	BrO_3^-	$+$	$2Cr^{3+}$	$+ 4H_2O \rightleftharpoons$	Br^-	$+$	$Cr_2O_7^{2-}$	$+$	$8H^+$
Initial concentration	0.010 0		0.010 0						1.00
Final concentration	0.010 0 − x		0.010 0 − 2x		x		x		1.00 + 8x

6-H. If each compound is dissolved in water, will the solution be acidic, basic, or neutral?

(a) Na^+Br^-
(b) $Na^+CH_3CO_2^-$
(c) $NH_4^+Cl^-$
(d) K_3PO_4
(e) $(CH_3)_4N^+Cl^-$
(f) $(CH_3)_4N^+$⟨benzene ring⟩$-CO_2^-$
(g) $Fe(NO_3)_3$

6-I. Succinic acid dissociates in two steps:

$$HOCCH_2CH_2COH \underset{}{\overset{K_1}{\rightleftharpoons}} HOCCH_2CH_2CO^- + H^+$$
$$K_1 = 6.2 \times 10^{-5}$$

$$HOCCH_2CH_2CO^- \underset{}{\overset{K_2}{\rightleftharpoons}} {}^-OCCH_2CH_2CO^- + H^+$$
$$K_2 = 2.3 \times 10^{-6}$$

Calculate K_{b1} and K_{b2} for the following reactions:

$${}^-OCCH_2CH_2CO^- + H_2O \overset{K_{b1}}{\rightleftharpoons} HOCCH_2CH_2CO^- + OH^-$$

$$HOCCH_2CH_2CO^- + H_2O \overset{K_{b2}}{\rightleftharpoons} HOCCH_2CH_2COH + OH^-$$

6-J. Histidine is a triprotic amino acid:

$$K_1 = 3 \times 10^{-2}$$
$$K_2 = 8.5 \times 10^{-7}$$
$$K_3 = 4.6 \times 10^{-10}$$

What is the value of the equilibrium constant for the reaction

HC—CH₂—(imidazole, NH_3^+) + H_2O ⇌ HC—CH₂—(imidazole, H^+, NH_3^+) + OH^-

6-K. (a) From K_w in Table 6-1, calculate the pH of pure water at $0°$, $20°$, and $40°C$.

(b) For the reaction $D_2O \rightleftharpoons D^+ + OD^-$, $K = [D^+][OD^-] = 1.35 \times 10^{-15}$ at $25°C$. In this equation, D stands for deuterium, which is the isotope ^2H. What is the pD $(= -\log[D^+])$ for neutral D_2O?

Problems

Equilibrium and Thermodynamics

6-1. To evaluate the equilibrium constant in Equation 6-2, we must express concentrations of solutes in mol/L, gases in bars, and omit solids, liquids, and solvents. Explain why.

6-2. Why do we say that the equilibrium constant for the reaction $H_2O \rightleftharpoons H^+ + OH^-$ (or any other reaction) is dimensionless?

6-3. Explain the statement that predictions about the direction of a reaction based on Gibbs free energy or Le Châtelier's principle are *thermodynamic*, not *kinetic*.

6-4. Write the expression for the equilibrium constant for each of the following reactions. Write the pressure of a gaseous molecule, X, as P_X.

(a) $3Ag^+(aq) + PO_4^{3-}(aq) \rightleftharpoons Ag_3PO_4(s)$
(b) $C_6H_6(l) + \frac{15}{2}O_2(g) \rightleftharpoons 3H_2O(l) + 6CO_2(g)$

6-5. For the reaction $2A(g) + B(aq) + 3C(l) \rightleftharpoons D(s) + 3E(g)$, the concentrations at equilibrium are found to be

A: 2.8×10^3 Pa   C: 12.8 M   E: 3.6×10^4 Torr
B: 1.2×10^{-2} M   D: 16.5 M

Find the numerical value of the equilibrium constant that would appear in a conventional table of equilibrium constants.

6-6. From the equations

$$HOCl \rightleftharpoons H^+ + OCl^- \qquad K = 3.0 \times 10^{-8}$$
$$HOCl + OBr^- \rightleftharpoons HOBr + OCl^- \qquad K = 15$$

find the value of K for the reaction $HOBr \rightleftharpoons H^+ + OBr^-$.

6-7. (a) A favorable entropy change occurs when ΔS is positive. Does the order of the system increase or decrease when ΔS is positive?

(b) A favorable enthalpy change occurs when ΔH is negative. Does the system absorb heat or give off heat when ΔH is negative?

(c) Write the relation between ΔG, ΔH, and ΔS. Use the results of parts (a) and (b) to state whether ΔG must be positive or negative for a spontaneous change.

6-8. For the reaction $HCO_3^- \rightleftharpoons H^+ + CO_3^{2-}$, $\Delta G° = +59.0$ kJ/mol at 298.15 K. Find the value of K for the reaction.

6-9. The formation of tetrafluoroethylene from its elements is highly exothermic:

$$2F_2(g) + 2C(s) \rightleftharpoons F_2C{=}CF_2(g)$$
Fluorine   Graphite   Tetrafluoroethylene

(a) If a mixture of F_2, graphite, and C_2F_4 is at equilibrium in a closed container, will the reaction go to the right or to the left if F_2 is added?

(b) Rare bacteria from the planet Teflon eat C_2F_4 and make Teflon for their cell walls. Will the reaction go to the right or to the left if these bacteria are added?

Teflon

(c) Will the reaction go right or left if solid graphite is added? (Neglect any effect of increased pressure due to the decreased volume in the vessel when solid is added.)

(d) Will the reaction go right or left if the container is crushed to one-eighth of its original volume?

(e) Does the equilibrium constant become larger or smaller if the container is heated?

6-10. $BaCl_2 \cdot H_2O(s)$ loses water when it is heated in an oven:

$$BaCl_2 \cdot H_2O(s) \rightleftharpoons BaCl_2(s) + H_2O(g)$$

$$\Delta H° = 63.11 \text{ kJ/mol at } 25°C$$

$$\Delta S° = +148 \text{ J/(K} \cdot \text{mol) at } 25°C$$

(a) Write the equilibrium constant for this reaction. Calculate the vapor pressure of gaseous H_2O (P_{H_2O}) above $BaCl_2 \cdot H_2O$ at 298 K.

(b) If $\Delta H°$ and $\Delta S°$ are not temperature dependent (a poor assumption), estimate the temperature at which P_{H_2O} above $BaCl_2 \cdot H_2O(s)$ will be 1 bar.

6-11. The equilibrium constant for the reaction of $NH_3(aq) + H_2O \rightleftharpoons NH_4^+ + OH^-$ is $K_b = 1.479 \times 10^{-5}$ at 5°C and 1.570×10^{-5} at 10°C.

(a) Assuming $\Delta H°$ and $\Delta S°$ are constant in the interval 5°–10°C (probably a good assumption for small ΔT), use Equation 6-8 to find $\Delta H°$ for the reaction in this temperature range.

(b) Describe how Equation 6-8 could be used to make a linear graph to determine $\Delta H°$, if $\Delta H°$ and $\Delta S°$ were constant over some temperature range.

6-12. For $H_2(g) + Br_2(g) \rightleftharpoons 2HBr(g)$, $K = 7.2 \times 10^{-4}$ at 1 362 K and $\Delta H°$ is positive. A vessel is charged with 48.0 Pa HBr, 1 370 Pa H_2, and 3 310 Pa Br_2 at 1 362 K.

(a) Will the reaction proceed to the left or the right to reach equilibrium?

(b) Calculate the pressure (in pascals) of each species in the vessel at equilibrium.

(c) If the mixture at equilibrium is compressed to half of its original volume, will the reaction proceed to the left or the right to reestablish equilibrium?

(d) If the mixture at equilibrium is heated from 1 362 to 1 407 K, will HBr be formed or consumed in order to reestablish equilibrium?

6-13. *Henry's law* states that the concentration of a gas dissolved in a liquid is proportional to the pressure of the gas. This law is a consequence of the equilibrium

$$X(g) \overset{K_h}{\rightleftharpoons} X(aq) \qquad K_h = \frac{[X]}{P_X}$$

where K_h is called the Henry's law constant. (The same law applies to other solvents, but the value of K_h is different for each solvent.) For the gasoline additive MTBE, $K_h = 1.71$ M/bar. Suppose we have a closed container with aqueous solution and air in equilibrium. If the concentration of MTBE in the water is 1.00×10^2 ppm ($= 100 \ \mu g$ MTBE/g solution $\approx 100 \ \mu g$/mL), what is the pressure of MTBE in the air?

$$CH_3-O-C(CH_3)_3 \qquad \text{Methyl-}t\text{-butylether (MTBE, FM 88.15)}$$

Solubility Product

6-14. Find the concentration of Cu^+ in equilibrium with $CuBr(s)$ and 0.10 M Br^-.

6-15. What concentration of $Fe(CN)_6^{4-}$ (ferrocyanide) is in equilibrium with 1.0 μM Ag^+ and $Ag_4Fe(CN)_6(s)$. Express your answer with a prefix from Table 1-3.

6-16. Find $[Cu^{2+}]$ in a solution saturated with $Cu_4(OH)_6(SO_4)$ if $[OH^-]$ is *fixed* at 1.0×10^{-6} M. Note that $Cu_4(OH)_6(SO_4)$ gives 1 mol of SO_4^{2-} for 4 mol of Cu^{2+}.

$$Cu_4(OH)_6(SO_4)(s) \rightleftharpoons 4Cu^{2+} + 6OH^- + SO_4^{2-}$$
$$K_{sp} = 2.3 \times 10^{-69}$$

6-17. (a) From the solubility product of zinc ferrocyanide, $Zn_2Fe(CN)_6$, calculate the concentration of $Fe(CN)_6^{4-}$ in 0.10 mM $ZnSO_4$ saturated with $Zn_2Fe(CN)_6$. Assume that $Zn_2Fe(CN)_6$ is a negligible source of Zn^{2+}.

(b) What concentration of $K_4Fe(CN)_6$ should be in a suspension of solid $Zn_2Fe(CN)_6$ in water to give $[Zn^{2+}] = 5.0 \times 10^{-7}$ M?

6-18. Solubility products predict that cation A^{3+} can be 99.999% separated from cation B^{2+} by precipitation with anion X^-. When the separation is tried, we find 0.2% contamination of $AX_3(s)$ with B^{2+}. Explain what might be happening.

6-19. A solution contains 0.050 0 M Ca^{2+} and 0.030 0 M Ag^+. Can 99% of Ca^{2+} be precipitated by sulfate without precipitating Ag^+? What will be the concentration of Ca^{2+} when Ag_2SO_4 begins to precipitate?

6-20. A solution contains 0.010 M Ba^{2+} and 0.010 M Ag^+. Can 99.90% of either ion be precipitated by chromate (CrO_4^{2-}) without precipitating the other metal ion?

6-21. If a solution containing 0.10 M Cl^-, Br^-, I^-, and CrO_4^{2-} is treated with Ag^+, in what order will the anions precipitate?

Complex Formation

6-22. Explain why the total solubility of lead in Figure 6-3 first decreases and then increases as $[I^-]$ increases. Give an example of the chemistry in each of the two domains.

6-23. Identify the Lewis acids in the following reactions:

(a) $BF_3 + NH_3 \rightleftharpoons F_3\bar{B}-\overset{+}{N}H_3$

(b) $F^- + AsF_5 \rightleftharpoons AsF_6^-$

6-24. The cumulative formation constant for $SnCl_2(aq)$ in 1.0 M $NaNO_3$ is $\beta_2 = 12$. Find the concentration of $SnCl_2(aq)$ for a solution in which the concentrations of Sn^{2+} and Cl^- are both somehow fixed at 0.20 M.

6-25. Given the following equilibria, calculate the concentration of each zinc species in a solution saturated with $Zn(OH)_2(s)$ and containing $[OH^-]$ at a fixed concentration of 3.2×10^{-7} M.

$Zn(OH)_2(s)$	$K_{sp} = 3 \times 10^{-16}$
$Zn(OH)^+$	$\beta_1 = 1 \times 10^4$
$Zn(OH)_2(aq)$	$\beta_2 = 2 \times 10^{10}$
$Zn(OH)_3^-$	$\beta_3 = 8 \times 10^{13}$
$Zn(OH)_4^{2-}$	$\beta_4 = 3 \times 10^{15}$

6-26. Although KOH, RbOH, and CsOH have little association between metal and hydroxide in aqueous solution, Li^+ and Na^+ do form complexes with OH^-:

$$Li^+ + OH^- \rightleftharpoons LiOH(aq) \qquad K_1 = \frac{[LiOH(aq)]}{[Li^+][OH^-]} = 0.83$$

$$Na^+ + OH^- \rightleftharpoons NaOH(aq) \qquad K_1 = 0.20$$

Prepare a table like the one in Exercise 6-B showing initial and final concentrations of Na^+, OH^-, and $NaOH(aq)$ in 1 F NaOH solution. Calculate the fraction of sodium in the form $NaOH(aq)$ at equilibrium.

6-27. In Figure 6-3, the concentration of $PbI_2(aq)$ is independent of $[I^-]$. Use any of the equilibrium constants for Reactions 6-11 through 6-15 to find the equilibrium constant for the reaction $PbI_2(s) \rightleftharpoons PbI_2(aq)$, which is equal to the concentration of $PbI_2(aq)$.

Acids and Bases

6-28. Distinguish Lewis acids and bases from Brønsted-Lowry acids and bases. Give an example of each.

6-29. Fill in the blanks:

(a) The product of a reaction between a Lewis acid and a Lewis base is called _____.

(b) The bond between a Lewis acid and a Lewis base is called _____ or _____.

(c) Brønsted-Lowry acids and bases related by gain or loss of one proton are said to be _____.

(d) A solution is *acidic* if _____. A solution is *basic* if _____.

6-30. Why is the pH of distilled water usually <7? How can you prevent this from happening?

6-31. Gaseous SO_2 is created by combustion of sulfur-containing fuels, especially coal. Explain how SO_2 in the atmosphere makes acidic rain.

6-32. Use electron dot structures to show why tetramethylammonium hydroxide, $(CH_3)_4N^+OH^-$, is an ionic compound. That is, show why hydroxide is not covalently bound to the rest of the molecule.

6-33. Identify the Brønsted-Lowry acids among the reactants in the following reactions:

(a) $KCN + HI \rightleftharpoons HCN + KI$

(b) $PO_4^{3-} + H_2O \rightleftharpoons HPO_4^{2-} + OH^-$

6-34. Write the autoprotolysis reaction of H_2SO_4.

6-35. Identify the conjugate acid-base pairs in the following reactions:

(a) $H_3NCH_2CH_2NH_3^+ + H_2O \rightleftharpoons H_3NCH_2CH_2NH_2^+ + H_3O^+$

(b)

Benzoic acid   Pyridine   Benzoate   Pyridinium

pH

6-36. Calculate $[H^+]$ and pH for the following solutions:

(a) 0.010 M HNO_3
(b) 0.035 M KOH
(c) 0.030 M HCl
(d) 3.0 M HCl
(e) 0.010 M $[(CH_3)_4N^+]OH^-$
     Tetramethylammonium hydroxide

6-37. Use Table 6-1 to calculate the pH of pure water at (a) 25°C and (b) 100°C.

6-38. The equilibrium constant for the reaction $H_2O \rightleftharpoons H^+ + OH^-$ is 1.0×10^{-14} at 25°C. What is the value of K for the reaction $4H_2O \rightleftharpoons 4H^+ + 4OH^-$?

6-39. An acidic solution containing 0.010 M La^{3+} is treated with NaOH until $La(OH)_3$ precipitates. At what pH does this occur?

6-40. Use Le Châtelier's principle and K_w in Table 6-1 to decide whether the autoprotolysis of water is endothermic or exothermic at (a) 25°C; (b) 100°C; (c) 300°C.

Strengths of Acids and Bases

6-41. Make a list of the common strong acids and strong bases. Memorize this list.

6-42. Write the formulas and names for three classes of weak acids and two classes of weak bases.

6-43. Write the K_a reaction for trichloroacetic acid, Cl_3CCO_2H, for anilinium ion, , and for lanthanum ion, La^{3+}.

6-44. Write the K_b reactions for pyridine and for sodium 2-mercaptoethanol.

Pyridine   Sodium 2-mercaptoethanol

6-45. Write the K_a and K_b reactions of $NaHCO_3$.

6-46. Write the stepwise acid-base reactions for the following ions in water. Write the correct symbol (for example, K_{b1}) for the equilibrium constant for each reaction.

(a) $H_3NCH_2CH_2NH_3^+$
    Ethylenediammonium ion

(b)

    Malonate ion

6-47. Which is a stronger acid, (a) or (b)?

(a)

Dichloroacetic acid
$K_a = 8 \times 10^{-2}$

(b)

Chloroacetic acid
$K_a = 1.36 \times 10^{-3}$

Which is a stronger base, (c) or (d)?

(c) H_2NNH_2
    Hydrazine
    $K_b = 1.1 \times 10^{-6}$

(d)

    Urea
    $K_b = 1.5 \times 10^{-14}$

6-48. Write the K_b reaction of CN^-. Given that the K_a value for HCN is 6.2×10^{-10}, calculate K_b for CN^-.

6-49. Write the K_{a2} reaction of phosphoric acid (H_3PO_4) and the K_{b2} reaction of disodium oxalate ($Na_2C_2O_4$).

6-50. From the K_b values for phosphate in Equations 6-32 through 6-34, calculate the three K_a values of phosphoric acid.

6-51. From the following equilibrium constants, calculate the equilibrium constant for the reaction $HO_2CCO_2H \rightleftharpoons 2H^+ + C_2O_4^{2-}$.

$K_1 = 5.4 \times 10^{-2}$

$K_2 = 5.4 \times 10^{-5}$

6-52. (a) Using only K_{sp} from Table 6-3, calculate how many moles of $Ca(OH)_2$ will dissolve in 1.00 L of water.

(b) How will the solubility calculated in part (a) be affected by the K_1 reaction in Table 6-3?

6-53. The planet Aragonose (which is made mostly of the mineral aragonite, whose composition is $CaCO_3$) has an atmosphere containing methane and carbon dioxide, each at a pressure of 0.10 bar. The oceans are saturated with aragonite and have a concentration of H^+ equal to 1.8×10^{-7} M. Given the following equilibria, calculate how many grams of calcium are contained in 2.00 L of Aragonose seawater.

$$CaCO_3 \, (s, \text{aragonite}) \rightleftharpoons Ca^{2+}(aq) + CO_3^{2-}(aq) \quad K_{sp} = 6.0 \times 10^{-9}$$
$$CO_2(g) \rightleftharpoons CO_2(aq) \quad K_{CO_2} = 3.4 \times 10^{-2}$$
$$CO_2(aq) + H_2O(l) \rightleftharpoons HCO_3^-(aq) + H^+(aq) \quad K_1 = 4.5 \times 10^{-7}$$
$$HCO_3^-(aq) \rightleftharpoons H^+(aq) + CO_3^{2-}(aq) \quad K_2 = 4.7 \times 10^{-11}$$

Don't panic! Reverse the first reaction, add all the reactions together, and see what cancels.

7 Activity and the Systematic Treatment of Equilibrium

Estimated number of waters of hydration

Molecule	Tightly bound H_2O
$CH_3CH_2CH_3$	0
C_6H_6	0
CH_3CH_2Cl	0
CH_3CH_2SH	0
$CH_3{-}O{-}CH_3$	1
CH_3CH_2OH	1
$(CH_3)_2C{=}O$	1.5
$CH_3CH{=}O$	1.5
CH_3CO_2H	2
$CH_3C{\equiv}N$	3
$CH_3\overset{\overset{O}{\|}}{C}NHCH_3$	4
CH_3NO_2	5
$CH_3CO_2^-$	5
CH_3NH_2	6
CH_3SO_3H	7
NH_3	9
$CH_3SO_3^-$	10
NH_4^+	12

From S. Fu and C. A. Lucy, "Prediction of Electrophoretic Mobilities," Anal. Chem. 1998, 70, 173.

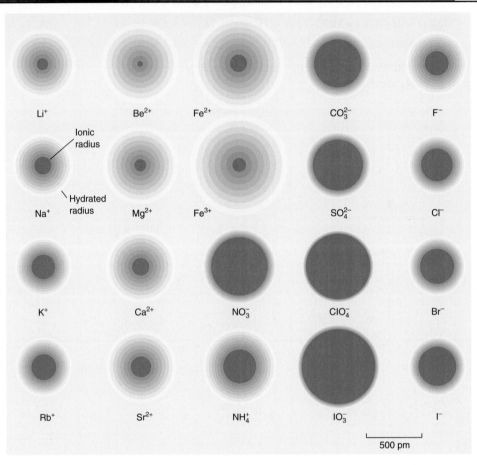

Ionic and hydrated radii of several ions. Smaller, more highly charged ions bind water molecules more tightly and behave as larger hydrated species.[3]

Ions and molecules in solution are surrounded by an organized sheath of solvent molecules. The oxygen atom of H_2O has a partial negative charge, and each hydrogen atom has half as much positive charge.

Water binds to cations through the oxygen atom. The first coordination sphere of Li^+, for example, is composed of ~4 H_2O molecules.[1] Cl^- binds ~6 H_2O molecules through hydrogen atoms.[1,2] H_2O exchanges rapidly between bulk solvent and ion-coordination sites.

Ionic radii in the figure are measured by X-ray diffraction of ions in crystals. Hydrated radii are estimated from diffusion coefficients of ions in solution and from the mobilities of aqueous ions in an electric field.[3,4] Smaller, more highly charged ions bind more water molecules and behave as larger species in solution. The *activity* of aqueous ions, which we study in this chapter, is related to the size of the hydrated species.

In Chapter 6, we wrote the equilibrium constant for a reaction in the form

$$Fe^{3+} + SCN^- \rightleftharpoons Fe(SCN)^{2+} \qquad K = \frac{[Fe(SCN)^{2+}]}{[Fe^{3+}][SCN^-]} \qquad (7\text{-}1)$$

Pale yellow Colorless Red

Figure 7-1, Demonstration 7-1, and Color Plate 2 show that the concentration quotient in Equation 7-1 decreases if you add the "inert" salt KNO_3 to the solution. That is, the equilibrium "constant" is not really constant. This chapter explains why concentrations are replaced by *activities* in the equilibrium constant and how activities are used.

FIGURE 7-1 Student data showing that the equilibrium quotient of concentrations for the reaction $Fe^{3+} + SCN^- \rightleftharpoons Fe(SCN)^{2+}$ decreases as potassium nitrate is added to the solution. Color Plate 2 shows the fading of the red color of $Fe(SCN)^{2+}$ after KNO_3 has been added. Problem 12-11 gives more information on this chemical system. [From R. J. Stolzberg, "Discovering a Change in Equilibrium Constant with Change in Ionic Strength," *J. Chem. Ed.* **1999**, *76*, 640.]

7-1 The Effect of Ionic Strength on Solubility of Salts

Consider a saturated solution of $CaSO_4$ in distilled water.

$$CaSO_4(s) \rightleftharpoons Ca^{2+} + SO_4^{2-} \qquad K_{sp} = 2.4 \times 10^{-5} \qquad (7\text{-}2)$$

Figure 6-1 showed that the solubility is 0.015 M. The dissolved species are mainly 0.010 M Ca^{2+}, 0.010 M SO_4^{2-}, and 0.005 M $CaSO_4(aq)$ (an ion pair).

Now an interesting effect is observed when a salt such as KNO_3 is added to the solution. Neither K^+ nor NO_3^- reacts with either Ca^{2+} or SO_4^{2-}. Yet, when 0.050 M KNO_3 is added to the saturated solution of $CaSO_4$, more solid dissolves until the concentrations of Ca^{2+} and SO_4^{2-} have each increased by about 30%.

In general, adding an "inert" salt (KNO_3) to a sparingly soluble salt ($CaSO_4$) increases the solubility of the sparingly soluble salt. "Inert" means that KNO_3 has no chemical reaction with $CaSO_4$. When we add salt to a solution, we say that the *ionic strength* of the solution increases. The definition of ionic strength will be given shortly.

Addition of an "inert" salt increases the solubility of an ionic compound.

The Explanation

Why does the solubility increase when salts are added to the solution? Consider one particular Ca^{2+} ion and one particular SO_4^{2-} ion in solution. The SO_4^{2-} ion is surrounded by cations (K^+, Ca^{2+}) and anions (NO_3^-, SO_4^{2-}) in solution. However, for the average anion, there will be more cations than anions nearby because cations are attracted to the anion, but anions are repelled. These interactions create a region of net positive charge around any particular anion. We call this region the **ionic atmosphere** (Figure 7-2). Ions continually diffuse into and out of the ionic atmosphere. The net charge in the atmosphere, averaged over time, is less than the charge of the anion at the center. Similarly, an atmosphere of negative charge surrounds any cation in solution.

The ionic atmosphere attenuates (decreases) the attraction between ions. The cation plus its negative atmosphere has less positive charge than the cation alone. The anion plus its ionic atmosphere has less negative charge than the anion alone. The net attraction between the cation with its ionic atmosphere and the anion with its ionic atmosphere is smaller than it would be between pure cation and anion in the absence of ionic atmospheres. *The greater the ionic strength of a solution, the higher the charge in the ionic atmosphere. Each ion-plus-atmosphere contains less net charge and there is less attraction between any particular cation and anion.*

An anion is surrounded by excess cations. A cation is surrounded by excess anions.

FIGURE 7-2 An ionic atmosphere, shown as a spherical cloud of charge $\delta+$ or $\delta-$, surrounds ions in solution. The charge of the atmosphere is less than the charge of the central ion. The greater the ionic strength of the solution, the greater the charge in each ionic atmosphere.

DEMONSTRATION 7-1   Effect of Ionic Strength on Ion Dissociation[5]

This experiment demonstrates the effect of ionic strength on the dissociation of the red iron(III) thiocyanate complex:

$$Fe(SCN)^{2+} \rightleftharpoons Fe^{3+} + SCN^-$$

Red   Pale yellow   Colorless

Prepare a solution of 1 mM $FeCl_3$ by dissolving 0.27 g of $FeCl_3 \cdot 6H_2O$ in 1 L of water containing 3 drops of 15 M (concentrated) HNO_3. The acid slows the precipitation of $Fe(OH)_3$, which occurs within a few days and necessitates the preparation of fresh solution for this demonstration.

To demonstrate the effect of ionic strength on the dissociation reaction, mix 300 mL of 1 mM $FeCl_3$ with 300 mL of 1.5 mM NH_4SCN or $KSCN$. Divide the pale red solution into two equal portions and add 12 g of KNO_3 to one of them to increase the ionic strength to 0.4 M. As KNO_3 dissolves, the red $Fe(SCN)^{2+}$ complex dissociates and the color fades (Color Plate 2).

Add a few crystals of NH_4SCN or $KSCN$ to either solution to drive the reaction toward formation of $Fe(SCN)^{2+}$, thereby intensifying the red color. This reaction demonstrates Le Châtelier's principle—adding a product creates more reactant.

FIGURE 7-3 Solubility of potassium hydrogen tartrate increases when the salts $MgSO_4$ or NaCl are added. There is no effect when the neutral compound glucose is added. Addition of KCl decreases the solubility. (*Why?*) [From C. J. Marzzacco, "Effect of Salts and Nonelectrolytes on the Solubility of Potassium Bitartrate," *J. Chem. Ed.* **1998,** *75,* 1628.]

Electrolyte	Molarity	Ionic strength
1:1	M	M
2:1	M	3M
3:1	M	6M
2:2	M	4M

Increasing ionic strength therefore reduces the attraction between any particular Ca^{2+} ion and any SO_4^{2-} ion, relative to their attraction for each other in distilled water. The effect is to reduce their tendency to come together, thereby *increasing* the solubility of $CaSO_4$.

Increasing ionic strength promotes dissociation into ions. Thus, each of the following reactions is driven to the right if the ionic strength is raised from, say, 0.01 to 0.1 M:

$$Fe(SCN)^{2+} \rightleftharpoons Fe^{3+} + SCN^-$$
Thiocyanate

$$\text{⬡—OH} \rightleftharpoons \text{⬡—O}^- + H^+$$
Phenol   Phenolate

$$\underset{\text{Potassium hydrogen tartrate}}{HO_2CCHCHCO_2K(s)} \rightleftharpoons HO_2CCHCHCO_2^- + K^+$$

Figure 7-3 shows the effect of added salt on the solubility of potassium hydrogen tartrate.

What Do We Mean by "Ionic Strength"?

Ionic strength, μ, is a measure of the total concentration of ions in solution. The more highly charged an ion, the more it is counted.

Ionic strength: $$\mu = \frac{1}{2}(c_1 z_1^2 + c_2 z_2^2 + \cdots) = \frac{1}{2}\sum_i c_i z_i^2 \qquad (7\text{-}3)$$

where c_i is the concentration of the *i*th species and z_i is its charge. The sum extends over *all* ions in solution.

EXAMPLE   **Calculation of Ionic Strength**

Find the ionic strength of **(a)** 0.10 M $NaNO_3$; **(b)** 0.010 M Na_2SO_4; and **(c)** 0.020 M KBr plus 0.010 M Na_2SO_4.

Solution

(a) $\mu = \frac{1}{2}\{[Na^+] \cdot (+1)^2 + [NO_3^-] \cdot (-1)^2\}$

$\qquad = \frac{1}{2}\{0.10 \cdot 1 + 0.10 \cdot 1\} = 0.10$ M

(b) $\mu = \frac{1}{2}\{[Na^+] \cdot (+1)^2 + [SO_4^{2-}] \cdot (-2)^2\}$

$\qquad = \frac{1}{2}\{(0.020 \cdot 1) + (0.010 \cdot 4)\} = 0.030$ M

Note that $[Na^+] = 0.020$ M because there are two moles of Na^+ per mole of Na_2SO_4.

(c) $\mu = \frac{1}{2}\{[K^+] \cdot (+1)^2 + [Br^-] \cdot (-1)^2 + [Na^+] \cdot (+1)^2 + [SO_4^{2-}] \cdot (-2)^2\}$

$\qquad = \frac{1}{2}\{(0.020 \cdot 1) + (0.020 \cdot 1) + (0.020 \cdot 1) + (0.010 \cdot 4)\} = 0.050$ M

Test Yourself   What is the ionic strength of 1 mM $CaCl_2$? (*Answer:* 3 mM)

$NaNO_3$ is called a 1:1 electrolyte because the cation and the anion both have a charge of 1. For 1:1 electrolytes, ionic strength equals molarity. For other stoichiometries (such as the 2:1 electrolyte Na_2SO_4), ionic strength is greater than molarity.

Computing the ionic strength of any but the most dilute solutions is complicated because salts with ions of charge ≥ 2 are not fully dissociated. In Box 7-1, we find that, at a formal concentration of 0.025 M $MgSO_4$, 35% of Mg^{2+} is bound in the ion pair, $MgSO_4(aq)$. The higher the concentration and the higher the ionic charge, the more the ion pairing. There is no simple way to find the ionic strength of 0.025 M $MgSO_4$.

BOX 7-1 **Salts with Ions of Charge ≥|2| Do Not Fully Dissociate[6]**

Salts composed of cations and anions with charges of ±1 dissociate almost completely at concentrations <0.1 M in water. Salts containing ions with a charge ≥2 are less dissociated, even in dilute solution. Appendix J gives formation constants for *ion pairing*:

Ion-pair formation constant:

$$M^{n+}(aq) + L^{m-}(aq) \rightleftharpoons M^{n+}L^{m-}(aq)$$

Ion pair

$$K = \frac{[ML]\gamma_{ML}}{[M]\gamma_M[L]\gamma_L}$$

where the γ_i are activity coefficients. With constants from Appendix J, activity coefficients from Equation 7-6, craft, and persistence, you might calculate the following percentages of ion pairing in 0.025 F solutions:

Percentage of metal ion bound as ion pair in 0.025 F M_xL_y solution[a]

M \ L	Cl^-	SO_4^{2-}
Na^+	0.6%	4%
Mg^{2+}	8%	35%

a. The size of ML was taken as 500 pm to compute its activity coefficient.

The table tells us that 0.025 F NaCl is only 0.6% associated as $Na^+Cl^-(aq)$ and Na_2SO_4 is 4% associated as $NaSO_4^-(aq)$. For $MgSO_4$, 35% is ion paired. A solution of 0.025 F $MgSO_4$ contains 0.016 M Mg^{2+}, 0.016 M SO_4^{2-}, and 0.009 M $MgSO_4(aq)$. The ionic strength of 0.025 F $MgSO_4$ is not 0.10 M, but just 0.065 M. Problem 7-28 provides an example of the type of calculation in this box.

7-2 Activity Coefficients

Equation 7-1 does not predict any effect of ionic strength on a chemical reaction. To account for the effect of ionic strength, concentrations are replaced by **activities**:

Activity of C:

$$\mathcal{A}_C = [C]\gamma_C \qquad (7\text{-}4)$$

Activity   Concentration   Activity coefficient
of C   of C   of C

The activity of species C is its concentration multiplied by its **activity coefficient**. The activity coefficient measures the deviation of behavior from ideality. If the activity coefficient were 1, then the behavior would be ideal and the form of the equilibrium constant in Equation 7-1 would be correct.

The correct form of the equilibrium constant is

General form of equilibrium constant:

$$K = \frac{\mathcal{A}_C^c \, \mathcal{A}_D^d}{\mathcal{A}_A^a \, \mathcal{A}_B^b} = \frac{[C]^c\gamma_C^c[D]^d\gamma_D^d}{[A]^a\gamma_A^a [B]^b\gamma_B^b} \qquad (7\text{-}5)$$

> Do not confuse the terms *activity* and *activity coefficient*.

> Equation 7-5 is the "real" equilibrium constant. Equation 6-2, the concentration quotient, K_c, did not include activity coefficients:
>
> $$K_c = \frac{[C]^c[D]^d}{[A]^a[B]^b} \qquad (6\text{-}2)$$

Equation 7-5 allows for the effect of ionic strength on a chemical equilibrium because the activity coefficients depend on ionic strength.

For Reaction 7-2, the equilibrium constant is

$$K_{sp} = \mathcal{A}_{Ca^{2+}}\mathcal{A}_{SO_4^{2-}} = [Ca^{2+}]\gamma_{Ca^{2+}}[SO_4^{2-}]\gamma_{SO_4^{2-}}$$

If the concentrations of Ca^{2+} and SO_4^{2-} are to *increase* when a second salt is added to increase ionic strength, the activity coefficients must *decrease* with increasing ionic strength.

At low ionic strength, activity coefficients approach unity, and the thermodynamic equilibrium constant (7-5) approaches the "concentration" equilibrium constant (6-2). One way to measure a thermodynamic equilibrium constant is to measure the concentration ratio (6-2) at successively lower ionic strengths and extrapolate to zero ionic strength. Commonly, tabulated equilibrium constants are not thermodynamic constants but just the concentration ratio (6-2) measured under a particular set of conditions.

EXAMPLE Exponents of Activity Coefficients

Write the solubility product expression for $La_2(SO_4)_3$ with activity coefficients.

Solution Exponents of activity coefficients are the same as exponents of concentrations:

$$K_{sp} = \mathcal{A}_{La^{3+}}^2\mathcal{A}_{SO_4^{2-}}^3 = [La^{3+}]^2\gamma_{La^{3+}}^2[SO_4^{2-}]^3\gamma_{SO_4^{2-}}^3$$

Activity Coefficients of Ions

The ionic atmosphere model leads to the **extended Debye-Hückel equation**, relating activity coefficients to ionic strength:

Extended Debye-Hückel equation:
$$\log \gamma = \frac{-0.51z^2 \sqrt{\mu}}{1 + (\alpha\sqrt{\mu}/305)} \qquad \text{(at 25°C)} \qquad (7\text{-}6)$$

1 pm (picometer) $= 10^{-12}$ m

In Equation 7-6, γ is the activity coefficient of an ion of charge $\pm z$ and size α (picometers, pm) in an aqueous solution of ionic strength μ. The equation works fairly well for $\mu \leq 0.1$ M. To find activity coefficients for ionic strengths above 0.1 M (up to molalities of 2–6 mol/kg for many salts), more complicated *Pitzer equations* are usually used.[7]

Table 7-1 lists sizes (α) and activity coefficients of many ions. All ions of the same size and charge appear in the same group and have the same activity coefficients. For example, Ba^{2+} and succinate ion $[^-O_2CCH_2CH_2CO_2^-$, listed as $(CH_2CO_2^-)_2]$, each have a size of 500 pm and are listed among the charge ± 2 ions. At an ionic strength of 0.001 M, both of these ions have an activity coefficient of 0.868.

The ion size α in Equation 7-6 is an empirical parameter that provides agreement between measured activity coefficients and ionic strength up to $\mu \approx 0.1$ M. In theory, α is the diameter of the hydrated ion.[8] However, sizes in Table 7-1 cannot be taken literally. For example, the diameter of Cs^+ ion in crystals is 340 pm. The hydrated Cs^+ ion must be larger than the ion in the crystal, but the size of Cs^+ in Table 7-1 is only 250 pm.

Ionic and hydrated ion sizes are shown at the opening of this chapter.

Even though ion sizes in Table 7-1 are empirical parameters, trends among sizes are sensible. Small, highly charged ions bind solvent more tightly and have larger effective sizes than do larger or less highly charged ions. For example, the order of sizes in Table 7-1 is $Li^+ > Na^+ > K^+ > Rb^+$, even though crystallographic radii are $Li^+ < Na^+ < K^+ < Rb^+$.

Effect of Ionic Strength, Ion Charge, and Ion Size on the Activity Coefficient

Over the range of ionic strengths from 0 to 0.1 M, the effect of each variable on activity coefficients is as follows:

1. As ionic strength increases, the activity coefficient decreases (Figure 7-4). The activity coefficient (γ) approaches unity as the ionic strength (μ) approaches 0.
2. As the magnitude of the charge of the ion increases, the departure of its activity coefficient from unity increases. Activity corrections are more important for ions with a charge of ± 3 than for ions with a charge of ± 1 (Figure 7-4).
3. The smaller the ion size (α), the more important activity effects become.

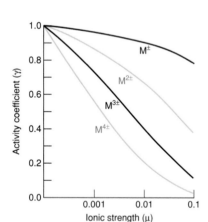

FIGURE 7-4 Activity coefficients for differently charged ions with a constant ionic size (α) of 500 pm. At zero ionic strength, $\gamma = 1$. The greater the charge of the ion, the more rapidly γ decreases as ionic strength increases. Note that the abscissa is logarithmic.

EXAMPLE Using Table 7-1

Find the activity coefficient of Ca^{2+} in a solution of 3.3 mM $CaCl_2$.

Solution The ionic strength is

$$\mu = \frac{1}{2}\{[Ca^{2+}] \cdot 2^2 + [Cl^-] \cdot (-1)^2\}$$

$$= \frac{1}{2}\{(0.003\,3) \cdot 4 + (0.006\,6) \cdot 1\} = 0.010 \text{ M}$$

In Table 7-1, Ca^{2+} is listed under the charge ± 2 and has a size of 600 pm. Thus $\gamma = 0.675$ when $\mu = 0.010$ M.

Test Yourself Find γ for Cl^- in 0.33 mM $CaCl_2$. (**Answer:** 0.964)

TABLE 7-1 Activity coefficients for aqueous solutions at 25°C

Ion	Ion size (α, pm)	Ionic strength (μ, M)				
		0.001	0.005	0.01	0.05	0.1
Charge = ±1		*Activity coefficient (γ)*				
H^+	900	0.967	0.933	0.914	0.86	0.83
$(C_6H_5)_2CHCO_2^-$, $(C_3H_7)_4N^+$	800	0.966	0.931	0.912	0.85	0.82
$(O_2N)_3C_6H_2O^-$, $(C_3H_7)_3NH^+$, $CH_3OC_6H_4CO_2^-$	700	0.965	0.930	0.909	0.845	0.81
Li^+, $C_6H_5CO_2^-$, $HOC_6H_4CO_2^-$, $ClC_6H_4CO_2^-$, $C_6H_5CH_2CO_2^-$, $CH_2{=}CHCH_2CO_2^-$, $(CH_3)_2CHCH_2CO_2^-$, $(CH_3CH_2)_4N^+$, $(C_3H_7)_2NH_2^+$	600	0.965	0.929	0.907	0.835	0.80
$Cl_2CHCO_2^-$, $Cl_3CCO_2^-$, $(CH_3CH_2)_3NH^+$, $(C_3H_7)NH_3^+$	500	0.964	0.928	0.904	0.83	0.79
Na^+, $CdCl^+$, ClO_2^-, IO_3^-, HCO_3^-, $H_2PO_4^-$, HSO_3^-, $H_2AsO_4^-$, $Co(NH_3)_4(NO_2)_2^+$, $CH_3CO_2^-$, $ClCH_2CO_2^-$, $(CH_3)_4N^+$, $(CH_3CH_2)_2NH_2^+$, $H_2NCH_2CO_2^-$	450	0.964	0.928	0.902	0.82	0.775
$^+H_3NCH_2CO_2H$, $(CH_3)_3NH^+$, $CH_3CH_2NH_3^+$	400	0.964	0.927	0.901	0.815	0.77
OH^-, F^-, SCN^-, OCN^-, HS^-, ClO_3^-, ClO_4^-, BrO_3^-, IO_4^-, MnO_4^-, HCO_2^-, $H_2citrate^-$, $CH_3NH_3^+$, $(CH_3)_2NH_2^+$	350	0.964	0.926	0.900	0.81	0.76
K^+, Cl^-, Br^-, I^-, CN^-, NO_2^-, NO_3^-	300	0.964	0.925	0.899	0.805	0.755
Rb^+, Cs^+, NH_4^+, Tl^+, Ag^+	250	0.964	0.924	0.898	0.80	0.75
Charge = ±2		*Activity coefficient (γ)*				
Mg^{2+}, Be^{2+}	800	0.872	0.755	0.69	0.52	0.45
$CH_2(CH_2CH_2CO_2^-)_2$, $(CH_2CH_2CH_2CO_2^-)_2$	700	0.872	0.755	0.685	0.50	0.425
Ca^{2+}, Cu^{2+}, Zn^{2+}, Sn^{2+}, Mn^{2+}, Fe^{2+}, Ni^{2+}, Co^{2+}, $C_6H_4(CO_2^-)_2$, $H_2C(CH_2CO_2^-)_2$, $(CH_2CH_2CO_2^-)_2$	600	0.870	0.749	0.675	0.485	0.405
Sr^{2+}, Ba^{2+}, Cd^{2+}, Hg^{2+}, S^{2-}, $S_2O_4^{2-}$, WO_4^{2-}, $H_2C(CO_2^-)_2$, $(CH_2CO_2^-)_2$, $(CHOHCO_2^-)_2$	500	0.868	0.744	0.67	0.465	0.38
Pb^{2+}, CO_3^{2-}, SO_3^{2-}, MoO_4^{2-}, $Co(NH_3)_5Cl^{2+}$, $Fe(CN)_5NO^{2-}$, $C_2O_4^{2-}$, $Hcitrate^{2-}$	450	0.867	0.742	0.665	0.455	0.37
Hg_2^{2+}, SO_4^{2-}, $S_2O_3^{2-}$, $S_2O_6^{2-}$, $S_2O_8^{2-}$, SeO_4^{2-}, CrO_4^{2-}, HPO_4^{2-}	400	0.867	0.740	0.660	0.445	0.355
Charge = ±3		*Activity coefficient (γ)*				
Al^{3+}, Fe^{3+}, Cr^{3+}, Sc^{3+}, Y^{3+}, In^{3+}, lanthanides[a]	900	0.738	0.54	0.445	0.245	0.18
$citrate^{3-}$	500	0.728	0.51	0.405	0.18	0.115
PO_4^{3-}, $Fe(CN)_6^{3-}$, $Cr(NH_3)_6^{3+}$, $Co(NH_3)_6^{3+}$, $Co(NH_3)_5H_2O^{3+}$	400	0.725	0.505	0.395	0.16	0.095
Charge = ±4		*Activity coefficient (γ)*				
Th^{4+}, Zr^{4+}, Ce^{4+}, Sn^{4+}	1 100	0.588	0.35	0.255	0.10	0.065
$Fe(CN)_6^{4-}$	500	0.57	0.31	0.20	0.048	0.021

a. Lanthanides are elements 57–71 in the periodic table.
SOURCE: J. Kielland, J. Am. Chem. Soc. **1937**, 59, 1675.

How to Interpolate

If you need to find an activity coefficient for an ionic strength that is between values in Table 7-1, you can use Equation 7-6. Alternatively, in the absence of a spreadsheet, it not hard to interpolate in Table 7-1. In *linear interpolation*, we assume that values between two entries of a table lie on a straight line. For example, consider a table in which $y = 0.67$ when $x = 10$ and $y = 0.83$ when $x = 20$. What is the value of y when $x = 16$?

Interpolation is the estimation of a number that lies *between* two values in a table. Estimating a number that lies *beyond* values in a table is called *extrapolation*.

x value	y value
10	0.67
16	?
20	0.83

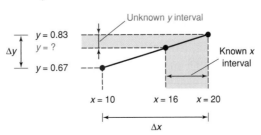

To interpolate a value of y, we can set up a proportion:

This calculation is equivalent to saying:

"16 is 60% of the way from 10 to 20. Therefore the y value will be 60% of the way from 0.67 to 0.83."

Interpolation:

$$\frac{\text{Unknown } y \text{ interval}}{\Delta y} = \frac{\text{known } x \text{ interval}}{\Delta x} \qquad (7\text{-}7)$$

$$\frac{0.83 - y}{0.83 - 0.67} = \frac{20 - 16}{20 - 10} \Rightarrow y = 0.76_6$$

For $x = 16$, our estimate of y is 0.76_6.

EXAMPLE **Interpolating Activity Coefficients**

Calculate the activity coefficient of H^+ when $\mu = 0.025$ M.

Solution H^+ is the first entry in Table 7-1.

μ	γ for H^+
0.01	0.914
0.025	?
0.05	0.86

The linear interpolation is set up as follows:

$$\frac{\text{Unknown } \gamma \text{ interval}}{\Delta \gamma} = \frac{\text{known } \mu \text{ interval}}{\Delta \mu}$$

$$\frac{0.86 - \gamma}{0.86 - 0.914} = \frac{0.05 - 0.025}{0.05 - 0.01}$$

$$\gamma = 0.89_4$$

Another Solution A better and slightly more tedious calculation uses Equation 7-6, with the ion size $\alpha = 900$ pm listed for H^+ in Table 7-1:

$$\log \gamma_{H^+} = \frac{(-0.51)(1^2)\sqrt{0.025}}{1 + (900\sqrt{0.025}/305)} = -0.054_{98}$$

$$\gamma_{H^+} = 10^{-0.054_{98}} = 0.88_1$$

Test Yourself By interpolation, find γ for H^+ when $\mu = 0.06$ M. (**Answer:** 0.85_4)

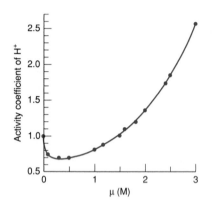

FIGURE 7-5 Activity coefficient of H^+ in solutions containing 0.010 0 M $HClO_4$ and varying amounts of $NaClO_4$. [Derived from L. Pezza, M. Molina, M. de Moraes, C. B. Melios, and J. O. Tognolli, *Talanta* 1996, *43*, 1689.] The authoritative source on electrolyte solutions is H. S. Harned and B. B. Owen, *The Physical Chemistry of Electrolyte Solutions* (New York: Reinhold, 1958 ed.).

Activity Coefficients of Nonionic Compounds

For neutral species, $\mathcal{A}_C \approx [C]$. A more accurate relation is $\log \gamma = k\mu$, where $k \approx 0$ for ion pairs, $k \approx 0.11$ for NH_3 and CO_2, and $k \approx 0.2$ for organic molecules. For an ionic strength of $\mu = 0.1$ M, $\gamma \approx 1.00$ for ion pairs, $\gamma \approx 1.03$ for NH_3, and $\gamma \approx 1.05$ for organic molecules.

Neutral molecules, such as benzene and acetic acid, have no ionic atmosphere because they have no charge. To a good approximation, their activity coefficients are unity when the ionic strength is less than 0.1 M. In this book, we set $\gamma = 1$ for neutral molecules. That is, *the activity of a neutral molecule will be assumed to be equal to its concentration.*

For gases such as H_2, the activity is written

$$\mathcal{A}_{H_2} = P_{H_2}\gamma_{H_2}$$

For gases, $\mathcal{A} \approx P$ (bar).

where P_{H_2} is pressure in bars. The activity of a gas is called its *fugacity*, and the activity coefficient is called the *fugacity coefficient*. Deviation of gas behavior from the ideal gas law results in deviation of the fugacity coefficient from unity. For gases at or below 1 bar, $\gamma \approx 1$. Therefore, for gases, *we will set* $\mathcal{A} = P$ *(bar)*.

High Ionic Strengths

At high ionic strength, γ increases with increasing μ.

Above an ionic strength of ~ 1 M, activity coefficients of most ions increase, as shown for H^+ in $NaClO_4$ solutions in Figure 7-5. We should not be too surprised that activity coefficients in concentrated salt solutions are not the same as those in dilute aqueous solution. The "solvent" is no longer H_2O but, rather, a mixture of H_2O and $NaClO_4$. Hereafter, we limit our attention to dilute aqueous solutions.

EXAMPLE Using Activity Coefficients

Find the concentration of Ca^{2+} in equilibrium with 0.050 M NaF saturated with CaF_2. The solubility of CaF_2 is small, so the concentration of F^- is 0.050 M from NaF.

Solution We find $[Ca^{2+}]$ from the solubility product expression, including activity coefficients. The ionic strength of 0.050 M NaF is 0.050 M. At $\mu = 0.050\ 0$ M in Table 7-1, we find $\gamma_{Ca^{2+}} = 0.485$ and $\gamma_{F^-} = 0.81$.

$$K_{sp} = [Ca^{2+}]\gamma_{Ca^{2+}}[F^-]^2\gamma_{F^-}^2$$
$$3.2 \times 10^{-11} = [Ca^{2+}](0.485)(0.050)^2(0.81)^2$$
$$[Ca^{2+}] = 4.0 \times 10^{-8}\ M$$

K_{sp} comes from Appendix F. Note that γ_{F^-} is squared.

Test Yourself Find $[Hg_2^{2+}]$ in equilibrium with 0.010 M KCl saturated with Hg_2Cl_2. (*Answer:* 2.2×10^{-14} M)

7-3 pH Revisited

The definition pH $\approx -\log[H^+]$ in Chapter 6 is not exact. A better definition is

$$\mathbf{pH} = -\log \mathcal{A}_{H^+} = -\log[H^+]\gamma_{H^+} \tag{7-8}$$

When we measure pH with a pH meter, we are measuring the negative logarithm of the hydrogen ion *activity*, not its concentration.

For a discussion of what pH really means and how the pH of primary standard solutions is measured, see B. Lunelli and F. Scagnolari, "pH Basics," *J. Chem. Ed.* **2009**, *86*, 246.

EXAMPLE pH of Pure Water at 25°C

Let's calculate the pH of pure water by using activity coefficients.

Solution The relevant equilibrium is

$$H_2O \xrightleftharpoons{K_w} H^+ + OH^- \tag{7-9}$$
$$K_w = \mathcal{A}_{H^+}\mathcal{A}_{OH^-} = [H^+]\gamma_{H^+}[OH^-]\gamma_{OH^-} \tag{7-10}$$

H^+ and OH^- are produced in a 1:1 mole ratio, so their concentrations must be equal. Calling each concentration x, we write

$$K_w = 1.0 \times 10^{-14} = (x)\gamma_{H^+}(x)\gamma_{OH^-}$$

But the ionic strength of pure water is so small that it is reasonable to guess that $\gamma_{H^+} = \gamma_{OH^-} = 1$. Using these values in the preceding equation gives

$$1.0 \times 10^{-14} = (x)(1)(x)(1) = x^2 \Rightarrow x = 1.0 \times 10^{-7}\ M$$

The concentrations of H^+ and OH^- are both 1.0×10^{-7} M. The ionic strength is 1.0×10^{-7} M, so each activity coefficient is very close to 1.00. The pH is

$$pH = -\log[H^+]\gamma_{H^+} = -\log(1.0 \times 10^{-7})(1.00) = 7.00$$

EXAMPLE pH of Water Containing a Salt

Now let's calculate the pH of water containing 0.10 M KCl at 25°C.

Solution Reaction 7-9 tells us that $[H^+] = [OH^-]$. However, the ionic strength of 0.10 M KCl is 0.10 M. The activity coefficients of H^+ and OH^- in Table 7-1 are 0.83 and 0.76, respectively, when $\mu = 0.10$ M. Putting these values into Equation 7-10 gives

$$K_w = [H^+]\gamma_{H^+}[OH^-]\gamma_{OH^-}$$
$$1.0 \times 10^{-14} = (x)(0.83)(x)(0.76)$$
$$x = 1.26 \times 10^{-7}\ M$$

The concentrations of H^+ and OH^- are equal and are both greater than 1.0×10^{-7} M. The activities of H^+ and OH^- are not equal in this solution:

$$\mathcal{A}_{H^+} = [H^+]\gamma_{H^+} = (1.26 \times 10^{-7})(0.83) = 1.05 \times 10^{-7}$$
$$\mathcal{A}_{OH^-} = [OH^-]\gamma_{OH^-} = (1.26 \times 10^{-7})(0.76) = 0.96 \times 10^{-7}$$

Finally, we calculate $pH = -\log \mathcal{A}_{H^+} = -\log(1.05 \times 10^{-7}) = 6.98$.

Test Yourself   Find $[H^+]$ and the pH of 0.05 M $LiNO_3$. (**Answer:** $1.2_0 \times 10^{-7}$ M, 6.99)

The pH of water changes from 7.00 to 6.98 when we add 0.10 M KCl. KCl is not an acid or a base. The pH changes because KCl affects the activities of H^+ and OH^-. The pH change of 0.02 units lies at the limit of accuracy of pH measurements and is hardly important. However, the *concentration* of H^+ in 0.10 M KCl (1.26×10^{-7} M) is 26% greater than the concentration of H^+ in pure water (1.00×10^{-7} M).

7-4   Systematic Treatment of Equilibrium

The *systematic treatment of equilibrium* is a way to deal with all types of chemical equilibria, regardless of their complexity. After setting up general equations, we often introduce specific conditions or judicious approximations that allow simplification. Even simplified calculations are usually very tedious, so we make liberal use of spreadsheets for numerical solutions. When you have mastered the systematic treatment of equilibrium, you should be able to explore the behavior of complex systems.

The systematic procedure is to write as many independent algebraic equations as there are unknowns (species) in the problem. The equations are generated by writing all the chemical equilibrium conditions plus two more: the balances of charge and of mass. There is only one charge balance in a given system, but there could be several mass balances.

Charge Balance

The margin note: **Solutions must have zero total charge.**

The **charge balance** is an algebraic statement of electroneutrality: *The sum of the positive charges in solution equals the sum of the negative charges in solution.*

Suppose that a solution contains the following ionic species: H^+, OH^-, K^+, $H_2PO_4^-$, HPO_4^{2-}, and PO_4^{3-}. The charge balance is

$$[H^+] + [K^+] = [OH^-] + [H_2PO_4^-] + 2[HPO_4^{2-}] + 3[PO_4^{3-}] \tag{7-11}$$

The margin note: **The coefficient of each term in the charge balance equals the magnitude of the charge on each ion.**

This statement says that the total charge contributed by H^+ and K^+ equals the magnitude of the charge contributed by all of the anions on the right side of the equation. *The coefficient in front of each species always equals the magnitude of the charge on the ion.* This statement is true because a mole of, say, PO_4^{3-} contributes three moles of negative charge. If $[PO_4^{3-}] = 0.01$ M, the negative charge is $3[PO_4^{3-}] = 3(0.01) = 0.03$ M.

Equation 7-11 appears unbalanced to many people. "The right side of the equation has much more charge than the left side!" you might think. But you would be wrong.

For example, consider a solution prepared by weighing out 0.025 0 mol of KH_2PO_4 plus 0.030 0 mol of KOH and diluting to 1.00 L. The concentrations of the species at equilibrium are calculated to be

$$[H^+] = 5.1 \times 10^{-12} \text{ M} \qquad [H_2PO_4^-] = 1.3 \times 10^{-6} \text{ M}$$
$$[K^+] = 0.055\ 0 \text{ M} \qquad [HPO_4^{2-}] = 0.022\ 0 \text{ M}$$
$$[OH^-] = 0.002\ 0 \text{ M} \qquad [PO_4^{3-}] = 0.003\ 0 \text{ M}$$

This calculation, which you should be able to do when you have finished studying acids and bases, takes into account the reaction of OH^- with $H_2PO_4^-$ to produce HPO_4^{2-} and PO_4^{3-}.

Are the charges balanced? Yes, indeed. Plugging into Equation 7-11, we find

$$[H^+] + [K^+] = [OH^-] + [H_2PO_4^-] + 2[HPO_4^{2-}] + 3[PO_4^{3-}]$$
$$5.1 \times 10^{-12} + 0.055\ 0 = 0.002\ 0 + 1.3 \times 10^{-6} + 2(0.022\ 0) + 3(0.003\ 0)$$
$$0.055\ 0 \text{ M} = 0.055\ 0 \text{ M}$$

The total positive charge is 0.055 0 M, and the total negative charge also is 0.055 0 M (Figure 7-6). Charges must balance in every solution. Otherwise, a beaker with excess positive charge would glide across the lab bench and smash into a beaker with excess negative charge.

FIGURE 7-6 Charge contributed by each ion in 1.00 L of solution containing 0.025 0 mol KH_2PO_4 plus 0.030 0 mol KOH. The total positive charge equals the total negative charge.

The general form of the charge balance for any solution is

Charge balance: $\qquad n_1[C_1] + n_2[C_2] + \cdots = m_1[A_1] + m_2[A_2] + \cdots \qquad$ (7-12)

where [C] is the concentration of a cation, n is the charge of the cation, [A] is the concentration of an anion, and m is the magnitude of the charge of the anion.

Σ[positive charges] = Σ[negative charges]

Activity coefficients do not appear in the charge balance. The charge contributed by 0.1 M H^+ is exactly 0.1 M. Think about this.

EXAMPLE   **Writing a Charge Balance**

Write the charge balance for a solution containing H_2O, H^+, OH^-, ClO_4^-, $Fe(CN)_6^{3-}$, CN^-, Fe^{3+}, Mg^{2+}, CH_3OH, HCN, NH_3, and NH_4^+.

Solution Neutral species (H_2O, CH_3OH, HCN, and NH_3) contribute no charge, so the charge balance is

$$[H^+] + 3[Fe^{3+}] + 2[Mg^{2+}] + [NH_4^+] = [OH^-] + [ClO_4^-] + 3[Fe(CN)_6^{3-}] + [CN^-]$$

Test Yourself   What would be the charge balance if you add $MgCl_2$ to the solution and it dissociates into $Mg^{2+} + 2Cl^-$? (*Answer:* $[H^+] + 3[Fe^{3+}] + 2[Mg^{2+}] + [NH_4^+] = [OH^-] + [ClO_4^-] + 3[Fe(CN)_6^{3-}] + [CN^-] + [Cl^-]$)

Mass Balance

The **mass balance**, also called the *material balance*, is a statement of the conservation of matter. The mass balance states that *the quantity of all species in a solution containing a particular atom (or group of atoms) must equal the amount of that atom (or group) delivered to the solution.* It is easier to see this relation through examples than by a general statement.

    Suppose that a solution is prepared by dissolving 0.050 mol of acetic acid in water to give a total volume of 1.00 L. Acetic acid partially dissociates into acetate:

$$\underset{\text{Acetic acid}}{CH_3CO_2H} \rightleftharpoons \underset{\text{Acetate}}{CH_3CO_2^-} + H^+$$

The mass balance is a statement of the conservation of matter. It really refers to conservation of atoms, not to mass.

The mass balance states that the quantity of dissociated and undissociated acetic acid in the solution must equal the amount of acetic acid put into the solution.

Mass balance for acetic acid in water: $\qquad \underset{\substack{\text{What we put into} \\ \text{the solution}}}{0.050\ M} = \underset{\substack{\text{Undissociated} \\ \text{product}}}{[CH_3CO_2H]} + \underset{\substack{\text{Dissociated} \\ \text{product}}}{[CH_3CO_2^-]}$

    When a compound dissociates in several ways, the mass balance must include all the products. Phosphoric acid (H_3PO_4), for example, dissociates to $H_2PO_4^-$, HPO_4^{2-}, and PO_4^{3-}. The mass balance for a solution prepared by dissolving 0.025 0 mol of H_3PO_4 in 1.00 L is

$$0.025\ 0\ M = [H_3PO_4] + [H_2PO_4^-] + [HPO_4^{2-}] + [PO_4^{3-}]$$

Activity coefficients do not appear in the mass balance. The concentration of each species counts exactly the number of atoms of that species.

EXAMPLE   **Mass Balance When the Total Concentration Is Known**

Write the mass balances for K^+ and for phosphate in a solution prepared by mixing 0.025 0 mol KH_2PO_4 plus 0.030 0 mol KOH and diluting to 1.00 L.

Solution The total K^+ is 0.025 0 M + 0.030 0 M, so one mass balance is

$$[K^+] = 0.055\ 0\ M$$

The total of *all forms* of phosphate is 0.025 0 M, so the mass balance for phosphate is

$$[H_3PO_4] + [H_2PO_4^-] + [HPO_4^{2-}] + [PO_4^{3-}] = 0.025\ 0\ M$$

Test Yourself   Write two mass balances for a 1.00-L solution containing 0.100 mol of sodium acetate. (*Answer:* $[Na^+] = 0.100$ M; $[CH_3CO_2H] + [CH_3CO_2^-] = 0.100$ M)

Now consider a solution prepared by dissolving $La(IO_3)_3$ in water.

$$La(IO_3)_3(s) \xrightleftharpoons{K_{sp}} La^{3+} + 3IO_3^-$$
$$\text{Iodate}$$

We do not know how much La^{3+} or IO_3^- is dissolved, but we do know that there must be three iodate ions for each lanthanum ion dissolved. That is, the iodate concentration must be three times the lanthanum concentration. If La^{3+} and IO_3^- are the only species derived from $La(IO_3)_3$, then the mass balance is

$$[IO_3^-] = 3[La^{3+}]$$

If the solution also contains the ion pair $LaIO_3^{2+}$ and the hydrolysis product $LaOH^{2+}$, the mass balance would be

$$[\text{Total iodate}] = 3[\text{total lanthanum}]$$
$$[IO_3^-] + [LaIO_3^{2+}] = 3\{[La^{3+}] + [LaIO_3^{2+}] + [LaOH^{2+}]\}$$

EXAMPLE **Mass Balance When the Total Concentration Is Unknown**

Write the mass balance for a saturated solution of the slightly soluble salt Ag_3PO_4, which produces PO_4^{3-} and $3Ag^+$ when it dissolves.

Solution If the phosphate in solution remained as PO_4^{3-}, we could write

$$[Ag^+] = 3[PO_4^{3-}]$$

because three silver ions are produced for each phosphate ion. However, phosphate reacts with water to give HPO_4^{2-}, $H_2PO_4^-$, and H_3PO_4, so the mass balance is

$$[Ag^+] = 3\{[PO_4^{3-}] + [HPO_4^{2-}] + [H_2PO_4^-] + [H_3PO_4]\}$$

> Atoms of Ag = 3(atoms of P)

That is, the number of atoms of Ag^+ must equal three times the total number of atoms of phosphorus, regardless of how many species contain phosphorus.

> Box 7-2 illustrates the operation of a mass balance in natural waters.

Test Yourself Write the mass balance for a saturated solution of $Ba(HSO_4)_2$ if the species in solution are Ba^{2+}, $BaSO_4(aq)$, HSO_4^-, SO_4^{2-}, and $BaOH^+$. (***Answer:*** 2 × total barium = total sulfate, or $2\{[Ba^{2+}] + [BaSO_4(aq)] + [BaOH^+]\} = [SO_4^{2-}] + [HSO_4^-] + [BaSO_4(aq)]$)

Systematic Treatment of Equilibrium

Now that we have considered the charge and mass balances, we are ready for the systematic treatment of equilibrium.[11] Here is the general prescription:

Step 1   Write the *pertinent reactions*.
Step 2   Write the *charge balance* equation.
Step 3   Write *mass balance* equations. There may be more than one.
Step 4   Write the *equilibrium constant expression* for each chemical reaction. This step is the only one in which activity coefficients appear.
Step 5   *Count the equations and unknowns.* There should be as many equations as unknowns (chemical species). If not, you must either find more equilibria or fix some concentrations at known values.
Step 6   *Solve* for all unknowns.

Steps 1 and 6 are the heart of the problem. Guessing what chemical equilibria exist in a given solution requires a fair degree of chemical intuition. In this book, you will usually be given help with step 1. Unless we know all the relevant equilibria, it is not possible to calculate the composition of a solution correctly. Because we do not know all the chemical reactions, we undoubtedly oversimplify many equilibrium problems. Step 6—solving the equations—is likely to be your biggest challenge. With n equations and n unknowns, the problem can be solved, at least in principle. In the simplest cases, you can do this by hand, but, for most problems, approximations are made or a spreadsheet is employed.

BOX 7-2 Calcium Carbonate Mass Balance in Rivers

Ca^{2+} is the most common cation in rivers and lakes. It comes from dissolution of the mineral calcite by the action of CO_2 to produce 2 moles of HCO_3^- for each mole of Ca^{2+}:

$$CaCO_3(s) + CO_2(aq) + H_2O \rightleftharpoons Ca^{2+} + 2HCO_3^- \quad \textbf{(A)}$$

Calcite   Bicarbonate

Near neutral pH, most of the product is bicarbonate, not CO_3^{2-} or H_2CO_3. The mass balance for the dissolution of calcite is therefore $[HCO_3^-] \approx 2[Ca^{2+}]$. Indeed, measurements of Ca^{2+} and HCO_3^- in many rivers conform to this mass balance, shown by the straight line on the graph. Rivers such as the Danube, the Mississippi, and the Congo, which lie on the line $[HCO_3^-] = 2[Ca^{2+}]$, appear to be saturated with calcium carbonate. If the river water were in equilibrium with atmospheric CO_2 ($P_{CO_2} = 10^{-3.4}$ bar), the concentration of Ca^{2+} would be 20 mg/L (see Problem 7-30). Rivers with more than 20 mg of Ca^{2+} per liter have a higher concentration of dissolved CO_2 produced by respiration or from inflow of groundwaters with a high CO_2 content. Rivers such as the Nile, the Niger, and the Amazon, for which $2[Ca^{2+}] < [HCO_3^-]$, are not saturated with $CaCO_3$.

Between 1960 and 2008, atmospheric CO_2 increased by 20% (Figure 0-4, page 4)—mostly from our burning of fossil fuel. This increase drives Reaction A to the right and threatens the existence of coral reefs,[9] which are living structures consisting largely of $CaCO_3$. Coral reefs are a unique habitat for many aquatic species. Continued buildup of atmospheric CO_2 threatens certain plankton with $CaCO_3$ shells,[10] a loss that, in turn, threatens higher members of the food chain.

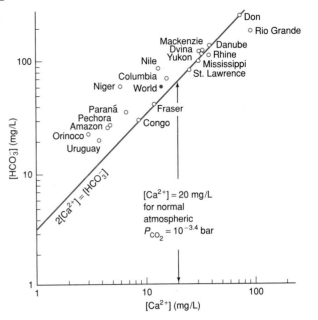

Concentrations of bicarbonate and calcium in many rivers conform to the mass balance $[HCO_3^-] \approx 2[Ca^{2+}]$. [Data from W. Stumm and J. J. Morgan, *Aquatic Chemistry*, 3rd ed. (New York: Wiley-Interscience, 1996), p. 189; and H. D. Holland, *The Chemistry of the Atmosphere and Oceans* (New York: Wiley-Interscience, 1978).]

7-5 Applying the Systematic Treatment of Equilibrium

Now we examine some problems to learn to apply the systematic treatment of equilibrium and to illustrate what can be done by hand and when a spreadsheet is really helpful.

A Solution of Ammonia

Let's find the concentrations of species in an aqueous solution containing 0.010 0 mol NH_3 in 1.000 L. The primary equilibrium is

$$NH_3 + H_2O \xrightleftharpoons{K_b} NH_4^+ + OH^- \qquad K_b = 1.76 \times 10^{-5} \text{ at } 25°C \qquad \textbf{(7-13)}$$

A second equilibrium in every aqueous solution is

$$H_2O \xrightleftharpoons{K_w} H^+ + OH^- \qquad K_w = 1.0 \times 10^{-14} \text{ at } 25°C \qquad \textbf{(7-14)}$$

Our goal is to find $[NH_3]$, $[NH_4^+]$, $[H^+]$, and $[OH^-]$.

Step 1   Pertinent reactions. They are 7-13 and 7-14.

Step 2   Charge balance. The sum of positive charge equals the sum of negative charge:

$$[NH_4^+] + [H^+] = [OH^-] \qquad \textbf{(7-15)}$$

Step 3   Mass balance. All of the ammonia delivered to the solution is either in the form NH_3 or NH_4^+. These two must add up to 0.010 0 M.

$$[NH_3] + [NH_4^+] = 0.010\ 0\ M \equiv F \qquad \textbf{(7-16)}$$

The symbol ≡ means "is defined as."

where F stands for formal concentration.

Step 4   Equilibrium expressions.

$$K_b = \frac{[NH_4^+]\gamma_{NH_4^+}[OH^-]\gamma_{OH^-}}{[NH_3]\gamma_{NH_3}} = 1.76 \times 10^{-5} \qquad \textbf{(7-17)}$$

$$K_w = [H^+]\gamma_{H^+}[OH^-]\gamma_{OH^-} = 1.0 \times 10^{-14} \qquad \textbf{(7-18)}$$

This is the only step in which activity coefficients enter the problem.

Step 5 Count equations and unknowns. We have four equations, 7-15 to 7-18, and four unknowns ($[NH_3]$, $[NH_4^+]$, $[H^+]$, and $[OH^-]$). We have enough information to solve the problem.

Step 6 Solve.

This "simple" problem is complicated. Let's start by ignoring activity coefficients; we will return to them later. Our approach is to eliminate one variable at a time until only one unknown is left. For an acid-base problem, I choose to express each concentration in terms of $[H^+]$. A substitution we can always make is $[OH^-] = K_w/[H^+]$. Putting this expression for $[OH^-]$ into the charge balance 7-15 gives

$$[NH_4^+] + [H^+] = \frac{K_w}{[H^+]}$$

from which we can solve for $[NH_4^+]$:

$$[NH_4^+] = \frac{K_w}{[H^+]} - [H^+] \tag{7-19}$$

The mass balance tells us that $[NH_3] = F - [NH_4^+]$. We can substitute the expression for $[NH_4^+]$ from Equation 7-19 into the mass balance to express $[NH_3]$ in terms of $[H^+]$:

$$[NH_3] = F - [NH_4^+] = F - \left(\frac{K_w}{[H^+]} - [H^+]\right) \tag{7-20}$$

Equation 7-19 gives $[NH_4^+]$ in terms of $[H^+]$. Equation 7-20 gives $[NH_3]$ in terms of $[H^+]$.

We can generate an equation in which the only unknown is $[H^+]$ by substituting our expressions for $[NH_4^+]$, $[NH_3]$, and $[OH^-]$ into the K_b equilibrium (still ignoring activity coefficients):

$$K_b = \frac{[NH_4^+][OH^-]}{[NH_3]} = \frac{\left(\dfrac{K_w}{[H^+]} - [H^+]\right)\left(\dfrac{K_w}{[H^+]}\right)}{\left(F - \dfrac{K_w}{[H^+]} + [H^+]\right)} \tag{7-21}$$

Equation 7-21 is horrible, but $[H^+]$ is the only unknown. One way you could solve for $[H^+]$ is to *guess* a value of $[H^+]$, put it into the right side of the equation, and see if the right side is equal to K_b on the left side. Of course, your guess would not make the right side of the equation equal to K_b. So guess a little smaller value of $[H^+]$ and see if you come closer to K_b. If your second answer is better than the first one, continue guessing smaller values of $[H^+]$ until you have solved the equation. If your second answer was worse than the first one, guess larger values of $[H^+]$ until you have solved the equation.

Excel has a procedure called Goal Seek, which follows a systematic guessing process to find the answer. We will learn to use Goal Seek later in this chapter.

After solving for $[H^+]$, we could calculate $[OH^-] = K_w/[H^+]$. We could find $[NH_4^+]$ from Equation 7-19 and we could compute $[NH_3]$ from the K_b equilibrium expression. Then we would have all four concentrations.

What could we do about the activity coefficients? From $[NH_4^+]$, $[H^+]$, and $[OH^-]$, we could calculate the ionic strength and then find the activity coefficients with Table 7-1 or the extended Debye-Hückel equation. Then we could put these coefficients into Equations 7-17 and 7-18 and solve the problem all over again. The new concentrations would be different from what we found without activity coefficients. We could use the new concentrations to find a new ionic strength and new activity coefficients and another set of new concentrations. We could repeat this process until differences between successive answers become negligible (say, less than 1%). At that point, we have solved the problem.

You should now appreciate that applying the systematic treatment of equilibrium to even the simplest of problems is not simple. In most equilibrium problems, we will make simplifying approximations to reach a good answer with a reasonable effort. After solving a problem, we must always verify that our approximations are valid.

Here is an approximation to simplify the ammonia problem. Ammonia is a base, so we expect that $[OH^-] \gg [H^+]$. For example, suppose the pH comes out to 9. Then $[H^+] = 10^{-9}$ M and $[OH^-](= K_w/[H^+]) = 10^{-14}/10^{-9} = 10^{-5}$ M. That is, $[OH^-] \gg [H^+]$. In the first term of the numerator of Equation 7-21, we can neglect $[H^+]$ in comparison with $K_w/[H^+]$.

In the denominator, we can also neglect $[H^+]$ in comparison with $K_w/[H^+]$. With these approximations, Equation 7-21 becomes

$$K_b = \frac{\left(\dfrac{K_w}{[H^+]} - \cancel{[H^+]}\right)\left(\dfrac{K_w}{[H^+]}\right)}{\left(F - \dfrac{K_w}{[H^+]} + \cancel{[H^+]}\right)} = \frac{\left(\dfrac{K_w}{[H^+]}\right)\left(\dfrac{K_w}{[H^+]}\right)}{\left(F - \dfrac{K_w}{[H^+]}\right)} = \frac{[OH^-]^2}{F - [OH^-]} \qquad (7\text{-}22)$$

Equation 7-22 is quadratic with one variable $[OH^-]$. We can solve for $[OH^-]$ by algebra. We will deal with equations of this type extensively in the next chapter on acids and bases.

Next, we take two examples all the way through to numerical answers.

Solubility of Calcium Sulfate

Let's find the concentrations of the major species in a saturated solution of $CaSO_4$.

Step 1 Pertinent reactions. Even in such a simple system, there are quite a few reactions:

$$CaSO_4(s) \xrightleftharpoons{K_{sp}} Ca^{2+} + SO_4^{2-} \qquad K_{sp} = 2.4 \times 10^{-5} \qquad (7\text{-}23)$$

Caveat Emptor! In all equilibrium problems, we are limited by how much of the system's chemistry is understood. Without all relevant equilibria, it is impossible to calculate the composition.

$$CaSO_4(s) \xrightleftharpoons{K_{\text{ion pair}}} CaSO_4(aq) \qquad K_{\text{ion pair}} = 5.0 \times 10^{-3} \qquad (7\text{-}24)$$

$$Ca^{2+} + H_2O \xrightleftharpoons{K_{\text{acid}}} CaOH^+ + H^+ \qquad K_{\text{acid}} = 2.0 \times 10^{-13} \qquad (7\text{-}25)$$

$$SO_4^{2-} + H_2O \xrightleftharpoons{K_{\text{base}}} HSO_4^- + OH^- \qquad K_{\text{base}} = 9.8 \times 10^{-13} \qquad (7\text{-}26)$$

$$H_2O \xrightleftharpoons{K_w} H^+ + OH^- \qquad K_w = 1.0 \times 10^{-14} \qquad (7\text{-}27)$$

There is no way you can be expected to come up with all of these reactions, so you will be given help with this step.

Step 2 Charge balance. Equating positive and negative charges gives

$$2[Ca^{2+}] + [CaOH^+] + [H^+] = 2[SO_4^{2-}] + [HSO_4^-] + [OH^-] \qquad (7\text{-}28)$$

Multiply $[Ca^{2+}]$ and $[SO_4^{2-}]$ by 2 because 1 mol of each ion has 2 mol of charge.

Step 3 Mass balance. Reaction 7-23 produces 1 mole of sulfate for each mole of calcium. No matter what happens to these ions, the total concentration of all species with sulfate must equal the total concentration of all species with calcium:

$$[\text{Total calcium}] = [\text{total sulfate}]$$

$$[Ca^{2+}] + [CaSO_4(aq)] + [CaOH^+] = [SO_4^{2-}] + [HSO_4^-] + [CaSO_4(aq)] \qquad (7\text{-}29)$$

Step 4 Equilibrium constant expressions. There is one for each chemical reaction.

$$K_{sp} = [Ca^{2+}]\gamma_{Ca^{2+}}[SO_4^{2-}]\gamma_{SO_4^{2-}} = 2.4 \times 10^{-5} \qquad (7\text{-}30)$$

$$K_{\text{ion pair}} = [CaSO_4(aq)] = 5.0 \times 10^{-3} \qquad (7\text{-}31)$$

$$K_{\text{acid}} = \frac{[CaOH^+]\gamma_{CaOH^+}[H^+]\gamma_{H^+}}{[Ca^{2+}]\gamma_{Ca^{2+}}} = 2.0 \times 10^{-13} \qquad (7\text{-}32)$$

$$K_{\text{base}} = \frac{[HSO_4^-]\gamma_{HSO_4^-}[OH^-]\gamma_{OH^-}}{[SO_4^{2-}]\gamma_{SO_4^{2-}}} = 9.8 \times 10^{-13} \qquad (7\text{-}33)$$

$$K_w = [H^+]\gamma_{H^+}[OH^-]\gamma_{OH^-} = 1.0 \times 10^{-14} \qquad (7\text{-}34)$$

Step 4 is the only one where activity coefficients come in.

Step 5 Count equations and unknowns. There are seven equations (7-28 through 7-34) and seven unknowns: $[Ca^{2+}]$, $[SO_4^{2-}]$, $[CaSO_4(aq)]$, $[CaOH^+]$, $[HSO_4^-]$, $[H^+]$, and $[OH^-]$.

Step 6 Solve. Well, this is not easy! We don't know the ionic strength, so we cannot evaluate activity coefficients. Also, where do we start when there are seven unknowns?

Just when the world looks grim, a Good Chemist comes galloping down the hill on her white stallion to rescue us with some hints. "First of all," she says, "Reactions 7-25 and 7-26 have small equilibrium constants, so they are not very important compared with Reactions 7-23 and 7-24. As for activity coefficients, you can begin by setting them all equal to 1 and solving for all concentrations. Then you can calculate the ionic strength and find values for the activity coefficients. With these coefficients, you can solve the problem a second time and find a new set of concentrations. You can repeat this process a few times until you reach a constant answer."

On the advice of the Good Chemist, we neglect the acid-base reactions 7-25 and 7-26. The remaining reactions of calcium sulfate, 7-23 and 7-24, are not reactions with H_2O. Therefore, in our approximation, Reaction 7-27 is independent of the calcium sulfate chemistry. Reaction 7-27 produces $[H^+]$ and $[OH^-]$, which we know are 1×10^{-7} M. These concentrations of H^+ and OH^- are not exactly correct, because we have neglected Reactions 7-25 and 7-26, which do play a role in determining $[H^+]$ and $[OH^-]$.

These approximations leave just Reactions 7-23 and 7-24. In the charge balance 7-28, $CaOH^+$ and HSO_4^- are neglected. Also, $[H^+] = [OH^-]$, so they cancel. The charge balance becomes $2[Ca^{2+}] = 2[SO_4^{2-}]$. In the mass balance 7-29, we discard $CaOH^+$ and HSO_4^- because we neglect reactions 7-25 and 7-26. The term $[CaSO_4(aq)]$ cancels, leaving $[Ca^{2+}] = [SO_4^{2-}]$, which is the same as the charge balance. We have three unknowns ($[Ca^{2+}]$, $[SO_4^{2-}]$, $[CaSO_4(aq)]$) and three equations:

The activity coefficient of the neutral species $CaSO_4(aq)$ is 1.

$$K_{sp} = [Ca^{2+}]\gamma_{Ca^{2+}}[SO_4^{2-}]\gamma_{SO_4^{2-}} = 2.4 \times 10^{-5} \qquad (7\text{-}30)$$

$$K_{ion\ pair} = [CaSO_4(aq)] = 5.0 \times 10^{-3} \qquad (7\text{-}31)$$

$$[Ca^{2+}] = [SO_4^{2-}] \qquad (7\text{-}35)$$

Equation 7-31 says $[CaSO_4(aq)] = 5.0 \times 10^{-3}$ M, so $[CaSO_4(aq)]$ is known.

The simplified problem is reduced to Equations 7-30 and 7-35. We find a first approximate solution by setting activity coefficients to 1:

$$[Ca^{2+}]_1\gamma_{Ca^{2+}}[SO_4^{2-}]_1\gamma_{SO_4^{2-}} = 2.4 \times 10^{-5} \qquad (7\text{-}36)$$

$$[Ca^{2+}]_1(1)[SO_4^{2-}]_1(1) = 2.4 \times 10^{-5}$$

$$[Ca^{2+}]_1(1)[Ca^{2+}]_1(1) = 2.4 \times 10^{-5} \Rightarrow [Ca^{2+}]_1 = 4.9 \times 10^{-3}\ M$$

We are carrying out a method of *successive approximations*. Each cycle is called an *iteration*.

where the subscript 1 means it is our first approximation. If $[Ca^{2+}] = [SO_4^{2-}] = 4.9 \times 10^{-3}$ M, then the ionic strength is $\mu = 4(4.9 \times 10^{-3}\ M) = 0.020$ M. Interpolating in Table 7-1, we find the activity coefficients $\gamma_{Ca^{2+}} = 0.628$ and $\gamma_{SO_4^{2-}} = 0.606$. Putting these coefficients back into Equation 7-36 provides a second approximation:

$$[Ca^{2+}]_2(0.628)[Ca^{2+}]_2(0.606) = 2.4 \times 10^{-5} \Rightarrow [Ca^{2+}]_2 = 7.9 \times 10^{-3}\ M$$

$$\Rightarrow \mu = 0.032\ M$$

Repeating this process gives the following results:

Iteration	$\gamma_{Ca^{2+}}$	$\gamma_{SO_4^{2-}}$	$[Ca^{2+}]$ (M)	μ (M)
1	1	1	0.004 9	0.020
2	0.628	0.606	0.007 9	0.032
3	0.570	0.542	0.008 8	0.035
4	0.556	0.526	0.009 1	0.036
5	0.551	0.520	0.009 2	0.037
6	0.547	0.515	0.009 2	0.037

Total concentration of dissolved sulfate
$= [SO_4^{2-}] + [CaSO_4(aq)]$
$= 0.009\ 2 + 0.005\ 0 = 0.014\ 2$ M

which is near the measured value in Figure 6-1. With the extended Debye-Hückel equation, instead of interpolation, total dissolved sulfate = 0.014 7 M.

The sixth iteration gives the same concentration as the fifth, so we have reached a constant answer.

With the advice of the Good Chemist, we simplified the problem tremendously. Now we need to know if her advice was good. From $[Ca^{2+}] = [SO_4^{2-}] = 0.009\ 2$ M and $[H^+] = [OH^-] = 1.0 \times 10^{-7}$ M, we can estimate $[CaOH^+]$ and $[HSO_4^-]$ to see if they are negligible. We are just looking for orders of magnitude, so we ignore activity coefficients. From Equation 7-32, we have

$$\frac{[CaOH^+][H^+]}{[Ca^{2+}]} = 2.0 \times 10^{-13}$$

$$[CaOH^+] \approx \frac{(2.0 \times 10^{-13})[Ca^{2+}]}{[H^+]} = \frac{(2.0 \times 10^{-13})[0.009\ 2]}{[1.0 \times 10^{-7}]} = 2 \times 10^{-8}\ M$$

From Equation 7-33, we estimate

$$[HSO_4^-] \approx \frac{(9.8 \times 10^{-13})[SO_4^{2-}]}{[OH^-]} = \frac{(9.8 \times 10^{-13})[0.009\ 2]}{[1.0 \times 10^{-7}]} = 9 \times 10^{-8}\ M$$

Both $[CaOH^+]$ and $[HSO_4^-]$ are $\sim 10^5$ times less than $[Ca^{2+}]$ and $[SO_4^{2-}]$, so it was reasonable to neglect $[CaOH^+]$ and $[HSO_4^-]$ in the charge and mass balances.

We simplified the problem to the point where we could assume that the pH is close to 7.00. If you need a better answer, Section 12-3 provides a method to find the pH using all five chemical reactions 7-23 through 7-27. The result in Problem 12-13 is pH = 7.06.

We Will Usually Omit Activity Coefficients

Although it is proper to write equilibrium constants in terms of activities, the complexity of manipulating activity coefficients is a nuisance. Most of the time, we will omit activity coefficients unless there is a particular point to be made. Occasional problems will remind you how to use activities.

Solubility of Magnesium Hydroxide

Let's find the concentrations of species in a saturated solution of $Mg(OH)_2$, given the following chemistry. For simplicity, we ignore activity coefficients.

$$Mg(OH)_2(s) \overset{K_{sp}}{\rightleftharpoons} Mg^{2+} + 2OH^- \qquad K_{sp} = [Mg^{2+}][OH^-]^2 = 7.1 \times 10^{-12} \quad (7\text{-}37)$$

$$Mg^{2+} + OH^- \overset{K_1}{\rightleftharpoons} MgOH^+ \qquad K_1 = \frac{[MgOH^+]}{[Mg^{2+}][OH^-]} = 3.8 \times 10^2 \quad (7\text{-}38)$$

$$H_2O \overset{K_w}{\rightleftharpoons} H^+ + OH^- \qquad K_w = [H^+][OH^-] = 1.0 \times 10^{-14} \quad (7\text{-}39)$$

Step 1 Pertinent reactions are listed above.

Step 2 Charge balance:

$$2[Mg^{2+}] + [MgOH^+] + [H^+] = [OH^-] \qquad (7\text{-}40)$$

Step 3 Mass balance. This is a little tricky. From Reaction 7-37, we could say that the concentrations of all species containing OH^- equal two times the concentrations of all magnesium species. However, Reaction 7-39 also creates 1 OH^- for each H^+. The mass balance accounts for both sources of OH^-:

$$\underbrace{[OH^-] + [MgOH^+]}_{\text{Species containing } OH^-} = 2\underbrace{\{[Mg^{2+}] + [MgOH^+]\}}_{\text{Species containing } Mg^{2+}} + [H^+] \qquad (7\text{-}41)$$

After all this work, Equation 7-41 is equivalent to Equation 7-40.

Step 4 Equilibrium constant expressions are in Equations 7-37 through 7-39.

Step 5 Count equations and unknowns. We have four equations (7-37 to 7-40) and four unknowns: $[Mg^{2+}]$, $[MgOH^+]$, $[H^+]$, and $[OH^-]$.

Step 6 Solve.

Before hitting the algebra, we can make one simplification. The solution must be basic because we made it from $Mg(OH)_2$. In basic solution, $[OH^-] \gg [H^+]$, so we can neglect $[H^+]$ on the left side of Equation 7-40 in comparison with $[OH^-]$ on the right side. The charge balance simplifies to

$$2[Mg^{2+}] + [MgOH^+] = [OH^-] \qquad (7\text{-}42)$$

From the K_1 expression 7-38, we write $[MgOH^+] = K_1[Mg^{2+}][OH^-]$. Substituting this expression for $[MgOH^+]$ into Equation 7-42 gives

$$2[Mg^{2+}] + K_1[Mg^{2+}][OH^-] = [OH^-]$$

which we solve for $[Mg^{2+}]$:

$$[Mg^{2+}] = \frac{[OH^-]}{2 + K_1[OH^-]}$$

Substituting this expression for $[Mg^{2+}]$ into the solubility product reduces the equation to a single variable:

$$K_{sp} = [Mg^{2+}][OH^-]^2 = \left(\frac{[OH^-]}{2 + K_1[OH^-]}\right)[OH^-]^2 = \frac{[OH^-]^3}{2 + K_1[OH^-]} \qquad (7\text{-}43)$$

	A	B	C	D
1	Mg(OH)$_2$ Solubility			
2				
3	K$_{sp}$ =		[OH$^-$]$_{guess}$ =	[OH$^-$]^3/(2 + K$_1$[OH$^-$]) =
4	7.1E-12		0.0002459	7.1000E-12
5	K$_1$ =			
6	3.8E+02		[Mg^{2+}] =	[MgOH$^+$] =
7			0.0001174	0.0000110
8				
9	D4 = C4^3/(2+A6*C4)			
10	C7 = A4/C4^2			
11	D8 = A6*C7*C4			

FIGURE 7-7 Spreadsheet for solving Equation 7-43.

We are down to solving the ugly Equation 7-43 for [OH$^-$]. Just when the world looks grim again, the Good Chemist says, "Use a spreadsheet to vary [OH$^-$] until Equation 7-43 is satisfied." We do this in Figure 7-7, where we *guess* a value of [OH$^-$] in cell C4 and evaluate the right side of Equation 7-43 in cell D4. When we have guessed the correct value of [OH$^-$], cell D4 is equal to K_{sp}. Guessing is done by systematically changing [OH$^-$] in cell C4 until cell D4 has the desired value of 7.1×10^{-12}.

Solving Equations with Excel Goal Seek®

A better procedure to solve Equation 7-43 is to use Excel Goal Seek to vary cell C4 until cell D4 is equal to K_{sp}. Put any guess for [OH$^-$], such as 0.01, into cell C4 in Figure 7-7. Before using Goal Seek in Excel 2007, click the Microsoft Office button at the upper left of the spreadsheet. At the bottom of the Office window, click Excel Options. At the left side of the next window, select Formulas. In Calculation Options, set Maximum Change to 1e-24, which enables Excel to look for differences smaller than the target value of 7.1×10^{-12}. Click OK. Return to the spreadsheet and select the Data ribbon. In Data Tools, click What-If Analysis and select Goal Seek. When the window at the right of Figure 7-7 appears, fill in the underlined values: Set cell <u>D4</u> To value <u>7.1e-12</u> by changing cell <u>C4</u>. Click OK and Excel varies cell C4 until cell D4 is equal to 7.1E-12. We find [OH$^-$] = 2.459×10^{-4}M. Figure 7-7 also computes [Mg^{2+}] and [MgOH$^+$] from Equations 7-37 and 7-38. The results confirm the approximation we made at the outset: [H$^+$] = K_w/[OH$^-$] = 4.1×10^{-11} M $\ll$ [OH$^-$].

In earlier versions of Excel, before using Goal Seek, go to the Tools menu and select Options. Select the Calculations tab and set Maximum change to 1e-24. Go back to Tools, select Goal Seek, and proceed as described above.

Excel Goal Seek

Terms to Understand

activity	extended Debye-Hückel	ionic strength
activity coefficient	equation	mass balance
charge balance	ionic atmosphere	pH

Summary

The thermodynamic equilibrium constant for the reaction $aA + bB \rightleftharpoons cC + dD$ is $K = \mathcal{A}_C^c \mathcal{A}_D^d/(\mathcal{A}_A^a \mathcal{A}_B^b)$, where $\mathcal{A}_i$ is the activity of the ith species. The activity is the product of the concentration (c) and the activity coefficient (γ): $\mathcal{A}_i = c_i\gamma_i$. For nonionic compounds and gases, $\gamma_i \approx 1$. For ionic species, the activity coefficient depends on the ionic strength, defined as $\mu = \frac{1}{2}\Sigma c_i z_i^2$, where z_i is the charge of an ion. The activity coefficient decreases as ionic strength increases, at least for low ionic strengths (≤ 0.1 M). Dissociation of ionic compounds increases with ionic strength because the ionic atmosphere of each ion diminishes the attraction of ions for one another. You should be able to estimate activity coeffi-

cients by interpolation in Table 7-1. pH is defined in terms of the activity of H$^+$: pH = $-\log \mathcal{A}_{H^+}$ = $-\log[H^+]\gamma_{H^+}$.

In the systematic treatment of equilibrium, we write pertinent equilibrium expressions, as well as the charge and mass balances. The charge balance states that the sum of all positive charges in solution equals the sum of all negative charges. The mass balance states that the moles of all forms of an element in solution must equal the moles of that element delivered to the solution. We make certain that we have as many equations as unknowns and then solve for the concentrations by using algebra, approximations, spreadsheets, magic, or anything else.

Exercises

7-A. Assuming complete dissociation of the salts, calculate the ionic strength of (a) 0.2 mM KNO_3; (b) 0.2 mM Cs_2CrO_4; (c) 0.2 mM $MgCl_2$ plus 0.3 mM $AlCl_3$.

7-B. Find the activity (not the activity coefficient) of the $(C_3H_7)_4N^+$ (tetrapropylammonium) ion in a solution containing 0.005 0 M $(C_3H_7)_4N^+Br^-$ plus 0.005 0 M $(CH_3)_4N^+Cl^-$.

7-C. Using activities, find $[Ag^+]$ in 0.060 M KSCN saturated with AgSCN(s).

7-D. Using activities, calculate the pH and concentration of H^+ in 0.050 M LiBr at 25°C.

7-E. A 40.0-mL solution of 0.040 0 M $Hg_2(NO_3)_2$ was titrated with 60.0 mL of 0.100 M KI to precipitate Hg_2I_2 ($K_{sp} = 4.6 \times 10^{-29}$).

(a) Show that 32.0 mL of KI are needed to reach the equivalence point.

(b) When 60.0 mL of KI have been added, virtually all Hg_2^{2+} has precipitated, along with 3.20 mmol of I^-. Considering all ions remaining in the solution, calculate the ionic strength when 60.0 mL of KI have been added.

(c) Using activities, calculate $pHg_2^{2+}(= -\log \mathcal{A}_{Hg_2^{2+}})$ for part (b).

7-F. (a) Write the mass balance for $CaCl_2$ in water if the species are Ca^{2+} and Cl^-.

(b) Write the mass balance if the species are Ca^{2+}, Cl^-, $CaCl^+$, and $CaOH^+$.

(c) Write the charge balance for part (b).

7-G. Write the charge and mass balances for dissolving CaF_2 in water if the reactions are

$$CaF_2(s) \rightleftharpoons Ca^{2+} + 2F^-$$
$$Ca^{2+} + H_2O \rightleftharpoons CaOH^+ + H^+$$
$$Ca^{2+} + F^- \rightleftharpoons CaF^+$$
$$CaF_2(s) \rightleftharpoons CaF_2(aq)$$
$$F^- + H^+ \rightleftharpoons HF(aq)$$
$$HF(aq) + F^- \rightleftharpoons HF_2^-$$

7-H. Write charge and mass balances for aqueous $Ca_3(PO_4)_2$ if the species are Ca^{2+}, $CaOH^+$, $CaPO_4^-$, PO_4^{3-}, HPO_4^{2-}, $H_2PO_4^-$, and H_3PO_4.

7-I. 🖳 (*Warning: long problem!*) Using activities, find the concentrations of the major species in 0.10 M $NaClO_4$ saturated with $Mn(OH)_2$. Take the ionic strength to be 0.10 M and suppose that the ion size of $MnOH^+$ is the same as Mn^{2+}. Consider just the following chemistry:

$$Mn(OH)_2(s) \overset{K_{sp}}{\rightleftharpoons} Mn^{2+} + 2OH^- \qquad K_{sp} = 1.6 \times 10^{-13}$$
$$Mn^{2+} + OH^- \overset{K_1}{\rightleftharpoons} MnOH^+ \qquad K_1 = 2.5 \times 10^3$$
$$H_2O \overset{K_w}{\rightleftharpoons} H^+ + OH^- \qquad K_w = 1.0 \times 10^{-14}$$

Problems

Activity Coefficients

7-1. Explain why the solubility of an ionic compound increases as the ionic strength of the solution increases (at least up to ~0.5 M).

7-2. Which statements are true? In the ionic strength range 0–0.1 M, activity coefficients decrease with (a) increasing ionic strength; (b) increasing ionic charge; (c) decreasing hydrated radius.

7-3. Calculate the ionic strength of (a) 0.008 7 M KOH and (b) 0.000 2 M $La(IO_3)_3$ (assuming complete dissociation at this low concentration and no hydrolysis reaction to make $LaOH^{2+}$).

7-4. Find the activity coefficient of each ion at the indicated ionic strength:

(a) SO_4^{2-} ($\mu = 0.01$ M)
(b) Sc^{3+} ($\mu = 0.005$ M)
(c) Eu^{3+} ($\mu = 0.1$ M)
(d) $(CH_3CH_2)_3NH^+$ ($\mu = 0.05$ M)

7-5. Interpolate in Table 7-1 to find the activity coefficient of H^+ when $\mu = 0.030$ M.

7-6. Calculate the activity coefficient of Zn^{2+} when $\mu = 0.083$ M by using (a) Equation 7-6; (b) linear interpolation in Table 7-1.

7-7. Calculate the activity coefficient of Al^{3+} when $\mu = 0.083$ M by linear interpolation in Table 7-1.

7-8. The equilibrium constant for dissolution in water of a nonionic compound, such as diethyl ether ($CH_3CH_2OCH_2CH_3$), can be written

$$ether(l) \rightleftharpoons ether(aq) \qquad K = [ether(aq)]\gamma_{ether}$$

At low ionic strength, $\gamma \approx 1$ for neutral compounds. At high ionic strength, most neutral molecules can be *salted out* of aqueous solution. That is, when a high concentration (typically >1 M) of a salt such as NaCl is added to an aqueous solution, neutral molecules usually become *less* soluble. Does the activity coefficient, γ_{ether}, increase or decrease at high ionic strength?

7-9. Including activity coefficients, find $[Hg_2^{2+}]$ in saturated Hg_2Br_2 in 0.001 00 M KBr.

7-10. Including activity coefficients, find the concentration of Ba^{2+} in a 0.100 M $(CH_3)_4NIO_3$ solution saturated with $Ba(IO_3)_2$.

7-11. Find the activity coefficient of H^+ in a solution containing 0.010 M HCl plus 0.040 M $KClO_4$. What is the pH of the solution?

7-12. Using activities, calculate the pH of a solution containing 0.010 M NaOH plus 0.012 0 M $LiNO_3$. What would be the pH if you neglected activities?

7-13. The temperature-dependent form of the extended Debye-Hückel equation 7-6 is

$$\log \gamma = \frac{(-1.825 \times 10^6)(\varepsilon T)^{-3/2}z^2\sqrt{\mu}}{1 + \alpha\sqrt{\mu}/(2.00\sqrt{\varepsilon T})}$$

where ε is the (dimensionless) dielectric constant[†] of water, T is temperature (K), z is the charge of the ion, μ is ionic strength (mol/L), and α is the ion size parameter (pm). The dependence of ε on temperature is

$$\varepsilon = 79.755 e^{(-4.6 \times 10^{-3})(T - 293.15)}$$

Calculate the activity coefficient of SO_4^{2-} at 50.00°C when $\mu = 0.100$ M. Compare your value with the one in Table 7-1.

7-14. 🖩 *Extended Debye-Hückel equation.* Use Equation 7-6 to calculate the activity coefficient (γ) as a function of ionic strength (μ) for $\mu = 0.000\ 1$, $0.000\ 3$, 0.001, 0.003, 0.01, 0.03, and 0.1 M.

(a) For an ionic charge of ± 1 and a size $\alpha = 400$ pm, make a table of γ (= $10^{\wedge}(\log \gamma)$) for each value of μ.

(b) Do the same for ionic charges of ± 2, ± 3, and ± 4.

(c) Plot γ versus $\log \mu$ to obtain a graph similar to Figure 7-4.

7-15. *Activity coefficient of a neutral molecule.* We use the approximation that the activity coefficient (γ) of neutral molecules is 1.00. A more accurate relation is $\log \gamma = k\mu$, where μ is ionic strength and $k \approx 0.11$ for NH_3 and CO_2 and $k \approx 0.2$ for organic molecules. With activity coefficients for HA, A^-, and H^+, predict the value of the quotient below for benzoic acid (HA $\equiv C_6H_5CO_2H$). The observed quotient is 0.63 ± 0.03.[12]

$$\text{Concentration quotient} = \dfrac{\dfrac{[H^+][A^-]}{[HA]}\,(\text{at } \mu = 0)}{\dfrac{[H^+][A^-]}{[HA]}\,(\text{at } \mu = 0.1 \text{ M})}$$

Systematic Treatment of Equilibrium

7-16. State the meaning of the charge and mass balance equations.

7-17. Why do activity coefficients not appear in the charge and mass balance equations?

7-18. Write the charge balance for a solution containing H^+, OH^-, Ca^{2+}, HCO_3^-, CO_3^{2-}, $Ca(HCO_3)^+$, $Ca(OH)^+$, K^+, and ClO_4^-.

7-19. Write the charge balance for a solution of H_2SO_4 in water if H_2SO_4 ionizes to HSO_4^- and SO_4^{2-}.

7-20. Write the charge balance for an aqueous solution of arsenic acid, H_3AsO_4, in which the acid can dissociate to $H_2AsO_4^-$, $HAsO_4^{2-}$, and AsO_4^{3-}. Look up the structure of arsenic acid in Appendix G and write the structure of $HAsO_4^{2-}$.

7-21. (a) Write the charge and mass balances for a solution made by dissolving $MgBr_2$ to give Mg^{2+}, Br^-, $MgBr^+$, and $MgOH^+$.

(b) Modify the mass balance if the solution was made by dissolving 0.2 mol $MgBr_2$ in 1 L.

[†] The dimensionless *dielectric constant*, ε, measures how well a solvent can separate oppositely charged ions. The force of attraction (newtons) between ions of charge q_1 and q_2 (coulombs) separated by distance r (meters) is

$$\text{Force} = -(8.988 \times 10^9)\dfrac{q_1 q_2}{\varepsilon r^2}$$

The larger the value of ε, the smaller the attraction between ions. Water, with $\varepsilon \approx 80$, separates ions very well. Here are some values of ε: methanol, 33; ethanol, 24; benzene, 2; vacuum and air, 1. Ionic compounds dissolved in solvents less polar than water may exist predominantly as ion pairs, not separate ions.

7-22. What would happen if charge balance did not exist in a solution? The force between two charges is given in the footnote on this page. Find the force between two beakers separated by 1.5 m of air if one contains 250 mL with 1.0×10^{-6} M excess negative charge and the other has 250 mL with 1.0×10^{-6} M excess positive charge. There are 9.648×10^4 coulombs per mole of charge. Convert force from N into pounds (0.224 8 pounds/N). Could two elephants hold the beakers apart?

7-23. For a 0.1 M aqueous solution of sodium acetate, $Na^+CH_3CO_2^-$, one mass balance is simply $[Na^+] = 0.1$ M. Write a mass balance involving acetate.

7-24. Consider the dissolution of the compound X_2Y_3, which gives $X_2Y_2^{2+}$, X_2Y^{4+}, $X_2Y_3(aq)$, and Y^{2-}. Use the mass balance to find an expression for $[Y^{2-}]$ in terms of the other concentrations. Simplify your answer as much as possible.

7-25. Write a mass balance for a solution of $Fe_2(SO_4)_3$ if the species are Fe^{3+}, $Fe(OH)^{2+}$, $Fe(OH)_2^+$, $Fe_2(OH)_2^{4+}$, $FeSO_4^+$, SO_4^{2-}, and HSO_4^-.

7-26. (a) Following the example of ammonia in Section 7-5, write the equilibria and charge and mass balances needed to find the composition of 0.01 M sodium acetate, which you should abbreviate as Na^+A^-. *Include activity coefficients* where appropriate.

(b) Neglect activity coefficients for this part and call the formal concentration F. Following the example of ammonia, reduce the equations to a single equation with $[H^+]$ as the sole unknown. Do not try to simplify the equation.

(c) 🖩 Use Excel Goal Seek to solve for $[H^+]$. Use the value $K_b = 5.7 \times 10^{-10}$. Before using Goal Seek in Excel 2007, click the Microsoft Office button, then Excel Options, and select Formulas. In Calculation Options, set Maximum Change to 1e-14. In earlier versions of Excel, go to Tools and Options. Select the Calculations tab and set Maximum change to 1e-14. From $[H^+]$, find $[OH^-]$, $[A^-]$, and [HA].

7-27. (a) Following the example of $Mg(OH)_2$ in Section 7-5, write the equations needed to find the solubility of $Ca(OH)_2$. *Include activity coefficients* where appropriate. Equilibrium constants are in Appendixes F and I.

(b) Neglecting activity coefficients, compute the concentrations of all species and compute the solubility of $Ca(OH)_2$ in g/L.

7-28. Look up the equilibrium constant for the *ion-pairing* reaction $Zn^{2+} + SO_4^{2-} \rightleftharpoons ZnSO_4(aq)$ in Appendix J.

(a) Use the systematic treatment of equilibrium to find $[Zn^{2+}]$ in 0.010 F $ZnSO_4$. Neglect activity coefficients and any other reactions.

(b) Use the answer from part (a) to compute the ionic strength and activity coefficients of Zn^{2+} and SO_4^{2-}. Then repeat the calculation using activity coefficients. Repeat the procedure two more times to find a good estimate of $[Zn^{2+}]$. What percentage is ion paired? What is the ionic strength of the solution?

(c) Two possibly important reactions that we did not consider are acid hydrolysis of Zn^{2+} and base hydrolysis of SO_4^{2-}. Write these two reactions and find their equilibrium constants in Appendixes I and G.

7-29. *Finding solubility by iteration.* Use the systematic treatment of equilibrium to find the concentrations of the major species in a saturated aqueous solution of LiF. Consider these reactions:

$$LiF(s) \rightleftharpoons Li^+ + F^- \qquad K_{sp} = [Li^+]\gamma_{Li^+}[F^-]\gamma_{F^-} = 0.001\ 7$$

$$LiF(s) \rightleftharpoons LiF(aq) \qquad K_{ion\ pair} = [LiF(aq)]\gamma_{LiF(aq)} = 0.002\ 9$$

(a) Initially, set the ionic strength to 0 and solve for all concentrations. Then compute the ionic strength and activity coefficients and find new concentrations. Use several iterations to home in on the correct solution.

(b) ▦ In the systematic treatment, you will find that the calculation simplifies to $[Li^+] = [F^-] = \sqrt{K_{sp}/(\gamma_{Li^+}\gamma_{F^-})}$. Set up the following spreadsheet, in which ionic strength in cell B4 is initially given the value 0. Activity coefficients in cells B6 and B8 are from the extended Debye-Hückel equation. Cell B10 computes $[Li^+] = [F^-] = \sqrt{K_{sp}/(\gamma_{Li^+}\gamma_{F^-})}$. With 0 in cell B4, your spreadsheet should compute 1 in cells B6 and B8 and $[Li^+] = [F^-] = 0.041\ 23$ M in cell B10.

	A	B	C
1	Spreadsheet for iterative LiF solubility computation		
2			
3	Size (pm) of Li$^+$ =	Ionic strength =	
4	600	0.00000	
5	Size (pm) of F$^-$ =	Activity coeff Li$^+$ =	
6	350	1	
7	K$_{sp}$ =	Activity coeff F$^-$ =	
8	0.0017	1	
9		[Li$^+$] = [F$^-$] =	
10		0.04123	
11			
12	B6 = 10^((-0.51)*SQRT(B4)/(1+A4*SQRT(B4)/305))		
13	B8 = 10^((-0.51)*SQRT(B4)/(1+A6*SQRT(B4)/305))		
14	B10 = SQRT(A8/(B6*B8))		

Because the ionic strength of a 1:1 electrolyte is equal to the concentration, copy the value 0.041 23 from cell B10 into cell B4. (To transfer a numerical value, rather than a formula, copy cell B10 and then highlight cell B4. In Excel 2007, go to the Home ribbon, select Paste, and then click on Paste Values. The numerical value from B10 will be pasted into B4. In earlier versions of Excel, go to the Edit menu, select Paste Special, and then choose Value.) This procedure gives new activity coefficients in cells B6 and B8 and a new concentration in cell B10. Copy the new concentration from cell B10 into cell B4 and repeat this procedure several times until you have a constant answer.

(c) *Using circular references in Excel.* Take your spreadsheet and enter 0 in cell B4. The value 0.041 23 is computed in cell B10. Ideally, you would like to write "=B10" in cell B4 so that the value from B10 would be copied to B4. Excel gives a "circular reference" error message because cell B10 depends on B4 and cell B4 depends on B10. To get around this problem, in Excel 2007, click the Microsoft Office button at the upper left of the spreadsheet. Click on Excel Options at the bottom of the window. Select Formulas. In Calculation Options, check Enable iterative calculation and set Maximum Change to 0.000 01. Click OK and Excel merrily iterates between cells B10 and B4 until the two values agree within 0.000 01. In earlier versions of Excel, go to the Tools menu and choose Options. Select Calculation and choose Iteration. Set Maximum change to 0.000 01.

7-30. *Heterogeneous equilibria and calcite solubility.* If river water in Box 7-2 is saturated with calcite ($CaCO_3$), $[Ca^{2+}]$ is governed by the following equilibria:

$$CaCO_3(s) \rightleftharpoons Ca^{2+} + CO_3^{2-} \qquad K_{sp} = 4.5 \times 10^{-9}$$
$$CO_2(g) \rightleftharpoons CO_2(aq) \qquad K_{CO_2} = 0.032$$
$$CO_2(aq) + H_2O \rightleftharpoons HCO_3^- + H^+ \qquad K_1 = 4.46 \times 10^{-7}$$
$$HCO_3^- \rightleftharpoons CO_3^{2-} + H^+ \qquad K_2 = 4.69 \times 10^{-11}$$

(a) From these reactions, find the equilibrium constant for the reaction

$$CaCO_3(s) + CO_2(aq) + H_2O \rightleftharpoons Ca^{2+} + 2HCO_3^- \qquad K = ? \quad (A)$$

(b) The mass balance for Reaction A is $[HCO_3^-] = 2[Ca^{2+}]$. Find $[Ca^{2+}]$ (in mol/L and in mg/L) in equilibrium with atmospheric CO_2 if $P_{CO_2} = 3.8 \times 10^{-4}$ bar. Locate this point on the line in Box 7-2.

(c) The concentration of Ca^{2+} in the Don River is 80 mg/L. What effective P_{CO_2} is in equilibrium with this much Ca^{2+}? How can the river have this much CO_2?

8 | Monoprotic Acid-Base Equilibria

MEASURING pH INSIDE CELLULAR COMPARTMENTS

(a) Mouse macrophage engulfs two foreign red blood cells as phagocytosis begins. [From J. P. Revel in B. Alberts, D. Bray, J. Lewis, M. Raff, K. Roberts, and J. D. Watson, *Molecular Biology of the Cell*, 2nd ed. (New York: Garland Publishing, 1989.]
(b) Macrophages with ingested 1.6-μm-diameter fluorescent beads.
(c) Fluorescence image of panel *b*. [From K. P. McNamara, T. Nguyen, G. Dumitrascu, J. Ji, N. Rosenzweig, and Z. Rosenzweig, "Synthesis, Characterization, and Application of Fluorescence Sensing Lipobeads for Intracellular pH Measurements," *Anal. Chem.* 2001, *73*, 3240.]

(a)

(b)

(c)

Macrophages are white blood cells that fight infection by ingesting and dissolving foreign cells—a process called *phagocytosis*. The compartment containing the ingested foreign cell merges with compartments called *lysosomes*, which contain digestive enzymes that are most active in acid. Low enzyme activity above pH 7 protects the cell from enzymes that leak into the cell.

One way to measure pH inside the compartment containing the ingested particle and digestive enzymes is to present macrophages with polystyrene beads coated with a lipid membrane to which fluorescent (light-emitting) dyes are covalently bound. Panel *d* shows that fluorescence intensity from the dye fluorescein depends on pH, but fluorescence from tetramethylrhodamine does not. The ratio of emission from the dyes is a measure of pH. Panel *e* shows the fluorescence intensity ratio changing in 3 s as the bead is ingested and the pH around the bead drops from 7.3 to 5.7 to allow digestion to commence.

(d) Fluorescence spectra of lipobeads in solutions at pH 5–8. (e) pH change during phagocytosis of a single bead by a macrophage. [From McNamara et al., ibid.]

(d)

(e)

cids and bases are essential to virtually every application of chemistry and for the intelligent use of analytical procedures such as chromatography and electrophoresis. It would be difficult to have a meaningful discussion of, say, protein purification or the weathering of rocks without understanding acids and bases. This chapter covers acid-base equilibria and buffers. Chapter 9 treats polyprotic systems involving two or more acidic protons. Nearly every biological macromolecule is polyprotic. Chapter 10 describes acid-base titrations. Now is the time to review fundamentals of acids and bases in Sections 6-5 through 6-7.

8-1 Strong Acids and Bases

What could be easier than calculating the pH of 0.10 M HBr? HBr is a **strong acid,** so the reaction

$$HBr + H_2O \rightarrow H_3O^+ + Br^-$$

goes to completion, and the concentration of H_3O^+ is 0.10 M. It will be our custom to write H^+ instead of H_3O^+, and we can say

$$pH = -\log[H^+] = -\log(0.10) = 1.00$$

EXAMPLE   Activity Coefficient in a Strong-Acid Calculation

Calculate the pH of 0.10 M HBr, using activity coefficients.

Solution   The ionic strength of 0.10 M HBr is $\mu = 0.10$ M, at which the activity coefficient of H^+ is 0.83 (Table 7-1). Remember that pH is $-\log \mathcal{A}_{H^+}$, not $-\log[H^+]$:

$$pH = -\log[H^+]\gamma_{H^+} = -\log(0.10)(0.83) = 1.08$$

Test Yourself   Calculate the pH of 0.010 M HBr in 0.090 M KBr. (**Answer:** 2.08)

Table 6-2 gave a list of strong acids and bases that you must memorize.

Equilibrium constants for the reaction[1]
$HX(aq) + H_2O \rightleftharpoons H_3O^+ + X^-$

HCl	$K_a = 10^{3.9}$
HBr	$K_a = 10^{5.8}$
HI	$K_a = 10^{10.4}$
HNO_3	$K_a = 10^{1.4}$

HNO_3 is discussed in Box 8-1.

BOX 8-1   Concentrated HNO_3 Is Only Slightly Dissociated[2]

Strong acids in dilute solution are essentially completely dissociated. As concentration increases, the degree of dissociation decreases. The figure shows a *Raman spectrum* of solutions of nitric acid of increasing concentration. The spectrum measures scattering of light whose energy corresponds to vibrational energies of molecules. The sharp signal at $1\,049\ cm^{-1}$ in the spectrum of 5.1 M $NaNO_3$ is characteristic of free NO_3^- anion.

A 10.0 M HNO_3 solution has a strong NO_3^- signal at $1\,049\ cm^{-1}$ from dissociated acid. Bands denoted by asterisks arise from *undissociated* HNO_3. As concentration increases, the $1\,049\ cm^{-1}$ signal disappears and signals attributed to undissociated HNO_3 increase. The graph shows the fraction of dissociation deduced from spectroscopic measurements. It is instructive to realize that, in 20 M HNO_3, there are fewer H_2O molecules than there are molecules of HNO_3. Dissociation decreases because there is not enough solvent to stabilize the free ions.

Theoretical studies indicate that dilute HNO_3 at a water-air *interface* is also a weak acid because there are not enough H_2O molecules to solvate the free ions.[3] This finding has implications for atmospheric chemistry at the surface of microscopic droplets in clouds.

Raman spectrum of aqueous HNO_3 at 25°C. Signals at 1 360, 1 049, and 720 cm^{-1} arise from NO_3^- anion. Signals denoted by asterisks are from undissociated HNO_3. The wavenumber unit, cm^{-1}, is 1/wavelength.

Temperature (°C)	Acid dissociation constant (K_a)
0	46.8
25	26.8
50	14.9

Now that we have reminded you of activity coefficients, you can breathe a sigh of relief because we will neglect activity coefficients unless there is a specific point to be made.

How do we calculate the pH of 0.10 M KOH? KOH is a **strong base** (completely dissociated), so $[OH^-] = 0.10$ M. Using $K_w = [H^+][OH^-]$, we write

From $[OH^-]$, you can always find $[H^+]$:

$$[H^+] = \frac{K_w}{[OH^-]}$$

$$[H^+] = \frac{K_w}{[OH^-]} = \frac{1.0 \times 10^{-14}}{0.10} = 1.0 \times 10^{-13}\,M$$

$$pH = -\log[H^+] = 13.00$$

Finding the pH of other concentrations of KOH is pretty trivial:

$[OH^-]$ (M)	$[H^+]$ (M)	pH
$10^{-3.00}$	$10^{-11.00}$	11.00
$10^{-4.00}$	$10^{-10.00}$	10.00
$10^{-5.00}$	$10^{-9.00}$	9.00

A generally useful relation is

The temperature dependence of K_w was given in Table 6-1.

Relation between pH and pOH:

$$pH + pOH = -\log K_w = 14.00 \text{ at } 25°C \qquad (8\text{-}1)$$

The Dilemma

Well, life seems simple enough so far. Now we ask, "What is the pH of 1.0×10^{-8} M KOH?" Applying our usual reasoning, we calculate

Adding base to water cannot *lower* the pH. (Lower pH is more *acidic*.) There must be something wrong.

$$[H^+] = K_w/(1.0 \times 10^{-8}) = 1.0 \times 10^{-6}\,M \Rightarrow pH = 6.00$$

How can the base KOH produce an acidic solution (pH < 7)? It's impossible.

The Cure

Clearly, there is something wrong with our calculation. In particular, we have not considered the contribution of OH^- from the ionization of water. In pure water, $[OH^-] = 1.0 \times 10^{-7}$ M, which is greater than the amount of KOH added to the solution. To handle this problem, we resort to the systematic treatment of equilibrium.

Step 1 *Pertinent reactions.* The only one is $H_2O \xrightleftharpoons{K_w} H^+ + OH^-$

Step 2 *Charge balance.* The species in solution are K^+, OH^-, and H^+. So,

$$[K^+] + [H^+] = [OH^-] \qquad (8\text{-}2)$$

Step 3 *Mass balance.* All K^+ comes from the KOH, so $[K^+] = 1.0 \times 10^{-8}$ M

If we had been using activities, step 4 is the only point at which activity coefficients would have entered.

Step 4 *Equilibrium constant.* $K_w = [H^+][OH^-] = 1.0 \times 10^{-14}$

Step 5 *Count.* There are three equations and three unknowns ($[H^+]$, $[OH^-]$, $[K^+]$), so we have enough information to solve the problem.

Step 6 *Solve.* We seek the pH, so let's set $[H^+] = x$. Writing $[K^+] = 1.0 \times 10^{-8}$ M in Equation 8-2, we get

$$[OH^-] = [K^+] + [H^+] = 1.0 \times 10^{-8} + x$$

Putting $[OH^-] = 1.0 \times 10^{-8} + x$ into the K_w equilibrium enables us to solve the problem:

$$[H^+][OH^-] = K_w$$

$$(x)(1.0 \times 10^{-8} + x) = 1.0 \times 10^{-14}$$

$$x^2 + (1.0 \times 10^{-8})x - (1.0 \times 10^{-14}) = 0$$

Solution of a quadratic equation:

$$ax^2 + bx + c = 0$$

$$x = \frac{-b \pm \sqrt{b^2 - 4ac}}{2a}$$

Retain all digits in your calculator because b^2 is sometimes nearly equal to $4ac$. If you round off before computing $b^2 - 4ac$, your answer may be garbage.

$$x = \frac{-1.0 \times 10^{-8} \pm \sqrt{(1.0 \times 10^{-8})^2 - 4(1)(-1.0 \times 10^{-14})}}{2(1)}$$

$$= 9.6 \times 10^{-8}\,M \quad \text{or} \quad -1.1 \times 10^{-7}\,M$$

Rejecting the negative concentration, we conclude that

$$[H^+] = 9.6 \times 10^{-8}\,M \Rightarrow pH = -\log[H^+] = 7.02$$

This pH is eminently reasonable, because 10^{-8} M KOH should be very slightly basic.

Figure 8-1 shows the pH calculated for different concentrations of strong base or strong acid in water. There are three regions:

1. When the concentration is "high" ($\geq 10^{-6}$ M), pH is calculated by just considering the added H^+ or OH^-. That is, the pH of $10^{-5.00}$ M KOH *is* 9.00.
2. When the concentration is "low" ($\leq 10^{-8}$ M), the pH is 7.00. We have not added enough acid or base to change the pH of the water itself.
3. At intermediate concentrations of 10^{-6} to 10^{-8} M, the effects of water ionization and the added acid or base are comparable. Only in this region is a systematic equilibrium calculation necessary.

Region 1 is the only practical case. Unless you were to protect 10^{-7} M KOH from the air, the pH would be governed by dissolved CO_2, not KOH.

Water Almost Never Produces 10^{-7} M H^+ and 10^{-7} M OH^-

The misconception that dissociation of water always produces 10^{-7} M H^+ and 10^{-7} M OH^- is true *only* in pure water with no added acid or base. In 10^{-4} M HBr, for example, the pH is 4. The concentration of OH^- is $[OH^-] = K_w/[H^+] = 10^{-10}$ M. But the only source of $[OH^-]$ is dissociation of water. If water produces 10^{-10} M OH^-, it must also produce 10^{-10} M H^+ because it makes one H^+ for every OH^-. In 10^{-4} M HBr solution, water dissociation produces only 10^{-10} M OH^- and 10^{-10} M H^+.

8-2 Weak Acids and Bases

Let's review the meaning of the **acid dissociation constant,** K_a, for the acid HA:

Weak-acid equilibrium:

$$HA \xrightleftharpoons{K_a} H^+ + A^- \qquad K_a = \frac{[H^+][A^-]}{[HA]} \qquad (8\text{-}3)$$

A **weak acid** is one that is not completely dissociated. That is, Reaction 8-3 does not go to completion. For a base, B, the **base hydrolysis constant,** K_b, is defined by the reaction

Weak-base equilibrium:

$$B + H_2O \xrightleftharpoons{K_b} BH^+ + OH^- \qquad K_b = \frac{[BH^+][OH^-]}{[B]} \qquad (8\text{-}4)$$

A **weak base** is one for which Reaction 8-4 does not go to completion.

pK is the negative logarithm of an equilibrium constant:

$$pK_w = -\log K_w$$
$$pK_a = -\log K_a$$
$$pK_b = -\log K_b$$

As K increases, pK decreases, and vice versa. Comparing formic and benzoic acids, we see that formic acid is stronger, with a larger K_a and smaller pK_a, than benzoic acid.

$$\underset{\text{Formic acid}}{\overset{O}{\underset{\|}{HCOH}}} \rightleftharpoons H^+ + \underset{\text{Formate}}{\overset{O}{\underset{\|}{HCO^-}}} \qquad \begin{array}{l} K_a = 1.80 \times 10^{-4} \\ pK_a = 3.744 \end{array}$$

$$\underset{\text{Benzoic acid}}{\overset{O}{\underset{\|}{\text{⬡—COH}}}} \rightleftharpoons \underset{\text{Benzoate}}{\overset{O}{\underset{\|}{\text{⬡—CO}^-}}} + H^+ \qquad \begin{array}{l} K_a = 6.28 \times 10^{-5} \\ pK_a = 4.202 \end{array}$$

The acid HA and its corresponding base, A^-, are said to be a **conjugate acid-base pair,** because they are related by the gain or loss of a proton. Similarly, B and BH^+ are a conjugate pair. The important relation between K_a and K_b for a conjugate acid-base pair is

Relation between K_a and K_b for conjugate pair:

$$K_a \cdot K_b = K_w \qquad (8\text{-}5)$$

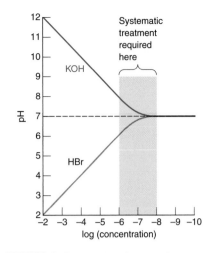

FIGURE 8-1 Calculated pH as a function of concentration of strong acid or strong base in water.

Any acid or base suppresses water ionization, as predicted by Le Châtelier's principle.

Question What concentrations of H^+ and OH^- are produced by H_2O dissociation in 0.01 M NaOH?

Of course, you know that K_a should really be expressed in terms of activities, not concentrations:

$$K_a = \mathcal{A}_{H^+}\mathcal{A}_{A^-}/\mathcal{A}_{HA}$$

Hydrolysis is reaction with water.

As K_a increases, pK_a decreases. Smaller pK_a means stronger acid.

HA and A^- are a *conjugate acid-base pair.* B and BH^+ also are conjugate.

Weak Is Conjugate to Weak

The conjugate base of a weak acid is a weak base. The conjugate acid of a weak base is a weak acid. Consider a weak acid, HA, with $K_a = 10^{-4}$. The conjugate base, A^-, has $K_b = K_w/K_a = 10^{-10}$. That is, if HA is a weak acid, A^- is a weak base. If K_a were 10^{-5}, then K_b would be 10^{-9}. As HA becomes a weaker acid, A^- becomes a stronger base (but never a strong base). Conversely, the greater the acid strength of HA, the less the base strength of A^-. However, if either A^- or HA is weak, so is its conjugate. If HA is strong (such as HCl), its conjugate base (Cl^-) is *so* weak that it is not a base at all in water.

Using Appendix G

Appendix G lists acid dissociation constants. Each compound is shown in its *fully protonated form*. Diethylamine, for example, is shown as $(CH_3CH_2)_2NH_2^+$, which is really the diethylammonium ion. The value of K_a (1.0×10^{-11}) given for diethylamine is actually K_a for the diethylammonium ion. To find K_b for diethylamine, we write $K_b = K_w/K_a = 1.0 \times 10^{-14}/1.0 \times 10^{-11} = 1.0 \times 10^{-3}$.

For polyprotic acids and bases, several K_a values are given. Pyridoxal phosphate is given in its fully protonated form as follows:[4]

Pyridoxal phosphate
(a derivative of vitamin B_6)

pK_a	K_a
1.4 (POH)	0.04
3.51 (OH)	3.1×10^{-4}
6.04 (POH)	9.1×10^{-7}
8.25 (NH)	5.6×10^{-9}

pK_1 (1.4) is for dissociation of one of the phosphate protons, and pK_2 (3.51) is for the hydroxyl proton. The third most acidic proton is the other phosphate proton, for which $pK_3 = 6.04$, and the NH^+ group is the least acidic ($pK_4 = 8.25$).

Species drawn in Appendix G are fully protonated. If a structure in Appendix G has a charge other than 0, it is not the structure that belongs with the name in the appendix. *Names refer to neutral molecules.* The neutral molecule pyridoxal phosphate is not the species drawn above, which has a +1 charge. The neutral molecule pyridoxal phosphate is

We took away a POH proton, not the NH^+ proton, because POH is the most acidic group in the molecule ($pK_a = 1.4$).

As another example, consider the molecule piperazine:

Structure shown for piperazine in Appendix G

Actual structure of piperazine, which *must be neutral*

Appendix G gives pK_a for ionic strengths of 0 and 0.1 M, when available. We will use pK_a for $\mu = 0$ unless there is no value listed or we need $\mu = 0.1$ M for a specific purpose. For pyridoxal phosphate, we used values for $\mu = 0.1$ M because none are listed for $\mu = 0$.

K_a at $\mu = 0$ is the *thermodynamic acid dissociation constant* that applies at any ionic strength when activity coefficients are used for that ionic strength:

$$K_a = \frac{\mathscr{A}_{H^+}\mathscr{A}_{A^-}}{\mathscr{A}_{HA}} = \frac{[H^+]\gamma_{H^+}[A^-]\gamma_{A^-}}{[HA]\gamma_{HA}}$$

K_a at $\mu = 0.1$ M is the quotient of concentrations when the ionic strength is 0.1 M:

$$K_a (\mu = 0.1 \text{ M}) = \frac{[H^+][A^-]}{[HA]}$$

8-3 Weak-Acid Equilibria

Let's compare the ionization of *ortho*- and *para*-hydroxybenzoic acids:

o-Hydroxybenzoic acid
(salicylic acid)
$pK_a = 2.97$

p-Hydroxybenzoic acid
$pK_a = 4.54$

Why is the *ortho* isomer 30 times more acidic than the *para* isomer? Any effect that increases the stability of the product of a reaction drives the reaction forward. In the *ortho* isomer, the product of the acid dissociation reaction can form a strong, internal hydrogen bond.

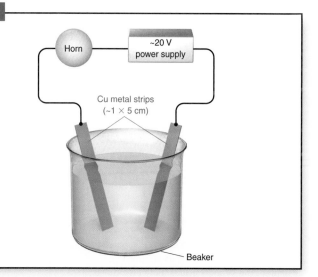

The *para* isomer cannot form such a bond because the —OH and —CO$_2^-$ groups are too far apart. By stabilizing the product, the internal hydrogen bond is thought to make *o*-hydroxybenzoic acid more acidic than *p*-hydroxybenzoic acid.

A Typical Weak-Acid Problem

The problem is to find the pH of a solution of the weak acid HA, given the formal concentration of HA and the value of K_a.[5] Let's call the formal concentration F and use the systematic treatment of equilibrium:

Reactions:

$$HA \xrightleftharpoons{K_a} H^+ + A^- \qquad H_2O \xrightleftharpoons{K_w} H^+ + OH^-$$

Charge balance:

$$[H^+] = [A^-] + [OH^-] \qquad (8\text{-}6)$$

Mass balance:

$$F = [A^-] + [HA] \qquad (8\text{-}7)$$

Equilibrium expressions:

$$K_a = \frac{[H^+][A^-]}{[HA]} \qquad (8\text{-}8)$$

$$K_w = [H^+][OH^-]$$

> *Formal concentration* is the total number of moles of a compound dissolved in a liter. The formal concentration of a weak acid is the total amount of HA placed in the solution, regardless of the fact that some has changed into A$^-$.

There are four equations and four unknowns ([A$^-$], [HA], [H$^+$], [OH$^-$]), so the problem is solved if we can just do the algebra.

But it's not so easy to solve these simultaneous equations. If you combine them, you will discover that a cubic equation results. At this point, the Good Chemist rides down again from the mountain on her white stallion to rescue us and cries, "Wait! There is no reason to solve a cubic equation. We can make an excellent, simplifying approximation. (Besides, I have trouble solving cubic equations.)"

For any respectable weak acid, [H$^+$] from HA will be much greater than [H$^+$] from H$_2$O. When HA dissociates, it produces A$^-$. When H$_2$O dissociates, it produces OH$^-$. If dissociation of HA is much greater than dissociation of H$_2$O, then [A$^-$] $\gg$ [OH$^-$], and Equation 8-6 reduces to

$$[H^+] \approx [A^-] \qquad (8\text{-}9)$$

DEMONSTRATION 8-1   Conductivity of Weak Electrolytes[6]

The relative conductivity of strong and weak acids is related to their different degrees of dissociation in aqueous solution. To demonstrate conductivity, we use a Radio Shack piezo alerting buzzer, but any kind of buzzer or light bulb could easily be substituted. The voltage required will depend on the buzzer or light chosen.

When a conducting solution is placed in the beaker, the horn sounds. First show that distilled water and sucrose solution are nonconductive. Solutions of the strong electrolytes NaCl or HCl are conductive. Compare strong and weak electrolytes by demonstrating that 1 mM HCl gives a loud sound, whereas 1 mM acetic acid gives little or no sound. With 10 mM acetic acid, the strength of the sound varies noticeably as the electrodes are moved away from each other in the beaker.

When CO$_2$ is absorbed by pure water, the conductivity increases from dissociation of the H$_2$CO$_3$ (carbonic acid) formed. Atmospheric CO$_2$ can be measured by conductivity.[7]

To solve the problem, first set $[H^+] = x$. Equation 8-9 says that $[A^-]$ also is equal to x. Equation 8-7 says that $[HA] = F - [A^-] = F - x$. Putting these expressions into Equation 8-8 gives

$$K_a = \frac{[H^+][A^-]}{[HA]} = \frac{(x)(x)}{F - x}$$

Setting $F = 0.050\ 0$ M and $K_a = 1.0_7 \times 10^{-3}$ for o-hydroxybenzoic acid, we can solve the equation, because it is just a quadratic equation.

$$\frac{x^2}{0.050\ 0 - x} = 1.0_7 \times 10^{-3}$$

$$x^2 = (1.0_7 \times 10^{-3})(0.050\ 0 - x)$$

$$x^2 + (1.07 \times 10^{-3})x - 5.35 \times 10^{-5} = 0$$

$$x = 6.8_0 \times 10^{-3}\ \text{M (negative root rejected)}$$

$$[H^+] = [A^-] = x = 6.8_0 \times 10^{-3}\ \text{M}$$

$$[HA] = F - x = 0.043_2\ \text{M}$$

$$pH = -\log x = 2.17$$

Was the approximation $[H^+] \approx [A^-]$ justified? The calculated pH is 2.17, which means that $[OH^-] = K_w/[H^+] = 1.5 \times 10^{-12}$ M.

$$[A^-](\text{from HA dissociation}) = 6.8 \times 10^{-3}\ \text{M}$$

$$\Rightarrow [H^+]\ \text{from HA dissociation} = 6.8 \times 10^{-3}\ \text{M}$$

$$[OH^-](\text{from H}_2\text{O dissociation}) = 1.5 \times 10^{-12}\ \text{M}$$

$$\Rightarrow [H^+]\ \text{from H}_2\text{O dissociation} = 1.5 \times 10^{-12}\ \text{M}$$

The assumption that H^+ is derived mainly from HA is excellent.

Fraction of Dissociation

The **fraction of dissociation, α,** is defined as the fraction of the acid in the form A^-:

Fractions of dissociation of an acid:
$$\alpha = \frac{[A^-]}{[A^-] + [HA]} = \frac{x}{x + (F - x)} = \frac{x}{F} \qquad (8\text{-}10)$$

For 0.050 0 M o-hydroxybenzoic acid, we find

$$\alpha = \frac{6.8 \times 10^{-3}\ \text{M}}{0.050\ 0\ \text{M}} = 0.14$$

That is, the acid is 14% dissociated at a formal concentration of 0.050 0 M.

The variation of α with formal concentration is shown in Figure 8-2. **Weak electrolytes** (compounds that are only partially dissociated) dissociate more as they are diluted. o-Hydroxybenzoic acid is more dissociated than p-hydroxybenzoic acid at the same formal concentration because the *ortho* isomer is a stronger acid. Demonstration 8-1 and Box 8-2 illustrate weak-acid properties.

The Essence of a Weak-Acid Problem

When faced with finding the pH of a weak acid, you should immediately realize that $[H^+] = [A^-] = x$ and proceed to set up and solve the equation

Equation for weak acids:
$$\frac{[H^+][A^-]}{[HA]} = \frac{x^2}{F - x} = K_a \qquad (8\text{-}11)$$

where F is the formal concentration of HA. The approximation $[H^+] = [A^-]$ would be poor only if the acid were too dilute or too weak, neither of which constitutes a practical problem.

EXAMPLE **A Weak-Acid Problem**

Find the pH of 0.050 M trimethylammonium chloride.

Trimethylammonium chloride

Margin notes (left column):

$x = [H^+]$ in weak-acid problems.

For uniformity, we express pH to the 0.01 decimal place, regardless of the number of places justified by significant figures. Ordinary pH measurements are not more accurate than ±0.02 pH units.

In a solution of a weak acid, H^+ is derived almost entirely from HA, not from H_2O.

α is the fraction of HA that has dissociated:
$$\alpha = \frac{[A^-]}{[A^-] + [HA]}$$

FIGURE 8-2 Fraction of dissociation of a weak electrolyte increases as electrolyte is diluted. The stronger acid is more dissociated than the weaker acid at all concentrations.

Solution We assume that ammonium halide salts are completely dissociated to give $(CH_3)_3NH^+$ and Cl^-.[†] We then recognize that trimethylammonium ion is a weak acid, being the conjugate acid of trimethylamine, $(CH_3)_3N$, a weak base. Cl^- has no basic or acidic properties and should be ignored. In Appendix G, we find trimethylammonium ion listed as trimethylamine but drawn as trimethylammonium ion, with $pK_a = 9.799$, at an ionic strength of $\mu = 0$. So,

> Cl^- has no acidic or basic properties because it is the conjugate base of the strong acid, HCl. If Cl^- had appreciable basicity, HCl would not be completely dissociated.

$$K_a = 10^{-pK_a} = 10^{-9.799} = 1.59 \times 10^{-10}$$

From here, everything is downhill.

$$(CH_3)_3NH^+ \xrightleftharpoons{\;K_a\;} (CH_3)_3N + H^+$$
$$F - x \qquad\qquad x \qquad x$$

$$\frac{x^2}{0.050 - x} = 1.59 \times 10^{-10} \qquad\qquad (8\text{-}12)$$

$$x = 2.8 \times 10^{-6}\,M \Rightarrow pH = 5.55$$

Test Yourself Find the pH of 0.050 M triethylammonium bromide. (*Answer:* 6.01)

BOX 8-2   Dyeing Fabrics and the Fraction of Dissociation[8]

Cotton fabrics are largely cellulose, a polymer with repeating units of the sugar glucose:

Structure of cellulose. Hydrogen bonding between glucose units helps make the structure rigid.

Dyes are colored molecules that can form covalent bonds to fabric. For example, Procion Brilliant Blue M-R is a dye with a blue *chromophore* (the colored part) attached to a reactive dichlorotriazine ring:

Oxygen atoms of the —CH_2OH groups on cellulose can replace Cl atoms of the dye to form covalent bonds that fix the dye permanently to the fabric:

After the fabric has been dyed in cold water, excess dye is removed with a hot wash. During the hot wash, the second Cl group of the dye is replaced by a second cellulose or by water (giving dye—OH).

The chemically reactive form of cellulose is the conjugate base:

$$\text{Cellulose—}CH_2OH \xrightleftharpoons{\;K_a \approx 10^{-15}\;} \text{cellulose—}CH_2O^- + H^+$$
$$\phantom{\text{Cellulose—}}ROH \qquad\qquad RO^-$$
$$\text{(conjugate base)}$$

To promote dissociation of the cellulose—CH_2OH proton, dyeing is carried out in sodium carbonate solution with a pH around 10.6. The fraction of reactive cellulose is given by the fraction of dissociation of the weak acid at pH 10.6:

$$\text{Fraction of dissociation} = \frac{[RO^-]}{[ROH] + [RO^-]} \approx \frac{[RO^-]}{[ROH]}$$

Because the fraction of dissociation of the very weak acid is so small, $[ROH] \gg [RO^-]$ in the denominator, which is therefore approximately just $[ROH]$. The quotient $[RO^-]/[ROH]$ can be calculated from K_a and the pH:

$$K_a = \frac{[RO^-][H^+]}{[ROH]} \Rightarrow \frac{[RO^-]}{[ROH]} = \frac{K_a}{[H^+]} \approx \frac{10^{-15}}{10^{-10.6}}$$

$$= 10^{-4.4} \approx \text{fraction of dissociation}$$

Only about one cellulose—CH_2OH group in 10^4 is in the reactive form at pH 10.6.

[†]$R_4N^+X^-$ salts are not completely dissociated, because there are some *ion pairs*, $R_4N^+X^-(aq)$ (Box 7-1). Equilibrium constants for $R_4N^+ + X^- \rightleftharpoons R_4N^+X^-(aq)$ are given below. For 0.050 F solutions, the fraction of ion pairing, calculated with activity coefficients, is 4% in $(CH_3)_4N^+Br^-$, 7% in $(CH_3CH_2)_4N^+Br^-$, and 9% in $(CH_3CH_2CH_2)_4N^+Br^-$.

R_4N^+	X^-	$K_{\text{ion pair}}$ ($\mu = 0$)	R_4N^+	X^-	$K_{\text{ion pair}}$ ($\mu = 0$)
Me_4N^+	Cl^-	1.1	Me_4N^+	I^-	2.0
Bu_4N^+	Cl^-	2.5	Et_4N^+	I^-	2.9
Me_4N^+	Br^-	1.4	Pr_4N^+	I^-	4.6
Et_4N^+	Br^-	2.4	Bu_4N^+	I^-	6.0
Pr_4N^+	Br^-	3.1			

$Me = CH_3—$, $Et = CH_3CH_2—$, $Pr = CH_3CH_2CH_2—$, $Bu = CH_3CH_2CH_2CH_2—$

A handy tip: Equation 8-11 can always be solved with the quadratic formula. However, an easier method worth trying first is to neglect x in the denominator. If x comes out much smaller than F, then your approximation was good and you need not use the quadratic formula. For Equation 8-12, the approximation works like this:

Approximation Neglect x in denominator. If x comes out to be less than 1% of F, accept the approximation.

$$\frac{x^2}{0.050 - x} \approx \frac{x^2}{0.050} = 1.59 \times 10^{-10} \Rightarrow x = \sqrt{(0.050)(1.59 \times 10^{-10})} = 2.8 \times 10^{-6} \text{ M}$$

The approximate solution ($x \approx 2.8 \times 10^{-6}$) is much smaller than the term 0.050 in the denominator of Equation 8-12. Therefore, the approximate solution is fine. A reasonable rule of thumb is to accept the approximation if x comes out to be less than 1% of F.

8-4 Weak-Base Equilibria

The treatment of weak bases is almost the same as that of weak acids.

As K_b increases, pK_b decreases and the base becomes stronger.

$$B + H_2O \xrightleftharpoons{K_b} BH^+ + OH^- \qquad K_b = \frac{[BH^+][OH^-]}{[B]}$$

We suppose that nearly all OH^- comes from the reaction of $B + H_2O$, and little comes from dissociation of H_2O. Setting $[OH^-] = x$, we must also set $[BH^+] = x$, because one BH^+ is produced for each OH^-. Calling the formal concentration of base F ($= [B] + [BH^+]$), we write

$$[B] = F - [BH^+] = F - x$$

Plugging these values into the K_b equilibrium expression, we get

A weak-base problem has the same algebra as a weak-acid problem, except $K = K_b$ and $x = [OH^-]$.

Equation for weak base: $\qquad \dfrac{[BH^+][OH^-]}{[B]} = \dfrac{x^2}{F - x} = K_b \qquad\qquad$ (8-13)

which looks a lot like a weak-acid problem, except that now $x = [OH^-]$.

A Typical Weak-Base Problem

Consider the commonly occurring weak base, cocaine.

If the formal concentration is 0.037 2 M, the problem is formulated as follows:

$$B + H_2O \rightleftharpoons BH^+ + OH^-$$
$$0.037\ 2 - x \qquad\qquad x \qquad x$$

$$\frac{x^2}{0.037\ 2 - x} = 2.6 \times 10^{-6} \Rightarrow x = 3.1 \times 10^{-4} \text{ M}$$

Because $x = [OH^-]$, we can write

Question What concentration of OH^- is produced by H_2O dissociation in this solution? Was it justified to neglect water dissociation as a source of OH^-?

$$[H^+] = K_w/[OH^-] = 1.0 \times 10^{-14}/3.1 \times 10^{-4} = 3.2 \times 10^{-11} \text{ M}$$

$$pH = -\log[H^+] = 10.49$$

This is a reasonable pH for a weak base.

What fraction of cocaine has reacted with water? We can formulate α for a base, called the **fraction of association:**

For a base, α is the fraction that has reacted with water.

Fraction of association of a base: $\qquad \alpha = \dfrac{[BH^+]}{[BH^+] + [B]} = \dfrac{x}{F} = 0.008\ 3 \qquad$ (8-14)

Only 0.83% of the base has reacted.

Conjugate Acids and Bases—Revisited

Earlier, we noted that **the conjugate base of a weak acid is a weak base,** and **the conjugate acid of a weak base is a weak acid.** We also derived an exceedingly important relation between the equilibrium constants for a conjugate acid-base pair: $K_a \cdot K_b = K_w$.

In Section 8-3, we considered *o*- and *p*-hydroxybenzoic acids, designated HA. Now consider their conjugate bases. For example, the salt sodium *o*-hydroxybenzoate dissolves to give Na^+ (which has no acid-base chemistry) and *o*-hydroxybenzoate, which is a weak base.

The acid-base chemistry is the reaction of *o*-hydroxybenzoate with water:

HA and A$^-$ are a conjugate acid-base pair. So are BH$^+$ and B.

In aqueous solution,

$$\frac{x^2}{F - x} = K_b$$

From K_a for each isomer, we calculate K_b for the conjugate base.

Isomer of hydroxybenzoic acid	K_a	$K_b = K_w/K_a$
ortho	$1.0_7 \times 10^{-3}$	9.3×10^{-12}
para	2.9×10^{-5}	3.5×10^{-10}

Using each value of K_b and letting F = 0.050 0 M, we find

> pH of 0.050 0 M *o*-hydroxybenzoate = 7.83
>
> pH of 0.050 0 M *p*-hydroxybenzoate = 8.62

These are reasonable pH values for solutions of weak bases. Furthermore, as expected, the conjugate base of the stronger acid is the weaker base.

EXAMPLE **A Weak-Base Problem**

Find the pH of 0.10 M ammonia.

Solution When ammonia is dissolved in water, its reaction is

$$NH_3 + H_2O \xrightarrow{K_b} NH_4^+ + OH^-$$

In Appendix G, we find ammonium ion, NH_4^+, listed next to ammonia. pK_a for ammonium ion is 9.245. Therefore, K_b for NH_3 is

$$K_b = \frac{K_w}{K_a} = \frac{10^{-14.00}}{10^{-9.245}} = 1.76 \times 10^{-5}$$

To find the pH of 0.10 M NH_3, we set up and solve the equation

$$\frac{[NH_4^+][OH^-]}{[NH_3]} = \frac{x^2}{0.10 - x} = K_b = 1.76 \times 10^{-5}$$

$$x = [OH^-] = 1.3_2 \times 10^{-3}\,M$$

$$[H^+] = \frac{K_w}{[OH^-]} = 7.6 \times 10^{-12}\,M \Rightarrow pH = -\log[H^+] = 11.12$$

Test Yourself Find the pH of 0.10 M methylamine. (**Answer:** 11.80)

8-5 Buffers

A buffered solution resists changes in pH when acids or bases are added or when dilution occurs. The **buffer** is a mixture of an acid and its conjugate base. There must be comparable amounts of the conjugate acid and base (within a factor of ~10) to exert significant buffering.

The importance of buffers in all areas of science is immense. At the outset of this chapter, we saw that digestive enzymes in lysosomes operate best in acid, a constraint that allows a cell

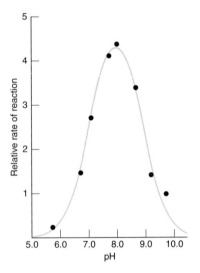

FIGURE 8-3 pH dependence of the rate of cleavage of an amide bond by the enzyme chymotrypsin, which helps digest proteins in your intestine. [M. L. Bender, G. E. Clement, F. J. Kézdy, and H. A. Heck, "The Correlation of the pH (pD) Dependence and the Stepwise Mechanism of α-Chymotrypsin-Catalyzed Reactions," *J. Am. Chem. Soc.* 1964, *86*, 3680.]

$$\begin{array}{c} O \\ \parallel \\ RC - NHR' \end{array}$$

Amide bond

When you mix a weak acid with its conjugate base, you get what you mix!

The approximation that the concentrations of HA and A⁻ remain unchanged breaks down for dilute solutions or at extremes of pH. We test the validity of the approximation on page 180.

$\log xy = \log x + \log y$

L. J. Henderson was a physician who wrote $[H^+] = K_a[\text{acid}]/[\text{salt}]$ in a physiology article in 1908, a year before the word "buffer" and the concept of pH were invented by the biochemist S. P. L. Sørensen. Henderson's contribution was the approximation of setting [acid] equal to the concentration of HA placed in solution and [salt] equal to the concentration of A⁻ placed in solution. In 1916, K. A. Hasselbalch wrote what we call the Henderson-Hasselbalch equation in a biochemical journal.[9]

to protect itself from its own enzymes. If enzymes leak into the buffered, neutral cytoplasm, they have low reactivity and do less damage to the cell than they would at their optimum pH. Figure 8-3 shows the pH dependence of an enzyme-catalyzed reaction that is fastest near pH 8.0. For an organism to survive, it must control the pH of each subcellular compartment so that each reaction proceeds at the proper rate.

Mixing a Weak Acid and Its Conjugate Base

If you mix A moles of a weak acid with B moles of its conjugate base, the moles of acid remain close to A and the moles of base remain close to B. Little reaction occurs to change either concentration.

To understand why this should be so, look at the K_a and K_b reactions in terms of Le Châtelier's principle. Consider an acid with $pK_a = 4.00$ and its conjugate base with $pK_b = 10.00$. Let's calculate the fraction of acid that dissociates in a 0.10 M solution of HA.

$$\underset{0.10-x}{HA} \rightleftharpoons \underset{x}{H^+} + \underset{x}{A^-} \qquad pK_a = 4.00$$

$$\frac{x^2}{F - x} = K_a \Rightarrow x = 3.1 \times 10^{-3}\,M$$

$$\text{Fraction of dissociation} = \alpha = \frac{x}{F} = 0.031$$

The acid is only 3.1% dissociated under these conditions.

In a solution containing 0.10 mol of A⁻ dissolved in 1.00 L, the extent of reaction of A⁻ with water is even smaller.

$$\underset{0.10-x}{A^-} + H_2O \rightleftharpoons \underset{x}{HA} + \underset{x}{OH^-} \qquad pK_b = 10.00$$

$$\frac{x^2}{F - x} = K_b \Rightarrow x = 3.2 \times 10^{-6}$$

$$\text{Fraction of association} = \alpha = \frac{x}{F} = 3.2 \times 10^{-5}$$

HA dissociates very little, and adding extra A⁻ to the solution makes HA dissociate even less. Similarly, A⁻ does not react much with water, and adding extra HA makes A⁻ react even less. If 0.050 mol of A⁻ plus 0.036 mol of HA are added to water, there will be close to 0.050 mol of A⁻ and close to 0.036 mol of HA in the solution at equilibrium.

Henderson-Hasselbalch Equation

The central equation for buffers is the **Henderson-Hasselbalch equation,** which is merely a rearranged form of the K_a equilibrium expression.

$$K_a = \frac{[H^+][A^-]}{[HA]}$$

$$\log K_a = \log \frac{[H^+][A^-]}{[HA]} = \log[H^+] + \log \frac{[A^-]}{[HA]}$$

$$\underbrace{-\log[H^+]}_{pH} = \underbrace{-\log K_a}_{pK_a} + \log \frac{[A^-]}{[HA]}$$

Henderson-Hasselbalch equation for an acid:

$$HA \xrightleftharpoons{K_a} H^+ + A^-$$

$$\boxed{pH = pK_a + \log \frac{[A^-]}{[HA]}} \qquad (8\text{-}16)$$

The Henderson-Hasselbalch equation tells us the pH of a solution, provided we know the ratio of the concentrations of conjugate acid and base, as well as pK_a for the acid. If a solution is prepared from the weak base B and its conjugate acid, the analogous equation is

Henderson-Hasselbalch equation for a base:

$$BH^+ \xrightleftharpoons{K_a = K_w/K_b} B + H^+$$

$$\boxed{pH = pK_a + \log \frac{[B]}{[BH^+]}} \qquad \begin{array}{l} pK_a \text{ applies to} \\ \checkmark \ this \text{ acid} \end{array} \qquad (8\text{-}17)$$

where pK_a is the acid dissociation constant of the weak acid BH^+. The important features of Equations 8-16 and 8-17 are that the base (A^- or B) appears in the numerator of both equations, and the equilibrium constant is K_a of the acid in the denominator.

Challenge Show that, when activities are included, the Henderson-Hasselbalch equation is

$$pH = pK_a + \log \frac{[A^-]\gamma_{A^-}}{[HA]\gamma_{HA}} \qquad (8\text{-}18)$$

The Henderson-Hasselbalch equation is not an approximation. It is just a rearranged form of the equilibrium expression. The approximations we make are the values of $[A^-]$ and $[HA]$. In most cases, it is valid to assume that what we mix is what we get in solution. At the end of this chapter, we treat the case in which what we mix is not what we get because the solution is too dilute or the acid is too strong.

Properties of the Henderson-Hasselbalch Equation

In Equation 8-16, we see that, if $[A^-] = [HA]$, then $pH = pK_a$.

$$pH = pK_a + \log \frac{[A^-]}{[HA]} = pK_a + \log 1 = pK_a$$

Regardless of how complex a solution may be, whenever $pH = pK_a$, for a particular acid, $[A^-]$ must equal $[HA]$ for that acid.

All equilibria must be satisfied simultaneously in any solution at equilibrium. If there are 10 different acids and bases in the solution, the 10 forms of Equation 8-16 will have 10 different quotients $[A^-]/[HA]$, but all 10 equations must give the same pH, because **there can be only one concentration of H^+ in a solution**.

Another feature of the Henderson-Hasselbalch equation is that, for every power-of-10 change in the ratio $[A^-]/[HA]$, the pH changes by one unit (Table 8-1). As the base (A^-) increases, the pH goes up. As the acid (HA) increases, the pH goes down. For any conjugate acid-base pair, you can say, for example, that if $pH = pK_a - 1$, there must be 10 times as much HA as A^-. Ten-elevenths is in the form HA and one-eleventh is in the form A^-.

| | EXAMPLE | **Using the Henderson-Hasselbalch Equation** |

Sodium hypochlorite (NaOCl, the active ingredient of almost all bleaches) was dissolved in a solution buffered to pH 6.20. Find the ratio $[OCl^-]/[HOCl]$ in this solution.

Solution In Appendix G, we find that $pK_a = 7.53$ for hypochlorous acid, HOCl. The pH is known, so the ratio $[OCl^-]/[HOCl]$ can be calculated from the Henderson-Hasselbalch equation.

$$HOCl \rightleftharpoons H^+ + OCl^-$$

$$pH = pK_a + \log \frac{[OCl^-]}{[HOCl]}$$

$$6.20 = 7.53 + \log \frac{[OCl^-]}{[HOCl]}$$

$$-1.33 = \log \frac{[OCl^-]}{[HOCl]}$$

$$10^{-1.33} = 10^{\log([OCl^-]/[HOCl])} = \frac{[OCl^-]}{[HOCl]}$$

$$0.047 = \frac{[OCl^-]}{[HOCl]}$$

The ratio $[OCl^-]/[HOCl]$ is set by pH and pK_a. We do not need to know how much NaOCl was added, or the volume.

Test Yourself Find $[OCl^-]/[HOCl]$ if pH is raised by one unit to 7.20. (***Answer:*** 0.47)

Equations 8-16 and 8-17 are only sensible when the base (A^- or B) is in the *numerator*. When the concentration of base increases, the log term increases and the pH increases.

When $[A^-] = [HA]$, $pH = pK_a$.

TABLE 8-1 Effect of $[A^-]/[HA]$ on pH

$[A^-]/[HA]$	pH
100:1	$pK_a + 2$
10:1	$pK_a + 1$
1:1	pK_a
1:10	$pK_a - 1$
1:100	$pK_a - 2$

$10^{\log z} = z$

A Buffer in Action

For illustration, we choose a widely used buffer called "tris," which is short for tris(hydroxymethyl)aminomethane.

In Appendix G, we find $pK_a = 8.072$ for the conjugate acid of tris. An example of a salt containing the BH^+ cation is tris hydrochloride, which is BH^+Cl^-. When BH^+Cl^- is dissolved in water, it dissociates to BH^+ and Cl^-.

EXAMPLE   A Buffer Solution

Find the pH of a 1.00-L aqueous solution prepared with 12.43 g of tris (FM 121.135) plus 4.67 g of tris hydrochloride (FM 157.596).

Solution   The concentrations of B and BH^+ added to the solution are

$$[B] = \frac{12.43 \text{ g/L}}{121.135 \text{ g/mol}} = 0.102\ 6 \text{ M} \qquad [BH^+] = \frac{4.67 \text{ g/L}}{157.596 \text{ g/mol}} = 0.029\ 6 \text{ M}$$

Assuming that what we mixed stays in the same form, we plug these concentrations into the Henderson-Hasselbalch equation to find the pH:

$$pH = pK_a + \log \frac{[B]}{[BH^+]} = 8.072 + \log \frac{0.102\ 6}{0.029\ 6} = 8.61$$

Test Yourself   Find the pH if we add another 1.00 g of tris hydrochloride. (***Answer:*** 8.53)

The pH of a buffer is nearly independent of volume.

Notice that *the volume of solution is irrelevant*, because volume cancels in the numerator and denominator of the log term:

$$pH = pK_a + \log \frac{\text{moles of B/L of solution}}{\text{moles of } BH^+/\text{L of solution}}$$

$$= pK_a + \log \frac{\text{moles of B}}{\text{moles of } BH^+}$$

BOX 8-3   Strong Plus Weak Reacts Completely

Strong acid reacts with a weak base essentially "completely" because the equilibrium constant is large.

$$\underset{\substack{\text{Weak} \\ \text{base}}}{B} + \underset{\substack{\text{Strong} \\ \text{acid}}}{H^+} \rightleftharpoons BH^+ \qquad K = \frac{1}{K_a \text{ (for } BH^+)}$$

If B is tris(hydroxymethyl)aminomethane, the equilibrium constant for reaction with HCl is

$$K = \frac{1}{K_a} = \frac{1}{10^{-8.072}} = 1.2 \times 10^8$$

Strong base reacts "completely" with a weak acid because the equilibrium constant is, again, very large.

$$\underset{\substack{\text{Strong} \\ \text{base}}}{OH^-} + \underset{\substack{\text{Weak} \\ \text{acid}}}{HA} \rightleftharpoons A^- + H_2O \qquad K = \frac{1}{K_b \text{ (for } A^-)}$$

If HA is acetic acid, then the equilibrium constant for reaction with NaOH is

$$K = \frac{1}{K_b} = \frac{K_a \text{ (for HA)}}{K_w} = 1.7 \times 10^9$$

The reaction of a strong acid with a strong base is even more complete than a strong plus weak reaction:

$$\underset{\substack{\text{Strong} \\ \text{acid}}}{H^+} + \underset{\substack{\text{Strong} \\ \text{base}}}{OH^-} \rightleftharpoons H_2O \qquad K = \frac{1}{K_w} = 10^{14}$$

If you mix a strong acid, a strong base, a weak acid, and a weak base, the strong acid and strong base will neutralize each other until one is used up. The remaining strong acid or strong base will then react with the weak base or weak acid.

EXAMPLE Effect of Adding Acid to a Buffer

If we add 12.0 mL of 1.00 M HCl to the solution in the previous example, what will be the new pH?

Solution The key is to realize that, *when a strong acid is added to a weak base, they react completely to give* BH^+ (Box 8-3). We are adding 12.0 mL of 1.00 M HCl, which contains $(0.012\ 0\ L)(1.00\ mol/L) = 0.012\ 0$ mol of H^+. This much H^+ consumes 0.012 0 mol of B to create 0.012 0 mol of BH^+:

	B Tris	+	H^+ From HCl	→	BH^+
Initial moles	0.102 6		0.012 0		0.029 6
Final moles	0.090 6		—		0.041 6
	$(0.102\ 6 - 0.012\ 0)$				$(0.029\ 6 + 0.012\ 0)$

Information in the table allows us to calculate the pH:

$$pH = pK_a + \log \frac{\text{moles of B}}{\text{moles of BH}^+}$$

$$= 8.072 + \log \frac{0.090\ 6}{0.041\ 6} = 8.41$$

The volume of the solution is irrelevant.

Test Yourself Find the pH if only 6.0 instead of 12.0 mL HCl were added. (*Answer:* 8.51)

We see that *the pH of a buffer does not change very much when a limited amount of strong acid or base is added*. Addition of 12.0 mL of 1.00 M HCl changed the pH from 8.61 to 8.41. Addition of 12.0 mL of 1.00 M HCl to 1.00 L of unbuffered solution would have lowered the pH to 1.93.

But *why* does a buffer resist changes in pH? It does so because the strong acid or base is consumed by B or BH^+. If you add HCl to tris, B is converted into BH^+. If you add NaOH, BH^+ is converted into B. As long as you don't use up B or BH^+ by adding too much HCl or NaOH, the log term of the Henderson-Hasselbalch equation does not change much and pH does not change much. Demonstration 8-2 illustrates what happens when buffer does get used up. The buffer has its maximum capacity to resist changes of pH when $pH = pK_a$. We return to this point later.

(Right margin) A buffer resists changes in pH . . .

. . . because the buffer consumes the added acid or base.

EXAMPLE Calculating How to Prepare a Buffer Solution

How many milliliters of 0.500 M NaOH should be added to 10.0 g of tris hydrochloride to give a pH of 7.60 in a final volume of 250 mL?

Solution The moles of tris hydrochloride are $(10.0\ g)/(157.596\ g/mol) = 0.063\ 5$ mol. We can make a table to help solve the problem:

Reaction with OH^-:	BH^+	+	OH^-	→	B
Initial moles	0.063 5		x		—
Final moles	$0.063\ 5 - x$		—		x

The Henderson-Hasselbalch equation allows us to find x, because we know pH and pK_a.

$$pH = pK_a + \log \frac{\text{mol B}}{\text{mol BH}^+}$$

$$7.60 = 8.072 + \log \frac{x}{0.063\ 5 - x}$$

$$-0.472 = \log \frac{x}{0.063\,5 - x}$$

$$10^{-0.472} = \frac{x}{0.063\,5 - x} \Rightarrow x = 0.016\,0 \text{ mol}$$

This many moles of NaOH is contained in

$$\frac{0.016\,0 \text{ mol}}{0.500 \text{ mol/L}} = 0.032\,0 \text{ L} = 32.0 \text{ mL}$$

Test Yourself How many mL of 0.500 M NaOH should be added to 10.0 g of tris hydrochloride to give a pH of 7.40 in a final volume of 500 mL? (**Answer:** 22.3 mL)

DEMONSTRATION 8-2 **How Buffers Work**

A buffer resists changes in pH because added acid or base is consumed by buffer. As buffer is used up, it becomes less resistant to changes in pH.

In this demonstration,[10] a mixture containing approximately a 10:1 mole ratio of $HSO_3^- : SO_3^{2-}$ is prepared. Because pK_a for HSO_3^- is 7.2, the pH should be approximately

$$pH = pK_a + \log \frac{[SO_3^{2-}]}{[HSO_3^-]} = 7.2 + \log \frac{1}{10} = 6.2$$

When formaldehyde is added, the net reaction is the consumption of HSO_3^-, but not of SO_3^{2-}:

$$H_2C{=}O + HSO_3^- \rightarrow H_2C{\Big\langle}^{O^-}_{SO_3H} \rightarrow H_2C{\Big\langle}^{OH}_{SO_3^-} \quad (A)$$

Formaldehyde   Bisulfite

$$H_2C{=}O + SO_3^{2-} \rightarrow H_2C{\Big\langle}^{O^-}_{SO_3^-} \quad (B)$$

Sulfite

$$H_2C{\Big\langle}^{O^-}_{SO_3^-} + HSO_3^- \rightarrow H_2C{\Big\langle}^{OH}_{SO_3^-} + SO_3^{2-}$$

(In sequence A, bisulfite is consumed directly. In sequence B, the net reaction is destruction of HSO_3^-, with no change in the SO_3^{2-} concentration.)

We can prepare a table showing how the pH should change as the HSO_3^- reacts.

Percentage of reaction completed	$[SO_3^{2-}] : [HSO_3^-]$	Calculated pH
0	1 : 10	6.2
90	1 : 1	7.2
99	1 : 0.1	8.2
99.9	1 : 0.01	9.2
99.99	1 : 0.001	10.2

Through 90% completion, the pH should rise by just 1 unit. In the next 9% of the reaction, the pH will rise by another unit. At the end of the reaction, the change in pH is very abrupt.

In the formaldehyde *clock reaction*,[11] formaldehyde is added to a solution containing HSO_3^-, SO_3^{2-}, and phenolphthalein indicator. Phenolphthalein is colorless below pH 8 and red above this pH. The

solution remains colorless for more than a minute. Suddenly pH shoots up and the liquid turns pink. Monitoring pH with a glass electrode gave the results in the graph.

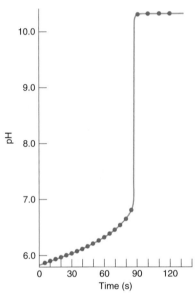

Graph of pH versus time in the formaldehyde clock reaction.

Procedure: Solutions must be fresh. Prepare a solution of formaldehyde by diluting 9 mL of 37 wt% formaldehyde to 100 mL. Dissolve 1.4 g of $Na_2S_2O_5$ (sodium metabisulfite)[12] and 0.18 g of Na_2SO_3 in 400 mL of water, and add ~1 mL of phenolphthalein solution (Table 10-3). Add 23 mL of formaldehyde solution to the well-stirred solution to initiate the reaction. Reaction time can be adjusted by changing temperature, concentrations, or volume.

A less toxic variant of this demonstration uses glyoxal

$$(HC{-}CH) \overset{O\ \ O}{\overset{\|\ \ \|}{}}$$ in place of formaldehyde.[13] A day before the demonstration, dilute 2.9 g 40 wt% glyoxal (20.0 mmol) to 25 mL. Dissolve 0.90 g $Na_2S_2O_5$ (4.7 mmol), 0.15 g Na_2SO_3 (1.2 mmol), and 0.18 g $Na_2EDTA \cdot 2H_2O$ (0.48 mmol, to protect sulfite from metal-catalyzed air oxidation) in 50 mL. A mole of $Na_2S_2O_5$ makes 2 moles of HSO_3^- by reaction with H_2O. For the demonstration, add 0.5 mL phenol red indicator (Table 10-3) to 400 mL H_2O plus 5.0 mL sulfite solution. Add 2.5 mL glyoxal solution to the well-stirred sulfite solution to start the clock reaction.

Preparing a Buffer in Real Life!

If you really wanted to prepare a tris buffer of pH 7.60, you would *not* do it by calculating what to mix. Suppose that you wish to prepare 1.00 L of buffer containing 0.100 M tris at a pH of 7.60. You have available solid tris hydrochloride and approximately 1 M NaOH. Here's how I would do it:

1. Weigh out 0.100 mol of tris hydrochloride and dissolve it in a beaker containing about 800 mL of water.
2. Place a calibrated pH electrode in the solution and monitor the pH.
3. Add NaOH solution until the pH is exactly 7.60.
4. Transfer the solution to a volumetric flask and wash the beaker a few times. Add the washings to the volumetric flask.
5. Dilute to the mark and mix.

You do not simply add the calculated quantity of NaOH because it would not give exactly the desired pH. The reason for using 800 mL of water in the first step is so that the volume will be reasonably close to the final volume during pH adjustment. Otherwise, the pH will change slightly when the sample is diluted to its final volume and the ionic strength changes.

Buffer Capacity[14]

Buffer capacity, β, is a measure of how well a solution resists changes in pH when strong acid or base is added. Buffer capacity is defined as

Buffer capacity:
$$\beta = \frac{dC_b}{d\text{pH}} = -\frac{dC_a}{d\text{pH}} \qquad (8\text{-}19)$$

where C_a and C_b are the number of moles of strong acid and strong base per liter needed to produce a unit change in pH. The greater the buffer capacity, the more resistant the solution is to pH change.

Figure 8-4a shows C_b versus pH for a solution containing 0.100 F HA with $pK_a = 5.00$. The ordinate (C_b) is the formal concentration of strong base needed to be mixed with 0.100 F HA to give the indicated pH. For example, a solution containing 0.050 F OH^- plus 0.100 F HA would have a pH of 5.00 (neglecting activities).

Figure 8-4b, which is the derivative of the upper curve, shows buffer capacity for the same system. Buffer capacity reaches a maximum when pH = pK_a. That is, *a buffer is most effective in resisting changes in pH when pH = pK_a* (that is, when [HA] = [A^-]).

In choosing a buffer, *seek one whose pK_a is as close as possible to the desired pH. The useful pH range of a buffer is usually considered to be $pK_a \pm 1$ pH unit.* Outside this range, there is not enough of either the weak acid or the weak base to react with added base or acid. Buffer capacity can be increased by increasing the concentration of the buffer.

Buffer capacity in Figure 8-4b continues upward at high pH (and at low pH, which is not shown) simply because there is a high concentration of OH^- at high pH (or H^+ at low pH). Addition of a small amount of acid or base to a large amount of OH^- (or H^+) will not have a large effect on pH. A solution of high pH is buffered by the H_2O/OH^- conjugate acid-conjugate base pair. A solution of low pH is buffered by the H_3O^+/H_2O conjugate acid-conjugate base pair.

Buffer pH Depends on Ionic Strength and Temperature

The correct Henderson-Hasselbalch equation, 8-18, includes activity coefficients. The most important reason why the calculated pH of a buffer is not equal to the observed pH is because the ionic strength is not 0, so activity coefficients are not 1. Table 8-2 lists pK_a values for common buffers that are widely used in biochemistry. Values are listed for ionic strengths of 0 and 0.1 M. If a buffer solution has an ionic strength closer to 0.1 M than to 0, it is sensible to use pK_a for $\mu = 0.1$ to obtain a more realistic calculation of pH.

If we mix 0.200 mol boric acid with 0.100 mol NaOH in 1.00 L, we generate a 1:1 mixture of boric acid and its conjugate base with an ionic strength of 0.10 M:

$$\underset{\text{Boric acid (HA)}}{B(OH)_3} + OH^- \rightarrow \underset{\text{Borate (A}^-\text{)}}{(HO)_2BO^-} + H_2O$$

For boric acid in Table 8-2 we find $pK_a = 9.24$ at $\mu = 0$ and $pK_a = 8.98$ at $\mu = 0.1$ M. We predict that the pH of a 1:1 mixture of boric acid and borate will have pH near $pK_a = 9.24$ at low ionic strength and near 8.98 at $\mu = 0.1$ M. As another example of ionic strength effects,

---

Reasons why a calculation could be wrong:

1. Activity coefficients were not considered.

2. The temperature might not be 25°C, for which pK_a is listed.

3. The approximations [HA] = F_{HA} and [A^-] = F_{A^-} could be in error.

4. pK_a listed for tris in your favorite table is probably not exactly what you would measure.

5. Other ions besides the acid and base species affect pH by ion-pairing reactions with acid and base species.

Choose a buffer whose pK_a is close to the desired pH.

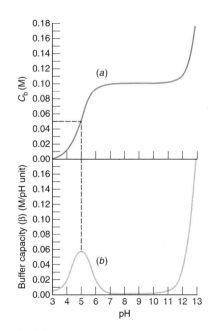

FIGURE 8-4 (*a*) C_b versus pH for a solution containing 0.100 F HA with $pK_a = 5.00$. (*b*) Buffer capacity versus pH for the same system reaches a maximum when pH = pK_a. The lower curve is the derivative of the upper curve.

TABLE 8-2 Structures and pK_a values for common buffers[a,b,c,d]

Name	Structure	pK_a^e $\mu = 0$	pK_a^e $\mu = 0.1$ M	Formula mass	$\Delta(pK_a)/\Delta T$ (K^{-1})
N-2-Acetamidoiminodiacetic acid (ADA)	O $H_2NCCH_2\overset{+}{N}H$ CH_2CO_2H / CH_2CO_2H	— (CO_2H)	1.59	190.15	—
N-Tris(hydroxymethyl)methylglycine (TRICINE)	$(HOCH_2)_3CNH_2CH_2CO_2H$	2.02 (CO_2H)	—	179.17	−0.003
Phosphoric acid	H_3PO_4	2.15 (pK_1)	1.92	98.00	0.005
N,N-Bis(2-hydroxyethyl)glycine (BICINE)	$(HOCH_2CH_2)_2\overset{+}{N}HCH_2CO_2H$	2.23 (CO_2H)	—	163.17	—
ADA	(see above)	2.48 (CO_2H)	2.31	190.15	—
Piperazine-N,N'-bis(2-ethanesulfonic acid) (PIPES)	$^-O_3SCH_2CH_2\overset{+}{N}H$ $H\overset{+}{N}CH_2CH_2SO_3^-$	— (pK_1)	2.67	302.37	—
Citric acid	OH / $HO_2CCH_2CCH_2CO_2H$ / CO_2H	3.13 (pK_1)	2.90	192.12	−0.002
Glycylglycine	O $H_3\overset{+}{N}CH_2CNHCH_2CO_2H$	3.14 (CO_2H)	3.11	132.12	0.000
Piperazine-N,N'-bis(3-propanesulfonic acid) (PIPPS)	$^-O_3S(CH_2)_3\overset{+}{N}H$ $H\overset{+}{N}(CH_2)_3SO_3^-$	— (pK_1)	3.79	330.42	—
Piperazine-N,N'-bis(4-butanesulfonic acid) (PIPBS)	$^-O_3S(CH_2)_4\overset{+}{N}H$ $H\overset{+}{N}(CH_2)_4SO_3^-$	— (pK_1)	4.29	358.47	—
N,N'-Diethylpiperazine dihydrochloride (DEPP·2HCl)	$CH_3CH_2\overset{+}{N}H$ $H\overset{+}{N}CH_2CH_3 \cdot 2Cl^-$	— (pK_1)	4.48	215.16	—
Citric acid	(see above)	4.76 (pK_2)	4.35	192.12	−0.001
Acetic acid	CH_3CO_2H	4.76	4.62	60.05	0.000
N,N'-Diethylethylenediamine-N,N'-bis(3-propanesulfonic acid) (DESPEN)	$^-O_3S(CH_2)_3\overset{+}{N}HCH_2CH_2H\overset{+}{N}(CH_2)_3SO_3^-$ / CH_3CH_2 CH_2CH_3	— (pK_1)	5.62	360.49	—
2-(N-Morpholino)ethanesulfonic acid (MES)	O $\overset{+}{N}HCH_2CH_2SO_3^-$	6.27	6.06	195.24	−0.009
Citric acid	(see above)	6.40 (pK_3)	5.70	192.12	0.002
N,N,N',N'-Tetraethylethylenediamine dihydrochloride (TEEN·2HCl)	$Et_2\overset{+}{N}HCH_2CH_2H\overset{+}{N}Et_2 \cdot 2Cl^-$	— (pK_1)	6.58	245.23	—
1,3-Bis[tris(hydroxymethyl)methyl-amino] propane hydrochloride (BIS-TRIS propane-2HCl)	$(HOCH_2)_3C\overset{+}{N}H_2(CH_2)_3\overset{+}{N}H_2 \cdot 2Cl^-$ / $(HOCH_2)_3C$	6.65 (pK_1)	—	355.26	—
ADA	(see above)	6.84 (NH)	6.67	190.15	−0.007

a. The protonated form of each molecule is shown. Acidic hydrogen atoms are shown in **bold** type. pK_a is for 25°C.

b. Many buffers in this table are widely used in biomedical research because of their weak metal binding and physiologic inertness (C. L. Bering, J. Chem. Ed. **1987**, 64, 803). In one study, where MES and MOPS had no discernible affinity for Cu^{2+}, a minor impurity in HEPES and HEPPS had a strong affinity for Cu^{2+} and MOPSO bound Cu^{2+} stoichiometrically (H. E. Marsh, Y.-P. Chin, L. Sigg, R. Hari, and H. Xu, Anal. Chem. **2003**, 75, 671). ADA, BICINE, ACES, and TES have some metal-binding ability (R. Nakon and C. R. Krishnamoorthy, Science **1983**, 221, 749). Lutidine buffers for the pH range 3 to 8 with limited metal-binding power have been described by U. Bips, H. Elias, M. Hauröder, G. Kleinhans, S. Pfeifer, and K. J. Wannowius, Inorg. Chem. **1983**, 22, 3862.

c. Some data from R. N. Goldberg, N. Kishore, and R. M. Lennen, J. Phys. Chem. Ref. Data **2002**, 31, 231. This paper gives the temperature dependence of pK_a.

d. Temperature and ionic strength dependence of pK_a for buffers: HEPES—D. Feng, W. F. Koch, and Y. C. Wu, Anal. Chem. **1989**, 61, 1400; MOPSO—Y. C. Wu, P. A. Berezansky, D. Feng, and W. F. Koch, Anal. Chem. **1993**, 65, 1084; ACES and CHES—R. N. Roy, J. Bice, J. Greer, J. A. Carlsten, J. Smithson, W. S. Good, C. P. Moore, L. N. Roy, and K. M. Kuhler, J. Chem. Eng. Data **1997**, 42, 41; TEMN, TEEN, DEPP, DESPEN, PIPES, PIPPS, PIPBS, MES, MOPS, and MOBS—A. Kandegedara and D. B. Rorabacher, Anal. Chem. **1999**, 71, 3140. This last set of buffers was specifically developed for low metal-binding ability (Q. Yu, A. Kandegedara, Y. Xu, and D. B. Rorabacher, Anal. Biochem. **1997**, 253, 50).

e. See marginal note on page 166 for the distinction between pK_a at $\mu = 0$ and at $\mu = 0.1$ M.

Name	Structure	pK_a $\mu = 0$	pK_a $\mu = 0.1$ M	Formula mass	$\Delta(pK_a)/\Delta T$ (K^{-1})
N-2-Acetamido-2-aminoethane-sulfonic acid (ACES)	$H_2NCCH_2\overset{+}{N}H_2CH_2CH_2SO_3^-$	6.85	6.75	182.20	−0.018
3-(N-Morpholino)-2-hydroxy-propanesulfonic acid (MOPSO)	$O\overset{+}{N}HCH_2CHCH_2SO_3^-$ (OH)	6.90	—	225.26	−0.015
Imidazole hydrochloride	$H\overset{+}{N}$ ring $\cdot Cl^-$	6.99	7.00	104.54	−0.022
PIPES	(see above)	7.14 (pK_2)	6.93	302.37	−0.007
3-(N-Morpholino)propanesulfonic acid (MOPS)	$O\overset{+}{N}HCH_2CH_2CH_2SO_3^-$	7.18	7.08	209.26	−0.012
Phosphoric acid	H_3PO_4	7.20 (pK_2)	6.71	98.00	−0.002
4-(N-Morpholino)butanesulfonic acid (MOBS)	$O\overset{+}{N}HCH_2CH_2CH_2CH_2SO_3^-$	—	7.48	223.29	—
N-Tris(hydroxymethyl)methyl-2-aminoethanesulfonic acid (TES)	$(HOCH_2)_3C\overset{+}{N}H_2CH_2CH_2SO_3^-$	7.55	7.60	229.25	−0.019
N-2-Hydroxyethylpiperazine-N'-2-ethanesulfonic acid (HEPES)	$HOCH_2CH_2N\overset{+}{N}HCH_2CH_2SO_3^-$	7.56	7.49	238.30	−0.012
PIPPS	(see above)	— (pK_2)	7.97	330.42	—
N-2-Hydroxyethylpiperazine-N'-3-propanesulfonic acid (HEPPS)	$HOCH_2CH_2N\overset{+}{N}H(CH_2)_3SO_3^-$	7.96	7.87	252.33	−0.013
Glycine amide hydrochloride	$H_3\overset{+}{N}CH_2CNH_2\cdot Cl^-$	—	8.04	110.54	—
Tris(hydroxymethyl)aminomethane hydrochloride (TRIS·HCl)	$(HOCH_2)_3C\overset{+}{N}H_3\cdot Cl^-$	8.07	8.10	157.60	−0.028
TRICINE	(see above)	8.14 (NH)	—	179.17	−0.018
Glycylglycine	(see above)	8.26 (NH)	8.09	132.12	−0.026
BICINE	(see above)	8.33 (NH)	8.22	163.17	−0.015
PIPBS	(see above)	— (pK_2)	8.55	358.47	—
DEPP·2HCl	(see above)	— (pK_2)	8.58	207.10	—
DESPEN	(see above)	— (pK_2)	9.06	360.49	—
BIS-TRIS propane·2HCl	(see above)	9.10 (pK_2)	—	355.26	—
Ammonia	NH_4^+	9.24	—	17.03	−0.031
Boric acid	$B(OH)_3$	9.24 (pK_1)	8.98	61.83	−0.008
Cyclohexylaminoethanesulfonic acid (CHES)	$⬡\overset{+}{N}H_2CH_2CH_2SO_3^-$	9.39	—	207.29	−0.023
TEEN·2HCl	(see above)	— (pK_2)	9.88	245.23	—
3-(Cyclohexylamino)propanesulfonic acid (CAPS)	$⬡\overset{+}{N}H_2CH_2CH_2CH_2SO_3^-$	10.50	10.39	221.32	−0.028
N,N,N',N'−Tetraethylmethylene-diamine·2HCl (TEMN·2HC1)	$Et_2\overset{+}{N}HCH_2\overset{+}{H}NEt_2\cdot 2Cl^-$	— (pK_2)	11.01	231.21	—
Phosphoric acid	H_3PO_4	12.35 (pK_3)	11.52	98.00	−0.009
Boric acid	$B(OH)_3$	12.74 (pK_2)	—	61.83	—

when a 0.5 M stock solution of phosphate buffer at pH 6.6 is diluted to 0.05 M, the pH rises to 6.9—a rather significant effect.

For almost all problems in this book we use pK_a for $\mu = 0$. As a practical matter, when there is no value of pK_a listed for $\mu = 0$, use pK_a for $\mu = 0.1$ M.

Buffer pK_a depends on temperature, as indicated in the last column of Table 8-2. Tris has an exceptionally large dependence, -0.028 pK_a units per degree, near room temperature. A solution of tris with pH 8.07 at 25°C will have pH ≈ 8.7 at 4°C and pH ≈ 7.7 at 37°C.

When What You Mix Is Not What You Get

In dilute solution or at extremes of pH, the concentrations of HA and A^- are not equal to their formal concentrations. Suppose we mix F_{HA} moles of HA and F_{A^-} moles of the salt Na^+A^-. The mass and charge balances are

Mass balance: $\qquad\qquad\qquad\qquad F_{HA} + F_{A^-} = [HA] + [A^-]$

Charge balance: $\qquad\qquad\qquad\quad [Na^+] + [H^+] = [OH^-] + [A^-]$

The substitution $[Na^+] = F_{A^-}$ and a little algebra leads to the equations

$$[HA] = F_{HA} - [H^+] + [OH^-] \qquad\qquad (8\text{-}20)$$

$$[A^-] = F_{A^-} + [H^+] - [OH^-] \qquad\qquad (8\text{-}21)$$

So far we have assumed that $[HA] \approx F_{HA}$ and $[A^-] \approx F_{A^-}$, and we used these values in the Henderson-Hasselbalch equation. A more rigorous procedure is to use Equations 8-20 and 8-21. If F_{HA} or F_{A^-} is small, or if $[H^+]$ or $[OH^-]$ is large, then the approximations $[HA] \approx F_{HA}$ and $[A^-] \approx F_{A^-}$ are not good. In acidic solutions, $[H^+] \gg [OH^-]$, so $[OH^-]$ can be ignored in Equations 8-20 and 8-21. In basic solutions, $[H^+]$ can be neglected.

EXAMPLE   **A Dilute Buffer Prepared from a Moderately Strong Acid**

What will be the pH if 0.010 0 mol of HA (with $pK_a = 2.00$) and 0.010 0 mol of A^- are dissolved in water to make 1.00 L of solution?

Solution   The solution is acidic (pH $\approx pK_a = 2.00$), so we neglect $[OH^-]$ in Equations 8-20 and 8-21. Setting $[H^+] = x$ in Equations 8-20 and 8-21, we use the K_a equation to find $[H^+]$:

$$\begin{array}{ccccc} HA & \rightleftharpoons & H^+ & + & A^- \\ 0.010\,0 - x & & x & & 0.010\,0 + x \end{array}$$

$$K_a = \frac{[H^+][A^-]}{[HA]} = \frac{(x)(0.010\,0 + x)}{(0.010\,0 - x)} = 10^{-2.00} \qquad (8\text{-}22)$$

$$\Rightarrow x = 0.004\,14 \text{ M} \quad \Rightarrow \quad pH = -\log[H^+] = 2.38$$

The pH is 2.38 instead of 2.00. [HA] and $[A^-]$ are not what we mixed:

$$[HA] = F_{HA} - [H^+] = 0.005\,86 \text{ M}$$

$$[A^-] = F_{A^-} + [H^+] = 0.014\,1 \text{ M}$$

In this example, HA is too strong and the concentrations are too low for HA and A^- to be equal to their formal concentrations.

Test Yourself   Find pH if $pK_a = 3.00$ instead of 2.00. Does the answer make sense? (**Answer:** 3.07)

The Henderson-Hasselbalch equation (with activity coefficients) is *always* true, because it is just a rearrangement of the K_a equilibrium expression. Approximations that are not always true are the statements $[HA] \approx F_{HA}$ and $[A^-] \approx F_{A^-}$.

In summary, a buffer consists of a mixture of a weak acid and its conjugate base. The buffer is most useful when pH $\approx pK_a$. Over a reasonable range of concentration, the pH of a buffer is nearly independent of concentration. A buffer resists changes in pH because it reacts with added acids or bases. If too much acid or base is added, the buffer will be consumed and will no longer resist changes in pH.

We saw at the end of Chapter 7 that Goal Seek solves numerical equations. In setting up Equation 8-22, we made the (superb) approximation $[H^+] \gg [OH^-]$ and neglected $[OH^-]$. With Goal Seek, it is easy to use Equations 8-20 and 8-21 without approximations:

$$K_a = \frac{[H^+][A^-]}{[HA]} = \frac{[H^+](F_{A^-} + [H^+] - [OH^-])}{F_{HA} - [H^+] + [OH^-]} \qquad (8\text{-}23)$$

The spreadsheet illustrates Goal Seek and the naming of cells to make formulas more meaningful. In column A, enter labels for K_a, K_w, F_{HA}, F_A ($= F_{A^-}$), H ($= [H^+]$), and OH ($= [OH^-]$). Write numerical values for K_a, K_w, F_{HA}, and F_{A^-} in B1:B4. In cell B5, enter a *guess* for $[H^+]$.

	A	B	C	D	E
1	Ka =	0.01		Reaction quotient	
2	Kw =	1.00E-1.4		for Ka =	
3	FHA =	0.01		[H+][A–]/[HA] =	
4	FA =	0.01		0.001222222	
5	H =	1.000E-03		<–Vary H with Goal Seek until D4 = Ka	
6	OH = Kw/H =	1E-11		D4 = H*(FA + H – OH)/(FHA – H + OH)	
7	pH = –log(H) =	3.00			

Now we want to name cells B1 through B6. In Excel 2007, select cell B1, go to the Formula ribbon, and click on Define Name. The dialog box will ask if you want to use the name "Ka" that appears in cell A1. If you like this name, click OK. Now when you select cell B1, the name box at the upper left of the spreadsheet displays Ka instead of B1. In earlier versions of Excel, select cell B1, go to the Insert menu, select Name and then Define. By this procedure, name the other cells in column B "Kw", "FHA", "FA", "H", and "OH". Now when you write a formula referring to cell B2, you can write Kw instead of B2. Kw is an *absolute reference* to cell $B$2.

In cell B6 enter the formula "=Kw/H" and Excel returns the value 1E-11 for $[OH^-]$. The beauty of naming cells is that "=Kw/H" is easier to understand than "=$B$2/$B$5". In cell B7, enter the formula "= –log(H)" for the pH.

In cell D4, write "=H*(FA+H-OH)/(FHA-H+OH)", which is the quotient in Equation 8-23. Excel returns the value 0.001 222 based on the guess $[H^+] = 0.001$ in cell B5.

Now use Goal Seek to vary $[H^+]$ in cell B5 until the reaction quotient in cell D4 equals 0.01, which is the value of K_a. Before using Goal Seek in Excel 2007, click the Microsoft Office button at the top left of the spreadsheet, click on Excel Options and then on Formulas. Set Maximum Change to a small number such as 1e-20 to find an answer with high precision. To execute Goal Seek, go to the Data ribbon, click on What-If Analysis and then on Goal Seek. In the dialog box, Set cell <u>D4</u> To value <u>0.01</u> By changing cell <u>B5</u>. Click OK and Excel varies cell B5 until the value $[H^+] = 4.142 \times 10^{-3}$ gives a reaction quotient of 0.01 in cell D4. Different guesses for H might give negative solutions or might not reach a solution. Only one positive value of H satisfies Equation 8-23. For earlier versions of Excel, in the Tools menu, select Options and go to Calculation. Set Maximum Change to 1e-20. In the Tools menu, select Goal Seek and proceed as described above.

Test Yourself Find H if $K_a = 0.001$. (***Answer:*** H = 8.44×10^{-4}, pH = 3.07)

Terms to Understand

acid dissociation constant, K_a	fraction of association, α	pK	weak base
base hydrolysis constant, K_b	(of a base)	strong acid	weak electrolyte
buffer	fraction of dissociation, α	strong base	
buffer capacity	(of an acid)	weak acid	
conjugate acid-base pair	Henderson-Hasselbalch equation		

Summary

Strong acids or bases. For practical concentrations ($\geq 10^{-6}$ M), pH or pOH is obtained directly from the formal concentration of acid or base. When the concentration is near 10^{-7} M, we use the systematic treatment of equilibrium to calculate pH. At still lower concentrations, the pH is 7.00, set by autoprotolysis of the solvent.

Weak acids. For the reaction $HA \rightleftharpoons H^+ + A^-$, we set up and solve the equation $K_a = x^2/(F - x)$, where $[H^+] = [A^-] = x$, and $[HA] = F - x$. The fraction of dissociation is given by $\alpha = [A^-]/([HA] + [A^-]) = x/F$. The term pK_a is defined as $pK_a = -\log K_a$.

Weak bases. For the reaction $B + H_2O \rightleftharpoons BH^+ + OH^-$, we set up and solve the equation $K_b = x^2/(F - x)$, where $[OH^-] = [BH^+] = x$, and $[B] = F - x$. The conjugate acid of a weak base is a weak acid, and the conjugate base of a weak acid is a weak base. For a conjugate acid-base pair, $K_a \cdot K_b = K_w$.

Buffers. A buffer is a mixture of a weak acid and its conjugate base. It resists changes in pH because it reacts with added acid or base. The pH is given by the Henderson-Hasselbalch equation:

$$pH = pK_a + \log \frac{[A^-]}{[HA]}$$

where pK_a applies to the species in the denominator. The concentrations of HA and A^- are essentially unchanged from those used to prepare the solution. The pH of a buffer is nearly independent of dilution, but buffer capacity increases as the concentration of buffer increases. The maximum buffer capacity is at $pH = pK_a$, and the useful range is $pH = pK_a \pm 1$.

The conjugate base of a weak acid is a weak base. The weaker the acid, the stronger the base. However, if one member of a conjugate pair is weak, so is its conjugate. The relation between K_a for an acid and K_b for its conjugate base in aqueous solution is $K_a \cdot K_b = K_w$. When a strong acid (or base) is added to a weak base (or acid), they react nearly completely.

Exercises

8-A. Using activity coefficients correctly, find the pH of 1.0×10^{-2} M NaOH.

8-B. (Without activities), calculate the pH of

(a) 1.0×10^{-8} M HBr

(b) 1.0×10^{-8} M H_2SO_4 (H_2SO_4 dissociates completely to $2H^+$ plus SO_4^{2-} at this low concentration.)

8-C. What is the pH of a solution prepared by dissolving 1.23 g of 2-nitrophenol (FM 139.11) in 0.250 L?

8-D. The pH of 0.010 M *o*-cresol is 6.16. Find pK_a for this weak acid.

o-Cresol

8-E. Calculate the limiting value of the fraction of dissociation (α) of a weak acid ($pK_a = 5.00$) as the concentration of HA approaches 0. Repeat the same calculation for $pK_a = 9.00$.

8-F. Find the pH of 0.050 M sodium butanoate (the sodium salt of butanoic acid, also called butyric acid).

8-G. The pH of 0.10 M ethylamine is 11.82.

(a) Without referring to Appendix G, find K_b for ethylamine.

(b) Using results from part (a), calculate the pH of 0.10 M ethylammonium chloride.

8-H. Which of the following bases would be most suitable for preparing a buffer of pH 9.00? (i) NH_3 (ammonia, $K_b = 1.76 \times 10^{-5}$); (ii) $C_6H_5NH_2$ (aniline, $K_b = 3.99 \times 10^{-10}$); (iii) H_2NNH_2 (hydrazine, $K_b = 1.05 \times 10^{-6}$); (iv) C_5H_5N (pyridine, $K_b = 1.58 \times 10^{-9}$)

8-I. A solution contains 63 different conjugate acid-base pairs. Among them is acrylic acid and acrylate ion, with the equilibrium ratio [acrylate]/[acrylic acid] = 0.75. What is the pH of the solution?

$$H_2C=CHCO_2H \qquad pK_a = 4.25$$
Acrylic acid

8-J. (a) Find the pH of a solution prepared by dissolving 1.00 g of glycine amide hydrochloride (Table 8-2) plus 1.00 g of glycine amide in 0.100 L.

Glycine amide
$C_2H_6N_2O$
FM 74.08

(b) How many grams of glycine amide should be added to 1.00 g of glycine amide hydrochloride to give 100 mL of solution with pH 8.00?

(c) What would be the pH if the solution in part (a) were mixed with 5.00 mL of 0.100 M HCl?

(d) What would be the pH if the solution in part (c) were mixed with 10.00 mL of 0.100 M NaOH?

(e) What would be the pH if the solution in part (a) were mixed with 90.46 mL of 0.100 M NaOH? (This is exactly the quantity of NaOH required to neutralize the glycine amide hydrochloride.)

8-K. A solution with an ionic strength of 0.10 M containing 0.010 0 M phenylhydrazine has a pH of 8.13. Using activity coefficients correctly, find pK_a for the phenylhydrazinium ion found in phenylhydrazine hydrochloride. Assume that $\gamma_{BH^+} = 0.80$.

Phenylhydrazine
B

Phenylhydrazine hydrochloride
BH^+Cl^-

8-L. [spreadsheet icon] Use the Goal Seek spreadsheet at the end of the chapter to find the pH of 1.00 L of solution containing 0.030 mol HA ($pK_a = 2.50$) and 0.015 mol NaA. What would the pH be with the approximations [HA] = 0.030 and $[A^-] = 0.015$?

Problems

Strong Acids and Bases

8-1. Why doesn't water dissociate to produce 10^{-7} M H^+ and 10^{-7} M OH^- when some HBr is added?

8-2. Calculate the pH of **(a)** 1.0×10^{-3} M HBr; **(b)** 1.0×10^{-2} M KOH.

8-3. Calculate the pH of 5.0×10^{-8} M $HClO_4$. What fraction of the total H^+ in this solution is derived from dissociation of water?

8-4. (a) The measured pH of 0.100 M HCl at 25°C is 1.092. From this information, calculate the activity coefficient of H^+ and compare your answer with that in Table 7-1.

(b) The measured pH of 0.010 0 M HCl + 0.090 0 M KCl at 25°C is 2.102. From this information, calculate the activity coefficient of H^+ in this solution.

(c) The ionic strengths of the solutions in parts **(a)** and **(b)** are the same. What can you conclude about the dependence of activity coefficients on the particular ions in a solution?

Weak-Acid Equilibria

8-5. Write the chemical reaction whose equilibrium constant is

(a) K_a for benzoic acid, $C_6H_5CO_2H$

(b) K_b for benzoate ion, $C_6H_5CO_2^-$

(c) K_b for aniline, $C_6H_5NH_2$

(d) K_a for anilinium ion, $C_6H_5NH_3^+$

8-6. Find the pH and fraction of dissociation (α) of a 0.100 M solution of the weak acid HA with $K_a = 1.00 \times 10^{-5}$.

8-7. $BH^+ClO_4^-$ is a salt formed from the base B ($K_b = 1.00 \times 10^{-4}$) and perchloric acid. It dissociates into BH^+, a weak acid, and ClO_4^-, which is neither an acid nor a base. Find the pH of 0.100 M $BH^+ClO_4^-$.

8-8. Find the pH and concentrations of $(CH_3)_3N$ and $(CH_3)_3NH^+$ in a 0.060 M solution of trimethylammonium chloride.

8-9. Use the reaction quotient, Q, to explain why the fraction of dissociation of weak acid, HA, increases when the solution is diluted by a factor of 2.

8-10. *When is a weak acid weak and when is a weak acid strong?* Show that the weak acid HA will be 92% dissociated when dissolved in water if the formal concentration is one-tenth of K_a (F = $K_a/10$). Show that the fraction of dissociation is 27% when F = $10K_a$. At what formal concentration will the acid be 99% dissociated? Compare your answer with the left-hand curve in Figure 8-2.

8-11. A 0.045 0 M solution of benzoic acid has a pH of 2.78. Calculate pK_a for this acid.

8-12. A 0.045 0 M solution of HA is 0.60% dissociated. Calculate pK_a for this acid.

8-13. Barbituric acid dissociates as follows:

Barbituric acid
HA A^-

(a) Calculate the pH and fraction of dissociation of $10^{-2.00}$ M barbituric acid.

(b) Calculate the pH and fraction of dissociation of $10^{-10.00}$ M barbituric acid.

8-14. Using activity coefficients, find the pH and fraction of dissociation of 50.0 mM hydroxybenzene (phenol) in 0.050 M LiBr. Take the size of $C_6H_5O^-$ to be 600 pm.

8-15. Cr^{3+} is acidic by virtue of the hydrolysis reaction

$$Cr^{3+} + H_2O \xrightleftharpoons{K_{a1}} Cr(OH)^{2+} + H^+$$

[Further reactions produce $Cr(OH)_2^+$, $Cr(OH)_3$, and $Cr(OH)_4^-$.] Find the value of K_{a1} in Figure 6-8. Considering only the K_{a1} reaction, find the pH of 0.010 M $Cr(ClO_4)_3$. What fraction of chromium is in the form $Cr(OH)^{2+}$?

8-16. From the dissociation constant of HNO_3 at 25°C in Box 8-1, find the percent dissociated in 0.100 M HNO_3 and in 1.00 M HNO_3.

8-17. ▦ *Excel Goal Seek.* Solve the equation $x^2/(F - x) = K$ by using Goal Seek. Guess a value of x in cell A4 and evaluate $x^2/(F - x)$ in cell B4. Use Goal Seek to vary the value of x until $x^2/(F - x)$ is equal to K. Use your spreadsheet to check your answer to Problem 8-6.

	A	B
1	Using Excel Goal Seek	
2		
3	x =	x^2/(F-x) =
4	0.01	1.1111E-03
5	F =	
6	0.1	

Weak-Base Equilibria

8-18. Covalent compounds generally have higher vapor pressure than ionic compounds. The "fishy" smell of fish arises from amines in the fish. Explain why squeezing lemon (which is acidic) onto fish reduces the fishy smell (and taste).

8-19. Find the pH and fraction of association (α) of a 0.100 M solution of the weak base B with $K_b = 1.00 \times 10^{-5}$.

8-20. Find the pH and concentrations of $(CH_3)_3N$ and $(CH_3)_3NH^+$ in a 0.060 M solution of trimethylamine.

8-21. Find the pH of 0.050 M NaCN.

8-22. Calculate the fraction of association (α) for 1.00×10^{-1}, 1.00×10^{-2}, and 1.00×10^{-12} M sodium acetate. Does α increase or decrease with dilution?

8-23. A 0.10 M solution of a base has pH = 9.28. Find K_b.

8-24. A 0.10 M solution of a base is 2.0% hydrolyzed ($\alpha = 0.020$). Find K_b.

8-25. Show that the limiting fraction of association of a base in water, as the concentration of base approaches 0, is $\alpha = 10^7 K_b/(1 + 10^7 K_b)$. Find the limiting value of α for $K_b = 10^{-4}$ and for $K_b = 10^{-10}$.

Buffers

8-26. Describe how to prepare 100 mL of 0.200 M acetate buffer, pH 5.00, starting with pure liquid acetic acid and solutions containing ~3 M HCl and ~3 M NaOH.

8-27. Why is the pH of a buffer nearly independent of concentration?

8-28. Why does buffer capacity increase as the concentration of buffer increases?

8-29. Why does buffer capacity increase as a solution becomes very acidic (pH ≈ 1) or very basic (pH ≈ 13)?

8-30. Why does the buffer capacity reach a maximum when $pH = pK_a$?

8-31. Explain the following statement: The Henderson-Hasselbalch equation (with activity coefficients) is *always* true; what may not be correct are the values of [A$^-$] and [HA] that we choose to use in the equation.

8-32. Which of the following acids would be most suitable for preparing a buffer of pH 3.10? (i) hydrogen peroxide; (ii) propanoic acid; (iii) cyanoacetic acid; (iv) 4-aminobenzenesulfonic acid.

8-33. A buffer was prepared by dissolving 0.100 mol of the weak acid HA ($K_a = 1.00 \times 10^{-5}$) plus 0.050 mol of its conjugate base Na$^+$A$^-$ in 1.00 L. Find the pH.

8-34. Write the Henderson-Hasselbalch equation for a solution of formic acid. Calculate the quotient [HCO$_2^-$]/[HCO$_2$H] at **(a)** pH 3.000; **(b)** pH 3.744; **(c)** pH 4.000.

8-35. Calculate the quotient [HCO$_2^-$]/[HCO$_2$H] at pH 3.744 if the ionic strength is 0.1 M by using the effective equilibrium constant listed for $\mu = 0.1$ in Appendix G.

8-36. Given that pK_b for nitrite ion (NO$_2^-$) is 10.85, find the quotient [HNO$_2$]/[NO$_2^-$] in a solution of sodium nitrite at **(a)** pH 2.00; **(b)** pH 10.00.

8-37. **(a)** Would you need NaOH or HCl to bring the pH of 0.050 0 M HEPES (Table 8-2) to 7.45?

(b) Describe how to prepare 0.250 L of 0.050 0 M HEPES, pH 7.45.

8-38. How many milliliters of 0.246 M HNO$_3$ should be added to 213 mL of 0.006 66 M 2,2'-bipyridine to give a pH of 4.19?

8-39. **(a)** Write the chemical reactions whose equilibrium constants are K_b and K_a for imidazole and imidazole hydrochloride, respectively.

(b) Calculate the pH of a solution prepared by mixing 1.00 g of imidazole with 1.00 g of imidazole hydrochloride and diluting to 100.0 mL.

(c) Calculate the pH of the solution if 2.30 mL of 1.07 M HClO$_4$ are added.

(d) How many milliliters of 1.07 M HClO$_4$ should be added to 1.00 g of imidazole to give a pH of 6.993?

8-40. Calculate the pH of a solution prepared by mixing 0.080 0 mol of chloroacetic acid plus 0.040 0 mol of sodium chloroacetate in 1.00 L of water.

(a) First do the calculation by assuming that the concentrations of HA and A$^-$ equal their formal concentrations.

(b) Then do the calculation, using the real values of [HA] and [A$^-$] in the solution.

(c) Using first your head, and then the Henderson-Hasselbalch equation, find the pH of a solution prepared by dissolving all the following compounds in one beaker containing a total volume of 1.00 L: 0.180 mol ClCH$_2$CO$_2$H, 0.020 mol ClCH$_2$CO$_2$Na, 0.080 mol HNO$_3$, and 0.080 mol Ca(OH)$_2$. Assume that Ca(OH)$_2$ dissociates completely.

8-41. Calculate how many milliliters of 0.626 M KOH should be added to 5.00 g of MOBS (Table 8-2) to give a pH of 7.40.

8-42. **(a)** Use Equations 8-20 and 8-21 to find the pH and concentrations of HA and A$^-$ in a solution prepared by mixing 0.002 00 mol of acetic acid plus 0.004 00 mol of sodium acetate in 1.00 L of water.

(b) 🖿 After working part **(a)** by hand, use Excel Goal Seek to find the same answers.

8-43. **(a)** Calculate the pH of a solution prepared by mixing 0.010 0 mol of the base B ($K_b = 10^{-2.00}$) with 0.020 0 mol of BH$^+$Br$^-$ and diluting to 1.00 L. First calculate the pH by assuming [B] = 0.010 0 and [BH$^+$] = 0.020 0 M. Compare this answer with the pH calculated without making such an assumption.

(b) 🖿 After working part **(a)** by hand, use Excel Goal Seek to find the same answers.

8-44. *Effect of ionic strength on pK$_a$.* K_a for the H$_2$PO$_4^-$/HPO$_4^{2-}$ buffer is

$$K_a = \frac{[\text{HPO}_4^{2-}][\text{H}^+]\gamma_{\text{HPO}_4^{2-}}\gamma_{\text{H}^+}}{[\text{H}_2\text{PO}_4^-]\gamma_{\text{H}_2\text{PO}_4^-}} = 10^{-7.20}$$

If you mix a 1:1 mole ratio of H$_2$PO$_4^-$ and HPO$_4^{2-}$ at 0 ionic strength, the pH is 7.20. Using activity coefficients from Table 7-1, calculate the pH of a 1:1 mixture of H$_2$PO$_4^-$ and HPO$_4^{2-}$ at an ionic strength of 0.10. Remember that pH = $-\log \mathcal{A}_{\text{H}^+}$ = $-\log[\text{H}^+]\gamma_{\text{H}^+}$.

8-45. *Systematic treatment of equilibrium.* The acidity of Al^{3+} is determined by the following reactions. Write the equations needed to find the pH of Al(ClO$_4$)$_3$ at a formal concentration F.

$$\text{Al}^{3+} + \text{H}_2\text{O} \xrightleftharpoons{\beta_1} \text{AlOH}^{2+} + \text{H}^+$$

$$\text{Al}^{3+} + 2\text{H}_2\text{O} \xrightleftharpoons{\beta_2} \text{Al(OH)}_2^+ + 2\text{H}^+$$

$$2\text{Al}^{3+} + 2\text{H}_2\text{O} \xrightleftharpoons{K_{22}} \text{Al}_2(\text{OH})_2^{4+} + 2\text{H}^+$$

$$\text{Al}^{3+} + 3\text{H}_2\text{O} \xrightleftharpoons{\beta_3} \text{Al(OH)}_3(aq) + 3\text{H}^+$$

$$\text{Al}^{3+} + 4\text{H}_2\text{O} \xrightleftharpoons{\beta_4} \text{Al(OH)}_4^- + 4\text{H}^+$$

$$3\text{Al}^{3+} + 4\text{H}_2\text{O} \xrightleftharpoons{K_{43}} \text{Al}_3(\text{OH})_4^{5+} + 4\text{H}^+$$

9 Polyprotic Acid-Base Equilibria

(a) Myoglobin backbone

(b) Heme structure

(c) Space-filling model of myoglobin

(a) Amino acid backbone of the protein myoglobin, which stores oxygen in muscle tissue. Substituents (R groups from Table 9-1) are omitted for clarity. The flat *heme* group at the right side of the protein contains an iron atom that can bind O_2, CO, and other small molecules. [From M. F. Perutz, "The Hemoglobin Molecule." Copyright © 1964 by Scientific American, Inc.] (b) Structure of heme. (c) Space-filling model of myoglobin, with charged acidic and basic amino acids in dark color and *hydrophobic* (nonpolar, water-repelling) amino acids in light color. White amino acids are *hydrophilic* (polar, water-loving), but not charged. The surface of this water-soluble protein is dominated by charged and hydrophilic groups. [From J. M. Berg, J. L. Tymoczko, and L. Stryer, *Biochemistry*, 5th ed. (New York: Freeman, 2002).]

Proteins perform biological functions such as structural support, catalysis of chemical reactions, immune response to foreign substances, transport of molecules across membranes, and control of genetic expression. The three-dimensional structure and function of a protein is determined by the sequence of *amino acids* from which the protein is made. The diagram below shows how amino acids are connected to make a *polypeptide*. Of the 20 common amino acids, three have basic substituents and four have acidic substituents. Myoglobin, shown above, folds into several helical (spiral) regions that control access of oxygen and other small molecules to the heme group, whose function is to store O_2 in muscle cells. Of the 153 amino acids in sperm-whale myoglobin, 35 have basic side groups and 23 are acidic.

$$H_3\overset{+}{N}-\underset{\underset{R_1}{|}}{\overset{\overset{H}{|}}{C}}-CO_2^- \ + \ H_3\overset{+}{N}-\underset{\underset{R_2}{|}}{\overset{\overset{H}{|}}{C}}-CO_2^- \ + \ H_3\overset{+}{N}-\underset{\underset{R_3}{|}}{\overset{\overset{H}{|}}{C}}-CO_2^- \quad \text{Amino acids}$$

A side group (also called a substituent)

$-2H_2O$

$$H_3\overset{+}{N}-\underset{\underset{R_1}{|}}{\overset{\overset{H}{|}}{C}}-\overset{\overset{O}{\|}}{C}-\underset{\underset{H}{|}}{N}-\underset{\underset{R_2}{|}}{\overset{\overset{H}{|}}{C}}-\overset{\overset{O}{\|}}{C}-\underset{\underset{H}{|}}{N}-\underset{\underset{R_3}{|}}{\overset{\overset{H}{|}}{C}}-CO_2^-$$

A polypeptide (A long polypeptide is called a protein.)

N-terminal residue

Peptide bond

C-terminal residue

Polyprotic acids and bases are those that can donate or accept more than one proton. After we have studied **diprotic** systems (with two acidic or basic sites), the extension to three or more acidic sites is straightforward. Then we step back and take a qualitative look at the big picture and think about which species are dominant at any given pH.

9-1 Diprotic Acids and Bases

Amino acids from which proteins are built have an acidic carboxylic acid group, a basic amino group, and a variable substituent designated R. The carboxyl group is a stronger acid than the ammonium group, so the nonionized form rearranges spontaneously to the **zwitterion,** which has both positive and negative sites:

A *zwitterion* is a molecule with both positive and negative charges.

Amino group ⟶ H_2N

Carboxylic acid ⟶ HO—C

CH—R ⟶

$H_3\overset{+}{N}$ ← Ammonium group

CH—R

^{-}O—C ← Carboxyl group

O

O

Zwitterion

pK_a values of amino acids in living cells are somewhat different from those in Table 9-1 because physiologic temperature is not 25°C and the ionic strength is not 0.

At low pH, both the ammonium group and the carboxyl group are protonated. At high pH, neither is protonated. Acid dissociation constants of amino acids are listed in Table 9-1, where each compound is drawn in its fully protonated form.

Zwitterions are stabilized in solution by interactions of $-NH_3^+$ and $-CO_2^-$ with water. The zwitterion is also the stable form of the amino acid in the solid state, where hydrogen bonding from $-NH_3^+$ to $-CO_2^-$ of neighboring molecules occurs. In the gas phase, there are no neighbors to stabilize the charges, so the nonionized structure in Figure 9-1, with intramolecular hydrogen bonding from $-NH_2$ to a carboxyl oxygen, predominates.

Our discussion will focus on the amino acid leucine, designated HL.

The substituent R in leucine is an isobutyl group: $(CH_3)_2CHCH_2-$

$$H_3\overset{+}{N}CHCO_2H \overset{pK_{a1} = 2.328}{\rightleftharpoons} H_3\overset{+}{N}CHCO_2^- \overset{pK_{a2} = 9.744}{\rightleftharpoons} H_2\overset{}{N}CHCO_2^-$$

H_2L^+ Leucine HL L^-

The equilibrium constants refer to the following reactions:

We customarily omit the subscript "a" in K_{a1} and K_{a2}. We will always write the subscript "b" in K_{b1} and K_{b2}.

Diprotic acid:

$$H_2L^+ \rightleftharpoons HL + H^+ \qquad K_{a1} \equiv K_1 \qquad (9\text{-}1)$$
$$HL \rightleftharpoons L^- + H^+ \qquad K_{a2} \equiv K_2 \qquad (9\text{-}2)$$

Diprotic base:

$$L^- + H_2O \rightleftharpoons HL + OH^- \qquad K_{b1} \qquad (9\text{-}3)$$
$$HL + H_2O \rightleftharpoons H_2L^+ + OH^- \qquad K_{b2} \qquad (9\text{-}4)$$

Recall that the relations between the acid and base equilibrium constants are

Relations between K_a and K_b:

$$K_{a1} \cdot K_{b2} = K_w \qquad (9\text{-}5)$$

$$K_{a2} \cdot K_{b1} = K_w \qquad (9\text{-}6)$$

We now set out to calculate the pH and composition of individual solutions of 0.050 0 M H_2L^+, 0.050 0 M HL, and 0.050 0 M L^-. Our methods are general. They do not depend on the charge type of the acids and bases. That is, *we would use the same procedure to find the pH of the diprotic H_2A, where A is anything, or H_2L^+, where HL is leucine.*

270 pm

288 pm

FIGURE 9-1 Gas-phase structure of alanine, determined by microwave spectroscopy. [From S. Blanco, A. Lesarri, J. C. López, and J. L. Alonso, "The Gas-Phase Structure of Alanine," *J. Am. Chem. Soc.* 2004, *126*, 11675.]

The Acidic Form, H_2L^+

Leucine hydrochloride contains the protonated species, H_2L^+, which can dissociate twice (Reactions 9-1 and 9-2). Because $K_1 = 4.70 \times 10^{-3}$, H_2L^+ is a weak acid. HL is an even weaker acid, because $K_2 = 1.80 \times 10^{-10}$. It appears that the H_2L^+ will dissociate only partly, and the resulting HL will hardly dissociate at all. For this reason, we make the (superb) approximation that a solution of H_2L^+ behaves as a monoprotic acid, with $K_a = K_1$.

With this approximation, finding the pH of 0.050 0 M H_2L^+ is easy.

H_2L^+ can be treated as monoprotic, with $K_a = K_{a1}$.

$$H_3\overset{+}{N}CHCO_2H \overset{K_a = K_{a1} = K_1}{\rightleftharpoons} H_3\overset{+}{N}CHCO_2^- + H^+$$

H_2L^+ HL H^+

0.050 0 − x   x   x

$$K_a = K_1 = 4.70 \times 10^{-3}$$

$$\frac{x^2}{F - x} = K_a \Rightarrow x = 1.32 \times 10^{-2}\ M$$ Solve for x with the quadratic equation.

$$[HL] = x = 1.32 \times 10^{-2}\ M$$
$$[H^+] = x = 1.32 \times 10^{-2}\ M \Rightarrow pH = 1.88$$
$$[H_2L^+] = F - x = 3.68 \times 10^{-2}\ M$$

TABLE 9-1   Acid dissociation constants of amino acids

Amino acid[a]	Substituent[a]	Carboxylic acid[b] pK_a	Ammonium[b] pK_a	Substituent[b] pK_a	Formula mass
Alanine (A)	$-CH_3$	2.344	9.868		89.09
Arginine (R)	$-CH_2CH_2CH_2NHC\overset{\overset{+}{N}H_2}{\underset{NH_2}{\|}}$	1.823	8.991	(12.1[c])	174.20
Asparagine (N)	$-CH_2\overset{O}{\overset{\|}{C}}NH_2$	2.16[c]	8.73[c]		132.12
Aspartic acid (D)	$-CH_2CO_2H$	1.990	10.002	3.900	133.10
Cysteine (C)	$-CH_2SH$	(1.7)	10.74	8.36	121.16
Glutamic acid (E)	$-CH_2CH_2CO_2H$	2.16	9.96	4.30	147.13
Glutamine (Q)	$-CH_2CH_2\overset{O}{\overset{\|}{C}}NH_2$	2.19[c]	9.00[c]		146.15
Glycine (G)	$-H$	2.350	9.778		75.07
Histidine (H)	$-CH_2\!\!\begin{smallmatrix}\text{imidazole}\end{smallmatrix}$	(1.6)	9.28	5.97	155.16
Isoleucine (I)	$-CH(CH_3)(CH_2CH_3)$	2.318	9.758		131.17
Leucine (L)	$-CH_2CH(CH_3)_2$	2.328	9.744		131.17
Lysine (K)	$-CH_2CH_2CH_2CH_2NH_3^+$	(1.77)	9.07	10.82	146.19
Methionine (M)	$-CH_2CH_2SCH_3$	2.18[c]	9.08[c]		149.21
Phenylalanine (F)	$-CH_2\!\!-\!\!\bigcirc$	2.20	9.31		165.19
Proline (P)	(Structure of entire amino acid)	1.952	10.640		115.13
Serine (S)	$-CH_2OH$	2.187	9.209		105.09
Threonine (T)	$-CH(CH_3)(OH)$	2.088	9.100		119.12
Tryptophan (W)	$-CH_2\!\!\begin{smallmatrix}\text{indole}\end{smallmatrix}$	2.37[c]	9.33[c]		204.23
Tyrosine (Y)	$-CH_2\!\!-\!\!\bigcirc\!\!-OH$	2.41[c]	8.67[c]	11.01[c]	181.19
Valine (V)	$-CH(CH_3)_2$	2.286	9.719		117.15

a. The acidic protons are shown in **bold** type. Each amino acid is written in its fully protonated form. Standard abbreviations are shown in parentheses.
b. pK_a values refer to 25°C and zero ionic strength unless marked by c. Values considered to be uncertain are enclosed in parentheses. Appendix G gives pK_a for $\mu = 0.1$ M.
c. For these entries, the ionic strength is 0.1 M, and the constant refers to a product of concentrations instead of activities.

SOURCE: A. E. Martell and R. J. Motekaitis, NIST Database 46 (Gaithersburg, MD: National Institute of Standards and Technology, 2001).

What is the concentration of L^- in the solution? We have already assumed that it is very small, but it cannot be 0. We can calculate $[L^-]$ from the K_2 equation, with the concentrations of HL and H^+ that we just computed.

$$K_2 = \frac{[H^+][L^-]}{[HL]} \Rightarrow [L^-] = \frac{K_2[HL]}{[H^+]} \tag{9-7}$$

$$[L^-] = \frac{(1.80 \times 10^{-10})(1.32 \times 10^{-2})}{(1.32 \times 10^{-2})} = 1.80 \times 10^{-10} \text{ M } (= K_2)$$

The approximation $[H^+] \approx [HL]$ reduces Equation 9-7 to $[L^-] = K_2$.

Our approximation is confirmed by this last result. The concentration of L^- is about eight orders of magnitude smaller than that of HL. The dissociation of HL is indeed negligible relative to the dissociation of H_2L^+. For most diprotic acids, K_1 is sufficiently larger than K_2 for this approximation to be valid. Even if K_2 were just 10 times less than K_1, $[H^+]$ calculated by ignoring the second ionization would be in error by only 4%. The error in pH would be only 0.01 pH unit. In summary, *a solution of a diprotic acid behaves like a solution of a monoprotic acid, with $K_a = K_1$.*

Dissolved carbon dioxide is one of the most important diprotic acids in Earth's ecosystem. Box 9-1 describes imminent danger to the entire ocean food chain as a result of increasing atmospheric CO_2 dissolving in the oceans. Reaction A in Box 9-1 lowers the concentration of CO_3^{2-} in the oceans. As a result, $CaCO_3$ shells and skeletons of creatures at the bottom of the food chain will dissolve by Reaction B. This effect is far more certain than the effects of atmospheric CO_2 on Earth's climate.

The Basic Form, L^-

The species L^-, found in a salt such as sodium leucinate, can be prepared by treating leucine (HL) with an equimolar quantity of NaOH. Dissolving sodium leucinate in water gives a solution of L^-, the fully basic species. K_b values for this dibasic anion are

$$L^- + H_2O \rightleftharpoons HL + OH^- \qquad K_{b1} = K_w/K_{a2} = 5.55 \times 10^{-5}$$

$$HL + H_2O \rightleftharpoons H_2L^+ + OH^- \qquad K_{b2} = K_w/K_{a1} = 2.13 \times 10^{-12}$$

Hydrolysis is the reaction of anything with water. Specifically, the reaction $L^- + H_2O \rightleftharpoons HL + OH^-$ is called hydrolysis.

K_{b1} tells us that L^- will not *hydrolyze* (react with water) very much to give HL. Furthermore, K_{b2} tells us that the resulting HL is such a weak base that hardly any further reaction to make H_2L^+ will occur.

We therefore treat L^- as a monobasic species, with $K_b = K_{b1}$. The results of this (fantastic) approximation are outlined as follows:

L^- can be treated as monobasic, with $K_b = K_{b1}$.

$$H_2NCHCO_2^- + H_2O \overset{K_b = K_{b1}}{\rightleftharpoons} H_3\overset{+}{N}CHCO_2^- + OH^-$$

$$\begin{array}{ccc} L^- & HL & OH^- \\ 0.050\ 0 - x & x & x \end{array}$$

$$K_b = K_{b1} = \frac{K_w}{K_{a2}} = 5.55 \times 10^{-5}$$

$$\frac{x^2}{F - x} = 5.55 \times 10^{-5} \Rightarrow x = 1.64 \times 10^{-3} \text{ M}$$

$$[HL] = x = 1.64 \times 10^{-3} \text{ M}$$

$$[H^+] = K_w/[OH^-] = K_w/x = 6.11 \times 10^{-12} \text{ M} \Rightarrow pH = 11.21$$

$$[L^-] = F - x = 4.84 \times 10^{-2} \text{ M}$$

The concentration of H_2L^+ can be found from the K_{b2} (or K_{a1}) equilibrium.

$$K_{b2} = \frac{[H_2L^+][OH^-]}{[HL]} = \frac{[H_2L^+]x}{x} = [H_2L^+]$$

We find that $[H_2L^+] = K_{b2} = 2.13 \times 10^{-12}$ M, and the approximation that $[H_2L^+]$ is insignificant relative to $[HL]$ is well justified. In summary, if there is any reasonable separation between K_{a1} and K_{a2} (and, therefore, between K_{b1} and K_{b2}), *the fully basic form of a diprotic acid can be treated as monobasic, with $K_b = K_{b1}$.*

BOX 9-1   Carbon Dioxide in the Air and Ocean

The graph on the opening page of Chapter 0 shows that atmospheric CO_2 has oscillated between about 180 and 280 ppm by volume ($\mu L/L$) for 800 000 years. Burning of fossil fuel and destruction of Earth's forests since 1800 caused an exponential increase in CO_2 that threatens to alter Earth's climate in your lifetime.

Increasing atmospheric CO_2 increases the concentration of CO_2 dissolved in the ocean, which consumes carbonate and lowers the pH:[1]

$$CO_2(aq) + H_2O + CO_3^{2-} \rightleftharpoons 2HCO_3^- \qquad \text{(A)}$$
$$\underset{\text{Carbonate}}{} \qquad \underset{\text{Bicarbonate}}{}$$

The pH of the ocean has already decreased from its preindustrial value of 8.16 to 8.04 today.[2] Without changes in our activities, the pH could be 7.7 by 2100.

Low carbonate concentration promotes dissolution of solid calcium carbonate:

$$CaCO_3(s) \rightleftharpoons Ca^{2+} + CO_3^{2-} \qquad \text{(B)}$$
$$\underset{\text{Calcium carbonate}}{}$$

Le Châtelier's principle tells us that decreasing $[CO_3^{2-}]$ draws the reaction to the right

If $[CO_3^{2-}]$ in the ocean decreases enough, organisms such as plankton and coral with $CaCO_3$ shells or skeletons will not survive.[3] Calcium carbonate has two crystalline forms called calcite and aragonite. Aragonite is more soluble than calcite. Aquatic organisms have either calcite or aragonite in their shells and skeletons.

Pteropods are a type of zooplankton also known as winged snails. When pteropods collected from the subarctic Pacific Ocean are kept in water that is less than saturated with aragonite, their shells begin to dissolve within 48 h. Animals such as the pteropod lie at the base of the food chain. Their destruction would reverberate throughout the entire ocean.

Today, ocean surface waters contain more than enough CO_3^{2-} to sustain aragonite and calcite. As atmospheric CO_2 inexorably increases during the twenty-first century, ocean surface waters will become undersaturated with respect to aragonite—killing off organisms that depend on this mineral for their structure. Polar regions will suffer this fate first because CO_2 is more soluble in cold water than in warm water and K_{a1} and K_{a2} at low temperature favor HCO_3^- and $CO_2(aq)$ relative to CO_3^{2-} (Problem 9-12).

Panel *b* shows the predicted concentration of CO_3^{2-} in polar ocean surface water as a function of atmospheric CO_2. The upper horizontal line is the concentration of CO_3^{2-} below which aragonite dissolves. Atmospheric CO_2 is presently near 400 ppm and $[CO_3^{2-}]$ is near 100 $\mu mol/kg$ of seawater—more than enough to precipitate aragonite or calcite. When atmospheric CO_2 reaches 600 ppm near the middle of the present century, $[CO_3^{2-}]$ will decrease to 60 $\mu mol/kg$ and creatures with aragonite structures will begin to disappear from polar waters. At still higher atmospheric CO_2 concentration, extinctions will move to lower latitudes and will overtake organisms with calcite structures as well as aragonite structures. *How long will we put CO_2 into the atmosphere to see if these predictions are borne out?*

(b)

Calculated $[CO_3^{2-}]$ in polar ocean surface waters as a function of atmospheric CO_2. When $[CO_3^{2-}]$ drops below the upper horizontal line, argonite dissolves. [Adapted from J. C. Orr et al., *Nature* 2005, *437*, 681.[4]] Reference 1 gives equations that allow you to calculate the curve in this figure.

(a)

Pteropods. The shell of a live pteropod begins to dissolve after 48 h in water that is undersaturated with aragonite. [David Wrobel/Visuals Unlimited.]

Nature counteracts some changes to mitigate their effects. For example, phytoplankton called coccolithophores are marine organisms with a $CaCO_3$ skeleton several micrometers in diameter. These organisms produce about one-third of all $CaCO_3$ in the ocean. In the past 220 years, as atmospheric CO_2 increased, the average mass of the coccolithophore species *Emiliania huxleyi* has increased by 40%, thereby removing some CO_2 from the ocean.[5] Coccolithophores can mitigate CO_2 increases up to a point. It is not likely that any *calcifying* ($CaCO_3$-producing) marine organism can survive if CO_2 increases to the level where $CaCO_3$ is no longer thermodynamically stable.

The Intermediate Form, HL

A solution prepared from leucine, HL, is more complicated than one prepared from either H_2L^+ or L^-, because HL is both an acid and a base.

$$HL \rightleftharpoons H^+ + L^- \qquad K_a = K_{a2} = 1.80 \times 10^{-10} \qquad (9\text{-}8)$$

$$HL + H_2O \rightleftharpoons H_2L^+ + OH^- \qquad K_b = K_{b2} = 2.13 \times 10^{-12} \qquad (9\text{-}9)$$

A molecule that can both donate and accept a proton is said to be **amphiprotic.** The acid dissociation reaction (9-8) has a larger equilibrium constant than the base hydrolysis reaction (9-9), so we expect that a solution of leucine will be acidic.

However, we cannot simply ignore Reaction 9-9, even if K_a and K_b differ by several orders of magnitude. Both reactions proceed to nearly equal extent, because H^+ produced in Reaction 9-8 reacts with OH^- from Reaction 9-9, thereby driving Reaction 9-9 to the right.

To treat this case, we resort to the systematic treatment of equilibrium. The procedure is applied to leucine, whose intermediate form (HL) has no net charge. However, the results apply to the intermediate form of *any* diprotic acid, regardless of its charge.

For Reactions 9-8 and 9-9, the charge balance is

$$[H^+] + [H_2L^+] = [L^-] + [OH^-] \quad \text{or} \quad [H_2L^+] - [L^-] + [H^+] - [OH^-] = 0$$

From the acid dissociation equilibria, we replace $[H_2L^+]$ with $[HL][H^+]/K_1$, and $[L^-]$ with $[HL]K_2/[H^+]$. Also, we can always write $[OH^-] = K_w/[H^+]$. Putting these expressions into the charge balance gives

$$\frac{[HL][H^+]}{K_1} - \frac{[HL]K_2}{[H^+]} + [H^+] - \frac{K_w}{[H^+]} = 0$$

which can be solved for $[H^+]$. First, multiply all terms by $[H^+]$:

$$\frac{[HL][H^+]^2}{K_1} - [HL]K_2 + [H^+]^2 - K_w = 0$$

Then rearrange and factor out $[H^+]^2$:

$$[H^+]^2 \left(\frac{[HL]}{K_1} + 1 \right) = K_2[HL] + K_w$$

$$[H^+]^2 = \frac{K_2[HL] + K_w}{\dfrac{[HL]}{K_1} + 1}$$

Now multiply the numerator and denominator by K_1 and take the square root:

$$[H^+] = \sqrt{\frac{K_1K_2[HL] + K_1K_w}{K_1 + [HL]}} \qquad (9\text{-}10)$$

Up to this point, we have made no approximations, except to neglect activity coefficients. We solved for $[H^+]$ in terms of known constants plus the single unknown, $[HL]$. Where do we proceed from here?

Just as we are feeling desperate, the Good Chemist again gallops down from the mountains to provide the missing insight: "The major species is HL, because it is both a weak acid and a weak base. Neither Reaction 9-8 nor Reaction 9-9 goes very far. For [HL] in Equation 9-10, you can simply substitute the formal concentration, 0.050 0 M."

Taking the Good Chemist's advice, we write Equation 9-10 in its most useful form.

Intermediate form of diprotic acid: $\qquad [H^+] \approx \sqrt{\dfrac{K_1K_2F + K_1K_w}{K_1 + F}} \qquad (9\text{-}11)$

where F is the formal concentration of HL ($= 0.050\ 0$ M in the present case).

At long last, we can calculate the pH of 0.050 0 M leucine:

$$[H^+] = \sqrt{\frac{(4.70 \times 10^{-3})(1.80 \times 10^{-10})(0.050\ 0) + (4.70 \times 10^{-3})(1.0 \times 10^{-14})}{4.70 \times 10^{-3} + 0.050\ 0}}$$

$$= 8.80 \times 10^{-7}\ M \Rightarrow pH = 6.06$$

The concentrations of H_2L^+ and L^- can be found from the K_1 and K_2 equilibria, using $[H^+] = 8.80 \times 10^{-7}$ M and $[HL] = 0.050\ 0$ M.

$$[H_2L^+] = \frac{[H^+][HL]}{K_1} = \frac{(8.80 \times 10^{-7})(0.050\ 0)}{4.70 \times 10^{-3}} = 9.36 \times 10^{-6}\ M$$

$$[L^-] = \frac{K_2[HL]}{[H^+]} = \frac{(1.80 \times 10^{-10})(0.050\ 0)}{8.80 \times 10^{-7}} = 1.02 \times 10^{-5}\ M$$

Was the approximation $[HL] \approx 0.050\ 0$ M a good one? It certainly was, because $[H_2L^+]$ $(= 9.36 \times 10^{-6}$ M) and $[L^-]$ $(= 1.02 \times 10^{-5}$ M) are small in comparison with $[HL]$ $(\approx 0.050\ 0$ M). Nearly all the leucine remained in the form HL. Note also that $[H_2L^+]$ is nearly equal to $[L^-]$. This result confirms that Reactions 9-8 and 9-9 proceed equally, even though K_a is 84 times bigger than K_b for leucine.

A summary of results for leucine is given here. Notice the relative concentrations of H_2L^+, HL, and L^- in each solution and notice the pH of each solution.

If $[H_2L^+]$ + $[L^-]$ is not much less than $[HL]$ and if you wish to refine your values of $[H_2L^+]$ and $[L^-]$, try the method in Box 9-2.

Solution	pH	$[H^+]$ (M)	$[H_2L^+]$ (M)	$[HL]$ (M)	$[L^-]$ (M)
0.050 0 M H_2A	1.88	1.32×10^{-2}	3.68×10^{-2}	1.32×10^{-2}	1.80×10^{-10}
0.050 0 M HA^-	6.06	8.80×10^{-7}	9.36×10^{-6}	5.00×10^{-2}	1.02×10^{-5}
0.050 0 M HA^{2-}	11.21	6.11×10^{-12}	2.13×10^{-12}	1.64×10^{-3}	4.84×10^{-2}

BOX 9-2   Successive Approximations

The method of *successive approximations* is a good way to deal with difficult equations that do not have simple solutions. For example, Equation 9-11 is not a good approximation when the concentration of the intermediate species of a diprotic acid is not close to F, the formal concentration. This situation arises when K_1 and K_2 are nearly equal and F is small. Consider a solution of 1.00×10^{-3} M HM^-, the intermediate form of malic acid.

HO   CO₂H
         |
         | $K_1 = 3.5 \times 10^{-4}$
         | $pK_1 = 3.46$
    CO₂H

Malic acid
   H_2M

HO   CO₂⁻
         |
         |
    CO₂H

   HM^-

         $K_2 = 7.9 \times 10^{-6}$
         $pK_2 = 5.10$

HO   CO₂⁻
         |
         |
    CO₂⁻

   M^{2-}

For a first approximation, assume that $[HM^-] \approx 1.00 \times 10^{-3}$ M. Plugging this value into Equation 9-10, we calculate first approximations for $[H^+]$, $[H_2M]$, and $[M^{2-}]$.

$$[H^+]_1 = \sqrt{\frac{K_1K_2(0.001\ 00) + K_1K_w}{K_1 + (0.001\ 00)}} = 4.53 \times 10^{-5}\ M$$

$$\Rightarrow [H_2M]_1 = \frac{[H^+][HM^-]}{K_1} = \frac{(4.53 \times 10^{-5})(1.00 \times 10^{-3})}{3.5 \times 10^{-4}}$$
$$= 1.29 \times 10^{-4}\ M$$

$$[M^{2-}]_1 = \frac{K_2[HM^-]}{[H^+]} = \frac{(7.9 \times 10^{-6})(1.00 \times 10^{-3})}{4.53 \times 10^{-5}}$$
$$= 1.75 \times 10^{-4}\ M$$

Clearly, $[H_2M]$ and $[M^{2-}]$ are not negligible relative to $F = 1.00 \times 10^{-3}$ M, so we need to revise our estimate of $[HM^-]$. The mass balance gives us a second approximation:

$$[HM^-]_2 = F - [H_2M]_1 - [M^{2-}]_1$$
$$= 0.001\ 00 - 0.000\ 129 - 0.000\ 175 = 0.000\ 696\ M$$

Inserting $[HM^-]_2 = 0.000\ 696$ into Equation 9-10 gives

$$[H^+]_2 = \sqrt{\frac{K_1K_2(0.000\ 696) + K_1K_w}{K_1 + (0.000\ 696)}} = 4.29 \times 10^{-5}\ M$$

$$\Rightarrow [H_2M]_2 = 8.53 \times 10^{-5}\ M$$
$$[M^{2-}]_2 = 1.28 \times 10^{-4}\ M$$

$[H_2M]_2$ and $[M^{2-}]_2$ can be used to calculate a third approximation for $[HM^-]$:

$$[HM^-]_3 = F - [H_2M]_2 - [M^{2-}]_2 = 0.000\ 786\ M$$

Plugging $[HM^-]_3$ into Equation 9-10 gives

$$[H^+]_3 = 4.37 \times 10^{-5}\ M$$

and the procedure can be repeated to get

$$[H^+]_4 = 4.35 \times 10^{-5}\ M$$

We are homing in on an estimate of $[H^+]$ in which the precision is already less than 1%. The fourth approximation gives pH = 4.36, compared with pH = 4.34 from the first approximation and pH = 4.28 from the formula pH $\approx \frac{1}{2}(pK_1 + pK_2)$. Considering the uncertainty in pH measurements, all this calculation was hardly worth the effort. However, the concentration $[HM^-]_5$ is 0.000 768 M, which is 23% less than the original estimate of 0.001 00 M. Successive approximations can be carried out by hand, but the process is more easily and reliably performed with a spreadsheet.

Simplified Calculation for the Intermediate Form

Usually Equation 9-11 is a fair-to-excellent approximation. An even simpler form results from two conditions that usually exist. First, if $K_2F \gg K_w$, the second term in the numerator of Equation 9-11 can be dropped.

$$[H^+] \approx \sqrt{\frac{K_1K_2F + K_1K_w}{K_1 + F}}$$

Then, if $K_1 \ll F$, the first term in the denominator also can be neglected.

$$[H^+] \approx \sqrt{\frac{K_1K_2F}{K_1 + F}}$$

Canceling F in the numerator and denominator gives

$$[H^+] \approx \sqrt{K_1K_2}$$

or

$$\log[H^+] \approx \frac{1}{2}(\log K_1 + \log K_2)$$

$$-\log[H^+] \approx -\frac{1}{2}(\log K_1 + \log K_2)$$

Recall that

$$\log(x^{1/2}) = \frac{1}{2}\log x$$

$$\log(xy) = \log x + \log y$$

$$\log(x/y) = \log x - \log y$$

The pH of the intermediate form of a diprotic acid is close to midway between the two pK_a values and is almost independent of concentration.

Intermediate form of diprotic acid: $\quad pH \approx \frac{1}{2}(pK_1 + pK_2)$ (9-12)

Equation 9-12 is really good to keep in your head. It gives a pH of 6.04 for leucine, compared with pH = 6.06 from Equation 9-11. Equation 9-12 says that *the pH of the intermediate form of a diprotic acid is close to midway between pK_1 and pK_2, regardless of the formal concentration.*

EXAMPLE **pH of the Intermediate Form of a Diprotic Acid**

Potassium hydrogen phthalate, KHP, is a salt of the intermediate form of phthalic acid. Calculate the pH of 0.10 M and 0.010 M KHP.

Phthalic acid H_2P — $pK_1 = 2.950$ — Monohydrogen phthalate HP^- + H^+ — $pK_2 = 5.408$ — Phthalate P^{2-} + H^+

(Potassium hydrogen phthalate = K^+HP^-)

Solution With Equation 9-12, the pH of potassium hydrogen phthalate is estimated as $\frac{1}{2}(pK_1 + pK_2) = 4.18$, regardless of concentration. With Equation 9-11, we calculate pH = 4.18 for 0.10 M K^+HP^- and pH = 4.20 for 0.010 M K^+HP^-.

Test Yourself Find the pH of 0.002 M K^+HP^- with Equation 9-11. (***Answer:*** 4.28)

Advice When faced with the intermediate form of a diprotic acid, use Equation 9-11 to calculate pH. The answer should be close to $\frac{1}{2}(pK_1 + pK_2)$.

Summary of Diprotic Acid Calculations

Here is how we calculate the pH and composition of solutions prepared from different forms of a diprotic acid (H_2A, HA^-, or A^{2-}).

Solution of H_2A

1. Treat H_2A as a monoprotic acid with $K_a = K_1$ to find $[H^+]$, $[HA^-]$, and $[H_2A]$.

$$H_2A \overset{K_1}{\rightleftharpoons} H^+ + HA^- \qquad \frac{x^2}{F - x} = K_1$$
$$ F - x \qquad x \qquad x$$

2. Use the K_2 equilibrium to solve for $[A^{2-}]$.

$$[A^{2-}] = \frac{K_2[HA^-]}{[H^+]} = K_2$$

Solution of HA⁻

1. Use the approximation $[HA^-] \approx F$ and find the pH with Equation 9-11.

$$[H^+] = \sqrt{\frac{K_1 K_2 F + K_1 K_w}{K_1 + F}}$$

The pH should be close to $\frac{1}{2}(pK_1 + pK_2)$.

2. With $[H^+]$ from step 1 and $[HA^-] \approx F$, solve for $[H_2A]$ and $[A^{2-}]$, using the K_1 and K_2 equilibria.

$$[H_2A] = \frac{[HA^-][H^+]}{K_1} \qquad [A^{2-}] = \frac{K_2[HA^-]}{[H^+]}$$

Solution of A²⁻

1. Treat A^{2-} as monobasic, with $K_b = K_{b1} = K_w/K_{a2}$ to find $[A^{2-}]$, $[HA^-]$, and $[H^+]$.

$$\underset{F-x}{A^{2-}} + H_2O \xrightleftharpoons{K_{b1}} \underset{x}{HA^-} + \underset{x}{OH^-} \qquad \frac{x^2}{F-x} = K_{b1} = \frac{K_w}{K_{a2}}$$

$$[H^+] = \frac{K_w}{[OH^-]} = \frac{K_w}{x}$$

2. Use the K_1 equilibrium to solve for $[H_2A]$.

$$[H_2A] = \frac{[HA^-][H^+]}{K_{a1}} = \frac{[\cancel{HA^-}](K_w/[\cancel{OH^-}])}{K_{a1}} = K_{b2}$$

The calculations we have been doing are really important to understand and use. However, we should not get too cocky about our great powers, because there could be equilibria we have not considered. For example, Na^+ or K^+ in solutions of HA^- or A^{2-} form weak ion pairs that we have neglected:[6]

$$K^+ + A^{2-} \rightleftharpoons \{K^+A^{2-}\}$$
$$K^+ + HA^- \rightleftharpoons \{K^+HA^-\}$$

9-2 Diprotic Buffers

A buffer made from a diprotic (or polyprotic) acid is treated in the same way as a buffer made from a monoprotic acid. For the acid H_2A, we can write *two* Henderson-Hasselbalch equations, both of which are *always* true. If we happen to know $[H_2A]$ and $[HA^-]$, then we will use the pK_1 equation. If we know $[HA^-]$ and $[A^{2-}]$, we will use the pK_2 equation.

$$pH = pK_1 + \log\frac{[HA^-]}{[H_2A]} \qquad pH = pK_2 + \log\frac{[A^{2-}]}{[HA^-]}$$

All Henderson-Hasselbalch equations (with activity coefficients) are always true for a solution at equilibrium.

EXAMPLE A Diprotic Buffer System

Find the pH of a solution prepared by dissolving 1.00 g of potassium hydrogen phthalate and 1.20 g of disodium phthalate in 50.0 mL of water.

Solution Monohydrogen phthalate and phthalate were shown in the preceding example. The formula masses are $KHP = C_8H_5O_4K = 204.221$ and $Na_2P = C_8H_4O_4Na_2 = 210.094$. We know $[HP^-]$ and $[P^{2-}]$, so we use the pK_2 Henderson-Hasselbalch equation to find the pH:

$$pH = pK_2 + \log\frac{[P^{2-}]}{[HP^-]} = 5.408 + \log\frac{(1.20 \text{ g})/(210.094 \text{ g/mol})}{(1.00 \text{ g})/(204.221 \text{ g/mol})} = 5.47$$

K_2 is the acid dissociation constant of HP^-, which appears in the denominator of the log term. Notice that the volume of solution was not used to answer the question.

Test Yourself Find the pH with 1.50 g Na_2P instead of 1.20 g. (*Answer:* 5.57)

EXAMPLE Preparing a Buffer in a Diprotic System

How many milliliters of 0.800 M KOH should be added to 3.38 g of oxalic acid to give a pH of 4.40 when diluted to 500 mL?

$$\underset{\substack{\text{Oxalic acid} \\ (H_2Ox) \\ \text{Formula mass} = 90.035}}{\overset{\displaystyle \overset{O\ O}{\underset{}{\|\ \|}}}{HOCCOH}} \qquad \begin{aligned} pK_1 &= 1.250 \\ pK_2 &= 4.266 \end{aligned}$$

Solution The desired pH is above pK_2. We know that a 1:1 mole ratio of $HOx^- : Ox^{2-}$ would have pH $= pK_2 = 4.266$. If the pH is to be 4.40, there must be more Ox^{2-} than HOx^- present. We must add enough base to convert all of the H_2Ox into HOx^-, plus enough additional base to convert the right amount of HOx^- into Ox^{2-}.

$$H_2Ox + OH^- \rightarrow HOx^- + H_2O$$

$$pH \approx \frac{1}{2}(pK_1 + pK_2) = 2.76$$

$$HOx^- + OH^- \rightarrow Ox^{2-} + H_2O$$

A 1:1 mixture would have
pH $= pK_2 = 4.266$

In 3.38 g of H_2Ox, there are $0.037\,5_4$ mol. The volume of 0.800 M KOH needed to react with this much H_2Ox to make HOx^- is $(0.037\,5_4 \text{ mol})/(0.800 \text{ M}) = 46.9_3$ mL.

To produce a pH of 4.40 requires more OH^-:

	HOx^-	+	OH^-	$\rightarrow$	Ox^{2-}
Initial moles	$0.037\,5_4$		x		—
Final moles	$0.037\,5_4 - x$		—		x

$$pH = pK_2 + \log\frac{[Ox^{2-}]}{[HOx^-]}$$

$$4.40 = 4.266 + \log\frac{x}{0.037\,5_4 - x} \Rightarrow x = 0.021\,6_6 \text{ mol}$$

The volume of KOH needed to deliver $0.021\,6_6$ mole is $(0.021\,6_4 \text{ mol})/(0.800 \text{ M}) = 27.0_5$ mL. The total volume of KOH needed to bring the pH to 4.40 is $46.9_3 + 27.0_5 = 73.9_8$ mL.

Test Yourself What volume of KOH would bring the pH to 4.50? (**Answer:** 76.5_6 mL)

9-3 Polyprotic Acids And Bases

The treatment of diprotic acids and bases can be extended to polyprotic systems. By way of review, let's write the pertinent equilibria for a triprotic system.

$$H_3A \rightleftharpoons H_2A^- + H^+ \qquad K_{a1} = K_1$$
$$H_2A^- \rightleftharpoons HA^{2-} + H^+ \qquad K_{a2} = K_2$$
$$HA^{2-} \rightleftharpoons A^{3-} + H^+ \qquad K_{a3} = K_3$$
$$A^{3-} + H_2O \rightleftharpoons HA^{2-} + OH^- \qquad K_{b1} = K_w/K_{a3}$$
$$HA^{2-} + H_2O \rightleftharpoons H_2A^- + OH^- \qquad K_{b2} = K_w/K_{a2}$$
$$H_2A^- + H_2O \rightleftharpoons H_3A + OH^- \qquad K_{b3} = K_w/K_{a1}$$

We deal with triprotic systems as follows:

1. H_3A is treated as a monoprotic weak acid, with $K_a = K_1$.
2. H_2A^- is treated as the intermediate form of a diprotic acid.

The K values in Equations 9-13 and 9-14 are K_a values for the triprotic acid.

$$[H^+] \approx \sqrt{\frac{K_1K_2F + K_1K_w}{K_1 + F}} \qquad (9\text{-}13)$$

3. HA^{2-} is also treated as the intermediate form of a diprotic acid. However, HA^{2-} is "surrounded" by H_2A^- and A^{3-}, so the equilibrium constants to use are K_2 and K_3, instead of K_1 and K_2.

$$[H^+] \approx \sqrt{\frac{K_2K_3F + K_2K_w}{K_2 + F}} \qquad (9\text{-}14)$$

4. A^{3-} is treated as monobasic, with $K_b = K_{b1} = K_w / K_{a3}$.

CHAPTER 9 Polyprotic Acid-Base Equilibria

EXAMPLE **A Triprotic System**

Find the pH of 0.10 M H_3His^{2+}, 0.10 M H_2His^+, 0.10 M HHis, and 0.10 M His^-, where His stands for the amino acid histidine.

Solution *0.10 M H_3His^{2+}:* Treating H_3His^{2+} as a monoprotic acid, we write

$$H_3His^{2+} \rightleftharpoons H_2His^+ + H^+$$
$$\quad\quad F - x \quad\quad\quad x \quad\quad x$$

$$\frac{x^2}{F - x} = K_1 = 10^{-1.6} \Rightarrow x = 3._9 \times 10^{-2}\,M \Rightarrow pH = 1.41$$

0.10 M H_2His^+: Using Equation 9-13, we find

$$[H^+] = \sqrt{\frac{(10^{-1.6})(10^{-5.97})(0.10) + (10^{-1.6})(1.0 \times 10^{-14})}{10^{-1.6} + 0.10}}$$

$$= 1.4_7 \times 10^{-4}\,M \Rightarrow pH = 3.83$$

which is close to $\frac{1}{2}(pK_1 + pK_2) = 3.78$.

0.10 M HHis: Equation 9-14 gives

$$[H^+] = \sqrt{\frac{(10^{-5.97})(10^{-9.28})(0.10) + (10^{-5.97})(1.0 \times 10^{-14})}{10^{-5.97} + 0.10}}$$

$$= 2.3_7 \times 10^{-8}\,M \Rightarrow pH = 7.62$$

which is the same as $\frac{1}{2}(pK_2 + pK_3) = 7.62$.

0.10 M His^-: Treating His^- as monobasic, we can write

$$His^- + H_2O \rightleftharpoons HHis + OH^-$$
$$\quad F - x \quad\quad\quad\quad\quad x \quad\quad x$$

$$\frac{x^2}{F - x} = K_{b1} = \frac{K_w}{K_{a3}} = 1.9 \times 10^{-5} \Rightarrow x = 1.3_7 \times 10^{-3}\,M$$

$$pH = -\log\left(\frac{K_w}{x}\right) = 11.14$$

Test Yourself Compute the pH of 0.010 M HHis. (***Answer:*** 7.62)

We have reduced acid-base problems to just three types. When you encounter an acid or base, decide whether you are dealing with an *acidic*, *basic*, or *intermediate* form. Then do the appropriate arithmetic to answer the question at hand.

Three forms of acids and bases:
- acidic
- basic
- intermediate (amphiprotic)

9-4 Which Is the Principal Species?

We sometimes must identify which species of acid, base, or intermediate predominates under given conditions. For example, "What is the principal form of benzoic acid at pH 8?"

Benzoic acid $pK_a = 4.20$

An example of when you need to know principal species is when you design a chromatographic or electrophoretic separation. You would use different strategies for separating cations, anions, and neutral compounds.

$$pH = pK_a + \log \frac{[A^-]}{[HA]}$$

pH	Major species
$< pK_a$	HA
$> pK_a$	A^-

pH	Major species
$pH < pK_1$	H_2A
$pK_1 < pH < pK_2$	HA^-
$pH > pK_2$	A^{2-}

pK_a for benzoic acid is 4.20. So, at pH 4.20, there is a 1:1 mixture of benzoic acid (HA) and benzoate ion (A^-). At $pH = pK_a + 1$ (= 5.20), the quotient $[A^-]/[HA]$ is 10:1. At $pH = pK_a + 2$ (= 6.20), the quotient $[A^-]/[HA]$ is 100:1. As pH increases, the quotient $[A^-]/[HA]$ increases still further.

For a monoprotic system, the basic species A^- is the predominant form when $pH > pK_a$. The acidic species, HA, is the predominant form when $pH < pK_a$. The predominant form of benzoic acid at pH 8 is the benzoate anion, $C_6H_5CO_2^-$.

EXAMPLE **Principal Species—Which One and How Much?**

What is the predominant form of ammonia in a solution at pH 7.0? Approximately what fraction is in this form?

Solution In Appendix G, we find $pK_a = 9.24$ for the ammonium ion (NH_4^+, the conjugate acid of ammonia, NH_3). At $pH = 9.24$, $[NH_4^+] = [NH_3]$. Below pH 9.24, NH_4^+ will be the predominant form. Because $pH = 7.0$ is about 2 pH units below pK_a, the quotient $[NH_4^+]/[NH_3]$ will be about 100:1. More than 99% is in the form NH_4^+.

Test Yourself Approximately what fraction of ammonia is in the form NH_3 at pH 11? (**Answer:** somewhat less than 99% because pH is almost 2 units above pK_a)

For polyprotic systems, our reasoning is similar, but there are several values of pK_a. Consider oxalic acid, H_2Ox, with $pK_1 = 1.25$ and $pK_2 = 4.27$. At $pH = pK_1$, $[H_2Ox] = [HOx^-]$. At $pH = pK_2$, $[HOx^-] = [Ox^{2-}]$. The chart in the margin shows the major species in each pH region.

EXAMPLE **Principal Species in a Polyprotic System**

The amino acid arginine has the following forms:

$$H_3N^+ \quad \overset{\alpha}{\longleftarrow}$$

(chemical structures with)

H_3Arg^{2+} Substituent $pK_1 = 1.82$

H_2Arg^+ $pK_2 = 8.99$

HArg
This is the neutral molecule
called arginine $pK_3 = 12.1$ Arg^-

Appendix G tells us that the α-ammonium group (at the left) is more acidic than the substituent (at the right). What is the principal form of arginine at pH 10.0? Approximately what fraction is in this form? What is the second most abundant form at this pH?

Solution We know that at $pH = pK_2 = 8.99$, $[H_2Arg^+] = [HArg]$. At $pH = pK_3 = 12.1$, $[HArg] = [Arg^-]$. At $pH = 10.0$, the major species is HArg. Because pH 10.0 is about one pH unit higher than pK_2, we can say that $[HArg]/[H_2Arg^+] \approx 10:1$. About 90% of arginine is in the form HArg. The second most important species is H_2Arg^+, which makes up about 10% of the arginine.

Test Yourself What is the predominant form of arginine at pH 11? What is the second major species? (**Answer:** HArg, Arg^-)

EXAMPLE **More on Polyprotic Systems**

In the pH range 1.82 to 8.99, H_2Arg^+ is the principal form of arginine. Which is the second most prominent species at pH 6.0? at pH 5.0?

Solution We know that the pH of the pure intermediate (amphiprotic) species, H_2Arg^+, is

$$pH \text{ of } H_2Arg^+ \approx \frac{1}{2}(pK_1 + pK_2) = 5.40$$

Above pH 5.40 (and below pH = pK_2), HArg, the conjugate base of H_2Arg^+, will be the second most important species. Below pH 5.40 (and above pH = pK_1), H_3Arg^{2+} will be the second most important species.

Test Yourself At what pH does $[H_2Arg^+] = [Arg^-]$? (***Answer:*** 10.54)

Figure 9-2 summarizes how we think of a triprotic system. We determine the principal species by comparing the pH of the solution with the pK_a values.

Go back and read the Example, "Preparing a Buffer in a Diprotic System," on page 193. See if it makes more sense now.

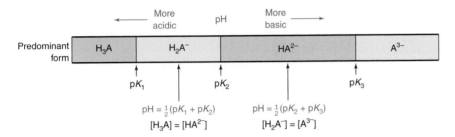

FIGURE 9-2 The predominant molecular form of a triprotic system (H_3A) in the various pH intervals.

Speciation describes the distribution of analyte among possible species. For an acid or base, speciation describes how much of each protonated form is present. When you ingest inorganic arsenic ($AsO(OH)_3$ and $As(OH)_3$) from drinking water, it is methylated to species such as $(CH_3)AsO(OH)_2$, $(CH_3)As(OH)_2$, $(CH_3)_2AsO(OH)$, $(CH_3)_2As(OH)$, $(CH_3)_3AsO$, and $(CH_3)_3As$. Speciation describes the forms and quantities that are present.

9-5 Fractional Composition Equations

We now derive equations that give the fraction of each species of acid or base at a given pH. These equations will be useful for acid-base and EDTA titrations, as well as for electrochemical equilibria. They are of key value in Chapter 12.

Monoprotic Systems

Our goal is to find an expression for the fraction of an acid in each form (HA and A^-) as a function of pH. We can do this by combining the equilibrium constant with the mass balance. Consider an acid with formal concentration F:

$$HA \overset{K_a}{\rightleftharpoons} H^+ + A^- \qquad K_a = \frac{[H^+][A^-]}{[HA]}$$

$$\text{Mass balance:} \quad F = [HA] + [A^-]$$

Rearranging the mass balance gives $[A^-] = F - [HA]$, which can be plugged into the K_a expression to give

$$K_a = \frac{[H^+](F - [HA])}{[HA]}$$

or, with a little algebra,

$$[HA] = \frac{[H^+]F}{[H^+] + K_a} \qquad (9-15)$$

The *fraction* of molecules in the form HA is called α_{HA}.

$$\alpha_{HA} = \frac{[HA]}{[HA] + [A^-]} = \frac{[HA]}{F} \qquad (9-16)$$

α_{HA} = fraction of species in the form HA
α_{A^-} = fraction of species in the form A^-

$$\boxed{\alpha_{HA} + \alpha_{A^-} = 1}$$

The fraction denoted here as α_{A^-} is the same thing we called the *fraction of dissociation* (α) previously.

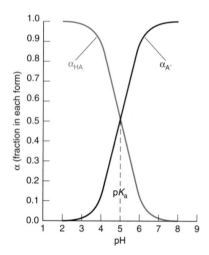

FIGURE 9-3 Fractional composition diagram of a monoprotic system with $pK_a = 5.00$. Below pH 5, HA is the dominant form, whereas, above pH 5, A^- dominates.

α_{H_2A} = fraction of species in the form H_2A
α_{HA^-} = fraction of species in the form HA^-
$\alpha_{A^{2-}}$ = fraction of species in the form A^{2-}

$$\alpha_{H_2A} + \alpha_{HA^-} + \alpha_{A^{2-}} = 1$$

The general form of α for the polyprotic acid H_nA is

$$\alpha_{H_nA} = \frac{[H^+]^n}{D}$$

$$\alpha_{H_{n-1}A} = \frac{K_1[H^+]^{n-1}}{D}$$

$$\alpha_{H_{n-j}A} = \frac{K_1K_2\ldots K_j[H^+]^{n-j}}{D}$$

where $D = [H^+]^n + K_1[H^+]^{n-1} + K_1K_2[H^+]^{n-2} + \cdots + K_1K_2K_3 \cdots K_n$.

How to apply fractional composition equations to bases

Dividing Equation 9-15 by F gives

Fraction in the form HA:
$$\alpha_{HA} = \frac{[HA]}{F} = \frac{[H^+]}{[H^+] + K_a} \qquad (9\text{-}17)$$

In a similar manner, the fraction in the form A^-, designated α_{A^-}, can be obtained:

Fraction in the form A^-:
$$\alpha_{A^-} = \frac{[A^-]}{F} = \frac{K_a}{[H^+] + K_a} \qquad (9\text{-}18)$$

Figure 9-3 shows α_{HA} and α_{A^-} for a system with $pK_a = 5.00$. At low pH, almost all of the acid is in the form HA. At high pH, almost everything is in the form A^-.

Diprotic Systems

The derivation of fractional composition equations for a diprotic system follows the same pattern used for the monoprotic system.

$$H_2A \overset{K_1}{\rightleftharpoons} H^+ + HA^-$$

$$HA^- \overset{K_2}{\rightleftharpoons} H^+ + A^{2-}$$

$$K_1 = \frac{[H^+][HA^-]}{[H_2A]} \Rightarrow [HA^-] = [H_2A]\frac{K_1}{[H^+]}$$

$$K_2 = \frac{[H^+][A^{2-}]}{[HA^-]} \Rightarrow [A^{2-}] = [HA^-]\frac{K_2}{[H^+]} = [H_2A]\frac{K_1K_2}{[H^+]^2}$$

Mass balance: $\quad F = [H_2A] + [HA^-] + [A^{2-}]$

$$F = [H_2A] + \frac{K_1}{[H^+]}[H_2A] + \frac{K_1K_2}{[H^+]^2}[H_2A]$$

$$F = [H_2A]\left(1 + \frac{K_1}{[H^+]} + \frac{K_1K_2}{[H^+]^2}\right) = [H_2A]\frac{[H^+]^2 + [H^+]K_1 + K_1K_2}{[H^+]^2}$$

For a diprotic system, we designate the fraction in the form H_2A as α_{H_2A}, the fraction in the form HA^- as α_{HA^-}, and the fraction in the form A^{2-} as $\alpha_{A^{2-}}$. From the definition of α_{H_2A}, we can write

Fraction in the form H_2A:
$$\alpha_{H_2A} = \frac{[H_2A]}{F} = \frac{[H^+]^2}{[H^+]^2 + [H^+]K_1 + K_1K_2} \qquad (9\text{-}19)$$

In a similar manner, we can derive the following equations:

Fraction in the form HA^-:
$$\alpha_{HA^-} = \frac{[HA^-]}{F} = \frac{K_1[H^+]}{[H^+]^2 + [H^+]K_1 + K_1K_2} \qquad (9\text{-}20)$$

Fraction in the form A^{2-}:
$$\alpha_{A^{2-}} = \frac{[A^{2-}]}{F} = \frac{K_1K_2}{[H^+]^2 + [H^+]K_1 + K_1K_2} \qquad (9\text{-}21)$$

Figure 9-4 shows the fractions α_{H_2A}, α_{HA^-}, and $\alpha_{A^{2-}}$ for fumaric acid, whose two pK_a values are only 1.46 units apart. α_{HA^-} rises only to 0.73 because the two pK values are so close together. There are substantial quantities of H_2A and A^{2-} in the region $pK_1 < pH < pK_2$.

Equations 9-19 through 9-21 apply equally well to B, BH^+, and BH_2^{2+} obtained from the base B. The fraction α_{H_2A} applies to the acidic form BH_2^{2+}. α_{HA^-} applies to BH^+, and $\alpha_{A^{2-}}$ applies to B. The constants K_1 and K_2 are the *acid* dissociation constants of BH_2^{2+} ($K_1 = K_w/K_{b2}$ and $K_2 = K_w/K_{b1}$).

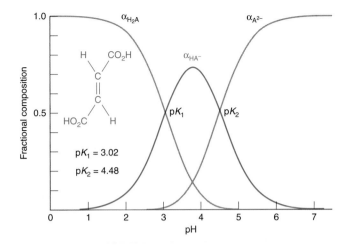

FIGURE 9-4 Fractional composition diagram for fumaric acid (*trans*-butenedioic acid). At low pH, H_2A is dominant. At intermediate pH, HA^- is dominant; and, at high pH, A^{2-} dominates. Because pK_1 and pK_2 are not separated very much, the fraction of HA^- never gets very close to unity.

9-6 Isoelectric and Isoionic pH

Biochemists speak of the isoelectric or isoionic pH of polyprotic molecules, such as proteins. These terms can be understood in terms of a diprotic system, such as the amino acid alanine.

$$\underset{\substack{\text{Alanine cation} \\ H_2A^+}}{\overset{\substack{CH_3 \\ | \\ H_3\overset{+}{N}CHCO_2H}}{}} \rightleftharpoons \underset{\substack{\text{Neutral zwitterion} \\ HA}}{\overset{\substack{CH_3 \\ | \\ H_3\overset{+}{N}CHCO_2^-}}{}} + H^+ \qquad pK_1 = 2.34$$

$$\overset{\substack{CH_3 \\ | \\ H_3\overset{+}{N}CHCO_2^-}}{} \rightleftharpoons \underset{\substack{\text{Alanine anion} \\ A^-}}{\overset{\substack{CH_3 \\ | \\ H_2NCHCO_2^-}}{}} + H^+ \qquad pK_2 = 9.87$$

The **isoionic point** (or isoionic pH) is the pH obtained when the pure, neutral polyprotic acid HA (the neutral zwitterion) is dissolved in water. The only ions are H_2A^+, A^-, H^+, and OH^-. Most alanine is in the form HA, and the concentrations of H_2A^+ and A^- are *not* equal to each other.

Isoionic pH is the pH of the pure, neutral, polyprotic acid.

The **isoelectric point** (or isoelectric pH) is the pH at which the *average* charge of the polyprotic acid is 0. Most of the molecules are in the uncharged form HA, and the concentrations of H_2A^+ and A^- *are* equal to each other. There is always some H_2A^+ and some A^- in equilibrium with HA.

Isoelectric pH is the pH at which average charge of the polyprotic acid is 0.

When alanine is dissolved in water, the pH of the solution, by definition, is the *isoionic* pH. Because alanine (HA) is the intermediate form of the diprotic acid, H_2A^+, $[H^+]$ is given by

Isoionic point:
$$[H^+] = \sqrt{\frac{K_1K_2F + K_1K_w}{K_1 + F}} \qquad (9\text{-}22)$$

Alanine is the intermediate form of a diprotic acid, so we use Equation 9-11 to find the pH.

where F is the formal concentration of alanine. For 0.10 M alanine, the isoionic pH is found from

$$[H^+] = \sqrt{\frac{K_1K_2(0.10) + K_1K_w}{K_1 + (0.10)}} = 7.7 \times 10^{-7}\,M \Rightarrow pH = 6.11$$

From $[H^+]$, K_1, and K_2, you could calculate $[H_2A^+] = 1.68 \times 10^{-5}$ M and $[A^-] = 1.76 \times 10^{-5}$ M for pure alanine in water (the *isoionic* solution). There is a slight excess of A^- because HA is a slightly stronger acid than it is a base. It dissociates to make A^- a little more than it reacts with water to make H_2A^+.

The *isoelectric* point is the pH at which $[H_2A^+] = [A^-]$, and, therefore, the average charge of alanine is 0. To go from the *isoionic* solution (pure HA in water) to the isoelectric solution, we could add just enough strong acid to decrease $[A^-]$ and increase $[H_2A^+]$ until they are equal. Adding acid necessarily lowers the pH. For alanine, the isoelectric pH must be lower than the isoionic pH.

We calculate the isoelectric pH by first writing expressions for $[H_2A^+]$ and $[A^-]$:

$$[H_2A^+] = \frac{[HA][H^+]}{K_1} \qquad [A^-] = \frac{K_2[HA]}{[H^+]}$$

Setting $[H_2A^+] = [A^-]$, we find

$$\frac{[HA][H^+]}{K_1} = \frac{K_2[HA]}{[H^+]} \Rightarrow [H^+] = \sqrt{K_1K_2}$$

which gives

The isoelectric point is midway between the two pK_a values "surrounding" the neutral, intermediate species.

Isoelectric point: $$pH = \frac{1}{2}(pK_1 + pK_2) \qquad (9\text{-}23)$$

For a diprotic amino acid, the isoelectric pH is halfway between the two pK_a values. The isoelectric pH of alanine is $\frac{1}{2}(2.34 + 9.87) = 6.10$.

The isoelectric and isoionic points for a polyprotic acid are almost the same. At the isoelectric pH, the average charge of the molecule is 0; thus $[H_2A^+] = [A^-]$ and $pH = \frac{1}{2}(pK_1 + pK_2)$. At the isoionic point, the pH is given by Equation 9-22, and $[H_2A^+]$ is not exactly equal to $[A^-]$.

For a protein, the *isoionic* pH is the pH of a solution of pure protein with no other ions except H^+ and OH^-. Proteins are usually isolated in a charged form together with counterions such as Na^+, NH_4^+, or Cl^-. When the protein is subjected to intensive *dialysis* (Demonstration 26-1) against pure water, the pH in the protein compartment approaches the isoionic point if

---

BOX 9-3 Isoelectric Focusing

At its *isoelectric point*, the average charge of all forms of a protein is 0. It will therefore not migrate in an electric field at its isoelectric pH. This effect is the basis of a technique of protein separation called **isoelectric focusing.** A protein mixture is subjected to a strong electric field in a medium specifically designed to have a pH gradient. Positively charged molecules move toward the negative pole and negatively charged molecules move toward the positive pole. Each protein migrates until it reaches the point where the pH is the same as its isoelectric pH. At this point, the protein has no net charge and no longer moves. Each protein in the mixture is focused in one region at its isoelectric pH.

Isoelectric focusing in a 6-mm-long × 100 µm-wide × 25-µm-deep capillary etched into silica glass is shown at the left below. Panel (i) shows fluorescent markers with known isoelectric pH (called pI) run as standards. Panels (ii) and (iii) show separations of fluorescence-labeled proteins. Proteins migrate until they reach their isoelectric pH and then stop migrating. If a molecule diffuses out of its isoelectric region, it becomes charged and immediately migrates back to its isoelectric zone. The graph shows measured pH versus distance in the capillary. Separations or reactions conducted in capillaries in glass or polymer chips are examples of *lab-on-a-chip* operations (Section 25-8).

The figure at the lower right shows separation of whole yeast cells at three different stages of growth (called early log, mid log, and stationary phase) by isoelectric focusing in a silica capillary tube. Acid-base properties (and, therefore, pI) of cell surfaces change during growth of the colony.

Lab-on-a-chip isoelectric focusing. (i) Fluorescent pI markers. (ii) and (iii) Separation of fluorescence-labeled proteins: (OVA) ovalbumin; (GFP) green fluorescent protein; (BSA) bovine serum albumin; (Tfer) transferrin; (CA) carbonic anhydrase; (PhB) phosphorylase B; and (Hb) hemoglobin. [G. J. Sommer, A. K. Singh, and A. V. Hatch, "On-Chip Isoelectric Focusing Using Photopolymerized Immobilized pH Gradients," *Anal. Chem.* 2008, 80, 3327.]

Capillary isoelectric focusing of whole yeast cells taken from three growth stages. After cells have been focused at their isoelectric pH, the inlet end of the capillary was elevated and liquid drained out of the capillary past an ultraviolet detector, creating three peaks observed here. The abscissa is the time required for the bands to reach the detector. [From R. Shen, S. J. Berger, and R. D. Smith, "Capillary Isoelectric Focusing of Yeast Cells," *Anal. Chem.* 2000, 72, 4603.]

---

the counterions are free to pass through the semipermeable dialysis membrane that retains the protein. The *isoelectric* point is the pH at which the protein has no net charge. Box 9-3 tells how proteins can be separated on the basis of their different isoelectric points.

Related properties—of interest to geology, environmental science, and ceramics—are the surface acidity of a solid[7] and the pH of zero charge.[8] Mineral, clay, and even organic surfaces behave as acids and bases. The silica (SiO_2) surface of sand or glass can be simplistically thought of as a diprotic acid:

$$\equiv Si-OH_2^+ \xrightleftharpoons{K_{a1}} \equiv Si-OH + H^+ \qquad K_{a1} = \frac{\{SiOH\}[H^+]}{\{SiOH_2^+\}} \qquad (9\text{-}24)$$

$$\equiv Si-OH \xrightleftharpoons{K_{a2}} \equiv Si-O^- + H^+ \qquad K_{a2} = \frac{\{SiO^-\}[H^+]}{\{SiOH\}} \qquad (9\text{-}25)$$

The notation $\equiv Si$ represents a surface silicon atom. Silanol groups ($\equiv Si-OH$) can donate or accept a proton to give the surface a negative or positive charge. In the equilibrium constants, the concentrations of the surface species $\{SiOH_2^+\}$, $\{SiOH\}$, and $\{SiO^-\}$ are measured in moles per gram of solid.

The *pH of zero charge* is the pH at which $\{SiOH_2^+\} = \{SiO^-\}$ and, therefore, the surface has no net charge. Like the isoelectric point of a diprotic acid, the pH of zero charge is $\frac{1}{2}(pK_{a1} + pK_{a2})$. *Colloidal particles* (those with diameters in the range 1–500 nm) tend to remain dispersed when they are charged, but they *flocculate* (come together and precipitate) near the pH of zero charge. In capillary electrophoresis (Chapter 25), the surface charge of the silica capillary governs the rate at which solvent moves through the capillary.

Terms to Understand

amino acid	hydrolysis	isoionic point	speciation
amphiprotic	isoelectric focusing	polyprotic acids and bases	zwitterion
diprotic acids and bases	isoelectric point		

Summary

Diprotic acids and bases fall into three categories:

1. The fully acidic form, H_2A, behaves as a monoprotic acid, $H_2A \rightleftharpoons H^+ + HA^-$, for which we solve the equation $K_{a1} = x^2/(F - x)$, where $[H^+] = [HA^-] = x$, and $[H_2A] = F - x$. From $[HA^-]$ and $[H^+]$, $[A^{2-}]$ can be found from the K_{a2} equilibrium.

2. The fully basic form, A^{2-}, behaves as a base, $A^{2-} + H_2O \rightleftharpoons HA^- + OH^-$, for which we solve the equation $K_{b1} = x^2/(F - x)$, where $[OH^-] = [HA^-] = x$, and $[A^{2-}] = F - x$. From these concentrations, $[H_2A]$ can be found from the K_{a1} or K_{b2} equilibrium.

3. The intermediate (amphiprotic) form, HA^-, is both an acid and a base. Its pH is given by

$$[H^+] = \sqrt{\frac{K_1 K_2 F + K_1 K_w}{K_1 + F}}$$

where K_1 and K_2 are acid dissociation constants for H_2A, and F is the formal concentration of the intermediate. In most cases, this equation reduces to the form $pH \approx \frac{1}{2}(pK_1 + pK_2)$, in which pH is independent of concentration.

In triprotic systems, there are two intermediate forms. The pH of each is found with an equation analogous to that for the intermediate

form of a diprotic system. Triprotic systems also have one fully acidic and one fully basic form; these species can be treated as monoprotic for the purpose of calculating pH. For polyprotic buffers, we write the Henderson-Hasselbalch equation connecting the two principal species in the system. pK_a in this equation is the one that applies to the acid in the denominator of the log term.

The principal species of a monoprotic or polyprotic system is found by comparing the pH with the pK_a values. For $pH < pK_1$, the fully protonated species, H_nA, is the predominant form. For $pK_1 < pH < pK_2$, the form $H_{n-1}A^-$ is favored; and, at each successive pK value, the next deprotonated species becomes principal. Finally, at pH values higher than the highest pK, the fully basic form (A^{n-}) is dominant. The fractional composition of a solution is expressed by α, given in Equations 9-17 and 9-18 for a monoprotic system and Equations 9-19 through 9-21 for a diprotic system.

The isoelectric pH of a polyprotic compound is that pH at which the average charge of all species is 0. For a diprotic amino acid whose amphiprotic form is neutral, the isoelectric pH is given by $pH = \frac{1}{2}(pK_1 + pK_2)$. The isoionic pH of a polyprotic species is the pH that would exist in a solution containing only ions derived from the neutral polyprotic species and from H_2O. For a diprotic amino acid whose amphiprotic form is neutral, the isoionic pH is found from $[H^+] = \sqrt{(K_1 K_2 F + K_1 K_w)/(K_1 + F)}$, where F is the formal concentration of the amino acid.

Exercises

9-A. Find the pH and the concentrations of H_2SO_3, HSO_3^-, and SO_3^{2-} in each solution: (a) 0.050 M H_2SO_3; (b) 0.050 M $NaHSO_3$; (c) 0.050 M Na_2SO_3.

9-B. (a) How many grams of $NaHCO_3$ (FM 84.007) must be added to 4.00 g of K_2CO_3 (FM 138.206) to give a pH of 10.80 in 500 mL of water?

(b) What will be the pH if 100 mL of 0.100 M HCl are added to the solution in part (a)?

(c) How many milliliters of 0.320 M HNO_3 should be added to 4.00 g of K_2CO_3 to give a pH of 10.00 in 250 mL?

9-C. How many milliliters of 0.800 M KOH should be added to 5.02 g of 1,5-pentanedioic acid ($C_5H_8O_4$, FM 132.11) to give a pH of 4.40 when diluted to 250 mL?

9-D. Calculate the pH of a 0.010 M solution of each amino acid in the form drawn here.

(a) $H_3\overset{+}{N}CHCO_2^-$ (b) $H_3\overset{+}{N}CHCO_2^-$ (c) $H_2NCHCO_2^-$
     Glutamine   Cysteine   Arginine

9-E. (a) Draw the structure of the predominant form (principal species) of 1,3-dihydroxybenzene at pH 9.00 and at pH 11.00.

(b) What is the second most prominent species at each pH?

(c) Calculate the percentage in the major form at each pH.

9-F. Draw the structures of the predominant forms of glutamic acid and tyrosine at pH 9.0 and pH 10.0. What is the second most abundant species at each pH?

9-G. Calculate the isoionic pH of 0.010 M lysine.

9-H. Neutral lysine can be written HL. The other forms of lysine are H_3L^{2+}, H_2L^+, and L^-. The isoelectric point is the pH at which the *average* charge of lysine is 0. Therefore, at the isoelectric point, $2[H_3L^{2+}] + [H_2L^+] = [L^-]$. Use this condition to calculate the isoelectric pH of lysine.

Problems

Diprotic Acids and Bases

9-1. Consider HA^-, the intermediate form of a diprotic acid. K_a for this species is 10^{-4} and K_b is 10^{-8}. Nonetheless, the K_a and K_b reactions proceed to nearly the same extent when NaHA is dissolved in water. Explain.

9-2. Write the general structure of an amino acid. Why do some amino acids in Table 9-1 have two pK values and others three?

9-3. Write the chemical reactions whose equilibrium constants are K_{b1} and K_{b2} for the amino acid proline. Find the values of K_{b1} and K_{b2}.

9-4. Consider the diprotic acid H_2A with $K_1 = 1.00 \times 10^{-4}$ and $K_2 = 1.00 \times 10^{-8}$. Find the pH and concentrations of H_2A, HA^-, and A^{2-} in (a) 0.100 M H_2A; (b) 0.100 M NaHA; (c) 0.100 M Na_2A.

9-5. We will abbreviate malonic acid, $CH_2(CO_2H)_2$, as H_2M. Find the pH and concentrations of H_2M, HM^-, and M^{2-} in (a) 0.100 M H_2M; (b) 0.100 M NaHM; (c) 0.100 M Na_2M.

9-6. Calculate the pH of 0.300 M piperazine. Calculate the concentration of each form of piperazine in this solution.

9-7. Use the method of Box 9-2 to calculate the concentrations of H^+, H_2A, HA^-, and A^{2-} in 0.001 00 M monosodium oxalate, NaHA.

9-8. *Activity.* In this problem, we calculate the pH of the intermediate form of a diprotic acid, taking activities into account.

(a) Including activity coefficients, derive Equation 9-11 for potassium hydrogen phthalate (K^+HP^- in the example following Equation 9-12).

(b) Calculate the pH of 0.050 M KHP, using the results in part (a). Assume that the sizes of both HP^- and P^{2-} are 600 pm.

9-9. 🖾 *Intermediate form of diprotic acid.* Use the method in Box 9-2 to find the pH and concentration of HA^- in a 0.01 F solution of the amphiprotic salt Na^+HA^- derived from the diprotic acid H_2A with $pK_1 = 4$ and (a) $pK_2 = 8$ or (b) $pK_2 = 5$.

9-10. *Derivation of Equation 9-10 for pH of the intermediate form Na^+HA^-.* The derivation in the chapter was for HL, the intermediate form of H_2L^+. Write the mass and charge balances for Na^+HA^- at formal concentration F. Substitute $[H_2A] + [HA^-] + [A^{2-}]$ for $[Na^+]$ in the charge balance. After canceling like terms on both sides of the charge balance, express all concentrations in terms of $[HA^-]$ and $[H^+]$. Solve for $[H^+]$ and show that you end up with the same expression as Equation 9-10.

9-11. *Heterogeneous equilibrium.* CO_2 dissolves in water to give "carbonic acid" (which is mostly dissolved CO_2, as described in Box 6-4).

$$CO_2(g) \rightleftharpoons CO_2(aq) \qquad K = 10^{-1.5}$$

(The equilibrium constant is called the *Henry's law constant* for carbon dioxide, because Henry's law states that the solubility of a gas in a liquid is proportional to the pressure of the gas.) The acid dissociation constants listed for "carbonic acid" in Appendix G apply to $CO_2(aq)$. Given that P_{CO_2} in the atmosphere is $10^{-3.4}$ atm, find the pH of water in equilibrium with the atmosphere.

9-12. *Effect of temperature on carbonic acid acidity and the solubility of $CaCO_3$.*[9] Box 9-1 states that marine life with $CaCO_3$ shells and skeletons will be threatened with extinction in cold polar waters before that will happen in warm tropical waters. The following equilibrium constants apply to seawater at 0° and 30°C, when concentrations are measured in moles per kilogram of seawater and pressure is in bars:

$$CO_2(g) \rightleftharpoons CO_2(aq) \tag{A}$$

$$K_H = \frac{[CO_2(aq)]}{P_{CO_2}} = 10^{-1.2073} \text{ mol kg}^{-1} \text{ bar}^{-1} \text{ at } 0°C$$

$$= 10^{-1.6048} \text{ mol kg}^{-1} \text{ bar}^{-1} \text{ at } 30°C$$

$$CO_2(aq) + H_2O \rightleftharpoons HCO_3^- + H^+ \qquad\qquad (B)$$

$$K_{a1} = \frac{[HCO_3^-][H^+]}{[CO_2(aq)]} = 10^{-6.1004} \ \text{mol kg}^{-1} \ \text{at } 0°C$$

$$= 10^{-5.8008} \ \text{mol kg}^{-1} \ \text{at } 30°C$$

$$HCO_3^- \rightleftharpoons CO_3^{2-} + H^+ \qquad\qquad (C)$$

$$K_{a2} = \frac{[CO_3^{2-}][H^+]}{[HCO_3^-]} = 10^{-9.3762} \ \text{mol kg}^{-1} \ \text{at } 0°C$$

$$= 10^{-8.8324} \ \text{mol kg}^{-1} \ \text{at } 30°C$$

$$CaCO_3(s, \textit{aragonite}) \rightleftharpoons Ca^{2+} + CO_3^{2-} \qquad\qquad (D)$$

$$K_{sp}^{arg} = [Ca^{2+}][CO_3^{2-}] = 10^{-6.1113} \ \text{mol}^2 \ \text{kg}^{-2} \ \text{at } 0°C$$

$$= 10^{-6.1391} \ \text{mol}^2 \ \text{kg}^{-2} \ \text{at } 30°C$$

$$CaCO_3(s, \textit{calcite}) \rightleftharpoons Ca^{2+} + CO_3^{2-} \qquad\qquad (E)$$

$$K_{sp}^{cal} = [Ca^{2+}][CO_3^{2-}] = 10^{-6.3652} \ \text{mol}^2 \ \text{kg}^{-2} \ \text{at } 0°C$$

$$= 10^{-6.3713} \ \text{mol}^2 \ \text{kg}^{-2} \ \text{at } 30°C$$

The first equilibrium constant is called K_H for Henry's law (Problem 9-11). Units are given to remind you what units you must use.

(a) Combine the expressions for K_H, K_{a1}, and K_{a2} to find an expression for $[CO_3^{2-}]$ in terms of P_{CO_2} and $[H^+]$.

(b) From the result of (a), calculate $[CO_3^{2-}]$ (mol kg^{-1}) at P_{CO_2} = 800 μbar and pH = 7.8 at temperatures of 0° (polar ocean) and 30°C (tropical ocean). These are conditions that could be reached around the year 2100 if we continue to release CO_2 at the present rate.

(c) The concentration of Ca^{2+} in the ocean is 0.010 M. Predict whether aragonite and calcite will dissolve under the conditions in (b).

Diprotic Buffers

9-13. How many grams of Na_2CO_3 (FM 105.99) should be mixed with 5.00 g of $NaHCO_3$ (FM 84.01) to produce 100 mL of buffer with pH 10.00?

9-14. How many milliliters of 0.202 M NaOH should be added to 25.0 mL of 0.023 3 M salicylic acid (2-hydroxybenzoic acid) to adjust the pH to 3.50?

9-15. Describe how you would prepare exactly 100 mL of 0.100 M picolinate buffer, pH 5.50. Possible starting materials are pure picolinic acid (pyridine-2-carboxylic acid, FM 123.11), 1.0 M HCl, and 1.0 M NaOH. Approximately how many milliliters of the HCl or NaOH will be required?

9-16. How many grams of Na_2SO_4 (FM 142.04) should be added to how many grams of sulfuric acid (FM 98.08) to give 1.00 L of buffer with pH 2.80 and a total sulfur (= SO_4^{2-} + HSO_4^- + H_2SO_4) concentration of 0.200 M?

Polyprotic Acids and Bases

9-17. Phosphate at 0.01 M is one of the main buffers in blood plasma, whose pH is 7.45. Would phosphate be as useful if the plasma pH were 8.5?

9-18. Starting with the fully protonated species, write the stepwise acid dissociation reactions of the amino acids glutamic acid and tyrosine. Be sure to remove the protons in the correct order. Which species are the neutral molecules that we call glutamic acid and tyrosine?

9-19. (a) Calculate the quotient $[H_3PO_4]/[H_2PO_4^-]$ in 0.050 0 M KH_2PO_4.

(b) Find the same quotient for 0.050 0 M K_2HPO_4.

9-20. (a) Which two of the following compounds would you mix to make a buffer of pH 7.45: H_3PO_4 (FM 98.00), NaH_2PO_4 (FM 119.98), Na_2HPO_4 (FM 141.96), and Na_3PO_4 (FM 163.94)?

(b) If you wanted to prepare 1.00 L of buffer with a total phosphate concentration of 0.050 0 M, how many grams of each of the two selected compounds would you mix together?

(c) If you did what you calculated in part (b), you would not get a pH of exactly 7.45. Explain how you would really prepare this buffer in the lab.

9-21. Find the pH and the concentration of each species of lysine in a solution of 0.010 0 M lysine · HCl, lysine monohydrochloride.

9-22. How many milliliters of 1.00 M KOH should be added to 100 mL of solution containing 10.0 g of histidine hydrochloride (His · HCl, FM 191.62) to get a pH of 9.30?

9-23. (a) Using activity coefficients, calculate the pH of a solution containing a 2.00:1.00 mole ratio of HC^{2-}:C^{3-} (H_3C = citric acid). The ionic strength is 0.010 M.

(b) What will be the pH if the ionic strength is raised to 0.10 M and the mole ratio HC^{2-}:C^{3-} is kept constant?

Which Is the Principal Species?

9-24. The acid HA has pK_a = 7.00.

(a) Which is the principal species, HA or A$^-$, at pH 6.00?

(b) Which is the principal species at pH 8.00?

(c) What is the quotient $[A^-]/[HA]$ at pH 7.00? at pH 6.00?

9-25. The diprotic acid H_2A has pK_1 = 4.00 and pK_2 = 8.00.

(a) At what pH is $[H_2A]$ = $[HA^-]$?

(b) At what pH is $[HA^-]$ = $[A^{2-}]$?

(c) Which is the principal species at pH 2.00: H_2A, HA^-, or A^{2-}?

(d) Which is the principal species at pH 6.00?

(e) Which is the principal species at pH 10.00?

9-26. The base B has pK_b = 5.00.

(a) What is the value of pK_a for the acid BH^+?

(b) At what pH is $[BH^+]$ = $[B]$?

(c) Which is the principal species at pH 7.00: B or BH^+?

(d) What is the quotient $[B]/[BH^+]$ at pH 12.00?

9-27. Draw the structure of the predominant form of pyridoxal-5-phosphate at pH 7.00.

Fractional Composition Equations

9-28. The acid HA has pK_a = 4.00. Use Equations 9-17 and 9-18 to find the fraction in the form HA and the fraction in the form A$^-$ at pH = 5.00. Does your answer agree with what you expect for the quotient $[A^-]/[HA]$ at pH 5.00?

9-29. A dibasic compound, B, has pK_{b1} = 4.00 and pK_{b2} = 6.00. Find the fraction in the form BH_2^{2+} at pH 7.00, using Equation 9-19. Note that K_1 and K_2 in Equation 9-19 are acid dissociation constants for BH_2^{2+} (K_1 = K_w/K_{b2} and K_2 = K_w/K_{b1}).

9-30. What fraction of ethane-1,2-dithiol is in each form (H_2A, HA^-, A^{2-}) at pH 8.00? at pH 10.00?

9-31. Calculate α_{H_2A}, α_{HA^-}, and $\alpha_{A^{2-}}$ for *cis*-butenedioic acid at pH 1.00, 1.92, 6.00, 6.27, and 10.00.

9-32. (a) Derive equations for α_{H_3A}, $\alpha_{H_2A^-}$, $\alpha_{HA^{2-}}$ and $\alpha_{A^{3-}}$ for a triprotic system.

(b) Calculate the values of these fractions for phosphoric acid at pH 7.00.

9-33. A solution containing acetic acid, oxalic acid, ammonia, and pyridine has a pH of 9.00. What fraction of ammonia is not protonated?

9-34. A solution was prepared from 10.0 mL of 0.100 M cacodylic acid and 10.0 mL of 0.080 0 M NaOH. To this mixture was added 1.00 mL of 1.27×10^{-6} M morphine. Calling morphine B, calculate the fraction of morphine present in the form BH^+.

O
||
$(CH_3)_2AsOH$ HO   O   OH

Cacodylic acid   Morphine

$K_a = 6.4 \times 10^{-7}$ $K_b = 1.6 \times 10^{-6}$

9-35. Fractional composition in a diprotic system. Create a spreadsheet with Equations 9-19 through 9-21 to compute the three curves in Figure 9-4. Plot the three curves in a beautifully labeled figure.

9-36. Fractional composition in a triprotic system. For a triprotic system, the fractional composition equations are

$$\alpha_{H_3A} = \frac{[H^+]^3}{D} \qquad \alpha_{HA^{2-}} = \frac{K_1 K_2 [H^+]}{D}$$

$$\alpha_{H_2A^-} = \frac{K_1 [H^+]^2}{D} \qquad \alpha_{A^{3-}} = \frac{K_1 K_2 K_3}{D}$$

where $D = [H^+]^3 + K_1[H^+]^2 + K_1 K_2[H^+] + K_1 K_2 K_3$. Use these equations to create a fractional composition diagram analogous to Figure 9-4 for the amino acid tyrosine. What is the fraction of each species at pH 10.00?

9-37. Fractional composition in a tetraprotic system. Prepare a fractional composition diagram analogous to Figure 9-4 for the tetraprotic system derived from hydrolysis of Cr^{3+}:

$$Cr^{3+} + H_2O \rightleftharpoons Cr(OH)^{2+} + H^+ \qquad K_{a1} = 10^{-3.80}$$

$$Cr(OH)^{2+} + H_2O \rightleftharpoons Cr(OH)_2^+ + H^+ \qquad K_{a2} = 10^{-6.40}$$

$$Cr(OH)_2^+ + H_2O \rightleftharpoons Cr(OH)_3(aq) + H^+ \qquad K_{a3} = 10^{-6.40}$$

$$Cr(OH)_3(aq) + H_2O \rightleftharpoons Cr(OH)_4^- + H^+ \qquad K_{a4} = 10^{-11.40}$$

(Yes, the values of K_{a2} and K_{a3} are equal.)

(a) Use these equilibrium constants to prepare a fractional composition diagram for this tetraprotic system.

(b) You should do this part with your head and your calculator, not your spreadsheet. The solubility of $Cr(OH)_3$ is given by

$$Cr(OH)_3(s) \rightleftharpoons Cr(OH)_3(aq) \qquad K = 10^{-6.84}$$

What concentration of $Cr(OH)_3(aq)$ is in equilibrium with solid $Cr(OH)_3(s)$?

(c) What concentration of $Cr(OH)^{2+}$ is in equilibrium with $Cr(OH)_3(s)$ if the solution pH is adjusted to 4.00?

Isoelectric and Isoionic pH

9-38. What is the difference between the isoelectric pH and the isoionic pH of a protein with many different acidic and basic substituents?

9-39. Explain what is wrong with the following statement: At its isoelectric point, the charge on all molecules of a particular protein is 0.

9-40. Calculate the isoelectric and isoionic pH of 0.010 M threonine.

9-41. Explain how isoelectric focusing works.

10 Acid-Base Titrations

ACID-BASE TITRATION OF A PROTEIN

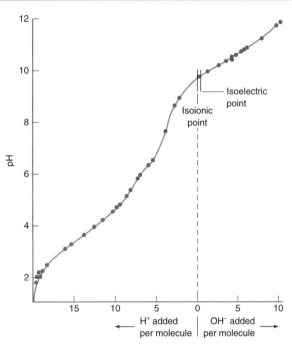

Acid-base titration of the enzyme ribonuclease. The isoionic point is the pH of the pure protein with no ions present except H^+ and OH^-. The isoelectric point is the pH at which the average charge on the protein is zero. [From C. T. Tanford and J. D. Hauenstein, "Hydrogen Ion Equilibria of Ribonuclease," *J. Am. Chem. Soc.* 1956, *78*, 5287.]

If tyrosine is buried deep inside the protein, it is not readily accessible and a high concentration of OH^- is required to remove the proton from the phenol group.

Ribonuclease is an enzyme with 124 amino acids. Its function is to cleave ribonucleic acid (RNA) into small fragments. A solution containing pure protein, with no other ions present except H^+ and OH^- derived from the protein and water, is said to be *isoionic*. From the isoionic point near pH 9.6 in the graph, the protein can be titrated with acid or base. Of the 124 amino acids, 16 can be protonated by acid and 20 can lose protons to added base. From the shape of the titration curve, it is possible to deduce the approximate pK_a for each titratable group.[1,2] This information provides insight into the environment of that amino acid in the protein. In ribonuclease, three tyrosine residues have "normal" values of pK_a (≈ 10) (Table 9-1) and three others have $pK_a > 12$. An interpretation is that three tyrosine groups are accessible to OH^-, and three are buried inside the protein where they cannot be easily titrated. The curve in the illustration is calculated from pK_a values for all titratable groups.

Theoretical titration curves for enzymes can be calculated from known crystal structures and first principles of electrostatics. Key amino acids at the active site where catalysis occurs have significantly perturbed pK_a values and unusual behavior in which they are partially protonated over a wide pH region.[3] In principle, such titration calculations can identify the active site of a protein whose structure is known, but whose function is not.

From an acid-base titration curve, we can deduce the quantities and pK_a values of acidic and basic substances in a mixture. In medicinal chemistry, the pK_a and *lipophilicity* of a candidate drug predict how easily it will cross cell membranes. From pK_a and pH, we can compute the charge of a polyprotic acid. Usually, the more highly charged a drug, the harder it is for that drug to cross a cell membrane. In this chapter, we learn to predict the shapes of titration curves and to find end points with electrodes and indicators.

10-1 Titration of Strong Base with Strong Acid

For each type of titration in this chapter, *our goal is to construct a graph showing how pH changes as titrant is added.* If you can do this, then you understand what is happening during the titration, and you will be able to interpret an experimental titration curve. pH is usually measured with a glass electrode whose operation is explained in Section 14-5.

First, write the reaction between *titrant* and *analyte*.

The first step in each case is to write the chemical reaction between titrant and analyte. Then use that reaction to calculate the composition and pH after each addition of titrant. As a simple example, let's focus on the titration of 50.00 mL of 0.020 00 M KOH with 0.100 0 M HBr. The chemical reaction between titrant and analyte is merely

The titration reaction.

$$H^+ + OH^- \rightarrow H_2O \qquad K = 1/K_w = 10^{14}$$

Because the equilibrium constant for this reaction is 10^{14}, it is fair to say that it "goes to completion." *Any amount of H^+ added will consume a stoichiometric amount of OH^-.*

It is useful to know the volume of HBr (V_e) needed to reach the equivalence point, which we find by equating moles of KOH being titrated to moles of added HBr:

$$\underbrace{(V_e\,(\text{L}))\left(0.100\,0\,\frac{\text{mol}}{\text{L}}\right)}_{\substack{\text{mol of HBr} \\ \text{at equivalence point}}} = \underbrace{(0.050\,00\,\text{L})\left(0.020\,00\,\frac{\text{mol}}{\text{L}}\right)}_{\substack{\text{mol of OH}^- \\ \text{being titrated}}} \Rightarrow V_e = 0.010\,00\,\text{L}$$

Instead of multiplying $\text{L} \times (\text{mol}/\text{L})$ to get mol, we frequently multiply $\text{mL} \times (\text{mol/L})$, which is the same as $\text{mL} \times (\text{mmol}/\text{mL}) = \text{mmol}$:

$$\underbrace{(V_e\,(\text{mL}))(0.100\,0\,\text{M})}_{\substack{\text{mmol of HBr} \\ \text{at equivalence point}}} = \underbrace{(50.00\,\text{mL})(0.020\,00\,\text{M})}_{\substack{\text{mmol of OH}^- \\ \text{being titrated}}} \Rightarrow V_e = 10.00\,\text{mL}$$

$$\text{mL} \times \frac{\text{mol}}{\text{L}} = \text{mL} \times \frac{\text{mmol}}{\text{mL}} = \text{mmol}$$

When 10.00 mL of HBr have been added, the titration is complete. Prior to this point, there is excess, unreacted OH^- present. After V_e, there is excess H^+ in the solution.

In the titration of any strong base with any strong acid, there are three regions of the titration curve that require different kinds of calculations:

1. Before the equivalence point, the pH is determined by excess OH^- in the solution.
2. At the equivalence point, H^+ is just sufficient to react with all OH^- to make H_2O. The pH is determined by dissociation of water.
3. After the equivalence point, pH is determined by excess H^+ in the solution.

We will do one sample calculation for each region. Complete results are shown in Table 10-1 and Figure 10-1. As a reminder, the *equivalence point* occurs when added titrant is exactly enough for stoichiometric reaction with the analyte. The equivalence point is the ideal result we seek in a titration. What we actually measure is the *end point*, which is marked by a sudden physical change, such as indicator color or an electrode potential.

Region 1: Before the Equivalence Point

Before the equivalence point, there is excess OH^-.

We will do this calculation once by the method you would have learned in general chemistry, and then we will streamline the process. When 3.00 mL of HBr have been added, the total volume is 53.00 mL. HBr is consumed by NaOH, leaving excess NaOH. The moles of added HBr are $(0.100\,0\,\text{M})(0.003\,00\,\text{L}) = 0.300 \times 10^{-3}$ mol HBr $= 0.300$ mmol HBr. The initial moles of NaOH were $(0.020\,00\,\text{M})(0.050\,00\,\text{L}) = 1.000 \times 10^{-3}$ mol NaOH $= 1.000$ mmol NaOH. Unreacted OH^- is the difference 1.000 mmol $- 0.300$ mmol $= 0.700$ mmol. The concentration of unreacted OH^- is $(0.700\,\text{mmol})/(53.00\,\text{mL}) = 0.013\,2$ M. So, $[H^+] = K_w/[OH^-] = 7.57 \times 10^{-13}$ M and pH $= -\log[H^+] = 12.12$.

In your sleep,

$$\frac{\text{mmol}}{\text{mL}} = \frac{\text{mol}}{\text{L}} = \text{M}$$

Now here is the streamlined calculation: When 3.00 mL of HBr have been added, the reaction is three-tenths complete because $V_e = 10.00$ mL. The fraction of OH^- left unreacted

TABLE 10-1 Calculation of the titration curve for 50.00 mL of 0.020 00 M KOH treated with 0.100 0 M HBr

	mL HBr added (V_a)	Concentration of unreacted OH⁻ (M)	Concentration of excess H⁺ (M)	pH
	0.00	0.020 0		12.30
	1.00	0.017 6		12.24
	2.00	0.015 4		12.18
	3.00	0.013 2		12.12
	4.00	0.011 1		12.04
Region 1	5.00	0.009 09		11.95
(excess OH⁻)	6.00	0.007 14		11.85
	7.00	0.005 26		11.72
	8.00	0.003 45		11.53
	9.00	0.001 69		11.22
	9.50	0.000 840		10.92
	9.90	0.000 167		10.22
	9.99	0.000 016 6		9.22
Region 2	10.00	—	—	7.00
	10.01		0.000 016 7	4.78
	10.10		0.000 166	3.78
	10.50		0.000 826	3.08
	11.00		0.001 64	2.79
Region 3	12.00		0.003 23	2.49
(excess H⁺)	13.00		0.004 76	2.32
	14.00		0.006 25	2.20
	15.00		0.007 69	2.11
	16.00		0.009 09	2.04

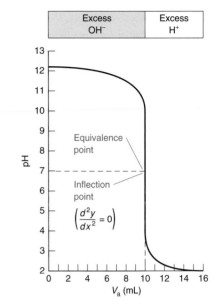

FIGURE 10-1 Calculated titration curve, showing how pH changes as 0.100 0 M HBr is added to 50.00 mL of 0.020 00 M KOH. The equivalence point is an inflection point at which the second derivative is zero.

is seven-tenths. The concentration of remaining OH⁻ is the product of the fraction remaining, the initial concentration, and a dilution factor:

$$[OH^-] = \left(\frac{10.00 - 3.00}{10.00}\right)(0.020\ 00\ M)\left(\frac{50.00}{50.00 + 3.00}\right) = 0.013\ 2\ M \qquad \text{(10-1)}$$

Fraction of OH⁻ remaining · Initial concentration of OH⁻ · Dilution factor · Initial volume of OH⁻ / Total volume of solution

$$[H^+] = \frac{K_w}{[OH^-]} = \frac{1.0 \times 10^{-14}}{0.013\ 2} = 7.5_7 \times 10^{-13}\ M \Rightarrow pH = 12.12$$

Equation 10-1 tells us that [OH⁻] is equal to a certain fraction of the initial concentration, with a correction for dilution. The dilution factor equals the initial volume of analyte divided by the total volume of solution.

In Table 10-1, the volume of acid added is designated V_a. pH is expressed to the 0.01 decimal place, regardless of what is justified by significant figures. We do this for the sake of consistency and also because 0.01 is near the limit of accuracy in pH measurements.

Region 2: At the Equivalence Point

Region 2 is the equivalence point, where just enough H⁺ has been added to consume OH⁻. We could prepare the same solution by dissolving KBr in water. pH is determined by dissociation of water:

$$H_2O \rightleftharpoons \underset{x}{H^+} + \underset{x}{OH^-}$$

$$K_w = x^2 \Rightarrow x = 1.00 \times 10^{-7}\ M \Rightarrow pH = 7.00$$

The pH at the equivalence point in the titration of any strong base (or acid) with strong acid (or base) will be 7.00 at 25°C.

Challenge Using a setup similar to Equation 10-1, calculate [OH⁻] when 6.00 mL of HBr have been added. Check your pH against the value in Table 10-1.

At the equivalence point, pH = 7.00, but *only* in a strong-acid–strong-base reaction.

We will soon discover that *the pH is not 7.00 at the equivalence point in the titration of weak acids or bases*. The pH is 7.00 only if the titrant and analyte are both strong.

Region 3: After the Equivalence Point

Beyond the equivalence point, we are adding excess HBr to the solution. The concentration of excess H^+ at, say, 10.50 mL is given by

After the equivalence point, there is excess H^+.

$$[H^+] = \underbrace{(0.100\ 0\ M)}_{\substack{\text{Initial} \\ \text{concentration} \\ \text{of } H^+}}\overbrace{\left(\frac{0.50}{\underbrace{50.00 + 10.50}_{\substack{\text{Total volume} \\ \text{of solution}}}}\right)}^{\substack{\text{Volume of} \\ \text{excess } H^+}} = 8.26 \times 10^{-4}\ M$$

Dilution factor

$$pH = -\log[H^+] = 3.08$$

At $V_a = 10.50$ mL, there is an excess of just $V_a - V_e = 10.50 - 10.00 = 0.50$ mL of HBr. That is the reason why 0.50 appears in the dilution factor.

The Titration Curve

The titration curve in Figure 10-1 exhibits a rapid change in pH near the equivalence point. The equivalence point is where the slope (dpH/dV_a) is greatest (and the second derivative is 0, which makes it an *inflection point*). To repeat an important statement, the pH at the equivalence point is 7.00 *only* in a strong-acid–strong-base titration. If one or both of the reactants is weak, the equivalence point pH is *not* 7.00.

10-2 Titration of Weak Acid with Strong Base

The titration of a weak acid with a strong base allows us to put all our knowledge of acid-base chemistry to work. The example we examine is the titration of 50.00 mL of 0.020 00 M MES with 0.100 0 M NaOH. MES is an abbreviation for 2-(*N*-morpholino)ethanesulfonic acid, which is a weak acid with $pK_a = 6.27$.

The *titration reaction* is

Always start by writing the titration reaction.

$$O\overset{+}{\diagdown}NHCH_2CH_2SO_3^- + OH^- \rightarrow O\diagdown NCH_2CH_2SO_3^- + H_2O \qquad (10\text{-}2)$$

$$\underset{\substack{\text{HA} \\ \text{MES, } pK_a = 6.27}}{} \qquad\qquad\qquad \underset{\text{A}^-}{}$$

Reaction 10-2 is the reverse of the K_b reaction for the base A^-. Therefore, the equilibrium constant for Reaction 10-2 is $K = 1/K_b = 1/(K_w/K_a \text{ (for HA))} = 5.4 \times 10^7$. The equilibrium constant is so large that we can say that the reaction goes "to completion" after each addition of OH^-. As we saw in Box 8-3, *strong plus weak react completely*.

Strong + weak → complete reaction

Let's first calculate the volume of base, V_b, needed to reach the equivalence point:

$$\underbrace{(V_b\ (\text{mL}))(0.100\ 0\ M)}_{\text{mmol of base}} = \underbrace{(50.00\ \text{mL})(0.020\ 00\ M)}_{\text{mmol of HA}} \Rightarrow V_b = 10.00\ \text{mL}$$

The titration calculations for this problem are of four types:

1. Before any base is added, the solution contains just HA in water. This is a weak acid whose pH is determined by the equilibrium

$$HA \overset{K_a}{\rightleftharpoons} H^+ + A^-$$

2. From the first addition of NaOH until immediately before the equivalence point, there is a mixture of unreacted HA plus the A^- produced by Reaction 10-2. *Aha! A buffer!* We can use the Henderson-Hasselbalch equation to find the pH.

3. At the equivalence point, "all" HA has been converted into A^-. The same solution could have been made by dissolving A^- in water. We have a weak base whose pH is determined by the reaction

$$A^- + H_2O \overset{K_b}{\rightleftharpoons} HA + OH^-$$

4. Beyond the equivalence point, excess NaOH is being added to a solution of A^-. To a good approximation, pH is determined by the strong base. We calculate the pH as if we had simply added excess NaOH to water. We neglect the tiny effect of A^-.

Region 1: Before Base Is Added

Before adding any base, we have a solution of 0.020 00 M HA with $pK_a = 6.27$. This is simply a weak-acid problem.

$$HA \rightleftharpoons H^+ + A^- \qquad K_a = 10^{-6.27}$$
$$ F - x \quad\ x \quad\ x$$

The initial solution contains just the *weak acid* HA.

$$\frac{x^2}{0.020\ 00 - x} = K_a \Rightarrow x = 1.03 \times 10^{-4} \Rightarrow pH = 3.99$$

Region 2: Before the Equivalence Point

Adding OH^- creates a mixture of HA and A^-. This mixture is a buffer whose pH can be calculated with the Henderson-Hasselbalch equation (8-16) from the quotient $[A^-]/[HA]$.

Suppose we wish to calculate $[A^-]/[HA]$ when 3.00 mL of OH^- have been added. Because $V_e = 10.00$ mL, we have added enough base to react with three-tenths of the HA. We can make a table showing the relative concentrations before and after the reaction:

Before the equivalence point, there is a mixture of HA and A^-, which is a *buffer*.

We only need *relative* concentrations because pH depends on the quotient $[A^-]/[HA]$.

Titration reaction:		HA	$+\ OH^-$	$\rightarrow A^-$	$+\ H_2O$
Relative initial quantities (HA $\equiv$ 1)		1	$\frac{3}{10}$	—	—
Relative final quantities		$\frac{7}{10}$	—	$\frac{3}{10}$	—

Once we know the *quotient* $[A^-]/[HA]$ in any solution, we know its pH:

$$pH = pK_a + \log\left(\frac{[A^-]}{[HA]}\right) = 6.27 + \log\left(\frac{3/10}{7/10}\right) = 5.90$$

The point at which the volume of titrant is $\frac{1}{2}V_e$ is a special one in any titration.

Titration reaction:		HA	$+\ OH^-$	$\rightarrow A^-$	$+\ H_2O$
Relative initial quantities		1	$\frac{1}{2}$	—	—
Relative final quantities		$\frac{1}{2}$	—	$\frac{1}{2}$	—

$$pH = pK_a + \log\left(\frac{1/2}{1/2}\right) = pK_a$$

When $V_b = \frac{1}{2}V_e$, pH $= pK_a$. This is a landmark in any titration.

When the volume of titrant is $\frac{1}{2}V_e$, pH $= pK_a$ for the acid HA (neglecting activity coefficients). From an experimental titration curve, you can find the approximate value of pK_a by reading the pH when $V_b = \frac{1}{2}V_e$, where V_b is the volume of added base. (To find the true value of pK_a requires activity coefficients.)

Advice As soon as you recognize a mixture of HA and A^- in any solution, *you have a buffer!* You can calculate the pH from the quotient $[A^-]/[HA]$.

$$pH = pK_a + \log\left(\frac{[A^-]}{[HA]}\right)$$

Learn to recognize buffers! They lurk in every corner of acid-base chemistry.

Region 3: At the Equivalence Point

At the equivalence point, there is exactly enough NaOH to consume HA.

At the equivalence point, HA has been converted into A^-, a *weak base*.

Titration reaction:		HA	$+\ OH^-$	$\rightarrow A^-$	$+\ H_2O$
Relative initial quantities		1	1	—	—
Relative initial quantities		—	—	1	—

The solution contains "just" A^-. We could have prepared the same solution by dissolving the salt Na^+A^- in distilled water. *A solution of Na^+A^- is merely a solution of a weak base.*

To compute the pH of a weak base, we write the reaction of the weak base with water:

$$A^- + H_2O \rightleftharpoons HA + OH^- \qquad K_b = \frac{K_w}{K_a}$$
$${}_{F-x} {}_x {}_x$$

The only tricky point is that the formal concentration of A^- is no longer 0.020 00 M, which was the initial concentration of HA. A^- has been diluted by NaOH from the buret:

$$F' = (0.020\ 00\ \text{M})\overbrace{\left(\underbrace{\frac{50.00}{50.00 + 10.00}}\right)}^{\text{Initial volume of HA}} = 0.016\ 7\ \text{M}$$

Initial concentration of HA   Dilution factor   Total volume of solution

With this value of F', we can solve the problem:

$$\frac{x^2}{F' - x} = K_b = \frac{K_w}{K_a} = 1.86 \times 10^{-8} \Rightarrow x = 1.76 \times 10^{-5}\ \text{M}$$

$$pH = -\log[H^+] = -\log\left(\frac{K_w}{x}\right) = 9.25$$

The pH is always higher than 7 at the equivalence point in the titration of a weak acid with a strong base.

The pH at the equivalence point in this titration is 9.25. **It is not 7.00.** The equivalence-point pH will *always* be above 7 for the titration of a weak acid, because the acid is converted into its conjugate base at the equivalence point.

Region 4: After the Equivalence Point

Here we assume that the pH is governed by the excess OH^-.

Now we are adding NaOH to a solution of A^-. The base NaOH is so much stronger than the base A^- that it is fair to say that the pH is determined by the excess OH^-.

Let's calculate the pH when $V_b = 10.10$ mL, which is just 0.10 mL past V_e. The concentration of excess OH^- is

Challenge Compare the concentration of OH^- from excess titrant at $V_b = 10.10$ mL to the concentration of OH^- from hydrolysis of A^-. Satisfy yourself that it is fair to neglect the contribution of A^- to the pH after the equivalence point.

$$[OH^-] = (0.100\ 0\ \text{M})\overbrace{\left(\underbrace{\frac{0.10}{50.00 + 10.10}}\right)}^{\text{Volume of excess } OH^-} = 1.66 \times 10^{-4}\ \text{M}$$

Initial concentration of OH^- Dilution factor   Total volume of solution

$$pH = -\log\left(\frac{K_w}{[OH^-]}\right) = 10.22$$

The Titration Curve

Landmarks in a titration:

At $V_b = V_e$, the curve is steepest.

At $V_b = \frac{1}{2}V_e$, pH = pK_a and the slope is minimal.

The *buffer capacity* measures the ability of the solution to resist changes in pH.

Calculations for the titration of MES with NaOH are shown in Table 10-2. The calculated titration curve in Figure 10-2 has two easily identified points. One is the equivalence point, which is the steepest part of the curve. The other landmark is the point where $V_b = \frac{1}{2}V_e$ and pH = pK_a. This latter point is also an inflection point, having the minimum slope.

If you look back at Figure 8-4b, you will note that the maximum *buffer capacity* occurs when pH = pK_a. That is, the solution is most resistant to pH changes when pH = pK_a (and $V_b = \frac{1}{2}V_e$); the slope (dpH/dV_b) is therefore at its minimum.

Figure 10-3 shows how the titration curve depends on the acid dissociation constant of HA and on the concentrations of reactants. As HA becomes a weaker acid, or as the concentrations of analyte and titrant decrease, the inflection near the equivalence point decreases, until the equivalence point becomes too shallow to detect. *It is not practical to titrate an acid or a base when its strength is too weak or its concentration too dilute.*

10-3 Titration of Weak Base with Strong Acid

The titration of a weak base with a strong acid is the reverse of the titration of a weak acid with a strong base. The *titration reaction* is

$$B + H^+ \rightarrow BH^+$$

TABLE 10-2 Calculation of the titration curve for 50.00 mL of 0.020 00 M MES treated with 0.100 0 M NaOH

	mL base added (V_b)	pH
Region 1 (weak acid)	0.00	3.99
	0.50	4.99
	1.00	5.32
	2.00	5.67
	3.00	5.90
	4.00	6.09
Region 2 (buffer)	5.00	6.27
	6.00	6.45
	7.00	6.64
	8.00	6.87
	9.00	7.22
	9.50	7.55
	9.90	8.27
Region 3 (weak base)	10.00	9.25
	10.10	10.22
	10.50	10.91
	11.00	11.21
Region 4 (excess OH$^-$)	12.00	11.50
	13.00	11.67
	14.00	11.79
	15.00	11.88
	16.00	11.95

FIGURE 10-2 Calculated titration curve for the reaction of 50.00 mL of 0.020 00 M MES with 0.100 0 M NaOH. Landmarks occur at half of the equivalence volume (pH = pK_a) and at the equivalence point, which is the steepest part of the curve.

Because the reactants are a weak base and a strong acid, the reaction goes essentially to completion after each addition of acid. There are four distinct regions of the titration curve:

1. Before acid is added, the solution contains just the weak base, B, in water. The pH is determined by the K_b reaction.

$$\underset{F-x}{B} + H_2O \overset{K_b}{\rightleftharpoons} \underset{x}{BH^+} + \underset{x}{OH^-}$$

When V_a (= volume of added acid) = 0, we have a *weak-base* problem.

2. Between the initial point and the equivalence point, there is a mixture of B and BH$^+$— *Aha! A buffer!* The pH is computed by using

$$pH = pK_a \text{ (for } BH^+) + \log\left(\frac{[B]}{[BH^+]}\right)$$

When $0 < V_a < V_e$, we have a *buffer*.

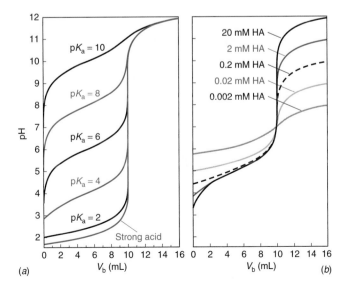

(a)

FIGURE 10-3 (*a*) Calculated curves showing the titration of 50.0 mL of 0.020 0 M HA with 0.100 M NaOH. (*b*) Calculated curves showing the titration of 50.0 mL of HA (pK_a = 5) with NaOH whose concentration is five times greater than that of HA. As the acid becomes weaker or more dilute, the end point becomes less distinct.

In adding acid (increasing V_a), we reach the special point where $V_a = \frac{1}{2}V_e$ and pH = pK_a (for BH^+). As before, pK_a can be determined easily from the titration curve.

3. At the equivalence point, B has been converted into BH^+, a weak acid. The pH is calculated by considering the acid dissociation reaction of BH^+.

When $V_a = V_e$, the solution contains the *weak acid* BH^+.

$$BH^+ \rightleftharpoons B + H^+ \qquad K_a = \frac{K_w}{K_b}$$
$$ F'-x \quad\; x \quad x$$

The formal concentration of BH^+, F', is not the original formal concentration of B because some dilution has occurred. The solution contains BH^+ at the equivalence point, so it is acidic. *The pH at the equivalence point must be below 7.*

For $V_a > V_e$, there is excess *strong acid*.

4. After the equivalence point, the excess strong acid determines the pH. We neglect the contribution of weak acid, BH^+.

EXAMPLE **Titration of Pyridine with HCl**

Consider the titration of 25.00 mL of 0.083 64 M pyridine with 0.106 7 M HCl.

$$\bigcirc N: \qquad K_b = 1.59 \times 10^{-9} \;\Rightarrow\; K_a = \frac{K_w}{K_b} = 6.31 \times 10^{-6} \qquad pK_a = 5.20$$

Pyridine

The titration reaction is

$$\bigcirc N: + H^+ \;\rightarrow\; \bigcirc NH^+$$

and the equivalence point occurs at 19.60 mL:

$$\underbrace{(V_e\,(mL))(0.106\ 7\ M)}_{\text{mmol of HCl}} = \underbrace{(25.00\ mL)(0.083\ 64\ M)}_{\text{mmol of pyridine}} \Rightarrow V_e = 19.60\ mL$$

Find the pH when $V_a = 4.63$ mL.

Solution Part of the pyridine has been neutralized, so there is a mixture of pyridine and pyridinium ion—*Aha! A buffer!* The fraction of pyridine that has been titrated is 4.63/19.60 = 0.236, because it takes 19.60 mL to titrate the whole sample. The fraction of pyridine remaining is $1 - 0.236 = 0.764$. The pH is

$$pH = pK_a + \log\left(\frac{[B]}{[BH^+]}\right)$$

$$= 5.20 + \log\frac{0.764}{0.236} = 5.71$$

Test Yourself Find the pH when $V_a = 14.63$ mL. (*Answer:* 4.73)

10-4 Titrations in Diprotic Systems

The principles developed for titrations of monoprotic acids and bases are readily extended to titrations of polyprotic acids and bases. We will examine two cases.

A Typical Case

The upper curve in Figure 10-4 is calculated for the titration of 10.0 mL of 0.100 M base (B) with 0.100 M HCl. The base is dibasic, with $pK_{b1} = 4.00$ and $pK_{b2} = 9.00$. The titration curve has reasonably sharp breaks at both equivalence points, corresponding to the reactions

$$B + H^+ \rightarrow BH^+$$
$$BH^+ + H^+ \rightarrow BH_2^{2+}$$

The volume at the first equivalence point is 10.00 mL because

$$\underbrace{(V_e\,(mL))(0.100\ M)}_{\text{mmol of HCl}} = \underbrace{(10.00\ mL)(0.100\ 0\ M)}_{\text{mmol of B}} \Rightarrow V_e = 10.00\ mL$$

FIGURE 10-4 (a) Titration of 10.0 mL of 0.100 M base (pK_{b1} = 4.00, pK_{b2} = 9.00) with 0.100 M HCl. The two equivalence points are C and E. Points B and D are the half-neutralization points, whose pH values equal pK_{a2} and pK_{a1}, respectively. (b) Titration of 10.0 mL of 0.100 M nicotine (pK_{b1} = 6.15, pK_{b2} = 10.85) with 0.100 M HCl. There is no sharp break at the second equivalence point, J, because the pH is too low.

The volume at the second equivalence point must be $2V_e$, because the second reaction requires the same number of moles of HCl as the first reaction.

$V_{e2} = 2V_{e1}$, always.

The pH calculations are similar to those for corresponding points in the titration of a monobasic compound. Let's examine points A through E in Figure 10-4.

Point A

Before acid is added, the solution contains just weak base, B, whose pH is governed by the reaction

$$B + H_2O \xrightleftharpoons{K_{b1}} BH^+ + OH^-$$
$$0.100 - x \qquad\qquad x \qquad x$$

$$\frac{x^2}{0.100 - x} = 1.00 \times 10^{-4} \Rightarrow x = 3.11 \times 10^{-3}$$

$$[H^+] = \frac{K_w}{x} \Rightarrow pH = 11.49$$

The fully basic form of a dibasic compound can be treated as if it were monobasic. (The K_{b2} reaction can be neglected.)

Point B

At any point between A (the initial point) and C (the first equivalence point), we have a buffer containing B and BH^+. Point B is halfway to the equivalence point, so [B] = [BH^+]. The pH is calculated from the Henderson-Hasselbalch equation *for the weak acid*, BH^+, whose acid dissociation constant is K_{a2} (for BH_2^{2+}) = K_w/K_{b1} = $10^{-10.00}$.

$$pH = pK_{a2} + \log \frac{[B]}{[BH^+]} = 10.00 + \log 1 = 10.00$$

Of course, you remember that

$$K_{a2} = \frac{K_w}{K_{b1}} \qquad K_{a1} = \frac{K_w}{K_{b2}}$$

So the pH at point B is just pK_{a2}.

To calculate the quotient [B]/[BH^+] at any point in the buffer region, just find what fraction of the way from point A to point C the titration has progressed. For example, if V_a = 1.5 mL, then

$$\frac{[B]}{[BH^+]} = \frac{8.5}{1.5}$$

because 10.0 mL are required to reach the equivalence point and we have added just 1.5 mL. The pH at V_a = 1.5 mL is

$$pH = 10.00 + \log \frac{8.5}{1.5} = 10.75$$

Point C

At the first equivalence point, *B has been converted into BH^+, the intermediate form of the diprotic acid, BH_2^{2+}. BH^+ is both an acid and a base.* From Equation 9-11, we know that

BH^+ is the *intermediate form* of a diprotic acid.

$$pH \approx \frac{1}{2}(pK_1 + pK_2)$$

$$[H^+] \approx \sqrt{\frac{K_1 K_2 F + K_1 K_w}{K_1 + F}} \qquad (10\text{-}3)$$

where K_1 and K_2 are the acid dissociation constants of BH_2^{2+}.

The formal concentration of BH^+ is calculated by considering dilution of the original solution of B.

$$F = (0.100 \text{ M})\left(\frac{10.0}{20.0}\right) = 0.050\ 0 \text{ M}$$

Initial volume of B

Total volume of solution

Original concentration of B   Dilution factor

Plugging all the numbers into Equation 10-3 gives

$$[H^+] = \sqrt{\frac{(10^{-5})(10^{-10})(0.050\ 0) + (10^{-5})(10^{-14})}{10^{-5} + 0.050\ 0}} = 3.16 \times 10^{-8}$$

$$pH = 7.50$$

Note that, in this example, $pH = \frac{1}{2}(pK_{a1} + pK_{a2})$.

Point C in Figure 10-4 shows where the intermediate form of a diprotic acid lies on a titration curve. This is the *least-buffered* point on the whole curve, because the pH changes most rapidly if small amounts of acid or base are added. There is a misconception that the intermediate form of a diprotic acid behaves as a buffer when, in fact, it is the *worst choice* for a buffer.

The intermediate form of a polyprotic acid is the worst possible choice for a buffer.

Point D

At any point between C and E, there is a buffer containing BH^+ (the base) and BH_2^{2+} (the acid). When $V_a = 15.0$ mL, $[BH^+] = [BH_2^{2+}]$ and

$$pH = pK_{a1} + \log\frac{[BH^+]}{[BH_2^{2+}]} = 5.00 + \log 1 = 5.00$$

Challenge Show that, if V_a were 17.2 mL, the ratio in the log term would be

$$\frac{[BH^+]}{[BH_2^{2+}]} = \frac{20.0 - 17.2}{17.2 - 10.0} = \frac{2.8}{7.2}$$

Point E

Point E is the second equivalence point, at which the solution is formally the same as one prepared by dissolving BH_2Cl_2 in water. The formal concentration of BH_2^{2+} is

$$F = (0.100 \text{ M})\left(\frac{10.0}{30.0}\right) = 0.033\ 3 \text{ M}$$

Original volume of B

Total volume of solution

The pH is determined by the acid dissociation reaction of BH_2^{2+}.

At the second equivalence point, we have made BH_2^{2+}, which can be treated as a monoprotic weak acid.

$$BH_2^{2+} \rightleftharpoons BH^+ + H^+ \qquad K_{a1} = \frac{K_w}{K_{b2}}$$

$$\phantom{BH_2^{2+} \rightleftharpoons} F - x \qquad\quad x \qquad x$$

$$\frac{x^2}{0.033\ 3 - x} = 1.0 \times 10^{-5} \Rightarrow x = 5.72 \times 10^{-4} \Rightarrow pH = 3.24$$

Beyond the second equivalence point ($V_a > 20.0$ mL), the pH of the solution can be calculated from the volume of strong acid added to the solution. For example, at $V_a = 25.00$ mL, there is an excess of 5.00 mL of 0.100 M HCl in a total volume of $10.00 + 25.00 = 35.00$ mL. The pH is found by writing

$$[H^+] = (0.100 \text{ M})\left(\frac{5.00}{35.00}\right) = 1.43 \times 10^{-2} \text{ M} \Rightarrow pH = 1.85$$

Blurred End Points

When the pH is too low or too high or when pK_a values are too close together, end points are obscured.

Titrations of many diprotic acids or bases show two clear end points, as in curve *a* in Figure 10-4. Some titrations do not show both end points, as illustrated by curve *b*, which is calculated for the titration of 10.0 mL of 0.100 M nicotine (pK_{b1} = 6.15, pK_{b2} = 10.85) with 0.100 M HCl. The two reactions are

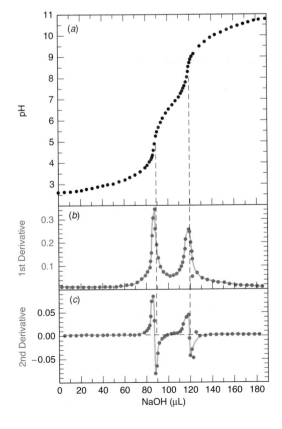

Nicotine (B) BH^+ BH_2^{2+}

There is no perceptible break at the second equivalence point (J), because BH_2^{2+} is too strong an acid (or, equivalently, BH^+ is too weak a base). As the titration approaches low pH($\lesssim$ 3), the approximation that HCl reacts completely with BH^+ to give BH_2^{2+} breaks down. To calculate pH between points I and J requires the systematic treatment of equilibrium. Later in this chapter, we describe how to calculate the whole curve with a spreadsheet.

In the titration of ribonuclease at the beginning of this chapter, there is a continuous change in pH, with no clear breaks. The reason why there are no breaks is that 29 groups are titrated in the pH interval shown. The 29 end points are so close together that a nearly uniform rise results. The curve can be analyzed to find the many pK_a values, but the individual pK_a values will not be determined very precisely.

10-5 Finding the End Point with a pH Electrode

Titrations are commonly performed to find out how much analyte is present or to measure equilibrium constants. We can obtain the information necessary for both purposes by monitoring pH during the titration. Figure 2-12 showed an *autotitrator*, which performs the entire operation automatically.[6] The instrument waits for pH to stabilize after each addition of titrant, before adding the next increment. The end point is computed automatically by finding the maximum slope in the titration curve.

Figure 10-5a shows experimental results for the manual titration of a hexaprotic weak acid, H_6A, with NaOH. Because the compound is difficult to purify, only a tiny amount was available for titration. Just 1.430 mg was dissolved in 1.000 mL of aqueous solution and titrated with microliter quantities of 0.065 92 M NaOH, delivered with a Hamilton syringe.

Box 10-1 illustrates an important application of acid-base titrations in environmental analysis.

FIGURE 10-5 (*a*) Experimental points in the titration of 1.430 mg of xylenol orange, a hexaprotic acid, dissolved in 1.000 mL of aqueous 0.10 M $NaNO_3$. The titrant was 0.065 92 M NaOH. (*b*) The first derivative, $\Delta pH/\Delta V$, of the titration curve. (*c*) The second derivative, $\Delta(\Delta pH/\Delta V)/\Delta V$, which is the derivative of the curve in panel *b*. Derivatives for the first end point are calculated in Figure 10-6. End points are taken as maxima in the derivative curve and zero crossings of the second derivative.

BOX 10-1   Alkalinity and Acidity

The *alkalinity* of a sample of natural water is defined as the moles of HCl equivalent to the excess moles of basic species in the sample from weak acids with $pK_a > 4.5$ at 25°C and zero ionic strength.[4] Alkalinity is approximately equal to the moles of HCl needed to bring 1 kg of water to pH 4.5, which is the second equivalence point in the titration of CO_3^{2-}. To a good approximation,

$$\text{Alkalinity} \approx [OH^-] + 2[CO_3^{2-}] + [HCO_3^-]$$

When water is titrated with HCl to pH 4.5, OH^-, CO_3^{2-}, and HCO_3^- will have reacted. Minor species that could contribute to alkalinity in natural waters include phosphate, borate, silicate, fluoride, ammonia, sulfide, and organic compounds. In oceanography, alkalinity is used to estimate anthropogenic (man-made) CO_2 penetration into the ocean and in measuring the marine $CaCO_3$ budget (sources and sinks for $CaCO_3$).[5] Oceanographers must account for salinity (ionic strength) and temperature in the measurement of alkalinity.[4]

Alkalinity and *hardness* (dissolved Ca^{2+} and Mg^{2+}, Box 11-3) are important characteristics of irrigation water. Alkalinity in excess of the $Ca^{2+} + Mg^{2+}$ content is called "residual sodium carbonate." Water with a residual sodium carbonate content equivalent to ≥ 2.5 mmol H^+/kg is not suitable for irrigation. Residual sodium carbonate between 1.25 and 2.5 mmol H^+/kg is marginal, whereas ≤ 1.25 mmol H^+/kg is suitable for irrigation.

Acidity of natural waters refers to the total acid content that can be titrated to pH 8.3 with NaOH. This pH is the second equivalence

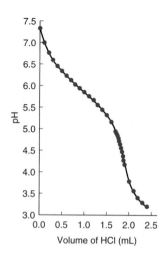

Alkalinity titration of 165.4 mL of saltwater at 20.05°C with 0.209 5 M HCl in a closed cell to prevent escape of CO_2. HCl contains NaCl, so ionic strength remains constant. [Data from Dickson.[4]]

point for titration of carbonic acid (H_2CO_3) with OH^-. Almost all weak acids in the water also will be titrated in this procedure. Acidity is expressed as millimoles of OH^- needed to bring 1 kg of water to pH 8.3.

Figure 10-5a shows two clear breaks, near 90 and 120 μL, which correspond to titration of the *third* and *fourth* protons of H_6A.

$$H_4A^{2-} + OH^- \rightarrow H_3A^{3-} + H_2O \qquad (\sim 90 \text{ μL equivalence point})$$

$$H_3A^{3-} + OH^- \rightarrow H_2A^{4-} + H_2O \qquad (\sim 120 \text{ μL equivalence point})$$

The first two and last two equivalence points are unrecognizable, because they occur at too low or too high a pH.

Using Derivatives to Find the End Point

End point: maximum slope second derivative = 0.

The end point is taken as the volume where the slope (dpH/dV) of the titration curve is greatest. The slope (first derivative) in Figure 10-5b is calculated in the spreadsheet in Figure 10-6.

	A	B	C	D	E	F
1	Derivatives of a Titration Curve					
2	Data		1st derivative		2nd derivative	
3	μL NaOH	pH	μL	$\Delta pH/\Delta \mu L$		$\Delta(\Delta pH/\Delta \mu L)$
4	85.0	4.245			μL	$\Delta \mu L$
5			85.5	0.155		
6	86.0	4.400			86.0	0.0710
7			86.5	0.226		
8	87.0	4.626			87.0	0.0810
9			87.5	0.307		
10	88.0	4.933			88.0	0.0330
11			88.5	0.340		
12	89.0	5.273			89.0	−0.0830
13			89.0	0.257		
14	90.0	5.530			90.0	−0.0680
15			90.5	0.189		
16	91.0	5.719			91.25	−0.0390
17			92.0	0.131		
18	93.0	5.980				
19	Representative formulas:					
20	C5 = (A6+A4)/2			E6 = (C7+C5)/2		
21	D5 = (B6−B4)/(A6−A4)			F6 = (D7−D5)/(C7−C5)		

FIGURE 10-6 Spreadsheet for computing first and second derivatives near 90 μL in Figure 10-5.

The first two columns contain experimental volume and pH. (The pH meter was precise to three digits, even though accuracy ends in the second decimal place.) To compute the first derivative, each pair of volumes is averaged and the quantity $\Delta pH/\Delta V$ is calculated. ΔpH is the change in pH between consecutive readings and ΔV is the change in volume between consecutive additions. Figure 10-5c and the last two columns of the spreadsheet give the second derivative, computed in an analogous manner. The end point is the volume at which the second derivative is 0. Figure 10-7 allows us to make good estimates of the end points.

FIGURE 10-7 Enlargement of the end-point regions in the second derivative curve shown in Figure 10-5c.

EXAMPLE   Computing Derivatives of a Titration Curve

Let's see how the first and second derivatives in Figure 10-6 are calculated.

Solution The first number in the third column, 85.5, is the average of the first two volumes (85.0 and 86.0) in the first column. The derivative $\Delta pH/\Delta V$ is calculated from the first two pH values and the first two volumes:

$$\frac{\Delta pH}{\Delta V} = \frac{4.400 - 4.245}{86.0 - 85.0} = 0.155$$

The coordinates ($x = 85.5$, $y = 0.155$) are one point in the graph of the first derivative in Figure 10-5b.

The second derivative is computed from the first derivative. The first entry in the fifth column of Figure 10-6 is 86.0, which is the average of 85.5 and 86.5. The second derivative is

$$\frac{\Delta(\Delta pH/\Delta V)}{\Delta V} = \frac{0.226 - 0.155}{86.5 - 85.5} = 0.071$$

The coordinates ($x = 86.0$, $y = 0.071$) are plotted in the second derivative graph in Figure 10-6c.

Test Yourself Verify the derivative in cell D7 of Figure 10-6.

Using a Gran Plot to Find the End Point[7, 8]

A problem with using derivatives to find the end point is that titration data are least accurate right near the end point, because buffering is minimal and electrode response is sluggish. A **Gran plot** uses data from before the end point (typically from 0.8 V_e or 0.9 V_e up to V_e) to locate the end point.

A related method uses data from the middle of the titration (not near the equivalence point) to deduce V_e and K_a.[9]

Consider the titration of a weak acid, HA:

$$HA \rightleftharpoons H^+ + A^- \qquad K_a = \frac{[H^+]\gamma_{H^+}[A^-]\gamma_{A^-}}{[HA]\gamma_{HA}}$$

It will be necessary to include activity coefficients in this discussion because a pH electrode responds to hydrogen ion *activity*, not concentration.

Prior to the equivalence point, it is a good approximation to say that each mole of NaOH converts 1 mol of HA into 1 mol of A^-. If we titrate V_a mL of HA (whose formal concentration is F_a) with V_b mL of NaOH (whose formal concentration is F_b), we can write

$$[A^-] = \frac{\text{moles of OH}^- \text{ delivered}}{\text{total volume}} = \frac{V_b F_b}{V_b + V_a}$$

$$[HA] = \frac{\text{original moles of HA} - \text{moles of OH}^-}{\text{total volume}} = \frac{V_a F_a - V_b F_b}{V_a + V_b}$$

Substitution of these values of $[A^-]$ and $[HA]$ into the equilibrium expression gives

$$K_a = \frac{[H^+]\gamma_{H^+} V_b F_b \gamma_{A^-}}{(V_a F_a - V_b F_b)\gamma_{HA}}$$

which can be rearranged to

$$\underbrace{V_b[H^+]\gamma_{H^+}}_{10^{-pH}} = \frac{\gamma_{HA}}{\gamma_{A^-}} K_a \left(\frac{V_a F_a - V_b F_b}{F_b} \right) \tag{10-4}$$

The term on the left is $V_b \cdot 10^{-pH}$, because $[H^+]\gamma_{H^+} = 10^{-pH}$. The term in parentheses on the right is

$$\frac{V_a F_a - V_b F_b}{F_b} = \frac{V_a F_a}{F_b} - V_b = V_e - V_b$$

Equation 10-4 can, therefore, be written in the form

Gran plot equation: $\qquad V_b \cdot 10^{-pH} = \dfrac{\gamma_{HA}}{\gamma_{A^-}} K_a (V_e - V_b) \tag{10-5}$

A graph of $V_b \cdot 10^{-pH}$ versus V_b is called a *Gran plot*. If γ_{HA}/γ_{A^-} is constant, the graph is a straight line with a slope of $-K_a \gamma_{HA}/\gamma_{A^-}$ and an *x*-intercept of V_e. Figure 10-8 shows a Gran plot for the titration in Figure 10-5. Any units can be used for V_b, but the same units must be used on both axes. In Figure 10-8, V_b is expressed in microliters on both axes.

The beauty of a Gran plot is that it enables us to use data taken *before* the end point to find the end point. The slope of the Gran plot enables us to find K_a. Although we derived the Gran function for a monoprotic acid, the same plot ($V_b 10^{-pH}$ versus V_b) applies to polyprotic acids (such as H_6A in Figure 10-5).

The Gran function, $V_b \cdot 10^{-pH}$, does not actually go to 0, because 10^{-pH} is never 0. The curve must be extrapolated to find V_e. The reason the function does not reach 0 is that we have used the approximation that every mole of OH^- generates 1 mol of A^-, which is not true as V_b approaches V_e. Only the linear portion of the Gran plot is used.

Another source of curvature in the Gran plot is changing ionic strength, which causes γ_{HA}/γ_{A^-} to vary. In Figure 10-8, this variation was avoided by maintaining nearly constant ionic strength with $NaNO_3$. Even without added salt, the last 10–20% of data before V_e gives a fairly straight line because the quotient γ_{HA}/γ_{A^-} does not change very much. The Gran plot in the acidic region gives accurate results even if CO_2 is dissolved in the strong-base titrant. The Gran plot in the basic region can be used to measure CO_2 in the strong base.[7]

Challenge Show that, when weak base, B, is titrated with a strong acid, the Gran function is

$$V_a \cdot 10^{+pH} = \left(\frac{1}{K_a} \cdot \frac{\gamma_B}{\gamma_{BH^+}} \right)(V_e - V_a) \tag{10-6}$$

where V_a is the volume of strong acid and K_a is the acid dissociation constant of BH^+. A graph of $V_a 10^{+pH}$ versus V_a should be a straight line with a slope of $-\gamma_B/(\gamma_{BH^+} K_a)$ and an *x*-intercept of V_e.

Margin notes:

Strong plus weak react completely.

$\mathcal{A}_{H^+} = [H^+]\gamma_{H^+} = 10^{-pH}$

$V_a F_a = V_e F_b \Rightarrow V_e = \dfrac{V_a F_a}{F_b}$

Gran plot:

Plot $V_b \cdot 10^{-pH}$ versus V_b

x-intercept = V_e

Slope = $-K_a \gamma_{HA}/\gamma_{A^-}$

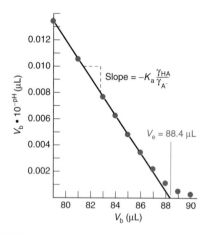

FIGURE 10-8 Gran plot for the first equivalence point of Figure 10-5. This plot gives an estimate of V_e that differs from that in Figure 10-7 by 0.2 μL (88.4 versus 88.2 μL). The last 10–20% of volume prior to V_e is normally used for a Gran plot.

10-6 Finding the End Point with Indicators

An acid-base **indicator** is itself an acid or base whose various protonated species have different colors. An example is thymol blue.

> An indicator is an acid or a base whose various protonated forms have different colors.

Red (R)
Thymol blue
Yellow (Y⁻)
Blue (B²⁻)

$pK_1 = 1.7$ $pK_2 = 8.9$

Below pH 1.7, the predominant species is red; between pH 1.7 and pH 8.9, the predominant species is yellow; and above pH 8.9, the predominant species is blue (Color Plate 3). For simplicity, we designate the three species R, Y⁻, and B²⁻.

The equilibrium between R and Y⁻ can be written

$$R \xrightleftharpoons{K_1} Y^- + H^+ \qquad pH = pK_1 + \log\frac{[Y^-]}{[R]} \tag{10-7}$$

pH	[Y⁻]:[R]	Color
0.7	1:10	red
1.7	1:1	orange
2.7	10:1	yellow

At pH 1.7 (= pK_1), there will be a 1:1 mixture of the yellow and red species, which appears orange. As a crude rule of thumb, we can say that the solution will appear red when $[Y^-]/[R]$ $\lesssim$ 1/10 and yellow when $[Y^-]/[R] \gtrsim$ 10/1. From Equation 10-7, we see that the solution will be red when pH ≈ pK_1 − 1 and yellow when pH ≈ pK_1 + 1. In tables of indicator colors, thymol blue is listed as red below pH 1.2 and yellow above pH 2.8. The pH values predicted by our rule of thumb are 0.7 and 2.7. Between pH 1.2 and 2.8, the indicator exhibits various shades of orange. The pH range (1.2 to 2.8) over which the color changes is called the **transition range.** Whereas most indicators have a single color change, thymol blue has another transition, from yellow to blue, between pH 8.0 and pH 9.6. In this range, various shades of green are seen.

Acid-base indicator color changes are featured in Demonstration 10-1. Box 10-2 shows how optical absorption by indicators allows us to measure pH.

Choosing an Indicator

A titration curve for which pH = 5.54 at the equivalence point is shown in Figure 10-9. An indicator with a color change near this pH would be useful for an end point. In Figure 10-9,

> One of the most common indicators is phenolphthalein, usually used for its colorless-to-pink transition at pH 8.0–9.6.

Colorless phenolphthalein
pH < 8.0

$2H^+$ $2OH^-$

Pink phenolphthalein
pH > 9.6

$+ 2H_2O$

> In strong acid, the colorless form of phenolphthalein turns orange-red. In strong base, the red species loses its color.[10]

Orange-red
(in 65–98% H_2SO_4)

Colorless
pH > 11

FIGURE 10-9 Calculated titration curve for the reaction of 100 mL of 0.010 0 M base (pK_b = 5.00) with 0.050 0 M HCl.

BOX 10-2 **What Does a Negative pH Mean?**

In the 1930s, Louis Hammett and his students measured the strengths of very weak acids and bases, using a weak reference base (B), such as *p*-nitroaniline ($pK_a = 0.99$), whose base strength could be measured in aqueous solution.

p-Nitroanilinium ion
BH$^+$

$pK_a = 0.99$

p-Nitroaniline
B

Suppose that some *p*-nitroaniline and a second base, C, are dissolved in a strong acid, such as 2 M HCl. The pK_a of CH$^+$ can be measured relative to that of BH$^+$ by first writing a Henderson-Hasselbalch equation for each acid:

$$pH = pK_a \text{ (for BH}^+\text{)} + \log \frac{[B]\gamma_B}{[BH^+]\gamma_{BH^+}}$$

$$pH = pK_a \text{ (for CH}^+\text{)} + \log \frac{[C]\gamma_C}{[CH^+]\gamma_{CH^+}}$$

Setting the two equations equal (because there is only one pH) gives

$$\underbrace{pK_a(\text{for CH}^+) - pK_a(\text{for BH}^+)}_{\Delta pK_a} = \log \frac{[B][CH^+]}{[C][BH^+]} + \log \frac{\gamma_B \gamma_{CH^+}}{\gamma_C \gamma_{BH^+}}$$

The ratio of activity coefficients is close to unity, so the second term on the right is close to 0. Neglecting this last term gives an operationally useful result:

$$\Delta pK_a \approx \log \frac{[B][CH^+]}{[C][BH^+]}$$

That is, if you have a way to find the concentrations of B, BH$^+$, C, and CH$^+$ and if you know pK_a for BH$^+$, then you can find pK_a for CH$^+$.

Concentrations can be measured with a spectrophotometer[12] or by nuclear magnetic resonance,[13] so pK_a for CH$^+$ can be determined. Then, with CH$^+$ as the reference, the pK_a for another compound, DH$^+$, can be measured. This procedure can be extended to measure the strengths of successively weaker bases (such as nitrobenzene, $pK_a = -11.38$), far too weak to be protonated in water.

The acidity of a solvent that protonates the weak base, B, is defined as the **Hammett acidity function:**

Hammett acidity function:
$$H_0 = pK_a(\text{for BH}^+) + \log \frac{[B]}{[BH^+]}$$

For dilute aqueous solutions, H_0 approaches pH. For concentrated acids, H_0 is a measure of the acid strength. The weaker a base, B, the stronger the acidity of the solvent must be to protonate the base. Acidity of strongly acidic solvents is now measured more conveniently by electrochemical methods.[14]

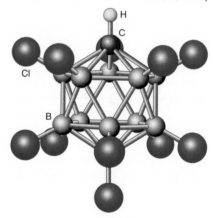

Hammett acidity function, H_0, for aqueous solutions of acids. [Data from R. A. Cox and K. Yates, "Acidity Functions," *Can. J. Chem.* **1983**, *61*, 2225.]

Icosahedral carborane anion of $[CHB_{11}Cl_{11}]^-H^+$, the strongest known acid.[15] Icosahedral $H_2[B_{12}Cl_{12}]$ is the strongest known diprotic acid.

When we refer to *negative* pH, we usually mean H_0 values. For example, as measured by its ability to protonate very weak bases, 8 M HClO$_4$ has a "pH" close to -4. The graph shows that HClO$_4$ is a stronger acid than other mineral acids. Values of H_0 for several powerfully acidic solvents are shown in the table. The strongest known acid is $[CHB_{11}Cl_{11}]^-H^+$, in which the icosahedral carborane cage has no significant affinity for H$^+$.

Acid	Name	H_0
H$_2$SO$_4$ (100%)	sulfuric acid	-11.93
H$_2$SO$_4 \cdot$ SO$_3$	fuming sulfuric acid (oleum)	-14.14
HSO$_3$F	fluorosulfuric acid	-15.07
HSO$_3$F + 10% SbF$_5$	"super acid"	-18.94
HSO$_3$F + 7% SbF$_5 \cdot$ 3SO$_3$	—	-19.35

This one is just plain fun.[11] Place 900 mL of water and a magnetic stirring bar in each of two 1-L graduated cylinders. Add 10 mL of 1 M NH_3 to each. Then put 2 mL of phenolphthalein indicator solution in one and 2 mL of bromothymol blue indicator solution in the other. Both indicators will have the color of their basic species.

Drop a few chunks of Dry Ice (solid CO_2) into each cylinder. As the CO_2 bubbles through each cylinder, the solutions become more acidic. First the pink phenolphthalein color disappears. After some time, the pH drops low enough for bromothymol blue to change from blue to green, but not all the way to yellow.

Add 20 mL of 6 M HCl to *the bottom* of each cylinder, using a length of Tygon tubing attached to a funnel. Then stir each solution for a few seconds on a magnetic stirrer. Explain what happens. The sequence of events is shown in Color Plate 4.

---

the pH drops steeply (from 7 to 4) over a small volume interval. Therefore, any indicator with a color change in this pH interval would provide a fair approximation to the equivalence point. The closer the point of color change is to pH 5.54, the more accurate will be the end point. The difference between the observed end point (color change) and the true equivalence point is called the **indicator error.**

If you dump half a bottle of indicator into your reaction, you will introduce a different indicator error. Because indicators are acids or bases, they react with analyte or titrant. The moles of indicator must be negligible relative to the moles of analyte. Never use more than a few drops of dilute indicator solution.

Several of the indicators in Table 10-3 would be useful for the titration in Figure 10-9. If bromocresol purple were used, we would use the purple-to-yellow color change as the end point. The last trace of purple should disappear near pH 5.2, which is quite close to the true equivalence point in Figure 10-9. If bromocresol green were used as the indicator, a color change from blue to green (= yellow + blue) would mark the end point.

TABLE 10-3 Common indicators

Indicator	Transition range (pH)	Acid color	Base color	Preparation
Methyl violet	0.0–1.6	Yellow	Violet	0.05 wt% in H_2O
Cresol red	0.2–1.8	Red	Yellow	0.1 g in 26.2 mL 0.01 M NaOH. Then add ~225 mL H_2O.
Thymol blue	1.2–2.8	Red	Yellow	0.1 g in 21.5 mL 0.01 M NaOH. Then add ~225 mL H_2O.
Cresol purple	1.2–2.8	Red	Yellow	0.1 g in 26.2 mL 0.01 M NaOH. Then add ~225 mL H_2O.
Erythrosine, disodium	2.2–3.6	Orange	Red	0.1 wt% in H_2O
Methyl orange	3.1–4.4	Red	Yellow	0.01 wt% in H_2O
Congo red	3.0–5.0	Violet	Red	0.1 wt% in H_2O
Ethyl orange	3.4–4.8	Red	Yellow	0.1 wt% in H_2O
Bromocresol green	3.8–5.4	Yellow	Blue	0.1 g in 14.3 mL 0.01 M NaOH. Then add ~225 mL H_2O.
Methyl red	4.8–6.0	Red	Yellow	0.02 g in 60 mL ethanol. Then add 40 mL H_2O.
Chlorophenol red	4.8–6.4	Yellow	Red	0.1 g in 23.6 mL 0.01 M NaOH. Then add ~225 mL H_2O.
Bromocresol purple	5.2–6.8	Yellow	Purple	0.1 g in 18.5 mL 0.01 M NaOH. Then add ~225 mL H_2O.
p-Nitrophenol	5.6–7.6	Colorless	Yellow	0.1 wt% in H_2O
Litmus	5.0–8.0	Red	Blue	0.1 wt% in H_2O
Bromothymol blue	6.0–7.6	Yellow	Blue	0.1 g in 16.0 mL 0.01 M NaOH. Then add ~225 mL H_2O.
Phenol red	6.4–8.0	Yellow	Red	0.1 g in 28.2 mL 0.01 M NaOH. Then add ~225 mL H_2O.
Neutral red	6.8–8.0	Red	Yellow	0.01 g in 50 mL ethanol. Then add 50 mL H_2O.
Cresol red	7.2–8.8	Yellow	Red	See above.
α-Naphtholphthalein	7.3–8.7	Pink	Green	0.1 g in 50 mL ethanol. Then add 50 mL H_2O.
Cresol purple	7.6–9.2	Yellow	Purple	See above.
Thymol blue	8.0–9.6	Yellow	Blue	See above.
Phenolphthalein	8.0–9.6	Colorless	Pink	0.05 g in 50 mL ethanol. Then add 50 mL H_2O.
Thymolphthalein	8.3–10.5	Colorless	Blue	0.04 g in 50 mL ethanol. Then add 50 mL H_2O.
Alizarin yellow	10.1–12.0	Yellow	Orange-red	0.01 wt% in H_2O
Nitramine	10.8–13.0	Colorless	Orange-brown	0.1 g in 70 mL ethanol. Then add 30 mL H_2O.
Tropaeolin O	11.1–12.7	Yellow	Orange	0.1 wt% in H_2O

TABLE 10-4   **Primary standards**

Compound	Density (g/mL) for buoyancy corrections	Notes
ACIDS		
Potassium hydrogen phthalate FM 204.221	1.64	The pure commercial material is dried at 105°C and used to standardize base. A phenolphthalein end point is satisfactory.
HCl Hydrochloric acid FM 36.461	—	HCl and water distill as an *azeotrope* (a mixture), whose composition (~6 M) depends on pressure and is tabulated as a function of the pressure during distillation. See Problem 10-56 for more information.
$KH(IO_3)_2$ Potassium hydrogen iodate FM 389.912	—	This is a strong acid, so any indicator with an end point between ~5 and ~9 is adequate.
Benzoic acid FM 122.121	1.27	Primary standard for nonaqueous titrations in solvents such as ethanol. A glass electrode is used to find the end point.
Sulfosalicylic acid double salt FM 550.639	—	1 mol of commercial grade sulfosalicylic acid is combined with 0.75 mol of reagent-grade $KHCO_3$, recrystallized several times from water, and dried at 110°C to produce the double salt with 3 K^+ ions and one titratable H^+.[16] Phenolphthalein is used as the indicator for titration with NaOH.
$H_3\overset{+}{N}SO_3^-$ Sulfamic acid FM 97.094	2.15	Sulfamic acid is a strong acid with one acidic proton, so any indicator with an end point between ~5 and ~9 is suitable.
BASES		
$H_2NC(CH_2OH)_3$ Tris(hydroxymethyl)aminomethane (also called tris or tham) FM 121.135	1.33	The pure commercial material is dried at 100°–103°C and titrated with strong acid. The end point is in the range pH 4.5–5. $$H_2NC(CH_2OH)_3 + H^+ \rightarrow H_3\overset{+}{N}C(CH_2OH)_3$$
HgO Mercuric oxide FM 216.59	11.1	Pure HgO is dissolved in a large excess of I^- or Br^-, whereupon 2 OH^- are liberated: $$HgO + 4I^- + H_2O \rightarrow HgI_4^{2-} + 2OH^-$$ The base is titrated, using an indicator end point.
Na_2CO_3 Sodium carbonate FM 105.988	2.53	Primary-standard-grade Na_2CO_3 is commercially available. Alternatively, recrystallized $NaHCO_3$ can be heated for 1 h at 260°–270°C to produce pure Na_2CO_3. Sodium carbonate is titrated with acid to an end point of pH 4–5. Just before the end point, the solution is boiled to expel CO_2.
$Na_2B_4O_7 \cdot 10H_2O$ Borax FM 381.372	1.73	The recrystallized material is dried in a chamber containing an aqueous solution saturated with NaCl and sucrose. This procedure gives the decahydrate in pure form.[17] The standard is titrated with acid to a methyl red end point. $$\text{"}B_4O_7 \cdot 10H_2O^{2-}\text{"} + 2H^+ \rightarrow 4B(OH)_3 + 5H_2O$$

In general, *we seek an indicator whose transition range overlaps the steepest part of the titration curve as closely as possible.* The steepness of the titration curve near the equivalence point in Figure 10-9 ensures that the indicator error caused by the noncoincidence of the end point and equivalence point will be small. If the indicator end point were at pH 6.4 (instead of 5.54), the error in V_e would be only 0.25% in this particular case. You can estimate the indicator error by calculating the volume of titrant required to attain pH 6.4 instead of pH 5.54.

Choose an indicator whose transition range overlaps the steepest part of the titration curve.

10-7 Practical Notes

Acids and bases in Table 10-4 can be obtained pure enough to be *primary standards.*[18] NaOH and KOH are not primary standards because they contain carbonate (from reaction with atmospheric CO_2) and adsorbed water. Solutions of NaOH and KOH must be standardized against a primary standard such as potassium hydrogen phthalate. Solutions of NaOH for titrations are prepared by diluting a stock solution of 50 wt% aqueous NaOH. Sodium carbonate is insoluble in this stock solution and settles to the bottom.

Procedures for preparing standard acid and base are given at the end of this chapter.

Alkaline solutions (for example, 0.1 M NaOH) must be protected from the atmosphere; otherwise, they absorb CO_2:

$$OH^- + CO_2 \rightarrow HCO_3^-$$

CO_2 changes the concentration of strong base over a period of time and decreases the extent of reaction near the end point in the titration of weak acids. If solutions are kept in tightly capped polyethylene bottles, they can be used for about a week with little change.

Standard solutions are commonly stored in high-density polyethylene bottles with screw caps. Evaporation from the bottle slowly changes the reagent concentration. The chemical supplier Sigma-Aldrich reports that an aqueous solution stored in a tightly capped bottle became 0.2% more concentrated in 2 years at 23°C and 0.5% more concentrated in 2 years at 30°C. Enclosing the bottle in a sealed, aluminized bag reduced evaporation by a factor of 10. The lesson is that a standard solution has a finite shelf life.

Strongly basic solutions attack glass and are best stored in plastic containers. Such solutions should not be kept in a buret longer than necessary. Boiling 0.01 M NaOH in a flask for 1 h decreases the molarity by 10% because OH^- reacts with glass.[19]

10-8 Kjeldahl Nitrogen Analysis

Developed in 1883, the **Kjeldahl nitrogen analysis** is one of the most widely used methods for determining nitrogen in organic substances. Protein is the main nitrogen-containing constituent of food. Most proteins contain close to 16 wt% nitrogen, so measuring nitrogen is a surrogate for measuring protein (Box 10-3). The other common way to measure nitrogen in food is by combustion analysis (Section 26-4).

In the Kjeldahl method, a sample is first *digested* (decomposed and dissolved) in boiling sulfuric acid, which converts amine and amide nitrogen into ammonium ion, NH_4^+, and oxidizes other elements present:[23]

Kjeldahl digestion: $\qquad$ organic C, H, N $\xrightarrow[H_2SO_4]{boiling}$ $NH_4^+ + CO_2 + H_2O$ $\qquad$ (10-8)

Each atom of nitrogen in starting material is converted into one NH_4^+ ion.

Mercury, copper, and selenium compounds catalyze the digestion. To speed the reaction, the boiling point of concentrated (98 wt%) sulfuric acid (338°C) is raised by adding K_2SO_4. Digestion is carried out in a long-neck *Kjeldahl flask* (Figure 10-10) that prevents loss of sample from spattering. Alternative digestion procedures employ H_2SO_4 plus H_2O_2 or $K_2S_2O_8$ plus NaOH[24] in a microwave bomb (a pressurized vessel shown in Figure 27-7).

After digestion is complete, the solution containing NH_4^+ is made basic, and the liberated NH_3 is steam distilled (with a large excess of steam) into a receiver containing a known amount of HCl[25] (Figure 10-11). Excess, unreacted HCl is then titrated with standard NaOH to determine how much HCl was consumed by NH_3.

BOX 10-3 Kjeldahl Nitrogen Analysis Behind the Headlines

In 2007, dogs and cats in North America suddenly began to die from kidney failure. The illness was traced to animal food containing ingredients from China. Melamine, used to make plastics, had been deliberately added to food ingredients "in a bid to meet the contractual demand for the amount of protein in the products."[20] Cyanuric acid, used to disinfect swimming pools, was also found in the food. Melamine alone does not cause kidney failure, but the combination of melamine and cyanuric acid does.

	Melamine	Cyanuric acid
	(66.6 wt% nitrogen)	(32.6 wt% nitrogen)

What do these compounds have to do with protein? Nothing— except that they are high in nitrogen. Protein, which contains ~16 wt% nitrogen, is the main source of nitrogen in food. Kjeldahl nitrogen analysis is a surrogate measurement for protein in food. For example, if food contains 10 wt% protein, it will contain ~16% of 10% = 1.6 wt% N. If you measure 1.6 wt% N in food, you could conclude that the food contains ~10 wt% protein. Melamine contains 66.6 wt% N, which is four times more than protein. Adding 1 wt% melamine to food makes it appear that the food contains an additional 4 wt% protein.

Protein source	Weight % nitrogen
Meat	16.0
Blood plasma	15.3
Milk	15.6
Flour	17.5
Egg	14.9

SOURCE: D. J. Holme and H. Peck, Analytical Biochemistry, 3rd ed. (New York: Addison Wesley Longman, 1998), p. 388.

Incredibly, in the summer of 2008, approximately 300 000 Chinese babies became sick and some suffered kidney failure.[21] Many Chinese companies had diluted milk with water and added melamine to make the protein content appear normal. Poisoned milk products were sold in the domestic and export markets. In a response to the appearance of melamine in food, at least one company developed a colorimetric assay that distinguishes protein nitrogen from nonprotein nitrogen.[22]

Another means to measure nitrogen in food is the *Dumas method*. Organic material mixed with CuO is heated in CO_2 at 650°–700°C to produce CO_2, H_2O, N_2, and nitrogen oxides. Products are carried by a stream of CO_2 through hot Cu to convert nitrogen oxides to N_2. The gases are bubbled through concentrated KOH(*aq*) to capture CO_2. The volume of N_2 is measured in a gas buret. This method does not distinguish protein from melamine.

Neutralization of NH_4^+:	$NH_4^+ + OH^- \rightarrow NH_3(g) + H_2O$	(10-9)
Distillation of NH_3 into standard HCl:	$NH_3 + H^+ \rightarrow NH_4^+$	(10-10)
Titration of unreacted HCl with NaOH:	$H^+ + OH^- \rightarrow H_2O$	(10-11)

An alternative to a titration is to neutralize the acid and raise the pH with a buffer, followed by addition of reagents that form a colored product with NH_3.[26] The absorbance of the colored product gives the concentration of NH_3 from the digestion.

(a)   (b)

FIGURE 10-10 (a) Kjeldahl digestion flask with long neck to minimize loss from spattering. (b) Six-port manifold for multiple samples provides for exhaust of fumes. [Courtesy Labconco Corp.]

EXAMPLE   Kjeldahl Analysis

A typical protein contains 16.2 wt% nitrogen. A 0.500-mL aliquot of protein solution was digested, and the liberated NH_3 was distilled into 10.00 mL of 0.021 40 M HCl. Unreacted HCl required 3.26 mL of 0.019 8 M NaOH for complete titration. Find the concentration of protein (mg protein/mL) in the original sample.

Solution The initial amount of HCl in the receiver was (10.00 mL)(0.021 40 mmol/mL) = 0.214 0 mmol. The NaOH required for titration of unreacted HCl in Reaction 10-11 was (3.26 mL)(0.019 8 mmol/mL) = 0.064 5 mmol. The difference, 0.214 0 − 0.064 5 = 0.149 5 mmol, must be the quantity of NH_3 produced in Reaction 10-9 and distilled into the HCl.

Because 1 mol of N in the protein produces 1 mol of NH_3, there must have been 0.149 5 mmol of N in the protein, corresponding to

$$(0.149\ 5\ \text{mmol})\left(14.006\ 74\ \frac{\text{mg N}}{\text{mmol}}\right) = 2.093\ \text{mg N}$$

If the protein contains 16.2 wt% N, there must be

$$\frac{2.093\ \text{mg N}}{0.162\ \text{mg N/mg protein}} = 12.9\ \text{mg protein} \Rightarrow \frac{12.9\ \text{mg protein}}{0.500\ \text{mL}} = 25.8\ \frac{\text{mg protein}}{\text{mL}}$$

Test Yourself Find mg protein/mL if 3.00 mL of NaOH were required. (**Answer: 26.7 mg/mL**)

FIGURE 10-11 Original apparatus used by Dutch chemist J. Kjeldahl (1849–1900). [From D. T. Burns, "Kjeldahl, the Man, the Method and the Carlsberg Laboratory," *Anal. Proc.* (Royal Society of Chemistry) 1984, *21*, 210.]

10-9 The Leveling Effect

The strongest acid that can exist in water is H_3O^+ and the strongest base is OH^-. If an acid stronger than H_3O^+ is dissolved in water, it protonates H_2O to make H_3O^+. If a base stronger than OH^- is dissolved in water, it deprotonates H_2O to make OH^-. Because of this **leveling effect,** $HClO_4$ and HCl behave as if they had the same acid strength; both are *leveled* to H_3O^+:

$$HClO_4 + H_2O \rightarrow H_3O^+ + ClO_4^-$$
$$HCl + H_2O \rightarrow H_3O^+ + Cl^-$$

In acetic acid solvent, which is less basic than H_2O, $HClO_4$ and HCl are not leveled to the same strength:

$$HClO_4 + CH_3CO_2H \rightleftharpoons CH_3CO_2H_2^+ + ClO_4^- \qquad K = 1.3 \times 10^{-5}$$
$$\text{Acetic acid solvent}$$
$$HCl + CH_3CO_2H \rightleftharpoons CH_3CO_2H_2^+ + Cl^- \qquad K = 2.8 \times 10^{-9}$$

The equilibrium constants show that $HClO_4$ is a stronger acid than HCl in acetic acid solvent.

Figure 10-12 shows a titration curve for a mixture of five acids titrated with 0.2 M tetra-butylammonium hydroxide in methyl isobutyl ketone solvent. This solvent is not protonated

In acetic acid solution, $HClO_4$ is a stronger acid than HCl; but, in aqueous solution, both are leveled to the strength of H_3O^+.

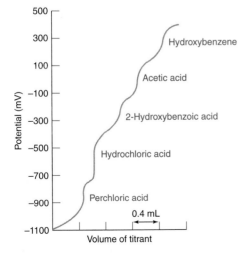

FIGURE 10-12 Titration of a mixture of acids with tetrabutylammonium hydroxide in methyl isobutyl ketone solvent shows that the order of acid strength is $HClO_4$ > HCl > 2-hydroxybenzoic acid > acetic acid > hydroxybenzene. Measurements were made with a glass electrode and a platinum reference electrode. The ordinate is proportional to pH, with increasing pH as the potential becomes more positive. [From D. B. Bruss and G. E. A. Wyld, "Methyl Isobutyl Ketone as a Wide-Range Solvent for Titration of Acid Mixtures and Nitrogen Bases," *Anal. Chem.* 1957, *29*, 232.]

Question Where do you think the end point for the acid $H_3O^+ClO_4^-$ would come in Figure 10-12?

to a great extent by any of the acids. We see that perchloric acid is a stronger acid than HCl in this solvent as well.

Now consider a base such as urea, $(H_2N)_2C\!=\!O$ ($K_b = 1.3 \times 10^{-14}$), that is too weak to give a distinct end point when titrated with a strong acid in water.

Titration with $HClO_4$ in H_2O: $\qquad\qquad\qquad B + H_3O^+ \rightleftharpoons BH^+ + H_2O$

The end point cannot be recognized because the equilibrium constant for the titration reaction is not large enough. If an acid stronger than H_3O^+ were available, the titration reaction might have an equilibrium constant large enough to give a distinct end point. If the same base were dissolved in acetic acid and titrated with $HClO_4$ in acetic acid, a clear end point might be observed. The reaction

Titration with $HClO_4$ in CH_3CO_2H: $\qquad\qquad B + HClO_4 \rightleftharpoons \underbrace{BH^+ClO_4^-}_{\text{An ion pair}}$

might have a large equilibrium constant, because $HClO_4$ is a much stronger acid than H_3O^+. (The product in this reaction is written as an ion pair because the *dielectric constant* of acetic acid is too low to allow ions to separate extensively.) Titrations that are not feasible in water might be feasible in other solvents.[27]

In *electrophoresis* (Chapter 25), ions are separated by their different mobilities in an electric field. Compounds shown in the margin are such weak bases that they cannot be protonated in aqueous solution and, therefore, cannot be made into charged species for aqueous electrophoresis. However, in anhydrous acetonitrile solvent, they are protonated by $HClO_4$ in anhydrous acetic acid and can be separated as cations.[28]

A base too weak to be titrated by H_3O^+ in water might be titrated by $HClO_4$ in acetic acid solvent.

Dielectric constant is discussed in Problem 7-13.

Compounds that can be protonated in acetonitrile by perchloric acid plus acetic acid, $CH_3C(OH)_2^+ClO_4^-$:

H₃C—C(=S)—NH₂ Thioacetamide

O₂N—C₆H₄—C(=O)—NH₂ 4-Nitrobenzamide

10-10 Calculating Titration Curves with Spreadsheets

This chapter has been critical for developing your understanding of the chemistry that occurs during titrations. However, the approximations we used are of limited value when concentrations are too dilute or equilibrium constants are not of the right magnitude or K_a values are too closely spaced, like those in a protein. This section develops equations to deal with titrations in a general manner, using spreadsheets.[29]

Experiment 10, "Fitting a Titration Curve with Excel Solver," at www.whfreeman.com/qca, applies equations developed in this section.

Titrating a Weak Acid with a Strong Base

Consider the titration of a volume V_a of acid HA (initial concentration C_a) with a volume V_b of NaOH of concentration C_b. The charge balance for this solution is

Charge balance: $\qquad\qquad [H^+] + [Na^+] = [A^-] + [OH^-]$

and the concentration of Na^+ is just

$$[Na^+] = \frac{C_b V_b}{V_a + V_b}$$

because we have diluted $C_b V_b$ moles of NaOH to a total volume of $V_a + V_b$. Similarly, the formal concentration of the weak acid is

$$F_{HA} = [HA] + [A^-] = \frac{C_a V_a}{V_a + V_b}$$

because we have diluted $C_a V_a$ moles of HA to a total volume of $V_a + V_b$.

Now we use the fractional composition equations from Section 9-5. The concentration of A^- can be written in terms of α_{A^-} defined in Equation 9-18:

α_{A^-} = fraction of acid in the form A^-:

$$\alpha_{A^-} = \frac{[A^-]}{F_{HA}}$$

$$[A^-] = \alpha_{A^-} \cdot F_{HA} = \frac{\alpha_{A^-} \cdot C_a V_a}{V_a + V_b} \qquad\qquad (10\text{-}12)$$

where $\alpha_{A^-} = K_a/([H^+] + K_a)$ and K_a is the acid dissociation constant of HA. Substituting for $[Na^+]$ and $[A^-]$ in the charge balance gives

$$[H^+] + \frac{C_b V_b}{V_a + V_b} = \frac{\alpha_{A^-} \cdot C_a V_a}{V_a + V_b} + [OH^-]$$

which you can rearrange to the form

Fraction of titration for
weak acid by strong base:

$$\phi \equiv \frac{C_b V_b}{C_a V_a} = \frac{\alpha_{A^-} - \dfrac{[H^+] - [OH^-]}{C_a}}{1 + \dfrac{[H^+] - [OH^-]}{C_b}} \quad (10\text{-}13)$$

$\phi = C_b V_b / C_a V_a$ is the fraction of the way to the equivalence point:

ϕ	Volume of base
0.5	$V_b = \frac{1}{2} V_e$
1	$V_b = V_e$
2	$V_b = 2 V_e$

At last! Equation 10-13 is really useful. It relates the volume of titrant (V_b) to the pH and a bunch of constants. The quantity ϕ, which is the quotient $C_b V_b / C_a V_a$, is the fraction of the way to the equivalence point, V_e. When $\phi = 1$, the volume of base added, V_b, is equal to V_e. Equation 10-13 works backward from the way you are accustomed to thinking, because you need to put in pH on the right to get out volume on the left. Let me say that again: We *put in a concentration of H^+* and *get out the volume of titrant* that produces that concentration.

Let's use Equation 10-13 to calculate the titration curve for 50.00 mL of the weak acid 0.020 00 M MES with 0.100 0 M NaOH, which was shown in Figure 10-2 and Table 10-2. The equivalence volume is $V_e = 10.00$ mL. Quantities in Equation 10-13 are

$$C_b = 0.1 \text{ M} \qquad\qquad [H^+] = 10^{-pH}$$

$$C_a = 0.02 \text{ M} \qquad\qquad [OH^-] = K_w/[H^+]$$

$$V_a = 50 \text{ mL}$$

$$K_a = 5.3_7 \times 10^{-7} \qquad \alpha_{A^-} = \frac{K_a}{[H^+] + K_a}$$

$$K_w = 10^{-14}$$

$$\boxed{\text{pH is the input}} \qquad \boxed{V_b = \frac{\phi C_a V_a}{C_b} \text{ is the output}}$$

The input to the spreadsheet in Figure 10-13 is pH in column B and the output is V_b in column G. From the pH, the values of $[H^+]$, $[OH^-]$, and α_{A^-} are computed in columns C, D, and E. Equation 10-13 is used in column F to find the fraction of titration, ϕ. From this value, we calculate the volume of titrant, V_b, in column G.

How do we know what pH values to put in? Trial-and-error allows us to find the starting pH, by putting in a pH and seeing whether V_b is positive or negative. In a few tries, it is easy to home in on the pH at which $V_b = 0$. In Figure 10-13, a pH of 3.90 is too low, because ϕ and V are both negative. Input values of pH are spaced as closely as you like, so that you can generate a smooth titration curve. To save space, we show only a few points in Figure 10-13, including the

2-(*N*-Morpholino)ethanesulfonic acid
MES, $pK_a = 6.27$

In Figure 10-13, you could use Excel Goal Seek (page 158) to vary pH in cell B5 until V_b in cell G5 is 0.

	A	B	C	D	E	F	G
1	Titration of weak acid with strong base						
2							
3	C_b =	pH	$[H^+]$	$[OH^-]$	$\alpha(A^-)$	ϕ	V_b (mL)
4	0.1	3.90	1.26E-04	7.94E-11	0.004	-0.002	-0.020
5	C_a =	3.99	1.02E-04	9.77E-11	0.005	0.000	0.001
6	0.02	5.00	1.00E-05	1.00E-09	0.051	0.050	0.505
7	V_a =	6.00	1.00E-06	1.00E-08	0.349	0.349	3.493
8	50	6.27	5.37E-07	1.86E-08	0.500	0.500	5.000
9	K_a =	7.00	1.00E-07	1.00E-07	0.843	0.843	8.430
10	5.37E-07	8.00	1.00E-08	1.00E-06	0.982	0.982	9.818
11	K_w =	9.00	1.00E-09	1.00E-05	0.998	0.999	9.987
12	1.E-14	9.25	5.62E-10	1.78E-05	0.999	1.000	10.000
13		10.00	1.00E-10	1.00E-04	1.000	1.006	10.058
14		11.00	1.00E-11	1.00E-03	1.000	1.061	10.606
15		12.00	1.00E-12	1.00E-02	1.000	1.667	16.667
16							
17	C4 = 10^-B4			F4 = (E4-(C4-D4)/\$A\$6)/(1+(C4-D4)/\$A\$4)			
18	D4 = \$A\$12/C4			G4 = F4*\$A\$6*\$A\$8/\$A\$4			
19	E4 = \$A\$10/(C4+\$A\$10)						

FIGURE 10-13 Spreadsheet uses Equation 10-13 to calculate the titration curve for 50 mL of the weak acid, 0.02 M MES ($pK_a = 6.27$) treated with 0.1 M NaOH. We provide pH as input in column B, and the spreadsheet tells us what volume of base is required to generate that pH.

TABLE 10-5   Titration equations for spreadsheets

CALCULATION OF ϕ

Titrating strong acid with strong base

$$\phi = \frac{C_b V_b}{C_a V_a} = \frac{1 - \dfrac{[H^+] - [OH^-]}{C_a}}{1 + \dfrac{[H^+] - [OH^-]}{C_b}}$$

Titrating strong base with strong acid

$$\phi = \frac{C_a V_a}{C_b V_b} = \frac{1 + \dfrac{[H^+] - [OH^-]}{C_b}}{1 - \dfrac{[H^+] - [OH^-]}{C_a}}$$

Titrating weak acid (HA) with weak base (B)

$$\phi = \frac{C_b V_b}{C_a V_a} = \frac{\alpha_{A^-} - \dfrac{[H^+] - [OH^-]}{C_a}}{\alpha_{BH^+} + \dfrac{[H^+] - [OH^-]}{C_b}}$$

Titrating H$_2$A with strong base ($\rightarrow \rightarrow A^{2-}$)

$$\phi = \frac{C_b V_b}{C_a V_a} = \frac{\alpha_{HA^-} + 2\alpha_{A^{2-}} - \dfrac{[H^+] - [OH^-]}{C_a}}{1 + \dfrac{[H^+] - [OH^-]}{C_b}}$$

Titrating dibasic B with strong acid ($\rightarrow \rightarrow BH_2^{2+}$)

$$\phi = \frac{C_a V_a}{C_b V_b} = \frac{\alpha_{BH^+} + 2\alpha_{BH_2^{2+}} + \dfrac{[H^+] - [OH^-]}{C_b}}{1 - \dfrac{[H^+] - [OH^-]}{C_a}}$$

Titrating weak acid (HA) with strong base

$$\phi = \frac{C_b V_b}{C_a V_a} = \frac{\alpha_{A^-} - \dfrac{[H^+] - [OH^-]}{C_a}}{1 + \dfrac{[H^+] - [OH^-]}{C_b}}$$

Titrating weak base (B) with strong acid

$$\phi = \frac{C_a V_a}{C_b V_b} = \frac{\alpha_{BH^+} + \dfrac{[H^+] - [OH^-]}{C_b}}{1 - \dfrac{[H^+] - [OH^-]}{C_a}}$$

Titrating weak base (B) with weak acid (HA)

$$\phi = \frac{C_a V_a}{C_b V_b} = \frac{\alpha_{BH^+} + \dfrac{[H^+] - [OH^-]}{C_b}}{\alpha_{A^-} - \dfrac{[H^+] - [OH^-]}{C_a}}$$

Titrating H$_3$A with strong base ($\rightarrow \rightarrow \rightarrow A^{3-}$)

$$\phi = \frac{C_b V_b}{C_a V_a} = \frac{\alpha_{H_2A^-} + 2\alpha_{HA^{2-}} + 3\alpha_{A^{3-}} - \dfrac{[H^+] - [OH^-]}{C_a}}{1 + \dfrac{[H^+] - [OH^-]}{C_b}}$$

Titrating tribasic B with strong acid ($\rightarrow \rightarrow \rightarrow BH_3^{3+}$)

$$\phi = \frac{C_a V_a}{C_b V_b} = \frac{\alpha_{BH^+} + 2\alpha_{BH_2^{2+}} + 3\alpha_{BH_3^{3+}} + \dfrac{[H^+] - [OH^-]}{C_b}}{1 - \dfrac{[H^+] - [OH^-]}{C_a}}$$

SYMBOLS

ϕ = fraction of the way to the first equivalence point
C_a = initial concentration of acid
C_b = initial concentration of base

α = fraction of dissociation of acid or fraction of association of base
V_a = volume of acid
V_b = volume of base

CALCULATION OF α

Monoprotic systems

$$\alpha_{HA} = \frac{[H^+]}{[H^+] + K_a} \qquad \alpha_{A^-} = \frac{K_a}{[H^+] + K_a}$$

$$\alpha_{BH^+} = \frac{[H^+]}{[H^+] + K_{BH^+}} \qquad \alpha_B = \frac{K_{BH^+}}{[H^+] + K_{BH^+}}$$

SYMBOLS
K_a = acid dissociation constant of HA
K_{BH^+} = acid dissociation constant of BH$^+$ ($= K_w/K_b$)

Diprotic systems

$$\alpha_{H_2A} = \alpha_{BH_2^{2+}} = \frac{[H^+]^2}{[H^+]^2 + [H^+]K_1 + K_1K_2} \qquad \alpha_{HA^-} = \alpha_{BH^+} = \frac{[H^+]K_1}{[H^+]^2 + [H^+]K_1 + K_1K_2} \qquad \alpha_{A^{2-}} = \alpha_B = \frac{K_1K_2}{[H^+]^2 + [H^+]K_1 + K_1K_2}$$

SYMBOLS
K_1 and K_2 for the acid are the acid dissociation constants of H$_2$A and HA$^-$, respectively.
K_1 and K_2 for the base are the acid dissociation constants of BH$_2^{2+}$ and BH$^+$, respectively: $K_1 = K_w/K_{b2}$; $K_2 = K_w/K_{b1}$.

Triprotic systems

$$\alpha_{H_3A} = \frac{[H^+]^3}{[H^+]^3 + [H^+]^2K_1 + [H^+]K_1K_2 + K_1K_2K_3} \qquad \alpha_{H_2A^-} = \frac{[H^+]^2K_1}{[H^+]^3 + [H^+]^2K_1 + [H^+]K_1K_2 + K_1K_2K_3}$$

$$\alpha_{HA^{2-}} = \frac{[H^+]K_1K_2}{[H^+]^3 + [H^+]^2K_1 + [H^+]K_1K_2 + K_1K_2K_3} \qquad \alpha_{A^{3-}} = \frac{K_1K_2K_3}{[H^+]^3 + [H^+]^2K_1 + [H^+]K_1K_2 + K_1K_2K_3}$$

midpoint (pH $6.27 \Rightarrow V_b = 5.00$ mL) and the end point (pH $9.25 \Rightarrow V_b = 10.00$ mL). This spreadsheet reproduces Table 10-2 without approximations other than neglecting activity coefficients. It gives correct results even when the approximations used in Table 10-2 fail.

Titrating a Weak Acid with a Weak Base

Now consider the titration of V_a mL of acid HA (initial concentration C_a) with V_b mL of base B whose concentration is C_b. Let the acid dissociation constant of HA be K_a and the acid dissociation constant of BH^+ be K_{BH^+}. The charge balance is

Charge balance: $\qquad [H^+] + [BH^+] = [A^-] + [OH^-]$

As before, we can say that $[A^-] = \alpha_{A^-} \cdot F_{HA}$, where $\alpha_{A^-} = K_a/([H^+] + K_a)$ and $F_{HA} = C_aV_a/(V_a + V_b)$.

We can write an analogous expression for $[BH^+]$, which is a monoprotic weak acid. If the acid were HA, we would use Equation 9-17 to say

$$[HA] = \alpha_{HA} \cdot F_{HA} \qquad \alpha_{HA} = \frac{[H^+]}{[H^+] + K_a}$$

where K_a applies to the acid HA. For the weak acid BH^+, we write

$$[BH^+] = \alpha_{BH^+} \cdot F_B \qquad \alpha_{BH^+} = \frac{[H^+]}{[H^+] + K_{BH^+}}$$

where the formal concentration of base is $F_B = C_bV_b/(V_a + V_b)$.

Substituting for $[BH^+]$ and $[A^-]$ in the charge balance gives

$$[H^+] + \frac{\alpha_{BH^+} \cdot C_bV_b}{V_a + V_b} = \frac{\alpha_{A^-} \cdot C_aV_a}{V_a + V_b} + [OH^-]$$

which can be rearranged to the useful result

Fraction of titration for *weak acid by weak base:* $\qquad \phi = \dfrac{C_bV_b}{C_aV_a} = \dfrac{\alpha_{A^-} - \dfrac{[H^+] - [OH^-]}{C_a}}{\alpha_{BH^+} + \dfrac{[H^+] - [OH^-]}{C_b}}$ $\qquad$ (10-14)

α_{HA} = fraction of acid in the form HA:

$$\alpha_{HA} = \frac{[HA]}{F_{HA}}$$

α_{BH^+} = fraction of base in the form BH^+:

$$\alpha_{BH^+} = \frac{[BH^+]}{F_B}$$

Equation 10-14 for a weak base looks just like Equation 10-13 for a strong base, except that α_{BH^+} replaces 1 in the denominator.

Table 10-5 gives useful equations derived by writing a charge balance and substituting fractional compositions for various concentrations. For titration of the diprotic acid, H_2A, ϕ is the fraction of the way to the first equivalence point. When $\phi = 2$, we are at the second equivalence point. It should not surprise you that, when $\phi = 0.5$, pH $\approx pK_1$, and when $\phi = 1.5$, pH $\approx pK_2$. When $\phi = 1$, we have the intermediate HA^- and pH $\approx \frac{1}{2}(pK_1 + pK_2)$.

Terms to Understand

Gran plot	indicator	Kjeldahl nitrogen analysis	transition range
Hammett acidity function	indicator error	leveling effect	

Summary

Key equations used to calculate titration curves:

Strong acid/strong base titration

$$H^+ + OH^- \rightarrow H_2O$$
pH is determined by the concentration of excess unreacted H^+ or OH^-

Weak acid titrated with OH^-

$$HA + OH^- \rightarrow A^- + H_2O \quad (V_e = \text{equivalence volume})$$
$$(V_b = \text{volume of added base})$$

$V_b = 0$: pH determined by K_a ($HA \overset{K_a}{\rightleftharpoons} H^+ + A^-$)
$0 < V_b < V_e$: pH = pK_a + log([A^-]/[HA])
pH = pK_a when $V_b = \frac{1}{2}V_e$ (neglecting activity)

At V_e: pH governed by K_b ($A^- + H_2O \overset{K_b}{\rightleftharpoons} HA + OH^-$)
After V_e: pH is determined by excess OH^-

Weak base titrated with H^+

$$B + H^+ \rightarrow BH^+ \quad (V_e = \text{equivalence volume})$$
$$(V_a = \text{volume of added acid})$$

$V_a = 0$: pH determined by K_b ($B + H_2O \overset{K_b}{\rightleftharpoons} BH^+ + OH^-$)
$0 < V_a < V_e$: pH = pK_{BH^+} + log([B]/[BH^+])
pH = pK_{BH^+} when $V_a = \frac{1}{2}V_e$

At V_e: pH governed by K_{BH^+} ($BH^+ \overset{K_{BH^+}}{\rightleftharpoons} B + H^+$)
After V_e: pH is determined by excess H^+

H_2A titrated with OH^-

$$H_2A \xrightarrow{OH^-} HA^- \xrightarrow{OH^-} A^{2-}$$
Equivalence volumes: $V_{e2} = 2V_{e1}$

$V_b = 0$: pH determined by K_1 ($H_2A \overset{K_1}{\rightleftharpoons} H^+ + HA^-$)

$0 < V_b < V_{e1}$: pH = pK_1 + log([HA^-]/[H_2A])
pH = pK_1 when $V_b = \frac{1}{2}V_{e1}$

At V_{e1}: $[H^+] = \sqrt{\dfrac{K_1 K_2 F' + K_1 K_w}{K_1 + F'}}$

$\Rightarrow$ pH $\approx \frac{1}{2}$(pK_1 + pK_2)
F' = formal concentration of HA^-
$V_{e1} < V_b < V_{e2}$: pH = pK_2 + log([A^{2-}]/[HA^-])
pH = pK_2 when $V_b = \frac{3}{2}V_{e1}$
At V_{e2}: pH governed by K_{b1}

$$(A^{2-} + H_2O \overset{K_{b1}}{\rightleftharpoons} HA^- + OH^-)$$
After V_{e2}: pH is determined by excess OH^-

Behavior of derivatives at the equivalence point
First derivative: ΔpH/ΔV has greatest magnitude
Second derivative: $\Delta(\Delta$pH/$\Delta V))/\Delta V = 0$

Gran plot
Plot $V_b \cdot 10^{-pH}$ versus V_b
x-intercept = V_e; slope = $-K_a \gamma_{HA}/\gamma_{A^-}$
K_a = acid dissociation constant
γ = activity coefficient

Choosing an indicator: Color transition range should match pH at V_e. Preferably the color change should occur entirely within the steep portion of the titration curve.

Kjeldahl nitrogen analysis: A nitrogen-containing organic compound is digested with a catalyst in boiling H_2SO_4. Nitrogen is converted into NH_4^+, which is converted into NH_3 with base and distilled into standard HCl. The excess, unreacted HCl tells us how much nitrogen was present in the original analyte.

Exercises

10-A. Calculate the pH at each of the following points in the titration of 50.00 mL of 0.010 0 M NaOH with 0.100 M HCl. Volume of acid added: 0.00, 1.00, 2.00, 3.00, 4.00, 4.50, 4.90, 4.99, 5.00, 5.01, 5.10, 5.50, 6.00, 8.00, and 10.00 mL. Make a graph of pH versus volume of HCl added.

10-B. Calculate the pH at each point listed for the titration of 50.0 mL of 0.050 0 M formic acid with 0.050 0 M KOH. The points to calculate are V_b = 0.0, 10.0, 20.0, 25.0, 30.0, 40.0, 45.0, 48.0, 49.0, 49.5, 50.0, 50.5, 51.0, 52.0, 55.0, and 60.0 mL. Draw a graph of pH versus V_b.

10-C. Calculate the pH at each point listed for the titration of 100.0 mL of 0.100 M cocaine (Section 8-4, $K_b = 2.6 \times 10^{-6}$) with 0.200 M HNO_3. The points to calculate are V_a = 0.0, 10.0, 20.0, 25.0, 30.0, 40.0, 49.0, 49.9, 50.0, 50.1, 51.0, and 60.0 mL. Draw a graph of pH versus V_a.

10-D. Consider the titration of 50.0 mL of 0.050 0 M malonic acid with 0.100 M NaOH. Calculate the pH at each point listed and sketch the titration curve: V_b = 0.0, 8.0, 12.5, 19.3, 25.0, 37.5, 50.0, and 56.3 mL.

10-E. Write the chemical reactions (including structures of reactants and products) that occur when the amino acid histidine is titrated with perchloric acid. (Histidine is a molecule with no net charge.) A solution containing 25.0 mL of 0.050 0 M histidine was titrated with

0.050 0 M $HClO_4$. Calculate the pH at the following values of V_a: 0, 4.0, 12.5, 25.0, 26.0, and 50.0 mL.

10-F. Select indicators from Table 10-3 that would be useful for the titrations in Figures 10-1 and 10-2 and the pK_a = 8 curve in Figure 10-3. Select a different indicator for each titration and state what color change you would use as the end point.

10-G. When 100.0 mL of a weak acid were titrated with 0.093 81 M NaOH, 27.63 mL were required to reach the equivalence point. The pH at the equivalence point was 10.99. What was the pH when only 19.47 mL of NaOH had been added?

10-H. A 0.100 M solution of the weak acid HA was titrated with 0.100 M NaOH. The pH measured when $V_b = \frac{1}{2}V_e$ was 4.62. Using activity coefficients, calculate pK_a. The size of the A^- anion is 450 pm.

10-I. [icon] *Finding the end point from pH measurements.* Here are data points around the second apparent end point in Figure 10-5:

V_b (μL)	pH	V_b (μL)	pH
107.0	6.921	117.0	7.878
110.0	7.117	118.0	8.090
113.0	7.359	119.0	8.343
114.0	7.457	120.0	8.591
115.0	7.569	121.0	8.794
116.0	7.705	122.0	8.952

(a) Prepare a spreadsheet or table analogous to Figure 10-6, showing the first and second derivatives. Plot both derivatives versus V_b and locate the end point in each plot.

(b) Prepare a Gran plot analogous to Figure 10-8. Use the least-squares procedure to find the best straight line and find the end point. You will have to use your judgment as to which points lie on the "straight" line.

10-J. *Indicator error.* Consider the titration in Figure 10-2 in which the equivalence-point pH in Table 10-2 is 9.25 at a volume of 10.00 mL.

(a) Suppose you used the yellow-to-blue transition of thymol blue indicator to find the end point. According to Table 10-3, the last trace of green disappears near pH 9.6. What volume of base is required to reach pH 9.6? The difference between this volume and 10 mL is the indicator error.

(b) If you used cresol red, with a color change at pH 8.8, what would be the indicator error?

10-K. *Spectrophotometry with indicators.*[†] Acid-base indicators are themselves acids or bases. Consider an indicator, HIn, which dissociates according to the equation

$$HIn \overset{K_a}{\rightleftharpoons} H^+ + In^-$$

The molar absorptivity, ε, is 2 080 $M^{-1} cm^{-1}$ for HIn and 14 200 $M^{-1} cm^{-1}$ for In$^-$, at a wavelength of 440 nm.

(a) Write an expression for the absorbance of a solution containing HIn at a concentration [HIn] and In$^-$ at a concentration [In$^-$] in a cell of pathlength 1.00 cm. The total absorbance is the sum of absorbances of each component.

(b) A solution containing indicator at a formal concentration of 1.84 $\times 10^{-4}$ M is adjusted to pH 6.23 and found to exhibit an absorbance of 0.868 at 440 nm. Calculate pK_a for this indicator.

[†]This problem is based on Beer's law in Section 17-2.

Problems

Titration of Strong Acid with Strong Base

10-1. Distinguish the terms *end point* and *equivalence point*.

10-2. Consider the titration of 100.0 mL of 0.100 M NaOH with 1.00 M HBr. Find the pH at the following volumes of acid added and make a graph of pH versus V_a: $V_a = 0, 1, 5, 9, 9.9, 10, 10.1,$ and 12 mL.

10-3. Why does an acid-base titration curve (pH versus volume of titrant) have an abrupt change at the equivalence point?

Titration of Weak Acid with Strong Base

10-4. Sketch the general appearance of the curve for the titration of a weak acid with a strong base. Explain (in words) what chemistry governs the pH in each of the four distinct regions of the curve.

10-5. Why is it not practical to titrate an acid or a base that is too weak or too dilute?

10-6. A weak acid HA ($pK_a = 5.00$) was titrated with 1.00 M KOH. The acid solution had a volume of 100.0 mL and a molarity of 0.100 M. Find the pH at the following volumes of base added and make a graph of pH versus V_b: $V_b = 0, 1, 5, 9, 9.9, 10, 10.1,$ and 12 mL.

10-7. Consider the titration of the weak acid HA with NaOH. At what fraction of V_e does pH = pK_a − 1? At what fraction of V_e does pH = pK_a + 1? Use these two points, plus $V_b = 0, \frac{1}{2}V_e, V_e,$ and $1.2V_e$ to sketch the titration curve for the reaction of 100 mL of 0.100 M anilinium bromide ("aminobenzene · HBr") with 0.100 M NaOH.

10-8. What is the pH at the equivalence point when 0.100 M hydroxyacetic acid is titrated with 0.050 0 M KOH?

10-9. Find the equilibrium constant for the reaction of MES (Table 8-2) with NaOH.

10-10. When 22.63 mL of aqueous NaOH were added to 1.214 g of cyclohexylaminoethanesulfonic acid (FM 207.29, structure in Table 8-2) dissolved in 41.37 mL of water, the pH was 9.24. Calculate the molarity of the NaOH.

10-11. *Use activity coefficients* to calculate the pH after 10.0 mL of 0.100 M trimethylammonium bromide were titrated with 4.0 mL of 0.100 M NaOH.

Titration of Weak Base with Strong Acid

10-12. Sketch the general appearance of the curve for the titration of a weak base with a strong acid. Explain (in words) what chemistry governs the pH in each of the four distinct regions of the curve.

10-13. Why is the equivalence-point pH necessarily below 7 when a weak base is titrated with strong acid?

10-14. A 100.0-mL aliquot of 0.100 M weak base B ($pK_b = 5.00$) was titrated with 1.00 M HClO$_4$. Find the pH at the following volumes of acid added and make a graph of pH versus V_a: $V_a = 0, 1, 5, 9, 9.9, 10, 10.1,$ and 12 mL.

10-15. At what point in the titration of a weak base with a strong acid is the maximum buffer capacity reached? This is the point at which a given small addition of acid causes the least pH change.

10-16. What is the equilibrium constant for the reaction between benzylamine and HCl?

10-17. A 50.0-mL solution of 0.031 9 M benzylamine was titrated with 0.050 0 M HCl. Calculate the pH at the following volumes of added acid: $V_a = 0, 12.0, \frac{1}{2}V_e, 30.0, V_e,$ and 35.0 mL.

10-18. Calculate the pH of a solution made by mixing 50.00 mL of 0.100 M NaCN with

(a) 4.20 mL of 0.438 M HClO$_4$

(b) 11.82 mL of 0.438 M HClO$_4$

(c) What is the pH at the equivalence point with 0.438 M HClO$_4$?

Titrations in Diprotic Systems

10-19. Sketch the general appearance of the curve for the titration of a weak diprotic acid with NaOH. Explain (in words) what chemistry governs the pH in each distinct region of the curve.

10-20. The opening page of this chapter shows the titration curve for an enzyme. Is the average charge of the protein positive, negative, or neutral at its isoionic point? How do you know?

10-21. The base Na^+A^-, whose anion is dibasic, was titrated with HCl to give curve b in Figure 10-4. Is the first equivalence point (H) the isoelectric point or the isoionic point?

10-22. The figure compares the titration of a monoprotic weak acid with a monoprotic weak base and the titration of a diprotic acid with strong base.

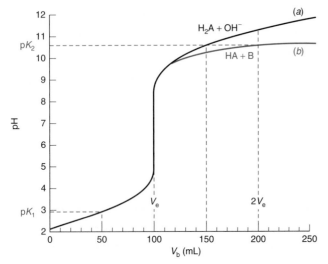

(a) Titration of 100 mL of 0.050 M H_2A ($pK_1 = 2.86$, $pK_2 = 10.64$) with 0.050 M NaOH. (b) Titration of 100 mL of the weak acid HA (0.050 M, $pK_a = 2.86$) with the weak base B (0.050 M, $pK_b = 3.36$).

(a) Write the reaction between the weak acid and the weak base and show that the equilibrium constant is $10^{7.78}$. This large value means that the reaction goes "to completion" after each addition of reagent.

(b) Why does pK_2 intersect the upper curve at $\frac{3}{2}V_e$ and the lower curve at $2V_e$? On the lower curve, "pK_2" is pK_a for the acid, BH^+.

10-23. The dibasic compound B ($pK_{b1} = 4.00$, $pK_{b2} = 8.00$) was titrated with 1.00 M HCl. The initial solution of B was 0.100 M and had a volume of 100.0 mL. Find the pH at the following volumes of acid added and make a graph of pH versus V_a: $V_a = 0, 1, 5, 9, 10, 11, 15, 19, 20,$ and 22 mL.

10-24. A 100.0-mL aliquot of 0.100 M diprotic acid H_2A ($pK_1 = 4.00$, $pK_2 = 8.00$) was titrated with 1.00 M NaOH. Find the pH at the following volumes of base added and make a graph of pH versus V_b: $V_b = 0, 1, 5, 9, 10, 11, 15, 19, 20,$ and 22 mL.

10-25. Calculate the pH at 10.0-mL intervals (from 0 to 100 mL) in the titration of 40.0 mL of 0.100 M piperazine with 0.100 M HCl. Make a graph of pH versus V_a.

10-26. Calculate the pH when 25.0 mL of 0.020 0 M 2-aminophenol have been titrated with 10.9 mL of 0.015 0 M $HClO_4$.

10-27. Consider the titration of 50.0 mL of 0.100 M sodium glycinate, $H_2NCH_2CO_2Na$, with 0.100 M HCl.

(a) Calculate the pH at the second equivalence point.

(b) Show that our approximate method of calculations gives incorrect (physically unreasonable) values of pH at $V_a = 90.0$ and $V_a = 101.0$ mL.

10-28. A solution containing 0.100 M glutamic acid (the molecule with no net charge) was titrated to its first equivalence point with 0.025 0 M RbOH.

(a) Draw the structures of reactants and products.

(b) Calculate the pH at the first equivalence point.

10-29. Find the pH of the solution when 0.010 0 M tyrosine is titrated to the equivalence point with 0.004 00 M $HClO_4$.

10-30. This problem deals with the amino acid cysteine, which we will abbreviate H_2C.

(a) A 0.030 0 M solution was prepared by dissolving dipotassium cysteine, K_2C, in water. Then 40.0 mL of this solution were titrated with 0.060 0 M $HClO_4$. Calculate the pH at the first equivalence point.

(b) Calculate the quotient $[C^{2-}]/[HC^-]$ in a solution of 0.050 0 M cysteinium bromide (the salt $H_3C^+Br^-$).

10-31. How many grams of dipotassium oxalate (FM 166.22) should be added to 20.0 mL of 0.800 M $HClO_4$ to give a pH of 4.40 when the solution is diluted to 500 mL?

10-32. When 5.00 mL of 0.103 2 M NaOH were added to 0.112 3 g of alanine (FM 89.093) in 100.0 mL of 0.10 M KNO_3, the measured pH was 9.57. *Use activity coefficients* to find pK_2 for alanine. Consider the ionic strength of the solution to be 0.10 M and consider each ionic form of alanine to have an activity coefficient of 0.77.

Finding the End Point with a pH Electrode

10-33. What is a Gran plot used for?

10-34. Data for the titration of 100.00 mL of a weak acid by NaOH are given below. Find the end point by preparing a Gran plot, using the last 10% of the volume prior to V_e.

mL NaOH	pH	mL NaOH	pH	mL NaOH	pH
0.00	4.14	20.75	6.09	22.70	6.70
1.31	4.30	21.01	6.14	22.76	6.74
2.34	4.44	21.10	6.15	22.80	6.78
3.91	4.61	21.13	6.16	22.85	6.82
5.93	4.79	21.20	6.17	22.91	6.86
7.90	4.95	21.30	6.19	22.97	6.92
11.35	5.19	21.41	6.22	23.01	6.98
13.46	5.35	21.51	6.25	23.11	7.11
15.50	5.50	21.61	6.27	23.17	7.20
16.92	5.63	21.77	6.32	23.21	7.30
18.00	5.71	21.93	6.37	23.30	7.49
18.35	5.77	22.10	6.42	23.32	7.74
18.95	5.82	22.27	6.48	23.40	8.30
19.43	5.89	22.37	6.53	23.46	9.21
19.93	5.95	22.48	6.58	23.55	9.86
20.48	6.04	22.57	6.63		

10-35. Prepare a second derivative graph to find the end point from the following titration data.

mL NaOH	pH	mL NaOH	pH	mL NaOH	pH
10.679	7.643	10.725	6.222	10.750	4.444
10.696	7.447	10.729	5.402	10.765	4.227
10.713	7.091	10.733	4.993		
10.721	6.700	10.738	4.761		

Finding the End Point with Indicators

10-36. Explain the origin of the rule of thumb that indicator color changes occur at $pK_{HIn} \pm 1$.

10-37. Why does a properly chosen indicator change color near the equivalence point in a titration?

10-38. The pH of microscopic vesicles (compartments) in living cells can be estimated by infusing an indicator (HIn) into the compartment and measuring the quotient $[In^-]/[HIn]$ from the spectrum of the indicator inside the vesicle. Explain how this tells us the pH.

10-39. Write the formula of a compound with a negative pK_a.

10-40. Consider the titration in Figure 10-2, for which the pH at the equivalence point is calculated to be 9.25. If thymol blue is used as an indicator, what color will be observed through most of the titration prior to the equivalence point? At the equivalence point? After the equivalence point?

10-41. What color do you expect to observe for cresol purple indicator (Table 10-3) at the following pH values? **(a)** 1.0; **(b)** 2.0; **(c)** 3.0

10-42. Cresol red has *two* transition ranges listed in Table 10-3. What color would you expect it to be at the following pH values? **(a)** 0; **(b)** 1; **(c)** 6; **(d)** 9

10-43. Would the indicator bromocresol green, with a transition range of pH 3.8–5.4, ever be useful in the titration of a weak acid with a strong base?

10-44. (a) What is the pH at the equivalence point when 0.030 0 M NaF is titrated with 0.060 0 M $HClO_4$?

(b) Why would an indicator end point probably not be useful in this titration?

10-45. A titration curve for $NaCO_3$ titrated with HCl is shown here. Suppose that *both* phenolphthalein and bromocresol green are present in the titration solution. State what colors you expect to observe at the following volumes of added HCl: **(a)** 2 mL; **(b)** 10 mL; **(c)** 19 mL.

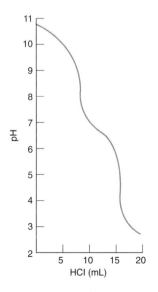

10-46. In the Kjeldahl nitrogen determination, the final product is a solution of NH_4^+ in HCl solution. It is necessary to titrate the HCl without titrating NH_4^+.

(a) Calculate the pH of pure 0.010 M NH_4Cl.

(b) Select an indicator that would allow you to titrate HCl but not NH_4^+.

10-47. A 10.231-g sample of window cleaner containing ammonia was diluted with 39.466 g of water. Then 4.373 g of solution were titrated with 14.22 mL of 0.106 3 M HCl to reach a bromocresol green end point. Find the weight percent of NH_3 (FM 17.031) in the cleaner.

10-48. A procedure to measure the alkalinity (Box 10-1) of home swimming pool water is to titrate a fixed volume of pool water by counting the number of drops of standard H_2SO_4 to reach the bromocresol green end point.[30] Explain what is measured in this titration and why bromocresol green was chosen.

Practical Notes, Kjeldahl Analysis, and Leveling Effect

10-49. Give the name and formula of a primary standard used to standardize **(a)** HCl and **(b)** NaOH.

10-50. Why is it more accurate to use a primary standard with a high equivalent mass (the mass required to provide or consume 1 mol of H^+) than one with a low equivalent mass?

10-51. Explain how to use potassium hydrogen phthalate to standardize a solution of NaOH.

10-52. A solution was prepared from 1.023 g of the primary standard tris (Table 10-4) plus 99.367 g of water; 4.963 g of the solution were titrated with 5.262 g of aqueous HNO_3 to reach the methyl red end point. Calculate the concentration of the HNO_3 (expressed as mol HNO_3/kg solution).

10-53. The balance says that you have weighed out 1.023 g of tris to standardize a solution of HCl. Use the buoyancy correction in Section 2-3 and the density in Table 10-4 to determine how many grams you have really weighed out. The volume of HCl required to react with the tris was 28.37 mL. Does the buoyancy correction introduce a random or a systematic error into the calculated molarity of HCl? What is the magnitude of the error expressed as a percentage? Is the calculated molarity of HCl higher or lower than the true molarity?

10-54. A solution was prepared by dissolving 0.194 7 g of HgO (Table 10-4) in 20 mL of water containing 4 g of KBr. Titration with HCl required 17.98 mL to reach a phenolphthalein end point. Calculate the molarity of the HCl.

10-55. How many grams of potassium hydrogen phthalate should be weighed into a flask to standardize ~0.05 M NaOH if you wish to use ~30 mL of base for the titration?

10-56. Constant-boiling aqueous HCl can be used as a primary standard for acid-base titrations. When ~20 wt% HCl (FM 36.461) is distilled, the composition of the distillate varies in a regular manner with the barometric pressure:

P (Torr)	HCl^a (g/100 g solution)
770	20.196
760	20.220
750	20.244
740	20.268
730	20.292

*a. The composition of distillate is from C. W. Foulk and M. Hollingsworth, J. Am. Chem. Soc. **1923**, 45, 1223, with numbers corrected for the current values of atomic masses.*

(a) Make a graph of the data in the table to find the weight percent of HCl collected at 746 Torr.

(b) What mass of distillate (weighed in air, using weights whose density is 8.0 g/mL) should be dissolved in 1.000 0 L to give 0.100 00 M HCl? The density of distillate over the whole range in the table is close to 1.096 g/mL. You will need this density to change the mass measured in vacuum to mass measured in air. See Section 2-3 for buoyancy corrections.

10-57. (a) *Uncertainty in formula mass.* In an extremely-high-precision gravimetric titration, the uncertainty in formula mass of the primary standard could contribute to the uncertainty of the result. Review Section 3-5 and express the formula mass of potassium hydrogen phthalate, $C_8H_5O_4K$, with its proper uncertainty based on a rectangular distribution of uncertainty of atomic mass.

(b) *Systematic uncertainty in reagent purity.* The manufacturer of potassium hydrogen phthalate states that the purity is $1.000\ 00 \pm 0.000\ 05$. In the absence of further information, we assume that the distribution of uncertainty is rectangular. What standard uncertainty would you use for the purity of this reagent?

10-58. The Kjeldahl procedure was used to analyze 256 μL of a solution containing 37.9 mg protein/mL. The liberated NH_3 was collected in 5.00 mL of 0.033 6 M HCl, and the remaining acid required 6.34 mL of 0.010 M NaOH for complete titration. What is the weight percent of nitrogen in the protein?

10-59. What is meant by the leveling effect?

10-60. Considering the following pK_a values,[31] explain why dilute sodium methoxide ($NaOCH_3$) and sodium ethoxide ($NaOCH_2CH_3$) are leveled to the same base strength in aqueous solution. Write the chemical reactions that occur when these bases are added to water.

CH_3OH	$pK_a = 15.54$
CH_3CH_2OH	$pK_a = 16.0$
HOH	$pK_a = 15.74$ (for $K_a = [H^+][OH^-]/[H_2O]$)

10-61. The base B is too weak to titrate in aqueous solution.

(a) Which solvent, pyridine or acetic acid, would be more suitable for the titration of B with $HClO_4$? Why?

(b) Which solvent would be more suitable for the titration of a very weak acid with tetrabutylammonium hydroxide? Why?

10-62. Explain why sodium amide ($NaNH_2$) and phenyl lithium (C_6H_5Li) are leveled to the same base strength in aqueous solution. Write the chemical reactions that occur when these reagents are added to water.

10-63. Pyridine is half protonated in aqueous phosphate buffer at pH 5.2. If you mix 45 mL of phosphate buffer with 55 mL of methanol, the buffer must have a pH of 3.2 to half protonate pyridine. Suggest a reason why.

Calculating Titration Curves with Spreadsheets

10-64. Derive the following equation for the titration of potassium hydrogen phthalate (K^+HP^-) with NaOH:

$$\phi = \frac{C_bV_b}{C_aV_a} = \frac{\alpha_{HP^-} + 2\alpha_{P^{2-}} - 1 - \dfrac{[H^+] - [OH^-]}{C_a}}{1 + \dfrac{[H^+] - [OH^-]}{C_b}}$$

10-65. [icon] *Effect of pK_a in the titration of weak acid with strong base.* Use Equation 10-13 to compute and plot the family of curves at the left side of Figure 10-3. For a strong acid, choose a large K_a, such as $K_a = 10^2$ or $pK_a = -2$.

10-66. [icon] *Effect of concentration in the titration of weak acid with strong base.* Use your spreadsheet from Problem 10-65 to prepare a family of titration curves for $pK_a = 6$, with the following combinations of concentrations: (a) $C_a = 20$ mM, $C_b = 100$ mM; (b) $C_a = 2$ mM, $C_b = 10$ mM; (c) $C_a = 0.2$ mM, $C_b = 1$ mM.

10-67. [icon] *Effect of pK_b in the titration of weak base with strong acid.* Using the appropriate equation in Table 10-5, compute and plot a family of curves analogous to the left part of Figure 10-3 for the titration of 50.0 mL of 0.020 0 M B ($pK_b = -2.00$, 2.00, 4.00, 6.00, 8.00, and 10.00) with 0.100 M HCl. (The value $pK_b = -2.00$ represents a strong base.) In the expression for α_{BH^+}, $K_{BH^+} = K_w/K_b$.

10-68. [icon] *Titrating weak acid with weak base.*

(a) Prepare a family of graphs for the titration of 50.0 mL of 0.020 0 M HA ($pK_a = 4.00$) with 0.100 M B ($pK_b = 3.00$, 6.00, and 9.00).

(b) Write the acid-base reaction that occurs when acetic acid and sodium benzoate (the salt of benzoic acid) are mixed, and find the equilibrium constant for the reaction. Find the pH of a solution prepared by mixing 212 mL of 0.200 M acetic acid with 325 mL of 0.050 0 M sodium benzoate.

10-69. [icon] *Titrating diprotic acid with strong base.* Prepare a family of graphs for the titration of 50.0 mL of 0.020 0 M H_2A with 0.100 M NaOH. Consider the following cases: (a) $pK_1 = 4.00$, $pK_2 = 8.00$; (b) $pK_1 = 4.00$, $pK_2 = 6.00$; (c) $pK_1 = 4.00$, $pK_2 = 5.00$.

10-70. [icon] *Titrating nicotine with strong acid.* Prepare a spreadsheet to reproduce the lower curve in Figure 10-4.

10-71. [icon] *Titrating triprotic acid with strong base.* Prepare a spreadsheet to graph the titration of 50.0 mL of 0.020 0 M histidine · 2HCl with 0.100 M NaOH. Treat histidine · 2HCl with the triprotic acid equation in Table 10-5.

10-72. [icon] *A tetraprotic system.* Write an equation for the titration of tetrabasic base with strong acid (B + $H^+ \rightarrow \rightarrow \rightarrow \rightarrow BH_4^{4+}$). You can do this by inspection of Table 10-5 or you can derive it from the charge balance for the titration reaction. Graph the titration of 50.0 mL of 0.020 0 M sodium pyrophosphate ($Na_4P_2O_7$) with 0.100 M $HClO_4$. Pyrophosphate is the anion of pyrophosphoric acid.

Using Beer's Law with Indicators[†]

10-73. Spectrophotometric properties of a particular indicator are given below:

$$HIn \underset{}{\overset{pK_a = 7.95}{\rightleftharpoons}} In^- + H^+$$

$\lambda_{max} = 395$ nm $\lambda_{max} = 604$ nm
$\varepsilon_{395} = 1.80 \times 10^4\ M^{-1}\ cm^{-1}$ $\varepsilon_{604} = 4.97 \times 10^4\ M^{-1}\ cm^{-1}$
$\varepsilon_{604} = 0$

A solution with a volume of 20.0 mL containing 1.40×10^{-5} M indicator plus 0.050 0 M benzene-1,2,3-tricarboxylic acid was treated with 20.0 mL of aqueous KOH. The resulting solution had an absorbance at 604 nm of 0.118 in a 1.00-cm cell. Calculate the molarity of the KOH solution.

[†]These problems are based on Beer's law in Section 17-2.

10-74. A certain acid-base indicator exists in three colored forms:

$$H_2In \overset{pK_1=1.00}{\rightleftharpoons} HIn^- \overset{pK_2=7.95}{\rightleftharpoons} In^{2-}$$

$\lambda_{max} = 520$ nm $\qquad$ $\lambda_{max} = 435$ nm $\qquad$ $\lambda_{max} = 572$ nm
$\varepsilon_{520} = 5.00 \times 10^4$ $\qquad$ $\varepsilon_{435} = 1.80 \times 10^4$ $\qquad$ $\varepsilon_{572} = 4.97 \times 10^4$
Red $\qquad\qquad$ Yellow $\qquad\qquad$ Red

$\varepsilon_{435} = 1.67 \times 10^4$ $\qquad$ $\varepsilon_{520} = 2.13 \times 10^3$ $\qquad$ $\varepsilon_{520} = 2.50 \times 10^4$

$\varepsilon_{572} = 2.03 \times 10^4$ $\qquad$ $\varepsilon_{572} = 2.00 \times 10^2$ $\qquad$ $\varepsilon_{435} = 1.15 \times 10^4$

The units of molar absorptivity, ε, are M^{-1} cm^{-1}. A solution containing 10.0 mL of 5.00×10^{-4} M indicator was mixed with 90.0 mL of 0.1 M phosphate buffer (pH 7.50). Calculate the absorbance of this solution at 435 nm in a 1.00-cm cell.

Reference Procedure: Preparing Standard Acid and Base

Standard 0.1 M NaOH

1. Prepare 50 wt% aqueous NaOH solution in advance and allow the Na_2CO_3 precipitate to settle overnight. (Na_2CO_3 is insoluble in this solution.) Store the solution in a tightly sealed polyethylene bottle and avoid disturbing the precipitate when supernate is taken. The density is close to 1.50 g of solution per milliliter.

2. Dry primary-standard-grade potassium hydrogen phthalate for 1 h at 110°C and store it in a desiccator.

Potassium hydrogen
phthalate FM 204.221

3. Boil 1 L of water for 5 min to expel CO_2. Pour the water into a polyethylene bottle, which should be tightly capped whenever possible. Calculate the volume of 50 wt% NaOH needed (~5.3 mL) to produce 1 L of ~0.1 M NaOH. Use a graduated cylinder to transfer this much NaOH to the bottle of water. Mix well and allow the solution to cool to room temperature (preferably overnight).

4. Weigh out four ~0.51-g portions of potassium hydrogen phthalate and dissolve each in ~25 mL of distilled water in a 125-mL flask. Each sample should require ~25 mL of 0.1 M NaOH. Add 3 drops of phenolphthalein indicator (Table 10-3) to each, and titrate one of them rapidly to find the approximate end point. The buret should have a loosely fitted cap to minimize entry of CO_2.

5. Calculate the volume of NaOH required for each of the other three samples and titrate them carefully. During each titration, periodically tilt and rotate the flask to wash liquid from the walls into the solution. When very near the end, deliver less than 1 drop of titrant at a time. To do this, carefully suspend a fraction of a drop from the buret tip, touch it to the inside wall of the flask, wash it into the bulk solution by careful tilting, and swirl the solution. The end point is the first appearance of pink color that persists for 15 s. The color will slowly fade as CO_2 from air dissolves in the solution.

6. Calculate the average molarity ($\bar{x}$), the standard deviation (s), and the relative standard deviation ($s/\bar{x}$). If you have used some care, the relative standard deviation should be <0.2%.

Standard 0.1 M HCl

1. The inside cover of this book tells us that 8.2 mL of ~37 wt% HCl should be added to 1 L of water to produce ~0.1 M HCl. Prepare this solution in a capped polyethylene bottle, using a graduated cylinder to deliver the HCl.

2. Dry primary-standard-grade Na_2CO_3 for 1 h at 110°C and cool it in a desiccator.

3. Weigh four samples containing enough Na_2CO_3 to react with ~25 mL of 0.1 M HCl from a buret and place each in a 125-mL flask. When ready to titrate each one, dissolve it in ~25 mL of distilled water.

$$2HCl + Na_2CO_3 \rightarrow CO_2 + 2NaCl + H_2O$$

FM 105.988

Add 3 drops of bromocresol green indicator (Table 10-3) and titrate one sample rapidly to a green color to find the approximate end point.

4. Carefully titrate each of the other samples until it just turns from blue to green. Then boil the solution to expel CO_2. The solution should return to a blue color. Carefully add HCl from the buret until the solution turns green again.

5. Titrate one blank prepared from 3 drops of indicator plus 50 mL of 0.05 M NaCl. Subtract the blank volume of HCl from that required to titrate Na_2CO_3.

6. Calculate the mean HCl molarity, standard deviation, and relative standard deviation.

11 | EDTA Titrations

Protein ion pore
spans cell membrane

Side view of
Gramicidin A dimer

View down
ion channel

Left and center: Two gramicidin A molecules associate to span a cell membrane. *Right:* Axial view showing ion channel.
[Structure from B. Roux, "Computational Studies of the Gramicidin Channel," *Acc. Chem. Res.* 2002, *35*, 366, based on solid-state nuclear magnetic resonance. Schematic at left from L. Stryer, *Biochemistry*, 4th ed. (New York: W. H. Freeman and Company, 1995).]

Amide
group

Gramicidin A is an antibiotic that kills cells by making their membranes permeable to Na^+ and K^+. Gramicidin A is made of 15 amino acids wound into a helix with a 0.4-nm-diameter channel through the middle. The channel is lined by polar amide groups, and the outside of gramicidin is covered by nonpolar hydrocarbon substituents (Table 9-1). *Polar* groups have positive and negative regions that attract neighboring molecules by electrostatic forces. *Nonpolar* groups have little charge separation and are soluble inside the nonpolar cell membrane.

Metal cations dissolve in water and are said to be *hydrophilic* ("water loving"). Cell membranes exclude water and are described as *hydrophobic* ("water hating"). Gramicidin A lodges in the cell membrane because the outside of the molecule is hydrophobic. Na^+ and K^+ pass through each hydrophilic pore at a rate of 10^7 ions/s. The pore is selective for monovalent cations; it excludes anions and more highly charged cations.

Part of the Nobel Prize in Chemistry in 2003 was awarded to Roderick MacKinnon for elucidating the structure of potassium channels that selectively permit K^+ to pass through membranes of cells such as nerves.[1] Unlike gramicidin A channels, the potassium channels are selective for K^+ over Na^+. Amide oxygen atoms of the protein backbone in the channel are spaced just right to replace waters of hydration from $K(H_2O)_6^+$. There is little change in energy when hydrated K^+ sheds H_2O and binds inside the channel. The spacing of amide oxygens is too great by 0.04 nm to displace H_2O from $Na(H_2O)_6^+$. Hydrated Na^+ remains outside the channel, whereas hydrated K^+ sheds H_2O and binds inside the channel. K^+ passes at a rate of 10^8 ions/s per channel—100 times faster than Na^+.

EDTA is a merciful abbreviation for *ethylenediaminetetraacetic acid,* a compound that forms strong 1:1 complexes with most metal ions (Figure 11-1) and finds wide use in quantitative analysis. EDTA plays a larger role as a strong metal-binding agent in industrial processes and in products such as detergents, cleaning agents, and food additives that prevent metal-catalyzed oxidation of food. EDTA is an emerging player in environmental chemistry.[2] For example, the majority of nickel discharged into San Francisco Bay and a significant fraction of the iron, lead, copper, and zinc are EDTA complexes that pass unscathed through wastewater treatment plants.

Dissolved nickel in the south San Francisco Bay reaches 110 nM in the summer.

FIGURE 11-1 EDTA forms strong 1:1 complexes with most metal ions, binding through four oxygen and two nitrogen atoms. The six-coordinate geometry of Mn^{2+}-EDTA found in the compound $KMnEDTA \cdot 2H_2O$ was deduced from X-ray crystallography. [From J. Stein, J. P. Fackler, Jr., G. J. McClune, J. A. Fee, and L. T. Chan, "Reactions of Mn-EDTA and MnCyDTA Complexes with O_2^-: X-Ray Structure of KMnEDTA $\cdot$ $2H_2O$," *Inorg. Chem.* **1979**, *18*, 3511.]

11-1 Metal-Chelate Complexes

Metal ions are **Lewis acids,** accepting electron pairs from electron-donating ligands that are **Lewis bases.** Cyanide (CN^-) is called a **monodentate** ligand because it binds to a metal ion through only one atom (the carbon atom). Most transition metal ions bind six ligand atoms. A ligand that attaches to a metal ion through more than one ligand atom is said to be **multidentate** ("many toothed"), or a **chelating ligand** (pronounced KEE-late-ing).[3]

A simple chelating ligand is 1,2-diaminoethane $H_2NCH_2CH_2NH_2$, also called ethylenediamine), whose binding to a metal ion is shown in the margin. We say that ethylenediamine is *bidentate* because it binds to the metal through two ligand atoms.

The **chelate effect** is the ability of multidentate ligands to form more stable metal complexes than those formed by similar monodentate ligands.[4,5] For example, the reaction of $Cd(H_2O)_6^{2+}$ with two molecules of ethylenediamine is more favorable than its reaction with four molecules of methylamine:

$$K \equiv \beta_2 = 8 \times 10^9 \qquad (11\text{-}1)$$

$$K \equiv \beta_4 = 4 \times 10^6 \qquad (11\text{-}2)$$

At pH 12 in the presence of 2 M ethylenediamine and 4 M methylamine, the quotient $[Cd(ethylenediamine)_2^{2+}]/[Cd(methylamine)_4^{2+}]$ is 30.

An important *tetradentate* ligand is adenosine triphosphate (ATP), which binds to divalent metal ions (such as Mg^{2+}, Mn^{2+}, Co^{2+}, and Ni^{2+}) through four of their six coordination positions (Figure 11-2). The fifth and sixth positions are occupied by water molecules. The biologically active form of ATP is generally the Mg^{2+} complex.

Metal-chelate complexes are ubiquitous in biology. Bacteria such as *Escherichia coli* and *Salmonella enterica* in your gut excrete a powerful iron chelator called enterobactin (Figure 11-3) to scavenge iron that is essential for bacterial growth. The iron-enterobactin complex is recognized at specific sites on the bacterial cell surface and taken into the cell. Iron is then released inside the bacterium by enzymatic disassembly of the chelator. To fight bacterial infection, your immune system produces a protein called siderocalin to sequester and inactivate enterobactin.[9] An important medical application of chelates is described in Box 11-1.

$Ag^+ \;+\; 2\,:\!\bar{C}\!\!\equiv\!\!N\!: \;\rightleftharpoons$
Lewis acid / Lewis base
(electron pair acceptor) / (electron pair donor)

$(:\!N\!\!\equiv\!\!C\!-\!Ag\!-\!C\!\!\equiv\!\!N\!:)^-$
Complex ion

Ethylenediamine

Bidentate coordination

Nomenclature for formation constants (K and β) was discussed in Box 6-2.

We have drawn *trans* isomers of the octahedral complexes (with H_2O ligands at opposite poles), but *cis* isomers are also possible (with H_2O ligands adjacent to each other).

BOX 11-1 Chelation Therapy and Thalassemia

Oxygen is carried in blood by the iron-containing protein hemoglobin, which consists of two pairs of subunits, designated α and β. β-Thalassemia major is a genetic disease in which the β subunits of hemoglobin are not synthesized in adequate quantities. Children afflicted with this disease survive only with frequent transfusions of normal red blood cells. The problem with this treatment is that patients accumulate 4–8 g of iron per year from the hemoglobin in the transfused cells. The body has no mechanism for excreting large quantities of iron, and patients eventually die from the toxic effects of iron overload.

To enhance iron excretion, intensive chelation therapy is used. The most successful drug is *desferrioxamine B*, a powerful Fe^{3+} chelator produced by the microbe *Streptomyces pilosus*.[6] The formation constant for the Fe(III) complex, called ferrioxamine B, is $10^{30.6}$. Used in conjunction with ascorbic acid—vitamin C, a reducing agent that reduces Fe^{3+} to the more soluble Fe^{2+}—desferrioxamine clears several grams of iron per year from an overloaded patient. The ferrioxamine complex is excreted in the urine.

Desferrioxamine reduces the incidence of heart and liver disease in thalassemia patients. In patients for whom desferrioxamine effectively controls iron overload, there is a 91% rate of cardiac disease-free survival after 15 years of therapy.[7] There are negative effects of desferrioxamine treatment; for example, too high a dose stunts a child's growth.

Desferrioxamine is expensive and must be taken by overnight subcutaneous infusion five to seven nights per week. It is not absorbed through the intestine. Many potent iron chelators have been tested to find an effective one that can be taken orally, but few have entered clinical practice.[8] The orally administered chelator deferiprone, introduced in 1987, is used with positive effect in more than 50 countries (but is not licensed in the United States and Canada). Combined use of desferrioxamine and deferiprone increases survival and reduces the incidence of cardiac disease, in comparison with desferrioxamine treatment only. An orally administered chelator called deferasirox was approved for use in the United States in 2005. The continued search for new chelators indicates that no current treatment is fully effective. In the long term, bone marrow transplants or gene therapy might cure the disease.

Structure of the iron complex ferrioxamine B and crystal structure of a related compound, ferrioxamine E, in which the chelate has a cyclic structure. Graph shows success of transfusions and transfusions plus chelation therapy. [Crystal structure kindly provided by M. Neu, Los Alamos National Laboratory, based on D. Van der Helm and M. Poling, *J. Am. Chem. Soc.* **1976**, *98*, 82. Graph from P. S. Dobbin and R. C. Hider, "Iron Chelation Therapy," *Chem. Br.* **1990**, *26*, 565.]

FIGURE 11-2 (*a*) Structure of adenosine triphosphate (ATP), with ligand atoms shown in color. (*b*) Possible structure of a metal-ATP complex; the metal, M, has four bonds to ATP and two bonds to H_2O ligands.

(*a*)   (*b*)

Aminocarboxylic acids in Figure 11-4 are synthetic chelating agents. Amine N atoms and carboxylate O atoms are potential ligand atoms in these molecules (Figures 11-5 and 11-6). When these molecules bind to a metal ion, the ligand atoms lose their protons. A medical application of the ligand DTPA in Figure 11-4 is illustrated by the tightly bound complex Gd^{3+}-DTPA, which is injected into humans at a concentration of ~ 0.5 mM to provide contrast in magnetic resonance imaging.[10]

A titration based on complex formation is called a **complexometric titration.** Ligands other than NTA in Figure 11-4 form strong 1:1 complexes with all metal ions except univalent ions such as Li^+, Na^+, and K^+. *The stoichiometry is 1:1 regardless of the charge on the ion.*

FIGURE 11-3 Iron(III)-enterobactin complex. Certain bacteria secrete enterobactin to capture iron and bring it into the cell. Enterobactin is one of several known chelates—designated *siderophores*—released by microbes to capture iron to be used by the cell. [From R. J. Abergel, J. A. Warner, D. K. Shuh, and K. N. Raymond, "Enterobactin Protonation and Iron Release," *J. Am. Chem. Soc.* **2006**, *128*, 8920.]

NTA
Nitrilotriacetic acid

EDTA
Ethylenediaminetetraacetic acid
(also called ethylenedinitrilotetraacetic acid)

DCTA
trans-1,2-Diaminocyclohexanetetraacetic acid

DTPA
Diethylenetriaminepentaacetic acid

EGTA
bis(Aminoethyl)glycolether-*N,N,N',N'*-tetraacetic acid

FIGURE 11-4 Structures of analytically useful chelating agents. Nitrilotriacetic acid (NTA) tends to form 2:1 (ligand:metal) complexes with metal ions, whereas the others form 1:1 complexes.

FIGURE 11-5 Structure of $Fe(NTA)_2^{3-}$ in the salt $Na_3[Fe(NTA)_2] \cdot 5H_2O$. The ligand at the right binds to Fe through three O atoms and one N atom. The other ligand uses two O atoms and one N atom. Its third carboxylate group is uncoordinated. The Fe atom is seven coordinate. [From W. Clegg, A. K. Powell, and M. J. Ware, "Structure of $Na_3[Fe(NTA)_2] \cdot 5H_2O$," *Acta Crystallogr.* **1984**, *C40*, 1822.]

Axial ligand

Axial ligand

Fe
C
O
N

FIGURE 11-6 Structure of $Fe(DTPA)^{2-}$ found in the salt $Na_2[Fe(DTPA)] \cdot 2H_2O$. The seven-coordinate pentagonal bipyramidal coordination environment of the iron atom features 3 N and 2 O ligands in the equatorial plane (dashed lines) and two axial O ligands. The axial Fe—O bond lengths are 11 to 19 pm shorter than those of the more crowded equatorial Fe—O bonds. One carboxyl group of the ligand is uncoordinated. [From D. C. Finnen, A. A. Pinkerton, W. R. Dunham, R. H. Sands, and M. O. Funk, Jr., "Structures and Spectroscopic Characterization of Fe(III)-DTPA Complexes," *Inorg. Chem.* **1991**, *30*, 3960.]

One mole of EDTA reacts with one mole of metal ion.

EDTA is, by far, the most widely used chelator in analytical chemistry. By direct titration or through an indirect sequence of reactions, virtually every element of the periodic table can be measured with EDTA.

Acid-Base Properties

EDTA is a hexaprotic system, designated H_6Y^{2+}. The highlighted, acidic hydrogen atoms are the ones that are lost upon metal-complex formation.

$$HO_2CCH_2 \diagdown \quad \overset{+}{H}NCH_2CH_2\overset{+}{N}H \diagup CH_2CO_2H$$
$$HO_2CCH_2 \diagup \qquad\qquad \diagdown CH_2CO_2H$$
$$H_6Y^{2+}$$

$pK_1 = 0.0 \ (CO_2H)$	$pK_4 = 2.69 \ (CO_2H)$
$pK_2 = 1.5 \ (CO_2H)$	$pK_5 = 6.13 \ (NH^+)$
$pK_3 = 2.00 \ (CO_2H)$	$pK_6 = 10.37 \ (NH^+)$

pK applies at 25°C and $\mu = 0.1$ M, except pK_1 applies at $\mu = 1$ M

The first four pK values apply to carboxyl protons, and the last two are for the ammonium protons. The neutral acid is tetraprotic, with the formula H_4Y.

H_4Y can be dried at 140°C for 2 h and used as a primary standard. It can be dissolved by adding NaOH solution from a plastic container. NaOH solution from a glass bottle should not be used because it contains alkaline earth metals leached from the glass. Reagent-grade $Na_2H_2Y \cdot 2H_2O$ contains ~0.3% excess water. It may be used in this form with suitable correction for the mass of excess water or dried to the composition $Na_2H_2Y \cdot 2H_2O$ at 80°C.[11] The certified reference material $CaCO_3$ can be used to standardize EDTA or to verify the composition of standard EDTA.

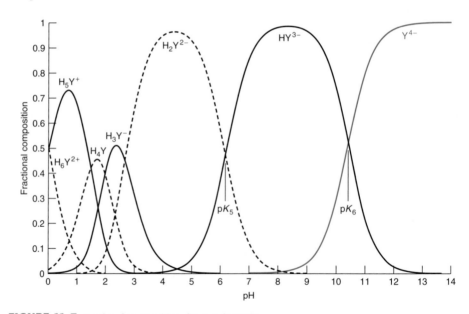

FIGURE 11-7 Fractional composition diagram for EDTA.

The fraction of EDTA in each of its protonated forms is plotted in Figure 11-7. As in Section 9-5, we can define α for each species as the fraction of EDTA in that form. For example, $\alpha_{Y^{4-}}$ is defined as

Fraction of EDTA in the form Y^{4-}:

$$\alpha_{Y^{4-}} = \frac{[Y^{4-}]}{[H_6Y^{2+}] + [H_5Y^+] + [H_4Y] + [H_3Y^-] + [H_2Y^{2-}] + [HY^{3-}] + [Y^{4-}]}$$

$$\alpha_{Y^{4-}} = \frac{[Y^{4-}]}{[EDTA]} \tag{11-3}$$

where [EDTA] is the total concentration of all *free* EDTA species in the solution. By "free," we mean EDTA not complexed to metal ions. Following the derivation in Section 9-5, we can show that $\alpha_{Y^{4-}}$ is given by

$$\alpha_{Y^{4-}} = \frac{K_1 K_2 K_3 K_4 K_5 K_6}{D} \qquad (11\text{-}4)$$

where $D = [H^+]^6 + [H^+]^5 K_1 + [H^+]^4 K_1 K_2 + [H^+]^3 K_1 K_2 K_3 + [H^+]^2 K_1 K_2 K_3 K_4 + [H^+] K_1 K_2 K_3 K_4 K_5 + K_1 K_2 K_3 K_4 K_5 K_6$. Table 11-1 gives values for $\alpha_{Y^{4-}}$ as a function of pH.

EXAMPLE What does $\alpha_{Y^{4-}}$ mean?

The fraction of all free EDTA in the form Y^{4-} is called $\alpha_{Y^{4-}}$. At pH 6.00 and a formal concentration of 0.10 M, the composition of an EDTA solution is

$[H_6Y^{2+}] = 8.9 \times 10^{-20}$ M $\quad$ $[H_5Y^+] = 8.9 \times 10^{-14}$ M $\quad$ $[H_4Y] = 2.8 \times 10^{-7}$ M

$[H_3Y^-] = 2.8 \times 10^{-5}$ M $\quad$ $[H_2Y^{2-}] = 0.057$ M $\quad$ $[HY^{3-}] = 0.043$ M

$[Y^{4-}] = 1.8 \times 10^{-6}$ M

Find $\alpha_{Y^{4-}}$.

Solution $\alpha_{Y^{4-}}$ is the fraction in the form Y^{4-}:

$$\alpha_{Y^{4-}} = \frac{[Y^{4-}]}{[H_6Y^{2+}] + [H_5Y^+] + [H_4Y] + [H_3Y^-] + [H_2Y^{2-}] + [HY^{3-}] + [Y^{4-}]}$$

$$= \frac{1.8 \times 10^{-6}}{(8.9 \times 10^{-20}) + (8.9 \times 10^{-14}) + (2.8 \times 10^{-7}) + (2.8 \times 10^{-5}) + (0.057) + (0.043) + (1.8 \times 10^{-6})}$$

$$= 1.8 \times 10^{-5}$$

Test Yourself At what pH does $\alpha_{Y^{4-}} = 0.50$? (*Answer:* pH = $pK_6 = 10.37$)

EDTA Complexes

The equilibrium constant for the reaction of a metal with a ligand is called the **formation constant, K_f**, or the **stability constant:**

Formation constant: $\quad M^{n+} + Y^{4-} \rightleftharpoons MY^{n-4} \quad\quad K_f = \dfrac{[MY^{n-4}]}{[M^{n+}][Y^{4-}]} \qquad (11\text{-}5)$

Note that K_f for EDTA is defined in terms of the species Y^{4-} reacting with the metal ion. The equilibrium constant could have been defined for any of the other six forms of EDTA in the solution. Equation 11-5 should not be interpreted to mean that only Y^{4-} reacts with metal ions. Table 11-2 shows that formation constants for most EDTA complexes are large and tend to be larger for more positively charged cations.

In many transition metal complexes, EDTA engulfs the metal ion, forming the six-coordinate species in Figure 11-1. If you try to build a space-filling model of a six-coordinate metal-EDTA complex, you will find strain in the chelate rings. This strain is relieved when the O ligands are drawn back toward the N atoms. Such distortion opens up a seventh coordination position, which can be occupied by H_2O, as in Figure 11-8. In some complexes, such as $Ca(EDTA)(H_2O)_2^{2-}$, the metal ion is so large that it accommodates eight ligand atoms.[12] Larger metal ions require more ligand atoms. Even if H_2O is attached to the metal ion, the formation constant is still given by Equation 11-5. The relation remains true because solvent (H_2O) is omitted from the reaction quotient.

Lanthanide and actinide elements typically have a coordination number of 9, with the shape of a tricapped trigonal prism (Figure 11-9).[13] Pu(IV) forms a complex with one EDTA plus three H_2O molecules to complete the coordination sphere with nine ligand atoms.[14] Eu(III) forms mixed complexes of the type Eu(EDTA)(NTA) in which EDTA provides six ligand atoms and NTA provides three ligand atoms (Figure 11-4).[15]

pH	$\alpha_{Y^{4-}}$
0	1.3×10^{-23}
1	1.4×10^{-18}
2	2.6×10^{-14}
3	2.1×10^{-11}
4	3.0×10^{-9}
5	2.9×10^{-7}
6	1.8×10^{-5}
7	3.8×10^{-4}
8	4.2×10^{-3}
9	0.041
10	0.30
11	0.81
12	0.98
13	1.00
14	1.00

Question From Figure 11-7, which species has the greatest concentration at pH 6? At pH 7? At pH 11?

Fe O C N

FIGURE 11-8 Seven-coordinate geometry of Fe(EDTA)(H$_2$O)$^-$. Other metal ions that form seven-coordinate EDTA complexes include Fe^{2+}, Mg^{2+}, Cd^{2+}, Co^{2+}, Mn^{2+}, Ru^{3+}, Cr^{3+}, Co^{3+}, V^{3+}, Ti^{3+}, In^{3+}, Sn^{4+}, Os^{4+}, and Ti^{4+}. Some of these same ions also form six-coordinate EDTA complexes. Eight-coordinate complexes are formed by Ca^{2+}, Er^{3+}, Yb^{3+}, and Zr^{4+}. [From T. Mizuta, J. Wang, and K. Miyoshi, "A 7-Coordinate Structure of Fe(III)-EDTA," *Bull. Chem. Soc. Japan* 1993, *66*, 2547.]

Top view

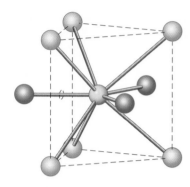

FIGURE 11-9 Tricapped trigonal prismatic structure of many Ln(III) and An(III) complexes, where Ln is a lanthanide element and An is an actinide element. In $M(H_2O)_9^{3+}$, bonds from the metal to the 6 O atoms at the corners of the prism are shorter than bonds from the metal to the 3 O atoms projecting out from the rectangular faces.

TABLE 11-2 Formation constants for metal-EDTA complexes

Ion	$\log K_f$	Ion	$\log K_f$	Ion	$\log K_f$
Li^+	2.95	V^{3+}	25.9^a	Tl^{3+}	35.3
Na^+	1.86	Cr^{3+}	23.4^a	Bi^{3+}	27.8^a
K^+	0.8	Mn^{3+}	25.2	Ce^{3+}	15.93
Be^{2+}	9.7	Fe^{3+}	25.1	Pr^{3+}	16.30
Mg^{2+}	8.79	Co^{3+}	41.4	Nd^{3+}	16.51
Ca^{2+}	10.65	Zr^{4+}	29.3	Pm^{3+}	16.9
Sr^{2+}	8.72	Hf^{4+}	29.5	Sm^{3+}	17.06
Ba^{2+}	7.88	VO^{2+}	18.7	Eu^{3+}	17.25
Ra^{2+}	7.4	VO_2^+	15.5	Gd^{3+}	17.35
Sc^{3+}	23.1^a	Ag^+	7.20	Tb^{3+}	17.87
Y^{3+}	18.08	Tl^+	6.41	Dy^{3+}	18.30
La^{3+}	15.36	Pd^{2+}	25.6^a	Ho^{3+}	18.56
V^{2+}	12.7^a	Zn^{2+}	16.5	Er^{3+}	18.89
Cr^{2+}	13.6^a	Cd^{2+}	16.5	Tm^{3+}	19.32
Mn^{2+}	13.89	Hg^{2+}	21.5	Yb^{3+}	19.49
Fe^{2+}	14.30	Sn^{2+}	18.3^b	Lu^{3+}	19.74
Co^{2+}	16.45	Pb^{2+}	18.0	Th^{4+}	23.2
Ni^{2+}	18.4	Al^{3+}	16.4	U^{4+}	25.7
Cu^{2+}	18.78	Ga^{3+}	21.7		
Ti^{3+}	21.3	In^{3+}	24.9		

NOTE: *The stability constant is the equilibrium constant for the reaction* $M^{n+} + Y^{4-} \rightleftharpoons MY^{n-4}$. *Values in table apply at 25°C and ionic strength 0.1 M unless otherwise indicated.*

a. 20°C, ionic strength = 0.1 M.   b. 20°C, ionic strength = 1 M.

SOURCE: *A. E. Martell, R. M. Smith, and R. J. Motekaitis, NIST Critically Selected Stability Constants of Metal Complexes, NIST Standard Reference Database 46, Gaithersburg, MD, 2001.*

Conditional Formation Constant

The formation constant $K_f = [MY^{n-4}]/[M^{n+}][Y^{4-}]$ describes the reaction between Y^{4-} and a metal ion. As you see in Figure 11-7, most EDTA is not Y^{4-} below pH 10.37. The species HY^{3-}, H_2Y^{2-}, and so on, predominate at lower pH. From the definition $\alpha_{Y^{4-}} = [Y^{4-}]/[EDTA]$, we can express the concentration of Y^{4-} as

$$[Y^{4-}] = \alpha_{Y^{4-}}[EDTA]$$

where [EDTA] is the total concentration of all EDTA species not bound to metal ion.

The formation constant can now be rewritten as

Only some of the free EDTA is in the form Y^{4-}.

$$K_f = \frac{[MY^{n-4}]}{[M^{n+}][Y^{4-}]} = \frac{[MY^{n-4}]}{[M^{n+}]\alpha_{Y^{4-}}[EDTA]}$$

If the pH is fixed by a buffer, then $\alpha_{Y^{4-}}$ is a constant that can be combined with K_f:

Conditional formation constant: $$K_f' = \alpha_{Y^{4-}}K_f = \frac{[MY^{n-4}]}{[M^{n+}][EDTA]} \qquad (11\text{-}6)$$

The number $K_f' = \alpha_{Y^{4-}}K_f$ is called the **conditional formation constant,** or the *effective formation constant.* It describes the formation of MY^{n-4} at any particular pH. After we learn to use Equation 11-6, we will modify it to allow for the possibility that not all metal ion is in the form M^{n+}.

The conditional formation constant allows us to look at EDTA complex formation as if the uncomplexed EDTA were all in one form:

With the conditional formation constant, we can treat EDTA complex formation as if all free EDTA were in one form.

$$M^{n+} + EDTA \rightleftharpoons MY^{n-4} \qquad K_f' = \alpha_{Y^{4-}}K_f$$

At any given pH, we can find $\alpha_{Y^{4-}}$ and evaluate K_f'.

EXAMPLE Using the Conditional Formation Constant

The formation constant in Table 11-2 for CaY^{2-} is $10^{10.65}$. Calculate the concentration of free Ca^{2+} in a solution of 0.10 M CaY^{2-} at pH 10.00 and at pH 6.00.

Solution The complex formation reaction is

$$Ca^{2+} + EDTA \rightleftharpoons CaY^{2-} \qquad K_f' = \alpha_{Y^{4-}} K_f$$

where EDTA on the left side of the equation refers to all forms of unbound EDTA (Y^{4-}, HY^{3-}, H_2Y^{2-}, H_3Y^-, and so on). Using $\alpha_{Y^{4-}}$ from Table 11-1, we find

At pH 10.00: $K_f' = (0.30)(10^{10.65}) = 1.3_4 \times 10^{10}$

At pH 6.00: $K_f' = (1.8 \times 10^{-5})(10^{10.65}) = 8.0 \times 10^5$

Dissociation of CaY^{2-} must produce equal quantities of Ca^{2+} and EDTA, so we can write

	Ca^{2+}	+	EDTA	$\rightleftharpoons$	CaY^{2-}
Initial concentration (M)	0		0		0.10
Final concentration (M)	x		x		$0.10 - x$

$$\frac{[CaY^{2-}]}{[Ca^{2+}][EDTA]} = \frac{0.10 - x}{x^2} = K_f' = 1.3_4 \times 10^{10} \quad \text{at pH 10.00}$$

$$= 8.0 \times 10^5 \quad \text{at pH 6.00}$$

Solving for x (= $[Ca^{2+}]$), we find $[Ca^{2+}] = 2.7 \times 10^{-6}$ M at pH 10.00 and 3.5×10^{-4} M at pH 6.00. *Using the conditional formation constant at a fixed pH, we treat the dissociated EDTA as if it were a single species.*

Test Yourself Find $[Ca^{2+}]$ in 0.10 M CaY^{2-} at pH 8.00. (***Answer:*** 2.3×10^{-5} M)

FIGURE 11-10 Titration of Ca^{2+} with EDTA as a function of pH. As the pH is lowered, the end point becomes less distinct. The potential was measured with mercury and calomel electrodes, as described in Exercise 14-B. [From C. N. Reilley and R. W. Schmid, "Chelometric Titration with Potentiometric End Point Detection. Mercury as a pM Indicator Electrode," *Anal. Chem.* **1958**, *30*, 947.]

You can see from the example that a metal-EDTA complex becomes less stable at lower pH. For a titration reaction to be effective, it must go "to completion" (say, 99.9%), which means that the equilibrium constant is large—the analyte and titrant are essentially completely reacted at the equivalence point. Figure 11-10 shows how pH affects the titration of Ca^{2+} with EDTA. Below pH ≈ 8, the end point is not sharp enough to allow accurate determination. The conditional formation constant for CaY^{2-} is just too small for "complete" reaction at low pH.

pH can be used to select which metals will be titrated by EDTA and which will not. Metals with higher formation constants can be titrated at lower pH. If a solution containing both Fe^{3+} and Ca^{2+} is titrated at pH 4, Fe^{3+} is titrated without interference from Ca^{2+}.

11-3 EDTA Titration Curves

In this section, we calculate the concentration of free M^{n+} during its titration with EDTA.[16] The titration reaction is

$$M^{n+} + EDTA \rightleftharpoons MY^{n-4} \qquad K_f' = \alpha_{Y^{4-}} K_f \qquad (11-7)$$

If K_f' is large, we can consider the reaction to be complete at each point in the titration.

The titration curve is a graph of pM (= $-\log[M^{n+}]$) versus the volume of added EDTA. The curve is analogous to plotting pH versus volume of titrant in an acid-base titration. There are three natural regions of the titration curve in Figure 11-11.

K_f' is the effective formation constant at the fixed pH of the solution.

Region 1: Before the Equivalence Point

In this region, there is excess M^{n+} left in solution after the EDTA has been consumed. The concentration of free metal ion is equal to the concentration of excess, unreacted M^{n+}. The dissociation of MY^{n-4} is negligible.

Region 2: At the Equivalence Point

There is exactly as much EDTA as metal in the solution. We can treat the solution as if it had been made by dissolving pure MY^{n-4}. Some free M^{n+} is generated by the slight dissociation of MY^{n-4}:

$$MY^{n-4} \rightleftharpoons M^{n+} + EDTA$$

In this reaction, EDTA represents free EDTA in all its forms. At the equivalence point, $[M^{n+}] = [EDTA]$.

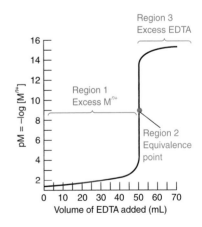

FIGURE 11-11 Three regions in an EDTA titration illustrated for reaction of 50.0 mL of 0.050 0 M M^{n+} with 0.050 0 M EDTA, assuming $K_f' = 1.15 \times 10^{16}$. The concentration of free M^{n+} decreases as the titration proceeds.

Region 3: After the Equivalence Point

Now there is excess EDTA, and virtually all the metal ion is in the form MY^{n-4}. The concentration of free EDTA can be equated to the concentration of excess EDTA added after the equivalence point.

Titration Calculations

Let's calculate the shape of the titration curve for the reaction of 50.0 mL of 0.040 0 M Ca^{2+} (buffered to pH 10.00) with 0.080 0 M EDTA:

$$Ca^{2+} + EDTA \rightarrow CaY^{2-}$$

$$K_f' = \alpha_{Y^{4-}} K_f = (0.30)(10^{10.65}) = 1.3_4 \times 10^{10}$$

The value of $\alpha_{Y^{4-}}$ comes from Table 11-1.

Because K_f' is large, it is reasonable to say that the reaction goes to completion with each addition of titrant. We want to make a graph in which pCa^{2+} $(= -\log[Ca^{2+}])$ is plotted versus milliliters of added EDTA. The equivalence volume is 25.0 mL.

Region 1: Before the Equivalence Point

Before the equivalence point, there is excess unreacted Ca^{2+}.

Consider the addition of 5.0 mL of EDTA. Because the equivalence point requires 25.0 mL of EDTA, one-fifth of the Ca^{2+} will be consumed and four-fifths remains.

$$[Ca^{2+}] = \underbrace{\left(\frac{25.0 - 5.0}{25.0}\right)}_{\substack{\text{Fraction} \\ \text{remaining} \\ (= 4/5)}} \underbrace{(0.040\ 0\ M)}_{\substack{\text{Original} \\ \text{concentration} \\ \text{of } Ca^{2+}}} \underbrace{\left(\frac{50.0}{55.0}\right)}_{\substack{\text{Dilution} \\ \text{factor}}}$$

(with labels: Initial volume of Ca^{2+} = 50.0; Total volume of solution = 55.0)

$$= 0.029\ 1\ M \Rightarrow pCa^{2+} = -\log[Ca^{2+}] = 1.54$$

In a similar manner, we could calculate pCa^{2+} for any volume of EDTA less than 25.0 mL.

Region 2: At the Equivalence Point

At the equivalence point, the major species is CaY^{2-}, in equilibrium with small, equal amounts of free Ca^{2+} and EDTA.

Virtually all the metal is in the form CaY^{2-}. With negligible dissociation, the concentration of CaY^{2-} is equal to the original concentration of Ca^{2+}, with a correction for dilution.

$$[CaY^{2-}] = \underbrace{(0.040\ 0\ M)}_{\substack{\text{Original} \\ \text{concentration} \\ \text{of } Ca^{2+}}} \underbrace{\left(\frac{50.0}{75.0}\right)}_{\substack{\text{Dilution} \\ \text{factor}}} = 0.026\ 7\ M$$

(with labels: Initial volume of Ca^{2+} = 50.0; Total volume of solution = 75.0)

The concentration of free Ca^{2+} is small and unknown. We can write

	Ca^{2+}	+	EDTA	$\rightleftharpoons$	CaY^{2-}
Initial concentration (M)	—		—		0.026 7
Final concentration (M)	x		x		0.026 7 $- x$

[EDTA] refers to the total concentration of all forms of EDTA not bound to metal.

$$\frac{[CaY^{2-}]}{[Ca^{2+}][EDTA]} = K_f' = 1.3_4 \times 10^{10}$$

$$\frac{0.026\ 7 - x}{x^2} = 1.3_4 \times 10^{10} \Rightarrow x = 1.4 \times 10^{-6}\ M$$

$$pCa^{2+} = -\log[Ca^{2+}] = -\log x = 5.85$$

Region 3: After the Equivalence Point

After the equivalence point, virtually all the metal is in the form CaY^{2-}. There is a known excess of EDTA. A small amount of free Ca^{2+} exists in equilibrium with CaY^{2-} and EDTA.

In this region, virtually all the metal is in the form CaY^{2-}, and there is excess, unreacted EDTA. The concentrations of CaY^{2-} and excess EDTA are easily calculated. For example, at 26.0 mL, there is 1.0 mL of excess EDTA.

$$[EDTA] = \underbrace{(0.080\ 0\ M)}_{\substack{\text{Original} \\ \text{concentration} \\ \text{of EDTA}}} \underbrace{\left(\frac{1.0}{76.0}\right)}_{\substack{\text{Dilution} \\ \text{factor}}} = 1.05 \times 10^{-3}\ M$$

(with labels: Volume of excess EDTA = 1.0; Total volume of solution = 76.0)

CHAPTER 11 EDTA Titrations

$$[CaY^{2-}] = (0.040\ 0\ M)\underbrace{\left(\frac{50.0}{76.0}\right)}_{\substack{\text{Dilution}\\ \text{factor}}} = 2.63 \times 10^{-2}\ M$$

Original volume of Ca^{2+} · Original concentration of Ca^{2+} · Total volume of solution

The concentration of Ca^{2+} is governed by

$$\frac{[CaY^{2-}]}{[Ca^{2+}][EDTA]} = K'_f = 1.3_4 \times 10^{10}$$

$$\frac{[2.63 \times 10^{-2}]}{[Ca^{2+}](1.05 \times 10^{-3})} = 1.3_4 \times 10^{10}$$

$$[Ca^{2+}] = 1.9 \times 10^{-9}\ M \Rightarrow pCa^{2+} = 8.73$$

The same sort of calculation can be used for any volume past the equivalence point.

The Titration Curve

Calculated titration curves for Ca^{2+} and Sr^{2+} in Figure 11-12 show a distinct break at the equivalence point, where the slope is greatest. The Ca^{2+} end point is more distinct than the Sr^{2+} end point because the conditional formation constant, $\alpha_{Y^{4-}}K_f$, for CaY^{2-} is greater than that of SrY^{2-}. If the pH is lowered, the conditional formation constant decreases (because $\alpha_{Y^{4-}}$ decreases), and the end point becomes less distinct, as we saw in Figure 11-10. The pH cannot be raised arbitrarily high, because metal hydroxide might precipitate.

11-4 🖿 Do It with a Spreadsheet

Let's see how to reproduce the EDTA titration curves in Figure 11-12 by using one equation that applies to the entire titration. Because the reactions are carried out at fixed pH, the equilibria and mass balances are sufficient to solve for all unknowns.

Consider the titration of metal ion M (initial concentration = C_M, volume = V_M) with a solution of ligand L (concentration = C_L, volume added = V_L) to form a 1:1 complex:

$$M + L \rightleftharpoons ML \qquad K_f = \frac{[ML]}{[M][L]} \Rightarrow [ML] = K_f[M][L] \qquad (11\text{-}8)$$

The mass balances for metal and ligand are

Mass balance for M: $\qquad [M] + [ML] = \dfrac{C_M V_M}{V_M + V_L}$

Mass balance for L: $\qquad [L] + [ML] = \dfrac{C_L V_L}{V_M + V_L}$

Substituting $K_f[M][L]$ (from Equation 11-8) for [ML] in the mass balances gives

$$[M](1 + K_f[L]) = \frac{C_M V_M}{V_M + V_L} \qquad (11\text{-}9)$$

$$[L](1 + K_f[M]) = \frac{C_L V_L}{V_M + V_L} \Rightarrow [L] = \frac{\dfrac{C_L V_L}{V_M + V_L}}{1 + K_f[M]} \qquad (11\text{-}10)$$

Now substitute the expression for [L] from Equation 11-10 back into Equation 11-9

$$[M]\left(1 + K_f \frac{\dfrac{C_L V_L}{V_M + V_L}}{1 + K_f[M]}\right) = \frac{C_M V_M}{V_M + V_L}$$

and do about five lines of algebra to solve for the fraction of titration, ϕ:

Spreadsheet equation for titration of M with L:
$$\phi = \frac{C_L V_L}{C_M V_M} = \frac{1 + K_f[M] - \dfrac{[M] + K_f[M]^2}{C_M}}{K_f[M] + \dfrac{[M] + K_f[M]^2}{C_L}} \qquad (11\text{-}11)$$

As in acid-base titrations in Table 10-5, ϕ is the fraction of the way to the equivalence point. When $\phi = 1$, $V_L = V_e$. When $\phi = \frac{1}{2}$, $V_L = \frac{1}{2}V_e$. And so on.

FIGURE 11-12 Theoretical titration curves for the reaction of 50.0 mL of 0.040 0 M metal ion with 0.080 0 M EDTA at pH 10.00.

Total metal concentration

$$= \frac{\text{initial moles of metal}}{\text{total volume}}$$

$$= \frac{C_M V_M}{V_M + V_L}$$

Total ligand concentration

$$= \frac{\text{moles of ligand added}}{\text{total volume}}$$

$$= \frac{C_L V_L}{V_M + V_L}$$

Replace K_f by K'_f if L = EDTA.

FIGURE 11-13 Spreadsheet for the titration
of 50.0 mL of 0.040 0 M Ca^{2+} with 0.080 0 M
EDTA at pH 10.00. This spreadsheet reproduces
calculations of Section 11-3. pM was varied by
trial and error to find volumes of 5.00, 25.00,
and 26.00 mL used in the preceding section.
Better yet, use Goal Seek (page 158) to vary pM
in cell B9 until the volume in cell E9 is 25.000 mL.

	A	B	C	D	E
1	Titration of 50 mL of 0.04 M Ca^{2+} with 0.08 M EDTA				
2					
3	C_M =	pM	M	Phi	V(ligand)
4	0.04	1.398	4.00E-02	0.000	0.002
5	V_M =	1.537	2.90E-02	0.201	5.026
6	50	2.00	1.00E-02	0.667	16.667
7	C(ligand) =	3.00	1.00E-03	0.963	24.074
8	0.08	4.00	1.00E-04	0.996	24.906
9	K_f' =	5.85	1.41E-06	1.000	25.0000
10	1.34E+10	7.00	1.00E-07	1.001	25.019
11		8.00	1.00E-08	1.007	25.187
12		8.73	1.86E-09	1.040	26.002
13	C4 = 10^-B4				
14	Equation 11-11:				
15	D4 = (1+$A$10*C4-(C4+C4*C4*$A$10)/$A$4)/				
16		(C4*$A$10+(C4+C4*C4*$A$10)/$A$8)			
17	E4 = D4*$A$4*$A$6/$A$8				

For a titration with EDTA, you can follow the derivation through and find that the formation constant, K_f, should be replaced in Equation 11-11 by the conditional formation constant, K_f', which applies at the fixed pH of the titration. Figure 11-13 shows a spreadsheet in which Equation 11-11 is used to calculate the Ca^{2+} titration curve in Figure 11-12. As in acid-base titrations, your input in column B is pM = $-\log[Ca^{2+}]$ and the output in column E is volume of titrant. To find the initial point, vary pM until V_1 is close to 0.

If you reverse the process and titrate ligand with metal ion, the fraction of the way to the equivalence point is the inverse of the fraction in Equation 11-11:

Spreadsheet equation for titration of L with M:

$$\phi = \frac{C_M V_M}{C_L V_L} = \frac{K_f[M] + \dfrac{[M] + K_f[M]^2}{C_L}}{1 + K_f[M] - \dfrac{[M] + K_f[M]^2}{C_M}} \quad (11\text{-}12)$$

Replace K_f by K_f' if L = EDTA.

HO   OH
HO_2C CO_2H
Tartaric acid

HO CO_2H
HO_2C CO_2H
Citric acid

$N(CH_2CH_2OH)_3$
Triethanolamine

11-5 Auxiliary Complexing Agents

EDTA titration conditions in this chapter were selected to prevent metal hydroxide precipitation at the chosen pH. To permit many metals to be titrated in alkaline solutions with EDTA, we use an **auxiliary complexing agent.** This reagent is a ligand, such as ammonia, tartrate, citrate, or triethanolamine, that binds the metal strongly enough to prevent metal hydroxide from precipitating but weakly enough to give up the metal when EDTA is added. Zn^{2+} is usually titrated in ammonia buffer, which fixes the pH and complexes the metal ion to keep it in solution. Let's see how this works.

Metal-Ligand Equilibria[17]

Consider a metal ion that forms two complexes with the auxiliary complexing ligand L:

$$M + L \rightleftharpoons ML \qquad \beta_1 = \frac{[ML]}{[M][L]} \quad (11\text{-}13)$$

$$M + 2L \rightleftharpoons ML_2 \qquad \beta_2 = \frac{[ML_2]}{[M][L]^2} \quad (11\text{-}14)$$

The equilibrium constants, β_i, are called *overall* or **cumulative formation constants.** The fraction of metal ion in the uncomplexed state, M, can be expressed as

$$\alpha_M = \frac{[M]}{M_{tot}} \quad (11\text{-}15)$$

where M_{tot} is the total concentration of all forms of M (= M, ML, and ML_2 in this case).

Now let's find a useful expression for α_M. The mass balance for metal is

$$M_{tot} = [M] + [ML] + [ML_2]$$

Equations 11-13 and 11-14 allow us to say $[ML] = \beta_1[M][L]$ and $[ML_2] = \beta_2[M][L]^2$. Therefore,

$$M_{tot} = [M] + \beta_1[M][L] + \beta_2[M][L]^2$$
$$= [M]\{1 + \beta_1[L] + \beta_2[L]^2\}$$

Substituting this last result into Equation 11-15 gives the desired result.

Fraction of free metal ion:
$$\alpha_M = \frac{[M]}{[M]\{1 + \beta_1[L] + \beta_2[L]^2\}} = \frac{1}{1 + \beta_1[L] + \beta_2[L]^2} \quad \text{(11-16)}$$

If the metal forms more than two complexes, Equation 11-16 takes the form

$$\alpha_M = \frac{1}{1 + \beta_1[L] + \beta_2[L]^2 + \cdots + \beta_n[L]^n}$$

EXAMPLE **Ammonia Complexes of Zinc**

Zn^{2+} and NH_3 form the complexes $Zn(NH_3)^{2+}$, $Zn(NH_3)_2^{2+}$, $Zn(NH_3)_3^{2+}$, and $Zn(NH_3)_4^{2+}$. If the concentration of free, *unprotonated* NH_3 is 0.10 M, find the fraction of zinc in the form Zn^{2+}. (At any pH, there will also be some NH_4^+ in equilibrium with NH_3.)

Solution Appendix I gives formation constants for the complexes $Zn(NH_3)^{2+}$ ($\beta_1 = 10^{2.18}$), $Zn(NH_3)_2^{2+}$ ($\beta_2 = 10^{4.43}$), $Zn(NH_3)_3^{2+}$ ($\beta_3 = 10^{6.74}$), and $Zn(NH_3)_4^{2+}$ ($\beta_4 = 10^{8.70}$). The appropriate form of Equation 11-16 is

$$\alpha_{Zn^{2+}} = \frac{1}{1 + \beta_1[L] + \beta_2[L]^2 + \beta_3[L]^3 + \beta_4[L]^4} \quad \text{(11-17)}$$

Equation 11-17 gives the fraction of zinc in the form Zn^{2+}. Putting in [L] = 0.10 M and the four values of β_i gives $\alpha_{Zn^{2+}} = 1.8 \times 10^{-5}$, which means there is very little free Zn^{2+} in the presence of 0.10 M NH_3.

Test Yourself Find $\alpha_{Zn^{2+}}$ if free, unprotonated $[NH_3] = 0.02$ M. (***Answer:*** 0.007 2)

BOX 11-2 **Metal Ion Hydrolysis Decreases the Effective Formation Constant for EDTA Complexes**

Equation 11-18 states that the effective (conditional) formation constant for an EDTA complex is the product of the formation constant, K_f, times the fraction of metal in the form M^{m+} times the fraction of EDTA in the form Y^{4-}: $K_f'' = \alpha_{M^{m+}}\alpha_{Y^{4-}}K_f$. Table 11-1 told us that $\alpha_{Y^{4-}}$ increases with pH until it levels off at 1 near pH 11.

In Section 11-3, we had no auxiliary complexing ligand and we implicitly assumed that $\alpha_{M^{m+}} = 1$. In fact, metal ions react with water to form $M(OH)_n$ species. Combinations of pH and metal ion in Section 11-3 were selected so that hydrolysis to $M(OH)_n$ is negligible. We can find such conditions for most M^{2+} ions, but not for M^{3+} or M^{4+}. Even in acidic solution, Fe^{3+} hydrolyzes to $Fe(OH)^{2+}$ and $Fe(OH)_2^{+}$.[18] (Appendix I gives formation constants for hydroxide complexes.) The graph shows that $\alpha_{Fe^{3+}}$ is close to 1 between pH 1 and 2 ($\log \alpha_{Fe^{3+}} \approx 0$), but then drops as hydrolysis occurs. At pH 5, the fraction of Fe(III) in the form Fe^{3+} is $\sim 10^{-5}$.

The effective formation constant for FeY^- in the graph has three contributions:

$$K_f''' = \frac{\alpha_{Fe^{3+}}\,\alpha_{Y^{4-}}}{\alpha_{FeY^-}}K_f$$

As pH increases, $\alpha_{Y^{4-}}$ increases, so K_f''' increases. As pH increases, metal hydrolysis occurs, so $\alpha_{Fe^{3+}}$ decreases. The increase in $\alpha_{Y^{4-}}$ is canceled by the decrease in $\alpha_{Fe^{3+}}$, so K_f''' is nearly constant above pH 3. The third contribution to K_f''' is α_{FeY^-}, which is the fraction of the EDTA complex in the form FeY^-. At low pH, some of the complex gains a proton to form FeHY, which decreases α_{FeY^-} near pH 1.

In the pH range 2 to 5, α_{FeY^-} is nearly constant at 1. In neutral and basic solution, complexes such as $Fe(OH)Y^{2-}$ and $[Fe(OH)Y]_2^{4-}$ are formed and α_{FeY^-} decreases.

Take-home message: In this book, we restrict ourselves to cases in which there is no hydrolysis and $\alpha_{M^{m+}}$ is controlled by a deliberately added auxiliary ligand. In reality, hydrolysis of M^{m+} and MY influences most EDTA titrations and makes the theoretical analysis more complicated than we pretend in this chapter.

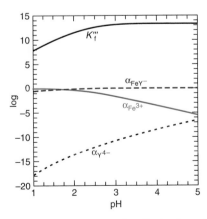

Contributions of $\alpha_{Y^{4-}}$, $\alpha_{Fe^{3+}}$, and α_{FeY^-} to the effective formation constant, K_f''', for FeY^-. Curves were calculated by considering the species H_6Y^{2+}, H_5Y^+, H_4Y, H_3Y^-, H_2Y^{2-}, HY^{3-}, Y^{4-}, Fe^{3+}, $Fe(OH)^{2+}$, $Fe(OH)_2^+$, FeY^-, and FeHY.

EDTA Titration with an Auxiliary Complexing Agent

Now consider a titration of Zn^{2+} by EDTA in the presence of NH_3. The extension of Equation 11-6 requires a new conditional formation constant to account for the fact that only some of the EDTA is in the form Y^{4-} and only some of the zinc not bound to EDTA is in the form Zn^{2+}:

$$K_f'' = \alpha_{Zn^{2+}}\alpha_{Y^{4-}}K_f \qquad (11\text{-}18)$$

In the margin: *K_f'' is the effective formation constant at a fixed pH and fixed concentration of auxiliary complexing agent. Box 11-2 describes the influence of metal ion hydrolysis on the effective formation constant.*

In this expression, $\alpha_{Zn^{2+}}$ is given by Equation 11-17 and $\alpha_{Y^{4-}}$ is given by Equation 11-4. For particular values of pH and $[NH_3]$, we can compute K_f'' and proceed with titration calculations as in Section 11-3, substituting K_f'' for K_f'. An assumption in this process is that EDTA is a much stronger complexing agent than ammonia, so essentially all EDTA binds Zn^{2+} until the metal ion is consumed.

EXAMPLE EDTA Titration in the Presence of Ammonia

Consider the titration of 50.0 mL of 1.00×10^{-3} M Zn^{2+} with 1.00×10^{-3} M EDTA at pH 10.00 in the presence of 0.10 M NH_3. (This is the concentration of NH_3. There is also NH_4^+ in the solution.) The equivalence point is at 50.0 mL. Find pZn^{2+} after addition of 20.0, 50.0, and 60.0 mL of EDTA.

Solution In Equation 11-17, we found that $\alpha_{Zn^{2+}} = 1.8 \times 10^{-5}$. Table 11-1 tells us that $\alpha_{Y^{4-}} = 0.30$. With K_f from Table 11-2, the conditional formation constant is

$$K_f'' = \alpha_{Zn^{2+}}\alpha_{Y^{4-}}K_f = (1.8 \times 10^{-5})(0.30)(10^{16.5}) = 1.7 \times 10^{11}$$

(a) *Before the equivalence point—20.0 mL:* Because the equivalence point is 50.0 mL, the fraction of Zn^{2+} remaining is 30.0/50.0. The dilution factor is 50.0/70.0. Therefore, the concentration of zinc not bound to EDTA is

$$C_{Zn^{2+}} = \left(\frac{30.0}{50.0}\right)(1.00 \times 10^{-3}\ \text{M})\left(\frac{50.0}{70.0}\right) = 4.3 \times 10^{-4}\ \text{M}$$

However, nearly all zinc not bound to EDTA is bound to NH_3. The concentration of free Zn^{2+} is

$$[Zn^{2+}] = \alpha_{Zn^{2+}}C_{Zn^{2+}} = (1.8 \times 10^{-5})(4.3 \times 10^{-4}) = 7.7 \times 10^{-9}\ \text{M}$$
$$\Rightarrow pZn^{2+} = -\log[Zn^{2+}] = 8.11$$

In the margin: *The relation $[Zn^{2+}] = \alpha_{Zn^{2+}}C_{Zn^{2+}}$ follows from Equation 11-15.*

Let's try a reality check. The product $[Zn^{2+}][OH^-]^2$ is $[10^{-8.11}][10^{-4.00}]^2 = 10^{-16.11}$, which does not exceed the solubility product of $Zn(OH)_2$ ($K_{sp} = 10^{-15.52}$).

(b) *At the equivalence point—50.0 mL:* At the equivalence point, the dilution factor is 50.0/100.0, so $[ZnY^{2-}] = (50.0/100.0)(1.00 \times 10^{-3}\ \text{M}) = 5.00 \times 10^{-4}\ \text{M}$. We then create a table of concentrations:

	$C_{Zn^{2+}}$	+ EDTA	$\rightleftharpoons$	ZnY^{2-}
Initial concentration (M)	0	0		5.00×10^{-4}
Final concentration (M)	x	x		$5.00 \times 10^{-4} - x$

$$K_f'' = 1.7 \times 10^{11} = \frac{[ZnY^{2-}]}{[C_{Zn^{2+}}][EDTA]} = \frac{5.00 \times 10^{-4} - x}{x^2}$$

$$\Rightarrow x = C_{Zn^{2+}} = 5.4 \times 10^{-8}\ \text{M}$$

$$[Zn^{2+}] = \alpha_{Zn^{2+}}C_{Zn^{2+}} = (1.8 \times 10^{-5})(5.4 \times 10^{-8}) = 9.7 \times 10^{-13}\ \text{M}$$

$$\Rightarrow pZn^{2+} = -\log[Zn^{2+}] = 12.01$$

(c) *After the equivalence point—60.0 mL:* Almost all zinc is in the form ZnY^{2-}. With a dilution factor of 50.0/110.0 for zinc, we find

$$[ZnY^{2-}] = \left(\frac{50.0}{110.0}\right)(1.00 \times 10^{-3}\ \text{M}) = 4.5 \times 10^{-4}\ \text{M}$$

We also know the concentration of excess EDTA, whose dilution factor is 10.0/110.0:

$$[EDTA] = \left(\frac{10.0}{110.0}\right)(1.00 \times 10^{-3}\ \text{M}) = 9.1 \times 10^{-5}\ \text{M}$$

Once we know $[ZnY^{2-}]$ and [EDTA], we can use the equilibrium constant to find $[Zn^{2+}]$:

$$\frac{[ZnY^{2-}]}{[Zn^{2+}][EDTA]} = \alpha_{Y^{4-}} K_f = K'_f = (0.30)(10^{16.5}) = 9.5 \times 10^{15}$$

$$\frac{[4.5 \times 10^{-4}]}{[Zn^{2+}][9.1 \times 10^{-5}]} = 9.5 \times 10^{15} \Rightarrow [Zn^{2+}] = 5.3 \times 10^{-16} \text{ M}$$

$$\Rightarrow pZn^{2+} = 15.28$$

After the equivalence point, the problem does not depend on the presence of NH_3, because we know both $[ZnY^{2-}]$ and [EDTA].

Test Yourself Find pZn^{2+} after adding 30.0 and 51.0 mL of EDTA. (***Answer:*** 8.35, 14.28)

Figure 11-14 compares the calculated titration curves for Zn^{2+} in the presence of different concentrations of auxiliary complexing agent. The greater the concentration of NH_3, the smaller the change of pZn^{2+} near the equivalence point. The auxiliary ligand must be kept below the level that would obliterate the end point of the titration. Color Plate 5 shows the appearance of a Cu^{2+}-ammonia solution during an EDTA titration.

11-6 Metal Ion Indicators

The most common way to detect the end point in EDTA titrations is with a metal ion indicator. Alternatives include a mercury electrode (Figure 11-10 and Exercise 14-B) and an ion-selective electrode (Section 14-6). A pH electrode will follow the course of the titration in unbuffered solution, because H_2Y^{2-} releases $2H^+$ when it forms a metal complex.

Metal ion indicators (Table 11-3) are compounds that change color when they bind to a metal ion. *Useful indicators must bind metal less strongly than EDTA does.*

A typical titration is illustrated by the reaction of Mg^{2+} with EDTA at pH 10 with Calmagite indicator.

$$\underset{\text{Red}}{MgIn} + \underset{\text{Colorless}}{EDTA} \rightarrow \underset{\text{Colorless}}{MgEDTA} + \underset{\text{Blue}}{In} \qquad (11\text{-}19)$$

At the start of the experiment, a small amount of indicator (In) is added to the colorless solution of Mg^{2+} to form a red complex. As EDTA is added, it reacts first with free, colorless Mg^{2+}. When free Mg^{2+} is used up, the last EDTA added before the equivalence point displaces indicator from the red MgIn complex. The change from the red MgIn to blue unbound In signals the end point of the titration (Demonstration 11-1).

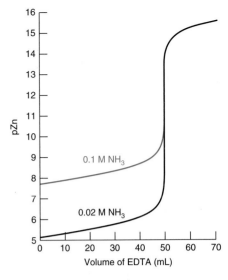

FIGURE 11-14 Titration curves for the reaction of 50.0 mL of 1.00×10^{-3} M Zn^{2+} with 1.00×10^{-3} M EDTA at pH 10.00 in the presence of two different concentrations of NH_3.

End-point detection methods:

1. Metal ion indicators
2. Mercury electrode
3. Ion-selective electrode
4. Glass (pH) electrode

The indicator must release its metal to EDTA.

DEMONSTRATION 11-1 Metal Ion Indicator Color Changes

This demonstration illustrates the color change associated with Reaction 11-19 and shows how a second dye can be added to produce a more easily detected color change.

STOCK SOLUTIONS

Buffer (pH 10.0): Add 142 mL of concentrated (14.5 M) aqueous ammonia to 17.5 g of ammonium chloride and dilute to 250 mL with water.

$MgCl_2$: 0.05 M

EDTA: 0.05 M $Na_2H_2EDTA \cdot 2H_2O$

Prepare a solution containing 25 mL of $MgCl_2$, 5 mL of buffer, and 300 mL of water. Add six drops of Eriochrome black T or Calmagite indicator (Table 11-3) and titrate with EDTA. Note the color change from wine red to pale blue at the end point (Color Plate 6a). The spectroscopic change accompanying the color change is shown in the figure.

For some observers, the change of indicator color is not as sharp as desired. The colors can be affected by adding an "inert"

dye whose color alters the appearance of the solution before and after the titration. Adding 3 mL of methyl red (Table 10-3) (or many other yellow dyes) produces an orange color prior to the end point and a green color after it. This sequence of colors is shown in Color Plate 6b.

Visible spectra of Mg^{2+}-Calmagite and free Calmagite at pH 10 in ammonia buffer. [From C. E. Dahm, J. W. Hall, and B. E. Mattioni, "A Laser Pointer-Based Spectrometer for Endpoint Detection of EDTA Titrations," *J. Chem. Ed.* 2004, *81*, 1787.]

TABLE 11-3 Common metal ion indicators

Name	Structure	pK_a	Color of free indicator	Color of metal ion complex
Calmagite	(H$_2$In$^-$) structure shown	$pK_2 = 8.1$ $pK_3 = 12.4$	H$_2$In$^-$ red HIn^{2-} blue In^{3-} orange	Wine red
Eriochrome black T	(H$_2$In$^-$) structure shown	$pK_2 = 6.3$ $pK_3 = 11.6$	H$_2$In$^-$ red HIn^{2-} blue In^{3-} orange	Wine red
Murexide	(H$_4$In$^-$) structure shown	$pK_2 = 9.2$ $pK_3 = 10.9$	H$_4$In$^-$ red-violet H$_3$In^{2-} violet H$_2$In^{3-} blue	Yellow (with Co^{2+}, Ni^{2+}, Cu^{2+}); red with Ca^{2+}
Xylenol orange	(H$_3$In^{3-}) structure shown	$pK_2 = 2.32$ $pK_3 = 2.85$ $pK_4 = 6.70$ $pK_5 = 10.47$ $pK_6 = 12.23$	H$_5$In$^-$ yellow H$_4$In^{2-} yellow H$_3$In^{3-} yellow H$_2$In^{4-} violet HIn^{5-} violet In^{6-} violet	Red
Pyrocatechol violet	(H$_3$In$^-$) structure shown	$pK_1 = 0.2$ $pK_2 = 7.8$ $pK_3 = 9.8$ $pK_4 = 11.7$	H$_4$In red H$_3$In$^-$ yellow H$_2$In^{2-} violet HIn^{3-} red-purple	Blue

PREPARATION AND STABILITY:

Calmagite: 0.05 g/100 mL H$_2$O; solution is stable for a year in the dark.

Eriochrome black T: Dissolve 0.1 g of the solid in 7.5 mL of triethanolamine plus 2.5 mL of absolute ethanol; solution is stable for months; best used for titrations above pH 6.5.

Murexide: Grind 10 mg of murexide with 5 g of reagent NaCl in a clean mortar; use 0.2–0.4 g of the mixture for each titration.

Xylenol orange: 0.5 g/100 mL H$_2$O; solution is stable indefinitely.

Pyrocatechol violet: 0.1 g/100 mL; solution is stable for several weeks.

Most metal ion indicators are also acid-base indicators, with pK_a values listed in Table 11-3. Because the color of free indicator is pH dependent, most indicators can be used only in certain pH ranges. For example, xylenol orange (pronounced ZY-leen-ol) changes from yellow to red when it binds to a metal ion at pH 5.5. This color change is easy to observe. At pH 7.5, the change is from violet to red and difficult to see. A spectrophotometer can measure the color change, but it is more convenient if we can see it. Figure 11-15 shows pH ranges in which many metals can be titrated and indicators that are useful in different ranges.

The indicator must give up its metal ion to the EDTA. If a metal does not freely dissociate from an indicator, the metal is said to **block** the indicator. Eriochrome black T is blocked by Cu^{2+}, Ni^{2+}, Co^{2+}, Cr^{3+}, Fe^{3+}, and Al^{3+}. It cannot be used for the *direct titration* of any of these metals. It can be used for a *back titration*, however. For example, excess standard EDTA can be added to Cu^{2+}. Then indicator is added and the excess EDTA is back-titrated with Mg^{2+}.

Question What color change occurs when the back titration is performed?

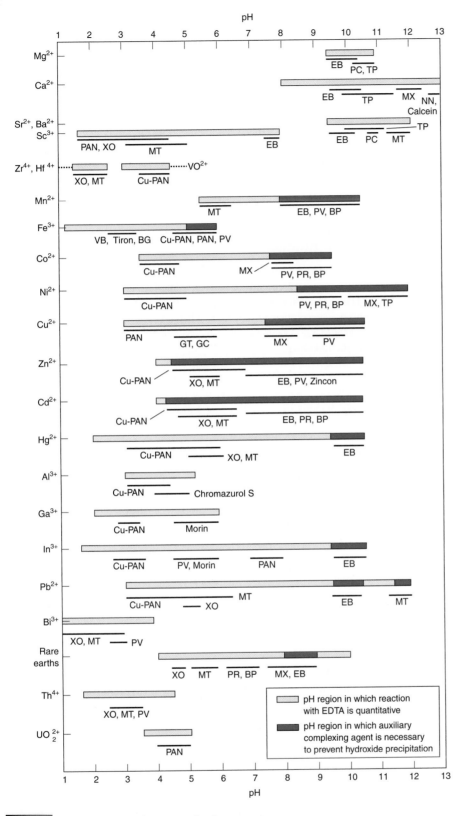

FIGURE 11-15 Guide to EDTA titrations of common metals. Light color shows pH region in which reaction with EDTA is quantitative. Dark color shows pH region in which auxiliary complexing agent is required to prevent metal from precipitating. Calmagite is more stable than Eriochrome black T (EB) and can be substituted for EB. [Adapted from K. Ueno, "Guide for Selecting Conditions of EDTA Titrations," *J. Chem. Ed.* 1965, *42*, 432.]

Abbreviations for indicators:

BG, Bindschedler's green leuco base
BP, Bromopyrogallol red
EB, Eriochrome black T
GC, Glycinecresol red
GT, Glycinethymol blue
MT, Methylthymol blue
MX, Murexide
NN, Patton & Reeder's dye
PAN, Pyridylazonaphthol
Cu-PAN, PAN plus Cu-EDTA
PC, *o*-Cresolphthalein complexone
PR, Pyrogallol red
PV, Pyrocatechol violet
TP, Thymolphthalein complexone
VB, Variamine blue B base
XO, Xylenol orange

$pK_4(OH) = 12.5$
$pK_3(OH) = 7.6$

"Tiron"
1,2-dihydroxybenzene-
3,5-disulfonic acid,
disodium salt

Tiron is an indicator for EDTA titration of Fe(III) at pH 2–3 at 40°C. Color change is from blue to pale yellow.

11-7 EDTA Titration Techniques

Because so many elements can be analyzed with EDTA, there is extensive literature dealing with many variations of the basic procedure.[16,21]

Direct Titration

In a **direct titration,** analyte is titrated with standard EDTA. Analyte is buffered to a pH at which the conditional formation constant for the metal-EDTA complex is large and the color of the free indicator is distinctly different from that of the metal-indicator complex.

Phytoremediation.[19,20] An approach to removing toxic metals from contaminated soil is to grow plants that accumulate 1–15 g metal/g dry mass of plant. The plant is harvested to recover metals such as Pb, Cd, and Ni. Phytoremediation is enhanced by adding EDTA to mobilize insoluble metals. Unfortunately, rain spreads soluble metal-EDTA complexes through the soil, so phytoremediation is limited to locations where the connection with groundwater is blocked or where leaching is not important. The natural chelate, EDDS mobilizes metal and biodegrades before it can spread very far.

S,S-Ethylenediaminedisuccinic acid (EDDS)

Auxiliary complexing agents such as NH_3, tartrate, citrate, or triethanolamine may be employed to prevent metal ion from precipitating in the absence of EDTA. For example, Pb^{2+} is titrated in NH_3 buffer at pH 10 in the presence of tartrate, which complexes Pb^{2+} and does not allow $Pb(OH)_2$ to precipitate. The lead-tartrate complex must be less stable than the lead-EDTA complex, or the titration would not be feasible.

Back Titration

In a **back titration,** a known excess of EDTA is added to the analyte. Excess EDTA is then titrated with a standard solution of a second metal ion. A back titration is necessary if analyte precipitates in the absence of EDTA, if it reacts too slowly with EDTA, or if it blocks the indicator. The metal ion for the back titration must not displace analyte from EDTA.

EXAMPLE **A Back Titration**

Ni^{2+} can be analyzed by a back titration using standard Zn^{2+} at pH 5.5 with xylenol orange indicator. A solution containing 25.00 mL of Ni^{2+} in dilute HCl is treated with 25.00 mL of 0.052 83 M Na_2EDTA. The solution is neutralized with NaOH, and the pH is adjusted to 5.5 with acetate buffer. The solution turns yellow when a few drops of indicator are added. Titration with 0.022 99 M Zn^{2+} requires 17.61 mL to reach the red end point. What is the molarity of Ni^{2+} in the unknown?

Solution The unknown was treated with 25.00 mL of 0.052 83 M EDTA, which contains (25.00 mL)(0.052 83 M) = 1.320 8 mmol of EDTA. Back titration required (17.61 mL) × (0.022 99 M) = 0.404 9 mmol of Zn^{2+}. Because 1 mol of EDTA reacts with 1 mol of any metal ion, there must have been

$$1.320\ 8\ \text{mmol EDTA} - 0.404\ 9\ \text{mmol Zn}^{2+} = 0.915\ 9\ \text{mmol Ni}^{2+}$$

The concentration of Ni^{2+} is 0.915 9 mmol/25.00 mL = 0.036 64 M.

Test Yourself If back titration required 13.00 mL Zn^{2+}, what was the original concentration of Ni^{2+}? (***Answer:*** 0.040 88 M)

Back titration prevents precipitation of analyte. For example, $Al(OH)_3$ precipitates at pH 7 in the absence of EDTA. An acidic solution of Al^{3+} can be treated with excess EDTA, adjusted to pH 7–8 with sodium acetate, and boiled to ensure complete formation of stable, soluble $Al(EDTA)^-$. The solution is then cooled; Calmagite indicator is added; and back titration with standard Zn^{2+} is performed.

Displacement Titration

Hg^{2+} does not have a satisfactory indicator, but a **displacement titration** is feasible. Hg^{2+} is treated with excess $Mg(EDTA)^{2-}$ to displace Mg^{2+}, which is titrated with standard EDTA.

$$M^{n+} + MgY^{2-} \rightarrow MY^{n-4} + Mg^{2+} \qquad (11\text{-}20)$$

The conditional formation constant for $Hg(EDTA)^{2-}$ must be greater than K_f' for $Mg(EDTA)^{2-}$, or else Mg^{2+} will not be displaced from $Mg(EDTA)^{2-}$.

There is no suitable indicator for Ag^+. However, Ag^+ will displace Ni^{2+} from tetracyanonickelate(II) ion:

$$2Ag^+ + Ni(CN)_4^{2-} \rightarrow 2Ag(CN)_2^- + Ni^{2+}$$

The liberated Ni^{2+} can then be titrated with EDTA to find out how much Ag^+ was added.

Indirect Titration

Anions that precipitate with certain metal ions can be analyzed with EDTA by **indirect titration.** For example, sulfate can be analyzed by precipitation with excess Ba^{2+} at pH 1. The $BaSO_4(s)$ is washed and then boiled with excess EDTA at pH 10 to bring Ba^{2+} back into solution as $Ba(EDTA)^{2-}$. Excess EDTA is back-titrated with Mg^{2+}.

BOX 11-3 **Water Hardness**

Hardness is the total concentration of alkaline earth (Group 2) ions, which are mainly Ca^{2+} and Mg^{2+}, in water. Hardness is commonly expressed as the equivalent number of milligrams of $CaCO_3$ per liter. Thus, if $[Ca^{2+}] + [Mg^{2+}] = 1$ mM, we would say that the hardness is 100 mg $CaCO_3$ per liter because 100 mg $CaCO_3$ = 1 mmol $CaCO_3$. Water of hardness less than 60 mg $CaCO_3$ per liter is considered to be "soft." If the hardness is above 270 mg/L, the water is considered to be "hard."

Hard water reacts with soap to form insoluble curds:

$$Ca^{2+} + 2RCO_2^- \rightarrow Ca(RCO_2)_2(s) \qquad (A)$$

Soap   Precipitate
R is a long-chain hydrocarbon such as $C_{17}H_{35}$—

Enough soap to consume Ca^{2+} and Mg^{2+} must be used before soap is available for cleaning. Hard water leaves solid deposits called *scale* on pipes when it evaporates, but hard water is not known to be unhealthy. Hardness is beneficial in irrigation water because alkaline earth ions tend to *flocculate* (cause to aggregate) colloidal particles in soil and thereby increase the permeability of the soil to water. Soft water etches concrete, plaster, and grout.

To measure hardness, the sample is treated with ascorbic acid (or hydroxylamine) to reduce Fe^{3+} to Fe^{2+} and Cu^{2+} to Cu^+ and then with cyanide to mask Fe^{2+}, Cu^+, and several other minor metal ions. Titration with EDTA at pH 10 in NH_3 buffer then gives the total concentrations of Ca^{2+} and Mg^{2+}. Ca^{2+} can be determined separately if the titration is carried out at pH 13 without ammonia. At this pH, $Mg(OH)_2$ precipitates and is inaccessible to EDTA. Interference by many metal ions can be reduced by the right choice of indicators.[23]

Insoluble carbonates are converted into soluble bicarbonates by excess carbon dioxide:

$$CaCO_3(s) + CO_2 + H_2O \rightarrow Ca(HCO_3)_2(aq) \qquad (B)$$

Heat converts bicarbonate into carbonate (driving off CO_2) and precipitates $CaCO_3$ scale that clogs boiler pipes. The fraction of hardness due to $Ca(HCO_3)_2(aq)$ is called *temporary hardness* because this calcium is lost (by precipitation of $CaCO_3$) upon heating. Hardness arising from other salts (mainly dissolved $CaSO_4$) is called *permanent hardness* because it is not removed by heating.

Alternatively, an anion can be precipitated with excess standard metal ion. The precipitate is filtered and washed, and excess metal in the filtrate is titrated with EDTA. Anions such as CO_3^{2-}, CrO_4^{2-}, S^{2-}, and SO_4^{2-} can be determined by indirect titration with EDTA.[22]

Masking

A **masking agent** is a reagent that protects some component of the analyte from reaction with EDTA. For example, Al^{3+} in a mixture of Mg^{2+} and Al^{3+} can be measured by first masking the Al^{3+} with F^-, thereby leaving only the Mg^{2+} to react with EDTA.

Cyanide masks Cd^{2+}, Zn^{2+}, Hg^{2+}, Co^{2+}, Cu^+, Ag^+, Ni^{2+}, Pd^{2+}, Pt^{2+}, Fe^{2+}, and Fe^{3+}, but not Mg^{2+}, Ca^{2+}, Mn^{2+}, or Pb^{2+}. When cyanide is added to a solution containing Cd^{2+} and Pb^{2+}, only Pb^{2+} reacts with EDTA. (*Caution:* Cyanide forms toxic gaseous HCN below pH 11. Cyanide solutions should be strongly basic and only handled in a hood.) Fluoride masks Al^{3+}, Fe^{3+}, Ti^{4+}, and Be^{2+}. (*Caution:* HF formed by F^- in acidic solution is extremely hazardous and should not contact skin and eyes. It may not be immediately painful, but the affected area should be flooded with water and then treated with calcium gluconate gel that you have on hand *before* the accident. First aid providers must wear rubber gloves to protect themselves.) Triethanolamine masks Al^{3+}, Fe^{3+}, and Mn^{2+}; 2,3-dimercapto-1-propanol masks Bi^{3+}, Cd^{2+}, Cu^{2+}, Hg^{2+}, and Pb^{2+}.

Demasking releases metal ion from a masking agent. Cyanide complexes can be demasked with formaldehyde:

$$M(CN)_m^{n-m} + mH_2CO + mH^+ \rightarrow mH_2C \begin{smallmatrix} OH \\ | \\ \\ CN \end{smallmatrix} + M^{n+}$$

Formaldehyde

Thiourea masks Cu^{2+} by reducing it to Cu^+ and complexing the Cu^+. Copper can be liberated from thiourea by oxidation with H_2O_2. Selectivity afforded by masking, demasking, and pH control allows individual components of complex mixtures of metal ions to be analyzed by EDTA titration.

Masking prevents one species from interfering in the analysis of another. Masking is not restricted to EDTA titrations. Box 11-3 gives an important application of masking.

SH
|
$HOCH_2CHCH_2SH$
2,3-Dimercapto-1-propanol

Terms to Understand

Summary

In a complexometric titration, analyte and titrant form a complex ion, and the equilibrium constant is called the formation constant, K_f. Chelating (multidentate) ligands form more stable complexes than do monodentate ligands. Synthetic aminocarboxylic acids such as EDTA have large metal-binding constants.

Formation constants for EDTA are expressed in terms of $[Y^{4-}]$, even though there are six protonated forms of EDTA. Because the fraction ($\alpha_{Y^{4-}}$) of free EDTA in the form Y^{4-} depends on pH, we define a conditional (or effective) formation constant as $K_f' = \alpha_{Y^{4-}} K_f = [MY^{n-4}]/[M^{n+}][EDTA]$. This constant describes the hypothetical reaction $M^{n+} + EDTA \rightleftharpoons MY^{n-4}$, where EDTA refers to all forms of EDTA not bound to metal ion. Titration calculations fall into three categories. When excess unreacted M^{n+} is present, pM is calculated directly from $pM = -\log[M^{n+}]$. When excess EDTA is present, we know both $[MY^{n-4}]$ and [EDTA], so $[M^{n+}]$ can be calculated from the conditional formation constant. At the equivalence point, the condition $[M^{n+}] = [EDTA]$ allows us to solve for $[M^{n+}]$. A single spreadsheet equation applies in all three regions of the titration curve.

The greater the effective formation constant, the sharper is the EDTA titration curve. Addition of auxiliary complexing agents, which compete with EDTA for the metal ion and thereby limit the sharpness of the titration curve, is often necessary to keep the metal in solution. Calculations for a solution containing EDTA and an auxiliary complexing agent utilize the conditional formation constant $K_f'' = \alpha_M \alpha_{Y^{4-}} K_f$, where α_M is the fraction of free metal ion not complexed by the auxiliary ligand.

For end-point detection, we commonly use metal ion indicators, a glass electrode, an ion-selective electrode, or a mercury electrode. When a direct titration is not suitable, because the analyte is unstable, reacts slowly with EDTA, or has no suitable indicator, a back titration of excess EDTA or a displacement titration of $Mg(EDTA)^{2-}$ may be feasible. Masking prevents interference by unwanted species. Indirect EDTA titrations are available for many anions and other species that do not react directly with the reagent.

Exercises

11-A. Potassium ion in a 250.0 (± 0.1) mL water sample was precipitated with sodium tetraphenylborate:

$$K^+ + (C_6H_5)_4B^- \rightarrow KB(C_6H_5)_4(s)$$

The precipitate was filtered, washed, dissolved in an organic solvent, and treated with excess $Hg(EDTA)^{2-}$:

$$4HgY^{2-} + (C_6H_5)_4B^- + 4H_2O \rightarrow$$
$$H_3BO_3 + 4C_6H_5Hg^+ + 4HY^{3-} + OH^-$$

The liberated EDTA was titrated with 28.73 (± 0.03) mL of 0.043 7 ($\pm 0.000\ 1$) M Zn^{2+}. Find $[K^+]$ (and its absolute uncertainty) in the original sample.

11-B. A 25.00-mL sample containing Fe^{3+} and Cu^{2+} required 16.06 mL of 0.050 83 M EDTA for complete titration. A 50.00-mL sample of the unknown was treated with NH_4F to protect the Fe^{3+}. Then Cu^{2+} was reduced and masked by thiourea. Addition of 25.00 mL of 0.050 83 M EDTA liberated Fe^{3+} from its fluoride complex to form an EDTA complex. The excess EDTA required 19.77 mL of 0.018 83 M Pb^{2+} to reach a xylenol orange end point. Find $[Cu^{2+}]$ in the unknown.

11-C. Calculate pCu^{2+} (to the 0.01 decimal place) at each of the following points in the titration of 50.0 mL of 0.040 0 M EDTA with 0.080 0 M $Cu(NO_3)_2$ at pH 5.00: 0.1, 5.0, 10.0, 15.0, 20.0, 24.0, 25.0, 26.0, and 30.0 mL. Make a graph of pCu^{2+} versus volume of titrant.

11-D. Calculate the concentration of H_2Y^{2-} at the equivalence point in Exercise 11-C.

11-E. Suppose that 0.010 0 M Mn^{2+} is titrated with 0.005 00 M EDTA at pH 7.00.

(a) What is the concentration of free Mn^{2+} at the equivalence point?

(b) What is the quotient $[H_3Y^-]/[H_2Y^{2-}]$ in the solution when the titration is just 63.7% of the way to the equivalence point?

11-F. A solution containing 20.0 mL of 1.00×10^{-3} M Co^{2+} in the presence of 0.10 M $C_2O_4^{2-}$ at pH 9.00 was titrated with 1.00×10^{-2} M EDTA. Using formation constants from Appendix I for $Co(C_2O_4)$ and $Co(C_2O_4)_2^{2-}$, calculate pCo^{2+} for the following volumes of added EDTA: 0, 1.00, 2.00, and 3.00 mL. Consider the concentration of $C_2O_4^{2-}$ to be fixed at 0.10 M. Sketch a graph of pCo^{2+} versus milliliters of added EDTA.

11-G. Iminodiacetic acid forms 2:1 complexes with many metal ions:

$$H_2\overset{+}{N}\diagup\diagdown \begin{matrix} CH_2CO_3H \\ CH_2CO_3H \end{matrix} \equiv H_3X^+$$

$$\alpha_{X^{2-}} = \frac{[X^{2-}]}{[H_3X^+] + [H_2X] + [HX^-] + [X^{2-}]}$$

$$Cu^{2+} + 2X^{2-} \rightleftharpoons CuX_2^{2-} \qquad K = \beta_2 = 3.5 \times 10^{16}$$

A 25.0 mL solution containing 0.120 M iminodiacetic acid buffered to pH 7.00 was titrated with 25.0 mL of 0.050 0 M Cu^{2+}. Given that $\alpha_{X^{2-}} = 4.6 \times 10^{-3}$ at pH 7.00, calculate $[Cu^{2+}]$ in the resulting solution.

Problems

EDTA

11-1. What is the chelate effect?

11-2. State (in words) what $\alpha_{Y^{4-}}$ means. Calculate $\alpha_{Y^{4-}}$ for EDTA at (a) pH 3.50 and (b) pH 10.50.

11-3. (a) Find the conditional formation constant for $Mg(EDTA)^{2-}$ at pH 9.00.

(b) Find the concentration of free Mg^{2+} in 0.050 M $Na_2[Mg(EDTA)]$ at pH 9.00.

11-4. *Metal ion buffers.* By analogy to a hydrogen ion buffer, a metal ion buffer tends to maintain a particular metal ion concentration in solution. A mixture of the acid HA and its conjugate base A^- maintains $[H^+]$ defined by the equation $K_a = [A^-][H^+]/[HA]$. A mixture of CaY^{2-} and Y^{4-} serves as a Ca^{2+} buffer governed by the equation $1/K_f' = [EDTA][Ca^{2+}]/[CaY^{2-}]$. How many grams of $Na_2EDTA \cdot 2H_2O$ (FM 372.23) should be mixed with 1.95 g of $Ca(NO_3)_2 \cdot 2H_2O$ (FM 200.12) in a 500-mL volumetric flask to give a buffer with $pCa^{2+} = 9.00$ at pH 9.00?

11-5. *Purification by reprecipitation and predominant species of polyprotic acids.* To measure oxygen isotopes in SO_4^{2-} for geologic studies, SO_4^{2-} was precipitated with excess Ba^{2+}.[24] In the presence of HNO_3, $BaSO_4$ precipitate is contaminated by NO_3^-. The solid can be purified by washing, redissolving in the absence of HNO_3, and reprecipitating. For purification, 30 mg of $BaSO_4$ were dissolved in 15 mL of 0.05 M DTPA (Figure 11-4) in 1 M NaOH with vigorous shaking at 70°C. $BaSO_4$ was reprecipitated by adding 10 M HCl dropwise to obtain pH 3–4 and allowing the mixture to stand for 1 h. The solid was washed twice by centrifuging, removing the liquid, and resuspending in deionized water. The molar ratio NO_3^-/SO_4^{2-} was reduced from 0.25 in the original precipitate to 0.001 after two cycles of dissolution and reprecipitation. What are the predominant species of sulfate and DTPA at pH 14 and pH 3? Explain why $BaSO_4$ dissolves in DTPA in 1 M NaOH and then reprecipitates when the pH is lowered to 3–4.

EDTA Titration Curves

11-6. A 100.0 mL solution of 0.050 0 M M^{n+} buffered to pH 9.00 was titrated with 0.050 0 M EDTA.

(a) What is the equivalence volume, V_e, in milliliters?

(b) Calculate the concentration of M^{n+} at $V = \frac{1}{2}V_e$.

(c) What fraction ($\alpha_{Y^{4-}}$) of free EDTA is in the form Y^{4-} at pH 9.00?

(d) The formation constant (K_f) is $10^{12.00}$. Calculate the value of the conditional formation constant K_f' ($= \alpha_{Y^{4-}} K_f$)

(e) Calculate the concentration of M^{n+} at $V = V_e$.

(f) What is the concentration of M^{n+} at $V = 1.100\ V_e$?

11-7. Calculate pCo^{2+} at each of the following points in the titration of 25.00 mL of 0.020 26 M Co^{2+} by 0.038 55 M EDTA at pH 6.00: (a) 12.00 mL; (b) V_e; (c) 14.00 mL.

11-8. Consider the titration of 25.0 mL of 0.020 0 M $MnSO_4$ with 0.010 0 M EDTA in a solution buffered to pH 8.00. Calculate pMn^{2+} at the following volumes of added EDTA and sketch the titration curve:

(a) 0 mL   (d) 49.0 mL   (g) 50.1 mL

(b) 20.0 mL   (e) 49.9 mL   (h) 55.0 mL

(c) 40.0 mL   (f) 50.0 mL   (i) 60.0 mL

11-9. For the same volumes used in Problem 11-8, calculate pCa^{2+} for the titration of 25.00 mL of 0.020 00 M EDTA with 0.010 00 M $CaSO_4$ at pH 10.00.

11-10. Calculate $[HY^{3-}]$ in a solution prepared by mixing 10.00 mL of 0.010 0 M $VOSO_4$, 9.90 mL of 0.010 0 M EDTA, and 10.0 mL of buffer with a pH of 4.00.

11-11. 🖳 *Titration of metal ion with EDTA.* Use Equation 11-11 to compute curves (pM versus mL of EDTA added) for the titration of 10.00 mL of 1.00 mM M^{2+} ($= Cd^{2+}$ or Cu^{2+}) with 10.0 mM EDTA at pH 5.00. Plot both curves on one graph.

11-12. 🖳 *Effect of pH on the EDTA titration.* Use Equation 11-11 to compute curves (pCa^{2+} versus mL of EDTA added) for the titration of 10.00 mL of 1.00 mM Ca^{2+} with 1.00 mM EDTA at pH 5.00, 6.00, 7.00, 8.00, and 9.00. Plot all curves on one graph and compare your results with Figure 11-10.

11-13. 🖳 *Titration of EDTA with metal ion.* Use Equation 11-12 to reproduce the results of Exercise 11-C.

Auxiliary Complexing Agents

11-14. State the purpose of an auxiliary complexing agent and give an example of its use.

11-15. According to Appendix I, Cu^{2+} forms two complexes with acetate:

$$Cu^{2+} + CH_3CO_2^- \rightleftharpoons Cu(CH_3CO_2)^+ \qquad \beta_1 (= K_1)$$
$$Cu^{2+} + 2CH_3CO_2^- \rightleftharpoons Cu(CH_3CO_2)_2(aq) \qquad \beta_2$$

(a) Referring to Box 6-2, find K_2 for the reaction

$$Cu(CH_3CO_2)^+ + CH_3CO_2^- \rightleftharpoons Cu(CH_3CO_2)_2(aq) \qquad K_2$$

(b) Consider 1.00 L of solution prepared by mixing 1.00×10^{-4} mol $Cu(ClO_4)_2$ and 0.100 mol CH_3CO_2Na. Use Equation 11-16 to find the fraction of copper in the form Cu^{2+}.

11-16. Calculate pCu^{2+} at each of the following points in the titration of 50.00 mL of 0.001 00 M Cu^{2+} with 0.001 00 M EDTA at pH 11.00 in a solution with $[NH_3]$ *fixed* at 1.00 M:

(a) 0 mL   (c) 45.00 mL   (e) 55.00 mL

(b) 1.00 mL   (d) 50.00 mL

11-17. Consider the derivation of the fraction α_M in Equation 11-16.

(a) Derive the following expressions for the fractions α_{ML} and α_{ML_2}:

$$\alpha_{ML} = \frac{\beta_1[L]}{1 + \beta_1[L] + \beta_2[L]^2} \qquad \alpha_{ML_2} = \frac{\beta_2[L]^2}{1 + \beta_1[L] + \beta_2[L]^2}$$

(b) Calculate the values of α_{ML} and α_{ML_2} for the conditions in Problem 11-15.

11-18. *Microequilibrium constants for binding of metal to a protein.* The iron-transport protein, transferrin, has two distinguishable metal-binding sites, designated a and b. The *microequilibrium* formation constants for each site are defined as follows:

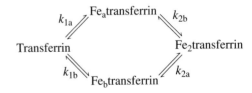

For example, the formation constant k_{1a} refers to the reaction $Fe^{3+} + transferrin \rightleftharpoons Fe_a transferrin$, in which the metal ion binds to site a:

$$k_{1a} = \frac{[Fe_a transferrin]}{[Fe^{3+}][transferrin]}$$

(a) Write the chemical reactions corresponding to the conventional macroscopic formation constants, K_1 and K_2.

(b) Show that $K_1 = k_{1a} + k_{1b}$ and $K_2^{-1} = k_{2a}^{-1} + k_{2b}^{-1}$.

(c) Show that $k_{1a}k_{2b} = k_{1b}k_{2a}$. This expression means that, if you know any three of the microequilibrium constants, you automatically know the fourth one.

(d) *A challenge to your sanity.* From the equilibrium constants below, *find the equilibrium fraction of each of the four species in the diagram above in circulating blood that is 40% saturated with iron* (that is, Fe/transferrin = 0.80, because each protein can bind 2 Fe).

Effective formation constants for blood plasma at pH 7.4

$k_{1a} = 6.0 \times 10^{22}$	$k_{2a} = 2.4 \times 10^{22}$
$k_{1b} = 1.0 \times 10^{22}$	$k_{2b} = 4.2 \times 10^{21}$
$K_1 = 7.0 \times 10^{22}$	$K_2 = 3.6 \times 10^{21}$

The binding constants are so large that you may assume that there is negligible free Fe^{3+}. To get started, let's use the abbreviations [T] = [transferrin], [FeT] = [Fe$_a$T] + [Fe$_b$T], and [Fe$_2$T] = [Fe$_2$transferrin]. Now we can write

Mass balance for protein: $[T] + [FeT] + [Fe_2T] = 1$ (A)

Mass balance for iron:

$$\frac{[FeT] + 2[Fe_2T]}{[T] + [FeT] + [Fe_2T]} = [FeT] + 2[Fe_2T] = 0.8 \quad \text{(B)}$$

Combined equilibria: $\dfrac{K_1}{K_2} = 19_{.44} = \dfrac{[FeT]^2}{[T][Fe_2T]}$ (C)

Now you have three equations with three unknowns and should be able to tackle this problem.

11-19. *Spreadsheet equation for auxiliary complexing agent.* Consider the titration of metal M (initial concentration = C_M, initial volume = V_M) with EDTA (concentration = C_{EDTA}, volume added = V_{EDTA}) in the presence of an auxiliary complexing ligand (such as ammonia). Follow the derivation in Section 11-4 to show that the master equation for the titration is

$$\phi = \frac{C_{EDTA}V_{EDTA}}{C_M V_M} = \frac{1 + K_f''[M]_{free} - \dfrac{[M]_{free} + K_f''[M]_{free}^2}{C_M}}{K_f''[M]_{free} + \dfrac{[M]_{free} + K_f''[M]_{free}^2}{C_{EDTA}}}$$

where K_f'' is the conditional formation constant in the presence of auxiliary complexing agent at the fixed pH of the titration (Equation 11-18) and $[M]_{free}$ is the total concentration of metal not bound to EDTA. $[M]_{free}$ is the same as M_{tot} in Equation 11-15. The result is equivalent to Equation 11-11, with [M] replaced by $[M]_{free}$ and K_f replaced by K_f''.

11-20. *Auxiliary complexing agent.* Use the equation derived in Problem 11-19.

(a) Prepare a spreadsheet to reproduce the 20-, 50-, and 60-mL points in the EDTA titration of Zn^{2+} in the presence of NH_3 in the example on pages 248–249.

(b) Use your spreadsheet to plot the curve for the titration of 50.00 mL of 5.00 mM Ni^{2+} by 10.00 mM EDTA at pH 11.00 in the presence of a fixed oxalate concentration of 0.100 M.

11-21. *Spreadsheet equation for formation of the complexes ML and* ML_2. Consider the titration of metal M (initial concentration = C_M, initial volume = V_M) with ligand L (concentration = C_L, volume added = V_L), which can form 1:1 and 2:1 complexes:

$$M + L \rightleftharpoons ML \qquad \beta_1 = \frac{[ML]}{[M][L]}$$

$$M + 2L \rightleftharpoons ML_2 \qquad \beta_2 = \frac{[ML_2]}{[M][L]^2}$$

Let α_M be the fraction of metal in the form M, α_{ML} be the fraction in the form ML, and α_{ML_2} be the fraction in the form ML_2. Following the derivation in Section 11-5, you could show that these fractions are given by

$$\alpha_M = \frac{1}{1 + \beta_1[L] + \beta_2[L]^2} \qquad \alpha_{ML} = \frac{\beta_1[L]}{1 + \beta_1[L] + \beta_2[L]^2}$$

$$\alpha_{ML_2} = \frac{\beta_2[L]^2}{1 + \beta_1[L] + \beta_2[L]^2}$$

The concentrations of ML and ML_2 are

$$[ML] = \alpha_{ML}\frac{C_M V_M}{V_M + V_L} \qquad [ML_2] = \alpha_{ML_2}\frac{C_M V_M}{V_M + V_L}$$

because $C_M V_M/(V_M + V_L)$ is the total concentration of all metal in the solution. The mass balance for ligand is

$$[L] + [ML] + 2[ML_2] = \frac{C_L V_L}{V_M + V_L}$$

By substituting expressions for [ML] and [ML$_2$] into the mass balance, show that the master equation for a titration of metal by ligand is

$$\phi = \frac{C_L V_L}{C_M V_M} = \frac{\alpha_{ML} + 2\alpha_{ML_2} + ([L]/C_M)}{1 - ([L]/C_L)}$$

11-22. *Titration of M with L to form ML and* ML_2. Use the equation from Problem 11-21, where M is Cu^{2+} and L is acetate. Consider adding 0.500 M acetate to 10.00 mL of 0.050 0 M Cu^{2+} at pH 7.00 (so that all ligand is present as $CH_3CO_2^-$, not CH_3CO_2H). Formation constants for $Cu(CH_3CO_2)^+$ and $Cu(CH_3CO_2)_2$ are given in Appendix I. Construct a spreadsheet in which the input is pL and the output is (a) [L]; (b) V_L; (c) [M]; (d) [ML]; (e) [ML$_2$]. Prepare a graph showing concentrations of L, M, ML, and ML_2 as V_L ranges from 0 to 3 mL.

Metal Ion Indicators

11-23. Explain why the change from red to blue in Reaction 11-19 occurs suddenly at the equivalence point instead of gradually throughout the entire titration.

11-24. List four methods for detecting the end point of an EDTA titration.

11-25. Calcium ion was titrated with EDTA at pH 11, using Calmagite as indicator (Table 11-3). Which is the principal species of Calmagite at pH 11? What color was observed before the equivalence point? After the equivalence point?

11-26. Pyrocatechol violet (Table 11-3) is to be used as a metal ion indicator in an EDTA titration. The procedure is as follows:

1. Add a known excess of EDTA to the unknown metal ion.
2. Adjust the pH with a suitable buffer.
3. Back-titrate the excess chelate with standard Al^{3+}.

From the following available buffers, select the best buffer, and then state what color change will be observed at the end point. Explain your answer.

(a) pH 6–7 (b) pH 7–8 (c) pH 8–9 (d) pH 9–10

EDTA Titration Techniques

11-27. Give three circumstances in which an EDTA back titration might be necessary.

11-28. Describe what is done in a displacement titration and give an example.

11-29. Give an example of the use of a masking agent.

11-30. What is meant by water hardness? Explain the difference between temporary and permanent hardness.

11-31. How many milliliters of 0.050 0 M EDTA are required to react with 50.0 mL of 0.010 0 M Ca^{2+}? With 50.0 mL of 0.010 0 M Al^{3+}?

11-32. A 50.0-mL sample containing Ni^{2+} was treated with 25.0 mL of 0.050 0 M EDTA to complex all the Ni^{2+} and leave excess EDTA in solution. The excess EDTA was then back-titrated, requiring 5.00 mL of 0.050 0 M Zn^{2+}. What was the concentration of Ni^{2+} in the original solution?

11-33. A 50.0-mL aliquot of solution containing 0.450 g of $MgSO_4$ (FM 120.37) in 0.500 L required 37.6 mL of EDTA solution for titration. How many milligrams of $CaCO_3$ (FM 100.09) will react with 1.00 mL of this EDTA solution?

11-34. Cyanide solution (12.73 mL) was treated with 25.00 mL of Ni^{2+} solution (containing excess Ni^{2+}) to convert the cyanide into tetracyanonickelate(II):

$$4CN^- + Ni^{2+} \rightarrow Ni(CN)_4^{2-}$$

Excess Ni^{2+} was then titrated with 10.15 mL of 0.013 07 M EDTA. $Ni(CN)_4^{2-}$ does not react with EDTA. If 39.35 mL of EDTA were required to react with 30.10 mL of the original Ni^{2+} solution, calculate the molarity of CN^- in the 12.73-mL sample.

11-35. A 1.000-mL sample of unknown containing Co^{2+} and Ni^{2+} was treated with 25.00 mL of 0.038 72 M EDTA. Back titration with 0.021 27 M Zn^{2+} at pH 5 required 23.54 mL to reach the xylenol orange end point. A 2.000-mL sample of unknown was passed through an ion-exchange column that retards Co^{2+} more than Ni^{2+}. The Ni^{2+} that passed through the column was treated with 25.00 mL of 0.038 72 M EDTA and required 25.63 mL of 0.021 27 M Zn^{2+} for back titration. The Co^{2+} emerged from the column later. It, too, was treated with 25.00 mL of 0.038 72 M EDTA. How many milliliters of 0.021 27 M Zn^{2+} will be required for back titration?

11-36. A 50.0-mL solution containing Ni^{2+} and Zn^{2+} was treated with 25.0 mL of 0.045 2 M EDTA to bind all the metal. The excess unreacted EDTA required 12.4 mL of 0.012 3 M Mg^{2+} for complete reaction. An excess of the reagent 2,3-dimercapto-1-propanol was then added to displace the EDTA from zinc. Another 29.2 mL of Mg^{2+} were required for reaction with the liberated EDTA. Calculate the molarity of Ni^{2+} and Zn^{2+} in the original solution.

11-37. Sulfide ion was determined by indirect titration with EDTA. To a solution containing 25.00 mL of 0.043 32 M $Cu(ClO_4)_2$ plus 15 mL of 1 M acetate buffer (pH 4.5) were added 25.00 mL of unknown sulfide solution with vigorous stirring. The CuS precipitate was filtered and washed with hot water. Then ammonia was added to the filtrate (which contained excess Cu^{2+}) until the blue color of $Cu(NH_3)_4^{2+}$ was observed. Titration of the filtrate with 0.039 27 M EDTA required 12.11 mL to reach the murexide end point. Calculate the molarity of sulfide in the unknown.

11-38. *Indirect EDTA determination of cesium.* Cesium ion does not form a strong EDTA complex, but it can be analyzed by adding a known excess of $NaBiI_4$ in cold concentrated acetic acid containing excess NaI. Solid $Cs_3Bi_2I_9$ is precipitated, filtered, and removed. The excess yellow BiI_4^- is then titrated with EDTA. The end point occurs when the yellow color disappears. (Sodium thiosulfate is used in the reaction to prevent the liberated I^- from being oxidized to yellow aqueous I_2 by O_2 from the air.) The precipitation is fairly selective for Cs^+. The ions Li^+, Na^+, K^+, and low concentrations of Rb^+ do not interfere, although Tl^+ does. Suppose that 25.00 mL of unknown containing Cs^+ were treated with 25.00 mL of 0.086 40 M $NaBiI_4$ and the unreacted BiI_4^- required 14.24 mL of 0.043 7 M EDTA for complete titration. Find the concentration of Cs^+ in the unknown.

11-39. The sulfur content of insoluble sulfides that do not readily dissolve in acid can be measured by oxidation with Br_2 to SO_4^{2-}.[25] Metal ions are then replaced with H^+ by an ion-exchange column, and sulfate is precipitated as $BaSO_4$ with a known excess of $BaCl_2$. The excess Ba^{2+} is then titrated with EDTA to determine how much was present. (To make the indicator end point clearer, a small, known quantity of Zn^{2+} also is added. The EDTA titrates both the Ba^{2+} and the Zn^{2+}.) Knowing the excess Ba^{2+}, we can calculate how much sulfur was in the original material. To analyze the mineral sphalerite (ZnS, FM 97.46), 5.89 mg of powdered solid were suspended in a mixture of CCl_4 and H_2O containing 1.5 mmol Br_2. After 1 h at 20°C and 2 h at 50°C, the powder dissolved and the solvent and excess Br_2 were removed by heating. The residue was dissolved in 3 mL of water and passed through an ion-exchange column to replace Zn^{2+} with H^+. Then 5.000 mL of 0.014 63 M $BaCl_2$ were added to precipitate all sulfate as $BaSO_4$. After the addition of 1.000 mL of 0.010 00 M $ZnCl_2$ and 3 mL of ammonia buffer, pH 10, the excess Ba^{2+} and Zn^{2+} required 2.39 mL of 0.009 63 M EDTA to reach the Calmagite end point. Find the weight percent of sulfur in the sphalerite. What is the theoretical value?

12 Advanced Topics in Equilibrium

St. Paul's Cathedral, London. [Pictor International/ Picture Quest.]

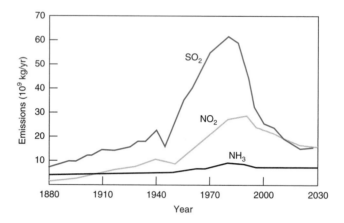

Estimated emissions over Europe. [From R. F. Wright, T. Larssen, L. Camarero, B. J. Crosby, R. C. Ferrier, R. Helliwell, M. Forsius, A. Jenkins, J. Kopáček, V. Majer, F. Moldan, M. Posch, M. Rogora, and W. Schöpp, "Recovery of Acidified European Surface Waters," *Environ. Sci. Technol.* **2005**, *39*, 64A.]

Limestone and marble are building materials whose main constituent is calcite, the common crystalline form of calcium carbonate. This mineral is not very soluble in neutral or basic solution ($K_{sp} = 4.5 \times 10^{-9}$), but it dissolves in acid by virtue of two **coupled equilibria,** in which the reactions have a species in common—carbonate, in this case:

$$CaCO_3(s) \rightleftharpoons Ca^{2+} + CO_3^{2-}$$

Calcite · · · · · · · · · · · Carbonate

$$CO_3^{2-} + H^+ \rightleftharpoons HCO_3^-$$

Bicarbonate

Carbonate produced in the first reaction is protonated to form bicarbonate in the second reaction. Le Châtelier's principle tells us that, if we remove a product of the first reaction, we will draw the reaction to the right, making calcite more soluble. This chapter deals with coupled equilibria in chemical systems.

Between 1980 and 1990, half a millimeter of the thickness of the external stone walls of St. Paul's Cathedral in London was dissolved by acidic rainfall. A corner of the building facing a power station dissolved at 10 times the rate of the rest of the building until the station was closed. The power station and other industries that burn coal emit SO_2, which is a major source of acid rain (described in Box 14-1). Loss of heavy industry and laws limiting emissions decreased atmospheric SO_2 from as high as 100 ppb in the 1970s to 10 ppb in 2000. Only a quarter of a millimeter of St. Paul's external stone disappeared between 1990 and 2000.[1,2]

This optional chapter provides tools to compute the concentrations of species in systems with many simultaneous equilibria.[3] The most important tool is the systematic treatment of equilibrium from Chapter 7. The other tool is a spreadsheet for numerical solution of the equilibrium equations. We will also see how to incorporate activity coefficients into equilibrium calculations. Later chapters in this book do not depend on Chapter 12.

12-1 General Approach to Acid-Base Systems

We first illustrate a general approach to find the concentrations of species in mixtures of acids and bases. Consider a solution made by dissolving 20.0 mmol sodium tartrate (Na^+HT^-), 15.0 mmol pyridinium chloride (PyH^+Cl^-), and 10.0 mmol KOH in a volume of 1.00 L. The problem is to find the pH and concentrations of all species in the solution.

The approach to equilibrium problems in this chapter is adapted from Julian Roberts, University of Redlands.

D-Tartaric acid
H_2T
$pK_1 = 3.036$, $pK_2 = 4.366$

Pyridinium chloride
PyH^+Cl^-
$pK_a = 5.20$

For this example, we designate the two acid dissociation constants of H_2T as K_1 and K_2. We designate the acid dissociation constant of PyH^+ as K_a.

The chemical reactions and equilibrium constants at 0 ionic strength are

$$H_2T \rightleftharpoons HT^- + H^+ \qquad K_1 = 10^{-3.036} \qquad (12\text{-}1)$$

$$HT^- \rightleftharpoons T^{2-} + H^+ \qquad K_2 = 10^{-4.366} \qquad (12\text{-}2)$$

$$PyH^+ \rightleftharpoons Py + H^+ \qquad K_a = 10^{-5.20} \qquad (12\text{-}3)$$

$$H_2O \rightleftharpoons H^+ + OH^- \qquad K_w = 10^{-14.00} \qquad (12\text{-}4)$$

The charge balance is

$$[H^+] + [PyH^+] + [Na^+] + [K^+] = [OH^-] + [HT^-] + 2[T^{2-}] + [Cl^-] \quad (12\text{-}5)$$

There is a factor of 2 in front of $[T^{2-}]$ because the ion has a charge -2. 1 M T^{2-} contributes a charge of 2 M.

and there are several mass balances:

$$[Na^+] = 0.020\,0\,M \qquad [K^+] = 0.010\,0\,M \qquad [Cl^-] = 0.015\,0\,M$$

$$[H_2T] + [HT^-] + [T^{2-}] = 0.020\,0\,M \qquad [PyH^+] + [Py] = 0.015\,0\,M$$

There are 10 independent equations and 10 species, so we have enough information to solve for all the concentrations.

There is a systematic way to handle this problem without algebraic gymnastics.

"Independent" equations cannot be derived from one another. As a trivial example, the equations $a = b + c$ and $2a = 2b + 2c$ are not independent. The three equilibrium expressions for K_a, K_b, and K_w for a weak acid and its conjugate base provide only two independent equations because we can derive K_b from K_a and K_w: $K_b = K_w/K_a$.

Step 1 Write a *fractional composition equation* from Section 9-5 for each acid or base that appears in the charge balance.

Step 2 Substitute the fractional composition expressions into the charge balance and enter known values for $[Na^+]$, $[K^+]$, and $[Cl^-]$. Also, write $[OH^-] = K_w/[H^+]$. At this point, you will have a complicated equation in which the only variable is $[H^+]$.

Step 3 Use your trusty spreadsheet to solve for $[H^+]$.

Here is a recap of the fractional composition equations from Section 9-5 for *any* monoprotic acid HA and *any* diprotic acid H_2A.

Monoprotic system:
$$[HA] = \alpha_{HA}F_{HA} = \frac{[H^+]F_{HA}}{[H^+] + K_a} \qquad (12\text{-}6a)$$

$F_{HA} = [HA] + [A^-]$

$$[A^-] = \alpha_{A^-}F_{HA} = \frac{K_aF_{HA}}{[H^+] + K_a} \qquad (12\text{-}6b)$$

Diprotic system:
$$[H_2A] = \alpha_{H_2A}F_{H_2A} = \frac{[H^+]^2F_{H_2A}}{[H^+]^2 + [H^+]K_1 + K_1K_2} \qquad (12\text{-}7a)$$

$F_{H_2A} = [H_2A] + [HA^-] + [A^{2-}]$

$$[HA^-] = \alpha_{HA^-}F_{H_2A} = \frac{K_1[H^+]F_{H_2A}}{[H^+]^2 + [H^+]K_1 + K_1K_2} \qquad (12\text{-}7b)$$

$$[A^{2-}] = \alpha_{A^{2-}}F_{H_2A} = \frac{K_1K_2F_{H_2A}}{[H^+]^2 + [H^+]K_1 + K_1K_2} \qquad (12\text{-}7c)$$

Table 10-5 gave fractional composition equations for H_3A.

In each equation, α_i is the fraction in each form. For example, $\alpha_{A^{2-}}$ is the fraction of diprotic acid in the form A^{2-}. When we multiply $\alpha_{A^{2-}}$ times F_{H_2A} (the total or formal concentration of H_2A), the product is the concentration of A^{2-}.

Applying the General Procedure

Now let's apply the general procedure to the mixture of 0.020 0 M sodium tartrate (Na^+HT^-), 0.015 0 M pyridinium chloride (PyH^+Cl^-), and 0.010 0 M KOH. We designate the formal concentrations as $F_{H_2T} = 0.020\ 0$ M and $F_{PyH^+} = 0.015\ 0$ M.

Step 1 Write a *fractional composition equation* for each acid or base that appears in the charge balance.

$$[PyH^+] = \alpha_{PyH^+}F_{PyH^+} = \frac{[H^+]F_{PyH^+}}{[H^+] + K_a} \tag{12-8}$$

$$[HT^-] = \alpha_{HT^-}F_{H_2T} = \frac{K_1[H^+]F_{H_2T}}{[H^+]^2 + [H^+]K_1 + K_1K_2} \tag{12-9}$$

$$[T^{2-}] = \alpha_{T^{2-}}F_{H_2T} = \frac{K_1K_2F_{H_2T}}{[H^+]^2 + [H^+]K_1 + K_1K_2} \tag{12-10}$$

All quantities on the right side of these expressions are known, except for $[H^+]$.

Step 2 Substitute the fractional composition expressions into the charge balance 12-5. Enter values for $[Na^+]$, $[K^+]$, and $[Cl^-]$, and write $[OH^-] = K_w/[H^+]$.

$$[H^+] + [PyH^+] + [Na^+] + [K^+] = [OH^-] + [HT^-] + 2[T^{2-}] + [Cl^-] \tag{12-5}$$

K_a, K_1, K_2, and $[H^+]$ are contained in the α expressions. The only variable in Equation 12-11 is $[H^+]$.

$$[H^+] + \alpha_{PyH^+}F_{PyH^+} + [0.020\ 0] + [0.010\ 0]$$
$$= \frac{K_w}{[H^+]} + \alpha_{HT^-}F_{H_2T} + 2\alpha_{T^{2-}}F_{H_2T} + [0.015\ 0] \tag{12-11}$$

Step 3 The spreadsheet in Figure 12-1 solves Equation 12-11 for $[H^+]$.

In Figure 12-1, shaded boxes contain input. Everything else is computed by the spreadsheet. Values for F_{H_2T}, pK_1, pK_2, F_{PyH^+}, pK_a, and $[K^+]$ were given in the problem. The initial value of pH in cell H13 is a *guess*. We will use Excel Solver to vary pH until the sum of charges in cell E15 is 0. Species in the charge balance are in cells B10:E13. $[H^+]$ in cell B10 is

Key step: *Guess* a value for $[H^+]$ and use Excel Solver to vary $[H^+]$ until it satisfies the charge balance.

	A	B	C	D	E	F	G	H	I
1	Mixture of 0.020 M Na⁺HT⁻, 0.015 M PyH⁺Cl⁻, and 0.010 M KOH								
2									
3	F_{H2T} =	0.020		F_{PyH+} =	0.015		$[K^+]$ =	0.010	
4	pK_1 =	3.036		pK_a =	5.20		K_w =	1.00E-14	
5	pK_2 =	4.366		K_a =	6.31E-06				
6	K_1 =	9.20E-04							
7	K_2 =	4.31E-05							
8									
9	Species in charge balance:						Other concentrations:		
10	$[H^+]$ =	1.00E-06		$[OH^-]$ =	1.00E-08		$[H_2T]$ =	4.93E-07	
11	$[PyH^+]$ =	2.05E-03		$[HT^-]$ =	4.54E-04		$[Py]$ =	1.29E-02	
12	$[Na^+]$ =	0.020		$[T^{2-}]$ =	1.95E-02				
13	$[K^+]$ =	0.010		$[Cl^-]$ =	0.015		pH =	6.000	← initial value
14									is a guess
15	Positive charge minus negative charge =				-2.25E-02	← vary pH in H13 with Solver to make this 0			
16					E15 = B10+B11+B12+B13-E10-E11-2*E12-E13				
17	Check: $[PyH^+]$ + $[Py]$ =			0.01500	(= B11+H11)				
18	Check: $[H_2T]$ + $[HT^-]$ + $[T^{2-}]$ =			0.02000	(= H10+E11+E12)				
19									
20	Formulas:								
21	B6 = 10^-B4		B7 = 10^-B5		E5 = 10^-E4		E10 = H4/B10		
22	B10 = 10^-H13		B12 = B3		B13 = H3		E13 = E3		
23	E11 = B6*B10*B3/(B10^2+B10*B6+B6*B7)						B11 = B10*E3/(B10+E5)		
24	E12 = B6*B7*B3/(B10^2+B10*B6+B6*B7)						H11 = E5*E3/(B10+E5)		
25	H10 = B10^2*B3/(B10^2+B10*B6+B6*B7)								

FIGURE 12-1 Spreadsheet for mixture of acids and bases uses Solver to find the value of pH in cell H13 that satisfies the charge balance in cell E15. The sums $[PyH^+]$ + $[Py]$ in cell D17 and $[H_2T]$ + $[HT^-]$ + $[T^{2-}]$ in cell D18 are computed to verify that the formulas for each of the species do not have mistakes. These sums are independent of pH.

computed from the pH we guessed in cell H13. $[PyH^+]$ in cell B11 is computed with Equation 12-8. Known values are entered for $[Na^+]$, $[K^+]$, and $[Cl^-]$. $[OH^-]$ is computed from $K_w/[H^+]$. $[HT^-]$ and $[T^{2-}]$ in cells E11 and E12 are computed with Equations 12-9 and 12-10.

The sum of charges, $[H^+] + [PyH^+] + [Na^+] + [K^+] - [OH^-] - [HT^-] - 2[T^{2-}] - [Cl^-]$, is computed in cell E15. If we had guessed the correct pH in cell H13, the sum of the charges would be 0. Instead, the sum is -2.25×10^{-2} M. We use Excel Solver to vary pH in cell H13 until the sum of charges in cell E15 is 0.

Using Excel Solver

In Excel 2007, in the Data ribbon, select Solver, and a window similar to Figure 18-4 appears. (If you don't see Solver in the Analysis section of the ribbon, click the Microsoft Office button at the top left of the spreadsheet. Click Excel Options and select Add-Ins. Highlight Solver Add-In, click OK, and Solver is loaded.) In the Data ribbon, click on Solver. In the Solver window, Set Target Cell E15 Equal To Value of 0 By Changing Cells H13. Then click Solve. Solver will vary the pH in cell H13 to make the net charge in cell E15 equal to 0. Starting with a pH of 6 in cell H13, Solver returns a net charge of $\sim 10^{-6}$ in cell E15 by adjusting the pH in cell H13 to 4.298. (In earlier versions of Excel, select Solver in the Tools menu and proceed as described above. If you don't see Solver in the Tools menu, select Add-Ins and click on Solver in the Add-Ins window. Click OK and Solver appears in the Tools menu.)

The reason the charge is 10^{-6} instead of 0 is that the default precision of Solver is 10^{-6}. To get a charge closer to 0, click on Options in the Solver window. Precision will probably have the default setting of 0.000 001. Enter 1e-16 for Precision and click OK. Run Solver again. This time, the charge in cell E15 is reduced to $\sim 10^{-16}$. The pH in cell H13 is still 4.298 (to three decimal places). The difference in pH required to reduce the net charge from 10^{-6} to 10^{-16} is not perceptible in the third decimal place of pH. Many chemistry problems involve very large or very small numbers for which you might need to adjust the precision of Solver. The concentrations after executing Solver are

	A	B	C	D	E	F	G	H
9	Species in charge balance:						Other concentrations:	
10	$[H^+]$ =	5.04E-05		$[OH^-]$ =	1.99E-10		$[H_2T]$ =	5.73E-04
11	$[PyH^+]$ =	1.33E-02		$[HT^-]$ =	1.05E-02		$[Py]$ =	1.67E-03
12	$[Na^+]$ =	0.020		$[T^{2-}]$ =	8.95E-03			
13	$[K^+]$ =	0.010		$[Cl^-]$ =	0.015		pH =	4.298
14								
15	Positive charge minus negative charge =				9.71E-17			

Ignorance Is Bliss: A Complication of Ion Pairing

We should not get too cocky with our newfound power to handle complex problems, because we have oversimplified the actual situation. For one thing, we have not included activity coefficients, which ordinarily affect the answer by a few tenths of a pH unit. In Section 12-2, we will show how to incorporate activity coefficients.

Even with activity coefficients, we are always limited by chemistry that we do not know. In the mixture of sodium hydrogen tartrate (Na^+HT^-), pyridinium chloride (PyH^+Cl^-), and KOH, several possible ion-pair equilibria are

$$Na^+ + T^{2-} \rightleftharpoons NaT^- \qquad K_{NaT^-} = \frac{[NaT^-]}{[Na^+][T^{2-}]} = 8 \qquad (12\text{-}12)$$

$$Na^+ + HT^- \rightleftharpoons NaHT \qquad K_{NaHT} = \frac{[NaHT]}{[Na^+][[HT^-]} = 1.6 \qquad (12\text{-}13)$$

$$Na^+ + Py \rightleftharpoons PyNa^+ \qquad K_{PyNa^+} = 1.0 \qquad (12\text{-}14)$$

$$K^+ + T^{2-} \rightleftharpoons KT^- \qquad K_{KT^-} = 3$$

$$K^+ + HT^- \rightleftharpoons KHT \qquad K_{KHT} = ?$$

$$PyH^+ + Cl^- \rightleftharpoons PyH^+Cl^- \qquad K_{PyHCl} = ?$$

$$PyH^+ + T^{2-} \rightleftharpoons PyHT^- \qquad K_{PyHT^-} = ?$$

Equilibrium constants from A. E. Martell, R. M. Smith, and R. J. Motekaitis, *NIST Standard Reference Database 46*, Version 6.0, 2001.

Some equilibrium constants at 0 ionic strength are listed above. Values for the other reactions are not available, but there is no reason to believe that the reactions do not occur.

How could we add ion pairing to our spreadsheet? For simplicity, we outline how to add just Reactions 12-12 and 12-13. With these reactions, the mass balance for sodium is

$$[Na^+] + [NaT^-] + [NaHT] = F_{Na} = F_{H_2T} = 0.020\ 0\ M \qquad (12\text{-}15)$$

From the ion-pair equilibria, we write $[NaT^-] = K_{NaT^-}[Na^+][T^{2-}]$ and $[NaHT] = K_{NaHT}[Na^+][HT^-]$. With these substitutions for $[NaT^-]$ and $[NaHT]$ in the mass balance for sodium, we find an expression for $[Na^+]$:

$$[Na^+] = \frac{F_{H_2T}}{1 + K_{NaT^-}[T^{2-}] + K_{NaHT}[HT^-]} \qquad (12\text{-}16)$$

A bigger nuisance when considering ion pairs is that the fractional composition equations for $[H_2T]$, $[HT^-]$, and $[T^{2-}]$ also are changed because the mass balance for H_2T now has five species in it instead of three:

$$F_{H_2T} = [H_2T] + [HT^-] + [T^{2-}] + [NaT^-] + [NaHT] \qquad (12\text{-}17)$$

We must derive new equations analogous to 12-9 and 12-10 from the mass balance 12-17.

The new fractional composition equations are messy, so we reserve this case for Problem 12-19. The end result is that ion-pair equilibria 12-12 and 12-13 change the calculated pH from 4.30 to 4.26. This change is not large, so neglecting ion pairs with small equilibrium constants does not lead to serious error. We find that 7% of sodium is tied up in ion pairs. *Our ability to compute the distribution of species in a solution is limited by our knowledge of relevant equilibria.*

12-2 Activity Coefficients

Even if we know all reactions and equilibrium constants for a given system, we cannot compute concentrations accurately without activity coefficients. Chapter 7 gave the extended Debye-Hückel equation 7-6 for activity coefficients with size parameters in Table 7-1. Many ions of interest are not in Table 7-1, and we do not know their size parameter. Therefore, we introduce the *Davies equation*, which has no size parameter:

Davies equation: $$\log \gamma = -0.51z^2\left(\frac{\sqrt{\mu}}{1 + \sqrt{\mu}} - 0.3\mu\right) \qquad (\text{at } 25°C) \qquad (12\text{-}18)$$

where γ is the activity coefficient for an ion of charge z at ionic strength μ. Equation 12-18 is used up to $\mu \approx 0.5$ M (Figure 12-2) but is more accurate at lower ionic strength. For best accuracy, Pitzer's equations are used (Chapter 7, reference 7).

Now consider a primary standard buffer containing 0.025 0 m KH_2PO_4 and 0.025 0 m Na_2HPO_4. Its pH at 25°C is 6.865 ± 0.006.[4] The concentration unit, m, is *molality*, which means moles of solute per kilogram of solvent. For precise chemical measurements, concentrations are expressed in molality, rather than molarity, because molality is independent of temperature. Tabulated equilibrium constants usually apply to molality, not molarity. Uncertainties in equilibrium constants are usually sufficiently great to render the ~0.3% difference between molality and molarity of dilute solutions unimportant.

Acid-base equilibrium constants for H_3PO_4 at $\mu = 0$ at 25°C are

$$H_3PO_4 \rightleftharpoons H_2PO_4^- + H^+ \qquad K_1 = \frac{[H_2PO_4^-]\gamma_{H_2PO_4^-}[H^+]\gamma_{H^+}}{[H_3PO_4]\gamma_{H_3PO_4}} = 10^{-2.148} \quad (12\text{-}19)$$

$$H_2PO_4^- \rightleftharpoons HPO_4^{2-} + H^+ \qquad K_2 = \frac{[HPO_4^{2-}]\gamma_{HPO_4^{2-}}[H^+]\gamma_{H^+}}{[H_2PO_4^-]\gamma_{H_2PO_4^-}} = 10^{-7.198} \quad (12\text{-}20)$$

$$HPO_4^{2-} \rightleftharpoons PO_4^{3-} + H^+ \qquad K_3 = \frac{[PO_4^{3-}]\gamma_{PO_4^{3-}}[H^+]\gamma_{H^+}}{[HPO_4^{2-}]\gamma_{HPO_4^{2-}}} = 10^{-12.375} \quad (12\text{-}21)$$

Equilibrium constants can be determined by measuring concentration quotients at several low ionic strengths and extrapolating to 0 ionic strength.

For $\mu \neq 0$, we rearrange the equilibrium constant expression to incorporate activity coefficients into an effective equilibrium constant, K' at the given ionic strength.

$$K_1' = K_1\left(\frac{\gamma_{H_3PO_4}}{\gamma_{H_2PO_4^-}\gamma_{H^+}}\right) = \frac{[H_2PO_4^-][H^+]}{H_3PO_4} \qquad (12\text{-}22)$$

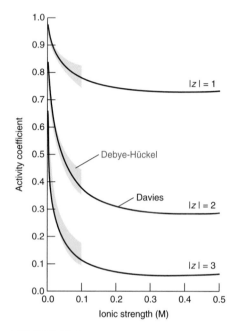

FIGURE 12-2 Activity coefficients from extended Debye-Hückel and Davies equations. Shaded areas give Debye-Hückel activity coefficients for the range of ion sizes in Table 7-1.

A 0.5 wt% solution of K_2HPO_4 has a *molarity* of 0.028 13 mol/L and a *molality* of 0.028 20 mol/kg. The difference is 0.25%.

$$K_2' = K_2\left(\frac{\gamma_{H_2PO_4^-}}{\gamma_{HPO_4^{2-}}\gamma_{H^+}}\right) = \frac{[HPO_4^{2-}][H^+]}{H_2PO_4^-} \tag{12-23}$$

$$K_3' = K_3\left(\frac{\gamma_{HPO_4^{2-}}}{\gamma_{PO_4^{3-}}\gamma_{H^+}}\right) = \frac{[PO_4^{3-}][H^+]}{[HPO_4^{2-}]} \tag{12-24}$$

K_2' gives the *concentration quotient*

$$\frac{[HPO_4^{2-}][H^+]}{[H_2PO_4^-]}$$

at a specified ionic strength.

For ionic species, we compute activity coefficients with the Davies equation 12-18. For the neutral species H_3PO_4, we assume that $\gamma \approx 1.00$.

Now we remind ourselves of the equations for water ionization

$$H_2O \rightleftharpoons H^+ + OH^- \qquad K_w = [H^+]\gamma_{H^+}[OH^-]\gamma_{OH^-} = 10^{-13.995}$$

Values of K_w are found in Table 6-1.

$$K_w' = \frac{K_w}{\gamma_{H^+}\gamma_{OH^-}} = [H^+][OH^-] \Rightarrow [OH^-] = K_w'/[H^+] \tag{12-25}$$

$$pH = -\log([H^+]\gamma_{H^+}) \tag{12-26}$$

Here is how we use effective equilibrium constants:

Step 1 Solve the acid-base problem with the constants K_1, K_2, and K_3, which apply at $\mu = 0$. Activity coefficients are considered to be unity in this first step.

Step 2 From the results of step 1, compute the ionic strength. Then use the Davies equation to compute activity coefficients. With these activity coefficients, compute the effective equilibrium constants, K_1', K_2', and K_3', and K_w'.

Step 3 Solve the acid-base problem again, this time with K_1', K_2', and K_3', and K_w'.

Step 4 From the results of step 3, compute a new ionic strength and a new set of K' values. Repeat the process several times until the ionic strength is constant.

Let's find the pH of 0.025 0 m KH_2PO_4 plus 0.025 0 m Na_2HPO_4. The chemical reactions are 12-19 through 12-21 plus water ionization. Mass balances are $[K^+] = 0.025\ 0\ m$, $[Na^+] = 0.050\ 0\ m$, and total phosphate $\equiv F_{H_3P} = 0.050\ 0\ m$. The charge balance is

$$[Na^+] + [K^+] + [H^+] = [H_2PO_4^-] + 2[HPO_4^{2-}] + 3[PO_4^{3-}] + [OH^-] \tag{12-27}$$

Our strategy is to substitute expressions into the charge balance to obtain an equation in which the only variable is $[H^+]$. For this purpose, we use the fractional composition equations for the triprotic acid, H_3PO_4, which we abbreviate as H_3P:

$$[P^{3-}] = \alpha_{P^{3-}}F_{H_3P} = \frac{K_1'K_2'K_3'F_{H_3P}}{[H^+]^3 + [H^+]^2K_1' + [H^+]K_1'K_2' + K_1'K_2'K_3'} \tag{12-28}$$

$$[HP^{2-}] = \alpha_{HP^{2-}}F_{H_3P} = \frac{[H^+]K_1'K_2'F_{H_3P}}{[H^+]^3 + [H^+]^2K_1' + [H^+]K_1'K_2' + K_1'K_2'K_3'} \tag{12-29}$$

$$[H_2P^-] = \alpha_{H_2P^-}F_{H_3P} = \frac{[H^+]^2K_1'F_{H_3P}}{[H^+]^3 + [H^+]^2K_1' + [H^+]K_1'K_2' + K_1'K_2'K_3'} \tag{12-30}$$

$$[H_3P] = \alpha_{H_3P}F_{H_3P} = \frac{[H^+]^3F_{H_3P}}{[H^+]^3 + [H^+]^2K_1' + [H^+]K_1'K_2' + K_1'K_2'K_3'} \tag{12-31}$$

Figure 12-3 puts everything together in a spreadsheet. Input values for $F_{KH_2PO_4}$, $F_{Na_2HPO_4}$, pK_1, pK_2, pK_3, and pK_w are in the shaded cells. We *guess* a value for pH in cell H15 and write the initial ionic strength of 0 in cell C19. Cells A9:H10 compute activities with the Davies equation. With $\mu = 0$, all activity coefficients are 1. Cells A13:H16 compute concentrations. $[H^+]$ in cell B13 is $(10^{-pH})/\gamma_{H^+} = (10^{\wedge}\text{-H15})/B9$. Cell E18 computes the sum of charges.

An initial guess of pH = 7 in cell H15 gives a net charge of 0.005 6 m in cell E18, which is not shown in Figure 12-3. Excel Solver was then used to vary pH in cell H15 to produce a net charge near 0 in cell E18. For this purpose, the precision was set to 1e-16 in Solver Options. Figure 12-3 shows that executing Solver gives pH = 7.198 in cell H15 and a net charge of -2×10^{-17} m in cell E18. The calculated ionic strength in cell C20 is 0.100 m.

For the second iteration, we write the ionic strength 0.100 m in cell C19 of Figure 12-4. This ionic strength automatically gives new activity coefficients in cells A9:H10 and new effective equilibrium constants in cells H3:H6. The sum of charges in cell E18 is no longer near zero. Application of Solver to vary pH in cell H15 to try to get 0 net charge produces a net charge of 9×10^{-17} m in cell E18. The new ionic strength in Cell C20 remains 0.100 m, so

12-2 Activity Coefficients   **263**

FIGURE 12-3 Spreadsheet solved for the system 0.025 0 m KH_2PO_4 plus 0.025 0 m Na_2HPO_4 with initial ionic strength = 0 and activity coefficients = 1.

	A	B	C	D	E	F	G	H
1	Mixture of KH_2PO_4 and Na_2HPO_4 including activity coefficients from Davies equation							
2								
3	F_{KH2PO4} =	0.0250		pK_1 =	2.148		K_1' =	7.11E-03
4	$F_{Na2HPO4}$ =	0.0250		pK_2 =	7.198		K_2' =	6.34E-08
5	F_{H3P} =	0.0500	(=B3+B4)	pK_3 =	12.375		K_3' =	4.22E-13
6				pK_w =	13.995		K_w' =	1.01E-14
7								
8	Activity coefficients:							
9	H^+ =	1.00		H_3P =	1.00	(fixed at 1)	HP^{2-} =	1.00
10	OH^- =	1.00		H_2P^- =	1.00		P^{3-} =	1.00
11								
12	Species in charge balance:						Other concentrations:	
13	$[H^+]$ =	6.34E-08		$[OH^-]$ =	1.60E-07		$[H_3P]$ =	2.23E-07
14	$[Na^+]$ =	0.050000		$[H_2P^-]$ =	2.50E-02			
15	$[K^+]$ =	0.025000		$[HP^{2-}]$ =	2.50E-02		pH =	7.198
16				$[P^{3-}]$ =	1.66E-07		↑ initial value is a guess	
17								
18	Positive charge minus negative charge =				-2.27E-17			
19	Ionic strength =		0.0000	←initial value is 0				
20	New ionic strength =		0.1000	←substitute this value into cell C19 for next iteration				
21								
22	Formulas:							
23	H3 = 10^-E3*E9/(E10*B9)				H4 = 10^-E4*E10/(H9*B9)			
24	H5 = 10^-E5*H9/(H10*B9)				H6 = 10^-E6/(B9*B10)			
25	B9 = B10 = E10 = 10^(-0.51*1^2*(SQRT($C$19)/(1+SQRT($C$19))-0.3*$C$19))							
26	H9 = 10^(-0.51*2^2*(SQRT($C$19)/(1+SQRT($C$19))-0.3*$C$19))							
27	H10 = 10^(-0.51*3^2*(SQRT($C$19)/(1+SQRT($C$19))-0.3*$C$19))							
28	B13 = (10^-H15)/B9		B14 = 2*B4		B15 = B3		E13 = H6/(B13)	
29	E14 = B13^2*H3*B5/(B13^3+B13^2*H3+B13*H3*H4+H3*H4*H5)							
30	E15 = B13*H3*H4*B5/(B13^3+B13^2*H3+B13*H3*H4+H3*H4*H5)							
31	E16 = H3*H4*H5*B5/(B13^3+B13^2*H3+B13*H3*H4+H3*H4*H5)							
32	H13 = B13^3*B5/(B13^3+B13^2*H3+B13*H3*H4+H3*H4*H5)							
33	E18 = B13+B14+B15-E13-E14-2*E15-3*E16							
34	C20 = 0.5*(B13+B14+B15+E13+E14+4*E15+9*E16)							

FIGURE 12-4 Second iteration of spreadsheet solved for the system 0.025 0 m KH_2PO_4 plus 0.025 0 m Na_2HPO_4 with ionic strength = 0.100 from the first iteration.

	A	B	C	D	E	F	G	H
1	Mixture of KH_2PO_4 and Na_2HPO_4 including activity coefficients from Davies equation							
2								
3	F_{KH2PO4} =	0.0250		pK_1 =	2.148		K_1' =	1.17E-02
4	$F_{Na2HPO4}$ =	0.0250		pK_2 =	7.198		K_2' =	1.70E-07
5	F_{H3P} =	0.0500	(=B3+B4)	pK_3 =	12.375		K_3' =	1.86E-12
6				pK_w =	13.995		K_w' =	1.66E-14
7								
8	Activity coefficients:							
9	H^+ =	0.78		H_3P =	1.00	(fixed at 1)	HP^{2-} =	0.37
10	OH^- =	0.78		H_2P^- =	0.78		P^{3-} =	0.11
11								
12	Species in charge balance:						Other concentrations:	
13	$[H^+]$ =	1.70E-07		$[OH^-]$ =	9.74E-08		$[H_3P]$ =	3.65E-07
14	$[Na^+]$ =	0.050000		$[H_2P^-]$ =	2.50E-02			
15	$[K^+]$ =	0.025000		$[HP^{2-}]$ =	2.50E-02		pH =	6.876
16				$[P^{3-}]$ =	2.73E-07		↑ initial value is a guess	
17								
18	Positive charge minus negative charge =				8.56E-17			
19	Ionic strength =		0.1000	← initial value is 0				
20	New ionic strength =		0.1000	← substitute this value into cell C19 for next iteration				

we are done. When the new ionic strength in cell C20 equals the old ionic strength in cell C19 (to three decimal places), there is no need for further iteration.

The final pH in cell H15 of Figure 12-4 is 6.876, which differs from the certified pH of 6.865 by 0.011. This difference is about as close an agreement between a measured and computed pH as you will encounter. If we had used extended Debye-Hückel activity coefficients for $\mu = 0.1 \, m$ in Table 7-1, the computed pH would be 6.859, differing from the stated pH by just 0.006.

Sometimes Solver cannot find a solution if Precision is set too small in the Options window. You can make the Precision larger and see if Solver can find a solution. You can also try a different initial guess for pH.

Back to Basics

A spreadsheet operating on the charge balance to reduce the net charge to zero is an excellent, general method for solving complex equilibrium problems. However, we did learn how to find the pH of a mixture of KH_2PO_4 plus Na_2HPO_4 back in Chapter 8 by a simple, less rigorous method. Recall that, when we mix a weak acid ($H_2PO_4^-$) and its conjugate base (HPO_4^{2-}), *what we mix is what we get*. The pH is estimated from the Henderson-Hasselbalch equation 8-18 with activity coefficients:

$$pH = pK_a + \log\frac{[A^-]\gamma_{A^-}}{[HA]\gamma_{HA}} = pK_2 + \log\frac{[HPO_4^{2-}]\gamma_{HPO_4^{2-}}}{[H_2PO_4^-]\gamma_{H_2PO_4^-}} \qquad (8\text{-}18)$$

For 0.025 m KH_2PO_4 plus 0.025 m Na_2HPO_4, the ionic strength is

$$\mu = \frac{1}{2}\sum_i c_i z_i^2 = \frac{1}{2}([K^+]\cdot(+1)^2 + [H_2PO_4^-]\cdot(-1)^2 + [Na^+]\cdot(+1)^2 + [HPO_4^{2-}]\cdot(-2)^2)$$

$$= \frac{1}{2}([0.025]\cdot 1 + [0.025]\cdot 1 + [0.050]\cdot 1 + [0.025]\cdot 4) = 0.100 \, m$$

In Table 7-1, the activity coefficients at $\mu = 0.1 \, m$ are 0.775 for $H_2PO_4^-$ and 0.355 for HPO_4^{2-}. Inserting these values into Equation 8-18 gives

$$pH = 7.198 + \log\frac{[0.025]0.355}{[0.025]0.775} = 6.859$$

$pK_a = pK_2$ in Equation 8-18 applies at $\mu = 0$.

The answer is the same that we generate with the spreadsheet because the approximation that what we mix is what we get is excellent in this case.

So you already knew how to find the pH of this buffer by a simple calculation. The value of the general method with the charge balance in the spreadsheet is that it applies even when what you mix is not what you get, because the concentrations are very low or K_2 is not so small or there are additional equilibria.

Ignorance Is Still Bliss

Even in such a simple solution as KH_2PO_4 plus Na_2HPO_4, for which we were justifiably proud of computing an accurate pH, we have overlooked numerous ion-pair equilibria:

$$PO_4^{3-} + Na^+ \rightleftharpoons NaPO_4^{2-} \qquad K = 27 \qquad HPO_4^{2-} + Na^+ \rightleftharpoons NaHPO_4^- \quad K = 12$$
$$H_2PO_4^- + Na^+ \rightleftharpoons NaH_2PO_4 \quad K = 2 \qquad NaPO_4^{2-} + Na^+ \rightleftharpoons Na_2PO_4^- \quad K = 14$$
$$Na_2PO_4^- + H^+ \rightleftharpoons Na_2HPO_4 \qquad K = 5.4 \times 10^{10}$$

There is an analogous set of reactions for K^+, whose equilibrium constants are similar to those for Na^+.

The reliability of the calculated concentrations depends on knowing all relevant equilibria and having the fortitude to include them all in the computation, which is by no means trivial.

The effective pK_2 for H_3PO_4 listed in the *NIST Critically Selected Stability Constants Database 46* (2001) for an ionic strength of 0.1 M has the following values: 6.71 for Na^+ background ions, 6.75 for K^+ background ions, and 6.92 for unspecified tetraalkylammonium background ions. The dependence of the effective pK on the nature of the background ions strongly suggests that ion-pairing reactions play an observable role in the solution chemistry.

12-3 Dependence of Solubility on pH

An important example of the effect of pH on solubility is tooth decay. Tooth enamel contains the mineral hydroxyapatite, which is insoluble near neutral pH, but dissolves in acid because both phosphate and hydroxide in the hydroxyapatite react with H^+:

$$\underset{\text{Calcium hydroxyapatite}}{Ca_{10}(PO_4)_6(OH)_2(s)} + 14H^+ \rightleftharpoons 10Ca^{2+} + 6H_2PO_4^- + 2H_2O$$

H₃C, H
HO CO₂H
L-Lactic acid

Bacteria on the surface of our teeth metabolize sugars to produce lactic acid, which lowers the pH enough to slowly dissolve tooth enamel. Fluoride inhibits tooth decay because it forms fluorapatite, $Ca_{10}(PO_4)_6F_2$, which is more acid resistant than hydroxyapatite.

Solubility of CaF₂

The mineral *fluorite*, CaF_2, in Figure 12-5 has a cubic crystal structure and often cleaves to form nearly perfect octahedra (eight-sided solids with equilateral triangular faces). Depending on impurities, the mineral takes on a variety of colors and may fluoresce when irradiated with an ultraviolet lamp.

The solubility of CaF_2 is governed by K_{sp} for the salt, hydrolysis of F^- and of Ca^{2+}, and by ion pairing between Ca^{2+} and F^-:

Solubility products are in Appendix F. The acid dissociation constant of HF is from Appendix G. The hydrolysis constant for Ca is the inverse of the formation constant of $CaOH^+$ in Appendix I. The ion-pair formation constant for CaF^+ is listed in Appendix J.

$$CaF_2(s) \rightleftharpoons Ca^{2+} + 2F^- \qquad K_{sp} = [Ca^{2+}][F^-]^2 = 3.2 \times 10^{-11} \qquad (12\text{-}32)$$

$$HF \rightleftharpoons H^+ + F^- \qquad K_{HF} = \frac{[H^+][F^-]}{[HF]} = 6.8 \times 10^{-4} \qquad (12\text{-}33)$$

$$Ca^{2+} + H_2O \rightleftharpoons CaOH^+ + H^+ \qquad K_a = \frac{[CaOH^+][H^+]}{[Ca^{2+}]} = 2 \times 10^{-13} \qquad (12\text{-}34)$$

$$Ca^{2+} + F^- \rightleftharpoons CaF^+ \qquad K_{ip} = \frac{[CaF^+]}{[Ca^{2+}][F^-]} = 4.3 \qquad (12\text{-}35)$$

$$H_2O \rightleftharpoons H^+ + OH^- \qquad K_w = [H^+][OH^-] = 1.0 \times 10^{-14} \qquad (12\text{-}36)$$

The charge balance is

Charge balance: $[H^+] + 2[Ca^{2+}] + [CaOH^+] + [CaF^+] = [OH^-] + [F^-]$ (12-37)

To find the mass balance, we need to realize that all calcium and fluoride species come from CaF_2. Therefore, the total fluoride equals two times the total calcium:

$$2[\text{total calcium species}] = [\text{total fluoride species}]$$

$$2\{[Ca^{2+}] + [CaOH^+] + [CaF^+]\} = [F^-] + [HF] + [CaF^+]$$

Mass balance: $2[Ca^{2+}] + 2[CaOH^+] + [CaF^+] = [F^-] + [HF]$ (12-38)

There are seven independent equations and seven unknowns, so we have enough information.

How to use activity coefficients

We omit activity coefficients here, but you do know how to use them. You would solve the problem with activity coefficients equal to 1, find the ionic strength, and then compute activity coefficients with the Davies equation. Then you would compute effective equilibrium constants incorporating activity coefficients and solve the problem again. After each iteration, you would find a new ionic strength and a new set of activity coefficients. Repeat the process until ionic strength is constant. Wow! You are smart!

We want to reduce seven equations with seven unknowns to one equation with one unknown—but this is not easy. However, we can express all concentrations in terms of $[H^+]$

FIGURE 12-5 Crystals of the mineral fluorite, CaF_2. Each Ca^{2+} ion is surrounded by eight F^- ions at the corners of a cube. Each F^- ion is surrounded by four Ca^{2+} ions at the corners of a tetrahedron. If you imagine the next unit cell above this one, you should see that the Ca^{2+} ion at the center of the top face of the cell shown here is adjacent to four F^- ions in this cell and four F^- ions in the next cell above it.

(a) (b)

0.546 nm

F

Ca

and [F$^-$], so we can reduce the mass and charge balance to two equations with two unknowns. Here are the substitutions from the equilibrium expressions:

$$[Ca^{2+}] = K_{sp}/[F^-]^2 \qquad\qquad [HF] = [H^+][F^-]/K_{HF}$$

$$[CaOH^+] = \frac{K_a[Ca^{2+}]}{[H^+]} = \frac{K_aK_{sp}}{[H^+][F^-]^2} \qquad [CaF^+] = K_{ip}[Ca^{2+}][F^-] = \frac{K_{ip}K_{sp}}{[F^-]}$$

The first expression for [CaOH$^+$] is from Equation 12-34. The second expression is obtained by substituting $[Ca^{2+}] = K_{sp}/[F^-]^2$ into the first equation.

Putting these expressions into the charge balance gives

$$[H^+] + 2[Ca^{2+}] + [CaOH^+] + [CaF^+] - [OH^-] - [F^-] = 0 \qquad (12\text{-}39a)$$

$$[H^+] + \frac{2K_{sp}}{[F^-]^2} + \frac{K_aK_{sp}}{[H^+][F^-]^2} + \frac{K_{ip}K_{sp}}{[F^-]} - \frac{K_w}{[H^+]} - [F^-] = 0 \qquad (12\text{-}39b)$$

Substitution in the mass balance gives

$$2[Ca^{2+}] + 2[CaOH^+] + [CaF^+] - [F^-] - [HF] = 0$$

$$\frac{2K_{sp}}{[F^-]^2} + \frac{2K_aK_{sp}}{[H^+][F^-]^2} + \frac{K_{ip}K_{sp}}{[F^-]} - [F^-] - \frac{[H^+][F^-]}{K_{HF}} = 0 \qquad (12\text{-}40)$$

We could do some messy algebra to solve Equation 12-39b for [H$^+$] and substitute this expression for [H$^+$] into Equation 12-40 to get an equation with [F$^-$] as the only unknown.[5]

Instead, we will use a spreadsheet for a numerical solution. *Suppose that the pH is somehow fixed by a buffer.* The mass balance relating calcium to fluoride is still correct. However, the charge balance is no longer valid, because the buffer introduces additional ions into the solution. We will find a way around the loss of the charge balance shortly.

When buffer is added, the mass balance still applies but the original charge balance is no longer correct.

In the spreadsheet in Figure 12-6, enter pH in column A. Compute $[H^+] = 10^{-pH}$ in column B and *guess* a value for [F$^-$] in column C. With these values of [H$^+$] and [F$^-$],

	A	B	C	D	E	F	G	H	I	J
1	Finding the concentrations of species in saturated calcium fluoride solution									
2										
3	K_{sp} =	3.2E-11		Mass balance:						
4	K_{HF} =	6.8E-04		$\dfrac{2K_{sp}}{[F^-]^2} + \dfrac{2K_aK_{sp}}{[H^+][F^-]^2} + \dfrac{K_{ip}K_{sp}}{[F^-]} - [F^-] - \dfrac{[H^+][F^-]}{K_{HF}} = 0$						
5	K_a =	2.E-13								
6	K_{ip} =	4.3								
7	K_w =	1.0E-14		Mass						
8	Input		[F$^-$] found by	balance						Sum of
9	pH	[H$^+$]	SOLVER	sum	[Ca^{2+}]	[CaOH$^+$]	[CaF$^+$]	[HF]	[OH$^-$]	charges
10	0	1.E+00	3.517E-05	5.6E-17	2.6E-02	5.17E-15	3.9E-06	5.17E-02	1.0E-14	1.1E+00
11	1	1.E-01	7.561E-05	0.0E+00	5.6E-03	1.12E-14	1.8E-06	1.11E-02	1.0E-13	1.1E-01
12	2	1.E-02	1.597E-04	-7.6E-17	1.3E-03	2.51E-14	8.6E-07	2.35E-03	1.0E-12	1.2E-02
13	3	1.E-03	2.960E-04	-2.8E-17	3.7E-04	7.31E-14	4.6E-07	4.35E-04	1.0E-11	1.4E-03
14	4	1.E-04	3.822E-04	-1.2E-17	2.2E-04	4.38E-13	3.6E-07	5.62E-05	1.0E-10	1.6E-04
15	5	1.E-05	3.982E-04	-1.5E-17	2.0E-04	4.04E-12	3.5E-07	5.86E-06	1.0E-09	1.6E-05
16	6	1.E-06	3.999E-04	-7.8E-18	2.0E-04	4.00E-11	3.4E-07	5.88E-07	1.0E-08	1.6E-06
17	7	1.E-07	4.001E-04	-1.0E-17	2.0E-04	4.00E-10	3.4E-07	5.88E-08	1.0E-07	5.8E-08
18	8	1.E-08	4.001E-04	-1.0E-17	2.0E-04	4.00E-09	3.4E-07	5.88E-09	1.0E-06	-9.9E-07
19	9	1.E-09	4.001E-04	-1.1E-17	2.0E-04	4.00E-08	3.4E-07	5.88E-10	1.0E-05	-1.0E-05
20	10	1.E-10	4.004E-04	-2.5E-17	2.0E-04	3.99E-07	3.4E-07	5.89E-11	1.0E-04	-1.0E-04
21	11	1.E-11	4.028E-04	-8.7E-18	2.0E-04	3.95E-06	3.4E-07	5.92E-12	1.0E-03	-1.0E-03
22	12	1.E-12	4.252E-04	-2.5E-17	1.8E-04	3.54E-05	3.2E-07	6.25E-13	1.0E-02	-1.0E-02
23	13	1.E-13	5.770E-04	-5.5E-17	9.6E-05	1.92E-04	2.4E-07	8.48E-14	1.0E-01	-1.0E-01
24	14	1.E-14	1.104E-03	-2.0E-17	2.6E-05	5.25E-04	1.2E-07	1.62E-14	1.0E+00	-1.0E+00
25										
26	B10 = 10^-A10			E10 = $B$3/C10^2			F10 = $B$5*$B$3/(B10*C10^2)			
27	G10 = $B$6*$B$3/C10			H10 = B10*C10/$B$4			I10 = $B$7/B10			
28	D10 = 2*$B$3/C10^2+2*$B$5*$B$3/(B10*C10^2)+$B$6*$B$3/C10-C10-B10*C10/$B$4									
29	J10 = B10+2*E10+F10+G10-I10-C10									

FIGURE 12-6 Spreadsheet using Solver for saturated CaF$_2$ at fixed pH values.

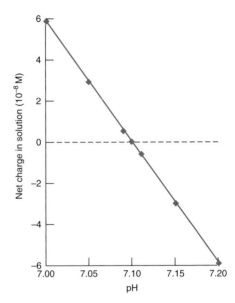

FIGURE 12-7 pH dependence of species in a saturated solution of CaF_2. As pH is lowered, H^+ reacts with F^- to make HF, and $[Ca^{2+}]$ increases. Note the logarithmic ordinate.

FIGURE 12-8 Net charge in a saturated solution of CaF_2 due to H^+, Ca^{2+}, $CaOH^+$, CaF^+, OH^-, and F^- as a function of pH. The net charge is 0 in unbuffered solution at pH 7.10.

compute the mass balance 12-40 in column D. Then use Solver to vary $[F^-]$ in column C to make the mass balance 0 in column D.

Each row of the spreadsheet must be dealt with separately. For example, in row 10, the pH was set to 0 in cell A10 and the initial *guessed* value of $[F^-]$ in cell C10 was 0.000 1 M. Before executing Solver, Precision was set to 1e-16 in Solver Options. In Solver, Set Target Cell <u>D10</u> Equal To Value of <u>0</u> By Changing Cells <u>C10</u>. Solver changes the value of $[F^-]$ in cell C10 to 3.517E-5 to satisfy the mass balance in cell D10. With the correct value of $[F^-]$ in cell C10, the concentrations of $[Ca^{2+}]$, $[CaOH^+]$, $[CaF^+]$, [HF], and $[OH^-]$ in columns E through I must be correct.

Figure 12-7 shows how the concentrations vary with pH. At low pH, H^+ reacts with F^- to produce HF and increases the solubility of CaF_2. The species CaF^+ and $CaOH^+$ are minor at most pH values, but $CaOH^+$ becomes the major form of calcium above pH 12.7, which is pK_a for Reaction 12-34. A reaction that we did not consider was precipitation of $Ca(OH)_2(s)$. Comparison of the product $[Ca^{2+}][OH^-]^2$ with K_{sp} for $Ca(OH)_2$ indicates that $Ca(OH)_2$ should precipitate at a pH between 13 and 14.

What would be the pH if we had not added buffer? Column J in Figure 12-6 gives the net charge from Equation 12-39a, which includes all ions other than buffer. The pH of unbuffered solution is the pH at which column J is 0. The net charge goes through 0 near pH 7. Finer computations in Figure 12-8 show that the net charge is 0 at pH 7.10. A saturated solution of CaF_2 is predicted to have a pH of 7.10 (ignoring activity coefficients).

Acid Rain Dissolves Minerals and Creates Environmental Hazards

In general, salts of basic ions such as F^-, OH^-, S^{2-}, CO_3^{2-}, $C_2O_4^{2-}$, and PO_4^{3-} have increased solubility at low pH, because the anions react with H^+. Figure 12-9 shows that marble, which is largely $CaCO_3$, dissolves more readily as the acidity of rain increases. Much of the acid in rain comes from SO_2 emissions from combustion of fuels containing sulfur and from nitrogen oxides produced by all types of combustion. SO_2, for example, reacts in the air to make sulfuric acid ($SO_2 + H_2O \rightarrow H_2SO_3 \xrightarrow{\text{oxidation}} H_2SO_4$), which finds its way back to the ground in rainfall.

Aluminum is the third most abundant element on Earth (after oxygen and silicon), but it is tightly locked into insoluble minerals such as kaolinite ($Al_2(OH)_4Si_2O_5$) and bauxite (AlOOH). Acid rain from human activities is a recent change for our planet, and it is introducing soluble forms of aluminum (and lead and mercury) into the environment.[6] Figure 12-10 shows that, below pH 5, aluminum is mobilized from minerals and its concentration in lake water rises rapidly. At a concentration of 130 μg/L, aluminum kills fish. In humans, high concentrations of aluminum cause dementia, softening of bones, and anemia. Aluminum is suspected

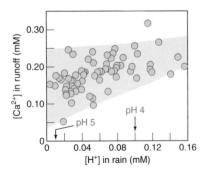

FIGURE 12-9 Measured calcium in acid rain runoff from marble stone (which is largely $CaCO_3$) roughly increases as $[H^+]$ in the rain increases. [Data from P. A. Baedecker and M. M. Reddy, "The Erosion of Carbonate Stone by Acid Rain," *J. Chem. Ed.* **1993**, *70*, 104.]

FIGURE 12-10 Relation of total aluminum (including dissolved and suspended species) in 1 000 Norwegian lakes as a function of the pH of the lake water. The more acidic the water, the greater the aluminum concentration. [From G. Howells, *Acid Rain and Acid Waters*, 2nd ed. (Hertfordshire: Ellis Horwood, 1995).]

as a possible cause of Alzheimer's disease. Although metallic elements from minerals are liberated by acid, the concentration and availability of metal ions in the environment tend to be regulated by organic matter that binds metal ions.[7]

Solubility of Barium Oxalate

Now consider the dissolution of $Ba(C_2O_4)$, whose anion is *dibasic* and whose cation is a weak acid.[8] The chemistry in this system is

$$Ba(C_2O_4)(s) \rightleftharpoons Ba^{2+} + C_2O_4^{2-} \qquad K_{sp} = [Ba^{2+}][C_2O_4^{2-}] = 1.0 \times 10^{-6} \qquad (12\text{-}41)$$
Oxalate

$$H_2C_2O_4 \rightleftharpoons HC_2O_4^- + H^+ \qquad K_1 = \frac{[H^+][HC_2O_4^-]}{[H_2C_2O_4]} = 5.4 \times 10^{-2} \qquad (12\text{-}42)$$

$$HC_2O_4^- \rightleftharpoons C_2O_4^{2-} + H^+ \qquad K_2 = \frac{[H^+][C_2O_4^{2-}]}{[HC_2O_4^-]} = 5.42 \times 10^{-5} \qquad (12\text{-}43)$$

$$Ba^{2+} + H_2O \rightleftharpoons BaOH^+ + H^+ \qquad K_a = \frac{[H^+][BaOH^+]}{[Ba^{2+}]} = 4.4 \times 10^{-14} \qquad (12\text{-}44)$$

Ba^{2+} is a weak acid.

$$Ba^{2+} + C_2O_4^{2-} \rightleftharpoons Ba(C_2O_4)(aq) \qquad K_{ip} = \frac{[Ba(C_2O_4)(aq)]}{[Ba^{2+}][C_2O_4^{2-}]} = 2.1 \times 10^2 \qquad (12\text{-}45)$$

Ion pairing

The value of K_{sp} applies at $\mu = 0.1$ M and 20°C. K_{ip} is for $\mu = 0$ and 18°C. K_1, K_2, and K_a apply at $\mu = 0$ and 25°C. For lack of better information, we will use this mixed set of equilibrium constants.

The charge balance is

Charge balance: $[H^+] + 2[Ba^{2+}] + [BaOH^+] = [OH^-] + [HC_2O_4^-] + 2[C_2O_4^{2-}]$ (12-46)

The mass balance states that the total moles of barium equal the total moles of oxalate:

$$[\text{Total barium}] = [\text{total oxalate}]$$

$$[Ba^{2+}] + [BaOH^+] + [\cancel{Ba(C_2O_4)(aq)}] = [H_2C_2O_4] + [HC_2O_4^-] + [C_2O_4^{2-}] + [\cancel{Ba(C_2O_4)(aq)}]$$

Mass balance: $\underbrace{[Ba^{2+}] + [BaOH^+]}_{F_{Ba}} = \underbrace{[H_2C_2O_4] + [HC_2O_4^-] + [C_2O_4^{2-}]}_{F_{H_2Ox}}$ (12-47)

We are defining F_{Ba} and F_{H_2O} to *exclude* the ion pair $Ba(C_2O_4)(aq)$.

There are eight unknowns and eight independent equations (including $[OH^-] = K_w/[H^+]$), so we have enough information to find the concentrations of all species.

We deal with ion pairing by adding Reactions 12-41 and 12-45 to find

$$Ba(C_2O_4)(s) \rightleftharpoons Ba(C_2O_4)(aq) \quad K = [Ba(C_2O_4)(aq)] = K_{sp}K_{ip} = 2.1 \times 10^{-4} \qquad (12\text{-}48)$$

Therefore, $[Ba(C_2O_4)(aq)] = 2.1 \times 10^{-4}$ M as long as undissolved $Ba(C_2O_4)(s)$ is present.

The ion pair $Ba(C_2O_4)(aq)$ has a constant concentration in this system.

HO, O
C—C
O, OH
Oxalic acid

Now for our old friends, the fractional composition equations. Abbreviating oxalic acid as H_2Ox, we can write

$$F_{H_2Ox} = [H_2C_2O_4] + [HC_2O_4^-] + [C_2O_4^{2-}]$$

$$[H_2Ox] = \alpha_{H_2Ox}F_{H_2Ox} = \frac{[H^+]^2F_{H_2Ox}}{[H^+]^2 + [H^+]K_1 + K_1K_2} \tag{12-49}$$

$$[HOx^-] = \alpha_{HOx^-}F_{H_2Ox} = \frac{K_1[H^+]F_{H_2Ox}}{[H^+]^2 + [H^+]K_1 + K_1K_2} \tag{12-50}$$

$$[Ox^{2-}] = \alpha_{Ox^{2-}}F_{H_2Ox} = \frac{K_1K_2F_{H_2Ox}}{[H^+]^2 + [H^+]K_1 + K_1K_2} \tag{12-51}$$

Also, Ba^{2+} and $BaOH^+$ are a conjugate acid-base pair. Ba^{2+} behaves as a monoprotic acid HA, and $BaOH^+$ is the conjugate base A^-.

$$F_{Ba} = [Ba^{2+}] + [BaOH^+]$$

$$[Ba^{2+}] = \alpha_{Ba^{2+}}F_{Ba} = \frac{[H^+]F_{Ba}}{[H^+] + K_a} \tag{12-52}$$

$$[BaOH^+] = \alpha_{BaOH^+}F_{Ba} = \frac{K_aF_{Ba}}{[H^+] + K_a} \tag{12-53}$$

Suppose that the pH is fixed by adding a buffer (and therefore the charge balance 12-46 is no longer valid). From K_{sp}, we can write

$$K_{sp} = [Ba^{2+}][C_2O_4^{2-}] = \alpha_{Ba^{2+}}F_{Ba}\alpha_{Ox^{2-}}F_{H_2Ox}$$

But the mass balance 12-47 told us that $F_{Ba} = F_{H_2Ox}$. Therefore,

$$K_{sp} = \alpha_{Ba^{2+}}F_{Ba}\alpha_{Ox^{2-}}F_{H_2Ox} = \alpha_{Ba^{2+}}F_{Ba}\alpha_{Ox^{2-}}F_{Ba}$$

$$\Rightarrow F_{Ba} = \sqrt{\frac{K_{sp}}{\alpha_{Ba^{2+}}\alpha_{Ox^{2-}}}} \tag{12-54}$$

In the spreadsheet in Figure 12-11, pH is specified in column A. From this pH, plus K_1 and K_2, the fractions α_{H_2Ox}, α_{HOx^-}, and $\alpha_{Ox^{2-}}$ are computed with Equations 12-49 through 12-51 in columns C, D, and E. From the pH and K_a, the fractions $\alpha_{Ba^{2+}}$ and α_{BaOH^+} are computed with Equations 12-52 and 12-53 in columns F and G. The total concentrations of barium and oxalate, F_{Ba} and F_{H_2Ox} are equal and calculated in column H from Equation 12-54. In a real spreadsheet, we would have continued to the right with column I. To fit on this page, the spreadsheet was continued in row 18. In this lower section, the concentrations of $[Ba^{2+}]$ and $[BaOH^+]$ are computed with Equations 12-52 and 12-53. $[H_2C_2O_4]$, $[HC_2O_4^-]$, and $[C_2O_4^{2-}]$ are found with Equations 12-49 through 12-51.

The net charge ($= [H^+] + 2[Ba^{2+}] + [BaOH^+] - [OH^-] - [HC_2O_4^-] - 2[C_2O_4^{2-}]$) is computed beginning in cell H19. If we had not added buffer to fix the pH, the net charge would be 0. Net charge goes from positive to negative between pH 6 and 8. By using Solver, we find the pH in cell A11 that makes the net charge 0 in cell H23 (with Solver Precision = 1e-16). This pH, 7.64, is the pH of unbuffered solution.

Results in Figure 12-12 show that the solubility of barium oxalate is steady at 10^{-3} M in the middle pH range. Solubility increases below pH 5 because $C_2O_4^{2-}$ reacts with H^+ to make $HC_2O_4^-$. Solubility increases above pH 13 because Ba^{2+} reacts with OH^- to make $BaOH^+$.

A final point is to see that the solubility of $Ba(OH)_2(s)$ is not exceeded. Evaluation of the product $[Ba^{2+}][OH^-]^2$ shows that $K_{sp} = 3 \times 10^{-4}$ is not exceeded below pH 13.9. We predict that $Ba(OH)_2(s)$ will begin to precipitate at pH 13.9.

12-4 Analyzing Acid-Base Titrations with Difference Plots[9]

A *difference plot*, also called a *Bjerrum plot*, is an excellent means to extract metal-ligand formation constants or acid dissociation constants from titration data obtained with electrodes. We will apply the difference plot to an acid-base titration curve.

We derive the key equation for a diprotic acid, H_2A, and extend it to a general acid, H_nA. The mean fraction of protons bound to H_2A ranges from 0 to 2 and is defined as

Niels Bjerrum (1879–1958) was a Danish physical chemist who made fundamental contributions to inorganic coordination chemistry and is responsible for much of our understanding of acids and bases and titration curves.[10]

$$\bar{n}_H = \frac{\text{moles of bound } H^+}{\text{total moles of weak acid}} = \frac{2[H_2A] + [HA^-]}{[H_2A] + [HA^-] + [A^{2-}]} \tag{12-55}$$

	A	B	C	D	E	F	G	H
1	Finding the concentrations of species in saturated barium oxalate solution							
2								
3	K_{sp} =	1.0E-06		K_1 =	5.4E-02		K_{ip} =	2.1E+02
4	K_a =	4.4E-14		K_2 =	5.42E-05		K_w =	1.0E-14
5								F_{Ba}
6	pH	[H^+]	$\alpha(H_2Ox)$	$\alpha(HOx^-)$	$\alpha(Ox^{2-})$	$\alpha(Ba^{2+})$	$\alpha(BaOH^+)$	= F_{H2Ox}
7	0	1.E+00	9.5E-01	5.1E-02	2.8E-06	1.0E+00	4.4E-14	6.0E-01
8	2	1.E-02	1.6E-01	8.4E-01	4.6E-03	1.0E+00	4.4E-12	1.5E-02
9	4	1.E-04	1.2E-03	6.5E-01	3.5E-01	1.0E+00	4.4E-10	1.7E-03
10	6	1.E-06	3.4E-07	1.8E-02	9.8E-01	1.0E+00	4.4E-08	1.0E-03
11	7.643	2.E-08	1.8E-10	4.2E-04	1.0E+00	1.0E+00	1.9E-06	1.0E-03
12	8	1.E-08	3.4E-11	1.8E-04	1.0E+00	1.0E+00	4.4E-06	1.0E-03
13	10	1.E-10	3.4E-15	1.8E-06	1.0E+00	1.0E+00	4.4E-04	1.0E-03
14	12	1.E-12	3.4E-19	1.8E-08	1.0E+00	9.6E-01	4.2E-02	1.0E-03
15	14	1.E-14	3.4E-23	1.8E-10	1.0E+00	1.9E-01	8.1E-01	2.3E-03
16								
17								net
18	pH	[Ba^{2+}]	[$BaOH^+$]	[H_2Ox]	[HOx^-]	[Ox^{2-}]	[OH^-]	charge
19	0	6.0E-01	2.6E-14	5.7E-01	3.1E-02	1.7E-06	1.0E-14	2.2E+00
20	2	1.5E-02	6.5E-14	2.3E-03	1.2E-02	6.7E-05	1.0E-12	2.7E-02
21	4	1.7E-03	7.4E-13	2.0E-06	1.1E-03	5.9E-04	1.0E-10	1.2E-03
22	6	1.0E-03	4.4E-11	3.4E-10	1.8E-05	9.9E-04	1.0E-08	1.9E-05
23	7.643	1.0E-03	1.9E-09	1.8E-13	4.2E-07	1.0E-03	4.4E-07	6.9E-18
24	8	1.0E-03	4.4E-09	3.4E-14	1.8E-07	1.0E-03	1.0E-06	-8.1E-07
25	10	1.0E-03	4.4E-07	3.4E-18	1.8E-09	1.0E-03	1.0E-04	-1.0E-04
26	12	9.8E-04	4.3E-05	3.5E-22	1.9E-11	1.0E-03	1.0E-02	-1.0E-02
27	14	4.3E-04	1.9E-03	7.9E-26	4.3E-13	2.3E-03	1.0E+00	-1.0E+00
28								
29	B7 = 10^-A7						B19 = F7*H7	
30	C7 = $B7^2/($B7^2+$B7*$E$3+$E$3*$E$4)						C19 = G7*H7	
31	D7 = $B7*$E$3/($B7^2+$B7*$E$3+$E$3*$E$4)						D19 = C7*H7	
32	E7 $E$3*$E$4/($B7^2+$B7*$E$3+$E$3*$E$4)						E19 = D7*H7	
33	F7 = B7/(B7+$B$4)						F19 = E7*H7	
34	G7 = $B$4/(B7+$B$4)						G19 = $H$4/B7	
35	H7 = SQRT($B$3/(E7*F7))		H19 = B7+2*B19+C19-G19-E19-2*F19					

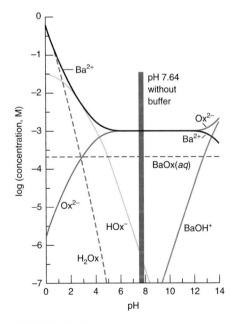

FIGURE 12-11 Spreadsheet for saturated BaC_2O_4. Solver was used to find the pH in cell A11 necessary to make the net charge 0 in cell H23.

FIGURE 12-12 pH dependence of the concentrations of species in saturated BaC_2O_4. As pH is lowered, H^+ reacts with $C_2O_4^{2-}$ to make $HC_2O_4^-$ and $H_2C_2O_4$, and the concentration of Ba^{2+} increases.

We can measure $\bar{n}_H$ by a titration beginning with a mixture of A mmol of H_2A and C mmol of HCl in V_0 mL. We add HCl to increase the degree of protonation of H_2A, which is partially dissociated in the absence of HCl. We titrate the solution with standard NaOH at a concentration of C_b mol/L. After adding v mL of NaOH, the number of mmol of Na^+ in the solution is $C_b v$.

To maintain a nearly constant ionic strength, the solution of H_2A plus HCl contains 0.10 M KCl, and the concentrations of H_2A and HCl are much less than 0.10 M. NaOH is sufficiently concentrated so that the added volume is small relative to V_0.

The charge balance for the titration solution is

$$[H^+] + [Na^+] + [K^+] = [OH^-] + [Cl^-]_{HCl} + [Cl^-]_{KCl} + [HA^-] + 2[A^{2-}]$$

where $[Cl^-]_{HCl}$ is from HCl and $[Cl^-]_{KCl}$ is from KCl. But $[K^+] = [Cl^-]_{KCl}$, so we cancel these terms. The net charge balance is

$$[H^+] + [Na^+] = [OH^-] + [Cl^-]_{HCl} + [HA^-] + 2[A^{2-}] \tag{12-56}$$

The denominator of Equation 12-55 is $F_{H_2A} = [H_2A] + [HA^-] + [A^{2-}]$. The numerator can be written as $2F_{H_2A} - [HA^-] - 2[A^{2-}]$. Therefore

$$\bar{n}_H = \frac{2F_{H_2A} - [HA^-] - 2[A^{2-}]}{F_{H_2A}} \tag{12-57}$$

From Equation 12-56, we can write: $-[HA^-] - 2[A^{2-}] = [OH^-] + [Cl^-]_{HCl} - [H^+] - [Na^+]$. Substituting this expression into the numerator of Equation 12-57 gives

$$\bar{n}_H = \frac{2F_{H_2A} + [OH^-] + [Cl^-]_{HCl} - [H^+] - [Na^+]}{F_{H_2A}} = 2 + \frac{[OH^-] + [Cl^-]_{HCl} - [H^+] - [Na^+]}{F_{H_2A}}$$

For the general polyprotic acid H_nA, the mean fraction of bound protons turns out to be

$$\bar{n}_H = n + \frac{[OH^-] + [Cl^-]_{HCl} - [H^+] - [Na^+]}{F_{H_nA}} \qquad (12\text{-}58)$$

Each term on the right side of Equation 12-58 is known during the titration. From the reagents that were mixed, we can say

$$F_{H_2A} = \frac{\text{mmol } H_2A}{\text{total volume}} = \frac{A}{V_0 + v} \qquad [Cl^-]_{HCl} = \frac{\text{mmol HCl}}{\text{total volume}} = \frac{C}{V_0 + v}$$

$$[Na^+] = \frac{\text{mmol NaOH}}{\text{total volume}} = \frac{C_b v}{V_0 + v}$$

$[H^+]$ and $[OH^-]$ are measured with a pH electrode and calculated as follows: Let the effective value of K_w that applies at $\mu = 0.10$ M be $K'_w = K_w/(\gamma_{H^+}\gamma_{OH^-}) = [H^+][OH^-]$ (Equation 12-25). Remembering that $pH = -\log([H^+]\gamma_{H^+})$, we write

$$[H^+] = \frac{10^{-pH}}{\gamma_{H^+}} \qquad [OH^-] = \frac{K'_w}{[H^+]} = 10^{(pH - pK'_w)} \cdot \gamma_{H^+}$$

Substituting into Equation 12-58 gives the measured fraction of bound protons:

Experimental fraction of protons bound to polyprotic acid

$$\bar{n}_H(\text{measured}) = n + \frac{10^{(pH - pK'_w)} \cdot \gamma_{H^+} + C/(V_0 + v) - (10^{-pH})/\gamma_{H^+} - C_b v/(V_0 + v)}{A/(V_0 + v)} \qquad (12\text{-}59)$$

In acid-base titrations, a **difference plot,** or *Bjerrum plot,* is a graph of the mean fraction of protons bound to an acid versus pH. The mean fraction is $\bar{n}_H$ calculated with Equation 12-59. For complex formation, the difference plot gives the mean number of ligands bound to a metal versus pL ($= -\log[\text{ligand}]$).

Equation 12-59 gives the measured value of $\bar{n}_H$. What is the theoretical value? For a diprotic acid, the theoretical mean fraction of bound protons is

$$\bar{n}_H(\text{theoretical}) = 2\alpha_{H_2A} + \alpha_{HA^-} \qquad (12\text{-}60)$$

If Equation 12-60 is not obvious, you can derive it from Equation 12-55 by writing the right-hand side as

$$\frac{2[H_2A] + [HA^-]}{[H_2A] + [HA^-] + [A^{2-}]}$$

$$= \frac{2[H_2A] + [HA^-]}{F_{H_2A}}$$

$$= \frac{2[H_2A]}{F_{H_2A}} + \frac{[HA^-]}{F_{H_2A}}$$

$$= 2\alpha_{H_2A} + \alpha_{HA^-}$$

where α_{H_2A} is the fraction of acid in the form H_2A and α_{HA^-} is the fraction in the form HA^-. You should be able to write expressions for α_{H_2A} and α_{HA^-} in your sleep by now.

$$\alpha_{H_2A} = \frac{[H^+]^2}{[H^+]^2 + [H^+]K_1 + K_1K_2} \qquad \alpha_{HA^-} = \frac{[H^+]K_1}{[H^+]^2 + [H^+]K_1 + K_1K_2} \qquad (12\text{-}61)$$

We extract K_1 and K_2 from an experimental titration by constructing a difference plot with Equation 12-59. This plot is a graph of $\bar{n}_H(\text{measured})$ versus pH. We then fit the theoretical curve (Equation 12-60) to the experimental curve by the method of least squares to find the values of K_1 and K_2 that minimize the sum of the squares of the residuals:

Best values of K_1 and K_2 minimize the sum of the squares of the residuals.

$$\sum(\text{residuals})^2 = \sum[\bar{n}_H(\text{measured}) - \bar{n}_H(\text{theoretical})]^2 \qquad (12\text{-}62)$$

Experimental data for a titration of the amino acid glycine are given in Figure 12-13. The initial 40.0-mL solution contained 0.190 mmol of glycine plus 0.232 mmol of HCl to increase the fraction of fully protonated $^+H_3NCH_2CO_2H$. Aliquots of 0.490 5 M NaOH were added and the pH was measured after each addition. Volumes and pH are listed in columns A and B beginning in row 16. pH was precise to the 0.001 decimal place, but the accuracy of pH measurement is, at best, ±0.02.

$H_3NCH_2CO_2H$
  Glycine

$pK_1 = 2.35$ at $\mu = 0$

$pK_2 = 9.78$ at $\mu = 0$

Input values of concentration, volume, and moles are in cells B3:B6 in Figure 12-13. Cell B7 has the value 2 to indicate that glycine is a diprotic acid. Cell B8 has the activity coefficient of H^+ computed with the Davies equation, 12-18. Cell B9 begins with the effective value of $pK'_w = 13.797$ in 0.1 M KCl.[11] We allowed the spreadsheet to vary pK'_w for best fit of the experimental data, giving 13.807 in cell B9. Cells B10 and B11 began with estimates of pK_1 and pK_2 for glycine. We used the values 2.35 and 9.78 from Table 9-1, which apply at $\mu = 0$. As explained in the next section, we use Solver to vary pK_1, pK_2, and pK'_w for best fit of the experimental data, giving 2.312 and 9.625 in cells B10 and B11.

	A	B	C	D	E	F	G	H	I
1	Difference plot for glycine								
2				C16 = 10^-B16/$B$8					
3	Titrant NaOH =	0.4905	C_b (M)	D16 = 10^-$B$9/C16					
4	Initial volume =	40	V_0 (mL)	E16 = $B$7+($B$6-$B$3*A16-(C16-D16)*($B$4+A16))/$B$5					
5	Glycine =	0.190	L (mmol)	F16 + $C16^2/($C16^2+$C16*$E$10+$E$10*$E$11)					
6	HCl added =	0.232	A (mmol)	G16 = $C16*$E$10/($C16^2+$C16*$E$10+$E$10*$E$11)					
7	Number of H+ =	2	n	H16 = 2*F16+G16					
8	Activity coeff =	0.78	γ_H	I16 = (E16-H16)^2					
9	pK_w' =	13.807							
10	pK_1 =	2.312		K_1 =	0.0048713	= 10^-B10			
11	pK_2 =	9.625		K_2 =	2.371E-10	= 10^-B11			
12	$\Sigma(resid)^2$ =	0.0048	= sum of column I						
13									
14	v	pH	[H+] =	[OH-] =	Measured			Theoretical	$(residuals)^2$ =
15	mL NaOH		$(10^{-pH})/\gamma_H$	$(10^{-pK_w})/[H^+]$	n_H	α_{H2A}	α_{HA^-}	n_H	$(n_{meas} - n_{theor})^2$
16	0.00	2.234	7.48E-03	2.08E-12	1.646	0.606	0.394	1.606	0.001656
17	0.02	2.244	7.31E-03	2.13E-12	1.630	0.600	0.400	1.600	0.000879
18	0.04	2.254	7.14E-03	2.18E-12	1.612	0.595	0.405	1.595	0.000319
19	0.06	2.266	6.95E-03	2.24E-12	1.601	0.588	0.412	1.588	0.000174
20	0.08	2.278	6.76E-03	2.30E-12	1.589	0.581	0.419	1.581	0.000056
21	0.10	2.291	6.56E-03	2.38E-12	1.578	0.574	0.426	1.574	0.000020
22	:								
23	0.50	2.675	2.71E-03	5.75E-12	1.353	0.357	0.643	1.357	0.000022
24	:								
25	1.56	11.492	4.13E-12	3.77E-03	0.016	0.000	0.017	0.017	0.000000
26	1.58	11.519	3.88E-12	4.01E-03	0.018	0.000	0.016	0.016	0.000004
27	1.60	11.541	3.69E-12	4.22E-03	0.015	0.000	0.015	0.015	0.000000

FIGURE 12-13 Spreadsheet for difference plot of the titration of 0.190 mmol glycine plus 0.232 mmol HCl in 40.0 mL with 0.490 5 M NaOH. Cells A16:B27 give only a fraction of the experimental data. [Complete data listed in Problem 12-16 provided by A. Kraft, Heriot-Watt University.]

The spreadsheet in Figure 12-13 computes $[H^+]$ and $[OH^-]$ in columns C and D beginning in row 16. The mean fraction of protonation, $\bar{n}_H$(measured) from Equation 12-59, is in column E. The Bjerrum difference plot in Figure 12-14 shows $\bar{n}_H$(measured) versus pH. Values of α_{H_2A} and α_{HA^-} from Equations 12-61 are computed in columns F and G and $\bar{n}_H$(theoretical) was computed with Equation 12-60 in column H. Column I contains the squares of the residuals, $[\bar{n}_H(\text{measured}) - \bar{n}_H(\text{theoretical})]^2$. The sum of the squares of the residuals is in cell B12.

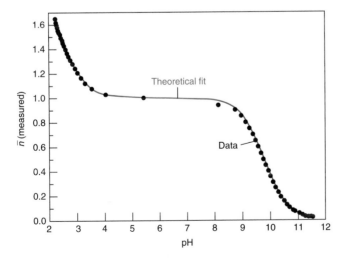

FIGURE 12-14 Bjerrum difference plot for the titration of glycine. Many experimental points are omitted from the figure for clarity.

Using Excel Solver to Optimize More Than One Parameter

We want values of pK_w', pK_1, and pK_2 that minimize the sum of squares of residuals in cell B12. Select Solver. In the Solver window, Set Target Cell B12 Equal To <u>Min</u> By Changing Cells <u>B9, B10, B11</u>. Then click Solve, and Solver finds the best values in cells B9, B10, and B11 to minimize the sum of squares of residuals in cell B12. Starting with 13.797, 2.35, and 9.78 in cells B9, B10, and B11 gives a sum of squares of residuals equal to 0.110 in cell B12.

After Solver is executed, cells B9, B10, and B11 become 13.807, 2.312, and 9.625. The sum in cell B12 is reduced to 0.004 8. When you use Solver to optimize several parameters at once, it is a good idea to try different starting values to see if the same solution is reached. Sometimes a local minimum can be reached that is not as low as another that might be reached elsewhere.

The theoretical curve $\bar{n}_H(\text{theoretical}) = 2\alpha_{H_2A} + \alpha_{HA^-}$ based on the results of Solver are contained in columns F, G, and H in Figure 12-13 and shown by the solid curve in Figure 12-14. The curve fits the experimental data well, a result suggesting that we have found reliable values of pK_1 and pK_2.

It may seem inappropriate to allow pK'_w to vary, because we claimed to know pK'_w at the outset. The change of pK'_w from 13.797 to 13.807 significantly improved the fit. The value 13.797 gave values of $\bar{n}_H(\text{measured})$ that became level near 0.04 at the end of the titration in Figure 12-14. This behavior is qualitatively incorrect, because $\bar{n}_H$ must approach 0 at high pH. A small change in pK'_w markedly improves the fit as $\bar{n}_H$ approaches 0.

Strictly speaking, pK_1 and pK_2 should have primes to indicate that they apply in 0.10 M KCl. We left off the primes to avoid complicating the symbols. We did distinguish K_w, which applies at $\mu = 0$, from K'_w, which applies at $\mu = 0.10$ M.

Terms to Understand

coupled equilibria   difference plot

Summary

Coupled equilibria are reversible reactions that have a species in common. Therefore, each reaction has an effect on the other.

The general treatment of acid-base systems begins with charge and mass balances and equilibrium expressions. There should be as many independent equations as chemical species. Substitute a fractional composition equation for each acid or base into the charge balance. After you enter known concentrations of species such as Na^+ and Cl^-, and substitute $K_w/[H^+]$ for $[OH^-]$, the remaining variable should be $[H^+]$. Use Excel Solver to find $[H^+]$ and then solve for all other concentrations from $[H^+]$. If there are equilibria in addition to acid-base reactions, such as ion pairing, then you need the full systematic treatment of equilibrium. Make maximum use of fractional composition equations to simplify the problem.

To use activity coefficients, first solve the equilibrium problem with all activity coefficients equal to unity. From the resulting concentrations, compute the ionic strength and use the Davies equation to find activity coefficients. With activity coefficients, calculate the effective equilibrium constant K' for each chemical reaction. K' is the equilibrium quotient of concentrations at a particular ionic strength. Solve the problem again with K' values and find a new ionic strength. Repeat the cycle until the concentrations reach constant values.

We considered solubility problems in which the cation and anion could each undergo one or more acid-base reactions and in which ion pairing could occur. Substitute fractional composition expressions for all acid-base species into the mass balance. In some systems, such as barium oxalate, the resulting equation contains the formal concentrations of anion and cation and $[H^+]$. The solubility product provides a relation between the formal concentrations of anion and cation, so you can eliminate one of them from the mass balance. By assuming a value for $[H^+]$, you can solve for the remaining formal concentration and, therefore, for all concentrations. By this means, find the composition as a function of pH. The pH of the unbuffered solution is the pH at which the charge balance is satisfied.

In the more difficult calcium fluoride solubility problem, ion pairing prevented us from reducing the equations to one unknown. Instead, $[H^+]$ and $[F^-]$ were both unknown. However, for a fixed value of $[H^+]$, we could use Solver to solve for $[F^-]$. The correct pH is the one at which the charge balance is satisfied.

To extract acid dissociation constants from an acid-base titration curve, we can construct a difference plot, or Bjerrum plot, which is a graph of the mean fraction of bound protons, $\bar{n}_H$, versus pH. This mean fraction can be measured from the quantities of reagents that were mixed and the measured pH. The theoretical shape of the difference plot is an expression in terms of fractional compositions. Use Excel Solver to vary equilibrium constants to obtain the best fit of the theoretical curve to the measured points. This process minimizes the sum of squares $[\bar{n}_H(\text{measured}) - \bar{n}_H(\text{theoretical})]^2$.

Exercises

Instructors: Most of these problems are quite long. Please be kind when you assign them.

12-A. Neglecting activity coefficients and ion pairing, find the pH and concentrations of species in 1.00 L of solution containing 0.010 mol hydroxybenzene (HA), 0.030 mol dimethylamine (B), and 0.015 mol HCl.

12-B. Repeat Exercise 12-A with activity coefficients from the Davies equation.

12-C. (a) Neglecting activity coefficients and ion pairing, find the pH and concentrations of species in 1.00 L of solution containing 0.040 mol 2-aminobenzoic acid (a neutral molecule, HA), 0.020 mol dimethylamine (B), and 0.015 mol HCl.

(b) What fraction of HA is in each of its three forms? What fraction of B is in each of its two forms? Compare your answers with what you would find if HCl reacted with B and then excess B reacted with HA. What pH do you predict from this simple estimate?

12-D. Include activity coefficients from the Davies equation to find the pH and concentrations of species in the mixture of sodium tartrate, pyridinium chloride, and KOH in Section 12-1. Consider only Reactions 12-1 through 12-4.

12-E. ▦ (a) Using the ion-pair equilibrium constant from Appendix J, with activity coefficients = 1, find the concentrations of species in 0.025 M $MgSO_4$. Hydrolysis of the cation and anion near neutral pH is negligible. Only consider ion-pair formation. You can solve this problem exactly with a quadratic equation. Alternatively, if you use Solver, set Precision to 1e-6 (not 1e-16) in the Solver Options. If Precision is much smaller, Solver does not find a satisfactory solution. The success of Solver in this problem depends on how close your initial guess is to the correct answer.

(b) Compute a new ionic strength and repeat part (a) with new activity coefficients from the Davies equation. Perform several iterations until the ionic strength is constant. The fraction of ion pairing that you find should be close to that in Box 7-1, which was derived with Debye-Hückel activity coefficients.

(c) We naively assigned the ionic strength of 0.025 M $MgSO_4$ to be 0.10 M. What is the actual ionic strength of this solution?

12-F. ▦ (a) Find the concentrations of species in saturated CaF_2 as a function of pH by using Reactions 12-32 through 12-36 and adding the following reaction:

$$HF(aq) + F^- \rightleftharpoons HF_2^- \qquad K_{HF_2^-} = 10^{0.58}$$

Do not include activity coefficients. Produce a graph similar to Figure 12-7.

(b) Find the pH and concentrations of species in saturated CaF_2.

12-G. ▦ Make a graph of $[Ag^+]$, $[AgOH(aq)]$, $[CN^-]$, and $[HCN]$ as a function of pH in a saturated solution of AgCN. Consider the following equilibria and do not consider activity coefficients. Find the pH if no buffer were added.

$$AgCN(s) \rightleftharpoons Ag^+ + CN^- \qquad pK_{sp} = 15.66$$
$$HCN(aq) \rightleftharpoons CN^- + H^+ \qquad pK_{HCN} = 9.21$$
$$Ag^+ + H_2O \rightleftharpoons AgOH(aq) + H^+ \qquad pK_{Ag} = 12.0$$

12-H. ▦ *Difference plot.* A solution containing 3.96 mmol acetic acid plus 0.484 mmol HCl in 200 mL of 0.10 M KCl was titrated with 0.490 5 M NaOH to measure K_a for acetic acid.

(a) Write expressions for the experimental mean fraction of protonation, $\bar{n}_H$(measured), and the theoretical mean fraction of protonation, $\bar{n}_H$(theoretical).

(b) From the following data, prepare a graph of $\bar{n}_H$(measured) versus pH. Find the best values of pK_a and pK'_w by minimizing the sum of the squares of the residuals, $\sum[\bar{n}_H(\text{measured}) - \bar{n}_H(\text{theoretical})]^2$.

V (mL)	pH	V (mL)	pH	V (mL)	pH	V (mL)	pH
0.00	2.79	2.70	4.25	5.40	4.92	8.10	5.76
0.30	2.89	3.00	4.35	5.70	4.98	8.40	5.97
0.60	3.06	3.30	4.42	6.00	5.05	8.70	6.28
0.90	3.26	3.60	4.50	6.30	5.12	9.00	7.23
1.20	3.48	3.90	4.58	6.60	5.21	9.30	10.14
1.50	3.72	4.20	4.67	6.90	5.29	9.60	10.85
1.80	3.87	4.50	4.72	7.20	5.38	9.90	11.20
2.10	4.01	4.80	4.78	7.50	5.49	10.20	11.39
2.40	4.15	5.10	4.85	7.80	5.61	10.50	11.54

Data from A. Kraft, J. Chem. Ed. **2003,** *80, 554.*

Problems

Instructors: Most of these problems are quite long. Please be kind when you assign them.

12-1. Why does the solubility of a salt of a basic anion increase with decreasing pH? Write chemical reactions for the minerals galena (PbS) and cerussite ($PbCO_3$) to explain how acid rain mobilizes traces of metal from relatively inert forms into the environment, where the metals can be taken up by plants and animals.

12-2. ▦ (a) Considering just acid-base chemistry, not ion pairing and not activity coefficients, use the systematic treatment of equilibrium to find the pH of 1.00 L of solution containing 0.010 0 mol hydroxybenzene (HA) and 0.005 0 mol KOH.

(b) What pH would you have predicted from your knowledge of Chapter 8?

(c) Find the pH if [HA] and [KOH] were both reduced by a factor of 100.

12-3. ▦ Repeat part (a) of Problem 12-2 with Davies activity coefficients. Remember that pH = $-\log([H^+]\gamma_{H^+})$.

12-4. From pK_1 and pK_2 for glycine at $\mu = 0$ in Table 9-1, compute pK'_1 and pK'_2 that apply at $\mu = 0.1$ M. Use the Davies equation for activity coefficients. Compare your answer with experimental values in cells B10 and B11 of Figure 12-13.

12-5. ▦ Considering just acid-base chemistry, not ion pairing and not activity coefficients, use the systematic treatment of equilibrium to find the pH and concentrations of species in 1.00 L of solution

containing 0.100 mol ethylenediamine and 0.035 mol HBr. Compare the pH to that found by the methods of Chapter 10.

12-6. ▦ Considering just acid-base chemistry, not ion pairing and not activity coefficients, find the pH and concentrations of species in 1.00 L of solution containing 0.040 mol benzene-1,2,3-tricarboxylic acid (H_3A), 0.030 mol imidazole (a neutral molecule, HB), and 0.035 mol NaOH.

12-7. ▦ Considering just acid-base chemistry, not ion pairing and not activity coefficients, find the pH and concentrations of species in 1.00 L of solution containing 0.020 mol arginine, 0.030 mol glutamic acid, and 0.005 mol KOH.

12-8. ▦ Solve Problem 12-7 by using Davies activity coefficients.

12-9. ▦ A solution containing 0.008 695 m KH_2PO_4 and 0.030 43 m Na_2HPO_4 is a primary standard buffer with a stated pH of 7.413 at 25°C. Calculate the pH of this solution by using the systematic treatment of equilibrium with activity coefficients from (a) the Davies equation and (b) the extended Debye-Hückel equation.

12-10. ▦ Considering just acid-base chemistry, not ion pairing and not activity coefficients, find the pH and composition of 1.00 L of solution containing 0.040 mol H_4EDTA (EDTA = ethylenedinitrilotetraacetic acid ≡ H_4A), 0.030 mol lysine (neutral molecule ≡ HL), and 0.050 mol NaOH.

12-11. ▦ The solution containing no added KNO_3 for Figure 7-1 contains 5.0 mM $Fe(NO_3)_3$, 5.0 μM NaSCN, and 15 mM HNO_3. We will use Davies activity coefficients to find the concentrations of all species in the following reactions:

$$Fe^{3+} + SCN^- \rightleftharpoons Fe(SCN)^{2+} \qquad \log \beta_1 = 3.03 \; (\mu = 0)$$

$$Fe^{3+} + 2SCN^- \rightleftharpoons Fe(SCN)_2^+ \qquad \log \beta_2 = 4.6 \; (\mu = 0)$$

$$Fe^{3+} + H_2O \rightleftharpoons FeOH^{2+} + H^+ \qquad pK_a = 2.195 \; (\mu = 0)$$

This problem is harder than you might think, so we will guide you through the steps.

(a) Write the four equilibrium expressions (including K_w). Express the effective equilibrium constants in terms of equilibrium constants and activity coefficients. For example, $K_w' = K_w/\gamma_{H^+}\gamma_{OH^-}$. Write expressions for $[Fe(SCN)^{2+}]$, $[Fe(SCN)_2^+]$, $[FeOH^{2+}]$, and $[OH^-]$ in terms of $[Fe^{3+}]$, $[SCN^-]$, and $[H^+]$.

(b) Write the charge balance.

(c) Write mass balances for iron, thiocyanate, Na^+, and NO_3^-.

(d) Substitute expressions from part (a) into the mass balance for thiocyanate to find an expression for $[Fe^{3+}]$ in terms of $[SCN^-]$.

(e) Substitute expressions from part (a) into the mass balance for iron to find an expression for $[H^+]$ in terms of $[Fe^{3+}]$ and $[SCN^-]$.

(f) In a spreadsheet, guess a value for $[SCN^-]$. From this value, compute $[Fe^{3+}]$ from part (d). From $[SCN^-]$ and $[Fe^{3+}]$, compute $[H^+]$ from part (e).

(g) Compute the concentrations of all other species from $[SCN^-]$, $[Fe^{3+}]$, and $[H^+]$.

(h) Compute the net charge. We want to vary $[SCN^-]$ to reduce the net charge to 0. When the net charge is 0, you have the correct concentrations. Solver did not work well for these equations. You have several recourses. One is to use Goal Seek, which is similar to Solver but only handles one unknown. Start with $\mu = 0$ and guess $[SCN^-] = 1e-6$. In Excel 2007, click on the Microsoft Office button at the upper left of the spreadsheet. Click on Excel Options and select Formulas. In Calculation Options, set Maximum Change to 1e-16 and click OK. On the Data ribbon, click on What-If Analysis and then Goal Seek. In the Goal Seek dialog box, set Maximum Change to 1e-16. Set cell {net charge} To value 0 By changing cell {[SCN^-]}. Click OK and Goal Seek finds a solution. Calculate the new ionic strength and activity coefficients from the Davies equation. Calculate effective equilibrium constants. Repeat the process until μ is constant to three significant digits. If Goal Seek fails to work at some stage, vary $[SCN^-]$ by hand to reduce the net charge to near 0. When you are close to a correct answer, Goal Seek can take you the rest of the way. When you have a solution, verify that mass balances are satisfied and compute $pH = -\log([H^+]\gamma_{H^+})$.

In earlier versions of Excel, go to the Tools menu, select Options and Calculation. Set Maximum Change to 1e-16. Then select Goal Seek from the Tools menu and continue as described above.

(i) Find the quotient $[Fe(SCN)^{2+}]/(\{[Fe^{3+}] + [FeOH^{2+}]\}[SCN^-])$. This is the point for $[KNO_3] = 0$ in Figure 7-1. Compare your answer with Figure 7-1. The ordinate of Figure 7-1 is labeled $[Fe(SCN)^{2+}]/([Fe^{3+}][SCN^-])$, but $[Fe^{3+}]$ really refers to the total concentration of iron not bound to thiocyanate.

(j) Find the quotient in part (i) when the solution also contains 0.20 M KNO_3. Compare your answer with Figure 7-1.

12-12. ▦ (a) Follow the steps of Problem 12-11 to solve this one. From the following equilibria, find the concentrations of species and the pH of 1.0 mM $La_2(SO_4)_3$. Use Davies activity coefficients. You may wish to use Goal Seek in place of Solver as described in Problem 12-11.

$$La^{3+} + SO_4^{2-} \rightleftharpoons La(SO_4)^+ \qquad \beta_1 = 10^{3.64} \; (\mu = 0)$$

$$La^{3+} + 2SO_4^{2-} \rightleftharpoons La(SO_4)_2^- \qquad \beta_2 = 10^{5.3} \; (\mu = 0)$$

$$La^{3+} + H_2O \rightleftharpoons LaOH^{2+} + H^+ \qquad K_a = 10^{-8.5} \; (\mu = 0)$$

(b) If $La_2(SO_4)_3$ were a strong electrolyte, what would be the ionic strength of 1.0 mM $La_2(SO_4)_3$? What is the actual ionic strength of this solution?

(c) What fraction of lanthanum is La^{3+}?

(d) Why did we not consider hydrolysis of SO_4^{2-} to give HSO_4^-?

(e) Will $La(OH)_3(s)$ precipitate in this solution?

12-13. ▦ Find the pH and composition of saturated $CaSO_4$ in water, considering Reactions 7-23 to 7-27 and using Davies activity coefficients. *Suggested procedure:* Use activity coefficients in all equilibrium expressions. Let the initial ionic strength be 0. (i) In the mass balance 7-29, express $[CaOH^+]$ in terms of $[Ca^{2+}]$ and $[H^+]$. Express $[HSO_4^-]$ in terms of $[SO_4^{2-}]$ and $[OH^-]$. (ii) From the K_{sp} equilibrium expression, express $[SO_4^{2-}]$ in terms of $[Ca^{2+}]$. From the K_w equilibrium, express $[OH^-]$ in terms of $[H^+]$. The mass balance now contains only $[Ca^{2+}]$, $[H^+]$, and equilibrium constants. (iii) Solve for $[Ca^{2+}]$ in terms of $[H^+]$. (iv) Set up a spreadsheet in which pH is the input. Guess an initial value pH = 7.00. Compute $[H^+] = 10^{-pH}/\gamma_{H^+}$. (v) From $[H^+]$, compute $[OH^-]$, $[Ca^{2+}]$, $[CaOH^+]$, $[SO_4^{2-}]$, and $[HSO_4^-]$. (vi) Compute the net charge of the solution from charge balance 7-28. (vii) Use Solver or Goal Seek to vary pH until the net charge is 0 ($<1 \times 10^{-18}$ M). (viii) Compute the ionic strength. (ix) With the new ionic strength, use Solver again to find the pH that reduces the net charge to $<1 \times 10^{-18}$ M. Perform several iterations until ionic strength no longer changes.

12-14. ▦ Find the composition of a saturated solution of AgCN containing 0.10 M KCN adjusted to pH 12.00 with NaOH. Consider the following equilibria and use Davies activity coefficients.

$$AgCN(s) \rightleftharpoons Ag^+ + CN^- \qquad pK_{sp} = 15.66$$

$$HCN(aq) \rightleftharpoons CN^- + H^+ \qquad pK_{HCN} = 9.21$$

$$Ag^+ + H_2O \rightleftharpoons AgOH(aq) + H^+ \qquad pK_{Ag} = 12.0$$

$$Ag^+ + CN^- + OH^- \rightleftharpoons Ag(OH)(CN)^- \qquad \log K_{AgOHCN} = 13.22$$

$$Ag^+ + 2CN^- \rightleftharpoons Ag(CN)_2^- \qquad \log \beta_2 = 20.48$$

$$Ag^+ + 3CN^- \rightleftharpoons Ag(CN)_3^{2-} \qquad \log \beta_3 = 21.7$$

Suggested procedure: Let $[CN^-]$ be the master variable. We know $[H^+]$ from the pH, and we can find $[Ag^+] = K_{sp}'/[CN^-]$. (i) Use equilibrium expressions to find $[OH^-]$, $[HCN]$, $[AgOH]$, $[Ag(OH)(CN)^-]$, $[Ag(CN)_2^-]$, and $[Ag(CN)_3^{2-}]$ in terms of $[CN^-]$, $[Ag^+]$, and $[H^+]$. (ii) Key mass balance: {total silver} + $[K^+]$ = {total cyanide}. (iii) Guess that ionic strength = 0.10 M and compute activity coefficients. (iv) Compute $[H^+] = 10^{-pH}/\gamma_{H^+}$. (v) Guess a value of $[CN^-]$ and compute the concentrations of all species in the mass balance. (vi) Use Solver to vary $[CN^-]$ to satisfy the mass balance. (vii) Compute $[Na^+]$ from the charge balance. (viii) Compute the ionic strength and perform several iterations of the entire process until the ionic strength no longer changes.

12-15. ▦ Consider the reactions of Fe^{2+} with the amino acid glycine:

$$Fe^{2+} + G^- \rightleftharpoons FeG^+ \qquad \log \beta_1 = 4.31$$

$$Fe^{2+} + 2G^- \rightleftharpoons FeG_2(aq) \qquad \log \beta_2 = 7.65$$

$$Fe^{2+} + 3G^- \rightleftharpoons FeG_3^- \qquad \log \beta_3 = 8.87$$

$$Fe^{2+} + H_2O \rightleftharpoons FeOH^+ + H^+ \qquad pK_a = 9.4$$

$$^+H_3NCH_2CO_2H \quad \text{glycine, } H_2G^+ \qquad pK_1 = 2.350, pK_2 = 9.778$$

Suppose that 0.050 mol of FeG_2 is dissolved in 1.00 L and enough HCl is added to adjust the pH to 8.50. Use Davies activity coefficients to find the composition of the solution. What fraction of iron is in each of its forms, and what fraction of glycine is in each of its forms? From the distribution of species, explain the principal chemistry that requires addition of HCl to obtain a pH of 8.50.

Suggested procedure: (i) Use equilibrium expressions to write all concentrations in terms of $[G^-]$, $[Fe^{2+}]$, and $[H^+]$. (ii) Write one mass balance for iron and one for glycine. (iii) Substitute expressions from step (i) into the mass balance for iron to find $[Fe^{2+}]$ in terms of $[G^-]$ and $[H^+]$. (iv) Guess an ionic strength of 0.01 M and compute activity coefficients. (v) Compute $[H^+] = 10^{-pH}/\gamma_{H^+}$. (vi) Guess a value of $[G^-]$ and compute all concentrations in the mass balance for glycine. (vii) Use Solver to vary $[G^-]$ to satisfy the mass balance for glycine. (viii) Compute $[Cl^-]$ from the charge balance. (ix) Compute the ionic strength and perform several iterations of the entire process until the ionic strength no longer changes.

12-16. 🖿 Data for the glycine difference plot in Figure 12-14 are given below.

v (mL)	pH	v (mL)	pH	v (mL)	pH	v (mL)	pH
0.00	2.234	0.40	2.550	0.80	3.528	1.20	10.383
0.02	2.244	0.42	2.572	0.82	3.713	1.22	10.488
0.04	2.254	0.44	2.596	0.84	4.026	1.24	10.595
0.06	2.266	0.46	2.620	0.86	5.408	1.26	10.697
0.08	2.278	0.48	2.646	0.88	8.149	1.28	10.795
0.10	2.291	0.50	2.675	0.90	8.727	1.30	10.884
0.12	2.304	0.52	2.702	0.92	8.955	1.32	10.966
0.14	2.318	0.54	2.736	0.94	9.117	1.34	11.037
0.16	2.333	0.56	2.768	0.96	9.250	1.36	11.101
0.18	2.348	0.58	2.802	0.98	9.365	1.38	11.158
0.20	2.363	0.60	2.838	1.00	9.467	1.40	11.209
0.22	2.380	0.62	2.877	1.02	9.565	1.42	11.255
0.24	2.397	0.64	2.920	1.04	9.660	1.44	11.296
0.26	2.413	0.66	2.966	1.06	9.745	1.46	11.335
0.28	2.429	0.68	3.017	1.08	9.830	1.48	11.371
0.30	2.448	0.70	3.073	1.10	9.913	1.50	11.405
0.32	2.467	0.72	3.136	1.12	10.000	1.52	11.436
0.34	2.487	0.74	3.207	1.14	10.090	1.54	11.466
0.36	2.506	0.76	3.291	1.16	10.183	1.56	11.492
0.38	2.528	0.78	3.396	1.18	10.280	1.58	11.519
						1.60	11.541

Data from A. Kraft, J. Chem. Ed. **2003**, *80, 554.*

(a) Reproduce the spreadsheet in Figure 12-13 and show that you get the same values of pK_1 and pK_2 in cells B10 and B11 after executing Solver. Start with different values of pK_1 and pK_2 and see if Solver finds the same solutions.

(b) Use Solver to find the best values of pK_1 and pK_2 while fixing pK'_w at its expected value of 13.797. Describe how $\bar{n}_H$(measured) behaves when pK'_w is fixed.

12-17. 🖿 *Difference plot.* A solution containing 0.139 mmol of the triprotic acid tris(2-aminoethyl)amine · 3HCl plus 0.115 mmol HCl in 40 mL of 0.10 M KCl was titrated with 0.490 5 M NaOH to measure acid dissociation constants.

$$N(CH_2CH_2NH_3^+)_3 \cdot 3Cl^-$$

Tris(2-aminoethyl)amine · 3HCl
$H_3A^{3+} \cdot 3Cl^-$

(a) Write expressions for the experimental mean fraction of protonation, $\bar{n}_H$(measured), and the theoretical mean fraction of protonation, $\bar{n}_H$(theoretical).

(b) From the following data, prepare a graph of $\bar{n}_H$(measured) versus pH. Find the best values of pK_1, pK_2, pK_3, and pK'_w by minimizing the sum of the squares of the residuals, $\Sigma[\bar{n}_H$(measured) $- \bar{n}_H$(theoretical)$]^2$.

v (mL)	pH	v (mL)	pH	v (mL)	pH	v (mL)	pH
0.00	2.709	0.36	8.283	0.72	9.687	1.08	10.826
0.02	2.743	0.38	8.393	0.74	9.748	1.10	10.892
0.04	2.781	0.40	8.497	0.76	9.806	1.12	10.955
0.06	2.826	0.42	8.592	0.78	9.864	1.14	11.019
0.08	2.877	0.44	8.681	0.80	9.926	1.16	11.075
0.10	2.937	0.46	8.768	0.82	9.984	1.18	11.128
0.12	3.007	0.48	8.851	0.84	10.042	1.20	11.179
0.14	3.097	0.50	8.932	0.86	10.106	1.22	11.224
0.16	3.211	0.52	9.011	0.88	10.167	1.24	11.268
0.18	3.366	0.54	9.087	0.90	10.230	1.26	11.306
0.20	3.608	0.56	9.158	0.92	10.293	1.28	11.344
0.22	4.146	0.58	9.231	0.94	10.358	1.30	11.378
0.24	5.807	0.60	9.299	0.96	10.414	1.32	11.410
0.26	6.953	0.62	9.367	0.98	10.476	1.34	11.439
0.28	7.523	0.64	9.436	1.00	10.545	1.36	11.468
0.30	7.809	0.66	9.502	1.02	10.615	1.38	11.496
0.32	8.003	0.68	9.564	1.04	10.686	1.40	11.521
0.34	8.158	0.70	9.626	1.06	10.756		

Data from A. Kraft, J. Chem. Ed. **2003**, *80, 554.*

(c) Create a fractional composition graph showing the fractions of H_3A^{3+}, H_2A^{2+}, HA^+, and A as a function of pH.

12-18. 🖿 (a) For the following reactions, prepare a diagram showing log(concentration) versus pH for all species in the pH range 2 to 12 for a solution made by dissolving 0.025 mol $CuSO_4$ in 1.00 L. Equilibrium constants apply at $\mu = 0.1$ M, which you should assume is constant. Do not use activity coefficients, because the equilibrium constants already apply at $\mu = 0.1$ M.

$Cu^{2+} + SO_4^{2-} \rightleftharpoons CuSO_4(aq)$	$\log K'_{ip} = 1.26$
$HSO_4^- \rightleftharpoons SO_4^{2-} + H^+$	$\log K'_a = -1.54$
$Cu^{2+} + OH^- \rightleftharpoons CuOH^+$	$\log \beta'_1 = 6.1$
$Cu^{2+} + 2OH^- \rightleftharpoons Cu(OH)_2(aq)$	$\log \beta'_2 = 11.2*$
$Cu^{2+} + 3OH^- \rightleftharpoons Cu(OH)_3^-$	$\log \beta'_3 = 14.5*$
$Cu^{2+} + 4OH^- \rightleftharpoons Cu(OH)_4^{2-}$	$\log \beta'_4 = 15.6*$
$2Cu^{2+} + OH^- \rightleftharpoons Cu_2(OH)^{3+}$	$\log \beta'_{12} = 8.2*$
$2Cu^{2+} + 2OH^- \rightleftharpoons Cu_2(OH)_2^{2+}$	$\log \beta'_{22} = 16.8$
$3Cu^{2+} + 4OH^- \rightleftharpoons Cu_3(OH)_4^{2+}$	$\log \beta'_{43} = 33.5$
$H_2O \rightleftharpoons H^+ + OH^-$	$\log K'_w = -13.79$

*Asterisk indicates that equilibrium constant was estimated for $\mu = 0.1$ M from data reported for a different ionic strength.

Remember that pH $= -\log([H^+]\gamma_{H^+})$ and $[OH^-] = K'_w/[H^+]$. Use $\gamma_{H^+} = 0.78$ for this problem.

Recommended procedure: From the mass balance for sulfate, find an expression for $[SO_4^{2-}]$ in terms of $[Cu^{2+}]$ and $[H^+]$. Set up a spreadsheet with pH values between 2 and 12 in column A. From pH, compute $[H^+]$ and $[OH^-]$ in columns B and C. *Guess* a value for $[Cu^{2+}]$ in column D. From $[H^+]$ and $[Cu^{2+}]$, calculate $[SO_4^{2-}]$ from the equation derived from the mass balance for sulfate. From $[H^+]$, $[OH^-]$, $[Cu^{2+}]$, and $[SO_4^{2-}]$, calculate all other concentrations by using equilibrium expressions. Find the total concentration of Cu by adding the concentrations of all the species. Use Solver to vary $[Cu^{2+}]$ in column D so that the total concentration of copper is 0.025 M. You need to use Solver for each line in the spreadsheet with a different pH.

(b) Find the pH of 0.025 M $CuSO_4$ if nothing is added to adjust the pH. This is the pH at which the net charge is 0.

(c) At what pH does each of the following salts precipitate from 0.025 M $CuSO_4$?

$$Cu(OH)_{1.5}(SO_4)_{0.25}(s) \rightleftharpoons Cu^{2+} + 1.5OH^- + 0.25SO_4^{2-}$$
$$\log K'_{sp}(OH\text{-}SO_4) = -16.68$$

$$Cu(OH)_2(s) \rightleftharpoons Cu^{2+} + 2OH^- \qquad \log K'_{sp}(OH) = -18.7*$$

$$CuO(s) + H_2O \rightleftharpoons Cu^{2+} + 2OH^- \qquad \log K'_{sp}(O) = -19.7*$$

---

*Asterisk indicates that equilibrium constant was estimated for $\mu = 0.1$ M from data reported for a different ionic strength.

12-19. ⊞ *Ion-pairing in acid-base systems.* This problem incorporates ion-pair equilibria 12-12 and 12-13 into the acid-base chemistry of Section 12-1.

(a) From mass balance 12-15, derive Equation 12-16.

(b) Substitute equilibrium expressions into mass balance 12-17 to derive an expression for $[T^{2-}]$ in terms of $[H^+]$, $[Na^+]$, and various equilibrium constants.

(c) With the same approach used in part (b), derive expressions for $[HT^-]$ and $[H_2T]$.

(d) Add the species $[NaT^-]$ and $[NaHT]$ to the spreadsheet in Figure 12-1 and compute the composition and pH of the solution. Compute $[Na^+]$ with Equation 12-16. Compute $[H_2T]$, $[HT^-]$, and $[T^{2-}]$ from the expressions derived in parts (b) and (c). Excel will indicate a *circular reference* problem because, for example, the formula for $[Na^+]$ depends on $[T^{2-}]$ and the formula for $[T^{2-}]$ depends on $[Na^+]$. To handle this circular reference in Excel 2007, click the Microsoft Office button at the upper left of the spreadsheet. Click on Excel Options at the bottom of the window. Select Formulas. In Calculation Options, check Enable iterative calculation. Click OK and then use Solver to find the pH in cell H13 that reduces the net charge in cell E15 to (near) zero. In earlier versions of Excel, select the Tools menu and go to Options. Select the Calculations tab and choose Iteration. Click OK and then run Solver.

Fundamentals of Electrochemistry

LITHIUM-ION BATTERY

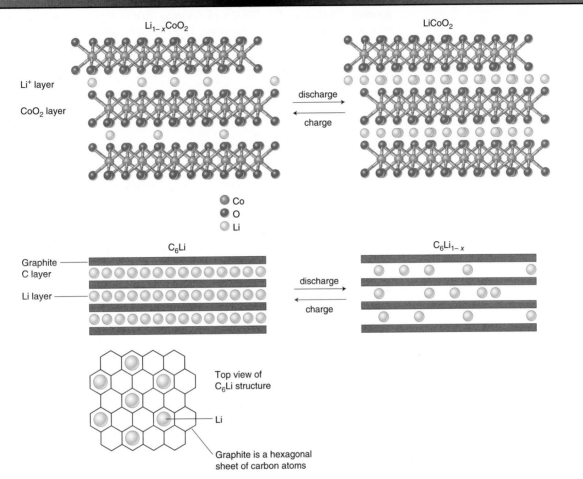

$Li_{1-x}CoO_2$

$LiCoO_2$

Li⁺ layer

CoO₂ layer

discharge
charge

- Co
- O
- Li

C_6Li

C_6Li_{1-x}

Graphite
C layer

Li layer

discharge
charge

Top view of
C_6Li structure

Li

Graphite is a hexagonal
sheet of carbon atoms

Lithium-ion
battery

Graphite layer

Porous separator plus Li⁺
salt in organic solvent

Cobalt oxide layer

Direction of electron flow from lithium battery
to laptop computer.

High-capacity, rechargeable lithium-ion batteries, such as those in cell phones and laptop computers, are a shining example of the fruits of materials chemistry research. The idealized chemistry is

$$C_6Li + Li_{1-x}CoO_2 \underset{\text{charge}}{\overset{\text{discharge}}{\rightleftharpoons}} C_6Li_{1-x} + LiCoO_2$$

In C_6Li, lithium atoms reside between layers of carbon in graphite. Atoms or molecules located between layers of a structure are said to be *intercalated*. During operation of the battery, lithium ions migrate from graphite to cobalt oxide. Lithium atoms leave electrons behind in the graphite, and the resulting Li^+ ions become intercalated between CoO_2 layers. To go from graphite to cobalt oxide, Li^+ passes through an electrolyte consisting of a lithium salt dissolved in a high-boiling organic solvent. A porous polymer separator between graphite and cobalt oxide is an electrical insulator that permits Li^+ ions to pass. Electrons travel through the external circuit to reach the cobalt oxide and maintain electroneutrality.

A single-cell lithium-ion battery produces ~3.7 volts. These batteries store twice as much energy per unit mass as the nickel-metal hydride batteries they replaced. Ongoing research is aimed at improved materials and high-area microstructures for the electrodes and the separator layer. Goals include higher energy density, longer life, and safer operation. Batteries are *galvanic cells*, which are the subject of this chapter. A galvanic cell uses a favorable chemical reaction to produce electricity.

Electrochemistry is a major branch of analytical chemistry that uses electrical measurements of chemical systems for analytical purposes.[1,2] For example, the opening of Chapter 1 showed an electrode being used to measure neurotransmitter molecules released by a nerve cell. Electrochemistry also refers to the use of electricity to drive a chemical reaction or to the use of a chemical reaction to produce electricity.

13-1 Basic Concepts

A **redox reaction** involves transfer of electrons from one species to another. A species is said to be **oxidized** when it *loses electrons*. It is **reduced** when it *gains electrons*. An **oxidizing agent,** also called an **oxidant,** takes electrons from another substance and becomes reduced. A **reducing agent,** also called a **reductant,** gives electrons to another substance and is oxidized in the process. In the reaction

$$Fe^{3+} + V^{2+} \rightarrow Fe^{2+} + V^{3+} \qquad (13\text{-}1)$$

Oxidizing agent   Reducing agent

Fe^{3+} is the oxidizing agent because it takes an electron from V^{2+}. V^{2+} is the reducing agent because it gives an electron to Fe^{3+}. Fe^{3+} is reduced, and V^{2+} is oxidized as the reaction proceeds from left to right. Appendix D reviews oxidation numbers and balancing of redox equations.

Chemistry and Electricity

When electrons from a redox reaction flow through an electric circuit, we can learn something about the reaction by measuring current and voltage. Electric current is proportional to the rate of reaction, and the cell voltage is proportional to the free energy change for the electrochemical reaction. In techniques such as voltammetry, the voltage can be used to identify reactants.

Electric Charge

Electric charge (q) is measured in **coulombs** (C). The magnitude of the charge of a single electron is 1.602×10^{-19} C, so a mole of electrons has a charge of $(1.602 \times 10^{-19}$ C$)(6.022 \times 10^{23}$ mol$^{-1}) = 9.649 \times 10^{4}$ C, which is called the **Faraday constant,** F.

Relation between charge and moles:

$$q = n \cdot F \qquad (13\text{-}2)$$

Coulombs   mol e$^-$ $\dfrac{\text{Coulombs}}{\text{mol e}^-}$

where n is the number moles of electrons transferred.

> **EXAMPLE**   Relating Coulombs to Quantity of Reaction
>
> If 5.585 g of Fe^{3+} were reduced in Reaction 13-1, how many coulombs of charge must have been transferred from V^{2+} to Fe^{3+}?
>
> **Solution** First, we find that 5.585 g of Fe^{3+} equal 0.100 0 mol of Fe^{3+}. Because each Fe^{3+} ion requires one electron in Reaction 13-1, 0.100 0 mol of electrons must have been transferred. Using the Faraday constant, we find that 0.100 0 mol of electrons corresponds to
>
> $$q = nF = (0.100\ 0 \text{ mol e}^-)\left(9.649 \times 10^4 \frac{\text{C}}{\text{mol e}^-}\right) = 9.649 \times 10^3 \text{ C}$$
>
> **Test Yourself** How many moles of Sn^{4+} are reduced to Sn^{2+} by 1.00 C of electric charge? (**Answer:** 5.18 μmol)

Electric Current

The quantity of charge flowing each second through a circuit is called the **current.** The unit of current is the **ampere,** abbreviated A. A current of 1 ampere represents a charge of 1 coulomb per second flowing past a point in a circuit.

Oxidation: loss of electrons
Reduction: gain of electrons
Oxidizing agent: takes electrons
Reducing agent: gives electrons

$Fe^{3+} + e^- \rightarrow Fe^{2+}$
$V^{2+} \rightarrow V^{3+} + e^-$

Faraday constant $(F) = \dfrac{\text{coulombs}}{\text{mole of electrons}}$

1 A = 1 C/s

EXAMPLE Relating Current to Rate of Reaction

Suppose that electrons are forced into a platinum wire immersed in a solution containing Sn^{4+} (Figure 13-1), which is reduced to Sn^{2+} at a constant rate of 4.24 mmol/h. How much current passes through the solution?

Solution *Two* electrons are required to reduce *one* Sn^{4+} ion:

$$Sn^{4+} + 2e^- \rightarrow Sn^{2+}$$

Electrons flow at a rate of (2 mmol e^-/mmol Sn^{4+})(4.24 mmol Sn^{4+}/h) = 8.48 mmol e^-/h, which corresponds to

$$\frac{8.48 \text{ mmol } e^-/h}{3\,600 \text{ s/h}} = 2.356 \times 10^{-3} \frac{\text{mmol } e^-}{s} = 2.356 \times 10^{-6} \frac{\text{mol } e^-}{s}$$

To find the current, we convert moles of electrons per second into coulombs per second:

$$\text{Current} = \frac{\text{charge}}{\text{time}} = \frac{\text{coulombs}}{\text{second}} = \frac{\text{moles } e^-}{\text{second}} \cdot \frac{\text{coulombs}}{\text{mole}}$$

$$= \left(2.356 \times 10^{-6} \frac{\text{mol}}{s}\right)\left(9.649 \times 10^4 \frac{C}{\text{mol}}\right)$$

$$= 0.227 \text{ C/s} = 0.227 \text{ A}$$

Test Yourself What current reduces Sn^{4+} at a rate of 1.00 mmol/h? (*Answer:* 53.6 mA)

FIGURE 13-1 Electrons flowing into a coil of Pt wire at which Sn^{4+} ions in solution are reduced to Sn^{2+}. This process could not happen by itself, because there is no complete circuit. If Sn^{4+} is to be reduced at this Pt electrode, some other species must be oxidized at some other place.

In Figure 13-1, we encountered a Pt **electrode,** which conducts electrons into or out of a chemical species in the redox reaction. Platinum is a common *inert* electrode. It does not participate in the redox chemistry except as a conductor of electrons.

Voltage, Work, and Free Energy

The difference in **electric potential,** E, between two points is the work needed (or that can be done) when moving an electric charge from one point to the other. *Potential difference* is measured in **volts** (V). The greater the potential difference between two points, the stronger will be the "push" on a charged particle traveling between those points.

A good analogy for understanding current and potential is to think of water flowing through a garden hose (Figure 13-2). Current is the electric charge flowing past a point in a wire each second. Current is analogous to the volume of water flowing past a point in the hose each second. The difference in electric potential is a measure of the force pushing on the electrons. The greater the force, the more current flows. Potential difference is analogous to the pressure on the water in the hose. The greater the pressure, the faster the water flows.

When a charge, q, moves through a potential difference, E, the work done is

It takes work to move like charges toward one another. Work can be done when opposite charges move toward one another.

Relation between work and voltage:

$$\underset{\text{Joules}}{\text{Work}} = \underset{\text{Volts}}{E} \cdot \underset{\text{Coulombs}}{q} \qquad (13\text{-}3)$$

Work has the dimensions of energy, whose units are **joules** (J). One *joule* of energy is gained or lost when 1 *coulomb* of charge moves between points whose potentials differ by 1 *volt*. Equation 13-3 tells us that the dimensions of volts are joules per coulomb (1 V = 1 J/C).

EXAMPLE Electrical Work

How much work can be done if 2.4 mmol of electrons fall through a potential difference of 0.27 V?

Solution To use Equation 13-3, we must convert moles of electrons into coulombs of charge. The relation is

$$q = nF = (2.4 \times 10^{-3} \text{ mol})(9.649 \times 10^4 \text{ C/mol}) = 2.3 \times 10^2 \text{ C}$$

The work that could be done is

$$\text{Work} = E \cdot q = (0.27 \text{ V})(2.3 \times 10^2 \text{ C}) = 62 \text{ J}$$

Test Yourself What must be the potential drop (V) for 1.00 μmol e^- to do 1.00 J of work? (*Answer:* 10.4 V)

Electric *current* is analogous to *volume* of water per second flowing out of a hose.

High pressure   Low pressure

Electric *potential difference* is analogous to the hydrostatic *pressure* pushing water through a hose. High pressure gives high flow.

FIGURE 13-2 Analogy between the flow of water through a hose and the flow of electricity through a wire.

In the garden hose analogy, suppose that one end of a hose is raised 1 m above the other end and a volume of 1 L of water flows through the hose. The flowing water goes through a mechanical device to do a certain amount of work. If one end of the hose is raised 2 m above the other, the amount of work that can be done by the falling water is twice as great. The elevation difference between the ends of the hose is analogous to electric potential difference and the volume of water is analogous to electric charge. The greater the electric potential difference between two points in a circuit, the more work can be done by the charge flowing between those two points.

Section 6-2 gave a brief discussion of ΔG.

The free energy change, ΔG, for a chemical reaction conducted reversibly at constant temperature and pressure equals the maximum possible electrical work that can be done by the reaction on its surroundings:

$$\text{Work done on surroundings} = -\Delta G \qquad (13\text{-}4)$$

The negative sign in Equation 13-4 indicates that the free energy of a system decreases when the work is done on the surroundings.

Combining Equations 13-2, 13-3, and 13-4 produces a relation of utmost importance:

$$\Delta G = -\text{work} = -E \cdot q$$

$q = nF$

Relation between free energy difference and electric potential difference: $\boxed{\Delta G = -nFE}$ (13-5)

Equation 13-5 relates the free energy change of a chemical reaction to the electrical potential difference (that is, the voltage) that can be generated by the reaction.

Ohm's Law

Ohm's law states that current, I, is directly proportional to the potential difference (voltage) across a circuit and inversely proportional to the **resistance, R,** of the circuit.

The greater the voltage, the more current will flow. The greater the resistance, the less current will flow.

Box 13-1 shows measurements of the resistance of single molecules by measuring current and voltage and applying Ohm's law.

Ohm's law: $\boxed{I = \dfrac{E}{R}}$ (13-6)

Units of resistance are **ohms,** assigned the Greek symbol Ω (omega). A current of 1 ampere flows through a circuit with a potential difference of 1 volt if the resistance is 1 ohm. From Equation 13-6, the unit ampere (A) is equivalent to V/Ω.

Power

Power, P, is the work done per unit time. The SI unit of power is J/s, better known as the **watt** (W).

$$P = \frac{\text{work}}{\text{s}} = \frac{E \cdot q}{\text{s}} = E \cdot \frac{q}{\text{s}} \qquad (13\text{-}7)$$

Power (watts) = work per second
$P = E \cdot I = (IR) \cdot I = I^2 R$

Because q/s is the current, I, we can write

$$P = E \cdot I \qquad (13\text{-}8)$$

A cell capable of delivering 1 ampere at a potential difference of 1 volt has a power output of 1 watt.

FIGURE 13-3 A circuit with a battery and a resistor. Benjamin Franklin investigated static electricity in the 1740s.[6] He thought electricity was a fluid that flows from a silk cloth to a glass rod when the rod is rubbed with the cloth. We now know that electrons flow from glass to silk. However, Franklin's convention for the direction of electric current has been retained, so we say that current flows from positive to negative—in the opposite direction of electron flow.

EXAMPLE   **Using Ohm's Law**

In the circuit in Figure 13-3, the battery generates a potential difference of 3.0 V, and the resistor has a resistance of 100 Ω. We assume that the wire connecting the battery and the resistor has negligible resistance. Find the current and power delivered by the battery.

Solution The current is

$$I = \frac{E}{R} = \frac{3.0\ \text{V}}{100\ \Omega} = 0.030\ \text{A} = 30\ \text{mA}$$

The power produced by the battery is

$$P = E \cdot I = (3.0\ \text{V})(0.030\ \text{A}) = 90\ \text{mW}$$

Test Yourself   What voltage is required to produce 180 mW of power? (***Answer:*** 4.24 V)

BOX 13-1 **Ohm's Law, Conductance, and Molecular Wire[3]**

The electrical conductance of a single molecule suspended between two gold electrodes is known from measurement of voltage and current by applying Ohm's law. Conductance is 1/resistance, so it has the units 1/ohm ≡ siemens (S).

To make molecular junctions, a gold scanning tunneling microscope tip was moved in and out of contact with a gold substrate in the presence of a solution containing a test molecule terminated by thiol (—SH) groups. Thiols spontaneously bind to gold, forming bridges such as that shown here. Nanoampere currents were observed with a potential difference of 0.1 V between the gold surfaces.

$$Au-S-CH_2-\overset{H}{\underset{O}{N}}-C-S-Au$$

The graph shows four observations of conductance as the scanning tunneling microscope tip was pulled away from the Au substrate. Plateaus are observed at multiples of 19 nS. An interpretation is that a single molecule connecting two Au surfaces has a conductance of 19 nS (or a resistance of 50 MΩ). If two molecules form parallel bridges, conductance increases to 38 nS. Three molecules give a conductance of 57 nS. If there are three bridges and the electrodes are pulled apart, one of the bridges breaks and conductance drops to 38 nS. When the second bridge breaks, conductance drops to 19 nS. The exact conductance varies because the environment of each molecule on the Au surface is not identical. A histogram of >500 observations in the inset shows peaks at 19, 38, and 57 nS.

Change in conductance as Au scanning tunneling microscope tip immersed in dithiol solution is withdrawn from Au substrate. [From X. Xiao, B. Xu, and N. Tao, "Conductance Titration of Single-Peptide Molecules," *J. Am. Chem. Soc.* 2004, *126*, 5370.]

Alkane hydrocarbons can be thought of as prototypical electrical insulators. The conductance of alkane dithiols decreases exponentially as chain length increases:[4]

$$HS(CH_2)_8SH \qquad conductance = 16.1\ nS$$

$$HS(CH_2)_{10}SH \qquad conductance = 1.37\ nS$$

$$HS(CH_2)_{12}SH \qquad conductance = 0.35\ nS$$

The conductance of a chain of Cr atoms with pyridylamine ligands is greater than that of alkanes.[5] The chart shows that conductance decreases exponentially with chain length, but at a lower rate than for the alkanes.

Cr₃ chain ($m = 0$)   conductance = 1 100 nS
Cr₅ chain ($m = 1$)   conductance = 310 nS
Cr₇ chain ($m = 2$)   conductance = 140 nS

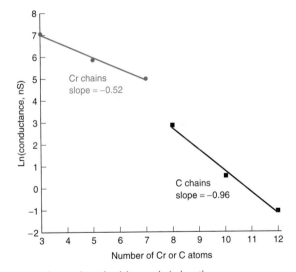

Dependence of conductivity on chain length.

What happens to the power generated by the circuit? *The energy appears as heat in the resistor.* The power (90 mW) equals the rate at which heat is produced in the resistor.

Here is a summary of symbols, units, and relations from the last few pages:

Relation between charge and moles:

$$q = n \cdot F$$

Charge (coulombs, C) Moles of e⁻   C/mole

Relation between work and voltage:

$$Work = E \cdot q$$

Joules (J)   Volts (V)   Coulombs (C)

Relation between free energy difference
and electric potential difference:

$$\Delta G \underset{\text{Joules}}{=} -nFE$$

Ohm's law:

$$I \underset{\substack{\text{Current} \\ \text{(A)}}}{=} E \underset{\substack{\text{Volts} \\ \text{(V)}}}{/} R \underset{\substack{\text{Resistance} \\ \text{(ohms, }\Omega\text{)}}}{}$$

Electric power:

$$P \underset{\substack{\text{Power} \\ \text{(watts, W)}}}{=} \frac{\text{work}}{\text{s}} \underset{\text{J/s}}{=} E \underset{\text{Volts}}{\cdot} I \underset{\text{Amperes}}{}$$

13-2 Galvanic Cells

> A galvanic cell uses a spontaneous chemical reaction to generate electricity.

A **galvanic cell** (also called a *voltaic cell*) uses a *spontaneous* chemical reaction to generate electricity. To accomplish this, one reagent must be oxidized and another must be reduced. The two cannot be in contact, or electrons would flow directly from the reducing agent to the oxidizing agent. Instead, oxidizing and reducing agents are physically separated, and electrons are forced to flow through an external circuit to go from one reactant to the other. Batteries[7] and fuel cells[8] are galvanic cells that consume their reactants to generate electricity. A battery has a static compartment filled with reactants. In a fuel cell, fresh reactants flow past the electrodes and products are continuously flushed from the cell.

A Cell in Action

Figure 13-4 shows a galvanic cell with two electrodes suspended in a solution of $CdCl_2$. One electrode is cadmium; the other is metallic silver coated with solid AgCl. The reactions are

Reduction: $\quad 2AgCl(s) + 2e^- \rightleftharpoons 2Ag(s) + 2Cl^-(aq)$

Oxidation: $\quad\quad\quad Cd(s) \rightleftharpoons Cd^{2+}(aq) + 2e^-$

Net reaction: $\quad Cd(s) + 2AgCl(s) \rightleftharpoons Cd^{2+}(aq) + 2Ag(s) + 2Cl^-(aq)$

The net reaction is composed of a reduction and an oxidation, each of which is called a **half-reaction.** The two half-reactions are written with equal numbers of electrons so that their sum includes no free electrons.

> Recall that ΔG is *negative* for a spontaneous reaction.

Oxidation of Cd metal to $Cd^{2+}(aq)$ provides electrons that flow through the circuit to the Ag electrode in Figure 13-4. At the Ag surface, Ag^+ (from AgCl) is reduced to Ag(s). Chloride from AgCl goes into solution. The free energy change for the net reaction, -150 kJ per mole of Cd, provides the driving force that pushes electrons through the circuit.

FIGURE 13-4 A simple galvanic cell. The **potentiometer** measures voltage. It has positive and negative terminals. When electrons flow into the negative terminal, as in this illustration, the voltage is positive.

EXAMPLE	Voltage Produced by a Chemical Reaction

Calculate the voltage that would be measured by the potentiometer in Figure 13-4.

Solution Because $\Delta G = -150$ kJ per mol of Cd, we can use Equation 13-5 (where n is moles of electrons transferred in the balanced net reaction) to write

$$E = -\frac{\Delta G}{nF} = -\frac{-150 \times 10^3 \text{ J}}{(2 \text{ mol})\left(9.649 \times 10^4 \dfrac{\text{C}}{\text{mol}}\right)}$$

$$= +0.777 \text{ J/C} = +0.777 \text{ V}$$

Reminder: 1 J/C = 1 volt

A spontaneous chemical reaction (negative ΔG) produces a *positive voltage*.

Test Yourself Find E if $\Delta G = +150$ kJ and $n = 1$ mol. (*Answer:* -1.55 V)

Chemists define the electrode at which *reduction* occurs as the **cathode.** The **anode** is the electrode at which *oxidation* occurs. In Figure 13-4, Ag is the cathode because reduction takes place at its surface ($2AgCl + 2e^- \rightarrow 2Ag + 2Cl^-$), and Cd is the anode because it is oxidized ($Cd \rightarrow Cd^{2+} + 2e^-$).

Cathode: where reduction occurs
Anode: where oxidation occurs

Salt Bridge

Consider the cell in Figure 13-5, in which the reactions are intended to be

Cathode:	$2Ag^+(aq) + 2e^- \rightleftharpoons 2Ag(s)$	
Anode:	$Cd(s) \rightleftharpoons Cd^{2+}(aq) + 2e^-$	
Net reaction:	$Cd(s) + 2Ag^+(aq) \rightleftharpoons Cd^{2+}(aq) + 2Ag(s)$	(13-9)

The net reaction is spontaneous, but little current flows through the circuit because Ag^+ is not forced to be reduced at the Ag electrode. Aqueous Ag^+ can react directly at the $Cd(s)$ surface, giving the same net reaction with no flow of electrons through the external circuit.

The cell in Figure 13-5 is *short-circuited.*

We can separate the reactants into two *half-cells*[11] if we connect the two halves with a **salt bridge,** as shown in Figure 13-6. The salt bridge is a U-shaped tube filled with a gel containing a high concentration of KNO_3 (or other electrolyte that does not affect the cell reaction). The ends of the bridge are porous glass disks that allow ions to diffuse but minimize mixing of solutions inside and outside the bridge. When the galvanic cell is operating, K^+ from the bridge migrates into the cathode compartment and a small amount of NO_3^- migrates from the cathode into the bridge. Ion migration offsets the charge buildup that would otherwise occur as electrons

A salt bridge maintains electroneutrality (no charge buildup) throughout the cell. See Demonstration 13-1.

FIGURE 13-5 A cell that will not work. The solution contains $Cd(NO_3)_2$ and $AgNO_3$.

FIGURE 13-6 A cell that works—thanks to the salt bridge!

flow into the silver electrode. The migration of ions out of the bridge is greater than the migration of ions into the bridge because the salt concentration in the bridge is much higher than the salt concentration in the half-cells. At the left side of the salt bridge, NO_3^- migrates into the anode compartment and a little Cd^{2+} migrates into the bridge to prevent buildup of positive charge.

For reactions that do not involve Ag^+ or other species that react with Cl^-, the salt bridge usually contains KCl electrolyte. A typical salt bridge is prepared by heating 3 g of agar with 30 g of KCl in 100 mL of water until a clear solution is obtained. The solution is poured into the U-tube and allowed to gel. The bridge is stored in saturated aqueous KCl.

Line Notation

Electrochemical cells are described by a notation employing just two symbols:

| phase boundary   ‖ salt bridge

The salt bridge symbol ‖ represents the two phase boundaries on either side of the bridge.

The cell in Figure 13-4 is represented by the *line diagram*

$$Cd(s)\,|\,CdCl_2(aq)\,|\,AgCl(s)\,|\,Ag(s)$$

Each phase boundary is indicated by a vertical line. The electrodes are shown at the extreme left- and right-hand sides of the line diagram. The cell in Figure 13-6 is

$$Cd(s)\,|\,Cd(NO_3)_2(aq)\,\|\,AgNO_3(aq)\,|\,Ag(s)$$

DEMONSTRATION 13-1   The Human Salt Bridge

A salt bridge is an ionic medium with a *semipermeable* barrier on each end. Small molecules and ions can cross a semipermeable barrier, but large molecules cannot. Demonstrate a "proper" salt bridge by filling a U-tube with agar and KCl as described in the text and construct the cell shown here.

The pH meter is a potentiometer whose negative terminal is the reference electrode socket.

Write the half-reactions for this cell and use the Nernst equation to calculate the theoretical voltage. Measure the voltage with a conventional salt bridge. Then replace the salt bridge with filter paper soaked in NaCl solution and measure the voltage again. Finally, replace the filter-paper salt bridge with two fingers and measure the voltage again. A human is just a bag of salt housed in a semipermeable membrane. Small differences in voltage observed when the salt bridge is replaced can be attributed to the junction potential discussed in Section 14-3. To prove that it is hard to distinguish a chemistry instructor from a hot dog, use a hot dog as a salt bridge[9] and measure the voltage again.

Challenge   One hundred eighty students at Virginia Tech made a salt bridge by holding hands.[10] Their resistance was lowered from $10^6\ \Omega$ per student to $10^4\ \Omega$ per student by wetting everyone's hands. Can your class beat this record?

13-3 Standard Potentials

The voltage measured in Figure 13-6 is the difference in electric potential between the Ag electrode on the right and the Cd electrode on the left. Voltage tells us how much work can be done by electrons flowing from one side to the other (Equation 13-3). The potentiometer (voltmeter) indicates a positive voltage when electrons flow into the negative terminal, as in Figure 13-6. If electrons flow the other way, the voltage is negative.

Sometimes the negative terminal of a voltmeter is labeled "common." It may be colored black and the positive terminal red. When a pH meter with a BNC socket is used as a potentiometer, the center wire is the positive input and the outer connection is the negative input. In older pH meters, the negative terminal is the narrow receptacle to which the reference electrode is connected.

To predict the voltage that will be observed when different half-cells are connected to each other, the **standard reduction potential,** $E°$, for each half-cell is measured by an experiment shown in an idealized form in Figure 13-7. The half-reaction of interest in this diagram is

$$Ag^+ + e^- \rightleftharpoons Ag(s) \tag{13-10}$$

which occurs in the half-cell at the right connected to the *positive* terminal of the potentiometer. *Standard* means that activities of all species are unity. For Reaction 13-10 under standard conditions, $\mathcal{A}_{Ag^+} = 1$ and, by definition, the activity of Ag(s) also is unity.

The left half-cell, connected to the *negative* terminal of the potentiometer, is called the **standard hydrogen electrode** (S.H.E.). It consists of a catalytic Pt surface in contact with an acidic solution in which $\mathcal{A}_{H^+} = 1$. A stream of $H_2(g)$ bubbled through the electrode saturates the solution with $H_2(aq)$. The activity of $H_2(g)$ is unity if the pressure of $H_2(g)$ is 1 bar. The reaction that comes to equilibrium at the surface of the Pt electrode is

$$\text{S.H.E. half-reaction:} \quad H^+(aq, \mathcal{A} = 1) + e^- \rightleftharpoons \frac{1}{2}H_2(g, \mathcal{A} = 1) \tag{13-11}$$

We *arbitrarily* assign a potential of 0 to the standard hydrogen electrode at 25°C. The voltage measured by the meter in Figure 13-7 can therefore be *assigned* to Reaction 13-10, which occurs in the right half-cell. The measured value $E° = +0.799$ V is the standard reduction potential for Reaction 13-10. The positive sign tells us that electrons flow from left to right through the meter.

We can arbitrarily *assign* a potential to Reaction 13-11 because it serves as a reference point from which we can measure other half-cell potentials. An analogy is the arbitrary assignment of 0°C to the freezing point of water. Relative to this freezing point, hexane boils at 69° and benzene boils at 80°. The difference between boiling points is $80° - 69° = 11°$. If we were to assign the freezing point of water to be 200°C instead of 0°C, hexane would boil at 269° and benzene would boil at 280°. The difference is still 11°. Regardless of where we set zero on the scale, differences between points remain constant.

> Positive terminal is the wire at the center inside connector
>
> **BNC connector**
>
> Reference (– terminal)
> pH (+ terminal)
>
> **U.S. standard connector**

> **Question** What is the pH of the standard hydrogen electrode?

> We will write all half-reactions as *reductions*. By convention, $E° = 0$ for S.H.E.

> Walther Nernst appears to have been the first to assign the potential of the hydrogen electrode as 0 in 1897.[12]

$$Pt(s) \,|\, H_2(g, \mathcal{A} = 1) \,|\, H^+(aq, \mathcal{A} = 1) \,\|\, Ag^+(aq, \mathcal{A} = 1) \,|\, Ag(s)$$

Standard hydrogen electrode
(S.H.E.)

FIGURE 13-7 Cell used to measure the standard potential of the reaction $Ag^+ + e^- \rightleftharpoons Ag(s)$. This cell is hypothetical because it is usually not possible to adjust the activity of a species to 1.

TABLE 13-1 Ordered standard reduction potentials

	Oxidizing agent	Reducing agent	$E°$(V)	
⬆ Oxidizing power increases	$F_2(g) + 2e^- \rightleftharpoons 2F^-$		2.890	Reducing power increases ⬇
	$O_3(g) + 2H^+ + 2e^- \rightleftharpoons O_2(g) + H_2O$		2.075	
	$MnO_4^- + 8H^+ + 5e^- \rightleftharpoons Mn^{2+} + 4H_2O$		1.507	
	$Ag^+ + e^- \rightleftharpoons Ag(s)$		0.799	
	$Cu^{2+} + 2e^- \rightleftharpoons Cu(s)$		0.339	
	$2H^+ + 2e^- \rightleftharpoons H_2(g)$		0.000	
	$Cd^{2+} + 2e^- \rightleftharpoons Cd(s)$		−0.402	
	$K^+ + e^- \rightleftharpoons K(s)$		−2.936	
	$Li^+ + e^- \rightleftharpoons Li(s)$		−3.040	

Question The potential for the reaction $K^+ + e^- \rightleftharpoons K(s)$ is −2.936 V. This means that K^+ is a very poor oxidizing agent. (It does not readily accept electrons.) Does this imply that K^+ is therefore a good reducing agent?

Answer: No! To be a good reducing agent, K^+ would have to give up electrons easily (forming K^{2+}), which it cannot do. (*But*, the large negative reduction potential does imply that $K(s)$ is a good reducing agent.)

The line notation for the cell in Figure 13-7 is

$$Pt(s)\,|\,H_2(g, \mathcal{A} = 1)\,|\,H^+(aq, \mathcal{A} = 1)\,\|\,Ag^+(aq, \mathcal{A} = 1)\,|\,Ag(s)$$

or

$$S.H.E.\,\|\,Ag^+(aq, \mathcal{A} = 1)\,|\,Ag(s)$$

By convention, the left-hand electrode (Pt) is attached to the negative (reference) terminal of the potentiometer and the right-hand electrode is attached to the positive terminal. A standard reduction potential is really a potential *difference* between the potential of the reaction of interest and the potential of S.H.E, which we have arbitrarily set to 0.

To measure the standard potential of the half-reaction

$$Cd^{2+} + 2e^- \rightleftharpoons Cd(s) \tag{13-12}$$

we construct the cell

$$S.H.E.\,\|\,Cd^{2+}(aq, \mathcal{A} = 1)\,|\,Cd(s)$$

with the cadmium half-cell connected to the positive terminal of the potentiometer. In this case, we observe a *negative* voltage of −0.402 V. The negative sign means that electrons flow from Cd to Pt, a direction opposite that of the cell in Figure 13-7.

Appendix H contains standard reduction potentials arranged alphabetically by element. If the half-reactions were arranged according to descending value of $E°$ (as in Table 13-1), we would find the strongest oxidizing agents at the upper left and the strongest reducing agents at the lower right. If we connected the two half-cells represented by Reactions 13-10 and 13-12, Ag^+ would be reduced to Ag(s) as Cd(s) is oxidized to Cd^{2+}.

We always attach the left-hand electrode to the negative terminal of the potentiometer and the right-hand electrode to the positive terminal. The voltage on the meter is the difference:

> voltage = right-hand electrode potential − left-hand electrode potential

Challenge Draw a picture of the cell S.H.E $\|$ Cd^{2+} (aq, $\mathcal{A}$ = 1) | Cd(s) and show the direction of electron flow.

13-4 Nernst Equation

Le Châtelier's principle tells us that increasing reactant concentrations drives a reaction to the right and increasing the product concentrations drives a reaction to the left. The net driving force for a reaction is expressed by the **Nernst equation**, whose two terms include the driving force under standard conditions ($E°$, which applies when all activities are unity) and a term showing the dependence on reagent concentrations.

A reaction is spontaneous if ΔG is negative and E is positive. $\Delta G°$ and $E°$ refer to the free energy change and potential when the activities of reactants and products are unity. $\Delta G° = -nFE°$.

Nernst Equation for a Half-Reaction

For the half-reaction

$$aA + ne^- \rightleftharpoons bB$$

the Nernst equation giving the half-cell potential, E, is

Challenge Show that Le Châtelier's principle requires a negative sign in front of the reaction quotient term in the Nernst equation. *Hint:* The more favorable a reaction, the more positive is E.

Nernst equation:

$$E = E° - \frac{RT}{nF} \ln \frac{\mathcal{A}_B^b}{\mathcal{A}_A^a} \tag{13-13}$$

where E° = standard reduction potential ($\mathcal{A}_A = \mathcal{A}_B = 1$)

R = gas constant (8.314 J/(K · mol) = 8.314(V · C)/(K · mol))

T = temperature (K)

n = number of electrons in the half-reaction

F = Faraday constant (9.649×10^4 C/mol)

$\mathcal{A}_i$ = activity of species i

The logarithmic term in the Nernst equation is the **reaction quotient, Q.**

$$Q = \mathcal{A}_B^b / \mathcal{A}_A^a \qquad (13\text{-}14)$$

Q has the same form as the equilibrium constant, but the activities need not have their equilibrium values. Pure solids, pure liquids, and solvents are omitted from Q because their activities are unity (or close to unity). Concentrations of solutes are expressed as moles per liter and concentrations of gases are expressed as pressures in bars. When all activities are unity, $Q = 1$ and $\ln Q = 0$, thus giving $E = E^\circ$.

Converting the natural logarithm in Equation 13-13 into the base 10 logarithm and inserting T = 298.15 K (25.00°C) gives the most useful form of the Nernst equation:

Appendix A shows that log x = (ln x)/(ln 10) = (ln x)/2.303.

Nernst equation at 25°C: $\qquad E = E^\circ - \dfrac{0.059\ 16\ \text{V}}{n} \log\dfrac{\mathcal{A}_B^b}{\mathcal{A}_A^a} \qquad (13\text{-}15)$

The potential changes by 59.16/n mV for each factor-of-10 change in Q.

EXAMPLE Writing the Nernst Equation for a Half-Reaction

Let's write the Nernst equation for the reduction of white phosphorus to phosphine gas:

$$\frac{1}{4}P_4(s,\text{ white}) + 3H^+ + 3e^- \rightleftharpoons PH_3(g) \qquad E^\circ = -0.046\ \text{V}$$

Phosphine

Phosphine is a highly poisonous gas with the odor of decaying fish.

Solution We omit solids from the reaction quotient, and the concentration of a gas is expressed as the pressure of the gas. Therefore, the Nernst equation is

$$E = -0.046 - \frac{0.059\ 16}{3} \log\frac{P_{PH_3}}{[H^+]^3}$$

Test Yourself With E° from Appendix H, write the Nernst equation for $ZnS(s) + 2e^- \rightleftharpoons Zn(s) + S^{2-}$. (**Answer:** $E = -1.405 - \dfrac{0.059\ 16}{2} \log[S^{2-}]$)

EXAMPLE Multiplying a Half-Reaction Does Not Change E°

If we multiply a half-reaction by any factor, E° does not change. However, the factor n before the log term and the form of the reaction quotient, Q, do change. Let's write the Nernst equation for the reaction in the preceding example, multiplied by 2:

$$\frac{1}{2}P_4(s,\text{ white}) + 6H^+ + 6e^- \rightleftharpoons 2PH_3(g) \qquad E^\circ = -0.046\ \text{V}$$

Solution

$$E = -0.046 - \frac{0.059\ 16}{6} \log\frac{P_{PH_3}^2}{[H^+]^6}$$

Even though this Nernst equation does not look like the one in the preceding example, Box 13-2 shows that the numerical value of E is unchanged. The squared term in the reaction quotient cancels the doubled value of n in front of the log term.

Test Yourself Write the Nernst equation for $P_4 + 12H^+ + 12e^- \rightleftharpoons 4PH_3$. From Box 13-2, show that E is the same as if the reaction were written with $\frac{1}{2}P_4$ or $\frac{1}{4}P_4$.

Multiplying a half-reaction by any number does not change the standard reduction potential, $E°$. The potential difference between two points is the work done *per coulomb of charge* carried through that potential difference (E = work/q). Work per coulomb is the same whether 0.1, 2.3, or 10^4 coulombs have been transferred. The total work is different in each case, but work per coulomb is constant. Therefore, we do not double $E°$ if we multiply a half-reaction by 2.

Multiplying a half-reaction by any number does not change the half-cell potential, E. Consider a half-cell reaction written with either one or two electrons:

$$Ag^+ + e^- \rightleftharpoons Ag(s) \qquad E = E° - 0.059\ 16 \log \frac{1}{[Ag^+]}$$

$$2Ag^+ + 2e^- \rightleftharpoons 2Ag(s) \qquad E = E° - \frac{0.059\ 16}{2} \log \frac{1}{[Ag^+]^2}$$

The two expressions are equal because $\log a^b = b \log a$:

$$\frac{0.059\ 16}{2} \log \frac{1}{[Ag^+]^2} = \frac{\cancel{2} \times 0.059\ 16}{\cancel{2}} \log \frac{1}{[Ag^+]}$$

$$= 0.059\ 16 \log \frac{1}{[Ag^+]}$$

The exponent in the log term is canceled by the factor $1/n$ preceding the log term. Cell voltage is a measurable quantity that cannot depend on how you write the reaction.

Nernst Equation for a Complete Reaction

In Figure 13-6, the measured voltage is the difference between the potentials of the two electrodes:

Nernst equation for a complete cell: $\qquad \boxed{E = E_+ - E_-}$ $\qquad\qquad$ (13-16)

where E_+ is the potential of the electrode attached to the positive terminal of the potentiometer and E_- is the potential of the electrode attached to the negative terminal. The potential of each half-reaction (*written as a reduction*) is governed by a Nernst equation, and the voltage for the complete reaction is the difference between the two half-cell potentials.

Here is a procedure for writing a net cell reaction and finding its voltage:

Step 1 Write *reduction* half-reactions for both half-cells and find $E°$ for each in Appendix H. Multiply the half-reactions as necessary so that they both contain the same number of electrons. When you multiply a reaction, you *do not* multiply $E°$.

Step 2 Write a Nernst equation for the right half-cell, which is attached to the positive terminal of the potentiometer. This is E_+.

Step 3 Write a Nernst equation for the left half-cell, which is attached to the negative terminal of the potentiometer. This is E_-.

Step 4 Find the net cell voltage by subtraction: $E = E_+ - E_-$.

Step 5 Write the net cell reaction by subtracting the left half-reaction from the right half-reaction. (Subtraction is equivalent to reversing the left half-reaction and adding.)

$E > 0$: net cell reaction goes $\rightarrow$
$E < 0$: net cell reaction goes $\leftarrow$

If the net cell voltage, E ($= E_+ - E_-$), is positive, then the net cell reaction is spontaneous in the forward direction. If the net cell voltage is negative, then the reaction is spontaneous in the reverse direction.

EXAMPLE Nernst Equation for a Complete Reaction

Find the voltage of the cell in Figure 13-6 if the right half-cell contains 0.50 M $AgNO_3(aq)$ and the left half-cell contains 0.010 M $Cd(NO_3)_2(aq)$. Write the net cell reaction and state whether it is spontaneous in the forward or reverse direction.

Solution

Step 1 Right electrode: $\qquad 2Ag^+ + 2e^- \rightleftharpoons 2Ag(s) \qquad\qquad E_+° = 0.799$ V

$\qquad\qquad$ Left electrode: $\qquad Cd^{2+} + 2e^- \rightleftharpoons Cd(s) \qquad\qquad E_-° = -0.402$ V

Step 2 Nernst equation for right electrode:

Pure solids, pure liquids, and solvents are omitted from Q.

$$E_+ = E_+° - \frac{0.059\ 16}{2} \log \frac{1}{[Ag^+]^2} = 0.799 - \frac{0.059\ 16}{2} \log \frac{1}{[0.50]^2} = 0.781$$ V

Step 3 Nernst equation for left electrode:

$$E_- = E^\circ_- - \frac{0.059\ 16}{2} \log\frac{1}{[Cd^{2+}]} = -0.402 - \frac{0.059\ 16}{2} \log\frac{1}{[0.010]} = -0.461 \text{ V}$$

Step 4 Cell voltage: $\quad E = E_+ - E_- = 0.781 - (-0.461) = +1.242 \text{ V}$

Step 5 Net cell reaction:

$$\begin{array}{r} 2Ag^+ + 2e^- \rightleftharpoons 2Ag(s) \\ - \quad \underline{Cd^{2+} + 2e^- \rightleftharpoons Cd(s)} \\ Cd(s) + 2Ag^+ \rightleftharpoons Cd^{2+} + 2Ag(s) \end{array}$$

Subtracting a reaction is the same as reversing the reaction and adding.

The voltage is positive, so the reaction is spontaneous in the forward direction. $Cd(s)$ is oxidized and Ag^+ is reduced. Electrons flow from the left-hand electrode to the right-hand electrode.

Test Yourself Is the reaction spontaneous if the cells contain 5.0 μM AgNO$_3$ and 1.0 M Cd(NO$_3$)$_2$? (*Answer:* $E_+ = 0.485$ V, $E_- = -0.402$ V, $E = +0.887$ V, spontaneous)

What if you had written the Nernst equation for the right half-cell with just one electron instead of two?

$$Ag^+ + e^- \rightleftharpoons Ag(s)$$

Would the net cell voltage be different from what we calculated? It better not be, because the chemistry is still the same. Box 13-2 shows that *neither E° nor E depend on how we write the reaction*. Box 13-3 shows how to derive standard reduction potentials for half-reactions that are the sum of other half-reactions.

An Intuitive Way to Think About Cell Potentials[2]

In the preceding example, we found that E for the silver half-cell was 0.781 V and E for the cadmium half-cell was −0.461 V. Place these values on the number line in Figure 13-8 and note that *electrons always flow toward more positive potential*. Therefore, electrons in the circuit flow from cadmium to silver. The separation of the two half-cells is 1.242 V. This diagram works the same way even if both half-cell potentials are positive or both are negative. Electrons always flow toward more positive potential.

Electrons flow toward more positive potential.

Different Descriptions of the Same Reaction

In Figure 13-4, the right half-reaction can be written

$$AgCl(s) + e^- \rightleftharpoons Ag(s) + Cl^- \qquad E^\circ_+ = 0.222 \text{ V} \qquad (13\text{-}17)$$

$$E_+ = E^\circ_+ - 0.059\ 16 \log[Cl^-] = 0.222 - 0.059\ 16 \log(0.033\ 4) = 0.309_3 \text{ V} \quad (13\text{-}18)$$

The Cl^- in the silver half-reaction was derived from 0.016 7 M CdCl$_2$(aq).

Suppose that a different, less handsome, author had written this book and had chosen to describe the half-reaction differently:

$$Ag^+ + e^- \rightleftharpoons Ag(s) \qquad E^\circ_+ = 0.799 \text{ V} \qquad (13\text{-}19)$$

This description is just as valid as the previous one. In both cases, Ag(I) is reduced to Ag(0).

Electrons flow from Cd to Ag

Cd^{2+} | Cd
$E = -0.461$ V

Ag$^+$ | Ag
$E = 0.781$ V

E(cell) = 0.781 − (−0.461)
= 1.242 V

FIGURE 13-8 Electrons always flow from the more negative to the more positive electrode. That is, they always flow to the right in this diagram.[2]

BOX 13-3 Latimer Diagrams: How to Find $E°$ for a New Half-Reaction

A **Latimer diagram** displays standard reduction potentials, $E°$, connecting various oxidation states of an element.[13] For example, in acid solution, the following standard reduction potentials are observed:

The notation $IO_3^- \xrightarrow{+1.154} HOI$ stands for the balanced equation

$$IO_3^- + 5H^+ + 4e^- \rightleftharpoons HOI + 2H_2O \qquad E° = +1.154 \text{ V}$$

We can derive reduction potentials for arrows that are not shown in the diagram by using $\Delta G°$. For example, the reaction shown by the dashed line in the Latimer diagram is

$$IO_3^- + 6H^+ + 6e^- \rightleftharpoons I^- + 3H_2O$$

To find $E°$ for this reaction, express the reaction as a sum of reactions whose potentials are known.

The standard free energy change, $\Delta G°$, for a reaction is given by

$$\Delta G° = -nFE°$$

When two reactions are added, $\Delta G°$ is the sum of $\Delta G°$ values for each reaction.

To apply free energy to our problem, we write reactions whose sum is the desired reaction:

$$IO_3^- + 6H^+ + 5e^- \xrightarrow{E_1° = 1.210} \tfrac{1}{2}I_2(s) + 3H_2O \quad \Delta G_1° = -5F(1.210)$$

$$\tfrac{1}{2}I_2(s) + e^- \xrightarrow{E_2° = 0.535} I^- \qquad\qquad \Delta G_2° = -1F(0.535)$$

$$IO_3^- + 6H^+ + 6e^- \xrightarrow{E_3° = ?} I^- + 3H_2O \qquad \Delta G_3° = -6FE_3°$$

But, because $\Delta G_1° + \Delta G_2° = \Delta G_3°$, we can solve for $E_3°$:

$$\Delta G_3° = \Delta G_1° + \Delta G_2°$$

$$-6FE_3° = -5F(1.210) - 1F(0.535)$$

$$E_3° = \frac{5(1.210) + 1(0.535)}{6} = 1.098 \text{ V}$$

If the two descriptions are equally valid, then they should predict the same voltage. The Nernst equation for Reaction 13-19 is

$$E_+ = 0.799 - 0.059\,16 \log \frac{1}{[Ag^+]}$$

To find the concentration of Ag^+, we use the solubility product for AgCl. Because the cell contains 0.033 4 M Cl^- and solid AgCl, we can say

$K_{sp} = [Ag^+][Cl^-]$

$$[Ag^+] = \frac{K_{sp}\,(\text{for AgCl})}{[Cl^-]} = \frac{1.8 \times 10^{-10}}{0.033\,4} = 5.4 \times 10^{-9} \text{ M}$$

Putting this value into the Nernst equation gives

$$E_+ = 0.799 - 0.059\,16 \log \frac{1}{5.4 \times 10^{-9}} = 0.309_9 \text{ V}$$

which differs slightly from the value in Equation 13-18 because of the accuracy of K_{sp} and the neglect of activity coefficients. Reactions 13-17 and 13-19 give the same voltage in the Nernst equation because they describe the same cell.

The cell voltage cannot depend on how we write the reaction!

Advice for Finding Relevant Half-Reactions

When faced with a cell drawing or a line diagram, first write reduction reactions for each half-cell. To do this, *look in the cell for an element in two oxidation states.* For the cell

How to figure out the half-cell reactions

$$Pb(s)|PbF_2(s)|F^-(aq)\,\|\,Cu^{2+}(aq)|Cu(s)$$

we see Pb in two oxidation states, as $Pb(s)$ and $PbF_2(s)$, and Cu in two oxidation states, as Cu^{2+} and $Cu(s)$. Thus, the half-reactions are

Right half-cell: $\qquad\qquad\qquad Cu^{2+} + 2e^- \rightleftharpoons Cu(s)$

Left half-cell: $\qquad\qquad\qquad PbF_2(s) + 2e^- \rightleftharpoons Pb(s) + 2F^-$ $\qquad$ (13-20)

You might have chosen to write the Pb half-reaction as

Left half-cell: $\qquad\qquad\qquad Pb^{2+} + 2e^- \rightleftharpoons Pb(s)$ $\qquad\qquad$ (13-21)

because you know that if $PbF_2(s)$ is present, there must be some Pb^{2+} in the solution. Reactions 13-20 and 13-21 are equally valid and must predict the same cell voltage. Your choice of reactions depends on whether the F^- or Pb^{2+} concentration is easier to figure out.

We described the left half-cell in terms of a redox reaction involving Pb because Pb is the element that appears in two oxidation states. We would not write a reaction such as $F_2(g) + 2e^- \rightleftharpoons 2F^-$ because $F_2(g)$ is not shown in the line diagram of the cell.

Don't invent species not shown in the cell. Use what is shown in the line diagram to select the half-reactions.

The Nernst Equation Is Used in Measuring Standard Reduction Potentials

The standard reduction potential would be observed if the half-cell of interest (with unit activities) were connected to a standard hydrogen electrode, as it is in Figure 13-7. It is nearly impossible to construct such a cell, because we have no way to adjust concentrations and ionic strength to give unit activities. In reality, activities less than unity are used in each half-cell, and the Nernst equation is employed to extract the value of $E°$ from the cell voltage.[14] In the hydrogen electrode, standard buffers with known pH (Table 14-3) are used to obtain known activities of H^+.

Problem 13-20 gives an example of the use of the Nernst equation to find $E°$.

13-5 $E°$ and the Equilibrium Constant

A galvanic cell produces electricity because the cell reaction is not at equilibrium. The potentiometer allows negligible current (Box 13-4), so concentrations in the cell remain unchanged. If we replaced the potentiometer with a wire, there would be much more current and concentrations would change until the cell reached equilibrium. At that point, nothing would drive the reaction, and E would be 0. When a battery (which is a galvanic cell) runs down to 0 V, the chemicals inside have reached equilibrium and the battery is "dead."

At equilibrium, E (not $E°$) = 0.

Now let's relate E for a whole cell to the reaction quotient, Q, for the net cell reaction. For the two half-reactions

Right electrode:	$aA + ne^- \rightleftharpoons cC$	$E°_+$
Left electrode:	$dD + ne^- \rightleftharpoons bB$	$E°_-$

Net cell reaction:

$aA + bB \rightleftharpoons cC + dD$

the Nernst equation looks like this:

$$E = E_+ - E_- = E°_+ - \frac{0.059\,16}{n} \log \frac{\mathcal{A}_C^c}{\mathcal{A}_A^a} - \left(E°_- - \frac{0.059\,16}{n} \log \frac{\mathcal{A}_B^b}{\mathcal{A}_D^d} \right)$$

$$E = \underbrace{(E°_+ - E°_-)}_{E°} - \frac{0.059\,16}{n} \log \underbrace{\frac{\mathcal{A}_C^c \mathcal{A}_D^d}{\mathcal{A}_A^a \mathcal{A}_B^b}}_{Q} = E° - \frac{0.059\,16}{n} \log Q \quad (13\text{-}22)$$

$\log a + \log b = \log ab$

Equation 13-22 is true at any time. In the special case when the cell is at equilibrium, $E = 0$ and $Q = K$, the equilibrium constant. Therefore, Equation 13-22 is transformed into these most important forms at equilibrium:

Finding $E°$ from K:

$$E° = \frac{0.059\,16}{n} \log K \quad \text{(at 25°C)} \quad (13\text{-}23)$$

Finding K from $E°$:

$$K = 10^{nE°/0.059\,16} \quad \text{(at 25°C)} \quad (13\text{-}24)$$

To go from Equation 13-23 to 13-24:

$$\frac{0.059\,16}{n} \log K = E°$$

$$\log K = \frac{nE°}{0.059\,16}$$

$$10^{\log K} = 10^{nE°/0.059\,16}$$

$$K = 10^{nE°/0.059\,16}$$

Equation 13-24 allows us to deduce the equilibrium constant from $E°$. Alternatively, we can find $E°$ from K with Equation 13-23.

BOX 13-4   Concentrations in the Operating Cell

Why doesn't operation of a cell change the concentrations in the cell? Cell voltage is measured under conditions of *negligible current flow*. The resistance of a high-quality pH meter is $10^{13}\ \Omega$. If a cell produces 1 V, the current through the circuit is

$$I = \frac{E}{R} = \frac{1\ V}{10^{13}\ \Omega} = 10^{-13}\ A$$

The electron flow is

$$\frac{10^{-13}\ C/s}{9.649 \times 10^4\ C/mol} = 10^{-18}\ \text{mol e}^-/s$$

which produces negligible oxidation and reduction of reagents in the cell. *The meter measures the voltage of the cell without affecting concentrations in the cell.*

If a salt bridge were left in a real cell for very long, concentrations and ionic strength would change because of diffusion between each compartment and the salt bridge. Cells should be set up for such a short time that mixing is insignificant.

EXAMPLE Using $E°$ to Find the Equilibrium Constant

Find the equilibrium constant for the reaction

$$Cu(s) + 2Fe^{3+} \rightleftharpoons 2Fe^{2+} + Cu^{2+}$$

Solution Divide the reaction into two half-reactions found in Appendix H:

$$
\begin{array}{ll}
2Fe^{3+} + 2e^- \rightleftharpoons 2Fe^{2+} & E_+^{\circ} = 0.771 \text{ V} \\
Cu^{2+} + 2e^- \rightleftharpoons Cu(s) & E_-^{\circ} = 0.339 \text{ V} \\
\hline
Cu(s) + 2Fe^{3+} \rightleftharpoons 2Fe^{2+} + Cu^{2+} &
\end{array}
$$

Then find $E°$ for the net reaction

$$E° = E_+^{\circ} - E_-^{\circ} = 0.771 - 0.339 = 0.432 \text{ V}$$

and compute the equilibrium constant with Equation 13-24:

$$K = 10^{(2)(0.432)/(0.059\,16)} = 4 \times 10^{14}$$

A modest value of $E°$ produces a large equilibrium constant. The value of K is correctly expressed with one significant figure, because $E°$ has three digits. Two are used for the exponent (14), and one is left for the multiplier (4).

Test Yourself Find K for the reaction $Cu(s) + 2Ag^+ \rightleftharpoons 2Ag(s) + Cu^{2+}$. (**Answer:** $E° = 0.460$ V, $K = 4 \times 10^{15}$)

Finding K for Net Reactions That Are Not Redox Reactions

Consider the following half-reactions whose difference is the solubility reaction for iron(II) carbonate (which is not a redox reaction):

$$
\begin{array}{ll}
FeCO_3(s) + 2e^- \rightleftharpoons Fe(s) + CO_3^{2-} & E_+^{\circ} = -0.756 \text{ V} \\
Fe^{2+} + 2e^- \rightleftharpoons Fe(s) & E_-^{\circ} = -0.44 \text{ V} \\
\hline
\underset{\substack{\text{Iron(II)} \\ \text{carbonate}}}{FeCO_3(s)} \rightleftharpoons Fe^{2+} + CO_3^{2-} & E° = -0.756 - (-0.44) = -0.31_6 \text{ V}
\end{array}
$$

$$K = K_{sp} = 10^{(2)(-0.31_6)/(0.059\,16)} = 10^{-11}$$

From $E°$ for the net reaction, we can compute K_{sp} for iron(II) carbonate. Potentiometric measurements allow us to find equilibrium constants that are too small or too large to measure by determining concentrations of reactants and products directly.

"Wait!" you protest. "How can there be a redox potential for a reaction that is not a redox reaction?" Box 13-3 shows that the redox potential is just another way of expressing the free energy change of the reaction. The more energetically favorable the reaction (the more negative $\Delta G°$), the more positive is $E°$.

The general form of a problem involving the relation between $E°$ values for half-reactions and K for a net reaction is

$$
\begin{array}{lll}
\text{Half-reaction:} & E_+^{\circ} & \\
\text{Half-reaction:} & E_-^{\circ} & \\
\hline
\text{Net reaction:} & E° = E_+^{\circ} - E_-^{\circ} & K = 10^{nE°/0.059\,16}
\end{array}
$$

If you know E_-° and E_+°, you can find $E°$ and K for the net cell reaction. Alternatively, if you know $E°$ and either E_-° or E_+°, you can find the missing standard potential. If you know K, you can calculate $E°$ and use it to find either E_-° or E_+°, provided you know one of them.

Possible structure of Ni(glycine)$_2$

EXAMPLE Relating $E°$ and K

From the formation constant of Ni(glycine)$_2$ plus $E°$ for the $Ni^{2+} | Ni(s)$ couple,

$$
\begin{array}{ll}
Ni^{2+} + 2 \text{ glycine}^- \rightleftharpoons Ni(glycine)_2 & K \equiv \beta_2 = 1.2 \times 10^{11} \\
Ni^{2+} + 2e^- \rightleftharpoons Ni(s) & E° = -0.236 \text{ V}
\end{array}
$$

deduce the value of $E°$ for the reaction

$$Ni(glycine)_2 + 2e^- \rightleftharpoons Ni(s) + 2\,glycine^- \qquad (13\text{-}25)$$

Solution We need to see the relations among the three reactions:

$$
\begin{array}{lll}
& Ni^{2+} + 2e^- \rightleftharpoons Ni(s) & E°_+ = -0.236\ V \\
- & \underline{Ni(glycine)_2 + 2e^- \rightleftharpoons Ni(s) + 2\,glycine^-} & E°_- = ? \\
& Ni^{2+} + 2\,glycine^- \rightleftharpoons Ni(glycine)_2 & E° = ? \quad K = 1.2 \times 10^{11}
\end{array}
$$

We know that $E°_+ - E°_-$ must equal $E°$, so we can deduce the value of $E°_-$ if we can find $E°$. But $E°$ can be determined from the equilibrium constant for the net reaction:

$$E° = \frac{0.059\ 16}{n}\log K = \frac{0.059\ 16}{2}\log(1.2 \times 10^{11}) = 0.328\ V$$

Hence, the standard reduction potential for half-reaction 13-25 is

$$E°_- = E°_+ - E° = -0.236 - 0.328 = -0.564\ V$$

Test Yourself Select half-reactions in Appendix H to find the formation constant β_2 for $Cu^+ + 2\ ethylenediamine \rightleftharpoons Cu(ethylenediamine)_2^+$. (**Answer:** $E° = 0.637\ V$, $\beta_2 = 6 \times 10^{10}$)

The more negative potential of −0.564 V for reducing Ni(glycine)$_2$ compared to −0.236 V for reducing Ni^{2+} tells us that it is harder to reduce Ni(glycine)$_2$ than Ni^{2+}. Ni^{2+} is stabilized with respect to reduction by complexation with glycine.

13-6 Cells as Chemical Probes[15]

It is essential to distinguish two classes of equilibria associated with galvanic cells:

1. Equilibrium *between* the two half-cells
2. Equilibrium *within* each half-cell

If a galvanic cell has a nonzero voltage, then the net cell reaction is not at equilibrium. We say that equilibrium *between* the two half-cells has not been established.

 *We allow half-cells to stand long enough to come to chemical equilibrium **within** each half-cell.* For example, in the right-hand half-cell in Figure 13-9, the reaction

$$AgCl(s) \rightleftharpoons Ag^+(aq) + Cl^-(aq)$$

is at equilibrium. It is not part of the net cell reaction. It simply occurs when AgCl(s) is in contact with water. In the left half-cell, the reaction

$$CH_3CO_2H \rightleftharpoons CH_3CO_2^- + H^+$$

has also come to equilibrium. Neither reaction is part of the net cell redox reaction.

A chemical reaction that occurs *within one half-cell* will reach equilibrium and is assumed to remain at equilibrium. Such a reaction is not the net cell reaction.

Pt(s) | H$_2$(1.00 bar) | CH$_3$CO$_2$H(0.050 M), CH$_3$CO$_2$Na(0.005 0 M) ‖ Cl$^-$(0.10 M) | AgCl(s) | Ag(s)

Aha! A buffer!

FIGURE 13-9 This galvanic cell can be used to measure the pH of the left half-cell.

The redox reaction for the right half-cell of Figure 13-9 is

$$AgCl(s) + e^- \rightleftharpoons Ag(s) + Cl^-(aq, 0.10\ M) \qquad E_+^\circ = 0.222\ V$$

But what is the reaction in the left half-cell? The only element we find in two oxidation states is hydrogen. We see that $H_2(g)$ bubbles into the cell, and we also realize that every aqueous solution contains H^+. Therefore, hydrogen is present in two oxidation states, and the half-reaction can be written as

$$2H^+(aq, ?\ M) + 2e^- \rightleftharpoons H_2(g, 1.00\ bar) \qquad E_-^\circ = 0$$

The net cell reaction is not at equilibrium, because the measured voltage is 0.503 V, not 0 V. The Nernst equation for the net cell reaction is

$$E = E_+ - E_- = (0.222 - 0.059\ 16\ \log[Cl^-]) - \left(0 - \frac{0.059\ 16}{2} \log \frac{P_{H_2}}{[H^+]^2}\right)$$

After inserting the known quantities, we discover that the only unknown is $[H^+]$. *The measured voltage therefore allows us to find $[H^+]$ in the left half-cell:*

$$0.503 = (0.222 - 0.059\ 16\ \log[0.10]) - \left(0 - \frac{0.059\ 16}{2} \log \frac{1.00}{[H^+]^2}\right)$$

$$\Rightarrow [H^+] = 1.8 \times 10^{-4}\ M$$

This, in turn, allows us to evaluate the equilibrium constant for the acid-base reaction that has come to equilibrium in the left half-cell:

$$K_a = \frac{[CH_3CO_2^-][H^+]}{[CH_3CO_2H]} = \frac{(0.005\ 0)(1.8 \times 10^{-4})}{0.050} = 1.8 \times 10^{-5}$$

Question Why can we assume that the concentrations of acetic acid and acetate ion are equal to their initial (formal) concentrations?

The cell in Figure 13-9 acts as a *probe* to measure $[H^+]$ in the left half-cell. Using this type of cell, we could determine the equilibrium constant for acid dissociation or base hydrolysis in the left half-cell.

Survival Tips

Problems in this chapter include some brainbusters designed to bring together your knowledge of electrochemistry, chemical equilibrium, solubility, complex formation, and acid-base chemistry. They require you to find the equilibrium constant for a reaction that occurs in only one half-cell. The reaction of interest is not the net cell reaction and is not a redox reaction. Here is a good approach:

Half-reactions that you write *must* involve species that appear in two oxidation states in the cell.

Step 1 Write the two half-reactions and their standard potentials. If you choose a half-reaction for which you cannot find E°, then find another way to write the reaction.

Step 2 Write a Nernst equation for the net reaction and put in all the known quantities. If all is well, there will be only one unknown in the equation.

Step 3 Solve for the unknown concentration and use that concentration to solve the chemical equilibrium problem that was originally posed.

EXAMPLE Analyzing a Very Complicated Cell

The cell in Figure 13-10 measures the formation constant (K_f) of $Hg(EDTA)^{2-}$. The right-hand compartment contains 0.500 mmol of Hg^{2+} and 2.00 mmol of EDTA in 0.100 L buffered to pH 6.00. The voltage is +0.342 V. Find the value of K_f for $Hg(EDTA)^{2-}$.

Solution

Step 1 The left half-cell is a standard hydrogen electrode for which we can say $E_- = 0$. In the right half-cell, mercury is in two oxidation states. So let's write the half-reaction

$$Hg^{2+} + 2e^- \rightleftharpoons Hg(l) \qquad E_+^\circ = 0.852\ V$$

$$E_+ = 0.852 - \frac{0.059\ 16}{2} \log\left(\frac{1}{[Hg^{2+}]}\right)$$

+0.342 V

H₂

Solution prepared from
50.0 mL of 0.010 0 M HgCl₂
40.0 mL of 0.050 0 M EDTA
10.0 mL of buffer, pH 6.00

Salt bridge

Pt

Hg(*l*)

Pt contact

Standard hydrogen
electrode

S.H.E. $\|$ Hg(EDTA)$^{2-}$(*aq*, 0.005 00 M), EDTA(*aq*, 0.015 0 M) $|$ Hg(*l*)

In the right half-cell, the reaction between Hg^{2+} and EDTA is

$$Hg^{2+} + Y^{4-} \overset{K_f}{\rightleftharpoons} HgY^{2-}$$

Because we expect K_f to be large, we assume that virtually all the Hg^{2+} has reacted to make HgY^{2-}. Therefore, the concentration of HgY^{2-} is 0.500 mmol/100 mL = 0.005 00 M. The remaining EDTA has a total concentration of (2.00 − 0.50) mmol/100 mL = 0.015 0 M. The right-hand compartment therefore contains 0.005 00 M HgY^{2-}, 0.015 0 M EDTA, and a small, unknown concentration of Hg^{2+}.

The formation constant for HgY^{2-} can be written

$$K_f = \frac{[HgY^{2-}]}{[Hg^{2+}][Y^{4-}]} = \frac{[HgY^{2-}]}{[Hg^{2+}]\alpha_{Y^{4-}}[EDTA]}$$

Recall that $[Y^{4-}] = \alpha_{Y^{4-}}[EDTA]$.

where [EDTA] is the formal concentration of EDTA not bound to metal. In this cell, [EDTA] = 0.015 0 M. The fraction of EDTA in the form Y^{4-} is $\alpha_{Y^{4-}}$ (Section 11-2). Because we know that $[HgY^{2-}] = 0.005\ 00$ M, all we need to find is $[Hg^{2+}]$ in order to evaluate K_f.

Step 2 The Nernst equation for the net cell reaction is

$$E = 0.342 = E_+ - E_- = \left[0.852 - \frac{0.059\ 16}{2}\log\left(\frac{1}{[Hg^{2+}]}\right)\right] - (0)$$

in which the only unknown is $[Hg^{2+}]$.

Step 3 Now we solve the Nernst equation to find $[Hg^{2+}] = 5._7 \times 10^{-18}$ M, and this value of $[Hg^{2+}]$ allows us to evaluate the formation constant for HgY^{2-}:

$$K_f = \frac{[HgY^{2-}]}{[Hg^{2+}]\alpha_{Y^{4-}}[EDTA]} = \frac{(0.005\ 00)}{(5._7 \times 10^{-18})(1.8 \times 10^{-5})(0.015\ 0)}$$

$$= 3 \times 10^{21}$$

$\alpha_{Y^{4-}}$ comes from Table 11-1.

The mixture of EDTA plus $Hg(EDTA)^{2-}$ in the cathode serves as a mercuric ion "buffer" that fixes the concentration of Hg^{2+}. This concentration, in turn, determines the cell voltage.

Test Yourself Find K_f if the cell voltage had been 0.300 V. (***Answer:*** 8×10^{22})

13-7 Biochemists Use $E^{\circ\prime}$

In respiration, molecules from food are oxidized by O_2 to yield energy or metabolic intermediates. The standard reduction potentials that we have been using so far apply to systems in which all activities of reactants and products are unity. If H^+ is involved in the reaction,

TABLE 13-2 Reduction potentials of biological interest

Reaction	$E°$ (V)	$E°'$ (V)
$O_2 + 4H^+ + 4e^- \rightleftharpoons 2H_2O$	+1.229	+0.815
$Fe^{3+} + e^- \rightleftharpoons Fe^{2+}$	+0.771	+0.771
$I_2 + 2e^- = 2I^-$	+0.535	+0.535
Cytochrome a (Fe^{3+}) + e^- $\rightleftharpoons$ cytochrome a (Fe^{2+})	+0.290	+0.290
$O_2(g) + 2H^+ + 2e^- \rightleftharpoons H_2O_2$	+0.695	+0.281
Cytochrome c (Fe^{3+}) + e^- $\rightleftharpoons$ cytochrome c (Fe^{2+})	—	+0.254
2,6-Dichlorophenolindophenol $+ 2H^+ + 2e^- \rightleftharpoons$ reduced 2,6-dichlorophenolindophenol	—	+0.22
Dehydroascorbate $+ 2H^+ + 2e^- \rightleftharpoons$ ascorbate + H_2O	+0.390	+0.058
Fumarate $+ 2H^+ + 2e^- \rightleftharpoons$ succinate	+0.433	+0.031
Methylene blue $+ 2H^+ + 2e^- \rightleftharpoons$ reduced product	+0.532	+0.011
Glyoxylate $+ 2H^+ + 2e^- \rightleftharpoons$ glycolate	—	−0.090
Oxaloacetate $+ 2H^+ + 2e^- \rightleftharpoons$ malate	+0.330	−0.102
Pyruvate $+ 2H^+ + 2e^- \rightleftharpoons$ lactate	+0.224	−0.190
Riboflavin $+ 2H^+ + 2e^- \rightleftharpoons$ reduced riboflavin	—	−0.208
$FAD + 2H^+ + 2e^- \rightleftharpoons FADH_2$	—	−0.219
$(Glutathione-S)_2 + 2H^+ + 2e^- \rightleftharpoons$ 2 glutathione-SH	—	−0.23
Safranine T $+ 2e^- \rightleftharpoons$ leucosafranine T	−0.235	−0.289
$(C_6H_5S)_2 + 2H^+ + 2e^- \rightleftharpoons 2C_6H_5SH$	—	−0.30
$NAD^+ + H^+ + 2e^- \rightleftharpoons NADH$	−0.105	−0.320
$NADP^+ + H^+ + 2e^- \rightleftharpoons NADPH$	—	−0.324
Cystine $+ 2H^+ + 2e^- \rightleftharpoons$ 2 cysteine	—	−0.340
Acetoacetate $+ 2H^+ + 2e^- \rightleftharpoons$ L-β-hydroxybutyrate	—	−0.346
Xanthine $+ 2H^+ + 2e^- \rightleftharpoons$ hypoxanthine + H_2O	—	−0.371
$2H^+ + 2e^- \rightleftharpoons H_2$	0.000	−0.414
Gluconate $+ 2H^+ + 2e^- \rightleftharpoons$ glucose + H_2O	—	−0.44
$SO_4^{2-} + 2e^- + 2H^+ \rightleftharpoons SO_3^{2-} + H_2O$	—	−0.454
$2SO_3^{2-} + 2e^- + 4H^+ \rightleftharpoons S_2O_4^{2-} + 2H_2O$	—	−0.527

$E°$ applies when pH = 0 ($\mathcal{A}_{H^+} = 1$). *Whenever H^+ appears in a redox reaction, or whenever reactants or products are acids or bases, reduction potentials are pH dependent.*

Because the pH inside a plant or animal cell is about 7, reduction potentials that apply at pH 0 are not particularly appropriate. For example, at pH 0, ascorbic acid (vitamin C) is a more powerful reducing agent than succinic acid. However, at pH 7, this order is reversed. It is the reducing strength at pH 7, not at pH 0, that is relevant to a living cell.

The *standard potential* for a redox reaction is defined for a galvanic cell in which all activities are unity. The **formal potential** is the reduction potential that applies under a *specified* set of conditions (including pH, ionic strength, and concentration of complexing agents). Biochemists call the formal potential at pH 7 $E°'$ (read "*E* zero prime"). Table 13-2 lists $E°'$ values for some biological redox couples.

The formal potential at pH = 7 is called $E°'$.

Relation Between $E°$ and $E°'$

Consider the half-reaction

$$a\text{A} + ne^- \rightleftharpoons b\text{B} + m\text{H}^+ \qquad E°$$

in which A is an oxidized species and B is a reduced species. Both A and B might be acids or bases, as well. The Nernst equation for this half-reaction is

$$E = E° - \frac{0.059\ 16}{n}\log\frac{[\text{B}]^b[\text{H}^+]^m}{[\text{A}]^a}$$

To find $E°'$, we rearrange the Nernst equation to a form in which the log term contains only the *formal concentrations* of A and B raised to the powers a and b, respectively.

Recipe for $E^{\circ\prime}$:

$$E = \underbrace{E^\circ + \text{other terms}}_{\substack{\text{All of this is called } E^{\circ\prime} \\ \text{when pH} = 7}} - \frac{0.059\ 16}{n} \log \frac{F_B^b}{F_A^a} \quad (13\text{-}26)$$

If we had included activity coefficients, they would appear in $E^{\circ\prime}$ also.

The entire collection of terms over the brace, evaluated at pH = 7, is called $E^{\circ\prime}$.

To convert [A] or [B] into F_A or F_B, we use fractional composition equations (Section 9-5), which relate the formal (that is, total) concentration of *all* forms of an acid or a base to its concentration in a *particular* form:

Monoprotic system:

$$[HA] = \alpha_{HA}F = \frac{[H^+]F}{[H^+] + K_a} \quad (13\text{-}27)$$

$$[A^-] = \alpha_{A^-}F = \frac{K_aF}{[H^+] + K_a} \quad (13\text{-}28)$$

For a monoprotic acid:

$$F = [HA] + [A^-]$$

For a diprotic acid:

$$F = [H_2A] + [HA^-] + [A^{2-}]$$

Diprotic system:

$$[H_2A] = \alpha_{H_2A}F = \frac{[H^+]^2F}{[H^+]^2 + [H^+]K_1 + K_1K_2} \quad (13\text{-}29)$$

$$[HA^-] = \alpha_{HA^-}F = \frac{K_1[H^+]F}{[H^+]^2 + [H^+]K_1 + K_1K_2} \quad (13\text{-}30)$$

$$[A^{2-}] = \alpha_{A^{2-}}F = \frac{K_1K_2F}{[H^+]^2 + [H^+]K_1 + K_1K_2} \quad (13\text{-}31)$$

where F is the formal concentration of HA or H_2A, K_a is the acid dissociation constant for HA, and K_1 and K_2 are the acid dissociation constants for H_2A.

One way to *measure $E^{\circ\prime}$* is to establish a half-cell in which the formal concentrations of the oxidized and reduced species are equal and the pH is adjusted to 7. Then the log term in Equation 13-26 is zero and the measured potential (versus S.H.E.) is $E^{\circ\prime}$.

EXAMPLE   **Calculating the Formal Potential**

Find $E^{\circ\prime}$ for the reaction

Dehydroascorbic acid (oxidized) $+ 2H^+ + 2e^- \rightleftharpoons$ Ascorbic acid (vitamin C) (reduced) $+ H_2O$ $E^\circ = 0.390$ V   (13-32)

$pK_1 = 4.10$ $pK_2 = 11.79$

Solution Abbreviating dehydroascorbic acid[16] as D, and ascorbic acid as H_2A, we rewrite the reduction as

$$D + 2H^+ + 2e^- \rightleftharpoons H_2A + H_2O$$

for which the Nernst equation is

$$E = E^\circ - \frac{0.059\ 16}{2} \log \frac{[H_2A]}{[D][H^+]^2} \quad (13\text{-}33)$$

D is not an acid or a base, so its formal concentration equals its molar concentration: $F_D = [D]$. For the diprotic acid H_2A, we use Equation 13-29 to express $[H_2A]$ in terms of F_{H_2A}:

$$[H_2A] = \frac{[H^+]^2 F_{H_2A}}{[H^+]^2 + [H^+]K_1 + K_1K_2}$$

Putting these values into Equation 13-33 gives

$$E = E° - \frac{0.059\ 16}{2} \log\left(\frac{\dfrac{[H^+]^2 F_{H_2A}}{[H^+]^{2+}[H^+]K_1 + K_1K_2}}{F_D[H^+]^2}\right)$$

which can be rearranged to the form

$$E = E° - \underbrace{\frac{0.059\ 16}{2} \log \frac{1}{[H^+]^{2+}[H^+]K_1 + K_1K_2}}_{\substack{\text{Formal potential } (= E°'\text{ if pH} = 7) \\ = +0.062\ \text{V}}} - \frac{0.059\ 16}{2} \log \frac{F_{H_2A}}{F_D} \qquad (13\text{-}34)$$

Putting the values of $E°$, K_1, and K_2 into Equation 13-34 and setting $[H^+] = 10^{-7.00}$, we find $E°' = +0.062$ V.

Test Yourself Compute $E°'$ for the reaction $O_2 + 4H^+ + 4e^- \rightleftharpoons 2H_2O$. (*Answer:* 0.815 V)

Curve *a* in Figure 13-11 shows how the calculated formal potential for Reaction 13-32 depends on pH. The potential decreases as the pH increases, until pH $\approx$ pK_2. Above pK_2, A^{2-} is the dominant form of ascorbic acid, and no protons are involved in the net redox reaction. Therefore, the potential becomes independent of pH.

A biological example of $E°'$ is the reduction of Fe(III) in the protein transferrin. This protein has two Fe(III)-binding sites, one in each half of the molecule designated C and N for the carboxyl and amino terminals of the peptide chain. Transferrin carries Fe(III) through the blood to cells that require iron. Membranes of these cells have a receptor that binds Fe(III)-transferrin and takes it into a compartment called an endosome into which H^+ is pumped to lower the pH to ~5.8. Iron is released from transferrin in the endosome and continues into the cell as Fe(II) attached to an intracellular metal-transport protein.

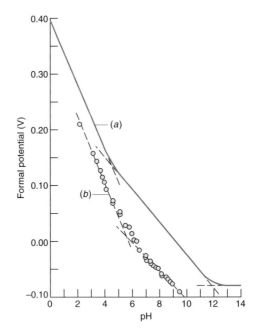

FIGURE 13-11 Reduction potential of ascorbic acid, showing its dependence on pH. (*a*) Graph of the function labeled formal potential in Equation 13-34. (*b*) Experimental polarographic half-wave reduction potential of ascorbic acid in a medium of ionic strength = 0.2 M. The half-wave potential (Chapter 16) is nearly the same as the formal potential. At high pH (>12), the half-wave potential does not level off to a slope of 0, as Equation 13-34 predicts. Instead, a hydrolysis reaction of ascorbic acid occurs and the chemistry is more complex than Reaction 13-32. [From J. J. Ruiz, A. Aldaz, and M. Dominguez, *Can. J. Chem.* **1977**, *55*, 2799; ibid. **1978**, *56*, 1533.]

The entire cycle of transferrin uptake, metal removal, and transferrin release back to the bloodstream takes 1–2 min. The time required for Fe(III) to dissociate from transferrin at pH 5.8 is ~6 min, which is too long to account for release in the endosome. The reduction potential of Fe(III)-transferrin at pH 5.8 is $E^{\circ\prime} = -0.52$ V, which is too low for physiologic reductants.

The mystery of how Fe(III) is released from transferrin in the endosome was solved by measuring $E^{\circ\prime}$ for the Fe(III)-transferrin-receptor complex at pH 5.8. To simplify the chemistry, transferrin was cleaved and only the C-terminal half of the protein (designated Trf_C) was used. Figure 13-12 shows measurements of $\log\{[\text{Fe(III)Trf}_C]/[\text{Fe(II)Trf}_C]\}$ for free protein and for the protein-receptor complex. In Equation 13-26, $E = E^{\circ\prime}$ when the log term is zero (that is, when $[\text{Fe(III)Trf}_C] = [\text{Fe(II)Trf}_C]$). Figure 13-12 shows that $E^{\circ\prime}$ for Fe(III)Trf$_C$ is near –0.50 V, but $E^{\circ\prime}$ for the Fe(III)Trf$_C$-receptor complex is –0.29 V. The reducing agents NADH and NADPH in Table 13-2 are strong enough to reduce Fe(III)Trf$_C$ bound to its receptor at pH 5.8, but not strong enough to reduce free Fe(III)-transferrin.

FIGURE 13-12 Spectroscopic measurement of $\log\{[\text{Fe(III)Trf}_C]/[\text{Fe(II)Trf}_C]\}$ versus potential at pH 5.8. [From S. Dhungana, C. H. Taboy, O. Zak, M. Larvie, A. L. Crumbliss, and P. Aisen, "Redox Properties of Human Transferrin Bound to Its Receptor," *Biochemistry* **2004**, *43*, 205.]

Terms to Understand

ampere	Faraday constant	oxidant	reduction
anode	formal potential	oxidation	resistance
cathode	galvanic cell	oxidizing agent	salt bridge
coulomb	half-reaction	potentiometer	standard hydrogen electrode
current	joule	power	standard reduction potential
$E^{\circ\prime}$	Latimer diagram	reaction quotient	volt
electric potential	Nernst equation	redox reaction	watt
electrochemistry	ohm	reducing agent	
electrode	Ohm's law	reductant	

Summary

Work done when a charge of q coulombs passes through a potential difference of E volts is work = $E \cdot q$. The maximum work that can be done on the surroundings by a spontaneous chemical reaction is related to the free energy change for the reaction: work = $-\Delta G$. If the chemical change produces a potential difference, E, the relation between free energy and the potential difference is $\Delta G = -nFE$. Ohm's law ($I = E/R$) describes the relation between current, voltage, and resistance in an electric circuit. It can be combined with the definitions of work and power (P = work per second) to give $P = E \cdot I = I^2R$.

A galvanic cell uses a spontaneous redox reaction to produce electricity. The electrode at which oxidation occurs is the anode, and the electrode at which reduction occurs is the cathode. Two half-cells are usually separated by a salt bridge that allows ions to migrate from one side to the other to maintain charge neutrality but prevents reactants in the two half-cells from mixing. The standard reduction potential of a half-reaction is measured by pairing that half-reaction with a standard hydrogen electrode. The term "standard" means that activities of reactants and products are unity. If several half-reactions are added to give another half-reaction, the standard potential of the net half-reaction can be found by equating the free energy of the net half-reaction to the sum of free energies of the component half-reactions.

The voltage for a complete reaction is the difference between the potentials of the two half-reactions: $E = E_+ - E_-$, where E_+ is the potential of the half-cell connected to the positive terminal of the potentiometer and E_- is the potential of the half-cell connected to the negative terminal. The potential of each half-reaction is given by the Nernst equation: $E = E^{\circ} - (0.059\ 16/n) \log Q$ (at 25°C), where each reaction is written as a reduction and Q is the reaction quotient. The reaction quotient has the same form as the equilibrium constant, but it is evaluated with concentrations existing at the time of interest. Electrons flow through the circuit from the electrode with the more negative potential to the electrode with the more positive potential.

Complex equilibria can be studied by making them part of an electrochemical cell. If we measure the voltage and know the concentrations (activities) of all but one of the reactants and products, the Nernst equation allows us to compute the concentration of the unknown species. The electrochemical cell serves as a probe for that species.

Biochemists use the formal potential of a half-reaction at pH 7 ($E^{\circ\prime}$) instead of the standard potential (E°), which applies at pH 0. $E^{\circ\prime}$ is found by writing the Nernst equation for the half-reaction and grouping together all terms except the logarithm containing the formal concentrations of reactant and product. The combination of terms, evaluated at pH 7, is $E^{\circ\prime}$.

Exercises

13-A. In olden days, mercury cells with the following chemistry were used to power heart pacemakers:

$$Zn(s) + HgO(s) \rightarrow ZnO(s) + Hg(l) \qquad E° = 1.35 \text{ V}$$

What is the cell voltage? If the power required to operate the pacemaker is 0.010 0 W, how many kilograms of HgO (FM 216.59) will be consumed in 365 days? How many pounds of HgO is this? (1 pound = 453.6 g)

13-B. Calculate $E°$ and K for each of the following reactions.

(a) $I_2(s) + 5Br_2(aq) + 6H_2O \rightleftharpoons 2IO_3^- + 10Br^- + 12H^+$

(b) $Cr^{2+} + Fe(s) \rightleftharpoons Fe^{2+} + Cr(s)$

(c) $Mg(s) + Cl_2(g) \rightleftharpoons Mg^{2+} + 2Cl^-$

(d) $5MnO_2(s) + 4H^+ \rightleftharpoons 3Mn^{2+} + 2MnO_4^- + 2H_2O$

(e) $Ag^+ + 2S_2O_3^{2-} \rightleftharpoons Ag(S_2O_3)_2^{3-}$

(f) $CuI(s) \rightleftharpoons Cu^+ + I^-$

13-C. Calculate the voltage of each of the following cells. With the reasoning in Figure 13-8, state the direction of electron flow.

(a) $Fe(s) \mid FeBr_2(0.010 \text{ M}) \parallel NaBr(0.050 \text{ M}) \mid Br_2(l) \mid Pt(s)$

(b) $Cu(s) \mid Cu(NO_3)_2(0.020 \text{ M}) \parallel Fe(NO_3)_2(0.050 \text{ M}) \mid Fe(s)$

(c) $Hg(l) \mid Hg_2Cl_2(s) \mid KCl(0.060 \text{ M}) \parallel KCl(0.040 \text{ M}) \mid$ $Cl_2(g, 0.50 \text{ bar}) \mid Pt(s)$

13-D. The left half-reaction of the cell drawn here can be written in *either* of two ways:

$$AgI(s) + e^- \rightleftharpoons Ag(s) + I^- \qquad (1)$$

$$Ag^+ + e^- \rightleftharpoons Ag(s) \qquad (2)$$

The right half-cell reaction is

$$H^+ + e^- \rightleftharpoons \frac{1}{2}H_2(g) \qquad (3)$$

$Ag(s) \mid AgI(s) \mid NaI (0.10 \text{ M}) \parallel HCl (0.10 \text{ M}) \mid H_2(g, 0.20 \text{ bar}) \mid Pt(s)$

(a) Using Reactions 2 and 3, calculate $E°$ and write the Nernst equation for the cell.

(b) Use the value of K_{sp} for AgI to compute $[Ag^+]$ and find the cell voltage. By the reasoning in Figure 13-8, in which direction do electrons flow?

(c) Suppose, instead, that you wish to describe the cell with Reactions 1 and 3. We know that the cell voltage (E, not $E°$) must be the same, no matter which description we use. Write the Nernst equation for Reactions 1 and 3 and use it to find $E°$ in Reaction 1. Compare your answer with the value in Appendix H.

13-E. Calculate the voltage of the cell

$$Cu(s) \mid Cu^{2+}(0.030 \text{ M}) \parallel K^+Ag(CN)_2^-(0.010 \text{ M}),$$
$$HCN(0.10 \text{ F}), \text{ buffer to pH } 8.21 \mid Ag(s)$$

by considering the following reactions:

$$Ag(CN)_2^- + e^- \rightleftharpoons Ag(s) + 2CN^- \qquad E° = -0.310 \text{ V}$$

$$HCN \rightleftharpoons H^+ + CN^- \qquad pK_a = 9.21$$

By the reasoning in Figure 13-8, in which direction do electrons flow?

13-F. (a) Write a balanced equation for the reaction $PuO_2^+ \rightarrow Pu^{4+}$ and calculate $E°$ for the reaction.

$$PuO_2^{2+} \xrightarrow{+0.966} PuO_2^+ \xrightarrow{?} Pu^{4+} \xrightarrow{+1.006} Pu^{3+}$$
$$\underset{1.021}{\underbrace{\phantom{PuO_2^{2+} \xrightarrow{+0.966} PuO_2^+ \xrightarrow{?}}}}$$

(b) Predict whether an equimolar mixture of PuO_2^{2+} and PuO_2^+ will oxidize H_2O to O_2 at a pH of 2.00 and $P_{O_2} = 0.20$ bar. Will O_2 be liberated at pH 7.00?

13-G. Calculate the voltage of the following cell, in which KHP is potassium hydrogen phthalate, the monopotassium salt of phthalic acid. By the reasoning in Figure 13-8, in which direction do electrons flow?

$$Hg(l) \mid Hg_2Cl_2(s) \mid KCl(0.10 \text{ M}) \parallel KHP(0.050 \text{ M}) \mid$$
$$H_2(g, 1.00 \text{ bar}) \mid Pt(s)$$

13-H. The following cell has a voltage of 0.083 V:

$$Hg(l) \mid Hg(NO_3)_2(0.001 \text{ 0 M}), KI(0.500 \text{ M}) \parallel S.H.E.$$

From this voltage, calculate the equilibrium constant for the reaction

$$Hg^{2+} + 4I^- \rightleftharpoons HgI_4^{2-}$$

In 0.5 M KI, virtually all the mercury is present as HgI_4^{2-}.

13-I. The formation constant for $Cu(EDTA)^{2-}$ is 6.3×10^{18}, and $E°$ is $+0.339$ V for the reaction $Cu^{2+} + 2e^- \rightleftharpoons Cu(s)$. From this information, find $E°$ for the reaction

$$CuY^{2-} + 2e^- \rightleftharpoons Cu(s) + Y^{4-}$$

13-J. On the basis of the following reaction, state which compound, $H_2(g)$ or glucose, is the more powerful reducing agent at pH = 0.00.

CO$_2$H   CHO

$$
\begin{array}{c}
CO_2H \\
|\\
HCOH \\
|\\
HOCH \\
|\\
HCOH \\
|\\
HCOH \\
|\\
CH_2OH \\
\text{Gluconic acid} \\
pK_a = 3.56
\end{array}
\quad + 2H^+ + 2e^- \underset{\xleftarrow{}}{\overset{E°' = -0.45 \text{ V}}{\xrightarrow{}}} \quad
\begin{array}{c}
CHO \\
|\\
HCOH \\
|\\
HOCH + H_2O \\
|\\
HCOH \\
|\\
HCOH \\
|\\
CH_2OH \\
\text{Glucose} \\
\text{(no acidic protons)}
\end{array}
$$

13-K. Living cells convert energy derived from sunlight or combustion of food into energy-rich ATP (adenosine triphosphate) molecules. For ATP synthesis, $\Delta G = +34.5$ kJ/mol. This energy is then made available to the cell when ATP is hydrolyzed to ADP (adenosine diphosphate). In animals, ATP is synthesized when protons pass through a complex enzyme in the mitochondrial membrane.[17] Two factors account for the movement of protons through this enzyme into the mitochondrion (see the figure): (1) $[H^+]$ is higher outside the mitochondrion than inside because protons are *pumped* out of the mitochondrion by enzymes that catalyze the oxidation of food. (2) The inside of the mitochondrion is negatively charged with respect to the outside.

(a) The synthesis of one ATP molecule requires $2H^+$ to pass through the phosphorylation enzyme. The difference in free energy when a molecule travels from a region of high activity to a region of low activity is

$$\Delta G = -RT \ln \frac{\mathcal{A}_{high}}{\mathcal{A}_{low}}$$

How big must the pH difference be (at 298 K) if the passage of two protons is to provide enough energy to synthesize one ATP molecule?

(b) pH differences this large have not been observed in mitochondria. How great an electric potential difference between inside and outside is necessary for the movement of two protons to provide energy to synthesize ATP? In answering this question, neglect any contribution from the pH difference.

(c) The energy for ATP synthesis is thought to be provided by *both* the pH difference and the electric potential. If the pH difference is 1.00 pH unit, what is the magnitude of the potential difference?

Problems

Basic Concepts

13-1. Explain the difference between electric charge (q, coulombs), electric current (I, amperes), and electric potential (E, volts).

13-2. (a) How many electrons are in one coulomb?

(b) How many coulombs are in one mole of charge?

13-3. The basal rate of consumption of O_2 by a 70-kg human is about 16 mol of O_2 per day. This O_2 oxidizes food and is reduced to H_2O, providing energy for the organism:

$$O_2 + 4H^+ + 4e^- \rightleftharpoons 2H_2O$$

(a) To what current (in amperes = C/s) does this respiration rate correspond? (Current is defined by the flow of electrons from food to O_2.)

(b) Compare your answer in part (a) with the current drawn by a refrigerator using 5.00×10^2 W at 115 V. Remember that power (in watts) = work/s = $E \cdot I$.

(c) If the electrons flow from nicotinamide adenine dinucleotide (NADH) to O_2, they experience a potential drop of 1.1 V. What is the power output (in watts) of our human friend?

13-4. A 6.00-V battery is connected across a 2.00-kΩ resistor.

(a) How many electrons per second flow through the circuit?

(b) How many joules of heat are produced for each electron?

(c) If the circuit operates for 30.0 min, how many moles of electrons will have flowed through the resistor?

(d) What voltage would the battery need to deliver for the power to be 1.00×10^2 W?

13-5. Consider the redox reaction

$$I_2 + 2S_2O_3^{2-} \rightleftharpoons 2I^- + S_4O_6^{2-}$$

Thiosulfate   Tetrathionate

(a) Identify the oxidizing agent on the left side of the reaction and write a balanced oxidation half-reaction.

(b) Identify the reducing agent of the left side of the reaction and write a balanced reduction half-reaction.

(c) How many coulombs of charge are passed from reductant to oxidant when 1.00 g of thiosulfate reacts?

(d) If the rate of reaction is 1.00 g of thiosulfate consumed per minute, what current (in amperes) flows from reductant to oxidant?

13-6. The space shuttle's expendable booster engines derive their power from solid reactants:

$$6NH_4^+ClO_4^-(s) + 10Al(s) \rightarrow 3N_2(g) + 9H_2O(g)$$

FM 117.49 $+ 5Al_2O_3(s) + 6HCl(g)$

(a) Find the oxidation numbers of the elements N, Cl, and Al in reactants and products. Which reactants act as reducing agents and which act as oxidants?

(b) The heat of reaction is −9 334 kJ for every 10 mol of Al consumed. Express this as heat released per gram of total reactants.

Galvanic Cells

13-7. Explain how a galvanic cell uses a spontaneous chemical reaction to generate electricity.

Galvanic cells for Problem 13.8

13-8. Write a line notation and two reduction half-reactions for each cell pictured above.

13-9. Draw a picture of the following cell and write reduction half-reactions for each electrode:

$Pt(s) | Fe^{3+}(aq), Fe^{2+}(aq) || Cr_2O_7^{2-}(aq), Cr^{3+}(aq), HA(aq) | Pt(s)$

13-10. Consider the rechargeable battery:

$Zn(s) | ZnCl_2(aq) || Cl^-(aq) | Cl_2(l) | C(s)$

(a) Write reduction half-reactions for each electrode. From which electrode will electrons flow from the battery into a circuit if the electrode potentials are not too different from $E°$ values?

(b) If the battery delivers a constant current of 1.00×10^3 A for 1.00 h, how many kilograms of Cl_2 will be consumed?

13-11. *Lithium-ion battery.*

(a) Ideal formulas for the electrodes of the Li^+-ion battery described in the chapter opener are C_6Li (FM 79.01) and $LiCoO_2$ (FM 97.87). When the battery operates, C_6Li is consumed and $LiCoO_2$ is formed. Write a half-reaction for each electrode, assuming that $x = 1$ in the reaction $C_6Li + Li_{1-x}CoO_2 \rightleftharpoons C_6Li_{1-x} + LiCoO_2$. (In fact, $x \neq 1$ in real cells.) Which is the anode and which is the cathode?

(b) Charge capacity of an electrode in a battery is expressed as mA·h/g, which is the number of milliamperes delivered by 1 g of material for 1 hour. How many coulombs are in 1 mA·h?

(c) Show that the theoretical capacity of the battery is 274 mA·h/g $LiCoO_2$.

(d) A Li^+-ion battery can deliver 140 mA·h/g $LiCoO_2$. What fraction of Li in the formula $LiCoO_2$ is available?

(e) Energy stored by a battery per unit mass of an electrode material is expressed as W·h/g. A Li^+-ion battery delivers 140 mA·h/g $LiCoO_2$ at 3.7 V. Express the energy storage as W·h/g $LiCoO_2$.

Standard Potentials

13-12. Which will be the strongest oxidizing agent under standard conditions (that is, all activities = 1): HNO_2, Se, UO_2^{2+}, Cl_2, H_2SO_3, or MnO_2?

13-13. (a) Cyanide ion causes $E°$ for Fe(III) to decrease:

$$Fe^{3+} + e^- \rightleftharpoons Fe^{2+} \qquad E° = 0.771 \text{ V}$$
Ferric $\qquad$ Ferrous

$$Fe(CN)_6^{3-} + e^- \rightleftharpoons Fe(CN)_6^{4-} \qquad E° = 0.356 \text{ V}$$
Ferricyanide $\qquad$ Ferrocyanide

Which ion, Fe(III) or Fe(II), is stabilized more by complexing with CN^-?

(b) Using Appendix H, answer the same question when the ligand is phenanthroline instead of cyanide.

Phenanthroline

Nernst Equation

13-14. What is the difference between E and $E°$ for a redox reaction? Which one runs down to 0 when the complete cell comes to equilibrium?

13-15. (a) Use the Nernst equation to write the spontaneous chemical reaction that occurs in the cell in Demonstration 13-1.

(b) If you use your fingers as a salt bridge in Demonstration 13-1, will your body take in Cu^{2+} or Zn^{2+}?

13-16. Write the Nernst equation for the following half-reaction and find E when pH = 3.00 and P_{AsH_3} = 1.0 mbar.

$$As(s) + 3H^+ + 3e^- \rightleftharpoons AsH_3(g) \qquad E° = -0.238 \text{ V}$$
Arsine

13-17. (a) Write the line notation for the following cell.

(b) Calculate the potential of each half-cell and the cell voltage, E. In which direction will electrons flow through the circuit? Write the spontaneous net cell reaction.

(c) The left half-cell was loaded with 14.3 mL of $Br_2(l)$ (density = 3.12 g/mL). The aluminum electrode contains 12.0 g of Al. Which element, Br_2 or Al, is the limiting reagent? (That is, which reagent will be used up first?)

(d) If the cell is somehow operated under conditions in which it produces a constant voltage of 1.50 V, how much electrical work will have been done when 0.231 mL of $Br_2(l)$ has been consumed?

(e) If the potentiometer is replaced by a 1.20-kΩ resistor and if the heat dissipated by the resistor is 1.00×10^{-4} J/s, at what rate (grams per second) is Al(s) dissolving? (In this question, the voltage is not 1.50 V.)

13-18. A nickel-metal hydride rechargeable battery formerly used in laptop computers is based on the following chemistry:

Cathode:

$$NiOOH(s) + H_2O + e^- \underset{\text{charge}}{\overset{\text{discharge}}{\rightleftharpoons}} Ni(OH)_2(s) + OH^-$$

Anode:

$$MH(s) + OH^- \underset{\text{charge}}{\overset{\text{discharge}}{\rightleftharpoons}} M(s) + H_2O + e^-$$

The anode material, MH, is a transition metal hydride or rare earth alloy hydride. Explain why the voltage remains nearly constant during the entire discharge cycle.

13-19. Suppose that the concentrations of NaF and KCl were each 0.10 M in the cell

$$Pb(s)|PbF_2(s)|F^-(aq)\|Cl^-(aq)|AgCl(s)|Ag(s)$$

(a) Using the half-reactions $2AgCl(s) + 2e^- \rightleftharpoons 2Ag(s) + 2Cl^-$ and $PbF_2(s) + 2e^- \rightleftharpoons Pb(s) + 2F^-$, calculate the cell voltage.

(b) By the reasoning in Figure 13-8, in which direction do electrons flow?

(c) Now calculate the cell voltage by using the reactions $2Ag^+ + 2e^- \rightleftharpoons 2Ag(s)$ and $Pb^{2+} + 2e^- \rightleftharpoons Pb(s)$. For this part, you will need the solubility products for PbF_2 and AgCl.

13-20. The following cell was set up to measure the standard reduction potential of the Ag^+ | Ag couple:

$$Pt(s)|HCl(0.010\ 00\ M), H_2(g)\|AgNO_3(0.010\ 00\ M)|Ag(s)$$

The temperature was 25°C (the standard condition) and atmospheric pressure was 751.0 Torr. Because the vapor pressure of water is 23.8 Torr at 25°C, P_{H_2} in the cell was 751.0 − 23.8 = 727.2 Torr. The Nernst equation for the cell, including activity coefficients, is derived as follows:

Right electrode: $\quad Ag^+ + e^- \rightleftharpoons Ag(s) \qquad\qquad E_+^\circ = E_{Ag^+|Ag}^\circ$

Left electrode: $\quad H^+ + e^- \rightleftharpoons \dfrac{1}{2}H_2(g) \qquad\qquad E_-^\circ = 0\ V$

$$E_+ = E_{Ag^+|Ag}^\circ - 0.059\ 16 \log\left(\frac{1}{[Ag^+]\gamma_{Ag^+}}\right)$$

$$E_- = 0 - 0.059\ 16 \log\left(\frac{P_{H_2}^{1/2}}{[H^+]\gamma_{H^+}}\right)$$

$$E = E_+ - E_- = E_{Ag^+|Ag}^\circ - 0.059\ 16 \log\left(\frac{[H^+]\gamma_{H^+}}{P_{H_2}^{1/2}[Ag^+]\gamma_{Ag^+}}\right)$$

Given a measured cell voltage of +0.798 3 V and using activity coefficients from Table 7-1, find $E_{Ag^+|Ag}^\circ$. Be sure to express P_{H_2} in bar in the reaction quotient.

13-21. Write a balanced chemical equation (in acidic solution) for the reaction represented by the question mark on the lower arrow.[18] Calculate E° for the reaction.

13-22. What must be the relation between E_1° and E_2° if the species X^+ is to disproportionate spontaneously under standard conditions to X^{3+} and $X(s)$? Write a balanced equation for the disproportionation.

$$\overset{E_1^\circ}{}\quad\overset{E_2^\circ}{}$$
$$X^{3+} \rightarrow X^+ \rightarrow X(s)$$

13-23. *Including activities,* calculate the voltage of the cell Ni(s) | $NiSO_4(0.002\ 0\ M)$ ‖ $CuCl_2(0.003\ 0\ M)$ | Cu(s). Assume that the salts are completely dissociated (that is, neglect ion-pair formation). By the reasoning in Figure 13-8, in which direction do electrons flow?

Relation of E° and the Equilibrium Constant

13-24. For the reaction $CO + \frac{1}{2}O_2 \rightleftharpoons CO_2$, $\Delta G^\circ = -257$ kJ per mole of CO at 298 K. Find E° and the equilibrium constant for the reaction.

13-25. Calculate E°, ΔG°, and K for the following reactions.
(a) $4Co^{3+} + 2H_2O \rightleftharpoons 4Co^{2+} + O_2(g) + 4H^+$
(b) $Ag(S_2O_3)_2^{3-} + Fe(CN)_6^{4-} \rightleftharpoons Ag(s) + 2S_2O_3^{2-} + Fe(CN)_6^{3-}$

13-26. A solution contains 0.100 M Ce^{3+}, 1.00×10^{-4} M Ce^{4+}, 1.00×10^{-4} M Mn^{2+}, 0.100 M MnO_4^-, and 1.00 M $HClO_4$.

(a) Write a balanced net reaction that can occur between species in this solution.

(b) Calculate ΔG° and K for the reaction.

(c) Calculate E for the conditions given.

(d) Calculate ΔG for the conditions given.

(e) At what pH would the given concentrations of Ce^{4+}, Ce^{3+}, Mn^{2+}, and MnO_4^- be in equilibrium at 298 K?

13-27. For the cell Pt(s) | $VO^{2+}(0.116$ M$)$, $V^{3+}(0.116$ M$)$, $H^+(1.57$ M$)$ ‖ $Sn^{2+}(0.031\ 8$ M$)$, $Sn^{4+}(0.031\ 8$ M$)$ | Pt(s), E (not E°) = −0.289 V. Write the net cell reaction and calculate its equilibrium constant. Do not use E° values from Appendix H to answer this question.

13-28. Calculate E° for the half-reaction $Pd(OH)_2(s) + 2e^- \rightleftharpoons Pd(s) + 2OH^-$ given that K_{sp} for $Pd(OH)_2$ is 3×10^{-28} and $E^\circ = 0.915$ V for the reaction $Pd^{2+} + 2e^- \rightleftharpoons Pd(s)$.

13-29. From the standard potentials for reduction of $Br_2(aq)$ and $Br_2(l)$ in Appendix H, calculate the solubility of Br_2 in water at 25°C. Express your answer as g/L.

13-30. Given the following information, calculate the standard potential for the reaction $FeY^- + e^- \rightleftharpoons FeY^{2-}$, where Y is EDTA.

$$FeY^- + e^- \rightleftharpoons Fe^{2+} + Y^{4-} \qquad\qquad E^\circ = -0.730\ V$$
$$FeY^{2-}: \qquad K_f = 2.1 \times 10^{14}$$
$$FeY^-: \qquad K_f = 1.3 \times 10^{25}$$

13-31. For modest temperature excursions away from 25°C, the change in $E°$ for a half-reaction can be written in the form

$$E°(T) = E° + \left(\frac{dE°}{dT}\right)\Delta T$$

where $E°(T)$ is the standard reduction potential at temperature T (°C), and ΔT is $(T - 25)$. For the reaction $Al^{3+} + 3e^- \rightleftharpoons Al(s)$, $dE°/dT = 0.533$ mV/K near 25°C. Find $E°$ for this half-reaction at 50°C.

13-32. This problem is slightly tricky. Calculate $E°$, $\Delta G°$, and K for the reaction

which is the sum of *three* half-reactions listed in Appendix H. Use $\Delta G° (= -nFE°)$ for each of the half-reactions to find $\Delta G°$ for the net reaction. Note that, if you reverse the direction of a reaction, you reverse the sign of $\Delta G°$.

13-33. *Thermodynamics of a solid-state reaction.* The following electrochemical cell is reversible at 1 000 K in an atmosphere of flowing $O_2(g)$:[19]

```
                    +0.152 9 V
            ┌─────────┤ V ├─────────┐
            │       −       +        │
  ┌─────────┴──────────────────────┴─────────┐
  │ ▊                              ▊          │
  │ ▊  Intimate mixture   CaF₂(s)  ▊ Intimate mixture │
  │ ▊  of MgF₂(s) and              ▊ of MgF₂(s),       │
  │ ▊  MgO(s)                      ▊ MgAl₂O₄(s),       │
  │ ▊                              ▊ and Al₂O₃(s)      │
  └────────────────────────────────────────────┘
        Pt mesh     F⁻ ion diffuses
        electrode   through CaF₂(s)
                    at 1 000 K
```

Left half-cell: $MgF_2(s) + \frac{1}{2}O_2(g) + 2e^- \rightleftharpoons MgO(s) + 2F^-$

Right half-cell: $MgF_2(s) + Al_2O_3(s) + \frac{1}{2}O_2(g) + 2e^- \rightleftharpoons$

$$MgAl_2O_4(s) + 2F^-$$

(a) Write a Nernst equation for each half-cell. Write the net reaction and its Nernst equation. The activity of $O_2(g)$ is the same on both sides, and the activity of F^- is the same on both sides, governed by F^- ions diffusing through $CaF_2(s)$. Show that the observed voltage is $E°$ for the net reaction.

(b) From the relation $\Delta G° = -nFE°$, find $\Delta G°$ for the net reaction. Note that 1 V = 1 J/C.

(c) The cell voltage in the temperature range $T = 900$ to 1 250 K is $E(V) = 0.122\ 3 + 3.06 \times 10^{-5}\ T$. Assuming that $\Delta H°$ and $\Delta S°$ are constant, find $\Delta H°$ and $\Delta S°$ from the relation $\Delta G° = \Delta H° - T\Delta S°$.

Using Cells as Chemical Probes

13-34. With Figure 13-10 as an example, explain what we mean when we say that there is equilibrium *within* each half-cell but not necessarily *between* the two half-cells.

13-35. The cell $Pt(s) \mid H_2(g, 1.00\ bar) \mid H^+(aq, pH = 3.60) \parallel Cl^-(aq, x\ M) \mid AgCl(s) \mid Ag(s)$ can be used as a probe to find the concentration of Cl^- in the right compartment.

(a) Write reactions for each half-cell, a balanced net cell reaction, and the Nernst equation for the net cell reaction.

(b) Given a measured cell voltage of 0.485 V, find $[Cl^-]$ in the right compartment.

13-36. The quinhydrone electrode was introduced in 1921 as a means of measuring pH.[20]

$Pt(s) \mid 1:1$ mole ratio of quinone(aq) and hydroquinone(aq),
unknown pH $\parallel Cl^-(aq, 0.50\ M) \mid Hg_2Cl_2(s) \mid Hg(l) \mid Pt(s)$

The solution whose pH is to be measured is placed in the left half-cell, which also contains a 1:1 mole ratio of quinone and hydroquinone. The half-cell reaction is

$$O=\!\!\!\bigcirc\!\!\!=O + 2H^+ + 2e^- \rightleftharpoons HO-\!\!\!\bigcirc\!\!\!-OH$$
Quinone   Hydroquinone

(a) Write half-reactions and Nernst equations for each half-cell.

(b) Ignoring activities, rearrange the Nernst equation for the net reaction to the form $E(\text{cell}) = A + (B \cdot pH)$, where A and B are constants. Calculate A and B at 25°C.

(c) If the pH were 4.50, in which direction would electrons flow through the potentiometer?

13-37. The voltage for the following cell is 0.490 V. Find K_b for the organic base RNH_2.

$Pt(s) \mid H_2(1.00\ bar) \mid$
$RNH_2(aq, 0.10\ M), RNH_3^+Cl^-(aq, 0.050\ M) \parallel S.H.E.$

13-38. The voltage of the cell shown here is −0.246 V. The right half-cell contains the metal ion, M^{2+}, whose standard reduction potential is −0.266 V.

$$M^{2+} + 2e^- \rightleftharpoons M(s) \qquad E° = -0.266\ V$$

Calculate K_f for the metal-EDTA complex.

28.0 mL of 0.010 0 M pyrophosphoric acid   28.0 mL of 0.010 0 M M²⁺
72.0 mL of 0.010 0 M KOH   72.0 mL of 0.010 0 M EDTA
                                               buffered to pH 8.00

13-39. The following cell was constructed to find the difference in K_{sp} between two naturally occurring forms of $CaCO_3(s)$, called *calcite* and *aragonite*.[21]

$$\text{Pb}(s)\,|\,\text{PbCO}_3(s)\,|\,\text{CaCO}_3(s, \text{calcite})\,|\,\text{buffer(pH 7.00)}\,\|$$
$$\text{buffer(pH 7.00)}\,|\,\text{CaCO}_3(s, \text{aragonite})\,|\,\text{PbCO}_3(s)\,|\,\text{Pb}(s)$$

Each compartment of the cell contains a mixture of solid PbCO_3 ($K_{sp} = 7.4 \times 10^{-14}$) and either calcite or aragonite, both of which have $K_{sp} \approx 5 \times 10^{-9}$. Each solution was buffered to pH 7.00 with an inert buffer, and the cell was completely isolated from atmospheric CO_2. The measured cell voltage was –1.8 mV. Find the ratio of solubility products, K_{sp} (for calcite)/K_{sp} (for aragonite).

13-40. *Do not ignore activity coefficients in this problem.* If the voltage for the following cell is 0.512 V, find K_{sp} for $\text{Cu(IO}_3)_2$. Neglect any ion pairing.

$$\text{Ni}(s)\,|\,\text{NiSO}_4(0.002\ 5\ \text{M})\,\|\,\text{KIO}_3(0.10\ \text{M})\,|\,\text{Cu(IO}_3)_2(s)\,|\,\text{Cu}(s)$$

Biochemists Use $E^{\circ\prime}$

13-41. Explain what $E^{\circ\prime}$ is and why it is preferred over E° in biochemistry.

13-42. We are going to find $E^{\circ\prime}$ for the reaction $\text{C}_2\text{H}_2(g) + 2\text{H}^+ + 2\text{e}^- \rightleftharpoons \text{C}_2\text{H}_4(g)$.

(a) Write the Nernst equation for the half-reaction, using E° from Appendix H.

(b) Rearrange the Nernst equation to the form

$$E = E^{\circ} + \text{other terms} - \frac{0.059\ 16}{2} \log \left(\frac{P_{\text{C}_2\text{H}_4}}{P_{\text{C}_2\text{H}_2}} \right)$$

(c) The quantity (E° + other terms) is $E^{\circ\prime}$. Evaluate $E^{\circ\prime}$ for pH = 7.00.

13-43. Evaluate $E^{\circ\prime}$ for the half-reaction $(\text{CN})_2(g) + 2\text{H}^+ + 2\text{e}^- \rightleftharpoons 2\text{HCN}(aq)$.

13-44. Calculate $E^{\circ\prime}$ for the reaction

$$\underset{\text{Oxalic acid}}{\text{H}_2\text{C}_2\text{O}_4} + 2\text{H}^+ + 2\text{e}^- \rightleftharpoons \underset{\text{Formic acid}}{2\text{HCO}_2\text{H}} \qquad E^{\circ} = 0.204\ \text{V}$$

13-45. HOx is a monoprotic acid with $K_a = 1.4 \times 10^{-5}$ and H_2Red^- is a diprotic acid with $K_1 = 3.6 \times 10^{-4}$ and $K_2 = 8.1 \times 10^{-8}$. Find E° for the reaction

$$\text{HOx} + \text{e}^- \rightleftharpoons \text{H}_2\text{Red}^- \qquad E^{\circ\prime} = 0.062\ \text{V}$$

13-46. Given the following information, find K_a for nitrous acid, HNO_2.

$$\text{NO}_3^- + 3\text{H}^+ + 2\text{e}^- \rightleftharpoons \text{HNO}_2 + \text{H}_2\text{O} \qquad \begin{aligned} E^{\circ} &= 0.940\ \text{V} \\ E^{\circ\prime} &= 0.433\ \text{V} \end{aligned}$$

13-47. Using the reaction

$$\text{HPO}_4^{2-} + 2\text{H}^+ + 2\text{e}^- \rightleftharpoons \text{HPO}_3^{2-} + \text{H}_2\text{O} \qquad E^{\circ} = -0.234\ \text{V}$$

and acid dissociation constants from Appendix G, calculate E° for the reaction

$$\text{H}_2\text{PO}_4^- + \text{H}^+ + 2\text{e}^- \rightleftharpoons \text{HPO}_3^{2-} + \text{H}_2\text{O}$$

13-48. This problem requires Beer's law from Chapter 17. The oxidized form (Ox) of a flavoprotein that functions as a one-electron reducing agent has a molar absorptivity (ε) of $1.12 \times 10^4\ \text{M}^{-1}\text{cm}^{-1}$ at 457 nm at pH 7.00. For the reduced form (Red), $\varepsilon = 3.82 \times 10^3\ \text{M}^{-1}\text{cm}^{-1}$ at 457 nm at pH 7.00.

$$\text{Ox} + \text{e}^- \rightleftharpoons \text{Red} \qquad E^{\circ\prime} = -0.128\ \text{V}$$

The substrate (S) is the molecule reduced by the protein.

$$\text{Red} + \text{S} \rightleftharpoons \text{Ox} + \text{S}^-$$

Both S and S^- are colorless. A solution at pH 7.00 was prepared by mixing enough protein plus substrate (Red + S) to produce initial concentrations [Red] = [S] = 5.70×10^{-5} M. The absorbance at 457 mm was 0.500 in a 1.00-cm cell.

(a) Calculate the concentrations of Ox and Red from the absorbance data.

(b) Calculate the concentrations of S and S^-.

(c) Calculate the value of $E^{\circ\prime}$ for the reaction $\text{S} + \text{e}^- \rightleftharpoons \text{S}^-$.

14 | Electrodes and Potentiometry

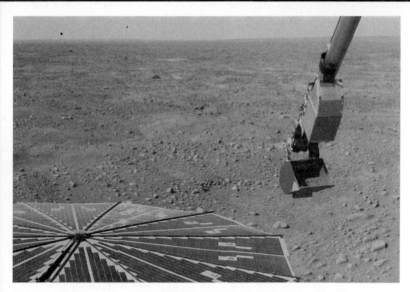

Robotic arm of *Phoenix Mars Lander* scoops up soil for chemical analysis on Mars.
[NASA photograph courtesy S. Kounaves, Tufts University.]

One of four Wet Chemistry Laboratory cells for soil analysis on *Phoenix*. [NASA photograph courtesy S. Kounaves, Tufts University.]

Sam Kounaves and his students at Tufts University had the thrill of a lifetime in 2008 when their Wet Chemistry Laboratory on the *Phoenix Mars Lander* began sending back data on soil scooped up near the north pole of Mars. The objective was to learn the chemical composition of salts that could be leached from the soil by stirring it with water. Measurements of dissolved salts were made by 23 electrochemical sensors, of which 15 were ion-selective electrodes similar to those discussed in this chapter. Sensors were embedded in the walls of four 40-mL epoxy plastic "beakers" used for the leaching experiments.[1]

The temperature on Mars is well below 0°C, so each operation began by using precious electric power for up to 90 min to melt an aqueous solution. A 25-mL solution containing 10^{-5} M ionic standards was added to the beaker and sensor responses were monitored for 15 min to obtain a first calibration point for each sensor. A crucible containing known masses of solid salts was then dropped into the beaker and the contents dissolved by stirring. Sensors were monitored for 30 min to obtain a second calibration point.

Earlier, the robotic arm scooped up soil and dumped it through a sieve to reject particles larger than 2 mm. Soil particles fell into a 1-mL drawer, which was later retracted to drop the soil into the leaching solution in the beaker. Each day, scientists on Earth wrote a series of commands to be carried out by the robotic arm. On the next day, photos transmitted by the Lander showed what the arm did and what the soil looked like. From the photos, scientists would plan commands for the following day.

Sensor response was monitored for 200 min while salts from the soil dissolved. The mixture was allowed to freeze during the Martian night and was thawed and measured the next day to observe changes induced by the freeze-thaw cycle. Then 4 mg of 2-nitrobenzoic acid were added to see what changes would be induced by weak acidification for 90 min. Finally, the solution was titrated by three additions of solid $BaCl_2$ to measure SO_4^{2-} by precipitating $BaSO_4$. Even on Mars in 2008, a precipitation titration has its place.

The sequence of operations measured Ca^{2+}, Mg^{2+}, K^+, NO_3^-, NH_4^+, SO_4^{2-}, Cl^-, Br^-, I^-, and pH with ion-selective electrodes. Other electrodes measured conductivity, reduction potential, redox couples, and reducible metals including Cu^{2+}, Cd^{2+}, Pb^{2+}, Fe^{2+}, Fe^{3+}, and Hg^{2+}. The biggest discovery was a completely unexpected, high concentration of ClO_4^- sensed by the nitrate electrode (Box 14-3). A second surprise was that sulfate was not observed.

Clever chemists have designed electrodes that respond selectively to specific analytes in solution or in the gas phase. Typical ion-selective electrodes are about the size of your pen. Really clever chemists created ion-sensing field effect transistors that are just hundreds of micrometers in size and can be inserted into a blood vessel. The use of electrodes to measure voltages that provide chemical information is called **potentiometry.**

In the simplest case, analyte is an *electroactive species* that is part of a galvanic cell. An **electroactive species** is one that can donate or accept electrons at an electrode. We turn the unknown solution into a half-cell by inserting an electrode, such as a Pt wire, that can transfer electrons to or from the analyte. Because this electrode responds to analyte, it is called the **indicator electrode.** We connect this half-cell to a second half-cell by a salt bridge. The second half-cell has a fixed composition, so it has a constant potential. Because of its constant potential, the second half-cell is called a **reference electrode.** The cell voltage is the difference between the variable potential of the analyte half-cell and the constant potential of the reference electrode.

Indicator electrode: responds to analyte activity

Reference electrode: maintains a fixed (reference) potential

14-1 Reference Electrodes

Suppose you want to measure the relative amounts of Fe^{2+} and Fe^{3+} in a solution. You can make this solution part of a galvanic cell by inserting a Pt wire and connecting the cell to a constant-potential half-cell by a salt bridge, as shown in Figure 14-1.

The two half-reactions (written as *reductions*) are

Right electrode: $\qquad Fe^{3+} + e^- \rightleftharpoons Fe^{2+} \qquad\qquad E_+^\circ = 0.771 \text{ V}$

Left electrode: $\qquad AgCl(s) + e^- \rightleftharpoons Ag(s) + Cl^- \qquad E_-^\circ = 0.222 \text{ V}$

The electrode potentials are

$$E_+ = 0.771 - 0.059\,16 \log\left(\frac{[Fe^{2+}]}{[Fe^{3+}]}\right)$$

$$E_- = 0.222 - 0.059\,16 \log[Cl^-]$$

and the cell voltage is the difference $E_+ - E_-$:

$$E = \left\{0.771 - 0.059\,16 \log\left(\frac{[Fe^{2+}]}{[Fe^{3+}]}\right)\right\} - \{0.222 - 0.059\,16 \log[Cl^-]\}$$

But $[Cl^-]$ in the left half-cell is constant, fixed by the solubility of KCl, with which the solution is saturated. Therefore, the cell voltage changes only when the quotient $[Fe^{2+}]/[Fe^{3+}]$ changes.

E_+ is the potential of the electrode attached to the positive input of the potentiometer.
E_- is the potential of the electrode attached to the negative input of the potentiometer.

The voltage really tells us the quotient of activities, $\mathcal{A}_{Fe^{2+}}/\mathcal{A}_{Fe^{3+}}$. We will neglect activity coefficients and write the Nernst equation with concentrations instead of activities.

Anode: $Ag + Cl^- \rightleftharpoons AgCl + e^-$ Cathode: $Fe^{3+} + e^- \rightleftharpoons Fe^{2+}$

$Ag(s)\,|\,AgCl(s)\,|\,Cl^-(aq)\,\|\,Fe^{2+}(aq), Fe^{3+}(aq)\,|\,Pt(s)$

FIGURE 14-1 A galvanic cell that can be used to measure the quotient $[Fe^{2+}]/[Fe^{3+}]$ in the right half-cell. The Pt wire is the *indicator electrode,* and the entire left half-cell plus salt bridge (enclosed by the dashed line) can be considered a *reference electrode.*

FIGURE 14-2 Another view of Figure 14-1. The contents of the dashed box in Figure 14-1 are now considered to be a reference electrode dipped into the analyte solution.

The half-cell on the left in Figure 14-1 acts as a *reference electrode*. We can picture the cell and salt bridge enclosed by the dashed line as a single unit dipped into the analyte solution, as shown in Figure 14-2. The Pt wire is the indicator electrode, whose potential responds to the quotient $[Fe^{2+}]/[Fe^{3+}]$. The reference electrode completes the redox reaction and provides a *constant potential* to the left side of the potentiometer. Changes in the cell voltage result from changes in the quotient $[Fe^{2+}]/[Fe^{3+}]$.

Silver-Silver Chloride Reference Electrode[2]

The half-cell enclosed by the dashed line in Figure 14-1 is called a **silver-silver chloride electrode.** Figure 14-3 shows how the electrode is reconstructed as a thin tube that can be dipped into analyte solution. Figure 14-4 shows a *double-junction electrode* that minimizes contact between analyte solution and KCl from the electrode. Silver-silver chloride and calomel reference electrodes (described next) are used because they are convenient. A standard hydrogen electrode (S.H.E.) is difficult to use because it requires H_2 gas and a freshly prepared catalytic Pt surface that is easily poisoned in many solutions.

FIGURE 14-3 Silver-silver chloride reference electrode.

Wire lead

Air inlet to allow electrolyte to drain slowly through porous plug

Ag wire bent into a loop

Aqueous solution saturated with KCl and AgCl

AgCl paste

Solid KCl plus some AgCl

Porous plug for contact with external solution (salt bridge)

Inner electrode   Outer compartment

Porous plugs

FIGURE 14-4 Double-junction reference electrode. The inner electrode is the same as the one in Figure 14-3. The solution in the outer compartment is compatible with analyte solution. For example, if you do not want Cl^- to contact the analyte, the outer electrode can be filled with KNO_3 solution. The inner and outer solutions slowly mix, so the outer compartment must be refilled periodically with fresh KNO_3 solution.
[Courtesy Fisher Scientific, Pittsburgh, PA.]

The standard reduction potential for the AgCl | Ag couple is +0.222 V at 25°C. This would be the potential of a silver-silver chloride electrode if $\mathcal{A}_{Cl^-}$ were unity. But the activity of Cl^- in a saturated solution of KCl at 25°C is not unity, and the potential of the electrode in Figure 14-3 is +0.197 V with respect to S.H.E. at 25°C.

Ag | AgCl electrode: $\quad AgCl(s) + e^- \rightleftharpoons Ag(s) + Cl^- \qquad E° = +0.222 \text{ V}$

$$E(\text{saturated KCl}) = +0.197 \text{ V}$$

A problem with reference electrodes is that porous plugs become clogged, thus causing sluggish, unstable electrical response. Some designs incorporate a free-flowing capillary in place of the porous plug. Other designs allow you to force fresh solution from the electrode through the electrode-analyte junction prior to a measurement.

Calomel Electrode

The **calomel electrode** in Figure 14-5 is based on the reaction

Calomel electrode: $\quad \frac{1}{2}Hg_2Cl_2(s) + e^- \rightleftharpoons Hg(l) + Cl^- \qquad E° = +0.268 \text{ V}$

Mercury(I) chloride
(Calomel)

$$E(\text{saturated KCl}) = +0.241 \text{ V}$$

The standard potential for this reaction is +0.268 V. If the cell is saturated with KCl at 25°C, the potential is +0.241 V. A calomel electrode saturated with KCl is called a **saturated calomel electrode,** abbreviated **S.C.E.** The advantage in using saturated KCl is that $[Cl^-]$ does not change if some liquid evaporates.

Wire lead

Pt wire

Hole to allow drainage through porous plug

Hg(*l*)

Hg, Hg_2Cl_2 + KCl

Glass wool

Opening

Saturated KCl solution

KCl(*s*)

Glass wall

Porous plug (salt bridge)

FIGURE 14-5 A saturated calomel electrode (S.C.E.).

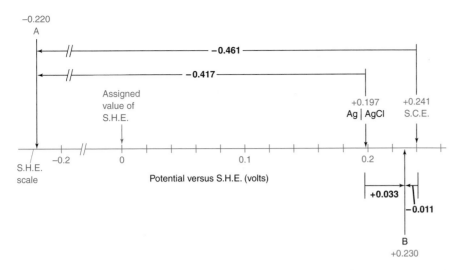

FIGURE 14-6 A diagram that helps us convert electrode potential between different reference scales.

Voltage Conversions Between Different Reference Scales

If an electrode has a potential of −0.461 V with respect to a calomel electrode, what is the potential with respect to a silver-silver chloride electrode? What would be the potential with respect to the standard hydrogen electrode?

To answer these questions, Figure 14-6 shows the positions of the calomel and silver-silver chloride electrodes with respect to the standard hydrogen electrode. We see that point A, which is −0.461 V from S.C.E., is −0.417 V from the silver-silver chloride electrode and −0.220 V with respect to S.H.E. Point B, whose potential is +0.033 V from silver-silver chloride, is −0.011 V from S.C.E. and +0.230 V from S.H.E. By keeping this diagram in mind, you can convert potentials from one scale into another.

14-2 Indicator Electrodes

We will study two broad classes of indicator electrodes. *Metal electrodes* described in this section develop an electric potential in response to a redox reaction at the metal surface. *Ion-selective electrodes* described later are not based on redox processes. Instead, selective binding of one type of ion to a membrane generates an electric potential.

The most common metal indicator electrode is platinum, which is relatively *inert*—it does not participate in many chemical reactions. Its purpose is simply to transmit electrons to or from species in solution. Gold electrodes are even more inert than Pt. Various types of carbon are used as indicator electrodes because the rates of many redox reactions on the carbon surface are fast. A metal electrode works best when its surface is large and clean. To clean the electrode, dip it in hot 8 M HNO_3 in a fume hood and rinse with distilled water.

Figure 14-7 shows how a silver electrode can be used with a reference electrode to measure Ag^+ concentration.[3] The reaction at the Ag indicator electrode is

$$Ag^+ + e^- \rightleftharpoons Ag(s) \qquad E^\circ_+ = 0.799 \text{ V}$$

The calomel reference half-cell reaction is

$$Hg_2Cl_2(s) + 2e^- \rightleftharpoons 2Hg(l) + 2Cl^- \qquad E_- = 0.241 \text{ V}$$

The reference potential (E_-, not E°_-) is fixed at 0.241 V because the reference cell is saturated with KCl. The Nernst equation for the entire cell is therefore

$$E = E_+ - E_- = \underbrace{\left\{ 0.799 - 0.059\ 16 \log\left(\frac{1}{[Ag^+]} \right) \right\}}_{\text{Potential of Ag}\,|\,Ag^+ \text{ indicator electrode}} - \underbrace{\{0.241\}}_{\substack{\text{Potential of} \\ \text{S.C.E. reference electrode}}}$$

$$E = 0.558 + 0.059\ 16 \log[Ag^+] \qquad (14\text{-}1)$$

FIGURE 14-7 Use of Ag and calomel electrodes to measure $[Ag^+]$. The calomel electrode has a double junction, like that in Figure 14-4. The outer compartment of the electrode is filled with KNO_3, so there is no direct contact between Cl^- in the inner compartment and Ag^+ in the beaker.

That is, the voltage of the cell in Figure 14-7 provides a measure of $[Ag^+]$. Ideally, the voltage changes by 59.16 mV (at 25°C) for each factor-of-10 change in $[Ag^+]$.

EXAMPLE   **Potentiometric Precipitation Titration**

A 100.0-mL solution containing 0.100 0 M NaCl was titrated with 0.100 0 M AgNO₃, and the voltage of the cell shown in Figure 14-7 was monitored. The equivalence volume is $V_e = 100.0$ mL. Calculate the voltage after the addition of **(a)** 65.0 and **(b)** 135.0 mL of AgNO₃.

Solution   The titration reaction is

$$Ag^+ + Cl^- \rightarrow AgCl(s)$$

(a) At 65.0 mL, 65.0% of Cl^- has precipitated and 35.0% remains in solution:

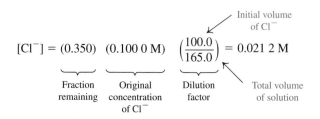

$$[Cl^-] = \underbrace{(0.350)}_{\substack{\text{Fraction} \\ \text{remaining}}} \; \underbrace{(0.100\ 0\ M)}_{\substack{\text{Original} \\ \text{concentration} \\ \text{of } Cl^-}} \; \underbrace{\left(\frac{100.0}{165.0}\right)}_{\substack{\text{Dilution} \\ \text{factor}}} = 0.021\ 2\ M$$

To find the cell voltage in Equation 14-1, we need to know $[Ag^+]$:

$$[Ag^+][Cl^-] = K_{sp} \Rightarrow [Ag^+] = \frac{K_{sp}}{[Cl^-]} = \frac{1.8 \times 10^{-10}}{0.021\ 2\ M} = 8.5 \times 10^{-9}\ M$$

The cell voltage is therefore

$$E = 0.558 + 0.059\ 16 \log(8.5 \times 10^{-9}) = 0.081\ V$$

(b) At 135.0 mL, there is an excess of 35.0 mL of AgNO₃ = 3.50 mmol Ag^+ in a total volume of 235.0 mL. Therefore, $[Ag^+] = (3.50\ \text{mmol})/(235.0\ \text{mL}) = 0.014\ 9\ M$. The cell voltage is

$$E = 0.558 + 0.059\ 16 \log(0.014\ 9) = 0.450\ V$$

Test Yourself   Find the voltage after addition of 99.0 mL of AgNO₃. (***Answer:*** 0.177 V)

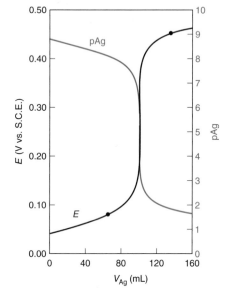

FIGURE 14-8 Titration curve computed for the addition of 0.100 0 M Ag^+ to 100.0 mL of 0.100 0 M Cl^- with the electrodes in Figure 14-7. Points computed at 65.0 and 135.0 mL are shown. Colored line is pAg = $-\log[Ag^+]$.

The cell responds to a change in $[Cl^-]$, which necessarily changes $[Ag^+]$ because $[Ag^+][Cl^-] = K_{sp}$.

Demonstration 14-1 is a great example of indicator and reference electrodes.

Figure 14-8 shows the titration curve for the preceding example. There is a strong analogy to acid-base titrations, with Ag^+ replacing H^+ and Cl^- acting as a base being titrated. As the acid-base titration proceeds, $[H^+]$ increases and pH decreases. As the Ag^+/Cl^- titration proceeds, $[Ag^+]$ increases and pAg ($\equiv -\log[Ag^+]$) decreases. The silver electrode measures pAg, which you can see by substituting pAg = $-\log[Ag^+]$ into Equation 14-1:

$$E = 0.558 - 0.059\ 16\ \text{pAg} \qquad (14\text{-}2)$$

A silver electrode is also a halide electrode, if solid silver halide is present.[4] If the solution contains AgCl(s), we substitute $[Ag^+] = K_{sp}/[Cl^-]$ into Equation 14-1 to find an expression relating the cell voltage to $[Cl^-]$:

$$E = 0.558 + 0.059\ 16 \log\left(\frac{K_{sp}}{[Cl^-]}\right) \qquad (14\text{-}3)$$

Metals including Ag, Cu, Zn, Cd, and Hg can be used as indicator electrodes for their aqueous ions. Most metals, however, are unsuitable for this purpose because the equilibrium $M^{n+} + ne^- \rightleftharpoons M$ is not readily established at the metal surface.

The Belousov-Zhabotinskii reaction is a cerium-catalyzed oxidation of malonic acid by bromate, in which the quotient $[Ce^{3+}]/[Ce^{4+}]$ oscillates by a factor of 10 to 100.[6]

$$3CH_2(CO_2H)_2 + 2BrO_3^- + 2H^+ \rightarrow$$
Malonic acid   Bromate

$$2BrCH(CO_2H)_2 + 3CO_2 + 4H_2O$$
Bromomalonic acid

When the Ce^{4+} concentration is high, the solution is yellow. When Ce^{3+} predominates, the solution is colorless. With redox indicators (Section 15-2), this reaction oscillates through a sequence of colors.[7]

Oscillation between yellow and colorless is set up in a 300-mL beaker with the following solutions:

160 mL of 1.5 M H_2SO_4
40 mL of 2 M malonic acid
30 mL of 0.5 M $NaBrO_3$ (or saturated $KBrO_3$)
4 mL of saturated ceric ammonium sulfate,
    $(Ce(SO_4)_2 \cdot 2(NH_4)_2SO_4 \cdot 2H_2O)$

After an induction period of 5 to 10 min with magnetic stirring, oscillations can be initiated by adding 1 mL of ceric ammonium sulfate solution. The reaction may need more Ce^{4+} over a 5-min period to initiate oscillations.

A galvanic cell is built around the reaction as shown in the figure. The quotient $[Ce^{3+}]/[Ce^{4+}]$ is monitored by Pt and calomel electrodes. You should be able to write the cell reactions and a Nernst equation for this experiment.

Apparatus used to monitor the quotient $[Ce^{3+}]/[Ce^{4+}]$ for an oscillating reaction. [The idea for this demonstration came from George Rossman, California Institute of Technology.]

In place of a potentiometer (a pH meter), use a computer or recorder to show the oscillations. Because the potential oscillates over a range of ~100 mV but is centered near ~1.2 V, the cell voltage is offset by ~1.2 V with any available power supply.[8] Trace *a* shows what is usually observed. The potential changes rapidly during the abrupt colorless-to-yellow change and gradually during the gentle yellow-to-colorless change. Trace *b* shows two different cycles superimposed in the same solution. This unusual event occurred in a reaction that had been oscillating normally for about 30 min.[9]

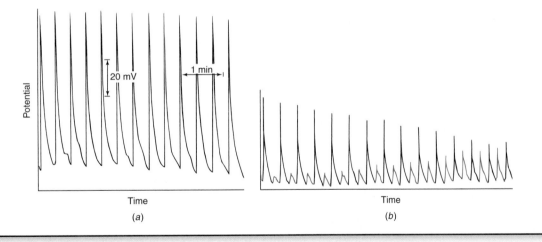

(a)   (b)

14-3 What Is a Junction Potential?

Whenever dissimilar electrolyte solutions are in contact, a voltage difference called the **junction potential** develops at their interface. This small voltage (usually a few millivolts) is found at each end of a salt bridge connecting two half-cells. *The junction potential puts a fundamental limitation on the accuracy of direct potentiometric measurements*, because we usually do not know the contribution of the junction to the measured voltage.

$$E_{observed} = E_{cell} + E_{junction}$$

Because the junction potential is usually unknown, E_{cell} is uncertain.

TABLE 14-1 Mobilities of ions in water at 25°C

Ion	Mobility $[m^2/(s \cdot V)]^a$
H^+	36.30×10^{-8}
Rb^+	7.92×10^{-8}
K^+	7.62×10^{-8}
NH_4^+	7.61×10^{-8}
La^{3+}	7.21×10^{-8}
Ba^{2+}	6.59×10^{-8}
Ag^+	6.42×10^{-8}
Ca^{2+}	6.12×10^{-8}
Cu^{2+}	5.56×10^{-8}
Na^+	5.19×10^{-8}
Li^+	4.01×10^{-8}
OH^-	20.50×10^{-8}
$Fe(CN)_6^{4-}$	11.45×10^{-8}
$Fe(CN)_6^{3-}$	10.47×10^{-8}
SO_4^{2-}	8.27×10^{-8}
Br^-	8.13×10^{-8}
I^-	7.96×10^{-8}
Cl^-	7.91×10^{-8}
NO_3^-	7.40×10^{-8}
ClO_4^-	7.05×10^{-8}
F^-	5.70×10^{-8}
HCO_3^-	4.61×10^{-8}
$CH_3CO_2^-$	4.24×10^{-8}

a. The mobility of an ion is the terminal velocity that the particle achieves in an electric field of 1 V/m. Mobility = velocity/field. The units of mobility are therefore $(m/s)/(V/m) = m^2/(s \cdot V)$.

TABLE 14-2 Liquid junction potentials at 25°C

Junction	Potential (mV)
0.1 M NaCl | 0.1 M KCl	−6.4
0.1 M NaCl | 3.5 M KCl	−0.2
1 M NaCl | 3.5 M KCl	−1.9
0.1 M HCl | 0.1 M KCl	+27
0.1 M HCl | 3.5 M KCl	+3.1

NOTE: A positive sign means that the right side of the junction becomes positive with respect to the left side.

Hydrophobic: "water hating" (does not mix with water)

FIGURE 14-9 Development of the junction potential caused by unequal mobilities of Na^+ and Cl^-.

To see why the junction potential occurs, consider a solution of NaCl in contact with distilled water (Figure 14-9). Na^+ and Cl^- ions begin to diffuse from the NaCl solution into the water. However, Cl^- ion has a greater **mobility** than Na^+. That is, Cl^- diffuses faster than Na^+. As a result, a region rich in Cl^-, with excess negative charge, develops at the front. Behind it is a positively charged region depleted of Cl^-. The result is an electric potential difference at the junction of the NaCl and H_2O phases. The junction potential opposes the movement of Cl^- and accelerates the movement of Na^+. The steady-state junction potential represents a balance between the unequal mobilities that create a charge imbalance and the tendency of the resulting charge imbalance to retard the movement of Cl^-.

Table 14-1 shows mobilities of several ions and Table 14-2 lists some liquid junction potentials. Saturated KCl is used in a salt bridge because K^+ and Cl^- have similar mobilities. Junction potentials at the two interfaces of a KCl salt bridge are slight.

Nonetheless, the junction potential of 0.1 M HCl | 3.5 M KCl is 3.1 mV. A pH electrode has a response of 59 mV per pH unit. A pH electrode dipped into 0.1 M HCl will have a junction potential of ~3 mV, or an error of 0.05 pH units (12% error in $[H^+]$).

EXAMPLE Junction Potential

A 0.1 M NaCl solution was placed in contact with a 0.1 M $NaNO_3$ solution. Which side of the junction is positive?

Solution Because $[Na^+]$ is equal on both sides, there is no net diffusion of Na^+ across the junction. However, Cl^- diffuses into $NaNO_3$, and NO_3^- diffuses into NaCl. The mobility of Cl^- is greater than that of NO_3^-, so the NaCl region will be depleted of Cl^- faster than the $NaNO_3$ region will be depleted of NO_3^-. The $NaNO_3$ side will become negative, and the NaCl side will become positive.

Test Yourself Which side of the 0.05 M NaCl | 0.05 M LiCl junction will be positive? (**Answer:** LiCl)

14-4 How Ion-Selective Electrodes Work[10]

Ion-selective electrodes discussed in the remainder of this chapter respond selectively to one ion. These electrodes are fundamentally different from metal electrodes in that ion-selective electrodes do not involve redox processes. The key feature of an ideal ion-selective electrode is a thin membrane capable of binding only the intended ion.

Consider the *liquid-based ion-selective electrode* in Figure 14-10a. The electrode is said to be "liquid based" because the ion-selective membrane is a *hydrophobic* organic polymer impregnated with a viscous organic solution containing an ion exchanger and, sometimes, a ligand that selectively binds the analyte cation, C^+. The inside of the electrode contains filling solution with the ions $C^+(aq)$ and $B^-(aq)$. The outside of the electrode is immersed in analyte solution containing $C^+(aq)$, $A^-(aq)$, and, perhaps, other ions. Ideally, it does not matter what A^- and B^- and the other ions are. The electric potential difference (the voltage) across the ion-selective membrane is measured by two reference electrodes, which might be Ag | AgCl. If the concentration (really, the activity) of C^+ in the analyte solution changes, the voltage measured between the two reference electrodes also changes. By using a calibration curve, the voltage tells us the activity of C^+ in the analyte solution.

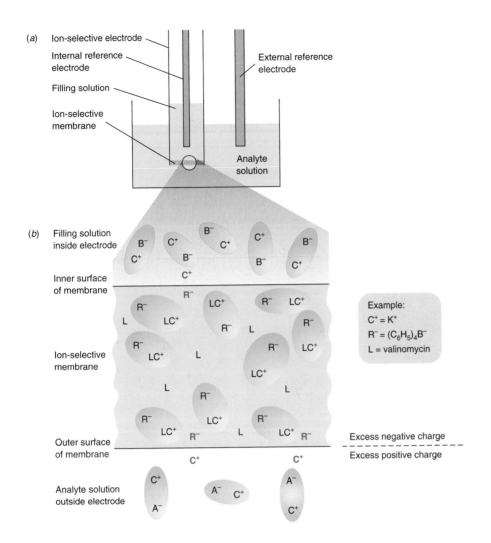

(a)
Ion-selective electrode
Internal reference electrode
External reference electrode
Filling solution
Ion-selective membrane
Analyte solution

(b) Filling solution inside electrode

Inner surface of membrane

Ion-selective membrane

Example:
$C^+ = K^+$
$R^- = (C_6H_5)_4B^-$
L = valinomycin

Outer surface of membrane

Excess negative charge
Excess positive charge

Analyte solution outside electrode

FIGURE 14-10 (*a*) Ion-selective electrode immersed in aqueous solution containing analyte cation, C^+. Typically, the membrane is made of poly(vinyl chloride) impregnated with the *plasticizer* dioctyl sebacate, a nonpolar liquid that softens the membrane and dissolves the ion-selective ionophore (L), the complex (LC^+), and a hydrophobic anion (R^-). (*b*) Close-up of membrane. Ellipses encircling pairs of ions are a guide for the eye to count the charge in each phase. Bold colored ions represent excess charge in each phase. The electric potential difference across each surface of the membrane depends on the activity of analyte ion in the aqueous solution contacting the membrane.

K O C N

Valinomycin-K^+ complex has six carbonyl oxygen atoms in octahedral coordination around K^+. [From L. Stryer, *Biochemistry*, 4th ed. (New York: W. H. Freeman and Company, 1995), p. 273.]

L has some ability to bind other ions besides C^+, so those other ions interfere to some extent with the measurement of C^+. An ion-selective electrode uses a ligand with a strong preference to bind the desired ion.

Example of hydrophobic anion, R^-:

Tetraphenylborate, $(C_6H_5)_4B^-$

Figure 14-10b shows how the electrode works. The key in this example is the ligand, L (called an *ionophore*), which is soluble inside the membrane and selectively binds analyte ion. In a potassium ion-selective electrode, for example, L could be valinomycin, a natural antibiotic secreted by certain microorganisms to carry K^+ ion across cell membranes. *The ligand, L, is chosen to have a high affinity for analyte cation, C^+, and low affinity for other ions.* In an ideal electrode, L binds only C^+. Real electrodes always have some affinity for other cations, so these cations interfere to some degree with the measurement of C^+. For charge neutrality, the membrane also contains a hydrophobic anion, R^-, such as tetraphenylborate, $(C_6H_5)_4B^-$, that is soluble in the membrane and poorly soluble in water.

Almost all analyte ion inside the membrane in Figure 14-10b is bound in the complex LC^+, which is in equilibrium with a small amount of free C^+ in the membrane. The membrane also contains excess free L. C^+ can diffuse across the interface. In an ideal electrode, R^- cannot leave the membrane, because it is not soluble in water, and the aqueous anion A^- cannot enter the membrane, because it is not soluble in the organic phase. As soon as a few C^+ ions diffuse from the membrane into the aqueous phase, there is excess positive charge in the aqueous phase. This imbalance creates an electric potential difference that opposes diffusion of more C^+ into the aqueous phase. The region of charge imbalance extends just a few nanometers into the membrane and into the neighboring solution.

When C^+ diffuses from a region of activity $\mathcal{A}_m$ in the membrane to a region of activity $\mathcal{A}_o$ in the outer solution, the free energy change is

$$\Delta G = \underbrace{\Delta G_{\text{solvation}}}_{\substack{\Delta G \text{ due to change} \\ \text{in solvent}}} - \underbrace{RT \ln\left(\frac{\mathcal{A}_m}{\mathcal{A}_o}\right)}_{\substack{\Delta G \text{ due to} \\ \text{change in activity} \\ \text{(concentration)}}}$$

where R is the gas constant and T is temperature (K). $\Delta G_{solvation}$ is the change in solvation energy when the environment around C^+ changes from the organic liquid in the membrane to the aqueous solution outside the membrane. The term $-RT \ln(\mathcal{A}_m/\mathcal{A}_o)$ gives the free energy change when a species diffuses between regions of different activities (concentrations). In the absence of a phase boundary, ΔG would always be negative when a species diffuses from a region of high activity to one of lower activity.

The driving force for diffusion of C^+ from the membrane to the aqueous solution is the favorable solvation of the ion by water. As C^+ diffuses from the membrane into the water, there is a buildup of positive charge in the water immediately adjacent to the membrane. The charge separation creates an electric potential difference (E_{outer}) across the membrane. The free energy difference for C^+ in the two phases is $\Delta G = -nFE_{outer}$, where F is the Faraday constant and n is the charge of the ion. At equilibrium, the net change in free energy for diffusion of C^+ across the membrane boundary must be 0:

$$\underbrace{\Delta G_{solvation} - RT \ln\left(\frac{\mathcal{A}_m}{\mathcal{A}_o}\right)}_{\substack{\Delta G \text{ due to transfer between phases} \\ \text{and activity difference}}} + \underbrace{(-nFE_{outer})}_{\substack{\Delta G \text{ due to} \\ \text{charge imbalance}}} = 0$$

Solving for E_{outer}, we find that the electric potential difference across the boundary between the membrane and the outer aqueous solution in Figure 14-10b is

Electric potential difference across phase boundary between membrane and analyte:
$$E_{outer} = \frac{\Delta G_{solvation}}{nF} - \left(\frac{RT}{nF}\right) \ln\left(\frac{\mathcal{A}_m}{\mathcal{A}_o}\right) \qquad (14\text{-}4)$$

There is also a potential difference E_{inner} at the boundary between the inner filling solution and the membrane, with terms analogous to those in Equation 14-4.

The potential difference between the outer analyte solution and the inner filling solution is the difference $E = E_{outer} - E_{inner}$. In Equation 14-4, E_{outer} depends on the activities of C^+ in the analyte solution and in the membrane near its outer surface. E_{inner} is constant because the activity of C^+ in the filling solution is constant.

But the activity of C^+ in the membrane ($\mathcal{A}_m$) is very nearly constant for the following reason: The high concentration of LC^+ in the membrane is in equilibrium with free L and a small concentration of free C^+ in the membrane. The hydrophobic anion R^- is poorly soluble in water and therefore cannot leave the membrane. *Very little* C^+ can diffuse out of the membrane because each C^+ that enters the aqueous phase leaves behind one R^- in the membrane. (This separation of charge is the source of the potential difference at the phase boundary.) As soon as a tiny fraction of C^+ diffuses from the membrane into solution, further diffusion is prevented by excess positive charge in the solution near the membrane.

So the potential difference between the outer and the inner solutions is

$$E = E_{outer} - E_{inner} = \frac{\Delta G_{solvation}}{nF} - \left(\frac{RT}{nF}\right) \ln\left(\frac{\mathcal{A}_m}{\mathcal{A}_o}\right) - E_{inner}$$

$$\ln\frac{x}{y} = \ln x - \ln y$$

$$E = \underbrace{\frac{\Delta G_{solvation}}{nF}}_{\text{Constant}} + \left(\frac{RT}{nF}\right) \ln \mathcal{A}_o - \underbrace{\left(\frac{RT}{nF}\right) \ln \mathcal{A}_m}_{\text{Constant}} - \underbrace{E_{inner}}_{\text{Constant}}$$

Combining the constant terms, we find that the potential difference across the membrane depends only on the activity of analyte in the outer solution:

$$E = \text{constant} + \left(\frac{RT}{nF}\right) \ln \mathcal{A}_o$$

From Appendix A, $\ln x = (\ln 10)(\log x) = 2.303 \log x$

Converting $\ln$ into $\log$ and inserting values of R, T, and F gives a useful expression for the potential difference across the membrane:

The number 0.059 16 V is $\frac{RT\ln 10}{F}$ at 25°C.

Electric potential difference for ion-selective electrode:
$$E = \text{constant} + \frac{0.059\ 16}{n} \log \mathcal{A}_o \text{ (volts at 25°C)} \qquad (14\text{-}5)$$

where n is the charge of the analyte ion and $\mathcal{A}_o$ is its activity in the outer (unknown) solution. Equation 14-5 applies to any ion-selective electrode, including a glass pH electrode. If the analyte is an anion, the sign of n is negative. Later, we will modify the equation to account for interfering ions.

A difference of 59.16 mV (at 25°C) builds up across a glass pH electrode for every factor-of-10 change in activity of H^+ in the analyte solution. Because a factor-of-10 difference in activity of H^+ is 1 pH unit, a difference of, say, 4.00 pH units would lead to a potential difference of $4.00 \times 59.16 = 237$ mV. The charge of a calcium ion is $n = 2$, so a potential difference of $59.16/2 = 29.58$ mV is expected for every factor-of-10 change in activity of Ca^{2+} in the analyte measured with a calcium ion-selective electrode.

14-5 pH Measurement with a Glass Electrode

The **glass electrode** used to measure pH is the most common *ion-selective electrode*. A typical pH **combination electrode,** incorporating both glass and reference electrodes in one body, is shown in Figure 14-11. A line diagram of this cell can be written as follows:

Glass membrane selectively binds H^+

$$Ag(s) \mid AgCl(s) \mid Cl^-(aq) \parallel H^+(aq, \text{outside}) \vdots H^+(aq, \text{inside}), Cl^-(aq) \mid AgCl(s) \mid Ag(s)$$

| Outer reference electrode | H^+ outside glass electrode (analyte solution) | H^+ inside glass electrode | Inner reference electrode |

The pH-sensitive part of the electrode is the thin glass bulb or cone at the bottom of the electrodes in Figures 14-11 and 14-12. The reference electrode at the left of the preceding

M. Cremer at the Institute of Physiology at Munich discovered in 1906 that a potential difference of 0.2 V developed across a glass membrane with acid on one side and neutral saline solution on the other. The student Klemensiewicz, working with F. Haber in Karlsruhe in 1908, improved the glass electrode and carried out the first acid-base titration to be monitored with a glass electrode.[11]

FIGURE 14-11 Diagram of a glass combination electrode with a silver-silver chloride reference electrode. The glass electrode is immersed in a solution of unknown pH so that the porous plug on the lower right is below the surface of the liquid. The two Ag | AgCl electrodes measure the voltage across the glass membrane.

FIGURE 14-12 (a) Glass-body combination electrode with pH-sensitive glass bulb at the bottom. The porous ceramic plug (the salt bridge) connects analyte solution to the reference electrode. Two silver wires coated with AgCl are visible inside the electrode. [Courtesy Fisher Scientific, Pittsburgh PA.] (b) A pH electrode with a platinum diaphragm (a bundle of Pt wires), which is said to be less prone to clogging than a ceramic plug. [From W. Knappek, *Am. Lab. News Ed.,* July 2003, p. 14.]

FIGURE 14-13 Schematic structure of glass, which consists of an irregular network of SiO_4 tetrahedra connected through oxygen atoms. Cations such as Li^+, Na^+, K^+, and Ca^{2+} are coordinated to the oxygen atoms. The silicate network is not planar. This diagram is a projection of each tetrahedron onto the plane of the page. [Adapted from G. A. Perley, "Glasses for Measurement of pH," *Anal. Chem.* **1949**, *21*, 394.]

○ = O   ● = Si   ● = cation

So little current flows across a glass electrode that it was not practical when discovered in 1906. One of the first people to use a vacuum tube amplifier to measure pH with a glass electrode was an undergraduate, W. H. Wright at the University of Illinois in 1928, who knew about electronics from amateur radio. Arnold Beckman at Caltech invented a portable, rugged, vacuum tube pH meter in 1935, a device that revolutionized chemical instrumentation.[12]

Electrodes

Beckman's pH meter

line diagram is the coiled Ag | AgCl electrode in the combination electrode in Figure 14-11. The reference electrode at the right side of the line diagram is the straight Ag | AgCl electrode at the center of the electrode in Figure 14-11. The two reference electrodes measure the electric potential difference across the glass membrane. The salt bridge in the line diagram is the porous plug at the bottom right side of the combination electrode in Figure 14-11.

Figure 14-13 shows the irregular structure of the silicate lattice in glass. Negatively charged oxygen atoms in glass can bind to cations of suitable size. Monovalent cations, particularly Na^+, can move sluggishly through the silicate lattice. A schematic cross section of the glass membrane of a pH electrode is shown in Figure 14-14. The two surfaces swell as they absorb water. Metal ions in these *hydrated gel* regions of the membrane diffuse out of the glass and into solution. H^+ can diffuse into the membrane to replace the metal ions. The reaction in which H^+ replaces cations in the glass is an **ion-exchange equilibrium** (Figure 14-15). A pH electrode responds selectively to H^+ because H^+ is the only ion that binds significantly to the hydrated gel layer.

To perform an electrical measurement, at least some tiny current must flow through the entire circuit—even across the glass pH electrode membrane. Studies with tritium (radioactive ^3H) show that H^+ does not cross the glass membrane. However, Na^+ sluggishly crosses the membrane. The H^+-sensitive membrane may be thought of as two surfaces electrically connected by Na^+ transport. The membrane's resistance is typically 10^8 Ω, so little current actually flows across it.

The potential difference between the inner and outer silver-silver chloride electrodes in Figure 14-11 depends on the chloride concentration in each electrode compartment and on the potential difference across the glass membrane. Because $[Cl^-]$ is fixed in each compartment and because $[H^+]$ is fixed on the inside of the glass membrane, the only variable is the pH of analyte solution outside the glass membrane. Equation 14-5 states that *the voltage of the ideal pH electrode changes by 59.16 mV for every pH-unit change of analyte activity at 25°C*.

The response of real glass electrodes is described by the Nernst-like equation

The number 0.059 16 V is (RT/F) ln 10, where R is the gas constant, T is temperature, and F is the Faraday constant.

Response of glass electrode:

$$E = \text{constant} + \beta(0.059\ 16) \log \mathcal{A}_{H^+}(\text{outside})$$

$$E = \text{constant} - \beta(0.059\ 16)\ \text{pH(outside)} \quad \text{(at 25°C)} \quad (14\text{-}6)$$

10 nm   0.1 mm (10^5 nm)   10 nm

Internal solution ($\mathcal{A}_{H^+} \approx 0.1$)

External solution ($\mathcal{A}_{H^+}$ is variable)

Hydrated gel layer

Dry glass layer

Hydrated gel layer

(Exchange sites occupied by H^+ and Na^+)

(All sites occupied by Na^+)

(Exchange sites occupied by H^+ and Na^+)

FIGURE 14-14 Schematic cross section of the glass membrane of a pH electrode.

H^+ ^-O O^- H^+

Glass

Internal solution

External solution

FIGURE 14-15 Ion-exchange equilibria on surfaces of a glass membrane: H^+ replaces metal ions bound to the negatively charged oxygen atoms. The pH of the internal solution is fixed. As the pH of the external solution (the sample) changes, the electric potential difference across the glass membrane changes.

The value of β, the *electromotive efficiency*, is close to 1.00 (typically >0.98). We measure the constant and β when we calibrate the electrode in solutions of known pH.

Calibrating a Glass Electrode

A pH electrode should be calibrated with two (or more) standard buffers selected so that the pH of the unknown lies within the range of the standards. Standards in Table 14-3 are accurate to ±0.01 pH unit.[13]

When you calibrate an electrode with standard buffers, you measure a voltage with the electrode in each buffer (Figure 14-16). The pH of buffer S1 is pH_{S1} and the measured electrode potential in this buffer is E_{S1}. The pH of buffer S2 is pH_{S2} and the measured electrode potential is E_{S2}. The equation of the line through the two standard points is

$$\frac{E - E_{S1}}{pH - pH_{S1}} = \frac{E_{S2} - E_{S1}}{pH_{S2} - pH_{S1}} \qquad (14\text{-}7)$$

The slope of the line is $\Delta E/\Delta pH = (E_{S2} - E_{S1})/(pH_{S2} - pH_{S1})$, which is 59.16 mV/pH unit at 25°C for an ideal electrode and $\beta(59.16)$ mV/pH unit for a real electrode, where β is the correction factor in Equation 14-6.

To measure the pH of an unknown, measure the potential of the unknown with the calibrated electrode and find pH by substitution in Equation 14-7:

$$\frac{E_{unknown} - E_{S1}}{pH_{unknown} - pH_{S1}} = \frac{E_{S2} - E_{S1}}{pH_{S2} - pH_{S1}} \qquad (14\text{-}8)$$

Alas, modern pH meters are "black boxes" that do these calculations for us by applying Equations 14-7 and 14-8 and automatically displaying pH.

Before using a pH electrode, be sure that the air inlet near the upper end of the electrode in Figure 14-11 is not capped. (This hole is capped during storage to prevent evaporation of the reference electrode filling solution.) Wash the electrode with distilled water and gently *blot* it dry with a tissue. Do not *wipe* it, because this action might produce a static charge on the glass.

To calibrate the electrode, dip it in a standard buffer whose pH is near 7 and allow the electrode to equilibrate with stirring for at least a minute. Following the manufacturer's instructions, press a key that might say "Calibrate" or "Read" on a microprocessor-controlled meter or adjust the reading of an analog meter to indicate the pH of the standard buffer. Then wash the electrode with water, blot it dry, and immerse it in a second standard whose pH is further from 7 than the pH of the first standard. Enter the second buffer on the meter. Finally, dip the electrode in the unknown, stir the liquid, allow the reading to stabilize, and read the pH.

Store a glass electrode in aqueous solution to prevent dehydration of the glass. Ideally, the solution should be similar to that inside the reference compartment of the electrode. If the electrode has dried, recondition it in dilute acid for several hours. If the electrode is to be used above pH 9, soak it in a high-pH buffer. (The field effect transistor pH electrode in Section 14-8 is stored dry. Prior to use, scrub it gently with a soft brush and soak it in pH 7 buffer for 10 min.)

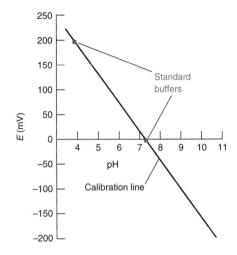

FIGURE 14-16 Two-point calibration of a pH electrode.

Do not leave a glass electrode out of water (or in a nonaqueous solvent) any longer than necessary.

If electrode response becomes sluggish or if an electrode cannot be calibrated properly, try soaking it in 6 M HCl, followed by water. As a last resort, soak the electrode in 20 wt% aqueous ammonium bifluoride, NH_4HF_2, for 1 min in a plastic beaker. This reagent dissolves glass and exposes fresh surface. Wash the electrode with water and try calibrating it again. Avoid contact with ammonium bifluoride, which produces a painful HF burn.

Errors in pH Measurement

1. *Standards*. A pH measurement cannot be more accurate than our standards, which are typically ±0.01 pH unit.
2. *Junction potential*. A *junction potential* exists at the porous plug near the bottom of the electrode in Figure 14-11. If the ionic composition of the analyte solution is different from that of the standard buffer, the junction potential will change *even if the pH of the two solutions is the same* (Box 14-1, page 322). This effect gives an uncertainty of at least ~0.01 pH unit.
3. *Junction potential drift*. Most combination electrodes have a Ag | AgCl reference electrode containing saturated KCl solution. More than 350 mg Ag per liter dissolve in the KCl, mainly as $AgCl_4^{3-}$ and $AgCl_3^{2-}$. In the porous plug, KCl is diluted and AgCl can precipitate. If analyte solution contains a reducing agent, Ag(*s*) also can precipitate

The apparent pH will change if the ionic composition of the analyte changes, even when the actual pH is constant.

TABLE 14-3 pH values of National Institute of Standards and Technology buffers

Temperature (°C)	0.05 m potassium tetroxalate (1)	Saturated (25°C) potassium hydrogen tartrate (2)	0.05 m potassium dihydrogen citrate (3)	0.05 m potassium hydrogen phthalate (4)	0.08 m MOPSO 0.08 m NaMOPSO 0.08 m NaCl (5)
0	1.667	—	3.863	4.003	7.268
5	1.666	—	3.840	3.999	7.182
10	1.665	—	3.820	3.998	7.098
15	1.669	—	3.802	3.999	7.018
20	1.672	—	3.788	4.002	6.940
25	1.677	3.557	3.776	4.008	6.865
30	1.681	3.552	3.766	4.015	6.792
35	1.688	3.549	3.759	4.024	6.722
37	—	3.548	3.756	4.028	6.695
40	1.694	3.547	3.753	4.035	6.654
45	1.699	3.547	3.750	4.047	6.588
50	1.706	3.549	3.749	4.060	6.524
55	1.713	3.554	—	4.075	—
60	1.722	3.560	—	4.091	—
70	—	3.580	—	4.126	—
80	—	3.609	—	4.164	—
90	—	3.650	—	4.205	—
95	—	3.674	—	4.227	—

NOTE: *The designation m stands for molality. Masses in the buffer recipes below are apparent masses measured in air.*

In the buffer solution preparations, it is essential to use high-purity materials and freshly distilled or deionized water of resistivity greater than 2 000 ohm · m. Solutions having pH 6 or above should be stored in plastic containers, preferably ones with an NaOH trap to prevent ingress of atmospheric carbon dioxide. They can normally be kept for 2–3 weeks, or slightly longer in a refrigerator. Buffer materials in this table are available as Standard Reference Materials from the National Institute of Standards and Technology. http://ts.nist.gov/srm. pH standards for D₂O and aqueous-organic solutions can be found in P. R. Mussini, T Mussini, and S. Rondinini, Pure Appl. Chem. 1997, 69, 1007.

1. 0.05 m potassium tetroxalate, $KHC_2O_4 \cdot H_2C_2O_4$. Dissolve 12.71 g of potassium tetroxalate dehydrate (Standard Reference Material used without drying) in 1 kg of water. pH values from P. M. Juusola, J. I. Partanen, K. P. Vahteristo, P. O. Minkkinen, and A. K. Covington, J. Chem. Eng. Data 2007, 52, 973.

2. Saturated (25°C) potassium hydrogen tartrate, $KHC_4H_4O_6$. An excess of the salt is shaken with water, and it can be stored in this way. Before use, it should be filtered or decanted at a temperature between 22° and 28°C.

3. 0.05 m potassium dihydrogen citrate, $KH_2C_6H_5O_7$. Dissolve 11.41 g of the salt in 1 L of solution at 25°C.

4. 0.05 m potassium hydrogen phthalate. Although this is not usually essential, the crystals may be dried at 100°C for 1 h, then cooled in a desiccator. At 25°C, 10.12 g $C_6H_4(CO_2H)(CO_2K)$ are dissolved in water, and the solution made up to 1 L.

5. 0.08 m MOPSO ((3-N-morpholino)-2-hydroxypropanesulfonic acid, Table 8-2), 0.08 m sodium salt of MOPSO, 0.08 m NaCl. Buffers 5 and 7 are recommended for 2-point standardization of electrodes for pH measurements of physiologic fluids. MOPSO is crystallized twice from 70 wt% ethanol and dried in vacuum at 50°C for 24 h. NaCl is dried at 110°C for 4 h. Na^+MOPSO^- may be prepared by neutralization of MOPSO with standard NaOH. The sodium salt is also available as a Standard Reference Material. Dissolve 18.00 g MOPSO, 19.76 g Na^+MOPSO^-, and 4.674 g NaCl in 1.000 kg H_2O.

> **Challenge** Use Equation 14-6 to show that the potential of the glass electrode changes by 1.3 mV when $\mathcal{A}_{H^+}$ changes by 5.0%. Show that 1.3 mV = 0.02 pH unit.
>
> *Moral:* A small uncertainty in voltage (1.3 mV) or pH (0.02 unit) corresponds to a large uncertainty (5%) in analyte concentration. Similar uncertainties arise in other potentiometric measurements.

in the plug. Both effects change the junction potential, causing a slow drift of the pH reading (solid colored circles in Figure 14-17). You can compensate for this error by recalibrating the electrode every 2 h.

4. *Sodium error.* When [H$^+$] is very low and [Na$^+$] is high, the electrode responds to Na$^+$ and the apparent pH is lower than the true pH. This is called the **sodium error** or *alkaline error* (Figure 14-18).

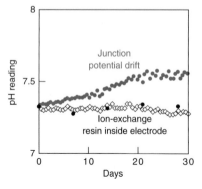

FIGURE 14-17 Solid colored circles show the drift in apparent pH of a low-conductivity industrial water supply measured continuously by a single electrode. Individual measurements with a freshly calibrated electrode (black circles) demonstrate that the pH is not drifting. Drift is attributed to slow clogging of the electrode's porous plug with AgCl(s). When a cation-exchange resin was placed inside the reference electrode near the porous plug, Ag(I) was bound by the resin and did not precipitate. This electrode gave the drift-free, continuous reading shown by open diamonds. [From S. Ito, H. Hachiya, K. Baba, Y. Asano, and H. Wada, "Improvement of the Ag | AgCl Reference Electrode and Its Application to pH Measurement," *Talanta* **1995**, *42*, 1685.]

0.025 *m* potassium dihydrogen phosphate 0.025 *m* disodium hydrogen phosphate (6)	0.08 *m* HEPES 0.08 *m* NaHEPES 0.08 *m* NaCl (7)	0.008 695 *m* potassium dihydrogen phosphate 0.030 43 *m* disodium hydrogen phosphate (8)	0.01 *m* borax (9)	0.025 *m* sodium bicarbonate 0.025 *m* sodium carbonate (10)
6.984	7.853	7.534	9.464	10.317
6.951	7.782	7.500	9.395	10.245
6.923	7.713	7.472	9.332	10.179
6.900	7.645	7.448	9.276	10.118
6.881	7.580	7.429	9.225	10.062
6.865	7.516	7.413	9.180	10.012
6.853	7.454	7.400	9.139	9.966
6.844	7.393	7.389	9.102	9.925
6.840	7.370	7.385	9.088	9.910
6.838	7.335	7.380	9.068	9.889
6.834	7.278	7.373	9.038	9.856
6.833	7.223	7.367	9.011	9.828
6.834	—	—	8.985	—
6.836	—	—	8.962	—
6.845	—	—	8.921	—
6.859	—	—	8.885	—
6.877	—	—	8.850	—
6.886	—	—	8.833	—

6. 0.025 m disodium hydrogen phosphate, 0.025 m potassium dihydrogen phosphate. The anhydrous salts are best; each should be dried for 2 h at 120°C and cooled in a desiccator, because they are slightly hygroscopic. Higher drying temperatures should be avoided to prevent formation of condensed phosphates. Dissolve 3.53 g Na_2HPO_4 and 3.39 g KH_2PO_4 in water to give 1 L of solution at 25°C.

7. 0.08 m HEPES (N-2-hydroxyethylpiperazine-N'-2-ethanesulfonic acid, Table 8-2), 0.08 m sodium salt of HEPES, 0.08 m NaCl. Buffers 5 and 7 are recommended for 2-point standardization of electrodes for pH measurements of physiologic fluids. HEPES is crystallized twice from 80 wt% ethanol and dried in vacuum at 50°C for 24 h. NaCl is dried at 110°C for 4 h. Na^+HEPES^- may be prepared by neutralization of HEPES with standard NaOH. The sodium salt is also available as a Standard Reference Material. Dissolve 19.04 g HEPES, 20.80 g Na^+HEPES^-, and 4.674 g NaCl in 1.000 kg H_2O.

8. 0.008 695 m potassium dihydrogen phosphate, 0.030 43 m disodium hydrogen phosphate. Prepare like Buffer 6; dissolve 1.179 g KH_2PO_4 and 4.30 g Na_2HPO_4 in water to give 1 L of solution at 25°C.

9. 0.01 m sodium tetraborate decahydrate. Dissolve 3.80 g $Na_2B_4O_7 \cdot 10H_2O$ in water to give 1 L of solution. This borax solution is particularly susceptible to pH change from carbon dioxide absorption, and it should be correspondingly protected.

10. 0.025 m sodium bicarbonate, 0.025 m sodium carbonate. Primary standard grade Na_2CO_3 is dried at 250°C for 90 min and stored over $CaCl_2$ and Drierite. Reagent-grade $NaHCO_3$ is dried over molecular sieves and Drierite for 2 days at room temperature. Do not heat $NaHCO_3$, or it may decompose to Na_2CO_3. Dissolve 2.092 g of $NaHCO_3$ and 2.640 g of Na_2CO_3 in 1 L of solution at 25°C.

*SOURCES: R. G. Bates, J. Res. Natl. Bureau Stds. **1962**, 66A, 179; B. R. Staples and R. G. Bates, J. Res. Natl. Bureau Stds. **1969**, 73A, 37. Data on HEPES and MOPSO are from Y. C Wu, P. A. Berezansky, D. Feng, and W. F. Koch, Anal. Chem. **1993**, 65, 1084, and D. Feng, W. F Koch, and Y. C Wu, Anal. Chem. **1989**, 61, 1400. Instructions for preparing some of these solutions are from G. Mattock in C. N. Reilley, ed., Advances in Analytical Chemistry and Instrumentation (New York: Wiley, 1963), Vol. 2, p. 45. See also R. G. Bates, Determination of pH: Theory and Practice, 2nd ed. (New York: Wiley, 1973), Chap. 4.*

5. *Acid error.* In strong acid, the measured pH is higher than the actual pH (Figure 14-18), perhaps because the glass is saturated with H^+ and cannot be further protonated.

6. *Equilibration time.* It takes time for an electrode to equilibrate with a solution. A well-buffered solution requires ~30 s with adequate stirring. A poorly buffered solution (such as one near the equivalence point of a titration) needs many minutes.

7. *Hydration of glass.* A dry electrode requires several hours of soaking before it responds to H^+ correctly.

8. *Temperature.* A pH meter should be calibrated at the same temperature at which the measurement will be made.

9. *Cleaning.* If an electrode has been exposed to a hydrophobic liquid, such as oil, it should be cleaned with a solvent that will dissolve the liquid and then conditioned well in aqueous solution. The reading of an improperly cleaned electrode can drift for hours while the electrode re-equilibrates with aqueous solution.

Errors 1 and 2 limit the accuracy of pH measurement with the glass electrode to ± 0.02 pH unit, at best. Measurement of pH *differences* between solutions can be accurate to about ± 0.002 pH unit, but knowledge of the true pH will still be at least an order of magnitude more uncertain. An uncertainty of ± 0.02 pH unit corresponds to an uncertainty of $\pm 5\%$ in $\mathcal{A}_{H^+}$.

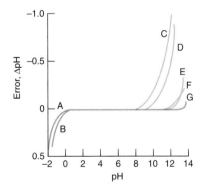

FIGURE 14-18 Acid and alkaline errors of some glass electrodes. A: Corning 015, H_2SO_4. B: Corning 015, HCl. C: Corning 015, 1 M Na^+. D: Beckman-GP, 1 M Na^+. E: L & N Black Dot, 1 M Na^+. F: Beckman Type E, 1 M Na^+. G: Ross electrode.[19] [From R. G. Bates, *Determination of pH: Theory and Practice,* 2nd ed. (New York: Wiley, 1973). Ross electrode data is from Orion *Ross pH Electrode Instruction Manual.*]

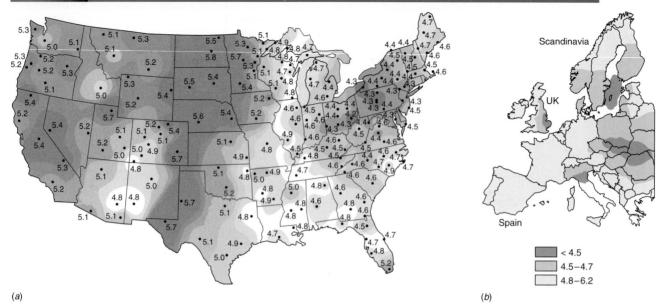

(a)

(b)

> < 4.5
> 4.5–4.7
> 4.8–6.2

(a) pH of precipitation in the United States in 2001. The lower the pH, the more acidic the water. [From National Atmospheric Deposition Program (NRSP-3)/National Trends Network (2002). Illinois State Water Survey, 2204 Griffith Dr., Champaign, IL 61820. See also http://nadp.sws.uiuc.edu and www.epa.gov/acidrain]

(b) pH of rain in Europe. No values are reported for Italy and Greece. [From H. Rodhe, F. Dentener, and M. Schulz, *Environ. Sci. Technol.* 2002, *36*, 4382.]

Combustion products from automobiles and factories include nitrogen oxides and sulfur dioxide, which react with water in the atmosphere to produce acids.[14]

$$SO_2 + H_2O \rightarrow H_2SO_3 \xrightarrow{\text{oxidation}} H_2SO_4$$
$$\underset{\text{Sulfurous acid}}{} \quad \underset{\text{Sulfuric acid}}{}$$

Acid rain in North America is most severe in the East, downwind of many coal-fired power plants. In the three-year period 1995–1997, after SO_2 emissions were limited by a new law, concentrations of SO_4^{2-} and H^+ in precipitation decreased by 10–25% in the eastern United States.[15]

Acid rain threatens lakes and forests throughout the world. Monitoring the pH of rainwater is a critical component of programs to measure and reduce the production of acid rain.

To identify systematic errors in the measurement of pH of rainwater, a careful study was conducted with 17 laboratories.[16] Eight samples were provided to each laboratory, along with instructions on how to conduct the measurements. Each laboratory used two buffers to standardize pH meters. Sixteen laboratories successfully measured the pH of Unknown A (within ±0.02 pH unit), which was 4.008 at 25°C. One lab whose measurement was 0.04 pH unit low had a faulty commercial standard buffer.

Panel (c) shows typical results for the pH of rainwater. The average of the 17 measurements is given by the horizontal line at pH 4.14, and the letters s, t, u, v, w, x, y, and z identify types of pH electrodes. Types s and w had relatively large systematic errors. The type s electrode was a combination electrode (Figure 14-11) containing a reference electrode liquid junction with an exceptionally large area. Electrode type w had a reference electrode filled with a gel.

A hypothesis was that variations in the liquid junction potential (Section 14-3) led to variations among the pH measurements. Standard buffers have ionic strengths of 0.05 to 0.1 M, whereas rainwater samples have ionic strengths two or more orders of magnitude lower. To see whether junction potential caused systematic errors, 2×10^{-4} M HCl was used as a pH standard in place of high ionic strength buffers. Panel (d) shows good results from all but the first lab. The standard deviation of 17 measurements was reduced from 0.077 pH unit (with standard buffer) to 0.029 pH unit (with HCl standard). It was concluded that junction potential caused most of the variability between labs and that a low ionic strength standard is appropriate for rainwater pH measurements.[17,18]

(c)

(c) pH of rainwater from identical samples measured at 17 different labs using standard buffers for calibration. Letters designate different types of pH electrodes.

(d)

(d) Rainwater pH measured by using low ionic strength HCl for calibration.

Not All pH Electrodes Are Glass

Glass electrodes are the most common, but not the only means, for measuring pH. Solid-state pH electrodes based on the field effect transistor are described at the end of this chapter. Liquid-based ion-selective electrodes for H^+ are described in Section 14-6.

An anhydrous IrO_2 layer formed by oxidation of iridium wire responds to pH by a half-reaction that might be[20]

$$IrO_2(s) + H^+ + e^- \rightleftharpoons IrOOH(s)$$

$$E = E° - 0.059\ 16 \log\left(\frac{1}{[H^+]}\right) = E° - 0.059\ 16\ pH$$

Other metal oxide electrodes have been used under extreme conditions. For example, a ZrO_2 electrode can measure pH up to 300°C.[21]

The *Phoenix Mars Lander* described at the opening of this chapter had two liquid-based ion-selective electrodes in each Wet Chemistry Lab to measure the pH of Martian soil suspended in water. It was not certain that these electrodes would survive the temperatures and pressures encountered during the mission, so a rugged IrO_2 pH electrode was added. The IrO_2 electrode remains accurate at pH > 9, a condition at which ion-selective electrodes were nonresponsive.

14-6 Ion-Selective Electrodes[22, 23]

A critically ill patient is wheeled into the emergency room, and the doctor needs blood chemistry information quickly to help her make a diagnosis. Analytes in Table 14-4 are part of the critical care profile of blood chemistry. Every analyte in the table can be measured by electrochemical means. Ion-selective electrodes are the method of choice for Na^+, K^+, Cl^-, pH, and P_{CO_2}. The "Chem 7" test constitutes up to 70% of tests performed in the hospital lab. It measures Na^+, K^+, Cl^-, total CO_2, glucose, urea, and creatinine, four of which are analyzed with ion-selective electrodes.

Most ion-selective electrodes fall into one of the following classes:

1. *Glass membranes* for H^+ and certain monovalent cations
2. *Solid-state electrodes* based on inorganic crystals or, recently, conductive polymers
3. *Liquid-based electrodes* using a hydrophobic polymer membrane saturated with a hydrophobic liquid ion exchanger
4. *Compound electrodes* with an analyte-selective electrode enclosed by a membrane that separates analyte from other species or that generates analyte in a chemical reaction.

Reminder: How Ion-Selective Electrodes Work

In Figure 14-10, analyte ions equilibrate with ion-exchange sites at the outer surface of the ion-selective membrane. Diffusion of analyte ions out of the membrane creates a slight charge imbalance (an electric potential difference) across the interface between the membrane and the analyte solution. Changes in analyte ion concentration in the solution change the potential difference across the outer boundary of the ion-selective membrane. A calibration curve relates the potential difference to analyte concentration.

An ion-selective electrode responds to the activity of *free analyte*, not complexed analyte. For example, when Pb^{2+} in tap water at pH 8 was measured with an ion-selective electrode, the result was $[Pb^{2+}] = 2 \times 10^{-10}$ M.[24] When lead in the same tap water was measured by inductively coupled plasma–mass spectrometry (Section 20-6), the result was more than 10 times greater: 3×10^{-9} M. The discrepancy arose because the inductively coupled plasma measures *all* lead and the ion-selective electrode measures *free* Pb^{2+}. In tap water at pH 8, much of the lead is complexed by CO_3^{2-}, OH^-, and other anions. When the pH of tap water was adjusted to 4, Pb^{2+} dissociated from its complexes and the concentration indicated by the ion-selective electrode was 3×10^{-9} M—the same value found by inductively coupled plasma.

Selectivity Coefficient

No electrode responds exclusively to one kind of ion, but the glass pH electrode is among the most selective. Sodium ion is the principal interfering species. Its effect on the pH reading is only significant when $[H^+] \lesssim 10^{-12}$ M and $[Na^+] \gtrsim 10^{-2}$ M (Figure 14-18).

TABLE 14-4 Critical care profile

Function	Analyte
Conduction	K^+, Ca^{2+}
Contraction	Ca^{2+}, Mg^{2+}
Energy level	Glucose, P_{O_2}, lactate, hematocrit
Ventilation	P_{O_2}, P_{CO_2}
Perfusion	Lactate, $SO_2\%$, hematocrit
Acid-base	pH, P_{CO_2}, HCO_3^-
Osmolality	Na^+, glucose
Electrolyte balance	Na^+, K^+, Ca^{2+}, Mg^{2+}
Renal function	Blood urea nitrogen, creatinine

SOURCE: C. C. Young, "Evolution of Blood Chemistry Analyzers Based on Ion Selective Electrodes," J. Chem. Ed. *1997*, 74, 177.

The United States conducts more than 200 million clinical assays of K^+ each year with ion-selective electrodes.

Analyte ions establish an ion-exchange equilibrium at the surface of the ion-selective membrane. Other ions that bind to the same sites interfere.

The ion-selective electrode responds to Pb^{2+} with little response to $Pb(OH)^+$ or $Pb(CO_3)(aq)$.

An electrode intended to measure ion A also responds to ion X. The **selectivity coefficient** gives the relative response to different species with the same charge:

Selectivity coefficient: $$K_{A,X}^{Pot} = \frac{\text{response to X}}{\text{response to A}} \qquad (14\text{-}9)$$

The superscript "Pot" for "potentiometric" is customary in the chemical literature. The smaller the selectivity coefficient, the less the interference by X. A K^+ ion-selective electrode that uses the chelator valinomycin as the liquid ion exchanger has selectivity coefficients $K_{K^+,Na^+}^{Pot} = 1 \times 10^{-5}$, $K_{K^+,Cs^+}^{Pot} = 0.44$, and $K_{K^+,Rb^+}^{Pot} = 2.8$. These coefficients tell us that Na^+ hardly interferes with the measurement of K^+, but Cs^+ and Rb^+ interfere strongly. In fact, the electrode responds more to Rb^+ than to K^+.

If the response to each ion is Nernstian, then the response of an ion-selective electrode to its primary ion (A) and to interfering ions of the *same charge* (X) is[10,25]

Response of ion-selective electrode: $$E = \text{constant} \pm \frac{0.059\,16}{z_A} \log\left[\mathcal{A}_A + \sum_X K_{A,X}^{Pot} \mathcal{A}_X \right] \qquad (14\text{-}10)$$

where z_A is the magnitude of the charge of A, $\mathcal{A}_A$ and $\mathcal{A}_X$ are activities, and $K_{A,X}^{Pot}$ is the selectivity coefficient for each interfering ion. If the ion-selective electrode is connected to the positive terminal of the potentiometer, the sign of the log term is positive if A is a cation and negative if A is an anion. Box 14-2 describes how selectivity coefficients are measured.

BOX 14-2 Measuring Selectivity Coefficients for an Ion-Selective Electrode

When measuring selectivity coefficients, you must *demonstrate* that the response of the electrode to each interfering ion follows the Nernst equation.[26,27,28] This is not as simple as it sounds. An ion-selective membrane that has equilibrated with its primary ion may become kinetically insensitive to weakly bound interfering ions.

The graph shows the *separate solution method* for measuring selectivity coefficients. In this method, a calibration curve is

Measurement of selectivity coefficients of Na^+ ion-selective electrode. Activities on the abscissa were calculated from concentrations and activity coefficients. [Adapted from E. Bakker, "Determination of Unbiased Selectivity Coefficients of Neutral Carrier-Based Cation-Selective Electrodes," *Anal. Chem.* 1997, *69*, 1061.]

constructed for each ion by itself. Other common procedures are the *fixed interference method* and the *matched potential method*.[26]

The graph shows the response of a sodium ion-selective electrode to the interfering ions K^+, Ca^{2+}, and Mg^{2+}. To obtain a Nernstian response to interfering ions, the electrode was prepared in the absence of Na^+. The electrode was filled with 0.01 M KCl and soaked in 0.01 M KCl overnight to condition the ion-selective membrane prior to measurements. After measuring K^+, Ca^{2+}, and Mg^{2+}, Na^+ was measured. For subsequent use to measure Na^+, the inner filling solution is replaced by 0.01 M NaCl.

The data demonstrate a near-Nernstian response for each ion. At the laboratory temperature of 21.5°C, Nernstian response would be $(RT \ln 10)/zF = 58.5/z$ mV per decade (factor of 10) of ion activity, where z is the charge of the ion. Measured slopes are 61.3 ± 1.5 mV for Na^+, 56.3 ± 0.6 mV for K^+, 26.0 ± 1.0 mV for Mg^{2+}, and 31.2 ± 0.7 mV for Ca^{2+}. The deviation of Ca^{2+} from a straight line above an activity of $10^{-2.5}$ is attributed to Na^+ impurity in high-purity $CaCl_2$. Electrode response to Na^+ is much greater than to Ca^{2+}, so a little Na^+ has a large effect.

To find the selectivity coefficient, we measure the difference between the Na^+ calibration line and the line for an interfering ion at any chosen activity and apply the formula

$$\log K_{A,X}^{Pot} = \frac{z_A F(E_X - E_A)}{RT \ln 10} + \log\left(\frac{\mathcal{A}_A}{(\mathcal{A}_X)^{z_A/z_X}} \right) \qquad (14\text{-}11)$$

where A = Na^+ with charge $z_A = 1$ and X is an interfering ion of charge z_X. At an activity of 10^{-3}, the dashed line shows a difference of $E_{Ca^{2+}} - E_{Na^+} = -363$ mV. The selectivity coefficient is

$$\log K_{Na^+,Ca^{2+}}^{Pot} = \frac{(+1)F(-0.363\ \text{V})}{RT \ln 10} + \log\left(\frac{10^{-3}}{(10^{-3})^{1/2}} \right) = -7.0$$

We could choose a different activity to measure $E_{Ca^{2+}} - E_{Na^+}$ and should find the same $K_{Na^+,Ca^{2+}}^{Pot}$. The other lines in the graph give $\log K_{Na^+,Mg^{2+}}^{Pot} = -8.0$ and $\log K_{Na^+,K^+}^{Pot} = -4.9$.

Problem 14-43 provides a formula for estimating the error in measuring primary ion A caused by interfering ion X, which need not have the same charge as A.

Using the Selectivity Coefficient

A fluoride ion-selective electrode has a selectivity coefficient $K^{Pot}_{F^-,OH^-} = 0.1$. What will be the change in electrode potential when 1.0×10^{-4} M F^- at pH 5.5 is raised to pH 10.5?

Solution From Equation 14-10, we find the potential with negligible OH^- at pH 5.5:

$$E = constant - 0.059\ 16 \log[1.0 \times 10^{-4}] = constant + 236.6\ mV$$

At pH 10.50, $[OH^-] = 3.2 \times 10^{-4}$ M, so the electrode potential is

$$E = constant - 0.059\ 16 \log[1.0 \times 10^{-4} + (0.1)(3.2 \times 10^{-4})]$$

$$= constant + 229.5\ mV$$

The change is $229.5 - 236.6 = -7.1$ mV, which is quite significant. If you didn't know about the pH change, you would think that the concentration of F^- had increased by 32%.

Test Yourself Find the change in potential when 1.0×10^{-4} M F^- at pH 5.5 is raised to pH 9.5. (*Answer:* -0.8 mV)

Solid-State Electrodes

A **solid-state ion-selective electrode** based on an inorganic crystal is shown in Figure 14-19. A common electrode of this type is the fluoride electrode, employing a crystal of LaF_3 doped with Eu^{2+}. *Doping* means adding a small amount of Eu^{2+} in place of La^{3+}. The filling solution contains 0.1 M NaF and 0.1 M NaCl. Fluoride electrodes are used to monitor the fluoridation of municipal water supplies.

To conduct a tiny electric current, F^- migrates across the LaF_3 crystal, as shown in Figure 14-20. Anion vacancies are created within the crystal when we dope LaF_3 with EuF_2. An adjacent fluoride ion can jump into the vacancy, leaving a new vacancy behind. In this manner, F^- diffuses from one side to the other.

By analogy with the pH electrode, the response of the F^- electrode is

Response of F^- electrode: $\quad E = constant - \beta(0.059\ 16) \log \mathcal{A}_{F^-}(\text{outside})$ $\quad$ (14-12)

where β is close to 1.00. Equation 14-12 has a negative sign before the log term because fluoride is an anion. The F^- electrode has a nearly Nernstian response over a F^- concentration range from about 10^{-6} M to 1 M (Figure 14-21). The electrode is more responsive to F^- than to most other ions by $>1\ 000$. The only interfering species is OH^-, for which the selectivity coefficient is $K^{Pot}_{F^-,OH^-} = 0.1$. At low pH, F^- is converted to HF ($pK_a = 3.17$), to which the electrode is insensitive.

A routine procedure for measuring F^- is to dilute the unknown in a high ionic strength buffer containing acetic acid, sodium citrate, NaCl, and NaOH to adjust the pH to 5.5. The

FIGURE 14-19 Diagram of an ion-selective electrode employing an inorganic salt crystal as the ion-selective membrane.

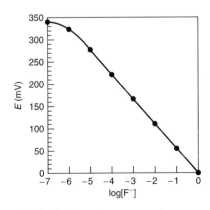

FIGURE 14-21 Calibration curve for fluoride ion-selective electrode. [Data from M. S. Frant and J. W. Ross, Jr., "Electrode for Sensing Fluoride Ion Activity in Solution," *Science* 1966, *154*, 1553.]

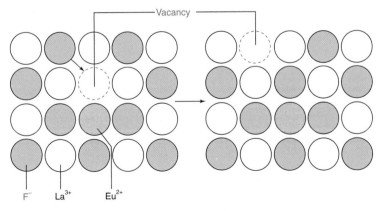

FIGURE 14-20 Migration of F^- through LaF_3 doped with EuF_2. Because Eu^{2+} has less charge than La^{3+}, an anion vacancy occurs for every Eu^{2+}. A neighboring F^- can jump into the vacancy, thereby moving the vacancy to another site. Repetition of this process moves F^- through the lattice.

buffer keeps all standards and unknowns at a constant ionic strength, so the fluoride activity coefficient is constant in all solutions (and can therefore be ignored).

$$E = \text{constant} - \beta(0.059\ 16)\log[F^-]\gamma_{F^-}$$

$$= \underbrace{\text{constant} - \beta(0.059\ 16)\log \gamma_{F^-}}_{\substack{\text{This expression is constant because} \\ \gamma_{F^-} \text{ is constant at constant ionic strength}}} - \beta(0.059\ 16)\log[F^-]$$

At pH 5.5, there is no interference by OH^- and little conversion of F^- to HF. Citrate complexes Fe^{3+} and Al^{3+}, which would otherwise bind F^- and interfere with the analysis.

EXAMPLE   **Response of an Ion-Selective Electrode**

When a fluoride electrode was immersed in standard solutions (maintained at a constant ionic strength of 0.1 M with $NaNO_3$), the following potentials (versus S.C.E.) were observed:

$[F^-]$ (M)	E (mV)
1.00×10^{-5}	100.0
1.00×10^{-4}	41.5
1.00×10^{-3}	-17.0

Because the ionic strength is constant, the response should depend on the logarithm of the F^- *concentration*. Find $[F^-]$ in an unknown that gave a potential of 0.0 mV.

Solution We fit the calibration data with Equation 14-12:

$$E = m \underbrace{\log[F^-]}_{x} + \underset{y}{b}$$

Plotting E versus $\log[F^-]$ gives a straight line with a slope $m = -58.5$ mV and a y-intercept $b = -192.5$ mV. Setting $E = 0.0$ mV, we solve for $[F^-]$:

$$0.0\ \text{mV} = (-58.5\ \text{mV})\log[F^-] - 192.5\ \text{mV} \Rightarrow [F^-] = 5.1 \times 10^{-4}\ \text{M}$$

Test Yourself   Find $[F^-]$ if $E = 81.2$ mV. Is the calibration curve valid for $E = 110.7$ mV? (***Answer:*** 2.1×10^{-4} M; no, because calibration points do not go above 100 mV)

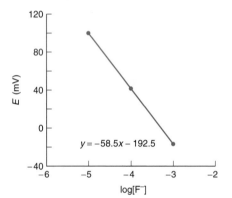

Another common inorganic crystal electrode uses Ag_2S for the membrane. This electrode responds to Ag^+ and to S^{2-}. If we dope the electrode with CuS, CdS, or PbS, we can prepare electrodes sensitive to Cu^{2+}, Cd^{2+}, or Pb^{2+}, respectively (Table 14-5).

Figure 14-22 illustrates the mechanism by which a CdS crystal responds selectively to certain ions. The crystal can be cleaved to expose planes of Cd atoms or S atoms. The Cd plane in Figure 14-22a selectively adsorbs HS^- ions, whereas an S plane does not interact strongly with HS^-. Figure 14-22b shows strong response of the exposed Cd face to HS^-, but only weak response when the S face is exposed. The opposite behavior is observed in response to Cd^{2+} ions. The partial response of the S face to HS^- in the upper curve is attributed to about 10% of the exposed atoms actually being Cd instead of S.

TABLE 14-5   **Properties of solid-state ion-selective electrodes**

Ion	Concentration range (M)	Membrane material	pH range	Interfering species
F^-	10^{-6}–1	LaF_3	5–8	OH^-(0.1 M)
Cl^-	10^{-4}–1	AgCl	2–11	$CN^-, S^{2-}, I^-, S_2O_3^{2-}, Br^-$
Br^-	10^{-5}–1	AgBr	2–12	CN^-, S^{2-}, I^-
I^-	10^{-6}–1	AgI	3–12	S^{2-}
SCN^-	10^{-5}–1	AgSCN	2–12	$S^{2-}, I^-, CN^-, Br^-, S_2O_3^{2-}$
CN^-	10^{-6}–10^{-2}	AgI	11–13	S^{2-}, I^-
S^{2-}	10^{-5}–1	Ag_2S	13–14	

(a)　　　　　　　　　　　　　　　　　(b)

FIGURE 14-22 (a) Crystal structure of hexagonal CdS, showing alternating planes of Cd and S along the vertical axis (the crystal c-axis). HS^- is shown adsorbed to the uppermost Cd plane. (b) Potentiometric response of exposed crystal faces to HS^-. [From K. Uosaki, Y. Shigematsu, H. Kita, Y. Umezawa, and R. Souda, "Crystal-Face-Specific Response of a Single-Crystal Cadmium Sulfide Based Ion-Selective Electrode," *Anal. Chem.* **1989**, *61*, 1980.]

Liquid-Based Ion-Selective Electrodes

A **liquid-based ion-selective electrode** is similar to the solid-state electrode in Figure 14-19, except that the liquid-based electrode has a hydrophobic membrane impregnated with a hydrophobic ion exchanger (called an *ionophore*) that is selective for analyte ion (Figure 14-23). The response of a Ca^{2+} ion-selective electrode is given by

Response of Ca^{2+} electrode:　　$E = \text{constant} + \beta\left(\dfrac{0.059\ 16}{2}\right)\log \mathcal{A}_{Ca^{2+}}(\text{outside})$　　**(14-13)**

where β is close to 1.00. Equations 14-13 and 14-12 have different signs before the log term because one involves an anion and the other a cation. Note also that the charge of Ca^{2+} requires a factor of 2 in the denominator before the logarithm.

The membrane at the bottom of the electrode in Figure 14-23 is made from poly(vinyl chloride) impregnated with ion exchanger. One particular Ca^{2+} liquid ion exchanger consists

To potentiometer

Filling solution (e.g., 0.01 M $CaCl_2$ saturated with AgCl for Ca^{2+} ion-selective electrode

Silver-silver chloride electrode

Hydrophobic polymer membrane containing hydrophobic ion exchanger and ion-selective ionophore

FIGURE 14-23 Calcium ion-selective electrode with a liquid ion exchanger.

Hydrophobic anion (R^-)
Tetrakis[3,5-bis(trifluoromethyl)phenyl]borate

Hydrophobic liquid solvent
2-Nitrophenyl octyl ether

Hydrophobic Ca^{2+}-binding ligand (L)
N, *N*-Dicyclohexyl-*N'*,*N'*-dioctadecyl-3-oxapentanediamide

Poly(vinyl chloride)

FIGURE 14-24 Membrane components of a Ca^{2+} ion-selective electrode. Ligand L is an ionophore that selectively binds Ca^{2+}.

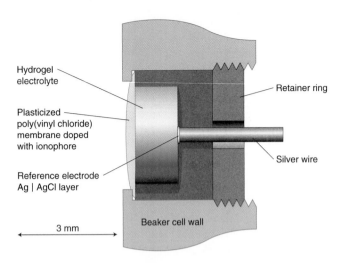

FIGURE 14-25 Liquid-based ion-selective electrodes built into walls of Wet Chemistry Lab beakers of *Phoenix Mars Lander*. Hydrogel electrolyte is a poly(2-hydroxyethyl methacrylate) gel that retains aqueous phase containing 1 mM M^+Cl^-, where M^+ is the cation being measured. [From S. P. Kounaves et al., "The 2007 Mars Scout Lander MECA Wet Chemistry Laboratory," *J. Geophys. Res.* **2009**, *113*, E00A19.]

FIGURE 14-26 Ionophore ETH 2418 for liquid-based H^+ ion-selective electrodes. ETH stands for the Swiss Federal Institute of Technology (Eidgenössische Technische Hochschule Zürich) where many ionophores were synthesized.

of the neutral hydrophobic ligand (L), called an *ionophore*, and a salt of the hydrophobic anion (Na^+R^-) dissolved in a hydrophobic liquid (Figure 14-24) in the poly(vinyl chloride) membrane. The most serious interference comes from Sr^{2+}. The selectivity coefficient in Equation 14-9 is $K_{Ca^{2+},Sr^{2+}}^{Pot} = 0.13$, which means that the response to Sr^{2+} is 13% as great as the response to an equal concentration of Ca^{2+}. For most cations, $K_{Ca^{2+},X}^{Pot} < 10^{-3}$.

The *Phoenix Mars Lander* at the opening of this chapter had 11 liquid-based ion-selective electrodes (Figure 14-25) and 4 solid-state electrodes built into the walls of each Wet Chemistry Laboratory beaker. The liquid-based ion-selective electrodes for H^+ used an ionophore called ETH 2418 (Figure 14-26). This ionophore responds over the pH range 1 to 9 and has selectivity coefficients $K_{H^+,Na^+}^{Pot} = 10^{-8.6}$, $K_{H^+,K^+}^{Pot} = 10^{-9.7}$, and $K_{H^+,Ca^{2+}}^{Pot} = 10^{-7.8}$. Box 14-3 tells how electrode interference led to the discovery of perchlorate on Mars.[29]

Breakthrough in Ion-Selective Electrode Detection Limits[24]

The black curve in Figure 14-27 is typical of many liquid-based ion-selective electrodes. The electrode detects changes in Pb^{2+} concentration above 10^{-6} M but is unresponsive for $[Pb^{2+}] < 10^{-6}$ M. The solution in the internal electrode compartment contains 0.5 mM $PbCl_2$.

The colored curve in Figure 14-27 was obtained with the same electrode, but the internal filling solution was replaced by a *metal ion buffer* (Section 14-7) that fixes $[Pb^{2+}]$ at 10^{-12} M. Now the electrode responds to changes in $[Pb^{2+}]$ down to ~10^{-11} M.

The sensitivity of liquid-based ion-selective electrodes has been limited by leakage of the primary ion (Pb^{2+}, in this case) from the internal filling solution through the ion-exchange membrane. Leakage provides a substantial concentration of primary ion at the external surface of the membrane. If analyte concentration is below 10^{-6} M, leakage from the electrode maintains an effective concentration near 10^{-6} M at the outer surface of the electrode. If we can lower the concentration of primary ion inside the electrode, the concentration of leaking

FIGURE 14-27 Response of Pb^{2+} liquid-based ion-selective electrode with (*black curve*) conventional filling solution containing 0.5 mM Pb^{2+} or (*colored curve*) metal ion buffer filling solution in which $[Pb^{2+}] = 10^{-12}$ M. [From T. Sokalski, A. Ceresa, T. Zwickl, and E. Pretsch, "Large Improvement of the Lower Detection Limit of Ion-Selective Polymer Membrane Electrodes," *J. Am. Chem. Soc.* **1997**, *119*, 11347.]

BOX 14-3   How Was Perchlorate Discovered on Mars?[29]

Nobody expected perchlorate (ClO_4^-) to be abundant on Mars, so the *Phoenix Mars Lander* Wet Chemistry Laboratory was not designed to look for ClO_4^-. However, the nitrate ion-selective electrode sent to Mars was 1 000 times *more sensitive* to ClO_4^- than to NO_3^-. That is, $K_{NO_3^-,ClO_4^-}^{Pot} = 10^3$. Liquid used to leach ions from the soil had a constant NO_3^- background near 1 mM. Nitrate would only be detected if it were present at concentrations above 1 mM.

When salts were leached from soil in the Wet Chemistry Laboratory, the NO_3^- electrode potential changed by 200 mV,

corresponding to an apparent NO_3^- concentration above 1 M, which would have required more NO_3^- than the mass of soil that was analyzed. However, 4–6 mg of ClO_4^- in 1 g of soil would have produced the observed response. Heating the soil released a product at 400°–600°C with a molecular mass of 32 (likely O_2), consistent with thermal decomposition of ClO_4^-. Perchlorate occurs at similar levels on Earth in arid regions including the Atacama Desert. On Earth, ClO_4^- is thought to arise from photochemical reactions of ozone (O_3) with chlorine species in the atmosphere.

TABLE 14-6 Detection limits and selectivity coefficients for liquid-based ion-selective electrodes operating without leakage of primary ion

Primary ion (A)	Detection limit for A (nM)	Selectivity coefficient for some interfering ions (X) $\log K_{A,X}^{Pot}$ (Equation 14-9)
Na^+	30	H^+, -4.8; K^+, -2.7; Ca^{2+}, -6.0
K^+	5	Na^+, -4.2; Mg^{2+}, -7.6; Ca^{2+}, -6.9
NH_3	20	
Cs^+	8	Na^+, -4.7; Mg^{2+}, -8.7; Ca^{2+}, -8.5
Ca^{2+}	0.1	H^+, -4.9; Na^+, -4.8; Mg^{2+}, -5.3
Ag^+	0.03	H^+, -10.2; Na^+, -10.3; Ca^{2+}, -11.3
Pb^{2+}	0.06	H^+, -5.6; Na^+, -5.6; Mg^{2+}, -13.8
Cd^{2+}	0.1	H^+, -6.7; Na^+, -8.4; Mg^{2+}, -13.4
Cu^{2+}	2	H^+, -0.7; Na^+, <-5.7; Mg^{2+}, <-6.9
ClO_4^-	20	OH^-, -5.0; Cl^-, -4.9; NO_3^-, -3.1
I^-	2	OH^-, -1.7

SOURCE: E. Bakker and E. Pretsch, "Modern Potentiometry," *Angew. Chem. Int. Ed.* **2007**, *46*, 5660.

ion outside the membrane is reduced by orders of magnitude and the detection limit is correspondingly reduced. The sensitivity of a solid-state electrode cannot be lowered by changing the filling solution, because analyte concentration is governed by the solubility of the inorganic salt crystal forming the ion-sensitive membrane.

The response of the electrode with 10^{-12} M Pb^{2+} in the filling solution is limited by interference from Na^+ in the internal solution, which contains 0.05 M Na_2EDTA from the metal ion buffer. When the filling solution is buffered to 10^{-12} M Pb^{2+}, the apparent selectivity coefficient is decreased by one to five orders of magnitude for most interfering cations. Not only is the detection limit for Pb^{2+} improved by 10^5, but the selectivity for Pb^{2+} over other cations increases by several orders of magnitude. Table 14-6 gives detection limits and selectivity coefficients for ion-selective electrodes in which precautions are taken to prevent leakage of the primary ion.

Compound Electrodes

Compound electrodes contain a conventional electrode surrounded by a membrane that isolates (or generates) the analyte to which the electrode responds. The CO_2 gas-sensing electrode in Figure 14-28 consists of an ordinary glass pH electrode surrounded by a thin layer of electrolyte solution enclosed in a semipermeable membrane made of rubber, Teflon,

A new direction in ion-selective electrodes is to eliminate the internal filling solution entirely. Ionophore is coated onto or dissolved in an electrically conducting polymer that serves as an ion-to-electron transducer.[30] A nitrate-selective all-solid-state electrode based on the conductive polymer polypyrrole can be made in a student experiment.[31]

Metal

Conductive polymer

Ion-selective membrane

To potentiometer

0.1 M HCl

Internal silver-silver chloride electrode

O-ring

Silver-silver chloride electrode in KCl electrolyte

Spacer

0.1 M KCl electrolyte with weak bicarbonate buffer

Glass electrode membrane

CO_2 permeable membrane

FIGURE 14-28 CO_2 gas-sensing electrode. The membrane is stretched taut, and there is a thin layer of electrolyte between the membrane and the glass bulb.

or polyethylene.[32] A Ag | AgCl reference electrode is immersed in the electrolyte solution. When CO_2 diffuses through the semipermeable membrane, it lowers the pH in the electrolyte. The response of the glass electrode to the change in pH is a measure of the CO_2 concentration outside the electrode. Other acidic or basic gases, including NH_3, SO_2, H_2S, NO_x (nitrogen oxides), and HN_3 (hydrazoic acid) can be detected in the same manner. These electrodes can be used to measure gases in solution or *in the gas phase.*

It is not necessary to use the compound electrode for CO_2 and NH_3 in aqueous solution. Ionophores are available to measure CO_3^{2-} and NH_4^+ with conventional liquid-based ion-selective electrodes.[33]

Numerous ingenious compound electrodes use *enzymes.*[34] These devices contain a conventional electrode coated with an enzyme that catalyzes a reaction of the analyte. The product of the reaction is detected by the electrode.

14-7 Using Ion-Selective Electrodes

Ion-selective electrodes respond linearly to the logarithm of analyte activity over four to six orders of magnitude. Electrodes do not consume unknowns, and they introduce negligible contamination. Response time is seconds or minutes, so electrodes are used to monitor flow streams in industrial applications. Color and turbidity do not hinder electrodes. Microelectrodes can be used inside living cells.

Precision is rarely better than 1% and is usually worse. Electrodes can be fouled by proteins or other organic solutes and the fouling leads to sluggish, drifting response. Certain ions interfere with or poison particular electrodes. Some electrodes are fragile and have limited shelf life.

Electrodes respond to the *activity* of *uncomplexed* analyte ion. Therefore, ligands must be absent or masked. Because we usually wish to know concentrations, not activities, an inert salt is often used to bring all standards and samples to a high, constant ionic strength. If activity coefficients are constant, the electrode potential gives concentrations directly.

Human blood plasma has eight major calcium-containing species that can be separated by capillary electrophoresis and measured by inductively coupled plasma atomic emission spectrometry (Figure 14-29) You will learn about these techniques later in this book. Of the eight species, one with a concentration of 1.05 mM was identified as free Ca^{2+}. In the other seven species, with a total concentration of 1.21 mM, Ca^{2+} is bound to proteins or other ligands. When Ca^{2+} in blood was measured with an ion-selective electrode, only free Ca^{2+} was observed.[35] Calcium bound to ligands is invisible to an ion-selective electrode.

Standard Addition with Ion-Selective Electrodes

When ion-selective electrodes are used, it is important that the composition of the standard solution closely approximates the composition of the unknown. The medium in which the analyte exists is called the **matrix.** For complex or unknown matrixes, the **standard addition** method (Section 5-3) can be used. In this technique, the electrode is immersed in unknown and the potential is recorded. Then a small volume of standard solution is added, so as not to perturb the ionic strength of the unknown. The change in potential tells how the electrode responds to analyte and, therefore, how much analyte was in the unknown. It is best to add several successive aliquots and use a graphical procedure to extrapolate back to the concentration of unknown. Standard addition is best if the additions increase analyte to 1.5 to 3 times its original concentration.

The graphical procedure is based on the equation for the response of the ion-selective electrode, which we will write in the form

$$E = k + \beta\left(\frac{RT \ln 10}{nF}\right)\log[X] \qquad (14\text{-}14)$$

where E is the meter reading in volts and $[X]$ is the concentration of analyte. This reading is the difference in potential of the ion-selective electrode and the reference electrode. The constants k and β depend on the particular ion-selective electrode. The factor $(RT/F) \ln 10$ is 0.059 16 V at 298.15 K. If $\beta = 1$, then the response is Nernstian. We abbreviate the term $(\beta RT/nF) \ln 10$ as S.

Let the initial volume of unknown be V_0 and the initial concentration of analyte be c_X. Let the volume of added standard be V_S and the concentration of standard be c_S. Then the total concentration of analyte after standard is added is $(V_0 c_X + V_S c_S)/(V_0 + V_S)$. Substituting this expression for $[X]$ in Equation 14-14 and doing some rearrangement gives

Advantages of ion-selective electrodes:

- linear response to log $\mathscr{A}$ over a wide range
- nondestructive
- noncontaminating
- short response time
- unaffected by color or turbidity

A 1-mV error in potential corresponds to a 4% error in monovalent ion activity. A 5-mV error corresponds to a 22% error. The relative error *doubles* for divalent ions and *triples* for trivalent ions.

FIGURE 14-29 Separation of calcium-containing species in human blood plasma. The largest peak is free Ca^{2+}. Other peaks are proteins or small molecules to which Ca^{2+} is bound. The detector measures calcium. [From B. Deng, P. Zhu, Y. Wang, J. Feng, X. Li, X. Xu, H. Lu, and Q. Xu, "Determination of Free Calcium and Calcium-Containing Species in Human Plasma by Capillary Electrophoresis-Inductively Coupled Plasma Optical Emission Spectrometry," *Anal. Chem.* **2008,** *80,* 5721.]

R = gas constant
T = temperature (K)
n = charge of the ion being detected
F = Faraday constant

Standard addition plot
for ion-selective electrode:

$$\underbrace{(V_0 + V_S) \, 10^{E/S}}_{y} = \underbrace{10^{k/S} \, V_0 c_X}_{b} + \underbrace{10^{k/S} \, c_S V_S}_{m \quad x} \quad \text{(14-15)}$$

A graph of $(V_0 + V_S)10^{E/S}$ on the y-axis versus V_S on the x-axis has a slope $m = 10^{k/S} c_S$ and a y-intercept of $10^{k/S} V_0 c_X$ (Figure 14-30). The x-intercept is found by setting $y = 0$:

$$x\text{-intercept} = -\frac{b}{m} = -\frac{10^{k/S} V_0 c_X}{10^{k/S} c_S} = -\frac{V_0 c_X}{c_S} \quad \text{(14-16)}$$

Equation 14-16 gives us the concentration of unknown, c_X, from V_0, c_S, and the x-intercept.

A weakness of standard addition with ion-selective electrodes is that we cannot measure β in Equation 14-14 in the unknown matrix. We could measure β in a series of standard solutions (not containing unknown) and use this value to compute S in the function $(V_0 + V_S)10^{E/S}$ in Equation 14-15. Another procedure is to add a concentrated, known matrix to the unknown and to all standards so that the matrix is essentially the same in all solutions.

Metal Ion Buffers

It is pointless to dilute $CaCl_2$ to 10^{-6} M for standardizing an ion-selective electrode. At this low concentration, Ca^{2+} might be lost by adsorption on glass or reaction with impurities.

An alternative is to prepare a **metal ion buffer** from the metal and a suitable ligand. For example, consider the reaction of Ca^{2+} with EDTA at pH 6.00 at which the fraction of EDTA in the form Y^{4-} is $\alpha_{Y^{4-}} = 1.8 \times 10^{-5}$ (Table 11-1):

$$Ca^{2+} + Y^{4-} \rightleftharpoons CaY^{2-}$$

$$K_f = 10^{10.65} = \frac{[CaY^{2-}]}{[Ca^{2+}]\alpha_{Y^{4-}}[EDTA]} \quad \text{(14-17)}$$

If equal concentrations of CaY^{2-} and EDTA are present in a solution,

$$[Ca^{2+}] = \frac{[CaY^{2-}]}{K_f \alpha_{Y^{4-}}[EDTA]} = \frac{[\cancel{CaY^{2-}}]}{(10^{10.65})(1.8 \times 10^{-5})[\cancel{EDTA}]} = 1.2 \times 10^{-6} \, M$$

EXAMPLE   Preparing a Metal Ion Buffer

What concentration of EDTA should be added to 0.010 M CaY^{2-} at pH 6.00 to give $[Ca^{2+}] = 1.0 \times 10^{-6}$ M?

Solution From Equation 14-17, we write

$$[EDTA] = \frac{[CaY^{2-}]}{K_f \alpha_{Y^{4-}}[Ca^{2+}]} = \frac{0.010}{(10^{10.65})(1.8 \times 10^{-5})(1.00 \times 10^{-6})} = 0.012_4 \, M$$

These are practical concentrations of CaY^{2-} and of EDTA.

Test Yourself   What concentration of EDTA should be added to 0.010 M CaY^{2-} at pH 6.00 to give $[Ca^{2+}] = 1.0 \times 10^{-7}$ M? (**Answer:** 0.12_4 M)

A metal ion buffer is the only way to obtain $[Pb^{2+}] = 10^{-12}$ M for the electrode filling solution in Figure 14-27.

14-8   Solid-State Chemical Sensors[36]

Solid-state chemical sensors are fabricated by the same technology used for microelectronic chips. The field effect transistor (FET) is the heart of many sensors such as the pH electrode in Figure 14-31.

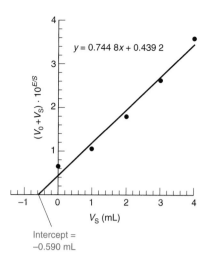

FIGURE 14-30 Standard addition graph for ion-selective electrode based on Equation 14-15. See Exercise 14-F. [Data from G. Li, B. J. Polk, L. A. Meazell, and D. W. Hatchett, "ISE Analysis of Hydrogen Sulfide in Cigarette Smoke," *J. Chem. Ed.* 2000, *77*, 1049.]

Plastic bottles are better than glass for very dilute solutions of metal salts because ions are adsorbed by glass.

[EDTA] = total concentration of all forms of EDTA not bound to metal ion.

$\alpha_{Y^{4-}}$ = fraction of unbound EDTA in the form Y^{4-}.

FIGURE 14-31 Combination pH electrode based on field effect transistor. The thermistor senses temperature and is used for automatic temperature compensation. [Courtesy SENTRON, Europe BV.]

Reference
electrode

pH FET
electrode
(1 mm wide)

Thermistor
measures
temperature

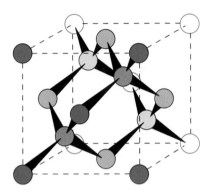

FIGURE 14-32 Diamondlike structure of silicon. Each atom is tetrahedrally bonded to four neighbors, with a Si—Si distance of 235 pm. All atoms in this diagram are Si. The darkest color is for atoms on the front face of the cube. Atoms with successively lighter shades are on planes that are successively further from the front face.

Semiconductors and Diodes

Semiconductors such as Si (Figure 14-32), Ge, and GaAs are materials whose electrical *resistivity*[37] lies between those of conductors and insulators. The four valence electrons of the pure materials are all involved in bonds between atoms (Figure 14-33a). A phosphorus impurity, with five valence electrons, provides one extra **conduction electron** that is free to move through the crystal (Figure 14-33b). An aluminum impurity has one less bonding electron than necessary, creating a vacancy called a **hole,** which behaves as a positive charge carrier. When a neighboring electron fills the hole, a new hole appears at an adjacent position (Figure 14-33c). A semiconductor with excess conduction electrons is called *n-type*, and one with excess holes is called *p-type*.

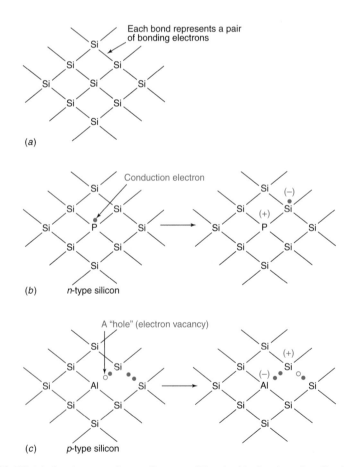

FIGURE 14-33 (*a*) The electrons of pure silicon are all involved in the sigma-bonding framework. (*b*) An impurity atom such as phosphorus adds one extra electron (•), which is relatively free to move through the crystal. (*c*) An aluminum impurity atom lacks one electron needed for the sigma-bonding framework. The hole (○) introduced by the Al atom can be occupied by an electron from a neighboring bond, effectively moving the hole to the neighboring bond.

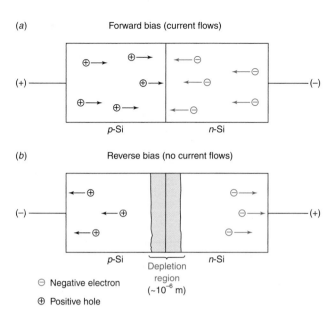

(a) Forward bias (current flows)

p-Si   *n*-Si

(b) Reverse bias (no current flows)

p-Si   *n*-Si

Depletion region (~10⁻⁶ m)

⊖ Negative electron
⊕ Positive hole

An *activation energy* is needed to get the charge carriers to move across the diode. For Si, ~0.6 V of forward bias is required before current will flow. For Ge, the requirement is ~0.2 V.

For moderate reverse bias voltages, no current flows. If the voltage is sufficiently negative, *breakdown* occurs and current flows in the reverse direction.

FIGURE 14-34 Behavior of a *pn* junction, showing that current (*a*) can flow under forward bias conditions, but (*b*) is prevented from flowing under reverse bias.

A **diode** is a *pn* junction (Figure 14-34a). If *n*-Si is made negative with respect to *p*-Si, electrons flow from the external circuit into the *n*-Si. At the *pn* junction, electrons and holes combine. As electrons move from the *p*-Si into the circuit, a fresh supply of holes is created in the *p*-Si. The net result is that current flows when *n*-Si is negative with respect to *p*-Si. The diode is said to be *forward biased*.

If the polarity is reversed (Figure 14-34b), electrons are drawn out of *n*-Si and holes are drawn out of *p*-Si, leaving a thin *depletion region* devoid of charge carriers near the *pn* junction. The diode is *reverse biased* and does not conduct current in the reverse direction.

Chemical-Sensing Field Effect Transistor

The *base* of the **field effect transistor** in Figure 14-35 is constructed of *p*-Si with two *n*-type regions called *source* and *drain*. An insulating surface layer of SiO$_2$ is overcoated by a conductive metal *gate* between source and drain. The source and the base are held at the same electric potential. When a voltage is applied between source and drain (Figure 14-35a), little current flows because the drain-base interface is a *pn* junction in reverse bias.

If the gate is made positive, electrons from the base are attracted toward the gate and form a conductive channel between source and drain (Figure 14-35b). Current increases as the gate is made more positive. *The gate potential regulates the current between source and drain.*

The more positive the gate, the more current can flow between source and drain.

FIGURE 14-35 Operation of a field effect transistor. (*a*) Nearly random distribution of holes and electrons in the base in the absence of gate potential. (*b*) Positive gate potential attracts electrons that form a conductive channel beneath the gate. Current can flow through this channel between source and drain.

FIGURE 14-36 Operation of a chemical-sensing field effect transistor. The transistor is coated with an insulating SiO_2 layer and a second layer of Si_3N_4 (silicon nitride), which is impervious to ions and improves electrical stability. The circuit at the lower left adjusts the potential difference between the reference electrode and the source in response to changes in the analyte solution, such that a constant drain-source current is maintained.

Circuit adjusts potential difference between the reference electrode and the source, in response to changing analyte concentration, to maintain constant drain-source current

This voltage sets the drain-source current at a desired level

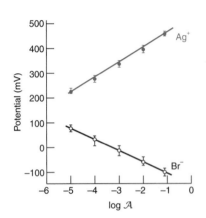

FIGURE 14-37 Response of a silver bromide-coated field effect transistor. Error bars are 95% confidence intervals for data obtained from 195 sensors prepared from different chips. [From R. P. Buck and D. E. Hackleman, "Field Effect Potentiometric Sensors," *Anal. Chem.* **1977**, *49*, 2315.]

The essential feature of the chemical-sensing field effect transistor in Figure 14-36 is the chemically sensitive layer over the gate. An example is a layer of AgBr. When exposed to silver nitrate solution, Ag^+ is adsorbed on the AgBr (Figure 26-2), thereby giving it a positive charge and increasing current between source and drain. *The voltage that must be applied by an external circuit to bring the current back to its initial value is the response to Ag^+.* Figure 14-37 shows that Ag^+ makes the gate more positive and Br^- makes the gate more negative. The response is close to 59 mV for a 10-fold concentration change. The transistor is smaller (Figure 14-31) and more rugged than ion-selective electrodes. The sensing surface is typically only 1 mm^2. Chemical-sensing field effect transistors have been designed for species such as H^+,[38] NH_4^+, Ca^{2+}, NO_3^-,[39] CO_2,[40] pesticides (detection limit = 10^{-12} M),[41,42] adrenaline,[43] and specific sequences of DNA.[44]

Terms to Understand

calomel electrode
combination electrode
compound electrode
conduction electron
diode
electroactive species
field effect transistor
glass electrode

hole
indicator electrode
ion-exchange equilibrium
ion-selective electrode
junction potential
liquid-based ion-selective
    electrode
matrix

metal ion buffer
mobility
potentiometry
reference electrode
saturated calomel electrode
    (S.C.E.)
selectivity coefficient
semiconductor

silver-silver chloride electrode
sodium error
solid-state ion-selective
    electrode
standard addition

Summary

In potentiometric measurements, the indicator electrode responds to changes in the activity of analyte, and the reference electrode is a self-contained half-cell with a constant potential. The most common reference electrodes are calomel and silver-silver chloride. Common indicator electrodes include (1) the inert Pt electrode, (2) a silver electrode responsive to Ag^+, halides, and other ions that react with Ag^+, and (3) ion-selective electrodes. Unknown junction potentials at liquid-liquid interfaces limit the accuracy of most potentiometric measurements.

Ion-selective electrodes, including the glass pH electrode, respond mainly to one ion that is selectively bound to the ion-exchange membrane of the electrode. The potential difference across the membrane, E, depends on the activity ($\mathcal{A}_o$) of the target ion in the external analyte solution. At 25°C, the ideal relation is $E(V) = \text{constant} + (0.059\ 16/n) \ln \mathcal{A}_o$, where n is the charge of the target ion. For interfering ions (X) with the same charge as the primary ion (A), the response of ion-selective electrodes is $E = \text{constant} \pm (0.059\ 16/n)\log[\mathcal{A}_A + \Sigma K_{A,X}^{Pot} \mathcal{A}_X]$, where $K_{A,X}^{Pot}$

is the selectivity coefficient for each species. Common ion-selective electrodes can be classified as solid state, liquid based, and compound. Quantitation with ion-selective electrodes is usually done with a calibration curve or by the method of standard addition. Metal ion buffers are appropriate for establishing and maintaining low concentrations of ions. A chemical-sensing field effect transistor is a solid-state device that uses a chemically sensitive coating to alter the electrical properties of a semiconductor in response to changes in the chemical environment.

Exercises

14-A. The apparatus in Figure 14-7 was used to monitor the titration of 50.0 mL of 0.100 M $AgNO_3$ with 0.200 M NaBr. Calculate the cell voltage at each volume of NaBr, and sketch the titration curve: 1.0, 12.5, 24.0, 24.9, 25.1, 26.0, and 35.0 mL.

14-B. The apparatus in the figure can follow the course of an EDTA titration and was used to generate the curves in Figure 11-10. The heart of the cell is a pool of liquid Hg in contact with the solution and with a Pt wire. A small amount of HgY^{2-} added to the analyte equilibrates with a very tiny amount of Hg^{2+}:

$$Hg^{2+} + Y^{4-} \rightleftharpoons HgY^{2-}$$

$$K_f = \frac{[HgY^{2-}]}{[Hg^{2+}][Y^{4-}]} = 10^{21.5} \qquad (A)$$

The redox equilibrium $Hg^{2+} + 2e^- \rightleftharpoons Hg(l)$ is established rapidly at the surface of the Hg electrode, so the Nernst equation for the cell

can be written in the form

$$E = E_+ - E_- = \left(0.852 - \frac{0.059\ 16}{2}\log\left(\frac{1}{[Hg^{2+}]}\right)\right) - E_- \quad (B)$$

where E_- is the constant potential of the reference electrode. From Equation A, $[Hg^{2+}] = [HgY^{2-}] / K_f[Y^{4-}]$, and this can be substituted into Equation B to give

$$E = 0.852 - \frac{0.059\ 16}{2}\log\left(\frac{[Y^{4-}]K_f}{[HgY^{2-}]}\right) - E_-$$

$$= 0.852 - E_- - \frac{0.059\ 16}{2}\log\left(\frac{K_f}{[HgY^{2-}]}\right) - \frac{0.059\ 16}{2}\log[Y^{4-}]$$
$$(C)$$

where K_f is the formation constant for HgY^{2-}. This apparatus thus responds to the changing EDTA concentration during an EDTA titration.

(a) (b)

(a) Apparatus for Exercise 14-B. (b) Enlarged view of mercury electrode.

Suppose that you titrate 50.0 mL of 0.010 0 M $MgSO_4$ with 0.020 0 M EDTA at pH 10.0, using the apparatus in the figure with an S.C.E. reference electrode. Analyte contains 1.0×10^{-4} M $Hg(EDTA)^{2-}$ added at the beginning of the titration. Calculate the cell voltage at the following volumes of added EDTA, and draw a graph of millivolts versus milliliters: 0, 10.0, 20.0, 24.9, 25.0, and 26.0 mL.

14-C. A solid-state fluoride ion-selective electrode responds to F^- but not to HF. It also responds to hydroxide ion at high concentration when $[OH^-] \approx [F^-]/10$. Suppose that such an electrode gave a potential of +100 mV (versus S.C.E.) in 10^{-5} M NaF and +41 mV in 10^{-4} M NaF. Sketch qualitatively how the potential would vary if the electrode were immersed in 10^{-5} M NaF and the pH ranged from 1 to 13.

14-D. One glass-membrane sodium ion-selective electrode has a selectivity coefficient $K_{Na^+,H^+}^{Pot} = 36$. When this electrode was immersed in 1.00 mM NaCl at pH 8.00, a potential of -38 mV (versus S.C.E.) was recorded.

(a) Neglecting activity coefficients, calculate the potential with Equation 14-10 if the electrode were immersed in 5.00 mM NaCl at pH 8.00.

(b) What would the potential be for 1.00 mM NaCl at pH 3.87? You can see that pH is a critical variable for the sodium electrode.

14-E. An ammonia gas-sensing electrode gave the following calibration points when all solutions contained 1 M NaOH.

NH_3 (M)	E (mV)	NH_3 (M)	E (mV)
1.00×10^{-5}	268.0	5.00×10^{-4}	368.0
5.00×10^{-5}	310.0	1.00×10^{-3}	386.4
1.00×10^{-4}	326.8	5.00×10^{-3}	427.6

A dry food sample weighing 312.4 mg was digested by the Kjeldahl procedure (Section 10-8) to convert all nitrogen into NH_4^+. The digestion solution was diluted to 1.00 L, and 20.0 mL were transferred to a 100-mL volumetric flask. The 20.0-mL aliquot was treated with 10.0 mL of 10.0 M NaOH plus enough NaI to complex the Hg catalyst from the digestion and diluted to 100.0 mL. When measured with the ammonia electrode, this solution gave a reading of 339.3 mV. Calculate the wt% nitrogen in the food sample.

14-F. H_2S from cigarette smoke was collected by bubbling smoke through aqueous NaOH and measured with a sulfide ion-selective electrode. Standard additions of volume V_S containing Na_2S at concentration $c_S = 1.78$ mM were then made to $V_0 = 25.0$ mL of unknown and the electrode response, E, was measured.

V_S (mL)	E (V)	V_S (mL)	E (V)
0	0.046 5	3.00	0.030 0
1.00	0.040 7	4.00	0.026 5
2.00	0.034 4		

From a separate calibration curve, it was found that $\beta = 0.985$ in Equation 14-14. Using $T = 298.15$ K and $n = -2$ (the charge of S^{2-}), prepare a standard addition graph with Equation 14-15 and find the concentration of sulfide in the unknown.

Problems

Reference Electrodes

14-1. **(a)** Write the half-reactions for the silver-silver chloride and calomel reference electrodes.

(b) Predict the voltage for the following cell.

Saturated silver-silver chloride electrode ‖ Saturated calomel electrode

14-2. Convert the following potentials. The $Ag | AgCl$ and calomel reference electrodes are saturated with KCl.

(a) 0.523 V versus S.H.E. = ? versus $Ag | AgCl$

(b) -0.111 V versus $Ag | AgCl$ = ? versus S.H.E.

(c) -0.222 V versus S.C.E. = ? versus S.H.E.

(d) 0.023 V versus $Ag | AgCl$ = ? versus S.C.E.

(e) -0.023 V versus S.C.E. = ? versus $Ag | AgCl$

14-3. Suppose that the silver-silver chloride electrode in Figure 14-2 is replaced by a saturated calomel electrode. Calculate the cell voltage if $[Fe^{2+}] / [Fe^{3+}] = 2.5 \times 10^{-3}$.

14-4. From the following potentials, calculate the *activity* of Cl^- in 1 M KCl.

$$E°(\text{calomel electrode}) = 0.268 \text{ V}$$

$$E(\text{calomel electrode, 1 M KCl}) = 0.280 \text{ V}$$

14-5. For a silver-silver chloride electrode, the following potentials are observed:

$$E° = 0.222 \text{ V} \qquad E(\text{saturated KCl}) = 0.197 \text{ V}$$

From these potentials, find the activity of Cl^- in saturated KCl. Calculate E for a calomel electrode saturated with KCl, given that $E°$ for the calomel electrode is 0.268 V. (Your answer will not be exactly the value 0.241 used in this book.)

Indicator Electrodes

14-6. A cell was prepared by dipping a Cu wire and a saturated calomel electrode into 0.10 M $CuSO_4$ solution. The Cu wire was attached to the positive terminal of a potentiometer and the calomel electrode was attached to the negative terminal.

(a) Write a half-reaction for the Cu electrode.

(b) Write the Nernst equation for the Cu electrode.

(c) Calculate the cell voltage.

14-7. Explain why a silver electrode can be an indicator electrode for Ag^+ and for halides.

14-8. A 10.0-mL solution of 0.050 0 M $AgNO_3$ was titrated with 0.025 0 M NaBr in the cell

$$\text{S.C.E.} ‖ \text{titration solution} | Ag(s)$$

Find the cell voltage for 0.1 and 30.0 mL of titrant.

14-9. A solution containing 50.0 mL of 0.100 M EDTA buffered to pH 10.00 was titrated with 50.0 mL of 0.020 0 M $Hg(ClO_4)_2$ in the cell shown in Exercise 14-B:

$$\text{S.C.E.} ‖ \text{titration solution} | Hg(l)$$

From the cell voltage $E = -0.027$ V, find the formation constant of $Hg(EDTA)^{2-}$.

14-10. Consider the cell S.C.E. ‖ cell solution | $Pt(s)$, whose voltage is -0.126 V. The cell solution contains 2.00 mmol of $Fe(NH_4)_2(SO_4)_2$, 1.00 mmol of $FeCl_3$, 4.00 mmol of Na_2EDTA, and lots of buffer, pH 6.78, in a volume of 1.00 L.

(a) Write a reaction for the right half-cell.

(b) Find the quotient $[Fe^{2+}]/[Fe^{3+}]$ in the cell solution. (This expression gives the ratio of *uncomplexed* ions.)

(c) Find the quotient of formation constants: (K_f for $FeEDTA^-$)/ (K_f for $FeEDTA^{2-}$).

14-11. *An equilibrium challenge:* Here's a cell you'll really like:

$$Ag(s) \mid AgCl(s) \mid KCl(aq, \text{ saturated}) \parallel \text{cell solution} \mid Cu(s)$$

The cell solution was made by mixing

25.0 mL of 4.00 mM KCN
25.0 mL of 4.00 mM KCu(CN)$_2$
25.0 mL of 0.400 M acid, HA, with pK_a = 9.50
25.0 mL of KOH solution

The measured voltage was -0.440 V. *Calculate the molarity of the KOH solution.* Assume that essentially all copper(I) is Cu(CN)$_2^-$. A little HCN comes from the reaction of KCN with HA. Neglect the small amount of HA consumed by reaction with HCN. For the right half-cell, the reaction is $Cu(CN)_2^- + e^- \rightleftharpoons Cu(s) + 2CN^-$. *Suggested procedure:* From E, find [CN$^-$]. From [CN$^-$], find pH. From pH, figure out how much OH$^-$ had been added.

Junction Potential

14-12. What causes a junction potential? How does this potential limit the accuracy of potentiometric analyses? Identify a cell in the illustrations in Section 13-2 that has no junction potential.

14-13. Why is the 0.1 M HCl $\mid$ 0.1 M KCl junction potential of opposite sign and greater magnitude than the 0.1 M NaCl $\mid$ 0.1 M KCl potential in Table 14-2?

14-14. Which side of the liquid junction 0.1 M KNO$_3$ $\mid$ 0.1 M NaCl will be negative?

14-15. Refer to the footnote in Table 14-1. How many seconds will it take for **(a)** H$^+$ and **(b)** NO$_3^-$ to migrate a distance of 12.0 cm in a field of 7.80×10^3 V/m?

14-16. Suppose that an ideal hypothetical cell such as that in Figure 13-7 were set up to measure $E°$ for the half-reaction $Ag^+ + e^- \rightleftharpoons Ag(s)$.
(a) Calculate the equilibrium constant for the net cell reaction.
(b) If there were a junction potential of +2 mV (increasing E from 0.799 to 0.801 V), by what percentage would the calculated equilibrium constant increase?
(c) Answer parts **(a)** and **(b)**, using 0.100 V instead of 0.799 V for the value of $E°$ for the silver reaction.

14-17. Explain how the cell Ag(s) $\mid$ AgCl(s) $\mid$ 0.1 M HCl $\mid$ 0.1 M KCl $\mid$ AgCl(s) $\mid$ Ag(s) can be used to measure the 0.1 M HCl $\mid$ 0.1 M KCl junction potential.

14-18. 🖩 *Henderson equation.* The junction potential, E_j, between solutions α and β can be estimated with the Henderson equation:

$$E_j \approx \frac{\sum_i \dfrac{|z_i|u_i}{z_i}[C_i(\beta) - C_i(\alpha)]}{\sum_i |z_i|u_i[C_i(\beta) - C_i(\alpha)]} \frac{RT}{F} \ln \frac{\sum_i |z_i|u_i\, C_i(\alpha)}{\sum_i |z_i|u_i\, C_i(\beta)}$$

where z_i is the charge of species i, u_i is the *mobility* of species i (Table 14-1), $C_i(\alpha)$ is the concentration of species i in phase α, and $C_i(\beta)$ is the concentration in phase β. (Activity coefficients are neglected in this equation.)

(a) Using your calculator, show that the junction potential of 0.1 M HCl $\mid$ 0.1 M KCl is 26.9 mV at 25°C. (Note that $(RT/F) \ln x = 0.059\ 16 \log x$.)

(b) Set up a spreadsheet to reproduce the result in part **(a)**. Then use your spreadsheet to compute and plot the junction potential for 0.1 M HCl $\mid$ x M KCl, where x varies from 1 mM to 4 M.

(c) Use your spreadsheet to explore the behavior of the junction potential for y M HCl $\mid$ x M KCl, where $y = 10^{-4}, 10^{-3}, 10^{-2}$, and 10^{-1} M and $x = 1$ mM or 4 M.

pH Measurement with a Glass Electrode

14-19. Describe how you would calibrate a pH electrode and measure the pH of blood (~7.5) *at 37°C*. Use the standard buffers in Table 14-3.

14-20. List the sources of error associated with pH measurement with the glass electrode.

14-21. If electrode C in Figure 14-18 were placed in a solution of pH 11.0, what would the pH reading be?

14-22. Which National Institute of Standards and Technology buffer(s) would you use to calibrate an electrode for pH measurements in the range 3–4?

14-23. Why do glass pH electrodes tend to indicate a pH lower than the actual pH in strongly basic solution?

14-24. Suppose that the Ag $\mid$ AgCl outer electrode in Figure 14-11 is filled with 0.1 M NaCl instead of saturated KCl. Suppose that the electrode is calibrated in a dilute buffer containing 0.1 M KCl at pH 6.54 at 25°C. The electrode is then dipped in a second buffer *at the same pH* and same temperature, but containing 3.5 M KCl. Use Table 14-2 to estimate how much the indicated pH will change.

14-25. (a) When the difference in pH across the membrane of a glass electrode at 25°C is 4.63 pH units, how much voltage is generated by the pH gradient?
(b) What would the voltage be for the same pH difference at 37°C?

14-26. When calibrating a glass electrode, 0.025 m potassium dihydrogen phosphate/0.025 m disodium hydrogen phosphate buffer (Table 14-3) gave a reading of -18.3 mV at 20°C and 0.05 m potassium hydrogen phthalate buffer gave a reading of +146.3 mV. What is the pH of an unknown giving a reading of +50.0 mV? What is the slope of the calibration curve (mV/pH unit) and what is the theoretical slope at 20°C? Find the value of β in Equation 14-6.

14-27. *Activity problem.* The 0.025 0 m KH$_2$PO$_4$/0.025 0 m Na$_2$HPO$_4$ buffer (6) in Table 14-3 has a pH of 6.865 at 25°C.
(a) Show that the ionic strength of the buffer is $\mu = 0.100\ m$.
(b) From the pH and K_2 for phosphoric acid, find the quotient of activity coefficients, $\gamma_{HPO_4^{2-}}/\gamma_{H_2PO_4^-}$, at $\mu = 0.100\ m$.
(c) You have the urgent need to prepare a pH 7.000 buffer to be used as a calibration standard.[45] You can use the activity coefficient ratio from part **(b)** to accurately prepare such a buffer if the ionic strength is kept at 0.100 m. What molalities of KH$_2$PO$_4$ and Na$_2$HPO$_4$ should be mixed to obtain a pH of 7.000 and $\mu = 0.100\ m$?

Ion-Selective Electrodes

14-28. Explain the principle of operation of ion-selective electrodes. How does a compound electrode differ from a simple ion-selective electrode?

14-29. What does the selectivity coefficient tell us? Is it better to have a large or a small selectivity coefficient?

14-30. What makes a liquid-based ion-selective electrode specific for one analyte?

14-31. Why is it preferable to use a metal ion buffer to achieve $pM = 8$ rather than just dissolving enough M to give a 10^{-8} M solution?

14-32. To determine the *concentration* of a dilute analyte with an ion-selective electrode, why do we use standards with a constant, high concentration of an inert salt?

14-33. A cyanide ion-selective electrode obeys the equation

$$E = \text{constant} - 0.059\,16\,\log[CN^-]$$

The potential was -0.230 V when the electrode was immersed in 1.00 mM NaCN.

(a) Evaluate the constant in the equation.

(b) Using the result from part (a), find $[CN^-]$ if $E = -0.300$ V.

(c) Without using the constant from part (a), find $[CN^-]$ if $E = -0.300$ V.

14-34. By how many volts will the potential of an ideal Mg^{2+} ion-selective electrode change if the electrode is removed from 1.00×10^{-4} M $MgCl_2$ and placed in 1.00×10^{-3} M $MgCl_2$ at 25°C?

14-35. When measured with a F^- ion-selective electrode with a Nernstian response at 25°C, the potential due to F^- in unfluoridated groundwater in Foxboro, Massachusetts, was 40.0 mV more positive than the potential of tap water in Providence, Rhode Island. Providence maintains its fluoridated water at the recommended level of 1.00 ± 0.05 mg F^-/L. What is the concentration of F^- in mg/L in groundwater in Foxboro? (Disregard the uncertainty.)

14-36. The selectivities of a Li^+ ion-selective electrode are indicated on the following diagram. Which alkali metal (Group 1) ion causes the most interference? Which alkaline earth (Group 2) ion causes the most interference? How much greater must be $[K^+]$ than $[Li^+]$ for the two ions to give equal response?

14-37. A metal ion buffer was prepared from 0.030 M ML and 0.020 M L, where ML is a metal-ligand complex and L is free ligand.

$$M + L \rightleftharpoons ML \qquad K_f = 4.0 \times 10^8$$

Calculate the concentration of free metal ion, M, in this buffer.

14-38. ▦ *Calibration curve and propagation of uncertainty for exponents.* The following data were obtained when a Ca^{2+}

ion-selective electrode was immersed in standard solutions whose ionic strength was constant at 2.0 M.

Ca^{2+} (M)	E (mV)
3.38×10^{-5}	-74.8
3.38×10^{-4}	-46.4
3.38×10^{-3}	-18.7
3.38×10^{-2}	$+10.0$
3.38×10^{-1}	$+37.7$

(a) Prepare a calibration curve and find the least-squares slope and intercept and their standard deviations.

(b) Calculate the value of β in Equation 14-13.

(c) For a measured potential, the calibration curve gives us $\log[Ca^{2+}]$. We can compute $[Ca^{2+}] = 10^{\log[Ca^{2+}]}$. Using rules for propagation of uncertainty in Table 3-1, calculate $[Ca^{2+}]$ and its associated uncertainty of a sample that gave a reading of $-22.5\,(\pm 0.3)$ mV in four replicate measurements.

14-39. The selectivity coefficient, $K^{Pot}_{Li^+,H^+}$, for a Li^+ ion-selective electrode is 4×10^{-4}. When this electrode is placed in 3.44×10^{-4} M Li^+ solution at pH 7.2, the potential is -0.333 V versus S.C.E. What would be the potential if the pH were lowered to 1.1 and the ionic strength were kept constant?

14-40. *Standard addition.* A particular CO_2 compound electrode like the one in Figure 14-28 obeys the equation $E = \text{constant} - [\beta RT\,(\ln 10)/2F]\log[CO_2]$, where R is the gas constant, T is temperature (303.15 K), F is the Faraday constant, and $\beta = 0.933$ (measured from a calibration curve). $[CO_2]$ represents all forms of dissolved carbon dioxide at the pH of the experiment, which was 5.0. Standard additions of volume V_S containing a standard concentration $c_S = 0.020\,0$ M $NaHCO_3$ were made to an unknown solution whose initial volume was $V_0 = 55.0$ mL.

V_S (mL)	E (V)	V_S (mL)	E (V)
0	0.079 0	0.300	0.058 8
0.100	0.072 4	0.800	0.050 9
0.200	0.065 3		

Prepare a graph with Equation 14-15 and find $[CO_2]$ in the unknown.

14-41. *Standard addition with confidence interval.* Ammonia in seawater was measured with an ammonia-selective electrode. A 100.0 mL aliquot of seawater was treated with 1.00 mL of 10 M NaOH to convert NH_4^+ to NH_3. Therefore, $V_0 = 101.0$ mL. A reading was then taken with the electrode. Then a series of 10.00 mL aliquots of standard $NH_4^+Cl^-$ were added with results in the table.

V_S (mL)	E (V)	V_S (mL)	E (V)
0	$-0.084\,4$	30.00	$-0.039\,4$
10.00	$-0.058\,1$	40.00	$-0.034\,7$
20.00	$-0.046\,9$		

SOURCE: Data derived from H. Van Ryswyk, E. W. Hall, S. J. Petesch, and A. E. Wiedeman, "Extending the Marine Microcosm Laboratory," *J. Chem. Ed.* **2007**, *84*, 306.

The standard contains 100.0 ppm (mg/L) of nitrogen in the form of $NH_4^+Cl^-$. A separate experiment determined that the electrode slope $\beta RT\,(\ln 10)/F$ is 0.056 6 V.

(a) Prepare a standard addition graph. Find the concentration and 95% confidence interval for ammonia nitrogen (ppm) in the 100.0 mL of seawater.

(b) Standard addition is best if the additions increase analyte to 1.5 to 3 times its original concentration. Does this experiment fall in that range? A criticism of this experiment is that too much added standard creates error because the standards contribute too much to the computed result and the reading from the initial solution is not weighted heavily enough.

14-42. Data below come from the graph in Box 14-2, for which the separate solutions method was used to measure selectivity coefficients for a sodium ion-selective electrode at 21.5°C. Use Equation 14-11 to calculate $\log K^{Pot}$ for each line below.

$(E_{Mg^{2+}} - E_{Na^+})$ at $\mathcal{A} = 10^{-3} = -0.385 \text{ V} \Rightarrow \log K^{Pot}_{Na^+,Mg^{2+}} = ?$

$(E_{Mg^{2+}} - E_{Na^+})$ at $\mathcal{A} = 10^{-2} = -0.418 \text{ V} \Rightarrow \log K^{Pot}_{Na^+,Mg^{2+}} = ?$

$(E_{K^+} - E_{Na^+})$ at $\mathcal{A} = 10^{-3} = -0.285 \text{ V} \Rightarrow \log K^{Pot}_{Na^+,K^+} = ?$

$(E_{K^+} - E_{Na^+})$ at $\mathcal{A} = 10^{-1.5} = -0.285 \text{ V} \Rightarrow \log K^{Pot}_{Na^+,K^+} = ?$

14-43. The H^+ ion-selective electrode on the *Phoenix Mars Lander* has selectivity coefficients $K^{Pot}_{H^+,Na^+} = 10^{-8.6}$ and $K^{Pot}_{H^+,Ca^{2+}} = 10^{-7.8}$. Let A be the primary ion sensed by the electrode and let its charge be z_A. Let X be an interfering ion with charge z_X. The relative error in primary ion activity due to an interfering ion is[46]

$$\text{Error in } \mathcal{A}_A \, (\%) = \frac{(K^{Pot}_{A,X})^{z_X/z_A} \, \mathcal{A}_X}{\mathcal{A}_A^{z_X/z_A}} \times 100$$

This expression applies for errors less than ~10%. If the pH is 8.0 ($\mathcal{A}_{H^+} = 10^{-8.0}$) and $\mathcal{A}_{Na^+} = 10^{-2.0}$, what is the relative error in measuring $\mathcal{A}_{H^+}$? If the pH is 8.0 and $\mathcal{A}_{Ca^{2+}} = 10^{-2.0}$, what is the relative error in measuring $\mathcal{A}_{H^+}$?

14-44. A Ca^{2+} ion-selective electrode was calibrated in metal ion buffers with ionic strength fixed at 0.50 M. Using the following electrode readings, write an equation for the response of the electrode to Ca^{2+} and Mg^{2+}.

$[Ca^{2+}]$ (M)	$[Mg^{2+}]$ (M)	mV
1.00×10^{-6}	0	-52.6
2.43×10^{-4}	0	$+16.1$
1.00×10^{-6}	3.68×10^{-3}	-38.0

14-45. The Pb^{2+} ion buffer used inside the electrode for the colored curve in Figure 14-27 was prepared by mixing 1.0 mL of 0.10 M $Pb(NO_3)_2$ with 100.0 mL of 0.050 M Na_2EDTA. At the measured pH of 4.34, $\alpha_{Y^{4-}} = 1.46 \times 10^{-8}$ (Equation 11-4). Show that $[Pb^{2+}] = 1.4 \times 10^{-12}$ M.

14-46. Solutions with a wide range of Hg^{2+} concentrations were prepared to calibrate an experimental Hg^{2+} ion-selective electrode. For the range $10^{-5} < [Hg^{2+}] < 10^{-1}$ M, $Hg(NO_3)_2$ was used directly. The range $10^{-11} < [Hg^{2+}] < 10^{-6}$ M could be covered by the buffer system $HgCl_2(s) + KCl(aq)$ (based on pK_{sp} for $HgCl_2 = 13.16$). The

range $10^{-15} < [Hg^{2+}] < 10^{-11}$ M was obtained with $HgBr_2(s) + KBr(aq)$ (based on pK_{sp} for $HgBr_2 = 17.43$). The resulting calibration curve is shown in the figure. Calibration points for the $HgCl_2$ / KCl buffer are not in line with the other data. Suggest a possible explanation.

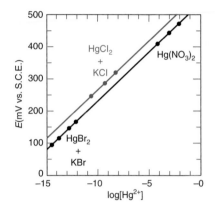

Hg^{2+} ion-selective electrode calibration curve from J. A. Shatkin, H. S. Brown, and S. Licht, "Composite Graphite Ion Selective Electrode Array Potentiometry for the Detection of Mercury and Other Relevant Ions in Aquatic Systems," *Anal. Chem.* **1995**, *67*, 1147. It was not stated in the paper, but we presume that all solutions had the same ionic strength.

14-47. *Activity problem.* Citric acid is a triprotic acid (H_3A) whose anion (A^{3-}) forms stable complexes with many metal ions.

$$Ca^{2+} + A^{3-} \overset{K_f}{\rightleftharpoons} CaA^-$$

When a Ca^{2+} ion-selective electrode with a slope of 29.58 mV was immersed in a solution having $\mathcal{A}_{Ca^{2+}} = 1.00 \times 10^{-3}$, the reading was +2.06 mV. Calcium citrate solution was prepared by mixing equal volumes of solutions 1 and 2.

Solution 1:
$[Ca^{2+}] = 1.00 \times 10^{-3}$ M, pH = 8.00, $\mu = 0.10$ M

Solution 2:
$[Citrate]_{total} = 1.00 \times 10^{-3}$ M, pH = 8.00, $\mu = 0.10$ M

The calcium citrate solution gave an electrode reading of -25.90 mV.

(a) Refer to the discussion with Figure B-2 in Appendix B. Calculate the activity of Ca^{2+} in the calcium citrate solution.

(b) Calculate the formation constant, K_f, for CaA^-. Assume that the size of CaA^- is 500 pm. At pH 8.00 and $\mu = 0.10$ M, the fraction of free citrate in the form A^{3-} is 0.998.

Solid-State Chemical Sensors

14-48. What does analyte do to a chemical-sensing field effect transistor to produce a signal related to the activity of analyte?

15 | Redox Titrations

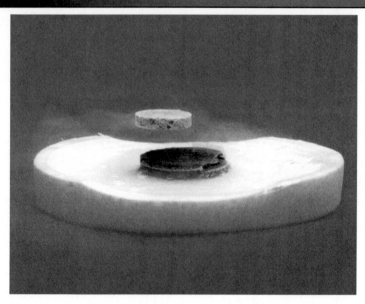

Permanent magnet levitates above superconducting disk cooled in a pool of liquid nitrogen. Redox titrations are crucial in measuring the chemical composition of a superconductor. [Photo courtesy D. Cornelius, Michelson Laboratory, with materials from T. Vanderah.]

Superconductors are materials that lose all electrical resistance when cooled below a critical temperature. Prior to 1987, all known superconductors required cooling to temperatures near that of liquid helium (4 K), a process that is costly and impractical for all but a few applications. In 1987, a giant step was taken when "high-temperature" superconductors that retain their superconductivity above the boiling point of liquid nitrogen (77 K) were discovered. The most startling characteristic of a superconductor is magnetic levitation, shown above. When a magnetic field is applied to a superconductor, current flows in the outer skin of the material such that the applied magnetic field is exactly canceled by the induced magnetic field, and the net field inside the specimen is zero. Expulsion of a magnetic field from a superconductor is called the *Meissner effect*.

A prototypical high-temperature superconductor is yttrium barium copper oxide, $YBa_2Cu_3O_7$, in which two-thirds of the copper is in the $+2$ oxidation state and one-third is in the unusual $+3$ state. Another example is $Bi_2Sr_2(Ca_{0.8}Y_{0.2})Cu_2O_{8.295}$, in which the average oxidation state of copper is $+2.105$ and the average oxidation state of bismuth is $+3.090$ (which is formally a mixture of Bi^{3+} and Bi^{5+}). The most reliable means to unravel these complex formulas is through "wet" oxidation-reduction titrations described in this chapter.

Iron and its compounds are environmentally acceptable redox agents that are finding increased use in remediating toxic waste in groundwaters:[1]

$$\underset{\substack{\text{Dissolved}\\\text{chromate}\\\text{(carcinogen)}}}{CrO_4^{2-}} + \underset{\substack{\text{Iron}\\\text{particles}\\\text{(reductant)}}}{Fe(0)} + 4H_2O \rightarrow$$

$$\underset{\substack{\text{Mixed hydroxide precipitate}\\\text{(relatively safe)}}}{\underline{Cr(OH)_3(s) + Fe(OH)_3(s)} + 2OH^-}$$

$$\underset{\text{Pollutant}}{3H_2S} + \underset{\substack{\text{Ferrate(VI)}\\\text{(oxidant)}}}{8HFeO_4^-} + 6H_2O \rightarrow$$

$$\underset{\text{(safe products)}}{\underline{8Fe(OH)_3(s) + 3SO_4^{2-}} + 2OH^-}$$

A redox titration is based on an oxidation-reduction reaction between analyte and titrant. In addition to the many common analytes in chemistry, biology, and environmental and materials science that can be measured by redox titrations, exotic oxidation states of elements in uncommon materials such as superconductors and laser materials are measured by redox titrations. For example, chromium added to laser crystals to increase their efficiency is found in the common oxidation states $+3$ and $+6$, and the unusual $+4$ state. A redox titration is a good way to unravel the nature of this complex mixture of chromium ions.

This chapter introduces the theory of redox titrations and discusses some common reagents. A few of the oxidants and reductants in Table 15-1 can be used as titrants.[2] Most reductants used as titrants react with O_2 and, therefore, require protection from air.

15-1 The Shape of a Redox Titration Curve

Consider the titration of iron(II) with standard cerium(IV), monitored potentiometrically with Pt and calomel electrodes as shown in Figure 15-1. The titration reaction is

Titration reaction:

$$Ce^{4+} + Fe^{2+} \rightarrow Ce^{3+} + Fe^{3+} \qquad (15\text{-}1)$$

Ceric   Ferrous   Cerous   Ferric
titrant   analyte

for which $K \approx 10^{16}$ in 1 M $HClO_4$. Each mole of ceric ion oxidizes 1 mol of ferrous ion rapidly and quantitatively. The titration reaction creates a mixture of Ce^{4+}, Ce^{3+}, Fe^{2+}, and Fe^{3+} in the beaker in Figure 15-1. Box 15-1 describes the likely mechanism of Reaction 15-1.

At the Pt indicator electrode, *two* reactions come to equilibrium:

Indicator half-reaction: $\quad Fe^{3+} + e^- \rightleftharpoons Fe^{2+} \qquad E° = 0.767$ V $\qquad (15\text{-}2)$

Indicator half-reaction: $\quad Ce^{4+} + e^- \rightleftharpoons Ce^{3+} \qquad E° = 1.70$ V $\qquad (15\text{-}3)$

The potentials are the formal potentials that apply in 1 M $HClO_4$. The Pt indicator electrode responds to the relative concentrations (really, activities) of Ce^{4+} and Ce^{3+} or Fe^{3+} and Fe^{2+}.

We now set out to calculate how the cell voltage changes as Fe^{2+} is titrated with Ce^{4+}. The titration curve has three regions.

Region 1: Before the Equivalence Point

As each aliquot of Ce^{4+} is added, titration reaction 15-1 consumes Ce^{4+} and creates an equal number of moles of Ce^{3+} and Fe^{3+}. Prior to the equivalence point, excess unreacted Fe^{2+} remains in solution. Therefore, we can find the concentrations of Fe^{2+} and Fe^{3+} without difficulty. On the other hand, we cannot find the concentration of Ce^{4+} without solving a fancy little equilibrium problem. Because the amounts of Fe^{2+} and Fe^{3+} are both

> The titration reaction goes to completion after each addition of titrant. The equilibrium constant is $K = 10^{nE°/0.059\ 16}$ at 25°C.

> Equilibria 15-2 and 15-3 are both established at the Pt electrode.

> We can use either Reaction 15-2 or Reaction 15-3 to describe the cell voltage at any time. However, because we know $[Fe^{2+}]$ and $[Fe^{3+}]$, it is more convenient for now to use Reaction 15-2.

TABLE 15-1 Oxidizing and reducing agents

Oxidants		Reductants	
BiO_3^-	Bismuthate		Ascorbic acid (vitamin C)
BrO_3^-	Bromate		
Br_2	Bromine		
Ce^{4+}	Ceric		
CH_3—〇—$SO_2NCl^-Na^+$	Chloramine T	BH_4^-	Borohydride
		Cr^{2+}	Chromous
Cl_2	Chlorine	$S_2O_4^{2-}$	Dithionite
ClO_2	Chlorine dioxide	Fe^{2+}	Ferrous
$Cr_2O_7^{2-}$	Dichromate	N_2H_4	Hydrazine
FeO_4^{2-}	Ferrate(VI)		
H_2O_2	Hydrogen peroxide	HO—〇—OH	Hydroquinone
$Fe^{2+} + H_2O_2$	Fenton reagent[3]	NH_2OH	Hydroxylamine
OCl^-	Hypochlorite	H_3PO_2	Hypophosphorous acid
IO_3^-	Iodate		
I_2	Iodine		Retinol (vitamin A)
$Pb(acetate)_4$	Lead(IV) acetate		
HNO_3	Nitric acid		
O	Atomic oxygen	Sn^{2+}	Stannous
O_2	Dioxygen (oxygen)	SO_3^{2-}	Sulfite
O_3	Ozone	SO_2	Sulfur dioxide
$HClO_4$	Perchloric acid	$S_2O_3^{2-}$	Thiosulfate
IO_4^-	Periodate		
MnO_4^-	Permanganate		α-Tocopherol (vitamin E)[4]
$S_2O_8^{2-}$	Peroxydisulfate		

known, it is *convenient* to calculate the cell voltage by using Reaction 15-2 instead of Reaction 15-3.

E_+ is the potential of the Pt electrode connected to the positive terminal of the potentiometer in Figure 15-1. E_- is the potential of the calomel electrode connected to the negative terminal.

$$E = E_+ - E_- \tag{15-4}$$

$$E = \left[\underset{\substack{\uparrow \\ \text{Formal potential for } Fe^{3+} \\ \text{reduction in} \\ 1 \text{ M HClO}_4}}{0.767} - 0.059\ 16 \log\left(\frac{[Fe^{2+}]}{[Fe^{3+}]}\right) \right] - \underset{\substack{\uparrow \\ \text{Potential of} \\ \text{saturated calomel} \\ \text{electrode}}}{0.241} \tag{15-5}$$

$$E = 0.526 - 0.059\ 16 \log\left(\frac{[Fe^{2+}]}{[Fe^{3+}]}\right) \tag{15-6}$$

For Reaction 15-2, $E_+ = E°(Fe^{3+} | Fe^{2+})$ when $V = \frac{1}{2}V_e$.

One special point is reached before the equivalence point. When the volume of titrant is one-half of the amount required to reach the equivalence point ($V = \frac{1}{2}V_e$), $[Fe^{3+}] =$

BOX 15-1   Many Redox Reactions Are Atom-Transfer Reactions

Reaction 15-1 appears as if an electron moves from Fe^{2+} to Ce^{4+} to give Fe^{3+} and Ce^{3+}. In fact, this reaction and many others are thought to proceed through atom transfer, not electron transfer.[5] In this case, a hydrogen atom (a proton plus an electron) could be transferred from aqueous Fe^{2+} to aqueous Ce^{4+} species:

"Ce⁴⁺"
$Ce(H_2O)_7(OH)^{3+}$

"Fe²⁺"
$Fe(H_2O)_6^{2+}$

"Ce³⁺"
$Ce(H_2O)_8^{3+}$

"Fe³⁺"
$Fe(H_2O)_5(OH)^{2+}$

Other common redox reactions between metallic species could proceed via transfer of oxygen atoms or halogen atoms to effect net electron transfer from one metal to another.

$[Fe^{2+}]$. In this case, the log term is 0, and $E_+ = E°$ for the $Fe^{3+} \mid Fe^{2+}$ couple. *The point at which $V = \frac{1}{2}V_e$ is analogous to the point at which $pH = pK_a$ when $V = \frac{1}{2}V_e$ in an acid-base titration.*

The voltage at zero titrant volume cannot be calculated, because we do not know how much Fe^{3+} is present. If $[Fe^{3+}] = 0$, the voltage calculated with Equation 15-6 would be $-\infty$. In fact, there must be some Fe^{3+} in each reagent, either as an impurity or from oxidation of Fe^{2+} by atmospheric oxygen. In any case, the voltage could never be lower than that needed to reduce the solvent ($H_2O + e^- \rightarrow \frac{1}{2}H_2 + OH^-$).

Region 2: At the Equivalence Point

Exactly enough Ce^{4+} has been added to react with all the Fe^{2+}. Virtually all cerium is in the form Ce^{3+}, and virtually all iron is in the form Fe^{3+}. Tiny amounts of Ce^{4+} and Fe^{2+} are present at equilibrium. From the stoichiometry of Reaction 15-1, we can say that

$$[Ce^{3+}] = [Fe^{3+}] \tag{15-7}$$

$$[Ce^{4+}] = [Fe^{2+}] \tag{15-8}$$

To understand why Equations 15-7 and 15-8 are true, imagine that *all* the cerium and the iron have been converted into Ce^{3+} and Fe^{3+}. Because we are at the equivalence point, $[Ce^{3+}] = [Fe^{3+}]$. Now let Reaction 15-1 come to equilibrium:

$$Fe^{3+} + Ce^{3+} \rightleftharpoons Fe^{2+} + Ce^{4+} \qquad \text{(reverse of Reaction 15-1)}$$

If a little bit of Fe^{3+} goes back to Fe^{2+}, an equal number of moles of Ce^{4+} must be made. So $[Ce^{4+}] = [Fe^{2+}]$.

At any time, Reactions 15-2 and 15-3 are *both* in equilibrium at the Pt electrode. At the equivalence point, it is *convenient* to use both reactions to describe the cell voltage. The Nernst equations for these reactions are

$$E_+ = 0.767 - 0.059\ 16 \log\left(\frac{[Fe^{2+}]}{[Fe^{3+}]}\right) \tag{15-9}$$

$$E_+ = 1.70 - 0.059\ 16 \log\left(\frac{[Ce^{3+}]}{[Ce^{4+}]}\right) \tag{15-10}$$

At the equivalence point, we use both Reactions 15-2 and 15-3 to calculate the cell voltage. This is strictly a matter of algebraic convenience.

Here is where we stand: Equations 15-9 and 15-10 are each a statement of algebraic truth. But neither one alone allows us to find E_+, because we do not know exactly what tiny concentrations of Fe^{2+} and Ce^{4+} are present. It is possible to solve the four simultaneous equations 15-7 through 15-10 by first *adding* Equations 15-9 and 15-10:

$$2E_+ = 0.767 + 1.70 - 0.059\ 16 \log\left(\frac{[Fe^{2+}]}{[Fe^{3+}]}\right) - 0.059\ 16 \log\left(\frac{[Ce^{3+}]}{[Ce^{4+}]}\right)$$

$\log a + \log b = \log ab$
$-\log a - \log b = -\log ab$

$$2E_+ = 2.46_7 - 0.059\ 16 \log\left(\frac{[Fe^{2+}][Ce^{3+}]}{[Fe^{3+}][Ce^{4+}]}\right)$$

But, because $[Ce^{3+}] = [Fe^{3+}]$ and $[Ce^{4+}] = [Fe^{2+}]$ at the equivalence point, the ratio of concentrations in the log term is unity. Therefore, the logarithm is 0 and

$$2E_+ = 2.46_7 \text{ V} \Rightarrow E_+ = 1.23 \text{ V}$$

The cell voltage is

$$E = E_+ - E(\text{calomel}) = 1.23 - 0.241 = 0.99 \text{ V} \tag{15-11}$$

In this particular titration, the equivalence-point voltage is independent of the concentrations and volumes of the reactants.

Region 3: After the Equivalence Point

Now virtually all iron atoms are Fe^{3+}. The moles of Ce^{3+} equal the moles of Fe^{3+}, and there is a known excess of unreacted Ce^{4+}. Because we know both $[Ce^{3+}]$ and $[Ce^{4+}]$, it is *convenient* to use Reaction 15-3 to describe the chemistry at the Pt electrode:

After the equivalence point, we use Reaction 15-3 because we can easily calculate $[Ce^{3+}]$ and $[Ce^{4+}]$. It is not convenient to use Reaction 15-2, because we do not know the concentration of Fe^{2+}, which has been "used up."

$$E = E_+ - E(\text{calomel}) = \left[1.70 - 0.059\ 16 \log\left(\frac{[Ce^{3+}]}{[Ce^{4+}]}\right)\right] - 0.241 \tag{15-12}$$

At the special point when $V = 2V_e$, $[Ce^{3+}] = [Ce^{4+}]$ and $E_+ = E°(Ce^{4+} \mid Ce^{3+}) = 1.70$ V.

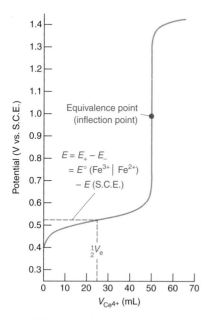

FIGURE 15-2 Theoretical curve for titration of 100.0 mL of 0.050 0 M Fe^{2+} with 0.100 M Ce^{4+} in 1 M $HClO_4$. You cannot calculate the potential for zero titrant, but you can start at a small volume such as $V_{Ce^{4+}} = 0.1$ mL.

Anyone with a serious need to calculate redox titration curves should use a spreadsheet with a more general set of equations than we use in this section.[7] The supplement at www.whfreeman.com/qca explains how to use spreadsheets to compute redox titration curves.

The curve in Figure 15-2 is essentially independent of the concentrations of analyte and titrant. The curve is symmetric near V_e because the stoichiometry is 1:1.

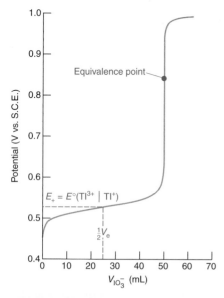

FIGURE 15-3 Theoretical curve for titration of 100.0 mL of 0.010 0 M Tl^+ with 0.010 0 M IO_3^- in 1.00 M HCl. The equivalence point at 0.842 V is not at the center of the steep part of the curve because the stoichiometry of the reaction is not 1:1.

Before V_e, the indicator electrode potential is fairly steady near $E°(Fe^{3+} \mid Fe^{2+}) = 0.77$ V.[6] After V_e, the indicator electrode potential levels off near $E°(Ce^{4+} \mid Ce^{3+}) = 1.70$ V. At V_e, there is a rapid rise in voltage.

EXAMPLE **Potentiometric Redox Titration**

Suppose that we titrate 100.0 mL of 0.050 0 M Fe^{2+} with 0.100 M Ce^{4+}, using the cell in Figure 15-1. The equivalence point occurs when $V_{Ce^{4+}} = 50.0$ mL. Calculate the cell voltage at 36.0, 50.0, and 63.0 mL.

Solution *At 36.0 mL:* This is 36.0/50.0 of the way to the equivalence point. Therefore, 36.0/50.0 of the iron is in the form Fe^{3+} and 14.0/50.0 is in the form Fe^{2+}. Putting $[Fe^{2+}]/[Fe^{3+}] = 14.0/36.0$ into Equation 15-6 gives $E = 0.550$ V.

At 50.0 mL: Equation 15-11 tells us that the cell voltage at the equivalence point is 0.99 V, regardless of the concentrations of reagents for this particular titration.

At 63.0 mL: The first 50.0 mL of cerium were converted into Ce^{3+}. There is an excess of 13.0 mL of Ce^{4+}, so $[Ce^{3+}]/[Ce^{4+}] = 50.0/13.0$ in Equation 15-12, and $E = 1.424$ V.

Test Yourself Find E at $V_{Ce^{4+}} = 20.0$ and 51.0 mL (**Answer:** 0.516, 1.358 V)

Shapes of Redox Titration Curves

The calculations above allow us to plot the titration curve in Figure 15-2, which shows potential as a function of the volume of added titrant. The equivalence point is marked by a steep rise in voltage. The value of E_+ at $\frac{1}{2}V_e$ is the formal potential of the $Fe^{3+} \mid Fe^{2+}$ couple, because the quotient $[Fe^{2+}]/[Fe^{3+}]$ is unity at this point. The voltage at any point in this titration depends only on the *ratio* of reactants; their *concentrations* do not figure in any calculations in this example. We expect, therefore, that the curve in Figure 15-2 will be independent of dilution. We should observe the same curve if both reactants were diluted by a factor of 10.

For Reaction 15-1, the titration curve in Figure 15-2 is symmetric near the equivalence point because the reaction stoichiometry is 1:1. Figure 15-3 shows the curve calculated for the titration of Tl^+ by IO_3^- in 1.00 M HCl.

$$IO_3^- + 2Tl^+ + 2Cl^- + 6H^+ \rightarrow ICl_2^- + 2Tl^{3+} + 3H_2O \qquad (15\text{-}13)$$

The curve is *not symmetric about the equivalence point* because the stoichiometry of reactants is 2:1, not 1:1. Still, the curve is so steep near the equivalence point that negligible error is introduced if the center of the steep part is taken as the end point. Demonstration 15-1 provides an example of an asymmetric titration curve whose shape also depends on the pH of the reaction medium.

The voltage change near the equivalence point increases as the difference between $E°$ of the two redox couples in the titration increases. The larger the difference in $E°$, the greater the equilibrium constant for the titration reaction. For Figure 15-2, half-reactions 15-2 and 15-3 differ by 0.93 V, and there is a large break at the equivalence point in the titration curve. For Figure 15-3, the half-reactions differ by 0.47 V, so there is a smaller break at the equivalence point.

$$IO_3^- + 2Cl^- + 6H^+ + 4e^- \rightleftharpoons ICl_2^- + 3H_2O \qquad E° = 1.24 \text{ V}$$
$$Tl^{3+} + 2e^- \rightleftharpoons Tl^+ \qquad E° = 0.77 \text{ V}$$

Clearest results are achieved with the strongest oxidizing and reducing agents. The same rule applies to acid-base titrations, where strong-acid or strong-base titrants give the sharpest break at the equivalence point.

15-2 Finding the End Point

As in acid-base titrations, indicators and electrodes are commonly used to find the end point of a redox titration.

This reaction illustrates many principles of potentiometric titrations:

$$MnO_4^- + 5Fe^{2+} + 8H^+ \rightarrow Mn^{2+} + 5Fe^{3+} + 4H_2O \qquad (A)$$

Titrant   Analyte

Dissolve 0.60 g of $Fe(NH_4)_2(SO_4)_2 \cdot 6H_2O$ (FM 392.13; 1.5 mmol) in 400 mL of 1 M H_2SO_4. Titrate the well-stirred solution with 0.02 M $KMnO_4$ ($V_e \approx 15$ mL), using Pt and calomel electrodes with a pH meter as a potentiometer. Before use, zero the meter by connecting the two inputs directly to each other and setting the millivolt scale to 0.

Calculate some points on the theoretical titration curve before performing the experiment. Then compare the theoretical and experimental results. Also note the coincidence of the potentiometric and visual end points.

Question Potassium permanganate is purple, and all the other species in this titration are colorless (or very faintly colored). What color change is expected at the equivalence point?

To calculate the theoretical curve, we use the following half-reactions:

$$Fe^{3+} + e^- \rightleftharpoons Fe^{2+} \qquad E^\circ = 0.68 \text{ V in 1 M } H_2SO_4 \qquad (B)$$

$$MnO_4^- + 8H^+ + 5e^- \rightarrow Mn^{2+} + 4H_2O \qquad E^\circ = 1.507 \text{ V} \qquad (C)$$

Prior to V_e, calculations are similar to those in Section 15-1 for the titration of Fe^{2+} by Ce^{4+}, but with $E^\circ = 0.68$ V. After V_e, you can find the potential by using Reaction C. For example, suppose that you titrate 0.400 L of 3.75 mM Fe^{2+} with 0.020 0 M $KMnO_4$. From the stoichiometry of Reaction A, $V_e = 15.0$ mL. When you have added 17.0 mL of $KMnO_4$, the concentrations of species in Reaction C are $[Mn^{2+}] = 0.719$ mM, $[MnO_4^-] = 0.095\ 9$ mM, and $[H^+] = 0.959$ M (neglecting the small quantity of H^+ consumed in the titration). The voltage should be

$$E = E_+ - E(\text{calomel})$$

$$= \left[1.507 - \frac{0.059\ 16}{5}\log\left(\frac{[Mn^{2+}]}{[MnO_4^-][H^+]^8}\right)\right] - 0.241$$

$$= \left[1.507 - \frac{0.059\ 16}{5}\log\left(\frac{7.19 \times 10^{-4}}{(9.59 \times 10^{-5})(0.959)^8}\right)\right] - 0.241$$

$$= 1.254 \text{ V}$$

To calculate the voltage at V_e, we add the Nernst equations for Reactions B and C, as we did for the cerium and iron reactions in Section 15-1. Before doing so, however, multiply the permanganate equation by 5 so that we can add the log terms:

$$E_+ = 0.68 - 0.059\ 16 \log\left(\frac{[Fe^{2+}]}{[Fe^{3+}]}\right)$$

$$5E_+ = 5\left[1.507 - \frac{0.059\ 16}{5}\log\left(\frac{[Mn^{2+}]}{[MnO_4^-][H^+]^8}\right)\right]$$

Now add the two equations to get

$$6E_+ = 8.215 - 0.059\ 16 \log\left(\frac{[Mn^{2+}][Fe^{2+}]}{[MnO_4^-][Fe^{3+}][H^+]^8}\right) \qquad (D)$$

The stoichiometry of the titration reaction A tells us that, at V_e, $[Fe^{3+}] = 5[Mn^{2+}]$ and $[Fe^{2+}] = 5[MnO_4^-]$. Substituting these values into Equation D gives

$$6E_+ = 8.215 - 0.059\ 16 \log\left(\frac{[Mn^{2+}](5[MnO_4^-])}{[MnO_4^-](5[Mn^{2+}])[H^+]^8}\right)$$

$$= 8.215 - 0.059\ 16 \log\left(\frac{1}{[H^+]^8}\right) \qquad (E)$$

Inserting the concentration of $[H^+]$, which is $(400/415)(1.00 \text{ M}) = 0.964$ M, we find

$$6E_+ = 8.215 - 0.059\ 16 \log\left(\frac{1}{(0.964)^8}\right) \quad \Rightarrow \quad E_+ = 1.368 \text{ V}$$

The predicted cell voltage at V_e is $E = E_+ - E(\text{calomel}) = 1.368 - 0.241 = 1.127$ V.

Redox Indicators

A **redox indicator** is a compound that changes color when it goes from its oxidized to its reduced state. The indicator ferroin changes from pale blue (almost colorless) to red.

Oxidized ferroin (pale blue) In(oxidized)   Reduced ferroin (red) In(reduced)

To predict the potential range over which the indicator color will change, we first write a Nernst equation for the indicator.

$$\text{In(oxidized)} + ne^- \rightleftharpoons \text{In(reduced)}$$

$$E = E^\circ - \frac{0.059\ 16}{n}\log\left(\frac{[\text{In(reduced)}]}{[\text{In(oxidized)}]}\right) \qquad (15\text{-}14)$$

As with acid-base indicators, the color of In(reduced) will be observed when

$$\frac{[\text{In(reduced)}]}{[\text{In(oxidized)}]} \gtrsim \frac{10}{1}$$

and the color of In(oxidized) will be observed when

$$\frac{[\text{In(reduced)}]}{[\text{In(oxidized)}]} \lesssim \frac{1}{10}$$

Putting these quotients into Equation 15-14 tells us that the color change will occur over the range

A redox indicator changes color over a range of $\pm(59/n)$ mV, centered at $E°$ for the indicator. n is the number of electrons in the indicator half-reaction.

$$E = \left(E° \ \pm \ \frac{0.059\ 16}{n} \right) \text{volts}$$

For ferroin, with $E° = 1.147$ V (Table 15-2), we expect the color change to occur in the approximate range 1.088 V to 1.206 V with respect to the standard hydrogen electrode. With a saturated calomel reference electrode, the indicator transition range will be

Figure 14-6 should help you understand Equation 15-15.

$$\begin{pmatrix} \text{Indicator transition} \\ \text{range versus calomel} \\ \text{electrode (S.C.E.)} \end{pmatrix} = \begin{pmatrix} \text{Transition range} \\ \text{versus standard hydrogen} \\ \text{electrode (S.H.E.)} \end{pmatrix} - E(\text{calomel}) \qquad \textbf{(15-15)}$$

$$= (1.088 \text{ to } 1.206) - (0.241)$$

$$= 0.847 \text{ to } 0.965 \text{ V (versus S.C.E.)}$$

The indicator transition range should overlap the steep part of the titration curve.

Ferroin would therefore be a useful indicator for the titrations in Figures 15-2 and 15-3.

The larger the difference in standard potential between titrant and analyte, the greater the break in the titration curve at the equivalence point. A redox titration is usually feasible if the difference between analyte and titrant is $\gtrsim 0.2$ V. However, the end point of such a titration is not very sharp and is best detected potentiometrically. If the difference in formal potentials is $\gtrsim 0.4$ V, then a redox indicator usually gives a satisfactory end point.

Gran Plot

With the apparatus in Figure 15-1, we measure electrode potential, E, versus volume of titrant, V, during a redox titration. The end point is the maximum of the first derivative, $\Delta E/\Delta V$, or the zero crossing of the second derivative, $\Delta(\Delta E/\Delta V)/\Delta V$ (Figure 10-5).

A more accurate way to use potentiometric data is to prepare a Gran plot[8,9] as we did for acid-base titrations in Section 10-5. The Gran plot uses data from well before the equivalence point (V_e) to locate V_e. Potentiometric data taken close to V_e are the least accurate because electrodes are slow to equilibrate with species in solution when one member of a redox couple is nearly used up.

TABLE 15-2 Redox indicators

Indicator	Color		$E°$
	Oxidized	Reduced	
Phenosafranine	Red	Colorless	0.28
Indigo tetrasulfonate	Blue	Colorless	0.36
Methylene blue	Blue	Colorless	0.53
Diphenylamine	Violet	Colorless	0.75
4′-Ethoxy-2,4-diaminoazobenzene	Yellow	Red	0.76
Diphenylamine sulfonic acid	Red-violet	Colorless	0.85
Diphenylbenzidine sulfonic acid	Violet	Colorless	0.87
Tris(2,2′-bipyridine)iron	Pale blue	Red	1.120
Tris(1,10-phenanthroline)iron (ferroin)	Pale blue	Red	1.147
Tris(5-nitro-1,10-phenanthroline)iron	Pale blue	Red-violet	1.25
Tris(2,2′-bipyridine)ruthenium	Pale blue	Yellow	1.29

For the oxidation of Fe^{2+} to Fe^{3+}, the potential prior to V_e is

$$E = \left[E^{\circ\prime} - 0.059\ 16 \log\left(\frac{[Fe^{2+}]}{[Fe^{3+}]}\right) \right] - E_{ref} \qquad (15\text{-}16)$$

where $E^{\circ\prime}$ is the formal potential for $Fe^{3+}\,|\,Fe^{2+}$ and E_{ref} is the potential of the reference electrode (which we have also been calling E_-). If the volume of titrant is V and if the reaction goes "to completion" with each addition of titrant, then $[Fe^{2+}]/[Fe^{3+}] = (V_e - V)/V$. Substituting this expression into Equation 15-16 and rearranging, we eventually stumble onto a linear equation:

$$\underbrace{V \cdot 10^{-nE/0.059\ 16}}_{y} = \underbrace{V_e \cdot 10^{-n(E_{ref} - E^{\circ\prime})/0.059\ 16}}_{b} - \underbrace{V}_{x} \cdot \underbrace{10^{-n(E_{ref} - E^{\circ\prime})/0.059\ 16}}_{m} \qquad (15\text{-}17)$$

where n is the number of electrons in the half-reaction at the indicator electrode.

A graph of $V \cdot 10^{-nE/0.059\ 16}$ versus V should be a straight line with x-intercept $= V_e$ (Figure 15-4). If the ionic strength of the reaction is constant, the activity coefficients are constant, and Equation 15-17 gives a straight line over a wide volume range. If ionic strength varies as titrant is added, we just use the last 10–20% of the data prior to V_e.

The Starch-Iodine Complex

Numerous analytical procedures are based on redox titrations involving iodine. Starch[10] is the indicator of choice for these procedures because it forms an intense blue complex with iodine. Starch is not a redox indicator; it responds specifically to the presence of I_2, not to a change in redox potential.

The active fraction of starch is amylose, a polymer of the sugar α-D-glucose, with the repeating unit shown in Figure 15-5. Small molecules can fit into the center of the coiled, helical polymer. In the presence of starch, iodine forms I_6 chains inside the amylose helix and the color turns dark blue.

I·I·I·I·I·I

Starch is readily biodegraded, so it should be freshly dissolved or the solution should contain a preservative, such as HgI_2 (~1 mg/100 mL) or thymol. A hydrolysis product of starch is glucose, which is a reducing agent. Therefore, partially hydrolyzed starch solution can be a source of error in a redox titration.

The constant 0.059 16 V is $(RT \ln 10)/nF$, where R is the gas constant, T is 298.15 K, F is the Faraday constant, and n is the number of electrons in the $Fe^{3+}\,|\,Fe^{2+}$ redox half-reaction ($n = 1$). For $T \neq 298.15$ K or $n \neq 1$, the number 0.059 16 will change.

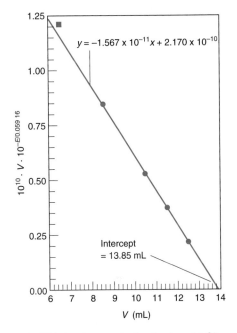

FIGURE 15-4 Gran plot for titration of Fe^{2+} by Ce^{4+} in Exercise 15-D.[8] The line was fit to the four points shown by circles. In the function on the ordinate, the value of n is 1. Numerical values were multiplied by 10^{10} for ease of display. Multiplication does not change the x-intercept.

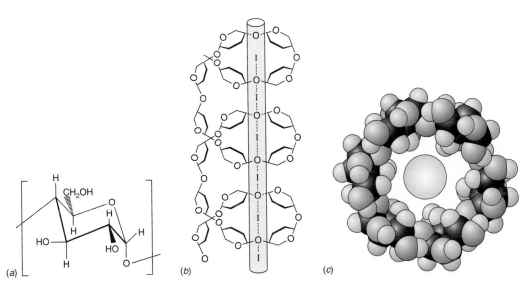

FIGURE 15-5 (a) Structure of the repeating unit of amylose. (b) Schematic structure of the starch-iodine complex. The amylose chain forms a helix around I_6 units. [Adapted from A. T. Calabrese and A. Khan, "Amylose-Iodine Complex Formation with KI: Evidence for Absence of Iodide Ions within the Complex," *J. Polymer Sci.* 1999, A37, 2711.] (c) View down the starch helix, showing iodine inside the helix.[10] [Figure kindly provided by R. D. Hancock, Power Engineering, Salt Lake City.]

FIGURE 15-6 A column filled with a solid reagent used for prereduction of analyte is called a *reductor*.

Analyte solution

Solid reductant

Sintered-glass disk

Reduced analyte

In *disproportionation*, a reactant is converted into products in higher and lower oxidation states. The compound oxidizes and reduces *itself*.

Challenge Write a half-reaction in which H_2O_2 behaves as an oxidant and a half-reaction in which it behaves as a reductant.

15-3 Adjustment of Analyte Oxidation State

Sometimes we need to adjust the oxidation state of analyte before it can be titrated. For example, Mn^{2+} can be **preoxidized** to MnO_4^- and then titrated with standard Fe^{2+}. Preadjustment must be quantitative, and you must eliminate excess preadjustment reagent so that it will not interfere in the subsequent titration.

Preoxidation

Several powerful oxidants can be easily removed after preoxidation. *Peroxydisulfate* ($S_2O_8^{2-}$, also called *persulfate*) is a strong oxidant that requires Ag^+ as a catalyst.

$$S_2O_8^{2-} + Ag^+ \rightarrow SO_4^{2-} + \underbrace{SO_4^- + Ag^{2+}}_{\substack{\text{Two powerful} \\ \text{oxidants}}}$$

Excess reagent is destroyed by boiling the solution after oxidation of analyte is complete.

$$2S_2O_8^{2-} + 2H_2O \xrightarrow{\text{boiling}} 4SO_4^{2-} + O_2 + 4H^+$$

The $S_2O_8^{2-}$ and Ag^+ mixture oxidizes Mn^{2+} to MnO_4^-, Ce^{3+} to Ce^{4+}, Cr^{3+} to $Cr_2O_7^{2-}$, and VO^{2+} to VO_2^+.

Silver(II) oxide (AgO) dissolves in concentrated mineral acids to give Ag^{2+}, with oxidizing power similar to $S_2O_8^{2-}$ plus Ag^+. Excess Ag^{2+} can be removed by boiling:

$$4Ag^{2+} + 2H_2O \xrightarrow{\text{boiling}} 4Ag^+ + O_2 + 4H^+$$

Solid *sodium bismuthate* ($NaBiO_3$) has an oxidizing strength similar to that of Ag^{2+} and $S_2O_8^{2-}$. Excess solid oxidant is removed by filtration.

Hydrogen peroxide is a good oxidant in basic solution. It can transform Co^{2+} into Co^{3+}, Fe^{2+} into Fe^{3+}, and Mn^{2+} into MnO_2. In acidic solution, it can *reduce* $Cr_2O_7^{2-}$ to Cr^{3+} and MnO_4^- to Mn^{2+}. Excess H_2O_2 spontaneously **disproportionates** in boiling water.

$$2H_2O_2 \xrightarrow{\text{boiling}} O_2 + 2H_2O$$

Prereduction

Stannous chloride ($SnCl_2$) will reduce Fe^{3+} to Fe^{2+} in hot HCl. Excess reductant is destroyed by adding excess $HgCl_2$:

$$Sn^{2+} + 2HgCl_2 \rightarrow Sn^{4+} + Hg_2Cl_2 + 2Cl^-$$

The Fe^{2+} is then titrated with an oxidant.

Chromous chloride is a powerful reductant sometimes used to **prereduce** analyte to a lower oxidation state. Excess Cr^{2+} is oxidized by atmospheric O_2. *Sulfur dioxide* and *hydrogen sulfide* are mild reducing agents that can be expelled by boiling an acidic solution after the reduction is complete.

An important prereduction technique uses a column packed with a solid reducing agent. Figure 15-6 shows the *Jones reductor*, which contains zinc coated with zinc *amalgam*. An **amalgam** is a solution of anything in mercury. The amalgam is prepared by mixing granular zinc with 2 wt% aqueous $HgCl_2$ for 10 min and then washing with water. You can reduce Fe^{3+} to Fe^{2+} by passage through a Jones reductor, using 1 M H_2SO_4 as solvent. Wash the column well with water and titrate the combined washings with standard MnO_4^-, Ce^{4+}, or $Cr_2O_7^{2-}$. Pass a solution containing only the matrix through the reductor in the same manner as the unknown for a blank determination.

Most reduced analytes are reoxidized by atmospheric oxygen. To prevent oxidation, the reduced analyte can be collected in an acidic solution of Fe^{3+}. Ferric ion is reduced to Fe^{2+}, which is stable in acid. The Fe^{2+} is then titrated with an oxidant. By this means, elements such as Cr, Ti, V, and Mo can be indirectly analyzed.

Zinc is a powerful reducing agent, with $E° = -0.764$ for the reaction $Zn^{2+} + 2e^- \rightleftharpoons Zn(s)$, so the Jones reductor is not very selective. The *Walden reductor*, filled with solid Ag and 1 M HCl, is more selective. The reduction potential for $Ag \mid AgCl$ (0.222 V) is high enough that species such as Cr^{3+} and TiO^{2+} are not reduced and therefore do not interfere in the analysis of a metal such as Fe^{3+}. Another selective reductor uses granular Cd metal. To determine levels of nitrogen oxides for air-pollution monitoring,[11] the gases are first converted into NO_3^-, which is not easy to analyze. Passing nitrate through a Cd-filled column reduces NO_3^- to NO_2^-, for which a convenient spectrophotometric analysis is available.

15-4 Oxidation with Potassium Permanganate

Potassium permanganate ($KMnO_4$) is a strong oxidant with an intense violet color. In strongly acidic solutions (pH $\leq$1), it is reduced to colorless Mn^{2+}.

$$\underset{\text{Permanganate}}{MnO_4^-} + 8H^+ + 5e^- \rightleftharpoons \underset{\text{Manganous}}{Mn^{2+}} + 4H_2O \qquad E° = 1.507 \text{ V}$$

In neutral or alkaline solution, the product is the brown solid, MnO_2.

$$MnO_4^- + 4H^+ + 3e^- \rightleftharpoons \underset{\substack{\text{Manganese} \\ \text{dioxide}}}{MnO_2(s)} + 2H_2O \qquad E° = 1.692 \text{ V}$$

In strongly alkaline solution (2 M NaOH), green manganate ion is produced.

$$MnO_4^- + e^- \rightleftharpoons \underset{\text{Manganate}}{MnO_4^{2-}} \qquad E° = 0.56 \text{ V}$$

Representative permanganate titrations are listed in Table 15-3. For titrations in strongly acidic solution, $KMnO_4$ serves as its own indicator because the product, Mn^{2+},

$KMnO_4$ serves as its own indicator in acidic solution.

TABLE 15-3 Analytical applications of permanganate titrations

Species analyzed	Oxidation reaction	Notes
Fe^{2+}	$Fe^{2+} \rightleftharpoons Fe^{3+} + e^-$	Fe^{3+} is reduced to Fe^{2+} with Sn^{2+} or a Jones reductor. Titration is carried out in 1 M H_2SO_4 or 1 M HCl containing Mn^{2+}, H_3PO_4, and H_2SO_4. Mn^{2+} inhibits oxidation of Cl^- by MnO_4^-. H_3PO_4 complexes Fe^{3+} to prevent formation of yellow Fe^{3+}-chloride complexes.
$H_2C_2O_4$	$H_2C_2O_4 \rightleftharpoons 2CO_2 + 2H^+ + 2e^-$	Add 95% of titrant at 25°C, then complete titration at 55°–60°C.
Br^-	$Br^- \rightleftharpoons \frac{1}{2}Br_2(g) + e^-$	Titrate in boiling 2 M H_2SO_4 to remove $Br_2(g)$.
H_2O_2	$H_2O_2 \rightleftharpoons O_2(g) + 2H^+ + 2e^-$	Titrate in 1 M H_2SO_4.
HNO_2	$HNO_2 + H_2O \rightleftharpoons NO_3^- + 3H^+ + 2e^-$	Add excess standard $KMnO_4$ and back-titrate after 15 min at 40°C with Fe^{2+}.
As^{3+}	$H_3AsO_3 + H_2O \rightleftharpoons H_3AsO_4 + 2H^+ + 2e^-$	Titrate in 1 M HCl with KI or ICl catalyst.
Sb^{3+}	$H_3SbO_3 + H_2O \rightleftharpoons H_3SbO_4 + 2H^+ + 2e^-$	Titrate in 2 M HCl.
Mo^{3+}	$Mo^{3+} + 2H_2O \rightleftharpoons MoO_2^{2+} + 4H^+ + 3e^-$	Reduce Mo in a Jones reductor, and run the Mo^{3+} into excess Fe^{3+} in 1 M H_2SO_4. Titrate the Fe^{2+} formed.
W^{3+}	$W^{3+} + 2H_2O \rightleftharpoons WO_2^{2+} + 4H^+ + 3e^-$	Reduce W with Pb(Hg) at 50°C and titrate in 1 M HCl.
U^{4+}	$U^{4+} + 2H_2O \rightleftharpoons UO_2^{2+} + 4H^+ + 2e^-$	Reduce U to U^{3+} with a Jones reductor. Expose to air to produce U^{4+}, which is titrated in 1 M H_2SO_4.
Ti^{3+}	$Ti^{3+} + H_2O \rightleftharpoons TiO^{2+} + 2H^+ + e^-$	Reduce Ti to Ti^{3+} with a Jones reductor, and run the Ti^{3+} into excess Fe^{3+} in 1 M H_2SO_4. Titrate the Fe^{2+} that is formed.
$Mg^{2+}, Ca^{2+}, Sr^{2+},$ $Ba^{2+}, Zn^{2+}, Co^{2+},$ $La^{3+}, Th^{4+}, Pb^{2+},$ Ce^{3+}, BiO^+, Ag^+	$H_2C_2O_4 \rightleftharpoons 2CO_2 + 2H^+ + 2e^-$	Precipitate the metal oxalate. Dissolve in acid and titrate the $H_2C_2O_4$.
$S_2O_8^{2-}$	$S_2O_8^{2-} + 2Fe^{2+} + 2H^+ \rightleftharpoons 2Fe^{3+} + 2HSO_4^-$	Peroxydisulfate is added to excess standard Fe^{2+} containing H_3PO_4. Unreacted Fe^{2+} is titrated with MnO_4^-.
PO_4^{3-}	$Mo^{3+} + 2H_2O \rightleftharpoons MoO_2^{2+} + 4H^+ + 3e^-$	$(NH_4)_3PO_4 \cdot 12MoO_3$ is precipitated and dissolved in H_2SO_4. The Mo(VI) is reduced (as above) and titrated.

is colorless (Color Plate 7). The end point is taken as the first persistent appearance of pale pink MnO_4^-. If the titrant is too dilute to be seen, an indicator such as ferroin can be used.

Preparation and Standardization

KMnO₄ is not a primary standard.

Potassium permanganate is not a primary standard because traces of MnO_2 are invariably present. In addition, distilled water usually contains enough organic impurities to reduce some freshly dissolved MnO_4^- to MnO_2. To prepare a 0.02 M stock solution, dissolve $KMnO_4$ in distilled water, boil it for an hour to hasten the reaction between MnO_4^- and organic impurities, and filter the resulting mixture through a clean, sintered-glass filter to remove precipitated MnO_2. Do not use filter paper (organic matter!) for the filtration. Store the reagent in a dark glass bottle. Aqueous $KMnO_4$ is unstable by virtue of the reaction

$$4MnO_4^- + 2H_2O \rightarrow 4MnO_2(s) + 3O_2 + 4OH^-$$

which is slow in the absence of MnO_2, Mn^{2+}, heat, light, acids, and bases. Permanganate should be standardized often for the most accurate work. Prepare and standardize fresh dilute solutions from 0.02 M stock solution, using water distilled from alkaline $KMnO_4$.

Potassium permanganate can be standardized by titration of sodium oxalate ($Na_2C_2O_4$) by Reaction 1-7 or pure electrolytic iron wire. Dissolve dry (105°C, 2 h) sodium oxalate (available as a 99.9–99.95% pure primary standard) in 1 M H_2SO_4 and treat it with 90–95% of the required $KMnO_4$ solution at room temperature. Then warm the solution to 55–60°C and complete the titration by slow addition of $KMnO_4$. Subtract a blank value to account for the quantity of titrant (usually one drop) needed to impart a pink color to the solution.

If pure Fe wire is used as a standard, dissolve it in warm 1.5 M H_2SO_4 under N_2. The product is Fe^{2+}, and the cooled solution can be used to standardize $KMnO_4$ (or other oxidants) with no special precautions. Adding 5 mL of 86 wt% H_3PO_4 per 100 mL of solution masks the yellow color of Fe^{3+} and makes the end point easier to see. Ferrous ammonium sulfate, $Fe(NH_4)_2(SO_4)_2 \cdot 6H_2O$, and ferrous ethylenediammonium sulfate, $Fe(H_3NCH_2CH_2NH_3)(SO_4)_2 \cdot 2H_2O$, are sufficiently pure to be standards for most purposes.

15-5 Oxidation with Ce⁴⁺

Reduction of Ce^{4+} to Ce^{3+} proceeds cleanly in acidic solutions. The aquo ion, $Ce(H_2O)_n^{4+}$, probably does not exist in any of these solutions, because Ce(IV) binds anions (ClO_4^-, SO_4^{2-}, NO_3^-, Cl^-). Variation of the $Ce^{4+} \mid Ce^{3+}$ formal potential with the medium is indicative of these interactions:

Different formal potentials imply that different cerium species are present in each solution.

$$Ce^{4+} + e^- \rightleftharpoons Ce^{3+} \qquad \text{Formal potential} \begin{cases} 1.70 \text{ V in 1 F } HClO_4 \\ 1.61 \text{ V in 1 F } HNO_3 \\ 1.47 \text{ V in 1 F } HCl \\ 1.44 \text{ V in 1 F } H_2SO_4 \end{cases}$$

Ce^{4+} is yellow and Ce^{3+} is colorless, but the color change is not distinct enough for cerium to be its own indicator. Ferroin and other substituted phenanthroline redox indicators (Table 15-2) are well suited to titrations with Ce^{4+}.

Ce^{4+} can be used in place of $KMnO_4$ in most procedures. In the oscillating reaction in Demonstration 14-1, Ce^{4+} oxidizes malonic acid to CO_2 and formic acid:

$$CH_2(CO_2H)_2 + 2H_2O + 6Ce^{4+} \rightarrow 2CO_2 + HCO_2H + 6Ce^{3+} + 6H^+$$

Malonic acid   Formic acid

This reaction can be used for quantitative analysis of malonic acid by heating a sample in 4 M $HClO_4$ with excess standard Ce^{4+} and back-titrating unreacted Ce^{4+} with Fe^{2+}. Analogous procedures are available for many alcohols, aldehydes, ketones, and carboxylic acids.

Preparation and Standardization

Primary-standard-grade ammonium hexanitratocerate(IV), $(NH_4)_2Ce(NO_3)_6$, can be dissolved in 1 M H_2SO_4 and used directly. Although the oxidizing strength of Ce^{4+} is greater in $HClO_4$ or HNO_3, these solutions undergo slow photochemical decomposition with concomitant oxidation of water. Ce^{4+} in H_2SO_4 is stable indefinitely, despite the fact that the reduction potential of 1.44 V is great enough to oxidize H_2O to O_2. Reaction with water is slow, even though it is thermodynamically favorable. Solutions in HCl are unstable because Cl^- is oxidized to Cl_2—rapidly, when the solution is hot. Sulfuric acid solutions of Ce^{4+} can be used for titrations of unknowns in HCl because the reaction with analyte is fast, whereas reaction with Cl^- is slow. Less expensive salts, including $Ce(HSO_4)_4$, $(NH_4)_4Ce(SO_4)_4 \cdot 2H_2O$, and $CeO_2 \cdot xH_2O$ (also called $Ce(OH)_4$), are adequate for preparing titrants that are subsequently standardized with $Na_2C_2O_4$ or Fe as described for MnO_4^-.

15-6 Oxidation with Potassium Dichromate

In acidic solution, orange dichromate ion is a powerful oxidant that is reduced to chromic ion:

$$Cr_2O_7^{2-} + 14H^+ + 6e^- \rightleftharpoons 2Cr^{3+} + 7H_2O \qquad E° = 1.36 \ V$$

Dichromate   Chromic

Cr(VI) waste is carcinogenic and should not be poured down the drain. See page 31 for disposal method.

In 1 M HCl, the formal potential is just 1.00 V and, in 2 M H_2SO_4, it is 1.11 V; so dichromate is a less powerful oxidizing agent than MnO_4^- or Ce^{4+}. In basic solution, $Cr_2O_7^{2-}$ is converted into yellow chromate ion (CrO_4^{2-}), whose oxidizing power is nil:

$$CrO_4^{2-} + 4H_2O + 3e^- \rightleftharpoons Cr(OH)_3(s, \text{hydrated}) + 5OH^- \qquad E° = -0.12 \ V$$

Potassium dichromate, $K_2Cr_2O_7$, is a primary standard. Its solutions are stable, and it is cheap. Dichromate is orange and Cr^{3+} complexes range from green to violet, so indicators with distinctive color changes, such as diphenylamine sulfonic acid or diphenylbenzidine sulfonic acid, are used to find a dichromate end point. Alternatively, reactions can be monitored with Pt and calomel electrodes.

$K_2Cr_2O_7$ is not as strong an oxidant as $KMnO_4$ or Ce^{4+}. It is employed chiefly for the determination of Fe^{2+} and, indirectly, for species that will oxidize Fe^{2+} to Fe^{3+}. For indirect analyses, the unknown is treated with a measured excess of Fe^{2+}. Then unreacted Fe^{2+} is titrated with $K_2Cr_2O_7$. For example, ClO_3^-, NO_3^-, MnO_4^-, and organic peroxides can be analyzed this way. Box 15-2 describes the use of dichromate in water pollution analysis.

Diphenylbenzidine sulfonate (reduced, colorless)

Diphenylbenzidine sulfonate (oxidized, violet)

$+ 2H^+ + 2e^-$

15-7 Methods Involving Iodine

When a reducing analyte is titrated with iodine (to produce I^-), the method is called *iodimetry*. In *iodometry*, an oxidizing analyte is added to excess I^- to produce iodine, which is then titrated with standard thiosulfate solution.

Iodimetry: titration *with* iodine
Iodometry: titration *of* iodine produced by a chemical reaction

Molecular iodine is only slightly soluble in water (1.3×10^{-3} M at 20°C), but its solubility is enhanced by complexation with iodide.

$$I_2(aq) + I^- \rightleftharpoons I_3^- \qquad K = 7 \times 10^2$$

Iodine   Iodide   Triiodide

A solution made from 1.5 mM I_2 + 1.5 mM KI in water contains[20]

0.9 mM I_2	5 μM I_5^-
0.9 mM I^-	40 nM I_6^{2-}
0.6 mM I_3^-	0.3 μM HOI

A typical 0.05 M solution of I_3^- for titrations is prepared by dissolving 0.12 mol of KI plus 0.05 mol of I_2 in 1 L of water. When we speak of using iodine as a titrant, we almost always mean that we are using a solution of I_2 plus excess I^-.

Use of Starch Indicator

As described in Section 15-2, starch is used as an indicator for iodine. In a solution with no other colored species, it is possible to see the color of ~5 μM I_3^-. With starch, the limit of detection is extended by about a factor of 10.

In iodimetry (titration *with* I_3^-), starch can be added at the beginning of the titration. The first drop of excess I_3^- after the equivalence point causes the solution to turn dark blue.

BOX 15-2 **Environmental Carbon Analysis and Oxygen Demand**

Drinking water and industrial waste streams are partially characterized and regulated on the basis of their carbon content and oxygen demand.[12] *Inorganic carbon* (IC) is the $CO_2(g)$ liberated when water is acidified to pH < 2 with H_3PO_4 and purged with Ar or N_2. IC corresponds to CO_3^{2-} and HCO_3^- in the sample. After inorganic carbon is removed, *total organic carbon* (TOC) is the CO_2 produced by oxidizing the remaining organic matter:

TOC analysis: Organic carbon $\xrightarrow[\text{metal catalyst}]{O_2/\sim700°C}$ CO_2

Total carbon (TC) is defined as the sum TC = TOC + IC.

Different oxidation techniques produce different values for TOC, because not all organic matter is oxidized by each technique. The state of the art is such that TOC is defined by the result obtained with a particular instrument.

Commercial instruments that measure TOC by thermal oxidation have detection limits of 4–50 ppb (4–50 μg C/L). A typical 20-μL water sample is analyzed in 3 min with infrared absorption used to measure CO_2. Other instruments oxidize organic matter by irradiating a suspension of solid TiO_2 catalyst (0.2 g/L) in water at pH 3.5 with ultraviolet light.[13] Light creates electron-hole pairs (Section 14-8) in the TiO_2. Holes oxidize H_2O to hydroxyl radical (HO·), a powerful oxidant that converts organic carbon into CO_2, which is measured by the electrical conductivity of carbonic acid.[14] Color Plate 8 shows an instrument in which $K_2S_2O_8$ in acid is exposed to ultraviolet radiation to generate sulfate radical (·SO_4^-), which oxidizes organic matter to CO_2. (Pure TiO_2 hardly absorbs visible light, so it cannot use sunlight efficiently. By doping TiO_2 with ~1 wt% carbon, the efficiency of using visible light is markedly increased.[15])

Ultraviolet radiation absorbed by TiO_2 creates an electron-hole pair. The hole oxidizes H_2O to the powerful oxidant HO•. The electron reduces O_2 to H_2O by a chain of reactions. TiO_2 is a catalyst and O_2 is consumed in the net reaction: organic C + O_2 → CO_2.

TOC is widely used to determine compliance with discharge laws. Municipal and industrial wastewater typically has TOC > 1 mg C/mL. Tap water TOC is 50–500 ng C/mL. High-purity water for the electronics industry has TOC < 1 ng C/mL.

Total oxygen demand (TOD) tells us how much O_2 is required for complete combustion of pollutants in a waste stream. A volume of N_2 containing a known quantity of O_2 is mixed with the sample and complete combustion is carried out. The remaining O_2 is measured by a potentiometric sensor. Different species in the waste stream consume different amounts of O_2. For example, urea consumes five times as much O_2 as formic acid does. Species such as NH_3 and H_2S also contribute to TOD.

Pollutants can be oxidized by refluxing with dichromate ($Cr_2O_7^{2-}$). *Chemical oxygen demand* (COD) is defined as the O_2 that is chemically equivalent to the $Cr_2O_7^{2-}$ consumed in this process. Each $Cr_2O_7^{2-}$ consumes 6e$^-$ (to make $2Cr^{3+}$) and each O_2 consumes 4e$^-$ (to make H_2O). Therefore, 1 mol of $Cr_2O_7^{2-}$ is chemically equivalent to 1.5 mol

of O_2 for this computation. COD analysis is carried out by refluxing polluted water for 2 h with excess standard $Cr_2O_7^{2-}$ in H_2SO_4 solution containing Ag^+ catalyst. Unreacted $Cr_2O_7^{2-}$ is measured by titration with standard Fe^{2+} or by spectrophotometry. Permits for industry may include COD limits for waste streams. "Oxidizability," which is used in Europe, is analogous to COD. Oxidizability is measured by refluxing with permanganate in acid solution at 100°C for 10 min. Each MnO_4^- consumes five electrons and is chemically equivalent to 1.25 O_2. Electrochemical methods based on photooxidation with TiO_2 could replace cumbersome refluxing with $Cr_2O_7^{2-}$ or MnO_4^-. Problem 16-21 describes one proven method.

Biological oxygen demand (BOD) is the O_2 required for biological degradation of organic matter by microorganisms. The procedure calls for incubating a sealed container of wastewater with no extra air space for 5 days at 20°C in the dark while microbes metabolize organic compounds in the waste. Dissolved O_2 is measured before and after incubation. The difference is BOD.[17] BOD also measures species such as HS^- and Fe^{2+} that may be in the water. Inhibitors are added to prevent oxidation of nitrogen species such as NH_3. There is interest in developing a rapid analysis to provide information equivalent to BOD. This goal could be achieved by substituting ferricyanide ($Fe(CN)_6^{3-}$) for O_2 as the electron sink for bacterial degradation of organic matter. Ferricyanide requires just 3 h, and results are similar to that of the 5-day standard procedure.[18]

Bound nitrogen includes all nitrogen-containing compounds, except N_2, dissolved in water. Kjeldahl nitrogen analysis described in Section 10-8 is excellent for amines and amides but fails to respond to many other forms of nitrogen. Combustion can convert most forms of nitrogen in aqueous samples into NO, which can be measured by chemiluminescence after reaction with ozone:[19]

Bound nitrogen analysis: Nitrogen compounds $\xrightarrow[\text{catalyst}]{O_2/\sim1\,000°C}$ NO

$$NO + O_3 \rightarrow NO_2^*$$
Nitrous oxide   Ozone   Nitric oxide Excited electronic state

$$NO_2^* \rightarrow NO_2 + h\nu$$
Characteristic light emission

Azide (N_3^-) and hydrazines ($RNHNH_2$) are not quantitatively converted into NO by combustion. Bound nitrogen measurements are required for compliance with wastewater discharge regulations.

TiO_2-blended PVC before irradiation   After irradiation for 20 days

Here is a "green" idea: TiO_2 can be blended into poly(vinyl chloride) (PVC) plastic so that the plastic is degraded by sunlight.[16] Ordinary PVC lasts many years in municipal landfills after it is discarded. TiO_2-blended PVC would decompose in a short time. [Courtesy H. Hidaka and S. Horikoshi, Meisei University, Tokyo.]

In iodometry (titration *of* I_3^-), I_3^- is present throughout the reaction up to the equivalence point. *Starch should not be added until immediately before the equivalence point* (as detected visually, by fading of the I_3^-; Color Plate 9). Otherwise some iodine tends to remain bound to starch particles after the equivalence point is reached.

Starch-iodine complexation is temperature dependent. At 50°C, the color is only one-tenth as intense as at 25°C. If maximum sensitivity is required, cooling in ice water is recommended.[22] Organic solvents decrease the affinity of iodine for starch and markedly reduce the utility of the indicator.

An alternative to using starch is to add a few milliliters of *p*-xylene to the vigorously stirred titration vessel. After each addition of reagent near the end point, stop stirring long enough to examine the color of the organic phase. I_2 is 400 times more soluble in *p*-xylene than in water, and its color is readily detected in the organic phase.[21]

Preparation and Standardization of I_3^- Solutions

Triiodide (I_3^-) is prepared by dissolving solid I_2 in excess KI. Sublimed I_2 is pure enough to be a primary standard, but it is seldom used as a standard because it evaporates while it is being weighed. Instead, the approximate amount is rapidly weighed, and the solution of I_3^- is standardized with a pure sample of analyte or $Na_2S_2O_3$.

Acidic solutions of I_3^- are unstable because the excess I^- is slowly oxidized by air:

$$6I^- + O_2 + 4H^+ \rightarrow 2I_3^- + 2H_2O$$

In neutral solutions, oxidation is insignificant in the absence of heat, light, and metal ions. At $pH \gtrsim 11$, triiodide disproportionates to hypoiodous acid, iodate, and iodide.

An excellent way to prepare standard I_3^- is to add a weighed quantity of the primary standard potassium iodate (KIO_3) to a small excess of KI.[23] Then add excess strong acid (giving $pH \approx 1$) to produce I_3^- by quantitative reverse disproportionation:

$$IO_3^- + 8I^- + 6H^+ \rightleftharpoons 3I_3^- + 3H_2O \qquad (15\text{-}18)$$

Freshly acidified iodate plus iodide can be used to standardize thiosulfate. I_3^- must be used immediately or it will be oxidized by air. The disadvantage of KIO_3 is its low molecular mass relative to the number of electrons it accepts, giving a larger-than-desirable relative weighing error in preparing solutions.

There is a significant vapor pressure of toxic I_2 above solid I_2 and aqueous I_3^-. Vessels containing I_2 or I_3^- should be capped and kept in a fume hood. Waste solutions of I_3^- should not be dumped into a sink in the open lab.

HOI   hypoiodous acid
IO_3^- iodate

Use of Sodium Thiosulfate

Sodium thiosulfate is the almost universal titrant for triiodide. In neutral or acidic solution, triiodide oxidizes thiosulfate to tetrathionate:

$$I_3^- + 2S_2O_3^{2-} \rightleftharpoons 3I^- + O=\overset{\displaystyle O}{\overset{\displaystyle \|}{\underset{\displaystyle \underset{\displaystyle O^-}{|}}{S}}}-S-S-\overset{\displaystyle O}{\overset{\displaystyle \|}{\underset{\displaystyle \underset{\displaystyle O^-}{|}}{S}}}=O \qquad (15\text{-}19)$$

Thiosulfate   Tetrathionate

In basic solution, I_3^- disproportionates to I^- and HOI, which can oxidize $S_2O_3^{2-}$ to SO_4^{2-}. Reaction 15-19 should be carried out below pH 9. The common form of thiosulfate, $Na_2S_2O_3 \cdot 5H_2O$, is not pure enough to be a primary standard. Instead, thiosulfate is usually standardized by reaction with a fresh solution of I_3^- prepared from KIO_3 plus KI.

A stable solution of $Na_2S_2O_3$ can be prepared by dissolving the reagent in high-quality, freshly boiled distilled water. Dissolved CO_2 makes the solution acidic and promotes disproportionation of $S_2O_3^{2-}$:

$$S_2O_3^{2-} + H^+ \rightleftharpoons HSO_3^- + S(s) \qquad (15\text{-}20)$$
Bisulfite   Sulfur

Anhydrous, primary standard $Na_2S_2O_3$ can be prepared from the pentahydrate.[24]

and metal ions catalyze atmospheric oxidation of thiosulfate:

$$2Cu^{2+} + 2S_2O_3^{2-} \rightarrow 2Cu^+ + S_4O_6^{2-}$$

$$2Cu^+ + \frac{1}{2}O_2 + 2H^+ \rightarrow 2Cu^{2+} + H_2O$$

Thiosulfate solutions should be stored in the dark. Addition of 0.1 g of sodium carbonate per liter maintains the pH in an optimum range for stability of the solution. Three drops of chloroform should also be added to each bottle of thiosulfate solution to help prevent bacterial growth. An acidic solution of thiosulfate is unstable, but the reagent can be used to titrate I_3^- in acidic solution because the reaction with triiodide is faster than Reaction 15-20.

Analytical Applications of Iodine

Reducing agent $+ I_3^- \rightarrow 3I^-$

Reducing agents can be titrated directly with standard I_3^- in the presence of starch, until reaching the intense blue starch-iodine end point (Table 15-4). An example is the iodimetric determination of vitamin C:

Ascorbic acid
(vitamin C)

Dehydroascorbic
acid[25]

Oxidizing agent $+ 3I^- \rightarrow I_3^-$

Oxidizing agents can be treated with excess I^- to produce I_3^- (Table 15-5, Box 15-3). The iodometric analysis is completed by titrating the liberated I_3^- with standard thiosulfate. Starch is not added until just before the end point.

TABLE 15-4 Titrations with standard triiodide (iodimetric titrations)

Species analyzed	Oxidation reaction	Notes
As^{3+}	$H_3AsO_3 + H_2O \rightleftharpoons H_3AsO_4 + 2H^+ + 2e^-$	Titrate directly in $NaHCO_3$ solution with I_3^-.
Sn^{2+}	$SnCl_4^{2-} + 2Cl^- \rightleftharpoons SnCl_6^{2-} + 2e^-$	Sn(IV) is reduced to Sn(II) with granular Pb or Ni in 1 M HCl and titrated in the absence of oxygen.
N_2H_4	$N_2H_4 \rightleftharpoons N_2 + 4H^+ + 4e^-$	Titrate in $NaHCO_3$ solution.
SO_2	$SO_2 + H_2O \rightleftharpoons H_2SO_3$ $H_2SO_3 + H_2O \rightleftharpoons SO_4^{2-} + 4H^+ + 2e^-$	Add SO_2 (or H_2SO_3 or HSO_3^- or SO_3^{2-}) to excess standard I_3^- in dilute acid and back-titrate unreacted I_3^- with standard thiosulfate.
H_2S	$H_2S \rightleftharpoons S(s) + 2H^+ + 2e^-$	Add H_2S to excess I_3^- in 1 M HCl and back-titrate with thiosulfate.
$Zn^{2+}, Cd^{2+}, Hg^{2+}, Pb^{2+}$	$M^{2+} + H_2S \rightarrow MS(s) + 2H^+$ $MS(s) \rightleftharpoons M^{2+} + S + 2e^-$	Precipitate and wash metal sulfide. Dissolve in 3 M HCl with excess standard I_3^- and back-titrate with thiosulfate.
Cysteine, glutathione, thioglycolic acid, mercaptoethanol	$2RSH \rightleftharpoons RSSR + 2H^+ + 2e^-$	Titrate the sulfhydryl compound at pH 4–5 with I_3^-.
HCN	$I_2 + HCN \rightleftharpoons ICN + I^- + H^+$	Titrate in carbonate-bicarbonate buffer, using *p*-xylene as an extraction indicator.
$H_2C{=}O$	$H_2CO + 3OH^- \rightleftharpoons HCO_2^- + 2H_2O + 2e^-$	Add excess I_3^- plus NaOH to the unknown. After 5 min, add HCl and back-titrate with thiosulfate.
Glucose (and other reducing sugars)	$\overset{\displaystyle O}{\overset{\|}{R}}CH + 3OH^- \rightleftharpoons RCO_2^- + 2H_2O + 2e^-$	Add excess I_3^- plus NaOH to the sample. After 5 min, add HCl and back-titrate with thiosulfate.
Ascorbic acid (vitamin C)	Ascorbate $+ H_2O \rightleftharpoons$ dehydroascorbate $+ 2H^+ + 2e^-$	Titrate directly with I_3^-.
H_3PO_3	$H_3PO_3 + H_2O \rightleftharpoons H_3PO_4 + 2H^+ + 2e^-$	Titrate in $NaHCO_3$ solution.

BOX 15-3 **Iodometric Analysis of High-Temperature Superconductors**

An important application of superconductors is in powerful electromagnets needed for medical magnetic resonance imaging. Ordinary conductors in such magnets require a huge amount of electric power. Because electricity moves through a superconductor with no resistance, the voltage can be removed from the electromagnetic coil once the current has started. Current continues to flow with no power consumption.

A breakthrough in superconductor technology came with the discovery[26] of yttrium barium copper oxide, $YBa_2Cu_3O_7$, whose crystal structure is shown here. When heated, the material readily loses oxygen atoms from the Cu—O chains, and any composition between $YBa_2Cu_3O_7$ and $YBa_2Cu_3O_6$ is observable.

When high-temperature superconductors were discovered, the oxygen content in the formula $YBa_2Cu_3O_x$ was unknown. $YBa_2Cu_3O_7$ represents an unusual set of oxidation states. Common states of yttrium and barium are Y^{3+} and Ba^{2+}, and the common states

of copper are Cu^{2+} and Cu^+. If all the copper were Cu^{2+}, the formula of the superconductor would be $(Y^{3+})(Ba^{2+})_2(Cu^{2+})_3(O^{2-})_{6.5}$, with a cation charge of $+13$ and an anion charge of -13. The composition $YBa_2Cu_3O_7$ requires Cu^{3+}, which is rather rare. Formally, $YBa_2Cu_3O_7$ can be thought of as $(Y^{3+})(Ba^{2+})_2(Cu^{2+})_2(Cu^{3+})(O^{2-})_7$ with a cation charge of $+14$ and an anion charge of -14.

Redox titrations proved to be the most reliable way to measure the oxidation state of copper and thereby deduce the oxygen content of $YBa_2Cu_3O_x$.[27] An iodometric method includes two experiments.

In *Experiment A*, $YBa_2Cu_3O_x$ is dissolved in dilute acid, in which Cu^{3+} is converted into Cu^{2+}. For simplicity, we write reactions for $YBa_2Cu_3O_7$, but you could balance these equations for $x \neq 7$.[28]

$$YBa_2Cu_3O_7 + 13H^+ \rightarrow$$

$$Y^{3+} + 2Ba^{2+} + 3Cu^{2+} + \frac{13}{2}H_2O + \frac{1}{4}O_2 \quad (1)$$

The total copper content is measured by treatment with iodide

$$Cu^{2+} + \frac{5}{2}I^- \rightarrow CuI(s) + \frac{1}{2}I_3^- \quad (2)$$

and titration of the liberated triiodide with standard thiosulfate by Reaction 15-19. Each mole of Cu in $YBa_2Cu_3O_7$ is equivalent to 1 mol of $S_2O_3^{2-}$ in Experiment A.

In *Experiment B*, $YBa_2Cu_3O_x$ is dissolved in dilute acid containing I^-. Each mole of Cu^{2+} produces 0.5 mol of I_3^- by Reaction (2) and each mole of Cu^{3+} produces 1 mol of I_3^-:

$$Cu^{3+} + 4I^- \rightarrow CuI(s) + I_3^- \quad (3)$$

The moles of $S_2O_3^{2-}$ required in Experiment A equal the total moles of Cu in the superconductor. The difference in $S_2O_3^{2-}$ required between Experiments B and A gives the Cu^{3+} content. From this difference, you can find x in the formula $YBa_2Cu_3O_x$.[29]

Although we can balance cation and anion charges in the formula $YBa_2Cu_3O_7$ by including Cu^{3+} in the formula, there is no evidence for discrete Cu^{3+} ions in the crystal. There is also no evidence that some of the oxygen is in the form of peroxide, O_2^{2-}, which also would balance the cation and anion charges. The best description of the valence state in the solid crystal involves electrons and holes delocalized in the Cu—O planes and chains. Nonetheless, the formal designation of Cu^{3+} and Reactions 1 through 3 accurately describe the redox chemistry of $YBa_2Cu_3O_7$. Problem 15-35 describes titrations that separately measure oxidation numbers of Cu and Bi in the superconductor $Bi_2Sr_2(Ca_{0.8}Y_{0.2})Cu_2O_{8.295}$.

Cu—O chains

Cu—O planes

Y · Cu · Ba · O

Structure of $YBa_2Cu_3O_7$. One-dimensional Cu—O chains (shown in color) run along the crystallographic b-axis and two-dimensional Cu—O sheets lie in the a–b plane. Loss of colored oxygen atoms from the chains at elevated temperature results in $YBa_2Cu_3O_6$. [From G. F. Holland and A. M. Stacy, "Physical Properties of the Quaternary Oxide Superconductor $YBa_2Cu_3O_x$," *Acc. Chem. Res.* 1988, *21*, 8.]

TABLE 15-5   Titration of I_3^- produced by analyte (iodometric titrations)

Species analyzed	Reaction	Notes
Cl_2	$Cl_2 + 3I^- \rightleftharpoons 2Cl^- + I_3^-$	Reaction in dilute acid.
$HOCl$	$HOCl + H^+ + 3I^- \rightleftharpoons Cl^- + I_3^- + H_2O$	Reaction in 0.5 M H_2SO_4.
Br_2	$Br_2 + 3I^- \rightleftharpoons 2Br^- + I_3^-$	Reaction in dilute acid.
BrO_3^-	$BrO_3^- + 6H^+ + 9I^- \rightleftharpoons Br^- + 3I_3^- + 3H_2O$	Reaction in 0.5 M H_2SO_4.
IO_3^-	$2IO_3^- + 16I^- + 12H^+ \rightleftharpoons 6I_3^- + 6H_2O$	Reaction in 0.5 M HCl.
IO_4^-	$2IO_4^- + 22I^- + 16H^+ \rightleftharpoons 8I_3^- + 8H_2O$	Reaction in 0.5 M HCl.
O_2	$O_2 + 4Mn(OH)_2 + 2H_2O \rightleftharpoons 4Mn(OH)_3$ $2Mn(OH)_3 + 6H^+ + 3I^- \rightleftharpoons 2Mn^{2+} + I_3^- + 6H_2O$	The sample is treated with Mn^{2+}, NaOH, and KI. After 1 min, it is acidified with H_2SO_4, and the I_3^- is titrated.
H_2O_2	$H_2O_2 + 3I^- + 2H^+ \rightleftharpoons I_3^- + 2H_2O$	Reaction in 1 M H_2SO_4 with NH_4MoO_3 catalyst.
$O_3{}^a$	$O_3 + 3I^- + 2H^+ \rightleftharpoons O_2 + I_3^- + H_2O$	O_3 is passed through neutral 2 wt% KI solution. Add H_2SO_4 and titrate.
NO_2^-	$2HNO_2 + 2H^+ + 3I^- \rightleftharpoons 2NO + I_3^- + 2H_2O$	The nitric oxide is removed (by bubbling CO_2 generated in situ) prior to titration of I_3^-.
As^{5+}	$H_3AsO_4 + 2H^+ + 3I^- \rightleftharpoons H_3AsO_3 + I_3^- + H_2O$	Reaction in 5 M HCl.
$S_2O_8^{2-}$	$S_2O_8^{2-} + 3I^- \rightleftharpoons 2SO_4^{2-} + I_3^-$	Reaction in neutral solution. Then acidify and titrate.
Cu^{2+}	$2Cu^{2+} + 5I^- \rightleftharpoons 2CuI(s) + I_3^-$	NH_4HF_2 is used as a buffer.
$Fe(CN)_6^{3-}$	$2Fe(CN)_6^{3-} + 3I^- \rightleftharpoons 2Fe(CN)_6^{4-} + I_3^-$	Reaction in 1 M HCl.
MnO_4^-	$2MnO_4^- + 16H^+ + 15I^- \rightleftharpoons 2Mn^{2+} + 5I_3^- + 8H_2O$	Reaction in 0.1 M HCl.
MnO_2	$MnO_2(s) + 4H^+ + 3I^- \rightleftharpoons Mn^{2+} + I_3^- + 2H_2O$	Reaction in 0.5 M H_3PO_4 or HCl.
$Cr_2O_7^{2-}$	$Cr_2O_7^{2-} + 14H^+ + 9I^- \rightleftharpoons 2Cr^{3+} + 3I_3^- + 7H_2O$	Reaction in 0.4 M HCl requires 5 min for completion and is particularly sensitive to air oxidation.
Ce^{4+}	$2Ce^{4+} + 3I^- \rightleftharpoons 2Ce^{3+} + I_3^-$	Reaction in 1 M H_2SO_4.

a. The pH must be $\gtrsim 7$ when O_3 is added to I^-. In acidic solution, each O_3 produces $1.25\ I_3^-$, not $1\ I_3^-$. [N. V. Klassen, D. Marchington, and H. C. E. McGowan, Anal. Chem. **1994**, 66, 2921.]

Terms to Understand

amalgam
disproportionation

preoxidation
prereduction

redox indicator
redox titration

Summary

Redox titrations are based on an oxidation-reduction reaction between analyte and titrant. Sometimes a quantitative chemical preoxidation (with reagents such as $S_2O_8^{2-}$, AgO, $NaBiO_3$, or H_2O_2) or prereduction (with reagents such as $SnCl_2$, $CrCl_2$, SO_2, H_2S, or a metallic reductor column) is necessary to adjust the oxidation state of the analyte prior to analysis. The end point of a redox titration is commonly detected by potentiometry or with a redox indicator. A useful indicator must have a transition range ($= E°(\text{indicator}) \pm 0.059\ 16/n$ V) that overlaps the abrupt change in potential of the titration curve.

The greater the difference in reduction potential between analyte and titrant, the sharper will be the end point. Plateaus before and after the equivalence point are centered near $E°(\text{analyte})$ and $E°(\text{titrant})$. Prior to the equivalence point, the half-reaction involving analyte is used to find the voltage because the concentrations of both the oxidized and the reduced forms of analyte are known. After the equivalence point, the half-reaction involving titrant is employed. At the equivalence point, both half-reactions are used to find the voltage.

Common oxidizing titrants include $KMnO_4$, Ce^{4+}, and $K_2Cr_2O_7$. Many procedures are based on oxidation with I_3^- or titration of I_3^- liberated in a chemical reaction.

Exercises

15-A. A 20.0-mL solution of 0.005 00 M Sn^{2+} in 1 M HCl was titrated with 0.020 0 M Ce^{4+} to give Sn^{4+} and Ce^{3+}. Calculate the potential (versus S.C.E.) at the following volumes of Ce^{4+}: 0.100, 1.00, 5.00, 9.50, 10.00, 10.10, and 12.00 mL. Sketch the titration curve.

15-B. Would indigo tetrasulfonate be a suitable redox indicator for the titration of $Fe(CN)_6^{4-}$ with Tl^{3+} in 1 M HCl? (*Hint:* The potential at the equivalence point must be between the potentials for each redox couple.)

15-C. Compute the titration curve for Demonstration 15-1, in which 400.0 mL of 3.75 mM Fe^{2+} are titrated with 20.0 mM MnO_4^- at a *fixed pH* of 0.00 in 1 M H_2SO_4. Calculate the potential versus S.C.E. at titrant volumes of 1.0, 7.5, 14.0, 15.0, 16.0, and 30.0 mL and sketch the titration curve.

15-D. A titration of 50.0 mL of unknown Fe^{2+} with 0.100 M Ce^{4+} at 25°C, monitored with Pt and calomel electrodes, gave data in the following table.[8] Prepare a Gran plot and decide which data lie on a straight line. Find the x-intercept of this line, which is the equivalence volume. Calculate the molarity of Fe^{2+} in the unknown.

Titrant volume, V (mL)	E (volts)
6.50	0.635
8.50	0.651
10.50	0.669
11.50	0.680
12.50	0.696

15-E. A solid mixture weighing 0.054 85 g contained only ferrous ammonium sulfate and ferrous chloride. The sample was dissolved in 1 M H_2SO_4, and the Fe^{2+} required 13.39 mL of 0.012 34 M Ce^{4+} for complete oxidation to Fe^{3+}. Calculate the weight percent of Cl in the original sample.

$FeSO_4 \cdot (NH_4)_2SO_4 \cdot 6H_2O$	$FeCl_2 \cdot 6H_2O$
Ferrous ammonium sulfate	Ferrous chloride
FM 392.13	FM 234.84

Problems

Shape of a Redox Titration Curve

15-1. Consider the titration in Figure 15-2.

(a) Write a balanced titration reaction.

(b) Write two different half-reactions for the indicator electrode.

(c) Write two different Nernst equations for the cell voltage.

(d) Calculate E at the following volumes of Ce^{4+}: 10.0, 25.0, 49.0, 50.0, 51.0, 60.0, and 100.0 mL. Compare your results with Figure 15-2.

15-2. Consider the titration of 100.0 mL of 0.010 0 M Ce^{4+} in 1 M $HClO_4$ by 0.040 0 M Cu^+ to give Ce^{3+} and Cu^{2+}, using Pt and saturated Ag | AgCl electrodes to find the end point.

(a) Write a balanced titration reaction.

(b) Write two different half-reactions for the indicator electrode.

(c) Write two different Nernst equations for the cell voltage.

(d) Calculate E at the following volumes of Cu^+: 1.00, 12.5, 24.5, 25.0, 25.5, 30.0, and 50.0 mL. Sketch the titration curve.

15-3. Consider the titration of 25.0 mL of 0.010 0 M Sn^{2+} by 0.050 0 M Tl^{3+} in 1 M HCl, using Pt and saturated calomel electrodes to find the end point.

(a) Write a balanced titration reaction.

(b) Write two different half-reactions for the indicator electrode.

(c) Write two different Nernst equations for the cell voltage.

(d) Calculate E at the following volumes of Tl^{3+}: 1.00, 2.50, 4.90, 5.00, 5.10, and 10.0 mL. Sketch the titration curve.

15-4. Ascorbic acid (0.010 0 M) was added to 10.0 mL of 0.020 0 M Fe^{3+} in a solution buffered to pH 0.30, and the potential was monitored with Pt and saturated Ag | AgCl electrodes.

$$\text{Dehydroascorbic acid} + 2H^+ + 2e^- \rightleftharpoons \text{ascorbic acid} + H_2O$$
$$E° = 0.390 \text{ V}$$

(a) Write a balanced equation for the titration reaction.

(b) Using $E° = 0.767$ V for the $Fe^{3+} | Fe^{2+}$ couple, calculate the cell voltage when 5.0, 10.0, and 15.0 mL of ascorbic acid have been added. (*Hint:* Refer to the calculations in Demonstration 15-1.)

15-5. Consider the titration of 25.0 mL of 0.050 0 M Sn^{2+} with 0.100 M Fe^{3+} in 1 M HCl to give Fe^{2+} and Sn^{4+}, using Pt and calomel electrodes.

(a) Write a balanced titration reaction.

(b) Write two half-reactions for the indicator electrode.

(c) Write two Nernst equations for the cell voltage.

(d) Calculate E at the following volumes of Fe^{3+}: 1.0, 12.5, 24.0, 25.0, 26.0, and 30.0 mL. Sketch the titration curve.

Finding the End Point

15-6. Select indicators from Table 15-2 that would be suitable for finding the end point in Figure 15-3. What color changes would be observed?

15-7. Would tris(2,2′-bipyridine)iron be a useful indicator for the titration of Sn^{2+} with $Mn(EDTA)^-$? (*Hint:* The potential at the equivalence point must be between the potentials for each redox couple.)

Adjustment of Analyte Oxidation State

15-8. Explain what we mean by *preoxidation* and *prereduction*. Why is it important to be able to destroy the reagents used for these purposes?

15-9. Write balanced reactions for the destruction of $S_2O_8^{2-}$, Ag^{2+}, and H_2O_2 by boiling.

15-10. What is a Jones reductor and what is it used for?

15-11. Why don't Cr^{3+} and TiO^{2+} interfere in the analysis of Fe^{3+} when a Walden reductor, instead of a Jones reductor, is used for prereduction?

Redox Reactions of $KMnO_4$, Ce(IV), and $K_2Cr_2O_7$

15-12. From information in Table 15-3, explain how you would use $KMnO_4$ to find the content of $(NH_4)_2S_2O_8$ in a solid mixture with $(NH_4)_2SO_4$. What is the purpose of phosphoric acid in the procedure?

15-13. Write balanced half-reactions in which MnO_4^- acts as an oxidant at (a) pH = 0; (b) pH = 10; (c) pH = 15.

15-14. When 25.00 mL of unknown were passed through a Jones reductor, molybdate ion (MoO_4^{2-}) was converted into Mo^{3+}. The filtrate required 16.43 mL of 0.010 33 M $KMnO_4$ to reach the purple end point.

$$MnO_4^- + Mo^{3+} \rightarrow Mn^{2+} + MoO_2^{2+}$$

A blank required 0.04 mL. Balance the reaction and find the molarity of molybdate in the unknown.

15-15. A 25.00-mL volume of commercial hydrogen peroxide solution was diluted to 250.0 mL in a volumetric flask. Then 25.00 mL of the diluted solution were mixed with 200 mL of water and 20 mL of 3 M H_2SO_4 and titrated with 0.021 23 M $KMnO_4$. The first pink color was observed with 27.66 mL of titrant. A blank prepared from water in place of H_2O_2 required 0.04 mL to give visible pink color. Using the H_2O_2 reaction in Table 15-3, find the molarity of the commercial H_2O_2.

15-16. Two possible reactions of MnO_4^- with H_2O_2 to produce O_2 and Mn^{2+} are

$$\text{Scheme 1:} \qquad MnO_4^- \rightarrow Mn^{2+}$$
$$H_2O_2 \rightarrow O_2$$
$$\text{Scheme 2:} \qquad MnO_4^- \rightarrow O_2 + Mn^{2+}$$
$$H_2O_2 \rightarrow H_2O$$

(a) Complete the half-reactions for both schemes by adding e^-, H_2O, and H^+ and write a balanced net equation for each scheme.

(b) Sodium peroxyborate tetrahydrate, $NaBO_3 \cdot 4H_2O$ (FM 153.86), produces H_2O_2 when dissolved in acid: $BO_3^- + 2H_2O \rightarrow H_2O_2 + H_2BO_3^-$. To decide whether Scheme 1 or Scheme 2 occurs, students at the U.S. Naval Academy[30] weighed 1.023 g $NaBO_3 \cdot 4H_2O$ into a 100-mL volumetric flask, added 20 mL of 1 M H_2SO_4, and diluted to the mark with H_2O. Then they titrated 10.00 mL of this solution with 0.010 46 M $KMnO_4$ until the first pale pink color persisted. How many mL of $KMnO_4$ are required in Scheme 1 and in Scheme 2? (The Scheme 1 stoichiometry was observed.)

15-17. A 50.00-mL sample containing La^{3+} was treated with sodium oxalate to precipitate $La_2(C_2O_4)_3$, which was washed, dissolved in acid, and titrated with 18.04 mL of 0.006 363 M $KMnO_4$. Calculate the molarity of La^{3+} in the unknown.

15-18. An aqueous glycerol solution weighing 100.0 mg was treated with 50.0 mL of 0.083 7 M Ce^{4+} in 4 M $HClO_4$ at 60°C for 15 min to oxidize the glycerol to formic acid:

$$\begin{array}{ccc} CH_2 - CH - CH_2 \\ | \quad\ | \quad\ | \\ OH \quad OH \quad OH \end{array} \qquad HCO_2H$$

Glycerol
FM 92.095

Formic acid

The excess Ce^{4+} required 12.11 mL of 0.044 8 M Fe^{2+} to reach a ferroin end point. What is the weight percent of glycerol in the unknown?

15-19. Nitrite (NO_2^-) can be determined by oxidation with excess Ce^{4+}, followed by back titration of the unreacted Ce^{4+}. A 4.030-g sample of solid containing only $NaNO_2$ (FM 68.995) and $NaNO_3$ was dissolved in 500.0 mL. A 25.00-mL sample of this solution was treated with 50.00 mL of 0.118 6 M Ce^{4+} in strong acid for 5 min, and the excess Ce^{4+} was back-titrated with 31.13 mL of 0.042 89 M ferrous ammonium sulfate.

$$2Ce^{4+} + NO_2^- + H_2O \rightarrow 2Ce^{3+} + NO_3^- + 2H^+$$
$$Ce^{4+} + Fe^{2+} \rightarrow Ce^{3+} + Fe^{3+}$$

Calculate the weight percent of $NaNO_2$ in the solid.

15-20. Calcium fluorapatite ($Ca_{10}(PO_4)_6F_2$, FM 1 008.6) laser crystals were doped with chromium to improve their efficiency. It was suspected that the chromium could be in the +4 oxidation state.

1. To measure the total oxidizing power of chromium in the material, a crystal was dissolved in 2.9 M $HClO_4$ at 100°C, cooled to

20°C, and titrated with standard Fe^{2+}, using Pt and Ag | AgCl electrodes to find the end point. Chromium oxidized above the +3 state should oxidize an equivalent amount of Fe^{2+} in this step. That is, Cr^{4+} would consume one Fe^{2+}, and each atom of Cr^{6+} in $Cr_2O_7^{2-}$ would consume three Fe^{2+}:

$$Cr^{4+} + Fe^{2+} \rightarrow Cr^{3+} + Fe^{3+}$$
$$\frac{1}{2} Cr_2O_7^{2-} + 3Fe^{2+} \rightarrow Cr^{3+} + 3Fe^{3+}$$

2. In a second step, the total chromium content was measured by dissolving a crystal in 2.9 M $HClO_4$ at 100°C and cooling to 20°C. Excess $S_2O_8^{2-}$ and Ag^+ were then added to oxidize all chromium to $Cr_2O_7^{2-}$. Unreacted $S_2O_8^{2-}$ was destroyed by boiling, and the remaining solution was titrated with standard Fe^{2+}. In this step, each Cr in the original unknown reacts with three Fe^{2+}.

$$Cr^{x+} \xrightarrow{S_2O_8^{2-}} Cr_2O_7^{2-}$$
$$\frac{1}{2} Cr_2O_7^{2-} + 3Fe^{2+} \rightarrow Cr^{3+} + 3Fe^{3+}$$

In step 1, 0.437 5 g of laser crystal required 0.498 mL of 2.786 mM Fe^{2+} (prepared by dissolving $Fe(NH_4)_2(SO_4)_2 \cdot 6H_2O$ in 2 M $HClO_4$). In step 2, 0.156 6 g of crystal required 0.703 mL of the same Fe^{2+} solution. Find the average oxidation number of Cr in the crystal and find the total micrograms of Cr per gram of crystal.

15-21. Primary-standard-grade arsenic(III) oxide (As_4O_6) is a useful (but carcinogenic) reagent for standardizing oxidants including MnO_4^- and I_3^-. To standardize MnO_4^-, As_4O_6 is dissolved in base and then titrated with MnO_4^- in acid. A small amount of iodide (I^-) or iodate (IO_3^-) catalyzes the reaction between H_3AsO_3 and MnO_4^-.

$$As_4O_6 + 8OH^- \rightleftharpoons 4HAsO_3^{2-} + 2H_2O$$
$$HAsO_3^{2-} + 2H^+ \rightleftharpoons H_3AsO_3$$
$$5H_3AsO_3 + 2MnO_4^- + 6H^+ \rightarrow 5H_3AsO_4 + 2Mn^{2+} + 3H_2O$$

(a) A 3.214-g aliquot of $KMnO_4$ (FM 158.034) was dissolved in 1.000 L of water, heated to cause any reactions with impurities to occur, cooled, and filtered. What is the theoretical molarity of this solution if no MnO_4^- was consumed by impurities?

(b) What mass of As_4O_6 (FM 395.68) would be just sufficient to react with 25.00 mL of the $KMnO_4$ solution in part (a)?

(c) It was found that 0.146 8 g of As_4O_6 required 29.98 mL of $KMnO_4$ solution for the faint color of unreacted MnO_4^- to appear. In a blank titration, 0.03 mL of MnO_4^- was required to produce enough color to be seen. Calculate the molarity of the permanganate solution.

Methods Involving Iodine

15-22. Why is iodine almost always used in a solution containing excess I^-?

15-23. State two ways to make standard triiodide solution.

15-24. In which technique, iodimetry or iodometry, is starch indicator not added until just before the end point? Why?

15-25. (a) Potassium iodate solution was prepared by dissolving 1.022 g of KIO_3 (FM 214.00) in a 500-mL volumetric flask. Then 50.00 mL of the solution were pipetted into a flask and treated with excess KI (2 g) and acid (10 mL of 0.5 M H_2SO_4). How many moles of I_3^- are created by the reaction?

(b) The triiodide from part (a) reacted with 37.66 mL of $Na_2S_2O_3$ solution. What is the concentration of the $Na_2S_2O_3$ solution?

(c) A 1.223-g sample of solid containing ascorbic acid and inert ingredients was dissolved in dilute H_2SO_4 and treated with 2 g of KI and 50.00 mL of KIO_3 solution from part (a). Excess triiodide required 14.22 mL of $Na_2S_2O_3$ solution from part (b). Find the weight percent of ascorbic acid (FM 176.13) in the unknown.

(d) Does it matter whether starch indicator is added at the beginning or near the end point in the titration in part (c)?

15-26. A 3.026-g portion of a copper(II) salt was dissolved in a 250-mL volumetric flask. A 50.0-mL aliquot was analyzed by adding 1 g of KI and titrating the liberated iodine with 23.33 mL of 0.046 68 M $Na_2S_2O_3$. Find the weight percent of Cu in the salt. Should starch indicator be added to this titration at the beginning or just before the end point?

15-27. H_2S was measured by slowly adding 25.00 mL of aqueous H_2S to 25.00 mL of acidified standard 0.010 44 M I_3^- to precipitate elemental sulfur. (If $[H_2S] > 0.01$ M, then precipitated sulfur traps some I_3^- solution, which is not subsequently titrated.) The remaining I_3^- was titrated with 14.44 mL of 0.009 336 M $Na_2S_2O_3$. Find the molarity of the H_2S solution. Should starch indicator be added to this titration at the beginning or just before the end point?

15-28. From the following reduction potentials

$$I_2(s) + 2e^- \rightleftharpoons 2I^- \qquad E° = 0.535 \text{ V}$$
$$I_2(aq) + 2e^- \rightleftharpoons 2I^- \qquad E° = 0.620 \text{ V}$$
$$I_3^- + 2e^- \rightleftharpoons 3I^- \qquad E° = 0.535 \text{ V}$$

(a) Calculate the equilibrium constant for $I_2(aq) + I^- \rightleftharpoons I_3^-$.

(b) Calculate the equilibrium constant for $I_2(s) + I^- \rightleftharpoons I_3^-$.

(c) Calculate the solubility (g/L) of $I_2(s)$ in water.

15-29. The Kjeldahl analysis in Section 10-8 is used to measure the nitrogen content of organic compounds, which are digested in boiling sulfuric acid to decompose to ammonia, which, in turn, is distilled into standard acid. The remaining acid is then back-titrated with base. Kjeldahl himself had difficulty in 1880 discerning by lamplight the methyl red indicator end point in the back titration. He could have refrained from working at night, but instead he chose to complete the analysis differently. After distilling the ammonia into standard sulfuric acid, he added a mixture of KIO_3 and KI to the acid. The liberated iodine was then titrated with thiosulfate, using starch for easy end-point detection—even by lamplight.[31] Explain how the thiosulfate titration is related to the nitrogen content of the unknown. Derive a relationship between moles of NH_3 liberated in the digestion and moles of thiosulfate required for titration of iodine.

15-30. Some people have an allergic reaction to the food preservative sulfite (SO_3^{2-}). Sulfite in wine was measured by the following procedure: To 50.0 mL of wine were added 5.00 mL of solution containing (0.804 3 g KIO_3 + 5 g KI)/100 mL. Acidification with 1.0 mL of 6.0 M H_2SO_4 quantitatively converted IO_3^- into I_3^-. The I_3^- reacted with SO_3^{2-} to generate SO_4^{2-}, leaving excess I_3^- in solution. The excess I_3^- required 12.86 mL of 0.048 18 M $Na_2S_2O_3$ to reach a starch end point.

(a) Write the reaction that occurs when H_2SO_4 is added to KIO_3 + KI and explain why 5 g KI were added to the stock solution. Is it necessary to measure out 5 g very accurately? Is it necessary to measure 1.0 mL of H_2SO_4 very accurately?

(b) Write a balanced reaction between I_3^- and sulfite.

(c) Find the concentration of sulfite in the wine. Express your answer in mol/L and in mg SO_3^{2-} per liter.

(d) *t test.* Another wine was found to contain 277.7 mg SO_3^{2-}/L with a standard deviation of ± 2.2 mg/L for three determinations by the iodimetric method. A spectrophotometric method gave 273.2 ± 2.1 mg/L in three determinations. Are these results significantly different at the 95% confidence level?

15-31. Potassium bromate, $KBrO_3$, is a primary standard for the generation of Br_2 in acidic solution:

$$BrO_3^- + 5Br^- + 6H^+ \rightleftharpoons 3Br_2(aq) + 3H_2O$$

The Br_2 can be used to analyze many unsaturated organic compounds. Al^{3+} was analyzed as follows: An unknown was treated with 8-hydroxyquinoline (oxine) at pH 5 to precipitate aluminum oxinate, $Al(C_9H_6ON)_3$. The precipitate was washed, dissolved in warm HCl containing excess KBr, and treated with 25.00 mL of 0.020 00 M $KBrO_3$.

The excess Br_2 was reduced with KI, which was converted into I_3^-. The I_3^- required 8.83 mL of 0.051 13 M $Na_2S_2O_3$ to reach a starch end point. How many milligrams of Al were in the unknown?

15-32. *Iodometric analysis of high-temperature superconductor.* The procedure in Box 15-3 was carried out to find the effective copper oxidation state, and therefore the number of oxygen atoms, in the formula $YBa_2Cu_3O_{7-z}$, where z ranges from 0 to 0.5.

(a) In Experiment A of Box 15-3, 1.00 g of superconductor required 4.55 mmol of $S_2O_3^{2-}$. In Experiment B, 1.00 g of superconductor required 5.68 mmol of $S_2O_3^{2-}$. Calculate the value of z in the formula $YBa_2Cu_3O_{7-z}$ (FM 666.246 – 15.999 4z).

(b) *Propagation of uncertainty.* In several replications of Experiment A, the thiosulfate required was 4.55 (± 0.10) mmol of $S_2O_3^{2-}$ per gram of $YBa_2Cu_3O_{7-z}$. In Experiment B, the thiosulfate required was 5.68 (± 0.05) mmol of $S_2O_3^{2-}$ per gram. Calculate the uncertainty of x in the formula $YBa_2Cu_3O_x$.

15-33. Here is a description of an analytical procedure for superconductors containing unknown quantities of Cu(I), Cu(II), Cu(III), and peroxide (O_2^{2-}):[32] "The possible trivalent copper and/or peroxide-type oxygen are reduced by Cu(I) when dissolving the sample (*ca.* 50 mg) in deoxygenated HCl solution (1 M) containing a known excess of monovalent copper ions (*ca.* 25 mg CuCl). On the other hand, if the sample itself contained monovalent copper, the amount of Cu(I) in the solution would increase upon dissolving the sample. The excess Cu(I) was then determined by coulometric back-titration . . . in an argon atmosphere." The abbreviation "*ca.*" means "approximately." *Coulometry* is an electrochemical method in which the electrons liberated in the reaction $Cu^+ \rightarrow Cu^{2+} + e^-$ are measured from the charge flowing through an electrode. Explain with your own words and equations how this analysis works.

15-34. $Li_{1+y}CoO_2$ is an anode for lithium batteries. Cobalt is present as a mixture of Co(III) and Co(II). Most preparations also contain inert lithium salts and moisture. To find the stoichiometry, Co was measured by atomic absorption and its average oxidation state was measured by a potentiometric tritration.[33] For the titration, 25.00 mg of solid were dissolved under N_2 in 5.000 mL containing 0.100 0 M Fe^{2+} in 6 M H_2SO_4 plus 6 M H_3PO_4 to give a clear pink solution:

$$Co^{3+} + Fe^{2+} \rightarrow Co^{2+} + Fe^{3+}$$

Unreacted Fe^{2+} required 3.228 mL of 0.015 93 M $K_2Cr_2O_7$ for complete titration.

(a) How many mmol of Co^{3+} are contained in 25.00 mg of the material?

(b) Atomic absorption found 56.4 wt% Co in the solid. What is the average oxidation state of Co?

(c) Find y in the formula $Li_{1+y}CoO_2$.

(d) What is the theoretical quotient wt% Li/wt% Co in the solid? The observed quotient, after washing away inert lithium salts, was $0.138\ 8 \pm 0.000\ 6$. Is the observed quotient consistent with the average cobalt oxidation state?

15-35. *Warning! The Surgeon General has determined that this problem is hazardous to your health.* The oxidation numbers of Cu and Bi in high-temperature superconductors of the type $Bi_2Sr_2(Ca_{0.8}Y_{0.2})Cu_2O_x$ (which could contain Cu^{2+}, Cu^{3+}, Bi^{3+}, and Bi^{5+}) can be measured by the following procedure.[34] In Experiment A, the superconductor is dissolved in 1 M HCl containing excess 2 mM CuCl. Bi^{5+} (written as BiO_3^-) and Cu^{3+} consume Cu^+ to make Cu^{2+}:

$$BiO_3^- + 2Cu^+ + 4H^+ \rightarrow BiO^+ + 2Cu^{2+} + 2H_2O \quad (1)$$

$$Cu^{3+} + Cu^+ \rightarrow 2Cu^{2+} \quad (2)$$

The excess, unreacted Cu^+ is then titrated by *coulometry* (described in Chapter 16). In Experiment B, the superconductor is dissolved in 1 M HCl containing excess 1 mM $FeCl_2 \cdot 4H_2O$. Bi^{5+} reacts with the Fe^{2+} but Cu^{3+} does not react with Fe^{2+}.[35]

$$BiO_3^- + 2Fe^{2+} + 4H^+ \rightarrow BiO^+ + 2Fe^{3+} + 2H_2O \quad (3)$$

$$Cu^{3+} + \frac{1}{2}H_2O \rightarrow Cu^{2+} + \frac{1}{4}O_2 + H^+ \quad (4)$$

The excess, unreacted Fe^{2+} is then titrated by coulometry. The total oxidation number of Cu + Bi is measured in Experiment A, and the oxidation number of Bi is determined in Experiment B. The difference gives the oxidation number of Cu.

(a) In Experiment A, a sample of $Bi_2Sr_2CaCu_2O_x$ (FM 760.37 + 15.999 4x) (containing no yttrium) weighing 102.3 mg was dissolved in 100.0 mL of 1 M HCl containing 2.000 mM CuCl. After reaction with the superconductor, coulometry detected 0.108 5 mmol of unreacted Cu^+ in the solution. In Experiment B, 94.6 mg of superconductor were dissolved in 100.0 mL of 1 M HCl containing 1.000 mM $FeCl_2 \cdot 4H_2O$. After reaction with the superconductor, coulometry detected 0.057 7 mmol of unreacted Fe^{2+}. Find the average oxidation numbers of Bi and Cu in the superconductor and the oxygen stoichiometry coefficient, x.

(b) Find the uncertainties in the oxidation numbers and x if the quantities in Experiment A are 102.3 ($\pm$0.2) mg and 0.108 5 ($\pm$0.000 7) mmol and the quantities in Experiment B are 94.6 ($\pm$0.2) mg and 0.057 7 ($\pm$0.000 7) mmol. Assume negligible uncertainty in other quantities.

16 Electroanalytical Techniques

(a) Electrochemical detector measures sugars emerging from a chromatography column. Sugars are oxidized at the Cu electrode, whose potential is regulated with respect to the Ag | AgCl reference electrode. Reduction ($H_2O + e^- \rightarrow \frac{1}{2}H_2 + OH^-$) occurs at the stainless steel exit arm, and electric current is measured between Cu and steel. [Adapted from Bioanalytical Systems, West Lafayette, IN.] (b) Anion-exchange separation of sugars in 0.1 M NaOH with CarboPac PA1 column. Upper chromatogram shows a standard mixture of (1) fucose, (2) methylglucose, (3) arabinose, (4) glucose, (5) fructose, (6) lactose, (7) sucrose, and (8) cellobiose. Lower chromatogram was obtained with Bud Dry beer diluted by a factor of 100 with water and filtered through a 0.45-μm membrane to remove particles. [From P. Luo, M. Z. Luo, and R. P. Baldwin, "Determination of Sugars in Food Products," J. Chem. Ed. 1993, 70, 679.]

(a)   (b)

You can measure sugars in your favorite beverage by separating the sugars by anion-exchange chromatography (Chapter 25) in strongly basic solution and detecting them with an electrode as they emerge from the column.[1] The —OH groups of sugars such as fructose, whose structure is shown in the chromatogram, partially dissociate to —O⁻ anions in 0.1 M NaOH. Anions are separated from one another when they pass through a column that has fixed positive charges. As sugars emerge from the column, they are detected by oxidation at a Cu electrode poised at a potential of +0.55 V versus Ag | AgCl. The chromatogram is a graph of detector current versus time. Each sugar gives a peak whose area is proportional to the moles exiting the column.

Brand	Sugar concentration (g/L)			
	Glucose	Fructose	Lactose	Maltose
Budweiser	0.54	0.26	0.84	2.05
Bud Dry	0.14	0.29	0.46	—
Coca-Cola	45.1	68.4	—	1.04
Pepsi	44.0	42.9	—	1.06
Diet Pepsi	0.03	0.01	—	—

Charles Martin Hall. [Courtesy of Alcoa.]

Convention: *cathodic current is considered positive.*

Previous chapters dealt with *potentiometry*—in which voltage was measured in the absence of significant current. Now we consider electroanalytical methods in which current is essential.[3] Techniques in this chapter are all examples of **electrolysis**—the process in which a chemical reaction is forced to occur at an electrode by an imposed voltage (Demonstration 16-1). The home glucose monitor described in this chapter, with sales of more than $3 billion in 2007, is the single largest electroanalytical application.

16-1 Fundamentals of Electrolysis

Suppose we dip Cu and Pt electrodes into a solution of Cu^{2+} and force electric current through to deposit copper metal at the cathode and to liberate O_2 at the anode.

Cathode: $$Cu^{2+} + 2e^- \rightleftharpoons Cu(s)$$

Anode: $$H_2O \rightleftharpoons \frac{1}{2}O_2(g) + 2H^+ + 2e^-$$

Net reaction: $$H_2O + Cu^{2+} \rightleftharpoons Cu(s) + \frac{1}{2}O_2(g) + 2H^+ \qquad \text{(16-1)}$$

Figure 16-1 shows how we might conduct the experiment. The potentiometer measures the voltage applied by the power source. The ammeter measures the current flowing through the circuit.

The electrode at which the reaction of interest occurs is called the **working electrode.** In Figure 16-1, we happen to be interested in reduction of Cu^{2+}, so Cu is the working electrode. The other electrode is called the *counter electrode*. We adopt the convention that *current is positive if reduction occurs at the working electrode.*

FIGURE 16-1 Electrolysis experiment. The power supply is a variable voltage source. The potentiometer measures voltage and the ammeter measures current.

Current Measures the Rate of Reaction

If a current I flows for a time t, the charge q passing any point in the circuit is

Relation of charge to current and time: $$q = I \cdot t \qquad \text{(16-2)}$$
Coulombs   Amperes · Seconds

An *ampere* is an electric current of 1 coulomb per second.

A *coulomb* contains $6.241\ 5 \times 10^{18}$ electrons.

The number of moles of electrons is

$$\text{Moles of e}^- = \frac{\text{coulombs}}{\text{coulombs/mole}} = \frac{I \cdot t}{F}.$$

If a reaction requires n electrons per molecule, the quantity reacting in time t is

Relation of moles to current and time:

$$\text{Moles reacted} = \frac{I \cdot t}{nF} \tag{16-3}$$

Faraday constant:

$F = 9.648\ 5 \times 10^4$ C/mol

$$\text{moles of electrons} = \frac{I \cdot t}{F}$$

EXAMPLE **Relating Current, Time, and Amount of Reaction**

If a current of 0.17 A flows for 16 min through the cell in Figure 16-1, how many grams of Cu(s) will be deposited?

Solution We first calculate the moles of e$^-$ flowing through the cell:

$$\text{Moles of e}^- = \frac{I \cdot t}{F} = \frac{\left(0.17\,\frac{C}{s}\right)(16\ \text{min})\left(60\,\frac{s}{min}\right)}{96\ 485\left(\frac{C}{mol}\right)} = 1.6_9 \times 10^{-3}\ \text{mol}$$

The cathode half-reaction requires 2e$^-$ for each Cu deposited. Therefore,

$$\text{Moles of Cu}(s) = \frac{1}{2}(\text{moles of e}^-) = 8.4_5 \times 10^{-4}\ \text{mol}$$

The mass of Cu(s) deposited is $(8.4_5 \times 10^{-4}\ \text{mol})(63.546\ \text{g/mol}) = 0.054$ g.

Test Yourself A *monolayer* (single layer of atoms) of Cu on the crystal face shown in the margin has 1.53×10^{15} atoms/cm$^2 = 2.54 \times 10^{-9}$ mol/cm^2. What current can deposit one layer of Cu atoms on 1 cm^2 in 1 s? (*Answer:* 0.490 mA)

(100) crystal face has 1.53×10^{15} atoms/cm^2

← 0.361 nm →

Face-centered cubic crystal

DEMONSTRATION 16-1 **Electrochemical Writing[4]**

Approximately 7% of electric power in the United States goes into electrolytic chemical production. The electrolysis apparatus pictured here consists of a sheet of Al foil taped or cemented to a wood surface. Any size will work, but an area of 15 cm on a side is convenient for a classroom demonstration. Tape to the metal foil (at one edge only) a sandwich consisting of filter paper, printer paper, and another sheet of filter paper. Make a stylus from Cu wire (18 gauge or thicker) looped at one end and passed through a length of glass tubing.

Front view   Side view

Prepare a fresh solution from 1.6 g of KI, 20 mL of water, 5 mL of 1 wt% starch solution, and 5 mL of phenolphthalein indicator solution. (If the solution darkens after several days, decolorize it with a few drops of dilute Na$_2$S$_2$O$_3$.) Soak the three layers of paper with the KI-starch-phenolphthalein solution. Connect the stylus and foil to a 12 V DC power source, and write on the paper with the stylus.

When the stylus is the cathode, water is reduced to H$_2$ plus OH$^-$ and pink color appears from the reaction of OH$^-$ with phenolphthalein.

$$\text{Cathode:} \quad H_2O + e^- \rightarrow \frac{1}{2}H_2(g) + OH^-$$

When the polarity is reversed and the stylus is the anode, I$^-$ is oxidized to I$_2$; a black (very dark blue) color appears from the reaction of I$_2$ with starch.

$$\text{Anode:} \quad I^- \rightarrow \frac{1}{2}I_2 + e^-$$

Pick up the top sheet of filter paper and the printer paper, and you will discover that the writing appears in the opposite color on the bottom sheet of filter paper (Color Plate 10).

Voltage Changes When Current Flows

Figure 16-1 is drawn with the same conventions as Figures 13-4 and 13-6. The cathode—where reduction occurs—is at the right side of the figure. The positive terminal of the potentiometer is on the right-hand side.

If electric current is negligible, the cell voltage is

$$E = E(\text{cathode}) - E(\text{anode}) \qquad (16\text{-}4)$$

To use $E = E(\text{cathode}) - E(\text{anode})$, you must write both reactions as *reductions*. $E(\text{cathode}) - E(\text{anode})$ is the *open-circuit* voltage. It is the voltage measured with negligible current flowing between cathode and anode.

The *cathode* is the electrode connected to the *negative* terminal of the power supply.

In Chapter 13, we wrote $E = E_+ - E_-$, where E_+ is the potential of the electrode attached to the positive terminal of the potentiometer and E_- is the potential of the electrode attached to the negative terminal of the potentiometer. Equation 16-4 is equivalent to $E = E_+ - E_-$. The polarity of the potentiometer in Figure 16-1 is the same as in Figures 13-4 and 13-6. In an electrolysis, electrons come from the negative terminal of the power supply into the cathode of the electrolysis cell. $E(\text{cathode})$ is the potential of the electrode connected to the negative terminal of the power supply, and $E(\text{anode})$ is the potential of the electrode connected to the positive terminal of the power supply.

If Reaction 16-1 contains 0.20 M Cu^{2+} and 1.0 M H^+ and liberates O_2 at a pressure of 1.0 bar, we find

$$E = \underbrace{\left\{ 0.339 - \frac{0.059\ 16}{2} \log\left(\frac{1}{[Cu^{2+}]} \right) \right\}}_{E\ (\text{cathode})} - \underbrace{\left\{ 1.229 - \frac{0.059\ 16}{2} \log\left(\frac{1}{P_{O_2}^{1/2}[H^+]^2} \right) \right\}}_{E\ (\text{anode})}$$

$$= \left\{ 0.339 - \frac{0.059\ 16}{2} \log\left(\frac{1}{[0.20]} \right) \right\} - \left\{ 1.229 - \frac{0.059\ 16}{2} \log\left(\frac{1}{(1.0)^{1/2}[1.0]^2} \right) \right\}$$

$$= 0.318 - 1.229 = -0.911 \text{ V}$$

Free energy change for Reaction 16-1:

$\Delta G = -nFE = -nF(-0.911 \text{ V})$

$\quad = -(2)\left(96\ 485\ \dfrac{C}{\text{mol}} \right)(-0.911 \text{ V})$

$\quad = +1.76 \times 10^5 \text{ C} \cdot \text{V/mol}$

$\quad = +1.76 \times 10^5 \text{ J/mol} = 176 \text{ kJ/mol}$

Note that $C \times V = J$

This voltage would be read on the potentiometer in Figure 16-1 if there were negligible current. The voltage is negative because the positive terminal of the potentiometer is connected to the negative side of the power supply. The free energy change computed in the margin is positive because the reaction is not spontaneous. We need the power supply to force the reaction to occur. If current is not negligible, *overpotential*, *ohmic potential*, and *concentration polarization* can change the voltage required to drive the reaction.

Overpotential is the voltage required to overcome the *activation energy* for a reaction at an electrode (Figure 16-2).[5] The faster you wish to drive the reaction, the greater the overpotential that must be applied. Electric current measures the rate of electron transfer. Applying a greater overpotential will sustain a higher *current density* (current per unit area of electrode surface, A/m^2). Table 16-1 shows that the overpotential for liberation of H_2 at a Cu surface must be increased from 0.479 to 1.254 V to increase the current density from 10 A/m^2 to 10 000 A/m^2. Activation energy depends on the nature of the surface. H_2 is evolved at a Pt surface with little overpotential, whereas a Hg surface requires ~1 V to drive the reaction.

FIGURE 16-2 (*a*) Schematic energy profile for electron transfer from a metal to H_3O^+, leading to liberation of H_2. (*b*) Applying a potential to the metal raises the energy of the electron in the metal and decreases the activation energy for electron transfer.

Ohmic potential is the voltage needed to overcome electrical resistance, R, of the solution in the electrochemical cell when a current, I, is flowing:

Ohmic potential: $\qquad\qquad\qquad E_{\text{ohmic}} = IR \qquad (16\text{-}5)$

TABLE 16-1 Overpotential (V) for gas evolution at various current densities (A/m²) at 25°C

Electrode	10 A/m²		100 A/m²		1 000 A/m²		10 000 A/m²	
	H_2	O_2	H_2	O_2	H_2	O_2	H_2	O_2
Platinized Pt	0.015 4	0.398	0.030 0	0.521	0.040 5	0.638	0.048 3	0.766
Smooth Pt	0.024	0.721	0.068	0.85	0.288	1.28	0.676	1.49
Cu	0.479	0.422	0.584	0.580	0.801	0.660	1.254	0.793
Ag	0.475 1	0.580	0.761 8	0.729	0.874 9	0.984	1.089 0	1.131
Au	0.241	0.673	0.390	0.963	0.588	1.244	0.798	1.63
Graphite	0.599 5		0.778 8		0.977 4		1.220 0	
Pb	0.52		1.090		1.179		1.262	
Zn	0.716		0.746		1.064		1.229	
Hg	0.9		1.0		1.1		1.1	

SOURCE: *International Critical Tables*, **1929**, 6, 339. This reference also gives overpotentials for Cl_2, Br_2, and I_2.

If the cell has a resistance of 2 ohms and a current of 20 mA is flowing, the voltage required to overcome the resistance is $E = (2 \; \Omega)(20 \; mA) = 0.040$ V.

Concentration polarization occurs when the concentrations of reactants or products are not the same at the surface of the electrode as they are in bulk solution. For Reaction 16-1, the Nernst equation should be written

$$E(\text{cathode}) = 0.339 - \frac{0.059\ 16}{2} \log\left(\frac{1}{[Cu^{2+}]_s}\right)$$

where $[Cu^{2+}]_s$ is the concentration in the solution *at the surface of the electrode*. If reduction of Cu^{2+} occurs rapidly, $[Cu^{2+}]_s$ could be very small because Cu^{2+} cannot diffuse to the electrode as fast as it is consumed. As $[Cu^{2+}]_s$ decreases, $E(\text{cathode})$ becomes more negative.

Overpotential, ohmic potential, and concentration polarization make electrolysis more difficult. They drive the cell voltage more negative, requiring more voltage from the power supply in Figure 16-1 to drive the reaction forward.

$$E = \underbrace{E(\text{cathode}) - E(\text{anode})}_{\substack{\text{These terms include the} \\ \text{effects of concentration polarization}}} - IR - \text{overpotentials} \qquad (16\text{-}6)$$

There can be concentration polarization and overpotential at both the cathode and the anode.

Resistance is measured in ohms, whose symbol is capital Greek omega, Ω.

Electrodes respond to concentrations of reactants and products adjacent to the electrode, not to concentrations in the bulk solution.

If $[Cu^{2+}]_s$ were reduced from 0.2 M to 2 μM, $E(\text{cathode})$ would change from 0.318 to 0.170 V.

EXAMPLE   Effects of Ohmic Potential, Overpotential, and Concentration Polarization

Suppose we wish to electrolyze I^- to I_3^- in a 0.10 M KI solution containing 3.0×10^{-5} M I_3^- at pH 10.00 with P_{H_2} fixed at 1.00 bar.

$$3I^- + 2H_2O \rightarrow I_3^- + H_2(g) + 2OH^-$$

(a) Find the cell voltage if no current is flowing. **(b)** Then suppose that electrolysis increases $[I_3^-]_s$ to 3.0×10^{-4} M, but other concentrations are unaffected. Suppose that the cell resistance is 2.0 Ω, the current is 63 mA, the cathode overpotential is 0.382 V, and the anode overpotential is 0.025 V. What voltage is needed to drive the reaction?

Solution **(a)** The open-circuit voltage is $E(\text{cathode}) - E(\text{anode})$:

Cathode: $2H_2O + 2e^- \rightarrow H_2(g) + 2OH^-$ $E° = -0.828$ V
Anode: $I_3^- + 2e^- \rightarrow 3I^-$ $E° = 0.535$ V

$$E(\text{cathode}) = -0.828 - \frac{0.059\ 16}{2}\log(P_{H_2}[OH^-]^2)$$

$$= -0.828 - \frac{0.059\ 16}{2}\log[(1.00)(1.0 \times 10^{-4})^2] = -0.591 \text{ V}$$

$$E(\text{anode}) = 0.535 - \frac{0.059\,16}{2}\log\left(\frac{[I^-]^3}{[I_3^-]}\right)$$

$$= 0.535 - \frac{0.059\,16}{2}\log\left(\frac{[0.10]^3}{[3.0 \times 10^{-5}]}\right) = 0.490 \text{ V}$$

$$E = E(\text{cathode}) - E(\text{anode}) = -1.081 \text{ V}$$

We would have to apply -1.081 V to force the reaction to occur.
(b) Now $E(\text{cathode})$ is unchanged but $E(\text{anode})$ changes because $[I_3^-]_s$ is different from $[\,I_3^-]$ in bulk solution.

$$E(\text{anode}) = 0.535 - \frac{0.059\,16}{2}\log\left(\frac{[0.10]^3}{[3.0 \times 10^{-4}]}\right) = 0.520 \text{ V}$$

$$E = E(\text{cathode}) - E(\text{anode}) - IR - \text{overpotentials}$$

$$= -0.591 \text{ V} - 0.520 \text{ V} - (2.0\,\Omega)(0.063\text{ A}) - 0.382 \text{ V} - 0.025 \text{ V}$$

$$= -1.644 \text{ V}$$

Instead of -1.081 V, we need to apply -1.644 V to drive the reaction.

Test Yourself Find the voltage in part **(b)** if $[I^-]_s = 0.01$ M. (***Answer:*** -1.732 V)

Controlled-Potential Electrolysis with a Three-Electrode Cell

An **electroactive species** is one that can be oxidized or reduced at an electrode. We regulate the potential of the working electrode to control which species react and which do not. Metal electrodes are said to be **polarizable,** which means that their potentials are easily changed when small currents flow. A reference electrode such as calomel or Ag | AgCl is said to be **nonpolarizable,** because its potential does not vary much unless a significant current is flowing. Ideally, we want to measure the potential of a polarizable working electrode with respect to a nonpolarizable reference electrode. How can we do this if there is to be significant current at the working electrode and negligible current at the reference electrode?

The answer is to introduce a third electrode (Figure 16-3). The **working electrode** is the one at which the reaction of interest occurs. A calomel or other **reference electrode** is used to measure the potential of the working electrode. The **auxiliary electrode** (also called the *counter electrode*) is the current-supporting partner of the working electrode. Current flows between the working and the auxiliary electrodes. Negligible current flows through the reference electrode, so its potential is unaffected by ohmic potential, concentration polarization, and overpotential. It truly maintains a constant reference potential. In **controlled-potential electrolysis,** the voltage difference between the working and reference electrodes in a three-electrode cell is regulated by an electronic device called a **potentiostat.**

Concentration polarization and overpotential can both occur at the working and auxiliary electrodes. There is an ohmic potential drop between the working and auxiliary electrodes. To obtain the best measurement of the working electrode potential, the reference electrode should be placed as close as possible to the working electrode (Figure 16-4).

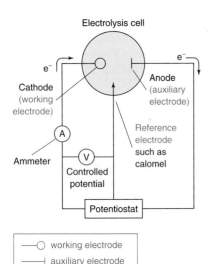

FIGURE 16-3 Circuit used for controlled-potential electrolysis with a three-electrode cell.

Working electrode: where the analytical reaction occurs

Auxiliary electrode: the other electrode needed for current flow

Reference electrode: used to measure the potential of the working electrode

The chromatographic detector at the opening of this chapter has a Cu working electrode, a stainless steel auxiliary electrode, and a Ag | AgCl reference electrode.

FIGURE 16-4 Use of a Luggin capillary to position a reference electrode as close as possible to the working electrode (shown as a dropping mercury electrode in this illustration). The capillary, with an opening of ~0.2 mm, is filled with the same electrolyte that is in the analyte solution. The reference electrode is in contact with the capillary solution. There is negligible current in the capillary, so there is negligible ohmic loss between the tip of the capillary and the reference electrode.

Color Plate 1 HCl Fountain (Demonstration 6-2) (*a*) Basic indicator solution in beaker. (*b*) Indicator is drawn into flask and changes to acidic color. (*c*) Solution levels at end of experiment.

(*a*)   (*b*)   (*c*)

(*a*)   (*b*)

Color Plate 2 Effect of Ionic Strength on Ionic Dissociation (Demonstration 7-1) (*a*) Two beakers containing identical solutions with $Fe(SCN)^{2+}$, Fe^{3+}, and SCN^-. (*b*) The red color of $Fe(SCN)^{2+}$ fades when KNO_3 is added to the right-hand beaker because the equilibrium $Fe^{3+} + SCN^- \rightleftharpoons Fe(SCN)^{2+}$ shifts to the left.

pH: 1   2   3   4   5   6   7   8   9   10   11

Color Plate 3 Thymol Blue (Section 10-6) Acid-base indicator thymol blue between pH 1 and 11. The pK values are 1.7 and 8.9.

(a)

(b)

(c)

(d)

(e)

Color Plate 4 Indicators and Acidity of CO_2 (Demonstration 10-1) (*a*) Cylinders before adding Dry Ice. Ethanol indicator solutions of phenolphthalein (left) and bromothymol blue (right) have not yet mixed with entire cylinder. (*b*) Adding Dry Ice causes bubbling and mixing. (*c*) Further mixing. (*d*) Phenolphthalein changes to its colorless acidic form. Color of bromothymol blue is due to mixture of acidic and basic forms. (*e*) After addition of HCl and stirring of right-hand cylinder, bubbles of CO_2 can be seen leaving solution, and indicator changes completely to its acidic color.

Color Plate 5 Titration of Cu(II) with EDTA, Using Auxiliary Complexing Agent (Section 11-5) 0.02 M $CuSO_4$ before titration (left). Color of Cu(II)-ammonia complex after adding ammonia buffer, pH 10 (center). End-point color when all ammonia ligands have been displaced by EDTA (right).

(a)

(b)

Color Plate 6 Titration of Mg^{2+} by EDTA, Using Eriochrome Black T Indicator (Demonstration 11-1) (a) Before (left), near (center), and after (right) equivalence point. (b) Same titration with methyl red added as inert dye to alter colors.

Color Plate 7 Titration of VO^{2+} with Potassium Permanganate (Section 15-4) Blue VO^{2+} solution prior to titration (left). Mixture of blue VO^{2+} and yellow VO_2^+ observed during titration (center). Dark color of MnO_4^- at end point (right).

Color Plate 8 Photolytic Environmental Carbon Analyzer (Box 15-2) A measured water sample is injected into the chamber at the left, where it is acidified with H_3PO_4 and sparged (bubbled with Ar or N_2) to remove CO_2 derived from HCO_3^- and CO_3^{2-}. The CO_2 is measured by its infrared absorbance. The sample is then forced into the digestion chamber, where $S_2O_8^{2-}$ is added and the sample is exposed to ultraviolet radiation from an immersion lamp (the coil at the center of the photo). Sulfate radicals (SO_4^-) formed by irradiation oxidize most organic compounds to CO_2, which is measured by infrared absorbance. The U-tube at the right contains Sn and Cu granules to scavenge volatile acids, such as HCl and HBr, liberated in the digestion. [Photo courtesy Ed Urbansky, U.S. Environmental Protection Agency, Cincinnati, OH.]

Color Plate 9 **Iodometric Titration (Section 15-7)**
I_3^- solution (left). I_3^- solution before end point in titration
with $S_2O_3^{2-}$ (left center). I_3^- solution immediately before
end point with starch indicator present (right center). At
the end point (right).

(a)

(b)

(c)

Color Plate 10 **Electrochemical Writing (Demonstration 16-1)** (a) Stylus used
as cathode. (b) Stylus used as anode. (c) The polarity of the foil backing is opposite that
of the stylus and produces reverse color on the bottom sheet of paper.

(a)

(b)

Color Plate 11 **Formation of Diffusion Layer During Electrolysis (Box 16-3)** (a) Cu
electrode (flat plate, left) and Pt electrode (mesh basket, right) immersed in solution containing
KI and starch, with no electric current. (b) Starch-iodine complex forms at surface of Pt anode
when current flows.

Color Plate 12 **Grating Dispersion (Section 17-2)** Visible spectrum produced by a grating inside a spectrophotometer.

Color Plate 13 **Beer's Law (Section 17-2)** Fe(phenanthroline)$_3^{2+}$ standards for spectrophotometric analysis. Volumetric flasks contain Fe(phenanthroline)$_3^{2+}$ with Fe concentrations ranging from 1 mg/L (left) to 10 mg/L (right). The absorbance, as evidenced by the intensity of the color, is proportional to the iron concentration.

(*a*)   (*b*)

Color Plate 14 **Absorption Spectra (Demonstration 17-1)** (*a*) Projected visible spectra of (from top to bottom) white light, potassium dichromate, bromophenol blue, and phenolphthalein. (*b*) Visible absorption spectra of the same compounds recorded with a spectrophotometer.

Color Plate 15 Absorption Spectra and Color (Section 17-2 and Problem 17-9) Flasks contain suspensions of silver nanoparticles whose color depends on the size and shape of the particles, which are approximately triangular plates with edge lengths of ~50–100 nm. The visible absorption spectrum of each suspension is shown in the graph. Stable suspensions of nanoparticles are called *colloids* (Demonstration 26-1). [From D. M. Ledwith, A. M. Whelan, and J. M. Kelly, *J. Mater. Chem.* **2007,** *17,* 2459. Courtesy J. M. Kelly and D. Ledwith, Trinity College, University of Dublin.]

(*a*)   (*b*)

Color Plate 16 Luminescence (Section 17-7) (*a*) Green crystal of yttrium aluminum garnet containing a small amount of Cr^{3+}. (*b*) When irradiated with high-intensity blue light from a laser at the right side, Cr^{3+} absorbs blue light and emits lower energy red light. When the laser is removed, the crystal appears green again. [Courtesy M. Seltzer, Michelson Laboratory, China Lake, CA.]

Color Plate 17 Quenching of Ru(II) Luminescence by O_2 (Section 18-6) *Left:* Orange-red luminescence from ~5 μM (bipyridyl)$_3$RuCl$_2$ in methanol after most of the air was removed by bubbling with Dry Ice. *Right:* Luminescence is quenched (decreased) after bubbling O_2 through the solution for 30 s.

Oregon green reference dye—
insensitive to O_2

Ru(II) dye—
sensitive to O_2

Color Plate 18 Fluorescence from O_2-indicator beads inside living cells. (Section 18-6) Green light emitted from the dye Oregon Green is independent of local O_2 concentration. Red light emitted from tris(4,7-diphenyl-1,10-phenanthroline)ruthenium(II) chloride decreases in the presence of O_2. The ratio of intensities at the red and green wavelengths tells us the concentration of O_2 inside the cell. [Courtesy R. Kopelman and E. Monson, University of Michigan. From H. Xu, J. W. Aylott, R. Kopelman, T. J. Miller, and M. A. Philbert, *Anal. Chem.* **2001**, *73*, 4124.]

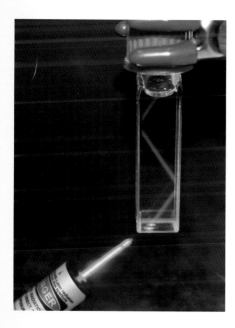

Color Plate 19 Upconversion (Box 18-2) Low-energy green light from a 5-mW laser is converted to high-energy blue fluorescence. Does this violate the conservation of energy? Box 18-2 provides an answer. [Courtesy F. N. Castellano and T. N. Singh-Rachford, Bowling Green State University. See R. R. Islangulov, D. V. Kozlov, and F. N. Castellano, *Chem. Commun.* **2005**, 3776.]

$n = -1$

Incident beam $n = 1$ $n = 2$
(a)

$n = 1$ $n = 2$
(b)

Color Plate 20 Laser Diffraction from a Compact Disk (Section 19-2) The grooves in an audio compact disk or a computer compact disk have a spacing of 1.6 μm. (a) When a red laser strikes disk at normal incidence ($\theta = 0$ in Figure 19-7 and Equation 19-2), three diffracted beams with orders $n = +1, +2$, and -1 are observed. (b) Red and green lasers strike the disk at normal incidence. Green light has a shorter wavelength than red light, and so, according to Equation 19-2, green light is diffracted at a smaller angle (ϕ). Beams have been made visible by "fog" from liquid nitrogen. [Courtesy J. Tellinghuisen, Vanderbilt University. See J. Tellinghuisen, *J. Chem. Ed.* **2002**, *79*, 703; and F. Wakabayashi and K. Hamada, *J. Chem. Ed.* **2006**, *83*, 56.]

(a)   (b)

Color Plate 21 **Transmission, Reflection, Refraction, Absorption, and Luminescence (Section 19-4)** (a) Blue-green laser is directed into a crystal of yttrium aluminum garnet doped with Er^{3+}, which emits yellow light. Light entering the crystal from the right is refracted (bent) and partially reflected at the right-hand surface of the crystal. The laser beam appears yellow inside the crystal because of luminescence from Er^{3+}. As it exits the crystal at the left side, the laser beam is refracted again, and partially reflected back into the crystal. (b) Same experiment, but with blue light instead of blue-green light. Blue light is absorbed by Er^{3+} and does not penetrate very far into the crystal. [Courtesy M. Seltzer, Michelson Laboratory, China Lake, CA.]

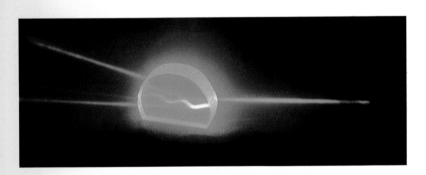

Color Plate 22 **Multiple Internal Reflections in a Bulk Crystal (Section 19-4)** Multiple internal reflections observed when blue laser light enters a crystal of yttrium aluminum garnet doped with Ho^{3+}. The beam entering from the right is mostly reflected back into the crystal at each face, creating a zigzag pattern inside the crystal. Part of the light is transmitted out of the crystal at each face. In an optical fiber, the angle of incidence is such that the beam is totally reflected within the fiber. [Courtesy M. Seltzer, Michelson Laboratory, China Lake, CA.]

25 μm   10 μm

(a)   (b)

Color Plate 23 **Oxygen Optode (Section 19-4)** (a) Sensor prepared from a 100-μm-diameter optical fiber. The active layer at the end contains tris(1,10-phenanthroline)Ru(II) chloride dissolved in polyacrylamide that is covalently bound to the fiber. Light coming down the fiber excites the Ru compound, which emits characteristic orange-red light that is collected with a microscope. When immersed in a sample containing O_2, emission is decreased. The decrease is a measure of O_2 concentration. (b) An optode with a submicrometer tip pulled from a larger fiber. This fiber can detect 10 amol of O_2. [From Z. Rosenzweig and R. Kopelman, *Anal. Chem.* **1995**, *67*, 2650.]

Color Plate 24 **Polychromator for Inductively Coupled Plasma Atomic Emission Spectrometer with One Detector for Each Element (Section 20-4)** Light emitted by a sample in the plasma enters the polychromator at the right and is dispersed into its component wavelengths by a grating at the bottom of the diagram. Each different emission wavelength (shown schematically by colored lines) is diffracted at a different angle and directed to a different photomultiplier detector on the focal curve. Each detector sees only one preselected element, and all elements are measured simultaneously. [Courtesy TJA Solutions, Franklin, MA.]

Color Plate 25 **Polychromator for Inductively Coupled Plasma Atomic Emission Spectrometer with One Detector for All Elements (Section 20-4)** Light emitted by a sample in the plasma enters the polychromator at the upper right and is dispersed vertically by a prism and then horizontally by a grating. The resulting two-dimensional pattern of wavelengths from 165 to 1 000 nm is detected by a charge injection device detector with 262 000 pixels. All elements are detected simultaneously. [Courtesy TJA Solutions, Franklin, MA.]

CID detector

Plasma

Light

Entrance aperture

Echelle grating

Shutter

CID detector

Prism

Collimating mirror

Focusing mirror

30° C

Color Plate 26 **Low-Temperature Plasma Ionizes Substances from Surfaces for Mass Spectral Analysis (Section 21-5)** Room-temperature plasma is made by passing He, Ar, N_2, or ambient air through a glass tube with a coaxial grounded wire. The tube is wrapped on the outside with a copper cladding to which is applied a 3-kV alternating current at a frequency of 2.5 kHz and power of 1 W. Excited-state species in the plasma ionize and dislodge molecules from a surface such as skin. The surface would be placed adjacent to the inlet of a mass spectrometer to record a spectrum of the ions. There is no electrical shock to the skin. [Courtesy R. G. Cooks, Purdue University. From J. D. Harper, N. A. Charipar, C. C. Mulligan, X. Zhang, R. G. Cooks, and Z. Ouyang, *Anal. Chem.* **2008**, *80*, 9097.]

After addition of phase transfer agent

Before addition of phase transfer agent

Colored anion: CrO_4^{2-} Food color   Gatorade drink   Orange II MnO_4^- $Cr_2O_7^{2-}$ Congo red

Color Plate 27 **Addition of Phase Transfer Agent Extracts Colored Anions from Water into Ether (Box 22-1)** *Lower panel:* Vials with aqueous lower phase and diethyl ether upper phase. Colored anions are present in the aqueous phase. *Upper panel:* After adding trioctylmethylammonium chloride to each vial and mixing well, trioctylmethylammonium cation extracts colored anion into ether phase.
[Courtesy A. J. Pezhathinal and R. Chan-Yu-King, University of Science and Arts of Oklahoma. From A. J. Pezhathinal, K. Rocke, L. Susanto, D. Handke, R. Chan-Yu-King, and P. Gordon, *J. Chem. Ed.* **2006**, *83*, 1161.]

(*a*)   (*b*)

Color Plate 28 **Thin-Layer Chromatography (Section 24-1)** The mixture to be separated is placed in tiny spots near the base of a plastic or glass plate coated with an adsorptive stationary phase. When the plate is placed in a shallow pool of solvent in a closed chamber, liquid migrates *up* the plate by capillary action. Different components of the mixture are carried along by the solvent to different extents, depending on how strongly they are adsorbed on the stationary phase. The stronger the adsorption, the slower a component travels. (*a*) Solvent ascends past a mixture of dyes near the bottom of the plate. (*b*) Separation achieved after solvent has ascended most of the way up the plate.

(a)

(b)

(c)

Color Plate 29 **Supercritical Carbon Dioxide (Box 24-3)**
(a) Liquid CO_2 in a 60-mL steel chamber at 30°C and 6.9 MPa. The red color is from a little I_2 added to the liquid to make it visible. (b) Beginning of the supercritical phase transition as the temperature is raised. (c) Single-phase supercritical carbon dioxide. [From H. Black, *Environ. Sci. Technol.* **1996**, *30*, 124A. Photos courtesy D. Pesiri and W. Tumas, Los Alamos National Laboratory.]

0

0

66

66

165

165

Hydrodynamic flow →
100-μm-diameter capillary

Electroosmotic flow →
75-μm-diameter capillary

Color Plate 30 **Velocity Profiles for Hydrodynamic and Electroosmotic Flow (Section 25-6)** A fluorescent dye was imaged inside a capillary tube at times 0, 66, and 165 ms after initiating flow. The highest concentration of dye is represented by blue and the lowest concentration is red in these images in which different colors are assigned to different fluorescence intensities. [From P. H. Paul, M. G. Garguilo, and D. J. Rakestraw, *Anal. Chem.* **1998**, *70*, 2459. See also D. Ross, T. J. Johnson, and L. E. Locascio, *Anal. Chem.* **2001**, *73*, 2509.]

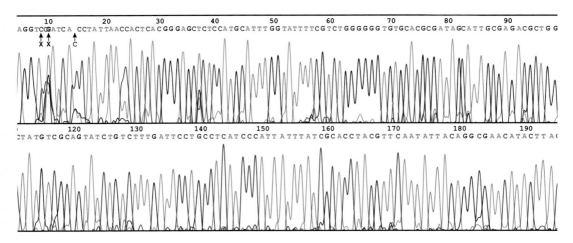

10   20   30   40   50   60   70   80   90
AGGTCCGATCA CCTA TTAACCACTCAC GGG GAGCTCTCCATGCATTT GG TATTTTCG TCT GGGGGG TGTGCACGCGATAGCATT GCGA GACGCT G G
X X   C

120   130   140   150   160   170   180   190
CTATGTCGCAG TATCTG TCTTTGATTCCT GCCTCATCCCATT ATTTATCGCACCTACGTT CAATATTACAGG CGAACATACTTA C

Color Plate 31 **DNA Sequencing by Capillary Gel Electrophoresis with Fluorescence Detection (Section 25-7)** Portion of DNA nucleotide base sequence made with a microfluidic device ("lab-on-a-chip") capable of reading a length of 365 bases with 99% accuracy. Successive peaks correspond to lengths of DNA with one more base. DNA strands terminated in each of the four different bases A, T, C, and G are tagged with a different fluorescent label that identifies the terminal base when it passes through a fluorescence detector. Different lengths of DNA are separated by sieving through an 18-cm-long electrophoresis channel containing polyacrylamide gel with 6 M urea to stabilize single strands. Injected sample volume is 30 nL containing 100 amol (60 million molecules) of DNA. [From R. G. Blazej, P. Kumaresan, S. A. Cronier, and R. A. Mathies, "Inline Injection Microdevice for Attomole-Scale Sanger DNA Sequencing," *Anal. Chem.* **2007,** *79,* 4499.]

(a)   *(b)*   *(c)*

Color Plate 32 **Colloids and Dialysis (Demonstration 26-1)** (*a*) Colloidal Fe(III) (left) and ordinary aqueous Fe(III) (right). (*b*) Dialysis bags containing colloidal Fe(III) (left) and a solution of Cu(II) (right) immediately after placement in flasks of water. (*c*) After 24 h of dialysis, the Cu(II) has diffused out and is dispersed uniformly between the bag and the flask, but the colloidal Fe(III) remains inside the bag.

Color Plate 33 **Fajans Titration of Cl⁻ with AgNO₃, Using Dichlorofluorescein (Demonstration 26-2)** (*a*) Indicator before beginning titration. (*b*) AgCl precipitate before end point. (*c*) Indicator adsorbed on precipitate after end point.

(a)   *(b)*   *(c)*

16-2 Electrogravimetric Analysis

In **electrogravimetric analysis,** analyte is quantitatively deposited on an electrode by electrolysis. The electrode is weighed before and after deposition. The increase in mass tells us how much analyte was deposited. We can measure Cu^{2+} in a solution by reducing it to $Cu(s)$ on a clean Pt gauze cathode with a large surface area (Figure 16-5). O_2 is liberated at the counter electrode.

How do you know when electrolysis is complete? One way is to observe the disappearance of color in a solution from which a colored species such as Cu^{2+} or Co^{2+} is removed. Another way is to expose most, but not all, of the surface of the cathode to the solution during electrolysis. To test whether or not the reaction is complete, raise the beaker or add water so that fresh surface of the cathode is exposed to the solution. After an additional period of electrolysis (15 min, say), see if the newly exposed electrode surface has a deposit. If it does, repeat the procedure. If not, the electrolysis is done. A third method is to remove a drop of solution and perform a qualitative test for analyte.

Tests for completion of the deposition:
- disappearance of color
- no deposition on freshly exposed electrode surface
- qualitative test for analyte in solution

(a)   (b)   (c)

FIGURE 16-5 (a) Electrogravimetric analysis. Analyte is deposited on the large Pt gauze electrode. If analyte is to be oxidized, rather than reduced, the polarity of the power supply is reversed so that deposition still occurs on the large electrode. (b) Outer Pt gauze electrode. (c) Optional inner Pt gauze electrode designed to be spun by a motor in place of magnetic stirring.

In the preceding section, we calculated that -0.911 V needs to be applied between the electrodes to deposit $Cu(s)$ on the cathode. The actual behavior of the electrolysis in Figure 16-6 shows that nothing special happens at -0.911 V. Near -2 V, the reaction begins in earnest. At low voltage, a small *residual current* is observed from reduction at the cathode and an equal amount of oxidation at the anode. Reduction might involve traces of dissolved O_2, impurities such as Fe^{3+}, or surface oxide on the electrode.

Table 16-1 shows that an overpotential of ~1 V is required for O_2 formation at the smooth Pt anode. Overpotential is the main reason why not much happens in Figure 16-6 until -2 V is applied. Beyond -2 V, the rate of reaction (the current) increases steadily. Around -4.6 V, the current increases more rapidly with the onset of reduction of H_3O^+ to H_2. Gas bubbles at the working electrode interfere with deposition of solids.

The voltage between the two electrodes is

$$E = E(\text{cathode}) - E(\text{anode}) - IR - \text{overpotentials} \qquad (16\text{-}6)$$

Suppose we hold the applied potential at $E = -2.0$ V. As Cu^{2+} is used up, the current decreases and both ohmic and overpotentials decrease in magnitude. $E(\text{anode})$ is fairly

16-2 Electrogravimetric Analysis   **367**

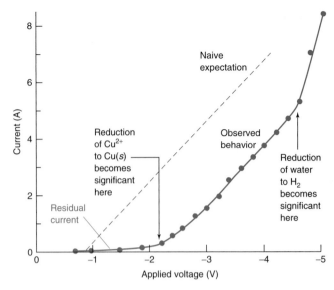

FIGURE 16-6 Observed current-voltage relation for electrolysis of 0.2 M $CuSO_4$ in 1 M $HClO_4$ under N_2, using the apparatus in Figure 16-5.

constant because of the high concentration of solvent being oxidized at the anode ($H_2O \rightarrow \frac{1}{2}O_2 + 2H^+ + 2e^-$). If E and E(anode) are constant and if IR and overpotentials decrease in magnitude, then *E(cathode) must become more negative* to maintain the algebraic equality in Equation 16-6. E(cathode) drops in Figure 16-7 to -0.4 V, at which H_3O^+ is reduced to H_2. As E(cathode) falls from $+0.3$ V to -0.4 V, other ions such as Co^{2+}, Sn^{2+}, and Ni^{2+} can be reduced. In general, then, *when the applied voltage is constant, the cathode potential drifts to more negative values and other solutes might be reduced.*

To prevent the cathode potential from becoming so negative that unintended ions are reduced, a cathodic **depolarizer** such as NO_3^- can be added to the solution. The cathodic depolarizer is more easily reduced than H_3O^+:

A *cathodic depolarizer* is reduced in preference to solvent. For *oxidation* reactions, *anodic depolarizers* include N_2H_4 (hydrazine) and NH_2OH (hydroxylamine).

$$NO_3^- + 10H^+ + 8e^- \rightarrow NH_4^+ + 3H_2O$$

Alternatively, we can use a three-electrode cell (Figure 16-3) with a potentiostat to control the cathode potential and prevent unwanted side reactions.

EXAMPLE **Controlled-Potential Electrolysis**

What cathode potential is required to reduce 99.99% of 0.10 M Cu^{2+} to $Cu(s)$? Is it possible to remove this Cu^{2+} without reducing 0.10 M Sn^{2+} in the same solution?

$$Cu^{2+} + 2e^- \rightleftharpoons Cu(s) \qquad E° = 0.339 \text{ V} \qquad (16\text{-}7)$$

$$Sn^{2+} + 2e^- \rightleftharpoons Sn(s) \qquad E° = -0.141 \text{ V} \qquad (16\text{-}8)$$

Solution If 99.99% of Cu^{2+} were reduced, the concentration of remaining Cu^{2+} would be 1.0×10^{-5} M, and the required cathode potential would be

$$E(\text{cathode}) = 0.339 - \frac{0.059\ 16}{2} \log\left(\frac{1}{\underbrace{1.0 \times 10^{-5}}_{[Cu^{2+}]}}\right) = 0.19 \text{ V}$$

The cathode potential required to reduce Sn^{2+} is

$$E(\text{cathode, for reduction of } Sn^{2+}) = -0.141 - \frac{0.059\ 16}{2} \log\left(\frac{1}{\underbrace{0.10}_{[Sn^{2+}]}}\right) = -0.17 \text{ V}$$

We do not expect reduction of Sn^{2+} at a cathode potential more positive than -0.17 V. The reduction of 99.99% of Cu^{2+} without reducing Sn^{2+} appears feasible.

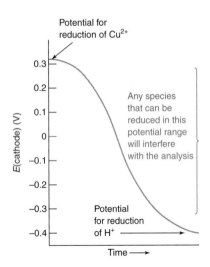

FIGURE 16-7 E(cathode) becomes more negative with time when electrolysis is conducted in a two-electrode cell with a constant voltage between the electrodes.

Test Yourself Will E(cathode) $= 0.19$ V reduce 0.10 M SbO^+ at pH 2 by the reaction $SbO^+ + 2H^+ + 3e^- \rightleftharpoons Sb(s) + H_2O$, $E° = 0.208$ V? (***Answer:*** E(cathode) for $SbO^+ = 0.11$ V, so reduction should not occur at 0.19 V)

Underpotential Deposition

When Sn^{2+} in 1 M HCl is electrolyzed at a gold working electrode, reduction is observed at $E(\text{cathode}) = -0.18$ V by the technique of cyclic voltammetry, which we study later in this chapter. From what you know so far, you should not expect that potentials more positive than -0.18 V will reduce Sn^{2+}. Yet, a small current is observed at $E(\text{cathode}) = +0.12$ V. The coulombs required at -0.18 V increase in proportion to $[Sn^{2+}]$. The coulombs required at $+0.12$ V are just enough to produce 8.7×10^{-10} mol $Sn(s)$ per square centimeter of gold electrode surface.[6] Then no more current flows at $E(\text{cathode}) = +0.12$ V.

Reduction at $+0.12$ V is called **underpotential deposition.** It occurs at a potential that is not predicted to reduce Sn^{2+} to bulk $Sn(s)$. It is explained by the reaction

$$Sn^{2+} + 2e^- \rightleftharpoons Sn(\textit{monolayer on Au}) \qquad (16\text{-}9)$$

in which the product is a one-atom-thick layer of tin on gold. It is thermodynamically more favorable to deposit a layer of tin atoms on gold than it is to deposit a fresh layer of tin on bulk tin metal. Therefore, the potential for Reaction 16-9 is more positive than the potential for Reaction 16-8.

Monolayer of Sn on Au

16-3 Coulometry

Coulometry is a chemical analysis based on counting the electrons used in a reaction. For example, cyclohexene can be titrated with Br_2 generated by electrolytic oxidation of Br^-:

$$2Br^- \rightarrow Br_2 + 2e^- \qquad (16\text{-}10)$$

$$Br_2 \ + \ \bigcirc \ \rightarrow \ \text{(trans-1,2-dibromocyclohexane structure)} \qquad (16\text{-}11)$$

Cyclohexene   *trans*-1,2-Dibromocyclohexane

Coulometric methods are based on measuring the number of electrons that participate in a chemical reaction.

The initial solution contains an unknown quantity of cyclohexene and a large amount of Br^-. When Reaction 16-10 has generated just enough Br_2 to react with all the cyclohexene, the moles of electrons liberated in Reaction 16-10 are equal to twice the moles of Br_2 and therefore twice the moles of cyclohexene.

The reaction is carried out *at a constant current* with the apparatus in Figure 16-8. Br_2 generated at the Pt anode at the left reacts with cyclohexene. When cyclohexene is consumed, the concentration of Br_2 suddenly rises, signaling the end of the reaction.

The rise in Br_2 concentration is detected by measuring the current between the two detector electrodes at the right in Figure 16-8. A voltage of 0.25 V applied between these two electrodes is not enough to electrolyze any solute, so only a tiny current of <1 μA flows through

To constant-current coulometric power supply

Port for adding reagents, with ground-glass stopper

Microammeter

Voltmeter

Pt generator anode

Pt cathode

Pt detector electrodes

Magnetic stirring bar

Amperometric end-point detector circuit

FIGURE 16-8 Apparatus for coulometric titration of cyclohexene with Br_2. The solution contains cyclohexene, 0.15 M KBr, and 3 mM mercuric acetate in a mixed solvent of acetic acid, methanol, and water. Mercuric acetate catalyzes the addition of Br_2 to the olefin.
[Adapted from D. H. Evans, "Coulometric Titration of Cyclohexene with Bromine," *J. Chem. Ed.* **1968**, *45*, 88.]

the microammeter. At the equivalence point, cyclohexene is consumed, $[Br_2]$ suddenly increases, and detector current flows by virtue of the reactions:

Both Br_2 and Br^- must be present for the detector half-reactions to occur. Prior to the equivalence point, there is Br^-, but virtually no Br_2.

Detector anode:	$2Br^- \rightarrow Br_2 + 2e^-$
Detector cathode:	$Br_2 + 2e^- \rightarrow 2Br^-$

In practice, enough Br_2 is first generated in the absence of cyclohexene to give a detector current of 20.0 μA. When cyclohexene is added, the current decreases to a tiny value because Br_2 is consumed. Br_2 is then generated by the coulometric circuit, and the end point is taken when the detector again reaches 20.0 μA. Because the reaction is begun with Br_2 present, impurities that can react with Br_2 before analyte is added are eliminated.

The electrolysis current (not to be confused with the detector current) for the Br_2-generating electrodes is controlled by a hand-operated switch. As the detector current approaches 20.0 μA, you close the switch for shorter and shorter intervals. This practice is analogous to adding titrant dropwise from a buret near the end of a titration. The switch in the coulometer circuit serves as a "stopcock" for addition of Br_2 to the reaction. Modern instruments automate the entire procedure.

EXAMPLE Coulometric Titration

A 2.000-mL volume containing 0.611 3 mg of cyclohexene/mL is to be titrated in Figure 16-8. How much time is required for titration at a constant current of 4.825 mA?

Solution The moles of cyclohexene are

$$\frac{(2.000 \text{ mL})(0.611\ 3 \text{ mg/mL})}{(82.146 \text{ mg/mmol})} = 0.014\ 88 \text{ mmol}$$

In Reactions 16-10 and 16-11, each mole of cyclohexene requires 1 mol of Br_2, which requires 2 mol of electrons. For 0.014 88 mmol of cyclohexene to react, 0.029 76 mmol of electrons must flow. From Equation 16-3,

$$\text{Moles of } e^- = \frac{I \cdot t}{F} \Rightarrow t = \frac{(\text{moles of } e^-)F}{I}$$

$$t = \frac{(0.029\ 76 \times 10^{-3} \text{mol})(96\ 485 \text{ C/mol})}{(4.825 \times 10^{-3} \text{ C/s})} = 595.1 \text{ s}$$

It will require just under 10 min to complete the reaction.

Test Yourself How much time is required to titrate 1.000 mg of cyclohexene at 4.000 mA? (**Answer:** 587.3 s)

Advantages of coulometry:

- precision
- sensitivity
- generation of unstable reagents in situ

The Latin *in situ* means "in place." Reagent is used right where it is generated.

Commercial coulometers deliver electrons with an accuracy of ~0.1%. With extreme care, the Faraday constant has been measured to within several parts per million by coulometry.[7] Automated coulometers commonly generate H^+, OH^-, Ag^+, and I_2 to titrate analytes including CO_2, sulfites in food, and sulfide in wastewater.[8] Unstable reagents such as Ag^{2+}, Cu^+, Mn^{3+}, and Ti^{3+} can be generated and used in situ.

In Figure 16-8, the reactive species (Br_2) is generated at the anode. The cathode products (H_2 from solvent and Hg from the catalyst) do not interfere with the reaction of Br_2 and cyclohexene. In some cases, however, H_2 or Hg could react with analyte. Then it is desirable to separate the counter electrode from the analyte, using the cell in Figure 16-9. H_2 bubbles innocuously out of the cathode chamber without mixing with the bulk solution.

Types of Coulometry

Coulometry employs either a *constant current* or a *controlled potential*. Constant-current methods, like the preceding Br_2/cyclohexene example, are called **coulometric titrations.** If we know the current and the time of reaction, we know how many coulombs have been delivered from Equation 16-2: $q = I \cdot t$.

Controlled-potential coulometry in a three-electrode cell is more selective than constant-current coulometry. Because the working electrode potential is constant, current decreases exponentially as analyte concentration decreases. Charge is measured by integrating current over the time of the reaction:

The number of coulombs is equal to the area under a curve of current versus time. Problem 16-21 provides an example.

$$q = \int_0^t I\, dt \qquad (16\text{-}12)$$

To power supply

Gas outlet

Inert electrolyte

Counter electrode

Fritted-glass disk

Analyte solution

Generator electrode

Magnetic stirring bar

FIGURE 16-9 Isolating the counter electrode from analyte. Ions can flow through the porous fritted-glass disk. The liquid level in the counter electrode compartment should be higher than the liquid in the reactor so that analyte solution does not flow into the compartment.

In controlled-potential coulometry, current decays exponentially. You can approach the equivalence point by letting the current decay to a set value. For example, the current (*above* the residual current) will ideally be 0.1% of its initial value when 99.9% of the analyte has been consumed. Alternatively, the current versus time curve can be extrapolated according to its theoretical behavior after measuring its actual behavior for some time.

16-4 Amperometry

In **amperometry,** we measure the electric current between a pair of electrodes that are driving an electrolysis reaction. One reactant is the intended analyte, and the measured current is proportional to the concentration of analyte. The measurement of dissolved O_2 with the **Clark electrode** in Box 16-1 is based on amperometry. A different kind of sensor based on conductivity—the "electronic nose"—is described in Box 16-2.

Amperometry: Electric current is proportional to the concentration of analyte.

Coulometry: Total number of electrons used for a reaction tells us how much analyte is present.

Box 16-1 Clark Oxygen Electrode

The Clark oxygen electrode[9] is widely used in medicine and biology to measure dissolved oxygen by amperometry. Leland Clark, who invented the electrode, also invented the glucose monitor and the heart-lung machine.

The glass body of the microelectrode in the figure is drawn to a fine point with a 5-μm opening at the base. Inside the opening is a 10- to 40-μm-long plug of silicone rubber, which is permeable to O_2. Oxygen diffuses into the electrode through the rubber and is reduced at the Au tip on the Pt wire, which is held at -0.75 V with respect to the Ag | AgCl reference electrode:

Pt | Au cathode: $O_2 + 4H^+ + 4e^- \rightarrow 2H_2O$

Ag | AgCl anode: $4Ag + 4Cl^- \rightarrow 4AgCl + 4e^-$

A Clark electrode is calibrated by placing it in solutions of known O_2 concentration, and a graph of current versus $[O_2]$ is constructed. The electrode also contains a silver *guard electrode* extending most of the way to the bottom. The guard electrode is kept at a negative potential so that any O_2 diffusing in from the top of the electrode is reduced and does not interfere with measurement of O_2 diffusing in through the silicone membrane at the bottom. Similar electrodes have been designed for detection of NO, H_2S, and CO.[10]

Coaxial Cu wire to Pt cathode

Ag | AgCl anode

Glass tubes drawn to points

Pt wire

Filling solution: 0.5 M KCl 0.5 M carbonate buffer, pH 10.5

Ag guard electrode

Gold button cathode electroplated onto Pt wire

Silicone rubber membrane

Clark oxygen microelectrode used to measure dissolved O_2 in small volumes. The tip of the cathode is plated with Au, which is less prone than Pt to fouling by adsorption of species from the test solution. [Adapted from N. P. Revsbech, "An Oxygen Microsensor with a Guard Column," *Limnol. Oceanogr.* 1989, *34*, 474.]

Box 16-2 **What Is An "Electronic Nose"?**

In the "old days," chemists prided themselves on their ability to identify compounds by odor. Smelling unknown chemicals is a bad idea because some vapors are toxic. Chemists are developing "electronic noses" that recognize odors to assess the freshness of meat, to find out if fruit is internally bruised, and to detect adulteration of food products.[11]

One approach to recognizing vapors is to coat interdigitated electrodes with an electronically conductive polymer, such as a derivative of polypyrrole.

Polypyrrole

When gaseous odor molecules are absorbed by the polymer, the electrical conductivity of the polymer changes. Different gases affect conductivity in different ways. Other sensor coatings are polymers containing conductive particles of silver or graphite. When the polymer absorbs small molecules, it swells and the conductivity decreases.

Pt/Au interdigitated electrodes

Interdigitated electrodes coated with conductive polymer to create an electronic nose. The conductivity of the polymer changes when it absorbs odor molecules. The spacing between "fingers" is ~0.25 mm.

One commercial "nose" has 32 sets of electrodes, each coated with a different polymer. The sensor yields 32 different responses when exposed to a vapor. The 32 changes are a "fingerprint" of the vapor. The electronic nose must be "trained" by pattern recognition algorithms to recognize an odor by its characteristic fingerprint. Other electronic noses are based on changes in optical absorption of polymers at the tips of optical fibers and on changes at the gates of field effect transistors (Section 14-8).

Enzyme: A protein that catalyzes a biochemical reaction. The enzyme increases the rate of reaction by many orders of magnitude.

Antibody: A protein that binds to a specific target molecule called an *antigen*. Antibodies bind to foreign cells that infect your body to initiate their destruction or identify them for attack by immune system cells.

Biosensors[12,13] use biological components such as *enzymes*, *antibodies*, or DNA for a highly selective response to one analyte. Biosensors that generate electric or optical signals are most common. Examples of amperometric biosensors are those that measure perchlorate in groundwater,[14] adenosine triphosphate (ATP) in biological tissue,[15] attomolar levels of genes,[16] and antibodies at femtomolar levels.[17] We now describe blood glucose monitors, which are, by far, the most widely used biosensors.

Blood Glucose Monitor

Diabetes afflicts approximately 5% of the population. Many people with diabetes must monitor their blood sugar (glucose) several times a day to control their disease through diet and insulin injections. Figure 16-10 shows a home glucose monitor featuring a disposable test strip with two carbon working electrodes and a Ag | AgCl reference electrode. As little as 4 μL of blood applied in the circular opening at the right of the figure is wicked over all three electrodes by a thin *hydrophilic* ("water loving") mesh. A 20-s measurement begins when the liquid reaches the reference electrode.

(a) (b)

FIGURE 16-10 (a) Personal glucose monitor used by diabetics to measure blood sugar level. (b) Details of disposable test strip to which a drop of blood is applied. [Courtesy Abbott Laboratories MediSense Products, Bedford, MA.]

Working electrode 1 is coated with the enzyme *glucose oxidase* and a *mediator*, which we describe soon. The enzyme catalyzes the reaction of glucose with O_2:

Reaction in coating above working electrode 1:

$$\text{Glucose} + O_2 \xrightarrow{\text{glucose oxidase}} \text{Gluconolactone} + H_2O_2 \quad (16\text{-}13)$$

Glucose   Oxygen   Gluconolactone   Hydrogen peroxide

In the absence of enzyme, the oxidation of glucose is negligible.

Early glucose monitors measured H_2O_2 from Reaction 16-13 by oxidation at a single working electrode, which was held at +0.6 V versus Ag | AgCl:

Reaction at working electrode 1:

$$H_2O_2 \rightarrow O_2 + 2H^+ + 2e^- \quad (16\text{-}14)$$

The current is proportional to the concentration of H_2O_2, which, in turn, is proportional to the glucose concentration in blood (Figure 16-11).

One problem with early glucose monitors is that their response depended on the concentration of O_2 in the enzyme layer, because O_2 participates in Reaction 16-13. If the O_2 concentration is low, the monitor responds as though the glucose concentration were low.

A good way to reduce O_2 dependence is to incorporate into the enzyme layer a species that substitutes for O_2 in Reaction 16-13. A substance that transports electrons between the analyte (glucose, in this case) and the electrode is called a **mediator.**

Reaction in coating above working electrode 1:

$$\text{Glucose} + 2 \left[\text{Fe} \right]^+ \xrightarrow{\text{glucose oxidase}} \text{gluconolactone} + 2 \text{ Fe} + 2H^+ \quad (16\text{-}15)$$

1,1'-Dimethylferricinium cation   1,1'-Dimethylferrocene
Mediator

The mediator consumed in Reaction 16-15 is then regenerated at the working electrode:

Reaction at working electrode 1:

$$\text{Fe} \xrightarrow[{-e^-}]{\text{working electrode}} \left[\text{Fe} \right]^+ \quad (16\text{-}16)$$

The current at the working electrode is proportional to the concentration of ferrocene, which, in turn, is proportional to the concentration of glucose in the blood.

One problem with glucose monitors is that species such as ascorbic acid (vitamin C), uric acid, and acetaminophen (Tylenol) found in blood can be oxidized at the same potential required to oxidize the mediator in Reaction 16-16. To correct for this interference, the test strip in Figure 16-10 has a second indicator electrode coated with mediator, *but not with glucose oxidase.* Interfering species that are reduced at electrode 1 are also reduced at electrode 2. The current due to glucose is the current at electrode 1 *minus* the current at electrode 2 (both measured with respect to the reference electrode). Now you see why the test strip has two working electrodes.

A challenge is to manufacture glucose monitors in such a reproducible manner that they do not require calibration. A user expects to add a drop of blood to the test strip and get a reliable reading without first constructing a calibration curve from known concentrations of glucose in blood. Each lot of test strips is highly reproducible and calibrated at the factory.

FIGURE 16-11 Response of an amperometric glucose electrode with dissolved O_2 concentration corresponding to an oxygen pressure of 0.027 bar, which is 20% lower than the typical concentration in subcutaneous tissue. [Data from S.-K. Jung and G. W. Wilson, "Polymeric Mercaptosilane-Modified Platinum Electrodes for Elimination of Interferents in Glucose Biosensors," *Anal. Chem.* **1996**, *68*, 591.]

A *mediator* transports electrons between the analyte and the working electrode. The mediator undergoes no net reaction itself.

Ferrocene contains flat, five-membered rings, similar to benzene. Each ring formally carries one negative charge, so the oxidation state of Fe, which sits between the rings, is +2. This molecule is called a *sandwich complex.*

The mediator lowers the required working electrode potential from 0.6 V to 0.2 V versus Ag | AgCl, thereby improving the stability of the sensor and eliminating some interference by other species in the blood.

A modified sensor measures glucose at a concentration of 2 fM in a 30-μL volume containing just 36 000 molecules of glucose.[18]

"Electrical Wiring" of Enzymes and Mediators for Blood Glucose Monitor

Demand for glucose monitors provides an economic stimulus for research that continues to improve the performance of home glucose monitors.[19] Noteworthy advances include (1) monitoring the reaction by coulometry instead of amperometry, (2) using a different enzyme to catalyze glucose oxidation, and (3) "electrical wiring" to increase the rate of reaction and to prevent reactants from diffusing away from the working electrode.

In *amperometry*, current flowing during glucose oxidation is measured. In *coulometry*, the number of electrons required to oxidize the glucose in a blood sample is counted. Amperometry measures the *rate of oxidation*. Coulometry measures the *number of molecules that have been oxidized*. The rate of reaction, and therefore the current, depends on temperature, but total charge transferred during oxidation is independent of temperature. Therefore, the coulometric measurement is independent of temperature. Total charge transferred is also insensitive to the activity of the enzyme (how quickly it works) and the mobility of the mediator, both of which affect current. Current is also affected by depletion of glucose during the measurement. In coulometry, the goal is to use up all the glucose.

Replacing the enzyme glucose oxidase with *glucose dehydrogenase* eliminates O_2 as a reactant. A *cofactor* called PQQ, which is bound to glucose dehydrogenase, receives $2H^+ + 2e^-$ during the oxidation.

A *cofactor* is a small, nonprotein molecule that is bound to an enzyme and is necessary for the activity of the enzyme.

Glucose

Pyrroloquinoline quinone (PQQ)
cofactor in glucose dehydrogenase

glucose dehydrogenase

Gluconolactone

$PQQH_2$

(16-17)

Unlike Reaction 16-13, O_2 is not involved in Reaction 16-17. Therefore, there is no dependence of the response to dissolved O_2.

In an "electrically wired" polymer gel on the surface of a carbon electrode (Figure 16-12), the enzyme and an osmium mediator are tethered to a polymer backbone. The $PQQH_2$ product of Reaction 16-17 is oxidized back to $PQQ + 2H^+$ by a nearby Os^{3+}. Os^{3+} is reduced to Os^{2+} in the process. Os^{2+} can exchange an electron with another Os^{3+}. Electrons are transported rapidly from Os to Os until they reach a carbon anode. Electrons then flow through a circuit to the Ag | AgCl counter electrode where AgCl is reduced to $Ag + Cl^-$.

"Electrical wiring" of the enzyme and the osmium mediators increases the current by a factor of 10 to 100 compared with an enzyme/mediator layer deposited onto an electrode. Increased current provides larger signal and a faster measurement. Covalent attachment of osmium to the polymer prevents the mediator from diffusing to the counter electrode, where it would react and create a large background current. Ligands for osmium are chosen so that the mildest possible potential (+0.1 V versus Ag | AgCl) can be applied to the electrode to oxidize glucose. At this potential, common oxidizable interferents produce acceptably small errors in glucose measurement.

The newest glucose monitor test strips require only 0.3 μL of blood for a measurement, significantly reducing the pain experienced by people who need to measure their glucose several times each day. Glucose in the entire volume is oxidized in about 1 min, and current is measured as a function of time. Integrating current versus time (Equation 16-12) gives the total charge required to reduce glucose.

Rotating Disk Electrode

A molecule has three ways to reach the surface of an electrode: (1) *diffusion* through a concentration gradient; (2) *convection*, which is the movement of bulk fluid by physical means

FIGURE 16-12 "Electrically wired" glucose dehydrogenase. The enzyme catalyzes the oxidation of glucose, reducing PQQ to PQQH$_2$. PQQH$_2$ is oxidized back to PQQ + 2H$^+$ by Os^{3+}. Electrons move through successive osmium atoms until they reach the carbon anode. All members of the redox chain are bound to a polymer backbone.

such a stirring or boiling; and (3) *migration*, which is the attraction or repulsion of an ion by a charged surface. A common working electrode for amperometry is the **rotating disk electrode,** for which convection and diffusion control the flux of analyte to the electrode.[20]

When the electrode in Figure 16-13a is spun at ~1 000 revolutions per minute, a vortex is established that brings analyte near the electrode rapidly by convection. If the potential is great enough, analyte reacts rapidly at the electrode, thereby reducing the concentration of analyte at the surface to near 0. The resulting concentration gradient is shown schematically in Figure 16-13b. Analyte must traverse the final, short distance (~10–100 μm) by diffusion alone.

Three ways for analyte to reach an electrode:
- diffusion
- convection
- migration

(a)

(b)

FIGURE 16-13 (*a*) Rotating disk electrode. Only the polished bottom surface of the electrode, which is typically 5 mm in diameter, contacts the solution. (*b*) Schematic concentration profile of analyte near the surface of the rotating disk electrode when the potential is great enough to reduce the concentration of analyte to 0 at the electrode surface.

The rate at which analyte diffuses from bulk solution to the surface of the electrode is proportional to the concentration difference between the two regions:

$$\text{Current} \propto \text{rate of diffusion} \propto [C]_0 - [C]_s \qquad \text{(16-18)}$$

where $[C]_0$ is the concentration in bulk solution and $[C]_s$ is the concentration at the surface of the electrode. At sufficiently great potential, the rate of reaction at the electrode is so fast that $[C]_s \ll [C]_0$ and Equation 16-18 reduces to

$$\text{Limiting current} = \text{diffusion current} \propto [C]_0 \qquad \text{(16-19)}$$

The limiting current is called the **diffusion current** because it is governed by the rate at which analyte can diffuse to the electrode. The proportionality of diffusion current to bulk-solute concentration is the basis for quantitative analysis by amperometry and, in the next section, voltammetry.

The faster a rotating disk electrode spins, the thinner is the diffusion layer in Figure 16-13b and the greater is the diffusion current. A rapidly rotating Pt electrode can measure 20 nM H_2O_2 in rainwater.[21] H_2O_2 is oxidized to O_2 at +0.4 V (versus S.C.E.) at the Pt surface, and the current is proportional to $[H_2O_2]$ in the rainwater.

16-5 Voltammetry

Voltammetry is a collection of techniques in which the relation between current and voltage is observed during electrochemical processes.[22] The **voltammogram** in Figure 16-14a is a graph of current versus working electrode potential for a mixture of ferricyanide and ferrocyanide being oxidized or reduced at a rotating disk electrode. By convention, current is positive when analyte is reduced at the working electrode. The limiting (diffusion) current for oxidation of $Fe(CN)_6^{4-}$ is observed at potentials above +0.5 V (versus S.C.E.).

$$Fe(CN)_6^{4-} \rightarrow Fe(CN)_6^{3-} + e^-$$

Ferrocyanide	Ferricyanide
Fe(II)	Fe(III)

In this region, current is governed by the rate at which $Fe(CN)_6^{4-}$ diffuses to the electrode. Diffusion current is proportional to the bulk concentration of $Fe(CN)_6^{4-}$ (Figure 16-14b) just as it is for the rotating disk electrode in Equation 16-19. Below 0 V, there is another plateau corresponding to the diffusion current for reduction of $Fe(CN)_6^{3-}$, whose concentration is constant in all the solutions.

Polarography

Voltammetry conducted with a **dropping-mercury electrode** is called **polarography** (Figure 16-15). The dispenser suspends one drop of mercury from the bottom of the capillary. After current and voltage are measured, the drop is mechanically dislodged. Then a

(a)

(b)

FIGURE 16-14 (a) Voltammograms for a mixture of 10 mM $K_3Fe(CN)_6$ and 20–60 mM $K_4Fe(CN)_6$ in 0.1 M Na_2SO_4 at a glassy carbon rotating electrode. Rotation speed = 2 000 revolutions/min and voltage sweep rate = 5 mV/s. (b) Dependence of limiting current on $K_4Fe(CN)_6$ concentration. [From J. Nikolic, E. Expósito, J. Iniesta, J. González-Garcia, and V. Montiel, "Theoretical Concepts and Applications of a Rotating Disk Electrode," *J. Chem. Ed.* 2000, 77, 1191.]

Electronically controlled Hg drop dispenser

Potentiostat

V A

Capillary tube

N₂ in

N₂ outlet

Platinum auxiliary electrode

Analyte solution

Dropping-mercury working electrode

Mercury drop

Calomel reference electrode

Pool of used mercury

FIGURE 16-15 Polarography apparatus featuring a dropping-mercury working electrode. Polarography was invented by J. Heyrovský in 1922, for which he received the Nobel Prize in 1959.

fresh drop is suspended and the next measurement is made. Freshly exposed Hg yields reproducible current-potential behavior. The current for other electrodes, such as Pt, depends on surface condition.

Most reactions studied with the Hg electrode are reductions. At a Pt surface, reduction of H^+ competes with reduction of many analytes:

$$2H^+ + 2e^- \rightarrow H_2(g) \qquad E° = 0$$

Table 16-1 showed that there is a large *overpotential* for reduction of H^+ at the Hg surface. Reactions that are thermodynamically less favorable than reduction of H^+ can be carried out without competitive reduction of H^+. In neutral or basic solutions, even alkali metal (Group 1) cations are reduced more easily than H^+. Furthermore, reduction of a metal into a mercury *amalgam* is more favorable than reduction to the solid state:

$$K^+ + e^- \rightarrow K(s) \qquad E° = -2.936 \text{ V}$$
$$K^+ + e^- + Hg \rightarrow K(in\ Hg) \qquad E° = -1.975 \text{ V}$$

Mercury is not useful for studying oxidations because Hg is oxidized in noncomplexing media near +0.25 V (versus S.C.E.). If the concentration of Cl^- is 1 M, Hg is oxidized near 0 V because Hg(II) is stabilized by Cl^-:

$$Hg(l) + 4Cl^- \rightleftharpoons HgCl_4^{2-} + 2e^-$$

For oxidations, Pt, Au, C, or diamond working electrodes in appropriate solvents provide a range of accessible potentials (Table 16-2). Boron-doped chemical-vapor-deposited diamond (Figure 16-16) is an exceptionally inert carbon electrode with a wide potential window, low background current,[24] and visible and infrared transparency.

An **amalgam** is anything dissolved in Hg.

5 μm

FIGURE 16-16 Boron-doped diamond coating on Pt electrode. [From J. Cvačka et al., *Anal. Chem.* **2003**, *75*, 2678. Courtesy G. M. Swain, Michigan State University.]

TABLE 16-2 Approximate working electrode potential range in 1 M H₂SO₄

Electrode	Potential range (V vs. S.C.E.)
Pt	−0.2 to +0.9 V
Au	−0.3 to +1.4 V
Hg	−1.3 to +0.1 V
Glassy carbon	−0.8 to +1.1 V
B-doped diamond[a]	−1.5 to +1.7 V
Fluorinated B-doped diamond[b]	−2.5 to +2.5 V

a. A. E. Fischer, Y. Show, and G. M. Swain, "Electrochemical Performance of Diamond Thin-Film Electrodes from Different Commercial Sources," *Anal. Chem.* **2004**, *76*, 2553; Y. Dai, G. M. Swain, M. D. Porter, and J. Zak, "Optically Transparent Carbon Electrodes," *Anal. Chem.* **2008**, *80*, 14; J. Stotter, Y. Show, S. Wang, and G. Swain, "Comparison of Electrical, Optical, and Electrochemical Properties of Diamond and Indium Tin Oxide Thin-Film Electrodes," *Chem. Mater.* **2005**, *17*, 4880.

b. S. Ferro and A. De Battisti, "The 5-V Window of Polarizability of Fluorinated Diamond Electrodes in Aqueous Solution," *Anal. Chem.* **2003**, *75*, 7040.

FIGURE 16-17 *Staircase voltage profile* used in *sampled current polarography*. Current is measured only during the intervals shown by heavy, colored lines. Potential is scanned toward more negative values as the experiment progresses. Lower graph shows that charging current decays more rapidly than faradaic current after each voltage step.

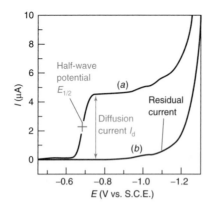

FIGURE 16-18 *Sampled current polarograms* of (*a*) 5 mM Cd^{2+} in 1 M HCl and (*b*) 1 M HCl alone.

FIGURE 16-19 Sampled current polarogram of 0.1 M KCl, saturated with air or purged (bubbled) with N_2 to remove O_2.

Polarograms in the older literature have large oscillations superimposed on the wave in Figure 16-18a. For the first 50 years of polarography, current was measured continuously as Hg flowed from a capillary tube. Each drop grew until it fell off and was replaced by a new drop. The current oscillated from a low value when the drop was small to a high value when the drop was big.

Faradaic current: due to redox reaction at the electrode

Charging current: due to electrostatic attraction or repulsion of ions in solution and electrons in the electrode

Faradaic current is the signal of interest. Charging current obscures the signal of interest, so we seek to minimize charging current.

One way to conduct a measurement is by **sampled current polarography** with the *staircase voltage ramp* in Figure 16-17. After each drop of Hg is dispensed, the potential is made more negative by 4 mV. After waiting almost 1 s, current is measured during the last 17 ms of the life of each Hg drop. The **polarographic wave** in Figure 16-18a results from reduction of Cd^{2+} analyte to form an amalgam:

$$Cd^{2+} + 2e^- \rightarrow Cd(in\ Hg)$$

The potential at which half the maximum current is reached in Figure 16-18a, called the **half-wave potential** ($E_{1/2}$), is characteristic of a given analyte in a given medium and can be used for qualitative analysis. For electrode reactions in which reactants and products are both in solution, such as $Fe^{3+} + e^- \rightleftharpoons Fe^{2+}$, $E_{1/2}$ (expressed with respect to S.H.E.) is nearly equal to $E°$ for the half-reaction.

For quantitative analysis, the *diffusion current* in the plateau region is proportional to the concentration of analyte. Diffusion current is measured from the baseline recorded without analyte in Figure 16-18b. **Residual current** in the absence of analyte is due to reduction of impurities in solution and on the surface of the electrodes. Near -1.2 V in Figure 16-18, current increases rapidly as reduction of H^+ to H_2 commences.

For quantitative analysis, the limiting current should be controlled by the rate at which analyte can diffuse to the electrode. We minimize convection by using an unstirred solution. We minimize migration (electrostatic attraction of analyte) by using a high concentration of *supporting electrolyte*, such as 1 M HCl in Figure 16-18.

Oxygen must be absent because O_2 gives two polarographic waves when it is reduced to H_2O_2 and then to H_2O (Figure 16-19). Typically, N_2 is bubbled through analyte solution for 10 min to remove O_2.[25] Then N_2 flow in the gas phase is continued to keep O_2 out. The liquid should not be purged with N_2 during a measurement, because we do not want convection of analyte to the electrode.

Faradaic and Charging Currents

The current we seek to measure in voltammetry is **faradaic current** due to reduction or oxidation of analyte at the working electrode. In Figure 16-18a, faradaic current comes from reduction of Cd^{2+} at the Hg electrode. Another current, called **charging current** (or *capacitor current* or *condenser current*) interferes with every measurement. We step the working electrode to a more negative potential by forcing electrons into the electrode from the potentiostat. In response, cations in solution flow toward the electrode, and anions flow away from the electrode (Box 16-3). This flow of ions and electrons, called *charging current*, is not from redox reactions. We try to minimize charging current, which obscures faradaic current. Charging current usually controls the detection limit in voltammetry.

The bottom of Figure 16-17 shows the behavior of faradaic and charging currents after each potential step. Faradaic current decays because analyte cannot diffuse to the electrode

BOX 16-3 The Electric Double Layer

When a power supply pumps electrons into or out of an electrode, the charged surface of the electrode attracts ions of opposite charge. The charged electrode and the oppositely charged ions next to it constitute the **electric double layer.**

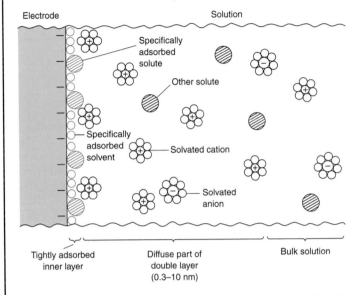

Electrode-solution interface. The tightly adsorbed inner layer (also called the *compact, Helmholtz,* or *Stern* layer) may include solvent and any solute molecules. Cations in the inner layer do not completely balance the charge of the electrode. Therefore, excess cations are required in the *diffuse part of the double layer* for charge balance.

A given solution has a *potential of zero charge* at which there is no excess charge on the electrode. This potential is -0.58 V (versus a calomel electrode containing 1 M KCl) for a mercury electrode immersed in 0.1 M KBr. It shifts to -0.72 V for the same electrode in 0.1 M KI.

The first layer of molecules at the surface of the electrode is *specifically adsorbed* by van der Waals and electrostatic forces. The adsorbed solute could be neutral molecules, anions, or cations. Iodide is more strongly adsorbed than bromide, so the potential of zero charge for KI is more negative than for KBr: A more negative potential is required to expel adsorbed iodide from the electrode surface.

The next layer beyond the specifically adsorbed layer is rich in cations attracted by the negative electrode. The excess of cations decreases with increasing distance from the electrode. This region, whose composition is different from that of bulk solution, is called the *diffuse part of the double layer* and is typically 0.3–10 nm thick. The thickness is controlled by the balance between attraction toward the electrode and randomization by thermal motion.

When a species is created or destroyed by an electrochemical reaction, its concentration near the electrode is different from its concentration in bulk solution (Figure 16-13b and Color Plate 11). The region containing excess product or decreased reactant is called the *diffusion layer* (not to be confused with the diffuse part of the double layer).

fast enough to sustain the high reaction rate. Charging current decays even faster because ions near the electrode redistribute themselves rapidly. A second after each potential step, faradaic current is still significant and charging current is small.

Square Wave Voltammetry

The most efficient voltage profile for voltammetry, called **square wave voltammetry,** uses the waveform in Figure 16-20, which consists of a square wave superimposed on a staircase.[26] During each cathodic pulse, analyte is reduced at the electrode surface. During the anodic

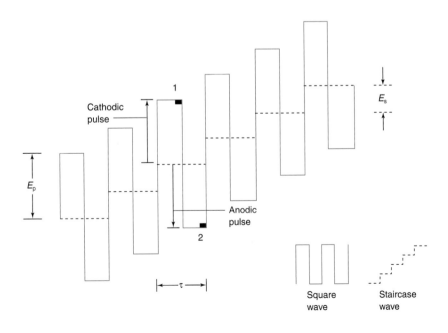

FIGURE 16-20 Waveform for square wave voltammetry. Typical parameters are pulse height (E_p) = 25 mV, step height (E_s) = 10 mV, and pulse period (τ) = 5 ms. Current is measured in regions 1 and 2. Optimum values are E_p = 50/n mV and E_s = 10/n mV, where n is the number of electrons in the half-reaction.

FIGURE 16-21 Comparison of polarograms of 5 mM Cd^{2+} in 1 M HCl. Waveforms are shown in Figures 16-17 and 16-20. Sampled current: drop time = 1 s, step height = 4 mV, sampling time = 17 ms. Square wave: drop time = 1 s, step height = 4 mV, pulse period = 67 ms, pulse height = 25 mV, sampling time = 17 ms.

pulse, analyte that was just reduced is reoxidized. The square wave polarogram in Figure 16-21 is the *difference* in current between intervals 1 and 2 in Figure 16-20. Electrons flow from the electrode to analyte at point 1 and in the reverse direction at point 2. Because the two currents have opposite signs, their difference is larger than either current alone. When the difference is plotted, the shape of the square wave polarogram in Figure 16-21 is essentially the derivative of the sampled current polarogram.

The signal in square wave voltammetry is increased relative to a sampled current voltammogram and the wave becomes peak shaped. Signal is increased because reduced product from each cathodic pulse is right at the surface of the electrode waiting to be oxidized by the next anodic pulse. Each anodic pulse provides a high concentration of reactant at the surface of the electrode for the next cathodic pulse. The detection limit is reduced from $\sim 10^{-5}$ M for sampled current voltammetry to $\sim 10^{-7}$ M in square wave voltammetry. Because it is easier to resolve neighboring peaks than neighboring waves, square wave voltammetry can resolve species whose half-wave potentials differ by ~ 0.05 V, whereas potentials must differ by ~ 0.2 V to be resolved in sampled current voltammetry. Square wave voltammetry is faster than other voltammetric techniques. The square wave polarogram in Figure 16-21 was recorded in one-fifteenth of the time required for the sampled current polarogram. In principle, the shorter the pulse period, τ, in Figure 16-20, the greater the current that will be observed. With $\tau = 5$ ms (a practical lower limit) and $E_s = 10$ mV, an entire square wave polarogram with a 1-V width is obtained with *one drop of Hg in 0.5 s*. Such rapid sweeps allow voltammograms to be recorded on individual components as they emerge from a chromatography column.

Stripping Analysis

In **stripping analysis,** analyte from a dilute solution is concentrated into a thin film of Hg or other electrode material, usually by electroreduction. The electroactive species is then *stripped* from the electrode by reversing the direction of the voltage sweep. The potential becomes more *positive, oxidizing* the species back into solution. Peak current measured during oxidation is proportional to the quantity of analyte that was deposited. Figure 16-22 shows an anodic stripping voltammogram of Cd, Pb, and Cu from honey.

Stripping is the most sensitive voltammetric technique (Table 16-3), because analyte is concentrated from a dilute solution. The longer the period of concentration, the more sensitive

Advantages of square wave voltammetry:

- increased signal
- derivative (peak) shape permits better resolution of neighboring signals
- faster measurement

Anodic stripping analysis:

1. Concentrate analyte onto electrode by reduction
2. Reoxidize analyte by scanning to more positive potential
3. Peak current during oxidation is proportional to analyte concentration

FIGURE 16-22 (a) Anodic stripping voltammogram of honey dissolved in water and acidified to pH 1.2 with HCl. Cd, Pb, and Cu were reduced from solution into a thin film of Hg for 5 min at −1.4 V (versus S.C.E.) prior to recording the voltammogram. (b) Voltammogram obtained without 5-min reduction step. The concentrations of Cd and Pb in the honey were 7 and 27 ng/g (ppb), respectively. Precision was 2–4%. [From Y. Li, F. Wahdat, and R. Neeb, "Digestion-Free Determination of Heavy Metals in Honey," *Fresenius J. Anal. Chem.* **1995**, *351*, 678.]

TABLE 16-3 Detection limits for stripping analysis

Analyte	Stripping mode	Detection limit
Ag^+	Anodic	2×10^{-12} M[a]
Testosterone	Anodic	2×10^{-10} M[b]
I^-	Cathodic	1×10^{-10} M[c]
DNA or RNA	Cathodic	2–5 pg/mL[d]
Fe^{3+}	Cathodic	1×10^{-11} M[e]

a. *S. Dong and Y. Wang, Anal. Chim. Acta* **1988**, *212*, 341.

b. *J. Wang, "Adsorptive Stripping Voltammetry." EG&G Princeton Applied Research Application Note A-7 (1985).*

c. *G. W. Luther III, C. Branson Swartz, and W. J. Ullman, Anal. Chem.* **1988**, *60*, 1721. I^- is deposited onto the mercury drop by anodic oxidation: $Hg(l) + I^- \rightleftharpoons \frac{1}{2}Hg_2I_2(adsorbed\ on\ Hg) + e^-$.

d. *S. Reher, Y. Lepka, and G. Schwedt, Fresenius J. Anal. Chem.* **2000**, *368*, 720; *J. Wang, Anal. Chim. Acta* **2003**, *500*, 247.

e. *C. M. G. van den Berg, Anal. Chem.* **2006**, *78*, 156.

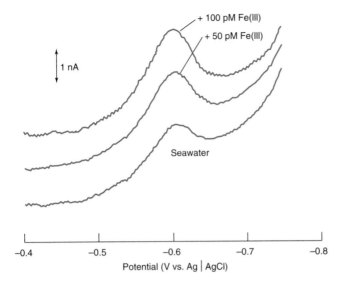

+ 100 pM Fe(III)

+ 50 pM Fe(III)

1 nA

Seawater

-0.4   -0.5   -0.6   -0.7   -0.8

Potential (V vs. Ag | AgCl)

FIGURE 16-23 Sampled current cathodic stripping voltammogram of Fe(III) in seawater plus two standard additions of 50 pM Fe(III). [From H. Obata and C. M. G. van den Berg, "Determination of Picomolar Levels of Iron in Seawater Using Catalytic Cathodic Stripping Voltammetry," *Anal. Chem.* **2001**, *73*, 2522. See also C. M. G. van den Berg, "Chemical Speciation of Iron in Seawater by Cathodic Stripping Voltammetry with Dihydroxynaphthalene," *Anal. Chem.* **2006**, *78*, 156.]

is the analysis. Only a fraction of analyte from the solution is deposited, so deposition must be done for a reproducible time (such as 5 min) with reproducible stirring.

The detection limit for Fe(III) in seawater can be reduced to 10^{-11} M with a *catalytic stripping* process. First, 20 μM 2,3-dihydroxynaphthalene (L), 20 mM bromate (BrO_3^-) and pH 8.0 buffer are added to seawater that has been purged with N_2 to remove O_2. Dihydroxynaphthalene forms a complex, $L_nFe(III)$, which adsorbs onto a mercury drop electrode poised at -0.1 V versus Ag | AgCl during 60 s of vigorous stirring. After stirring is stopped and the solution becomes stationary, the potential is scanned from -0.1 to -0.8 V to give the lower trace in Figure 16-23. At -0.6 V, Fe(III) is reduced to Fe(II), which begins to diffuse away from the electrode. Before Fe(II) goes very far, BrO_3^- oxidizes Fe(II) back to Fe(III), which is readsorbed and available to be reduced again. The cathodic stripping current is 290 times greater in the presence of 20 mM BrO_3^- than without BrO_3^-. Fe(II) is a catalyst for the net reduction of BrO_3^-.

2,3-Dihydroxynaphthalene

(Presumably, the polarizable pi electron cloud of naphthalene is bound to the polarizable Hg by van der Waals forces)

$$L_nFe(III)_{adsorbed} + e^- \xrightarrow[\text{stripping}]{\text{cathodic}} L_nFe(II)_{\text{in diffusion layer}}$$

$$\downarrow BrO_3^-$$

$$L_nFe(III)_{\text{in diffusion layer}}$$

Rigorous precautions must be taken to remove iron from reagents and equipment when measuring concentrations of 10^{-11} M. For example, 3 M KCl in the salt bridge had to be purified, and the bridge itself was made of Teflon instead of glass.

Electrodes other than mercury can be used for stripping analysis if there is a means to adsorb analyte on the electrode. Traces of As(III) can be measured by codeposition with gold on a boron-doped diamond electrode. Pure diamond is an electrical insulator. Conductive boron-doped diamond can be grown on Si by gas-phase decomposition of 0.5 vol% CH_4 with 10 ppm B_2H_6 (diborane) in a 0.06-bar H_2 atmosphere at 800°C in a microwave plasma.

Figure 16-24a shows the bare surface of boron-doped diamond. Figure 16-24b shows the surface after 90 s of deposition at -0.45 V (vs. Ag | AgCl) from 10 ppm As(III) + 10 ppm

(a)

(b)

FIGURE 16-24 Scanning electron micrographs of (a) bare boron-doped diamond electrode surface and (b) diamond surface with Au_xAs_y nanoparticles deposited at -0.45 V in 90 s from 10 ppm As(III) + 10 ppm Au(III) in 1 M HCl. [From Y. Song and G. M. Swain, "Development of a Method for Total Inorganic Arsenic Analysis Using Anodic Stripping Voltammetry and a Au-coated, Diamond Thin-Film Electrode," *Anal. Chem.* **2007**, *79*, 2412.]

FIGURE 16-25 Anodic stripping of As(III) from a boron-doped diamond electrode. Au$_x$As$_y$ was deposited from a solution containing 100 ppm Au(III) in 1 M HCl. Peak current is proportional to the concentration of As(III) in the deposition solution. [From Y. Song and G. M. Swain, "Development of a Method for Total Inorganic Arsenic Analysis Using Anodic Stripping Voltammetry and a Au-coated, Diamond Thin-Film Electrode," *Anal. Chem.* **2007**, *79*, 2412.]

Au(III) in 1 M HCl. An intermetallic Au$_x$As$_y$ compound has been deposited as particles with a mean diameter of 22 nm. No As is deposited in the absence of gold. After deposition, the potential is scanned to positive values to oxidize As back to As(III) (Figure 16-25). The peak current observed at +0.16 V is proportional to the As(III) analyte concentration in the original solution. Finally, the potential is held at +0.6 V to oxidize all As and Au off the electrode before beginning a new analysis.

The detection limit for As(III) is 0.005 ppb = 7×10^{-11} M. As(V) cannot be measured directly, but it can be quantitatively reduced to As(III) with sodium sulfite (Na$_2$SO$_3$) in HCl. The resulting As(III) can be measured by stripping analysis. A solution containing both As(III) and As(V) can be analyzed by first measuring As(III) on one aliquot. Then the As(V) is reduced in a second aliquot, and total As in the second aliquot is measured by stripping.

Boron-doped diamond is an especially useful electrode because its surface is stable over a wide potential range and is not altered from one analysis to the next. It does not adsorb polar molecules that might interfere with deposition of analyte. Diamond has an exceptionally wide potential limit (-1.5 to $+1.7$ V vs. S.C.E.) and gives exceptionally low background currents in the absence of analyte.

Cyclic Voltammetry

In **cyclic voltammetry,** we apply the triangular waveform in Figure 16-26 to the working electrode. After the application of a linear voltage ramp between times t_0 and t_1 (typically a few seconds), the ramp is reversed to bring the potential back to its initial value at time t_2. The cycle might be repeated many times.

The initial portion of the cyclic voltammogram in Figure 16-27, beginning at t_0, exhibits a *cathodic wave*. Instead of leveling off at the top of the wave, current decreases at more negative potential because analyte becomes depleted near the electrode. Diffusion is too slow to replenish analyte near the electrode. At the time of peak voltage (t_1), the cathodic current has decayed to a small value. After t_1, the potential is reversed and, eventually, reduced product near the electrode is oxidized. This reaction produces an *anodic wave*. Finally, as the reduced product is depleted, the anodic current decays back toward its initial value at t_2.

Figure 16-27a illustrates a *reversible* reaction that is fast enough to maintain equilibrium concentrations of reactant and product *at the electrode surface*. The peak anodic and peak cathodic currents have equal magnitudes in a reversible process, and

$$E_{pa} - E_{pc} = \frac{2.22RT}{nF} = \frac{57.0}{n}(mV) \quad (at\ 25°C) \qquad (16\text{-}20)$$

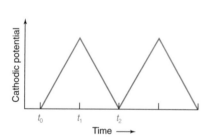

FIGURE 16-26 Waveform for cyclic voltammetry. Corresponding times are indicated in Figure 16-27.

FIGURE 16-27 Cyclic voltammograms of (a) 1 mM O$_2$ in acetonitrile with 0.10 M (C$_2$H$_5$)$_4$N$^+$ClO$_4^-$ electrolyte and (b) 0.060 mM 2-nitropropane in acetonitrile with 0.10 M (*n*-C$_7$H$_{15}$)$_4$N$^+$ClO$_4^-$ electrolyte. The reaction in curve *a* is

$$O_2 + e^- \rightleftharpoons O_2^-$$
$$\text{Superoxide}$$

Working electrode, Hg; reference electrode, Ag | 0.001 M AgNO$_3$(*aq*) | 0.10 M (C$_2$H$_5$)$_4$N$^+$ClO$_4^-$ in acetonitrile; scan rate = 100 V/s. I_{pa} is the peak anodic current and I_{pc} is the peak cathodic current. E_{pa} and E_{pc} are the potentials at which these currents are observed. [From D. H. Evans, K. M. O'Connell, R. A. Petersen, and M. J. Kelly, "Cyclic Voltammetry," *J. Chem. Ed.* **1983**, *60*, 290.]

C_{60}

(a)

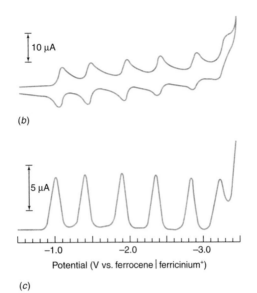

(b)

(c)

FIGURE 16-28 (*a*) Structure of C_{60} (buckminsterfullerene). (*b*) Cyclic voltammetry and (*c*) polarogram of 0.8 mM C_{60}, showing six waves for reduction to C_{60}^{-}, C_{60}^{2-}, ..., C_{60}^{6-}. The acetonitrile/toluene solution was at $-10°C$ with $(n\text{-}C_4H_9)_4N^+PF_6^-$ supporting electrolyte. The reference electrode contains the ferrocene | ferricinium$^+$ redox couple. Ferrocene is $(C_5H_5)_2Fe$ and ferricinium cation is $(C_5H_5)_2Fe^+$. The structure of ferrocene was shown in Reaction 16-15. [From Q. Xie, E. Pérez-Cordero, and L. Echegoyen, "Electrochemical Detection of C_{60}^{6-} and C_{70}^{6-}," *J. Am. Chem. Soc.* **1992**, *114*, 3978.]

where E_{pa} and E_{pc} are the potentials at which the *peak anodic* and *peak cathodic* currents are observed and n is the number of electrons in the half-reaction. The half-wave potential, $E_{1/2}$, lies midway between the two peak potentials. Figure 16-27b is the cyclic voltammogram of an *irreversible* reaction. The cathodic and anodic peaks are broader and more separated. At the limit of irreversibility, where oxidation is very slow, no anodic peak is seen.

For a reversible reaction, the peak current (I_{pc}, amperes) for the forward sweep of the first cycle is proportional to the concentration of analyte and the square root of sweep rate:

$$I_{pc} = (2.69 \times 10^8)n^{3/2}ACD^{1/2}v^{1/2} \qquad \text{(at 25°C)} \qquad (16\text{-}21)$$

where n is the number of electrons in the half-reaction, A is the area of the electrode (m^2), C is the concentration (mol/L), D is the diffusion coefficient of the electroactive species (m^2/s), and v is sweep rate (V/s). The faster the sweep rate, the greater the peak current, as long as the reaction remains reversible. If the electroactive species is adsorbed on the electrode, the peak current is proportional to v rather than $\sqrt{v}$.

For catalytic stripping in Figure 16-23, current was proportional to $\sqrt{v}$, consistent with the rate-limiting step being diffusion of BrO_3^- to the electrode. If the rate-limiting step were reduction of Fe(III) adsorbed on the electrode, the peak current would have been proportional to v.

Cyclic voltammetry is used to characterize the redox behavior of compounds such as C_{60} in Figure 16-28 and to elucidate the kinetics of electrode reactions.[27]

Microelectrodes

Microelectrodes have working dimensions from a few tens of micrometers down to nanometers.[28] The surface area of the electrode is small, so the current is tiny. With a small current, the ohmic drop ($= IR$) in a highly resistive medium is small, thus allowing microelectrodes to be used in poorly conducting nonaqueous media (Figure 16-29). The electric

Advantages of ultramicroelectrodes:

- Fit into small places
- Useful in resistive, nonaqueous media (because of small ohmic losses)
- Rapid voltage scans (possible because of small double-layer capacitance) allow short-lived species to be studied
- Detection limits increased by orders of magnitude because of low charging current

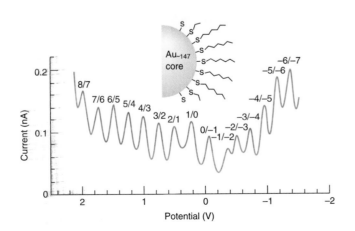

FIGURE 16-29 Voltammogram of gold nanoparticles ($Au_{\sim 147}$ capped with ~50 hexanethiol molecules) in 1,2-dichloroethane solution recorded with 25-μm-diameter Pt working electrode. The nanoparticle exhibits oxidation states from -7 to $+8$ over the potential range of this scan. Supporting electrolyte: 10 mM $[(C_6H_5)_3P=N=P(C_6H_5)_3]^+[(C_6F_5)_4B]^-$. Potential measured versus "quasi-reference" electrode—a silver wire whose potential is ~0.1 V versus Ag | AgCl. [From B. M. Quinn, P. Liljeroth, V. Ruiz, T. Laaksonen, and K. Kontturi, "Electrochemical Resolution of 15 Oxidation States for Monolayer Protected Gold Nanoparticles," *J. Am. Chem. Soc.* **2003**, *125*, 6644.]

$$-[(CF_2CF_2)_n-CFCF_2]_x-$$

Nafion membrane permits cations to pass, but not anions

O
|
CF_2
|
$CFCF_3$
|
O
|
$CF_2CF_2SO_3^-Na^+$

Dopamine

$+ 2H^+ + 2e^-$

FIGURE 16-30 Electron micrograph of the tip of a Nafion-coated carbon-fiber electrode. The carbon inside the electrode has a diameter of 10 μm. Nafion permits cations to pass but excludes anions. [Photo courtesy R. M. Wightman. From R. M. Wightman, L. J. May, and A. C. Michael, "Detection of Dopamine Dynamics in the Brain," *Anal. Chem.* **1988,** *60,* 769A.]

40 μm

capacitance of the double layer (Box 16-3) of a microelectrode is very small. Low capacitance gives low background charging current relative to the faradaic current of a redox reaction. The result is a lowering of the detection limit by as much as three orders of magnitude over conventional electrodes. Low capacitance also enables the potential to be varied at rates up to 500 kV/s, thus allowing short-lived species with lifetimes of less than 1 μs to be studied.

Sufficiently small electrodes fit inside a living cell. A carbon fiber coated with insulating polymer provides a 1-μm-diameter working electrode at the exposed tip. Carbon is fouled by adsorption of organic molecules inside living cells, so a thin layer of Pt or Au is electrolytically plated onto the exposed carbon. The metallized electrode is cleaned in situ (inside the cell during an experiment) by an anodic voltage pulse that desorbs surface-bound species and produces an oxide layer on the metal, followed by a cathodic pulse to reduce the oxide.[29]

Figure 16-30 shows a carbon fiber coated with a cation-exchange membrane called Nafion, which has fixed negative charges. Cations diffuse through the membrane, but anions are excluded. The electrode can measure the cationic neurotransmitter, dopamine, in a rat brain.[30] Negatively charged ascorbate, which ordinarily interferes with dopamine analysis, is excluded by Nafion. The response to dopamine is 1 000 times higher than the response to ascorbate.

Figure 16-31 shows an application of a microelectrode array in biology. Cells from a line called pheochromocytoma PC 12, derived from a tumor of the adrenal gland, release dopamine when stimulated by K^+. Neurotransmitters are contained in small intracellular compartments called *vesicles*. To release their content outside the cell, vesicles fuse with the cell membrane and open up. This process is called *exocytosis*. The microelectrode array consists of 5-μm-diameter carbon fibers in a seven-barrel glass capillary pulled to a fine point. Each carbon fiber in Figure 16-31a is an independent working electrode that can sample a region approximately 5 μm in diameter. Figure 16-31b shows the electrode pressed against a single cell, distorting the cell into a crescent shape. When 0.1 M KCl from the micropipette at the left in Figure 16-31b is injected into the medium, the cell releases dopamine in discrete exocytotic events. Figure 16-31c shows traces from electrodes A–G over 16 min during which KCl is injected every 45 s. All traces are different, indicating that different patches of cell surface respond differently. For example, patches near electrodes F and G are "cold" for the first 8 min—with little release of dopamine. Something changes after 8 min, and these two patches become "hot."

(a)

Arrow points to cell pressed against electrode

Micropipet delivers KCl

Microelectrode array

10 μm

(b)

(c)

FIGURE 16-31 (a) Carbon-fiber microelectrode array with seven electrodes. (b) Microelectrode array (right) pressed against a PC 12 cell distorted into a crescent shape by the electrode. Micropipette at left is used to deliver 0.1 M KCl to stimulate dopamine release from the cell. A Ag|AgCl reference/auxiliary electrode is not visible. (c) Amperometric traces from all seven electrodes during exocytotic release of neurotransmitter. [From B. Zhang, K. L. Adams, S. J. Luber, D. J. Eves, M. L. Heien, and A. G. Ewing, "Spatially and Temporally Resolved Single-Cell Exocytosis Utilizing Individually Addressable Carbon Microelectrode Arrays," *Anal. Chem.* **2008,** *80,* 1394.]

16-6 Karl Fischer Titration of H₂O

The **Karl Fischer titration,**[31] which measures traces of water in transformer oil, solvents, foods, polymers, and other substances, is performed half a million times each day.[32] The titration is usually performed by delivering titrant from an automated buret or by coulometric generation of titrant. The volumetric procedure tends to be appropriate for larger amounts of water (but can go as low as ~1 mg H_2O), and the coulometric procedure tends to be appropriate for smaller amounts of water.

We illustrate the coulometric procedure in Figure 16-32, in which the main compartment contains anode solution plus unknown. The smaller compartment at the left has an internal Pt electrode immersed in cathode solution and an external Pt electrode immersed in the anode solution of the main compartment. The two compartments are separated by an ion-permeable membrane. Two Pt electrodes are used for end-point detection.

Anode solution contains an alcohol, a base, SO_2, I^-, and possibly another organic solvent. Methanol and diethylene glycol monomethyl ether ($CH_3OCH_2CH_2OCH_2CH_2OH$) are typical alcohols. Typical bases are imidazole and diethanolamine. The organic solvent may contain chloroform, formamide, or other solvents. The trend is to avoid chlorinated solvents because of their environmental hazards. When analyzing nonpolar substances such as transformer oil, sufficient solvent, such as chloroform, should be used to make the reaction homogeneous. Otherwise, moisture trapped in oily emulsions is inaccessible. (An *emulsion* is a fine suspension of liquid-phase droplets in another liquid.)

The anode at the lower left in Figure 16-32 generates I_2 by oxidation of I^-. In the presence of H_2O, reactions occur between the alcohol (ROH), base (B), SO_2, and I_2.

$$ROH + SO_2 + B \rightarrow BH^+ + ROSO_2^- \qquad (16\text{-}22)$$

$$H_2O + I_2 + ROSO_2^- + 2B \rightarrow ROSO_3^- + 2BH^+I^- \qquad (16\text{-}23)$$

The net reaction is oxidation of SO_2 by I_2, with formation of $ROSO_3^-$. One mole of I_2 is consumed for each mole of H_2O when the solvent is methanol. In other solvents, the stoichiometry can be more complex.[32]

In a typical procedure, the main compartment in Figure 16-32 is filled with anode solution and the coulometric generator is filled with cathode solution that may contain reagents designed to be reduced at the cathode. Current is run until moisture in the main compartment is consumed, as indicated by the end-point detection system described after the Example. An unknown is injected through the septum, and the coulometer is run again until moisture has been consumed. Two moles of electrons correspond to 1 mol of H_2O if the I_2:H_2O stoichiometry is 1:1.

Imidazole   Diethanolamine

Maintain pH in the range 4 to 7. Above pH 8, nonstoichiometric side reactions occur. Below pH 3, the reaction is slow.

FIGURE 16-32 Apparatus for coulometric Karl Fischer titration.

Pt electrodes for coulometric generation of I₂

Pt electrodes for bipotentiometric end-point detection

Cathode solution

Cathode of generator

Ion-permeable membrane

Anode where I₂ is generated

Septum for sample injection

Anode solution to which unknown is added

Magnetic stirring bar

CH₃CH₂CH₂ structure label area:

CH$_3$CH$_2$CH$_2$
CH$_3$
N
H H
H
HO—CH
H
CONH—CH
H H
HO
O
H
H
OH H
+HCl
+H$_2$O
H
OH H
H OH
SCH$_3$

Lincomycin hydrochloride monohydrate
C$_{18}$H$_{37}$N$_2$O$_7$SCl, FM 461.01
3.91 wt% H$_2$O

EXAMPLE **Standardization and Blank Correction in Karl Fischer Titration**

It is routine to standardize Karl Fischer reagents, or even a coulometer, with a standard such as lincomycin hydrochloride monohydrate, which contains 3.91 wt% H_2O. The coulometer is run until the end point is reached, indicating that the Karl Fischer reagent is dry. A port is opened briefly to add solid lincomycin, which is then titrated to the same end point. Then an unknown is added and titrated in the same manner. Find the wt% H_2O in the unknown.

Milligrams lincomycin	μg H_2O observed	μg H_2O theoretical	Difference (μg) = blank correction
3.89	172.4	152.1	172.4 − 152.1 = 20.3
13.64	556.3	533.3	556.3 − 533.3 = 23.0
19.25	771.4	752.7	771.4 − 752.7 = 18.7
			Average = 20.7

Milligrams unknown	μg H_2O observed	μg H_2O corrected (= observed − 20.7)	wt% H_2O in unknown
24.17	540.8	520.1	520.1 μg/24.17 mg = 2.15%
17.08	387.6	366.9	366.9 μg/17.08 mg = 2.15%

SOURCE: *Data from W. C. Schinzer, Pfizer Co., Michigan Pharmaceutical Sciences, Portage, MI.*

Solution For lincomycin, we observe ~20.7 μg more H_2O than expected, *independent of the sample size*. Excess H_2O comes from the atmosphere when the port is opened to add solid. To determine moisture in unknowns, subtract this blank from the total moisture titrated. This procedure can generate very reproducible data.

Test Yourself Observed H_2O in 20.33 mg of unknown was 888.8 μg. Apply the correction and find wt% H_2O in the unknown. (***Answer:*** 4.27%)

A **bipotentiometric** measurement is the most common way to detect the end point of a Karl Fischer titration. The detector circuit maintains a *constant current* (usually 5 or 10 μA) between the two detector electrodes at the right in Figure 16-32 while measuring the voltage needed to sustain the current. Prior to the equivalence point, the solution contains I^-, but little I_2 (which is consumed in Reaction 16-23 as fast as it is generated). To maintain a current of 10 μA, the cathode potential must be negative enough to reduce some component of the solvent system. At the equivalence point, excess I_2 suddenly appears and current can be carried at very low voltage by Reactions 16-24 and 16-25. The abrupt voltage drop marks the end point.

Cathode: $I_3^- + 2e^- \rightarrow 3I^-$ (16-24)

Anode: $3I^- \rightarrow I_3^- + 2e^-$ (16-25)

A trend in Karl Fischer coulometric instrumentation is to eliminate the separate cathode compartment in Figure 16-32 to reduce conditioning time required before samples can be analyzed and to eliminate clogging of the membrane.[33] The challenge is to minimize interference by products of the cathodic reaction.

End points in Karl Fischer titrations tend to drift because of slow reactions and water leaking into the cell from the air. Some instruments measure the rate at which I_2 must be generated to maintain the end point and then compare this rate with that measured before sample was added. Other instruments allow you to set a "persistence of end point" time, typically 5 to 60 s, during which the detector voltage must be stable to define the end point.

A round robin study of accuracy and precision of the coulometric procedure identified sources of systematic error.[34] In some labs, either the instruments were inaccurate or workers did not measure the quantity of standards accurately. In other cases, the solvent was not appropriate. Commercial reagents are designed for Karl Fischer analysis. Reagents recommended by the instrument manufacturer should be used with each instrument.

Terms to Understand

Summary

In electrolysis, a chemical reaction is forced to occur by the flow of electricity through a cell. The moles of electrons flowing through the cell are It/F, where I is current, t is time, and F is the Faraday constant. The magnitude of the voltage that must be applied to an electrolysis cell is $E = E(\text{cathode}) - E(\text{anode}) - IR -$ overpotentials.

1. Overpotential is the voltage required to overcome the activation energy of an electrode reaction. A greater overpotential is required to drive a reaction at a faster rate.
2. Ohmic potential ($= IR$) is that voltage needed to overcome internal resistance of the cell.
3. Concentration polarization occurs when the concentration of electroactive species near an electrode is not the same as its concentration in bulk solution. Concentration polarization is embedded in the terms $E(\text{cathode})$ and $E(\text{anode})$.

Overpotential, ohmic potential, and concentration polarization always oppose the desired reaction and require a greater voltage to be applied for electrolysis.

Controlled-potential electrolysis is conducted in a three-electrode cell in which the potential of the working electrode is measured with respect to a reference electrode to which negligible current flows. Current flows between the working and auxiliary electrodes.

In electrogravimetric analysis, analyte is deposited on an electrode, whose increase in mass is then measured. With a constant voltage in a two-electrode cell, electrolysis is not very selective, because the working electrode potential changes as the reaction proceeds.

In coulometry, the moles of electrons needed for a chemical reaction are measured. In a coulometric (constant-current) titration, the time needed for complete reaction measures the number of electrons consumed. Controlled-potential coulometry is more selective than constant-current coulometry, but slower. Electrons consumed in the reaction are measured by integrating the current-versus-time curve.

In amperometry, current at the working electrode is proportional to analyte concentration. The amperometric glucose monitor generates H_2O_2 by enzymatic oxidation of glucose, and the H_2O_2 is measured by oxidation at an electrode. A mediator is employed to rapidly shuttle electrons between electrode and analyte. A coulometric glucose monitor counts electrons released by oxidation of all glucose in a small blood sample. "Electrical wiring" of the enzyme and mediator in a glucose monitor increases signal from the desired reaction and decreases background current from mediator diffusing to the auxiliary electrode.

Voltammetry is a collection of methods in which the dependence of current on the applied potential of the working electrode is observed. Polarography is voltammetry with a dropping-mercury working electrode. This electrode gives reproducible results because fresh surface is always exposed. Hg is useful for reductions because the high overpotential for H^+ reduction on Hg prevents interference by H^+ reduction. Oxidations are usually studied with other electrodes because Hg is readily oxidized. For quantitative analysis, the diffusion current is proportional to analyte concentration if there is a sufficient concentration of supporting electrolyte. The half-wave potential is characteristic of a particular analyte in a particular medium.

Sampled current polarography uses a staircase voltage profile for measurements with successive, static drops of Hg. One second after each voltage step, charging current is nearly 0, but there is still substantial faradaic current from the redox reaction.

Square wave polarography achieves increased sensitivity and a derivative peak shape by applying a square wave superimposed on a staircase voltage ramp. With each cathodic pulse, there is a rush of analyte to be reduced at the electrode surface. During the anodic pulse, reduced analyte is reoxidized. The polarogram is the difference between the cathodic and anodic currents. Square wave polarography permits fast, real-time measurements not possible with other electrochemical methods.

Stripping is the most sensitive form of voltammetry. In anodic stripping polarography, analyte is concentrated into a single drop or thin film of mercury by reduction at a fixed voltage for a fixed time. The potential is then made more positive, and current is measured as analyte is reoxidized. In cyclic voltammetry, a triangular waveform is applied, and cathodic and anodic processes are observed in succession. Microelectrodes fit into small places, and their low current allows them to be used in resistive, nonaqueous media. Their low capacitance increases sensitivity by reducing charging current and permits rapid voltage scanning, which allows very short-lived species to be studied.

The Karl Fischer titration of water uses a buret to deliver reagent or coulometry to generate reagent. In bipotentiometric end-point detection, the voltage needed to maintain a constant current between two Pt electrodes is measured. Voltage changes abruptly at the equivalence point, when one member of a redox couple is either created or destroyed.

Exercises

16-A. A dilute Na_2SO_4 solution is to be electrolyzed with a pair of smooth Pt electrodes at a current density of 100 A/m^2 and a current of 0.100 A. The products are $H_2(g)$ and $O_2(g)$ at 1.00 bar. Calculate the required voltage if the cell resistance is 2.00 Ω and there is no concentration polarization. What voltage would be required if the Pt electrodes were replaced by Au electrodes?

16-B. (a) At what cathode potential will Sb(s) deposition commence from 0.010 M SbO$^+$ solution at pH 0.00? Express this potential versus S.H.E. and versus Ag | AgCl.

$$SbO^+ + 2H^+ + 3e^- \rightleftharpoons Sb(s) + H_2O \qquad E^\circ = 0.208 \text{ V}$$

(b) What percentage of 0.10 M Cu^{2+} could be reduced electrolytically to Cu(s) before 0.010 M SbO$^+$ in the same solution begins to be reduced at pH 0.00?

16-C. Calculate the cathode potential (versus S.C.E.) needed to reduce cobalt(II) to 1.0 μM in each of the following solutions. In each case, Co(s) is the product of the reaction.

(a) 0.10 M HClO$_4$

(b) 0.10 M C$_2$O$_4^{2-}$ (Find the potential at which $[Co(C_2O_4)_2^{2-}] = 1.0$ μM.)

$$Co(C_2O_4)_2^{2-} + 2e^- \rightleftharpoons Co(s) + 2C_2O_4^{2-} \qquad E^\circ = -0.474 \text{ V}$$

(c) 0.10 M EDTA at pH 7.00 (Find the potential at which $[Co(EDTA)^{2-}] = 1.0$ μM.)

16-D. Ions that react with Ag$^+$ can be determined electrogravimetrically by deposition on a silver working anode:

$$Ag(s) + X^- \rightarrow AgX(s) + e^-$$

(a) What will be the final mass of a silver anode used to electrolyze 75.00 mL of 0.023 80 M KSCN if the initial mass of the anode is 12.463 8 g?

(b) At what electrolysis voltage (versus S.C.E.) will AgBr(s) be deposited from 0.10 M Br$^-$? (Consider negligible current flow, so that there is no ohmic potential, concentration polarization, or overpotential.)

(c) Is it theoretically possible to separate 99.99% of 0.10 M KI from 0.10 M KBr by controlled-potential electrolysis?

16-E. Chlorine has been used for decades to disinfect drinking water. An undesirable side effect of this treatment is reaction with organic impurities to create organochlorine compounds, some of which could be toxic. Monitoring total organic halide (designated TOX) is required for many water providers. A standard procedure for TOX is to pass water through activated charcoal, which adsorbs organic compounds. Then the charcoal is combusted to liberate hydrogen halides:

$$\text{Organic halide (RX)} \xrightarrow{O_2/800^\circ C} CO_2 + H_2O + HX$$

HX is absorbed into aqueous solution and measured by coulometric titration with a silver anode:

$$X^-(aq) + Ag(s) \rightarrow AgX(s) + e^-$$

When 1.00 L of drinking water was analyzed, a current of 4.23 mA was required for 387 s. A blank prepared by oxidizing charcoal required 6 s at 4.23 mA. Express the TOX of the drinking water as μmol halogen/L. If all halogen is chlorine, express the TOX as μg Cl/L.

16-F. Cd^{2+} was used as an internal standard in the analysis of Pb^{2+} by square wave polarography. Cd^{2+} gives a reduction wave at -0.60 (±0.02) V and Pb^{2+} gives a reduction wave at -0.40 (±0.02) V. It was first verified that the ratio of peak heights is proportional to the ratio of concentrations over the whole range employed in the experiment. Here are results for known and unknown mixtures.

Analyte	Concentration (M)	Current (μA)
Known		
Cd^{2+}	$3.23\ (\pm0.01) \times 10^{-5}$	1.64 (±0.03)
Pb^{2+}	$4.18\ (\pm0.01) \times 10^{-5}$	1.58 (±0.03)
Unknown + Internal Standard		
Cd^{2+}	?	2.00 (±0.03)
Pb^{2+}	?	3.00 (±0.03)

The unknown mixture was prepared by mixing 25.00 (±0.05) mL of unknown (containing only Pb^{2+}) plus 10.00 (±0.05) mL of 3.23 (±0.01) $\times$ 10^{-4} M Cd^{2+} and diluting to 50.00 (±0.05) mL.

(a) Disregarding uncertainties, find [Pb^{2+}] in the undiluted unknown.

(b) Find the absolute uncertainty for the answer to part (a).

16-G. Consider the cyclic voltammogram of the Co^{3+} compound Co(B$_9$C$_2$H$_{11}$)$_2^-$. Suggest a chemical reaction to account for each wave. Are the reactions reversible? How many electrons are involved in each step? Sketch the sampled current and square wave polarograms expected for this compound.

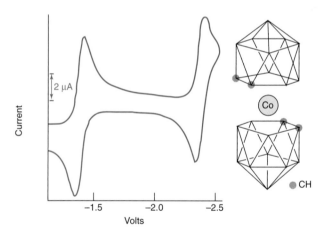

Cyclic voltammogram of Co(B$_9$C$_2$H$_{11}$)$_2^-$. [From W. E. Geiger, Jr., W. L. Bowden, and N. El Murr, "An Electrochemical Study of the Protonation Site of the Cobaltocene Anion and of Cyclopentadienylcobalt(I) Dicarbollides," *Inorg. Chem.* **1979**, *18*, 2358.]

$E_{1/2}$ (V vs. S.C.E)	I_{pa}/I_{pc}	$E_{pa} - E_{pc}$ (mV)
-1.38	1.01	60
-2.38	1.00	60

16-H. In a coulometric Karl Fischer water analysis, 25.00 mL of pure "dry" methanol required 4.23 C to generate enough I$_2$ to react with the residual H$_2$O in the methanol. A suspension of 0.847 6 g of finely ground polymeric material in 25.00 mL of the same "dry" methanol required 63.16 C. Find the weight percent of H$_2$O in the polymer.

Problems

Fundamentals of Electrolysis

16-1. The figure shows the behavior of Pt and Ag cathodes at which reduction of H_3O^+ to $H_2(g)$ occurs. Explain why the two curves are not superimposed.

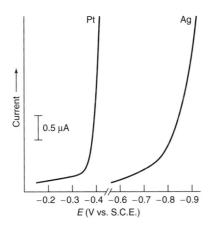

Current versus voltage for Pt and Ag electrodes in O_2-free, aqueous H_2SO_4 adjusted to pH 3.2. [From D. Marín, F. Mendicuti, and C. Teijeiro, "An Electrochemistry Experiment: Hydrogen Evolution Reaction on Different Electrodes," *J. Chem. Ed.* **1994**, *71*, A277.]

16-2. How many hours are required for 0.100 mol of electrons to flow through a circuit if the current is 1.00 A?

16-3. The standard free energy change for the formation of $H_2(g) + \frac{1}{2}O_2(g)$ from $H_2O(l)$ is $\Delta G° = +237.13$ kJ. The reactions are

Cathode: $\quad 2H_2O + 2e^- \rightleftharpoons H_2(g) + 2OH^-$

Anode: $\quad H_2O \rightleftharpoons \frac{1}{2}O_2(g) + 2H^+ + 2e^-$

Calculate the standard voltage ($E°$) needed to decompose water into its elements by electrolysis. What does the word *standard* mean in this question?

16-4. Consider the following electrolysis reactions.

Cathode: $\quad H_2O(l) + e^- \rightleftharpoons \frac{1}{2}H_2(g, 1.0\text{ bar}) + OH^+(aq, 0.10\text{ M})$

Anode: $\quad Br^-(aq, 0.10\text{ M}) \rightleftharpoons \frac{1}{2}Br_2(l) + e^-$

(a) Calculate the voltage needed to drive the net reaction if current is negligible.

(b) Suppose that the cell has a resistance of 2.0 Ω and a current of 100 mA. How much voltage is needed to overcome the cell resistance? This is the ohmic potential.

(c) Suppose that the anode reaction has an overpotential of 0.20 V and that the cathode overpotential is 0.40 V. What voltage is needed to overcome these effects combined with those of parts (a) and (b)?

(d) Suppose that concentration polarization occurs. $[OH^-]_s$ at the cathode surface increases to 1.0 M and $[Br^-]_s$ at the anode surface decreases to 0.010 M. What voltage is needed to overcome these effects combined with those of (b) and (c)?

16-5. Which voltage, V_1 or V_2, in the diagram (see top of next column), is constant in controlled-potential electrolysis? Which are the working, auxiliary, and reference electrodes in the diagram?

16-6. The Weston cell is a stable voltage standard formerly used in potentiometers. (The potentiometer compares an unknown voltage with that of the standard. In contrast with the conditions of this problem, very little current may be drawn from the cell if it is to be a voltage standard.)

$$Hg_2SO_4 + Cd(in\ Hg) \rightleftharpoons 2Hg + CdSO_4$$

(a) How much work (J) can be done by the Weston cell if the voltage is 1.02 V and 1.00 mL of Hg (density = 13.53 g/mL) is deposited?

(b) If the cell passes current through a 100-Ω resistor that dissipates heat at a rate of 0.209 J/min, how many grams of Cd are oxidized each hour? (This question is not meant to be consistent with part (a). The voltage is no longer 1.02 volts.)

16-7. The chlor-alkali process,[35] in which seawater is electrolyzed to produce Cl_2 and NaOH, is the second most important commercial electrolysis, behind production of aluminum.

Anode: $\quad Cl^- \to \frac{1}{2}Cl_2 + e^-$

Hg cathode: $\quad Na^+ + H_2O + e^- \to NaOH + \frac{1}{2}H_2$

The Nafion membrane (page 384) used to separate the anode and cathode compartments resists chemical attack. Its anionic side chains permit conduction of Na^+, but not anions. The cathode compartment contains pure water, and the anode compartment contains seawater from which Ca^{2+} and Mg^{2+} have been removed. Explain how the membrane allows NaOH to be formed free of NaCl.

16-8. A lead-acid battery in a car has six cells in series, each delivering close to 2.0 V for a total of 12 V when the battery is discharging. Recharging requires ~2.4 V per cell, or ~14 V for the entire battery.[36] Explain these observations in terms of Equation 16-6.

Electrogravimetric Analysis

16-9. A 0.326 8-g unknown containing $Pb(CH_3CHOHCO_2)_2$ (lead lactate, FM 385.3) plus inert material was electrolyzed to produce 0.111 1 g of PbO_2 (FM 239.2). Was the PbO_2 deposited at the anode or at the cathode? Find the weight percent of lead lactate in the unknown.

16-10. A solution of Sn^{2+} is to be electrolyzed to reduce the Sn^{2+} to $Sn(s)$. Calculate the cathode potential (versus S.H.E.) needed to reduce $[Sn^{2+}]$ to 1.0×10^{-8} M if no concentration polarization occurs. What would be the potential versus S.C.E. instead of S.H.E.? Would the potential be more positive or more negative if concentration polarization occurred?

16-11. What cathode potential (versus S.H.E.) is required to reduce 99.99% of Cd(II) from a solution containing 0.10 M Cd(II) in 1.0 M ammonia if there is negligible current? Consider the following reactions and assume that nearly all Cd(II) is in the form $Cd(NH_3)_4^{2+}$.

$$Cd^{2+} + 4NH_3 \rightleftharpoons Cd(NH_3)_4^{2+} \qquad \beta_4 = 3.6 \times 10^6$$
$$Cd^{2+} + 2e^- \rightleftharpoons Cd(s) \qquad E° = -0.402 \text{ V}$$

16-12. *Electroplating efficiency.*[37] Nickel was electrolytically plated onto a carbon electrode from a bath containing 290 g/L $NiSO_4 \cdot 6H_2O$, 30 g/L $B(OH)_3$, and 8 g/L NaCl at -1.2 V vs. Ag | AgCl. The most important side reaction is reduction of H^+ to H_2. In one experiment, a carbon electrode weighing 0.477 5 g before deposition weighed 0.479 8 g after 8.082 C had passed through the circuit. What percentage of the current went into the reaction $Ni^{2+} + 2e^- \rightarrow Ni(s)$?

Coulometry

16-13. Explain how the amperometric end-point detector in Figure 16-8 operates.

16-14. What does a mediator do?

16-15. The sensitivity of a coulometer is governed by the delivery of its minimum current for its minimum time. Suppose that 5 mA can be delivered for 0.1 s.

(a) How many moles of electrons are delivered by 5 mA for 0.1 s?

(b) How many milliliters of a 0.01 M solution of a two-electron reducing agent are required to deliver the same number of electrons?

16-16. The experiment in Figure 16-8 required 5.32 mA for 964 s for complete reaction of a 5.00-mL aliquot of unknown cyclohexene solution.

(a) How many moles of electrons passed through the cell?

(b) How many moles of cyclohexene reacted?

(c) What was the molarity of cyclohexene in the unknown?

16-17. $H_2S(aq)$ can be analyzed by titration with coulometrically generated I_2.

$$H_2S + I_2 \rightarrow S(s) + 2H^+ + 2I^-$$

To 50.00 mL of sample were added 4 g of KI. Electrolysis required 812 s at 52.6 mA. Calculate the concentration of H_2S (μg/mL) in the sample.

16-18. Ti^{3+} is to be generated in 0.10 M $HClO_4$ solution for coulometric reduction of azobenzene.

$$TiO^{2+} + 2H^+ + e^- \rightleftharpoons Ti^{3+} + H_2O \qquad E° = 0.100 \text{ V}$$
$$4Ti^{3+} + C_6H_5N{=}NC_6H_5 + 4H_2O \rightarrow 2C_6H_5NH_2 + 4TiO^{2+} + 4H^+$$

$\qquad$ Azobenzene $\qquad\qquad\qquad$ Aniline

At the counter electrode, water is oxidized, and O_2 is liberated at a pressure of 0.20 bar. Both electrodes are made of smooth Pt, and each has a total surface area of 1.00 cm^2. The rate of reduction of the azobenzene is 25.9 nmol/s, and the resistance of the solution between the generator electrodes is 52.4 Ω.

(a) Calculate the current density (A/m^2) at the electrode surface. Use Table 16-1 to estimate the overpotential for O_2 liberation.

(b) Calculate the cathode potential (versus S.H.E.) assuming that $[TiO^{2+}]_{surface} = [TiO^{2+}]_{bulk} = 0.050$ M and $[Ti^{3+}]_{surface} = 0.10$ M.

(c) Calculate the anode potential (versus S.H.E.).

(d) What should the applied voltage be?

16-19. In an extremely accurate measurement of the Faraday constant, a pure silver anode was oxidized to Ag^+ with a constant current of 0.203 639 0 ($\pm 0.000\ 000\ 4$) A for 18 000.075 (± 0.010) s to give a mass loss of 4.097 900 ($\pm 0.000\ 003$) g from the anode. Given that the atomic mass of Ag is 107.868 2 ($\pm 0.000\ 2$), find the value of the Faraday constant and its uncertainty.

16-20. *Coulometric titration of sulfite in wine.*[38] Sulfur dioxide is added to many foods as a preservative. In aqueous solution, the following species are in equilibrium:

$$\underset{\text{Sulfur dioxide}}{SO_2} \rightleftharpoons \underset{\text{Sulfurous acid}}{H_2SO_3} \rightleftharpoons \underset{\text{Bisulfite}}{HSO_3^-} \rightleftharpoons \underset{\text{Sulfite}}{SO_3^{2-}} \qquad \text{(A)}$$

Bisulfite reacts with aldehydes in food near neutral pH:

$\qquad\qquad$ Aldehyde $\qquad\qquad\qquad\qquad$ Adduct

Sulfite is released from the adduct in 2 M NaOH and can be analyzed by its reaction with I_3^- to give I^- and sulfate. Excess I_3^- must be present for quantitative reaction.

$\qquad$ Here is a coulometric procedure for analysis of total sulfite in white wine. Total sulfite means all species in Reaction A and the adduct in Reaction B. We use white wine so that we can see the color of a starch-iodine end point.

1. Mix 9.00 mL of wine plus 0.8 g NaOH and dilute to 10.00 mL. The NaOH releases sulfite from its organic adducts.
2. Generate I_3^- at the working electrode (the anode) by passing a known current for a known time through the cell in Figure 16-9. The cell contains 30 mL of 1 M acetate buffer (pH 3.7) plus 0.1 M KI. In the cathode compartment, H_2O is reduced to $H_2 + OH^-$. The frit retards diffusion of OH^- into the main compartment, where it would react with I_3^- to give IO^-.
3. Generate I_3^- at the anode with a current of 10.0 mA for 4.00 min.
4. Inject 2.000 mL of the wine/NaOH solution into the cell, where the sulfite reacts with I_3^-, leaving excess I_3^-.
5. Add 0.500 mL of 0.050 7 M thiosulfate to consume I_3^- by Reaction 15-19 and leave excess thiosulfate.
6. Add starch indicator to the cell and generate fresh I_3^- with a constant current of 10.0 mA. A time of 131 s was required to consume excess thiosulfate and reach the starch end point.

(a) In what pH range is each form of sulfurous acid predominant?

(b) Write balanced half-reactions for the anode and cathode.

(c) At pH 3.7, the dominant form of sulfurous acid is HSO_3^- and the dominant form of sulfuric acid is HSO_4^{2-}. Write balanced reactions between I_3^- and HSO_3^- and between I_3^- and thiosulfate.

(d) Find the concentration of total sulfite in undiluted wine.

16-21. *Chemical oxygen demand by coulometry.* An electrochemical device incorporating photooxidation on a TiO_2 surface could replace refluxing with $Cr_2O_7^{2-}$ to measure chemical oxygen demand (Box 15-2). The diagram below shows a working electrode held at $+0.30$ V versus Ag │AgCl and coated with nanoparticles of TiO_2. Upon ultraviolet irradiation, electrons and holes are generated in the TiO_2. Holes oxidize organic matter at the surface. Electrons reduce H_2O at the auxiliary electrode in a compartment connected to the working compartment by a salt bridge. The sample compartment is only 0.18 mm thick with a volume of 13.5 μL. It requires ~1 min for all organic matter to diffuse to the TiO_2 surface and be exhaustively oxidized.

The blank curve in the graph below shows the response when the sample compartment contains just electrolyte. Before irradiation, no current is observed. Ultraviolet radiation causes a spike in the current, followed by a decrease to a steady level near 40 μA. This current arises from oxidation of water at the TiO_2 surface under ultraviolet exposure. The upper curve shows the same experiment, but with wastewater in the sample compartment. The increased current arises from oxidation of organic matter. When the organic matter is consumed, the current decreases to the blank level. The area between the two curves tells us how many electrons flow from oxidation of organic matter in the sample.

Photocurrent response for sample and blank. Both solutions contain 2 M $NaNO_3$. [From H. Zhao, D. Jiang, S. Zhang, K. Catterall, and R. John, "Development of a Direct Photoelectrochemical Method for Determination of Chemical Oxygen Demand," *Anal. Chem.* **2004**, *76*, 155.]

(a) Balance the oxidation half-reaction that occurs in this cell:

$$C_cH_hO_oN_nX_x + AH_2O \rightarrow BCO_2 + CX^- + DNH_3 + EH^+ + Fe^-$$

where X is any halogen. Express the stoichiometry coefficients A, B, C, D, E, and F in terms of c, h, o, n, and x.

(b) How many molecules of O_2 are required to balance the half-reaction in part **(a)** by reduction of oxygen ($O_2 + 4H^+ + 4e^- \rightarrow 2H_2O$)?

(c) The area between the two curves in the graph is $\int_0^\infty (I_{sample} - I_{blank})dt = 9.43$ mC. This is the number of electrons liberated by complete oxidation of the sample. How many moles of O_2 would be required for the same oxidation?

(d) Chemical oxygen demand (COD) is expressed as mg of O_2 required to oxidize 1 L of sample. Find the COD for this sample.

(e) If the only oxidizable substance in the sample were $C_9H_6NO_2ClBr_2$, what is its concentration in mol/L?

Amperometry

16-22. What is a Clark electrode, and how does it work?

16-23. (a) How does the amperometric glucose monitor in Figure 16-10 work?

(b) Why is a mediator advantageous in the glucose monitor?

(c) How does the coulometric glucose monitor in Figure 16-12 work?

(d) Why does the signal in the amperometric measurement depend on the temperature of the blood sample, whereas the signal in coulometry is independent of temperature? Do you expect the signal to increase or decrease with increasing temperature in amperometry?

(e) Glucose ($C_6H_6O_{12}$, FM 180.16) is present in normal human blood at a concentration near 1 g/L. How many microcoulombs are required for complete oxidation of glucose in 0.300 μL of blood in a home glucose monitor if the concentration is 1.00 g/L?

16-24. For a rotating disk electrode operating at sufficiently great potential, the redox reaction rate is governed by the rate at which analyte diffuses through the diffusion layer to the electrode (Figure 16-13b). The thickness of the diffusion layer is

$$\delta = 1.61 D^{1/3} \nu^{1/6} \omega^{-1/2}$$

where D is the diffusion coefficient of reactant (m²/s), ν is the kinematic viscosity of the liquid (= viscosity/density = m²/s), and ω is the rotation rate (radians/s) of the electrode. There are 2π radians in a circle. The current density (A/m²) is

$$\text{Current density} = 0.62nFD^{2/3}\nu^{-1/6}\omega^{1/2}C_0$$

where n is the number of electrons in the half-reaction, F is the Faraday constant, and C_0 is the concentration of the electroactive species in bulk solution (mol/m³, not mol/L). Consider the oxidation of $Fe(CN)_6^{4-}$ in a solution of 10.0 mM $K_3Fe(CN)_6$ + 50.0 mM $K_4Fe(CN)_6$ at +0.90 V (versus S.C.E.) at a rotation speed of 2.00×10^3 revolutions per minute.[20] The diffusion coefficient of $Fe(CN)_6^{4-}$ is 2.5×10^{-9} m²/s, and the kinematic viscosity is 1.1×10^{-6} m²/s. Calculate the thickness of the diffusion layer and the current density. If you are careful, the current density should look like the value in Figure 16-14b.

Voltammetry

16-25. In 1 M NH_3/1 M NH_4Cl solution, Cu^{2+} is reduced to Cu^+ near -0.3 (versus S.C.E.), and Cu^+ is reduced to Cu(*in Hg*) near -0.6 V.

(a) Sketch a qualitative sampled current polarogram for a solution of Cu^+.

(b) Sketch a polarogram for a solution of Cu^{2+}.

(c) Suppose that Pt, instead of Hg, were used as the working electrode. Which, if any, reduction potential would you expect to change?

16-26. (a) Explain the difference between charging current and faradaic current.

(b) What is the purpose of waiting 1 s after a voltage pulse before measuring current in sampled current voltammetry?

(c) Why is square wave voltammetry more sensitive than sampled current voltammetry?

16-27. Suppose that the diffusion current in a polarogram for reduction of Cd^{2+} at a mercury electrode is 14 µA. If the solution contains 25 mL of 0.50 mM Cd^{2+}, what percentage of Cd^{2+} is reduced in the 3.4 min required to scan from -0.6 to -1.2 V?

16-28. The drug Librium gives a polarographic wave with $E_{1/2} = -0.265$ V (versus S.C.E.) in 0.05 M H_2SO_4. A 50.0-mL sample containing Librium gave a wave height of 0.37 µA. When 2.00 mL of 3.00 mM Librium in 0.05 M H_2SO_4 were added to the sample, the wave height increased to 0.80 µA. Find the molarity of Librium in the unknown.

16-29. Explain what is done in anodic stripping voltammetry. Why is stripping the most sensitive voltammetric technique?

16-30. The figure below shows a series of standard additions of Cu^{2+} to acidified tap water measured by anodic stripping voltammetry at an iridium electrode. The unknown and all standard additions were made up to the same final volume.

(a) What chemical reaction occurs during the concentration stage of the analysis?

(b) What chemical reaction occurs during the stripping stage of the analysis?

(c) Find the concentration of Cu^{2+} in the tap water.

Anodic stripping voltammograms of tap water and five standard additions of 100 ppb Cu^{2+}. [From M. A. Nolan and S. P. Kounaves, "Microfabricated Array of Ir Microdisks for Determination of Cu^{2+} or Hg^{2+} Using Square Wave Stripping Voltammetry," *Anal. Chem.* **1999**, *71*, 3567.]

16-31. From the two standard additions of 50 pm Fe(III) in Figure 16-23, find the concentration of Fe(III) in the seawater. Estimate where the baseline should be drawn for each trace and measure the peak height from the baseline. Consider the volume to be constant for all three solutions.

16-32. The cyclic voltammogram of the antibiotic chloramphenicol (abbreviated RNO_2) is shown here. The scan was started at 0 V, and potential was swept toward negative voltage. The first cathodic wave, A, is from the reaction $RNO_2 + 4e^- + 4H^+ \rightarrow RNHOH + H_2O$. Explain what happens at peaks B and C by using the reaction $RNO + 2e^- + 2H^+ \rightleftharpoons RNHOH$. Why was peak C not seen in the initial scan?

Cyclic voltammogram of 3.7×10^{-4} chloramphenicol in 0.1 M acetate buffer, pH 4.62. The voltage of the carbon paste working electrode was scanned at a rate of 350 mV/s. [From P. T. Kissinger and W. R. Heineman, "Cyclic Voltammetry," *J. Chem. Ed.* **1983**, *60*, 702.]

16-33. Peak current (I_p) and scan rate (v) are listed for cyclic voltammetry (Fe(II) → Fe(III)) of a water-soluble ferrocene derivative in 0.1 M NaCl.[39]

Scan rate (V/s)	Peak anodic current (µA)
0.019 2	2.18
0.048 9	3.46
0.075 1	4.17
0.125	5.66
0.175	6.54
0.251	7.55

If a graph of I_p versus $\sqrt{v}$ gives a straight line, then the reaction is diffusion controlled. Prepare such a graph and use it to find the diffusion coefficient of the reactant for this one-electron oxidation. The area of the working electrode is 0.020 1 cm^2, and the concentration of reactant is 1.00 mM.

16-34. What are the advantages of using a microelectrode for voltammetric measurements?

16-35. What is the purpose of the Nafion membrane in Figure 16-30?

Karl Fischer Titration

16-36. Write the chemical reactions that show that 1 mol of I_2 is required for 1 mol of H_2O in a Karl Fischer titration.

16-37. Explain how the end point is detected in a Karl Fischer titration in Figure 16-32.

17 | Fundamentals of Spectrophotometry

THE OZONE HOLE[1]

Spectrum of ozone, showing maximum absorption of ultraviolet radiation at a wavelength near 260 nm. At this wavelength, a layer of ozone is more opaque than a layer of gold of the same mass. [Adapted from R. P. Wayne, *Chemistry of Atmospheres* (Oxford: Clarendon Press, 1991).]

Spectroscopically measured concentrations of O_3 and ClO (measured in ppb = nL/L) in the stratosphere near the South Pole in 1987. The loss of O_3 at latitudes where ClO has a high concentration is consistent with the known chemistry of catalytic destruction of O_3 by halogen radicals. [From J. G. Anderson, W. H. Brune, and M. H. Proffitt, *J. Geophys. Res.* 1989, *94D*, 11465.]

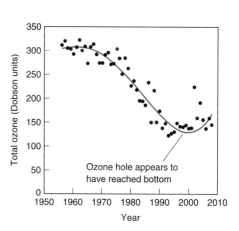

Mean atmospheric O_3 at Halley in Antarctica in October. *Dobson units* are defined in Problem 17-14. [From J. D. Shanklin, British Antarctic Survey, http://www.antarctica.ac.uk/met/jds/ozone/.]

Ozone, formed at altitudes of 20 to 40 km by the action of solar ultraviolet radiation ($h\nu$) on O_2, absorbs ultraviolet radiation that causes sunburn and skin cancer.

$$O_2 \xrightarrow{h\nu} 2O \qquad\qquad O + O_2 \rightarrow O_3$$
$$\text{Ozone}$$

In 1985, the British Antarctic Survey reported that the total ozone in the Antarctic stratosphere had decreased by 50% in early spring, relative to levels observed in the preceding 20 years. Ground, airborne, and satellite observations have since shown that this "ozone hole" occurs only in early spring (Figure 1-1) and continued to deepen until the year 2000.

An explanation begins with chlorofluorocarbons such as Freon-12 (CCl_2F_2), formerly used in refrigerators and air conditioners. These long-lived compounds, which are not found in nature,[2] diffuse to the stratosphere, where they catalyze ozone decomposition.

(1) $CCl_2F_2 \xrightarrow{h\nu} CClF_2 + Cl$ Photochemical Cl formation

(2) $Cl + O_3 \rightarrow ClO + O_2$

(3) $O_3 \xrightarrow{h\nu} O + O_2$ Net reaction of (2)–(4): catalytic O_3 destruction

(4) $O + ClO \rightarrow Cl + O_2$ $2O_3 \rightarrow 3O_2$

Cl produced in step 4 reacts again in step 2, so a single Cl atom can destroy $>10^5$ molecules of O_3. The chain is terminated when Cl or ClO reacts with hydrocarbons or NO_2 to form HCl or $ClONO_2$.

Stratospheric clouds[3] formed during the Antarctic winter catalyze the reaction of HCl with $ClONO_2$ to form Cl_2, which is split by sunlight into Cl atoms to initiate O_3 destruction:

$$HCl + ClONO_2 \xrightarrow[\text{polar clouds}]{\text{surface of}} Cl_2 + HNO_3 \qquad Cl_2 \xrightarrow{h\nu} 2Cl$$

Polar stratospheric clouds require winter cold to form. Only when the sun is rising in September and October, and clouds are still present, are conditions right for O_3 destruction.

To protect life from ultraviolet radiation, international treaties now ban or phase out chlorofluorocarbons, and there is an effort to find safe substitutes.

Spectrophotometry is any technique that uses light to measure chemical concentrations. A procedure based on absorption of visible light is called *colorimetry*. The most-cited article in the journal *Analytical Chemistry* from 1945 to 1999 describes a colorimetric method by which biochemists measure sugars.[4] Chapter 17 is intended to give a stand-alone overview of spectrophotometry sufficient for introductory purposes. Chapter 18 goes further into applications, and Chapter 19 discusses instrumentation.

Following the discovery of the Antarctic ozone "hole" in 1985, atmospheric chemist Susan Solomon led the first expedition in 1986 specifically intended to make chemical measurements of the Antarctic atmosphere by using balloons and ground-based spectroscopy. The expedition discovered that ozone depletion occurred after polar sunrise and that the concentration of chemically active chlorine in the stratosphere was ~100 times greater than had been predicted from gas-phase chemistry. Solomon's group identified chlorine as the culprit in ozone destruction and polar stratospheric clouds as the catalytic surface for the release of so much chlorine.

17-1 Properties of Light

It is convenient to describe light in terms of both particles and waves. Light waves consist of perpendicular, oscillating electric and magnetic fields. For simplicity, a *plane-polarized* wave is shown in Figure 17-1. In this figure, the electric field is in the *xy* plane, and the magnetic field is in the *xz* plane. **Wavelength**, λ, is the crest-to-crest distance between waves. **Frequency**, ν, is the number of complete oscillations that the wave makes each second. The unit of frequency is 1/second. One oscillation per second is called one **hertz** (Hz). A frequency of 10^6 s^{-1} is therefore said to be 10^6 Hz, or 1 *megahertz* (MHz).

The relation between frequency and wavelength is

Relation between frequency and wavelength: $\qquad \nu\lambda = c \qquad\qquad$ (17-1)

where c is the speed of light (2.998×10^8 m/s in vacuum). In a medium other than vacuum, the speed of light is c/n, where n is the **refractive index** of that medium. For visible wavelengths in most substances, $n > 1$, so visible light travels more slowly through matter than through vacuum. When light moves between media with different refractive indexes, the frequency remains constant but the wavelength changes.

With regard to energy, it is more convenient to think of light as particles called **photons.** Each photon carries the energy, E, which is given by

Relation between energy and frequency: $\qquad E = h\nu \qquad\qquad$ (17-2)

where h is *Planck's constant* ($= 6.626 \times 10^{-34}$ J·s).

Equation 17-2 states that energy is proportional to frequency. Combining Equations 17-1 and 17-2, we can write

$$E = \frac{hc}{\lambda} = hc\tilde{\nu} \qquad\qquad (17\text{-}3)$$

where $\tilde{\nu}(=1/\lambda)$ is called **wavenumber.** Energy is inversely proportional to wavelength and directly proportional to wavenumber. Red light, with a longer wavelength than blue light, is less energetic than blue light. The most common unit of wavenumber in the literature is cm^{-1}, read "reciprocal centimeters" or "wavenumbers."

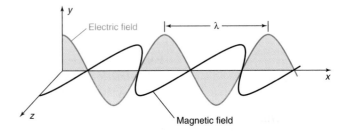

FIGURE 17-1 Plane-polarized electromagnetic radiation of wavelength λ, propagating along the *x*-axis. The electric field of plane-polarized light is confined to a single plane. Ordinary, unpolarized light has electric field components in all planes.

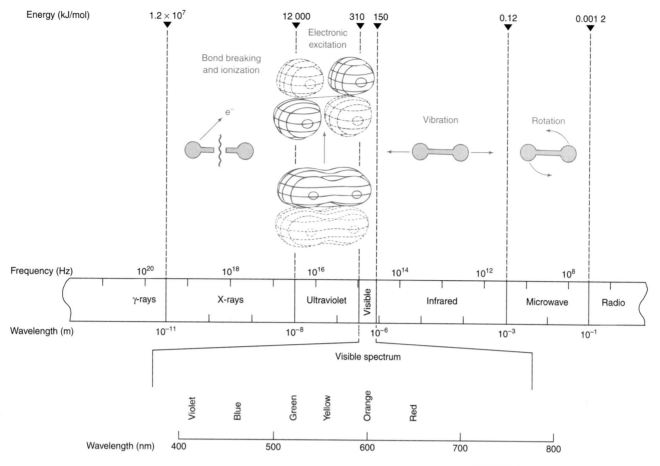

FIGURE 17-2 Electromagnetic spectrum, showing representative molecular processes that occur when light in each region is absorbed. The visible spectrum spans the wavelength range 380–780 nanometers (1 nm = 10^{-9} m).

The names of regions of the **electromagnetic spectrum** in Figure 17-2 are historical. There are no abrupt changes in characteristics as we go from one region to the next, such as visible to infrared. Visible light—the kind of electromagnetic radiation we see—represents only a small fraction of the electromagnetic spectrum.

17-2 Absorption of Light

When a molecule absorbs a photon, the energy of the molecule increases. We say that the molecule is promoted to an **excited state** (Figure 17-3). If a molecule emits a photon, the energy of the molecule is lowered. The lowest energy state of a molecule is called the **ground state.** Figure 17-2 indicates that microwave radiation stimulates rotation of molecules when it is absorbed. Infrared radiation stimulates vibrations. Visible and ultraviolet radiation promote electrons to higher energy orbitals. X-rays and short-wavelength ultraviolet radiation break chemical bonds and ionize molecules. Medical X-rays damage the human body, so your exposure should be minimized.

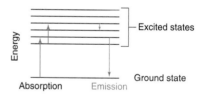

FIGURE 17-3 Absorption of light increases the energy of a molecule. Emission of light decreases its energy.

EXAMPLE   Photon Energies

By how many kilojoules per mole is the energy of O_2 increased when it absorbs ultraviolet radiation with a wavelength of 147 nm? How much is the energy of CO_2 increased when it absorbs infrared radiation with a wavenumber of 2 300 cm^{-1}?

Solution For the ultraviolet radiation, the energy increase is

$$\Delta E = h\nu = h\frac{c}{\lambda}$$

$$= (6.626 \times 10^{-34}\text{ J} \cdot s)\left[\frac{(2.998 \times 10^{8}\text{ m/s})}{(147\text{ nm})(10^{-9}\text{ m/nm})}\right] = 1.35 \times 10^{-18}\text{ J/molecule}$$

$$(1.35 \times 10^{-18}\text{ J/molecule})(6.022 \times 10^{23}\text{ molecules/mol}) = 814\text{ kJ/mol}$$

This is enough energy to break the O=O bond in oxygen. For CO_2, the energy increase is

$$\Delta E = h\nu = h\frac{c}{\lambda} = hc\tilde{\nu} \qquad \left(\text{recall that } \tilde{\nu} = \frac{1}{\lambda}\right)$$

$$= (6.626 \times 10^{-34} \text{ J} \cdot \text{s})(2.998 \times 10^8 \text{ m/s})(2\,300 \text{ cm}^{-1})(100 \text{ cm/m})$$

$$= 4.6 \times 10^{-20} \text{ J/molecule} = 28 \text{ kJ/mol}$$

Infrared absorption increases the amplitude of the vibrations of the CO_2 bonds.

Test Yourself   What is the wavelength, wavenumber, and name of radiation with an energy of 100 kJ/mol? (**Answer:** 1.20 μm, 8.36×10^3 cm^{-1}, infrared)

When light is absorbed by a sample, the *irradiance* of the beam of light is decreased. **Irradiance,** P, is the energy per second per unit area of the light beam. A rudimentary spectrophotometric experiment is illustrated in Figure 17-4. Light is passed through a **monochromator** (a prism, a grating, or even a filter) to select one wavelength (Color Plate 12). Light with a very narrow range of wavelength is said to be **monochromatic** ("one color.") The monochromatic light, with irradiance P_0, strikes a sample of length b. The irradiance of the beam emerging from the other side of the sample is P. Some of the light may be absorbed by the sample, so $P \leq P_0$.

Irradiance is the energy per unit time per unit area in the light beam (watts per square meter, W/m^2). The terms *intensity* or *radiant power* have been used for this same quantity.

Monochromatic light consists of "one color" (one wavelength). The better the monochromator, the narrower is the range of wavelengths in the emerging beam.

FIGURE 17-4 Schematic diagram of a single-beam spectrophotometric experiment. P_0, irradiance of beam entering sample; P, irradiance of beam emerging from sample; b, length of path through sample.

Transmittance, T, is defined as the fraction of the original light that passes through the sample.

Transmittance:
$$T = \frac{P}{P_0} \tag{17-4}$$

Therefore, T has the range 0 to 1. The *percent transmittance* is simply $100T$ and ranges between 0 and 100%. **Absorbance** is defined as

Relation between transmittance and absorbance:

P/P_0	% T	A
1	100	0
0.1	10	1
0.01	1	2

Absorbance:
$$A = \log\left(\frac{P_0}{P}\right) = -\log T \tag{17-5}$$

When no light is absorbed, $P = P_0$ and $A = 0$. If 90% of the light is absorbed, 10% is transmitted and $P = P_0/10$. This ratio gives $A = 1$. If only 1% of the light is transmitted, $A = 2$. Absorbance is sometimes called *optical density.*

Absorbance is so important because it is directly proportional to the concentration, c, of the light-absorbing species in the sample (Color Plate 13).

Box 17-1 explains why absorbance, not transmittance, is directly proportional to concentration.

Beer's law:
$$A = \varepsilon b c \tag{17-6}$$

Equation 17-6, which is the heart of spectrophotometry as applied to analytical chemistry, is called the *Beer-Lambert law,*[6] or simply **Beer's law.** Absorbance is dimensionless, but some people write "absorbance units" after absorbance. The concentration of the sample, c, is usually given in units of moles per liter (M). The pathlength, b, is commonly expressed in centimeters. The quantity ε (epsilon) is called the **molar absorptivity** (or *extinction coefficient* in the older literature) and has the units M^{-1} cm^{-1} to make the product $\varepsilon b c$ dimensionless. Molar absorptivity is the characteristic of a substance that tells how much light is absorbed at a particular wavelength.

EXAMPLE   **Absorbance, Transmittance, and Beer's Law**

Find the absorbance and transmittance of a 0.002 40 M solution of a substance with a molar absorptivity of 313 M^{-1} cm^{-1} in a cell with a 2.00-cm pathlength.

Beer's law, Equation 17-6, states that *absorbance* is proportional to the concentration of the absorbing species. The fraction of light passing through a sample (the *transmittance*) is related logarithmically, not linearly, to the sample concentration. Why should this be?

Imagine light of irradiance P passing through an *infinitesimally thin* layer of solution whose thickness is dx. A physical model of the absorption process suggests that, within the infinitesimally thin layer, decrease in power (dP) ought to be proportional to the incident power (P), to the concentration of absorbing species (c), and to the thickness of the section (dx):

$$dP = -\beta Pc\, dx \qquad (A)$$

where β is a constant of proportionality and the minus sign indicates a decrease in P as x increases. The rationale for saying that the decrease in power is proportional to the incident power may be understood from a numerical example. If 1 photon out of 1 000 incident photons is absorbed in a thin layer, we would expect that 2 out of 2 000 incident photons would be absorbed. The decrease in photons (power) is proportional to the incident flux of photons (power).

Equation A can be rearranged and integrated to find an expression for P:

$$-\frac{dP}{P} = \beta c\, dx \quad \Rightarrow \quad -\int_{P_0}^{P}\frac{dP}{P} = \beta c\int_{0}^{b} dx$$

The limits of integration are $P = P_0$ at $x = 0$ and $P = P$ at $x = b$.

$$-\ln P - (-\ln P_0) = \beta cb \quad \Rightarrow \quad \ln\!\left(\frac{P_0}{P}\right) = \beta cb$$

Finally, converting ln into log, using the relation $\ln z = (\ln 10)(\log z)$, gives Beer's law:

$$\underbrace{\log\!\left(\frac{P_0}{P}\right)}_{\text{Absorbance}} = \underbrace{\left(\frac{\beta}{\ln 10}\right)}_{\text{Constant} \equiv \varepsilon} cb \quad \Rightarrow \quad A = \varepsilon cb$$

The logarithmic relation of P_0/P to concentration arises because, in each infinitesimal portion of the total volume, *the decrease in power is proportional to the power incident upon that section*. As light travels through the sample, the power loss in each succeeding layer decreases, because the magnitude of the incident power that reaches each layer is decreasing. Molar absorptivity ranges from 0 (if the probability for photon absorption is 0) to approximately $10^5\ \mathrm{M^{-1}\ cm^{-1}}$ (when the probability for photon absorption approaches unity).

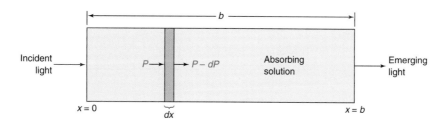

Incident light $P \longrightarrow$ $\longrightarrow P - dP$ Absorbing solution   Emerging light

$x = 0$ dx $x = b$

Solution   Equation 17-6 gives us the absorbance.

$$A = \varepsilon bc = (313\ \mathrm{M^{-1}\ cm^{-1}})(2.00\ \text{cm})(0.002\ 40\ \text{M}) = 1.50$$

Transmittance is obtained from Equation 17-5 by raising 10 to the power equal to the expression on each side of the equation:

$$\log T = -A$$
$$T = 10^{\log T} = 10^{-A} = 10^{-1.50} = 0.031\ 6$$

Just 3.16% of the incident light emerges from this solution.

If $x = y$, $10^x = 10^y$.

Test Yourself   The transmittance of a 0.010 M solution of a compound in a 0.100-cm-pathlength cell is $T = 8.23\%$. Find the absorbance (A) and the molar absorptivity (ε). (**Answer:** 1.08, $1.08 \times 10^3\ \mathrm{M^{-1}\ cm^{-1}}$)

Equation 17-6 could be written

$$A_\lambda = \varepsilon_\lambda bc$$

because A and ε depend on the wavelength of light. The quantity ε is simply a coefficient of proportionality between absorbance and the product bc. The greater the molar absorptivity, the greater the absorbance. An **absorption spectrum** (Demonstration 17-1) is a graph showing how A (or ε) varies with wavelength.

The plural of "spectrum" is "spectra."

The part of a molecule responsible for light absorption is called a **chromophore.** Any substance that absorbs visible light appears colored when white light is transmitted through it or reflected from it. (White light contains all the colors in the visible spectrum.) The substance absorbs certain wavelengths of the white light, and our eyes detect the wavelengths that are

The spectrum of visible light can be projected on a screen in a darkened room in the following manner:[7] Mount four layers of plastic diffraction grating[†] on a cardboard frame having a square hole large enough to cover the lens of an overhead projector. Tape the assembly over the projector lens facing the screen. Place an opaque cardboard surface with two 1 × 3 cm slits on the working surface of the projector.

When the lamp is turned on, the white image of each slit is projected on the center of the screen. A visible spectrum appears on either side of each image. By placing a beaker of colored solution over one slit, you can see its color projected on the screen where the white image previously appeared. The spectrum beside the colored image loses intensity at wavelengths absorbed by the colored species.

Color Plate 14a shows the spectrum of white light and the spectra of three different colored solutions. Potassium dichromate, which appears orange or yellow, absorbs blue wavelengths. Bromophenol blue absorbs orange wavelengths and appears blue to our eyes. Phenolphthalein absorbs the center of the visible spectrum. For comparison, spectra of these three solutions recorded with a spectrophotometer are shown in Color Plate 14b.

This same setup can be used to demonstrate fluorescence and the properties of colors.[7] Other demonstrations of absorption and emission spectra[8] and decomposition of spectra into chromaticity coordinates have been described.[9]

(a) Overhead projector. (b) Diffraction grating mounted on cardboard. (c) Mask for working surface.

[†]Edmund Scientific Co., www.edmundoptics.com, catalog no. NT40-267.

The color of a solution is the complement of the color of the light that it absorbs. The color we perceive depends not only on the wavelength of light, but on its intensity.

not absorbed. Table 17-1 gives a rough guide to colors.[10] The observed color is called the *complement* of the absorbed color. For example, bromophenol blue has maximum absorbance at 591 nm and its observed color is blue. Color Plate 15 displays absorption spectra and observed colors.

TABLE 17-1 Colors of visible light

Wavelength of maximum absorption (nm)	Color absorbed	Color observed
380–420	Violet	Green-yellow
420–440	Violet-blue	Yellow
440–470	Blue	Orange
470–500	Blue-green	Red
500–520	Green	Purple
520–550	Yellow-green	Violet
550–580	Yellow	Violet-blue
580–620	Orange	Blue
620–680	Red	Blue-green
680–780	Red	Green

When Beer's Law Fails

Beer's law states that absorbance is proportional to the concentration of the absorbing species. It applies to *monochromatic* radiation,[11] and it works very well for dilute solutions ($\lesssim 0.01$ M) of most substances.

In concentrated solutions, solute molecules influence one another as a result of their proximity. When solute molecules get close to one another, their properties (including molar absorptivity) change somewhat. At very high concentration, the solute *becomes* the solvent.

Properties of a molecule are not exactly the same in different solvents. Nonabsorbing solutes in a solution can also interact with the absorbing species and alter the absorptivity.

If the absorbing molecule participates in a concentration-dependent chemical equilibrium, the absorptivity changes with concentration. For example, in concentrated solution, a weak acid, HA, may be mostly undissociated. As the solution is diluted, dissociation increases. If the absorptivity of A^- is not the same as that of HA, the solution will appear not to obey Beer's law as it is diluted.

Beer's law works for monochromatic radiation passing through a dilute solution in which the absorbing species is not participating in a concentration-dependent equilibrium.

17-3 Measuring Absorbance

The minimum requirements for a spectrophotometer (a device to measure absorbance of light) were shown in Figure 17-4. Light from a continuous source is passed through a monochromator, which selects a narrow band of wavelengths from the incident beam. This "monochromatic" light travels through a sample of pathlength b, and the irradiance of the emergent light is measured.

For visible and ultraviolet spectroscopy, a liquid sample is usually contained in a cell called a **cuvet** that has flat, fused-silica (SiO_2) faces (Figure 17-5). Glass is suitable for visible, but not ultraviolet, spectroscopy because it absorbs ultraviolet radiation. The most common cuvets have a 1.000-cm pathlength and are sold in matched sets for sample and reference.

For infrared measurements, cells are commonly constructed of NaCl or KBr. For the 400 to 50 cm^{-1} far-infrared region, polyethylene is a transparent window. Solid samples are commonly ground to a fine powder, which can be added to mineral oil (a viscous hydrocarbon also called Nujol) to give a dispersion that is called a *mull* and is pressed between two KBr plates. The analyte spectrum is obscured in a few regions in which the mineral oil absorbs infrared radiation. Alternatively, a 1 wt% mixture of solid sample with KBr can be ground to a fine powder and pressed into a translucent pellet at a pressure of ~60 MPa (600 bar). Solids and powders can also be examined by *diffuse reflectance*, in which reflected infrared radiation, instead of transmitted infrared radiation, is observed. Wavelengths absorbed by the sample are not reflected as well. This technique is sensitive only to the surface of the sample.

Gases are more dilute than liquids and require cells with longer pathlengths, typically ranging from 10 cm to many meters. A pathlength of many meters is obtained by reflecting light so that it traverses the sample many times before reaching the detector.

Figure 17-4 outlines a *single-beam* instrument, which has only one beam of light. We do not measure the incident irradiance, P_0, directly. Rather, the irradiance of light passing through a reference cuvet containing pure solvent (or a reagent blank) is *defined* as P_0. This cuvet is then removed and replaced by an identical one containing sample. The irradiance of light striking the detector after passing through the sample is the quantity P. Knowing both P and P_0 allows T or A to be determined. The reference cuvet compensates for reflection,

Approximate low-energy cutoff for common infrared windows:

sapphire (Al_2O_3)	1 500 cm^{-1}
NaCl	650 cm^{-1}
KBr	350 cm^{-1}
AgCl	350 cm^{-1}
CsBr	250 cm^{-1}
CsI	200 cm^{-1}

FIGURE 17-5 Common cuvets for visible and ultraviolet spectroscopy. Flow cells permit continuous flow of solution through the cell. In the thermal cell, liquid from a constant-temperature bath flows through the cell jacket to maintain a desired temperature. [Courtesy A. H. Thomas Co., Philadelphia, PA.]

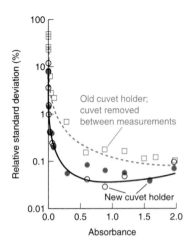

FIGURE 17-6 Precision of replicate absorbance measurements with a diode array spectrometer at 350 nm with a dichromate solution. Filled circles are from replicate measurements in which the sample was not removed from the cuvet holder between measurements. Open circles are from measurements in which the sample was removed and then replaced in the cuvet holder between measurements. Best reproducibility is observed at intermediate absorbance ($A \approx 0.3$ to 2). Note logarithmic ordinate. Lines are least-squares fit of data to theoretical equations. [Data from J. Galbán, S. de Marcos, I. Sanz, C. Ubide, and J. Zuriarrain, "Uncertainty in Modern Spectrophotometers," *Anal. Chem.* **2007**, *79*, 4763.]

Don't touch the clear faces of a cuvet—fingerprints scatter and absorb light.

This example illustrates the measurement of molar absorptivity from a single solution. It is better to measure several concentrations to obtain a more reliable absorptivity and to demonstrate that Beer's law is obeyed.

scattering, and absorption by the cuvet and solvent. A *double-beam* instrument, which splits the light to pass alternately between sample and reference cuvets, is described in Chapter 19.

In recording an absorbance spectrum, first record a *baseline spectrum* with reference solutions (pure solvent or a reagent blank) in both cuvets. If the instrument were perfect, the baseline would be 0 everywhere. In our imperfect world, the baseline usually exhibits small positive and negative absorbance. We subtract the baseline absorbance from the sample absorbance to obtain the true absorbance at each wavelength.

For spectrophotometric analysis, we normally choose the wavelength of maximum absorbance for two reasons: (1) The sensitivity of the analysis is greatest at maximum absorbance; that is, we get the maximum response for a given concentration of analyte. (2) The curve is relatively flat at the maximum, so there is little variation in absorbance if the monochromator drifts a little or if the width of the transmitted band changes slightly. Beer's law is obeyed when the absorbance is constant across the selected waveband.

Modern spectrophotometers are most precise (reproducible) at intermediate levels of absorbance ($A \approx 0.3$ to 2). If too little light gets through the sample (high absorbance), intensity is hard to measure. If too much light gets through (low absorbance), it is hard to distinguish transmittance of the sample from that of the reference. It is desirable to adjust sample concentration so that absorbance falls in this intermediate range. Compartments must be tightly closed to avoid stray light, which leads to false readings.

Figure 17-6 shows the relative standard deviation of replicate measurements made at 350 nm with a diode array spectrometer. Filled circles are from replicate measurements in which the sample was not removed from the cuvet holder between measurements. Open circles are from measurements in which the sample was removed and then replaced in the cuvet holder between measurements. Relative standard deviation is below 0.1% in both cases for absorbance in the range 0.3 to 2. Data points represented by open squares were obtained when a 10-year-old cuvet holder was used and the sample was removed and replaced in the holder between measurements. Variability in the position of the cuvet more than doubles the relative standard deviation. The conclusion is that modern spectrometers with modern cell holders provide excellent reproducibility. Precision was degraded when an old cell holder was used and the sample was removed and inserted between measurements.

Always keep containers covered to exclude dust, which scatters light and appears to the spectrophotometer to be absorbance. Filtering the final solution through a very fine filter may be necessary in critical work. Handle cuvets with a tissue to avoid putting fingerprints on the faces, and keep cuvets scrupulously clean.

Slight mismatch between sample and reference cuvets leads to systematic errors in spectrophotometry. Place cuvets in the spectrophotometer as reproducibly as possible. Variation in apparent absorbance arises from slightly misplacing the cuvet in its holder, or turning a flat cuvet around by 180°, or rotating a circular cuvet.

17-4 Beer's Law in Chemical Analysis

For a compound to be analyzed by spectrophotometry, it must absorb light, and this absorption should be distinguishable from that due to other substances in the sample. Because most compounds absorb ultraviolet radiation, measurements in this region of the spectrum tend to be inconclusive, and analysis is usually restricted to the visible spectrum. If there are no interfering species, however, ultraviolet absorbance is satisfactory. Proteins are normally assayed in the ultraviolet region at 280 nm because aromatic groups present in virtually every protein have an absorbance maximum at 280 nm.

> **EXAMPLE** Measuring Benzene in Hexane
>
> (a) Pure hexane has negligible ultraviolet absorbance above a wavelength of 200 nm. A solution prepared by dissolving 25.8 mg of benzene (C_6H_6, FM 78.11) in hexane and diluting to 250.0 mL had an absorption peak at 256 nm and an absorbance of 0.266 in a 1.000-cm cell. Find the molar absorptivity of benzene at this wavelength.
>
> **Solution** The concentration of benzene is
>
> $$[C_6H_6] = \frac{(0.025\,8\ \text{g})/(78.11\ \text{g/mol})}{0.250\,0\ \text{L}} = 1.32_1 \times 10^{-3}\ \text{M}$$

We find the molar absorptivity from Beer's law:

$$\text{Molar absorptivity} = \varepsilon = \frac{A}{bc} = \frac{(0.266)}{(1.00\ \text{cm})(1.32_1 \times 10^{-3}\ \text{M})} = 201._3\ \text{M}^{-1}\ \text{cm}^{-1}$$

(b) A sample of hexane contaminated with benzene had an absorbance of 0.070 at 256 nm in a cuvet with a 5.000-cm pathlength. Find the concentration of benzene in mg/L.

Solution Using Beer's law with the molar absorptivity from part (a), we find

$$[C_6H_6] = \frac{A}{\varepsilon b} = \frac{0.070}{201._3\ \text{M}^{-1}\ \text{cm}^{-1}(5.00\ \text{cm})} = 6.9_5 \times 10^{-5}\ \text{M}$$

$$[C_6H_6] = \left(6.9_5 \times 10^{-5}\ \frac{\text{mol}}{\text{L}}\right)\left(78.11 \times 10^3\ \frac{\text{mg}}{\text{mol}}\right) = 5.4\ \frac{\text{mg}}{\text{L}}$$

Test Yourself 0.10 mM KMnO$_4$ has an absorbance maximum of 0.26 near 525 nm in a 1.000-cm cell. Find the molar absorptivity and the concentration of a solution whose absorbance is 0.52 at 525 nm in the same cell. (**Answer:** 2 600 M^{-1} cm^{-1}, 0.20 mM)

Serum Iron Determination

Iron for biosynthesis is transported through the bloodstream by the protein *transferrin*, whose Fe^{3+}-binding site is shown in Figure 17-7. The following procedure measures the Fe content of transferrin in blood.[12] This analysis requires only about 1 µg of Fe for an accuracy of 2–5%. Human blood usually contains about 45 vol% cells and 55 vol% *plasma* (liquid). If blood is collected without an anticoagulant, the blood clots; the liquid that remains is called *serum*. Serum normally contains about 1 µg of Fe/mL attached to transferrin.

The serum iron determination has three steps:

Step 1 Reduce Fe^{3+} in transferrin to Fe^{2+}, which is released from the protein. Commonly employed reducing agents are hydroxylamine hydrochloride (NH$_3$OH$^+$Cl$^-$), thioglycolic acid, or ascorbic acid.

$$2Fe^{3+} + 2HSCH_2CO_2H \rightarrow 2Fe^{2+} + HO_2CCH_2S\!-\!SCH_2CO_2H + 2H^+$$

Thioglycolic acid

Step 2 Add trichloroacetic acid (Cl$_3$CCO$_2$H) to precipitate proteins, leaving Fe^{2+} in solution. Centrifuge the mixture to remove the precipitate. If protein were left in the solution, it would partially precipitate in the final solution. Light scattered by the precipitate would be mistaken for absorbance.

$$\text{Protein}(aq) \xrightarrow{\text{CCl}_3\text{CO}_2\text{H}} \text{protein}(s)$$

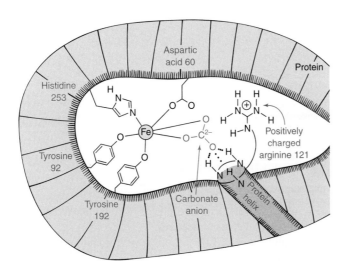

Hydroxylamine hydrochloride

Ascorbic acid
(vitamin C)

FIGURE 17-7 Each of the two Fe-binding sites of transferrin is located in a cleft in the protein. Each site has one nitrogen ligand from the amino acid histidine and three oxygen ligands from tyrosine and aspartic acid. Two oxygen ligands come from a carbonate anion (CO$_3^{2-}$) anchored by electrostatic attraction to positively charged arginine and by hydrogen bonding to the protein helix. When transferrin is taken up by a cell, it is brought into a compartment whose pH is lowered to 5.5. H$^+$ reacts with carbonate to make HCO$_3^-$, thereby releasing Fe^{3+} from the protein. [Adapted from E. N. Baker, B. F. Anderson, H. M. Baker, M. Haridas, G. E. Norris, S. V. Rumball, and C. A. Smith, "Metal and Anion Binding Sites in Lactoferrin and Related Proteins," *Pure Appl. Chem.* **1990**, *62*, 1067.]

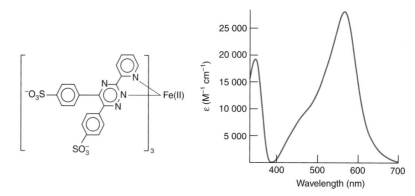

FIGURE 17-8 Visible absorption spectrum of the complex (ferrozine)$_3$Fe(II) used in the colorimetric analysis of iron.

Supernate is the liquid layer above the solid that collects at the bottom of a tube during centrifugation.

Step 3 Transfer a measured volume of supernatant liquid from step 2 to a fresh vessel. Add buffer plus excess ferrozine to form a purple complex. Measure the absorbance at the 562-nm peak (Figure 17-8). Buffer sets a pH at which complex formation is complete.

$$Fe^{2+} + 3\ ferrozine^{2-} \rightarrow (ferrozine)_3Fe^{4-}$$
$$\text{Purple complex}$$
$$\lambda_{max} = 562\ nm$$

The blank described here contains all sources of absorbance other than analyte. An alternative blank for some analyses contains analyte but no color-forming reagent. The choice of blank depends on which species interfere at the analytical wavelength.

In most spectrophotometric analyses, it is important to prepare a **reagent blank** containing all reagents, but with analyte replaced by distilled water. Any absorbance of the blank is due to the color of uncomplexed ferrozine plus the color caused by the iron impurities in the reagents and glassware. *Subtract the blank absorbance from the absorbance of samples and standards before doing any calculations.*

Use a series of iron standards for a *calibration curve* (Figure 17-9) to show that Beer's law is obeyed. Standards are prepared by the same procedure as unknowns. The absorbance of the unknown should fall within the region covered by the standards. Pure iron wire with a shiny, rust-free surface is dissolved in acid to prepare standards (see Appendix K). Ferrous ammonium sulfate (Fe(NH$_4$)$_2$(SO$_4$)$_2$ · 6H$_2$O) and ferrous ethylenediammonium sulfate (Fe(H$_3$NCH$_2$CH$_2$NH$_3$)(SO$_4$)$_2$ · 4H$_2$O) are suitable standards for less accurate work.

If unknowns and standards are prepared with identical volumes, then the quantity of iron in the unknown can be calculated from the least-squares equation for the calibration line. For example, in Figure 17-9, if the unknown has an absorbance of 0.357 (after subtracting the absorbance of the blank), the sample contains 3.59 μg of iron.

In the procedure we just described, results would be about 10% high because serum copper also forms a colored complex with ferrozine. Interference is eliminated if neocuproine or thiourea is added. These reagents **mask** Cu$^+$ by forming strong complexes that prevent Cu$^+$ from reacting with ferrozine.

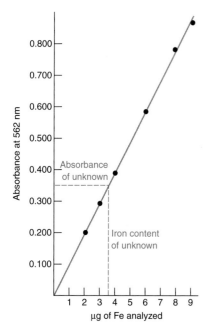

FIGURE 17-9 Calibration curve showing the validity of Beer's law for the (ferrozine)$_3$Fe(II) complex used in serum iron determination. Each sample was diluted to a final volume of 5.00 mL. Therefore, 1.00 μg of iron is equivalent to a concentration of 3.58×10^{-6} M.

Neocuproine
A **masking agent** for copper

Tightly bound complex with low absorbance at 562 nm

EXAMPLE **Serum Iron Analysis**

Serum iron and standard iron solutions were analyzed as follows:

Step 1 To 1.00 mL of sample, add 2.00 mL of reducing agent and 2.00 mL of acid to reduce and release Fe from transferrin.

Step 2 Precipitate proteins with 1.00 mL of 30 wt% trichloroacetic acid. Centrifuge the mixture to remove protein.

Step 3 Transfer 4.00 mL of supernatant liquid to a fresh test tube and add 1.00 mL of solution containing ferrozine and buffer. Measure the absorbance after 10 min.

Step 4 To establish each point on the calibration curve in Figure 17-9, use 1.00 mL of standard containing 2–9 μg Fe in place of serum.

The blank absorbance was 0.038 at 562 nm in a 1.000-cm cell. A serum sample had an absorbance of 0.129. After the blank was subtracted from each standard absorbance, the points in Figure 17-9 were obtained. The least-squares line through the standard points is

$$\text{Absorbance} = 0.067_0 \times (\text{μg Fe in initial sample}) + 0.001_5$$

According to Beer's law, the intercept should be 0, not 0.001_5. We will use the small, observed intercept for our analysis. Find the concentration of iron in the serum.

Solution Rearranging the least-squares equation of the calibration line and inserting the corrected absorbance (observed − blank = 0.129 − 0.038 = 0.091) of unknown, we find

$$\text{μg Fe in unknown} = \frac{\text{absorbance} - 0.001_5}{0.067_0} = \frac{0.091 - 0.001_5}{0.067_0} = 1.33_6 \text{ μg}$$

To find the uncertainty in μg Fe, use Equation 4-27.

To find the uncertainty in μg Fe, use Equation 4-27.

The concentration of Fe in the serum is

$$[\text{Fe}] = \text{moles of Fe/liters of serum}$$
$$= \left(\frac{1.33_6 \times 10^{-6} \text{ g Fe}}{55.845 \text{ g Fe/mol Fe}}\right)/(1.00 \times 10^{-3} \text{ L}) = 2.39 \times 10^{-5} \text{ M}$$

Test Yourself If the observed absorbance is 0.200 and the blank absorbance is 0.049, what is the concentration of Fe (μg/mL) in the serum? (***Answer:*** 2.23 μg/mL)

17-5 Spectrophotometric Titrations

In a **spectrophotometric titration,** we monitor changes in absorbance during a titration to tell when the equivalence point has been reached. A solution of the iron-transport protein, transferrin (Figure 17-7), can be titrated with iron to measure the transferrin content. Transferrin without iron, called apotransferrin, is colorless. Each molecule, with a molecular mass of 81 000, binds two Fe^{3+} ions. When Fe^{3+} binds to the protein, a red color with an absorbance maximum at a wavelength of 465 nm develops. The absorbance is proportional to the concentration of iron bound to the protein. Therefore, absorbance may be used to follow the course of a titration of an unknown amount of apotransferrin with a standard solution of Fe^{3+}.

$$\text{Apotransferrin} + 2Fe^{3+} \rightarrow (Fe^{3+})_2\text{transferrin}$$
(colorless)   (red)

This titration works well for a solution of purified transferrin, but it is not very good for serum because of the background color in serum.

Figure 17-10 shows the titration of 2.000 mL of apotransferrin with 1.79 mM ferric nitrilotriacetate. As iron is added to the protein, red color develops and absorbance increases. When protein is saturated with iron, no more iron can bind and the curve levels off. The extrapolated intersection of the two straight portions of the titration curve at 203 μL in Figure 17-10 is taken as the end point. Absorbance rises slowly after the equivalence point because ferric nitrilotriacetate has some absorbance at 465 nm.

The quantity of Fe^{3+} required for complete reaction in Figure 17-10 is $(203 \times 10^{-6} \text{ L}) \times (1.79 \times 10^{-3} \text{ mol/L}) = 0.363$ μmol. Each protein molecule binds 2 Fe^{3+} ions, so the moles of protein in the sample must be $\frac{1}{2}(0.363$ μmol$) = 0.182$ μmol.

To construct the graph in Figure 17-10, dilution must be considered because the volume is different at each point. Each point on the graph represents the absorbance that would be observed *if the solution had not been diluted from its original volume of 2.000 mL.*

$$\text{Corrected absorbance} = \left(\frac{\text{total volume}}{\text{initial volume}}\right)(\text{observed absorbance}) \quad (17\text{-}7)$$

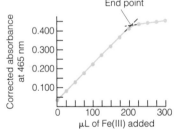

FIGURE 17-10 Spectrophotometric titration of transferrin with ferric nitrilotriacetate. Absorbance is corrected as if no dilution had taken place. The initial absorbance of the solution, before iron is added, is due to a colored impurity.

Ferric nitrilotriacetate is soluble at neutral pH. In the absence of nitrilotriacetate, Fe^{3+} precipitates as $Fe(OH)_3$ in neutral solution. Nitrilotriacetate binds Fe^{3+} through four atoms shown in **bold** type:

Nitrilotriacetate anion

Correcting Absorbance for the Effect of Dilution

The absorbance measured after adding 125 μL (= 0.125 mL) of ferric nitrilotriacetate to 2.000 mL of apotransferrin was 0.260. Calculate the corrected absorbance that should be plotted in Figure 17-10.

Solution The total volume was 2.000 + 0.125 = 2.125 mL. If the volume had been 2.000 mL, the absorbance would have been greater than 0.260 by a factor of 2.125/2.000.

$$\text{Corrected absorbance} = \left(\frac{2.125 \text{ mL}}{2.000 \text{ mL}}\right)(0.260) = 0.276$$

The absorbance plotted in Figure 17-10 is 0.276.

Test Yourself In a different titration, the absorbance after adding 75 μL of ferric nitrilotriacetate to 1.500 mL of apotransferrin was 0.222. Find the corrected absorbance. (*Answer:* 0.233)

17-6 What Happens When a Molecule Absorbs Light?

When a molecule absorbs a photon, the molecule is promoted to a more energetic *excited state* (Figure 17-3). Conversely, when a molecule emits a photon, the energy of the molecule falls by an amount equal to the energy of the photon.

For example, consider formaldehyde in Figure 17-11a. In its ground state, the molecule is planar, with a double bond between carbon and oxygen. From the electron dot description of formaldehyde, we expect two pairs of nonbonding electrons to be localized on the oxygen atom. The double bond consists of a sigma bond between carbon and oxygen and a pi bond made from the $2p_y$ (out-of-plane) atomic orbitals of carbon and oxygen.

Electronic States of Formaldehyde

Molecular orbitals describe the distribution of electrons in a molecule, just as *atomic orbitals* describe the distribution of electrons in an atom. In the molecular orbital diagram for formaldehyde in Figure 17-12, one of the nonbonding orbitals of oxygen is mixed with the three sigma bonding orbitals. These four orbitals, labeled σ_1 through σ_4, are each occupied by a pair of electrons with opposite spin (spin quantum numbers $= +\frac{1}{2}$ and $-\frac{1}{2}$). At higher energy is an occupied pi bonding orbital (π), made of the p_y atomic orbitals of carbon and oxygen. The highest energy occupied orbital is the nonbonding orbital (n), composed principally of the oxygen $2p_x$ atomic orbital. The lowest energy unoccupied orbital is the pi antibonding orbital (π^*). Electrons in this orbital produce repulsion, rather than attraction, between the carbon and oxygen atoms.

In an **electronic transition,** an electron from one molecular orbital moves to another orbital, with a concomitant increase or decrease in the energy of the molecule. The lowest energy electronic transition of formaldehyde promotes a nonbonding (n) electron to the antibonding pi orbital (π^*).[13] There are two possible transitions, depending on the spin quantum numbers in the excited state (Figure 17-13). The state in which the spins are opposed is called a **singlet state.** If the spins are parallel, we have a **triplet state.**

The lowest energy excited singlet and triplet states are called S_1 and T_1, respectively. In general, T_1 has lower energy than S_1. In formaldehyde, the transition $n \rightarrow \pi^*(T_1)$ requires absorption of visible light with a wavelength of 397 nm. The $n \rightarrow \pi^*(S_1)$ transition occurs when ultraviolet radiation with a wavelength of 355 nm is absorbed.

With an electronic transition near 397 nm, you might expect from Table 17-1 that formaldehyde appears green-yellow. In fact, formaldehyde is colorless, because the probability of undergoing a transition between singlet and triplet states (such as $n(S_0) \rightarrow \pi^*(T_1)$) is exceedingly small. Formaldehyde absorbs so little light at 397 nm that our eyes do not detect any absorbance at all. Singlet-to-singlet transitions such as $n(S_0) \rightarrow \pi^*(S_1)$ are much more probable, so the ultraviolet absorption is more intense.

Although formaldehyde is planar in its ground state, it has a pyramidal structure in both the S_1 (Figure 17-11b) and T_1 excited states. Promotion of a nonbonding electron to an antibonding C—O orbital lengthens the C—O bond and changes the molecular geometry.

FIGURE 17-11 Geometry of formaldehyde. (*a*) Ground state. (*b*) Lowest excited singlet state.

A triplet state splits into three slightly different energy levels in a magnetic field, but a singlet state is not split by a magnetic field.

The shorter the wavelength of electromagnetic radiation, the greater the energy.

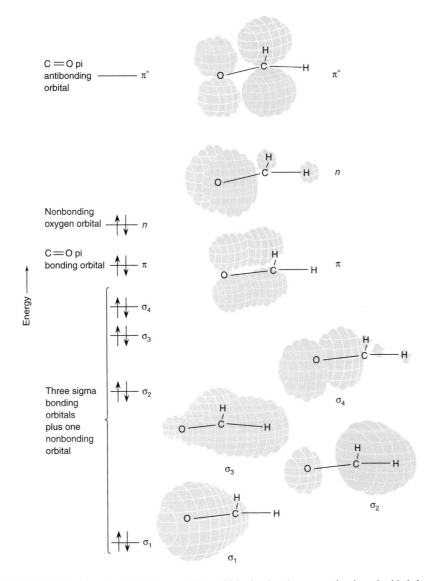

C=O pi antibonding orbital ——— π^*

π^*

n

Nonbonding oxygen orbital —‖— n

C=O pi bonding orbital —‖— π

π

Energy

σ_4

σ_3

Three sigma bonding orbitals plus one nonbonding orbital

σ_2

σ_4

σ_3

σ_2

σ_1

σ_1

FIGURE 17-12 Molecular orbital diagram of formaldehyde, showing energy levels and orbital shapes. The coordinate system of the molecule was shown in Figure 17-11. [From W. L. Jorgensen and L. Salem, *The Organic Chemist's Book of Orbitals* (New York: Academic Press, 1973).]

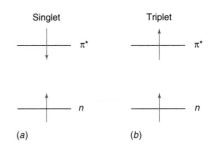

Singlet — π^* Triplet — π^*

n n

(a)   (b)

FIGURE 17-13 Two possible electronic states arise from an $n \rightarrow \pi^*$ transition. (*a*) Excited singlet state, S_1. (*b*) Excited triplet state, T_1.

Symmetric C–H stretching 2 766 cm^{-1}

Asymmetric C–H stretching 2 843 cm^{-1}

C–O stretching 1 746 cm^{-1}

Symmetric bending 1 500 cm^{-1}

Asymmetric bending 1 251 cm^{-1}

Out-of-plane bending 1 167 cm^{-1}

FIGURE 17-14 The six modes of vibrations of formaldehyde. The wavenumber of infrared radiation needed to stimulate each kind of motion is given in units of reciprocal centimeters, cm^{-1}.

A nonlinear molecule with n atoms has $3n - 6$ vibrational modes and three rotations. A linear molecule can rotate about only two axes; it therefore has $3n - 5$ vibrational modes and two rotations.

The C—O stretching vibration of formaldehyde is reduced from 1 746 cm^{-1} in the S_0 state to 1 183 cm^{-1} in the S_1 state because the strength of the C—O bond decreases when the antibonding π^* orbital is populated.

Your microwave oven heats food by transferring rotational energy to water molecules in the food.

Vibrational and Rotational States of Formaldehyde

Absorption of visible or ultraviolet radiation promotes electrons to higher energy orbitals in formaldehyde. Infrared and microwave radiation are not energetic enough to induce electronic transitions, but they can change the vibrational or rotational motion of the molecule.

Each of the four atoms of formaldehyde can move along three axes in space, so the entire molecule can move in $4 \times 3 = 12$ different ways. Three of these motions correspond to translation of the entire molecule in the x, y, and z directions. Another three motions correspond to rotation about x-, y-, and z-axes placed at the center of mass of the molecule. The remaining six motions are vibrations shown in Figure 17-14.

When formaldehyde absorbs an infrared photon with a wavenumber of 1 251 cm^{-1} (= 14.97 kJ/mol), the asymmetric bending vibration in Figure 17-14 is stimulated: Oscillations of the atoms increase in amplitude, and the energy of the molecule increases.

Spacings between rotational energy levels of a molecule are smaller than vibrational energy spacings. A molecule in the rotational ground state could absorb microwave photons with energies of 0.029 07 or 0.087 16 kJ/mol (wavelengths of 4.115 or 1.372 mm) to be promoted to the two lowest excited states. Absorption of microwave radiation causes the molecule to rotate faster than it does in its ground state.

Combined Electronic, Vibrational, and Rotational Transitions

Vibrational transitions usually involve simultaneous rotational transitions. Electronic transitions usually involve simultaneous vibrational and rotational transitions.

In general, when a molecule absorbs light having sufficient energy to cause an electronic transition, **vibrational** and **rotational transitions**—that is, changes in the vibrational and rotational states—occur as well. Formaldehyde can absorb one photon with just the right energy to cause the following simultaneous changes: (1) a transition from the S_0 to the S_1 electronic state; (2) a change in vibrational energy from the ground vibrational state of S_0 to an excited vibrational state of S_1; and (3) a transition from one rotational state of S_0 to a different rotational state of S_1.

Electronic absorption bands are usually broad (Figure 17-8) because many different vibrational and rotational levels are available at slightly different energies. A molecule can absorb photons with a wide range of energies and still be promoted from the ground electronic state to one particular excited electronic state.

What Happens to Absorbed Energy?

Suppose that absorption promotes a molecule from the ground electronic state, S_0, to a vibrationally and rotationally excited level of the excited electronic state S_1 (Figure 17-15). Usually, the first process after absorption is *vibrational relaxation* to the lowest vibrational level of S_1. In this *radiationless* transition, labeled R_1 in Figure 17-15, vibrational energy is transferred to other molecules (solvent, for example) through collisions, not by emission of a photon. The net effect is to convert part of the energy of the absorbed photon into heat spread throughout the entire medium.

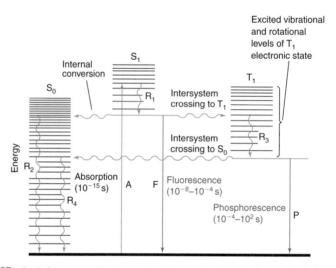

FIGURE 17-15 Physical processes that can occur after a molecule absorbs an ultraviolet or visible photon. S_0 is the ground electronic state. S_1 and T_1 are the lowest excited singlet and triplet electronic states. Straight arrows represent processes involving photons, and wavy arrows are radiationless transitions. R denotes vibrational relaxation. Absorption could terminate in any of the vibrational levels of S_1, not just the one shown. Fluorescence and phosphorescence can terminate in any of the vibrational levels of S_0.

Internal conversion is a radiationless transition between states with the same spin quantum numbers (e.g., $S_1 \rightarrow S_0$).

At the S_1 level, several events can happen. The molecule could enter a highly excited vibrational level of S_0 having the same energy as S_1. This is called *internal conversion* (IC). From this excited state, the molecule can relax back to the ground vibrational state and transfer its energy to neighboring molecules through collisions. This radiationless process is labeled R_2. If a molecule follows the path A–R_1–IC–R_2 in Figure 17-15, the entire energy of the photon will have been converted into heat.

Intersystem crossing is a radiationless transition between states with different spin quantum numbers (e.g., $T_1 \rightarrow S_0$).

Alternatively, the molecule could cross from S_1 into an excited vibrational level of T_1. Such an event is known as *intersystem crossing* (ISC). After the radiationless vibrational relaxation R_3, the molecule finds itself at the lowest vibrational level of T_1. From here, the molecule might undergo a second intersystem crossing to S_0, followed by the radiationless relaxation R_4. All processes mentioned so far simply convert light into heat.

Fluorescence: emission of a photon during a transition between states with the same spin quantum numbers (e.g., $S_1 \rightarrow S_0$).

Phosphorescence: emission of a photon during a transition between states with different spin quantum numbers (e.g., $T_1 \rightarrow S_0$).

A molecule could also relax from S_1 or T_1 to S_0 by emitting a photon. The radiational transition $S_1 \rightarrow S_0$ is called **fluorescence** (Box 17-2), and the radiational transition $T_1 \rightarrow S_0$ is called **phosphorescence.** The relative rates of internal conversion, intersystem crossing, fluorescence, and phosphorescence depend on the molecule, the solvent, and conditions such as temperature and pressure. The energy of phosphorescence is less than the energy of fluorescence, so phosphorescence comes at longer wavelengths than fluorescence (Figure 17-16).

Fluorescence and phosphorescence are relatively rare. Molecules generally decay from the excited state by radiationless transitions. The *lifetime* of fluorescence is 10^{-8} to 10^{-4} s.

FIGURE 17-16 Emission spectra showing that phosphorescence comes at lower energy than fluorescence from the same molecule. The phosphorescence is ~10 times weaker than the fluorescence and is only observed when the sample is cooled. [Data from J. C. Fister, III, J. M. Harris, D. Rank, and W. Wacholtz, "Molecular Photophysics of Acridine Yellow Studied by Phosphorescence and Delayed Fluorescence," *J. Chem. Ed.* **1997**, *74*, 1208.]

The lifetime of phosphorescence is longer (10^{-4} to 10^{2} s) because phosphorescence involves a change in spin quantum numbers (from two unpaired electrons to no unpaired electrons), which is an improbable event. A few materials, such as strontium aluminate doped with europium and dysprosium ($SrAl_2O_4$:Eu:Dy) phosphoresce for *hours* after exposure to light.[15] One application for this material is in signs leading to emergency exits when power is lost.

The *lifetime* of a state is the time needed for the population of that state to decay to 1/e of its initial value, where e is the base of natural logarithms.

BOX 17-2   Fluorescence All Around Us

A fluorescent lamp is a glass tube filled with Hg vapor. The inner walls are coated with a blend of red and green *phosphors* (luminescent substances). The red phosphor is Eu^{3+} doped into Y_2O_3. The green phosphor is Tb^{3+} doped into $CeMgAl_{11}O_{19}$. Hg vapor in the lamp is excited by an electric current and emits ultraviolet radiation at 185 and 254 nm, and a series of visible lines in panel *a*. Hg emission by itself looks blue to our eyes. When the ultraviolet radiation is absorbed by the phosphors, Eu^{3+} emits red light at 612 nm and Tb^{3+} emits green light at 542 nm. The combination of blue, red, and green emissions appears white to our eyes.

(*a*) Atomic emission spectrum of Hg vapor. [From S. R. Goode and L. A. Metz, "Emission Spectroscopy in the Undergraduate Laboratory," *J. Chem. Ed.* **2003**, *80*, 1455.]
(*b*) Emission spectra of phosphors scraped from the inside of a compact fluorescent lamp. Tb^{3+} is selectively excited at 280 nm and Eu^{3+} is selectively excited at 240 nm. [From C. Degli Esposti and L. Bizzochi, "The Radiative Decay of Green and Red Photoluminescent Phosphors," *J. Chem. Ed.* **2008**, *85*, 839.]

A 13-W fluorescent light that fits in a standard screw-in socket provides the same light as the 60-W incandescent bulb it replaces. The expected lifetime of the fluorescent lamp is 10 000 h, whereas that of the incandescent bulb is 750 h. The fluorescent light is more expensive than the incandescent bulb but saves a great deal of electricity and money over its lifetime.

Most white fabrics also fluoresce. Just for fun, turn on an ultraviolet lamp in a darkened room containing several people (*but do not look directly at the lamp*). You will discover emission from white fabrics (shirts, pants, shoelaces, unmentionables) containing fluorescent compounds to enhance whiteness. You might see fluorescence from teeth and from recently bruised areas of skin that show no surface damage. Many demonstrations of fluorescence and phosphorescence have been described.[14]

A fluorescent whitener from laundry detergent

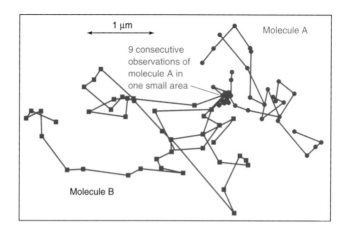

FIGURE 17-17 Tracks of two molecules of 20 pM Rhodamine 6G in silica gel observed by fluorescence integrated over 0.20-s periods at 0.78-s intervals. Some points are not connected, because the molecule disappeared above or below the focal plane in the 0.45-μm thick film and was not observed in a particular observation interval. In the nine periods when molecule A was in one location, it might have been adsorbed to a particle of silica. An individual molecule emits thousands of photons in 0.2 s as the molecule cycles between the ground and excited state. Only a fraction of these photons reach the detector, which generates a burst of ~10–50 electrons. [From K. S. McCain, D. C. Hanley, and J. M. Harris, "Single-Molecule Fluorescence Trajectories for Investigating Molecular Transport in Thin Silica Sol-Gel Films," *Anal. Chem.* **2003**, *75*, 4351.]

Rhodamine 6G

17-7 Luminescence

Fluorescence and phosphorescence are examples of **luminescence,** which is emission of light from an excited state of a molecule. Luminescence is inherently more sensitive than absorption. Imagine yourself in a stadium at night; the lights are off, but each of the 50 000 raving fans is holding a lighted candle. If 500 people blow out their candles, you will hardly notice the difference. Now imagine that the stadium is completely dark; then 500 people suddenly light their candles. In this case, the effect is dramatic. The first example is analogous to changing transmittance from 100% to 99%, which is equivalent to an absorbance of $-\log 0.99 = 0.004\ 4$. It is hard to measure such a small absorbance because the background is so bright. The second example is analogous to observing fluorescence from 1% of the molecules in a sample. Against a black background, fluorescence is striking.

Luminescence is sensitive enough to observe *single molecules*.[16] Figure 17-17 shows *observed* tracks of two molecules of the highly fluorescent Rhodamine 6G at 0.78-s intervals in a thin layer of silica gel. These observations confirm the "random walk" of diffusing molecules postulated by Albert Einstein in 1905.

Relation Between Absorption and Emission Spectra

Figure 17-15 shows that fluorescence and phosphorescence come at lower energy than absorption (the *excitation* energy). That is, molecules emit radiation at longer wavelengths than the radiation they absorb. Examples are shown in Figure 17-18 and Color Plate 16.

Figure 17-19 explains why emission comes at lower energy than absorption and why the emission spectrum is roughly the mirror image of the absorption spectrum. In absorption, wavelength λ_0 corresponds to a transition from the ground vibrational level of S_0 to the lowest vibrational level of S_1. Absorption maxima at higher energy (shorter wavelength) correspond to the S_0 to S_1 transition accompanied by absorption of one or more quanta of vibrational

FIGURE 17-18 Absorption (black line) and emission (colored line) of bis(benzylimido)-perylene in dichloromethane solution, illustrating the approximate mirror image relation between absorption and emission. The 10^{-11} M solution used for emission had an average of just 10 analyte molecules in the volume probed by the 514-nm excitation laser. [From P. J. G. Goulet, N. P. W. Pieczonka, and R. F. Aroca, "Overtones and Combinations in Single-Molecule Surface-Enhanced Resonance Raman Scattering Spectra," *Anal. Chem.* **2003**, *75*, 1918.]

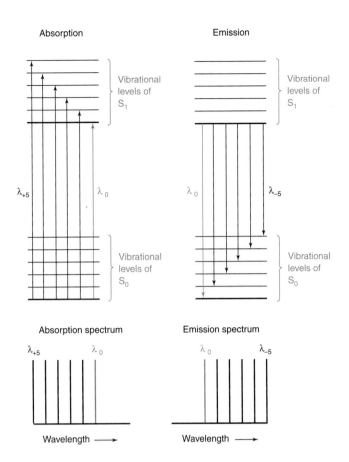

Absorption

Emission

Vibrational
levels of
S_1

Vibrational
levels of
S_1

λ_{+5} λ_0 λ_0 λ_{-5}

Vibrational
levels of
S_0

Vibrational
levels of
S_0

Absorption spectrum

Emission spectrum

λ_{+5} λ_0 λ_0 λ_{-5}

Wavelength ⟶   Wavelength ⟶

FIGURE 17-19 Energy-level diagram showing why structure is seen in the absorption and emission spectra and why the spectra are roughly mirror images of each other. In absorption, wavelength λ_0 comes at lowest energy, and λ_{+5} is at highest energy. In emission, wavelength λ_0 comes at highest energy, and λ_{-5} is at lowest energy.

energy. In polar solvents, vibrational structure is usually broadened beyond recognition, and only a broad envelope of absorption is observed. In less polar or nonpolar solvents, vibrational structure is observed.

After absorption, the vibrationally excited S_1 molecule relaxes back to the lowest vibrational level of S_1 prior to emitting any radiation. Emission from S_1 in Figure 17-19 can go to any vibrational level of S_0. The highest energy transition comes at wavelength λ_0, with a series of peaks following at longer wavelength. Absorption and emission spectra will have an approximate mirror image relation if the spacings between vibrational levels are roughly equal and if the transition probabilities are similar.

The λ_0 transitions in Figure 17-18 (and later in Figure 17-22) do not exactly overlap. In the emission spectrum, λ_0 comes at slightly lower energy than in the absorption spectrum. The reason is seen in Figure 17-20. A molecule absorbing radiation is initially in its electronic ground state, S_0. This molecule possesses a certain geometry and solvation. Suppose that the excited state is S_1. The electronic transition is faster than the vibrational motion of atoms or the translational motion of solvent molecules. When radiation is first absorbed, the excited S_1 molecule still possesses its S_0 geometry and solvation. Shortly after excitation, the geometry and solvation change to their most favorable values for the S_1 state. This rearrangement lowers the energy of the excited molecule. When an S_1 molecule fluoresces, it returns to the S_0 state with S_1 geometry and solvation. This unstable configuration must have a higher energy than that of an S_0 molecule with S_0 geometry and solvation. The net effect in Figure 17-20 is that the λ_0 emission energy is less than the λ_0 excitation energy.

Electronic transitions are fast, relative to nuclear motion, so each atom has nearly the same position and momentum before and after a transition. This is called the *Franck-Condon principle.*

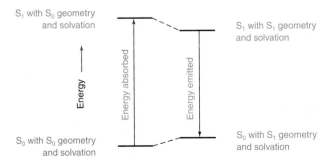

S_1 with S_0 geometry
and solvation

S_1 with S_1 geometry
and solvation

Energy ⟶

Energy absorbed

Energy emitted

S_0 with S_0 geometry
and solvation

S_0 with S_1 geometry
and solvation

FIGURE 17-20 Diagram showing why the λ_0 transitions do not exactly overlap in Figures 17-18 and 17-22.

Solution-phase spectra are broadened because absorbing molecules are surrounded by solvent molecules with a variety of orientations that create slightly different energy levels for different absorbing molecules. Simple gas-phase molecules, which are not in close contact with neighbors and which have a limited number of energy levels, have extremely sharp absorptions. Individual rotational transitions of gaseous $H_2{}^{16}O$, $H_2{}^{17}O$, and $H_2{}^{18}O$ with line widths of 0.02 cm^{-1} are readily distinguished even though they are just 0.2 cm^{-1} apart.

Excitation and Emission Spectra

An emission experiment is shown in Figure 17-21. An excitation wavelength (λ_{ex}) is selected by one monochromator, and luminescence is observed through a second monochromator, usually positioned at 90° to the incident light to minimize the intensity of scattered light reaching the detector. If we hold the excitation wavelength fixed and scan through the emitted radiation, an **emission spectrum** such as Figure 17-16 is produced. An emission spectrum is a graph of emission intensity versus emission wavelength.

FIGURE 17-21 Essentials of a luminescence experiment. The sample is irradiated at one wavelength and emission is observed over a range of wavelengths. The excitation monochromator selects the excitation wavelength (λ_{ex}) and the emission monochromator selects one wavelength at a time (λ_{em}) to observe.

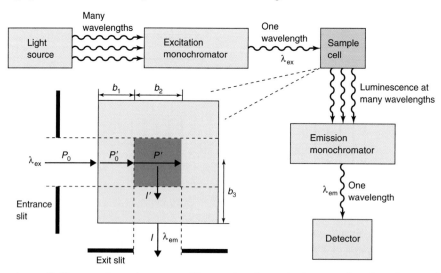

Emission spectrum: constant λ_{ex} and variable λ_{em}

Excitation spectrum: variable λ_{ex} and constant λ_{em}

An **excitation spectrum** is measured by varying the excitation wavelength and measuring emitted light at one particular wavelength (λ_{em}). An excitation spectrum is a graph of emission intensity versus excitation wavelength (Figure 17-22). *An excitation spectrum looks very much like an absorption spectrum* because, the greater the absorbance at the excitation wavelength, the more molecules are promoted to the excited state and the more emission will be observed.

In emission spectroscopy, we measure emitted radiation, rather than the fraction of incident radiation striking the detector. Detector response varies with wavelength, so the recorded emission spectrum is not a true profile of emitted irradiance versus emission wavelength. For analytical measurements employing a single emission wavelength, this effect is inconsequential. If a true profile is required, it is necessary to calibrate the detector response at each wavelength.

FIGURE 17-22 Excitation and emission spectra of anthracene have the same mirror image relationship as the absorption and emission spectra in Figure 17-18. An excitation spectrum is nearly the same as an absorption spectrum. [From C. M. Byron and T. C. Werner, "Experiments in Synchronous Fluorescence Spectroscopy for the Undergraduate Instrumental Chemistry Course," *J. Chem. Ed.* **1991,** *68,* 433.]

BOX 17-3   Rayleigh and Raman Scattering

Emission spectra can exhibit confusing features in addition to fluorescence and phosphorescence. The chart shows an emission spectrum of aqueous dichlorofluorescein (colored line) and, for reference, an emission spectrum of pure water (black line). The excitation wavelength is 400 nm. The only difference between the two traces is fluorescence from dichlorofluorescein with a peak at 522 nm.

Upper trace: Emission spectrum of aqueous dichlorofluorescein. *Lower trace:* Spectrum observed from pure water. [Courtesy Kris Varazo, Francis Marion University. See R. J. Clark and A. Oprysa, "Fluorescence and Light Scattering," *J. Chem. Ed.* **2004**, *81*, 705.]

What are the other peaks? The strongest peak, which is off-scale, is observed at the excitation wavelength of 400 nm. It is called **Rayleigh scattering** after the same Lord Rayleigh (J. W. Strutt) who discovered argon (page 76). The oscillating electromagnetic field of the excitation light source causes electrons in water molecules to oscillate at the same frequency as the incident radiation. Oscillating electrons emit this same frequency of radiation in all directions. The time required for scattering is essentially the period of one oscillation of the incoming electromagnetic wave, which is ~10^{-15} s for 400-nm light. By comparison, the time for fluorescence is ~10^{-8} to 10^{-4} s. Rayleigh scattering is always present and is usually filtered out so that it is not displayed in the emission spectrum.

The second strongest peak in this example comes at 800 nm, which is exactly twice the excitation wavelength. It is an artifact of the monochromator. Grating monochromators designed to pass wavelength λ also pass integer fractions $\lambda/2$, $\lambda/3$, and so on, with decreasing efficiency. When the emission monochromator in Figure 17-21 is set to pass 800 nm, it also passes some light at 400 nm. Rayleigh scattering at 400 nm passes through the monochromator set to 800 nm. We call this *second-order diffraction* from the monochromator. If we used a filter to block 400-nm light between the sample cell and the emission monochromator in Figure 17-21, there would be no peak at 800 nm in the emission spectrum.

A weak, but reproducible, peak is observed in both water and dichlorofluorescein solution at 462 nm. The difference in energy between incident light at 400 nm and the peak at 462 nm corresponds to a vibrational energy of H_2O. The peak at 462 nm is called **Raman scattering** after the Indian physicist C. V. Raman, who discovered this phenomenon in 1928 and was awarded the Nobel Prize in 1930. In this type of scattering, which also occurs in a time frame of ~10^{-15} s, a small fraction of incident photons gives up one quantum of molecular vibrational energy to H_2O. Scattered radiation emerges with less energy than the excitation energy. Vibrational energy is customarily expressed as the wavenumber (cm^{-1}) of a photon with that energy. Liquid H_2O has a broad range of vibrational energies centered near 3 404 cm^{-1}. The wavenumber of the exciting radiation is 1/wavelength = 1/400 nm = 25 000 cm^{-1}. In Raman scattering, an incident photon with energy of 25 000 cm^{-1} gives up 3 404 cm^{-1} and emerges at (25 000 – 3 404) = 21 596 cm^{-1}. The wavelength is 1/(21 596 cm^{-1}) $\approx$ 463 nm. The observed peak is at 462 nm.

What are the lessons from this example? First, compare the spectrum of pure solvent to the spectrum of a sample under study so that you can disregard peaks due to solvent. Second, fluorescence comes at a fixed location, such as 522 nm for dichlorofluorescein. The wavelength of scattered radiation varies with incident wavelength. If we used an excitation wavelength of 410 nm instead of 400, the second-order grating line would be seen at 820 nm and the water Raman line would be at an energy that is 3 404 cm^{-1} less than the exciting light, or 477 nm. Scattered radiation shifts with the incident wavelength, but fluorescence and phosphorescence do not.

Box 17-3 describes common forms of light scattering that can be confused with emission when you are interpreting an experimental spectrum.

Luminescence Intensity

A simplified view of processes occurring during a luminescence measurement is shown in the enlarged sample cell at the lower left of Figure 17-21. We expect emission to be proportional to the irradiance absorbed by the sample. In Figure 17-21, the irradiance (W/m^2) incident on the sample cell is P_0. Some is absorbed over the pathlength b_1, so the irradiance striking the central region of the cell is

$$\text{Irradiance striking central region} = P_0' = P_0 \cdot 10^{-\varepsilon_{ex}b_1c} \qquad (17\text{-}8)$$

where ε_{ex} is the molar absorptivity at the wavelength λ_{ex} and c is the concentration of analyte. The irradiance of the beam when it has traveled the additional distance b_2 is

$$P' = P_0' \cdot 10^{-\varepsilon_{ex}b_2c} \qquad (17\text{-}9)$$

Emission intensity I is proportional to irradiance absorbed in the central region of the cell:

$$\text{Emission intensity} = I' = k'(P_0' - P') \qquad (17\text{-}10)$$

Equation 17-8 comes from Equations 17-5 and 17-6. If species other than analyte absorbed at the wavelengths of interest, we would have to include them.

where k' is a constant of proportionality. Not all radiation emitted from the center of the cell in the direction of the exit slit is observed. Some is absorbed between the center and the edge of the cell. Emission intensity I emerging from the cell is given by Beer's law:

$$I = I' \cdot 10^{-\varepsilon_{em}b_3c} \tag{17-11}$$

where ε_{em} is the molar absorptivity at the emission wavelength and b_3 is the distance from the center to the edge of the cell.

Combining Equations 17-10 and 17-11 gives an expression for emission intensity:

$$I = k'(P_0' - P')10^{-\varepsilon_{em}b_3c}$$

Substituting expressions for P_0' and P' from Equations 17-8 and 17-9, we obtain a relation between incident irradiance and emission intensity:

$$I = k'(P_0 \cdot 10^{-\varepsilon_{ex}b_1c} - P_0 \cdot 10^{-\varepsilon_{ex}b_1c} \cdot 10^{-\varepsilon_{ex}b_2c})10^{-\varepsilon_{ex}b_3c}$$
$$= k'P_0 \cdot \underbrace{10^{-\varepsilon_{ex}b_1c}}_{} \underbrace{(1 - 10^{-\varepsilon_{ex}b_2c})}_{} \underbrace{10^{-\varepsilon_{em}b_3c}}_{} \tag{17-12}$$

Loss of intensity Absorption Loss of intensity
in region 1   in region 2   in region 3

Consider the limit of low concentration, which means that the exponents $\varepsilon_{ex}b_1c$, $\varepsilon_{ex}b_2c$, and $\varepsilon_{em}b_3c$ are all very small. The terms $10^{-\varepsilon_{ex}b_1c}$, $10^{-\varepsilon_{ex}b_2c}$, and $10^{-\varepsilon_{em}b_3c}$ are all close to unity. We can replace $10^{-\varepsilon_{ex}b_1c}$ and $10^{-\varepsilon_{em}b_3c}$ with 1 in Equation 17-12. We cannot replace $10^{-\varepsilon_{ex}b_2c}$ with 1, because we are subtracting this term from 1 and would be left with 0. Instead, we expand $10^{-\varepsilon_{ex}b_2c}$ in a power series:

The series 17-13 follows from the relation $10^{-A} = (e^{\ln 10})^{-A} = e^{-A \ln 10}$ and the expansion of e^x:

$$e^x = 1 + \frac{x^1}{1!} + \frac{x^2}{2!} + \frac{x^3}{3!} + \cdots$$

$$10^{-\varepsilon_{ex}b_2c} = 1 - \varepsilon_{ex}b_2c \ln 10 + \frac{(\varepsilon_{ex}b_2c \ln 10)^2}{2!} - \frac{(\varepsilon_{ex}b_2c \ln 10)^3}{3!} + \cdots \tag{17-13}$$

Each term of 17-13 becomes smaller and smaller, so we just keep the first two terms. The central factor in Equation 17-12 becomes $(1 - 10^{-\varepsilon_{ex}b_2c}) = (1 - [1 - \varepsilon_{ex}b_2c \ln 10]) = \varepsilon_{ex}b_2c \ln 10$, so the entire equation can be written

Emission intensity at low concentration: $k'P_0(\varepsilon_{ex}b_2c \ln 10) =$ $\boxed{I = kP_0c}$ (17-14)

where $k = k'\varepsilon_{ex}b_2 \ln 10$ is a constant.

Equation 17-14 says that, *at low concentration, emission intensity is proportional to analyte concentration.* Data for anthracene in Figure 17-23 are linear below 10^{-6} M. Blank samples invariably scatter light and must be run in every analysis. Equation 17-14 tells us that doubling the incident irradiance (P_0) will double the emission intensity (up to a point). In contrast, doubling P_0 has no effect on absorbance, which is a *ratio* of two intensities. The sensitivity of a luminescence measurement can be increased by more than a factor of 3 by the simple expedient of using a mirror coating on the two walls of the sample cell opposite the slits in Figure 17-21.[17]

For higher concentrations, we need all the terms in Equation 17-12, or we need an even more accurate equation.[18] As concentration increases, maximum emission is reached. Then emission decreases because absorption increases more rapidly than the emission. We say the emission is *quenched* (decreased) by **self-absorption,** which is the absorption of excitation or emission energy by analyte molecules in the solution. Quenching by self-absorption is also called the *inner filter effect.* At high concentration, even the *shape* of the emission spectrum can change, because absorption and emission both depend on wavelength.

Example: Fluorimetric Assay of Selenium in Brazil Nuts

Selenium is a trace element essential to life. For example, the selenium-containing enzyme glutathione peroxidase catalyzes the destruction of peroxides (ROOH) that are harmful to cells. Conversely, at high concentration, selenium can be toxic.

To measure selenium in Brazil nuts, 0.1 g of nut is digested with 2.5 mL of 70 wt% HNO_3 in a Teflon *bomb* (Figure 27-7) in a microwave oven. Hydrogen selenate (H_2SeO_4) in the digest is reduced to hydrogen selenite (H_2SeO_3) with hydroxylamine (NH_2OH). Selenite is then **derivatized** to form a fluorescent product that is extracted into cyclohexane.

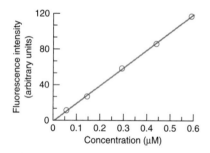

FIGURE 17-23 Linear calibration curve for fluorescence of anthracene measured at the wavelength of maximum fluorescence in Figure 17-22. [From C. M. Byron and T. C. Werner, "Experiments in Synchronous Fluorescence Spectroscopy for the Undergraduate Instrumental Chemistry Course," *J. Chem. Ed.* **1991,** *68,* 433.]

Derivatization is the chemical alteration of analyte so that it can be detected easily or separated easily from other species.

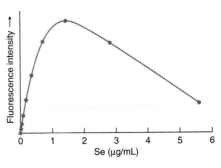

$$\text{2,3-Diaminonaphthalene} + H_2SeO_3 \xrightarrow[50°C]{pH = 2} \text{Fluorescent product} + 3H_2O \quad \text{(17-15)}$$

Maximum response of the fluorescent product was observed with an excitation wavelength of 378 nm and an emission wavelength of 518 nm in Figure 17-24. Emission is proportional to concentration only up to ~0.1 μg Se/mL. Beyond 0.1 μg Se/mL, the response becomes curved, eventually reaches a maximum, and finally *decreases* with increasing concentration as self-absorption dominates. This behavior is predicted by Equation 17-12.

FIGURE 17-24 Fluorescence calibration curve for the selenium-containing product in Reaction 17-15. The curvature and maximum are due to self-absorption. [From M.-C. Sheffield and T. M. Nahir, "Analysis of Selenium in Brazil Nuts by Microwave Digestion and Fluorescence Detection," *J. Chem. Ed.* **2002**, *79*, 1345.]

Luminescence in Analytical Chemistry[19]

Some analytes, such as riboflavin (vitamin B_2)[20] and polycyclic aromatic compounds (an important class of carcinogens), are naturally fluorescent and can be analyzed directly. Most compounds are not luminescent. However, coupling to a fluorescent moiety provides a route to sensitive analyses. *Fluorescein* is a strongly fluorescent compound that can be coupled to many molecules for analytical purposes. Fluorescent labeling of fingerprints is a powerful tool in forensic analysis.[21] Sensor molecules whose luminescence responds selectively to a variety of simple cations and anions are available.[22] Ca^{2+} can be measured from the fluorescence of a complex it forms with a derivative of fluorescein called calcein.

Fluorescein

Chelating aminodiacetate group

Calcein

Molecular biologists use *DNA microarrays* ("gene chips") to monitor gene expression and mutations and to detect and identify pathological microorganisms.[23] A single chip can contain thousands of known single-strand DNA sequences in known locations. The chip is incubated with unknown single-strand DNA that has been tagged with fluorescent labels. After the unknown DNA has bound to its complementary strands on the chip, the amount bound to each spot on the chip is measured by fluorescence intensity.

Light from a firefly or light stick[24] is an example of **chemiluminescence**—emission of light from a chemical reaction.[25] Chemiluminescence detectors for sulfur and nitrogen in organic compounds are employed in gas chromatography (Section 23-3). Nitric oxide (NO), which transmits signals between living cells, can be measured at parts per billion levels by its chemiluminescent reaction with the compound luminol.[26] Other biological analytes measurable by chemiluminescence include Ca^{2+} in mitochondria[27] and endocrine-disrupting compounds in municipal wastewater.[28]

Luminol
(5-amino-2,3-dihydro-1,4-phthalazinedione)

oxidant (such as NO or H_2O_2)

OH^-, metal catalyst

$+ N_2 +$ blue light

Terms to Understand

absorbance	excitation spectrum	molecular orbital	self-absorption
absorption spectrum	excited state	monochromatic light	singlet state
Beer's law	fluorescence	monochromator	spectrophotometric titration
chemiluminescence	frequency	phosphorescence	spectrophotometry
chromophore	ground state	photon	transmittance
cuvet	hertz	Raman scattering	triplet state
derivatization	irradiance	Rayleigh scattering	vibrational transition
electromagnetic spectrum	luminescence	reagent blank	wavelength
electronic transition	masking	refractive index	wavenumber
emission spectrum	molar absorptivity	rotational transition	

Summary

Light can be thought of as waves whose wavelength (λ) and frequency (ν) have the important relation $\lambda\nu = c$, where c is the speed of light. Alternatively, light may be viewed as consisting of photons whose energy (E) is given by $E = h\nu = hc/\lambda = hc\tilde{\nu}$, where h is Planck's constant and $\tilde{\nu}(= 1/\lambda)$ is the wavenumber. Absorption of light is commonly measured by absorbance (A) or transmittance (T), defined as $A = \log(P_0/P)$ and $T = P/P_0$, where P_0 is the incident irradiance and P is the exiting irradiance. Absorption spectroscopy is useful in quantitative analysis because absorbance is proportional to the concentration of the absorbing species in dilute solution (Beer's law): $A = \varepsilon bc$. In this equation, b is pathlength, c is concentration, and ε is the molar absorptivity (a constant of proportionality).

Basic components of a spectrophotometer include a radiation source, a monochromator, a sample cell, and a detector. To minimize errors in spectrophotometry, samples should be free of particles, cuvets must be clean, and they should be positioned reproducibly in the sample holder. Measurements should be made at a wavelength of maximum absorbance. Instrument errors tend to be minimized if the absorbance falls in the range $A \approx 0.3$–2.

In a spectrophotometric titration, absorbance is monitored as titrant is added. For many reactions, there is an abrupt change in slope when the equivalence point is reached.

When a molecule absorbs light, it is promoted to an excited state from which it returns to the ground state by radiationless processes or by fluorescence (singlet → singlet emission) or phosphorescence (triplet → singlet emission). Emission intensity is proportional to concentration at low concentration. At sufficiently high concentration, emission decreases because of self-absorption by the analyte. An excitation spectrum (a graph of emission intensity versus excitation wavelength) is similar to an absorption spectrum (a graph of absorbance versus wavelength). An emission spectrum (a graph of emission intensity versus emission wavelength) is observed at lower energy than the absorption spectrum and tends to be the mirror image of the absorption spectrum. A molecule that is not fluorescent can be analyzed by attaching a fluorescent group to it. Light emitted by a chemical reaction—chemiluminescence—is also used for quantitative analysis.

Exercises

17-A. (a) What value of absorbance corresponds to 45.0% T?

(b) If a 0.010 0 M solution exhibits 45.0% T at some wavelength, what will be the percent transmittance for a 0.020 0 M solution of the same substance?

17-B. (a) A 3.96×10^{-4} M solution of compound A exhibited an absorbance of 0.624 at 238 nm in a 1.000-cm cuvet; a blank solution containing only solvent had an absorbance of 0.029 at the same wavelength. Find the molar absorptivity of compound A.

(b) The absorbance of an unknown solution of compound A in the same solvent and cuvet was 0.375 at 238 nm. Find the concentration of A in the unknown.

(c) A concentrated solution of compound A in the same solvent was diluted from an initial volume of 2.00 mL to a final volume of 25.00 mL and then had an absorbance of 0.733. What is the concentration of A in the concentrated solution?

17-C. Ammonia can be determined spectrophotometrically by reaction with phenol in the presence of hypochlorite (OCl^-):

2 [phenol]—OH + NH₃ →(OCl⁻)→ O=[ring]=N—[ring]—O⁻

Phenol (colorless)   Ammonia (colorless)   Blue product, $\lambda_{max} = 625$ nm

A 4.37-mg sample of protein was chemically digested to convert its nitrogen into ammonia and then diluted to 100.0 mL. Then 10.0 mL of the solution were placed in a 50-mL volumetric flask and treated with 5 mL of phenol solution plus 2 mL of sodium hypochlorite solution. The sample was diluted to 50.0 mL, and the absorbance at 625 nm was measured in a 1.00-cm cuvet after 30 min. For reference, a standard solution was prepared from 0.010 0 g of NH₄Cl (FM 53.49) dissolved in 1.00 L of water. Then 10.0 mL of this standard were placed in a 50-mL volumetric flask and analyzed in the same manner as the unknown. A reagent blank was prepared by using distilled water in place of unknown.

Sample	Absorbance at 625 nm
Blank	0.140
Reference	0.308
Unknown	0.592

(a) Calculate the molar absorptivity of the blue product.

(b) Calculate the weight percent of nitrogen in the protein.

17-D. Cu^+ reacts with neocuproine to form the colored complex (neocuproine)$_2$Cu$^+$, with an absorption maximum at 454 nm. Neocuproine is particularly useful because it reacts with few other metals. The copper complex is soluble in 3-methyl-1-butanol (isoamyl alcohol), an organic solvent that does not dissolve appreciably in water. In other words, when isoamyl alcohol is added to water, a two-layered mixture results, with the denser water layer at the bottom. If (neocuproine)$_2$Cu$^+$ is present, virtually all of it goes into the organic phase. For the purpose of this problem, assume that the isoamyl alcohol does not dissolve in the water at all and that all of the colored complex will be in the organic phase. Suppose that the following procedure is carried out:

1. A rock containing copper is pulverized, and all metals are extracted from it with strong acid. The acidic solution is neutralized with base and made up to 250.0 mL in flask A.

2. Next, 10.00 mL of the solution are transferred to flask B and treated with 10.00 mL of reducing agent to convert Cu^{2+} to Cu^+. Then 10.00 mL of buffer are added so that the pH is suitable for complex formation with neocuproine.

3. 15.00 mL of this solution are withdrawn and placed in flask C. To the flask are added 10.00 mL of an aqueous solution containing neocuproine and 20.00 mL of isoamyl alcohol. After the mixture

Transfer all liquid to flask   Transfer 10.00 mL   Transfer 15.00 mL

Rock

Grind rock to fine powder and extract Cu from powder with strong acid

Flask A 250.0 mL

Add 20.00 mL of reagents

Flask B

Add 10.00 mL of aqueous reagent + 20.00 mL organic solvent

Flask C

20.00 mL organic phase containing all Cu-neocuproine

25.00 mL aqueous phase containing no Cu

has been shaken well and the phases allowed to separate, all (neocuproine)$_2$Cu$^+$ is in the organic phase.

4. A few milliliters of the upper layer are withdrawn, and the absorbance at 454 nm is measured in a 1.00-cm cell. A blank carried through the same procedure gives an absorbance of 0.056.

(a) Suppose that the rock contained 1.00 mg of Cu. What will be the concentration of Cu (mol/L) in the isoamyl alcohol phase?

(b) If the molar absorptivity of (neocuproine)$_2$Cu$^+$ is 7.90×10^3 M^{-1} cm^{-1}, what will be the observed absorbance? Remember that a blank carried through the same procedure gave an absorbance of 0.056.

(c) A rock is analyzed and found to give a final absorbance of 0.874 (uncorrected for the blank). How many milligrams of Cu are in the rock?

17-E. Semi-xylenol orange is a yellow compound at pH 5.9 but turns red when it reacts with Pb^{2+}. A 2.025-mL sample of semi-xylenol orange at pH 5.9 was titrated with 7.515×10^{-4} M Pb(NO$_3$)$_2$, with the following results:

Total μL Pb^{2+} added	Absorbance at 490-nm wavelength	Total μL Pb^{2+} added	Absorbance at 490-nm wavelength
0.0	0.227	42.0	0.425
6.0	0.256	48.0	0.445
12.0	0.286	54.0	0.448
18.0	0.316	60.0	0.449
24.0	0.345	70.0	0.450
30.0	0.370	80.0	0.447
36.0	0.399		

Make a graph of absorbance versus microliters of Pb^{2+} added. Be sure to correct the absorbances for dilution. Corrected absorbance is what would be observed if the volume were not changed from its initial value of 2.025 mL. Assuming that the reaction of semi-xylenol orange with Pb^{2+} has a 1:1 stoichiometry, find the molarity of semi-xylenol orange in the original solution.

Problems

Properties of Light

17-1. Fill in the blanks.

(a) If you double the frequency of electromagnetic radiation, you _____ the energy.

(b) If you double the wavelength, you _____ the energy.

(c) If you double the wavenumber, you _____ the energy.

17-2. (a) How much energy (in kilojoules) is carried by one mole of photons of red light with $\lambda = 650$ nm?

(b) How many kilojoules are carried by one mole of photons of violet light with $\lambda = 400$ nm?

17-3. Calculate the frequency (Hz), wavenumber (cm^{-1}), and energy (J/photon and J/[mol of photons]) of visible light with a wavelength of 562 nm.

17-4. Which molecular processes correspond to the energies of microwave, infrared, visible, and ultraviolet photons?

17-5. Characteristic orange light produced by sodium in a flame is due to an intense emission called the sodium D line, which is actually a doublet, with wavelengths (measured in vacuum) of 589.157 88 and 589.755 37 nm. The index of refraction of air at a wavelength near 589 nm is 1.000 292 6. Calculate the frequency, wavelength, and wavenumber of each component of the D line, measured in air.

Absorption of Light and Measuring Absorbance

17-6. Explain the difference between transmittance, absorbance, and molar absorptivity. Which one is proportional to concentration?

17-7. What is an absorption spectrum?

17-8. Why does a compound whose visible absorption maximum is at 480 nm (blue-green) appear to be red?

17-9. *Color and absorption spectra.* Color Plate 15 shows colored solutions and their spectra. From Table 17-1, predict the color of each solution from the wavelength of maximum absorption. Do observed colors agree with predicted colors?

17-10. Why is it most accurate to measure absorbances in the range $A = 0.3$–2?

17-11. The absorbance of a 2.31×10^{-5} M solution of a compound is 0.822 at a wavelength of 266 nm in a 1.00-cm cell. Calculate the molar absorptivity at 266 nm.

17-12. What color would you expect to observe for a solution of Fe(ferrozine)$_3^{4-}$, which has a visible absorbance maximum at 562 nm?

17-13. When I was a boy, Uncle Wilbur let me watch as he analyzed the iron content of runoff from his banana ranch. A 25.0-mL sample was acidified with nitric acid and treated with excess KSCN to form a red complex. (KSCN itself is colorless.) The solution was then diluted to 100.0 mL and put in a variable-pathlength cell. For comparison, a 10.0-mL reference sample of 6.80×10^{-4} M Fe^{3+} was treated with HNO_3 and KSCN and diluted to 50.0 mL. The reference was placed in a cell with a 1.00-cm light path. The runoff sample exhibited the same absorbance as the reference when the pathlength of the runoff cell was 2.48 cm. What was the concentration of iron in Uncle Wilbur's runoff?

17-14. The *absorption cross section* on the ordinate of the ozone absorption spectrum at the beginning of this chapter is defined by the relation

$$\text{Transmittance } (T) = e^{-n\sigma b}$$

where n is the number of absorbing molecules per cubic centimeter, σ is the absorption cross section (cm^2), and b is the pathlength (cm). The total ozone in the atmosphere is approximately 8×10^{18} molecules above each square centimeter of Earth's surface (from the surface up to the top of the atmosphere). If this were compressed into a 1-cm-thick layer, the concentration would be 8×10^{18} molecules/cm^3.

(a) Using the ozone spectrum at the beginning of the chapter, estimate the transmittance and absorbance of this 1-cm^3 sample at 325 and 300 nm.

(b) Sunburns are caused by radiation in the 295- to 310-nm region. At the center of this region, the transmittance of atmospheric ozone is 0.14. Calculate the absorption cross section for $T = 0.14$, $n = 8 \times 10^{18}$ molecules/cm^3, and $b = 1$ cm. By what percentage does the transmittance increase if the ozone concentration decreases by 1% to 7.92×10^{18} molecules/cm^3?

(c) Atmospheric O_3 is measured in *Dobson units* (1 unit = 2.69×10^{16} molecules O_3 above each cm^2 of Earth's surface). (Dobson unit $\equiv$ thickness [in hundredths of a millimeter] that the O_3 column would occupy if it were compressed to 1 atm at 0°C.) The graph shows variations in O_3 concentration as a function of latitude and season. Using an absorption cross section of 2.5×10^{-19} cm^2, calculate the transmittance in the winter and in the summer at 30°–50° N latitude, at which O_3 varies from 290 to 350 Dobson units. By what percentage is the ultraviolet transmittance greater in winter than in summer?

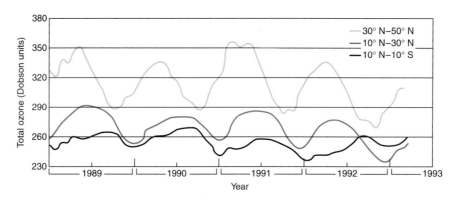

Variation in atmospheric ozone at different latitudes. [From P. S. Zurer, *Chem. Eng. News*, 24 May 1993, p. 8.]

Beer's Law in Chemical Analysis

17-15. What is the purpose of neocuproine in the serum iron analysis?

17-16. A compound with molecular mass 292.16 g/mol was dissolved in a 5-mL volumetric flask. A 1.00-mL aliquot was withdrawn, placed in a 10-mL volumetric flask, and diluted to the mark. The absorbance at 340 nm was 0.427 in a 1.000-cm cuvet. The molar absorptivity at 340 nm is $\varepsilon_{340} = 6\ 130$ $M^{-1}\ cm^{-1}$.

(a) Calculate the concentration of compound in the cuvet.

(b) What was the concentration of compound in the 5-mL flask?

(c) How many milligrams of compound were used to make the 5-mL solution?

17-17. If a sample for spectrophotometric analysis is placed in a 10-cm cell, the absorbance will be 10 times greater than the absorbance in a 1-cm cell. Will the absorbance of the reagent-blank solution also be increased by a factor of 10?

17-18. You have been sent to India to investigate the occurrence of goiter disease attributed to iodine deficiency. As part of your investigation, you must make field measurements of traces of iodide (I^-) in groundwater. The procedure is to oxidize I^- to I_2 and convert the I_2 into an intensely colored complex with the dye brilliant green in the organic solvent toluene.

(a) A 3.15×10^{-6} M solution of the colored complex exhibited an absorbance of 0.267 at 635 nm in a 1.000-cm cuvet. A blank solution made from distilled water in place of groundwater had an absorbance of 0.019. Find the molar absorptivity of the colored complex.

(b) The absorbance of an unknown solution prepared from groundwater was 0.175. Find the concentration of the unknown.

17-19. Nitrite ion, NO_2^-, is a preservative for bacon and other foods, but it is potentially carcinogenic. A spectrophotometric determination of NO_2^- makes use of the following reactions:

$$HO_3S-\langle\bigcirc\rangle-NH_2 + NO_2^- + 2H^+ \longrightarrow$$

Sulfanilic acid

$$HO_3S-\langle\bigcirc\rangle-\overset{+}{N}\equiv N + 2H_2O$$

$$HO_3S-\langle\bigcirc\rangle-\overset{+}{N}\equiv N + \langle\bigcirc\rangle-NH_2 \longrightarrow$$

1-Aminonaphthalene

$$HO_3S-\langle\bigcirc\rangle-N=N-\langle\bigcirc\rangle-NH_2 + H^+$$

Colored product, $\lambda_{max} = 520$ nm

Here is an abbreviated procedure for the determination:

1. To 50.0 mL of unknown solution containing nitrite is added 1.00 mL of sulfanilic acid solution.

2. After 10 min, 2.00 mL of 1-aminonaphthalene solution and 1.00 mL of buffer are added.

3. After 15 min, the absorbance is read at 520 nm in a 5.00-cm cell.

The following solutions were analyzed:

A. 50.0 mL of food extract known to contain no nitrite (that is, a negligible amount); final absorbance = 0.153.

B. 50.0 mL of food extract suspected of containing nitrite; final absorbance = 0.622.

C. Same as B, but with 10.0 μL of 7.50×10^{-3} M $NaNO_2$ added to the 50.0-mL sample; final absorbance = 0.967.

(a) Calculate the molar absorptivity, ε, of the colored product. Remember that a 5.00-cm cell was used.

(b) How many micrograms of NO_2^- were present in 50.0 mL of food extract?

Spectrophotometric Titrations

17-20. A 2.00-mL solution of apotransferrin was titrated as illustrated in Figure 17-10. It required 163 μL of 1.43 mM ferric nitrilotriacetate to reach the end point.

(a) Why does the slope of the absorbance-versus-volume graph change abruptly at the equivalence point?

(b) How many moles of Fe(III) (= ferric nitrilotriacetate) were required to reach the end point?

(c) Each apotransferrin molecule binds two ferric ions. Find the concentration of apotransferrin in the 2.00-mL solution.

17-21. The iron-binding site of transferrin in Figure 17-7 can accommodate certain other metal ions besides Fe^{3+} and certain other anions besides CO_3^{2-}. Data are given in the table for the titration of transferrin (3.57 mg in 2.00 mL) with 6.64 mM Ga^{3+} solution in the presence of the anion oxalate, $C_2O_4^{2-}$, and in the absence of a suitable anion. Prepare a graph similar to Figure 17-10, showing both sets of data. Indicate the theoretical equivalence point for the binding of one and two Ga^{3+} ions per molecule of protein and the observed end point. How many Ga^{3+} ions are bound to transferrin in the presence and in the absence of oxalate?

Titration in presence of $C_2O_3^{2-}$		Titration in absence of anion	
Total μL Ga^{3+} added	Absorbance at 241 nm	Total μL Ga^{3+} added	Absorbance at 241 nm
0.0	0.044	0.0	0.000
2.0	0.143	2.0	0.007
4.0	0.222	6.0	0.012
6.0	0.306	10.0	0.019
8.0	0.381	14.0	0.024
10.0	0.452	18.0	0.030
12.0	0.508	22.0	0.035
14.0	0.541	26.0	0.037
16.0	0.558		
18.0	0.562		
21.0	0.569		
24.0	0.576		

17-22. Gold nanoparticles (Figure 16-29) can be titrated with the oxidizing agent TCNQ in the presence of excess of Br^- to oxidize Au(0) to AuBr in deaerated toluene. Gold atoms in the interior of the particle are Au(0). Gold atoms bound to $C_{12}H_{25}S$— (dodecanethiol) ligands on the surface of the particle are Au(I) and are not titrated.

$$Au(0) + \text{TCNQ} + 2Br^- \rightarrow AuBr_2^- + TCNQ^{\overline{\cdot}}$$

TCNQ
Tetracyanoquinodimethane

Reduced
radical anion
λ_{max} = 856 nm

Reduced $TCNQ^{\overline{\cdot}}$ has a low-energy electronic absorption peak at 856 nm.

The table gives the absorbance at 856 nm as 0.700 mL of 1.00×10^{-4} M TCNQ + 0.05 M $(C_8H_{17})_4N^+Br^-$ in toluene is titrated with gold nanoparticles (1.43 g/L in toluene) from a microsyringe with a Teflon-coated needle. Absorbance in the table has already been corrected for dilution.

Total μL nanoparticles	Absorbance at 856 nm	Total μL nanoparticles	Absorbance at 856 nm
4.9	0.208	19.1	0.706
8.0	0.301	22.0	0.770
11.0	0.405	25.0	0.784
13.7	0.502	30.0	0.785
16.2	0.610	35.9	0.784

SOURCE: G. Zotti, B. Vercelli, and A. Berlin, "Reaction of Gold Nanoparticles with Tetracyanoquinoidal Molecules," *Anal. Chem.* **2008**, *80*, 815.

(a) Make a graph of absorbance versus volume of titrant and estimate the equivalence point. Calculate the mmol of Au(0) in 1.00 g of nanoparticles.

(b) From other analyses of similarly prepared nanoparticles, it is estimated that 25 wt% of the mass of the particle is dodecanethiol ligand ($C_{12}H_{25}S$—, FM 201.40). Calculate mmol of $C_{12}H_{25}S$ in 1.00 g of nanoparticles.

(c) The Au(I) content of 1.00 g of nanoparticles should be 1.00 − mass of Au(0) − mass of $C_{12}H_{25}S$. Calculate the micromoles of Au(I) in 1.00 g of nanoparticles and the mole ratio Au(I):$C_{12}H_{25}S$. In principle, this ratio should be 1:1. The difference is most likely because $C_{12}H_{25}S$ was not measured for this specific nanoparticle preparation.

Luminescence

17-23. In formaldehyde, the transition $n \rightarrow \pi^*(T_1)$ occurs at 397 nm, and the $n \rightarrow \pi^*(S_1)$ transition comes at 355 nm. What is the difference in energy (kJ/mol) between the S_1 and T_1 states? This difference is due to the different electron spins in the two states.

17-24. What is the difference between fluorescence and phosphorescence?

17-25. What is the difference between luminescence and chemiluminescence?

17-26. Explain what happens in Rayleigh scattering and Raman scattering. How much faster is scattering of visible light than fluorescence?

17-27. Consider a molecule that can fluoresce from the S_1 state and phosphoresce from the T_1 state. Which is emitted at longer wavelength, fluorescence or phosphorescence? Make a sketch showing absorption, fluorescence, and phosphorescence on a single spectrum.

17-28. What is the difference between a fluorescence excitation spectrum and a fluorescence emission spectrum? Which one resembles an absorption spectrum?

17-29. The wavelengths of maximum absorption and emission of anthracene in Figure 17-22 are approximately 357 and 402 nm. Molar absorptivities at these wavelengths are $\varepsilon_{ex} = 9.0 \times 10^3 \, M^{-1} \, cm^{-1}$ and $\varepsilon_{em} = 5 \times 10^1 \, M^{-1} \, cm^{-1}$. Consider a fluorescence experiment in Figure 17-21 with cell dimensions $b_1 = 0.30$ cm, $b_2 = 0.40$ cm, and $b_3 = 0.5$ cm. Calculate the relative fluorescence intensity with Equation 17-12 as a function of concentration over the range 10^{-8} to 10^{-3} M. Explain the shape of the curve. Up to approximately what concentration is fluorescence proportional to concentration (within 5%)? Is the calibration range in Figure 17-23 sensible?

17-30. *Standard addition.* Selenium from 0.108 g of Brazil nuts was converted into the fluorescent product in Reaction 17-15, which was extracted into 10.0 mL of cyclohexane. Then 2.00 mL of the cyclohexane solution were placed in a cuvet for fluorescence measurement. Standard additions of fluorescent product containing 1.40 μg Se/mL are given in the table. Construct a standard addition graph to find the concentration of Se in the 2.00-mL unknown solution. Find the wt% of Se in the nuts and its uncertainty and 95% confidence interval.

Volume of standard added (μL)	Fluorescence intensity (arbitrary units)
0	41.4
10.0	49.2
20.0	56.4
30.0	63.8
40.0	70.3

Applications of Spectrophotometry

FLUORESCENCE RESONANCE ENERGY TRANSFER BIOSENSOR

- Radiant energy absorber (donor)
- Radiant energy emitter (acceptor)

Response of biosensor to TNT is a *decrease* in fluorescence with increasing concentration of analyte. [From I. L. Medintz, E. R. Goldman, M. E. Lassman, A. Hayhurst, A. W. Kusterbeck, and J. R. Deschamps, "Self-Assembled TNT Biosensor Based on Modular Multifunctional Surface-Tethered Components," *Anal. Chem.* 2005, *77*, 365.]

A *biosensor* is a device that uses biological components such as enzymes, antibodies, or DNA, in combination with electric, optical, or other signals, to achieve a selective response to one analyte.[1] The biosensor in the diagram consists of two surface-tethered components. The biorecognition element can be an antibody, DNA, RNA, or carbohydrate with specific affinity for an analyte. A structural analog of the analyte is bound to a flexible arm adjacent to the recognition element. In the absence of analyte, the tethered analog binds to the recognition element.

A chromophore that efficiently absorbs radiant energy is attached to the recognition element adjacent to the recognition site. The fluorescent chromophore is attached to the flexible arm adjacent to the analyte structural analog. At the left side of the figure, absorbing and emitting chromophores are close to each other. Light with a wavelength of 510 nm is absorbed by and efficiently transferred through space to , which fluoresces strongly at 600 nm. This *fluorescence resonance energy transfer* decreases with the sixth power of the distance between the donor and acceptor.[2,3]

When analyte is added, it displaces the tethered analog from the biorecognition element. The higher the concentration of analyte, the more displacement occurs. When the analog is displaced from the binding site, and are no longer close enough for energy transfer, and fluorescence is decreased. In the graph, analyte is trinitrotoluene (the explosive, TNT) and the detection limit is 0.1 mg/L (0.1 ppm). After a measurement, the sensor is washed to remove analyte and then reused for more analyses.

This chapter describes applications of absorption and emission of electromagnetic radiation in chemical analysis. *Flow injection analysis* is introduced as an important analytical method that commonly uses absorption or emission to measure analytical response. We use Excel Solver and spreadsheet matrix operations as powerful tools for numerical analysis.

18-1 Analysis of a Mixture

The absorbance of a solution at any wavelength is the sum of absorbances of all the species in the solution.

Absorbance is additive.

Absorbance of a mixture:
$$A = \varepsilon_X b[X] + \varepsilon_Y b[Y] + \varepsilon_Z b[Z] + \cdots \qquad (18\text{-}1)$$

where ε is the molar absorptivity of each species at the wavelength in question and b is the cell pathlength (Figure 17-4). If we know the spectra of the pure components, we can

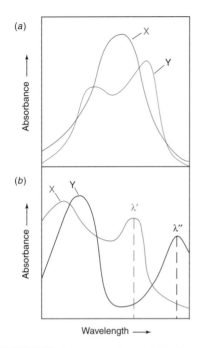

(a)

(b)

FIGURE 18-1 Two cases for analysis of a mixture. (*a*) Spectra of the pure components have substantial overlap. (*b*) Regions exist in which each component makes the major contribution.

mathematically disassemble the spectrum of a mixture into those of its components. In acid-base chemistry, such a procedure allows us to measure the concentrations of the acidic and basic forms of an indicator. This information, combined with the Henderson-Hasselbalch equation (8-16), provides a precise measurement of pH by spectrophotometry.[4]

For a mixture of compounds X and Y, two cases are distinguished. In Figure 18-1a, the absorption bands of pure X and pure Y overlap significantly everywhere. This case is best treated by a spreadsheet procedure that uses measurements at many wavelengths. In Figure 18-1b, X and Y have relatively little overlap in some regions. We can analyze this case with a pair of measurements at wavelength λ', where X makes the major contribution, and wavelength λ'', where Y makes the major contribution.

What to Do When the Individual Spectra Overlap

We will apply a least-squares analysis of Equation 18-1 to the spectrum of an unknown mixture of H_2O_2 complexes of Ti(IV) and V(V) in Figure 18-2. The figure also shows spectra of standard 1.32 mM Ti(IV) and 1.89 mM V(V).

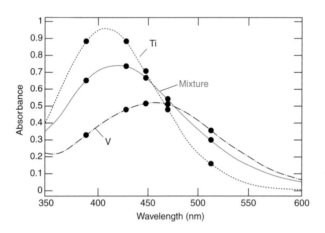

FIGURE 18-2 Visible spectra of 1.32 mM Ti(IV), 1.89 mM V(V), and an unknown mixture containing both ions. All solutions contain 0.5 wt% H_2O_2 and 0.01 M H_2SO_4. Absorbance at points shown by dots is listed in Figure 18-3. [From M. Blanco, H. Iturriaga, S. Maspoch, and P. Tarín, "A Simple Method for Spectrophotometric Determination of Two Components with Overlapped Spectra," *J. Chem. Ed.* **1989**, *66*, 178.]

Measure absorbance at more wavelengths than there are components in the mixture. More wavelengths increase the accuracy.

First, we measure the absorbance of each standard at several wavelengths shown by dots in Figure 18-2. Results are listed in columns A–C of Figure 18-3. Concentrations of the standards are entered in cells A14 and A16 and the pathlength (1.000 cm) is indicated in cell A19.

	A	B	C	D	E	F	G	H
1	Analysis of a Mixture When You Have More Data Points Than Components of the Mixture							
2				Measured				
3				Absorbance			Calculated	
4	Wave-	Absorbance of Standard:		of Mixture	Molar Absorptivity		Absorbance	
5	length	Titanium	Vanadium	Am	Titanium	Vanadium	Acalc	[Acalc-Am]^2
6	390	0.895	0.326	0.651	678.0	172.5	0.8505	3.981E-02
7	430	0.884	0.497	0.743	669.7	263.0	0.9327	3.597E-02
8	450	0.694	0.528	0.665	525.8	279.4	0.8051	1.963E-02
9	470	0.481	0.512	0.547	364.4	270.9	0.6353	7.796E-03
10	510	0.173	0.374	0.314	131.1	197.9	0.3289	2.233E-04
11							sum =	1.034E-01
12	Standards		Concentrations in the mixture					
13	[Ti](M) =		(to be found by Solver)					
14	0.00132		[Ti] =	0.001000				
15	[V](M) =		[V] =	0.001000				
16	0.00189							
17	Pathlength		E6 = B6/($A$19*$A$14)					
18	(cm) =		F6 = C6/($A$19*$A$16)					
19	1.000		G6 = E6*$A$19*$D$14+F6*$A$19*$D$15					
20			H6 = (G6-D6)^2					

FIGURE 18-3
Spreadsheet using Solver to analyze the mixture in Figure 18-2.

Calling the two components X (= Ti) and Y (= V), we find the molar absorptivity of each component at each wavelength from Beer's law:

$$\varepsilon_X = \frac{A_{X_s}}{b[X]_s} \qquad \varepsilon_Y = \frac{A_{Y_s}}{b[Y]_s} \qquad (18\text{-}2)$$

where A_{X_s} is the absorbance of the standard and $[X]_s$ is the concentration of the standard. This computation is shown in columns E and F of Figure 18-3. The measured absorbance of the unknown mixture at each wavelength, A_m, is listed in column D. At each wavelength, this absorbance is the sum of absorbances of the components:

$$A_m = \varepsilon_X b[X] + \varepsilon_Y b[Y] \qquad (18\text{-}3)$$

However, we do not know the concentrations [X] and [Y] in the mixture.

To find [X] and [Y], we begin by *guessing* concentrations and inserting them in cells D14 and D15. The guesses do not have to be close to correct values. We arbitrarily chose 0.001 M for both guesses. The calculated absorbance of the mixture is then computed in column G from the equation

$$A_{calc} = \varepsilon_X b[X]_{guess} + \varepsilon_Y b[Y]_{guess} \qquad (18\text{-}4)$$

For example, A_{calc} in cell G6 = (678.0)(1.000)[0.001] + (172.5)(1.000)[0.001]. Column H gives the square of the difference between calculated and measured absorbance.

Column H contains $(A_{calc} - A_m)^2$

The least-squares condition is to minimize the sum of squares $(A_{calc} - A_m)^2$ by varying the concentrations $[X]_{guess}$ and $[Y]_{guess}$. The "best" values of $[X]_{guess}$ and $[Y]_{guess}$ in cells D14 and D15 are those that minimize the sum of squares in cell H11.

Excel has a powerful tool called Solver that carries out the minimization for us. In Excel 2007, you will find Solver in the Data ribbon in the Analysis section. If you don't see Solver, click the Microsoft Office button at the top left of the spreadsheet. Click Excel Options and select Add-Ins. Highlight Solver Add-In, click OK, and Solver is loaded. In earlier versions of Excel, Solver appears in the Tools menu. If it is not there, go to Tools, Add-Ins and select Solver Add-In. Click OK and Solver will be loaded and should appear in the Tools menu.

Highlight cell H11 in Figure 18-3 and select Solver. The window in Figure 18-4 appears. Enter "H11" in Set Target Cell. Then select the button that says Min. Enter "D14,D15" in By Changing Cells. We just told Solver to minimize cell H11 by changing cells D14 and D15. Click Solve. After a little work, Solver finds the values 0.000 670 in cell D14 and 0.001 123 in cell D15. The sum of squares in cell H11 is reduced from 0.103 to 0.000 028. Cells D14 and D15 now tell us that [Ti(IV)] = 0.670 mM and [V(V)] = 1.123 mM in the mixture.

FIGURE 18-4 Solver window in Excel.

This procedure is readily extended to mixtures containing more than two components. The more points you measure, the more accurate the result is likely to be.

What to Do When the Individual Spectra Are Well Resolved

If spectra of the individual components of a mixture are moderately resolved from one another, as at wavelengths λ' and λ'' in Figure 18-1b, we can solve two simultaneous equations to find the concentrations in the mixture. The absorbance of the mixture at any wavelength is the sum of absorbances of each component at that wavelength. For wavelengths λ' and λ'',

$$A' = \varepsilon_X' b[X] + \varepsilon_Y' b[Y]$$
$$A'' = \varepsilon_X'' b[X] + \varepsilon_Y'' b[Y] \qquad (18\text{-}5)$$

where the ε values apply to each species at each wavelength. The absorptivities of X and Y at wavelengths λ' and λ'' must be measured in separate experiments.

$$\begin{vmatrix} a & b \\ c & d \end{vmatrix} = (a \times d) - (b \times c)$$

We can solve the Equations 18-5 for the two unknowns [X] and [Y]. The result is

Analysis of a mixture when spectra are resolved:

$$[X] = \dfrac{\begin{vmatrix} A' & \varepsilon_Y' b \\ A'' & \varepsilon_Y'' b \end{vmatrix}}{\begin{vmatrix} \varepsilon_X' b & \varepsilon_Y' b \\ \varepsilon_X'' b & \varepsilon_Y'' b \end{vmatrix}} \qquad [Y] = \dfrac{\begin{vmatrix} \varepsilon_X' b & A' \\ \varepsilon_X'' b & A'' \end{vmatrix}}{\begin{vmatrix} \varepsilon_X' b & \varepsilon_Y' b \\ \varepsilon_X'' b & \varepsilon_Y'' b \end{vmatrix}} \qquad (18\text{-}6)$$

In Equations 18-6, each symbol $\begin{vmatrix} a & b \\ c & d \end{vmatrix}$ is a *determinant*. It is a shorthand way of writing the product $(a \times d) - (b \times c)$. Thus, $\begin{vmatrix} 1 & 2 \\ 3 & 4 \end{vmatrix}$ means $(1 \times 4) - (2 \times 3) = -2$.

EXAMPLE **Analysis of a Mixture, Using Equations 18-6**

The molar absorptivities of compounds X and Y were measured with pure samples of each:

	$\varepsilon \, (M^{-1} \, cm^{-1})$	
λ (nm)	X	Y
272	16 400	3 870
327	3 990	6 420

A mixture of compounds X and Y in a 1.000-cm cell had an absorbance of 0.957 at 272 nm and 0.559 at 327 nm. Find the concentrations of X and Y in the mixture.

Solution Using Equations 18-6 and setting $b = 1.000$, we find

$$[X] = \dfrac{\begin{vmatrix} 0.957 & 3\,870 \\ 0.559 & 6\,420 \end{vmatrix}}{\begin{vmatrix} 16\,400 & 3\,870 \\ 3\,990 & 6\,420 \end{vmatrix}} = \dfrac{(0.957)(6\,420) - (3\,870)(0.559)}{(16\,400)(6\,420) - (3\,870)(3\,990)} = 4.43 \times 10^{-5} \, M$$

$$[Y] = \dfrac{\begin{vmatrix} 16\,400 & 0.957 \\ 3\,990 & 0.559 \end{vmatrix}}{\begin{vmatrix} 16\,400 & 3\,870 \\ 3\,990 & 6\,420 \end{vmatrix}} = 5.95 \times 10^{-5} \, M$$

Test Yourself Find the concentration of [X] if the absorbances are 0.700 at 272 nm and 0.550 at 327 nm. (*Answer:* 2.63×10^{-5} M)

To analyze a mixture of two compounds, it is necessary to measure absorbance at two wavelengths and to know ε at each wavelength for each compound. Similarly, a mixture of n components can be analyzed by making n absorbance measurements at n wavelengths.

Solving Simultaneous Linear Equations with Excel

Excel solves systems of linear equations with a single statement. If the following matrix mathematics below is unfamiliar to you, disregard it. The important result is the template in Figure 18-5 for solving simultaneous equations. You can use this template by following the instructions in the last paragraph of this section, even if the math is not familiar.

The simultaneous equations of the preceding example are

$$A' = \varepsilon_X' b[X] + \varepsilon_Y' b[Y] \qquad\qquad 0.957 = 16\,440[X] + 3\,870[Y]$$
$$\Rightarrow$$
$$A'' = \varepsilon_X'' b[X] + \varepsilon_Y'' b[Y] \qquad\qquad 0.559 = 3\,990[X] + 6\,420[Y]$$

which can be written in matrix notation in the form

$$\begin{bmatrix} 0.957 \\ 0.559 \end{bmatrix} = \begin{bmatrix} 16\,400 & 3\,870 \\ 3\,990 & 6\,420 \end{bmatrix} \begin{bmatrix} [X] \\ [Y] \end{bmatrix} \qquad (18\text{-}7)$$

$$\mathbf{A} \quad = \quad\quad \mathbf{K} \quad\quad\quad \mathbf{C}$$

K is the *matrix* of molar absorptivity times pathlength, εb. **A** is the matrix of absorbance of the unknown. A matrix, such as **A**, with only one column or one row is called a *vector*. **C** is the vector of unknown concentrations.

	A	B	C	D	E	F	G
1	Solving Simultaneous Linear Equations with Excel Matrix Operations						
2							
3	Wavelength	Coefficient Matrix		Absorbance		Concentrations	
4				of unknown		in mixture	
5	272	16400	3870	0.957		4.4304E-05	← [X]
6	327	3990	6420	0.559		5.9537E-05	← [Y]
7		K		A		C	
8							
9	1. Enter matrix of coefficients εb in cells B5:C6						
10	2. Enter absorbance of unknown at each wavelength (cells D5:D6)						
11	3. Highlight block of blank cells required for solution (F5 and F6)						
12	4. Type the formula "= MMULT(MINVERSE(B5:C6),D5:D6)"						
13	5. Press CONTROL+SHIFT+ENTER on a PC or COMMAND+RETURN on a Mac						
14	6. Behold! The answer appears in cells F5 and F6						

FIGURE 18-5 Solving simultaneous linear equations with Excel.

A matrix K^{-1}, called the *inverse* of **K,** is such that the products KK^{-1} or $K^{-1}K$ are equal to a unit matrix with 1's on the diagonal and 0 elsewhere.[5] We can solve Equation 18-7 for the concentration vector, **C,** by multiplying both sides of the equation by K^{-1}:

$$KC = A$$
$$\underbrace{K^{-1}KC}_{C} = K^{-1}A$$

The product $K^{-1}KC$ is just **C**:

$$\underbrace{\begin{bmatrix} 1 & 0 \\ 0 & 1 \end{bmatrix}}_{K^{-1}K} \underbrace{\begin{bmatrix} [X] \\ [Y] \end{bmatrix}}_{C} = \underbrace{\begin{bmatrix} [X] \\ [Y] \end{bmatrix}}_{C}$$

To solve simultaneous equations, find the inverse matrix K^{-1} and multiply it times A. The product is C, the concentrations in the unknown mixture.

In Figure 18-5, we enter the wavelengths in column A just to keep track of information. We will not use these wavelengths for computation. Enter the products εb for pure X in column B and εb for pure Y in column C. The array in cells B5:C6 is the matrix **K.** The Excel function MINVERSE(B5:C6) gives the inverse matrix, K^{-1}. The function MMULT(matrix 1, matrix 2) gives the product of two matrices (or a matrix and a vector). The concentration vector, **C,** is equal to $K^{-1}A$, which we get with the single statement

$$=MMULT(\underbrace{MINVERSE(B5:C6)}_{K^{-1}}, \underbrace{D5:D6}_{A})$$

To use the template in Figure 18-5, enter the coefficients, εb, measured from the pure compounds, in cells B5:C6. Enter the absorbance of the unknown in cells D5:D6. Highlight cells F5:F6 and type the formula "=MMULT(MINVERSE(B5:C6), D5:D6)". Press CONTROL+SHIFT+ENTER on a PC or COMMAND(⌘)+RETURN on a Mac. The concentrations [X] and [Y] in the mixture now appear in cells F5:F6.

Procedure for solving simultaneous equations with Excel

Isosbestic Points

Often one absorbing species, X, is converted into another absorbing species, Y, in the course of a chemical reaction. This transformation leads to an obvious behavior shown in Figure 18-6. If the spectra of pure X and pure Y cross each other at any wavelength, then every spectrum recorded during this chemical reaction will cross at that same point, called an **isosbestic point.** *An isosbestic point is good evidence that only two principal species are present.*[6]

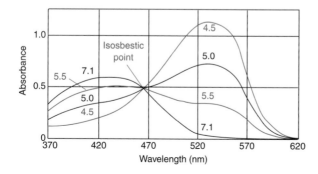

FIGURE 18-6 Absorption spectrum of 3.7×10^{-4} M methyl red between pH 4.5 and 7.1. [From E. J. King, *Acid-Base Equilibria* (Oxford: Pergamon Press, 1965).]

Consider the indicator methyl red, which changes between red (HIn) and yellow (In$^-$) near pH 5.1:

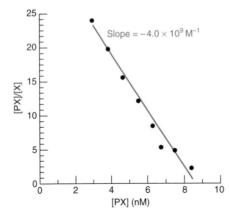

$$pK_2 \approx 5.1$$

HIn
(red)

In$^-$
(yellow)

Because the spectra of HIn and In$^-$ (at the same concentration) happen to cross at 465 nm, all spectra in Figure 18-6 cross at this point. (If the spectra of HIn and In$^-$ crossed at several points, there would be several isosbestic points.)

To see why there is an isosbestic point, we write an equation for the absorbance of the solution at 465 nm:

$$A^{465} = \varepsilon_{HIn}^{465} b[HIn] + \varepsilon_{In^-}^{465} b[In^-] \tag{18-8}$$

But because the spectra of pure HIn and pure In$^-$ (at the same concentration) cross at 465 nm, ε_{HIn}^{465} must be equal to $\varepsilon_{In^-}^{465}$. Setting $\varepsilon_{HIn}^{465} = \varepsilon_{In^-}^{465} = \varepsilon^{465}$, we can factor Equation 18-8:

$$A^{465} = \varepsilon^{465} b([HIn] + [In^-]) \tag{18-9}$$

An isosbestic point occurs when $\varepsilon_X = \varepsilon_Y$ and $[X] + [Y]$ is constant.

In Figure 18-6, all solutions contain the same total concentration of methyl red (= [HIn] + [In$^-$]). Only the pH varies. Therefore, the sum of concentrations in Equation 18-9 is constant, and A^{465} is constant.

18-2 Measuring an Equilibrium Constant: The Scatchard Plot

To measure an equilibrium constant, we measure concentrations (actually activities) of species at equilibrium. This section shows how spectrophotometry can be used for this purpose.

Absorbance is proportional to *concentration* (not activity), so concentrations must be converted into activities to get true equilibrium constants.

Let's examine the equilibrium in which the species P and X react to form PX.

$$P + X \rightleftharpoons PX \tag{18-10}$$

Neglecting activity coefficients, we can write

$$K = \frac{[PX]}{[P][X]} \tag{18-11}$$

Consider a series of solutions in which increments of X are added to a constant amount of P. Letting P_0 be the total concentration of P (in the forms P and PX), we can write

Clearing the cobwebs from your brain, you realize that Equation 18-12 is a mass balance.

$$[P] = P_0 - [PX] \tag{18-12}$$

Now the equilibrium expression, Equation 18-11, can be rearranged as follows:

$$\frac{[PX]}{[X]} = K[P] = K(P_0 - [PX]) \tag{18-13}$$

A *Scatchard plot* is a graph of [PX]/[X] versus [PX]. The slope is $-K$.

A graph of [PX]/[X] versus [PX] has a slope of $-K$ and is called a **Scatchard plot.**[7] It is widely used in biochemistry to measure equilibrium constants (Figure 18-7).

Slope = -4.0×10^9 M^{-1}

[PX]/[X]

[PX] (nM)

FIGURE 18-7 Scatchard plot for binding of antigen (X) to antibody (P). The antibody binds the explosive, trinitrotoluene (TNT). The antigen is a fluorescent analog of TNT. From the slope, the binding constant for the reaction P + X $\rightleftharpoons$ PX is $K = 4.0 \times 10^9$ M^{-1}. [Derived from Figure 4 of A. Bromberg and R. A. Mathies, "Homogeneous Immunoassay for Detection of TNT on a Capillary Electrophoresis Chip," *Anal. Chem.* 2003, *75*, 1188.]

If we know [PX], we can find [X] with the mass balance

$$X_0 = [\text{total X}] = [\text{PX}] + [\text{X}]$$

To measure [PX], we might use spectrophotometric absorbance. Suppose that P and PX each have some absorbance at wavelength λ, but X has no absorbance at this wavelength. For simplicity, let all measurements be made in a cell of pathlength 1.000 cm so that we can omit b (= 1.000 cm) when writing Beer's law.

The absorbance at some wavelength is the sum of absorbances of PX and P:

$$A = \varepsilon_{PX}[\text{PX}] + \varepsilon_{P}[\text{P}]$$

Substituting $[\text{P}] = P_0 - [\text{PX}]$, we can write

$$A = \varepsilon_{PX}[\text{PX}] + \underbrace{\varepsilon_{P}P_0}_{A_0} - \varepsilon_{P}[\text{PX}] \tag{18-14}$$

But $\varepsilon_{P}P_0$ is A_0, the initial absorbance before any X is added. Therefore,

$$A = [\text{PX}](\varepsilon_{PX} - \varepsilon_{P}) + A_0 \Rightarrow [\text{PX}] = \frac{\Delta A}{\Delta \varepsilon} \tag{18-15}$$

where $\Delta \varepsilon = \varepsilon_{PX} - \varepsilon_{P}$ and $\Delta A \ (= A - A_0)$ is the observed absorbance after each addition of X minus the initial absorbance.

Substituting [PX] from Equation 18-15 into Equation 18-13 gives

Scatchard equation:

$$\frac{\Delta A}{\text{X}} = K\Delta \varepsilon P_0 - K\Delta A \tag{18-16}$$

A graph of $\Delta A/[\text{X}]$ versus ΔA should be a straight line with a slope of $-K$. Absorbance measured while P is titrated with X can be used to find K for the reaction of X with P.

Two cases commonly arise in the application of Equation 18-16. If K is small, then large concentrations of X are needed to produce PX. Therefore, $X_0 \gg P_0$, and $[\text{X}] \approx X_0$. Alternatively, if K is not small, then $[\text{X}] \neq X_0$, and [X] must be measured, either at another wavelength or by measuring a different physical property.

Errors in a Scatchard plot could be substantial. We define the fraction of saturation as

$$\text{Fraction of saturation} = S = \frac{[\text{PX}]}{P_0} \tag{18-17}$$

The most accurate data are obtained for $0.2 \lesssim S \lesssim 0.8$.[8] A range representing ~75% of the total saturation curve should be measured before concluding that equilibrium (18-10) is obeyed. People have made mistakes by exploring too little of the binding curve.

Problem 18-13 gives an alternate way to find K by using Excel Solver.

Equation 18-13 can be recast as
$S/[\text{X}] = K(1 - S)$.
Plot $S/[\text{X}]$ versus S.

18-3 The Method of Continuous Variation

Suppose that several complexes can form between species P and X:

$$\text{P} + \text{X} \rightleftharpoons \text{PX} \tag{18-18}$$

$$\text{P} + 2\text{X} \rightleftharpoons \text{PX}_2 \tag{18-19}$$

$$\text{P} + 3\text{X} \rightleftharpoons \text{PX}_3 \tag{18-20}$$

If one complex (say, PX_2) predominates, the **method of continuous variation** (also called *Job's method*)[9] allows us to identify the stoichiometry of the predominant complex.

The classical procedure is to mix P and X and dilute to constant volume so that the total concentration [P] + [X] is constant. For example, 2.50 mM solutions of P and X could be mixed as shown in Table 18-1 to give various X:P ratios, but constant total concentration. The absorbance of each solution is measured, typically at λ_{max} for the complex, and a graph is made, showing *corrected* absorbance (defined in Equation 18-21) versus mole fraction of X. *Maximum absorbance is reached at the composition corresponding to the stoichiometry of the predominant complex.*

Corrected absorbance is the measured absorbance minus the absorbance that would be produced by free P and free X alone:

$$\text{Corrected absorbance} = \text{measured absorbance} - \varepsilon_{P}bP_{T} - \varepsilon_{X}bX_{T} \tag{18-21}$$

For the reaction $\text{P} + n\text{X} \rightleftharpoons \text{PX}_n$, you could show that $[\text{PX}_n]$ reaches a maximum when the initial concentrations have the ratio $[\text{X}]_0 = n[\text{P}]_0$. To do this, write

$$K = \frac{[\text{PX}_n]}{([\text{P}]_0 - [\text{PX}_n])([\text{X}]_0 - n[\text{PX}_n])}$$

and set the partial derivatives $\partial[\text{PX}_n]/\partial[\text{P}]_0$ and $\partial[\text{PX}_n]/\partial[\text{X}]_0$ equal to 0.

TABLE 18-1 Solutions for the method of continuous variation

mL of 2.50 mM P	mL of 2.50 mM X	Mole ratio (X:P)	Mole fraction of X $\left(\dfrac{\text{mol X}}{\text{mol X + mol P}}\right)$
1.00	9.00	9.00:1	0.900
2.00	8.00	4.00:1	0.800
2.50	7.50	3.00:1	0.750
3.33	6.67	2.00:1	0.667
4.00	6.00	1.50:1	0.600
5.00	5.00	1.00:1	0.500
6.00	4.00	1:1.50	0.400
6.67	3.33	1:2.00	0.333
7.50	2.50	1:3.00	0.250
8.00	2.00	1:4.00	0.200
9.00	1.00	1:9.00	0.100

NOTE: All solutions are diluted to a total volume of 25.0 mL with a buffer.

where ε_P and ε_X are the molar absorptivities of pure P and pure X, b is the pathlength, and P_T and X_T are the total concentrations of P and X in the solution. For the first solution in Table 18-1, $P_T = (1.00/25.0)(2.50 \text{ mM}) = 0.100 \text{ mM}$ and $X_T = (9.00/25.0)(2.50 \text{ mM}) = 0.900 \text{ mM}$. If P and X do not absorb at the wavelength of interest, no absorbance correction is needed.

Maximum absorbance occurs at the mole fraction of X corresponding to the stoichiometry of the complex (Figure 18-8). If the predominant complex is PX_2, the maximum occurs at mole fraction of X = $2/(2 + 1) = 0.667$.

Method of continuous variation:

$$P + nX \rightleftharpoons PX_n$$

Maximum absorbance occurs when (mole fraction of X) = $n/(n + 1)$.

$$\text{Mole fraction of X in } P_aX_b = \frac{b}{b + a} \ (= 0.667 \text{ when } b = 2 \text{ and } a = 1)$$

If P_3X were predominant, the maximum would occur at (mole fraction of X) = $1/(1 + 3) = 0.250$.

Here are some precautions for applying the method of continuous variation:

1. Verify that the complex follows Beer's law.
2. Use constant ionic strength and pH, if applicable.
3. Take readings at more than one wavelength; the maximum should occur at the same mole fraction for each wavelength.
4. Do experiments at different total concentrations of P + X. If a second set of solutions were prepared in the proportions given in Table 18-1, but from stock concentrations of 5.00 mM, the maximum should still occur at the same mole fraction.

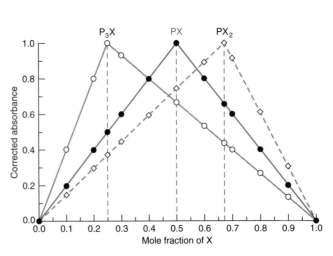

FIGURE 18-8 Ideal behavior of Job plots for formation of the complexes P_3X, PX, and PX_2.

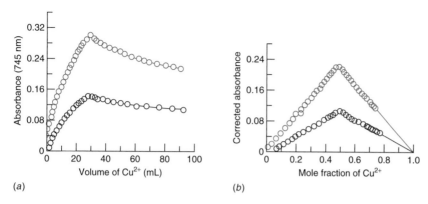

(a)   (b)

FIGURE 18-9 (a) Spectrophotometric titration of 30.0 mL of EDTA in acetate buffer with $CuSO_4$ in the same buffer. *Upper curve:* [EDTA] = [Cu^{2+}] = 5.00 mM. *Lower curve:* [EDTA] = [Cu^{2+}] = 2.50 mM. The absorbance has not been "corrected" in any way. (b) Transformation of data into mole fraction format. The absorbance of free $CuSO_4$ at the same formal concentration has been subtracted from each point in panel a. EDTA is transparent at this wavelength. [From Z. D. Hill and P. MacCarthy, "Novel Approach to Job's Method," *J. Chem. Ed.* **1986**, *63*, 162.]

The method of continuous variation can be carried out with many separate solutions, as in Table 18-1. However, a titration is more sensible. Figure 18-9a shows a titration of EDTA with Cu^{2+}. In Figure 18-9b, the abscissa has been transformed into mole fraction of Cu^{2+} (= [moles of Cu^{2+}]/[moles of Cu^{2+} + moles of EDTA]) instead of volume of Cu^{2+}. The sharp maximum at a mole fraction of 0.5 indicates formation of a 1:1 complex. If the equilibrium constant is not large, the maximum is more curved than in Figure 18-9b. The curvature can be used to estimate the equilibrium constant.[10]

18-4 Flow Injection Analysis and Sequential Injection

In **flow injection analysis,** a liquid sample is injected into a *continuously flowing* liquid carrier containing a reagent that reacts with the sample.[11–15] Additional reagents might be added further downstream. As sample flows from the injector to the detector, the sample zone broadens and reacts with reagent to form a product to which the detector responds. Advantages of flow injection over "batch processes" in which individual samples are separately analyzed include speed, automation of solution handling, and low cost. It is normal for flow injection to handle 100 samples per hour. Autosamplers holding hundreds of samples are essential for automated analysis. Flow injection is a workhorse in many soil and water analysis labs, which routinely measure large numbers of samples.

Figure 18-10 shows a representative analysis for traces of the herbicide acetochlor in food.[16] The sample is prepared by homogenizing a food such as grain or flour, extracting

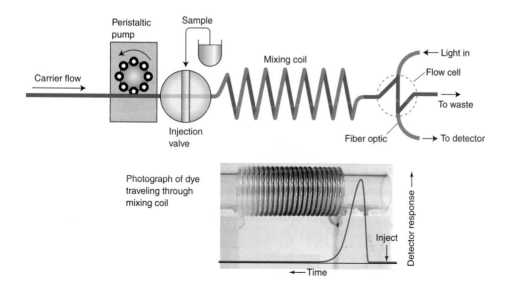

FIGURE 18-10 Schematic diagram of flow injection analysis in which sample is injected into carrier flow containing reagents that form a color with the analyte. The peristaltic pump pushes liquid through flexible tubing by the action of eight rollers along the tubing. Photograph and graph show dispersion of a dye injected into the carrier. [Adapted from tutorial by J. Ruzicka, *Flow Injection Analysis*, 4th ed., 2009, available free from www.flowinjection.com/freecd.aspx]

herbicide from the food with an organic solvent, and hydrolyzing the extract. Aqueous hydrolysis product is injected into the carrier stream for flow injection analysis in Figure 18-10.

Acetochlor herbicide → (0.2 M HCl, 100°C) → Hydrolysis product

The carrier stream contains a diazonium salt (the *reagent*) that reacts with the sample to give a colored product, which can be measured by its visible absorbance at 400 nm.

Diazonium reagent in carrier stream + Hydrolysis product in sample → (analysis) → Colored product (an azo dye)

Figure 18-11 shows dispersion and reaction of the sample after it has been injected into the carrier stream. If there were no reagent, the sample would just begin to spread out (disperse) as it travels downstream. When liquid flows relatively slowly through a cylindrical tube, the flow is *laminar*. Friction with the walls of the tube slows the flow to zero at the walls. The flow at the center of the channel is twice as fast as the average flow. Between the center and walls of the tube, there is a parabolic velocity profile. The top panel in Figure 18-11 shows curved leading and trailing edges of the sample zone. Liquid near the walls mixes with liquid in the bulk by radial diffusion. The narrower the tube, the faster is radial mixing. With a typical tube diameter of 0.5 to 0.75 mm, diffusion of liquid away from the walls is significant in a few seconds. In Figure 18-10, the bulk of the path between injection and detection is a helical mixing coil. Curvature and sharp bends in the flow path create turbulence, which promotes mixing.

As the sample plug is transported through the mixing coil, reaction with reagent in the carrier stream occurs from the leading and trailing edges of the sample zone. Turbulence is required for good mixing of the reagent with the sample because the length of the sample zone is much greater than the diameter of the tubing. Formation of product depends on the rate of the chemical reaction, as well as the rate of mixing of the zones. Typical times between injection and detection are just tens of seconds. In contrast to most methods of chemical analysis, the mixing of analyte with reagents is incomplete, and chemical equilibrium is not reached in flow injection analysis.

The key to analytical precision is the repeatability of flow injection. The concentration profile of product passing through the detector depends on many conditions, including volume of sample, flow rate, reaction rate, and temperature. Reproducible conditions provide a reproducible response. For replicate injections of the herbicide acetochlor, the standard deviation was 1.6% when measuring 1 ppm of the herbicide in food.

The detector flow cell in Figure 18-10 has a Z-shaped liquid path. Monochromatic light is brought by way of a fiber optic. Light that has passed through the flow cell travels to the detector via another fiber optic. Common flow cells have a pathlength of 10 mm with a volume of 60 µL.

Flow injection is a dynamic process in which equilibrium is not reached. Reproducibility is obtained by repeating the same conditions in every run.

FIGURE 18-11 Dispersion and reaction of sample as it travels downstream after injection into the carrier. [Adapted from tutorial by J. Ruzicka, *Flow Injection Analysis*, 4th ed., 2009, www.flowinjection.com.]

CHAPTER 18 Applications of Spectrophotometry

FIGURE 18-12 Flow injection analysis with two reagent channels for continuous analysis of nanomolar concentrations of NH_3 in seawater. [Adapted from N. Amornthammarong and J.-Z. Zhang, "Shipboard Fluorometric Flow Analyzer for High-Resolution Underway Measurement of Ammonium in Seawater," *Anal. Chem.* **2008**, *80*, 1019.]

Typical volumes of injected samples are tens of microliters. Peak height, rather than peak area, is commonly taken as the analytical signal in flow injection.

Figure 18-12 shows a slightly more complicated flow injection setup designed for continuous shipboard analysis of nanomolar concentrations of ammonia in seawater. The peristaltic pump feeds three liquids into the reaction coil, each at a rate of 160 μL/min. One liquid is seawater collected 1 m below the ocean surface. Reagent 1 (25 mM *o*-phthaldialdehyde) and reagent 2 (10 mM sodium sulfite plus 5 mM formaldehyde) mix with the seawater before the stream enters the 1.0-mm-diameter × 2-m-long reaction coil, which is kept at 65°C to speed the reaction to produce a fluorescent product. In the detector, the solution is irradiated with 365-nm ultraviolet light brought through a fiber optic. Fluorescence at 423 nm is collected by a second fiber optic at a right angle to the excitation beam. Calibration in Figure 18-13 is carried out by periodically replacing seawater with 0.958-mL volumes of NH_4Cl standards through the injection valve. The standard deviation of replicate injections was 2.2% for 200 nM NH_4Cl and 6.7% for 1.0 nM NH_4Cl.

Analysis time in Figure 18-13 is unusually long. Large-volume standards run at slow flow to boost sensitivity and reduce reagent consumption were measured at a rate of 8 per hour. By contrast, continuous measurement of seawater produced 3 600 readings per hour. Flow injection typically measures 100 samples per hour.

The terms "mixing coil" and "reaction coil" are used interchangeably.

FIGURE 18-13 Response to replicate injections of NH_4Cl standards. [From N. Amornthammarong and J.-Z. Zhang, "Shipboard Fluorometric Flow Analyzer for High-Resolution Underway Measurement of Ammonium in Seawater," *Anal. Chem.* **2008**, *80*, 1019.]

Three-way valve

Temperature-controlled holding coil

Six-way valve

Carrier buffer

Syringe pump

Flow cell

To waste

Fiber optic

Samples

Peristaltic pump

To waste

Reagent 1   Reagent 2

FIGURE 18-14 Schematic diagram of sequential injection equipment. Rotation of the valve can connect central port C to any of the ports 1 through 6. [Adapted from tutorial by J. Ruzicka, *Flow Injection Analysis*, 4th ed., 2009, www.flowinjection.com.]

Sequential Injection[11,17]

Sequential injection is distinguished from flow injection by *flow programming* and *flow reversal* under computer control. Flow is *not continuous*. Smaller volumes of reagents are required and less waste is generated than in flow injection. Sequential injection has been miniaturized to the point where it is also called a "lab-on-a-valve." Miniaturization and discontinuous flow reduce the consumption of expensive reagents such as enzymes and antibodies for biochemical assays. Sequential injection is used for continuous monitoring of environmental and industrial processes, even at remote locations.

The central feature of the sequential injection apparatus in Figure 18-14 is a six-way valve. In this example, ports 2, 3, 4, and 5 are used. The figure shows a connection between port 5 and central port C. Liquid samples are brought to port 5 by an auxiliary peristaltic pump. Rotation of the valve by computer control can connect any one of the other ports to C. Near the upper left is a stepper-motor-controlled syringe pump, which moves precise volumes of liquid forward or backward under computer control. The valve at the top of the syringe pump can connect the syringe to a reservoir of carrier buffer or to the holding coil.

Figure 18-15 shows how a reaction of a sample with one reagent could be conducted. Sample from port 5 of Figure 18-14 is first taken into the holding coil. Then the six-way valve is rotated to take reagent 1 from port 3 into the holding coil. Flow is then stopped to allow sample and reagent to mix and react in the holding coil. After a selected time, flow is reversed and the liquid is sent out via port 2 through the flow cell in which absorbance is monitored at

FIGURE 18-15 Mixing and reaction of sample with reagent in holding coil in sequential injection. After taking in sample (A) and reagent (B), flow is stopped (C) to allow product to form. Flow is then reversed (D) to pass product through the detector (E). [Adapted from tutorial by J. Ruzicka, *Flow Injection Analysis*, 4th ed., 2009, www.flowinjection.com.]

Holding coil   Sample   Valve   Detector

A   carrier

B   carrier   Reagent

C   Product

Flow stopped for mixing and reaction

D

E

Syringe pump

Holding coil

Optical fiber to spectrophotometer

Six-way valve

Optical fiber from light source

Detector flow cell

FIGURE 18-16 Sequential injection apparatus with spectrophotometric detection. The central feature of the equipment is the six-way valve, so sequential injection is sometimes called "lab-on-a-valve." [From tutorial by J. Ruzicka, *Flow Injection Analysis*, 4th ed., 2009, www.flowinjection.com.]

a selected wavelength. Volumes of each reagent and time in the mixing coil are under computer control. In Figure 18-14, two different reagents could be mixed with the sample. A powerful variant of the procedure is to stop the flow when the product zone reaches the detector and to measure the change of absorbance versus time as more product forms in the detector. Figure 18-16 shows sequential injection equipment, which can be smaller than a laptop computer. Figure 19-23 shows another example of sequential injection.

18-5 Immunoassays and Aptamers

An important application of fluorescence is in **immunoassays,** which employ antibodies to detect analyte. An *antibody* is a protein produced by the immune system of an animal in response to a foreign molecule called an *antigen*. The antibody recognizes the antigen that stimulated synthesis of the antibody. The formation constant for the antibody-antigen complex is large, whereas the binding of the antibody to other molecules is weak.

Figure 18-17 illustrates the principle of an *enzyme-linked immunosorbent assay*, abbreviated ELISA in biochemical literature. Antibody 1, which is specific for the analyte of interest (the antigen), is bound to a polymeric support. In steps 1 and 2, analyte is incubated with the polymer-bound antibody to form a complex. The fraction of antibody sites that bind analyte is proportional to the concentration of analyte in the unknown. The surface is then

Rosalyn Yalow received the Nobel Prize in medicine in 1977 for developing immunoassay techniques in the 1950s, using proteins labeled with radioactive ^{131}I to enable their detection.[18] Yalow, a physicist, worked with Solomon Berson, a medical doctor, in this pioneering effort.

Antibody 1

Polymer support

1. Add sample containing analyte
2. Wash to remove unbound molecules

Analyte protein bound to antibody 1

3. Add enzyme-labeled antibody 2
4. Wash to remove unbound antibody

Enzyme covalently bound to antibody 2

Antibody 2

FIGURE 18-17 Enzyme-linked immunosorbent assay. Antibody 1, which is specific for the analyte of interest, is bound to a polymer support and treated with unknown. After excess, unbound molecules have been washed away, the analyte remains bound to antibody 1. The bound analyte is then treated with antibody 2, which recognizes a different site on the analyte and to which an enzyme is covalently attached. After unbound material has been washed away, each molecule of analyte is coupled to an enzyme that will be used in Figure 18-18.

FIGURE 18-18 Enzyme bound to antibody 2 can catalyze reactions that produce colored or fluorescent products. Each molecule of analyte bound in the immunoassay leads to many molecules of colored or fluorescent product that are easily measured.

washed to remove unbound substances. In steps 3 and 4, the antibody-antigen complex is treated with antibody 2, which recognizes a different region of the analyte. Antibody 2 was specially prepared for the immunoassay by covalent attachment of an enzyme that will be used later in the process. Again, excess unbound substances are washed away.

The enzyme attached to antibody 2 is critical for quantitative analysis. Figure 18-18 shows two ways in which the enzyme can be used. The enzyme can transform a colorless reactant into a colored product. Because one enzyme molecule catalyzes the same reaction many times, many molecules of colored product are created for each analyte molecule. The enzyme thereby *amplifies* the signal in the chemical analysis. The higher the concentration of analyte in the original unknown, the more enzyme is bound and the greater the extent of the enzyme-catalyzed reaction. Alternatively, the enzyme can convert a nonfluorescent reactant into a fluorescent product. Colorimetric and fluorometric enzyme-linked immunosorbent assays are sensitive to less than a nanogram of analyte. Pregnancy tests are based on the immunoassay of a placental protein in urine.

Immunoassays in Environmental Analysis

Commercial immunoassay kits are available for screening and analysis of pesticides, industrial chemicals, explosives, and microbial toxins at the parts per trillion to parts per million levels in groundwater, soil, and food.[19] An advantage of screening in the field is that uncontaminated regions that require no further attention are readily identified. An immunoassay can be 20–40 times less expensive than a chromatographic analysis and can be completed in 0.3–3 h in the field, using 1-mL samples. Chromatography generally must be done in a lab and might require several days, because analyte must first be extracted or concentrated from liter-quantity samples to obtain a sufficient concentration.

Time-Resolved Fluorescence Immunoassays[20]

The sensitivity of fluorescence immunoassays can be enhanced by a factor of 100 (to detect 10^{-13} M analyte) with time-resolved measurements of luminescence from the lanthanide ion Eu^{3+}. Organic chromophores such as fluorescein are plagued by background fluorescence at 350–600 nm from solvent, solutes, and particles. This background decays to a negligible level within 100 μs after excitation. Sharp luminescence at 615 nm from Eu^{3+}, however, has a much longer lifetime, falling to 1/e ($= 37\%$) of its initial intensity in approximately 700 μs. In a time-resolved fluorescence measurement (Figure 18-19), luminescence is measured between 200 and 600 μs after a brief laser pulse at 340 nm. The next pulse is flashed at 1 000 μs, and the cycle is repeated approximately 1 000 times per second. Rejecting emission within 200 μs of the excitation eliminates most of the background fluorescence.

Figure 18-20 shows how Eu^{3+} can be incorporated into an immunoassay. A chelating group that binds lanthanide ions is attached to antibody 2 in Figure 18-17. While bound to the antibody, Eu^{3+} has weak luminescence. After all steps in Figure 18-17 have been completed, the pH of the solution is lowered in the presence of a soluble chelator that extracts Eu^{3+} into solution. Strong luminescence from the soluble metal ion is then easily detected by a time-resolved measurement.

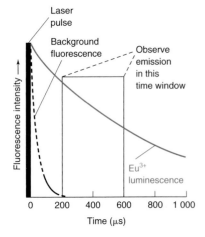

FIGURE 18-19 Emission intensity in a time-resolved fluorescence experiment.

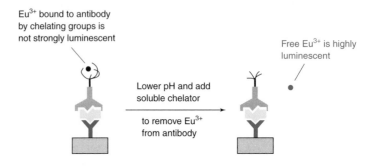

FIGURE 18-20 Antibody 2 in the immunosorbent assay of Figure 18-17 can be labeled with Eu^{3+}, which is not strongly luminescent when bound to the antibody. To complete the analysis, the pH of the solution is lowered to liberate strongly luminescent Eu^{3+} ion.

Aptamer

Flat aromatic molecule can be inserted between nucleotide base pairs of DNA

Citrulline enmeshed within aptamer

FIGURE 18-21 Aptamer specifically binds citrulline within a pocket in a short stretch of RNA. Long straight lines represent hydrogen-bonded nucleotide bases. The three-dimensional structure was deduced from nuclear magnetic resonance. [From M. Famulok, G. Mayer, and M. Blind, "Nucleic Acid Aptamers—From Selection in Vitro to Applications in Vivo," *Acc. Chem. Res.* **2000**, *33*, 591.]

Citrulline

Aptamers: Synthetic Nucleic Acid "Antibodies"

Aptamers are ~15–40-base pieces of DNA (deoxyribonucleic acid) or RNA (ribonucleic acid) that strongly and selectively bind to a specific molecule[21] or the surface of one type of living cell.[22] An aptamer for a desired target molecule is chosen from a pool of random DNA or RNA sequences by successive cycles of binding to the target, removing unbound material, and replicating the bound nucleic acid. Once the sequence of nucleic acids in an aptamer for a specific target is known, that aptamer can be synthesized in large quantities. The aptamer behaves as a custom-made, synthetic "antibody." An aptamer can bind to a section of a macromolecule, such as a protein, or it can engulf a small molecule, as in Figure 18-21.

Figure 18-21 also indicates that flat, aromatic molecules can be inserted between the flat, hydrogen-bonded nucleotide base pairs of DNA. We say that the flat molecule is *intercalated* between the base pairs. One of many analytical applications of aptamers utilizes a "molecular light switch" intercalated in the aptamer.

The ruthenium complex in the margin is slightly luminescent in aqueous solution but is strongly luminescent when intercalated in DNA. To measure the protein immunoglobulin E (IgE), an aptamer that binds IgE is selected. Ruthenium complex added to the aptamer is highly luminescent. When IgE is added to the intercalated aptamer, the aptamer binds to IgE and expels the ruthenium complex. Each "light switch" molecule that is expelled loses most of its luminescence. Loss of luminescence is proportional to IgE concentration in the range 0.1 to 1 nM.[23]

Unlike antibodies, which are fragile proteins that must be refrigerated for storage, aptamers have a long shelf life at room temperature. Aptamers have tremendous potential for use in highly specific chemical sensors.

Flat aromatic structure can intercalate into DNA

"Molecular light switch"

18-6 Sensors Based on Luminescence Quenching

When a molecule absorbs a photon, it is promoted to an excited state from which it can lose energy as heat or it can emit a lower energy photon (Figure 17-15). Box 18-1 describes how absorbed light can be converted into electricity. In this section, we discuss how excited-state molecules can serve as chemical sensors (Figure 18-22).

Luminescence Quenching

Suppose that the molecule M absorbs light and is promoted to the excited state M*:

Absorption: $\qquad$ $M + h\nu \rightarrow M^*$ $\qquad$ Rate $= \dfrac{d[M^*]}{dt} = k_a[M]$

The rate at which M* is created, $d[M^*]/dt$, is proportional to the concentration of M. The rate constant, k_a, depends on the intensity of illumination and the absorptivity of M. The more intense the light and the more efficiently it is absorbed, the faster M* will be created.

Following absorption, M* can emit a photon and return to the ground state:

Emission: $\qquad$ $M^* \rightarrow M + h\nu$ $\qquad$ Rate $= -\dfrac{d[M^*]}{dt} = k_e[M^*]$

FIGURE 18-22 Fiber-optic sensor measures O_2 by its ability to quench the luminescence of Ru(II) at the tip of the fiber. A blue-light-emitting diode provides excitation energy. Inset shows several probes. [Courtesy Ocean Optics, Dunedin, FL.]

BOX 18-1   Converting Light into Electricity

Earth's deserts receive 250–300 W/m^2 of solar irradiance. If solar energy could be used with 10% efficiency, 5% of the sunlight falling on the deserts would provide all the energy used in the world in 2007. The solar cell[24] shown below has a conversion efficiency close to 7%.

Sunlight enters the cell from the left through a transparent electrode made of fluorine-doped tin oxide. The electrode is coated with a 10-μm-thick layer of nanometer-size TiO$_2$ particles that are each coated with a *sensitizer*. The sensitizer is a Ru(II) complex that absorbs a large fraction of visible light. A layer of sensitizer coated on a planar electrode is so thin that it would absorb just 1% of the light. Nano-TiO$_2$ particles have so much surface area that there is enough sensitizer to absorb 99% of the light entering the cell.

When Ru(II) sensitizer absorbs light, it is promoted to an excited state from which an electron is injected into the *conduction band* of the semiconductor TiO$_2$ within ~50 fs. Instead of flowing back from TiO$_2$ to Ru(III), most electrons flow through an external circuit (where they can do useful work) to the Pt electrode at the right side of the cell. At the Pt surface, I$_3^-$ is reduced to I$^-$. The cycle is completed when I$^-$ reduces Ru(III) back to Ru(II):

$$Ru(II) + h\nu \quad \rightarrow \quad Ru(II)^* \qquad (*denotes\ excited\ state)$$
$$Ru(II)^* \quad \rightarrow \quad Ru(III) + e^- \ injected\ into\ TiO_2$$

$$e^-\ flows\ through\ circuit\ from\ tin\ oxide\ electrode\ to\ Pt$$

at Pt electrode:
$$I_3^- + 2e^- \quad \rightarrow \quad 3I^-$$
$$3I^- + 2Ru(III) \quad \rightarrow \quad I_3^- + 2Ru(II)$$

The graph shows the efficiency with which photons incident on the left side of the cell are converted into electrons in the circuit. The sensitizer absorbs much of the solar spectrum. The cell retains 92% of its initial efficiency after 1 000 h at 80°C. Dye-sensitized photocells suitable for external use on buildings are commercially available.[25] The sunlight-to-electric power conversion efficiency is 8 to 11%. Photovoltaic cells made of silicon are more efficient at converting sunlight into electricity, but dye-sensitized TiO$_2$ photocells have the possibility of generating electricity at lower cost. One focus of current research is to synthesize dyes that have higher solar absorptivity and that block I$_3^-$ from reduction at the TiO$_2$ surface.[26]

Scanning electron micrograph of sintered, nanocrystalline TiO$_2$. [From A. Hagfeldt and M. Grätzel, "Molecular Photovoltaics," *Acc. Chem. Res.* **2000**, *33*, 269.]

Photocell based on sensitizer-coated nano-TiO$_2$ particles.

The rate at which M* is lost is proportional to the concentration of M*. Alternatively, the excited molecule can lose energy in the form of heat:

Deactivation: M* → M + heat   Rate $= -\dfrac{d[M^*]}{dt} = k_d[M^*]$

Quenching is the process whereby emission from an excited molecule is decreased by energy transfer to another molecule (the quencher).

Still another possibility is that the excited molecule can transfer energy to a different molecule, called a *quencher* (Q), to promote the quencher to an excited state (Q*):

Quenching: M* + Q → M + Q*   Rate $= -\dfrac{d[M^*]}{dt} = k_q[M^*][Q]$

The excited quencher can then lose its energy by a variety of processes.

Under constant illumination, the system soon reaches a steady state in which the concentrations of M* and M remain constant. In the steady state, the rate of appearance of M* must equal the rate of destruction of M*. The rate of appearance is

$$Rate\ of\ appearance\ of\ M^* = \frac{d[M^*]}{dt} = k_a[M]$$

Photoaction spectrum showing the efficiency with which photons incident on the solar cell are converted into electrons in the circuit. [From P. Wang, C. Klein, R. Humphry-Baker, S. M. Zakeeruddin, and M. Grätzel, "A High Molar Extinction Coefficient Sensitizer for Stable Dye-Sensitized Solar Cells," *J. Am. Chem. Soc.* 2005, *127*, 808]

Converting sunlight directly into electricity is good, but converting sunlight into a fuel such as H_2 or CH_3OH would be even better. The fuel could be used when it is needed, not just when the sun is shining. Green leaves of plants use sunlight to reduce CO_2 to carbohydrates, a fuel that is oxidized back to CO_2 by animals and plants to provide energy.

We are rapidly exhausting Earth's supply of fossil fuels (coal, oil, and gas), which are also raw materials for plastics, fabrics, and many essential items. We face a major problem when the supply of raw materials is gone. Moreover, burning fossil fuels increases CO_2 in the atmosphere, which threatens to alter the climate.

Why do we burn irreplaceable raw materials? The shortsighted answer is that electricity from fossil-fuel-burning generators is relatively inexpensive and reliable. The charts show estimated costs of generating electricity in the year 2000 and estimated CO_2 emissions from each source of power. Fossil fuels produce the most CO_2. We will probably consume fossil fuel until this increasingly scarce resource becomes more expensive than the cost of renewable energy—which is likely to happen in your lifetime.

Nuclear energy is cost competitive with fossil fuel,[27] has very low greenhouse gas emission, and creates far less air pollution. Fear of accidents and intractable issues of waste containment have prevented construction of nuclear power plants in the United States for three decades.

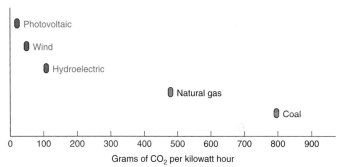

Estimated electricity costs and CO_2 emissions. Solar thermal electric power refers to sunlight concentrated by mirrors to produce high-temperature steam for generating electricity. [Cost is from L. Glickman, "Energy Efficiency in the Built Environment," *Physics Today*, July 2008, p. 35. CO_2 emissions from S. Pacca and A. Horvath, "Greenhouse Gas Emissions from Building and Operating Electric Power Plants in the Upper Colorado River Basin," *Environ. Sci. Technol.* 2002, *36*, 3194.]

The rate of disappearance is the sum of the rates of emission, deactivation, and quenching:

$$\text{Rate of disappearance of M*} = k_e[M^*] + k_d[M^*] + k_q[M^*][Q]$$

Setting the rates of appearance and disappearance equal to each other gives

$$k_a[M] = k_e[M^*] + k_d[M^*] + k_q[M^*][Q] \tag{18-22}$$

The **quantum yield** for a photochemical process is the fraction of absorbed photons that produce a desired result. If the result occurs every time a photon is absorbed, then the quantum yield is unity. The quantum yield is a number between 0 and 1.

The quantum yield for emission from M* is the rate of emission divided by the rate of absorption. In the absence of quencher, we designate this quantum yield Φ_0:

$$\Phi_0 = \frac{\text{photons emitted per second}}{\text{photons absorbed per second}} = \frac{\text{emission rate}}{\text{absorption rate}} = \frac{k_e[M^*]}{k_a[M]}$$

Substituting in the value of $k_a[M]$ from Equation 18-22 and setting $[Q] = 0$ gives an expression for the quantum yield of emission in the steady state:

Quantum yield for emission in the absence of quenching (Φ_0).

$$\Phi_0 = \frac{k_e[M^*]}{k_e[M^*] + k_d[M^*] + k_q[M^*][0]} = \frac{k_e}{k_e + k_d} \qquad (18\text{-}23)$$

If $[Q] \neq 0$, then the quantum yield for emission (Φ_Q) is

Quenching reduces the quantum yield for emission ($\Phi_Q < \Phi_0$).

$$\Phi_Q = \frac{k_e[M^*]}{k_e[M^*] + k_d[M^*] + k_q[M^*][Q]} = \frac{k_e}{k_e + k_d + k_q[Q]} \qquad (18\text{-}24)$$

In luminescence-quenching experiments, we measure emission in the absence and presence of a quencher. Equations 18-23 and 18-24 tell us that the relative yields are

Graph of Stern-Volmer equation:

Stern-Volmer equation:
$$\frac{\Phi_0}{\Phi_Q} = \frac{k_e + k_d + k_q[Q]}{k_e + k_d} = 1 + \left(\frac{k_q}{k_e + k_d}\right)[Q] \qquad (18\text{-}25)$$

The *Stern-Volmer equation* says that, if we measure relative emission (Φ_0/Φ_Q) as a function of quencher concentration and plot this quantity versus $[Q]$, we should observe a straight line. The quantity Φ_0/Φ_Q in Equation 18-25 is equivalent to I_0/I_Q, where I_0 is the emission intensity in the absence of quencher and I_Q is the intensity in the presence of quencher.

A Luminescent Intracellular O_2 Sensor

We illustrate our discussion with Ru(II) complexes, which strongly absorb visible light, efficiently emit light at significantly longer wavelengths than they absorb, are stable for long periods, and have a long-lived excited state whose emission is quenched by O_2 (Color Plate 17).[28] A widely used luminescent Ru(II) complex is $Ru(dpp)_3^{2+} \cdot 2Cl^-$.

$$\left(\begin{array}{c} C_6H_5 \\ \vdots \\ C_6H_5 \end{array}\right)_3 Ru(II) \equiv (dpp)_3Ru(II)$$

dpp = 4,7-diphenyl-1,10-phenanthroline

Oxygen is a good quencher because its ground state has two unpaired electrons—it is a *triplet* state designated $^3O_2$. O_2 has a low-lying *singlet* state with no unpaired electrons. Figure 17-15 showed that the lowest excited state of many molecules is a triplet. This triplet excited state $^3M^*$ can transfer energy to $^3O_2$ to produce a ground-state singlet molecule and excited $^1O_2^*$.

The ground state of Ru(II) is a singlet, and the lowest excited state is a triplet. When Ru(II) absorbs visible light, the excited singlet relaxes to the luminescent triplet state. O_2 quenches the luminescence by providing a radiationless pathway by which the triplet is converted into the ground-state singlet.

$$^3M^* \quad + \quad ^3O_2 \quad \rightarrow \quad ^1M \quad + \quad ^1O_2^*$$

| Excited state | Ground state | Ground state | Excited state |

There are two electron spins up and two down in both the reactants and products. This energy transfer therefore conserves the net spin and is more rapid than processes that change the spin.

Color Plate 18 shows tiny dots of light from fluorescent silica beads shot into living cells by a "gene-gun" that is usually used to infect cells with DNA-coated particles. Two dyes are bound inside each porous bead whose size is in the range 100 to 600 nm.

When illuminated with blue light, one dye emits green light near 525 nm and the other emits red light near 610 nm (Figure 18-23). The green dye is not affected by O_2, but the red ruthenium dye is affected. From the ratio of red-to-green emission intensity, we can calculate the concentration of O_2 in the vicinity of the beads. Results show that more than three-fourths of available intracellular O_2 is consumed by the cells within 2 min at 21°C after removing O_2 from the external liquid medium. The sensitivity of intracellular O_2 measurement has been increased further by using luminescent Pt-based dyes.[29]

Conditions	Intracellular O_2 (ppm)
Air-saturated buffer solution	8.8 ± 0.8
Cells in air-saturated buffer	7.9 ± 2.1
Cells in buffer 25 s after removing O_2	6.5 ± 1.7
Cells in buffer 120 s after removing O_2	≤ 1.5

Red dye,
sensitive to O_2

Green dye,
not sensitive
to O_2

Solution purged
with N_2

Purged
with air

Purged
with O_2

Fluorescence intensity →

Wavelength (nm)

500   550   600   650   700

FIGURE 18-23 (*a*) Fluorescence from O_2-indicator beads showing constant intensity near 525 nm and variable intensity near 610 nm. The ratio of emission intensity at these two wavelengths is related to O_2 concentration. [From H. Xu, J. W. Aylott, R. Kopelman, T. J. Miller, and M. A. Philbert, "Real-Time Method for Determination of O_2 Inside Living Cells Using Optical Nanosensors," *Anal. Chem.* **2001**, *73*, 4124.]

We end this chapter with Color Plate 19, which shows blue fluorescence from a solution irradiated with green light. Blue photons carry more energy than green photons, so how could this happen? You now possess knowledge to understand the explanation in Box 18-2.

BOX 18-2   Upconversion

Fluorescence and phosphorescence always come at lower energy than the excitation energy, as in Figure 17-18, because part of the excitation energy is converted to heat by vibrational relaxation such as R1 and R3 in Figure 17-15. Color Plate 19 shows *green* laser light shining into a solution that is emitting *blue* fluorescence.[30] Blue photons carry *more* energy than green photons. This *upconversion*, which creates high-energy photons from low-energy photons, does not violate conservation of energy because it requires *two* green photons to create *one* blue photon.

The solution contains a ruthenium(II) complex plus 9,10-diphenylanthracene in a deaerated organic solvent.

$Ru(II)$ ≡ L_3Ru
L = 4,4' dimethyl-
2,2'-bipyridine

≡ A
9,10-diphenyl-
anthracene

Absorption of green laser light promotes the ground state (S_0) Ru complex to its excited singlet state (S_1), which decays to the triplet state (T_1).

$$L_3Ru(S_0) \xrightarrow{\text{green photon}} L_3Ru^*(S_1) \xrightarrow{-\text{ heat}} L_3Ru^*(T_1)$$

We denote an excited state with an asterisk. The excited triplet $L_3Ru^*(T_1)$ is relatively long lived in the absence of O_2. It can transfer its excitation energy to ground-state anthracene to make an excited triplet state of anthracene.

$$L_3Ru^*(T_1) + A(S_0) \longrightarrow L_3Ru(S_0) + A^*(T_1)$$

The reaction of $L_3Ru^*(T_1)$ with $A(S_0)$ conserves electron-spin angular momentum, with two spins up and two spins down in reactants and products. In general, reactions that conserve spin angular momentum are faster than reactions in which spin changes. All reactions below conserve spin angular momentum.

Triplet anthracene survives long enough to allow two triplets to collide. One is promoted to the excited singlet state S_1 and the other is demoted to the ground state S_0.

$$A^*(T_1) + A^*(T_1) \longrightarrow A^*(S_1) + A(S_0)$$

Finally, excited singlet anthracene can emit a blue photon to return to the ground state.

$$A^*(S_1) \xrightarrow{\text{blue fluorescence}} A(S_0)$$

The net result is the conversion of two green photons absorbed by $L_3Ru(II)$ into one blue photon emitted by anthracene from its excited singlet state. This rare combination of cleverly selected reactions converts green light into blue light.

The measured *quantum yield* for this process is 3.3%. That is, for every 100 green photons absorbed, 3.3 blue photons are emitted.

Terms to Understand

aptamer

flow injection analysis

immunoassay

isosbestic point

method of continuous variation

quantum yield

quenching

Scatchard plot

sequential injection

Summary

The absorbance of a mixture is the sum of absorbances of the individual components. At a minimum, you should be able to find the concentrations of two species in a mixture by writing and solving two simultaneous equations for absorbance at two wavelengths. This procedure is most accurate if the two absorption spectra have regions where they do not overlap very much. With a spreadsheet, you should be able to use matrix operations to solve n simultaneous Beer's law equations for n components in a solution, with measurements at n wavelengths. You should be able to use Excel Solver to decompose a spectrum into a sum of spectra of the components by minimizing the function $(A_{calc} - A_m)^2$.

Isosbestic (crossing) points are observed when a solution contains variable proportions of two components with a constant total concentration. A Scatchard plot is used to measure an equilibrium constant, and the method of continuous variation allows us to determine the stoichiometry of a complex.

In flow injection analysis, sample injected into a flowing carrier stream is mixed with a color-forming reagent and passed into a flow-through detector. Analyte disperses and reacts with reagent without reaching equilibrium. Precision depends on carrying out the process reproducibly. In sequential injection, sample and reagent are separately aspirated into the carrier by a computer-controlled syringe pump. After being stationary in the reaction coil, analyte, product, and reagent are pushed through the detector.

Immunoassays use antibodies to detect the analyte of interest. In an enzyme-linked immunosorbent assay, signal is amplified by coupling analyte to an enzyme that catalyzes many cycles of a reaction producing either a colored or fluorescent product. Time-resolved fluorescence measurements provide sensitivity by separating analyte fluorescence in time and wavelength from background fluorescence. Aptamers are short pieces of DNA or RNA that are selected to bind tightly to a target molecule, which can be small or large. Once an aptamer is identified for a particular target, it can be synthesized and used in place of antibodies in chemical analyses.

Luminescence intensity is proportional to the concentration of the emitting species if the concentration is low enough. We can measure some analytes, such as O_2, by their ability to quench (decrease) the luminescence of another compound.

Exercises

18-A. 🖩 This problem can be worked with Equations 18-6 on a calculator or with the spreadsheet in Figure 18-5. Transferrin is the iron-transport protein found in blood. It has a molecular mass of 81 000 and carries two Fe^{3+} ions. Desferrioxamine B is a chelator used to treat patients with iron overload (Box 11-1). It has a molecular mass of about 650 and can bind one Fe^{3+}. Desferrioxamine can take iron from many sites within the body and is excreted (with its iron) through the kidneys. Molar absorptivities of these compounds (saturated with iron) at two wavelengths are given in the table. Both compounds are colorless (no visible absorption) in the absence of iron.

	$\varepsilon(M^{-1} cm^{-1})$	
$\lambda(nm)$	Transferrin	Desferrioxamine
428	3 540	2 730
470	4 170	2 290

(a) A solution of transferrin exhibits an absorbance of 0.463 at 470 nm in a 1.000-cm cell. Calculate the concentration of transferrin in milligrams per milliliter and the concentration of bound iron in micrograms per milliliter.

(b) After adding desferrioxamine (which dilutes the sample), the absorbance at 470 nm was 0.424, and the absorbance at 428 nm was 0.401. Calculate the fraction of iron in transferrin and the fraction in desferrioxamine. Remember that transferrin binds two iron atoms and desferrioxamine binds only one.

18-B. 🖩 The spreadsheet lists molar absorptivities of three dyes and the absorbance of a mixture of the dyes at visible wavelengths.

Use the least-squares procedure in Figure 18-3 to find the concentration of each dye in the mixture.

	A	B	C	D	E
1	Mixture of Dyes				
2					Absorbance
3	Wavelength	——— Molar Absorptivity ———			of mixture
4	(nm)	Tartrazine	Sunset Yellow	Ponceau 4R	Am
5	350	6.229E+03	2.019E+03	4.172E+03	0.557
6	375	1.324E+04	4.474E+03	2.313E+03	0.853
7	400	2.144E+04	7.403E+03	3.310E+03	1.332
8	425	2.514E+04	8.551E+03	4.534E+03	1.603
9	450	2.200E+04	1.275E+04	6.575E+03	1.792
10	475	1.055E+04	1.940E+04	1.229E+04	2.006
11	500	1.403E+03	1.869E+04	1.673E+04	1.821
12	525	0.000E+00	7.641E+03	1.528E+04	1.155
13	550	0.000E+00	3.959E+02	9.522E+03	0.445
14	575	0.000E+00	0.000E+00	1.814E+03	0.084

SOURCE: *J. J. B. Nevado, J. R. Flores, and M. J. V. Llerena, "Simultaneous Spectrophotometric Determination of Tartrazine, Sunset Yellow, and Ponceau 4R in Commercial Products," Fresenius J. Anal. Chem.* **1998,** *361, 465.*

18-C. Compound P, which absorbs light at 305 nm, was titrated with X, which does not absorb at this wavelength. The product, PX, also absorbs at 305 nm. Absorbance was measured in a 1.000-cm cell, and the concentration of free X was determined independently, with results in the table. Prepare a Scatchard plot and find the equilibrium constant for the reaction $X + P \rightleftharpoons PX$.

CHAPTER 18 Applications of Spectrophotometry

Experiment	P_0 (M)	X_0 (M)	A	[X] (M)
0	0.010 0	0	0.213	0
1	0.010 0	0.001 00	0.303	4.42×10^{-6}
2	0.010 0	0.002 00	0.394	9.10×10^{-6}
3	0.010 0	0.003 00	0.484	1.60×10^{-5}
4	0.010 0	0.004 00	0.575	2.47×10^{-5}
5	0.010 0	0.005 00	0.663	3.57×10^{-5}
6	0.010 0	0.006 00	0.752	5.52×10^{-5}
7	0.010 0	0.007 00	0.840	8.20×10^{-5}
8	0.010 0	0.008 00	0.926	1.42×10^{-4}
9	0.010 0	0.009 00	1.006	2.69×10^{-4}
10	0.010 0	0.010 00	1.066	5.87×10^{-4}
11	0.010 0	0.020 00	1.117	9.66×10^{-3}

18-D. Complex formation by 3-aminopyridine and picric acid in chloroform solution gives a yellow product with an absorbance maximum at 400 nm. Neither starting material absorbs significantly at this wavelength. Stock solutions containing 1.00×10^{-4} M of each compound were mixed as indicated, and the absorbances of the mixtures were recorded. Prepare a graph of absorbance versus mole fraction of 3-aminopyridine and find the stoichiometry of the complex.

Picric acid (mL)	3-Aminopyridine (mL)	Absorbance at 400 nm
2.70	0.30	0.106
2.40	0.60	0.214
2.10	0.90	0.311
1.80	1.20	0.402
1.50	1.50	0.442
1.20	1.80	0.404
0.90	2.10	0.318
0.60	2.40	0.222
0.30	2.70	0.110

SOURCE: E. Bruneau, D. Lavabre, G. Levy, and J. C. Micheau, "Quantitative Analysis of Continuous-Variation Plots with a Comparison of Several Methods," J. Chem. Ed. **1992**, 69, 833.

Problems

Analysis of a Mixture

18-1. 🖩 This problem can be worked by calculator or with the spreadsheet in Figure 18-5. Consider compounds X and Y in the example labeled "Analysis of a Mixture, Using Equations 18-6" in Section 18-1. Find [X] and [Y] in a solution whose absorbance is 0.233 at 272 nm and 0.200 at 327 nm in a 0.100-cm cell.

18-2. 🖩 The figure shows spectra of 1.00×10^{-4} M MnO_4^-, 1.00×10^{-4} M $Cr_2O_7^{2-}$, and an unknown mixture of both, all in 1.000-cm-pathlength cells. Absorbances are given in the table. Use the least-squares procedure in Figure 18-3 to find the concentration of each species in the mixture.

Wavelength (nm)	MnO_4^- standard	$Cr_2O_7^{2-}$ standard	Mixture
266	0.042	0.410	0.766
288	0.082	0.283	0.571
320	0.168	0.158	0.422
350	0.125	0.318	0.672
360	0.056	0.181	0.366

18-3. When are isosbestic points observed and why?

18-4. The metal ion indicator xylenol orange (Table 11-3) is yellow at pH 6 ($\lambda_{max} = 439$ nm). The spectral changes that occur as VO^{2+} is added to the indicator at pH 6 are shown here. The mole ratio VO^{2+}/xylenol orange at each point is

Trace	Mole ratio	Trace	Mole ratio	Trace	Mole ratio
0	0	6	0.60	12	1.3
1	0.10	7	0.70	13	1.5
2	0.20	8	0.80	14	2.0
3	0.30	9	0.90	15	3.1
4	0.40	10	1.0	16	4.1
5	0.50	11	1.1		

Suggest a sequence of chemical reactions to explain the spectral changes, especially the isosbestic points at 457 and 528 nm.

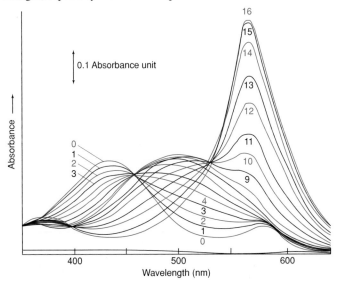

Absorption spectra for the reaction of xylenol orange with VO^{2+} at pH 6.0.
[From D. C. Harris and M. H. Gelb, "Binding of Xylenol Orange to Transferrin: Demonstration of Metal-Anion Linkage," Biochim. Biophys. Acta **1980**, 623, 1.]

Visible spectrum of MnO_4^-, $Cr_2O_7^{2-}$, and an unknown mixture containing both ions. [From M. Blanco, H. Iturriaga, S. Maspoch, and P. Tarín, "A Simple Method for Spectrophotometric Determination of Two-Components with Overlapped Spectra," J. Chem. Ed. **1989**, 66, 178.]

18-5. Infrared spectra are customarily recorded on a transmittance scale so that weak and strong bands can be displayed on the same scale. The region near $2\,000\ cm^{-1}$ in the infrared spectra of compounds A and B is shown in the figure. Note that absorption corresponds to a downward peak on this scale. The spectra were recorded from a $0.010\,0$ M solution of each, in cells with $0.005\,00$-cm pathlengths. A mixture of A and B in a $0.005\,00$-cm cell gave a transmittance of 34.0% at $2\,022\ cm^{-1}$ and 38.3% at $1\,993\ cm^{-1}$. Find the concentrations of A and B.

Wavenumber	Pure A	Pure B
$2\,022\ cm^{-1}$	31.0% T	97.4% T
$1\,993\ cm^{-1}$	79.7% T	20.0% T

18-6. Spectroscopic data for the indicators thymol blue (TB), semithymol blue (STB), and methylthymol blue (MTB) are shown in the table. A solution of TB, STB, and MTB in a 1.000-cm cuvet had absorbances of 0.412 at 455 nm, 0.350 at 485 nm, and 0.632 at 545 nm. Modify the spreadsheet in Figure 18-5 to handle three simultaneous equations and find [TB], [STB], and [MTB] in the mixture.

	ε $(M^{-1}\ cm^{-1})$		
λ (nm)	TB	STB	MTB
455	4 800	11 100	18 900
485	7 350	11 200	11 800
545	36 400	13 900	4 450

SOURCE: S. Kiciak, H. Gontarz, and E. Krzyżanowska, "Monitoring the Synthesis of Semimethylthymol Blue and Methylthymol Blue," *Talanta* **1995**, *42*, 1245.

18-7. The spreadsheet gives the product εb for four pure compounds and a mixture at infrared wavelengths. Modify Figure 18-5 to solve four equations and find the concentration of each compound. You can treat the coefficient matrix as if it were molar absorptivity because the pathlength was constant (but unknown) for all measurements.

Wavelength	Coefficient matrix (εb)				Absorbance
(μm)	p-xylene	m-xylene	o-xylene	ethylbenzene	of unknown
12.5	1.5020	0.0514	0	0.0408	0.1013
13.0	0.0261	1.1516	0	0.0820	0.09943
13.4	0.0342	0.0355	2.532	0.2933	0.2194
14.3	0.0340	0.0684	0	0.3470	0.03396

SOURCE: Z. Zdravkovski, "Mathcad in Chemistry Calculations. II. Arrays," *J. Chem. Ed.* **1992**, *69*, 242A.

18-8. A solution was prepared by mixing 25.00 mL of $0.080\,0$ M aniline, 25.00 mL of $0.060\,0$ M sulfanilic acid, and 1.00 mL of 1.23×10^{-4} M HIn and then diluting to 100.0 mL. (HIn stands for protonated indicator.)

Anilinium ion
$pK_a = 4.601$

Sulfanilic acid
$pK_a = 3.232$

$$HIn \rightleftharpoons H^+ + In^-$$

$\varepsilon_{325} = 2.45 \times 10^4\ M^{-1}\ cm^{-1}$ $\varepsilon_{325} = 4.39 \times 10^3\ M^{-1}\ cm^{-1}$
$\varepsilon_{550} = 2.26 \times 10^4\ M^{-1}\ cm^{-1}$ $\varepsilon_{550} = 1.53 \times 10^4\ M^{-1}\ cm^{-1}$

The absorbance measured at 550 nm in a *5.00-cm* cell was 0.110. Find the concentrations of HIn and In$^-$ and pK_a for HIn.

18-9. *Chemical equilibrium and analysis of a mixture. (Warning! This is a long problem.)* A remote optical sensor for CO_2 in the ocean was designed to operate without the need for calibration.[31]

$CO_2(g)$

$CO_2(aq) \rightleftharpoons CO_2(aq)$

$$CO_2(aq) + H_2O \overset{K_1}{\rightleftharpoons} HCO_3^- + H^+$$

$$HCO_3^- \overset{K_2}{\rightleftharpoons} CO_3^{2-} + H^+$$

$$HIn^- \overset{K_{In}}{\rightleftharpoons} In^{2-} + H^+$$

ε^{HIn}_{620} ε^{In}_{620}
ε^{HIn}_{434} ε^{In}_{434}

Seawater   Inside sensor compartment

Silicone membrane allows CO_2, but not other species, to diffuse across

The sensor compartment is separated from seawater by a silicone membrane through which CO_2, but not dissolved ions, can diffuse. Inside the sensor, CO_2 equilibrates with HCO_3^- and CO_3^{2-}. For each measurement, the sensor is flushed with fresh solution containing 50.0 μM bromothymol blue indicator (NaHIn) and 42.0 μM NaOH. All indicator is in the form HIn^- or In^{2-} near neutral pH, so we can write two mass balances: (1) $[HIn^-] + [In^{2-}] = F_{In} = 50.0\ \mu$M and (2) $[Na^+] = F_{Na} = 50.0\ \mu$M + 42.0 μM = 92.0 μM. HIn^- has an absorbance maximum at 434 nm and In^{2-} has a maximum at 620 nm. The sensor measures the absorbance *ratio* $R_A = A_{620}/A_{434}$ reproducibly without need for calibration. From this ratio, we can find $[CO_2(aq)]$ in the seawater as outlined here:

(a) From Beer's law for the mixture, write equations for $[HIn^-]$ and $[In^{2-}]$ in terms of the absorbances at 620 and 434 nm (A_{620} and A_{434}). Then show that

$$\frac{[In^{2-}]}{[HIn^-]} = \frac{R_A \varepsilon^{HIn^-}_{434} - \varepsilon^{HIn^-}_{620}}{\varepsilon^{In^{2-}}_{620} - R_A \varepsilon^{In^{2-}}_{434}} \equiv R_{In} \qquad (A)$$

(b) From the mass balance (1) and the acid dissociation constant K_{In}, show that

$$[HIn^-] = \frac{F_{In}}{R_{In} + 1} \qquad (B)$$

$$[In^{2-}] = \frac{K_{In}F_{In}}{[H^+](R_{In} + 1)} \qquad (C)$$

(c) Show that $[H^+] = K_{In}/R_{In}$. $\qquad (D)$

(d) From the carbonic acid dissociation equilibria, show that

$$[HCO_3^-] = \frac{K_1[CO_2(aq)]}{[H^+]} \tag{E}$$

$$[CO_3^{2-}] = \frac{K_1 K_2[CO_2(aq)]}{[H^+]^2} \tag{F}$$

(e) Write the charge balance for the solution in the sensor compartment. Substitute in expressions B, C, E, and F for $[HIn^-]$, $[In^{2-}]$, $[HCO_3^-]$, and $[CO_3^{2-}]$.

(f) Suppose that the various constants have the following values:

$$\varepsilon_{434}^{HIn^-} = 8.00 \times 10^3 \text{ M}^{-1} \text{ cm}^{-1} \qquad K_1 = 3.0 \times 10^{-7}$$

$$\varepsilon_{620}^{HIn^-} = 0 \qquad K_2 = 3.3 \times 10^{-11}$$

$$\varepsilon_{434}^{In^{2-}} = 1.90 \times 10^3 \text{ M}^{-1} \text{ cm}^{-1} \qquad K_{In} = 2.0 \times 10^{-7}$$

$$\varepsilon_{620}^{In^{2-}} = 1.70 \times 10^4 \text{ M}^{-1} \text{ cm}^{-1} \qquad K_w = 6.7 \times 10^{-15}$$

From the measured absorbance ratio $R_A = A_{620}/A_{434} = 2.84$, find $[CO_2(aq)]$ in the seawater.

(g) Approximately what is the ionic strength inside the sensor compartment? Were we justified in neglecting activity coefficients in working this problem?

Measuring an Equilibrium Constant

18-10. Figure 18-7 is a Scatchard plot for the addition of 0–20 nM antigen X to a fixed concentration of antibody P ($P_0 = 10$ nM). Prepare a Scatchard plot from the data in the table and find K for the reaction $P + X \rightleftharpoons PX$. The table gives measured concentrations of unbound X and the complex PX. It is recommended that the fraction of saturation should span the range ~0.2–0.8. What is the range of the fraction of saturation for the data?

[X] (nM)	[PX] (nM)	[X] (nM)	[PX] (nM)
0.12_0	2.8_7	0.73_1	6.2_9
0.19_2	3.8_0	1.2_2	6.7_7
0.29_6	4.6_6	1.5_0	7.5_2
0.45_0	5.5_4	3.6_1	8.4_5

18-11. 🖳 *Scatchard plot for binding of estradiol to albumin.* Data in the table come from a student experiment to measure the binding constant of the radioactively labeled hormone estradiol (X) to the protein, bovine serum albumin (P). Estradiol (7.5 nM) was equilibrated with various concentrations of albumin for 30 min at 37°C. A small fraction of unbound estradiol was removed by solid-phase microextraction (Section 23-4) and measured by liquid scintillation counting. Albumin is present in large excess, so its concentration in any given solution is essentially equal to its initial concentration in that solution. Call the initial concentration of estradiol $[X]_0$ and the final concentration of unbound estradiol $[X]$. Then bound estradiol is $[X]_0 - [X]$ and the equilibrium constant is

$$X + P \rightleftharpoons PX \qquad K = \frac{[PX]}{[X][P]} = \frac{[X]_0 - [X]}{[X][P]}$$

which you can rearrange to

$$\frac{[X]_0}{[X]} = K[P] + 1$$

A graph of $[X]_0/[X]$ versus $[P]$ should be a straight line with a slope of K. The quotient $[X]_0/[X]$ is equal to the counts of radioactive

estradiol extracted from a solution without albumin divided by the counts of estradiol extracted from a solution with estradiol.

(a) Prepare a graph of $[X]_0/[X]$ versus $[P]$ and find K. From the standard deviation of the slope, find the 95% confidence interval for K.

(b) What fraction of estradiol is bound to albumin at the first and last points?

$[X]_0/[X]$	$[P]$ (μM)	$[X]_0/[X]$	$[P]$ (μM)
1.26	6.3	3.33	50.0
1.62	10.0	4.19	60.0
2.16	20.0	4.13	70.0
2.51	30.0	4.36	80.0
3.34	40.0		

SOURCE: P. Liang, B. Adhyaru, W. L. Pearson, and K. R. Williams, "The Binding Constant of Estradiol to Bovine Serum Albumin," *J. Chem. Ed.* **2006**, *83*, 294.

18-12. Iodine reacts with mesitylene to form a complex with an absorption maximum at 332 nm in CCl_4 solution:

$$I_2 + \text{[mesitylene]} \rightleftharpoons \text{[complex]} \qquad K = \frac{[\text{complex}]}{[I_2][\text{mesitylene}]}$$

Iodine   Mesitylene   Complex
$\varepsilon_{332} \approx 0$ $\varepsilon_{332} \approx 0$ $\lambda_{max} = 332$ nm

(a) Given that the product absorbs at 332 nm, but neither reactant has significant absorbance at this wavelength, use the equilibrium constant, K, and Beer's law to show that

$$\frac{A}{[\text{mesitylene}][I_2]_{tot}} = K\varepsilon - \frac{KA}{[I_2]_{tot}}$$

where A is the absorbance at 332 nm, ε is the molar absorptivity of the complex at 332 nm, [mesitylene] is the concentration of free mesitylene, and $[I_2]_{tot}$ is the total concentration of iodine in the solution ($= [I_2] + [\text{complex}]$). The cell pathlength is 1.000 cm.

(b) Spectrophotometric data for this reaction are shown in the table. Because $[\text{mesitylene}]_{tot} \gg [I_2]$, we can say that $[\text{mesitylene}] \approx [\text{mesitylene}]_{tot}$. Prepare a graph of $A/([\text{mesitylene}][I_2]_{tot})$ versus $A/[I_2]_{tot}$ and find the equilibrium constant and molar absorptivity of the complex.

$[\text{Mesitylene}]_{tot}$ (M)	$[I_2]_{tot}$ (M)	Absorbance at 332 nm
1.690	7.817×10^{-5}	0.369
0.921 8	2.558×10^{-4}	0.822
0.633 8	3.224×10^{-4}	0.787
0.482 9	3.573×10^{-4}	0.703
0.390 0	3.788×10^{-4}	0.624
0.327 1	3.934×10^{-4}	0.556

SOURCE: P. J. Ogren and J. R. Norton, "Applying a Simple Linear Least-Squares Algorithm to Data with Uncertainties in Both Variables," *J. Chem. Ed.* **1992**, *69*, A130.

18-13. 🖳 Now we use Solver to find K for the previous problem. The only absorbing species at 332 nm is the complex, so, from Beer's law, $[\text{complex}] = A/\varepsilon$ (because pathlength = 1.000 cm). I_2 is either free or bound in the complex, so $[I_2] = [I_2]_{tot} - [\text{complex}]$. There is a huge excess of mesitylene, so $[\text{mesitylene}] \approx [\text{mesitylene}]_{tot}$.

$$K = \frac{[\text{complex}]}{[I_2][\text{mesitylene}]} = \frac{A/\varepsilon}{([I_2]_{tot} - A/\varepsilon)[\text{mesitylene}]_{tot}}$$

The spreadsheet shows some of the data. You will need to use all the data. Column A contains [mesitylene] and column B contains $[I_2]_{tot}$.

Column C lists the measured absorbance. *Guess* a value of the molar absorptivity of the complex, ε, in cell A7. Then compute the concentration of the complex ($= A/\varepsilon$) in column D. The equilibrium constant in column E is given by E2 $=$ [complex]/($[I_2]$[mesitylene]) $=$ (D2)/((B2-D2)*A2).

	A	B	C	D	E
1	[Mesitylene]	[I2]tot	A	[Complex] = A/ε	Keq
2	1.6900	7.817E-05	0.369	7.380E-05	9.99282
3	0.9218	2.558E-04	0.822	1.644E-04	1.95128
4	0.6338	3.224E-04	0.787	1.574E-04	1.50511
5				Average =	4.48307
6	Guess for ε:			Standard Dev =	4.77680
7	5.000E+03			Stdev/Average =	1.06552

What should we minimize with Solver? We want to vary ε in cell A7 until the values of K in column E are as constant as possible. We would like to minimize a function like $\Sigma(K_i - K_{average})^2$, where K_i is the value in each line of the table and $K_{average}$ is the average of all computed values. The problem with $\Sigma(K_i - K_{average})^2$ is that we can minimize this function simply by making K_i very small, but not necessarily constant. What we really want is for all the K_i to be clustered around the mean value. A good way to do this is to minimize the *relative standard deviation* of the K_i, which is (standard deviation)/average. In cell E5, we compute the average value of K and in cell E6 the standard deviation. Cell E7 contains the relative standard deviation. Use Solver to minimize cell E7 by varying cell A7. Compare your answer with that of Problem 18-12.

Method of Continuous Variation

18-14. *Method of continuous variation.* Make a graph of absorbance versus mole fraction of thiocyanate from the data in the table.

mL Fe³⁺ solution[a]	mL SCN⁻ solution[b]	Absorbance at 455 nm
30.00	0	0.001
27.00	3.00	0.122
24.00	6.00	0.226
21.00	9.00	0.293
18.00	12.00	0.331
15.00	15.00	0.346
12.00	18.00	0.327
9.00	21.00	0.286
6.00	24.00	0.214
3.00	27.00	0.109
0	30.00	0.002

a. Fe^{3+} solution: 1.00 mM $Fe(NO_3)_3$ + 10.0 mM HNO_3.

b. SCN^- solution: 1.00 mM KSCN + 15.0 mM HCl.

SOURCE: Z. D. Hill and P. MacCarthy, "Novel Approach to Job's Method," *J. Chem. Ed.* **1986**, *63*, 162.

(a) What is the stoichiometry of the predominant $Fe(SCN)_n^{3-n}$ species?

(b) Why is the peak not as sharp as those in Figure 18-8?

(c) Why does one solution contain 10.0 mM acid and the other 15.0 mM acid?

18-15. The indicator xylenol orange (Table 11-3) forms a complex with Zr(IV) in HCl solution. Prepare a Job plot from the data in the table and suggest the stoichiometry of the complex (xylenol orange)$_x$Zr$_z$.

Mole fraction xylenol orange	Relative absorbance at $\lambda_{max} = 550$ nm	Mole fraction xylenol orange	Relative absorbance at $\lambda_{max} = 550$ nm
0.100	0.110	0.600	0.236
0.200	0.235	0.700	0.145
0.300	0.352	0.800	0.088
0.400	0.412	0.900	0.045
0.500	0.348		

SOURCE: L. Perring and B. Bourqui, "An Alternative Method for Fluoride Determination by Ion Chromatography Using Post-Column Reaction," *Am. Lab. News Ed.*, March 2002, p. 28.

18-16. *Simulating a Job's plot.* Consider the reaction $A + 2B \rightleftharpoons AB_2$, for which $K = [AB_2]/[A][B]^2$. Suppose that the following mixtures of A and B at a fixed total concentration of 10^{-4} M are prepared:

$[A]_{total}$ (M)	$[B]_{total}$ (M)	$[A]_{total}$ (M)	$[B]_{total}$ (M)
1.00×10^{-5}	9.00×10^{-5}	5.00×10^{-5}	5.00×10^{-5}
2.00×10^{-5}	8.00×10^{-5}	6.00×10^{-5}	4.00×10^{-5}
2.50×10^{-5}	7.50×10^{-5}	7.00×10^{-5}	3.00×10^{-5}
3.00×10^{-5}	7.00×10^{-5}	8.00×10^{-5}	2.00×10^{-5}
3.33×10^{-5}	6.67×10^{-5}	9.00×10^{-5}	1.00×10^{-5}
4.00×10^{-5}	6.00×10^{-5}		

(a) Prepare a spreadsheet to find the concentration of AB_2 for each mixture, for equilibrium constants of $K = 10^6$, 10^7, and 10^8. One way to do this is to enter the values of $[A]_{total}$ and $[B]_{total}$ in columns A and B, respectively. Then put a trial (guessed) value of $[AB_2]$ in column C. From the mass balances $[A]_{total} = [A] + [AB_2]$ and $[B]_{total} = [B] + 2[AB_2]$, we can write $K = [AB_2]/[A][B]^2 = [AB_2]/\{([A]_{total} - [AB_2])([B]_{total} - 2[AB_2])^2\}$. In column D, enter the reaction quotient $[AB_2]/[A][B]^2$. For example, cell D2 has the formula "=C2/((A2-C2)(B2-2*C2)^2)". Then vary the value of $[AB_2]$ in cell C2 with Solver until the reaction quotient in cell D2 is equal to the desired equilibrium constant (such as 10^8).

(b) Prepare a graph by the method of continuous variation in which you plot $[AB_2]$ versus mole fraction of A for each equilibrium constant. Explain the shapes of the curves.

18-17. A study was conducted with derivatives of the DNA nucleotide bases adenine and thymine bound inside micelles (Box 25-1) in aqueous solution.

Adenine derivative   Thymine derivative   Sodium dodecyl sulfate

Sodium dodecyl sulfate forms micelles with the hydrocarbon tails pointed inward and ionic headgroups exposed to water. It was hypothesized that the bases would form a 1:1 hydrogen-bonded complex inside the micelle as they do in DNA:

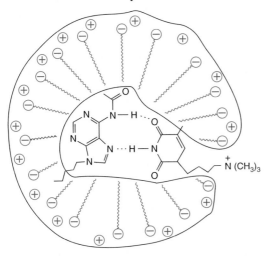

Hydrogen-bonded base pair inside micelle, with hydrocarbon tails anchoring the bases to the micelle.

To test the hypothesis, aliquots of 5.0 mM adenine derivative were mixed with aliquots of 5.0 mM thymine derivative in proportions shown in the table. Each solution also contained 20 mM sodium dodecyl sulfate. The concentration of product measured by nuclear magnetic resonance also is shown in the table. Are the results consistent with formation of a 1:1 complex? Explain your answer.

Adenine volume (mL)	Thymine volume (mL)	Concentration of product (mM)
0.450	0.050	0.118 ± 0.009
0.400	0.100	0.202 ± 0.038
0.350	0.150	0.265 ± 0.021
0.300	0.200	0.307 ± 0.032
0.250	0.250	0.312 ± 0.060
0.200	0.300	0.296 ± 0.073
0.150	0.350	0.260 ± 0.122
0.100	0.400	0.187 ± 0.110
0.050	0.450	0.103 ± 0.104

Flow Injection, Luminescence, and Immunoassay

18-18. Explain what is done in flow injection analysis and sequential injection. What is the principal difference between the two techniques? Which one is called "lab-on-a-valve"?

18-19. Explain how signal amplification is achieved in enzyme-linked immunosorbent assays.

18-20. What is the advantage of a time-resolved emission measurement with Eu^{3+} versus measurement of fluorescence from organic chromophores?

18-21. Here is an *immunoassay* to measure explosives such as trinitrotoluene (TNT) in organic solvent extracts of soil. The assay employs a *flow cytometer*, which counts small particles (such as living cells) flowing through a narrow tube past a detector. The cytometer in this experiment irradiates the particles with a green laser and measures fluorescence from each particle as it flows past the detector.

1. Antibodies that bind TNT are chemically attached to 5-μm-diameter latex beads.

2. The beads are incubated with a fluorescent derivative of TNT to saturate the antibodies, and excess TNT derivative is removed. The beads are resuspended in aqueous detergent.

Trinitrotoluene (TNT)

Fluorescently labeled TNT

3. 5 μL of the suspension are added to 100 μL of sample or standard. TNT in the sample or standard displaces some derivatized TNT from bound antibodies. The higher the concentration of TNT, the more derivatized TNT is displaced.

4. An aliquot is injected into the flow cytometer, which measures fluorescence of individual beads as they pass the detector. The figure shows median fluorescence intensity ± standard deviation. TNT can be quantified in the ppb to ppm range.

Fluorescence of TNT-antibody-beads versus TNT concentration. [From G. P. Anderson, S. C. Moreira, P. T. Charles, I. L. Medintz, E. R. Goldman, M. Zeinali, and C. R. Taitt, "TNT Detection Using Multiplexed Liquid Array Displacement Immunoassays," *Anal. Chem.* **2006**, *78*, 2279.]

Draw pictures showing the state of the beads in steps 1, 2, and 3 and explain how this method works.

18-22. The graph shows the effect of pH on quenching of luminescence of tris(2,2'-bipyridine)Ru(II) by 2,6-dimethylphenol. The ordinate, K_{SV}, is the collection of constants, $k_q/(k_e + k_d)$, in the Stern-Volmer equation 18-25. The greater K_{SV}, the greater the quenching. Suggest an explanation for the shape of the graph and estimate pK_a for 2,6-dimethylphenol.

Tris(2,2'-bipyridine)ruthenium(II)

2,6-Dimethylphenol

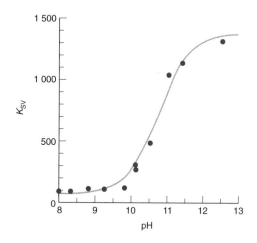

Quenching of tris(2,2'-bipyridine)Ru(II) by 2,6-dimethylphenol as a function of pH. [Data from H. Gsponer, G. A. Argüello, and G. A. Argüello, "Determinations of pK_a from Luminescence Quenching Data," *J. Chem. Ed.* **1997**, *74*, 968.]

18-23. *Fluorescence quenching in micelles.* Consider an aqueous solution with a high concentration of micelles (Box 25-1) and relatively low concentrations of the fluorescent molecule pyrene and a quencher (cetylpyridinium chloride, designated Q), both of which dissolve in the micelles.

Pyrene
(fluorescent species)

$CH_2(CH_2)_{14}CH_3$

N^+ Cl^-

Cetylpyridinium chloride
(quencher, Q)

Quenching occurs if pyrene and Q are in the same micelle. Let the total concentration of quencher be [Q] and the concentration of micelles be [M]. The average number of quenchers per micelle is $\overline{Q}$ = [Q]/[M]. If Q is randomly distributed among the micelles, then the probability that a particular micelle has *n* molecules of Q is given by the *Poisson distribution*:[32]

$$\text{Probability of } n \text{ molecules of Q in micelle} \equiv P_n = \frac{\overline{Q}^n}{n!}e^{-\overline{Q}} \quad (1)$$

where *n*! is *n* factorial (= *n*[*n* − 1][*n* − 2] ... [1]). The probability that there are no molecules of Q in a micelle is

$$\text{Probability of } 0 \text{ molecules of Q in micelle} = \frac{\overline{Q}^0}{0!}e^{-\overline{Q}} = e^{-\overline{Q}} \quad (2)$$

because 0! ≡ 1.

Let I_0 be the fluorescence intensity of pyrene in the absence of Q and let I_Q be the intensity in the presence of Q (both measured at the same concentration of micelles). The quotient I_Q/I_0 must be $e^{-\overline{Q}}$, which is the probability that a micelle does not possess a quencher molecule. Substituting $\overline{Q}$ = [Q]/[M] gives

$$I_Q/I_0 = e^{-\overline{Q}} = e^{-[Q]/[M]} \quad (3)$$

Micelles are made of the surfactant molecule, sodium dodecyl sulfate (shown in Problem 18-17). When surfactant is added to a solution, no micelles form until a minimum concentration called the *critical micelle concentration* (CMC) is attained. When the total concentration of surfactant, [S], exceeds the critical concentration, then the surfactant found in micelles is [S] − [CMC]. The molar concentration of micelles is

$$[M] = \frac{[S] - [CMC]}{N_{av}} \quad (4)$$

where N_{av} is the average number of molecules of surfactant in each micelle.

Combining Equations 3 and 4 gives an expression for fluorescence as a function of total quencher concentration, [Q]:

$$\ln\frac{I_0}{I_Q} = \frac{[Q]N_{av}}{[S] - [CMC]} \quad (5)$$

By measuring fluorescence intensity as a function of [Q] at fixed [S], we can find the average number of molecules of S per micelle if we know the critical micelle concentration (which is independently measured in solutions of S). The table gives data for 3.8 μM pyrene in a micellar solution with a total concentration of sodium dodecyl sulfate [S] = 20.8 mM.

Q (μM)	I_0/I_Q	Q (μM)	I_0/I_Q	Q (μM)	I_0/I_Q
0	1	158	2.03	316	4.04
53	1.28	210	2.60	366	5.02
105	1.61	262	3.30	418	6.32

SOURCE: *M. F. R. Prieto, M. C. R. Rodríguez, M. M. González, A. M. R. Rodríguez, and J. C. M. Fernández, "Fluorescene Quenching in Microheterogeneous Media," J. Chem. Ed.* **1995**, *72*, 662.

(a) If micelles were not present, quenching would be expected to follow the Stern-Volmer equation 18-25. Show that the graph of I_0/I_Q versus [Q] is not linear.

(b) The critical micelle concentration is 8.1 mM. Prepare a graph of $\ln(I_0/I_Q)$ versus [Q]. Use Equation 5 to find N_{av}, the average number of sodium dodecyl sulfate molecules per micelle.

(c) Find the concentration of micelles, [M], and the average number of molecules of Q per micelle, $\overline{Q}$, when [Q] = 0.200 mM.

(d) Compute the fractions of micelles containing 0, 1, and 2 molecules of Q when [Q] = 0.200 mM.

19 Spectrophotometers

Cavity ring-down spectrum of ~3 mbar of CO_2, which is similar to the concentration in human breath. [From E. R. Crosson, K. N. Ricci, B. A. Richman, F. C. Chilese, T. G. Owano, R. A. Provencal, M. W. Todd, J. Glasser, A. A. Kachanov, B. A. Paldus, T. G. Spence, and R. N. Zare, "Stable Isotope Ratios Using Cavity Ring-Down Spectroscopy: Determination of $^{13}C/^{12}C$ for Carbon Dioxide in Human Breath," *Anal. Chem.* **2002,** *74,* 2003.]

Cavity ring-down can measure absorbance as low as ~10^{-6} and has the potential to provide sensitive detectors for chromatography.[1] In panel *a,* a laser pulse is directed into a cavity with mirrors on both ends. If the mirror reflectivity is 99.98%, then 0.02% of the power penetrates mirror M_1 and enters the cavity. The laser is shut off and light inside the cavity bounces back and forth, losing 0.02% of its intensity each time it strikes a mirror. A detector outside mirror M_2 measures light leaking through M_2. Graph *b* shows decay of the detector signal from a cavity containing a nonabsorbing liquid. If an absorbing species is present, decay is faster because signal is lost by absorption during each pass between the mirrors. The difference in signal decay time with and without absorber provides a measure of absorbance. The effective pathlength is ~10^3 times the length of the cavity because light makes ~10^3 passes between mirrors during the measurement. A commercial cavity ring-down spectrometer for measurement of isotopes of gases such as CO_2, CH_4, NH_3, H_2S, HF, H_2CO, and C_2H_4 has an effective pathlength of nearly 20 km.[2]

Spectrum *c* shows absorbance measured for $CO_2(g)$ with the natural mixture of 98.9% ^{12}C and 1.1% ^{13}C. Peaks arise from transitions between rotational levels of two vibrational states. The spectral region was chosen to include a strong absorption of $^{13}CO_2$ and a weak absorption of $^{12}CO_2$, so the intensities of the isotopic peaks are similar. Each point in the spectrum was obtained by varying the laser frequency.

The areas of the $^{13}CO_2$ and $^{12}CO_2$ peaks from human breath were used to determine whether a patient was infected with *Helicobacter pylori,* a bacterium that causes ulcers. After a patient ingests ^{13}C-urea, *H. pylori* converts ^{13}C-urea into $^{13}CO_2$, which appears in the patient's breath. The ratio $^{13}C/^{12}C$ in the breath of an infected person increases by 1–5%, whereas the ratio $^{13}C/^{12}C$ from an uninfected person is constant to within 0.1%.

Figure 17-4 showed the essential features of a *single-beam spectrophotometer.* Light from a *source* is separated into narrow bands of wavelength by a *monochromator,* passed through a sample, and measured by a *detector.* We measure the *irradiance* (P_0, watts/m^2) striking the detector with a *reference* cell (a solvent blank or reagent blank) in the sample compartment. When the reference is replaced by the sample of interest, some radiation is usually absorbed and the irradiance striking the detector, P, is smaller than P_0. The quotient P/P_0, which is a number between 0 and 1, is the *transmittance, T. Absorbance,* which is proportional to concentration, is $A = \log P_0/P = -\log T$.

A single-beam spectrophotometer is inconvenient because the sample and reference must be placed alternately in the beam. For measurements at multiple wavelengths, the reference

Transmittance: $T = P/P_0$

Absorbance: $A = -\log T = \varepsilon bc$

ε = molar absorptivity
b = pathlength
c = concentration

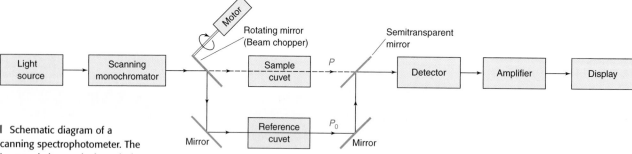

FIGURE 19-1 Schematic diagram of a double-beam scanning spectrophotometer. The incident beam is passed alternately through the sample and reference cuvets by the rotating beam chopper.

must be run at each wavelength. A single-beam instrument is poorly suited to measuring absorbance as a function of time, as in a kinetics experiment, because both the source intensity and the detector response slowly drift.

Figure 19-1 shows a *double-beam spectrophotometer*, in which light alternately passes through the sample and the reference (blank), directed by a rotating mirror (the *chopper*) into and out of the light path. When light passes through the sample, the detector measures irradiance P. When the chopper diverts the beam through the reference cuvet, the detector measures P_0. The beam is chopped several times per second, and the circuitry automatically compares P and P_0 to obtain transmittance and absorbance. This procedure provides automatic correction for changes of source intensity and detector response with time and wavelength, because the power emerging from the two samples is compared so frequently. Most research-quality spectrophotometers provide automatic wavelength scanning and continuous recording of absorbance versus wavelength.

It is routine to first record a *baseline* spectrum with reference solution in both cuvets. The baseline absorbance at each wavelength is then subtracted from the measured absorbance of the sample to obtain the true absorbance of the sample at each wavelength.

Components of a double-beam ultraviolet-visible spectrophotometer are shown in Figure 19-2. Visible light comes from a quartz-halogen lamp (like that in an automobile

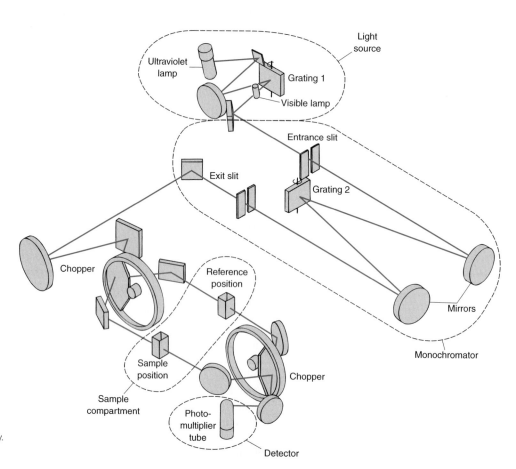

FIGURE 19-2 Optical train of Varian Cary 3E ultraviolet-visible double-beam spectrophotometer. [Courtesy Varian Australia Pty. Ltd., Victoria, Australia.]

(a)

FIGURE 19-3 (a) Thermo Scientific Evolution 600 ultraviolet-visible double-beam spectrophotometer. (b) Optical train of Evolution 600, showing layout of components. [Courtesy Thermo Fisher Scientific, Madison, WI.]

(b)

headlight), and the ultraviolet source is a deuterium arc lamp that emits in the range 200 to 400 nm. Only one lamp is used at a time. Grating 1 selects a narrow band of wavelengths to enter the monochromator, which in turn selects an even narrower band to pass through the sample. After being chopped and transmitted through the sample and reference cells, the signal is detected by a *photomultiplier tube*, which creates an electric current proportional to irradiance. Figure 19-3 shows a research-quality double-beam spectrophotometer. Now we describe the components of the spectrophotometer in more detail.

19-1 Lamps and Lasers: Sources of Light

A *tungsten lamp* is an excellent source of continuous visible and near-infrared radiation. A typical tungsten filament operates at a temperature near 3 000 K and produces useful radiation in the range 320 to 2 500 nm (Figure 19-4). This range covers the entire visible region and parts of the infrared and ultraviolet regions as well. Ultraviolet spectroscopy normally employs a *deuterium arc lamp* in which a controlled electric discharge causes D_2 to dissociate and emit ultraviolet radiation from 110 to 400 nm (Figure 19-4). In a typical ultraviolet-visible spectrophotometer, a switch is made between deuterium and tungsten lamps when passing through 360 nm, so that the source with the highest intensity is always employed. For selected visible and ultraviolet frequencies, electric discharge arc lamps filled with mercury vapor (Box 17-2) or xenon gas are widely used. Light-emitting diodes provide narrow bands of visible and near-infrared (close to visible) radiation.[3]

Infrared radiation in the range 4 000 to 200 cm^{-1} is commonly obtained from a silicon carbide *globar*, heated to near 1 500 K by an electric current. The globar emits radiation with approximately the same spectrum as a *blackbody* at 1 000 K (Box 19-1).

Lasers provide isolated lines of a single wavelength for many applications. A laser with a wavelength of 3 μm might have a **bandwidth** (range of wavelengths) of 3×10^{-14} to 3×10^{-8} μm. The bandwidth is measured where the radiant power falls to half of its maximum value. The brightness of a low-power laser at its output wavelength is 10^{13} times greater than that of the sun at its brightest (yellow) wavelength. (Of course, the sun emits all wavelengths, whereas the laser emits only a narrow band. The total brightness of the sun is much greater than that of the laser.) The angular divergence of the laser beam from its direction of travel is typically less than 0.05°, a property that allows us to illuminate a small target. Laser light is typically *plane polarized*, with the electric field

FIGURE 19-4 Intensities of a tungsten filament at 3 200 K and a deuterium lamp.

Caution Ultraviolet radiation is harmful to the naked eye. Do not view an ultraviolet source without protection.

Properties of laser light:

monochromatic:	one wavelength
extremely bright:	high power at one wavelength
collimated:	parallel rays
polarized:	electric field oscillates in one plane
coherent:	all waves in phase

BOX 19-1 **Blackbody Radiation and the Greenhouse Effect**

When an object is heated, it emits radiation—it glows. Even at room temperature, objects emit infrared radiation. Imagine a hollow sphere whose inside surface is perfectly black. That is, the surface absorbs all radiation striking it. If the sphere is at constant temperature, it must emit as much radiation as it absorbs, or else its temperature would rise. If a small hole were made in the wall, we would observe that the escaping radiation has a continuous spectral distribution. The object is called a *blackbody*, and the radiation is called **blackbody radiation.** Emission from real objects such as the tungsten filament of a light bulb resembles that from an ideal blackbody.

The power per unit area radiating from the surface of an object is called the *exitance* (or *emittance*), *M*. For a blackbody, exitance is given by

Exitance from blackbody: $\qquad M = \sigma T^4$

where σ is the Stefan-Boltzmann constant ($5.669\ 8 \times 10^{-8}$ W/($m^2 \cdot K^4$)). A blackbody whose temperature is 1 000 K radiates 5.67×10^4 watts per square meter of surface area. If the temperature is doubled, the exitance increases by a factor of $2^4 = 16$.

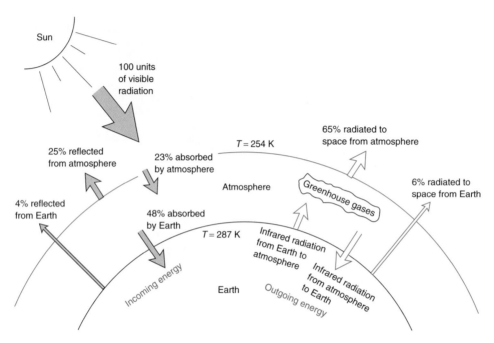

Balance between energy reaching Earth from the sun and energy reradiated to space. Exchange of infrared radiation between Earth and its atmosphere keeps Earth's surface 33 K warmer than the upper atmosphere.

oscillating in one plane perpendicular to the direction of travel (Figure 17-1). Laser light is *coherent*, which means that all waves emerging from the laser oscillate in phase with one another.

A necessary condition for lasing is *population inversion*, in which a higher energy state has a greater population, *n*, than a lower energy state in the lasing medium. In Figure 19-5a, this condition occurs when the population of state E_2 exceeds that of E_1. Molecules in ground state E_0 of the lasing medium are *pumped* to excited state E_3 by broadband radiation from a powerful lamp or by an electric discharge. Molecules in state E_3 rapidly relax to E_2, which has a relatively long lifetime. After a molecule in E_2 decays to E_1, it rapidly relaxes to the ground state, E_0 (thereby keeping the population of E_2 greater than the population of E_1).

A photon with an energy that exactly spans two states can be absorbed to raise a molecule to an excited state. Alternatively, that same photon can stimulate the excited molecule to emit a photon and return to the lower state. This is called *stimulated emission*. When a photon emitted by a molecule falling from E_2 to E_1 strikes another molecule in E_2, a second photon can be emitted with the same phase and polarization as the incident photon. If there is a

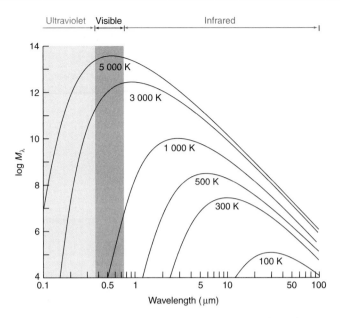

Spectral distribution of blackbody radiation. Both axes are logarithmic. The family of curves is called the *Planck distribution* after the German physicist, Max Planck (1858–1947). His derivation of the law governing blackbody radiation in 1900 relied on the hypothesis that electromagnetic energy could only be emitted in discrete quanta. Planck received the Nobel Prize in 1918.

The graph shows that maximum blackbody emission for an object near 300 K occurs at infrared wavelengths (~10 μm). The surface of the sun behaves like a blackbody with a temperature near 5 800 K, emitting mainly *visible* light (~0.5 μm = 500 nm).

Of the solar flux of 1 368 W/m^2 reaching Earth's upper atmosphere, 23% is absorbed by the atmosphere and 25% is reflected back into space. Earth absorbs 48% of the solar flux and reflects 4%. Radiation reaching Earth should be just enough to keep the surface temperature at 254 K, which would not support life as we know it. Why does the average temperature of Earth's surface stay at a comfortable 287 K?

The blackbody curves tell us that Earth radiates mainly *infrared* radiation, rather than visible light. Although the atmosphere is transparent to incoming visible light, it absorbs outgoing infrared radiation. The main absorbers, called *greenhouse gases*, are water[4] and CO_2 and, to lesser extents, O_3, CH_4, chlorofluorocarbons, and N_2O. Radiation emitted from Earth is absorbed by the atmosphere and part of it is reradiated back to Earth. The atmosphere behaves like an insulating blanket, maintaining Earth's surface temperature 33 K warmer than the temperature of the upper atmosphere.[5]

Man's activities since the dawn of the Industrial Revolution have increased atmospheric CO_2 through the burning of fossil fuel. Relative to preindustrial conditions, CO_2 in 2005 was estimated to contribute an additional 1.7 W/m^2 of radiative heating of Earth's surface.[6] CO_2 is the largest anthropogenic source of radiative heating. The chart shows that Earth's surface was ~0.8°C warmer in 2000–2005 than it was in 1850–1899.[6] Will there be disastrous climatic changes? Will there be compensating responses that lead to little temperature change? We cannot accurately answer these questions, but prudence suggests that we should avoid making such large relative changes in our atmosphere.

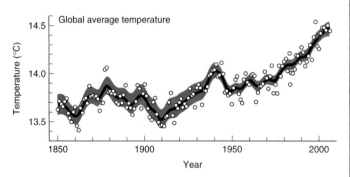

Estimate of global temperature based on proxy data such as tree rings and isotope ratios in sediments and ice cores. The 1990s was the warmest decade in 2 000 years. [From Report of the Intergovernmental Panel on Climate Change, 2007.[6] See also M. E. Mann and P. D. Jones, "Global Surface Temperatures over the Past Two Millennia," *Geophys. Res. Lett.* 2003, *30*, 1820.]

population inversion ($n_2 > n_1$), one photon stimulates the emission of many photons as it travels through the laser.

Figure 19-5b shows essential components of a laser. Pump energy directed through the side of the lasing medium creates the population inversion. One end of the laser cavity is a mirror that reflects all light (0% transmittance). The other end is a partially transparent mirror that reflects most light (1% transmittance). Photons with energy $E_2 - E_1$ bouncing back and forth between the mirrors stimulate an avalanche of new photons. The small fraction of light passing through the partially transparent mirror at the right is the useful output of the laser.

A helium-neon laser is a common source of red light with a wavelength of 632.8 nm and an output power of 0.1–25 mW. An electric discharge pumps helium atoms to state E_3 in Figure 19-5. The excited helium transfers energy by colliding with a neon atom, raising the neon to state E_2. The high concentration of helium and intense electric pumping create a population inversion among neon atoms.

In a *laser diode*, population inversion of charge carriers in a semiconductor is achieved by a high electric field across a *pn* junction in GaAs,[7] GaN,[8] or compositions such as $Al_xGa_{1-x}N$. Commonly available laser diodes cover the range 360–1 550 nm.

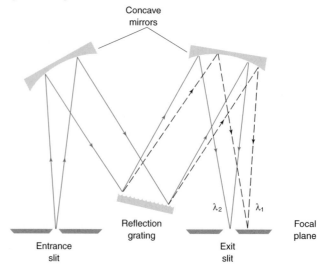

FIGURE 19-5 (*a*) Energy-level diagram illustrating the principle of operation of a laser. (*b*) Basic components of a laser. The population inversion is created in the lasing medium. Pump energy might be derived from either intense lamps or an electric discharge.

19-2 Monochromators

A **monochromator** disperses light into its component wavelengths and selects a narrow band of wavelengths to pass on to the sample or detector. The monochromator in Figure 19-2 consists of entrance and exit slits, mirrors, and a *grating* to disperse the light. *Prisms* were used instead of gratings in older instruments.

Gratings[9]

Grating: optical element with closely spaced lines
Diffraction: bending of light by a grating
Refraction: bending of light by a lens or prism

A **grating** is a reflective or transmissive optical component with a series of closely ruled lines. When light is reflected from or transmitted through the grating, each line behaves as a separate source of radiation. Different wavelengths of light are reflected or transmitted at different angles from the grating (Color Plate 12). The bending of light rays by a grating is called **diffraction.** (The bending of light rays by a prism or lens, which is called *refraction*, is discussed in Section 19-4.)

Polychromatic: many wavelengths
Monochromatic: one wavelength

In the grating monochromator in Figure 19-6, *polychromatic* radiation from the entrance slit is *collimated* (made into a beam of parallel rays) by a concave mirror. These rays fall on a reflection grating, whereupon different wavelengths are diffracted at different angles. The light strikes a second concave mirror, which focuses each wavelength at a different point on the focal plane. The orientation of the reflection grating directs only one narrow band of wavelengths to the exit slit of the monochromator. Rotation of the grating allows different wavelengths to pass through the exit slit.

FIGURE 19-6 Czerny-Turner grating monochromator.

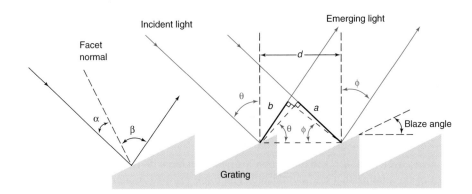

FIGURE 19-7 Principle of a reflection grating.

The reflection grating in Figure 19-7 is ruled with a series of closely spaced, parallel grooves with a repeat distance d. The grating is coated with aluminum to make it reflective. A thin layer of silica (SiO_2) on top of the aluminum protects the metal from oxidation, which would reduce its reflectivity. When light is reflected from the grating, each groove behaves as a source of radiation. When adjacent light rays are in phase, they reinforce one another. When they are not in phase, they partially or completely cancel one another (Figure 19-8).

Consider the incident and emerging light rays shown in Figure 19-7. Fully constructive interference occurs when the difference in length of the two paths is an integral multiple of the wavelength of light. The difference in pathlength is $a - b$ in Figure 19-7. Constructive interference occurs if

$$n\lambda = a - b \tag{19-1}$$

where the diffraction order $n = \pm 1, \pm 2, \pm 3, \pm 4, \ldots$. The interference maximum for which $n = \pm 1$ is called *first-order diffraction*. When $n = \pm 2$, we have *second-order diffraction*, and so on.

In Figure 19-7, the incident angle θ is defined to be positive. The diffraction angle ϕ goes in the opposite direction, so, by convention, ϕ is negative. It is possible for ϕ to be on the same side of the facet normal as θ, in which case ϕ would be positive. In Figure 19-7, $a = d\sin\theta$ and $b = -d\sin\phi$ (because ϕ is negative and $\sin\phi$ is negative). Substituting into Equation 19-1 gives the condition for constructive interference:

Grating equation: $$n\lambda = d(\sin\theta + \sin\phi) \tag{19-2}$$

where d is the distance between adjacent grooves. For each incident angle, θ, there is a series of reflection angles, ϕ, at which a given wavelength will produce maximum constructive interference (Color Plate 20).

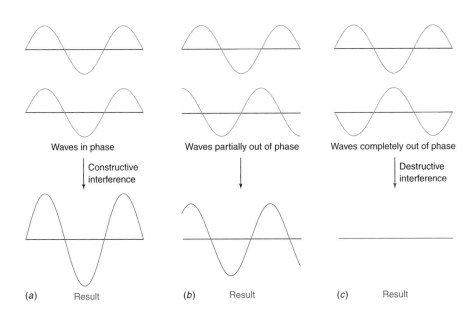

(a) Waves in phase — Constructive interference — Result

(b) Waves partially out of phase — Result

(c) Waves completely out of phase — Destructive interference — Result

FIGURE 19-8 Interference of adjacent waves that are (a) 0°, (b) 90°, and (c) 180° out of phase.

Resolution, Dispersion, and Efficiency of a Grating

Resolution measures the ability to separate two closely spaced peaks. The greater the resolution, the smaller is the difference ($\Delta\lambda$) between two wavelengths that can be distinguished from each other. The precise definition (which is beyond the scope of this discussion) means that the valley between the two peaks is about three-fourths of the height of the peaks when they are just barely resolved. The resolution of a grating is given by

Resolution of grating:
$$\frac{\lambda}{\Delta\lambda} = nN \qquad (19\text{-}3)$$

where λ is wavelength, n is the diffraction order in Equation 19-1, and N is the number of grooves of the grating that are illuminated. The more grooves in a grating, the better the resolution between closely spaced wavelengths. If we need to resolve lines that are 0.05 nm apart at a wavelength of 500 nm, the required resolution is $\lambda/\Delta\lambda = 500$ nm/0.05 nm $= 10^4$. Equation 19-3 tells us that, if we desire a first-order resolution of 10^4, there must be 10^4 grooves in the grating. If the grating has a ruled length of 10 cm, we require 10^3 grooves/cm.

Dispersion measures the ability to separate wavelengths differing by $\Delta\lambda$ through the difference in angle, $\Delta\phi$ (radians). For the grating in Figure 19-7, the dispersion is

Dispersion of grating:
$$\frac{\Delta\phi}{\Delta\lambda} = \frac{n}{d\cos\phi} \qquad (19\text{-}4)$$

where n is the diffraction order. Dispersion and resolution both increase with decreasing groove spacing. Equation 19-4 tells us that a grating with 10^3 grooves/cm provides a resolution of 0.102 radians (5.8°) per micrometer of wavelength if $n = 1$ and $\phi = 10°$. Wavelengths differing by 1 μm would be separated by an angle of 5.8°.

To select a narrower band of wavelengths from the monochromator, we decrease the exit slit width in Figure 19-6. Decreasing the exit slit width decreases the energy reaching the detector. Thus, *resolution of closely spaced absorption bands is achieved at the expense of decreased signal-to-noise ratio.* For quantitative analysis, a monochromator bandwidth that is ≲1/5 of the width of the absorption band (measured at half the peak height) is reasonable.

The relative *efficiency* of a grating (which is typically 45–80%) is defined as

$$\text{Relative efficiency} = \frac{E_\lambda^n(\text{grating})}{E_\lambda(\text{mirror})} \qquad (19\text{-}5)$$

where E_λ^n (grating) is the irradiance at a particular wavelength diffracted in the order of interest, n, and E_λ(mirror) is the irradiance at the same wavelength that would be reflected by a mirror with the same coating as the grating. Efficiency is partially controlled by the *blaze angle* at which the grooves are cut in Figure 19-7. Reflection from a flat surface is maximum when the angle of incidence (α) is equal to the angle of reflection (β). These angles are measured with respect to the facet normal in Figure 19-7. To optimize diffraction for a certain wavelength, the blaze angle is chosen to satisfy the two conditions $n\lambda = d(\sin\theta + \sin\phi)$ and $\alpha = \beta$. Each grating is optimized for a limited range of wavelengths (Figure 19-9), so a spectrophotometer may require several different gratings to scan through its entire spectral range.

Resolution: ability to distinguish two closely spaced peaks

Dispersion: ability to produce angular separation of adjacent wavelengths

Trade-off between resolution and signal: The narrower the exit slit, the greater the resolution and the noisier the signal.

Specular reflection: For a flat reflective surface, such as a mirror, the angle of incidence (α in Figure 19-7) is equal to the angle of reflection (β).

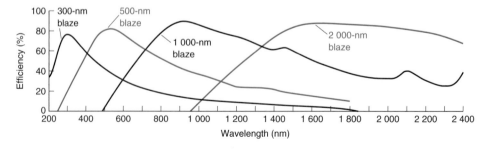

FIGURE 19-9 Efficiencies of diffraction gratings with 3 000 grooves/cm and blaze angles optimized for different wavelengths. [Courtesy Princeton Instruments, Trenton, NJ.]

Choosing the Monochromator Bandwidth

The wider the exit slit in Figure 19-6, the wider the band of wavelengths selected by the monochromator. We usually measure slit width in terms of the bandwidth of radiation selected by the slit. Instead of saying that a slit is 0.3 mm wide, we might say that the *bandwidth* getting through the slit is 1.0 nm.

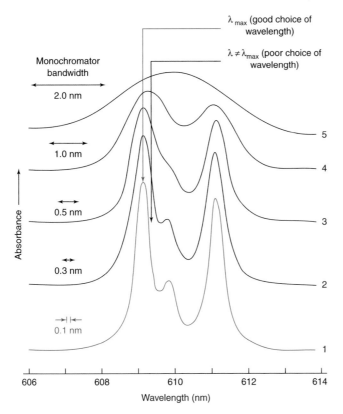

FIGURE 19-10 Increasing monochromator bandwidth broadens the bands and decreases the apparent absorbance of Pr^{3+} in a crystal of yttrium aluminum garnet (a laser material). [Courtesy M. D. Seltzer, Michelson Laboratory, China Lake, CA.]

A wide slit increases the energy reaching the detector and gives a high signal-to-noise ratio, leading to good precision in measuring absorbance. However, Figure 19-10 shows that, if the bandwidth is large relative to the width of the peak being measured, peak shape is distorted. We choose a bandwidth as wide as the spectrum permits to allow the most possible light to reach the detector. A monochromator bandwidth that is 1/5 as wide as the absorption peak generally gives acceptably small distortion of the peak shape.[10]

Monochromator bandwidth should be as large as possible, but small compared with the width of the peak being measured.

Stray Light

In every instrument, some **stray light** (wavelengths outside the bandwidth expected from the monochromator) reaches the detector. Stray light coming through the monochromator from the light source arises from diffraction into unwanted orders and angles and unintended scatter from optical components and walls. Stray light can also come from outside the instrument if the sample compartment is not perfectly sealed. Entry holes for tubing or electrical wires required at the sample for some experiments should be sealed to reduce stray light. Error from stray light is most serious when the sample absorbance is high (Figure 19-11) because the stray light constitutes a large fraction of the light reaching the detector. Stray light is expressed as a percentage of P_0, which is the irradiance reaching the detector in the absence of the sample.

High-quality spectrometers could have two monochromators in series (called a *double monochromator*) to reduce stray light. Unwanted radiation that passes through the first monochromator is rejected by the second monochromator.

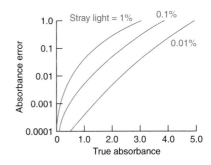

EXAMPLE Stray Light

If the true absorbance of a sample is 2.00 and there is 1.0% stray light, find the apparent absorbance.

Solution A true absorbance of 2.00 means that the true transmittance is $T = 10^{-A} = 10^{-2.00} = 0.010 = 1.0\%$. Transmittance is the irradiance passing through the sample, P, divided by the irradiance passing through the reference, P_0: $T = P/P_0$. If stray light with irradiance S passes through both the sample and the reference, the apparent transmittance is

$$\text{Apparent transmittance} = \frac{P + S}{P_0 + S} \qquad (19\text{-}6)$$

FIGURE 19-11 Absorbance error introduced by different levels of stray light. Stray light is expressed as a percentage of the irradiance incident on the sample. [From M. R. Sharp, "Stray Light in UV-VIS Spectrophotometers," *Anal. Chem.* 1984, *53*, 339A.]

If $P/P_0 = 0.010$ and there is 1.0% stray light, then $S = 0.010$ and the apparent transmittance is

$$\text{Apparent transmittance} = \frac{P + S}{P_0 + S} = \frac{0.010 + 0.010}{1 + 0.010} = 0.019_8$$

The apparent absorbance is $-\log T = -\log(0.019_8) = 1.70$, instead of 2.00.

Test Yourself What level of stray light gives an absorbance error of 0.01 at an absorbance of 2? That is, what value of S gives an apparent absorbance of 1.99? (**Answer:** $S = 0.000\ 23_{54} = 0.024\%$)

Stray light in research-quality instruments can be 0.01% to 0.000 1%, or even less.

Table 19-1 gives the absorbance of a solution that you can prepare to test the accuracy of absorbance measurements on your spectrophotometer. Absorbance accuracy is affected by all components of the spectrophotometer, as well as stray light. Two wavelength accuracy standards are a solution of holmium oxide[11] for absorption and a mercury-argon discharge lamp for emission.[12]

TABLE 19-1 Calibration standard for ultraviolet absorbance

Wavelength (nm)	Absorbance of $K_2Cr_2O_7$ (60.06 mg/L) in 5.0 mM H_2SO_4 in 1-cm cell
235	0.748 ± 0.010
257	0.865 ± 0.010
313	0.292 ± 0.010
350	0.640 ± 0.010

SOURCE: S. Ebel, "Validation of Analysis Methods," *Fresenius J. Anal. Chem.* **1992**, *342*, 769.

Filters

It is frequently necessary to *filter* (remove) bands of radiation from a signal. For example, the grating monochromator in Figure 19-6 directs first-order diffraction of a small wavelength band to the exit slit. (By "first order," we mean $n = 1$ in Equation 19-2.) Let λ_1 be the wavelength whose first-order diffraction reaches the exit slit. Equation 19-2 shows that, if $n = 2$, the wavelength $\frac{1}{2}\lambda_1$ also reaches the same exit slit because $\frac{1}{2}\lambda_1$ gives constructive interference at the same angle as λ_1. For $n = 3$, the wavelength $\frac{1}{3}\lambda_1$ also reaches the slit. One solution for selecting just λ_1 is to place a filter in the beam, so that wavelengths $\frac{1}{2}\lambda_1$ and $\frac{1}{3}\lambda_1$ will be blocked. To scan a wide range of wavelengths, it may be necessary to use several filters and to change them as the wavelength changes.

The simplest filter is colored glass, which absorbs a broad portion of the spectrum and transmits other portions. For finer control, *interference filters* and *holographic filters* are constructed to pass radiation in the region of interest and reflect other wavelengths (Figure 19-12). These devices derive their performance from constructive or destructive interference of light waves within the filter. Some holographic notch filters have such a sharp cutoff that it is possible to attenuate the Rayleigh line in Raman spectroscopy (Box 17-3) by a factor of 10^6 while allowing signals to be observed 100 cm^{-1} away from the Rayleigh line.

19-3 Detectors

A detector produces an electric signal when it is struck by photons. For example, a **phototube** emits electrons from a photosensitive, negatively charged surface (the cathode) when struck by visible light or ultraviolet radiation. Electrons flow through a vacuum to a positively charged collector whose current is proportional to the radiation intensity.

Figure 19-13 shows that detector response depends on the wavelength of the incident photons. For example, for a given radiant power (W/m^2) of 420-nm light, the S-20 photomultiplier produces a current about four times greater than the current produced for the same radiant power of 300-nm radiation. The response below 280 nm and above 800 nm is essentially 0. In a single-beam spectrophotometer, the 100% transmittance control must be readjusted

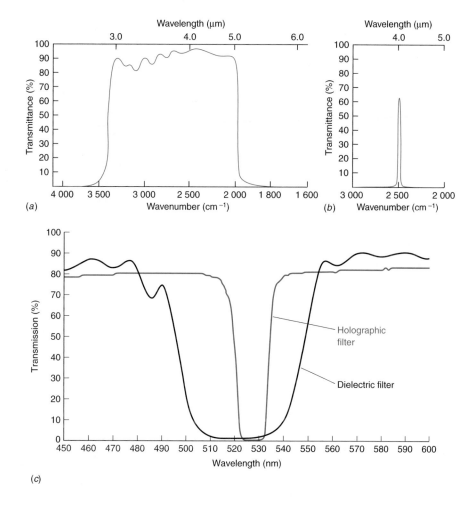

(a)

(b)

(c)

FIGURE 19-12 Transmission of filters. (a) Wide band-pass dielectric interference filter has ~90% transmission in the 3- to 5-μm wavelength range but <0.01% transmittance outside this range. (b) Narrow band-pass filter dielectric interference has a transmission width of 0.1 μm centered at 4 μm. [Courtesy Barr Associates, Westford, MA.] (c) Holographic interference filter provides greater attenuation and narrower band-pass than dielectric filter. [From H. Owen, "The Impact of Volume Phase Holographic Filters and Gratings on the Development of Raman Instrumentation," J. Chem. Ed. 2007, 84, 61.]

each time the wavelength is changed. This calibration adjusts the spectrophotometer to the maximum detector output that can be obtained at each wavelength. Subsequent readings are scaled to the 100% reading.

FIGURE 19-13 Detector response. The greater the sensitivity, the greater the current or voltage produced by the detector for a given incident irradiance (W/m^2) of photons. Each curve is normalized to a maximum value of 1. In$_x$Ga$_{1-x}$As response can be shifted to longer or shorter wavelength by varying the composition. [Courtesy Barr Associates, Westford, MA. GaN data from APA Optics, Blaine, MN.]

Photomultiplier Tube

A **photomultiplier tube** (Figure 19-14) is a very sensitive device in which electrons emitted from the photosensitive surface strike a second surface, called a *dynode*, which is positive with respect to the photosensitive emitter. Electrons are accelerated and strike the dynode with more than their original kinetic energy. Each energetic electron knocks more than one electron from the dynode. These new electrons are accelerated toward a second dynode, which is more positive than the first dynode. Upon striking the second dynode, even more electrons are knocked off and accelerated toward a third dynode. This process is repeated several times, so more than 10^6 electrons are finally collected for each photon striking the first surface. Extremely low light intensities are translated into measurable electric signals. As sensitive as a photomultiplier is, your eye is even more sensitive (Box 19-2).

FIGURE 19-14 Photomultiplier tube with nine dynodes. Amplification of the signal occurs at each dynode, which is approximately 90 volts more positive than the previous dynode. [Photo by David J. Green/Alamy.]

Many electrons emitted from dynode 1 for each electron striking dynode 1

Photoelectrons emitted from cathode

Transparent window

Incident radiation ($h\nu$)

Dynode

Grill

Photoemissive cathode

Anode, > 10^6 electrons for each photon

(a)

(b)

All photodetectors produce a small response in the absence of light. This *dark current* could arise from spontaneous emission of electrons from the cathode of a photomultiplier tube or spontaneous generation of electrons and holes in a semiconductor device. For example, atomic vibrations can provide enough energy to an electron for it to escape from the cathode. The higher the temperature of the cathode, the greater the dark current.

BOX 19-2   The Most Important Photoreceptor

The retina at the back of your eye contains photosensitive cells, called *rods* and *cones*, that are sensitive to levels of light varying over seven orders of magnitude. Light impinging on these cells is translated into nerve impulses that are transmitted by the optic nerve to the brain. Rod cells detect the dimmest light but cannot distinguish colors. Cone cells operate in bright light and give us color vision.

A stack of about 1 000 *disks* in each rod cell contains the light-sensing protein *rhodopsin*,[13] in which the chromophore 11-*cis*-retinal (from vitamin A) is attached to the protein *opsin*. When light is absorbed by rhodopsin, a series of rapid transformations releases all-*trans*-retinal. At this stage, the pigment is *bleached* (loses all color) and cannot respond to more light until retinal isomerizes back to the 11-*cis* form and recombines with the protein.

In the dark, there is a continuous flow of 10^9 Na^+ ions per second out of the rod cell's inner segment, through the adjoining medium, and into the cell's outer segment. An energy-dependent process using adenosine triphosphate (ATP) and oxygen pumps Na^+ out of the cell. Another process involving a molecule called cyclic GMP keeps the gates of the outer segment open for ions to flow back into the cell. When light is absorbed and rhodopsin is bleached, a series of reactions leads to destruction of cyclic GMP and shutdown of the channels through which Na^+ flows into the cell. A single photon reduces the ion current by 3%—corresponding to a decreased current of 3×10^7 ions per second. This *amplification* is greater than that of a photomultiplier tube, which is one of the most sensitive man-made photodetectors. The ion current returns to its dark value as the protein and retinal recombine and cyclic GMP is restored to its initial concentration.

Na$^+$

Disks

Plasma membrane

Outer segment

Cytoplasm

Na$^+$

Inner segment

Nucleus

Synapse

Rod cell

cis double bond

NH—Protein

Chromophore

Rhodopsin

Light

Opsin +

All-*trans*-retinal

Bleached products

For a refresher on semiconductors, see Section 14-8.

FIGURE 19-15 (*a*) Schematic cross-sectional view of photodiode array. (*b*) Photograph of array with 1 024 elements, each 25 μm wide and 2.5 mm high. The central black rectangle is the photosensitive area. The entire chip is 5 cm in length. [Courtesy Oriel Corporation, Stratford, CT.]

Photodiode Array

Spectrophotometers with a photomultiplier tube detector scan slowly through a spectrum one wavelength at a time. A photodiode array spectrophotometer records the entire spectrum at once in a fraction of a second. One application of rapid scanning is chromatography, in which the full spectrum of a compound is recorded as it emerges from the column.

At the heart of rapid spectroscopy is the **photodiode array** shown in Figure 19-15 (or the charge coupled device described later). Rows of *p*-type silicon on a substrate (the underlying body) of *n*-type silicon create a series of *pn* junction diodes. A reverse bias is applied to each diode, drawing electrons and holes away from the junction. In the depletion region at each junction, there are few electrons and holes. The junction acts as a capacitor, with charge stored on either side of the depletion region. At the beginning of the measurement cycle, each diode is fully charged.

When radiation strikes the semiconductor, free electrons and holes are created and migrate to regions of opposite charge, partially discharging the capacitor. The more radiation that strikes each diode, the less charge remains at the end of the measurement. The longer the array is irradiated between readings, the more each capacitor is discharged. The state of each capacitor is determined at the end of the cycle by measuring the current needed to recharge the capacitor.

In a *dispersive spectrometer* (Figure 19-1), only one narrow band of wavelengths reaches the detector at any time. In a *photodiode array spectrophotometer* (Figure 19-16), all wavelengths are recorded simultaneously, allowing more rapid acquisition of the spectrum or higher signal-to-noise ratio, or some combination of both. In the photodiode array spectrophotometer, *white light* (with all wavelengths) passes through the sample. The light then enters a **polychromator,** which disperses the light into its component wavelengths and directs the light at the diode array. Each diode receives a *different wavelength*, and all wavelengths are measured simultaneously. Resolution depends on how closely spaced the diodes are and how much dispersion is produced by the polychromator.

A photodiode array spectrophotometer measures all wavelengths at once, giving faster acquisition and higher signal-to-noise ratio.

Photodiode arrays allow faster spectral acquisition (<1 s) than dispersive instruments (which require several minutes). Photodiode array instruments have almost no moving parts, so they are more rugged than dispersive instruments that must rotate the grating and change filters to scan through the spectrum. The resolution of ~0.1 nm attainable with a dispersive instrument and the wavelength accuracy are better than those of a photodiode array (~0.5–1.5 nm resolution). Stray light is less in a dispersive instrument than in a photodiode instrument, thus giving the dispersive instrument a greater dynamic range for measuring high absorbance. Stray light in a photodiode array instrument is not substantially increased when the sample compartment is open. In a dispersive instrument, the compartment must be tightly closed.

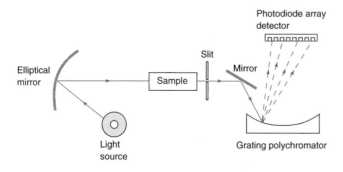

FIGURE 19-16 Design of photodiode array spectrophotometer.

Charge Coupled Device[14]

A **charge coupled device** is an extremely sensitive detector that stores photo-generated charge in a two-dimensional array. A charge coupled device can produce a higher signal-to-noise ratio than can be obtained with a photomultiplier tube. The device in Figure 19-17a is constructed of p-doped Si on an n-doped substrate. The structure is capped with an insulating layer of SiO_2, on top of which is placed a pattern of conducting Si electrodes. When light is absorbed in the p-doped region, an electron is introduced into the conduction band and a hole is left in the valence band. The electron is attracted to the region beneath the positive electrode, where it is stored. The hole migrates to the n-doped substrate, where it combines with an electron. Each electrode can store $\sim 10^5$ electrons before electrons spill out into adjacent elements.

The charge coupled device is a two-dimensional array, as shown in Figure 19-17b. After the desired observation time, electrons stored in each *pixel* (picture element) of the top row are moved into the serial register at the top and then moved, one pixel at a time, to the top right position, where the charge is read out. Then the next row is moved up and read out, and the sequence is repeated until the entire array has been read. The transfer of stored charges is carried out by an array of electrodes considerably more complex than we have indicated in

(a)　　　(b)

(c)

FIGURE 19-17 Schematic representation of a charge coupled device. (*a*) Cross-sectional view, indicating charge generation and storage in each pixel. (*b*) Top view, showing two-dimensional nature of an array. An actual array is about the size of a postage stamp. (*c*) Effect of multiplier stage on signal-to-noise ratio of a signal that is too weak to be seen without multiplication. [From C. G. Coates, "A Sensitive Detector of Ultralow-Light Imaging," *Am. Lab.*, August 2004, p. 13. Courtesy C. G. Coates, Andor Technology, Belfast, Ireland.]

Signal acquisition time (s)	Photodiode array		Photomultiplier tube		Charge coupled device	
	Ultraviolet	Visible	Ultraviolet	Visible	Ultraviolet	Visible
1	6 000	3 300	30	122	31	17
10	671	363	6.3	26	3.1	1.7
100	112	62	1.8	7.3	0.3	0.2

SOURCE: R. B. Bilhorn, J. V. Sweedler, P. M. Epperson, and M. B. Denton, "Charge Transfer Device Detectors for Analytical Optical Spectroscopy," Appl. Spectros. *1987, 41, 1114.*

Figure 19-17a. Charge transfer from one pixel to the next is extremely efficient, with a loss of approximately five of every million electrons.

The minimum detectable signal for visible light in Table 19-2 is 17 photons/s. The sensitivity of the charge coupled device is derived from its high *quantum efficiency* (electrons generated per incident photon), low background electrical noise (thermally generated free electrons), and low noise associated with readout.

The most sensitive charge coupled devices have a "multiplication stage," which multiplies the signal by $\sim 10^2$ to 10^3 between the serial register and the output amplifier. Noise that occurs during signal collection is also multiplied, but noise associated with readout is not. For the weakest signals in which readout is the dominant source of noise, multiplication increases the signal-to-noise ratio (Figure 19-17c). For cases in which the dominant noise occurs during signal collection, multiplication cannot improve the signal-to-noise ratio, but multiplication does decrease the time needed to collect the signal.

Infrared Detectors

Detectors for visible and ultraviolet radiation rely on incoming photons to eject electrons from a photosensitive surface or to promote electrons from the valence band of silicon to the conduction band. Infrared photons do not have sufficient energy to generate a signal in either kind of detector. Therefore, other kinds of devices are used for infrared detection.

A **thermocouple** is a junction between two different electrical conductors. Electrons have lower free energy in one conductor than in the other, so they flow from one to the other until the resulting voltage difference prevents further flow. The junction potential is temperature dependent because electrons flow back to the high-energy conductor at higher temperature. If a thermocouple is blackened to absorb radiation, its temperature (and hence voltage) becomes sensitive to radiation. A typical sensitivity is 6 V per watt of radiation absorbed.

A **ferroelectric material,** such as deuterated triglycine sulfate, has a permanent electric polarization because of alignment of the molecules in the crystal. One face of the crystal is positively charged and the opposite face is negative. The polarization is temperature dependent, and its variation with temperature is called the *pyroelectric effect*. When the crystal absorbs infrared radiation, its temperature and polarization change. The voltage change is the signal in a pyroelectric detector. Deuterated triglycine sulfate is a common detector in Fourier transform spectrometers described later in this chapter.

A **photoconductive detector** is a semiconductor whose conductivity increases when infrared radiation excites electrons from the valence band to the conduction band. **Photovoltaic detectors** contain *pn* junctions, across which an electric field exists. Absorption of infrared radiation creates electrons and holes, which are attracted to opposite sides of the junction and which change the voltage across the junction. Mercury cadmium telluride ($Hg_{1-x}Cd_xTe$, $0 < x < 1$) is a detector material whose sensitivity to different wavelengths is affected by the stoichiometry coefficient, x. Photoconductive and photovoltaic devices can be cooled to 77 K (liquid nitrogen temperature) to reduce thermal electrical noise by more than an order of magnitude.

The infrared spectrophotometer set up by Charles David Keeling (Section 0-1) to measure atmospheric CO_2 on Mauna Loa in Hawaii in 1958 used a simple detector that is very different from any we have described. This instrument, described in Box 19-3, operated for 48 years before it was retired.

Calibrating Detector Response for Luminescence Measurements

Figure 19-13 showed that each kind of detector has a different spectral response. For the same number of input photons at different wavelengths, a particular detector will generate different

In a *ferroelectric material,* dipole moments of molecules remain aligned in the absence of an external field. This alignment gives the material a permanent electric polarization.

Infrared radiation promotes electrons from the valence band of silicon to the conduction band. Semiconductors that are used as infrared detectors have smaller band gaps than silicon.

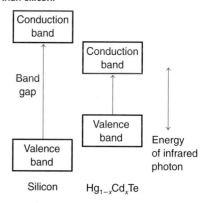

BOX 19-3 **Nondispersive Infrared Measurement of CO$_2$ on Mauna Loa**

When Charles David Keeling proposed continuous monitoring of atmospheric CO$_2$ at Mauna Loa in Hawaii in 1956, he chose a *nondispersive* infrared analyzer that is very different from today's instruments.[15] "Nondispersive" means that there was no prism or grating to spread the radiation into its component wavelengths.

The infrared source is a resistively heated wire at ~525°C. Radiation is split into two beams and "chopped" (interrupted) at 20 Hz by a rotating wheel. The sample cell contains dry air pumped in from outside the observatory. The reference cell contains dry, CO$_2$-free air. CO$_2$ in the sample cell absorbs some infrared radiation, but gas in the reference cell does not. Detector cells contain CO$_2$ in Ar. CO$_2$ in the detector absorbs infrared radiation, so the detector gas alternately warms (expands) and cools (contracts) at a frequency of 20 Hz. Warming and cooling create a pressure oscillation at 20 Hz that is detected by a microphone in each detector. The recorder displays the difference in the responses of the two detectors. The more CO$_2$ in the sample, the less radiation reaches the detector, and the larger the difference between the responses of the two cells.

The observatory on Mauna Loa in Hawaii at an altitude of 3.4 km is intended to measure pristine air over the Pacific Ocean. Four air intakes located 90° apart are each 7 m above the ground and 175 m from the observatory. The two upwind intakes are selected for monitoring. The spectrometer monitors air from one intake for 10 min, then monitors the other for 10 min, and then measures a reference gas for 10 min.

A strip chart recorder shows the average difference between air and reference gas from four air measurements each hour. Sometimes readings are steady and other times they vary when CO$_2$ is emitted from volcanic vents on Mauna Loa. Data representing pristine air was obtained by rejecting readings for any hour when the variation in CO$_2$ was more than 0.5 ppm. A reading for a given day required that there be at least 6 consecutive hours of steady data from which to compose an average. If readings varied too much, no value was reported for that day.

The key to accuracy was the reference gas measured by manometry (Box 3-2) in Keeling's lab in California. Experimental uncertainty for CO$_2$ in air was estimated as ±0.2 ppm for levels of 300–400 ppm. Annually increasing CO$_2$ since 1957 in Figure 0-4 has awakened us to our influence on Earth's atmosphere.

output signals. This variation in response is not a problem when measuring transmittance, which is the quotient of transmitted irradiance (P) divided by incident irradiance (P_0). The ratio at a given wavelength does not depend on how sensitive the detector is at that wavelength. Spectral response of the detector is also not a problem for quantitative analysis using luminescence at a single wavelength, as in Figure 17-23.

However, if you want to measure the true *shape* of a luminescence band, you must know how your detector responds at different wavelengths. Figure 19-18 shows fluorescence from one solution measured by two fluorometers. The band measured by instrument A lies at shorter wavelength than the same fluorescence measured by instrument B. The detector in instrument B is more sensitive at long wavelength than is the detector in instrument A.

To obtain the true shape of a luminescence spectrum, we must measure the relative response of the detector at each wavelength. Calibration can be done with certified luminescence standards whose fluorescence has been measured with calibrated detectors.[16] Apparent fluorescence measured by your instrument is compared with the known fluorescence to obtain a calibration factor at each wavelength. When a measured spectrum is multiplied by the calibration factor at each wavelength, a true spectral shape is obtained. The corrected spectra in Figure 19-18 coincide—as they must—because they come from the same solution.

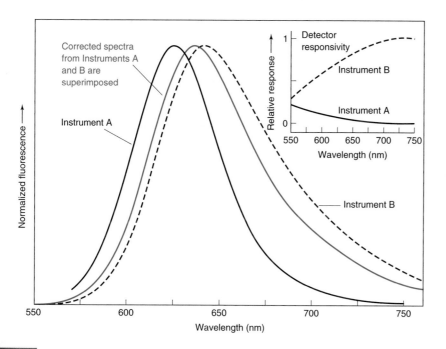

Corrected spectra from Instruments A and B are superimposed

Instrument A

Instrument B

Normalized fluorescence

Wavelength (nm)

Detector responsivity

Instrument B

Instrument A

Relative response

Wavelength (nm)

FIGURE 19-18 Uncorrected fluorescence spectra of the same solution recorded with two different spectrophotometers are different from each other. After correcting for response of each detector, the spectra are superimposed. [From U. Resch-Genger and P. Nording, Sigma-Aldrich Certified Luminescence Standards application note. Courtesy Sigma-Aldrich, St. Louis, MO.]

19-4 Optical Sensors

An *optode* is a chemical sensor based on an *optical fiber.*[17] To understand how optodes work, we first need to know a little about refraction of light.

Refraction

The speed of light in a medium of **refractive index** n is c/n, where c is the speed of light in vacuum. That is, for vacuum, $n = 1$. The refractive index of a liquid is commonly reported for 20°C at the wavelength of the sodium D line ($\lambda = 589.3$ nm). The frequency of light, ν, inside a medium does not change from the frequency in vacuum. The wavelength of light in matter decreases from that in vacuum because $\lambda \nu = c/n$.

When light is reflected, the angle of reflection is equal to the angle of incidence (Figure 19-19). When light passes from one medium into another, its path is bent (Color Plate 21). This bending, called **refraction,** is described by **Snell's law:**

Snell's law:
$$n_1 \sin \theta_1 = n_2 \sin \theta_2 \qquad (19\text{-}7)$$

where n_1 and n_2 are the refractive indexes of the two media and θ_1 and θ_2 are angles defined in Figure 19-19.

Refractive index at sodium D line:

vacuum	1
air (0°, 1 bar)	1.000 29
water	1.33
fused silica	1.46
benzene	1.50
bromine	1.66
iodine	3.34

EXAMPLE Refraction of Light by Water

Visible light travels from air (medium 1) into water (medium 2) at a 45° angle (θ_1 in Figure 19-19). At what angle, θ_2, does the light ray pass through the water?

Solution The refractive index is close to 1 for air and 1.33 for water. From Snell's law,

$$(1.00)(\sin 45°) = (1.33)(\sin \theta_2) \Rightarrow \theta_2 = 32°$$

If your radians and inverse trigonometric functions are rusty, here is how to solve for θ_2: $\sin \theta_2 = (\sin 45°)/1.33 = 0.531\ 7$, so $\theta_2 = \sin^{-1}(0.531\ 7) \equiv \arcsin(0.513\ 7)$. In Excel, the inverse sine function is ASIN and angles are expressed in radians. The Excel function ASIN (0.531 7) returns a value of 0.560 6 radians. Degrees = $180 \times \dfrac{\text{radians}}{\pi} = 180 \times \dfrac{0.560\ 6}{\pi} = 32°$.

What is θ_2 if the incident ray is perpendicular to the surface (that is, $\theta_1 = 0°$)?
$$(1.00)(\sin 0°) = (1.33)(\sin \theta_2) \Rightarrow \theta_2 = 0°$$

A perpendicular ray is not refracted.

Test Yourself Visible light strikes the surface of benzene at a 45° angle. At what angle does the light ray pass through the benzene? (***Answer:*** 28°)

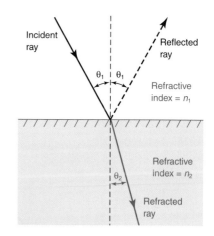

Incident ray

Reflected ray

θ_1 | θ_1

Refractive index = n_1

θ_2

Refractive index = n_2

Refracted ray

FIGURE 19-19 *Snell's law:* $n_1 \sin \theta_1 = n_2 \sin \theta_2$. When light passes from air into any medium, the greater the refractive index of the medium, the smaller is θ_2.

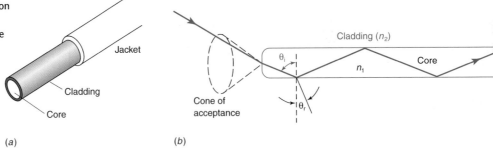

FIGURE 19-20 (a) Optical fiber construction and (b) principle of operation. Any light ray entering within the cone of acceptance will be totally reflected at the wall of the fiber.

Optical Fibers

Optical fibers carry light by *total internal reflection*. Optical fibers are replacing electrical wires for communication because fibers are immune to electrical noise, transmit data at a higher rate, and can handle more signals. Optical fibers can bring an optical signal from inside a chemical reactor out to a spectrophotometer for process monitoring.

A flexible optical fiber has a high-refractive-index, transparent core enclosed in a lower-refractive-index, transparent cladding (Figure 19-20a). The cladding is enclosed in a protective plastic jacket. The core and coating can be made from glass or polymer.

Consider the light ray striking the wall of the core in Figure 19-20b at the angle of incidence, θ_i. Part of the ray is reflected inside the core, and part might be transmitted into the cladding at the angle of refraction, θ_r (Color Plate 22). If the index of refraction of the core is n_1 and the index of the cladding is n_2, Snell's law (Equation 19-7) tells us that

$$n_1 \sin \theta_i = n_2 \sin \theta_r \Rightarrow \sin \theta_r = \frac{n_1}{n_2} \sin \theta_i \qquad (19\text{-}8)$$

> Light traveling from a region of high refractive index (n_1) to a region of low refractive index (n_2) is totally reflected if the angle of incidence exceeds the *critical angle* given by $\sin \theta_{critical} = n_2/n_1$.

If $(n_1/n_2) \sin \theta_i$ is greater than 1, no light is transmitted into the cladding because $\sin \theta_r$ cannot be greater than 1. In such case, θ_i exceeds the *critical angle* for total internal reflection. *If $n_1/n_2 > 1$, there is a range of angles θ_i in which essentially all light is reflected at the walls of the core, and a negligible amount enters the cladding.* All rays entering one end of the fiber within a certain cone of acceptance emerge from the other end of the fiber with little loss.

Optodes

We can create optical sensors for specific analytes by placing a chemically sensitive layer at the end of the fiber. An optical fiber sensor is called an **optode** (or *optrode*), derived from the words "optical" and "electrode." Optodes have been designed to respond to analytes such as dissolved O_2 and pH, Li^+ in blood, NO in cells, and sulfites in food.[18]

The end of the O_2 optode in Color Plate 23 is coated with a Ru(II) complex in a layer of polymer. Luminescence from Ru(II) is *quenched* (decreased) by O_2, as discussed in Section 18-6. The optode is inserted into a liquid sample as small as 100 fL (100×10^{-15} L) on the stage of a microscope. The degree of quenching tells us the concentration of O_2. The detection limit is 10 amol of O_2. By incorporating the enzyme glucose oxidase (Reaction 16-13), the optode becomes a glucose sensor with a detection limit of 1 fmol of glucose.[19] An optode for measuring biochemical oxygen demand (Box 15-2) uses yeast cells immobilized in a membrane to consume O_2 and Ru(II) luminescence for measuring O_2.[20]

> **Question** How many molecules are in 10 amol? (**Answer:** 6 million)

Fiber-Optic Spectrophotometer

You are now ready to appreciate the fiber-optic spectrophotometer used in flow injection and sequential injection analysis. Figure 18-10 showed the main features of flow injection, including a flow cell to measure absorbance. Sample flows through a cylindrical cell that is irradiated at one end with visible light delivered by a fiber optic from a tungsten-halogen lamp. On the other end of the flow cell, a fiber optic delivers transmitted light to a spectrophotometer. Figure 18-16 shows a sequential analyzer with a flow cell. Figure 18-12 employs a fiber optic spectrophotometer to measure fluorescence.

Figure 19-21 shows the optical train of the spectrophotometer. An optical fiber from the flow cell delivers light through connector 1 and entrance slit 2. The entrance slit width determines how many lines of the diffraction grating will be illuminated and, therefore, the resolution of the monochromator (Equation 19-3). Filter 3 allows only a limited band of wavelengths to enter the spectrophotometer. Mirror 4 collimates the beam so that all rays are

Mirror (6)   Mirror (4)   Lens (7)   Filter (9)   CCD detector (8)

Connector (1)   Entrance slit (2)   Grating (5)
                and filter (3)

FIGURE 19-21 Optical train of Ocean Optics USB 4000 fiber-optic spectrophotometer used in flow injection and sequential injection analysis. This miniature spectrophotometer fits in your hand and weighs 190 grams. [Courtesy Ocean Optics, Dunedin, FL.]

parallel. Grating 5 disperses light into its component wavelengths. Mirror 6 focuses diffracted light onto the cylindrical collection lens 7, which directs the light onto the 3 648-pixel linear charge coupled device detector 8. Each pixel, whose physical size is 8 μm wide and 200 μm tall, receives a range of wavelengths. Filter 9, located between lens 7 and detector 8, blocks second- and third-order ($n = 2$ and $n = 3$) diffracted radiation.

When a customer orders the spectrophotometer, she specifies the wavelength range of interest for her application. The manufacturer installs the grating and filters at the factory for the desired wavelength range. The spectrometer is capable of operating in parts of the 200 to 1 100 nm range with different gratings and filters. Visible stray light is ~0.05 to 0.1%. The instrument requires just 4 ms to measure a spectrum. Signal can be integrated for as long as 10 s to improve the signal-to-noise ratio (Section 19-6). There are no moving parts in this rugged and relatively inexpensive instrument.

Attenuated Total Reflectance

Figure 19-20 showed total internal reflection of a light ray traveling through an optical fiber. The same behavior is observed in a flat layer of material whose refractive index, n_1, is greater than the refractive index of the surroundings, n_2. A planar layer in which light is totally reflected is called a **waveguide.** A chemical sensor can be fabricated by placing a chemically sensitive layer on a waveguide.[21]

When light in Figure 19-20 strikes the wall, the ray is totally reflected if θ_i exceeds the critical angle given by $\sin \theta_{critical} = n_2/n_1$. Even though light is totally reflected, the electric field of the light penetrates the cladding to some extent. Figure 19-22 shows that the field dies out exponentially inside the cladding. The part of the light that penetrates the wall of an optical fiber or waveguide is called an *evanescent wave.*

Evanescent means "vanishing" or "fleeting." Light "escapes" from the waveguide, but it vanishes over a short distance.

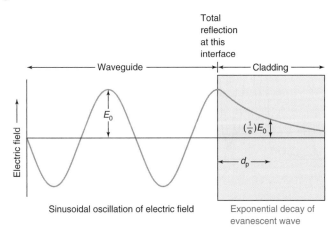

Total reflection at this interface

Waveguide   Cladding

E_0

$(\frac{1}{e})E_0$

d_p

Sinusoidal oscillation of electric field   Exponential decay of evanescent wave

FIGURE 19-22 Behavior of electromagnetic wave when it strikes a surface from which it is totally reflected. The field penetrates the reflective barrier and dies out exponentially.

The electric field, E, of the evanescent wave in Figure 19-22 decays as

$$\frac{E}{E_0} = e^{-x/d_p} \qquad \left(d_p = \frac{\lambda/n_1}{2\pi\sqrt{\sin^2\theta_i - (n_2/n_1)^2}} \right) \qquad (19\text{-}9)$$

where E_0 is the magnitude of the field at the reflective interface, x is the distance into the cladding, and λ is the wavelength of light in a vacuum. The *penetration depth*, d_p, is the distance at which the evanescent field dies down to $1/e$ of its value at the interface. Consider a waveguide with $n_1 = 1.70$ and $n_2 = 1.45$, giving a critical angle of 58.5°. If light with a wavelength of 590 nm has a 70° angle of incidence, then the penetration depth is 140 nm. This depth is great enough to allow the light to interact with many layers of large molecules, such as proteins, whose dimensions are on the order of 10 nm.

Figure 19-23a shows an **attenuated total reflectance** infrared sensor for measuring caffeine in soft drinks. "Attenuated" means "decreased." The diamond crystal at the right side of the diagram acts as a waveguide; it has a circular upper surface whose diameter is 3 mm. When light passes through the crystal, it is totally reflected three times. The upper surface is in contact with a liquid channel carved into a block of poly(tetrafluoroethylene) (Teflon®). At the center of the channel are 5 mg of hydrophobic sorbent beads made from polystyrene-divinylbenzene (Figure 25-1) held in place by cotton wool. When soft drink flows through the channel, caffeine is adsorbed by the beads and becomes concentrated in the beads. Sugar and caramel colorant present in the soft drink are not retained by the beads.

FIGURE 19-23 (*a*) Sequential injection apparatus for measuring caffeine in soft drinks by attenuated total reflectance from analyte retained by sorbent beads. (*b*) Attenuated total reflectance spectrum of caffeine adsorbed from 3 mL of standard containing 50 μg caffeine/mL. [Adapted from M. C. Alcudia-León, R. Lucena, S. Cárdenas, and M. Valcárcel, "Characterization of an Attenuated Total Reflection-Based Sensor for Integrated Solid-Phase Extraction and Infrared Detection," *Anal. Chem.* **2008**, *80*, 1146.]

The evanescent wave from infrared radiation traveling through the diamond waveguide extends into the sorbent beads. When caffeine is present, wavelengths absorbed by caffeine are attenuated. The resulting spectrum is shown in Figure 19-23b. Each peak in the spectrum arises from a vibrational mode of caffeine. The integrated area beneath the entire spectrum is taken as the signal in this experiment.

Sequential injection apparatus (Section 18-4) in Figure 19-23a carries out the following operations. First, 2 mL of water are pumped through the detector cell and a background spectrum is recorded. Then 3 mL of soft drink or standard are pumped through the cell. Caffeine is retained by the beads while sugar and caramel colorant in the soft drink are not retained. After washing the beads with 2 mL of water to remove sugar and colorant, a second spectrum is recorded. The background spectrum is subtracted from the second spectrum to obtain a spectrum whose integrated intensity is proportional to the concentration of caffeine in the soft drink. Finally, the beads are washed with 0.25 mL of acetonitrile, which quantitatively removes caffeine. The procedure can be repeated indefinitely to measure more unknowns and standards.

Surface Plasmon Resonance[22]

Conduction electrons in a metal are nearly free to move within the metal in response to an applied electric field. A *surface plasma wave*, also called a *surface plasmon*, is an electromagnetic wave that propagates along the boundary between a metal and a *dielectric* (an electrical insulator). The electromagnetic field decreases exponentially into both layers but is concentrated in the dielectric layer.

Figure 19-24a shows essentials of one common **surface plasmon resonance** measurement. Monochromatic light whose electric field oscillates in the plane of the page is directed into a prism whose bottom face is coated with ~50 nm of gold. The bottom surface of the gold is coated with a chemical layer (~2–20 nm) that selectively binds an analyte of interest. At low angles of incidence, θ, much, but not all, of the light is reflected by the gold. When θ increases to the critical angle for total internal reflection, the reflectivity is ideally 100%. As θ increases further, a surface plasmon (oscillating electron cloud) is set up, absorbing energy from the incident light. Because some of the energy is absorbed in the gold layer, the reflectivity decreases from 100%. There is a small range of angles at which the plasmon is in resonance with the incident light, a condition creating the sharp dip in the curve in Figure 19-24b. As θ increases beyond the resonance condition, less energy is absorbed and reflectivity increases.

The angle at which reflectivity is minimum depends on the refractive indexes of all the layers in Figure 19-24a. The angle for minimum reflectivity in Figure 19-24b changes by ~0.1° when a thin layer of polymer is coated onto the gold. When analyte binds to a chemically sensitive layer on the gold, the refractive index of that layer changes slightly and the angle for minimum reflectivity changes slightly. Commercial instruments can measure changes in the surface plasmon resonance angle with a precision of ~10^{-4} to 10^{-5} degrees. For biosensors, the chemically sensitive layer might contain an antibody or antigen, DNA or RNA, a protein, or a carbohydrate that has a selective interaction with some analyte.

Figures 19-25 shows a scheme by which surface plasmon resonance can detect a specific sequence of DNA. A gold coating on a prism is treated with a synthetic peptide nucleic acid and mercaptohexanol. *Peptide nucleic acid* (abbreviated PNA) has the same bases as DNA (abbreviated A, T, C, and G), but the deoxyribose sugar-phosphate backbone of DNA is

Peptide nucleic acid (PNA)

Base =
A-Adenine
T-Thymine
C-Cytosine
G-Guanine

Mercaptohexanol

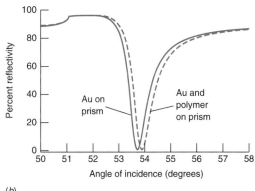

FIGURE 19-24 (*a*) Essentials of a surface plasmon resonance measurement. (*b*) Reflectivity versus angle, θ. [Adapted from J. M. Brockman, B. P. Nelson, and R. M. Corn, "Surface Plasmon Resonance Imaging Measurements of Ultrathin Organic Films," *Annu. Rev. Phys. Chem.* **2000**, *51*, 41.]

FIGURE 19-25 Scheme for detection of specific DNA sequence by surface plasmon resonance.

1. Add DNA-biotin

DNA   Biotin

2. Wash away noncomplementary DNA

Mercapto-hexanol

PNA (peptide nucleic acid)

Au

DNA-PNA hybrid

3. Add streptavidin

Biotin

FIGURE 19-26 A small shift in surface plasmon resonance angle occurs when a small strand of DNA binds to complementary PNA on a gold surface. A large shift occurs when the protein streptavidin binds to biotin at the end of the DNA strand. [From J. Liu, S. Tian, L. Tiefenauer, P. E. Nielsen, and W. Knoll, "Simultaneously Amplified Electrochemical and Surface Plasmon Detection of DNA Hybridization Based on Ferrocene-Streptavidin Conjugates," *Anal. Chem.* **2005**, *77*, 2756.]

replaced by a peptide backbone. A sulfur atom at one end of the PNA chain binds to the gold surface. Mercaptohexanol occupies binding sites on gold not occupied by PNA. The complex (called a *hybrid*) between PNA and a complementary strand of DNA is stronger than the complex between two strands of complementary DNA. The sensor is designed to detect a specific, short sequence of DNA that is complementary to the synthetic PNA. When a mixture of DNA strands is incubated with PNA, only complementary DNA binds tightly. Noncomplementary DNA can be washed away. Hybrids of DNA to PNA are more stable to variations in temperature and ionic strength than are hybrids of DNA to DNA.

Prior to exposing unknown DNA to the Au-PNA surface, *biotin* is attached to one end of each DNA strand. Biotin is an essential B vitamin. The protein *streptavidin* is a 53 000-dalton tetrameric protein isolated from the bacterium *Streptomyces avidinii*. It has a strong, specific affinity for biotin, with a formation constant of 10^{15}. When biotin-DNA is exposed to streptavidin, the protein binds tightly to biotin (Figure 19-25).

The solid line in Figure 19-26 is the surface plasmon resonance of the gold film with bound PNA and mercaptohexanol. When complementary DNA labeled with biotin binds to the PNA, the resonance shown by the dotted line is almost imperceptibly shifted because the

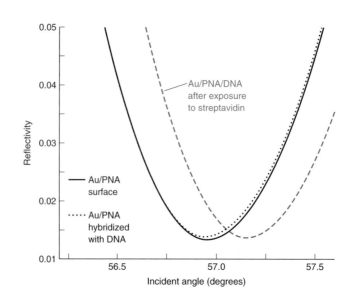

Au/PNA/DNA after exposure to streptavidin

—— Au/PNA surface

······ Au/PNA hybridized with DNA

Reflectivity

0.05

0.04

0.03

0.02

0.01

56.5   57.0   57.5

Incident angle (degrees)

amount of DNA that is bound is very small and the refractive index of the thin surface layer is barely changed. When streptavidin is added, the large protein binds to the biotin on the end of every DNA. The massive protein changes the refractive index of the organic film enough to shift the resonance position by 0.3° (dashed line in Figure 19-26). The signal in Figure 19-26 arises from ~1 × 10^{12} PNA molecules per square centimeter of gold surface. DNA with 12 nucleotide bases and one mismatch with the PNA does not bind to the PNA.

19-5 Fourier Transform Infrared Spectroscopy[23]

A photodiode array or charge coupled device can measure an entire spectrum at once. The spectrum is spread into its component wavelengths, and each wavelength is directed onto one detector element. For the infrared region, the most important method for observing the entire spectrum at once is *Fourier transform spectroscopy*.

Fourier Analysis

Fourier analysis is a procedure in which a curve is decomposed into a sum of sine and cosine terms, called a *Fourier series*. To analyze the curve in Figure 19-27, which spans the interval $x_1 = 0$ to $x_2 = 10$, the Fourier series has the form

Fourier series: $y = a_0 \sin(0\omega x) + b_0 \cos(0\omega x) + a_1 \sin(1\omega x) + b_1 \cos(1\omega x)$

$$+ a_2 \sin(2\omega x) + b_2 \cos(2\omega x) + \ldots$$

$$= \sum_{n=0}^{\infty} [a_n \sin(n\omega x) + b_n \cos(n\omega x)] \qquad (19\text{-}10)$$

where

$$\omega = \frac{2\pi}{x_2 - x_1} = \frac{2\pi}{10 - 0} = \frac{\pi}{5} \qquad (19\text{-}11)$$

Equation 19-10 says that the value of y for any value of x can be expressed by an infinite sum of sine and cosine waves. Successive terms correspond to waves with increasing frequency.

Figure 19-28 shows how sequences of three, five, or nine sine and cosine waves give better and better approximations to the curve in Figure 19-27. The coefficients a_n and b_n required to construct the curves in Figure 19-28 are given in Table 19-3.

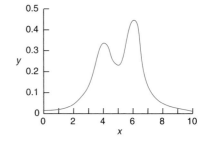

FIGURE 19-27 A curve to be decomposed into a sum of sine and cosine terms by Fourier analysis.

TABLE 19-3 Fourier coefficients for Figure 19-28

n	a_n	b_n
0	0	0.136 912
1	−0.006 906	−0.160 994
2	0.015 185	0.037 705
3	−0.014 397	0.024 718
4	0.007 860	−0.043 718
5	0.000 089	0.034 864
6	−0.004 813	−0.018 858
7	0.006 059	0.004 580
8	−0.004 399	0.003 019

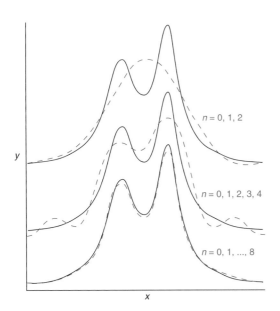

FIGURE 19-28 Fourier series reconstruction of the curve in Figure 19-27. Solid line is the original curve and dashed lines are made from a series of $n = 0$ to $n = 2, 4$, or 8 in Equation 19-10. Coefficients a_n and b_n are given in Table 19-3.

Interferometry

The heart of a Fourier transform infrared spectrophotometer is the **interferometer** in Figure 19-29. Radiation from the source at the left strikes a *beamsplitter*, which transmits some light and reflects some light. For ease of explanation, consider a beam of monochromatic radiation. (In fact, the Fourier transform spectrophotometer uses a continuum source of infrared radiation, not a monochromatic source.) Suppose that the beamsplitter reflects half of the light and transmits half. When light strikes the beamsplitter at point O, some is reflected to a stationary

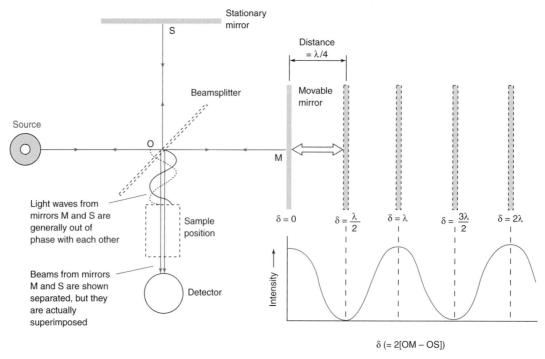

FIGURE 19-29 Schematic diagram of Michelson interferometer. Detector response as a function of retardation (= 2[OM − OS]) is shown for monochromatic incident radiation of wavelength λ.

Albert Michelson developed the interferometer around 1880 and conducted the Michelson-Morley experiment in 1887, in which it was found that the speed of light is independent of the motion of the source and the observer. This crucial experiment led Einstein to the theory of relativity. Michelson also used the interferometer to create the predecessor of today's length standard based on the wavelength of light. He received the Nobel Prize in 1907 "for precision optical instruments and the spectroscopic and metrological investigations carried out with their aid."

mirror at a distance OS and some is transmitted to a movable mirror at a distance OM. Rays reflected by the mirrors travel back to the beamsplitter, where half of each ray is transmitted and half is reflected. One recombined ray travels in the direction of the detector, and another heads back to the source.

In general, the paths OM and OS are not equal, so the two waves reaching the detector are not in phase. If the two waves are in phase, they interfere constructively to give a wave with twice the amplitude, as shown in Figure 19-8. If the waves are one-half wavelength (180°) out of phase, they interfere destructively and cancel. For any intermediate phase difference, there is partial cancellation.

The difference in pathlength followed by the two waves in Figure 19-29 is 2(OM − OS). This difference is called the *retardation*, δ. Constructive interference occurs whenever δ is an integer multiple of the wavelength, λ. A minimum appears when δ is a half-integer multiple of λ. If mirror M moves away from the beamsplitter at a constant speed, light reaching the detector goes through a sequence of maxima and minima as the interference alternates between constructive and destructive phases.

A graph of output light intensity versus retardation, δ, is called an **interferogram.** If the light from the source is monochromatic, the interferogram is a simple cosine wave:

$$I(\delta) = B(\tilde{\nu}) \cos\left(\frac{2\pi\delta}{\lambda}\right) = B(\tilde{\nu}) \cos(2\pi\tilde{\nu}\delta) \qquad (19\text{-}12)$$

where $I(\delta)$ is the intensity of light reaching the detector and $\tilde{\nu}$ is the wavenumber ($=1/\lambda$) of the light. Clearly, I is a function of the retardation, δ. $B(\tilde{\nu})$ is a constant that accounts for the intensity of the light source, efficiency of the beamsplitter (which never gives exactly 50% reflection and 50% transmission), and response of the detector. All these factors depend on $\tilde{\nu}$. For monochromatic light, there is only one value of $\tilde{\nu}$.

Figure 19-30a shows the interferogram produced by monochromatic radiation of wavenumber $\tilde{\nu}_0 = 2 \text{ cm}^{-1}$. The wavelength (repeat distance) of the interferogram can be seen in the figure to be λ = 0.5 cm, which is equal to $1/\tilde{\nu}_0 = 1/(2 \text{ cm}^{-1})$. Figure 19-30b shows the interferogram that results from a source with two monochromatic waves ($\tilde{\nu}_0 = 2$ and $\tilde{\nu}_0 = 8 \text{ cm}^{-1}$) with relative intensities 1:1. A short wave oscillation (λ = $\frac{1}{8}$ cm) is superimposed on a long wave oscillation (λ = $\frac{1}{2}$ cm). The interferogram is a sum of two terms:

$$I(\delta) = B_1 \cos(2\pi\tilde{\nu}_1\delta) + B_2 \cos(2\pi\tilde{\nu}_2\delta) \qquad (19\text{-}13)$$

where $B_1 = 1$, $\tilde{\nu}_1 = 2 \text{ cm}^{-1}$, $B_2 = 1$, and $\tilde{\nu}_2 = 8 \text{ cm}^{-1}$.

Spectrum

Interferogram

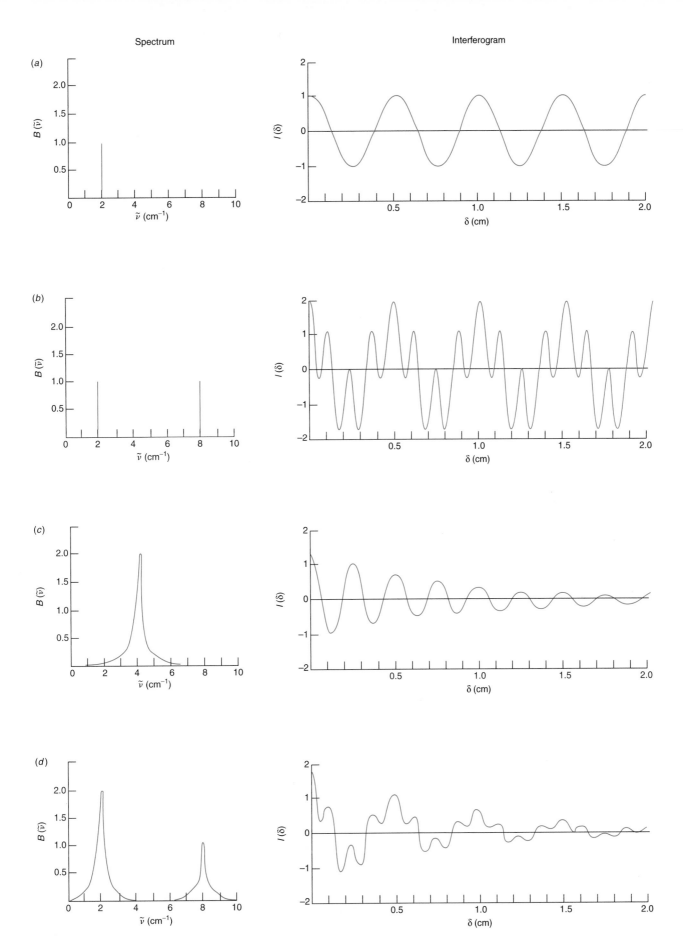

FIGURE 19-30 Interferograms produced by different spectra.

Fourier analysis of the interferogram gives back the spectrum from which the interferogram is made. *The spectrum is the Fourier transform of the interferogram.*

Fourier analysis decomposes a curve into its component wavelengths. Fourier analysis of the interferogram in Figure 19-30a gives the (trivial) result that the interferogram is made from a single wavelength function, with $\lambda = \frac{1}{2}$ cm. Fourier analysis of the interferogram in Figure 19-30b gives the slightly more interesting result that the interferogram is composed of two wavelengths ($\lambda = \frac{1}{2}$ and $\lambda = \frac{1}{8}$ cm) with relative contributions 1:1. We say that the spectrum is the *Fourier transform* of the interferogram.

The interferogram in Figure 19-30c is derived from a spectrum with an absorption band centered at $\tilde{\nu}_0 = 4$ cm^{-1}. The interferogram is the sum of contributions from all source wavelengths. The Fourier transform of the interferogram in Figure 19-30c is indeed the third spectrum in Figure 19-30c. That is, decomposition of the interferogram into its component wavelengths gives back the band centered at $\tilde{\nu}_0 = 4$ cm^{-1}. *Fourier analysis of the interferogram gives back the intensities of its component wavelengths.*

The interferogram in Figure 19-30d comes from two absorption bands in the spectrum at the left. The Fourier transform of this interferogram gives back the spectrum to its left.

Fourier Transform Spectroscopy

The interferogram loses intensity at those wavelengths absorbed by the sample.

In a Fourier transform spectrometer, the sample is usually placed between the interferometer and the detector, as in Figures 19-29 and 19-31. Because the sample absorbs certain wavelengths, *the interferogram contains the spectrum of the source minus the spectrum of the sample.* An interferogram of a reference sample containing the cell and solvent is first recorded and transformed into a spectrum. Then the interferogram of a sample in the same solvent is recorded and transformed into a spectrum. The quotient of the sample spectrum divided by the reference spectrum is the transmission spectrum of the sample (Figure 19-32). The quotient is the same as computing P/P_0 to find transmittance. P_0 is the irradiance received at the detector through the reference, and P is the irradiance received after passage through the sample.

resolution $\approx (1/\Delta)$ cm^{-1}
Δ = maximum retardation (cm)

The interferogram is recorded at discrete intervals. The *resolution* of the spectrum (ability to discern closely spaced peaks) is approximately equal to $(1/\Delta)$ cm^{-1}, where Δ is the maximum retardation. If the mirror travel is ± 2 cm, the retardation is ± 4 cm and the resolution is $1/(4 \text{ cm}) = 0.25$ cm^{-1}.

For a spectral range of $\Delta\tilde{\nu}$ cm^{-1}, points must be taken at retardation intervals of $1/(2\Delta\tilde{\nu})$.

Capital delta has two different purposes in this section:

$\Delta\tilde{\nu}$ = spectral range (cm^{-1})
Δ = maximum retardation (cm)

The wavenumber range of the spectrum is determined by how the interferogram is sampled. The closer the spacing between data points, the greater the range. Covering a range of $\Delta\tilde{\nu}$ wavenumbers requires sampling the interferogram at retardation intervals of $\delta = 1/(2\Delta\tilde{\nu})$. If $\Delta\tilde{\nu}$ is 4 000 cm^{-1}, sampling must occur at intervals of $\delta = 1/(2 \cdot 4\,000 \text{ cm}^{-1}) = 1.25 \times 10^{-4}$ cm $= 1.25\ \mu$m. This sampling interval corresponds to a mirror motion of 0.625 μm. For every centimeter of mirror travel, 1.6×10^4 data points must be collected. If the mirror moves at a rate of 2 mm per second, the data collection rate would be 3 200 points per second.

FIGURE 19-31 Layout of Fourier transform infrared spectrometer. [Courtesy Nicolet, Madison, WI.]

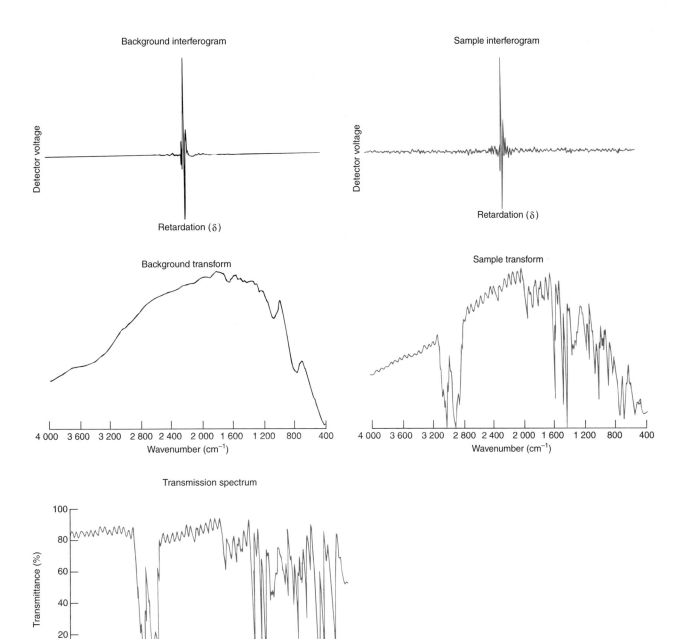

FIGURE 19-32 Fourier transform infrared spectrum of polystyrene film. The Fourier transform of the background interferogram gives a spectrum determined by the source intensity, beamsplitter efficiency, detector response, and absorption by traces of H_2O and CO_2 in the atmosphere. The sample compartment is purged with dry N_2 to reduce the levels of H_2O and CO_2. The transform of the sample interferogram is a measure of all the instrumental factors, plus absorption by the sample. The transmission spectrum is obtained by dividing the sample transform by the background transform. Each interferogram is an average of 32 scans and contains 4 096 data points, giving a resolution of 4 cm^{-1}. The mirror velocity was 0.693 cm/s. [Courtesy M. P. Nadler, Michelson Laboratory, China Lake, CA.]

The source, beamsplitter, and detector each limit the usable wavelength range. Clearly, the instrument cannot respond to a wavelength that is absorbed by the beamsplitter or a wavelength to which the detector does not respond. The beamsplitter for the mid-infrared region (~4 000 to 400 cm^{-1}) is typically a layer of germanium evaporated onto a KBr plate. For longer wavelengths ($\tilde{\nu} < 400$ cm^{-1}), a film of the organic polymer Mylar is a suitable beamsplitter.

To control the sampling interval for the interferogram, a monochromatic visible laser beam is passed through the interferometer along with the polychromatic infrared light (Figure 19-31). The laser gives destructive interference whenever the retardation is a half-integer multiple of the laser wavelength. These zeros in the laser signal, observed with a visible detector, are used to control sampling of the infrared interferogram. For example, an infrared data point might be taken at every second zero point of the visible-light interferogram. The precision with which the laser frequency is known gives an accuracy of 0.01 cm^{-1} in the infrared spectrum, which is a 100-fold improvement over the accuracy of dispersive (grating) instruments.

Advantages of Fourier Transform Spectroscopy

Compared with dispersive instruments, the Fourier transform spectrometer offers improved signal-to-noise ratio at a given resolution, better frequency accuracy, speed, and built-in data-handling capabilities. Signal-to-noise improvement comes mainly because the Fourier transform spectrometer uses energy from the entire spectrum, instead of analyzing a sequence of small wavebands available from a monochromator. Precise reproduction of wavenumber position from one spectrum to the next allows Fourier transform instruments to average signals from multiple scans to further improve the signal-to-noise ratio. Wavenumber precision and low noise levels allow spectra with slight differences to be subtracted from each other to expose those differences. Fourier transform instruments are not as accurate as dispersive spectrometers for measuring transmittance.

Advantages of Fourier transform infrared spectrometers are so great that it is nearly impossible to purchase a dispersive infrared spectrometer. Fourier transform visible and ultraviolet spectrometers are not commercially available, because of the requirement to sample the interferometer at intervals of $\delta = 1/(2\Delta\tilde{\nu})$. For visible spectroscopy, $\Delta\tilde{\nu}$ could be $25\ 000$ cm^{-1} (corresponding to 400 nm), giving $\delta = 0.2$ μm and a mirror movement of 0.1 μm between data points. Such fine control over significant ranges of mirror motion is not feasible.

19-6 Dealing with Noise[24]

An advantage of Fourier transform spectroscopy is that the entire interferogram is recorded in a few seconds and stored in a computer. The signal-to-noise ratio can be improved by collecting tens or hundreds of interferograms and averaging them.

Signal Averaging

Signal averaging can improve the quality of data, as illustrated in Figure 19-33.[25] The lowest trace contains a great deal of noise. A simple way to estimate the noise level is to measure the maximum amplitude of the noise in a region free of signal. The signal is measured from the middle of the baseline noise to the middle of the noisy peak. By this criterion, the lowest trace of Figure 19-33 has a signal-to-noise ratio of $14/9 = 1.6$.

A more common measurement of noise, which requires a digitized signal, is the **root-mean-square (rms) noise,** defined as

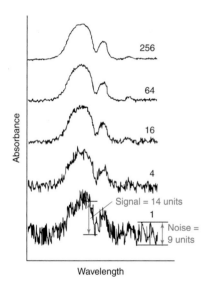

Absorbance

256

64

16

4

Signal = 14 units

1

Noise = 9 units

Wavelength

FIGURE 19-33 Effect of signal averaging on a simulated noisy spectrum. Labels refer to number of scans averaged. [From R. Q. Thompson, "Experiments in Software Data Handling," *J. Chem. Ed.* 1985, *62*, 866.]

To improve the signal-to-noise ratio by a factor of n requires averaging n^2 spectra.

Question By what factor should the signal-to-noise ratio be improved when 16 spectra are averaged? Measure the noise level in Figure 19-33 to test your prediction.

Root-mean-square noise:
$$\text{rms noise} = \sqrt{\frac{\sum_i (A_i - \overline{A})^2}{n}} \tag{19-14}$$

where A_i is the measured signal for the ith data point, $\overline{A}$ is the mean signal, and n is the number of data points. For a large number of data points, the rms noise is the standard deviation of the noise. It is best to apply Equation 19-14 where the signal is flat, as it is at the left and right sides of Figure 19-33. If you must use data at the center of Figure 19-33, the average $\overline{A}$ must be fitted to the continuously rising or falling signal. The rms noise is ~5 times less than the peak-to-peak noise. If we used rms noise instead of peak-to-peak noise, we would say that the noise in the bottom spectrum of Figure 19-33 is $9/5 = 1.8$ and the signal-to-noise ratio is $14/1.8 = 7.8$. Clearly, the signal-to-noise ratio depends on how you define noise.

Consider what happens if you record the spectrum twice and add the results. The signal is the same in both spectra and adds to give twice the value of each spectrum. If n spectra are added, the signal will be n times as large as in the first spectrum. Noise is random, so it may be positive or negative at any point. It turns out that, if n spectra are added, the noise increases in proportion to $\sqrt{n}$. Because the signal increases in proportion to n, the signal-to-noise ratio increases in proportion to $n/\sqrt{n} = \sqrt{n}$.

When we average n spectra, the signal-to-noise ratio is improved by $\sqrt{n}$. To improve the signal-to-noise ratio by a factor of 2 requires averaging four spectra. To improve the signal-to-noise ratio by a factor of 10 requires averaging 100 spectra. Spectroscopists record as many as 10^4 to 10^5 scans to observe weak signals. It is rarely possible to do better than this because instrumental instabilities cause drift in addition to random noise.

Types of Noise

Figure 19-34 shows three common types of noise in electrical instruments.[26] The upper trace is *white noise*, also called *Gaussian noise*. One source of white noise, called *Johnson noise*, is random fluctuations of electrons in an electronic device. Lowering the operating temperature is one way to reduce Johnson noise. *Shot noise* is another form of white noise attributable to the quantized nature of charge carriers and photons. At low signal levels, noise arises from random variation in the small number of photons reaching a detector or the small number of electrons and holes generated in a semiconductor.

The second trace in Figure 19-34 shows *1/f noise*, also called *drift*, which is greatest at zero frequency and decreases in proportion to 1/frequency. An example of low-frequency noise in laboratory instruments is flickering or drifting of a light source in a spectrophotometer or a flame in atomic spectroscopy. Drift arises from causes such as slow changes in instrument components with temperature and age or variation of power-line voltage to an instrument. The classical way to detect and account for drift is to periodically measure standards and correct the instrument reading for any observed change.[27]

The bottom trace of Figure 19-34 shows *line noise* (also called *interference* or *whistle noise*) at discrete frequencies such as the 60-Hz transmission-line frequency or the 0.2-Hz vibrational frequency when elephants walk through the basement of your building. Electrical shielding and grounding the shielding and instrument to the same ground point help reduce line noise.

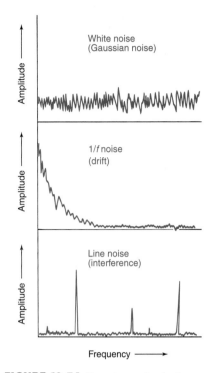

FIGURE 19-34 Three types of noise in electrical instruments. White noise is always present. A beam chopping frequency can be selected to reduce 1/f noise and line noise to insignificant values.

Beam Chopping

Spectrophotometers in Figures 19-1 through 19-3 have a rotating mirror called a *chopper* that alternately sends light through sample and reference cells. Chopping allows both cells to be sampled almost continuously, and it provides a means of noise reduction. **Beam chopping** moves the analytical signal from zero frequency to the frequency of the chopper. The chopping frequency can be selected so that 1/f noise and line noise are minimal. High-frequency detector circuits are required to take advantage of beam chopping.

A Low-Noise Spectrophotometer

Spectrophotometer noise is broadly attributable to (i) sources that are independent of the light level, (ii) sources that are proportional to photo-generated current, and (iii) variation of the intensity of the light source.[28] For several decades, precision has been governed by variation of intensity of the light source, which produces a root-mean-square noise equivalent to an apparent absorbance of ~0.000 03 at visible wavelengths. That is, absorbance below ~0.000 03 cannot be measured because it is hidden in the noise.

Figure 19-35 shows a system designed to cancel the noise generated by variation of the light source. Visible light from a tungsten-halogen lamp goes through a monochromator and is then split into two beams that pass through sample and reference cells. Light from the sample cell goes to a photodetector that generates current I_{sam}. Light from the reference cell goes to a photodetector that generates current I_{ref}. Electronics then convert the currents to voltages V_{sam} and V_{ref} and the *difference voltage* $V_{dif} = V_{sam} - V_{ref}$. Noise from variation of the light source intensity affects V_{sam} and V_{ref} equally. That is, if the light momentarily increases by 0.1%, the voltages recorded by each detector increase by 0.1%. The difference V_{dif} should be *zero* in the absence of sample absorption. The traces at the right of Figure 19-35 are simulations showing a weak absorption superimposed on strong noise. The signal is too weak to see in the sample spectrum (V_{sam}), but is loud and clear in the difference spectrum (V_{dif}).

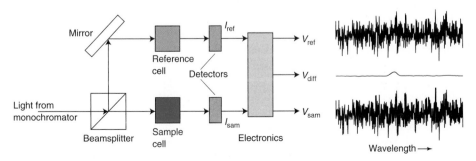

FIGURE 19-35 Principle of low-noise spectrophotometer. [Adapted from Z. Xu and D. W. Larsen, "Development of Ultra-Low-Noise Spectrophotometry for Analytical Applications," *Anal. Chem.* **2005**, *77*, 6463.]

FIGURE 19-36 Spectrum of 2 nM Nile Blue in methanol recorded with (*a*) commercial spectrophotometer and (*b*) same instrument modified to cancel light source variation. [From Z. Xu and D. W. Larsen, "Development of Ultra-Low-Noise Spectrophotometry for Analytical Applications," *Anal. Chem.* **2005**, *77*, 6463.]

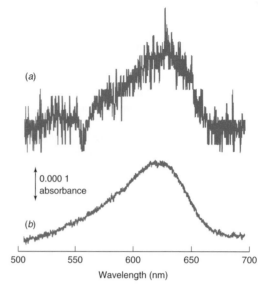

Figure 19-36 is an experimental implementation of the scheme to cancel source intensity noise. The upper trace is the absorption spectrum of a 9-nM dye solution with a peak absorbance of 0.000 2. The lower spectrum was obtained from the same instrument modified to display absorbance computed from the difference voltage V_{dif}.[29] The signal-to-noise ratio is improved by a factor of 10. Hardware modifications are expected to enable an additional factor-of-10 improvement for a total signal-to-noise ratio reduction of 100.

Terms to Understand

attenuated total reflectance	Fourier analysis	photoconductive detector	resolution
bandwidth	grating	photodiode array	root-mean-square (rms) noise
beam chopping	interferogram	photomultiplier tube	signal averaging
blackbody radiation	interferometer	phototube	Snell's law
charge coupled device	laser	photovoltaic detector	stray light
diffraction	monochromator	polychromator	surface plasmon resonance
dispersion	optical fiber	refraction	thermocouple
ferroelectric material	optode	refractive index	waveguide

Summary

Spectrophotometer components include the source, sample cell, monochromator, and detector. Tungsten and deuterium lamps provide visible and ultraviolet radiation; a silicon carbide globar is a good infrared source. Tungsten and silicon carbide lamps behave approximately as blackbodies, which are objects that absorb all light striking them. Emission of radiant energy from the surface of the blackbody is proportional to the fourth power of temperature and shifts to shorter wavelengths as temperature increases. Lasers provide high-intensity, coherent, monochromatic radiation by stimulated emission from a medium in which an excited state has been pumped to a higher population than that of a lower state. Sample cells must be transparent to the radiation of interest. A reference sample compensates for reflection and scattering by the cell and solvent. A grating monochromator disperses light into its component wavelengths. The finer a grating is ruled, the higher the resolution and the greater the dispersion of wavelengths over angles. Narrow slits improve resolution but increase noise, because less light reaches the detector. A bandwidth that is 1/5 of the width of the spectral peak is a good compromise between maximizing signal-to-noise ratio and minimizing peak distortion. Stray light introduces absorbance errors that are most serious when the transmittance of a sample is very low. Filters pass wide bands of wavelength and reject other bands.

A photomultiplier tube is a sensitive detector of visible and ultraviolet radiation; photons cause electrons to be ejected from a metallic cathode. The signal is amplified at each successive dynode on which the photoelectrons impinge. Photodiode arrays and charge coupled devices are solid-state detectors in which photons create electrons and holes in semiconductor materials. Coupled to a polychromator, these devices can record all wavelengths of a spectrum simultaneously, with resolution limited by the number and spacing of detector elements. Common infrared detectors include thermocouples, ferroelectric materials, and photoconductive and photovoltaic devices.

When light passes from a region of refractive index n_1 to a region of refractive index n_2, the angle of refraction (θ_2) is related to the angle of incidence (θ_1) by Snell's law: $n_1 \sin \theta_1 = n_2 \sin \theta_2$. Optical fibers and flat waveguides transmit light by a series of total internal reflections. Optodes are sensors based on optical fibers. Some optodes have a layer of material whose absorbance or fluorescence changes in the presence of analyte. Light can be carried to and from the tip by the optical fiber. When light is transmitted through a fiber optic or a waveguide by total internal reflectance, some light, called the evanescent wave, penetrates through the reflective interface during each reflection. In attenuated total reflectance devices, the waveguide is coated with a substance that absorbs light in the presence of analyte. In a surface plasmon resonance sensor, we measure the change in the angle of minimum reflectance from a gold film coated with a chemically sensitive layer on the back face of a prism.

Fourier analysis decomposes a signal into its component wavelengths. An interferometer contains a beamsplitter, a stationary mirror, and a movable mirror. Reflection of light from the two mirrors creates an interferogram. Fourier analysis of the interferogram tells us what frequencies went into the interferogram. In a Fourier transform spectrophotometer, the interferogram of the source is first measured without sample. Then sample is placed in the beam and a second interferogram is recorded. The transforms of the interferograms tell what amount of light at each frequency reaches the detector with and without the sample. The quotient of the two transforms is the transmission spectrum. The resolution of a Fourier transform spectrum is approximately $1/\Delta$, where Δ is the maximum retardation. To cover a wavenumber range $\Delta\tilde{\nu}$ requires sampling the interferogram at intervals of $\delta = 1/(2\Delta\tilde{\nu})$.

If you average n scans, the signal-to-noise ratio should increase by $\sqrt{n}$. White noise, which is independent of frequency, arises from sources such as random fluctuations of electrons in components (Johnson noise) and the discrete nature of charge carriers and photons (shot noise). $1/f$ noise decreases with increasing frequency. Drift and flicker of the intensity of a light source or brightness of a flame in atomic spectroscopy are sources of $1/f$ noise. Line noise occurs at discrete frequencies such as the 60-Hz line frequency of a power supply. Beam chopping in a dual-beam spectrophotometer reduces $1/f$ and line noise. A spectrometer modified to record the difference between the sample and reference signals can reduce noise from lamp flicker by at least an additional factor of 10.

Exercises

19-A. (a) If a diffraction grating has a resolution of 10^4, is it possible to distinguish two spectral lines with wavelengths of 10.00 and 10.01 μm?

(b) With a resolution of 10^4, how close in wavenumbers (cm^{-1}) is the closest line to 1 000 cm^{-1} that can barely be resolved?

(c) Calculate the resolution of a 5.0-cm-long grating ruled at 250 lines/mm for first-order $(n = 1)$ diffraction and tenth-order $(n = 10)$ diffraction.

(d) Find the angular dispersion $(\Delta\phi$, in radians and degrees) between light rays with wavenumbers of 1 000 and 1 001 cm^{-1} for second-order diffraction $(n = 2)$ from a grating with 250 lines/mm and $\phi = 30°$.

19-B. The true absorbance of a sample is 1.000, but the monochromator passes 1.0% stray light. Add the light coming through the sample to the stray light to find the apparent transmittance of the sample. Convert this answer back into absorbance and find the relative error in the calculated concentration of the sample.

19-C. Refer to the Fourier transform infrared spectrum in Figure 19-32.

(a) The interferogram was sampled at retardation intervals of 1.2660×10^{-4} cm. What is the theoretical wavenumber range (0 to ?) of the spectrum?

(b) A total of 4 096 data points were collected from $\delta = -\Delta$ to $\delta = +\Delta$. Compute the value of Δ, the maximum retardation.

(c) Calculate the approximate resolution of the spectrum.

(d) The interferometer mirror velocity is given in the figure caption. How many microseconds elapse between each datum?

(e) How many seconds were required to record each interferogram once?

(f) What kind of beamsplitter is typically used for the region 400 to 4 000 cm^{-1}? Why is the region below 400 cm^{-1} not observed?

19-D. The table shows signal-to-noise ratios recorded in a nuclear magnetic resonance experiment. Construct graphs of (a) signal-to-noise ratio versus n and (b) signal-to-noise ratio versus $\sqrt{n}$, where n is the number of scans. Draw error bars corresponding to the standard deviation at each point. Is the signal-to-noise ratio proportional to $\sqrt{n}$? Find the 95% confidence interval for each row of the table.

Signal-to-noise ratio at the aromatic protons of 1% ethylbenzene in CCl_4

Number of experiments	Number of accumulations	Signal-to-noise ratio	Standard deviation
8	1	18.9	1.9
6	4	36.4	3.7
6	9	47.3	4.9
8	16	66.7	7.0
6	25	84.6	8.6
6	36	107.2	10.7
6	49	130.3	13.3
4	64	143.2	15.1
4	81	146.2	15.0
4	100	159.4	17.1

source: M. Henner, P. Levior, and B. Ancian, "An NMR Spectrometer-Computer Interface Experiment," J. Chem. Ed. 1979, 56, 685.

Problems

The Spectrophotometer

19-1. Describe the role of each component of the spectrophotometer in Figure 19-1.

19-2. Explain how a laser generates light. List important properties of laser light.

19-3. Would you use a tungsten or a deuterium lamp as a source of 300-nm radiation? What kind of lamp provides radiation at 4-μm wavelength?

19-4. Which variables increase the resolution of a grating? Which variables increase the dispersion of the grating? How is the blaze angle chosen to optimize a grating for a particular wavelength?

19-5. What is the role of a filter in a grating monochromator?

19-6. What are the advantages and disadvantages of decreasing monochromator slit width?

19-7. Explain how the following detectors of visible radiation work: (a) photomultiplier tube, (b) photodiode array, and (c) charge coupled device.

19-8. Deuterated triglycine sulfate (abbreviated DTGS) is a common ferroelectric infrared detector material. Explain how it works.

19-9. Consider a reflection grating operating with an incident angle of 40° in Figure 19-7.

(a) How many lines per centimeter should be etched in the grating if the first-order diffraction angle for 600-nm (visible) light is to be $-30°$?

(b) Answer the same question for 1 000 cm^{-1} (infrared) light.

19-10. Show that a grating with 10^3 grooves/cm provides a dispersion of 5.8° per μm of wavelength if $n = 1$ and $\phi = 10°$ in Equation 19-4.

19-11. (a) What resolution is required for a diffraction grating to resolve wavelengths of 512.23 and 512.26 nm?

(b) With a resolution of 10^4, how close in nm is the closest line to 512.23 nm that can barely be resolved?

(c) Calculate the fourth-order resolution of a grating that is 8.00 cm long and is ruled at 185 lines/mm.

(d) Find the angular dispersion ($\Delta\phi$) between light rays with wavelengths of 512.23 and 512.26 nm for first-order diffraction ($n = 1$) and thirtieth-order diffraction from a grating with 250 lines/mm and $\phi = 3.0°$.

19-12. (a) The true absorbance of a sample is 1.500, but 0.50% stray light reaches the detector. Find the apparent transmittance and apparent absorbance of the sample.

(b) How much stray light can be tolerated if the absorbance error is not to exceed 0.001 at a true absorbance of 2?

(c) A research-quality spectrophotometer has a stray light level of <0.000 05% at 340 nm. What will be the maximum absorbance error for a sample with a true absorbance of 2? Of 3?

19-13. The pathlength of a cell for infrared spectroscopy can be measured by counting *interference fringes* (ripples in the transmission spectrum). The following spectrum shows 30 interference maxima between 1 906 and 698 cm^{-1} obtained by placing an empty KBr cell in a spectrophotometer.

The fringes arise because light reflected from the cell compartment interferes constructively or destructively with the unreflected beam.

If the reflected beam travels an extra distance λ, it will interfere constructively with the unreflected beam. If the reflection pathlength is λ/2, destructive interference occurs. Peaks therefore arise when $m\lambda = 2b$ and troughs occur when $m\lambda/2 = 2b$, where m is an integer. If the medium between the KBr plates has refractive index n, the wavelength in the medium is λ/n, so the equations become $m\lambda/n = 2b$ and $m\lambda/2n = 2b$. The cell pathlength can be shown to be given by

$$b = \frac{N}{2n} \cdot \frac{\lambda_1\lambda_2}{\lambda_2 - \lambda_1} = \frac{N}{2n} \cdot \frac{1}{\tilde{\nu}_1 - \tilde{\nu}_2}$$

where N maxima occur between wavelengths λ_1 and λ_2. Calculate the pathlength of the cell that gave the interference fringes shown earlier.

19-14. Calculate the power per unit area (the exitance, W/m^2) radiating from a blackbody at 77 K (liquid nitrogen temperature) and at 298 K (room temperature).

19-15. The exitance (power per unit area per unit wavelength) from a blackbody (Box 19-1) is given by the *Planck distribution*:

$$M_\lambda = \frac{2\pi hc^2}{\lambda^5}\left(\frac{1}{e^{hc/\lambda kT} - 1}\right)$$

where λ is wavelength, T is temperature (K), h is Planck's constant, c is the speed of light, and k is Boltzmann's constant. The area under each curve between two wavelengths in the blackbody graph in Box 19-1 is equal to the power per unit area (W/m^2) emitted between those two wavelengths. We find the area by integrating the Planck function between wavelengths λ_1 and λ_2:

$$\text{Power emitted} = \int_{\lambda_1}^{\lambda_2} M_\lambda \, d\lambda$$

For a narrow wavelength range, $\Delta\lambda$, the value of M_λ is nearly constant and the power emitted is simply the product $M_\lambda\Delta\lambda$.

(a) Evaluate M_λ at $\lambda = 2.00$ μm and at $\lambda = 10.00$ μm at $T = 1\,000$ K.

(b) Calculate the power emitted per square meter at 1 000 K in the interval $\lambda = 1.99$ μm to $\lambda = 2.01$ μm by evaluating the product $M_\lambda\Delta\lambda$, where $\Delta\lambda = 0.02$ μm.

(c) Repeat part **(b)** for the interval 9.99 to 10.01 μm.

(d) The quantity $[M_\lambda (\lambda = 2\ \mu m)]/[M_\lambda (\lambda = 10\ \mu m)]$ is the relative exitance at the two wavelengths. Compare the relative exitance at these two wavelengths at 1 000 K with the relative exitance at 100 K. What does your answer mean?

19-16. In the cavity ring-down measurement at the opening of this chapter, absorbance is given by

$$A = \frac{L}{c \ln 10}\left(\frac{1}{\tau} - \frac{1}{\tau_0}\right)$$

where L is the length of the cavity between mirrors, c is the speed of light, τ is the ring-down lifetime with sample in the cavity, and τ_0 is the ring-down lifetime with no sample in the cavity. Ring-down lifetime is obtained by fitting the observed ring-down signal intensity I to an exponential decay of the form $I = I_0 e^{-t/\tau}$, where I_0 is the initial intensity and t is time. A measurement of CO_2 is made at a wavelength absorbed by the molecule. The ring-down lifetime for a 21.0-cm-long empty cavity is 18.52 μs and 16.06 μs for a cavity containing CO_2. Find the absorbance of CO_2 at this wavelength.

Optical Sensors

19-17. Light passes from benzene (medium 1) to water (medium 2) in Figure 19-19 at **(a)** $\theta_1 = 30°$ or **(b)** $\theta_1 = 0°$. Find the angle θ_2 in each case.

19-18. Explain how an optical fiber works. Why does the fiber still work when it is bent?

19-19. The photograph of upconversion in Color Plate 19 shows total internal reflection of the blue ray inside the cuvet. The angle of incidence of the blue ray on the wall of the cuvet is ~55°. We estimate that the refractive index of the organic solvent is 1.50 and the refractive index of the fused-silica cuvet is 1.46. Calculate the critical angle for total internal reflection at the solvent/silica interface and at the silica/air interface. From your calculation, which interface is responsible for total internal reflection in the photo?

19-20. Explain how the attenuated total reflection sensor in Figure 19-23 works.

19-21. Consider a planar waveguide used for attenuated total reflection measurement of a film coated on one surface of the waveguide. For a given angle of incidence, the sensitivity of attenuated total reflectance increases as the thickness of the waveguide decreases. Explain why. (The waveguide can be less than 1 µm thick.)

19-22. (a) Find the critical value of θ_i in Figure 19-20 beyond which there is total internal reflection in a ZrF_4-based infrared optical fiber whose core refractive index is 1.52 and whose cladding refractive index is 1.50.

(b) The loss of radiant power (due to absorption and scatter) in an optical fiber of length ℓ is expressed in decibels per meter (dB/m), defined as

$$\frac{\text{Power out}}{\text{Power in}} = 10^{-\ell(\text{dB/m})/10}$$

Calculate the quotient power out/power in for a 20.0-m-long fiber with a loss of 0.010 0 dB/m.

19-23. Find the minimum angle θ_i for total reflection in the optical fiber in Figure 19-20 if the index of refraction of the cladding is 1.400 and the index of refraction of the core is **(a)** 1.600 or **(b)** 1.800.

19-24. The prism shown here is used to totally reflect light at a 90° angle. No surface of this prism is silvered. Use Snell's law to explain why total reflection occurs. What is the minimum refractive index of the prism for total reflection?

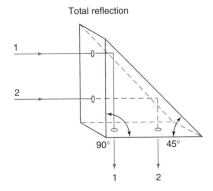

19-25. Here is an extremely sensitive method for measuring nitrite (NO_2^-) down to 1 nM in natural waters. The water sample is treated

with sulfanilamide and *N*-(1-naphthylethylenediamine) in acid solution to produce a colored product with a molar absorptivity of $4.5 \times 10^4 \ M^{-1} \ cm^{-1}$ at 540 nm. The colored solution is pumped into a 4.5-meter-long, coiled Teflon tube whose fluorocarbon wall has a refractive index of 1.29. The aqueous solution inside the tube has a refractive index near 1.33. The colored solution is pumped through the coiled tube. An optical fiber delivers white light into one end of the tube, and an optical fiber at the other end leads to a polychromator and detector.

Long-pathlength spectrometer. [Adapted from W. Yao, R. H. Byrne, and R. D. Waterbury, "Determination of Nanomolar Concentrations of Nitrite and Nitrate Using Long Path Length Absorbance Spectroscopy," *Environ. Sci. Technol.* **1998**, *32*, 2646.]

(a) Explain the purpose of the coiled Teflon tube and explain how it functions.

(b) What is the critical angle of incidence for total internal reflection at the Teflon/water interface?

(c) What is the predicted absorbance of a 1.0-nM solution of colored reagent?

19-26. (a) A particular silica glass waveguide is reported to have a loss coefficient of 0.050 dB/cm (power out/power in, defined in Problem 19-22) for 514-nm-wavelength light. The thickness of the waveguide is 0.60 µm and the length is 3.0 cm. The angle of incidence (θ_i in the figure) is 70°. What fraction of the incident radiant intensity is transmitted through the waveguide?

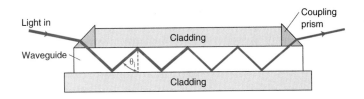

(b) If the index of refraction of the waveguide material is 1.5, what is the wavelength of light inside the waveguide? What is the frequency?

19-27. 🖳 The variation of refractive index, n, with wavelength for fused silica is given by

$$n^2 - 1 = \frac{(0.696\ 166\ 3)\lambda^2}{\lambda^2 - (0.068\ 404\ 3)^2} + \frac{(0.407\ 942\ 6)\lambda^2}{\lambda^2 - (0.116\ 241\ 4)^2}$$
$$+ \frac{(0.897\ 479\ 4)\lambda^2}{\lambda^2 - (9.896\ 161)^2}$$

where λ is expressed in µm.

(a) Make a graph of n versus λ with points at the following wavelengths: 0.2, 0.4, 0.6, 0.8, 1, 2, 3, 4, 5, and 6 µm.

(b) The ability of a prism to spread apart (disperse) neighboring wavelengths increases as the slope $dn/d\lambda$ increases. Is the dispersion of fused silica greater for blue light or red light?

Fourier Transform Spectroscopy

19-28. The interferometer mirror of a Fourier transform infrared spectrophotometer travels ±1 cm.

(a) How many centimeters is the maximum retardation, Δ?

(b) State what is meant by resolution.

(c) What is the approximate resolution (cm^{-1}) of the instrument?

(d) At what retardation interval, δ, must the interferogram be sampled (converted into digital form) to cover a spectral range of 0 to 2 000 cm^{-1}?

19-29. Explain why the transmission spectrum in Figure 19-32 is calculated from the quotient (sample transform)/(background transform) instead of the difference (sample transform) − (background transform).

Dealing with Noise

19-30. Describe three general types of noise that have a different dependence on frequency. Give an example of the source of each kind of noise.

19-31. Explain how beam chopping reduces line noise and noise from source drift.

19-32. Explain how the difference voltage in Figure 19-35 reduces noise from source flicker.

19-33. A spectrum has a signal-to-noise ratio of 8/1. How many spectra must be averaged to increase the signal-to-noise ratio to 20/1?

19-34. A measurement with a signal-to-noise ratio of 100/1 can be thought of as a signal, S, with 1% uncertainty, e. That is, the measurement is $S \pm e = 100 \pm 1$.

(a) Use the rules for propagation of uncertainty to show that, if you add two such signals, the result is total signal = $200 \pm \sqrt{2}$, giving a signal-to-noise ratio of $200/\sqrt{2} = 141/1$.

(b) Show that if you add four such measurements, the signal-to-noise ratio increases to 200/1.

(c) Show that averaging n measurements increases the signal-to-noise ratio by a factor of $\sqrt{n}$ compared with the value for one measurement.

19-35. Results of an electrochemical experiment are shown in the figure. In each case, a voltage is applied between two electrodes at time = 20 s and the absorbance of a solution decreases until the voltage is stepped back to its initial value at time = 60 s. The upper traces show the average results for 100, 300, and 1 000 repetitions of the experiment. The measured signal-to-rms noise ratio in the upper trace is 60.0. Predict the signal-to-noise ratios expected for 300, 100, and 1 cycle and compare your answers with the observed values in the figure.

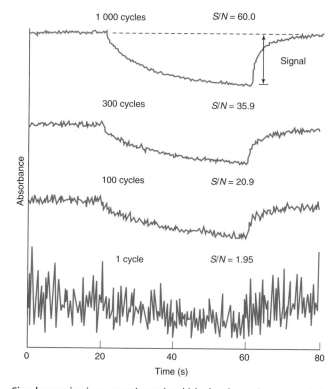

Signal averaging in an experiment in which absorbance is measured after the electric potential is changed at 20 s. [From A. F. Slaterbeck, T. H. Ridgeway, C. J. Seliskar, and W. R. Heineman, "Spectroelectrochemical Sensing Based on Multimode Selectivity Simultaneously Achievable in a Single Device," *Anal. Chem.* **1999**, *71*, 1196.]

20 | Atomic Spectroscopy

Trace element profile of a tooth from a modern man and from a person who lived in Scandinavia 200 years ago. [From A. Cox, F. Keenan, M. Cooke, and J. Appleton, "Trace Element Profiling of Dental Tissues Using Laser Ablation Inductively Coupled Plasma–Mass Spectrometry," *Fresenius J. Anal. Chem.* **1996**, *354*, 254.]

In atomic spectroscopy, a substance is decomposed into atoms in a flame, furnace, or *plasma*. (A **plasma** is a gas that is hot enough to contain ions and free electrons.) Each element is measured by absorption or emission of ultraviolet or visible radiation by the gaseous atoms. To measure trace elements in a tooth, tiny portions of the tooth are vaporized (*ablated*) by a laser pulse[1] and swept into a plasma. The plasma ionizes some of the atoms, which pass into a mass spectrometer that separates ions by their mass and measures their quantity.

Elements are incorporated into teeth from the diet or by inhalation. The figure shows trace element profiles measured by laser ablation–plasma ionization–mass spectrometry of the dentine of teeth from a modern person and one who lived in Scandinavia about A.D. 1800. The contrast is striking. The old tooth contains significant amounts of tin and bismuth, which are nearly absent in the modern tooth. The old tooth contains more lead and antimony than the modern tooth. Tin and lead are constituents of pewter, which was used for cooking vessels and utensils. Bismuth and antimony also might come from pewter.

Even more striking in the old tooth is the abundance of rare earths (dysprosium, holmium, erbium, thulium, ytterbium, and lutetium) and the elements tantalum, tungsten, gold, thorium, and uranium. Rare earth minerals are found in Scandinavia (in fact, many rare earth elements were discovered there), but what were they used for? Did people prepare food with them? Did they somehow get into the food chain?

Inductively coupled argon plasma atomizes substances at 6 000 K.

I n *atomic spectroscopy,* samples vaporized at 2 000–8 000 K decompose into atoms. Concentrations of atoms in the vapor are measured by emission or absorption of characteristic wavelengths of radiation. Because of its high sensitivity, its ability to distinguish one element from another in a complex sample, its ability to perform simultaneous multielement analyses, and the ease with which many samples can be automatically analyzed, atomic spectroscopy is a principal tool of analytical chemistry.[2,3] Ions in the vapor can also be measured in a mass spectrometer, which we describe in this chapter. Equipment for atomic spectroscopy is expensive, but widely available.

Analyte is measured at parts per million (μg/g) to parts per trillion (pg/g) levels. To analyze major constituents, the sample must be diluted. Trace constituents can be measured directly without *preconcentration*. The precision of atomic spectroscopy is typically a few percent (depending on the type of sample and matrix), which is not as good as that of some wet chemical methods. Done in the most careful manner with suitable samples, inductively coupled plasma–atomic emission spectroscopy has an accuracy and precision on the order of 0.1% and can be used to certify DNA reference materials on the basis of phosphorus content.[4]

20-1 An Overview

Three forms of atomic spectroscopy are based on absorption, emission, and fluorescence (Figure 20-1).[5] In **atomic absorption** in Figure 20-2, a liquid sample is aspirated (sucked) into a flame whose temperature is 2 000–3 000 K. Liquid evaporates and the remaining solid is **atomized** (broken into atoms) in the flame, which replaces the cuvet of conventional spectrophotometry. The pathlength of the flame is typically 10 cm. The *hollow-cathode lamp* at the left in Figure 20-2 has an iron cathode. When the cathode is bombarded with energetic Ne^+ or Ar^+ ions, excited Fe atoms vaporize and emit light with the same frequencies absorbed by analyte Fe in the flame. At the right side of Figure 20-2, a detector measures the amount of light that passes through the flame.

An important difference between atomic and molecular spectroscopy is the width of absorption or emission bands. Absorption and emission spectra of liquids and solids typi-

Preconcentration: concentrating a dilute analyte to a level high enough to be analyzed

Types of atomic spectroscopy:
- *emission* from a thermally populated excited state
- *absorption* of sharp lines from hollow-cathode lamp
- *fluorescence* following absorption of laser radiation

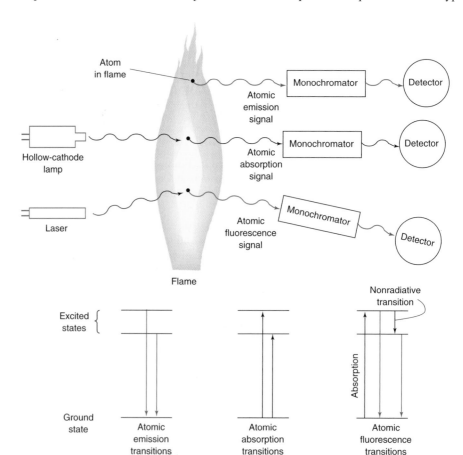

FIGURE 20-1 Emission, absorption, and fluorescence by atoms in a flame. In atomic absorption, atoms absorb part of the light from the source and the remainder of the light reaches the detector. Atomic emission comes from atoms that are in an excited state because of the high thermal energy of the flame. To observe atomic fluorescence, atoms are excited by an external lamp or laser. An excited atom can fall to a lower state and emit radiation.

FIGURE 20-2 Atomic absorption experiment. As in Figure 17-4, transmittance $= T = P/P_0$ and absorbance $= A = -\log T$. In practice, P_0 is the irradiance reaching the detector when no sample is going into the flame, and P is measured while sample is present.

cally have bandwidths of ~10 to 100 nm, as in Figures 17-8 and 17-18. In contrast, spectra of gaseous atoms consist of sharp lines with widths of ~0.001 nm (Figure 20-3). Lines are so sharp that there is usually little overlap between the spectra of different elements in the same sample. Therefore, some instruments can measure more than 70 elements simultaneously. We will see later that sharp analyte absorption lines require that the light source also has sharp lines.

Figure 20-1 also illustrates an **atomic fluorescence** experiment. Atoms in the flame are irradiated by a laser to promote them to an excited electronic state from which they can fluoresce to return to the ground state. Figure 20-4 shows atomic fluorescence from 2 ppb of lead in tap water. Atomic fluorescence is potentially a thousand times more sensitive than atomic absorption, but equipment for atomic fluorescence is not common. An important example of atomic fluorescence is in the analysis of mercury (Box 20-1).

Fluorescence is more sensitive than absorption because we can observe a weak fluorescence signal above a dark background. In absorption, we are looking for small differences between large amounts of light reaching the detector.

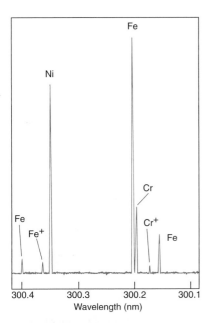

FIGURE 20-3 A portion of the emission spectrum of a steel hollow-cathode lamp, showing lines from gaseous Fe, Ni, and Cr atoms and weak lines from Cr^+ and Fe^+ ions. The monochromator resolution is 0.001 nm, which is comparable to the true linewidths. [From A. P. Thorne, "Fourier Transform Spectrometry in the Ultraviolet," *Anal. Chem.* **1991,** *63*, 57A.]

FIGURE 20-4 Atomic fluorescence from Pb at 405.8 nm. Water containing parts per billion of colloidal $PbCO_3$ was ejected from a capillary tube and exposed to a 6-ns pulse of 1 064-nm laser radiation focused on the drop. This pulse created a plume of vapor moving toward the laser. After 2.5 μs, the plume was exposed to a 193-nm laser pulse, creating excited Pb atoms whose fluorescence was measured for 0.1 μs with an optical system whose resolution was 0.2 nm. The figure shows a calibration curve constructed from colloidal $PbCO_3$ standards and the signal from tap water containing 2 ppb Pb. [From S. K. Ho and N. H. Cheung, "Sub-Part-per-Billion Analysis of Aqueous Lead Colloids by ArF Laser Induced Atomic Fluorescence," *Anal. Chem.* **2005,** *77*, 193.]

BOX 20-1 Mercury Analysis by Cold Vapor Atomic Fluorescence

Mercury is a volatile toxic pollutant. The map shows Hg(0) concentrations in the air near Earth's surface. Mercury is also found as Hg(II)(*aq*) in clouds and on particles in the atmosphere. Approximately two-thirds of atmospheric mercury comes from human activities, including coal burning, waste incineration, and Cl_2 production by the chlor-alkali process (Problem 16-7).

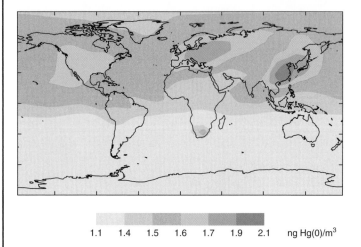

Global annual average surface concentration of Hg(0). [From C. Seigneur, K. Vijayaraghavan, P. Karamchandani, and C. Scott, "Global Source Attribution of Mercury Deposition in the United States," *Environ. Sci. Technol.* 2004, *38*, 555.]

A sensitive method to measure mercury in matrices such as water, soil, and fish generates Hg(*g*), which is measured by atomic absorption or fluorescence. For most environmental samples, automated digestion/analysis equipment is available.[6,7] For water analysis by one standard method, all mercury is first oxidized to Hg(II) with BrCl in the purge flask in the drawing at the right. Halogens are reduced with hydroxylamine (NH_2OH), and Hg(II) is reduced to Hg(0) with $SnCl_2$. Hg(0) is then purged from solution by bubbling

purified Ar or N_2. Hg(0) is collected at room temperature in the sample trap, which contains gold-coated silica sand. Hg binds to Au while other gases in the purge stream pass through. The sample trap is then heated to 450°C to release Hg(*g*), which is caught in the analytical trap at room temperature. Two traps are used so that all other gaseous impurities are removed prior to analysis. Hg(*g*) is then released from the analytical trap by heating and flows into the fluorescence cell. Fluorescence intensity strongly depends on gaseous impurities that can quench the emission from Hg.

The lower limit of quantitation is ~0.5 ng/L (parts per trillion). To measure such small quantities requires extraordinary care at every stage of analysis to prevent contamination. Mercury amalgam fillings in a worker's teeth can contaminate samples exposed to exhaled breath.

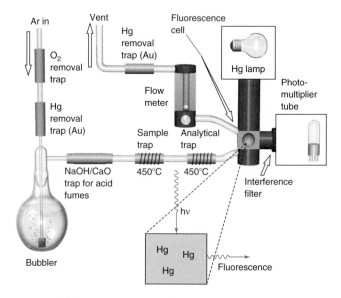

Mercury analysis by U.S. Environmental Protection Agency Method 1631.

By contrast, **atomic emission** (Figure 20-1) is widely used.[8] Collisions in the hot *plasma* promote some atoms to excited electronic states from which they can emit photons to return to lower energy states. No lamp is required. Emission intensity is proportional to the concentration of the element in the sample. Emission from atoms in a plasma is now the dominant form of atomic spectroscopy.

20-2 Atomization: Flames, Furnaces, and Plasmas

In atomic spectroscopy, analyte is *atomized* in a flame, an electrically heated furnace, or a plasma. Flames were used for decades, but they have been replaced by the inductively coupled plasma and the graphite furnace. We begin our discussion with flames because they are still common in teaching labs.

Flames

Most flame spectrometers use a **premix burner,** such as that in Figure 20-5, in which fuel, oxidant, and sample are mixed before introduction into the flame. Sample solution is drawn into the *pneumatic nebulizer* by the rapid flow of oxidant (usually air) past the tip of the sample capillary. Liquid breaks into a fine mist as it leaves the capillary. The spray is directed against a glass bead, upon which the droplets break into smaller particles. The formation of small droplets is termed **nebulization.** A fine suspension of liquid (or solid) particles in a gas is called an **aerosol.** The nebulizer creates an aerosol from the liquid sample. The mist, oxidant, and fuel flow past baffles that promote further mixing and block large droplets of liquid.

Organic solvents with surface tensions lower than that of water are excellent for atomic spectroscopy because they form smaller droplets, thus leading to more efficient atomization.

(a)

Flame

Burner head

Spray chamber

Fuel

Oxidant

Sample

Nebulizer

To drain

Baffles

Glass bead

(b)

Outer cone

Interconal layer

Preheating region

Blue cone

Burner head

(c)

Volume percent

10.8

5.4

0

6   12

Diameter (μm)

FIGURE 20-5 (*a*) Premix burner. (*b*) End view of flame. The slot in the burner head is about 0.5 mm wide. (*c*) Distribution of droplet sizes produced by a particular nebulizer. [From R. H. Clifford, I. Ishii, A. Montaser, and G. A. Meyer, "Droplet-Size and Velocity Distributions of Aerosols from Commonly Used Nebulizers," *Anal. Chem.* **1990**, *62*, 390.]

Excess liquid collects at the bottom of the spray chamber and flows out to a drain. Aerosol reaching the flame contains only about 5% of the initial sample.

The most common fuel-oxidizer combination is acetylene and air, which produces a flame temperature of 2 400–2 700 K (Table 20-1). If a hotter flame is required to atomize high-boiling elements (called *refractory* elements), acetylene and nitrous oxide are usually used. In the flame profile in Figure 20-5b, gas entering the preheating region is heated by conduction and radiation from the primary reaction zone (the blue cone in the flame). Combustion is completed in the outer cone, where surrounding air is drawn into the flame. Flames emit light that must be subtracted from the total signal to obtain the analyte signal.

Droplets entering the flame evaporate; then the remaining solid vaporizes and decomposes into atoms. Many elements form oxides and hydroxides in the outer cone. Molecules do not have the same spectra as atoms, so the atomic signal is lowered. Molecules also emit broad radiation that must be subtracted from the sharp atomic signals. If the flame is relatively rich in fuel (a "rich" flame), excess carbon tends to reduce metal oxides and hydroxides and thereby increases sensitivity. A "lean" flame, with excess oxidant, is hotter. Different elements require either rich or lean flames for best analysis. The height in the flame at which maximum atomic absorption or emission is observed depends on the element being measured and the flow rates of sample, fuel, and oxidizer.[9]

TABLE 20-1   Maximum flame temperatures

Fuel	Oxidant	Temperature (K)
Acetylene, HC≡CH	Air	2 400–2 700
Acetylene	Nitrous oxide, N_2O	2 900–3 100
Acetylene	Oxygen	3 300–3 400
Hydrogen	Air	2 300–2 400
Hydrogen	Oxygen	2 800–3 000
Cyanogen, N≡C—C≡N	Oxygen	4 800

Furnaces[10]

An electrically heated **graphite furnace** is more sensitive than a flame and requires less sample. From 1 to 100 μL of sample are injected into the furnace through the hole at the center of Figure 20-6. Light from a hollow-cathode lamp travels through windows at each end of the graphite tube. To prevent oxidation of the graphite, Ar gas is passed over the furnace and the maximum recommended temperature is 2 550°C for not more than 7 s.

In flame spectroscopy, the *residence time* of analyte in the optical path is <1 s as it rises through the flame. A graphite furnace confines the atomized sample in the optical path for several seconds, thereby affording higher sensitivity. Whereas 1–2 mL is the minimum volume of solution necessary for flame analysis, as little as 1 μL is adequate for a furnace. Precision is rarely better than 5–10% with manual sample injection, but automated injection improves reproducibility to ~1%.

Graphite furnace

FIGURE 20-6 A 38-mm-long, electrically heated graphite furnace for atomic spectroscopy. [Courtesy Instrumentation Laboratory, Wilmington, MA.]

You must determine reasonable time and temperature for each stage of the analysis. Once a program is established, it can be applied to a large number of similar samples.

(a)

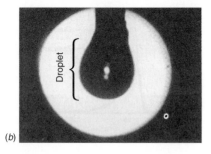

(b)

FIGURE 20-7 (*a*) Correct position for injecting sample into a graphite furnace deposits the droplet in a small volume on the floor of the furnace. (*b*) If injection is too high, sample splatters and precision is poor. [From P. K. Booth, "Improvements in Method Development for Graphite Furnace Atomic Absorption Spectrometry," *Am. Lab.*, February **1995**, p. 48X.]

When you inject sample, the droplet should contact the floor of the furnace and remain in a small area (Figure 20-7a). If you inject the droplet too high (Figure 20-7b), it splashes and spreads, leading to poor precision. In the worst case, the drop adheres to the tip of the pipet and is finally deposited around the injector hole when the pipet is withdrawn.

Compared with flames, furnaces require more operator skill to find proper conditions for each type of sample. The furnace is heated in three or more steps to properly atomize the sample. To measure Fe in the iron-storage protein ferritin, 10 μL of sample containing ~0.1 ppm Fe are injected into the furnace at ~90°C. The furnace is programmed to *dry* the sample at 125°C for 20 s to remove solvent. Drying is followed by 60 s of *charring* at 1 400°C to destroy organic matter. Charring is also called *pyrolysis*, which means decomposing with heat. Charring creates smoke that would interfere with the Fe determination. After charring, the sample is atomized at 2 100°C for 10 s. Absorbance reaches a maximum and then decreases as Fe evaporates from the oven. The analytical signal is the time-integrated absorbance (the peak area) during atomization. After atomization, the furnace is heated to 2 500°C for 3 s to clean out remaining residue.

The furnace is purged with Ar or N_2 during each step except atomization to remove volatile material. Gas flow is halted during atomization to avoid blowing analyte out of the furnace. When developing a method for a new kind of sample, it is important to record the signal as a function of time, because signals are also observed from smoke during charring and from the glow of the red-hot furnace in the last part of atomization. A skilled operator interprets which signal is due to analyte so that the right peak is integrated.

The furnace in Figure 20-8a performs better than a simple graphite tube. Sample is injected onto a platform that is heated by radiation from the furnace wall, so its temperature lags behind that of the wall. Analyte does not vaporize until the wall reaches constant temperature (Figure 20-8b). At constant furnace temperature, the area under the absorbance peak in Figure 20-8b is a reliable measure of the analyte. A heating rate of 2 000 K/s rapidly dissociates molecules and increases the concentration of free atoms in the furnace.

The furnace in Figure 20-8a is heated *transversely* (from side to side) to provide nearly uniform temperature over the whole furnace. In furnaces with *longitudinal* (end-to-end) heating, the center of the furnace is hotter than the ends. Atoms from the central region condense at the ends, where they can vaporize during the next sample run. Interference from previous runs, called a *memory effect*, is reduced in a transversely heated furnace. To further reduce memory effects, ordinary graphite is coated with a dense layer of *pyrolytic graphite* formed by thermal decomposition of an organic vapor. The coating seals the relatively porous graphite, so it cannot absorb foreign atoms.

A sample can be *preconcentrated* by injecting and evaporating multiple aliquots in the graphite furnace prior to analysis.[11] To measure traces of As in drinking water, a 30-μL aliquot of water plus matrix modifier was injected and evaporated. The procedure was

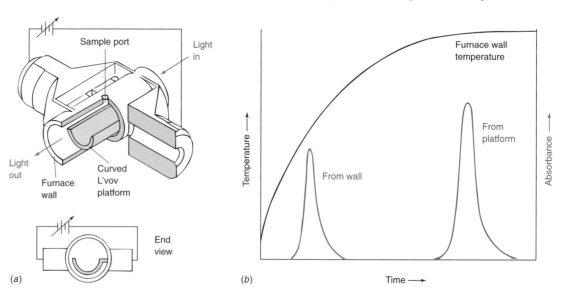

(a)

(b)

FIGURE 20-8 (*a*) Transversely heated graphite furnace maintains nearly constant temperature over its whole length, thereby reducing memory effect from previous runs. The *L'vov platform* is uniformly heated by radiation from the outer wall, not by conduction. The platform is attached to the wall by one small connection that is hidden from view. [Courtesy Perkin-Elmer Corp., Norwalk, CT.] (**b**) Heating profiles comparing analyte evaporation from wall and from platform. [From W. Slavin, "Atomic Absorption Spectroscopy," *Anal. Chem.* **1982**, *54*, 685A.]

repeated five more times so that the total sample volume was 180 μL. The detection limit for As was 0.3 μg/L (parts per billion). Without preconcentration, the detection limit would have been 1.8 μg/L. This increased capability is critical because As is considered to be a health hazard at concentrations of just a few parts per billion.

Liquid samples are ordinarily used in furnaces. However, in *direct solid sampling*, a solid is analyzed without sample preparation (Figure 20-9). For example, trace impurities in tungsten powder used to make components for industry can be analyzed by weighing 0.1 to 100 mg of powder onto a graphite platform.[12] The platform is transferred into the furnace and heated to 2 600°C to atomize impurities in the tungsten, but not the tungsten itself, which melts at 3 410°C. After several runs, residual tungsten is scraped off the platform, which could be reused 400 times. Because so much more sample is analyzed when solid is injected than when liquid is injected, detection limits can be 100 times lower than those obtained for liquid injection. For example, Zn could be detected at a level of 10 pg/g (10 parts per trillion) in 100 mg of tungsten. Calibration curves are obtained by injecting standard solutions of the trace elements and analyzing them as conventional liquids. Results obtained from direct solid sampling are in good agreement with results obtained by laboriously dissolving the solid. Other solids that have been analyzed by direct solid sampling include graphite, silicon carbide, cement, river sediments, hair, and vegetable matter.[13]

FIGURE 20-9 Direct solid sampling showing end view of furnace.

Matrix Modifiers for Furnaces

Everything in a sample other than analyte is called the **matrix.** Ideally, the matrix decomposes and vaporizes during the charring step. A **matrix modifier** is a substance added to the sample to reduce the loss of analyte during charring by making the matrix more volatile or the analyte less volatile.

The matrix modifier ammonium nitrate can be added to seawater to increase the volatility of the matrix NaCl. Figure 20-10a shows a graphite furnace heating profile used to analyze Mn in seawater. When 0.5 M NaCl solution is subjected to this profile, signals are observed at the analytical wavelength of Mn (Figure 20-10b). Much of the apparent absorbance is probably due to optical scatter by smoke created by heating NaCl. Absorption at the start of the atomization step interferes with the measurement of Mn. Adding NH_4NO_3 to the sample in Figure 20-10c reduces matrix absorption peaks. NH_4NO_3 plus NaCl give NH_4Cl and $NaNO_3$, which cleanly evaporate instead of making smoke.

The matrix modifier $Pd(NO_3)_2$ is added to seawater to decrease the volatility of the analyte Sb. In the absence of modifier, 90% of Sb is lost during charring at 1 250°C. With the modifier, seawater matrix can be largely evaporated at 1 400°C without loss of Sb.[14]

Matrix modifier: increases volatility of matrix or decreases volatility of analyte to obtain a cleaner separation between matrix and analyte.

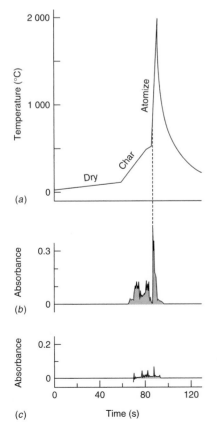

FIGURE 20-10 Reduction of interference by using a matrix modifier. (*a*) Graphite furnace temperature profile for analysis of Mn in seawater. (*b*) Absorbance profile from 10 μL of 0.5 M reagent-grade NaCl subjected to temperature profile in panel *a*. Absorbance is monitored at the Mn wavelength of 279.5 nm with a bandwidth of 0.5 nm. (*c*) Reduced absorbance from 10 μL of 0.5 M NaCl plus 10 μL of 50 wt% NH_4NO_3 matrix modifier. [From M. N. Quigley and F. Vernon, "Matrix Modification Experiment for Electrothermal Atomic Absorption Spectrophotometry," *J. Chem. Ed.* **1996**, *73*, 980.]

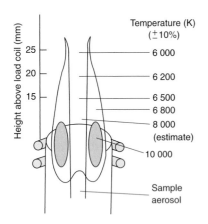

FIGURE 20-11 Temperature profile of inductively coupled plasma. [From V. A. Fassel, "Simultaneous or Sequential Determination of the Elements at All Concentration Levels," *Anal. Chem.* 1979, *51*, 1290A.]

FIGURE 20-12 Inductively coupled plasma burner. [From R. N. Savage and G. M. Hieftje, "Miniature Inductively Coupled Plasma Source for Atomic Emission Spectrometry," *Anal. Chem.* 1979, *51*, 408.]

The matrix modifier $Mg(NO_3)_2$ raises the temperature for atomization of Al analyte.[15] At high temperature, $Mg(NO_3)_2$ decomposes to $MgO(g)$ and Al is converted into Al_2O_3. At sufficiently high temperature, Al_2O_3 decomposes to Al and O, and Al evaporates. Evaporation of Al is retarded by $MgO(g)$ by virtue of the reaction

$$3MgO(g) + 2Al(s) \rightleftharpoons 3Mg(g) + Al_2O_3(s) \qquad (20\text{-}1)$$

When MgO has evaporated, Reaction 20-1 no longer occurs and Al_2O_3 decomposes and evaporates. A matrix modifier that raises the boiling temperature of analyte allows a higher charring temperature to be used to remove matrix without losing analyte.

It is important to monitor the absorption signal from a graphite furnace as a function of time, as in Figure 20-10b. Peak shapes help you decide how to adjust time and temperature for each step to obtain a clean signal from analyte. Also, a graphite furnace has a finite lifetime. Degradation of peak shape or a change in the slope of the calibration curve tell you that it is time to change the furnace.

Inductively Coupled Plasmas

The **inductively coupled plasma**[16] shown at the beginning of the chapter is twice as hot as a combustion flame (Figure 20-11). The high temperature, stability, and relatively inert Ar environment eliminate much of the interference encountered with flames. Simultaneous multi-element analysis described in Section 20-4 is routine for inductively coupled plasma–atomic emission spectroscopy, which has replaced flame atomic absorption. The plasma instrument costs more to purchase and operate than a flame instrument.

The cross-sectional view of an inductively coupled plasma burner in Figure 20-12 shows two turns of a 27- or 41-MHz induction coil wrapped around the upper opening of the quartz apparatus. High-purity Ar gas is fed through the plasma gas inlet. After a spark from a Tesla coil ionizes Ar, free electrons are accelerated by the radio-frequency field. Electrons collide with atoms and transfer energy to the entire gas, maintaining a temperature of 6 000 to 10 000 K. The quartz torch is protected from overheating by Ar coolant gas.

The concentration of analyte needed for adequate signal is reduced by an order of magnitude with an *ultrasonic nebulizer* (Figure 20-13), in which sample solution is directed onto a

FIGURE 20-13 (*a*) Ultrasonic nebulizer lowers detection limit for most elements by an order of magnitude. (*b*) Mist created when sample is sprayed against vibrating crystal. [Courtesy Cetac Technologies, Omaha, NB.]

TABLE 20-2 Comparison of detection limits for Ni$^+$ ion at 231 nm

Technique[a]	Detection limits for different instruments (ng/g)
ICP/atomic emission (pneumatic nebulizer)	3–50
ICP/atomic emission (ultrasonic nebulizer)	0.3–4
Graphite furnace/atomic absorption	0.02–0.06
ICP/mass spectrometry	0.001–0.2

a. ICP = inductively coupled plasma.

SOURCE: J. M. Mermet and E. Poussel, "ICP Emission Spectrometers: Analytical Figures of Merit," Appl. Spectros. **1995,** 49, 12A.

piezoelectric crystal oscillating at 1 MHz. The vibrating crystal creates a fine aerosol that is carried by an Ar stream through a heated tube where solvent evaporates. In the next refrigerated zone, solvent condenses and is removed. Then the stream enters a desolvator containing a microporous polytetrafluoroethylene membrane in a chamber maintained at 160°C. Remaining solvent vapor diffuses through the membrane and is swept away by flowing Ar. Analyte reaches the plasma flame as an aerosol of dry, solid particles. Plasma energy is not needed to evaporate solvent, so more energy is available for atomization. Also, a larger fraction of the sample reaches the plasma than with a conventional nebulizer.

Sensitivity with an inductively coupled plasma is further enhanced by a factor of 3 to 10 by observing emission along the length of the plasma (axial view) instead of across the diameter of the plasma. Additional sensitivity is obtained by detecting ions with a mass spectrometer instead of by optical emission (Table 20-2), as described in Section 20-6.

> A *piezoelectric crystal* is one whose dimensions change in an applied electric field. A sinusoidal voltage applied between two faces of the crystal causes it to oscillate. Quartz is the most common piezoelectric material.

20-3 How Temperature Affects Atomic Spectroscopy

Temperature determines the degree to which a sample breaks down into atoms and the extent to which a given atom is found in its ground, excited, or ionized states. Each of these effects influences the strength of the signal we observe.

The Boltzmann Distribution

Consider an atom with energy levels E_0 and E^* separated by ΔE (Figure 20-14). An atom (or molecule) may have more than one state at a given energy. Figure 20-14 shows three states at E^* and two at E_0. The number of states at each energy is called the *degeneracy*. We will call the degeneracies g_0 and g^*.

The **Boltzmann distribution** describes the relative populations of different states at thermal equilibrium. If equilibrium exists (which is not true in the blue cone of a flame but is probably true above the blue cone), the relative population (N^*/N_0) of any two states is

FIGURE 20-14 Two energy levels with different degeneracies. Ground-state atoms that absorb light are promoted to the excited state. Excited-state atoms can emit light to return to the ground state.

Boltzmann distribution:
$$\frac{N^*}{N_0} = \frac{g^*}{g_0}e^{-\Delta E/kT} \qquad (20\text{-}2)$$

where T is temperature (K) and k is Boltzmann's constant (1.381×10^{-23} J/K).

> The Boltzmann distribution applies to a system at thermal equilibrium.

Effect of Temperature on Excited-State Population

The lowest excited state of a sodium atom lies 3.371×10^{-19} J/atom above the ground state. The degeneracy of the excited state is 2, whereas that of the ground state is 1. The fraction of Na in the excited state in an acetylene-air flame at 2 600 K is, from Equation 20-2,

$$\frac{N^*}{N_0} = \left(\frac{2}{1}\right)e^{-(3.371\times10^{-19}\,\text{J})/[(1.381\times10^{-23}\,\text{J/K})(2\,600\,\text{K})]} = 1.67 \times 10^{-4}$$

That is, less than 0.02% of the atoms are in the excited state.

If the temperature were 2 610 K, the fraction of atoms in the excited state would be

$$\frac{N^*}{N_0} = \left(\frac{2}{1}\right)e^{-(3.371\times10^{-19}\,\text{J})/[(1.381\times10^{-23}\,\text{J/K})(2\,610\,\text{K})]} = 1.74 \times 10^{-4}$$

The fraction of atoms in the excited state is still less than 0.02%, but that fraction has increased by $100(1.74 - 1.67)/1.67 = 4\%$.

> A 10-K temperature rise changes the excited-state population by 4% in this example.

The Effect of Temperature on Absorption and Emission

We see that more than 99.98% of the sodium atoms are in their ground state at 2 600 K. *Varying the temperature by 10 K hardly affects the ground-state population and would not noticeably affect the signal in atomic absorption.*

How would emission intensity be affected by a 10 K rise in temperature? In Figure 20-14, absorption arises from ground-state atoms, but emission arises from excited-state atoms. Emission intensity is proportional to the population of the excited state. *Because the excited-state population changes by 4% when the temperature rises 10 K, emission intensity rises by 4%.* It is critical in atomic *emission* spectroscopy that the flame be very stable, or emission intensity will vary significantly. In atomic *absorption* spectroscopy, temperature variation is important but not as critical.

Almost all atomic emission is carried out with an inductively coupled plasma, whose temperature is more stable than that of a flame. Plasma is normally used for emission, not absorption, because it is so hot that there is a substantial population of excited-state atoms and ions. Table 20-3 compares excited-state populations for a flame at 2 500 K and a plasma at 6 000 K. Although the fraction of excited atoms is small, each atom emits many photons per second because it is rapidly promoted back to the excited state by collisions.

TABLE 20-3 Effect of energy difference and temperature on population of excited states

Wavelength difference of states (nm)	Energy difference of states (J/atom)	Excited-state fraction $(N^*/N_0)^a$	
		2 500 K	6 000 K
250	7.95×10^{-19}	1.0×10^{-10}	6.8×10^{-5}
500	3.97×10^{-19}	1.0×10^{-5}	8.3×10^{-3}
750	2.65×10^{-19}	4.6×10^{-4}	4.1×10^{-2}

a. Based on the equation $N^*/N_0 = (g^*/g_0)e^{-\Delta E/kT}$ in which $g^* = g_0 = 1$.

Energy levels of halogen atoms (F, Cl, Br, I) are so high that they emit ultraviolet radiation below 200 nm. This spectral region is called *vacuum ultraviolet* because radiation below 200 nm is absorbed by O_2, so spectrometers for the far-ultraviolet were customarily evacuated. Some plasma emission spectrometers are purged with N_2 to exclude air so that the region 130 to 200 nm is accessible and Cl, Br, I, P, and S can be analyzed.[17] In another application, nitrogen in fertilizers is measured along with other major elements in the fertilizer. The torch is designed to be purged with Ar to exclude N_2 from air. Unknowns are purged with He to remove dissolved air. Emission from N is observed near 174 nm.

20-4 Instrumentation

Fundamental requirements for an atomic absorption experiment are shown in Figure 20-2. Principal differences between atomic and ordinary molecular spectroscopy lie in the light source (or lack of a light source in atomic emission), the sample container (the flame, furnace, or plasma), and the need to subtract background emission.

Atomic Linewidths[18]

Beer's law requires that the linewidth of the radiation source should be substantially narrower than the linewidth of the absorbing sample. Otherwise, the measured absorbance will not be proportional to the sample concentration. Atomic absorption lines are very sharp, with an intrinsic width of only $\sim 10^{-4}$ nm.

Linewidth is governed by the **Heisenberg uncertainty principle,** which says that the shorter the lifetime of the excited state, the more uncertain is its energy:

Heisenberg uncertainty principle: $$\delta E \delta t \gtrsim \frac{h}{4\pi} \tag{20-3}$$

where δE is the uncertainty in the energy difference between ground and excited states, δt is the lifetime of the excited state before it decays to the ground state, and h is Planck's constant. Equation 20-3 says that the uncertainty in the energy difference between two states multiplied by the lifetime of the excited state is at least $h/4\pi$. If δt decreases, then δE increases. The

---

Margin notes:

Atomic absorption is not as sensitive to temperature as atomic emission, which is exponentially sensitive to temperature.

The linewidth of the source must be narrower than the linewidth of the atomic vapor for Beer's law to be obeyed. "Linewidth" and "bandwidth" are used interchangeably, but "lines" are narrower than "bands."

ΔE in Equation 20-2 is the difference in energy between ground and excited states. δE in Equation 20-3 is the uncertainty in ΔE. δE is a small fraction of ΔE.

lifetime of an excited state of an isolated gaseous atom is $\sim 10^{-9}$ s. Therefore, the uncertainty in its energy is

$$\delta E \gtrsim \frac{h}{4\pi \delta t} = \frac{6.6 \times 10^{-34}\,\text{J} \cdot \text{s}}{4\pi(10^{-9}\,\text{s})} \approx 10^{-25}\,\text{J}$$

Suppose that the energy difference (ΔE) between the ground and the excited state of an atom corresponds to visible light with a wavelength of $\lambda = 500$ nm. This energy difference is $\Delta E = hc/\lambda = 4.0 \times 10^{-19}$ J (Equation 17-3, c is the speed of light). The relative uncertainty in the energy difference is $\delta E/\Delta E \gtrsim (10^{-25}\,\text{J}) / (4.0 \times 10^{-19}\,\text{J}) \approx 2 \times 10^{-7}$. The relative uncertainty in wavelength ($\delta\lambda/\lambda$) is the same as the relative uncertainty in energy:

$$\frac{\delta\lambda}{\lambda} = \frac{\delta E}{\Delta E} \gtrsim 2 \times 10^{-7} \quad \Rightarrow \quad \delta\lambda \gtrsim (2 \times 10^{-7})(500\,\text{nm}) = 10^{-4}\,\text{nm} \qquad \text{(20-4)}$$

The inherent linewidth of an atomic absorption or emission signal is $\sim 10^{-4}$ nm because of the short lifetime of the excited state.

Two mechanisms broaden the lines to 10^{-3} to 10^{-2} nm in atomic spectroscopy. One is the **Doppler effect.** An atom moving toward the radiation source experiences more oscillations of the electromagnetic wave in a given time period than one moving away from the source (Figure 20-15). That is, an atom moving toward the source "sees" higher frequency light than that encountered by one moving away. In the laboratory frame of reference, the atom moving toward the source absorbs lower frequency light than that absorbed by the one moving away. The linewidth, $\delta\lambda$, due to the Doppler effect, is

Doppler linewidth: $\qquad\qquad \delta\lambda \approx \lambda(7 \times 10^{-7})\sqrt{\dfrac{T}{M}} \qquad\qquad$ (20-5)

where T is temperature (K) and M is the mass of the atom in atomic mass units. For an emission line near $\lambda = 300$ nm from Fe ($M = 56$ atomic mass units) at 2 500 K in Figure 20-3, the Doppler linewidth is $(300\,\text{nm})(7 \times 10^{-7})\sqrt{2\,500/56} = 0.001\,4$ nm, which is an order of magnitude greater than the natural linewidth.

Linewidth is also affected by **pressure broadening** from collisions between atoms. Collisions shorten the lifetime of the excited state. Uncertainty in the frequency of atomic absorption and emission lines is roughly numerically equal to the collision frequency between atoms and is proportional to pressure. The Doppler effect and pressure broadening are similar in magnitude and yield linewidths of 10^{-3} to 10^{-2} nm in atomic spectroscopy.

Hollow-Cathode Lamp

Monochromators generally cannot isolate lines narrower than 10^{-3} to 10^{-2} nm. To produce narrow lines of the correct frequency, we use a **hollow-cathode lamp** containing a vapor of the same element as that being analyzed.

The hollow-cathode lamp in Figure 20-16 is filled with Ne or Ar at a pressure of $\sim 130–700$ Pa (1–5 Torr). The cathode is made of the element whose emission lines we want. When ~ 500 V is applied between the anode and the cathode, gas is ionized and positive ions are accelerated toward the cathode. After ionization occurs, the lamp is maintained at a constant current of 2–30 mA by a lower voltage. Cations strike the cathode with enough energy to "sputter" metal atoms from the cathode into the gas phase. Gaseous atoms excited by collisions with high-energy electrons emit photons. This atomic radiation has the same frequency absorbed by analyte in the flame or furnace. Atoms in the lamp are cooler than atoms in a flame, so lamp emission is sufficiently narrower than the absorption bandwidth of atoms in the flame to be nearly "monochromatic" (Figure 20-17). The purpose of a monochromator in

FIGURE 20-16 A hollow-cathode lamp.

Insulating disk
Quartz or glass window
(−)
(+)
Hollow cathode
Anode

$\delta\lambda$ is the width of an absorption or emission line measured at half the height of the peak.

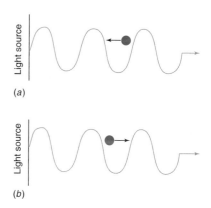

(a)

(b)

FIGURE 20-15 The Doppler effect. A molecule moving (a) toward the radiation source "feels" the electromagnetic field oscillate more often than one moving (b) away from the source.

Doppler and pressure effects broaden the atomic lines by one to two orders of magnitude relative to their inherent linewidths.

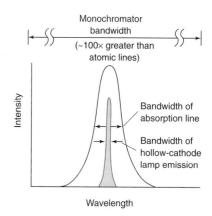

Monochromator bandwidth
(~100× greater than atomic lines)

Bandwidth of absorption line

Bandwidth of hollow-cathode lamp emission

Wavelength

FIGURE 20-17 Relative bandwidths of hollow-cathode emission, atomic absorption, and a monochromator. Linewidths are measured at half the signal height. The linewidth from the hollow cathode is relatively narrow because gas temperature in the lamp is lower than flame temperature (so there is less Doppler broadening) and pressure in the lamp is lower than pressure in a flame (so there is less pressure broadening).

atomic spectroscopy is to select one line from the hollow-cathode lamp and to reject as much emission from the flame or furnace as possible. A different lamp is usually required for each element, although some lamps are made with more than one element in the cathode.

Multielement Detection with the Inductively Coupled Plasma

An inductively coupled plasma emission spectrometer does not require any lamps and can measure as many as 70 elements simultaneously. Color Plates 24 and 25 illustrate two designs for multielement analysis. In Color Plate 24, atomic emission enters the polychromator and is dispersed into its component wavelengths by the grating at the bottom. One photomultiplier detector (Figure 19-14) is required at the correct position for each element.

In Color Plate 25, atomic emission entering from the top right is reflected by a collimating mirror (which makes light rays parallel), dispersed in the vertical plane by a prism, and then dispersed in the horizontal plane by a grating. Dispersed radiation lands on a *charge injection device* (CID) detector, which is related to the charge coupled device (CCD) in Figure 19-17. Different wavelengths are spread over the 262 000 pixels of the CID shown at the upper left of Color Plate 25. In a CCD detector, each pixel must be read one at a time in row-by-row order. Each pixel of a CID detector can be read individually at any time. Selectively reading relevant pixels eliminates time spent reading pixels that are of no interest. A given pixel can be monitored and read before it becomes full. The charge in the pixel is then neutralized to reset the pixel to zero. The pixel can then accumulate more charge and be read several times while other pixels are filling up at a slower pace. This process increases the dynamic range of the detector by allowing large signals to be measured in some pixels while small signals are measured in other pixels. Another advantage of the CID detector over the CCD detector is that strong signals in one pixel are less prone to spill over into neighboring pixels (a process called *blooming* in CCD detectors). CID detectors can therefore measure weak emission signals adjacent to strong signals. Figure 20-18 shows an actual spectrum as seen by a CID detector.

The spectrometer in Color Plate 25 is purged with N_2 or Ar to exclude O_2, thereby allowing ultraviolet wavelengths in the 100–200-nm range to be observed. This spectral region permits more sensitive detection of some elements that are normally detected at longer wavelengths and allows halogens, P, S, and N to be measured (with poor detection limits of tens of

Capabilities of CID detector:
- pixels are individually addressed
- rapidly filling pixel can be read, re-zeroed, and read again
- filled pixel does not bloom into neighbors

FIGURE 20-18 "Constellation image" of inductively coupled plasma emission from 200 µg Fe/mL seen by charge injection detector. Almost all peaks are from iron. Horizontally blurred "galaxies" near the top are Ar plasma emission. A prism spreads wavelengths of 200–400 nm over most of the detector. Wavelengths >400 nm are bunched together at the top. A grating provides high resolution in the horizontal direction. Selected peaks are labeled with wavelength (in nanometers) and diffraction order (*n* in Equation 19-1) in parentheses. Two Fe peaks labeled in color at the lower left and lower right are both the same wavelength (238.204 nm) diffracted into different orders by the grating. [Courtesy M. D. Seltzer, Michelson Laboratory, China Lake, CA.]

FIGURE 20-19 Graphite furnace absorption spectrum of bronze dissolved in HNO_3. [From B. T. Jones, B. W. Smith, and J. D. Winefordner, "Continuum Source Atomic Absorption Spectrometry in a Graphite Furnace with Photodiode Array Detection," *Anal. Chem.* **1989**, *61*, 1670.]

parts per million). These nonmetallic elements cannot be observed at wavelengths above 200 nm. The photomultiplier spectrometer in Color Plate 24 is more expensive and complicated than the CID spectrometer in Color Plate 25 but provides lower detection limits because a photomultiplier tube is more sensitive than a CID detector.

Background Correction

Atomic spectroscopy must provide **background correction** to distinguish analyte signal from absorption, emission, and optical scattering of the sample matrix, the flame, plasma, or red-hot graphite furnace. Figure 20-19 shows the spectrum of a sample analyzed in a graphite furnace. Sharp atomic signals with a maximum absorbance near 1.0 are superimposed on a broad background with an absorbance of 0.3. If we did not subtract the background absorbance, significant errors would result. Background correction is critical for graphite furnaces, which tend to contain residual smoke from charring. Optical scatter from smoke must be distinguished from optical absorption by analyte.

Figure 20-20 shows how background is subtracted in an emission spectrum collected with a charge injection device detector. The figure shows 15 pixels from one row of the detector centered on an analytical peak. (The spectrum is manipulated by a computer algorithm to make it look smooth, even though the original data consist of a single reading for each pixel and would look like a bar graph.) Pixels 7 and 8 were selected to represent the peak. Pixels 1 and 2 represent the baseline at the left and pixels 14 and 15 represent the baseline at the right. The mean baseline is the average of pixels 1, 2, 14, and 15. The mean peak amplitude is the average of pixels 7 and 8. The corrected peak height is the mean peak amplitude minus the mean baseline amplitude.

For atomic absorption, *beam chopping* or electrical *modulation* of the hollow-cathode lamp (pulsing it on and off) can distinguish the signal of the flame from the atomic line at the same wavelength. Figure 20-21 shows light from the lamp being periodically blocked by a rotating chopper. Signal reaching the detector while the beam is blocked must be from flame emission. Signal reaching the detector when the beam is not blocked is from the lamp and the flame. The difference between the two signals is the desired analytical signal.

Beam chopping corrects for flame emission but not for scattering. Most spectrometers provide an additional means to correct for scattering and broad background absorption. Deuterium lamps and Zeeman correction systems are most common.

For *deuterium lamp background correction*, broad emission from a D_2 lamp (Figure 19-4) is passed through the flame in alternation with that from the hollow cathode. The monochromator bandwidth is so wide that a negligible fraction of D_2 radiation is absorbed by the analyte atomic absorption line. Light from the hollow-cathode lamp is absorbed by analyte

Background signal arises from absorption, emission, or scatter by everything in the sample besides analyte (the *matrix*), as well as from absorption, emission, or scatter by the flame, the plasma, or the furnace.

Background correction methods:
- adjacent pixels of CID display
- beam chopping
- D_2 lamp
- Zeeman

FIGURE 20-21 Beam chopper for flame background correction. (*a*) Lamp and flame emission reach detector. (*b*) Only flame emission reaches detector. (*c*) Resulting square wave signal.

FIGURE 20-20 Data from charge injection detector illustrating baseline correction in plasma emission spectrometry. The mean value of pixels on either side of a peak is subtracted from the mean value of pixels under the peak. [Courtesy M. D. Seltzer, Michelson Laboratory, China Lake, CA.]

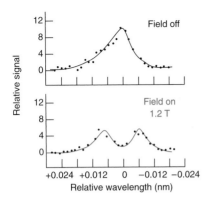

FIGURE 20-22 Zeeman effect on Co fluorescence in a graphite furnace with excitation at 301 nm and detection at 341 nm. The magnetic field strength for the lower spectrum is 1.2 tesla. [From J. P. Dougherty, F. R. Preli, Jr., J. T. McCaffrey, M. D. Seltzer, and R. G. Michel, "Instrumentation for Zeeman Electrothermal Atomizer Laser Excited Atomic Fluorescence Spectrometry," *Anal. Chem.* **1987**, *59*, 1112.]

FIGURE 20-23 Measurement of peak-to-peak noise level and signal level. The signal is measured from its base at the midpoint of the noise along the slightly slanted baseline. This sample exhibits a signal-to-noise ratio of 2.4.

and absorbed and scattered by background. Light from the D_2 lamp is absorbed and scattered only by background. The difference between absorbance measured with the hollow-cathode lamp and absorbance measured with the D_2 lamp is the absorbance of analyte.

An excellent, but expensive, background correction technique for a graphite furnace for many elements relies on the *Zeeman effect* (pronounced ZAY-mon). When a magnetic field is applied parallel to the light path through a furnace, the absorption (or emission) line of analyte atoms is split into three components. Two are shifted to slightly lower and higher wavelengths (Figure 20-22), and one component is unshifted. The unshifted component does not have the correct electromagnetic polarization to absorb light traveling parallel to the magnetic field and is therefore "invisible."

To use the Zeeman effect for background correction, a strong magnetic field is pulsed on and off. Sample and background are observed when the field is off. Background alone is observed when the field is on. The difference is the corrected signal.

The advantage of Zeeman background correction is that it operates at the analytical wavelength. In contrast, D_2 background correction is made over a broad band. A structured or sloping background is averaged by this process, potentially misrepresenting the true background signal at the analytical wavelength. Background correction in Figure 20-20 is similar to D_2 background correction, but the wavelength range in Figure 20-20 is restricted to the immediate vicinity of the analytical peak.

Detection Limits

One definition of **detection limit** is the concentration of an element that gives a signal equal to two times the peak-to-peak noise level of the baseline (Figure 20-23). The baseline noise level should be measured for a blank sample.[19]

Figure 20-24 compares detection limits for flame, furnace, and inductively coupled plasma analyses on instruments from one manufacturer. The detection limit for furnaces is typically two orders of magnitude lower than that observed with a flame because the sample is confined in the small volume of the furnace for a relatively long time. For instruments in Figure 20-24, detection limits for the inductively coupled plasma are intermediate between the flame and the furnace. With ultrasonic nebulization and axial plasma viewing, the sensitivity of the inductively coupled plasma is close to that of the graphite furnace.

Commercial standard solutions for flame atomic absorption are not necessarily suitable for plasma and furnace analyses. The latter methods require purer grades of water and acids for standard solutions and, especially, for dilutions. For the most sensitive analyses, solutions are prepared in a dust-free clean room or hood with a filtered air supply to reduce background contamination that *will be* detected by your instruments.

Any standard solution has a limited shelf life. High-density polyethylene plastic bottles containing standards for atomic spectroscopy are packaged in sealed, aluminized bags to retard evaporation. In one study, the concentration of standard not stored in an aluminized bag increased by 0.11% per year at 23°C and 0.26% per year at 30°C because water evaporated

FIGURE 20-24 Flame, furnace, and inductively coupled plasma emission and inductively coupled plasma–mass spectrometry detection limits (ng/g = ppb) with instruments from GBC Scientific Equipment, Australia. Accurate quantitative analysis requires concentrations 10–100 times greater than the detection limit. [Flame, furnace, ICP from R. J. Gill, *Am. Lab.*, November 1993, p. 24F. ICP–MS from T. T. Nham, *Am. Lab.*, August 1998, 17A. Data for Cl, Br, and I are from reference 17.]

Detection limits (ng/g) — For each element the four values are, top to bottom: Inductively coupled plasma emission, Flame atomic absorption, Graphite furnace atomic absorption, Inductively coupled plasma–mass spectrometry.

Element	ICP emission	Flame AA	Graphite furnace AA	ICP–MS
Li	0.7	2	0.1	0.0002
Be	0.07	1	0.02	0.0009
B	1	500	15	0.0008
C	10	—		
Na	3	0.2	0.005	0.0002
Mg	0.08	0.3	0.004	0.0003
Al	2	30	0.01	0.0002
Si	5	100	0.1	<0.0001
P	7	40 000	30	<0.0001
S	3	—	—	0.0001
Cl	60	—	—	—
K	20	3	0.1	0.0002
Ca	0.07	0.5	0.01	0.007
Sc	0.3	40	—	0.0002
Ti	0.4	70	0.5	0.004
V	0.7	50	0.2	0.0003
Cr	2	3	0.01	0.0002
Mn	0.2	2	0.01	0.0002
Fe	0.7	5	0.02	0.008
Co	1	4	0.02	0.0002
Ni	3	90	0.1	0.001
Cu	0.9	1	0.02	0.0005
Zn	0.6	0.5	0.001	0.003
Ga	10	60	0.5	0.002
Ge	20	200	—	0.003
As	7	200	0.2	0.003
Se	10	250	0.5	0.05
Br	150	—	—	0.02
Rb	1	7	0.05	0.0003
Sr	0.2	2	0.1	0.0003
Y	0.6	200	—	0.0003
Zr	2	1000	—	0.0006
Nb	5	2000	—	0.0008
Mo	3	20	0.02	0.002
Tc				
Ru	10	60	1	0.001
Rh	20	4	—	0.0003
Pd	4	10	0.3	0.001
Ag	0.8	1	0.005	0.0007
Cd	0.5	0.4	0.003	0.0008
In	20	40	1	0.0003
Sn	9	30	0.2	0.0009
Sb	9	40	0.1	0.001
Te	4	30	0.1	0.02
I	40	—	—	0.002
Cs	40 000	4	0.2	0.0003
Ba	0.6	10	0.04	0.0003
La	1	2000	—	0.0003
Hf	4	2000	—	0.0008
Ta	10	2000	—	0.0005
W	8	1000	—	0.002
Re	3	600	—	0.0007
Os	0.2	100	—	0.002
Ir	7	400	—	0.0004
Pt	7	100	—	0.0009
Au	2	10	0.2	0.0009
Hg	7	150	0.1	0.0004
Tl	10	20	0.1	0.0006
Pb	10	10	0.05	0.0006
Bi	7	40	0.1	0.0005
Ce	2	—	—	0.0003
Pr	9	6000	—	0.0002
Nd	10	1000	—	0.001
Sm	10	1000	—	0.001
Eu	0.9	20	—	0.0004
Gd	5	2000	—	0.0002
Tb	6	500	—	0.0002
Dy	2	30	—	0.0009
Ho	2	40	—	0.0002
Er	0.7	30	2	0.0007
Tm	2	900	—	0.0002
Yb	0.3	4	—	0.001
Lu	0.3	300	—	0.0002
Th	7	—	—	0.0003
U	60	40 000	—	0.0005

Shaded cells: Requires N_2O/C_2H_2 flame and is therefore better analyzed by inductively coupled plasma

Best analyzed by emission

from the sealed bottle.[20] Evaporation from bottles sealed in aluminized bags was 0.01% per year at 23°C and 0.08% per year at 30°C.

Standards evaporate—even from sealed bottles

20-5 Interference

Interference is any effect that changes the signal while analyte concentration remains unchanged. Interference can be corrected by removing the source of interference or by preparing standards that exhibit the same interference.

Types of Interference

Spectral interference refers to the overlap of analyte signal with signals due to other elements or molecules in the sample or with signals due to the flame or furnace. Interference from the flame can be subtracted by using D_2 or Zeeman background correction. The best means of dealing with overlap between lines of different elements in the sample is to choose another wavelength for analysis. High-resolution spectrometers eliminate interference from other elements by resolving closely spaced lines (Figure 20-25).

A single element such as Nb gives rise to over 1 000 discrete emission lines, so some lines from different elements overlap—even at high resolution. Software in one commercial instrument measures each element from several different spectral lines.[21] It converts each intensity to a concentration, using a different calibration curve for every line. If there were no spectral interference, all lines for one element would give the same concentration. Results that deviate more than a little from most results are attributed to spectral interference and are discarded.

Elements that form very stable diatomic oxides are incompletely atomized at the temperature of a flame or furnace. The spectrum of a molecule is much broader and more complex than that of an atom, because vibrational and rotational transitions are combined with electronic transitions (Section 17-6). The broad spectrum leads to spectral interference at many wavelengths. Figure 20-26 shows a plasma containing Y and Ba atoms as well as YO molecules. Note how broad the molecular emission is relative to the atomic emission.

When trace impurities in tungsten powder are analyzed by graphite furnace atomic absorption using direct solid sampling (Figure 20-9), WO_3 from the surface of the powder sublimes and fills the furnace with vapor, creating spectral interference throughout the visible and ultraviolet regions. Heating the powder under H_2 at 1 000°–1 200°C in the furnace prior to atomization reduces WO_3 to metallic tungsten and eliminates the interference.[12]

Chemical interference is caused by any component of the sample that decreases the extent of atomization of analyte. For example, SO_4^{2-} and PO_4^{3-} hinder the atomization of Ca^{2+}, perhaps by forming nonvolatile salts. **Releasing agents** are chemicals added to a sample to decrease chemical interference. EDTA and 8-hydroxyquinoline protect Ca^{2+} from interference by SO_4^{2-} and PO_4^{3-}. La^{3+} is a releasing agent, apparently because it preferentially reacts with PO_4^{3-} and frees the Ca^{2+}. A fuel-rich flame reduces certain oxidized analyte species that would otherwise hinder atomization. Higher flame temperatures eliminate many kinds of chemical interference.

Ionization interference can be a problem in the analysis of alkali metals at relatively low temperature and in the analyses of other elements at higher temperature. For any element, we can write a gas-phase ionization reaction:

$$M(g) \rightleftharpoons M^+(g) + e^-(g) \qquad K = \frac{[M^+][e^-]}{[M]} \qquad (20\text{-}6)$$

Because alkali metals have low ionization potentials, they are most extensively ionized. At 2 450 K and a pressure of 0.1 Pa, sodium is 5% ionized. With its lower ionization potential, potassium is 33% ionized. Ions have energy levels different from those of neutral atoms, so the desired signal is decreased. If there is a strong signal from the ion, you could use the ion signal rather than the atomic signal.

An **ionization suppressor** decreases the extent of ionization of analyte. In the analysis of potassium, it is recommended that solutions contain 1 000 ppm of CsCl, because cesium is more easily ionized than potassium. By producing a high concentration of electrons in the flame, ionization of Cs suppresses ionization of K. Ionization suppression is desirable in a low-temperature flame in which we want to observe neutral atoms.

The *method of standard addition* (Section 5-3) compensates for many types of interference by adding known quantities of analyte to the unknown in its complex matrix. For example, Figure 20-27 shows the analysis of strontium in aquarium water by standard addition. The

Types of interference:

- *spectral:* unwanted signals overlapping analyte signal
- *chemical:* chemical reactions decreasing the concentration of analyte atoms
- *ionization:* ionization of analyte atoms decreasing the concentration of neutral atoms

FIGURE 20-25 A Cd line at 228.802 nm causes spectral interference with the As line at 228.812 nm in most spectrometers. With sufficiently high resolution, peaks are separated and there is no interference. The instrument used for this spectrum has a 1-m Czerny-Turner monochromator (Figure 19-6) with a resolution of 0.005 nm from 160 to 320 nm and 0.010 nm from 320 to 800 nm. [Courtesy Jobin Yvon Horiba Group, Longjumeau Cedex, France.]

FIGURE 20-26 Emission from a plasma produced by laser irradiation of the high-temperature superconductor $YBa_2Cu_3O_7$. Solid is vaporized by the laser, and excited atoms and molecules such as YO emit light at characteristic wavelengths. [From W. A. Weimer, "Plasma Emission from Laser Ablation of $YBa_2Cu_3O_7$," *Appl. Phys. Lett.* 1988, *52,* 2171.]

Le Châtelier's principle tells us that adding electrons to the right side of Reaction 20-6 drives the reaction back to the left.

FIGURE 20-27 Atomic absorption calibration curve for Sr added to distilled water and standard addition of Sr to aquarium water. All solutions are made up to a constant volume, so the ordinate is the concentration of added Sr. [Data from L. D. Gilles de Pelichy, C. Adams, and E. T. Smith, "Analysis of Sr in Marine Aquariums by Atomic Absorption Spectroscopy," *J. Chem. Ed.* **1997**, *74*, 1192.]

slope of the standard addition curve is 0.018 8 absorbance units/ppm. If, instead, Sr is added to distilled water, the slope is 0.030 8 absorbance units/ppm. That is, in distilled water, the absorbance increases 0.030 8/0.018 8 = 1.64 times more than it does in aquarium water for each addition of standard Sr. We attribute the lower response in aquarium water to interference by other species. The absolute value of the *x*-intercept of the standard addition curve, 7.41 ppm, is a reliable measure of Sr in the aquarium.

Virtues of the Inductively Coupled Plasma

An inductively coupled argon plasma eliminates many common interferences.[22] The plasma is twice as hot as a conventional flame, and the residence time of analyte in the plasma is about twice as long. Therefore, atomization is more complete and signal is enhanced. Formation of analyte oxides and hydroxides is negligible. The plasma is remarkably free of background radiation where sample emission is observed 15–35 mm above the load coil.

Concentrations of excited atoms in the cooler, outer part of the flame are lower than in the warmer, central part of the flame. Emission from the central region is absorbed in the outer region. This **self-absorption** increases with increasing concentration of analyte and gives nonlinear calibration curves. In a plasma, the temperature is more uniform, and self-absorption is not nearly so important. Plasma emission calibration curves are linear over five orders of magnitude. In flames and furnaces, the linear range is two orders of magnitude. For inductively coupled plasma–mass spectrometry, the linear range is eight orders of magnitude (Table 20-4).

TABLE 20-4 Comparison of atomic analysis methods

	Flame absorption	Furnace absorption	Plasma emission	Plasma–mass spectrometry
Detection limits (ng/g)	10–1 000	0.01–1	0.1–10	0.000 01–0.000 1
Linear range	10^2	10^2	10^5	10^8
Precision				
short term (5–10 min)	0.1–1%	0.5–5%	0.1–2%	0.5–2%
long term (hours)	1–10%	1–10%	1–5%	<5%
Interferences				
spectral	very few	very few	many	few
chemical	many	very many	very few	some
mass	—	—	—	many
Sample throughput	10–15 s/element	3–4 min/element	6–60 elements/min	all elements in 2–5 min
Dissolved solid	0.5–5%	>20% slurries and solids	1–20%	0.1–0.4%
Sample volume	large	very small	medium	medium
Purchase cost	1	2	4–9	10–15

SOURCE: Adapted from TJA Solutions, Franklin, MA.

Sampling by Laser Ablation

At the opening of this chapter, we saw an example of *laser ablation–inductively coupled plasma–mass spectrometry*[23] for the analysis of teeth. In laser **ablation,** a pulsed laser beam is focused onto a microscopic spot on a solid sample, creating an explosion of particles, atoms, electrons, and ions into the gas phase. Most commonly, a Nd:YAG (neodymium-doped yttrium aluminum garnet) laser with a wavelength of 1.064 μm is used. A 10-ns pulse with 10 mJ of energy focused onto a spot diameter of 50 μm delivers a power of 50 GW/cm². Material is typically removed to a depth of 0.02 to 5 μm by each pulse (Figure 20-28). Each pulse ablates just nanograms of material, making the method almost nondestructive. Ablation product generated in a sealed chamber is swept by Ar or He through a Teflon-coated tube into the plasma for analysis by mass spectrometry or atomic emission. *Depth profiling* can be done by successive pulses probing deeper and deeper to measure elemental concentrations as a function of depth.

If the plasma is analyzed by observing atomic emission, the method is called **laser-induced breakdown spectroscopy.** Each laser pulse produces a short-lived plasma with a temperature of 10 000 to 20 000 K in which material dissociates into excited atoms and ions. Initial optical emission from the plasma is a continuum with little useful information. The plasma expands supersonically and cools to temperatures at which discrete atomic emission lines are observed after ~1–10 μs. The detector is *gated* to record atomic emission several microseconds after the laser pulse. One commercial system observes plasma emission through seven optical fibers directed to seven polychromators, each containing a linear 2 048-pixel charge coupled device to view a different spectral region. Spectral resolution is 0.1 nm over the range 200–980 nm.

A major challenge for quantitative analysis by laser ablation is that elements can be selectively ablated, selectively transported to the plasma, or selectively atomized in the plasma. Therefore, the relative numbers of ions detected are not necessarily equal to relative quantities in the solid sample. The most reliable—but usually unattainable—calibration is a standard sample containing elements of interest in the same matrix as the unknown. Mussel shell in Figure 20-28 is composed principally of $CaCO_3$. A calibration standard was made by dissolving known quantities of metals with a large excess of Ca^{2+} in acid and precipitating everything with CO_3^{2-}. The carbonate precipitate was washed, dried, and pressed into a dense pellet whose ablation behavior is similar to that of mussel shell. In the absence of *matrix-matched standards*, results from laser ablation can be compared with results obtained by completely digesting a material and analyzing the homogeneous solution. The difficulty of calibration renders laser ablation methods semiquantitative, with accuracies of ±10% or worse.

FIGURE 20-28 Microscopic crater ablated into a mussel shell by 10 pulses from a 266-nm laser with a beam energy of 4.5 mJ per 10-ns pulse and a repetition rate of 10 Hz. [From V. R. Bellotto and N. Miekely, "Improvements in Calibration Procedures for the Quantitative Determination of Trace Elements in Carbonate Material (Mussel Shells) by Laser Ablation ICP–MS," *Fresenius J. Anal. Chem.* **2000,** *367,* 635.]

20-6 Inductively Coupled Plasma–Mass Spectrometry

The ionization energy of Ar is 15.8 electron volts (eV), which is higher than those of all elements except He, Ne, and F. In an Ar plasma, analyte elements can be ionized by collisions with Ar^+, excited Ar atoms, or energetic electrons. The plasma can be directed into a mass spectrometer (Chapter 21), which separates and measures ions according to their mass-to-charge ratio.[24] For the most accurate measurements of isotope ratios, the mass spectrometer has one detector for each desired isotope.[25]

The trace element profile of teeth at the opening of this chapter was obtained by inductively coupled plasma–mass spectrometry. Figure 20-29 shows an example in which coffee beans were extracted with trace-metal-grade nitric acid and the aqueous extract was analyzed by inductively coupled plasma–mass spectrometry. Coffee brewed from either Hawaiian or Cuban beans contains ~15 ng Pb/mL. However, the Cuban beans also contain Hg at a concentration similar to that of Pb.

The difficulty in sampling anything with a mass spectrometer is that the spectrometer requires high vacuum to avoid collisions between ions and background gas molecules that divert the ions from their trajectory in a magnetic field. Figure 20-30 shows an interface between a horizontal Ar plasma and a mass spectrometer. The plasma at the left is directed onto a water-cooled Ni sampling cone with a 1-mm-diameter orifice through which a fraction of the plasma can pass. Behind the sampling cone is a water-cooled skimmer cone with an even smaller orifice. The extraction lens behind the skimmer cone has a high negative potential to attract positive ions from the plasma. Pressure is reduced in each successive section of the instrument. From the skimmer cone, ions enter a collision cell that might contain H_2 or He or both. The collision cell guides ions to the entrance of the mass separator and reduces the

FIGURE 20-29 Partial elemental profile of coffee beans by inductively coupled plasma–mass spectrometry. The Cuban beans have a much higher Hg content than the Hawaiian beans. A blank has not been subtracted from either spectrum, so the small amount of Hg in the upper spectrum could be in the blank. [Courtesy G. S. Ostrom and M. D. Seltzer, Michelson Laboratory, China Lake, CA.]

spread of ion kinetic energies by a factor of 10. Following the collision cell, ions are separated by a mass spectrometer. Ions of selected mass-to-charge ratio are deflected into the detector (at the right of the diagram), where they are counted. Photons from the plasma do not hit the detector, or they would generate a signal.

Part per trillion detection limits for inductively coupled plasma–mass spectrometry are so low (Figure 20-24 and Table 20-2) as to tax the cleanliness of reagents, glassware, and procedures. Solutions must be made from extremely pure water and trace-metal-grade HNO_3 in Teflon or polyethylene vessels protected from dust. HCl and H_2SO_4 are avoided because they create *isobaric interferences* discussed below. The plasma–mass spectrometer interface cannot tolerate high concentrations of dissolved solids that clog the orifice of the sampling cone. The plasma reduces organic matter to carbon that can clog the orifice. Organic material can be analyzed if some O_2 is fed into the plasma to oxidize the carbon.

Matrix effects on the yield of ions in the plasma are important, so calibration standards should be in the same matrix as the unknown. Standard addition is highly appropriate to correct for matrix effects. For example, when trace elements in blood were measured by inductively coupled plasma–mass spectrometry, matrix interference produced values for $^{75}As^+$ and $^{78}Se^+$

FIGURE 20-30 Interface between inductively coupled plasma and mass spectrometer. [Courtesy TJA Solutions, Franklin, MA.] Chapter 21 discusses mass spectrometry.

that were almost a factor of 2 above certified values for standard reference materials.[26] Calibration by standard addition reduced the errors to 0–4%.

Alternatively, internal standards can be used if they have nearly the same ionization energy as the analyte. For example, Tm can be used as an internal standard for U. The ionization energies of these two elements are 5.81 and 6.08 eV, respectively, so they should ionize to nearly the same extent in different matrices. If possible, internal standards with just one major isotope should be selected for maximum response.

Isobaric Interference

Ar is an "inert" gas with virtually no chemistry. However, Ar^+ has the same electronic configuration as Cl, and their chemistry is analogous. The inductively couple plasma is a rich source of Ar^+ and ions such as ArH^+, ArC^+, ArN^+, $ArNH^+$, ArO^+, $ArCl^+$, and Ar_2^+. These ions interfere with the measurement of analyte ions of the same mass-to-charge ratio.

For example, $^{40}Ar^{16}O^+$ has nearly the same mass as $^{56}Fe^+$, and $^{40}Ar_2^+$ has nearly the same mass as $^{80}Se^+$. Interference by ions of similar mass-to-charge ratio is called **isobaric interference.** Doubly ionized $^{138}Ba^{2+}$ interferes with $^{69}Ga^+$ because each has nearly the same mass-to-charge ratio (138/2 = 69/1). High-resolution mass spectrometers[27] eliminate interference by resolving species such as $^{40}Ar^{16}O^+$ and $^{56}Fe^+$, which differ by 0.02 atomic mass units, but most systems do not have high resolution.

For elements with multiple isotopes, you can check for isobaric interference by measuring isotope ratios. For example, if the ratio of Se isotopes agrees with those found in nature ($^{74}Se{:}^{76}Se{:}^{77}Se{:}^{78}Se{:}^{80}Se{:}^{82}Se$ = 0.008 7:0.090:0.078:0.235:0.498:0.092), then it is unlikely that there is interference at any of these masses.

In the collision cell in Figure 20-30, collisions of high-velocity ions such as ArO^+ with H_2 or He redistribute the kinetic energy of the ion into vibrational energy, which can cause ArO^+ to dissociate. *Endothermic* bond breaking derives its energy from the initial kinetic energy (the high velocity) of ArO^+. However, the concentration of ArO^+ is not sufficiently reduced to eliminate isobaric interference.

A *dynamic reaction cell*, which can replace the collision cell in Figure 20-30, uses *thermodynamically favorable* reactions to reduce isobaric interference.[29] The dynamic reaction cell contains a reactive gas such as NH_3, CH_4, N_2O, CO, or O_2, and its electric field is configured to select lower and upper masses of ions to pass through the cell. Plasma species that

Ar^+ is similar to Cl in its chemical reactivity.

Box 20-2 gives an important environmental application of the dynamic reaction cell.

BOX 20-2 GEOTRACES

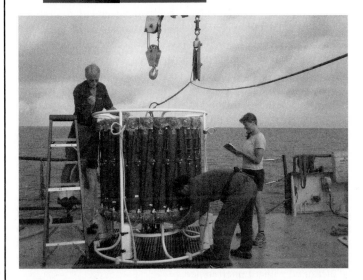

A carousel of 12-L plastic bottles lowered on a Kevlar line from a research vessel is used to collect ocean water without introducing trace metal contamination during collection. Samples are removed from the bottles in a plastic-lined clean room inside a cargo container on the deck of the ship.
[Courtesy Gregory A. Cutter, Old Dominion University, Norfolk, VA.]

GEOTRACES is an international program to characterize geographic and depth distributions of trace metals and their isotopes including Al, Mn, Fe, Cu, Zn, Cd, and a host of other elements at part per trillion levels (ng/kg) in ocean water around the globe. The goal is to establish a baseline and then monitor responses to changing environmental conditions such as global warming and ocean acidification, as well as reveal the chemical and biological processes affecting trace metals. Models suggest that Fe is a growth-limiting nutrient in the ocean, Al is a tracer for mineral dust, and Mn responds to redox processes. Measuring part per trillion levels requires *preconcentration* of the elements from seawater, described in Section 25-1.[28] ^{55}Mn, ^{54}Fe, and ^{56}Fe are affected by isobaric interference from $^{40}Ar^{14}NH^+$, $^{40}Ar^{14}N^+$, and $^{40}Ar^{16}O^+$ in inductively coupled plasma–mass spectrometry analysis. Therefore, a dynamic reaction cell with NH_3 gas is turned on for the analysis of Mn and Fe and then turned off for the analysis of other trace metals in the same sample. Detection limits for Al, Mn, Fe, Co, Ni, Cu, Zn, Cd, and Pb range from 0.1 to 7 ng/kg. Concentrations of these metals in the open ocean range from 1 ng/kg for Co to 500 ng/kg for Ni. Another means used to overcome isobaric interference is to employ high resolution mass spectrometry.

interfere with some elements can be reduced by as many as nine orders of magnitude by reactions such as

Electron transfer from NH₃:

$$^{40}Ar^+ + NH_3 \rightarrow NH_3^+ + Ar \qquad\qquad ^{40}Ar^{14}N^+ + NH_3 \rightarrow NH_3^+ + Ar + N$$

$$^{40}Ar^{12}C^+ + NH_3 \rightarrow NH_3^+ + Ar + C \qquad\qquad ^{40}Ar^{16}O^+ + NH_3 \rightarrow NH_3^+ + Ar + O$$

Proton transfer to NH₃:

$$^{40}ArH^+ + NH_3 \rightarrow NH_4^+ + Ar \qquad\qquad ^{35,37}Cl^{16}OH^+ + NH_3 \rightarrow NH_4^+ + ClO$$

$$^{40}Ar^{14}NH^+ + NH_3 \rightarrow NH_4^+ + Ar + N$$

For example, $^{40}Ar^{12}C^+$ interferes with $^{52}Cr^+$ and $^{35}Cl^{17}OH^+$ interferes with $^{53}Cr^+$. A dynamic reaction cell with NH_3 permits the measurement of Cr by removing ArC^+ and $ClOH^+$. $^{40}Ar^{16}O^+$ interferes with $^{56}Fe^+$. Either ArO^+ can be removed by reaction with NH_3 or Fe^+ can be shifted to a different mass by reaction with N_2O:

$$Fe^+ + N_2O \rightarrow FeO^+ + N_2$$

In natural waters with traces of iron and high concentrations of calcium, $^{40}Ca^{16}O^+$ interferes with $^{56}Fe^+$. CaO^+ can be removed by reaction with CO:

$$CaO^+ + CO \rightarrow Ca^+ + CO_2$$

Terms to Understand

ablation	chemical interference	ionization interference	piezoelectric crystal
aerosol	detection limit	ionization suppressor	plasma
atomic absorption	Doppler effect	isobaric interference	premix burner
atomic emission	graphite furnace	laser-induced breakdown	pressure broadening
atomic fluorescence	Heisenberg uncertainty	spectroscopy	releasing agent
atomization	principle	matrix	self-absorption
background correction	hollow-cathode lamp	matrix modifier	spectral interference
Boltzmann distribution	inductively coupled plasma	nebulization	

Summary

In atomic spectroscopy, absorption, emission, or fluorescence from gaseous atoms is measured. Liquids may be atomized by a plasma, a furnace, or a flame. Flame temperatures are usually in the range 2 300–3 400 K. The choice of fuel and oxidant determines the temperature of the flame and affects the extent of spectral, chemical, or ionization interference that will be encountered. Temperature instability affects atomization in atomic absorption and has an even larger effect on atomic emission, because the excited-state population is exponentially sensitive to temperature. An electrically heated graphite furnace requires less sample than a flame and has a lower detection limit. In an inductively coupled plasma, a radio-frequency induction coil heats Ar^+ ions to 6 000–10 000 K. At this high temperature, emission is observed from electronically excited atoms and ions. There is little chemical interference in an inductively coupled plasma, the temperature is very stable, and little self-absorption is observed.

Plasma emission spectroscopy does not require a light source and is capable of measuring ~70 elements simultaneously with a charge injection device detector. Background correction for a given emission peak is based on subtracting the intensity of neighboring pixels in the detector. The lowest detection limits are obtained by directing the plasma into a mass spectrometer that separates and measures ions from the plasma. In flame and furnace atomic absorption spectroscopy, a hollow-cathode lamp made of the analyte element provides spectral lines sharper than those of the atomic vapor. The inherent linewidth of atomic lines is limited by the Heisenberg uncertainty principle. Lines in a flame, furnace, or plasma are broadened by a factor of 10–100 by the Doppler effect and by atomic collisions. Correction for background emission from the flame is possible by electrically pulsing the lamp on and off or mechanically chopping the beam. Light scattering and spectral background can be subtracted by measuring absorption with a deuterium lamp or by Zeeman background correction, in which the atomic energy levels are alternately shifted in and out of resonance with the lamp frequency by a magnetic field. Chemical interference can be reduced by addition of releasing agents, which prevent the analyte from reacting with interfering species. Ionization interference in flames is suppressed by adding easily ionized elements such as Cs.

In inductively coupled plasma–mass spectrometry, isobaric interference occurs between species with the same mass and charge. Interference can be eliminated if mass spectral resolution is sufficiently great, but this is not often the case. A dynamic reaction cell reduces isobaric interference by using an exothermic reaction of a gas such as NH_3, N_2O, or CO to remove interfering molecular ions such as ArO^+ or to transform analyte into a molecular ion that can be measured without interference.

For qualitative and semiquantitative analysis, solids and liquids can be sampled by laser ablation. Ablated material can be swept through an inductively coupled plasma into a mass spectrometer. In laser-induced breakdown spectroscopy, plasma emission is used to quantify elements. Matrix-matched standards are necessary for semiquantitative analysis.

Exercises

20-A. Li was determined by atomic emission with the method of standard addition. Prepare a standard addition graph (Section 5-3) to find the concentration of Li and its uncertainty in pure unknown. The Li standard contained 1.62 μg Li/mL.

Unknown (mL)	Standard (mL)	Final volume (mL)	Emission intensity (arbitrary units)
10.00	0.00	100.0	309
10.00	5.00	100.0	452
10.00	10.00	100.0	600
10.00	15.00	100.0	765
10.00	20.00	100.0	906

20-B. Mn was used as an internal standard for measuring Fe by atomic absorption. A standard mixture containing 2.00 μg Mn/mL and 2.50 μg Fe/mL gave a quotient (Fe signal/Mn signal) = 1.05/1.00. A mixture with a volume of 6.00 mL was prepared by mixing 5.00 mL of unknown Fe solution with 1.00 mL containing 13.5 μg Mn/mL. The absorbance of this mixture at the Mn wavelength was 0.128, and the absorbance at the Fe wavelength was 0.185. Find the molarity of the unknown Fe solution.

20-C. (a) The atomic absorption signal shown here was obtained with 0.048 5 μg Fe/mL in a graphite furnace. Estimate the detection limit for Fe, defined for this problem as the concentration of Fe that gives a signal-to-noise ratio of 2.

(b) Seven replicate measurements of a standard containing 1.00 ng Hg/L gave readings of 0.88, 1.48, 0.94, 1.12, 1.03, 1.40, and 1.14 ng/L in cold vapor atomic absorption (Box 20-1). From Equations 5-5 and 5-6, estimate the detection and quantitation limits. (Note that in Equations 5-5 and 5-6, the quotient s/m is the standard deviation in concentration.)

20-D. The measurement of Li in brine (salt water) is used by geochemists to help determine the origin of this fluid in oil fields. Flame atomic emission and absorption of Li are subject to interference by scattering, ionization, and overlapping spectral emission from other elements. Atomic absorption analysis of replicate samples of a marine sediment gave results in the table.

Sample and treatment	Li found (μg/g)	Analytical method	Flame type
1. None	25.1	standard curve	air/C_2H_2
2. Dilute to 1/10 with H_2O	64.8	standard curve	air/C_2H_2
3. Dilute to 1/10 with H_2O	82.5	standard addition	air/C_2H_2
4. None	77.3	standard curve	N_2O/C_2H_2
5. Dilute to 1/10 with H_2O	79.6	standard curve	N_2O/C_2H_2
6. Dilute to 1/10 with H_2O	80.4	standard addition	N_2O/C_2H_2

SOURCE: *B. Baraj, L. F. H. Niencheski, R. D. Trapaga, R. G. França, V. Cocoli, and D. Robinson, "Interference in the Flame Atomic Absorption Determination of Li," Fresenius J. Anal. Chem.* **1999**, *364, 678.*

(a) Suggest a reason for the increasing apparent concentration of Li in samples 1 through 3.

(b) Why do samples 4 through 6 give an almost constant result?

(c) What value would you recommend for reporting the real concentration of Li in the sample?

Problems

Techniques of Atomic Spectroscopy

20-1. In which technique, atomic absorption or atomic emission, is flame temperature stability more critical? Why?

20-2. State the advantages and disadvantages of a furnace compared with a flame in atomic absorption spectroscopy.

20-3. Figure 20-10 shows a temperature profile for a furnace atomic absorption experiment. Explain the purpose of each different part of the heating profile.

20-4. State the advantages and disadvantages of the inductively coupled plasma compared with a flame in atomic spectroscopy.

20-5. Explain what is meant by the Doppler effect. Rationalize why Doppler broadening increases with increasing temperature and decreasing mass in Equation 20-5.

20-6. Explain how the following background correction techniques work: (a) beam chopping; (b) deuterium lamp; (c) Zeeman.

20-7. Explain what is meant by spectral, chemical, ionization, and isobaric interference.

20-8. Bone consists of the protein collagen and the mineral hydroxyapatite, $Ca_{10}(PO_4)_6(OH)_2$. The Pb content of archaeological human skeletons measured by graphite furnace atomic absorption sheds light on customs and economic status of individuals in historical times.[30] Explain why La^{3+} is added to bone samples to suppress matrix interference in Pb analysis.

20-9. (a) Explain the purpose of the collision cell in Figure 20-30. (b) Explain the purpose of a dynamic reaction cell, which can replace the collision cell.

(c) In geologic strontium isotopic analysis, there is isobaric interference between $^{87}Rb^+$ and $^{87}Sr^+$. A collision cell with CH_3F converts Sr^+ to SrF^+ but does not convert Rb^+ to RbF^+. How does this reaction eliminate interference?

20-10. What is the purpose of a matrix-matched standard in laser-induced breakdown spectroscopy?

20-11. Laser atomic fluorescence excitation and emission spectra of sodium in an air-acetylene flame are shown here. In the *excitation* spectrum, the laser (bandwidth = 0.03 nm) was scanned through various wavelengths, while the detector monochromator (bandwidth = 1.6 nm) was held fixed near 589 nm. In the *emission* spectrum, the laser was fixed at 589.0 nm, and the detector monochromator wavelength was varied. Explain why the emission spectrum gives one broad band, whereas the excitation spectrum gives two sharp lines. How can the excitation linewidths be much narrower than the detector monochromator bandwidth?

(a) Wavelength (nm)   (b) Wavelength (nm)

Fluorescence excitation and emission spectra of the two sodium D lines in an air-acetylene flame. (a) In the excitation spectrum, the laser was scanned. (b) In the emission spectrum, the monochromator was scanned. The monochromator slit width was the same for both spectra. [From S. J. Weeks, H. Haraguchi, and J. D. Winefordner, "Improvement of Detection Limits in Laser-Excited Atomic Fluorescence Flame Spectrometry," *Anal. Chem.* **1978**, *50*, 360.]

20-12. Concentrations (pg per g of snow) of metals by atomic fluorescence in the Agassiz Ice Cap in Greenland for the period 1988–1992 are[31] Pb, 1.0_4 ($\pm 0.1_7$) $\times 10^2$; Tl, $0.43 \pm 0.08_7$; Cd, $3.5 \pm 0.8_7$; Zn, 1.7_4 ($\pm 0.2_6$) $\times 10^2$; and Al, $6._1$ ($\pm 1._7$) $\times 10^3$. The mean annual snowfall was 11.5 g/cm^2. Calculate the mean annual flux of each metal in units of ng/cm^2. *Flux* means how much metal falls on each cm^2.

20-13. Calculate the emission wavelength (nm) of excited atoms that lie 3.371×10^{-19} J per molecule above the ground state.

20-14. Derive the entries for 500 nm in Table 20-3. Find N^*/N_0 at 6 000 K if $g^* = 3$ and $g_0 = 1$?

20-15. Calculate the Doppler linewidth for the 589-nm line of Na and for the 254-nm line of Hg, both at 2 000 K.

20-16. The first excited state of Ca is reached by absorption of 422.7-nm light.

(a) Find the energy difference (kJ/mol) between ground and excited states.

(b) The degeneracies are $g^*/g_0 = 3$ for Ca. Find N^*/N_0 at 2 500 K.

(c) By what percentage will N^*/N_0 change with a 15-K rise in temperature?

(d) Find N^*/N_0 at 6 000 K.

20-17. An *electron volt* (eV) is the energy change of an electron moved through a potential difference of 1 volt: eV = (1.602 $\times$ 10^{-19} C)(1 V) = 1.602×10^{-19} J per electron = 96.49 KJ per mole of electrons. Use the Boltzmann distribution to fill in the table and explain why Br is not readily observed in atomic absorption or atomic emission.

	Na	Cu	Br
Excited-state energy (eV)	2.10	3.78	8.04
Wavelength (nm)			
Degeneracy ratio (g^*/g_0)	3	3	2/3
N^*/N_0 at 2 600 K in flame			
N^*/N_0 at 6 000 K in plasma			

20-18. MgO prevents premature evaporation of Al in a furnace by maintaining the aluminum as Al_2O_3. Another type of matrix modifier prevents loss of signal from the atom X that readily forms the molecular carbide XC in a graphite furnace (a source of carbon). For example, adding yttrium to a sample containing barium increases the Ba signal by 30%. The bond dissociation energy of YC is greater than that of BaC. Explain what is happening to increase the Ba signal.

20-19. The 20-μm-radius laser ablation pit in Figure 20-28 was created by a laser pulse with a duration of 10 ns and an energy of 2.4 mJ. Express the laser power density in units of W/cm^2. Recall that 1 W = 1 J/s. If the depth of the pit is 1 μm and the density of the material is 4 g/mL, how much mass is removed by one pulse?

Quantitative Analysis by Atomic Spectroscopy

20-20. Why is an internal standard most appropriate for quantitative analysis when unavoidable sample losses are expected during sample preparation?

20-21. *Standard addition*. To measure Ca in breakfast cereal, 0.521 6 g of crushed Cheerios was ashed in a crucible at 600°C in air for 2 h.[32] The residue was dissolved in 6 M HCl, quantitatively transferred to a volumetric flask, and diluted to 100.0 mL. Then 5.00-mL aliquots were transferred to 50-mL volumetric flasks. Each was treated with standard Ca^{2+} (containing 20.0 μg/mL), diluted to volume with H_2O, and analyzed by flame atomic absorption. Construct a standard addition graph and use the method of least squares to find the x-intercept and its uncertainty. Find wt% Ca in Cheerios and its uncertainty.

Ca^{2+} standard (mL)	Absorbance	Ca^{2+} standard (mL)	Absorbance
0	0.151	8.00	0.388
1.00	0.185	10.00	0.445
3.00	0.247	15.00	0.572
5.00	0.300	20.00	0.723

20-22. *Internal standard.* A solution was prepared by mixing 10.00 mL of unknown (X) with 5.00 mL of standard (S) containing 8.24 μg S/mL and diluting the mixture to 50.0 mL. The measured signal quotient was (signal due to X/signal due to S) = 1.690/1.000.

(a) In a separate experiment, in which the concentrations of X and S were equal, the quotient was (signal due to X/signal due to S) = 0.930/1.000. What is the concentration of X in the unknown?

(b) Answer the same question if, in a separate experiment in which the concentration of X was 3.42 times the concentration of S, the quotient was (signal due to X/signal due to S) = 0.930/1.000.

20-23. ▦ Potassium standards gave the following emission intensities at 404.3 nm. Emission from the unknown was 417. Find $[K^+]$ and its uncertainty in the unknown.

Sample (μg K/mL):	0	5.00	10.00	20.00	30.00
Relative emission:	0	124	243	486	712

20-24. *Quality assurance.* Tin is leached (dissolved) into canned foods from the tin-plated steel can.[33] For analysis by inductively coupled plasma–atomic emission, food is digested by microwave heating in a Teflon bomb (Figure 27-7) in three steps with HNO_3, H_2O_2, and HCl.

(a) CsCl is added to the final solution at a concentration of 1 g/L. What is the purpose of the CsCl?

(b) ▦ Calibration data are shown in the table. Find the slope and intercept and their standard deviations and R^2, which is a measure of the goodness of fit of the data to a line. Draw the calibration curve.

Sn (μg/L)	Emission at 189.927 nm	Sn (μg/L)	Emission at 189.927 nm
0	4.0	40.0	31.1
10.0	8.5	60.0	41.7
20.0	19.6	100.0	78.8
30.0	23.6	200.0	159.1

(c) Interference by high concentrations of other elements was assessed at different emission lines of Sn. Foods containing little tin were digested and spiked with Sn at 100.0 μg/L. Then other elements were deliberately added. The table shows selected results. Which elements interfere at each of the two wavelengths? Which wavelength is preferred for the analysis?

Element added at 50 mg/L	Sn found (μg/L) with 189.927-nm emission line	Sn found (μg/L) with 235.485-nm emission line
None	100.0	100.0
Ca	96.4	104.2
Mg	98.9	92.6
P	106.7	104.6
Si	105.7	102.9
Cu	100.9	116.2
Fe	103.3	intense emission
Mn	99.5	126.3
Zn	105.3	112.8
Cr	102.8	76.4

(d) *Limits of detection and quantitation.* The slope of the calibration curve in part **(b)** is 0.782 units per (μg/L) of Sn. Food containing little Sn gave a mean signal of 5.1 units for seven replicates. Food spiked with 30.0 μg Sn/L gave a mean signal of 29.3 units with a standard deviation of 2.4 units for seven replicates. Use Equations 5-5 and 5-6 to estimate the limits of detection and quantitation.

(e) A 2.0-g food sample was digested and eventually diluted to 50 mL for analysis. Express the limit of quantitation from part **(d)** in terms of mg Sn/kg food.

20-25. Titanocene dichloride, $(\pi\text{-}C_5H_5)_2TiCl_2$, is a potential antitumor drug thought to be carried to cancer cells by the protein transferrin (Figure 17-7). To measure the Ti(IV) binding capacity of transferrin, the protein was treated with excess titanocene dichloride. After allowing time for Ti(IV) binding to the protein, excess small molecules were removed by dialysis (Demonstration 26-1). The protein was then digested with 2 M NH_3 and used to prepare a series of solutions with standard additions for chemical analysis. All solutions were made to the same total volume. Titanium and sulfur in each solution were measured by inductively coupled plasma–atomic emission spectrometry, with results in the table. Each transferrin molecule contains 39 sulfur atoms. Find the molar ratio Ti/transferrin in the protein.

Added Ti (mg/L)	ICP-AES signal	Added S (mg/L)	ICP-AES signal
0	0.86	0	0.017 4
3.00	1.10	37.0	0.022 1
6.00	1.34	74.0	0.026 8
12.0	1.82	148.0	0.036 2

Data derived from A. Cardona and E. Meléndez, "Determination of the Titanium Content of Human Transferrin by Inductively-Coupled Plasma-Atomic-Emission Spectroscopy," Anal. Bioanal. Chem. **2006**, *386, 1689.*

21 Mass Spectrometry

Deflection and disintegration of liquid droplets falling past a wire held at +6 000 V. [From D. B. Hager and N. J. Dovichi, "Behavior of Microscopic Liquid Droplets Near a Strong Electrostatic Field: Droplet Electrospray," *Anal. Chem.* **1994**, *66*, 1593. See also D. B. Hager, N. J. Dovichi, J. Klassen, and P. Kebarle, "Droplet Electrospray Mass Spectrometry," *Anal. Chem.* **1994**, *66*, 3944.]

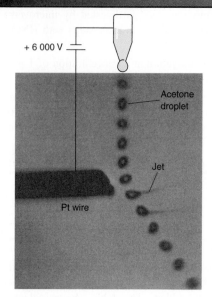

One method for expelling charged protein molecules into the gas phase for mass spectrometry is called *electrospray*. In the experiment shown here, droplets of acetone with a diameter of 16 μm fall past a Pt wire held at +6 000 V with respect to the nozzle from which the droplets came. High voltage creates a glowing electric corona discharge (a plasma containing electrons and positive ions) around the wire, but the discharge is not visible in this photograph. Droplets falling through the discharge become positively charged and are repelled by the wire, thus deflecting their path to the right. When positively charged droplets come close to the wire, we see a fine stream of liquid jetting away from the positively charged wire. Microscopic droplets in the fine spray rapidly evaporate. If the liquid had been an aqueous protein solution, the water would evaporate, thereby leaving charged protein molecules in the gas phase.

Francis W. Aston (1877–1945) developed a "mass spectrograph" in 1919 that could separate ions differing in mass by 1% and focus them onto a photographic plate. Aston immediately found that neon consists of two isotopes (^{20}Ne and ^{22}Ne) and went on to discover 212 of the 281 naturally occurring isotopes. He received the Nobel Prize for chemistry in 1922.

*M*ass spectrometry has long been used to measure isotopes and decipher organic structures. Hydrogen and oxygen isotopes in ice cores record the history of Earth's climate (Figure 0-6). The constancy of the $^{18}O/^{16}O$ ratio in certain dinosaur bones strongly suggests that these species were warm blooded.[1] Mass spectrometry can elucidate the amino acid sequence in a protein,[2] the sequence of nucleic acids in DNA, the structure of a complex carbohydrate, and the types of lipids in a single organism. Mass spectrometry can measure masses of individual cells[3] and viruses.[4] Mass spectrometry is the most powerful detector for chromatography, offering both qualitative and quantitative information, providing high sensitivity, and distinguishing different substances with the same retention time.

21-1 What Is Mass Spectrometry?

Mass spectrometry is a technique for studying the masses of atoms or molecules or fragments of molecules.[5,6] To obtain a mass spectrum, gaseous species desorbed from condensed phases are ionized, the ions are accelerated by an electric field and then separated according to their mass-to-charge ratio, m/z. If all charges are +1, then m/z is numerically equal to the mass. If an ion has a charge of +2, for example, then m/z is 1/2 of the mass. The **mass spectrum** in Figure 21-1 displays detector response versus m/z, showing four natural isotopes of Pb^+ ions. The area of each peak is proportional to the abundance of each isotope. Box 21-1 defines *nominal mass*, which is the mass we usually speak of in this chapter.

Figure 21-2 shows a **magnetic sector mass spectrometer,** which uses a magnetic field to allow ions of a selected m/z to pass from the ion source to the detector.[7] Gaseous molecules entering at the upper left are converted into ions (usually positive ions), accelerated by an electric field, and expelled into the analyzer tube, where they encounter a magnetic field perpendicular to their direction of travel. The tube is maintained under high vacuum ($\sim 10^{-5}$ Pa, $\sim 10^{-7}$ Torr) so that ions are not deflected by collision with background gas molecules. The magnet deflects ions toward the detector at the far end of the tube (see Box 21-2). Heavy ions are not deflected enough and light ones are deflected too much to reach the detector. The spectrum of masses is obtained by varying the magnetic field strength.

At the *electron multiplier* detector[8] in Figure 21-2, each arriving ion starts a cascade of electrons, just as a photon starts a cascade of electrons in a photomultiplier tube (Figure 19-14). A series of dynodes multiplies the number of electrons by $\sim 10^5$ before they reach the anode where current is measured. The mass spectrum shows detector current as a function of m/z selected by the magnetic field.

Mass spectrometers can measure negative ions by reversing voltages where the ions are formed and detected. To detect negative ions, a *conversion dynode* with a positive potential is placed before the electron multiplier. When bombarded by negative ions, the dynode liberates positive ions that are accelerated into the electron multiplier, which amplifies the signal.

Electron Ionization

Molecules entering the ion source in Figure 21-2 are converted into ions by **electron ionization.** Electrons emitted from a hot filament (like the one in a light bulb) are accelerated through 70 V before interacting with incoming molecules. Some ($\sim 0.01\%$) molecules (M)

FIGURE 21-1 Mass spectrum showing natural isotopes of Pb observed as an impurity in brass. The variability of isotopic abundances in Pb from natural sources creates a large uncertainty in the atomic mass (207.2 ± 0.1) in the periodic table. [From Y. Su, Y. Duan, and Z. Jin, "Development and Evaluation of a Glow Discharge Microwave-Induced Plasma Tandem Source for Time-of-Flight Mass Spectrometry," *Anal. Chem.* 2000, 72, 5600.]

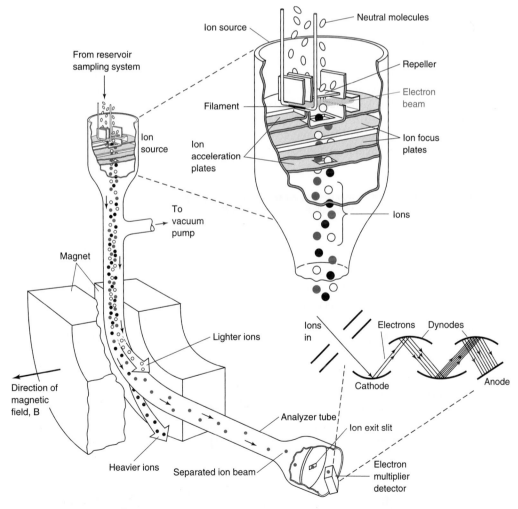

FIGURE 21-2 Magnetic sector mass spectrometer. [Adapted from F. W. McLafferty, *Interpretation of Mass Spectra* (New York: Benjamin, 1966).]

Atomic mass is the weighted average of the masses of the isotopes of an element. Bromine consists of 50.69% ^{79}Br with a mass of 78.918 34 Da (daltons) and 49.31% ^{81}Br with a mass of 80.916 29 Da. Therefore, its atomic mass is $(0.506\ 9)(78.918\ 34) + (0.493\ 1)(80.916\ 29) = 79.904$ Da. The unit of atomic mass is the dalton, Da, defined as 1/12 of the mass of ^{12}C. Mass spectrometrists prefer "u" for "unified atomic mass unit." Da and u are synonymous.

The **molecular mass** of a molecule or an ion is the sum of atomic masses listed in the periodic table. For bromoethane,

C_2H_5Br, the molecular mass is $(2 \times 12.010\ 7) + (5 \times 1.007\ 94) + (1 \times 79.904) = 108.965$.

The **nominal mass** of a molecule or ion is the *integer* mass of the species with the most abundant isotope of each of the constituent atoms. For carbon, hydrogen, and bromine the most abundant isotopes are ^{12}C, ^{1}H, and ^{79}Br. Therefore, the nominal mass of C_2H_5Br is $(2 \times 12) + (5 \times 1) + (1 \times 79) = 108$.

FIGURE 21-3 Ionization energies (I.E.) of valence electrons in formaldehyde. [Data from C. R. Brundle, M. B. Robin, N. A. Kuebler, and H. Basch, "Perfluoro Effects in Photoelectron Spectroscopy," *J. Am. Chem. Soc.* 1972, *94*, 1451.] Here are other first ionization energies:

$CH_3CH_2CH_2CH_3$	10.6 eV (sigma)
$CH_2{=}CHCH_2CH_3$	9.6 eV (pi)
$(CH_3CH_2)_2\ddot{O}$	9.6 eV (nonbonding)
⬡NH	8.6 eV (nonbonding)
⬡	9.2 eV (pi)

absorb as much as 12–15 electron volts (1 eV = 96.5 kJ/mol), which is enough for ionization:

$$M \quad + \quad e^- \quad \rightarrow \quad M^{+\bullet} \quad + \quad e^- \quad + \quad e^-$$
$$\ \ 70\ eV \quad\quad\ \text{Molecular ion} \quad {\sim}55\ eV \quad\ \ 0.1\ eV$$

Almost all stable molecules have an even number of electrons. When one electron is lost, the resulting cation with one unpaired electron is designated $M^{+\bullet}$, the **molecular ion.** After ionization, $M^{+\bullet}$ usually has enough internal energy (~1 eV) to break into fragments.

A small positive potential on the repeller plate of the ion source pushes ions toward the analyzer tube, and a small potential on the ion focus plates creates a focused beam. High voltage (~1 000–10 000 V) between the ion acceleration plates imparts a high velocity to ions as they are expelled from the bottom of the ion source.

The electron kinetic energy of 70 eV is much greater than the ionization energy of molecules. Consider formaldehyde in Figure 21-3, whose molecular orbitals were shown in Figure 17-12. The most easily lost electron comes from a nonbonding ("lone pair") orbital centered on oxygen, with an ionization energy of 11.0 eV. To remove a pi bonding electron from neutral formaldehyde requires 14.1 eV, and to remove the highest energy sigma bonding electron from the neutral molecule requires 15.9 eV.

Interaction with a 70-eV electron most likely removes the electron with lowest ionization energy. The resulting molecular ion, $M^{+\bullet}$, can have so much extra energy that it breaks into fragments. There might be so little $M^{+\bullet}$ that its peak is small or absent in the mass spectrum.

Electron ionization in the ion source of the mass spectrometer in Figure 21-2 creates positive ions, M^{z+}, with different masses. Let the mass of an ion be m and let its charge be $+ze$ (where e is the magnitude of the charge of an electron). When the ion is accelerated through a potential difference V by the ion acceleration plates, it acquires a kinetic energy equal to the electric potential difference:

$$\underbrace{\frac{1}{2}mv^2}_{\substack{\text{Kinetic energy}\\(v = \text{velocity})}} = \underbrace{zeV}_{\substack{\text{Potential energy}}} \Rightarrow v = \sqrt{\frac{2zeV}{m}} \quad (A)$$

An ion with charge ze and velocity v traveling perpendicular to a magnetic field B experiences a force $zevB$ that is perpendicular to both the velocity vector and the magnetic field vector. This force deflects the ion through a circular path of radius r. The centripetal force (mv^2/r) required to deflect the particle is provided by the magnetic field.

$$\underbrace{\frac{mv^2}{r}}_{\text{Centripetal force}} = \underbrace{zevB}_{\text{Magnetic force}} \Rightarrow v = \frac{zeBr}{m} \quad (B)$$

Equating velocities from Equations A and B gives

$$\frac{zeBr}{m} = \sqrt{\frac{2zeV}{m}} \quad \Rightarrow \quad \boxed{\frac{m}{z} = \frac{eB^2r^2}{2V}} \quad (C)$$

Equation C tells us the radius of curvature of the path traveled by an ion with mass m and charge z. The radius of curvature is fixed by the geometry of the hardware. Ions can be selected to reach the detector by adjusting the magnetic field, B, or the accelerating voltage, V. Normally, B is varied to select ions and V is fixed near 3 000 V. Transmission of ions and detector response both decrease when V is decreased.

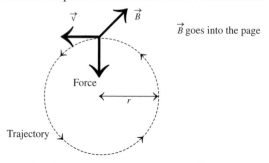

$\vec{B}$ goes into the page

FIGURE 21-4 Mass spectra of the sedative pentobarbital from electron ionization (*left*) or chemical ionization (*right*). The molecular ion ($M^{+\bullet}$, m/z 226) is not evident with electron ionization. The dominant ion from chemical ionization is MH^+. The peak at m/z 255 in the chemical ionization spectrum is from $M(C_2H_5)^+$. [Courtesy Varian Associates, Sunnyvale, CA.]

The electron ionization mass spectrum at the left side of Figure 21-4 does not exhibit an $M^{+\bullet}$ peak that would be at m/z 226. Instead, there are peaks at m/z 197, 156, 141, 112, 98, 69, and 55, arising from fragmentation of $M^{+\bullet}$. These peaks provide clues about the structure of the molecule. A computer search is commonly used to match the spectrum of an unknown to similar spectra in a library.[10]

If you lower the kinetic energy of electrons in the ionization source to, say, 20 eV, there will be a lower yield of ions and less fragmentation. You would likely observe a greater abundance of molecular ions. We customarily use 70 eV because it gives reproducible fragmentation patterns that can be compared with library spectra.

The most intense peak in a mass spectrum is called the **base peak.** Intensities of other peaks are expressed as a percentage of the base peak intensity. In the electron ionization spectrum in Figure 21-4, the base peak is at m/z 141.

Electrons with an energy near 70 eV almost exclusively create cationic molecular products. If the electron energy is lower, it is possible to form negative ions from molecules with a sufficiently great electron affinity:

Resonance capture: $M \quad + \quad e^- \quad \rightarrow \quad M^{-\bullet}$

             <0.1 eV

Dissociative capture: $XY \quad + \quad e^- \quad \rightarrow \quad X \quad + \quad Y^{-\bullet}$

          0.1–10 eV

Chemical ionization produces less fragmentation than electron ionization. For chemical ionization, the ionization source is filled with a *reagent gas* such as methane, isobutane, or ammonia, at a pressure of ~100 Pa (~1 mbar, ~1 Torr). Energetic electrons (100–200 eV) convert CH_4 into a variety of reactive products:

$$CH_4 + e^- \rightarrow CH_4^{+\bullet} + 2e^-$$

$$CH_4^{+\bullet} + CH_4 \rightarrow CH_5^+ + {}^\bullet CH_3$$

$$CH_4^{+\bullet} \rightarrow CH_3^+ + H^\bullet$$

$$CH_3^+ + CH_4 \rightarrow C_2H_5^+ + H_2$$

CH_5^+ is a potent proton donor that reacts with analyte to give the *protonated molecule*, MH^+, which is usually the most abundant ion in the chemical ionization mass spectrum.

$$CH_5^+ + M \rightarrow CH_4 + MH^+$$

In the chemical ionization spectrum in Figure 21-4, MH^+ at m/z 227 is the second strongest peak and there are fewer fragments than in the electron ionization spectrum.

Ammonia or isobutane are used in place of CH_4 to reduce the fragmentation of MH^+. These reagents bind H^+ more strongly than CH_4 does and impart less energy to MH^+ when the proton is transferred to M. Another mild, versatile ionization reagent is NO^+ generated from NO by radioactive ^{210}Po.[12]

A reasonable match of the experimental spectrum to one in the computer library is **not** proof of molecular structure—it is just a clue.[9] You must be able to explain all major peaks (and even minor peaks at high m/z) in the spectrum in terms of the proposed structure, and you should obtain a matching spectrum from an authentic sample before reaching a conclusion. The authentic sample must have the same chromatographic retention time as the proposed unknown. Many isomers produce nearly identical mass spectra.

CH_5^+ is described as a CH_3 tripod with an added H_2 unit. [H–C–H] is held together by two electrons distributed over three atoms. Atoms of the H_2 unit rapidly exchange with atoms of the CH_3 unit.[11]

The *molecular ion*, $M^{+\bullet}$, can be formed by reactions such as

$$CH_4^{+\bullet} + M \rightarrow CH_4 + M^{+\bullet}$$

MH^+ is the *protonated molecule,* not the molecular ion.

FIGURE 21-5 Resolving power. (*a*) By one definition, the resolving power is $m/\Delta m = 500/1 = 500$. (*b*) By a second definition, the resolving power for the same pair of peaks is $m/m_{1/2} = 500/0.481 = 1\,040$. (*c*) With the second definition, two peaks at *m/z* 500 and 501 are just barely discernible if the resolving power is 500.

Negative chemical ionization reagents include O_2^-, F^-, and SF_6^-.[12] Anions such as Cl^- and OH^- in solution can produce negative ions in the mass spectrum by reactions such as

$$M + Cl^- \rightarrow (M + Cl)^-$$

$$M + OH^- \rightarrow (M - H)^- + H_2O$$

Resolving Power

Mass spectra in Figure 21-4 are computer-generated bar graphs. In contrast, Figure 21-1 shows the actual detector signal. Each mass spectral peak has a width that limits how closely two peaks could be spaced and still be resolved. If peaks are too close, they appear to be a single peak.

The higher the **resolving power** of a mass spectrometer, the better it is able to separate two peaks with similar mass.

The expression $m/m_{1/2}$ gives a value twice as great as $m/\Delta m$ for resolving power.

$$\text{Resolving power} = \frac{m}{\Delta m} \quad \text{or} \quad \frac{m}{m_{1/2}} \qquad (21\text{-}1)$$

$$\underset{\text{Figure 21-5a}}{} \qquad \underset{\text{Figure 21-5b}}{}$$

Resolution is the smallest difference in *m/z* values that can be detected as separate peaks and should be reported with the *m/z* value where it is measured. Examples: "$\Delta m = 0.1$ at $m/z = 1\,000$" or "$\Delta m = 0.1$ throughout the *m/z* scale." Resolving power is a big number; resolution is a small number.

where *m* is the smaller value of *m/z*. The denominator is defined in two different ways. In Figure 21-5a, the denominator is the separation of the two peaks (Δm) when the overlap at their base is 10% of the maximum peak height. The resolving power in Figure 21-5a is $m/\Delta m = 500/1.00 = 5.00 \times 10^2$. In Figure 21-5b, the denominator is taken as $m_{1/2}$, the width of the peak at half the maximum height, which is 0.481 Da (dalton, see Box 21-1). By this definition, the resolving power is $m/m_{1/2} = 500/0.481 = 1.04 \times 10^3$ for the same two peaks. Figure 21-5c shows that, when the expression $m/m_{1/2}$ gives a resolving power of 5.00×10^2, two peaks at *m/z* 500 and 501 are barely discernible. Specify which definition you use when expressing resolving power.

> **EXAMPLE** Resolving Power
>
> Using the ^{208}Pb peak in Figure 21-1, find the resolving power from the expression $m/m_{1/2}$.
>
> **Solution** The width at half-height is 0.146 *m/z* units. Therefore,
>
> $$\text{Resolving power} = \frac{m}{m_{1/2}} = \frac{208}{0.146} = 1.42 \times 10^3$$
>
> An instrument with a resolving power of 1.42×10^3 separates peaks well near *m/z* 200 but gives a barely discernible separation at *m/z* 1 420.
>
> **Test Yourself** Measure the resolving power $m/m_{1/2}$ from the ^{31}P peak in Figure 21-9. (***Answer:*** $m/m_{1/2} = 30.974$ Da/0.003 3 Da $= 9\,400$)

21-2 Oh, Mass Spectrum, Speak to Me!*

*An expression borrowed from O. David Sparkman, a master teacher.

Each mass spectrum has a story to tell. The molecular ion, $M^{+\bullet}$, tells us the molecular mass of an unknown. Unfortunately, with electron ionization, some compounds do not exhibit a molecular ion, because $M^{+\bullet}$ breaks apart so efficiently. However, fragments provide clues to the structure of an unknown. To find the molecular mass, we can obtain a chemical ionization mass spectrum, which usually has a strong peak for MH^+.

The **nitrogen rule** helps us propose compositions for molecular ions: If a compound has an odd number of nitrogen atoms—in addition to any number of C, H, halogens, O, S, Si and P—then $M^{+\bullet}$ has an odd nominal mass. For a compound with an even number of nitrogen atoms (0, 2, 4, . . .), $M^{+\bullet}$ has an even nominal mass. A molecular ion at m/z 128 can have 0 or 2 N atoms, but it cannot have 1 N atom.

Molecular Ion and Isotope Patterns

Electron ionization of aromatic compounds (those containing benzene rings) usually gives significant intensity for $M^{+\bullet}$. $M^{+\bullet}$ is the base peak (most intense) in the spectra of benzene and biphenyl in Figure 21-6.

The next higher mass peak, M+1, provides information on elemental composition. Table 21-1 lists the natural abundance of several isotopes. For carbon, 98.93% of atoms are ^{12}C and 1.07% are ^{13}C. The ratio of the two isotopes is $^{13}C/^{12}C = 1.07/98.93 = 0.010\ 8$. Nearly all hydrogen is ^1H, with 0.012% ^2H. Applying the factors in Table 21-2 to C_nH_m, the intensity of the M+1 peak should be

Although we write the molecular ion as $M^{+\bullet}$, we use the notation M+1 and M−29 for other peaks, without indicating positive charge. M+1 refers to an ion with a mass one unit greater than that of $M^{+\bullet}$.

Intensity of M+1 relative to molecular ion for C_nH_m:

$$\text{Intensity} = \underbrace{n \times 1.08\%}_{\text{From }^{13}C} + \underbrace{m \times 0.012\%}_{\text{From }^2H} \qquad (21\text{-}2)$$

For benzene, C_6H_6, $M^{+\bullet}$ is observed at m/z 78. The predicted intensity at m/z 79 is $6 \times 1.08 + 6 \times 0.012 = 6.55\%$ of the abundance of $M^{+\bullet}$. The observed intensity in Figure 21-6 is 6.5%. An intensity within $\pm 10\%$ of the expected value (5.9 to 7.2, in this case) is within the precision of ordinary mass spectrometers. For biphenyl ($C_{12}H_{10}$), we predict that M+1 should have $12 \times 1.08 + 10 \times 0.012 = 13.1\%$ of the intensity of $M^{+\bullet}$. The observed value is 12.9%.

FIGURE 21-6 Electron ionization (70 eV) mass spectra of molecular ion region of benzene (C_6H_6) and biphenyl ($C_{12}H_{10}$). [From NIST/EPA/NIH Mass Spectral Database.[10]]

EXAMPLE   Listening to a Mass Spectrum

In the chemical ionization mass spectrum of pentobarbital in Figure 21-4, the peak with the most significant intensity at the high end of the mass spectrum at m/z 227 is suspected to be MH^+. If this is so, then the nominal mass of M is 226. The nitrogen rule tells us that a molecule with an even mass must have an even number of nitrogen atoms. If you know from elemental analysis that the compound contains only C, H, N, and O, how many atoms of carbon would you suspect are in the molecule?

Solution The m/z 228 peak has 12.0% of the height of the m/z 227 peak. Table 21-2 tells us that n carbon atoms will contribute $n \times 1.08\%$ intensity at m/z 228 from ^{13}C. Contributions from ^2H and ^{17}O are small. ^{15}N makes a larger contribution, but there are probably few atoms of nitrogen in the compound. Our first guess is

TABLE 21-1 Isotopes of selected elements

Element	Mass number	Mass (Da)[a]	Abundance (atom%)[b]	Element	Mass number	Mass (Da)[a]	Abundance (atom%)[b]
Proton	1	1.007 276 467	—	Cl	35	34.968 85	75.78
Neutron	1	1.008 664 916	—		37	36.965 90	24.22
Electron	—	0.000 548 580	—	Ar	36	35.967 55	0.336
H	1	1.007 825	99.988		38	37.962 73	0.063
	2	2.014 10	0.012		40	39.962 38	99.600
B	10	10.012 94	19.9	Fe	54	53.939 61	5.845
	11	11.009 31	80.1		56	55.934 94	91.754
C	12	12(exact)	98.93		57	56.935 40	2.119
	13	13.003 35	1.07		58	57.933 28	0.282
N	14	14.003 07	99.632	Br	79	78.918 34	50.69
	15	15.000 11	0.368		81	80.916 29	49.31
O	16	15.994 91	99.757	I	127	126.904 47	100
	17	16.999 13	0.038	Hg	196	195.965 81	0.15
	18	17.999 16	0.205		198	197.966 75	9.97
F	19	18.998 40	100		199	198.968 26	16.87
Si	28	27.976 93	92.230		200	199.968 31	23.10
	29	28.976 49	4.683		201	200.970 29	13.18
	30	29.973 77	3.087		202	201.970 63	29.86
P	31	30.973 76	100		204	203.973 48	6.87
S	32	31.972 07	94.93	Pb	204	203.973 03	1.4
	33	32.971 46	0.76		206	205.974 45	24.1
	34	33.967 87	4.29		207	206.975 88	22.1
	36	35.967 08	0.02		208	207.976 64	52.4

a. 1 dalton (Da) $\equiv$ 1/12 of the mass of ^{12}C = 1.660 538 782 (83) $\times 10^{-27}$ kg (from http://physics.nist.gov/constants). Nuclide masses from G. Audi, A. H. Wapsta, and C. Thibault, Nucl. Phys. **2003**, A729, 337, found at www.nndc.bnl.gov/masses. This source provides more significant figures for atomic mass than are cited in this table.

b. Abundance is representative of what is found in nature. Significant variations are observed. For example, ^{18}O in natural substances has been found in the range 0.188 to 0.222 atom%. The latest list of isotope abundances, which is slightly different from this table, is found in J. K. Böhlke et al., "Isotopic Compositions of the Elements, 2001," J. Phys. Chem. Ref. Data **2005**, 34, 57.

$$\text{Number of C atoms} = \frac{\text{observed peak intensity for M} + 1}{\text{contribution per carbon atom}} = \frac{12.0\%}{1.08\%} = 11.1 \approx 11$$

The actual composition of MH^+ at m/z 227 is $C_{11}H_{19}O_3N_2$. From the factors in Table 21-2, the theoretical intensity at m/z 228 is

$$\text{Intensity} = \underbrace{11 \times 1.08\%}_{^{13}C} + \underbrace{19 \times 0.012\%}_{^{2}H} + \underbrace{3 \times 0.038\%}_{^{17}O} + \underbrace{2 \times 0.369\%}_{^{15}N} = 13.0\%$$

The theoretical intensity is within the 10% uncertainty of the observed value of 12.0%.

TABLE 21-2 Isotope abundance factors (%) for interpreting mass spectra

Element	X+1	X+2	X+3	X+4	X+5	X+6
H	0.012n					
C	1.08n	0.005 8n(n − 1)				
N	0.369n					
O	0.038n	0.205n				
F	0					
Si	5.08n	3.35n	0.170n(n − 1)	0.056n(n − 1)		
P	0					
S	0.801n	4.52n	0.036n(n − 1)	0.102n(n − 1)		
Cl	—	32.0n	—	5.11n(n − 1)	—	0.544n(n − 1)(n − 2)
Br	—	97.3n	—	47.3n(n − 1)	—	15.3n(n − 1)(n − 2)
I	0					

EXAMPLE: For a peak at m/z = X containing n carbon atoms, the intensity from carbon at X+1 is n × 1.08% of the intensity at X. The intensity at X + 2 is n(n − 1) × 0.005 8% of the intensity at X. Contributions from isotopes of other atoms in the ion are additive.

Test Yourself A compound with a parent ion at m/z 117 has a peak at m/z 118 with a relative intensity of 9.3%. Is there an odd or even number of N atoms? How many carbons are in the formula? (*Answer:* odd, 8C. It cannot be 9C because mass of C_9N is 122 Da, which is >117 Da.)

FIGURE 21-7 Calculated isotopic patterns for species containing Cl and Br.

FIGURE 21-8 Electron ionization mass spectrum (70 eV) of 1-bromobutane. [From A. Illies, P. B. Shevlin, G. Childers, M. Peschke, and J. Tsai, "Mass Spectrometry for Large Undergraduate Laboratory Sections," *J. Chem. Ed.* **1995**, *72*, 717. Referee from Maddy Harris.]

Ions containing Cl or Br have distinctive isotopic peaks shown in Figure 21-7.[16] In the mass spectrum of 1-bromobutane in Figure 21-8, two nearly equal peaks at m/z 136 and 138 are a strong indication that the molecular ion contains one Br atom. The fragment at m/z 107 has a nearly equal partner at m/z 109, strongly suggesting that this fragment ion contains Br. Box 21-3 describes other information available from isotope ratios.

BOX 21-3 Isotope Ratio Mass Spectrometry

The opening of Chapter 23 describes the analysis of cholesterol from ancient human bones by *isotope ratio mass spectrometry*.[13] A compound eluted from a gas chromatography column is passed through a combustion furnace loaded with a metal catalyst (such as CuO/Pt at 820°C) to oxidize organic compounds to CO_2. The H_2O combustion by-product is removed by passage through a Nafion fluorocarbon tube (page 384). Water diffuses through the membrane, but other combustion products are retained. CO_2 then enters a mass spectrometer that monitors m/z 44 ($^{12}CO_2$) and 45 ($^{13}CO_2$). One detector is dedicated to each ion.

Natural carbon is composed of 98.9% ^{12}C and 1.1% ^{13}C. The chart shows consistent, small variations in ^{13}C from natural sources.

The standard used to measure carbon isotope ratios is calcium carbonate from the Pee Dee belemnite (fossil shell) formation in South Carolina (designated PDB) with a ratio $R_{PDB} = {}^{13}C/{}^{12}C = 0.011\ 23_{72}$. (The composition is accurate to four significant figures, but small differences are precise to six significant figures.) The $\delta^{13}C$ scale expresses small variations in isotopic compositions:

$$\delta^{13}C \text{ (parts per thousand, \%)} = 1\ 000 \left(\frac{R_{sample} - R_{PDB}}{R_{PDB}} \right)$$

$\delta^{13}C$ of natural materials provides information about their biological and geographic origins.[14,15]

Variation in ^{13}C from natural sources. C_3, C_4, and CAM are types of plants with distinct metabolic pathways leading to different ^{13}C incorporation. [Adapted from W. Meier-Augenstein, *LCGC* **1997**, *15*, 244.]

FIGURE 21-9 High-resolution double-focusing sector-field mass spectrum distinguishing $^{31}P^+$, $^{15}N^{16}O^+$, and $^{14}N^{16}OH^+$ with a resolution of $m/\Delta m = 4\,000$. The spectrum is a bar graph of detector counts/s at each increment of m/z.
[From J. S. Becker, S. F. Boulyga, C. Pickhardt, J. Becker, S. Buddrus, and M. Przybylski, "Determination of Phosphorus in Small Amounts of Protein Samples by ICP-MS," *Anal. Bioanal. Chem.* **2003**, *375*, 561.]

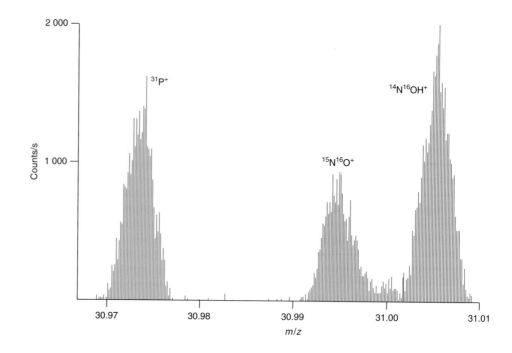

Predicted masses from Table 21-1 and masses observed by high-resolution mass spectrometry:

$C_5H_8O^{+\bullet}$:

5C	$5 \times 12.000\ 00$
8H	$+8 \times 1.007\ 825$
1O	$+1 \times 15.994\ 91$
$-e^-$	$-1 \times 0.000\ 55$
	$84.056\ 96$
observed:[17]	$84.059\ 1$

$C_6H_{12}^{+\bullet}$:

6C	$6 \times 12.000\ 00$
12H	$12 \times 1.007\ 825$
$-e^-$	$-1 \times 0.000\ 55$
	$84.093\ 35$
observed:[17]	$84.093\ 9$

If P makes more than three bonds or if S makes more than two bonds, then Equation 21-3 does not include these extra bonds. Examples that violate Equation 21-3:

High-Resolution Mass Spectrometry

An ion at m/z 84 could have a variety of elemental compositions, such as $C_5H_8O^+ = 84.056\ 96$ Da or $C_6H_{12}^+ = 84.093\ 35$ Da. We can determine which composition is correct if the spectrometer can distinguish sufficiently small differences in mass. With a *double-focusing mass spectrometer*, a *time-of-flight mass spectrometer*, an *orbitrap mass spectrometer*, or a *Fourier transform ion cyclotron resonance mass spectrometer* (Section 21-3), we can resolve differences of 0.001 or less at m/z 100. Figure 21-9 shows a high-resolution spectrum that distinguishes $^{31}P^+$, $^{15}N^{16}O^+$, and $^{14}N^{16}OH^+$—all with a nominal mass of 31.

To ensure accurate mass measurement, spectrometers are calibrated with compounds such as perfluorokerosene ($CF_3(CF_2)_nCF_3$) or perfluorotributylamine (($CF_3CF_2CF_2CF_2)_3N$). In high-resolution spectra, the exact masses of fluorocarbon fragments are slightly lower than those of ions containing C, H, O, N, and S. For high-resolution work, standards should be run with the unknown.

Rings + Double Bonds

If we know the composition of a molecular ion, and we want to propose its structure, a handy equation that gives the number of **rings + double bonds** (R + DB) is

Rings + double bonds formula:

$$R + DB = c - h/2 + n/2 + 1 \qquad (21\text{-}3)$$

where c is the number of Group 14 atoms (C, Si, and so on, which all make four bonds), h is the number of (H + halogen) atoms (which make one bond), and n is the number of Group 15 atoms (N, P, As, and so on, which make three bonds). Group 16 atoms (O, S, and so on, which make two bonds) do not enter the formula. Here is an example:

$$R + DB = c - h/2 + n/2 + 1$$

$$R + DB = (14 + 1) - \frac{22 + 1 + 1}{2} + \frac{1 + 1}{2} + 1 = 5$$

The molecule has one ring + four double bonds.
Note that c includes C + Si, h includes H + Cl + Br, and n includes N + As.

$C_{14}SiH_{22}ClBrNAsO_3S$

FIGURE 21-10 Electron ionization (70 eV) mass spectra of isomeric ketones with the composition $C_6H_{12}O$. [From NIST/EPA/NIH Mass Spectral Database.[10]]

Identifying the Molecular Ion Peak

Figure 21-10 shows electron ionization mass spectra of isomers with the elemental composition $C_6H_{12}O$. $M^{+\bullet}$ is marked by the solid triangle at m/z 100.

If these spectra had been obtained from unknowns, how would we know that the peak at m/z 100 represents the molecular ion? Here are some guidelines:

1. $M^{+\bullet}$ will be at the highest m/z value of any of the "significant" peaks in the spectrum that cannot be attributed to isotopes or background. "Background" arises from sources such as pump oil in the spectrometer and stationary phase from a gas chromatography column. Experience is required to recognize these signals. With electron ionization, the molecular ion peak intensity is often no greater than 5–20% of the base peak and might not represent more than 1% of all ions.
2. Intensities of isotopic peaks at M+1, M+2, and so on, must be consistent with the proposed formula.
3. The peak for the heaviest fragment ion should not correspond to an improbable mass loss from $M^{+\bullet}$. It is rare to find a mass loss in the range 3–14 or 21–25 Da. Common losses include 15 (CH_3), 17 (OH or NH_3), 18 (H_2O), 29 (C_2H_5), 31 (OCH_3), 43 Da (CH_3CO or C_3H_7), and many others. If you think that m/z 150 represents $M^{+\bullet}$, but there is a significant peak at m/z 145, then $M^{+\bullet}$ is not assigned correctly, because a mass loss of 5 Da is improbable. Alternatively, the two peaks represent ions from different compounds or both peaks represent fragments from a compound whose mass is greater than 150 Da.
4. If a fragment ion is known to contain, say, three atoms of element X, then there must be at least three atoms of X in the molecular ion.

In both spectra in Figure 21-10, the peak at highest m/z with "significant" intensity is m/z 100. The next highest significant peak is m/z 85, representing a loss of 15 Da (probably CH_3). Peaks at m/z 85 and 100 are consistent with m/z 100 representing $M^{+\bullet}$. If $M^{+\bullet}$ has an even mass, the nitrogen rule says there must be an even number of N atoms (which could be 0) in the molecule.

In both spectra, M+1 has 6% of the intensity of $M^{+\bullet}$, with just one significant digit in the measurement. From the intensity of M+1, we estimate the number of C atoms:

$$\text{Number of C atoms} = \frac{\text{observed (M+1)/M}^{+\bullet} \text{ intensity}}{\text{contribution per carbon atom}} = \frac{6\%}{1.08\%} = 5._6 \approx 6$$

If there are six C atoms and no N atoms, a possible composition is $C_6H_{12}O$. The expected intensity of M+1 is

$$\text{Intensity} = \underbrace{6 \times 1.08\%}_{^{13}C} + \underbrace{12 \times 0.012\%}_{^2H} + \underbrace{1 \times 0.038\%}_{^{17}O} = 6.7\% \text{ of M}^{+\bullet}$$

The observed intensity of 6% for M+1 is within experimental error of 6.7%, so $C_6H_{12}O$ is consistent with the data so far. $C_6H_{12}O$ requires one ring or double bond.

The small peak at m/z 86 is not a loss of 14 Da from $M^{+\bullet}$. It is the isotopic partner of m/z 85 containing ^{13}C.

We see no evidence at M+1, M+2, and M+3 for the elements Cl, Br, Si, or S.

Challenge What intensities would you see at M+2 and M+3 if there were one Cl, one Br, one Si, or one S in the molecule?

R + DB = $c - h/2 + n/2 + 1$
= $6 - 12/2 + 0 + 1 = 1$

FIGURE 21-11 Four possible fragmentation pathways for the molecular ion of 2-hexanone.

⌢ is transfer of one electron

⌒ is transfer of two electrons

Types of bond breaking:

- *Homolytic cleavage:* 1 electron remains with each fragment
- *Heterolytic cleavage:* Both electrons stay with one fragment

In general, the most intense peaks correspond to the most stable fragments.

$CH_3CH_2CH_2CH_2$—C(=O)—CH_3 M⁺·
→ A homolytic cleavage →
$CH_3CH_2CH_2CH_2^+$:$\overset{+}{O}$≡C—CH_3 m/z = 43

$CH_3CH_2CH_2CH_2$—C(=$\overset{+}{O}$)—CH_3
→ B homolytic cleavage →
$CH_3CH_2CH_2CH_2$—C≡$\overset{+}{O}$:   ·CH_3 m/z = 85

$CH_3CH_2CH_2CH_2$—C(=$\overset{+}{O}$)—CH_3
→ C heterolytic cleavage →
$CH_3CH_2CH_2CH_2$—C·(=O)   ⁺CH_3 m/z = 15

CH_3 γ H +O: β α C CH_3
→ D McLafferty rearrangement →
CH_3 CH_2 Propene (neutral molecule)   H—$\overset{+}{O}$: H_2C—C—CH_3 m/z = 58

Bond cleavages of 4-methyl-2-pentanone leading to m/z 15, 85, 43, and 57:

57 57
43 43 85
15

Challenge Draw a rearrangement like Reaction D in Figure 21-11 to show how m/z 58 arises.

Interpreting Fragmentation Patterns

Consider how the 2-hexanone molecular ion can break apart to give the many peaks in Figure 21-10. Reactions A and B in Figure 21-11 show $M^{+\cdot}$ derived from loss of a nonbonding electron from oxygen, which has the lowest ionization energy. In Reaction A, the C—C bond adjacent to C=O splits so that one electron goes to each C atom. The products are a neutral butyl radical (·C_4H_9) and $CH_3CO^{+\cdot}$. Only the ion is detected by the mass spectrometer, giving the base peak at m/z 43. Cleavage of the C—C bond in Reaction B gives an ion with m/z 85, corresponding to loss of ·CH_3 from the molecular ion. Two other major peaks in the spectrum arise from C_4—C_5 bond cleavage to give $CH_3CH_2^+$ (m/z 29) and $^+CH_2CH_2COCH_3$ (m/z 71). The nitrogen rule told us that molecules containing only C, H, halogens, O, S, Si, P, and an even number of N atoms (such as 0) have an even mass. A fragment of a neutral molecule that is missing an H atom must have an odd mass.

We have yet to account for the second-tallest peak at m/z 58, which is special because it has an *even* mass. The molecular ion has an even mass (100 Da). Radical fragments, such as $CH_3CH_2^+$, have an *odd* mass. All of the fragments discussed so far have an odd mass. The peak at m/z 58 results from the loss of a *neutral* molecule with an *even* mass of 42 Da.

Reaction D in Figure 21-11 shows a common rearrangement that leads to loss of a neutral molecule with even mass. In ketones with an H atom on the γ carbon atom, the H atom can be transferred to O^+. Concomitantly, the C_α—C_β bond cleaves and a neutral molecule of propene ($CH_3CH=CH_2$, 42 Da) is lost. The resulting ion has a mass of 58 Da.

The spectra in Figure 21-10 allow us to distinguish one isomer of $C_6H_{12}O$ from the other. The main difference between the spectra is a peak at m/z 71 for 2-hexanone that is absent in 4-methyl-2-pentanone. The peak at m/z 71 results from loss of ethyl radical, $CH_3CH_2^+$, from $M^{+\cdot}$. Ethyl radical is derived from carbon atoms 5 and 6 of 2-hexanone. There is no simple way for an ethyl radical to cleave from 4-methyl-2-pentanone. The diagram in the margin shows how peaks at m/z 15, 85, 43, and 57 can arise from breaking bonds in 4-methyl-2-pentanone. A rearrangement like that at the bottom of Figure 21-11 accounts for m/z 58.

Other major peaks in Figure 21-10 might be $CH_2=C=O^{+\cdot}$ (m/z 42), $C_3H_5^+$ (m/z 41), $C_3H_3^+$ (m/z 39), $C_2H_5^+$ or $HC\equiv O^+$ (m/z 29), and $C_2H_3^+$ (m/z 27). Small fragments are common in many spectra and not very useful for structure determination.

Interpretation of mass spectra to elucidate molecular structure is an important field.[18] Fragmentation patterns can even unravel the structures of large biological molecules.

21-3 Types of Mass Spectrometers

Figure 21-2 shows a *magnetic sector mass spectrometer* in which ions with different mass, but constant kinetic energy, are separated by their trajectories in a magnetic field. The kinetic energy is imparted to the ions by the voltage between the acceleration plates in the ion source.

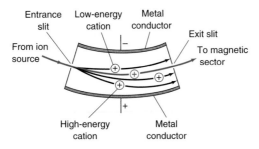

FIGURE 21-12 Electric sector of a double-focusing mass spectrometer. Positive ions are attracted toward the negative plate. Trajectories of high-energy ions are changed less than trajectories of low-energy ions. Ions reaching the exit slit have a narrow range of kinetic energies.

Ions are created from the neutral molecule with a small spread of kinetic energies and are accelerated to different extents, depending on where in the ion source they were formed.

The resolving power of a mass spectrometer is limited by the variation in kinetic energy of ions emerging from the source, which is typically ~0.1%. This variation limits the resolving power to ~1 000, corresponding to a resolution of 0.1 at m/z 100. In a **double-focusing mass spectrometer,** ions ejected from the source pass through an electric sector as well as a magnetic sector (Figure 21-12). With both sectors in series, it is possible to achieve a resolving power of ~10^5, corresponding to a resolution of 0.001 at m/z 100.

Transmission Quadrupole Mass Spectrometer

Magnetic sector and double-focusing mass spectrometers are not detectors of choice for chromatography. Figure 21-13 shows a **transmission quadrupole mass spectrometer**[19] connected to an open tubular gas chromatography column to record multiple spectra from each component as it is eluted. Species exiting the chromatography column pass through a heated connector into the electron ionization chamber, which is pumped to maintain a pressure of ~10^{-4} Pa (~10^{-9} bar, ~10^{-6} Torr), using a high-speed turbomolecular or oil diffusion pump. Ions are accelerated through a potential of 5–15 V before entering the quadrupole filter.

The quadrupole is a common mass separator because of its low cost. It consists of four parallel metal rods to which are applied both a constant voltage and a radio-frequency oscillating voltage. The electric field deflects ions in complex trajectories as they migrate from the ionization chamber toward the detector, allowing only ions with one particular mass-to-charge ratio to reach the detector. Other ions (nonresonant ions) collide with the rods and

FIGURE 21-13 Quadrupole mass spectrometer. Ideally, the rods should have a hyperbolic cross section on the surfaces that face one another.

FIGURE 21-14 Continuous-dynode electron multiplier, also known as a Channeltron®. For each incident electron, ~10⁵ electrons are detected at the far end of the horn. [Adapted from J. T. Watson and O. D. Sparkman, *Introduction to Mass Spectrometry*, 4th ed. (Chichester: Wiley, 2007).]

Ideally, the kinetic energy of an ion expelled from the source is zeV, where ze is the charge of the ion and V is the voltage on the backplate.

are lost before they reach the detector. Rapidly varying voltages select ions of different masses to reach the detector. Transmission quadrupoles can record 2–8 spectra per second, covering a range as high as 4 000 *m/z* units. They typically can resolve peaks separated by *m/z* 0.3.

Magnetic sector and double-focusing instruments, as well as time-of-flight instruments described next, operate at *constant resolving power*, which means that the separation between peaks decreases as *m/z* increases. Transmission quadrupoles (and other quadrupole instruments described later) operate at *constant resolution*. That is, ions of *m/z* 100 and 101 are separated to the same extent as ions of *m/z* 500 and 501.

In place of the *discrete-dynode* detector in Figure 21-2, we show a *high-energy dynode* and a *continuous-dynode electron multiplier* detector in Figure 21-13. A high-energy dynode is used with quadrupole mass separators and ion traps (described below) to ensure that all ions produce a similar electrical response at the detector; otherwise, the mass spectrum may exhibit *mass discrimination* in which ions of different *m/z* produce different responses. Cations emerging from the quadrupole mass analyzer are attracted to the high-energy dynode held at −10 kV. Each cation striking the dynode liberates electrons, which are accelerated toward the mouth of the continuous-dynode electron multiplier. The electrically resistive lead-doped glass wall of the horn-shaped electron multiplier is at −2 kV at the mouth and at ground potential at the narrow end (Figure 21-14). An electron striking the wall of the electron multiplier liberates several electrons, which are accelerated toward more positive potential deeper into the horn. After many rebounds, each incident electron is multiplied to ~10⁵ electrons at the narrow end of the horn.

Time-of-Flight Mass Spectrometer

The principle of the **time-of-flight mass spectrometer** is shown in Figure 21-15.[20] The ion source is at the upper left. About 3 000 to 20 000 times per second, a voltage of 5 000 V is applied to the backplate to accelerate ions toward the right and expel them from the ion source into the drift region, where there is no electric or magnetic field and no further acceleration. Ideally, all ions have the same kinetic energy, which is $\frac{1}{2}mv^2$, where m is the mass of the ion and v is its velocity. If ions have the same kinetic energy, but different masses, the lighter ones travel faster than the heavier ones. In its simplest incarnation, the time-of-flight mass spectrometer is just a long, straight, evacuated tube with the source at one end and the detector at the other end. Ions expelled from the source drift to the detector in order of increasing mass, because the lighter ones travel faster.

The instrument in Figure 21-15 is designed for improved resolving power. The main limitation on resolving power is that all ions do not emerge from the source with the same kinetic energy. An ion formed close to the backplate is accelerated through a higher voltage difference than one formed near the grid, so the ion near the backplate gains more kinetic energy. Also, there is some distribution of kinetic energies among the ions, even in the absence of the accelerating voltage. Heavier ions with more-than-average kinetic energy reach the detector at the same time as lighter ions with less-than-average energy.

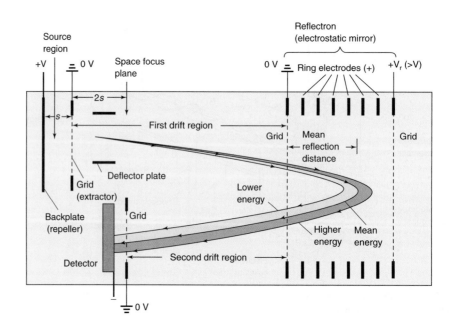

FIGURE 21-15 Time-of-flight mass spectrometer. Positive ions are accelerated out of the source by voltage +V periodically applied to the backplate. Light ions travel faster and reach the detector sooner than heavier ions.

Consider two ions expelled from the source with different speeds at different times. The ion formed close to the grid has less kinetic energy but is expelled first. The ion formed close to the backplate has more kinetic energy but is expelled later. Eventually, the faster ion catches up to the slower ion at the *space focus plane* at a distance $2s$ from the grid (where s is the distance between the backplate and the grid). It can be shown that all ions of a given mass reach the space focus plane at the same time. After the space focus plane, ions diverge again, with faster ions overtaking slower ones. In the absence of countermeasures, ions would be badly spread by the time they reach the detector and resolving power would be low.

To improve resolving power, ions are turned around by the "reflectron" at the right side of Figure 21-15. The reflectron is a series of hollow rings held at increasingly positive potential, terminated by a grid whose potential is more positive than the accelerating potential on the backplate of the source. Ions entering the reflectron are slowed down, stopped, and reflected back to the left. The more kinetic energy an ion had when it entered the reflectron, the further it penetrates before it is turned around. Reflected ions reach a new space focus plane at the grid in front of the detector. *All ions of the same mass reach this grid at the same time, regardless of their initial kinetic energies.*

Resolving power can be as high as 1 000 to 25 000, and m/z accuracy is ~0.001. Other advantages of the time-of-flight spectrometer are its high acquisition rate (10^2 to 10^4 spectra/s) and its capability for measuring very high masses ($m/z \approx 10^6$). The time-of-flight mass spectrometer requires a lower operating pressure than do the transmission quadrupole and magnetic sector instruments (10^{-7} Pa versus 10^{-4} Pa or 10^{-9} Torr versus 10^{-6} Torr).

Three-Dimensional Quadrupole Ion-Trap Mass Spectrometer

The **three-dimensional quadrupole ion-trap mass spectrometer**[21] is a compact device that is well suited as a chromatography detector. In Figure 21-16, substances emerging from the chromatography column enter the cavity of the three-dimensional ion-trap mass analyzer from the lower left through a heated transfer line. In the implementation of the ion trap shown in Figure 21-16, the gate electrode periodically admits electrons from the filament at the top into the cavity through holes in the end cap. Molecules undergo electron ionization in the cavity formed by the two end caps and a ring electrode, all of which are electrically isolated from one another. Alternatively, chemical ionization is achieved by adding a reagent gas such as methane to the cavity, ionizing it with the electron beam to form reagent ions, and then allowing reagent ions to react with analyte molecules. Some gas chromatography–mass spectrometry systems and all liquid chromatography–mass spectrometry systems produce ions outside the ion trap and inject them into the trap.

A constant-frequency radio-frequency voltage applied to the central ring electrode causes ions to circulate in stable, three-dimensional orbits in the cavity, with the lowest m/z in the outermost orbits. Increasing the amplitude of the radio-frequency voltage destabilizes the orbits of ions of one m/z value at a time, sending them flying out of the two end caps. Ions expelled through the lower end cap in Figure 21-16b are detected by the electron multiplier with high sensitivity (~1–10 pg). Scans from m/z 10 to 650 can be conducted eight times per second. Resolving power is 1 000–4 000, m/z accuracy is 0.1, and the maximum m/z is ~20 000.

At the University of Bonn, W. Paul showed in the 1950s that ions could be manipulated by quadrupole electric fields. He received the Nobel Prize for physics in 1989.

(a)

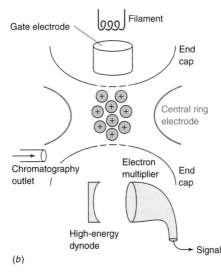

(b)

FIGURE 21-16 Three-dimensional quadrupole ion-trap mass spectrometer. (*a*) Mass analyzer consists of two end caps (left and right) and central ring electrode. (*b*) **Schematic diagram.** [Courtesy Varian Associates, Sunnyvale, CA.]

In other mass analyzers, a small fraction of ions reaches the detector. With a three-dimensional ion trap, half of the ions reach the detector, giving this device 10 to 100 times more sensitivity than the transmission quadrupole.

The quadrupole ion trap contains He gas at 10^{-1} Pa (10^{-3} Torr), which is much higher than the allowed pressure in a transmission quadrupole (10^{-4} Pa or 10^{-6} Torr) or a time-of-flight instrument (10^{-7} Pa, 10^{-9} Torr). Gas in the ion trap cools the ions by collisions, which absorb excess vibrational and rotational energy from the ions. Cool ions have less random displacement from the ideal orbits established by the electric field of the quadrupole. Lower pressure is required in other mass filters to increase the mean free path of ions so that they are not diverted from the intended trajectory by collisions with background gas.

The capacity of the three-dimensional quadrupole to hold ions is limited by the small volume enclosed by the electrodes and by the total charge of the ions, which can alter the electric field inside the quadrupole. The field created by the ions is called the *space charge*. When the space charge is too great, performance of the ion trap is degraded. Mass spectrometers have feed-back controls that limit the space charge to tolerable levels by controlling the number of ions allowed in the trap.

Linear Quadrupole Ion-Trap Mass Spectrometer

The **linear quadrupole ion-trap mass spectrometer,** which has higher trapping efficiency and higher storage capacity than the three-dimensional quadrupole ion trap, is finding its way into many instruments. The quadrupole mass filter in Figure 21-13 used a direct current volt-age and a radio-frequency voltage to select ions of a particular m/z to be transmitted through the filter. The linear ion trap in Figure 21-17 adds sections at each end of the quadrupole to create a potential well. If the ends are sufficiently positive with respect to the center section, cations become trapped in the center section. Ions are confined in the radial direction (the xy plane) by a radio-frequency field applied to the central section. By manipulating the voltages, ions of a specific m/z value can be expelled through slits in the x direction or out the ends in the z direction to one or more detectors.

Three-dimensional quadrupole ion traps have low detection limits, but they are limited by the number of ions they can hold without exceeding the allowed space charge. Also, they only trap ~5% of injected ions. The linear ion trap has even lower detection limits because it can hold ~30 times more ions and can be more than 10 times as efficient at trapping injected ions.

Linear and three-dimensional quadrupole ion traps can trade spectral acquisition rate for resolution. They are commonly operated at unit resolution with scan rates on the order of 11 000 m/z units per second. At the expense of slowing the scan rate to 27 m/z units per second, a resolution of 0.05 m/z units can be attained.

(a)   (b)

FIGURE 21-17 Linear ion trap. [Drawing from Z. Ouyang, G. Wu, Y. Song, H. Li, W. R. Plass, and R. G. Cooks, "Rectilinear Ion Trap: Concepts, Calculations, and Analytical Performance of a New Mass Analyzer," *Anal. Chem.* **2004,** *76*, 4595. Photograph of LTQ XL linear ion trip courtesy Thermo Fisher Scientific, San Jose, CA.]

Orbitrap Mass Spectrometer

The orbitrap is a high-resolution mass analyzer that does not require a magnetic field or a radio-frequency field. In its commercial realization, the **orbitrap mass spectrometer** provides a resolving power of ~150 000, m/z accuracy of 2–5 ppm with external calibration, upper m/z limit of 6 000, and a dynamic range of several thousand. With internal calibration standards,

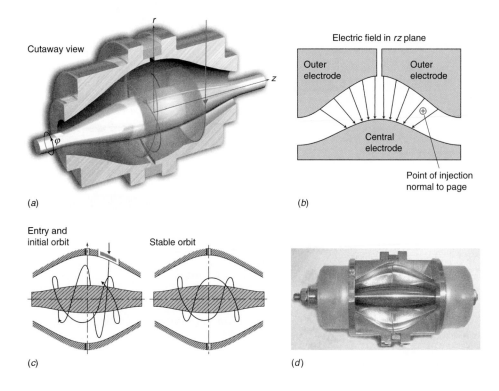

(a)

Cutaway view

Electric field in *rz* plane

Outer electrode

Outer electrode

Central electrode

Point of injection normal to page

(b)

Entry and initial orbit

Stable orbit

(c)

(d)

FIGURE 21-18 (*a*) **Cutaway drawing of orbitrap.** [From A. Makarov, "Electrostatic Axially Harmonic Orbital Trapping: A High-Performance Technique of Mass Analysis," *Anal. Chem.* **2000,** *72,* 1156.] (*b*) **Electric field in one longitudinal plane of orbitrap.** (*c*) **Initial path of ion entering the orbitrap and stable path for successive orbits.** (*d*) **Photograph of orbitrap with half of outer sheath removed. Inner diameter of outer electrodes is 30 mm and length of exposed section is 4 cm.** [Courtesy Thermo Fisher Scientific, San Jose, CA.]

sub-part-per-million *m/z* accuracy is attainable. The cutaway view in Figure 21-18a shows precisely machined central and outer electrodes. The central electrode is held at −5 kV while the two outer electrodes (Figure 21-18b) are close to ground potential and electrically isolated from each other. Figure 21-18b shows electric field vectors perpendicular to the axis of symmetry (*z*) near the center of the orbitrap, but increasingly angled toward the center of the orbitrap at increasing distance from the center.

A packet of ions is introduced perpendicular to the plane of Figure 21-18b at the indicated point. The electric field pushes ions into an orbit toward the center of the orbitrap in Figure 21-18c. Initial momentum carries ions from the right side of the orbitrap to the left side until the electric field is strong enough to push the ions back to the right. When ions have traveled far enough to the right, they are pushed back to the left. They circle back and forth around the central electrode as long as they do not strike another molecule. Maintaining the unperturbed orbit long enough requires the best vacuum of any mass spectrometer at $\sim 10^{-8}$ Pa ($\sim 10^{-10}$ Torr)—providing a mean free path of 100 km. The frequency of oscillation of an ion between the right and left halves of the orbitrap is proportional to $1/\sqrt{m/z}$.

Ions oscillating between the two halves of the orbitrap induce an opposite charge called the *image charge* in the outer electrode. A packet of cations in the right half of the orbitrap attracts electrons in the outer right electrode. A packet of cations in the left half of the orbitrap attracts electrons in the outer left electrode. An amplifier connected to the two halves of the split outer electrode measures the image current oscillating in synchrony with the ions. The orbitrap contains ions with different *m/z* values, each creating a component of current with a different frequency. The observed signal is the sum of currents from all *m/z* values. After recording the current for a predetermined time (~0.1 to 1.5 s), a computer decomposes the current into its component frequencies—and hence component *m/z* values—through a Fourier transform (Section 19-5).

Ions must be injected into the orbitrap in one small packet. Figure 21-19 shows how this can be done. Ions from an electrospray source (Section 21-4) are accumulated in a linear ion trap and then transferred in one batch to a C-trap, which electrodynamically squeezes them into a small packet. The packet is ejected through ion optics into the orbitrap. During the ~0.1-ms injection period, the electric field in the orbitrap is ramped up in such a manner that the ions begin to orbit the central electrode. After the voltage on the central electrode reaches its constant level for stable orbits, detection begins.

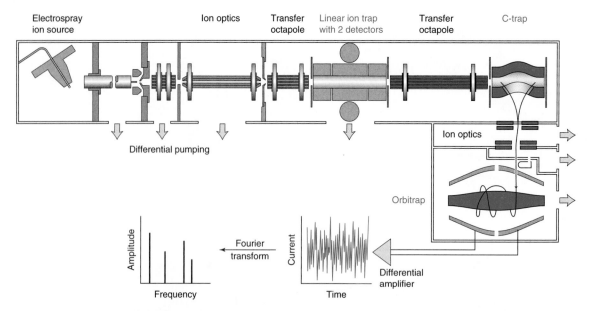

Electrospray ion source   Ion optics   Transfer octapole   Linear ion trap with 2 detectors   Transfer octapole   C-trap

Differential pumping

Ion optics

Orbitrap

Amplitude / Frequency

Fourier transform

Current / Time

Differential amplifier

FIGURE 21-19 Arrangement of linear ion trap and C-trap for collecting ions, compressing them into a small packet, and feeding the packet into the orbitrap. [From A. Makarov, E. Denisov, A. Kholomeev, W. Balschun, O. Lange, K. Strupat, and S. Horning, "Performance Evaluation of a Hybrid Linear Ion Trap/Orbitrap Mass Spectrometer," *Anal. Chem.* **2006**, *78*, 2113.]

Ion Mobility Spectrometer[22]

More than 10^4 **ion mobility spectrometers** are deployed at airport security checkpoints to detect explosives, and perhaps 10^5 hand-portable devices are used by military and civil defense personnel. Although functionally similar to mass spectrometers, ion mobility spectrometers are operated in air at ambient pressure, and ion mobility spectrometry is *not* a form of mass spectrometry. Ion mobility spectrometry does not measure molecular mass and provides no structural information. However, it is so widely used that we introduce it here.

Electrophoresis, which is discussed in Chapter 25, is the migration of ions in solution under the influence of an electric field. Ion mobility spectrometry is *gas-phase electrophoresis*, which separates ions according to their size-to-charge ratio. Unlike mass spectrometry, ion mobility spectrometry is capable of separating isomers. Ion mobility can be used to study large protein assemblies with masses up to 50 MDa.[23]

At first sight, the ion mobility spectrometer in Figure 21-20a reminds us of a time-of-flight spectrometer. In portable units, the drift tube is 5 to 10 cm long. Typically, sample adsorbed on a cotton swab is placed in the heated anvil at the left to desorb analyte vapor. Dry air doped with a chemical ionization reagent (such as Cl_2 for anions and acetone or NH_3 for cations) sweeps the vapor through a tube containing 10 millicuries of radioactive ^{63}Ni. Reagent gas ionized by β-emission from ^{63}Ni reacts with analyte to generate analyte ions.

A spectral scan is initiated when a ~250-μs voltage pulse on a gating grid admits a packet of ions into the drift tube. An electric field of 200–300 V/cm in the drift region is established by potential differences on the drift rings. The field causes either cations or anions to drift to the right at ~1 to 2 m/s. Ions are retarded by collisions with gas molecules at atmospheric pressure. Each ion travels at its own speed equal to KE, where E is the electric field and K is the *mobility*. Small ions have greater mobility than large ions of the same charge because large ions experience more drag.

The ion mobility "spectrum" in Figure 21-20b is a graph of detector response versus drift time for several explosives. Peak area is proportional to the number of ions. Peaks are identified by their mobility, which is reproducibly measured relative to that of an internal standard. Runs are repeated ~20 times per second. The displayed spectrum averages many runs and requires 2–5 s to obtain.

Detection limits are 0.1 to 1 pg for compounds with favorable ionization chemistry. Mobility spectrometers have limited resolution, but false positives are minimized by combining mobility determinations with selective ionization. Resolution can be improved by operating in a closed chamber with pressures of 2–4 bar.[24] Ion mobility separations are now being combined with mass spectrometry to identify the separated components.

Small ions drift faster than large ions. For a description of ion mobility spectrometry in electrophoretic terms, see Problem 25-48.

(a)

(b)

FIGURE 21-20 (a) Schematic illustration of ion mobility spectrometer. The aperture grid prevents serious line broadening. Ions approaching a bare detector plate induce current, which appears as signal before the arrival of the ions. The aperture grid protects the detector from induced current until ions are between the grid and the detector. (b) Negative ion mobility spectrum of explosives designated RDX, TNT, PETN, and so on. The calibrant 4-nitrobenzonitrile is an internal mobility standard. Cl_2 is the reagent gas for generating anions by chemical ionization.

[Courtesy W. R. Stott, Smiths Detection, Toronto.]

21-4 **Chromatography–Mass Spectrometry**

Mass spectrometry is widely used as the detector in chromatography to provide both qualitative and quantitative information. The spectrometer can be highly selective for the analyte of interest. This selectivity eases the requirements for sample preparation or complete chromatographic separation of components in a mixture, and it increases the signal-to-noise ratio.

Mass spectrometry requires high vacuum to prevent molecular collisions during ion separation. Chromatography is inherently a high-pressure technique. The problem in marrying the two techniques is to remove the huge excess of matter between the chromatograph and the spectrometer. Gas chromatography felicitously evolved to employ narrow capillary columns whose eluate does not overwhelm the pumping capabilities of the mass spectrometer vacuum system. The capillary column is connected directly to the inlet of the mass spectrometer through a heated transfer line, as in Figures 21-13 and 21-16.[25]

Liquid chromatography creates a huge volume of gas when solvent vaporizes at the interface between the column and the mass spectrometer.[26] Most of this gas must be removed prior to ion separation. Nonvolatile mobile-phase additives (such as phosphate buffer), which are common in chromatography, must be avoided for mass spectrometry. *Electrospray ionization* and *atmospheric pressure chemical ionization* are dominant methods for introducing eluate from liquid chromatography into a mass spectrometer.

Electrospray Ionization

Electrospray ionization,[28] also called *ion spray*, is illustrated in Figure 21-21a. Liquid from the chromatography column enters the steel nebulizer capillary at the upper left, along with a coaxial flow of N_2 gas. For positive ion mass spectrometry, the nebulizer is held at 0 V and

Volatile buffer components and additives for liquid chromatography that are compatible with mass spectrometry include NH_3, HCO_2H, CH_3CO_2H, CCl_3CO_2H, $(CH_3)_3N$, and $(C_2H_5)_3N$. Avoid additive concentrations >20 mM and detergent concentrations >10 μM.

J. B. Fenn received part of the Nobel Prize in chemistry in 2002[27] for electrospray ionization. K. Tanaka received part of the same prize for matrix-assisted laser desorption/ionization, described in Box 21-4.

(a)

(b)

FIGURE 21-21 (a) Pneumatically assisted electrospray interface for mass spectrometry. (b) Gas-phase ion formation. [Adapted from E. C. Huang, T. Wachs, J. J. Conboy, and J. D. Henion, "Atmospheric Pressure Ionization Mass Spectrometry," *Anal. Chem.* **1990**, *62*, 713A and P. Kebarle and L. Tang, "From Ions in Solution to Ions in the Gas Phase," *Anal. Chem.* **1993**, *65*, 972A.] (c) Electrospray from a silica capillary. [Courtesy R. D. Smith, Pacific Northwest Laboratory, Richland, WA.]

(c)

the spray chamber is held at $-3\,500$ V. For negative ion mass spectrometry, all voltages are reversed. The strong electric field at the nebulizer outlet, combined with the coaxial flow of N_2 gas, creates a fine aerosol of charged particles.

Typically, but not always, *ions that vaporize from aerosol droplets were already in solution in the chromatography column*. For example, protonated bases (BH^+) and ionized acids (A^-) can be observed. Other gas-phase ions arise from complexation between analyte, M (which could be neutral or charged), and stable ions from the solution. Examples include

$$MH^+ (mass = M + 1) \qquad M(K^+)(mass = M + 39)$$

$$M(NH_4^+)(mass = M + 18) \qquad M(HCO_2^-)(mass = M + 45)$$

$$M(Na^+)(mass = M + 23) \qquad M(CH_3CO_2^-)(mass = M + 59)$$

For protein electrospray, it is common to find multiply charged ions such as $[M + nH]^{n+}$ and, sometimes, $[M + nNa]^{n+}$ or $[M + nNH_4]^{n+}$. There is little fragmentation in electrospray.

Positive ions from the aerosol are attracted toward the glass capillary leading into the mass spectrometer by an even more negative potential of $-4\,500$ V. Gas flowing from atmospheric pressure in the spray chamber transports ions to the right through the capillary to the exit, where the pressure is reduced to ~300 Pa (~2 Torr) by a vacuum pump.

Figure 21-21b provides more detail on ionization. Voltage imposed between the steel nebulizer capillary and the inlet to the mass spectrometer creates excess charge in the liquid by redox reactions. If the nebulizer is positively biased, oxidation enriches the liquid in positive ions by reactions such as

$$Fe(s) \rightarrow Fe^{2+} + 2e^-$$

$$H_2O(l) \rightarrow \frac{1}{2}O_2 + 2H^+ + 2e^-$$

Electrons from the oxidation flow through the external circuit and eventually neutralize gaseous positive ions at the inlet to the mass spectrometer. It is possible for analyte to be chemically altered by species such as $HO^\bullet$ generated during electrospray.[29]

Charged liquid exiting the capillary forms a cone and then a fine filament and finally breaks into a spray of fine droplets (see Figure 21-21c and the opening of this chapter). It is thought that a droplet shrinks to ~1 μm by solvent evaporation until the repulsive force of the excess charge equals the cohesive force of surface tension. At that point, the droplet breaks up by ejecting tiny droplets with diameters of ~10 nm. They evaporate, leaving their cargo of ions in the gas phase. Aerodynamic forces might also contribute to droplet breakup.[30]

In electrospray, little fragmentation of analyte occurs and mass spectra are simple. Fragmentation can be intentionally increased by **collisionally activated dissociation** (also referred to as *collision-induced dissociation*) in the region between the glass capillary and the skimmer cone in Figure 21-21a. The pressure in this region is ~300 Pa (~2 Torr) and the background gas is mainly N_2. In Figure 21-21a, the outlet of the glass capillary is coated with a metal layer that is held at $+40$ V. The potential difference between the metal skimmer cone and the capillary is $-20 - (40) = -60$ V. Positive ions accelerated through 60 V collide with N_2 molecules with enough energy to break into a few fragments. Adjusting the skimmer cone voltage controls the degree of fragmentation. A small voltage difference favors molecular ions, whereas a large voltage difference creates fragments that aid in identification of analyte. Collisionally activated dissociation also tends to break apart complexes such as $M(Na^+)$.

For example, with a cone voltage difference of -20 V, the positive ion spectrum of the drug acetaminophen exhibits a base peak at m/z 152 for the protonated molecule, $[M+H]^+$ (colored species in the margin). A smaller peak at m/z 110 probably corresponds to the fragment shown in the margin. When the cone voltage difference is -50 V, collisionally activated dissociation decreases the peak at m/z 152 and increases the fragment peak at m/z 110. The negative ion spectrum has a large peak at m/z 150 for $[M-H]^-$. As the cone voltage difference is raised from $+20$ V to $+50$ V, this peak decreases and a fragment at m/z 107 increases.

Figure 21-22 shows an electrospray interface for capillary electrophoresis. The silica capillary is contained in a stainless steel capillary held at the required outlet potential for electrophoresis. The steel makes electrical contact with the liquid inside the silica by a liquid sheath flowing between the capillaries. The sheath liquid, which is typically a mixed organic/aqueous solvent, constitutes ~90% of the aerosol.

You, the chemist, need to adjust the pH of the chromatography solvent to favor BH^+ or A^- for mass spectrometric detection.

Electrospray requires low buffer concentration so that buffer ions do not overwhelm analyte ions in the mass spectrum. Low-surface-tension organic solvent is better than water. In reversed-phase chromatography (Section 24-1), it is good to use a stationary phase that strongly retains analyte so that a high fraction of organic solvent can be used. A flow rate of 0.05 to 0.4 mL/min is best for electrospray.

Collisionally activated dissociation of acetaminophen:

$[M+H]^+$
$m/z = 152$

$[M-H]^-$
$m/z = 150$

collisionally activated dissociation

$m/z = 110$
positive ions

$m/z = 107$
negative ions

FIGURE 21-22 Electrospray interface for capillary electrophoresis/mass spectrometry.

Liquid sheath

Electrospray

Stainless steel capillary (+3–6 kV with respect to mass spectrometer inlet)

Fused silica capillary

Unlike electrospray, atmospheric pressure chemical ionization *creates gaseous ions from neutral analyte molecules.* Analyte must have some volatility. For nonvolatile molecules such as sugars and proteins, electrospray can be used.

Atmospheric Pressure Chemical Ionization

In **atmospheric pressure chemical ionization,** heat and a coaxial flow of N_2 convert eluate into a fine mist from which solvent and analyte evaporate (Figure 21-23). Like chemical ionization in the ion source of a mass spectrometer, atmospheric pressure chemical ionization creates new ions from gas-phase reactions between ions and molecules. The distinguishing feature of this technique is that a high voltage is applied to a metal needle in the path of the aerosol. An electric *corona* (a plasma containing charged particles) forms around the needle, injecting electrons into the aerosol and creating ions. For example, protonated analyte, MH^+, can be formed in the following manner:

$$N_2 + e^- \rightarrow N_2^{+\bullet} + 2e^-$$
$$N_2^{+\bullet} + 2N_2 \rightarrow N_4^{+\bullet} + N_2$$
$$N_4^{+\bullet} + H_2O \rightarrow H_2O^{+\bullet} + 2N_2$$
$$H_2O^{+\bullet} + H_2O \rightarrow H_3O^+ + {}^\bullet OH$$
$$H_3O^+ + nH_2O \rightarrow H_3O^+(H_2O)_n$$
$$H_3O^+(H_2O)_n + M \rightarrow MH^+ + (n+1)H_2O$$

Analyte M might also form a negative ion by electron capture:

$$M + e^- \rightarrow M^{-\bullet}$$

A molecule, X—Y, in the eluate might create a negative ion by the reaction

$$X{-}Y + e^- \rightarrow X^\bullet + Y^-$$

The species Y^- could abstract a proton from a weakly acidic analyte, AH:

$$AH + Y^- \rightarrow A^- + HY$$

Atmospheric pressure chemical ionization handles a variety of analytes and accepts chromatography flow rates up to 2 mL/min. Generally, to be observed, analyte M must be capable of forming the protonated ion, MH^+. Atmospheric pressure chemical ionization tends to produce single-charge ions and is unsuitable for macromolecules such as proteins. There is little fragmentation, but the skimmer cone voltage difference can be adjusted to favor *collisionally activated dissociation* into a small number of fragments.

At least one commercial detector responds to an increased range of analytes by simultaneous electrospray and atmospheric pressure chemical ionization. Ions produced by electrospray are routed directly to the mass analyzer. Neutral molecules left after electrospray then undergo atmospheric pressure chemical ionization.[31]

Selected Ion Monitoring and Extracted Ion Monitoring

Selected ion monitoring and *extracted ion monitoring* increase the selectivity of mass spectrometry for individual analytes and decrease the response to everything else (decreasing

FIGURE 21-24 Chromatograms of herbicides (designated 1–6) spiked into river water at a level near 1 ppb demonstrate increased signal-to-noise ratio in selected ion monitoring. (*a*) Ultraviolet detection at 240 nm. (*b*) Electrospray reconstructed total ion chromatogram. (*c*) Electrospray selected ion monitoring at *m/z* 312. [From A. Laganà, G. Fago, and A. Marino, "Simultaneous Determination of Imidazolinone Herbicides from Soil and Natural Waters," *Anal. Chem.* **1998,** *70,* 121.]

background noise). Figure 21-24a shows a liquid chromatogram with ultraviolet absorbance detection of a mixture of herbicides (designated 1–6) spiked at a level near 1 ppb into river water. (*Spiked* means deliberately added.) The broad hump underlying the analytes represents many natural substances in the river water. The simplest way to use a mass spectrometer as a chromatographic detector is to substitute it for the ultraviolet monitor and add up the total current of all ions of all masses detected above a selected value. This **reconstructed total ion chromatogram** is shown in Figure 21-24b, which is just as congested as the ultraviolet chromatogram because all substances emerging at any moment contribute to the signal. This chromatogram is "reconstructed" by a computer from individual mass spectra recorded during chromatography.

To be more selective, we use **selected ion monitoring** in which the mass spectrometer is set to monitor just a few values of *m/z* (never more than four or five in any time interval). Figure 21-24c shows the **selected ion chromatogram** for which just *m/z* 312 is monitored. The signal corresponds to MH$^+$ from herbicide 6, which is imazaquin. The signal-to-noise ratio in selected ion monitoring is greater than the signal-to-noise ratio in chromatograms *a* or *b* because (1) most of the spectral acquisition time is spent collecting data in a small mass range and (2) little but the intended analyte gives a signal at *m/z* 312.

An **extracted ion chromatogram** is like a selected ion chromatogram, but the extracted ion chromatogram does not benefit from taking all available time to measure just one or a few mass spectral peaks. To create an extracted ion chromatogram, the entire mass spectrum is

FIGURE 21-25 (*a*) Total ion chromatogram of pesticides from fruit-based soft drinks from London's Gatwick Airport. (*b*) Extracted ion chromatogram set for thiabendazole with *m/z* window 202.043 ± 0.01. Inset shows mass spectrum of thiabendazole peak at 9.31 min. [From J. F. García-Reyes, B. Gilbert-López, A. Molina-Díaz, and A. R. Fernández-Alba, "Determination of Pesticide Residues in Fruit-Based Soft Drinks," *Anal. Chem.* **2008**, *80*, 8966.]

recorded repeatedly during chromatography. Then one value of *m/z* is taken from each spectrum for display. For example, the chromatogram might display the intensity of *m/z* 312 observed as a function of time. In selected ion monitoring, only the *m/z* 312 signal would be collected. For the extracted ion chromatogram, all values of *m/z* are measured, but only *m/z* 312 intensity is displayed. You should collect the entire mass spectrum when you do not know what you are looking for and you want to observe everything.

Figure 21-25 shows an important example of an extracted ion chromatogram. By using liquid chromatography–high resolution time-of-flight mass spectrometry, it is possible to search for 100 pesticides at once in food extracts. The time-of-flight spectrometer provides nearly unique identification of small molecules, such as pesticides, by exact mass measurement with uncertainties of ~1–2 parts per million in *m/z*. Components of fruit-based soft drinks isolated by solid-phase extraction (Section 27-3) were separated by liquid chromatography to obtain the complex total ion chromatogram in the upper part of Figure 21-25. When the extracted ion chromatogram window was set for *m/z* 202.043 ± 0.01 to find the pesticide thiabendazole, one major peak at 9.31 min was observed. Combined levels of several pesticides found in the majority of tested soft drinks from several countries exceeded European allowed maximum residue levels for drinking water by factors of 10 to 35.

Selected Reaction Monitoring

Selectivity and signal-to-noise ratio in a chromatogram is increased markedly by **selected reaction monitoring,** illustrated in Figure 21-26 with a *triple quadrupole mass spectrometer.* A mixture of ions enters quadrupole Q1, which passes just one selected **precursor ion** to the second stage, Q2. The second stage passes all ions of all masses straight on to the third stage, Q3. However, while inside Q2, which is called a *collision cell,* the precursor ion collides with N_2 or Ar at a pressure of $\sim 10^{-3}$ to 10^{-1} Pa ($\sim 10^{-5}$ to 10^{-3} Torr) and breaks into fragments called **product ions.** Quadrupole Q3 allows only specific product ions to reach the detector.

Selected reaction monitoring is highly selective for the analyte of interest. An example is provided by the analysis of human estrogens in sewage at parts per trillion levels (ng/L).

Selected reaction monitoring is one of several techniques carried out by consecutive mass filters. These techniques are collectively called *tandem mass spectrometry* or *mass spectrometry/mass spectrometry* or simply *MS/MS.*

(b)

FIGURE 21-26 (a) Principle of selected reaction monitoring, also called tandem mass spectrometry, mass spectrometry/mass spectrometry, or MS/MS. (b) Cutaway view of triple quadrupole mass spectrometer. [Courtesy Agilent Technologies, Santa Clara, CA.]

Estrogens are hormones in the ovarian cycle. The synthetic estrogen, 17α-ethinylestradiol (designated EE_2) is a contraceptive. Even at parts per trillion levels, some estrogens can provoke reproductive disturbances in fish.

A project in Italy measured the estrogens entering the aquatic environment from human waste. Now think about this problem. Sewage contains thousands of organic compounds—many at high concentrations. To measure nanograms of one analyte was beyond the capabilities of analytical chemistry until recently. Some *sample preparation* was necessary to remove polar compounds from the less polar analyte and to *preconcentrate* analyte. Raw sewage (150 mL) was filtered to remove particles >1.5 μm and then passed through a *solid-phase extraction* cartridge (Section 27-3) containing carbon adsorbent that retained analyte. The cartridge was washed with polar solvents to remove polar materials. Estrogens were washed from the cartridge by a mixture of dichloromethane and methanol. Solvent was evaporated and the residue was dissolved in 200 μL of aqueous solution containing another estrogen as internal standard. A volume of 50 μL was injected for chromatography.

Figure 21-27a shows the collisionally activated dissociation mass spectrum of the deprotonated molecule of estrogen EE_2. The *precursor ion* $[M - H]^-$ (*m/z* 295) obtained by electrospray was isolated by mass separator Q1 in Figure 21-26 and sent into Q2 for *collisionally activated dissociation*. Then all fragments >*m/z* 140 were analyzed by Q3 to give the mass spectrum in Figure 21-27a. For subsequent *selected reaction monitoring*, only the *product ions* at *m/z* 159 and 145 would be selected by Q3. The chromatogram in Figure 21-27b shows the signal from these product ions when *m/z* 295 was selected by Q1. From the area of the peak for EE_2, its concentration in the sewage is calculated to be 3.6 ng/L. Amazingly, there are other compounds that contribute mass spectral signals for this same set of masses (295 → 159 + 145) at elution times of 15–18 min. EE_2 was identified by its retention time and its complete mass spectrum.

Since the pioneering study, the widespread problem of pharmaceuticals and illegal drugs in municipal wastewater has been recognized.[32] Wastes that we generate find their way from wastewater into drinking water and the environment.

Other combinations of mass separators can be used for selected reaction monitoring. For example, in Figure 21-19, the linear ion trap can select the precursor ion and carry out collisionally activated dissociation. Product ions would be analyzed in the orbitrap.

The process in Q2 is called *collisionally activated dissociation*.

In *preconcentration*, analyte collected from 150 mL was dissolved in 200 μL. The concentration increased by a factor of

$$\frac{150 \times 10^{-3} \text{ L}}{200 \times 10^{-6} \text{ L}} = 750$$

3.6 ng/L in sewage became 750 × 3.6 ng/L = 2.7 μg/L for chromatography.

(a)

(b)

FIGURE 21-27 (a) Electrospray tandem mass spectrum of pure estrogen EE_2. The ion $[M-H]^-$ (m/z 295) was selected by quadrupole Q1 in Figure 21-26 and dissociated in Q2; the full spectrum of fragments was measured by Q3. (b) Selected reaction monitoring chromatogram showing the elution of 3.6 ng/L of estrogen EE_2 extracted from sewage. The signal is the sum of m/z 159 + 145 from Q3 when m/z 295 was selected by Q1. [From C. Baronti, R. Curini, G. D'Ascenzo, A. di Corcia, A. Gentili, and R. Samperi, "Monitoring Estrogens at Activated Sludge Sewage Treatment Plants and in a Receiving River Water," *Environ. Sci. Tech.* **2000**, *34*, 5059.]

The three-dimensional quadrupole ion-trap mass spectrometer in Figure 21-16 can perform selected reaction monitoring without additional hardware. After a collection of ions of different m/z is injected into the ion trap, all but one m/z are intentionally ejected. Precursor ions with the single m/z remaining in the ion trap are then given increased kinetic energy by increasing the amplitude of the applied radio-frequency electric field. Increased energy promotes collisional dissociation with background He atoms in the ion trap. Product ions are retained in the trap but do not have enough kinetic energy for further collisional dissociation. After a period of dissociation of the precursor ion, product ions are expelled to the detector to record a mass spectrum.

This process can be repeated by selecting a product ion for further dissociation. The repeated process is called MS^n, which denotes multiple repetitions of selected reaction monitoring. The beauty of MS^n with a three-dimensional quadrupole ion trap is that the entire process takes place in one piece of hardware under software control. To conduct more cycles of selected reaction monitoring with the equipment in Figure 21-26, it would be necessary to add another Q2 and Q3 chamber for each cycle, which is not practical.

MS^n: successive cycles of selected reaction monitoring. The product ion of one cycle becomes the precursor ion for the next cycle.

Box 21-4 describes MALDI, the other most useful way to introduce proteins into a mass spectrometer.

Electrospray of Proteins

Electrospray is well suited for the study of charged macromolecules such as proteins.[35,36] It has been used to study intact viruses with masses up to 40 MDa.[37] A typical protein has carboxylic acid and amine side chains (Table 9-1) that give it a net positive or negative charge, depending on pH. Electrospray (Figure 21-21) ejects preexisting ions from solution into the gas phase.

Each peak in the mass spectrum of the protein transferrin in Figure 21-28 arises from molecules with different numbers of protons, MH_n^{n+}.[38] Although we have already labeled charges

FIGURE 21-28 Electrospray time-of-flight mass spectrum of one anion-exchange chromatographic peak containing the protein transferrin with one particular set of carbohydrate substituents. Peaks arise from species with different numbers of protons, MH_n^{n+}. [From M. E. Del Castillo Busto, M. Montes-Bayón, E. Blanco-González, J. Meija, and A. Sanz-Medel, "Strategies to Study Human Serum Transferrin Isoforms Using Integrated Liquid Chromatography ICPMS, MALDI-TOF, and ESI-Q-TOF Detection: Application to Chronic Alcohol Abuse," *Anal. Chem.* **2005**, *77*, 5615.]

BOX 21-4 **Matrix-Assisted Laser Desorption/Ionization**

Major methods for introducing proteins and other macromolecules into mass spectrometers are electrospray and **matrix-assisted laser desorption/ionization (MALDI).**[20,33] Most often, MALDI is used with a time-of-flight mass spectrometer, which can measure m/z up to 10^6. Typically, 1 μL of a 10-μM solution of analyte is mixed with 1 μL of a 1- to 100-mM solution of an ultraviolet-absorbing compound such as 2,5-dihydroxybenzoic acid (the *matrix*) directly on a probe that fits into the source of the spectrometer. Evaporation of the liquid leaves an intimate mixture of fine crystals of matrix plus analyte.

To introduce ions into the gas phase for mass spectrometry, a brief infrared or ultraviolet pulse (600 ps) from a laser is directed onto the sample. The matrix vaporizes and expands into the gas phase, carrying analyte along with it. The high matrix/sample ratio inhibits association between analyte molecules and provides protonated or ionic species that transfer charge to analyte, much of which carries a single charge. Shortly after ions expand into the source, a voltage pulse applied to the backplate expels ions into the spectrometer. The resolving power is 10^3–10^4, and mass accuracy can be 0.005–0.05%. The spectrum below shows proteins from milk that has not undergone any sample preparation except for mixing with the matrix. It is possible to map differences in chemical composition in different regions by directing a laser at different parts of fixed cells, such as neurons.[34]

Sequence of events in matrix-assisted laser desorption/ionization. (*a*) Dried mixture of analyte and matrix on sample probe inserted into backplate of ion source. (*b1*) Enlarged view of laser pulse striking sample. (*b2*) Matrix is ionized and vaporized by laser and transfers some charge to analyte. (*b3*) Vapor expands in a supersonic plume.

Partial mass spectrum of cow's milk (containing 2% milk fat) observed by MALDI/time-of-flight mass spectrometry. [From R. M. Whittal and L. Li, "Time-Lag Focusing MALDI-TOF Mass Spectrometry," *Am. Lab.*, December 1997, p. 30.]

on several peaks, we do not know what these charges are before analyzing the spectrum. If we can find the charge for each species, we can find the molecular mass, M, of the neutral protein.

To find the charge, consider a peak with $m/z = m_n$ derived from the neutral molecule plus n protons:

$$m_n = \frac{\text{mass}}{\text{charge}} = \frac{M + n(1.008)}{n} = \frac{M}{n} + 1.008 \quad \Rightarrow \quad \boxed{m_n - 1.008 = \frac{M}{n}} \quad (21\text{-}4)$$

Mass = mass of protein (M) + mass of n atoms of H ($n \times 1.008$).

The next peak at lower m/z should have $n + 1$ protons and a charge of $n + 1$. For this peak,

$$m_{n+1} = \frac{M + (n+1)(1.008)}{n+1} = \frac{M}{n+1} + 1.008 \quad \Rightarrow \quad \boxed{m_{n+1} - 1.008 = \frac{M}{n+1}} \quad (21\text{-}5)$$

TABLE 21-3 Analysis of electrospray mass spectrum of tetrasialo-transferrin in Figure 21-28

Observed $m/z \equiv m_n$	$m_{n+1} - 1.008$	$m_n - m_{n+1}$	Charge $= n =$ $\dfrac{m_{n+1} - 1.008}{m_n - m_{n+1}}$	Molecular mass $=$ $n \times (m_n - 1.008)$
2 652.722 7	2 566.212 5	85.502 2	$30.013 \approx 30$	79 551.44
2 567.220 5	2 485 975 6	80.236 9	$30.983 \approx 31$	79 552.59
2 486.983 6	2 410.652 6	75.323 0	$32.004 \approx 32$	79 551.22
2 411.660 6	2 339.748 3	70.904 3	$32.999 \approx 33$	79 551.54
2 340.756 3	2 272.901 2	66.847 1	$34.001 \approx 34$	79 551.44
2 273.909 2	2 209.768 9	63.132 3	$35.002 \approx 35$	79 551.54
2 210.776 9	2 150.042 8	59.726 1	$35.998 \approx 36$	79 551.68
2 151.050 8	2 093.482 2	56.560 6	$37.013 \approx 37$	79 551.58
2 094.490 2	2 039.806 3	53.675 9	$38.002 \approx 38$	79 552.32
2 040.814 3	1 988.798 0	51.008 3	$38.990 \approx 39$	79 552.45
1 989.806 0	1 894.143 4		40	79 551.92
				mean $= 79\ 551.78 \pm 0.48$

SOURCE: *M. E. Del Castillo Busto, M. Montes-Bayón, E. Blanco-González, J. Meija, and A. Sanz-Medel, "Strategies to Study Human Serum Transferrin Isoforms Using Integrated Liquid Chromatography ICPMS, MALDI-TOF, and ESI-Q-TOF Detection: Application to Chronic Alcohol Abuse," Anal. Chem.* **2005,** *77, 5615.*

Forming the quotient of the expressions in the boxes, we get

$$\frac{m_n - 1.008}{m_{n+1} - 1.008} = \frac{M/n}{M/(n+1)} = \frac{n+1}{n} \qquad (21\text{-}6)$$

Solving Equation 21-6 for n gives the charge on peak m_n:

$$n = \frac{m_{n+1} - 1.008}{m_n - m_{n+1}} \qquad (21\text{-}7)$$

The fourth column in Table 21-3 shows the charge of each peak, n, calculated with Equation 21-7. The charge of the peak at m/z 2 652.722 7 is $n = 30$. We assign this peak as MH_{30}^{30+}. The next peak at m/z 2 567.220 5 is MH_{31}^{31+}, and so on. We find highly protonated species because the chromatography solvent was acidic (95 vol% acetonitrile + 5 vol% H_2O + 0.2 vol% formic acid). Even without adding formic acid, electrolysis in positive-ion electrospray produces H^+ by oxidation of H_2O.

From any peak, we can find the mass of the neutral molecule by rearranging the right side of Equation 21-4:

$$M = n \times (m_n - 1.008) \qquad (21\text{-}8)$$

Reproducible masses computed with Equation 21-8 appear in the last column of Table 21-3. A limitation on the accuracy of molecular mass determination is the accuracy of the m/z scale. For this work, m/z was calibrated with an external standard of poly(propylene glycol).

Electron-Transfer Dissociation for Protein Sequencing

We saw at the opening of Chapter 9 that proteins are chains of amino acids held together by amide (also called peptide) bonds. Proteins are synthesized on *ribosomes*, which are assemblies of RNA and protein whose job is to translate the sequence of DNA into a corresponding sequence of amino acids. After synthesis, some proteins are specifically modified by enzymes to add groups such as acetate, phosphate, carbohydrates, and lipids to specific amino acid side chains. A branch of biochemistry called *proteomics* seeks to characterize the structure and function of the full complement of proteins in an organism.

Mass spectrometry is a principal tool for deducing the sequence of amino acids in a protein. Protein is digested into shorter chains by enzymatic cleavage. Individual chains are cleaved to make fragments of all possible lengths. High-resolution mass spectrometry allows us to deduce which amino acids are in each fragment. A computer can use this information to reconstruct the sequence of amino acids.

Electron-transfer dissociation is a selective way to cleave polypeptides into fragments in a mass spectrometer.[39] This process involves the exothermic transfer of an electron from a gas-phase anion to a gas-phase polypeptide cation with concomitant bond breaking. In the chain of amino acids below,

cleavage occurs at each indicated position to form 10 possible charged fragments designated c_1 to c_5 and z_1 to z_5. In the c fragments, charge resides on the fragment to the left of the broken bond. In the z fragments, charge resides on the fragment to the right of the broken bond. Unlike collisionally activated dissociation, electron-transfer dissociation does not break other bonds in the peptide chain or the side groups, including groups such as phosphate or carbohydrates bound to side chains.

The mass spectrometer in Figure 21-19, when equipped with two electrospray sources, can carry out electron-transfer dissociation for polypeptide sequencing.[40] Polypeptide cations produced by electrospray for 0.2 s are collected in the linear ion trap and stored in the downstream section by application of appropriate voltages. The polypeptide source is then turned off and, after 0.4 s, 9-anthracenecarboxylic acid solution is electrosprayed from a second source for 0.2 s. Anthracenecarboxylate anions are collected in the upstream section of the linear ion trap. Collisionally activated decarboxylation in the ion trap produces anion A^-:

9-Anthracenecarboxylate A^-

Potentials are then adjusted to allow anions and cations that had been at opposite ends of the ion trap to mix with one another. A^- transfers an electron to a polypeptide P^{n+}, thereby inducing electron-transfer dissociation of one peptide bond:

$$A^- + P^{n+} \rightarrow A + P^{(n-1)+} \rightarrow \text{cleavage of one poptide bond}$$

Different bonds are cleaved in individual molecules to make all possible c and z fragments drawn above. The reaction is quenched by ejecting anions from the ion trap. Finally, peptide cations are expelled and exact m/z values are measured with the orbitrap in Figure 21-19. With m/z accuracy of 1 part per million, the majority of electron-transfer dissociation product ions with $m/z < 1000$ can be uniquely assigned and identified as N- or C-terminal peptides.[41]

21-5 Open-Air Sampling for Mass Spectrometry

Since 2004, new mass spectral sampling techniques have been introduced to vaporize and ionize analytes directly from the surface of an object in the open air with little harm to the object. This sampling capability opens up new vistas for mass spectrometry in qualitative analysis.

Direct Analysis in Real Time (DART)

A **direct analysis in real time** (DART) source produces electronically excited He or vibrationally excited N_2, which are directed at the surface of an object to be sampled in open air. In Figure 21-29, heated gas flows through a needle electrode held at $+1$ to $+5$ kV with respect to a grounded counterelectrode in the form of a perforated disk. The glow discharge plasma contains electrons, ions, and excited neutral species. Electrodes 1 and 2 are held at positive potentials for positive ion mass spectrometry and at negative potentials for negative ion mass spectrometry. At positive potential, electrodes 1 and 2 prevent cations from exiting the DART source. At negative potential, anions and electrons are retained.

The DART gun is directed at an object to be sampled. With a helium source, excited He atoms with an energy of 19.8 eV (designated 2^3S and shown as He* in the margin) react with atmospheric water vapor to create protonated water clusters. These clusters can react with analyte M on the surface of an object to create MH^+. Other chemistry can produce $(M-H)^-$, M^-, or adducts such as $(M+NH_4)^+$ or $(M+Cl)^-$.

$He^* + H_2O \rightarrow H_2O^+ + He + e^-$
$H_2O^+ + H_2O \rightarrow H_3O^+ + OH$
$H_3O^+ + (n-1)H_2O \rightarrow (H_2O)_nH^+$

FIGURE 21-29 Direct Analysis in Real Time (DART) source. [Adapted from R. B. Cody, J. A. Laramée, and H. D. Durst, "Versatile New Ion Source for the Analysis of Materials in Open Air Under Ambient Conditions," *Anal. Chem.* **2005,** *77,* 2297 and JEOL USA, Peabody, MA.]

If the sample is a poppy seed from a bagel, two major species observed by a high-resolution time-of-flight mass spectrometer are the protonated molecules, (morphine)H$^+$ (C$_{17}$H$_{19}$NO$_3$H$^+$ at *m/z* 286.144 3) and (codeine)H$^+$ (C$_{18}$H$_{21}$NO$_3$H$^+$ at *m/z* 300.161 1).[42] Separate quantitative analysis shows that poppy seeds contain ~33 and ~14 µg/g (ppm) of morphine and codeine, respectively.

Low-Temperature Plasma

In a technique related to DART, a low-temperature plasma is created by passing He, Ar, N$_2$, or ambient air through a glass tube with a grounded wire at the center (Figure 21-30 and Color Plate 26). The tube is wrapped on the outside with a copper sheath to which is applied a 3-kV alternating current. Excited-state species in the plasma ionize and dislodge molecules from a surface such as human skin into the source of a mass spectrometer. There is no electrical shock to the surface under study.

FIGURE 21-30 Low-temperature plasma for sampling a surface in the open air. [From J. D. Harper, N. A. Charipar, C. C. Mulligan, X. Zhang, R. G. Cooks, and Z. Ouyang, "Low-Temperature Plasma Probe for Ambient Desorption Ionization," *Anal. Chem.* **2008,** *80,* 9097.]

Desorption Electrospray Ionization (DESI)

In **desorption electrospray ionization** (DESI), micron-sized, charged droplets created by electrospray of analyte-free solvent (Figure 21-21) are directed onto the surface of an object under study.[43] Analyte on the surface dissolves in the droplets. Further bombardment knocks droplets into the air toward a mass spectrometer inlet. As in conventional electrospray, it is common to observe multiply charged ions and alkali metal adducts in the mass spectrum.

Figure 21-31 shows DESI mapping of inks on a page. A mixture of methanol and water is electrosprayed toward the paper, just 2 mm away from the spray tip. The mass spectrum (*m/z* 150–600) of rebounded droplets is recorded every 0.67 s. The paper is translated in small steps to make a two-dimensional map. The mass spectral signal observed at a blank area of the paper is subtracted from the signal observed from an inked area to obtain the spectrum of the ink.

In this example, the date "1432" was written with a blue pen. Then a second blue pen was used to change 4 to 9 and 3 to 8, so that the date would be "1982." The second pen had a blue ink different from that in the first pen, but the difference is subtle. The objective of the demonstration was to show that the original date had been altered.

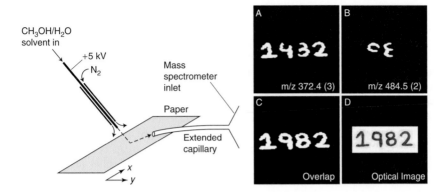

FIGURE 21-31 Desorption ElectroSpray Ionization (DESI) mapping of inks. Images A, B, and C at the right are maps made by scanning the page in the *x* and *y* directions beneath the electrospray source. D is an optical image of the surface. [From D. R. Ifa, L. M. Gumaelius, L. S. Eberlin, N. E. Manicke, and R. G. Cooks, "Forensic Analysis of Inks by Imaging Desorption Electrospray Ionization (DESI) Mass Spectrometry," *Analyst* **2007**, *132*, 461.]

Basic Violet 3
$^{12}C_{25}\,^{1}H_{30}\,^{14}N_3^+$ *m/z* 372.243

Solvent Blue 2
$^{12}C_{34}\,^{1}H_{34}\,^{14}N_3^+$ *m/z* 484.275

The first ink was Basic Violet 3, with a parent ion M^+ at *m/z* 372.243. The second ink was Solvent Blue 2, with a parent ion at *m/z* 484.275. Map A in Figure 21-31 shows extracted-ion monitoring of *m/z* 372.4 with unit mass resolution. White shows where *m/z* 372.4 is maximal and black shows where *m/z* 372.4 is minimal. Therefore, map A shows where Basic Violet 3 ink is located. Map B is made from extracted-ion monitoring of *m/z* 484.5 with unit resolution to show where Solvent Blue 2 ink is located. We see that the second ink was written over the first ink to change "1432" to "1982." Map C is the sum of maps A and B. Frame D is an optical image of the writing.

Terms to Understand

atomic mass
atmospheric pressure chemical
  ionization
base peak
chemical ionization
collisionally activated dissociation
desorption electrospray
  ionization (DESI)
direct analysis in real time
  (DART)
double-focusing mass
  spectrometer

electron ionization
electron-transfer dissociation
electrospray ionization
extracted ion chromatogram
ion mobility spectrometer
linear quadrupole ion-trap mass
  spectrometer
magnetic sector mass
  spectrometer
mass spectrometry
mass spectrum

matrix-assisted laser desorption/
  ionization (MALDI)
molecular ion
molecular mass
nitrogen rule
nominal mass
orbitrap mass spectrometer
precursor ion
product ion
reconstructed total ion
  chromatogram

resolving power
rings + double bonds formula
selected ion chromatogram
selected ion monitoring
selected reaction monitoring
three-dimensional quadrupole
  ion-trap mass spectrometer
time-of-flight mass spectrometer
transmission quadrupole mass
  spectrometer

Summary

Ions are created or desorbed in the ion source of a mass spectrometer. Neutral molecules are converted into ions by electron ionization (which produces a molecular ion, $M^{+\bullet}$, and many fragments) or by chemical ionization (which tends to create MH^+ and few fragments). A magnetic sector mass spectrometer separates gaseous ions by accelerating them in an electric field and deflecting ions of different mass-to-charge (*m/z*) ratio through different arcs. Ions are detected by an electron multiplier, which works like a photomultiplier tube. The mass spectrum is a graph of detector response versus *m/z* value. A double-focusing mass spectrometer attains high resolution by employing an electric sector with the magnetic sector to select ions with a narrow range of kinetic energy. Other mass separators include the transmission quadrupole mass spectrometer, the time-of-flight mass spectrometer, the three-dimensional quadrupole ion trap, the linear ion trap, and the orbitrap.

The time-of-flight instrument is capable of high acquisition rates and has a nearly unlimited upper mass range. Resolving power is defined as $m/\Delta m$ or $m/m_{1/2}$, where m is the mass being measured, Δm is the difference in mass between two peaks that are separated with a 10% valley between them, and $m_{1/2}$ is the width of a peak at half-height. Time-of-flight and orbitrap spectrometers can provide high-resolution spectra. An ion mobility spectrometer separates gas-phase ions by their different mobilities in an electric field at atmospheric pressure.

In a mass spectrum, the molecular ion is found from the highest *m/z* value of any "significant" peak that cannot be attributed to isotopes or background signals. For a given composition, you should be able to predict the relative intensities of the isotopic peaks at M+1, M+2, and so on. Among common elements, Cl and Br have particularly diagnostic isotope patterns. From a molecular composition, the rings + double bonds equation helps us propose structures. An organic compound with an odd number of nitrogen atoms will have an odd mass. Fragment ions arising from bond cleavage and rearrangements provide clues to molecular structure.

Gas emerging from a capillary gas chromatography column can go directly into the ion source of a well-pumped mass spectrometer to provide qualitative and quantitative information about the components of a mixture. For liquid chromatography, atmospheric pressure chemical ionization utilizes a corona discharge needle to create a variety of gaseous ions. Alternatively, electrospray employs high voltage at the exit of the column, combined with coaxial N_2 gas flow, to create a fine aerosol containing charged species that were already present in the liquid phase. Analyte is often associated with other ions to give species such as $[MNa]^+$ or $[M(CH_3CO_2)]^-$. Control of pH helps ensure that selected analytes are in anionic or cationic form. Both atmospheric pressure chemical ionization and electrospray tend to produce unfragmented ions. Collisionally activated dissociation to produce fragment ions is controlled by the cone voltage at the mass spectrometer inlet. Electrospray of proteins typically creates an array of highly charged ions such as MH_n^{n+}. Matrix-assisted laser desorp-

tion/ionization is a gentle way to produce predominantly singly charged, intact protein ions for mass spectrometry.

A reconstructed total ion chromatogram shows the signal from all ions above a chosen m/z emerging from chromatography as a function of time. An extracted ion chromatogram shows the signal for one ion taken from the complete mass spectrum. Selected ion monitoring of one or a few values of m/z improves signal-to-noise because all of the time is spent measuring just that one or a few ions. In selected reaction monitoring, a precursor ion isolated by one mass filter passes into a collision cell in which it breaks into products. One (or more) product ion is then selected by a second mass filter for passage to the detector. This process is extremely selective for just one analyte and vastly increases the signal-to-noise ratio for this analyte. Electron-transfer dissociation is employed in protein sequencing to break amide bonds in a polypeptide without breaking other bonds.

Several methods can ionize molecules from the surface of an object in ambient atmosphere. Direct analysis in real time (DART) and a low-temperature plasma use excited-state helium or nitrogen to ionize analytes. Desorption electrospray ionization (DESI) directs electrosprayed solvent onto a surface to dislodge ions.

Exercises

21-A. Measure the width at half-height of the peak at m/z 53 and calculate the resolving power of the spectrometer from the expression $m/m_{1/2}$. Would you expect to be able to resolve two peaks at 100 and 101 Da?

Mass spectrum. [From V. J. Angelico, S. A. Mitchell, and V. H. Wysocki, "Low-Energy Ion–Surface Reactions of Pyrazine with Two Classes of Self-Assembled Monolayers," *Anal. Chem.* **2000**, *72*, 2603.]

21-B. What resolving power is required to distinguish $CH_3CH_2^+$ from $HC\equiv O^+$?

21-C. *Isotope patterns.* Consider an element with two isotopes whose natural abundances are a and b ($a + b = 1$). If there are n atoms of the element in a compound, the probability of finding each combination of isotopes is derived from the expansion of the binomial $(a + b)^n$. For carbon, the abundances are $a = 0.989\ 3$ for ^{12}C and $b = 0.010\ 7$ for ^{13}C. The probability of finding 2 ^{12}C atoms in acetylene, $HC\equiv CH$, is given by the first term of the expansion of $(a + b)^2 = a^2 + 2ab + b^2$. The value of a^2 is $(0.989\ 3)^2 = 0.978\ 7$, so the probability of finding two ^{12}C atoms in acetylene is $0.978\ 7$. The probability of finding 1 ^{12}C + 1 ^{13}C is $2ab = 2(0.989\ 3)(0.010\ 7) = 0.021\ 2$. The probability of finding 2 ^{13}C is $(0.010\ 7)^2 = 0.000\ 114$. The molecular ion, by definition, contains 2 ^{12}C atoms. The M+1 peak contains 1 ^{12}C + 1 ^{13}C. The intensity of M+1 relative to $M^{+\cdot}$ will be $(0.021\ 2)/(0.978\ 7) = 0.021\ 7$. (We are ignoring ^2H because its natural abundance is small.) Predict the relative amounts of $C_6H_4{}^{35}Cl_2$, $C_6H_4{}^{35}Cl^{37}Cl$, and $C_6H_4{}^{37}Cl_2$ in 1,2-dichlorobenzene. Draw a stick diagram of the distribution, like Figure 21-6.

21-D. (a) Find the number of rings plus double bonds in a molecule with the composition $C_{14}H_{12}$ and draw one plausible structure.

(b) For an ion or radical, the rings + double bonds formula gives noninteger answers because the formula is based on valences in neutral molecules with all electrons paired. How many rings plus double bonds are predicted for $C_4H_{10}NO^+$? Draw one structure for $C_4H_{10}NO^+$.

21-E. (a) Spectra A and B belong to the isomers of $C_6H_{12}O$ below. Explain how you can tell which isomer goes with each spectrum.

Mass spectra of isomeric ketones with the composition $C_6H_{12}O$. [From NIST/EPA/NIH Mass Spectral Database.[10]]

(b) The intensity of the M+1 peak at m/z 101 must be incorrect in both spectra. It is entirely missing in spectrum A and too intense (15.6% of intensity of $M^{+\cdot}$) in spectrum B. What should be the intensity of M+1 relative to $M^{+\cdot}$ for the composition $C_6H_{12}O$?

21-F. (This is a long exercise suitable for group work.) Relative intensities for the molecular ion region of several compounds are listed in parts (a)–(d) and shown in the figure. Suggest a composition for each molecule and calculate the expected isotopic peak intensities.

(a) m/z (intensity): 94 (999), 95 (68), 96 (3)

(b) m/z (intensity): 156 (566), 157 (46), 158 (520), 159 (35)

(c) m/z (intensity): 224 (791), 225 (63), 226 (754), 227 (60), 228 (264), 229 (19), 230 (29)

(d) m/z (intensity): 154 (122), 155 (9), 156 (12) (*Hint:* Contains sulfur.)

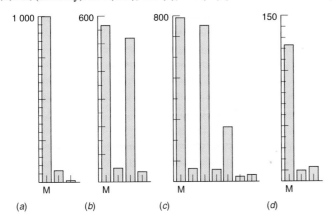

Mass spectra. [From NIST/EPA/NIH Mass Spectral Database.[10]]

21-G. *Protein molecular mass from electrospray.* The enzyme lysozyme[44] exhibits MH_n^{n+} peaks at $m/z = 1\,789.1, 1\,590.4, 1\,431.5, 1\,301.5$, and $1\,193.1$. Follow the procedure of Table 21-3 to find the mean molecular mass and its standard deviation.

21-H. *Quantitative analysis by selected ion monitoring.* Caffeine in beverages and urine can be measured by adding caffeine-D_3 as an internal standard and using selected ion monitoring to measure each compound by gas chromatography. The figure shows mass chromatograms of caffeine (m/z 194) and caffeine-D_3 (m/z 197), which have nearly the same retention time.

Selected ion monitoring mass chromatogram showing caffeine and caffeine-D_3 eluted from a capillary gas chromatography column. [From D. W. Hill, B. T. McSharry, and L. S. Trzupek, "Quantitative Analysis by Isotopic Dilution Using Mass Spectrometry," *J. Chem. Ed.* **1988,** *65,* 907.]

Caffeine (M = 194 Da)   Caffeine-D_3 (M = 197 Da)

Suppose that the following data were obtained for standard mixtures:

Caffeine (mg/L)	Caffeine-D_3 (mg/L)	Caffeine peak area	Caffeine-D_3 peak area
13.60×10^2	3.70×10^2	11 438	2 992
6.80×10^2	3.70×10^2	6 068	3 237
3.40×10^2	3.70×10^2	2 755	2 819

NOTE: *Injected volume was different in all three runs.*

(a) Compute the mean response factor in the equation

$$\frac{\text{Area of analyte signal}}{\text{Area of standard signal}} = F\left(\frac{\text{concentration of analyte}}{\text{concentration of standard}}\right)$$

(b) For analysis of a cola beverage, 1.000 mL of beverage was treated with 50.0 μL of standard solution containing 1.11 g/L caffeine-D_3 in methanol. The combined solution was passed through a solid-phase extraction cartridge that retains caffeine. Polar solutes were washed off with water. Then the caffeine was washed off the cartridge with an organic solvent and the solvent was evaporated to dryness. The residue was dissolved in 50 μL of methanol for gas chromatography. Peak areas were 1 144 for m/z 197 and 1 733 for m/z 194. Find the concentration of caffeine (mg/L) in the beverage.

Problems

What Is Mass Spectrometry?

21-1. Briefly describe how a magnetic sector mass spectrometer works.

21-2. How are ions created for each of the mass spectra in Figure 21-4? Why are the two spectra so different?

21-3. Define the unit *dalton*. From this definition, compute the mass of 1 Da in grams. The mean of 60 measurements of the mass of individual *E. coli* cells vaporized by MALDI and measured with a quadrupole ion trap was 5.03 ($\pm$0.14) $\times 10^{10}$ Da.[3] Express this mass in femtograms.

21-4. Nickel has two major and three minor isotopes. For the purpose of this problem, suppose that the *only* isotopes are ^{58}Ni and ^{60}Ni. The atomic mass of ^{58}Ni is 57.935 3 Da and the mass of ^{60}Ni is 59.933 2 Da. From the amplitude of the peaks in the following spectrum, calculate the atomic mass of Ni and compare your answer with the value in the periodic table.

Mass spectrum. [From Y. Su, Y. Duan, and Z. Jin, "Helium Plasma Source Time-of-Flight Mass Spectrometry: Off-Cone Sampling for Elemental Analysis," *Anal. Chem.* **2000**, *72*, 2455.]

21-5. Measure the width at half-height of the tallest peak in the spectrum below and calculate the resolving power of the spectrometer from the expression $m/m_{1/2}$. Would you expect to be able to distinguish two peaks at 10 000 and 10 001 Da?

MALDI mass spectrum of the peptide melittin. [From P. B. O'Connor and C. E. Costello, "Application of Multishot Acquisition in Fourier Transform Mass Spectrometry," *Anal. Chem.* **2000**, *72*, 5125.]

21-6. The two peaks near m/z 31.00 in Figure 21-9 differ in mass by 0.010 Da. *Estimate* the resolving power of the spectrometer from the expression $m/\Delta m$ without making any measurements in the figure.

21-7. The highest resolution mass spectra are obtained by Fourier transform ion cyclotron resonance mass spectrometry.[45] Molecular ions of two peptides (chains of seven amino acids) differing in mass by 0.000 45 Da were separated with a 10% valley between them. The ions each have a mass of 906.49 Da and a width at half-height of 0.000 27 Da. Compute the resolving power by the 10% valley formula and by the half-width formula, both in Equation 21-1. Compare the difference in mass of these two compounds with the mass of an electron.

The Mass Spectrum

21-8. The mass of a fragment ion in a high-resolution spectrum is 83.086 5 Da. Which composition, $C_5H_7O^+$ or $C_6H_{11}^+$, better matches the observed mass?

21-9. Calculate the theoretical masses of the species in Figure 21-9 and compare your answers with the values observed in the figure.

21-10. *Isotope patterns.* Referring to Exercise 21-C, predict the relative amounts of $C_2H_2{}^{79}Br_2$, $C_2H_2{}^{79}Br^{81}Br$, and $C_2H_2{}^{81}Br_2$ in 1,2-dibromoethylene. Compare your answer with Figure 21-7.

21-11. *Isotope patterns.* Referring to Exercise 21-C, predict the relative abundances of $^{10}B_2H_6$, $^{10}B^{11}BH_6$, and $^{11}B_2H_6$ for diborane (B_2H_6).

21-12. Find the number of rings plus double bonds in molecules with the following compositions and draw one plausible structure for each: **(a)** $C_{11}H_{18}N_2O_3$; **(b)** $C_{12}H_{15}BrNPOS$; **(c)** fragment in a mass spectrum with the composition $C_3H_5^+$.

21-13. (Each part of this problem is quite long and best worked by groups of students.) Peak intensities of the molecular ion region are listed in parts **(a)**–**(g)** and shown in the figure. Identify which peak represents the molecular ion, suggest a composition for it, and calculate the expected isotopic peak intensities. Restrict your attention to elements in Table 21-1.

(a) m/z (intensity): 112 (999), 113 (69), 114 (329), 115 (21)

(b) m/z (intensity): 146 (999), 147 (56), 148 (624), 149 (33), 150 (99), 151 (5)

(c) m/z (intensity): 90 (2), 91 (13), 92 (96), 93 (999), 94 (71), 95 (2)

(d) m/z (intensity): 226 (4), 227 (6), 228 (130), 229 (215), 230 (291), 231 (168), 232 (366), 233 (2), 234 (83) (Calculate expected intensities from isotopes of the major element present.)

(e) m/z (intensity): 172 (531), 173 (12), 174 (999), 175 (10), 176 (497)

(f) m/z (intensity): 177 (3), 178 (9), 179 (422), 180 (999), 181 (138), 182 (9)

(g) m/z (intensity): 182 (4), 183 (1), 184 (83), 185 (16), 186 (999), 187 (132), 188 (10)

Mass spectra for Problem 21-13. [From NIST/EPA/NIH Mass Spectral Database.[10]]

21-14. Suggest a composition for the halogen compound whose mass spectrum is shown in the figure. Assign each of the major peaks.

Mass spectrum for Problem 21-14. [From NIST/EPA/NIH Mass Spectral Database.[10]]

21-15. Box 21-3 (page 509) shows that CO_2 in human breath in the United States has a value of $\delta^{13}C$ different from that of CO_2 in human breath in mainland Europe. Suggest an explanation.

21-16. **(a)** The mass of ^1H in Table 21-1 is 1.007 825 Da. Compare it with the sum of the masses of a proton and an electron given in the table.

(b) ^2H (deuterium) contains one proton, one neutron, and one electron. Compare the sum of the masses of these three particles with the mass of ^2H.

(c) The discrepancy in part **(b)** comes from the conversion of mass into binding energy that holds the nucleus together. The relation of mass, m, to energy, E, is $E = mc^2$, where c is the speed of light. From the discrepancy in part **(b)**, calculate the binding energy of ^2H in joules and in kJ/mol. (1 Da $= 1.660\ 5 \times 10^{-27}$ kg)

(d) The binding energy (ionization energy) of the electron in a hydrogen or deuterium atom is 13.6 eV. Use Table 1-4 to convert this number into kJ/mol and compare it with the binding energy of the ^2H nucleus.

(e) A typical bond dissociation energy in a molecule is 400 kJ/mol. How many times larger is the nuclear binding energy of ^2H than a bond energy?

21-17. *Isotope patterns.* From the natural abundance of ^{79}Br and ^{81}Br, predict the relative amounts of $CH^{79}Br_3$, $CH^{79}Br_2{}^{81}Br$, $CH^{79}Br^{81}Br_2$, and $CH^{81}Br_3$. As in Exercise 21-C, the fraction of each isotopic molecule comes from the expansion of $(a + b)^3$, where a is the abundance of ^{79}Br and b is the abundance of ^{81}Br. Note that $(a + b)^n =$

$$a^n - na^{n-1}b + \frac{n(n-1)}{2!} a^{n-2}b^2 + \frac{n(n-1)(n-2)}{3!} a^{n-3}b^3 + \cdots.$$

Compare your answer with Figure 21-7.

21-18. 🖳 *Isotope patterns.* (*Caution:* This problem could lead to serious brain injury.) For an element with three isotopes with abundances a, b, and c, the distribution of isotopes in a molecule with n atoms is based on the expansion of $(a + b + c)^n$. Predict what the mass spectrum of Si_2 will look like.

Mass Spectrometers

21-19. Explain how a double-focusing mass spectrometer achieves high resolution.

21-20. A limitation on how many spectra per second can be recorded by a time-of-flight mass spectrometer is the time it takes the slowest ions to go from the source to the detector. Suppose we want to scan up to m/z 500. Calculate the speed of this heaviest ion if it is accelerated through 5.00 kV in the source. How long would it take to drift 2.00 m through a spectrometer? At what frequency could you record spectra if a new extraction cycle were begun each

time this heaviest ion reached the detector? What would be the frequency if you wanted to scan up to m/z 1 000?

21-21. What is the purpose of the reflectron in a time-of-flight mass spectrometer?

21-22. **(a)** The *mean free path* is the average distance a molecule travels before colliding with another molecule. The mean free path, λ, is given by $\lambda = kT/(\sqrt{2}\sigma P)$, where k is Boltzmann's constant, T is temperature (K), P is pressure (Pa), and σ is the collision cross section. For a molecule with a diameter d, the collision cross section is πd^2. The collision cross section is the area swept out by the molecule within which it will strike any other molecule it encounters. The magnetic sector mass spectrometer is maintained at a pressure of ~10^{-5} Pa (~10^{-7} Torr) so that ions do not collide with (and deflect) one another as they travel through the mass analyzer. What is the mean free path of a molecule with a diameter of 1 nm at 300 K in the mass analyzer?

(b) The vacuum in an orbitrap mass separator is ~10^{-8} Pa (~10^{-13} bar, ~10^{-10} Torr). Find the mean free path in the orbitrap for the same conditions as **(a)**.

Chromatography/Mass Spectrometry

21-23. Which liquid chromatography/mass spectrometry interface, atmospheric pressure chemical ionization or electrospray, requires analyte ions to be in solution prior to the interface? How does the other interface create gaseous ions from neutral species in solution?

21-24. What is collisionally activated dissociation? At what points in a mass spectrometer does it occur?

21-25. What is the difference between a reconstructed total ion chromatogram, an extracted ion chromatogram, and a selected ion chromatogram?

21-26. What is selected reaction monitoring? Why is it also called MS/MS? Why does it improve the signal/noise ratio for a particular analyte?

21-27. **(a)** To detect the drug ibuprofen by liquid chromatography/mass spectrometry, would you choose the positive or negative ion mode for the spectrometer? Would you choose acidic or neutral chromatography solvent? State your reasons.

Ibuprofen (FM 206)

(b) If the unfragmented ion has an intensity of 100, what should be the intensity of M+1?

21-28. An electrospray/transmission quadrupole mass spectrum of the α chain of hemoglobin from acidic solution exhibits nine peaks corresponding to MH_n^{n+}. Find the charge, n, for peaks A–I. Calculate the molecular mass of the neutral protein, M, from peaks A, B, G, H, and I, and find the mean value.

Peak	m/z	Amplitude	Peak	m/z	Amplitude
A	1 261.5	0.024	F	not stated	1.000
B	1 164.6	0.209	G	834.3	0.959
C	not stated	0.528	H	797.1	0.546
D	not stated	0.922	I	757.2	0.189
E	not stated	0.959			

21-29. The molecular ion region in the mass spectrum of a large molecule, such as a protein, consists of a cluster of peaks differing by 1 Da. This pattern occurs because a molecule with many atoms has a high probability of containing one or several atoms of ^{13}C, ^{15}N, ^{18}O, ^{2}H, and ^{34}S. In fact, the probability of finding a molecule with only ^{12}C, ^{14}N, ^{16}O, ^{1}H, and ^{32}S may be so small that the nominal molecular ion is not observed. The electrospray mass spectrum of the rat protein interleukin-8 consists of a series of clusters of peaks arising from intact molecular ions with different charges. One cluster has peaks at m/z 1 961.12, 1 961.35, 1 961.63, 1 961.88, 1 962.12 (tallest peak), 1 962.36, 1 962.60, 1 962.87, 1 963.10, 1 963.34, 1 963.59, 1 963.85, and 1 964.09. These peaks correspond to isotopic ions differing by 1 Da. From the observed peak separation, find the charge of the ions in this cluster. From m/z of the tallest peak, estimate the molecular mass of the protein.

21-30. Phytoplankton at the ocean surface maintain the fluidity of their cell membranes by altering their lipid (fat) composition when the temperature changes. When the ocean temperature is high, plankton synthesize relatively more 37:2 than 37:3.[46]

O
‖
⟍(CH₂)₁₁⟍⟍(CH₂)₅⟍⟍(CH₂)₁₃CH₃

$$37:2 = C_{37}H_{70}O$$

O
‖
⟍(CH₂)₄⟍⟍(CH₂)₅⟍⟍(CH₂)₅⟍⟍(CH₂)₁₃CH₃

$$37:3 = C_{37}H_{68}O$$

After they die, plankton sink to the ocean floor and end up buried in sediment. The deeper we sample a sediment, the further back into time we delve. By measuring the relative quantities of cell-membrane compounds at different depths in the sediment, we can infer the temperature of the ocean long ago.

The molecular ion regions of the chemical ionization mass spectra of 37:2 and 37:3 are listed in the table. Predict the expected intensities of M, M+1, and M+2 for each of the four species listed. Include contributions from C, H, O, and N, as appropriate. Compare your predictions with the observed values. Discrepant intensities in these data are typical unless care is taken to obtain high-quality data.

Compound	Species in mass spectrum	M	M+1	M+2
37:3	$[MNH_4]^+$ (m/z 546)[a]	100	35.8	7.0
37:3	$[MH]^+$ (m/z 529)[b]	100	23.0	8.0
37:2	$[MNH_4]^+$ (m/z 548)[a]	100	40.8	3.7
37:2	$[MH]^+$ (m/z 531)[b]	100	33.4	8.4

a. Chemical ionization with ammonia.
b. Chemical ionization with isobutane.

21-31. Chlorate (ClO_3^-), chlorite (ClO_2^-), bromate (BrO_3^-), and iodate (IO_3^-) can be measured in drinking water at the 1-ppb level with 1% precision by selected reaction monitoring.[47] Chlorate and chlorite arise from ClO_2 used as a disinfectant. Bromate and iodate can be formed from Br^- or I^- when water is disinfected with ozone (O_3). For the highly selective measurement of chlorate, the negative ion selected by Q1 in Figure 21-26 is m/z 83 and the negative ion selected by Q3 is m/z 67. Explain how this measurement works and how it distinguishes ClO_3^- from ClO_2^-, BrO_3^-, and IO_3^-.

21-32. *Quantitative analysis by isotope dilution.* In *isotope dilution*, a known amount of an unusual isotope (called the *spike*) is added to an unknown as an internal standard for quantitative analysis. After the mixture has been homogenized, some of the element of interest must be isolated. The ratio of the isotopes is then measured. From this ratio, the quantity of the element in the original unknown can be calculated.

Natural vanadium has atom fractions $^{51}V = 0.997\ 5$ and $^{50}V = 0.002\ 5$. The atom fraction is defined as

$$\text{Atom fraction of } ^{51}V = \frac{\text{atoms of } ^{51}V}{\text{atoms of } ^{50}V + \text{atoms of } ^{51}V}$$

A spike enriched in ^{50}V has atom fractions $^{51}V = 0.639\ 1$ and $^{50}V = 0.360\ 9$.

(a) Let isotope A be ^{51}V and isotope B be ^{50}V. Let A_x be the atom fraction of isotope A (= atoms of A/[atoms of A + atoms of B]) in an unknown. Let B_x be the atom fraction of B in an unknown. Let A_s and B_s be the corresponding atom fractions in a spike. Let C_x be the total concentration of all isotopes of vanadium (μmol/g) in the unknown and let C_s be the concentration in the spike. Let m_x be the mass of unknown and m_s be the mass of spike. After m_x grams of unknown are mixed with m_s grams of spike, the ratio of isotopes in the mixture is found to be R. Show that

$$R = \frac{\text{mol A}}{\text{mol B}} = \frac{A_x C_x m_x + A_s C_s m_s}{B_x C_x m_x + B_s C_s m_s} \quad \text{(A)}$$

(b) Solve Equation A for C_x to show that

$$C_x = \left(\frac{C_s m_s}{m_x}\right)\left(\frac{A_s - RB_s}{RB_x - A_x}\right) \quad \text{(B)}$$

(c) A 0.401 67-g sample of crude oil containing an unknown concentration of natural vanadium was mixed with 0.419 46 g of spike containing 2.243 5 μmol V/g enriched with ^{50}V (atom fractions: $^{51}V = 0.639\ 1$, $^{50}V = 0.360\ 9$).[48] After dissolution and equilibration of the oil and the spike, some of the vanadium was isolated by ion-exchange chromatography. The measured isotope ratio in the isolated vanadium was $R = {}^{51}V/{}^{50}V = 10.545$. Find the concentration of vanadium (μmol/g) in the crude oil.

(d) Examine the calculation in part (c) and express the answer with the correct number of significant figures.

21-33. *A literature project.* Until the 1960s, dinosaurs were thought to be cold-blooded animals, which means they could not regulate their body temperature. Reference 1 describes how the $^{18}O/^{16}O$ ratio in dinosaur bones suggests that some species were warm blooded. Find reference 1, preferably at http://pubs.acs.org/ac if your institution has an electronic subscription to *Analytical Chemistry*. Explain how the $^{18}O/^{16}O$ ratio implies that an animal is warm or cold blooded. Explain the criteria that were used to determine the likelihood that $^{18}O/^{16}O$ in bone phosphate was altered after the dinosaur died. Describe how bone samples were prepared for analysis of oxygen isotopes and state the results of the measurements.

MEASURING SILICONES LEAKING FROM BREAST IMPLANTS

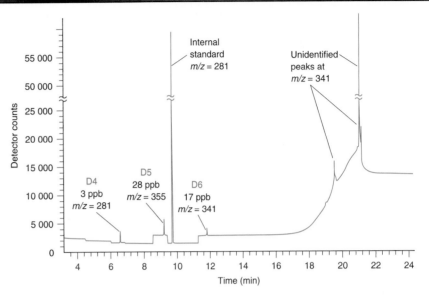

Selected ion monitoring gas chromatogram of plasma extract showing traces of siloxanes 5 years after 5-year-old breast implants were removed. [From D. Flassbeck, B. Pfleiderer, R. Grumping, and A. V. Hirner, "Determination of Low Molecular Weight Silicones in Women After Exposure to Breast Implants by GC/MS," *Anal. Chem.* **2001**, *73*, 606.]

Name:	D3	D4	D5	D6
Molecular mass:	222	296	370	444
Main fragments:	**207**	**281**	73, 267, **355**	73, **341**, 429

High-molecular-mass poly(dimethylsiloxane), $[(CH_3)_2SiO]_n$, is used as a stationary phase in gas chromatography (Table 23-1) and as the gel in breast implants. Approximately 1–2% of silicones in breast implants are low-molecular-mass materials that can leak from intact implants and travel through the circulatory and lymph systems to take up residence in lipid-rich tissues.

Gas chromatography with selected ion mass spectrometric detection (Section 21-4) provides a specific, sensitive means to measure silicones. Analytes were extracted with hexane from 1 mL of blood plasma to which the internal standard, $[(CH_3)_3SiO]_4Si$, was added. Each analyte was monitored at the mass of its most abundant fragment. The chromatogram on this page would be enormously more complex if detection were not by selected ion monitoring. By setting the mass spectrometer to observe only each intended analyte near its known retention time, everything else eluted from the column becomes invisible.

The observation of silicones at part per billion levels in human tissue does not necessarily imply a health risk. Risk must be assessed in medical studies that make use of the analytical data.

Two liquids are **miscible** if they form a single phase when they are mixed in any ratio. *Immiscible* liquids remain in separate phases. Organic solvents with low polarity are generally immiscible with water, which is highly polar.

$[S]_2$ Phase 2

$[S]_1$ Phase 1

FIGURE 22-1 Partitioning of a solute between two liquid phases.

For simplicity, we assume that the two phases are not soluble in each other. A more realistic treatment considers that most liquids are partially soluble in each other.[1]

The larger the partition coefficient, the less solute remains in phase 1.

Example: If $q = \frac{1}{4}$, then $\frac{1}{4}$ of the solute remains in phase 1 after one extraction. A second extraction reduces the concentration to $\left(\frac{1}{4}\right)\left(\frac{1}{4}\right) = \frac{1}{16}$ of initial concentration.

In the vast majority of real analytical problems, we must separate, identify, and measure one or more components from a complex mixture. This chapter discusses fundamentals of analytical separations, and the next three chapters describe specific methods.

22-1 Solvent Extraction

Extraction is the transfer of a solute from one phase to another. Common reasons to carry out an extraction in analytical chemistry are to isolate or concentrate the desired analyte or to separate it from species that would interfere in the analysis. The most common case is the extraction of an aqueous solution with an organic solvent. Diethyl ether, toluene, and hexane are common solvents that are *immiscible* with and less dense than water. They form a separate phase that floats on top of the aqueous phase. Chloroform, dichloromethane, and carbon tetrachloride are common solvents that are denser than water.[†] In the two-phase mixture, one phase is predominantly water and the other phase is predominantly organic.

Suppose that solute S is partitioned between phases 1 and 2, as depicted in Figure 22-1. The **partition coefficient,** K, is the equilibrium constant for the reaction

$$S \text{ (in phase 1)} \rightleftharpoons S \text{ (in phase 2)}$$

Partition coefficient:
$$K = \frac{\mathcal{A}_{S_2}}{\mathcal{A}_{S_1}} \approx \frac{[S]_2}{[S]_1} \tag{22-1}$$

where $\mathcal{A}_{S_1}$ refers to the activity of solute in phase 1. Lacking knowledge of the activity coefficients, we will write the partition coefficient in terms of concentrations.

Suppose that solute S in V_1 mL of solvent 1 (water) is extracted with V_2 mL of solvent 2 (toluene). Let m be the moles of S in the system and let q be the fraction of S remaining in phase 1 at equilibrium. The molarity in phase 1 is therefore qm/V_1. The fraction of total solute transferred to phase 2 is $(1 - q)$, and the molarity in phase 2 is $(1 - q)m/V_2$. Therefore,

$$K = \frac{[S]_2}{[S]_1} = \frac{(1 - q)m/V_2}{qm/V_1}$$

from which we can solve for q:

$$\begin{matrix} \text{Fraction remaining in phase 1} \\ \text{after 1 extraction} \end{matrix} = q = \frac{V_1}{V_1 + KV_2} \tag{22-2}$$

Equation 22-2 says that the fraction of solute remaining in the water (phase 1) depends on the partition coefficient and the volumes. If the phases are separated and fresh toluene (solvent 2) is added, the fraction of solute remaining in the water at equilibrium will be

$$\begin{matrix} \text{Fraction remaining in phase 1} \\ \text{after 2 extractions} \end{matrix} = q \cdot q = \left(\frac{V_1}{V_1 + KV_2}\right)^2$$

After n extractions, each with volume V_2, the fraction remaining in the water is

$$\begin{matrix} \text{Fraction remaining in phase 1} \\ \text{after } n \text{ extractions} \end{matrix} = q^n = \left(\frac{V_1}{V_1 + KV_2}\right)^n \tag{22-3}$$

EXAMPLE   Extraction Efficiency

Solute A has a partition coefficient of 3 between toluene and water, with three times as much in the toluene phase. Suppose that 100 mL of a 0.010 M aqueous solution of A are extracted with toluene. What fraction of A remains in the aqueous phase **(a)** if one extraction with 500 mL is performed or **(b)** if five extractions with 100 mL are performed?

Solution **(a)** With water as phase 1 and toluene as phase 2, Equation 22-2 says that, after a 500-mL extraction, the fraction remaining in the aqueous phase is

$$q = \frac{100}{100 + (3)(500)} = 0.062 \approx 6\%$$

[†]Whenever a choice exists between $CHCl_3$ and CCl_4, the less toxic $CHCl_3$ should be chosen. Hexane and toluene are greatly preferred over benzene, which is a carcinogen.

(b) With five 100-mL extractions, the fraction remaining is given by Equation 22-3:

$$\text{Fraction remaining} = \left(\frac{100}{100 + (3)(100)}\right)^5 = 0.00098 \approx 0.1\%$$

It is more efficient to do several small extractions than one big extraction.

Test Yourself If the partition coefficient is 10, what fraction remains in 100 mL of water after 1 and 5 extractions with 20 mL of toluene? (**Answer:** 33%, 0.41%)

pH Effects

If a solute is an acid or base, its charge changes as the pH is changed. Usually, a neutral species is more soluble in an organic solvent and a charged species is more soluble in aqueous solution. Consider a basic amine whose neutral form, B, has partition coefficient K between aqueous phase 1 and organic phase 2. Suppose that the conjugate acid, BH^+, is soluble *only* in aqueous phase 1. Let's denote its acid dissociation constant as K_a. The **distribution coefficient, D,** is defined as

Distribution coefficient:
$$D = \frac{\text{total concentration in phase 2}}{\text{total concentration in phase 1}} \qquad (22\text{-}4)$$

which becomes

$$D = \frac{[B]_2}{[B]_1 + [BH^+]_1} \qquad (22\text{-}5)$$

Substituting $K = [B]_2/[B]_1$ and $K_a = [H^+][B]_1/[BH^+]_1$ into Equation 22-5 leads to

Distribution of base between two phases:
$$D = \frac{K \cdot K_a}{K_a + [H^+]} = K \cdot \alpha_B \qquad (22\text{-}6)$$

where α_B is the fraction of weak base in the neutral form, B, in the aqueous phase. *The distribution coefficient D is used in place of the partition coefficient K in Equation 22-2 when dealing with a species that has more than one chemical form, such as B and BH^+.*

Charged species tend to be more soluble in water than in organic solvent. To extract a base into water, use a pH low enough to convert B into BH^+ (Figure 22-2). To extract the acid HA into water, use a pH high enough to convert HA into A^-.

> *Challenge* Suppose that the acid HA (with dissociation constant K_a) is partitioned between aqueous phase 1 and organic phase 2. Calling the partition coefficient K for HA and assuming that A^- is not soluble in the organic phase, show that the distribution coefficient is given by
>
> *Distribution of acid between two phases:*
> $$D = \frac{K \cdot [H^+]}{[H^+] + K_a} = K \cdot \alpha_{HA} \qquad (22\text{-}7)$$
>
> where α_{HA} is the fraction of weak acid in the form HA in the aqueous phase.

EXAMPLE Effect of pH on Extraction

Suppose that the partition coefficient for an amine, B, is $K = 3.0$ and the acid dissociation constant of BH^+ is $K_a = 1.0 \times 10^{-9}$. If 50 mL of 0.010 M aqueous amine are extracted with 100 mL of solvent, what will be the formal concentration remaining in the aqueous phase **(a)** at pH 10.00 and **(b)** at pH 8.00?

Solution **(a)** At pH 10.00, $D = KK_a/(K_a + [H^+]) = (3.0)(1.0 \times 10^{-9})/(1.0 \times 10^{-9} + 1.0 \times 10^{-10}) = 2.73$. *Using D in place of K,* Equation 22-2 says that the fraction remaining in the aqueous phase is

$$q = \frac{50}{50 + (2.73)(100)} = 0.15 \Rightarrow 15\% \text{ left in water}$$

The concentration of amine in the aqueous phase is 15% of 0.010 M = 0.001 5 M.

Many small extractions are more effective than a few large extractions.

The limit for extracting solute S from phase 1 (volume V_1) into phase 2 (volume V_2) is attained by dividing V_2 into an infinite number of infinitesimally small portions for extraction. With $K = [S]_2/[S]_1$, the limiting fraction of solute remaining in phase 1 is[2]

$$q_{\text{limit}} = e^{-(V_2/V_1)K}$$

$$\alpha_B = \frac{[B]_{aq}}{[B]_{aq} + [BH^+]_{aq}}$$

α_B is the same as α_{A^-} in Equation 9-18.

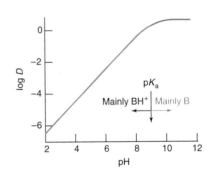

FIGURE 22-2 Effect of pH on the distribution coefficient for the extraction of a base into an organic solvent. In this example, $K = 3.0$ and pK_a for BH^+ is 9.00.

(b) At pH 8.00, $D = (3.0)(1.0 \times 10^{-9})/(1.0 \times 10^{-9} + 1.0 \times 10^{-8}) = 0.273$. Therefore,

$$q = \frac{50}{50 + (0.273)(100)} = 0.65 \Rightarrow 65\% \text{ left in water}$$

The concentration in the aqueous phase is 0.006 5 M. At pH 10, the base is predominantly in the form B and is extracted into the organic solvent. At pH 8, it is in the form BH^+ and remains in the water.

Test Yourself Consider an acid HA with $K = 3.0$ and $K_a = 1.0 \times 10^{-9}$. What will be the formal concentration remaining in the aqueous phase at pH 10.00 and at pH 8.00? Explain why the acid and base have opposite answers. (***Answer:*** 65% at pH 10 and 15% at pH 8; neutral species HA or B are more soluble in organic phase)

Extraction with a Metal Chelator

Most complexes that can be extracted into organic solvents are neutral. Charged complexes, such as $Fe(EDTA)^-$ or $Fe(1,10\text{-phenanthroline})_3^{2+}$, are not very soluble in organic solvents. One scheme for separating metal ions from one another is to selectively complex one ion with an organic ligand and extract it into an organic solvent. Ligands such as dithizone (Demonstration 22-1), 8-hydroxyquinoline, and cupferron are commonly employed. Each is a weak acid, HL, which loses a proton when it binds to a metal ion through atoms shown in **bold** type.

8-Hydroxyquinoline
(oxine)

Cupferron

β is the overall formation constant defined in Box 6-2.

$$HL(aq) \rightleftharpoons H^+(aq) + L^-(aq) \qquad K_a = \frac{[H^+]_{aq}[L^-]_{aq}}{[HL]_{aq}} \qquad (22\text{-}8)$$

$$nL^-(aq) + M^{n+}(aq) \rightleftharpoons ML_n(aq) \qquad \beta = \frac{[ML_n]_{aq}}{[M^{n+}]_{aq}[L^-]_{aq}^n} \qquad (22\text{-}9)$$

DEMONSTRATION 22-1 **Extraction with Dithizone**

Dithizone (diphenylthiocarbazone) is a green compound that is soluble in nonpolar organic solvents and insoluble in water below pH 7.[3] In alkaline aqueous solution, it forms a soluble yellow ion. It forms red, hydrophobic complexes with di- and trivalent metals. Dithizone is used for analytical extractions and for colorimetric determinations of metal ions.

Dithizone
(green)   (colorless)
$pK_a \approx 4.5$

$\beta_2 = 10^{16}$

Metal complex
(red)

$+2H^+$

You can demonstrate the equilibrium between green ligand and red complex by using three large test tubes sealed with tightly fitting rubber stoppers. Place some hexane plus a few milliliters of dithizone solution (prepared by dissolving 1 mg of dithizone in 100 mL of $CHCl_3$) in each test tube. Add distilled water to tube A, tap water to tube B, and 2 mM $Pb(NO_3)_2$ to tube C. After shaking and settling, tubes B and C contain a red upper phase, whereas A remains green.

Proton equilibrium in the dithizone reaction is shown by adding a few drops of 1 M HCl to tube C. After shaking, the dithizone turns green again. Competition with a stronger ligand is shown by adding a few drops of 0.05 M EDTA solution to tube B. Again, shaking causes a reversion to the green color.

Practicing "Green" Chemistry

Chemical procedures that produce less waste or less hazardous waste are said to be "green" because they reduce harmful environmental effects. In chemical analyses with dithizone, you can substitute aqueous micelles (Box 25-1) for the organic phase (which has traditionally been chloroform, $CHCl_3$) to eliminate chlorinated solvent and the tedious extraction.[4] For example, a solution containing 5.0 wt% of the micelle-forming surfactant Triton X-100 dissolves 8.3×10^{-5} M dithizone at 25°C and pH < 7. The concentration of dithizone inside the micelles, which constitute a small fraction of the volume of solution, is much greater than 8.3×10^{-5} M. Aqueous micellar solutions of dithizone can be used for the spectrophotometric analysis of metals such as Zn(II), Cd(II), Hg(II), Cu(II), and Pb(II) with results comparable to those obtained with an organic solvent.

Each ligand can react with many different metal ions, but some selectivity is achieved by controlling the pH.

Let's derive an equation for the distribution coefficient of a metal between two phases when essentially all the metal in the aqueous phase (*aq*) is in the form M^{n+} and all the metal in the organic phase (*org*) is in the form ML_n (Figure 22-3). We define the partition coefficients for ligand and complex as follows:

M^{n+} is in the aqueous phase and ML_n is in the organic phase.

$$HL(aq) \rightleftharpoons HL(org) \qquad K_L = \frac{[HL]_{org}}{[HL]_{aq}} \qquad (22\text{-}10)$$

$$ML_n(aq) \rightleftharpoons ML_n(org) \qquad K_M = \frac{[ML_n]_{org}}{[ML_n]_{aq}} \qquad (22\text{-}11)$$

The distribution coefficient we seek is

$$D = \frac{[\text{total metal}]_{org}}{[\text{total metal}]_{aq}} \approx \frac{[ML_n]_{org}}{[M^{n+}]_{aq}} \qquad (22\text{-}12)$$

From Equations 22-11 and 22-9, we can write

$$[ML_n]_{org} = K_M[ML_n]_{aq} = K_M\beta[M^{n+}]_{aq}[L^-]^n_{aq}$$

Using $[L^-]_{aq}$ from Equation 22-8 gives

$$[ML_n]_{org} = \frac{K_M\beta[M^{n+}]_{aq}K_a^n[HL]^n_{aq}}{[H^+]^n_{aq}}$$

Putting this value of $[ML_n]_{org}$ into Equation 22-12 gives

$$D \approx \frac{K_M\beta K_a^n[HL]^n_{aq}}{[H^+]^n_{aq}}$$

Because most HL is in the organic phase, we substitute $[HL]_{aq} = [HL]_{org}/K_L$ to produce the most useful expression for the distribution coefficient:

Distribution of metal-chelate complex between phases: $\qquad D \approx \frac{K_M\beta K_a^n}{K_L^n} \frac{[HL]^n_{org}}{[H^+]^n_{aq}} \qquad (22\text{-}13)$

We see that the distribution coefficient for metal ion extraction depends on pH and ligand concentration. It is often possible to select a pH where D is large for one metal and small for another. For example, Figure 22-4 shows that Cu^{2+} could be separated from Pb^{2+} and Zn^{2+} by extraction with dithizone at pH 5. Demonstration 22-1 illustrates the pH dependence of an extraction. Box 22-1 describes *crown ethers* and *phase transfer agents* used to extract polar reagents into nonpolar solvents for chemical reactions.

You can select a pH to bring the metal into either phase.

FIGURE 22-3 Extraction of a metal ion with a chelator. The predominant form of metal in the aqueous phase is M^{n+}, and the predominant form in the organic phase is ML_n.

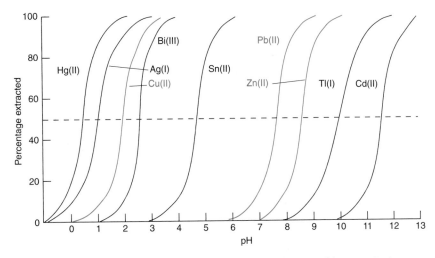

FIGURE 22-4 Extraction of metal ions by dithizone into CCl_4. At pH 5, Cu^{2+} is completely extracted into CCl_4, whereas Pb^{2+} and Zn^{2+} remain in the aqueous phase. [Adapted from G. H. Morrison and H. Freiser in C. L. Wilson and D. Wilson, eds., *Comprehensive Analytical Chemistry*, Vol. IA (New York: Elsevier, 1959).]

BOX 22-1   Crown Ethers and Phase Transfer Agents

Crown ethers are synthetic compounds that envelop metal ions (especially alkali metal cations) in a pocket of oxygen ligands. Crown ethers are used as *phase transfer catalysts* because they can extract water-soluble ionic reagents into nonpolar solvents, where reaction with hydrophobic compounds can occur. In the potassium complex of dibenzo-30-crown-10, K^+ is engulfed by 10 oxygen atoms, with K—O distances averaging 288 pm. Only the hydrophobic outside of the complex is exposed to solvent.

Hydrophobic cations and anions can function as phase transfer agents to bring ions of the opposite charge into organic solvents. For example, Color Plate 27 shows the extraction of colored anions from the lower aqueous phase into the upper diethyl ether layer when

trioctylmethylammonium chloride is added and the mixture is stirred. For $KMnO_4$ in the aqueous phase, the equilibrium below lies well to the right:

$$K^+(aq) + MnO_4^-(aq) + (C_8H_{17})_3\overset{+}{N}CH_3\,Cl^-(org) \rightleftharpoons$$
$$\text{Ion pair}$$

$$K^+(aq) + Cl^-(aq) + (C_8H_{17})_3\overset{+}{N}CH_3\,MnO_4^-(org)$$
$$\text{Ion pair}$$

An important application of this reaction is to bring permanganate into the organic phase to oxidize an organic compound. A phase transfer, analytical separation of hafnium from tungsten is used in geochemical studies to detect debris from supernova explosions.[5]

(a)   (b)   ● C   ● O   ○ K

(a) Molecular structure of dibenzo-30-crown-10. (b) Three-dimensional structure of its K^+ complex.
[Adapted from M. A. Bush and M. R. Truter, "Crystal Structures of Alkali-Metal Complexes with Cyclic Polyethers," *J. Chem. Soc. Chem. Commun.* **1970**, 1439.]

22-2   What Is Chromatography?

In 1903 in Warsaw, the botanist M. Tswett invented adsorption chromatography to separate plant pigments, using a hydrocarbon solvent and inulin powder (a carbohydrate) as stationary phase. The separation of colored bands led to the name *chromatography*, from the Greek *chromatos* ("color") and *graphein* ("to write")—"color writing." Tswett later found that $CaCO_3$ or sucrose could also be used as stationary phases.[6]

Chromatography lay dormant until Tswett's methods were applied, beginning in 1931, to biochemical separations by E. Lederer and R. Kuhn in Heidelberg, P. Karrer in Zurich, and L. Zechmeister in Hungary.[7] During the 1930s, adsorption chromatography became an established tool in biochemistry.

Chromatography operates on the same principle as extraction, but one phase is held in place while the other moves past it.[8,9] Figure 22-5 shows a solution containing solutes A and B placed on top of a column packed with solid particles and filled with solvent. When the outlet is opened, solutes A and B flow down into the column. Fresh solvent is then applied to the top of the column and the mixture is washed down the column by continuous solvent flow. If solute A is more strongly adsorbed than solute B on the solid particles, then solute A spends a smaller fraction of the time free in solution. Solute A moves down the column more slowly than solute B and emerges at the bottom after solute B. We have just separated a mixture into its components by *chromatography*.

The **mobile phase** (the solvent moving through the column) in chromatography is either a liquid or a gas. The **stationary phase** (the one that stays in place inside the column) is most commonly a viscous liquid chemically bonded to the inside of a capillary tube or onto the surface of solid particles packed in the column. Alternatively, as in Figure 22-5, the solid particles themselves may be the stationary phase. In any case, the partitioning of solutes between mobile and stationary phases gives rise to separation.

Fluid entering the column is called **eluent.** Fluid emerging from the end of the column is called **eluate:**

elu*ent*—in
elu*ate*—out

$$\text{eluent in} \quad \rightarrow \quad \boxed{\text{COLUMN}} \quad \rightarrow \quad \text{eluate out}$$

The process of passing liquid or gas through a chromatography column is called **elution.**

FIGURE 22-5 The idea behind chromatography: solute A, with a greater affinity than solute B for the stationary phase, remains on the column longer. Panel *f* is a reconstruction of the separation of pigments from red paprika skin from the work of L. Zechmeister in the 1930s. Bands marked by horizontal lines are different pigments. The lower stationary phase is $Ca(OH)_2$ and the upper stationary phase is $CaCO_3$.
[Panel *f* from L. S. Ettre, "The Rebirth of Chromatography 75 Years Ago," *LCGC* **2007**, *25*, 640.]

Columns are either **packed** or **open tubular.** A packed column is filled with particles of stationary phase, as in Figure 22-5. An open tubular column is a narrow, hollow capillary with stationary phase coated on the inside walls.

Types of Chromatography

Chromatography is divided into categories on the basis of the mechanism of interaction of the solute with the stationary phase, as shown in Figure 22-6.

Adsorption chromatography. A solid stationary phase and a liquid or gaseous mobile phase are used. Solute is adsorbed on the surface of the solid particles. The more strongly a solute is adsorbed, the slower it travels through the column.

Partition chromatography. A liquid stationary phase is bonded to a solid surface, which is typically the inside of the silica (SiO_2) chromatography column in gas chromatography. Solute equilibrates between the stationary liquid and the mobile phase, which is a flowing gas in gas chromatography.

Ion-exchange chromatography. Anions such as $—SO_3^-$ or cations such as $—N(CH_3)_3^+$ are covalently attached to the stationary solid phase, usually a *resin*. Solute ions of the opposite charge are attracted to the stationary phase. The mobile phase is a liquid.

Molecular exclusion chromatography. Also called *size exclusion*, **gel filtration,** or **gel permeation** chromatography, this technique separates molecules by size, with the larger solutes passing through most quickly. In the ideal case of molecular exclusion, there is no attractive interaction between the stationary phase and the solute. Rather, the liquid or gaseous mobile phase passes through a porous gel. The pores are small enough to exclude large solute molecules but not small ones. Large molecules stream past without entering the pores. Small molecules take longer to pass through the column because they enter the gel and therefore must flow through a larger volume before leaving the column.

Affinity chromatography. This most selective kind of chromatography employs specific interactions between one kind of solute molecule and a second molecule that is covalently attached (immobilized) to the stationary phase. For example, the immobilized molecule might be an antibody to a particular protein. When a mixture containing a thousand proteins is passed through the column, only the one protein that reacts with the antibody binds to the column. After all other solutes have been washed from the column, the desired protein is dislodged by changing the pH or ionic strength.

This form of chromatography was invented by Tswett in 1903.

For their pioneering work on liquid-liquid partition chromatography in 1941, A. J. P. Martin and R. L. M. Synge received a Nobel Prize in 1952.

B. A. Adams and E. L. Holmes developed the first synthetic ion-exchange resins in 1935. *Resins* are relatively hard, amorphous organic solids. *Gels* are relatively soft.

Large molecules pass through the column *faster* than small molecules.

FIGURE 22-6 Major types of chromatography.

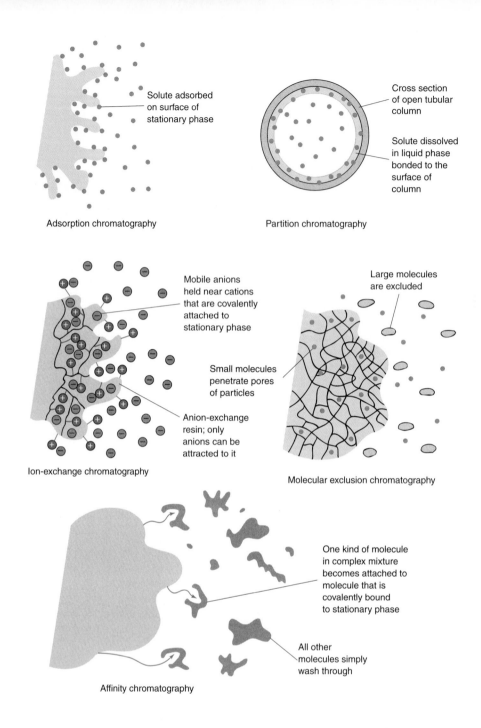

Adsorption chromatography

Partition chromatography

Ion-exchange chromatography

Molecular exclusion chromatography

Affinity chromatography

22-3 A Plumber's View of Chromatography

The speed of the mobile phase passing through a chromatography column is expressed either as a volume flow rate or as a linear flow rate. Consider a liquid chromatography experiment in which the column has an inner diameter of 0.60 cm (radius $\equiv r = 0.30$ cm) and the mobile phase occupies 20% of the column volume. Each centimeter of column length has a volume of $\pi r^2 \times \text{length} = \pi(0.30 \text{ cm})^2(1 \text{ cm}) = 0.283$ mL, of which 20% ($= 0.056\ 5$ mL) is mobile phase (solvent). The **volume flow rate,** such as 0.30 mL/min, tells how many milliliters of solvent per minute travel through the column. The **linear flow rate** tells how many centimeters are traveled in 1 min by the solvent. Because 1 cm of column length contains 0.056 5 mL of mobile phase, 0.30 mL would occupy (0.30 mL)/(0.056 5 mL/cm) = 5.3 cm of column length. The linear flow rate corresponding to 0.30 mL/min is 5.3 cm/min.

The Chromatogram

Solutes eluted from a chromatography column are observed with detectors described in later chapters. A **chromatogram** is a graph showing the detector response as a function of elution

Volume flow rate = volume of solvent per unit time traveling through column.

Linear flow rate = distance per unit time traveled by solvent.

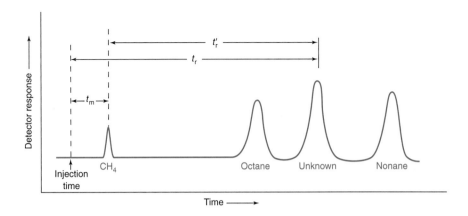

time. Figure 22-7 shows what might be observed when a mixture of octane, nonane, and an unknown are separated by gas chromatography, which is described in Chapter 23. The **retention time,** t_r, for each component is the time that elapses between injection of the mixture onto the column and the arrival of that component at the detector. **Retention volume,** V_r, is the volume of mobile phase required to elute a particular solute from the column.

   Mobile phase or an unretained solute travels through the column in the minimum possible time, t_m. The **adjusted retention time,** t_r', for a retained solute is the additional time required to travel the length of the column, beyond that required by solvent:

Adjusted retention time: $$t_r' = t_r - t_m$$ (22-14)

In gas chromatography, t_m is usually taken as the time needed for CH_4 to travel through the column (Figure 22-7).

   For two components 1 and 2, the **relative retention,** α (also called *separation factor*), is the ratio of their adjusted retention times:

Relative retention: $$\alpha = \frac{t_{r2}'}{t_{r1}'}$$ (22-15)

where $t_{r2}' > t_{r1}'$, so $\alpha > 1$. The greater the relative retention, the greater the separation between two components. Relative retention is fairly independent of flow rate and can therefore be used to help identify peaks when the flow rate changes.

   For component 2 eluted after component 1, the *unadjusted relative retention,* γ, is the ratio of their unadjusted retention times:

Unadjusted relative retention: $$\gamma = \frac{t_{r2}}{t_{r1}}$$ (22-16)

Unadusted relative retention
$$= \frac{\text{retention time of component 2}}{\text{retention time of component 1}}$$
$$= \frac{\text{speed of component 1}}{\text{speed of component 2}}$$

The unadjusted relative retention is the inverse of the ratio of the speeds at which the two components travel.

   For each peak in the chromatogram, the **retention factor,** k, is the time required to elute that peak minus the time t_m required for mobile phase to pass through the column, expressed in multiples of t_m.

Retention factor: $$k = \frac{t_r - t_m}{t_m}$$ (22-17)

Retention factor is also called *capacity factor, capacity ratio,* or *partition ratio* and was formerly written as k' instead of k.

The longer a component is retained by the column, the greater is the retention factor. It takes volume V_m to push solvent from the beginning of the column to the end of the column. If it takes an additional volume $3V_m$ to elute a solute, then the retention factor for that solute is 3.

EXAMPLE   **Retention Parameters**

A mixture of benzene, toluene, and methane was injected into a gas chromatograph. Methane gave a sharp spike in 42 s, whereas benzene required 251 s and toluene was eluted in 333 s. Find the adjusted retention time and retention factor for each solute, the relative retention, and the unadjusted relative retention.

Solution The adjusted retention times are

Benzene: $t'_r = t_r - t_m = 251 - 42 = 209$ s   Toluene: $t'_r = 333 - 42 = 291$ s

The retention factors are

Benezene: $k = \dfrac{t_r - t_m}{t_m} = \dfrac{251 - 42}{42} = 5.0$ Toluene: $k = \dfrac{333 - 42}{42} = 6.9$

The relative retention is expressed as a number greater than unity:

$$\alpha = \frac{t'_r(\text{toluene})}{t'_r(\text{benzene})} = \frac{333 - 42}{251 - 42} = 1.39$$

The unadjusted relative retention is

$$\gamma = \frac{t_r(\text{toluene})}{t_r(\text{benzene})} = \frac{333}{251} = 1.33$$

Test Yourself Ethylbenzene was eluted at 350 s. Find its retention factor and the relative retention and unadjusted relative retention for ethylbenzene and toluene. (**Answer:** 7.33, 1.058, 1.051)

Relation Between Retention Time and the Partition Coefficient

The retention factor in Equation 22-17 is equivalent to

$$k = \frac{\text{time solute spends in stationary phase}}{\text{time solute spends in mobile phase}} \tag{22-18a}$$

Let's see why this is true. If the solute spends all its time in the mobile phase and none in the stationary phase, it would be eluted in time t_m. Putting $t_r = t_m$ into Equation 22-17 gives $k = 0$, because solute spends no time in the stationary phase. Suppose that solute spends equal time in the stationary and mobile phases. The retention time would then be $t_r = 2t_m$ and $k = (2t_m - t_m)/t_m = 1$. If solute spends three times as much time in the stationary phase as in the mobile phase, $t_r = 4t_m$ and $k = (4t_m - t_m)/t_m = 3$.

If solute spends three times as much time in the stationary phase as in the mobile phase, there will be three times as many moles of solute in the stationary phase as in the mobile phase at any time. The quotient in Equation 22-18a is equivalent to

$$\frac{\text{Time solute spends in stationary phase}}{\text{Time solute spends in mobile phase}} = \frac{\text{moles of solute in stationary phase}}{\text{moles of solute in mobile phase}}$$

$$k = \frac{c_s V_s}{c_m V_m} \tag{22-18b}$$

where c_s is the concentration of solute in the stationary phase, V_s is the volume of the stationary phase, c_m is the concentration of solute in the mobile phase, and V_m is the volume of the mobile phase.

The quotient c_s/c_m is the ratio of concentrations of solute in the stationary and mobile phases. If the column is run slowly enough to be at equilibrium, the quotient c_s/c_m is the *partition coefficient*, K, introduced in connection with solvent extraction. Therefore, we cast Equation 22-18b in the form

Partition coefficient $= K = \dfrac{c_s}{c_m}$

Relation of retention time to partition coefficient: $k = K\dfrac{V_s}{V_m} \overset{\text{Eq. 22.17}}{=} \dfrac{t_r - t_m}{t_m} = \dfrac{t'_r}{t_m}$ $\tag{22-19}$

which relates retention time to the partition coefficient and the volumes of stationary and mobile phases. Because $t'_r \propto k \propto K$, relative retention can also be expressed as

Relative retention: $\alpha = \dfrac{t'_{r2}}{t'_{r1}} = \dfrac{k_2}{k_1} = \dfrac{K_2}{K_1}$ $\tag{22-20}$

Physical basis of chromatography:

The greater the ratio of partition coefficients between mobile and stationary phases, the greater the separation between two components of a mixture.

That is, the relative retention of two solutes is proportional to the ratio of their partition coefficients. This relation is the physical basis of chromatography.

In the preceding example, methane gave a sharp spike in 42 s, whereas benzene required 251 s. The open tubular chromatography column has an inner diameter of 250 μm and is coated on the inside with a layer of stationary phase 1.0 μm thick. Estimate the partition coefficient ($K = c_s/c_m$) for benzene between stationary and mobile phases and state what fraction of the time benzene spends in the mobile phase.

Solution We need to calculate the relative volumes of the stationary and mobile phases. The column is an open tube with a thin coating of stationary phase on the inside wall.

Radius of hollow cavity: $r_1 = 124$ μm
Radius to middle of stationary phase:
$r_2 = 124.5$ μm

Cross-sectional area of column $= \pi r_1^2$
$= \pi(124$ μm$)^2 = 4.83 \times 10^4$ μm^2

Cross-sectional area of coating $\approx 2\pi r_2 \times$ thickness
$= 2\pi(124.5$ μm$)(1.0$ μm$) = 7.8 \times 10^2$ μm^2

The relative volumes of the phases are proportional to the relative cross-sectional areas of the phases. Therefore, $V_s/V_m = (7.8 \times 10^2$ μm$^2)/(4.83 \times 10^4$ μm$^2) = 0.016\ 1$. In the preceding example, we found that the retention factor for benzene is

$$k = \frac{t_r - t_m}{t_m} = \frac{251 - 42}{42} = 5.0$$

Substituting this value into Equation 22-19 gives the partition coefficient:

$$k = K\frac{V_s}{V_m} \Rightarrow 5.0 = K(0.016\ 1) \Rightarrow K = 310$$

To find the fraction of time spent in the mobile phase, we use Equations 22-17 and 22-18a:

$$k = \frac{\text{time in stationary phase}}{\text{time in mobile phase}} = \frac{t_r - t_m}{t_m} = \frac{t_s}{t_m} \Rightarrow t_s = kt_m$$

where t_s is the time in the stationary phase. The fraction of time in the mobile phase is

$$\text{Fraction of time in mobile phase} = \frac{t_m}{t_s + t_m} = \frac{t_m}{kt_m + t_m} = \frac{1}{k + 1} = \frac{1}{5.0 + 1} = 0.17$$

Test Yourself Find the partition coefficient for toluene ($t_r = 333$ s) and state what fraction of the time it spends in the mobile phase. (**Answer:** 430, 0.13)

FIGURE 22-8 Industrial-scale preparative chromatography column can purify a kilogram of material. The column volume is 300 L.
[Courtesy Prochrom, Inc., Indianapolis, IN.]

Retention volume, V_r, is the volume of mobile phase required to elute a particular solute from the column:

Retention volume: $V_r = t_r \cdot u_v$ (22-21)

where u_v is the volume flow rate (volume per unit time) of the mobile phase. The retention volume of a particular solute is constant over a range of flow rates.

> Volume is proportional to time, so any ratio of times can be written as the corresponding ratio of volumes. If V_m is the elution volume for unretained solute,
>
> $$k = \frac{t_r - t_m}{t_m} = \frac{V_r - V_m}{V_m}$$
>
> where V_r is the retention volume for solute.

Scaling Up

We normally carry out chromatography for *analytical* purposes (to separate and identify or measure the components of a mixture) or for *preparative* purposes (to purify a significant quantity of a component of a mixture). Analytical chromatography is usually performed with thin columns that provide good separation. For preparative chromatography, we use fatter columns that can handle more load (Figure 22-8).[10] Preparative chromatography is especially important in the pharmaceutical industry, which can afford the high cost of separating compounds such as *optical isomers* of drugs (Box 23-1).

If you have developed a procedure to separate 2 mg of a mixture on a column with a diameter of 1.0 cm, what size column should you use to separate 20 mg of the mixture? The most straightforward way to scale up is to maintain the same column length and to increase the cross-sectional area to maintain a constant ratio of sample mass to column volume. Cross-sectional area is πr^2, where r is the column radius, so the desired diameter is given by

Scaling rules:

- Keep column length constant
- Cross-sectional area of column $\propto$ mass of analyte:

$$\frac{Mass_2}{Mass_1} = \left(\frac{radius_2}{radius_1}\right)^2$$

(The symbol $\propto$ means "is proportional to.")

- Maintain constant linear flow rate:

$$\frac{Volume\ flow_2}{Volume\ flow_1} = \left(\frac{radius_2}{radius_1}\right)^2$$

- Sample volume applied to column $\propto$ mass of analyte
- If you change column length, mass of sample can be increased in proportion to total length

Scaling equation:

$$\frac{\text{Large mass}}{\text{Small mass}} = \left(\frac{\text{large column radius}}{\text{small column radius}}\right)^2 \qquad (22\text{-}22)$$

$$\frac{20\ mg}{2\ mg} = \left(\frac{\text{large column radius}}{0.50\ cm}\right)^2$$

Large column radius = 1.58 cm

A column with a diameter near 3 cm would be appropriate.

To reproduce the conditions of the smaller column in the larger column, the *linear flow rate* (not the volume flow rate) should be kept constant. Because the area (and hence volume) of the large column is 10 times greater than that of the small column in this example, the volume flow rate should be 10 times greater to maintain a constant linear flow rate. If the small column had a volume flow rate of 0.3 mL/min, the large column should be run at 3 mL/min.

The mass of sample (g) that can be run in preparative chromatography on a reversed-phase (page 603) silica-based column is *roughly*

$$\text{Column capacity (g)} \approx (2.2 \times 10^{-7})Ld_c^2\sigma_g$$

where L is column length in mm, d_c is column diameter in mm, and σ_g is the surface area (m²) per gram of stationary phase.[11] For L = 250 mm, d_c = 50 mm, and σ_g = 200 m²/g, we estimate the column capacity as $(2.2 \times 10^{-7})(250)(50)^2(200)$ = 28 g. This calculation presumes that the band will occupy the entire volume of the column, so it is surely an upper limit estimate. If you want the band to occupy just 20% of the column, so there is room for chromatography, the mass of sample would be (0.2)(28 g) = 5.6 g.

22-4 Efficiency of Separation

Two factors contribute to how well compounds are separated by chromatography. One is the difference in elution times between peaks: The farther apart, the better their separation. The other factor is how broad the peaks are: the wider the peaks, the poorer their separation. This section discusses how we measure the efficiency of a separation.

Resolution

Solute moving through a chromatography column tends to spread into a Gaussian shape with standard deviation σ (Figure 22-9). The longer a solute resides in a column, the broader the band becomes. Common measures of breadth are (1) the width $w_{1/2}$ measured at a height

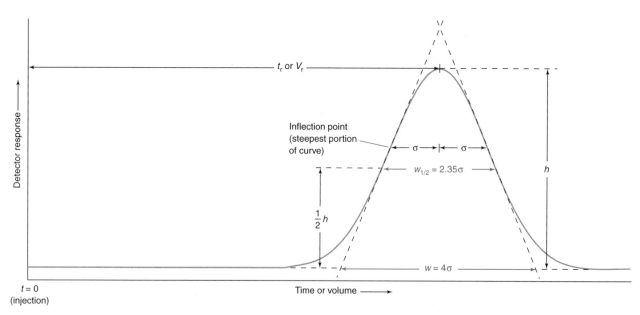

FIGURE 22-9 Idealized Gaussian chromatogram showing how w and $w_{1/2}$ are measured. The value of w is obtained by extrapolating the tangents to the inflection points down to the baseline.

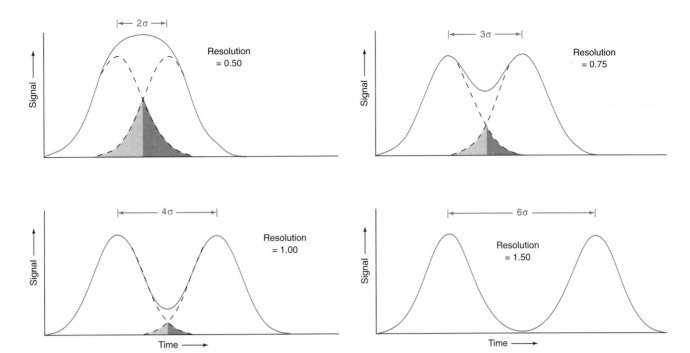

FIGURE 22-10 Resolution of Gaussian peaks of equal area and amplitude. Dashed lines show individual peaks, and solid lines are the sum of two peaks. Overlapping area is shaded.

equal to half of the peak height and (2) the width w at the baseline between tangents drawn to the steepest parts of the peak. From Equation 4-3 for a Gaussian peak, it is possible to show that $w_{1/2} = 2.35\sigma$ and $w = 4\sigma$.

In chromatography, the **resolution** of two peaks from each other is defined as

Resolution:
$$\text{Resolution} = \frac{\Delta t_r}{w_{av}} = \frac{\Delta V_r}{w_{av}} = \frac{0.589\Delta t_r}{w_{1/2av}} \qquad (22\text{-}23)$$

where Δt_r or ΔV_r is the separation between peaks (in units of time or volume) and w_{av} is the average width of the two peaks in corresponding units. (Peak width is measured at the base, as shown in Figure 22-9.) Alternatively, the last expression in Equation 22-23 uses $w_{1/2av}$, the width at half-height of Gaussian peaks. The width at half-height is usually used because it is easiest to measure. Figure 22-10 shows the overlap of two peaks with different degrees of resolution. For quantitative analysis, a resolution >1.5 is highly desirable.

EXAMPLE **Measuring Resolution**

A peak with a retention time of 407 s has a width at half-height of 7.6 s. A neighboring peak is eluted 17 s later with $w_{1/2} = 9.4$ s. Find the resolution for these two components.

Solution

$$\text{Resolution} = \frac{0.589\Delta t_r}{w_{1/2av}} = \frac{0.589(17 \text{ s})}{\frac{1}{2}(7.6 \text{ s} + 9.4 \text{ s})} = 1.18$$

Test Yourself What difference in retention times is required for an adequate resolution of 1.5? (**Answer:** 21.6 s)

Diffusion

A band of solute broadens as it moves through a chromatography column (Figure 22-11). Ideally, an infinitely narrow band applied to the inlet of the column emerges with a Gaussian shape at the outlet. In less ideal circumstances, the band becomes asymmetric.

One main cause of band spreading is **diffusion,** which is the net transport of a solute from a region of high concentration to a region of low concentration caused by the random

(a)

(b)

(c)

FIGURE 22-11 (*a*) Schematic representation of broadening of an initially sharp band of solute as it moves through a chromatography column. (*b*) Experimentally observed diffusional broadening of a band after 2 and 26 min in a capillary electrophoresis column. (c) Expanded view of Gaussian bandshape at 26 min. [From M. U. Musheev, S. Javaherian, V. Okhonin, and S. N. Krylov, "Diffusion as a Tool of Measuring Temperature Inside a Capillary," *Anal. Chem.* **2008**, *80*, 6752.]

movement of molecules. Figure 22-12 lets us visualize the random motion of molecules through the *Brownian motion* of a fluorescent bead inside a microscopic water droplet. The bead is pushed by water molecules moving with random directions and speeds. Changes in the x and y coordinates of the bead in successive intervals follow a Gaussian distribution.

The **diffusion coefficient** measures the rate at which molecules move randomly from a region of high concentration to a region of low concentration. Figure 22-13 shows motion of solute across a plane with a concentration gradient dc/dx. The number of moles crossing each square meter per second, called the *flux*, *J*, is proportional to the concentration gradient:

Equation 22-24 is called *Fick's first law of diffusion*. If concentration is expressed as mol/m³, the units of *D* are m²/s.

Definition of diffusion coefficient:
$$\text{Flux}\left(\frac{\text{mol}}{\text{m}^2 \cdot \text{s}}\right) \equiv J = -D\frac{dc}{dx} \qquad (22\text{-}24)$$

The constant of proportionality, *D*, is the diffusion coefficient, and the negative sign is necessary because the net flux is from the region of high concentration to the region of low concentration.

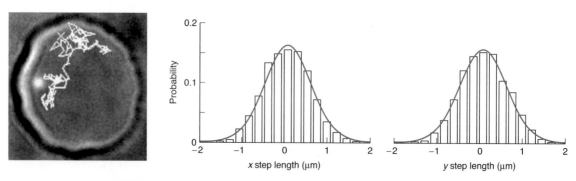

FIGURE 22-12 Brownian motion of a 290-nm-diameter fluorescent bead in a 20-μm-diameter water drop at 155-ms intervals. Histograms show Δx and Δy for each step in 800 photographs. Smooth curve is Gaussian fit. [From J. C. Gadd, C. L. Kuyper, B. S. Fujimoto, R. W. Allen, and D. T. Chiu, "Sizing Subcellular Organelles and Nanoparticles Confined Within Aqueous Droplets," *Anal. Chem.* **2008**, *80*, 3450.]

TABLE 22-1   **Representative diffusion coefficients at 298 K**

Solute	Solvent	Diffusion coefficient (m^2/s)
H_2O	H_2O	2.3×10^{-9}
Sucrose	H_2O	0.52×10^{-9}
Glycine	H_2O	1.1×10^{-9}
CH_3OH	H_2O	1.6×10^{-9}
Ribonuclease (FM 13 700)	H_2O (293 K)	0.12×10^{-9}
Serum albumin (FM 66 000)	H_2O (293 K)	0.059×10^{-9}
I_2	Hexane	4.0×10^{-9}
CCl_4	Heptane	3.2×10^{-9}
N_2	CCl_4	3.4×10^{-9}
$CS_2(g)$	Air (293 K)	1.0×10^{-5}
$O_2 (g)$	Air (273 K)	1.8×10^{-5}
H^+	H_2O	9.3×10^{-9}
OH^-	H_2O	5.3×10^{-9}
Li^+	H_2O	1.0×10^{-9}
Na^+	H_2O	1.3×10^{-9}
K^+	H_2O	2.0×10^{-9}
Cl^-	H_2O	2.0×10^{-9}
I^-	H_2O	2.0×10^{-9}

FIGURE 22-13 The flux of molecules diffusing across a plane of unit area is proportional to the concentration gradient and to the diffusion coefficient: $J = -D(dc/dx)$.

Table 22-1 shows that diffusion in liquids is 10^4 times slower than diffusion in gases. Macromolecules such as ribonuclease and albumin diffuse 10 to 100 times slower than small molecules.

If solute begins its journey through a column in an infinitely sharp layer with m moles per unit cross-sectional area of the column and spreads by diffusion as it travels, then the Gaussian profile of the band is described by

Broadening of chromatography band by diffusion:

$$c = \frac{m}{\sqrt{4\pi Dt}} e^{-x^2/4Dt} \qquad (22\text{-}25)$$

where c is concentration (mol/m^3), t is time, and x is the distance along the column from the current center of the band. (The band center is always $x = 0$ in this equation.) Comparison of Equations 22-25 and 4-3 shows that the standard deviation of the band is

Standard deviation of band: $\sigma = \sqrt{2Dt}$ (22-26)

Plate Height: A Measure of Column Efficiency

Equation 22-26 tells us that the standard deviation for diffusive band spreading is $\sqrt{2Dt}$. If solute has traveled a distance x at the linear flow rate u_x (m/s), then the time it has been on the column is $t = x/u_x$. Therefore,

$$\sigma^2 = 2Dt = 2D\frac{x}{u_x} = \underbrace{\left(\frac{2D}{u_x}\right)}_{\text{Plate height} \equiv H}x = Hx$$

Plate height: $H = \sigma^2/x$ (22-27)

Plate height, H, is the constant of proportionality between the variance, σ^2, of the band and the distance it has traveled, x. The name came from the theory of distillation in which separation could be performed in discrete stages called plates. Plate height is also called the *height equivalent to a theoretical plate*. It is approximately the length of column required for one equilibration of solute between mobile and stationary phases. We explore this concept later in Box 22-2. *The smaller the plate height, the narrower the bandwidth.*

The ability of a column to separate components of a mixture is improved by decreasing plate height. An efficient column has more theoretical plates than an inefficient column.

Bandwidth $\propto \sqrt{t}$. If elution time increases by a factor of 4, diffusion will broaden the band by a factor of 2.

u_x = linear flow rate (distance/time)
u_v = volume flow rate (volume/time)

As a teenager, A. J. P. Martin, coinventor of partition chromatography, built distillation columns in discrete sections from coffee cans. (We don't know what he was distilling!) When he formulated the theory of partition chromatography, he adopted terms from distillation theory.

Different solutes passing through the same column have different plate heights because they have different diffusion coefficients. Plate heights are ~0.1 to 1 mm in gas chromatography, ~10 μm in high-performance liquid chromatography, and <1 μm in capillary electrophoresis.

Plate height is the length σ^2/x, where σ is the standard deviation of the Gaussian band in Figure 22-9 and x is the distance traveled. For solute emerging from a column of length L, the number of plates, N, in the entire column is the length L divided by the plate height:

$$N = \frac{L}{H} = \frac{Lx}{\sigma^2} = \frac{L^2}{\sigma^2} = \frac{16L^2}{w^2}$$

because $x = L$ and $\sigma = w/4$. In this expression, w has units of length and the number of plates is dimensionless. If we express L and w (or σ) in units of time instead of length, N is still dimensionless. We obtain a useful expression for N by writing

Number of plates on column:
$$N = \frac{16t_r^2}{w^2} = \frac{t_r^2}{\sigma^2}$$
(22-28a)

where t_r is the retention time of the peak and w is the width at the base in Figure 22-9 *in units of time*. If we use the width at half-height, also called the *half-width*, instead of the width at the base, we get

Number of plates on column:
$$N = \frac{5.55t_r^2}{w_{1/2}^2}$$
(22-28b)

EXAMPLE Measuring Plates

A solute with a retention time of 407 s has a width at the base of 13.0 s on a column 12.2 m long. Find the number of plates and plate height.

Solution

$$N = \frac{16t_r^2}{w^2} = \frac{16 \cdot 407^2}{13.0^2} = 1.57 \times 10^4$$

$$H = \frac{L}{N} = \frac{12.2 \text{ m}}{1.57 \times 10^4} = 0.78 \text{ mm}$$

Test Yourself The half-width of the same peak is 7.6 s. Find the plate height. (**Answer:** 0.77 mm)

To estimate the number of theoretical plates for the asymmetric peak in Figure 22-14, draw a horizontal line across the band at 1/10 of the maximum height. The quantities A and B can then be measured, and the number of plates is[12]

$$N \approx \frac{41.7(t_r/w_{0.1})^2}{A/B + 1.25}$$
(22-29)

where $w_{0.1}$ is the width ($= A + B$) at 1/10 height.

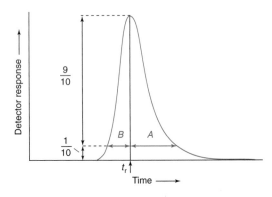

FIGURE 22-14 Asymmetric peak showing parameters used to estimate the number of theoretical plates.

Small plate height ⇒ narrow peaks ⇒ better separations

Choose a peak with a retention factor greater than 5 when you measure plate height for a column.

Challenge If N is constant, show that the width of a chromatographic peak increases in proportion to retention time. That is, successive peaks in a chromatogram should be increasingly broad.

All quantities must be measured in the same units, such as time or length.

Factors Affecting Resolution

For two closely spaced, symmetric peaks, the relation between plates and resolution is[13]

$$\text{Resolution} = \frac{\sqrt{N}}{4}(\gamma - 1) \qquad (22\text{-}30)$$

where N is the number of theoretical plates and γ is the *unadjusted relative retention* in Equation 22-16. For retention times $t_A = 341$ s and $t_B = 348$ s, $\gamma = 348/341 = 1.021$.

One important feature of Equation 22-30 is that resolution is proportional to $\sqrt{N}$. Therefore, *doubling the column length increases resolution by* $\sqrt{2}$. Figure 22-15 shows the effect of column length on the separation of L-phenylalanine from L-phenylalanine-D_5, in which the phenyl ring bears five deuterium atoms. The mixture was cycled repeatedly through a pair of chromatography columns by an ingenious valving system. After 1 pass in Figure 22-15, the peaks are barely resolved. By 15 passes, baseline separation has been achieved. The inset shows that the square of resolution is proportional to the number of passes, as predicted by Equation 22-30.

Unadjusted relative retention: $\gamma = \dfrac{u_A}{u_B} = \dfrac{t_B}{t_A}$

u_A, u_B = linear velocities of components A and B
t_A, t_B = retention times of components A and B

Resolution $\propto \sqrt{N} \propto \sqrt{L}$

FIGURE 22-15 Separation of 0.5 mM L-phenylalanine and 0.5 mM L-phenylalanine-D_5 by repeated passes through a pair of Hypersil C8 liquid chromatography columns (25 cm × 4.6 mm) eluted with 10:90 acetonitrile:water containing 25 mM Na_2SO_4 and 0.1% trifluoroacetic acid in the water. The relative retention on the first pass is $\alpha = 1.03$. [From K. Lan and J. W. Jorgensen, "Pressure-Induced Retention Variations in Reversed-Phase Alternate-Pumping Recycle Chromatography," *Anal. Chem.* **1998**, *70*, 2773.]

Equation 22-30 also tells us that resolution increases as the unadjusted relative retention γ increases. The unadjusted relative retention is the relative velocity of the two components through the column. The way to change relative velocity is to change the stationary phase in gas chromatography or either the stationary or the mobile phase in liquid chromatography. Important equations from chromatography are summarized in Table 22-2.

EXAMPLE   Plates Needed for Desired Resolution

Two solutes have an unadjusted relative retention of $\gamma = 1.06$. How many plates are required to give a resolution of 1.5? Of 3.0? If the plate height is 0.5 mm in gas chromatography, how long must the column be for a resolution of 1.5?

Solution We use Equation 22-30:

$$\text{Resolution} = 1.5 = \frac{\sqrt{N}}{4}(\gamma - 1) \Rightarrow N = \left(\frac{4(1.5)}{1.06 - 1}\right)^2 = 1.0 \times 10^4 \text{ plates}$$

To double the resolution to 3.0 requires four times as many plates $= 4.0 \times 10^4$ plates. For resolution $= 1.5$, the required length is (0.5 mm/plate)(1.0×10^4 plates) $= 5$ m.

Test Yourself If $\gamma = 1.06$ and $H = 3$ μm in liquid chromatography, what length of column in cm gives a resolution of 1.5? (**Answer:** 3.0 cm)

TABLE 22-2 Summary of chromatography equations

Quantity	Equation	Parameters
Partition coefficient	$K = c_s/c_m$	c_s = concentration of solute in stationary phase
		c_m = concentration of solute in mobile phase
Adjusted retention time	$t_r' = t_r - t_m$	t_r = retention time of solute of interest
		t_m = retention time of unretained solute
Retention volume	$V_r = t_r \cdot u_v$	u_v = volume flow rate = volume/unit time
Retention factor	$k = t_r'/t_m = KV_s/V_m$	V_s = volume of stationary phase
		V_m = volume of mobile phase
	$k = \dfrac{t_s}{t_m}$	t_s = time solute spends in stationary phase
		t_m = time solute spends in mobile phase
Relative retention	$\alpha = \dfrac{t_{r2}'}{t_{r1}'} = \dfrac{k_2}{k_1} = \dfrac{K_2}{K_1}$	Subscripts 1 and 2 refer to two solutes
Unadjusted relative retention	$\gamma = t_2/t_1 \; (\gamma > 1)$	t_2 = retention time of solute 2
		t_1 = retention time of solute 1
		w = width at base
Number of plates	$N = \dfrac{16t_r^2}{w^2} = \dfrac{5.55t_r^2}{w_{1/2}^2}$	$w_{1/2}$ = width at half-height
Plate height	$H = \dfrac{\sigma^2}{x} = \dfrac{L}{N}$	σ = standard deviation of band
		x = distance traveled by center of band
		L = length of column
		N = number of plates on column
Resolution	$\text{Resolution} = \dfrac{\Delta t_r}{w_{av}} = \dfrac{\Delta V_r}{w_{av}}$	Δt_r = difference in retention times
		ΔV_r = difference in retention volumes
		w_{av} = average width measured at baseline in same units as numerator (time or volume)
	$\text{Resolution} = \dfrac{\sqrt{N}}{4}(\gamma - 1)$	N = number of plates
		γ = unadjusted relative retention

22-5 Why Bands Spread[14]

A band of solute invariably spreads as it travels through a chromatography column (Figure 22-11) and emerges at the detector with a standard deviation σ. Each individual mechanism contributing to broadening produces a standard deviation σ_i. The observed variance (σ_{obs}^2) is the sum of variances from all contributing mechanisms:

Variance is additive, but standard deviation is not.

Variance is additive:
$$\sigma_{obs}^2 = \sigma_1^2 + \sigma_2^2 + \sigma_3^2 + \cdots = \sum \sigma_i^2 \qquad (22\text{-}31)$$

Broadening Outside the Column

Solute cannot be applied to the column in an infinitesimally thin zone, so the band has a finite width even before it enters the column. If the band is applied as a plug of width Δt (measured in units of time), the contribution to the variance of the final bandwidth is

Variance due to injection or detection:
$$\sigma_{injection}^2 = \sigma_{detector}^2 = \dfrac{(\Delta t)^2}{12} \qquad (22\text{-}32)$$

The same relation holds for broadening in a detector that requires a time Δt for the sample to pass through. Sometimes on-column detection is possible, thereby eliminating the problem of band spreading in a detector.

EXAMPLE Band Spreading Before and After the Column

A band from a column eluted at a rate of 1.35 mL/min has a width at half-height of 16.3 s. The sample was applied as a sharp plug with a volume of 0.30 mL, and the detector volume is 0.20 mL. Find the variances introduced by injection and detection. What would $w_{1/2}$ be if broadening occurred only on the column?

Solution Figure 22-9 tells us that the width at half-height is $w_{1/2} = 2.35\sigma$. Therefore, the observed total variance is

$$\sigma_{obs}^2 = \left(\frac{w_{1/2}}{2.35}\right)^2 = \left(\frac{16.3}{2.35}\right)^2 = 48.11 \text{ s}^2$$

Injection time is $\Delta t_{injection} = (0.30 \text{ mL})/(1.35 \text{ mL/min}) = 0.222 \text{ min} = 13.3 \text{ s}$. Therefore,

$$\sigma_{injection}^2 = \frac{\Delta t_{injection}^2}{12} = \frac{13.3^2}{12} = 14.81 \text{ s}^2$$

The time spent in the detector is $\Delta t_{detector} = (0.20 \text{ mL})/(1.35 \text{ mL/min}) = 8.89 \text{ s}$, so $\sigma_{detector}^2 = \Delta t_{detector}^2/12 = 6.58 \text{ s}^2$. The observed variance is

$$\sigma_{obs}^2 = \sigma_{column}^2 + \sigma_{injection}^2 + \sigma_{detector}^2$$

$$48.11 = \sigma_{column}^2 + 14.81 + 6.58 \quad \Rightarrow \quad \sigma_{column} = 5.17 \text{ s}$$

The width due to column broadening alone is $w_{1/2} = 2.35\sigma_{column} = 12.1 \text{ s}$, which is about three-fourths of the observed width.

Test Yourself Predict $w_{1/2}$ if the injected volume were decreased to 0.15 mL?
(*Answer:* 14.3 s)

The worst possible band spreading occurs in the large dead space beneath some rudimentary benchtop chromatography columns where each new drop exiting the column mixes with a significant volume of eluate already in the dead space. To minimize band spreading, dead spaces and tubing lengths should be minimized. Sample should be applied uniformly in a narrow zone and allowed to enter the column before mixing with eluent.

Plate Height Equation

Plate height, H, is proportional to the variance of a chromatographic band (Equation 22-27): the smaller the plate height, the narrower the band. The **van Deemter equation** tells us how the column and flow rate affect plate height:

van Deemter equation for plate height:

$$H \approx A + \frac{B}{u_x} + Cu_x \qquad (22\text{-}33)$$

$$\underset{\text{Multiple paths}}{} \quad \underset{\text{Longitudinal diffusion}}{} \quad \underset{\text{Equilibration time}}{}$$

where u_x is the linear flow rate and A, B, and C are constants for a given column and stationary phase. Changing the column and stationary phase changes A, B, and C. The van Deemter equation says there are band-broadening mechanisms that are proportional to flow rate, inversely proportional to flow rate, and independent of flow rate (Figure 22-16).

In packed columns, all three terms contribute to band broadening. For open tubular columns, the multiple path term, A, is 0, so bandwidth decreases and resolution increases. In capillary electrophoresis (Chapter 25), both A and C go to 0, thereby reducing plate height to submicron values and providing extraordinary separation powers.

Longitudinal Diffusion

If you could apply a thin, disk-shaped band of solute to the center of a column, the band would slowly broaden as molecules diffuse from the high concentration within the band to regions of lower concentration on the edges of the band. Diffusional broadening of a band, called **longitudinal diffusion** because it takes place along the axis of the column, occurs while the band is transported along the column by the flow of solvent (Figure 22-17).

The term B/u_x in Equation 22-33 arises from longitudinal diffusion. The faster the linear flow, the less time is spent in the column and the less diffusional broadening occurs. Equation 22-26 told us that the variance resulting from diffusion is

$$\sigma^2 = 2D_m t = \frac{2D_m L}{u_x}$$

Plate height due to longitudinal diffusion:

$$H_D = \frac{\sigma^2}{L} = \frac{2D_m}{u_x} \equiv \frac{B}{u_x} \qquad (22\text{-}34)$$

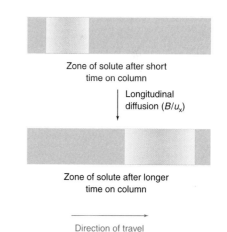

FIGURE 22-16 Application of van Deemter equation to gas chromatography: $A = 1.65$ mm, $B = 25.8$ mm · mL/min, and $C = 0.023\ 6$ mm · min/mL. [Experimental points from H. W. Moody, "The Evaluation of the Parameters in the van Deemter Equation," *J. Chem. Ed.* **1982**, *59*, 290.]

Packed columns: $A, B, C \neq 0$
Open tubular columns: $A = 0$
Capillary electrophoresis: $A = C = 0$

Zone of solute after short time on column

Longitudinal diffusion (B/u_x)

Zone of solute after longer time on column

Direction of travel

FIGURE 22-17 Longitudinal diffusion gives rise to B/u_x in the van Deemter equation. Solute continuously diffuses away from the concentrated center of its zone. The faster the flow, the less time is spent on the column and the less longitudinal diffusion occurs.

Longitudinal diffusion in a gas is much faster than diffusion in a liquid, so the optimum linear flow rate in gas chromatography is higher than in liquid chromatography.

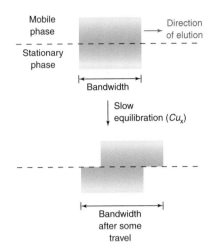

FIGURE 22-18 The finite time required for solute to equilibrate between mobile and stationary phases gives rise to Cu_x in the van Deemter equation. The slower the linear flow, the more complete equilibration is and the less zone broadening occurs.

Mobile phase

Direction of elution

Stationary phase

Bandwidth

Slow equilibration (Cu_x)

Bandwidth after some travel

The term A was formerly called the *eddy diffusion* term.

where D_m is the diffusion coefficient of solute in the mobile phase, t is time, and H_D is the plate height due to longitudinal diffusion. The time needed to travel the length of the column is L/u_x, where L is the column length and u_x is the linear flow rate.

Finite Equilibration Time Between Phases

The term Cu_x in Equation 22-33 comes from the finite time required for solute to equilibrate between mobile and stationary phases.[15] Although some solute is stuck in the stationary phase, the remainder in the mobile phase moves forward, spreading the overall zone of solute (Figure 22-18).

Plate height from finite equilibration time, also called the *mass transfer term*, is

Plate height due to finite equilibration time:
$$H_{\text{mass transfer}} = Cu_x = (C_s + C_m)u_x \qquad (22\text{-}35)$$

where C_s describes the rate of mass transfer through the stationary phase and C_m describes mass transfer through the mobile phase. Specific equations for C_s and C_m depend on the type of chromatography.

For gas chromatography in an open tubular column, the terms are

Mass transfer in stationary phase:
$$C_s = \frac{2k}{3(k+1)^2} \frac{d^2}{D_s} \qquad (22\text{-}35a)$$

Mass transfer in mobile phase:
$$C_m = \frac{1 + 6k + 11k^2}{24(k+1)^2} \frac{r^2}{D_m} \qquad (22\text{-}35b)$$

where k is the retention factor, d is the thickness of stationary phase, D_s is the diffusion coefficient of solute in the stationary phase, r is the column radius, and D_m is the diffusion coefficient of solute in the mobile phase. Decreasing stationary phase thickness, d, reduces plate height and increases efficiency because solute can diffuse faster from the farthest depths of the stationary phase into the mobile phase. Decreasing column radius, r, reduces plate height by decreasing the distance through which solute must diffuse to reach the stationary phase.

Mass transfer plate height is also decreased by increasing temperature, which increases the diffusion coefficient of solute in the stationary phase. In Figure 22-19, raising the temperature allows the linear flow rate to be increased by a factor of 5 while maintaining acceptable resolution. Resolution is maintained because of the increased rate of mass transfer between phases at elevated temperature. Many common silica-based stationary phases for liquid chromatography are not stable at elevated temperature. The zirconia (ZrO_2)-based material in Figure 22-19 is used because it is stable.

Multiple Flow Paths

The term A in the van Deemter equation (22-33) arises from multiple effects for which the theory is murky. Figure 22-20 is a pictorial explanation of one effect. Because some flow paths are longer than others, molecules entering the column at the same time on the left are

FIGURE 22-19 Liquid chromatography showing decreased analysis time when temperature is raised from 30° to 100°C. 1, uracil; 2, *p*-nitroaniline; 3, methyl benzoate; 4, phenetole; 5, toluene. The 4.6-mm-diameter × 100-cm-long column was packed with 4.5-μm-diameter zirconia (ZrO_2) coated with 2.1 wt% polybutadiene and eluted with 20 vol% acetonitrile in water. [From J. Li, Y. Hu, and P. W. Carr, "Fast Separations at Elevated Temperatures on Polybutadiene-Coated Zirconia Reversed-Phase Material," *Anal. Chem.* **1997**, *69*, 3884.] For a silica-based stationary phase, temperature is usually kept below 60°C to prevent hydrolysis of the silica.

Time ⟶

FIGURE 22-20 Band spreading from multiple flow paths. The smaller the stationary phase particles, the less serious this problem is. This process is absent in an open tubular column. [Adapted from H. M. McNair and E. J. Bonelli, *Basic Gas Chromatography* (Palo Alto, CA: Varian Instrument Division, 1968).]

eluted at different times on the right. For simplicity, we approximate many different effects by the constant A in Equation 22-33.

Advantages of Open Tubular Columns

In gas chromatography, we have a choice of using open tubular columns or packed columns. For similar analysis times, open tubular columns provide higher resolution and increased sensitivity to small quantities of analyte. Open tubular columns have small sample capacity, so they are not useful for preparative separations.

Particles in a packed column resist flow of the mobile phase, so the linear flow rate cannot be very fast. For the same length of column and applied pressure, the linear flow rate in an open tubular column is much higher than that of a packed column. Therefore, the open tubular column can be made 100 times longer than the packed column and still achieve a similar pressure drop and linear flow rate. If plate height is the same, the longer column provides 100 times more theoretical plates, yielding $\sqrt{100} = 10$ times more resolution.

Plate height is reduced in an open tubular column because band spreading by multiple flow paths (Figure 22-20) cannot occur. In the van Deemter curve for the packed column in Figure 22-16, the A term accounts for half of the plate height at the most efficient flow rate (minimum H) near 30 mL/min. If A were deleted, the number of plates on the column would be doubled. To obtain high performance from an open tubular column, the radius of the column must be small and the stationary phase must be as thin as possible to ensure rapid exchange of solute between mobile and stationary phases.

Table 22-3 compares the performances of packed and open tubular gas chromatography columns with the same stationary phase. For similar analysis times, the open tubular column gives resolution seven times better (10.6 versus 1.5) than that of the packed column. Alternatively, speed could be traded for resolution. If the open tubular column were reduced to 5 m in length, the same solutes could be separated with a resolution of 1.5, but the time would be reduced from 38.5 to 0.83 min.

A Touch of Reality: Asymmetric Bandshapes

A Gaussian bandshape results when the partition coefficient, K $(= c_s/c_m)$, is independent of the concentration of solute on the column. In real columns, K changes as the concentration of solute increases, and bandshapes are skewed.[16] A graph of c_s versus c_m (at a given temperature) is called an *isotherm*. Three common isotherms and their resulting bandshapes are shown in Figure 22-21. The ideal center isotherm gives a symmetric peak.

The upper isotherm in Figure 22-21 arises from an *overloaded* column in which too much solute has been applied to the column. As the concentration of solute increases, the solute becomes more and more soluble in the stationary phase. There is so much solute in the stationary phase that the stationary phase begins to resemble solute. (There is a rule of thumb in chemistry that "like dissolves like.") The front of an overloaded peak has gradually increasing concentration.

Compared with packed columns, open tubular columns can provide

- higher resolution
- shorter analysis time
- increased sensitivity
- lower sample capacity

For a given pressure, linear flow rate is proportional to cross-sectional area of the column and inversely proportional to column length:

$$u_x \propto \frac{\text{area}}{\text{length}}$$

Compared with packed columns, open tubular columns allow

- increased linear flow rate or a longer column or both
- decreased plate height, which means higher resolution

c_s = concentration of solute in stationary phase
c_m = concentration of solute in mobile phase

TABLE 22-3 Comparison of packed and wall-coated open tubular column performance[a]

Property	Packed	Open tubular
Column length, L	2.4 m	100 m
Linear gas velocity	8 cm/s	16 cm/s
Plate height for methyl oleate	0.73 mm	0.34 mm
Retention factor, k, for methyl oleate	58.6	2.7
Theoretical plates, N	3 290	294 000
Resolution of methyl stearate and methyl oleate	1.5	10.6
Retention time of methyl oleate	29.8 min	38.5 min

a. Methyl stearate ($CH_3(CH_2)_{16}CO_2CH_3$) and methyl oleate (cis-$CH_3(CH_2)_7CH{=}CH(CH_2)_7CO_2CH_3$) were separated on columns with poly(diethylene glycol succinate) stationary phase at 180°C.

SOURCE: L. S. Ettre. *Introduction to Open Tubular Columns* (Norwalk, CT: Perkin-Elmer Corp., 1979), p. 26.

Overloading produces a gradual rise and an abrupt fall of the chromatographic peak.

A long tail occurs when some sites retain solute more strongly than other sites.

As the concentration increases, the band becomes overloaded. The solute is so soluble in the overloaded zone that little solute trails behind the peak. The band emerges gradually from the column but ends abruptly.

The lower isotherm in Figure 22-21 arises when small quantities of solute are retained more strongly than large quantities. It leads to a long "tail" of gradually decreasing concentration after the peak.

Sites that bind solute strongly cause tailing. Silica surfaces of columns and stationary phase particles have hydroxyl groups that form hydrogen bonds with polar solutes, thereby leading to serious tailing. **Silanization** reduces tailing by blocking the hydroxyl groups with nonpolar trimethylsilyl groups:

$$
\begin{array}{c}
\overset{\displaystyle OH}{|} \qquad \overset{\displaystyle OH}{|} \\
-Si-O-Si- \; + \; (CH_3)_3SiNHSi(CH_3)_3 \; \rightarrow \;
\overset{\displaystyle (CH_3)_3SiO}{\underset{}{|}} \quad \overset{\displaystyle OSi(CH_3)_3}{\underset{}{|}} \\
\qquad\qquad -Si-O-Si- \; + \; NH_3
\end{array}
$$

(22-36)

Solid phase with exposed —OH groups   Hexamethyldisilazane   Protected surface

Glass and silica columns used for gas and liquid chromatography can also be silanized to minimize interaction of the solute with active sites on the walls.

Now that you have been exposed to many concepts, you might want to read about a microscopic model of chromatography in Box 22-2.

BOX 22-2   Microscopic Description of Chromatography

A *stochastic* theory provides a model for chromatography.[17] The term "stochastic" implies the presence of a random variable. The model supposes that as a molecule travels through a column, it spends an average time τ_m in the mobile phase between adsorption events. The time between desorption and the next adsorption is random, but the *average* time is τ_m. The average time spent adsorbed to the stationary phase between one adsorption and one desorption is τ_s. While the molecule is adsorbed on the stationary phase, it does not move. When the molecule is in the mobile phase, it moves with the speed u_x of the mobile phase. The probability that an adsorption or desorption occurs in a given time follows the Poisson distribution, which was described briefly in Problem 18-23.

We assume that all molecules spend total time t_m in the mobile phase. This is the retention time of unretained solute. Important results of the stochastic model are

- A solute molecule is adsorbed and desorbed an average of n times as it flows through the column, where $n = t_m/\tau_m$.
- The adjusted retention time for a solute is

$$t'_r = n\tau_s \qquad (A)$$

This is the average time that the solute is bound to the stationary phase during its transit through the column.
- The width (standard deviation) of a peak due to effects of the stationary phase is

$$\sigma = \tau_s\sqrt{2n} \qquad (B)$$

Consider the idealized chromatogram in the illustration, with one unretained component and two retained substances, A and B. The chromatographic parameters are representative of a high-performance liquid chromatography separation on a 15-cm-long × 0.39-cm-diameter column packed with 5-μm-diameter spherical particles of C_{18}-silica (Section 24-1). With a volume flow rate of 1.0 mL/min, the linear velocity is $u_x = 2.4$ mm/s. From the measured width at half-height ($w_{1/2}$), the standard deviation (σ) of a Gaussian peak is computed from $w_{1/2} = 2.35\sigma$ (Figure 22-9). The plate number for components A and B, computed with Equation 22-28, is $N = (t_r/\sigma)^2 = 1.00 \times 10^4$.

The stochastic model applies to processes involving the stationary phase. To analyze the chromatogram, we need to subtract contributions to peak broadening from dispersion in the mobile phase and extra-column effects such as finite injection width and finite detector volume. These effects account for the width of the unretained peak. To subtract the unwanted effects, we write

$$\sigma^2_{observed} = \sigma^2_{stationary\ phase} + \sigma^2_{unretained\ peak}$$

$$\sigma^2_{stationary\ phase} = \sigma^2_{observed} - \sigma^2_{unretained\ peak}$$

For component A, $\sigma^2_{stationary\ phase} = \sigma^2_{observed} - \sigma^2_{unretained\ peak} = (3.6\ s)^2 - (1.5\ s)^2 \Rightarrow \sigma_{stationary\ phase} = 3.27$ s. For component B, we find $\sigma_{stationary\ phase} = 5.81$ s. The adjusted retention time for component A is $t'_r = t_r - t_m = 360 - 60 = 300$ s. For component B, $t'_r = 600 - 60 = 540$ s.

Now we use t'_r and σ ($= \sigma_{stationary\ phase}$) for each component to find physically meaningful parameters. Combining Equations A and B above, we find

$$n = 2\left(\frac{t'_r}{\sigma}\right)^2 \qquad \tau_s = \frac{\sigma^2}{2t'_r}$$

and we already knew that $\tau_m = t_m/n$. From the parameters in the illustration, we compute the results in the table below.

	Component A	Component B
n	16 800	17 300
τ_s	17.8 ms	31.2 ms
τ_m	3.6 ms	3.5 ms
Distance between adsorptions ($= u_x\tau_m$)	8.6 μm	8.4 μm

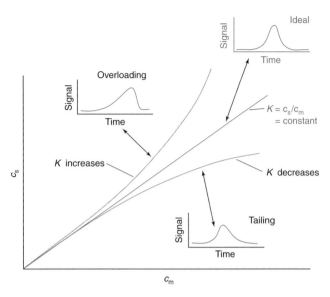

FIGURE 22-21 Common isotherms and their resulting chromatographic bandshapes.

Both components spend nearly the same time (~3.5 ms) in the mobile phase between adsorption events. Component A spends an average of 17.8 ms bound to the stationary phase each time it is adsorbed and component B spends an average of 31.2 ms. This difference in τ_s is the reason why A and B are separated from each other.

During its transit through the column, each substance becomes adsorbed $n \approx 17\,000$ times. The distance traveled between adsorptions is ~8.6 µm. The chromatogram was simulated for a column with $N = 10\,000$ theoretical plates. The plate height is 15 cm/(10 000 plates) = 15 µm. In Section 22-4, we stated that plate height is approximately the length of column required for one equilibration of solute between mobile and stationary phases. From the stochastic theory in this example, we find that there are approximately two equilibrations with the stationary phase in each length corresponding to the plate height.

The time required for a solute to flow past a stationary-phase particle whose diameter is $d = 5$ µm is $t = (5\ \mu\text{m})/(2.4\ \text{mm/s}) =$

2.1 ms. The stochastic theory predicts that the fraction of time that a molecule in the mobile phase will travel *less than* distance d is $1 - e^{-t/\tau_m} = 1 - e^{-(2.1\ \text{ms})/(3.5\ \text{ms})} = 0.55$. That is, approximately half of the time, a solute molecule does not travel as far as the next particle of stationary phase before becoming adsorbed again to the same particle from which it just desorbed. If we lined up spherical particles of stationary phase, it would take 30 000 particles to cover the 15-cm length of the column. Each solute molecule binds ~17 000 times as it transits the column, and half of those binding steps are to the same particle from which it just desorbed.

The model provides insight into microscopic events that occur during chromatography. The model omits some phenomena that occur in real columns. For example, in a porous stationary phase, the mobile phase could be stagnant inside the pores. When a molecule enters such a pore, it will adsorb and desorb many times from the same particle before escaping from the pore.

Idealized liquid chromatographic separation of 3 components.

Terms to Understand

adjusted retention time	elution	molecular exclusion	retention factor
adsorption chromatography	extraction	chromatography	retention time
affinity chromatography	gel filtration chromatography	open tubular column	retention volume
chromatogram	gel permeation chromatography	packed column	silanization
diffusion	ion-exchange chromatography	partition chromatography	stationary phase
diffusion coefficient	linear flow rate	partition coefficient	van Deemter equation
distribution coefficient	longitudinal diffusion	plate height	volume flow rate
eluate	miscible	relative retention	
eluent	mobile phase	resolution	

Summary

A solute can be extracted from one phase into another in which it is more soluble. The ratio of solute concentrations in each phase at equilibrium is called the partition coefficient. If more than one form of the solute exists, we use a distribution coefficient instead of a partition coefficient. We derived equations relating the fraction of solute extracted to the partition or distribution coefficient, volumes, and pH. Many small extractions are more effective than a few large extractions. A metal chelator, soluble only in organic solvents, can extract metal ions from aqueous solutions, with selectivity achieved by adjusting pH. Crown ethers and salts containing a hydrophobic ion act as phase transfer agents to bring a hydrophilic ion from an aqueous phase into an organic phase.

In adsorption and partition chromatography, a continuous equilibration of solute between mobile and stationary phases occurs. Eluent goes into a column and eluate comes out. Columns may be packed with stationary phase or may be open tubular, with stationary phase bonded to the inner wall. In ion-exchange chromatography, the solute is attracted to the stationary phase by coulombic forces. In molecular exclusion chromatography, the fraction of stationary phase volume available to solute decreases as the size of the solute molecules increases. Affinity chromatography relies on specific, noncovalent interactions between the stationary phase and one solute in a complex mixture.

The relative retention of two components is the quotient of their adjusted retention times. The retention factor for a single component is the adjusted retention time divided by the elution time for solvent. Retention factor gives the ratio of time spent by solute in the stationary phase to time spent in the mobile phase. When a separation is scaled up from a small load to a large load, linear flow rate is held constant and the cross-sectional area of the column should be increased in proportion to the loading.

Plate height ($H = \sigma^2/x$) is related to the breadth of a band emerging from the column. The smaller the plate height, the sharper the band. The number of plates for a Gaussian peak is $N = 5.55\, t_r^2/w_{1/2}^2$. Plate height is approximately the length of column required for one equilibration of solute between mobile and stationary phases. Resolution of neighboring peaks is the difference in retention time divided by the average width (measured at the baseline, $w = 4\sigma$). Resolution is proportional to $\sqrt{N}$ and increases with the unadjusted relative retention, γ, which is the quotient of linear velocities of two components. Doubling the length of a column increases resolution by $\sqrt{2}$.

The standard deviation of a diffusing band of solute is $\sigma = \sqrt{2Dt}$, where D is the diffusion coefficient and t is time. The van Deemter equation describes band broadening on the chromatographic column: $H \approx A + B/u_x + Cu_x$, where H is plate height, u_x is linear flow rate, and A, B, and C are constants. The first term represents irregular flow paths, the second, longitudinal diffusion, and the third, the finite rate of transfer of solute between mobile and stationary phases. The optimum flow rate, which minimizes plate height, is faster for gas chromatography than for liquid chromatography. The number of plates and the optimal flow rate increase as the stationary phase particle size decreases. In gas chromatography, open tubular columns can provide higher resolution or shorter analysis times than packed columns. Bands spread during injection and detection, as well as during passage through the column. The observed variance of the band is the sum of the variances for all mechanisms of spreading. Overloading and tailing can be corrected by using smaller samples and by masking strong adsorption sites on the stationary phase.

Exercises

22-A. Consider a chromatography experiment in which two components with retention factors $k_1 = 4.00$ and $k_2 = 5.00$ are injected into a column with $N = 1.00 \times 10^3$ theoretical plates. The retention time for the less-retained component is $t_{r1} = 10.0$ min.

(a) Calculate t_m and t_{r2}. Find $w_{1/2}$ (width at half-height) and w (width at the base) for each peak.

(b) Using graph paper, sketch the chromatogram analogous to Figure 22-7, supposing that the two peaks have the same amplitude (height). Draw the half-widths accurately.

(c) Calculate the resolution of the two peaks and compare this value with those drawn in Figure 22-10.

22-B. A solute with a partition coefficient of 4.0 is extracted from 10 mL of phase 1 into phase 2.

(a) What volume of phase 2 is needed to extract 99% of the solute in one extraction?

(b) What is the total volume of solvent 2 needed to remove 99% of the solute in three equal extractions instead?

22-C. (a) Find the retention factors for octane and nonane in Figure 22-7.

(b) Find the ratio

$$\frac{\text{Time octane spends in stationary phase}}{\text{Total time octane spends on column}}$$

(c) Find the relative retention for octane and nonane.

(d) Find the partition coefficient for octane by assuming that the volume of the stationary phase equals half the volume of the mobile phase.

22-D. A gas chromatogram of a mixture of toluene and ethyl acetate is shown here.

(a) Use the width of each peak (measured at the base) to calculate the number of theoretical plates in the column. Estimate all lengths to the nearest 0.1 mm.

(b) Using the width of the toluene peak at its base, calculate the width expected at half-height. Compare the measured and calculated values. When the thickness of the pen line is significant relative to the length being measured, it is important to take the pen line width into account. You can measure from the edge of one line to the corresponding edge of the other line, as shown here.

22-E. The three chromatograms shown below were obtained with 2.5, 1.0, and 0.4 μL of ethyl acetate injected on the same column under the same conditions. Explain why the peak becomes less symmetric with increasing sample size.

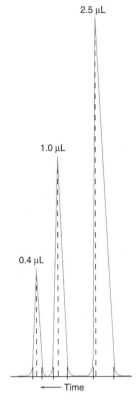

22-F. The relative retention for two compounds in gas chromatography is 1.068 on a column with a plate height of 0.520 mm. The retention factor for compound 1 is 5.16.

(a) Find the unadjusted relative retention (γ) for the two compounds.

(b) What length of column will separate the compounds with a resolution of 1.00?

(c) The retention time for air (t_m) is 2.00 min. If the number of plates is the same for both compounds, find t_r and $w_{1/2}$ for each peak.

(d) If the ratio of stationary phase to mobile phase is 0.30, find the partition coefficient for compound 1.

Problems

Solvent Extraction

22-1. If you are extracting a substance from water into ether, is it more effective to do one extraction with 300 mL of ether or three extractions with 100 mL?

22-2. If you wish to extract aqueous acetic acid into hexane, is it more effective to adjust the aqueous phase to pH 3 or pH 8?

22-3. (a) Why is it difficult to extract the EDTA complex of aluminum into an organic solvent but easy to extract the 8-hydroxyquinoline complex?

(b) If you need to bring the EDTA complex into the organic solvent, should you add a phase transfer agent with a hydrophobic cation or a hydrophobic anion?

22-4. Why is the extraction of a metal ion into an organic solvent with 8-hydroxyquinoline more complete at higher pH?

22-5. The distribution coefficient for extraction of a metal complex from aqueous to organic solvents is $D = [\text{total metal}]_{\text{org}}/[\text{total metal}]_{\text{aq}}$. Give physical reasons why β and K_a appear in the numerator of Equation 22-13, but K_L and $[\text{H}^+]_{\text{aq}}$ appear in the denominator.

22-6. Give a physical interpretation of Equations 22-6 and 22-7 in terms of the fractional composition equations for a monoprotic acid discussed in Section 9-5.

22-7. Solute S has a partition coefficient of 4.0 between water (phase 1) and chloroform (phase 2) in Equation 22-1.

(a) Calculate the concentration of S in chloroform if [S(aq)] is 0.020 M.

(b) If the volume of water is 80.0 mL and the volume of chloroform is 10.0 mL, find the quotient (mol S in chloroform)/(mol S in water).

22-8. The solute in Problem 22-7 is initially dissolved in 80.0 mL of water. It is then extracted six times with 10.0-mL portions of chloroform. Find the fraction of solute remaining in the aqueous phase.

22-9. The weak base B ($K_b = 1.0 \times 10^{-5}$) equilibrates between water (phase 1) and benzene (phase 2).

(a) Define the distribution coefficient, D, for this system.

(b) Explain the difference between D and K, the partition coefficient.

(c) Calculate D at pH 8.00 if $K = 50.0$.

(d) Will D be greater or less at pH 10 than at pH 8? Explain why.

22-10. Consider the extraction of M^{n+} from aqueous solution into organic solution by reaction with protonated ligand, HL:

$$M^{n+}(aq) + nHL(org) \rightleftharpoons ML_n(org) + nH^+(aq)$$

$$K_{extraction} = \frac{[ML_n]_{org}[H^+]_{aq}^n}{[M^{n+}]_{aq}[HL]_{org}^n}$$

Rewrite Equation 22-13 in terms of $K_{extraction}$ and express $K_{extraction}$ in terms of the constants in Equation 22-13. Give a physical reason why each constant increases or decreases $K_{extraction}$.

22-11. Butanoic acid has a partition coefficient of 3.0 (favoring benzene) when distributed between water and benzene. Find the formal concentration of butanoic acid in each phase when 100 mL of 0.10 M aqueous butanoic acid is extracted with 25 mL of benzene (a) at pH 4.00 and (b) at pH 10.00.

22-12. For a given value of $[HL]_{org}$ in Equation 22-13, over what pH range (how many pH units) will D change from 0.01 to 100 if $n = 2$?

22-13. For the extraction of Cu^{2+} by dithizone in CCl_4, $K_L = 1.1 \times 10^4$, $K_M = 7 \times 10^4$, $K_a = 3 \times 10^{-5}$, $\beta = 5 \times 10^{22}$, and $n = 2$.

(a) Calculate the distribution coefficient for extraction of 0.1 μM Cu^{2+} into CCl_4 by 0.1 mM dithizone at pH 1.0 and at pH 4.0.

(b) If 100 mL of 0.1 μM aqueous Cu^{2+} are extracted once with 10 mL of 0.1 mM dithizone at pH 1.0, what fraction of Cu^{2+} remains in the aqueous phase?

22-14. Consider the extraction of 100.0 mL of $M^{2+}(aq)$ by 2.0 mL of 1×10^{-5} M dithizone in $CHCl_3$, for which $K_L = 1.1 \times 10^4$, $K_M = 7 \times 10^4$, $K_a = 3 \times 10^{-5}$, $\beta = 5 \times 10^{18}$, and $n = 2$.

(a) Derive an expression for the fraction of metal ion extracted into the organic phase, in terms of the distribution coefficient and volumes of the two phases.

(b) 📊 Prepare a graph of the percentage of metal ion extracted over the pH range 0 to 5.

22-15. 📊 The theoretical limit for extracting solute S from phase 1 (volume V_1) into phase 2 (volume V_2) is attained by dividing V_2 into an infinite number of infinitesimally small portions and conducting an infinite number of extractions. With a partition coefficient $K = [S]_2/[S]_1$, the limiting fraction of solute remaining in phase 1 is $q_{limit} = e^{-(V_2/V_1)K}$. Let $V_1 = V_2 = 50$ mL and let $K = 2$. Let volume V_2 be divided into n equal portions to conduct n extractions. Find the fraction of S extracted into phase 2 for $n = 1, 2, 10$ extractions. How many portions are required to attain 95% of the theoretical limit?

A Plumber's View of Chromatography

22-16. Match the terms in the first list with the characteristics in the second list.

1. adsorption chromatography

2. partition chromatography

3. ion-exchange chromatography

4. molecular exclusion chromatography

5. affinity chromatography

A. Ions in mobile phase are attracted to counterions covalently attached to stationary phase.

B. Solute in mobile phase is attracted to specific groups covalently attached to stationary phase.

C. Solute equilibrates between mobile phase and surface of stationary phase.

D. Solute equilibrates between mobile phase and film of liquid attached to stationary phase.

E. Different-sized solutes penetrate voids in stationary phase to different extents. Largest solutes are eluted first.

22-17. The partition coefficient for a solute in chromatography is $K = c_s/c_m$, where c_s is the concentration in the stationary phase and c_m is the concentration in the mobile phase. The larger the partition coefficient, the longer it takes a solute to be eluted. Explain why.

22-18. (a) Write the meaning of the retention factor, k, in terms of time spent by solute in each phase.

(b) Write an expression in terms of k for the fraction of time spent by a solute molecule in the mobile phase.

(c) The *retention ratio* in chromatography is defined as

$$R = \frac{\text{time for solvent to pass through column}}{\text{time for solute to pass through column}} = \frac{t_m}{t_r}$$

Show that R is related to the retention factor by the equation $R = 1/(k + 1)$.

22-19. (a) A chromatography column with a length of 10.3 cm and inner diameter of 4.61 mm is packed with a stationary phase that occupies 61.0% of the volume. If the volume flow rate is 1.13 mL/min, find the linear flow rate in cm/min.

(b) How long does it take for solvent (which is the same as unretained solute) to pass through the column?

(c) Find the retention time for a solute with a retention factor of 10.0.

22-20. An open tubular column is 30.1 m long and has an inner diameter of 0.530 mm. It is coated on the inside wall with a layer of stationary phase that is 3.1 μm thick. Unretained solute passes through in 2.16 min, whereas a particular solute has a retention time of 17.32 min.

(a) Find the linear and volume flow rates.

(b) Find the retention factor for the solute and the fraction of time spent in the stationary phase.

(c) Find the partition coefficient, c_s/c_m, for this solute.

22-21. A chromatographic procedure separates 4.0 mg of unknown mixture on a column with a length of 40 cm and a diameter of 0.85 cm.

(a) What size column would you use to separate 100 mg of the same mixture?

(b) If the flow is 0.22 mL/min on the small column, what volume flow rate should be used on the large column?

(c) If mobile phase occupies 35% of the column volume, calculate the linear flow rate for the small column and the large column.

22-22. Solvent passes through a column in 3.0 min but solute requires 9.0 min.

(a) Calculate the retention factor, k.

(b) What fraction of time is the solute in the mobile phase in the column?

(c) The volume of stationary phase is 1/10 of the volume of the mobile phase in the column ($V_s = 0.10V_m$). Find the partition coefficient, K, for this system.

22-23. Solvent occupies 15% of the volume of a chromatography column whose inner diameter is 3.0 mm. If the volume flow rate is 0.2 mL/min, find the linear flow rate.

22-24. Consider a chromatography column in which $V_s = V_m/5$. Find the retention factor if $K = 3$ and if $K = 30$.

22-25. The retention volume of a solute is 76.2 mL for a column with $V_m = 16.6$ mL and $V_s = 12.7$ mL. Calculate the retention factor and the partition coefficient for this solute.

22-26. An open tubular column has an inner diameter of 207 μm and the thickness of the stationary phase on the inner wall is 0.50 μm. Unretained solute passes through in 63 s and a particular solute emerges in 433 s. Find the partition coefficient for this solute and find the fraction of time spent in the stationary phase.

22-27. Isotopic compounds are separated in Figure 22-15 by repeated passage through a pair of columns. Each cycle in the figure represents one pass through length $L = 50$ cm containing N theoretical plates. The unadjusted relative retention is $\gamma = 1.018$.

(a) The observed resolution after 10 cycles is 1.60. Calculate the number of theoretical plates, N, in column length L. The mixture has passed through length $10L$ in 10 cycles.

(b) Find the plate height in μm.

(c) Predict the resolution expected from two cycles. The observed value is 0.71.

Efficiency and Band Spreading

22-28. Chromatograms of compounds A and B were obtained at the same flow rate with two columns of equal length.

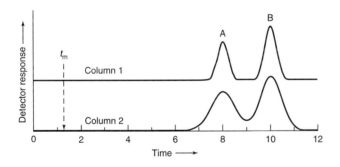

(a) Which column has more theoretical plates?

(b) Which column has a larger plate height?

(c) Which column gives higher resolution?

(d) Which column gives a greater relative retention?

(e) Which compound has a higher retention factor?

(f) Which compound has a greater partition coefficient?

(g) What is the numerical value of the unadjusted relative retention?

22-29. Why does plate height depend on linear flow rate, not volume flow rate?

22-30. Which column is more efficient: plate height = (a) 0.1 mm or (b) 1 mm?

22-31. Why is longitudinal diffusion a more serious problem in gas chromatography than in liquid chromatography?

22-32. In chromatography, why is the optimal flow rate greater if the stationary phase particle size is smaller?

22-33. What is the optimal flow rate in Figure 22-16 for best separation of solutes?

22-34. Explain how silanization reduces tailing of chromatographic peaks.

22-35. Describe how nonlinear partition isotherms lead to non-Gaussian bandshapes. Draw the bandshape produced by an overloaded column and a column with tailing.

22-36. A separation of 2.5 mg of an unknown mixture has been optimized on a column of length L and diameter d. Explain why you might not achieve the same resolution for 5.0 mg on a column of length $2L$ and diameter d.

22-37. An infinitely sharp zone of solute is placed at the center of a column at time $t = 0$. After diffusion for time t_1, the standard deviation of the Gaussian band is 1.0 mm. After 20 min more, at time t_2, the standard deviation is 2.0 mm. What will be the width after another 20 min, at time t_3?

22-38. A chromatogram with ideal Gaussian bands has $t_r = 9.0$ min and $w_{1/2} = 2.0$ min.

(a) How many theoretical plates are present?

(b) Find the plate height if the column is 10 cm long.

22-39. (a) The asymmetric chromatogram in Figure 22-14 has a retention time equal to 15 min, and the values of A and B are 33 and 11 s, respectively. Find the number of theoretical plates.

(b) The width of the Gaussian peak in Figure 22-9 at a height equal to 1/10 of the peak height is 4.297σ. Suppose that the peak in part (a) is symmetric with $A = B = 22$ s. Use Equations 22-28 and 22-29 to find the plate number.

22-40. Two chromatographic peaks with widths, w, of 6 min are eluted at 24 and 29 min. Which diagram in Figure 22-10 will most closely resemble the chromatogram?

22-41. A chromatographic band has a width, w, of 4.0 mL and a retention volume of 49 mL. What width is expected for a band with a retention volume of 127 mL? Assume that the only band spreading occurs on the column itself.

22-42. A band from a column eluted at 0.66 mL/min has a width at half-height, $w_{1/2}$, of 39.6 s. The sample was applied as a sharp plug with a volume of 0.40 mL, and the detector volume is 0.25 mL. Find the variances introduced by injection and detection. What would $w_{1/2}$ be if the only broadening occurred on the column?

22-43. Two compounds with partition coefficients of 15 and 18 are to be separated on a column with $V_m/V_s = 3.0$ and $t_m = 1.0$ min. Calculate the number of theoretical plates needed to produce a resolution of 1.5.

22-44. (a) Calculate the number of theoretical plates needed to achieve a resolution of 2.0 for compounds with unadjusted relative retention times of $t_2/t_1 = 1.01$, 1.05, or 1.10.

(b) How can you increase N and $\gamma = t_2/t_1$ in a chromatography experiment?

22-45. Consider the peaks for pentafluorobenzene and benzene in the chromatogram below. The elution time for unretained solute is 1.06 min. The open tubular column is 30.0 m in length and 0.530 mm in diameter, with a layer of stationary phase 3.0 μm thick on the inner wall.

(a) Find the adjusted retention times and retention factors for both compounds.

(b) Find the relative retention, α.

(c) Find the unadjusted relative retention, γ.

(d) Measuring $w_{1/2}$ on the chromatogram, find the number of plates, N_1 and N_2, and the plate height for these two compounds.

(e) Measuring the width, w, at the baseline on the chromatogram, find the number of plates for these two compounds.

(f) Use your answer to part (e) to find the resolution between the two peaks.

(g) Using the number of plates [$N = \sqrt{N_1 N_2}$, with values from part (e)] and the observed unadjusted relative retention, calculate what the resolution should be and compare your answer with the measured resolution in part (f).

22-46. A layer with negligible thickness containing 10.0 nmol of methanol ($D = 1.6 \times 10^{-9}$ m²/s) was placed in a tube of water 5.00 cm in diameter and allowed to spread by diffusion. Using Equation 22-25, prepare a graph showing the Gaussian concentration profile of the methanol zone after 1.00, 10.0, and 100 min. Prepare a second graph showing the same experiment with the enzyme ribonuclease ($D = 0.12 \times 10^{-9}$ m²/s).

22-47. A 0.25-mm-diameter open tubular gas chromatography column is coated with stationary phase that is 0.25 μm thick. The diffusion coefficient for a compound with a retention factor $k = 10$ is $D_m = 1.0 \times 10^{-5}$ m²/s in the gas phase and $D_s = 1.0 \times 10^{-9}$ m²/s in the stationary phase. Consider longitudinal diffusion and finite equilibration time in the mobile and stationary phases as sources of broadening. Prepare a graph showing the plate height from each of these three sources and the total plate height as a function of linear flow rate (from 2 cm/s to 1 m/s). Then change the stationary phase thickness to 2 μm and repeat the calculations. Explain the difference in the two results.

22-48. Consider two Gaussian peaks with relative areas of 4:1. Construct a set of graphs to show the overlapping peaks if the resolution is 0.5, 1, or 2.

23 Gas Chromatography

Gas chromatogram of cholesterol and other lipids extracted from bones and derivatized with trimethylsilyl $((CH_3)_3Si—)$ groups to increase volatility for chromatography. Bone contains 2 to 50 μg cholesterol/gram of dry bone. [From A. W. Stott and R. P. Evershed, "$\delta^{13}C$ Analysis of Cholesterol Preserved in Archaeological Bones and Teeth," *Anal. Chem.* **1996**, *68*, 4402. Excellent questions for student interpretation of this article can be found in L. Roecker, "Introducing Students to the Scientific Literature," *J. Chem. Ed.* **2007**, *84*, 1380.]

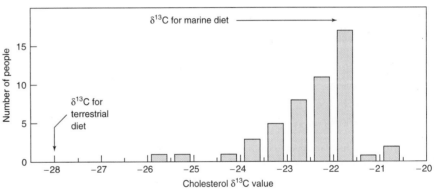

^{13}C content of cholesterol from bones of 50 people who lived on British coast in the years A.D. 500–1800. $\delta^{13}C$ is the deviation of the atomic ratio $^{13}C/^{12}C$ from that of a standard material, measured in parts per thousand. [Data from A. W. Stott and R. P. Evershed, "$\delta^{13}C$ Analysis of Cholesterol Preserved in Archaeological Bones and Teeth," *Anal. Chem.* **1996**, *68*, 4402.]

The ^{13}C content of cholesterol preserved in ancient bones provides information on the diets of people who lived long ago. Approximately 1.1% of natural carbon is ^{13}C and 98.9% is ^{12}C. Different types of plants and animals have consistent, slightly different ratios of $^{13}C/^{12}C$, which reflect their biosynthetic pathways.

To learn whether people from the ancient British coastal town of Barton-on-Humber ate mainly plants or fish, cholesterol from the bones of 50 people was extracted with organic solvent, isolated by gas chromatography, combusted to convert it into CO_2, and its $^{13}C/^{12}C$ ratio measured by mass spectrometry. Observed $^{13}C/^{12}C$ ratios differ from that of a standard material by about -21 to -24 parts per thousand. A diet of local plants yields a $\delta^{13}C$ value (defined in Box 21-3) in cholesterol of -28 parts per thousand. Values more positive than -28 parts per thousand are indicative of a marine diet. It appears that the population got much of its food from the sea.

Chapter 22 gave a foundation for understanding chromatographic separations. Chapters 23 through 25 discuss specific methods and instrumentation. The goal is for you to understand how chromatographic methods work and what parameters you can control for best results.[1]

23-1 The Separation Process in Gas Chromatography

In **gas chromatography**,[2,3,4] gaseous analyte is transported through the column by a gaseous mobile phase, called the **carrier gas.** In *gas-liquid partition chromatography*, the stationary phase is a nonvolatile liquid bonded to the inside of the column or to a fine solid support

Gas chromatography:

mobile phase: gas

stationary phase: usually a nonvolatile liquid, but sometimes a solid

analyte: gas or volatile liquid

FIGURE 23-1 Schematic diagram of a gas chromatograph.

(Figure 22-6, upper right). In *gas-solid adsorption chromatography*, analyte is adsorbed directly on solid particles of stationary phase (Figure 22-6, upper left).

In the schematic *gas chromatograph* in Figure 23-1, volatile liquid or gaseous sample is injected through a **septum** (a rubber disk) into a heated port, in which it rapidly evaporates. Vapor is swept through the column by He, N_2, or H_2 carrier gas, and separated analytes flow through a detector, whose response is displayed on a computer. The column must be hot enough to provide sufficient vapor pressure for analytes to be eluted in a reasonable time. The detector is maintained at a higher temperature than the column so analytes will be gaseous.

Open Tubular Columns

The vast majority of analyses use long, narrow **open tubular columns** (Figure 23-2) made of fused silica (SiO_2) and coated with polyimide (a plastic capable of withstanding 350°C) for support and protection from atmospheric moisture.[5] As discussed in Section 22-5, open tubular columns offer higher resolution, shorter analysis time, and greater sensitivity than packed columns, but they have less sample capacity.

The *wall-coated* column in Figure 23-2c features a 0.1- to 5-μm-thick film of stationary liquid phase on the inner wall of the column. A *support-coated* column has solid particles coated with stationary liquid phase and attached to the inner wall. In the *porous-layer* column in Figure 23-3, solid particles *are* the active stationary phase. With their high surface area, support-coated columns can handle larger samples than can wall-coated columns. The performance of support-coated columns is intermediate between those of wall-coated columns and packed columns.

The choice of carrier gas depends on the detector and the desired separation efficiency and speed.

Compared with packed columns, open tubular columns offer
- higher resolution
- shorter analysis time
- greater sensitivity
- lower sample capacity

Wall-coated open tubular column (WCOT): liquid stationary phase on inside wall of column

Support-coated open tubular column (SCOT): liquid stationary phase coated on solid support attached to inside wall of column

Porous-layer open tubular column (PLOT): solid stationary phase on inside wall of column

FIGURE 23-2 (*a*) Typical dimensions of open tubular gas chromatography column. (*b*) Fused-silica column with a cage diameter of 0.2 m and column length of 15–100 m. (*c*) Cross-sectional view of wall-coated, support-coated, and porous-layer columns.

1 Acetaldehyde
2 Methanol
3 Ethanol
4 Ethyl acetate
5 2-Methyl-2-propanol
6 1-Butanol
7 2-Methyl-1-butanol
8 3-Methyl-1-butanol
9 Isoamyl acetate
10 Ethyl caproate
11 2-Phenylethanol

(a)

(b)

FIGURE 23-3 (a) Porous carbon stationary phase (2 μm thick) on inside wall of fused-silica open tubular column. (b) Chromatogram of vapors from the headspace of a beer can, obtained with 0.25-mm-diameter × 30-m-long porous carbon column operated at 30°C for 2 min and then ramped up to 160°C at 20°/min. [Courtesy Alltech Associates, State College, PA.]

Column inner diameters are typically 0.10 to 0.53 mm and lengths are 15 to 100 m, with 30 m being common. Narrow columns provide higher resolution than wider columns (Figure 23-4 and Equation 22-35b) but require higher operating pressure and have less sample capacity. Diameters ≥0.32 mm tend to overload the vacuum system of a mass spectrometer, so the gas stream must be split and only a fraction sent to the spectrometer. The number of theoretical plates, N, on a column is proportional to length. In Equation 22-30, resolution is proportional to $\sqrt{N}$ and, therefore, to the square root of column length (Figure 23-5).

At the constant linear velocity in Figure 23-6, increasing the thickness of the stationary phase increases retention time and sample capacity and increases resolution of early-eluting peaks with a *retention factor* (Equation 22-17) of $k \lesssim 5$. Thick films of stationary phase can shield analytes from the silica surface and reduce *tailing* (Figure 22-21), but

Equation 22-30:

$$\text{Resolution} = \frac{\sqrt{N}}{4}(\gamma - 1)$$

N = plate number
γ = unadjusted relative retention

Equation 22-17:

$$k = \frac{t_r - t_m}{t_m}$$

t_r = retention time of solute
t_m = transit time of solvent

1. 1,3-Dichlorobenzene
2. 1,4-Dichlorobenzene
3. *sec*-Butylbenzene
4. 1,2-Dichlorobenzene

0.32-mm inner diameter

0.25-mm inner diameter

FIGURE 23-4 Effect of open tubular column inner diameter on resolution. Narrower columns provide higher resolution. Notice the increased resolution of peaks 1 and 2 in the narrow column. Conditions: DB-1 stationary phase (0.25 μm thick) in 15-m wall-coated column operated at 95°C with He linear velocity of 34 cm/s. [Courtesy J&W Scientific, Folsom, CA.]

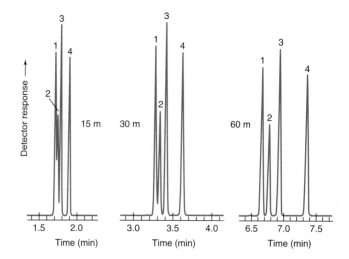

FIGURE 23-5 Resolution increases in proportion to the square root of column length. Notice the increased resolution of peaks 1 and 2 as length is increased. Conditions: DB-1 stationary phase (0.25 μm thick) in 0.32-mm-diameter wall-coated column operated at 95°C with He linear velocity of 34 cm/s. Compounds 1–4 are the same as in Figure 23-4. [Courtesy J&W Scientific, Folsom, CA.]

(Diphenyl)(dimethyl)-polysiloxane

Arylene polysiloxane

they can also increase bleed (decomposition and evaporation) of the stationary phase at elevated temperature. A thickness of 0.25 μm is standard, but thicker films are used for volatile analytes.

The choice of liquid stationary phase (Table 23-1) is based on the rule "like dissolves like." Nonpolar columns are best for nonpolar solutes (Table 23-2). Columns of intermediate polarity are best for intermediate polarity solutes, and strongly polar columns are best for strongly polar solutes. Box 23-1 describes *chiral* (optically active) bonded phases for separating optical isomers.

As a column ages, stationary phase can be lost, surface silanol groups (Si—O—H) are exposed, and tailing increases. To reduce the tendency of stationary phase to bleed from the column at high temperature, it is usually *bonded* (covalently attached) to the silica surface and covalently *cross-linked* to itself. To monitor column performance, it is good practice to periodically measure the retention factor of a standard (Equation 22-17), the number of plates (Equation 22-28), and peak asymmetry (Figure 22-14). Changes in these parameters indicate degradation of the column.

At upper operating temperatures, stationary phases decompose, giving a slow "bleed" of decomposition products from the column. These products produce elevated background signal in most detectors, reducing the signal-to-noise for analyte and potentially contaminating the detector. Arylene stationary phases have increased thermal stability, bleed less at high temperature, and are especially suitable for gas chromatography–mass spectrometry. Compared with (diphenyl)(dimethyl)polysiloxanes, arylene phases produce some differences in relative retention of different compounds.

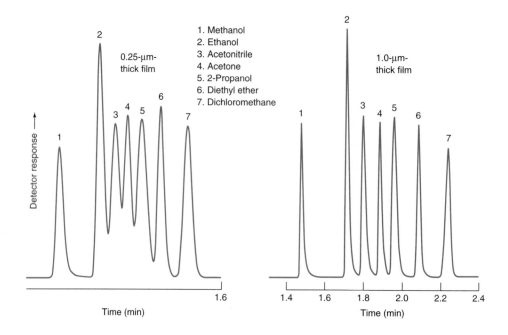

FIGURE 23-6 Effect of stationary phase thickness on open tubular column performance. Increasing thickness increases retention time and increases resolution of early-eluting peaks. Conditions: DB-1 stationary phase in 15-m-long × 0.32-mm-diameter wall-coated column operated at 40°C with He linear velocity of 38 cm/s. [Courtesy J&W Scientific, Folsom, CA.]

1. Methanol
2. Ethanol
3. Acetonitrile
4. Acetone
5. 2-Propanol
6. Diethyl ether
7. Dichloromethane

TABLE 23-1   Common stationary phases in capillary gas chromatography

Structure	Polarity	Temperature range (°C)
(Diphenyl)$_x$(dimethyl)$_{1-x}$ polysiloxane	$x = 0$ Nonpolar $x = 0.05$ Nonpolar $x = 0.35$ Intermediate polarity $x = 0.65$ Intermediate polarity	$-60°\text{–}320°$ $-60°\text{–}320°$ $0°\text{–}300°$ $50°\text{–}370°$
(Cyanopropylphenyl)$_{0.14}$ (dimethyl)$_{0.86}$ polysiloxane	Intermediate polarity	$-20°\text{–}280°$
$\text{+CH}_2\text{CH}_2\text{—O}\text{}_n$ Carbowax (poly(ethylene glycol))	Strongly polar	$40°\text{–}250°$
(Biscyanopropyl)$_{0.9}$ (cyanopropylphenyl)$_{0.1}$ polysiloxane	Strongly polar	$0°\text{–}275°$

TABLE 23-2   Polarity of solutes

Nonpolar	Weak intermediate polarity
Saturated hydrocarbons	Ethers
Olefinic hydrocarbons	Ketones
Aromatic hydrocarbons	Aldehydes
Halocarbons	Esters
Mercaptans	Tertiary amines
Sulfides	Nitro compounds (without α-H atoms)
CS_2	Nitriles (without α-atoms)

Strong intermediate polarity	Strongly polar
Alcohols	Polyhydroxyalcohols
Carboxylic acids	Amino alcohols
Phenols	Hydroxy acids
Primary and secondary amines	Polyprotic acids
Oximes	Polyphenols
Nitro compounds (with α-H atoms)	
Nitriles (with α-H atoms)	

SOURCE: *Adapted from H. M. McNair and E. J. Bonelli, Basic Gas Chromatography (Palo Alto, CA: Varian Instrument Division, 1968).*

BOX 23-1 **Chiral Phases for Separating Optical Isomers**

Optical isomers—also called *enantiomers*—are mirror image compounds that cannot be superimposed. For example, the natural amino acid building blocks of proteins are L-amino acids.

Mirror plane

L-amino acid

D-amino acid

Enantiomers of an amino acid

Volatile derivative for gas chromatography

Chromatography with a *chiral* (optically active) stationary phase is one of the few ways to separate enantiomers. We can estimate ages of fossils up to 500 million years old by measuring the fraction of amino acid that has transformed into the D enantiomer in a fossil.[6,7] Amino acids do not have enough vapor pressure for gas chromatography. A volatile derivative suitable for gas chromatography is shown in the figure above.[8]

Common chiral stationary phases for gas chromatography have *cyclodextrins* bonded to a conventional polysiloxane stationary phase.[9,10] Cyclodextrins are naturally occurring cyclic sugars. β-Cyclodextrin has a 0.78-nm-diameter opening into a chiral, hydrophobic cavity. The hydroxyls are capped with alkyl groups to decrease the polarity of the faces.

Primary alcohol

Secondary alcohol

(a)

Chiral hydrophobic cavity

0.78 nm

Hydrophilic entrance

Secondary face (14 OH groups)

Primary face (7 OH groups)

(b)

(*a*) Structure of β-cyclodextrin, a cyclic sugar made of seven glucose molecules. (α-Cyclodextrin contains six monomers and γ-cyclodextrin contains eight.) (*b*) Primary hydroxyl groups lie on one face and the secondary hydroxyl groups lie on the other face.

Enantiomers have different affinities for the cyclodextrin cavity, so they separate as they travel through the chromatography column. The chromatogram shows a chiral separation of a by-product found in pesticides.

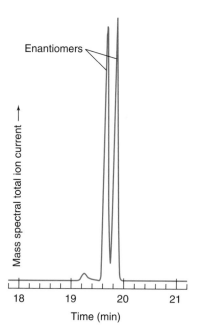

Enantiomers

Programmed temperature (120°–200°C) chiral separation on a 0.25-mm × 25-m open tubular column with a 0.25-μm-thick stationary phase containing 10 wt% fully methylated β-cyclodextrin chemically bonded to poly(dimethylsiloxane). [From W. Vetter and W. Jun, "Elucidation of a Polychlorinated Bipyrrole Structure Using Enantioselective GC," *Anal. Chem.* **2002**, *74*, 4287.]

Mirror plane

Chlorinated pesticide impurity. The two rings are perpendicular to each other. The mirror images are not superimposable, because there is no free rotation about the C—N bond between the rings.

To reduce interference from column bleeding, use the thinnest possible stationary phase and the narrowest and shortest column that provides adequate separation. Oxidation of the stationary phase by O_2 is also a major source of bleed. High-purity carrier gas should be used, and it should be passed through an O_2 scrubber before the column. Even 1 ppb of O_2 slowly degrades the column. To a lesser extent, H_2O can break down the stationary phase by hydrolysis. To minimize bleed, manufacturers modify the silica surface of the capillary to eliminate exposed silanol groups (Si—OH), which can initiate breakdown of the stationary phase.

Ionic liquids are the newest type of stationary phase for gas chromatography. They melt below room temperature and have a wide liquid range with low volatility at elevated temperature. Ionic liquids have the potential to offer novel selectivities for polar analytes and increased operating temperature with low bleed.

1,9-Di(3-vinylimidazolium)nonane bis(trifluoromethyl)sulfonylimidate
(Supelco SP-IL 100 stationary phase)

Among the solids used for porous-layer open tubular columns, porous polymers, high-surface-area carbon (Figure 23-3), and *alumina* (Al_2O_3) can separate hydrocarbons in gas-solid adsorption chromatography. **Molecular sieves** (Figure 23-7) are inorganic or organic materials with cavities into which small molecules enter and are partially retained.[11] Molecules such as H_2, O_2, N_2, CO_2, and CH_4 can be separated from one another. Gases can be dried by passage through traps containing molecular sieves because water is strongly retained. Inorganic sieves can be regenerated (dried) by heating to 300°C in vacuum or under flowing N_2.

Cubo-octahedron

(a) ◯ Oxygen ⬤ Si or Al (b) Open cavity

FIGURE 23-7 Structure of the molecular sieve $Na_{12}(Al_{12}Si_{12}O_{48}) \cdot 27H_2O$. (*a*) Aluminosilicate framework of one cubo-octahedron of a mineral class called *zeolites*. (*b*) Interconnection of eight cubo-octahedra to produce a cavity into which small molecules can enter.

Packed Columns

Packed columns contain fine particles of solid support coated with nonvolatile liquid stationary phase, or the solid itself may be the stationary phase. Compared with open tubular columns, packed columns provide greater sample capacity but give broader peaks, longer retention times, and less resolution. (Compare Figure 23-8 with Figure 23-3.) Despite their inferior resolution, packed columns are used for preparative separations, which require a great deal of stationary phase, or to separate gases that are poorly retained. Packed columns are usually made of stainless steel or glass and are typically 3–6 mm in diameter and 1–5 m in length. The solid support is often silica that is *silanized* (Reaction 22-36) to reduce hydrogen bonding to polar solutes. For tenaciously binding solutes, Teflon is a useful support, but it is limited to <200°C.

In a packed column, uniform particle size decreases the multiple path term in the van Deemter equation (22-33), thereby reducing plate height and increasing resolution. Small particle

Teflon is a chemically inert polymer with the structure —CF_2—CF_2—CF_2—CF_2—.

FIGURE 23-8 Chromatogram of alcohol mixture at 40°C using packed column (2 mm inner diameter × 76 cm long) containing 20% Carbowax 20M on Gas-Chrom R support and flame ionization detector. [Courtesy Norman Pearson.]

size decreases the time required for solute equilibration, thereby improving column efficiency. However, the smaller the particle size, the less space between particles and the more pressure required to force mobile phase through the column. Particle size is expressed in micrometers or as a *mesh size*, which refers to the size of screens through which the particles are passed or retained (Table 27-2). A 100/200 mesh particle passes through a 100 mesh screen, but not through a 200 mesh screen. The mesh number equals the number of openings per linear inch of screen.

Retention Index

Figure 23-9 illustrates how the relative retention times of polar and nonpolar solutes change as the polarity of the stationary phase changes. In Figure 23-9a, 10 compounds are eluted nearly in order of increasing boiling point from a nonpolar stationary phase. The principal determinant of retention on this column is the volatility of the solutes. In Figure 23-9b, the strongly polar stationary phase retains the strongly polar solutes. The three alcohols are the last to be eluted, following the three ketones, which follow four alkanes. Hydrogen bonding

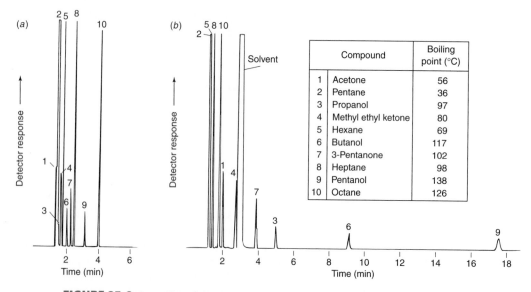

	Compound	Boiling point (°C)
1	Acetone	56
2	Pentane	36
3	Propanol	97
4	Methyl ethyl ketone	80
5	Hexane	69
6	Butanol	117
7	3-Pentanone	102
8	Heptane	98
9	Pentanol	138
10	Octane	126

FIGURE 23-9 Separation of 10 compounds on (*a*) nonpolar poly(dimethylsiloxane) and (*b*) strongly polar poly(ethylene glycol) 1-μm-thick stationary phases in 0.32-mm-diameter × 30-m-long open tubular columns at 70°C. [Courtesy Restek Co., Bellefonte, PA.]

to the stationary phase is probably the strongest force leading to retention. Dipole interactions of the ketones are the second strongest force.

The Kovats **retention index,** I, for a linear alkane equals 100 times the number of carbon atoms. For octane, $I = 800$; and for nonane, $I = 900$. A compound eluted between octane and nonane (Figure 22-7) has a retention index between 800 and 900 computed by the formula

Retention index:
$$I = 100\left[n + (N - n)\frac{\log t_r'(\text{unknown}) - \log t_r'(n)}{\log t_r'(N) - \log t_r'(n)} \right] \tag{23-1}$$

Retention index relates the retention time of a solute to the retention times of linear alkanes.

where n is the number of carbon atoms in the *smaller* alkane; N is the number of carbon atoms in the *larger* alkane; $t_r'(n)$ is the adjusted retention time of the *smaller* alkane; and $t_r'(N)$ is the adjusted retention time of the *larger* alkane.

Adjusted retention time $= t_r' = t_r - t_m$
t_r = retention time for solute
t_m = time for unretained solute (CH_4) to pass through column

EXAMPLE   Retention Index

If the retention times in Figure 22-7 are $t_r(CH_4) = 0.5$ min, $t_r(\text{octane}) = 14.3$ min, $t_r(\text{unknown}) = 15.7$ min, and $t_r(\text{nonane}) = 18.5$ min, find the retention index for the unknown.

Solution The index is computed with Equation 23-1:

$$I = 100\left[8 + (9 - 8)\frac{\log 15.2 - \log 13.8}{\log 18.0 - \log 13.8} \right] = 836$$

Test Yourself Where would an unknown with a retention index of 936 be eluted in Figure 22-7? (***Answer:*** after nonane)

The retention index of 657 for benzene on poly(dimethylsiloxane) in Table 23-3 means that benzene is eluted between hexane ($I \equiv 600$) and heptane ($I \equiv 700$) from this nonpolar stationary phase. Nitropropane is eluted just after heptane on the same column. As we go down the table, the stationary phases become more polar. For (biscyanopropyl)$_{0.9}$-(cyanopropylphenyl)$_{0.1}$polysiloxane at the bottom of the table, benzene is eluted after decane, and nitropropane is eluted after n-$C_{14}H_{30}$.

When identifying an eluted compound by comparing its mass spectrum with a mass spectral library, false matches are frequent. If retention index is used as a second characteristic, false matches are reduced.

Temperature and Pressure Programming

A large fraction of all gas chromatography is run with **temperature programming,** in which the temperature of the column is raised *during* the separation to increase analyte vapor pressure

Raising column temperature:
• decreases retention time
• sharpens peaks

TABLE 23-3   Retention indexes for several compounds on common stationary phases

	Retention index[a]				
Phase	**Benzene** b.p. 80°C	**Butanol** b.p. 117°C	**2-Pentanone** b.p. 102°C	**1-Nitropropane** b.p. 132°C	**Pyridine** b.p. 116°C
Poly(dimethylsiloxane)	657	648	670	708	737
(Diphenyl)$_{0.05}$(dimethyl)$_{0.95}$-polysiloxane	672	664	691	745	761
(Diphenyl)$_{0.35}$ (dimethyl)$_{0.65}$-polysiloxane	754	717	777	871	879
(Cyanopropylphenyl)$_{0.14}$-(dimethyl)$_{0.86}$ polysiloxane	726	773	784	880	852
(Diphenyl)$_{0.65}$(dimethyl)$_{0.35}$-polysiloxane	797	779	824	941	943
Poly(ethylene glycol)	956	1 142	987	1 217	1 185
(Biscyanopropyl)$_{0.9}$-(cyanopropylphenyl)$_{0.1}$-polysiloxane	1 061	1 232	1 174	1 409	1 331

a. For reference, boiling points (b.p.) for various alkanes are hexane, 69°C; heptane, 98°C; octane, 126°C; nonane, 151°C; decane, 174°C; undecane, 196°C. Retention indexes for the straight-chain alkanes are fixed values and do not vary with the stationary phase: hexane, 600; heptane, 700; octane, 800; nonane, 900; decane, 1 000; undecane, 1 100.

SOURCE: *Restek Chromatography Products Catalog, 1993–94, Bellefonte, PA.*

FIGURE 23-10 Programmed-temperature chromatography of linear alkanes at three different heating rates on a 3.0-m-long poly(dimethylpolylsiloxane) column etched in a silicon chip. [From S. Reidy, D. George, M. Agah, and R. Sacks, "Temperature-Programmed GC Using Silicon Microfabricated Columns with Integrated Heaters and Temperature Sensors," *Anal. Chem.* **2007,** *79,* 2911.]

and decrease retention times of late-eluting components. Figure 23-10 shows the elution of C_5 to C_{15} linear alkanes from a 3-m-long column etched in a silicon chip described in Box 23-2. At a constant temperature of 30°C (not shown), heavier alkanes take so long to be eluted and emerge over such a long time that they would not be detected. The three traces in Figure 23-10 show what happens when column temperature is raised from 30° to 150°C at rates of (a) 20°C/min, (b) 40°C/min, and (c) 60°C/min. Broad, late-eluting peaks can be sharpened and eluted in less time with temperature programming. To maintain adequate resolution of earlier eluting peaks, programs often include a period of time at constant, low temperature prior to raising the temperature.

Most gas chromatography columns come with a label showing two temperatures limits. The lower one is the isothermal temperature limit at which the column can be kept for a long time. The upper one is the programmed temperature limit to which the column should only be exposed for a few minutes at a time at the end of a programmed temperature run. High temperatures decompose the stationary phase and cause column "bleeding." An increase in baseline signal at low temperature is an indicator of column degradation. Other signs of column degradation are peak broadening, tailing, and changing retention times.

Many chromatographs are equipped with electronic pressure control of the carrier gas. Increasing the inlet pressure increases the flow of mobile phase and decreases retention time. In some cases, programmed pressure can be used instead of programmed temperature to reduce retention times of late-eluting components. At the end of a run, the pressure can be rapidly reduced to its initial value for the next run. Time is not wasted waiting for a hot column to cool before the next injection. Programmed pressure is useful for analytes that cannot tolerate high temperature.

Carrier Gas

Helium is the most common carrier gas and is compatible with most detectors. For a flame ionization detector, N_2 gives a lower detection limit than He. Figure 23-11 shows that H_2, He, and N_2 give essentially the same optimal plate height (0.3 mm) at significantly different flow rates. Optimal flow rate increases in the order $N_2 < He < H_2$. Fastest separations can be achieved with H_2 as carrier gas, and H_2 can be run much faster than its optimal velocity with little penalty in resolution.[12] Figure 23-12 shows the effect of carrier gas on the separation of two compounds on the same column with the same temperature program.

The main reason H_2 was not used more often in the past is that concentrations >4 vol% in air are explosive. Flow rates in capillary chromatography are unlikely to create a dangerous concentration of H_2. Electrolytic generators produce high-purity H_2 and eliminate the need for tanks of compressed H_2. For gas chromatography–mass spectrometry, H_2 reduces the

FIGURE 23-11 van Deemter curves for gas chromatography of n-$C_{17}H_{36}$ at 175°C, using N_2, He, or H_2 in a 0.25-mm-diameter × 25-m-long wall-coated column with OV-101 stationary phase. [From R. R. Freeman, ed., *High Resolution Gas Chromatography* (Palo Alto, CA: Hewlett Packard Co., 1981).]

FIGURE 23-12 Separation of two polyaromatic hydrocarbons on a wall-coated open tubular column with different carrier gases. Resolution, R, increases and analysis time decreases as we change from N_2 to He to H_2 carrier gas. [Courtesy J&W Scientific, Folsom, CA.]

efficiency of a turbomolecular vacuum pump but has little effect on a diffusion pump.[13] It is possible for H_2 to react with unsaturated compounds on metal surfaces.

H_2 and He give better resolution (smaller plate height) than N_2 at high flow rate because solutes diffuse more rapidly through H_2 and He than through N_2. The more rapidly a solute diffuses between phases, the smaller is the mass transfer (Cu_x) term in the van Deemter equation (22-33). Equations 22-35a and 22-35b describe the effects of the finite rate of mass transfer in an open tubular column. If the stationary phase is thin enough ($\lesssim 0.5$ μm), mass transfer is dominated by slow diffusion through the *mobile phase* rather than through the *stationary phase*. That is, $C_s \ll C_m$ in Equations 22-35a and 22-35b. For a column of a given radius, r, and a solute of a given retention factor, k, the only variable affecting the rate of mass transfer in the mobile phase (Equation 22-35b) is the diffusion coefficient of solute through the mobile phase. Diffusion coefficients follow the order $H_2 > He > N_2$.

Most analyses are run at carrier gas velocities that are 1.5 to 2 times greater than the optimum velocity at the minimum of the van Deemter curve. The higher velocity is chosen to give maximum efficiency (most theoretical plates) per unit time. A decrease in resolution is tolerated in return for faster analyses.

Gas flow through a narrow column may be too low for best detector performance, so extra *makeup gas* is sometimes added between the column and the detector. Makeup gas that is optimum for detection can be a different gas from that used in the column.

Impurities in carrier gas degrade the stationary phase. High-quality gas should be used, and it should be passed through purifiers to remove traces of O_2, H_2O, and organic compounds. An oxygen indicator trap should be in line after the main purifier. Steel or copper tubing, rather than plastic or rubber, should be used for gas lines because metals are less permeable to air and do not release volatile contaminants into the gas stream. As with thermal degradation, symptoms of oxidative degradation of the stationary phase include increased baseline signal at low temperature, peak broadening and tailing, and altered retention times.

Guard Columns and Retention Gaps

In gas chromatography, a *guard column* and a *retention gap* are both typically a 3- to 10-m length of empty capillary in front of the capillary chromatography column. The capillary is silanized so that solutes are not retained by the bare silica wall. Physically, the guard column and the retention gap are identical, but they are employed for different purposes.

The purpose of a **guard column** is to accumulate nonvolatile substances that would otherwise contaminate the chromatography column and degrade its performance. Periodically,

van Deemter equation:

$$H \approx A + \frac{B}{u_x} + Cu_x$$

Multiple Longitudinal Equilibration
paths   diffusion   time

Guard column: accumulates nonvolatile substances that would contaminate chromatography column

BOX 23-2 **Chromatography Column on a Chip**

The column for Figure 23-10 is part of a microfabricated instrument being developed for environmental monitoring, medical diagnosis, and forensic science. A 3-m-long square spiral channel was etched into silicon, whose edge dimension is 3.2 cm. In panel *a*, gas flows into the white spiral and out from the colored spiral. The connection between the two paths at the center of the structure is shown in panel *b*. A glass plate bonded to the top of the silicon creates a gas-tight channel, which is coated with a ~0.15-μm-thick layer of cross-linked poly(dimethylsiloxane) (Table 23-1). The carrier gas for a chip designed for field deployment is air that is filtered to remove water and organic vapors.

Efficiency of the column (plate number) increases as the channels are made narrower. Panel *c* shows the measured effect of channel width on optimum plate height. The smaller the plate height, the greater the plate number. The narrower the channels, the less time is needed for solute to diffuse from the gas phase to the mobile phase and the more rapidly equilibration occurs between phases.

In addition to band broadening by longitudinal diffusion and finite equilibration time (the terms B/u_x and Cu_x in the van Deemter equation 22-33), broadening is created by right-angle corners where gas at the outside of the corner travels farther than gas at the inside of the corner. This problem was reduced in later designs by making curved bends instead of right-angle bends. Parts of the system outside the column create band broadening that is proportional to the square of the linear velocity—giving an additional term Du_x^2 in the van Deemter equation. Major challenges in creating a small, autonomous chromatograph include making low-power miniature pumps, injectors, detectors, and a gas sampling system to run on battery power.

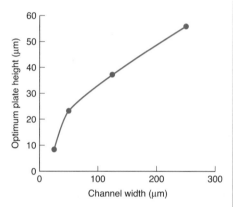

(*a*) *Layout of square spiral.* (*b*) Gas flow turnaround at center of chip and etched-channel structure (150-μm-wide × 240-μm-deep channels). [From G. Lambertus, A. Elstro, K. Sensenig, J. Potkay, M. Agah, S. Scheuering, K. Wise, F. Dorman, and R. Sacks, "Design, Fabrication, and Evaluation of Microfabricated Columns for Gas Chromatography," *Anal. Chem.* **2004**, *76*, 2629. See also S. Reidy, G. Lambertus, J. Reece, and R. Sacks, "High-Performance, Static-Coated Silicon Microfabricated Columns for Gas Chromatography," *Anal. Chem.* **2006**, *78*, 2623.]

(*c*) Optimum plate height decreases (column efficiency increases) as channel width decreases. This graph applies to a chip with round bends, not 90° bends. [Data from M. A. Zareian-Jahromi, M. Ashraf-Khorassani, L. T. Taylor, and M. Agah, "Design, Modeling, and Fabrication of MEMS-Based Multicapillary Gas Chromatographic Columns," *J. Microelectromech. Syst.* **2009**, *18*, 28.]

the beginning of the guard column should be cut off to eliminate nonvolatile residues. Trim the guard column when you observe irregular peak shapes from a column that had been producing symmetric peaks. It is a good idea to trim 10–20 cm off the guard column every time the injection inlet liner is changed. With electronic pneumatically controlled chromatographs, be sure to enter the new length of the guard column in the control software.

A **retention gap** is used to improve peak shapes under certain conditions. If you introduce a large volume of sample (>2 μL) by *splitless* or *on-column* injection (described in the next section), microdroplets of liquid solvent can persist inside the column for the first few meters. Solutes dissolved in the droplets are carried along with them and give rise to a series of ragged bands. The retention gap allows solvent to evaporate prior to entering the chromatography column. Use at least 1 m of retention gap per microliter of solvent. Even small volumes of solvent with very different polarity from the stationary phase can cause irregular peak shapes. The retention gap helps separate solvent from solute to improve peak shapes.

We calculate plate number, N, with Equation 22-28 by using retention time and peak width. Plate height, H, is the length of the column, L, divided by N. Do not include the retention gap or guard column as part of L for calculating H.[14] For peaks with a retention factor $k < 5$, plate height might not be meaningful when a guard column or retention gap is used.

Retention gap: improves peak shape by separating volatile solvent from less volatile solutes prior to chromatography

Glass barrel

Air   Sample   Air   Solvent   Air

FIGURE 23-13 "Sandwich" injection technique. [Adapted from J. T. Watson, *Introduction to Mass Spectrometry*, 3rd ed. (Philadelphia: Lippincott-Raven, 1997).]

23-2 Sample Injection[15]

Figure 23-13 shows a good technique for using a syringe to inject liquid sample into a gas chromatograph. After cleaning the syringe several times with solvent, take up air, then solvent, then air, then sample, and then more air. When the needle is inserted through the rubber septum into the heated injection port of the chromatograph, sample does not immediately evaporate because there is no sample in the needle. If there were sample in the needle, the most volatile components would begin to evaporate and would be depleted before the sample is injected. The air bubble behind the sample plug prevents sample and solvent from mixing. The solvent plug washes sample out of the needle, and the final air plug expels solvent from the needle. Many autosamplers are capable of this "sandwich" injection.

An injection port with a silanized glass liner is shown in Figure 23-14. Carrier gas sweeps vaporized sample from the port into the chromatography column. For analytical chromatography, the injected volume is typically 0.1–2 μL of liquid sample. Gases are injected by a gas-tight syringe into the same kind of sample loop used in liquid chromatography (Figure 24-18). Decomposed sample, nonvolatile components, and septum debris accumulate in the glass liner, which is periodically replaced. Glass wool is used near the bottom of some liners to trap particles and pyrolysis products to prevent them from reaching the column. The liner must seal properly or carrier gas will bypass the liner. The lifetime of the rubber septum could be as little as 20 manual injections or ~100 autosampler injections.

Different injection port liners are designed for split, splitless, and on-column injection and for use with solid-phase microextraction.

Split Injection

If analytes of interest constitute >0.1% of the sample, **split injection** is usually preferred. For high-resolution work, best results are obtained with the smallest amount of sample (≤1 μL) that can be adequately detected—preferably containing ≤1 ng of each component. A complete injection contains too much material for a 0.32-mm-diameter or smaller column. A split injection delivers only 0.2–2% of the sample to the column. In Figure 23-14, sample is injected rapidly (<1 s) through the septum into the evaporation zone. The injector temperature is kept high (for example, 350°C) to promote fast evaporation. A brisk flow of carrier gas sweeps the sample through the *mixing chamber*, where complete vaporization and good mixing occur. At the split point, a small fraction of vapor enters the chromatography column, but most passes through needle valve 2 to a waste vent. The pressure regulator leading to needle valve 2 controls the fraction of sample discarded. The proportion of sample that does not reach the column is called the *split ratio* and typically ranges from 50:1 to 600:1. After sample has been flushed from the injection port (~30 s), needle valve 2 is closed and carrier gas

Injection into open tubular columns:

split: routine for introducing small sample volume into open tubular column

splitless: best for trace levels of high-boiling solutes in low-boiling solvents

on-column: best for thermally unstable solutes and high-boiling solvents; best for quantitative analysis

To vent

Needle valve

Sample evaporation zone   Heated metal wall   Silanized glass liner   Mixing chamber

Split point

Inject here

Septum

Flow controller

Pressure regulator

Chromatography column

Carrier gas in   Molecular sieves   Hydrocarbon trap   Oxygen scrubber

Needle valve

To vent

FIGURE 23-14 Injection port for split injection into an open tubular column. The glass liner is slowly contaminated by nonvolatile and decomposed samples and must be replaced periodically. For splitless injection, the glass liner is a straight tube with no mixing chamber. For dirty samples, split injection is used and a packing material can be placed inside the liner to adsorb undesirable components of the sample. Glass wool might be placed at the near end of the liner so that liquid on the outside of the syringe needle gets wiped off by the wool before the needle is withdrawn.

flow is correspondingly reduced. Quantitative analysis with split injection can be inaccurate because the split ratio is not reproducible from run to run.

A 1-μL liquid injection creates roughly 0.5 mL of gas volume, which can fill the glass liner in Figure 23-14. Some vapor can escape backward toward the septum. Lower boiling components evaporate first and are more likely to escape than components with higher boiling points. The injection port temperature should be high enough to minimize this fractionation of the sample. However, if the injector temperature is too high, decomposition can occur. During injection and chromatography, *septum purge* gas flow through needle valve 1 in Figure 23-14 is run at ~1 mL/min to remove excess sample vapor and gas that bleeds from the hot rubber septum.

Splitless Injection

For trace analysis[16] of analytes that are less than 0.01% of the sample, **splitless injection** is appropriate. The same port shown for split injection in Figure 23-14 is used. However, the glass liner is a straight, empty tube with no mixing chamber, as shown in Figure 23-15. A large volume (~2 μL) of dilute solution in a low-boiling solvent is injected slowly (~2 s) into the liner, with the split vent closed. Slow flow through the septum purge is maintained during injection and chromatography to remove any vapors that escape from the injection liner. Injector temperature for splitless injection is lower (~220°C) than that for split injection, because the sample spends more time in the port and we do not want it to decompose. The residence time of the sample in the glass liner is ~1 min, because carrier gas flows through the liner at the column flow rate, which is ~1 mL/min. In splitless injection, ~80% of the sample is applied to the column, and little fractionation occurs during injection.

The initial column temperature is set 40°C *below* the boiling point of the solvent, which therefore condenses at the beginning of the column. As solutes catch up with the condensed plug of solvent, they are trapped in the solvent in a narrow band at the beginning of the column. This **solvent trapping** leads to sharp chromatographic peaks. Without solvent trapping, the bands could not be sharper than the 1-min injection time. Chromatography is initiated by raising the column temperature to vaporize the solvent trapped at the head of the column.

An alternative means of condensing solutes in a narrow band at the beginning of the column is called **cold trapping.** The initial column temperature is 150°C lower than the boiling points of the solutes of interest. Solvent and low-boiling components are eluted rapidly, but high-boiling solutes remain in a narrow band at the beginning of the column. The column is then rapidly warmed to initiate chromatography of the high-boiling solutes. For low-boiling solutes, *cryogenic focusing* is required, with an initial column temperature below room temperature.

Figure 23-16 shows effects of operating parameters in split and splitless injections. Experiment A is a standard split injection with brisk flow through the split vent in Figure 23-15. The column was kept at 75°C. The injection liner was purged rapidly by carrier gas, and peaks are quite sharp. Experiment B shows the same sample injected in the same way, except the split vent was closed. Then the injection liner was purged slowly, and sample was applied to the column over a long time. Peaks are broad, and they tail badly because fresh carrier gas continuously mixes with vapor in the injector, making it more and more dilute but never completely flushing the

For solvent trapping, sample should contain 10^4 times as much solvent as analyte and column temperature should be 40°C below the solvent's boiling point.

For cold trapping, stationary phase film thickness must be ≥2 μm.

FIGURE 23-15 Representative injection conditions for split, splitless, and on-column injection into an open tubular column.

FIGURE 23-16 Split and splitless injections of a solution containing 1 vol% methyl isobutyl ketone (b.p. 118°C) and 1 vol% *p*-xylene (b.p. 138°C) in dichloromethane (b.p. 40°C) on a BP-10 moderately polar cyanopropylphenyl methyl silicone open tubular column (0.22 mm diameter × 10 m long, film thickness = 0.25 μm, column temperature = 75°C). Vertical scale is the same for A–C. In D, signal heights should be multiplied by 2.33 to be on the same scale as A–C. [From P. J. Marriott and P. D. Carpenter, "Capillary Gas Chromatography Injection," *J. Chem. Ed.* **1996,** *73,* 96.]

sample from the injector. Peak areas in B are much greater than those in A because the entire sample reaches the column in B, whereas only a small fraction of sample reaches the column in A.

Experiment C is the same as B, but the split vent was opened after 30 s to rapidly purge all vapors from the injection liner. The bands in chromatogram C would be similar to those in B, but the bands are truncated after 30 s. Experiment D was the same as C, except that the column was initially cooled to 25°C to trap solvent and solutes at the beginning of the column. This is the correct condition for splitless injection. Solute peaks are sharp because the solutes were applied to the column in a narrow band of trapped solvent. Detector response in D is different from A–C. Actual peak areas in D are greater than those in A because most of the sample is applied to the column in D, but only a small fraction is applied in A. To make experiment D a proper splitless injection, the sample would need to be much more dilute.

On-Column Injection

On-column injection is used for samples that decompose above their boiling points and is preferred for quantitative analysis. Solution is injected directly into the column, without going through a hot injector (Figure 23-15). The initial column temperature is low enough to condense solutes in a narrow zone. Warming the column initiates chromatography. Samples are subjected to the lowest possible temperature in this procedure, and little loss of any solute occurs. The needle of a standard microliter syringe fits inside a 0.53-mm-diameter column, but this column does not give the best resolution. For 0.20- to 0.32-mm-diameter columns, which give better resolution, special syringes with thin silica needles are required.

23-3 Detectors

For *qualitative analysis,* a mass spectrometer (Chapter 21) can identify a chromatographic peak by comparing its spectrum with a library of spectra. For mass spectral identification, sometimes two prominent mass spectral peaks are selected in the electron ionization spectrum. The *quantitation ion* is used for quantitative analysis. The *confirmation ion* is used for qualitative identification. For example, the confirmation ion might be expected to be 65% as abundant as the quantitation ion. If the observed abundance is not close to 65%, then we suspect that the compound is misidentified.

Another method to identify a peak is to compare its retention time with that of an authentic sample of the suspected compound. The most reliable way to compare retention times is by **spiking,** also called *co-chromatography*, in which an authentic compound is added to the unknown. If the added compound is identical with a component of the unknown, then the relative area of that one peak will increase. Identification is tentative when carried out with one column but it is firmer when carried out on several columns with different stationary phases.

TABLE 23-4 Detection limits and linear ranges of gas chromatography detectors

Detector	Approximate detection limit	Linear range
Thermal conductivity	400 pg/mL (propane)	$>10^5$
Flame ionization	2 pg/s	$>10^7$
Electron capture	As low as 5 fg/s	10^4
Flame photometric	<1 pg/s (phosphorus)	$>10^4$
	<10 pg/s (sulfur)	$>10^3$
Nitrogen-phosphorus	100 fg/s	10^5
Sulfur chemiluminescence	100 fg/s (sulfur)	10^5
Photoionization	25 pg to 50 pg (aromatics)	$>10^5$
Fourier transform infrared	200 pg to 40 ng	10^4
Mass spectrometric	25 fg to 100 pg	10^5

SOURCE: *Most data are from D. G. Westmoreland and G. R. Rhodes, "Detectors for Gas Chromatography," Pure Appl. Chem.* **1989**, *61, 1147.*

Linear response means that peak area is proportional to analyte concentration. For very narrow peaks, peak height is often substituted for peak area.

Quantitative analysis with internal standard:

$$\frac{A_X}{[X]} = F\left(\frac{A_S}{[S]}\right)$$

A_X = area of analyte signal
A_S = area of internal standard
[X] = concentration of analyte
[S] = concentration of standard
F = response factor

Quantitative analysis is based on the area of a chromatographic peak. In the *linear response* concentration range, *the area of a peak is proportional to the quantity of that component.* In most instruments, peak area is reported by a computer. Judgment is required to draw baselines beneath peaks and to decide on the boundaries of the area to be measured.[17] If area must be measured by hand, and if the peak is Gaussian, then the area is

$$\text{Area of Gaussian peak} = 1.064 \times \text{peak height} \times w_{1/2} \qquad (23\text{-}2)$$

where $w_{1/2}$ is the width at half-height (Figure 22-9). Quantitative analysis is almost always performed by adding a known quantity of *internal standard* to the unknown (Section 5-4). After measurement of the *response factor* with standard mixtures, the equation in the margin is used to measure the quantity of unknown.

Thermal Conductivity Detector

In the past, **thermal conductivity detectors** were common in gas chromatography because they are simple and *universal*—responding to all analytes. Thermal conductivity is useful for packed columns. It is less sensitive than other detectors used with open tubular columns (Table 23-4).

Thermal conductivity measures the ability of a substance to transport heat from a hot region to a cold region (Table 23-5). Helium is the carrier gas commonly used with a thermal conductivity detector. Helium has the second highest thermal conductivity (after H_2), so any analyte mixed with helium lowers the conductivity of the gas stream. In Figure 23-17, eluate from the chromatography column flows over a hot tungsten-rhenium filament. When analyte emerges from the column, the conductivity of the gas stream decreases, the filament gets hotter, its electrical resistance increases, and the voltage across the filament changes. The detector measures the change in voltage.

TABLE 23-5 Thermal conductivity at 273 K and 1 atm

Gas	Thermal conductivity J/(K · m · s)
H_2	0.170
He	0.141
NH_3	0.021 5
N_2	0.024 3
C_2H_4	0.017 0
O_2	0.024 6
Ar	0.016 2
C_3H_8	0.015 1
CO_2	0.014 4
Cl_2	0.007 6

The energy per unit area per unit time flowing from a hot region to a cold region is given by

$$\text{Energy flux } (J/m^2 \cdot s) = -\kappa(dT/dx)$$

where κ is the thermal conductivity [units = J/(K · m · s)] and dT/dx is the temperature gradient (K/m). Thermal conductivity is to energy flux as the diffusion coefficient is to mass flux.

FIGURE 23-17 Thermal conductivity detectors. [Adapted from J. V. Hinshaw, "The Thermal Conductivity Detector," *LCGC* **2006**, *24*, 38.]

It is common to split the carrier gas into two streams, sending part through the analytical column and part through a matched reference column. Each stream is passed over a different filament or alternated over a single filament. Resistance of the sample filament is measured with respect to that of the reference filament. The reference column minimizes flow differences when temperature is changed. Sensitivity increases with the square of the filament current. However, to avoid burning out the filament, the maximum recommended current should not be exceeded. The filament should be off when carrier gas is not flowing.

The sensitivity of a thermal conductivity detector (but *not* that of the flame ionization detector described next) is inversely proportional to flow rate: It is more sensitive at a lower flow rate. Sensitivity also increases with increasing temperature differences between the filament and the surrounding block in Figure 23-17. The block should therefore be maintained at the lowest temperature that allows all solutes to remain gaseous.

Flame Ionization Detector

In the **flame ionization detector** in Figure 23-18, eluate is burned in a mixture of H_2 and air.[18] Carbon atoms (except carbonyl and carboxyl carbons) produce CH radicals, which are thought to produce CHO^+ ions and electrons in the flame.

$$CH + O \rightarrow CHO^+ + e^-$$

Only about 1 in 10^5 carbon atoms produces an ion, but ion production is proportional to the number of susceptible carbon atoms entering the flame. In the absence of analyte, $\sim 10^{-14}$ A flows between the flame tip and the collector, which is held at +200 to 300 V with respect to the flame tip. Eluted analytes produce a current of $\sim 10^{-12}$ A, which is converted to voltage, amplified, filtered to remove high-frequency noise, and finally converted to a digital signal.

Response to organic compounds is proportional to solute mass over seven orders of magnitude. The detection limit is ~ 100 times smaller than that of the thermal conductivity detector (Table 23-4) and is reduced by 50% when N_2 carrier gas is used instead of He. For open tubular columns, N_2 makeup gas is added to the H_2 or He eluate before it enters the detector. The flame ionization detector is sensitive enough for narrow-bore open tubular columns. It responds to most hydrocarbons and is insensitive to nonhydrocarbons such as H_2, He, N_2, O_2, CO, CO_2, H_2O, NH_3, NO, H_2S, and SiF_4.

Electron Capture Detector

Most detectors other than flame ionization and thermal conductivity respond to limited classes of analytes. The **electron capture detector** (Figure 23-19) is particularly sensitive to halogen-containing molecules, conjugated carbonyls, nitriles, nitro compounds, and organometallic compounds but relatively insensitive to hydrocarbons, alcohols, and ketones. The carrier or makeup gas must be either N_2 or 5% methane in Ar. Moisture decreases the sensitivity. Gas entering the detector is ionized by high-energy electrons ("β-rays") emitted from a foil containing radioactive ^{63}Ni. Electrons in the plasma thus formed are attracted to an anode, producing a small current that is maintained at a constant level by variable frequency pulses applied between the cathode and anode. When analytes with a high electron affinity enter the detector, they capture some electrons and decrease the conductivity of the plasma. The detector responds by varying the frequency of voltage pulses to maintain a constant current. The frequency of the pulses is the detector signal. The electron capture detector is extremely sensitive (Table 23-4), with a detection limit comparable to that of mass spectrometric selected ion monitoring.

Other Detectors

The *nitrogen-phosphorus detector*, also called an *alkali flame detector*, is a modified flame ionization detector that is especially sensitive to compounds containing N and P.[19] Its response to N and P is 10^4–10^6 times greater than the response to carbon. It is particularly important for drug, pesticide, and herbicide analyses. The detector features a Rb_2SO_4-containing glass bead at the burner tip. Ions such as NO_2^-, CN^-, and PO_2^- produced by these elements when they contact the bead carry the current that is measured. N_2 from air is inert to this detector and does not interfere. The bead must be replaced periodically because Rb_2SO_4 is consumed. Figure 23-26 (in the next section) shows a chromatogram from a nitrogen-phosphorus detector.

A *flame photometric detector* measures optical emission from phosphorus, sulfur, lead, tin, or other selected elements. When eluate passes through a H_2-air flame, as in the flame

Thermal conductivity detector:

- 10^5 linear response range
- H_2 and He give lowest detection limit
- sensitivity increases with
    increasing filament current
    decreasing flow rate
    lower detector block temperature

Flame ionization detector:

- N_2 gives best detection limit
- signal proportional to number of susceptible carbon atoms
- 100-fold better detection than thermal conductivity
- 10^7 linear response range

FIGURE 23-18 Flame ionization detector.
[Courtesy Varian Associates, Palo Alto, CA.]

FIGURE 23-19 Electron capture detector.

FIGURE 23-20 Gas chromatograms showing sulfur compounds in natural gas: (a) flame ionization detector response and (b) sulfur chemiluminescence detector response. The organosulfur compounds are too dilute to be seen in flame ionization, and the sulfur chemiluminescence is insensitive to hydrocarbons. [From N. G. Johansen and J. W. Birks, "Determination of Sulfur Compounds in Difficult Matrices," *Am. Lab.,* February 1991, p. 112.]

(a)

(b)

Other gas chromatography detectors:

electron capture: halogens, conjugated $C=O$, $-C\equiv N$, $-NO_2$

nitrogen-phosphorus: highlights P, N

flame photometer: individual selected elements, such as P, S, Sn, Pb

photoionization: aromatics, unsaturated compounds

sulfur chemiluminescence: S

nitrogen chemiluminescence: N

atomic emission: most elements (selected individually)

mass spectrometer: most analytes

infrared spectrometer: most analytes

Reactions thought to give sulfur chemiluminescence:

$$sulfur\ compound \xrightarrow{\text{H}_2\text{-O}_2\ \text{flame}} SO + products$$

$$SO + O_3 \rightarrow SO_2^* + O_2 \quad (SO_2^* = \text{excited state})$$

$$SO_2^* \rightarrow SO_2 + h\nu$$

ionization detector, excited atoms emit characteristic light. Phosphorus emission at 536 nm or sulfur emission at 394 nm can be isolated by a narrow-band interference filter and detected with a photomultiplier tube.

A *photoionization detector* uses a vacuum ultraviolet source to ionize aromatic and unsaturated compounds; it has little response to saturated hydrocarbons or halocarbons. Electrons produced by the ionization are collected and measured.

A *sulfur chemiluminescence detector* takes exhaust from a flame ionization detector, in which sulfur has been oxidized to SO, and mixes it with ozone (O_3) to form an excited state of SO_2 that emits blue light and ultraviolet radiation. Emission intensity is proportional to the mass of sulfur eluted, regardless of the source, and the response to S is 10^7 times greater than the response to C (Figure 23-20). A *nitrogen chemiluminescence detector* is analogous. Combustion of eluate at 1 800°C converts nitrogen into NO, which reacts with O_3 to create a chemiluminescent product. The response to N is 10^7 times greater than the response to C.

Gas Chromatography–Mass Spectrometry

Mass spectrometry is a sensitive detector that provides both qualitative and quantitative information. With *selected ion monitoring* or *selected reaction monitoring* (Section 21-4), we can measure one component in a complex chromatogram of poorly separated compounds. Selected ion monitoring lowers the detection limit by a factor of 10^2–10^3 compared with m/z scanning, because more time is spent just collecting ions of interest in selected ion monitoring.

Figure 23-21 illustrates **selected ion monitoring.** The *reconstructed total ion chromatogram* in trace *a* was obtained from a portable gas chromatograph–mass spectrometer designed for identification of spills at accident sites. A total of 1 072 spectra of eluate were recorded at equal time intervals between 1 and 10 min. The ordinate in the reconstructed total ion chromatogram is the sum of detector signal for all m/z above a selected cutoff. It measures everything eluted from the column. The selected ion chromatogram in trace *b* is obtained by parking the detector at m/z 78 and only measuring this one mass. By spending all the time measuring just one ion, the signal-to-noise ratio increases and the

FIGURE 23-21 Selected ion monitoring in gas chromatography–mass spectrometry. (*a*) Reconstructed total ion chromatogram of automobile exhaust with electron ionization. (*b*) Selected ion monitoring at *m/z* 78. (*c*) and (*d*) Quantitative analysis of benzene after adding an internal standard with a prominent ion at *m/z* 69. [Courtesy Inficon, Syracuse, NY.]

chromatogram is simplified. A peak is observed for benzene (nominal mass 78 Da) and minor peaks are observed for benzene derivatives at 7–9 min. For quantitative analysis, an internal standard with a signal at *m/z* 69 is added to the mixture. Even though this internal standard overlaps the congested part of the chromatogram near a retention time of 2 min, the selected ion chromatogram for *m/z* 69 has a single peak in trace *c*. To measure benzene, the area of the *m/z* 78 peak in trace *d* is compared with the area of *m/z* 69 in trace *c*.

Selected reaction monitoring is illustrated in Figure 23-22. Trace *a* is the reconstructed total ion chromatogram of extract from an orange peel. To make the analysis selective for the pesticide fensulfothion, the precursor ion *m/z* 293 selected by mass filter Q1 in Figure 21-26 was passed to collision cell Q2, in which it broke into fragments with a prominent ion at *m/z* 264. Trace *b* in Figure 23-22 shows the detector signal at *m/z* 264 from mass filter Q3. Only

FIGURE 23-22 Selected reaction monitoring in gas chromatography–mass spectrometry. (*a*) Reconstructed total ion chromatogram of extract from orange peel with electron ionization. (*b*) Selected reaction monitoring with the precursor ion *m/z* 293 selected by mass filter Q1 in Figure 21-26 and product ion *m/z* 264 selected by mass filter Q3. The chromatogram is a graph of intensity at *m/z* 264 from Q3 versus time. [Courtesy Thermo Finnigan GC and GC/MS Division, San Jose, CA.]

FIGURE 23-23 Extracted element chromatograms produced by gas chromatography–inductively coupled plasma–mass spectrometry. Each trace responds to just one element. [From D. Profrock, P. Leonhard, S. Wilbur, and A. Prange, "Sensitive, Simultaneous Determination of P, S, Cl, Br, and I Containing Pesticides in Environmental Samples by GC Hyphenated with Collision-Cell ICP-MS," *J. Anal. Atom. Spectros.* **2004,** *19,* 1.]

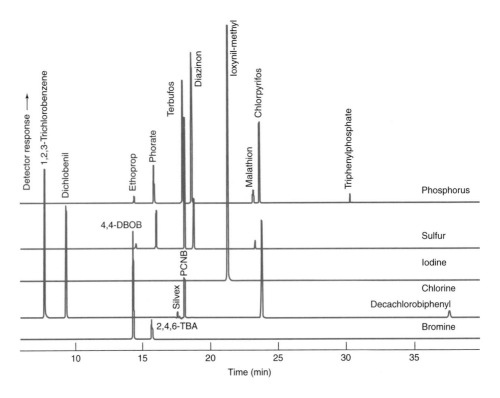

one peak is observed because few compounds other than fensulfothion give rise to an ion at *m/z* 293 that produces a fragment at *m/z* 264. Selected reaction monitoring increases the signal-to-noise ratio in chromatographic analysis and eliminates much interference.

Fensulfothion
(nominal mass 308 Da)

$-e^-$
$-CH_3$

M − 15 (*m/z* 293)
Selected by Q1

collision
cell, Q2

M − 15 − 29 (*m/z* 264)
Selected by Q3

Element-Specific Plasma Detectors

Eluate from a chromatography column can be passed through a plasma to atomize and ionize its components and measure selected elements by atomic emission spectroscopy or mass spectrometry. An *atomic emission detector* directs eluate through a helium plasma in a microwave cavity. Every element of the periodic table produces characteristic emission that can be detected by a photodiode array polychromator (Figure 19-16). Sensitivity for sulfur can be 10 times better than the sensitivity of a flame photometric detector.

The extremely sensitive inductively coupled plasma–mass spectrometer was described in Section 20-6. Figure 23-23 shows 15 pesticides measured by gas chromatography–inductively coupled plasma–mass spectrometry. Eluate was atomized and ionized in the plasma. Ions were measured by a mass spectrometer that could monitor any set of *m/z* values. The figure shows traces for P, S, I, Cl, and Br.

23-4 Sample Preparation

Sample preparation is the process of transforming a sample into a form that is suitable for analysis. This process might entail extracting analyte from a complex matrix, *preconcentrating* very dilute analytes to get a concentration high enough to measure, removing or masking interfering species, or chemically transforming (*derivatizing*) analyte into a more convenient or more easily detected form. Chapter 27 addresses sample preparation, so we now describe only techniques that are especially applicable to gas chromatography.

Solid-phase microextraction extracts compounds from liquids, air, or even sludge without using any solvent.[20] The key component is a fused silica fiber coated with a 10- to 100-μm-thick film of stationary phase similar to those used in gas chromatography.

Example of derivatization:

O
‖
RCOH
Nonvolatile
carboxylic acid

O
‖
RCOSi(CH₃)₃
Volatile trimethylsilyl
ester derivative

Figure 23-24 shows the fiber attached to the base of a syringe with a fixed metal needle. The fiber can be extended from the needle or retracted inside the needle. Figure 23-25 demonstrates the procedure of exposing the fiber to a sample solution (or the gaseous headspace above the liquid) for a fixed time while stirring and, perhaps, heating. It is best to determine by experiment how much time is required for the fiber to become saturated with analyte and to allow this much time for extraction. If you use shorter times, the concentration of analyte in the fiber is likely to vary from sample to sample. Only a fraction of the analyte in the sample is extracted into the fiber. When you extract the headspace, liquid sample should occupy about two-thirds of the vial. Too much headspace volume reduces extraction efficiency.

After sample collection, retract the fiber and insert the syringe into a gas chromatograph equipped with a 0.7-mm-inner diameter injection port liner. Extend the fiber into the hot injection liner, where analyte is thermally desorbed from the fiber in splitless mode for a fixed time. The narrow port maintains desorbed analyte in a narrow band. Collect desorbed analyte by *cold trapping* (page 578) at the head of the column prior to chromatography. If there will be a long time between sampling and injection, insert the needle into a septum to seal the fiber from the atmosphere. Figure 23-26 shows a chromatogram of chemical warfare nerve agents isolated from seawater by solid-phase microextraction and detected with a nitrogen-phosphorus detector. The chromatogram is deceptively simple because the detector responds only to compounds containing N and P.

In solid-phase microextraction, the mass of analyte (m, μg) absorbed in the coated fiber is

Mass of analyte extracted:
$$m = \frac{KV_f c_0 V_s}{KV_f + V_s} \qquad (23\text{-}3)$$

where V_f is the volume of film on the fiber, V_s is the volume of solution being extracted, and c_0 is the initial concentration (μg/mL) of analyte in the solution being extracted. K is the partition coefficient for solute between the film and the solution: $K = c_f/c_s$, where c_f is the concentration of analyte in the film and c_s is the concentration of analyte in the solution. If you extract a large volume of solution such that $V_s \gg KV_f$, then Equation 23-3 reduces to $m = KV_f c_0$. That is, the mass extracted is proportional to the concentration of analyte in solution. For quantitative analysis, you can construct a calibration curve by extracting known solutions. Alternatively, internal standards and standard additions are both useful for solid-phase microextraction.

FIGURE 23-24 Syringe for solid-phase microextraction. The fused silica fiber is withdrawn inside the steel needle after sample collection and when the syringe is used to pierce a septum.

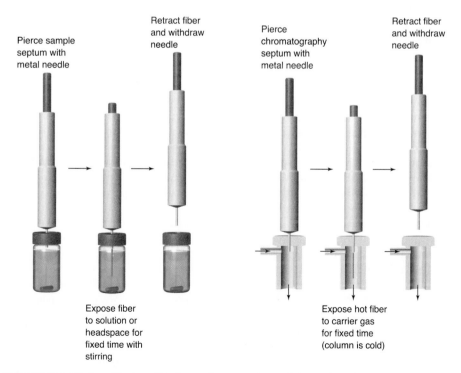

FIGURE 23-25 Sampling by solid-phase microextraction and desorption of analyte from the coated fiber into a gas chromatograph. [Adapted from Supelco Chromatography Products catalog, Bellefonte, PA.]

Stir-bar sorption

[Courtesy Gerstel, Inc., Linthicum, MD.]

FIGURE 23-26 Gas chromatogram of nerve agents sampled by solid-phase microextraction for 30 min from seawater spiked with 60 nL of each agent per liter (60 ppb by volume). The fiber had a 65-μm-thick coating of copoly(dimethylsiloxane/divinylbenzene). The nitrogen-phosphorus detector had a detection limit of 0.05 ppb. Analytes were desorbed from the fiber for 2 min at 250°C in splitless mode in the injection port. The column temperature was 30°C during desorption, and then it was ramped up at 10°C/min during chromatography. The column was 0.32 mm × 30 m with a 1-μm coating of $(phenyl)_{0.05}(methyl)_{0.95}$polysiloxane. Soman appears as a split peak because it has two isomers. [From H.-Å. Lakso and W. F. Ng, "Determination of Chemical Warfare Agents in Natural Water Samples by Solid-Phase Microextraction," *Anal. Chem.* **1997**, *69*, 1866.]

Tenax

You need to establish the time and temperature required to purge 100% of the analyte from the sample in separate control experiments.

Stir-bar sorptive extraction is closely related to solid-phase microextraction, but it is ~100 times more sensitive for trace analysis.[21] A magnetic stirring bar enclosed in a thin glass jacket is coated with a 0.5- to 1-mm-thick layer of sorbent such as poly(dimethylsiloxane)—the same as the stationary phase in nonpolar gas chromatography columns. The stirring bar is placed in an aqueous liquid sample such as fruit juice, wine, urine, or blood plasma and stirred for 0.5–4 h to absorb hydrophobic analytes. The mass of analyte extracted is given by Equation 23-3, but the volume of sorbent (V_f) is increased from ~0.5 μL in solid-phase microextraction to 25–125 μL in stir-bar sorptive extraction. Therefore, 50 to 250 times more analyte is extracted with the stir bar. After extraction, the stir bar is touched to a tissue to remove water droplets, possibly rinsed with a few milliliters of water, and then placed in a *thermal desorption tube*. Desorption is typically conducted by heating the tube to 250°C for 5 min in flowing carrier gas. Volatile analytes are collected by cold trapping and then separated by gas chromatography. Analytes at part per billion or lower concentration can be measured by using standard addition, isotopic internal standards, or a calibration curve constructed with the same matrix.

Purge and trap is a method for removing volatile analytes from liquids or solids (such as groundwater or soil), concentrating the analytes, and introducing them into a gas chromatograph. In contrast with solid-phase microextraction, which removes only a portion of analyte from the sample, the goal in purge and trap is to remove 100% of analyte from the sample. Quantitative removal of polar analytes from polar matrices can be difficult.

Figure 23-27 shows apparatus for measuring volatile flavor components in carbonated beverages. Helium purge gas from a stainless steel needle is bubbled through the cola in the sample vial, which is heated to 50°C to aid evaporation of analytes. Purge gas exiting the sample vial passes through an adsorption tube containing three layers of adsorbent compounds with increasing adsorbent strength. For example, the moderate adsorbent could be a nonpolar phenylmethylpolysiloxane, the stronger adsorbent could be the polymer Tenax, and the strongest adsorbent could be carbon molecular sieves.[22]

During the purge and trap process, gas flows through the adsorbent tube from end A to end B in Figure 23-27. After purging all analyte from the sample into the adsorption tube, reverse the gas flow to go from B to A and purge the trap at 25°C to remove as much water or other solvent as possible from the adsorbents. Then connect outlet A of the adsorption tube to the injection port of a gas chromatograph operating in splitless mode and heat the trap to ~200°C. Desorbed analytes flow into the chromatography column where they are concentrated by cold trapping. After complete desorption from the trap, warm the chromatography column to initiate the separation.

Thermal desorption is a method for releasing volatile compounds from solid samples. A weighed sample is placed in a steel or glass tube and held in place with glass wool. The sample is purged with carrier gas to remove O_2, which is vented to the air, not into the chromatography column. The desorption tube is then connected to the chromatography column and heated to release volatile substances, which are collected by *cold trapping* at the beginning of the column. The column is then heated rapidly to initiate chromatography.

23-5 Method Development in Gas Chromatography

With the bewildering choice of parameters in gas chromatography, is there a rational way to choose a procedure for a particular problem? In general, there are many satisfactory solutions. As a broad guide to method selection, consider the following topics in this order: (1) the goal of the analysis, (2) sample preparation, (3) detector, (4) column, and (5) injection.[23]

Order of decisions:

1. goal of analysis
2. sample preparation
3. detector
4. column
5. injection

Goal of the Analysis

What is required from the analysis? Is it qualitative identification of components in a mixture? Will you require high-resolution separation of everything or do you just need good resolution in a portion of the chromatogram? Can you sacrifice resolution to shorten the analysis time? Do you need quantitative analysis of one or many components? Do you need high precision? Will analytes be present in adequate concentration or do you need preconcentration or a very sensitive detector for ultratrace analysis? How much can the analysis cost? Each of these factors creates trade-offs in selecting techniques.

Sample Preparation

The key to successful chromatography of a complex sample is to clean it up before it ever sees the column. In Section 23-4, we described solid-phase microextraction, stir-bar sorptive extraction, purge and trap, and thermal desorption to isolate volatile components from complex matrices. Other methods include liquid extraction, supercritical fluid extraction, and solid-phase extraction, most of which are described in Chapter 27. These techniques isolate desired analytes from interfering substances, and they can concentrate dilute analytes up to detectable levels. If you do not clean up your samples, chromatograms could contain a broad "forest" of unresolved peaks, and nonvolatile substances will ruin the expensive chromatography column.

Garbage in—garbage out!

Choosing the Detector

The next step is to choose a detector. Do you need information about everything in the sample or do you want to detect a particular element or class of compounds?

The most general purpose detector for open tubular chromatography is a mass spectrometer. Flame ionization is probably the most popular detector, but it responds to hydrocarbons; and Table 23-4 shows that it is not as sensitive as electron capture, nitrogen-phosphorus, or chemiluminescence detectors. The flame ionization detector requires the sample to contain ≥ 10 ppm of each analyte for split injection. The thermal conductivity detector responds to all classes of compounds, but it is not very sensitive.

Sensitive detectors for ultratrace analysis each respond to a limited class of analytes. The electron capture detector is specific for halogen-containing molecules, nitriles, nitro compounds, and conjugated carbonyls. For split injection, the sample should contain ≥ 100 ppb of each analyte if an electron capture detector is to be used. A photoionization detector can be specific for aromatic and unsaturated compounds. The nitrogen-phosphorus detector has enhanced response to compounds containing either of these two elements, but it also responds to hydrocarbons. Sulfur and nitrogen chemiluminescence detectors each respond to just one element. Flame photometric detectors are specific for selected elements, such as S, P, Pb, or Sn. A selective detector might be chosen to simplify the chromatogram by not responding to everything that is eluted (as in Figure 23-23). Mass spectrometry with selected reaction monitoring (Figure 23-22) is an excellent way to monitor one analyte of interest in a complex sample.

If qualitative information is required to identify eluates, then mass spectral or infrared detectors are good choices. The infrared detector, like the thermal conductivity detector, is not sensitive enough for high-resolution, narrow-bore, open tubular columns.

Selecting the Column

The basic choices are stationary phase, column diameter and length, and thickness of the stationary phase. A nonpolar stationary phase from Table 23-1 is most useful. An intermediate polarity stationary phase will handle most separations that the nonpolar column cannot. For highly polar compounds, a strongly polar column might be necessary. Optical isomers and closely related geometric isomers require special stationary phases for separation.

Table 23-6 shows that there are only a few sensible combinations of column diameter and film thickness. Highest resolution is afforded by the narrowest columns. Thin-film, narrow-bore columns are especially good for separating mixtures of high-boiling-point compounds that are retained too strongly on thick-film columns. Short retention times provide high-speed analyses. However, thin-film, narrow-bore columns have very low sample capacity, require

FIGURE 23-27 Purge and trap apparatus for extracting volatile substances from a liquid or solid by flowing gas.

TABLE 23-6　Gas chromatography column comparisons[23]

	⭘	⚪	⬤
Description	Thin-film narrow-bore	Thick-film narrow-bore	Thick-film wide-bore
Inner diameter	0.10–0.32 mm	0.25–0.32 mm	0.53 mm
Film thickness	~0.2 μm	~1–2 μm	~2–5 μm
Advantages	High resolution Trace analysis Fast separations Low temperatures Elute high-b.p. compounds	Good capacity Good resolution 　(4 000 plates/m) Easy to use Retains volatile compounds Good for mass spectrometry	High capacity (100 ng/ 　solute) Good for thermal conductivity 　and infrared detectors Simple injection techniques
Disadvantages	Low capacity 　(≤1 ng/solute) Requires high-sensitivity 　detector (not mass spectrometry) Surface activity of exposed silica	Moderate resolution Long retention time for 　high-b.p. compounds	Low resolution 　(500–2 000 plates/m) Long retention time for 　high-b.p. compounds

high-sensitivity detectors (flame ionization might not be adequate), do not retain low-boiling-point compounds well, and could suffer from exposure of active sites on the silica.

Thick-film, narrow-bore columns in Table 23-6 provide a good compromise between resolution and sample capacity. They can be used with most detectors (but usually not thermal conductivity detectors) and with compounds of high volatility. Retention times are longer than those of thin-film columns. Thick-film, wide-bore columns are required for use with thermal conductivity and infrared detectors. They have high sample capacity and can handle highly volatile compounds, but they give low resolution and have long retention times.

If a particular column meets most of your requirements but does not provide sufficient resolution, then a narrower column of the same type could be used (Equation 22-35b). To obtain similar retention times for the same length of column, the thickness of the stationary phase should be decreased in proportion to the diameter. A column diameter of 0.15 mm is sensible for maximizing resolution without requiring a different gas chromatograph designed for narrower columns.[24]

Doubling the length of a column doubles the number of plates and, according to Equation 22-30, increases resolution by $\sqrt{2}$. Doubling the column length is not the best way to increase resolution because it doubles retention time. A narrower column increases resolution with no penalty in retention time. Selecting another stationary phase changes the unadjusted relative retention (γ in Equation 22-30) and might resolve components of interest.

To increase the speed of analysis without decreasing resolution, a shorter, narrower column can be chosen. Another way to decrease retention time without sacrificing resolution is to switch carrier gas from He to H_2 and increase flow rate by a factor of 1.5 to 2 (Figure 23-11).

If you have measured resolution of a few key components of a mixture under a small number of conditions, commercial software is available to optimize conditions (such as temperature and pressure programming) for the best separation.[25]

Figure 23-28 suggests approximate upper and lower analyte mass limits for different columns and detectors. The abscissa shows the mass of a given analyte that reaches the column or detector. A typical volume of injected liquid is 1 μL, with a mass of 1 mg. If analyte concentration is 1 ppm, the mass of analyte in 1 μL is 10^{-9} g = 1 ng. A vertical line at 10^{-9} g falls in the operating range of all columns and four of the detectors. The sample mass is too small for thermal conductivity, too great for electron capture, and near the upper limit for selected ion monitoring. For a 100:1 split injection, the mass of analyte introduced onto the column would be 100 times smaller, or 10^{-11} g. This mass is in range for all but the thermal conductivity detector. For columns, the region to the left of 10^{-11} g is shaded because chromatography becomes progressively more problematic as the

To improve resolution, use a
- longer column
- narrower column
- different stationary phase

FIGURE 23-28 Approximate injected analyte mass limits for wall-coated open tubular gas chromatography columns and detectors. [Adapted from J. V. Hinshaw, "Setting Realistic Expectations for GC Optimization," *LCGC* **2006**, *24*, 1194.]

mass gets smaller. Analyte masses less than 10^{-11} g could be lost to adsorption or decomposition in the injector and column. Polar solutes might be lost to adsorption at even higher levels. At the upper end of the column scale, a dividing line recommends where to change from a thin-film stationary phase to a thick film to have adequate sample capacity.

Choosing the Injection Method

The last major decision is how to inject the sample. *Split injection* is best for high concentrations of analyte or gas analysis. Quantitative analysis is very poor. Less volatile components can be lost during injection. Split injection offers high resolution and can handle dirty samples if an adsorbent packing is added to the injection liner. Thermally unstable compounds can decompose during the high-temperature injection.

Splitless injection is required for very dilute solutions. It offers high resolution but is poor for quantitative analysis because less volatile compounds can be lost during injection. It is better than split injection for compounds of moderate thermal stability because the injection temperature is lower. Splitless injection introduces sample onto the column slowly, so solvent trapping or cold trapping is required. Therefore, splitless injection cannot be used for isothermal chromatography. Samples containing less than 100 ppm of each analyte can be analyzed with a column film thickness of <1 μm and splitless injection. Samples containing 100–1 000 ppm of each analyte require a column film thickness ≥1 μm.

On-column injection is best for quantitative analysis and for thermally sensitive compounds. It is strictly a low-resolution technique and cannot be used with columns whose inner diameter is less than 0.2 mm. It can handle dilute or concentrated solutions and relatively large or small volumes. Other column requirements are the same as in splitless injection.

Split injection:
- concentrated sample
- high resolution
- dirty samples (use packed liner)
- could cause thermal decomposition

Splitless injection:
- dilute sample
- high resolution
- requires solvent trapping or cold trapping

On-column injection:
- best for quantitative analysis
- thermally sensitive compounds
- low resolution

Terms to Understand

carrier gas	on-column injection	selected ion monitoring	splitless injection
cold trapping	open tubular column	selected reaction monitoring	stir-bar sorptive extraction
electron capture detector	packed column	septum	temperature programming
flame ionization detector	purge and trap	solid-phase microextraction	thermal conductivity detector
gas chromatography	retention gap	solvent trapping	thermal desorption
guard column	retention index	spiking	
molecular sieve	sample preparation	split injection	

Summary

In gas chromatography, a volatile liquid or gaseous solute is carried by a gaseous mobile phase over a stationary phase on the inside of an open tubular column or on a solid support. Long, narrow, fused-silica open tubular columns have low capacity but give excellent separation. They can be wall coated, support coated, or porous layer. Packed columns provide high capacity but poor resolution. Each liquid stationary phase most strongly retains solutes in its own polarity class ("like dissolves like"). Solid stationary phases include porous carbon, alumina, and molecular sieves. The retention index measures elution times in relation to those of linear alkanes. Temperature or pressure programming reduce elution times of strongly retained components. Without compromising separation efficiency, the linear flow rate may be increased when H_2 or He, instead of N_2, is used as carrier gas. Split injection provides high-resolution separations of relatively concentrated samples. Splitless injection of very dilute samples requires solvent trapping or cold trapping to concentrate solutes at the start of the column (to give sharp bands). On-column injection is best for quantitative analysis and for thermally unstable solutes.

Quantitative analysis is usually done with internal standards in gas chromatography. Co-elution of an unknown peak with a spike of a known compound on several different columns is useful for qualitative identification of the unknown peak. Mass spectral and infrared detectors provide qualitative information to help identify unknowns.

The mass spectrometer is more sensitive and less subject to interference when selected ion monitoring or selected reaction monitoring are employed. Thermal conductivity detection has universal response but is not very sensitive. Flame ionization detection is sensitive enough for most columns and responds to most organic compounds. Electron capture, nitrogen-phosphorus, flame photometry, photoionization, chemiluminescence, and atomic emission detectors are specific for certain classes of compounds or individual elements.

You need to define the goal of an analysis before developing a chromatographic method. The key to successful chromatography is a clean sample. Solid-phase microextraction, stir-bar sorptive extraction, purge and trap, and thermal desorption can isolate volatile components from complex matrices. After sample preparation, the remaining decisions for method development are to select a detector, a column, and the injection method, in that order.

Exercises

23-A. (a) When a solution containing 234 mg of butanol (FM 74.12) and 312 mg of hexanol (FM 102.17) in 10.0 mL was separated by gas chromatography, the relative peak areas were butanol:hexanol = 1.00:1.45. Considering butanol to be the internal standard, find the response factor for hexanol.

(b) Use Equation 23-2 to estimate the areas of the peaks for butanol and hexanol in Figure 23-8.

(c) The solution from which the chromatogram was generated contained 112 mg of butanol. What mass of hexanol was in the solution?

(d) What is the largest source in uncertainty in this problem? How great is this uncertainty?

23-B. When 1.06 mmol of 1-pentanol and 1.53 mmol of 1-hexanol were separated by gas chromatography, they gave relative peak areas of 922 and 1 570 units, respectively. When 0.57 mmol of pentanol was added to an unknown containing hexanol, the relative chromatographic peak areas were 843:816 (pentanol:hexanol). How much hexanol did the unknown contain?

23-C. (a) In Table 23-3, 2-pentanone has a retention index of 987 on a poly(ethylene glycol) column (also called Carbowax). Between which two straight-chain hydrocarbons is 2-pentanone eluted?

(b) An unretained solute is eluted from a certain column in 1.80 min. Decane ($C_{10}H_{22}$) is eluted in 15.63 min and undecane ($C_{11}H_{24}$) is eluted in 17.22 min. What is the retention time of a compound whose retention index is 1 050?

23-D. For isothermal elution of a homologous series of compounds (those with similar structures, but differing by the number of CH_2 groups in a chain), log t_r' is usually a linear function of the number of carbon atoms. A compound was known to be a member of the family

$$(CH_3)_2CH(CH_2)_nCH_2OSi(CH_3)_3$$

(a) From the gas chromatographic retention times given here, prepare a graph of log t_r' versus n and estimate the value of n in the chemical formula.

$n = 7$	4.0 min	CH_4	1.1 min
$n = 8$	6.5 min	unknown	42.5 min
$n = 14$	86.9 min		

(b) Calculate the retention factor for the unknown.

23-E. Resolution of two peaks (Equation 22-30) depends on the column plate number, N, and the unadjusted relative retention, γ. Suppose you have two peaks with a resolution of 1.0 and you want to increase the resolution to 1.5 for a baseline separation for quantitative analysis (Figure 22-10).

(a) You can increase the resolution to 1.5 by increasing column length. By what factor must the column length be increased? If flow rate is constant, how much more time will the separation require when the length is increased?

(b) You might change relative retention by choosing a different stationary phase. If γ was 1.013, to what value must it be increased to obtain a resolution of 1.5? If you were separating two alcohols with a (diphenyl)$_{0.05}$(dimethyl)$_{0.95}$ polysiloxane stationary phase (Table 23-1), what stationary phase would you choose to increase γ? Will this change affect the time required for chromatography?

Problems

23-1. (a) What is the advantage of temperature programming in gas chromatography?

(b) What is the advantage of pressure programming?

23-2. (a) What are the relative advantages and disadvantages of packed and open tubular columns in gas chromatography?

(b) Explain the difference between wall-coated, support-coated, and porous-layer open tubular columns.

(c) What is the advantage of bonding (covalently attaching) the stationary phase to the column wall or cross-linking the stationary phase to itself?

23-3. (a) Why do open tubular columns provide greater resolution than packed columns in gas chromatography?

(b) Why do H_2 and He allow more rapid linear flow rates in gas chromatography than N_2 does, without loss of column efficiency (Figure 23-12)?

23-4. (a) When would you use split, splitless, or on-column injection in gas chromatography?

(b) Explain how solvent trapping and cold trapping work in splitless injection.

23-5. To which kinds of analytes do the following gas chromatography detectors respond?

(a) thermal conductivity

(f) photoionization

(b) flame ionization

(g) sulfur chemiluminescence

(c) electron capture

(h) atomic emission

(d) flame photometric

(i) mass spectrometer

(e) nitrogen-phosphorus

23-6. Why does a thermal conductivity detector respond to all analytes except the carrier gas? Why isn't the flame ionization detector universal?

23-7. Explain what is displayed in a reconstructed total ion chromatogram, selected ion monitoring, and selected reaction monitoring. Which technique is most selective and which is least selective and why?

23-8. Use Table 23-3 to predict the elution order of the following compounds from columns containing (a) poly(dimethylsiloxane), (b) (diphenyl)$_{0.35}$(dimethyl)$_{0.65}$polysiloxane, and (c) poly(ethylene glycol): hexane, heptane, octane, benzene, butanol, 2-pentanone.

23-9. Use Table 23-3 to predict the elution order of the following compounds from columns containing (a) poly(dimethylsiloxane), (b) (diphenyl)$_{0.35}$(dimethyl)$_{0.65}$polysiloxane, and (c) poly(ethylene glycol):

1. 1-pentanol (n-C$_5$H$_{11}$OH, b.p. 138°C)

2. 2-hexanone (CH$_3$C(=O)C$_4$H$_9$, b.p. 128°C)

3. heptane (n-C$_7$H$_{16}$, b.p. 98°C)

4. octane (n-C$_8$H$_{18}$, b.p. 126°C)

5. nonane (n-C$_9$H$_{20}$, b.p. 151°C)

6. decane (n-C$_{10}$H$_{22}$, b.p. 174°C)

23-10. This problem reviews concepts from Chapter 22. An unretained solute passes through a chromatography column in 3.7 min and analyte requires 8.4 min.

(a) Find the adjusted retention time and retention factor for the analyte.

(b) The volume of the mobile phase is 1.4 times the volume of the stationary phase. Find the partition coefficient for the analyte.

23-11. If retention times in Figure 22-7 are 1.0 min for CH$_4$, 12.0 min for octane, 13.0 min for unknown, and 15.0 min for nonane, find the Kovats retention index for the unknown.

23-12. Retention time depends on temperature, T, according to the equation $\log t_r' = (a/T) + b$, where a and b are constants for a specific compound on a specific column. A compound is eluted from a gas chromatography column at an adjusted retention time $t_r' = 15.0$ min when the column temperature is 373 K. At 363 K, $t_r' = 20.0$ min. Find the parameters a and b and predict t_r' at 353 K.

23-13. What is the purpose of derivatization in chromatography? Give an example.

23-14. (a) Explain how solid-phase microextraction works. Why is cold trapping necessary during injection with this technique? Is all the analyte in an unknown extracted into the fiber in solid-phase microextraction?

(b) Explain the differences between stir-bar sorptive extraction and solid-phase microextraction. Which is more sensitive and why?

23-15. Why is splitless injection used with purge and trap sample preparation?

23-16. State the order of decisions in method development for gas chromatography.

23-17. (a) Why is it illogical to use a thin stationary phase (0.2 μm) in a wide-bore (0.53-mm) open tubular column?

(b) Consider a narrow-bore (0.25 mm diameter), thin-film (0.10 μm) column with 5 000 plates per meter. Consider also a wide-bore (0.53 mm diameter), thick-film (5.0 μm) column with 1 500 plates per meter. The density of stationary phase is approximately 1.0 g/mL. What mass of stationary phase is in each column in a length equivalent to one theoretical plate? How many nanograms of analyte can be injected into each column if the mass of analyte is not to exceed 1.0% of the mass of stationary phase in one theoretical plate?

23-18. How can you improve the resolution between two closely spaced peaks in gas chromatography?

23-19. The graph shows van Deemter curves for n-nonane at 70°C in the 3.0-m-long microfabricated column in Box 23-2 with a 1- to 2-μm-thick stationary phase.

van Deemter curves. [From G. Lambertus, A. Elstro, K. Sensenig, J. Potkay, M. Agah, S. Scheuering, K. Wise, F. Dorman, and R. Sacks, "Design, Fabrication, and Evaluation of Microfabricated Columns for Gas Chromatography," *Anal. Chem.* **2004**, *76*, 2629.]

(a) Why would air be chosen as the carrier gas? What is the danger of using air as carrier gas?

(b) Measure the optimum velocity and plate height for air and for H$_2$ carriers.

(c) How many plates are there in the 3-m-long column for each carrier at optimum flow rate?

(d) How long does unretained gas take to travel through the column at optimum velocity for each gas?

(e) If stationary phase is sufficiently thin with respect to column diameter, which of the two mass transfer terms (22-35a and 22-35b) becomes negligible? Why?

(f) Why is the loss of column efficiency at high flow rates less severe for H$_2$ than for air carrier gas?

23-20. (a) When a solution containing 234 mg of pentanol (FM 88.15) and 237 mg of 2,3-dimethyl-2-butanol (FM 102.17) in 10.0 mL was separated, the relative peak areas were pentanol:2,3-dimethyl-2-butanol = 0.913:1.00. Considering pentanol to be the internal standard, find the response factor for 2,3-dimethyl-2-butanol.

(b) Use Equation 23-2 to find the areas for pentanol and 2,3-dimethyl-2-butanol in Figure 23-8.

(c) The concentration of pentanol internal standard in the unknown solution was 93.7 mM. What was the concentration of 2,3-dimethyl-2-butanol?

23-21. A standard solution containing 6.3×10^{-8} M iodoacetone and 2.0×10^{-7} M p-dichlorobenzene (an internal standard) gave peak areas of 395 and 787, respectively, in a gas chromatogram. A 3.00-mL unknown solution of iodoacetone was treated with 0.100 mL of 1.6×10^{-5} M p-dichlorobenzene and the mixture was diluted to 10.00 mL. Gas chromatography gave peak areas of 633 and 520 for iodoacetone and p-dichlorobenzene, respectively. Find the concentration of iodoacetone in the 3.00 mL of original unknown.

23-22. Heptane, decane, and an unknown had adjusted retention times of 12.6 min (heptane), 22.9 min (decane), and 20.0 min (unknown). The retention indexes for heptane and decane are 700 and 1 000, respectively. Find the retention index for the unknown.

23-23. The gasoline additive methyl *t*-butyl ether (MTBE) has been leaking into groundwater ever since its introduction in the 1990s. MTBE can be measured at parts per billion levels by solid-phase microextraction from groundwater to which 25% (wt/vol) NaCl has been added (*salting out*, Problem 7-8). After microextraction, analytes are thermally desorbed from the fiber in the port of a gas chromatograph. The figure below shows a reconstructed total ion chromatogram and selected ion monitoring of substances desorbed from the extraction fiber.

Methyl *t*-butyl ether	Ethyl *t*-butyl ether	*t*-Amyl methyl ether
MTBE	ETBE	TAME

Nominal mass:	88	102	102

(a) What is the purpose of adding NaCl prior to extraction?

(b) What nominal mass is being observed in selected ion monitoring? Why are only three peaks observed?

(c) Here is a list of major ions above *m/z* 50 in the mass spectra. The base (tallest) peak is marked by an asterisk. Given that MTBE and TAME have an intense peak at *m/z* 73 and there is no significant peak at *m/z* 73 for ETBE, suggest a structure for *m/z* 73. Suggest structures for all ions listed in the table.

MTBE	ETBE	TAME
73*	87	87
57	59*	73*
	57	71
		55

Reconstructed total ion chromatogram and selected ion monitoring of solid-phase microextract of groundwater. Chromatography conditions: 0.32-mm × 30-m column with 5-μm film of poly(dimethylsiloxane). Temperature = 50°C for 4 min, then raised 20°C/min to 90°C, held for 3 min, and then raised 40°C/min to 200°C. [From D. A. Cassada, Y. Zhang, D. D. Snow, and R. F. Spalding, "Trace Analysis of Ethanol, MTBE, and Related Oxygenate Compounds in Water Using Solid-Phase Microextraction and Gas Chromatography/Mass Spectrometry," *Anal. Chem.* **2000**, *72*, 4654.]

23-24. Here is a student procedure to measure nicotine in urine. A 1.00-mL sample of biological fluid was placed in a 12-mL vial containing 0.7 g Na_2CO_3 powder. After 5.00 μg of the internal standard 5-aminoquinoline were injected, the vial was capped with a Teflon-coated silicone rubber septum. The vial was heated to 80°C for 20 min and then a solid-phase microextraction needle was passed through the septum and left in the headspace for 5.00 min. The fiber was retracted and inserted into a gas chromatograph. Volatile substances were desorbed from the fiber at 250°C for 9.5 min in the injection port while the column was at 60°C. The column temperature was then raised to 260°C at 25°C/min and eluate was monitored by electron ionization mass spectrometry with selected ion monitoring at *m/z* 84 for nicotine and *m/z* 144 for internal standard. Calibration data from replicate standard mixtures taken through the same procedure are given in the table.

Nicotine in urine (μg/L)	Area ratio *m/z* 84/144
12	$0.05_6, 0.05_9$
51	$0.40_2, 0.39_1$
102	$0.68_4, 0.66_9$
157	$1.01_1, 1.06_3$
205	$1.27_8, 1.35_5$

SOURCE: A. E. Wittner, D. M. Klinger, X. Fan, M. Lam, D. T. Mathers, and S. A. Mabury, "*Quantitative Determination of Nicotine and Cotinine in Urine and Sputum Using a Combined SPME-GC/MS Method*," J. Chem. Ed. **2002**, 79, 1257.

(a) Why was the vial heated to 80°C before and during extraction?

(b) Why was the chromatography column kept at 60°C during thermal desorption of the extraction fiber?

(c) Suggest a structure for *m/z* 84 from nicotine. What is the *m/z* 144 ion from the internal standard, 5-aminoquinoline?

Nicotine $C_{10}H_{14}N_2$	5-Aminoquinoline $C_9H_8N_2$

(d) Urine from an adult female nonsmoker had an area ratio *m/z* 84/144 = 0.51 and 0.53 in replicate determinations. Urine from a nonsmoking girl whose parents are heavy smokers had an area ratio 1.18 and 1.32. Find the nicotine concentration (μg/L) and its uncertainty in the urine of each person.

23-25. Nitric oxide (NO) is a cell-signaling agent in physiologic processes including vasodilation, inhibition of clotting, and inflammation. A sensitive chromatography–mass spectrometry method was developed to measure two of its metabolites, nitrite (NO_2^-) and nitrate (NO_3^-), in biological fluids. Internal standards, $^{15}NO_2^-$ and $^{15}NO_3^-$, were added to the fluid at concentrations of 80.0 and 800.0 μM, respectively. Natural $^{14}NO_2^-$ and $^{14}NO_3^-$ plus the internal standards were then converted to volatile derivatives:

Pentafluorobenzyl bromide

$O-NO_2$ *m/z* 62 for ^{14}N *m/z* 63 for ^{15}N

$O-NO$ *m/z* 46 for ^{14}N *m/z* 47 for ^{15}N

Because biological fluids are so complex, the derivatives were first isolated by high-performance liquid chromatography. For quantitative analysis, liquid chromatography peaks corresponding to the two products were injected into a gas chromatograph, ionized by *negative ion* chemical ionization (giving major peaks for NO_2^- and NO_3^-) and the products measured by selected ion monitoring. Results are shown in the figure. If the ^{15}N internal standards undergo the same reactions and same separations at the same rate as the ^{14}N analytes, then the concentrations of analytes are simply

$$[^{14}NO_x^-] = [^{15}NO_x^-](R - R_{blank})$$

where R is the measured peak area ratio (m/z 46/47 for nitrite and m/z 62/63 for nitrate) and R_{blank} is the measured ratio of peak areas in a blank prepared from the same buffers and reagents with no added nitrate or nitrite. The ratios of peak areas are m/z 46/47 = 0.062 and m/z 62/63 = 0.538. The ratios for the blank were m/z 46/47 = 0.040 and m/z 62/63 = 0.058. Find the concentrations of nitrite and nitrate in the urine.

Selected ion chromatogram showing *negative ions* at m/z 46, 47, 62, and 63 obtained by derivatizing nitrite and nitrate plus internal standards ($^{15}NO_2^-$ and $^{15}NO_3^-$) in urine. [From D. Tsikas, "Derivatization and Quantification of Nitrite and Nitrate in Biological Fluids by Gas Chromatography/Mass Spectrometry," *Anal. Chem.* **2000**, *72*, 4064.]

23-26. *van Deemter equation for open tubular column.* Equation 22-33 contains terms (A, B, and C) describing three band-broadening mechanisms.

(a) Which term is 0 for an open tubular column? Why?

(b) Express the value of B in terms of measurable physical properties.

(c) Express the value of C in terms of measurable physical quantities.

(d) The linear flow rate that produces minimum plate height is found by setting the derivative $dH/du_x = 0$. Find an expression of the minimum plate height in terms of the measurable physical quantities used to answer parts (b) and (c).

23-27. *Theoretical performance in gas chromatography.* As the inside radius of an open tubular column is decreased, the maximum possible column efficiency increases and sample capacity decreases. For a thin stationary phase that equilibrates rapidly with analyte, the minimum theoretical plate height is given by

$$\frac{H_{min}}{r} = \sqrt{\frac{1 + 6k + 11k^2}{3(1 + k)^2}}$$

where r is the inside radius of the column and k is the retention factor.

(a) Find the limit of the square-root term as $k \to 0$ (unretained solute) and as $k \to \infty$ (infinitely retained solute).

(b) If the column radius is 0.10 mm, find H_{min} for the two cases in (a).

(c) What is the maximum number of theoretical plates in a 50-m-long column with a 0.10-mm radius if k = 5.0?

(d) The relation between retention factor k and partition coefficient K (Equation 22-19) can also be written $k = 2tK/r$, where t is the thickness of the stationary phase in a wall-coated column and r is the inside radius of the column. Derive the equation $k = 2tK/r$ and find k if K = 1 000, t = 0.20 µm, and r = 0.10 mm.

23-28. Consider the chromatography of n-$C_{12}H_{26}$ on a 25-m-long × 0.53-mm-diameter open tubular column of 5% phenyl–95% methyl polysiloxane with a stationary-phase thickness of 3.0 µm and He carrier gas at 125°C. The observed retention factor for n-$C_{12}H_{26}$ is 8.0. Measurements were made of plate height, H, at various values of linear velocity, u_x (m/s). A least-squares curve through the data points is given by

$$H(m) = (6.0 \times 10^{-5}\ m^2/s)/u_x + (2.09 \times 10^{-3}\ s)u_x$$

From the coefficients of the van Deemter equation, find the diffusion coefficient of n-$C_{12}H_{26}$ in the mobile and stationary phases. Why is one of these diffusion coefficients so much greater than the other?

23-29. 🔲 *Efficiency of solid-phase microextraction.* Equation 23-3 gives the mass of analyte extracted into a solid-phase microextraction fiber as a function of the partition coefficient between the fiber coating and the solution.

(a) A commercial fiber with a 100-µm-thick coating has a film volume of 6.9×10^{-4} mL.[26] Suppose that the initial concentration of analyte in solution is c_0 = 0.10 µg/mL (100 ppb). Use a spreadsheet to prepare a graph showing the mass of analyte extracted into the fiber as a function of solution volume for partition coefficients of 10 000, 5 000, 1 000, and 100. Let the solution volume vary from 0 to 100 mL.

(b) Evaluate the limit of Equation 23-3 as V_s gets big relative to KV_f. Does the extracted mass in your graph approach this limit?

(c) What percentage of the analyte from 10.0 mL of solution is extracted into the fiber when K = 100 and when K = 10 000?

23-30. *Mass spectral interpretation.* Box 23-1 shows the separation of enantiomers with the formula $C_9H_4N_2Cl_6$.

(a) Verify that the formula for rings + double bonds (21-3) agrees with the structure.

(b) Find the nominal mass of $C_9H_4N_2Cl_6$ (Box 21-1)

(c) The high-mass region of the electron impact mass spectrum of one enantiomer is shown on page 594. Suggest an assignment for m/z 350, 315, 280, 245, and 210.

(d) 🔲 The relative abundance of ^{35}Cl and ^{37}Cl in a molecule containing n Cl atoms is given by the terms of the binomial expansion

$$(a + b)^n = a^n + na^{n-1}b + \frac{n(n-1)}{2!}a^{n-2}b^2 + \frac{n(n-1)(n-2)}{3!}$$

$a^{n-3}b^3 + \cdots$, where a is the natural abundance of ^{35}Cl (0.757 7) and b is the natural abundance of ^{37}Cl (0.242 3). The first term gives

Mass spectrum of one enantiomer of $C_9H_4N_2Cl_6$ from Box 23-1. [From W. Vetter and W. Jun, "Elucidation of a Polychlorinated Bipyrrole Structure Using Enantioselective GC," *Anal. Chem.* **2002,** *74,* 4287.]

the abundance of ${}^{35}Cl_n{}^{37}Cl_0$, the second term gives ${}^{35}Cl_{n-1}{}^{37}Cl_1$, and so forth. The spreadsheet below shows how to compute the terms of $(a + b)^6$ with Excel in cells D8:D14. In cell D8, the function is "=BINOMDIST(A8,\$B\$5,\$B\$3,FALSE)" = BINOMDIST $(6,6,0.7577,\text{FALSE})$, which translates into D8 = $(0.7577)^6$. When you highlight cell D8 and Fill Down, the function in cell D9 is "=BINOMDIST (A9,\$B\$5,\$B\$3,FALSE)" = BINOMDIST $(5,6,0.7577,\text{FALSE})$, which translates into D9 = $6(0.7577)^5$ $(0.2423)^1$—the second term of the expansion. Column E normalizes the abundances in column D so that the most intense peak is 100. The spreadsheet predicts that a molecule with 6 Cl atoms will have a ratio of intensities $M:M+2:M+4:M+6:M+8:M+10:M+12$ $= 52.12:100:79.95:34.09:8.18:1.05:0.06$ (if there are no other significant isotopes of other elements in the molecule). Compute the expected abundances of Cl isotopes for species with 5, 4, 3, and 2 Cl atoms and compare your results with the observed clusters in the mass spectrum.

Spreadsheet illustrating use of binomial distribution function.

	A	B	C	D	E
1	Isotopic abundance from binomial distribution				
2					
3	${}^{35}Cl =$	0.7577	natural abundance		
4	${}^{37}Cl =$	0.2423	natural abundance		
5	$n =$	6			
6					Relative
7	${}^{35}Cl$	Formula	Mass	Abundance	abundance
8	6	${}^{35}Cl_6{}^{37}Cl_0$	M	0.18923	52.12
9	5	${}^{35}Cl_5{}^{37}Cl_1$	M+2	0.36307	100.00
10	4	${}^{35}Cl_4{}^{37}Cl_2$	M+4	0.29026	79.95
11	3	${}^{35}Cl_3{}^{37}Cl_3$	M+6	0.12376	34.09
12	2	${}^{35}Cl_2{}^{37}Cl_4$	M+8	0.02968	8.18
13	1	${}^{35}Cl_1{}^{37}Cl_5$	M+10	0.00380	1.05
14	0	${}^{35}Cl_0{}^{37}Cl_6$	M+12	0.00020	0.06
15					
16	D8 = BINOMDIST (A8,\$B\$5,\$B\$3,FALSE)				
17	E8 = 100*D8/MAX(\$D\$8:\$D\$14)				

24 | High-Performance Liquid Chromatography

Liquid chromatogram of archaeal cell membrane lipids extracted from sea floor sediment. Compounds 4 and 4′ are isomers with an unknown relationship. Chromatography was run on a (2.1 × 150 mm) column containing 3 μm Prevail Cyano phase. [From S. Schouten, C. Huguet, E. C. Hopmans, M. V. Kienhuis, and J. S. Sinninghe Damsté, "Analytical Methodology for TEX$_{86}$ Paleothermometry by High-Performance Liquid Chromatography/Atmospheric Pressure Chemical Ionization-Mass Spectrometry," *Anal. Chem.* 2007, *79*, 2940.]

Correlation of today's sea surface temperature with TEX$_{86}$ in 223 core-top sediments from the Atlantic, Pacific, and Indian Oceans. [Data from J.-H. Kim, S. Schouten, E. C. Hopmans, B. Donner, and J. S. Sinninghe Damsté, "Global Sediment Core-top Calibration of the TEX$_{86}$ Paleothermometer in the Ocean," *Geochim. Cosmochim. Acta* 2008, *72*, 1154.]

Archaea are single-cell organisms that constitute a substantial fraction of life in the ocean. They manufacture cell membrane lipids, which can be measured by liquid chromatography. Lipids 0 and 4 are the major components. Of the minor components, type 1 is predominant in archaea living at low temperature (near 0°C). Types 2, 3, and 4′ are manufactured in increasing amounts at warmer ocean temperatures, presumably giving membranes proper fluidity. When archaea die, they fall to the ocean floor, where their lipids remain in the sediment. The surface of the sediment contains remains of the most recent archaea. The deeper we dig into the sediment, the older are the remains.

The graph shows the correlation between today's sea surface temperature and membrane lipid types found in the topmost sediment. The function in the graph is the sum of lipids 2, 3, and 4′ expressed as a fraction of lipids 1, 2, 3, and 4′:

$$TEX_{86} \equiv \frac{[2] + [3] + [4']}{[1] + [2] + [3] + [4']} \text{ (measured by chromatography)}$$

Empirical correlation: Sea surface temperature (°C) = −10.78 + 56.2(TEX$_{86}$)

By measuring archaeal membrane lipids as a function of depth in ocean sediment and in sedimentary rock on land, it is possible to construct a history of ocean temperature and to compare the membrane lipid ratio to other indicators of climate.[1]

FIGURE 24-1 High-performance liquid chromatography (HPLC) equipment including a mass spectrometer for detection. The column is enlarged at the right. In operation, the door to the column oven would be closed to maintain constant temperature. [Courtesy E. Erickson, Michelson Laboratory, China Lake, CA.]

The pioneer of high-performance liquid chromatography was C. Horváth at Yale University in 1965.

FIGURE 24-2 (*a* and *b*) Chromatograms of the same sample run at the same linear velocity on 5.0-cm-long columns packed with C_{18}-silica. (*c*) A stronger solvent was used to elute solutes more rapidly from the column in panel *b*. [From Y. Yang and C. C. Hodges, "Assay Transfer from HPLC to UPLC for Higher Analysis Throughput," *LCGC Supplement*, May 2005, p. 31.]

High-performance liquid chromatography (HPLC) uses high pressure to force solvent through closed columns containing fine particles that give high-resolution separations.[2–6] The HPLC system in Figure 24-1 consists of an autosampler, a solvent delivery system, a sample injection valve, a high-pressure chromatography column, and a mass spectrometer, which serves as a detector. Not shown in the photograph are solvent reservoirs, a photodiode array absorbance detector, and a computer to control the hardware and display results. The column is housed in an oven whose door is normally closed to keep the column at constant temperature. In this chapter, we discuss liquid-liquid partition and liquid-solid adsorption chromatography. Chapter 25 deals with ion-exchange, molecular exclusion, affinity, and hydrophobic interaction chromatography.

Chromatographers generally choose gas chromatography over liquid chromatography when there is a choice, because gas chromatography is normally less expensive and generates much less waste. Liquid chromatography is important because most compounds are not sufficiently volatile for gas chromatography.

24-1 The Chromatographic Process

Increasing the rate at which solute equilibrates between stationary and mobile phases increases the efficiency of chromatography. For gas chromatography with an open tubular column, rapid equilibration is accomplished by reducing the diameter of the column so that molecules can diffuse quickly between the channel and the stationary phase on the wall. Diffusion in liquids is 100 times slower than diffusion in gases. Therefore, in liquid chromatography, it is not generally feasible to use open tubular columns, because the diameter of the solvent channel is too great to be traversed by a solute molecule in a short time. Liquid chromatography is conducted with packed columns so that a solute molecule does not have to diffuse very far to encounter the stationary phase.

Small Particles Give High Efficiency but Require High Pressure

The efficiency of a packed column increases as the size of the stationary phase particles decreases. Typical particle sizes in HPLC are 1.7 to 5 μm. Figure 24-2a and b illustrate the increased resolution afforded by decreasing particle size. Plate number increased from 2 000 to 7 500 when the particle size decreased, so the peaks are sharper with the smaller particle size. In Figure 24-2c, a stronger solvent was used to elute the peaks in less time. Decreasing particle size permits us to improve resolution or to maintain the same resolution while decreasing run time.

EXAMPLE Scaling Relations Between Columns

Commonly, silica particles occupy ~40% of the column volume and solvent occupies ~60% of the column volume, regardless of particle size. The column used in Figure 24-2a has an inside diameter of 4.6 mm and was run at a volume flow rate (u_v) of 3.0 mL/min with a sample size of 20 μL. The column used in Figure 24-2b has a diameter of $d_c = 2.1$ mm. What flow rate should be used in trace *b* to achieve the same linear velocity (u_x) as in trace *a*? What sample volume should be injected?

Solution Column volume is proportional to the square of column diameter. Changing the diameter from 4.6 to 2.1 mm, reduces volume by a factor of $(2.1/4.6)^2 = 0.208$. Therefore, u_v should be reduced by a factor of 0.208 to maintain the same linear velocity.

$$u_v(\text{small column}) = 0.208 \times u_v(\text{large column}) = (0.208)(3.0 \text{ mL/min}) = 0.62 \text{ mL/min}$$

To maintain the same ratio of injected sample to column volume,

Injection volume in small column $= 0.208 \times$ (injection volume in large column)
$$= (0.208)(20 \text{ μL}) = 4.2 \text{ μL}$$

Test Yourself What should be the volume flow rate and injected volume for a 1.5-mm-diameter column? (***Answer:*** 0.32 mL/min, 2.1 μL)

van Deemter plots of plate height versus linear flow rate in Figure 24-3 show that small particles reduce plate height and that plate height is not very sensitive to increased flow rate when the particles are small. At the optimum flow rate for each column (minimum plate height in Figure 24-3), the number of theoretical plates in a column of length L (cm) is *approximately*[7]

$$N \approx \frac{3\,000\,L\,(\text{cm})}{d_p\,(\text{μm})} \qquad (24\text{-}1)$$

where d_p is the particle diameter in μm. The 5.0-cm-long column in Figure 24-2a with 4.0-μm-diameter particles is predicted to provide $\sim(3\,000)(5.0)/4.0 = 3\,800$ plates. The observed plate number for the second peak is 2 000. Perhaps the column was not run at optimum flow rate. When the stationary phase particle diameter is reduced to 1.7 μm, the optimum plate number is expected to be $\sim(3\,000)(5.0)/1.7 = 8\,800$. The observed value is 7 500.

One reason why small particles give better resolution is that they provide more uniform flow through the column, thereby reducing the multiple path term, A, in the van Deemter equation (22-33). A second reason is that the distance through which solute must diffuse in the mobile and stationary phases is on the order of the particle size. The smaller the particles, the less distance solute must diffuse. This effect decreases the C term in the van Deemter equation for finite equilibration time. Optimum flow rate for small particles is faster than for large particles because solutes diffuse through smaller distances.

Increasing efficiency is equivalent to decreasing plate height, H, in the van Deemter equation (22-33):

$$H \approx A + \frac{B}{u_x} + Cu_x$$

u_x = linear flow rate

FIGURE 24-3 van Deemter curves: Plate height as a function of *linear flow rate* (mm/s) for microporous (Figure 24-5) stationary phase particle diameters of 5.0, 3.5, and 1.8 μm, as well as *superficially porous* particles (Figure 24-9) with a diameter of 2.7 μm (0.5 μm porous layer thickness). Measurements for naphthalene eluted from C_{18}-silica (50 mm long × 4.6 mm diameter) with 60 vol% acetonitrile/40 vol% H_2O at 24°C. [Courtesy MAC-MOD Analytical, Chadds Ford, PA.]

Smaller particle size leads to
- higher plate number
- higher pressure
- shorter optimum run time
- lower detection limit

An added benefit of small particle size, coupled with a narrow column and higher flow is that analyte is not diluted so much as it travels through the column. The limit of quantitation for conditions in Figure 24-2c (50 µg/L) is four times lower than the limit of quantitation in Figure 24-2a (200 µg/L).

The penalty for small particle size is resistance to solvent flow. The pressure required to drive solvent through a column is

Viscosity measures resistance of a fluid to flow. The more viscous a liquid, the slower it flows at a given pressure.

Column pressure:
$$P = f\frac{u_x \eta L}{\pi r^2 d_p^2}$$
(24-2)

where u_x is linear flow rate, η is the viscosity of the solvent, L is the length of the column, r is column radius, and d_p is the particle diameter. The factor f depends on particle shape and particle packing. The physical significance of Equation 24-2 is that pressure in HPLC is proportional to flow rate and column length and inversely proportional to the square of column radius (or diameter) and the square of particle size. The difference between traces *a* and *b* in Figure 24-2 is that particle size was decreased from 4.0 µm to 1.7 µm and column diameter was decreased from 4.6 mm to 2.1 mm. Therefore, the required pressure increases by a factor of $(4.6 \text{ mm}/2.1 \text{ mm})^2(4.0 \text{ µm}/1.7 \text{ µm})^2 = 27$. That is, *27 times more pressure* is required to operate the column in Figure 24-2b.

Analogy of fluid flow with electric current:

Electric power (W) = current × voltage

Chromatographic heat generation:

Power (W) = volume flow rate × pressure drop

$\underbrace{}_{m^3/s}$ $\underbrace{}_{Pa = kg/(m \cdot s^2)}$

Until recently, HPLC operated at pressures of ~7–40 MPa (70–400 bar, 1 000–6 000 pounds/inch^2) to attain flow rates of ~0.5–5 mL/min. In 2004, commercial equipment became available to employ 1.5- to 2-µm-diameter particles at pressures up to 100 MPa (1 000 bar, 15 000 pounds/inch^2). These instruments provide substantially increased resolution or decreased run time. Table 24-1 shows theoretical performance for different particle sizes; such performance was realized in research with ultrahigh-pressure equipment. Chromatography with 1.5- to 2-µm-diameter particles at high pressure is commonly called UPLC (Ultra Performance Liquid Chromatography), which is a trademark of Waters Corporation. Peaks eluted from a UPLC column could be so narrow that there is not enough time for mass spectral detection.

Another penalty of small particle size is increased frictional heating as solvent is forced through the particle bed.[8] The center of a column is warmer than the outer wall, and the outlet is warmer than the inlet. A 100-mm-long × 2.1-mm-diameter column containing 1.7-µm particles eluted with acetonitrile generates a temperature difference of ~10°C from the inlet to the outlet at a flow rate of 1.0 mL/min. The centerline of the column can be ~2°C warmer than the wall. To avoid undue band broadening from temperature differences, column diameter should be ≤2.1 mm for 1.7-µm particles.

The Column

Columns are expensive and easily degraded by dust or particles in the sample or solvent and by irreversible adsorption of impurities from the sample or solvent. To avoid introducing particulate matter into the column, samples should be centrifuged and/or filtered through a 0.5-µm filter before they are loaded into vials for an autosampler or taken into a syringe for manual injection. An in-line 0.5-µm filter should be installed immediately downsteam of the autosampler.

The HPLC equipment in Figure 24-1 uses steel or plastic columns that are 5–30 cm in length, with an inner diameter of 1–5 mm (Figure 24-4). The entrance to the main column is protected by a short **guard column** containing the same stationary phase as the main column.

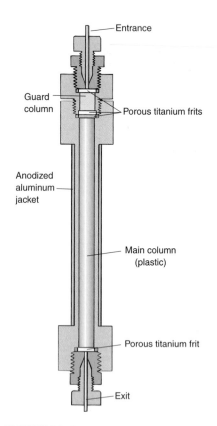

FIGURE 24-4 HPLC column with replaceable guard column to collect irreversibly adsorbed impurities. Titanium frits contain the stationary phase and distribute liquid evenly over the diameter of the column. [Courtesy Upchurch Scientific, Oak Harbor, WA.] In Figure 24-1, flow direction is bottom to top, which is opposite the flow direction in this figure.

TABLE 24-1 **Performance as a function of particle diameter**

Particle size d_p (µm)	Retention time (min)	Plate number (N)	Required pressure (bar)
5.0	30	25 000	19
3.0	18	42 000	87
1.5	9	83 000	700
1.0	6	125 000	2 300

NOTE: *Theoretical performance of 33-µm-diameter × 25-cm-long capillary for minimum plate height for solute with retention factor k = 2 and diffusion coefficient = 6.7 × 10^{-10} m^2/s in water-acetonitrile eluent.*

SOURCE: *J. E. MacNair, K. D. Patel, and J. W. Jorgenson, "Ultrahigh-Pressure Reversed-Phase Capillary Liquid Chromatography with 1.0-µm Particles," Anal. Chem.* **1999,** *71, 700.*

CHAPTER 24 High-Performance Liquid Chromatography

Fine particles and strongly adsorbed solutes are retained in the guard column, which is periodically replaced when column pressure increases or after a set number of injections or time in service. While guard columns make sense with 10- to 30-cm-long chromatography columns, many people do not consider a guard column to be cost effective for a 5-cm column.

Heating a chromatography column[9] usually decreases the viscosity of the solvent, thereby reducing the required pressure or permitting faster flow. Increased temperature decreases retention times (Figure 22-19) and improves resolution by hastening diffusion of solutes. However, increased temperature can degrade the stationary phase and decrease column lifetime. When column temperature is not controlled, it fluctuates with the ambient temperature. Using a column heater set 10°C above room temperature improves the reproducibility of retention times and the precision of quantitative analysis. Some chromatographers routinely conduct all separations at 50° or 60°C. For a heated column, the mobile phase should be passed through a preheating metal coil between the injector and the column so that the solvent and column are at the same temperature. If their temperatures are different, peaks become distorted and retention times change.

In the recent past, the most common HPLC column diameter was 4.6 mm. Now, 2.1 mm is becoming most common. The narrow column is more compatible with mass spectrometry, which requires low solvent flow. The narrow column requires less sample and produces less waste. Instruments that used the 4.6-mm column also operate with the 2.1-mm column. Columns narrower than 2.1 mm require specially designed instruments to reduce band broadening outside the column. Capillary columns as narrow as 25 μm can be used.

The Stationary Phase

The most common support is highly pure, spherical, **microporous particles** of silica (Figure 24-5) that are permeable to solvent and have a surface area of several hundred square meters per gram. Most silica cannot be used above pH 8, because it dissolves in base. Figure 24-6 shows the structure of ordinary silica and silica with ethylene bridges, which resist hydrolysis up to pH 12.[10] For separation of basic compounds at pH 8–12, ethylene-bridged silica or polymeric supports such as polystyrene (Figure 25-1) can be used. Stationary phase is covalently attached to the polymer.

A silica surface (Figure 24-6) has up to 8 μmol of silanol groups (Si—OH) per square meter. Silanol groups are protonated at pH ~2–3. They dissociate to negative Si—O⁻ over a

FIGURE 24-5 Scanning electron micrograph of 4.4-μm-diameter microporous silica chromatography particle from an experimental batch made by K. Wyndham at Waters Corporation. [Photo kindly provided by J. Jorgensen, University of North Carolina.]

Hydrogen-bonded vicinal silanol groups

Isolated silanol groups

Siloxane bonds

Geminal silanol groups

(a)

(b)

FIGURE 24-6 (a) Schematic structure of silica particle. [From R. E. Majors, *LCGC*, May 1997, p. S8.] (b) Base-hydrolysis-resistant silica incorporates ethylene bridges in place of oxide bridges between some silicon atoms. Ethylene-bridged structure is more rigid and well suited for particles that are <2 μm in diameter, which must withstand high pressure.

FIGURE 24-7 Tailing of amine bases on silica: (*a*) Type A silica support gives distorted peaks. (*b*) Less acidic Type B silica with fewer Si—OH groups and less metallic impurity gives symmetric peaks with shorter retention time. In both cases, chromatography was performed with a 0.46 × 15 cm column eluted at 1.0 mL/min at 40°C with 30 vol% acetonitrile/70 vol% sodium phosphate buffer (pH 2.5) containing 0.2 wt% triethylamine and 0.2 wt% trifluoroacetic acid. The detector measured ultraviolet absorbance at 254 nm. Additives such as triethylamine and trifluoroacetic acid are often used to mask strong adsorption sites and thereby reduce tailing. [From J. J. Kirkland, *Am. Lab.*, June 1994, p. 28K.]

broad pH range above 3. In the old Type A silica, exposed Si—O⁻ groups strongly retain protonated bases (for example, RNH_3^+) and lead to tailing (Figure 24-7). Metallic impurities in Type A silica also cause tailing. Type B silica in Figure 24-7, which has fewer exposed silanol groups and fewer metallic impurities, is the most common form used today. Type C silica causes even less tailing because 90% of the Si—OH groups are replaced by Si—H bonds, which do not retain solutes by hydrogen bonding.

Bare silica can be used as the stationary phase for adsorption chromatography. Most commonly, liquid-liquid partition chromatography is conducted with a **bonded stationary phase** covalently attached to the silica surface by reactions such as

Residual silanol groups on the silica surface are capped with trimethylsilyl groups by reaction with $ClSi(CH_3)_3$ to eliminate polar adsorption sites that cause tailing.

Common polar phases

R = $(CH_2)_3NH_2$	amino
R = $(CH_2)_3C{\equiv}N$	cyano
R = $(CH_2)_2OCH_2CH(OH)CH_2OH$	diol

Common nonpolar phases

R = $(CH_2)_{17}CH_3$	octadecyl
R = $(CH_2)_7CH_3$	octyl
R = $(CH_2)_3C_6H_5$	phenyl
R = $(CH_2)_3C_6F_5$	F₅-phenyl

There are ~4 μmol of R groups per square meter of support surface area, with little bleeding of the stationary phase from the column during chromatography.

The octadecyl (C_{18}) stationary phase (often abbreviated ODS) is by far the most common in HPLC. Retention factors for a given solute on columns with the same nominal stationary phase from different manufacturers are quite variable. Different surface areas of different columns account for much of the variation.[11] The nonpolar perfluorophenyl (F₅-phenyl) phase provides selectivities different from those of the octadecyl phase and can be especially useful for separating aromatic compounds. The separation of cell membrane lipids at the opening of this chapter was done with a polar cyano column.

The siloxane (Si—O—SiR) bond hydrolyzes below pH 2, so HPLC with a bonded phase on a silica support is generally limited to the pH range 2–8. If bulky isobutyl groups are attached to the silicon atom of the bonded phase (Figure 24-8), the stationary phase is protected from attack by H_3O^+ and is stable for long periods at low pH, even at elevated temperature (for example, pH 0.9 at 90°C). Box 24-1 describes *monolithic* stationary phases.

Bidentate C_{18} stationary phase provides increased stability above pH 8:

BOX 24-1 Monolithic Silica Columns

Monolithic silica rods.

Structure of rod showing ~2-μm pores.

Enlarged view. Invisible 13-nm pores are located within the silica skeleton.

[Photos courtesy D. Cunningham, Merck KGaA, Darmstadt, Germany.]

Time is money. In commercial laboratories, the faster an analysis can be done, the less it costs. *Monolithic silica columns* enable us to increase the flow rate in liquid chromatography while retaining good separation.[12]

Each column in the left-hand photograph is a single, porous silica rod polymerized from liquid precursors. The adjacent micrographs show the silica skeleton with a network of ~2-μm pores. The inside of the skeleton contains a finer network of ~13-nm pores that are too small to be seen in the micrographs. Approximately 80% of the volume of the rod is empty space. The surface area is 300 m^2/g, which compares favorably with that of excellent stationary phase materials. C_{18} or other bonded phases are attached to the silica for reversed-phase chromatography. After fabrication, the silica rod is tightly encased in a chemically resistant plastic tube made of polyether ether ketone (PEEK).

Solvent flows through the open, rigid structure of the monolithic column with relatively low resistance. The same pressure required to obtain a flow rate of 1 mL/min with 3.5-μm spherical particles provides a flow rate of 9 mL/min in the monolithic column. At 9 mL/min, the plate height in the monolithic column is only 50% greater than the minimum plate height observed at 2 mL/min.

With low resistance to solvent flow, monolithic columns can be made very long to achieve exquisite separations at ordinary HPLC pressures. The chromatogram shows the resolution of deuterated benzene and toluene molecules achieved with a 440-cm long × 0.1 mm-diameter-column run by ordinary HPLC hardware at a pressure of 396 bar.

Separation of isotopic molecules on a 440-cm-long monolithic C_{18}-silica column eluted with CH_3CN/H_2O (30:70 vol/vol) at 30°C. [From K. Miyamoto, T. Hara, H. Kobayashi, H. Morisaka, D. Tokuda, K. Horie, K. Koduki, S. Makino, O. Núñez, C. Yang, T. Kawabe, T. Ikegami, H. Takubo, Y. Ishihama, and N. Tanaka, "High-Efficiency Liquid Chromatographic Separation Utilizing Long Monolithic Silica Capillary Columns," *Anal. Chem.* 2008, *80*, 8741.]

FIGURE 24-8 Bulky isobutyl groups protect siloxane bonds from hydrolysis at low pH. [From J. J. Kirkland, *Am. Lab.*, June 1994, p. 28K.]

FIGURE 24-9 Rapid separation of proteins on superficially porous C_{18}-silica in 2.1 × 75 mm column containing Poroshell 300SB-C18. Mobile phase A: 0.1 wt% trifluoroacetic acid in H_2O. Mobile phase B: 0.07 wt% trifluoroacetic acid in acetonitrile. Solvent was changed continuously from 95 vol% A/5 vol% B to 100% B over 1 min. Flow = 3 mL/min at 70°C at 26 MPa (260 bar) with ultraviolet detection at 215 nm. Peaks: 1, angiotensin II; 2, neurotensin; 3, ribonuclease; 4, insulin; 5, lysozyme; 6, myoglobin; 7, carbonic anhydrase; 8, ovalbumin. [From R. E. Majors, *LCGC Column Technology Supplement*, June 2004, p. 8K. Courtesy Agilent Technologies.]

Nonpolar bonded phase with embedded polar amide group offers different selectivity from C_{18}, has improved peak shape for bases, and tolerates 100% aqueous eluent.

Another type of nonpolar stationary phase has a *polar embedded group*. The example in the margin consists of a long hydrocarbon chain with a polar amide group near its base. Embedded polar groups provide alternate selectivities from C_{18} stationary phases, improved peak shapes for bases, and compatibility with 100% aqueous phase. Other nonpolar stationary phases should not be exposed to 100% aqueous phase because they become very difficult to re-equilibrate with organic phase.

Figure 24-9 shows a rapid separation of proteins on **superficially porous particles** (also called *fused-core* particles), which consist of a 0.25-μm-thick porous silica layer on a 5-μm-diameter nonporous silica core. A stationary phase such as C_{18} is bonded throughout the thin, porous outer layer. Mass transfer of solute into a 0.25-μm-thick layer is 10 times faster than mass transfer into fully porous particles with a radius of 2.5 μm, thus enabling high efficiency at high flow rate. Superficially porous particles are especially suitable for separation of macromolecules such as proteins, which diffuse more slowly than small molecules. Figure 24-3 showed that the van Deemter curve for superficially porous particles with a total diameter of 2.7 μm and a porous layer thickness of 0.5 μm is similar to that of a conventional, totally porous particle with a diameter of 1.8 μm. The superficially porous particle enables separations similar to those achieved with 1.8-μm totally porous particles without requiring such high pressure.

Porous graphitic carbon deposited on silica[13] is a stationary phase that exhibits increased retention of nonpolar compounds relative to retention by C_{18}. Graphite has high affinity for polar compounds and separates isomers that cannot be separated on C_{18}. The stationary phase is stable in 10 M acid and 10 M base.

Pharmaceutical companies often separate the two enantiomers (mirror image isomers) of a drug because each enantiomer has a different pharmacological effect. To resolve enantiomers, optically active bonded phases, such as those shown in Figure 24-10 and Exercise 24-B, are used.[14] Figure 24-10 shows the calculated geometry of the chiral drug naproxen binding to one enantiomer of the stationary phase. Mirror image forms of the drug are designated *R* and *S*. Mirror image forms of the stationary phase are designated (*R*,*R*) and (*S*,*S*). Binding of (*S*)-naproxen to (*S*,*S*)-stationary phase is stronger than binding of (*R*)-naproxen to (*S*,*S*)-stationary phase. Therefore, (*R*)-naproxen is eluted before (*S*)-naproxen from (*S*,*S*)-stationary phase. Some other chiral stationary phases are based on substituted cellulose, on cyclic peptides with sugar substituents, and on cyclodextrins (Box 23-1).

The Elution Process

In *adsorption chromatography*, solvent molecules compete with solute molecules for sites on the stationary phase (Figure 24-11 and Color Plate 28). The relative abilities of different solvents to elute a given solute from the adsorbent are nearly independent of the nature of the solute. Elution occurs when solvent displaces solute from the stationary phase.

Chiral stationary phase Whelk-O 1

(R,R)   (S,S)

(S)-Naproxen   (R)-Naproxen

Interaction of (R)- and (S)-naproxen with (S,S) stationary phase

To silica

Naphthalene group

Dinitrophenyl group

(S)-Naproxen
(S,S)
More stable adsorbate

(R)-Naproxen   (S,S)
Less stable adsorbate

FIGURE 24-10 Interaction of enantiomers of the drug naproxen with (S,S) chiral stationary phase Whelk-O 1. (S)-Naproxen is adsorbed more strongly and is therefore retained longer by the column. [From S. Ahuja, "A Strategy for Developing HPLC Methods for Chiral Drugs," *LCGC* 2007, *25*, 1112.]

An *eluotropic series* ranks solvents by their relative abilities to displace solutes from a given adsorbent. The **eluent strength, ε°,** is a measure of the solvent adsorption energy, with the value for pentane defined as 0 for adsorption on bare silica. Table 24-2 ranks solvents by their eluent strength on bare silica. The more polar the solvent, the greater is its eluent strength for adsorption chromatography with bare silica. The greater the eluent strength, the more rapidly will solutes be eluted from the column.

Adsorption chromatography on bare silica is an example of **normal-phase chromatography,** in which we use a polar stationary phase and a less polar solvent. *A more polar solvent has a higher eluent strength.* **Reversed-phase chromatography** is the more common scheme in which the stationary phase is nonpolar or weakly polar and the solvent is more polar. *A less polar solvent has a higher eluent strength.* In general, eluent strength is increased by making the mobile phase more like the stationary phase. Reversed-phase chromatography eliminates peak tailing because the stationary phase has few sites that can strongly adsorb a solute to cause tailing (Figure 22-21). Normal-phase chromatography is sensitive to small amounts of

Normal-phase chromatography:
- polar stationary phase
- more polar solvent has higher eluent strength

Reversed-phase chromatography:
- nonpolar stationary phase
- less polar solvent has higher eluent strength

Solvent

Flow

Desorbed solute

Solute adsorbed on stationary phase

Stationary phase

Solvent adsorbed on stationary phase

(a)   (b)

FIGURE 24-11 Solvent molecules compete with solute molecules for binding sites on the stationary phase. The greater the eluent strength of the solvent, the more easily it displaces the solute.

Solvent	Eluent strength ($\varepsilon°$)	Ultraviolet cutoff (nm)
Pentane	0.00	190
Hexane	0.01	195
Heptane	0.01	200
Trichlorotrifluoroethane	0.02	231
Toluene	0.22	284
Chloroform	0.26	245
Dichloromethane	0.30	233
Diethyl ether	0.43	215
Ethyl acetate	0.48	256
Methyl t-butyl ether	0.48	210
Dioxane	0.51	215
Acetonitrile	0.52	190
Acetone	0.53	330
Tetrahydrofuran	0.53	212
2-Propanol	0.60	205
Methanol	0.70	205

NOTE: The ultraviolet cutoff for water is 190 nm.

SOURCES: L. R. Snyder, in High-Performance Liquid Chromatography (C. Horváth, ed.), Vol. 3 (New York: Academic Press, 1983); Burdick & Jackson Solvent Guide, 3rd ed. (Muskegon, MI: Burdick & Jackson Laboratories, 1990).

Isocratic elution: one solvent

Gradient elution: continuous change of solvent composition to increase eluent strength. Gradient elution in HPLC is analogous to temperature programming in gas chromatography. Increased eluent strength is required to elute more strongly retained solutes.

water in the eluent, but reversed–phase chromatography is not. Box 24-2 describes the solvent–stationary phase interface in reversed-phase chromatography.

Isocratic and Gradient Elution

Isocratic elution is performed with a single solvent (or constant solvent mixture). If one solvent does not provide sufficiently rapid elution of all components, then **gradient elution** can be used. In this case, increasing amounts of solvent B are added to solvent A to create a continuous gradient.

BOX 24-2   Structure of the Solvent–Bonded Phase Interface

Structured layers of solvent at the interface between the stationary and mobile phases influence the separation in liquid chromatography. The diagrams show solvent layers at the surface of C_{18}-silica in the presence of methanol/water (40:60 vol:vol) and acetonitrile/water (30:70 vol:vol), as deduced from adsorption studies of phenol and caffeine. The 40:60 and 30:70 compositions were chosen because they have nearly the same eluent strength.

Methanol forms a *monolayer* (one molecule thick) of adsorbed solvent on the C_{18} surface with a thickness of 0.25 nm. Acetonitrile

forms a pool of adsorbed solvent with a thickness of 1.3 nm and a high capacity for dissolved solute. Adsorbed phenol and caffeine molecules can reside inside the C_{18} layer or on the outer surface of the C_{18} layer, or they can be dissolved in the CH_3CN pool. Each type of adsorption site has a different binding energy and binding capacity and affects the shape of the eluted solute band differently at different solute concentrations. The pool of adsorbed CH_3CN is a stronger eluent than the bulk organic/aqueous solution and explains why acetonitrile is a stronger eluent than methanol.

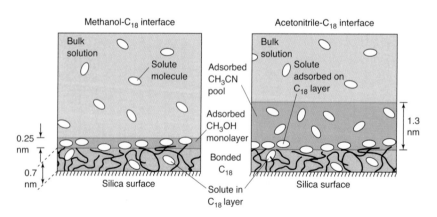

Structure of adsorbed solvent on C_{18} bonded phase in reversed-phase chromatography. [Adapted from F. Gritti and G. Guiochon, "Adsorption Mechanism in RPLC. Effect of the Nature of the Organic Modifier," *Anal. Chem.* 2005, 77, 4257.]

Figure 24-12 shows the effect of increasing eluent strength in the isocratic elution of eight compounds from a reversed-phase column. In a reversed-phase separation, eluent strength *decreases* as the solvent becomes *more* polar. The first chromatogram (upper left) was obtained with a solvent consisting of 90 vol% acetonitrile and 10 vol% aqueous buffer. Acetonitrile has a high eluent strength, and all compounds are eluted rapidly. Only

- Aqueous buffer for HPLC is prepared and the pH adjusted *prior* to mixing with organic solvent.[15]
- Ultrapure water for HPLC should be freshly prepared by a purification train or by distillation. Water extracts impurities from polyethylene or glass after storage for a few hours.
- To prepare 70% B, for example, mix 70 mL of B with 30 mL of A. *The result is different* from placing 70 mL of B in a volumetric flask and diluting to 100 mL with A because there is a volume change when A and B are mixed.

General elution problem: For a complex mixture, isocratic conditions can often be found to produce adequate separation of early-eluting peaks or late-eluting peaks, but not both. This problem drives us to use gradient elution.

FIGURE 24-12 Isocratic HPLC separation of a mixture of aromatic compounds at 1.0 mL/min on a 0.46 × 25 cm Hypersil ODS column (C_{18} on 5-μm silica) at ambient temperature (~22°C): (1) benzyl alcohol; (2) phenol; (3) 3′,4′-dimethoxyacetophenone; (4) benzoin; (5) ethyl benzoate; (6) toluene; (7) 2,6-dimethoxytoluene; (8) o-methoxybiphenyl. Eluent consisted of aqueous buffer (designated A) and acetonitrile (designated B). The notation "90% B" in the first chromatogram means 10 vol% A and 90 vol% B. The buffer contained 25 mM KH_2PO_4 plus 0.1 g/L sodium azide adjusted to pH 3.5 with HCl.

three peaks are observed because of overlap. It is customary to call the aqueous solvent A and the organic solvent B. The first chromatogram was obtained with 90% B. When eluent strength is reduced by changing the solvent to 80% B, there is slightly more separation and five peaks are observed. At 60% B, we begin to see a sixth peak. At 40% B, there are eight clear peaks, but compounds 2 and 3 are not fully resolved. At 30% B, all peaks would be resolved, but the separation takes too long. Backing up to 35% B (the bottom trace) separates all peaks in a little over 2 h (which is still too long for many purposes).

From the isocratic elutions in Figure 24-12, the gradient in Figure 24-13 was selected to resolve all peaks in 38 min. First, 30% B (B = acetonitrile) was run for 8 min to separate components 1, 2, and 3. Then eluent strength was increased over 5 min to 45% B and held there for 15 min to elute peaks 4 and 5. Finally, the solvent was changed to 80% B over 2 min and held there to elute the last peaks.

Box 24-3 describes gradient elution in *supercritical fluid chromatography*.

Hydrophilic: "water loving"—soluble in water, surface is wetted by water

Hydrophobic: "water hating"—insoluble in water, surface is not wetted by water

Hydrophilic Interaction Chromatography (HILIC)

Hydrophilic substances are soluble in water or attract water to their surfaces. Polar organic molecules have hydrophilic regions. **Hydrophilic interaction chromatography** is most useful

---

BOX 24-3   "Green" Technology: Supercritical Fluid Chromatography

In the phase diagram for carbon dioxide, solid CO_2 (Dry Ice) is in equilibrium with gaseous CO_2 at a temperature of $-78.7°C$ and a pressure of 1.00 bar. The solid *sublimes* without turning into liquid. Above the *triple point* at $-56.6°C$, liquid and vapor coexist as separate phases. For example, at 0°C, liquid is in equilibrium with gas at 34.9 bar. Moving up the liquid-gas boundary, we see that two phases always exist until the *critical point* is reached at 31.3°C and 73.9 bar. *Above this temperature, only one phase exists, no matter what the pressure.* We call this phase a **supercritical fluid** (Color Plate 29). Its density and viscosity are between those of the gas and liquid, as is its ability to act as a solvent.

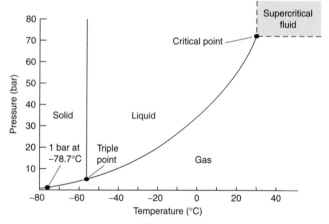

Phase diagram for CO_2.

An interesting supercritical fluid to demonstrate is SF_6.[16] The photographs at the right show different changes as fluid is warmed and cooled through its supercritical temperature.

Supercritical fluid chromatography with a mixture of CO_2 and organic solvent is a "green" technology that reduces organic solvent use by up to 90% for the separation of kilograms of compounds and enantiomers in the pharmaceutical industry.[17] The low viscosity of the supercritical fluid also permits faster flow to increase productivity. Though CO_2 is a weak solvent by itself, when mixed with some organic solvent, it is capable of dissolving a variety of compounds.

Supercritical fluid chromatography provides increased speed and resolution, relative to liquid chromatography, because of higher diffusion coefficients of solutes in supercritical fluids. Unlike gases, supercritical fluids can dissolve nonvolatile solutes. When pressure is released, the solvent turns to gas, leaving the solute in the gas phase for easy detection. Carbon dioxide is the supercritical fluid of choice for chromatography because it is compatible with flame ionization and ultraviolet detectors, it has a low critical temperature, and it is nontoxic.

Equipment for supercritical fluid chromatography is similar to that for HPLC with packed columns or open tubular columns. Eluent strength is increased in HPLC by gradient elution and in gas chromatography by raising the temperature. In supercritical fluid chromatography, eluent strength is increased by making the solvent *denser* by increasing the pressure. The chromatogram shows a density-gradient elution.

Critical constants

Compound	Critical temperature (°C)	Critical pressure (bar)	Critical density (g/mL)
Argon	-122.5	47	0.53
Carbon dioxide	31.3	73.9	0.448
Sulfur hexafluoride	45.6	37.0	0.755
Ammonia	132.2	113.0	0.24
Diethyl ether	193.6	36.8	0.267
Methanol	240.5	79.9	0.272
Water	374.4	229.8	0.344

for small molecules that are too polar to be retained by reversed-phase columns. Stationary phases for hydrophilic interaction chromatography, such as those in Figure 24-14, are strongly polar. The mobile phase typically contains (25–97 vol%) CH_3CN or other polar organic solvent mixed with aqueous buffer. Solutes equilibrate between the mobile phase and a layer of aqueous phase on the surface of the stationary phase. In biochemistry, HILIC is useful for separating peptides and saccharides (sugars).

In HILIC, eluent strength is increased by *increasing* the fraction of water in the mobile phase. Gradient elution goes from low aqueous content to high aqueous content. In normal-phase chromatography, the solvent is nonaqueous. To increase eluent strength, we increase the polarity of the nonaqueous solvent. In reversed-phase chromatography, the solvent is usually aqueous, and eluent strength is increased by *decreasing* the fraction of water in the mobile phase to increase the solubility of solutes in the mobile phase.

Selecting the Separation Mode

There can be many ways to separate components of a given mixture. Figure 24-15 is a decision tree for choosing a starting point. If the molecular mass of analyte is below 2 000, use the upper part of the figure; if the molecular mass is greater than 2 000, use the lower part. In

There are no firm rules in Figure 24-15. Methods in either part of the diagram might work for molecules whose size belongs to the other part.

Warming and cooling SF_6 through its supercritical temperature. Upon warming (*upper panel*), liquid boils and the meniscus rises as the liquid expands. Upon cooling (*lower panel*), droplets of liquid form throughout the fluid. Gravity pulls droplets down, creating a "storm" before separation into two distinct phases. [From P. Licence, D. Litchfield, M. P. Dellar, and M. Poliakoff, "'Supercriticality'; a Dramatic but Safe Demonstration of the Critical Point," *Green Chem.* 2004, *6*, 352. Courtesy Peter Licence, University of Nottingham.]

Open tubular capillary supercritical chromatogram of aromatic compounds with CO_2, using density-gradient elution at 140°C. [From R. D. Smith, B. W. Wright, and C. R. Yonker, "Supercritical Fluid Chromatography," *Anal. Chem.* 1988, *60*, 1323A.]

FIGURE 24-13 Gradient elution of the same mixture of aromatic compounds in Figure 24-12 with the same column, flow rate, and solvents. The upper trace is the *segmented gradient* profile, so named because it is divided into several different segments.

either part, the first question is whether the solutes dissolve in water or in organic solvents. Suppose we have a mixture of small molecules (molecular mass $<2\ 000$) soluble in dichloromethane. Table 24-2 is essentially a ranking of solvent polarity, with the most polar solvents at the bottom. The eluent strength of dichloromethane (0.30) is closer to that of chloroform (0.26) than it is to those of alcohols, acetonitrile, or ethyl acetate (≥ 0.48). Therefore, Figure 24-15 suggests that we try adsorption chromatography on silica. The decision path is highlighted in color.

FIGURE 24-14 Stationary phases for hydrophilic interaction chromatography (HILIC).

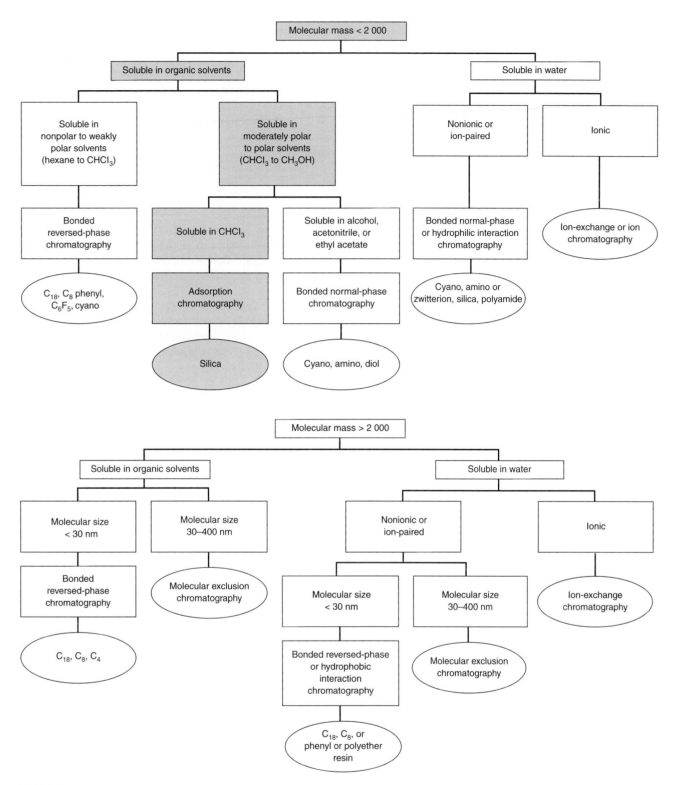

FIGURE 24-15 Guide to HPLC mode selection.

If solutes dissolve only in nonpolar or weakly polar solvents, the decision tree suggests that we try reversed-phase chromatography. Our choices include bonded phases containing octadecyl (C_{18}), octyl, phenyl, pentafluorophenyl, and cyano groups.

If molecular masses of solutes are >2 000 and if they are soluble in organic solvents and their molecular diameter is >30 nm, Figure 24-15 tells us to try molecular exclusion chromatography, described in Section 25-3. If molecular masses of solutes are >2 000, and they are soluble in water, but not ionic, and have diameters <30 nm, the decision tree says to use reversed-phase chromatography, or *hydrophobic interaction chromatography*, which is described in Section 25-5.

Solvents

Pure HPLC-grade (expensive) solvents are required to prevent degradation of costly columns with impurities and to minimize detector background signals from contaminants. Before use, solvents are purged with He or evacuated to remove dissolved air. Air bubbles create difficulties for pumps, columns, and detectors. Dissolved O_2 absorbs ultraviolet radiation in the 250- to 200-nm wavelength range,[18] interfering with ultraviolet detection. A filter is used on the intake tubing in the solvent reservoir to exclude >0.5-μm particles.

V_m = volume of solvent in column between stationary phase particles and inside pores of particles.

If time for mobile phase or unretained solute to transit the column is t_m, and flow rate is F, then $V_m = t_mF$.

If you don't know t_m, you can estimate

$$V_m(mL) \approx \frac{1}{2}d_c^2 L,$$ where d_c = column inner diameter (cm) and L = column length (cm).

Sample and solvent are passed through a short, expendable *guard column* (Figures 24-1 and 24-4) that has the same stationary phase as the analytical column and retains strongly adsorbed species. At the end of a reversed-phase separation, the column should be washed with 10–20 mobile phase volumes (V_m) of strong eluent to remove strongly retained solutes.[19] To clean the column after a series of runs, first replace aqueous buffer by water. For example, if the last eluent was acetonitrile-aqueous buffer (40:60 vol/vol), wash the column with 5–10 volumes (V_m) of acetonitrile-water (40:60 vol/vol). Then wash the column with 10–20 mobile phase volumes of strong solvent such as acetonitrile-water (95:5 vol/vol) and store the column with this solvent. This procedure is suitable for alkyl, aryl, cyano, and embedded polar-phase columns. A different procedure is recommended for normal-phase columns.[20]

Normal-phase separations are sensitive to water in the solvent. To speed the equilibration of stationary phase with changing eluents, organic solvents for normal-phase chromatography should be 50% saturated with water. Saturation can be achieved by adding a few milliliters of water to dry solvent and stirring. Then the wet solvent is separated from the excess water and mixed with an equal volume of dry solvent.

For gradient elution in reversed-phase separations, 10–20 volumes (V_m) of initial solvent should be passed through the column after a run to equilibrate the stationary phase with solvent for the next run. Equilibration can take as long as the separation. By adding 3 vol% 1-propanol to each solvent (so that there is always 3% 1-propanol at any point in the gradient), you can reduce the re-equilibration volume down to $3V_m$.[21] Propanol is thought to coat the stationary phase with a monolayer of alcohol that does not change much throughout the gradient.

Reduce Waste Solvent Without Sacrificing Resolution

- Use shorter columns with smaller diameter particles.
- Switch from 4.6-mm-diameter column to 2.1 mm.
- For isocratic separations, use an electronic recycler that recycles eluate when no peak is being eluted.

Maintaining Symmetric Bandshape

A standard mixture should be injected each day to evaluate the HPLC system. Changes in peak shapes or retention times alert you to a problem.

HPLC columns should provide narrow, symmetric peaks. If a new column does not reproduce the quality of the separation of a standard mixture whose chromatogram comes with the column, and if you have satisfied yourself that the problem is not in the rest of your system, return the column.

The asymmetry factor A/B in Figure 22-14 should rarely be outside the range 0.9–1.5. Tailing of amines (Figure 24-7) might be eliminated by adding 30 mM triethylamine to the mobile phase. The additive binds to sites on silica that would otherwise strongly bind analyte. Tailing of acidic compounds might be eliminated by adding 30 mM ammonium acetate. For unknown mixtures, 30 mM triethylammonium acetate is useful. If tailing persists, 10 mM dimethyloctylamine or dimethyloctylammonium acetate might be effective. A problem with additives is that they increase the equilibration time required when changing solvents. Improved grades of silica (types B and C) have reduced tailing and, therefore, reduced the need for additives.

Tailing or splitting of peaks can occur if the frit at the beginning of the column becomes clogged with particles.[22] You can try to unclog the frit by disconnecting and reversing the column and flushing it with 20–30 mL. The column should not be connected to the detector during reverse flushing. If peaks remain distorted, replace the column.

Doubled peaks or altered retention times (Figure 24-16) sometimes occur if the solvent in which sample is dissolved has much greater eluent strength than the mobile phase. Try to dissolve the sample in a solvent of lower eluent strength or in the mobile phase itself.

Overloading causes the distorted shape shown in Figure 22-21.[23] To see if overloading is occurring, reduce the sample mass by a factor of 10 and see if retention times increase or peaks become narrower. If either change occurs, reduce the mass again until the injection size does not affect retention time and peak shape. In general, reversed-phase columns can handle 1–10 μg of sample per gram of silica. A 4.6-mm-diameter column contains 1 g of silica in a length of 10 cm. To prevent peak broadening from too large an injection volume, the injected volume should be less than 15% of the peak volume measured at the baseline. For example, if a peak eluted at 1 mL/min has a width of 0.2 min, the peak volume is 0.2 mL. The injection volume should not exceed 15% of 0.2 mL, or 30 μL.

The volume of a chromatography system outside of the column from the point of injection to the point of detection is called the **dead volume,** or the *extra-column volume.* Excessive dead volume allows bands to broaden by diffusion or mixing. Use short, narrow tubing whenever possible, and be sure that connections are made with matched fittings to minimize dead volume and thereby minimize extra-column band spreading.

HPLC columns have a typical lifetime of 500 to 2 000 injections.[7] You can monitor the health of a column by keeping a record of pressure, resolution, and peak shape. The pressure required to maintain a given flow rate increases as a column ages. System wear becomes serious when pressure exceeds 17 MPa (2 500 pounds/inch^2). It is desirable to develop methods in which pressure does not exceed 14 MPa (2 000 pounds/inch^2). When the pressure reaches 17 MPa, the in-line 0.5-μm frit between the autosampler and the column should be replaced. If this does not help, it is probably time for a new column. If you use a column for repetitive analyses, replace the column when the required resolution is lost or when tailing becomes significant. Resolution and tailing criteria should be established during method development.

24-2 Injection and Detection in HPLC

We now consider the hardware required to inject sample and solvent onto the column and to detect compounds as they leave the column. Mass spectrometric detection, which is extremely powerful and important, was discussed in Section 21-4.

Pumps and Injection Valves

The quality of a pump for HPLC is measured by how steady and reproducible a flow it can produce. A fluctuating flow rate can create detector noise that obscures weak signals. Figure 24-17 shows a pump with two sapphire pistons that produce a programmable, constant flow rate up to 10 mL/min at pressures up to 40 MPa (400 bar, 6 000 pounds/inch^2). Gradients made from up to four solvents are constructed by proportioning the liquids through a four-way valve at low pressure and then pumping the mixture at high pressure into the column. The gradient is electronically controlled and programmable in 0.1 vol% increments.

The *injection valve* in Figure 24-18 has interchangeable sample loops, each of which holds a fixed volume. Loops of different sizes hold volumes that range from 2 to 1 000 μL. In the load position, a syringe is used to wash and load the loop with fresh sample at atmospheric pressure. High-pressure flow from the pump to the column passes through the lower left segment of the valve. When the valve is rotated 60° counterclockwise, the content of the sample loop is injected into the column at high pressure.

FIGURE 24-16 Effect of sample solvent on retention time and peak shape of *n*-butylaniline. Eluent (1 mL/min) is 90:10 (vol/vol) H$_2$O/acetonitrile with 0.1 wt% trifluoroacetic acid. Lower sample was dissolved in eluent. Upper sample was dissolved in methanol, which is a much stronger solvent than eluent. Column: 15 cm × 4.6 mm, 5-μm-diameter C$_{18}$-silica, 30°C. Injection: 10 μL containing 0.5 μg analyte. Ultraviolet detection at 254 nm. [Courtesy Supelco, Bellefonte, PA.]

FIGURE 24-17 High-pressure piston pump for HPLC. Solvent at the left passes through an electronic inlet valve synchronized with the large piston and designed to minimize the formation of solvent vapor bubbles during the intake stroke. The spring-loaded outlet valve maintains a constant outlet pressure, and the damper further reduces pressure surges. Pressure surges from the first piston are decreased in the damper that "breathes" against a constant outside pressure. Pressure surges are typically <1% of the operating pressure. As the large piston draws in liquid, the small piston propels liquid to the column. During the return stroke of the small piston, the large piston delivers solvent into the expanding chamber of the small piston. Part of the solvent fills the chamber while the remainder flows to the column. Delivery rate is controlled by the stroke volumes. [Adapted from Hewlett-Packard Co., Palo Alto, CA.]

- Pass samples through a 0.5-μm filter to remove particles prior to injection.
- The HPLC syringe needle is *blunt*, not pointed, to prevent damage to the injection port.

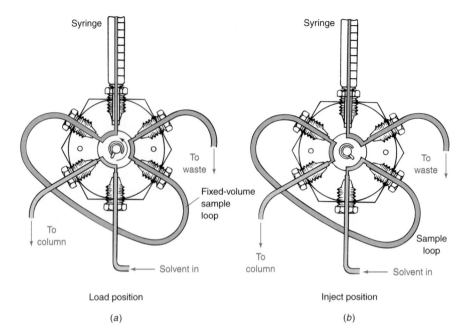

FIGURE 24-18 Injection valve for HPLC.

Linear range: analyte concentration range over which detector response is proportional to concentration

Dynamic range: range over which detector responds in any manner (not necessarily linearly) to changes in analyte concentration (see page 89)

Detection limit: concentration of analyte that gives a specified signal-to-noise ratio

FIGURE 24-19 Light path in a spectrophotometric micro flow cell. One common cell with a pathlength of 1 cm has a volume of 8 μL. Another cell with a 0.5-cm pathlength contains only 2.5 μL. In the future, it should be possible to lower detection limits by 1–2 orders of magnitude by placing mirrors on both ends of the cavity. Light transits the cell multiple times, with ~1% of the power leaking through the mirror to the detector on each pass.[24]

Spectrophotometric Detectors

An ideal detector of any type (Table 24-3) is sensitive to low concentrations of every analyte, provides linear response, and does not broaden the eluted peaks. It is also insensitive to changes in temperature and solvent composition. To prevent peak broadening, the detector volume should be less than 20% of the volume of the chromatographic band. Gas bubbles in the detector create noise, so back pressure may be applied to the detector to prevent bubble formation during depressurization of eluate.

An **ultraviolet detector** using a flow cell such as that in Figure 24-19 is the most common HPLC detector, because many solutes absorb ultraviolet light. Simple systems employ the intense 254-nm emission of a mercury vapor lamp or other discrete ultraviolet wavelengths from zinc or cadmium vapor lamps. More versatile instruments have broadband deuterium, xenon, or tungsten lamps and a monochromator, so you can choose an optimum wavelength for your analytes. The system in Figure 24-20 uses a *photodiode array* to record the spectrum of each solute as it is eluted. High-quality detectors provide full-scale absorbance ranges from 0.000 5 to 3 absorbance units, with a noise level near 1% of full scale. The linear range extends over five orders of magnitude of solute concentration (which is another way of saying that Beer's law is obeyed over this range). Ultraviolet detectors are

TABLE 24-3 Comparison of commercial HPLC detectors

Detector	Approximate limit of detection^a (ng)	Useful with gradient?
Ultraviolet	0.1–1	Yes
Refractive index	100–1 000	No
Evaporative light-scattering	0.1–1	Yes
Charged aerosol	1	Yes
Electrochemical	0.01–1	No
Fluorescence	0.001–0.01	Yes
Nitrogen ($N \xrightarrow{\text{combustion}} NO \xrightarrow{O_3} NO_2^* \to h\nu$)	0.3	Yes
Conductivity	0.5–1	No
Mass spectrometry	0.1–1	Yes
Fourier transform infrared	1 000	Yes

a. Most detection limits from E. W. Yeung and R. E. Synovec, "Detectors for Liquid Chromatography," *Anal. Chem.* **1986**, *58*, 1237A. Detection limit for charged aerosol detector from T. Górecki, F. Lynen, R. Szucs, and P. Sandra, "Universal Response in Liquid Chromatography Using Charged Aerosol Detection," *Anal. Chem.* **2006**, *78*, 3186.

FIGURE 24-20 Photodiode array ultraviolet detector for HPLC. (*a*) Dual-beam optical system uses grating polychromator, one diode array for the sample spectrum, and another diode array for the reference spectrum. Photodiode arrays are described in Section 19-3. (*b*) Reversed-phase chromatography (using C_{18}-silica) of sample containing 0.2 ng of anthracene, with detection at 250 nm. Full-scale absorbance is 0.001. (*c*) Spectrum of anthracene recorded as it emerged from the column. [Courtesy Perkin-Elmer Corp., Norwalk, CT.]

good for gradient elution with nonabsorbing solvents. Table 24-2 lists approximate cutoff wavelengths below which a solvent absorbs too strongly.

Fluorescence detectors excite the eluate with a laser and measure fluorescence (Figure 17-21). These detectors are very sensitive but respond only to the few analytes that fluoresce. To increase the utility of fluorescence and electrochemical detectors (described later), fluorescent or electroactive groups can be covalently attached to the analyte. This process of **derivatization** can be performed on the mixture prior to chromatography or by addition of reagents to the eluate between the column and the detector (called *post-column derivatization*). For example, Tb(EDTA)$^-$ can be added to the hormones epinephrine, norepinephrine, and dopamine emerging from a chromatography column.[25] These compounds form complexes with Tb(III), which can be excited near 300 nm and emit strongly at 500–600 nm. The detection limit is 10 to 100 nM with Tb fluorescence.

Evaporative Light-Scattering Detector

An **evaporative light-scattering detector** responds to any analyte that is significantly less volatile than the mobile phase.[26] In Figure 24-21, eluate enters the detector at the top. In the nebulizer, eluate is mixed with nitrogen gas and forced through a small-bore needle to form a uniform dispersion of droplets. Solvent evaporates from the droplets in the heated drift tube, leaving a fine mist of solid particles to enter the detection zone at the bottom. The particles are detected by the light that they scatter from a diode laser to a photodiode.

The evaporative light-scattering detector response is related to the mass of analyte, not to the structure or molecular mass of the analyte. If you see a large peak and a small peak, you can be pretty sure that there is less material in the small peak than in the large peak. With an ultraviolet detector, a small mass of strongly absorbing analyte can give a stronger signal than a large mass of weakly absorbing analyte. The evaporative light-scattering detector response is fit to the equation

$$\log A = a + b \log m \qquad (24\text{-}3)$$

FIGURE 24-21 (*a*) Operation of an evaporative light-scattering detector. [Courtesy Alltech Associates, Deerfield, IL.] (*b*) Comparison of ultraviolet absorbance and evaporative light-scattering detector response. Soluble components were extracted from a drug tablet and separated by reversed-phase liquid chromatography on a 4.6 × 150 mm column containing C_8-silica with a 30-nm pore size for separating polymers. Solvent A is 0.01 wt% trifluoroacetic acid in H_2O. Solvent B is 45 wt% propanol and 0.01 wt% trifluoroacetic acid in H_2O. A gradient was run from 10 vol% B up to 90 vol% B at 1 mL/min. Gradient time was not stated. [From L. A. Doshier, J. Hepp, and K. Benedek, "Method Development Tools for the Analysis of Complex Pharmaceutical Samples," *Am. Lab.*, December 2002, p. 18.]

where A is the area of the signal (in a graph of scattered light intensity versus time), m is the mass of analyte, and a and b are constants to fit the calibration curve.

The evaporative light-scattering detector is compatible with gradient elution. Also, there are no peaks associated with the solvent front, so there is no interference with early eluting peaks. What do we mean by solvent front? In Figure 24-13, you can see small positive and negative signals at 3–4 min. These signals arise from changes in the refractive index of the mobile phase due to solvent in which the sample was dissolved. This change displaces the ultraviolet detector signal at the time t_m required for unretained mobile phase to pass through the column. If a peak is eluted close to the time t_m, it can be distorted by the solvent front peaks. An evaporative light-scattering detector has no solvent front peaks.

If you use a buffer in the eluent, it must be volatile or else it will evaporate down to solid particles that scatter light and obscure the analyte. Low-concentration buffers made from acetic, formic, or trifluoroacetic acid, ammonium acetate, diammonium phosphate, ammonia, or triethylamine are suitable. Buffers for evaporative light scattering are the same as those for mass spectrometric detection.

Figure 24-21b compares evaporative light scattering with ultraviolet absorbance for the detection of soluble components of a drug tablet containing polymers and small-molecule active ingredients. Two components have absorbance at 232 nm. Four or five components are evident with 205-nm detection. The solvent gradient produces a sloping baseline at 205 nm. All components are observed by evaporative light scattering, and there is no slope to the baseline. Some broad peaks arise from polymers with a distribution of molecular masses.

Charged Aerosol Detector

The **charged aerosol detector** is a sensitive, almost universal detector with nearly equal response to equal masses of nonvolatile analytes.[27] Introduced in 2005, it already has a prominent place in industry. For example, it is useful for measuring the material balance in

FIGURE 24-22 Operation of the charged aerosol detector. [Courtesy ESA, Inc., Chelmsford MA.]

pharmaceutical preparations because the relative area of each peak in the chromatogram is nearly equal to the relative mass of that component in the preparation. Lipids and carbohydrates, which have little response from an ultraviolet detector, give the same response as other components of the mixture. A minor peak in the charged aerosol chromatogram must be a minor component of the mixture, whereas it could be a major component in the ultraviolet chromatogram.

At the top left of the charged aerosol nebulizer in Figure 24-22, eluate and N_2 gas enter a nebulizer similar to the premix burner in Figure 20-5. Fine mist from the nebulizer reaches the drying tube, while larger droplets fall to the drain. In the drying tube, solvent evaporates at ambient temperature, leaving an aerosol containing ~1% of the original analyte. Meanwhile, part of the N_2 stream passes over a Pt needle held at ~+10 kV with respect to the outer case of the corona charging chamber to form N_2^+. A chain of events, perhaps as written for atmospheric pressure chemical ionization on page 522, transfers positive charge to aerosol particles that flow out of the charging chamber through a small-ion trap. Charged plates of the trap attract small mobile ions. Aerosol particles are too massive to be deflected and pass through the trap to the collector. Total charge reaching the collector is measured by an electrometer, which produces the detector signal for the chromatogram.

The dynamic range of the detector spans 4–5 orders of magnitude in concentration. Response is approximately proportional to $\sqrt{m}$. Equal masses of different analytes give equal response within ~15% in isocratic elution. Response depends on solvent composition, with higher response for higher percentage of volatile organic solvent and lower response for water. As the organic composition of a gradient increases, so does the response. However, the detector can still be used with gradient elution.[28]

Electrochemical Detector

An **electrochemical detector** responds to analytes that can be oxidized or reduced, such as phenols, aromatic amines, peroxides, mercaptans, ketones, aldehydes, conjugated nitriles, aromatic halogen compounds, and aromatic nitro compounds. The opening of Chapter 16 shows a detector in which eluate is oxidized or reduced at the working electrode. The potential is maintained at a selected value with respect to the Ag|AgCl reference electrode, and current is measured between the working electrode and the stainless steel auxiliary electrode. For oxidizable solutes, copper or glassy carbon working electrodes are common. For reducible solutes, a drop of mercury is a good working electrode. Current is proportional to solute concentration over six orders of magnitude. Aqueous or other polar solvents containing dissolved electrolytes are required, and they must be rigorously freed of oxygen. Metal ions extracted from tubing can be masked by adding EDTA to the solvent. The detector is sensitive to flow rate and temperature changes.

FIGURE 24-23 Pulsed electrochemical detection of alcohols separated on Dionex AS-1 anion-exchange column with 0.05 M HClO₄. Peaks: 1, glycerol; 2, ethylene glycol; 3, propylene glycol; 4, methanol; 5, ethanol; 6, 2-propanol; 7, 1-propanol; 8, 2-butanol; 9, 2-methyl-1-propanol; 10, 1-butanol; 11, 3-methyl-1-butanol; 12, 1-pentanol; 13, cyclohexanol; 14, 1-hexanol; 15, diethylene glycol. [From D. C. Johnson and W. R. LaCourse, "Liquid Chromatography with Pulsed Electrochemical Detection at Gold and Platinum Electrodes," *Anal. Chem.* **1990**, *62*, 589A.]

Pulsed electrochemical measurements at Au or Pt working electrodes expand the classes of detectable compounds to include alcohols, carbohydrates, and sulfur compounds.[29] The electrode is held at +0.8 V (versus a saturated calomel electrode) for 120 ms to oxidatively desorb organic compounds from the electrode surface and to oxidize the metal surface. Then the electrode is brought to −0.6 V for 200 ms to reduce the oxide to pristine metal. The electrode is then brought to a constant working potential (typically in the range +0.4 to −0.4 V), at which analyte is oxidized or reduced. After a delay of 400 ms to allow the charging current (Figure 16-17) to decay to 0, current is integrated for the next 200 ms to measure analyte. The sequence of pulses just described is then repeated to measure successive data points as eluate emerges from the column. By this means, 10 ppb ethylene glycol (HOCH₂CH₂OH) in the chromatographic system in Figure 24-23 would give a signal-to-noise ratio of 3.

Refractive Index Detector

A **refractive index detector** responds to almost every solute, but its detection limit is about 1 000 times poorer than that of the ultraviolet detector. A deflection-type detector has two triangular 5- to 10-μL compartments through which pure solvent or eluate passes. Collimated (parallel) visible light, filtered to remove infrared radiation that would heat the sample, passes through the cell with pure solvent in both compartments and is directed to a photodiode array. When solute with a different refractive index enters the cell, the beam is deflected and different pixels of the array are irradiated.

A refractive index detector is useless in gradient elution because we cannot match the sample and reference while solvent composition is changing. Refractive index is sensitive to changes in pressure and temperature (~0.01°C). With its low sensitivity, a refractive index detector is not useful for trace analysis. It also has a small linear range, spanning only a factor of 500 in solute concentration. The primary appeal of this detector is its universal response to all solutes, including those that have little ultraviolet absorption.

Comments on Mass Spectrometric Detection in Liquid Chromatography

Atmospheric pressure chemical ionization (Figure 21-23) and electrospray ionization (Figure 21-21) are the most common liquid chromatography–mass spectrometry interfaces. Atmospheric pressure chemical ionization creates ions from analyte, as well as from other species in solution. Electrospray requires that analyte be charged in solution.

Volatile, low-concentration (≤10 mM) buffers used with mass spectrometers are made from formic acid, acetic acid, and ammonia. HCl should not be used because it corrodes the metal interface. Blank injections should be run to observe background ions generated in the absence of sample. Suitable organic solvents include acetonitrile, methanol, ethanol, propanol, and acetone. Tetrahydrofuran is less desirable because, for some types of samples, it can lead to large background signals in a total ion chromatogram and because it is not compatible with commonly used polyether ether ketone (PEEK) tubing with mass spectrometric detection. Ion-pair agents (Section 25-2) and surfactants should not be used because they create background signals and suppress electrospray signals. Triethylamine suppresses ionization and gives an intense MH^+ peak at m/z 102.

Mass spectral sensitivity is reduced by aqueous mobile phase. More organic phase can be added through a "T" connection between the column and the mass spectrometer. Alternatively, a different stationary phase can be chosen to require more organic phase. The pH of the mobile phase can be adjusted between the column and the detector so that acidic or basic analytes are ionized. Atmospheric pressure chemical ionization and electrospray with positive and negative ion modes can all be tried to obtain adequate sensitivity. Eluate emerging from the column prior to t_m can be diverted to waste to avoid introducing unnecessary salts into the mass spectrometer.

Run a *performance qualification* sample or an internal standard frequently to verify that the separation and the mass spectrometer are behaving normally. When a liquid chromatography–mass spectrometry method develops a problem, but the performance qualification check is normal, the problem is most likely in the chromatography of the unknown. If the performance qualification check fails, the problem is likely in the mass spectrometer. You can test the mass spectrometer alone by direct flow injection of a standard sample or calibration solution into the mass spectrometer.

24-3 Method Development for Reversed-Phase Separations[2,3,30]

Many separations encountered in industrial and research laboratories can be handled by reversed-phase chromatography. We now describe a general procedure for developing an isocratic separation of an unknown mixture with a reversed-phase column. The next section deals with gradient separations. In method development, the goals are to obtain adequate separation in a reasonable time. Ideally, the procedure should be *rugged*, which means that the separation will not be seriously degraded by gradual deterioration of the column, by *small* variations in solvent composition, pH, and temperature, or by use of a different batch of the same stationary phase, perhaps from a different manufacturer. If your column does not have temperature control, you can at least insulate it to reduce temperature fluctuations.

Reversed-phase chromatography is usually adequate to separate mixtures of low-molecular-mass neutral or charged organic compounds. If isomers do not separate well, normal-phase chromatography or porous graphitic carbon are recommended because solutes have stronger, more specific interactions with the stationary phase. For enantiomers, chiral stationary phases (Figure 24-10) are required. Separations of inorganic ions, polymers, and biological macromolecules are described in Chapter 25.

As in gas chromatography (Section 23-5), the first steps in method development are to (1) determine the goal of the analysis, (2) select a method of sample preparation to ensure a "clean" sample, and (3) choose a detector that allows you to observe the desired analytes in the mixture. The remainder of method development described in the following sections assumes that steps 1 through 3 have been carried out.

Criteria for an Adequate Separation

The *retention factor k* in Equation 22-17 measures corrected retention time, $t_r - t_m$, in units of the time t_m required for mobile phase to pass through the column. Reasonable separations demand that k for all peaks be in the range 0.5–20. If k is too small, the first peak is distorted by the solvent front. If k is too great, the run takes too long. In the lowest trace in Figure 24-12, t_m is the time when the first baseline disturbance is observed near 3 min. If you do not observe a baseline disturbance, you can estimate

Solvent front: $$V_m \approx \frac{Ld_c^2}{2} \quad \text{or, equivalently,} \quad t_m \approx \frac{Ld_c^2}{2F} \tag{24-4}$$

where V_m is the volume at which unretained solute is eluted ($=$ volume at which solvent front appears), L is column length (cm), d_c is column diameter (cm), and F is flow rate (mL/min). In reversed-phase chromatography, t_m could be measured by running the unretained solutes uracil (detected at 260 nm) or $NaNO_3$ (detected at 210 nm) through the column.

For quantitation, a minimum resolution (Figure 22-10) of 1.5 between the two closest peaks is desired. For ruggedness, a resolution of 2 is desirable so that resolution remains adequate despite small changes in conditions or slow deterioration of the column.

Desired attributes of a new chromatographic method:
- adequate resolution of desired analytes
- short run time
- rugged (not drastically affected by small variations in conditions)

Initial steps in method development:
1. determine goal
2. select method of sample preparation
3. choose detector

Retention factor: $k = \dfrac{t_r - t_m}{t_m}$

t_r = retention time of analyte
t_m = elution time for mobile phase or unretained solute

Attributes of a good separation:
- $0.5 \leq k \leq 20$
- resolution ≥ 2
- operating pressure ≤ 15 MPa
- $0.9 \leq$ asymmetry factor ≤ 1.5

A chromatographic method must not exceed the upper allowable operating pressure for the hardware. Keeping pressure below ~15 MPa (150 bar, 2 200 pounds/inch^2) prolongs the life of the pump, valves, seals, and autosampler. Pressure can double during the life of a column because of progressive clogging. Establishing an operating pressure of ≤15 MPa during method development allows for column degradation.

All peaks (certainly all peaks that need to be measured) should be symmetric, with an asymmetry factor A/B in Figure 22-14 in the range 0.9–1.5. Asymmetric peak shapes should be corrected as described at the end of Section 24-1 (page 610) before optimizing a separation.

Optimization with One Organic Solvent

Combinations of acetonitrile, methanol, and tetrahydrofuran with water (or aqueous buffer) provide a sufficient range of dipolar and hydrogen-bonding interactions with solutes to separate a vast number of compounds in reversed-phase chromatography. The first solvent mixture to try is acetonitrile and water. Acetonitrile has low viscosity, which allows relatively low operating pressure, and it permits ultraviolet detection down to 190 nm (Table 24-2). At 190 nm, many analytes have some absorbance. Methanol is the second choice for organic solvent because it has a higher viscosity and longer wavelength ultraviolet cutoff. Tetrahydrofuran is the third choice because it has less usable ultraviolet range, it is slowly oxidized,[32] and it equilibrates more slowly with stationary phase. Starting conditions for reversed-phase HPLC are listed in Table 24-4.

Choice of organic solvent:

1. acetonitrile
2. methanol
3. tetrahydrofuran

To avoid disposing of acetonitrile as hazardous waste, you can hydrolyze it to sodium acetate and flush it down the drain.[31]

TABLE 24-4   Starting conditions for reversed-phase chromatography

Stationary phase:	C_{18} or C_8 on 5- or 3-μm-diameter spherical type B silica for operation in the pH 2–7.5 range. For operation in the pH 8–12 range, use ethylene-bridged silica (Figure 24-6). For operation above 50°C, use sterically protected silica (Figure 24-8). Alternatively, polymer or zirconia stationary phases are stable over wide pH and temperature ranges.

Column:	0.21 × 15 cm column for 5-μm particles
	0.21 × 10 cm column for 3-μm particles (shorter run, similar resolution)
	    When changing particle size, maintain constant (column length/particle diameter) for
	    similar resolution at same linear flow rate.
	0.46 × 15 cm column for 5-μm particles
	0.46 × 10 cm column for 3-μm particles (shorter run, similar resolution)

Flow rate:	1.0 mL/min for 0.46-cm-diameter column $0.2 \text{ mL/min for 0.21-cm-diameter column} = \left(\dfrac{0.21 \text{ cm}}{0.46 \text{ cm}}\right)^2 \left(1.0 \dfrac{\text{mL}}{\text{min}}\right)$
Mobile phase:	CH_3CN/H_2O for neutral analytes CH_3CN/aqueous buffer[a] for ionic organic analytes 5 vol% CH_3CN in H_2O to 95 vol% CH_3CN in H_2O for gradient elution
Temperature:	30°–40°C with temperature control Some people routinely prefer 50°–60°C
Sample size:	0.21-cm-diameter column: 2 μL containing 25–50 μg of each analyte 0.46-cm-diameter column: 5–10 μL containing 25–50 μg of each analyte (Older ultraviolet absorbance detectors might require more sample.)
For mass spectrometry:	0.21 × 5 cm column with 3-μm particles of C_{18}-silica run at 30°–60°C Mobile phase is acetonitrile or methanol plus volatile, aqueous buffer. Aqueous buffer is made from 2 mM ammonium formate plus 1–10 g formic acid per L (giving pH ≈ 2.7–1.7). Ammonium formate buffer can be used for pH ≈ 2.8–4.8. Ammonium acetate buffer can be used for pH ≈ 3.8–5.8. Use NH_3, formic acid, or acetic acid to adjust pH. Ammonium carbonate can be used for pH ≈ 8–11. Atmospheric pressure chemical ionization or electrospray interface. Electrospray requires analyte to be charged in solution.

a. *Buffer is 25 mM phosphate/pH 2–3 made by treating H_3PO_4 with KOH. K^+ is more soluble than Na^+ in organic solvents and leads to less tailing. This concentration of phosphate is suitable for ultraviolet detection. Add 0.2 g sodium azide per liter as a preservative if buffer will not be consumed quickly.*

Adapted and modified from L. R. Synder, J. J Kirkland, and J. L. Glajch, Practical HPLC Method Development (New York: Wiley, 1997) and S. Needham, "HPLC Method Development for LC/MS," short course presented at Pittcon 2009.

Figure 24-12 showed a succession of experiments to establish that 35 vol% CH_3CN (designated B) plus 65 vol% buffer is a good solvent to separate the particular mixture of analytes. The initial experiment was done with a high concentration of CH_3CN (90% B) to ensure the elution of all components of the unknown. Then %B was successively lowered to separate all the components. Eluent containing 40% B did not separate peaks 2 and 3 adequately, and 30% B took too long to elute peak 8. Therefore, 35% B was selected.

With 35% B, peak 1 is eluted at 4.9 min and peak 8 is eluted at 125.2 min. The solvent front appears at $t_m = 2.7$ min. Therefore, k for peak 1 is $(4.9 - 2.7)/2.7 = 0.8$ and k for peak 8 is $(125.2 - 2.7)/2.7 = 45$. *If $k > 20$, gradient elution is suggested* (Section 24-4). If all peaks in Figure 24-12 could be resolved while maintaining $0.5 \leq k \leq 20$, then we would have a successful isocratic separation. If we were not concerned about measuring peaks 2 and 3, then 45% B would probably be a good choice.

Optimization with Two or Three Organic Solvents

Figures 24-24 and 24-25 illustrate a systematic process to develop a separation with combinations of solvents. Method development is finished as soon as the separation meets your criteria. There is a good chance of attaining adequate separation without going through all the steps.

Step 1   Optimize the separation with acetonitrile/buffer to generate chromatogram A in Figure 24-25.
Step 2   Optimize the separation with methanol/buffer to generate chromatogram B.
Step 3   Optimize the separation with tetrahydrofuran/buffer to generate chromatogram C.
Step 4   Mix the solvents used in A, B, and C, one pair at a time, in 1:1 proportion, to generate chromatograms D, E, and F.
Step 5   Construct a 1:1:1 mixture of the solvents for A, B, and C to generate chromatogram G.
Step 6   If some of the results A through G are almost good enough, select the two best points and mix the solvents to obtain points between those two.

Let's examine Figure 24-25 to see how the systematic procedure works. Step 1 is to generate chromatogram A by varying the proportions of acetonitrile and aqueous buffer (as in Figure 24-12) to obtain the best separation within the constraint that $0.5 \leq k \leq 20$. At the best composition, 30 vol% acetonitrile/70 vol% buffer, peaks 4 and 5 are not resolved adequately for quantitative analysis.

In HPLC, lowering the flow rate usually improves resolution. With the flow rate of 1.0 mL/min in Figure 24-25, we chose to keep $k < 10$ (instead of <20) to obtain a run time below ~25 min. In chromatogram A, $k = 1.1$ for peak 1 and $k = 8.1$ for peak 7. If the flow rate had been 2.0 mL/min, then $k < 20$ would give a run time below ~25 min.

Step 2 seeks a methanol/buffer solvent to obtain the best separation at point B. It is not necessary to start all over again with 90% methanol. Figure 24-26 allows us to select a methanol/water mixture that has approximately the same eluent strength as a particular acetonitrile/water mixture. A vertical line drawn at 30% acetonitrile (the composition used in chromatogram A) intersects the methanol line near 40%. Therefore, 40% methanol has about the same eluent strength as 30% acetonitrile. The first experiment carried out to establish point B in Figure 24-25 utilized 40% methanol. A little trial-and-error (with 45% methanol and 35% methanol) demonstrated that 40% methanol gave the best separation, but the separation is poor. In chromatogram B in Figure 24-25, the seven components give only five peaks. When we changed from acetonitrile to methanol, the order of elution of some compounds changed.

Step 3 generates chromatogram C in Figure 24-25, using tetrahydrofuran. Figure 24-26 tells us that 22% tetrahydrofuran has the same eluent strength as 30% acetonitrile. When 22% tetrahydrofuran was tried, elution times were too long. Trial-and-error demonstrated that 32% tetrahydrofuran was best. All seven compounds are cleanly separated in chromatogram C in an acceptable time. However, a baseline dip associated with the solvent front between peaks 3 and 1 interferes with quantitative analysis of compound 1. The order of elution with tetrahydrofuran is quite different from the order with acetonitrile. In general, changing solvent is a powerful way to change relative retention of different compounds.

Step 4 generates chromatograms D, E, and F. The composition at D is a 1:1 mixture of the solvents used in A and B. Because we used 30% acetonitrile at A and 40% methanol at B,

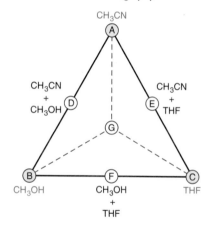

FIGURE 24-24 HPLC method development triangle. THF stands for tetrahydrofuran. Figure 24-25 shows how the procedure is applied to a real chromatographic separation.

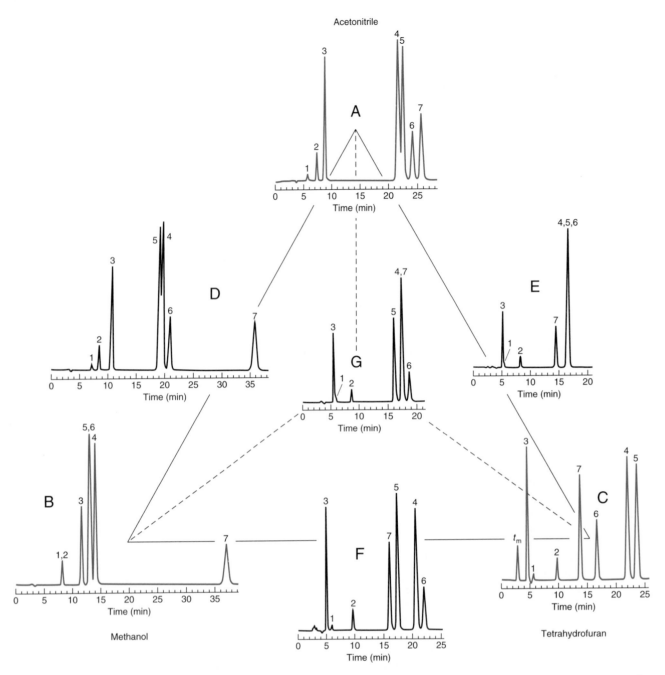

FIGURE 24-25 Application of the method development triangle to the separation of seven aromatic compounds by HPLC. Column: 0.46 × 25 cm Hypersil ODS (C$_{18}$ on 5-μm silica) at ambient temperature (~22°C). Elution rate was 1.0 mL/min with the following solvents: (A) 30 vol% acetonitrile/70 vol% buffer; (B) 40% methanol/60% buffer; (C) 32% tetrahydrofuran/68% buffer. The aqueous buffer contained 25 mM KH$_2$PO$_4$ plus 0.1 g/L NaN$_3$ adjusted to pH 3.5 with HCl. Points D, E, and F are midway between the vertices: (D) 15% acetonitrile/20% methanol/65% buffer; (E) 15% acetonitrile/16% tetrahydrofuran/69% buffer; (F) 20% methanol/16% tetrahydrofuran/64% buffer. Point G at the center of the triangle is an equal blend of A, B, and C with the composition 10% acetonitrile/13% methanol/11% tetrahydrofuran/66% buffer. The negative dip in C between peaks 3 and 1 is associated with the solvent front. Peak identities were tracked with a photodiode array ultraviolet spectrophotometer: (1) benzyl alcohol; (2) phenol; (3) 3′,4′-dimethoxyacetophenone; (4), *m*-dinitrobenzene; (5) *p*-dinitrobenzene; (6) *o*-dinitrobenzene; (7) benzoin.

D was obtained with 15% acetonitrile/20% methanol/65% buffer. Similarly, E was obtained with a 1:1 mixture of the solvents for A and C. F was obtained with a 1:1 mixture of the solvents for B and C.

Chromatogram D is not acceptable, because peaks 4 and 5 overlap. Chromatogram E is terrible because peaks 1 and 3 overlap and peaks 4, 5, and 6 overlap. But (hurrah!) chromatogram

CHAPTER 24 High-Performance Liquid Chromatography

FIGURE 24-26 Nomogram showing volume percentage of solvents having the same eluent strength. A vertical line intersects each solvent line at the same eluent strength. For example, 30 vol% acetonitrile/70 vol% water has about the same eluent strength as 40 vol% methanol or 22 vol% tetrahydrofuran. [From L. R. Snyder, J. J. Kirkland, and J. L. Glajch, *Practical HPLC Method Development* (New York: Wiley, 1997).]

F is what we have been looking for: All peaks are separated and the first peak (3) is adequately removed from the solvent dip. We are finished! The solvent composition at F (20% methanol/16% tetrahydrofuran/64% buffer) does the job. Peaks 4 and 6 have the minimum resolution of 1.8. (We would have preferred resolution >2.0.) All peaks are symmetric and within the range $k = 0.9$–7.5. The operating pressure remained reasonable.

If none of the trials had worked, step 5 would generate chromatogram G in Figure 24-25, using a 1:1:1 mixture of solvents from A, B, and C. For completeness, the result at G is shown. Peaks 1 and 3 overlap and peaks 4 and 7 overlap.

If some of the compositions A through G were almost good enough, compositions between the points might have been better. For example, if A and D were almost good enough, perhaps a mixture of the solvents for A and D might have been better.

Temperature as a Variable

Column temperature affects the relative retention of different compounds, and elevated temperature permits high-speed chromatography to be conducted.[33] Figure 24-27 suggests a systematic procedure for method development in which solvent composition and temperature are the two independent variables. For elevated temperature operation, pH should be below 6 to retard dissolution of silica. Alternatively, zirconia-based stationary phases work up to at least 200°C.

Choosing a Stationary Phase

C_{18}-silica is the most common stationary phase, and it separates a wide range of mixtures when the solvent is chosen carefully, as in Figure 24-25. However, it cannot achieve all separations. Table 24-5 is a guide to selecting other bonded phases. A quantitative "hydrophobic subtraction model" with a database of several hundred commercial HPLC columns is freely available to help analysts select columns for particular types of separations.[34]

$$\text{Resolution} = \frac{\Delta t_r}{w_{av}} = \frac{0.589\ \Delta t_r}{w_{1/2}}$$

Δt_r = separation between peaks
w_{av} = average width at baseline
$w_{1/2}$ = average width at half-height

After optimizing the solvent, you might still need to improve resolution. To increase resolution, you can

- decrease flow rate
- increase column length
- decrease particle size

FIGURE 24-27 Isocratic method development for HPLC, using solvent composition, %B, and temperature, T, as independent variables. %B and T are each varied between selected low and high values. From the appearance of chromatograms resulting from conditions A–D, we can select intermediate conditions to improve the separation.

TABLE 24-5 Selection of bonded stationary phases for HPLC

Bonded group	Polarity	Retention mechanisms	Comments
C_{18}, C_8, C_4	Nonpolar	van der Waals	C_8 does not retain hydrophobic compounds as strongly as C_{18}
Phenyl	Nonpolar	Hydrophobic and pi-pi	
Cyano	Intermediate	Hydrophobic, dipole-dipole, and pi-pi	Resolves polar organic compounds by reversed-phase or normal-phase chromatography
Amino	Polar (—NH$_2$) or ionic (—NH$_3^+$)	Dipole-dipole and H-bonding	Normal-phase or ion-exchange separations; separates carbohydrates, polar organic compounds, and inorganic ions; reacts with aldehydes and ketones
Bare silica	Very polar	H-bonding	Normal-phase separations

For a free column selection tool, see http://www.usp.org/USPNF/columnsDB.html.

SOURCE: *C. S. Young and R. J. Weigand, "An Efficient Approach to Column Selection in HPLC Method Development," LCGC 2002, 20, 464.*

EXAMPLE   Selecting a Bonded Phase

Suggest an approach to resolve the following compounds in a short time.

1. Rutin

2. Chatechin

3. Myricetin

4. Quercitin

5. Apigenin

Double bond

6. Naringenin

Single bond

Solution Compounds with the most similar structures are likely to be hardest to separate. Of the compounds in the mixture, **5** and **6** are most similar, differing by just one C=C bond. Table 24-5 suggests that a stationary phase with phenyl or cyano groups, which retain analytes by pi-pi interactions, might be good for distinguishing **5** and **6**. For rapid analysis but good resolution, select a short column with small particle size and run it at a rapid flow rate. Figure 24-28 shows that **5** and **6** are resolved by a phenyl column but not by a C_{18} column.

FIGURE 24-28 Separation of six compounds on (*a*) phenyl- and (*b*) C_{18}-silica columns with 3-μm particle size, using 35:65 (vol/vol) acetonitrile/0.2% aqueous trifluoroacetic acid. Column size: 7 × 53 mm; flow rate = 2.5 mL/min. [From C. S. Young and R. J. Weigand, "An Efficient Approach to Column Selection in HPLC Method Development," *LCGC* **2002**, *20*, 464. Courtesy Alltech Associates.]

Test Yourself What other stationary phase in Table 24-5 might separate the six compounds? (**Answer:** I would try bare silica to separate these polar compounds.)

We have discussed several ways to change the relative retention of two closely spaced peaks. Some methods are harder to implement than others. One suggested order of steps to try—from easiest to hardest—is (1) change the solvent strength by varying the fraction of each solvent; (2) change the temperature; (3) change the pH (in small steps); (4) use a different solvent; (5) use a different kind of stationary phase.[35]

24-4 Gradient Separations[3]

Figure 24-12 shows an isocratic separation of eight compounds that required a run time of more than 2 h. When eluent strength was low enough to resolve early peaks (2 and 3), the elution of later peaks was very slow. To retain the desired resolution but decrease the analysis time, the *segmented gradient* (a gradient with several distinct parts) in Figure 24-13 was selected. Peaks 1–3 were separated with a low eluent strength (30% B). Between 8 and 13 min, B was increased linearly from 30 to 45% to elute the middle peaks. Between 28 and 30 min, B was increased linearly from 45 to 80% to elute the final peaks.

Dwell Volume and Dwell Time

The volume between the point at which solvents are mixed and the beginning of the column is called the **dwell volume.** The *dwell time*, t_D, is the time required for the gradient to reach the column. Dwell volumes range from 0.5 to 10 mL in different systems. For Figure 24-13, the dwell volume is 5 mL, and the flow rate is 1.0 mL/min. Therefore, the dwell time is 5 min. A solvent change initiated at 8 min does not reach the column until 13 min.

$$t_D = \frac{\text{dwell volume (mL)}}{\text{flow rate (mL/min)}}$$

Differences in dwell volumes between different systems are an important reason why gradient separations on one chromatograph do not necessarily transfer to another. It is helpful to state the dwell volume for your system when you report a gradient separation. One way to compensate for dwell volume is to inject sample at the time t_D instead of at $t = 0$.

You can measure dwell volume by first disconnecting the column and connecting the inlet tube directly to the outlet tube. Place water in reservoirs A and B of the solvent delivery system. Add 0.1 vol% acetone to reservoir B. Program the gradient to go from 0 to 100% B in 20 min and begin the gradient at $t = 0$. With the detector set to 260 nm, the response will ideally look like that in Figure 24-29. The delay between the start of the gradient and the first response at the detector is the dwell time, t_D.

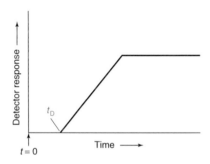

FIGURE 24-29 Measurement of dwell time, using nonabsorbing solvent in reservoir A and a weak absorber in reservoir B. The gradient from 0 to 100% B is begun at time $t = 0$ but does not reach the detector until time t_D. The column is removed from the system for this measurement. Real response will be rounded instead of having the sharp intersections shown in this illustration.

Begin Method Development with a Scouting Gradient

The quickest way to survey a new mixture to decide whether to use isocratic or gradient elution is to run a broad gradient, as in Figure 24-30a.[36] This figure shows how the sample mixture in Figure 24-12 is separated by a linear gradient from 10 to 90% acetonitrile in 40 min. The *gradient time*, t_G, is the time over which the solvent composition is changed (40 min). Let Δt be the difference in the retention time between the first and the last peak in the chromatogram. In Figure 24-30a, $\Delta t = 35.5 - 14.0 = 21.5$ min. The criterion to choose whether to use a gradient is

- The first run on a new mixture should be a gradient.
- If $\Delta t/t_G > 0.25$, use gradient elution.
- If $\Delta t/t_G < 0.25$, use isocratic elution.
- Isocratic solvent should have the composition applied to the column halfway through the period Δt.

Use a gradient if $\Delta t/t_G > 0.25$ Use isocratic elution if $\Delta t/t_G < 0.25$

If all peaks are eluted over a narrow solvent range, then isocratic elution is feasible. If a wide solvent range is required, then gradient elution is more practical. In Figure 24-30a, $\Delta t/t_G = 21.5/40 = 0.54 > 0.25$. Therefore, gradient elution is recommended. Isocratic elution is possible, but the time required in Figure 24-12 is impractically long.

If isocratic elution is indicated because $\Delta t/t_G < 0.25$, then a good starting solvent is the composition at the point halfway through the interval Δt. That is, if the first peak is eluted at 10 min and the last peak is eluted at 20 min, a reasonable isocratic solvent has the composition at 15 min in the gradient.

Steps in gradient method development:

1. Run a wide gradient (e.g., 5 to 95% B) over 40–60 min. From this run, decide whether gradient or isocratic elution is best.

2. If gradient elution is chosen, eliminate portions of the gradient prior to the first peak and following the last peak. Use the same gradient time as in step 1.

3. If the separation in step 2 is acceptable, try reducing the gradient time to reduce the run time.

Developing a Gradient Separation

The first run should survey a broad range of eluent strength such as 10 to 90% B in 40 min in Figure 24-30a. Because the dwell time was 5 min and the gradient began at $t = 0$, the gradient did not reach the column until $t = 5$ min. (It would have been better to inject the

FIGURE 24-30 Linear gradient separations of the mixture from Figure 24-12 in the same column and solvent system [buffer (solvent A) with acetonitrile (solvent B)] at a flow rate of 1.0 mL/min. Dwell time = 5 min.

sample at $t = 5$ min, but this was not done.) Serendipitously, the first run in Figure 24-30 separated all eight peaks. We could stop at this point if we were willing to settle for a 36-min run time.

The next step in developing a gradient method is to spread the peaks out by choosing a shallower gradient. For a dwell time of 5 min, the gradient profile for Figure 24-30a looks like Figure 24-31. Peak 1 was eluted at 14.0 min when the solvent was 28% B. Peak 8 was eluted near 35.5 min, when the solvent was 71% B. The portions of the gradient from 10 to 28% B and 71 to 90% B were not really needed. Therefore, the second run could be made with a

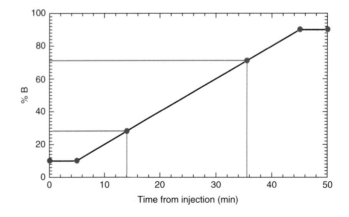

FIGURE 24-31 Solvent gradient for Figure 24-30a. The gradient was begun at the time of injection ($t = 0$), but the dwell time was 5 min. Therefore, the solvent was 10% B for the first 5 min. Then the composition increased linearly to 90% B over 40 min. After $t = 45$ min, the composition was constant at 90% B.

gradient from 28 to 71% B over the same t_G (40 min). Conditions chosen for the run in Figure 24-30b were 30 to 82% B in 40 min. This gradient spread the peaks out and reduced the run time slightly to 32 min.

In Figure 24-30c, we want to see whether a steeper gradient could be used to reduce the run time. The gradient limits were the same as in trace *b*, but t_G was reduced to 20 min. Peaks 6 and 7 are not fully resolved with the shorter gradient time. Trace *b* represents reasonable conditions for the gradient separation.

If the separation in Figure 24-30b were not acceptable, you could try to improve it by reducing the flow rate or going to a *segmented gradient*, as in Figure 24-13. The segmented gradient provides an appropriate solvent composition for each region of the chromatogram. It is easy to experiment with flow rate and gradient profiles. More difficult approaches to improve the separation are to change the solvent, use a longer column, use a smaller particle size, or change the stationary phase. Box 24-4 provides guidance on selecting gradient time and scaling gradients from one size column to another.

24-5 ▦ Do It with a Computer

Method development is simplified by computer simulations using commercial software[3,37] or your own spreadsheet. With data from a few experiments, you can predict the effects of solvent composition and temperature in isocratic or gradient separations. You can estimate what will be optimum conditions in hours of work instead of days of work.

The basis for most simulations of reversed-phase separations is the empirical *linear-solvent-strength model*, which supposes a logarithmic relationship between retention factor *k* for a given solute and the mobile phase composition Φ:

$$\log k \approx \log k_w - S\Phi \qquad (24\text{-}5)$$

An *empirical* model is based on observations, not on theory. Equation 24-5 is empirical in that it approximately describes the relation between *k* and Φ. The relationship is not predicted by a theory based on first principles.

BOX 24-4   Choosing Gradient Conditions and Scaling Gradients

We now discuss equations that allow you to select sensible linear gradient conditions and to scale gradients from one column to another. For gradient elution, the average retention factor $k*$ for each solute is the value of *k* when that solute is halfway through the column:

$$k* = \frac{t_G F}{\Delta \Phi V_m S} \qquad (24\text{-}10)$$

where t_G is gradient time (min), *F* is flow rate (mL/min), $\Delta\Phi$ is the change in solvent composition during the gradient, V_m is the volume of mobile phase in the column (mL), and *S* is the slope in the linear-solvent-strength model (Equation 24-5). We take a representative $S = 4$ for this discussion.

In isocratic elution, a retention factor $k \approx 5$ provides separation from the solvent front and does not require excessive time. For gradient elution, $k* \approx 5$ is a reasonable starting condition. Let's calculate a sensible gradient time for the experiment in Figure 24-30a, in which we chose a gradient from 10 to 90% B ($\Delta\Phi = 0.8$) in a 0.46×25 cm column eluted at 1.0 mL/min. From Equation 24-4, $V_m \approx L d_c^2/2 = (25\ \text{cm})(0.46\ \text{cm})^2/2 = 2.6_5$ mL. We calculate the required gradient time by rearranging Equation 24-10:

$$t_G = \frac{k* \Delta \Phi V_m S}{F} = \frac{(5)(0.8)(2.65\ \text{mL})(4)}{(1.0\ \text{mL/min})} = 42\ \text{min} \qquad (24\text{-}11)$$

A reasonable gradient time would be 42 min. In Figure 24-30a, t_G is 40 min, giving $k* = 4.7$. In Figure 24-30b, we changed the gradient to $\Delta\Phi = 0.52$, giving better separation:

$$k* = \frac{t_G F}{\Delta \Phi V_m S} = \frac{(40\ \text{min})(1.0\ \text{mL/min})}{(0.52)(2.65\ \text{mL})(4)} = 7.3$$

The separation is poorer in Figure 24-30c, for which $k* = 3.6$.

If you have a successful gradient separation and want to transfer it from column 1 to column 2, whose dimensions are different, the scaling relations are

$$\frac{F_2}{F_1} = \frac{m_2}{m_1} = \frac{d_2}{d_1} = \frac{V_2}{V_1} \qquad (24\text{-}12)$$

where *F* is volume flow rate (mL/min), *m* is the mass of sample, *d* is the dwell time before the gradient reaches the column, and *V* is total column volume. The gradient time, t_G should not be changed. In Figure 24-31, the dwell time $d = 5$ min is due to the dwell volume between the mixer and the column. Equation 24-12 tells us to change volume flow rate, sample mass, and dwell time in proportion to column volume. If dwell volume is small in comparison with the volume of solvent on the column, V_m, the dwell time *d* can be inconsequential. If dwell volume is large, it becomes an important factor over which you might have little control.

Suppose that you have optimized a gradient on a 0.46×25 cm column and you want to transfer it to a 0.21×10 cm column. The quotient V_2/V_1 is $(\pi r^2 L)_2/(\pi r^2 L)_1$, where *r* is column radius and *L* is column length. For these columns, $V_2/V_1 = 0.083$. Equation 24-12 tells us to decrease the volume flow rate, the sample mass, and the dwell time to 0.083 times the values used for the large column. The gradient time should not be changed.

When you make these changes, you will discover that $k*$ is the same for both columns. If you change a condition that affects $k*$, you should make a compensating change to restore $k*$. For example, Equation 24-10 tells us that, if we choose to double t_G, we could cut the flow rate in half so that the product $t_G F$ is constant and $k*$ remains constant.

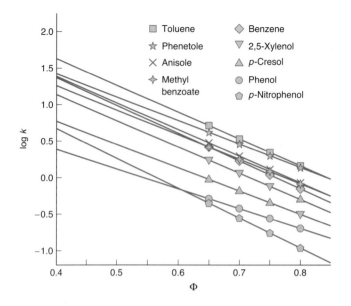

FIGURE 24-32 Linear-solvent-strength model. Graph of log k versus Φ for nine organic compounds eluted from a C_{18} column by methanol : water. [From R. A. Shalliker, S. Kayillo, and G. R. Dennis, "Optimizing Chromatographic Separation: An Experiment Using an HPLC Simulator," *J. Chem. Ed.* **2008**, *85*, 1265.]

where log k_w is the extrapolated retention factor for 100% aqueous eluent, Φ is the fraction of organic solvent ($\Phi = 0.4$ for 40 vol% organic solvent/60 vol% H_2O), and S is a constant for each compound, with a typical value of ~4 for small molecules. Figure 24-32 shows measurements for nine compounds in the range $\Phi = 0.65$ to 0.8 in methanol:water on a C_{18} column. The parameters S and log k_w are the slopes and y-intercepts of the lines in Figure 24-32.

Equation 24-5 is not accurate over wide ranges of solvent composition Φ. In Figure 24-32, the authors only varied Φ from 0.65 to 0.8. When we extrapolate beyond the measured range of Φ, we are in uncharted territory. Data for Figure 24-32 were collected at high values of Φ to keep the run time short. In the measured range of Φ, p-nitrophenol is eluted first and phenol is eluted next. The extrapolated lines cross at $\Phi = 0.61$. For $\Phi < 0.61$, we predict that phenol will be eluted before p-nitrophenol.

To simulate a chromatogram for an isocratic solvent composition such as $\Phi = 0.6$ for 60 vol% CH_3OH/40 vol% H_2O on a particular HPLC column, we begin with solvent transit time (t_m) = 1.85 min and plate number (N) = 7 000. Then, for a chosen value of solvent strength Φ, the retention factor k for each component is computed from

$$k = 10^{(\log k_w - S\Phi)} \tag{24-6}$$

Retention time t_r is found by rearranging Equation 22-17 to the form:

$$t_r = t_m(k + 1) \tag{24-7}$$

Assuming Gaussian peak shape, the standard deviation of the band in Figure 22-9 is found by rearranging Equation 22-28a:

$$\sigma = t_r / \sqrt{N} \tag{24-8}$$

The spreadsheet in Figure 24-33 simulates an isocratic separation. Highlighted cells require your input. The time step in cell C7 gives the interval between calculated points in the chromatogram. Relative areas in cells E14:E22 are arbitrary. You could set them all to 1 or you could try to vary them to match peak heights of an experimental chromatogram. The linear solvent strength parameters log k_w and S in cells C14:D22 are from experimental measurements in Figure 24-32. The spreadsheet computes k in cells F14:F22 with Equation 24-6. It computes t_r in cells G14:G22 with Equation 24-7 and the standard deviation of each Gaussian peak with Equation 24-8.

The shape of each Gaussian chromatographic peak is given by

$$\text{Detector signal } (y) = \frac{\text{relative area}}{\sigma\sqrt{2\pi}} e^{-(t - t_r)^2/2\sigma^2} \tag{24-9}$$

where relative areas are the numbers you specified in cells E:14:E22, t is time, t_r is the retention time in cells G14:G22, and σ is the standard deviation in cells H14:H22. The detector signal beginning in cell E30 is the sum of nine terms of the form of Equation 24-9—one term for each compound in the mixture. Each compound has its own σ, t_r, and relative area. The spreadsheet calculates detector signal for times beginning at $t = 0$ and proceeding past elution of the last peak.

	A	B	C	D	E	F	G	H
1	Chromatogram simulator - Gaussian peaks							
2	Data for methanol:water with Waters C-18 column - Shalliker et al., J. Chem. Ed. **2008**, _85_, 1265							
3								
4		constants						
5		t_m =	1.85	min	(time for mobile phase to transit the column)			
6		N =	7000	plates	(plate number for column)			
7		time step =	0.01	min	(time between calculated points)			
8		sqrt(2*pi) =	2.50663					
9		Φ =	0.56		(fraction of organic solvent)			
10								
11						k	Retention	σ (std dev
12		Compound			Relative	retention	time	peak width)
13	Number	Name	log k_w	S	area	factor	t_r (min)	(min)
14	1	p-nitrophenol	2.323	4.113	0.25	1.05	3.79	0.045
15	2	phenol	1.488	2.734	0.2	0.91	3.53	0.042
16	3	p-cresol	2.059	3.205	0.25	1.84	5.25	0.063
17	4	2,5-xylenol	2.591	3.619	0.4	3.67	8.63	0.103
18	5	benzene	2.895	3.806	0.8	5.80	12.59	0.150
19	6	methyl benzoate	2.617	3.392	0.8	5.22	11.50	0.137
20	7	anisole	2.840	3.646	0.4	6.28	13.48	0.161
21	8	phenetole	2.734	3.258	0.4	8.12	16.87	0.202
22	9	toluene	3.118	3.705	0.5	11.05	22.28	0.266

	C	D	E				
				F14 = 10^(C14-D14*$C$9)			
27			y	G14 = $C$5*(F14+1)			
28	time	time	detector	H14 = G14/SQRT($C$6)			
29	step	(min)	signal	Detector signal:			
30	0	0	0	E30 = ($E$14/($H$14*$C$8)*EXP(-((D30-$G$14)^2)/(2*$H$14^2)))			
31	1	0.01	0	+ ($E$15/($H$15*$C$8)*EXP(-((D30-$G$15)^2)/(2*$H$15^2)))			
32	2	0.02	0	+ an analogous term for each of the other compounds			

FIGURE 24-33 A spreadsheet to simulate isocratic chromatographic separation.

Figure 24-34 shows simulations done with the spreadsheet. At a solvent strength Φ = 0.75, all nine compounds are eluted within 6 min, but resolution of peaks 5, 6, and 7 is poor. At Φ = 0.60, peaks 1 and 2 overlap and peaks 5 and 6 have changed their order of elution. At Φ = 0.56, all peaks are resolved and the last peak is eluted in 22 min.

Solvent composition Φ = 0.56 (56 vol% methanol/44 vol% aqueous buffer) is not good enough to make the chromatographic method robust. A _robust_ separation is one that maintains adequate resolution despite _small_ changes in conditions such as Φ, pH, and temperature. _Resolution_ between closely spaced peaks is 1.5 for peaks 2 and 1, 1.8 for peaks 6 and 5, and 1.3 for peaks 5 and 7. For a robust separation, we desire a minimum resolution of 2.0, but this value cannot always be achieved. At solvent composition Φ = 0.54, resolution of peaks 2 and 1 and peaks 6 and 5 is above 2.0, but the resolution of peaks 5 and 7 has decreased to 1.2. Peak 9, which is off the chart, has a retention time of 26 min. The composition Φ = 0.56 appears to be about the best we can do with methanol and water on this column. To obtain better resolution, we could decrease flow rate, use smaller particles, increase column length, or change the temperature, solvent, or stationary phase.

$$\text{Resolution} = \frac{\Delta t_r}{w_{av}} \quad \text{(Equation 22-23)}$$

Δt_r = difference in retention time
w_{av} = average width at base of peak = 4σ

From a few experiments to find S and log k_w, we can use the spreadsheet to _estimate_ that Φ = 0.56 is the optimum condition to try in the lab. Solvent composition Φ = 0.56 is outside the measured range in Figure 24-32. The only way to know if the curves in Figure 24-32 remain linear down to Φ = 0.56 is to try the experiment. With only a little more complexity, the linear-solvent-strength model enables us to simulate and optimize gradient separations.[3]

With the tools in this chapter, you can usually find a way to separate the components of a mixture if it does not contain too many compounds. If reversed-phase chromatography fails, normal-phase chromatography or one of the methods in Chapter 25 could be appropriate. Method development is part science, part art, and part luck.

In supplementary topics at the Web site www.whfreeman.com/qca8e, you will find equations and a spreadsheet to simulate gradient and isocratic elution.

FIGURE 24-34 Chromatograms simulated with the spreadsheet in Figure 24-33. Peak numbers are *p*-nitrophenol (1), phenol (2), *p*-cresol (3), 2,5-xylenol (4), benzene (5), methyl benzoate (6), anisole (7), phenetole (8), and toluene (9).

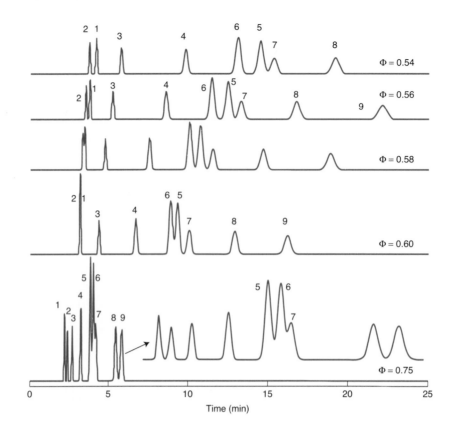

Terms to Understand

bonded stationary phase
charged aerosol detector
dead volume
derivatization
dwell volume
electrochemical detector
eluent strength

evaporative light-scattering
    detector
gradient elution
guard column
high-performance liquid
    chromatography

hydrophilic interaction
    chromatography
hydrophilic substance
isocratic elution
microporous particles
normal-phase chromatography

refractive index detector
reversed-phase
    chromatography
supercritical fluid
superficially porous particle
ultraviolet detector

Summary

In high-performance liquid chromatography (HPLC), solvent is pumped at high pressure through a column containing stationary phase particles with diameters of 1.5–5 μm. The smaller the particle size, the more efficient the column, but the greater the resistance to flow. Microporous silica particles with a covalently bonded liquid phase such as octadecyl groups ($-C_{18}H_{37}$) are most common. Eluent strength measures the ability of a given solvent to elute solutes from the column. In normal-phase chromatography, the stationary phase is polar and a less polar solvent is used. Eluent strength increases as the polarity of the solvent increases. Reversed-phase chromatography employs a nonpolar stationary phase and polar solvent. Eluent strength increases as the polarity of the solvent decreases. Most separations of organic compounds can be done on reversed-phase columns. Polar compounds can be separated by normal-phase chromatography or hydrophilic interaction chromatography. Normal-phase chromatography or porous graphitic carbon is good for separating isomers. Chiral phases are used for optical isomers. Techniques for separating inorganic ions, polymers, and biological macromolecules are described in Chapter 25.

If a solution of organic solvent and water is used in reversed-phase chromatography, eluent strength increases as the percentage of organic solvent increases. If the solvent has a fixed composition, the process is called isocratic elution. In gradient elution, eluent strength is increased during chromatography by increasing the percentage of strong solvent.

A short guard column containing the same stationary phase as the analytical column is placed before the analytical column to protect it from contamination with particles or irreversibly adsorbed solutes. A high-quality pump provides smooth solvent flow. The injection valve allows rapid, precise sample introduction. The column is best housed in an oven to maintain a reproducible temperature. Column efficiency increases at elevated temperature because the rate of mass transfer between phases is increased. Mass spectrometric detection provides quantitative and qualitative information for each substance eluted from the column. Ultraviolet detection is most common, and it can provide qualitative information if a photodiode array is used to record a full spectrum of each analyte. Refractive index detection has universal response but is not very sensitive. Evaporative light scattering and the charged aerosol detector respond to the mass of each nonvolatile solute. Electrochemical and fluorescence detectors are extremely sensitive but selective. In supercritical fluid chromatography, nonvolatile solutes are separated by a process whose efficiency, speed, and detectors more closely resemble those of gas chromatography than those of liquid chromatography.

Steps in method development: (1) determine the goal of the analysis, (2) select a method of sample preparation, (3) choose a detector, and (4) use a systematic procedure to select solvent for isocratic or gradient elution. Aqueous acetonitrile, methanol, and tetrahydrofuran are customary solvents for reversed-phase separations. A separation can be optimized by varying several solvents or by using one solvent and temperature as the principal variables. If further resolution is required, flow rate can be decreased, a longer column can be employed, and smaller particle size can be selected. Criteria for a successful separation are $0.5 \le k \le 20$, resolution ≥ 2.0, operating pressure ≤ 15 MPa (for conventional equipment), and asymmetry factor in the range 0.9–1.5. In gradient elution, solvent composition does not begin to change until the dwell volume has passed from the point of solvent mixing to the start of the column. A wide gradient is a good first choice to determine whether to use isocratic or gradient elution. Retention factors measured at a few solvent compositions can be fit to the linear-solvent-strength model to enable computer optimization of a separation.

Exercises

24-A. A known mixture of compounds A and B gave the following HPLC results:

Compound	Concentration (mg/mL in mixture)	Peak area (arbitrary units)
A	1.03	10.86
B	1.16	4.37

A solution was prepared by mixing 12.49 mg of B plus 10.00 mL of unknown containing just A and diluting to 25.00 mL. Peak areas of 5.97 and 6.38 were observed for A and B, respectively. Find the concentration of A (mg/mL) in the unknown.

24-B. A bonded stationary phase for the separation of optical isomers has the structure

Optically active center

To resolve the enantiomers of amines, alcohols, or thiols, the compounds are first derivatized with a nitroaromatic group that increases their interaction with the bonded phase and makes them observable with a spectrophotometric detector.

Aromatic derivative
(mixture of two enantiomers)

When the mixture is eluted with 20 vol% 2-propanol in hexane, the (R)-enantiomer is eluted before the (S)-enantiomer, with the following chromatographic parameters:

$$\text{Resolution} = \frac{\Delta t_r}{w_{av}} = 7.7 \qquad \text{Relative retention } (\alpha) = 4.53$$

$$k \text{ for } (R)\text{-isomer} = 1.35 \qquad t_m = 1.00 \text{ min}$$

where w_{av} is the average width of the two Gaussian peaks at their base.

(a) Find t_1, t_2, and w_{av}, with units of minutes.

(b) The width of a peak at half-height is $w_{1/2}$ (Figure 22-9). If the plate number for each peak is the same, find $w_{1/2}$ for each peak.

(c) The area of a Gaussian peak is $1.064 \times$ peak height $\times w_{1/2}$. Given that the areas under the two bands should be equal, find the relative peak heights (height$_R$/height$_S$).

24-C. Two peaks emerge from a chromatography column as sketched in the illustration.

According to Equation 22-30, resolution is given by

$$\text{Resolution} = \frac{\sqrt{N}}{4}(\gamma - 1)$$

where N is plate number and γ is the unadjusted relative retention (Equation 22-16).

(a) If you change the solvent or the stationary phase, you will change the relative retention. Sketch the chromatogram if γ increases but N is constant.

(b) If you increase column length or decrease particle size or flow rate, you can increase the plate number. Sketch the chromatogram if N increases but γ is constant.

24-D. After poisonous melamine and cyanuric acid appeared in milk in China (Box 10-3) in 2008, there was a flurry of activity to develop methods to measure these substances. An analytical method for milk is to treat 1 volume of milk with 9 volumes of $H_2O:CH_3CN$ (20:80 vol/vol) to precipitate proteins. The mixture is centrifuged for 5 min to remove precipitate. The supernatant liquid is filtered through a 0.5-μm filter and injected into a HILIC liquid chromatography column (TSKgel Amide-80 stationary phase) and products are measured by mass spectrometry with selected reaction monitoring (Section 21-4). Melamine is measured in positive ion mode with the transition m/z

$127 \rightarrow 85$. Cyanuric acid is measured in negative ion mode with the transition m/z $128 \rightarrow 42$.

(a) Write the formulas for the four ions and propose structures for all four ions.

(b) Even though milk is a complex substance, only one clean peak is observed for melamine and one for cyanuric acid spiked into milk. Explain why.

24-E. (a) Make a graph showing the retention time of each peak in Figure 24-25 in chromatograms A, D, and B as a function of position along the line AB. Predict the retention times for solvent compositions midway between A and D and midway between D and B. Draw a stick diagram (representing each peak as a vertical line) of the two predicted chromatograms.

(b) What would be the solvent compositions midway between A and D and midway between D and B?

Problems

High-Performance Liquid Chromatography

24-1. (a) Why does eluent strength increase as solvent becomes less polar in reversed-phase chromatography, whereas eluent strength increases as solvent becomes more polar in normal-phase chromatography?

(b) What kind of gradient is used in supercritical fluid chromatography?

24-2. Why are the relative eluent strengths of solvents in adsorption chromatography fairly independent of solute?

24-3. In hydrophilic interaction chromatography (HILIC), why is eluent strength increased by *increasing* the fraction of water in the mobile phase.

24-4. (a) Why is high pressure needed in HPLC?

(b) What is a bonded phase in liquid chromatography?

24-5. (a) Use Equation 24-1 to estimate the length of column required to achieve 1.0×10^4 plates if the stationary phase particle size is 10.0, 5.0, 3.0, or 1.5 μm.

(b) Why do smaller particles give better resolution?

24-6. If a 15-cm-long column has a plate height of 5.0 μm, what will be the half-width (in seconds) of a peak eluted at 10.0 min? If plate height = 25 μm, what will be $w_{1/2}$?

24-7. Why are silica stationary phases generally limited to operating in the pH range 2–8? Why does the silica in Figure 24-8 have improved stability at low pH?

24-8. How do additives such as triethylamine reduce tailing of certain solutes?

24-9. HPLC peaks should generally not have an asymmetry factor A/B in Figure 22-14 outside the range 0.9–1.5.

(a) Sketch the shape of a peak with an asymmetry of 1.8.

(b) What might you do to correct the asymmetry?

24-10. (a) Sketch a graph of the van Deemter equation (plate height versus linear flow rate). What would the curve look like if the multiple path term were 0? If the longitudinal diffusion term were 0? If the finite equilibration time term were 0?

(b) Explain why the van Deemter curve for 1.8-μm particles in Figure 24-3 is nearly flat at high flow rate. What can you say about each of the terms in the van Deemter equation for 1.8-μm particles?

(c) Explain why the 2.7-μm superficially porous particle enables separations similar to those achieved by 1.8-μm totally porous particles, but the superficially porous particle requires lower pressure.

24-11. The figure below shows the separation of two enantiomers on a chiral stationary phase.

L enantiomer
$t_r = 4.70$ min
$w_{1/2} = 0.28$ min

D enantiomer
$t_r = 5.37$ min
$w_{1/2} = 0.35$ min

Separation of enantiomers of Ritalin by HPLC with a chiral stationary phase.
[From R. Bakhtiar, L. Ramos, and F. L. S. Tse, "Quantification of Methylphenidate in Plasma Using Chiral Liquid-Chromatography/Tandem Mass Spectrometry: Application to Toxicokinetic Studies," *Anal. Chim. Acta* **2002**, *469*, 261.]

(a) From t_r and $w_{1/2}$, find N for each peak.

(b) From t_r and $w_{1/2}$, find the resolution.

(c) Use Equation 22-30 with the average N to predict the resolution.

24-12. (a) According to Equation 24-2, if all conditions are constant, but particle size is reduced from 3 μm to 0.7 μm, by what factor must pressure be increased to maintain constant linear velocity?

(b) If all conditions except pressure are constant, by what factor will linear velocity increase if column pressure is increased by a factor of 10?

(c) With 0.7-μm particles in a 50-μm-diameter × 9-cm-long column, increasing pressure from 70 MPa to 700 MPa decreased analysis time by approximately a factor of 10 while increasing plate count from 12 000 to 45 000.[38] Explain why small particles permit 10-fold faster flow without losing efficiency or, in this case, with improved efficiency.

24-13. Use Figure 24-15 to suggest which type of liquid chromatography you could use to separate compounds in each of the following categories:

(a) Molecular mass <2 000, soluble in octane

(b) Molecular mass <2 000, soluble in dioxane

(c) Molecular mass <2 000, ionic

(d) Molecular mass >2 000, soluble in water, nonionic, size 50 nm

(e) Molecular mass >2 000, soluble in water, ionic

(f) Molecular mass >2 000, soluble in tetrahydrofuran, size 50 nm

24-14. Microporous silica particles with a density of 2.2 g/mL and a diameter of 10 μm have a measured surface area of 300 m²/g. Calculate the surface area of the spherical silica as if it were simply solid particles. What does this calculation tell you about the shape or porosity of the particles?

24-15. (a) Nonpolar aromatic compounds were separated by HPLC on an octadecyl (C_{18}) bonded phase. The eluent was 65 vol% methanol in water. How would the retention times be affected if 90% methanol were used instead?

(b) Octanoic acid and 1-aminooctane were passed through the same column described in part (a), using an eluent of 20% methanol/80% buffer (pH 3.0). State which compound is expected to be eluted first and why.

$$CH_3CH_2CH_2CH_2CH_2CH_2CH_2CO_2H$$
Octanoic acid

$$CH_3CH_2CH_2CH_2CH_2CH_2CH_2CH_2NH_2$$
1-Aminooctane

24-16. Consider the chromatogram of deuterated benzenes in Box 24-1.

(a) Unretained thiourea, is eluted in 41.7 min. Find the linear velocity u_x (mm/s).

(b) Find the retention factor k for C_6D_6.

(c) Find the plate number N and plate height for C_6D_6.

(d) Assuming that the peak widths for C_6H_5D and C_6H_6 are the same as that of C_6D_6, find the resolution of C_6H_5D and C_6H_6.

(e) Retention times for C_6H_5D and C_6H_6 are 193.3 and 194.3 min, respectively. Find the relative retention (α) and unadjusted relative retention (γ) between C_6H_5D and C_6H_6.

(f) If we just increased the column length to increase N, what value of N and what column length would be required for a resolution of 1.000?

(g) Without increasing the length of the column, and without changing the stationary phase, how might you improve the resolution?

(h) When the solvent was changed from CH_3CN/H_2O (30:70 vol/vol) to $CH_3CN/CH_3OH/H_2O$ (10:5:85), the unadjusted relative retention changed from $\gamma = 1.005_2$ to 1.008_3. If the plate number were unchanged, what would be the resolution?

24-17. The antitumor drug gimatecan is available as nearly pure (*S*)-enantiomer. Neither pure (*R*)-enantiomer nor a *racemic* (equal) mixture of the two enantiomers are available. To measure small quantites of (*R*)-enantiomer in nearly pure (*S*)-gimatecan, a prepara-

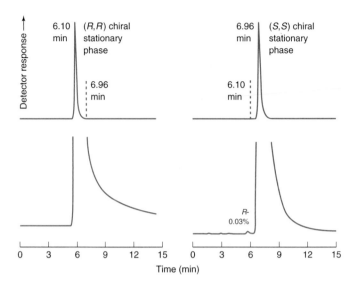

Chromatography of gimatecan on each enantiomer of a chiral stationary phase. Lower traces have enlarged vertical scale. [From E. Badaloni, W. Cabri, A. Ciogli, R. Deias, F. Gasparrini, F. Giorgi, A. Vigevani, and C. Villani, "Combination of HPLC 'Inverted Chirality Columns Approach' and MS/MS Detection for Extreme Enantiomeric Excess Determination Even in Absence of Reference Samples." *Anal. Chem.* **2007**, *79*, 6013.]

tion was subjected to normal-phase chromatography on each of the enantiomers of a commercial, chiral stationary phase designated (*S,S*)- and (*R,R*)-DACH-DNB. Chromatography on the (*R,R*)-stationary phase gave a slightly asymmetric peak at $t_r = 6.10$ min with retention factor $k = 1.22$. Chromatography on the (*S,S*)-stationary phase gave a slightly asymmetric peak at $t_r = 6.96$ min with $k = 1.50$. With the (*S,S*)-stationary phase, a small peak with 0.03% of the area of the main peak was observed at 6.10 min.

(a) Explain the appearance of the upper chromatograms. Dashed lines are position markers, not part of the chromatogram. What would the chromatogram of pure (*R*)-gimatecan look like on the same two stationary phases?

(b) Explain the appearance of the two lower chromatograms and why it can be concluded that the gimatecan contained 0.03% of the (*R*)-enantiomer. Why is the (*R*)-enantiomer not observed with the (*R,R*)-stationary phase?

(c) Find the adjusted relative retention (α) and unadjusted relative retention (γ) for the two enantiomers on the (*S,S*)-stationary phase.

(d) The column provides $N = 6\ 800$ plates. What would be the resolution between the two equal peaks in a racemic (equal) mixture of (*R*)- and (*S*)-gimatecan? If the peaks were symmetric, does this resolution provide "baseline" separation in which signal returns to baseline before the next peak begins?

24-18. Suppose that an HPLC column produces Gaussian peaks. The detector measures absorbance at 254 nm. A sample containing equal moles of compounds A and B was injected into the column. Compound A ($\varepsilon_{254} = 2.26 \times 10^4\ M^{-1}\ cm^{-1}$) has a height $h = 128$ mm and a half-width $w_{1/2} = 10.1$ mm. Compound B ($\varepsilon_{254} = 1.68 \times 10^4\ M^{-1}\ cm^{-1}$) has $w_{1/2} = 7.6$ mm. What is the height of peak B in millimeters?

24-19. Retention factors for three solutes separated on a C_8 nonpolar stationary phase are listed in the table on the next page. Eluent was a 70:30 (vol/vol) mixture of 50 mM citrate buffer (adjusted to pH with NH_3) plus methanol. Draw the dominant species of each compound at each pH in the table and explain the behavior of the retention factors.

Analyte	Retention factor		
	pH 3	pH 5	pH 7
Acetophenone	4.21	4.28	4.37
Salicylic acid	2.97	0.65	0.62
Nicotine	0.00	0.13	3.11

Acetophenone

Salicylic acid
pK_a = 2.97

Nicotine
pK_1 = 3.15
pK_2 = 7.85

24-20. Morphine and morphine 3-β-D-glucuronide were separated on two different 4.6-mm-diameter × 50-mm-long columns with 3-μm particles.[39] Column A was C_{18}-silica run at 1.4 mL/min, and column B was bare silica run at 2.0 mL/min.

Morphine

Morphine 3-β-D-glucuronide

(a) Estimate the volume, V_m, and time, t_m, at which unretained solute would emerge from each column. The observed times are 0.65 min for column A and 0.50 min for column B.

(b) Column A was eluted with 2 vol% acetonitrile in water containing 10 mM ammonium formate at pH 3. Morphine 3-β-D-glucuronide emerged at 1.5 min and morphine at 2.8 min. Explain the order of elution.

(c) Column B was eluted with a 5.0-min gradient beginning at 90 vol% acetonitrile in water and ending at 50 vol% acetonitrile in water. Both solvents contained 10 mM ammonium formate, pH 3. Morphine emerged at 1.3 min and morphine 3-β-D-glucuronide emerged at 2.7 min. Explain the order of elution. Why does the gradient go from high to low acetonitrile volume fraction?

(d) Find the retention factor k for each solute on column A, using t_m = 0.65 min.

(e) From Equation 24-10 in Box 24-4, estimate k^*, assuming $S = 4$ and with t_m = 0.50 min.

24-21. The rate at which heat is generated inside a chromatography column from friction of flowing liquid is power (watts, W = J/s) = volume flow rate (m^3/s) × pressure drop (pascals, Pa = kg/[m · s^2]).

(a) Explain the analogy between heat generated in a chromatography column and heat generated in an electric circuit (power = current × voltage).

(b) At what rate (watts = J/s) is heat generated for a flow of 1 mL/min with a pressure difference of 3 500 bar between the inlet and outlet? You will need to convert mL/min to m^3/s. Also 1 bar ≡ 10^5 Pa.

24-22. *Chromatography–mass spectrometry.* Cocaine metabolism in rats can be studied by injecting the drug and periodically withdrawing blood to measure levels of metabolites by HPLC–mass spectrometry. For quantitative analysis, isotopically labeled internal standards are mixed with the blood sample. Blood was analyzed by reversed-phase chromatography with an acidic eluent and atmospheric pressure chemical ionization mass spectrometry for detection. The mass spectrum of the collisionally activated dissociation products from the m/z 304 positive ion is shown in the figure. Selected reaction monitoring (m/z 304 from mass filter Q1 and m/z 182 from Q3 in Figure 21-26) gave a single chromatographic peak at 9.22 min for cocaine. The internal standard $^2H_5$-cocaine gave a single peak at 9.19 min for m/z 309 (Q1) → 182 (Q3).

(a) Draw the structure of the ion at m/z 304.

(b) Suggest a structure for the ion at m/z 182.

(c) The intense peaks at m/z 182 and 304 do not have ^{13}C isotopic partners at m/z 183 and 305. Explain why.

(d) Rat plasma is exceedingly complex. Why does the chromatogram show just one clean peak?

(e) Given that $^2H_5$-cocaine has only two major mass spectral peaks at m/z 309 and 182, which atoms are labeled with deuterium?

(f) Explain how you would use $^2H_5$-cocaine for measuring cocaine in blood.

Cocaine

Mass spectrum of m/z 304 after collisionally activated dissociation

182

82   105   150   304

50   100   150   200   250   300
m/z

9.22

Cocaine
m/z 304 → m/z 182

9.19

$^2H_5$-Cocaine
m/z 309 → m/z 182

9   10   11   12
Time (min)

For Problem 24-22.
Left: Mass spectrum of collisionally activated dissociation products from m/z 304 positive ion from atmospheric pressure chemical ionization mass spectrum of cocaine. *Right:* Chromatograms obtained by selected reaction monitoring. [From G. Singh, V. Arora, P. T. Fenn, B. Mets, and I. A. Blair, "Isotope Dilution Liquid Chromatography Tandem Mass Spectrometry Assay for Trace Analysis of Cocaine and Its Metabolites in Plasma," *Anal. Chem.* **1999**, *71*, 2021.]

24-23. *Chromatography–mass spectrometry.* HPLC separation of enantiomers of the drug Ritalin on a chiral stationary phase was shown in Problem 24-11.

Ritalin (methylphenidate)
$C_{14}H_{19}O_2N$
Nominal mass = 233

(a) Detection is by atmospheric pressure chemical ionization with selected reaction monitoring of the m/z 234 $\rightarrow$ 84 transition. Explain how this detection works and propose structures for m/z 234 and m/z 84.

(b) For quantitative analysis, the internal standard $^2H_3$-Ritalin with a deuterated methyl group was added. Deuterated enantiomers have the same retention times as unlabeled enantiomers. Which selected reaction monitoring transition should be monitored to produce a chromatogram of the internal standard in which unlabeled Ritalin will be invisible?

Method Development

24-24. (a) Explain how to measure k and resolution.

(b) State three methods for measuring t_m in reversed-phase chromatography.

(c) Estimate t_m for a 0.46 × 15 cm column containing 5-μm particles operating at a flow rate of 1.5 mL/min. Estimate t_m if the particle size were 3.5 μm instead.

24-25. What is the difference between dead volume and dwell volume? How do each of these volumes affect a chromatogram?

24-26. What does it mean for a separation procedure to be "rugged" and why is it desirable?

24-27. What are criteria for an adequate isocratic chromatographic separation?

24-28. Explain how to use a gradient for the first run to decide whether isocratic or gradient elution would be more appropriate.

24-29. What are the general steps in developing an isocratic separation for reversed-phase chromatography with one organic solvent?

24-30. What are the general steps in developing an isocratic separation for reversed-phase chromatography with two organic solvents?

24-31. What are the general steps in developing an isocratic separation for reversed-phase chromatography with one organic solvent and temperature as variables?

24-32. The "rule of three" states that the retention factor for a given solute decreases *approximately* threefold when the organic phase increases by 10%. In Figure 24-12, t_m = 2.7 min. Find k for peak 5 at 50% B. Predict the retention time for peak 5 at 40% B and compare the observed and predicted times.

24-33. (a) Make a graph showing retention times of peaks 6, 7, and 8 in Figure 24-12 as a function of %acetonitrile (%B). Predict the retention time of peak 8 at 45% B.

(b) *Linear-solvent-strength model:* In Figure 24-12, t_m = 2.7 min. Compute k for peaks 6, 7, and 8 as a function of %B. Make a graph of log k versus Φ, where Φ = %B/100. Find the equation of a straight line through a suitable linear range for peak 8. The slope is $-S$ and the intercept is log k_w. From the line, predict t_r for peak 8 at 45% B and compare your answer with part **(a)**.

24-34. (a) Make a graph showing the retention time of each peak in Figure 24-25 in chromatograms B, F, and C as a function of position along the line BC. Predict the retention times for solvent compositions midway between B and F and midway between F and C. Draw a stick diagram (representing each peak as a vertical line) of each of the two predicted chromatograms.

(b) What would be the solvent compositions midway between B and F and midway between F and C?

24-35. Suppose that in Figure 24-25 the optimum concentrations of solvents at points A, B, and C are 50% acetonitrile, 60% methanol, and 40% tetrahydrofuran, respectively. What will be the solvent compositions at points D, E, F, and G?

24-36. A reversed-phase separation of a reaction mixture calls for isocratic elution with 48% methanol/52% water. If you want to change the procedure to use acetonitrile/water, what is a good starting percentage of acetonitrile to try?

24-37. (a) When you try separating an unknown mixture by reversed-phase chromatography with 50% acetonitrile/50% water, the peaks are too close together and are eluted in the range k = 2–6. Should you use a higher or lower concentration of acetonitrile in the next run?

(b) When you try separating an unknown mixture by normal-phase chromatography with pure 50% hexane/50% methyl *t*-butyl ether, the peaks are too close together and are eluted in the range k = 2–6. Should you use a higher or lower concentration of hexane in the next run?

24-38. A mixture of 14 compounds was subjected to a reversed-phase gradient separation going from 5% to 100% acetonitrile with a gradient time of 60 min. The sample was injected at t = dwell time. All peaks were eluted between 22 and 41 min.

(a) Is the mixture more suitable for isocratic or gradient elution?

(b) If the next run is a gradient, select the starting and ending %acetonitrile and the gradient time.

24-39. (a) List ways in which the resolution between two closely spaced peaks might be changed.

(b) After optimization of an isocratic elution with several solvents, the resolution of two peaks is 1.2. How might you improve the resolution without changing solvents or the kind of stationary phase?

24-40. (a) You wish to use a wide gradient from 5 vol% to 95 vol% B for the first separation of a mixture of small molecules to decide whether to use gradient or isocratic elution. What should be the gradient time, t_G, for a 0.46 × 15 cm column containing 3-μm particles with a flow of 1.0 mL/min?

(b) You optimized the gradient separation going from 20 vol% to 34 vol% B in 11.5 min at 1.0 mL/min. Find k^* for this optimized separation. To scale up to a 1.0 × 15 cm column, what should be the gradient time and the volume flow rate? If the sample load on the small column was 1 mg, what sample load can be applied to the large column? Verify that k^* is unchanged.

24-41. 🖳 *Simulating a separation with a spreadsheet.* Use the spreadsheet in Figure 24-33 to simulate the chromatograms for Φ = 0.75 and Φ = 0.56 in Figure 24-34.

25 Chromatographic Methods and Capillary Electrophoresis

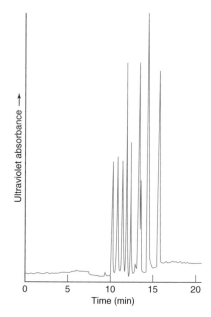

The inside of a monolithic electrochromatography column contains silicate "fingers" coated with stationary phase. Aromatic compounds were separated with a mean plate number of 80 000 in a 50-cm column with 15 kV applied voltage. [From J. D. Hayes and A. Malik, "Sol-Gel Monolithic Columns with Reversed Electroosmotic Flow for Capillary Electrochromatography," *Anal. Chem.* 2000, 72, 4090. Photo courtesy A. Malik, University of South Florida.]

Electrochromatography[1] uses an electric field, rather than pressure, to propel mobile phase through a capillary column with uniform, pluglike flow (Color Plate 30). The capillary in the photograph contains a monolithic silicate structure similar to that in Box 24-1. Polymerization of soluble precursors inside the capillary forms chemical structures such as that shown below. The surface is coated with covalently anchored, positively charged quaternary ammonium groups. Mobile anions in solution provide charge balance. A strong electric field draws anions toward the anode, pulling the entire solution in the capillary along with them.

C_{18} groups attached to the quaternary ammonium cations are the chromatographic stationary phase. Solutes passing through the capillary are separated when they partition between the mobile solvent and the stationary C_{18} phase.

This chapter continues our discussion of pressure-driven chromatographic methods including ion-exchange, molecular exclusion, affinity, and hydrophobic interaction chromatography. We then introduce separations driven by an electric field. In electrophoresis and electrochromatography, liquid moves through a capillary tube by *electroosmosis*. This process enables us to create miniature analytical chips in which fluids are driven through capillary channels etched into glass or plastic. Chemical reactions and chemical separations are carried out on these chips. In the future, people will carry a "lab-on-a-chip" into the field for investigations that require large instruments today.

25-1 Ion-Exchange Chromatography

In **ion-exchange chromatography,** retention is based on the attraction between solute ions and charged sites bound to the stationary phase (Figure 22-6). In **anion exchangers,** positively charged groups on the stationary phase attract solute <u>anions</u>. **Cation exchangers** contain covalently bound, negatively charged sites that attract solute <u>cations</u>.

Anion exchangers contain bound *positive* groups.

Cation exchangers contain bound *negative* groups.

Ion Exchangers

Resins are amorphous (noncrystalline) particles of organic material. *Polystyrene resins* for ion exchange are made by the copolymerization of styrene and divinylbenzene (Figure 25-1). Divinylbenzene content is varied from 1 to 16% to increase the extent of **cross-linking** of the insoluble hydrocarbon polymer. Benzene rings can be modified to produce a cation-exchange resin, containing sulfonate groups ($-SO_3^-$), or an anion-exchange resin, containing ammonium groups ($-NR_3^+$). If methacrylic acid is used in place of styrene, a polymer with carboxyl groups results.

Table 25-1 classifies ion exchangers as strongly or weakly acidic or basic. Sulfonate groups ($-SO_3^-$) of strongly acidic resins remain ionized even in strongly acidic solutions. Carboxyl groups ($-CO_2^-$) of the weakly acidic resins are protonated near pH 4 and lose their cation-exchange capacity. "Strongly basic" quaternary ammonium groups ($-CH_2NR_3^+$) (which are not really basic at all) remain cationic at all values of pH. Weakly basic tertiary

Methacrylic acid

Strongly acidic cation exchangers: RSO_3^-
Weakly acidic cation exchangers: RCO_2^-
"Strongly basic" anion exchangers: RNR'_3^+
Weakly basic anion exchangers: $RNR_2'H^+$

Cross-linked styrene-divinylbenzene copolymer

Strongly acidic cation-exchange resin

Strongly basic anion-exchange resin

FIGURE 25-1 Structures of styrene-divinylbenzene cross-linked ion-exchange resins.

TABLE 25-1   Ion-exchange resins

Resin type	Chemical constitution	Usual form as purchased	Common trade names		Selectivity	Thermal stability
			Rohm & Haas	Dow Chemical		
Strongly acidic cation exchanger	Sulfonic acid groups attached to styrene and divinylbenzene copolymer	$Aryl\!-\!SO_3^-H^+$	Amberlite IR-120	Dowex 50W	$Ag^+ > Rb^+ > Cs^+ >$ $K^+ > NH_4^+ > Na^+ >$ $H^+ > Li^+$ $Zn^{2+} > Cu^{2+} > Ni^{2+}$ $> Co^{2+}$	Good up to 150°C
Weakly acidic cation exchanger	Carboxylic acid groups attached to acrylic and divinylbenzene copolymer	$R\!-\!COO^-Na^+$	Amberlite IRC-50	—	$H^+ \gg Ag^+ > K^+ >$ $Na^+ > Li^+$ $H^+ \gg Fe^{2+} > Ba^{2+}$ $Sr^{2+} > Ca^{2+} > Mg^{2+}$	Good up to 100°C
Strongly basic anion exchanger	Quaternary ammonium groups attached to styrene and divinylbenzene copolymer	$Aryl\!-\!CH_2N(CH_3)_3^+Cl^-$	Amberlite IRA-400	Dowex 1	$I^- > phenolate^- >$ $HSO_4^- > ClO_3^- >$ $NO_3^- > Br^- > CN^-$ $> HSO_3^- > NO_2^- > Cl^- >$ $HCO_3^- > IO_3^-$ $> HCOO^- >$ $acetate^- > OH^- > F^-$	OH^- form fair up to 50°C; Cl^- and other forms good up to 150°C
Weakly basic anion exchanger	Polyalkylamine groups attached to styrene and divinylbenzene copolymer	$Aryl\!-\!NH(R)_2^+Cl^-$	Amberlite IR-45	Dowex 3	$Aryl\!-\!SO_3H > citric >$ $CrO_3 > H_2SO_4 >$ $tartaric > oxalic >$ $H_3PO_4 > H_3AsO_4 >$ $HNO_3 > HI > HBr$ $> HCl > HF >$ $HCO_2H >$ $CH_3CO_2H > H_2CO_3$	Extensive information not available; tentatively limited to 65°C

SOURCE: *Adapted from J. X. Khym, Analytical Ion-Exchange Procedures in Chemistry and Biology (Englewood Cliffs, NJ: Prentice Hall, 1974).*

Because they are much softer than polystyrene *resins,* dextran and its relatives are called **gels.**

Figure 25-15 shows the structure of polyacrylamide.

ammonium ($-CH_2NHR_2^+$) anion exchangers are deprotonated in moderately basic solution and lose their ability to bind anions.

The extent of cross-linking is indicated by the notation "-XN" after the name of the resin. For example, Dowex 1-X4 contains 4% divinylbenzene, and Bio-Rad AG 50W-X12 contains 12% divinylbenzene. The resin becomes more rigid and less porous as cross-linking increases. Lightly cross-linked resins permit rapid equilibration of solute between the inside and the outside of the particle. However, resins with little cross-linking swell in water. Hydration decreases both the density of ion-exchange sites and the selectivity of the resin for different ions. More heavily cross-linked resins exhibit less swelling and higher exchange capacity and selectivity, but they have longer equilibration times. The charge density of polystyrene ion exchangers is so great that highly charged macromolecules such as proteins may be irreversibly bound.

Cellulose and dextran ion exchangers, which are polymers of the sugar glucose, possess larger pore sizes and lower charge densities than those of polystyrene resins. They are well suited to ion exchange of macromolecules, such as proteins. Dextran, cross-linked by glycerin, is sold under the name Sephadex (Figure 25-2). Other macroporous ion exchangers are based on the polysaccharide agarose and on polyacrylamide. Table 25-2 lists charged functional groups used to derivatize polysaccharide hydroxyl groups. DEAE-Sephadex, for example, is an anion-exchange Sephadex containing diethylaminoethyl groups.

FIGURE 25-2 Structure of Sephadex, a cross-linked dextran originally sold by Pharmacia Fine Chemicals.

TABLE 25-2 Common active groups of ion-exchange gels

Type	Abbreviation	Name	Structure
Cation Exchangers			
Strong acid	SP	Sulfopropyl	$-OCH_2CH_2CH_2SO_3H$
	SE	Sulfoethyl	$-OCH_2CH_2SO_3H$
Intermediate acid	P	Phosphate	$-OPO_3H_2$
Weak acid	CM	Carboxymethyl	$-OCH_2CO_2H$
Anion Exchangers			
Strong base	TEAE	Triethylaminoethyl	$-OCH_2CH_2\overset{+}{N}(CH_2CH_3)_3$
	QAE	Diethyl(2-hydroxypropyl) quaternary amino	$-OCH_2CH_2\overset{+}{N}(CH_2CH_3)_2$ $\mid$ $CH_2CHOHCH_3$
Intermediate base	DEAE	Diethylaminoethyl	$-OCH_2CH_2N(CH_2CH_3)_2$
	ECTEOLA	Triethanolamine coupled to cellulose through glyceryl chains	
	BD	Benzoylated DEAE groups	
Weak base	PAB	*p*-Aminobenzyl	$-O-CH_2-\bigotimes-NH_2$

Ion-Exchange Selectivity

Consider the competition of Na^+ and Li^+ for sites on the cation-exchange resin R^-:

Selectivity coefficient: $R^-Na^+ + Li^+ \rightleftharpoons R^-Li^+ + Na^+$ $K = \dfrac{[R^-Li^+][Na^+]}{[R^-Na^+][Li^+]}$ (25-1)

The equilibrium constant is called the **selectivity coefficient,** because it describes the relative selectivity of the resin for Li^+ and Na^+. Selectivities of polystyrene resins in Table 25-3 tend to increase with the extent of cross-linking, because the pore size of the resin shrinks as cross-linking increases. Ions such as Li^+, with a large *hydrated radius* (Chapter 7 opener), do not have as much access to the resin as smaller ions, such as Cs^+, do.

TABLE 25-3 Relative selectivity coefficients of ion-exchange resins

| | Sulfonic acid cation-exchange resin | | | Quaternary ammonium anion-exchange resin | |
| | Relative selectivity for divinylbenzene content | | | | |
Cation	4%	8%	10%	Anion	Relative selectivity
Li^+	1.00	1.00	1.00	F^-	0.09
H^+	1.30	1.26	1.45	OH^-	0.09
Na^+	1.49	1.88	2.23	Cl^-	1.0
NH_4^+	1.75	2.22	3.07	Br^-	2.8
K^+	2.09	2.63	4.15	NO_3^-	3.8
Rb^+	2.22	2.89	4.19	I^-	8.7
Cs^+	2.37	2.91	4.15	ClO_4^-	10.0
Ag^+	4.00	7.36	19.4		
Tl^+	5.20	9.66	22.2		

SOURCE: *Amberlite Ion Exchange Resins—Laboratory Guide (Rohm & Haas Co., 1979).*

Polarizability measures the ability of an ion's electron cloud to be deformed by nearby charges. Deformation of the cloud induces a dipole in the ion. The attraction between the induced dipole and the nearby charge increases the binding of the ion to the resin.

In general, ion exchangers favor the binding of ions of higher charge, decreased hydrated radius, and increased *polarizability*. A fairly general order of selectivity for cations is

$$Pu^{4+} \gg La^{3+} > Ce^{3+} > Pr^{3+} > Eu^{3+} > Y^{3+} > Sc^{3+} > Al^{3+} \gg$$
$$Ba^{2+} > Pb^{2+} > Sr^{2+} > Ca^{2+} > Ni^{2+} > Cd^{2+} > Cu^{2+} >$$
$$Co^{2+} > Zn^{2+} > Mg^{2+} > UO_2^{2+} \gg Tl^+ > Ag^+ > Rb^+ > K^+ >$$
$$NH_4^+ > Na^+ > H^+ > Li^+$$

Reaction 25-1 can be driven in either direction, even though Na^+ is bound more tightly than Li^+. Washing a column containing Na^+ with a substantial excess of Li^+ will replace Na^+ with Li^+. Washing a column in the Li^+ form with Na^+ will convert it into the Na^+ form.

Na^+ has a *smaller* hydrated radius than Li^+.

Ion exchangers loaded with one kind of ion bind small amounts of a different ion nearly quantitatively. Na^+-loaded resin will bind small amounts of Li^+ nearly quantitatively, even though the selectivity is greater for Na^+. The same column binds large quantities of Ni^{2+} or Fe^{3+} because the resin has greater selectivity for these ions than for Na^+. Even though Fe^{3+} is bound more tightly than H^+, Fe^{3+} is removed from the resin by washing with excess acid.

Donnan Equilibrium

A phase containing bound charges tends to exclude electrolyte.

When an ion exchanger is placed in an electrolyte solution, *the concentration of electrolyte is higher outside the resin than inside it.* The equilibrium between ions in solution and ions inside the resin is called the **Donnan equilibrium.**

Consider a quaternary ammonium anion-exchange resin (R^+) in its Cl^- form immersed in a solution of KCl. Let the concentration of an ion inside the resin be $[X]_i$ and the concentration outside the resin be $[X]_o$. It can be shown from thermodynamics that the ion product inside the resin is approximately equal to the product outside the resin:

$$[K^+]_i[Cl^-]_i = [K^+]_o[Cl^-]_o \tag{25-2}$$

From considerations of charge balance, we know that

$$[K^+]_o = [Cl^-]_o \tag{25-3}$$

We are ignoring H^+ and OH^-, which are assumed to be negligible.

Inside the resin, there are three charged species, and the charge balance is

$$[R^+]_i + [K^+]_i = [Cl^-]_i \tag{25-4}$$

where $[R^+]$ is the concentration of quaternary ammonium ions attached to the resin. Substituting Equations 25-3 and 25-4 into Equation 25-2 gives

$$[K^+]_i([K^+]_i + [R^+]_i) = [K^+]_o^2 \tag{25-5}$$

which says that $[K^+]_o$ must be greater than $[K^+]_i$.

EXAMPLE   Exclusion of Cations by an Anion Exchanger

Suppose that the concentration of cationic sites in the resin is 6.0 M. When the Cl^- form of this resin is immersed in 0.050 M KCl, what will be the ratio $[K^+]_o/[K^+]_i$?

Solution Let us assume that $[K^+]_o$ remains 0.050 M. Equation 25-5 gives

$$[K^+]_i([K^+]_i + 6.0) = (0.050)^2 \implies [K^+]_i = 0.000\ 42\ M$$

The concentration of K^+ inside the resin is less than 1% of that outside the resin.

Test Yourself Find the $[K^+]_i$ if cationic sites in the resin were 3.0 M instead of 6.0 M. Does your answer make sense? (**Answer:** 0.000 83 M)

Ions with the *same* charge as the resin are excluded. (The quaternary ammonium resin excludes K^+.) The counterion, Cl^-, is *not excluded* from the resin. There is no electrostatic barrier to penetration of an anion into the resin. Anion exchange takes place freely in the quaternary ammonium resin even though cations are repelled from the resin.

The Donnan equilibrium is the basis of *ion-exclusion chromatography*. Because dilute electrolytes are excluded from the resin, they pass through a column faster than nonelectrolytes, such as sugar, which freely penetrates the resin. When a solution of NaCl and sugar is applied to an ion-exchange column, NaCl emerges from the column *before* sugar.

The high concentration of positive charges within the resin repels cations from the resin.

Conducting Ion-Exchange Chromatography

Ion-exchange *resins* are used for applications involving small molecules (FM $\lesssim$ 500), which can penetrate the small pores of the resin. A mesh size (Table 27-2) of 100/200 is suitable for most work. Higher mesh numbers (smaller particle size) lead to finer separations but slower column operation. For preparative separations, the sample may occupy 10 to 20% of the column volume. Ion-exchange *gels* are used for large molecules (such as proteins and nucleic acids), which cannot penetrate the pores of resins. Separations involving harsh chemical conditions (high temperature, high radiation levels, strongly basic solution, or powerful oxidizing agents) employ *inorganic ion exchangers*, such as hydrous oxides of Zr, Ti, Sn, and W.

Three classes of ion exchangers:
1. resins
2. gels
3. inorganic exchangers

Gradient elution with increasing ionic strength or changing pH is common in ion-exchange chromatography. Consider a column to which anion A^- is bound more tightly than anion B^-. We separate A^- from B^- by elution with C^-, which is less tightly bound than either A^- or B^-. As the concentration of C^- is increased, B^- is eventually displaced and moves down the column. At a still higher concentration of C^-, the anion A^- also is eluted.

In Figure 25-3, a gradient of ammonium acetate was used for an anion-exchange separation of differently charged forms of the protein transferrin at pH 6.0. Transferrin (Figure 17-7)

An ionic strength gradient is analogous to a solvent or temperature gradient.

FIGURE 25-3 Ionic-strength-gradient anion-exchange separation of *isoforms* S_2 to S_6 of the protein transferrin bearing two to six sialic acid groups. The 5 × 50 mm column contained Pharmacia Mono-Q resin, which consists of 10-μm-diameter beads with quaternary ammonium anion-exchange sites. [From S. A. Rodríguez, E. B. Gonzalez, G. A. Llamas, M. Montes-Bayón, and A. Sanz-Medel, "Detection of Transferrin Isoforms in Human Serum: Comparison of UV and ICP–MS Detection after CZE and HPLC Separations," *Anal. Bioanal. Chem.* **2005**, *383*, 390.]

has a molecular mass near 81 000 and can bind two Fe^{3+} ions for transport through the blood. Attached to the protein are two carbohydrate chains, each terminated in one to four negatively charged sialic acid sugars. The most common form, designated S_4 in Figure 25-3, contains four sialic acid groups. Transferrin molecules containing different numbers of sialic acid sugars are called *isoforms* and are labeled S_2 through S_6 in Figure 25-3. Relative amounts of different isoforms vary with certain pathological disorders. In Figure 25-3, transferrin was detected by directing eluate into an inductively coupled plasma–mass spectrometer to detect ^{56}Fe (Section 20-6). Serum species that do not contain iron are invisible to this detector.

Applications of Ion Exchange

Ion exchange can be used to convert one salt into another. For example, we can prepare tetrapropylammonium hydroxide from a tetrapropylammonium salt of some other anion:

$$(CH_3CH_2CH_2)_4N^+I^- \xrightarrow[OH^- \text{ form}]{\text{anion exchanger}} (CH_3CH_2CH_2)_4N^+OH^-$$

Tetrapropylammonium iodide   Tetrapropylammonium hydroxide

Preconcentration: process of concentrating trace components of a sample prior to their analysis

Ion exchange is used for **preconcentration** of trace components of a solution to obtain enough for analysis. For example, the GEOTRACES program (Box 20-2) is measuring Al, Mn, Fe, Cu, Zn, and Cd at part per trillion levels in the presence of orders-of-magnitude higher concentrations of Na^+, Ca^{2+}, and Mg^{2+} in the ocean. Trace elements are preconcentrated and separated from alkali (Group 1) and alkaline earth (Group 2) elements by passage through the Chelate-PA1 ion-exchange resin shown in Figure 25-4. Ethylenediaminetriacetic acid groups in the resin bind analyte elements quantitatively at pH 6 and above. Alkali and alkaline earth elements are weakly bound at pH 6. The procedure is to pass 125 g of seawater (adjusted to pH 6 with ammonium acetate buffer) through 0.5 g of resin and wash away loosely bound metals with 40 mL of pH 6 buffer. Strongly retained metal ions are then quantitatively eluted in the reverse direction with 15 mL of 1 M HNO_3. Trace metals are concentrated by a factor of 8 from 125 g of seawater into ~15 g of eluate, and >99.9% of alkali and alkaline earth metals are removed. Eluate is analyzed by inductively coupled plasma–mass spectrometry.

FIGURE 25-4 Nobias Chelate-PA1 resin used to preconcentrate trace metals from seawater contains ethylenediaminetriacetic acid and iminodiacetic acid groups covalently attached to hydrophilic methacrylate polymer resin. Graphs show fraction of metal ions retained by resin as a function of pH. Mn^{2+} and Al^{3+} are well retained at pH 6. Na^+ and Ca^{2+} are poorly retained at pH 6. [Data from Y. Sohrin, S. Urushihara, S. Nakatsuka, T. Kono, E. Higo, T. Minami, K. Norisuye, and S. Umetani, "Multielemental Determination of GEOTRACES Key Trace Metals in Seawater by ICPMS after Preconcentration Using an Ethylenediaminetriacetic Acid Chelating Resin," *Anal. Chem.* **2008**, *80*, 6267.]

Ion exchange is used to purify water. **Deionized water** is prepared by passing water through an anion-exchange resin in its OH⁻ form and a cation-exchange resin in its H⁺ form. Suppose, for example, that $Cu(NO_3)_2$ is present in the solution. The cation-exchange resin binds Cu^{2+} and replaces it with $2H^+$. The anion-exchange resin binds NO_3^- and replaces it with OH⁻. The eluate is pure water:

Water softeners use ion exchange to remove Ca^{2+} and Mg^{2+} from "hard" water (Box 11-3).

$$\left.\begin{array}{l} Cu^{2+} \xrightarrow{\text{H}^+ \text{ ion exchange}} 2H^+ \\[1em] 2NO_3^- \xrightarrow{\text{OH}^- \text{ ion exchange}} 2OH^- \end{array}\right\} \longrightarrow \text{pure } H_2O$$

In many laboratory buildings, tap water is purified, first, by passage through activated carbon to adsorb organic material, and then by *reverse osmosis*. In the latter process, water is forced by pressure through a membrane containing pores through which few molecules larger than H_2O can pass. Ions do not pass through the pores because their hydrated radius is too large. Reverse osmosis removes 95–99% of ions, organic molecules, bacteria, and particles from water.

"Water polishing" equipment is used in many laboratories after reverse osmosis. The water is passed through activated carbon and then through several ion-exchange cartridges that convert ions into H⁺ and OH⁻. The resulting high-purity water has a resistivity (Chapter 14, note 37) of 180 000 ohm · m (18 Mohm · cm), with concentrations of individual ions below 1 ng/mL (1 ppb).[2]

Cation-exchange chromatography has been used to separate enantiomers (mirror image isomers) of cationic metal complexes by elution with one enantiomer of tartrate anion.[3] Tartrate has a different ion-pair formation constant with each cation enantiomer and therefore elutes one cation enantiomer from the column before the other.

In pharmaceuticals, ion-exchange resins are used for drug stabilization and as aids for tablet disintegration. Ion exchangers are also used for taste masking, for sustained-release products, as topical products for application to skin, and for ophthalmic or nasal delivery.[4]

Simultaneous Separation of Anions and Cations on One Column

Section 24-1 described *hydrophilic interaction chromatography* (HILIC) with a polar stationary phase and a mixed organic-aqueous mobile phase. The zwitterionic bonded stationary phase in Figure 24-14 has fixed positive and negative charges and is useful for the simultaneous separation of cations and anions. HILIC utilizes a mixed aqueous-organic eluent. The gradient in Figure 25-5 goes from 85 vol% acetonitrile to 10 vol% acetonitrile. Eluent strength *increases* as the fraction of acetonitrile *decreases*.

FIGURE 25-5 Separation of cations and anions on 5-μm zwitterion stationary phase ZIC-HILIC® (Figure 24-14), using an evaporative light-scattering detector operating at 55°C and 3 bar N_2. Column: 4.6 × 250 mm eluted at 1 mL/min. Solvent A: 85 vol% acetonitrile-15% aqueous buffer (100 mM ammonium acetate at pH 5). Solvent B: 10% acetonitrile-90% buffer. Gradient: 0–2 min, 100% A; 2–22 min, linear gradient to 100% B. [From D. S. Risely and B. W. Pack, "Simultaneous Determination of Positive and Negative Counterions Using a Hydrophilic Interaction Chromatography Method," *LCGC* 2006, *24*, 776.]

Ion	Concentrations observed ($\mu g/L = ppb$)	
	Minimum	Maximum
F^-	0.10	6.20
Cl^-	25	40 100
Br^-	0.8	49.4
NO_3^-	8.6	354
SO_4^{2-}	10.6	4 020
$H_2PO_4^-$	1.8	49.0
HCO_2^-	1.1	45.7
$CH_3CO_2^-$	5.0	182
$CH_3SO_3^-$	1.1	281
NH_4^+	2.4	46.5
Na^+	15	17 050
K^+	3.1	740
Mg^{2+}	2.7	1 450
Ca^{2+}	12.6	1 010

SOURCE: R. Udisti, S. Bellandi, and G. Piccardi, "Analysis of Snow from Antarctica," *Fresenius J. Anal. Chem.* **1994,** *349,* 289.

25-2 Ion Chromatography

Ion chromatography, a high-performance version of ion-exchange chromatography, is generally the method of choice for anion analysis.[5] In the semiconductor industry, it is used to monitor anions and cations at 0.1-ppb levels in deionized water.

Suppressed-Ion Anion and Cation Chromatography

In **suppressed-ion anion chromatography** (Figure 25-6a), a mixture of anions is separated by ion exchange and detected by electrical conductivity. The key feature of suppressed-ion chromatography is removal of unwanted electrolyte prior to conductivity measurement.

For the sake of illustration, consider a sample containing $NaNO_3$ and $CaSO_4$ injected into the *separator column*—an anion-exchange column in the hydroxide form—followed by elution with KOH. NO_3^- and SO_4^{2-} equilibrate with the resin and are slowly displaced by OH^- eluent. Na^+ and Ca^{2+} cations are not retained and simply wash through. Eventually, KNO_3 and K_2SO_4 are eluted from the separator column, as shown in the upper graph of Figure 25-6a. These species cannot be easily detected, however, because eluate contains a high concentration of KOH, whose high conductivity obscures that of analyte.

To remedy this problem, the solution next passes through a *suppressor,* in which cations are replaced by H^+. H^+ exchanges with K^+, in this example, through a cation-exchange membrane in the suppressor. H^+ diffuses from high concentration outside the membrane to low concentration inside the membrane. K^+ diffuses from high concentration inside to low concentration outside. K^+ is carried away outside the membrane, so its concentration is always low on the outside. The net result is that KOH eluent, which has high conductivity, is converted into H_2O, which has low conductivity. When analyte is present, HNO_3 or H_2SO_4 with high conductivity is produced and detected.

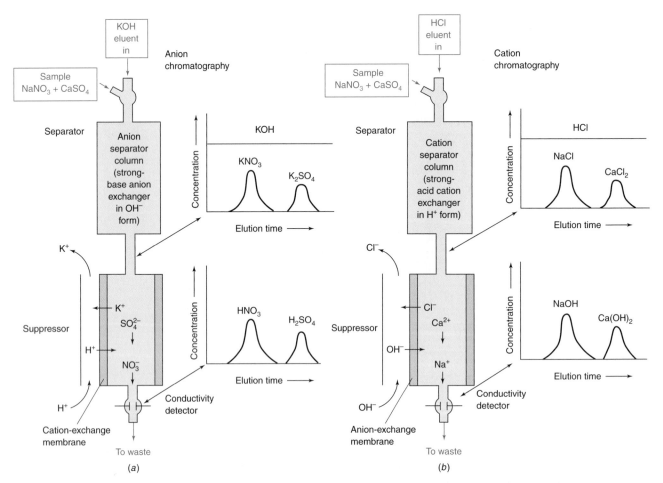

FIGURE 25-6 Schematic illustrations of (*a*) suppressed-ion anion chromatography and (*b*) suppressed-ion cation chromatography.

FIGURE 25-7 Ion chromatography of pond water. Upper chromatograms were obtained from mixtures of standards. Concentrations of ions in lower chromatograms from pond water are in units of μg/mL (ppm). Anion analysis was done with an IonPac AS14 column with 1.0 mM $NaHCO_3$/3.5 mM Na_2CO_3 eluent with ion suppression and conductivity detection. Cation analysis used an IonPac CS12A column with 11 mM H_2SO_4 eluent, ion suppression, and conductivity detection. [From K. Sinniah and K. Piers, "Ion Chromatography: Analysis of Ions in Pond Waters," *J. Chem. Ed.* 2001, *78*, 358.]

Suppressed-ion <u>cation</u> chromatography is conducted in a similar manner, but the suppressor replaces Cl^- from eluent with OH^- through an anion-exchange membrane. Figure 25-6b illustrates the separation of $NaNO_3$ and $CaSO_4$. With HCl eluent, NaCl and $CaCl_2$ emerge from the cation-exchange separator column, and NaOH and $Ca(OH)_2$ emerge from the suppressor column. HCl eluate is converted into H_2O in the suppressor.

Figure 25-7 illustrates a student experiment to measure ions in pond water. Eluent for the anion separation was $NaHCO_3$/Na_2CO_3 buffer. The product from eluent after passing through the suppressor is H_2CO_3, which has low conductivity.

In automated systems, H^+ and OH^- eluents and suppressors are generated by electrolysis of H_2O. Figure 25-8 shows a system that generates KOH for >1 000 hours of isocratic or gradient elution before it is necessary to refresh the reagents. Water in the reservoir of aqueous K_2HPO_4 reacts at the metal anode to produce H^+ and $O_2(g)$. H^+ combines with HPO_4^- to form $H_2PO_4^{2-}$. For each H^+ ion that is generated, one K^+ ion migrates through the cation-exchange barrier membrane, which transports K^+, but not anions, and allows negligible liquid to pass. The barrier membrane must withstand the high pressure of liquid in the KOH generation chamber destined for the chromatography column. For each H^+ generated at the

The separator column separates analytes, and the suppressor replaces ionic eluent with a nonionic species.

Benzene-1,4-diammonium cation is a stronger eluent that can be used instead of H^+ for suppressed-ion cation chromatography. After suppression, a neutral product is formed:

$$H_3\overset{+}{N}-\bigcirc-\overset{+}{N}H_3 \xrightarrow{OH^-} H_2N-\bigcirc-NH_2$$

Benzene-1,4-diammonium ion

FIGURE 25-8 Electrolytic KOH eluent generator for ion chromatography. [Adapted from Y. Liu, K. Srinivasan, C. Pohl, and N. Avdalovic, "Recent Developments in Electrolytic Devices for Ion Chromatography," *J. Biochem. Biophys. Methods* 2004, *60*, 205.]

FIGURE 25-9 Anion separation by ion chromatography with a gradient of electrolytically generated KOH and conductivity detection after suppression. Column: Dionex IonPac AS11; 4 mm diameter; flow = 2.0 mL/min, Eluent: 0.5 mM KOH for 2.5 min, 0.5 to 5.0 mM KOH from 2.5 to 6 min; 5.0 to 38.2 mM KOH from 6 to 18 min. Peaks: (1) quinate, (2) F^-, (3) acetate, (4) propanoate, (5) formate, (6) methylsulfonate, (7) pyruvate, (8) valerate, (9) chloroacetate, (10) BrO_3^-, (11) Cl^-, (12) NO_2^-, (13) trifluoroacetate, (14) Br^-, (15) NO_3^-, (16) ClO_3^-, (17) selenite, (18) CO_3^{2-}, (19) malonate, (20) maleate, (21) SO_4^{2-}, (22) $C_2O_4^{2-}$, (23) tungstate, (24) phthalate, (25) phosphate, (26) chromate, (27) citrate, (28) tricarballylate, (29) isocitrate, (30) *cis*-aconitate, (31) *trans*-aconitate. [Courtesy Dionex Corp., Sunnyvale CA.]

anode, one K^+ flows through the cation-exchange barrier and one OH^- is generated at the cathode. Liquid exiting the KOH generation chamber contains KOH and H_2. The stream passes through an anion trap that removes traces of anions such as carbonate and degradation products from the ion-exchange resin. The trap is continuously replenished with electrolytically generated OH^-, which is not shown in the diagram. After the trap, liquid flows through a polymeric capillary that is permeable to H_2, which diffuses into an external liquid stream and is removed. The concentration of KOH produced by the apparatus in Figure 25-8 is governed by the liquid flow rate and the electric current. With computer control of the power supply, a precise gradient can be generated.

In the past, KOH eluent was usually contaminated with CO_3^{2-}. When CO_3^{2-} passes through the suppressor after the ion chromatography column, it is converted into H_2CO_3, which has some electrical conductivity that interferes with detection of analytes. In gradient elution with increasing KOH, the concentration of H_2CO_3 also increases, and so does the background conductivity. The feedstock for the electrolytic generator is pure H_2O, and the product is aqueous KOH containing very little CO_3^{2-}. Figure 25-9 shows a separation of 31 anions with a hydroxide gradient.

Suppressors in Figure 25-6 have been replaced by electrolytic units such as that in Figure 25-10, which generates H^+ or OH^- to neutralize eluate and requires only H_2O as feedstock. With electrolytic eluent generation and electrolytic suppression, ion chromatography has been

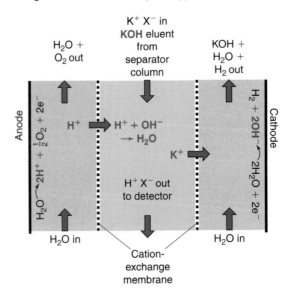

FIGURE 25-10 Electrolytic suppression for anion chromatography replaces KOH eluent with H_2O.

simplified and highly automated. Readily available software can be used to simulate and optimize ion chromatographic separations.[6]

Ion Chromatography Without Suppression

If the ion-exchange capacity of the separator column is sufficiently low and if dilute eluent is used, ion suppression is unnecessary. Also, anions of weak acids, such as borate, silicate, sulfide, and cyanide, cannot be determined with ion suppression, because these anions are converted into very weakly conductive products (such as H_2S).

For *nonsuppressed anion chromatography*, we use a resin with an exchange capacity near 5 μmol/g, with 10^{-4} M Na^+ or K^+ salts of benzoic, *p*-hydroxybenzoic, or phthalic acid as eluent. These eluents give a low background conductivity, and analyte anions are detected by a small *change* in conductivity as they emerge from the column. A judicious choice of pH produces an average eluent charge between 0 and −2 and provides control of eluent strength. Even dilute carboxylic acids, which are slightly ionized, are suitable eluents for some separations. *Nonsuppressed cation chromatography* is conducted with dilute HNO_3 eluent for monovalent ions and with ethylenediammonium salts ($^+H_2NCH_2CH_2NH_2^+$) for divalent ions.

Detectors

Conductivity detectors respond to all ions. In suppressed-ion chromatography, it is easy to measure analyte because eluent conductivity is reduced to near 0 by suppression. Suppression also allows us to use eluent concentration gradients.

In nonsuppressed anion chromatography, the conductivity of the analyte anion is higher than that of the eluent, so conductivity increases when analyte emerges from the column. Detection limits are normally in the mid-ppb to low-ppm range but can be lowered by a factor of 10 by using carboxylic acid eluents instead of carboxylate salts.

Benzoate or phthalate eluents enable sensitive (<1 ppm) **indirect detection** of anions. In Figure 25-11, eluate has strong, constant ultraviolet absorption. As each analyte emerges, nonabsorbing analyte anion replaces an equivalent amount of absorbing eluent anion. Absorbance therefore *decreases* when analyte appears. For cation chromatography, $CuSO_4$ is a suitable ultraviolet-absorbing eluent.

Ion-Pair Chromatography

Ion-pair chromatography uses a reversed-phase HPLC column instead of an ion-exchange column to separate polar or ionic compounds.[8] To separate cations (for example, protonated organic bases), an anionic *surfactant* (Box 25-1) such as $n\text{-}C_8H_{17}SO_3^-$ is added to the mobile

FIGURE 25-11 Indirect spectrophotometric detection of transparent ions. The column was eluted with 1 mM sodium phthalate plus 1 mM borate buffer, pH 10. The principle of indirect detection is illustrated in Figure 25-35. [Reproduced from H. Small, "Indirect Photometric Chromatography," *Anal. Chem.* **1982**, *54*, 462.]

BOX 25-1 Surfactants and Micelles

A **surfactant** is a molecule that accumulates at the interface between two phases and modifies the surface tension. (*Surface tension* is the energy per unit area needed to form a surface or interface.) Molecules with long hydrophobic tails and ionic headgroups are a common class of surfactants for aqueous solution:

Cetyltrimethylammonium bromide
$C_{16}H_{33}N(CH_3)_3^+Br^-$

A **micelle** is an aggregate of surfactants. In water, the hydrophobic tails form clusters that are, in effect, little oil drops insulated from the aqueous phase by the ionic headgroups. At low concentration, surfactant molecules do not form micelles. When their concentration exceeds the *critical micelle concentration*, spontaneous aggregation into micelles occurs.[7] Isolated surfactant molecules exist in equilibrium with micelles. Nonpolar organic solutes are soluble inside micelles. Cetyltrimethylammonium bromide forms micelles containing

~61 molecules (mass ≈ 22 000 Da) in water at 25°C at a critical micelle concentration of 0.9 mM.

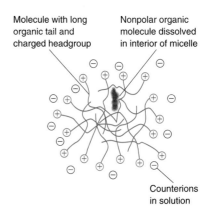

Structure of a micelle formed when ionic molecules with long, nonpolar tails aggregate in aqueous solution. The interior of the micelle resembles a nonpolar organic solvent, whereas the exterior charged groups interact strongly with water. [From F. M. Menger, R. Zana, and B. Lindman, "Portraying the Structure of Micelles," *J. Chem. Ed.* **1998**, *75*, 115.]

FIGURE 25-12 Ion-pair chromatography. The surfactant sodium octanesulfonate added to the mobile phase binds to the nonpolar stationary phase. Negative sulfonate groups protruding from the stationary phase then act as ion-exchange sites for analyte cations such as protonated organic bases, BH^+.

phase. The surfactant lodges in the stationary phase, effectively transforming the stationary phase into an ion exchanger (Figure 25-12). Analyte cations are attracted to the surfactant anions.[9] The retention mechanism is a mixture of reversed-phase and ion-exchange interactions. To separate anions, tetrabutylammonium salts can be added to the mobile phase as the ion-pair reagent (Figure 25-13).

Ion-pair chromatography is more complex than reversed-phase chromatography because equilibration of the surfactant with the stationary phase is slow, the separation is more sensitive to variations in temperature and pH, and the concentration of surfactant affects the separation. Methanol is the organic solvent of choice because ionic surfactants are more soluble in methanol/water mixtures than in acetonitrile/water mixtures. Strategies for method development analogous to the scheme in Figure 24-27 vary the pH and surfactant concentration with fixed methanol concentration and temperature.[10] Because of the slow equilibration of surfactant with the stationary phase, gradient elution is not recommended in ion-pair chromatography. Many ion-pair reagents have significant ultraviolet absorption, which makes ultraviolet detection of analytes problematic. Reversed-phase stationary phases with a *polar embedded group* (page 602) are a possible alternative to ion-pair chromatography for polar compounds.

FIGURE 25-13 Separation of carbohydrates by ion-pair chromatography. Carbohydrates were *derivatized* by covalently attaching p-aminobenzoate ($H_2N-C_6H_4-CO_2^-$), which changes carbohydrates into fluorescent anions. The anions were separated on a 0.30×25 cm column of AQUA® C_{18}-silica, using tetrabutylammonium cation as the ion-pair reagent. Eluent was a linear 60-min gradient starting with 20 mM aqueous (n-$C_4H_9)_4N^+HSO_4^-$, pH 2.0 (solvent A) and ending with 50:50 A:methanol. The method was used to measure carbohydrates at 10- to 100-ng/mL levels in water leaching from landfills.
[From A. Meyer, C. Raba, and K. Fischer, "Ion-Pair HPLC Determination of Sugars, Amino Sugars, and Uronic Acids," *Anal. Chem.* **2001**, *73*, 2377.]

25-3 Molecular Exclusion Chromatography

In **molecular exclusion chromatography** (also called *size exclusion* or **gel filtration** or *gel permeation chromatography*), molecules are separated according to size.[11] Small molecules penetrate the pores in the stationary phase, but large molecules do not (Figure 22-6). Because small molecules must pass through an effectively larger volume, *large molecules are eluted first* (Figure 25-14). This technique is widely used in biochemistry to purify macromolecules.

Salts of low molecular mass (or any small molecule) can be removed from solutions of large molecules by gel filtration because the large molecules are eluted first. This technique, called *desalting*, is useful for changing the buffer composition of a macromolecule solution.

Large molecules pass through the column faster than small molecules do.

Gel filtration: usually refers to a hydrophilic stationary phase and aqueous eluent

Gel permeation: usually refers to hydrophobic stationary phase and organic eluent

The Elution Equation

The *total* volume of mobile phase in a chromatography column is V_m, which includes solvent inside and outside the gel particles. The volume of mobile phase *outside* the gel particles is called the **void volume**, V_o. The volume of solvent *inside* the gel is therefore $V_m - V_o$. The quantity K_{av} (read "K average") is defined as

$$K_{av} = \frac{V_r - V_o}{V_m - V_o} \qquad (25\text{-}6)$$

where V_r is the retention volume for a solute. For a large molecule that does not penetrate the gel, $V_r = V_o$, and $K_{av} = 0$. For a small molecule that freely penetrates the gel, $V_r = V_m$, and $K_{av} = 1$. Molecules of intermediate size penetrate some gel pores, but not others, so K_{av} is between 0 and 1. Ideally, gel penetration is the only mechanism by which molecules are retained in this type of chromatography. In fact, there is always some adsorption, so K_{av} can be greater than 1.

Void volume is measured by passing a large, inert molecule through the column.[12] Its elution volume is defined as V_o. Blue Dextran 2000, a blue dye of molecular mass 2×10^6, is commonly used for this purpose. The volume V_m can be calculated from the measured column bed volume per gram of dry gel. For example, 1 g of dry Sephadex G-100 occupies 15 to 20 mL, when swollen with aqueous solution. The solid phase of the swollen gel occupies only $\sim$1 mL, so V_m is 14 to 19 mL, or 93–95% of the total column volume. Equal masses of different solid phases produce widely varying volumes when swollen with solvent.

In pure molecular exclusion, all molecules are eluted between $K_{av} = 0$ and $K_{av} = 1$.

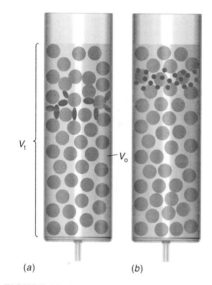

(a)   (b)

FIGURE 25-14 (*a*) Large molecules cannot penetrate the pores of the stationary phase. They are eluted by a volume of solvent equal to the volume of mobile phase. (*b*) Small molecules, which can be found inside or outside the gel, require a larger volume for elution. V_t is the total column volume occupied by gel plus solvent. V_o is the volume of solvent outside the gel particles. V_m is the total volume of solvent inside and outside the gel particles.

Stationary Phase[13]

Gels for open-column, preparative-scale molecular exclusion include Sephadex (Table 25-4), whose structure was given in Figure 25-2, and Bio-Gel P, which is a polyacrylamide cross-linked by N,N'-methylenebisacrylamide (Figure 25-15). The smallest pore sizes in highly cross-linked gels exclude molecules with a molecular mass $\geq$700, whereas the largest pore sizes exclude molecules with molecular mass $\geq 10^8$. The finer the particle size of the gel, the greater the resolution and the slower the flow rate of the column. Hydrophilic HPLC packings for molecular exclusion are made of poly(vinyl alcohol), polyacrylamide, and sulfonated polystyrene. Silica (Table 25-4) with controlled pore size provides 10 000–16 000 plates per meter. The silica is coated with a hydrophilic phase to minimize solute adsorption. A hydroxylated polyether resin with a well-defined pore size can be used over the pH range 2–12, whereas silica phases generally cannot be used above pH 8. Particles with different pore sizes can be mixed to give a wider molecular size separation range.

TABLE 25-4 Representative molecular exclusion media

Gel filtration in open columns		TSK SW silica for HPLC		
Name	Fractionation range for globular proteins (Da)	Name	Pore size (nm)	Fractionation range for globular proteins (Da)
Sephadex G-10	to 700	G2000SW	13	500–60 000
Sephadex G-25	1 000–5 000	G3000SW	24	1 000–300 000
Sephadex G-50	1 500–30 000	G4000SW	45	5 000–1 000 000
Sephadex G-75	3 000–80 000	G5000SW	100	>1 500 000
Sephadex G-100	4 000–150 000			
Sephadex G-200	5 000–600 000			

NOTE: *Sephadex is manufactured by GE Amersham Biosciences. TSK SW silica is manufactured by Tosoh Corp.*

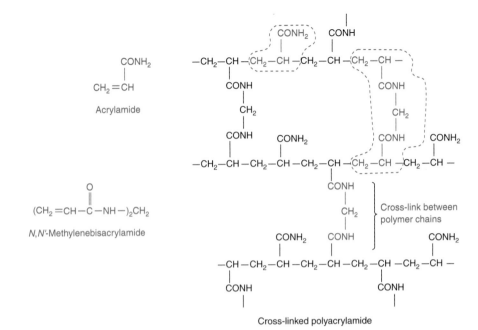

FIGURE 25-15 Structure of polyacrylamide.

For HPLC of hydrophobic polymers, cross-linked polystyrene spheres are available, with pore sizes ranging from 5 nm up to hundreds of nanometers. Particles with a 5-μm diameter yield up to 80 000 plates per meter of column length.

Molecular Mass Determination

Gel filtration is used mainly to separate molecules of significantly different molecular sizes (Figure 25-16). For each stationary phase, we construct a calibration curve, which is a graph of log(molecular mass) versus elution volume (Figure 25-17). We estimate the molecular mass of an unknown by comparing its elution volume with those of standards. We must exercise caution in interpreting results, however, because molecules with the same molecular mass but different shapes exhibit different elution characteristics. For proteins, it is important to use an ionic strength high enough (>0.05 M) to eliminate electrostatic adsorption of solute by occasional charged sites on the gel.

Nanoparticles can be separated by molecular exclusion chromatography just as proteins are separated. Figure 25-18 shows the relation between measured size and retention time of

1 Glutamate dehydrogenase (290 000)
2 Lactate dehydrogenase (140 000)
3 Enolase kinase (67 000)
4 Adenylate kinase (32 000)
5 Cytochrome *c* (12 400)

FIGURE 25-16 Separation of proteins by molecular exclusion chromatography with TSK 3000SW column. [Courtesy Varian Associates, Palo Alto, CA.]

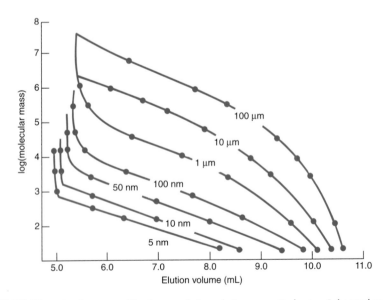

FIGURE 25-17 Molecular mass calibration graph for polystyrene on Beckman μSpherogel molecular exclusion column (0.77 × 30 cm). Resin pore size labeled on the lines ranges from 5 nm to 100 μm. [Courtesy Anspec Co., Ann Arbor, MI.]

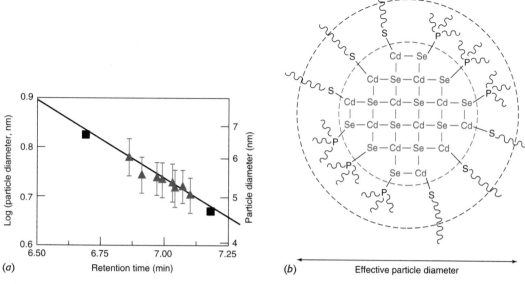

FIGURE 25-18 Larger CdSe quantum dots are eluted before smaller quantum dots by 0.1 M trioctylphosphine in toluene at 1.0 mL/min in size exclusion chromatography on a 7.5 × 300 mm cross-linked polystyrene column of Polymer Labs PLgel 5 μm with 100-nm pore size. Triangles are CdSe, and squares are polystyrene calibration standards. The size of the CdSe core was measured with a transmission electron microscope, and the length of 1-dodecanethiol endcaps (0.123 nm) was added to the radius. [Data from K. M. Krueger, A. M. Al-Somali, J. C. Falkner, and V. L. Colvin, "Characterization of Nanocrystalline CdSe by Size Exclusion Chromatography," *Anal. Chem.* **2005,** *77,* 3511.]

CdSe *quantum dots.* These are particles containing ~2 000 CdSe units in a dense, crystalline core capped by alkyl thiol (RS) groups on Cd and trialkylphosphine (R_3P) groups on Se. Quantum dots are useful because the wavelength of their visible emission depends on their size. Size is controlled during synthesis of the quantum dot by reaction time or other conditions. Quantum dots with different sizes can be used as spectroscopic labels in biological experiments.[14]

25-4 Affinity Chromatography

Affinity chromatography is used to isolate a single compound from a complex mixture. The technique is based on specific binding of that one compound to the stationary phase (Figure 22-6). When sample is passed through the column, only one solute is bound. After everything else has washed through, the one adhering solute is eluted by changing a condition such as pH or ionic strength to weaken its binding. Affinity chromatography is especially applicable in biochemistry and is based on specific interactions between enzymes and substrates, antibodies and antigens, or receptors and hormones.

Figure 25-19 shows the isolation of the protein immunoglobulin G (IgG) by affinity chromatography on a column containing covalently bound *protein A*. Protein A binds to one specific region of IgG at pH ≳ 7.2. When a crude mixture containing IgG and other proteins was passed through the column at pH 7.6, everything except IgG was eluted within 0.3 min. At 1 min, the eluent pH was lowered to 2.6 and IgG was cleanly eluted at 1.3 min.

Optical isomers of a drug can have completely different therapeutic effects. Affinity chromatography can be used to isolate individual optical isomers for drug evaluation.[15] The

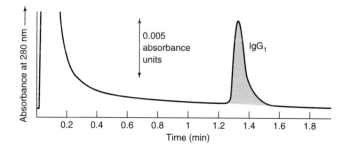

FIGURE 25-19 Purification of monoclonal antibody IgG by affinity chromatography on a 0.46 × 5 cm column containing protein A covalently attached to polymer support. Other proteins in the sample are eluted from 0 to 0.3 min at pH 7.6. When eluent pH is lowered to 2.6, IgG is freed from protein A and emerges from the column. [From B. J. Compton and L. Kreilgaard, "Chromatographic Analysis of Therapeutic Proteins," *Anal. Chem.* **1994,** *66,* 1175A.]

A **molecularly imprinted polymer** is synthesized in the presence of a template molecule to which components of the polymer have some affinity. When the template is removed, the polymer is "imprinted" with the shape of the template and with complementary functional groups that can bind the template. The template can be the analyte of interest, but it is better to use a structurally related molecule so that residual template in the polymer does not give false positive results when the polymer is employed. Imprinted polymer can be used as a stationary phase in affinity chromatography or a recognition element in a chemical sensor.

A molecularly imprinted polymer can be used to collect and preconcentrate penicillin antibiotics from river water for analysis. The figure shows a conceptual structure of a polymer pocket formed when monomers are polymerized with penicillin G as a template. After removing penicillin G by washing with methanol and HCl, the pocket retains its shape and arrangement of functional groups to bind similar molecules, such as penicillin V, which is shown in color.

When river water spiked with 30-ppb levels of eight different penicillin variants was passed through a column containing imprinted polymer, 90–99% of six of the penicillins were retained by the column. Two penicillin variants did not bind as well. Penicillins retained by the column were eluted by a small volume of 0.05 M tetrabutylammonium hydrogen sulfate in methanol and analyzed by HPLC.

Notional structure of pocket in a polymer imprinted to bind penicillin derivatives. [From J. L. Urraca, M. C. Moreno-Bondi, A. J. Hall, and B. Sellergren, "Direct Extraction of Penicillin G and Derivatives from Aqueous Samples Using a Stoichiometrically Imprinted Polymer," *Anal. Chem.* **2007**, *79*, 695.]

drug candidate in the margin has two chiral carbon atoms indicated by colored dots. With two possible geometries at each site, there are four stereoisomers. The mixture of isomers was covalently attached to protein and injected into mice to raise a mixture of antibodies to all stereoisomers. Antibodies are produced by B cells in the spleen. One B cell produces only one kind of antibody. By isolating individual B cells, it is possible to isolate the gene for the antibody to each of the stereoisomers. The gene can be transplanted into *E. coli* cells for mass production of a single kind of antibody, called a *monoclonal antibody*. When the mixture of stereoisomers is passed through a column to which just one kind of antibody is attached, only one of the four stereoisomers is retained. By lowering the pH, the retained isomer is eluted in pure form.

Box 25-2 shows how *molecularly imprinted polymers* can be used as affinity media. *Aptamers* (Section 18-5) are another class of chromatographically useful compounds with high affinity for a selected target.[17]

25-5 Hydrophobic Interaction Chromatography

Hydrophobic substances repel water, and their surfaces are not wetted by water. A protein can have *hydrophilic* regions that make it soluble in water and *hydrophobic* regions capable of interacting with a hydrophobic stationary chromatography phase. High concentrations of ammonium sulfate cause proteins to *salt out* of solution. Ammonium, sodium, and potassium salts of phosphate and sulfate decrease the solubility of proteins in water. Thiocyanate, iodide, and perchlorate salts have the opposite effect of dissolving proteins.

Hydrophobic interaction chromatography is used principally for protein purification.[19] A common stationary phase shown in Figure 25-20 has hydrophobic phenyl or alkyl groups attached to agarose gel (a polysaccharide) whose pore size is large enough for proteins to enter. When protein solution with a high concentration (such as 1 M) of ammonium sulfate is applied to the column, the salt induces the protein to bind to the hydrophobic surface of the stationary phase. Then a gradient of *decreasing salt concentration* is applied to *increase* the solubility of proteins in water and elute them from the column.

25-6 Principles of Capillary Electrophoresis[20,21]

Late in 2007, more than 200 people receiving the anticoagulant *heparin* suffered acute, allergic reactions and died.[22] Heparin is a complex mixture of sulfate-substituted polysaccharides that have molecular masses of 2 to 50 kDa and are isolated from pig intestines. As soon as the

Water does not *wet* a hydrophobic surface made of carbon nanotubes, so a drop remains almost spherical. The drop would flatten out on a hydrophilic surface, such as glass. [Courtesy Karen Gleason, Massachusetts Institute of Technology.[18]]

CHAPTER 25 Chromatographic Methods and Capillary Electrophoresis

problem was recognized in January 2008, U.S. distributors recalled heparin products and the U.S. Food and Drug Administration launched an investigation. Heparin is administered thousands of times every day to manage life-threatening conditions, so an immediate understanding and solution to the problem were required.

When exposed to the enzyme heparinase, heparin is cleaved into disaccharide units. Tainted heparin contained 20 to 50 wt% of macromolecular components that did not react with heparinase. *Capillary electrophoresis* proved to be the tool of choice to observe two contaminants (Figure 25-21).[23] One was dermatan sulfate, which was not known to cause allergic reactions. The other was identified by nuclear magnetic resonance as oversulfated chondroitin sulfate. An animal study verified that oversulfated chondroitin sulfate caused the allergic reaction. By March 2008, deaths from contaminated heparin had ceased and emergency regulations were issued to incorporate capillary electrophoresis and nuclear magnetic resonance into required testing of heparin imported into the U.S. Contaminated heparin had been prepared in China. Oversulfated chondroitin sulfate might have been added because it has anticoagulant activity and costs less than heparin.

Electrophoresis is the migration of ions in solution under the influence of an electric field. The technique was pioneered in the 1930s by the Swedish biophysical chemist A. Tiselius, who received the Nobel Prize in 1948 for his work on electrophoresis and "discoveries concerning the complex nature of serum proteins."

In **capillary electrophoresis,** shown in Figure 25-22, components of a solution are separated by applying a voltage of ~30 kV from end to end of a fused-silica (SiO_2) capillary tube that is 50 cm long and has an inner diameter of 25–75 μm. Different ions have different *mobilities* and migrate through the capillary at different speeds.[25] Modifications of this experiment described later allow neutral molecules, as well as ions, to be separated. Electrophoresis can separate whole cells for medical diagnosis and detection of food contamination.[26] Electrophoresis can analyze single cells, nuclei, vesicles, or mitochondria.[27] Single-cell enzyme assays have detection limits at the zeptomol (10^{-21} mol) level.[28]

Capillary electrophoresis provides high resolution. When we conduct chromatography in a packed column, peaks are broadened by three mechanisms in the van Deemter equation (22-33): multiple flow paths, longitudinal diffusion, and finite rate of mass transfer. An open tubular column eliminates multiple paths and thereby reduces plate height and improves resolution. Capillary electrophoresis reduces plate height further by knocking out the mass transfer term that comes from the finite time needed for solute to equilibrate between the mobile and stationary phases. In capillary electrophoresis, *there*

FIGURE 25-20 Stationary phase for hydrophobic interaction chromatography has ~10–20% as many bonded phenyl or alkyl groups per unit volume as a reversed-phase stationary phase.

Cations are attracted to the negative terminal (the cathode).

Anions are attracted to the positive terminal (the anode).

Electric potential difference = 30 kV

$$Electric\ field = \frac{30\ kV}{0.50\ m} = 60\ \frac{kV}{m}$$

FIGURE 25-21 Electropherogram of heparin (30 mg/mL) spiked with oversulfated chondroitin sulfate and dermatan sulfate. Tainted heparin had ~200 times more oversulfated chondroitin sulfate than shown here. Conditions: −16 kV, 20°C, 25 μm × 30 cm capillary, detector at 21.5 cm. Background buffer was made by adding 0.60 M H_3PO_4 to 0.60 M Li_3PO_4 to reach pH 2.8. [Courtesy Robert Weinberger, CE Technologies and Todd Wielgos, Baxter Healthcare. For details, see T. Wielgos, K. Havel, N. Ivanova, and R. Weinberger, "Determination of Impurities in Heparin by Capillary Electrophoresis using High Molarity Phosphate Buffers," *J. Pharma. Biomed. Anal.* **2009**, *49*, 319.]

Electrophoresis in glass capillaries was first described by J. W. Jorgenson in 1981.[24]

FIGURE 25-23 Comparison of peak widths for benzyl alcohol ($C_6H_5CH_2OH$) in capillary electrophoresis and HPLC. [From S. Fazio, R. Vivilecchia, L. Lesueur, and J. Sheridan, *Am. Biotech. Lab.*, January 1990, p. 10.]

We encountered *mobility* earlier in connection with junction potentials (Table 14-1).

is no stationary phase. The only fundamental source of broadening under ideal conditions is longitudinal diffusion:

$$ H \;=\; \cancel{A} \;+\; \frac{B}{u_x} \;+\; \cancel{C u_x} $$

(25-7)

Multiple path term eliminated by open tubular column

Mass transfer term eliminated because there is no stationary phase

(Other sources of broadening in real systems are mentioned later.) Capillary electrophoresis routinely produces 50 000–500 000 theoretical plates (Figure 25-23), a performance that is an order-of-magnitude better than that of chromatography.

Electrophoresis

When an ion with charge q (coulombs) is placed in an electric field E (V/m), the force on the ion is qE (newtons). In solution, the retarding frictional force is fu_{ep}, where u_{ep} is the velocity of the ion and f is the *friction coefficient*. The subscript "ep" stands for "electrophoresis." The ion quickly reaches a steady speed when the accelerating force equals the frictional force:

$$ qE \;=\; fu_{ep} $$

Accelerating force = Frictional force

Electrophoretic mobility:

$$ u_{ep} = \frac{q}{f}E \equiv \mu_{ep}E $$

(25-8)

Electrophoretic mobility

*Electrophoretic **mobility** (μ_{ep})* is the constant of proportionality between the speed of the ion and electric field strength. Mobility is proportional to the charge of the ion and inversely proportional to the friction coefficient. For molecules of similar size, mobility increases with charge:

$\mu_{ep} = -2.54 \times 10^{-8}\ \dfrac{m^2}{V \cdot s}$ $\mu_{ep} = -4.69 \times 10^{-8}\ \dfrac{m^2}{V \cdot s}$ $\mu_{ep} = -5.95 \times 10^{-8}\ \dfrac{m^2}{V \cdot s}$

(solvent is H_2O at 25°C)

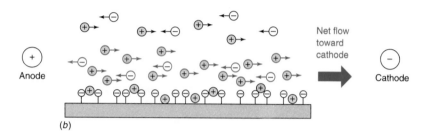

FIGURE 25-24 (a) Electric double layer created by negatively charged silica surface and nearby cations. (b) Predominance of cations in diffuse part of the double layer produces net electroosmotic flow toward the cathode when an external field is applied.

For a spherical particle of radius r moving through a fluid of *viscosity* η, the friction coefficient, f, is

Stokes equation: $$f = 6\pi\eta r \qquad (25\text{-}9)$$

Mobility is q/f, so the greater the radius, the lower the mobility. Most molecules are not spherical, but Equation 25-9 defines an effective *hydrodynamic radius* of a molecule, as if it were a sphere, based on its observed mobility.

Viscosity measures resistance to flow in a fluid. The units are kg m^{-1} s^{-1}. Relative to water, maple syrup is very viscous and hexane has low viscosity.

Electroosmosis

The inside wall of a fused-silica capillary is covered with silanol (Si—OH) groups, which have a negative charge (Si—O$^-$) above pH 3. Figure 25-24a shows the *electric double layer* (Box 16-3) at the capillary surface. The double layer consists of fixed negative charges on the wall and excess cations near the wall. A tightly adsorbed, immobile layer of cations partially neutralizes the negative charge. The remaining negative charge is neutralized by mobile cations in the *diffuse part of the double layer* in solution near the wall. The thickness of the diffuse part of the double layer ranges from ~10 nm when the ionic strength is 1 mM to ~0.3 nm when the ionic strength is 1 M.

In an electric field, cations are attracted to the cathode and anions are attracted to the anode (Figure 25-24b). Excess cations in the diffuse part of the double layer impart net momentum toward the cathode. This pumping action, called **electroosmosis** (or *electroendosmosis*), is driven by cations within ~10 nm of the walls and creates uniform pluglike *electroosmotic flow* of the entire solution toward the cathode (Figure 25-25a). This process is in sharp contrast with *hydrodynamic flow*, which is driven by a pressure difference. In hydrodynamic flow, the velocity profile through a cross section of the fluid is parabolic: It is fastest at the center and slows to 0 at the walls (Figure 25-25b and Color Plate 30).

Ions in the diffuse part of the double layer adjacent to the capillary wall are the "pump" that drives electroosmotic flow.

FIGURE 25-25 (a) Electroosmosis gives uniform flow over more than 99.9% of the cross section of the capillary. The speed decreases immediately adjacent to the capillary wall. (b) Parabolic velocity profile of hydrodynamic flow (also called *laminar flow*), with the highest velocity at the center of the tube and zero velocity at the walls. Experimentally observed velocity profiles are shown in Color Plate 30.

Electroosmotic velocity is measured by adding to the sample a *neutral* molecule to which the detector responds.

$$\text{Electroosmotic velocity} = \frac{\text{distance from injector to detector}}{\text{migration time of neutral molecule}}$$

The constant of proportionality between electroosmotic velocity, u_{eo}, and applied field is called *electroosmotic mobility*, μ_{eo}.

Electroosmotic mobility:
$$u_{eo} = \mu_{eo}E \tag{25-10}$$

$\uparrow$
Electroosmotic mobility
(units = m^2/[V · s])

Electroosmotic mobility is proportional to the surface charge density on the silica and inversely proportional to the square root of ionic strength. Electroosmosis decreases at low pH (Si—O$^-$ → Si—OH decreases surface charge density) and high ionic strength. At pH 9 in 20 mM borate buffer, electroosmotic flow is ~2 mm/s. At pH 3, flow is reduced by an order of magnitude.

Uniform electroosmotic flow contributes to the high resolution of capillary electrophoresis. Any effect that decreases uniformity creates band broadening and decreases resolution. The flow of ions in the capillary generates heat (called *Joule heating*) at a rate of I^2R joules per second, where I is current (A) and R is the resistance of the solution (ohms) (Section 13-1). Most of the capillary in Figure 25-22 is in a water-thermostated compartment necessary for temperature control inside the capillary.[29] Typically, the centerline of the capillary channel is 0.02 to 0.3 K hotter than the edge of the channel. Lower viscosity in the warmer region disturbs the flat electroosmotic profile of the fluid. Joule heating is not a serious problem in a capillary tube with a diameter of 50 μm, but the temperature gradient would be prohibitive if the diameter were $\geq$1 mm. Some instruments cool the capillary to reduce the electrical conductivity of solution inside the capillary and prevent runaway Joule heating.

The capillary must be thin enough to dissipate heat rapidly. Temperature gradients disturb the flow and reduce resolution.

Mobility

The *apparent* (or observed) *mobility*, μ_{app}, of an ion is the sum of the electrophoretic mobility of the ion plus the electroosmotic mobility of the solution.

Apparent mobility:
$$\mu_{app} = \mu_{ep} + \mu_{eo} \tag{25-11}$$

For an analyte *cation* moving in the same direction as the electroosmotic flow, μ_{ep} and μ_{eo} have the same sign, so μ_{app} is greater than μ_{ep}. Electrophoresis transports *anions* in the opposite direction from electroosmosis (Figure 25-24b), so for anions the two terms in Equation 25-11 have opposite signs. At neutral or high pH, brisk electroosmosis transports anions to the *cathode* because electroosmosis is usually faster than electrophoresis. At low pH, electroosmosis is weak and anions may never reach the detector. If you want to separate anions at low pH, you can reverse the polarity to make the sample end negative and the detector end positive.

The apparent mobility, μ_{app}, of a particular species is the net speed, u_{net}, of the species divided by the electric field, E:

$$\text{Speed} = \frac{\text{distance to detector}}{\text{migration time}} = \frac{L_d}{t}$$

$$\text{Electric field} = \frac{\text{applied voltage}}{\text{capillary length}} = \frac{V}{L_t}$$

Apparent mobility:
$$\mu_{app} = \frac{u_{net}}{E} = \frac{L_d/t}{V/L_t} \tag{25-12}$$

where L_d is the length of column from injection to the detector, L_t is the total length of the column from end to end, V is the voltage applied between the two ends, and t is the time required for solute to migrate from the injection end to the detector. Electroosmotic flow is measured by adding an ultraviolet-absorbing neutral solute to the sample and measuring its *migration time*, $t_{neutral}$, to the detector.

For quantitative analysis by electrophoresis, *normalized peak areas* are required. The normalized peak area is the measured peak area divided by the migration time. In chromatography, each analyte passes through the detector at the same rate, so peak area is proportional to the quantity of analyte. In electrophoresis, analytes with different apparent mobilities pass through the detector at different rates. The higher the apparent mobility, the shorter the migration time and the less time the analyte spends in the detector. To correct for time spent in the detector, divide the peak area for each analyte by its migration time.

Electroosmotic mobility is the speed of the neutral species, $u_{neutral}$, divided by the electric field:

For quantitative analysis, use

$$\frac{\text{Peak area}}{\text{Migration time}}$$

Electroosmotic mobility:
$$\mu_{eo} = \frac{u_{neutral}}{E} = \frac{L_d/t_{neutral}}{V/L_t} \tag{25-13}$$

The *electrophoretic mobility* of an analyte is the difference $\mu_{app} - \mu_{eo}$.

For maximum precision, mobilities are measured relative to an internal standard. Absolute variation from run to run should not affect relative mobilities, unless there are time-dependent (nonequilibrium) interactions of the solute with the wall.

For molecules of similar size, the magnitude of the electrophoretic mobility increases with charge. A protein "charge ladder" is a synthetic mixture made from a single protein with many different charges.[30] For example, we can obtain such a mixture by acetylating variable numbers of lysine amino acid side chains (Table 9-1) to reduce their charge from $+1$ ($R{-}NH_3^+$) to 0 ($R{-}NHC({=}O)CH_3$).

Lysine $-NH_2$ groups have $pK_a \approx 10.3$. At pH 8.3, 99% of these groups are protonated ($-NH_3^+$). Acetylation gives a mixture with every possible number of modified amino groups from 0 to the total number of lysine residues. This mixture gives the electropherogram in Figure 25-26 with a series of nearly evenly spaced peaks. Each molecule has approximately the same size and shape (and therefore nearly the same friction coefficient) but a different charge.

Problem 25-41 shows how to find the charge of unmodified protein from the charge ladder.

FIGURE 25-26 Protein charge ladder. Bovine carbonic anhydrase was acetylated to give species with charges of $n-$ (unacetylated), $(n + 1)-$, $(n + 2)-$, . . . , $(n + 18)-$ (fully acetylated). Electrophoresis was carried out at pH 8.3 at 2.50×10^4 V in a capillary with a length of 0.840 m and a distance to the detector of 0.640 m. The neutral, ultraviolet-absorbing marker used to measure electroosmotic flow was mesityl oxide, $(CH_3)_2C{=}CHC({=}O)CH_3$.
[From M. K. Menon and A. L. Zydney, "Determination of Effective Protein Charge by Capillary Electrophoresis," *Anal. Chem.* **2000**, *72*, 5714.]

EXAMPLE Mobilities in a Protein Charge Ladder

Bovine carbonic anhydrase is a protein with 18 lysine residues. The 19 peaks in Figure 25-26 arise from unmodified protein (P^{n-}) plus protein with every possible degree of acetylation: $P^{(n + 1)-}$, $P^{(n + 2)-}$, $P^{(n + 3)-}$, . . . , $P^{(n + 18)-}$. The voltage applied to the 0.840-m-long capillary is 2.50×10^4 V. A neutral marker molecule, carried by electroosmotic flow, requires 308 s to travel 0.640 m from the inlet to the detector. Migration times of P^{n-} and $P^{(n + 1)-}$ are 343 s and 355 s, respectively. Find the electroosmotic velocity and electroosmotic mobility. Find the apparent and electrophoretic mobilities of P^{n-} and $P^{(n + 1)-}$.

Solution Electroosmotic velocity, u_{eo}, is found from the migration time of the neutral marker:

$$\text{Electroosmotic velocity} = \frac{\text{distance to detector }(L_d)}{\text{migration time}} = \frac{0.640 \text{ m}}{308 \text{ s}} = 2.08 \text{ mm/s}$$

Electric field is the voltage divided by the total length, L_t, of the column: $E = 25\ 000$ V/0.840 m $= 2.98 \times 10^4$ V/m. Mobility is the constant of proportionality between velocity and electric field:

$$u_{eo} = \mu_{eo}E \implies \mu_{eo} = \frac{u_{eo}}{E} = \frac{0.002\ 08\ \text{m/s}}{2.98 \times 10^4\ \text{V/m}} = 6.98 \times 10^{-8}\ \frac{\text{m}^2}{\text{V} \cdot \text{s}}$$

The mobility of the neutral marker, which we just calculated, is the electroosmotic mobility for the entire solution.

The apparent mobility of P^{n-} is obtained from its migration time:

$$\mu_{app} = \frac{u_{net}}{E} = \frac{0.640\ \text{m/343 s}}{2.98 \times 10^4\ \text{V/m}} = 6.26 \times 10^{-8}\ \frac{\text{m}^2}{\text{V} \cdot \text{s}}$$

Electrophoretic mobility describes the response of the ion to the electric field. Subtract electroosmotic mobility from apparent mobility to find electrophoretic mobility:

$$\mu_{app} = \mu_{ep} + \mu_{eo} \implies \mu_{ep} = \mu_{app} - \mu_{eo} = (6.26 - 6.98) \times 10^{-8}\ \frac{\text{m}^2}{\text{V} \cdot \text{s}}$$

$$= -0.72 \times 10^{-8}\ \frac{\text{m}^2}{\text{V} \cdot \text{s}}$$

The electrophoretic mobility is negative because the protein has a negative charge and migrates in the opposite direction from electroosmotic flow. Electroosmotic flow at pH 8.3 is faster than electromigration, so the protein does get carried to the detector. Similar calculations for the modified protein $P^{(n+1)-}$ give $\mu_{app} = 6.05 \times 10^{-8}$ m^2/(V · s) and $\mu_{ep} = -0.93 \times 10^{-8}$ m^2/(V · s). The electrophoretic mobility of $P^{(n+1)-}$ is more negative than that of P^{n-} because the charge is more negative.

Test Yourself If electroosmotic mobility were 5.00×10^{-8} m^2/(V · s), what would be the migration times of neutral marker and P^{n-}? (**Answer:** 430 s, 502 s)

Theoretical Plates and Resolution

Consider a capillary of length L_d from the inlet to the detector. In Section 22-4, we defined the number of theoretical plates as $N = L_d^2/\sigma^2$, where σ is the standard deviation of the band. If the only mechanism of zone broadening is longitudinal diffusion, the standard deviation was given by Equation 22-26: $\sigma = \sqrt{2Dt}$, where D is the diffusion coefficient and t is the migration time ($= L_d/u_{net} = L_d/[\mu_{app}E]$). Combining these equations with the definition of electric field ($E = V/L_t$, where V is the applied voltage) gives an expression for the number of plates:

Number of plates:
$$N = \frac{\mu_{app}V}{2D} \frac{L_d}{L_t} \tag{25-14}$$

How many theoretical plates might we hope to attain? Using a typical value of $\mu_{app} = 2 \times 10^{-8}$ m^2/(V · s) (derived for a 10-min migration time in a capillary with $L_t = 60$ cm, $L_d = 50$ cm, and 25 kV) and using diffusion coefficients from Table 22-1, we find

For K$^+$: $\quad N = \dfrac{[2 \times 10^{-8}\ \text{m}^2/(\text{V} \cdot \text{s})][25\ 000\ \text{V}]}{2(2 \times 10^{-9}\ \text{m}^2/\text{s})} \dfrac{0.50\ \text{m}}{0.60\ \text{m}} = 1.0 \times 10^5$ plates

For serum albumin: $\quad N = \dfrac{[2 \times 10^{-8}\ \text{m}^2/(\text{V} \cdot \text{s})][25\ 000\ \text{V}]}{2(0.059 \times 10^{-9}\ \text{m}^2/\text{s})} \dfrac{0.50\ \text{m}}{0.60\ \text{m}} = 3.5 \times 10^6$ plates

For the small, rapidly diffusing K$^+$ ion, we expect 100 000 plates. For the slowly diffusing protein serum albumin (FM 66 000), we expect more than 3 million plates. High plate count means that bands are very narrow and resolution between adjacent bands is excellent.

In reality, additional sources of zone broadening include the finite width of the injected band (Equation 22-32), a parabolic flow profile from heating inside the capillary, adsorption of solute on the capillary wall (which acts as a stationary phase), the finite length of the detection zone, and mobility mismatch of solute and buffer ions that leads to nonideal electrophoretic behavior. If these factors are properly controlled, $\sim 10^5$ plates are routinely achieved.

Side notes (left margin):

Number of plates: $\quad N = \dfrac{L_d^2}{\sigma^2}$

L_d = distance to detector
σ = standard deviation of Gaussian band
L_t = total length of column

Under special conditions in which a reverse hydrodynamic flow was imposed to slow the passage of analytes through the capillary, up to 17 million plates were observed in the separation of small molecules![31]

FIGURE 25-27 Small region from the electropherogram of a complex mixture shows that increasing voltage increases resolution. All conditions are the same in both runs except voltage, which is ordinarily limited to ~30 kV. Special precautions were required to prevent electric arcing and overheating at 120 kV. [From K. M. Hutterer and J. W. Jorgenson, "Ultrahigh-Voltage Capillary Zone Electrophoresis," *Anal. Chem.* **1999**, *71*, 1293.]

Equation 25-14 says that, for constant L_d/L_t, plate count is independent of capillary length. In contrast with chromatography, longer capillaries in electrophoresis do not give higher resolution.

Equation 25-14 also tells us that, the higher the voltage, the greater the number of plates (Figure 25-27). Voltage is ultimately limited by capillary heating, which produces a parabolic temperature profile that gives band broadening. The optimum voltage is found by making an *Ohm's law plot* of current versus voltage with *background electrolyte* (also called *run buffer*) in the capillary. In the absence of overheating, this curve should be a straight line. The maximum allowable voltage is the value at which the curve deviates from linearity (by, say, 5%). Buffer concentration and composition, thermostat temperature, and active cooling all play roles in how much voltage can be tolerated. Up to a point, higher voltage gives better resolution and faster separations.

As in chromatography, resolution between closely spaced peaks A and B in an electropherogram is related to plate count, N, and separation factor, γ, by Equation 22-30: resolution = $(\sqrt{N}/4)(\gamma - 1)$. The *separation factor* ($\gamma = u_{net,A}/u_{net,B}$) is the quotient of migration times t_B/t_A. Increasing γ increases separation of peaks, and increasing N decreases their width.

25-7 Conducting Capillary Electrophoresis

Clever variations of electrophoresis allow us to separate neutral molecules as well as ions, to separate optical isomers, and to lower detection limits by up to 10^6.

Controlling the Environment Inside the Capillary

The inside capillary wall controls electroosmotic velocity and provides undesired adsorption sites for multiply charged molecules, such as proteins. A fused-silica capillary should be prepared for its *first* use by washing for 1 h with 1 M NaOH at a flow rate of ~4 column volumes/min, followed by 1 h with water, followed by 1 h with 6 M HCl, followed by 1 h with run buffer.[32] NaOH is thought to generate Si—OH groups on the silica surface and HCl removes metal ions from the surface. For subsequent use at high pH, wash for ~10 s with 0.1 M NaOH, followed by deionized water, and then by at least 5 min with run buffer.[33] If the capillary is being run with pH 2.5 phosphate buffer, wash between runs with 1 M phosphoric acid, deionized water, and run buffer.[34] When changing buffers, allow at least 5 min of flow for equilibration. For the pH range 4–6, at which equilibration of the wall with buffer is slow, the capillary needs frequent regeneration with 0.1 M NaOH if migration times become erratic. Buffer in both reservoirs should be replaced periodically because ions become depleted and because electrolysis raises the pH at the cathode and lowers the pH at the anode. The capillary inlet should be ~2 mm away from and below the electrode to minimize entry of electrolytically generated acid or base into the column.[35] Stored capillaries should be filled with distilled water.

Background electrolyte (the solution in the capillary and the electrode reservoirs) controls pH and electrolyte composition in the capillary.

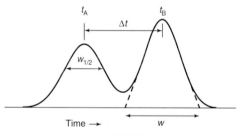

$$\text{Resolution} = \frac{\Delta t}{w_{av}} = \frac{0.589 \Delta t}{w_{1/2av}}$$

$$\text{Resolution} = \frac{\sqrt{N}}{4} (\gamma - 1)$$

N = plate number
γ = separation factor
$= \dfrac{\text{speed of faster species}}{\text{speed of slower species}}$

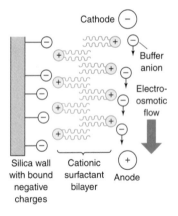

FIGURE 25-28 Effect of wall coating on electroosmotic mobility. Bare silica has little charge below pH 3 and high negative charge above pH 8. Polybrene cation embedded in silicate (lower structure) gives nearly constant positive charge to the wall. Anionic dextran sulfate adsorbed on polybrene (upper structure) gives constant negative charge to the wall. The bare silica wall goes from having little charge at pH 3 to full negative charge at pH 8. The wall behaves as if it has two populations of Si—OH groups with a total surface density of ~4.3 × 10^{-7} mol/m². One-fourth of the silanols have pK_a = 4.0 and three-fourths have pK_a = 5.5.[36] [From M. R. N. Monton, M. Tomita, T. Soga, and Y. Ishihama, "Polymer Entrapment in Polymerized Silicate for Preparing Highly Stable Capillary Coatings for CE and CE–MS," *Anal. Chem.* **2007**, *79*, 7838.]

Covalent coating helps prevent protein from sticking to capillary and provides reproducible migration times:

Different separations require more or less electroosmotic flow. Small anions with high mobility and highly negatively charged proteins require brisk electroosmotic flow or they will not travel toward the cathode. At pH 3, there is little charge on the silanol groups and little electroosmotic flow. At pH 8, the wall is highly charged and electroosmotic flow is strong. The graph in Figure 25-28 shows that electroosmotic mobility in a bare silica capillary is small and positive below pH 3. Mobility increases and reaches a high, steady value above pH 8.

Proteins with many positively charged substituents can bind tightly to negatively charged silica. To control this, 30–60 mM diaminopropane (which gives $^+H_3NCH_2CH_2CH_2NH_3^+$) may be added to the run buffer to neutralize charge on the wall. The wall charge can be reduced to near 0 by covalent attachment of silanes with neutral, hydrophilic substituents. However, many coatings are unstable under alkaline conditions.

You can reverse the direction of electroosmotic flow by adding a cationic surfactant, such as didodecyldimethylammonium bromide, to the run buffer.[37] This molecule has a positive charge at one end and two long hydrocarbon tails. The surfactant coats the negatively charged silica, with the tails pointing away from the surface (Figure 25-29). A second layer of surfactant orients itself in the opposite direction so that the tails form a nonpolar hydrocarbon layer. This *bilayer* adheres tightly to the wall of the capillary and effectively reverses the wall charge from negative to positive. Buffer anions create electroosmotic flow from cathode to anode when voltage is applied. Electroosmotic flow is in the direction opposite that shown in Figure 25-24. Best results are obtained when the capillary is freshly regenerated for each run.

Figure 25-28 shows a more stable cationic coating formed by embedding the cationic polymer polybrene in a silicate layer formed in situ on the capillary wall. The graph shows that electroosmotic flow is nearly constant in the pH range 2–11 and opposite that of bare silica. A stable, pH-independent negative surface can be made by adsorption of the anionic polymer dextran sulfate on the cationic polybrene surface.

FIGURE 25-29 Charge reversal created by a cationic surfactant bilayer coated on the capillary wall. The diffuse part of the double layer contains excess anions, and electroosmotic flow is opposite that shown in Figure 25-24. The surfactant is the didodecyldimethylammonium ion, $(n\text{-}C_{12}H_{25})N(CH_3)_2^+$, represented as in the illustration.

Sample Injection and Composition

Hydrodynamic injection uses pressure to force sample into the capillary (Figure 25-22). **Electrokinetic injection** uses the electric field to drive sample into the capillary. For hydrodynamic injection, the injected volume is

For quantitative analysis, it is critical to use an internal standard because the amount of sample injected into the capillary is not reproducible.

Hydrodynamic injection: $$\text{Volume} = \frac{\Delta P \pi d^4 t}{128 \eta L_t} \qquad (25\text{-}15)$$

where ΔP is the pressure difference between the ends of the capillary, d is the capillary inner diameter, t is injection time, η is sample viscosity, and L_t is the total length of the capillary.

EXAMPLE   Hydrodynamic Injection Time

How much time is required to inject a sample equal to 2.0% of the length of a 50-cm capillary if the diameter is 50 μm and the pressure difference is 2.0×10^4 Pa (0.20 bar)? Assume that the viscosity is 0.001 0 kg/(m · s), which is close to the viscosity of water.

Solution   The injection plug will be 1.0 cm long and occupy a volume of $\pi r^2 \times$ length = $\pi(25 \times 10^{-6}\text{ m})^2(1.0 \times 10^{-2}\text{ m}) = 1.9_6 \times 10^{-11}\text{ m}^3$. The required time is

$$t = \frac{128\eta L_t(\text{volume})}{\Delta P \pi d^4} = \frac{128[0.001\ 0\text{ kg/(m · s)}](0.50\text{ m})(1.9_6 \times 10^{-11}\text{ m}^3)}{(2.0 \times 10^4\text{ Pa})\pi(50 \times 10^{-6}\text{ m})^4} = 3.2\text{ s}$$

The units work out when we realize that Pa = force/area = (kg · m/s^2)/m^2 = kg/(m · s^2).

Test Yourself   How much time would be required to inject a 1.0-cm-long sample with twice the viscosity of water into a 40-cm-long column at the same ΔP? (***Answer:*** 5.1 s)

For electrokinetic injection, the capillary is dipped in the sample and a voltage is applied between the ends of the capillary. The moles of each ion taken into the capillary in t seconds are

Electrokinetic injection: $\text{Moles injected} = \mu_{\text{app}}\underbrace{\left(E\frac{\kappa_b}{\kappa_s}\right)}_{\text{Effective electric field} \equiv E_{\text{eff}}}t\pi r^2 C$ (25-16)

where μ_{app} is the apparent mobility of analyte (= $\mu_{\text{ep}} + \mu_{\text{eo}}$), E is the applied electric field (V/m), r is capillary radius, C is sample concentration (mol/m^3), and κ_b/κ_s is the ratio of conductivities of buffer and sample. Each analyte has a different mobility, so the injected sample does not have the same composition as the original sample. Electrokinetic injection is most useful for capillary gel electrophoresis (described later), in which liquid in the capillary is too viscous for hydrodynamic injection.

EXAMPLE   Electrokinetic Injection Time

How much time is required to inject a sample equal to 2.0% of the length of a 50-cm capillary if the diameter is 50 μm and the injection electric field is 10 kV/m? Assume that the sample has 1/10 of the conductivity of background electrolyte and $\mu_{\text{app}} = 2.0 \times 10^{-8}$ m^2/(V · s).

Solution   The factor κ_b/κ_s in Equation 25-16 is 10 in this case. The length of sample plug injected onto the column is (sample speed) $\times$ (time) = $\mu_{\text{app}}E_{\text{eff}}t$. The desired injection plug will be 1.0 cm long. The required time is

$$t = \frac{\text{plug length}}{\text{speed}} = \frac{\text{plug length}}{\mu_{\text{app}}\left(E\dfrac{\kappa_b}{\kappa_s}\right)} = \frac{0.010\text{ m}}{[2.0 \times 10^{-8}\text{ m}^2/(\text{V · s})](10\ 000\text{ V/m})(10)} = 5.0\text{ s}$$

Equation 25-16 multiplies the plug length times its cross-sectional area to find its volume and then multiplies by concentration to find the moles in that volume.

Test Yourself   What is the effect on injection time if you decrease the applied voltage by a factor of 2? (***Answer:*** injection time is doubled)

Conductivity Effects: Stacking and Skewed Bands

We choose conditions so that analyte is focused into narrow bands at the start of the capillary by a process called **stacking.** Without stacking, if you inject a zone with a length of 10 mm, no analyte band can be narrower than 10 mm when it reaches the detector.

Stacking depends on the relation between the electric field in the zone of injected sample and that in the background electrolyte on either side of the sample. Optimal buffer concentration in the sample solution is 1/10 of the background electrolyte concentration, and the sample concentration should be 1/500 of the background electrolyte concentration. If sample has a much lower ionic strength than the run buffer, the sample's conductivity is lower and its

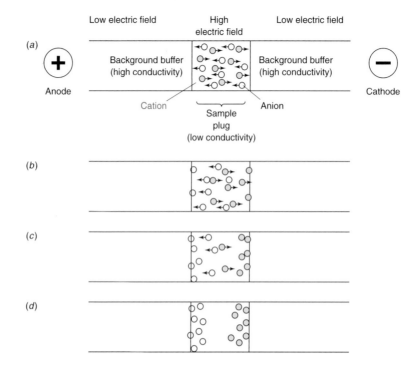

FIGURE 25-30 Stacking of anions and cations at opposite ends of low-conductivity sample plug occurs because the electric field in the sample plug is much higher than the field in background electrolyte. Time increases from panels *a* to *d*. Electroneutrality is maintained by migration of background electrolyte ions, which are not shown.

resistance is much greater. Electric field is inversely proportional to conductivity: the lower the conductivity, the greater the electric field. The electric field across the sample plug inside the capillary is higher than the electric field in the background electrolyte. Figure 25-30 shows ions in the sample plug migrating very fast, because the electric field is very high. When ions reach the zone boundary, they slow down because the field is lower outside the sample plug. This process of *stacking* continues until analyte cations are concentrated at one end of the sample plug and analyte anions are at the other end. The broad injection becomes concentrated into narrow bands of analyte cations or anions. Figure 25-31 shows signal enhancement by stacking.

If the conductivity of an analyte band is significantly different from the conductivity of the background electrolyte, peak distortion occurs. Figure 25-32 shows a band containing one analyte (as opposed to the sample plug in Figure 25-30, which contains all analytes in the entire injection). If background conductivity is greater than analyte conductivity ($\kappa_b > \kappa_a$), the electric field is lower outside the analyte band than inside the band. The band migrates to the right in Figure 25-32. An analyte molecule that diffuses past the front on the right suddenly encounters a lower electric field and it slows down. Soon, the analyte zone catches up

FIGURE 25-31 *Lower trace:* Sample injected electrokinetically for 2 s without stacking is limited in volume to prevent band broadening. *Upper trace:* With stacking, 15 times more sample could be injected (for 30 s), so the signal is 15 times stronger with no increase in bandwidth. [From Y. Zhao and C. E. Lunte, "pH-Mediated Field Amplification On-Column Preconcentration of Anions in Physiological Samples for Capillary Electrophoresis," *Anal. Chem.* **1999**, *71*, 3985.]

FIGURE 25-32 Irregular peak shapes arise when conductivity of analyte band, κ_a, is not the same as background conductivity, κ_b. Electropherogram shows cations extracted from the surface of a silicon semiconductor wafer. Background electrolyte contains imidazolium ion for *indirect* spectrophotometric detection, whose principle is shown in Figure 25-35. [From T. Ehmann, L. Fabry, L. Kotz, and S. Pahlke, "Monitoring of Ionic Contaminants on Silicon Wafer Surfaces using Capillary Electrophoresis," *Am. Lab.*, June 2002, p. 18.]

50 μm

150 μm

Light path

1

2

3

(a)

Light path

(b)

FIGURE 25-33 Capillary designs to increase pathlength for measuring ultraviolet absorption. (*a*) Bubble cell. "Plug" of solute zone is maintained as it passes through the bubble. (*b*) Right-angle bend. Light path is made of black fused silica to reduce stray light. Reflective interior serves as "light pipe" to maximize transmission. Detector response is linear up to 1.4 absorbance units. [Courtesy Agilent Technologies, Palo Alto, CA.]

with the molecule and it is back in the zone. A molecule that diffuses out of the zone on the left encounters a lower electric field and it also slows down. The analyte zone is moving faster than the wayward molecule and pulls away. This condition leads to a sharp front and a broad tail, as shown in the lower right electropherogram in Figure 25-32. When $\kappa_b < \kappa_a$, we observe the opposite electropherogram.

To minimize band distortion, sample concentration must be much less than the background electrolyte concentration. Otherwise, it is necessary to choose a buffer co-ion that has the same mobility as the analyte ion. (The *co-ion* is the buffer ion with the same charge as analyte. The *counterion* has the opposite charge.)

Detectors

Water is so transparent that *ultraviolet detectors* can operate at wavelengths as short as 185 nm, where most solutes have strong absorption. To take advantage of short-wavelength ultraviolet detection, background electrolyte must have very low absorption. Borate buffers are commonly used in electrophoresis because of their transparency.[38] Sensitivity is poor because the optical pathlength is only as wide as the capillary, which is 25–75 μm. Figure 25-33 shows a "bubble cell" that increases the absorbance signal-to-noise ratio by a factor of 3 to 5 and a right-angle bend that increases signal-to-noise by a factor of 10. However, the greater pathlength in the right-angle bend design leads to some band broadening. Successive peaks must be separated by 3 mm or they will overlap in the detector.

Fluorescence detection (shown later in Figure 25-42) is sensitive to naturally fluorescent analytes or to fluorescent derivatives. *Amperometric detection* is sensitive to analytes that can be oxidized or reduced at an electrode (Figure 25-34). *Conductivity detection* with ion-exchange

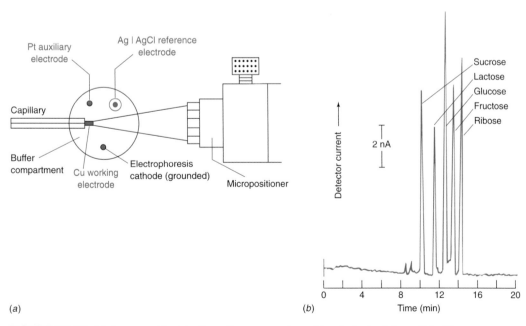

(a)

(b)

FIGURE 25-34 (*a*) Amperometric detection with macroscopic working electrode at the outlet of the capillary. (*b*) Electropherogram of sugars separated in 0.1 M NaOH, in which OH groups are partially ionized, thereby turning the molecules into anions. [From J. Ye and R. P. Baldwin, "Amperometric Detection in Capillary Electrophoresis with Normal Size Electrodes," *Anal. Chem.* **1993**, *65*, 3525.]

Bright solvent   Dark analyte   Bright solvent
zone   zone   zone

FIGURE 25-35 Principle of indirect detection. When analyte emerges from the capillary, the strong background signal decreases.

suppression of the background electrolyte (as in Figure 25-6) can detect small analyte ions at 1–10 ng/mL. *Electrospray mass spectrometry* (Figure 21-22) provides low detection limits and gives qualitative information about analytes.

Figure 25-35 shows the principle of *indirect detection*, which applies to fluorescence, absorbance, amperometry, conductivity, and other forms of detection. A substance with a steady background signal is added to the background electrolyte. In the analyte band, analyte molecules displace the chromophoric substance, so the detector signal *decreases* when analyte passes by. Figure 25-36 shows an impressive separation of Cl^- isotopes with indirect detection in the presence of the ultraviolet-absorbing anion chromate. Electroneutrality dictates that an analyte band containing Cl^- must have a lower concentration of CrO_4^{2-} than is found in the background electrolyte. With less CrO_4^{2-} to absorb ultraviolet radiation, a negative peak appears when Cl^- reaches the detector. Benzoate and phthalate are other anions useful for this purpose. Detection limits in capillary electrophoresis are generally about an order of magnitude higher than detection limits in ion chromatography but one to two orders of magnitude lower than detection limits for ion-selective electrodes.

Approximate detection limits (μM) for indirect detection in electrophoresis:

Ultraviolet absorption	1–100
Fluorescence	0.001–1
Chemiluminescence	0.001–0.01
Conductivity	0.01–100
Amperometry	10^{-5}–10
Mass spectrometry	0.001–0.01

SOURCE: Mostly from C. Vogt and G. L. Klunder, "Separation of Metal Ions by Capillary Electrophoresis—Diversity, Advantages, and Drawbacks of Detection Methods," *Fresenius J. Anal. Chem.* **2001**, *370*, 316.

FIGURE 25-36 Separation of natural isotopes of 0.56 mM Cl^- by capillary electrophoresis with indirect spectrophotometric detection at 254 nm. Background electrolyte contains 5 mM CrO_4^{2-} to provide absorbance at 254 nm and 2 mM borate buffer, pH 9.2. The capillary had a diameter of 75 μm, a total length of 47 cm (length to detector = 40 cm), and an applied voltage of 20 kV. The difference in electrophoretic mobility of $^{35}Cl^-$ and $^{37}Cl^-$ is just 0.12%. Conditions were adjusted so that electroosmotic flow was nearly equal to and opposite electrophoretic flow. The resulting near-zero net velocity gave the two isotopes maximum time to be separated by their slightly different mobilities. [From C. A. Lucy and T. L. McDonald, "Separation of Chloride Isotopes by Capillary Electrophoresis Based on the Isotope Effect on Ion Mobility," *Anal. Chem.* **1995**, *67*, 1074.]

Micellar Electrokinetic Chromatography[39]

The type of electrophoresis we have been discussing so far is called **capillary zone electrophoresis.** Separation is based on differences in electrophoretic mobility. If the capillary wall is negative, electroosmotic flow is toward the cathode (Figure 25-24) and the order of elution is cations before neutrals before anions. If the capillary wall charge is reversed by coating it with a cationic surfactant (Figure 25-29) and instrument polarity is reversed, then the order of elution is anions before neutrals before cations. Neither scheme separates neutral molecules from one another.

Micellar electrokinetic chromatography separates neutral molecules and ions. We illustrate a case in which the anionic surfactant sodium dodecyl sulfate is present above its *critical micelle concentration* (Box 25-1), so negatively charged micelles are formed.[41] In Figure 25-37, electroosmotic flow is to the right. Electrophoretic migration of the negatively charged micelles is to the left, but net motion is to the right because the electroosmotic flow dominates.

In the absence of micelles, all neutral molecules reach the detector in time t_0. Micelles injected with the sample reach the detector in time t_{mc}, which is longer than t_0 because the micelles migrate upstream. If a neutral molecule equilibrates between free solution and the

Normal order of elution in capillary zone electrophoresis:

1. cations (highest mobility first)
2. all neutrals (unseparated)
3. anions (highest mobility last)

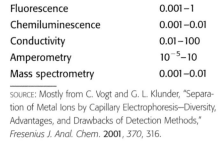

Sodium dodecyl sulfate (n-$C_{12}H_{25}OSO_3^-$ Na^+)

Micellar electrokinetic chromatography: The more time the neutral analyte spends inside the micelle, the longer is its migration time. This technique was introduced by S. Terabe in 1984.[40]

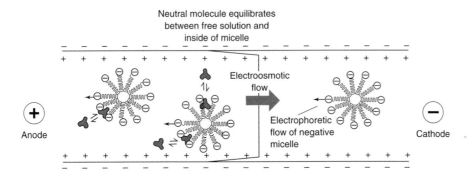

Neutral molecule equilibrates between free solution and inside of micelle

Electroosmotic flow

Electrophoretic flow of negative micelle

Anode

Cathode

FIGURE 25-37 Negatively charged sodium dodecyl sulfate micelles migrate against electroosmotic flow. Neutral molecules (solid color) are in dynamic equilibrium between free solution and the inside of the micelle. The more time spent in the micelle, the more the neutral molecule lags behind electroosmotic flow.

inside of the micelles, its migration time is increased, because it migrates at the slower rate of the micelle part of the time. The neutral molecule reaches the detector at a time between t_0 and t_{mc}. *The more time the neutral molecule spends inside the micelle, the longer is its migration time.* Migration times of cations and anions also are affected by micelles, because ions partition between the solution and the micelles and interact electrostatically with the micelles.

Micellar electrokinetic chromatography is a form of chromatography because the micelles behave as a *pseudostationary phase*. Separation of neutral molecules is based on partitioning between the solution and the pseudostationary phase. The mass transfer term Cu_x is no longer 0 in the van Deemter equation 25-7, but mass transfer into the micelles is fairly fast and band broadening is modest.

Imagination runs wild with the variables in micellar electrokinetic chromatography. We can add anionic, cationic, zwitterionic, and neutral surfactants to change partition coefficients of analytes. (Cationic surfactants also change the charge on the wall and the direction of electroosmotic flow.) We can add solvents such as acetonitrile or *N*-methylformamide to increase the solubility of organic analytes and to change the partition coefficient between the solution and the micelles.[42] We can add cyclodextrins (Box 23-1) to separate optical isomers that spend different fractions of the time associated with the cyclodextrins.[43] In Figure 25-38, chiral micelles were used to separate enantiomers of chiral drugs.

A method called *sweeping*, which we will not describe in detail, concentrates analyte by a factor of 10^3-10^5 for trace analysis. In sweeping, a migrating ionic reagent such as sodium dodecyl sulfate micelles or a chelator binds analyte and concentrates it into a narrow band. Procedures have been described for sweeping neutral analytes,[44] anions,[45] and cations.[46] Another method to concentrate analyte utilizes dynamic changes in pH inside the capillary.[47] A method called *isotachophoresis* stacks analytes to concentrate them by factors up to 10^5-10^6.[48]

Capillary electrochromatography, described at the opening of this chapter, differs from micellar electrokinetic chromatography by using a true stationary phase.[49] Solvent is driven by electroosmosis. Capillary electrochromatography provides about twice as many plates as HPLC for the same particle size and column length. Pressure is not used

Confusing terms:

Micellar electrokinetic chromatography: electrophoresis with micelles acting as pseudostationary phase

Capillary electrochromatography: similar to HPLC, except mobile phase is driven by electroosmosis instead of pressure

Mass spectral signal

Chiral center

Compound 5 (oxprenolol)

Time (min)

FIGURE 25-38 Separation of enantiomers of eight β-blocker drugs by micellar electrokinetic chromatography at pH 8.0 in a 120-cm capillary at 30 kV. The structure of one compound is shown. Micelles were formed by a polymer surfactant containing L-leucinate substituents for chiral recognition.
[From C. Akbay, S. A. A. Rizvi, and S. A. Shamsi, "Simultaneous Enantioseparation and Tandem UV-MS Detection of Eight β-Blockers in Micellar Electrokinetic Chromatography Using a Chiral Molecular Micelle," *Anal. Chem.* **2005,** *77,* 1672.]

Polymer chain

Covalent cross-link

Chemical gel
(has chemical bonds
(a)   between chains)

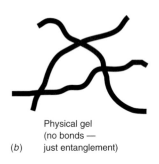

Physical gel
(no bonds —
(b)   just entanglement)

FIGURE 25-39 (*a*) A chemical gel contains covalent cross-links between different polymer chains. (*b*) A physical gel is not cross-linked but derives its properties from physical entanglement of the polymers.

to drive the mobile phase, so there is no pressure drop associated with the small particles. Capillary electrochromatography has a full range of applications, such as separation of enantiomers,[50] ion exchange,[51] and trace analysis.[52]

Capillary Gel Electrophoresis

Capillary gel electrophoresis is a variant of gel electrophoresis, which has been a primary tool in biochemistry for four decades. Slabs of polymer gel used to separate macromolecules according to size have customarily been chemical gels, in which chains are cross-linked by chemical bonds (Figure 25-39a). Chemical gels cannot be flushed from a capillary, so physical gels (Figure 25-39b) in which polymers are simply entangled are used. Physical gels can be flushed and reloaded to generate a fresh capillary for each separation.

Macromolecules are separated in a gel by *sieving*, in which smaller molecules migrate faster than large molecules through the entangled polymer network. Color Plate 31 shows part of a deoxyribonucleic acid (DNA) sequence analysis in which a mixture of fluorescence-labeled fragments with up to 400 nucleotides was separated in 32 minutes in a capillary containing 38 g/L of linear polyacrylamide (Figure 25-15, with no cross-links) and 6 M urea to stabilize single strands of DNA. Each strand, terminating in one of the four bases A, T, C, or G, is tagged with one of four different fluorescent labels that identify the terminal bases as they pass through a fluorescence detector. Capillary electrophoresis was an enabling technology for determining the sequence of nucleic acids in the human genome.[53]

Biochemists measure molecular mass of proteins by *sodium dodecyl sulfate (SDS)–gel electrophoresis*. Proteins are first *denatured* (unfolded to random coils) by reducing their disulfide bonds (—S—S—) with excess 2-mercaptoethanol (HSCH$_2$CH$_2$OH) and adding sodium dodecyl sulfate (C$_{12}$H$_{25}$OSO$_3^-$ Na$^+$). Dodecyl sulfate anion coats hydrophobic regions and gives the protein a large negative charge that is approximately proportional to the length of the protein. Denatured proteins are then separated by electrophoresis through a sieving gel. Large molecules are retarded more than small molecules, which is the opposite behavior from size exclusion chromatography. In Figure 25-40, the logarithm of molecular mass of the SDS-coated protein is proportional to 1/(migration time). Absolute migration times vary from run to run, so relative migration times are measured. The relative migration time is the migration time of a protein divided by the migration time of a fast-moving small dye molecule.

Method Development

Capillary electrophoresis is not used as much as liquid chromatography. Advantages of electrophoresis relative to chromatography include (1) higher resolution, (2) low waste production, and (3) generally simpler equipment. Drawbacks of electrophoresis include (1) higher limits of detection, (2) run-to-run irreproducibility of migration times, (3) insolubility of some analytes in common electrolyte solutions, and (4) inability to scale up to a preparative separation.

Liquid chromatography is two decades more mature than capillary electrophoresis. As trained "electropherographers" become more common, more separations will be handled

FIGURE 25-40 Calibration curve for protein molecular mass in sodium dodecyl sulfate–capillary gel electrophoresis. On the abscissa, t_{rel}, is the migration time of each protein divided by the migration time of a small dye molecule. [Data from J. K. Grady, J. Zang, T. M. Laue, P. Arosio, and N. D. Chasteen, "Characterization of the H- and L-Subunit Ratios in Ferritins by Sodium Dodecyl Sulfate–Capillary Gel Electrophoresis," *Anal. Biochem.* **2002**, *302*, 263.]

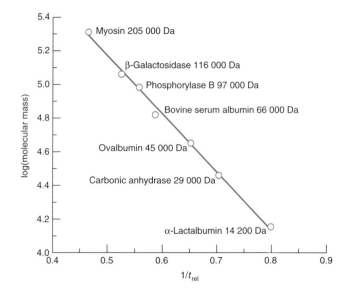

by electrophoresis. For example, electrophoresis displaced liquid chromatography as the preferred method for forensic analysis of alkaloids in opium and heroin.[54] The enabling technology for this application was dynamic coating of the capillary between runs to eliminate adsorption of analytes on the silica surface and decrease variations in migration times to less than 0.5%.

Method development for capillary electrophoresis addresses the following points:[55]

1. Select a detection method that provides the required limit of detection. For ultraviolet absorption, select the optimum wavelength. If necessary, use indirect detection or derivatization.

2. If there is a choice, separate analytes as anions, which do not stick to the negatively charged wall. If separating polycations, such as proteins at low pH, select additives to coat the walls or to reverse the wall charge.

3. Dissolve the entire sample. If the sample is not soluble in dilute aqueous buffer, try adding 6 M urea or surfactants. Acetate buffer tends to dissolve more organic solutes than does phosphate buffer. (If it is necessary to dissolve the sample, nonaqueous solvents can be used.[56] However, the nonaqueous solvent must be compatible with plastic parts in the system. Acetonitrile and methanol are recommended, but electric current should be kept low to minimize outgassing and evaporation. Aqueous micelles or cyclodextrin solve some solubility problems that would otherwise require organic solvent.)

4. Determine how many peaks are present. Assign each peak by using authentic samples of analytes and diode array ultraviolet detection or mass spectrometry.

5. For complex mixtures, use computerized experimental design to help optimize separation conditions.[57]

6. Use the short end of the capillary in Figure 25-22 for quick scouting runs to determine the direction of migration and whether broad peaks are present. Broad peaks indicate that wall effects are occurring and a coating could be required.

7. See if pH alone provides adequate separation. For acids, begin with 50 mM borate buffer, pH 9.3. For bases, try 50 mM phosphate, pH 2.5. If the separation is not adequate, try adjusting buffer pH close to the average pK_a of the solutes.

8. If pH does not provide adequate separation or if analytes are neutral, use surfactants for micellar electrokinetic capillary chromatography. For chiral solutes, try adding cyclodextrins.

9. Select a capillary wash procedure. If migration times are reproducible from run to run with no wash, then no wash is required. If migration times increase, wash for 5 to 10 s with 0.1 M NaOH, followed by 5 min with buffer. If migration times still drift, try increasing or decreasing the NaOH wash time by a few seconds. If migration times decrease, wash with 0.1 M H_3PO_4. If proteins or other cations stick to the walls, try washing with 0.1 M sodium dodecyl sulfate or use a commercial dynamic coating.

10. If necessary, select a sample cleanup method. Cleanup could be required if resolution is poor, if the salt content is high, or if the capillary fouls. Cleanup might involve solid-phase extraction (Section 27-3), protein precipitation, or dialysis (Demonstration 26-1).

11. If the detection limit is not sufficient, select a stacking or sweeping method to concentrate analyte in the capillary.

12. For quantitative analysis, determine the linear range required to measure the least concentrated and most concentrated analytes. If desired, select an internal standard. If the migration time or peak area of the standard changes, then you have an indication that some condition has drifted out of control.

25-8 Lab-on-a-Chip: Probing Brain Chemistry

One of the most exciting and rapidly developing areas of analytical chemistry is the "lab-on-a-chip."[59] Liquids can be moved by electroosmosis or by pressure through micron-size channels etched into glass or plastic chips that are about the size of a microscope slide. Chemical reactions can be conducted by moving picoliters of fluid from different reservoirs, mixing them, and subjecting the products to chemical analysis on the chip. Microreactors consuming micrograms of reactants can be used to optimize conditions for synthesis, exploring dozens of conditions in a short time. Chips that manipulate small volumes of liquid are called *microfluidic devices*. We describe one that is coupled to a microdialysis probe to monitor chemicals in the brain.[60]

Dialysis is the process in which small molecules diffuse across a *semipermeable membrane* with pores large enough to pass small molecules but not large molecules. The

The European Space Agency ExoMars mission will search for signs of life on Mars with a lab-on-a-chip to measure amines, amino acids, and enantiomers of amino acids.[58] Sub-part-per-trillion detection limits are 1 000 times lower than that of the *Viking* gas chromatography–mass spectrometer, which did not detect organic compounds on Mars in 1976.

(a) (b) (c)

FIGURE 25-41 (*a*) Diagram and (*b*) photo of microdialysis probe. Small molecules pass through the semipermeable membrane, but large molecules cannot. (*c*) Probe inserted into rat to sample chemicals in brain. [Courtesy R. T. Kennedy and Z. D. Sandlin, University of Michigan.]

Derivatization:

o-Phthaldialdehyde + H₂N — CHRCO₂⁻

o-Phthaldialdehyde Glutamate
(R = CH₂CH₂CO₂H)

+ HSCH₂CH₂OH $\xrightarrow{\text{pH 9.5}}$

Fluorescent product

microdialysis probe in Figure 25-41 has a rigid semipermeable tube that can be inserted into the brain of an anesthetized rat to collect neurotransmitter molecules. Fluid pumped through the probe at a rate of 3 μL/min transports small molecules that diffused into the probe. Fluid exiting the probe (*dialysate*) is routed to the sample introduction channel at the lower left of the microfluidic chip in Figure 25-42a.

Liquid in the chip in Figure 25-42a is connected to electrical ground (0 V) at the left and to −25 kV through the waste outlet at the right. Electroosmotic flow proceeds from left to right through the channels. Some liquid from the sample introduction channel is drawn toward the reaction channel by electroosmotic flow. *Derivatization* reagent that reacts with the sample to make it fluorescent is also drawn in and mixes with the sample. The reaction channel provides ~1 min for derivatization before liquid enters the separation channel. Dimensions of the channels determine the electric potential at different points and direct fluid along the desired path. When the high-voltage switch is in the closed position shown in Figure 25-42a, liquid from the reaction channel flows to the waste channel. When the high-voltage switch is momentarily opened, a small plug of solution from the reaction channel is diverted to the separation channel for electrophoretic analysis. Fluorescent products are observed by the detector in Figure 25-42b when they reach the detection zone. The 9-cm-long separation channel provides ~10^5 theoretical plates with a detection limit of 0.2 μM for the neurotransmitter glutamate. Dialysate analyzed in Figure 25-42c contained 3.3 μM glutamate. The system is capable of measuring changes in glutamate in response to physiologic stimulation with a time resolution of 2–4 min. Changes in glutamate concentration occurring in less than 2–4 min would not be resolved from one another.

The microfluidic chip simplifies the handling of dialysate, making microdialysis sampling accessible to more neuroscience laboratories. The chip could be modified to conduct multiple analyses by directing aliquots of dialysate to different chambers for different derivatizations. Microfluidic chips promise to make complex analyses available to more laboratories at lower cost.

(a)

(b) Detection of fluorescent product

(c)

FIGURE 25-42 Microfluidic chip to measure neurotransmitters in dialysate. (*a*) Layout of chip. Solid lines are channels for liquid. Dashed lines are electrical connections. Fuzzy blue circles are ports where liquid enters or exits the chip. Arrows denote electroosmotic flow. (*b*) Fluorescence detector. (*c*) Electropherogram of dialysate from rat brain showing the neurotransmitter glutamate. [From Z. D. Sandlin, M. Shou, J. G. Shackman, and R. T. Kennedy, "Microfluidic Electrophoresis Chip Coupled to Microdialysis for in Vivo Monitoring of Amino Acid Neurotransmitters," *Anal. Chem.* **2005**, *77*, 7702.]

Terms to Understand

affinity chromatography	electroosmosis	micellar electrokinetic chromatography
anion exchanger	electrophoresis	micelle
capillary electrochromatography	gel	mobility
capillary electrophoresis	gel filtration	molecular exclusion chromatography
capillary gel electrophoresis	gradient elution	molecularly imprinted polymer
capillary zone electrophoresis	hydrodynamic injection	preconcentration
cation exchanger	hydrophobic interaction chromatography	resin
cross-linking	hydrophobic substance	selectivity coefficient
deionized water	indirect detection	stacking
dialysis	ion chromatography	suppressed-ion chromatography
Donnan equilibrium	ion-exchange chromatography	surfactant
electrokinetic injection	ion-pair chromatography	void volume

Summary

Ion-exchange chromatography employs resins and gels with covalently bound charged groups that attract solute counterions (and that exclude ions having the same charge as the resin). Polystyrene resins are useful for small ions. Greater cross-linking of the resin increases the capacity, selectivity, and time needed for equilibration. Ion-exchange gels based on cellulose and dextran have large pore sizes and low charge densities, suitable for the separation of macromolecules. Certain inorganic solids have ion-exchange properties and are useful at extremes of temperature or radiation. Ion exchangers operate by the principle of mass action, with a gradient of increasing ionic strength most commonly used to effect a separation. Zwitterionic hydrophobic interaction chromatography can separate anions and cations on the same column.

In suppressed-ion chromatography, a separator column separates ions of interest, and a suppressor membrane converts eluent into a nonionic form so that analytes can be detected by their conductivity. Eluent and suppressor can be generated continuously by electrolysis. Alternatively, nonsuppressed ion chromatography uses an ion-exchange column and low-concentration eluent. If eluent absorbs light, indirect spectrophotometric detection is convenient and sensitive. Ion-pair chromatography utilizes an ionic surfactant in the eluent to make a reversed-phase column function as an ion-exchange column.

Molecular exclusion chromatography is used for separations based on size and for molecular mass determinations of macromolecules. Molecular exclusion is based on the inability of large molecules to enter small pores in the stationary phase. Small molecules enter these pores and therefore exhibit longer elution times than large molecules. In affinity chromatography, the stationary phase retains one particular solute in a complex mixture. After all other components have been eluted, the desired species is liberated by a change in conditions. In hydrophobic interaction chromatography, high concentrations of ammonium sulfate induce proteins to adhere to a hydrophobic stationary phase. A gradient of decreasing salt concentration is applied to increase the solubility of proteins in water and elute them from the column.

In capillary zone electrophoresis, ions are separated by differences in mobility in a strong electric field applied between the ends of a silica capillary tube. The greater the charge and the smaller the hydrodynamic radius, the greater the electrophoretic mobility. Normally, the capillary wall is negative, and solution is transported from anode to cathode by electroosmosis of cations in the electric double layer. Solute cations arrive first, followed by neutral species,
followed by solute anions (if electroosmosis is stronger than electrophoresis). Apparent mobility is the sum of electrophoretic mobility and electroosmotic mobility (which is the same for all species). Zone dispersion (broadening) arises mainly from longitudinal diffusion and the finite length of the injected sample. Stacking of solute ions in the capillary occurs when the sample has a low conductivity. Electroosmotic flow is reduced at low pH because surface $Si-O^-$ groups are protonated. $Si-O^-$ groups can be masked by polyamine cations, and the wall charge can be reversed by a cationic surfactant that forms a bilayer along the wall. Covalent coatings reduce electroosmosis and wall adsorption. Hydrodynamic sample injection uses pressure or siphoning; electrokinetic injection uses an electric field. Ultraviolet absorbance is commonly used for detection. Micellar electrophoretic chromatography uses micelles as a pseudostationary phase to separate neutral molecules and ions, which can be concentrated by stacking. Capillary electrochromatography is essentially the same as HPLC, but the mobile phase is driven by electroosmosis instead of pressure. Capillary gel electrophoresis separates macromolecules by sieving. In contrast with molecular exclusion chromatography, small molecules move fastest in gel electrophoresis. Microfluidic devices ("lab-on-a-chip") use electroosmotic or hydrodynamic flow in lithographically fabricated channels to conduct chemical reactions and chemical analysis.

Exercises

25-A. Vanadyl sulfate ($VOSO_4$, FM 163.00), as supplied commercially, is contaminated with H_2SO_4 (FM 98.08) and H_2O. A solution was prepared by dissolving 0.244 7 g of impure $VOSO_4$ in 50.0 mL of water. Spectrophotometric analysis indicated that the concentration of the blue VO^{2+} ion was 0.024 3 M. A 5.00-mL sample was passed through a cation-exchange column loaded with H^+. When VO^{2+} from the 5.00-mL sample became bound to the column, the H^+ released required 13.03 mL of 0.022 74 M NaOH for titration. Find the weight percent of each component ($VOSO_4$, H_2SO_4, and H_2O) in the vanadyl sulfate.

25-B. Blue Dextran 2000 was eluted during gel filtration in a volume of 36.4 mL from a 2.0 × 40 cm (diameter × length) column of Sephadex G-50, which fractionates molecules in the molecular mass range 1 500 to 30 000.

(a) At what retention volume would hemoglobin (molecular mass 64 000) be expected?

(b) Suppose that radioactive $^{22}NaCl$, which is not adsorbed on the column, is eluted in a volume of 109.8 mL. What would be the retention volume of a molecule with $K_{av} = 0.65$?

25-C. Consider a capillary electrophoresis experiment conducted near pH 9, at which the electroosmotic flow is stronger than the electrophoretic flow.

(a) Draw a picture of the capillary, showing the placement of the anode, cathode, injector, and detector. Show the direction of electroosmotic flow and the direction of electrophoretic flow of a cation and an anion. Show the direction of net flow.

(b) Using Table 14-1, explain why Cl^- has a shorter migration time than I^-. Predict whether Br^- will have a shorter migration time than Cl^- or a greater migration time than I^-.

(c) Why is the mobility of I^- greater than that of Cl^-?

Problems

Ion-Exchange and Ion Chromatography

25-1. State the purpose of the separator and suppressor in suppressed-ion chromatography. For cation chromatography, why is the suppressor an anion-exchange membrane?

25-2. State the effects of increasing cross-linking on an ion-exchange column.

25-3. What is deionized water? What kind of impurities are not removed by deionization?

25-4. The exchange capacity of an ion-exchange resin can be defined as the number of moles of charged sites per gram of dry resin. Describe how you would measure the exchange capacity of an anion-exchange resin by using standard NaOH, standard HCl, or any other reagent you wish.

25-5. Consider a negatively charged protein adsorbed on an anion-exchange gel at pH 8.

(a) How will a gradient of eluent pH (from pH 8 to some lower pH) be useful for eluting the protein? Assume that the ionic strength of the eluent is kept constant.

(b) How would a gradient of ionic strength (at constant pH) be useful for eluting the protein?

25-6. What does the designation 200/400 mesh mean on a bottle of chromatography stationary phase? What is the size range of such particles? (See Table 27-2.) Which particles are smaller, 100/200 or 200/400 mesh?

25-7. Propose a scheme for separating trimethylamine, dimethylamine, methylamine, and ammonia from one another by ion-exchange chromatography.

25-8. Suppose that an ion-exchange resin (R^-Na^+) is immersed in a solution of NaCl. Let the concentration of R^- in the resin be 3.0 M.

(a) What will the ratio $[Cl^-]_o/[Cl^-]_i$ be if $[Cl^-]_o$ is 0.10 M?

(b) What will the ratio $[Cl^-]_o/[Cl^-]_i$ be if $[Cl^-]_o$ is 1.0 M?

(c) Will the fraction of electrolyte inside the resin increase or decrease as the outside concentration of electrolyte increases?

25-9. *Material balance.* If you intend to measure all the anions and cations in an unknown, one sanity check on your results is that the total positive charge should equal the total negative charge. Concentrations of anions and cations in pond water in Figure 25-7 are expressed in $\mu g/mL$. Find the total concentration of negative and positive charge (mol/L) to assess the quality of the analysis. What do you conclude about this analysis?

25-10. (a) Explain how anions and cations can be separated by hydrophilic interaction chromatography in Figure 25-5, using the zwitterionic stationary phase in Figure 24-14. Why does the gradient go from high CH_3CN to low CH_3CN content?

(b) Isocratic elution with 20 vol% acetonitrile plus 80 vol% aqueous buffer did not separate Na^+ from Cl^-. Eluent with 40 vol% acetonitrile plus 60 vol% aqueous buffer gave a baseline separation. Suggest an explanation.

25-11. In the separation of proteins by hydrophobic interaction chromatograpy, why does eluent strength increase with *decreasing* salt concentration in the aqueous eluent?

25-12. In ion-exclusion chromatography, ions are separated from nonelectrolytes by an ion-exchange column. Nonelectrolytes penetrate the stationary phase, whereas half of the ions are repelled by the fixed charges. Because electrolytes have access to less of the column volume, they are eluted before nonelectrolytes. The chromatogram here shows the separation of trichloroacetic acid (TCA, $pK_a = 0.5$), dichloroacetic acid (DCA, $pK_a = 1.1$), and monochloroacetic acid (MCA, $pK_a = 2.86$) by passage through a cation-exchange resin eluted with 0.01 M HCl. Explain why the three acids are separated and why they emerge in the order shown.

Separation of acids on cation-exchange column. [From V. T. Turkelson and M. Richards, "Separation of the Citric Acid Cycle Acids by Liquid Chromatography," *Anal. Chem.* **1978**, *50*, 1420.]

25-13. 🖩 Norepinephrine (NE) in human urine can be assayed by ion-pair chromatography by using an octadecylsilane stationary phase and sodium octyl sulfate as the mobile-phase additive. Electrochemical detection (oxidation at 0.65 V versus Ag | AgCl) is used, with 2,3-dihydroxybenzylamine (DHBA) as internal standard.

Norepinephrine cation (NE)

2,3-Dihydroxybenzylamine cation (DHBA)

$$CH_3CH_2CH_2CH_2CH_2CH_2CH_2CH_2OSO_3^-\,Na^+$$
Sodium octyl sulfate

(a) Explain the physical mechanism by which an ion-pair separation works.

(b) A urine sample containing an unknown amount of NE and a fixed, added concentration of DHBA gave a detector peak height ratio NE/DHBA = 0.298. Then small standard additions of NE were made, with the following results:

Added concentration of NE (ng/mL)	Peak height ratio NE/DHBA
12	0.414
24	0.554
36	0.664
48	0.792

Using the graphical treatment shown in Section 5-3, find the original concentration of NE in the specimen.

25-14. Decomposition of dithionite ($S_2O_4^{2-}$) was studied by chromatography on an anion-exchange column eluted with 20 mM trisodium 1,3,6-naphthalenetrisulfonate in 90 vol% H_2O/10 vol% CH_3CN with ultraviolet detection at 280 nm. A solution of sodium dithionite stored for 34 days in the absence of air gave five peaks identified as SO_3^{2-}, SO_4^{2-}, $S_2O_3^{2-}$, $S_2O_4^{2-}$, and $S_2O_5^{2-}$. All the peaks had a *negative* absorbance. Explain why.

25-15. (a) Suppose that the reservoir in Figure 25-8 contains 1.5 L of 2.0 M K_2PO_4. For how many hours can the reservoir provide 20 mM KOH at a flow rate of 1.0 mL/min if 75% consumption of K^+ in the reservoir is feasible?

(b) What starting and ending current would be required to produce a gradient from 5.0 mM KOH to 0.10 M KOH at 1.0 mL/min flow rate?

25-16. The system in Figure 25-8 can be adapted to produce the strong acid eluent methanesulfonic acid ($CH_3SO_3^-H^+$). For this purpose, the polarity of the electrodes is reversed and the reservoir can contain $NH_4^+CH_3SO_3^-$. The barrier membrane and the resin bed at the bottom of the figure must both be anion-exchangers loaded with $CH_3SO_3^-$. Draw this system and write the chemistry that occurs in each part.

Molecular Exclusion and Affinity Chromatography

25-17. (a) How can molecular exclusion chromatography be used to measure the molecular mass of a protein?

(b) Which pore size in Figure 25-17 is most suitable for chromatography of molecules with molecular mass near 100 000?

25-18. A gel-filtration column has a radius, r, of 0.80 cm and a length, l, of 20.0 cm.

(a) Calculate the volume, V_t, of the column, which is equal to $\pi r^2 l$.

(b) The void volume, V_o, was 18.1 mL, and the total volume of mobile phase was 35.8 mL. Find K_{av} for a solute eluted at 27.4 mL.

25-19. Ferritin (molecular mass 450 000), transferrin (molecular mass 80 000), and ferric citrate were separated by molecular exclusion chromatography on Bio-Gel P-300. The column had a length of 37 cm and a 1.5-cm diameter. Eluate fractions of 0.65 mL were collected. The maximum of each peak came at the following fractions: ferritin, 22; transferrin, 32; and ferric citrate, 84. (That is, the ferritin peak came at an elution volume of $22 \times 0.65 = 14.3$ mL.) Assuming that ferritin is eluted at the void volume and that ferric citrate is eluted at V_m, find K_{av} for transferrin.

25-20. (a) The void volume in Figure 25-17 is the volume at which the curves rise vertically at the left. What is the smallest molecular mass of molecules excluded from the 10-nm-pore-size column?

(b) What is the molecular mass of molecules eluted at 6.5 mL from the 10-nm column?

25-21. A polystyrene resin molecular exclusion HPLC column has a diameter of 7.8 mm and a length of 30 cm. The solid portions of the gel particles occupy 20% of the volume, the pores occupy 40%, and the volume between particles occupies 40%.

(a) At what volume would totally excluded molecules be expected to emerge?

(b) At what volume would the smallest molecules be expected?

(c) A mixture of polyethylene glycols of various molecular masses is eluted between 23 and 27 mL. What does this imply about the retention mechanism for these solutes on the column?

25-22. 🔲 The following substances were separated on a gel filtration column. Estimate the molecular mass of the unknown.

Compound	V_r (mL)	Molecular mass (Da)
Blue Dextran 2000	17.7	2×10^6
Aldolase	35.6	158 000
Catalase	32.3	210 000
Ferritin	28.6	440 000
Thyroglobulin	25.1	669 000
Unknown	30.3	?

Capillary Electrophoresis

25-23. What is electroosmosis?

25-24. Electroosmotic velocities of buffered solutions are shown for a bare silica capillary and one with aminopropyl groups (silica—Si—$CH_2CH_2CH_2NH_2$) covalently attached to the wall. A positive sign means that flow is toward the cathode. Explain the signs and relative magnitudes of the velocities.

Capillary wall	Electroosmotic velocity (mm/s) for $E = 4.0 \times 10^4$ V/m	
	pH 10	pH 2.5
Bare silica	+3.1	+0.2
Aminopropyl-modified silica	+1.8	−1.3

SOURCE: K. Emoto, J. M. Harris, and M. Van Alstine, "Grafting Poly(ethylene glycol) Epoxide to Amino-Derivatized Quartz: Effect of Temperature and pH on Grafting Density," Anal. Chem. 1996, 68, 3751.

25-25. Fluorescent derivatives of amino acids separated by capillary zone electrophoresis had migration times with the following order: arginine (fastest) < phenylalanine < asparagine < serine < glycine (slowest). Explain why arginine has the shortest migration time.

25-26. What is the principal source of zone broadening in ideal capillary electrophoresis?

25-27. Consider the electrophoresis of heparin in Figure 25-21.

(a) Electrophoresis was carried out at pH 2.8, at which sulfate groups are negative. Why was reverse polarity (detector end positive) used?

(b) The ionic strength of 30 mg/mL heparin samples is higher than that of typical samples for electrophoresis. What is the benefit of high buffer concentration (0.6 M phosphate)?

(c) A narrow capillary (25 μm diameter) was chosen to be compatible with the high-ionic-strength buffer. What is the advantage of the narrow capillary?

(d) Li^+ has lower mobility than Na^+. Explain why lithium phosphate can be used at higher electric field strength than sodium phosphate to generate the same current. What is the advantage of higher field strength for this separation?

25-28. (a) An electrophoresis channel etched into a glass plate has a rectangular 12×50 μm cross section. How long (mm) is the sample zone for a 100-pL sample?

(b) If sample travels 24 mm in 8 s to reach the detector, what is the standard deviation in bandwidth contributed by the finite length of the injection zone? (*Hint:* See Equation 22-32.)

(c) If the diffusion coefficient of a solute is 1.0×10^{-8} m²/s, what is the diffusional contribution (standard deviation) to band broadening?

(d) What is the expected total bandwidth at the baseline ($w = 4\sigma$)?

25-29. State three different methods to reduce electroosmotic flow. Why does the direction of electroosmotic flow change when a silica capillary is washed with a cationic surfactant?

25-30. Explain how neutral molecules can be separated by micellar electrokinetic chromatography. Why is this a form of chromatography?

25-31. (a) What pressure difference is required to inject a sample equal to 1.0% of the length of a 60.0-cm capillary in 4.0 s if the diameter is 50 μm? Assume that the viscosity of the solution is 0.001 0 kg/(m · s).

(b) The pressure exerted by a column of water of height h is $h\rho g$, where ρ is the density of water and g is the acceleration of gravity (9.8 m/s²). To what height would you need to raise the sample vial to create the necessary pressure to load the sample in 4.0 s? Is it possible to raise the inlet of this column to this height? How could you obtain the desired pressure?

25-32. (a) How many moles of analyte are present in a 10.0-μM solution that occupies 1.0% of the length of a 25-μm $\times$ 60.0-cm capillary?

(b) What voltage is required to inject this many moles into a capillary in 4.0 s if the sample has 1/10 of the conductivity of background electrolyte, $\mu_{app} = 3.0 \times 10^{-8}$ m²/(V · s), and the sample concentration is 10.0 μM?

25-33. Measure the number of plates for the electrophoretic peak in Figure 25-23. Use Equation 22-29 for asymmetric peaks to find the number of plates for the chromatographic peak.

25-34. (a) A long thin molecule has a greater friction coefficient than a short fat molecule. Predict whether fumarate or maleate will have greater electrophoretic mobility.

Fumarate   Maleate

(b) Electrophoresis is run with the injection end positive and the detection end negative. At pH 8.5, both anions have a charge of -2. The electroosmotic flow from the positive terminal to the negative terminal is greater than the electrophoretic flow, so these two anions have a net migration from the positive to the negative end of the capillary in electrophoresis. From your answer to part (a), predict the order of elution of these two species.

(c) At pH 4.0, both anions have a charge close to -1, and the electroosmotic flow is weak. Therefore, electrophoresis is run with the injection end negative and the detection end positive. The anions migrate from the negative end of the capillary to the positive end. Predict the order of elution.

25-35. (a) A particular solution in a particular capillary has an electroosmotic mobility of 1.3×10^{-8} m^2/(V $\cdot$ s) at pH 2 and 8.1×10^{-8} m^2/(V $\cdot$ s) at pH 12. How long will it take a neutral solute to travel 52 cm from the injector to the detector if 27 kV is applied across the 62-cm-long capillary tube at pH 2? At pH 12?

(b) An analyte anion has an electrophoretic mobility of -1.6×10^{-8} m^2/(V $\cdot$ s). How long will it take to reach the detector at pH 2? At pH 12?

25-36. Figure 25-27 shows the effect on resolution of increasing voltage from 28 to 120 kV.

(a) What is the expected ratio of migration times $(t_{120\ kV}/t_{28\ kV})$ in the two experiments? Measure the migration times for peak 1 and find the observed ratio.

(b) What is the expected ratio of plates $(N_{120\ kV}/N_{28\ kV})$ in the two experiments?

(c) What is the expected ratio of bandwidths $(\sigma_{120\ kV}/\sigma_{28\ kV})$?

(d) What is the physical reason why increasing voltage decreases bandwidth and increases resolution?

25-37. The observed behavior of benzyl alcohol ($C_6H_5CH_2OH$) in capillary electrophoresis is given here. Draw a graph showing the number of plates versus the electric field and explain what happens as the field increases.

Electric field (V/m)	Number of plates
6 400	38 000
12 700	78 000
19 000	96 000
25 500	124 000
31 700	124 000
38 000	96 000

25-38. Measure the width of the $^{35}Cl^-$ peak at half-height in Figure 25-36 and calculate the plate number. The capillary was 40.0 cm in length. Find the plate height.

25-39. The migration time for Cl^- in a capillary zone electrophoresis experiment is 17.12 min and the migration time for I^- is 17.78 min. From mobilities in Table 14-1, predict the migration time of Br^-. (The observed value is 19.6 min.)

23-40. 🔳 *Molecular mass by sodium dodecylsulfate–gel electrophoresis.* Ferritin is a hollow iron-storage protein[61] consisting of 24 subunits that are a variable mixture of heavy (H) or light (L) chains, arranged in octahedral symmetry. The hollow center, with a diameter of 8 nm, can hold up to 4 500 iron atoms in the approximate form of the mineral ferrihydrite ($5Fe_2O_3 \cdot 9H_2O$). Iron(II) enters the protein through eight pores located on the threefold

symmetry axes of the octahedron. Oxidation to Fe(III) occurs at catalytic sites on the H chains. Other sites on the inside of the L chains appear to nucleate the crystallization of ferrihydrite.

Migration times for protein standards and the ferritin subunits are given in the table. Prepare a graph of log(molecular mass) versus 1/(relative migration time), where relative migration time = (migration time)/(migration time of marker dye). Compute the molecular mass of the ferritin light and heavy chains. The masses of the chains, computed from amino acid sequences, are 19 766 and 21 099 Da.

Protein	Molecular mass (Da)	Migration time (min)
Orange G marker dye	small	13.17
α-Lactalbumin	14 200	16.46
Carbonic anhydrase	29 000	18.66
Ovalbumin	45 000	20.16
Bovine serum albumin	66 000	22.36
Phosphorylase B	97 000	23.56
β-Galactosidase	116 000	24.97
Myosin	205 000	28.25
Ferritin light chain		17.07
Ferritin heavy chain		17.97

SOURCE: J. K. Grady, J. Zang, T. M. Laue, P. Arosio, and N. D. Chasteen, *"Characterization of the H- and L-Subunit Ratios in Ferritins by Sodium Dodecyl Sulfate–Capillary Gel Electrophoresis," Anal. Biochem. 2002, 302, 263.*

25-41. 🔳 *Protein charge ladder.* Electrophoretic mobility is proportional to charge. If members of a charge ladder (Figure 25-26) have the same friction coefficient (that is, the same size and shape), then the charge of the unmodified protein divided by its electrophoretic mobility, z_0/μ_0, is equal to the charge of the nth member divided by its electrophoretic mobility $(z_0 + \Delta z_n)/\mu_n$. Setting these two expressions equal to each other and rearranging gives

$$\Delta z_n = z_0\left(\frac{\mu_n}{\mu_0} - 1\right)$$

where z_0 is the charge of the unmodified protein, Δz_n is the charge difference between the nth modified protein and the unmodified protein, μ_n is the electrophoretic mobility of the nth modified protein, and μ_0 is the electrophoretic mobility of the unmodified protein. The migration time of the neutral marker molecule in Figure 25-26 is 308.5 s. The migration time of the unmodified protein is 343.0 s. Other members of the charge ladder have migration times of 355.4, 368.2, 382.2, 395.5, 409.1, 424.9, 438.5, 453.0, 467.0, 482.0, 496.4, 510.1, 524.1, 536.9, 551.4, 565.1, 577.4, and 588.5 s. Calculate the electrophoretic mobility of each protein and prepare a plot of Δz_n versus $(\mu_n/\mu_0) - 1$. If the points lie on a straight line, the slope is the charge of the unmodified protein, z_0. Prepare such a plot, suggest an explanation for its shape, and find z_0.

25-42. *Resolution.* Suppose that the electroosmotic mobility of a solution is $+1.61 \times 10^{-7}$ m^2/(V $\cdot$ s). How many plates are required to separate sulfate from bromide with a resolution of 2.0? Use Table 14-1 for mobilities and Equation 22-30 for resolution.

25-43. The water-soluble vitamins niacinamide (a neutral compound), riboflavin (a neutral compound), niacin (an anion), and thiamine (a cation) were separated by micellar electrokinetic chromatography in 15 mM borate buffer (pH 8.0) with 50 mM sodium dodecyl sulfate. The migration times were niacinamide (8.1 min), riboflavin (13.0 min), niacin (14.3 min), and thiamine (21.9 min). What would the order have been in the absence of sodium dodecyl sulfate? Which compound is most soluble in the micelles?

25-44. When the following three compounds are separated by micellar electrokinetic chromatography at pH 9.6, three peaks are observed. When 10 mM α-cyclodextrin is added to the run buffer, two of the three peaks split into two peaks, giving a total of five peaks. Explain this observation and predict which compound does not split.

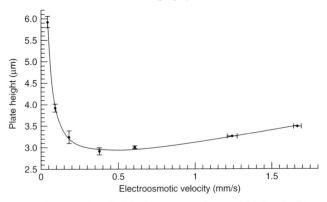

1	2	3
Cyclobarbital	Thiopental	Phenobarbital

25-45. A van Deemter plot for the separation of neutral dyes by micellar electrokinetic chromatography is shown below.[62]

[Plot: Plate height (μm) on y-axis ranging 2.5 to 6.0, Electroosmotic velocity (mm/s) on x-axis ranging 0 to 1.5]

(a) Explain why plate height increases at low and high velocities.

(b) The irregular flow path term, A, in the van Deemter equation should really be 0 for the ideal case of micellar electrokinetic chromatography. The observed value of A is 2.32 μm, which accounts for two-thirds of the band broadening at the optimum velocity. Suggest some reasons why A is not 0.

25-46. To obtain the best separation of two weak acids in capillary electrophoresis, it makes sense to use the pH at which their charge difference is greatest. Prepare a spreadsheet to examine the charges of malonic and phthalic acid as a function of pH. At what pH is the difference greatest?

25-47. *Optimizing a separation of acids.* Benzoic acid containing ^{16}O can be separated from benzoic acid containing ^{18}O by electrophoresis at a suitable pH because they have slightly different acid dissociation constants. The difference in mobility is caused by the different fraction of each acid in the anionic form, A^-. Calling this fraction α, we can write

$$H^{16}A \overset{^{16}K}{\rightleftharpoons} H^+ + {}^{16}A^- \qquad H^{18}A \overset{^{18}K}{\rightleftharpoons} H^+ + {}^{18}A^-$$

$$^{16}\alpha = \frac{^{16}K}{^{16}K + [H^+]} \qquad {}^{18}\alpha = \frac{^{18}K}{^{18}K + [H^+]}$$

where K is the equilibrium constant. The greater the fraction of acid in the form A^-, the faster it will migrate in the electric field.

It can be shown that, for electrophoresis, the maximum separation will occur when $\Delta\alpha/\sqrt{\overline{\alpha}}$ is a maximum. In this expression, $\Delta\alpha = {}^{16}\alpha - {}^{18}\alpha$, and $\overline{\alpha}$ is the average fraction of dissociation $[= \frac{1}{2}({}^{16}\alpha + {}^{18}\alpha)]$.

(a) Let us denote the ratio of acid dissociation constants as $R = {}^{16}K/{}^{18}K$. In general, R will be close to unity. For benzoic acid, $R = 1.020$. Abbreviate ^{16}K as K and write $^{18}K = K/R$. Derive an expression for $\Delta\alpha/\sqrt{\overline{\alpha}}$ in terms of K, $[H^+]$, and R. Because both equilibrium constants are nearly equal (R is close to unity), set $\overline{\alpha}$ equal to $^{16}\alpha$ in your expression.

(b) Find the maximum value of $\Delta\alpha/\sqrt{\overline{\alpha}}$ by taking the derivative with respect to $[H^+]$ and setting it equal to 0. Show that the maximum difference in mobility of isotopic benzoic acids occurs when $[H^+] = (K/2R)(1 + \sqrt{1 + 8R})$.

(c) Show that, for $R \approx 1$, this expression simplifies to $[H^+] = 2K$, or $pH = pK - 0.30$. That is, the maximum electrophoretic separation should occur when the column buffer has $pH = pK - 0.30$, regardless of the exact value of R.[63]

25-48. (a) *Ion mobility spectrometry* (Section 21-3) is *gas-phase electrophoresis.* Describe how ion mobility spectrometry works and state the analogies between this technique and capillary electrophoresis.

(b) As in electrophoresis, the velocity, u, of a gas-phase ion is $u = \mu E$, where μ is the mobility of the ion and E is the electric field ($E = V/L$, where V is the voltage difference across distance L). In ion mobility spectrometry, the time to go from the gate to the detector (Figure 21-20a) is called *drift time*, t_d. Drift time is related to voltage: $t_d = L/u = L/(\mu E) = L/(\mu(V/L)) = L^2/\mu V$. Plate number is $N = 5.55(t_d/w_{1/2})^2$, where $w_{1/2}$ is the width of the peak at half-height. Ideally, peak width depends only on the width of the gate pulse that admits ions to the drift tube and on diffusive broadening of ions while they migrate:[64]

$$\underset{\substack{\text{Initial width} \\ \text{from gate pulse}}}{w_{1/2}^2} = \underset{}{t_g^2} + \underset{\substack{\text{Diffusive} \\ \text{broadening}}}{\left(\frac{16kT \ln 2}{Vez}\right)t_d^2}$$

where t_g is the time that the ion gate is open, k is Boltzmann's constant, T is temperature, V is the potential difference from the gate to the detector, e is the elementary charge, and z is the charge of the ion. Prepare a graph of N versus V ($0 \le V \le 20\,000$) for an ion with $\mu = 8 \times 10^{-5}$ m²/(s · V), and $t_g = 0, 0.05,$ or 0.2 ms at 300 K. Let the length of the drift region be $L = 0.2$ m. Explain the shapes of the curves. What is the disadvantage of using short t_g?

(c) Why does decreasing T increase N?

(d) In a well-optimized ion mobility spectrometer, protonated arginine ion ($z = 1$) had a drift time of 24.925 ms and $w_{1/2} = 0.154$ ms at 300 K. Find N. For $V = 12\,500$ V and $t_g = 0.05$ ms, what is the theoretical plate number?

(e) Resolution is given by $R = (\sqrt{N}/4)(\gamma - 1)$, where γ is the ratio of drift times for two components. In a well-optimized ion mobility spectrometer, protonated leucine had $t_d = 22.5$ ms and protonated isoleucine had $t_d = 22.0$ ms. Both had $N \approx 80\,000$. What is the resolution of the two peaks?

26 | Gravimetric Analysis, Precipitation Titrations, and Combustion Analysis

Layers of rock exposed in the Grand Canyon by the erosive action of the Colorado River provide a window on a billion years of Earth's history. [*Left:* Adapted from F. Press, R. Siever, J. Gratzinger, and T. H. Jordan. *Understanding Earth*, 4th ed. (New York: W. H. Freeman and Company, 2004). *Right:* Carol Polich/Lonely Planet Images.]

In the 1800s, geologists understood that new layers (*strata*) of rock are deposited on top of older layers. Characteristic fossils in each layer helped geologists to identify strata from the same geologic era all around the world. However, the actual age of each layer was unknown.

Ernest Rutherford, Frederick Soddy, Bertram Boltwood, and Robert Strutt showed in the early 1900s that uranium decays to lead plus eight atoms of helium with a half-life of several billion years. Rutherford estimated the age of a rock from its U and He content. Boltwood obtained more reliable ages of minerals by measuring the U and Pb content.

In 1910, Arthur Holmes, a 20-year-old student of Strutt at Imperial College in London, became the first person to assign actual ages to minerals formed in specific geologic periods. Holmes conjectured that when certain U-containing minerals crystallized from hot magma, the crystals would be relatively free of impurities such as Pb. Once the mineral solidified, Pb would begin to accumulate. The ratio Pb/U would tell how long ago the mineral crystallized. Holmes measured U by the rate of production of radioactive Rn gas. To measure Pb, he dissolved each mineral in molten borax, dissolved the fused mass in acid, and quantitatively precipitated milligrams of $PbSO_4$. The nearly constant ratio Pb/U = 0.045 g/g in 15 minerals was consistent with the hypotheses that Pb is the end product of U decay and that little Pb had been present when the minerals crystallized. The calculated age of the "Devonian-age" minerals was 370 million years—four times older than the most accepted age of Earth at that time.

Geologic ages deduced by Holmes in 1911

Geologic period	Pb/U (g/g)	Millions of years	Today's accepted value
Carboniferous	0.041	340	330–362
Devonian	0.045	370	362–380
Silurian	0.053	430	418–443
Precambrian	0.125–0.20	1 025–1 640	900–2 500

SOURCE: *C. Lewis, The Dating Game (Cambridge: Cambridge University Press, 2000); A. Holmes, "The Association of Lead with Uranium in Rock-Minerals, and Its Application to the Measurement of Geological Time," Proc. R. Soc. Lond. A **1911**, 85, 248.*

Gravimetry was the main form of chemical analysis in the eighteenth and nineteenth centuries, but it is too tedious to be a method of choice today. However, gravimetry remains one of the most accurate methods. Standards used to calibrate instruments are frequently derived from gravimetric or titrimetric procedures.

In **gravimetric analysis,** the mass of a product is used to calculate the quantity of the original analyte (the species being analyzed). Careful gravimetric analysis by T. W. Richards and his colleagues early in the twentieth century determined the atomic masses of Ag, Cl, and N to six-figure accuracy.[1] This Nobel Prize–winning work allowed the accurate determination of atomic masses of many elements. In **precipitation titrations,** the quantity of titrant required for complete precipitation of analyte tells us how much analyte was present. In **combustion analysis,** a sample is burned in excess oxygen and products are measured. Combustion is typically used to measure C, H, N, S, and halogens in organic matter. To measure other elements, organic matter is burned in a closed system. Products and *ash* (solid residue) are then dissolved in acid or base and measured by inductively coupled plasma with atomic emission or mass spectrometry.

26-1 Examples of Gravimetric Analysis

Chloride can be measured by precipitating the anion with Ag^+ and finding the mass of AgCl.

$$Ag^+ + Cl^- \rightarrow AgCl(s) \tag{26-1}$$

EXAMPLE A Gravimetric Calculation

A 10.00-mL solution containing Cl^- was treated with excess $AgNO_3$ to precipitate 0.436 8 g of AgCl. What was the molarity of Cl^- in the unknown?

Solution The formula mass of AgCl is 143.321. Precipitate weighing 0.436 8 g contains

$$\frac{0.436\ 8\ \text{g AgCl}}{143.321\ \text{g AgCl/mol AgCl}} = 3.048 \times 10^{-3}\ \text{mol AgCl}$$

Because 1 mol of AgCl contains 1 mol of Cl^-, there must have been 3.048×10^{-3} mol of Cl^- in the unknown.

$$[Cl^-] = \frac{3.048 \times 10^{-3}\ \text{mol}}{0.010\ 00\ \text{L}} = 0.304\ 8\ \text{M}$$

Test Yourself How many grams of Br^- were in a sample that produced 1.000 g of AgBr precipitate (FM = 187.77)? (***Answer:*** 0.425 5 g)

Marie and Pierre Curie and Henri Becquerel shared the Nobel Prize in physics in 1903 for pioneering investigations of radioactivity. The Curies needed four years to isolate 100 mg of $RaCl_2$ from several tons of ore. Marie received the Nobel Prize in chemistry in 1911 for her isolation of metallic radium. Linus Pauling, John Bardeen, and Frederick Sanger are the only others who received two Nobel Prizes.

EXAMPLE Marie Curie's Measurement of the Atomic Mass of Radium

In her Ph.D. research (*Radioactive Substances*, 1903), Marie Curie measured the atomic mass of the element radium, which she discovered. She knew that radium is in the same family as barium, so the formula of radium chloride is $RaCl_2$. When 0.091 92 g of pure $RaCl_2$ was dissolved and treated with excess $AgNO_3$, 0.088 90 g of AgCl precipitated. How many moles of Cl^- were in the $RaCl_2$? From this measurement, find the atomic mass of Ra.

Solution AgCl precipitate weighing 0.088 90 g contains

$$\frac{0.088\ 90\ \text{g AgCl}}{143.321\ \text{g AgCl /mol AgCl}} = 6.202_9 \times 10^{-4}\ \text{mol AgCl}$$

Because 1 mol of AgCl contains 1 mol of Cl^-, there must have been $6.202_9 \times 10^{-4}$ mol of Cl^- in the $RaCl_2$. For 2 mol of Cl, there must be 1 mol of Ra, so

$$\text{mol radium} = \frac{6.202_9 \times 10^{-4}\ \text{mol Cl}}{2\ \text{mol Cl/mol Ra}} = 3.101_4 \times 10^{-4}\ \text{mol}$$

Let the formula mass of $RaCl_2$ be x. We found that 0.091 92 g $RaCl_2$ contains $3.101_4 \times 10^{-4}$ mol $RaCl_2$. Therefore,

$$3.101_4 \times 10^{-4} \text{ mol RaCl}_2 = \frac{0.091\ 92 \text{ g RaCl}}{x \text{ g RaCl}_2/\text{mol RaCl}_2}$$

$$x = \frac{0.091\ 92 \text{ g RaCl}}{3.101_4 \times 10^{-4} \text{ mol RaCl}_2} = 296.3_8 \text{ g/mol}$$

The atomic mass of Cl is 35.453, so the formula mass of $RaCl_2$ is

Formula mass of $RaCl_2$ = atomic mass of Ra + 2(35.453 g/mol) = 296.3_8 g/mol

$\Rightarrow$ atomic mass of Ra = 225.5 g/mol

The inside cover of this book lists the atomic number (the integer mass) of the long-lived isotope of Ra, which is 226.

Test Yourself   How many grams of AgBr would have been formed from 0.100 g of $RaBr_2$? (*Answer:* 0.097 g)

Representative analytical precipitations are listed in Table 26-1. A few common organic **precipitants** (agents that cause precipitation) are listed in Table 26-2. Conditions must be controlled to selectively precipitate one species. Potentially interfering substances may need to be removed prior to analysis.

TABLE 26-1   Representative gravimetric analyses

Species analyzed	Precipitated form	Form weighed	Interfering species
K^+	$KB(C_6H_5)_4$	$KB(C_6H_5)_4$	NH_4^+, Ag^+, Hg^{2+}, Tl^+, Rb^+, Cs^+
Mg^{2+}	$Mg(NH_4)PO_4 \cdot 6H_2O$	$Mg_2P_2O_7$	Many metals except Na^+ and K^+
Ca^{2+}	$CaC_2O_4 \cdot H_2O$	$CaCO_3$ or CaO	Many metals except Mg^{2+}, Na^+, K^+
Ba^{2+}	$BaSO_4$	$BaSO_4$	Na^+, K^+, Li^+, Ca^{2+}, Al^{3+}, Cr^{3+}, Fe^{3+}, Sr^{2+}, Pb^{2+}, NO_3^-
Ti^{4+}	TiO (5,7-dibromo-8-hydroxyquinoline)$_2$	Same	Fe^{3+}, Zr^{4+}, Cu^{2+}, $C_2O_4^{2-}$, citrate, HF
VO_4^{3-}	Hg_3VO_4	V_2O_5	Cl^-, Br^-, I^-, SO_4^{2-}, CrO_4^{2-}, AsO_4^{3-}, PO_4^{3-}
Cr^{3+}	$PbCrO_4$	$PbCrO_4$	Ag^+, NH_4^+
Mn^{2+}	$Mn(NH_4)PO_4 \cdot H_2O$	$Mn_2P_2O_7$	Many metals
Fe^{3+}	$Fe(HCO_2)_3$	Fe_2O_3	Many metals
Co^{2+}	$Co(1\text{-nitroso-2-naphtholate})_2$	$CoSO_4$ (by reaction with H_2SO_4)	Fe^{3+}, Pd^{2+}, Zr^{4+}
Ni^{2+}	$Ni(dimethylglyoximate)_2$	Same	Pd^{2+}, Pt^{2+}, Bi^{3+}, Au^{3+}
Cu^{2+}	CuSCN (after reduction of Cu^{2+} to Cu^+ with HSO_3^-)	CuSCN	NH_4^+, Pb^{2+}, Hg^{2+}, Ag^+
Zn^{2+}	$Zn(NH_4)PO_4 \cdot H_2O$	$Zn_2P_2O_7$	Many metals
Ce^{4+}	$Ce(IO_3)_4$	CeO_2	Th^{4+}, Ti^{4+}, Zr^{4+}
Al^{3+}	$Al(8\text{-hydroxyquinolate})_3$	Same	Many metals
Sn^{4+}	$Sn(cupferron)_4$	SnO_2	Cu^{2+}, Pb^{2+}, As(III)
Pb^{2+}	$PbSO_4$	$PbSO_4$	Ca^{2+}, Sr^{2+}, Ba^{2+}, Hg^{2+}, Ag^+, HCl, HNO_3
NH_4^+	$NH_4B(C_6H_5)_4$	$NH_4B(C_6H_5)_4$	K^+, Rb^+, Cs^+
Cl^-	AgCl	AgCl	Br^-, I^-, SCN^-, S^{2-}, $S_2O_3^{2-}$, CN^-
Br^-	AgBr	AgBr	Cl^-, I^-, SCN^-, S^{2-}, $S_2O_3^{2-}$, CN^-
I^-	AgI	AgI	Cl^-, Br^-, SCN^-, S^{2-}, $S_2O_3^{2-}$, CN^-
SCN^-	CuSCN	CuSCN	NH_4^+, Pb^{2+}, Hg^{2+}, Ag^+
CN^-	AgCN	AgCN	Cl^-, Br^-, I^-, SCN^-, S^{2-}, $S_2O_3^{2-}$
F^-	$(C_6H_5)_3SnF$	$(C_6H_5)_3SnF$	Many metals (except alkali metals), SiO_4^{4-}, CO_3^{2-}
ClO_4^-	$KClO_4$	$KClO_4$	
SO_4^{2-}	$BaSO_4$	$BaSO_4$	Na^+, K^+, Li^+, Ca^{2+}, Al^{3+}, Cr^{3+}, Fe^{3+}, Sr^{2+}, Pb^{2+}, NO_3^-
PO_4^{3-}	$Mg(NH_4)PO_4 \cdot 6H_2O$	$Mg_2P_2O_7$	Many metals except Na^+, K^+
NO_3^-	Nitron nitrate	Nitron nitrate	ClO_4^-, I^-, SCN^-, CrO_4^{2-}, ClO_3^-, NO_2^-, Br^-, $C_2O_4^{2-}$
CO_3^{2-}	CO_2 (by acidification)	CO_2	(The liberated CO_2 is trapped with Ascarite and weighed.)

TABLE 26-2 Common organic precipitating agents

Name	Structure	Ions precipitated
Dimethylglyoxime	(structure)	Ni^{2+}, Pd^{2+}, Pt^{2+}
Cupferron	(structure)	Fe^{3+}, VO_2^+, Ti^{4+}, Zr^{4+}, Ce^{4+}, Ga^{3+}, Sn^{4+}
8-Hydroxyquinoline (oxine)	(structure)	Mg^{2+}, Zn^{2+}, Cu^{2+}, Cd^{2+}, Pb^{2+}, Al^{3+}, Fe^{3+}, Bi^{3+}, Ga^{3+}, Th^{4+}, Zr^{4+}, UO_2^{2+}, TiO^{2+}
Salicylaldoxime	(structure)	Cu^{2+}, Pb^{2+}, Bi^{3+}, Zn^{2+}, Ni^{2+}, Pd^{2+}
1-Nitroso-2-naphthol	(structure)	Co^{2+}, Fe^{3+}, Pd^{2+}, Zr^{4+}
Nitron	(structure)	NO_3^-, ClO_4^-, BF_4^-, WO_4^{2-}
Sodium tetraphenylborate	$Na^+B(C_6H_5)_4^-$	K^+, Rb^+, Cs^+, NH_4^+, Ag^+, organic ammonium ions
Tetraphenylarsonium chloride	$(C_6H_5)_4As^+Cl^-$	$Cr_2O_7^{2-}$, MnO_4^-, ReO_4^-, MoO_4^{2-}, WO_4^{2-}, ClO_4^-, I_3^-

FIGURE 26-1 Particle-size distribution of colloids formed when $FeSO_4$ was oxidized to Fe^{3+} in 10^{-4} M OH^- in the presence of phosphate (PO_4^{3-}), silicate (SiO_4^{4-}), or no added anions. [From M. L. Magnuson, D. A. Lytle, C. M. Frietch, and C. A. Kelty, "Characterization of Submicron Aqueous Iron(III) Colloids by Sedimentation Field Flow Fractionation," *Anal. Chem.* **2001**, *73*, 4815.]

Supersaturation tends to decrease the particle size of a precipitate.

26-2 Precipitation

The ideal product of a gravimetric analysis should be pure, insoluble, and easily filterable, and should possess a known composition. Few substances meet these requirements, but appropriate techniques can help optimize properties of gravimetric precipitates.

Precipitated particles should not be so small that they clog or pass through the filter. Larger crystals have less surface area to which impurities can become attached. At the other extreme is a *colloidal suspension* of particles that have diameters in the approximate range 1–500 nm and pass through most filters (Figure 26-1 and Demonstration 26-1). Precipitation conditions determine the particle size.

Crystal Growth

Crystallization occurs in two phases: nucleation and particle growth. In **nucleation,** solutes are thought to form a disorganized cluster of sufficient size, which then reorganizes into an ordered structure capable of growing into larger particles.[5] Nucleation can occur on suspended impurity particles or scratches on a glass surface. When Fe(III) reacts with 0.1 M tetramethylammonium hydroxide at 25°C, nuclei of hydrated $Fe(OH)_3$ are 4 nm in diameter and contain ~50 Fe atoms.[6] In *particle growth*, molecules or ions condense onto the nucleus to form a larger crystal. $Fe(OH)_3$ nuclei grow into plates with lateral dimensions of ~30 × 7 nm after 15 min at 60°C.

A **supersaturated** solution contains more solute than should be present at equilibrium. Nucleation proceeds faster than particle growth in a highly supersaturated solution to produce tiny particles or, worse, a colloid. In a less supersaturated solution, nucleation is slower, and nuclei have a chance to grow into larger, more tractable particles.

Techniques that promote particle growth include

1. Raising the temperature to increase solubility and thereby decrease supersaturation
2. Adding precipitant slowly with vigorous mixing, to prevent a local, highly supersaturated condition where the stream of precipitant first enters the analyte
3. Using a large volume of solution so that concentrations of analyte and precipitant are low

Colloids are particles with diameters of ~1–100 nm. They are larger than most molecules but too small to precipitate. They remain in solution, suspended by the *Brownian motion* (random movement) of solvent molecules.[2]

To prepare colloidal iron(III) hydroxide, heat 200 mL of distilled water in a beaker to 70°–90°C and leave an identical beaker of water at room temperature. Add 1 mL of 1 M FeCl$_3$ to each beaker and stir. The warm solution turns brown-red in a few seconds, whereas the cold solution remains yellow (Color Plate 32). The yellow color is characteristic of low-molecular-mass Fe^{3+} compounds. The red color results from colloidal aggregates of Fe^{3+} ions held together by hydroxide, oxide, and some chloride ions. Each particle contains ~10^3 atoms of Fe with a molecular mass of 10^5 and a diameter of 10 nm.

You can demonstrate the size of colloidal particles by **dialysis,** a process in which two solutions are separated by a *semipermeable membrane* that has pores with diameters of 1–5 nm.[3] Small molecules diffuse through these pores, but large molecules (such as proteins or colloids) cannot. (*Microdialysis* in biology is shown in Figure 25-41.)

Pour some brown-red colloid into a dialysis tube knotted at one end; then tie off the other end. Drop the tube into a flask of distilled water to show that the color remains inside the bag after several days (Color Plate 32). For comparison, leave an identical bag containing dark blue 1 M CuSO$_4$ · 5H$_2$O in another flask of water. Cu^{2+} diffuses out and the internal and external solution will be light blue in 24 h. The yellow food coloring, tartrazine, can be used in place of Cu^{2+}. If dialysis is conducted in hot water, it is completed during one class period.[4]

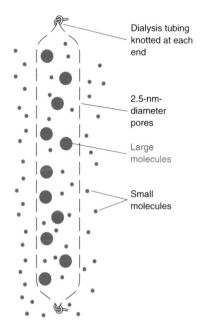

Dialysis tubing knotted at each end

2.5-nm-diameter pores

Large molecules

Small molecules

Large molecules remain trapped inside a dialysis bag, whereas small molecules diffuse through the membrane in both directions.

Dialysis is used to treat patients suffering from kidney failure. Blood is passed over a membrane through which metabolic waste products diffuse and are diluted into a large volume of liquid that is discarded. Protein molecules, which are a necessary part of the blood plasma, are too large to cross the membrane and are retained in the blood.

Homogeneous Precipitation

In **homogeneous precipitation,** the precipitant is generated slowly by a chemical reaction (Table 26-3). For example, urea decomposes slowly in boiling water to produce OH$^-$:

$$\underset{\text{Urea}}{H_2N-\overset{\overset{\displaystyle O}{\|}}{C}-NH_2} + 3H_2O \xrightarrow{\text{heat}} CO_2 + 2NH_4^+ + 2OH^- \qquad (26\text{-}2)$$

Gradual OH$^-$ production enhances the particle size of Fe(III) formate precipitate:

$$\underset{\text{Formic acid}}{H-\overset{\overset{\displaystyle O}{\|}}{C}-OH} + OH^- \rightarrow \underset{\text{Formate}}{HCO_2^-} + H_2O \qquad (26\text{-}3)$$

$$3HCO_2^- + Fe^{3+} \rightarrow \underset{\text{Fe(III) formate}}{Fe(HCO_2)_3 \cdot nH_2O(s)\downarrow} \qquad (26\text{-}4)$$

Precipitation in the Presence of Electrolyte

Ionic compounds are usually precipitated in the presence of an electrolyte. To understand why, we must discuss how tiny colloidal crystallites *coagulate* (come together) into larger crystals. We illustrate the case of AgCl, which is commonly formed in 0.1 M HNO$_3$.

Figure 26-2 shows a colloidal particle of AgCl growing in a solution containing excess Ag$^+$, H$^+$, and NO$_3^-$. The surface of the particle has excess positive charge due to the **adsorption** of extra silver ions on exposed chloride ions. (To be adsorbed means to be attached to the surface. In contrast, **absorption** means penetration beyond the surface, to the inside.) The

An *electrolyte* is a compound that dissociates into ions when it dissolves.

Although it is common to find the excess common ion adsorbed on the crystal surface, other ions can be selectively adsorbed. Citrate is bound in preference to sulfate on the surface of BaSO$_4$.

TABLE 26-3 Common reagents used for homogeneous precipitation

Precipitant	Reagent	Reaction	Some elements precipitated
OH^-	Urea	$(H_2N)_2CO + 3H_2O \rightarrow CO_2 + 2NH_4^+ + 2OH^-$	Al, Ga, Th, Bi, Fe, Sn
OH^-	Potassium cyanate	$HOCN + 2H_2O \rightarrow NH_4^+ + CO_2 + OH^-$ Hydrogen cyanate	Cr, Fe
S^{2-}	Thioacetamide[a]	$\underset{\substack{\|\| \\ S}}{CH_3CNH_2} + H_2O \rightarrow \underset{\substack{\|\| \\ O}}{CH_3CNH_2} + H_2S$	Sb, Mo, Cu, Cd
SO_4^{2-}	Sulfamic acid	$H_3\overset{+}{N}SO_3^- + H_2O \rightarrow NH_4^+ + SO_4^{2-} + H^+$	Ba, Ca, Sr, Pb
$C_2O_4^{2-}$	Dimethyl oxalate	$\underset{\substack{\|\| \| \| \\ OO}}{CH_3OCCOCH_3} + 2H_2O \rightarrow 2CH_3OH + C_2O_4^{2-} + 2H^+$	Ca, Mg, Zn
PO_4^{3-}	Trimethyl phosphate	$(CH_3O)_3P{=}O + 3H_2O \rightarrow 3CH_3OH + PO_4^{3-} + 3H^+$	Zr, Hf
CrO_4^{2-}	Chromic ion plus bromate	$2Cr^{3+} + BrO_3^- + 5H_2O \rightarrow 2CrO_4^{2-} + Br^- + 10H^+$	Pb
8-Hydroxyquinoline	8-Acetoxyquinoline	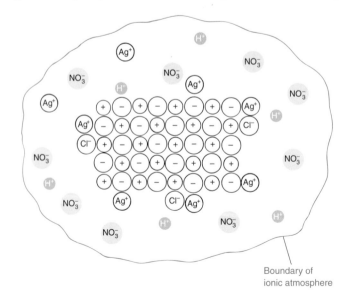	Al, U, Mg, Zn

*a. Hydrogen sulfide is volatile and toxic; it should be handled only in a well-vented hood. Thioacetamide is a carcinogen that should be handled with gloves. If thioacetamide contacts your skin, wash yourself thoroughly immediately. Leftover reagent is destroyed by heating at 50°C with 5 mol of NaOCl per mole of thioacetamide. [H. Elo, J. Chem. Ed. **1987**, 64, A144.]*

positively charged surface attracts anions and repels cations from the *ionic atmosphere* (Figure 26-2) surrounding the particle. The positively charged particle and the negatively charged ionic atmosphere together are called the **electric double layer.**

Colloidal particles must collide with one another to coalesce. However, the negatively charged ionic atmospheres of the particles repel one another. Particles must have enough kinetic energy to overcome electrostatic repulsion before they can coalesce.

Heating promotes coalescence by increasing the kinetic energy. Increasing electrolyte concentration (HNO_3 for AgCl) decreases the volume of the ionic atmosphere and allows particles to come closer together before electrostatic repulsion becomes significant. For this reason, most gravimetric precipitations are done in the presence of an electrolyte.

FIGURE 26-2 Colloidal particle of AgCl growing in a solution containing excess Ag^+, H^+, and NO_3^-. The particle has a net positive charge because of adsorbed Ag^+ ions. The region of solution surrounding the particle is called the *ionic atmosphere*. It has a net negative charge because the particle attracts anions and repels cations.

Digestion

Liquid from which a substance precipitates or crystallizes is called the *mother liquor*. After precipitation, most procedures call for a period of standing in the presence of the hot mother liquor. This treatment, called **digestion,** promotes slow recrystallization of the precipitate. Particle size increases and impurities tend to be expelled from the crystal.

Purity

Adsorbed impurities are bound to the surface of a crystal. *Absorbed* impurities (within the crystal) are classified as *inclusions* or *occlusions*. Inclusions are impurity ions that randomly occupy sites in the crystal lattice normally occupied by ions that belong in the crystal. Inclusions are more likely when the impurity ion has a size and charge similar to those of one of the ions that belongs to the product. Occlusions are pockets of impurity that are literally trapped inside the growing crystal.

Adsorbed, occluded, and included impurities are said to be **coprecipitated.** That is, the impurity is precipitated along with the desired product, even though the solubility of the impurity has not been exceeded (Figure 26-3).

Coprecipitation tends to be worst in colloidal precipitates such as $BaSO_4$, $Al(OH)_3$, and $Fe(OH)_3$, which all have a large surface area. Many procedures call for washing away the mother liquor, redissolving the precipitate, and *reprecipitating* the product. During the second precipitation, the concentration of impurities in solution is lower than during the first precipitation, and the degree of coprecipitation therefore tends to be lower.

Occasionally, a trace component is intentionally isolated by coprecipitation with a major component of the solution. The precipitate used to collect the trace component is said to be a *gathering agent*, and the process is called **gathering.** Natural arsenic in drinking water in Bangladesh is a significant health hazard. One way to remove arsenic is by coprecipitation with $Fe(OH)_3$.[7] Fe(II) or Fe(s) is added to the water and oxidized in air for several hours to precipitate $Fe(OH)_3$. After filtration through sand to remove solids, the water is drinkable.

Some impurities can be treated with a **masking agent** to prevent them from reacting with the precipitant. In the gravimetric analysis of Be^{2+}, Mg^{2+}, Ca^{2+}, or Ba^{2+} with the reagent *N-p*-chlorophenylcinnamohydroxamic acid, impurities such as Ag^+, Mn^{2+}, Zn^{2+}, Cd^{2+}, Hg^{2+}, Fe^{2+}, and Ga^{3+} are kept in solution by excess KCN. Pb^{2+}, Pd^{2+}, Sb^{3+}, Sn^{2+}, Bi^{3+}, Zr^{4+}, Ti^{4+}, V^{5+}, and Mo^{6+} are masked with a mixture of citrate and oxalate.

$$Ca^{2+} \ + \ 2RH \ \rightarrow \ CaR_2(s)\downarrow + \ 2H^+$$

Analyte   *N-p*-Chlorophenyl-   Precipitate
          cinnamohydroxamic acid

$$Mn^{2+} \ + \ 6CN^- \ \rightarrow \ Mn(CN)_6^{4-}$$

Impurity   Masking agent   Stays in solution

Even when a precipitate forms in a pure state, impurities might collect on the product while it is standing in the mother liquor. This is called *postprecipitation* and usually involves a supersaturated impurity that does not readily crystallize. An example is the crystallization of MgC_2O_4 on CaC_2O_4.

Washing a precipitate on a filter helps remove droplets of liquid containing excess solute. Some precipitates can be washed with water, but many require electrolyte to maintain coherence. For these precipitates, the ionic atmosphere is required to neutralize the surface charge of the tiny particles. If electrolyte is washed away with water, the charged solid particles repel one another and the product breaks up. This breaking up, called **peptization,** results in loss of product through the filter. AgCl will peptize if washed with water, so it is washed with dilute HNO_3 instead. Electrolyte used for washing must be volatile so that it will be lost during drying. Volatile electrolytes include HNO_3, HCl, NH_4NO_3, NH_4Cl, and $(NH_4)_2CO_3$.

Product Composition

The final product must have a known, stable composition. A **hygroscopic substance** is one that picks up water from the air and is therefore difficult to weigh accurately. Many precipitates contain a variable quantity of water and must be dried under conditions that give a known (possibly zero) stoichiometry of H_2O.

Ignition (strong heating) is used to change the chemical form of some precipitates. For example, igniting $Fe(HCO_2)_3 \cdot nH_2O$ at 850°C for 1 h gives Fe_2O_3, and igniting $Mg(NH_4)PO_4 \cdot 6H_2O$ at 1 100°C gives $Mg_2P_2O_7$.

FIGURE 26-3 Coprecipitation of phosphate with calcium carbonate in coral. Coprecipitated phosphate is proportional to phosphate concentration in seawater. By measuring P/Ca in ancient coral, we can infer that phosphate concentration in the western Mediterranean Sea 11 200 years ago was twice as high as current values. [Data from P. Montagna, M. McCulloch, M. Taviani, C. Mazzoli, and B. Vendrell, "Phosphorus in Cold-Water Corals as a Proxy for Seawater Nutrient Chemistry," *Science* **2006,** *312,* 1788.]

Removal of occluded NO_3^- from $BaSO_4$ by *reprecipitation*

	$[NO_3^-]/[SO_4^-]$ in precipitate
Initial precipitate	0.279
1st reprecipitation	0.028
2nd reprecipitation	0.001

Data from H. Bao, "Purifying Barite for Oxygen Isotope Measurement by Dissolution and Reprecipitation in a Chelating Solution," *Anal. Chem.* **2006,** *78,* 304.

N-p-Chlorophenylcinnamo-
hydroxamic acid (RH)
(ligand atoms are **bold**)

Ammonium chloride decomposes when heated:

$$NH_4Cl(s) \rightarrow NH_3(g) + HCl(g)$$

FIGURE 26-4 Thermogravimetric curve for calcium salicylate. [From G. Liptay, ed., *Atlas of Thermoanalytical Curves* (London: Heyden and Son, 1976).]

If you were performing this analysis, it would be important to determine that the impurities in the piperazine do not precipitate—otherwise the result will be high.

In **thermogravimetric analysis,** a substance is heated, and its mass is measured as a function of temperature. Figure 26-4 shows how the composition of calcium salicylate changes in four stages:

$$(26\text{-}5)$$

Calcium salicylate monohydrate

Calcium oxide   Calcium carbonate

The composition of the product depends on the temperature and duration of heating.

26-3 Examples of Gravimetric Calculations

We now examine some examples that illustrate how to relate the mass of a gravimetric precipitate to the quantity of the original analyte. The general approach is to relate the moles of product to the moles of reactant.

EXAMPLE   Relating Mass of Product to Mass of Reactant

The piperazine content of an impure commercial material can be determined by precipitating and weighing the diacetate:[8]

$$:NH\ HN: \ + \ 2CH_3CO_2H \ \rightarrow \ H_2\overset{+}{N} \quad \overset{+}{N}H_2(CH_3CO_2^-)_2 \qquad (26\text{-}6)$$

Piperazine   Acetic acid   Piperazine diacetate
FM 86.136   FM 60.052   FM 206.240

In one experiment, 0.312 6 g of sample was dissolved in 25 mL of acetone, and 1 mL of acetic acid was added. After 5 min, the precipitate was filtered, washed with acetone, dried at 110°C, and found to weigh 0.712 1 g. Find the wt% of piperazine in the sample.

Solution For each mole of piperazine in the impure material, 1 mol of product is formed.

$$\text{Moles of product} = \frac{0.712\ 1\ \text{g}}{206.240\ \text{g/mol}} = 3.453 \times 10^{-3}\ \text{mol}$$

This many moles of piperazine corresponds to

$$\text{Grams of piperazine} = (3.453 \times 10^{-3}\ \text{mol})\left(86.136\ \frac{\text{g}}{\text{mol}}\right) = 0.297\ 4\ \text{g}$$

which gives

$$\text{Percentage of piperazine in analyte} = \frac{0.297\ 4\ \text{g}}{0.312\ 6\ \text{g}} \times 100 = 95.14\%$$

An alternative (but equivalent) way to work this problem is to realize that 206.240 g (1 mol) of product will be formed for every 86.136 g (1 mol) of piperazine analyzed. Because 0.712 1 g of product was formed, the amount of reactant is given by

$$\frac{x\ \text{g piperazine}}{0.712\ 1\ \text{g product}} = \frac{86.136\ \text{g piperazine}}{206.243\ \text{g product}}$$

$$\Rightarrow x = \left(\frac{86.136\ \text{g piperazine}}{206.240\ \text{g product}}\right)0.712\ 1\ \text{g product} = 0.297\ 4\ \text{g piperazine}$$

The *gravimetric factor* relates mass of product to mass of analyte.

The quantity 86.136/206.240 is the *gravimetric factor* relating the mass of starting material to the mass of product.

Test Yourself   A 0.385 4 g sample gave 0.800 0 g of product. Find the wt% of piperazine in the sample. (***Answer:*** 86.69%)

For a reaction in which the stoichiometric relation between analyte and product is not 1:1, we must use the correct stoichiometry in formulating the gravimetric factor. For example, an unknown containing Mg^{2+} (atomic mass = 24.305 0) can be analyzed gravimetrically to produce magnesium pyrophosphate ($Mg_2P_2O_7$, FM 222.553). The gravimetric factor would be

$$\frac{\text{Grams of Mg in analyte}}{\text{Grams of } Mg_2P_2O_7 \text{ formed}} = \frac{2 \times (24.305\ 0)}{222.553}$$

because it takes two Mg^{2+} to make one $Mg_2P_2O_7$.

EXAMPLE Calculating How Much Precipitant to Use

(a) To measure the nickel content in steel, the alloy is dissolved in 12 M HCl and neutralized in the presence of citrate ion, which maintains iron in solution. The slightly basic solution is warmed, and dimethylglyoxime (DMG) is added to precipitate the red DMG-nickel complex quantitatively. The product is filtered, washed with cold water, and dried at 110°C.

$$Ni^{2+} + 2 \qquad \longrightarrow \qquad + 2H^+ \qquad (26\text{-}7)$$

DMG   Bis(dimethylglyoximate)nickel(II)
FM 58.69   FM 116.12   FM 288.91

If the nickel content is known to be near 3 wt% and you wish to analyze 1.0 g of steel, what volume of 1.0 wt% alcoholic DMG solution should be used to give a 50% excess of DMG for the analysis? Assume that the density of the alcohol solution is 0.79 g/mL.

Solution Because the Ni content is about 3%, 1.0 g of steel will contain about 0.03 g of Ni, which corresponds to

$$\frac{0.03\ \text{g Ni}}{58.69\ \text{g Ni/mol Ni}} = 5.11 \times 10^{-4}\ \text{mol Ni}$$

This amount of metal requires

$$2(5.11 \times 10^{-4}\ \text{mol Ni})(116.12\ \text{g DMG/mol Ni}) = 0.119\ \text{g DMG}$$

because 1 mol of Ni^{2+} requires 2 mol of DMG. A 50% excess of DMG would be $(1.5)(0.119\ \text{g}) = 0.178\ \text{g}$. This much DMG is contained in

$$\frac{0.178\ \text{g DMG}}{0.010\ \text{g DMG/g solution}} = 17.8\ \text{g solution}$$

which occupies a volume of

$$\frac{17.8\ \text{g solution}}{0.79\ \text{g solution/mL}} = 23\ \text{mL}$$

(b) If 1.163 4 g of steel gives 0.179 5 g of precipitate, what is the percentage of Ni in the steel?

Solution For each mole of Ni in the steel, 1 mol of precipitate will be formed. Therefore, 0.179 5 g of precipitate corresponds to

$$\frac{0.179\ 5\ \text{g Ni(DMG)}_2}{288.91\ \text{g Ni(DMG)}_2/\text{mol Ni(DMG)}_2} = 6.213 \times 10^{-4}\ \text{mol Ni(DMG)}_2$$

The mass of Ni in the steel is $(6.213 \times 10^{-4}\ \text{mol Ni})(58.69\ \text{g/mol Ni}) = 0.036\ 46\ \text{g}$, and the wt% Ni in steel is

$$\frac{0.036\ 46\ \text{g Ni}}{1.163\ 4\ \text{g steel}} \times 100 = 3.134\%$$

A slightly simpler way to approach this problem comes from realizing that 58.69 g of Ni (1 mol) would give 288.91 g (1 mol) of product. Calling the mass of Ni in the sample x, we can write

$$\frac{\text{Grams of Ni analyzed}}{\text{Grams of product formed}} = \frac{x}{0.179\ 5} = \frac{58.69}{288.91} \Rightarrow \text{Ni} = 0.036\ 46\ \text{g}$$

Test Yourself An alloy contains ~2.0 wt% Ni. What volume of 0.83 wt% DMG should be used to provide a 50% excess of DMG for the analysis of 1.8 g of steel? What mass of $\text{Ni}(\text{DMG})_2$ precipitate is expected? (***Answer:*** 33 mL, 0.18 g)

EXAMPLE **A Problem with Two Components**

A mixture of the 8-hydroxyquinoline complexes of Al and Mg weighed 1.084 3 g. When ignited in a furnace open to the air, the mixture decomposed, leaving a residue of Al_2O_3 and MgO weighing 0.134 4 g. Find the weight percent of $\text{Al}(\text{C}_9\text{H}_6\text{NO})_3$ in the original mixture.

AlQ$_3$
FM 459.43

MgQ$_2$
FM 312.61

FM 101.96   FM 40.304

Solution We will abbreviate the 8-hydroxyquinoline anion as Q. Letting the mass of AlQ$_3$ be x and the mass of MgQ$_2$ be y, we can write

$$\underset{\substack{\text{Mass of} \\ \text{AlQ}_3}}{x} + \underset{\substack{\text{Mass of} \\ \text{MgQ}_2}}{y} = 1.084\ 3\ \text{g}$$

The moles of Al are $x/459.43$, and the moles of Mg are $y/312.61$. The moles of Al_2O_3 are one-half of the total moles of Al, because it takes 2 mol of Al to make 1 mol of Al_2O_3.

$$\text{Moles of Al}_2\text{O}_3 = \left(\frac{1}{2}\right)\frac{x}{459.43}$$

The moles of MgO will equal the moles of Mg $= y/312.61$. Now we can write

$$\overset{\text{Mass of Al}_2\text{O}_3}{\overbrace{\underset{\substack{\text{mol Al}_2\text{O}_3 \quad \frac{\text{g Al}_2\text{O}_3}{\text{mol Al}_2\text{O}_3}}}{\left(\frac{1}{2}\right)\frac{x}{459.43}(101.96)}}} + \overset{\text{Mass of MgO}}{\overbrace{\underset{\substack{\text{mol MgO} \quad \frac{\text{g MgO}}{\text{mol MgO}}}}{\frac{y}{312.61}(40.304)}}} = 0.134\ 4\ \text{g}$$

Substituting $y = 1.084\ 3 - x$ into the preceding equation gives

$$\left(\frac{1}{2}\right)\left(\frac{x}{459.43}\right)(101.96) + \left(\frac{1.084\ 3 - x}{312.61}\right)(40.304) = 0.134\ 4\ \text{g}$$

from which we find $x = 0.300\ 3$ g, which is 27.70% of the original mixture.

Test Yourself If reproducibility is ± 0.5 mg, product mass might have been 0.133 9 to 0.134 9 g. Find wt% $\text{Al}(\text{C}_9\text{H}_6\text{NO})_3$ if the product weighed 0.133 9 g. (***Answer:*** 30.27%)

Please appreciate the huge uncertainty: A 0.5-mg difference in mass of product gives a 9% difference in calculated composition of the mixture.

26-4 Combustion Analysis

A historically important form of gravimetric analysis was *combustion analysis* to determine C and H in organic compounds burned in excess O_2 (Figure 26-5). Instead of weighing combustion products, modern instruments use thermal conductivity, infrared absorption, flame photometry (for S), and coulometry (for halogens) to measure products.

Gravimetric Combustion Analysis

In gravimetric combustion analysis, partially combusted product is passed through catalysts such as Pt gauze, CuO, PbO$_2$, or MnO$_2$ at elevated temperature to complete the oxidation to

FIGURE 26-5 Gravimetric combustion analysis for carbon and hydrogen.

Sample in Pt boat · Electric oven · Catalysts · O$_2$ in · Burner · P$_4$O$_{10}$ or Mg(ClO$_4$)$_2$ · NaOH on asbestos · P$_4$O$_{10}$ + NaOH/asbestos (guard tube) · O$_2$ out

CO_2 and H_2O. The combustion products are flushed through a chamber containing P_4O_{10} ("phosphorus pentoxide"), which absorbs water, and then through a chamber of Ascarite (NaOH on asbestos), which absorbs CO_2. The increase in mass of each chamber tells how much H and C were initially present. A guard tube prevents atmospheric H_2O or CO_2 from entering the chambers.

Asbestos would not be used today because it causes lung desease.

EXAMPLE Combustion Analysis Calculations

A compound weighing 5.714 mg produced 14.414 mg of CO_2 and 2.529 mg of H_2O upon combustion. Find the wt% of C and H in the sample.

Solution One mole of CO_2 contains 1 mol of carbon. Therefore,

Moles of C in sample = moles of CO_2 produced

$$= \frac{14.414 \times 10^{-3} \text{ g CO}_2}{44.010 \text{ g/mol CO}_2} = 3.275 \times 10^{-4} \text{ mol}$$

Mass of C in sample = $(3.275 \times 10^{-4} \text{ mol C})(12.010\ 7 \text{ g/mol C}) = 3.934$ mg

$$\text{wt% C} = \frac{3.934 \text{ mg C}}{5.714 \text{ mg sample}} \times 100 = 68.84\%$$

One mole of H_2O contains 2 mol of H. Therefore,

Moles of H in sample = 2(moles of H_2O produced)

$$= 2\left(\frac{2.529 \times 10^{-3} \text{ g H}_2\text{O}}{18.015 \text{ g/mol H}_2\text{O}}\right) = 2.808 \times 10^{-4} \text{ mol}$$

Mass of H in sample = $(2.808 \times 10^{-4} \text{ mol H})(1.007\ 9 \text{ g/mol H}) = 2.830 \times 10^{-4} \text{ g}$

$$\text{wt% H} = \frac{0.283\ 0 \text{ mg H}}{5.714 \text{ mg sample}} \times 100 = 4.95\%$$

Test Yourself A 6.234-mg sample produced 12.123 mg CO_2 and 2.529 mg of H_2O. Find the wt% of C and H in the sample. (**Answer:** 53.07, 4.54%)

Combustion Analysis Today[9]

Figure 26-6 shows an instrument that measures C, H, N, and S in a single operation. First, a ~2-mg sample is accurately weighed and sealed in a tin or silver capsule. The analyzer is swept with He gas that has been treated to remove traces of O_2, H_2O, and CO_2. At the start of a run, a measured excess of O_2 is added to the He stream. Then the capsule is dropped into a preheated ceramic crucible, where the capsule melts and sample is rapidly oxidized.

$$\text{C, H, N, S} \xrightarrow{1\ 050°\text{C} / \text{O}_2} CO_2(g) + H_2O(g) + N_2(g) + \underbrace{SO_2(g) + SO_3(g)}_{95\% \text{ SO}_2}$$

Oxygen

Autosampler

Encapsulated sample

WO₃ oxidation catalyst

Ceramic crucible

1 050°C

850°C

Cu reduction catalyst and O₂ scavenger

Thermal conductivity detector

Exit

N_2 CO_2 H_2O SO_2

Gas chromatograph

Data station

FIGURE 26-6 C,H,N,S elemental analyzer uses gas chromatography with thermal conductivity detection to measure N_2, CO_2, H_2O, and SO_2 combustion products. [Adapted from E. Pella, "Elemental Organic Analysis. 2. State of the Art," *Am. Lab.*, August 1990, p. 28.]

An *oxidation catalyst* completes the oxidation of sample, and a *reduction catalyst* carries out any required reduction and removes excess O_2.

The Sn capsule is oxidized to SnO_2, which

1. Liberates heat to vaporize and crack (decompose) sample
2. Uses available oxygen immediately
3. Ensures that sample oxidation occurs in gas phase
4. Acts as an oxidation catalyst

Injected O_2

Eluted N_2

Sample in

Flash combustion

Start

| 120 | 90 | 60 | 30 | 0 |

Time (s)

FIGURE 26-7 Sequence of events in dynamic flash combustion. [From E. Pella, "Elemental Organic Analysis. 1. Historical Developments," *Am. Lab.*, February 1990, p. 116.]

Products pass through hot WO_3 oxidation catalyst to complete the combustion of C to CO_2. In the next zone, metallic Cu at 850°C reduces SO_3 to SO_2 and removes excess O_2:

$$Cu + SO_3 \xrightarrow{850°C} SO_2 + CuO(s)$$

$$Cu + \frac{1}{2}O_2 \xrightarrow{850°C} CuO(s)$$

The mixture of CO_2, H_2O, N_2, and SO_2 is separated by gas chromatography, and each component is measured with a thermal conductivity detector (Figure 23-17). Alternatively, CO_2, H_2O, and SO_2 can be measured by infrared absorbance.

A key to elemental analysis is *dynamic flash combustion*, which creates a short burst of gaseous products, instead of slowly bleeding products out over several minutes. Chromatographic analysis requires that the whole sample be injected at once. Otherwise, the injection zone is so broad that the products cannot be separated.

In dynamic flash combustion, the tin-encapsulated sample is dropped into the preheated furnace shortly after the flow of a 50 vol% O_2/50 vol% He mixture is started (Figure 26-7). The Sn capsule melts at 235°C and is instantly oxidized to SnO_2, thereby liberating 594 kJ/mol, and heating the sample to 1 700°–1 800°C. By dropping the sample in before much O_2 is present, decomposition (cracking) occurs prior to oxidation, which minimizes the formation of nitrogen oxides. (Flammable liquid samples would be admitted prior to any O_2 to prevent explosions.)

Analyzers that measure C, H, and N, but not S, use better optimized catalysts. The oxidation catalyst is Cr_2O_3. The gas then passes through hot Co_3O_4 coated with Ag to absorb halogens and sulfur. A hot Cu column scavenges excess O_2.

Oxygen analysis requires a different strategy. The sample is thermally decomposed (a process called **pyrolysis**) in the absence of added O_2. Gaseous products are passed through nickelized carbon at 1 075°C to convert oxygen from the compound into CO (not CO_2). Other products include N_2, H_2, CH_4, and hydrogen halides. Acidic products are absorbed by NaOH, and the remaining gases are separated and measured by gas chromatography with a thermal conductivity detector.

For halogenated compounds, combustion gives CO_2, H_2O, N_2, and HX (X = halogen). The HX is trapped in aqueous solution and titrated with Ag^+ ions in a coulometer (Section 16-3). This instrument counts the electrons produced (one e^- for each Ag^+) during complete reaction with HX.

Table 26-4 shows representative results for two of seven compounds sent to more than 35 laboratories to compare their performance in combustion analysis. The accuracy for all seven compounds is excellent: Mean values of wt% C, H, N, and S for ~150 measurements of each

TABLE 26-4 Accuracy and precision of combustion analysis of pure compounds[a]

Substance	C	H	N	S
$C_7H_9NO_2S$ theoretical wt%	49.10	5.30	8.18	18.73
Toluene-4-sulfonamide	49.1 ± 0.63	5.3 ± 0.31	8.2 ± 0.38	18.7 ± 0.89
$C_4H_7NO_2S$ theoretical wt%	36.07	5.30	10.52	24.08
4-Thiazolidinecarboxylic acid	36.0 ± 0.33	5.3 ± 0.16	10.5 ± 0.16	24.0 ± 0.53
Mean uncertainty (wt%) for 7 different compounds	± 0.47	± 0.24	± 0.31	± 0.76

a. Results for two of seven pure compounds that were analyzed by 33–45 laboratories each year over six years. Each lab analyzed each compound at least five times during at least two days. For each substance, first row gives theoretical wt% and second row gives measured wt%. Uncertainties are 95% confidence intervals computed for all results after rejecting outliers at the 1% significance level.
SOURCE: R. Companyó, R. Rubio, A. Sahuquillo, R. Boqué, A. Maroto, and J. Riu, "Uncertainty Estimation in Organic Elemental Analysis Using Information from Proficiency Tests," *Anal. Bioanal. Chem.* **2008**, *392*, 1497.

compound are almost always within 0.1 wt% of theoretical values. Precision for all seven compounds is summarized at the bottom of the table. The mean 95% confidence interval for C is ± 0.47 wt%. For H, N, and S, the 95% confidence intervals are ± 0.24, ± 0.31, and ± 0.76 wt%, respectively. Chemists consider a result within ± 0.3 of the theoretical wt% to be good evidence that the compound has the expected formula. This criterion can be difficult to meet for C and S with a single analysis because the 95% confidence intervals are larger than ± 0.3 wt%.

Silicon compounds, such as SiC, Si_3N_4, and silicates (rocks) can be analyzed by reaction with elemental fluorine (F_2) in a Ni vessel to produce SiF_4 and fluorinated products of every element in the periodic table except O, N, He, Ne, Ar, and Kr.[10] Products can be measured by mass spectrometry. Nitrogen in Si_3N_4 and some other metal nitrides can be analyzed by heating to 3 000°C in an inert atmosphere to liberate nitrogen as N_2, which is measured by thermal conductivity.

F_2 is exceedingly reactive and dangerous. It must be handled only in systems designed for its use.

26-5 Precipitation Titration Curves

In gravimetric analysis, we could measure an unknown concentration of I^- by adding excess Ag^+ and weighing the AgI precipitate. In a *precipitation titration*, we monitor the course of the reaction between analyte (I^-) and titrant (Ag^+) to locate the **equivalence point** at which there is exactly enough titrant for stoichiometric reaction with the analyte. Knowing how much titrant was added tells us how much analyte was present. We seek the equivalence point in a titration, but we observe the **end point** at which there is an abrupt change in a physical property (such as an electrode potential) that is being measured. The physical property is chosen to make the end point as close as possible to the equivalence point.

The **titration curve** is a graph showing how the concentration of a reactant varies as titrant is added. We will derive equations that can be used to predict precipitation titration curves. One reason to calculate titration curves is to understand the chemistry that occurs during titrations. A second reason is to learn how experimental control can be exerted to influence the quality of an analytical titration. Concentrations of analyte and titrant and the size of the *solubility product* (K_{sp}) influence the sharpness of the end point.

Concentration varies over orders of magnitude, so it is useful to plot the p function:

The introduction to titrations in Section 1-5 and solubility products in Section 6-3 are prerequisites for studying precipitation titration curves.

p Function:
$$pX = -\log_{10}[X] \qquad (26\text{-}8)$$

The p function should really be written in terms of activity: $pX = -\log \mathcal{A}_X$. For simplicity, we use $pX = -\log[X]$ in this chapter.

where [X] is the concentration of X.

Consider the titration of 25.00 mL of 0.100 0 M I^- with 0.050 00 M Ag^+,

Titration reaction:
$$I^- + Ag^+ \rightarrow AgI(s) \qquad (26\text{-}9)$$

and suppose that we are monitoring $[Ag^+]$ with an electrode. Reaction 26-9 is the reverse of the dissolution of AgI(s), whose solubility product is rather small:

$$AgI(s) \rightleftharpoons Ag^+ + I^- \qquad K_{sp} = [Ag^+][I^-] = 8.3 \times 10^{-17} \qquad (26\text{-}10)$$

The equilibrium constant for the *titration reaction* 26-9 is large ($K = 1/K_{sp} = 1.2 \times 10^{16}$), so the equilibrium lies far to the right. Each aliquot of Ag^+ reacts nearly completely with I^-, leaving only a tiny amount of Ag^+ in solution. At the equivalence point, there will be a sudden increase in $[Ag^+]$ because there is no I^- left to consume the added Ag^+.

What volume of Ag^+ titrant is needed to reach the equivalence point? We calculate this volume, designated V_e, with the fact that 1 mol of Ag^+ reacts with 1 mol of I^-.

$$\underbrace{(0.025\ 00\ \text{L})(0.100\ 0\ \text{mol}\ I^-/\text{L})}_{\text{mol}\ I^-} = \underbrace{(V_e)(0.050\ 00\ \text{mol}\ Ag^+/\text{L})}_{\text{mol}\ Ag^+}$$

$$\Rightarrow V_e = 0.050\ 00\ \text{L} = 50.00\ \text{mL}$$

The titration curve has three distinct regions, depending on whether we are before, at, or after the equivalence point. Let's consider each region separately.

Before the Equivalence Point

Suppose that 10.00 mL of Ag^+ have been added. There are more moles of I^- than Ag^+ at this point, so virtually all Ag^+ is "used up" to make $AgI(s)$. We want to find the small concentration of Ag^+ remaining in solution after reaction with I^-. Imagine that Reaction 26-9 has gone to completion and that some AgI redissolves (Reaction 26-10). The solubility of Ag^+ is determined by the concentration of free I^- remaining in the solution:

$$[Ag^+] = \frac{K_{sp}}{[I^-]} \tag{26-11}$$

Free I^- is overwhelmingly from the I^- that has not been precipitated by 10.00 mL of Ag^+. By comparison, I^- from dissolution of $AgI(s)$ is negligible.

So let's find the concentration of unprecipitated I^-:

$$\begin{aligned}
\text{Moles of}\ I^- &= \text{original mols of}\ I^- - \text{moles of}\ Ag^+\ \text{added} \\
&= (0.025\ 00\ \text{L})(0.100\ \text{mol/L}) - (0.010\ 00\ \text{L})(0.050\ 00\ \text{mol/L}) \\
&= 0.002\ 000\ \text{mol}\ I^-
\end{aligned}$$

The volume is 0.035 00 L (25.00 mL + 10.00 mL), so the concentration is

$$[I^-] = \frac{0.002\ 000\ \text{mol}\ I^-}{0.035\ 00\ \text{L}} = 0.057\ 14\ \text{M} \tag{26-12}$$

The concentration of Ag^+ in equilibrium with this much I^- is

$$[Ag^+] = \frac{K_{sp}}{[I^-]} = \frac{8.3 \times 10^{-17}}{0.057\ 14} = 1.4_5 \times 10^{-15}\ \text{M} \tag{26-13}$$

Finally, the p function we seek is

$$pAg^+ = -\log[Ag^+] = -\log(1.4_5 \times 10^{-15}) = 14.84 \tag{26-14}$$

Two significant figures in K_{sp} provide two significant figures in $[Ag^+]$. The two figures in $[Ag^+]$ translate into two figures in the *mantissa* of the p function, which is written as 14.84.

The preceding step-by-step calculation is a tedious way to find the concentration of I^-. Here is a streamlined procedure that is well worth learning. Bear in mind that $V_e = 50.00$ mL. When 10.00 mL of Ag^+ have been added, the reaction is one-fifth complete because 10.00 mL out of the 50.00 mL of Ag^+ needed for complete reaction have been added. Therefore, four-fifths of the I^- is unreacted. If there were no dilution, $[I^-]$ would be four-fifths of its original value. However, the original volume of 25.00 mL has been increased to 35.00 mL. If no I^- had been consumed, the concentration would be the original value of $[I^-]$ times (25.00/35.00). Accounting for both the reaction and the dilution, we can write

$$[I^-] = \underbrace{\left(\frac{4.000}{5.000}\right)}_{\substack{\text{Fraction} \\ \text{remaining}}} \underbrace{(0.100\ 0\ \text{M})}_{\substack{\text{Original} \\ \text{concentration}}} \overbrace{\underbrace{\left(\frac{25.00}{35.00}\right)}_{\substack{\text{Dilution} \\ \text{factor}}}}^{\text{Original volume of}\ I^-}_{\text{Total volume of solution}} = 0.057\ 14\ \text{M}$$

This is the same result found in Equation 26-12.

V_e = volume of titrant at equivalence point

Eventually, we will derive a single, unified equation for a spreadsheet that treats all regions of the titration curve. To understand the chemistry, we break the curve into three regions described by approximate equations that are easy to use.

When $V < V_e$, the concentration of unreacted I^- regulates the solubility of AgI.

$\log(\underset{\substack{\text{Two significant} \\ \text{figures}}}{1.4_5} \times 10^{-15}) = 14.\underset{\substack{\text{Two digits} \\ \text{in mantissa}}}{84}$

Significant figures in logarithms were discussed in Section 3-2.

Streamlined calculation *well worth using.*

Let's calculate pAg^+ when V_{Ag^+} (the volume added from the buret) is 49.00 mL.

Solution Because $V_e = 50.00$ mL, the fraction of I^- reacted is 49.00/50.00, and the fraction remaining is 1.00/50.00. The total volume is $25.00 + 49.00 = 74.00$ mL.

$$[I^-] = \underbrace{\left(\frac{1.00}{50.00}\right)}_{\substack{\text{Fraction} \\ \text{remaining}}} \underbrace{(0.100\ 0\ \text{M})}_{\substack{\text{Original} \\ \text{concentration}}} \underbrace{\left(\frac{25.00}{74.00}\right)}_{\substack{\text{Dilution} \\ \text{factor}}} = 6.76 \times 10^{-4}\ \text{M}$$

$$[Ag^+] = K_{sp}/[I^-] = (8.3 \times 10^{-17})/(6.76 \times 10^{-4}) = 1.2_3 \times 10^{-13}\ \text{M}$$

$$pAg^+ = -\log[Ag^+] = 12.91$$

The concentration of Ag^+ is negligible compared with the concentration of unreacted I^-, even though the titration is 98% complete.

Test Yourself   Find pAg^+ at 49.1 mL. (**Answer:** 12.86)

At the Equivalence Point

Now we have added exactly enough Ag^+ to react with all the I^-. We can imagine that all AgI precipitates and some redissolves to give equal concentrations of Ag^+ and I^-. The value of pAg^+ is found by setting $[Ag^+] = [I^-] = x$ in the solubility product:

$$[Ag^+][I^-] = K_{sp}$$
$$(x)(x) = 8.3 \times 10^{-17} \Rightarrow x = 9.1 \times 10^{-9} \Rightarrow pAg^+ = -\log x = 8.04$$

This value of pAg^+ is independent of the original concentrations or volumes.

When $V = V_e$, $[Ag^+]$ is determined by the solubility of pure AgI. *This problem is the same as if we had just added AgI(s) to water.*

After the Equivalence Point

Virtually all Ag^+ added *before* the equivalence point has precipitated. The solution contains all of the Ag^+ added *after* the equivalence point. Suppose that $V_{Ag^+} = 52.00$ mL. The volume past the equivalence point is 2.00 mL. The calculation proceeds as follows:

$$\text{Moles of } Ag^+ = (0.002\ 00\ \text{L})(0.050\ 00\ \text{mol } Ag^+/\text{L}) = 0.000\ 100\ \text{mol}$$

$$[Ag^+] = (0.000\ 100\ \text{mol})/(0.077\ 00\ \text{L}) = 1.30 \times 10^{-3}\ \text{M} \Rightarrow pAg^+ = 2.89$$

Total volume = 77.00 mL

When $V > V_e$, $[Ag^+]$ is determined by the excess Ag^+ added from the buret.

We could justify three significant figures for the mantissa of pAg^+ because there are three significant figures in $[Ag^+]$. For consistency with earlier results, we retain only two figures.

For a streamlined calculation, the concentration of Ag^+ in the buret is 0.050 00 M, and 2.00 mL of titrant are being diluted to $(25.00 + 52.00) = 77.00$ mL. Hence, $[Ag^+]$ is

$$[Ag^+] = \underbrace{(0.050\ 00\ \text{M})}_{\substack{\text{Original} \\ \text{concentration} \\ \text{of } Ag^+}} \underbrace{\left(\frac{2.00}{77.00}\right)}_{\substack{\text{Dilution} \\ \text{factor}}} = 1.30 \times 10^{-3}\ \text{M}$$

Volume of excess Ag^+

Total volume of solution

The Shape of the Titration Curve

Titration curves in Figure 26-8 illustrate the effect of reactant concentration. The equivalence point is the steepest point of the curve. It is the point of maximum slope (a negative slope in this case) and is therefore an inflection point (at which the second derivative is 0):

Steepest slope: $\dfrac{dy}{dx}$ reaches its greatest value

Inflection point: $\dfrac{d^2y}{dx^2} = 0$

In titrations involving 1:1 stoichiometry of reactants, the equivalence point is the steepest point of the titration curve. For stoichiometries other than 1:1, such as $2Ag^+ + CrO_4^{2-} \rightarrow Ag_2CrO_4(s)$, the curve is not symmetric. The equivalence point is not at the center of the

FIGURE 26-8 Titration curves showing the effect of diluting the reactants.

Outer curve: 25.00 mL of 0.100 0 M I⁻ titrated with 0.050 00 M Ag⁺

Middle curve: 25.00 mL of 0.010 00 M I⁻ titrated with 0.005 000 M Ag⁺

Inner curve: 25.00 mL of 0.001 000 M I⁻ titrated with 0.000 500 0 M Ag⁺

At the equivalence point, the titration curve is steepest for the least soluble precipitate.

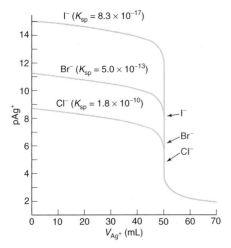

FIGURE 26-9 Titration curves showing the effect of K_{sp}. Each curve is calculated for 25.00 mL of 0.100 0 M halide titrated with 0.050 00 M Ag⁺. Equivalence points are marked by arrows.

steepest section of the curve, and it is not an inflection point. In practice, conditions are chosen such that titration curves are steep enough for the steepest point to be a good estimate of the equivalence point, regardless of the stoichiometry.

Figure 26-9 illustrates how K_{sp} affects the titration of halide ions. The least soluble product, AgI, gives the sharpest change at the equivalence point. However, even for AgCl, the curve is steep enough to locate the equivalence point accurately. The larger the equilibrium constant for a titration reaction, the more pronounced will be the change in concentration near the equivalence point.

EXAMPLE **Calculating Concentrations During a Precipitation Titration**

25.00 mL of 0.041 32 M $Hg_2(NO_3)_2$ were titrated with 0.057 89 M KIO_3.

$$Hg_2^{2+} + 2IO_3^- \rightarrow Hg_2(IO_3)_2(s)$$

Iodate

For $Hg_2(IO_3)_2$, $K_{sp} = 1.3 \times 10^{-18}$. Find $[Hg_2^{2+}]$ in the solution after addition of **(a)** 34.00 mL of KIO_3; **(b)** 36.00 mL of KIO_3; and **(c)** at the equivalence point.

Solution The volume of iodate needed to reach the equivalence point is found as follows:

$$\text{Moles of } IO_3^- = \left(\frac{2 \text{ mol } IO_3^-}{1 \text{ mol } Hg_2^{2+}}\right)(\text{moles of } Hg_2^{2+})$$

$$\underbrace{(V_e)(0.057\ 89\ M)}_{\text{Moles of } IO_3^-} = \underbrace{2(25.00\ mL)(0.041\ 32\ M)}_{\text{Moles of } Hg_2^{2+}} \Rightarrow V_e = 35.69\ mL$$

(a) When $V = 34.00$ mL, the precipitation of Hg_2^{2+} is not yet complete.

$$[Hg_2^{2+}] = \underbrace{\left(\frac{35.69 - 34.00}{35.69}\right)}_{\substack{\text{Fraction} \\ \text{remaining}}} \underbrace{(0.041\ 32\ M)}_{\substack{\text{Original} \\ \text{concentration} \\ \text{of } Hg_2^{2+}}} \underbrace{\left(\frac{25.00}{25.00 + 34.00}\right)}_{\substack{\text{Dilution} \\ \text{factor}}} = 8.29 \times 10^{-4}\ M$$

Original volume of Hg_2^{2+}

Total volume of solution

(b) When $V = 36.00$ mL, the precipitation is complete. We have gone $(36.00 - 35.69) = 0.31$ mL *past* the equivalence point. The concentration of excess IO_3^- is

$$[IO_3^-] = \underbrace{(0.057\ 89\ M)}_{\substack{\text{Original} \\ \text{concentration} \\ \text{of } IO_3^-}} \left(\underbrace{\frac{0.31}{25.00 + 36.00}}_{\substack{\text{Dilution} \\ \text{factor}}} \right) = 2.9_4 \times 10^{-4}\ M$$

Volume of excess IO_3^-

Total volume of solution

The concentration of Hg_2^{2+} in equilibrium with solid $Hg_2(IO_3)_2$ plus this much IO_3^- is

$$[Hg_2^{2+}] = \frac{K_{sp}}{[IO_3^-]^2} = \frac{1.3 \times 10^{-18}}{(2.9_4 \times 10^{-4})^2} = 1.5 \times 10^{-11}\ M$$

(c) At the equivalence point, there is exactly enough IO_3^- to react with all Hg_2^{2+}. We can imagine that all of the ions precipitate and then some $Hg_2(IO_3)_2(s)$ redissolves, giving two moles of iodate for each mole of mercurous ion:

$$Hg_2(IO_3)_2(s) \rightleftharpoons \underset{x}{Hg_2^{2+}} + \underset{2x}{2IO_3^-}$$

$$(x)(2x)^2 = K_{sp} \Rightarrow x = [Hg_2^{2+}] = 6.9 \times 10^{-7}\ M$$

Test Yourself Find $[Hg_2^{2+}]$ at 34.50 and 36.5 mL. (**Answer:** 5.79×10^{-4} M, 2.2×10^{-12} M)

Our calculations presume that the only chemistry that occurs is the reaction of anion with cation to precipitate solid salt. If other reactions occur, such as complex formation or ion-pair formation, we would have to modify the calculations.

26-6 Titration of a Mixture

If a mixture of two ions is titrated, the less soluble precipitate forms first. If the solubilities are sufficiently different, the first precipitation is nearly complete before the second commences.

Consider the addition of $AgNO_3$ to a solution containing KI and KCl. Because $K_{sp}(AgI) \ll K_{sp}(AgCl)$, AgI precipitates first. When precipitation of I^- is almost complete, the concentration of Ag^+ abruptly increases and AgCl begins to precipitate. When Cl^- is consumed, another abrupt increase in $[Ag^+]$ occurs. We expect two breaks in the titration curve, first at V_e for AgI and then at V_e for AgCl.

Figure 26-10 shows an experimental curve for this titration. The apparatus used to measure the curve is shown in Figure 26-11, and the theory of how this system measures Ag^+ concentration is discussed in Section 14-2.

A liquid containing suspended particles is said to be *turbid* because the particles scatter light.

The product with the smaller K_{sp} precipitates first, if the stoichiometry of the precipitates is the same. Precipitation of I^- and Cl^- with Ag^+ produces two breaks in the titration curve, first for I^- and then for Cl^-.

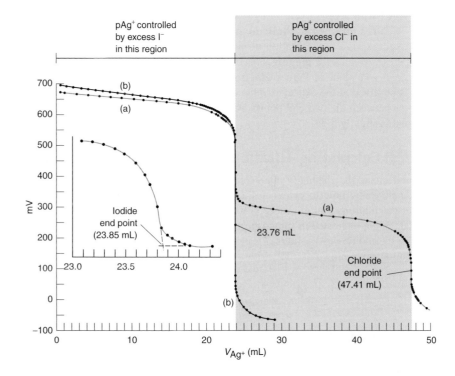

FIGURE 26-10 Experimental titration curves. (*a*) 40.00 mL of 0.050 2 M KI plus 0.050 0 M KCl titrated with 0.084 5 M $AgNO_3$. Inset expands the region near the first equivalence point. (*b*) 20.00 mL of 0.100 4 M I^- titrated with 0.084 5 M Ag^+.

FIGURE 26-11 Apparatus for measuring the titration curves in Figure 26-10. The silver electrode responds to changes in Ag^+ concentration, and the glass electrode provides a constant reference potential in this experiment. The measured voltage changes by approximately 59 mV for each factor-of-10 change in $[Ag^+]$. All solutions, including $AgNO_3$, were maintained at pH 2.0 by using 0.010 M sulfate buffer prepared from H_2SO_4 and KOH.

The I^- end point is taken as the intersection of the steep and nearly horizontal curves shown in the inset of Figure 26-10. Precipitation of I^- is not quite complete when Cl^- begins to precipitate. (The way we know that I^- precipitation is not complete is by a calculation. That's what these obnoxious calculations are for!) Therefore, the end of the steep portion (the intersection) is a better approximation of the equivalence point than is the middle of the steep section. The Cl^- end point is taken as the midpoint of the second steep section, at 47.41 mL. The moles of Cl^- in the sample equal the moles of Ag^+ delivered between the first and second end points. That is, it requires 23.85 mL of Ag^+ to precipitate I^-, and $(47.41 - 23.85) = 23.56$ mL of Ag^+ to precipitate Cl^-.

Comparison of the I^-/Cl^- and pure I^- titration curves in Figure 26-10 shows that the I^- end point is 0.38% too high in the I^-/Cl^- titration. We expect the first end point at 23.76 mL, but it is observed at 23.85 mL. Two factors contribute to this high value. One is experimental error, which is always present. This discrepancy is as likely to be positive as negative. However, the end point in some titrations, especially Br^-/Cl^- titrations, is systematically 0 to 3% high, depending on conditions. This error is attributed to *coprecipitation* of AgCl with AgBr. Even though the solubility of AgCl has not been exceeded, some Cl^- becomes attached to AgBr as it precipitates and carries down an equivalent amount of Ag^+. A high concentration of nitrate reduces coprecipitation, perhaps because NO_3^- competes with Cl^- for binding sites.

The second end point in Figure 26-10 corresponds to complete precipitation of both halides. It is observed at the expected value of V_{Ag^+}. The concentration of Cl^-, found from the *difference* between the two end points, will be slightly low in Figure 26-10, because the first end point is slightly high.

26-7 📊 Calculating Titration Curves with a Spreadsheet

Now you understand the chemistry that occurs at different stages of a precipitation titration, and you should know how to calculate the shape of a titration curve. We now introduce spreadsheet calculations that are more powerful than hand calculations and less prone to error. If you are not interested in spreadsheets at this time, you can skip this section.

Consider the addition of V_M liters of cation M^+ (whose initial concentration is C_M^0) to V_X^0 liters of solution containing anion X^- with a concentration C_X^0.

$$\underset{\substack{\text{Titrant} \\ C_M^0, V_M}}{M^+} + \underset{\substack{\text{Analyte} \\ C_X^0, V_X^0}}{X^-} \overset{K_{sp}}{\rightleftharpoons} MX(s) \qquad (26\text{-}15)$$

The total moles of added M $(= C_M^0 \cdot V_M)$ must equal the moles of M^+ in solution $(= [M^+] \cdot (V_M + V_X^0))$ plus the moles of precipitated MX(s). (This equality is called a *mass balance,*

	A	B	C	D	E
1	Titration of I- with Ag+				
2					
3	Ksp(AgI) =	pAg	[Ag+]	[I-]	Vm
4	8.30E-17	15.08	8.32E-16	9.98E-02	0.035
5	Vo =	15	1.00E-15	8.30E-02	3.195
6	25	14	1.00E-14	8.30E-03	39.322
7	Co(I) =	12	1.00E-12	8.30E-05	49.876
8	0.1	10	1.00E-10	8.30E-07	49.999
9	Co(Ag) =	8	1.00E-08	8.30E-09	50.000
10	0.05	6	1.00E-06	8.30E-11	50.001
11		4	1.00E-04	8.30E-13	50.150
12		3	1.00E-03	8.30E-14	51.531
13		2	1.00E-02	8.30E-15	68.750
14	C4 = 10^-B4				
15	D4 = $A$4/C4				
16	E4 = $A$6*($A$8+C4-D4)/($A$10-C4+D4)				

FIGURE 26-12 Spreadsheet for titration of 25 mL of 0.1 M I^- with 0.05 M Ag^+.

even though it is really a *mole balance*.) In a similar manner, we can write a mass balance for X.

Mass balance for M:
$$C_M^0 \cdot V_M = \underbrace{[M^+](V_M + V_X^0)}_{\text{Moles of M in solution}} + \underbrace{\text{mol MX}(s)}_{\text{Moles of M in precipitate}} \tag{26-16}$$

Total moles of added M

The *mass balance* states that the moles of an element in all species in a mixture equals the total moles of that element delivered to the solution.

Mass balance for X:
$$C_X^0 \cdot V_X^0 = \underbrace{[X^-](V_M + V_X^0)}_{\text{Moles of X in solution}} + \underbrace{\text{mol MX}(s)}_{\text{Moles of X in precipitate}} \tag{26-17}$$

Total moles of added X

Now equate mol MX(s) from Equation 26-16 with mol MX(s) from Equation 26-17:

$$C_M^0 \cdot V_M - [M^+](V_M + V_X^0) = C_X^0 \cdot V_X^0 - [X^-](V_M + V_X^0)$$

which can be rearranged to

Precipitation of X^- with M^+:
$$V_M = V_X^0 \left(\frac{C_X^0 + [M^+] - [X^-]}{C_M^0 - [M^+] + [X^-]} \right) \tag{26-18}$$

Equation 26-18 relates the volume of added M^+ to $[M^+]$, $[X^-]$, and the constants V_X^0, C_X^0, and C_M^0. To use Equation 26-18 in a spreadsheet, *enter values of pM and compute corresponding values of V_M*, as shown in Figure 26-12 for the iodide titration of Figure 26-9. This is the reverse of the way you normally calculate a titration curve in which V_M would be input and pM would be output. Column C of Figure 26-12 is calculated with the formula $[M^+] = 10^{-pM}$, and column D is given by $[X^-] = K_{sp}/[M^+]$. Column E is calculated from Equation 26-18. The first input value of pM (15.08) was selected by trial and error to produce a small V_M. You can start wherever you like. If your initial value of pM is before the true starting point, then V_M in column E will be negative. In practice, you will want more points than we have shown so that you can plot an accurate titration curve.

A supplementary section at www.whfreeman.com/qca8e derives a spreadsheet equation for the titration of a mixture, such as that in Figure 26-10.

26-8 End-Point Detection

Precipitation titration end points are commonly found with electrodes (Figure 26-11) or indicators. We now describe two indicator methods for the titration of Cl^- with Ag^+:

Titrations with Ag^+ are called **argentometric titrations.**

Volhard titration: formation of a soluble, colored complex at the end point
Fajans titration: adsorption of a colored indicator on the precipitate at the end point

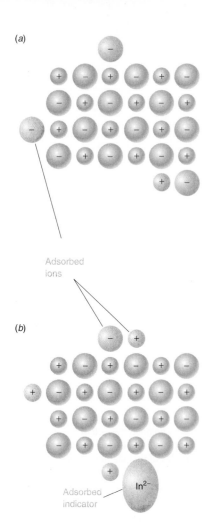

(a)

(b)

Adsorbed
ions

In^{2-}

Adsorbed
indicator

FIGURE 26-13 Ions from solution are adsorbed on the surface of a growing crystallite. (*a*) A crystal growing in the presence of excess lattice anions (anions that belong in the crystal) will have a slight negative charge because anions are predominantly adsorbed. (*b*) A crystal growing in the presence of excess lattice cations will have a slight positive charge and can therefore adsorb a negative indicator ion. Anions and cations in the solution that do not belong in the crystal lattice are less likely to be adsorbed than are ions belonging to the lattice. These diagrams omit other ions in solution. Overall, each solution plus its growing crystallites must have zero total charge.

Volhard Titration

The Volhard titration is a titration of Ag^+ in HNO_3 solution. For Cl^-, a back titration is necessary. First, Cl^- is precipitated by a known, excess quantity of standard $AgNO_3$.

$$Ag^+ + Cl^- \rightarrow AgCl(s)$$

The AgCl is filtered and washed, and excess Ag^+ in the combined filtrate is titrated with standard KSCN (potassium thiocyanate) in the presence of Fe^{3+}.

$$Ag^+ + SCN^- \rightarrow AgSCN(s)$$

When all Ag^+ has been consumed, SCN^- reacts with Fe^{3+} to form a red complex.

$$Fe^{3+} + SCN^- \rightarrow \underset{\text{Red}}{FeSCN^{2+}}$$

The appearance of red color is the end point. Knowing how much SCN^- was required for the back titration tells us how much Ag^+ was left over from the reaction with Cl^-. The total amount of Ag^+ is known, so the amount consumed by Cl^- can be calculated.

In the analysis of Cl^- by the Volhard method, the end point would slowly fade if the AgCl were not filtered off, because AgCl is more soluble than AgSCN. The AgCl slowly dissolves and is replaced by AgSCN. To eliminate this secondary reaction, we filter the AgCl and titrate only the Ag^+ in the filtrate. Br^- and I^-, whose silver salts are *less* soluble than AgSCN, can be titrated by the Volhard method without isolating the silver halide precipitate.

Fajans Titration

The Fajans titration uses an **adsorption indicator.** To see how this works, consider the electric charge at the surface of a precipitate. When Ag^+ is added to Cl^-, there is excess Cl^- in solution prior to the equivalence point. Some Cl^- is adsorbed onto the AgCl surface, imparting a negative charge to the crystal (Figure 26-13a). After the equivalence point, there is excess Ag^+ in solution. Adsorption of Ag^+ onto the AgCl surface places positive charge on the precipitate (Figure 26-13b). The abrupt change from negative to positive occurs at the equivalence point.

Common adsorption indicators are anionic dyes, which are attracted to positively charged particles produced immediately after the equivalence point. Adsorption of the negatively charged dye onto the positively charged surface changes the color of the dye. The color change is the end point in the titration. Because the indicator reacts with the precipitate surface, we want as much surface area as possible. To attain maximum surface area, we use conditions that keep the particles as small as possible, because small particles have more surface area than an equal volume of large particles. Low electrolyte concentration helps prevent coagulation of the precipitate and maintains small particle size.

The most common indicator for AgCl is dichlorofluorescein. This dye is greenish yellow in solution but turns pink when adsorbed onto AgCl (Demonstration 26-2 and Color Plate 33). The pH of the reaction must be controlled because the indicator is a weak acid and must be present in its anionic form. The dye eosin is useful in the titration of Br^-, I^-, and SCN^-. It gives a sharper end point than dichlorofluorescein and is more sensitive (that is, less halide can be titrated). It cannot be used for AgCl, because eosin is more strongly bound than Cl^- to AgCl. Eosin binds to AgCl crystallites even before the particles become positively charged.

In all argentometric titrations, but especially with adsorption indicators, strong light (such as daylight through a window) should be avoided. Light decomposes silver salts, and adsorbed indicators are especially light sensitive.

DEMONSTRATION 26-2 Fajans Titration

The Fajans titration of Cl^- with Ag^+ demonstrates indicator end points in precipitation titrations. Dissolve 0.5 g of NaCl plus 0.15 g of dextrin in 400 mL of water. Dextrin retards coagulation of the AgCl precipitate. Add 1 mL of dichlorofluorescein indicator solution containing either 1 mg/mL of dichlorofluorescein in 95% aqueous ethanol or 1 mg/mL of the sodium salt in water.

Titrate the NaCl solution with a solution containing 2 g of $AgNO_3$ in 30 mL H_2O. About 20 mL are required. Color Plate 33 shows yellow indicator prior to titration, milky white AgCl suspension during the titration, and pink suspension at the end point, when the anionic indicator becomes adsorbed on the cationic precipitate.

CHAPTER 26 Gravimetric Analysis, Precipitation Titrations, and Combustion Analysis

TABLE 26-5   Applications of precipitation titrations

Species analyzed	Notes
	Volhard Method
Br^-, I^-, SCN^-, CNO^-, AsO_4^{3-},	Precipitate removal is unnecessary.
Cl^-, PO_4^{3-}, CN^-, $C_2O_4^{2-}$, CO_3^{2-}, S^{2-}, CrO_4^{2-}	Precipitate removal required.
BH_4^-	Back titration of Ag^+ left after reaction with BH_4^-: $$BH_4^- + 8Ag^+ + 8OH^- \rightarrow 8Ag(s) + H_2BO_3^- + 5H_2O$$
K^+	K^+ is first precipitated with a known excess of $(C_6H_5)_4B^-$. Remaining $(C_6H_5)_4B^-$ is precipitated with a known excess of Ag^+. Unreacted Ag^+ is then titrated with SCN^-.
	Fajans Method
Cl^-, Br^-, I^-, SCN^-, $Fe(CN)_6^{4-}$	Titration with Ag^+. Detection with dyes such as fluorescein, dichlorofluorescein, eosin, bromophenol blue.
Zn^{2+}	Titration with $K_4Fe(CN)_6$ to produce $K_2Zn_3[Fe(CN)_6]_2$. End-point detection with diphenylamine.
SO_4^{2-}	Titration with $Ba(OH)_2$ in 50 vol% aqueous methanol using alizarin red S as indicator.
Hg_2^{2+}	Titration with NaCl to produce Hg_2Cl_2. End point detected with bromophenol blue.
PO_4^{3-}, $C_2O_4^{2-}$	Titration with $Pb(CH_3CO_2)_2$ to give $Pb_3(PO_4)_2$ or PbC_2O_4. End point detected with dibromofluorescein (PO_4^{3-}) or fluorescein ($C_2O_4^{2-}$).

Dichlorofluorescein

Tetrabromofluorescein (eosin)

Applications of precipitation titrations are listed in Table 26-5. Whereas the Volhard method is an argentometric titration, the Fajans method has wider applications. Because the Volhard titration is carried out in acidic solution (typically 0.2 M HNO_3), it avoids certain interferences that affect other titrations. Silver salts of CO_3^{2-}, $C_2O_4^{2-}$, and AsO_4^{3-} are soluble in acidic solution, so these anions do not interfere.

Because the Volhard method is a titration of Ag^+, it can be adapted for the determination of any anion that forms an insoluble silver salt.

Terms to Understand

absorption	dialysis	gravimetric analysis	precipitant
adsorption	digestion	homogeneous precipitation	precipitation titration
adsorption indicator	electric double layer	hygroscopic substance	pyrolysis
argentometric titration	end point	ignition	supersaturated solution
colloid	equivalence point	masking agent	thermogravimetric analysis
combustion analysis	Fajans titration	nucleation	titration curve
coprecipitation	gathering	peptization	Volhard titration

Summary

In gravimetric analysis, we relate the mass of a known product to the mass of analyte. Most commonly, analyte ion is precipitated by a suitable counterion. Measures taken to reduce supersaturation and promote the formation of large, easily filtered particles (as opposed to colloids) include (1) raising the temperature during precipitation, (2) slowly adding and vigorously mixing reagents, (3) maintaining a large sample volume, and (4) using homogeneous precipitation. Precipitates are usually digested in hot mother liquor to promote particle growth and recrystallization. Precipitates are then filtered and washed; some must be washed with a volatile electrolyte to prevent peptization. The product is heated to dryness or ignited to achieve a reproducible, stable composition. Gravimetric calculations relate moles of product to moles of analyte.

In combustion analysis for C, H, N, S, and halogens, an organic compound in a tin capsule is rapidly heated with excess O_2 to give predominantly CO_2, H_2O, N_2, SO_2, and HX (hydrogen halides). A hot oxidation catalyst completes the process, and hot Cu scavenges excess O_2. For sulfur analysis, hot copper also converts SO_3 into SO_2. Products may be separated by gas chromatography and measured by their thermal conductivity. Some instruments use infrared absorption to measure CO_2, H_2O, and SO_2. HX is trapped in aqueous solution and measured by coulometric titration (counting electrons) with electrolytically generated Ag^+. Oxygen in organic compounds can be measured by pyrolysis in the absence of added O_2 to convert oxygen from the compound into CO.

Concentrations of reactants and products during a precipitation titration are calculated in three regions. Before the equivalence point, there is excess analyte. Its concentration is the product (fraction remaining) × (original concentration) × (dilution factor). The concentration of titrant can be found from the solubility product of the precipitate and the known concentration of excess analyte. At the equivalence point, concentrations of both reactants are governed by the solubility product. After the equivalence point, the concentration of analyte can be determined from the solubility product of precipitate and the known concentration of excess titrant.

The Volhard titration measures Ag^+ by the reaction of Fe^{3+} with SCN^- after the precipitation of AgSCN is complete. Anions that precipitate with Ag^+ are measured by treating them with a known excess of Ag^+ and then back-titrating the excess Ag^+ by the Volhard titration. The Fajans titration applies to many precipitations that do not necessarily involve Ag^+. The end point is a color change that occurs when a charged indicator is adsorbed onto the charged surface of the precipitate after the equivalence point.

Exercises

26-A. An organic compound with a formula mass of 417 g/mol was analyzed for ethoxyl (CH_3CH_2O—) groups by the reactions

$$ROCH_2CH_3 + HI \rightarrow ROH + CH_3CH_2I$$

$$CH_3CH_2I + Ag^+ + OH^- \rightarrow AgI(s) + CH_3CH_2OH$$

A 25.42-mg sample of compound produced 29.03 mg of AgI. How many ethoxyl groups are there in each molecule?

26-B. A 0.649-g sample containing only K_2SO_4 (FM 174.27) and $(NH_4)_2SO_4$ (FM 132.14) was dissolved in water and treated with $Ba(NO_3)_2$ to precipitate all SO_4^{2-} as $BaSO_4$ (FM 233.39). Find the weight percent of K_2SO_4 in the sample if 0.977 g of precipitate was formed.

26-C. Consider a mixture of the two solids, $BaCl_2 \cdot 2H_2O$ (FM 244.26) and KCl (FM 74.551), in an unknown ratio. (The notation $BaCl_2 \cdot 2H_2O$ means that a crystal is formed with two water molecules for each $BaCl_2$.) When the unknown is heated to 160°C for 1 h, the water of crystallization is driven off:

$$BaCl_2 \cdot 2H_2O(s) \xrightarrow{160°C} BaCl_2(s) + 2H_2O(g)$$

A sample originally weighing 1.783 9 g weighed 1.562 3 g after heating. Calculate the weight percent of Ba, K, and Cl in the original sample.

26-D. A mixture containing only aluminum tetrafluoroborate, $Al(BF_4)_3$ (FM 287.39), and magnesium nitrate, $Mg(NO_3)_2$ (FM 148.31), weighed 0.282 8 g. It was dissolved in 1 wt% HF(aq) and treated with nitron solution to precipitate a mixture of nitron tetrafluoroborate and nitron nitrate weighing 1.322 g. Find the wt% Mg in the original solid mixture.

Nitron
$C_{20}H_{16}N_4$
FM 312.37

Nitron tetrafluoroborate
$C_{20}H_{17}N_4BF_4$
FM 400.18

Nitron nitrate
$C_{20}H_{17}N_5O_3$
FM 375.39

BF_4^-

NO_3^-

26-E. A 50.0-mL sample of 0.080 0 M KSCN is titrated with 0.040 0 M Cu^+. The solubility product of CuSCN is 4.8×10^{-15}. At each of the following volumes of titrant, calculate pCu^+, and construct a graph of pCu^+ versus milliliters of Cu^+ added: 0.10, 10.0, 25.0, 50.0, 75.0, 95.0, 99.0, 99.9, 100.0, 100.1, 101.0, 110.0 mL.

26-F. Consider the titration of 50.00 (± 0.05) mL of a mixture of I^- and SCN^- with 0.068 3 (± 0.000 1) M Ag^+. The first equivalence point is observed at 12.6 (± 0.4) mL, and the second occurs at 27.7 (± 0.3) mL.

(a) Find the molarity and the uncertainty in molarity of thiocyanate in the original mixture.

(b) Suppose that the uncertainties are all the same, except that the uncertainty of the first equivalence point (12.6 $\pm$? mL) is variable. What is the maximum uncertainty (milliliters) of the first equivalence point if the uncertainty in SCN^- molarity is to be $\leq 4.0\%$?

Problems

Gravimetric Analysis

26-1. (a) What is the difference between absorption and adsorption?

(b) How is an inclusion different from an occlusion?

26-2. State four desirable properties of a gravimetric precipitate.

26-3. Why is high relative supersaturation undesirable in a gravimetric precipitation?

26-4. What measures can be taken to decrease the relative supersaturation during a precipitation?

26-5. Why are many ionic precipitates washed with electrolyte solution instead of pure water?

26-6. Why is it less desirable to wash AgCl precipitate with aqueous $NaNO_3$ than with HNO_3 solution?

26-7. Why would a reprecipitation be employed in a gravimetric analysis?

26-8. Explain what is done in thermogravimetric analysis.

26-9. Explain how the quartz crystal microbalance at the opening of Chapter 2 measures small masses.

26-10. A 50.00-mL solution containing NaBr was treated with excess $AgNO_3$ to precipitate 0.214 6 g of AgBr (FM 187.772). What was the molarity of NaBr in the solution?

26-11. To find the Ce^{4+} content of a solid, 4.37 g were dissolved and treated with excess iodate to precipitate $Ce(IO_3)_4$. The precipitate was collected, washed well, dried, and ignited to produce 0.104 g of CeO_2 (FM 172.114). What was the weight percent of Ce in the original solid?

26-12. Marie Curie dissolved 0.091 92 g of $RaCl_2$ and treated it with excess $AgNO_3$ to precipitate 0.088 90 g of AgCl. In her time (1900), the atomic mass of Ag was known to be 107.8 and that of Cl was 35.4. From these values, find the atomic mass of Ra that Marie Curie would have calculated.

26-13. A 0.050 02-g sample of impure piperazine contained 71.29 wt% piperazine (FM 86.136). How many grams of product (FM 206.240) will be formed when this sample is analyzed by Reaction 26-6?

26-14. A 1.000-g sample of unknown gave 2.500 g of bis(dimethyl-glyoximate)nickel(II) (FM 288.91) when analyzed by Reaction 26-7. Find the weight percent of Ni in the unknown.

26-15. Name the product obtained in Figure 26-4 when calcium salicylate monohydrate is heated to 550°C and to 1 000°C. Using the formula masses of these products, calculate what mass is expected to remain when 0.635 6 g of calcium salicylate monohydrate is heated to 550°C and to 1 000°C.

26-16. A method to measure soluble organic carbon in seawater includes oxidation of the organic materials to CO_2 with $K_2S_2O_8$, followed by gravimetric determination of the CO_2 trapped by a column of NaOH-coated asbestos. A water sample weighing 6.234 g produced 2.378 mg of CO_2 (FM 44.010). Calculate the ppm carbon in the seawater.

26-17. How many milliliters of 2.15% alcoholic dimethylglyoxime should be used to provide a 50.0% excess for Reaction 26-7 with 0.998 4 g of steel containing 2.07 wt% Ni? Assume that the density of the dimethylglyoxime solution is 0.790 g/mL.

26-18. Twenty dietary iron tablets with a total mass of 22.131 g were ground and mixed thoroughly. Then 2.998 g of the powder were dissolved in HNO_3 and heated to convert all iron into Fe^{3+}. Addition of NH_3 precipitated $Fe_2O_3 \cdot xH_2O$, which was ignited to give 0.264 g of Fe_2O_3 (FM 159.69). What is the average mass of $FeSO_4 \cdot 7H_2O$ (FM 278.01) in each tablet?

26-19. Finely ground mineral (0.632 4 g) was dissolved in 25 mL of boiling 4 M HCl and diluted with 175 mL H_2O containing two drops of methyl red indicator. The solution was heated to 100°C, and 50 mL of warm solution containing 2.0 g $(NH_4)_2C_2O_4$ were slowly added to precipitate CaC_2O_4. Then 6 M NH_3 was added until the indicator changed from red to yellow, showing that the liquid was neutral or slightly basic. After slow cooling for 1 h, the liquid was decanted and the solid transferred to a filter crucible and washed with cold 0.1 wt% $(NH_4)_2C_2O_4$ solution five times until no Cl^- was detected in the filtrate upon addition of $AgNO_3$ solution. The crucible was dried at 105°C for 1 h and then at 500° ± 25°C in a furnace for 2 h.

$$Ca^{2+} + C_2O_4^{2-} \xrightarrow{105°C} CaC_2O_4 \cdot H_2O(s) \xrightarrow{500°C} CaCO_3(s)$$
$$\text{FM 40.078} \qquad\qquad\qquad\qquad\qquad\qquad \text{FM 100.087}$$

The mass of the empty crucible was 18.231 1 g, and the mass of the crucible with $CaCO_3(s)$ was 18.546 7 g.

(a) Find the wt% Ca in the mineral.

(b) Why is the unknown solution heated to boiling and the precipitant solution, $(NH_4)_2C_2O_4$, also heated before slowly mixing the two?

(c) What is the purpose of washing the precipitate with 0.1 wt% $(NH_4)_2C_2O_4$?

(d) What is the purpose of testing the filtrate with $AgNO_3$ solution?

26-20. *Man in the vat problem.*[11] Long ago, a workman at a dye factory fell into a vat containing hot, concentrated sulfuric and nitric acids. He dissolved completely! Because nobody witnessed the accident, it was necessary to prove that he fell in so that the man's wife could collect his insurance money. The man weighed 70 kg, and a human body contains ~6.3 parts per thousand (mg/g) phosphorus.

The acid in the vat was analyzed for phosphorus to see whether it contained a dissolved human.

(a) The vat contained 8.00×10^3 L of liquid, and a 100.0-mL sample was analyzed. If the man did fall into the vat, what is the expected quantity of phosphorus in 100.0 mL?

(b) The 100.0-mL sample was treated with a molybdate reagent that precipitated ammonium phosphomolybdate, $(NH_4)_3[P(Mo_{12}O_{40})] \cdot 12H_2O$. This substance was dried at 110°C to remove waters of hydration and heated to 400°C until it reached the constant composition $P_2O_5 \cdot 24MoO_3$, which weighed 0.371 8 g. When a fresh mixture of the same acids (not from the vat) was treated in the same manner, 0.033 1 g of $P_2O_5 \cdot 24MoO_3$ (FM 3 596.46) was produced. This *blank determination* gives the amount of phosphorus in the starting reagents. The $P_2O_5 \cdot 24MoO_3$ that could have come from the dissolved man is therefore $0.371\ 8 - 0.033\ 1 = 0.338\ 7$ g. How much phosphorus was present in the 100.0-mL sample? Is this quantity consistent with a dissolved man?

26-21. A 1.475-g sample containing NH_4Cl (FM 53.492), K_2CO_3 (FM 138.21), and inert ingredients was dissolved to give 0.100 L of solution. A 25.0-mL aliquot was acidified and treated with excess sodium tetraphenylborate, $Na^+B(C_6H_5)_4^-$, to precipitate K^+ and NH_4^+ ions completely:

$$(C_6H_5)_4B^- + K^+ \rightarrow (C_6H_5)_4BK(s)$$
$$\text{FM 358.33}$$

$$(C_6H_5)_4B^- + NH_4^+ \rightarrow (C_6H_5)_4BNH_4(s)$$
$$\text{FM 337.27}$$

The resulting precipitate amounted to 0.617 g. A fresh 50.0-mL aliquot of the original solution was made alkaline and heated to drive off all the NH_3:

$$NH_4^+ + OH^- \rightarrow NH_3(g) + H_2O$$

It was then acidified and treated with sodium tetraphenylborate to give 0.554 g of precipitate. Find the weight percent of NH_4Cl and K_2CO_3 in the original solid.

26-22. A mixture containing only Al_2O_3 (FM 101.96) and Fe_2O_3 (FM 159.69) weighs 2.019 g. When heated under a stream of H_2, Al_2O_3 is unchanged, but Fe_2O_3 is converted into metallic Fe plus $H_2O(g)$. If the residue weighs 1.774 g, what is the weight percent of Fe_2O_3 in the original mixture?

26-23. A solid mixture weighing 0.548 5 g contained only ferrous ammonium sulfate hexahydrate and ferrous chloride hexahydrate. The sample was dissolved in 1 M H_2SO_4, oxidized to Fe^{3+} with H_2O_2, and precipitated with cupferron. The ferric cupferron complex was ignited to produce 0.167 8 g of ferric oxide, Fe_2O_3 (FM 159.69). Calculate the weight percent of Cl in the original sample.

$FeSO_4 \cdot (NH_4)_2SO_4 \cdot 6H_2O$	$FeCl_2 \cdot 6H_2O$
Ferrous ammonium sulfate	Ferrous chloride
hexahydrate	hexahydrate
FM 392.13	FM 234.84

Cupferron
FM 155.16

26-24. *Propagation of error.* A mixture containing only silver nitrate and mercurous nitrate was dissolved in water and treated with excess sodium cobalticyanide, $Na_3[Co(CN)_6]$ to precipitate both cobalticyanide salts:

$AgNO_3$	FM 169.873
$Ag_3[Co(CN)_6]$	FM 538.643
$Hg_2(NO_3)_2$	FM 525.19
$(Hg_2)_3[Co(CN)_6]_2$	FM 1 633.62

(a) The unknown weighed 0.432 1 g and the product weighed 0.451 5 g. Find wt% $AgNO_3$ in the unknown. *Caution:* Keep all the digits in your calculator or else serious rounding errors may occur. Do not round off until the end.

(b) Even a skilled analyst is not likely to have less than a 0.3% error in isolating the precipitate. Suppose that there is negligible error in all quantities, except the mass of product, which has an uncertainty of 0.30%. Calculate the relative uncertainty in the mass of $AgNO_3$ in the unknown.

26-25. The thermogravimetric trace below shows mass loss by $Y_2(OH)_5Cl \cdot xH_2O$ upon heating. In the first step, waters of hydration are lost to give ~8.1% mass loss. After a second decomposition step, 19.2% of the original mass is lost. Finally, the composition stabilizes at Y_2O_3 above 800°C.

(a) Find x in the formula $Y_2(OH)_5Cl \cdot xH_2O$. Because the 8.1% mass loss is not accurately defined in the experiment, use the 31.8% total mass loss for your calculation.

(b) Suggest a formula for the material remaining at the 19.2% plateau. Be sure that the charges of all ions in your formula sum to 0. The cation is Y^{3+}.

Thermogravimetric analysis of $Y_2(OH)_5Cl \cdot xH_2O$. [From T. Hours, P. Bergez, J. Charpin, A. Larbot, C. Guizard, and L. Cot, "Preparation and Characterization of Yttrium Oxide by a Sol-Gel Process," *Ceramic Bull.* **1992**, *71*, 200.]

26-26. *Thermogravimetric analysis and propagation of error.*[12] Crystals of deuterated potassium dihydrogen phosphate, $K(D_xH_{1-x})_2PO_4$, are used in optics as a light valve, as a light deflector, and for frequency doubling of lasers. The optical properties are sensitive to the fraction of deuterium in the material. A publication states that deuterium content can be determined by measuring the mass lost by dehydration of the crystal after slow heating to 450°C in a Pt crucible under N_2.

$$K(D_xH_{1-x})_2PO_4(s) \xrightarrow{450°C} KPO_3(s) + (D_xH_{1-x})_2O(g)$$
$$\text{FM } 136.085\ 3 + 2.012\ 55x \qquad\qquad \text{FM } 118.070\ 3$$

(a) Let α be the mass of product divided by the mass of reactant:

$$\alpha = \frac{\text{mass of } KPO_3}{\text{mass of } K(D_xH_{1-x})_2PO_4}$$

Show that the coefficient x in $K(D_xH_{1-x})_2PO_4$ is related to α by the equation

$$x = \frac{58.667\ 0}{\alpha} - 67.618\ 3$$

What would be the value of α if starting material were 100% deuterated?

(b) One crystal measured in triplicate gave an average value of $\alpha = 0.856\ 7_7$. Find x for this crystal.

(c) From Equation C.1 in Appendix C, show that the uncertainty in x (e_x) is related to the uncertainty in α (e_α) by the equation

$$e_x = \frac{58.667\ 0\ e_\alpha}{\alpha^2}$$

(d) The uncertainty in deuterium:hydrogen stoichiometry is e_x. The authors estimate that their uncertainty in α is $e_\alpha = 0.000\ 1$. From e_α, compute e_x. Write the stoichiometry in the form $x \pm e_x$. If e_α were 0.001 (which is perfectly reasonable), what would e_x be?

26-27. When the *high-temperature superconductor* yttrium barium copper oxide (see Chapter 15 opening and Box 15-3) is heated under flowing H_2, the solid remaining at 1 000°C is a mixture of Y_2O_3, BaO, and Cu. The starting material has the formula $YBa_2Cu_3O_{7-x}$, in which the oxygen stoichiometry varies between 7 and 6.5 ($x = 0$ to 0.5).

$$YBa_2Cu_3O_{7-x}(s) + (3.5 - x)H_2(g) \xrightarrow{1\ 000°C}$$
$$\underset{\text{FM}}{} \quad 666.19 - 16.00x$$

$$\underbrace{\frac{1}{2}Y_2O_3(s) + 2BaO(s) + 3Cu(s)}_{YBa_2Cu_3O_{3.5}} + (3.5 - x)H_2O(g)$$

(a) *Thermogravimetric analysis.* When 34.397 mg of $YBa_2Cu_3O_{7-x}$ were subjected to this analysis, 31.661 mg of solid remained after heating to 1 000°C. Find the value of x in $YBa_2Cu_3O_{7-x}$.

(b) *Propagation of error.* Suppose that the uncertainty in each mass in part (a) is ±0.002 mg. Find the uncertainty in the value of x.

26-28. A reprecipitation was employed to remove occluded NO_3^- from $BaSO_4$ prior to isotopic analysis of oxygen for geologic studies.[13] Approximately 30 mg of $BaSO_4$ crystals were mixed with 15 mL of 0.05 M DTPA (Figure 11-4) in 1 M NaOH. After dissolving the solid at 70°C, it was reprecipitated by adding 10 M HCl dropwise to obtain pH 3–4 and allowing the mixture to stand for 1 h. The mixture was centrifuged, the mother liquor was removed, and the solid was resuspended in deionized water. Centrifugation and washing were repeated a second time to reduce the molar ratio $[NO_3^-]/[SO_4^{2-}]$ from 0.28 in the original precipitate to 0.001 in the purified material. What will be the predominant species of sulfate and DTPA at pH 14 and at pH 3? Explain why $BaSO_4$ dissolves in DTPA in 1 M NaOH and then reprecipitates when the pH is lowered to 3–4.

Combustion Analysis

26-29. What is the difference between combustion and pyrolysis?

26-30. What is the purpose of the WO_3 and Cu in Figure 26-6?

26-31. Why is tin used to encapsulate a sample for combustion analysis?

26-32. Why is sample dropped into the preheated furnace before the oxygen concentration reaches its peak in Figure 26-7?

26-33. Write a balanced equation for the combustion of benzoic acid, $C_6H_5CO_2H$, to give CO_2 and H_2O. How many milligrams of CO_2 and of H_2O will be produced by the combustion of 4.635 mg of benzoic acid?

26-34. Write a balanced equation for the combustion of $C_8H_7NO_2SBrCl$ in a C,H,N,S elemental analyzer.

26-35. Combustion analysis of a compound known to contain just C, H, N, and O demonstrated that it is 46.21 wt% C, 9.02 wt% H, 13.74 wt% N, and, by difference, $100 - 46.21 - 9.02 - 13.74 = 31.03\%$ O. This means that 100 g of unknown would contain 46.21 g of C, 9.02 g of H, and so on. Find the atomic ratio C:H:N:O and express it as the lowest reasonable integer ratio.

26-36. A mixture weighing 7.290 mg contained only cyclohexane, C_6H_{12} (FM 84.159), and oxirane, C_2H_4O (FM 44.053). When the mixture was analyzed by combustion analysis, 21.999 mg of CO_2 (FM 44.010) were produced. Find the weight percent of oxirane in the mixture.

26-37. Combustion analysis of an organic compound gave the composition 71.17 ± 0.41 wt% C, 6.76 ± 0.12 wt% H, and 10.34 ± 0.08 wt% N. Find the coefficients h and n and their uncertainties x and y in the formula $C_8H_{h \pm x}N_{n \pm y}$.

26-38. One way to determine sulfur is by combustion analysis. This procedure produces a mixture of SO_2 and SO_3 that can be passed through H_2O_2 to convert both into H_2SO_4, which is titrated with standard base. When 6.123 mg of a substance were burned, the H_2SO_4 required 3.01 mL of 0.015 76 M NaOH for titration. Find the weight percent of sulfur in the sample.

26-39. *Statistics of coprecipitation.*[14] In Experiment 1, 200.0 mL of solution containing 10.0 mg of SO_4^{2-} (from Na_2SO_4) were treated with excess $BaCl_2$ solution to precipitate $BaSO_4$ containing some coprecipitated Cl^-. To find out how much coprecipitated Cl^- was present, the precipitate was dissolved in 35 mL of 98 wt% H_2SO_4 and boiled to liberate HCl, which was removed by bubbling N_2 gas through the H_2SO_4. The HCl/N_2 stream was passed into a reagent solution that reacted with Cl^- to give a color that was measured. Ten replicate trials gave values of 7.8, 9.8, 7.8, 7.8, 7.8, 7.8, 13.7, 12.7, 13.7, and 12.7 μmol Cl^-. Experiment 2 was identical to the first one, except that the 200.0-mL solution also contained 6.0 g of Cl^- (from NaCl). Ten replicate trials gave 7.8, 10.8, 8.8, 7.8, 6.9, 8.8, 15.7, 12.7, 13.7, and 14.7 μmol Cl^-.

(a) Find the mean, standard deviation, and 95% confidence interval for Cl^- in each experiment.

(b) Is there a significant difference between the two experiments? What does your answer mean?

(c) If there were no coprecipitate, what mass of $BaSO_4$ (FM 233.39) would be expected?

(d) If the coprecipitate is $BaCl_2$ (FM 208.23), what is the average mass of precipitate ($BaSO_4 + BaCl_2$) in Experiment 1? By what percentage is the mass greater than the mass in part (c)?

Shape of a Precipitation Curve

26-40. Describe the chemistry that occurs in each of the following regions in curve (a) in Figure 26-10: (i) before the first equivalence point; (ii) at the first equivalence point; (iii) between the first and second equivalence points; (iv) at the second equivalence point; and

(v) past the second equivalence point. For each region except (ii), write the equation that you would use to calculate [Ag^+].

26-41. Consider the titration of 25.00 mL of 0.082 30 M KI with 0.051 10 M $AgNO_3$. Calculate pAg$^+$ at the following volumes of added $AgNO_3$: (a) 39.00 mL; (b) V_e; (c) 44.30 mL.

26-42. The text claims that I^- precipitation is not complete before Cl^- begins to precipitate in Figure 26-10. Calculate [Ag^+] at the equivalence point in the titration of I^- alone. Show that this concentration of Ag^+ will precipitate Cl^-.

26-43. A 25.00-mL solution containing 0.031 10 M $Na_2C_2O_4$ was titrated with 0.025 70 M $Ca(NO_3)_2$ to precipitate calcium oxalate: $Ca^{2+} + C_2O_4^{2-} \rightarrow CaC_2O_4(s)$. Find pCa^{2+} at the following volumes of $Ca(NO_3)_2$: (a) 10.00; (b) V_e; (c) 35.00 mL.

26-44. In precipitation titrations of halides by Ag^+, the ion pair AgX(aq) (X = Cl, Br, I) is in equilibrium with the precipitate. Use Appendix J to find the concentrations of AgCl(aq), AgBr(aq), and AgI(aq) during the precipitations.

Titration of a Mixture

26-45. A 0.238 6-g sample contained only NaCl and KBr. It was dissolved in water and required 48.40 mL of 0.048 37 M $AgNO_3$ for complete titration of both halides [giving AgCl(s) and AgBr(s)]. Calculate the weight percent of Br in the solid sample.

26-46. A procedure[15] for determining halogens in organic compounds uses an argentometric titration. To 50 mL of anhydrous ether is added a carefully weighed sample (10–100 mg) of unknown, plus 2 mL of sodium dispersion and 1 mL of methanol. (Sodium dispersion is finely divided solid sodium suspended in oil. With methanol, it makes sodium methoxide, $CH_3O^-Na^+$, which attacks the organic compound, liberating halides.) Excess sodium is destroyed by slow addition of 2-propanol, after which 100 mL of water are added. (Sodium should not be treated directly with water because the H_2 produced can explode in the presence of O_2: $2Na + 2H_2O \rightarrow 2NaOH + H_2$.) This procedure gives a two-phase mixture, with an ether layer floating on top of the aqueous layer that contains halide salts. The aqueous layer is adjusted to pH 4 and titrated with Ag^+, using the electrodes in Figure 26-11. How much 0.025 70 M $AgNO_3$ solution will be required to reach each equivalence point when 82.67 mg of 1-bromo-4-chlorobutane ($BrCH_2CH_2CH_2CH_2Cl$; FM 171.46) are analyzed?

Using Spreadsheets

26-47. Derive an expression analogous to Equation 26-18 for the titration of M^+ (concentration = C_M^0, volume = V_M^0) with X^- (titrant concentration = C_X^0). Your equation allows you to compute the volume of titrant (V_X) as a function of [X^-].

26-48. [spreadsheet icon] Use Equation 26-18 to reproduce the curves in Figure 26-9. Plot your results on a single graph.

26-49. Consider precipitation of X^{x-} with M^{m+}:

$$x M^{m+} + m X^{x-} \rightleftharpoons M_x X_m(s) \qquad K_{sp} = [M^{m+}]^x[X^{x-}]^m$$

Write mass balance equations for M and X and derive the equation

$$V_M = V_X^0 \left(\frac{x C_X^0 + m[M^{m+}] - x[X^{x-}]}{m C_M^0 - m[M^{m+}] + x[X^{x-}]} \right)$$

where $[X^{x-}] = (K_{sp}/[M^{m+}]^x)^{1/m}$.

26-50. 🖳 Use the equation in the Problem 26-49 to calculate the titration curve for 10.0 mL of 0.100 M CrO_4^{2-} titrated with 0.100 M Ag^+ to produce $Ag_2CrO_4(s)$.

End-Point Detection

26-51. Why does the surface charge of a precipitate change sign at the equivalence point?

26-52. Examine the procedure in Table 26-5 for the Fajans titration of Zn^{2+}. Do you expect the charge on the precipitate to be positive or negative after the equivalence point?

26-53. Describe how to analyze a solution of NaI by using the Volhard titration.

26-54. A 30.00-mL solution of I^- was treated with 50.00 mL of 0.365 0 M $AgNO_3$. $AgI(s)$ was filtered off, and the filtrate (plus Fe^{3+}) was titrated with 0.287 0 M KSCN. When 37.60 mL had been added, the solution turned red. How many milligrams of I^- were in the original solution?

COCAINE USE? ASK THE RIVER

Cocaine

Benzoylecgonine

Map of Italy, showing where the Po River was sampled. [From E. Zuccato, C. Chiabrando, S. Castiglioni, D. Calamari, R. Bagnati, S. Schiarea, and R. Fanelli, "Cocaine in Surface Waters: A New Evidence-Based Tool to Monitor Community Drug Use," *Environ. Health* **2005**, *4*, 14, available at www.ehjournal.net/content/4/1/14]

How honest do you expect people to be when questioned about illegal drug use? In Italy in 2001, 1.1% of people aged 15 to 34 years old acknowledged using cocaine "at least once in the preceding month." Researchers studying the occurrence of therapeutic drugs in sewage realized that they had a tool to measure illegal drug use.

After ingestion, cocaine is largely converted to benzoylecgonine before being excreted in urine. Scientists collected representative composite samples of water from the Po River and samples of wastewater entering treatment plants serving four Italian cities. They **preconcentrated** minute quantities of benzoylecgonine from large volumes of water by solid-phase extraction, described on pages 713–714. Extracted chemicals were washed from the solid phase by a small quantity of solvent, separated by liquid chromatography, and measured by mass spectrometry. Cocaine use was estimated from the concentration of benzoylecgonine, the volume of water flowing in the river, and the fact that 5.4 million people live upstream of the collection site.

Benzoylecgonine in the Po River corresponded to 27 ± 5 100-mg doses of cocaine per 1 000 people per day by the 15- to 34-year-old population. Similar results were observed in water from four treatment plants. Cocaine use is much higher than people admit in a survey.

Heterogeneous: different composition from place to place in a material

Homogeneous: same composition everywhere

A chemical analysis is meaningless unless you begin with a meaningful sample. To measure cholesterol in a dinosaur skeleton or herbicide in a truckload of oranges, you must have a strategy for selecting a *representative sample* from a *heterogeneous* material. Figure 27-1 shows that the concentration of nitrate in sediment beneath a lake drops by two orders of magnitude in the first 3 mm below the surface. If you want to measure nitrate in sediment, it makes an enormous difference whether you select a core sample that is 1 m deep or skim the top 2 mm of sediment for analysis. **Sampling** is the process of collecting a representative sample for analysis.[1] Real samples generally require some degree of *sample preparation* to remove substances that interfere in the analysis of the desired analyte and, perhaps, to convert analyte into a form suitable for analysis.[2]

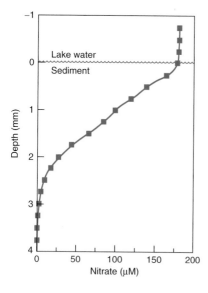

FIGURE 27-1 Depth profile of nitrate in sediment from freshwater Lake Søbygård in Denmark. A similar profile was observed in saltwater sediment. Measurements were made with a *biosensor* containing live bacteria that convert NO_3^- into N_2O, which was measured amperometrically by reduction at a silver cathode. [From L. H. Larsen, T. Kjær, and N. P. Revsbech, "A Microscale NO_3^- Biosensor for Environmental Applications," *Anal. Chem.* **1997**, *69*, 3527.]

FIGURE 27-2 *Sampling* is the process of selecting a representative bulk sample from the lot. *Sample preparation* is the process that converts a bulk sample into a homogeneous laboratory sample. Sample preparation also refers to steps that eliminate interfering species or that concentrate the analyte.

TABLE 27-1 Manganese concentration of serum stored in washed and unwashed polyethylene containers

Container[a]	Mn (ng/mL)
Unwashed	0.85
Unwashed	0.55
Unwashed	0.20
Unwashed	0.67
Average	0.57 ± 0.27
Washed	0.096
Washed	0.018
Washed	0.12
Washed	0.10
Average	0.084 ± 0.045

a. Washed containers were rinsed with water distilled twice from fused-silica vessels, which introduce less contamination into water than does glass.

SOURCE: *J. Versieck, Trends Anal. Chem.* **1983**, *2*, 110.

The terminology of sampling and sample preparation is shown in Figure 27-2. A *lot* is the total material (dinosaur skeleton or truckload of oranges) from which samples are taken. A *bulk sample* (also called a *gross sample*) is taken from the lot for analysis or *archiving* (storing for future reference). The bulk sample must be representative of the lot, and the choice of bulk sample is critical to producing a valid analysis. Box 0-1 gave a strategy for sampling heterogeneous material.

From the representative bulk sample, a smaller, homogeneous *laboratory sample* is formed that must have the same composition as the bulk sample. For example, we might obtain a laboratory sample by grinding the entire solid bulk sample to fine powder, mixing thoroughly, and keeping one bottle of powder for testing. Small portions (called *aliquots*) of the laboratory sample are used for individual analyses. *Sample preparation* is the series of steps that convert a representative bulk sample into a form suitable for analysis.

Besides choosing a sample judiciously, we must be careful about storing the sample. The composition may change with time after collection because of chemical changes, reaction with air, or interaction of the sample with its container. Glass is a notorious ion exchanger that alters the concentrations of trace ions in solution. Therefore, plastic (especially Teflon) collection bottles are frequently employed. Even these materials can absorb trace levels of analytes. For example, a 0.2-µM $HgCl_2$ solution lost 40–95% of its concentration in 4 h in polyethylene bottles. A 2-µM Ag^+ solution in a Teflon bottle lost 2% of its concentration in a day and 28% in a month.[3]

Plastic containers must be washed before use. Table 27-1 shows that manganese in blood serum samples increased by a factor of seven when stored in unwashed polyethylene containers prior to analysis. In the most demanding *trace analysis* of lead at 1 pg/g in polar ice cores, polyethylene containers contributed a measurable flux of 1 fg of lead per cm^2 per day even after they had been soaked in acid for seven months.[4] Steel needles are an avoidable source of metal contamination in biochemical analysis.

A study of mercury in Lake Michigan found levels near 1.6 pM (1.6×10^{-12} M), which is *two orders of magnitude* below concentrations observed in earlier studies.[5] Previous investigators apparently unknowingly contaminated their samples. A study of handling techniques for the analysis of lead in rivers investigated variations in sample collection, sample containers, protection during transportation, filtration, preservatives, and preconcentration procedures.[6] Each step that deviated from best practice *doubled* the apparent concentration of lead in stream water. Clean rooms with filtered air supplies are essential in trace analysis. Even with the best precautions, the precision of trace analysis becomes poorer as the concentration of analyte decreases (Box 5-2).

"Unless the complete history of any sample is known with certainty, the analyst is well advised not to spend his [or her] time in analyzing it."[7] Your laboratory notebook should describe how a sample was collected and stored and exactly how it was handled, as well as stating how it was analyzed.

27-1 Statistics of Sampling[8]

For random errors, the overall *variance*, s_o^2, is the sum of the variance of the analytical procedure, s_a^2, and the variance of the sampling operation, s_s^2:

Additivity of variance: $$s_o^2 = s_a^2 + s_s^2 \tag{27-1}$$

variance = (standard deviation)^2

$$\text{total variance} = \text{analytical variance} + \text{sampling variance}$$

If either s_a or s_s is sufficiently smaller than the other, there is little point in trying to reduce the smaller one. For example, if s_s is 10% and s_a is 5%, the overall standard deviation is 11% ($\sqrt{0.10^2 + 0.05^2} = 0.11$). A more expensive and time-consuming analytical procedure that reduces s_a to 1% only improves s_o from 11 to 10% ($\sqrt{0.10^2 + 0.01^2} = 0.10$).

Origin of Sampling Variance

To understand the nature of the uncertainty in selecting a sample for analysis, consider a random mixture of two kinds of solid particles. The theory of probability allows us to state the likelihood that a randomly drawn sample has the same composition as the bulk sample. It might surprise you to learn how large a sample is required for accurate sampling.[9]

Suppose that the mixture contains n_A particles of type A and n_B particles of type B. The probabilities of drawing A or B from the mixture are

$$p = \text{probability of drawing A} = \frac{n_A}{n_A + n_B} \tag{27-2}$$

$$q = \text{probability of drawing B} = \frac{n_B}{n_A + n_B} = 1 - p \tag{27-3}$$

If n particles are drawn at random, the expected number of particles of type A are np and the standard deviation of many drawings is known from the binomial distribution to be

Standard deviation in sampling operation: $$s_n = \sqrt{npq} \tag{27-4}$$

EXAMPLE Statistics of Drawing Particles

A mixture contains 1% KCl particles and 99% KNO_3 particles. If 10^4 particles are taken, what is the expected number of KCl particles, and what will be the standard deviation if the experiment is repeated many times?

Solution The expected number is

$$\text{Expected number of KCl particles} = np = (10^4)(0.01) = 100 \text{ particles}$$

and the standard deviation will be

$$\text{Standard deviation} = \sqrt{npq} = \sqrt{(10^4)(0.01)(0.99)} = 9.9$$

The standard deviation $\sqrt{npq}$ applies to both kinds of particles. The standard deviation is 9.9% of the expected number of KCl particles, but only 0.1% of the expected number of KNO_3 particles ($nq = 9\,900$). To measure nitrate, this sample is probably sufficient. For chloride, 9.9% uncertainty may not be acceptable.

Test Yourself It 10^5 particles are taken, what is the relative standard deviation of each measurement? (***Answer:*** 3% for KCl and 0.03% for KNO_3)

How much sample corresponds to 10^4 particles? Suppose that the particles are 1-mm-diameter spheres. The volume of a 1-mm-diameter sphere is $\frac{4}{3}\pi(0.5 \text{ mm})^3 = 0.524 \text{ μL}$. The density of KCl is 1.984 g/mL and that of KNO_3 is 2.109 g/mL, so the average density of the mixture is $(0.01)(1.984) + (0.99)(2.109) = 2.108$ g/mL. The mass of mixture containing 10^4 particles is $(10^4)(0.524 \times 10^{-3} \text{ mL})(2.108 \text{ g/mL}) = 11.0$ g. *If you take 11.0-g test portions from a larger laboratory sample, the expected sampling standard deviation for chloride is 9.9%.* The sampling standard deviation for nitrate will be only 0.1%.

How can you prepare a mixture of 1-mm-diameter particles? You might make such a mixture by grinding larger particles and passing them through a 16 mesh sieve, whose screen openings are squares with sides of length 1.18 mm (Table 27-2). Particles that pass through the screen are then passed through a 20 mesh sieve, whose openings are 0.85 mm, and material that does not pass is

Wow! I'm surprised that an 11.0-g sample has such a large standard deviation.

TABLE 27-2 Standard test sieves

Sieve number	Screen opening (mm)	Sieve number	Screen opening (mm)
5	4.00	45	0.355
6	3.35	50	0.300
7	2.80	60	0.250
8	2.36	70	0.212
10	2.00	80	0.180
12	1.70	100	0.150
14	1.40	120	0.125
16	1.18	140	0.106
18	1.00	170	0.090
20	0.850	200	0.075
25	0.710	230	0.063
30	0.600	270	0.053
35	0.500	325	0.045
40	0.425	400	0.038

EXAMPLE: Particles designated 50/100 mesh pass through a 50 mesh sieve but are retained by a 100 mesh sieve. Their size is in the range 0.150–0.300 mm.

retained for your sample. This procedure gives particles whose diameters are in the range 0.85–1.18 mm. We refer to the size range as 16/20 *mesh*.

Suppose that much finer particles of 80/120 mesh size (average diameter = 152 μm, average volume = 1.84 nL) were used instead. Now the mass containing 10^4 particles is reduced from 11.0 to 0.038 8 g. We could analyze a larger sample to reduce the sampling uncertainty for chloride.

EXAMPLE **Reducing Sample Uncertainty with a Larger Test Portion**

How many grams of 80/120 mesh sample are required to reduce the chloride sampling uncertainty to 1%?

Solution We are looking for a standard deviation of 1% of the number of KCl particles (= 1% of *np*):

$$\sigma_n = \sqrt{npq} = (0.01)np$$

Using $p = 0.01$ and $q = 0.99$, we find $n = 9.9 \times 10^5$ particles. With a particle volume of 1.84 nL and an average density of 2.108 g/mL, the mass required for 1% chloride sampling uncertainty is

$$\text{Mass} = (9.9 \times 10^5 \text{ particles})\left(1.84 \times 10^{-6}\frac{\text{mL}}{\text{particle}}\right)\left(2.108\frac{\text{g}}{\text{mL}}\right) = 3.84 \text{ g}$$

Even with an average particle diameter of 152 μm, we must analyze 3.84 g to reduce the sampling uncertainty to 1%. There is no point using an expensive analytical method with a precision of 0.1%, because the overall uncertainty will still be 1% from sampling.

Test Yourself What mass of 170/200 mesh particles reduces the KCl sampling uncertainty to 1%? (*Answer:* particle diameter = 0.082_5 mm, 9.9×10^5 particles, 0.61 g)

There is no advantage to reducing the analytical uncertainty if the sampling uncertainty is high, and vice versa.

TABLE 27-3 Nickel content of crushed ore

Particle mesh size	Ni content (wt%)
<230	13.52 ± 0.69
120/230	13.20 ± 0.74
25/120	13.22 ± 0.49
10/25	10.54 ± 0.84
>10	9.08 ± 0.69

NOTE: Uncertainty is ±1 standard deviation.

SOURCE: J. G. Dunn, D. N. Phillips, and W. van Bronswijk. "An Exercise to Illustrate the Importance of Sample Preparation in Analytical Chemistry," J. Chem. Ed. **1997,** *74, 1188.*

Sampling uncertainty arises from the random nature of drawing particles from a mixture. If the mixture is a liquid and the particles are molecules, there are about 10^{22} particles/mL. It will not require much volume of homogeneous liquid solution to reduce the sampling error to a negligible value. Solids, however, must be ground to very fine dimensions, and large quantities must be used to ensure a small sampling variance. Grinding invariably contaminates the sample with material from the grinding apparatus.

Table 27-3 illustrates another problem with heterogeneous materials. Nickel ore was crushed into small particles that were sieved and analyzed. Parts of the ore that are deficient in nickel are relatively resistant to fracture, so the larger particles do not have the same chemical composition as the smaller particles. It is necessary to grind the entire ore to a fine powder to have any hope of obtaining a representative sample.

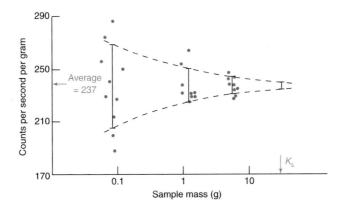

FIGURE 27-3 Sampling diagram of experimental results for ^{24}Na in liver homogenate. Dots are experimental points, and error bars extend ±1 standard deviation about the mean. Abscissa is logarithmic. [From B. Kratochvil and J. K. Taylor, "Sampling for Chemical Analysis," *Anal. Chem.* **1981**, *53*, 925A; National Bureau of Standards Internal Report 80-2164, 1980, p. 66.]

Choosing a Sample Size

A well-mixed powder containing KCl and KNO_3 is an example of a heterogeneous material in which the variation from place to place is random. *How much of a random mixture should be analyzed to reduce the sampling variance for one analysis to a desired level?*

To answer this question, consider Figure 27-3, which shows results for sampling the radioisotope ^{24}Na in human liver. The tissue was "homogenized" in a blender but was not truly homogeneous because it was a suspension of small particles in water. The average number of radioactive counts per second per gram of sample was about 237. When the mass of sample for each analysis was about 0.09 g, the standard deviation (shown by the error bar at the left in the diagram) was ±31 counts per second per gram of homogenate, which is ±13.1% of the mean value (237). When the sample size was increased to about 1.3 g, the standard deviation decreased to ±13 counts/s/g, or ±5.5% of the mean. For a sample size near 5.8 g, the standard deviation was reduced to ±5.7 counts/s/g, or ±2.4% of the mean.

Equation 27-4 told us that, when n particles are drawn from a mixture of two kinds of particles (such as liver tissue particles and droplets of water), the sampling standard deviation will be $\sigma_n = \sqrt{npq}$, where p and q are the fraction of each kind of particle present. The relative standard deviation is $\sigma_n/n = \sqrt{npq}/n = \sqrt{pq/n}$. The relative variance is

$$\text{Relative variance} \equiv R^2 = \left(\frac{\sigma_n}{n}\right)^2 = \frac{pq}{n} \quad \Rightarrow \quad nR^2 = pq \quad (27\text{-}5)$$

Noting that the mass of sample drawn, m, is proportional to the number of particles drawn, we can rewrite Equation 27-5 in the form

Sampling constant: $\qquad\qquad mR^2 = K_s \qquad\qquad (27\text{-}6)$

in which R is the relative standard deviation (expressed as a percentage) due to sampling and K_s is called the *sampling constant*. K_s is the mass of sample producing a relative sampling standard deviation of 1%.

Let's see if Equation 27-6 describes Figure 27-3. Table 27-4 shows that mR^2 is approximately constant for large samples, but agreement is poor for the smallest sample. Attributing the poor agreement at low mass to random sampling variation, we assign $K_s \approx 36$ g in Equation 27-6. This is the average from the 1.3- and 5.8-g samples in Table 27-4.

TABLE 27-4 Calculation of sampling constant for Figure 27-3

Sample mass, m (g)	Relative standard deviation (%)	mR^2 (g)
0.09	13.1	15.4
1.3	5.5	39.3
5.8	2.4	33.4

EXAMPLE Mass of Sample Required to Produce a Given Sampling Variance

What mass in Figure 27-3 will give a sampling standard deviation of ±7%?

Solution With the sampling constant $K_s \approx 36$ g, the answer is

$$m = \frac{K_s}{R^2} = \frac{36 \text{ g}}{7^2} = 0.73 \text{ g}$$

A 0.7-g sample should give ~7% sampling standard deviation. This is strictly a sampling standard deviation. The net variance will be the sum of variances from sampling and from analysis (Equation 27-1).

Test Yourself By what factor must the mass increase to reduce the sampling standard deviation by a factor of 2? (*Answer:* mass must be 4× greater)

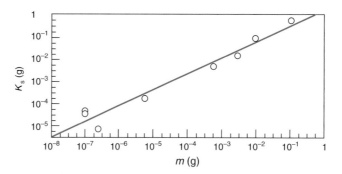

FIGURE 27-4 Data for Mn in powdered algae show that the sampling constant, K_s, in Equation 27-6 is approximately proportional to sample mass, m, over six orders of magnitude of mass. [From M. Rossbach and E. Zeiller, "Assessment of Element-Specific Homogeneity in Reference Materials Using Microanalytical Techniques," *Anal. Bioanal. Chem.* **2003**, *377*, 334.]

Figure 27-4 shows an example in which Equation 27-6 is approximately valid over six orders of magnitude of sample mass.

Choosing the Number of Replicate Analyses

We just saw that a single 0.7-g sample is expected to give a sampling standard deviation of ±7%. *How many 0.7-g samples must be analyzed to give 95% confidence that the mean is known to within ±4%?* The meaning of 95% confidence is that there is only a 5% chance that the true mean lies more than ±4% away from the measured mean. The question we just asked refers only to sampling uncertainty and assumes that analytical uncertainty is negligible.

Rearranging Student's t equation 4-6 allows us to answer the question:

Required number of replicate analyses:
$$\mu - \bar{x} = \underbrace{\frac{ts_s}{\sqrt{n}}}_{e} \Rightarrow n = \frac{t^2 s_s^2}{e^2} \tag{27-7}$$

> The sampling contribution to the overall uncertainty can be reduced by analyzing more samples.

in which μ is the true population mean, $\bar{x}$ is the measured mean, n is the number of samples needed, s_s^2 is the sampling variance, and e is the sought-for uncertainty. Both s_s and e must be expressed as absolute uncertainties or both must be expressed as relative uncertainties. Student's t is taken from Table 4-2 for 95% confidence at $n - 1$ degrees of freedom. Because n is not yet known, the value of t for $n = \infty$ can be used to estimate n. After a value of n is calculated, repeat the process a few times until n becomes constant.

EXAMPLE Sampling a Random Bulk Material

How many 0.7-g samples must be analyzed to give 95% confidence that the mean is known to within ±4%?

Solution A 0.7-g sample gives $s_s = 7\%$, and we are seeking $e = 4\%$. We will express both uncertainties in relative form. Taking $t = 1.960$ (from Table 4-2 for 95% confidence and ∞ degrees of freedom) as a starting value, we find

$$n \approx \frac{(1.960)^2(0.07)^2}{(0.04)^2} = 11.8 \approx 12$$

For $n = 12$, there are 11 degrees of freedom, so a second trial value of Student's t (interpolated from Table 4-2) is 2.209. A second cycle of calculation gives

$$n \approx \frac{(2.209)^2(0.07)^2}{(0.04)^2} = 14.9 \approx 15$$

For $n = 15$, there are 14 degrees of freedom and $t = 2.150$, which gives

$$n \approx \frac{(2.150)^2(0.07)^2}{(0.04)^2} = 14.2 \approx 14$$

For $n = 14$, there are 13 degrees of freedom and $t = 2.170$, which gives

$$n \approx \frac{(2.170)^2(0.07)^2}{(0.04)^2} = 14.4 \approx 14$$

The calculations reach a constant value near $n \approx 14$, so we need 14 samples of 0.7-g size to determine the mean value to within 4% with 95% confidence.

Test Yourself How many 2.8-g samples must be analyzed to give 95% confidence that the mean is known to within ±4%? (*Answer:* 6)

For the preceding calculations, we needed prior knowledge of the standard deviation. A preliminary study of the sample must be made before the remainder of the analysis can be planned. If there are many similar samples to be analyzed, a thorough analysis of one sample might allow you to plan less thorough—but adequate—analyses of the remainder.

27-2 Dissolving Samples for Analysis[10]

Once a *bulk sample* is selected, a *laboratory sample* must be prepared for analysis (Figure 27-2). A coarse solid sample should be ground and mixed so that the lab sample has the same composition as the bulk. Solids are typically dried at 110°C at atmospheric pressure to remove adsorbed water prior to analysis. Temperature-sensitive samples may simply be stored in an environment that brings them to a constant, reproducible moisture level.

The laboratory sample is usually dissolved for analysis. It is important to dissolve the entire sample, or else we cannot be sure that all of the analyte dissolved. If the sample does not dissolve under mild conditions, *acid digestion* or *fusion* may be used. Organic material may be destroyed by *combustion* (also called *dry ashing*) or *wet ashing* (oxidation with liquid reagents) to place inorganic elements in suitable form for analysis.

Grinding

Solids can be ground in a **mortar and pestle** like those shown in Figure 27-5. The steel mortar (also called a percussion mortar or "diamond" mortar) is a hardened steel tool into which the sleeve and pestle fit snugly. Ores and minerals can be crushed by striking the pestle lightly with a hammer. The *agate* mortar (or similar ones made of porcelain, mullite, or alumina) is designed for grinding small particles into a fine powder. Less expensive mortars tend to be more porous and more easily scratched, which leads to contamination of the sample with mortar material or portions of previously ground samples. A ceramic mortar can be cleaned by wiping with a wet tissue and washing with distilled water. Difficult residues can be removed by grinding with 4 M HCl in the mortar or by grinding an abrasive cleaner (such as Ajax), followed by washing with HCl and water. A *boron carbide* mortar and pestle is five times harder than agate and less likely to contaminate the sample.

Agate

Steel

Boron carbide

FIGURE 27-5 Steel, agate, and boron carbide mortars and pestles. The mortar is the bowl and the pestle is the grinding tool. The boron carbide mortar is a hemispheric shell enclosed in a plastic or aluminum body. The pestle has a boron carbide button at the tip of a plastic handle. [Courtesy Thomas Scientific, Swedesboro, NJ; and Spex Industries, Edison, NJ.]

A **ball mill** is a grinding device in which steel or ceramic balls are rotated inside a drum to crush the sample to a fine powder. A Wig-L-Bug® pulverizes a sample by shaking it in a vial with a ball that moves back and forth. For soft materials, plastic vials and balls are appropriate. For harder materials, steel, agate, and tungsten carbide are used. A Shatterbox laboratory mill spins a puck and ring inside a grinding container at 825 revolutions per minute to pulverize up to 100 g of material (Figure 27-6). Tungsten carbide and zirconia containers are used for very hard samples.

Shatterbox grinding action

FIGURE 27-6 Shatterbox laboratory mill spins a puck and ring inside a container at high speed to grind up to 100 mL of sample to a fine powder. [Courtesy Spex Industries, Edison, NJ.]

Dissolving Inorganic Materials with Acids

Table 27-5 lists acids commonly used for dissolving inorganic materials. The nonoxidizing acids HCl, HBr, HF, H_3PO_4, dilute H_2SO_4, and dilute $HClO_4$ dissolve metals by the redox reaction

$$M + nH^+ \rightarrow M^{n+} + \frac{n}{2}H_2 \qquad (27\text{-}8)$$

Metals with negative reduction potentials should dissolve, although some, such as Al, form a protective oxide coat that inhibits dissolution. Volatile species formed by protonation of anions such as carbonate ($CO_3^{2-} \rightarrow H_2CO_3 \rightarrow CO_2$), sulfide ($S^{2-} \rightarrow H_2S$), phosphide ($P^{3-} \rightarrow PH_3$), fluoride ($F^- \rightarrow HF$), and borate ($BO_3^{3-} \rightarrow H_3BO_3$) will be lost from hot acids in open vessels. Volatile metal halides such as $SnCl_4$ and $HgCl_2$ and some molecular oxides such as OsO_4 and RuO_4 also can be lost. Hot hydrofluoric acid is especially useful for dissolving silicates. Glass or platinum vessels can be used for HCl, HBr, H_2SO_4, H_3PO_4, and

HF causes excruciating burns. Exposure of just 2% of your body to concentrated (48 wt%) HF can kill you. Flood the affected area with water for 5 min and then coat the skin with 2.5% calcium gluconate gel kept in the lab for this purpose, and seek medical help. If the gel is not available, use whatever calcium salt is handy. HF damage can continue to develop days after exposure.

TABLE 27-5 Acids for sample dissolution

Acid	Typical composition (wt% and density)	Notes
HCl	37% 1.19 g/mL	Nonoxidizing acid useful for many metals, oxides, sulfides, carbonates, and phosphates. Constant boiling composition at 109°C is 20% HCl. As, Sb, Ge, and Pb form volatile chlorides that may be lost from an open vessel.
HBr	48–65%	Similar to HCl in solvent properties. Constant boiling composition at 124°C is 48% HBr.
H_2SO_4	95–98% 1.84 g/mL	Good solvent at its boiling point of 338°C. Attacks metals. Dehydrates and oxidizes organic compounds.
H_3PO_4	85% 1.70 g/mL	Hot acid dissolves refractory oxides insoluble in other acids. Becomes anhydrous above 150°C. Dehydrates to pyrophosphoric acid (H_2PO_3—O—PO_3H_2) above 200°C and dehydrates further to metaphosphoric acid ($[HPO_3]_n$) above 300°C.
HF	50% 1.16 g/mL	Used primarily to dissolve silicates, making volatile SiF_4. This product and excess HF are removed by adding H_2SO_4 or $HClO_4$ and heating. Constant boiling composition at 112°C is 38% HF. Used in Teflon, silver, or platinum containers. Extremely harmful upon contact or inhalation. Fluorides of As, B, Ge, Se, Ta, Nb, Ti, and Te are volatile. LaF_3, CaF_2, and YF_3 precipitate. F^- is removed by adding H_3BO_3 and taking to dryness with H_2SO_4 present.
$HClO_4$	60–72% 1.54–1.67 g/mL	Cold and dilute acid are not oxidizing, but hot, concentrated acid is an extremely powerful, explosive oxidant, especially useful for organic matter that has already been partially oxidized by hot HNO_3. Constant boiling composition at 203°C is 72%. **Before using $HClO_4$, evaporate the sample to near dryness several times with hot HNO_3 to destroy as much organic material as possible.**

FIGURE 27-7 Microwave digestion bomb lined with Teflon. The outer container retains strength to 150°C but rarely reaches 50°C.
[Courtesy Parr Instrument Co., Moline, IL.]

$HClO_4$. HF should be used in Teflon, polyethylene, silver, or platinum vessels. The highest quality acids must be used to minimize contamination by the concentrated reagent.

Substances that do not dissolve in nonoxidizing acids may dissolve in the oxidizing acids HNO_3, hot, concentrated H_2SO_4, or hot, concentrated $HClO_4$. Nitric acid attacks most metals, but not Au and Pt, which dissolve in the 3:1 (vol/vol) mixture of $HCl:HNO_3$ called **aqua regia.** Strong oxidants such as Cl_2 or $HClO_4$ in HCl dissolve difficult materials such as Ir at elevated temperature. A mixture of HNO_3 and HF attacks the refractory carbides, nitrides, and borides of Ti, Zr, Ta, and W. A powerful oxidizing solution known as "piranha solution" is a mixture of concentrated (30 wt%) H_2O_2 plus concentrated (98 wt%) H_2SO_4 variously described as 1:1 or 3:7 (vol/vol) mixtures. Hot, concentrated $HClO_4$ (described later for organic substances) is a DANGEROUS, powerful oxidant whose oxidizing power is increased by adding concentrated H_2SO_4 and catalysts such as V_2O_5 or CrO_3.

Digestion is conveniently carried out in a Teflon-lined **bomb** (a sealed vessel) heated in a microwave oven.[11] The vessel in Figure 27-7 has a volume of 23 mL and digests up to 1 g of inorganic material in up to 15 mL of concentrated acid or digests 0.1 g of organic material, which releases a great deal of $CO_2(g)$. Microwave energy heats the contents to 200°C in a minute. To prevent explosions, the lid releases gas from the vessel if the internal pressure exceeds 8 MPa (80 bar). The bomb cannot be made of metal, which absorbs microwaves. The bomb is cooled prior to opening to prevent loss of volatile products.

An example of a complex sample that can be digested by microwave heating is mixed electronic waste.[12] Components such as circuit boards are cut into small pieces and mixed thoroughly. Then 0.1 g is digested with 6 mL 70 wt% HNO_3/2 mL 30 wt% H_2O_2/1 mL 49 wt% HF. Raw plastics can be digested with 9 mL 70 wt% HNO_3. Power was ramped from 0 to 600 W over 5 min, held for 15 min, ramped to 1 400 W over 5 min, and held for 20 min.

Dissolving Inorganic Materials by Fusion

Substances that will not dissolve in acid can usually be dissolved by a hot, molten inorganic **flux** (Table 27-6). Finely powdered unknown is mixed with 2 to 20 times its mass of solid flux, and **fusion** (melting) is carried out in a platinum-gold alloy crucible at 300° to 1 200°C in a furnace or over a burner. Apparatus in Figure 27-8 fuses three samples at once over propane burners while rotating the crucibles to homogenize the melt. Then the molten flux is poured into beakers containing 10 wt% aqueous HNO_3 to dissolve the product.

Most fusions use lithium tetraborate ($Li_2B_4O_7$, m.p. 930°C), lithium metaborate ($LiBO_2$, m.p. 845°C), or a mixture of the two. A nonwetting agent such as KI can be added to prevent the flux from sticking to the crucible. For example, 0.2 g of cement might be fused with 2 g of $Li_2B_4O_7$ and 30 mg of KI.

A disadvantage of a flux is that impurities are introduced by the large mass of solid reagent. If part of the unknown can be dissolved with acid prior to fusion, it should be dissolved. Then the insoluble component is dissolved with less flux and the two portions are combined for analysis.

Basic fluxes in Table 27-6 ($LiBO_2$, Na_2CO_3, NaOH, KOH, and Na_2O_2) are best used to dissolve acidic oxides of Si and P. Acidic fluxes ($Li_2B_4O_7$, $Na_2B_4O_7$, $K_2S_2O_7$, and B_2O_3) are most suitable for basic oxides (including cements and ores) of the alkali metals, alkaline earths, lanthanides, and Al. KHF_2 is useful for lanthanide oxides. Sulfides and some oxides,

FIGURE 27-8 Automated apparatus that fuses three samples at once over propane burners. The Pt/Au crucibles rotate while they are inclined. [Courtesy Claisse, Quebec.]

Fusion is a last resort because the flux can introduce impurities.

TABLE 27-6 Fluxes for sample dissolution

Flux	Crucible	Uses
Na_2CO_3	Pt	For dissolving silicates (clays, rocks, minerals, glasses), refractory oxides, insoluble phosphates, and sulfates.
$Li_2B_4O_7$ or $LiBO_2$ or $Na_2B_4O_7$	Pt, graphite Au-Pt alloy, Au-Rh-Pt alloy	Individual or mixed borates are used to dissolve aluminosilicates, carbonates, and samples with high concentrations of basic oxides. $B_4O_7^{2-}$ is called tetraborate and BO_2^- is metaborate.
NaOH or KOH	Au, Ag	Dissolves silicates and SiC. Frothing occurs when H_2O is eliminated from flux, so it is best to prefuse flux and then add sample. Analytical capabilities are limited by impurities in NaOH and KOH.
Na_2O_2	Zr, Ni	Strong base and powerful oxidant good for silicates not dissolved by Na_2CO_3. Useful for iron and chromium alloys. Because it slowly attacks crucibles, a good procedure is to coat the inside of a Ni crucible with molten Na_2CO_3, cool, and add Na_2O_2. The peroxide melts at lower temperature than the carbonate, which shields the crucible from the melt.
$K_2S_2O_7$	Porcelain, SiO_2, Au, Pt	Potassium pyrosulfate ($K_2S_2O_7$) is prepared by heating $KHSO_4$ until all water is lost and foaming ceases. Alternatively, potassium persulfate ($K_2S_2O_8$) decomposes to $K_2S_2O_7$ upon heating. Good for refractory oxides, not silicates.
B_2O_3	Pt	Useful for oxides and silicates. Principal advantage is that flux can be completely removed as volatile methyl borate ($[CH_3O]_3B$) by several treatments with HCl in methanol.
$Li_2B_4O_7$ + Li_2SO_4 (2:1 wt/wt)	Pt	Example of a powerful mixture for dissolving refractory silicates and oxides in 10–20 min at 1 000°C. One gram of flux dissolves 0.1 g of sample. The solidified melt dissolves readily in 20 mL of hot 1.2 M HCl.

some iron and platinum alloys, and some silicates require an oxidizing flux for dissolution. For this purpose, pure Na_2O_2 may be suitable or oxidants such as KNO_3, $KClO_3$, or Na_2O_2 can be added to Na_2CO_3. Boric oxide can be converted into $B(OCH_3)_3$ after fusion and completely evaporated. Solidified flux is treated with 100 mL of CH_3OH saturated with HCl gas and heated gently. The procedure is repeated several times to remove all of the boron.

Platinum crucibles are expensive and should be heated in air, not a reducing atmosphere. Hot Pt should only be handled with Pt-tipped tongs. You can put Pt foil on the tip of ordinary tongs to handle Pt crucibles. The hot crucible should only be placed on a clean, inert surface, or in a Pt/Ir triangle. Carbon from smoky flames can embrittle Pt. Other elements including Sb, As, Pb, P, Se, Te, and Zn also embrittle Pt. Molten Ag, Au, and most base metals dissolve Pt. Oxides of Fe and Pb harm Pt above 1 000°C, as will silicates under reducing conditions.

Decomposition of Organic Substances

Digestion of organic material is classified as either **dry ashing,** when the procedure does not include liquid, or **wet ashing,** when liquid is used. Occasionally, fusion with Na_2O_2 (called

Parr oxidation) or alkali metals may be carried out in a sealed bomb. Section 26-4 discussed *combustion analysis*, in which C, H, N, S, and halogens are measured.

One form of *dry ashing* is microwave-induced combustion, an example of which is the analysis of halogens in coal.[13] Pellets of coal weighing 50 to 500 mg were wrapped in low-ash filter paper and placed on a quartz holder in a quartz vessel containing 6 mL of 50 mM $(NH_4)_2CO_3$ solution at the bottom. After wetting the filter paper with 50 mL of 6 M NH_4NO_3 (an oxidizer), the vessel was capped and pressurized with 20 bar of O_2. Application of 1 400 W of microwave power initiated combustion in which the coal reached a temperature of 1 400°C. Halides released by combustion dissolved in the $(NH_4)_2CO_3$ solution and were measured by ion chromatography.

Convenient *wet ashing* procedures include microwave digestion with acid in a Teflon bomb (Figure 27-7). For example, 0.25 g of animal tissue can be digested for metal analysis by placing the sample in a 60-mL Teflon vessel containing 1.5 mL of high-purity 70% HNO_3 plus 1.5 mL of high-purity 96% H_2SO_4 and heating it in a 700-W kitchen microwave oven for 1 min.[14] Teflon bombs fitted with temperature and pressure sensors allow safe, programmable control of digestion conditions. An important wet-ashing process is *Kjeldahl digestion* with H_2SO_4 for nitrogen analysis (Section 10-8).

In the *Carius method*, digestion is performed with fuming HNO_3 (which contains excess dissolved NO_2) in a sealed, heavy-walled glass tube at 200°–300°C. For safety, the glass Carius tube should be contained in a steel vessel pressurized to approximately the same pressure expected inside the glass tube.[15] For trace analysis, sample should be placed inside a fused silica tube inside the glass tube. Silica provides as little as 1–10% as much extractable metal as glass.[16]

Figure 27-9 shows microwave wet-ashing apparatus. Sulfuric acid or a mixture of H_2SO_4 and HNO_3 (~15 mL of acid per gram of unknown) are added to an organic substance in a glass digestion tube fitted with the reflux cap. In the first step, sample is *carbonized* for 10 to 20 min at gentle reflux until all particles have dissolved and the solution has a uniform black appearance. Power is turned off and the sample is allowed to cool for 1–2 min. Next, *oxidation* is carried out by adding H_2O_2 or HNO_3 through the reflux cap until the color disappears or the solution is just barely tinted. If the solution is not homogeneous, the power is turned up and the sample is heated to bring all solids into solution. Repeated cycles of oxidation and dissolution may be required. Once conditions for a particular type of material are worked out, the procedure is automated, with power levels and reagent delivery (by the peristaltic pump) programmed into the controller.

| Digestion tubes | Microwave heater and controller | Peristaltic pump | Reflux cap |

FIGURE 27-9 Microwave apparatus for digesting organic materials by wet ashing. [Courtesy Spex Industries, Edison, NJ.]

FIGURE 27-10 High-pressure autoclave allows digestion up to 270°C without H_2SO_4 in open vessels inside autoclave. [Adapted from B. Maichin, M. Zischka, and G. Knapp, "Pressurized Wet Digestion in Open Vessels," *Anal. Bioanal. Chem.* 2003, *376*, 715.]

Pressure-relief valve

Loose Teflon lid

140 bar

Steel autoclave

Fluoropolymer vessel

Teflon liner

Titanium rack

Fused silica vessel

Water bath with 5 vol% H_2O_2

Digestion sample

Thermocouple

Heater

Lip

Hole

Spout

FIGURE 27-11 Reflux cap for wet ashing in an Erlenmeyer flask. The hole allows vapor to escape. The curved spout touches the inside of the flask. [From D. D. Siemer and H. G. Brinkley, "Erlenmeyer Flask-Reflux Cap for Acid Sample Decomposition," *Anal. Chem.* 1981, *53*, 750.]

$HClO_4$ with organic material is an extreme **explosion hazard.** Always oxidize first with HNO_3. Always use a blast shield for $HClO_4$.

Fenton's reagent generates OH· radical and, possibly, $Fe^{II}OOH$ as the active species.[18]

Section 23-4 described *solid-phase microextraction, purge and trap,* and *thermal desorption*—sample preparation methods that are especially useful for gas chromatography.

The high-pressure asher in Figure 27-10 uses a resistive heating element inside a sealed chamber for digestion at temperatures up to 270°C under a pressure up to 140 bar. High pressure allows acids to be heated to high temperature without boiling. At high temperature, HNO_3 oxidizes organic matter without assistance from H_2SO_4, which is not as pure as HNO_3 and is therefore less suitable for trace analysis. Silica or fluoropolymer vessels inside the sealed chamber are loosely sealed by Teflon caps that permit evolved gases to escape. The bottom of the vessel is filled with 5 vol% H_2O_2 in H_2O. Hydrogen peroxide reduces nitrogen oxides generated by digestion of organic matter. As an example, a 1-g sample of animal tissue could be digested in a 50-mL fused-silica vessel containing 5 mL of high-purity 70 vol% HNO_3 plus 0.2 mL of high-purity 37 vol% HCl. Metallic elements in the digestion solution could be measured at part per billion to part per million levels by inductively coupled plasma–atomic emission.

Wet ashing with refluxing HNO_3-$HClO_4$ (Figure 27-11) is a widely applicable, but hazardous, procedure.[17] **Perchloric acid has caused numerous explosions.** Use a good blast shield in a metal-lined fume hood designed for $HClO_4$. First, heat the sample slowly to boiling with HNO_3 but *without* $HClO_4$. Boil to near dryness to destroy easily oxidized material that might explode in the presence of $HClO_4$. Add fresh HNO_3 and repeat the evaporation several times. After the sample cools to room temperature, add $HClO_4$ and heat again. If possible, HNO_3 should be present during the $HClO_4$ treatment. A large excess of HNO_3 should be present when oxidizing organic materials.

Bottles of $HClO_4$ should not be stored on wooden shelves, because acid spilled on wood can form explosive cellulose perchlorate esters. Perchloric acid also should not be stored near organic reagents or reducing agents. A reviewer of this book once wrote, "I have seen someone substitute perchloric acid for sulfuric acid in a Jones reductor experiment with spectacular results—no explosion but the tube melted!"

The combination of Fe^{2+} and H_2O_2, called *Fenton's reagent*, oxidizes organic material in dilute aqueous solutions.[19] For example, organic components of urine could be destroyed in 30 min at 50°C to release traces of mercury for analysis.[20] To do so, a 50-mL sample was adjusted to pH 3–4 with 0.5 M H_2SO_4. Then 50 µL of saturated aqueous ferrous ammonium sulfate, $Fe(NH_4)_2(SO_4)_2$, were added, followed by 100 µL of 30% H_2O_2.

27-3 Sample Preparation Techniques

Sample preparation is the series of steps required to transform a sample so that it is suitable for analysis. Sample preparation may include dissolving the sample, extracting analyte from a complex matrix, concentrating a dilute analyte to a level that can be measured, chemically converting analyte into a detectable form, and removing or masking interfering species.

Liquid Extraction Techniques

In **extraction,** analyte is dissolved in a solvent that does not necessarily dissolve the entire sample and does not decompose the analyte. In a typical *microwave-assisted extraction* of pesticides from soil, a mixture of soil plus acetone and hexane is placed in a Teflon-lined

FIGURE 27-12 Extraction vessels in a microwave oven that processes up to 12 samples in under 30 min. Each 100-mL vessel has a vent tube that releases vapor if the pressure exceeds 14 bar. Vapors from the chamber are ultimately vented to a fume hood. The temperature inside each vessel can be monitored and used to control the microwave power. [Courtesy CEM Corp., Matthews, NC.]

bomb (Figures 27-7 and 27-12) and heated by microwaves to 150°C. This temperature is 50° to 100° higher than the boiling points of solvents at atmospheric pressure. Pesticides dissolve, but the soil remains behind. The liquid is then analyzed by chromatography.

Acetone absorbs microwaves, so it can be heated in a microwave oven. Hexane does not absorb microwaves. To perform an extraction with pure hexane, the liquid is placed in a fluoropolymer insert inside the Teflon vessel in Figure 27-7.[21] The walls of the insert contain carbon black, which absorbs microwaves and heats the solvent.

Supercritical fluid extraction uses a *supercritical fluid* (Box 24-3) as the extraction solvent.[23] CO_2 is the most common supercritical fluid because it is inexpensive and it eliminates the need for costly disposal of waste organic solvents. Addition of a second solvent such as methanol increases the solubility of polar analytes. Nonpolar substances, such as petroleum hydrocarbons, can be extracted with supercritical argon.[24] The extraction process can be monitored by infrared spectroscopy because Ar has no infrared absorption.

Figure 27-13a shows how a supercritical fluid extraction can be carried out. Pressurized fluid is pumped through a heated extraction vessel. Fluid can be left in contact with the sample

Some chelators can extract metal ions into supercritical CO_2 (containing small quantities of methanol or water). The ligand below dissolves lanthanides and actinides:[22]

FIGURE 27-13 (*a*) Apparatus for supercritical fluid extraction. (*b*) Vessel for extracting house dust at 50°C with 20 mol% methanol/80 mol% CO_2 at 24.0 MPa (240 bar). (*c*) Gas chromatogram of CH_2Cl_2 solution of extract using a 30-m × 0.25-mm diphenyl$_{0.05}$dimethyl$_{0.95}$siloxane column (1-μm film thickness) with a temperature ranging from 40° to 280°C and flame ionization detection. [From T. S. Reighard and S. V. Olesik, "Comparison of Supercritical Fluids and Enhanced-Fluidity Liquids for the Extraction of Phenolic Pollutants from House Dust," *Anal. Chem.* **1996,** *68,* 3612.]

FIGURE 27-14 Continuous liquid-liquid extraction apparatus used when extraction solvent is (*a*) denser than the liquid being extracted or (*b*) lighter than the liquid being extracted.

Labels in figure (a): Condenser; Extraction solvent; Vapors from boiling solvent; Liquid being extracted; Extraction solvent returns to reservoir; Solvent return tube; Extraction solvent reservoir; Heat

Labels in figure (b): Liquid being extracted; Frit

for some time or it can be pumped through continuously. At the outlet of the extraction vessel, the fluid flows through a capillary tube to release pressure. Exiting CO_2 evaporates, leaving extracted analyte in the collection vessel. Alternatively, the CO_2 can be bubbled through a solvent in the collection vessel to leave a solution of analyte.

Figure 27-13b shows the extraction of organic compounds from dust collected with a vacuum cleaner from doormats at the chemistry building of Ohio State University. The chromatogram of the extract in Figure 27-13c exhibits myriad organic compounds that you and I inhale in every breath. In another study, polybrominated diphenyl ether flame retardants were found in house dust.[25] Levels in the U.S. were an order of magnitude higher than levels in Europe. Ingested dust was estimated to provide 0.1–6 μg of flame retardants per day to toddlers aged 1–4.

Figure 27-14 shows glassware for continuous *liquid-liquid extraction* of a nonvolatile analyte. In Figure 27-14a, the extracting solvent is denser than the liquid being extracted. Solvent boils from the flask and condenses into the extraction vessel. Dense droplets of solvent falling through the liquid column extract the analyte. When the liquid level is high enough, extraction solvent is pushed through the return tube to the solvent reservoir. By this means, analyte is slowly transferred from the light liquid at the left into the dense liquid in the reservoir. Figure 27-14b shows the procedure when the extraction solvent is less dense than the liquid being extracted.

One way to reduce solvent use in liquid-liquid extraction is *solid-supported liquid-liquid extraction* (Figure 27-15).[26] Solid-supported liquid-liquid extraction is conveniently carried out with a 96-well plate described in the next section. In a representative procedure for the extraction of prescription drugs, human plasma was diluted with an equal volume of 0.5 M NH_3. Then 200 μL of diluted plasma were applied to a small column of microporous diatomaceous earth with suction for 10 s. After allowing 5 min for the aqueous phase to disperse through the solid phase, the column was washed with 1 mL of immiscible organic solvent (hexane:2-methyl-1-butanol, 98:2 vol/vol), which flowed for 5 min by gravity. Suction was applied for 2 min to complete the elution. After evaporating the solvent to dryness, the residue was dissolved in mobile phase for liquid chromatography. In addition to using small volumes of liquid, solid-supported liquid-liquid extraction does not create emulsions, which plague ordinary liquid-liquid extraction.

Add aqueous sample such as urine, plasma, or beverage

Add organic solvent to extract analyte from adsorbed aqueous phase

High-purity diatomaceous earth

Allow 5 min for liquid to disperse in solid

Collect extract in organic solvent

(a)   (b)   (c)

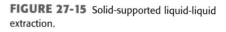

FIGURE 27-15 Solid-supported liquid-liquid extraction.

Solid-Phase Extraction[27]

Solid-phase extraction uses a small volume of a chromatographic stationary phase or molecular imprinted polymer[28] (Box 25-2) to isolate analytes from a sample. The extraction removes much of the sample matrix to simplify the analysis.

Figure 27-16 shows typical steps in the solid-phase extraction of 10 ng/mL of steroids from urine. First, a syringe containing 1 mL of C_{18}-silica is conditioned with 2 mL of methanol to remove adsorbed organic material. Then the column is washed with 2 mL of water. When the 10-mL urine sample is applied, nonpolar components adhere to the C_{18}-silica, and polar components pass through. The column is then rinsed with 4 mL of 25 mM borate buffer at pH 8 to remove polar substances (Figure 27-16c). Then rinses with 4 mL of 40 vol% methanol/60% water and 4 mL of 20% acetone/80% water remove less polar substances. Finally, elution with two 0.5-mL aliquots of 73% methanol/27% water washes the steroids from the column.

Figure 27-17 compares chromatograms of the drug Naproxen in blood serum with or without sample cleanup by solid-phase extraction. Without cleanup, serum proteins overlap and obscure the signal from Naproxen. Solid-phase extraction removes most of the protein.

Solid-phase extraction was used to *preconcentrate* and partially purify traces of cocaine and benzoylecgonine at the opening of this chapter. A 500-mL volume of river water was filtered, spiked with 10 ng of internal standard, and acidified to pH 2.0 with HCl. A solid-phase cation-exchange extraction cartridge containing 60 mg of resin was conditioned before use by washing with 6 mL of CH_3OH, 3 mL of deionized H_2O, and 3 mL of H_2O acidified to pH 2.0 with HCl. River water was sucked through the cartridge at 20 mL/min. Liquid was removed from the cartridge by suction for 5 min. Analytes were then eluted from the cartridge with 2 mL of CH_3OH followed by 2 mL of 2% ammonia solution in CH_3OH. This process preconcentrates the sample by a factor of 500 mL/4 mL = 125.

Desired analyte

1. MeOH
2. H_2O

Weak solvent

Stronger solvent

Still stronger solvent

Condition column

Apply crude sample

Elute weakly bound solutes

Elute desired analyte

(a)   (b)   (c)   (d)   (e)

FIGURE 27-16 Steps in solid-phase extraction.

Direct injection without cleanup

Proteins and naproxen

Nitrobenzene internal standard

With solid-phase extraction cleanup

Proteins

Naproxen

Nitrobenzene internal standard

Absorbance →

0   1   2   3   4   5

Time (min)

FIGURE 27-17 HPLC of Naproxen in blood serum with no cleanup (*upper trace*) or with prior sample cleanup (*lower trace*) by solid-phase extraction on C_8-silica. [From R. E. Majors and A. D. Broske, "New Directions in Solid-Phase Extraction Particle Design," *Am Lab.*, February 2002, p. 22.]

In Figure 27-18, the notation 1 meq/g means 1 milliequivalent of ion-exchange sites per gram of resin. A milliequivalent is one millimole of charged sites.

Mixed-mode Cation eXchange (MCX) reversed-phase sorbent
$pK_a < 1$; 1 meq/g

Mixed-mode Anion eXchange (MAX) reversed-phase sorbent
0.25 meq/g

Hydrophilic pyrrolidone

Lipophilic benzene

Hydrophilic-Lipophilic Balanced (HLB) reversed-phase sorbent
Stable pH 0–14
No silanol interactions

Mixed-mode **Weak Cation eXchange (WCX)** reversed-phase sorbent
$pK_a \sim 5$; 0.75 meq/g

Mixed-mode **Weak Anion eXchange (WAX)** reversed-phase sorbent
$pK_a \sim 6$; 0.6 meq/g

FIGURE 27-18 Water-wettable, hydrophobic ion-exchange Oasis® polymer sorbents for solid-phase extraction. [From Waters Corporation, Milford, MA.]

The solid-phase extraction cartridge used to preconcentrate cocaine from river water is the mixed-mode cation-exchange reversed-phase sorbent at the upper left in Figure 27-18. This is one of a family of resins whose backbone contains lipophilic benzene rings and hydrophilic pyrrolidone rings. The resins are wettable by water and have affinity for both polar and nonpolar substances. The four ion-exchange derivatives are useful for retaining and then releasing different types of analytes when conditions such as pH, solvent, and ionic strength are changed.

Preconcentration of cocaine from 500 mL of river water was done with just 60 mg of resin in a single syringe. For multiple samples or exploratory research, a *96-well plate* such as those shown in Figure 27-19 can be used. The conventional plate at the upper right has eight rows of 12 syringe-like wells, each of which can contain 5 to 60 mg of resin. The μElution® plate at the upper left has 96 Pasteur-pipet-like wells with a small volume, which can be eluted by 25-50 μL of solvent.

Solid-phase extractions reduce solvent consumption in analytical chemistry. One standard procedure for the analysis of pesticides in wastewater requires 200 mL of dichloromethane for the liquid-liquid extraction of 1 L of water. The same analytes can be isolated by solid-phase extraction on C_{18}-silica disks. The pesticides are recovered from the disks by supercritical fluid extraction with CO_2 that is finally vented into a small volume of hexane. This one kind of analysis can save 10^5 kg of CH_2Cl_2 per year.[29]

Ion-exchange resins can capture basic or acidic gases. Carbonate liberated as CO_2 from $(ZrO)_2CO_3(OH)_2 \cdot xH_2O$ used in nuclear fuel reprocessing can be measured by placing a known amount of powdered solid in the test tube in Figure 27-20 and adding 3 M HNO_3. When the solution is purged with N_2, CO_2 is captured quantitatively by moist anion-exchange resin in the sidearm:

$$CO_2 + H_2O \rightarrow H_2CO_3$$

$$2\,resin^+OH^- + H_2CO_3 \rightarrow (resin^+)_2CO_3^{2-} + 2H_2O$$

FIGURE 27-19 96-Well Plate and 96-well μElution Plate® for solid-phase extraction. [Courtesy Waters Corporation, Milford, MA.]

FIGURE 27-20 Apparatus for trapping basic or acidic gases by ion exchange. [From D. D. Siemer, "Ion Exchange Resins for Trapping Gases: Carbonate Determination," *Anal. Chem.* **1987**, *59*, 2439.]

Carbonate is eluted from the resin with 1 M $NaNO_3$ and measured by titration with acid. Table 27-7 gives other applications of this technique.

Derivatization

Derivatization is a procedure in which analyte is chemically modified to make it easier to detect or separate. For example, formaldehyde and other aldehydes and ketones in air, breath, or cigarette smoke[30] can be trapped and derivatized by passing air through a tiny cartridge containing 0.35 g of silica coated with 0.3 wt% 2,4-dinitrophenylhydrazine. Carbonyls are converted into the 2,4-dinitrophenylhydrazone derivative, which is eluted with 5 mL of acetonitrile and analyzed by HPLC. The products are readily detected by their strong ultraviolet absorbance near 360 nm.

TABLE 27-7   Use of ion-exchange resin for trapping gases

Gas	Species trapped	Eluent	Analytical method
CO_2	CO_3^{2-}	1 M $NaNO_3$	Titrate with acid
H_2S	S^{2-}	0.5 M $Na_2CO_3 + H_2O_2$	S^{2-} is oxidized to SO_4^{2-} by H_2O_2. The sulfate is measured by ion chromatography.
SO_2	SO_3^{2-}	0.5 M $Na_2CO_3 + H_2O_2$	SO_3^{2-} is oxidized to SO_4^{2-} by H_2O_2. The sulfate is measured by ion chromatography.
HCN	CN^-	1 M Na_2SO_4	Titration of CN^- with hypobromite: $$CN^- + OBr^- \rightarrow CNO^- + Br^-$$
NH_3	NH_4^+	1 M $NaNO_3$	Colorimetric assay with Nessler's reagent: $$2K_2HgI_4 + 2NH_3 \rightarrow NH_2Hg_2I_3 + 4KI + NH_4I$$ $\quad$ Nessler's reagent $\qquad\qquad$ Absorbs strongly at 400–425 nm

SOURCE: D. D. Siemer, "Ion Exchange Resins for Trapping Gases: Carbonate Determination," *Anal. Chem.* **1987**, *59*, 2439.

Terms to Understand

aqua regia	dry ashing	mortar and pestle	solid-phase extraction
ball mill	extraction	preconcentration	supercritical fluid extraction
bomb	flux	sample preparation	wet ashing
derivatization	fusion	sampling	

Summary

The variance of an analysis is the sum of the sampling variance and the analytical variance. Sampling variance can be understood in terms of the statistics of selecting particles from a heterogeneous mixture. If the probabilities of selecting two kinds of particles from a two-particle mixture are p and q, the standard deviation in selecting n particles is $\sqrt{npq}$. You should be able to use this relation to estimate how large a sample is required to reduce the sampling variance to a desired level. Student's t can be used to estimate how many repetitions of the analysis are required to reach a level of confidence in the final result.

Many inorganic materials can be dissolved in strong acids with heating. Glass vessels are often useful, but Teflon, platinum, or silver are required for HF, which dissolves silicates. If a nonoxidizing acid is insufficient, aqua regia or other oxidizing acids may do the job. A Teflon-lined bomb heated in a microwave oven is a convenient means of dissolving difficult samples. If acid digestion fails, fusion in a molten salt will usually work, but the large quantity of flux adds trace impurities. Organic materials are decomposed by wet ashing with hot concentrated acids or by dry ashing with heat.

Analytes can be separated from complex matrices by sample preparation techniques that include liquid extraction, supercritical fluid extraction, and solid-phase extraction. Derivatization transforms the analyte into a more easily detected or separated form.

Exercises

27-A. A box contains 120 000 red marbles and 880 000 yellow marbles.

(a) If you draw a random sample of 1 000 marbles from the box, what are the expected numbers of red and yellow marbles?

(b) Now put those marbles back in the box and repeat the experiment. What will be the absolute and relative standard deviations for the numbers in part **(a)** after many drawings of 1 000 marbles?

(c) What will be the absolute and relative standard deviations after many drawings of 4 000 marbles?

(d) If you quadruple the size of the sample, you decrease the sampling standard deviation by a factor of _____. If you increase the sample size by a factor of n, you decrease the sampling standard deviation by a factor of _____.

(e) What sample size is required to reduce the sampling standard deviation of red marbles to ±2%?

27-B. **(a)** What mass of sample in Figure 27-3 is expected to give a sampling standard deviation of ±10%?

(b) With the mass from part **(a)**, how many samples should be taken to assure 95% confidence that the mean is known to within ±20 counts per second per gram?

27-C. A soil sample contains some acid-soluble inorganic matter, some organic material, and some minerals that do not dissolve in any combination of hot acids that you try. Suggest a procedure for dissolving the entire sample.

Problems

Statistics of Sampling

27-1. Explain what is meant by the statement, "Unless the complete history of any sample is known with certainty, the analyst is well advised not to spend his or her time in analyzing it."

27-2. Explain what is meant by "analytical quality" and "data quality" in the following quotation: "We need to update the environmental data quality model to explicitly distinguish *analytical quality* from *data quality*. We need to begin spending as much effort ensuring sample and subsample representativeness for heterogeneous matrices as we spend overseeing the analysis of an extract. We need to stop acting like data variability stemming from laboratory analysis is all-important while variability stemming from the sampling process can be ignored . . . ."[31]

27-3. **(a)** In the analysis of a barrel of powder, the standard deviation of the sampling operation is ±4% and the standard deviation of the analytical procedure is ±3%. What is the overall standard deviation?

(b) To what value must the sampling standard deviation be reduced so that the overall standard deviation is ±4%?

27-4. What mass of sample in Figure 27-3 is expected to give a sampling standard deviation of ±6%?

27-5. Explain how to prepare a powder with an average particle diameter near 100 μm by using sieves from Table 27-2. How would such a particle mesh size be designated?

27-6. An example of a mixture of 1-mm-diameter particles of KCl and KNO_3 in a number ratio 1:99 follows Equation 27-4. A sample containing 10^4 particles weighs 11.0 g. What is the expected number and relative standard deviation of KCl particles in a sample weighing 11.0×10^2 g?

27-7. When you flip a coin, the probability of its landing on each side is $p = q = \frac{1}{2}$ in Equations 27-2 and 27-3. If you flip it n times, the expected number of heads equals the expected number of tails $= np = nq = \frac{1}{2}n$. The expected standard deviation for n flips is $\sigma_n = \sqrt{npq}$. From Table 4-1, we expect that 68.3% of the results will lie within $\pm 1\sigma_n$ and 95.5% of the results will lie within $\pm 2\sigma_n$.

(a) Find the expected standard deviation for the number of heads in 1 000 coin flips.

(b) By interpolation in Table 4-1, find the value of z that includes 90% of the area of the Gaussian curve. We expect that 90% of the results will lie within this number of standard deviations from the mean.

(c) If you repeat the 1 000 coin flips many times, what is the expected range for the number of heads that includes 90% of the results? (For example, your answer might be, "The range 490 to 510 will be observed 90% of the time.")

27-8. In analyzing a lot with random sample variation, you find a sampling standard deviation of ±5%. Assuming negligible error in the analytical procedure, how many samples must be analyzed to give 95% confidence that the error in the mean is within ±4% of the true value? Answer the same question for a confidence level of 90%.

27-9. In an experiment analogous to that in Figure 27-3, the sampling constant is found to be $K_s = 20$ g.

(a) What mass of sample is required for a ±2% sampling standard deviation?

(b) How many samples of the size in part **(a)** are required to produce 90% confidence that the mean is known to within 1.5%?

27-10. $^{87}Sr/^{86}Sr$ isotope ratios were measured in polar ice cores of varying sample size. The 95% confidence interval, expressed as a percentage of the mean $^{87}Sr/^{86}Sr$ isotope ratio decreased with increasing quantity of Sr measured:

pg Sr	Confidence interval	pg Sr	Confidence interval
57	±0.057%	506	±0.035%
68	±0.069%	515	±0.027%
110	±0.049%	916	±0.018%
110	±0.045%	955	±0.022%

SOURCE: *G. R. Burton, V. I. Morgan, C. F. Boutron, and K. J. R. Rosman, "High-Sensitivity Measurements of Strontium Isotopes in Polar Ice," Anal. Chim. Acta **2002**, 469, 225.*

We postulate that the confidence interval is related to the mass of sample by the relation $mR^2 = K_s$, where m is the mass of Sr in picograms, R is the confidence interval expressed as a percentage of the isotope ratio ($R = 0.022$ for the 955-pg sample), and K_s is a constant with units of picograms. Find the average value of K_s and its standard deviation. Justify why we expect $mR^2 = K_s$ to hold if all measurements are made with the same number of replications.

27-11. Consider a random mixture containing 4.00 g of Na_2CO_3 (density 2.532 g/mL) and 96.00 g of K_2CO_3 (density 2.428 g/mL) with a uniform spherical particle radius of 0.075 mm.

(a) Calculate the mass of a single particle of Na_2CO_3 and the number of particles of Na_2CO_3 in the mixture. Do the same for K_2CO_3.

(b) What is the expected number of particles in 0.100 g of the mixture?

(c) Calculate the relative sampling standard deviation in the number of particles of each type in a 0.100-g sample of the mixture.

Sample Preparation

27-12. From their standard reduction potentials, which of the following metals would you expect to dissolve in HCl by the reaction $M + nH^+ \rightarrow M^{n+} \rightarrow \frac{n}{2}H_2$: Zn, Fe, Co, Al, Hg, Cu, Pt, Au? (When the potential predicts that the element will not dissolve, it probably will not. If it is expected to dissolve, it may dissolve if some other process does not interfere. Predictions based on standard reduction potentials at 25°C are only tentative, because the potentials and activities in hot, concentrated solutions vary widely from those in the table of standard potentials.)

27-13. The following wet-ashing procedure was used to measure arsenic in organic soil samples by atomic absorption spectroscopy: A 0.1- to 0.5-g sample was heated in a 150-mL Teflon bomb in a microwave oven for 2.5 min with 3.5 mL of 70% HNO_3. After the sample cooled, a mixture containing 3.5 mL of 70% HNO_3, 1.5 mL of 70% $HClO_4$, and 1.0 mL of H_2SO_4 was added and the sample was reheated for three 2.5-min intervals with 2-min unheated periods in between. The final solution was diluted with 0.2 M HCl for analysis. Explain why $HClO_4$ was not introduced until the second heating.

27-14. Barbital can be isolated from urine by solid-phase extraction with C_{18}-silica. The barbital is then eluted with 1:1 vol/vol acetone:chloroform. Explain how this procedure works.

Barbital

27-15. To preconcentrate cocaine and benzoylecgonine from river water at the opening of this chapter, solid-phase extraction was carried out at pH 2 using the mixed-mode cation-exchange resin in Figure 27-18. After passing 500 mL of river water through 60 mg of resin, the retained analytes were eluted first with 2 mL of CH_3OH and then with 2 mL of 2% ammonia solution in CH_3OH. Explain the purpose of using pH 2 for retention and dilute ammonia for elution.

27-16. Referring to Table 27-7, explain how an anion-exchange resin can be used for absorption and analysis of SO_2 released by combustion.

27-17. Why is it advantageous to use large particles (50 μm) for solid-phase extraction, but small particles (5 μm) for chromatography?

27-18. In 2002, workers at the Swedish National Food Administration discovered that heated, carbohydrate-rich foods, such as french fries, potato chips, and bread, contain alarming levels (0.1 to 4 μg/g) of acrylamide, a known carcinogen.[32]

Acrylamide

After the discovery, simplified methods were developed to measure ppm levels of acrylamide in food. In one procedure, 10 g of pulverized, frozen french fries were mixed for 20 min with 50 mL of H_2O to extract acrylamide, which is very soluble in water (216 g/100 mL). The liquid was decanted and centrifuged to remove suspended

Chromatograms of acrylamide extract after passage through solid-phase extraction column. *Left:* Phenomenex Synergi Polar-RP 4-μm column eluted with 96:4 (vol/vol) H_2O:CH_3CN. *Right:* Phenomenex Synergi Hydro-RP 4-μm column eluted with 96:4:0.1 (vol/vol/vol) H_2O:CH_3OH:HCO_2H.

[From L. Peng, T. Farkas, L. Loo, J. Teuscher, and K. Kallury, "Rapid and Reproducible Extraction of Acrylamide in French Fries Using a Single Solid-Phase Sorbent," *Am. Lab. News Ed.*, October 2003, p. 10.]

matter. The internal standard $^2H_3$-acrylamide was added to 1 mL of extract. A solid-phase extraction column containing 100 mg of cation-exchange polymer with protonated sulfonic acid groups ($-SO_3H$) was washed twice with 1-mL portions of methanol and twice with 1-mL portions of water. The aqueous food extract (1 mL) was then passed through the column to bind protonated acrylamide ($-NH_3^+$) to sulfonate ($-SO_3^-$) on the column. The column was dried for 30 s at 0.3 bar and then acrylamide was eluted with 1 mL of H_2O. Eluate was analyzed by liquid chromatography with a polar bonded phase. The chromatograms show the results monitored by ultraviolet absorbance or by mass spectrometry. The retention time of acrylamide is different on the two columns because they have different dimensions and different flow rates.

(a) What is the purpose of solid-phase extraction prior to chromatography? How does the ion-exchange sorbent retain acrylamide?

(b) Why are there many peaks when chromatography is monitored by ultraviolet absorbance?

(c) Mass spectral detection used selected reaction monitoring (Figure 21-26) with the m/z $72 \rightarrow 55$ transition for acrylamide and $75 \rightarrow 58$ for $^2H_3$-acrylamide. Explain how this detection method works and suggest structures for the ions with m/z 72 and 55 from acrylamide.

(d) Why does mass spectral detection give just one major peak?

(e) How is the internal standard used for quantitation with mass spectral detection?

(f) Where does $^2H_3$-acrylamide appear with ultraviolet absorbance? With mass spectral selected reaction monitoring?

(g) Why does the mass spectral method give quantitative results even though retention of acrylamide by the ion-exchange sorbent is not quantitative and elution of acrylamide from the sorbent by 1 mL of water might not be quantitative?

27-19. Many metals in seawater can be preconcentrated for analysis by coprecipitation with $Ga(OH)_3$. A 200-μL HCl

Depth profile of elements in seawater near hydrothermal vents. [From T. Akagi and H. Haraguchi, "Simultaneous Multielement Determination of Trace Metals Using 10 mL of Seawater by Inductively Coupled Plasma Atomic Emission Spectrometry with Gallium Coprecipitation and Microsampling Technique," *Anal. Chem.* **1990,** *62,* 81.]

solution containing 50 μg of Ga^{3+} is added to 10.00 mL of the seawater. When the pH is brought to 9.1 with NaOH, a jellylike precipitate forms. After centrifugation to pack the precipitate, the water is removed and the gel is washed with water. Then the gel is dissolved in 50 μL of 1 M HNO_3 and aspirated into an inductively coupled plasma for atomic emission analysis. The preconcentration factor is 10 mL/50 μL = 200. The figure shows elemental concentrations in seawater as a function of depth near hydrothermal vents.

(a) What is the atomic ratio (Ga added):(Ni in seawater) for the sample with the highest concentration of Ni?

(b) The results given by gray lines were obtained with seawater samples that were not filtered prior to coprecipitation. The colored lines refer to filtered samples. The results for Ni do not vary between the two procedures, but the results for Fe do vary. Explain what this means.

27-20. Barium titanate, a ceramic used in electronics, was analyzed by the following procedure: Into a Pt crucible was placed 1.2 g of Na_2CO_3 and 0.8 g of $Na_2B_4O_7$ plus 0.314 6 g of unknown. After fusion at 1 000°C in a furnace for 30 min, the cooled solid was extracted with 50 mL of 6 M HCl, transferred to a 100-mL volumetric flask, and diluted to the mark. A 25.00-mL aliquot was treated with 5 mL of 15% tartaric acid (which complexes Ti^{4+} and keeps it in aqueous solution) and 25 mL of ammonia buffer, pH 9.5. The solution was treated with organic reagents that complex Ba^{2+}, and the Ba complex was extracted into CCl_4. After acidification (to release the Ba^{2+} from its organic complex), the Ba^{2+} was back-extracted into 0.1 M HCl. The final aqueous sample was treated with ammonia buffer and methylthymol blue (a metal ion indicator) and titrated with 32.49 mL of 0.011 44 M EDTA. Find the weight percent of Ba in the ceramic.

27-21. Acid-base equilibria of Cr(III) were summarized in Problem 9-37. Cr(VI) in aqueous solution above pH 6 exists as the yellow tetrahedral chromate ion, CrO_4^{2-}. Between pH 2 and 6, Cr(VI) exists as an equilibrium mixture of $HCrO_4^-$ and orange-red dichromate, $Cr_2O_7^{2-}$. Cr(VI) is a carcinogen, but Cr(III) is not considered to be as harmful. The following procedure was used to measure Cr(VI) in airborne particulate matter in workplaces.

1. Particles were collected by drawing a known volume of air through a polyvinyl chloride filter with 5-μM pore size.

2. The filter was placed in a centrifuge tube and 10 mL of 0.05 M $(NH_4)_2SO_4/0.05$ M NH_3 buffer, pH 8, were added. The immersed filter was agitated by ultrasonic vibration for 30 min at 35°C to extract all Cr(III) and Cr(VI) into solution.

3. A measured volume of extract was passed through a "strongly basic" anion exchanger (Table 25-1) in the Cl^- form. Then the resin was washed with distilled water. Liquid containing Cr(III) from the extract, and the wash was discarded.

4. Cr(VI) was then eluted from the column with 0.5 M $(NH_4)_2SO_4/0.05$ M NH_3 buffer, pH 8, and collected in a vial.

5. The eluted Cr(VI) solution was acidified with HCl and treated with a solution of 1,5-diphenylcarbazide, a reagent that forms a colored complex with Cr(VI). The concentration of the complex was measured by its visible absorbance.

(a) What are the dominant species of Cr(VI) and Cr(III) at pH 8?

(b) What is the purpose of the anion exchanger in step 3?

(c) Why is a "strongly basic" anion exchanger used instead of a "weakly basic" exchanger?

(d) Why is Cr(VI) eluted in step 4 but not step 3?

27-22. The county landfill in the diagram was monitored to verify that toxic compounds were not leaching into the local water supply. Wells drilled at 21 locations were monitored over a year and pollutants were observed only at sites 8, 11, 12, and 13. Monitoring all 21 sites each month is very expensive. Suggest a strategy to use *composite samples* (Box 0-1) made from more than one well at a time to reduce the cost of routine monitoring. How will your scheme affect the minimum detectable level for pollutants at a particular site?

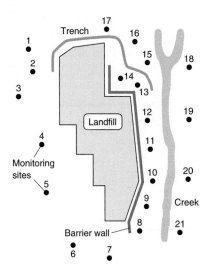

Diagram of county landfill, showing the location of wells used to monitor groundwater. [Adapted from P.-C. Li and R. Rajagopal, *Am. Environ. Lab.*, October 1994, p. 37.]

Chapter 0

1. C. D. Keeling, "Rewards and Penalties of Monitoring the Earth," *Ann. Rev. Energy Environ.* **1998**, *23*, 25–82. This spellbinding autobiography can be downloaded free of charge from http://scrippsco2.ucsd.edu/publications/keeling_autobiography.pdf.

2. J. C. Orr et al., "Anthropogenic Ocean Acidification over the Twenty-first Century and Its Impact on Calcifying Organisms," *Nature* **2005**, *437*, 681.

3. S. P. Beckett, *The Science of Chocolate*, 2nd ed. (Cambridge: Royal Society of Chemistry, 2008); G. Tannenbaum, "Chocolate: A Marvelous Natural Product of Chemistry," *J. Chem. Ed.* **2004**, *81*, 1131.

4. T. J. Wenzel, "A New Approach to Undergraduate Analytical Chemistry," *Anal. Chem.* **1995**, *67*, 470A. See also T. J. Wenzel, "The Lecture as a Learning Device," *Anal. Chem.* **1999**, *71*, 817A; T. J. Wenzel, "Cooperative Student Activities as Learning Devices," *Anal. Chem.* **2000**, *72*, 293A; T. J. Wenzel, "Practical Tips for Cooperative Learning," *Anal. Chem.* **2000**, *72*, 359A; T. J. Wenzel, "Undergraduate Research as a Capstone Learning Experience," *Anal. Chem.* **2000**, *72,* 547A.

5. W. R. Kreiser and R. A. Martin, Jr., *J. Assoc. Off. Anal. Chem.* **1978**, *61*, 1424; W. R. Kreiser and R. A. Martin, Jr., *J. Assoc. Off. Anal. Chem.* **1980**, *63*, 591. Today you would find many more recent references on caffeine.

6. A good source for many well-tested analytical procedures is W. Horwitz, *Official Methods of Analysis of AOAC International*, 18th ed. (Gaithersburg, MD: AOAC International, 2007). Can be found from http://my.aoac.org/scriptcontent/index.cfm.

7. W. Fresenius, "The Position of the Analyst as Expert: Yesterday and Today," *Fresenius J. Anal. Chem.* **2000**, *368*, 548.

Chapter 1

1. J. R. de Laeter and H. S. Peiser, "A Century of Progress in the Sciences Due to Atomic Weight and Isotopic Composition Measurements," *Anal. Bioanal. Chem.* **2003**, *375*, 62.

2. *Reagent Chemicals*, 10th ed. (Washington, DC: American Chemical Society, 2008). http://pubs.acs.org/reagents/index.html.

3. R. W. Ramette, "In Support of Weight Titrations," *J. Chem. Ed.* **2004**, *81*, 1715.

4. J. L. Sarmiento and N. Gruber, "Sinks for Anthropogenic Carbon," *Physics Today*, August 2002, p. 30.

5. U. Shahin, S.-M. Yi, R. D. Paode, and T. M. Holsen, "Long-Term Elemental Dry Deposition Fluxes Measured Around Lake Michigan," *Environ. Sci. Tech.* **2000**, *34*, 1887.

Chapter 2

1. V. Tsionsky, "The Quartz-Crystal Microbalance in an Undergraduate Laboratory Experiment," *J. Chem. Ed.* **2007**, *84*, 1334, 1337, 1340.

2. A $GaPO_4$ crystal microbalance has better properties than quartz for variable- and high-temperature measurements. (J. W. Elam and M. J. Pellin, "$GaPO_4$ Sensors for Gravimetric Monitoring during Atomic Layer Deposition at High Temperature," *Anal. Chem.* **2005**, *77*, 3531.)

3. A vibrating cantilever is 10^7 times more sensitive than a quartz crystal microbalance and can measure 1 fg (fg = femtogram = 10^{-15} g). (D. Maraldo, K. Rijal, G. Campbell, and R. Mutharasan, "Method for Label-Free Detection of Femtogram Quantities of Biologics in Flowing Liquid Samples," *Anal. Chem.* **2007**, *79*, 2762.)

4. The change in frequency for a given mass loading goes as the square of the vibrational frequency. Careful design and micromachining can produce a 62-MHz quartz oscillator. Commercial quartz microbalances oscillate at 5–10 MHz. The mass sensitivity of the custom oscillator is a factor of at least $(62/10)^2 = 38$ greater. (P. Kao, A. Patwardham, D. Allara, and S. Tadigadapa, "Human Serum Albumin Adsorption Study on 62-MHz Miniaturized Quartz Gravimetric Sensors," *Anal. Chem.* **2008**, *80*, 5930.)

5. Excellent training in basic laboratory techniques is available from http://jchemed.chem.wisc.edu/ and from www.academysavant.com.

6. R. J. Lewis, Sr., *Hazardous Chemicals Desk Reference*, 5th ed. (New York: Wiley, 2002); P. Patnaik, *A Comprehensive Guide to the Hazardous Properties of Chemical Substances*, 2nd ed. (New York: Wiley, 1999); G. Lunn and E. B. Sansone, *Destruction of Hazardous Chemicals in the Laboratory* (New York: Wiley, 1994); and M. A. Armour, *Hazardous Laboratory Chemical Disposal Guide*, 2nd ed. (Boca Raton, FL: CRC Press, 1996).

7. To recover gold from electronics, see J. W. Hill and T. A. Lear, "Recovery of Gold from Electronic Scrap," *J. Chem. Ed.* **1988**, *65*, 802. To remove Hg from gold, soak it in a 1:1 mixture of 0.01 M $(NH_4)_2S_2O_8$ and 0.01 M HNO_3; see T. Nomura and M. Fujisawa, "Electrolytic Determination of Mercury(II) in Water with a Piezoelectric Quartz Crystal," *Anal. Chim. Acta* **1986**, *182*, 267.

8. P. T. Anastas and J. C. Warner, *Green Chemistry: Theory and Practice* (New York: Oxford University Press, 1998); M. C. Cann and M. E. Connelly, *Real-World Cases in Green Chemistry* (Washington, DC: American Chemical Society, 2000); M. Lancaster, *Green Chemistry: An Introductory Text* (Cambridge: Royal Society of Chemistry, 2002); C. Baird and M. Cann, *Environmental Chemistry*, 3rd ed. (New York: W. H. Freeman and Company, 2005); J. E. Girard, *Principles of Environmental Chemistry* (Sudbury, MA: Bartlett, 2005); B. Braun, R. Charney, A. Clarens, J. Farrugia, C. Kitchens, C. Lisowski, D. Naistat, and A. O'Neil, *J. Chem. Ed.* **2006**, *83*, 1126.

9. J. M. Bonicamp, "Weigh This Way," *J. Chem. Ed.* **2002**, *79*, 476.

10. B. B. Johnson and J. D. Wells, "Cautions Concerning Electronic Analytical Balances," *J. Chem. Ed.* **1986**, *63*, 86.

11. For a demonstration of buoyancy, see K. D. Pinkerton, "Sink or Swim: The Cartesian Diver," *J. Chem. Ed.* **2001**, *78*, 200A (JCE Classroom Activity #33).

12. R. Batting and A. G. Williamson, "Single-Pan Balances, Buoyancy, and Gravity or 'A Mass of Confusion,'" *J. Chem. Ed.* **1984**, *61*, 51; J. E. Lewis and L. A. Woolf, "Air Buoyancy Corrections for Single-Pan Balances," *J. Chem. Ed.* **1971**, *48*, 639; F. F. Cantwell, B. Kratochvil, and W. E. Harris, "Air Buoyancy Errors and the Optical Scale of a Constant-Load Balance," *Anal. Chem.* **1978**, *50*, 1010; G. D. Chapman, "Weighing with Electronic Balances," National Research Council of Canada, Report NRCC 38659 (1996).

13. Air density (g/L) = $(0.003\ 485\ B - 0.001\ 318\ v)/T$, where B is barometric pressure (Pa), v is the vapor pressure of water in the air (Pa), and T is air temperature (K).

14. U. Henriksson and J. C. Eriksson, "Thermodynamics of Capillary Rise: Why Is the Meniscus Curved?" *J. Chem. Ed.* **2004**, *81*, 150.

15. Cleaning solution is prepared by dissolving 36 g of ammonium peroxydisulfate, $(NH_4)_2S_2O_8$, in a *loosely stoppered* 2.2-L ("one gallon") bottle of 98 wt% sulfuric acid (H. M. Stahr, W. Hyde, and L. Sigler, "Oxidizing Acid Baths—without Chromate Hazards," *Anal. Chem.* **1982**, *54*, 1456A). Add $(NH_4)_2S_2O_8$ every few weeks to maintain oxidizing power. Keep it loosely stoppered to prevent gas buildup (P. S. Surdhar, "Laboratory Hazard," *Anal. Chem.* **1992**, *64*, 310A). The commercial cleaning solution EOSULF (containing the metal binder, EDTA, and a sulfonate detergent) is an alternative for removing baked-on lipid or protein from glassware. (P. L. Manske, T. M. Stimpfel, and E. L. Gershey, "A Less Hazardous Chromic Acid Substitute for Cleaning Glassware," *J. Chem. Ed.* **1990**, *67*, A280.) Another powerful, oxidizing cleaning solution, called "piranha solution," is a 1:1 (vol/vol) mixture of 30 wt% H_2O_2 plus 98 wt% H_2SO_4.

16. W. B. Guenther, "Supertitrations: High-Precision Methods," *J. Chem. Ed.* **1988**, *65*, 1097; D. D. Siemer, S. D. Reeder, and M. A. Wade, "Syringe Buret Adaptor," *J. Chem. Ed.* **1988**, *65*, 467.

17. M. M. Singh, C. McGowan, Z. Szafran, and R. M. Pike, "A Modified Microburet for Microscale Titration," *J. Chem. Ed.* **1998**, *75*, 371; "A Comparative Study of Microscale and Standard Burets," *J. Chem. Ed.* **2000**, *77*, 625.

18. D. R. Burfield and G. Hefter, "Oven Drying of Volumetric Glassware," *J. Chem. Ed.* **1987**, *64*, 1054.

19. R. H. Obenauf and N. Kocherlakota, "Identifying Contamination in Trace Metal Laboratories," *Spectroscopy Applications Supplement*, March 2006, p. 12.

20. W. Vaccaro, "Minimizing Liquid Delivery Risk: Operators as Sources of Error," *Am. Lab. News Ed.* September 2007, p. 16; A. B. Carle, "Minimizing Liquid Delivery Risk: Barometric Pressure and Thermal Disequilibrium," *Am. Lab. News Ed.* January 2008, p. 8.

21. K. J. Albert, "Minimizing Liquid Delivery Risk: Automated Liquid Handlers as Sources of Error," *Am. Lab. News Ed.* June/July 2007, p. 8.

22. M. Connors and R. Curtis, "Pipetting Error," *Am. Lab. News Ed.* June 1999, p. 20; ibid. December 1999, p. 12; R. H. Curtis and G. Rodrigues, ibid. February 2004, p. 12.

23. R. Curtis, "Minimizing Liquid Delivery Risk: Pipets as Sources of Error," *Am. Lab. News Ed.* March 2007, p. 8.

24. B. Kratochvil and N. Motkosky, "Precision and Accuracy of Mechanical-Action Micropipets," *Anal. Chem.* **1987**, *59*, 1064. A colorimetric calibration kit is available from Artel, Inc., Westbrook, ME, www.artel-usa.com/.

25. E. J. Billo, *Microsoft Excel for Chemists*, 2nd ed. (New York: Wiley, 2001); R. de Levie, *How to Use Excel® in Analytical Chemistry and in General Scientific Data Analysis* (Cambridge: Cambridge University Press, 2001); E. J. Billo, *Excel for Scientists and Engineers: Numerical Methods* (New York: Wiley, 2007); R. de Levie, *Advanced Excel for Scientific Data Analysis,* 2nd ed. (Oxford: Oxford University Press, 2008).

26. D. Bohrer, P. Cícero do Nascimento, P. Martins, and R. Binotto, "Availability of Aluminum from Glass on an Al Form Ion Exchanger in the Presence of Complexing Agents and Amino Acids," *Anal. Chim. Acta* **2002**, *459*, 267.

Chapter 3

1. A catalog of Standard Reference Materials is available from SRMINFO@enh.nist.gov. European Reference Materials are available from http://www.erm-crm.org.

2. J. R. Taylor, *An Introduction to Error Analysis*, 2nd ed. (Sausalito, CA: University Science Books, 1997). An outstandingly readable book.

3. Readable treatments of propagation of uncertainty, which go beyond what we can cover in this book include B. Wampfler, M. Rösslein, and H. Felber, "The New Measurement Concept Explained by Using an Introductory Example," *J. Chem. Ed.* **2006**, *83*, 1382; EURACHEM/CITAG Guide CG 4, *Quantifying Uncertainty in Analytical Measurements*, 2nd ed., http://www.measurementuncertainty.org/mu/QUAM2000-1.pdf; *The NIST Reference on Constants, Units, and Uncertainty*, http://physics.nist.gov/cuu/.

4. P. De Bièvre, S. Valkiers, and P. D. P. Taylor, "The Importance of Avogadro's Constant for Amount-of-Substance Measurements," *Fresenius J. Anal. Chem.* **1998**, *361*, 227.

Chapter 4

1. Excellent, readable sources on statistics are D. B. Hibbert and J. J. Gooding, *Data Analysis for Chemistry* (Oxford: Oxford University Press, 2006); J. C. Miller and J. N. Miller, *Statistics and Chemometrics for Analytical Chemistry*, 5th ed. (Harlow, UK: Pearson Prentice Hall, 2005); and P. C. Meier and R. E. Zünd, *Statistical Methods in Analytical Chemistry*, 2nd ed. (New York: Wiley, 2000).

2. L. H. Keith, W. Crummett, J. Deegan, Jr., R. A. Libby, J. K. Taylor, and G. Wentler, "Principles of Environmental Analysis," *Anal. Chem.* **1983**, *55*, 2210.

3. When $t_{\text{calculated}}$ from Equation 4-8 is less than t_{table}, we conclude that two means are not statistically significantly different at a chosen confidence level. This test does not provide the same confidence that two means *are equal*. The two one-sided *t* test (TOST) provides a way to show that two means are equivalent. See S. E. Lewis and J. E. Lewis, "The Same or Not the Same: Equivalence as an Issue in Educational Research," *J. Chem. Ed.* **2005**, *82*, 1408, and G. B. Limentani, M. C. Ringo, F. Ye, M. L. Bergquist, and E. O. McSorley, "Beyond the *t*-Test: Statistical Equivalence Testing," *Anal. Chem.* **2005**, *77*, 221A.

4. *NIST/SEMATECH e-Handbook of Statistical Methods*, http://www.itl.nist.gov/div898/handbook/prc/section3/prc31.htm. Equation 4-9a is called the Welch-Satterthwaite approximation.

5. S. A. Lee, R. K. Ross, and M. C. Pike, "An Overview of Menopausal Oestrogen–Progestin Hormone Therapy and Breast Cancer Risk," *Br. J. Cancer* **2005**, *92*, 2049.

6. For a comprehensive approach to least-squares fitting of nonlinear curves, including analysis of uncertainty, see J. Tellinghuisen, "Understanding Least Squares through Monte Carlo Calculations," *J. Chem. Ed.* **2005**, *82*, 157; P. Ogren, B. Davis, and N. Guy, "Curve Fitting, Confidence Intervals and Correlations, and Monte Carlo Visualizations for Multilinear Problems in Chemistry: A General Spreadsheet Approach," *J. Chem. Ed.* **2001**, *78*, 827; see also D. C. Harris, "Nonlinear Least-Squares Curve Fitting with Microsoft Excel Solver," *J. Chem. Ed.* **1998**, *75*, 119; C. Salter and R. de Levie, "Nonlinear Fits of Standard Curves: A Simple Route to Uncertainties in Unknowns," *J. Chem. Ed.* **2002**, *79*, 268; R. de Levie, "Estimating Parameter Precision in Nonlinear Least Squares with Excel's Solver," *J. Chem. Ed.* **1999**, *76*, 1594; S. E. Feller and C. F. Blaich, "Error Estimates for Fitted Parameters," *J. Chem. Ed.* **2001**, *78*, 409; R. de Levie, "When, Why, and How to Use Weighted Least Squares," *J. Chem. Ed.* **1986**, *63*, 10; P. J. Ogren and J. R. Norton, "Applying a Simple Linear Least-Squares Algorithm to Data with Uncertainties in Both Variables," *J. Chem. Ed.* **1992**, *69*, A130.

7. In this book, we plot analytical response on the *y*-axis versus concentration on the *x*-axis. The inverse calibration (y = concentration, x = response) provides a more precise estimate of concentration from a measured response. The advantage of the inverse calibration increases as the noise in the response increases. There are cases, such as some spectrophotometric measurements, in which uncertainty in response (absorbance) is smaller than uncertainty in concentration. In such case, you should plot response on the *x*-axis and concentration on the *y*-axis. See J. Tellinghuisen, "Inverse vs Classical Calibration for Small Data Sets," *Fresenius J. Anal. Chem.* **2000**, *368*, 585; V. Centner, D. L. Massart, and S. de Jong, "Inverse Calibration Predicts Better Than Classical Calibration," *Fresenius J. Anal. Chem.* **1998**, *361*, 2; D. Grientschnig, "Relation Between Prediction Errors of Inverse and Classical Calibration," *Fresenius J. Anal. Chem.* **2000**, *367*, 497.

8. K. Danzer and L. A. Currie, "Guidelines for Calibration in Analytical Chemistry," *Pure Appl. Chem.* **1998**, *70*, 993.

9. C. Salter, "Error Analysis Using the Variance-Covariance Matrix," *J. Chem. Ed.* **2000**, *77*, 1239. Salter's Equation 8 is equivalent to Equation 4-27, although this equivalence is not obvious.

10. N. J. Lawryk and C. P. Weisel, "Concentration of Volatile Organic Compounds in the Passenger Compartments of Automobiles," *Environ. Sci. Tech.* **1996**, *30*, 810.

Chapter 5

1. C. Hogue, "Ferreting Out Erroneous Data," *Chem. Eng. News*, 1 April 2002, p. 49.

2. D. B. Hibbert, *Quality Assurance for the Analytical Chemistry Laboratory* (Oxford: Oxford University Press, 2007); W. Funk, V. Dammann, and G. Donnevert, *Quality Assurance in Analytical Chemistry* (Hoboken, NJ: Wiley, 2006); B. W. Wenclawiak, M. Koch, and E. Hadjiscostas, eds., *Quality Assurance in Analytical Chemistry* (Heidelberg: Springer-Verlag, 2004); E. Mullins, *Statistics for the Quality Control Chemistry Laboratory* (Cambridge: Royal Society of Chemistry, 2003); P. Quevauviller, *Quality Assurance for Water Analysis* (Chichester: Wiley, 2002); M. Valcárcel, *Principles of Analytical Chemistry* (Berlin: Springer-Verlag, 2000).

3. K. M. Phillips, K. Y. Patterson, A. S. Rasor, J. Exler, D. B. Haytowitz, J. M. Holden, and P. R. Pehrsson, "Quality-Control Material in the USDA National Food and Nutrient Analysis Program," *Anal. Bioanal. Chem.* **2006**, *384*, 1341.

4. C. C. Chan, H. Lam, Y. C. Lee, X.-M. Zhang, eds., *Analytical Method Validation and Instrument Performance Verification* (New York: Wiley, 2004); J. M. Green, "A Practical Guide to Analytical Method Validation," *Anal. Chem.* **1996**, *68*, 305A; M. Swartz and I. S. Krull, "Validation of Bioanalytical Methods—Highlights of FDA's Guidance," *LCGC* **2003**, *21*, 136; J. D. Orr, I. S. Krull, and M. E. Swartz, "Validation of Impurity Methods," *LCGC* **2003**, *21*, 626 and 1146.

5. R. de Levie, "Two Linear Correlation Coefficients," *J. Chem. Ed.* **2003**, *80*, 1030.

6. W. Horwitz, L. R. Kamps, and K. W. Boyer, *J. Assoc. Off. Anal. Chem.* **1980**, *63*, 1344; W. Horwitz, "Evaluation of Analytical Methods Used for Regulation of Foods and Drugs," *Anal. Chem.* **1982**, *54*, 67A; P. Hall and B. Selinger, "A Statistical Justification Relating Interlaboratory Coefficients of

Variation with Concentration Levels," *Anal. Chem.* **1989**, *61*, 1465; R. Albert and W. Horwitz, "A Heuristic Derivation of the Horwitz Curve," *Anal. Chem.* **1997**, *69*, 789.

7. J. Vial and A. Jardy, "Experimental Comparison of the Different Approaches to Estimate LOD and LOQ of an HPLC Method," *Anal. Chem.* **1999**, *71*, 2672; G. L. Long and J. D. Winefordner, "Limit of Detection," *Anal. Chem.* **1983**, *55*, 713A; W. R. Porter, "Proper Statistical Evaluation of Calibration Data," *Anal. Chem.* **1983**, *55*, 1290A; S. Geiß and J. W. Einmax, "Comparison of Detection Limits in Environmental Analysis," *Fresenius J. Anal. Chem.* **2001**, *370*, 673; M. E. Zorn, R. D. Gibbons, and W. C. Sonzogni, "Evaluation of Approximate Methods for Calculating the Limit of Detection and Limit of Quantitation," *Environ. Sci. Technol.* **1999**, *33*, 2291; J. D. Burdge, D. L. MacTaggart, and S. O. Farwell, "Realistic Detection Limits from Confidence Bands," *J. Chem. Ed.* **1999**, *76*, 434.

8. The procedure in the text leading to Equation 5-5 is the one most recommended for determining the detection limit. If you do not have replicate determinations of the blank and a low-concentration sample, but you do have a linear calibration curve such as the one in Figure 4-13, you can use the least-squares parameters to estimate a limit of detection for analyte at a desired confidence level. The formula below comes from ISO 11843-2:2000 (International Organization for Standardization, Geneva, www.iso.org). Suppose you measure I calibration standards (including the blank), with J replicates at each concentration. Then you make K replicate measurements of unknown. The detection limit is

$$\text{Detection limit} = \frac{2ts_y}{m}\sqrt{\frac{1}{K} + \frac{1}{I \times J} + \frac{\bar{x}^2}{J\sum(x_i - \bar{x})^2}} \quad \text{(A)}$$

where s_y is the standard deviation of y (Equation 4-20), m is the slope (Equation 4-16), and $\bar{x}$ is the mean value of x for the standards (including the blank). Student's t is selected from Table 4-2 for $(I \times J) - 2$ degrees of freedom. Column headings in Table 4-2 are for a two-tailed distribution. The required value of t in Equation A is for a one-tailed distribution. The equation gives the concentration of analyte that will lead with a probability $(1 - \beta)$ to the conclusion that the concentration of analyte in the unknown is larger than that of a blank. For 95% confidence, $\beta = 0.05$. In this case select t from the column labeled 90% confidence. For 99% confidence, $\beta = 0.01$, and you select t from the column labeled 98% confidence.

Example: Consider the calibration data in Problem 4-33, for which $m = 869.1$ mV/vol%, $s_y = 18.05$ mV, $\bar{x} = 0.544$ vol%, and $\sum(x_i - \bar{x})^2 = 2.878$ vol%². There are seven calibration points, including the blank, so $I = 7$ and degrees of freedom = $7 - 2 = 5$. There is one measurement at each calibration concentration, so $J = 1$. There are four replicate measurements of unknown, so $K = 4$. Suppose we want a 99% confidence level detection limit. Therefore, select $t = 3.365$ from Table 4-2 for 98% confidence and 5 degrees of freedom.

$$\text{Detection limit} = \frac{2(3.365)(18.05\ \text{mV})}{(869.1\ \text{mV/vol%})}\sqrt{\frac{1}{4} + \frac{1}{7 \times 1} + \frac{(0.544\ \text{vol%})^2}{(1)(2.878\ \text{vol%})^2}}$$

$$= (0.140)\sqrt{0.250 + 0.143 + 0.0357} = 0.092\ \text{vol%}$$

If more replicate measurements of unknown are made, the first term in the square root becomes smaller and the detection limit would decrease.

9. M. Bader, "A Systematic Approach to Standard Addition Methods in Instrumental Analysis," *J. Chem. Ed.* **1980**, *57*, 703.

10. W. R. Kelly, B. S. MacDonald, and W. F. Guthrie, "Gravimetric Approach to the Standard Addition Method in Instrumental Analysis," *Anal. Chem.* **2008**, *80*, 6154.

11. G. R. Bruce and P. S. Gill, "Estimates of Precision in a Standard Additions Analysis," *J. Chem. Ed.* **1999**, *76*, 805.

12. R. G. Brereton, *Applied Chemometrics for Scientists* (Chichester: Wiley, 2007); M. Otto, *Chemometrics* (Wenheim: Wiley-VCH, 2007); D. Montgomery, *Design and Analysis of Experiments*, 5th ed., (New York: Wiley, 2001); C. F. Wu and M. Hamada, *Experiments: Planning, Analysis, and Parameter Design Optimization* (New York: Wiley, 2000); M. Anderson and P. Whitcomb, *DoE Simplified: Practical Tools for Effective Experimentation* (Portland, OR: Productivity, Inc., 2000); G. E. P. Box, W. G. Hunter, and J. S. Hunter, *Statistics for Experimenters: An Introduction to Design Data Analysis and Model Building* (New York: Wiley, 1978); R. S. Strange, "Introduction to Experimental Design for Chemists," *J. Chem. Ed.* **1990**, *67*, 113;

J. M. Gozálvez and J. C. García-Díaz, "Mixture Design Experiments Applied to the Formulation of Colorant Solutions," *J. Chem. Ed.* **2006**, *83*, 647.

13. S. N. Deming and S. L. Morgan, "Simplex Optimization of Variables in Analytical Chemistry," *Anal. Chem.* **1973**, *45*, 278A; D. J. Leggett, "Instrumental Simplex Optimization," *J. Chem. Ed.* **1983**, *60*, 707; S. Srijaranai, R. Burakham, T. Khammeng, and R. L. Deming, "Use of the Simplex Method to Optimize the Mobile Phase for the Micellar Chromatographic Separation of Inorganic Anions," *Anal. Bioanal. Chem.* **2002**, *374*, 145; D. Betteridge, A. P. Wade, and A. G. Howard, "Reflections on the Modified Simplex," *Talanta* **1985**, *32*, 709, 723.

14. P. de B. Harrington, E. Kolbrich, and J. Cline, "Experimental Design and Multiplexed Modeling Using Titrimetry and Spreadsheets," *J. Chem. Ed.* **2002**, *79*, 863.

15. J. A. Day, M. Montes-Bayón, A. P. Vonderheide, and J. A. Caruso, "A Study of Method Robustness for Arsenic Speciation in Drinking Water Samples by Anion Exchange HPLC-ICP-MS," *Anal. Bioanal. Chem.* **2002**, *373*, 664.

16. X. Zhao and C. D. Metcalf, "Characterizing and Compensating for Matrix Effects Using Atmospheric Pressure Chemical Ionization Liquid Chromatography–Tandem Mass Spectrometry: Analysis of Neutral Pharmaceuticals in Municipal Wastewater," *Anal. Chem.* **2008**, *80*, 2010.

Chapter 6

1. D. P. Sheer and D. C. Harris, "Acidity Control in the North Branch Potomac," *J. Water Pollution Control Federation* **1982**, *54*, 1441.

2. R. E. Weston, Jr., "Climate Change and Its Effect on Coral Reefs," *J. Chem. Ed.* **2000**, *77*, 1574.

3. P. D. Thacker, "Global Warming's Other Effects on the Oceans," *Environ. Sci. Technol.* **2005**, *39*, 10A.

4. J. K. Baird, "A Generalized Statement of the Law of Mass Action," *J. Chem. Ed.* **1999**, *76*, 1146; R. de Levie, "What's in a Name?" *J. Chem. Ed.* **2000**, *77*, 610.

5. For thermodynamic data, see N. Jacobson, "Use of Tabulated Thermochemical Data for Pure Compounds," *J. Chem. Ed.* **2001**, *78*, 814; http://webbook.nist.gov/chemistry/ and http://www.crct.polymtl.ca/fact/websites.htm; M. W. Chase, Jr., *NIST-JANAF Thermochemical Tables*, 4th ed; *J. Phys. Chem. Ref. Data*: Monograph 9 (New York: American Chemical Society and American Physical Society, 1998).

6. The solubility of most ionic compounds increases with temperature, despite the fact that the standard heat of solution ($\Delta H°$) is negative for about half of them. Discussions of this seeming contradiction can be found in G. M. Bodner, "On the Misuse of Le Châtelier's Principle for the Prediction of the Temperature Dependence of the Solubility of Salts," *J. Chem. Ed.* **1980**, *57*, 117, and R. S. Treptow, "Le Châtelier's Principle Applied to the Temperature Dependence of Solubility," *J. Chem. Ed.* **1984**, *61*, 499.

7. A. K. Sawyer, "Solubility and K_{sp} of Calcium Sulfate: A General Chemistry Laboratory Experiment," *J. Chem. Ed.* **1983**, *60*, 416.

8. A really good book to read about solubility and all types of equilibrium calculations is W. B. Guenther, *Unified Equilibrium Calculations* (New York: Wiley, 1991).

9. E. Koubek, "Demonstration of the Common Ion Effect," *J. Chem. Ed.* **1993**, *70*, 155.

10. For many *great* chemical demonstrations, see B. Z. Shakhashiri, *Chemical Demonstrations: A Handbook for Teachers of Chemistry* (Madison, WI: University of Wisconsin Press, 1983–1992), 4 volumes. See also L. E. Summerlin and J. L. Ealy, Jr., *Chemical Demonstrations: A Sourcebook for Teachers*, 2nd ed. (Washington, DC: American Chemical Society, 1988).

11. A demonstration of selective precipitation by addition of Pb^{2+} to a solution containing CO_3^{2-} and I^- is described by T. P. Chirpich, "A Simple, Vivid Demonstration of Selective Precipitation," *J. Chem. Ed.* **1988**, *65*, 359.

12. Classroom demonstration of complex equilibria: A. R. Johnson, T. M. McQueen, and K. T. Rodolfa, "Species Distribution Diagrams in the Copper-Ammonia System," *J. Chem. Ed.* **2005**, *82*, 408.

13. A computer database of critically selected equilibrium constants is found in R. M. Smith, A. E. Martell, and R. J. Motekaitis, *NIST Critical Stability Constants of Metal Complexes Database 46* (Gaithersburg, MD: National Institute of Standards and Technology, 2001). Measurement of equilibrium constants is described in A. Martell and R. Motekaitis, *Determination and Use*

of *Stability Constants* (New York: VCH Publishers, 1992); K. A. Conners, *Binding Constants: The Measurement of Molecular Complex Stability* (New York: Wiley, 1987); and D. J. Leggett, ed., *Computational Methods for the Determination of Formation Constants* (New York: Plenum Press, 1985).

14. P. A. Giguère, "The Great Fallacy of the H^+ Ion," *J. Chem. Ed.* **1979**, *56*, 571; P. A. Giguère and S. Turrell, "The Nature of Hydrofluoric Acid: A Spectroscopic Study of the Proton-Transfer Complex, $H_3O^+ \cdot F^-$," *J. Am. Chem. Soc.* **1980**, *102*, 5473.

15. Z. Xie, R. Bau, and C. A. Reed, "A Crystalline $[H_9O_4]^+$ Hydronium Salt with a Weakly Coordinating Anion," *Inorg. Chem.* **1995**, *34*, 5403.

16. F. A. Cotton, C. K. Fair, G. E. Lewis, G. N. Mott, K. K. Ross, A. J. Schultz, and J. M. Williams, "X-Ray and Neutron Diffraction Studies of $[V(H_2O)_6][H_5O_2][CF_3SO_3]_4$," *J. Am. Chem. Soc.* **1984**, *106*, 5319.

17. J. M. Headrick, E. G. Diken, R. W. Walters, N. I. Hammer, R. A. Christie, J. Cui, E. M. Myshakin, M. A. Duncan, M. A. Johnson, and K. D. Jordan, "Spectral Signatures of Hydrated Proton Vibrations in Water Clusters," *Science* **2005**, *308*, 1765.

18. S. Wei, Z. Shi, and A. W. Castleman, Jr., "Mixed Cluster Ions as a Structure Probe: Experimental Evidence for Clathrate Structure of $(H_2O)_{20}H^+$ and $(H_2O)_{21}H^+$," *J. Chem. Phys.* **1991**, *94*, 3268.

19. K. Abu-Dari, K. N. Raymond, and D. P. Freyberg, "The Bihydroxide $(H_3O_2^-)$ Anion," *J. Am. Chem. Soc.* **1979**, *101*, 3688.

20. W. B. Jensen, "The Symbol for pH," *J. Chem. Ed.* **2004**, *81*, 21.

21. V. Buch, A. Milet, R. Vácha, P. Jungwirth, and J. P. Devlin, "Water Surface Is Acidic," *Proc. Natl. Acad. Sci. USA* **2007**, *104*, 7342.

22. D. K. Nordstrom, C. N. Alpers, C. J. Ptacek, and D. W. Blowes, "Negative pH and Extremely Acidic Mine Waters from Iron Mountain, California," *Environ. Sci. Technol.* **2000**, *34*, 254.

23. For a CO_2 fountain, see S.-J. Kang and E.-H. Ryu, "Carbon Dioxide Fountain," *J. Chem. Ed.* **2007**, *84*, 1671. For NH_3 fountains, see N. C. Thomas, S. Faulk, and R. Sullivan, "A Hand-Held Ammonia Fountain," *J. Chem. Ed.* **2008**, *85*, 1063; M. D. Alexander, "The Ammonia Smoke Fountain," *J. Chem. Ed.* **1999**, *76*, 210; N. C. Thomas, "A Chemiluminescent Ammonia Fountain," *J. Chem. Ed.* **1990**, *67*, 339; and N. Steadman, "Ammonia Fountain Improvements," *J. Chem. Ed.* **1992**, *66*, 764.

24. L. M. Schwartz, "Ion-pair Complexation in Moderately Strong Aqueous Acids," *J. Chem. Ed.* **1995**, *72*, 823. Even though it is not "free," H_3O^+ in ion pairs with certain anions such as $CF_3CO_2^-$ and $CCl_3CO_2^-$ appears to participate in ionic conductance (R. I. Gelb and J. S. Alper, "Anomalous Conductance in Electrolyte Solutions," *Anal. Chem.* **2000**, *72*, 1322).

25. Z. Tian, B. Chan, M. B. Sullivan, L. Radom, and S. R. Kass, "Lithium Monoxide Anion: A Ground-State Triplet with the Strongest Base to Date," *Proc. Natl. Acad. Sci. USA* **2008**, *105*, 7647.

26. S. J. Hawkes, "All Positive Ions Give Acid Solutions in Water," *J. Chem. Ed.* **1996**, *73*, 516.

27. M. Kern, "The Hydration of Carbon Dioxide," *J. Chem. Ed.* **1960**, *37*, 14. Great demonstrations with CO_2, including one with carbonic anhydrase, are described by J. A. Bell, "Every Year Begins a Millennium," *J. Chem. Ed.* **2000**, *77*, 1098.

28. J. A. Tossell, "H_2CO_3 and Its Oligomers: Structures, Stabilities, Vibrational and NMR Spectra, and Acidities," *Inorg. Chem.* **2006**, *45*, 5961.

Chapter 7

1. H. Ohtaki and T. Radnal, "Structure and Dynamics of Hydrated Ions," *Chem. Rev.* **1993**, *93*, 1157.

2. A. G. Sharpe, "The Solvation of Halide Ions and Its Chemical Significance," *J. Chem. Ed.* **1990**, *67*, 309.

3. E. R. Nightingale, Jr., "Phenomenological Theory of Ion Solvation. Effective Radii of Hydrated Ions," *J. Phys. Chem.* **1959**, *63*, 1381.

4. K. H. Stern and E. S. Amis, "Ionic Size," *Chem. Rev.* **1959**, *59*, 1.

5. D. R. Driscol, "Invitation to Enquiry: The Fe^{3+}/CNS^- Equilibrium," *J. Chem. Ed.* **1979**, *56*, 603.

6. S. J. Hawkes, "Salts are Mostly NOT Ionized," *J. Chem. Ed.* **1996**, *73*, 421; S. O. Russo and G. I. H. Hanania, "Ion Association, Solubilities, and Reduction Potentials in Aqueous Solution," *J. Chem. Ed.* **1989**, *66*, 148.

7. K. S. Pitzer, *Activity Coefficients in Electrolyte Solutions*, 2nd ed. (Boca Raton, FL: CRC Press, 1991); B. S. Krumgalz, R. Pogorelskii, A. Sokolov, and K. S. Pitzer, "Volumetric Ion Interaction Parameters for Single-Solute Aqueous Electrolyte Solutions at Various Temperatures," *J. Phys. Chem. Ref. Data* **2000**, *29*, 1123.

8. J. Kielland, "Individual Activity Coefficients of Ions in Aqueous Solutions," *J. Am. Chem. Soc.* **1937**, *59*, 1675.

9. R. E. Weston, Jr., "Climate Change and Its Effect on Coral Reefs," *J. Chem. Ed.* **2000**, *77*, 1574.

10. R. A. Feely, C. L. Sabine, K. Lee, W. Berelson, J. Kleypas, V. J. Fabry, and F. J. Millero, "Impact of Anthropogenic CO_2 on the $CaCO_3$ System in the Oceans," *Science* **2004**, *305*, 362.

11. For more about equilibrium calculations, see W. B. Guenther, *Unified Equilibrium Calculations* (New York: Wiley, 1991); J. N. Butler, *Ionic Equilibrium: Solubility and pH Calculations* (New York: Wiley, 1998); and M. Meloun, *Computation of Solution Equilibria* (New York: Wiley, 1988). For equilibrium calculation software, see http://www.micromath.com/ and http://www.acadsoft.co.uk/

12. E. Koort, P. Gans, K. Herodes, V. Pihl, and I. Leito, "Acidity Constants in Different Media (I = 0 and I = 0.1 M KCl) from the Uncertainty Perspective," *Anal. Bioanal. Chem.* **2006**, *385*, 1124.

Chapter 8

1. R. Schmid and A. M. Miah, "The Strength of the Hydrohalic Acids," *J. Chem. Ed.* **2001**, *78*, 116.

2. T. F. Young, L. F. Maranville, and H. M. Smith, "Raman Spectral Investigations of Ionic Equilibria in Solutions of Strong Electrolytes" in W. J. Hamer, ed., *The Structure of Electrolytic Solutions* (New York: Wiley, 1959).

3. E. S. Shamay, V. Buch, M. Parrinello, and G. L. Richmond, "At the Water's Edge: Nitric Acid as a Weak Acid," *J. Am. Chem. Soc.* **2007**, *129*, 12910.

4. Acid dissociation constants do not tell us which protons dissociate in each step. Assignments for pyridoxal phosphate come from nuclear magnetic resonance spectroscopy (B. Szpoganicz and A. E. Martell, "Thermodynamic and Microscopic Equilibrium Constants of Pyridoxal 5′-Phosphate," *J. Am. Chem. Soc.* **1984**, *106*, 5513).

5. For an alternative treatment approach, see H. L. Pardue, I. N. Odeh, and T. M. Tesfai, "Unified Approximations: A New Approach for Monoprotic Weak Acid-Base Equilibria," *J. Chem. Ed.* **2004**, *81*, 1367.

6. R. T. da Rocha, I. G. R. Gutz, and C. L. do Lago, "A Low-Cost and High-Performance Conductivity Meter," *J. Chem. Ed.* **1997**, *74*, 572; G. Berenato and D. F. Maynard, "A Simple Audio Conductivity Device," *J. Chem. Ed.* **1997**, *74*, 415; S. K. S. Zawacky, "A Cheap, Semiquantitative Hand-Held Conductivity Tester," *J. Chem. Ed.* **1995**, *72*, 728; T. R. Rettich, "An Inexpensive and Easily Constructed Device for Quantitative Conductivity Experiments," *J. Chem. Ed.* **1989**, *66*, 168; and D. A. Katz and C. Willis, "Two Safe Student Conductivity Apparatus," *J. Chem. Ed.* **1994**, *71*, 330.

7. L. R. Kuck, R. D. Godec, P. P. Kosenka, and J. W. Birks, "High-Precision Conductometric Detector for the Measurement of Atmospheric CO_2," *Anal. Chem.* **1998**, *70*, 4678.

8. M. C. Bonneau, "The Chemistry of Fabric Reactive Dyes," *J. Chem. Ed.* **1995**, *72*, 724.

9. H. N. Po and N. M. Senozan, "The Henderson-Hasselbalch Equation: Its History and Limitations," *J. Chem. Ed.* **2001**, *78*, 1499; R. de Levie, "The Henderson-Hasselbalch Equation: Its History and Limitations," *J. Chem. Ed.* **2003**, *80*, 146.

10. F. B. Dutton and G. Gordon in H. N. Alyea and F. B. Dutton, eds., *Tested Demonstrations in Chemistry*, 6th ed. (Easton, PA: Journal of Chemical Education, 1965), p. 147; R. L. Barrett, "The Formaldehyde Clock Reaction," *J. Chem. Ed.* **1955**, *32*, 78. See also J. J. Fortman and J. A. Schreier, "Some Modified Two-Color Formaldehyde Clock Salutes for Schools with Colors of Gold and Green or Gold and Red," *J. Chem. Ed.* **1991**, *68*, 324; M. G. Burnett, "The Mechanism of the Formaldehyde Clock Reaction," *J. Chem. Ed.* **1982**, *59*, 160; and P. Warneck, "The Formaldehyde-Sulfite Clock Reaction Revisited," *J. Chem. Ed.* **1989**, *66*, 334.

11. Many other clock reactions are described in the literature. For a summary, see A. P. Oliveira and R. B. Faria, "The Chlorate-Iodine Clock Reaction," *J. Am. Chem. Soc.* **2005**, *127*, 18022.

12. The chemical we call sodium bisulfite ($NaHSO_3$) is apparently not the solid in the reagent bottle, which is reported to be sodium metabisulfite ($Na_2S_2O_5$) (D. Tudela, "Solid $NaHSO_3$ Does Not Exist," *J. Chem. Ed.* **2000**, *77*, 830; see also H. D. B. Jenkins and D. Tudela, "New Methods to Estimate Lattice Energies: Application to Bisulfite and Metabisulfite," *J. Chem. Ed.* **2003**, *80*, 1482). $NaHSO_3$ is produced when $Na_2S_2O_5$ reacts with H_2O. A reagent bottle I use for the formaldehyde clock reaction is labeled "sodium bisulfite," but no formula is given. The label gives the reagent assay "as SO_2: minimum 58.5%." Pure $NaHSO_3$ is equivalent to 61.56 wt% SO_2 and pure $Na_2S_2O_5$ is equivalent to 67.40 wt% SO_2.

13. J. B. Early, A. R. Negron, J. Stephens, R. Stauffer, and S. D. Furrow, "The Glyoxal Clock Reaction," *J. Chem. Ed.* **2007**, *84*, 1965.

14. E. T. Urbansky and M. R. Schock, "Understanding, Deriving, and Computing Buffer Capacity," *J. Chem. Ed.* **2000**, *77*, 1640.

Chapter 9

1. B. J. Bozlee, M. Janebo, and G. Jahn, "A Simplified Model to Predict the Effect of Increasing Atmospheric CO_2 on Carbonate Chemistry in the Ocean," *J. Chem. Ed.* **2008**, *85*, 213.

2. P. D. Thacker, "Global Warming's Other Effects on the Oceans," *Environ. Sci. Technol.* **2005**, *39*, 10A.

3. R. E. Weston, Jr., "Climate Change and Its Effect on Coral Reefs," *J. Chem. Ed.* **2000**, *77*, 1574.

4. J. C. Orr, V. J. Fabry, O. Aumont, L. Bopp, S. C. Doney, R. A. Feely, A. Gnanadesikan, N. Gruber, A. Ishida, F. Joos, R. M. Key, K. Lindsay, E. Maier-Reimer, R. Matear, P. Monfray, A. Mouchet, R. G. Najjar, G.-K. Plattner, K. B. Rodgers, C. L. Sabine, J. L. Sarmiento, R. Schlitzer, R. D. Slater, I. J. Totterdell, M.-F. Weirig, Y. Yamanaka, and A. Yool, "Anthropogenic Ocean Acidification over the Twenty-first Century and Its Impact on Calcifying Organisms," *Nature* **2005**, *437*, 681.

5. M. D. Iglesias-Rodriguez, P. R. Halloran, R. E. M. Rickaby, I. R. Hall, E. Colmenero-Hidalgo, J. R. Gittins, D. R. H. Green, T. Tyrrell, S. J. Gibbs, P. von Dassow, E. Rehm, E. V. Armbrust, and K. P. Boessenkool, "Phytoplankton Calcification in a High-CO_2 World," *Science* **2008**, *320*, 336.

6. P. G. Daniele, A. De Robertis, C. De Stefano, S. Sammartano, and C. Rigano, "Na^+, K^+, and Ca^{2+} Complexes of Low Molecular Weight Ligands in Aqueous Solution," *J. Chem. Soc. Dalton Trans.* **1985**, 2353.

7. Experiment on surface acidity of a solid: L. Tribe and B. C. Barja, "Adsorption of Phosphate on Goethite," *J. Chem. Ed.* **2004**, *81*, 1624.

8. Experiment on pH of zero charge: M Davranche, S. Lacour, F. Bordas, and J.-C. Bollinger, "Determination of the Surface Chemical Properties of Natural Solids," *J. Chem. Ed.* **2003**, *80*, 76.

9. W. Stumm and J. J. Morgan, *Aquatic Chemistry*, 3rd ed. (New York: Wiley, 1996), pp. 343–348; F. J. Millero, "Thermodynamics of the Carbon Dioxide System in the Oceans," *Geochim. Cosmochim. Acta* **1995**, *59*, 661; Ocean carbon thermodynamics: http://cdiac.esd.ornl.gov/oceans/glodap/cther.htm.

Chapter 10

1. B. Mörnstam, K.-G. Wahlund, and B. Jönsson, "Potentiometric Acid-Base Titration of a Colloidal Solution," *Anal. Chem.* **1997**, *69*, 5037. For titration of entire cell surfaces, see I. Sokolov, D. S. Smith, G. S. Henderson, Y. A. Gorby, and F. G. Ferris, "Cell Surface Electrochemical Heterogeneity of the Fe(III)-Reducing Bacteria *Shewanella putrefaciens*," *Environ. Sci. Technol.* **2001**, *35*, 341.

2. A method to measure total charge on a protein as it binds to assorted ions is described by M. K. Menon and A. L. Zydney, "Measurement of Protein Charge and Ion Binding Using Capillary Electrophoresis," *Anal. Chem.* **1998**, *70*, 1581.

3. M. J. Ondrechen, J. G. Clifton, and D. Ringe, "THEMATICS: A Simple Computational Predictor of Enzyme Function from Structure," *Proc. Natl. Acad. Sci. USA* **2001**, *98*, 12473.

4. A. G. Dickson, http://andrew.ucsd.edu/co2qc/handbook/sop03.pdf.

5. T. R. Martz, A. G. Dickson, and M. D. DeGrandpre, "Tracer Monitored Titrations: Measurement of Total Alkalinity," *Anal. Chem.* **2006**, *78*, 1817.

6. K. R. Williams, "Automatic Titrators in the Analytical and Physical Chemistry Laboratories," *J. Chem. Ed.* **1998**, *75*, 1133; K. L. Headrick,

T. K. Davies, and A. N. Haegele, "A Simple Laboratory-Constructed Automatic Titrator," *J. Chem. Ed.* **2000**, *77*, 389.

7. M. Inoue and Q. Fernando, "Effect of Dissolved CO_2 on Gran Plots," *J. Chem. Ed.* **2001**, *78*, 1132; G. Gran, "Equivalence Volumes in Potentiometric Titrations," *Anal. Chim. Acta* **1988**, *206*, 111; F. J. C. Rossotti and H. Rossotti, "Potentiometric Titrations Using Gran Plots," *J. Chem. Ed.* **1965**, *42*, 375; L. M. Schwartz, "Uncertainty of a Titration Equivalence Point," *J. Chem. Ed.* **1992**, *69*, 879; L. M. Schwartz, "Advances in Acid-Base Gran Plot Methodology," *J. Chem. Ed.* **1987**, *64*, 947.

8. M. Rigobello-Masini and J. C. Masini, "Application of Modified Gran Functions and Derivative Methods to Potentiometric Acid Titration Studies of the Distribution of Inorganic Carbon Species in Cultivation Medium of Marine Microalgae," *Anal. Chim. Acta* **2001**, *448*, 239.

9. G. Papanastasiou and I. Ziogas, "Simultaneous Determination of Equivalence Volumes and Acid Dissociation Constants from Potentiometric Titration Data," *Talanta* **1995**, *42*, 827.

10. G. Wittke, "Reactions of Phenolphthalein at Various pH Values," *J. Chem. Ed.* **1983**, *60*, 239.

11. Demonstrations with universal indicator (a mixed indicator with many color changes) are described in J. T. Riley, "Flashy Solutions," *J. Chem. Ed.* **1977**, *54*, 29.

12. T. A. Canada, L. R. Allain, D. B. Beach, and Z. Xue, "High-Acidity Determination in Salt-Containing Acids by Optical Sensors," *Anal. Chem.* **2002**, *74*, 2535.

13. D. Fárcasiu and A. Ghenciu, "Acidity Functions from ^{13}C-NMR," *J. Am. Chem. Soc.* **1993**, *115*, 10901.

14. B. Hammouti, H. Oudda, A. El Maslout, and A. Benayada, "A Sensor for the In Situ Determination of Acidity Levels in Concentrated Sulfuric Acid," *Fresenius J. Anal. Chem.* **1999**, *365*, 310. For use of glass electrodes to measure pH as low as -4, see D. K. Nordstrom, C. N. Alpers, C. J. Ptacek, and D. W. Blowes, "Negative pH and Extremely Acidic Mine Waters from Iron Mountain, California," *Environ. Sci. Technol.* **2000**, *34*, 254.

15. M. Juhasz, S. Hoffmann, E. Stoyanov, K.-C. Kim, and C. A. Reed, "The Strongest Isolable Acid," *Angew. Chem. Int. Ed.* **2004**, *43*, 5352; A. Avelar, F. S. Tham, and C. A. Reed, "Superacidity of Boron Acids $H_2(B_{12}X_{12})$ (X = Cl, Br)," *Angew. Chem. Int. Ed.* **2009**, *48*, 3491.

16. R. A. Butler and R. G. Bates, "Double Potassium Salt of Sulfosalicylic Acid in Acidimetry and pH Control," *Anal. Chem.* **1976**, *48*, 1669.

17. Borax goes down to the pentahydrate upon standing: R. Naumann, C. Alexander-Weber, and F. G. K. Baucke, "Limited Stability of the pH Reference Material Sodium Tetraborate Decahydrate ("Borax")," *Fresenius J. Anal. Chem.* **1994**, *350*, 119.

18. Instructions for purifying and using primary standards can be found in the following books: J. A. Dean, *Analytical Chemistry Handbook* (New York: McGraw-Hill, 1995), pp. 3-28 to 3-30; J. Bassett, R. C. Denney, G. H. Jeffery, and J. Mendham, *Vogel's Textbook of Quantitative Inorganic Analysis*, 4th ed. (Essex: Longman, 1978), pp. 296–306; I. M. Kolthoff and V. A. Stenger, *Volumetric Analysis*, Vol. 2 (New York: Wiley-Interscience, 1947).

19. A. A. Smith, "Consumption of Base by Glassware," *J. Chem. Ed.* **1986**, *63*, 85; G. Perera and R. H. Doremus, "Dissolution Rates of Commercial Soda-Lime and Pyrex Borosilicate Glasses," *J. Am. Ceramic Soc.* **1991**, *74*, 1554.

20. D. Lee, "Plant Linked to Pet Deaths Had History of Polluting," *Los Angeles Times*, 9 May 2007, p. C1; B. Puschner, R. H. Poppenga, L. J. Lowenstine, M. S. Filigenzi, and P. A. Pesavento, "Assessment of Melamine and Cyanuric Acid Toxicity in Cats," *J. Vet. Diagn. Invest.* **2007**, *19*, 616.

21. D. Lee and A. Goldman, "Anguished Chinese Flood Hospitals," *Los Angeles Times*, 19 September 2008, p. A3; R. M. Baum, *Chem. Eng. News*, 13 October 2008, p. 3. In December 2008, the Chinese government estimated that 294 000 babies had become sick.

22. J. J. Urh, "Protein Testing Enters the 21st Century: Innovative Protein Analyzer Not Affected by Melamine," *Am. Lab.*, October 2008, p. 18.

23. The Kjeldahl digestion captures amine ($—NR_2$) or amide ($—C[=O]NR_2$) nitrogens (where R can be H or an organic group), but not oxidized nitrogen such as nitro ($—NO_2$) or azo ($—N=N—$) groups, which must be reduced first to amines or amides.

24. W. Maher, F. Krikowa, D. Wruck, H. Louie, T. Nguyen, and W. Y. Huang, "Determination of Total Phosphorus and Nitrogen in Turbid Waters by Oxidation with Alkaline Potassium Peroxodisulfate," *Anal. Chim. Acta* **2002**, *463*, 283.

25. An alternative to using standard HCl is to use ~4 wt% aqueous boric acid. This reagent captures ammonia as ammonium borate, which can be titrated with standard acid. (F. M. Scales and A P. Harrison, "Boric Acid Modification of the Kjeldahl Method for Crop and Soil Analysis," *J. Ind. Eng. Chem.* **1920**, *12*, 350.)

26. http://www.epa.gov/grtlakes/lmmb/methods/tknalr2.pdf http://www.flowinjection.com/methods/tkn.aspx

27. J. S. Fritz, *Acid-Base Titrations in Nonaqueous Solvents* (Boston: Allyn and Bacon, 1973); J. Kucharsky and L. Safarik, *Titrations in Non-Aqueous Solvents* (New York: Elsevier, 1963); W. Huber, *Titrations in Nonaqueous Solvents* (New York: Academic Press, 1967); I. Gyenes, *Titration in Non-Aqueous Media* (Princeton, NJ: Van Nostrand, 1967).

28. S. P. Porras, "Capillary Zone Electrophoresis of Some Extremely Weak Bases in Acetonitrile," *Anal. Chem.* **2006**, *78*, 5061.

29. R. de Levie, "A General Simulator for Acid-Base Titrations," *J. Chem. Ed.* **1999**, *76*, 987; R. de Levie, "Explicit Expressions of the General Form of the Titration Curve in Terms of Concentration," *J. Chem. Ed.* **1993**, *70*, 209; R. de Levie, "General Expressions for Acid-Base Titrations of Arbitrary Mixtures," *Anal. Chem.* **1996**, *68*, 585; R. de Levie, *Principles of Quantitative Chemical Analysis* (New York: McGraw-Hill, 1997); J. Burnett and W. A. Burns, "Using a Spreadsheet to Fit Experimental pH Titration Data to a Theoretical Expression: Estimation of Analyte Concentration and K_a," *J. Chem. Ed.* **2006**, *83*, 1190.

30. C. Salter and D. L. Langhus, "The Chemistry of Swimming Pool Maintenance," *J. Chem. Ed.* **2007**, *84*, 1124.

31. P. Ballinger and F. A. Long, "Acid Ionization Constants of Alcohols," *J. Am. Chem. Soc.* **1960**, *82*, 795.

Chapter 11

1. R. MacKinnon, "Potassium Channels and the Atomic Basis of Selective Ion Conduction" (Nobel Lecture), *Angew. Chem. Int. Ed.* **2004**, *43*, 4265.

2. W. D. Bedsworth and D. L. Sedlak, "Sources and Environmental Fate of Strongly Complexed Nickel in Estuarine Waters," *Environ. Sci. Technol.* **1999**, *33*, 926; B. Nowack, "Environmental Chemistry of Aminopolycarboxylate Chelating Agents," *Environ. Sci. Technol.* **2002**, *36*, 4009.

3. D. T. Haworth, "Some Linguistic Details on Chelation," *J. Chem. Ed.* **1998**, *75*, 47.

4. Classroom demonstration: D. C. Bowman, "A Colorful Look at the Chelate Effect," *J. Chem. Ed.* **2006**, *83*, 1158.

5. The chelate effect is often attributed to a favorable entropy change for multidentate binding. Recent work does not support this explanation: V. Vallet, U. Wahlgren, and I. Grenthe, "Chelate Effect and Thermodynamics of Metal Complex Formation in Solution: A Quantum Chemical Study," *J. Am. Chem. Soc.* **2003**, *125*, 14941.

6. Z. Hou, K. N. Raymond, B. O'Sullivan, T. W. Esker, and T. Nishio, "Microbial Macrocyclic Dihydroxamate Chelating Agents," *Inorg. Chem.* **1998**, *37*, 6630. Ferrioxamines found in the ocean at concentrations of 0.1–10 pM are presumably excreted by microorganisms to enable them to accumulate scarce iron from the ocean. (E. Mawji, M. Gledhill, J. A. Milton, G. A. Tarran, S. Ussher, A. Thompson, G. A. Wolff, P. J. Worsfold, and E. P. Achterberg, "Hydroxamate Siderophores: Occurrence and Importance in the Atlantic Ocean," *Environ. Sci. Technol.* **2008**, *42*, 8675.)

7. N. F. Olivieri and G. M. Brittenham, "Iron-Chelating Therapy and the Treatment of Thalassemia," *Blood* **1997**, *89*, 739.

8. E. J. Neufeld, "Oral Chelators Deferasirox and Deferiprone for Transfusional Iron Overload in Thalassemia Major: New Data, New Questions," *Blood* **2006**, *107*, 3436; K. Farmaki, "Reversal of Complications Following Intensive Combined Chelation in Beta Thalassemia Major Patients," Abstract LB4, 49th American Society of Hematology Annual Meeting, Atlanta, GA, December 2007.

9. R. J. Abergel, E. G. Moore, R. K. Strong, and K. N. Raymond, "Microbial Evasion of the Immune System: Structural Modifications of Enterobactin Impair Siderocalin Recognition," *J. Am. Chem. Soc.* **2006**, *128*, 10998.

10. J. Künnemeyer, L. Terborg, S. Nowak, L. Telgmann, F. Tokmak, B. K. Krämer, A. Günsel, G. A. Wiesmüller, J. Waldeck, C. Bremer, and U. Karst, "Analysis of the Contrast Agent Magnevist and Its Transmetalation Products in Blood Plasma by Capillary Electrophoresis/Electrospray Ionization Time-of-Flight Mass Spectrometry," *Anal. Chem.* **2009**, *81*, 3600.

11. W. J. Blaedel and H. T. Knight, "Purification and Properties of Disodium Salt of Ethylenediaminetetraacetic Acid as a Primary Standard," *Anal. Chem.* **1954**, *26*, 741.

12. R. L. Barnett and V. A. Uchtman, "Crystal Structures of Ca(CaEDTA) · 7H$_2$O and NaCaNTA," *Inorg. Chem.* **1979**, *18*, 2674.

13. P. Lindqvist-Reis, C. Apostolidis, J. Rebizant, A. Morgenstern, R. Klenze, O. Walter, T. Fanghänel, and R. G. Haire, "The Structures and Optical Spectra of Hydrated Transplutonium Ions in the Solid State and in Solution," *Angew. Chem. Int. Ed.* **2007**, *46*, 919; S. Skanthakumar, M. R. Antonio, R. E. Wilson, and L. Soderholm, "The Curium Aqua Ion," *Inorg. Chem.* **2007**, *46*, 3485.

14. S. G. John, C. E. Ruggiero, L. E. Hersman, C.-S. Tung, and M. P. Neu, "Siderophore Mediated Plutonium Accumulation by *Microbacterium flavescens*," *Environ. Sci. Technol.* **2001**, *35*, 2942.

15. J. N. Mathur, P. Thakur, C. J. Dodge, A. J. Francis, and G. R. Choppin, "Coordination Modes in the Formation of the Ternary Am(III), Cm(III), and Eu(III) Complexes with EDTA and NTA," *Inorg. Chem.* **2006**, *456*, 8026.

16. A definitive reference for the theory of EDTA titration curves is A. Ringbom, *Complexation in Analytical Chemistry* (New York: Wiley, 1963).

17. For discussion of metal-ligand equilibria with numerous examples, see P. Letkeman, "Computer-Modeling of Metal Speciation in Human Blood Serum," *J. Chem. Ed.* **1996**, *73*, 165; A. Rojas-Hernández, M. T. Ramírez, I. González, and J. G. Ibanez, "Predominance-Zone Diagrams in Solution Chemistry," *J. Chem. Ed.* **1995**, *72*, 1099; and A. Bianchi and E. Garcia-España, "Use of Calculated Species Distribution Diagrams to Analyze Thermodynamic Selectivity," *J. Chem. Ed.* **1999**, *76*, 1727.

18. W. N. Perara and G. Hefter, "Mononuclear Cyano- and Hydroxo-Complexes of Iron(III)," *Inorg. Chem.* **2003**, *42*, 5917.

19. S. Tandy, K. Bossart, R. Mueller, J. Ritschel, L. Hauser, R. Schulin, and B. Nowack, "Extraction of Heavy Metals from Soils Using Biodegradable Chelating Agents," *Environ. Sci. Technol.* **2004**, *38*, 937; B. Kos and D. Leštan, "Induced Phytoextraction/Soil Washing of Lead Using Biodegradable Chelate and Permeable Barriers," *Environ. Sci. Technol.* **2003**, *37*, 624; S. V. Sahi, N. L. Bryant, N. C. Sharma, and S. R. Singh, "Characterization of a Lead Hyperaccumulator Shrub," *Environ. Sci. Technol.* **2002**, *36*, 4676.

20. B. Nowack, R. Schulin, and B. H. Robinson, "Critical Assessment of Chelant-Enhanced Metal Phytoextraction," *Environ. Sci. Technol.* **2006**, *40*, 5225.

21. G. Schwarzenbach and H. Flaschka, *Complexometric Titrations*, H. M. N. H. Irving, trans. (London: Methuen, 1969); H. A. Flaschka, *EDTA Titrations* (New York: Pergamon Press, 1959); J. A. Dean, *Analytical Chemistry Handbook* (New York: McGraw-Hill, 1995); A. E. Martell and R. D. Hancock, *Metal Complexes in Aqueous Solution* (New York: Plenum Press, 1996).

22. Indirect determinations of monovalent *cations* are described by I. M. Yurist, M. M. Talmud, and P. M. Zaitsev, "Complexometric Determination of Monovalent Metals," *J. Anal. Chem. USSR* **1987**, *42*, 911.

23. D. P. S. Rathore, P. K. Bhargava, M. Kumar, and R. K. Talra, "Indicators for the Titrimetric Determination of Ca and Total Ca + Mg with EDTA," *Anal. Chim. Acta* **1993**, *281*, 173.

24. H. Bao, "Purifying Barite for Oxygen Isotope Measurement by Dissolution and Reprecipitation in a Chelating Solution," *Anal. Chem.* **2006**, *78*, 304.

25. T. Darjaa, K. Yamada, N. Sato, T. Fujino, and Y. Waseda, "Determination of Sulfur in Metal Sulfides by Bromine Water-CCl$_4$ Oxidative Dissolution and Modified EDTA Titration," *Fresenius J. Anal. Chem.* **1998**, *361*, 442.

Chapter 12

1. J. Gorman in *Science News*, 9 September 2000, p. 165.

2. R. F. Wright et al., "Recovery of Acidified European Surface Waters," *Environ. Sci. Technol.* **2005**, *39*, 64A.

3. Books about equilibrium calculations: W. B. Guenther, *Unified Equilibrium Calculations* (New York: Wiley, 1991); J. N. Butler, *Ionic Equilibrium: Solubility and pH Calculations* (New York: Wiley, 1998); and M. Meloun, *Computation of Solution Equilibria* (New York: Wiley, 1988). Software for equilibrium calculations: http://www.micromath.com/ and http://www.acadsoft.co.uk/

4. R. G. Bates, *Determination of pH*, 2nd ed. (New York: Wiley, 1973), p. 86, is the authoritative reference on pH. The pH uncertainty of primary standards could be greater than ± 0.006 at temperatures other than 25°C.

5. Excel might be used to solve for two unknown concentrations simultaneously: R. de Levie, "How to Compute Labile Metal-Ligand Equilibria," *J. Chem. Ed.* **2007**, *84*, 136.

6. R. B. Martin, "Aluminum: A Neurotoxic Product of Acid Rain," *Acc. Chem. Res.* **1994**, *27*, 204.

7. R. Jugdaohsingh, M. M. Campbell, R. P. H. Thompson, C. R. McCrohan, K. N. White, and J. J. Powell, "Mucus Secretion by the Freshwater Snail *Lymnaea stagnalis* Limits Aluminum Concentrations of the Aqueous Environment," *Environ. Sci. Technol.* **1998**, *32*, 2591; M. Ravichandran, G. R. Aiken, M. M. Reddy, and J. N. Ryan, "Enhanced Dissolution of Cinnabar (Mercuric Sulfide) by Dissolved Organic Matter Isolated from the Florida Everglades," *Environ. Sci. Technol.* **1998**, *32*, 3305; S. Sauvé, M. McBride, and W. Hendershot, "Lead Phosphate Solubility in Water and Soil Suspensions," *Environ. Sci. Technol.* **1998**, *32*, 388.

8. Our approach is similar to one by J. L. Guiñón, J. Garcia-Antón, and V. Pérez-Herranz, "Spreadsheet Techniques for Evaluating the Solubility of Sparingly Soluble Salts of Weak Acids," *J. Chem. Ed.* **1999**, *76*, 1157.

9. A. Kraft, "Determination of the pK_a of Multiprotic, Weak Acids by Analyzing Potentiometric Acid-Base Titration Data with Difference Plots," *J. Chem. Ed.* **2003**, *80*, 554.

10. G. B. Kauffman, "Niels Bjerrum: A Centennial Evaluation," *J. Chem. Ed.* **1980**, *57*, 779, 863.

11. Table 6-1 gives $pK_w = 13.995$ at $\mu = 0$ at 25°C. The expression of K_w to which this value applies is given in terms of molalities, m:

$$K_w = \frac{m_{H^+}\gamma_{H^+} m_{OH^-}\gamma_{OH^-}}{\mathcal{A}_{H_2O}} = 10^{-13.995}$$

We want to evaluate K'_w for 0.1 M KCl. The factor for converting molality into molarity in 0.1 M K is 0.994 in Table 12-1-1A of H. S. Harned and B. B. Owen, *Physical Chemistry of Electrolyte Solutions*, 3rd ed. (New York: Reinhold, 1958), p. 725. The factor $\gamma_{H^+}\gamma_{OH^-}/\mathcal{A}_{H_2O}$ is 0.626 in 0.10 M KCl, interpolated from Table 15-2-1A of Harned and Owen, p. 752. K'_w is the product of concentrations $[H^+][OH^-]$:

$$[H^+][OH^-] = \frac{m_{H^+}(0.994)\gamma_{H^+} m_{OH^-}(0.994)\gamma_{OH^-}}{\mathcal{A}_{H_2O}} \frac{\mathcal{A}_{H_2O}}{\gamma_{H^+}\gamma_{OH^-}}$$

$$= 10^{-13.995}(0.994^2)\left(\frac{1}{0.626}\right) = 10^{-13.797}$$

Chapter 13

1. General treatments of electrochemistry: A. Hamnett, C. H. Hamann, and W. Vielstich, *Electrochemistry* (New York: Wiley, 1998); Z. Galus, *Fundamentals of Electrochemical Analysis* (New York: Ellis Horwood, 1994); C. M. A. Brett and A. M. O. Brett, *Electrochemistry* (Oxford: Oxford University Press, 1993); and H. B. Oldham and J. C. Myland, *Fundamentals of Electrochemical Science* (San Diego: Academic Press, 1993).

2. K. Rajeshwar and J. G. Ibanez, *Environmental Electrochemistry* (San Diego: Academic Press, 1997).

3. N. J. Tao, "Measurement and Control of Single Molecule Conductance," *J. Mater. Chem.* **2005**, *15*, 3260; N. Tao, "Electrochemical Fabrication of Metallic Quantum Wires," *J. Chem. Ed.* **2005**, *82*, 720; S. Lindsay, "Single-Molecule Electronic Measurements with Metal Electrodes," *J. Chem. Ed.* **2005**, *82*, 727; R. A. Wassel and C. B. Gorman, "Establishing the Molecular Basis for Molecular Electronics," *Angew. Chem. Int. Ed.* **2004**, *43*, 5120.

4. T. Morita and S. Lindsay, "Determination of Single Molecule Conductances of Alkanedithiols by Conducting-Atomic Force Microscopy with Large Gold Nanoparticles," *J. Am. Chem. Soc.* **2007**, *129*, 7262.

5. I.-W. P. Chen, M.-D. Fu, W.-H. Tseng, J.-Y. Yu, S.-H. Wu, C.-J. Ku, C.-H. Chen, and S.-M. Peng, "Conductance and Stochastic Switching of Ligand-Supported Linear Chains of Metal Atoms," *Angew. Chem. Int. Ed.* **2006**, *45*, 5814.

6. S. Weinberg, *The Discovery of Subatomic Particles* (Cambridge: Cambridge University Press, 2003), pp. 13–16. A wonderful book by a Nobel Prize winner.

7. M. J. Smith, A. M. Fonseca, and M. M. Silva, "The Lead-Lead Oxide Secondary Cell as a Teaching Resource," *J. Chem. Ed.* **2009**, *86*, 357; M. J. Smith and C. A. Vincent, "Structure and Content of Some Primary Batteries," *J. Chem. Ed.* **2001**, *78*, 519; R. S. Treptow, "The Lead-Acid Battery: Its Voltage in Theory and in Practice," *J. Chem. Ed.* **2002**, *79*, 334; M. J. Smith and C. A. Vincent, "Why Do Some Batteries Last Longer Than Others?" *J. Chem. Ed.* **2002**, *79*, 851; M. Tamez and J. H. Yu, "Aluminum-Air Battery," *J. Chem. Ed.* **2007**, *84*, 1936A; H. Goto, H. Yoneyama, F. Togashi, R. Ohta, A. Tsujimoto, E. Kita, K. Ohshima, and D. Rosenberg, "Preparation of Conducting Polymers by Electrochemical Methods and Demonstration of a Polymer Battery," *J. Chem. Ed.* **2008**, *85*, 1067.

8. O. Zerbinati, "A Direct Methanol Fuel Cell," *J. Chem. Ed.* **2002**, *79*, 829; J. M. Ogden, "Hydrogen: The Fuel of the Future?" *Physics Today,* April 2002, p. 69.

9. P. Krause and J. Manion, "A Novel Approach to Teaching Electrochemical Principles," *J. Chem. Ed.* **1996**, *73*, 354.

10. L. P. Silverman and B. B. Bunn, "The World's Longest Human Salt Bridge," *J. Chem. Ed.* **1992**, *69*, 309.

11. Classroom demonstrations: J. D. Ciparick, "Half Cell Reactions: Do Students Ever See Them?" *J. Chem. Ed.* **1991**, *68*, 247; P.-O. Eggen, T. Grønneberg, and L. Kvittengen, "Small-Scale and Low-Cost Galvanic Cells," *J. Chem. Ed.* **2006**, *83*, 1201.

12. A. W. von Smolinski, C. E. Moore, and B. Jaselskis, "The Choice of the Hydrogen Electrode as the Base for the Electromotive Series" in *Electrochemistry, Past and Present*, ACS Symposium Series 390, J. T. Stock and M. V. Orna, eds. (Washington, DC: American Chemical Society, 1989), Chap. 9.

13. H. Frieser, "Enhanced Latimer Potential Diagrams Via Spreadsheets," *J. Chem. Ed.* **1994**, *71*, 786.

14. A. Arévalo and G. Pastor, "Verification of the Nernst Equation and Determination of a Standard Electrode Potential," *J. Chem. Ed.* **1985**, *62*, 882.

15. For a classroom demonstration using a cell as a chemical probe, see R. H. Anderson, "An Expanded Silver Ion Equilibria Demonstration: Including Use of the Nernst Equation and Calculation of Nine Equilibrium Constants," *J. Chem. Ed.* **1993**, *70*, 940.

16. Structure of dehydroascorbic acid: R. C. Kerber, "'As Simple as Possible, But Not Simpler'—The Case of Dehydroascorbic Acid," *J. Chem. Ed.* **2008**, *85*, 1237.

17. J. E. Walker, "ATP Synthesis by Rotary Catalysis (Nobel Lecture)," *Angew. Chem. Int. Ed.* **1998**, *37*, 2309; P. D. Boyer, "Energy, Life, and ATP (Nobel Lecture)," *Angew. Chem. Int. Ed.* **1998**, *37*, 2297; W. S. Allison, "F₁-ATPase," *Acc. Chem. Res.* **1998**, *31*, 819.

18. For an advanced equilibrium problem based on the bromine Latimer diagram, see T. Michalowski, "Calculation of pH and Potential *E* for Bromine Aqueous Solution," *J. Chem. Ed.* **1994**, *71*, 560.

19. K. T. Jacob, K. P. Jayadevan, and Y. Waseda, "Electrochemical Determination of the Gibbs Energy of Formation of $MgAl_2O_4$," *J. Am. Ceram. Soc.* **1998**, *81*, 209.

20. J. T. Stock, "Einar Biilmann (1873–1946): pH Determination Made Easy," *J. Chem. Ed.* **1989**, *66*, 910.

21. The cell in this problem would not give an accurate result because of the *junction potential* at each liquid junction (Section 14-3). A cell without liquid junctions is described by P. A. Rock, "Electrochemical Double Cells," *J. Chem. Ed.* **1975**, *52*, 787.

Chapter 14

1. S. P. Kounaves, M. H. Hecht, S. J. West, J.-M. Morookian, S. M. M. Young, R. Quinn, P. Grunthaner, X. Wen, M. Weilert, C. A. Cable, A. Fisher, K. Gospodinova, J. Kapit, S. Stroble, P.-C. Hsu, B. C. Clark, D. W. Ming, and P. H. Smith, "The 2007 Mars Scout Lander MECA Wet Chemistry Laboratory," *J. Geophys. Res.* **2009**, *113*, E00A19.

2. Practical aspects of electrode fabrication are discussed in D. T. Sawyer, A. Sobkowiak, and J. L. Roberts, Jr., *Electrochemistry for Chemists*, 2nd ed.

(New York: Wiley, 1995); G. A. East and M. A. del Valle, "Easy-to-Make Ag/AgCl Reference Electrode," *J. Chem. Ed.* **2000**, *77*, 97.

3. A demonstration of potentiometry with a silver electrode (or a microscale experiment for general chemistry) is described by D. W. Brooks, D. Epp, and H. B. Brooks, "Small-Scale Potentiometry and Silver One-Pot Reactions," *J. Chem. Ed.* **1995**, *72*, A162.

4. D. Dobčnik, J. Stergulec, and S. Gomišček, "Preparation of an Iodide Ion-Selective Electrode by Chemical Treatment of a Silver Wire," *Fresenius J. Anal. Chem.* **1996**, *354*, 494.

5. I. R. Epstein and J. A. Pojman, *An Introduction to Nonlinear Chemical Dynamics: Oscillations, Waves, Patterns, and Chaos* (New York: Oxford University Press, 1998); I. R. Epstein, K. Kustin, P. De Kepper, and M. Orbán, *Scientific American*, March 1983, p. 112; and H. Degn, "Oscillating Chemical Reactions in Homogeneous Phase," *J. Chem. Ed.* **1972**, *49*, 302.

6. Mechanisms of oscillating reactions are discussed by O. Benini, R. Cervellati, and P. Fetto, "The BZ Reaction: Experimental and Model Studies in the Physical Chemistry Laboratory," *J. Chem. Ed.* **1996**, *73*, 865; R. J. Field and F. W. Schneider, "Oscillating Chemical Reactions and Nonlinear Dynamics," *J. Chem. Ed.* **1989**, *66*, 195; R. M. Noyes, "Some Models of Chemical Oscillators," *J. Chem. Ed.* **1989**, *66*, 190; P. Ruoff, M. Varga, and E. Körös, "How Bromate Oscillators Are Controlled," *Acc. Chem. Res.* **1988**, *21*, 326; M. M. C. Ferriera, W. C. Ferriera, Jr., A. C. S. Lino, and M. E. G. Porto, "Uncovering Oscillations, Complexity, and Chaos in Chemical Kinetics Using *Mathematica*," *J. Chem. Ed.* **1999**, *76*, 861; G. Schmitz, L. Kolar-Anić, S. Anić, and Z. Čupić, "The Illustration of Multistability," *J. Chem. Ed.* **2000**, *77*, 1502.

7. H. E. Prypsztejn, "Chemiluminescent Oscillating Demonstrations: The Chemical Buoy, The Lighting Wave, and the Ghostly Cylinder," *J. Chem. Ed.* **2005**, *82*, 53; D. Kolb, "Overhead Projector Demonstrations," *J. Chem. Ed.* **1988**, *65*, 1004; R. J. Field, "A Reaction Periodic in Time and Space," *J. Chem. Ed.* **1972**, *49*, 308; J. N. Demas and D. Diemente, "An Oscillating Chemical Reaction with a Luminescent Indicator," *J. Chem. Ed.* **1973**, *50*, 357; J. F. Lefelhocz, "The Color Blind Traffic Light," *J. Chem. Ed.* **1972**, *49*, 312: P. Aroca, Jr., and R. Aroca, "Chemical Oscillations: A Microcomputer-Controlled Experiment," *J. Chem. Ed.* **1987**, *64*, 1017: J. Amrehn, P. Resch, and F. W. Schneider, "Oscillating Chemiluminescence with Luminol in the Continuous Flow Stirred Tank Reactor," *J. Phys. Chem.* **1988**, *92*, 3318; D. Avnir, "Chemically Induced Pulsations of Interfaces: The Mercury Beating Heart," *J. Chem. Ed.* **1989**, *66*, 211; K. Yoshikawa, S. Nakata, M. Yamanaka, and T. Waki, "Amusement with a Salt-Water Oscillator," *J. Chem. Ed.* **1989**, *66*, 205; L. J. Soltzberg, M. M. Boucher, D. M. Crane, and S. S. Pazar, "Far from Equilibrium—The Flashback Oscillator," *J. Chem. Ed.* **1987**, *64*, 1043; S. M. Kaushik, Z. Yuan, and R. M. Noyes, "A Simple Demonstration of a Gas Evolution Oscillator," *J. Chem. Ed.* **1986**, *63*, 76; R. F. Melka, G. Olsen, L. Beavers, and J. A. Draeger, "The Kinetics of Oscillating Reactions," *J. Chem. Ed.* **1992**, *69*, 596; J. M. Merino, "A Simple, Continuous-Flow Stirred-Tank Reactor for the Demonstration and Investigation of Oscillating Reactions," *J. Chem. Ed.* **1992**, *69*, 754.

8. T. Kappes and P. C. Hauser, "A Simple Supplementary Offset Device for Data Acquisition Systems," *J. Chem. Ed.* **1999**, *76*, 1429.

9. [Br⁻] also oscillates in this experiment. For an [I⁻] oscillator, see T. S. Briggs and W. C. Rauscher, "An Oscillating Iodine Clock," *J. Chem. Ed.* **1973**, *50*, 496.

10. E. Bakker, P. Bühlmann, and E. Pretsch, "Carrier-Based Ion-Selective Electrodes and Bulk Optodes," *Chem. Rev.* **1997**, *97*, 3083; ibid. **1998**, *98*, 1593.

11. C. E. Moore, B. Jaselskis, and A. von Smolinski, "Development of the Glass Electrode" in *Electrochemistry, Past and Present*, ACS Symposium Series 390, J. T. Stock and M. V. Orna, eds. (Washington, DC: American Chemical Society, 1989), Chap. 19.

12. B. Jaselskis, C. E. Moore, and A. von Smolinski, "Development of the pH Meter" in *Electrochemistry, Past and Present*, ACS Symposium Series 390, J. T. Stock and M. V. Orna, eds. (Washington, DC: American Chemical Society, 1989), Chap. 18.

13. R. P. Buck, S. Rondinini, A. K. Covington, F. G. K. Baucke, C. M. A. Brett, M. F. Camoes, M. J. T. Milton, T. Mussini, R. Naumann, K. W. Pratt, P. Spitzer, and G. S. Wilson, "Measurement of pH. Definitions, Standards, and Procedures," *Pure Appl. Chem.* **2002**, *74*, 2169; B. Lunelli and F. Scagnolari, "pH Basics," *J. Chem. Ed.* **2009**, *86*, 246.

14. L. M. Goss, "A Demonstration of Acid Rain and Lake Acidification: Wet Deposition of Sulfur Dioxide," *J. Chem. Ed.* **2003**, *80*, 39.

15. J. A. Lynch, V. C. Bowersox, and J. W. Grimm, "Acid Rain Reduced in Eastern United States," *Environ. Sci. Technol.* **2000**, *34*, 940; R. E. Baumgardner, Jr., T. F. Lavery, C. M. Rogers, and S. S. Isil, "Estimates of the Atmospheric Deposition of Sulfur and Nitrogen Species: Clean Air Status and Trends Network, 1990–2000," *Environ. Sci. Technol.* **2002**, *36*, 2614; www.epa.gov/acidrain

16. W. F. Koch, G. Marinenko, and R. C. Paule, "An Interlaboratory Test of pH Measurements in Rainwater," *J. Res. Natl. Bur. Stand.* **1986**, *91*, 23.

17. *Free diffusion junction* electrodes are designed to minimize junction potentials. The junction is a Teflon capillary tube containing electrolyte that is periodically renewed by syringe.

18. Spectrophotometry with acid-base indictors is another means for measuring the pH of low ionic strength natural waters. (C. R. French, J. J. Carr, E. M. Dougherty, L. A. K. Eidson, J. C. Reynolds, and M. D. DeGrandpre, "Spectrophotometric pH Measurements of Freshwater," *Anal. Chim. Acta* **2002**, *453*, 13.)

19. The reference electrode in the Ross combination electrode is Pt | I₂, I⁻. This electrode is claimed to give improved precision and accuracy over conventional pH electrodes (R. C. Metcalf, *Analyst* **1987**, *112*, 1573).

20. A. N. Bezbaruah and T. C. Zhang, "Fabrication of Anodically Electrodeposited Iridium Oxide Film pH Microelectrodes for Microenvironmental Studies," *Anal. Chem.* **2002**, *74*, 5726. Iridium oxide electrodes are commercially available or can be prepared by methods such as those described by J.-P. Ndobo-Epoy, E. Lesniewska, and J.-P. Guicquero, "Nano-pH Sensor for the Study of Reactive Materials," *Anal. Chem.* **2007**, *79*, 7560 or R.-G. Du, R.-G. Hu, R.-S. Huang, and C.-J. Lin, "In Situ Measurement of Cl⁻ Concentrations and pH at the Reinforcing Steel/Concrete Interface by Combination Sensors," *Anal. Chem.* **2006**, *78*, 3179.

21. L. W. Niedrach, "Electrodes for Potential Measurements in Aqueous Systems at High Temperatures and Pressures," *Angew. Chem.* **1987**, *26*, 161.

22. History of ion-selective electrodes: M. S. Frant, "Where Did Ion Selective Electrodes Come From?" *J. Chem. Ed.* **1997**, *74*, 159; J. Ruzicka, "The Seventies: Golden Age for Ion Selective Electrodes," *J. Chem. Ed.* **1997**, *74*, 167; T. S. Light, "Industrial Use and Applications of Ion Selective Electrodes," *J. Chem. Ed.* **1997**, *74*, 171; and C. C. Young, "Evolution of Blood Chemistry Analyzers Based on Ion Selective Electrodes," *J. Chem. Ed.* **1997**, *74*, 177; R. P. Buck and E. Lindner, "Tracing the History of Selective Ion Sensors," *Anal. Chem.* **2001**, *73*, 88A.

23. E. Bakker and E. Pretsch, "Modern Potentiometry," *Angew. Chem. Int. Ed.* **2007**, *46*, 5660.

24. E. Bakker and E. Pretsch, "The New Wave of Ion-Selective Electrodes," *Anal. Chem.* **2002**, *74*, 420A.

25. For interfering ions X with charge different from the primary ion A, you will find the incorrect, empirical Nicolsky-Eisenman equation in the literature:

$$E = \text{constant} \pm \frac{0.059\,16}{z_A} \log \left[\mathcal{A}_A + \sum_X K_{A,X}^{Pot} \mathcal{A}_X^{(z_A/z_X)} \right]$$

where z_A is the charge of primary ion A and z_X is the charge of interfering ion X. This equation should not be used. (Y. Umezawa, K. Umezawa, and H. Sato, "Selectivity Coefficients for Ion-Selective Electrodes: Recommended Methods for Reporting $K_{A,X}^{Pot}$ Values," *Pure Appl. Chem.* **1995**, *67*, 507.) Correct equations for interfering ions with charges different from that of the primary ion are complicated. You can find them in E. Bakker, R. Meruva, E. Pretsch, and M. Meyerhoff, "Selectivity of Polymer Membrane-Based Ion-Selective Electrodes: Self-Consistent Model Describing the Potentiometric Response in Mixed Ion Solutions of Different Charge," *Anal. Chem.* **1994**, *66*, 3021, and N. Nägele, E. Bakker, and E. Pretsch, "General Description of the Simultaneous Response of Potentiometric Ionophore-Based Sensors to Ions of Different Charge," *Anal. Chem.* **1999**, *71*, 1041.

26. E. Bakker, E. Pretsch, and P. Bühlmann, "Selectivity of Potentiometric Ion Sensors," *Anal. Chem.* **2000**, *72*, 1127; E. Bakker, "Determination of Unbiased Selectivity Coefficients of Neutral Carrier-Based Cation-Selective Electrodes," *Anal. Chem.* **1997**, *69*, 1061.

27. Y. Umezawa, K. Umezawa, and H. Sato, "Selectivity Coefficients for Ion-Selective Electrodes: Recommended Methods for Reporting $K_{A,X}^{Pot}$ Values," *Pure Appl. Chem.* **1995**, *67*, 507.

28. K. Ren, "Selectivity Problems of Membrane Ion-Selective Electrodes," *Fresenius J. Anal. Chem.* **1999**, *365*, 389.

29. M. H. Hecht, S. P. Kounaves, R. C. Quinn, S. J. West, S. M. M. Young, D. W. Ming, D. C. Catling, B. C. Clark, W. V. Boynton, J. Hoffman, L. P. DeFlores, K. Gospodinova, J. Kapit, and P. H. Smith, "Detection of Perchlorate and the Soluble Chemistry of Martian Soil at the Phoenix Lander Site," *Science* **2009**, *325*, 64.

30. A. Michalska, "Optimizing the Analytical Performance and Construction of Ion-Selective Electrodes with Conducting Polymer-Based Ion-to-Electron Transducers," *Anal. Bioanal. Chem.* **2006**, *384*, 391; A. Rzewuska, M. Wojciechowski, E. Bulska, E. A. H. Hall, K. Maksymuik, and A. Michalska, "Composite Polyacrylate-Poly(3,4-ethylenedioxythiophene) Membranes for Improved All-Solid-State Ion-Selective Sensors," *Anal. Chem.* **2008**, *80*, 321.

31. T. A. Bendikov and T. C. Harmon, "A Sensitive Nitrate Ion-Selective Electrode from a Pencil Lead," *J. Chem. Ed.* **2005**, *82*, 439.

32. Other kinds of CO_2 electrodes have been described (e.g., J. H. Shin, J. S. Lee, S. H. Choi, D. K. Lee, H. Nam, and G. S. Cha, "A Planar pCO_2 Sensor with Enhanced Electrochemical Properties," *Anal. Chem.* **2000**, *72*, 4468).

33. Y. S. Choi, L. Lvova, J. H. Shin, S. H. Oh, C. S. Lee, B. H. Kim, G. S. Cha, and H. Nam, "Determination of Oceanic Carbon Dioxide Using a Carbonate-Selective Electrode," *Anal. Chem.* **2002**, *74*, 2435.

34. J. D. Czaban, "Electrochemical Sensors in Clinical Chemistry," *Anal. Chem.* **1985**, *57*, 345A.

35. M. Umemoto, W. Tani, K. Kuwa, and Y. Ujihira, "Measuring Calcium in Plasma," *Anal. Chem.* **1994**, *66*, 352A.

36. P. Gründler, *Chemical Sensors: An Introduction for Scientists and Engineer*s (New York: Springer, 2007).

37. *Resistivity*, ρ, measures how well a substance retards the flow of electric current when an electric field is applied: $J = E/\rho$, where J is current density (current flowing through a unit cross section of the material, A/m^2) and E is electric field (V/m). Units of resistivity are $V \cdot m/A$ or $\Omega \cdot m$, because $\Omega = V/A$, where $\Omega =$ ohm. Conductors have resistivities near $10^{-8} \Omega \cdot m$, semiconductors have resistivities of 10^{-4} to $10^7 \Omega \cdot m$, and insulators have resistivities of 10^{12} to $10^{20} \Omega \cdot m$. The reciprocal of resistivity is *conductivity*. Resistivity does not depend on the dimensions of the substance. Resistance, R, is related to resistivity by the equation $R = \rho l/A$, where l is the length and A is the cross-sectional area of the conducting substance.

38. S.-S. Jan, J.-L. Chiang, Y.-C. Chen, J.-C. Chou, and C.-C Cheng, "Characteristics of the Hydrogen Ion-Sensitive Field Effect Transistors with Sol-Gel-Derived Lead Titanate Gate," *Anal. Chim. Acta* **2002**, *469*, 205.

39. C. Jiménez, I. Marqués, and J. Bartrolí, "Continuous-Flow Systems for On-Line Water Monitoring Using Back-Side Contact ISFET-Based Sensors," *Anal. Chem.* **1996**, *68*, 3801.

40. J. H. Shin, H. J. Lee, C. Y. Kim, B. K. Oh, K. L. Rho, H. Nam, and G. S. Cha, "ISFET-Based Differential pCO_2 Sensors Employing a Low-Resistance Gas-Permeable Membrane," *Anal. Chem.* **1996**, *68*, 3166.

41. A. M. Nyamsi Hendji, N. Jaffrezic-Renault, C. Martelet, P. Clechet, A. A. Shul'ga, V. I. Strikha, L. I. Netchiporuk, A. P. Soldatkin, and W. B. Wlodarski, "Sensitive Detection of Pesticides Using Differential ISFET-Based System with Immobilized Cholinesterases," *Anal. Chim. Acta* **1993**, *281*, 3.

42. M. Lahav, A. B. Kharitonov, O. Katz, T. Kunitake, and I. Willner, "Tailored Chemosensors for Chloroaromatic Acids Using Molecular Imprinted TiO_2 Thin Films on Ion-Sensitive Field-Effect Transistor," *Anal. Chem.* **2001**, *73*, 720.

43. A. B. Kharitonov, A. N. Shipway, and I. Willner, "An Au Nanoparticle/Bisbipyridinium Cyclophane-Functionalized Ion-Sensitive Field-Effect Transistor for the Sensing of Adrenaline," *Anal. Chem.* **1999**, *71*, 5441.

44. T. Uno, H. Tabata, and T. Kawai, "Peptide–Nucleic Acid-Modified Ion-Sensitive Field-Effect Transistor-Based Biosensor for Direct Detection of DNA Hybridization," *Anal. Chem.* **2007**, *79*, 52.

45. D. C. Jackman, "A Recipe for the Preparation of a pH 7.00 Calibration Buffer," *J. Chem. Ed.* **1993**, *70*, 853.

46. Equation 46 in E. Bakker, R. Meruva, E. Pretsch, and M. Meyerhoff, "Selectivity of Polymer Membrane-Based Ion-Selective Electrodes: Self-Consistent Model Describing the Potentiometric Response in Mixed Ion Solutions of Different Charge," *Anal. Chem.* **1994**, *66*, 3021.

Chapter 15

1. T. Astrup, S. L. S. Stipp, and T. H. Christensen, "Immobilization of Chromate from Coal Fly Ash Leachate Using an Attenuating Barrier Containing Zero-Valent Iron," *Environ. Sci. Technol.* **2000**, *34*, 4163; S. H. Joo, A. J. Feitz, and T. D. Waite, "Oxidative Degradation of the Carbothiolate Herbicide, Molinate, Using Nanoscale Zero-Valent Iron," *Environ. Sci. Technol.* **2004**, *38*, 2242; R. Miehr, P. G. Tratnyek, J. Z. Bandstra, M. M. Scherer, M. J. Alowitz, and E. U. Bylaska, "Diversity of Contaminant Reduction Reactions by Zerovalent Iron: Role of the Reductant," *Environ. Sci. Technol.* **2004**, *38*, 139; V. K. Sharma, C. R. Burnett, D. B. O'Connor, and D. Cabelli, "Iron(VI) and Iron(V) Oxidation of Thiocyanate," *Environ. Sci. Technol.* **2002**, *36*, 4182.

2. Information on redox titrations: J. Bassett, R. C. Denney, G. H. Jeffery, and J. Mendham, *Vogel's Textbook of Inorganic Analysis*, 4th ed. (Essex, UK: Longman, 1978); H. A. Laitinen and W. E. Harris, *Chemical Analysis*, 2nd ed. (New York: McGraw-Hill, 1975); I. M. Kolthoff, R. Belcher, V. A. Stenger, and G. Matsuyama, *Volumetric Analysis*, Vol. 3 (New York: Wiley, 1957); A. Berka, J. Vulterin, and J. Zýka, *Newer Redox Titrants*, H. Weisz, trans. (Oxford: Pergamon, 1965).

3. J. Ermírio, F. Moraes, F. H. Quina, C. A. O. Nascimento, D. N. Silva, and O. Chiavone-Filho, "Treatment of Saline Wastewater Contaminate with Hydrocarbons by the Photo-Fenton Process," *Environ. Sci. Technol.* **2004**, *38*, 1183; B. Gözmen, M. A. Oturan, N. Oturan, and O. Erbatur, "Indirect Electrochemical Treatment of Bisphenol A in Water via Electrochemically Generated Fenton's Reagent," *Environ. Sci. Technol.* **2003**, *37*, 3716. $(H_2O)_5Fe^{IV}{=}O^{2+}$ is not the Fenton intermediate in acidic and neutral aqueous solution: O. Pestovsky, S. Stoian, E. L. Bominaar, X. Shan, E. Münck, L. Que, Jr., and A. Bakac, "Aqueous $Fe^{IV}{=}O$: Spectroscopic Identification and Oxo-Group Exchange," *Angew. Chem. Int. Ed.* **2005**, *44*, 6871.

4. R. D. Webster, "New Insights into the Oxidative Electrochemistry of Vitamin E," *Acc. Chem. Res.* **2007**, *40*, 251.

5. D. T. Sawyer, "Conceptual Considerations in Molecular Science," *J. Chem. Ed.* **2005**, *82*, 985.

6. Equations 15-9 and 15-10 are analogous to the Henderson-Hasselbalch equation of acid-base buffers. Prior to the equivalence point, the redox titration is *buffered* to a potential near $E_+ =$ formal potential for $Fe^{3+} | Fe^{2+}$ by the presence of Fe^{3+} and Fe^{2+}. After the equivalence point, the reaction is *buffered* to a potential near $E_+ =$ formal potential for $Ce^{4+} | Ce^{3+}$. (R. de Levie, "Redox Buffer Strength," *J. Chem. Ed.* **1999**, *76*, 574.)

7. D. W. King, "A General Approach for Calculating Speciation and Poising Capacity of Redox Systems with Multiple Oxidation States: Application to Redox Titrations and the Generation of pε-pH Diagrams," *J. Chem. Ed.* **2002**, *79*, 1135.

8. T. J. MacDonald, B. J. Barker, and J. A. Caruso, "Computer Evaluation of Titrations by Gran's Method," *J. Chem. Ed.* **1972**, *49*, 200.

9. M. da Conceição Silva Barreto, L. de Lucena Medieros, and P. C. de Holanda Furtado, "Indirect Potentiometric Titration of Fe(III) with Ce(IV) by Gran's Method," *J. Chem. Ed.* **2001**, *78*, 91.

10. R. D. Hancock and B. J. Tarbet, "The Other Double Helix: The Fascinating Chemistry of Starch," *J. Chem. Ed.* **2000**, *77*, 988.

11. J. H. Margeson, J. C. Suggs, and M. R. Midgett, "Reduction of Nitrate to Nitrite with Cadmium," *Anal. Chem.* **1980**, *52*, 1955.

12. E. T. Urbansky, "Total Organic Carbon Analyzers as Tools for Measuring Carbonaceous Matter in Natural Waters," *J. Environ. Monit.* **2001**, *3*, 102. General references on environmental analysis: M. Radojevic and V. N. Bashkin, *Practical Environmental Analysis* (Cambridge: Royal Society of Chemistry, 1999) and D. Perez-Bendito and S. Rubio, *Environmental Analytical Chemistry* (Amsterdam: Elsevier, 1998).

13. L. J. Stolzberg and V. Brown, "Note on Photocatalytic Destruction of Organic Wastes: Methyl Red as a Substrate," *J. Chem. Ed.* **2005**, *82*, 526; J. A. Poce-Fatou, M. L. A. Gil, R. Alcántara, C. Botella, and J. Martin, "Photochemical Reactor for the Study of Kinetics and Adsorption Phenomena,"

J. Chem. Ed. **2004**, *81*, 537; J. C. Yu and L. Y. L. Chan, "Photocatalytic Degradation of a Gaseous Organic Pollutant," *J. Chem. Ed.* **1998**, *75*, 750.

14. R. Dunn, "New Developments in Membrane-Selective Conductometric Instruments for Total Organic Carbon Determination in Water," *Am. Lab.*, September 2004, p. 22.

15. S. Sakthivel and H. Kisch, "Daylight Photocatalysis by Carbon-Modified Titanium Dioxide," *Angew. Chem. Int. Ed.* **2003**, *42*, 4908.

16. S. Horikoshi, N. Serpone, Y. Hisamatsu, and H. Hidaka, "Photocatalyzed Degradation of Polymers in Aqueous Semiconductor Suspensions," *Environ. Sci. Technol.* **1998**, *32*, 4010.

17. BOD and COD procedures are described in *Standard Methods for the Examination of Wastewater*, 21st ed. (Washington, DC: American Public Health Association, 2005), which is the standard reference for water analysis.

18. K. Catterall, H. Zhao, N. Pasco, and R. John, "Development of a Rapid Ferricyanide-Mediated Assay for Biochemical Oxygen Demand Using a Mixed Microbial Consortium," *Anal. Chem.* **2003**, *75*, 2584.

19. B. Wallace and M. Purcell, "The Benefits of Nitrogen and Total Organic Carbon Determination by High-Temperature Combustion," *Am. Lab. News Ed.*, February 2003, p. 58.

20. W. Gottardi, "Redox-Potentiometric/Titrimetric Analysis of Aqueous Iodine Solutions," *Fresenius J. Anal. Chem.* **1998**, *362*, 263.

21. S. C. Petrovic and G. M. Bodner, "An Alternative to Halogenated Solvents for Halogen/Halide Extractions," *J. Chem. Ed.* **1991**, *68*, 509.

22. G. L. Hatch, "Effect of Temperature on the Starch-Iodine Spectrophotometric Calibration Line," *Anal. Chem.* **1982**, *54*, 2002.

23. Y. Xie, M. R. McDonald, and D. W. Margerum, "Mechanism of the Reaction Between Iodate and Iodide Ions in Acid Solutions," *Inorg. Chem.* **1999**, *38*, 3938.

24. Prepare anhydrous $Na_2S_2O_3$ by refluxing 21 g of $Na_2S_2O_3 \cdot 5H_2O$ with 100 mL of methanol for 20 min. Then filter the anhydrous salt, wash with 20 mL of methanol, and dry at 70°C for 30 min. (A. A. Woolf, "Anhydrous Sodium Thiosulfate as a Primary Iodometric Standard," *Anal. Chem.* **1982**, *54*, 2134.)

25. J. Hvoslef and B. Pedersen, "The Structure of Dehydroascorbic Acid in Solution," *Acta Chem. Scand.* **1979**, *B33*, 503; D. T. Sawyer, G. Chiericato, Jr., and T. Tsuchiya, "Oxidation of Ascorbic Acid and Dehydroascorbic Acid by Superoxide in Aprotic Media," *J. Am. Chem. Soc.* **1982**, *104*, 6273; R. C. Kerber, "As Simple As Possible, But Not Simpler—The Case of Dehydroascorbic Acid," *J. Chem. Ed.* **2008**, *85*, 1237.

26. R. J. Cava, "Oxide Superconductors," *J. Am. Ceram. Soc.* **2000**, *83*, 5.

27. D. C. Harris, M. E. Hills, and T. A. Hewston, "Preparation, Iodometric Analysis, and Classroom Demonstration of Superconductivity in $YBa_2Cu_3O_{8-x}$," *J. Chem. Ed.* **1987**, *64*, 847; D. C. Harris, "Oxidation State Chemical Analysis," in T. A. Vanderah, ed., *Chemistry of Superconductor Materials* (Park Ridge, NJ: Noyes, 1992); B. D. Fahlman, "Superconductor Synthesis: An Improvement," *J. Chem. Ed.* **2001**, *78*, 1182. Superconductor demonstration kits can be purchased from several vendors, including Sargent-Welch, 7400 N. Linder Ave., Skokie, IL 60077-1026.

28. Experiments with an ^{18}O-enriched superconductor show that the O_2 evolved in Reaction 1 is all derived from the solid, not from the solvent (M. W. Shafer, R. A. de Groot, M. M. Plechaty, G. J. Scilla, B. L. Olson, and E. I. Cooper, "Evolution and Chemical State of Oxygen Upon Acid Dissolution of $YBa_2Cu_3O_{6.98}$," *Mater. Res. Bull.* **1989**, *24*, 687; P. Salvador, E. Fernandez-Sanchez, J. A. Garcia Dominguez, J. Amdor, C. Cascales, and I. Rasines, "Spontaneous O_2 Release from $SmBa_2Cu_3O_{7-x}$ High T_c Superconductor in Contact with Water," *Solid State Commun.* **1989**, *70*, 71).

29. A more sensitive and elegant iodometric procedure is described by E. H. Appelman, L. R. Morss, A. M. Kini, U. Geiser, A. Umezawa, G. W. Crabtree, and K. D. Carlson, "Oxygen Content of Superconducting Perovskites $La_{2-x}Sr_xCuO_y$ and $YBa_2Cu_3O_y$," *Inorg. Chem.* **1987**, *26*, 3237. This method can be modified by adding standard Br_2 to analyze superconductors with oxygen in the range 6.0–6.5, in which there is formally Cu^+ and Cu^{2+}. The use of electrodes instead of starch to find the end point in iodometric titrations of superconductors is recommended (P. Phinyocheep and I. M. Tang, "Determination of the Hole Concentration (Copper Valency) in the High T_c Superconductors," *J. Chem. Ed.* **1994**, *71*, A115).

30. C. L. Copper and E. Koubek, "Analysis of an Oxygen Bleach," *J. Chem. Ed.* **2001**, *78*, 652.

31. M. T. Garrett, Jr., and J. F. Stehlik, "Classical Analysis," *Anal. Chem.* **1992**, *64*, 310A.

32. K. Peitola, K. Fujinami, M. Karppinen, H. Yamauchi, and L. Niiniströ, "Stoichiometry and Copper Valence in the $Ba_{1-y}CuO_{2+\delta}$ System," *J. Mater. Chem.* **1999**, *9*, 465.

33. S. Scaccia and M. Carewska, "Determination of Stoichiometry of $Li_{1+y}CoO_2$ Materials by Flame Atomic Absorption Spectrometry and Automated Potentiometric Titration," *Anal. Chim. Acta* **2002**, *453*, 35.

34. M. Karppinen, A. Fukuoka, J. Wang, S. Takano, M. Wakata, T. Ikemachi, and H. Yamauchi, "Valence Studies on Various Superconducting Bismuth and Lead Cuprates and Related Materials," *Physica* **1993**, *C208*, 130.

35. The oxygen liberated in Reaction 4 of this problem may be derived from the superconductor, not from solvent water. In any case, BiO_3^- reacts with Fe^{2+}, and Cu^{3+} does not, when the sample is dissolved in acid.

Chapter 16

1. T. R. I. Cataldi, C. Campa, and G. E. De Benedetto, "Carbohydrate Analysis by High-Performance Anion-Exchange Chromatography with Pulsed Amperometric Detection," *Fresenius J. Anal. Chem.* **2000**, *368*, 739.

2. W. E. Haupin, "Electrochemistry of the Hall-Heroult Process for Aluminum Smelting," *J. Chem. Ed.* **1983**, *60*, 279; N. C. Craig, "Charles Martin Hall: The Young Man, His Mentor, and His Metal," *J. Chem. Ed.* **1986**, *63*, 557.

3. A. J. Bard and L. R. Faulkner, *Electrochemical Methods and Applications*, 2nd ed. (New York: Wiley, 2001); J. Wang, *Analytical Electrochemistry*, 3rd ed. (New York: Wiley-VCH, 2006); J. O'M. Bockris, and A. K. N. Reddy, *Modern Electrochemistry*, 2nd ed. (Dordrecht, Netherlands: Kluwer, 1998–2001, 3 vols.); F. Scholz, ed., *Electroanalytical Methods* (Berlin: Springer-Verlag, 2002); A. J. Bard and M. Stratmann, eds., *Encyclopedia of Electrochemistry* (New York: Wiley-VCH, 2002).

4. E. C. Gilbert in H. N. Alyea and F. B. Dutton, eds., *Tested Demonstrations in Chemistry* (Easton, PA: Journal of Chemical Education, 1965), p. 145.

5. J. O'M. Bockris, "Overpotential: A Lacuna in Scientific Knowledge," *J. Chem. Ed.* **1971**, *48*, 352.

6. Z. Qiao, W. Shang, X. Zhang, and C. Wang, "Underpotential Deposition of Tin(II) on a Gold Disk Electrode and Determination of Tin in a Tin Plate Sample," *Anal. Bioanal. Chem.* **2005**, *381*, 1467.

7. D. N. Craig, J. I. Hoffman, C. A. Law, and W. J. Hamer, "Determination of the Value of the Faraday with a Silver-Perchloric Acid Coulometer," *J. Res. Natl. Bur. Stand.* **1960**, *64A*, 381; H. Diehl, "High-Precision Coulometry and the Value of the Faraday," *Anal. Chem.* **1979**, *51*, 318A.

8. J. Greyson and S. Zeller, "Analytical Coulometry in Monier-Williams Sulfite-in-Food Determinations," *Am. Lab.*, July 1987, p. 44; D. T. Pierce, M. S. Applebee, C. Lacher, and J. Bessie, "Low Parts Per Billion Determination of Sulfide by Coulometric Argentometry," *Environ. Sci. Technol.* **1998**, *32*, 1734.

9. L. C. Clark, R. Wolf, D. Granger, and A. Taylor, "Continuous Recording of Blood Oxygen Tension by Polarography," *J. Appl. Physiol.* **1953**, *6*, 189.

10. F. Schreiber, L. Polerecky, and D. De Beer, "Nitric Oxide Microsensor for High Spatial Resolution Measurements in Biofilms and Sediments," *Anal. Chem.* **2008**, *80*, 1152; Y. Lee and J. Kim, "Simultaneous Electrochemical Detection of Nitric Oxide and Carbon Monoxide Generated from Mouse Kidney Organ Tissues," *Anal. Chem.* **2008**, *80*, 7669.

11. J. Yinon, "Detection of Explosives by Electronic Noses," *Anal. Chem.* **2003**, *75*, 99A; M. C. C. Oliveros, J. L. P. Pavón, C. G. Pinto, M. E. F. Laespada, B. M. Cordero, and M. Forina, "Electronic Nose Based on Metal Oxide Semiconductor Sensors as a Fast Alternative for the Detection of Adulteration of Virgin Olive Oils," *Anal. Chim. Acta* **2002**, *459*, 219; C. L. Honeybourne, "Organic Vapor Sensors for Food Quality Assessment," *J. Chem. Ed.* **2000**, *77*, 338; E. Zubritsky, "E-Noses Keep an Eye on the Future," *Anal. Chem.* **2000**, *72*, 421A.

12. R. Renneberg and F. Lisdat, eds., *Biosensing for the 21st Century* (Heidelberg: Springer, 2008); J. M. Cooper and A. E. G. Cass, eds., *Biosensors*, 2nd ed. (Oxford: Oxford University Press, 2004); A. Mulchandani and O. A. Sadik, eds., *Chemical and Biological Sensors for Environmental Monitoring Biosensors* (Washington, DC: American Chemical Society, 2000)); E. Palaček and M. Fojta, "Detecting DNA Hybridization and Damage," *Anal. Chem.* **2001**, *73*, 75A.

13. Examples of biosensors: J. H. Thomas, S. K. Kim, P. J. Hesketh, H. B. Halsall, and W. R. Heineman, "Bead-Based Electrochemical Immunoassay for Bacteriophage MS2," *Anal. Chem.* **2004**, *76*, 2700; Y. Zhang, H.-H. Kim, and A. Heller, "Enzyme-Amplified Amperometric Detection of 3000 Copies of DNA in a 10-μL Droplet at 0.5 fM Concentration," *Anal. Chem.* **2003**, *75*, 3267; A. S. Mittelmann, E. Z. Ron, and J. Rishpon, "Amperometric Quantification of Total Coliforms and Specific Detection of *E. coli*," *Anal. Chem.* **2002**, *74*, 903; A. Avramescu, S. Andreescu, T. Noguer, C. Bala, D. Andreescu, and J.-L. Marty, "Biosensors Designed for Environmental and Food Quality Control," *Anal. Bioanal. Chem.* **2002**, *374*, 25; T. M. O'Regan, L. J. O'Riordan, M. Pravda, C. K. O'Sullivan, and G. G. Guilbault, "Direct Detection of Myoglobin in Whole Blood Using a Disposable Amperometric Immunosensor," *Anal. Chim. Acta* **2002**, *460*, 141.

14. B. C. Okeke, G. Ma, Q. Cheng, M. E. Losi, and W. T. Frankenberger, Jr., "Development of Perchlorate Reductase Based Biosensor for Real Time Analysis of Perchlorate in Water," *J. Microbiol. Meth.* **2007**, *68*, 69.

15. E. Liaudet, S. Hatz, M. Droniou, and N. Dale, "Microelectrode Biosensor for Real-Time Measurement of ATP in Biological Tissue," *Anal. Chem.* **2005**, *77*, 3267.

16. O. A. Loaiza, S. Campuzano, M. Pedrero, M. I. Pividori, P. García, and J. M. Pingarrón, "Disposable Magnetic DNA Sensors for the Determination at the Attomolar Level of a Specific *Enterobacteriaceae* Family Gene," *Anal. Chem.* **2008**, *80*, 8239.

17. Y. Zhang and A. Heller, "Reduction of the Nonspecific Binding of a Target Antibody and of Its Enzyme-Labeled Detection Probe Enabling Electrochemical Immunoassay of an Antibody Through the 7 pg/mL–100 ng/mL (40 fM–400 pM) Range," *Anal. Chem.* **2005**, *77*, 7758.

18. N. Mano and A. Heller, "Detection of Glucose at 2 fM Concentration," *Anal. Chem.* **2005**, *77*, 729.

19. A. Heller and B. Feldman, "Electrochemical Glucose Sensors and Their Applications in Diabetes Management," *Chem. Rev.* **2008**, *108*, 2482.

20. J. Nikolic, E. Expósito, J. Iniesta, J. González-Garcia, and V. Montiel, "Theoretical Concepts and Applications of a Rotating Disk Electrode," *J. Chem. Ed.* **2000**, *77*, 1191.

21. J. Lagrange and P. Lagrange, "Voltammetric Method for the Determination of H_2O_2 in Rainwater," *Fresenius J. Anal. Chem.* **1991**, *339*, 452.

22. A. J. Bard, "The Rise of Voltammetry: From Polarography to the Scanning Electrochemical Microscope," *J. Chem. Ed.* **2007**, *84*, 644; A. J. Bard and C. G. Zoski, "Voltammetry Retrospective," *Anal. Chem.* **2000**, *72*, 346A; A. M. Bond, *Broadening Electrochemical Horizons* (Oxford: Oxford University Press, 2002).

23. To clean a Hg spill, consolidate the droplets with a piece of cardboard. Use an evacuated filter flask to suck up Hg with a Pasteur pipet attached to a hose. To remove residual Hg, sprinkle elemental zinc powder on the surface and dampen the powder with 5% aqueous H_2SO_4 to make a paste. Mercury dissolves in the zinc. After working the paste into contaminated areas with a sponge or brush, allow the paste to dry and sweep it up. Discard the powder as contaminated Hg waste. This procedure is better than sprinkling sulfur on the spill. Sulfur coats Hg but does not react with the bulk of the droplet [D. N. Easton, "Management and Control of Hg Exposure," *Am. Lab.*, July 1988, p. 66].

24. Low background current in diamond is attributed to (i) the absence of redox-active surface groups and (ii) low capacitance due to the absence of ionizable surface groups and low internal charge carrier concentration. (A. E. Fischer, Y. Show, and G. M. Swain, "Electrochemical Performance of Diamond Thin-Film Electrodes from Different Commercial Sources," *Anal. Chem.* **2004**, *76*, 2553.)

25. To remove traces of O_2 from N_2, bubble the gas through two consecutive columns of liquid. The first column removes O_2 by reaction with V^{2+}, and the second saturates the gas stream with water at the same vapor pressure as that in the voltammetry cell. Fill the second column with the same supporting electrolyte solution used for voltammetry. Prepare the first column by boiling 2 g NH_4VO_3 (ammonium metavanadate) with 25 mL 12 M HCl and reduce VO_3^- to V^{2+} with zinc amalgam. (Amalgam is prepared by covering granulated Zn with 2 wt% $HgCl_2$ solution and stirring for 10 min to reduce Hg^{2+} to Hg, which coats the Zn. The liquid is decanted, and the amalgam is washed three times with water by decantation. Amalgamation increases the overpotential for H^+ reduction at the Zn surface, so the Zn is not wasted by reaction with acid.) The blue or green oxidized vanadium solution turns violet upon reduction. When the violet color is exhausted through use, it can be regenerated by adding more zinc amalgam or HCl or both. Two V^{2+} bubble tubes can be used in series (in addition to a third tube with supporting electrolyte). When V^{2+} in the first tube is expended, the second tube is still effective.

26. J. G. Osteryoung and R. A. Osteryoung, "Square Wave Voltammetry," *Anal. Chem.* **1985**, *57*, 101A; J. G. Osteryoung, "Voltammetry for the Future," *Acc. Chem. Res.* **1993**, *26*, 77.

27. P. Zanello, *Inorganic Electrochemistry: Theory, Practice and Application* (Cambridge: Royal Society of Chemistry, 2003); G. A. Mabbott, "An Introduction to Cyclic Voltammetry," *J. Chem. Ed.* **1983**, *60*, 697; P. T. Kissinger and W. R. Heineman, "Cyclic Voltammetry," *J. Chem. Ed.* **1983**, *60*, 702; D. H. Evans, K. M. O'Connell, R. A. Petersen, and M. J. Kelly, "Cyclic Voltammetry," *J. Chem. Ed.* **1983**, *60*, 290; H. H. Thorp, "Electrochemistry of Proton-Coupled Redox Reactions," *J. Chem. Ed.* **1992**, *69*, 251.

28. J. J. Watkins, B. Zhang, and H. S. White, "Electrochemistry at Nanometer-Scaled Electrodes," *J. Chem. Ed.* **2005**, *82*, 713; R. J. Forster, "Microelectrodes: New Dimensions in Electrochemistry," *Chem. Soc. Rev.* **1994**, 289; S. Ching, R. Dudek, and E. Tabet, "Cyclic Voltammetry with Ultramicroelectrodes," *J. Chem. Ed.* **1994**, *71*, 602; E. Howard and J. Cassidy, "Analysis with Microelectrodes Using Microsoft Excel Solver," *J. Chem. Ed.* **2000**, *77*, 409.

29. T. K. Chen, Y. Y. Lau, D. K. Y. Wong, and A. G. Ewing, "Pulse Voltammetry in Single Cells Using Platinum Microelectrodes," *Anal. Chem.* **1992**, *64*, 1264.

30. A. J. Cunningham and J. B. Justice, Jr., "Approaches to Voltammetric and Chromatographic Monitoring of Neurochemicals in Vivo," *J. Chem. Ed.* **1987**, *64*, A34. Another Nafion-coated electrode can detect 10^{-20} mol of the neurotransmitter nitric oxide within a single cell (T. Malinski and Z. Taha, "Nitric Oxide Release from a Single Cell Measured *in Situ* by a Porphyrinic-Based Microsensor," *Nature* **1992**, *358*, 676).

31. S. K. MacLeod, "Moisture Determination Using Karl Fischer Titrations," *Anal. Chem.* **1991**, *63*, 557A.

32. S. Grünke and G. Wünsch, "Kinetics and Stoichiometry in the Karl Fischer Solution," *Fresenius J. Anal. Chem.* **2000**, *368*, 139.

33. A. Cedergren and S. Jonsson, "Progress in Karl Fischer Coulometry Using Diaphragm-Free Cells," *Anal. Chem.* **2001**, *73*, 5611.

34. S. A. Margolis and J. B. Angelo, "Interlaboratory Assessment of Measurement Precision and Bias in the Coulometric Karl Fischer Determination of Water," *Anal. Bioanal. Chem.* **2002**, *374*, 505.

35. C. M. Sánchez-Sánchez, E. Expósito, A. Frías-Ferrer, J. González-García, V. Montiel, and A. Aldaz, "Chlor-Alkali Industry: A Laboratory Scale Approach," *J. Chem. Ed.* **2004**, *81*, 698; D. J. Wink, "The Conversion of Chemical Energy," *J. Chem. Ed.* **1992**, *69*, 108; S. Venkatesh and B. V. Tilak, "Chlor-Alkali Technology," *J. Chem. Ed.* **1983**, *60*, 276. Nafion is a trademark of DuPont Co.

36. R. S. Treptow, "The Lead-Acid Battery: Its Voltage in Theory and Practice," *J. Chem. Ed.* **2002**, *79*, 334. Includes *activity coefficients* of electrolyte in the battery.

37. Y. Chyan and O. Chyan, "Metal Electrodeposition on an Integrated, Screen-Printed Electrode Assembly," *J. Chem. Ed.* **2008**, *85*, 565.

38. D. Lowinsohn and M. Bertotti, "Coulometric Titrations in Wine Samples: Determination of S(IV) and the Formation of Adducts," *J. Chem. Ed.* **2002**, *79*, 103. Some species in wine in addition to sulfite react with I_3^-. A blank titration to correct for such reactions is described in this article.

39. M. E. Gomez and A. E. Kaifer, "Voltammetric Behavior of a Ferrocene Derivative," *J. Chem. Ed.* **1992**, *69*, 502.

Chapter 17

1. R. S. Stolarski, "The Antarctic Ozone Hole," *Scientific American,* January 1988. The 1995 Nobel Prize in Chemistry was shared by Paul Crutzen, Mario Molina, and F. Sherwood Rowland for "their work in atmospheric chemistry, particularly concerning the formation and decomposition of ozone." Their Nobel lectures can be found in P. J. Crutzen, "My Life With O_3, NO_x, and Other YZO_x Compounds," *Angew. Chem. Int. Ed. Engl.* **1996,** *35*, 1759;

M. J. Molina, "Polar Ozone Depletion," ibid., 1779; F. S. Rowland, "Stratospheric Ozone Depletion by Chlorofluorocarbons," ibid., 1787.

2. J. H. Butler, M. Battle, M. L. Bender, S. A. Montzka, A. D. Clarke, E. S. Saltzman, C. M. Sucher, J. P. Severinghaus, and J. W. Elkins, "A Record of Atmospheric Halocarbons During the Twentieth Century from Polar Firn Air," *Nature* **1999**, *399*, 749.

3. O. B. Toon and R. P. Turco, "Polar Stratospheric Clouds and Ozone Depletion," *Scientific American,* June 1991; A. J. Prenni and M. A. Tolbert, "Studies of Polar Stratospheric Cloud Formation," *Acc. Chem. Res.* **2001**, *34*, 545.

4. M. Dubois, K. A. Gilles, J. K. Hamilton, P. A. Rebers, and F. Smith, "Colorimetric Method for Determination of Sugars and Related Substances," *Anal. Chem.* **1956**, *28*, 350. (J. Riordon, E. Zubritsky, and A. Newman, "Top 10 Articles," *Anal. Chem.* **2000**, *72*, 324A.)

5. Classroom exercise to "derive" Beer's law: R. W. Ricci, M. A. Ditzler, and L. P. Nestor, "Discovering the Beer-Lambert Law," *J. Chem. Ed.* **1994,** *71*, 983. An alternate derivation: W. D. Bare, "A More Pedagogically Sound Treatment of Beer's Law: A Derivation Based on a Corpuscular-Probability Model," *J. Chem. Ed.* **2000**, *77*, 929.

6. D. R. Malinin and J. H. Yoe, "Development of the Laws of Colorimetry: A Historical Sketch," *J. Chem. Ed.*, **1961**, *38*, 129. The equation that we call "Beer's Law" embodies contributions by P. Bouguer (1698–1758), J. H. Lambert (1728–1777), and A. Beer (1825–1863). Beer published his work in 1852, and similar conclusions were independently reached and published within a few months by F. Bernard.

7. D. H. Alman and F. W. Billmeyer, Jr., "A Simple System for Demonstrations in Spectroscopy," *J. Chem. Ed.* **1976**, *53*, 166. For another approach, see F. H. Juergens, "Spectroscopy in Large Lecture Halls," *J. Chem. Ed.* **1988**, *65*, 266.

8. B. K. Niece, "Simultaneous Display of Spectral Images and Graphs Using a Web Camera and Fiber-Optic Spectrophotometer," *J. Chem. Ed.* **2006**, *83*, 761.

9. D. J. Williams, T. J. Flaherty, C. L. Jupe, S. A. Coleman, K. A. Marquez, and J. H. Stanton, "Beyond λ_{max}: Transforming Visible Spectra into 24-Bit Color Values," *J. Chem. Ed.* **2007**, *84*, 1873.

10. D. J. Williams, T. J. Flaherty, and B. K. Alnasleh, "Beyond λ_{max} Part 2: Predicting Molecular Color," *J. Chem. Ed.* **2009**, *86*, 333.

11. For Beer's law, "monochromatic" means that the bandwidth of the light must be substantially smaller than the width of the absorption band in the spectrum of the chromophore (W. E. Wentworth, "Dependence of the Beer-Lambert Absorption Law on Monochromatic Radiation," *J. Chem. Ed.* **1966**, *43*, 262).

12. D. C. Harris, "Serum Iron Determination: A Sensitive Colorimetric Experiment," *J. Chem. Ed.* **1978**, *55*, 539.

13. A pictorial description of the dynamics of the $n \rightarrow \pi^*$ and $\pi \rightarrow \pi^*$ transitions is given by G. Henderson, "A New Look at Carbonyl Electronic Transitions," *J. Chem. Ed.* **1990**, *67*, 392.

14. R. B. Weinberg, "An Iodine Fluorescence Quenching Clock Reaction," *J. Chem. Ed.* **2007**, *84*, 797; R. B. Weinberg, "How Does Your Laundry Glow?" *J. Chem. Ed.* **2007**, *84*, 800A; J. P. Blitz, D. J. Sheeran, and T. L. Becker, "Classroom Demonstrations of Concepts in Molecular Fluorescence," *J. Chem. Ed.* **2006**, *83*, 758; J. W. Bozzelli, "A Fluorescence Lecture Demonstration," *J. Chem. Ed.* **1982**, *59*, 787; G. L. Goe, "A Phosphorescence Demonstration," *J. Chem. Ed.* **1972**, *49*, 412; E. M. Schulman, "Room Temperature Phosphorescence," *J. Chem. Ed.* **1976**, *53*, 522; F. B. Bramwell and M. L. Spinner, "Phosphorescence: A Demonstration," *J. Chem. Ed.* **1977**, *54*, 167; S. Roalstad, C. Rue, C. B. LeMaster, and C. Lasko, "A Room-Temperature Emission Lifetime Experiment for the Physical Chemistry Laboratory," *J. Chem. Ed.* **1997**, *74*, 853.

15. J.-S. Filhol, D. Zitoun, L. Bernaud, and A. Manteghetti, "Microwave Synthesis of a Long-Lasting Phosphor," *J. Chem. Ed.* **2009**, *86*, 72.

16. C. Gell, D. Brockwell, and A. Smith, *Handbook of Single Molecule Fluorescence Spectroscopy* (Oxford: Oxford University Press, 2006); C. Zander, J. Enderlein, and R. Keller, eds., *Single Molecule Detection in Solution* (New York: Wiley, 2002); R. A. Keller, W. P. Ambrose, A. A. Arias, H. Cai, S. R. Emory, P. M. Goodwin, and J. H. Jett, "Analytical Applications of Single-Molecule Detection," *Anal. Chem.* **2002**, *74*, 317A; J. Zimmermann, A. van Dorp, and A. Renn, "Fluorescence Microscopy of Single Molecules," *J. Chem. Ed.* **2004**, *81*, 553; T. A Byassee, W. C. W. Chan, and S. Nie, "Probing Single Molecules in Single Living Cells," *Anal. Chem.* **2000**, *72*, 5606.

17. B. Fanget, O. Devos, and M. Draye, "Correction of Inner Filter Effect in Mirror Coating Cells for Trace Level Fluorescence Measurements," *Anal. Chem.* **2003**, *75*, 2790.

18. Q. Gu and J. E. Kenny, "Improvement of Inner Filter Effect Correction Based on Determination of Effective Geometric Parameters Using a Conventional Fluorimeter," *Anal. Chem.* **2009**, *81*, 420.

19. J. R. Albani, *Principles and Applications of Fluorescence Spectroscopy* (Oxford: Blackwell, 2007).

20. D. S. Chatellier and H. B. White, III, "What Color Is Egg White? A Biochemical Demonstration of the Formation of a Vitamin-Protein Complex Using Fluorescence Quenching," *J. Chem. Ed.* **1988**, *65*, 814.

21. E. R. Menzel, "Detection of Latent Fingerprints by Laser-Excited Luminescence," *Anal. Chem.* **1989**, *61*, 557A.

22. S. O. Obare and C. J. Murphy, "A Two-Color Fluorescent Lithium Ion Sensor," *Inorg. Chem.* **2001**, *40*, 6080; L. Fabbrizzi, N. Marcotte, F. Stomeo, and A. Taglietti, "Pyrophosphate Detection in Water by Fluorescence Competition Assays," *Angew. Chem. Int. Ed.* **2002**, *41*, 3811.

23. J. B. Rampal, ed., *DNA Arrays: Methods and Protocols* (Totowa, NJ: Humana Press, 2001); D. Gerion, F. Chen, B. Kannan, A. Fu, W. J. Parak, D. J. Chen, A. Majumdar, and A. P. Alivisatos, "Room-Temperature Single-Nucleotide Polymorphism and Multiallele DNA Detection Using Fluorescent Nanocrystals and Microarrays," *Anal. Chem.* **2003**, *75*, 4766.

24. T. S. Kuntzleman, A. E. Comfort, and B. W. Baldwin, "Glowmatography," *J. Chem. Ed.* **2009**, *86*, 65; C. Salter, K. Range, and G. Salter, "Laser-Induced Fluorescence of Lightsticks," *J. Chem. Ed.* **1999**, *76*, 84; E. Wilson, "Light Sticks," *Chem. Eng. News*, 18 January 1999, p. 65.

25. A. M. Garcia-Campana and W. R. G. Baeyens, eds., *Chemiluminescence in Analytical Chemistry* (New York: Marcel Dekker, 2001); L. J. Kricka, "Clinical Applications of Chemiluminescence," *Anal. Chim. Acta* **2003**, *500*, 279.

26. J. K. Robinson, M. J. Bollinger, and J. W. Birks, "Luminol/H_2O_2 Chemiluminescence Detector for the Analysis of NO in Exhaled Breath," *Anal. Chem.* **1999**, *71*, 5131. Many substances can be analyzed by coupling their chemistry to luminol oxidation. See, for example, O. V. Zui and J. W. Birks, "Trace Analysis of Phosphorus in Water by Sorption Preconcentration and Luminol Chemiluminescence," *Anal. Chem.* **2000**, *72*, 1699.

27. R. Rizzuto, A. W. M. Simpson, M. Brini, and T. Pozzan, "Rapid Changes of Mitochondrial Ca^{2+} Revealed by Specifically Targeted Recombinant Aequorin," *Nature* **1992**, *358*, 325; A. Toda, P. Pasini, M. Guardigli, M. Baraldini, M. Musiani, and M. Mirasoli, "Bio- and Chemiluminescence in Bioanalysis," *Fresenius J. Anal. Chem.* **2000**, *366*, 752.

28. A. Roda, M. Marasoli, E. Michelini, M. Magliulo, P. Simoni, M. Guardigli, R. Curini, M. Sergi, and A. Marino, "Analytical Approach for Monitoring Endocrine-Disrupting Compounds in Urban Waste Water Treatment Plants," *Anal. Bioanal. Chem.* **2006**, *385*, 742.

Chapter 18

1. F. S. Ligler and C. A. R. Taitt, eds., *Optical Biosensors: Present and Future* (Saint Louis: Elsevier, 2002); F. S. Ligler, "Perspective on Optical Biosensors and Integrated Sensor Systems," *Anal. Chem.* **2009**, *81*, 519.

2. L. Stryer, "Fluorescence Energy Transfer as a Spectroscopic Ruler," *Annu. Rev. Biochem.* **1978**, *47*, 819; C. Berney and G. Danuser, "FRET or No FRET: A Quantitative Comparison," *Biophys. J.* **2003**, *84*, 3992; http://www.probes.com/handbook/

3. K. M. Sanchez, D. E. Schlamadinger, J. E. Gable, and J. E. Kim, "Förster Resonance Energy Transfer and Conformational Stability of Proteins," *J. Chem. Ed.* **2008**, *85*, 1253.

4. X. Liu, Z. A. Wang, R. H. Byrne, E. A. Kaltenbacher, and R. E. Bernstein, "Spectrophotometric Measurements of pH in Situ: Laboratory and Field Evaluations of Instrumental Performance," *Environ. Sci. Tech.* **2006**, *40*, 5036; T. R. Martz, J. J. Carr, C. R. French, and M. D. DeGrandpre, "A Submersible Autonomous Sensor for Spectrophotometric pH Measurements of Natural Waters," *Anal. Chem.* **2003**, *75*, 1844; W. Yao and R. H. Byrne, "Spectrophotometric Determination of Freshwater pH Using Bromocresol Purple and Phenol Red," *Environ. Sci. Technol.* **2001**, *35*, 1197; H. Yamazaki, R. P. Sperline, and H. Freiser, "Spectrophotometric Determination of pH and Its

Applications to Determination of Thermodynamic Equilibrium Constants," *Anal. Chem.* **1992**, *64*, 2720.

5. To multiply a matrix times a vector, multiply each row of the matrix times each element of the vector as follows:

$$\begin{bmatrix} a_1 & a_2 \\ b_1 & b_2 \end{bmatrix} \begin{bmatrix} X \\ Y \end{bmatrix} = \begin{bmatrix} a_1X + a_2Y \\ b_1X + b_2Y \end{bmatrix}$$

The product is a vector. The product of a matrix times a matrix is another matrix obtained by multiplying rows times columns:

$$\begin{bmatrix} a_1 & a_2 \\ b_1 & b_2 \end{bmatrix} \begin{bmatrix} c_1 & c_2 \\ d_1 & d_2 \end{bmatrix} = \begin{bmatrix} row\ 1 \times column\ 1 & row\ 1 \times column\ 2 \\ row\ 2 \times column\ 1 & row\ 2 \times column\ 2 \end{bmatrix}$$

$$= \begin{bmatrix} a_1c_1 + a_2d_1 & a_1c_2 + a_2d_2 \\ b_1c_1 + b_2d_1 & b_1c_2 + b_2d_2 \end{bmatrix}$$

The matrix **B** below is the inverse of **A** because their product is the unit matrix:

$$\underset{\mathbf{A}}{\begin{bmatrix} 1 & 2 \\ 3 & 4 \end{bmatrix}} \underset{\mathbf{B}}{\begin{bmatrix} -2 & 1 \\ \frac{3}{2} & -\frac{1}{2} \end{bmatrix}} = \begin{bmatrix} 1 \cdot -2 + 2 \cdot \frac{3}{2} & 1 \cdot 1 + 2 \cdot -\frac{1}{2} \\ 3 \cdot -2 + 4 \cdot \frac{3}{2} & 3 \cdot 1 + 4 \cdot -\frac{1}{2} \end{bmatrix} = \underset{\text{Unit matrix}}{\begin{bmatrix} 1 & 0 \\ 0 & 1 \end{bmatrix}}$$

6. Under certain conditions, it is possible for a solution with more than two principal species to exhibit an isosbestic point. See D. V. Stynes, "Misinterpretation of Isosbestic Points: Ambident Properties of Imidazole," *Inorg. Chem.* **1975**, *14*, 453.

7. G. Scatchard, *Ann. N. Y. Acad. Sci.* **1949**, *51*, 660.

8. D. A. Deranleau, "Theory of the Measurement of Weak Molecular Complexes," *J. Am. Chem. Soc.* **1969**, *91*, 4044.

9. J. Job, "Formation and Stability of Inorganic Complexes in Solution," *Ann. Chim.* **1928**, *9*, 113.

10. E. Bruneau, D. Lavabre, G. Levy, and J. C. Micheau, "Quantitative Analysis of Continuous-Variation Plots with a Comparison of Several Methods," *J. Chem. Ed.* **1992**, *69*, 833; V. M. S. Gil and N. C. Oliveira, "On the Use of the Method of Continuous Variation," *J. Chem. Ed.* **1990**, *67*, 473; Z. D. Hill and P. MacCarthy, "Novel Approach to Job's Method," *J. Chem. Ed.* **1986**, *63*, 162.

11. J. Ruzicka, *Flow Injection Analysis*, 4th ed., 2009. (Tutorial available free of charge from www.flowinjection.com.)

12. J. Ruzicka and E. H. Hansen, *Flow Injection Analysis*, 2nd ed. (New York: Wiley, 1988); J. Ruzicka and E. H. Hansen, "Flow Injection Analysis: From Beaker to Microfluidics," *Anal. Chem.* **2000**, *72*, 212A; J. Ruzicka and L. Scampavia, "From Flow Injection to Bead Injection," *Anal. Chem.* **1999**, *71*, 257A.

13. M. Valcarcel and M. D. Luque de Castro, *Flow Injection Analysis: Principles and Applications* (Chichester: Ellis Horwood, 1987).

14. M. Trojanowicz, *Flow Injection Analysis: Instrumentation and Applications* (Singapore: World Scientific, 2000).

15. S. Kolev and Ian McKelvie, eds., *Advances in Flow Injection Analysis and Related Techniques* (New York: Elsevier Science, 2008).

16. J. Shah, M. Rasul Jan, and N. Bashir, "Flow Injection Spectrophotometric Determination of Acetochlor in Food Samples," *Am. Lab. News Ed.*, March 2008, p. 12.

17. C. E. Lenehan, N. W. Barnett, and S. W. Lewis, "Sequential Injection Analysis," *Analyst* **2002**, *127*, 997.

18. R. S. Yalow, "Development and Proliferation of Radioimmunoassay Technology," *J. Chem. Ed.* **1999**, *76*, 767; R. P. Ekins, "Immunoassay, DNA Analysis, and Other Ligand Binding Assay Techniques," *J. Chem. Ed.* **1999**, *76*, 769; E. F. Ullman, "Homogeneous Immunoassays," *J. Chem. Ed.* **1999**, *76*, 781; E. Straus, "Radioimmunoassay of Gastrointestinal Hormones," *J. Chem. Ed.* **1999**, *76*, 788.

19. S. J. Gee, B. D. Hammock, and J. M. Van Emon, *Environmental Immunochemical Analysis for Detection of Pesticides and Other Chemicals: A User's Guide* (Westwood, NJ: Noyes, 1997); R. Petkewich, "Sleuthing out Contamination," *Chem. Eng. News,* 1 September 2008, p. 49; E. M. Brun, E. Bonet, R. Puchades, and A. Maquieira, "Selective Enzyme-Linked Immunosorbent Assay for Triclosan. Application to Wastewater Treatment Plant Effluents," *Environ. Sci. Technol.* **2008**, *42*, 1665; M. Farré, R. Brix, M. Kuster, F. Rubio, Y. Goda, M. J. López de Alda, and D. Barceló, "Evaluation of Commercial Immunoassays for the Detection of Estrogens in Water by Comparison with High-Performance Liquid Chromatography

Tandem Mass Spectrometry HPLC-MS/MS (QqQ)," *Anal. Bioanal. Chem.* **2006**, *385*, 1001.

20. E. P. Diamandis and T. K. Christopoulos, "Europium Chelate Labels in Time-Resolved Fluorescence Immunoassays and DNA Hybridization Assays," *Anal. Chem.* **1990**, *62*, 1149A.

21. M. Mascini, *Aptamers in Bioanalysis* (Hoboken, NJ: Wiley, 2009); T. Mairal, V. C. Özlap, P. L. Sanchez, M. Mir, I. Katakis, and C. K. O'Sullivan, "Aptamers: Molecular Tools for Analytical Applications," *Anal. Bioanal. Chem.* **2008**, *390*, 989; G. Mayer, "The Chemical Biology of Aptamers," *Angew. Chem. Int. Ed.* **2009**, *48*, 2672.

22. R. Mukhopadhyay, "Aptamers are Ready for the Spotlight," *Anal. Chem.* **2005**, *77*, 114A; S. D. Mendonsa and M. T. Bowser, "In Vitro Selection of High-Affinity DNA Ligands for Human IgE Using Capillary Electrophoresis," *Anal. Chem.* **2004**, *76*, 5387.

23. Y. Jiang, X. Fang, and C. Bai, "Signaling Aptamer/Protein Binding by a Molecular Light Switch Complex," *Anal. Chem.* **2004**, *76*, 5230.

24. G. J. Meyer, "Efficient Light-to-Electrical Energy Conversion: Nanocrystalline TiO$_2$ Films Modified with Inorganic Sensitizers," *J. Chem. Ed.* **1997**, *74*, 652. Student experiment to construct a photocell: G. P. Smestad and M. Grätzel, "Demonstrating Electron Transfer and Nanotechnology: A Natural Dye-Sensitized Nanocrystalline Energy Converter," *J. Chem. Ed.* **1998**, *75*, 752. Solar cell kits for students are available from http://ice.chem.wisc.edu/Catalog.html

25. http://www.greatcell.com

26. C. Lee, J.-H. Yum, H. Choi, S. O. Kang, J. Ko, R. Humphry-Baker, M. Grätzel, and Md. K. Nazeeruddin, "Phenomenally High Molar Extinction Coefficient Sensitizer with 'Donor-Acceptor' Ligands for Dye-Sensitized Solar Cell Applications," *Inorg. Chem.* **2008**, *47*, 2267.

27. F. A. Settle, "Uranium to Electricity: The Chemistry of the Nuclear Fuel Cycle," *J. Chem. Ed.* **2009**, *86*, 316.

28. K. A. Kneas, W. Xu, J. N. Demas, and B. A. DeGraff, "Dramatic Demonstration of Oxygen Sensing by Luminescence Quenching," *J. Chem. Ed.* **1997**, *74*, 696.

29. Y.-E. L. Koo, Y. Cao, R. Kopelman, S. M. Koo, M. Brasuel, and M. A. Philbert, "Real-Time Measurements of Dissolved Oxygen Inside Living Cells by Organically Modified Silicate Fluorescent Nanosensors," *Anal. Chem.* **2004**, *76*, 2498.

30. R. R. Islangulov, D. V. Kozlov, and F. N. Castellano, "Low Power Upconversion Using MLCT Sensitizers," *Chem. Commun.* **2005**, 3776. For red-to-yellow and red-to-green upconversion, see T. N. Singh-Rachford, A. Haefele, R. Ziessel, and F. N. Castellano, *J. Am. Chem. Soc.* **2008**, *130*, 16164.

31. M. D. DeGrandpre, M. M. Baehr, and T. R. Hammar, "Calibration-Free Optical Chemical Sensors," *Anal. Chem.* **1999**, *71*, 1152.

32. A Poisson distribution is valid when (a) all possible outcomes are random and independent of one another, (b) the maximum possible value of n is a large number, and (c) the average value of n is a small fraction of the maximum possible value.

Chapter 19

1. K. L. Bechtel, R. N. Zare, A. A. Kachanov, S. S. Sanders, and B. A. Paldus, "Cavity Ring-Down Spectroscopy for HPLC," *Anal. Chem.* **2005**, *77*, 1177; B. Bahnev, L. van der Sneppen, A. E. Wiskerke, F. Ariese, C. Gooijer, and W. Ubachs, "Miniaturized Cavity Ring-Down Detection in a Liquid Flow Cell," *Anal. Chem.* **2005**, *77*, 1188.

2. www.picarro.com

3. A. Bergh, G. Craford, A. Duggal, and R. Haitz, "The Promise and Challenge of Solid-State Lighting," *Physics Today,* December 2001, p. 42; H. Sevian, S. Müller, H. Rudmann, and M. F. Rubner, "Using Organic Light-Emitting Electrochemical Thin-Film Devices to Teach Materials Science," *J. Chem. Ed.* **2004**, *81*, 1620.

4. J. M. Kauffman, "Water in the Atmosphere," *J. Chem. Ed.* **2004**, *81*, 1229.

5. S. K. Lower, "Thermal Physics (and Some Chemistry) of the Atmosphere," *J. Chem. Ed.* **1998**, *75*, 837; W. C. Trogler, "Environmental Chemistry of Trace Atmospheric Gases," *J. Chem. Ed.* **1995**, *72*, 973.

6. S. Solomon, D. Qin, M. Manning, Z. Chen, M. Marquis, K. B. Averyt, M. Tignor, and H. L. Miller, eds., "IPCC, 2007: Summary for Policymakers," in

Climate Change 2007: The Physical Science Basis. Contribution of Working Group I to the Fourth Assessment Report of the Intergovernmental Panel on Climate Change (Cambridge: Cambridge University Press, 2007).

7. M. G. D. Baumann, J. C. Wright, A. B. Ellis, T. Kuech, and G. C. Lisensky, "Diode Lasers," *J. Chem. Ed.* **1992**, *69*, 89; T. Imasaka and N. Ishibashi, "Diode Lasers and Practical Trace Analysis," *Anal. Chem.* **1990**, *62*, 363A.

8. S. Nakamura, S. Pearton, and G. Fasol, *The Blue Laser Diode: The Complete Story* (Heidelberg: Springer, 2000).

9. W. E. L. Grossman, "The Optical Characteristics and Production of Diffraction Gratings," *J. Chem. Ed.* **1993**, *70*, 741.

10. G. C.-Y. Chan and W. T. Chan, "Beer's Law Measurements Using Non-Monochromatic Light Sources — A Computer Simulation," *J. Chem. Ed.* **2001**, *78*, 1285.

11. J. C. Travis et al., "Intrinsic Wavelength Standard Absorption Bands in Holmium Oxide Solution for UV/Visible Molecular Absorption Spectrophotometry," *J. Phys. Chem. Ref. Data* **2005**, *34*, 41.

12. www.oceanoptics.com. Look at spectrometers and then standards.

13. K. Palczewski et al., "Crystal Structure of Rhodopsin," *Science* **2000**, *289*, 739.

14. S. C. Denson, C. J. S. Pommier, and M. B. Denton, "The Impact of Array Detectors on Raman Spectroscopy," *J. Chem. Ed.* **2007**, *84*, 67; J. M. Harnly and R. E. Fields, "Solid-State Array Detectors for Analytical Spectrometry," *Appl. Spectros.* **1997**, *51*, 334A; Q. S. Hanley, C. W. Earle, F. M. Pennebaker, S. P. Madden, and M. B. Denton, "Charge-Transfer Devices in Analytical Instrumentation," *Anal. Chem.* **1996**, *68*, 661A; J. V. Sweedler, K. L. Ratzlaff, and M. B. Denton, eds., *Charge Transfer Devices in Spectroscopy* (New York: VCH, 1994).

15. J. C. Pales and C. D. Keeling, "The Concentration of Atmospheric Carbon Dioxide in Hawaii," *J. Geophys. Res.* **1965**, *70*, 6053; V. N. Smith, "A Recording Infrared Analyzer," *Instruments* **1953**, *9*, 421.

16. U. Resch-Genger, D. Pfeifer, C. Monte, W. Pilz, A. Hoffmann, M. Spieles, K. Rurack, J. Hollandt, D. Taubert, B. Schönenberger, and P. Nording, "Traceability in Fluorometry: Part II. Spectral Fluorescence Standards," *J. Fluoresc.* **2005**, *15*, 314.

17. F. Baldini, A. N. Chester, J. Homola, and S. Martellucci, *Optical Chemical Sensors* (Dordrecht, Netherlands: Springer, 2006).

18. A. S. Kocincova, S. M. Borisov, C. Krause, and O. S. Wolfbeis, "Fiber-Optic Microsensors for Simultaneous Sensing of Oxygen and pH, and of Oxygen and Temperature," *Anal. Chem.* **2007**, *79*, 8486; D. Citterio, J. Takeda, M. Kosugi, H. Hisamoto, S.-I. Sasaki, H. Komatsu, and K. Suzuki, "pH-Independent Fluorescent Chemosensor for Highly Selective Lithium Ion Sensing," *Anal. Chem.* **2007**, *79*, 1237; W. Tan, R. Kopenman, S. L. R. Barker, and M. T. Miller, "Ultrasmall Optical Sensors for Cellular Measurements," *Anal. Chem.* **1999**, *71*, 606A; S. L. R. Barker, Y. Zhao, M. A. Marletta, and R. Kopelman, "Cellular Applications of a Fiber-Optic Biosensor Based on a Dye-Labeled Guanylate Cyclase," *Anal. Chem.* **1999**, *71*, 2071; M. Kuratli and E. Pretsch, "SO_2-Selective Optodes," *Anal. Chem.* **1994**, *66*, 85.

19. Z. Rosenzweig and R. Kopelman, "Analytical Properties and Sensor Size Effects of a Micrometer-Sized Optical Fiber Glucose Biosensor," *Anal. Chem.* **1996**, *68*, 1408.

20. C. Preininger, I. Klimant, and O. S. Wolfbeis, "Optical Fiber Sensor for Biological Oxygen Demand," *Anal. Chem.* **1994**, *66*, 1841.

21. J. T. Bradshaw, S. B. Mendes, and S. S. Saavedra, "Planar Integrated Optical Waveguide Spectroscopy," *Anal. Chem.* **2005**, *77*, 28A.

22. J. Homola, ed., *Surface Plasmon Resonance Based Sensors* (New York: Springer, 2006); D. J. Campbell and Y. Xia, "Plasmons: Why Should We Care?" *J. Chem. Ed.* **2007**, *84*, 91; J. Homola, "Present and Future of Surface Plasmon Resonance Biosensors," *Anal. Bioanal. Chem.* **2003**, *377*, 528.

23. P. R. Griffiths and J. A. de Haseth, *Fourier Transform Infrared Spectrometry*, 2nd ed. (Hoboken, NJ: Wiley, 2007) and W. D. Perkins, "Fourier Transform-Infrared Spectrometry," *J. Chem. Ed.* **1986**, *63*, A5; ibid. **1987**, *64*, A269, A296.

24. Digital and electronic techniques for improving signal-to-noise ratio: T. C. O'Haver, "An Introduction to Signal Processing in Chemical Measurement,"

J. Chem. Ed. **1991**, *68*, A147; M. G. Prais, "Spreadsheet Exercises for Instrumental Analysis," *J. Chem. Ed.* **1992**, *69*, 488; R. Q. Thompson, "Experiments in Software Data Handling," *J. Chem. Ed.* **1985**, *62*, 866; B. H. Vassos and L. López, "Signal-to-Noise Improvement," *J. Chem. Ed.* **1985**, *62*, 542; K. Overway, "FT Digital Filtering: Simulating Fourier Transform Apodization via Excel," *J. Chem. Ed.* **2008**, *85*, 1151; M. P. Eastman, G. Kostal, and T. Mayhew, "An Introduction to Fast Fourier Transforms Through the Study of Oscillating Reactions," *J. Chem. Ed.* **1986**, *63*, 453.

25. G. M. Hieftje, "Signal-to-Noise Enhancement Through Instrumental Techniques," *Anal. Chem.* **1972**, *44*, 81A [No. 6], 69A [No. 7]; D. C. Tardy, "Signal Averaging," *J. Chem. Ed.* **1986**, *63*, 648; see also T. Kaneta, "Hadamard Transform CE," *Anal. Chem.* **2001**, *73*, 540A.

26. N. N. Sesi, M. W. Borer, T. K. Starn, and G. M. Hieftje, "A Standard Approach to Collecting and Calculating Noise Amplitude Spectra," *J. Chem. Ed.* **1998**, *75*, 788.

27. M. L. Salit and G. C. Turk, "A Drift Correction Procedure," *Anal. Chem.* **1998**, *70*, 3184.

28. L. D. Rothman, S. R. Crouch, and J. D. Ingle, Jr., "Theoretical and Experimental Investigation of Factors Affecting Precision in Molecular Absorption Spectrophotometry," *Anal. Chem.* **1975**, *47*, 1226.

29. Let voltages V_{dif}^0 and V_{ref}^0 be recorded with pure solvent in both cells. Let voltages V_{dif} and V_{ref} be recorded with sample in one cell and solvent in the other cell. Absorbance is related to these voltages by $A = -\log[(1 + V_{dif}/V_{ref})/(1 + V_{dif}^0/V_{ref}^0)]$.

Chapter 20

1. E. R. Denoyer, K. J. Fredeen, and J. W. Hager, "Laser Solid Sampling for Inductively Coupled Plasma Mass Spectrometry," *Anal. Chem.* **1991**, *63*, 445A; K. Niemax, "Laser Ablation—Reflections on a Very Complex Technique for Solid Sampling," *Fresenius J. Anal. Chem.* **2001**, *370*, 332. A major challenge for laser ablation–mass spectrometry is quantitative analysis. One scheme achieves semiquantitative analysis without standards by comparing the signal from each element with the total mass spectrometric signal: A. M. Leach and G. M. Hieftje, "Standardless Semiquantitative Analysis of Metals Using Single-Shot Laser Ablation Inductively Coupled Plasma Time-of-Flight Mass Spectrometry," *Anal. Chem.* **2001**, *73*, 2959.

2. L. H. J. Lajunen and P. Perämäki, *Spectrochemical Analysis by Atomic Absorption* (Cambridge: Royal Society of Chemistry, 2004); M. Cullen, *Atomic Spectroscopy in Elemental Analysis* (Oxford: Blackwell, 2003); J. R. Dean and D. J. Ando, *Atomic Absorption and Plasma Spectroscopy* (New York: Wiley, 2002); J. A. C. Broekaert, *Analytical and Atomic Spectrometry with Flames and Plasmas* (Weinheim: Wiley-VCH, 2002); L. Ebdon, E. H. Evans, A. S. Fisher, and S. J. Hill, *An Introduction to Analytical Atomic Spectrometry* (Chichester: Wiley, 1998).

3. Excellent training modules are available from www.academysavant.com

4. M. J. Holden, S. A. Rabb, Y. B. Tewari, and M. R. Winchester, "Traceable Phosphorus Measurements by ICP-OES and HPLC for the Quantitation of DNA," *Anal. Chem.* **2007**, *79*, 1536.

5. History: A. Walsh, "The Development of Atomic Absorption Methods of Elemental Analysis 1952–1962," *Anal. Chem.* **1991**, *63*, 933A; B. V. L'vov, "Graphite Furnace Atomic Absorption Spectrometry," *Anal. Chem.* **1991**, *63*, 924A.

6. D. L. Pfeil and A. Reed, "Automating the Digestion and Determination of Mercury in a Variety of Environmental Sample Matrices," *Am. Lab.*, March 2002, p. 26.

7. D Pfeil, "When U.S. EPA Method 245.7 (Determination of Hg by Atomic Fluorescence) Is Approved, Will You Be Ready?" *Am. Lab.*, August 2005, p. 26.

8. History: R. F. Jarrell, "A Brief History of Atomic Emission Spectrochemical Analysis, 1666–1950," *J. Chem. Ed.* **2000**, *77*, 573; R. F. Jarrell, F. Brech, and M. J. Gustafson, "A History of Thermo Jarrell Ash Corporation and Spectroscopist Richard F. Jarrell," *J. Chem. Ed.* **2000**, *77*, 592; G. M. Hieftje, "Atomic Emission Spectroscopy—It Lasts and Lasts and Lasts," *J. Chem. Ed.* **2000**, *77*, 577.

9. R. J. Stolzberg, "Optimizing Signal-to-Noise Ratio in Flame Atomic Absorption Using Sequential Simplex Optimization," *J. Chem. Ed.* **1999**, *76*, 834; C. R. Dockery, M. J. Blew, and S. R. Goode, "Visualizing the Solute

Vaporization Interference in Flame Atomic Absorption Spectroscopy," *J. Chem. Ed.* **2008**, *85*, 854.

10. D. J. Butcher and J. Sneddon, *A Practical Guide to Graphite Furnace Atomic Absorption Spectrometry* (New York: Wiley, 1998).

11. J. B. Voit, "Low-Level Determination of Arsenic in Drinking Water," *Am. Lab. News Ed.*, February 2002, p. 62.

12. M. Hornung and V. Krivan, "Determination of Trace Impurities in Tungsten by Direct Solid Sampling Using a Transversely Heated Graphite Tube," *Anal. Chem.* **1998**, *70*, 3444.

13. U. Schäffer and V. Krivan, "Analysis of High Purity Graphite and Silicon Carbide by Direct Solid Sampling Electrothermal Atomic Absorption Spectrometry," *Fresenius J. Anal. Chem.* **2001**, *371*, 859; R. Nowka and H. Müller, "Direct Analysis of Solid Samples by Graphite Furnace Atomic Absorption Spectrometry," *Fresenius J. Anal. Chem.* **1997**, *359*, 132; J. Štupar and F. Dolinšek, "Determination of Chromium, Manganese, Lead, and Cadmium in Biological Samples Including Hair using Direct Electrothermal Atomic Absorption Spectrometry," *Spectrochim. Acta* **1996**, *B51*, 665.

14. J. Y. Cabon, "Influence of Experimental Parameters on the Determination of Antimony in Seawater by Atomic Absorption Spectrometry," *Anal. Bioanal. Chem.* **2003**, *374*, 1282.

15. D. L. Styris and D. A. Redfield, "Mechanisms of Graphite Furnace Atomization of Aluminum by Molecular Beam Sampling Mass Spectrometry," *Anal. Chem.* **1987**, *59*, 2891.

16. S. J. Hill, *Inductively Coupled Plasma Spectrometry and Its Applications* (Chichester: Wiley, 2007); J. R. Dean, *Practical Inductively Coupled Plasma Spectroscopy* (Chichester: Wiley, 2005); S. Greenfield, "Invention of the Annular Inductively Coupled Plasma as a Spectroscopic Source," *J. Chem. Ed.* **2000**, *77*, 584.

17. V. B. E. Thomsen, G. J. Roberts, and D. A. Tsourides, "Vacuumless Spectrochemistry in the Vacuum Ultraviolet," *Am. Lab.*, August 1997, p. 18H.

18. V. B. E. Thomsen, "Why Do Spectral Lines Have a Linewidth?" *J. Chem. Ed.* **1995**, *72*, 616.

19. A better measure of noise is the root-mean-square noise (Equation 19-14), which is ~5 times less than the peak-to-peak noise. Therefore, 2 times the peak-to-peak noise level is ~10 times the root-mean-square noise. The detection limit that is 2 times the peak-to-peak noise level is close to the limit of quantitation in Equation 5-6. The lesson is that you should define how you express a detection limit when you report one.

20. Sigma-Aldrich data. M. Weber and J. Wüthrich, www.sigma-aldrich.com/tracecert

21. J.-M. Mermet, A. Cosnier, S. Vélasquez, and S. Lebouil, "Efficient Use of Spectral Information Through Multiline Analysis in ICP-AES," *Am. Lab. News Ed.*, August 2007, p. 16.

22. A diagnostic for whether emission from an Ar plasma will be sensitive to matrix effects is the ratio of emission from Mg^+ and Mg atoms (at 280.270 and 285.213 nm, respectively). When the relative intensity is above 10, the plasma is not sensitive to variations in the sample matrix. When the ratio is below 4, there is high sensitivity to matrix effects. See J.-M. Mermet, "Mg as a Test Element for ICP Atomic Emission Diagnostics," *Anal. Chim. Acta* **1991**, *250*, 85 and J.-M. Mermet and E. Poussel, "ICP Emission Spectrometers: Analytical Figures of Merit," *Appl. Spectros.* **1995**, *49*[10], 12A. Intensities of the two lines must be corrected for different grating diffraction efficiencies at the two wavelengths. (J. W. Olesik, J. A. Kinser, and B. Harkleroad, "Inductively Coupled Plasma Optical Emission Spectrometry Using Nebulizers with Widely Different Sample Consumption Rates," *Anal. Chem.* **1994**, *66*, 2022.)

23. B. Hattendorf, C. Latkoczy, and D. Günther, "Laser Ablation–ICPMS," *Anal. Chem.* **2003**, *75*, 341A; R. E. Russo, X. Mao, and S. S. Mao, "The Physics of Laser Ablation in Microchemical Analysis," *Anal. Chem.* **2002**, *74*, 71A.

24. J. S. Becker, *Inorganic Mass Spectrometry: Principles and Applications* (Chichester: Wiley, 2008); R. Thomas, *Practical Guide to ICP-MS* (New York: Marcel Dekker, 2004); H. E. Taylor, *Inductively Coupled Plasma-Mass Spectrometry* (San Diego: Academic Press, 2001); C. M. Barshick, D. C. Duckworth, and D. H. Smith, eds., *Inorganic Mass Spectrometry* (New York: Marcel Dekker, 2000); S. J. Hill, ed., *Inductively Coupled Plasma Spectrometry and Its Applications* (Sheffield, England: Sheffield Academic Press, 1999); A. Montaser, ed., *Inductively Coupled Plasma Mass Spectrometry* (New York: Wiley, 1998); G. Holland and S. D. Tanner, eds., *Plasma Source Mass Spectrometry* (Cambridge: Royal Society of Chemistry, 1997, 1999).

25. M. J. Felton, "Plasma Opens New Doors in Isotope Ratio MS," *Anal. Chem.* **2003**, *75*, 119A.

26. R. Wahlen, L. Evans, J. Turner, and R. Hearn, "The Use of Collision/Reaction Cell ICP-MS for the Determination of Elements in Blood and Serum Samples," *Spectroscopy*, December 2005, p. 84.

27. L. Moens and N. Jakubowski, "Double-Focusing Mass Spectrometers in ICPMS," *Anal. Chem.* **1998**, *70*, 251A; F. A. M. Planchon, C. F. Boutron, C. Barbante, E. W. Wolff, G. Cozzi, V. Gaspari, C. P. Ferrari, and P. Cescon, "Ultrasensitive Determination of Heavy Metals at the Sub-pg/g Level in Ultraclean Antarctic Snow Samples by Inductively Coupled Plasma Sector Field Mass Spectrometry," *Anal. Chim. Acta* **2001**, *450*, 193.

28. Y. Sohrin, S. Urushihara, S. Nakatsuka, T. Kono, E. Higo, T. Minami, K. Norisuye, and S. Umetani, "Multielemental Determination of GEOTRACES Key Trace Metals in Seawater by ICPMS after Preconcentration Using an Ethylenediaminetriacetic Acid Chelating Resin," *Anal. Chem.* **2008**, *80*, 6267; Y. Nakagawa, M. L. Firdaus, K. Norisuye, Y. Sohrin, K. Irisawa, and T. Hirata, "Precise Isotopic Analysis of Mo in Seawater Using Multiple Collector-Inductively Coupled Mass Spectrometry Coupled with a Chelating Resin Column Preconcentration Method," *Anal. Chem.* **2008**, *80*, 9213.

29. D. R. Bandura, V. I. Baranov, and S. D. Tanner, "Reaction Chemistry and Collisional Processes in Multipole Devices for Resolving Isobaric Interferences in ICP-MS," *Fresenius J. Anal. Chem.* **2001**, *370*, 454; D. R. Bandura, V. I. Baranov, and S. D. Tanner, "Inductively Coupled Plasma Mass Spectrometer with Axial Field in a Quadrupole Reaction Cell," *J. Am. Soc. Mass Spec.* **2002**, *13*, 1176.

30. L. Wittmers, Jr., A. Aufderheide, G. Rapp, and A. Alich, "Archaeological Contributions of Skeletal Lead Analysis," *Acc. Chem. Res.* **2002**, *35*, 669.

31. V. Cheam, G. Lawson, I. Lechner, and R. Desrosiers, "Recent Metal Pollution in Agassiz Ice Cap," *Environ. Sci. Technol.* **1998**, *32*, 3974.

32. A. Bazzi, B. Kreuz, and J. Fischer, "Determination of Calcium in Cereal with Flame Atomic Absorption Spectroscopy," *J. Chem. Ed.* **2004**, *81*, 1042.

33. L. Perring and M. Basic-Dvorzak, "Determination of Total Tin in Canned Food Using Inductively Coupled Plasma Atomic Emission Spectroscopy," *Anal. Bioanal. Chem.* **2002**, *374*, 235.

Chapter 21

1. W. J. Showers, R. Barrick, and B. Genna, "Isotopic Analysis of Dinosaur Bones," *Anal. Chem.* **2002**, *74*, 142A.

2. F. Klink, *Introduction to Protein and Peptide Analysis with Mass Spectrometry* (Fullerton, CA: Academy Savant, 2004), computer training Program CMSP-10.

3. W.-P. Peng, Y.-C. Yang, M.-W. Kang, Y. T. Lee, and H.-C. Chang, "Measuring Masses of Single Bacterial Whole Cells with a Quadrupole Ion Trap," *J. Am. Chem. Soc.* **2004**, *126*, 11766–11767; J. J. Jones, M. J. Stump, R. C. Fleming, J. O. Lay, Jr., and C. L. Wilkins, "Investigation of MALDI-TOF and FT-MS Techniques for Analysis of *Escherichia coli* Whole Cells," *Anal. Chem.* **2003**, *75*, 1340.

4. Z. Nie, Y.-K. Tzeng, H.-C. Chang, C.-C. Chiu, C.-Y. Chang, C.-M. Chang, and M.-H. Tao, "Microscopy-Based Mass Measurement of a Single Whole Virus in a Cylindrical Ion Trap," *Angew. Chem. Int. Ed.* **2006**, *45*, 8131.

5. J. T. Watson and O. D. Sparkman, *Introduction to Mass Spectrometry*, 4th ed. (Chichester: Wiley, 2007); E. de Hoffmann and V. Stroobant, *Mass Spectrometry: Principles and Applications*, 3rd ed. (Chichester: Wiley, 2007); C. Dass, *Fundamentals of Contemporary Mass Spectrometry* (Wenheim: Wiley-VCH, 2007); J. H. Gross, *Mass Spectrometry: A Textbook* (Berlin: Springer-Verlag, 2004); K. Doward, *Mass Spectrometry: A Foundation Course* (Cambridge; Royal Society of Chemistry, 2004); C. G. Herbert and R. A. W. Johnstone, *Mass Spectrometry Basics* (Boca Raton, FL: CRC Press, 2002); R. K. Boyd, C. Basic, and R. A. Bethem, *Trace Quantitative Analysis by Mass Spectrometry* (Chichester: Wiley, 2008); C. Dass, *Principles and Practice of Biological Mass Spectrometry* (New York: Wiley, 2001).

6. Training modules in mass spectrometry and chromatography–mass spectrometry are available from www.academysavant.com

7. For a demonstration of mass spectrometry, see N. C. Grim and J. L. Sarquis, "Mass Spectrometry Analogy on the Overhead Projector," *J. Chem. Ed.* **1995**, *72*, 930.

8. D. W. Koppenaal, C. J. Barinaga, M. B. Denton, R. P. Sperline, G. M. Hieftje, G. D. Schilling, F. J. Andrade, and J. H. Barnes, IV, "MS Detectors," *Anal. Chem.* **2005**, *77*, 419A.

9. O. D. Sparkman, "Evaluating Electron Ionization Mass Spectral Library Search Results," *J. Am. Soc. Mass Spectrom.* **1996**, *7*, 313.

10. Electron ionization mass spectra of many compounds can be viewed at http://webbook.nist.gov/chemistry. Information for ordering the NIST/EPA/NIH mass spectral library can be found at www.nist.gov/srd/nist1a.htm. Purchasing information for the *Wiley Registry of Mass Spectral Data* can be found at www.wiley.com/WileyCDA.

11. O. Asvany, P. Kumar P, B. Redlich, I. Hegemann, S. Schlemmer, and D. Marx, "Understanding the Infrared Spectrum of Bare CH_5^+," *Science* **2005**, *309*, 1219.

12. J. D. Hearn and G. D. Smith, "A Chemical Ionization Mass Spectrometry Method for the Online Analysis of Organic Aerosols," *Anal. Chem.* **2004**, *76*, 2820.

13. I. T. Platzner, *Modern Isotope Ratio Mass Spectrometry* (New York: Wiley, 1997).

14. W. Chen and M. V. Orna, "Recent Advances in Archaeological Chemistry," *J. Chem. Ed.* **1996**, *73*, 485.

15. A. M. Pollard and C. Heron, *Archaeological Chemistry*, 2nd ed. (Cambridge: Royal Society of Chemistry, 2008).

16. R. A. Gross, Jr., "A Mass Spectral Chlorine Rule for Use in Structure Determinations in Sophomore Organic Chemistry," *J. Chem. Ed.* **2004**, *81*, 1161.

17. J. T. Watson and K. Biemann, "High-Resolution Mass Spectra of Compounds Emerging from a Gas Chromatograph," *Anal. Chem.* **1964**, *36*, 1135. A classic paper on gas chromatography/mass spectrometry.

18. R. M. Smith, *Understanding Mass Spectra: A Basic Approach*, 2nd ed. (Hoboken, NJ: Wiley, 2004).

19. P. E. Miller and M. B. Denton, "The Quadrupole Mass Filter: Basic Operating Concepts," *J. Chem. Ed.* **1986**, *63*, 617; M. Henchman and C. Steel, "Design and Operation of a Portable Quadrupole Mass Spectrometer for the Undergraduate Curriculum," *J. Chem. Ed.* **1998**, *75*, 1042; C. Steel and M. Henchman, "Understanding the Quadrupole Mass Filter through Computer Simulation," *J. Chem. Ed.* **1998**, *75*, 1049; J. J. Leary and R. L. Schidt, "Quadrupole Mass Spectrometers: An Intuitive Look at the Math," *J. Chem. Ed.* **1996**, *73*, 1142.

20. R. J. Cotter, *Time-of-Flight Mass Spectrometry* (Washington, DC: American Chemical Society, 1997); R. J. Cotter, "The New Time-of-Flight Mass Spectrometry," *Anal. Chem.* **1999**, *71*, 445A.

21. Z. Ziegler, "Ion Traps Come of Age," *Anal. Chem.* **2002**, *74*, 489A; C. M. Henry, "The Incredible Shrinking Mass Spectrometers," *Anal. Chem.* **1999**, *71*, 264A; R. G. Cooks and R. E. Kaiser, Jr., "Quadrupole Ion Trap Mass Spectrometry," *Acc. Chem. Res.* **1990**, *23*, 213.

22. G. A. Eiceman and J. A. Stone, "Ion Mobility Spectrometers in National Defense," *Anal. Chem.* **2004**, *76*, 392A; H. H. Hill, W. F. Siems, R. H. St. Louis, and D. G. McMinn, "Ion Mobility Spectrometry," *Anal. Chem.* **1990**, *62*, 1201A; G. A. Eiceman and Z. Karpas, *Ion Mobility Spectrometry*, 2nd ed. (Boca Raton, FL: Taylor & Francis CRC Press, 2005).

23. C. S. Kaddis and J. A. Loo, "Native Protein MS and Ion Mobility: Large Flying Proteins with ESI," *Anal. Chem.* **2007**, *79*, 1778.

24. E. J. Davis, P. Dwivedi, M. Tam, W. F. Siems, and H. H. Hill, "High-Pressure Ion Mobility Spectrometry," *Anal. Chem.* **2009**, *81*, 3270.

25. H.-J. Hübschmann, *Handbook of GC/MS: Fundamentals and Applications* (Weinheim: Wiley-VCH, 2001); W. M. A. Niessen, ed., *Current Practice of Gas Chromatography–Mass Spectrometry* (Boca Raton, FL: CRC Press, 2001); M. Oehme, *Practical Introduction to GC-MS Analysis with Quadrupoles* (Heidelberg: Hüthig Verlag, 1998); F. G. Kitson, B. S. Larsen, and C. N. McEwen, *Gas Chromatography and Mass Spectrometry: A Practical Guide* (San Diego, CA: Academic Press, 1996).

26. B. Ardrey, *Liquid Chromatography–Mass Spectrometry: An Introduction* (Chichester, UK: Wiley, 2003); W. M. A. Niessen, *Liquid Chromatography–Mass*

Spectrometry, 3rd ed. (Boca Raton, FL: Taylor & Francis, 2006); M. S. Lee, *LC/MS Applications in Drug Development* (New York: Wiley: 2002); J. Abian, "Historical Feature: The Coupling of Gas and Liquid Chromatography with Mass Spectrometry," *J. Mass Spectrom.* **1999**, *34*, 157.

27. M. M. Vestling, "Using Mass Spectrometry for Proteins," *J. Chem. Ed.* **2002**, *80*, 122; C. M. Henry, "Winning Ways," *Chem. Eng. News*, 18 November 2002, p. 62.

28. R. B. Cole, ed., *Electrospray Ionization Mass Spectrometry: Fundamentals, Instrumentation, and Applications* (New York: Wiley, 1997).

29. G. J. van Berkel and V. Kertesz, "Using the Electrochemistry of the Electrospray Ion Source," *Anal. Chem.* **2007**, *79*, 5510; S. Liu, W. J. Griffiths, and J. Sjövall, "On-Column Electrochemical Reactions Accompanying the Electrospray Process," *Anal. Chem.* **2003**, *75*, 1022.

30. S. R. Mabbett, L. W. Zilch, J. T. Maze, J. W. Smith, and M. F. Jarrold, "Pulsed Acceleration Charge Detection Mass Spectrometry: Application to Weighing Electrosprayed Droplets," *Anal. Chem.* **2007**, *79*, 8431.

31. W. P. Duncan and P. D. Perkins, "LC-MS with Simultaneous Electrospray and Atmospheric Pressure Chemical Ionization," *Am. Lab.*, March 2005, p. 28.

32. A. R. Batt, M. S. Kostich, and J. M. Lazorchak, "Analysis of Ecologically Relevant Pharmaceuticals in Wastewater and Surface Water Using Selective Solid-Phase Extraction and UPLC-MS/MS," *Anal. Chem.* **2008**, *80*, 5021; M. Schultz and E. T. Furlong, "Trace Analysis of Antidepressant Pharmaceuticals and Their Select Degradates in Aquatic Matrixes by LC/ESI/MS/MS," *Anal. Chem.* **2008**, *80*, 1756; M. J. M. Bueno, A. Agüera, M. J. Gómez, M. D. Hernando, J. F. García-Reyes, and A. R. Fernández-Alba, "Application of Liquid Chromatography/Quadrupole-Linear Ion Trap Mass Spectrometry and Time-of-Flight Mass Spectrometry to the Determination of Pharmaceuticals and Related Contaminants in Wastewater," *Anal. Chem.* **2007**, *79*, 9372; C. Postigo, M. J. Lopez de Alda, and D. Barceló, "Fully Automated Determination in the Low Nanogram per Liter Level of Different Classes of Drugs of Abuse in Sewage Water by On-Line Solid-Phase Extraction-Liquid Chromatography–Electrospray-Tandem Mass Spectrometry," *Anal. Chem.* **2008**, *80*, 3123.

33. C. Fenselau, "MALDI MS and Strategies for Protein Analysis," *Anal. Chem.* **1997**, *69*, 661A; R. W. Nelson, D. Nedelkov, and K. A. Tubbs, "Biomolecular Interaction Analysis Mass Spectrometry," *Anal. Chem.* **2000**, *72*, 405A; J. J. Thomas, R. Bakhtiar, and G. Siuzdak, "Mass Spectrometry in Viral Proteomics," *Acc. Chem. Res.* **2000**, *33*, 179; A. P. Snyder, *Interpreting Protein Mass Spectra* (Washington, DC: American Chemical Society, 2000); S. C. Moyer and R. J. Cotter, "Atmospheric Pressure MALDI," *Anal. Chem.* **2002**, *74*, 469A.

34. S. S. Rubakhin, W. T. Greenough, and J. V. Sweedler, "Spatial Profiling with MALDI MS: Distribution of Neuropeptides Within Single Neurons," *Anal. Chem.* **2003**, *75*, 5374.

35. S. A. Hofstadler, R. Bakhtiar, and R. D. Smith, "Electrospray Ionization Mass Spectrometry: Instrumentation and Spectral Interpretation," *J. Chem. Ed.* **1996**, *73*, A82.

36. R. Bakhtiar, R. Hofstadler, and R. D. Smith, "Electrospray Ionization Mass Spectrometry: Characterization of Peptides and Proteins," *J. Chem. Ed.* **1996**, *73*, A118; C. E. C. A. Hop and R. Bakhtiar, "Electrospray Ionization Mass Spectrometry: Applications in Inorganic Chemistry and Synthetic Polymer Chemistry," *J. Chem. Ed.* **1996**, *73*, A162.

37. S. D. Fuerstenau, W. H. Benner, J. J. Thomas, C. Brugidou, B. Bothner, and G. Siuzdak, "Mass Spectrometry of an Intact Virus," *Angew. Chem. Int. Ed.* **2001**, *40*, 542.

38. Multiply charged ions can be converted into singly charged ions by passage through a corona discharge between the electrospray nozzle and the mass analyzer. With a time-of-flight spectrometer to measure high *m/z*, singly charged ions simplify protein analysis. D. D. Ebeling, M. S. Westphall, M. Scalf, and L. M. Smith, "Corona Discharge in Charge Reduction Electrospray Mass Spectrometry," *Anal. Chem.* **2000**, *72*, 5158.

39. J. J. Coon, "Collisions or Electrons? Protein Sequence Analysis in the 21st Century," *Anal. Chem.* **2009**, *81*, 3208.

40. G. C. McAlister, D. Phanstiel, D. M. Good, W. T. Berggren, and J. J. Coon, "Implementation of Electron-Transfer Dissociation on a Hybrid Linear Ion Trap–Orbitrap Mass Spectrometer," *Anal. Chem.* **2007**, *79*, 3525.

An alternative way to use the linear ion trap for electron-transfer dissociation is described by X. Liang, J. W. Hager, and S. A. McLuckey, "Transmission Mode Ion-Ion Electron-Transfer Dissociation in a Linear Ion Trap," *Anal. Chem.* **2007**, *79*, 3363.

41. S. L. Hubler, A. Jue, J. Keith, G. C. McAlister, G. Craciun, and J. J. Coon, "Valence Parity Renders *z*-Type Ions Chemically Distinct," *J. Am. Chem. Soc.* **2008**, *130*, 6388.

42. *JEOL DART Applications Notebook*, www.jeolusa.com

43. Z. Tokáts, J. M. Wiseman, B. Gologan, and R. G. Cooks, "Mass Spectrometry Sampling Under Ambient Conditions with Desorption Electrospray Ionization," *Science* **2004**, *306*, 471.

44. S. D. Maleknia and K. M. Downard, "Charge Ratio Analysis Method: Approach for the Deconvolution of Electrospray Mass Spectra," *Anal. Chem.* **2005**, *77*, 111.

45. F. He, C. L. Hendrickson, and A. G. Marshall, "Baseline Mass Resolution of Peptide Isobars: A Record for Molecular Mass Resolution," *Anal. Chem.* **2001**, *73*, 647.

46. R. Chaler, J. O. Grimalt, C. Pelejero, and E. Calvo, "Sensitivity Effects in $U_{37}^{k'}$ Paleotemperature Estimation by Chemical Ionization Mass Spectrometry," *Anal. Chem.* **2000**, *72*, 5892.

47. L. Charles and D. Pépin, "Electrospray Ion Chromatography–Tandem Mass Spectrometry of Oxyhalides at Sub-ppb Levels," *Anal. Chem.* **1998**, *70*, 353.

48. J. D. Fassett and P. J. Paulsen, "Isotope Dilution Mass Spectrometry for Accurate Elemental Analysis," *Anal. Chem.* **1989**, *61*, 643A.

Chapter 22

1. T. Michalowski, "Effect of Mutual Solubility of Solvents in Multiple Liquid-Liquid Extraction," *J. Chem. Ed.* **2002**, *79*, 1267.

2. Derivation from Joshua Erickson, a student at Utah Valley State College.

3. K. Ueno, T. Imamura, and K. L. Cheng, *CRC Handbook of Organic Analytical Reagents*, 2nd ed. (Boca Raton, FL: CRC Press, 1992), pp. 431–444.

4. R. P. Paradkar and R. R. Williams, "Micellar Colorimetric Determination of Dithizone Metal Chelates," *Anal. Chem.* **1994**, *66*, 2752.

5. S. Maji, S. Lahiri, B. Wierczinski, and G. Korschinek, "Separation of Trace Level Hafnium from Tungsten: A Step Toward Solving an Astronomical Puzzle," *Anal. Chem.* **2006**, *78*, 2303.

6. H. H. Strain and J. Sherma, "M. Tswett: Adsorption Analysis and Chromatographic Methods," *J. Chem. Ed.* **1967**, *44*, 238 (a translation of Tswett's original article); H. H. Strain and J. Sherma, "Michael Tswett's Contributions to Sixty Years of Chromatography," *J. Chem. Ed.* **1967**, *44*, 235; L. S. Ettre, "M. S. Tswett and the Invention of Chromatography," *LCGC* **2003**, *21*, 458; L. S. Ettre, "The Birth of Partition Chromatography," *LCGC* **2001**, *19*, 506; L. S. Ettre, "The Story of Thin-Layer Chromatography," *LCGC* **2001**, *19*, 712; L. S. Ettre, "A. A. Zhuykhovitskii—A Russian Pioneer of Gas Chromatography," *LCGC* **2000**, *18*, 1148; V. R. Meyer, "Michael Tswett and His Method," *Anal. Chem.* **1997**, *69*, 284A; Ya. I. Yashin, "History of Chromatography (1903–1993)," *Russ. J. Anal. Chem.* **1994**, *49*, 939; K. I. Sakodynskii, "Discovery of Chromatography by M. S. Tsvet," *Russ. J. Anal. Chem.* **1994**, *48*, 897.

7. L. S. Ettre, "The Rebirth of Chromatography 75 Years Ago," *LCGC* **2007**, *25*, 640.

8. C. F. Poole, *The Essence of Chromatography* (Amsterdam: Elsevier, 2003) (graduate-level treatment of theory and specific techniques); J. M. Miller, *Chromatography: Concepts and Contrasts*, 2nd ed. (Hoboken, NJ: Wiley, 2005).

9. L. V. Heumann, "Colorful Column Chromatography: A Classroom Demonstration of a Three-Component Separation," *J. Chem. Ed.* **2008**, *85*, 524; M. J. Samide, "Separation Anxiety: An In-Class Game Designed to Help Students Discover Chromatography," *J. Chem. Ed.* **2008**, *85*, 1512; C. A. Smith and F. W. Villaescusa, "Simulating Chromatographic Separations in the Classroom," *J. Chem. Ed.* **2003**, *80*, 1023.

10. H. Schmidt-Traub, ed., *Preparative Chromatography of Fine Chemicals and Pharmaceutical Agents* (Weinheim: Wiley-VCH, 2005); S. Miller, "Prep LC Systems for Chemical Separations," *Anal. Chem.* **2003**, *75*, 477A;

R. E. Majors, "The Role of the Column in Preparative HPLC," *LCGC* **2004**, *22*, 416; A. Brandt and S. Kueppers, "Practical Aspects of Preparative HPLC in Pharmaceutical Development and Production," *LCGC* **2002**, *20*, 14.

11. L. R. Snyder and J. W. Dolan, *High-Performance Gradient Elution* (Hoboken, NJ: Wiley, 2007), Equation V.15.

12. J. P. Foley and J. G. Dorsey, "Equations for Calculation of Chromatographic Figures of Merit for Ideal and Skewed Peaks," *Anal. Chem.* **1983**, *55*, 730; B. A. Bidlingmeyer and F. V. Warren, Jr., "Column Efficiency Measurement," *Anal. Chem.* **1984**, *56*, 1583A.

13. M. T. Bowser, G. M. Bebault, X. Peng, and D. D. Y. Chen, "Redefining the Separation Factor: Pathway to a Unified Separation Science," *Electrophoresis* **1997**, *18*, 2928. The conventional equation is resolution $= \frac{\sqrt{N}}{4}\left(\frac{\alpha - 1}{\alpha}\right)\left(\frac{k_2'}{1 + k_{\mathrm{av}}'}\right)$, where α is relative retention, k_2' is the retention factor for the more retained component, and k_{av}' is the average retention factor for the two components. This expression is equivalent to $\frac{\sqrt{N}}{4}(\gamma - 1)$ for closely spaced peaks for which $k_{\mathrm{A}}' \approx k_{\mathrm{B}}' \approx k_{\mathrm{av}}'$.

14. J. C. Giddings, *Unified Separation Science* (New York: Wiley, 1991); S. J. Hawkes, "Modernization of the van Deemter Equation for Chromatographic Zone Dispersion," *J. Chem. Ed.* **1983**, *60*, 393.

15. G. Desmet and K. Broeckhoven, "Equivalance of the Different C_{m}- and C_{s}-Term Expressions Used in Liquid Chromatography and a Geometrical Model Uniting Them," *Anal. Chem.* **2008**, *80*, 8076.

16. For numerical simulation of skewed bandshapes, see S. Sugata and Y. Abe, "An Analogue Column Model for Nonlinear Isotherms: The Test Tube Model," *J. Chem. Ed.* **1997**, *74*, 406 and B. R. Sundheim, "Column Operations: A Spreadsheet Model," *J. Chem. Ed.* **1992**, *69*, 1003.

17. A. Felinger, "Molecular Movement in an HPLC Column: A Stochastic Analysis," *LCGC* **2004**, *22*, 642; J. C. Giddings, *Dynamics of Chromatography* (New York: Marcel Dekker, 1965).

Chapter 23

1. Excellent training modules are available from www.academysavant.com. For instructive exercises to complement lab experience in chromatography, see D. C. Stone, "Teaching Chromatography Using Virtual Laboratory Exercises," *J. Chem. Ed.* **2007**, *84*, 1488.

2. J. V. Hinshaw and L. S. Ettre, *Introduction to Open Tubular Gas Chromatography* (Cleveland, OH: Advanstar Communications, 1994); H. M. McNair and J. M. Miller, *Basic Gas Chromatography* (New York: Wiley, 1998); R. L. Grob and E. F. Barry, eds., *Modern Practice of Gas Chromatography* (Hoboken, NJ: Wiley, 2004); E. F. Barry and R. L. Grob, *Columns for Gas Chromatography* (Hoboken, NJ: Wiley, 2007).

3. E. Cremer and F. Prior first explored gas-solid adsorption chromatography in the mid-1940s at the University of Innsbruck in Austria. Gas-liquid partition chromatography is attributed to A. J. P. Martin and A. T. James in 1950–1952 at the British National Institute of Medical Research. See L. S. Ettre, "The Beginnings of Gas Adsorption Chromatography 60 Years Ago," *LCGC* **2008**, *26*, 48.

4. A. Wollrab, "Lecture Experiments in Gas-Liquid Chromatography," *J. Chem. Ed.* **1982**, *59*, 1042; C. E. Bricker, M. A. Taylor, and K. E. Kolb, "Simple Classroom Demonstration of Gas Chromatography," *J. Chem. Ed.* **1981**, *58*, 41.

5. R. E. Majors, "The Art and Science of GC Capillary Column Production," *LCGC* **2007**, *25*, 616; L. S. Ettre, "Evolution of Capillary Columns for Gas Chromatography," *LCGC* **2001**, *19*, 48.

6. V. R. Meyer, "Amino Acid Racemization: A Tool for Fossil Dating," *Chemtech*, July 1992, p. 412.

7. A. M. Pollard and C. Heron, *Archaeological Chemistry* (Cambridge: Royal Society of Chemistry, 1996)—a very good book.

8. P. A. Levkin, A. Levkina, and V. Schurig, "Combining the Enantioselectivities of L-Valine Diamide and Permethylated β-Cyclodextrin in One Gas Chromatographic Chiral Stationary Phase," *Anal. Chem.* **2006**, *78*, 5143; I. Molnár-Perl, *Quantitation of Amino Acids and Amines by Chromatography* (New York: Elsevier Science, 2005).

9. T. Cserháti and E. Forgács, *Cyclodextrins in Chromatography* (Cambridge: Royal Society of Chemistry, 2003); B. Haldar, A. Mallick, and N. Chattopadhyay, "Supramolecular Inclusion in Cyclodextrins: A Pictorial Spectroscopic Demonstration," *J. Chem. Ed.* **2008**, *85*, 429; J. Hernández-Benito, M. P. Garcia-Santos, E. O'Brien, E. Calle, and J. Casado, "A Practical Integrated Approach to Supramolecular Chemistry III. Thermodynamics of Inclusion Phenomena," *J. Chem. Ed.* **2004**, *81*, 540; B. D. Wagner, P. J. MacDonald, and M. Wagner, "Visual Demonstration of Supramolecular Chemistry: Fluorescence Enhancement upon Host-Guest Inclusion," *J. Chem. Ed.* **2000**, *77*, 178; D. Díaz, I. Vargas-Baca, and J. Graci-Mora, "β-Cyclodextrin Inclusion Complexes with Iodine," *J. Chem. Ed.* **1994**, *71*, 708.

10. A. Berthod, "Chiral Recognition Mechanisms," *Anal. Chem.* **2006**, *78*, 2093; T. J. Ward, "Chiral Separations," *Anal. Chem.* **2002**, *74*, 2863; A. Berthod, W. Li, and D. W. Armstrong, "Multiple Enantioselective Retention Mechanisms on Derivatized Cyclodextrin Gas Chromatographic Chiral Stationary Phases," *Anal. Chem.* **1992**, *64*, 873; K. Bester, "Chiral Analysis for Environmental Applications," *Anal. Bioanal. Chem.* **2003**, *376*, 302.

11. E. N. Coker and P. J. Davis, "Experiments with Zeolites at the Secondary-School Level," *J. Chem. Ed.* **1999**, *76*, 1417.

12. In programmed temperature runs with a constant inlet pressure, the flow rate decreases during the run because the viscosity of the carrier gas increases as temperature increases. The effect can be significant (e.g., 30% decrease in linear velocity for a 200° temperature increase), so it is a good idea to set the initial linear velocity *above* the optimum value so that it does not decrease too far below the optimum. Equations for calculating flow rates as a function of temperature and pressure are given by J. V. Hinshaw and L. S. Ettre, *Introduction to Open Tubular Gas Chromatography* (Cleveland, OH: Advanstar Communications, 1994) and L. S. Ettre and J. V. Hinshaw, *Basic Relationships of Gas Chromatography* (Cleveland, OH: Advanstar Communications, 1993).

13. R. Stevenson and D. Sparkman, "Chromatography in Times of Shortage: Points to Consider in Changing to Hydrogen for Your Carrier Gas," *Am. Lab.*, April 2009, p. 30.

14. J. V. Hinshaw, "The Retention Gap Effect," *LCGC* **2004**, *22*, 624.

15. K. Grob, *Split and Splitless Injection for Quantitative Gas Chromatography* (New York: Wiley, 2001).

16. Techniques and hardware for *large volume injections* for trace analysis are described by S. M. Song, P. Marriott, D. Ryan, and P. Wynne, "Analytical Limbo: How Low Can You Go?" *LCGC* **2006**, *24*, 1012.

17. A. N. Papas and M. F. Delaney, "Evaluation of Chromatographic Integrators and Data Systems," *Anal. Chem.* **1987**, *59*, 55A.

18. J. V. Hinshaw, "The Flame Ionization Detector," *LCGC* **2005**, *23*, 1262.

19. B. Erickson, "Measuring Nitrogen and Phosphorus in the Presence of Hydrocarbons," *Anal. Chem.* **1998**, *70*, 599A.

20. S. Risticevic, V. H. Niri, D. Vuckovic, and J. Pawliszyn, "Recent Developments in Solid-Phase Microextraction," *Anal. Bioanal. Chem.* **2009**, *393*, 781; F. Pragst, "Application of Solid-Phase Microextraction in Analytical Toxicology," *Anal. Bioanal. Chem.* **2007**, *388*, 1393; J. B. Quintana and I. Rodríguez, "Strategies for the Microextraction of Polar Organic Contaminants in Water Samples," *Anal. Bioanal. Chem.* **2006**, *384*, 1447; P. Mayer, J. Tolls, J. C. M. Hermens, and D. MacKay, "Equilibrium Sampling Devices," *Environ. Sci. Technol.* **2003**, *37*, 185A.

21. C. Bicci, E. Liberto, C. Cordero, B. Sgorbini, and P. Rubiolo, "Stir-Bar Sorptive Extraction and Headspace Sorptive Extraction: An Overview," *LCGC* **2009**, *27*, 376.

22. K. Dettmer and W. Engewald, "Absorbent Materials Commonly Used in Air Analysis for Adsorptive Enrichment and Thermal Desorption of Volatile Organic Compounds," *Anal. Bioanal. Chem.* **2002**, *373*, 490; M. Schneider and K.-U. Goss, "Systematic Investigation of the Sorption Properties of Tenax TA, Chromosorb 106, Porapak N, and Carbopak F," *Anal. Chem.* **2009**, *813*, 3017.

23. S. Cram, "How to Develop, Validate, and Troubleshoot Capillary GC Methods," American Chemical Society Short Course, 1996.

24. J. De Zeeuw and M. Barnes, "Fast, Practical GC and GC-MS," *Am. Lab. News Ed.*, January 2007, p. 26.

25. For software to optimize liquid and gas chromatography, see www.molnar-institut.com; J. V. Hinshaw, "Strategies for GC Optimization: Software," *LCGC*, **2000**, *18*, 1040.

26. J. J. Langenfeld, S. B. Hawthorne, and D. J. Miller, "Quantitative Analysis of Fuel-Related Hydrocarbons in Surface Water and Wastewater by Solid-Phase Microextraction," *Anal. Chem.* **1996**, *68*, 144.

Chapter 24

1. C. E. Burgess, P. N. Pearson, C. H. Lear1, H. E. G. Morgans, L. Handley, R. D. Pancost, and S. Schouten, "Middle Eocene Climate Cyclicity in the Southern Pacific: Implications for Global Ice Volume, *Geology* **2008**, *36*, 651; P. N. Pearson, B. E. van Dongen, C. J. Nicholas, R. D. Pancost, S. Schouten, J. M. Singano, and B. S. Wade, "Stable Warm Tropical Climate Through the Eocene Epoch," *Geology* **2007**, *35*, 211; C. Huguet, J.-H. Kim, J. S. Sinninghe Damsté, and S. Schouten, "Reconstruction of Sea Surface Temperature Variations in the Arabian Sea of the Last 23 kyr Using Organic Proxies TEX_{86} and $U_{37}^{K'}$," *Paleooceanography* **2006**, *21*, PA3003.

2. L. R. Snyder, J. J. Kirkland, and J. L. Glajch, *Practical HPLC Method Development* (New York: Wiley, 1997). References 2 and 3 are definitive resources for liquid chromatography.

3. L. R. Snyder and J. W. Dolan, *High-Performance Gradient Elution* (Hoboken, NJ: Wiley, 2007).

4. S. Kromidas, *More Practical Problem Solving in HPLC* (Weinheim: Wiley-VCH, 2005); S. Kromidas, *Practical Problem Solving in HPLC* (New York: Wiley, 2000); V. R. Meyer, *Practical High-Performance Liquid Chromatography*, 4th ed. (Chichester: Wiley, 2004); T. Hanai and R. M. Smith, *HPLC: A Practical Guide* (New York: Springer-Verlag, 1999); V. R. Meyer, *Pitfalls and Errors of HPLC in Pictures* (New York: Wiley, 1998); R. Eksteen, P. Schoenmakers, and N. Miller, eds., *Handbook of HPLC* (New York: Marcel Dekker, 1998); U. D. Neue, *HPLC Columns: Theory, Technology, and Practice* (New York: Wiley, 1997); A. Weston and P. R. Brown, *HPLC and CE* (San Diego: Academic Press, 1997); L. R. Snyder, "HPLC—Past and Present," *Anal. Chem.* **2000**, *72*, 412A.

5. For exercises to complement laboratory experience, see D. C. Stone, "Teaching Chromatography Using Virtual Laboratory Exercises," *J. Chem. Ed.* **2007**, *84*, 1488.

6. For excellent training modules, see www.academysavant.com.

7. J. W. Dolan, "LC Troubleshooting," *LCGC* **2003**, *21*, 888.

8. F. Gritti, M. Martin, and G. Guiochon, "Influence of Viscous Friction Heating on the Efficiency of Columns Operated under Very High Pressures," *Anal. Chem.* **2009**, *81*, 3365; F. Gritti and G. Guiochon, "Heat Exchanges in Fast, High-Performance Liquid Chromatography. A Complete Thermodynamic Study," *Anal. Chem.* **2008**, *80*, 6488; F. Gritti and G. Guiochon, "Complete Temperature Profiles in Ultra-High-Pressure Liquid Chromatography Columns," *Anal. Chem.* **2008**, *80*, 5009; J. R. Mazzeo, U. D. Neue, M. Kele, and R. S. Plumb "Advancing LC Performance with Smaller Particles and Higher Pressure," *Anal. Chem.* **2005**, *77*, 460A.

9. J. W. Dolan, "The Importance of Temperature," *LCGC* **2002**, *20*, 524; Y. Yang and D. R. Lynch, Jr., "Stationary Phases for High-Temperature LC Separations," *LCGC Supplement*, June 2004, p. 34.

10. For high-pH operation of any silica-based stationary phase, temperature should not exceed 40°C, organic buffers should be used instead of phosphate or carbonate, and methanol instead of acetonitrile should be the organic solvent. (J. J. Kirkland, J. D. Martosella, J. W. Henderson, C. H. Dilks, Jr., and J. B. Adams, Jr., "HPLC of Basic Compounds at High pH with a Silica-Based Bidentate-C18 Bonded-Phase Column," *Am. Lab.*, November 1999, p. 22; E. D. Neue, "HPLC Troubleshooting," *Am. Lab.*, February 2002, p. 72.)

11. A Giaquinto, Z. Liu, A. Bach, and Y. Kazakevich, "Surface Area of Reversed-Phase HPLC Columns," *Anal. Chem.* **2008**, *80*, 6358.

12. N. Tanaka, H. Kobayashi, K. Nakanishi, H. Minakuchi, and N. Ishizuka, "Monolithic LC Columns," *Anal. Chem.* **2001**, *73*, 421A; N. Tanaka and H. Kobayashi, "Monolithic Columns for Liquid Chromatogaphy," *Anal. Bioanal. Chem.* **2003**, *376*, 298; F. Svec, T. B. Tennikova, and Z. Deyl, eds., *Monolithic Materials: Preparation, Properties and Applications* (Amsterdam: Elsevier, 2003).

13. P. Ross, "Porous Graphitic Carbon in HPLC," *LCGC* **2000**, *18*, 18; S. Mazan, G. Crétier, N. Gilon, J.-M. Mermet, and J.-L. Rocca, "Porous

Graphitic Carbon as Stationary Phase for LC-ICPMS Separation of Arsenic Compounds," *Anal. Chem.* **2002**, *74*, 1281. For another robust stationary phase, see C. J. Dunlap, C. V. McNeff, D. Stoll, and P. W. Carr, "Zirconia Stationary Phases for Extreme Separations," *Anal. Chem.* **2001**, *73*, 599A.

14. D. W. Armstrong and B. Zhang, "Chiral Stationary Phases for HPLC," *Anal. Chem.* **2001**, *73*, 557A; S. Ahuja, "A Strategy for Developing HPLC Methods for Chiral Drugs," *LCGC* **2007**, *25*, 1112.

15. A. P. Schellinger and P. W. Carr, "Solubility of Buffers in Aqueous-Organic Eluents for Reversed-Phase Liquid Chromatography," *LCGC* **2004**, *22*, 544; D. Sykora, E. Tesarova, and D. W. Armstrong, "Practical Considerations of the Influence of Organic Modifiers on the Ionization of Analytes and Buffers in Reversed-Phase LC," *LCGC* **2002**, *20*, 974; G. W. Tindall, "Mobile-Phase Buffers. I. The Interpretation of pH in Partially Aqueous Mobile Phases," *LCGC* **2002**, *20*, 102; S. Espinosa, E. Bosch, and M. Rosés, "Acid-Base Constants of Neutral Bases in Acetonitrile-Water Mixtures," *Anal. Chim. Acta* **2002**, *454*, 157.

16. SF_6 demonstrations: P. Licence, D. Litchfield, M. P. Dellar, and M. Poliakoff, "'Supercriticality'; a Dramatic but Safe Demonstration of the Critical Point," *Green Chem.* **2004**, *6*, 352; R. Chang and J. F. Skinner, "A Lecture Demonstration of the Critical Phenomenon," *J. Chem. Ed.* **1992**, *69*, 158. CO_2 demonstration: V. T. Lieu, "Simple Experiment for Demonstration of Phase Diagram of Carbon Dioxide," *J. Chem. Ed.* **1996**, *73*, 837. H_2O + isobutyric acid: M. R. Johnson, "A Demonstration of the Continuous Phase (Second-Order) Transition of a Binary Liquid System in the Region Around Its Critical Point," *J. Chem. Ed.* **2006**, *83*, 1014.

17. C. J. Welch, W. R. Leonard, Jr., O. DaSilva, M. Biba, J. Albaneze-Walker, D. W. Henderson, B. Laing, and D. J. Mathre, "Preparative Chiral SFC as a Green Technology for Rapid Access to Enantiopurity in Pharmaceutical Process Research," *LCGC* **2005**, *23*, 16A; K. Anton and C. Berger, eds., *Supercritical Fluid Chromatography with Packed Columns—Techniques and Applications* (New York: Marcel Dekker, 1998).

18. Y. Egi and A. Ueyanagi, "Ghost Peaks and Aerated Sample Solvent," *LCGC* **1998**, *16*, 112.

19. J. W. Dolan, "Column Care," *LCGC* **2008**, *26*, 692.

20. *Washing normal-phase column:* Remove the guard column so that impurities from the guard column are not washed into the analytical column. Bare silica and diol-bonded phases can be washed (in order) with heptane, chloroform, ethyl acetate, acetone, ethanol, and water. Then the order is reversed, using dried solvents, to reactivate the column. Use 15 mobile phase volumes ($15V_m$) of each solvent. Amino-bonded phases are washed in the same manner as silica, but 0.5 M ammonia is used after water. [F. Rabel and K. Palmer, *Am. Lab.*, August 1992, p. 65.] Between uses, normal-phase columns can be stored in 2-propanol or hexane.

21. D. L. Warner and J. G. Dorsey, "Reduction of Total Analysis Time in Gradient Elution, Reversed Phase Liquid Chromatography," *LCGC* **1997**, *15*, 254.

22. C. Hawkins and J. W. Dolan, "Understanding Split Peaks," *LCGC* **2003**, *21*, 1134.

23. J. W. Dolan, "How Much Is Too Much?" *LCGC* **1999**, *17*, 508.

24. L. N. Seetohul, Z. Ali, and M. Islam, "Broadband Cavity Enhanced Absorption Spectroscopy as a Detector for HPLC," *Anal. Chem.* **2009**, *81*, 4106.

25. M. A. Fotopoulou and P. C. Ioannou, "Post-Column Terbium Complexation and Sensitized Fluorescence for the Determination of Norepinephrine, Epinephrine and Dopamine using High-Performance Liquid Chromatography," *Anal. Chim. Acta* **2002**, *462*, 179.

26. J. A. Koropchak, S. Sadain, X. Yang, L.-E. Magnusson, M. Heybroek, M. Anisimov, and S. L. Kaufman, "Nanoparticle Detection Technology," *Anal. Chem.* **1999**, *71*, 386A; C. S. Young and J. W. Dolan, "Success with Evaporative Light-Scattering Detection," *LCGC* **2003**, *21*, 120.

27. P. H. Gamache, R. S. McCarthy, S. M. Freeto, D. J. Asa, M. J. Woodcock, K. Laws, and R. O. Cole, "HPLC Analysis of Nonvolatile Analytes Using Charged Aerosol Detection," *LCGC* **2005**, *23*, 150.

28. The effect of the gradient on the response of the charged aerosol detector can be cancelled by delivering an exact inverse gradient with a second pump and mixing the inverse gradient with eluate prior to the detector. (T. Górecki, F. Lynen, R. Szucs, and P. Sandra, "Universal Response in Liquid Chromatography Using Charged Aerosol Detection," *Anal. Chem.* **2006**, *78*, 3186.)

29. W. R. LaCourse, *Pulsed Electrochemical Detection in High-Performance Liquid Chromatography* (New York: Wiley, 1997).

30. For a concise, readable, and expert guide to method development for reversed-phase separations, see J. W. Dolan, "The Perfect Method," *LCGC*, **2007**, *25*, 546, 632, 704, 944, 1014, 1094, 1178.

31. Acetonitrile can be hydrolyzed to sodium acetate and poured down the drain for disposal: $CH_3CN + NaOH + H_2O \rightarrow CH_3CO_2Na + NH_3$. Dilute the chromatography eluate to 10 vol% CH_3CN with water. To 1.0 L of 10 vol% CH_3CN, add 475 mL of 10 M NaOH. The solution can be left at 20°C for 25 days in a hood or heated to 80°C for 70 min to reduce the CH_3CN concentration to 0.025 vol%. Mix the hydrolysate with waste acid so that it is approximately neutral before disposal. (K. Gilomen, H. P. Stauffer, and V. R. Meyer, *LCGC* **1996**, *14*, 56.)

32. Tetrahydrofuran can be stored for at least half a year without oxidation by adding 25 vol% H_2O. (J. Zhao and P. W. Carr, "The Magic of Water in Tetrahydrofuran—Preventing Peroxide Formation," *LCGC* **1999**, *17*, 346.)

33. J. D. Thompson and P. W. Carr, "High-Speed Liquid Chromatography by Simultaneous Optimization of Temperature and Eluent Composition," *Anal. Chem.* **2002**, *74*, 4150.

34. L. R. Snyder, J. W. Dolan, and P. W. Carr, "A New Look at the Selectivity of RPC Columns," *Anal. Chem.* **2007**, *79*, 3255; See http://www.usp.org/USPNF/columnsDB.html for the column selection tool and http://www.usp.org/pdf/EN/USPNF/HPLCclassification.pdf for an explanation.

35. J. W. Dolan, "Resolving Minor Peaks," *LCGC* **2002**, *20*, 594.

36. For more on developing gradient separations, see Reference 3 and J. W. Dolan, "The Scouting Gradient Alternative," *LCGC* **2000**, *18*, 478.

37. Available software includes *JCE Software* **2003**, SP28 (www.jce.divched.org/JCE-Soft/programs/Collections/ACC/index.html); DryLab, Molnar Institute, Berlin (www.molnar-institut.com/cd/indexe.htm); ChromSword, Merck, Darmstadt (www.chromsword.net); Simulator, Advanced Chemistry Development, Toronto (www.acdlabs.com/products/chrom_lab/); Virtual Column (http://www.virtualcolumn.com).

38. J. M. Cintrón and L. A. Colón, "Organo-Silica Nano-Particles Used in Ultrahigh-Pressure Liquid Chromatography," *Analyst* **2002**, *127*, 705.

39. E. S. Grumbach, D. M. Wagrowski-Diehl, J. R. Mazzeo, B. Alden, and P. C. Iraneta, "Hydrophilic Interaction Chromatography Using Silica Columns for the Retention of Polar Analytes and Enhanced ESI-MS Sensitivity," *LCGC* **2004**, *22*, 1010.

Chapter 25

1. K. K. Unger, M. Huber, K. Walhagen, T. P. Hennessy, and M. T. W. Hearn, "A Critical Appraisal of Capillary Electrochromatography," *Anal. Chem.* **2002**, *74*, 200A.

2. B. A. Kyereboah-Taylor, "Ultrapure Water for Ion Chromatography," *Am. Lab.*, August 1995, p. 24; B. M. Stewart and D. Darbouret, "Ultrapure Water for ICP-MS," *Am. Lab. News Ed.*, April 1998, p. 36; A. de Chatellus, "Purification Media," *Am. Lab. News Ed.*, January 1998, p. 8.

3. A mixture of enantiomers of cationic metal complexes was applied to a cation-exchange column and eluted by one enantiomer of tartrate anion. The tartrate has a different ion-pair formation constant with each enantiomer of the metal complex and therefore removes one metal enantiomer from the column before the other metal enantiomer. (M. Cantuel, G. Bernardinelli, G. Muller, J. P. Riehl, and C. Piguet, "The First Enantiomerically Pure Helical Noncovalent Tripod for Assembling Nine-Coordinate Lanthanide(III) Podates," *Inorg. Chem.* **2004**, *43*, 1840; Y. Yoshikawa and K. Yamasaki, "Chromatographic Resolution of Metal Complexes on Sephadex Ion Exchangers," *Coord. Chem. Rev.* **1979**, *28*, 2005.)

4. D. P. Elder, "Pharmaceutical Applications of Ion-Exchange Resins," *J. Chem. Ed.* **2005**, *82*, 575.

5. J. S. Fritz and D. T. Gjerde, *Ion Chromatography*, 3rd ed. (New York: Wiley-VCH, 2000); J. Weiss, *Handbook of Ion Chromatography* (Weinheim: Wiley-VCH, 2004); P. R. Haddad and P. E. Jackson, *Ion Chromatography: Principles and Applications* (New York: Elsevier, 1990); H. Small, *Ion Chromatography* (New York: Plenum Press, 1989); P. R. Haddad, "Ion Chromatography Retrospective," *Anal. Chem.* **2001**, *73*, 266A; H. Small,

"Ion Chromatography: An Account of Its Conception and Early Development," *J. Chem. Ed.* **2004**, *81*, 1277; B. Evans, "The History of Ion Chromatography: The Engineering Perspective," *J. Chem. Ed.* **2004**, *81*, 1285.

6. J. E. Madden, M. J. Shaw, G. W. Dicinoski, and P. R. Haddad, "Simulation and Optimization of Retention in Ion Chromatography Using Virtual Column 2 Software," *Anal. Chem.* **2002**, *74*, 6023; P. R. Haddad, M. J. Shaw, J. E. Madden, and G. W. Dicinoski, "Computer-Based Undergraduate Exercise Using Internet-Accessible Simulation Software for the Study of Retention Behavior and Optimization of Separation Conditions in Ion Chromatography," *J. Chem. Ed.* **2004**, *81*, 1293; http://www.virtualcolumn.com

7. P. C. Schulz and D. Clausse, "An Undergraduate Physical Chemistry Experiment on Surfactants: Electrochemical Study of Commercial Soap," *J. Chem. Ed.* **2003**, *80*, 1053; A. Domínguez, A. Fernández, N. González, E. Iglesias, and L. Montenegro, "Determination of Critical Micelle Concentration of Some Surfactants by Three Techniques," *J. Chem. Ed.* **1997**, *74*, 1227; K. R. Williams and L. H. Tennant, "Micelles in the Physical/Analytical Chemistry Laboratory. Acid Dissociation of Neutral Red Indicator," *J. Chem. Ed.* **2001**, *78*, 349; S. A. Tucker, V. L. Amszi, and W. E. Acree, Jr., "Studying Acid-Base Equilibria in Two-Phase Solvent Media," *J. Chem. Ed.* **1993**, *70*, 80.

8. J. W. Dolan, "Ion Pairing—Blessing or Curse?" *LCGC* **2008**, *26*, 170.

9. M. A. Hervas and C. E. Fabara, "A Simple Demonstration of the Ion-Pairing Effect on the Solubility of Charged Molecules," *J. Chem. Ed.* **1995**, *72*, 437.

10. L. R. Snyder, J. J. Kirkland, and J. L. Glajch, *Practical HPLC Method Development* (New York: Wiley, 1997); J. W. Dolan, "Improving an Ion-Pairing Method," *LCGC* **1996**, *14*, 768.

11. G. A. Marson and B. Baptista, *Principles of Gel Permeation Chromatography: Interactive Software* (J. Chem. Ed. WebWare, http://www.jce.divched.org/JCEDLib/WebWare/).

12. T. W. Perkins, T. W. Root, and E. N. Lightfoot, "Measuring Column Void Volumes with NMR," *Anal. Chem.* **1997**, *69*, 3293.

13. C.-S. Wu, ed., *Handbook of Size Exclusion Chromatography*, 2nd ed. (New York: Marcel Dekker, 2004); S. Mori and H. G. Barth, *Size Exclusion Chromatography* (Berlin: Springer, 1999).

14. R. Gill, M. Zayats, and I. Willner, "Semiconductor Quantum Dots for Bioanalysis," *Angew. Chem. Int. Ed.* **2008**, *47*, 7602; E. R. Goldman, I. L. Medintz, and H. Mattoussi, "Luminescent Quantum Dots in Immunoassays," *Anal. Bioanal. Chem.* **2006**, *384*, 560; B. M. Hutchins, T. T. Morgan, M. G. Ucak-Astarlioglu, and M. E. Williams, "Optical Properties of Fluorescent Mixtures: Comparing Quantum Dots to Organic Dyes," *J. Chem. Ed.* **2007**, *84*, 1301; E. M. Boatman, G. C. Lisensky, and K. J. Nordell, "A Safer, Easier, Faster Synthesis for CdSe Quantum Dot Nanocrystals," *J. Chem. Ed.* **2005**, *82*, 1697; L. D. Winkler, J. F. Arceo, W. C. Hughes, B. A. DeGraff, and B. H. Augustine, "Quantum Dots: An Experiment for Physical or Materials Chemistry," *J. Chem. Ed.* **2005**, *82*, 1700.

15. T. K. Nevanen, H. Simolin, T. Suortti, A. Koivula, and H. Söderlund, "Development of a High-Throughput Format for Solid-Phase Extraction of Enantiomers Using an Immunosorbent in 384-Well Plates," *Anal. Chem.* **2005**, *77*, 3038.

16. X. Jiang, N. Jiang, H. Zhang, and M. Liu, "Small Organic Molecular Imprinted Materials: Their Preparation and Application," *Anal. Bioanal. Chem.* **2007**, *389*, 355; F. B. Kaabi and V. Pichon, "Different Approaches to Synthesizing Molecularly Imprinted Polymers for Solid-Phase Extraction," *LCGC* **2007**, *25*, 732.

17. G. Zhao, X.-F. Li, and X. Chris Le, "Aptamer-Modified Monolithic Capillary Chromatography for Protein Separation and Detection," *Anal. Chem.* **2008**, *80*, 3915.

18. K. K. S. Lau, J. Bico, K. B. K. Teo, M. Chhowalla, G. A. J. Amaratunga, W. I. Milne, G. H. McKinley, and K. K. Gleason, "Superhydrophobic Carbon Nanotube Forrests," *Nano Lett.* **2003**, *3*, 1701.

19. Amersham Pharmacia Biotech, *Hydrophobic Interaction Chromatography: Principles and Methods,* freely available from http://teachline.ls.huji.ac.il/72682/Booklets/AMERSHAM_hydrophobic_interactionManual.pdf

20. R. Weinberger, *Practical Capillary Electrophoresis*, 2nd ed. (San Diego: Academic Press, 2000); P. Landers, ed., *Handbook of Capillary and Microchip Electrophoresis and Associated Microtechniques*, 3rd ed. (Boca Raton, FL: CRC Press, 2008); T. Wehr, R. Rodríguez-Diaz, and M. Zhu, *Capillary Electrophoresis of Proteins* (New York: Marcel Dekker, 1999); M. G. Khaledi, ed., *High Performance Capillary Electrophoresis: Theory, Techniques, and Applications* (New York: Wiley, 1998); P. Camilleri, ed., *Capillary Electrophoresis: Theory and Practice* (Boca Raton, FL: CRC Press, 1998); J. R. Petersen and A. A. Mohammad, eds., *Clinical and Forensic Applications of Capillary Electrophoresis* (Totowa, NJ: Humana Press, 2001).

21. C. L. Cooper, "Capillary Electrophoresis: Theoretical and Experimental Background," *J. Chem. Ed.* **1998**, *75*, 343; C. L. Cooper and K. W. Whitaker, "Capillary Electrophoresis: Applications," *J. Chem. Ed.* **1998**, *75*, 347.

22. J. Kemsley, "Heparin Undone," *Chem. Eng. News*, 12 May 2008, p. 38.

23. M. Guerrini, D. Beccati, Z. Shriver, A. Naggi, K. Viswanathan, A. Bisio, I. Capila, J. C. Lansing, S. Guglieri, B. Fraser, A. Al-Hakim, N. S. Gunay, Z. Zhang, L. Robinson, L. Buhse, M. Nasr, J. Woodcock, R. Langer, G. Venkataraman, R. J. Linhardt, B. Casu, G. Torri, and R. Sasisekharan, "Oversulfated Chondroitin Sulfate is a Contaminant in Heparin Associated with Adverse Clinical Events," *Nature Biotech.* **2008**, *26*, 669.

24. J. W. Jorgensen and K. D. Lukacs, "Zone Electrophoresis in Open Tubular Glass Capillaries," *Anal. Chem.* **1981**, *53*, 1298.

25. For a demonstration of electrophoresis, see J. G. Ibanez, M. M. Singh, R. M. Pike, and Z. Szafran, "Microscale Electrokinetic Processing of Soils," *J. Chem. Ed.* **1998**, *75*, 634.

26. E. Klodzińska and B. Buszewski, "Electrokinetic Detection and Characterization of Intact Microorganisms," *Anal. Chem.* **2009**, *81*, 8.

27. I. G. Arcibal, M. F. Santillo, and A. G. Ewing, "Recent Advances in Capillary Electrophoretic Analysis of Individual Cells," *Anal. Bioanal. Chem.* **2007**, *387*, 51; A. K. Price and C. T. Culbertson, "Chemical Analysis of Single Mammalian Cells with Microfluidics," *Anal. Chem.* **2007**, *79*, 2615; P. B. Allen, B. R. Doepker, and D. T. Chiu, "High-Throughput Capillary-Electrophoresis Analysis of the Contents of a Single Mitochondria [sic]," *Anal. Chem.* **2009**, *81*, 3784.

28. G. K. Shoemaker, J. Lorieau, L. H. Lau, C. S. Gillmor, and M. M. Palcic, "Multiple Sampling in Single-Cell Enzyme Assays," *Anal. Chem.* **2005**, *77*, 3132; X. Sun and W. Jin, "Catalysis-Electrochemical Determination of Zeptomol Enzyme and Its Application for Single-Cell Analysis," *Anal. Chem.* **2003**, *75*, 6050.

29. C. J. Evenhuis, R. M. Guijt, M. Macka, P. J. Marriott, and P. R. Haddad, "Temperature Profiles and Heat Dissipation in Capillary Electrophoresis," *Anal. Chem.* **2006**, *78*, 2684; M. U. Musheev, S. Javaherian, V. Okhonin, and S. N. Krylov, "Diffusion as a Tool of Measuring Temperature Inside a Capillary," *Anal. Chem.* **2008**, *80*, 6752.

30. I. Gitlin, J. D. Carbeck, and G. M. Whitesides, "Why are Proteins Charged? Networks of Charge–Charge Interactions in Proteins Measured by Charge Ladders and Capillary Electrophoresis," *Angew. Chem. Int. Ed.* **2006**, *45*, 3022.

31. C. T. Culbertson and J. W. Jorgenson, "Flow Counterbalanced Capillary Electrophoresis," *Anal. Chem.* **1994**, *66*, 955.

32. J. E. Gomez and J. E. Sandoval, "The Effect of Conditioning of Fused-Silica Capillaries on Their Electrophoretic Performance," *Electrophoresis* **2008**, *29*, 381; R. Weinberger, "Capillary Conditioning Revisited," *Am. Lab.*, August 2008, p. 28.

33. R. Weinberger, "Capillary Electrophoresis," *Am. Lab.*, April 2005, p. 25.

34. H. Watley, "Making CE Work—Points to Consider," *LCGC* **1999**, *17*, 426.

35. M. Macka, P. Andersson, and P. R. Haddad, "Changes in Electrolyte pH Due to Electrolysis During Capillary Zone Electrophoresis," *Anal. Chem.* **1998**, *70*, 743.

36. H. Hijnen, "A Theoretical Analysis of the Influence of Electroosmosis on the Effective Ionic Mobility in Capillary Zone Electrophoresis," *J. Chem. Ed.* **2009**, *86*, 869.

37. N. E. Baryla, J. E. Melanson, M. T. McDermott, and C. A. Lucy, "Characterization of Surfactant Coatings in Capillary Electrophoresis by Atomic Force Microscopy," *Anal. Chem.* **2001**, *73*, 4558; M. M. Yassine and C. A. Lucy, "Enhanced Stability Self-Assembled Coatings for Protein Separations by Capillary Zone Electrophoresis Through the Use of Long-Chained Surfactants," *Anal. Chem.* **2005**, *77*, 62.

38. At a wavelength of 190 nm, phosphate at pH 7.2 has about three times the absorbance of borate at pH 9.2. Glycine, citrate, HEPES, and TRIS buffers have significant absorption near 210 nm. Borate buffer should be prepared from sodium tetraborate (borax, $Na_2B_4O_7 \cdot 10H_2O$), not from boric acid ($B(OH)_3$), which has somewhat different acid-base chemistry. An ultraviolet-absorbing impurity in borate buffer can be removed by passage through a C_{18} solid-phase extraction column.

39. A. Berethod and C. García-Alvarez-Coque, *Micellar Liquid Chromatography* (New York: Marcel Dekker, 2000); S. Terabe, "Micellar Electrokinetic Chromatography," *Anal. Chem.* **2004**, *76*, 240A; U. Pyell, "Micellar Electrokinetic Chromatography—From Theoretical Concepts to Real Samples (Review)," *Fresenius J. Anal. Chem.* **2001**, *371*, 691.

40. S. Terabe, K. Otsuka, K. Ichikawa, A. Tsuchiya, and T. Ando, "Electrokinetic Separations with Micellar Solutions and Open Tubular Capillaries," *Anal. Chem.* **1984**, *56*, 111; ibid., **1985**, *57*, 834.

41. Demonstrations: N. Gani and J. Khanam, "Are Surfactant Molecules Really Oriented in the Interface?" *J. Chem. Ed.* **2002**, *79*, 332; C. J. Marzzacco, "The Effect of SDS Micelle on the Rate of a Reaction," *J. Chem. Ed.* **1992**, *69*, 1024; C. P. Palmer, "Demonstrating Chemical and Analytical Concepts Using Electrophoresis and Micellar Electrokinetic Chromatography," *J. Chem. Ed.* **1999**, *76*, 1542.

42. Polar organic solvents with electrolytes such as sodium *p*-toluenesulfonate are compatible with capillary electrophoresis. Background electrolyte need not be an aqueous solution. (P. B. Wright, A. S. Lister, and J. G. Dorsey, "Behavior and Use of Nonaqueous Media without Supporting Electrolyte in Capillary Electrophoresis and Capillary Electrochromatography," *Anal. Chem.* **1997**, *69*, 3251; I. E. Valkó, H. Sirén, and M.-L. Riekkola, "Capillary Electrophoresis in Nonaqueous Media: An Overview," *LCGC* **1997**, *15*, 560; S. Li and S. G. Wever, "Separation of Neutral Compounds in Nonaqueous Solvents by Capillary Zone Electrophoresis," *J. Am. Chem. Soc.* **2000**, *122*, 3787.)

43. S. Conradi, C. Vogt, and E. Rohde, "Separation of Enantiomeric Barbiturates by Capillary Electrophoresis Using a Cyclodextrin-Containing Run Buffer," *J. Chem. Ed.* **1997**, *74*, 1122; F.-T. Chen and R. A. Evangelista, "Highly Sulfated Cyclodextrins: The Solution for Chiral Analysis," *Am. Lab.*, August 2002, p. 30; M. A. Schwarz and P. C. Hauser, "Chiral On-Chip Separations of Neurotransmitters," *Anal. Chem.* **2000**, *752*, 4691; W. Zhu and G. Vigh, "A Family of Single-Isomer, Sulfated γ-Cyclodextrin Chiral Resolving Agents for Capillary Electrophoresis," *Anal. Chem.* **2000**, *72*, 310; S. A. A. Rizvi, J. Zheng, R. P. Apkarian, S. N. Dublin, and S. A. Shamsi, "Polymeric Sulfated Amino Acid Surfactants: A Class of Versatile Chiral Selectors for Micellar Electrokinetic Chromatography (MEKC) and MEKC–MS," *Anal. Chem.* **2007**, *79*, 879.

44. J. Palmer, N. J. Munro, and J. P. Landers, "A Universal Concept for Stacking Neutral Analytes in Micellar Capillary Electrophoresis," *Anal. Chem.* **1999**, *71*, 1679; J. Palmer, D. S. Burji, N. J. Munro, and J. P. Landers, "Electrokinetic Injection for Stacking Neutral Analytes in Capillary and Microchip Electrophoresis," *Anal. Chem.* **2001**, *73*, 725; J. P. Quirino, S. Terabe, and P. Bocek, "Sweeping of Neutral Analytes in Electrokinetic Chromatography with High-Salt-Containing Matrixes," *Anal. Chem.* **2000**, *72*, 1934.

45. M. Gong, K. R. Wehmeyer, P. A. Limbach, and W. R. Heineman, "Unlimited-Volume Electrokinetic Stacking Injection in Sweeping Capillary Electrophoresis using a Cationic Surfactant," *Anal. Chem.* **2006**, *78*, 6035; L. Zhu, C. Tu, and H. K. Lee, "On-Line Concentration of Acidic Compounds by Anion-Selective Exhaustive Injection-Sweeping-Micellar Electrokinetic Chromatography," *Anal. Chem.* **2002**, *74*, 5820.

46. K. Isoo and S. Terabe, "Analysis of Metal Ions by Sweeping via Dynamic Complexation and Cation-Selective Exhaustive Injection in Capillary Electrophoresis," *Anal. Chem.* **2003**, *75*, 6789.

47. W. Wei, G. Zue, and E. S. Yeung, "One-Step Concentration of Analytes Based on Dynamic Change in pH in Capillary Zone Electrophoresis," *Anal. Chem.* **2002**, *74*, 934; P. Britz-McKibbin, K. Otsuka, and S. Terabe, "On-Line Focusing of Flavin Derivatives Using Dynamic pH Junction-Sweeping Capillary Electrophoresis with Laser-Induced Fluorescence Detection," *Anal. Chem.* **2002**, *74*, 3736.

48. B. Jung, R. Bharadwaj, and J. G. Santiago, "On-Chip Millionfold Sample Stacking using Transient Isotachophoresis," *Anal. Chem.* **2006**, *78*, 2319;

M. C. Breadmore and J. P. Quirino, "100 000-Fold Concentration of Anions in Capillary Zone Electrophoresis using Electroosmotic Flow Controlled Counterflow Isotachophoretic Stacking under Field Amplified Conditions," *Anal. Chem.* **2008**, *80*, 6373.

49. L. A. Colón, Y. Guo, and A. Fermier, "Capillary Electrochromatography," *Anal. Chem.* **1997**, *69*, 461A.

50. Y. Gong and H. K. Lee, "Application of Cyclam-Capped β-Cyclodextrin-Bonded Silica Particles as a Chiral Stationary Phase in Capillary Electrochromatography for Enantiomeric Separations," *Anal. Chem.* **2003**, *75*, 1348; M. W. Kamande, X. Zhu, C. Kapnissi-Christodoulou, and I. M. Warner, "Chiral Separations Using a Polypeptide and Polymeric Dipeptide Surfactant Polyelectrolyte Multilayer Coating in Open-Tubular Capillary Electrochromatography," *Anal. Chem.* **2004**, *76*, 6681.

51. J. R. Hutchinson, P. Zakaria, A. R. Bowie, M. Macka, N. Avdalovic, and P. R. Haddad, "Anion-Exchange Capillary Electrochromatography and In-Line Sample Preconcentration in Capillary Electrophoresis," *Anal. Chem.* **2005**, *77*, 407.

52. J. P. Quirino, M. T. Duylay, B. D. Bennett, and R. N. Zare, "Strategy for On-Line Preconcentration in Chromatographic Separations," *Anal. Chem.* **2001**, *73*, 5539; J. P. Quirino, M. T. Duylay, and R. N. Zare, "On-Line Preconcentration in Capillary Electrochromatography Using a Porous Monolith Together with Solvent Gradient and Sample Stacking," *Anal. Chem.* **2001**, *73*, 5557.

53. E. Zubritsky, "How Analytical Chemists Saved the Human Genome Project," *Anal. Chem.* **2002**, *74*, 23A.

54. R. Weinberger, "An Interview with Ira Lurie of the DEA," *Am. Lab.*, January 2005, p. 6; I. S. Lurie, P. A. Hays, A. D. Garcia, and S. Panicker, "Use of Dynamically Coated Capillaries for the Determination of Heroin, Basic Impurities and Adulterants with Capillary Electrophoresis," *J. Chromatogr. A* **2004**, *1034*, 227.

55. R. Weinberger, "Method Development for Capillary Electrophoresis," *Am. Lab.*, March 2003, p. 54.

56. S. P. Porras, M. L. Riekkola, and E. Kenndler, "The Principles of Migration and Dispersion in Capillary Zone Electrophoresis in Nonaqueous Solvents," *Electrophoresis* **2003**, *24*, 1485; M. L. Riekkola, "Recent Advances in Nonaqueous Capillary Electrophoresis," *Electrophoresis* **2002**, *23*, 3865; F. Steiner, and M. Hassel, "Non-aqueous Capillary Electrophoresis. A Versatile Completion of Electrophoretic Separation Techniques," *Electrophoresis* **2000**, *21*, 3994; R. Weinberger, "Nonaqueous Capillary Electrophoresis," *Am. Lab.*, February 2006, p. 49.

57. A. M. Siouffi and R. Phan-Tan-Luu, "Optimization Methods in Chromatography and Electrophoresis," *J. Chromatogr. A* **2000**, *892*, 75.

58. T. N. Chiesl, W. K. Chu, A. M. Stockton, X. Amashukeli, F. Grunthaner, and R. A. Mathies, "Enhanced Amine and Amino Acid Analysis Using Pacific Blue and the Mars Organic Analyzer Microchip Capillary Electrophoresis System," *Anal. Chem.* **2009**, *81*, 2537.

59. S. Hardt and F. Schönfeld, eds., *Microfluidic Technologies for Miniaturized Analysis Systems* (Heidelberg: Springer, 2007; H.-F. Li and J.-M. Lin, "Applications of Microfluidic Systems in Environmental Analysis," *Anal. Bioanal. Chem.* **2009**, *393*, 555; M. G. Roper, C. J. Easley, and J. P. Landers, "Advances in Polymerase Chain Reaction on Microfluidic Chips," *Anal. Chem.* **2005**, *77*, 3887; C.-C. Lee, G. Sui, A. Elizarov, C. J. Shu, Y.-S. Shin, A. N. Dooley, J. Huang, A. Daridon, P. Wyatt, D. Stout, H. C. Kolb, O. N. Witte, N. Satyamurthy, J. R. Heath, M. E. Phelps, S. R. Quake, and H.-R. Tseng, "Multistep Synthesis of a Radiolabeled Imaging Probe Using Integrated Microfluidics," *Science* **2005**, *310*, 1793; C. H. Legge, "Chemistry Under the Microscope—Lab-on-a-Chip Technologies," *J. Chem. Ed.* **2002**, *79*, 173.

60. C. J. Watson, B. J. Venton, and R. T. Kennedy, "In-Vivo Measurements of Neurotransmitters by Microdialysis Sampling," *Anal. Chem.* **2006**, *78*, 1391; M. Shou, C. R. Ferrario, K. N. Schultz, T. E. Robinson, and R. T. Kennedy, "Monitoring Dopamine In-Vivo by Microdialysis Sampling and On-Line CE–Laser-Induced Fluorescence," *Anal. Chem.* **2006**, *78*, 1391.

61. N. D. Chasteen and P. M. Harrison, "Mineralization in Ferritin: An Efficient Means of Iron Storage," *J. Struct. Biol.* **1999**, *126*, 182.

62. Data from A. W. Moore, Jr., S. C. Jacobson, and J. M. Ramsey, "Microchip Separations of Neutral Species via Micellar Electrokinetic Capillary Chromotography," *Anal. Chem.* **1995**, *67*, 4184.

63. The optimum pH for separating cations is pK + 0.30: K. K.-C. Yeung and C. A. Lucy, "Isotopic Separation of [^{14}N]- and [^{15}N]-Aniline by Capillary Electrophoresis Using Surfactant-Controlled Reversed Electroosmotic Flow," *Anal. Chem.* **1998**, *70*, 3286.

64. G. R. Asbury and H. H. Hill, Jr., "Evaluation of Ultrahigh Resolution Ion Mobility Spectrometry as an Analytical Separation Device in Chromatographic Terms," *J. Microcolumn Sep.* **2000**, *12*, 172; H. E. Revercomb and E. A. Mason, "Theory of Plasma Chromatography/Gaseous Electrophoresis," *Anal. Chem.* **1975**, *47*, 970.

Chapter 26

1. T. W. Richards, *Chem. Rev.* **1925**, *1*, 1; C. M. Beck, II, "Classical Analysis: A Look at the Past, Present, and Future," *Anal. Chem.* **1994**, *66*, 225A; I. M. Kolthoff, "Analytical Chemistry in the U.S.A. in the First Quarter of This Century," *Anal. Chem.* **1994**, *66*, 241A; D. T. Burns, "Highlights in the History of Quantitation in Chemistry," *Fresenius J. Anal. Chem.* **1990**, *337*, 205; L. Niiniströ, "Analytical Instrumentation in the 18th Century," *Fresenius J. Anal. Chem.* **1990**, *337*, 213.

2. Colloid experiment: C. D. Keating, M. D. Musick, M. H. Keefe, and M. J. Natan, "Kinetics and Thermodynamics of Au Colloid Monolayer Self Assembly," *J. Chem. Ed.* **1999**, *76*, 949.

3. Cellulose tubing such as catalog number 3787, sold by Thomas Scientific, www.thomassci.com, is adequate for this demonstration.

4. M. Suzuki, "The Movement of Molecules and Heat Energy: Two Demonstrative Experiments," *J. Chem. Ed.* **1993**, *70*, 821.

5. D. Erdemir, A. Y. Lee, and A. S. Myerson, "Nucleation of Crystals from Solution: Classical and Two-Step Models," *Acc. Chem. Res.* **2009**, *42*, 621.

6. A. Abou-Hassan, O. Sandre, S. Neveu, and V. Cabuil, "Synthesis of Goethite by Separation of the Nucleation and Growth Processes of Ferrihydrite Nanoparticles Using Microfluidics," *Angew. Chem. Int. Ed.* **2009**, *48*, 2342.

7. L. C. Roberts, S. J. Hug, T. Ruettimann, M. Billah, A. W. Khan, and M. T. Rahman, "Arsenic Removal with Iron(II) and Iron(III) Waters with High Silicate and Phosphate Concentrations," *Environ. Sci. Technol.* **2004**, *38*, 307. A more convenient method to remove arsenite and arsenate from drinking water is to pass the water through diatomaceous earth (silica skeletons of algae with 80–90% voids) containing hydrous ferric oxide. (M. Jang, S.-H. Min, T.-H. Kim, and J. K. Park, "Removal of Arsenite and Arsenate Using Hydrous Ferric Oxide Incorporated in Naturally Occurring Porous Diatomite," *Environ. Sci. Technol.* **2006**, *40*, 1636.)

8. G. W. Latimer, Jr., "Piperazine as the Diacetate," *J. Chem. Ed.* **1966**, *43*, 148; G. R. Bond, *Anal. Chem.* **1962**, *32*, 1332.

9. E. Pella, "Elemental Organic Analysis. 1. Historical Developments," *Am. Lab.*, February 1990, p. 116; "Elemental Organic Analysis. 2. State of the Art," *Am. Lab.*, August 1990, p. 28.

10. K. Russe, H. Kipphardt, and J. A. C. Broekaert, "Determination of Main and Minor Components of Silicon Based Materials by Combustion with F$_2$," *Anal. Chem.* **2000**, *72*, 3875.

11. R. W. Ramette, "Stoichiometry to the Rescue (A Calculation Challenge)," *J. Chem. Ed.* **1988**, *65*, 800.

12. G. Li, G. Su, X. Zhuang, Z. Li, and Y. He, "A New Method to Determine the Deuterium Content of DKDP Crystal with Thermo-Gravimetric Apparatus," *Optical Mater.* **2006**, *29*, 220.

13. H. Bao, "Purifying Barite for Oxygen Isotope Measurement by Dissolution and Reprecipitation in a Chelating Solution," *Anal. Chem.* **2006**, *78*, 304.

14. F. Torrades and M. Castellvi, "Spectrophotometric Determination of Cl$^-$ in BaSO$_4$ Precipitate," *Fresenius J. Anal. Chem.* **1994**, *349*, 734.

15. M. L. Ware, M. D. Argentine, and G. W. Rice, "Potentiometric Determination of Halogen Content in Organic Compounds Using Dispersed Sodium Reduction," *Anal. Chem.* **1988**, *60*, 383.

Chapter 27

1. E. P. Popek, *Sampling and Analysis of Environmental Chemical Pollutants* (Amsterdam: Academic Press, 2003); P. Gy, *Sampling for Analytical Purposes* (Chichester, UK: Wiley, 1998); B. B. Kebbekus and S. Mitra, *Environmental Chemical Analysis* (London: Blackie, 1998); N. T. Crosby and I. Patel, *General Principles of Good Sampling Practice* (Cambridge: Royal Society of Chemistry, 1995); R. F. Cross, "Reducing Sample Size and Obtaining Representative Samples," *LCGC* **2000**, *18*, 468.

2. S. Mitra, ed., *Sample Preparation Techniques in Analytical Chemistry* (Hoboken, NJ: Wiley, 2003).

3. D. T. Sawyer, A. Sobkowiak, and J. L. Roberts, Jr., *Electrochemistry for Chemists*, 2nd ed. (New York: Wiley, 1995), p. 262.

4. P. Vallelonga, K. Van de Velde, J. P. Candelone, C. Ly, K. J. R. Rosman, C. F. Boutron, V. I. Morgan, and D. J. Mackey, "Recent Advances in Measurement of Pb Isotopes in Polar Ice and Snow at Sub-Picogram per Gram Concentrations Using Thermal Ionisation Mass Spectrometry," *Anal. Chim. Acta* **2002**, *453*, 1.

5. R. P. Mason and K. A. Sullivan, "Mercury in Lake Michigan," *Environ. Sci. Technol.* **1997**, *31*, 942.

6. G. Benoit, K. S. Hunter, and T. F. Rozan, "Sources of Trace Metal Contamination Artifacts During Collection, Handling, and Analysis of Freshwaters," *Anal. Chem.* **1997**, *69*, 1006.

7. R. E. Thiers, *Methods of Biochemical Analysis* (D. Glick, ed.), Vol. 5 (New York: Interscience, 1957), p. 274.

8. B. Kratochvil and J. K. Taylor, "Sampling for Chemical Analysis," *Anal. Chem.* **1981**, *53*, 924A; H. A. Laitinen and W. E. Harris, *Chemical Analysis*, 2nd ed. (New York: McGraw-Hill, 1975), Chap. 27; S. K. Thompson, *Sampling* (New York: Wiley, 1992).

9. J. E. Vitt and R. C. Engstrom, "Effect of Sample Size on Sampling Error," *J. Chem. Ed.* **1999**, *76*, 99; R. D. Guy, L. Ramaley, and P. D. Wentzell, "Experiment in the Sampling of Solids for Chemical Analysis," *J. Chem. Ed.* **1998**, *75*, 1028. Demonstrations: L. S. Canaes, M. L. Brancalion, A. V. Rossi, and S. Rath, "Using Candy Samples to Learn About Sampling Techniques and Statistical Data Evaluation," *J. Chem. Ed.* **2008**, *85*, 1083; M. R. Ross, "A Classroom Exercise in Sampling Technique," *J. Chem. Ed.* **2000**, *77*, 1015; J. R. Hartman, "In-Class Experiment on the Importance of Sampling Techniques and Statistical Analysis of Data," *J. Chem. Ed.* **2000**, *77*, 1017.

10. D. C. Bogen, *Treatise on Analytical Chemistry*, 2nd ed. (P. J. Elving, E. Grushka, and I. M. Kolthoff, eds.), Part I, Vol. 5 (New York: Wiley, 1982), Chap. 1.

11. H. M. Kingston and S. J. Haswell, *Microwave-Enhanced Chemistry* (Washington, DC: American Chemical Society, 1997); B. D. Zehr, "Development of Inorganic Microwave Dissolutions," *Am. Lab.*, December 1992, p. 24; B. D. Zehr, J. P. VanKuren, and H. M. McMahon, "Inorganic Microwave Digestions Incorporating Bases," *Anal. Chem.* **1994**, *66*, 2194.

12. Z. Grosser, L. Thompson, and L. Davidowski, "Inorganic Analysis for Environmental RoHS Compliance," *Am. Lab.*, October 2007, p. 30.

13. E. M. M. Flores, M. F. Mesko, D. P. Moraes, J. S. F. Pereira, P. A. Mello, J. S. Barin, and G. Knapp, "Determination of Halogens in Coal after Digestion using the Microwave-Induced Combustion Technique," *Anal. Chem.* **2008**, *80*, 1865.

14. P. Aysola, P. Anderson, and C. H. Langford, "Wet Ashing in Biological Samples in a Microwave Oven under Pressure Using Poly(tetrafluoroethylene) Vessels," *Anal. Chem.* **1987**, *59*, 1582.

15. S. E. Long and W. R. Kelly, "Determination of Mercury in Coal by Isotope Dilution Cold-Vapor Generation Inductively Coupled Plasma Mass Spectrometry," *Anal. Chem.* **2002**, *74*, 1477.

16. M. Rehkämper, A. N. Halliday, and R. F. Wentz, "Low-Blank Digestion of Geological Samples for Pt-Group Analysis Using a Modified Carius Tube Design," *Fresenius J. Anal. Chem.* **1998**, *361*, 217.

17. A. A. Schilt, *Perchloric Acid and Perchlorates*, 2nd ed. (Powell, OH: GFS Chemicals, 2003).

18. C. Walling, "Intermediates in the Reactions of Fenton Type Reagents," *Acc. Chem. Res.* **1998**, *31*, 155; P. A. MacFaul, D. D. M. Wayner, and K. U. Ingold, "A Radical Account of 'Oxygenated Fenton Chemistry,'" *Acc. Chem. Res.* **1998**, *31*, 159; D. T. Sawyer, A. Sobkowiak, and T. Matsushita, "Oxygenated Fenton Chemistry," *Acc. Chem. Res.* **1996**, *29*, 409; O. Pestovsky, S. Stoian, E. L. Bominaar, X. Shan, E. Münck, L. Que, Jr., and A. Bakac, "Aqueous Fe^{IV}=O: Spectroscopic Identification and Oxo-Group Exchange," *Angew. Chem. Int. Ed.* **2005**, *44*, 6871.

19. D. C. Luehrs and A. E. Roher, "Demonstration of the Fenton Reaction," *J. Chem. Ed.* **2007**, *84*, 1290.

20. L. Ping and P. K. Dasgupta, "Determination of Total Mercury in Water and Urine by a Gold Film Sensor Following Fenton's Reagent Digestion," *Anal. Chem.* **1989**, *61*, 1230.

21. G. LeBlanc, "Microwave-Accelerated Techniques for Solid Sample Extraction," *LCGC* **1999**, *17*, S30 (June 1999 supplement).

22. Y. Lin and C. M. Wai, "Supercritical Fluid Extraction of Lanthanides with Fluorinated β-Diketones and Tributyl Phosphate," *Anal. Chem.* **1994**, *66*, 1971.

23. L. T. Taylor, *Supercritical Fluid Extraction* (New York: Wiley, 1996); C. L. Phelps, N. G. Smart, and C. M. Wai, "Past, Present, and Possible Future Applications of Supercritical Fluid Extraction Technology," *J. Chem. Ed.* **1996**, *73*, 1163; M. E. P. McNally, "Advances in Environmental SFE," *Anal. Chem.* **1995**, *67*, 308A; L. T. Taylor, "Strategies for Analytical SFE," *Anal. Chem.* **1995**, *67*, 364A.

24. S. Liang and D. C. Tilotta, "Extraction of Petroleum Hydrocarbons from Soil Using Supercritical Argon," *Anal. Chem.* **1998**, *70*, 616.

25. H. M. Stapleton, N. G. Dodder, J. H. Offenberg, M. M. Schantz, and S. A. Wise, "Polybrominated Diphenyl Ethers in House Dust and Clothes Dryer Lint," *Environ. Sci. Technol.* **2005**, *39*, 925. "Best Papers of 2005," *Environ. Sci. Technol.* **2006**, *40*, 2087.

26. www.biotage.com; L. Williams, H. Lodder, S. Merriman, A. Howells, S. Jordan, J. Labadie, M. Cleeve, C. Desbrow, R. Calverley, and M. Burke, "Extraction of Drugs from Plasma using ISOLUTE SLE+ Supported Liquid Extraction Plates," *LCGC Application Notebook*, February 2006, p. 12.

27. E. M. Thurman and M. S. Mills, *Solid-Phase Extraction: Principles and Practice* (New York: Wiley, 1998); N. J. K. Simpson, *Solid-Phase Extraction* (New York: Marcel Dekker, 2000).

28. J.-H. Zhang, M. Jiang, L. Zou, D. Shi, S.-R. Mei, Y.-X. Zhu, Y. Shi, K. Dai, and B. Lu, "Selective Solid-Phase Extraction of Bisphenol A using Molecularly Imprinted Polymers and its Application to Biological and Environmental Samples," *Anal. Bioanal. Chem.* **2006**, *385*, 780.

29. R. Hites and V. S. Ong, "Determination of Pesticides and Polychlorinated Biphenyls in Water: A Low-Solvent Method," *Environ. Sci. Technol.* **1995**, *29*, 1259.

30. J. W. Wong, K. K. Ngim, T. Shibamoto, S. A. Mabury, J. P. Eiserich, and H. C. H. Yeo, "Determination of Formaldehyde in Cigarette Smoke," *J. Chem. Ed.* **1997**, *74*, 1100. As another example, an analysis of nitrite (NO_2^-) in natural waters is based on its reaction with 2,4-dinitrophenylhydrazine to give an azide ($R-N_3$) that is measured by HPLC with ultraviolet detection at 307 nm. (R. J. Kieber and P. J. Seaton, "Determination of Subnanomolar Concentrations of Nitrite in Natural Waters," *Anal. Chem.* **1995**, *67*, 3261.)

31. D. M. Crumbling, Letter to Editor, *C&E News,* 12 August 2002, p. 4.

32. E. Tareke, P. Rydberg, P. Karlsson, S. Eriksson, and M. Tornqvist, "Analysis of Acrylamide, A Carcinogen Formed in Heated Foodstuffs," *J. Agric. Food Chem.* **2002**, *50*, 4998; B. E. Erickson, "Finding Acrylamide," *Anal. Chem.* **2004**, *76*, 247A. See collection of papers in *J. AOAC Int.*, **2005**, *88*, 227–330.

GLOSSARY

ablation   Vaporization of a small volume of material by a laser pulse.

abscissa   Horizontal *(x)* axis of a graph.

absolute uncertainty   An expression of the margin of uncertainty associated with a measurement. Absolute error also could refer to the difference between a measured value and the "true" value.

absorbance, A   Defined as $A = \log(P_0/P)$, where P_0 is the radiant power of light (power per unit area) striking the sample on one side and P is the radiant power emerging from the other side. Also called *optical density*.

absorptance, a   Fraction of incident radiant power absorbed by a sample.

absorption   Occurs when a substance is taken up *inside* another. See also *adsorption*.

absorption coefficient   Light absorbed by a sample is attenuated at the rate $P_2/P_1 = e^{-\alpha b}$, where P_1 is the initial radiant power, P_2 is the power after traversing a pathlength b, and α is called the absorption coefficient.

absorption spectrum   A graph of absorbance of light versus wavelength, frequency, or wavenumber.

accuracy   A measure of how close a measured value is to the "true" value.

acid   A substance that increases the concentration of H^+ when added to water.

acid-base titration   One in which the reaction between analyte and titrant is an acid-base reaction.

acid dissociation constant, K_a   Equilibrium constant for the reaction of an acid, HA, with H_2O:

$$HA + H_2O \overset{K_a}{\rightleftharpoons} A^- + H_3O^+ \qquad K_a = \frac{\mathcal{A}_{A^-}\mathcal{A}_{H_3O^+}}{\mathcal{A}_{HA}}$$

acid error   Systematic error that occurs in strongly acidic solutions, where glass electrodes tend to indicate a value of pH that is too high.

acidic solution   One in which the activity of H^+ is greater than the activity of OH^-.

acidity   In natural waters, the quantity of carbonic acid and other dissolved acids that react with strong base when the pH of the sample is raised to 8.3. Expressed as mmol OH^- needed to raise the pH of 1 L to pH 8.3.

acid wash   Procedure in which glassware is soaked in 3–6 M HCl for >1 h (followed by rinsing well with distilled water and soaking in distilled water) to remove traces of cations adsorbed on the surface of the glass and to replace them with H^+.

activation energy, E_a   Energy needed for a process to overcome a barrier that otherwise prevents the process from occurring.

activity, $\mathcal{A}$ The value that replaces concentration in a thermodynamically correct equilibrium expression. The activity of X is given by $\mathcal{A}_X = [X]\gamma_X$, where γ_X is the activity coefficient and [X] is the concentration.

activity coefficient, γ   The number by which the concentration must be multiplied to give activity.

adduct   Product formed when a Lewis base combines with a Lewis acid.

adjusted retention time, t'_r   In chromatography, this parameter is $t'_r = t_r - t_m$, where t_r is the retention time of a solute and t_m is the time needed for mobile phase to travel the length of the column.

adsorption   Occurs when a substance becomes attached to the *surface* of another substance. See also *absorption*.

adsorption chromatography   A technique in which solute equilibrates between the mobile phase and adsorption sites on the stationary phase.

adsorption indicator   Used for precipitation titrations, it becomes attached to a precipitate and changes color when the surface charge of the precipitate changes sign at the equivalence point.

aerosol   A suspension of very small liquid or solid particles in air or gas. Examples include fog and smoke.

affinity chromatography   A technique in which a particular solute is retained by a column by virtue of a specific interaction with a molecule covalently bound to the stationary phase.

aliquot   Portion.

alkali flame detector   Modified flame ionization detector that responds to N and P, which produce ions when they contact a Rb_2SO_4-containing glass bead in the flame. Also called *nitrogen-phosphorus detector*.

alkalimetric titration   With reference to EDTA, this technique involves titration of the protons liberated from EDTA upon binding to a metal.

alkaline error   See *sodium error*.

alkalinity   In natural water, the quantity of base (mainly HCO_3^-, CO_3^{2-}, and OH^-) that reacts with strong acid when the pH of the sample is lowered to 4.5. Expressed as mmol H^+ needed to lower the pH of 1 L to pH 4.5.

amalgam   A solution of anything in mercury.

amine   A compound with the general formula RNH_2, R_2NH, or R_3N, where R is any group of atoms.

amino acid   One of 20 building blocks of proteins, having the general structure

$$\overset{\displaystyle R}{\underset{}{H_3\overset{+}{N}CCO_2^-}}$$

where R is a different substituent for each acid.

ammeter   Device for measuring electric current.

ammonium ion   *The* ammonium ion is NH_4^+. *An* ammonium ion is any ion of the type RNH_3^+, $R_2NH_2^+$, R_3NH^+, or R_4N^+, where R is an organic substituent.

ampere, A   One ampere is the current that will produce a force of exactly 2×10^{-7} N/m when that current flows through two "infinitely" long, parallel conductors of negligible cross section, with a spacing of 1 m, in a vacuum.

amperometric detector   See *electrochemical detector*.

amperometric titration   One in which the end point is determined by monitoring the current passing between two electrodes immersed in the sample solution and maintained at a constant potential difference.

amperometry   Measurement of electric current for analytical purposes.

amphiprotic molecule   One that can act as both a proton donor and a proton acceptor. The intermediate species of polyprotic acids are amphiprotic.

analysis of variance   A statistical tool used to dissect the overall random error into contributions from several sources.

analyte   Substance being measured or detected.

analytical chromatography   Chromatography of small quantities of material conducted for the purpose of qualitative or quantitative analysis or both.

analytical concentration   See *formal concentration*.

anhydrous   Adjective describing a substance from which all water has been removed.

anion   A negatively charged ion.

anion exchanger   An ion exchanger with positively charged groups covalently attached to the support. It can reversibly bind anions.

anode   Electrode at which oxidation occurs. In electrophoresis, it is the positively charged electrode.

anodic depolarizer   A molecule that is easily oxidized, thereby preventing the anode potential of an electrochemical cell from becoming too positive.

anodic wave   In voltammetry, a flow of current due to oxidation at the working electrode.

anolyte   Solution present in the anode chamber of an electrochemical cell.

antibody   A protein manufactured by an organism to sequester foreign molecules and mark them for destruction.

antigen   A molecule that is foreign to an organism and causes the organism to make antibodies.

antilogarithm   The antilogarithm of a is b if $10^a = b$.

apparent mobility   Constant of proportionality, μ_{app}, between the net speed, u_{net}, of an ion in solution and the applied electric field, E: $u_{net} = \mu_{app}E$. Apparent mobility is the sum of the *electrophoretic* and *electroosmotic* *mobilities*: $\mu_{app} = \mu_{ep} + \mu_{eo}$.

aprotic solvent One that cannot donate protons (hydrogen ions) in an acid-base reaction.

aptamer A short (15–40 bases) length of single- or double-stranded DNA (deoxyribonucleic acid) or RNA (ribonucleic acid) that strongly binds to a selected molecule.

aqua regia A 3:1 (vol/vol) mixture of concentrated (37 wt%) HCl and concentrated (70 wt%) HNO_3.

aqueous In water (as an *aqueous* solution).

aquo ion The species $M(H_2O)_n^{m+}$, containing just the cation M and its tightly bound water ligands.

argentometric titration One using Ag^+ ion.

ashless filter paper Specially treated paper that leaves a negligible residue after ignition. It is used for gravimetric analysis.

assessment In quality assurance, the process of (1) collecting data to show that analytical procedures are operating within specified limits and (2) verifying that final results meet use objectives.

asymmetry potential When the activity of analyte is the same on the inside and outside of an ion-selective electrode, there should be no voltage across the membrane. In fact, the two surfaces are never identical, and some voltage (called the asymmetry potential) is usually observed.

atmosphere, atm One atm is defined as a pressure of 101 325 Pa. It is equal to the pressure exerted by a column of Hg 760 mm in height at Earth's surface.

atmospheric pressure chemical ionization A method to interface liquid chromatography to mass spectrometry. Liquid is nebulized into a fine aerosol by a coaxial gas flow and the application of heat. Electrons from a high-voltage corona discharge create cations and anions from analyte exiting the chromatography column. The most common species observed with this interface is MH^+, the protonated analyte, with little fragmentation.

atomic absorption spectroscopy A technique in which the absorption of light by free gaseous atoms in a plasma, flame, or furnace is used to measure the concentration of atoms.

atomic emission spectroscopy A technique in which the emission of light by thermally excited atoms in a flame or furnace is used to measure the concentration of atoms.

atomic fluorescence spectroscopy A technique in which electronic transitions of atoms in a flame, furnace, or plasma are excited by light, and the fluorescence is observed at a right angle to the incident beam.

atomic mass Number of grams of an element containing Avogadro's number of atoms.

atomization Process in which a compound is decomposed into its atoms at high temperature.

attenuated total reflection An analytical technique based on passage of light through a waveguide or optical fiber by total internal reflection. The absorption of the cladding is sensitive to the presence of analyte. Some of the evanescent wave is absorbed in the cladding during each reflection in the presence of analyte. The more analyte, the more signal is lost.

autoprotolysis The reaction in which two molecules of the same species transfer a proton from one to the other; e.g., $CH_3OH + CH_3OH \rightleftharpoons CH_3OH_2^+ + CH_3O^-$.

autoprotolysis constant The equilibrium constant for an autoprotolysis reaction.

autotitrator A device that dispenses measured amounts of titrant into a solution and monitors a property such as pH or electrode potential after each addition. The instrument performs the titration automatically and can determine the end point automatically. Data can be transferred to a spreadsheet for further interpretation.

auxiliary complexing agent A species, such as ammonia, that is added to a solution to stabilize a metal ion and keep that metal in solution. It binds loosely enough to be displaced by a titrant.

auxiliary electrode Current-carrying partner of the working electrode in an electrolysis. Also called *counter electrode*.

average The sum of several values divided by the number of values. Also called *mean*.

Avogadro's number The number of atoms in exactly 0.012 kg of ^{12}C, approximately 6.022×10^{23}.

azeotrope The constant-boiling distillate produced by two liquids. It is of constant composition, containing both substances.

background correction In atomic spectroscopy, a means of distinguishing signal due to analyte from signal due to absorption, emission, or scattering by the flame, furnace, plasma, or sample matrix.

background electrolyte In capillary electrophoresis, the buffer in which separation is carried out. Also called *run buffer*.

back titration One in which an excess of standard reagent is added to react with analyte. Then the excess reagent is titrated with a second reagent or with a standard solution of analyte.

ball mill A drum in which solid sample is ground to fine powder by tumbling with hard ceramic balls.

band gap Energy separating the valence band and conduction band in a semiconductor.

band-pass filter A filter that allows a band of wavelengths to pass through it while absorbing or reflecting other wavelengths.

bandwidth Usually, the range of wavelengths or frequencies of an absorption or emission band, typically measured at a height equal to half of the peak height. Also, the width of radiation emerging from the exit slit of a monochromator.

base A substance that decreases the concentration of H^+ when added to water.

base "dissociation" constant A misnomer for *base hydrolysis constant*, K_b.

base hydrolysis constant, K_b The equilibrium constant for the reaction of a base, B, with H_2O:

$$B + H_2O \overset{K_b}{\rightleftharpoons} BH^+ + OH^- \qquad K_b = \frac{\mathcal{A}_{BH^+}\mathcal{A}_{OH^-}}{\mathcal{A}_B}$$

base peak Most intense peak in a mass spectrum.

basic solution One in which the activity of OH^- is greater than the activity of H^+.

beam chopper A rotating mirror that directs light alternately through the sample and reference cells of a double-beam spectrophotometer.

beam chopping A technique using a rotating *beam chopper* to modulate the signal in a spectrophotometer at a frequency at which noise is reduced. In atomic absorption, periodic blocking of the beam allows a distinction to be made between light from the source and light from the flame.

beamsplitter A partially reflective, partially transparent plate that reflects some light and transmits some light.

Beer's law Relates the absorbance, A, of a sample to its concentration, c, pathlength, b, and molar absorptivity, ε: $A = \varepsilon bc$. More correctly called *Beer-Lambert-Bouguer law*.

biamperometric titration An amperometric titration conducted with two polarizable electrodes held at a constant potential difference.

bilayer Formed by a surfactant, a two-dimensional membrane structure in which polar or ionic headgroups are pointing outward and nonpolar tails are pointing inward.

biological oxygen demand, BOD In a water sample, the quantity of dissolved oxygen consumed by microorganisms during a 5-day incubation in a sealed vessel at 20°C. Oxygen consumption is limited by organic nutrients, so BOD is a measure of pollutant concentration.

biosensor Device that uses biological components such as enzymes, antibodies, or DNA, in combination with electric, optical, or other signals, to achieve a highly selective response to one analyte.

bipotentiometric titration A potentiometric titration in which a constant current is passed between two polarizable electrodes immersed in the sample solution. An abrupt change in potential characterizes the end point.

Bjerrum plot See *difference plot*.

blackbody An ideal surface that absorbs all photons striking it. If the blackbody is at constant temperature, it must emit as much radiant energy as it absorbs.

blackbody radiation Radiation emitted by a blackbody. The energy and spectral distribution of the emission depend only on the temperature of the blackbody.

blank solution   A solution not intended to contain analyte. It could be made from all reagents—except unknown—that would be used in an analytical procedure. Analyte signal measured with a blank solution could be due to impurities in the reagents or, possibly, interference.

blank titration   One in which a solution containing all reagents except analyte is titrated. The volume of titrant needed in the blank titration should be subtracted from the volume needed to titrate unknown.

blind sample   See *performance test sample*.

blocking   Occurs when a metal ion binds tightly to a metal ion indicator. A blocked indicator is unsuitable for a titration because no color change is observed at the end point.

Boltzmann distribution   Relative population of two states at thermal equilibrium:

$$\frac{N_2}{N_1} = \frac{g_2}{g_1} e^{-(E_2 - E_1)/kT}$$

where N_i is the population of the state, g_i is the degeneracy of the state, E_i is the energy of the state, k is Boltzmann's constant, and T is temperature in kelvins. *Degeneracy* refers to the number of states with the same energy.

bomb   Sealed vessel for conducting high-temperature, high-pressure reactions.

bonded stationary phase   In high-performance liquid chromatography, a stationary liquid phase covalently attached to the solid support.

Brønsted-Lowry acid   A proton (hydrogen ion) donor.

Brønsted-Lowry base   A proton (hydrogen ion) acceptor.

Brownian motion   Random motion of a small particle in a liquid caused by collisions with molecules moving with random speeds in random directions.

buffer   A mixture of a weak acid and its conjugate base. A buffered solution is one that resists changes in pH when acids or bases are added.

buffer capacity, β   A measure of the ability of a buffer to resist changes in pH. The larger the buffer capacity, the greater the resistance to pH change. The definition of buffer capacity is $\beta = dC_b/d\text{pH} = -dC_a/d\text{pH}$, where C_a and C_b are the number of moles of strong acid or base per liter needed to produce a unit change in pH. Also called *buffer intensity*.

buffer intensity   See *buffer capacity*.

bulk sample   Material taken from lot being analyzed—usually chosen to be representative of the entire lot. Also called *gross sample*.

bulk solution   Chemists' jargon referring to the main body of a solution. In electrochemistry, we distinguish properties of the bulk solution from properties that may be different in the immediate vicinity of an electrode.

buoyancy   Upward force exerted on an object in a liquid or gaseous fluid. An object weighed in air appears lighter than its actual mass by an amount equal to the mass of air that it displaces.

buret   A calibrated glass tube with a stopcock at the bottom. Used to deliver known volumes of liquid.

calibration   Process of relating the actual physical quantity (such as mass, volume, force, or electric current) to the quantity indicated on the scale of an instrument.

calibration check   In a series of analytical measurements, a calibration check is an analysis of a solution formulated by the analyst to contain a known concentration of analyte. It is the analyst's own check that procedures and instruments are functioning correctly.

calibration curve   A graph showing the value of some property versus concentration of analyte. When the corresponding property of an unknown is measured, its concentration can be determined from the graph.

calomel electrode   A common reference electrode based on the half-reaction $Hg_2Cl_2(s) + 2e^- \rightleftharpoons 2Hg(l) + 2Cl^-$. See also *S.C.E.*

candela, cd   Luminous intensity, in a given direction, of a source that emits monochromatic radiation of frequency 540 THz and that has a radiant intensity of 1/683 W/sr in that direction.

capacitance   The electric capacitance of two parallel, charged surfaces is the charge on each surface divided by the electric potential difference (volts) between the two surfaces.

capacitor current   See *charging current*.

capacity factor   See *retention factor*.

capacity ratio   See *retention factor*.

capillary electrochromatography   A version of high-performance liquid chromatography in which mobile phase is driven by electroosmosis instead of a pressure gradient. Solutes are separated by partitioning between the mobile and stationary phases.

capillary electrophoresis   Separation of a mixture into its components by a strong electric field imposed between the two ends of a narrow capillary tube filled with electrolyte solution. Unlike chromatography, there is no stationary phase in electrophoresis. Solutes are separated by differences in mobility.

capillary gel electrophoresis   A form of capillary electrophoresis in which the tube is filled with a polymer gel that serves as a sieve for macromolecules. The largest molecules have the slowest migration through the gel.

capillary zone electrophoresis   A form of capillary electrophoresis in which ionic solutes are separated because of differences in their electrophoretic mobility.

carboxylate anion   Conjugate base (RCO_2^-) of a carboxylic acid.

carboxylic acid   A molecule with the general structure RCO_2H, where R is any group of atoms.

carcinogen   A cancer-causing agent.

carrier gas   Mobile phase gas in gas chromatography.

cathode   Electrode at which reduction occurs. In electrophoresis, it is the negatively charged electrode.

cathodic depolarizer   A molecule that is easily reduced, thereby preventing the cathode potential of an electrochemical cell from becoming very low.

cathodic wave   In voltammetry, a flow of current due to reduction at the working electrode.

catholyte   Solution present in the cathode chamber of an electrochemical cell.

cation   A positively charged ion.

cation exchanger   An ion exchanger with negatively charged groups covalently attached to the support. It can reversibly bind cations.

certified reference material   Samples sold by national measurement institutes and containing known quantities of analytes to test accuracy of analytical procedures. The U.S. National Institute of Standards and Technology calls its certified materials *Standard Reference Materials*.

chain of custody   Trail followed by a sample from the time it is collected to the time it is analyzed and, possibly, archived.

characteristic   The part of a logarithm at the left of the decimal point.

charge balance   A statement that the sum of all positive charge in solution equals the magnitude of the sum of all negative charge in solution.

charge coupled device   An extremely sensitive detector in which light creates electrons and holes in a semiconductor material. The electrons are attracted to regions near positive electrodes, where the electrons are "stored" until they are ready to be counted. The number of electrons in each pixel (picture element) is proportional to the number of photons striking the pixel.

charged aerosol detector   Sensitive, nearly universal detector for liquid chromatography in which solvent is evaporated from eluate to leave an aerosol of fine particles of nonvolatile solute. Aerosol particles are charged by adsorption of N_2^+ ions and flow to a collector that measures total charge reaching the detector versus time.

charging current   Electric current arising from charging or discharging of the electric double layer at the electrode–solution interface. Also called *capacitor current* or *condenser current*.

charring   In a gravimetric analysis, the precipitate and filter paper are first *dried* gently. Then the filter paper is *charred* at intermediate temperature to destroy the paper without letting it inflame. Finally, precipitate is *ignited* at high temperature to convert it into its analytical form.

chelate effect   The observation that a single multidentate ligand forms metal complexes that are more stable than those formed by several individual ligands with the same ligand atoms.

chelating ligand   A ligand that binds to a metal through more than one atom.

chemical interference   In atomic spectroscopy, any chemical reaction that decreases the efficiency of atomization.

chemical ionization   A gentle method of producing ions for a mass spectrometer without extensive fragmentation of the analyte molecule, M. A reagent gas such as CH_4 is bombarded with electrons to make CH_5^+, which transfers H^+ to M, giving MH^+.

chemical oxygen demand, COD   In a natural water or industrial effluent sample, the quantity of O_2 equivalent to the quantity of $K_2Cr_2O_7$ consumed by refluxing the sample with a standard dichromate–sulfuric acid solution containing Ag^+ catalyst. Because 1 mol of $K_2Cr_2O_7$ consumes $6e^-$ ($Cr^{6+} \rightarrow Cr^{3+}$), it is equivalent to 1.5 mol of O_2 ($O \rightarrow O^{2-}$). See also *oxidizability*.

chemiluminescence   Emission of light by an excited-state product of a chemical reaction.

chiral molecule   One that is not superimposable on its mirror image in any accessible conformation. Also called an *optically active molecule*, a chiral molecule rotates the plane of polarization of light.

chromatogram   A graph showing chromatography detector response as a function of elution time or volume.

chromatograph   A machine used to perform chromatography.

chromatography   A technique in which molecules in a mobile phase are separated because of their different affinities for a stationary phase. The greater the affinity for the stationary phase, the longer a molecule is retained.

chromophore   The part of a molecule responsible for absorption of light of a particular frequency.

chronoamperometry   A technique in which the potential of a working electrode in an unstirred solution is varied rapidly while the current between the working and the auxiliary electrodes is measured. Suppose that the analyte is reducible and that the potential of the working electrode is made more negative. Initially, no reduction occurs. At a certain potential, the analyte begins to be reduced and the current increases. As the potential becomes more negative, the current increases further until the concentration of analyte at the surface of the electrode is sufficiently depleted. Then the current decreases, even though the potential becomes more negative. The maximum current is proportional to the concentration of analyte in bulk solution.

chronopotentiometry   A technique in which a constant current is forced to flow between two electrodes. The voltage remains fairly steady until the concentration of an electroactive species becomes depleted. Then the voltage changes rapidly as a new redox reaction assumes the burden of current flow. The elapsed time when the voltage suddenly changes is proportional to the concentration of the initial electroactive species in bulk solution.

cladding   Covering; overlay. Layer surrounding the core of an optical fiber.

Clark electrode   One that measures the concentration of dissolved oxygen by amperometry.

coagulation   With respect to gravimetric analysis, small crystallites coming together to form larger crystals.

co-chromatography   See *spike*.

coefficient of variation   The standard deviation, s, expressed as a percentage of the mean value $\bar{x}$: coefficient of variation = $100 \times s/\bar{x}$. Also called *relative standard deviation*.

coherence   Degree to which electromagnetic waves are in phase with one another. Laser light is highly coherent.

co-ion   An ion with the same charge as the ion of interest.

cold trapping   Splitless gas chromatography injection technique in which solute is condensed far below its boiling point in a narrow band at the start of the column.

collimated light   Light in which all rays travel in parallel paths.

collimation   Process of making light rays travel parallel to one another.

collisionally activated dissociation   Fragmentation of an ion in a mass spectrometer by high-energy collisions with gas molecules. In atmospheric pressure chemical ionization or electrospray interfaces, collisionally activated dissociation at the inlet to the mass filter can be promoted by varying the cone voltage. In tandem mass spectrometry, dissociation occurs in a

collision cell between the two mass separators. Also called *collision-induced dissociation*.

collision cell   Middle stage of a tandem mass spectrometer in which the precursor ion selected by the first stage is fragmented by collisions with gas molecules.

colloid   A dissolved particle with a diameter in the approximate range 1–500 nm. It is too large to be considered one molecule but too small to simply precipitate.

combination electrode   A glass electrode with a concentric reference electrode built on the same body.

combustion analysis   A technique in which a sample is heated in an atmosphere of O_2 to oxidize it to CO_2 and H_2O, which are collected and weighed or measured by gas chromatography. Modifications permit the simultaneous analysis of N, S, and halogens.

common ion effect   Occurs when a salt is dissolved in a solution already containing one of the ions of the salt. The salt is less soluble than it would be in a solution without that extra ion. An application of Le Châtelier's principle.

complex ion   Historical name for any ion containing two or more ions or molecules that are each stable by themselves; e.g., $CuCl_3^-$ contains $Cu^+ + 3Cl^-$.

complexometric titration   One in which the reaction between analyte and titrant involves complex formation.

composite sample   A representative sample prepared from a heterogeneous material. If the material consists of distinct regions, the composite is made of portions of each region, with relative amounts proportional to the size of each region.

compound electrode   An ion-selective electrode consisting of a conventional electrode surrounded by a barrier that is selectively permeable to the analyte of interest. Alternatively, the barrier region might convert external analyte into a different species, to which the inner electrode is sensitive.

concentration   An expression of the quantity per unit volume or unit mass of a substance. Common measures of concentration are molarity (mol/L) and molality (mol/kg of solvent).

concentration cell   A galvanic cell in which both half-reactions are the same, but the concentrations in each half-cell are not identical. The cell reaction increases the concentration of species in one half-cell and decreases the concentration in the other.

concentration polarization   Occurs when an electrode reaction occurs so rapidly that the concentration of solute near the surface of the electrode is not the same as the concentration in bulk solution.

condenser current   See *charging current*.

conditional formation constant, K_f'   Equilibrium constant for formation of a complex under a particular stated set of conditions, such as pH, ionic strength, and concentration of auxiliary complexing species. Also called *effective formation constant*.

conduction band   Energy levels containing conduction electrons in a semiconductor.

conduction electron   An electron free to move about within a solid and carry electric current. In a semiconductor, the energies of the conduction electrons are above those of the valence electrons that are localized in chemical bonds. The energy separating the valence and conduction bands is called the *band gap*.

conductivity, σ   Proportionality constant between electric current density, J (A/m^2), and electric field, E (V/m): $J = \sigma E$. Units are $\Omega^{-1}\,m^{-1}$. Conductivity is the reciprocal of *resistivity*.

cone voltage   Voltage applied between the *skimmer cone* and a nearby orifice through which gaseous ions flow into the mass separator of a mass spectrometer. The magnitude of the voltage can be increased to promote collisionally activated dissociation of ions prior to mass separation.

confidence interval   Range of values within which there is a specified probability that the true value lies.

conjugate acid-base pair   An acid and a base that differ only through the gain or loss of a single proton.

constant-current electrolysis   Electrolysis in which a constant current flows between working and auxiliary electrodes. As reactants are consumed, increasing voltage is required to keep the current flowing, so this is the least selective form of electrolysis.

constant mass   In gravimetric analysis, the product is heated and cooled to room temperature in a desiccator until successive weighings are "constant." There is no standard definition of "constant mass"; but, for ordinary work, it is usually taken to be about ± 0.3 mg. Constancy is usually limited by the irreproducible regain of moisture during cooling and weighing.

constant-voltage electrolysis   Electrolysis in which a constant voltage is maintained between working and auxiliary electrodes. This is less selective than controlled-potential electrolysis because the potential of the working electrode becomes more extreme as ohmic potential and overpotential change.

continuous-dynode electron multiplier   An electron detector that works like a photomultiplier tube. An electron striking the lead-doped glass wall of a horn-shaped tube liberates several electrons which are accelerated into the horn by increasingly positive potential. After many bounces, $\sim 10^5$ electrons reach the narrow end of the horn for each incident electron.

control chart   A graph in which successive observations of a process are recorded to determine whether the process is within specified control limits.

controlled-potential electrolysis   A technique for selective reduction (or oxidation), in which the voltage between working and reference electrodes is held constant.

convection   Process in which solute is carried from one place to another by bulk motion of the solution.

coprecipitation   Occurs when a substance whose solubility is not exceeded precipitates along with one whose solubility is exceeded.

correlation coefficient   The square of the correlation coefficient, R^2, is a measure of goodness of fit of data points to a straight line. The closer R^2 is to 1, the better the fit.

coulomb, C   Amount of charge per second that flows past any point in a circuit when the current is 1 ampere. There are approximately 96 485 coulombs in a mole of electrons.

coulometer   A device that generates a redox reagent for quantitative reaction with analyte and measures the number of electrons required to generate the redox reagent.

coulometric titration   One conducted with a constant current for a measured time.

coulometry   A technique in which the quantity of analyte is determined by measuring the number of coulombs needed for complete electrolysis.

counter electrode   See *auxiliary electrode*.

counterion   An ion with a charge opposite that of the ion of interest.

coupled equilibria   Reversible chemical reactions that have a species in common. For example, the product of one reaction could be a reactant in another reaction.

critical point   Critical temperature and pressure of a substance.

critical pressure   Pressure above which a fluid cannot be condensed to two phases (liquid and gas), no matter how low the temperature.

critical temperature   Temperature above which a fluid cannot be condensed to two phases (liquid and gas), no matter how great a pressure is applied.

cross-linking   Covalent linkage between different strands of a polymer.

cryogenic focusing   In gas chromatography, cold trapping of solutes below ambient temperature at the beginning of a column. A *cryogen* is a cold fluid such as liquid nitrogen used for cooling.

crystallization   Process in which a substance comes out of solution slowly to form a solid with a regular arrangement of atoms.

cumulative formation constant, β_n   Equilibrium constant for a reaction of the type $M + nX \rightleftharpoons MX_n$. Also called *overall formation constant*.

current, I   Amount of charge flowing through a circuit per unit time (A/s).

current density   Electric current per unit area (A/m^2).

cuvet   A cell with transparent walls used to hold samples for spectrophotometric measurements.

cyclic voltammetry   A polarographic technique with a triangular waveform. Both cathodic and anodic currents are observed for reversible reactions.

dalton, Da   Unit of atomic mass defined as 1/12 of the mass of ^{12}C.

dark current   Small current produced by a photodetector in the absence of light.

DART   See *direct analysis in real time*.

data quality objectives   Accuracy, precision, and sampling requirements for an analytical method.

dead volume   Volume of a chromatography system (not including the column) between the point of injection and the point of detection. Also called *extra-column volume*.

Debye-Hückel equation   Gives the activity coefficient, γ, as a function of ionic strength, μ. The *extended Debye-Hückel equation*, applicable to ionic strengths up to about 0.1 M, is $\log \gamma = [-0.51z^2\sqrt{\mu}]/[1 + (\alpha\sqrt{\mu}/305)]$, where z is the ionic charge and α is the effective hydrated radius in picometers.

decant   To pour liquid off a solid or, perhaps, a denser liquid. The denser phase is left behind.

decomposition potential   In an electrolysis, that voltage at which rapid reaction first begins.

degrees of freedom   In statistics, the number of observations minus the number of parameters estimated from the observations.

deionized water   Water that has been passed through a cation exchanger (in the H^+ form) and an anion exchanger (in the OH^- form) to remove ions from the solution.

deliquescent substance   A *hygroscopic* solid that spontaneously picks up so much water from the air that the substance completely dissolves.

demasking   Removal of a masking agent from the species protected by the masking agent.

density   Mass per unit volume.

depolarizer   A molecule that is oxidized or reduced at a modest potential. It is added to an electrolytic cell to prevent the cathode or anode potential from becoming too extreme.

derivatization   Chemical alteration to attach a group to a molecule so that the molecule can be detected conveniently. Alternatively, treatment can alter volatility or solubility to allow easier separation.

desalting   Removal of salts (or any small molecules) from a solution of macromolecules. Gel filtration or dialysis are used for desalting.

DESI   See *desorption electrospray ionization*.

desiccant   A drying agent.

desiccator   A sealed chamber in which samples can be dried in the presence of a desiccant or by vacuum pumping or both.

desorption   The release of an adsorbed substance from a surface.

desorption electrospray ionization, DESI   A solvent is electrosprayed onto a surface to dissolve analyte from the surface into aerosol microdroplets, which can be analyzed with a mass spectrometer.

detection limit   The smallest quantity of analyte that is "significantly different" from a blank. The detection limit is often taken as the mean signal for blanks plus three times the standard deviation of a low-concentration sample. Also called *lower limit of detection*.

determinant   The value of the two-dimensional determinant $\begin{vmatrix} a & b \\ c & d \end{vmatrix}$ is the difference $ad - bc$.

determinate error   See *systematic error*.

deuterium arc lamp   Source of broadband ultraviolet radiation. An electric discharge (a spark) in deuterium gas causes D_2 molecules to dissociate and emit many wavelengths of radiation.

dialysis   A technique in which solutions are placed on either side of a semipermeable membrane that allows small molecules, but not large molecules, to cross. Small molecules in the two solutions diffuse across and equilibrate between the two sides. Large molecules are retained on their original side.

dielectric constant, ε   The electrostatic force, F, between two charged particles is given by $F = kq_1q_2/\varepsilon r^2$, where k is a constant, q_1 and q_2 are the charges, r is the separation between particles, and ε is the dielectric constant of the medium. The higher the dielectric constant, the less force is exerted by one charged particle on another.

difference plot A graph of the mean fraction of protons bound to an acid versus pH. For complex formation, the difference plot gives the mean number of ligands bound to a metal versus pL (= −log[ligand concentration]). Also called *Bjerrum plot*.

diffraction Occurs when electromagnetic radiation passes through or is reflected from slits with a spacing comparable to the wavelength. Interference of waves from adjacent slits produces a spectrum of radiation, with each wavelength emerging at a different angle.

diffuse part of the double layer Region of solution near a charged surface in which excess counterions are attracted to the charge. The thickness of this layer is 0.3–10 nm.

diffuse reflection Occurs when a rough surface reflects light in all directions.

diffusion Net transport of a solute from a region of high concentration to a region of low concentration caused by the random movement of molecules in a liquid or gas (or, very slowly, in a solid).

diffusion coefficient, D Defined by Fick's first law of diffusion: $J = -D(dc/dx)$, where J is the rate at which molecules diffuse across a plane of unit area and dc/dx is the concentration gradient in the direction of diffusion.

diffusion current Current observed when the rate of electrolysis is limited by the rate of diffusion of analyte to the electrode. In polarography, diffusion current = limiting current − residual current.

diffusion layer Region near an electrode containing excess product or decreased reactant involved in the electrode reaction. The thickness of this layer can be hundreds of micrometers.

digestion Process in which a precipitate is left (usually warm) in the presence of mother liquor to promote particle recrystallization and growth. Purer, more easily filterable crystals result. Also used to describe any chemical treatment in which a substance is decomposed to transform the analyte into a form suitable for analysis.

dilution factor Factor (initial volume of reagent)/(total volume of solution) used to multiply the initial concentration of reagent to find the diluted concentration.

dimer A molecule made from two identical units.

diode A semiconductor device consisting of a *pn* junction through which current can pass in only one direction. Current flows when the *n*-type material is made negative and the *p*-type material is made positive. A voltage sufficient to overcome the activation energy for carrier movement must be supplied before any current flows. For silicon diodes, this voltage is ~0.6 V. If a sufficiently large voltage, called the *breakdown voltage*, is applied in the reverse direction, current will flow in the wrong direction through the diode.

diprotic acids and bases Compounds that can donate or accept two protons.

direct analysis in real time, DART A DART source produces excited He or N_2, which is directed at the surface of an object to be sampled in ambient atmosphere. The excited species react with ambient moisture to create protonated water clusters that react with analyte M to make MH^+. The MH^+ is measured by mass spectrometry.

direct current polarography Classical form of polarography, in which a linear voltage ramp is applied to the working electrode.

direct titration One in which the analyte is treated with titrant and the volume of titrant required for complete reaction is measured.

dispersion A measure of the ability of a monochromator to separate wavelengths differing by $\Delta\lambda$ through the angle $\Delta\phi$. The greater the dispersion, the greater the angle separating two closely spaced wavelengths. For a prism, dispersion refers to the rate of change of refractive index with wavelength, $dn/d\lambda$.

displacement titration An EDTA titration procedure in which analyte is treated with excess $MgEDTA^{2-}$ to displace Mg^{2+}: M^{n+} + $MgEDTA^{2-} \rightleftharpoons MEDTA^{n-4} + Mg^{2+}$. The liberated Mg^{2+} is then titrated with EDTA. This procedure is useful if there is no suitable indicator for direct titration of M^{n+}.

disproportionation A reaction in which an element in one oxidation state gives products containing that element in both higher and lower oxidation states; e.g., $2Cu^+ \rightleftharpoons Cu^{2+} + Cu(s)$.

distribution coefficient, D For a solute partitioned between two phases, the distribution coefficient is the total concentration of all forms of solute in phase 2 divided by the total concentration in phase 1.

Donnan equilibrium Ions of the same charge as those fixed on an ion-exchange resin are repelled from the resin. Thus, anions do not readily penetrate a cation-exchange resin, and cations are repelled from an anion-exchange resin.

dopant When small amounts of substance B are added to substance A, we call B a dopant and say that A is doped with B. Doping is done to alter the properties of A.

Doppler effect A molecule moving toward a source of radiation experiences a higher frequency than one moving away from the source.

double-focusing mass spectrometer A spectrometer that uses electric and magnetic sectors in series to obtain high resolution.

double-junction electrode An electrode with inner and outer compartments designed to minimize contact between analyte solution and the contents of the inner electrode. The outer compartment serves as a salt bridge with ions that are chemically compatible with the analyte.

double layer See *electric double layer*.

drift Slow change in the response of an instrument due to various causes such as changes in electrical components with temperature, variation in power-line voltage to an instrument, and aging of components within instruments. Same as *1/f noise* or *flicker noise*.

dropping-mercury electrode One that delivers fresh drops of Hg to a polarographic cell.

dry ashing Oxidation of organic matter with O_2 at high temperature to leave behind inorganic components for analysis.

dwell time In chromatography, time between mixing of solvents and when they reach the start of the column.

dwell volume In chromatography, volume between the point of mixing solvents and the start of the column.

dynamic range Range of analyte concentration over which a change in concentration gives a change in detector response.

dynode A metal surface that easily emits several electrons each time it is struck by one accelerated electron in a photomultiplier tube or an electron multiplier.

$E°$ Standard reduction potential.

$E°'$ Effective standard reduction potential at pH 7 (or at some other specified conditions).

EDTA (ethylenediaminetetraacetic acid) $(HO_2CCH_2)_2NCH_2CH_2N(CH_2CO_2H)_2$, the most widely used reagent for complexometric titrations. It forms 1:1 complexes with virtually all cations with a charge of 2 or more.

effective formation constant, K'_f See *conditional formation constant*.

effervescence Rapid release of gas with bubbling and hissing.

efflorescence Property by which the outer surface or entire mass of a substance turns into powder from loss of water of crystallization.

effluent See *eluate*.

einstein A mole of photons. The symbol of this unit is the same as its name, einstein.

electric charge, q Quantity of electricity, measured in coulombs.

electric discharge emission spectroscopy A technique in which atomization and excitation are stimulated by an electric arc, a spark, or microwave radiation.

electric double layer Region comprising the charged surface of an electrode or a particle plus the oppositely charged region of solution adjacent to the surface. Also called *double layer*.

electric potential, E The electric potential (in volts) at a point is the energy (in joules) needed to bring one coulomb of positive charge from infinity to that point. The *potential difference* between two points is the energy needed to transport one coulomb of positive charge from the negative point to the positive point.

electroactive species Any species that can be oxidized or reduced at an electrode.

electrocapillary maximum Potential at which the net charge on a mercury drop from a dropping-mercury electrode is 0 (and the surface tension of the drop is maximal).

electrochemical cell See *galvanic cell.*

electrochemical detector Liquid chromatography detector that measures current when an electroactive solute emerges from the column and passes over a working electrode held at a fixed potential with respect to a reference electrode. Also called *amperometric detector.*

electrochemistry Use of electrical measurements on a chemical system for analytical purposes. Also refers to use of electricity to drive a chemical reaction or use of a chemical reaction to produce electricity.

electrode An electrical conductor through which electrons flow into or out of chemical species involved in a redox reaction.

electroendosmosis See *electroosmosis.*

electrogravimetric analysis A technique in which the mass of an electrolytic deposit is used to quantify the analyte.

electrokinetic injection In capillary electrophoresis, the use of an electric field to inject sample into the capillary. Because different species have different mobilities, the injected sample does not have the same composition as the original sample.

electrolysis Process in which the passage of electric current causes a chemical reaction to occur.

electrolyte A substance that produces ions when dissolved.

electrolytic cell A cell in which a chemical reaction that would not otherwise occur is driven by a voltage applied between two electrodes.

electromagnetic spectrum The whole range of electromagnetic radiation, including visible light, radio waves, X-rays, etc.

electron capture detector Gas chromatography detector that is particularly sensitive to compounds with halogen atoms, nitro groups, and other groups with high electron affinity. Makeup gas (N_2 or 5% CH_4 in Ar) is ionized by β-rays from ^{63}Ni to liberate electrons that produce a small, steady current. High-electron-affinity analytes capture some of the electrons and reduce the detector current.

electronic balance A weighing device that uses an electromagnetic servomotor to balance the load on the pan. The mass of the load is proportional to the current needed to balance it.

electronic transition One in which an electron is promoted from one energy level to another.

electron ionization Interaction of analyte molecules (M) with high-energy electrons in the ion source of a mass spectrometer to give the cation radical, $M^{+\cdot}$, and fragments derived from $M^{+\cdot}$.

electron multiplier An ion detector that works like a photomultiplier tube. Cations striking a cathode liberate electrons. A series of *dynodes* multiplies the number of electrons by $\sim 10^5$ before they reach the anode.

electron-transfer dissociation Breaking of a chemical bond by the exothermic transfer of an electron from one species to another. This process is used in mass spectral sequencing of polypeptides because it cleaves peptide bonds without breaking other bonds in the molecule.

electroosmosis Bulk flow of fluid in a capillary tube induced by an electric field. Mobile ions in the diffuse part of the double layer at the wall of the capillary serve as the "pump." Also called *electroendosmosis.*

electroosmotic flow Uniform, pluglike flow of fluid in a capillary tube under the influence of an electric field. The greater the charge on the wall of the capillary, the greater the number of counterions in the double layer and the stronger the electroosmotic flow.

electroosmotic mobility, μ_{eo} Constant of proportionality between the electroosmotic speed, u_{eo}, of a fluid in a capillary and the applied electric field, E: $u_{eo} = \mu_{eo}E$. Also equal to the speed of a neutral species, $u_{neutral}$, divided by the electric field, E. See also *apparent mobility.*

electroosmotic velocity Speed with which solvent flows through a capillary electrophoresis column. It is measured by adding a detectable neutral molecule to the sample. Electroosmotic velocity is the distance from injector to detector divided by the time required for the neutral molecule to reach the detector.

electropherogram A graph of detector response versus time for electrophoresis.

electrophoresis Migration of ions in solution in an electric field. Cations move toward the cathode and anions move toward the anode. Ions can be separated from one another by their differing rates of migration in a strong electric field.

electrophoretic mobility, μ_{ep} Constant of proportionality between the electrophoretic speed, u_{ep}, of an ion in solution and the applied electric field, E: $u_{ep} = \mu_{ep}E$. See also *apparent mobility.*

electrospray ionization A method for interfacing liquid chromatography to mass spectrometry. A high potential applied to the liquid at the column exit creates charged droplets in a fine aerosol. Gaseous ions are derived from ions that were already in the mobile phase on the column. It is common to observe protonated bases (BH^+), ionized acids (A^-), and complexes formed between analyte, M (which could be neutral or charged), and stable ions such as NH_4^+, Na^+, HCO_2^-, or $CH_3CO_2^-$ that were already in solution.

eluate What comes out of a chromatography column. Also called *effluent.*

eluent Solvent applied to the beginning of a chromatography column.

eluent strength, $\varepsilon°$ A measure of the ability of a solvent to elute solutes from a chromatography column. Eluent strength is a measure of the adsorption energy of a solvent on the stationary phase in chromatography. Also called *solvent strength.*

eluotropic series Ranks solvents according to their ability to displace solutes from the stationary phase in adsorption chromatography.

elution Process of passing a liquid or a gas through a chromatography column.

emission spectrum A graph of luminescence intensity versus luminescence wavelength (or frequency or wavenumber), obtained with a fixed excitation wavelength.

emissivity A quotient given by the radiant emission from a real object divided by the radiant emission of a blackbody at the same temperature.

emulsion A stable dispersion of immiscible liquids, which might be made by vigorous shaking. Homogenized milk is an emulsion of cream in an aqueous phase. Emulsions usually require an emulsifying agent (a surfactant) for stability. The emulsifying agent stabilizes the interface between the two phases by its affinity for both phases.

enantiomers Mirror image isomers that cannot be superimposed on each other. Also called *optical isomers.*

endothermic reaction One for which ΔH is positive; heat must be supplied to reactants for them to react.

end point Point in a titration at which there is a sudden change in a physical property, such as indicator color, pH, conductivity, or absorbance. Used as a measure of the equivalence point.

energy The product force $\times$ distance.

enthalpy change, ΔH The heat absorbed or released when a reaction occurs at constant pressure.

enthalpy of hydration Heat liberated when a gaseous species is transferred to water.

entropy A measure of the "disorder" of a substance.

enzyme A protein that catalyzes a chemical reaction.

equilibrium State in which the forward and reverse rates of all reactions are equal, so the concentrations of all species remain constant.

equilibrium constant, K For the reaction $aA + bB \rightleftharpoons cC + dD$, $K = \mathcal{A}_C^c \mathcal{A}_D^d / \mathcal{A}_A^a \mathcal{A}_B^b$, where $\mathcal{A}_i$ is the activity of the ith species.

equimolar mixture of compounds One that contains an equal number of moles of each compound.

equivalence point Point in a titration at which the quantity of titrant is exactly sufficient for stoichiometric reaction with the analyte.

equivalent For a redox reaction, the amount of reagent that can donate or accept one mole of electrons. For an acid-base reaction, the amount of reagent that can donate or accept one mole of protons.

equivalent weight The mass of substance containing one equivalent.

error bar Graphical depiction of the uncertainty in a measurement.

evanescent wave   Light that penetrates the walls of an optical fiber or waveguide in which the light travels by total internal reflection.

evaporative light-scattering detector   A liquid chromatography detector that makes a fine mist of eluate and evaporates solvent from the mist in a heated zone. The remaining particles of liquid or solid solute flow past a laser beam and are detected by their ability to scatter the light.

excitation spectrum   A graph of luminescence (measured at a fixed wavelength) versus excitation frequency or wavelength. It closely corresponds to an absorption spectrum because the luminescence is generally proportional to the absorbance.

excited state   Any state of an atom or a molecule having more than the minimum possible energy.

exitance, M   Power per unit area radiating from the surface of an object.

exothermic reaction   One for which ΔH is negative; heat is liberated when products are formed.

extended Debye-Hückel equation   See *Debye-Hückel equation*.

extinction coefficient   See *molar absorptivity*.

extra-column volume   See *dead volume*.

extracted ion chromatogram   Chromatogram made by collecting consecutive full-range mass spectra, but selecting just one value of *m/z* for display. Most time is spent monitoring values of *m/z* that will not be displayed. A *selected ion chromatogram* provides greater signal-to-noise ratio than an extracted ion chromatogram because all of the time is taken to monitor just one or a few values of *m/z* in the selected ion chromatogram.

extraction   Process in which a solute is transferred from one phase to another. Analyte is sometimes removed from a sample by extraction into a solvent that dissolves the analyte.

extrapolation   Estimation of a value that lies beyond the range of measured data.

F test   For two variances, s_1^2 and s_2^2 (with s_1 chosen to be the larger of the two), the statistic F is defined as $F = s_1^2/s_2^2$. To decide whether s_1 is significantly greater than s_2, we compare F with the critical values in a table based on a certain confidence level. If the calculated value of F is greater than the value in the table, the difference is significant.

Fajans titration   A precipitation titration in which the end point is signaled by adsorption of a colored indicator on the precipitate.

false negative   A conclusion that the concentration of analyte is below a certain limit when, in fact, the concentration is above the limit.

false positive   A conclusion that the concentration of analyte exceeds a certain limit when, in fact, the concentration is below the limit.

farad, F   Unit of electrical capacitance; 1 farad of capacitance will store 1 coulomb of charge in a potential difference of 1 volt.

faradaic current   That component of current in an electrochemical cell due to oxidation and reduction reactions.

Faraday constant, F   The number of coulombs in a mole of elementary charges, approximately $9.648\,5 \times 10^4$ C/mol of charge.

Faraday cup   A mass spectrometric ion detector in which each arriving cation is neutralized by an electron. The current required to neutralize the ions is proportional to the number of cations arriving at the Faraday cup.

Faraday's laws   Two laws stating that the extent of an electrochemical reaction is directly proportional to the quantity of electricity that has passed through the cell. The mass of substance that reacts is proportional to its formula mass and inversely proportional to the number of electrons required in its half-reaction.

ferroelectric material   A solid with a permanent electric polarization (dipole) in the absence of an external electric field. The polarization results from alignment of molecules within the solid.

Fick's first law of diffusion   See *diffusion coefficient*.

field blank   A blank sample exposed to the environment at the sample collection site and transported in the same manner as other samples between the lab and the field.

field effect transistor   A semiconductor device in which the electric field between gate and base governs the flow of current between source and drain.

filtrate   Liquid that passes through a filter.

flame ionization detector   A gas chromatography detector in which solute is burned in an H_2-air flame to produce CHO^+ ions. The current carried through the flame by these ions is proportional to the concentration of susceptible species in the eluate.

flame photometer   A device that uses flame atomic emission and a filter photometer to quantify Li, Na, K, and Ca in liquid samples.

flame photometric detector   Gas chromatography detector that measures optical emission from S, P, Pb, Sn, or other elements in a H_2–O_2 flame.

flicker noise   See *drift*.

flow adaptor   An adjustable plungerlike device that may be used on either side of a chromatographic bed to support the bed and to minimize the dead space through which liquid can flow outside of the column bed.

flow injection analysis   Analytical technique in which sample is injected into flowing liquid carrier containing a reagent that reacts with the analyte. Additional reagents might be added further downstream. As sample flows from injector to detector, the sample zone broadens and reacts with reagent to form a product to which the detector responds.

fluorescence   Process in which a molecule emits a photon 10^{-8} to 10^{-4} s after absorbing a photon. It results from a transition between states of the same spin multiplicity (e.g., singlet → singlet).

fluorescence detector   Liquid chromatography detector that uses a strong light or laser to irradiate eluate emerging from a column and detects radiant emission from fluorescent solutes.

flux   In sample preparation, flux is the agent used as the medium for a fusion. In transport phenomena, flux is the quantity of whatever you like crossing each unit area in one unit of time. For example, the flux of diffusing molecules could be mol/(m² · s). Heat flux would be J/(m² · s).

force   The product mass × acceleration.

formal concentration, F   The molarity of a substance if it did not change its chemical form on being dissolved. It represents the total number of moles of substance dissolved in a liter of solution, regardless of any reactions that take place when the solute is dissolved. Also called *analytical concentration* or *formality*.

formality, F   See *formal concentration*.

formal potential   Potential of a half-reaction (relative to a standard hydrogen electrode) when the formal concentrations of reactants and products are unity. Any other conditions (such as pH, ionic strength, and concentrations of ligands) also must be specified.

formation constant, K_f   Equilibrium constant for the reaction of a metal with its ligands to form a metal-ligand complex. Also called *stability constant*. See also *conditional formation constant*.

formula mass, FM   The mass containing one mole of the indicated chemical formula of a substance. For example, the formula mass of $CuSO_4 \cdot 5H_2O$ is the sum of the masses of copper, sulfate, and five water molecules.

fortification   Same as a *spike*—a deliberate addition of analyte to a sample.

Fourier analysis   Process of decomposing a function into an infinite series of sine and cosine terms. Because each term represents a certain frequency or wavelength, Fourier analysis decomposes a function into its component frequencies or wavelengths.

Fourier series   Infinite sum of sine and cosine terms that add to give a particular function in a particular interval.

fraction of association, α   For the reaction of a base (B) with H_2O, the fraction of base in the form BH^+.

fraction of dissociation, α   For the dissociation of an acid (HA), the fraction of acid in the form A^-.

free energy   See *Gibbs free energy*.

frequency, ν   The number of cycles per unit time for a repetitive event.

friction coefficient   A molecule migrating through a solution is retarded by a force that is proportional to its speed. The constant of proportionality is the friction coefficient.

fugacity   The activity of a gas.

fugacity coefficient   Activity coefficient for a gas.

fused-core particle   See *superficially porous particle*.

fusion   Process in which an otherwise insoluble substance is dissolved in a molten salt such as Na_2CO_3, Na_2O_2, or KOH. Once the substance has dissolved, the melt is cooled, dissolved in aqueous solution, and analyzed.

galvanic cell   One that produces electricity by means of a spontaneous chemical reaction. Also called a *voltaic cell.*

gas chromatography   A form of chromatography in which the mobile phase is a gas.

gathering   A process in which a trace constituent of a solution is intentionally coprecipitated with a major constituent.

Gaussian distribution   Theoretical bell-shaped distribution of measurements when all error is random. The center of the curve is the mean, μ, and the width is characterized by the standard deviation, σ. A *normalized* Gaussian distribution, also called the *normal error curve,* has an area of unity and is given by

$$y = \frac{1}{\sigma\sqrt{2\pi}}\, e^{-(x-\mu)^2/2\sigma^2}$$

Gaussian noise   See *white noise.*

gel   Chromatographic stationary phase particles, such as Sephadex or polyacrylamide, which are soft and pliable.

gel filtration chromatography   See *molecular exclusion chromatography.*

gel permeation chromatography   See *molecular exclusion chromatography.*

geometric mean   For a series of n measurements with the values x_i, geometric mean $= \sqrt[n]{x_1 x_2 \ldots x_n}$.

Gibbs free energy, G   The change in Gibbs free energy, ΔG, for any process at constant temperature is related to the change in enthalpy, ΔH, and entropy, ΔS, by the equation $\Delta G = \Delta H - T\Delta S$, where T is temperature in kelvins. A process is spontaneous (thermodynamically favorable) if ΔG is negative.

glass electrode   One that has a thin glass membrane across which a pH-dependent voltage develops. The voltage (and hence pH) is measured by a pair of reference electrodes on either side of the membrane.

glassy carbon electrode   An inert carbon electrode, impermeable to gas, and especially well suited as an anode. The isotropic structure (same in all directions) is thought to consist of tangled ribbons of graphitelike sheets, with some cross-linking.

globar   An infrared radiation source made of a ceramic such as silicon carbide heated by passage of electricity.

Gooch filter crucible   A short, cup-shaped container with holes at the bottom, used for filtration and ignition of precipitates. For ignition, the crucible is made of porcelain or platinum and lined with a mat of ceramic fibers to retain the precipitate. For precipitates that do not need ignition, the crucible is made of glass and has a porous glass disk instead of holes at the bottom.

gradient elution   Chromatography in which the composition of the mobile phase is progressively changed to increase the eluent strength of the solvent.

graduated cylinder   A tube with volume calibrations along its length. Also called *graduate.*

gram-atom   The amount of an element containing Avogadro's number of atoms; it is the same as a mole of the element.

Gran plot   A graph such as the plot of $V_b \cdot 10^{-pH}$ versus V_b used to find the end point of a titration. V_b is the volume of base (titrant) added to an acid being titrated. The slope of the linear portion of the graph is related to the dissociation constant of the acid.

graphite furnace   A graphite tube that can be heated electrically to about 2 500 K to decompose and atomize a sample for atomic spectroscopy.

grating   Either a reflective or a transmitting surface etched with closely spaced lines; used to disperse light into its component wavelengths.

gravimetric analysis   Any analytical method that relies on measuring the mass of a substance (such as a precipitate) to complete the analysis.

gravimetric titration   A titration in which the mass of titrant is measured, instead of the volume. Titrant concentration is conveniently expressed as mol reagent/kg titrant solution. Gravimetric titrations can be more accurate and precise than volumetric titrations.

green chemistry   Principles intended to change our behavior in a manner that will help sustain the habitability of Earth. Green chemistry seeks to design chemical products and processes to reduce the use of resources and energy and the generation of hazardous waste.

gross sample   See *bulk sample.*

ground state   State of an atom or a molecule with the minimum possible energy.

Grubbs test   Statistical test used to decide whether to discard a datum that appears discrepant.

guard column   In high-performance liquid chromatography, a short column packed with the same material as the main column and placed between the injector and the main column. The guard column removes impurities that might irreversibly bind to the main column and degrade it. Also called *precolumn*. In gas chromatography, the guard column is a length of empty, silanized capillary ahead of the chromatography column. Nonvolatile residues accumulate in the guard column.

half-cell   Part of an electrochemical cell in which half of an electrochemical reaction (either the oxidation or the reduction reaction) occurs.

half-height   Half of the maximum amplitude of a signal.

half-reaction   Any redox reaction can be conceptually broken into two half-reactions, one involving only oxidation and one involving only reduction.

half-wave potential   Potential at the midpoint of the rise in the current of a polarographic wave.

half-width   Width of a signal at its half-height.

Hall-Héroult process   Electrolytic production of aluminum metal from a molten solution of Al_2O_3 and cryolite (Na_3AlF_6).

Hammett acidity function   The acidity of a solvent that protonates the weak base, B, is called the Hammett acidity function, H_0, and is given by

$$H_0 = pK_a \text{ (for BH}^+) + \log\frac{[B]}{[BH^+]}$$

For dilute aqueous solutions, H_0 approaches pH.

hanging-drop electrode   One with a stationary drop of Hg that is used for stripping analysis.

hardness   Total concentration of alkaline earth ions in natural water expressed as mg $CaCO_3$ per liter of water as if all of the alkaline earths present were $CaCO_3$. See also *permanent hardness* and *temporary hardness.*

Heisenberg uncertainty principle   Certain pairs of physical quantities cannot be known simultaneously with arbitrary accuracy. If δE is the uncertainty in the energy difference between two atomic states and δt is the lifetime of the excited state, their product cannot be known more accurately than $\delta E \delta t \gtrsim h/(4\pi)$, where h is Planck's constant. A similar relation holds between the position and the momentum of a particle. If position is known very accurately, then the uncertainty in momentum is large, and vice versa.

Henderson-Hasselbalch equation   A logarithmic rearranged form of the acid dissociation equilibrium equation:

$$pH = pK_a + \log\frac{[A^-]}{[HA]}$$

Henry's law   The partial pressure of a gas in equilibrium with gas dissolved in a solution is proportional to the concentration of dissolved gas: $P = k[\text{dissolved gas}]$. The constant k is called the *Henry's law constant*. It is a function of the gas, the liquid, and the temperature.

hertz, Hz   Unit of frequency, s^{-1}, also called *reciprocal seconds.*

heterogeneous   Not uniform throughout.

HETP, height equivalent to a theoretical plate   The length of a chromatography column divided by the number of theoretical plates in the column.

hexadentate ligand   One that binds to a metal atom through six ligand atoms.

high-performance liquid chromatography, HPLC A chromatographic technique using very small stationary phase particles and high pressure to force solvent through the column.

HILIC See *hydrophilic interaction chromatography*.

hole Absence of an electron in a semiconductor. When a neighboring electron moves into the hole, a new hole is created where the electron came from. By this means, a hole can move through a solid just as an electron can move through a solid.

hollow-cathode lamp One that emits sharp atomic lines characteristic of the element from which the cathode is made.

homogeneous Having the same composition everywhere.

homogeneous precipitation A technique in which a precipitating agent is generated slowly by a reaction in homogeneous solution, effecting a slow crystallization instead of a rapid precipitation of product.

hydrated radius Effective size of an ion or a molecule plus its associated water molecules in solution.

hydrodynamic flow Motion of fluid through a tube, driven by a pressure difference. Hydrodynamic flow is usually laminar, in which there is a parabolic profile of velocity vectors, with the highest velocity at the center of the stream and zero velocity at the walls.

hydrodynamic injection In capillary electrophoresis, the use of a pressure difference between the two ends of the capillary to inject sample into the capillary. Injection is achieved by applying pressure on one end, by applying suction on one end, or by siphoning.

hydrodynamic radius Effective radius of a molecule migrating through a fluid. It is defined by the *Stokes equation*, in which the friction coefficient is $6\pi\eta r$, where η is the viscosity of the fluid and r is the hydrodynamic radius of the molecule.

hydrolysis "Reaction with water." The reaction $B + H_2O \rightleftharpoons BH^+ + OH^-$ is often called hydrolysis of a base.

hydronium ion, H_3O^+ What we really mean when we write $H^+(aq)$.

hydrophilic interaction chromatography Chromatographic separation of polar solutes with a hydrophilic stationary phase using mixed organic-aqueous eluent. Eluent strength increases with decreasing organic solvent. Commonly called HILIC.

hydrophilic substance One that is soluble in water or attracts water to its surface.

hydrophobic interaction chromatography Chromatographic separation based on the interaction of a hydrophobic solute with a hydrophobic stationary phase.

hydrophobic substance One that is insoluble in water or repels water from its surface.

hygroscopic substance One that readily picks up water from the atmosphere.

ignition The heating to high temperature of some gravimetric precipitates to convert them into a known, constant composition that can be weighed.

immiscible liquids Two liquids that do not form a single phase when mixed together.

immunoassay An analytical measurement using antibodies.

inclusion An impurity that occupies lattice sites in a crystal.

indeterminate error See *random error*.

indicator A compound having a physical property (usually color) that changes abruptly near the equivalence point of a chemical reaction.

indicator electrode One that develops a potential whose magnitude depends on the activity of one or more species in contact with the electrode.

indicator error Difference between the indicator end point of a titration and the true equivalence point.

indirect detection Chromatographic detection based on the *absence* of signal from a background species. For example, in ion chromatography, a light-absorbing ionic species can be added to the eluent. Nonabsorbing analyte replaces an equivalent amount of light-absorbing eluent when analyte emerges from the column, thereby decreasing the absorbance of eluate.

indirect titration One that is used when the analyte cannot be directly titrated. For example, analyte A may be precipitated with excess reagent R. The product is filtered, and the excess R washed away. Then AR is dissolved in a new solution, and R can be titrated.

inductively coupled plasma A high-temperature plasma that derives its energy from an oscillating radio-frequency field. It is used to atomize a sample for atomic emission spectroscopy.

inflection point One at which the derivative of the slope is 0: $d^2y/dx^2 = 0$. That is, the slope reaches a maximum or minimum value.

injection precision See *instrument precision*.

inner filter effect See *self-absorption*.

inorganic carbon In a natural water or industrial effluent sample, the quantity of dissolved carbonate and bicarbonate.

instrument precision Reproducibility observed when the same quantity of one sample is repeatedly introduced into an instrument. Also called *injection precision*.

intensity Power per unit area of a beam of electromagnetic radiation (W/m^2). Also called *radiant power* or *irradiance*.

intercalation Binding of a flat, aromatic molecule between the flat, hydrogen-bonded base pairs in DNA or RNA.

intercept For a straight line whose equation is $y = mx + b$, b is the intercept. It is the value of y when $x = 0$.

interference A phenomenon in which the presence of one substance changes the response in the analysis of another substance.

interference filter A filter that transmits a particular band of wavelengths and reflects others. Transmitted light interferes constructively within the filter, whereas light that is reflected interferes destructively.

interferogram A graph of light intensity versus retardation (or time) for the radiation emerging from an *interferometer*.

interferometer A device with a beamsplitter, fixed mirror, and moving mirror that breaks input light into two beams that interfere with each other. The degree of interference depends on the difference in pathlength of the two beams.

interlaboratory precision The reproducibility observed when aliquots of the same sample are analyzed by different people in different laboratories.

intermediate precision Precision observed when an assay is performed by different people on different instruments on different days in the same lab. Also called *ruggedness*.

internal conversion A radiationless, isoenergetic, electronic transition between states of the same electron-spin multiplicity.

internal standard A known quantity of a compound other than analyte added to a solution containing an unknown quantity of analyte. The concentration of analyte is then measured relative to that of the internal standard.

interpolation Estimation of the value of a quantity that lies between two known values.

intersystem crossing A radiationless, isoenergetic, electronic transition between states of different electron-spin multiplicity.

intra-assay precision Precision observed when analyzing aliquots of a homogeneous material several times by one person on one day with the same equipment.

iodimetry Use of triiodide (or iodine) as a titrant.

iodometry A technique in which an oxidant is treated with I^- to produce I_3^-, which is then titrated (usually with thiosulfate).

ion chromatography High-performance liquid chromatography ion-exchange separation of ions. See also *suppressed-ion chromatography* and *single-column ion chromatography*.

ion-exchange chromatography A technique in which solute ions are retained by oppositely charged sites in the stationary phase.

ion-exchange equilibrium An equilibrium involving replacement of a cation by a different cation or replacement of an anion by a different anion. Usually the ions in these reactions are bound by electrostatic forces.

ion-exchange membrane Membrane containing covalently bound charged groups. Oppositely charged ions in solution penetrate the membrane freely, but similarly charged ions tend to be excluded from the membrane by the bound charges.

ion-exclusion chromatography A technique in which electrolytes are separated from nonelectrolytes by an ion-exchange resin.

ionic atmosphere The region of solution around an ion or a charged particle. It contains an excess of oppositely charged ions.

ionic liquid A salt that melts near or below room temperature and has a large temperature range over which it remains liquid.

ionic radius Effective size of an ion in a crystal.

ionic strength, μ Given by $\mu = \frac{1}{2} \sum_i c_i z_i^2$, where c_i is the concentration of the ith ion in solution and z_i is the charge on that ion. The sum extends over all ions in solution, including the ions whose activity coefficients are being calculated.

ionization interference In atomic spectroscopy, a lowering of signal intensity as a result of ionization of analyte atoms.

ionization suppressor An element used in atomic spectroscopy to decrease the extent of ionization of the analyte.

ion mobility spectrometer An instrument that measures the drift time of gaseous ions migrating in an electric field against a flow of gas. The "spectrum" of detector current versus drift time is really an electropherogram of a gas.

ionophore A molecule with a hydrophobic outside and a polar inside that can engulf an ion and carry the ion through a hydrophobic phase (such as a cell membrane).

ion pair A closely associated anion and cation, held together by electrostatic attraction. In solvents less polar than water, ions are usually found as ion pairs.

ion-pair chromatography Separation of ions on reversed-phase high-performance liquid chromatography column by adding to the eluent a hydrophobic counterion that pairs with analyte ion and is attracted to stationary phase.

ion-selective electrode One whose potential is selectively dependent on the concentration of one particular ion in solution.

ion spray See *electrospray ionization.*

irradiance Power per unit area (W/m^2) of a beam of electromagnetic radiation. Also called *radiant power* or *intensity.*

isobaric interference In mass spectrometry, overlap of two peaks with nearly the same mass. For example, $^{41}K^+$ and $^{40}ArH^+$ differ by 0.01 atomic mass unit and appear as a single peak unless the spectrometer resolution is great enough to separate them.

isocratic elution Chromatography using a single solvent for the mobile phase.

isoelectric buffer A neutral, polyprotic acid occasionally used as a low-conductivity "buffer" for capillary zone electrophoresis. For example, a solution of pure aspartic acid ($pK_1 = 1.99$, $pK_2 = 3.90$, $pK_3 = 10.00$) has $pH = \frac{1}{2}(pK_1 + pK_2) = 2.94$. Calling pure aspartic acid a "buffer" is an oxymoron, because the buffer capacity is a *minimum* at pH 2.94 and increases to maxima at pH 1.99 and 3.90. However, as the pH drifts away from 2.94, the solution gains significant buffer capacity. When electrophoresis is conducted in a background electrolyte of aspartic acid, the pH stays near 2.94 and conductivity remains very low, permitting a high electric field to be used, thus enabling rapid separations.

isoelectric focusing A technique in which a sample containing polyprotic molecules is subjected to a strong electric field in a medium with a pH gradient. Each species migrates until it reaches the region of its isoelectric pH. In that region, the molecule has no net charge, ceases to migrate, and remains focused in a narrow band.

isoelectric point That pH at which the average charge of a polyprotic species is 0. Same as *isoelectric pH.*

isoionic point The pH of a pure solution of a neutral, polyprotic molecule. The only ions present are H^+, OH^-, and those derived from the polyprotic species. Same as *isoionic pH.*

isosbestic point A wavelength at which the absorbance spectra of two species cross each other. The appearance of isosbestic points in a solution in which a chemical reaction is occurring is evidence that there are only two components present, with a constant total concentration.

isotherm Graph of C_s (mass transfer in stationary phase) versus C_m (mass transfer in the mobile phase) at a given temperature.

isotope ratio mass spectrometry A mass spectrometric technique designed to provide accurate measurements of the ratio of different ions of a selected element. The instrument has one detector dedicated to each isotope.

Job's method See *method of continuous variation.*

Johnson noise A form of white noise arising from random fluctuations of electrons in an electronic device. Lowering the temperature lowers the Johnson noise. Also called *Nyquist noise.*

Jones reductor A column packed with zinc amalgam. An oxidized analyte is passed through to reduce the analyte, which is then titrated with an oxidizing agent.

joule, J SI unit of energy. One joule is expended when a force of 1 N acts over a distance of 1 m. This energy is equivalent to that required to raise 102 g (about $\frac{1}{4}$ pound) by 1 m at sea level.

Joule heating Heat produced in an electric circuit by the flow of electricity. Power (J/s) $= I^2 R$, where I is the current (A) and R is the resistance (ohms).

junction potential An electric potential that exists at the junction (interface) between two different electrolyte solutions or substances. It arises in solutions as a result of unequal rates of diffusion of different ions.

Karl Fischer titration A sensitive technique for measuring traces of water, based on the reaction of H_2O with an amine, I_2, SO_2, and an alcohol.

kelvin, K Absolute unit of temperature defined such that the temperature of water at its triple point (where water, ice, and water vapor are at equilibrium) is 273.16 K and the absolute zero of temperature is 0 K.

kilogram, kg SI unit of mass equal to the mass of a particular Pt–Ir cylinder kept at the International Bureau of Weights and Measures, Sèvres, France.

Kjeldahl nitrogen analysis Procedure for the analysis of nitrogen in organic compounds. The compound is digested with boiling H_2SO_4 to convert nitrogen into NH_4^+, which is treated with base and distilled as NH_3 into a standard acid solution. The moles of acid consumed equal the moles of NH_3 liberated from the compound.

Kovats index See *retention index.*

laboratory sample Portion of bulk sample taken to the lab for analysis. Must have the same composition as the bulk sample.

laminar flow Motion with a parabolic velocity profile of fluid through a tube. Motion is fastest at the center and zero at the walls.

laser Source of intense, coherent monochromatic radiation. Light is produced by stimulated emission of radiation from a medium in which an excited state has been pumped to a high population. Coherence means that all light exiting the laser has the same phase.

laser-induced breakdown spectroscopy Semiquantitative measurement of elements in a surface by vaporizing a small patch with a short laser pulse and measuring atomic emission from the plasma above the surface.

Latimer diagram One that shows the reduction potentials connecting a series of species containing an element in different oxidation states.

law of mass action For the chemical reaction $aA + bB \rightleftharpoons cC + dD$, the condition at equilibrium is $K = \mathcal{A}_C^c \mathcal{A}_D^d / \mathcal{A}_A^a \mathcal{A}_B^b$, where A_i is the activity of the ith species. The law is usually used in approximate form, in which activities are replaced by concentrations.

least squares Process of fitting a mathematical function to a set of measured points by minimizing the sum of the squares of the distances from the points to the curve.

Le Châtelier's principle If a system at equilibrium is disturbed, the direction in which it proceeds back to equilibrium is such that the disturbance is partly offset.

leveling effect The strongest acid that can exist in solution is the protonated form of the solvent. A stronger acid will donate its proton to the solvent and be leveled to the acid strength of the protonated solvent. Similarly, the strongest base that can exist in a solvent is the deprotonated form of the solvent.

Lewis acid One that can form a chemical bond by sharing a pair of electrons donated by another species.

Lewis base One that can form a chemical bond by sharing a pair of its electrons with another species.

ligand An atom or a group attached to a central atom in a molecule. The term is often used to mean any group attached to anything else of interest.

limiting current In a polarographic experiment, the current that is reached at the plateau of a polarographic wave. See also *diffusion current*.

limit of quantitation The minimum signal that can be measured "accurately," often taken as the mean signal for blanks plus 10 times the standard deviation of a low-concentration sample.

linear flow rate In chromatography, the distance per unit time traveled by the mobile phase.

linear interpolation A form of interpolation in which the variation in some quantity is assumed to be linear. For example, to find the value of b when $a = 32.4$ in the following table,

$$a: \quad 32 \qquad 32.4 \qquad 33$$
$$b: \quad \ \ 12.85 \qquad x \qquad 17.96$$

you can set up the proportion

$$\frac{32.4 - 32}{33 - 32} = \frac{x - 12.85}{17.96 - 12.85}$$

which gives $x = 14.89$.

linear quadrupole ion-trap mass spectrometer An instrument that separates gaseous ions by trapping them in stable trajectories inside a linear quadrupole by use of radio-frequency fields. Ions can be expelled from the trap in order of increasing m/z for mass spectrometry.

linearity A measure of how well data in a graph follow a straight line, showing that response is proportional to the quantity of analyte.

linear range Concentration range over which the change in detector response is proportional to the change in analyte concentration.

linear response The case in which the analytical signal is directly proportional to the concentration of analyte.

linear-solvent-strength model In liquid chromatography, a model in which the retention factor k is related to mobile phase composition Φ by the empirical equation, $\log k \approx \log k_w - S\Phi$, where $\log k_w$ and S are constants.

linear voltage ramp The linearly increasing potential that is applied to the working electrode in polarography.

line noise Noise concentrated at discrete frequencies that come from sources external to an intended measuring system. Common sources include radiation emanating from the 60-Hz power line, vacuum-pump motors, and radio-frequency devices. Same as *whistle noise*.

lipid bilayer Double layer formed by molecules containing hydrophilic headgroup and hydrophobic tail. The tails of the two layers associate with each other and the headgroups face the aqueous solvent.

liquid-based ion-selective electrode One that has a hydrophobic membrane separating an inner reference electrode from the analyte solution. The membrane is saturated with a liquid ion exchanger dissolved in a nonpolar solvent. The ion-exchange equilibrium of analyte between the liquid ion exchanger and the aqueous solution gives rise to the electrode potential.

liquid chromatography A form of chromatography in which the mobile phase is a liquid.

liquid-liquid extraction Extraction of a solute from one liquid phase to another liquid phase.

liter, L Common unit of volume equal to exactly $1\ 000\ \text{cm}^3$.

logarithm The base 10 logarithm of n is a if $10^a = n$ (which means $\log n = a$). The natural logarithm of n is a if $e^a = n$ (which means $\ln n = a$). The number e ($= 2.718\ 28\ldots$) is called the base of the natural logarithm.

longitudinal diffusion Diffusion of solute molecules parallel to the direction of travel through a chromatography column.

lot Entire material that is to be analyzed. Examples are a bottle of reagent, a lake, or a truckload of gravel.

lower limit of detection See *detection limit*.

lower limit of quantitation Smallest amount of analyte that can be measured with reasonable accuracy. Usually taken as 10 times the standard deviation of a low-concentration sample. Also called *quantitation limit*.

luminescence Any emission of light by a molecule.

L'vov platform Platform on which sample is placed in a graphite-tube furnace for atomic spectroscopy to prevent sample vaporization before the walls reach constant temperature.

magnetic sector mass spectrometer A device that separates gaseous ions that have the same kinetic energy by passing them through a magnetic field perpendicular to their velocity. Trajectories of ions with a certain mass-to-charge ratio are bent exactly enough to reach the detector. Other ions are deflected too much or too little.

makeup gas Gas added to the exit stream from a gas chromatography column for the purpose of changing flow rate or gas composition to optimize detection of analyte.

MALDI See *matrix-assisted laser desorption/ionization*.

mantissa The part of a logarithm to the right of the decimal point.

masking Process of adding a chemical substance (a *masking agent*) to a sample to prevent one or more components from interfering in a chemical analysis.

masking agent A reagent that selectively reacts with one (or more) component(s) of a solution to prevent the component(s) from interfering in a chemical analysis.

mass balance A statement that the sum of the moles of any element in all of its forms in a solution must equal the moles of that element delivered to the solution.

mass chromatogram See *selected ion chromatogram*.

mass spectrometer An instrument that converts gaseous molecules into ions, accelerates them in an electric field, separates them according to their mass-to-charge ratio, and detects the amount of each species.

mass spectrometry A technique in which gaseous molecules are ionized, accelerated by an electric field, and then separated according to their mass.

mass spectrometry–mass spectrometry, MS–MS See *selected reaction monitoring*.

mass spectrum In mass spectrometry, a graph showing the relative abundance of each ion as a function of its mass-to-charge ratio.

mass titration One in which the mass of titrant, instead of the volume, is measured.

mass-to-charge ratio, m/z The mass of an ion in daltons divided by the charge of the ion measured in multiples of the elementary charge. For $^{23}\text{Na}^+$, for example, $m/z \approx 23/1 = 23$.

matrix The medium containing analyte. For many analyses, it is important that standards be prepared in the same matrix as the unknown.

matrix-assisted laser desorption/ionization, MALDI A gentle technique for introducing predominantly singly charged, intact macromolecular ions into the gas phase. An intimate solid mixture of analyte plus a large excess of a small, ultraviolet-absorbing molecule is irradiated by a pulse from an ultraviolet laser. The small molecule (the matrix) absorbs the radiation, becomes ionized, evaporates, and expands in a supersonic jet that carries analyte into the gas phase. Matrix ions apparently transfer charge to the analyte.

matrix effect A change in analytical signal caused by anything in the sample other than analyte.

matrix modifier Substance added to sample for atomic spectroscopy to make the matrix more volatile or the analyte less volatile so that the matrix evaporates before analyte does.

mean The sum of a set of results divided by the number of values in the set. Also called *average*.

mean activity coefficient For the salt (cation)$_m$(anion)$_n$, the mean activity coefficient, $\gamma_\pm$, is related to the individual ion activity coefficients (γ_+ and γ_-) by the equation $\gamma_\pm = (\gamma_+^m \gamma_-^n)^{1/(m+n)}$.

mechanical balance A balance having a beam that pivots on a fulcrum. Standard masses are used to measure the mass of an unknown.

median For a set of data, that value above and below which there are equal numbers of data.

mediator In electrolysis, a molecule that carries electrons between the electrode and the intended analyte. Used when the analyte cannot react directly at the electrode or when analyte concentration is so low that other reagents react instead. Mediator is recycled indefinitely by oxidation or reduction at the counter electrode.

memory effect Interference in a later analysis caused by substances retained in the instrument or apparatus from an earlier analysis.

meniscus Curved surface of a liquid.

mesh size The number of openings per linear inch in a standard screen used to sort particles. A 100/200 mesh particle passes through a 100 mesh screen but not through a 200 mesh screen.

metal ion buffer Consists of a metal-ligand complex plus excess free ligand. The two serve to fix the concentration of free metal ion through the reaction $M + nL \rightleftharpoons ML_n$.

metal ion indicator A compound that changes color when it binds to a metal ion.

meter, m SI unit of length defined as the distance that light travels in a vacuum during $\frac{1}{299\,792\,458}$ of a second.

method blank A sample without deliberately added analyte. The method blank is taken through all steps of a chemical analysis, including sample preparation.

method of continuous variation Procedure for finding the stoichiometry of a complex by preparing a series of solutions with different metal-to-ligand ratios. The ratio at which the extreme response (such as spectrophotometric absorbance) occurs corresponds to the stoichiometry of the complex. Also called *Job's method*.

method of least squares Process of fitting a mathematical function to a set of measured points by minimizing the sum of the squares of the distances from the points to the curve.

method validation Process of proving that an analytical method is acceptable for its intended purpose.

micellar electrokinetic capillary chromatography A form of capillary electrophoresis in which a micelle-forming surfactant is present. Migration times of solutes depend on the fraction of time spent in the micelles.

micelle An aggregate of molecules with ionic headgroups and long, non-polar tails. The inside of the micelle resembles hydrocarbon solvent, whereas the outside interacts strongly with aqueous solution.

microelectrode An electrode with a diameter on the order of 10 μm (or less). Microelectrodes fit into small places, such as living cells. Their small current gives rise to little ohmic loss, so they can be used in resistive, non-aqueous media. Small double-layer capacitance allows their voltage to be changed rapidly, permitting short-lived species to be studied.

microequilibrium constant An equilibrium constant that describes the reaction of a chemically distinct site in a molecule. For example, a base may be protonated at two distinct sites, each of which has a different equilibrium constant.

microporous particles Chromatographic stationary phase consisting of porous particles 1.5–10 μm in diameter, with high efficiency and high capacity for solute.

migration Electrostatically induced motion of ions in a solution under the influence of an electric field.

miscible liquids Two liquids that form a single phase when mixed in any ratio.

mobile phase In chromatography, the phase that travels through the column.

mobility The terminal velocity that an ion reaches in a field of 1 V/m. Velocity = mobility × field. See also *apparent mobility* and *electrophoretic mobility*.

modulation amplitude In polarography, the magnitude of the voltage pulse applied to the working electrode.

Mohr titration Argentometric titration conducted in the presence of chromate. The end point is signaled by the formation of red $Ag_2CrO_4(s)$.

molality, *m* A measure of concentration equal to the number of moles of solute per kilogram of solvent.

molar absorptivity, ε Constant of proportionality in Beer's law: $A = \varepsilon bc$, where A is absorbance, b is pathlength, and c is the molarity of the absorbing species. Also called *extinction coefficient*.

molarity, M A measure of concentration equal to the number of moles of solute per liter of solution.

mole, mol SI unit for the amount of substance that contains as many molecules as there are atoms in 12 g of ^{12}C. There are approximately 6.022×10^{23} molecules per mole.

molecular exclusion chromatography A technique in which the stationary phase has a porous structure into which small molecules can enter but large molecules cannot. Molecules are separated by size, with larger molecules moving faster than smaller ones. Also called *size exclusion, gel filtration,* or *gel permeation chromatography*.

molecular ion, M$^{+\bullet}$ In mass spectrometry, an ion that has not lost or gained any atoms during ionization.

molecularly imprinted polymer A polymer synthesized in the presence of a template molecule. After the template is removed, the polymer has a void with the right shape to hold the template, and polymer functional groups are positioned correctly to bind to template functional groups.

molecular mass The number of grams of a substance that contains Avogadro's number of molecules.

molecular orbital Describes the distribution of an electron within a molecule.

molecular sieve A solid particle with pores the size of small molecules. Zeolites (sodium aluminosilicates) are a common type.

mole fraction The number of moles of a substance in a mixture divided by the total number of moles of all components present.

monochromatic light Light with a very narrow range of wavelengths ("one color").

monochromator A device (usually a prism, grating, or filter) that disperses light into its component wavelengths and selects a narrow band of wavelengths to pass through the exit slit.

monodentate ligand One that binds to a metal ion through only one atom.

monolithic column Chromatographic column in which polymerization is conducted inside the column to fill the column with porous stationary phase. Monolithic columns allow faster flow rates because the pore structure is maintained at high pressure.

monoprotic acids and bases Compounds that can donate or accept one proton.

mortar and pestle A mortar is a hard ceramic or steel vessel in which a solid sample is ground with a hard tool called a pestle.

mother liquor Solution from which a substance has crystallized or precipitated.

MS^n Successive cycles of selected reaction monitoring. The product ion from one cycle becomes the precursor ion for the next cycle. This experiment can be conducted in a single three-dimensional quadrupole ion-trap mass spectrometer under software control.

mull A fine dispersion of a solid in an oil.

multidentate ligand One that binds to a metal ion through more than one atom.

m/z See *mass-to-charge ratio*.

natural logarithm The natural logarithm (ln) of a is b if $e^b = a$, where $e = 2.718\,28\ldots$. See also *logarithm*.

nebulization Process of breaking a liquid into a mist of fine droplets.

nebulizer In atomic spectroscopy, this device breaks the liquid sample into a mist of fine droplets.

needle valve A valve with a tapered plunger that fits into a small orifice to constrict flow.

nephelometry A technique in which the intensity of light scattered at 90° by a suspension is measured to determine the concentration of suspended particles. In a precipitation titration, the scattering increases until the equivalence point is reached, and then remains constant.

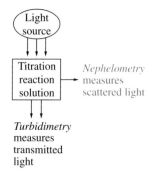

Nernst equation Relates the voltage of a cell, E, to the activities of reactants and products:

$$E = E° - \frac{RT}{nF} \ln Q$$

where R is the gas constant, T is temperature in kelvins, F is the Faraday constant, Q is the reaction quotient, and n is the number of electrons transferred in the balanced reaction. $E°$ is the cell voltage when all activities are unity.

neutralization Process in which a stoichiometric equivalent of acid is added to a base (or vice versa).

neutron-activation analysis A technique in which radiation is observed from a sample bombarded by slow neutrons. The radiation gives both qualitative and quantitative information about the sample composition.

newton, N SI unit of force. One newton will accelerate a mass of 1 kg by 1 m/s^2.

nitrogen chemiluminescence detector Gas chromatography detector in which combustion of eluate at 1 800°C converts nitrogen into NO, which reacts with O_3 to create a chemiluminescent product. Response to N is 10^7 times greater than response to C.

nitrogen-phosphorus detector See *alkali flame detector*.

nitrogen rule A compound with an odd number of nitrogen atoms—in addition to C, H, halogens, O, S, Si, and P—will have an odd nominal mass. A compound with an even number of nitrogen atoms (0, 2, 4, ...) will have an even nominal mass.

noise Signals originating from sources other than those intended to be measured. See also *line noise* and *white noise*.

nominal mass *Integer* mass of the species with the most abundant isotope of each of the constituent atoms. For C, H, and Br, the most abundant isotopes are ^{12}C, ^{1}H, and ^{79}Br. Therefore, the nominal mass of C_2H_5Br is $(2 \times 12) + (5 \times 1) + (1 \times 79) = 108$.

nonelectrolyte A substance that does not dissociate into ions when dissolved.

nonpolarizable electrode One whose potential remains nearly constant, even when current flows; e.g., a saturated calomel electrode.

nonpolar substance A substance, such as a hydrocarbon, with little charge separation within the molecule and no net ionic charge. Nonpolar substances interact with other substances by weak van der Waals forces and are generally not soluble in water.

normal error curve See *Gaussian distribution*.

normal hydrogen electrode, N.H.E. Same as *standard hydrogen electrode*.

normality n times the molarity of a redox reagent, where n is the number of electrons donated or accepted by that species in a particular chemical reaction. For acids and bases, it is also n times the molarity, but n is the number of protons donated or accepted by the species.

normal-phase chromatography A chromatographic separation utilizing a polar stationary phase and a less polar mobile phase.

nucleation Process whereby molecules in solution come together randomly to form small crystalline aggregates that can grow into larger crystals.

null hypothesis In statistics, the supposition that two quantities do not differ from each other or that two methods do not give different results.

Nyquist noise See *Johnson noise*.

occlusion An impurity that becomes trapped (sometimes with solvent) in a pocket within a growing crystal.

ohm, Ω SI unit of electrical resistance. A current of 1 A flows across a potential difference of 1 V if the resistance of the circuit is 1 Ω.

ohmic potential Voltage required to overcome the electrical resistance of an electrochemical cell.

Ohm's law States that the current, I, in a circuit is proportional to voltage, E, and inversely proportional to resistance, R: $I = E/R$.

Ohm's law plot In capillary electrophoresis, a graph of current versus applied voltage. The graph deviates from a straight line when Joule heating becomes significant.

on-column injection Technique used in gas chromatography to place a thermally unstable sample directly on the column without excessive heating in an injection port. Solute is condensed at the start of the column by low temperature, and then the temperature is raised to initiate chromatography.

$1/f$ noise See *drift*.

open tubular column In chromatography, a capillary column whose walls are coated with stationary phase.

optical density, OD See *absorbance*.

optical fiber Fiber that carries light by total internal reflection because the transparent core has a higher refractive index than the surrounding cladding.

optical isomers See *enantiomers*.

optode A sensor based on an optical fiber. Also called *optrode*.

orbitrap mass spectrometer A device that traps ions in stable orbits around a central electrode. Ions oscillate from one end of the trap to the other, inducing image currents in the outer electrodes. Fourier analysis of the image currents gives m/z for the oscillating ions.

order of magnitude A power of 10.

ordinate Vertical (y) axis of a graph.

osmolarity An expression of concentration that gives the total number of particles (ions and molecules) per liter of solution. For nonelectrolytes such as glucose, the osmolarity equals the molarity. For the strong electrolyte $CaCl_2$, the osmolarity is three times the molarity, because each mole of $CaCl_2$ provides 3 mol of ions ($Ca^{2+} + 2Cl^-$).

outlier A datum that lies far from the other data in a set of measurements.

overall formation constant, β_n Same as *cumulative formation constant*.

overpotential Voltage required to overcome the activation energy for a reaction at an electrode. The electrode potential is above that expected from the equilibrium potential, concentration polarization, and ohmic potential. Overpotential is 0 for a reversible reaction.

oxidant See *oxidizing agent*.

oxidation A loss of electrons or a raising of the oxidation state.

oxidation number See *oxidation state*.

oxidation state A bookkeeping device used to tell how many electrons have been gained or lost by a neutral atom when it forms a compound. Also called *oxidation number*.

oxidizability In a natural water or industrial effluent sample, the quantity of O_2 equivalent to the quantity of $KMnO_4$ consumed by refluxing the sample with standard permanganate. Each $KMnO_4$ consumes five electrons and is chemically equivalent to 1.25 mol of O_2. See also *chemical oxygen demand*.

oxidizing agent A substance that takes electrons in a chemical reaction. Also called *oxidant*.

packed column A chromatography column filled with stationary phase particles.

parallax Apparent displacement of an object when the observer changes position. Occurs when the scale of an instrument is viewed from a position that is not perpendicular to the scale. The apparent reading is not the true reading.

particle growth   Process in which molecules become attached to a crystal to form a larger crystal.

partition chromatography   A technique in which separation is achieved by equilibration of solute between two phases.

partition coefficient, K   The equilibrium constant for the reaction in which a solute is partitioned between two phases: solute (in phase 1) $\rightleftharpoons$ solute (in phase 2).

partition ratio   See *retention factor*.

parts per billion, ppb   An expression of concentration denoting nanograms (10^{-9} g) of solute per gram of solution.

parts per million, ppm   An expression of concentration denoting micrograms (10^{-6} g) of solute per gram of solution.

pascal, Pa   SI unit of pressure equal to 1 N/m^2. There are 10^5 Pa in 1 bar and 101 325 Pa in 1 atm.

pellicular particles   A type of stationary phase used in liquid chromatography. Contains a thin layer of liquid coated on a spherical bead. It has high efficiency (low plate height) but low capacity.

peptization   Occurs when washing some ionic precipitates with distilled water causes the ions that neutralize the charges of individual particles to be washed away. The particles then repel one another, disintegrate, and pass through the filter with the wash liquid.

performance test sample   In a series of analytical measurements, a performance test sample is inserted to see whether the procedure gives correct results when the analyst does not know the right answer. The performance test sample is formulated by someone other than the analyst to contain a known concentration of analyte. Also called a *quality control sample* or *blind sample*.

permanent hardness   Component of water hardness not due to dissolved alkaline earth bicarbonates. This hardness remains in the water after boiling. See also *hardness*.

p function   The negative logarithm (base 10) of a quantity: pX = $-\log$ X.

pH   Defined as pH = $-\log \mathcal{A}_{H^+}$ where $\mathcal{A}_{H^+}$ is the activity of H^+. In most approximate applications, the pH is taken as $-\log[H^+]$.

phase transfer agent   A compound such as a crown ether or a salt of a hydrophobic ion used to extract an ionic species from water into an organic solvent.

pH meter   A potentiometer that can measure voltage when extremely little current is flowing. It is used with a glass electrode to measure pH.

pH of zero charge   The pH at which the net charge on the surface of a solid is zero.

phospholipid   A molecule with a phosphate-containing polar headgroup and long hydrocarbon (lipid) tail.

phosphorescence   Emission of light during a transition between states of different spin multiplicity (e.g., triplet → singlet). Phosphorescence is slower than fluorescence, with emission occurring ~10^{-4} to 10^2 s after absorption of a photon.

photochemistry   Chemical reaction initiated by absorption of a photon.

photoconductive detector   A detector whose conductivity changes when light is absorbed by the detector material.

photodiode array   An array of semiconductor diodes used to detect light. The array is normally used to detect light that has been spread into its component wavelengths. One small band of wavelengths falls on each detector.

photoionization detector   A gas chromatography detector that uses vacuum ultraviolet radiation to ionize aromatic and unsaturated compounds; it has little response to saturated hydrocarbons or halocarbons. Electrons produced by the ionization are collected and measured.

photomultiplier tube   One in which the cathode emits electrons when struck by light. The electrons then strike a series of dynodes (plates that are positive with respect to the cathode), and more electrons are released each time a dynode is struck. As a result, more than 10^6 electrons may reach the anode for every photon striking the cathode.

photon   A "particle" of light with energy $h\nu$, where h is Planck's constant and ν is the frequency of the light.

phototube   A vacuum tube with a photoemissive cathode. The electric current flowing between the cathode and the anode is proportional to the intensity of light striking the cathode.

photovoltaic detector   A photodetector with a junction across which the voltage changes when light is absorbed by the detector material.

pH-stat   A device that maintains a constant pH in a solution by continually injecting (or electrochemically generating) acid or base to counteract pH changes.

piezoelectric crystal   A crystal that deforms when an electric field is applied.

piezoelectric effect   Development of electric charge on the surface of certain crystals when subjected to pressure. Conversely, application of an electric field deforms the crystal.

pipet   A glass tube calibrated to deliver a fixed or variable volume of liquid.

pK   The negative logarithm (base 10) of an equilibrium constant: pK = $-\log K$.

Planck distribution   Equation giving the spectral distribution of black-body radiation:

$$M_\lambda = \frac{2\pi hc^2}{\lambda^5}\left(\frac{1}{e^{hc/\lambda kT} - 1}\right)$$

where h is Planck's constant, c is the speed of light, λ is the wavelength of light, k is Boltzmann's constant, and T is temperature in kelvins. M_λ is the power (watts) per square meter of surface per meter of wavelength radiating from the surface. The integral $\int_{\lambda_1}^{\lambda_2} M_\lambda \, d\lambda$ gives the power emitted per unit area in the wavelength interval from λ_1 to λ_2.

Planck's constant   Fundamental constant of nature equal to the energy of light divided by its frequency: $h = E/\nu \approx 6.626 \times 10^{-34}$ J $\cdot$ s.

plane polarized light   Light whose electric field oscillates in one plane.

plasma   A gas that is hot enough to contain free ions and electrons, as well as neutral molecules.

plasmon   A collective oscillation of the free electrons in a metal.

plate height, H   Length of a chromatography column divided by the number of *theoretical plates* in the column. Calculated as the variance, σ^2, of the analyte band divided by the distance, x, it has traveled: $H = \sigma^2/x$.

polarizability   Proportionality constant relating the induced dipole to the strength of the electric field. When a molecule is placed in an electric field, a dipole is induced in the molecule by attraction of the electrons toward the positive pole and attraction of the nuclei toward the negative pole.

polarizable electrode   One whose potential can change readily when a small current flows. Examples are Pt or Ag wires used as indicator electrodes.

polarogram   A graph showing the relation between current and potential during a polarographic experiment.

polarograph   An instrument used to obtain and record a polarogram.

polarographic wave   S-shaped increase in current during a redox reaction in polarography.

polarography   A voltammetry experiment using a dropping-mercury electrode.

polar substance   A substance, such as an alcohol, that has positive and negative regions that attract neighboring molecules by electrostatic forces. Polar substances tend to be soluble in water and insoluble in nonpolar substances, such as hydrocarbons.

polychromatic light   Light of many wavelengths ("many colors").

polychromator   A device that spreads light into its component wavelengths and directs each small band of wavelengths to a different region where it is detected by a photodiode array.

polyprotic acid or base   Compound that can donate or accept more than one proton.

population inversion   A necessary condition for laser operation in which the population of an excited energy level is greater than that of a lower energy level.

porous-layer column   Gas chromatography column containing an adsorptive solid phase coated on the inside surface of its wall.

postprecipitation   Adsorption of otherwise soluble impurities on the surface of a precipitate after the precipitation is over.

potential   See *electric potential*.

potential difference   See *electric potential*.

potentiometer   A device that measures electric potential by balancing it with a known potential of the opposite sign. A potentiometer measures the same quantity as that measured by a voltmeter, but the potentiometer is designed to draw much less current from the circuit being measured.

potentiometry   An analytical method in which an electric potential difference (a voltage) of a cell is measured.

potentiostat   An electronic device that maintains a constant voltage between a pair of electrodes.

power   Energy expended (work done) per unit time. SI units are J/s = watts, W.

ppb, parts per billion   An expression of concentration denoting nanograms (10^{-9} g) of solute per gram of solution.

ppm, parts per million   An expression of concentration denoting micrograms (10^{-6} g) of solute per gram of solution.

precipitant   A substance that precipitates a species from solution.

precipitation   Occurs when a substance leaves solution rapidly (to form either microcrystalline or amorphous solid).

precipitation titration   One in which the analyte forms a precipitate with the titrant.

precision   How well replicate measurements agree with each other.

precolumn   See *guard column*.

preconcentration   Process of concentrating trace components of a mixture prior to their analysis.

precursor ion   In tandem mass spectrometry (selected reaction monitoring), the ion selected by the first mass separator for fragmentation in the collision cell.

premix burner   In atomic spectroscopy, one in which the sample is nebulized and simultaneously mixed with fuel and oxidant before being fed into the flame.

preoxidation   Process of oxidizing analyte prior to titrating it with a reducing agent.

preparative chromatography   Chromatography of large quantities of material conducted for the purpose of isolating pure material.

prereduction   Process of reducing an analyte to a lower oxidation state prior to performing a titration with an oxidizing agent.

pressure   Force per unit area, commonly measured in pascals (N/m) or bars.

pressure broadening   In spectroscopy, line broadening due to collisions between molecules.

primary standard   A reagent that is pure enough and stable enough to be used directly after weighing. The entire mass is considered to be pure reagent.

prism   A transparent, triangular solid. Each wavelength of light passing through the prism is bent (refracted) at a different angle.

product   The species created in a chemical reaction. Products appear on the right side of the chemical equation.

product ion   In tandem mass spectrometry (selected reaction monitoring), the fragment ion from the collision cell selected by the final mass separator for passage through to the detector.

protic solvent   One with an acidic hydrogen atom.

protocol   In quality assurance, written directions stating what must be documented and how the documentation is to be done.

proton   The ion H^+.

proton acceptor   A Brønsted-Lowry base: a molecule that combines with H^+.

protonated molecule   In mass spectrometry, the ion MH^+ resulting from addition of H^+ to the analyte.

proton donor   A Brønsted-Lowry acid: a molecule that can provide H^+ to another molecule.

purge   To force a fluid (usually gas) to flow through a substance or a chamber, usually to extract something from the substance being purged or to replace the fluid in the chamber with the purging fluid.

purge and trap   A method for removing volatile analytes from liquids or solids, concentrating the analytes, and introducing them into a gas chromatograph. A carrier gas bubbled through a liquid or solid extracts volatile analytes, which are then trapped in a tube containing adsorbent. After analyte has been collected, the adsorbent tube is heated and purged to desorb the analytes, which are collected by cold trapping at the start of a gas chromatography column.

pyroelectric effect   Variation with temperature in the electric polarization of a ferroelectric material.

pyrolysis   Thermal decomposition of a substance.

Q test   Statistical test used to decide whether to discard a datum that appears discrepant.

quadrupole ion-trap mass spectrometer   See *three-dimensional ion-trap mass spectrometer* and *linear quadrupole ion-trap mass spectrometer*.

qualitative analysis   Process of determining the identity of the constituents of a substance.

quality assurance   Quantitative indications that demonstrate whether data requirements have been met. Also refers to the broader process that includes quality control, quality assessment, and documentation of procedures and results designed to ensure adequate data quality.

quality control   Active measures taken to ensure the required accuracy and precision of a chemical analysis.

quality control sample   See *performance test sample*.

quantitation limit   See *lower limit of quantitation*.

quantitative analysis   Process of measuring how much of a constituent is present in a substance.

quantitative transfer   Transfer of the entire contents from one vessel to another. This process is usually accomplished by rinsing the first vessel several times with fresh liquid and pouring each rinse into the receiving vessel.

quantum yield   In photochemistry, the fraction of absorbed photons that produce a particular result. For example, if a molecule can isomerize from a *cis* to a *trans* isomer when light is absorbed, the quantum yield for isomerization is the number of molecules that isomerize divided by the number that absorb photons. Quantum yield is in the range 0 to 1.

quaternary ammonium ion   A cation containing four substituents attached to a nitrogen atom; e.g., $(CH_3CH_2)_4N^+$, the tetraethylammonium ion.

quenching   Process in which emission from an excited molecule is decreased by energy transfer to another molecule called a *quencher*.

radian, rad   SI unit of plane angle. There are 2π radians in a complete circle.

radiant power   Power per unit area (W/m^2) of a beam of electromagnetic radiation. Also called *irradiance* or *intensity*.

Raleigh scattering   Scattering of light in all directions by molecules or particles that are much smaller than the wavelength of the light. The wavelength of scattered light is the same as that of incident light. The intensity of scattered light increases as $1/\lambda^4$, where λ is the wavelength.

Raman scattering   Scattering of light in which the wavelength of scattered light is changed from that of incident light by an energy corresponding to vibrational energy of the molecule responsible for scattering. In Stokes Raman scattering, the molecule gains vibrational energy and scattered light has less energy than incident light. In anti–Stokes Raman scattering, an excited molecule loses vibrational energy and scattered light has more energy than incident light.

random error   A type of error, which can be either positive or negative and cannot be eliminated, based on the ultimate limitations on a physical measurement. Also called *indeterminate error*.

random heterogeneous material   A material in which there are differences in composition with no pattern or predictability and on a fine scale. When you collect a portion of the material for analysis, you obtain some of each of the different compositions.

random sample   Bulk sample constructed by taking portions of the entire lot at random.

range The difference between the highest and lowest values in a set of data. Also called *spread*. With respect to an analytical method, range is the concentration interval over which linearity, accuracy, and precision are all acceptable.

raw data Individual values of a measured quantity, such as peak areas from a chromatogram or volumes from a buret.

reactant The species consumed in a chemical reaction. It appears on the left side of a chemical equation.

reaction quotient, Q Expression having the same form as the equilibrium constant for a reaction. However, the reaction quotient is evaluated for a particular set of existing activities (concentrations), which are generally not the equilibrium values. At equilibrium, $Q = K$.

reagent blank A solution prepared from all of the reagents, but no analyte. The blank measures the response of the analytical method to impurities in the reagents or any other effects caused by any component other than the analyte. The reagent blank, unlike the *method blank,* is not subjected to all sample preparation steps before analysis.

reagent gas In a chemical ionization source for mass spectrometry, reagent gas (normally methane, isobutane, or ammonia at ~1 mbar) is converted into strongly proton donating species such as CH_5^+ by a process beginning with electron ionization. Protonated reagent gas reacts with analyte to create protonated analyte.

reagent grade chemical A high-purity chemical generally suitable for use in quantitative analysis and meeting purity requirements set by organizations such as the American Chemical Society.

reciprocal centimeter, cm^{-1} The most common unit of wavenumber, $1/\lambda$, where λ is wavelength in cm.

reconstructed total ion chromatogram In chromatography, a graph of the sum of intensities of all ions detected at all masses (above a selected cutoff) versus time.

redox couple A pair of reagents related by electron transfer; e.g., $Fe^{3+} \mid Fe^{2+}$ or $MnO_4^- \mid Mn^{2+}$.

redox indicator A compound used to find the end point of a redox titration because its various oxidation states have different colors. The standard potential of the indicator must be such that its color changes near the equivalence point of the titration.

redox reaction A chemical reaction in which electrons are transferred from one element to another.

redox titration One in which the reaction between analyte and titrant is an oxidation-reduction reaction.

reduced plate height In chromatography, the quotient plate height/d, where the numerator is the height equivalent to a theoretical plate and the denominator is the diameter of stationary phase particles.

reducing agent A substance that donates electrons in a chemical reaction. Also called *reductant*.

reductant See *reducing agent*.

reduction A gain of electrons or a lowering of the oxidation state.

reference electrode One that maintains a constant potential against which the potential of another half-cell may be measured.

reflectance Fraction of incident radiant power reflected by an object.

refraction Bending of light when it passes between media with different refractive indexes.

refractive index, n The speed of light in any medium is c/n, where c is the speed of light in vacuum and n is the refractive index of the medium. The refractive index also measures the angle at which a light ray is bent when it passes from one medium into another. Snell's law states that $n_1 \sin\theta_1 = n_2 \sin\theta_2$, where n_i is the refractive index for each medium and θ_i is the angle of the ray with respect to the normal between the two media.

refractive index detector Liquid chromatography detector that measures the change in refractive index of eluate as solutes emerge from the column.

relative retention, α In chromatography, the ratio of adjusted retention times for two components. If component 1 has an adjusted retention time of t'_{r1} and component 2 has an adjusted retention time of t'_{r2} ($>t'_{r1}$), the relative retention is $\alpha = t'_{r2}/t'_{r1}$. Also called *separation factor*. See also *unadjusted relative retention*, γ.

relative standard deviation See *coefficient of variation*.

relative supersaturation Defined as $(Q - S)/S$, where S is the concentration of solute in a saturated solution and Q is the concentration in a particular supersaturated solution.

relative uncertainty Uncertainty of a quantity divided by the value of the quantity. It is usually expressed as a percentage of the measured quantity.

releasing agent In atomic spectroscopy, a substance that prevents chemical interference.

replicate measurements Repeated measurements of the same quantity.

reporting limit Concentration below which regulations dictate that an analyte is reported as "not detected." The reporting limit is typically set 5 to 10 times higher than the detection limit.

reprecipitation Sometimes a gravimetric precipitate can be freed of impurities only by redissolving it in fresh solvent and reprecipitating it. The impurities are present at lower concentration during the second precipitation and are less likely to coprecipitate.

residual current The small current that is observed prior to the decomposition potential in an electrolysis.

resin Small, hard particles of an ion exchanger, such as polystyrene with ionic substituents.

resistance, R A measure of the retarding force opposing the flow of electric current. SI unit is ohm, Ω.

resistivity, ρ A measure of the ability of a material to retard the flow of electric current. $J = E/\rho$, where J is the current density (A/m^2) and E is electric field (V/m). Units of resistivity are $V \cdot m/A = ohm \cdot m = \Omega \cdot m$. The resistance, R, of a conductor with a given length and cross-sectional area is given by $R = \rho \cdot$ length/area. Resistivity is the reciprocal of *conductivity*.

resolution How close two bands in a spectrum or a chromatogram can be to each other and still be seen as two peaks. In chromatography, it is defined as the difference in retention times of adjacent peaks divided by their width. In mass spectrometry, resolution is the smallest difference in m/z values that can be detected as separate peaks and should be reported with the m/z value where it is measured.

resolving power In mass spectrometry, resolving power can be defined as $m/\Delta m$, where Δm is the separation of two peaks when the overlap at the base is 10% of the peak height and m is the smaller of the two m/z values. Alternatively, resolving power can be taken as $m/m_{1/2}$, where $m_{1/2}$ is the width of a peak at half the maximum height. In this case, the dip between two barely resolved peaks is 8% below the peak heights.

response factor, F Relative response of a detector to analyte (X) and internal standard (S): (signal from X)/[X] = F(signal from S)/[S]. Once you have measured F with a standard mixture, you can use it to find [X] in an unknown if you know [S] and the quotient (signal from X)/(signal from S).

results What we ultimately report after applying statistics to treated data.

retardation, δ Difference in pathlength between light striking the stationary and moving mirrors of an interferometer.

retention factor, k In chromatography, the adjusted retention time for a peak divided by the time for the mobile phase to travel through the column. Retention factor is also equal to the ratio of the time spent by the solute in the stationary phase to the time spent in the mobile phase. Also called *capacity factor*, *capacity ratio*, and *partition ratio*.

retention gap In gas chromatography, a 3- to 10-m length of empty, silanized capillary ahead of the chromatography column. The retention gap improves the peak shape of solutes that elute close to solvent when large volumes of solvent are injected or when the solvent has a very different polarity from that of the stationary phase.

retention index, I In gas chromatography, the Kovats retention index is a logarithmic scale that relates the retention time of a compound to those of linear alkanes. Pentane would be given an index of 500, hexane 600, heptane 700, and so on.

retention ratio In chromatography, the time required for solvent to pass through the column divided by the time required for solute to pass through the column.

retention time, t_r The time, measured from injection, needed for a solute to be eluted from a chromatography column.

retention volume, V_r The volume of solvent needed to elute a solute from a chromatography column.

reversed-phase chromatography A technique in which the stationary phase is less polar than the mobile phase.

rings + double bonds formula The number of rings + double bonds in a molecule with the formula $C_cH_hN_nO_x$ is $c - h/2 + n/2 + 1$, where c includes all Group 14 atoms (C, Si, Ge, Sn, Pb, which all make four bonds), h includes H + halogens (which make one bond), and n is the number of Group 15 atoms (N, P, As, Sb, Bi, which make three bonds). Group 16 atoms (which make two bonds) do not affect the result.

robustness Ability of an analytical method to be unaffected by small, deliberate changes in operating parameters.

root-mean-square (rms) noise Standard deviation of the noise in a region where the signal is flat:

$$\text{rms noise} = \sqrt{\frac{\sum_i (A_i - \overline{A})^2}{n}}$$

where A_i is the measured signal for the ith data point, $\overline{A}$ is the mean signal, and n is the number of data points.

rotating disk electrode A motor-driven electrode with a smooth flat face in contact with the solution. Rapid convection created by rotation brings fresh analyte to the surface of the electrode. A Pt electrode is especially suitable for studying anodic processes, in which a mercury electrode would be too easily oxidized.

rotational transition Occurs when a molecule changes its rotation energy.

rubber policeman A glass rod with a flattened piece of rubber on the tip. The rubber is used to scrape solid particles from glass surfaces in gravimetric analysis.

ruggedness See *intermediate precision*.

run buffer See *background electrolyte*.

salt An ionic solid.

salt bridge A conducting ionic medium in contact with two electrolyte solutions. It allows ions to flow without allowing immediate diffusion of one electrolyte solution into the other.

sample cleanup Removal of portions of the sample that do not contain analyte and may interfere with analysis.

sample preparation Transforming a sample into a state that is suitable for analysis. This process can include concentrating a dilute analyte and removing or masking interfering species.

sampling The process of collecting a representative sample for analysis.

sampling variance The square of the standard deviation arising from heterogeneity of the sample, not from the analytical procedure. For inhomogeneous materials, it is necessary to take larger portions or more portions to reduce the uncertainty of composition due to variation from one region to another. Total variance is the sum of variances from sampling and from analysis.

saturated calomel electrode, S.C.E. A calomel electrode saturated with KCl. The electrode half-reaction is $Hg_2Cl_2(s) + 2e^- \rightleftharpoons 2Hg(l) + 2Cl^-$.

saturated solution One that contains the maximum amount of a compound that can dissolve at equilibrium.

Scatchard plot A graph used to find the equilibrium constant for a reaction such as $X + P \rightleftharpoons PX$. It is a graph of $[PX]/[X]$ versus $[PX]$ or any functions proportional to these quantities. The magnitude of the slope of the graph is the equilibrium constant.

S.C.E. See *saturated calomel electrode*.

schlieren Streaks in a liquid mixture observed before the two phases have mixed. Streaks arise from regions that refract light differently.

Schottky noise See *white noise*.

second, s SI unit of time equal to the duration of 9 192 631 770 periods of the radiation corresponding to the transition between two hyperfine levels of the ground state of ^{133}Cs.

segregated heterogeneous material A material in which differences in composition are on a large scale. Different regions have obviously different composition.

selected ion chromatogram A graph of detector response versus time when a mass spectrometer monitors just one or a few species of selected mass-to-charge ratio, m/z, emerging from a chromatograph. Also called *mass chromatogram*.

selected ion monitoring Use of a mass spectrometer to monitor species with just one or a few mass-to-charge ratios, m/z.

selected reaction monitoring A technique in which the precursor ion selected by one mass separator passes through a collision cell in which the precursor breaks into several fragment ions (product ions). A second mass separator then selects one (or a few) of these ions for detection. Selected reaction monitoring improves chromatographic signal-to-noise ratio because it is insensitive to almost everything other than the intended analyte. Also called *mass spectrometry–mass spectrometry (MS–MS)* or *tandem mass spectrometry*.

selectivity See *specificity*.

selectivity coefficient With respect to an ion-selective electrode, a measure of the relative response of the electrode to two different ions. In ion-exchange chromatography, the selectivity coefficient is the equilibrium constant for displacement of one ion by another from the resin.

self-absorption In a luminescence measurement, a high concentration of analyte molecules can absorb excitation energy from excited analyte. Also called *inner filter effect*. If the absorbed energy is dissipated as heat instead of light, fluorescence does not increase in proportion to analyte concentration. Analyte concentration can be so high that fluorescence *decreases* with increasing concentration. In flame emission atomic spectroscopy, there is a lower concentration of excited-state atoms in the cool, outer part of the flame than in the hot, inner flame. The cool atoms can absorb emission from the hot ones and thereby decrease the observed signal.

semiconductor A material whose conductivity (10^{-7} to $10^4 \ \Omega^{-1} \cdot m^{-1}$) is intermediate between that of good conductors ($10^8 \ \Omega^{-1} \cdot m^{-1}$) and that of insulators (10^{-20} to $10^{-12} \ \Omega^{-1} \cdot m^{-1}$).

semipermeable membrane A thin layer of material that allows some substances, but not others, to pass across the material. A dialysis membrane allows small molecules to pass, but not large molecules.

sensitivity Capability of responding reliably and measurably to changes in analyte concentration. In quantitative terms, sensitivity is the amount of instrument response per unit change in concentration of analyte.

separation factor See *relative retention*.

separator column Ion-exchange column used to separate analyte species in ion chromatography.

septum A disk, usually made of silicone rubber, covering the injection port of a gas chromatograph. The sample is injected by syringe through the septum.

sequential injection analysis Analytical technique related to flow injection. Sample and reagents are taken into a holding coil through a multiport valve. After a suitable reaction time, flow is reversed and the zones of reagent, product, and sample are pushed through a detector to measure the amount of product. Flow is not continuous, so sequential injection consumes less reagents than does flow injection.

shot noise A form of white noise (Gaussian noise) arising from the quantized nature of charge carriers and photons. At low signal levels, shot noise arises from random variation in the small number of photons reaching a detector or the small number of electrons and holes generated in a semiconductor. Also called *Schottky noise*.

SI units International system of units based on the meter, kilogram, second, ampere, kelvin, candela, mole, radian, and steradian.

sieving In electrophoresis, the separation of macromolecules by migration through a polymer gel. Movement of the smallest molecules is fastest and that of the largest is slowest.

signal averaging Improvement of a signal by averaging successive scans. The signal increases in proportion to the number of scans accumulated. The noise increases in proportion to the square root of the number of scans. Therefore, the signal-to-noise ratio improves in proportion to the square root of the number of scans collected.

signal-to-noise ratio The height of a signal divided by the noise in the baseline around the signal. Noise is commonly expressed as root-mean-square noise. The higher the signal-to-noise ratio, the less uncertainty there is in the signal.

significant figure The number of significant digits in a quantity is the minimum number of digits needed to express the quantity in scientific notation without loss of precision. In experimental data, the first uncertain figure is the last significant figure.

silanization Treatment of a chromatographic solid support or glass column with hydrophobic silicon compounds that bind to the most reactive Si—OH groups. It reduces irreversible adsorption and tailing of polar solutes.

silver-silver chloride electrode A common reference electrode containing a silver wire coated with AgCl paste and dipped in a solution saturated with AgCl and (usually) KCl. The half-reaction is $AgCl(s) + e^- \rightleftharpoons Ag(s) + Cl^-$.

single-column ion chromatography Separation of ions on a low-capacity ion-exchange column, using low-ionic-strength eluent.

single-electrode potential Voltage measured when the electrode of interest is connected to the positive terminal of a potentiometer and a standard hydrogen electrode is connected to the negative terminal.

singlet state One in which all electron spins are paired.

size exclusion chromatography See *molecular exclusion chromatography*.

skimmer cone voltage See *cone voltage*.

slope For a straight line whose equation is $y = mx + b$, the value of m is the slope. It is the ratio $\Delta y/\Delta x$ for any segment of the line.

slurry A suspension of a solid in a solvent.

Smith-Hieftje background correction In atomic absorption spectroscopy, a method of distinguishing analyte signal from background signal, based on applying a periodic pulse of high current to the hollow-cathode lamp to distort the lamp signal. Signal detected during the current pulse is subtracted from signal detected without the pulse to obtain the corrected response.

smoothing Use of a mathematical procedure or electrical filtering to improve the quality of a signal.

Snell's law Relates angle of refraction, θ_2, to angle of incidence, θ_1, for light passing from a medium with refractive index n_1 to a medium of refractive index n_2: $n_1 \sin \theta_1 = n_2 \sin \theta_2$. Angles are measured with respect to the normal to the surface between the two media.

sodium error Systematic error that occurs when a glass pH electrode is placed in a strongly basic solution containing very little H^+ and a high concentration of Na^+. The electrode begins to respond to Na^+ as if it were H^+, so the pH reading is lower than the actual pH. Also called *alkaline error*.

solid-phase extraction Preconcentration procedure in which a solution is passed through a short column of chromatographic stationary phase, such as C_{18} on silica. Trace solutes adsorbed on the column can be eluted with a small volume of solvent of high eluent strength.

solid-phase microextraction Extraction of compounds from liquids or gases into a coated fiber dispensed from a syringe needle. After extraction, the fiber is withdrawn into the needle and the needle is inserted through the septum of a chromatograph. The fiber is extended inside the injection port and adsorbed solutes are desorbed by heating (for gas chromatography) or solvent (for liquid chromatography).

solid-state ion-selective electrode An ion-selective electrode that has a solid membrane made of an inorganic salt crystal. Ion-exchange equilibria between the solution and the surface of the crystal account for the electrode potential.

solid-supported liquid-liquid extraction A form of liquid-liquid extraction in which the liquid to be extracted is applied to a porous solid which retains the liquid. The second liquid is then passed through the porous solid to extract solute from the first liquid.

solubility product, K_{sp} Equilibrium constant for the dissociation of a solid salt to give its ions in solution. For the reaction $M_mN_n(s) \rightleftharpoons mM^{n+} + nN^{m-}$, $K_{sp} = \mathcal{A}_{M^{n+}}^m \cdot \mathcal{A}_{N^{m-}}^n$ where $\mathcal{A}$ is the activity of each species.

solute A minor component of a solution.

solution A homogeneous mixture of two or more substances.

solvation Interaction of solvent molecules with solute. Solvent molecules orient themselves around solute to minimize the energy through dipole and van der Waals forces.

solvent Major constituent of a solution.

solvent extraction A method in which a chemical species is transferred from one liquid phase to another. It is used to separate components of a mixture.

solvent strength See *eluent strength*.

solvent trapping Splitless gas chromatography injection technique in which solvent is condensed near its boiling point at the start of the column. Solutes dissolve in a narrow band in the condensed solvent.

speciation Describes the distribution of an element or compound among different chemical forms.

species Chemists refer to any element, compound, or ion of interest as a *species*. The word *species* is both singular and plural.

specific adsorption Process in which molecules are held tightly to a surface by van der Waals or electrostatic forces.

specifications In quality assurance, specifications are written statements describing how good analytical results need to be and what precautions are required in an analytical method.

specific gravity A dimensionless quantity equal to the mass of a substance divided by the mass of an equal volume of water at 4°C. Specific gravity is virtually identical with density in g/mL.

specificity Capability of an analytical method to distinguish analyte from other species in the sample. Also called *selectivity*.

spectral interference In atomic spectroscopy, any physical process that affects light intensity at the analytical wavelength. Created by substances that absorb, scatter, or emit light of the analytical wavelength.

spectrophotometer A device used to measure absorption of light. It includes a source of light, a wavelength selector (monochromator), and an electrical means of detecting light.

spectrophotometric analysis Any method in which light absorption, emission, reflection, or scattering is used to measure chemical concentrations.

spectrophotometric titration One in which absorption of light is used to monitor the progress of the titration reaction and to find the equivalence point.

spectrophotometry In a broad sense, any method using light to measure chemical concentrations.

specular reflection Reflection of light at an angle equal to the angle of incidence.

spike Addition of a known compound (usually at a known concentration) to an unknown. In isotope dilution mass spectrometry, the spike is the added, unusual isotope. *Spike* is a noun and a verb. In *co-chromatography*, spiking is the simultaneous chromatography of a known compound with an unknown. If a known and an unknown have the same retention time on several columns, they are probably identical. Also called *fortification*.

spike recovery The fraction of a spike eventually found by chemical analysis.

split injection Technique used in capillary gas chromatography to inject a small fraction of sample onto the column; the rest of the sample is blown out to waste.

splitless injection Technique used in capillary gas chromatography for trace analysis and quantitative analysis. The entire sample in a low-boiling solvent is directed to the column, where the sample is concentrated by *solvent trapping* (condensing the solvent below its boiling point) or *cold trapping* (condensing solutes far below their boiling range). The column is then warmed to initiate separation.

spontaneous process One that is energetically favorable. It will eventually occur, but thermodynamics makes no prediction about how long it will take.

spread See *range*.

square wave voltammetry A form of *voltammetry* (measurement of current versus potential in an electrochemical cell) in which the potential waveform consists of a square wave superimposed on a voltage staircase. The technique is faster and more sensitive than voltammetry with other waveforms.

stability constant   See *formation constant.*

stacking   In electrophoresis, the process of concentrating ions into a narrow band at the interface of electrolytes of low conductivity and high conductivity. Stacking occurs because the electric field in low-conductivity electrolyte is stronger than the field in high-conductivity electrolyte. Ions in the low-conductivity region migrate rapidly until they reach the interface, where the electric field is much smaller.

standard addition   A technique in which an analytical signal due to an unknown is first measured. Then a known quantity of analyte is added, and the increase in signal is recorded. From the response, it is possible to calculate what quantity of analyte was in the unknown.

standard curve   A graph showing the response of an analytical technique to known quantities of analyte.

standard deviation   A statistic measuring how closely data are clustered about the mean value. For a finite set of data, the standard deviation, s, is computed from the formula

$$s = \sqrt{\frac{\sum_i (x_i - \bar{x})^2}{n-1}} = \sqrt{\frac{\sum_i (x_i^2)}{n-1} - \frac{\left(\sum_i x_i\right)^2}{n(n-1)}}$$

where n is the number of results, x_i is an individual result, and $\bar{x}$ is the mean result. For a large number of measurements, s approaches σ, the true standard deviation of the population, and $\bar{x}$ approaches μ, the true population mean.

standard deviation of the mean, σ_n   The standard deviation of a set of measurements (σ) divided by the square root of the number of measurements (n) in the set: $\sigma/\sqrt{n}$.

standard hydrogen electrode, S.H.E.   One that contains $H_2(g)$ bubbling over a catalytic Pt surface immersed in aqueous H^+. The activities of H_2 and H^+ are both unity in the hypothetical standard electrode. The reaction is $H^+ + e^- \rightleftharpoons \frac{1}{2}H_2(g)$. Also called *normal hydrogen electrode (N.H.E.).*

standardization   Process of determining the concentration of a reagent by reaction with a known quantity of a second reagent.

standard operating procedure   A written procedure that must be rigorously followed to ensure the quality of a chemical analysis.

standard reduction potential, $E°$ The voltage that would be measured when a hypothetical cell containing the desired half-reaction (with all species present at unit activity) is connected to a standard hydrogen electrode anode.

Standard Reference Material   Same as *certified reference material.*

standard solution   A solution whose composition is known by virtue of the way that it was made from a reagent of known purity or by virtue of its reaction with a known quantity of a standard reagent.

standard state   The standard state of a solute is 1 M and the standard state of a gas is 1 bar. Pure solids and liquids are considered to be in their standard states. In equilibrium constants, dimensionless concentrations are expressed as a ratio of the concentration of each species to its concentration in its standard state.

stationary phase   In chromatography, the phase that does not move through the column.

stepwise formation constant, K_n   Equilibrium constant for a reaction of the type $ML_{n-1} + L \rightleftharpoons ML_n$.

steradian, sr   Unit of solid angle. There are 4π steradians in a complete sphere.

stimulated emission   Emission of a photon induced by the passage of another photon of the same wavelength.

stir-bar sorptive extraction   Sample preparation method similar to solid-phase microextraction, except the sorptive layer is coated on the outside of a stirring bar. Coating volume is greater than the fiber volume in solid-phase microextraction, so it provides $\sim 10^2$ times greater sensitivity for traces of analyte. Analyte is removed from the coating by thermal desorption for chromatography.

stoichiometry   Ratios of substances participating in a chemical reaction.

Stokes equation   The friction coefficient for a molecule migrating through solution is $6\pi\eta r$, where η is the viscosity of the fluid and r is the hydrodynamic radius (equivalent spherical radius) of the molecule.

stray light   In spectrophotometry, light reaching the detector that is not part of the narrow set of wavelengths expected from the monochromator.

stripping analysis   A sensitive polarographic technique in which analyte is concentrated from dilute solution by reduction into a drop (or a film) of Hg. It is then analyzed polarographically during an anodic redissolution process. Some analytes can be oxidatively concentrated onto an electrode other than Hg and stripped in a reductive process.

strong acids and bases   Those that are completely dissociated (to H^+ or OH^-) in water.

strong electrolyte   One that mostly dissociates into ions in solution.

Student's t   A statistical tool used to express confidence intervals and to compare results from different experiments.

sulfur chemiluminescence detector   Gas chromatography detector for the element sulfur. Exhaust from a flame ionization detector is mixed with O_3 to form an excited state of SO_2 that emits light, which is detected. Response to S is 10^7 times greater than response to C.

superconductor   A material that loses all electrical resistance when cooled below a critical temperature.

supercritical fluid   A fluid whose temperature is above its critical temperature and whose pressure is above its critical pressure. It has properties of both a liquid and a gas.

supercritical fluid chromatography   Chromatography using supercritical fluid as the mobile phase. Capable of highly efficient separations of nonvolatile solutes and able to use detectors suitable for gas or liquid.

supercritical fluid extraction   Extraction of compounds (usually from solids) with a supercritical fluid solvent.

superficially porous particle   A stationary phase particle for liquid chromatography containing a thin, porous outer layer and a dense, nonporous core. Mass transfer is faster in the superfically porous particle than in a fully porous particle of the same diameter. Also called *fused-core particle.*

supernatant liquid   Liquid remaining above the solid after a precipitation. Also called *supernate.*

supersaturated solution   One that contains more dissolved solute than would be present at equilibrium.

support-coated column   Open tubular gas chromatography column in which the stationary phase is coated on solid support particles attached to the inside wall of the column.

supporting electrolyte   An unreactive salt added in high concentration to solutions for voltammetric measurements (such as polarography). The supporting electrolyte carries most of the ion-migration current and therefore decreases the coulombic migration (drift of ions in the electric field) of electroactive species to a negligible level. The electrolyte also decreases the resistance of the solution.

suppressed-ion chromatography   Separation of ions by using an ion-exchange column followed by a suppressor (membrane or column) to remove ionic eluent.

suppressor   In ion chromatography, a device that transforms ionic eluent into a nonionic form.

surface plasmon resonance   A sensitive means to measure the binding of molecules to a thin gold layer on the underside of a prism. Light directed through the prism is reflected from the gold surface. There is one narrow range of angles at which reflection is nearly 0 because the gold absorbs the light to set up oscillations (called *plasmons*) of the electron cloud in the metal. When a thin layer of material (such as protein or DNA) binds to the side of the gold away from the prism, the electrical properties of the gold are changed and the reflectivity changes.

surfactant   A molecule with an ionic or polar headgroup and a long, nonpolar tail. Surfactants aggregate in aqueous solution to form micelles. Surfactants derive their name from the fact that they accumulate at boundaries between polar and nonpolar phases and modify the surface tension, which is the free energy of formation of the surface. Soaps are surfactants.

sweeping   In capillary electrophoresis, migration of a collector species such as a micelle or chelator to concentrate analyte into a narrow region at the front of the migrating collector species.

syringe A device having a calibrated barrel into which liquid is sucked by a plunger. The liquid is expelled through a needle by pushing on the plunger.

systematic error Error due to procedural or instrumental factors that cause a measurement to be consistently too large or too small. The error can, in principle, be discovered and corrected. Also called *determinate error.*

systematic treatment of equilibrium A method that uses the charge balance, mass balance(s), and equilibria to completely specify a system's composition.

t **test** Statistical test used to decide whether the results of two experiments are within experimental uncertainty of each other. The uncertainty must be specified to within a certain probability.

tailing Asymmetric chromatographic band in which the later part elutes very slowly. It often results from adsorption of a solute onto a few active sites on the stationary phase.

tandem mass spectrometry See *selected reaction monitoring.*

tare As a noun, *tare* is the mass of an empty vessel used to receive a substance to be weighed. As a verb, *tare* means setting the balance reading to 0 when an empty vessel or weighing paper is placed on the pan.

temperature programming Raising the temperature of a gas chromatography column during a separation to reduce the retention time of late-eluting components.

temporary hardness Component of water hardness due to dissolved alkaline earth bicarbonates. It is temporary because boiling causes precipitation of the carbonates.

test portion Part of the laboratory sample used for one analysis. Also called *aliquot.*

theoretical plate An imaginary construct in chromatography denoting a segment of a column in which one equilibration of solute occurs between stationary and mobile phases. The number of theoretical plates on a column with Gaussian bandshapes is defined as $N = t_r^2/\sigma^2$, where t_r is the retention time of a peak and σ is the standard deviation of the band. See also *plate height.*

thermal conductivity, κ Rate at which a substance transports heat (energy per unit time per unit area) through a temperature gradient (degrees per unit distance). Energy flow $[J/(s \cdot m^2)] = -\kappa(dT/dx)$, where κ is the thermal conductivity $[W/(m \cdot K)]$ and dT/dx is the temperature gradient (K/m).

thermal conductivity detector A device that detects substances eluted from a gas chromatography column by measuring changes in the thermal conductivity of the gas stream.

thermal desorption A sample preparation technique used in gas chromatography to release volatile substances from a solid sample by heating.

thermistor A device whose electrical resistance changes markedly with changes in temperature.

thermocouple An electrical junction across which a temperature-dependent voltage exists. Thermocouples are calibrated for measurement of temperature and usually consist of two dissimilar metals in contact with each other.

thermogravimetric analysis A technique in which the mass of a substance is measured as the substance is heated. Changes in mass indicate decomposition of the substance, often to well-defined products.

thermometric titration One in which the temperature is measured to determine the end point. Most titration reactions are exothermic, so the temperature rises during the reaction and suddenly stops rising when the equivalence point is reached.

thin-layer chromatography Liquid chromatography in which the stationary phase is coated on a flat glass or plastic plate. Solute is spotted near the bottom of the plate. The bottom edge of the plate is placed in contact with solvent, which creeps up the plate by capillary action.

three-dimensional quadrupole ion-trap mass spectrometer An instrument that separates gaseous ions by trapping them in stable three-dimensional trajectories inside a metal chamber to which a radio-frequency electric field is applied. Application of an oscillating electric field between the ends of the chamber destabilizes the trajectories of ions with a particular mass-to-charge ratio, expelling them from the cavity and into a detector.

time-of-flight mass spectrometer Ions of different mass accelerated through the same electric field have different velocities: The lighter ions move faster than the heavier ions. The time-of-flight spectrometer finds the mass-to-charge ratio by measuring the time that each group of ions requires to travel a fixed distance to the detector.

titer A measure of concentration, usually defined as how many milligrams of reagent B will react with 1 mL of reagent A. One milliliter of $AgNO_3$ solution with a titer of 1.28 mg NaCl/mL will be consumed by 1.28 mg NaCl in the reaction $Ag^+ + Cl^- \rightarrow AgCl(s)$. The same solution of $AgNO_3$ has a titer of 0.993 mg of KH_2PO_4/mL, because 1 mL of $AgNO_3$ solution will be consumed by 0.993 mg KH_2PO_4 to precipitate Ag_3PO_4.

titrant Substance added to the analyte in a titration.

titration A procedure in which one substance (titrant) is carefully added to another (analyte) until complete reaction has occurred. The quantity of titrant required for complete reaction tells how much analyte is present.

titration curve A graph showing how the concentration of a reactant or a physical property of the solution varies as one reactant (the titrant) is added to another (the analyte).

titration error Difference between the observed end point and the true equivalence point in a titration.

tolerance Manufacturer's stated uncertainty in the accuracy of a device such as a buret or volumetric flask. A 100-mL flask with a tolerance of ±0.08 mL may contain 99.92 to 100.08 mL and be within tolerance.

total carbon In a natural water or industrial effluent sample, the quantity of CO_2 produced when the sample is completely oxidized by oxygen at 900°C in the presence of a catalyst.

total ion chromatogram A graph of detector response versus time when a mass spectrometer monitors all ions above a selected *m/z* ratio emerging from a chromatograph.

total organic carbon In a natural water or industrial effluent sample, the quantity of CO_2 produced when the sample is first acidified and purged to remove carbonate and bicarbonate and then completely oxidized by oxygen at 900°C in the presence of a catalyst.

total oxygen demand In a natural water or industrial effluent sample, the quantity of O_2 required for complete oxidation of species in the water at 900°C in the presence of a catalyst.

trace analysis Chemical analysis of very low levels of analyte, typically ppm and lower.

transition range For an acid-base indicator, the pH range over which the color change occurs. For a redox indicator, the potential range over which the color change occurs.

transmission quadrupole mass spectrometer A mass spectrometer that separates ions by passing them between four metallic cylinders to which are applied direct current and oscillating electric fields. Resonant ions with the right mass-to-charge ratio pass through the chamber to the detector while nonresonant ions are deflected into the cylinders and are lost.

transmittance, *T* Defined as $T = P/P_0$, where P_0 is the radiant power of light striking the sample on one side and P is the radiant power of light emerging from the other side of the sample.

treated data Concentrations or amounts of analyte found from raw data with a calibration curve or some other calibration method.

triple point The one temperature and pressure at which the solid, liquid, and gaseous forms of a substance are in equilibrium with one another.

triplet state An electronic state in which there are two unpaired electrons.

tungsten lamp An ordinary light bulb in which electricity passing through a tungsten filament heats the wire and causes it to emit visible light.

turbidimetry A technique in which the decrease in intensity of light traveling through a turbid solution (a solution containing suspended particles) is measured. The greater the concentration of suspended particles, the less light is transmitted. See diagram under *nephelometry.*

turbidity   Light-scattering property associated with suspended particles in a liquid. A turbid solution appears cloudy.

turbidity coefficient   The transmittance of a turbid solution is given by $P/P_0 = e^{-\tau b}$, where P is the transmitted radiant power, P_0 is the incident radiant power, b is the pathlength, and τ is the turbidity coefficient.

ultraviolet detector   Liquid chromatography detector that measures ultraviolet absorbance of solutes emerging from the column.

unadjusted relative retention, γ   For components A and B separated by chromatography or electrophoresis, the unadjusted relative retention γ is the quotient of linear velocities or unadjusted retention times: $\gamma = u_A/u_B = t_B/t_A$, where u is linear velocity and t is retention time.

underpotential deposition   Reduction of M^{n+} to make an atomic monolayer of M on the surface of another material, such as gold. If the reduction potential to make the monolayer on gold is less negative than the potential required to reduce M^{n+} onto bulk metal M, we say that underpotential deposition occurs.

upconversion   A process in which a higher energy photon is created by the combination of energy of several lower energy photons.

use objectives   In quality assurance, use objectives are a written statement of how results will be used. Use objectives are required before specifications can be written for the method.

valence band   Energy levels containing valence electrons in a semiconductor. The electrons in these levels are localized in chemical bonds.

van Deemter equation   Describes the dependence of chromatographic plate height, H, on linear flow rate, u_x: $H = A + B/u_x + Cu_x$. The constant A depends on band-broadening processes such as multiple flow paths that are independent of flow rate. B depends on the rate of diffusion of solute in the mobile phase. C depends on the rate of mass transfer between the stationary and mobile phases.

variance, σ^2   The square of the standard deviation.

vibrational transition   Occurs when a molecule changes its vibrational energy.

viscosity   Resistance to flow in a fluid.

void volume, V_0   Volume of the mobile phase outside the gel particles in a molecular exclusion chromatography column.

volatile   Easily vaporized.

volatilization   Selective removal of a component from a mixture by transforming the component into a volatile (low-boiling) species and removing it by heating, pumping, or bubbling a gas through the mixture.

Volhard titration   Titration of Ag^+ with SCN^- in the presence of Fe^{3+}. Formation of red $Fe(SCN)^{2+}$ marks the end point.

volt, V   Unit of electric potential difference. If the potential difference between two points is one volt, then one joule of energy is required to move one coulomb of charge between the two points.

voltaic cell   See *galvanic cell.*

voltammetry   An analytical method in which the relation between current and voltage is observed during an electrochemical reaction.

voltammogram   A graph of current versus electrode potential in an electrochemical cell.

voltmeter   Device for measuring electric potential difference. See also *potentiometer.*

volume flow rate   In chromatography, the volume of mobile phase per unit time eluted from the column.

volume percent, vol%   Defined as (volume of solute/volume of solution) $\times$ 100.

volumetric analysis   A technique in which the volume of material needed to react with the analyte is measured.

volumetric flask   One having a tall, thin neck with a calibration mark. When the liquid level is at the calibration mark, the flask contains its specified volume of liquid at a specified temperature.

Walden reducer   A column packed with silver and eluted with HCl. Analyte is reduced during passage through the column. The reduced product is titrated with an oxidizing agent.

wall-coated column   Hollow chromatographic column in which the stationary phase is coated on the inside surface of the wall.

watt, W   SI unit of power equal to an energy flow of one joule per second. When an electric current of one ampere flows through a potential difference of one volt, the power is one watt.

waveguide   A thin layer or hollow structure in which electromagnetic radiation is totally reflected.

wavelength, λ   Distance between consecutive crests of a wave.

wavenumber, $\tilde{\nu}$ Reciprocal of the wavelength, $1/\lambda$, usually expressed in units of cm^{-1}.

weak acids and bases   Those whose dissociation constants are not large.

weak electrolyte   One that only partly dissociates into ions when it dissolves.

weighing paper   Paper on which to place a solid reagent on a balance. Weighing paper has a very smooth surface, from which solids fall easily for transfer to a vessel.

weight percent, wt%   (Mass of solute/mass of solution) $\times$ 100.

weight/volume percent   [(Mass of solute, g)/(volume of solution, mL)] $\times$ 100.

Weston cell   A stable voltage source based on the reaction $Cd(s) + HgSO_4(aq) \rightleftharpoons CdSO_4(aq) + Hg(l)$. It was formerly used to standardize a potentiometer.

wet ashing   Destruction of organic matter in a sample by a liquid reagent (such as boiling aqueous $HClO_4$) prior to analysis of an inorganic component.

whistle noise   See *line noise.*

white light   Light comprising all wavelengths.

white noise   Random noise, also called *Gaussian noise,* due to random movement of charge carriers in an electric circuit (called *thermal noise, Johnson noise,* or *Nyquist noise*) or from random arrival of photons or charge carriers to a detector (called *shot noise* or *Schottky noise*).

width at half-height, $w_{1/2}$ Width of a signal at half of its maximum height. Also called *half-width.*

work   Energy required or released when an object is moved from one point to another. Units of work are joules, J.

working electrode   One at which the reaction of interest occurs.

y-intercept   The value of y at which a line crosses the y-axis.

Zeeman background correction   Technique used in atomic spectroscopy in which analyte signals are shifted outside the detector monochromator range by applying a strong magnetic field to the sample. Signal that remains is the background.

Zeeman effect   Shifting of atomic energy levels in a magnetic field.

zwitterion   A molecule with a positive charge localized at one position and a negative charge localized at another position.

If a is the base 10 logarithm of n ($a = \log n$), then $n = 10^a$. On a calculator, you find the logarithm of a number by pressing the "log" button. If you know $a = \log n$ and you wish to find n, use the "antilog" button or raise 10 to the power a:

$$a = \log n$$

$$10^a = 10^{\log n} = n \ (\Rightarrow n = \text{antilog } a)$$

Natural logarithms (ln) are based on the number e ($= 2.718\ 281 \ldots$) instead of 10:

$$b = \ln n$$

$$e^b = e^{\ln n} = n$$

On a calculator, you find the ln of n with the "ln" button. To find n when you know $b = \ln n$, use the e^x key.

Here are some useful properties to know:

$$\log (a \cdot b) = \log a + \log b \qquad \log 10^a = a$$

$$\log \left(\frac{a}{b}\right) = \log a - \log b \qquad a^b \cdot a^c = a^{(b+c)}$$

$$\log (a^b) = b \log a \qquad \frac{a^b}{a^c} = a^{(b-c)}$$

Problems

Test yourself by simplifying each expression as much as possible:

(a) $e^{\ln a}$

(b) $10^{\log a}$

(c) $\log 10^a$

(d) $10^{-\log a}$

(e) $e^{-\ln a^3}$

(f) $e^{\ln a^{-3}}$

(g) $\log (10^{1/a^3})$

(h) $\log (10^{-a^2})$

(i) $\log (10^{a^2 - b})$

(j) $\log (2a^3\ 10^{b^2})$

(k) $e^{(a + \ln b)}$

(l) $10^{[(\log 3) - (4 \log 2)]}$

Solving a logarithmic equation: In working with the Nernst and Henderson-Hasselbalch equations, we will need to solve equations such as

$$a = b - c \log \frac{d}{gx}$$

for the variable, x. First isolate the log term:

$$\log \frac{d}{gx} = \frac{(b - a)}{c}$$

Then raise 10 to the value of each side of the equation:

$$10^{\log(d/gx)} = 10^{(b - a)/c}$$

But $10^{\log (d/gx)}$ is just d/gx, so

$$\frac{d}{gx} = 10^{(b - a)/c} \Rightarrow x = \frac{d}{g10^{(b - a)/c}}$$

Converting between ln x and log x: The relation between them is derived by writing $x = 10^{\log x}$ and taking ln of both sides:

$$\ln x = \ln(10^{\log x}) = (\log x)(\ln 10)$$

because $\ln a^b = b \ln a$.

Answers

(a) a

(b) a

(c) a

(d) $1/a$

(e) $1/a^3$

(f) $1/a^3$

(g) $1/a^3$

(h) $-a^2$

(i) $a^2 - b$

(j) $b^2 + \log(2a^3)$

(k) be^a

(l) $3/16$

The general form of the equation of a straight line is

$$y = mx + b$$

where $m = \text{slope} = \dfrac{\Delta y}{\Delta x} = \dfrac{y_2 - y_1}{x_2 - x_1}$

    $b = \text{intercept on } y\text{-axis}$

The meanings of slope and intercept are illustrated in Figure B-1.

FIGURE B-1 Parameters of a straight line.

If you know two points [(x_1, y_1) and (x_2, y_2)] that lie on the line, you can generate the equation of the line by noting that the slope is the same for every pair of points on the line. Calling some general point on the line (x, y), we can write

$$\frac{y - y_1}{x - x_1} = \frac{y_2 - y_1}{x_2 - x_1} = m \qquad \text{(B-1)}$$

which can be rearranged to the form

$$y - y_1 = \left(\frac{y_2 - y_1}{x_2 - x_1}\right)(x - x_1)$$

$$y = \underbrace{\left(\frac{y_2 - y_1}{x_2 - x_1}\right)}_{m} x + \underbrace{y_1 - \left(\frac{y_2 - y_1}{x_2 - x_1}\right)x_1}_{b}$$

When you have a series of experimental points that should lie on a line, the best line is generally obtained by the method of least squares, described in Chapter 4. This method gives the slope and the intercept directly. If, instead, you wish to draw the "best" line by eye, you can derive the equation of the line by selecting two points *that lie on the line* and applying Equation B-1.

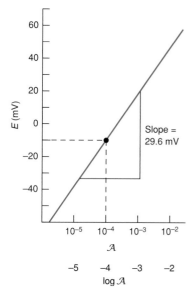

FIGURE B-2 A linear graph in which one axis is a logarithmic function.

Sometimes you are presented with a linear plot in which x or y or both are nonlinear functions. An example is shown in Figure B-2, in which the potential of an electrode is expressed as a function of the activity of analyte. Given that the slope is 29.6 mV and the line passes through the point ($\mathcal{A} = 10^{-4}$, $E = -10.2$), find the equation of the line. To do this, first note that the y-axis is linear but the x-axis is *logarithmic*. That is, the function E versus $\mathcal{A}$ is *not* linear, but E versus log $\mathcal{A}$ is linear. The form of the straight line should therefore be

$$E = (29.6) \underbrace{\log \mathcal{A}}_{} + b$$

$$\underset{y}{\uparrow} \qquad \underset{m}{\uparrow} \qquad \underset{x}{\uparrow}$$

To find b, we can use the coordinates of the one known point in Equation B-1:

$$\frac{y - y_1}{x - x_1} = \frac{E - E_1}{\log \mathcal{A} - \log \mathcal{A}_1} = \frac{E - (-10.2)}{\log \mathcal{A} - \log(10^{-4})} = m = 29.6$$

or

$$E + 10.2 = 29.6 \log \mathcal{A} + (29.6)(4)$$

$$E \,(\text{mV}) = 29.6 \,(\text{mV}) \log \mathcal{A} + 108.2 \,(\text{mV})$$

APPENDIX C Propagation of Uncertainty

The rules for propagation of uncertainty in Table 3-1 are special cases of a general formula. Suppose you wish to calculate the function, F, of several experimental quantities, $x, y, z, \ldots$. If the errors $(e_x, e_y, e_z, \ldots)$ in $x, y, z, \ldots$ are small, random, and independent of one another, then the uncertainty, e_F, in the function F is approximately:*

$$e_F = \sqrt{\left(\frac{\partial F}{\partial x}\right)^2 e_x^2 + \left(\frac{\partial F}{\partial y}\right)^2 e_y^2 + \left(\frac{\partial F}{\partial z}\right)^2 e_z^2 + \ldots} \quad \text{(C-1)}$$

The quantities in parentheses are partial derivatives, which are calculated in the same manner as ordinary derivatives, except that all but one variable are treated as constants. For example, if $F = 3xy^2$, then $\partial F/\partial x = 3y^2$ and $\partial F/\partial y = (3x)(2y) = 6xy$.

As an example of using Equation C-1, let's find the uncertainty in the function

$$F = x^y = (2.00 \pm 0.02)^{3.00 \pm 0.09}$$

The partial derivatives are

$$\frac{\partial F}{\partial x} = yx^{y-1} \qquad \frac{\partial F}{\partial y} = x^y \ln x$$

Putting these quantities into Equation C-1 gives

$$e_F = \sqrt{(yx^{y-1})^2 e_x^2 + (x^y \ln x)^2 e_y^2}$$
$$= \sqrt{y^2 x^{2y-2} e_x^2 + x^{2y} (\ln x)^2 e_y^2}$$
$$= \sqrt{y^2 x^{2y} \left(\frac{e_x}{x}\right)^2 + x^{2y} (\ln x)^2 e_y^2}$$

Multiplying and dividing the second term by y^2 allows us to rearrange to a more pleasant form:

$$e_F = \sqrt{y^2 x^{2y} \left(\frac{e_x}{x}\right)^2 + y^2 x^{2y} (\ln x)^2 \left(\frac{e_y}{y}\right)^2}$$

Removing $\sqrt{y^2 x^{2y}} = yF$ from both terms gives

$$e_F = yF \sqrt{\left(\frac{e_x}{x}\right)^2 + (\ln x)^2 \left(\frac{e_y}{y}\right)^2}$$

Now for the number crunching. Disregarding uncertainties for a moment, we know that $F = 2.00^{3.00} = 8.00 \pm ?$ The uncertainty is obtained from the equation above:

$$e_F = 3.00 \cdot 8.00 \sqrt{\left(\frac{0.02}{2.00}\right)^2 + (\ln 2.00)^2 \left(\frac{0.09}{3.00}\right)^2} = 0.55$$

Reasonable answers are $F = 8.0_0 \pm 0.5_5$ or 8.0 ± 0.6.

---

*A numerical recipe for evaluating Equation C-1 with a spreadsheet is given by R. de Levie, *J. Chem. Ed.* **2000,** *77,* 534.

Exercises

C-1. Verify the following calculations.

(a) $2.36^{4.39\pm0.08} = 43._4 \pm 3._0$
(b) $(2.36 \pm 0.06)^{4.39\pm0.08} = 43._4 \pm 5._7$

C-2. For $F = \sin(2\pi xy)$, show that

$$e_F = 2\pi xy \cos(2\pi xy)\sqrt{\left(\frac{e_x}{x}\right)^2 + \left(\frac{e_y}{y}\right)^2}$$

Covariance in Propagation of Uncertainty

Equation C-1 presumes that errors in x, y, and z are independent of one another. A common case in which this is not true is when we use the least-squares slope and intercept to compute a new quantity, such as the value of x from an observed value of y. In general, uncertainties in the slope and intercept are correlated, so they are not independent errors.

Let's restrict our attention to a function, F, of the two experimental parameters, m and b, whose uncertainties are s_m and s_b. If the uncertainties are correlated, the equation for propagation of uncertainty is[†]

$$e_F = \sqrt{\underbrace{\left(\frac{\partial F}{\partial m}\right)^2 s_m^2 + \left(\frac{\partial F}{\partial b}\right)^2 s_b^2}_{\substack{\text{Variance terms from}\\ \text{Equation C-1}}} + \underbrace{2\left(\frac{\partial F}{\partial m}\right)\left(\frac{\partial F}{\partial b}\right) s_{mb}}_{\substack{\text{Covariance accounts for}\\ \text{correlation of } m \text{ and } b}}} \quad \text{(C-2)}$$

The last term in Equation C-2 reflects the fact that uncertainties in m and b are not independent of each other. The quantity s_{mb} is called the *covariance* and it can be positive or negative.

In linear least-squares analysis, the variance and covariance are[‡]

Variance:

$$s_m^2 = \frac{s_y^2 n}{D} \qquad s_b^2 = \frac{s_y^2 \Sigma(x_i^2)}{D} \quad \text{(Equations 4-21 and 4-22)}$$

Covariance:

$$s_{mb} = \frac{-s_y^2 \Sigma(x_i)}{D} \quad \text{(C-3)}$$

where s_y^2 is the square of Equation 4-20, D is given by Equation 4-18, and n is the number of data points.

[†]E. F. Meyer, *J. Chem. Ed.* **1997,** *74,* 1339.
[‡]C. Salter, *J. Chem. Ed.* **2000,** *77,* 1239.

(Continued)

For the line $y = mx + b$, the *x*-intercept occurs when $y = 0$, or $x = -b/m$. Let's designate the *x*-intercept as the function $F = -b/m$. Find the *x*-intercept and its uncertainty for the least-squares line in Figure 4-11.

Solution   The following quantities are computed in Section 4-7:

$$m = 0.615\ 38 \quad s_m^2 = 0.002\ 958\ 6 \quad s_y^2 = 0.038\ 462 \quad \Sigma(x_i) = 14$$

$$b = 1.346\ 15 \quad s_b^2 = 0.045\ 859 \quad D = 52$$

The covariance in Equation C-3 is therefore

$$s_{mb} = \frac{-s_y^2\,\Sigma(x_i)}{D} = \frac{-(0.038\ 462)(14)}{52} = -0.010\ 355$$

The *x*-intercept is just $F = -b/m = -(1.346\ 15)/(0.615\ 38) = -2.187\ 5$.

To find the uncertainty in F, we use Equation C-2. The derivatives in C-2 are

$$\frac{\partial F}{\partial m} = \frac{\partial(-b/m)}{\partial m} = \frac{b}{m^2} = \frac{1.346\ 15}{0.615\ 38^2} = 3.554\ 7$$

$$\frac{\partial F}{\partial b} = \frac{\partial(-b/m)}{\partial b} = \frac{-1}{m} = \frac{-1}{0.615\ 38} = -1.625\ 0$$

Now we can evaluate the uncertainty with Equation C-2:

$$e_F = \sqrt{\left(\frac{\partial F}{\partial m}\right)^2 s_m^2 + \left(\frac{\partial F}{\partial b}\right)^2 s_b^2 + 2\left(\frac{\partial F}{\partial m}\right)\left(\frac{\partial F}{\partial b}\right)s_{mb}}$$

$$= \sqrt{\begin{array}{l}(3.554\ 7)^2(0.002\ 958\ 6)\ + \\ (-1.625\ 0)^2(0.045\ 859)\ + \\ 2(3.554\ 7)(-1.625\ 0)(-0.010\ 355)\end{array}}$$

$$= 0.527\ 36$$

The final answer can now be written with a reasonable number of digits:

$$F = -2.187\ 5 \pm 0.527\ 36 = -2.1_9 \pm 0.5_3$$

If we had used Equation C-1 and ignored the covariance term in Equation C-2, we would have computed an uncertainty of $\pm 0.4_0$.

To learn how to compute variance and covariance and to see how to include weighting factors in least-squares curve fitting, see J. Tellinghuisen, "Understanding Least Squares Through Monte Carlo Calculations," *J. Chem. Ed.* **2005,** *82,* 157.

The *oxidation number,* or *oxidation state,* is a bookkeeping device used to keep track of the number of electrons formally associated with a particular element. The oxidation number is meant to tell how many electrons have been lost or gained by a neutral atom when it forms a compound. Because oxidation numbers have no real physical meaning, they are somewhat arbitrary, and not all chemists will assign the same oxidation number to a given element in an unusual compound. However, there are some ground rules that provide a useful start.

1. The oxidation number of an element by itself—e.g., $Cu(s)$ or $Cl_2(g)$—is 0.
2. The oxidation number of H is almost always $+1$, except in metal hydrides—e.g., NaH—in which H is -1.
3. The oxidation number of oxygen is almost always -2. The only common exceptions are peroxides, in which two oxygen atoms are connected and each has an oxidation number of -1. Two examples are hydrogen peroxide (H—O—O—H) and its anion (H—O—O$^-$). The oxidation number of oxygen in gaseous O_2 is, of course, 0.
4. The alkali metals (Li, Na, K, Rb, Cs, Fr) almost always have an oxidation number of $+1$. The alkaline earth metals (Be, Mg, Ca, Sr, Ba, Ra) are almost always in the $+2$ oxidation state.
5. The halogens (F, Cl, Br, I) are usually in the -1 oxidation state. Exceptions occur when two different halogens are bound to each other or when a halogen is bound to more than one atom. When different halogens are bound to each other, we assign the oxidation number -1 to the more electronegative halogen.

The sum of the oxidation numbers of each atom in a molecule must equal the charge of the molecule. In H_2O, for example, we have

$$
\begin{aligned}
2 \text{ hydrogen} = 2(+1) &= +2 \\
\text{oxygen} &= \underline{-2} \\
\text{net charge} & \;\; 0
\end{aligned}
$$

In SO_4^{2-}, sulfur must have an oxidation number of $+6$ so that the sum of the oxidation numbers will be -2:

$$
\begin{aligned}
\text{oxygen} = 4(-2) &= \;\; -8 \\
\text{sulfur} &= \underline{+6} \\
\text{net charge} & \;\; -2
\end{aligned}
$$

In benzene (C_6H_6), the oxidation number of each carbon must be -1 if hydrogen is assigned the number $+1$. In cyclohexane (C_6H_{12}), the oxidation number of each carbon must be -2, for the same reason. The carbons in benzene are in a higher oxidation state than those in cyclohexane.

The oxidation number of iodine in ICl_2^- is $+1$. This is unusual, because halogens are usually -1. However, because chlorine is more electronegative than iodine, we assign Cl as -1, thereby forcing I to be $+1$.

The oxidation number of As in As_2S_3 is $+3$, and the value for S is -2. This is arbitrary but reasonable. Because S is more electronegative than As, we make S negative and As positive; and, because S is in the same family as oxygen, which is usually -2, we assign S as -2, thus leaving As as $+3$.

The oxidation number of S in $S_4O_6^{2-}$ (tetrathionate) is $+2.5$. The *fractional oxidation state* comes about because six O atoms contribute -12. Because the charge is -2, the four S atoms must contribute $+10$. The average oxidation number of S must be $+\frac{10}{4} = 2.5$.

The oxidation number of Fe in $K_3Fe(CN)_6$ is $+3$. To make this assignment, we first recognize cyanide (CN^-) as a common ion that carries a charge of -1. Six cyanide ions give -6, and three potassium ions (K^+) give $+3$. Therefore, Fe should have an oxidation number of $+3$ for the whole formula to be neutral. In this approach, it is not necessary to assign individual oxidation numbers to carbon and nitrogen, as long as we recognize that the charge of CN is -1.

Problems

Answers are given at the end of this appendix.

D-1. Write the oxidation state of the boldface atom in each of the following species.

(a) **Ag**Br
(b) **S**$_2$O$_3^{2-}$
(c) **Se**F$_6$
(d) H**S**$_2$O$_3^-$
(e) H**O**$_2$
(f) **N**O
(g) **Cr**$^{3+}$
(h) **Mn**O$_2$
(i) **Pb**(OH)$_3^-$
(j) **Fe**(OH)$_3$
(k) **Cl**O$^-$
(l) K$_4$**Fe**(CN)$_6$
(m) **Cl**O$_2$
(n) **Cl**O$_2^-$
(o) **Mn**(CN)$_6^{4-}$

(p) **N**$_2$
(q) **N**H$_4^+$
(r) **N**$_2$H$_5^+$
(s) H**As**O$_3^{2-}$
(t) **Co**$_2$(CO)$_8$ (CO group is neutral)
(u) (CH$_3$)$_4$**Li**$_4$
(v) **P**$_4$O$_{10}$
(w) **C**$_2$H$_6$O (ethanol, CH$_3$CH$_2$OH)
(x) **V**O(SO$_4$)
(y) **Fe**$_3$O$_4$
(z) **C**$_3$H$_3^+$

Structure:

D-2. Identify the oxidizing agent and the reducing agent on the left side of each of the following reactions.

(a) $Cr_2O_7^{2-} + 3Sn^{2+} + 14H^+ \rightarrow 2Cr^{3+} + 3Sn^{4+} + 7H_2O$

(b) $4I^- + O_2 + 4H^+ \rightarrow 2I_2 + 2H_2O$

(c) $5CH_3\overset{\displaystyle O}{\overset{\displaystyle \|}{C}}H + 2MnO_4^- + 6H^+ \rightarrow 5CH_3\overset{\displaystyle O}{\overset{\displaystyle \|}{C}}OH + 2Mn^{2+} + 3H_2O$

(d) $HOCH_2CHOHCH_2OH + 2IO_4^- \rightarrow$
     Glycerol $2H_2C{=}O + HCO_2H + 2IO_3^- + H_2O$
                   Formaldehyde   Formic acid

(e) $C_8H_8 + 2Na \rightarrow C_8H_8^{2-} + 2Na^+$
     C_8H_8 is cyclooctatetraene with the structure

(f) $I_2 + OH^- \rightarrow HOI + I^-$
     Hypoiodous
     acid

Balancing Redox Reactions

To balance a reaction involving oxidation and reduction, we must first identify which element is oxidized and which is reduced. We then break the net reaction into two imaginary *half-reactions,* one of which involves only oxidation and the other only reduction. Although free electrons never appear in a balanced net reaction, they do appear in balanced half-reactions. If we are dealing with aqueous solutions, we proceed to balance each half-reaction, using H_2O and either H^+ or OH^-, as necessary. *A reaction is balanced when the number of atoms of each element is the same on both sides and the net charge is the same on both sides.*[*]

---

[*]A completely different method for balancing complex redox equations by inspection has been described by D. Kolb, *J. Chem. Ed.* **1981,** *58,* 642. For some challenging problems in balancing redox equations, see R. Stout, *J. Chem. Ed.* **1995,** *72,* 1125.

(Continued)

Acidic Solutions

Here are the steps we will follow:

1. Assign oxidation numbers to the elements that are oxidized or reduced.
2. Break the reaction into two half-reactions, one involving oxidation and the other reduction.
3. For each half-reaction, balance the number of atoms that are oxidized or reduced.
4. Balance the electrons to account for the change in oxidation number by adding electrons to one side of each half-reaction.
5. Balance oxygen atoms by adding H_2O to one side of each half-reaction.
6. Balance the H atoms by adding H^+ to one side of each half-reaction.
7. Multiply each half-reaction by the number of electrons in the other half-reaction so that the number of electrons on each side of the total reaction will cancel. Then add the two half-reactions and simplify to the smallest integral coefficients.

EXAMPLE   Balancing a Redox Equation

Balance the following equation using H^+, but not OH^-:

$$Fe^{2+} + MnO_4^- \rightleftharpoons Fe^{3+} + Mn^{2+}$$
$$\quad +2 \qquad +7 \qquad\quad +3 \qquad +2$$
$$\qquad \text{Permanganate}$$

Solution

1. *Assign oxidation numbers.* They are assigned for Fe and Mn in each species in the above reaction.
2. *Break the reaction into two half-reactions.*

   Oxidation half-reaction: $Fe^{2+} \rightleftharpoons Fe^{3+}$
   $$\qquad\qquad\qquad\quad +2 \qquad +3$$

   Reduction half-reaction: $MnO_4^- \rightleftharpoons Mn^{2+}$
   $$\qquad\qquad\qquad\quad +7 \qquad\quad +2$$

3. *Balance the atoms that are oxidized or reduced.* Because there is only one Fe or Mn in each species on each side of the equation, the atoms of Fe and Mn are already balanced.
4. *Balance electrons.* Electrons are added to account for the change in each oxidation state.

   $$Fe^{2+} \rightleftharpoons Fe^{3+} + e^-$$
   $$MnO_4^- + 5e^- \rightleftharpoons Mn^{2+}$$

   In the second case, we need $5e^-$ on the left side to take Mn from $+7$ to $+2$.
5. *Balance oxygen atoms.* There are no oxygen atoms in the Fe half-reaction. There are four oxygen atoms on the left side of the Mn reaction, so we add four molecules of H_2O to the right side:

   $$MnO_4^- + 5e^- \rightleftharpoons Mn^{2+} + 4H_2O$$

6. *Balance hydrogen atoms.* The Fe equation is already balanced. The Mn equation needs $8H^+$ on the left.

   $$MnO_4^- + 5e^- + 8H^+ \rightarrow Mn^{2+} + 4H_2O$$

   At this point, each half-reaction must be completely balanced (the same number of atoms and charge on each side) *or you have made a mistake.*
7. *Multiply and add the reactions.* We multiply the Fe equation by 5 and the Mn equation by 1 and add:

   $$5Fe^{2+} \rightleftharpoons 5Fe^{3+} + 5e^-$$
   $$\underline{MnO_4^- + 5e^- + 8H^+ \rightleftharpoons Mn^{2+} + 4H_2O}$$
   $$5Fe^{2+} + MnO_4^- + 8H^+ \rightleftharpoons 5Fe^{3+} + Mn^{2+} + 4H_2O$$

The total charge on each side is $+17$, and we find the same number of atoms of each element on each side. The equation is balanced.

EXAMPLE   A Reverse Disproportionation

Now try the next reaction, which represents the reverse of a *disproportionation*. (In a disproportionation, an element in one oxidation state reacts to give the same element in higher and lower oxidation states.)

$$I_2 + IO_3^- + Cl^- \rightleftharpoons ICl_2^-$$
$$0 \quad +5 \qquad -1 \qquad +1 -1$$
$$\text{Iodine} \quad \text{Iodate}$$

Solution

1. The oxidation numbers are assigned above. Note that chlorine has an oxidation number of -1 on both sides of the equation. Only iodine is involved in electron transfer.
2. Oxidation half-reaction: $I_2 \rightleftharpoons ICl_2^-$
   $$\qquad\qquad\qquad\qquad 0 \qquad +1$$
   Reduction half-reaction: $IO_3^- \rightleftharpoons ICl_2^-$
   $$\qquad\qquad\qquad\qquad +5 \qquad +1$$
3. We need to balance I atoms in the first reaction and add Cl^- to each reaction to balance Cl.

   $$I_2 + 4Cl^- \rightleftharpoons 2ICl_2^-$$
   $$IO_3^- + 2Cl^- \rightleftharpoons ICl_2^-$$

4. Now add electrons to each.

   $$I_2 + 4Cl^- \rightleftharpoons 2ICl_2^- + 2e^-$$
   $$IO_3^- + 2Cl^- + 4e^- \rightleftharpoons ICl_2^-$$

   The first reaction needs $2e^-$ because there are two I atoms, each of which changes from 0 to $+1$.
5. The second reaction needs $3H_2O$ on the right side to balance oxygen atoms.

   $$IO_3^- + 2Cl^- + 4e^- \rightleftharpoons ICl_2^- + 3H_2O$$

6. The first reaction is balanced, but the second needs $6H^+$ on the left.

   $$IO_3^- + 2Cl^- + 4e^- + 6H^+ \rightleftharpoons ICl_2^- + 3H_2O$$

   As a check, the charge on each side of this half-reaction is -1, and all atoms are balanced.
7. Multiply and add.

   $$2(I_2 + 4Cl^- \qquad\qquad \rightleftharpoons 2ICl_2^- + 2e^-)$$
   $$\underline{IO_3^- + 2Cl^- + 4e^- + 6H^+ \rightleftharpoons ICl_2^- + 3H_2O}$$
   $$2I_2 + IO_3^- + 10Cl^- + 6H^+ \rightleftharpoons 5ICl_2^- + 3H_2O \qquad \text{(D-1)}$$

We multiplied the first reaction by 2 so that there would be the same number of electrons in each half-reaction. You could have multiplied the first reaction by 4 and the second by 2, but then all coefficients would simply be doubled. We customarily write the smallest coefficients.

Basic Solutions

The method many people prefer for basic solutions is to balance the equation first with H^+. The answer can then be converted into one in which OH^- is used instead. This is done by adding to each side of the equation a number of hydroxide ions equal to the number of H^+ ions appearing in the equation. For example, to balance Equation D-1 with OH^- instead of H^+, proceed as follows:

$$2I_2 + IO_3^- + 10Cl^- + 6H^+ \rightleftharpoons 5ICl_2 + 3H_2O$$
$$\qquad\qquad\qquad\qquad + 6OH^- \qquad\qquad\quad + 6OH^-$$
$$\overline{2I_2 + IO_3^- + 10Cl^- + \underbrace{6H^+ + 6OH^-}_{} \rightleftharpoons 5ICl_2 + 3H_2O + 6OH^-}$$
$$\qquad\qquad\qquad\qquad\quad 6H_2O$$
$$\qquad\qquad\qquad\qquad\quad \Downarrow$$
$$\qquad\qquad\qquad\qquad\quad 3H_2O$$

Realizing that $6H^+ + 6OH^- = 6H_2O$, and canceling $3H_2O$ on each side, gives the final result:

$$2I_2 + IO_3^- + 10Cl^- + 3H_2O \rightleftharpoons 5ICl_2 + 6OH^-$$

Problems

D-3. Balance the following reactions by using H^+ but not OH^-.

(a) $Fe^{3+} + Hg_2^{2+} \rightleftharpoons Fe^{2+} + Hg^{2+}$

(b) $Ag + NO_3^- \rightleftharpoons Ag^+ + NO$

(c) $VO^{2+} + Sn^{2+} \rightleftharpoons V^{3+} + Sn^{4+}$

(d) $SeO_4^{2-} + Hg + Cl^- \rightleftharpoons SeO_3^{2-} + Hg_2Cl_2$

(e) $CuS + NO_3^- \rightleftharpoons Cu^{2+} + SO_4^{2-} + NO$

(f) $S_2O_3^{2-} + I_2 \rightleftharpoons I^- + S_4O_6^{2-}$

(g) $ClO_3^- + As_2S_3 \rightleftharpoons Cl^- + H_2AsO_4^- + SO_4^{2-}$

(h) $Cr_2O_7^{2-} + CH_3CH\!\!\overset{O}{\overset{\|}{}} \rightleftharpoons CH_3C\!\!\overset{O}{\overset{\|}{}}OH + Cr^{3+}$

(i) $MnO_4^{2-} \rightleftharpoons MnO_2 + MnO_4^-$

(j) $Hg_2SO_4 + Ca^{2+} + S_8 \rightleftharpoons Hg_2^{2+} + CaS_2O_3$

(k) $ClO_3^- \rightleftharpoons Cl_2 + O_2$

D-4. Balance the following equations by using OH^- but not H^+.

(a) $PbO_2 + Cl^- \rightleftharpoons ClO^- + Pb(OH)_3^-$

(b) $HNO_2 + SbO^+ \rightleftharpoons NO + Sb_2O_5$

(c) $Ag_2S + CN^- + O_2 \rightleftharpoons S + Ag(CN)_2^- + OH^-$

(d) $HO_2^- + Cr(OH)_3^- \rightleftharpoons CrO_4^{2-} + OH^-$

(e) $ClO_2 + OH^- \rightleftharpoons ClO_2^- + ClO_3^-$

(f) $WO_3^- + O_2 \rightleftharpoons HW_6O_{21}^{5-} + OH^-$

(g) $Mn_2O_3 + CN^- \rightleftharpoons Mn(CN)_6^{4-} + (CN)_2$

(h) $Cu^{2+} + H_2 \rightleftharpoons Cu + H_2O$

(i) $BH_4^- + H_2O \rightleftharpoons H_3BO_3 + H_2$

(j) $Mn_2O_3 + Hg + CN^- \rightleftharpoons Mn(CN)_6^{4-} + Hg(CN)_2$

(k) $MnO_4^- + HC\!\!\overset{O}{\overset{\|}{}}CH_2CH_2OH \rightleftharpoons CH_2(CO_2^-)_2 + MnO_2$

(l) $K_3V_5O_{14} + HOCH_2CHOHCH_2OH \rightleftharpoons VO(OH)_2 + HCO_2^- + K^+$

Answers

D-1.

(a) $+1$	**(j)** $+3$	**(s)** $+3$
(b) $+2$	**(k)** $+1$	**(t)** 0
(c) $+6$	**(l)** $+2$	**(u)** -4
(d) $+2$	**(m)** $+4$	**(v)** $+5$
(e) $-\frac{1}{2}$	**(n)** $+3$	**(w)** -2
(f) $+2$	**(o)** $+2$	**(x)** $+4$
(g) $+3$	**(p)** 0	**(y)** $+8/3$
(h) $+4$	**(q)** -3	**(z)** $-2/3$
(i) $+2$	**(r)** -2	

D-2.

	Oxidizing agent	Reducing agent
(a)	$Cr_2O_7^{2-}$	Sn^{2+}
(b)	O_2	I^-
(c)	MnO_4^-	CH_3CHO
(d)	IO_4^-	Glycerol
(e)	C_8H_8	Na
(f)	I_2	I_2

Reaction **(f)** is called a *disproportionation*, because an element in one oxidation state is transformed into two different oxidation states—one higher and one lower than the original oxidation state.

D-3. **(a)** $2Fe^{3+} + Hg_2^{2+} \rightleftharpoons 2Fe^{2+} + 2Hg^{2+}$

(b) $3Ag + NO_3^- + 4H^+ \rightleftharpoons 3Ag^+ + NO + 2H_2O$

(c) $4H^+ + 2VO^{2+} + Sn^{2+} \rightleftharpoons 2V^{3+} + Sn^{4+} + 2H_2O$

(d) $2Hg + 2Cl^- + SeO_4^{2-} + 2H^+ \rightleftharpoons Hg_2Cl_2 + SeO_3^{2-} + H_2O$

(e) $3CuS + 8NO_3^- + 8H^+ \rightleftharpoons 3Cu^{2+} + 3SO_4^{2-} + 8NO + 4H_2O$

(f) $2S_2O_3^{2-} + I_2 \rightleftharpoons S_4O_6^{2-} + 2I^-$

(g) $14ClO_3^- + 3As_2S_3 + 18H_2O \rightleftharpoons$
$$14Cl^- + 6H_2AsO_4^- + 9SO_4^{2-} + 24H^+$$

(h) $Cr_2O_7^{2-} + 3CH_3CHO + 8H^+ \rightleftharpoons 2Cr^{3+} + 3CH_3CO_2H + 4H_2O$

(i) $4H^+ + 3MnO_4^{2-} \rightleftharpoons MnO_2 + 2MnO_4^- + 2H_2O$

(j) $2Hg_2SO_4 + 3Ca^{2+} + \frac{1}{8}S_8 + H_2O \rightleftharpoons 2Hg_2^{2+} + 3CaS_2O_3 + 2H^+$

(k) $2H^+ + 2ClO_3^- \rightleftharpoons Cl_2 + \frac{5}{2}O_2 + H_2O$
The balanced half-reaction for As_2S_3 in **(g)** is

$$As_2S_3 + 20H_2O \rightleftharpoons 2H_2AsO_4^- + 3SO_4^{2-} + 28e^- + 36H^+$$
$$\;+3\;-2\qquad\qquad\qquad +5\qquad\qquad +6$$

Because As_2S_3 is a single compound, we must consider the $As_2S_3 \to H_2AsO_4^-$ and $As_2S_3 \to SO_4^{2-}$ reactions together. The net change in oxidation number for the *two* As atoms is $2(5 - 3) = +4$. The net change in oxidation number for the *three* S atoms is $3[6 - (-2)] = +24$. Therefore, $24 + 4 = 28e^-$ are involved in the half-reaction.

D-4. **(a)** $H_2O + OH^- + PbO_2 + Cl^- \rightleftharpoons Pb(OH)_3^- + ClO^-$

(b) $4HNO_2 + 2SbO^- + 2OH^- \rightleftharpoons 4NO + Sb_2O_5 + 3H_2O$

(c) $Ag_2S + 4CN^- + \frac{1}{2}O_2 + H_2O \rightleftharpoons S + 2Ag(CN)_2^- + 2OH^-$

(d) $2HO_2^- + Cr(OH)_3^- \rightleftharpoons CrO_4^{2-} + OH^- + 2H_2O$

(e) $2ClO_2 + 2OH^- \rightleftharpoons ClO_2^- + ClO_3^- + H_2O$

(f) $12WO_3^- + 3O_2 + 2H_2O \rightleftharpoons 2HW_6O_{21}^{5-} + 2OH^-$

(g) $Mn_2O_3 + 14CN^- + 3H_2O \rightleftharpoons 2Mn(CN)_6^{4-} + (CN)_2 + 6OH^-$

(h) $Cu^{2+} + H_2 + 2OH^- \rightleftharpoons Cu + 2H_2O$

(i) $BH_4^- + 4H_2O \rightleftharpoons H_3BO_3 + 4H_2 + OH^-$

(j) $3H_2O + Mn_2O_3 + Hg + 14CN^- \rightleftharpoons$
$$2Mn(CN)_6^{4-} + Hg(CN)_2 + 6OH^-$$

(k) $2MnO_4^- + HC\!\!\overset{O}{\overset{\|}{}}CH_2CH_2OH \rightleftharpoons 2MnO_2 + 2H_2O + CH_2(CO_2^-)_2$

For **(k)**, the organic half-reaction is $8OH^- + C_3H_6O_2 \rightleftharpoons$
$$C_3H_2O_4^{2-} + 6e^- + 6H_2O.$$

(l) $32H_2O + 8K_3V_5O_{14} + 5HOCH_2CHOHCH_2OH \rightleftharpoons$
$$40VO(OH)_2 + 15HCO_2^- + 9OH^- + 24K^+$$

For **(l)**, the two half-reactions are $K_3V_5O_{14} + 9H_2O + 5e^- \rightleftharpoons 5VO(OH)_2 + 8OH^- + 3K^+$ and $C_3H_8O_3 + 11OH^- \rightleftharpoons 3HCO_2^- + 8e^- + 8H_2O.$

The *normality*, N, of a redox reagent is n times the molarity, where n is the number of electrons donated or accepted by that species in a chemical reaction.

$$N = nM \qquad \text{(E-1)}$$

For example, in the half-reaction

$$MnO_4^- + 8H^+ + 5e^- \rightleftharpoons Mn^{2+} + 4H_2O \qquad \text{(E-2)}$$

the normality of permanganate ion is five times its molarity, because each MnO_4^- accepts $5e^-$. If the molarity of permanganate is 0.1 M, the normality for the reaction

$$MnO_4^- + 5Fe^{2+} + 8H^+ \rightleftharpoons Mn^{2+} + 5Fe^{3+} + 4H_2O \qquad \text{(E-3)}$$

is $5 \times 0.1 = 0.5$ N (read "0.5 normal"). In this reaction, each Fe^{2+} ion donates one electron. The normality of ferrous ion *equals* the molarity of ferrous ion, even though it takes five ferrous ions to balance the reaction.

In the half-reaction

$$MnO_4^- + 4H^+ + 3e^- \rightleftharpoons MnO_2 + 2H_2O \qquad \text{(E-4)}$$

each MnO_4^- ion accepts only *three* electrons. The normality of permanganate for this reaction is equal to three times the molarity of permanganate. A 0.06 N permanganate solution for this reaction contains 0.02 M MnO_4^-.

The normality of a solution is a statement of the moles of "reacting units" per liter. One mole of reacting units is called one *equivalent*. Therefore, the units of normality are equivalents per liter (equiv/L). For redox reagents, *one equivalent is the amount of substance that can donate or accept one mole of electrons*. It is possible to speak of equivalents only with respect to a particular half-reaction. For example, in Reaction E-2, there are five equivalents per mole of MnO_4^-; but, in Reaction E-4, there are only three equivalents per mole of MnO_4^-. The mass of substance containing one equivalent is called the *equivalent mass*. The formula mass of $KMnO_4$ is 158.033 9. The equivalent mass of $KMnO_4$ for Reaction E-2 is $158.033\ 9/5 = 31.606\ 8$ g/equiv. The equivalent mass of $KMnO_4$ for Reaction E-4 is $158.033\ 9/3 = 52.678\ 0$ g/equiv.

EXAMPLE Finding Normality

Find the normality of a solution containing 6.34 g of ascorbic acid in 250.0 mL if the relevant half-reaction is

Ascorbic acid (vitamin C)   Dehydroascorbic acid

Solution The formula mass of ascorbic acid ($C_6H_8O_6$) is 176.124. In 6.34 g, there are $(6.34\text{ g})/(176.124\text{ g/mol}) = 3.60 \times 10^{-2}$ mol. Because each mole contains 2 equivalents in this example, $6.34\text{ g} = (2\text{ equiv/mol})(3.60 \times 10^{-2}\text{ mol}) = 7.20 \times 10^{-2}$ equivalent. The normality is $(7.20 \times 10^{-2}\text{ equiv})/(0.250\ 0\text{ L}) = 0.288$ N.

EXAMPLE Using Normality

How many grams of potassium oxalate should be dissolved in 500.0 mL to make a 0.100 N solution for titration of MnO_4^-?

$$5H_2C_2O_4 + 2MnO_4^- + 6H^+ \rightleftharpoons 2Mn^{2+} + 10CO_2 + 8H_2O \quad \text{(E-5)}$$

Solution It is first necessary to write the oxalic acid half-reaction:

$$H_2C_2O_4 \rightleftharpoons 2CO_2 + 2H^+ + 2e^-$$

It is apparent that there are two equivalents per mole of oxalic acid. Hence, a 0.100 N solution will be 0.050 0 M:

$$\frac{0.100 \text{ equiv/L}}{2 \text{ equiv/mol}} = 0.050\ 0 \text{ mol/L} = 0.050\ 0 \text{ M}$$

Therefore, we must dissolve $(0.050\ 0\text{ mol/L})(0.500\ 0\text{ L}) = 0.025\ 0$ mol in 500.0 mL. Because the formula mass of $K_2C_2O_4$ is 166.216, we should use $(0.025\ 0\text{ mol}) \times (166.216\text{ g/mol}) = 4.15$ g of potassium oxalate.

The utility of normality in volumetric analysis lies in the equation

$$N_1V_1 = N_2V_2 \qquad \text{(E-6)}$$

where N_1 is the normality of reagent 1, V_1 is the volume of reagent 1, N_2 is the normality of reagent 2, and V_2 is the volume of reagent 2. V_1 and V_2 may be expressed in any units, as long as the same units are used for both.

EXAMPLE Finding Normality

A solution containing 25.0 mL of oxalic acid required 13.78 mL of 0.041 62 N $KMnO_4$ for titration, according to Reaction E-5. Find the normality and molarity of the oxalic acid.

Solution Setting up Equation E-6, we write

$$N_1(25.0\text{ mL}) = (0.041\ 62\text{ N})(13.78\text{ mL})$$
$$N_1 = 0.022\ 94 \text{ equiv/L}$$

Because there are two equivalents per mole of oxalic acid in Reaction E-5,

$$M = \frac{N}{n} = \frac{0.022\ 94}{2} = 0.011\ 47 \text{ M}$$

Normality is sometimes used in acid-base or ion-exchange chemistry. With respect to acids and bases, the equivalent mass of a reagent is the amount that can donate or accept 1 mole of H^+. With respect to ion exchange, the equivalent mass is the mass of reagent containing 1 mole of charge.

Formula	pK_{sp}	K_{sp}	Formula	pK_{sp}	K_{sp}
Azides: L = N_3^-			**Chromates: L = CrO_4^{2-}**		
CuL	8.31	4.9×10^{-9}	BaL	9.67	2.1×10^{-10}
AgL	8.56	2.8×10^{-9}	CuL	5.44	3.6×10^{-6}
Hg_2L_2	9.15	7.1×10^{-10}	Ag_2L	11.92	1.2×10^{-12}
TlL	3.66	2.2×10^{-4}	Hg_2L	8.70	2.0×10^{-9}
$PdL_2(\alpha)$	8.57	2.7×10^{-9}	Tl_2L	12.01	9.8×10^{-13}
Bromates: L = BrO_3^-			**Cobalticyanides: L = $Co(CN)_6^{3-}$**		
$BaL \cdot H_2O$ (f)	5.11	7.8×10^{-6}	Ag_3L	25.41	3.9×10^{-26}
AgL	4.26	5.5×10^{-5}	$(Hg_2)_3L_2$	36.72	1.9×10^{-37}
TlL	3.78	1.7×10^{-4}	**Cyanides: L = CN^-**		
PbL_2	5.10	7.9×10^{-6}	AgL	15.66	2.2×10^{-16}
Bromides: L = Br^-			Hg_2L_2	39.3	5×10^{-40}
CuL	8.3	5×10^{-9}	ZnL_2 (h)	15.5	3×10^{-16}
AgL	12.30	5.0×10^{-13}	**Ferrocyanides: L = $Fe(CN)_6^{4-}$**		
Hg_2L_2	22.25	5.6×10^{-23}	Ag_4L	44.07	8.5×10^{-45}
TlL	5.44	3.6×10^{-6}	Zn_2L	15.68	2.1×10^{-16}
HgL_2 (f)	18.9	1.3×10^{-19}	Cd_2L	17.38	4.2×10^{-18}
PbL_2	5.68	2.1×10^{-6}	Pb_2L	18.02	9.5×10^{-19}
Carbonates: L = CO_3^{2-}			**Fluorides: L = F^-**		
MgL	7.46	3.5×10^{-8}	LiL	2.77	1.7×10^{-3}
CaL (calcite)	8.35	4.5×10^{-9}	MgL_2	8.13	7.4×10^{-9}
CaL (aragonite)	8.22	6.0×10^{-9}	CaL_2	10.50	3.2×10^{-11}
SrL	9.03	9.3×10^{-10}	SrL_2	8.58	2.6×10^{-9}
BaL	8.30	5.0×10^{-9}	BaL_2	5.82	1.5×10^{-6}
Y_2L_3	30.6	2.5×10^{-31}	LaL_3	18.7	2×10^{-19}
La_2L_3	33.4	4.0×10^{-34}	ThL_4	28.3	5×10^{-29}
MnL	9.30	5.0×10^{-10}	PbL_2	7.44	3.6×10^{-8}
FeL	10.68	2.1×10^{-11}			
CoL	9.98	1.0×10^{-10}	**Hydroxides: L = OH^-**		
NiL	6.87	1.3×10^{-7}	MgL_2 (amorphous)	9.2	6×10^{-10}
CuL	9.63	2.3×10^{-10}	MgL_2 (brucite crystal)	11.15	7.1×10^{-12}
Ag_2L	11.09	8.1×10^{-12}	CaL_2	5.19	6.5×10^{-6}
Hg_2L	16.05	8.9×10^{-17}	$BaL_2 \cdot 8H_2O$	3.6	3×10^{-4}
ZnL	10.00	1.0×10^{-10}	YL_3	23.2	6×10^{-24}
CdL	13.74	1.8×10^{-14}	LaL_3	20.7	2×10^{-21}
PbL	13.13	7.4×10^{-14}	CeL_3	21.2	6×10^{-22}
Chlorides: L = Cl^-			UO_2 ($\rightleftharpoons U^{4+} + 4OH^-$)	56.2	6×10^{-57}
CuL	6.73	1.9×10^{-7}	UO_2L_2 ($\rightleftharpoons UO_2^{2+} + 2OH^-$)	22.4	4×10^{-23}
AgL	9.74	1.8×10^{-10}	MnL_2	12.8	1.6×10^{-13}
Hg_2L_2	17.91	1.2×10^{-18}	FeL_2	15.1	7.9×10^{-16}
TlL	3.74	1.8×10^{-4}	CoL_2	14.9	1.3×10^{-15}
PbL_2	4.78	1.7×10^{-5}	NiL_2	15.2	6×10^{-16}

*The designations α, β, or γ after some formulas refer to particular crystalline forms (which are customarily identified by Greek letters). Data for salts except oxalates are taken mainly from A. E. Martell and R. M. Smith, *Critical Stability Constants, Vol. 4* (New York: Plenum Press, 1976). Data for oxalates are from L. G. Sillén and A. E. Martell, *Stability Constants of Metal-Ion Complexes,* Supplement No. 1 (London: The Chemical Society, Special Publication No. 25, 1971). Another source: R. M. H. Verbeeck et al., *Inorg. Chem.* **1984,** 23, 1922.

Conditions are 25°C and zero ionic strength unless otherwise indicated: (a) 19°C; (b) 20°C; (c) 38°C; (d) 0.1 M; (e) 0.2 M; (f) 0.5 M; (g) 1 M; (h) 3 M; (i) 4 M; (j) 5 M.

(Continued)

Formula	pK_{sp}	K_{sp}
CuL_2	19.32	4.8×10^{-20}
VL_3	34.4	4.0×10^{-35}
CrL_3 (d)	29.8	1.6×10^{-30}
FeL_3	38.8	1.6×10^{-39}
CoL_3 (a)	44.5	3×10^{-45}
$VOL_2 (\rightleftharpoons VO^{2+} + 2OH^-)$	23.5	3×10^{-24}
PdL_2	28.5	3×10^{-29}
ZnL_2 (amorphous)	15.52	3.0×10^{-16}
CdL_2 (β)	14.35	4.5×10^{-15}
HgO (red) $(\rightleftharpoons Hg^{2+} + 2OH^-)$	25.44	3.6×10^{-26}
$Cu_2O (\rightleftharpoons 2Cu^+ + 2OH^-)$	29.4	4×10^{-30}
$Ag_2O (\rightleftharpoons 2Ag^+ + 2OH^-)$	15.42	3.8×10^{-16}
AuL_3	5.5	3×10^{-6}
AlL_3 (α)	33.5	3×10^{-34}
GaL_3 (amorphous)	37	10^{-37}
InL_3	36.9	1.3×10^{-37}
$SnO (\rightleftharpoons Sn^{2+} + 2OH^-)$	26.2	6×10^{-27}
PbO (yellow) $(\rightleftharpoons Pb^{2+} + 2OH^-)$	15.1	8×10^{-16}
PbO (red) $(\rightleftharpoons Pb^{2+} + 2OH^-)$	15.3	5×10^{-16}

Iodates: $L = IO_3^-$

Formula	pK_{sp}	K_{sp}
CaL_2	6.15	7.1×10^{-7}
SrL_2	6.48	3.3×10^{-7}
BaL_2	8.81	1.5×10^{-9}
YL_3	10.15	7.1×10^{-11}
LaL_3	10.99	1.0×10^{-11}
CeL_3	10.86	1.4×10^{-11}
ThL_4 (f)	14.62	2.4×10^{-15}
$UO_2L_2 (\rightleftharpoons UO_2^{2+} + 2IO_3^-)$(e)	7.01	9.8×10^{-8}
CrL_3 (f)	5.3	5×10^{-6}
AgL	7.51	3.1×10^{-8}
Hg_2L_2	17.89	1.3×10^{-18}
TlL	5.51	3.1×10^{-6}
ZnL_2	5.41	3.9×10^{-6}
CdL_2	7.64	2.3×10^{-8}
PbL_2	12.61	2.5×10^{-13}

Iodides: $L = I^-$

Formula	pK_{sp}	K_{sp}
CuL	12.0	1×10^{-12}
AgL	16.08	8.3×10^{-17}
$CH_3HgL (\rightleftharpoons CH_3Hg^+ + I^-)$ (b, g)	11.46	3.5×10^{-12}
$CH_3CH_2HgL (\rightleftharpoons CH_3CH_2Hg^+ + I^-)$	4.11	7.8×10^{-5}
TlL	7.23	5.9×10^{-8}
Hg_2L_2	28.34	4.6×10^{-29}
SnL_2 (i)	5.08	8.3×10^{-6}
PbL_2	8.10	7.9×10^{-9}

Oxalates: $L = C_2O_4^{2-}$

Formula	pK_{sp}	K_{sp}
CaL (b, d)	7.9	1.3×10^{-8}
SrL (b, d)	6.4	4×10^{-7}
BaL (b, d)	6.0	1×10^{-6}
La_2L_3 (b, d)	25.0	1×10^{-25}
ThL_2 (g)	21.38	4.2×10^{-22}
$UO_2L (\rightleftharpoons UO_2^{2+} + C_2O_4^{2-})$ (b, d)	8.66	2.2×10^{-9}

Phosphates: $L = PO_4^{3-}$

Formula	pK_{sp}	K_{sp}
$MgHL \cdot 3H_2O (\rightleftharpoons Mg^{2+} + HL^{2-})$	5.78	1.7×10^{-6}
$CaHL \cdot 2H_2O (\rightleftharpoons Ca^{2+} + HL^{2-})$	6.58	2.6×10^{-7}
$SrHL (\rightleftharpoons Sr^{2+} + HL^{2-})$ (b)	6.92	1.2×10^{-7}
$BaHL (\rightleftharpoons Ba^{2+} + HL^{2-})$ (b)	7.40	4.0×10^{-8}
LaL (f)	22.43	3.7×10^{-23}
$Fe_3L_2 \cdot 8H_2O$	36.0	1×10^{-36}
$FeL \cdot 2H_2O$	26.4	4×10^{-27}
$(VO)_3L_2 (\rightleftharpoons 3VO^{2+} + 2L^{3-})$	25.1	8×10^{-26}
Ag_3L	17.55	2.8×10^{-18}
$Hg_2HL (\rightleftharpoons Hg_2^{2+} + HL^{2-})$	12.40	4.0×10^{-13}
$Zn_3L_2 \cdot 4H_2O$	35.3	5×10^{-36}
Pb_3L_2 (c)	43.53	3.0×10^{-44}
GaL (g)	21.0	1×10^{-21}
InL (g)	21.63	2.3×10^{-22}

Sulfates: $L = SO_4^{2-}$

Formula	pK_{sp}	K_{sp}
CaL	4.62	2.4×10^{-5}
SrL	6.50	3.2×10^{-7}
BaL	9.96	1.1×10^{-10}
RaL (b)	10.37	4.3×10^{-11}
Ag_2L	4.83	1.5×10^{-5}
Hg_2L	6.13	7.4×10^{-7}
PbL	6.20	6.3×10^{-7}

Sulfides: $L = S^{2-}$

Formula	pK_{sp}	K_{sp}
MnL (pink)	10.5	3×10^{-11}
MnL (green)	13.5	3×10^{-14}
FeL	18.1	8×10^{-19}
CoL (α)	21.3	5×10^{-22}
CoL (β)	25.6	3×10^{-26}
NiL (α)	19.4	4×10^{-20}
NiL (β)	24.9	1.3×10^{-25}
NiL (γ)	26.6	3×10^{-27}
CuL	36.1	8×10^{-37}
Cu_2L	48.5	3×10^{-49}
Ag_2L	50.1	8×10^{-51}
Tl_2L	21.2	6×10^{-22}
ZnL (α)	24.7	2×10^{-25}
ZnL (β)	22.5	3×10^{-23}
CdL	27.0	1×10^{-27}
HgL (black)	52.7	2×10^{-53}
HgL (red)	53.3	5×10^{-54}
SnL	25.9	1.3×10^{-26}
PbL	27.5	3×10^{-28}
In_2L_3	69.4	4×10^{-70}

Thiocyanates: $L = SCN^-$

Formula	pK_{sp}	K_{sp}
CuL (j)	13.40	4.0×10^{-14}
AgL	11.97	1.1×10^{-12}
Hg_2L_2	19.52	3.0×10^{-20}
TlL	3.79	1.6×10^{-4}
HgL_2	19.56	2.8×10^{-20}

Name	Structure*	Ionic strength (μ) = 0		μ = 0.1 M§
		pK_a†	K_a‡	pK_a†
Acetic acid (ethanoic acid)	CH_3CO_2H	4.756	1.75×10^{-5}	4.56
Alanine	NH_3^+ $\|$ $CHCH_3$ $\|$ CO_2H	2.344 (CO_2H) 9.868 (NH_3)	4.53×10^{-3} 1.36×10^{-10}	2.33 9.71
Aminobenzene (aniline)	C$_6$H$_5$—NH_3^+	4.601	2.51×10^{-5}	4.64
4-Aminobenzenesulfonic acid (sulfanilic acid)	^-O_3S—C$_6$H$_4$—NH_3^+	3.232	5.86×10^{-4}	3.01
2-Aminobenzoic acid (anthranilic acid)	C$_6$H$_4$ with NH_3^+ and CO_2H	2.08 (CO_2H) 4.96 (NH_3)	8.3×10^{-3} 1.10×10^{-5}	2.01 4.78
2-Aminoethanethiol (2-mercaptoethylamine)	$HSCH_2CH_2NH_3^+$	—— ——		8.21 (SH) 10.73 (NH_3)
2-Aminoethanol (ethanolamine)	$HOCH_2CH_2NH_3^+$	9.498	3.18×10^{-10}	9.52
2-Aminophenol	C$_6$H$_4$ with OH and NH_3^+	4.70 (NH_3) (20°) 9.97 (OH) (20°)	2.0×10^{-5} 1.05×10^{-10}	4.74 9.87
Ammonia	NH_4^+	9.245	5.69×10^{-10}	9.26
Arginine	NH_3^+ $\|$ $CHCH_2CH_2CH_2NHC$ with $=NH_2^+$ and NH_2 $\|$ CO_2H	1.823 (CO_2H) 8.991 (NH_3) —— (NH_2)	1.50×10^{-2} 1.02×10^{-9} ——	2.03 9.00 (12.1)
Arsenic acid (hydrogen arsenate)	O $\|\|$ HO—As—OH $\|$ OH	2.24 6.96 (11.50)	5.8×10^{-3} 1.10×10^{-7} 3.2×10^{-12}	2.15 6.65 (11.18)
Arsenious acid (hydrogen arsenite)	$As(OH)_3$	9.29	5.1×10^{-10}	9.14
Asparagine	NH_3^+ O $\|$ $\|\|$ $CHCH_2CNH_2$ $\|$ CO_2H	—— ——	—— ——	2.16 (CO_2H) 8.73 (NH_3)

*Each acid is written in its protonated form. The acidic protons are indicated in *bold* type.

†pK_a values refer to 25°C unless otherwise indicated. Values in parentheses are considered to be less reliable. Data are from A. E. Martell, R. M. Smith, and R. J. Motekaitis, *NIST Database 46* (Gaithersburg, MD: National Institute of Standards and Technology, 2001).

‡The accurate way to calculate K_b for the conjugate base is $pK_b = 13.995 - pK_a$ and $K_b = 10^{-pK_b}$.

§See marginal note on page 166 for distinction between pK_a at $\mu = 0$ and at $\mu = 0.1$ M.

(Continued)

Name	Structure	Ionic strength (μ) = 0		μ = 0.1 M
		pK_a	K_a	pK_a
Aspartic acid	NH$_3^+$ | CHCH$_2$CO$_2$H | α CO$_2$H (β)	1.990 (α-CO$_2$H) 3.900 (β-CO$_2$H) 10.002 (NH$_3$)	1.02×10^{-2} 1.26×10^{-4} 9.95×10^{-11}	1.95 3.71 9.96
Aziridine (dimethyleneimine)	NH$_2^+$	8.04	9.1×10^{-9}	——
Benzene-1,2,3-tricarboxylic acid (hemimellitic acid)	CO$_2$H / CO$_2$H / CO$_2$H	2.86 4.30 6.28	1.38×10^{-3} 5.0×10^{-5} 5.2×10^{-7}	2.67 3.91 5.50
Benzoic acid	CO$_2$H	4.202	6.28×10^{-5}	4.01
Benzylamine	CH$_2$NH$_3^+$	9.35	4.5×10^{-10}	9.40
2,2'-Bipyridine	N N H$^+$ $^+$H	—— 4.34	—— 4.6×10^{-5}	(1.3) 4.41
Boric acid (hydrogen borate)	B(OH)$_3$	9.237 (12.74) (20°) (13.80) (20°)	5.79×10^{-10} 1.82×10^{-13} 1.58×10^{-14}	8.98 —— ——
Bromoacetic acid	BrCH$_2$CO$_2$H	2.902	1.25×10^{-3}	2.71
Butane-2,3-dione dioxime (dimethylglyoxime)	HON NOH CH$_3$ CH$_3$	10.66 (12.0)	2.2×10^{-11} 1×10^{-12}	10.45 (11.9)
Butanoic acid	CH$_3$CH$_2$CH$_2$CO$_2$H	4.818	1.52×10^{-5}	4.62
cis-Butenedioic acid (maleic acid)	CO$_2$H / CO$_2$H	1.92 6.27	1.20×10^{-2} 5.37×10^{-7}	1.75 5.84
trans-Butenedioic acid (fumaric acid)	CO$_2$H / HO$_2$C	3.02 4.48	9.5×10^{-4} 3.3×10^{-5}	2.84 4.09
Butylamine	CH$_3$CH$_2$CH$_2$CH$_2$NH$_3^+$	10.640	2.29×10^{-11}	10.66
Carbonic acid* (hydrogen carbonate)	O || HO—C—OH	6.351 10.329	4.46×10^{-7} 4.69×10^{-11}	6.13 9.91
Chloroacetic acid	ClCH$_2$CO$_2$H	2.865	1.36×10^{-3}	2.69
3-Chloropropanoic acid	ClCH$_2$CH$_2$CO$_2$H	4.11	7.8×10^{-5}	3.92
Chlorous acid (hydrogen chlorite)	HOCl=O	1.96	1.10×10^{-2}	——

*The concentration of "carbonic acid" is considered to be the sum $[H_2CO_3] + [CO_2(aq)]$. See Box 6-4.

Name	Structure	Ionic strength (μ) = 0		$\mu = 0.1$ M
		pK_a	K_a	pK_a
Chromic acid (hydrogen chromate)	O ‖ HO—Cr—OH ‖ O	(−0.2) (20°) 6.51	1.6 3.1×10^{-7}	(−0.6) (20°C) 6.05
Citric acid (2-hydroxypropane-1,2,3-tricarboxylic acid)	CO$_2$H | HO$_2$CCH$_2$CCH$_2$CO$_2$H | OH	3.128 4.761 6.396	7.44×10^{-4} 1.73×10^{-5} 4.02×10^{-7}	2.90 4.35 5.70
Cyanoacetic acid	NCCH$_2$CO$_2$H	2.472	3.37×10^{-3}	——
Cyclohexylamine	(cyclohexyl)—NH$_3^+$	10.567	2.71×10^{-11}	10.62
Cysteine	NH$_3^+$ | CHCH$_2$SH | CO$_2$H	(1.7) (CO$_2$H) 8.36 (SH) 10.74 (NH$_3$)	2×10^{-2} 4.4×10^{-9} 1.82×10^{-11}	(1.90) 8.18 10.30
Dichloroacetic acid	Cl$_2$CHCO$_2$H	(1.1)	8×10^{-2}	(0.9)
Diethylamine	(CH$_3$CH$_2$)$_2$NH$_2^+$	11.00	1.0×10^{-11}	11.04
1,2-Dihydroxybenzene (catechol)	OH OH (benzene ring)	9.45 ——	3.5×10^{-10} ——	9.26 (13.3)
1,3-Dihydroxybenzene (resorcinol)	OH OH (benzene ring)	—— ——	—— ——	9.30 11.06
D-2,3-Dihydroxybutanedioic acid (D-tartaric acid)	OH | HO$_2$CCHCHCO$_2$H | OH	3.036 4.366	9.20×10^{-4} 4.31×10^{-5}	2.82 3.97
2,3-Dimercaptopropanol	HOCH$_2$CHCH$_2$SH | SH	—— ——	—— ——	8.63 10.65
Dimethylamine	(CH$_3$)$_2$NH$_2^+$	10.774	1.68×10^{-11}	10.81
2,4-Dinitrophenol	NO$_2$ O$_2$N—(benzene ring)—OH	4.114	7.69×10^{-5}	3.92
Ethane-1,2-dithiol	HSCH$_2$CH$_2$SH	—— ——	—— ——	8.85 (30°C) 10.43 (30°C)
Ethylamine	CH$_3$CH$_2$NH$_3^+$	10.673	2.12×10^{-11}	10.69
Ethylenediamine (1,2-diaminoethane)	H$_3^+$NCH$_2$CH$_2$N$^+$H$_3$	6.848 9.928	1.42×10^{-7} 1.18×10^{-10}	7.11 9.92

(*Continued*)

Name	Structure	Ionic strength (μ) = 0		μ = 0.1 M
		pK_a	K_a	pK_a
Ethylenedinitrilotetraacetic acid (EDTA)	$(HO_2CCH_2)_2\overset{+}{N}HCH_2CH_2\overset{+}{N}H(CH_2CO_2H)_2$	—— (CO₂H)		(0.0) (CO$_2$H) (μ = 1 M)
		—— (CO₂H)		(1.5) (CO$_2$H)
		—— (CO₂H)		2.00 (CO$_2$H)
		—— (CO₂H)		2.69 (CO$_2$H)
		6.273 (NH)	5.3×10^{-7}	6.13 (NH)
		10.948 (NH)	1.13×10^{-11}	10.37 (NH)
Formic acid (methanoic acid)	HCO_2H	3.744	1.80×10^{-4}	3.57
Glutamic acid	$\overset{NH_3^+}{\underset{\alpha\,CO_2H}{\overset{\mid}{CHCH_2CH_2CO_2H}}}\ ^\gamma$	2.160 (α-CO$_2$H)	6.92×10^{-3}	2.16
		4.30 (γ-CO$_2$H)	5.0×10^{-5}	4.15
		9.96 (NH$_3$)	1.10×10^{-10}	9.58
Glutamine	$\overset{NH_3^+}{\underset{CO_2H}{\overset{\mid}{CHCH_2CH_2\overset{O}{\overset{\parallel}{C}}NH_2}}}$	——	——	2.19 (CO$_2$H)
		——	——	9.00 (NH$_3$)
Glycine (aminoacetic acid)	$\overset{NH_3^+}{\underset{CO_2H}{\overset{\mid}{\underset{\mid}{CH_2}}}}$	2.350 (CO$_2$H)	4.47×10^{-3}	2.33
		9.778 (NH$_3$)	1.67×10^{-10}	9.57
Guanidine	$\overset{+NH_2}{\underset{}{H_2N-\overset{\parallel}{C}-NH_2}}$	——	——	(13.5) (μ = 1 M)
1,6-Hexanedioic acid (adipic acid)	$HO_2CCH_2CH_2CH_2CH_2CO_2H$	4.424	3.77×10^{-5}	4.26
		5.420	3.80×10^{-6}	5.04
Hexane-2,4-dione	$CH_3\overset{O}{\overset{\parallel}{C}}CH_2\overset{O}{\overset{\parallel}{C}}CH_2CH_3$	9.38	4.2×10^{-10}	9.11 (20°C)
Histidine	$\overset{NH_3^+}{\underset{CO_2H}{\overset{\mid}{CHCH_2}}}$ (imidazole ring)	(1.6) (CO$_2$H)	3×10^{-2}	(1.7)
		5.97 (NH)	1.07×10^{-6}	6.05
		9.28 (NH$_3$)	5.2×10^{-10}	9.10
Hydrazine	$H_3\overset{+}{N}-\overset{+}{N}H_3$	−0.99	1.0×10^{1}	(−0.21) (μ = 0.5 M)
		7.98	1.05×10^{-8}	8.07
Hydrazoic acid (hydrogen azide)	$HN=\overset{+}{N}=\overset{-}{N}$	4.65	2.2×10^{-5}	4.45
Hydrogen cyanate	$HOC\equiv N$	3.48	3.3×10^{-4}	——
Hydrogen cyanide	$HC\equiv N$	9.21	6.2×10^{-10}	9.04
Hydrogen fluoride	HF	3.17	6.8×10^{-4}	2.94
Hydrogen peroxide	$HOOH$	11.65	2.2×10^{-12}	——
Hydrogen sulfide	H_2S	7.02	9.5×10^{-8}	6.82
		14.0*	$1.0 \times 10^{-14*}$	——
Hydrogen thiocyanate	$HSC\equiv N$	(−1.1) (20°C)	1.3×10^{1}	——
Hydroxyacetic acid (glycolic acid)	$HOCH_2CO_2H$	3.832	1.48×10^{-4}	3.62

*D. J. Phillips and S. L. Phillips. "High Temperature Dissociation Constants of HS$^-$ and the Standard Thermodynamic Values for S^{2-}," J. Chem. Eng. Data **2000**, 45, 981.

Name	Structure	Ionic strength (μ) = 0		μ = 0.1 M
		pK_a	K_a	pK_a
Hydroxybenzene (phenol)	(benzene ring)—OH	9.997	1.01×10^{-10}	9.78
2-Hydroxybenzoic acid (salicylic acid)	(benzene ring with CO₂H and OH)	2.972 (CO$_2$H) (13.7) (OH)	1.07×10^{-3} 2×10^{-14}	2.80 (13.4)
L-Hydroxybutanedioic acid (malic acid)	$HO_2CCH_2CHCO_2H$ with OH	3.459 5.097	3.48×10^{-4} 8.00×10^{-6}	3.24 4.68
Hydroxylamine	$HON\overset{+}{H}_3$	5.96 (NH) (13.74) (OH)	1.10×10^{-6} 1.8×10^{-14}	5.96 ——
8-Hydroxyquinoline (oxine)	(quinoline structure) HO, N$^+$H	4.94 (NH) 9.82 (OH)	1.15×10^{-5} 1.51×10^{-10}	4.97 9.65
Hypobromous acid (hydrogen hypobromite)	HOBr	8.63	2.3×10^{-9}	——
Hypochlorous acid (hydrogen hypochlorite)	HOCl	7.53	3.0×10^{-8}	——
Hypoiodous acid (hydrogen hypoiodite)	HOI	10.64	2.3×10^{-11}	——
Hypophosphorous acid (hydrogen hypophosphite)	H_2POH with O double bond	(1.3)	5×10^{-2}	(1.1)
Imidazole (1,3-diazole)	(imidazole ring) —NH$^+$, N, H	6.993 (14.5)	1.02×10^{-7} 3×10^{-15}	7.00 ——
Iminodiacetic acid	$H_2\overset{+}{N}(CH_2CO_2H)_2$	(1.85) (CO$_2$H) 2.84 (CO$_2$H) 9.79 (NH$_2$)	1.41×10^{-2} 1.45×10^{-3} 1.62×10^{-10}	(1.77) 2.62 9.34
Iodic acid (hydrogen iodate)	$HOI=O$ with O double bond	0.77	0.17	——
Iodoacetic acid	ICH_2CO_2H	3.175	6.68×10^{-4}	2.98
Isoleucine	$N\overset{+}{H}_3$ / CHCH(CH$_3$)CH$_2$CH$_3$ / CO$_2$H	2.318 (CO$_2$H) 9.758 (NH$_3$)	4.81×10^{-3} 1.75×10^{-10}	2.26 9.60
Leucine	$N\overset{+}{H}_3$ / CHCH$_2$CH(CH$_3$)$_2$ / CO$_2$H	2.328 (CO$_2$H) 9.744 (NH$_3$)	4.70×10^{-3} 1.80×10^{-10}	2.32 9.58
Lysine	α N$\overset{+}{H}_3$ / CHCH$_2$CH$_2$CH$_2$CH$_2\overset{\epsilon}{N}\overset{+}{H}_3$ / CO$_2$H	(1.77) (CO$_2$H) 9.07 (α-NH$_3$) 10.82 (ϵ-NH$_3$)	1.7×10^{-2} 8.5×10^{-10} 1.51×10^{-11}	2.15 9.15 10.66

(Continued)

| Name | Structure | Ionic strength (μ) = 0 | | μ = 0.1 M |
		pK_a	K_a	pK_a
Malonic acid (propanedioic acid)	$HO_2CCH_2CO_2H$	2.847 5.696	1.42×10^{-3} 2.01×10^{-6}	2.65 5.27
Mercaptoacetic acid (thioglycolic acid)	$HSCH_2CO_2H$	3.64 (CO_2H) 10.61 (SH)	2.3×10^{-4} 2.5×10^{-11}	3.48 10.11
2-Mercaptoethanol	$HSCH_2CH_2OH$	9.72	1.9×10^{-10}	9.40
Methionine	NH_3^+ $\vert$ $CHCH_2CH_2SCH_3$ $\vert$ CO_2H	—— ——	—— ——	2.18 (CO_2H) 9.08 (NH_3)
2-Methoxyaniline (o-anisidine)		4.526	2.98×10^{-5}	——
4-Methoxyaniline (p-anisidine)	$CH_3O-\!\!\bigcirc\!\!-\overset{+}{N}H_3$	5.357	4.40×10^{-6}	5.33
Methylamine	$CH_3\overset{+}{N}H_3$	10.632	2.33×10^{-11}	10.65
2-Methylaniline (o-toluidine)		4.447	3.57×10^{-5}	——
4-Methylaniline (p-toluidine)	$CH_3-\!\!\bigcirc\!\!-\overset{+}{N}H_3$	5.080	8.32×10^{-6}	5.09
2-Methylphenol (o-cresol)		10.31	4.9×10^{-11}	10.09
4-Methylphenol (p-cresol)	$CH_3-\!\!\bigcirc\!\!-OH$	10.269	5.5×10^{-11}	10.04
Morpholine (perhydro-1,4-oxazine)		8.492	3.22×10^{-9}	——
1-Naphthoic acid		3.67	2.1×10^{-4}	——
2-Naphthoic acid		4.16	6.9×10^{-5}	——
1-Naphthol		9.416	3.84×10^{-10}	9.14
2-Naphthol		9.573	2.67×10^{-10}	9.31
Nitrilotriacetic acid	$H\overset{+}{N}(CH_2CO_2H)_3$	—— (CO_2H) 2.0 (CO_2H) (25°) 2.940 (CO_2H) (20°) 10.334 (NH) (20°)	—— 0.010 1.15×10^{-3} 4.63×10^{-11}	(1.0) 1.81 2.52 9.46

| Name | Structure | Ionic strength (μ) = 0 | | μ = 0.1 M |
		pK_a	K_a	pK_a
2-Nitrobenzoic acid	(structure: benzene ring with NO$_2$ and CO$_2$H)	2.185	6.53×10^{-3}	——
3-Nitrobenzoic acid	(structure: benzene ring with NO$_2$ and CO$_2$H)	3.449	3.56×10^{-4}	3.28
4-Nitrobenzoic acid	O_2N——⟨⟩——CO_2H	3.442	3.61×10^{-4}	3.28
Nitroethane	$CH_3CH_2NO_2$	8.57	2.7×10^{-9}	——
2-Nitrophenol	(structure: benzene ring with NO$_2$ and OH)	7.230	5.89×10^{-8}	7.04
3-Nitrophenol	(structure: benzene ring with NO$_2$ and OH)	8.37	4.3×10^{-9}	8.16
4-Nitrophenol	O_2N——⟨⟩——OH	7.149	7.10×10^{-8}	6.96
N-Nitrosophenylhydroxylamine (cupferron)	(structure: phenyl–N(NO)–OH)	——	——	4.16
Nitrous acid	$HON{=}O$	3.15	7.1×10^{-4}	——
Oxalic acid (ethanedioic acid)	HO_2CCO_2H	1.250 / 4.266	5.62×10^{-2} / 5.42×10^{-5}	(1.2) / 3.80
Oxoacetic acid (glyoxylic acid)	$HCCO_2H$ (with O double bond)	3.46	3.5×10^{-4}	3.05
Oxobutanedioic acid (oxaloacetic acid)	$HO_2CCH_2CCO_2H$ (with O double bond)	2.56 / 4.37	2.8×10^{-3} / 4.3×10^{-5}	2.26 / 3.90
2-Oxopentanedioic (α-ketoglutaric acid)	$HO_2CCH_2CH_2CCO_2H$ (with O double bond)	—— / ——	—— / ——	(1.9) (μ = 0.5 M) / 4.44 (μ = 0.5 M)
2-Oxopropanoic acid (pyruvic acid)	CH_3CCO_2H (with O double bond)	2.48	3.3×10^{-3}	2.26
1,5-Pentanedioic acid (glutaric acid)	$HO_2CCH_2CH_2CH_2CO_2H$	4.345 / 5.422	4.52×10^{-5} / 3.78×10^{-6}	4.19 / 5.06
Pentanoic acid (valeric acid)	$CH_3CH_2CH_2CH_2CO_2H$	4.843	1.44×10^{-5}	4.63 (18°C)
1,10-Phenanthroline	(structure: phenanthroline with NH HN$^+$)	—— / 4.91	—— / 1.23×10^{-5}	(1.8) / 4.92
Phenylacetic acid	⟨⟩——CH_2CO_2H	4.310	4.90×10^{-5}	4.11

(Continued)

Name	Structure	Ionic strength (μ) = 0		μ = 0.1 M
		pK_a	K_a	pK_a
Phenylalanine	NH$_3^+$ | CHCH$_2$—⬡ | CO$_2$H	2.20 (CO$_2$H) 9.31 (NH$_3$)	6.3×10^{-3} 4.9×10^{-10}	2.18 9.09
Phosphoric acid* (hydrogen phosphate)	O ‖ HO—P—OH | OH	2.148 7.198 12.375	7.11×10^{-3} 6.34×10^{-8} 4.22×10^{-13}	1.92 6.71 11.52
Phosphorous acid (hydrogen phosphite)	O ‖ HP—OH | OH	(1.5) 6.78	3×10^{-2} 1.66×10^{-7}	—— ——
Phthalic acid (benzene-1,2-dicarboxylic acid)	⬡—CO$_2$H / CO$_2$H	2.950 5.408	1.12×10^{-3} 3.90×10^{-6}	2.76 4.92
Piperazine (perhydro-1,4-diazine)	H$_2^+$N⬡NH$_2^+$	5.333 9.731	4.65×10^{-6} 1.86×10^{-10}	5.64 9.74
Piperidine	⬡NH$_2^+$	11.125	7.50×10^{-12}	11.08
Proline	⬠—CO$_2$H / N / $^+$H$_2$	1.952 (CO$_2$H) 10.640 (NH$_2$)	1.12×10^{-2} 2.29×10^{-11}	1.89 10.46
Propanoic acid	CH$_3$CH$_2$CO$_2$H	4.874	1.34×10^{-5}	4.69
Propenoic acid (acrylic acid)	H$_2$C=CHCO$_2$H	4.258	5.52×10^{-5}	——
Propylamine	CH$_3$CH$_2$CH$_2$NH$_3^+$	10.566	2.72×10^{-11}	10.64
Pyridine (azine)	⬡NH$^+$	5.20	6.3×10^{-6}	5.24
Pyridine-2-carboxylic acid (picolinic acid)	⬡NH$^+$ / CO$_2$H	(1.01) (CO$_2$H) 5.39 (NH)	9.8×10^{-2} 4.1×10^{-6}	(0.95) 5.21
Pyridine-3-carboxylic acid (nicotinic acid)	HO$_2$C / ⬡NH$^+$	2.03 (CO$_2$H) 4.82 (NH)	9.3×10^{-3} 1.51×10^{-5}	2.08 4.69
Pyridoxal-5-phosphate	O O=CH ‖ HO—POCH$_2$—⬡—OH | HO N—CH$_3$ $^+$H	—— —— —— ——	—— —— —— ——	(1.4) (POH) 3.51 (OH) 6.04 (POH) 8.25 (NH)
Pyrophosphoric acid (hydrogen diphosphate)	O O ‖ ‖ (HO)$_2$POP(OH)$_2$	(0.9) 2.28 6.70 9.40	0.13 5.2×10^{-3} 2.0×10^{-7} 4.0×10^{-10}	(0.8) (1.9) 5.94 8.25

*pK_3 from A. G. Miller and J. W. Macklin, Anal. Chem. **1983**, 55, 684.

Name	Structure	Ionic strength (μ) = 0		μ = 0.1 M
		pK_a	K_a	pK_a
Pyrrolidine	(cyclopentyl-NH_2^+)	11.305	4.95×10^{-12}	11.3
Serine	NH_3^+ | $CHCH_2OH$ | CO_2H	2.187 (CO_2H) 9.209 (NH_3)	6.50×10^{-3} 6.18×10^{-10}	2.16 9.05
Succinic acid (butanedioic acid)	$HO_2CCH_2CH_2CO_2H$	4.207 5.636	6.21×10^{-5} 2.31×10^{-6}	3.99 5.24
Sulfuric acid (hydrogen sulfate)	$HO-S-OH$ (with O above and O below, double bonds)	1.987 (pK_2)	1.03×10^{-2}	1.54
Sulfurous acid (hydrogen sulfite)	$HOSOH$ (with O double bond above)	1.857 7.172	1.39×10^{-2} 6.73×10^{-8}	1.66 6.85
Thiosulfuric acid (hydrogen thiosulfate)	$HOSSH$ (with O above and O below)	(0.6) (1.6)	0.3 0.03	—— (1.3)
Threonine	NH_3^+ | $CHCHOHCH_3$ | CO_2H	2.088 (CO_2H) 9.100 (NH_3)	8.17×10^{-3} 7.94×10^{-10}	2.20 8.94
Trichloroacetic acid	Cl_3CCO_2H	(−0.5)	3	——
Triethanolamine	$(HOCH_2CH_2)_3NH^+$	7.762	1.73×10^{-8}	7.85
Triethylamine	$(CH_3CH_2)_3NH^+$	10.72	1.9×10^{-11}	10.76
1,2,3-Trihydroxybenzene (pyrogallol)	(benzene ring with OH, OH, OH)	—— —— ——	—— —— ——	8.96 11.00 (14.0) (20°C)
Trimethylamine	$(CH_3)_3NH^+$	9.799	1.59×10^{-10}	9.82
Tris(hydroxymethyl)amino- methane (tris or tham)	$(HOCH_2)_3CNH_3^+$	8.072	8.47×10^{-9}	8.10
Tryptophan	NH_3^+ | $CHCH_2$ (indole) | CO_2H	—— ——	—— ——	2.37 (CO_2H) 9.33 (NH_3)
Tyrosine	NH_3^+ | $CHCH_2$—(benzene)—OH | CO_2H	—— —— ——	—— —— ——	2.41 (CO_2H) 8.67 (NH_3) 11.01 (OH)
Valine	NH_3^+ | $CHCH(CH_3)_2$ | CO_2H	2.286 (CO_2H) 9.719 (NH_3)	5.18×10^{-3} 1.91×10^{-10}	2.27 9.52
Water*	H_2O	13.997	1.01×10^{-14}	——

*The constant given for water is K_w.

Reaction	$E°$ (volts)	$dE°/dT$ (mV/K)
Aluminum		
$Al^{3+} + 3e^- \rightleftharpoons Al(s)$	−1.677	0.533
$AlCl^{2+} + 3e^- \rightleftharpoons Al(s) + Cl^-$	−1.802	
$AlF_6^{3-} + 3e^- \rightleftharpoons Al(s) + 6F^-$	−2.069	
$Al(OH)_4^- + 3e^- \rightleftharpoons Al(s) + 4OH^-$	−2.328	−1.13
Antimony		
$SbO^+ + 2H^+ + 3e^- \rightleftharpoons Sb(s) + H_2O$	0.208	
$Sb_2O_3(s) + 6H^+ + 6e^- \rightleftharpoons 2Sb(s) + 3H_2O$	0.147	−0.369
$Sb(s) + 3H^+ + 3e^- \rightleftharpoons SbH_3(g)$	−0.510	−0.030
Arsenic		
$H_3AsO_4 + 2H^+ + 2e^- \rightleftharpoons H_3AsO_3 + H_2O$	0.575	−0.257
$H_3AsO_3 + 3H^+ + 3e^- \rightleftharpoons As(s) + 3H_2O$	0.247 5	−0.505
$As(s) + 3H^+ + 3e^- \rightleftharpoons AsH_3(g)$	−0.238	−0.029
Barium		
$Ba^{2+} + 2e^- + Hg \rightleftharpoons Ba(in\ Hg)$	−1.717	
$Ba^{2+} + 2e^- \rightleftharpoons Ba(s)$	−2.906	−0.401
Beryllium		
$Be^{2+} + 2e^- \rightleftharpoons Be(s)$	−1.968	0.60
Bismuth		
$Bi^{3+} + 3e^- \rightleftharpoons Bi(s)$	0.308	0.18
$BiCl_4^- + 3e^- \rightleftharpoons Bi(s) + 4Cl^-$	0.16	
$BiOCl(s) + 2H^+ + 3e^- \rightleftharpoons Bi(s) + H_2O + Cl^-$	0.160	
Boron		
$2B(s) + 6H^+ + 6e^- \rightleftharpoons B_2H_6(g)$	−0.150	−0.296
$B_4O_7^{2-} + 14H^+ + 12e^- \rightleftharpoons 4B(s) + 7H_2O$	−0.792	
$B(OH)_3 + 3H^+ + 3e^- \rightleftharpoons B(s) + 3H_2O$	−0.889	−0.492
Bromine		
$BrO_4^- + 2H^+ + 2e^- \rightleftharpoons BrO_3^- + H_2O$	1.745	−0.511
$HOBr + H^+ + e^- \rightleftharpoons \frac{1}{2}Br_2(l) + H_2O$	1.584	−0.75
$BrO_3^- + 6H^+ + 5e^- \rightleftharpoons \frac{1}{2}Br_2(l) + 3H_2O$	1.513	−0.419
$Br_2(aq) + 2e^- \rightleftharpoons 2Br^-$	1.098	−0.499
$Br_2(l) + 2e^- \rightleftharpoons 2Br^-$	1.078	−0.611
$Br_3^- + 2e^- \rightleftharpoons 3Br^-$	1.062	−0.512
$BrO^- + H_2O + 2e^- \rightleftharpoons Br^- + 2OH^-$	0.766	−0.94
$BrO_3^- + 3H_2O + 6e^- \rightleftharpoons Br^- + 6OH^-$	0.613	−1.287
Cadmium		
$Cd^{2+} + 2e^- + Hg \rightleftharpoons Cd(in\ Hg)$	−0.380	
$Cd^{2+} + 2e^- \rightleftharpoons Cd(s)$	−0.402	−0.029
$Cd(C_2O_4)(s) + 2e^- \rightleftharpoons Cd(s) + C_2O_4^{2-}$	−0.522	
$Cd(C_2O_4)_2^{2-} + 2e^- \rightleftharpoons Cd(s) + 2C_2O_4^{2-}$	−0.572	
$Cd(NH_3)_4^{2+} + 2e^- \rightleftharpoons Cd(s) + 4NH_3$	−0.613	
$CdS(s) + 2e^- \rightleftharpoons Cd(s) + S^{2-}$	−1.175	
Calcium		
$Ca(s) + 2H^+ + 2e^- \rightleftharpoons CaH_2(s)$	0.776	
$Ca^{2+} + 2e^- + Hg \rightleftharpoons Ca(in\ Hg)$	−2.003	
$Ca^{2+} + 2e^- \rightleftharpoons Ca(s)$	−2.868	−0.186

*All species are aqueous unless otherwise indicated. The reference state for amalgams is an infinitely dilute solution of the element in Hg. The temperature coefficient, $dE°/dT$ allows us to calculate the standard potential, $E°(T)$, at temperature T: $E°(T) = E° + (dE°/dT)\Delta T$, where ΔT is $T − 298.15$ K. Note the units mV/K for $dE°/dT$. Once you know $E°$ for a net cell reaction at temperature T, you can find the equilibrium constant, K, for the reaction from the formula $K = 10^{nFE°/RT \ln 10}$, where n is the number of electrons in each half-reaction, F is the Faraday constant, and R is the gas constant.

SOURCES: The most authoritative source is S. G. Bratsch, J. Phys. Chem. Ref. Data **1989**, 18, 1. Additional data come from L. G. Sillén and A. E. Martell, Stability Constants of Metal-Ion Complexes (London: The Chemical Society, Special Publications Nos. 17 and 25. 1964 and 1971); G. Milazzo and S. Caroli, Tables of Standard Electrode Potentials (New York: Wiley, 1978); T. Mussini, P. Longhi, and S. Rondinini, Pure Appl. Chem. **1985**, 57, 169. Another good source is A. J. Bard, R. Parsons, and J. Jordan. Standard Potentials in Aqueous Solution (New York: Marcel Dekker, 1985). Reduction potentials for 1 200 free radical reactions are given by P. Wardman, J. Phys. Chem. Ref. Data **1989**, 18, 1637.

Reaction	$E°$ (volts)	$dE°/dT$ (mV/K)
$Ca(acetate)^+ + 2e^- \rightleftharpoons Ca(s) + acetate^-$	−2.891	
$CaSO_4(s) + 2e^- \rightleftharpoons Ca(s) + SO_4^{2-}$	−2.936	
$Ca(malonate)(s) + 2e^- \rightleftharpoons Ca(s) + malonate^{2-}$	−3.608	

Carbon

Reaction	$E°$ (volts)	$dE°/dT$ (mV/K)
$C_2H_2(g) + 2H^+ + 2e^- \rightleftharpoons C_2H_4(g)$	0.731	
$O=\!\!\bigcirc\!\!=O + 2H^+ + 2e^- \rightleftharpoons HO-\!\!\bigcirc\!\!-OH$	0.700	
$CH_3OH + 2H^+ + 2e^- \rightleftharpoons CH_4(g) + H_2O$	0.583	−0.039
Dehydroascorbic acid $+ 2H^+ + 2e^- \rightleftharpoons$ ascorbic acid $+ H_2O$	0.390	
$(CN)_2(g) + 2H^+ + 2e^- \rightleftharpoons 2HCN(aq)$	0.373	
$H_2CO + 2H^+ + 2e^- \rightleftharpoons CH_3OH$	0.237	−0.51
$C(s) + 4H^+ + 4e^- \rightleftharpoons CH_4(g)$	0.131 5	−0.209 2
$HCO_2H + 2H^+ + 2e^- \rightleftharpoons H_2CO + H_2O$	−0.029	−0.63
$CO_2(g) + 2H^+ + 2e^- \rightleftharpoons CO(g) + H_2O$	−0.103 8	−0.397 7
$CO_2(g) + 2H^+ + 2e^- \rightleftharpoons HCO_2H$	−0.114	−0.94
$2CO_2(g) + 2H^+ + 2e^- \rightleftharpoons H_2C_2O_4$	−0.432	−1.76

Cerium

Reaction	$E°$ (volts)		$dE°/dT$ (mV/K)
	1.72		1.54
	1.70	1 F HClO$_4$	
$Ce^{4+} + e^- \rightleftharpoons Ce^{3+}$	1.44	1 F H$_2$SO$_4$	
	1.61	1 F HNO$_3$	
	1.47	1 F HCl	
$Ce^{3+} + 3e^- \rightleftharpoons Ce(s)$	−2.336		0.280

Cesium

Reaction	$E°$ (volts)	$dE°/dT$ (mV/K)
$Cs^+ + e^- + Hg \rightleftharpoons Cs(in\ Hg)$	−1.950	
$Cs^+ + e^- \rightleftharpoons Cs(s)$	−3.026	−1.172

Chlorine

Reaction	$E°$ (volts)	$dE°/dT$ (mV/K)
$HClO_2 + 2H^+ + 2e^- \rightleftharpoons HOCl + H_2O$	1.674	0.55
$HClO + H^+ + e^- \rightleftharpoons \frac{1}{2}Cl_2(g) + H_2O$	1.630	−0.27
$ClO_3^- + 6H^+ + 5e^- \rightleftharpoons \frac{1}{2}Cl_2(g) + 3H_2O$	1.458	−0.347
$Cl_2(aq) + 2e^- \rightleftharpoons 2Cl^-$	1.396	−0.72
$Cl_2(g) + 2e^- \rightleftharpoons 2Cl^-$	1.360 4	−1.248
$ClO_4^- + 2H^+ + 2e^- \rightleftharpoons ClO_3^- + H_2O$	1.226	−0.416
$ClO_3^- + 3H^+ + 2e^- \rightleftharpoons HClO_2 + H_2O$	1.157	−0.180
$ClO_3^- + 2H^+ + e^- \rightleftharpoons ClO_2 + H_2O$	1.130	0.074
$ClO_2 + e^- \rightleftharpoons ClO_2^-$	1.068	−1.335

Chromium

Reaction	$E°$ (volts)	$dE°/dT$ (mV/K)
$Cr_2O_7^{2-} + 14H^+ + 6e^- \rightleftharpoons 2Cr^{3+} + 7H_2O$	1.36	−1.32
$CrO_4^{2-} + 4H_2O + 3e^- \rightleftharpoons Cr(OH)_3\ (s,\ hydrated) + 5OH^-$	−0.12	−1.62
$Cr^{3+} + e^- \rightleftharpoons Cr^{2+}$	−0.42	1.4
$Cr^{3+} + 3e^- \rightleftharpoons Cr(s)$	−0.74	0.44
$Cr^{2+} + 2e^- \rightleftharpoons Cr(s)$	−0.89	−0.04

Cobalt

Reaction	$E°$ (volts)		$dE°/dT$ (mV/K)
	1.92		1.23
$Co^{3+} + e^- \rightleftharpoons Co^{2+}$	1.817	8 F H$_2$SO$_4$	
	1.850	4 F HNO$_3$	
$Co(NH_3)_5(H_2O)^{3+} + e^- \rightleftharpoons Co(NH_3)_5(H_2O)^{2+}$	0.37	1 F NH$_4$NO$_3$	
$Co(NH_3)_6^{3+} + e^- \rightleftharpoons Co(NH_3)_6^{2+}$	0.1		
$CoOH^+ + H^+ + 2e^- \rightleftharpoons Co(s) + H_2O$	0.003		−0.04
$Co^{2+} + 2e^- \rightleftharpoons Co(s)$	−0.282		0.065
$Co(OH)_2(s) + 2e^- \rightleftharpoons Co(s) + 2OH^-$	−0.746		−1.02

Copper

Reaction	$E°$ (volts)	$dE°/dT$ (mV/K)
$Cu^+ + e^- \rightleftharpoons Cu(s)$	0.518	−0.754
$Cu^{2+} + 2e^- \rightleftharpoons Cu(s)$	0.339	0.011
$Cu^{2+} + e^- \rightleftharpoons Cu^+$	0.161	0.776
$CuCl(s) + e^- \rightleftharpoons Cu(s) + Cl^-$	0.137	
$Cu(IO_3)_2(s) + 2e^- \rightleftharpoons Cu(s) + 2IO_3^-$	−0.079	
$Cu(ethylenediamine)_2^+ + e^- \rightleftharpoons Cu(s) + 2$ ethylenediamine	−0.119	
$CuI(s) + e^- \rightleftharpoons Cu(s) + I^-$	−0.185	
$Cu(EDTA)^{2-} + 2e^- \rightleftharpoons Cu(s) + EDTA^{4-}$	−0.216	
$Cu(OH)_2(s) + 2e^- \rightleftharpoons Cu(s) + 2OH^-$	−0.222	
$Cu(CN)_2^- + e^- \rightleftharpoons Cu(s) + 2CN^-$	−0.429	
$CuCN(s) + e^- \rightleftharpoons Cu(s) + CN^-$	−0.639	

(Continued)

Reaction	$E°$ (volts)	$dE°/dT$ (mV/K)
Dysprosium		
$Dy^{3+} + 3e^- \rightleftharpoons Dy(s)$	-2.295	0.373
Erbium		
$Er^{3+} + 3e^- \rightleftharpoons Er(s)$	-2.331	0.388
Europium		
$Eu^{3+} + e^- \rightleftharpoons Eu^{2+}$	-0.35	1.53
$Eu^{3+} + 3e^- \rightleftharpoons Eu(s)$	-1.991	0.338
$Eu^{2+} + 2e^- \rightleftharpoons Eu(s)$	-2.812	-0.26
Fluorine		
$F_2(g) + 2e^- \rightleftharpoons 2F^-$	2.890	-1.870
$F_2O(g) + 2H^+ + 4e^- \rightleftharpoons 2F^- + H_2O$	2.168	-1.208
Gadolinium		
$Gd^{3+} + 3e^- \rightleftharpoons Gd(s)$	-2.279	0.315
Gallium		
$Ga^{3+} + 3e^- \rightleftharpoons Ga(s)$	-0.549	0.61
$GaOOH(s) + H_2O + 3e^- \rightleftharpoons Ga(s) + 3OH^-$	-1.320	-1.08
Germanium		
$Ge^{2+} + 2e^- \rightleftharpoons Ge(s)$	0.1	
$H_4GeO_4 + 4H^+ + 4e^- \rightleftharpoons Ge(s) + 4H_2O$	-0.039	-0.429
Gold		
$Au^+ + e^- \rightleftharpoons Au(s)$	1.69	-1.1
$Au^{3+} + 2e^- \rightleftharpoons Au^+$	1.41	
$AuCl_2^- + e^- \rightleftharpoons Au(s) + 2Cl^-$	1.154	
$AuCl_4^- + 2e^- \rightleftharpoons AuCl_2^- + 2Cl^-$	0.926	
Hafnium		
$Hf^{4+} + 4e^- \rightleftharpoons Hf(s)$	-1.55	0.68
$HfO_2(s) + 4H^+ + 4e^- \rightleftharpoons Hf(s) + 2H_2O$	-1.591	-0.355
Holmium		
$Ho^{3+} + 3e^- \rightleftharpoons Ho(s)$	-2.33	0.371
Hydrogen		
$2H^+ + 2e^- \rightleftharpoons H_2(g)$	$0.000\ 0$	0
$H_2O + e^- \rightleftharpoons \frac{1}{2}H_2(g) + OH^-$	$-0.828\ 0$	$-0.836\ 0$
Indium		
$In^{3+} + 3e^- + Hg \rightleftharpoons In(in\ Hg)$	-0.313	
$In^{3+} + 3e^- \rightleftharpoons In(s)$	-0.338	0.42
$In^{3+} + 2e^- \rightleftharpoons In^+$	-0.444	
$In(OH)_3(s) + 3e^- \rightleftharpoons In(s) + 3OH^-$	-0.99	-0.95
Iodine		
$IO_4^- + 2H^+ + 2e^- \rightleftharpoons IO_3^- + H_2O$	1.589	-0.85
$H_5IO_6 + 2H^+ + 2e^- \rightleftharpoons HIO_3 + 3H_2O$	1.567	-0.12
$HOI + H^+ + e^- \rightleftharpoons \frac{1}{2}I_2(s) + H_2O$	1.430	-0.339
$ICl_3(s) + 3e^- \rightleftharpoons \frac{1}{2}I_2(s) + 3Cl^-$	1.28	
$ICl(s) + e^- \rightleftharpoons \frac{1}{2}I_2(s) + Cl^-$	1.22	
$IO_3^- + 6H^+ + 5e^- \rightleftharpoons \frac{1}{2}I_2(s) + 3H_2O$	1.210	-0.367
$IO_3^- + 5H^+ + 4e^- \rightleftharpoons HOI + 2H_2O$	1.154	-0.374
$I_2(aq) + 2e^- \rightleftharpoons 2I^-$	0.620	-0.234
$I_2(s) + 2e^- \rightleftharpoons 2I^-$	0.535	-0.125
$I_3^- + 2e^- \rightleftharpoons 3I^-$	0.535	-0.186
$IO_3^- + 3H_2O + 6e^- \rightleftharpoons I^- + 6OH^-$	0.269	-1.163
Iridium		
$IrCl_6^{2-} + e^- \rightleftharpoons IrCl_6^{3-}$	1.026 1 F HCl	
$IrBr_6^{2-} + e^- \rightleftharpoons IrBr_6^{3-}$	0.947 2 F NaBr	
$IrCl_6^{2-} + 4e^- \rightleftharpoons Ir(s) + 6Cl^-$	0.835	
$IrO_2(s) + 4H^+ + 4e^- \rightleftharpoons Ir(s) + 2H_2O$	0.73	-0.36
$IrI_6^{2-} + e^- \rightleftharpoons IrI_6^{3-}$	0.485 1 F KI	
Iron		
$Fe(phenanthroline)_3^{3+} + e^- \rightleftharpoons Fe(phenanthroline)_3^{2+}$	1.147	
$Fe(bipyridyl)_3^{3+} + e^- \rightleftharpoons Fe(bipyridyl)_3^{2+}$	1.120	
$FeOH^{2+} + H^+ + e^- \rightleftharpoons Fe^{2+} + H_2O$	0.900	0.096
$FeO_4^{2-} + 3H_2O + 3e^- \rightleftharpoons FeOOH(s) + 5OH^-$	0.80	-1.59
$Fe^{3+} + e^- \rightleftharpoons Fe^{2+}$	$\begin{cases} 0.771 \\ 0.732 \ \ \text{1 F HCl} \\ 0.767 \ \ \text{1 F HClO}_4 \\ 0.746 \ \ \text{1 F HNO}_3 \\ 0.68 \ \ \ \text{1 F H}_2\text{SO}_4 \end{cases}$	1.175

Reaction	$E°$ (volts)	$dE°/dT$ (mV/K)
$FeOOH(s) + 3H^+ + e^- \rightleftharpoons Fe^{2+} + 2H_2O$	0.74	-1.05
$Ferricinium^+ + e^- \rightleftharpoons ferrocene$	0.400	
$Fe(CN)_6^{3-} + e^- \rightleftharpoons Fe(CN)_6^{4-}$	0.356	
$Fe(glutamate)^{3+} + e^- \rightleftharpoons Fe(glutamate)^{2+}$	0.240	
$FeOH^+ + H^+ + 2e^- \rightleftharpoons Fe(s) + H_2O$	-0.16	0.07
$Fe^{2+} + 2e^- \rightleftharpoons Fe(s)$	-0.44	0.07
$FeCO_3(s) + 2e^- \rightleftharpoons Fe(s) + CO_3^{2-}$	-0.756	-1.293

Lanthanum

Reaction	$E°$ (volts)	$dE°/dT$ (mV/K)
$La^{3+} + 3e^- \rightleftharpoons La(s)$	-2.379	0.242
$La(succinate)^+ + 3e^- \rightleftharpoons La(s) + succinate^{2-}$	-2.601	

Lead

Reaction	$E°$ (volts)	$dE°/dT$ (mV/K)
$Pb^{4+} + 2e^- \rightleftharpoons Pb^{2+}$	1.69 1 F HNO$_3$	
$PbO_2(s) + 4H^+ + SO_4^{2-} + 2e^- \rightleftharpoons PbSO_4(s) + 2H_2O$	1.685	
$PbO_2(s) + 4H^+ + 2e^- \rightleftharpoons Pb^{2+} + 2H_2O$	1.458	-0.253
$3PbO_2(s) + 2H_2O + 4e^- \rightleftharpoons Pb_3O_4(s) + 4OH^-$	0.269	-1.136
$Pb_3O_4(s) + H_2O + 2e^- \rightleftharpoons 3PbO(s, red) + 2OH^-$	0.224	-1.211
$Pb_3O_4(s) + H_2O + 2e^- \rightleftharpoons 3PbO(s, yellow) + 2OH^-$	0.207	-1.177
$Pb^{2+} + 2e^- \rightleftharpoons Pb(s)$	-0.126	-0.395
$PbF_2(s) + 2e^- \rightleftharpoons Pb(s) + 2F^-$	-0.350	
$PbSO_4(s) + 2e^- \rightleftharpoons Pb(s) + SO_4^{2-}$	-0.355	

Lithium

Reaction	$E°$ (volts)	$dE°/dT$ (mV/K)
$Li^+ + e^- + Hg \rightleftharpoons Li(in\ Hg)$	-2.195	
$Li^+ + e^- \rightleftharpoons Li(s)$	-3.040	-0.514

Lutetium

Reaction	$E°$ (volts)	$dE°/dT$ (mV/K)
$Lu^{3+} + 3e^- \rightleftharpoons Lu(s)$	-2.28	0.412

Magnesium

Reaction	$E°$ (volts)	$dE°/dT$ (mV/K)
$Mg^{2+} + 2e^- + Hg \rightleftharpoons Mg(in\ Hg)$	-1.980	
$Mg(OH)^+ + H^+ + 2e^- \rightleftharpoons Mg(s) + H_2O$	-2.022	0.25
$Mg^{2+} + 2e^- \rightleftharpoons Mg(s)$	-2.360	0.199
$Mg(C_2O_4)(s) + 2e^- \rightleftharpoons Mg(s) + C_2O_4^{2-}$	-2.493	
$Mg(OH)_2(s) + 2e^- \rightleftharpoons Mg(s) + 2OH^-$	-2.690	-0.946

Manganese

Reaction	$E°$ (volts)	$dE°/dT$ (mV/K)
$MnO_4^- + 4H^+ + 3e^- \rightleftharpoons MnO_2(s) + 2H_2O$	1.692	-0.671
$Mn^{3+} + e^- \rightleftharpoons Mn^{2+}$	1.56	1.8
$MnO_4^- + 8H^+ + 5e^- \rightleftharpoons Mn^{2+} + 4H_2O$	1.507	-0.646
$Mn_2O_3(s) + 6H^+ + 2e^- \rightleftharpoons 2Mn^{2+} + 3H_2O$	1.485	-0.926
$MnO_2(s) + 4H^+ + 2e^- \rightleftharpoons Mn^{2+} + 2H_2O$	1.230	-0.609
$Mn(EDTA)^- + e^- \rightleftharpoons Mn(EDTA)^{2-}$	0.825	-1.10
$MnO_4^- + e^- \rightleftharpoons MnO_4^{2-}$	0.56	-2.05
$3Mn_2O_3(s) + H_2O + 2e^- \rightleftharpoons 2Mn_3O_4(s) + 2OH^-$	0.002	-1.256
$Mn_3O_4(s) + 4H_2O + 2e^- \rightleftharpoons 3Mn(OH)_2(s) + 2OH^-$	-0.352	-1.61
$Mn^{2+} + 2e^- \rightleftharpoons Mn(s)$	-1.182	-1.129
$Mn(OH)_2(s) + 2e^- \rightleftharpoons Mn(s) + 2OH^-$	-1.565	-1.10

Mercury

Reaction	$E°$ (volts)	$dE°/dT$ (mV/K)
$2Hg^{2+} + 2e^- \rightleftharpoons Hg_2^{2+}$	0.908	0.095
$Hg^{2+} + 2e^- \rightleftharpoons Hg(l)$	0.852	-0.116
$Hg_2^{2+} + 2e^- \rightleftharpoons 2Hg(l)$	0.796	-0.327
$Hg_2SO_4(s) + 2e^- \rightleftharpoons 2Hg(l) + SO_4^{2-}$	0.614	
$Hg_2Cl_2(s) + 2e^- \rightleftharpoons 2Hg(l) + 2Cl^-$	$\begin{cases} 0.268 \\ 0.241 \text{ (saturated calomel electrode)} \end{cases}$	
$Hg(OH)_3^- + 2e^- \rightleftharpoons Hg(l) + 3OH^-$	0.231	
$Hg(OH)_2 + 2e^- \rightleftharpoons Hg(l) + 2OH^-$	0.206	-1.24
$Hg_2Br_2(s) + 2e^- \rightleftharpoons 2Hg(l) + 2Br^-$	0.140	
$HgO(s, yellow) + H_2O + 2e^- \rightleftharpoons Hg(l) + 2OH^-$	0.098 3	-1.125
$HgO(s, red) + H_2O + 2e^- \rightleftharpoons Hg(l) + 2OH^-$	0.097 7	$-1.120\ 6$

Molybdenum

Reaction	$E°$ (volts)	$dE°/dT$ (mV/K)
$MoO_4^{2-} + 2H_2O + 2e^- \rightleftharpoons MoO_2(s) + 4OH^-$	-0.818	-1.69
$MoO_4^{2-} + 4H_2O + 6e^- \rightleftharpoons Mo(s) + 8OH^-$	-0.926	-1.36
$MoO_2(s) + 2H_2O + 4e^- \rightleftharpoons Mo(s) + 4OH^-$	-0.980	-1.196

Neodymium

Reaction	$E°$ (volts)	$dE°/dT$ (mV/K)
$Nd^{3+} + 3e^- \rightleftharpoons Nd(s)$	-2.323	0.282

Neptunium

Reaction	$E°$ (volts)	$dE°/dT$ (mV/K)
$NpO_3^+ + 2H^+ + e^- \rightleftharpoons NpO_2^{2+} + H_2O$	2.04	
$NpO_2^{2+} + e^- \rightleftharpoons NpO_2^+$	1.236	0.058

(Continued)

Reaction	$E°$ (volts)	$dE°/dT$ (mV/K)
$NpO_2^+ + 4H^+ + e^- \rightleftharpoons Np^{4+} + 2H_2O$	0.567	-3.30
$Np^{4+} + e^- \rightleftharpoons Np^{3+}$	0.157	1.53
$Np^{3+} + 3e^- \rightleftharpoons Np(s)$	-1.768	0.18
Nickel		
$NiOOH(s) + 3H^+ + e^- \rightleftharpoons Ni^{2+} + 2H_2O$	2.05	-1.17
$Ni^{2+} + 2e^- \rightleftharpoons Ni(s)$	-0.236	0.146
$Ni(CN)_4^{2-} + e^- \rightleftharpoons Ni(CN)_3^{2-} + CN^-$	-0.401	
$Ni(OH)_2(s) + 2e^- \rightleftharpoons Ni(s) + 2OH^-$	-0.714	-1.02
Niobium		
$\frac{1}{2}Nb_2O_5(s) + H^+ + e^- \rightleftharpoons NbO_2(s) + \frac{1}{2}H_2O$	-0.248	-0.460
$\frac{1}{2}Nb_2O_5(s) + 5H^+ + 5e^- \rightleftharpoons Nb(s) + \frac{5}{2}H_2O$	-0.601	-0.381
$NbO_2(s) + 2H^+ + 2e^- \rightleftharpoons NbO(s) + H_2O$	-0.646	-0.347
$NbO_2(s) + 4H^+ + 4e^- \rightleftharpoons Nb(s) + 2H_2O$	-0.690	-0.361
Nitrogen		
$HN_3 + 3H^+ + 2e^- \rightleftharpoons N_2(g) + NH_4^+$	2.079	0.147
$N_2O(g) + 2H^+ + 2e^- \rightleftharpoons N_2(g) + H_2O$	1.769	-0.461
$2NO(g) + 2H^+ + 2e^- \rightleftharpoons N_2O(g) + H_2O$	1.587	-1.359
$NO^+ + e^- \rightleftharpoons NO(g)$	1.46	
$2NH_3OH^+ + H^+ + 2e^- \rightleftharpoons N_2H_5^+ + 2H_2O$	1.40	-0.60
$NH_3OH^+ + 2H^+ + 2e^- \rightleftharpoons NH_4^+ + H_2O$	1.33	-0.44
$N_2H_5^+ + 3H^+ + 2e^- \rightleftharpoons 2NH_4^+$	1.250	-0.28
$HNO_2 + H^+ + e^- \rightleftharpoons NO(g) + H_2O$	0.984	0.649
$NO_3^- + 4H^+ + 3e^- \rightleftharpoons NO(g) + 2H_2O$	0.955	0.028
$NO_3^- + 3H^+ + 2e^- \rightleftharpoons HNO_2 + H_2O$	0.940	-0.282
$NO_3^- + 2H^+ + e^- \rightleftharpoons \frac{1}{2}N_2O_4(g) + H_2O$	0.798	0.107
$N_2(g) + 8H^+ + 6e^- \rightleftharpoons 2NH_4^+$	0.274	-0.616
$N_2(g) + 5H^+ + 4e^- \rightleftharpoons N_2H_5^+$	-0.214	-0.78
$N_2(g) + 2H_2O + 4H^+ + 2e^- \rightleftharpoons 2NH_3OH^+$	-1.83	-0.96
$\frac{3}{2}N_2(g) + H^+ + e^- \rightleftharpoons HN_3$	-3.334	-2.141
Osmium		
$OsO_4(s) + 8H^+ + 8e^- \rightleftharpoons Os(s) + 4H_2O$	0.834	-0.458
$OsCl_6^{2-} + e^- \rightleftharpoons OsCl_6^{3-}$	0.85 1 F HCl	
Oxygen		
$OH + H^+ + e^- \rightleftharpoons H_2O$	2.56	-1.0
$O(g) + 2H^+ + 2e^- \rightleftharpoons H_2O$	2.430 1	$-1.148\ 4$
$O_3(g) + 2H^+ + 2e^- \rightleftharpoons O_2(g) + H_2O$	2.075	-0.489
$H_2O_2 + 2H^+ + 2e^- \rightleftharpoons 2H_2O$	1.763	-0.698
$HO_2 + H^+ + e^- \rightleftharpoons H_2O_2$	1.44	-0.7
$\frac{1}{2}O_2(g) + 2H^+ + 2e^- \rightleftharpoons H_2O$	1.229 1	$-0.845\ 6$
$O_2(g) + 2H^+ + 2e^- \rightleftharpoons H_2O_2$	0.695	-0.993
$O_2(g) + H^+ + e^- \rightleftharpoons HO_2$	-0.05	-1.3
Palladium		
$Pd^{2+} + 2e^- \rightleftharpoons Pd(s)$	0.915	0.12
$PdO(s) + 2H^+ + 2e^- \rightleftharpoons Pd(s) + H_2O$	0.79	-0.33
$PdCl_6^{4-} + 2e^- \rightleftharpoons Pd(s) + 6Cl^-$	0.615	
$PdO_2(s) + H_2O + 2e^- \rightleftharpoons PdO(s) + 2OH^-$	0.64	-1.2
Phosphorus		
$\frac{1}{4}P_4(s, white) + 3H^+ + 3e^- \rightleftharpoons PH_3(g)$	-0.046	-0.093
$\frac{1}{4}P_4(s, red) + 3H^+ + 3e^- \rightleftharpoons PH_3(g)$	-0.088	-0.030
$H_3PO_4 + 2H^+ + 2e^- \rightleftharpoons H_3PO_3 + H_2O$	-0.30	-0.36
$H_3PO_4 + 5H^+ + 5e^- \rightleftharpoons \frac{1}{4}P_4(s, white) + 4H_2O$	-0.402	-0.340
$H_3PO_3 + 2H^+ + 2e^- \rightleftharpoons H_3PO_2 + H_2O$	-0.48	-0.37
$H_3PO_2 + H^+ + e^- \rightleftharpoons \frac{1}{4}P_4(s) + 2H_2O$	-0.51	
Platinum		
$Pt^{2+} + 2e^- \rightleftharpoons Pt(s)$	1.18	-0.05
$PtO_2(s) + 4H^+ + 4e^- \rightleftharpoons Pt(s) + 2H_2O$	0.92	-0.36
$PtCl_4^{2-} + 2e^- \rightleftharpoons Pt(s) + 4Cl^-$	0.755	
$PtCl_6^{2-} + 2e^- \rightleftharpoons PtCl_4^{2-} + 2Cl^-$	0.68	
Plutonium		
$PuO_2^+ + e^- \rightleftharpoons PuO_2(s)$	1.585	0.39
$PuO_2^{2+} + 4H^+ + 2e^- \rightleftharpoons Pu^{4+} + 2H_2O$	1.000	$-1.615\ 1$
$Pu^{4+} + e^- \rightleftharpoons Pu^{3+}$	1.006	1.441
$PuO_2^{2+} + e^- \rightleftharpoons PuO_2^+$	0.966	0.03
$PuO_2(s) + 4H^+ + 4e^- \rightleftharpoons Pu(s) + 2H_2O$	-1.369	-0.38
$Pu^{3+} + 3e^- \rightleftharpoons Pu(s)$	-1.978	0.23

Reaction	$E°$ (volts)		$dE°/dT$ (mV/K)

Potassium

| $K^+ + e^- + Hg \rightleftharpoons K(in\ Hg)$ | -1.975 | | |
| $K^+ + e^- \rightleftharpoons K(s)$ | -2.936 | | -1.074 |

Praseodymium

| $Pr^{4+} + e^- \rightleftharpoons Pr^{3+}$ | 3.2 | | 1.4 |
| $Pr^{3+} + 3e^- \rightleftharpoons Pr(s)$ | -2.353 | | 0.291 |

Promethium

| $Pm^{3+} + 3e^- \rightleftharpoons Pm(s)$ | -2.30 | | 0.29 |

Radium

| $Ra^{2+} + 2e^- \rightleftharpoons Ra(s)$ | -2.80 | | -0.44 |

Rhenium

| $ReO_4^- + 2H^+ + e^- \rightleftharpoons ReO_3(s) + H_2O$ | 0.72 | | -1.17 |
| $ReO_4^- + 4H^+ + 3e^- \rightleftharpoons ReO_2(s) + 2H_2O$ | 0.510 | | -0.70 |

Rhodium

$Rh^{6+} + 3e^- \rightleftharpoons Rh^{3+}$	1.48	1 F $HClO_4$	
$Rh^{4+} + e^- \rightleftharpoons Rh^{3+}$	1.44	3 F H_2SO_4	
$RhCl_6^{2-} + e^- \rightleftharpoons RhCl_6^{3-}$	1.2		
$Rh^{3+} + 3e^- \rightleftharpoons Rh(s)$	0.76		0.4
$2Rh^{3+} + 2e^- \rightleftharpoons Rh_2^{4+}$	0.7		
$RhCl_6^{3-} + 3e^- \rightleftharpoons Rh(s) + 6Cl^-$	0.44		

Rubidium

| $Rb^+ + e^- + Hg \rightleftharpoons Rb(in\ Hg)$ | -1.970 | | |
| $Rb^+ + e^- \rightleftharpoons Rb(s)$ | -2.943 | | -1.140 |

Ruthenium

$RuO_4^- + 6H^+ + 3e^- \rightleftharpoons Ru(OH)_2^{2+} + 2H_2O$	1.53		
$Ru(dipyridyl)_3^{3+} + e^- \rightleftharpoons Ru(dipyridyl)_3^{2+}$	1.29		
$RuO_4(s) + 8H^+ + 8e^- \rightleftharpoons Ru(s) + 4H_2O$	1.032		-0.467
$Ru^{2+} + 2e^- \rightleftharpoons Ru(s)$	0.8		
$Ru^{3+} + 3e^- \rightleftharpoons Ru(s)$	0.60		
$Ru^{3+} + e^- \rightleftharpoons Ru^{2+}$	0.24		
$Ru(NH_3)_6^{3+} + e^- \rightleftharpoons Ru(NH_3)_6^{2+}$	0.214		

Samarium

| $Sm^{3+} + 3e^- \rightleftharpoons Sm(s)$ | -2.304 | | 0.279 |
| $Sm^{2+} + 2e^- \rightleftharpoons Sm(s)$ | -2.68 | | -0.28 |

Scandium

| $Sc^{3+} + 3e^- \rightleftharpoons Sc(s)$ | -2.09 | | 0.41 |

Selenium

$SeO_4^{2-} + 4H^+ + 2e^- \rightleftharpoons H_2SeO_3 + H_2O$	1.150		0.483
$H_2SeO_3 + 4H^+ + 4e^- \rightleftharpoons Se(s) + 3H_2O$	0.739		-0.562
$Se(s) + 2H^+ + 2e^- \rightleftharpoons H_2Se(g)$	-0.082		0.238
$Se(s) + 2e^- \rightleftharpoons Se^{2-}$	-0.67		-1.2

Silicon

$Si(s) + 4H^+ + 4e^- \rightleftharpoons SiH_4(g)$	-0.147		-0.196
$SiO_2(s, quartz) + 4H^+ + 4e^- \rightleftharpoons Si(s) + 2H_2O$	-0.990		-0.374
$SiF_6^{2-} + 4e^- \rightleftharpoons Si(s) + 6F^-$	-1.24		

Silver

$Ag^{2+} + e^- \rightleftharpoons Ag^+$	$\begin{cases} 2.000 & 4\ F\ HClO_4 \\ 1.989 & \\ 1.929 & 4\ F\ HNO_3 \end{cases}$		0.99
$Ag^{3+} + 2e^- \rightleftharpoons Ag^+$	1.9		
$AgO(s) + H^+ + e^- \rightleftharpoons \frac{1}{2}Ag_2O(s) + \frac{1}{2}H_2O$	1.40		
$Ag^+ + e^- \rightleftharpoons Ag(s)$	$0.799\ 3$		-0.989
$Ag_2C_2O_4(s) + 2e^- \rightleftharpoons 2Ag(s) + C_2O_4^{2-}$	0.465		
$AgN_3(s) + e^- \rightleftharpoons Ag(s) + N_3^-$	0.293		
$AgCl(s) + e^- \rightleftharpoons Ag(s) + Cl^-$	$\begin{cases} 0.222 & \\ 0.197 & \text{saturated KCl} \end{cases}$		
$AgBr(s) + e^- \rightleftharpoons Ag(s) + Br^-$	0.071		
$Ag(S_2O_3)_2^{3-} + e^- \rightleftharpoons Ag(s) + 2S_2O_3^{2-}$	0.017		
$AgI(s) + e^- \rightleftharpoons Ag(s) + I^-$	-0.152		
$Ag_2S(s) + H^+ + 2e^- \rightleftharpoons 2Ag(s) + SH^-$	-0.272		

Sodium

$Na^+ + e^- + Hg \rightleftharpoons Na(in\ Hg)$	-1.959		
$Na^+ + \frac{1}{2}H_2(g) + e^- \rightleftharpoons NaH(s)$	-2.367		-1.550
$Na^+ + e^- \rightleftharpoons Na(s)$	$-2.714\ 3$		-0.757

(Continued)

Reaction	$E°$ (volts)		$dE°/dT$ (mV/K)
Strontium			
$Sr^{2+} + 2e^- \rightleftharpoons Sr(s)$	−2.889		−0.237
Sulfur			
$S_2O_8^{2-} + 2e^- \rightleftharpoons 2SO_4^{2-}$	2.01		
$S_2O_6^{2-} + 4H^+ + 2e^- \rightleftharpoons 2H_2SO_3$	0.57		
$4SO_2 + 4H^+ + 6e^- \rightleftharpoons S_4O_6^{2-} + 2H_2O$	0.539		−1.11
$SO_2 + 4H^+ + 4e^- \rightleftharpoons S(s) + 2H_2O$	0.450		−0.652
$2H_2SO_3 + 2H^+ + 4e^- \rightleftharpoons S_2O_3^{2-} + 3H_2O$	0.40		
$S(s) + 2H^+ + 2e^- \rightleftharpoons H_2S(g)$	0.174		0.224
$S(s) + 2H^+ + 2e^- \rightleftharpoons H_2S(aq)$	0.144		−0.21
$S_4O_6^{2-} + 2H^+ + 2e^- \rightleftharpoons 2HS_2O_3^-$	0.10		−0.23
$5S(s) + 2e^- \rightleftharpoons S_5^{2-}$	−0.340		
$S(s) + 2e^- \rightleftharpoons S^{2-}$	−0.476		−0.925
$2S(s) + 2e^- \rightleftharpoons S_2^{2-}$	−0.50		−1.16
$2SO_3^{2-} + 3H_2O + 4e^- \rightleftharpoons S_2O_3^{2-} + 6OH^-$	−0.566		−1.06
$SO_3^{2-} + 3H_2O + 4e^- \rightleftharpoons S(s) + 6OH^-$	−0.659		−1.23
$SO_4^{2-} + 4H_2O + 6e^- \rightleftharpoons S(s) + 8OH^-$	−0.751		−1.288
$SO_4^{2-} + H_2O + 2e^- \rightleftharpoons SO_3^{2-} + 2OH^-$	−0.936		−1.41
$2SO_3^{2-} + 2H_2O + 2e^- \rightleftharpoons S_2O_4^{2-} + 4OH^-$	−1.130		−0.85
$2SO_4^{2-} + 2H_2O + 2e^- \rightleftharpoons S_2O_6^{2-} + 4OH^-$	−1.71		−1.00
Tantalum			
$Ta_2O_5(s) + 10H^+ + 10e^- \rightleftharpoons 2Ta(s) + 5H_2O$	−0.752		−0.377
Technetium			
$TcO_4^- + 2H_2O + 3e^- \rightleftharpoons TcO_2(s) + 4OH^-$	−0.366		−1.82
$TcO_4^- + 4H_2O + 7e^- \rightleftharpoons Tc(s) + 8OH^-$	−0.474		−1.46
Tellurium			
$TeO_3^{2-} + 3H_2O + 4e^- \rightleftharpoons Te(s) + 6OH^-$	−0.47		−1.39
$2Te(s) + 2e^- \rightleftharpoons Te_2^{2-}$	−0.84		
$Te(s) + 2e^- \rightleftharpoons Te^{2-}$	−0.90		−1.0
Terbium			
$Tb^{4+} + e^- \rightleftharpoons Tb^{3+}$	3.1		1.5
$Tb^{3+} + 3e^- \rightleftharpoons Tb(s)$	−2.28		0.350
Thallium			
$Tl^{3+} + 2e^- \rightleftharpoons Tl^+$	1.280		0.97
	0.77	1 F HCl	
	1.22	1 F H_2SO_4	
	1.23	1 F HNO_3	
	1.26	1 F HClO_4	
$Tl^+ + e^- + Hg \rightleftharpoons Tl(in\ Hg)$	−0.294		
$Tl^+ + e^- \rightleftharpoons Tl(s)$	−0.336		−1.312
$TlCl(s) + e^- \rightleftharpoons Tl(s) + Cl^-$	−0.557		
Thorium			
$Th^{4+} + 4e^- \rightleftharpoons Th(s)$	−1.826		0.557
Thulium			
$Tm^{3+} + 3e^- \rightleftharpoons Tm(s)$	−2.319		0.394
Tin			
$Sn(OH)_3^+ + 3H^+ + 2e^- \rightleftharpoons Sn^{2+} + 3H_2O$	0.142		
$Sn^{4+} + 2e^- \rightleftharpoons Sn^{2+}$	0.139	1 F HCl	
$SnO_2(s) + 4H^+ + 2e^- \rightleftharpoons Sn^{2+} + 2H_2O$	−0.094		−0.31
$Sn^{2+} + 2e^- \rightleftharpoons Sn(s)$	−0.141		−0.32
$SnF_6^{2-} + 4e^- \rightleftharpoons Sn(s) + 6F^-$	−0.25		
$Sn(OH)_6^{2-} + 2e^- \rightleftharpoons Sn(OH)_3^- + 3OH^-$	−0.93		
$Sn(s) + 4H_2O + 4e^- \rightleftharpoons SnH_4(g) + 4OH^-$	−1.316		−1.057
$SnO_2(s) + H_2O + 2e^- \rightleftharpoons SnO(s) + 2OH^-$	−0.961		−1.129
Titanium			
$TiO^{2+} + 2H^+ + e^- \rightleftharpoons Ti^{3+} + H_2O$	0.1		−0.6
$Ti^{3+} + e^- \rightleftharpoons Ti^{2+}$	−0.9		1.5
$TiO_2(s) + 4H^+ + 4e^- \rightleftharpoons Ti(s) + 2H_2O$	−1.076		0.365
$TiF_6^{2-} + 4e^- \rightleftharpoons Ti(s) + 6F^-$	−1.191		
$Ti^{2+} + 2e^- \rightleftharpoons Ti(s)$	−1.60		−0.16
Tungsten			
$W(CN)_8^{3-} + e^- \rightleftharpoons W(CN)_8^{4-}$	0.457		
$W^{6+} + e^- \rightleftharpoons W^{5+}$	0.26	12 F HCl	
$WO_3(s) + 6H^+ + 6e^- \rightleftharpoons W(s) + 3H_2O$	−0.091		−0.389

Reaction	$E°$ (volts)		$dE°/dT$ (mV/K)
$W^{5+} + e^- \rightleftharpoons W^{4+}$	-0.3	12 F HCl	
$WO_2(s) + 2H_2O + 4e^- \rightleftharpoons W(s) + 4OH^-$	-0.982		-1.197
$WO_4^{2-} + 4H_2O + 6e^- \rightleftharpoons W(s) + 8OH^-$	-1.060		-1.36
Uranium			
$UO_2^+ + 4H^+ + e^- \rightleftharpoons U^{4+} + 2H_2O$	0.39		-3.4
$UO_2^{2+} + 4H^+ + 2e^- \rightleftharpoons U^{4+} + 2H_2O$	0.273		-1.582
$UO_2^{2+} + e^- \rightleftharpoons UO_2^+$	0.16		0.2
$U^{4+} + e^- \rightleftharpoons U^{3+}$	-0.577		1.61
$U^{3+} + 3e^- \rightleftharpoons U(s)$	-1.642		0.16
Vanadium			
$VO_2^+ + 2H^+ + e^- \rightleftharpoons VO^{2+} + H_2O$	1.001		-0.901
$VO^{2+} + 2H^+ + e^- \rightleftharpoons V^{3+} + H_2O$	0.337		-1.6
$V^{3+} + e^- \rightleftharpoons V^{2+}$	-0.255		1.5
$V^{2+} + 2e^- \rightleftharpoons V(s)$	-1.125		-0.11
Xenon			
$H_4XeO_6 + 2H^+ + 2e^- \rightleftharpoons XeO_3 + 3H_2O$	2.38		0.0
$XeF_2 + 2H^+ + 2e^- \rightleftharpoons Xe(g) + 2HF$	2.2		
$XeO_3 + 6H^+ + 6e^- \rightleftharpoons Xe(g) + 3H_2O$	2.1		-0.34
Ytterbium			
$Yb^{3+} + 3e^- \rightleftharpoons Yb(s)$	-2.19		0.363
$Yb^{2+} + 2e^- \rightleftharpoons Yb(s)$	-2.76		-0.16
Yttrium			
$Y^{3+} + 3e^- \rightleftharpoons Y(s)$	-2.38		0.034
Zinc			
$ZnOH^+ + H^+ + 2e^- \rightleftharpoons Zn(s) + H_2O$	-0.497		0.03
$Zn^{2+} + 2e^- \rightleftharpoons Zn(s)$	-0.762		0.119
$Zn^{2+} + 2e^- + Hg \rightleftharpoons Zn(in\ Hg)$	-0.801		
$Zn(NH_3)_4^{2+} + 2e^- \rightleftharpoons Zn(s) + 4NH_3$	-1.04		
$ZnCO_3(s) + 2e^- \rightleftharpoons Zn(s) + CO_3^{2-}$	-1.06		
$Zn(OH)_3^- + 2e^- \rightleftharpoons Zn(s) + 3OH^-$	-1.183		
$Zn(OH)_4^{2-} + 2e^- \rightleftharpoons Zn(s) + 4OH^-$	-1.199		
$Zn(OH)_2(s) + 2e^- \rightleftharpoons Zn(s) + 2OH^-$	-1.249		-0.999
$ZnO(s) + H_2O + 2e^- \rightleftharpoons Zn(s) + 2OH^-$	-1.260		-1.160
$ZnS(s) + 2e^- \rightleftharpoons Zn(s) + S^{2-}$	-1.405		
Zirconium			
$Zr^{4+} + 4e^- \rightleftharpoons Zr(s)$	-1.45		0.67
$ZrO_2(s) + 4H^+ + 4e^- \rightleftharpoons Zr(s) + 2H_2O$	-1.473		-0.344

Reacting ions	log β_1	log β_2	log β_3	log β_4	log β_5	log β_6	Temperature (°C)	Ionic strength (μ, M)
Acetate, $CH_3CO_2^-$								
Ag^+	0.73	0.64					25	0
Ca^{2+}	1.24						25	0
Cd^{2+}	1.93	3.15					25	0
Cu^{2+}	2.23	3.63					25	0
Fe^{2+}	1.82						25	0.5
Fe^{3+}	3.38	7.1	9.7				20	0.1
Mg^{2+}	1.25						25	0
Mn^{2+}	1.40						25	0
Na^+	−0.18						25	0
Ni^{2+}	1.43						25	0
Zn^{2+}	1.28	2.09					20	0.1
Ammonia, NH_3								
Ag^+	3.31	7.23					25	0
Cd^{2+}	2.51	4.47	5.77	6.56			30	0
Co^{2+}	1.99	3.50	4.43	5.07	5.13	4.39	30	0
Cu^{2+}	3.99	7.33	10.06	12.03			30	0
Hg^{2+}	8.8	17.5	18.50	19.28			22	2
Ni^{2+}	2.67	4.79	6.40	7.47	8.10	8.01	30	0
Zn^{2+}	2.18	4.43	6.74	8.70			30	0
Cyanide, CN^-								
Ag^+		20	21				20	0
Cd^{2+}	5.18	9.60	13.92	17.11			25	?
Cu^+		24	28.6	30.3			25	0
Ni^{2+}				30			25	0
Tl^{3+}	13.21	26.50	35.17	42.61			25	4
Zn^{2+}		11.07	16.05	19.62			25	0
Ethylenediamine (1,2-diaminoethane), $H_2NCH_2CH_2NH_2$								
Ag^+	4.70	7.70	9.7				20	0.1
Cd^{2+}	5.69	10.36	12.80				25	0.5
Cu^{2+}	10.66	19.99					20	0
Hg^{2+}	14.3	23.3	23.2				25	0.1
Ni^{2+}	7.52	13.84	18.33				20	0
Zn^{2+}	5.77	10.83	14.11				20	0
Hydroxide, OH^-								
Ag^+	2.0	3.99					25	0
Al^{3+}	9.00	17.9	25.2	33.3			25	0
	log β_{22} = 20.3 log β_{43} = 42.1							
Ba^{2+}	0.64						25	0
Bi^{3+}	12.9	23.5	33.0	34.8			25	0
	log $\beta_{12\,6}$ = 165.3 (μ = 1)							
Be^{2+}	8.6	14.4	18.8	18.6			25	0
	log β_{12} = 10.82 (μ = 0.1) log β_{33} = 32.54 (μ = 0.1) log β_{65} = 66.24 (μ = 3) log β_{86} = 85 (μ = 0)							
Ca^{2+}	1.30						25	0
Cd^{2+}	3.9	7.7	10.3 (μ = 3)	12.0 (μ = 3)			25	0
	log β_{12} = 4.6 log β_{44} = 23.2							
Ce^{3+}	4.9						25	3
	log β_{22} = 12.4 log β_{53} = 35.1							
Co^{2+}	4.3	9.2	10.5	9.7			25	0
	log β_{12} = 3 log β_{44} = 25.5							
Co^{3+}	13.52						25	3

*The overall (cumulative) formation constant, β_n, is the equilibrium constant for the reaction $M + nL \rightleftharpoons ML_n$: $\beta_n = [ML_n]/([M][L]^n)$. β_n is related to stepwise formation constants (K_i) by $\beta_n = K_1 K_2 \ldots K_n$ (Box 6-2). β_{nm} is the cumulative formation constant for the reaction $mM + nL \rightleftharpoons M_m L_n$: $\beta_{nm} = [M_m L_n]/([M]^m[L]^n)$. The subscript n refers to the ligand and m refers to the metal. Data from L. G. Sillén and A. E. Martell, Stability Constants of Metal-Ion Complexes (London: The Chemical Society. Special Publications No. 17 and 25, 1964 and 1971); and A. E. Martell, R. M. Smith, and R. J. Motekaitis, NIST Critical Stability Constants of Metal Complexes Database 46 (Gaithersburg, MD: National Institute of Standards and Technology, 2001).

Reacting ions	$\log \beta_1$	$\log \beta_2$	$\log \beta_3$	$\log \beta_4$	$\log \beta_5$	$\log \beta_6$	Temperature (°C)	Ionic strength (μ, M)
Cr^{2+}	8.5						25	1
Cr^{3+}	10.34 (μ = 0.1)	17.3					25	0
	$\log \beta_{22} = 24.0 \ (\mu = 1)$	$\log \beta_{43} = 37.0 \ (\mu = 1)$	$\log \beta_{44} = 50.7 \ (\mu = 2)$					
Cu^{2+}	6.5	11.8	14.5 (μ = 1)	15.6 (μ = 1)			25	0
	$\log \beta_{12} = 8.2 \ (\mu = 3)$	$\log \beta_{22} = 17.4$	$\log \beta_{43} = 35.2$					
Fe^{2+}	4.6	7.5	13	10			25	0
Fe^{3+}	11.81	23.4		34.4			25	0
	$\log \beta_{22} = 25.14$	$\log \beta_{43} = 49.7$						
Ga^{3+}	11.4	22.1	31.7	39.4			25	0
Gd^{3+}	4.9						25	3
	$\log \beta_{22} = 14.14$							
Hf^{4+}	13.7				52.8		25	0
Hg_2^{2+}	8.7						25	0.5
	$\log \beta_{12} = 11.5 \ (\mu = 3)$	$\log \beta_{45} = 48.24 \ (\mu = 3)$						
Hg^{2+}	10.60	21.8	20.9				25	0
	$\log \beta_{12} = 10.7$	$\log \beta_{33} = 35.6$						
In^{3+}	10.1	20.2	29.5	33.8			25	0
	$\log \beta_{22} = 23.2 \ (\mu = 3)$	$\log \beta_{44} = 47.8 \ (\mu = 0.1)$	$\log \beta_{64} = 43.1 \ (\mu = 0.1)$					
La^{3+}	5.5						25	0
	$\log \beta_{22} = 10.7 \ (\mu = 3)$	$\log \beta_{95} = 38.4$						
Li^+	0.36						25	0
Mg^{2+}	2.6	−0.3 (μ = 3)					25	0
	$\log \beta_{44} = 18.1 \ (\mu = 3)$							
Mn^{2+}	3.4			7.7			25	0
	$\log \beta_{12} = 6.8$	$\log \beta_{32} = 18.1$						
Na^+	0.1						25	0
Ni^{2+}	4.1	9	12				25	0
	$\log \beta_{12} = 4.7 \ (\mu = 1)$	$\log \beta_{44} = 28.3$						
Pb^{2+}	6.4	10.9	13.9				25	0
	$\log \beta_{12} = 7.6$	$\log \beta_{43} = 32.1$	$\log \beta_{44} = 36.0$	$\log \beta_{86} = 68.4$				
Pd^{2+}	13.0	25.8					25	0
Rh^{3+}	10.67						25	2.5
Sc^{3+}	9.7	18.3	25.9	30			25	0
	$\log \beta_{22} = 22.0$	$\log \beta_{53} = 53.8$						
Sn^{2+}	10.6	20.9	25.4				25	0
	$\log \beta_{22} = 23.2$	$\log \beta_{43} = 49.1$						
Sr^{2+}	0.82						25	0
Th^{4+}	10.8	21.1		41.1 (μ = 3)			25	0
	$\log \beta_{22} = 23.6 \ (\mu = 3)$	$\log \beta_{32} = 33.8 \ (\mu = 3)$	$\log \beta_{53} = 53.7 \ (\mu = 3)$					
Ti^{3+}	12.7						25	0
	$\log \beta_{22} = 24.6 \ (\mu = 1)$							
Tl^+	0.79	−0.8 (μ = 3)					25	0
Tl^{3+}	13.4	26.6	38.7	41.0			25	0
U^{4+}	13.4						25	0
VO^{2+}	8.3						25	0
	$\log \beta_{22} = 21.3$							
Y^{3+}	6.3						25	0
	$\log \beta_{22} = 13.8$	$\log \beta_{53} = 38.4$						
Zn^{2+}	5.0	10.2	13.9	15.5			25	0
	$\log \beta_{12} = 5.5 \ (\mu = 3)$	$\log \beta_{44} = 27.9 \ (\mu = 3)$						
Zr^{4+}	14.3				54.0		25	0
	$\log \beta_{43} = 55.4$	$\log \beta_{84} = 106.0$						
Nitrilotriacetate, $N(CH_2CO_2^-)_3$								
Ag^+	5.16						20	0.1
Al^{3+}	9.5						20	0.1
Ba^{2+}	4.83						20	0.1
Ca^{2+}	6.46						20	0.1
Cd^{2+}	10.0	14.6					20	0.1

(Continued)

Reacting ions	$\log \beta_1$	$\log \beta_2$	$\log \beta_3$	$\log \beta_4$	$\log \beta_5$	$\log \beta_6$	Temperature (°C)	Ionic strength (μ, M)
Co^{2+}	10.0	13.9					20	0.1
Cu^{2+}	11.5	14.8					20	0.1
Fe^{3+}	15.91	24.61					20	0.1
Ga^{3+}	13.6	21.8					20	0.1
In^{3+}	16.9						20	0.1
Mg^{2+}	5.46						20	0.1
Mn^{2+}	7.4						20	0.1
Ni^{2+}	11.54						20	0.1
Pb^{2+}	11.47						20	0.1
Tl^{+}	4.75						20	0.1
Zn^{2+}	10.44						20	0.1

Oxalate, $^-O_2CCO_2^-$

Reacting ions	$\log \beta_1$	$\log \beta_2$	$\log \beta_3$	$\log \beta_4$	$\log \beta_5$	$\log \beta_6$	Temperature (°C)	Ionic strength (μ, M)
Al^{3+}			15.60				20	0.1
Ba^{2+}	2.31						18	0
Ca^{2+}	1.66	2.69					25	1
Cd^{2+}	3.71						20	0.1
Co^{2+}	4.69	7.15					25	0
Cu^{2+}	6.23	10.27					25	0
Fe^{3+}	7.54	14.59	20.00				?	0.5
Ni^{2+}	5.16	6.5					25	0
Zn^{2+}	4.85	7.6					25	0

1,10-Phenanthroline,

Reacting ions	$\log \beta_1$	$\log \beta_2$	$\log \beta_3$	$\log \beta_4$	$\log \beta_5$	$\log \beta_6$	Temperature (°C)	Ionic strength (μ, M)
Ag^{+}	5.02	12.07					25	0.1
Ca^{2+}	0.7						20	0.1
Cd^{2+}	5.17	10.00	14.25				25	0.1
Co^{2+}	7.02	13.72	20.10				25	0.1
Cu^{2+}	8.82	15.39	20.41				25	0.1
Fe^{2+}	5.86	11.11	21.14				25	0.1
Fe^{3+}			14.10				25	0.1
Hg^{2+}		19.65	23.4				20	0.1
Mn^{2+}	4.50	8.65	12.70				25	0.1
Ni^{2+}	8.0	16.0	23.9				25	0.1
Zn^{2+}	6.30	11.95	17.05				25	0.1

Logarithm of the Formation Constant for the Reaction $M(aq) + L(aq) \rightleftharpoons ML(aq)$*

M	F⁻	Cl⁻	Br⁻	I⁻	NO₃⁻	ClO₄⁻	IO₃⁻	SCN⁻	SO₄²⁻	CO₃²⁻
Li^+	0.23	—	—	—	—	—	—	—	0.64	—
Na^+	−0.2	−0.5	—	—	−0.55	−0.7	−0.4	—	0.72	1.27
K^+	−1.2[a]	−0.5	—	−0.4	−0.19	−0.03	−0.27	—	0.85	—
Rb^+	—	−0.4	—	0.04	−0.08	0.15	−0.19	—	0.60	—
Cs^+	—	−0.2	0.03	−0.03	−0.02	0.23	−0.11	—	0.3	—
Ag^+	0.4	3.31	4.6	6.6	−0.1	−0.1	0.63	4.8	1.3	—
$(CH_3)_4N^+$	—	0.04	0.16	0.31	—	0.27	—	—	—	—
Mg^{2+}	2.05	0.6	−1.4[d]	—	—	—	0.72	−0.9[d]	2.23	2.92
Ca^{2+}	0.63	0.2[b]	—	—	0.5	—	0.89	—	2.36	3.20
Sr^{2+}	0.14	−0.22[a]	—	—	0.6	—	1.00	—	2.2	2.81
Ba^{2+}	−0.20	−0.44[a]	—	—	0.7	—	1.10	—	2.2	2.71
Zn^{2+}	1.3	0.4	−0.07	−1.5[d]	0.4	—	—	1.33	2.34	4.76
Cd^{2+}	1.2	1.98	2.15	2.28	0.5	—	0.51[a]	1.98	2.46	3.49[b]
Hg_2^{2+}	—	—	—	—	0.08[f]	—	—	—	1.30[f]	—
Hg^{2+}	1.03[f]	7.30	9.07[f]	12.87[f]	0.11[d]	—	—	9.64	1.34[f]	11.0[f]
Sn^{2+}	—	1.64	1.16	0.70[e]	0.44[a]	—	—	0.83[a]	—	—
Y^{3+}	4.81	−0.1[a]	−0.15[a]	—	—	—	—	−0.07[f]	3.47	8.2
La^{3+}	3.60	−0.1[a]	—	—	0.1[a]	—	—	0.12[a]	3.64	5.6[d]
In^{3+}	4.65	2.32[c]	2.01[c]	1.64[c]	0.18	—	—	3.15	1.85[a]	—

*Unless otherwise indicated, conditions are 25°C and $\mu = 0$.

a. $\mu = 1$ M; b. $\mu = 0.1$ M; c. $\mu = 0.7$ M; d. $\mu = 3$ M; e. $\mu = 4$ M; f. $\mu = 0.5$ M.

SOURCE: A. E. Martell, R. M. Smith, and R. J. Motekaitis, NIST Critical Stability Constants of Metal Complexes Database 46 (Gaithersburg, MD: National Institute of Standards and Technology, 2001).

The table in this appendix recommends primary standards for many elements. An *elemental assay standard* must contain a known amount of the desired element. A *matrix matching standard* must contain extremely low concentrations of undesired impurities, such as the analyte. If you want to prepare 10 ppm Fe in 10% aqueous NaCl, the NaCl must not contain significant Fe impurity, because the impurity would then have a higher concentration than the deliberately added Fe.

Rather than using compounds in the table, many people purchase certified solutions whose concentrations are traceable to standards from the National Institute of Standards and Technology (NIST or other national institutes of standards). By *NIST traceable,* we mean that the solution has been prepared from a standard material certified by NIST or that it has been compared with an NIST standard by a reliable analytical procedure.

Manufacturers frequently indicate elemental purity by some number of 9s. This deceptive nomenclature is based on the measurement of certain impurities. For example, 99.999% (five 9s) pure Al is certified to contain $\leq 0.001\%$ *metallic* impurities, based on the analysis of other metals present. However, C, H, N, and O are not measured. The Al might contain 0.1% Al_2O_3 and still be "five 9s pure." For the most accurate work, the dissolved gas content in solid elements may also be a source of error.

Carbonates, oxides, and other compounds may not possess the expected stoichiometry. For example, TbO_2 will have a high Tb content if some Tb_4O_7 is present. Ignition in an O_2 atmosphere may be helpful, but the final stoichiometry is never guaranteed. Carbonates may contain traces of bicarbonate, oxide, and hydroxide. Firing in a CO_2 atmosphere may improve the stoichiometry. Sulfates may contain some HSO_4^-. Some chemical analysis may be required to ensure that you know what you are really working with.

Most metal standards dissolve in 6 M HCl or HNO_3 or a mixture of the two, possibly with heating. Frothing accompanies dissolution of metals or carbonates in acid, so vessels should be loosely covered by a watchglass or Teflon lid to prevent loss of material. Concentrated HNO_3 (16 M) may *passivate* some metals, forming an insoluble oxide coat that prevents dissolution. If you have a choice between using a bulk element or a powder as standards, the bulk form is preferred because it has a smaller surface area on which oxides can form and impurities can be adsorbed. After a pure metal to be used as a standard is cut, it should be etched ("pickled") in a dilute solution of the acid in which it will be dissolved to remove surface oxides and contamination from the cutter. The metal is then washed well with water and dried in a vacuum desiccator.

Dilute solutions of metals are best prepared in Teflon or plastic vessels, because glass is an ion exchanger that can replace analyte species. Specially cleaned glass vials are commercially available for trace organic analysis. Volumetric dilutions are rarely more accurate than 0.1%, so gravimetric dilutions are required for greater accuracy. Of course, weights should be corrected for buoyancy with Equation 2-1. Evaporation of standard solutions is a source of error that is prevented if the mass of the reagent bottle is recorded after each use. If the mass changes between uses, the contents are evaporating.

Calibration standards

Element	Source[a]	Purity	Comments[b]
Li	SRM 924 (Li_2CO_3)	100.05 ± 0.02%	E; dry at 200°C for 4 h.
	Li_2CO_3	five–six 9s	M; purity calculated from impurities. Stoichiometry unknown.
Na	SRM 919 or 2201 (NaCl)	99.9%	E; dry for 24 h over $Mg(ClO_4)_2$.
	Na_2CO_3	three 9s	M; purity based on metallic impurities.
K	SRM 918 (KCl)	99.9%	E; dry for 24 h over $Mg(ClO_4)_2$.
	SRM 999 (KCl)	52.435 ± 0.004% K	E; ignite at 500°C for 4 h.
	K_2CO_3	five–six 9s	M; purity based on metallic impurities.
Rb	SRM 984 (RbCl)	99.90 ± 0.02%	E; hygroscopic. Dry for 24 h over $Mg(ClO_4)_2$.
	Rb_2CO_3		M
Cs	Cs_2CO_3		M
Be	metal	three 9s	E, M; purity based on metallic impurities.
Mg	SRM 929	100.1 ± 0.4%	E; magnesium gluconate clinical standard.
		5.403 ± 0.022% Mg	Dry for 24 h over $Mg(ClO_4)_2$.
	metal	five 9s	E; purity based on metallic impurities.
Ca	SRM 915 ($CaCO_3$)	three 9s	E; use without drying.
	$CaCO_3$	five 9s	E, M; dry at 200°C for 4 h in CO_2. User must determine stoichiometry.
Sr	SRM 987 ($SrCO_3$)	99.8%	E; ignite to establish stoichiometry. Dry at 110°C for 1 h.
	$SrCO_3$	five 9s	M; up to 1% off stoichiometry. Ignite to establish stoichiometry. Dry at 200°C for 4 h.
Ba	$BaCO_3$	four–five 9s	M; dry at 200°C for 4 h.

Transition metals: Use pure metals (usually $\geq$ four 9s) for elemental and matrix standards. Assays are based on impurities and do not include dissolved gases.

Lanthanides: Use pure metals (usually $\geq$ four 9s) for elemental standards and oxides as matrix standards. Oxides may be difficult to dry and stoichiometry is not certain.

a. SRM is the National Institute of Standards and Technology designation for a Standard Reference Material.

b. E means elemental assay standard; M means matrix matching standard.

SOURCE: J. R. Moody, R. R. Greenberg, K. W. Pratt, and T. C. Rains, "Recommended Inorganic Chemicals for Calibration," Anal. Chem. **1988,** 60, 1203A.

Calibration standards (*continued*)

Element	Source[a]	Purity	Comments[b]
B	SRM 951 (H_3BO_3)	100.00 ± 0.01	E; expose to room humidity ($\sim$35%) for 30 min before use.
Al	metal	five 9s	E, M; SRM 1257 Al metal available.
Ga	metal	five 9s	E, M; SRM 994 Ga metal available.
In	metal	five 9s	E, M
Tl	metal	five 9s	E, M; SRM 997 Tl metal available.
C			No recommendation.
Si	metal	six 9s	E, M; SRM 990 SiO_2 available.
Ge	metal	five 9s	E, M
Sn	metal	six 9s	E, M; SRM 741 Sn metal available.
Pb	metal	five 9s	E, M; several SRMs available.
N	NH_4Cl	six 9s	E; can be prepared from $HCl + NH_3$.
	N_2	>three 9s	E
	HNO_3	six 9s	M; contaminated with NO_x. Purity based on impurities.
P	SRM 194 ($NH_4H_2PO_4$)	three 9s	E
	P_2O_5	five 9s	E, M; difficult to keep dry.
	H_3PO_4	four 9s	E; must titrate 2 hydrogens to be certain of stoichiometry.
As	metal	five 9s	E, M
	SRM 83d (As_2O_3)	$99.992\,6 \pm 0.003\,0\%$	Redox standard. As assay not ensured.
Sb	metal	four 9s	E, M
Bi	metal	five 9s	E, M
O	H_2O	eight 9s	E, M; contains dissolved gases.
	O_2	>four 9s	E
S	element	six 9s	E, M; difficult to dry. Other sources are H_2SO_4, Na_2SO_4, and K_2SO_4. Stoichiometry must be proved (e.g., no SO_3^{2-} present).
Se	metal	five 9s	E, M; SRM 726 Se metal available.
Te	metal	five 9s	E, M
F	NaF	four 9s	E, M; no good directions for drying.
Cl	NaCl	four 9s	E, M; dry for 24 h over $Mg(ClO_4)_2$. Several SRMs (NaCl and KCl) available.
Br	KBr	four 9s	E, M; need to dry and demonstrate stoichiometry.
	Br_2	four 9s	E
I	sublimed I_2	six 9s	E
	KI	three 9s	E, M
	KIO_3	three 9s	Stoichiometry not ensured.

SOLUTIONS TO EXERCISES

Chapter 1

1-A. **(a)** $\dfrac{(25.00\ \text{mL})(0.791\ 4\ \text{g/mL})/(32.042\ \text{g/mol})}{0.500\ 0\ \text{L}} = 1.235\ \text{M}$

(b) 500.0 mL of solution weighs $(1.454\ \text{g/mL}) \times (500.0\ \text{mL}) = 727.0\ \text{g}$ and contains 25.00 mL ($= 19.78\ \text{g}$) of methanol. The mass of chloroform in 500.0 mL must be $727.0 - 19.78 = 707.2\ \text{g}$. The molality of methanol is

$$\text{Molality} = \frac{\text{mol methanol}}{\text{kg chloroform}}$$
$$= \frac{(19.78\ \text{g})/(32.042\ \text{g/mol})}{0.707\ 2\ \text{kg}} = 0.872\ 9\ m$$

If you keep all the numbers in your calculator, the answer is $0.873\ 1\ m$. You will find small discrepancies due to intermediate round-offs in many answers in this book.

1-B. **(a)**

$$\left(\frac{48.0\ \text{g HBr}}{100.0\ \text{g solution}}\right)\left(1.50\ \frac{\text{g solution}}{\text{mL solution}}\right) = \left(\frac{0.720\ \text{g HBr}}{\text{mL solution}}\right) = \left(\frac{720\ \text{g HBr}}{\text{L solution}}\right)$$

$$\text{Formal concentration} = \frac{720\ \text{g HBr/L}}{80.912\ \text{g/mol}} = 8.90\ \text{M}$$

(b) $\dfrac{36.0\ \text{g HBr}}{0.480\ \text{g HBr/g solution}} = 75.0\ \text{g solution}$

(c) $233\ \text{mmol} = 0.233\ \text{mol}$

$\dfrac{0.233\ \text{mol}}{8.90\ \text{mol/L}} = 0.026\ 2\ \text{L} = 26.2\ \text{mL}$

(d) $M_{\text{conc}} \cdot V_{\text{conc}} = M_{\text{dil}} \cdot V_{\text{dil}}$

$(8.90\ \text{M}) \cdot (x\ \text{mL}) = (0.160\ \text{M}) \cdot (250\ \text{mL}) \Rightarrow x = 4.49\ \text{mL}$

1-C. Each mol of $Ca(NO_3)_2$ (FM 164.088) contains 2 mol NO_3^- (FM 62.005), so the fraction of mass that is nitrate is

$$\left(\frac{2\ \text{mol NO}_3^-}{\text{mol Ca(NO}_3)_2}\right)\left(\frac{62.005\ \text{g NO}_3^-/\text{mol NO}_3^-}{164.088\ \text{g Ca(NO}_3)_2/\text{mol Ca(NO}_3)_2}\right)$$
$$= 0.755\ 7\ \frac{\text{g NO}_3^-}{\text{g Ca(NO}_3)_2}$$

If the dissolved $Ca(NO_3)_2$ has a concentration of 12.6 ppm, the concentration of dissolved NO_3^- is $(0.755\ 7)(12.6\ \text{ppm}) = 9.52\ \text{ppm}$.

1-D. **(a)** Formula mass of ascorbic acid
= 6(atomic mass of C) + 8(atomic mass of H) + 6(atomic mass of O)
= $6(12.010\ 7) + 8(1.007\ 94) + 6(15.999\ 4) = 176.124\ \text{g/mol}$

(b) $\dfrac{0.197\ 0\ \text{g of ascorbic acid}}{176.124\ \text{g/mol}} = 1.118\ 5\ \text{mmol}$

Molarity of $I_3^- = 1.118\ 5\ \text{mmol}/29.41\ \text{mL}$
$= 0.038\ 03\ \text{M}$

(c) $31.63\ \text{mL of } I_3^- = 1.203\ \text{mmol of } I_3^-$
$= 1.203\ \text{mmol of ascorbic acid}$
$= 0.211\ 9\ \text{g} = 49.94\%$ of the tablet

1-E. **(a)** $\dfrac{0.824\ \text{g acid}}{204.221\ \text{g/mol}} = 4.03_{48}\ \text{mmol}$. This many mmol of NaOH is

contained in 0.038 314 kg of NaOH solution

$\Rightarrow \text{concentration} = \dfrac{4.03_{48} \times 10^{-3}\ \text{mol NaOH}}{0.038\ 314\ \text{kg solution}}$

$= 0.105_{31}\ \text{mol/kg solution}$

(b) mol NaOH $= (0.057\ 911\ \text{kg})(0.105_{31}\ \text{mol/kg}) = 6.09_{86}\ \text{mmol}$
Because 2 mol NaOH react with 1 mol H_2SO_4,

$$[H_2SO_4] = \frac{\frac{1}{2}(6.09_{86}\ \text{mmol})}{10.00\ \text{mL}} = 0.305\ \frac{\text{mmol}}{\text{mL}} = 0.305\ \text{M}$$

Chapter 2

2-A. **(a)** At 15°C, water density = 0.999 102 6 g/mL.

$$m = \frac{(5.397\ 4\ \text{g})\left(1 - \dfrac{0.001\ 2\ \text{g/mL}}{8.0\ \text{g/mL}}\right)}{\left(1 - \dfrac{0.001\ 2\ \text{g/mL}}{0.999\ 102\ 6\ \text{g/mL}}\right)} = 5.403\ 1\ \text{g}$$

(b) At 25°C, water density = 0.997 047 9 g/mL and $m = 5.403\ 1$ g.

2-B. Use Equation 2-1 with $m' = 0.296\ 1$ g, $d_a = 0.001\ 2$ g/mL, $d_w = 8.0$ g/mL, and $d = 5.24$ g/mL $\Rightarrow m = 0.296\ 3$ g.

2-C. $\dfrac{c'}{d'} = \dfrac{c}{d}$

Let the primes stand for 16°C:

$$\Rightarrow \frac{c'\ \text{at } 16°C}{0.998\ 946\ 0\ \text{g/mL}} = \frac{0.051\ 38\ \text{M}}{0.997\ 299\ 5\ \text{g/mL}}$$
$$\Rightarrow c'\ \text{at } 16°C = 0.051\ 46\ \text{M}$$

2-D. Column 3 of Table 2-7 tells us that water occupies 1.003 3 mL/g at 22°C. Therefore, $(15.569\ \text{g}) \times (1.003\ 3\ \text{mL/g}) = 15.620\ \text{mL}$.

Chapter 3

3-A. **(a)** $[12.41\ (\pm 0.09) \div 4.16\ (\pm 0.01)] \times 7.068\ 2\ (\pm 0.000\ 4)$

$= \dfrac{12.41\ (\pm 0.725\%) \times 7.068\ 2\ (\pm 0.005\ 7\%)}{4.16\ (\pm 0.240\%)}$

$= 21.086\ (\pm 0.764\%)$ (because $\sqrt{0.725^2 + 0.005\ 7^2 + 0.240^2} = 0.764$)

$= 21.0_9\ (\pm 0.1_6)$ or $21.1\ (\pm 0.2)$

Relative uncertainty $= \dfrac{0.1_6}{21.0_9} \times 100 = 0.8\%$

(b) $[3.26\ (\pm 0.10) \times 8.47\ (\pm 0.05)] - 0.18\ (\pm 0.06)$
$= [3.26\ (\pm 3.07\%) \times 8.47\ (\pm 0.59\%)] - 0.18\ (\pm 0.06)$
$= [27.612\ (\pm 3.13\%)] - 0.18\ (\pm 0.06)$
$= [27.612\ (\pm 0.864)] - 0.18\ (\pm 0.06)$
$= [27.4_3\ (\pm 0.8_7)]$ or $27.4\ (\pm 0.9)$; relative uncertainty $= 3._2\%$

(c) $6.843\ (\pm 0.008) \times 10^4 \div \underbrace{[2.09\ (\pm 0.04) - 1.63\ (\pm 0.01)]}_{\text{Combine absolute uncertainties}}$

$= 6.843\ (\pm 0.008) \times 10^4 \div \underbrace{[0.46\ (\pm 0.041\ 2)]}_{\text{Combine relative uncertainties}}$

$= 6.843\ (\pm 0.117\%) \times 10^4 \div [0.46\ (\pm 8.96\%)] = 1.49\ (\pm 8.96\%) \times 10^5$
$= 1.49\ (\pm 0.1_3) \times 10^5$; relative uncertainty $= 9._0\%$

(d) $\%e_y = \dfrac{1}{2}\%e_x = \dfrac{1}{2}\left(\dfrac{0.08}{3.24} \times 100\right) = 1.235\%$

$(3.24 \pm 0.08)^{1/2} = 1.80 \pm 1.235\%$
$= 1.80 \pm 0.02_2\ (\pm 1._2\%)$

(e) $\%e_y = 4\%e_x = 4\left(\dfrac{0.08}{3.24} \times 100\right) = 9.877\%$

$(3.24 \pm 0.08)^4 = 110.20 \pm 9.877\%$
$= 1.1_0\ (\pm 0.1_1) \times 10^2\ (\pm 9._9\%)$

(f) $e_y = 0.434\ 29\ \dfrac{e_x}{x} = 0.434\ 29\left(\dfrac{0.08}{3.24}\right) = 0.010\ 7$

$\log(3.24 \pm 0.08) = 0.510\ 5 \pm 0.010\ 7$
$= 0.51 \pm 0.01\ (\pm 2._1\%)$

(g) $\dfrac{e_y}{y} = 2.302\ 6\ e_x = 2.302\ 6\ (0.08) = 0.184$

$10^{3.24 \pm 0.08} = 1.74 \times 10^3 \pm 18.4\%$
$= 1.7_4\ (\pm 0.3_2) \times 10^3\ (\pm 18\%)$

3-B. **(a)** 2.000 L of 0.169 M NaOH (FM = 39.997) requires 0.338 mol = 13.52 g NaOH.

$$\frac{13.52 \text{ g NaOH}}{0.534 \text{ g NaOH/g solution}} = 25.32 \text{ g solution}$$

$$\frac{25.32 \text{ g solution}}{1.52 \text{ g solution/mL solution}} = 16.6_6 \text{ mL}$$

(b) Molarity =

$$\frac{[16.66\,(\pm0.10)\text{ mL}]\left[1.52\,(\pm0.01)\dfrac{\text{g solution}}{\text{mL}}\right]\left[0.534\,(\pm0.004)\dfrac{\text{g NaOH}}{\text{g solution}}\right]}{\left(39.997\,1\dfrac{\text{g NaOH}}{\text{mol}}\right)(2.000\text{ L})}$$

Because the relative errors in formula mass and final volume are negligible (≈ 0), we can write

$$\text{Relative error in molarity} = \sqrt{\left(\frac{0.10}{16.66}\right)^2 + \left(\frac{0.01}{1.52}\right)^2 + \left(\frac{0.004}{0.534}\right)^2} = 1.16\%$$

Molarity = 0.169 ($\pm$0.002)

3-C. $0.050\,0\,(\pm2\%)$ mol =

$$\frac{[4.18\,(\pm x)\text{ mL}]\left[1.18\,(\pm0.01)\dfrac{\text{g solution}}{\text{mL}}\right]\left[0.370\,(\pm0.005)\dfrac{\text{g NaCl}}{\text{g solution}}\right]}{\left(36.461\dfrac{\text{g HCl}}{\text{mol}}\right)}$$

Error analysis:

$$(0.02)^2 = \left(\frac{x}{4.18}\right)^2 + \left(\frac{0.01}{1.18}\right)^2 + \left(\frac{0.005}{0.370}\right)^2$$

$$x = 0.05 \text{ mL}$$

3-D. Atomic masses from periodic table:

N: $14.006\,7 \pm 0.000\,2$

H: $1.007\,94 \pm 0.000\,07$

Standard uncertainties from rectangular distribution:

N: $0.000\,2/\sqrt{3} = 0.000\,12$

H: $0.000\,07/\sqrt{3} = 0.000\,040$

$n \times$ (atomic mass $\pm$ standard uncertainty):

N: $14.006\,7 \pm 0.000\,12 = 14.006\,7 \pm 0.000\,12$

3H: $3(1.007\,94 \pm 0.000\,040) = 3.023\,82 \pm 0.000\,12$

NH_3: $17.030\,52 \pm \sqrt{0.000\,12^2 + 0.000\,12^2}$
$= 17.030\,52 \pm 0.000\,17 = 17.030\,5 \pm 0.000\,2$

$$\text{Percent relative uncertainty} = 100 \times \frac{0.000\,17}{17.030\,5} = 0.001\,0\%$$

Chapter 4

4-A. Mean = $\frac{1}{5}(116.0 + 97.9 + 114.2 + 106.8 + 108.3)$
$= 108.6_4$

$$\text{Standard deviation} = \sqrt{\frac{(116.0 - 108.6_4)^2 + \cdots + (108.3 - 108.6_4)^2}{5 - 1}}$$
$$= 7.1_4$$

Range = $116.0 - 97.9 = 18.1$

$$90\% \text{ confidence interval} = 108.6_4 \pm \frac{(2.132)(7.1_4)}{\sqrt{5}} = 108.6_4 \pm 6.8_1$$

$G_{\text{calculated}} = |97.9 - 108.6_4| / 7.1_4 = 1.50$
$G_{\text{table}} = 1.672$ for five measurements
Because $G_{\text{calculated}} < G_{\text{table}}$, we retain 97.9.

4-B.

	A	B	C	D
1	Computing standard deviation			
2				
3		Data = x	x − mean	(x-mean)^2
4		17.4	−0.44	0.1936
5		18.1	0.26	0.0676
6		18.2	0.36	0.1296
7		17.9	0.06	0.0036
8		17.6	−0.24	0.0576
9	sum =	89.2		0.452
10	mean =	17.84		
11	std dev =	0.3362		
12				
13	Formulas:	B9 = B4+B5+B6+B7+B8		
14		B10 = B9/5		
15		B11 = SQRT(D9/(5−1))		
16		C4 = B4−$B$10		
17		D4 = C4^2		
18		D9 = D4+D5+D6+D7+D8		
19				
20	Calculations using built-in functions:			
21	sum =	89.2		
22	mean =	17.84		
23	std dev =	0.3362		
24				
25	Formulas:	B21 = SUM(B4:B8)		
26		B22 = AVERAGE(B4:B8)		
27		B23 = STDEV(B4:B8)		

4-C. **(a)** We need to find the fraction of the area of the Gaussian curve between $x = -\infty$ and $x = 40\,860$ h. When $x = 40\,860$, $z = (40\,860 - 62\,700)/10\,400 = -2.100\,0$. The Gaussian curve is symmetric, so the area from $-\infty$ to $-2.100\,0$ is the same as the area from $2.100\,0$ to $+\infty$. Table 4-1 tells us that the area between $z = 0$ and $z = 2.1$ is 0.482 1. Because the area from $z = 0$ to $z = \infty$ is 0.500 0, the area from $z = 2.100\,0$ to $z = \infty$ is $0.500\,0 - 0.482\,1 = 0.017\,9$. The fraction of brakes expected to be 80% worn in less than 40 860 miles is 0.017 9, or 1.79%.
(b) At 57 500 miles, $z = (57\,500 - 62\,700)/10\,400 = -0.500\,0$. At 71 020 miles, $z = (71\,020 - 62\,700)/10\,400 = +0.800\,0$. The area under the Gaussian curve from $z = -0.500\,0$ to $z = 0$ is the same as the area from $z = 0$ to $z = +0.500\,0$, which is 0.191 5 in Table 4-1. The area from $z = 0$ to $z = +0.800\,0$ is 0.288 1. The total area from $z = -0.500\,0$ to $z = +0.800\,0$ is $0.191\,5 + 0.288\,1 = 0.479\,6$. The fraction of brakes expected to be 80% worn between 57 500 and 71 020 miles is 0.479 6, or 47.96%.

4-D. The answers in cells C4 and C9 of the following spreadsheet are **(a)** 0.052 and **(b)** 0.361.

	A	B	C
1	Mean =	Std dev =	
2	62700	10400	
3			
4	Area from −∞ to 45800 =		0.052081
5	Area from −∞ to 60000 =		0.397580
6	Area from −∞ to 70000 =		0.758637
7			
8	Area from 60000 to 70000		
9		= C6−C5 =	0.361056
10			
11	Formula:		
12	C4 = NORMDIST(45800,A2,B2,TRUE)		

4-E. For 117, 119, 111, 115, 120 μmol/100 mL, $\bar{x} = 116._4$ and $s = 3._{58}$. The 95% confidence interval for 4 degrees of freedom is

$$\bar{x} \pm \frac{ts}{\sqrt{n}} = 116._4 \pm \frac{(2.776)(3._{58})}{\sqrt{5}} = 116._4 \pm 4._4$$

$$= 112._0 \text{ to } 120._8 \ \mu\text{mol}/100 \text{ mL}$$

The 95% confidence interval does not include the accepted value of 111 μmol/100 mL, so the *difference is significant.*

4-F. **(a)** pg/g corresponds to 10^{-12} g/g, which is parts per trillion.
(b) $F_{\text{calculated}} = 4.6^2/3.6^2 = 1.6_3 < F_{\text{table}} = 5.05$ (for 5 degrees of freedom in both numerator and denominator). Standard deviations are not significantly different at 95% confidence level.
(c) Because $F_{\text{calculated}} < F_{\text{table}}$, we use Equations 4-8 and 4-9.

$$s_{\text{pooled}} = \sqrt{\frac{s_1^2(n_1 - 1) + s_2^2(n_2 - 1)}{n_1 + n_2 - 2}}$$

$$= \sqrt{\frac{4.6^2(6 - 1) + 3.6^2(6 - 1)}{6 + 6 - 2}} = 4.1_3$$

$$t_{\text{calculated}} = \frac{|\bar{x}_1 - \bar{x}_2|}{s_{\text{pooled}}}\sqrt{\frac{n_1 n_2}{n_1 + n_2}} = \frac{|51.1 - 34.4|}{4.1_3}\sqrt{\frac{6 \cdot 6}{6 + 6}} = 7.0_0$$

Because $t_{\text{calculated}} (7.0_0) > t_{\text{table}} (= 2.228 \text{ for } 10 \text{ degrees of freedom})$, the difference is significant at the 95% confidence level.
(d) $F_{\text{calculated}} = 3.6^2/1.2^2 = 9.0_0 > F_{\text{table}} = 5.05$. The standard deviations are significantly different at the 95% confidence level. Therefore, we use Equations 4-8a and 4-9a to compare the means:

$$\text{Degrees of freedom} = \frac{(s_1^2/n_1 + s_2^2/n_2)^2}{\dfrac{(s_1^2/n_1)^2}{n_1 - 1} + \dfrac{(s_2^2/n_2)^2}{n_2 - 1}}$$

$$= \frac{(3.6^2/6 + 1.2^2/6)^2}{\dfrac{(3.6^2/6)^2}{6 - 1} + \dfrac{(1.2^2/6)^2}{6 - 1}} = 6.10 \approx 6$$

$$t_{\text{calculated}} = \frac{|\bar{x}_1 - \bar{x}_2|}{\sqrt{s_1^2/n_1 + s_2^2/n_2}} = \frac{|34.4 - 42.9|}{\sqrt{3.6^2/6 + 1.2^2/6}} = 5.4_9$$

Because $t_{\text{calculated}} (= 5.4_9) > t_{\text{table}} (= 2.447 \text{ for } 6 \text{ degrees of freedom})$, the difference is significant at the 95% confidence level.

4-G. **(a)**

x_i	y_i	$x_i y_i$	x_i^2	d_i	d_i^2
0.00	0.466	0	0	$-0.004\,6$	2.12×10^{-5}
9.36	0.676	6.327	87.61	$+0.001\,6$	2.58×10^{-6}
18.72	0.883	16.530	350.44	$+0.004\,8$	2.31×10^{-5}
28.08	1.086	30.495	788.49	$+0.004\,0$	1.61×10^{-5}
37.44	1.280	47.923	1 401.75	$-0.005\,8$	3.34×10^{-5}
Sum: 93.60	4.391	101.275	2 628.29		9.64×10^{-5}

$$D = \begin{vmatrix} \Sigma(x_i^2) & \Sigma x_i \\ \Sigma x_i & n \end{vmatrix}$$

$$= (2\,628.29)(5) - (93.60)(93.60) = 4\,380.5$$

$$m = \begin{vmatrix} \Sigma(x_i y_i) & \Sigma x_i \\ \Sigma y_i & n \end{vmatrix} \div D$$

$$= \frac{(101.275)(5) - (93.60)(4.391)}{D}$$

$$= 95.377 \div 4\,380.5 = 0.021\,773$$

$$b = \begin{vmatrix} \Sigma(x_1^2) & \Sigma(x_i y_i) \\ \Sigma x_i & \Sigma y_i \end{vmatrix} \div D$$

$$= \frac{(2\,628.29)(4.391) - (101.275)(93.60)}{D}$$

$$= 2\,061.48 \div 4\,380.5 = 0.470\,60$$

$$s_y^2 = \frac{\Sigma(d_i^2)}{n - 2} = \frac{9.64 \times 10^{-5}}{3}$$

$$= 3.21 \times 10^{-5}; \ s_y = 0.005\,67$$

$$s_m = \sqrt{\frac{s_y^2 n}{D}} = \sqrt{\frac{(3.21 \times 10^{-5})5}{4\,380.5}} = 0.000\,191$$

$$s_b = \sqrt{\frac{s_y^2\,\Sigma(x_i^2)}{D}} = \sqrt{\frac{(3.21 \times 10^{-5})(2\,628.29)}{4\,380.5}} = 0.004\,39$$

	A	B	C	D	E	F	G	H	I
1	Least-Squares Spreadsheet								
2									
3		x	y						
4		0	0.466						
5		9.36	0.676						
6		18.72	0.883						
7		28.08	1.086						
8		37.44	1.280						
9									
10		LINEST output:							
11	m	0.02177	0.47060	b					
12	s_m	0.00019	0.00439	s_b					
13	R^2	0.99977	0.00567	s_y					
14									
15	n =		5	B15 = COUNT(B4:B8)					
16	Mean y =		0.878	B16 = AVERAGE(C4:C8)					
17	$\Sigma(x_i\text{-mean } x)^2$ =		876.096	B17 = DEVSQ(B4:B8)					
18									
19	Measured y =		0.973	Input					
20	Number of replicate measurements of y (k) =		1	Input					
21	Derived x		23.0739	B21 = (B19-C11)/B11					
22	s_x =		0.2878	B22 = (C13/B11)*SQRT((1/B20)+(1/B15)+((B19-B16)^2)/(B11^2*B17))					

Spreadsheet for Exercise 4-G.

Equation of the best line:

$$y = [0.021\ 8\ (\pm 0.000\ 2)]x + [0.471\ (\pm 0.004)]$$

(c) $x = \dfrac{y - b}{m} = \dfrac{0.973 - 0.471}{0.021\ 8} = 23.0\ \mu g.$

If you keep more digits for m and b, $x = 23.07\ \mu g$.
Uncertainty in x (s_x)

$$= \frac{s_y}{|m|}\sqrt{\frac{1}{k} + \frac{1}{n} + \frac{(y - \bar{y})^2}{m^2\ \Sigma(x_i - \bar{x})^2}}$$

$$= \frac{0.005\ 67}{|0.021\ 77|}\sqrt{\frac{1}{1} + \frac{1}{5} + \frac{(0.973 - 0.878\ 2)^2}{(0.021\ 77)^2\ (876.1)}} = 0.29\ \mu g$$

Final answer is $23.1 \pm 0.3\ \mu g$.

Chapter 5

5-A. (a) Standard deviation of 9 samples $= s = 0.000\ 6_{44}$
Mean blank $= y_{\text{blank}} = 0.001\ 1_{89}$
$y_{\text{dl}} = y_{\text{blank}} + 3s = 0.001\ 1_8 + (3)(0.000\ 6_{44}) = 0.003\ 1_{12}$

(b) Minimum detectable concentration $= \dfrac{3s}{m} = \dfrac{(3)(0.000\ 6_{44})}{2.24 \times 10^4\ \text{M}^{-1}}$

$= 8._6 \times 10^{-8}\ \text{M}$

(c) Lower limit of quantitation $= \dfrac{10s}{m} = \dfrac{(10)(0.000\ 6_{44})}{2.24 \times 10^4\ \text{M}^{-1}} = 2._9 \times 10^{-7}\ \text{M}$

5-B. (a) $[\text{Ni}^{2+}]_f = [\text{Ni}^{2+}]_i \dfrac{V_i}{V_f} = [\text{Ni}^{2+}]_i \left(\dfrac{25.0}{25.5}\right) = 0.980_4[\text{Ni}^{2+}]_i$

(b) $[\text{S}]_f = (0.028\ 7\ \text{M})\left(\dfrac{0.500}{25.5}\right) = 0.000\ 562_7\ \text{M}$

(c) $\dfrac{[\text{Ni}^{2+}]_i}{0.000\ 562\ 7 + 0.980\ 4[\text{Ni}^{2+}]_i} = \dfrac{2.36\ \mu A}{3.79\ \mu A}$

$\Rightarrow [\text{Ni}^{2+}]_i = 9.00 \times 10^{-4}\ \text{M}$

5-C. Use the standard mixture to find the response factor. We know that, when $[\text{X}] = [\text{S}]$, the ratio of signals A_X/A_S is 1.31.

$$\frac{A_X}{[\text{X}]} = F\left(\frac{A_S}{[\text{S}]}\right) \Rightarrow F = \frac{A_X/A_S}{[\text{X}]/[\text{S}]} = \frac{1.31}{1} = 1.31$$

In the mixture of unknown plus standard, the concentration of S is

$$[\text{S}] = \underbrace{(4.13\ \mu g/\text{mL})}_{\substack{\text{Initial}\\\text{concentration}}} \underbrace{\left(\frac{2.00}{10.0}\right)}_{\substack{\text{Dilution}\\\text{factor}}} = 0.826\ \mu g/\text{mL}$$

For the unknown mixture: $F = \dfrac{A_X/A_S}{[\text{X}]/[\text{S}]}$

$1.31 = \dfrac{0.808}{[\text{X}]/[0.826\ \mu g/\text{mL}]} \Rightarrow [\text{X}] = 0.509\ \mu g/\text{mL}$

Because X was diluted from 5.00 to 10.0 mL in the mixture with S, the original concentration of X was $(10.0/5.0)(0.509\ \mu g/\text{mL}) = 1.02\ \mu g/\text{mL}$.

5-D. There are 9 points, so there are $9 - 2 = 7$ degrees of freedom. For 90% confidence, $t = 1.895$, so the 90% confidence interval is $\pm(1.895)(0.098\ \text{mM}) = \pm 0.19\ \text{mM}$. For 99% confidence, $t = 3.500$, and the 99% confidence interval is $\pm(3.500)(0.098\ \text{mM}) = \pm 0.34\ \text{mM}$.

5-E. For the data in this problem, mean $= 0.84_{11}$ and standard deviation $= 0.18_{88}$. Stability criteria are

- There should be no observations outside the action lines—One observation (day 101) lies above the upper action line.
- There are not 2 out of 3 consecutive measurements between warning and action lines—OK.
- There are not 7 consecutive measurements all above or all below the center line—OK.
- There are not 6 consecutive measurements all steadily increasing or all steadily decreasing, wherever they are located—OK.
- There are not 14 consecutive points alternating up and down, regardless of where they are located—OK.
- There is no obvious nonrandom pattern—OK.

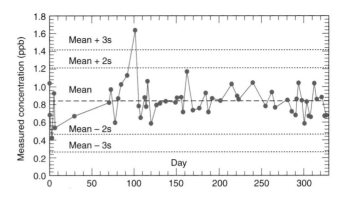

Chapter 6

6-A. (a) $\cancel{\text{Ag}^+} + \cancel{\text{Cl}^-} \rightleftharpoons \text{AgCl}(aq)$
$\text{AgCl}(s) \rightleftharpoons \cancel{\text{Ag}^+} + \cancel{\text{Cl}^-}$
$\overline{\text{AgCl}(s) \rightleftharpoons \text{AgCl}(aq)}$

$K_1 = 2.0 \times 10^3$
$K_2 = 1.8 \times 10^{-10}$
$\overline{K_3 = K_1K_2 = 3.6 \times 10^{-7}}$

(b) The answer to (a) tells us $[\text{AgCl}(aq)] = 3.6 \times 10^{-7}$

(c) $\text{AgCl}_2^- \rightleftharpoons \cancel{\text{AgCl}}(aq) + \text{Cl}^-$
$\text{Ag}^+ + \cancel{\text{Cl}^-} \rightleftharpoons \text{AgCl}(s)$
$\cancel{\text{AgCl}}(aq) \rightleftharpoons \cancel{\text{Ag}^+} + \cancel{\text{Cl}^-}$
$\overline{\text{AgCl}_2^- \rightleftharpoons \text{AgCl}(s) + \text{Cl}^-}$

$K_1 = 1/(9.3 \times 10^1)$
$K_2 = 1/(1.8 \times 10^{-10})$
$K_3 = 1/(2.0 \times 10^3)$
$\overline{K_4 = K_1K_2K_3 = 3.0 \times 10^4}$

6-B. (a) $\dfrac{(x)(x)(1.00 + 8x)^8}{(0.010\ 0 - x)(0.010\ 0 - 2x)^2} = 1 \times 10^{11}$

(b) $[\text{Br}^-]$ and $[\text{Cr}_2\text{O}_7^{2-}]$ will both be 0.005 00 M because Cr^{3+} is the *limiting reagent*. The reaction requires 2 moles of Cr^{3+} per mole of BrO_3^-. The Cr^{3+} will be used up first, making 1 mol Br^- and 1 mol $\text{Cr}_2\text{O}_7^{2-}$ per 2 mol Cr^{3+} consumed. To solve the preceding equation, we set $x = 0.005\ 00$ M in all terms except $[\text{Cr}^{3+}]$. The concentration of Cr^{3+} will be a small, unknown quantity.

$$\frac{(0.005\ 00)(0.005\ 00)[1.00 + 8(0.005\ 00)]^8}{(0.010\ 0 - 0.005\ 00)\ [\text{Cr}^{3+}]^2} = 1 \times 10^{11}$$

$[\text{Cr}^{3+}] = 2._6 \times 10^{-7}\ \text{M}$

$[\text{BrO}_3^-] = 0.010\ 0 - 0.005\ 00 = 0.005\ 00\ \text{M}$

6-C. K_{sp} for $\text{La}(\text{IO}_3)_3$ is small (1.0×10^{-11}), so we presume that the concentration of iodate will not be altered by the small amount of $\text{La}(\text{IO}_3)_3$ that dissolves.

$$[\text{La}^{3+}] = \frac{K_{\text{sp}}}{[\text{IO}_3^-]^3} = \frac{1.0 \times 10^{-11}}{(0.050)^3} = 8.0 \times 10^{-8}\ \text{M}$$

The answer agrees with the assumption that iodate from $\text{La}(\text{IO}_3)_3$ $\ll 0.050$ M.

6-D. We expect $\text{Ca}(\text{IO}_3)_2$ to be more soluble because its K_{sp} is larger and the two salts have the same stoichiometry. If the stoichiometry were not the same, we could not directly compare values of K_{sp}. Our prediction could be wrong if, for example, the barium salt formed a great deal of the ion pairs $\text{Ba}(\text{IO}_3)^+$ or $\text{Ba}(\text{IO}_3)_2(aq)$ and the calcium salt did not form ion pairs.

6-E. $[\text{Fe}^{3+}][\text{OH}^-]^3 = (10^{-10})[\text{OH}^-]^3 = 1.6 \times 10^{-39}$
$\Rightarrow [\text{OH}^-] = 2.5 \times 10^{-10}\ \text{M}$
$[\text{Fe}^{2+}][\text{OH}^-]^2 = (10^{-10})[\text{OH}^-]^2 = 7.9 \times 10^{-16}$
$\Rightarrow [\text{OH}^-] = 2.8 \times 10^{-3}\ \text{M}$

6-F. We want to reduce $[Ce^{3+}]$ to 1.0% of 0.010 M $= 0.000\ 10$ M. The concentration of oxalate in equilibrium with $0.000\ 10$ M Ce^{3+} is computed as follows:

$$[Ce^{3+}]^2[C_2O_4^{2-}]^3 = K_{sp} = 5.9 \times 10^{-30}$$

$$(0.000\ 10)^2[C_2O_4^{2-}]^3 = 5.9 \times 10^{-30}$$

$$[C_2O_4^{2-}] = \left(\frac{5.9 \times 10^{-30}}{(0.000\ 10)^2}\right)^{1/3} = 8.4 \times 10^{-8}\ \text{M}$$

To see if 8.4×10^{-8} M $C_2O_4^{2-}$ will precipitate 0.010 M Ca^{2+}, evaluate Q for CaC_2O_4:

$$Q = [Ca^{2+}][C_2O_4^{2-}] = (0.010)(8.4 \times 10^{-8}) = 8.4 \times 10^{-10}$$

Because $Q < K_{sp}$ for CaC_2O_4 ($= 1.3 \times 10^{-8}$), Ca^{2+} will not precipitate.

6-G. Assuming that all nickel is in the form $Ni(en)_3^{2+}$, $[Ni(en)_3^{2+}] = 1.00 \times 10^{-5}$ M. This uses up just 3×10^{-5} mol of en, which leaves the en concentration at 0.100 M. Adding the three equations gives

$$Ni^{2+} + 3en \rightleftharpoons Ni(en)_3^{2+}$$

$$K = K_1K_2K_3 = 2.1_4 \times 10^{18}$$

$$[Ni^{2+}] = \frac{[Ni(en)_3^{2+}]}{K[en]^3}$$

$$= \frac{(1.00 \times 10^{-5})}{(2.1_4 \times 10^{18})(0.100)^3} = 4.7 \times 10^{-21}\ \text{M}$$

Now we verify that $[Ni(en)^{2+}]$ and $[Ni(en)_2^{2+}] \ll 10^{-5}$ M:

$$[Ni(en)^{2+}] = K_1[Ni^{2+}][en] = 1.5 \times 10^{-14}\ \text{M}$$

$$[Ni(en)_2^{2+}] = K_2[Ni(en)^{2+}][en] = 3.2 \times 10^{-9}\ \text{M}$$

6-H. **(a)** Neutral—Neither Na^+ nor Br^- has any acidic or basic properties.
(b) Basic—$CH_3CO_2^-$ is the conjugate base of acetic acid, and Na^+ is neither acidic nor basic.
(c) Acidic—NH_4^+ is the conjugate acid of NH_3, and Cl^- is neither acidic nor basic.
(d) Basic—PO_4^{3-} is a base, and K^+ is neither acidic nor basic.
(e) Neutral—Neither ion is acidic nor basic.
(f) Basic—The quaternary ammonium ion is neither acidic nor basic, and the $C_6H_5CO_2^-$ anion is the conjugate base of benzoic acid.
(g) Acidic—Fe^{3+} is acidic, and nitrate is neither acidic nor basic.

6-I. $K_{b1} = K_w/K_{a2} = 4.3 \times 10^{-9}$
$K_{b2} = K_w/K_{a1} = 1.6 \times 10^{-10}$

6-J. $K \equiv K_{b2} = K_w/K_{a2} = 1.2 \times 10^{-8}$

6-K. **(a)** $[H^+][OH^-] = x^2 = K_w \Rightarrow x = \sqrt{K_w} \Rightarrow pH = -\log\sqrt{K_w} = 7.469$ at $0°C$, 7.082 at $20°C$, and 6.770 at $40°C$.
(b) Because $[D^+] = [OD^-]$ in pure D_2O, $K = 1.35 \times 10^{-15} = [D^+][OD^-] = [D^+]^2 \Rightarrow [D^+] = 3.67 \times 10^{-8}$ M $\Rightarrow$ pD $= 7.435$.

Chapter 7

7-A. **(a)** $\mu = \frac{1}{2}([K^+] \cdot 1^2 + [NO_3^-] \cdot (-1)^2) = 0.2$ mM
(b) $\mu = \frac{1}{2}([Cs^+] \cdot 1^2 + [CrO_4^{2-}] \cdot (-2)^2)$
$= \frac{1}{2}([0.4] \cdot 1 + [0.2] \cdot 4) = 0.6$ mM
(c) $\mu = \frac{1}{2}([Mg^{2+}] \cdot 2^2 + [Cl^-] \cdot (-1)^2 + [Al^{3+}] \cdot 3^2)$
$= \frac{1}{2}([0.2] \cdot 4 + [0.4 \quad + \quad 0.9] \cdot 1 + [0.3] \cdot 9) = 2.4$ mM
$\qquad\qquad \uparrow \qquad\qquad \uparrow$
$\qquad$ From $MgCl_2$ $\quad$ From $AlCl_3$

7-B. For $0.005\ 0$ M $(CH_3CH_2CH_2)_4N^+Br^-$ plus $0.005\ 0$ M $(CH_3)_4N^+Cl^-$, $\mu = 0.010$ M. The size of the ion $(CH_3CH_2CH_2)_4N^+$ is 800 pm. At $\mu = 0.01$ M, $\gamma = 0.912$ for an ion of charge ±1 with $\alpha = 800$ pm. $\mathcal{A} = (0.005\ 0)(0.912) = 0.004\ 6$.

7-C. $\mu = 0.060$ M from KSCN, assuming negligible solubility of AgSCN.
$K_{sp} = [Ag^+]\gamma_{Ag^+}[SCN^-]\gamma_{SCN^-} = 1.1 \times 10^{-12}$
The activity coefficients at $\mu = 0.060$ M are $\gamma_{Ag^+} = 0.79$ and $\gamma_{SCN^-} = 0.80$.
$K_{sp} = [Ag^+](0.79)[0.060](0.80) = 1.1 \times 10^{-12}$
$\Rightarrow [Ag^+] = 2.9 \times 10^{-11}$ M.

7-D. At an ionic strength of 0.050 M, $\gamma_{H^+} = 0.86$ and $\gamma_{OH^-} = 0.81$.
$[H^+]\gamma_{H^+}[OH^-]\gamma_{OH^-} = (x)(0.86)(x)(0.81) = 1.0 \times 10^{-14}$
$\Rightarrow x = [H^+] = 1.2 \times 10^{-7}$ M. pH $= -\log[(1.2 \times 10^{-7})(0.86)] = 6.99$.

7-E. **(a)** Moles of $I^- = 2$(moles of Hg_2^{2+})
$(V_e)(0.100\ \text{M}) = 2(40.\ 0\ \text{mL})(0.040\ 0\ \text{M}) \Rightarrow V_e = 32.0$ mL
(b) Virtually all the Hg_2^{2+} has precipitated, along with 3.20 mmol of I^-. The ions remaining in solution are

$$[NO_3^-] = \frac{3.20\ \text{mmol}}{100.0\ \text{mL}} = 0.032\ 0\ \text{M}$$

$$[I^-] = \frac{2.80\ \text{mmol}}{100.0\ \text{mL}} = 0.028\ 0\ \text{M}$$

$$[K^+] = \frac{6.00\ \text{mmol}}{100.0\ \text{mL}} = 0.060\ 0\ \text{M}$$

$$\mu = \frac{1}{2}\Sigma c_i z_i^2 = 0.060\ 0\ \text{M}$$

(c) $\mathcal{A}_{Hg_2^{2+}} = K_{sp}/\mathcal{A}_{I^-}^2 = K_{sp}/[I^-]^2\gamma_{I^-}^2$
$= 4.6 \times 10^{-29}/(0.028\ 0)^2(0.795)^2 = 9.3 \times 10^{-26}$
$\Rightarrow pHg_2^{2+} = -\log \mathcal{A}_{Hg_2^{2+}} = 25.03$

7-F. **(a)** $[Cl^-] = 2[Ca^{2+}]$
(b) $\underbrace{[Cl^-] + [CaCl^+]}_{\text{Species containing Cl}^-} = 2\{\underbrace{[Ca^{2+}] + [CaCl^+] + [CaOH^+]}_{\text{Species containing Ca}^{2+}}\}$
(c) $[Cl^-] + [OH^-] = 2[Ca^{2+}] + [CaCl^+] + [CaOH^+] + [H^+]$

7-G. Charge balance:
$[F^-] + [HF_2^-] + [OH^-] = 2[Ca^{2+}] + [CaOH^+] + [CaF^+] + [H^+]$
Mass balance: CaF_2 gives 2 mol F for each mol Ca.
$\underbrace{[F^-] + [CaF^+] + 2[CaF_2(aq)] + [HF] + 2[HF_2^-]}_{\text{Species containing F}^-}$
$= 2\{\underbrace{[Ca^{2+}] + [CaOH^+] + [CaF^+] + [CaF_2(aq)]}_{\text{Species containing Ca}^{2+}}\}$

7-H. Charge balance:
$2[Ca^{2+}] + [CaOH^+] + [H^+]$
$\quad = [CaPO_4^-] + 3[PO_4^{3-}] + 2[HPO_4^{2-}] + [H_2PO_4^-] + [OH^-]$
Mass balance: Equate 2(calcium species) = 3(phosphate species).
$2\{\underbrace{[Ca^{2+}] + [CaOH^+] + [CaPO_4^-]}_{\text{Species containing calcium}}\}$
$= 3\{\underbrace{[CaPO_4^-] + [PO_4^{3-}] + [HPO_4^{2-}] + [H_2PO_4^-] + [H_3PO_4]}_{\text{Species containing phosphate}}\}$

7-I. Pertinent reactions:

$$Mn(OH)_2(s) \overset{K_{sp}}{\rightleftharpoons} Mn^{2+} + 2OH^- \qquad K_{sp} = 1.6 \times 10^{-13}$$

$$Mn^{2+} + 2OH^- \overset{K_1}{\rightleftharpoons} MnOH^+ \qquad K_1 = 2.5 \times 10^3$$

$$H_2O \overset{K_w}{\rightleftharpoons} H^+ + OH^- \qquad K_w = 1.0 \times 10^{-14}$$

Charge balance: $2[Mn^{2+}] + [MnOH^+] + [H^+] = [OH^-]$
Mass balance: $\underbrace{[OH^-] + [MnOH^+]}_{\text{Species containing OH}^-} = 2\{\underbrace{[Mn^{2+}] + [MnOH^+]}_{\text{Species containing Mn}^{2+}}\} + [H^+]$

(Mass balance gives the same result as charge balance.)

Equilibrium constant expressions:
$K_{sp} = [Mn^{2+}]\gamma_{Mn^{2+}}[OH^-]^2\gamma_{OH^-}^2$
$K_1 = \dfrac{[MnOH^+]\gamma_{MnOH^+}}{[Mn^{2+}]\gamma_{Mn^{2+}}[OH^-]\gamma_{OH^-}}$
$K_w = [H^+]\gamma_{H^+}[OH^-]\gamma_{OH^-}$
From K_1, we write $[MnOH^+] = (K_1/\gamma_{MnOH^+})[Mn^{2+}]\gamma_{Mn^{2+}}[OH^-]\gamma_{OH^-}$.
Substitute for $[MnOH^+]$ in the charge balance:
$2[Mn^{2+}] + (K_1/\gamma_{MnOH^+})[Mn^{2+}]\gamma_{Mn^{2+}}[OH^-]\gamma_{OH^-} + [H^+] = [OH^-]$ (A)

	A	B	C	D
1	Mn(OH)$_2$ solubility with activity coefficients			
2				
3	K$_{sp}$ =		[OH$^-$]$_{guess}$ =	A10*[OH]^3*D10^2/(2+(K$_1$*A10/C10)[OH]*D10) =
4	1.6E-13		1.130E-04	1.6000E-13
5	K$_1$ =			
6	2.5E+03		[Mn^{2+}] =	[MnOH$^+$] =
7			5.358E-05	5.823E-06
8	Activity coefficients:			
9	Mn^{2+}		MnOH$^+$	OH$^-$
10	0.405		0.80	0.76
11				
12	D4 = (A10*C4^3*D10^2)/(2+(A6*A10/C10)*C4*D10)			
13	C7 = A4/(A10*C4^2*D10^2)			
14	D7 = (A6/C10)*C7*A10*C4*D10			

Spreadsheet for Exercise 7-I.

In a basic solution, neglect [H$^+$] in comparison with [OH$^-$] and solve (A) for [Mn^{2+}]:

$$[Mn^{2+}] = \frac{[OH^-]}{2 + (K_1\gamma_{Mn^{2+}}/\gamma_{MnOH^+})[OH^-]\gamma_{OH^-}} \quad (B)$$

Substitute (B) into K_{sp}:

$$K_{sp} = [Mn^{2+}]\gamma_{Mn^{2+}}[OH^-]^2\gamma_{OH^-}^2$$

$$K_{sp} = \left(\frac{\gamma_{Mn^{2+}}[OH^-]^3\gamma_{OH^-}^2}{2 + (K_1\gamma_{Mn^{2+}}/\gamma_{MnOH^+})[OH^-]\gamma_{OH^-}}\right) \quad (C)$$

We take the ion size of MnOH$^+$ to be the same as Mn^{2+}, which is 600 pm. (But the charge of MnOH$^+$ is +1, not +2.) For μ = 0.10 M, the activity coefficients are $\gamma_{Mn^{2+}}$ = 0.405, γ_{MnOH^+} = 0.80, and γ_{OH^-} = 0.76. In the spreadsheet, we use Goal Seek to find the value [OH$^-$] = 1.13 × 10^{-4} M in cell C4 that makes the right side of Equation (C) equal to 1.6 × 10^{-13} (= K_{sp}) in cell D4. With [OH$^-$] known, we use K_{sp} and K_1 to find:

$[Mn^{2+}] = K_{sp}/(\gamma_{Mn^{2+}}[OH^-]^2\gamma_{OH^-}^2) = 5.36 \times 10^{-5}$ M

$[MnOH^+] = (K_1/\gamma_{MnOH^+})[Mn^{2+}]\gamma_{Mn^{2+}}[OH^-]\gamma_{OH^-} = 5.82 \times 10^{-6}$ M

You can check that the charge balance $2[Mn^{2+}] + [MnOH^+] \approx [OH^-]$ is satisfied.

Chapter 8

8-A. pH = $-\log\mathcal{A}_{H^+}$. But $\mathcal{A}_{H^+}\mathcal{A}_{OH^-} = K_w \Rightarrow \mathcal{A}_{H^+} = K_w/\mathcal{A}_{OH^-}$. For 1.0 × 10^{-2} M NaOH, [OH$^-$] = 1.0 × 10^{-2} M and γ_{OH^-} = 0.900 (Table 7-1, with μ = 0.010 M).

$$\mathcal{A}_{H^+} = \frac{K_w}{[OH^-]\gamma_{OH^-}} = \frac{1.0 \times 10^{-14}}{(1.0 \times 10^{-2})(0.900)}$$

$= 1.11 \times 10^{-12} \Rightarrow$ pH = $-\log\mathcal{A}_{H^+}$ = 11.95

8-B. (a) Charge balance: [H$^+$] = [OH$^-$] + [Br$^-$]

Mass balance: [Br$^-$] = 1.0 × 10^{-8} M

Equilibrium: [H$^+$][OH$^-$] = K_w

Setting [H$^+$] = x and [Br$^-$] = 1.0 × 10^{-8} M, the charge balance tells us that [OH$^-$] = x − 1.0 × 10^{-8}. Putting this into the K_w equilibrium gives $(x)(x - 1.0 \times 10^{-8}) = 1.0 \times 10^{-14} \Rightarrow x = 1.0_5 \times 10^{-7}$ M $\Rightarrow$ pH = 6.98.

(b) Charge balance: [H$^+$] = [OH$^-$] + 2[SO$_4^{2-}$]

Mass balance: [SO$_4^{2-}$] = 1.0 × 10^{-8} M

Equilibrium: [H$^+$][OH$^-$] = K_w

As before, writing [H$^+$] = x and [SO$_4^{2-}$] = 1.0 × 10^{-8} M gives [OH$^-$] = x − 2.0 × 10^{-8} M and [H$^+$][OH$^-$] = $(x)[x - (2.0 \times 10^{-8})] = 1.0 \times 10^{-14}$ $\Rightarrow x = 1.10 \times 10^{-7}$ M $\Rightarrow$ pH = 6.96.

8-C.

2-Nitrophenol FM = 139.11
C$_6$H$_5$NO$_3$ K_a = 5.89 × 10^{-8}

$$F_{HA} \text{ (formal concentration)} = \frac{1.23 \text{ g}/(139.11 \text{ g/mol})}{0.250 \text{ L}}$$

$$= 0.035 \text{ 4 M}$$

$$\underset{F-x}{HA} \rightleftharpoons \underset{x}{H^+} + \underset{x}{A^-}$$

$$\frac{x^2}{0.035 \text{ 4} - x} = 5.89 \times 10^{-8} \text{ M}$$

$$\Rightarrow x = 4.56 \times 10^{-5} \text{ M} \Rightarrow \text{pH} = -\log x = 4.34$$

8-D.

But [H$^+$] = 10^{-pH} = 6.9 × 10^{-7} M $\Rightarrow$ [A$^-$] = 6.9 × 10^{-7} M and [HA] = 0.010 − [H$^+$] = 0.010.

$$K_a = \frac{[H^+][A^-]}{[HA]} = \frac{(6.9 \times 10^{-7})^2}{0.010} = 4.8 \times 10^{-11} \Rightarrow pK_a = 10.32$$

8-E. As [HA] → 0, pH → 7. If pH = 7,

$$\frac{[H^+][A^-]}{[HA]} = K_a \Rightarrow [A^-] = \frac{K_a}{[H^+]}[HA]$$

$$= \frac{10^{-5.00}}{10^{-7.00}}[HA] = 100[HA]$$

$$\alpha = \frac{[A^-]}{[HA] + [A^-]} = \frac{100[HA]}{[HA] + 100[HA]} = \frac{100}{101} = 99\%$$

If pK_a = 9.00, we find α = 0.99%.

8-F. $\underset{F-x}{CH_3CH_2CH_2CO_2^-} + H_2O \rightleftharpoons \underset{x}{CH_3CH_2CH_2CO_2H} + \underset{x}{OH^-}$

$$K_b = \frac{K_w}{K_a} = 6.58 \times 10^{-10}$$

$$\frac{x^2}{F - x} = K_b \Rightarrow x = 5.7_4 \times 10^{-6} \text{ M}$$

$$\text{pH} = -\log\left(\frac{K_w}{x}\right) = 8.76$$

8-G. (a) $\underset{F-x}{CH_3CH_2NH_2} + H_2O \rightleftharpoons \underset{x}{CH_3CH_2NH_3^+} + \underset{x}{OH^-}$

Because pH = 11.82, [OH$^-$] = $K_w/10^{-pH}$ = 6.6 × 10^{-3} M = [BH$^+$]. [B] = F − x = 0.093 M.

$$K_b = \frac{[BH^+][OH^-]}{[B]} = \frac{(6.6 \times 10^{-3})^2}{0.093} = 4.7 \times 10^{-4}$$

(b) $CH_3CH_2NH_3^+ \rightleftharpoons CH_3CH_2NH_2 + H^+$

$F-xxx$

$K_a = \dfrac{K_w}{K_b} = 2.1 \times 10^{-11}$

$\dfrac{x^2}{F-x} = K_a \Rightarrow x = 1.4_5 \times 10^{-6}\ M \Rightarrow pH = 5.84$

8-H.

Compound	pKa (for conjugate acid)	
Ammonia	9.24	← Most suitable, because pKa is closest to pH 9.00
Aniline	4.60	
Hydrazine	8.02	
Pyridine	5.20	

8-I. $pH = 4.25 + \log 0.75 = 4.13$

8-J. **(a)** $pH = pK_a + \log \dfrac{[B]}{[BH^+]}$

$= 8.04 + \log \dfrac{[(1.00\ g)/(74.08\ g/mol)]}{[(1.00\ g)/(110.54\ g/mol)]} = 8.21$

(b) $pH = pK_a + \log \dfrac{mol\ B}{mol\ BH^+}$

$8.00 = 8.04 + \log \dfrac{mol\ B}{(1.00\ g)/(110.54\ g/mol)}$

$\Rightarrow mol\ B = 0.008\ 25 = 0.611\ g$ of glycine amide

(c)

	B	+	H⁺	→	BH⁺
Initial moles:	0.013 499		0.000 500		0.009 046
Final moles:	0.012 999		—		0.009 546

$$pH = 8.04 + \left(\dfrac{0.012\ 999}{0.009\ 546}\right) = 8.17$$

(d)

	BH⁺	+	OH⁻	→	B
Initial moles:	0.009 546		0.001 000		0.012 999
Final moles:	0.008 546		—		0.013 999

$$pH = 8.04 + \log\left(\dfrac{0.013\ 999}{0.008\ 546}\right) = 8.25$$

(e) The solution in **(a)** contains 9.046 mmol glycine amide hydrochloride and 13.499 mmol glycine amide. Now we are adding 9.046 mmol of OH⁻, which will convert all of the glycine amide hydrochloride into glycine amide. The new solution contains 9.046 + 13.499 = 22.545 mmol of glycine amide in 190.46 mL. The concentration of glycine amide is $(22.545\ mmol)/(190.46\ mL) = 0.118_4\ M$. The pH is determined by hydrolysis of glycine amide:

$0.118_4 - x x x$

$\dfrac{x^2}{0.118_4 - x} = K_b = \dfrac{K_w}{K_a} = \dfrac{10^{-14.00}}{10^{-8.04}} = 1.10 \times 10^{-6}\ M$

$\Rightarrow x = 3.60 \times 10^{-4}\ M$

$pH = -\log(K_w/x) = 10.56$

8-K. The reaction of phenylhydrazine with water is
$B + H_2O \rightleftharpoons BH^+ + OH^- K_b$
We know that pH = 8.13, so we can find [OH⁻].

$[OH^-] = \dfrac{\mathscr{A}_{OH^-}}{\gamma_{OH^-}} = \dfrac{K_w/10^{-pH}}{\gamma_{OH^-}} = 1.78 \times 10^{-6}\ M$

(using $\gamma_{OH^-} = 0.76$ for $\mu = 0.10\ M$)

$K_b = \dfrac{[BH^+]\gamma_{BH^+}[OH^-]\gamma_{OH^-}}{[B]\gamma_B}$

$= \dfrac{(1.78 \times 10^{-6})(0.80)(1.78 \times 10^{-6})(0.76)}{[0.010 - (1.78 \times 10^{-6})](1.00)}$

$= 1.93 \times 10^{-10}$

$K_a = \dfrac{K_w}{K_b} = 5.19 \times 10^{-5} \Rightarrow pK_a = 4.28$

To find K_b, we made use of the equality $[BH^+] = [OH^-]$.

8-L. We use Goal Seek to vary cell B5 until cell D4 is equal to K_a. The spreadsheet shows that $[H^+] = 4.236 \times 10^{-3}$ (cell B5) and pH = 2.37 in cell B7.

	A	B	C	D
1	Ka = 10^-pKa =	0.00316228		Reaction quotient
2	Kw =	1.00E-14		for Ka =
3	FHA =	0.03		[H+][A-]/[HA] =
4	FA =	0.015		0.0031623
5	H =	4.236E-03		<-Goal Seek solution
6	OH = Kw/H =	2.3609E-12		D4=H*(FA+H-OH)/
7	pH = -logH =	2.37		(FHA-H+OH)

If we were doing this problem by hand with the approximation that what we mix is what we get, $[H^+] = K_a[HA]/[A^-] = 10^{-2.50}[0.030]/[0.015] = 0.006\ 32\ M \Rightarrow pH = 2.20$.

Chapter 9

9-A. **(a)** $H_2SO_3 \rightleftharpoons HSO_3^- + H^+$

$0.050 - x x x$

$\dfrac{x^2}{0.050 - x} = K_1 = 1.39 \times 10^{-2} \Rightarrow x = 2.03 \times 10^{-2}$

$[HSO_3^-] = [H^+] = 2.03 \times 10^{-2}\ M \Rightarrow pH = 1.69$

$[H_2SO_3] = 0.050 - x = 0.030\ M$

$[SO_3^{2-}] = \dfrac{K_2[HSO_3^-]}{[H^+]} = K_2 = 6.7 \times 10^{-8}\ M$

(b) $[H^+] = \sqrt{\dfrac{K_1K_2(0.050) + K_1K_w}{K_1 + (0.050)}}$

$= 2.70 \times 10^{-5}\ M \Rightarrow pH = 4.57$

$[H_2SO_3] = \dfrac{[H^+][HSO_3^-]}{K_1} = \dfrac{(2.70 \times 10^{-5})(0.050)}{1.39 \times 10^{-2}}$

$= 9.7 \times 10^{-5}\ M$

$[SO_3^{2-}] = \dfrac{K_2[HSO_3^-]}{[H^+]} = 1.2 \times 10^{-4}\ M$

$[HSO_3^-] = 0.050\ M$

(c) $SO_3^{2-} + H_2O \rightleftharpoons HSO_3^- + OH^-$

$0.050 - x x x$

$\dfrac{x^2}{0.050 - x} = K_{b1} = \dfrac{K_w}{K_{a2}} = 1.49 \times 10^{-7}$

$[HSO_3^-] = x = 8.6 \times 10^{-5}\ M$

$[H^+] = \dfrac{K_w}{x} = 1.16 \times 10^{-10}\ M \Rightarrow pH = 9.94$

$[SO_3^{2-}] = 0.050 - x = 0.050\ M$

$[H_2SO_3] = \dfrac{[H^+][HSO_3^-]}{K_1} = 7.2 \times 10^{-13}\ M$

9-B. **(a)** $pH = pK_2 \text{ (for } H_2CO_3) + \log \dfrac{[CO_3^{2-}]}{[HCO_3^-]}$

$10.80 = 10.329 + \log \dfrac{(4.00\ g)/(138.206\ g/mol)}{x/(84.007\ g/mol)}$

$\Rightarrow x = 0.822\ g$

(b)

	CO_3^{2-}	$+$	H^+	$\rightarrow$	HCO_3^-
Initial moles:	$0.028\,9_4$		$0.010\,0$		$0.009\,78$
Final moles:	$0.018\,9_4$		$—$		$0.019\,7_8$

$$pH = 10.329 + \log\frac{0.018\,9_4}{0.019\,7_8} = 10.31$$

(c)

	CO_3^{2-}	$+$	H^+	$\rightarrow$	HCO_3^-
Initial moles:	$0.028\,9_4$		x		$—$
Final moles:	$0.028\,9_4 - x$		$—$		x

$$10.00 = 10.329 + \log\frac{0.028\,9_4 - x}{x} \Rightarrow x = 0.019\,7\ \text{mol}$$

$$\Rightarrow \text{volume} = \frac{0.019\,7\ \text{mol}}{0.320\ \text{M}} = 61.6\ \text{mL}$$

9-C.

$\frac{1}{2}(pK_1 + pK_2) = 4.88$

pH 4.40 is above pK_1. At pH $= pK_1$, there would be a 1:1 mixture of H_2A and HA^-. We must add enough KOH to convert some H_2A to HA^- to create a mixture with pH $= 4.40$. $5.02\ \text{g}\ H_2A/(132.11\ \text{g/mol}) = 0.038\,0\ \text{mol}\ H_2A$.

	H_2A	$+$	OH^-	$\rightarrow$	HA^-
Initial moles:	$0.038\,0$		x		$—$
Final moles:	$0.038\,0 - x$		$—$		x

$$pH = pK_1 + \log\frac{[HA^-]}{[H_2A]}$$

$$4.40 = 4.345 + \log\frac{x}{0.038\,0 - x} \Rightarrow x = 0.020\,2\ \text{mol}$$

Volume of KOH $= (0.020\,2\ \text{mol})/(0.800\ \text{M}) = 25.2\ \text{mL}$.

9-D. **(a)** Call the three forms of glutamine H_2G^+, HG, and G^-. The form shown is HG.

$$[H^+] = \sqrt{\frac{K_1K_2(0.010) + K_1K_w}{K_1 + 0.010}} = 1.9_8 \times 10^{-6}\ \text{M} \Rightarrow pH = 5.70$$

(b) Call the four forms of cysteine H_3C^+, H_2C, HC^-, and C^{2-}. The form shown is HC^-.

$$[H^+] = \sqrt{\frac{K_2K_3(0.010) + K_2K_w}{K_2 + 0.010}} = 2.8_9 \times 10^{-10}\ \text{M} \Rightarrow pH = 9.54$$

(c) Call the four forms of arginine H_3A^{2+}, H_2A^+, HA, and A^-. The form shown is HA.

$$[H^+] = \sqrt{\frac{K_2K_3(0.010) + K_2K_w}{K_2 + 0.010}} = 4.2_8 \times 10^{-11}\ \text{M} \Rightarrow pH = 10.37$$

9-E.

	pH 9.00	pH 11.00
Principal species:	(OH, OH benzene)	(O⁻, OH benzene)
Secondary species:	(O⁻, OH benzene)	(O⁻, O⁻ benzene)
Percentage in major form:	66.5%	52.9%

The percentage in the major form was calculated with the formulas for α_{H_2A} (Equation 9-19 at pH 9.00) and α_{HA^-} (Equation 9-20 at pH 11.00).

9-F.

	pH 9.0	pH 10.0
Predominant form:	NH_3^+ CHCH$_2$CH$_2$CO$_2^-$ CO$_2^-$	NH_2 CHCH$_2$CH$_2$CO$_2^-$ CO$_2^-$
Secondary form:	NH_2 CHCH$_2$CH$_2$CO$_2^-$ CO$_2^-$	NH_3^+ CHCH$_2$CH$_2$CO$_2^-$ CO$_2^-$

	pH 9.0	pH 10.0
Predominant form:	NH_2 CHCH$_2$—⬡—OH CO$_2^-$	NH_2 CHCH$_2$—⬡—OH CO$_2^-$
Secondary form:	NH_3^+ CHCH$_2$—⬡—OH CO$_2^-$	NH_2 CHCH$_2$—⬡—O$^-$ CO$_2^-$

9-G. The isoionic pH is the pH of a solution of pure neutral lysine, which is

NH_2
$|$
CHCH$_2$CH$_2$CH$_2$CH$_2$NH$_3^+$
$|$
CO$_2^-$

$$[H^+] = \sqrt{\frac{K_2K_3F + K_2K_w}{K_2 + F}} \Rightarrow pH = 9.93$$

9-H. We know that the isoelectric point will be near $\frac{1}{2}(pK_2 + pK_3) \approx 9.95$. At this pH, the fraction of lysine in the form H_3L^{2+} is negligible. Therefore, the electroneutrality condition reduces to $[H_2L^+] = [L^-]$, for which the expression isoelectric pH $= \frac{1}{2}(pK_2 + pK_3) = 9.95$ applies.

Chapter 10

10-A. The titration reaction is $H^+ + OH^- \rightarrow H_2O$ and $V_e = 5.00\ \text{mL}$. Three representative calculations are given:

At 1.00 mL: $[OH^-] = \left(\frac{4.00}{5.00}\right)(0.010\,0\ \text{M})\left(\frac{50.00}{51.00}\right) = 0.007\,84\ \text{M}$

$$pH = -\log\left(\frac{K_w}{[OH^-]}\right) = 11.89$$

At 5.00 mL: $H_2O \rightleftharpoons \underset{x}{H^+} + \underset{x}{OH^-}$

$x^2 = K_w \Rightarrow x = 1.0 \times 10^{-7}\ \text{M}$

$pH = -\log x = 7.00$

At 5.01 mL: $[H^+] = \left(\frac{0.01}{55.01}\right)(0.100\ \text{M}) = 1.82 \times 10^{-5}\ \text{M} \Rightarrow pH = 4.74$

V_e (mL)	pH	V_e	pH	V_e	pH
0.00	12.00	4.50	10.96	5.10	3.74
1.00	11.89	4.90	10.26	5.50	3.05
2.00	11.76	4.99	9.26	6.00	2.75
3.00	11.58	5.00	7.00	8.00	2.29
4.00	11.27	5.01	4.74	10.00	2.08

10-B. The titration reaction is $HCO_2H + OH^- \rightarrow HCO_2^- + H_2O$ and V_e = 50.0 mL. For formic acid, $K_a = 1.80 \times 10^{-4}$. Four representative calculations are given:

At 0.0 mL: $HA \rightleftharpoons H^+ + A^-$
$x x$

$$\frac{x^2}{0.050\ 0 - x} = K_a \Rightarrow x = 2.91 \times 10^{-3}\ M \Rightarrow pH = 2.54$$

At 48.0 mL:

	HA	+	OH$^-$	$\rightarrow$	A$^-$	+	H$_2$O
Initial:	50		48		—		—
Final:	2		—		48		48

$$pH = pK_a + \log \frac{[A^-]}{[HA]} = 3.744 + \log \frac{48.0}{2.0} = 5.12$$

At 50.0 mL: $A^- + H_2O \overset{K_b}{\rightleftharpoons} HA + OH^-$
$F - x x x$

$$K_b = \frac{K_w}{K_a} \quad \text{and} \quad F = \left(\frac{50}{100}\right)(0.05\ M)$$

$$\frac{x^2}{0.025\ 0 - x} = 5.56 \times 10^{-11} \Rightarrow x = 1.18 \times 10^{-6}\ M$$

$$pH = -\log\left(\frac{K_w}{x}\right) = 8.07$$

At 60.0 mL: $[OH^-] = \left(\frac{10.0}{110.0}\right)(0.050\ 0\ M)$

$$= 4.55 \times 10^{-3}\ M \Rightarrow pH = 11.66$$

V_b (mL)	pH	V_b	pH	V_b	pH
0.0	2.54	45.0	4.70	50.5	10.40
10.0	3.14	48.0	5.12	51.0	10.69
20.0	3.57	49.0	5.43	52.0	10.99
25.0	3.74	49.5	5.74	55.0	11.38
30.0	3.92	50.0	8.07	60.0	11.66
40.0	4.35				

10-C. The titration reaction is $B + H^+ \rightarrow BH^+$ and V_e = 50.0 mL. Representative calculations:

At V_a = 0.0 mL: $B + H_2O \rightleftharpoons BH^+ + OH^-$
$0.100 - x x x$

$$\frac{x^2}{0.100 - x} = 2.6 \times 10^{-6} \Rightarrow x = 5.09 \times 10^{-4}\ M$$

$$pH = -\log\left(\frac{K_w}{x}\right) = 10.71$$

At V_a = 20.0 mL:

	B	+	H$^+$	$\rightarrow$	BH$^+$
Initial:	50.0		20.0		—
Final:	30.0		—		20.0

$$pH = pK_a\ (\text{for } BH^+) + \log \frac{[B]}{[BH^+]} = 8.41 + \log \frac{30.0}{20.0} = 8.59$$

At $V_a = V_e$ = 50.0 mL: All B has been converted into the conjugate acid, BH^+. The formal concentration of BH^+ is $(\frac{100}{150})(0.100\ M)$ = 0.066 7 M. The pH is determined by the reaction

$BH^+ \rightleftharpoons B + H^+$
$0.066\ 7 - x x x$

$$\frac{x^2}{0.066\ 7 - x} = K_a = \frac{K_w}{K_b} \Rightarrow x = 1.60 \times 10^{-5}\ M \Rightarrow pH = 4.80$$

At V_a = 51.0 mL: There is excess H^+:

$$[H^+] = \left(\frac{1.0}{151.0}\right)(0.200\ M) = 1.32 \times 10^{-3}\ M \Rightarrow pH = 2.88$$

V_a (mL)	pH	V_a	pH	V_a	pH
0.0	10.71	30.0	8.23	50.0	4.80
10.0	9.01	40.0	7.81	50.1	3.88
20.0	8.59	49.0	6.72	51.0	2.88
25.0	8.41	49.9	5.71	60.0	1.90

10-D. The titration reactions are

$$HO_2CCH_2CO_2H + OH^- \rightarrow {}^-O_2CCH_2CO_2H + H_2O$$
$${}^-O_2CCH_2CO_2H + OH^- \rightarrow {}^-O_2CCH_2CO_2^- + H_2O$$

and the equivalence points are 25.0 and 50.0 mL. Let's designate malonic acid as H_2M. Also, $K_{a1} = 1.42 \times 10^{-3}$ and $K_{a2} = 2.01 \times 10^{-6}$.

At 0.0 mL: $H_2M \rightleftharpoons H^+ + HM^-$
$0.050\ 0 - x x x$

$$\frac{x^2}{0.050\ 0 - x} = K_1 \Rightarrow x = 7.75 \times 10^{-3}\ M \Rightarrow pH = 2.11$$

At 8.0 mL:

	H$_2$M	+	OH$^-$	$\rightarrow$	HM$^-$	+	H$_2$O
Initial:	25		8		—		—
Final:	17		—		8		—

$$pH = pK_1 + \log \frac{[HM^-]}{[H_2M]} = 2.847 + \log \frac{8}{17} = 2.52$$

At 12.5 mL: $V_b = \frac{1}{2}V_e \Rightarrow pH = pK_1 = 2.85$

At 19.3 mL:

	H$_2$M	+	OH$^-$	$\rightarrow$	HM$^-$	+	H$_2$O
Initial:	25		19.3		—		—
Final:	5.7		—		19.3		—

$$pH = pK_1 + \log \frac{19.3}{5.7} = 3.38$$

At 25.0 mL: At V_e, H_2M has been converted into HM^-.

$$[H^+] = \sqrt{\frac{K_1K_2F + K_1K_w}{K_1 + F}}$$

where $F = \left(\frac{50}{75}\right)(0.050\ 0\ M) = 0.033\ 3\ M$.

$[H^+] = 5.23 \times 10^{-5}\ M \Rightarrow pH = 4.28$

At 37.5 mL: $V_b = \frac{3}{2}V_e \Rightarrow pH = pK_2 = 5.70$

At 50.0 mL: At $2V_e$, HM^- has been converted into M^{2-}:

$M^{2-} + H_2O \rightleftharpoons HM^- + OH^-$
$(\frac{50}{100})(0.050\ 0) - x x x$

$$\frac{x^2}{0.025\ 0 - x} = K_{b1} = \frac{K_w}{K_{a2}} \Rightarrow x = 1.12 \times 10^{-5}\ M$$

$$\Rightarrow pH = -\log\left(\frac{K_w}{x}\right) = 9.05$$

At 56.3 mL: There are 6.3 mL of excess NaOH.

$$[OH^-] = \left(\frac{6.3}{106.3}\right)(0.100\ M) = 5.93 \times 10^{-3}\ M \Rightarrow pH = 11.77$$

10-E.

The equivalence points occur at 25.0 and 50.0 mL.

At 0 mL: HHis is the second intermediate form derived from the triprotic acid H_3His^{2+}.

$$[H^+] = \sqrt{\frac{K_2K_3(0.050\ 0) + K_2K_w}{K_2 + (0.050\ 0)}}$$

$$= 2.37 \times 10^{-8}\ M \Rightarrow pH = 7.62$$

At 4.0 mL:

	HHis	+	H^+	$\rightarrow$	H_2His^+
Initial:	25		4		—
Final:	21		—		4

$$pH = pK_2 + \log\frac{21}{4} = 6.69$$

At 12.5 mL: $pH = pK_2 = 5.97$

At 25.0 mL: The histidine has been converted into H_2His^+ at the formal concentration $F = \left(\frac{25}{50}\right)(0.050\ 0\ M) = 0.025\ 0\ M$.

$$[H^+] = \sqrt{\frac{K_1K_2F + K_1K_w}{K_1 + F}} = 1.16 \times 10^{-4}\ M \Rightarrow pH = 3.94$$

At 26.0 mL:

	H_2His^+	+	H^+	$\rightarrow$	H_3His^{2+}
Initial:	25		1		—
Final:	24		—		1

$$pH = pK_1 + \log\frac{24}{1} = 2.98$$

The approximation that histidine reacts completely with HCl breaks down between 25 and 50 mL. If you use titration equations in Table 10-5, you would find pH = 3.28, instead of 2.98, at $V_a = 26.0$ mL.

At 50.0 mL: The histidine has been converted into H_3His^{2+} at the formal concentration $F = \left(\frac{25}{75}\right)(0.050\ 0\ M) = 0.016\ 7\ M$.

$$H_3His^{2+} \rightleftharpoons H_2His^+ + H^+$$
$$\phantom{H_3His^{2+}}\ \ 0.016\ 7 - x \quad\quad x \quad\quad\quad x$$

$$\frac{x^2}{0.016\ 7 - x} = K_1 \Rightarrow 0.011\ 5\ M \Rightarrow pH = 1.94$$

10-F.
Figure 10-1: bromothymol blue: blue $\rightarrow$ yellow

Figure 10-2: thymol blue: yellow $\rightarrow$ blue

Figure 10-3: thymolphthalein: colorless $\rightarrow$ blue

10-G.
The titration reaction is $HA + OH^- \rightarrow A^- + H_2O$. It requires 1 mol NaOH to react with 1 mol HA. Therefore, the formal concentration of A^- at the equivalence point is

$$\underbrace{\left(\frac{27.63}{127.63}\right)}_{\substack{\text{Dilution factor}\\\text{for NaOH}}} \times \underbrace{(0.093\ 81)}_{\substack{\text{Initial concentration}\\\text{of NaOH}}} = 0.020\ 31\ M$$

Because the pH is 10.99, $[OH^-] = 9.77 \times 10^{-4}$ M, and we can write

$$A^- + H_2O \rightleftharpoons HA + OH^-$$

$$K_b = \frac{[HA][OH^-]}{[A^-]} = \frac{(9.77 \times 10^{-4})^2}{0.020\ 31 - (9.77 \times 10^{-4})} = 4.94 \times 10^{-5}$$

$$K_a = \frac{K_w}{K_b} = 2.03 \times 10^{-10} \Rightarrow pK_a = 9.69$$

For the 19.47-mL point, we have

	HA	+	OH^-	$\rightarrow$	A^-	+	H_2O
Initial:	27.63		19.47		—		—
Final:	8.16		—		19.47		—

$$pH = pK_a + \log\frac{[A^-]}{[HA]} = 9.69 + \log\frac{19.47}{8.16} = 10.07$$

10-H.
When $V_b = \frac{1}{2}V_e$, $[HA] = [A^-] = 0.033\ 3$ M (using a correction for dilution by NaOH). $[Na^+] = 0.033\ 3$ M as well. Ionic strength = 0.033 3 M.

$$pK_a = pH - \log\frac{[A^-]\gamma_{A^-}}{[HA]\gamma_{HA}}$$

$$= 4.62 - \log\frac{(0.033\ 3)(0.854)}{(0.033\ 3)(1.00)} = 4.69$$

The activity coefficient of A^- was found by interpolation in Table 7-1.

10-I.
(a) The derivatives are shown in the spreadsheet below. In the first derivative graph, the maximum is near 119 mL. In Figure 10-7, the second derivative graph gives an end point of 118.9 μL.

	A	B	C	D	E	F	G
1	Derivatives of titration curve						
2							
3			First Derivative		Second Derivative		
4	μL NaOH	pH	μL	Derivative	μL	Derivative	Vb*10^-pH
5	107	6.921					
6	110	7.117	108.5	6.533E-02			
7	113	7.359	111.5	8.067E-02	110	5.11E-03	4.94E-06
8	114	7.457	113.5	9.800E-02	112.5	8.67E-03	3.98E-06
9	115	7.569	114.5	1.120E-01	114	1.40E-02	3.10E-06
10	116	7.705	115.5	1.360E-01	115	2.40E-02	2.29E-06
11	117	7.878	116.5	1.730E-01	116	3.70E-02	1.55E-06
12	118	8.090	117.5	2.120E-01	117	3.90E-02	9.59E-07
13	119	8.343	118.5	2.530E-01	118	4.10E-02	5.40E-07
14	120	8.591	119.5	2.480E-01	119	-5.00E-03	3.08E-07
15	121	8.794	120.5	2.030E-01	120	-4.50E-02	1.94E-07
16	122	8.952	121.5	1.580E-01	121	-4.50E-02	1.36E-07
17							
18	C6 = (A6+A5)/2			E7 = (C7+C6)/2		G7 = A7*10^-B7	
19	D6 = (B6-B5)/(A6-A5)			F7 = (D7-D6)/(C7-C6)			

(b) Column G in the spreadsheet gives $V_b(10^{-pH})$. In a graph of $V_b(10^{-pH})$ vs. V_b, the points from 113 to 117 µL give a straight line whose slope is -1.178×10^6 and whose intercept (end point) is 118.7 µL.

10-J. (a) pH 9.6 is past the equivalence point, so excess volume, V, is given by $[OH^-] = 10^{-4.4} = (0.100\ 0\ M)\dfrac{V}{50.00 + 10.00 + V} \Rightarrow V = 0.024$ mL.

(b) pH 8.8 is before the equivalence point:

$$8.8 = 6.27 + \log \frac{[A^-]}{[HA]} \Rightarrow \frac{[A^-]}{[HA]} = 339$$

Titration reaction:	HA	+	OH⁻	→	A⁻	+	H₂O
Relative initial quantities:	10		V		—		—
Relative final quantities:	$10 - V$		—		V		—

To attain a ratio $[A^-]/[HA] = 339$, we need $V/(10 - V) = 339 \Rightarrow V = 9.97$ mL. The indicator error is $10 - 9.97 = 0.03$ mL.

10-K. (a) $A = 2\ 080[HIn] + 14\ 200[In^-]$

(b) $[HIn] = x$; $[In^-] = 1.84 \times 10^{-4} - x$

$A = 0.868 = 2\ 080x + 14\ 200(1.84 \times 10^{-4} - x)$

$\Rightarrow x = 1.44 \times 10^{-4}$ M

$pK_a = pH - \log \dfrac{[In]}{[HIn]}$

$= 6.23 - \log \dfrac{(1.84 \times 10^{-4}) - (1.44 \times 10^{-4})}{(1.44 \times 10^{-4})} = 6.79$

Chapter 11

11-A. For every mole of K^+ entering the first reaction, 4 moles of EDTA are produced in the second reaction.

Moles of EDTA = moles of Zn^{2+} used in titration

$[K^+] = \dfrac{\frac{1}{4}(\text{moles of } Zn^{2+})}{\text{volume of original sample}}$

$= \dfrac{\frac{1}{4}[28.73\ (\pm 0.03)][0.043\ 7\ (\pm 0.000\ 1)]}{250.0\ (\pm 0.1)}$

$= \dfrac{[\frac{1}{4}\ (\pm 0\%)][28.73\ (\pm 0.104\%)][0.043\ 7\ (\pm 0.229\%)]}{250.0\ (\pm 0.040\ 0\%)}$

$= 1.256\ (\pm 0.255\%) \times 10^{-3}$ M $= 1.256\ (\pm 0.003)$ mM

11-B. Total $Fe^{3+} + Cu^{2+}$ in 25.00 mL = (16.06 mL) × (0.050 83 M) = 0.816 3 mmol.

Second titration:

millimoles EDTA used: (25.00 mL)(0.050 83 M) = 1.270 8

millimoles Pb^{2+} needed: (19.77 mL)(0.018 83 M) = 0.372 3

millimoles Fe^{3+} present: (difference)   0.898 5

Because 50.00 mL of unknown were used in the second titration, Fe^{3+} in 25.00 mL $= \frac{1}{2}$ (0.898 5 mmol) = 0.449 2 mmol. The millimoles of Cu^{2+} in 25.00 mL are 0.816 3 − 0.449 2 = 0.367 1 mmol/25.00 mL = 0.014 68 M.

11-C. Designating the total concentration of free EDTA as [EDTA], we can write

$K_f' = \dfrac{[CuY^{2-}]}{[Cu^{2+}][EDTA]} = \alpha_{Y^{4-}} K_f = (2.9 \times 10^{-7})(10^{18.78})$

$= 1.7_4 \times 10^{12}$

Representative calculations:

At 0.1 mL:

$[EDTA] = \left(\dfrac{25.0 - 0.1}{25.0}\right)(0.040\ 0\ M)\left(\dfrac{50.0}{50.1}\right) = 0.039\ 8$ M

$[CuY^{2-}] = \left(\dfrac{0.1}{50.1}\right)(0.080\ 0\ M) = 1.60 \times 10^{-4}$ M

$[Cu^{2+}] = \dfrac{[CuY^{2-}]}{K_f'[EDTA]} = \dfrac{(1.60 \times 10^{-4}\ M)}{(1.7_4 \times 10^{12})(0.039\ 8\ M)}$

$= 2.3 \times 10^{-15}$ M $\Rightarrow pCu^{2+} = 14.64$

At 25.0 mL:

Formal concentration of $CuY^{2-} = \left(\dfrac{25.0}{75.0}\right)(0.080\ 0\ M) = 0.026\ 7$ M

	Cu²⁺	+	EDTA	⇌	CuY²⁻
Initial concentration:	—		—		0.026 7
Final concentration:	x		x		$0.026\ 7 - x$

$\dfrac{0.0267 - x}{x^2} = 1.7_4 \times 10^{12} \Rightarrow [Cu^{2+}] = 1.2_4 \times 10^{-7}$ M $\Rightarrow pCu^{2+} = 6.91$

At 26.0 mL:

$[Cu^{2+}] = \left(\dfrac{1.0}{76.0}\right)(0.080\ 0\ M) = 1.05 \times 10^{-3}$ M $\Rightarrow pCu^{2+} = 2.98$

Volume (mL)	pCu²⁺	Volume	pCu²⁺	Volume	pCu²⁺
0.1	14.64	15.0	12.07	25.0	6.91
5.0	12.84	20.0	11.64	26.0	2.98
10.0	12.42	24.0	10.86	30.0	2.30

11-D. We seek a relationship between $[H_2Y^{2-}]$ and $[Y^{4-}]$, which we can relate to total EDTA:

$HY^{3-} \rightleftharpoons H^+ + Y^{4-} \qquad K_6$

$H_2Y^{2-} \rightleftharpoons H^+ + HY^{3-} \qquad K_5$

$H_2Y^{2-} \rightleftharpoons 2H^+ + Y^{4-} \qquad K = K_5K_6 = \dfrac{[H^+]^2[Y^{4-}]}{[H_2Y^{2-}]}$

$[H_2Y^{2-}] = \dfrac{[H^+]^2[Y^{4-}]}{K_5K_6} = \dfrac{[H^+]^2\alpha_{Y^{4-}}\ [EDTA]}{K_5K_6}$

Using the values $[H^+] = 10^{-5.00}$ M, $\alpha_{Y^{4-}} = 2.9 \times 10^{-7}$, and $[EDTA] = 1.2_4 \times 10^{-7}$ M gives $[H_2Y^{2-}] = 1.1 \times 10^{-7}$ M.

11-E. (a) One volume of Mn^{2+} requires two volumes of EDTA to reach the equivalence point. The formal concentration of MnY^{2-} at the equivalence point is $(\frac{1}{3})(0.010\ 0\ M) = 0.003\ 33$ M.

	Mn²⁺	+	EDTA	⇌	MnY²⁻
	x		x		$0.003\ 33 - x$

$\dfrac{0.003\ 33 - x}{x^2} = \alpha_{Y^{4-}} K_f = (3.8 \times 10^{-4})10^{13.89} = 2.9 \times 10^{10}$

$\Rightarrow x = [Mn^{2+}] = 3.4 \times 10^{-7}$ M

(b) pH is constant, so the *quotient* $[H_3Y^-]/[H_2Y^{2-}]$ is constant throughout the *entire* titration.

$\dfrac{[H_2Y^{2-}][H^+]}{[H_3Y^-]} = K_4 \Rightarrow \dfrac{[H_3Y^-]}{[H_2Y^{2-}]} = \dfrac{[H^+]}{K_4} = \dfrac{10^{-7.00}}{10^{-2.69}} = 4.9 \times 10^{-5}$

11-F. K_f for $CoY^{2-} = 10^{16.45} = 2.8 \times 10^{16}$

$\alpha_{Y^{4-}} = 0.041$ at pH 9.00

$\alpha_{Co^{2+}} = \dfrac{1}{1 + \beta_1[C_2O_4^{2-}] + \beta_2[C_2O_4^{2-}]^2} = 6.8 \times 10^{-6}$

(using $\beta_1 = K_1 = 10^{4.69}$ and $\beta_2 = K_1K_2 = 10^{7.15}$)

$K_f' = \alpha_{Y^{4-}} K_f = 1.1_6 \times 10^{15}$

$K_f'' = \alpha_{Co^{2+}} \alpha_{Y^{4-}} K_f = 7.9 \times 10^9$

At 0 mL:

$[Co^{2+}] = \alpha_{Co^{2+}}(1.00 \times 10^{-3} \text{ M}) = 6.8 \times 10^{-9} \text{ M} \Rightarrow pCo^{2+} = 8.17$

At 1.00 mL:

$$C_{Co^{2+}} = \underbrace{\left(\frac{1.00}{2.00}\right)}_{\substack{\text{Fraction} \\ \text{remaining}}} \underbrace{(1.00 \times 10^{-3} \text{ M})}_{\substack{\text{Initial} \\ \text{concentration}}} \underbrace{\left(\frac{20.00}{21.00}\right)}_{\substack{\text{Dilution} \\ \text{factor}}}$$

$$= 4.76 \times 10^{-4} \text{ M}$$

$[Co^{2+}] = \alpha_{Co^{2+}} C_{Co^{2+}} = 3.2 \times 10^{-9} \text{ M} \Rightarrow pCo^{2+} = 8.49$

At 2.00 mL: This is the equivalence point.

$$C_{Co^{2+}} \quad + \quad EDTA \quad \overset{K_f''}{\rightleftharpoons} \quad CoY^{2-}$$

$$x \qquad\qquad x \qquad\qquad \left(\frac{20.00}{22.00}\right)(1.00 \times 10^{-3} \text{ M}) - x$$

$K_f'' = \dfrac{9.09 \times 10^{-4} - x}{x^2} \Rightarrow x = 3.4 \times 10^{-7} \text{ M} = C_{Co^{2+}}$

$[Co^{2+}] = \alpha_{Co^{2+}} C_{Co^{2+}} = 2.3 \times 10^{-12} \text{ M} \Rightarrow pCo^{2+} = 11.64$

At 3.00 mL:

$[\text{excess EDTA}] = \dfrac{1.00}{23.00}(1.00 \times 10^{-2} \text{ M}) = 4.35 \times 10^{-4} \text{ M}$

$[CoY^{2-}] = \dfrac{20.00}{23.00}(1.00 \times 10^{-3} \text{ M}) = 8.70 \times 10^{-4} \text{ M}$

Knowing [EDTA] and $[CoY^{2-}]$, we use K_f' to find $[Co^{2+}]$:

$K_f' = \dfrac{[CoY^{2-}]}{[Co^{2+}][EDTA]} = \dfrac{[8.70 \times 10^{-4} \text{ M}]}{[Co^{2+}][4.35 \times 10^{-4} \text{ M}]}$

$\Rightarrow [Co^{2+}] = 1.7 \times 10^{-15} \text{ M} \Rightarrow pCo^{2+} = 14.76$

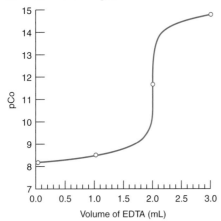

11-G. 25.0 mL of 0.120 M iminodiacetic acid = 3.00 mmol

25.0 mL of 0.050 0 M Cu^{2+} = 1.25 mmol

	Cu^{2+}	+	2 iminodiacetic acid	$\rightleftharpoons$	CuX_2^{2-}
Initial mmol:	1.25		3.00		—
Final mmol:	—		0.50		1.25

$\dfrac{[CuX_2^{2-}]}{[Cu^{2+}][X^{2-}]^2} = K_f$

$\dfrac{[1.25/50.0]}{[Cu^{2+}][(0.50/50.0)(4.6 \times 10^{-3})]^2} = 3.5 \times 10^{16}$

$\Rightarrow [Cu^{2+}] = 3.4 \times 10^{-10} \text{ M}$

Chapter 12

12-A. Hydroxybenzene = HA with $pK_{HA} = 9.997$

Dimethylamine = B from monoprotic BH^+ with $pK_{BH^+} = 10.774$

Mixture contains 0.010 mol HA, 0.030 mol B, and 0.015 mol HCl in 1.00 L

Chemical reactions:

$\quad HA \rightleftharpoons A^- + H^+ \qquad\qquad K_{HA} = 10^{-9.997}$

$\quad BH^+ \rightleftharpoons B + H^+ \qquad\qquad K_{BH^+} = 10^{-10.774}$

$\quad H_2O \rightleftharpoons H^+ + OH^- \qquad K_w = 10^{-14.00}$

Charge balance: $[H^+] + [BH^+] = [OH^-] + [A^-] + [Cl^-]$

Mass balances: $[Cl^-] = 0.015 \text{ M}$

$\qquad\qquad [BH^+] + [B] = 0.030 \text{ M} \equiv F_B$

$\qquad\qquad [HA] + [A^-] = 0.010 \text{ M} \equiv F_A$

We have seven equations and seven chemical species.

Fractional composition equations:

$[BH^+] = \alpha_{BH^+} F_B = \dfrac{[H^+]F_B}{[H^+] + K_{BH^+}}$

$[B] = \alpha_B F_B = \dfrac{K_{BH^+}F_B}{[H^+] + K_{BH^+}}$

$[HA] = \alpha_{HA} F_A = \dfrac{[H^+]F_A}{[H^+] + K_{HA}}$

$[A^-] = \alpha_{A^-} F_A = \dfrac{K_{HA}F_A}{[H^+] + K_{HA}}$

Substitute into charge balance:

$[H^+] + \alpha_{BH^+} F_B = K_w/[H^+] + \alpha_{A^-} F_A + [0.015 \text{ M}]$ \qquad (A)

We solve Equation A for $[H^+]$ by using Solver in the spreadsheet, with an initial guess of pH = 10 in cell H10. In the Data ribbon, select Solver and choose Options. Set Precision to 1e-16 and click OK. In the Solver

	A	B	C	D	E	F	G	H	I
1	Mixture of 0.010 M HA, 0.030 M B, and 0.015 M HCl								
2									
3	F_A =	0.010		F_B =	0.030		[Cl⁻] =	0.015	
4	pK_{HA} =	9.997		pK_{BH+} =	10.774		pK_w =	14.000	
5	K_{HA} =	1.01E-10		K_{BH+} =	1.68E-11		K_w =	1.00E-14	
6									
7	Species in charge balance:						Other concentrations:		
8	[H⁺] =	4.67E-11		[A⁻] =	6.83E-03		[HA] =	3.17E-03	
9	[BH⁺] =	2.20E-02		[Cl⁻] =	0.015		[B] =	7.95E-03	
10				[OH⁻] =	2.14E-04		pH =	10.331	
11							↑ initial value is a guess		
12	Positive charge minus negative charge				-4.92E-17	= B8+B9-E8-E9-E10			
13	Formulas:								
14	B5 = 10^-B4		B8 = 10^-H10				H5 = 10^-H4		
15	E5 = 10^-E4		E10 = H5/B8				E9 = H3		
16	B9 = B8*E3/(B8+E5)						E8 = B5*B3/(B8+B5)		
17	H9 = E5*E3/(B8+E5)						H8 = B8*B3/(B8+B5)		

Spreadsheet for Exercise 12-A.

window, Set Target Cell E12 Equal To Value of 0 By Changing Cells H10. Click Solve and Solver finds pH = 10.33 in cell H10, giving a net charge near 0 in cell E12.

12-B. We use effective equilibrium constants, K', as follows:

$$K_{HA} = \frac{[A^-]\gamma_{A^-}[H^+]\gamma_{H^+}}{[HA]\gamma_{HA}} = 10^{-9.997}$$

$$K'_{HA} = K_{HA}\left(\frac{\gamma_{HA}}{\gamma_{A^-}\gamma_{H^+}}\right) = \frac{[A^-][H^+]}{[HA]}$$

$$[HA] = \alpha_{HA}F_A = \frac{[H^+]F_A}{[H^+] + K'_{HA}}$$

$$[A^-] = \alpha_{A^-}F_A = \frac{K'_{HA}F_A}{[H^+] + K'_{HA}}$$

$$K_{BH^+} = \frac{[B]\gamma_B[H^+]\gamma_{H^+}}{[BH^+]\gamma_{BH^+}} = 10^{-10.774}$$

$$K'_{BH^+} = K_{BH^+}\left(\frac{\gamma_{BH^+}}{\gamma_B\gamma_{H^+}}\right) = \frac{[B][H^+]}{[BH^+]}$$

$$[BH^+] = \alpha_{BH^+}F_B = \frac{[H^+]F_B}{[H^+] + K'_{BH^+}}$$

$$[B] = \alpha_B F_B = \frac{K'_{BH^+}F_B}{[H^+] + K'_{BH^+}}$$

$$K_w = [H^+]\gamma_{H^+}[OH^-]\gamma_{OH^-} = 10^{-13.995}$$

$$K'_w = \frac{K_w}{\gamma_{H^+}\gamma_{OH^-}} = [H^+][OH^-]$$

$$[OH^-] = K'_w/[H^+]$$

$$pH = -\log([H^+]\gamma_{H^+})$$

$$[H^+] = (10^{-pH})/\gamma_{H^+}$$

The spreadsheet in Exercise 12-A is modified to add activity coefficients in cells A8:H9. Effective equilibrium constants are computed in row 5. Because we are going to the trouble of using activity coefficients, we use $pK_w = 13.995$ in cell H4 instead of the less accurate value of 14.00. With an initial ionic strength of 0 in cell C17 and a guess of pH = 10 in cell H14, the net charge in cell E16 is 0.005 56 m. Execute Solver to find the pH that reduces the net charge to near 0. Results are shown in the spreadsheet. The calculated ionic strength in cell C18 is 0.022 0 m. We write this number in cell C17 and use Solver again to find pH. Here are the results:

Iteration	Ionic strength	pH
1	0	10.331
2	0.022 0	10.357
3	0.022 5	10.358
4	0.022 5	10.358

12-C. **(a)** 2-Aminobenzoic acid = HA from diprotic H_2A^+, $pK_1 = 2.08$, $pK_2 = 4.96$

Dimethylamine = B from monoprotic BH^+, $pK_a = 10.774$

Mixture contains 0.040 mol HA, 0.020 mol B, and 0.015 mol HCl in 1.00 L.

Chemical reactions:

$$H_2A^+ \rightleftharpoons HA + H^+ \qquad K_1 = 10^{-2.08}$$
$$HA \rightleftharpoons A^- + H^+ \qquad K_2 = 10^{-4.96}$$
$$BH^+ \rightleftharpoons B + H^+ \qquad K_a = 10^{-10.774}$$
$$H_2O \rightleftharpoons H^+ + OH^- \qquad K_w = 10^{-14.00}$$

Charge balance: $\quad [H^+] + [H_2A^+] + [BH^+] = [OH^-] + [A^-] + [Cl^-]$

Mass balances: $\quad [Cl^-] = 0.015$ M

$$[BH^+] + [B] = 0.020 \text{ M} \equiv F_B$$
$$[H_2A^+] + [HA] + [A^-] = 0.040 \text{ M} \equiv F_A$$

We have eight equations and eight chemical species.

	A	B	C	D	E	F	G	H
1	Mixture of 0.010 M HA, 0.030 M B, and 0.015 M HCl							
2	With activities							
3	F_A =	0.010		F_B =	0.030		$[Cl^-]$ =	0.015
4	pK_{HA} =	9.997		pK_{BH^+} =	10.774		pK_w =	13.995
5	K_{HA}' =	1.01E-10		K_{BH^+}' =	1.68E-11		K_w' =	1.01E-14
6								
7	Activity coefficients:							
8	H^+ =	1.00		A^-	1.00		HA	1.00
9	OH^- =	1.00		BH^+	1.00		B	1.00
10								
11	Species in charge balance:						Other concentrations:	
12	$[H^+]$ =	4.67E-11		$[A^-]$ =	6.83E-03		[HA] =	3.17E-03
13	$[BH^+]$ =	2.20E-02		$[Cl^-]$ =	0.015		[B] =	7.95E-03
14				$[OH^-]$ =	2.17E-04		pH =	10.331
15							↑ initial value is a guess	
16	Positive charge minus negative charge =				-3.24E-17	= B12+B13-E12-E13-E14		
17	Ionic strength =		0.0000	← initial value is 0				
18	New ionic strength =		0.0220	← substitute this value into cell C17 for next iteration				
19								
20	Formulas:							
21	B5 = (10^-B4)*H8/(E8*B8)				H8 = H9 = 1			
22	E5 = (10^-E4)*E9/(H9*B8)				E13 = H3			
23	H5 = (10^-H4)/(B8*B9)							
24	B8 = B9 = E8 = E9 = 10^(-0.51*1^2*(SQRT($C$17)/(1+SQRT($C$17))-0.3*$C$17))							
25	B12 = (10^-H14)/B8				E14 = H5/B12			
26	B13 = B12*E3/(B12+E5)				H13 = E5*E3/(B12+E5)			
27	E12 = B5*B3/(B12+B5)				H12 = B12*B3/(B12+B5)			
28	C18 = 0.5*(B12+B13+E12+E13+E14)							

Spreadsheet for Exercise 12-B.

	A	B	C	D	E	F	G	H	I
1	Mixture of 0.040 M HA, 0.020 M B, and 0.015 M HCl								
2									
3	F_A =	0.040		F_B =	0.020		[Cl⁻] =	0.015	
4	pK_1 =	2.080		pK_a =	10.774		K_w =	1.00E-14	
5	pK_2 =	4.960		K_a =	1.68E-11				
6	K_1 =	8.32E-03							
7	K_2 =	1.10E-05							
8									
9	Species in charge balance:						Other concentrations:		
10	[H⁺] =	7.03E-05		[H₂A⁺] =	2.90E-04		[HA] =	3.43E-02	
11	[BH⁺] =	2.00E-02		[A⁻] =	5.36E-03		[B] =	4.79E-09	
12	[OH⁻] =	1.42E-10		[Cl⁻] =	0.015		pH =	4.153	
13							↑ initial value is a guess		
14	Positive charge minus negative charge				0.00E+00		= B10+B11+E10-B12-E11-E12		
15	Formulas:								
16	B16 = 10^-B4		B7 = 10^-B5				E5 = 10^-E4		
17	B10 =10^-H12		B12 =H4/B10				E12 = H3		
18	B11 = B10*E3/(B10+E5)								
19	E10 = B10^2*B3/(B10^2+B10*B6+B6*B7)								
20	E11 = B6*B7*B3/(B10^2+B10*B6+B6*B7)								
21	H10 = B10*B6*B3/(B10^2+B10*B6+B6*B7)								
22	H11 = E5*E3/(B10+E5)								

Spreadsheet for Exercise 12-C.

Fractional composition equations:

$$[BH^+] = \alpha_{BH^+}F_B = \frac{[H^+]F_B}{[H^+] + K_a}$$

$$[B] = \alpha_B\, F_B = \frac{K_a F_B}{[H^+] + K_a}$$

$$[H_2A^+] = \alpha_{H_2A^+}F_A = \frac{[H^+]^2 F_A}{[H^+]^2 + [H^+]K_1 + K_1K_2}$$

$$[HA] = \alpha_{HA}\, F_A = \frac{K_1[H^+]F_A}{[H^+]^2 + [H^+]K_1 + K_1K_2}$$

$$[A^-] = \alpha_A\cdot F_A = \frac{K_1K_2F_A}{[H^+]^2 + [H^+]K_1 + K_1K_2}$$

Substitute into charge balance:

$$[H^+] + \alpha_{H_2A^+}F_A + \alpha_{BH^+}F_B = K_w/[H^+] + \alpha_A F_A + [0.015\ M] \quad (A)$$

Solve Equation A for [H⁺] by using Solver in the spreadsheet, with an initial guess of pH = 7 in cell H12. In the Data ribbon, select Solver and choose Options. Set Precision to 1e-16 and click OK. In the Solver window, Set Target Cell E14 Equal To Value of 0 By Changing Cells H12. Click Solve and Solver finds pH = 4.15 in cell H12, giving a net charge near 0 in cell E14.

(b) From the concentrations in the spreadsheet, we find the following fractions of 2-aminobenzoic acid: H_2A^+ = 0.7%, HA = 85.9%, and A^- = 13.4%. The fractions of dimethylamine are BH^+ = 100.0% and B = 0.0%. The simple prediction is that HCl would consume B, giving 100% BH^+. The remaining 5 mmol B consumes 5 mmol HA, making 5 mmol A^- and leaving 35 mmol HA.

Predicted fractions: A^- = 5/40 = 12.5%, HA = 35/40 = 87.5%

Estimated pH = pK_2 + log([A⁻]/[HA]) = 4.96 + log(5/35) = 4.11

12-D. Effective equilibrium constants:

$$H_2T \overset{pK_1 = 3.036}{\rightleftharpoons} HT^- + H^+ \qquad K_1' = K_1\left(\frac{\gamma_{H_2T}}{\gamma_{HT^-}\gamma_{H^+}}\right) = \frac{[HT^-][H^+]}{[H_2T]}$$

$$HT^- \overset{pK_2 = 4.366}{\rightleftharpoons} T^{2-} + H^+ \qquad K_2' = K_2\left(\frac{\gamma_{HT^-}}{\gamma_{T^{2-}}\gamma_{H^+}}\right) = \frac{[T^{2-}][H^+]}{[HT^-]}$$

$$PyH^+ \overset{pK_a = 5.20}{\rightleftharpoons} Py + H^+ \qquad K_a' = K_a\left(\frac{\gamma_{PyH^+}}{\gamma_{Py}\gamma_{H^+}}\right) = \frac{[Py][H^+]}{[PyH^+]}$$

$$H_2O \overset{pK_w = 13.995}{\rightleftharpoons} H^+ + OH^- \qquad K_w' = \frac{K_w}{\gamma_{H^+}\gamma_{OH^-}} = [H^+][OH^-]$$

$$[OH^-] = K_w'/[H^+] \qquad\qquad pH = -\log([H^+]\gamma_{H^+})$$

The spreadsheet is modified from the one in the text by addition of activity coefficients in cells A10:H11. The ionic strength for computing activity coefficients with the Davies equation is in cell C20. The initial ionic strength is set to 0. The first guess for pH in cell H17 is 5. Then use Solver to find the pH in cell H17 that makes the net charge in cell E19 nearly 0. From this pH, all concentrations are computed and the new ionic strength is found in cell C21. This new ionic strength is typed into cell C20 for the next iteration. From the new ionic strength, new activity coefficients are computed and new values of K' are computed in cells B6, B7, E5, and H5. The pH satisfying the charge balance in cell E19 is then found again with Solver and the process is repeated until ionic strength reaches a constant value.

Iteration	Ionic strength	pH
1	0	4.298
2	0.052 3	4.116
3	0.053 9	4.114
4	0.054 0	4.114

	A	B	C	D	E	F	G	H
1	Mixture of 0.020 M Na$^+$HT$^-$, 0.015 M PyH$^+$Cl$^-$, and 0.010 M KOH - with activity							
2								
3	F_{H2T} =	0.020		F_{PyH^+} =	0.015		[K$^+$] =	0.010
4	pK$_1$ =	3.036		pK$_a$ =	5.20		pK$_w$ =	13.995
5	pK$_2$ =	4.366		K$_a$' =	6.31E-06		K$_w$' =	1.52E-14
6	K$_1$' =	1.38E-03						
7	K$_2$' =	9.67E-05						
8								
9	Activity coefficients from Davies equation:							
10	H$^+$ =	0.82		HT$^-$ =	0.82		OH$^-$ =	0.82
11	PyH$^+$ =	0.82		T^{2-} =	0.45			
12								
13	Species in charge balance:						Other concentrations:	
14	[H$^+$] =	9.42E-05		[OH$^-$] =	1.61E-10		[H$_2$T] =	6.52E-04
15	[PyH$^+$] =	1.41E-02		[HT$^-$] =	9.54E-03		[Py] =	9.42E-04
16	[Na$^+$] =	0.020		[T^{2-}] =	9.80E-03			
17	[K$^+$] =	0.010		[Cl$^-$] =	0.015		pH =	4.114
18							↑ initial value is a guess	
19	Positive charge minus negative charge =				2.78E-17		=B14+B15+B16+B17	
20	Ionic strength =		0.0540	← initial value is 0			-E14-E15-2*E16-E17	
21	New ionic strength =		0.0540	← substitute this value into cell C17 for next iteration				
22								
23	Formulas:							
24	B6 = 10^-B4*(1/(E10*B10))			B14 = (10^-H17)/B10			E14 = H5/B14	
25	B7 = 10^-B5*(E10/(E11*B10))			B16 = B3			E17 = E3	
26	E5 = 10^-E4*(B11/B10)			B17 = H3				
27	H5 = (10^-H4)/(B10*H10)							
28	B10=B11=E10=H10 = 10^(-0.51*1^2*(SQRT($C$20)/(1+SQRT($C$20))-0.3*$C$20))							
29	E11 = 10^(-0.51*2^2*(SQRT($C$20)/(1+SQRT($C$20))-0.3*$C$20))							
30	B15 = B14*E3/(B14+E5)							
31	E15 = B14*B6*B3/(B14^2+B14*B6+B6*B7)							
32	E16 = B6*B7*B3/(B14^2+B14*B6+B6*B7)							
33	H14 = B14^2*B3/(B14^2+B14*B6+B6*B7)							
34	H15 = E5*E3/(B14+E5)							
35	C21 = 0.5*(B14+B15+B16+B17+E14+E15+4*E16+E17)							

Spreadsheet for Exercise 12-D.

12-E. $Mg^{2+} + SO_4^{2-} \rightleftharpoons MgSO_4(aq)$ $K_{ip} = 10^{2.23} (=169.8)$

If hydrolysis is neglected, $[Mg^{2+}] = [SO_4^{2-}]$ and $[MgSO_4] = F - [Mg^{2+}]$, where F is the formal concentration (0.025 M). We will solve the equation

$$K'_{ip} = K_{ip}\left(\frac{\gamma_{Mg^{2+}}\gamma_{SO_4^{2-}}}{\gamma_{MgSO_4}}\right) = \frac{[MgSO_4(aq)]}{[Mg^{2+}][SO_4^{2-}]} = \frac{F - [Mg^{2+}]}{[Mg^{2+}]^2} \quad (A)$$

with $\gamma_{MgSO_4} = 1$ because it is neutral.

(a) First, set $\gamma_{Mg^{2+}} = \gamma_{SO_4^{2-}} = 1.00$ and use Solver (or the quadratic equation) to find $[Mg^{2+}]$ to satisfy Equation A. In the spreadsheet, guess an initial value of $[Mg^{2+}] = 0.012\ 5$ M in cell B11. Open Solver from the Data ribbon. Select Solver Options and set Precision = 1e-6. Click OK. In the Solver window, Set Target Cell F14 Equal To Value of 169.8 By Changing Cells B11. Solver finds $[Mg^{2+}] = 0.009\ 54$ M and $\mu = 0.038\ 2$ M.

(b) Write the value $\mu = 0.038\ 2$ M in cell F3. This ionic strength changes K'_{ip} from 169.8 to 40.7. Execute Solver again. This time Set Target Cell F14 Equal To Value of 40.7 By Changing Cells B11. Solver finds $[Mg^{2+}] = 0.015\ 4$ M and $\mu = 0.061\ 5$ M. Write 0.061 5 in cell F3 and repeat the process again. The succession of results is shown in the table. The fraction of ion pairing is $[MgSO_4(aq)]/F = 33\%$.

Iteration	Ionic strength (mM)	[Mg^{2+}] (mM)
1	0	9.54
2	38.2	15.4
3	61.5	16.5
4	66.0	16.7
5	66.7	16.7
6	66.8	16.7

(c) The ionic strength is 0.067 M, not 0.10 M.

12-F. **(a)** In addition to Reactions 12-32 through 12-36, we add the reaction

$$HF(aq) + F^- \rightleftharpoons HF_2^- \quad K_{HF_2^-} = 10^{0.58}$$

$$\Rightarrow [HF_2^-] = K_{HF_2^-}[F^-][HF] = K_{HF_2^-}[F^-]([H^+][F^-]/K_{HF})$$

$$= K_{HF_2^-}[F^-]^2[H^+]/K_{HF}$$

The charge balance becomes

$$[H^+] + 2[Ca^{2+}] + [CaOH^+] + [CaF^+] - [OH^-] - [F^-] - [HF_2^-] = 0 \quad (A)$$

and the mass balance is

$$2\{[Ca^{2+}] + [CaOH^+] + [CaF^+]\} = [F^-] + [HF] + [CaF^+] + 2[HF_2^-]$$

$$2[Ca^{2+}] + 2[CaOH^+] + [CaF^+] - [F^-] - [HF] - 2[HF_2^-] = 0$$

	A	**B**	**C**	**D**	**E**	**F**
1	Ion pairing of $MgSO_4$ with Davies activity					
2						
3	F =	0.025		Ionic strength =		0.0000
4	log K_{ip} =	2.23		New ionic strength =		0.0382
5	K_{ip}' =	169.8				
6						
7	Activity coefficients from Davies equation:					
8	γ_{Mg} =	1.00		γ_{SO4} =	1.00	
9						
10	Concentrations:					
11	$[Mg^{2+}]$ =	0.00954	← initial value is a guess			
12	$[MgSO_4(aq)]$ =	0.01546		$[SO_4^{2-}]$ =	0.00954	
13						
14	Reaction quotient = $[MgSO_4(aq)]/([Mg^{2+}][SO_4^{2-}])$ =					169.8
15	% ion pair =	61.8	= 100*B12/B3			
16						
17	Formulas:					
18	B5 = (10^B4)*B8*E8		B12 = B3-B11		E12 = B11	
19	B8 = E8 = 10^(-0.51*2^2*(SQRT($F$3)/(1+SQRT($F$3))-0.3*$F$3))					
20	F14 = B12/(B11*E12)			F4 = 0.5*(B11*4+E12*4)		

Spreadsheet for Exercise 12-E.

Substitute equilibrium expressions for the various species into the mass balance:

$$\frac{2K_{sp}}{[F^-]^2} + \frac{2K_aK_{sp}}{[H^+][F^-]^2} + \frac{K_{ip}K_{sp}}{[F^-]} - [F^-] - \frac{[H^+][F^-]}{K_{HF}} - \frac{2K_{HF_2^-}[F^-]^2[H^+]}{K_{HF}} = 0 \tag{B}$$

For a given value of $[H^+]$, use Solver to find the value of $[F^-]$ that makes the left side of Equation B equal to 0. From $[H^+]$ and $[F^-]$, compute the remaining concentrations with the equilibrium expressions. Results are shown in the graph.

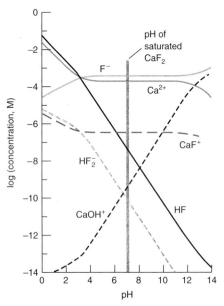

Graph for Exercise 12-F.

(b) To find the pH of unbuffered solution, find $[H^+]$ at which the charge balance (A) is also satisfied. Because $[HF_2^-]$ is negligible near neutral pH, the pH is unchanged from the case worked in the text. The pH of unbuffered solution is 7.10 and concentrations are

$[F^-] = 4.00 \times 10^{-4}$ M	$[Ca^{2+}] = 2.00 \times 10^{-4}$ M
$[HF] = 4.67 \times 10^{-8}$ M	$[CaOH^+] = 5.03 \times 10^{-10}$ M
$[HF_2^-] = 7.11 \times 10^{-11}$ M	$[CaF^+] = 3.44 \times 10^{-7}$ M

12-G. $AgCN(s) \rightleftharpoons Ag^+ + CN^-$ $pK_{sp} = 15.66$

$HCN(aq) \rightleftharpoons CN^- + H^+$ $pK_{HCN} = 9.21$

$Ag^+ + H_2O \rightleftharpoons AgOH(aq) + H^+$ $pK_{Ag} = 12.0$

Express all concentrations in terms of $[Ag^+]$ and $[H^+]$:

$$[CN^-] = K_{sp}/[Ag^+]$$

$$[HCN(aq)] = \frac{[H^+][CN^-]}{K_{HCN}} = \frac{[H^+]K_{sp}}{K_{HCN}[Ag^+]}$$

$$[AgOH(aq)] = \frac{K_{Ag}[Ag^+]}{[H^+]}$$

Mass balance: total silver = total cyanide

$$[Ag^+] + [AgOH(aq)] = [CN^-] + [HCN(aq)]$$

Substitute expressions for concentrations into the mass balance:

$$[Ag^+] + \frac{K_{Ag}[Ag^+]}{[H^+]} = \frac{K_{sp}}{[Ag^+]} + \frac{[H^+]K_{sp}}{K_{HCN}[Ag^+]} \tag{A}$$

Rearrange Equation A to solve for $[Ag^+]$ as a function of $[H^+]$ or use Solver to find $[Ag^+]$ as a function of $[H^+]$. We will use the algebraic

	A	B	C	D	E	F	G	H
1	Solubility of AgCN							
2								
3	pK_{sp} =	15.66		K_{sp} =	2.2E-16		K_w =	1.00E-14
4	pK_{HCN} =	9.21		K_{HCN} =	6.2E-10			
5	pK_{Ag} =	12.00		K_{Ag} =	1.0E-12			
6								Net
7	pH	[H$^+$]	[Ag$^+$]	[CN$^-$]	[HCN]	[AgOH]	[OH$^-$]	charge
8	0	1.0E+00	6.0E-04	3.7E-13	6.0E-04	6.0E-16	1.00E-14	1.0E+00
9	2	1.0E-02	6.0E-05	3.7E-12	6.0E-05	6.0E-15	1.00E-12	1.0E-02
10	4	1.0E-04	6.0E-06	3.7E-11	6.0E-06	6.0E-14	1.00E-10	1.1E-04
11	6	1.0E-06	6.0E-07	3.7E-10	6.0E-07	6.0E-13	1.00E-08	1.6E-06
12	7.28	5.3E-08	1.4E-07	1.6E-09	1.4E-07	2.6E-12	1.89E-07	3.7E-21
13	8	1.0E-08	6.1E-08	3.6E-09	5.8E-08	6.1E-12	1.00E-06	-9.3E-07
14	10	1.0E-10	1.6E-08	1.4E-08	2.2E-09	1.6E-10	1.00E-04	-1.0E-04
15	12	1.0E-12	1.0E-08	2.1E-08	3.4E-11	1.0E-08	1.00E-02	-1.0E-02
16	14	1.0E-14	1.5E-09	1.5E-07	2.4E-12	1.5E-07	1.00E+00	-1.0E+00
17								
18	B8 = 10^-A8			D8 = \$E\$3/C8		E8 = B8*D8/\$E\$4		F8 = \$E\$5*C8/B8
19	C8 = SQRT(\$E\$3*(\$E\$4+B8)*(B8)/(\$E\$4*(B8+\$E\$5)))							
20	G8 = \$H\$3/B8			H8 = B8+C8-D8-G8				

Spreadsheet for Exercise 12-G.

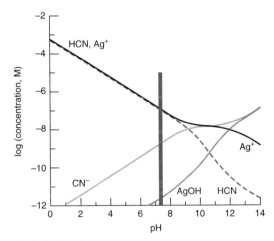

Graph for Exercise 12-G.

solution, which is easy for this exercise. Multiply both sides by [Ag$^+$] and solve:

$$[Ag^+]^2 + \frac{K_{Ag}[Ag^+]^2}{[H^+]} = K_{sp} + \frac{[H^+]K_{sp}}{K_{HCN}}$$

$$[Ag^+]^2 \left(\frac{[H^+] + K_{Ag}}{[H^+]} \right) = K_{sp} \left(\frac{K_{HCN} + [H^+]}{K_{HCN}} \right)$$

$$[Ag^+] = \sqrt{\frac{K_{sp}(K_{HCN} + [H^+])[H^+]}{K_{HCN}([H^+] + K_{Ag})}} \qquad (B)$$

The spreadsheet uses Equation B to find [Ag$^+$] in column C. pH is input in column A. To find the pH of unbuffered solution, we find the pH at which the net charge in column H is zero. We used Solver to find that pH = 7.28 in cell A12 makes the net charge in cell H12 equal to 0.

12-H. **(a)** $\bar{n}_H = \dfrac{\text{moles of bound H}^+}{\text{total moles of weak acid}} = \dfrac{[HA]}{[HA] + [A^-]}$

$$= \frac{[HA]}{F_{HA}} = \frac{F_{HA} - [A^-]}{F_{HA}} \qquad (A)$$

Charge balance: $[H^+] + [Na^+] = [OH^-] + [Cl^-]_{HCl} + [A^-]$

or $\quad -[A^-] = [OH^-] + [Cl^-]_{HCl} - [H^+] - [Na^+]$

Put this expression for $-[A^-]$ into numerator of (A):

$$\bar{n}_H(\text{measured}) = \frac{F_{HA} + [OH^-] + [Cl^-]_{HCl} - [H^+] - [Na^+]}{F_{HA}}$$

$$= 1 + \frac{[OH^-] + [Cl^-]_{HCl} - [H^+] - [Na^+]}{F_{HA}}$$

Making the same substitutions used in Section 12-4 produces Equation 12-59 with $n = 1$. The theoretical expression is $\bar{n}_H(\text{theoretical}) = \alpha_{HA} = [H^+]/([H^+] + K_a)$.

(b) Optimized values of pK'_w and pK_a in cells B9 and B10 in the spreadsheet are 13.869 and 4.726. These were obtained from initial guesses of pK'_w = 14 and pK_a = 5 after executing Solver to minimize the sum of squared residuals in cell B11. The NIST database lists pK_a = 4.757 at μ = 0 and pK_a = 4.56 at μ = 0.1 M. Our value of 4.726 at μ = 0.1 M suggests that the titration experiment was not very accurate.

	A	B	C	D	E	F	G
1	Difference plot for acetic acid						
2				C15 = 10^-B15/$B$8			
3	Titrant NaOH =	0.4905	C_b (M)	D15 = 10^-$B$9/C15			
4	Initial volume =	200	V_o (mL)	E15 = $B$7+($B$6-$B$3*A15			
5	Acetic acid =	3.96	L (mmol)	-(C15-D15)*($B$4+A15))/$B$5			
6	HCl added =	0.484	A (mmol)	F15 = $C15/($C15+$E$10)			
7	Number of H^+ =	1	n	G15 = (E15-F15)^2			
8	Activity coeff =	0.78	γ_H				
9	pK_w' =	13.869					
10	pK_a =	4.726		K_a =	1.881E-05	= 10^-B10	
11	$\Sigma(resid)^2$ =	0.0045	= sum of column G				
12							
13	v	pH	$[H^+]$ =	$[OH^-]$ =	Measured	Theoretical	$(residuals)^2$ =
14	mL NaOH		$(10^{-pH})/\gamma_H$	$(10^{-pKw})/[H^+]$	n_H	$n_H = \alpha_{HA}$	$(n_{meas} - n_{theor})^2$
15	0.00	2.79	2.08E-03	6.50E-12	1.017	0.991	0.000685
16	0.30	2.89	1.65E-03	8.19E-12	1.002	0.989	0.000163
17	:						
18	4.80	4.78	2.13E-05	6.36E-10	0.527	0.531	0.000018
19	5.10	4.85	1.81E-05	7.47E-10	0.490	0.491	0.000001
20	:						
21	10.20	11.39	5.22E-12	2.59E-03	-0.004	0.000	0.000014
22	10.50	11.54	3.70E-12	3.66E-03	0.016	0.000	0.000259

Spreadsheet for Exercise 12-H.

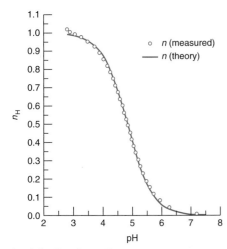

Graph for Exercise 12-H.

Chapter 13

13-A. The cell voltage will be 1.35 V because all activities are unity.

$I = P/E = 0.010\ 0\ W/1.35\ V = 7.41 \times 10^{-3}\ C/s$

$mol\ e^-/s = (7.41 \times 10^{-3}\ C/s)/(9.649 \times 10^4\ C/mol)$

$= 7.68 \times 10^{-8}\ mol\ e^-/s = 2.42\ mol\ e^-/365\ days$

$= 1.21\ mol\ HgO/365\ days = 0.262\ kg\ HgO$

$= 0.578\ lb$

13-B.

(a)
$$\underline{\begin{array}{l} 5Br_2(aq) + 10e^- \rightleftharpoons 10Br^- \\ 2IO_3^- + 12H^+ + 10e^- \rightleftharpoons I_2(s) + 6H_2O \end{array}}$$
$$I_2(s) + 5Br_2(aq) + 6H_2O \rightleftharpoons 2IO_3^- + 10Br^- + 12H^+$$
$E_+^\circ = 1.098\ V$

$\underline{E_-^\circ = 1.210\ V}$

$E^\circ = 1.098 - 1.210 = -0.112\ V$

$K = 10^{10(-0.112)/0.059\ 16} = 1 \times 10^{-19}$

(b)
$$\underline{\begin{array}{l} Cr^{2+} + 2e^- \rightleftharpoons Cr(s) \\ Fe^{2+} + 2e^- \rightleftharpoons Fe(s) \end{array}}$$
$$Cr^{2+} + Fe(s) \rightleftharpoons Cr(s) + Fe^{2+}$$
$E_+^\circ = -0.89\ V$

$\underline{E_-^\circ = -0.44\ V}$

$E^\circ = -0.89 - (-0.44) = -0.45\ V$

$K = 10^{2(-0.45)/0.059\ 16} = 6 \times 10^{-16}$

(c)
$$\underline{\begin{array}{l} Cl_2(g) + 2e^- \rightleftharpoons 2Cl^- \\ Mg^{2+} + 2e^- \rightleftharpoons Mg(s) \end{array}}$$
$$Mg(s) + Cl_2(g) \rightleftharpoons Mg^{2+} + 2Cl^-$$
$E_+^\circ = 1.360\ V$

$\underline{E_-^\circ = -2.360\ V}$

$E^\circ = 1.360 - (-2.360) = 3.720\ V$

$K = 10^{2(3.720)/0.059\ 16} = 6 \times 10^{125}$

(d)
$$\underline{\begin{array}{l} 3[MnO_2(s) + 4H^+ + 2e^- \rightleftharpoons Mn^{2+} + 2H_2O] \\ 2[MnO_4^- + 4H^+ + 3e^- \rightleftharpoons MnO_2(s) + 2H_2O] \end{array}}$$
$$5MnO_2(s) + 4H^+ \rightleftharpoons 2MnO_4^- + 3Mn^{2+} + 2H_2O$$
$E_+^\circ = 1.230\ V$

$\underline{E_-^\circ = 1.692\ V}$

$E^\circ = 1.230 - 1.692 = -0.462\ V$

$K = 10^{6(-0.462)/0.059\ 16} = 1 \times 10^{-47}$

An alternate way to answer **(d)** is
$$\underline{\begin{array}{l} 5[MnO_2(s) + 4H^+ + 2e^- \rightleftharpoons Mn^{2+} + 2H_2O] \\ 2[MnO_4^- + 8H^+ + 5e^- \rightleftharpoons Mn^{2+} + 4H_2O] \end{array}}$$
$$5MnO_2(s) + 4H^+ \rightleftharpoons 2MnO_4^- + 3Mn^{2+} + 2H_2O$$
$E_+^\circ = 1.230\ V$

$\underline{E_-^\circ = 1.507\ V}$

$E^\circ = 1.230 - 1.507 = -0.277\ V$

$K = 10^{10(-0.277)/0.059\ 16} = 2 \times 10^{-47}$

(e)

$$\underline{\quad} \quad Ag^+ + e^- \rightleftharpoons Ag(s)$$

$$\frac{Ag(S_2O_3)_2^{3-} + e^- \rightleftharpoons Ag(s) + 2S_2O_3^{2-}}{Ag^+ + 2S_2O_3^{2-} \rightleftharpoons Ag(S_2O_3)_2^{3-}}$$

$E_+^\circ = 0.799$ V

$\underline{E_-^\circ = 0.017 \text{ V}}$

$E^\circ = 0.799 - 0.017 = 0.782$ V

$K = 10^{0.782/0.059\,16} = 2 \times 10^{13}$

(f)

$$\underline{\quad} \quad CuI(s) + e^- \rightleftharpoons Cu(s) + I^-$$

$$\frac{Cu^+ + e^- \rightleftharpoons Cu(s)}{CuI(s) \rightleftharpoons Cu^+ + I^-}$$

$E_+^\circ = -0.185$ V

$\underline{E_-^\circ = 0.518 \text{ V}}$

$E^\circ = -0.185 - 0.518 = -0.703$ V

$K = 10^{-0.703/0.059\,16} = 1 \times 10^{-12}$

13-C. (a)

$\underline{\quad}$ $Br_2(l) + 2e^- \rightleftharpoons 2Br^-$	$E_+^\circ = 1.078$ V
$\dfrac{Fe^{2+} + 2e^- \rightleftharpoons Fe(s)}{Br_2(l) + Fe(s) \rightleftharpoons 2Br^- + Fe^{2+}}$	$E_-^\circ = -0.44$ V

$$E = \left\{1.078 - \frac{0.059\,16}{2}\log(0.050)^2\right\} - \left\{-0.44 - \frac{0.059\,16}{2}\log\frac{1}{0.010}\right\}$$

$$= 1.555 - (-0.50) = 1.65 \text{ V}$$

Electrons flow from the more negative Fe electrode (-0.50 V) to the more positive Pt electrode (1.155 V).

(b)

$\underline{\quad}$ $Fe^{2+} + 2e^- \rightleftharpoons Fe(s)$	$E_+^\circ = -0.44$ V
$\dfrac{Cu^{2+} + 2e^- \rightleftharpoons Cu(s)}{Fe^{2+} + Cu(s) \rightleftharpoons Fe(s) + Cu^{2+}}$	$E_-^\circ = 0.339$ V

$$E = \left\{-0.44 - \frac{0.059\,16}{2}\log\frac{1}{0.050}\right\} - \left\{0.339 - \frac{0.059\,16}{2}\log\frac{1}{0.020}\right\}$$

$$= -0.48 - (0.289) = -0.77 \text{ V}$$

Electrons flow from the more negative Fe electrode (-0.48 V) to the more positive Cu electrode (0.289 V).

(c)

$\underline{\quad}$ $Cl_2(g) + 2e^- \rightleftharpoons 2Cl^-$	$E_+^\circ = 1.360$ V
$\dfrac{Hg_2Cl_2(s) + 2e^- \rightleftharpoons 2Hg(l) + 2Cl^-}{Cl_2(g) + 2Hg(l) \rightleftharpoons Hg_2Cl_2(s)}$	$E_-^\circ = 0.268$ V

$$E = \left\{1.360 - \frac{0.059\,16}{2}\log\frac{(0.040)^2}{0.50}\right\} - \left\{0.268 - \frac{0.059\,16}{2}\log(0.060)^2\right\}$$

$$= 1.434 - (0.340) = 1.094 \text{ V}$$

Electrons flow from the more negative Hg electrode (0.340 V) to the more positive Pt electrode (1.434 V).

13-D. (a) $H^+ + e^- \rightleftharpoons \frac{1}{2}H_2(g)$ $E_+^\circ = 0$ V

$Ag^+ + e^- \rightleftharpoons Ag(s)$ $E_-^\circ = 0.799$ V

$E^\circ = E_+^\circ - E_-^\circ = -0.799$ V

$$E = \left\{0 - 0.059\,16\log\frac{P_{H_2}^{1/2}}{[H^+]}\right\} - \left\{0.799 - 0.059\,16\log\frac{1}{[Ag^+]}\right\}$$

(b) $[Ag^+] = \dfrac{K_{sp}}{[I^-]} = \dfrac{8.3 \times 10^{-17}}{0.10} = 8.3 \times 10^{-16}$ M

$$E = \left\{0 - 0.059\log\frac{\sqrt{0.20}}{0.10}\right\} - \left\{0.799 - 0.059\,16\log\frac{1}{8.3 \times 10^{-16}}\right\}$$

$$= -0.038 - (-0.093) = 0.055 \text{ V}$$

Electrons flow from the more negative Ag electrode (-0.093 V) to the more positive Pt electrode (-0.038 V).

(c) $H^+ + e^- \rightleftharpoons \frac{1}{2}H_2(g)$ $E_+^\circ = 0$ V

$AgI(s) + e^- \rightleftharpoons Ag(s) + I^-$ $E_-^\circ = ?$

$$0.055 = \left\{0 - 0.059\,16\log\frac{\sqrt{0.20}}{0.10}\right\} - \left\{E_-^\circ - 0.059\,16\log(0.10)\right\}$$

$$\Rightarrow E_-^\circ = -0.153 \text{ V}$$

(Appendix H gives $E_-^\circ = -0.152$ V.)

13-E. $Ag(CN)_2^- + e^- \rightleftharpoons Ag(s) + 2CN^-$ $E_+^\circ = -0.310$ V

$Cu^{2+} + 2e^- \rightleftharpoons Cu(s)$ $E_-^\circ = 0.339$ V

$$E = \left\{-0.310 - 0.059\,16\log\frac{[CN^-]^2}{[Ag(CN)_2^-]}\right\} - \left\{0.339 - \frac{0.059\,16}{2}\log\frac{1}{[Cu^{2+}]}\right\}$$

We know that $[Ag(CN)_2^-] = 0.010$ M and $[Cu^{2+}] = 0.030$ M. To find $[CN^-]$ at pH 8.21, we write

$$\frac{[CN^-]}{[HCN]} = \frac{K_a}{[H^+]} \Rightarrow [CN^-] = 0.10\,[HCN]$$

But, because $[CN^-] + [HCN] = 0.10$ M, $[CN^-] = 0.009\,1$ M. Putting this concentration into the Nernst equation gives $E = -0.187 - (0.294) = -0.481$ V. Electrons flow from the more negative Ag electrode (-0.187 V) to the more positive Cu electrode (0.294 V).

13-F. (a) $PuO_2^+ + e^- + 4H^+ \rightleftharpoons Pu^{4+} + 2H_2O$

$PuO_2^{2+} \rightarrow PuO_2^+$ $\Delta G = -1F\,(0.966)$

$PuO_2^+ \rightarrow Pu^{4+}$ $\Delta G = -1F\,E^\circ$

$\underline{Pu^{4+} \rightarrow Pu^{3+} \quad\quad \Delta G = -1F\,(1.006)}$

$PuO_2^{2+} \rightarrow Pu^{3+}$ $\Delta G = -3F\,(1.021)$

$-3F\,(1.021) = -1F\,(0.966) - 1F\,E^\circ - 1F\,(1.006) \Rightarrow E^\circ = 1.091$ V

(b) $2PuO_2^{2+} + 2e^- \rightleftharpoons 2PuO_2^+$ $E_+^\circ = 0.966$ V

$\dfrac{\frac{1}{2}O_2(g) + 2H^+ + 2e^- \rightleftharpoons H_2O}{2PuO_2^{2+} + H_2O \rightleftharpoons 2PuO_2^+ + \frac{1}{2}O_2(g) + 2H^+}$ $E_-^\circ = 1.229$ V

$$E = \left\{0.966 - \frac{0.059\,16}{2}\log\frac{[PuO_2^+]^2}{[PuO_2^{2+}]^2}\right\} - \left\{1.229 - \frac{0.059\,16}{2}\log\frac{1}{P_{O_2}^{1/2}[H^+]^2}\right\}$$

$[PuO_2^+]$ cancels $[PuO_2^{2+}]$ because they are equal. At pH 2.00, we insert $[H^+] = 10^{-2.00}$ and $P_{O_2} = 0.20$ to find $E = -0.134$ V. Because $E < 0$, the reaction is not spontaneous and water is not oxidized. At pH 7.00, we find $E = +0.161$ V, so water will be oxidized.

13-G. $2H^+ + 2e^- \rightleftharpoons H_2(g)$ $E_+^\circ = 0$ V

$Hg_2Cl_2(s) + 2e^- \rightleftharpoons 2Hg(l) + 2Cl^-$ $E_-^\circ = 0.268$ V

$$E = \left\{-\frac{0.059\,16}{2}\log\frac{P_{H_2}}{[H^+]^2}\right\} - \left\{0.268 - \frac{0.059\,16}{2}\log[Cl^-]^2\right\}$$

We find $[H^+]$ in the right half-cell by considering the acid-base chemistry of KHP, the intermediate form of a diprotic acid:

$$[H^+] = \sqrt{\frac{K_1K_2(0.050) + K_1K_w}{K_1 + 0.050}} = 6.5 \times 10^{-5} \text{ M}$$

$$E = \left\{-\frac{0.059\,16}{2}\log\frac{1}{(6.5 \times 10^{-5})^2}\right\} - \left\{0.268 - \frac{0.059\,16}{2}\log(0.10)^2\right\}$$

$$= -0.247_7 - 0.327_2 = -0.575 \text{ V}$$

Electrons flow from the more negative Pt electrode (-0.247_7 V) to the more positive Hg electrode (0.327_2 V).

13-H. $2H^+ + 2e^- \rightleftharpoons H_2(g)$ $E_+^\circ = 0$ V

$Hg^{2+} + 2e^- \rightleftharpoons Hg(l)$ $E_-^\circ = 0.852$ V

$$E = \left\{-\frac{0.059\,16}{2}\log\frac{P_{H_2}}{[H^+]^2}\right\} - \left\{0.852 - \frac{0.059\,16}{2}\log\frac{1}{[Hg^{2+}]}\right\}$$

$$0.083 = \left\{-\frac{0.059\,16}{2}\log\frac{1}{1^2}\right\} - \left\{0.852 - \frac{0.059\,16}{2}\log\frac{1}{[Hg^{2+}]}\right\}$$

$\Rightarrow [Hg^{2+}] = 2.5 \times 10^{-32}$ M

$[HgI_4^{2-}] = 0.001\,0$ M. To make this much HgI_4^{2-}, the concentration of I^- must have been reduced from 0.500 M to 0.496 M, because one Hg^{2+} ion reacts with four I^- ions.

$$K = \frac{[HgI_4^{2-}]}{[Hg^{2+}][I^-]^4} = \frac{(0.001\,0)}{(2.5 \times 10^{-32})(0.496)^4} = 7 \times 10^{29}$$

13-I.

$\underline{\quad}$ $CuY^{2-} + 2e^- \rightleftharpoons Cu(s) + Y^{4-}$	$E_+^\circ = ?$
$Cu^{2+} + 2e^- \rightleftharpoons Cu(s)$	$E_-^\circ = 0.339$ V
$CuY^{2-} \overset{1/K_f}{\rightleftharpoons} Cu^{2+} + Y^{4-}$	E°

$$E^\circ = \frac{0.059\,16}{2}\log\frac{1}{K_f} = -0.556 \text{ V}$$

$E_+^\circ = E^\circ + E_-^\circ = -0.556 + 0.339 = -0.217$ V

13-J. To compare glucose and H_2 at pH = 0, we need to know $E°$ for each. For H_2, $E° = 0$ V. For glucose, we find $E°$ from $E°'$:

$$HA + 2H^+ + 2e^- \rightleftharpoons G + H_2O$$

Gluconic acid   Glucose

$$E = E° - \frac{0.059\ 16}{2}\log\frac{[G]}{[HA][H^+]^2}$$

But $F_G = [G]$ and $[HA] = \dfrac{[H^+]F_{HA}}{[H^+] + K_a}$. Putting these into the Nernst equation gives

$$E = E° - \frac{0.059\ 16}{2}\log\frac{F_G}{\left(\dfrac{[H^+]F_{HA}}{[H^+]+K_a}\right)[H^+]^2}$$

$$= E° - \frac{0.059\ 16}{2}\log\frac{[H^+]+K_a}{[H^+]^3} - \frac{0.059\ 16}{2}\log\frac{F_G}{F_{HA}}$$

$\underbrace{\hspace{5cm}}$

This is $E°' = -0.45$ V when $[H^+] = 10^{-7}$

$$-0.45\ V = E° - \frac{0.059\ 16}{2}\log\frac{10^{-7.00} + 10^{-3.56}}{(10^{-7.00})^3}$$

$$\Rightarrow E° = +0.06_6 \text{ V for glucose}$$

Because $E°$ for H_2 is more negative than $E°$ for glucose, H_2 is the stronger reducing agent at pH 0.00.

13-K. **(a)** Each H^+ must provide $\frac{1}{2}(34.5\text{ kJ})$ when it passes from outside to inside.

$$\Delta G = -\frac{1}{2}(34.5 \times 10^3\text{ J}) = -RT\ln\frac{\mathcal{A}_{high}}{\mathcal{A}_{low}}$$

$$\frac{\mathcal{A}_{high}}{\mathcal{A}_{low}} = 1.05 \times 10^3 \Rightarrow \Delta pH = \log(1.05 \times 10^3) = 3.02 \text{ pH units}$$

(b) $\Delta G = -nFE$ (where n = charge of $H^+ = 1$)

$$-\frac{1}{2}(34.5 \times 10^3\text{ J}) = -1FE \Rightarrow E = 0.179 \text{ V}$$

(c) If $\Delta pH = 1.00$, $\mathcal{A}_{high}/\mathcal{A}_{low} = 10$.

$$\Delta G(pH) = -RT\ln 10 = -5.7 \times 10^3\text{ J}$$

$$\Delta G(\text{electric}) = \left[\frac{1}{2}(34.5) - 5.7\right]\text{ kJ} = 11.5 \text{ kJ}$$

$$E = \frac{\Delta G(\text{electric})}{F} = 0.120 \text{ V}$$

Chapter 14

14-A. The reaction at the silver electrode (written as a reduction) is $Ag^+ + e^- \rightleftharpoons Ag(s)$, and the cell voltage is written as

$$E = E_+ - E_- = E_+ - E(\text{S.C.E.}) = E_+ - 0.241$$

$$= \left(0.799 - 0.059\ 16\log\frac{1}{[Ag^+]}\right) - 0.241$$

$$= 0.558 + 0.059\ 16\log[Ag^+]$$

Titration reaction: $Br^- + Ag^+ \rightarrow AgBr(s)$ $K_{sp} = 5.0 \times 10^{-13}$

The equivalence point is $V_e = 25.0$. Between 0 and 25 mL, there is unreacted Ag^+ in the solution.

1.0 mL: $[Ag^+] = \left(\dfrac{24.0}{25.0}\right)(0.100\text{ M})\left(\dfrac{50.0}{51.0}\right) = 0.094\ 1$ M

Fraction of Ag^+ ↗ Initial concentration Dilution ↖
remaining   of Ag^+ factor

$$\Rightarrow E = 0.558 + 0.059\ 16\log[0.094\ 1] = 0.497 \text{ V}$$

12.5 mL: $[Ag^+] = \left(\dfrac{12.5}{25.0}\right)(0.100\text{ M})\left(\dfrac{50.0}{62.5}\right) = 0.040\ 0$ M

$$\Rightarrow E = 0.475 \text{ V}$$

24.0 mL: $[Ag^+] = \left(\dfrac{1.0}{25.0}\right)(0.100\text{ M})\left(\dfrac{50.0}{74.0}\right) = 0.002\ 70$ M

$$\Rightarrow E = 0.406 \text{ V}$$

24.9 mL: $[Ag^+] = \left(\dfrac{0.10}{25.0}\right)(0.100\text{ M})\left(\dfrac{50.0}{74.9}\right) = 2.67 \times 10^{-4}$ M

$$\Rightarrow E = 0.347 \text{ V}$$

Beyond 25 mL, all AgBr has precipitated and there is excess Br^- in solution.

25.1 mL: $[Br^-] = \left(\dfrac{0.1}{75.1}\right)(0.200\text{ M}) = 2._{67} \times 10^{-4}$ M

$$\Rightarrow [Ag^+] = K_{sp}/[Br^-] = (5.0 \times 10^{-13})/(2._{67} \times 10^{-4}) = 1._{88} \times 10^{-9}\text{ M}$$
$$\Rightarrow E = +0.042 \text{ V}$$

26.0 mL: $[Br^-] = \left(\dfrac{1.0}{76.0}\right)(0.200\text{ M}) = 2.6_6 \times 10^{-4}$ M

$$\Rightarrow [Ag^+] = 1.9_0 \times 10^{-10}\text{ M} \Rightarrow E = -0.017 \text{ V}$$

At 35.0 mL: $[Br^-] = \left(\dfrac{10.0}{85.0}\right)(0.200\text{ M}) = 0.023\ 5$ M

$$\Rightarrow [Ag^+] = 2.1_2 \times 10^{-11}\text{ M} \Rightarrow E = -0.073 \text{ V}$$

14-B. The cell voltage is given by Equation C, in which K_f is the formation constant for $Hg(EDTA)^{2-}$ ($= 10^{21.5}$). To find the voltage, we must calculate $[HgY^{2-}]$ and $[Y^{4-}]$ at each point. The concentration of HgY^{2-} is 1.0×10^{-4} M when $V = 0$ and is thereafter affected only by dilution because $K_f(HgY^{2-}) \gg K_f(MgY^{2-})$. The concentration of Y^{4-} is found from the Mg-EDTA equilibrium at all but the first point. At $V = 0$ mL, the Hg-EDTA equilibrium determines $[Y^{4-}]$.

0 mL: $\dfrac{HgY^{2-}}{[Hg^{2+}][EDTA]} = \alpha_{Y^{4-}} \cdot K_f \text{ (for } HgY^{4-}) = (0.30)(10^{21.5})$

$$\frac{1.0 \times 10^{-4} - x}{(x)(x)} = 9.5 \times 10^{20} \Rightarrow x = [EDTA] = 3.2 \times 10^{-13}\text{ M}$$

$$[Y^{4-}] = \alpha_{Y^{4-}}[EDTA] = 9.7 \times 10^{-14}\text{ M}$$

Using Equation C, we write

$$E = 0.852 - 0.241 - \frac{0.059\ 16}{2}\log\frac{10^{21.5}}{1.0 \times 10^{-4}}$$
$$- \frac{0.059\ 16}{2}\log(9.7 \times 10^{-14}) = 0.242 \text{ V}$$

10.0 mL: $V_e = 25.0$ mL, so $\frac{10}{25}$ of Mg^{2+} is in the form MgY^{2-}, and $\frac{15}{25}$ is in the form Mg^{2+}.

$$[Y^{4-}] = \frac{[MgY^{2-}]}{[Mg^{2+}]}/K_f \text{ (for } MgY^{2-}) = \left(\frac{10}{15}\right)/6.2 \times 10^8 = 1.08 \times 10^{-9}\text{ M}$$

$$[HgY^{2-}] = \left(\frac{50.0}{60.0}\right)(1.0 \times 10^{-4}\text{ M}) = 8.33 \times 10^{-5}\text{ M}$$

Dilution factor

$$E = 0.852 - 0.241 - \frac{0.059\ 16}{2}\log\frac{10^{21.5}}{8.33 \times 10^{-5}}$$
$$- \frac{0.059\ 16}{2}\log(1.08 \times 10^{-9}) = 0.120 \text{ V}$$

20.0 mL: $[Y^{4-}] = \left(\dfrac{20}{5}\right)/6.2 \times 10^8 = 6.45 \times 10^{-9}$ M

$$[HgY^{2-}] = \left(\frac{50.0}{70.0}\right)(1.0 \times 10^{-4}\text{ M}) = 7.14 \times 10^{-5}\text{ M}$$

$$\Rightarrow E = 0.095 \text{ V}$$

24.9 mL: $[Y^{4-}] = \left(\dfrac{24.9}{0.1}\right)/6.2 \times 10^8 = 4.02 \times 10^{-7}$ M

$[HgY^{2-}] = \left(\dfrac{50.0}{74.9}\right)(1.0 \times 10^{-4}$ M$) = 6.68 \times 10^{-5}$ M

$\Rightarrow E = 0.041$ V

25.0 mL: This is the equivalence point, at which $[Mg^{2+}] = [EDTA]$.

$\dfrac{[MgY^{2-}]}{[Mg^{2+}][EDTA]} = \alpha_{Y^{4-}} \cdot K_f$ (for MgY^{2-})

$\dfrac{\left(\dfrac{50.0}{75.0}\right)(0.010\ 0) - x}{x^2} = 1.85 \times 10^8 \Rightarrow x = 6.0 \times 10^{-6}$ M

$[Y^{4-}] = \alpha_{Y^4}\cdot(6.0 \times 10^{-6}$ M$) = 1.80 \times 10^{-6}$ M

$[HgY^{2-}] = \left(\dfrac{50.0}{75.0}\right)(1.0 \times 10^{-4}$ M$) = 6.67 \times 10^{-5}$ M

$\Rightarrow E = 0.021$ V

26.0 mL: Now there is excess EDTA in the solution.

$[Y^{4-}] = \alpha_{Y^4}\cdot[EDTA] = (0.30)\left[\left(\dfrac{1.0}{76.0}\right)(0.020\ 0$ M$)\right] = 7.89 \times 10^{-5}$ M

$[HgY^{2-}] = \left(\dfrac{50.0}{76.0}\right)(1.0 \times 10^{-4}$ M$) = 6.58 \times 10^{-5}$ M

$\Rightarrow E = -0.027$ V

14-C.   At intermediate pH, the voltage will be constant at 100 mV. When $[OH^-] \approx [F^-]/10 = 10^{-6}$ M (pH = 8), the electrode begins to respond to OH^- and the voltage will decrease (i.e., the electrode potential will change in the same direction as if more F^- were being added). Near pH = 3.17 (= pK_a for HF), F^- reacts with H^+ and the concentration of free F^- decreases. At pH = 1.17, $[F^-] \approx 1\%$ of 10^{-5} M $= 10^{-7}$ M, and $E \approx 100 + 2(59) = 218$ mV. A qualitative sketch of this behavior is shown here.

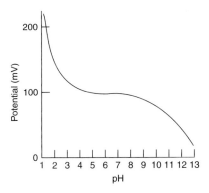

14-D.   **(a)** For 1.00 mM Na^+ at pH 8.00, we can write

$E = $ constant $+ 0.059\ 16 \log ([Na^+] + 36[H^+]) - 0.038$
$= $ constant $+ 0.059\ 16 \log [(1.00 \times 10^{-3}) + (36 \times 10^{-8})]$
$\Rightarrow$ constant $= +0.139$ V
For 5.00 mM Na^+ at pH 8.00, we have
$E = +0.139 + 0.059\ 16 \log [(5.00 \times 10^{-3}) + (36 \times 10^{-8})]$
$= 0.003$ V

(b) For 1.00 mM Na^+ at pH 3.87, we have
$E = +0.139 + 0.059\ 16 \log [(1.00 \times 10^{-3}) + (36 \times 10^{-3.87})]$
$= 0.007$ V

14-E.   A graph of E(mV) versus $\log[NH_3(M)]$ gives a straight line whose equation is $E = 563.4 + 59.05 \times \log [NH_3]$. For $E = 339.3$ mV, $[NH_3] = 1.60 \times 10^{-4}$ M. The sample analyzed contains $(100$ mL$)(1.60 \times 10^{-4}$ M$) = 0.016\ 0$ mmol of nitrogen. But this sample represents just 2.00% (20.0 mL/1.00 L) of the food sample. Therefore, the food contains $(0.016$ mmol N$)/0.020\ 0 = 0.800$ mmol N $= 11.2$ mg of N $= 3.59$ wt% N.

14-F.   The function to plot on the y-axis is $(V_0 + V_S)10^{E/S}$, where $S = (\beta RT/nF) \ln 10$. β is 0.985. Putting in $R = 8.314\ 5$ J/mol·K), $F = 96\ 485$ C/mol, $T = 298.15$ K, and $n = -2$ gives $S = -0.029\ 136$ J/C $= -0.029\ 136$ V. (Recall that joule/coulomb = volt.)

V_S (mL)	E (V)	y
0	0.046 5	0.633 8
1.00	0.040 7	1.042 5
2.00	0.034 4	1.781 1
3.00	0.030 0	2.615 2
4.00	0.026 5	3.571 7

The data are plotted in Figure 14-30, which has a slope of $m = 0.744\ 84$ and an intercept of $b = 0.439\ 19$, giving an x-intercept of $-b/m = -0.58_9\ _{65}$ mL. The concentration of original unknown is

$c_X = \dfrac{(x\text{-intercept})c_S}{V_0} = -\dfrac{(-0.58_9\ _{65}\ \text{mL})(1.78\ \text{mM})}{25.0\ \text{mL}} = 4.2 \times 10^{-5}$ M

(We decided that the last significant digit in the x-intercept was the 0.01 decimal place because the original data were only measured to the 0.01 decimal place.)

Chapter 15

15-A.   Titration: $Sn^{2+} + 2Ce^{4+} \rightarrow Sn^{4+} + 2Ce^{3+}$ $V_e = 10.0$ mL
Representative calculations:

0.100 mL: $E_+ = 0.139 - \dfrac{0.059\ 16}{2} \log \dfrac{[Sn^{2+}]}{[Sn^{4+}]}$

$= 0.139 - \dfrac{0.059\ 16}{2} \log \dfrac{9.90}{0.100} = 0.080$ V

$E = E_+ - E_- = 0.080 - 0.241 = -0.161$ V

10.00 mL: $2E_+ = 2(0.139) - 0.059\ 16 \log \dfrac{[Sn^{2+}]}{[Sn^{4+}]}$

$\underline{\qquad E_+ = 1.47 - 0.059\ 16 \log \dfrac{[Ce^{3+}]}{[Ce^{4+}]}\qquad}$

$3E_+ = 1.748 - 0.059\ 16 \log \dfrac{[Sn^{2+}][Ce^{3+}]}{[Sn^{4+}][Ce^{4+}]}$

At the equivalence point, $[Sn^{4+}] = \frac{1}{2}[Ce^{3+}]$ and $[Sn^{2+}] = \frac{1}{2}[Ce^{4+}]$, which makes the log term 0. Therefore, $3E_+ = 1.748$ and $E_+ = 0.583$ V. $E = E_+ - E_- = 0.583 - 0.241 = 0.342$ V

10.10 mL: $E_+ = 1.47 - 0.059\ 16 \log \dfrac{[Ce^{3+}]}{[Ce^{4+}]}$

$= 1.47 - 0.059\ 16 \log \dfrac{10.0}{0.10} = 1.35_2$ V

$E = E_+ - E_- = 1.35_2 - 0.241 = 1.11$ V

mL	E (V)	mL	E (V)
0.100	-0.161	10.00	0.342
1.00	-0.130	10.10	1.11
5.00	-0.102	12.00	1.19
9.50	-0.064		

15-B.   Standard potentials: indigo tetrasulfonate, 0.36 V; $Fe[CN]_6^{3-}$ | $Fe[CN]_6^{4-}$, 0.356 V; Tl^{3+} | Tl^+, 0.77 V. The end-point potential will be between 0.356 and 0.77 V. Indigo tetrasulfonate changes color near 0.36 V. Therefore, it will not be a useful indicator for this titration.

15-C.   Titration: $MnO_4^- + 5Fe^{2+} + 8H^+ \rightarrow Mn^{2+} + 5Fe^{3+} + 4H_2O$
$Fe^{3+} + e^- \rightleftharpoons Fe^{2+}$ $E° = 0.68$ V in 1 M H_2SO_4
$MnO_4^- + 8H^+ + 5e^- \rightarrow Mn^{2+} + 4H_2O$ $E° = 1.507$ V
The equivalence point comes at 15.0 mL. Before the equivalence point,

$E = E_+ - E_- = \left(0.68 - 0.059\ 16 \log \dfrac{[Fe^{2+}]}{[Fe^{3+}]}\right) - 0.241$

1.0 mL: $[Fe^{2+}]/[Fe^{3+}] = 14.0/1.0 \Rightarrow E = 0.371$ V

7.5 mL: $[Fe^{2+}]/[Fe^{3+}] = 7.5/7.5 \Rightarrow E = 0.439$ V

14.0 mL: $[Fe^{2+}]/[Fe^{3+}] = 1.0/14.0 \Rightarrow E = 0.507$ V

At the equivalence point, use Equation E of Demonstration 15-1:

$$6E_+ = 8.215 - 0.059\,16 \log \frac{1}{[\text{H}^+]^8} \overset{\text{pH}=0}{\Longrightarrow} E_+ = 1.369 \text{ V}$$

$$E = E_+ - E_- = 1.369 - 0.241 = 1.128 \text{ V}$$

After the equivalence point:

$$E = E_+ - E_- = \left(1.507 - \frac{0.059\,16}{5} \log \frac{[\text{Mn}^{2+}]}{[\text{MnO}_4^-][\text{H}^+]^8}\right) - 0241$$

16.0 mL: $[\text{Mn}^{2+}]/[\text{MnO}_4^-] = 15.0/1.0$ and $[\text{H}^+] = 1$ M

$\Rightarrow E = 1.252 \text{ V}$

30.0 mL: $[\text{Mn}^{2+}]/[\text{MnO}_4^-] = 15.0/15.0$ and $[\text{H}^+] = 1$ M

$\Rightarrow E = 1.266 \text{ V}$

15-D. The Gran plot of $V \cdot 10^{-E/0.059\,16}$ versus V is shown in Figure 15-4. The data from 8.5 to 12.5 mL appear to be on a straight line. The least-squares line through these four points has a slope of $m = -1.567\,3 \times 10^{-11}$ and an intercept of $b = 2.170\,2 \times 10^{-10}$. The x-intercept is $-b/m = 13.85$ mL. The amount of Ce^{4+} required to reach the equivalence point is (0.100 mmol/mL)(13.85 mL) = 1.385 mmol, and the concentration of unknown Fe^{2+} is 1.385 mmol/50.0 mL = 0.027 7 M.

Titrant volume, V (mL)	E (volts)	$V \cdot 10^{-E/0.059\,16}$
6.50	0.635	$1.200\,3 \times 10^{-10}$
8.50	0.651	$8.421\,0 \times 10^{-11}$
10.50	0.669	$5.162\,6 \times 10^{-11}$
11.50	0.680	$3.685\,1 \times 10^{-11}$
12.50	0.696	$2.148\,8 \times 10^{-11}$

15-E. Let x = mg of $\text{FeSO}_4 \cdot (\text{NH}_4)_2\text{SO}_4 \cdot 6\text{H}_2\text{O}$ and $(54.85 - x)$ = mg of $\text{FeCl}_2 \cdot 6\text{H}_2\text{O}$.

mmol of Ce^{4+} = mmol $\text{FeSO}_4 \cdot (\text{NH}_4)_2\text{SO}_4 \cdot 6\text{H}_2\text{O}$ + mmol $\text{FeCl}_2 \cdot 6\text{H}_2\text{O}$

$$(13.39 \text{ mL})(0.012\,34 \text{ M}) = \frac{x \text{ mg}}{392.13 \text{ mg/mmol}} + \frac{(54.85 - x)}{234.84 \text{ mg/mmol}}$$

$\Rightarrow x = 40.01$ mg $\text{FeSO}_4 \cdot (\text{NH}_4)_2\text{SO}_4 \cdot 6\text{H}_2\text{O}$

Mass of $\text{FeCl}_2 \cdot 6\text{H}_2\text{O}$ = 14.84 mg = 0.063 19 mmol = 4.48 mg Cl

$$\text{wt\% Cl} = \frac{4.48 \text{ mg}}{54.85 \text{ mg}} \times 100 = 8.17\%$$

Chapter 16

16-A. Cathode: $2\text{H}^+ + 2\text{e}^- \rightleftharpoons \text{H}_2(g)$ $E° = 0$ V
Anode (written as a reduction):

$\frac{1}{2}\text{O}_2(g) + 2\text{H}^+ + 2\text{e}^- \rightleftharpoons \text{H}_2\text{O}$ $E = 1.229$ V

$$E(\text{cathode}) = 0 - \frac{0.059\,16}{2} \log \frac{P_{\text{H}_2}}{[\text{H}^+]^2}$$

$$E(\text{anode}) = 1.229 - \frac{0.059\,16}{2} \log \frac{1}{[\text{H}^+]^2 P_{\text{O}_2}^{1/2}}$$

$E(\text{cell}) = E(\text{cathode}) - E(\text{anode})$

$$= -1.229 - \frac{0.059\,16}{2} \log P_{\text{H}_2} P_{\text{O}_2}^{1/2} = -1.229 \text{ V}$$

$E = E(\text{cell}) - I \cdot R - \text{overpotentials}$
$= -1.229 - (0.100 \text{ A})(2.00 \text{ Ω})$

$$\underbrace{-0.85 \text{ V}}_{\substack{\text{Anode} \\ \text{overpotential}}} \underbrace{- 0.068 \text{ V}}_{\substack{\text{Cathode} \\ \text{overpotential}}} = -2.35 \text{ V}$$

From Table 16-1

For Au electrodes, overpotentials of 0.963 and 0.390 V, give $E = -2.78$ V.

16-B. (a) To electrolyze 0.010 M SbO^+ requires a potential of

$$E(\text{cathode}) = 0.208 - \frac{0.059\,16}{3} \log \frac{1}{[\text{SbO}^+][\text{H}^+]^2}$$

$$= 0.208 - \frac{0.059\,16}{3} \log \frac{1}{(0.010)(1.0)^2} = 0.169 \text{ V}$$

$E(\text{cathode versus Ag} \mid \text{AgCl}) = E(\text{versus S.H.E.}) - E(\text{Ag} \mid \text{AgCl})$
$= 0.169 - 0.197 = -0.028$ V

(b) The concentration of Cu^{2+} in equilibrium with $\text{Cu}(s)$ at 0.169 V is
$\text{Cu}^{2+} + 2\text{e}^- \rightleftharpoons \text{Cu}(s)$ $E° = 0.339$

$$E(\text{cathode}) = 0.339 - \frac{0.059\,16}{2} \log \frac{1}{[\text{Cu}^{2+}]}$$

$$0.169 = 0.339 - \frac{0.059\,16}{2} \log \frac{1}{[\text{Cu}^{2+}]} \Rightarrow [\text{Cu}^{2+}] = 1.8 \times 10^{-6} \text{ M}$$

Percentage of Cu^{2+} not reduced $= \frac{1.8 \times 10^{-6}}{0.10} \times 100 = 1.8 \times 10^{-3}\%$

Percentage of Cu^{2+} reduced = 99.998%

16-C. (a) $\text{Co}^{2+} + 2\text{e}^- \rightleftharpoons \text{Co}(s)$ $E° = -0.282$ V

$$E(\text{cathode versus S.H.E.}) = -0.282 - \frac{0.059\,16}{2} \log \frac{1}{[\text{Co}^{2+}]}$$

Putting in $[\text{Co}^{2+}] = 1.0 \times 10^{-6}$ M gives $E = -0.459$ V and
$E(\text{cathode versus S.C.E.}) = -0.459 - \underbrace{0.241}_{E \text{ (S.C.E.)}} = -0.700 \text{ V}$

(b) $\text{Co}(\text{C}_2\text{O}_4)_2^{2-} + 2\text{e}^- \rightleftharpoons \text{Co}(s) + 2\text{C}_2\text{O}_4^{2-}$ $E° = -0.474$ V
$E(\text{cathode versus S.C.E.})$

$$= -0.474 - \frac{0.059\,16}{2} \log \frac{[\text{C}_2\text{O}_4^{2-}]^2}{[\text{Co}(\text{C}_2\text{O}_4)_2^{2-}]} - 0.241$$

For $[\text{C}_2\text{O}_4^{2-}] = 0.10$ M and $[\text{Co}(\text{C}_2\text{O}_4)_2^{2-}] = 1.0 \times 10^{-6}$ M, $E = -0.833$ V.

(c) We can think of the reduction as $\text{Co}^{2+} + 2\text{e}^- \rightleftharpoons \text{Co}(s)$, for which $E° = -0.282$ V. But $[\text{Co}^{2+}]$ is the tiny amount in equilibrium with 0.10 M EDTA plus 1.0×10^{-6} M $\text{Co}(\text{EDTA})^{2-}$. In Table 11-2, we find that the formation constant for $\text{Co}(\text{EDTA})^{2-}$ is $10^{16.45} = 2.8 \times 10^{16}$.

$$K_f = \frac{[\text{Co}(\text{EDTA})^{2-}]}{[\text{Co}^{2+}][\text{EDTA}^{4-}]} = \frac{[\text{Co}(\text{EDTA})^{2-}]}{[\text{Co}^{2+}]\alpha_{\text{Y}^{4-}}\text{F}}$$

where F is the formal concentration of EDTA (= 0.10 M) and $\alpha_{\text{Y}^{4-}} = 3.8 \times 10^{-4}$ at pH 7.00 (Table 11-1). Putting in $[\text{Co}(\text{EDTA})^{2-}] = 1.0 \times 10^{-6}$ M and solving for $[\text{Co}^{2+}]$ gives $[\text{Co}^{2+}] = 9.4 \times 10^{-19}$ M.

$$E = -0.282 - \frac{0.059\,16}{2} \log \frac{1}{9.4 \times 10^{-19}} - 0.241 = -1.056 \text{ V}$$

16-D. (a) 75.00 mL of 0.023 80 M KSCN = 1.785 mmol of SCN^-, which gives 1.785 mmol of AgSCN, containing 0.103 7 g of SCN. Final mass = 12.463 8 + 0.103 7 = 12.567 5 g.

(b) Anode: $\text{AgBr}(s) + \text{e}^- \rightleftharpoons \text{Ag}(s) + \text{Br}^-$ $E° = 0.071$ V
$E(\text{anode}) = 0.071 - 0.059\,16 \log[\text{Br}^-]$
$= 0.071 - 0.059\,16 \log [0.10] = 0.130$ V
$E(\text{cathode}) = E(\text{S.C.E.}) = 0.241$ V
$E = E(\text{cathode}) - E(\text{anode}) = 0.111$ V

(c) To remove 99.99% of 0.10 M KI will leave $[\text{I}^-] = 1.0 \times 10^{-5}$ M. The concentration of Ag^+ in equilibrium with this much I^- is

$[Ag^+] = K_{sp}/[I^-] = (8.3 \times 10^{-17})/(1.0 \times 10^{-5}) = 8.3 \times 10^{-12}$ M. The concentration of Ag^+ in equilibrium with 0.10 M Br^- is $[Ag^+] = K_{sp}/[Br^-] = (5.0 \times 10^{-13})/(0.10) = 5.0 \times 10^{-12}$ M. Therefore, 8.3×10^{-12} M Ag^+ will begin to precipitate 0.10 M Br^-. The separation is not possible.

16-E. The corrected coulometric titration time is $387 - 6 = 381$ s. $q = It/F = (4.23 \text{ mA})(381 \text{ s})/(96\ 485 \text{ C/mol}) = 16.7$ µmol e^-. Because $1e^-$ is equivalent to one X^-, the concentration of organohalide is 16.7 µM. If all halogen is Cl, this corresponds to 592 µg Cl/L.

16-F. **(a)** Use the internal standard equation with $X = Pb^{2+}$ and $S = Cd^{2+}$. From the standard mixture, we find the response factor, F:

$$\frac{Signal_X}{[X]} = F\left(\frac{signal_S}{[S]}\right)$$

$$\frac{1.58 \text{ µA}}{[41.8 \text{ µM}]} = F\left(\frac{1.64 \text{ µA}}{[32.3 \text{ µM}]}\right) \Rightarrow F = 0.744_5$$

$[Cd^{2+}]$ standard added to unknown

$$= \left(\frac{10.00}{50.00}\right)(3.23 \times 10^{-4} \text{ M}) = 6.46 \times 10^{-5} \text{ M}$$

For the unknown mixture, we can now say

$$\frac{Signal_X}{[X]} = F\left(\frac{signal_S}{[S]}\right)$$

$$\frac{3.00 \text{ µA}}{[Pb^{2+}]} = 0.744_5\left(\frac{2.00 \text{ µA}}{[64.6 \text{ µM}]}\right) \Rightarrow [Pb^{2+}] = 130._2 \text{ µM}$$

The concentration of Pb^{2+} in diluted unknown is 130._2 µM. In the undiluted unknown, the concentration is $\left(\frac{50.00}{25.00}\right)(130._2 \text{ µM}) = 2.60 \times 10^{-4}$ M.

(b) First find the relative uncertainty in the response factor:

$$F = \frac{(1.58 \pm 0.03)(32.3 \pm 0.1)}{(1.64 \pm 0.03)(41.8 \pm 0.1)} \Rightarrow F = 0.7445 \pm 0.019\ 9 \ (\pm 2.67\%)$$

Then find the uncertainty in $[Pb^{2+}]$:

$$[Pb^{2+}] = \frac{(3.00 \pm 0.03)\dfrac{(10.00 \pm 0.05)}{(50.00 \pm 0.05)}(3.23(\pm 0.01) \times 10^{-4})}{(2.00 \pm 0.03)(0.744\ 5 \pm 0.019\ 9)}$$

$$\Rightarrow [Pb^{2+}] = 2.60\ (\pm 0.09) \times 10^{-4} \text{ M}$$

16-G. We see two consecutive reductions. From the value of $E_{pa} - E_{pc}$, we find that one electron is involved in each reduction (using Equation 16-20). A possible sequence of reaction is

$$Co(III)(B_9C_2H_{11})_2^- \rightarrow Co(II)(B_9C_2H_{11})_2^{2-} \rightarrow Co(I)(B_9C_2H_{11})_2^{3-}$$

The equality of the anodic and cathodic peak heights suggests that the reactions are reversible. The expected sampled current (panel a) and square wave (panel b) polarograms are sketched below.

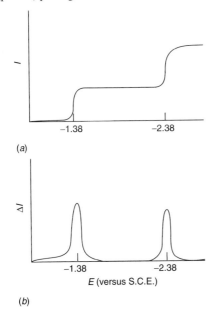

(a)

(b)

E (versus S.C.E.)

16-H. Electricity required for H_2O in 0.847 6 g of polymer $= (63.16 - 4.23) = 58.93$ C.

$$\frac{58.93 \text{ C}}{96\ 485 \text{ C/mol}} = 0.610\ 8 \text{ mmol of } e^- \text{ which corresponds to}$$

$\frac{1}{2}(0.610\ 8) = 0.305\ 4$ mmol of $I_2 = 0.305\ 4$ mmol of $H_2O = 5.502$ mg H_2O.

$$\text{Water content} = 100 \times \frac{5.502 \text{ mg } H_2O}{847.6 \text{ mg polymer}} = 0.649\ 1 \text{ wt\%}$$

Chapter 17

17-A. **(a)** $A = -\log P/P_0 = -\log T = -\log(0.45) = 0.347$

(b) Absorbance is proportional to concentration, so the absorbance will double to 0.694, giving $T = 10^{-A} = 10^{-0.694} = 0.202 \Rightarrow \%T = 20.2\%$.

17-B. **(a)** $\varepsilon = \dfrac{A}{cb} = \dfrac{0.624 - 0.029}{(3.96 \times 10^{-4} \text{ M})(1.000 \text{ cm})} = 1.50 \times 10^3 \text{ M}^{-1} \text{cm}^{-1}$

(b) $c = \dfrac{A}{\varepsilon b} = \dfrac{0.375 - 0.029}{(1.50 \times 10^3 \text{ M}^{-1} \text{ cm}^{-1})(1.000 \text{ cm})} = 2.31 \times 10^{-4} \text{ M}$

(c) $c = \underbrace{\left(\dfrac{25.00 \text{ mL}}{2.00 \text{ mL}}\right)}_{\text{Dilution factor}} \dfrac{0.733 - 0.029}{(1.50 \times 10^3 \text{ M}^{-1} \text{ cm}^{-1})(1.000 \text{ cm})} = 5.87 \times 10^{-3} \text{ M}$

17-C. **(a)** 1.00×10^{-2} g of NH_4Cl in 1.00 L $= 1.869 \times 10^{-4}$ M. In the colored solution, the concentration is $(\frac{10}{50})(1.869 \times 10^{-4} \text{ M}) = 3.739 \times 10^{-5}$ M. $\varepsilon = A/bc = (0.308 - 0.140)/[(1.00)(3.739 \times 10^{-5})] = 4.49_3 \times 10^3 \text{ M}^{-1} \text{ cm}^{-1}$.

(b) $\dfrac{\text{Absorbance of unknown}}{\text{Absorbance of reference}}$

$= \dfrac{0.592 - 0.140}{0.308 - 0.140} = \dfrac{\text{concentration of unknown}}{\text{concentration of reference}}$

$\Rightarrow$ concentration of NH_3 in unknown

$= \left(\dfrac{0.452}{0.168}\right)(1.869 \times 10^{-4}) = 5.028 \times 10^{-4} \text{ M}$

100.00 mL of unknown $= 5.028 \times 10^{-5}$ mol of N $= 7.043 \times 10^{-4}$ g of N

$\Rightarrow$ wt\% N $= (7.043 \times 10^{-4} \text{ g})/(4.37 \times 10^{-3} \text{ g}) = 16.1 \%$

17-D. **(a)** Cu in flask C $= (1.00 \text{ mg})\left(\dfrac{10}{250}\right)\left(\dfrac{15}{30}\right) = 0.020\ 0$ mg.

This entire quantity is in the isoamyl alcohol (20.00 mL), so the concentration is $(2.00 \times 10^{-5} \text{ g})/[(0.020\ 0 \text{ L})(63.546 \text{ g/mol})] = 1.57_4 \times 10^{-5}$ M.

(b) Observed absorbance

$$= \text{absorbance due to Cu in rock} + \text{blank absorbance} = \varepsilon bc + 0.056$$

$$= (7.90 \times 10^3)(1.00)(1.57_4 \times 10^{-5}) + 0.056 = 0.180$$

Note that the observed absorbance is equal to the absorbance from Cu in the rock *plus* the blank absorbance. In the lab, we measure the observed absorbance and subtract the blank absorbance from it to find the absorbance due to copper.

(c) $\dfrac{\text{Cu in unknown}}{\text{Cu in known}} = \dfrac{A \text{ of unknown}}{A \text{ of known}}$

$\dfrac{x \text{ mg}}{1.00 \text{ mg}} = \dfrac{0.874 - 0.056}{0.180 - 0.056} \Rightarrow x = 6.60 \text{ mg Cu}$

17-E. Absorbance is corrected by multiplying observed absorbance by (total volume/initial volume). For example, at 36.0 µL, $A(\text{corrected}) = (0.399)[2\ 025 + 36)/2\ 025] = 0.406$. A graph of corrected absorbance versus volume of Pb^{2+} (µL) is similar to Figure 17-10, with the end point at 46.7 µL. The moles of Pb^{2+} in this volume are $(46.7 \times 10^{-6} \text{ L})(7.515 \times 10^{-4} \text{ M}) = 3.510 \times 10^{-8}$ mol. The concentration of semi-xylenol orange is $(3.510 \times 10^{-8} \text{ mol})/(2.025 \times 10^{-3} \text{ L}) = 1.73 \times 10^{-5}$ M.

Chapter 18

18-A. **(a)** $c = A/\varepsilon b = 0.463/[(4\,170)(1.00)] = 1.110 \times 10^{-4}$ M $=$ 8.99 g/L $=$ 8.99 mg of transferrin/mL. The Fe concentration is 2.220×10^{-4} M $= 0.012\,4$ g/L $= 12.4$ µg/mL.

(b) $A_\lambda = \Sigma \varepsilon bc$

At 470 nm: $\quad 0.424 = 4\,170[T] + 2\,290[D]$

At 428 nm: $\quad 0.401 = 3\,540[T] + 2\,730[D]$

where [T] and [D] are the concentrations of transferrin and desferrioxamine, respectively. Solving the simultaneous equations gives [T] $= 7.30 \times 10^{-5}$ M and [D] $= 5.22 \times 10^{-5}$ M. The fraction of iron in transferrin (which binds two ferric ions) is $2[T]/(2[T] + [D]) = 73.7\%$. The fraction in desferrioxamine is 26.3%. The spreadsheet solution looks like this:

	A	B	C	D	E	F	G
1	Transferrin/Desferrioxamine mixture						
2							
3	Wavelength	Coefficient matrix		Absorbance		Concentrations	
4				of unknown		in mixture	
5	428	3540	2730	0.401		7.2992E-05	← [TRF]
6	470	4170	2290	0.424		5.2238E-05	← [DFO]
7			K	A		C	

Spreadsheet for Exercise 18-A.

18-B.

	A	B	C	D	E	F	G
1	Mixture of dyes						
2					Absorbance	Calculated	
3	Wavelength	Molar absorptivity			of mixture	absorbance	
4	(nm)	Tartrazine	Sunset yellow	Ponceau 4R	Am	Acalc	(Acalc-Am)^2
5	350	6.229E+03	2.019E+03	4.172E+03	0.557	0.536	0.0004
6	375	1.324E+04	4.474E+03	2.313E+03	0.853	0.837	0.0002
7	400	2.144E+04	7.403E+03	3.310E+03	1.332	1.343	0.0001
8	425	2.514E+04	8.551E+03	4.534E+03	1.603	1.600	0.0000
9	450	2.200E+04	1.275E+04	6.575E+03	1.792	1.801	0.0001
10	475	1.055E+04	1.940E+04	1.229E+04	2.006	1.999	0.0000
11	500	1.403E+03	1.869E+04	1.673E+04	1.821	1.834	0.0002
12	525	0.000E+00	7.641E+03	1.528E+04	1.155	1.130	0.0006
13	550	0.000E+00	3.959E+02	9.522E+03	0.445	0.474	0.0008
14	575	0.000E+00	0.000E+00	1.814E+03	0.084	0.086	0.0000
15		Least-squares				Sum =	0.0026
16		guessed concentrations					
17	Tartrazine	3.71E-05					
18	Sunset y	5.27E-05					
19	Ponceau 4R	4.76E-05					

Spreadsheet for Exercise 18-B.

18-C. The Scatchard plot is a graph of $\Delta A/[X]$ versus ΔA (Equation 18-16).

Experiment	ΔA	$\Delta A/[X]$
1	0.090	20 360
2	0.181	19 890
3	0.271	16 940
4	0.361	14 620
5	0.450	12 610
6	0.539	9 764
7	0.627	7 646
8	0.713	5 021
9	0.793	2 948
10	0.853	1 453
11	0.904	93.6

Points 2–11 lie on a reasonably straight line whose slope is -2.72×10^4 M^{-1}, giving $K = 2.72 \times 10^4$.

18-D.

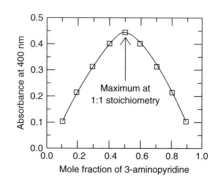

19-A.   (a) For $\lambda = 10.00\ \mu m$ and $\Delta\lambda = 0.01\ \mu m$, $\lambda/\Delta\lambda = 10.00/0.01 = 10^3$. The resolution is 10^4, so these lines will be resolved.

(b) $\lambda = \dfrac{1}{\tilde{v}} = \dfrac{1}{(1\,000\ cm^{-1})(10^{-4}\ cm/\mu m)} = 10\ \mu m$

$\Delta\lambda = \dfrac{\lambda}{10^4} = 10^{-3}\ \mu m$

$\Rightarrow$ 10.001 μm could be resolved from 10.000 μm

$\left.\begin{array}{l} 10.000\ \mu m = 1000.0\ cm^{-1} \\ 10.001\ \mu m = 999.9\ cm^{-1} \end{array}\right\}$ Difference $= 0.1\ cm^{-1}$

(c) $5.0\ cm \times 2\,500\ lines/cm = 12\,500\ lines$
Resolution $= 1 \cdot 12\,500 = 12\,500$ for $n = 1$
$\qquad\qquad = 10 \cdot 12\,500 = 125\,000$ for $n = 10$

(d) $\dfrac{\Delta\phi}{\Delta\lambda} = \dfrac{n}{d\cos\phi} = \dfrac{2}{\left(\dfrac{1\ mm}{250}\right)\cos 30°}$

$\qquad = 577\ \dfrac{radians}{mm} = 0.577\ \dfrac{radians}{\mu m}$

Convert radians to degrees $= \dfrac{radians}{\pi} \times 180$

$\Rightarrow \dfrac{\Delta\phi}{\Delta\lambda} = 33.1$ degrees/μm

The two wavelengths are $1\,000\ cm^{-1} = 10.00\ \mu m$ and $1\,001\ cm^{-1} = 9.99\ \mu m \Rightarrow \Delta\lambda = 0.01\ \mu m$.

$\Delta\phi = 0.577\ \dfrac{radians}{\mu m} \times 0.01\ \mu m$

$\qquad = 6 \times 10^{-3}\ radian = 0.3°$

19-B.   True transmittance $= 10^{-1.000} = 0.100$. With 1.0% stray light, the apparent transmittance is

Apparent transmittance $= \dfrac{P + S}{P_0 + S} = \dfrac{0.100 + 0.010}{1 + 0.010} = 0.109$

The apparent absorbance is $-\log T = -\log 0.109 = 0.963$.

Apparent concentration $= 96.3\%$ of true concentration $\Rightarrow$ error $= -3.7\%$

19-C.   (a) $\Delta\tilde{v} = 1/2\delta = 1/(2 \cdot 1.266\,0 \times 10^{-4}\ cm)$
$\qquad\qquad = 3\,949\ cm^{-1}$

(b) Each interval is $1.266\,0 \times 10^{-4}\ cm$. 4 096 intervals $= (4\,096)(1.266\,0 \times 10^{-4}\ cm) = 0.518\,6\ cm$. This is a range of $\pm\Delta$, so $\Delta = 0.259\,3\ cm$.

(c) Resolution $\approx 1/\Delta = 1/(0.259\,3\ cm) = 3.86\ cm^{-1}$

(d) Mirror velocity $= 0.693\ cm/s$

Interval $= \dfrac{1.2660 \times 10^{-4}\ cm}{0.693\ cm/s} = 183\ \mu s$

(e) (4 096 points) (183 μs/point) $= 0.748\ s$

(f) The beamsplitter is germanium on KBr. KBr absorbs light below $400\ cm^{-1}$, which the background transform shows clearly.

19-D.   The graphs show that signal-to-noise ratio is proportional to $\sqrt{n}$. The confidence interval is $\pm ts/\sqrt{n}$, where s is the standard deviation, n is the number of experiments, and t is Student's t from Table 4-2 for 95% confidence and $n - 1$ degrees of freedom. For the first line of the table, $n = 8$, $s = 1.9$, and $t = 2.365$ for 7 degrees of freedom. 95% confidence interval $= \pm(2.365)(1.9)/\sqrt{8} = \pm 1.6$. For remaining rows, 95% confidence interval $= 3.9, 5.1, 5.9, 9.0, 11.2, 14.0, 24.0, 23.9,$ and 27.2.

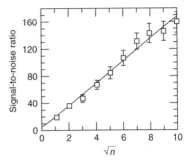

Chapter 20

20-A.   A graph of intensity versus concentration of added standard has an x-intercept of $-0.164 \pm 0.005\ \mu g/mL$. Because the sample was diluted by a factor of 10, the original concentration is $1.64 \pm 0.05\ \mu g/mL$.

	A	B	C	D
1	Standard Addition Constant Volume Least-Squares			
2	x = Added Li	y		
3	(g/mL)	Signal		
4	0.000	309		
5	0.081	452		
6	0.162	600		
7	0.243	765		
8	0.324	906		
9	B11:C13 = LINEST(B4:B8,A4:A8,TRUE,TRUE)			
10		LINEST output:		
11	m	1860.5	305.0	b
12	s_m	26.3	5.2	s_b
13	R^2	0.9994	6.7	s_y
14	x-intercept = -b/m =	-0.164		
15	n =	5	= COUNT(A4:A8)	
16	Mean y =	606.400	= AVERAGE(B4:B8)	
17	$\sum(x_i$ - mean x$)^2$ =	0.06561	= DEVSQ(A4:A8)	
18	Std deviation of			
19	x-intercept =	0.0049		
20	B19=(C13/ABS(B11))*SQRT((1/B15) + B16^2/(B11^2*B17))			

20-B.   The concentration of Mn in the unknown mixture is $(13.5\ \mu g/mL)(1.00/6.00) = 2.25\ \mu g/mL$.

Standard mixture:

$\dfrac{A_X}{[X]} = F\left(\dfrac{A_S}{[S]}\right)$

$\dfrac{1.05}{[2.50\ \mu g/mL]} = F\left(\dfrac{1.00}{[2.00\ \mu g/mL]}\right) \Rightarrow F = 0.840$

Unknown mixture:

$$\frac{A_X}{[X]} = F\left(\frac{A_S}{[S]}\right)$$

$$\frac{0.185}{[Fe]} = 0.840\left(\frac{0.128}{[2.25\ \mu g/mL]}\right) \Rightarrow [Fe] = 3.87\ \mu g/mL$$

The original concentration of Fe must have been

$$\frac{6.00}{5.00}(3.87\ \mu g/mL) = 4.65\ \mu g/mL = 8.33 \times 10^{-5}\ M$$

20-C. **(a)** The ratio of signal to peak-to-peak noise level is measured to be 17 in the figure. The concentration of Fe needed to give a signal-to-noise ratio of 2 is $(\frac{2}{17})(0.048\ 5\ \mu g/mL) = 0.005\ 7\ \mu g/mL$ (= 5.7 ppb).

(b) Standard deviation of 7 standards = 0.22_{48} ng/L ≡ s/m

Detection limit = 3s/m = 0.67 ng/L

Quantitation limit = 10s/m = 2.2 ng/L

20-D. **(a)** The higher result in experiment 2 compared with experiment 1 is probably the effect of diluting interfering species, so they do not interfere as much in experiment 2 as in experiment 1. Dilution lowers the concentration of species that might react with Li or make smoke that scatters light. In experiment 3, interference is present to the same extent as in experiment 2, but the standard addition procedure corrects for the interference. The whole point of standard addition is to measure the effect of the complex interfering matrix on the response to known quantities of analyte.

(b) Experiments 4–6 use a hotter flame than experiments 1–3. High temperature appears to eliminate most of the interference observed at lower temperature. Dilution has only a tiny effect on the results.

(c) Because it appears from experiments 1–3 that standard addition gives a true result, we surmise that experiments 3 and 6, and possibly 5, are within experimental error of each other. I would probably report the "true" value as the mean of experiments 3 and 6 (81.4 ppm). It might also be reasonable to take the average of experiments 3, 5, and 6 (80.8 ppm).

Chapter 21

21-A. **(a)** Resolving power $= \dfrac{m}{m_{1/2}} = \dfrac{53}{0.6_0} \approx 88$

We should be able to barely distinguish two peaks differing by 1 Da at a mass of 88 Da. We will probably not be able to distinguish two peaks at 100 and 101 Da.

21-B. $C_2H_5^+$:

	$2 \times 12.000\ 00$
	$+5 \times\ \ 1.007\ 825$
$-e^-$ mass	$-1 \times\ \ 0.000\ 55$
	$29.038\ 58$
HCO^+:	$1 \times 12.000\ 00$
	$1 \times\ \ 1.007\ 825$
	$+1 \times 15.994\ 91$
$-e^-$ mass	$-1 \times\ \ 0.000\ 55$
	$29.002\ 18$

We need to distinguish a mass difference of 29.038 58 − 29.002 18 = 0.036 40. The required resolving power is $m/\Delta m = 29.0/(0.036\ 40) \approx 800$.

21-C. ^{35}Cl abundance ≡ a = 0.757 8

^{37}Cl abundance ≡ b = 0.242 2

Relative abundance of $C_6H_4{}^{35}Cl_2 = a^2 = 0.574\ 2_6$

Relative abundance of $C_6H_4{}^{35}Cl^{37}Cl = 2ab = 0.367\ 0_8$

Relative abundance of $C_6H_4{}^{37}Cl_2 = b^2 = 0.058\ 66_1$

Relative abundances: $^{35}Cl_2 : {}^{35}Cl^{37}Cl : {}^{37}Cl_2 = 1 : 0.639\ 2 : 0.102\ 2$

Figure 21-7 shows the stick diagram.

21-D. **(a)** $C_{14}H_{12}$

$R + DB = c - h/2 + n/2 + 1 = 14 - 12/2 + 0/2 + 1 = 9$

A molecule with two rings + seven double bonds is *trans*-stilbene:

(b) $C_4H_{10}NO^+$

$R + DB = c - h/2 + n/2 + 1 = 4 - 10/2 + 1/2 + 1 = \dfrac{1}{2}$ Huh?

We came out with a fraction because the species is an ion in which at least one atom does not make its usual number of bonds. In the following structure, N makes four bonds instead of three:

A fragment in a mass spectrum

21-E. **(a)** The principal difference between the two spectra is the appearance of a significant peak at m/z 72 in A that is missing in B. This peak represents loss of a neutral molecule with an even mass of 28 Da from the molecular ion. The McLafferty rearrangement can split C_2H_4 from 3-methyl-2-pentanone but not from 3,3-dimethyl-2-butanone, which lacks a γ-CH group.

3-Methyl-2-pentanone   Ethylene   m/z = 72
                             28 Da

Spectrum A must be from 3-methyl-2-pentanone and spectrum B is from 3,3-dimethyl-2-butanone.

(b) Expected intensity of M+1 relative to $M^{+\bullet}$ for $C_6H_{12}O$:

Intensity = $\underbrace{6 \times 1.08\%}_{^{13}C} + \underbrace{12 \times 0.012\%}_{^2H} + \underbrace{1 \times 0.038\%}_{^{17}O} = 6.7\%$ of $M^{+\bullet}$

21-F. **(a)** —OH C_6H_6O: $M^{+\bullet} = 94$

Rings + double bonds = $c - h/2 + n/2 + 1 = 6 - 6/2 + 0/2 + 1 = 4$

Expected intensity of M+1 from Table 21-2:

$$\underset{\text{Carbon}}{1.08(6)} + \underset{\text{Hydrogen}}{0.012(6)} + \underset{\text{Oxygen}}{0.038(1)} = 6.59\%$$

Observed intensity of M+1 = 68/999 = 6.8%

Expected intensity of M+2 = 0.005 8(6)(5) + 0.205(1) = 0.38%

Observed intensity of M+2 = 0.3%

(b) —Br C_6H_5Br: $M^{+\bullet} = 156$

The two nearly equal peaks at m/z 156 and 158 scream out "bromine!"

Rings + double bonds = $c - h/2 + n/2 + 1 = 6 - 6/2 + 0/2 + 1 = 4$
                                          ↑
                                h includes H + Br

Expected intensity of M+1 = $\underset{\text{Carbon}}{1.08(6)} + \underset{\text{Hydrogen}}{0.012(5)} = 6.54\%$

Observed intensity of M+1 = 46/566 = 8.1%

Expected intensity of M+2 = $\underset{\text{Carbon}}{0.005\ 8(6)(5)} + \underset{\text{Bromine}}{97.3(1)} = 97.5\%$

Observed intensity of M+2 = 520/566 = 91.9%

The M+3 peak is the isotopic partner of the M+2 peak ($C_6H_5{}^{81}Br$). M+3 contains ^{81}Br plus either 1 ^{13}C or 1 ^2H. Therefore, the expected intensity of M+3 (relative to $C_6H_5{}^{81}Br$ at M+2) is 1.08(6) + 0.012(5) = 6.54% of predicted intensity of $C_6H_5{}^{81}Br$ at M+2 = (0.065 4)(97.3) = 6.4% of $M^{+\bullet}$. Observed intensity of M+3 is 35/566 = 6.2%

(c) [structure] —CO_2H $C_7H_3O_2Cl_3$: $M^{+\cdot} = 224$

From Figure 21-7, the M : M+2 : M+4 pattern looks like a molecule containing three chlorine atoms. The correct structure is shown here, but there is no way you could assign the isomeric structure from the data.

Rings + double bonds = $c - h/2 + n/2 + 1 = 7 - 6/2 + 0/2 + 1 = 5$

Expected intensity of M+1 from Table 21-2:

$$\underset{\text{Carbon}}{1.08(7)} + \underset{\text{Hydrogen}}{0.012(3)} + \underset{\text{Oxygen}}{0.038(2)} = 7.67\%$$

Observed intensity of M+1 = $63/791 = 8.0\%$

Expected intensity of M+2 = $0.005\,8(7)(6) + 0.205(2) + 32.0(3) = 96.7\%$

Observed intensity of M+2 = $754/791 = 95.4\%$

The M+3 peak is the isotopic partner of $C_7H_3O_2{}^{35}Cl_2{}^{37}Cl$ at M+2. M+3 contains one ^{37}Cl plus either 1 ^{13}C or 1 ^2H or 1 ^{17}O. Expected intensity of M+3 (relative to $C_7H_3O_2{}^{35}Cl_2{}^{37}Cl$ at M+2) = $1.08(7) + 0.012(3) + 0.038(2) = 7.67\%$ of predicted intensity of $C_7H_3O_2{}^{35}Cl_2{}^{37}Cl$ at M+2. The predicted intensity of $C_7H_3O_2{}^{35}Cl_2{}^{37}Cl$ is $32.0(3) = 96.0\%$ of $M^{+\cdot}$.

Expected intensity of M+3 = 7.67% of $96.0\% = 7.4\%$ of $M^{+\cdot}$

Observed intensity = $60/791 = 7.6\%$.

M+4 is composed mainly of $C_7H_3O_2{}^{35}Cl^{37}Cl_2$ plus a small amount of $C_7H_3{}^{16}O^{18}O^{35}Cl_2{}^{37}Cl$. Other formulas such as $^{12}C_6{}^{13}CH_2{}^{16}O^{17}O^{35}Cl_2{}^{37}Cl$ also add up to M+4, but they are even less likely to occur because they have two minor isotopes (^{13}C and ^{17}O). Expected intensity of M+4 from $C_7H_3O_2{}^{35}Cl^{37}Cl_2$ is $5.11(3)(2) = 30.7\%$ of $M^{+\cdot}$. The contribution from $C_7H_3{}^{16}O^{18}O^{35}Cl_2{}^{37}Cl$ is based on the predicted intensity of $C_7H_3O_2{}^{35}Cl_2{}^{37}Cl$ at M+2. The predicted intensity of $C_7H_3O_2{}^{35}Cl_2{}^{37}Cl$ is $32.0(3) = 96.0\%$ of $M^{+\cdot}$. The predicted intensity from $C_7H_3{}^{16}O^{18}O^{35}Cl_2{}^{37}Cl$ at M+4 is $0.205(2) = 0.410\%$ of $96.0\% = 0.4\%$. Total expected intensity of M+4 is $30.7\% + 0.4\% = 31.1\%$ of $M^{+\cdot}$.

Observed intensity = $264/791 = 33.4\%$

Expected intensity of M+5 from $^{12}C_6{}^{13}CH_3O_2{}^{35}Cl^{37}Cl_2$ and $^{12}C_7H_2{}^2HO_2{}^{35}Cl^{37}Cl_2$ and $C_7H_3{}^{16}O^{17}O^{35}Cl^{37}Cl_2$ is based on the predicted intensity of $C_7H_3O_2{}^{35}Cl^{37}Cl_2$ at M+4. M+5 should have $1.08(7) + 0.012(3) + 0.038(2) = 7.7\%$ of $C_7H_3O_2{}^{35}Cl^{37}Cl_2$ at M+4 = 7.7% of $30.7\% = 2.4\%$.

Observed intensity = $19/791 = 2.4\%$

Expected intensity of M+6 from $C_7H_3O_2{}^{37}Cl_3$ is $0.544(3)(2)(1) = 3.26\%$ of $M^{+\cdot}$. There will also be a small contribution from $C_7H_3{}^{16}O^{18}O^{35}Cl^{37}Cl_2$, which will be $0.205(2) = 0.410\%$ of predicted intensity of $C_7H_3{}^{16}O^{35}Cl^{37}Cl_2$ = 0.410% of 30.7% of $M^{+\cdot} = 0.13\%$ of $M^{+\cdot}$. The total expected intensity at M+6 is therefore $3.26 + 0.13 = 3.4\%$ of $M^{+\cdot}$.

Observed intensity = $29/791 = 3.7\%$

(d) HS [structure with OH, SH, OH] $C_4H_{10}O_2S_2$: $M^{+\cdot} = 154$

Sulfur gives a significant M+2 peak (4.52% of M per sulfur). The observed M+2 is $12/122 = 9.8\%$, which could represent two sulfur atoms. The composition $C_4H_{10}O_2S_2$ has two sulfur atoms and has a molecular mass of 154. The known structure is shown here, but you could not deduce the structure from the composition.

Rings + double bonds = $c - h/2 + n/2 + 1$

$$= 4 - 10/2 + 0/2 + 1 = 0$$

Expected intensity of M+1:

$$\underset{\text{Carbon}}{1.08(4)} + \underset{\text{Hydrogen}}{0.012(10)} + \underset{\text{Oxygen}}{0.038(2)} + \underset{\text{Sulfur}}{0.801(2)} = 6.12\%$$

Observed intensity of M+1 = $9/122 = 7.4\%$

Expected intensity of M+2 = $0.005\,8(4)(3) + 0.205(2) + 4.52(2) = 9.52\%$

Observed intensity of M+2 = $12/122 = 9.8\%$

21-G.

Analysis of electrospray mass spectrum of lysozyme

Observed $m/z \equiv m_n$	$m_{n+1} - 1.008$	$m_n - m_{n+1}$	Charge = n = $\dfrac{m_{n+1} - 1.008}{m_n - m_{n+1}}$	Molecular mass $= n \times (m_n - 1.008)$
1 789.1	1 589.39	198.7	$7.99 \approx 8$	14 304.7
1 590.4	1 430.49	158.9	$9.00 \approx 9$	14 304.5
1 431.5	1 300.49	130.0	$10.00 \approx 10$	14 304.9
1 301.5	1 192.09	108.4	$11.00 \approx 11$	14 305.4
1 193.1	—	—	12	14 305.1
			Mean =	14 304.9 (± 0.3)

21-H. **(a)** To find the response factor, we insert values from the first line of the table into the equation:

$$\frac{\text{Area of analyte signal}}{\text{Area of standard signal}} = F\left(\frac{\text{concentration of analyte}}{\text{concentration of standard}}\right)$$

$$\frac{11\,438}{2\,992} = F\left(\frac{13.60 \times 10^2}{3.70 \times 10^2}\right) = F = 1.04_0$$

For the next two sets of data, we find $F = 1.02_0$ and 1.06_4, giving a mean value $F = 1.04_1$.

(b) The concentration of internal standard in the mixture of caffeine-D_3 plus cola is

$$(1.11 \text{ g/L}) \times \frac{0.050\,0 \text{ mL}}{1.050 \text{ mL}} = 52.8_6 \text{ mg/L}$$

The concentration of caffeine in the chromatography solution is

$$\frac{\text{Area of analyte signal}}{\text{Area of standard signal}} = F\left(\frac{\text{concentration of analyte}}{\text{concentration of standard}}\right)$$

$$\frac{1\,733}{1\,144} = 1.04_1\left(\frac{[\text{caffeine}]}{52.8_6 \text{ mg/L}}\right) \Rightarrow [\text{caffeine}] = 76.9 \text{ mg/L}$$

The unknown beverage had been diluted from 1.000 to 1.050 mL when the standard was added, so the concentration of caffeine in the original beverage was $\left(\frac{1.050\,0 \text{ mL}}{1.00 \text{ mL}}\right)(76.9 \text{ mg/L}) = 80.8 \text{ mg/L}$.

Chapter 22

22-A. **(a)** $k_1 = \dfrac{t_{r1} - t_m}{t_m}$

$$\Rightarrow t_m = \frac{t_{r1}}{k_1 + 1} = \frac{10.0 \text{ min}}{5.00} = 2.00 \text{ min}$$

$$t_{r2} = t_m(k_2 + 1) = 2.00 \text{ min}(5.00 + 1) = 12.0 \text{ min}$$

$$\sigma_1 = \frac{t_{r1}}{\sqrt{N}} = \frac{10.0 \text{ min}}{\sqrt{1\,000}} = 0.316 \text{ min}$$

$$\Rightarrow = w_{1/2} \text{ (peak 1)} = 2.35\sigma_1 = 0.74 \text{ min}$$

$$w_1 = 4\sigma_1 = 1.26 \text{ min}$$

$$\sigma_2 = \frac{t_{r2}}{\sqrt{N}} = \frac{12.0 \text{ min}}{\sqrt{1\,000}} = 0.379 \text{ min}$$

$$\Rightarrow = w_{1/2} \text{ (peak 2)} = 2.35\sigma_2 = 0.89 \text{ min}$$

$$w_2 = 4\sigma_2 = 1.52 \text{ min}$$

(b)

(c) Resolution $= \dfrac{\Delta t_r}{w_{\text{av}}} = \dfrac{2 \text{ min}}{(1.26 \text{ min} + 1.52 \text{ min})/2} = 1.44$

22-B. (a) Fraction remaining $= q = \dfrac{V_1}{V_1 + KV_2}$

$0.01 = \dfrac{10}{10 + 4.0V_2} \Rightarrow V_2 = 248$ mL

(b) $q^3 = 0.01 = \left(\dfrac{10}{10 + 4.0V_2}\right)^3 \Rightarrow V_2 = 9.1$ mL

and total volume $= 27.3$ mL.

22-C. (a) Relative distances measured from Figure 22-7:

$t_m = 10._4$

$t_r' = 39._8$ for octane

$t_r' = 76._0$ for nonane

$k = t_r'/t_m = 3.8_3$ for octane and 7.3_1 for nonane

(b) Let $t_s =$ time in stationary phase, $t_m =$ time in mobile phase, and $t =$ total time on column. We know that $k = t_s/t_m$. But

$t = t_s + t_m = t_s + \dfrac{t_s}{k} = t_s\left(1 + \dfrac{1}{k}\right) = t_s\left(\dfrac{k+1}{k}\right)$

Therefore, $t_s/t = \dfrac{k}{k+1} = 3.8_3/4.8_3 = 0.79_3$

(c) $\alpha = t_r'$ (nonane)$/t_r'$(octane) $= 76._0/39._8 = 1.9_1$

(d) $K = kV_m/V_s = 3.8_3(V_m/\frac{1}{2}V_m) = 7.6_6$

22-D. (a) For ethyl acetate, I measure $t_r = 11.3$ mm and $w = 1.5$ mm. (You will have different numbers depending on the size of the figure in your book.) Therefore, $N = 16t_r^2/w^2 = 910$ plates. For toluene, the values are $t_r = 36.2$ mm, $w = 4.2$ mm, and $N = 1\,200$ plates.

(b) We expect $w_{1/2} = (2.35/4)w$. The measured value of $w_{1/2}$ is in good agreement with the calculated value.

22-E. The column is overloaded, causing a gradual rise and an abrupt fall of the peak. As the sample size is decreased, the overloading decreases and the peak becomes more symmetric.

22-F. (a) We know that $\alpha = 1.068$ and $k_1 = 5.16$. For resolution, we need to find the unadjusted relative retention $\gamma = t_2/t_1$, where t is retention time. The relation between retention factor and retention time is $k_1 = (t_1 - t_m)/t_m = t_1/t_m - 1 \Rightarrow t_1 = t_m(k_1 + 1) = t_m(6.16)$. Also, $k_2 = \alpha k_1 = (1.068)(5.16) = 5.51_1$. Therefore, $t_2 = t_m(k_2 + 1) = t_m(6.51_1)$.

$\dfrac{t_2}{t_1} = \dfrac{t_m(6.51_1)}{t_m(6.16)} = 1.057_0$

(b) Resolution $= \dfrac{\sqrt{N}}{4}(\gamma - 1)$

$1.00 = \dfrac{\sqrt{N}}{4}(1.057_0 - 1) \Rightarrow N = 4.92 \times 10^3$ plates

Required length $= (4.92 \times 10^3 \text{ plates})(0.520 \text{ mm/plate}) = 2.56$ m

(c) From (a), $t_{r1} = t_m(6.16) = (2.00 \text{ min})(6.16) = 12.32$ min

$t_{r2} = t_m(6.51_1) = (2.00 \text{ min})(6.51_1) = 13.02$ min

$w_{1/2} = \sqrt{\dfrac{5.55}{N}}\,t_r = \sqrt{\dfrac{5.55}{4.92 \times 10^3}}\,(12.32 \text{ min}) = 0.41$ min for component 1

$w_{1/2} = \sqrt{\dfrac{5.55}{4.92 \times 10^3}}\,(13.02 \text{ min}) = 0.44$ min for component 2

(d) $k = KV_s/V_m$

$5.16 = K(0.30) \Rightarrow K = 17._2$

Chapter 23

23-A. (a) S $= [\text{butanol}] = \dfrac{234 \text{ mg}/(74.12 \text{ g/mol})}{10.0 \text{ mL}} = 0.315_7$ M

X $= [\text{hexanol}] = \dfrac{312 \text{ mg}/(102.17 \text{ g/mol})}{10.0 \text{ mL}} = 0.305_4$ M

$\dfrac{A_X}{[X]} = F\left(\dfrac{A_S}{[S]}\right) \Rightarrow \dfrac{1.45}{[0.305_4 \text{ M}]} = F\left(\dfrac{1.00}{[0.315_7 \text{ M}]}\right) \Rightarrow F = 1.49_9$

(b) I estimate the areas by measuring the height and $w_{1/2}$ in millimeters. Your answer will be different from mine if the figure size in your book is different from that in my manuscript. However, the relative peak areas should be the same.

Butanol: Height $= 41.3$ mm; $w_{1/2} = 2.2$ mm;

Area $= 1.064 \times$ peak height $\times w_{1/2} = 96._7$ mm^2

Hexanol: Height $= 21.9$ mm; $w_{1/2} = 6.9$ mm;

Area $= 161$ mm^2

(c) The volume of solution is not stated, but the concentration is directly proportional to the number of moles. We can substitute moles for concentrations in the internal standard equation:

$\dfrac{A_X}{[X]} = F\left(\dfrac{A_S}{[S]}\right) \Rightarrow \dfrac{161 \text{ mm}^2}{\text{mg hexanol}/(102.17 \text{ mg/mmol})}$

$= 1.49_9\left(\dfrac{96.7 \text{ mm}^2}{112 \text{ mg}/(74.12 \text{ mg/mmol})}\right)$

$\Rightarrow$ hexanol $= 171$ mg

(d) The greatest uncertainty is in the width of the fairly narrow butanol peak. The uncertainty in width is ~5–10%.

23-B. S $= [\text{pentanol}]$; X $= [\text{hexanol}]$. We will substitute mmol for concentrations, because the volume is unknown and concentrations are proportional to mmol.

For the standard mixture, we can write

$\dfrac{A_X}{[X]} = F\left(\dfrac{A_S}{[S]}\right) \Rightarrow \dfrac{1\,570}{[1.53]} = F\left(\dfrac{922}{[1.06]}\right) \Rightarrow F = 1.18_0$

For the unknown mixture,

$\dfrac{816}{[X]} = 1.18_0\left(\dfrac{843}{[0.57]}\right) \Rightarrow [X] = 0.47$ mmol

23-C. (a) Between nonane (C_9H_{20}) and decane ($C_{10}H_{22}$).

(b) Adjusted retention times are 13.83 (C_{10}) and 15.42 min (C_{11})

$1\,050 = 100\left[10 + (11 - 10)\dfrac{\log t_r'(\text{unknown}) - \log 13.83}{\log 15.42 - \log 13.83}\right]$

$\Rightarrow t_r'$ (unknown) $= 14.60$ min $\Rightarrow t_r$ (unknown) $= 16.40$ min

23-D. (a) A plot of $\log t_r'$ versus (number of carbon atoms) should be a fairly straight line for a homologous series of compounds.

Peak	t_r'	$\log t_r'$
$n = 7$	2.9	0.46
$n = 8$	5.4	0.73
$n = 14$	85.8	1.93
Unknown	41.4	1.62

From a graph of $\log t_r'$ versus n, it appears that $n = 12$ for the unknown.

(b) $k = t_r'/t_m = 41.4/1.1 = 38$

23-E. (a) Plate number is proportional to column length. If everything is the same except for length, we can say from Equation 22-30 that

$\dfrac{1.5}{1.0} = \dfrac{R_2}{R_1} = \dfrac{\sqrt{N_2}}{\sqrt{N_1}} \Rightarrow N_2 = 2.25\,N_1$

The column must be 2.25 times longer to achieve the desired resolution, and the elution time will be 2.25 times longer.

(b) If everything is the same except for γ, we can say from Equation 22-30 that

$\dfrac{1.5}{1.0} = \dfrac{R_2}{R_1} = \dfrac{\gamma_2 - 1}{\gamma_1 - 1} = \dfrac{\gamma_2 - 1}{1.013 - 1} \Rightarrow \gamma_2 = 1.020$

Alcohols are polar, so we could probably increase the relative retention by choosing a more polar stationary phase. (Diphenyl)$_{0.05}$(dimethyl)$_{0.95}$ polysiloxane is listed as nonpolar. We could try an intermediate polarity phase such as (diphenyl)$_{0.35}$(dimethyl)$_{0.65}$polysiloxane. The more polar phase will probably retain alcohols more strongly and increase the retention time. We have no way to predict how much the retention time will increase.

24-A. $\dfrac{\text{Area}_A}{[A]} = F\left(\dfrac{\text{Area}_B}{[B]}\right) \Rightarrow \dfrac{10.86}{[1.03]} = F\left(\dfrac{4.37}{[1.16]}\right) \Rightarrow F = 2.79_9$

The concentration of internal standard (B) mixed with unknown (A) is
12.49 mg/25.00 mL = 0.499 6 mg/mL.

$\dfrac{5.97}{[A]} = 2.79_9\left(\dfrac{6.38}{[0.499\ 6]}\right) \Rightarrow [A] = 0.167_0$ mg/mL

[A] in original unknown $= \dfrac{25.00}{10.00}(0.167_0$ mg/mL$) = 0.418$ mg/mL

24-B. **(a)** Equation 22-17: $k = \dfrac{t_r - t_m}{t_m} \Rightarrow \dfrac{t_1 - 1.00}{1.00} = 1.35$

$\Rightarrow t_1 = 2.35$ min

Equation 22-15: $\alpha = \dfrac{t'_{r2}}{t'_{r1}} \Rightarrow 4.53 = \dfrac{t_2 - 1.00}{t_1 - 1.00} \Rightarrow t_2 = 7.12$ min

Equation 22-23: Resolution $= \dfrac{\Delta t_r}{w_{av}}$

$\Rightarrow 7.7 = \dfrac{7.12 - 2.35}{w_{av}} \Rightarrow w_{av} = 0.62$ min

(b) From Equation 22-28b, we know that $w_{1/2}$ is proportional to t_r if N is
constant. Therefore, $\dfrac{w_{1/2}(\text{peak 1})}{w_{1/2}(\text{peak 2})} = \dfrac{t_1}{t_2} = \dfrac{2.35}{7.12} = 0.330$. We know that

w_{av}, the average width at the base, is 0.62 min. For each peak, w = 4σ and
$w_{1/2} = 2.35\sigma$, so w = $1.70w_{1/2}$.

$w_{av} = 0.62 = \dfrac{1}{2}(w_1 + w_2)$

$\qquad = \dfrac{1}{2}[1.70w_{1/2}\,(\text{peak 1}) + 1.70w_{1/2}\,(\text{peak 2})]$

Substituting $w_{1/2}$ (peak 1) = $0.330w_{1/2}$ (peak 2) into the previous equation
gives $w_{1/2}$ (peak 2) = 0.54_8 min. Then $w_{1/2}$ (peak 1) = $0.330w_{1/2}$ (peak 2) =
0.18_1 min.

(c) Because the areas are equal, we can say

Height$_R$ × w_R = height$_S$ × w_S

$\Rightarrow \dfrac{\text{Height}_R}{\text{Height}_S} = \dfrac{w_S}{w_R} = \dfrac{0.54_8}{0.18_1} = 3.0$

24-C.

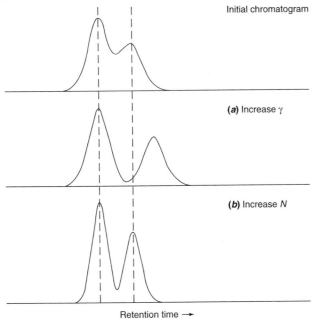

Initial chromatogram

(a) Increase γ

(b) Increase N

Retention time →

[Adapted from L. R. Snyder, J. J. Kirkland, and J. L. Glajch, *Practical HPLC Method
Development* (New York: Wiley, 1997).]

24-D. **(a)**

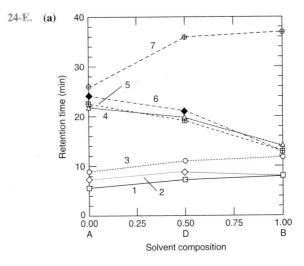

MelamineH$^+$
m/z 127

[Cyanuric acid-H]$^-$
m/z 128

When melamineH$^+$ decomposes to m/z 85, it loses a mass of 42 Da, which
is probably CN_2H_2. The remaining cation is $C_2N_4H_5^+$, which might have
the structure

$$H_2N{-}\underset{}{\overset{+}{C}}{\,}{\diagdown}_{N{-}C}{\overset{\nearrow\ NH_2}{\searrow_{NH}}}$$

[Cyanuric acid-H]$^-$ decomposes to m/z 42, which is most likely cyanate,
N≡C—O$^-$.

(b) In selected reaction monitoring for melamine, the mass spectrometer
isolates m/z 127$^+$ from the milk, and this ion then undergoes collisionally
activated dissociation to m/z 85$^+$. There are not many other molecular
species in the milk that provide the same two ions. For cyanuric acid, the
spectrometer isolates m/z 128$^-$, and this ion dissociates to m/z 42$^-$. Again,
there are not many other species in the milk that provide the same two ions.

24-E. **(a)**

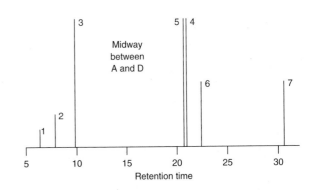

				Peaks			
Solvent composition	1	2	3	4	5	6	7
0.0	5.6	7.2	8.7	21.6	23.3	24.0	25.5
0.5	7.1	8.5	10.8	19.5	19.0	20.9	35.7
1.0	8.0	8.0	11.5	13.8	12.8	12.8	37.0
Predicted positions (by linear interpolation)							
0.25	6.35	7.85	9.75	21.05	20.65	22.45	30.60
0.75	7.55	8.25	11.15	16.65	15.90	16.85	36.35

Midway
between
A and D

Retention time

(b) A: 30% acetonitrile/70% buffer

B: 40% methanol/60% buffer

D: 14% acetonitrile/20% methanol/65% buffer

Between A and D: 22.5% acetonitrile/10% methanol/67.5% buffer

Between D and B: 7.5% acetonitrile/30% methanol/62.5% buffer

Chapter 25

25-A. 13.03 mL of 0.022 74 M NaOH = 0.296 3 mmol of OH^-, which must equal the total cation charge ($= 2[VO^{2+}] + 2[H_2SO_4]$) in the 5.00-mL aliquot. 50.0 mL therefore contains 2.963 mmol of cation charge. The VO^{2+} content is (50.0 mL)(0.024 3 M) = 1.215 mmol = 2.43 mmol of charge. The H_2SO_4 must therefore be $(2.963 - 2.43)/2 = 0.266_5$ mmol.

1.215 mmol $VOSO_4$ = 0.198 g $VOSO_4$ in 0.244 7 g sample = 80.9 wt%

0.266_5 mmol H_2SO_4 = 0.026 1 g H_2SO_4 in 0.244 7 g sample = 10.7 wt%

H_2O (by difference) = 8.4 wt%

25-B. **(a)** Fractionation range of Sephadex G-50 = 1 500–30 000, so hemoglobin should not be retained and ought to be eluted in a volume of 36.4 mL.

(b) The elution volume of ^{22}NaCl is V_m. Inserting $V_o = 36.4$ mL into the elution equation gives

$$K_{av} = \frac{V_r - V_o}{V_m - V_o} \Rightarrow 0.65 = \frac{V_r - 36.4}{109.8 - 36.4} \Rightarrow V_r = 84.1 \text{ mL}$$

25-C. **(a)**

Net flow of cations and anions is to the right, because electroosmotic flow is stronger than electrophoretic flow at high pH

(b) I^- has a greater mobility than Cl^-. Therefore, I^- swims upstream faster than Cl^- (because electrophoresis opposes electroosmosis) and is eluted later than Cl^-. The mobility of Br^- is greater than that of I^- in Table 14-1. Therefore, Br^- will have a longer migration time than I^-.

(c) Bare I^- is a larger ion than bare Cl^-, so the charge density in I^- is lower than the charge density in Cl^-. Therefore, I^- should have a smaller hydrated radius than Cl^-. This means that I^- has less friction than Cl^- and a greater mobility than Cl^-.

Chapter 26

26-A. 1 mol ethoxyl groups produces 1 mol AgI. 29.03 mg AgI = 0.123_5 mmol. The amount of compound analyzed is 25.42 mg/(417 g/mol) = 0.060_9_6 mmol. There are

$$\frac{0.123_5 \text{ mmol ethoxyl groups}}{0.060_9_6 \text{ mmol compound}} = 2.03 \ (= 2) \text{ ethoxyl groups/molecule}$$

26-B. There is 1 mol SO_4^{2-} in each mole of each reactant and of the product. Let x = g of K_2SO_4 and y = g of $(NH_4)_2SO_4$.

$$x + y = 0.649 \text{ g} \tag{1}$$

$$\underbrace{\frac{x}{174.27}}_{\substack{\text{Moles of} \\ K_2SO_4}} + \underbrace{\frac{y}{132.14}}_{\substack{\text{Moles of} \\ (NH_4)_2SO_4}} = \underbrace{\frac{0.977}{233.39}}_{\substack{\text{Moles of} \\ BaSO_4}} \tag{2}$$

Making the substitution $y = 0.649 - x$ in Equation 2 gives $x = 0.397$ g = 61.1% of the sample.

26-C. Formula and atomic masses: Ba(137.327), Cl(35.453), K(39.098), H_2O(18.015), KCl(74.551), $BaCl_2 \cdot 2H_2O$(244.26), H_2O lost = 1.783 9 − 1.562 3 = 0.221 6 g = $1.230 1 \times 10^{-2}$ mol of H_2O. For 2 mol H_2O lost, 1 mol $BaCl_2 \cdot 2H_2O$ must have been present. $\frac{1}{2}(1.230 1 \times 10^{-2} \text{ mol } H_2O \text{ lost}) = 6.150 4 \times 10^{-3}$ mol $BaCl_2 \cdot 2H_2O$ = 1.502 4 g. The Ba and Cl contents of the $BaCl_2 \cdot 2H_2O$ are

$$Ba = \left(\frac{137.33}{244.26}\right)(1.502 \text{ 4 g}) = 0.844 \text{ 69 g}$$

$$Cl = \left(\frac{2(35.453)}{244.26}\right)(1.502 \text{ 4 g}) = 0.436 \text{ 13 g}$$

Because the total sample weighs 1.783 9 g and contains 1.502 4 g of $BaCl_2 \cdot 2H_2O$, the sample must contain 1.783 9 − 1.502 4 = 0.281 5 g of KCl, which contains

$$K = \left(\frac{39.098}{74.551}\right)(0.281_5) = 0.147 \text{ 63 g}$$

$$Cl = \left(\frac{35.453}{74.551}\right)(0.281_5) = 0.133 \text{ 87 g}$$

Weight percent of each element:

$$Ba = \frac{0.844 \text{ 69}}{1.783 \text{ 9}} = 47.35\%$$

$$K = \frac{0.147 \text{ 63}}{1.783 \text{ 9}} = 8.28\%$$

$$Cl = \frac{0.436 \text{ 13} + 0.133 \text{ 87}}{1.783 \text{ 9}} = 31.95\%$$

26-D. Let x = mass of $Al(BF_4)_3$ and y = mass of $Mg(NO_3)_2$. We can say that $x + y = 0.282$ 8 g. We also know that

$$\text{Moles of nitron tetrafluoroborate} = 3(\text{moles of } Al(BF_4)_3) = \frac{3x}{287.39}$$

$$\text{Moles of nitron nitrate} = 2(\text{moles of } Mg(NO_3)_2) = \frac{2y}{148.31}$$

Equating the mass of product to the mass of nitron tetrafluoroborate plus the mass of nitron nitrate, we can write

$$\underbrace{1.322}_{\substack{\text{Mass of} \\ \text{product}}} = \underbrace{\left(\frac{3x}{287.39}\right)(400.18)}_{\substack{\text{Mass of nitron} \\ \text{tetrafluoroborate}}} + \underbrace{\left(\frac{2y}{148.31}\right)(375.39)}_{\substack{\text{Mass of nitron} \\ \text{nitrate}}}$$

Making the substitution $x = 0.282$ 8 − y allows us to find $y = 0.158$ 9 g of $Mg(NO_3)_2$ = 1.072 mmol of Mg = 0.026 05 g of Mg = 9.210% of the original solid sample.

26-E. Reaction: $SCN^- + Cu^+ \rightarrow CuSCN(s)$

At V_e, mol Cu^+ = mol $SCN^- \Rightarrow V_e = 100.0$ mL.

Before V_e, there is excess SCN^- in the solution. We calculate the molarity of SCN^- and then find $[Cu^+]$ from the relation $[Cu^+] = K_{sp}/[SCN^-]$. For example, when 0.10 mL of Cu^+ has been added:

$$[SCN^-] = \left(\frac{100.0 - 0.10}{100.0}\right)(0.080 \text{ 0 M})\left(\frac{50.0}{50.1}\right) = 7.98 \times 10^{-2} \text{ M}$$

$$[Cu^+] = 4.8 \times 10^{-15}/7.98 \times 10^{-2} = 6.0 \times 10^{-14} \text{ M}$$

$$pCu^+ = 13.22$$

At V_e, $[Cu^+][SCN^-] = x^2 = K_{sp}$
$$\Rightarrow x = [Cu^+] = 6.9 \times 10^{-8} \text{ M} \Rightarrow pCu^+ = 7.16$$

Past V_e, there is excess $[Cu^+]$. For example, when $V = 101.0$ mL,

$$[Cu^+] = (0.040\ 0\ M)\left(\frac{101.0 - 100.0}{151.0}\right) = 2.6 \times 10^{-4}\ M$$

$pCu^+ = 3.58$

mL	pCu	mL	pCu	mL	pCu
0.10	13.22	75.0	12.22	100.0	7.16
10.0	13.10	95.0	11.46	100.1	4.57
25.0	12.92	99.0	10.75	101.0	3.58
50.0	12.62	99.9	9.75	110.0	2.60

26-F. **(a)** 12.6 mL of Ag^+ are required to precipitate I^-.

$(27.7 - 12.6) = 15.1$ mL are required to precipitate SCN^-.

$$[SCN^-] = \frac{\text{moles of } Ag^+ \text{ needed to react with } SCN^-}{\text{original volume of } SCN^-}$$

$$= \frac{[27.7(\pm 0.3) - 12.6(\pm 0.4)][0.068\ 3\ (\pm 0.000\ 1)]}{50.00\ (\pm 0.05)}$$

$$= \frac{[15.1(\pm 0.5)][0.068\ 3\ (\pm 0.000\ 1)]}{50.00\ (\pm 0.05)}$$

$$= \frac{[15.1(\pm 3.31\%)][0.068\ 3\ (\pm 0.146\%)]}{50.00\ (\pm 0.100\%)}$$

$$= 0.020\ 6\ (\pm 0.000\ 7)\ M$$

(b) $[SCN^-](\pm 4.0\%) = \dfrac{[27.7(\pm 0.3) - 12.6(\pm ?)][0.068\ 3\ (\pm 0.000\ 1)]}{50.00(\pm 0.05)}$

Let the error in 15.1 mL be $y\%$:

$(4.0\%)^2 = (y\%)^2 + (0.146\%)^2 + (0.100\%)^2$

$\Rightarrow y = 4.00\% = 0.603$ mL

$27.7\ (\pm 0.3) - 12.6\ (\pm ?) = 15.1\ (\pm 0.603)$

$\Rightarrow 0.3^2 + ?^2 = 0.603^2 \Rightarrow ? = 0.5$ mL

Chapter 27

27-A. **(a)** Expected number of red marbles $= np_{red} = (1\ 000)(0.12) = 120$.
Expected number of yellow $= nq_{yellow} = (1\ 000)(0.88) = 880$.

(b) Absolute: $\sigma_{red} = \sigma_{yellow} = \sqrt{npq}$

$= \sqrt{(1\ 000)(0.12)(0.88)} = 10.28$

Relative: $\sigma_{red}/n_{red} = 10.28/120 = 8.56\%$

$\sigma_{yellow}/n_{yellow} = 10.28/880 = 1.17\%$

(c) For 4 000 marbles, $n_{red} = 480$ and $n_{yellow} = 3\ 520$.

$\sigma_{red} = \sigma_{yellow} = \sqrt{npq} = \sqrt{(4\ 000)(0.12)(0.88)} = 20.55$

$\sigma_{red}/n_{red} = 4.28\%$ $\sigma_{yellow}/n_{yellow} = 0.58\%$

(d) 2, $\sqrt{n}$

(e) $\dfrac{\sigma_{red}}{n_{red}} = 0.02 = \dfrac{\sqrt{n(0.12)(0.88)}}{(0.12)n}$

$\Rightarrow n = 1.83 \times 10^4$

27-B. **(a)** $mR^2 = K_s \Rightarrow m(10)^2 = 36 \Rightarrow m = 0.36$ g

(b) An uncertainty of ± 20 counts per second per gram is $100 \times 20/237 = 8.4\%$.

$n = \dfrac{t^2 s_s^2}{e^2} = \dfrac{(1.96)^2(0.10)^2}{(0.084)^2} = 5.4 \approx 5$

$\Rightarrow t = 2.776$ (4 degrees of freedom)

$n \approx \dfrac{(2.776)^2(0.10)^2}{(0.084)^2} = 10.9 \approx 11 \Rightarrow t = 2.228$

$n \approx \dfrac{(2.228)^2(0.10)^2}{(0.084)^2} = 7.0 \approx 7 \Rightarrow t = 2.447$

$n \approx \dfrac{(2.447)^2(0.10)^2}{(0.084)^2} = 8.5 \approx 8 \Rightarrow t = 2.365$

$n \approx \dfrac{(2.365)^2(0.10)^2}{(0.084)^2} = 7.9 \approx 8$

27-C. The acid-soluble inorganic matter and the organic material can probably be dissolved (and oxidized) together by wet ashing with $HNO_3 + H_2SO_4$ in a Teflon-lined bomb in a microwave oven. The insoluble residue should be washed well with water and the washings combined with the acid solution. After the residue has been dried, it can be fused with one of the fluxes in Table 27-6, dissolved in dilute acid, and combined with the previous solution.

Chapter 1

3. (a) milliwatt = 10^{-3} watt   (b) picometer = 10^{-12} meter
(c) kiloohm = 10^3 ohm   (d) microfarad = 10^{-6} farad
(e) terajoule = 10^{12} joule   (f) nanosecond = 10^{-9} second
(g) femtogram = 10^{-15} gram   (h) decipascal = 10^{-1} pascal

4. (a) 100 fJ or 0.1 pJ   (b) 43.172 8 nF   (c) 299.79 THz or
0.299 79 PHz   (d) 0.1 nm or 100 pm   (e) 21 TW
(f) 0.483 amol or 483 zmol

5. (a) 5.4×10^{12} kg of C   (b) 2.0×10^{13} kg CO_2
(c) 2.0×10^{10} ton CO_2 = 4 tons per person

6. 7.457×10^4 J/s, 6.416×10^7 cal/h

7. (a) 2.0 W/kg and 3.0 W/kg   (b) The person consumes 1.1×10^2 W.

8. 1.47×10^3 J/s, 1.47×10^3 W

9. (a) 0.621 37 miles/km   (b) 51 miles per gallon   (c) diesel produces
5.38 metric tons and gasoline produces 6.42 metric tons of CO_2

11. 6 tons/year

14. 1.10 M

15. 5.48 g

16. (a) 1.9×10^{-7} bar   (b) 11 nM

17. 10^{-3} g/L, 10^3 µg/L, 1 µg/mL, 1 mg/L

18. 7×10^{-10} M

19. 26.5 g $HClO_4$, 11.1 g H_2O

20. (a) 1 670 g solution   (b) 1.18×10^3 g $HClO_4$ (c) 11.7 mol

21. 1.51 m

22. (a) 6.0 amol/vesicle   (b) 3.6×10^6 molecules
(c) 3.35×10^{-20} m^3, 3.35×10^{-17} L   (d) 0.30 M

23. 4.4×10^{-3} M, 6.7×10^{-3} M

24. (a) 1 046 g, 376.6 g/L   (b) 9.07 m

25. Cal/g, Cal/ounce: Shredded Wheat (3.6, 102); doughnut (3.9, 111);
hamburger (2.8, 79); apple (0.48, 14)

26. 2.5×10^6 g F^-, 3.2×10^6 g H_2SiF_6

27. (a) 2.11×10^{-7} M   (b) Ar: 3.77×10^{-4} M; Kr: 4.60×10^{-8} M;
Xe: 3.5×10^{-9} M

28. 6.18 g in a 2-L volumetric flask

29. Dissolve 6.18 g $B(OH)_3$ in 2.00 kg H_2O.

30. 3.2 L

31. 8.0 g

32. (a) 55.6 mL   (b) 1.80 g/mL

33. 1.52 g/mL

34. 1.29 mL

35. 14.4 g

42. 32.0 mL

43. 43.20 mL $KMnO_4$, 270.0 mL $H_2C_2O_4$

44. 0.149 M

45. 0.100 3 M

46. 92.0 wt%

Chapter 2

3. Reducing Cr(VI) to Cr(III) decreases toxicity. Converting
Cr(III)(aq) to $Cr(OH)_3(s)$ decreases solubility. Evaporation minimizes
waste volume.

4. The upper "0" means that the reagent has no fire hazard. The right-hand
"0" indicates that the reagent is stable. The "3" tells us that the reagent is
corrosive or toxic and we should avoid skin contact or inhalation.

5. The lab notebook must (1) state what was done; (2) state what was
observed; and (3) be understandable to a stranger.

7. The buoyancy correction is 1 when the substance being weighed has
the same density as the weight used to calibrate the balance.

8. 14.85 g

9. smallest: PbO_2; largest: Li

10. 4.239 1 g, lower by 0.06%

11. (a) 0.000 164 g/mL   (b) 0.823 g

12. (a) 979 Pa   (b) 0.001 1 g/mL   (c) 1.001 0 g

13. 99.999 1 g

14. TD means "to deliver" and TC means "to contain."

15. Dissolve (0.250 0 L)(0.150 0 mol/L) = 0.037 50 mol K_2SO_4 (= 6.535 g,
FM 174.26 g/mol) in <250 mL H_2O in a 250-mL volumetric flask. Add
H_2O and mix. Dilute to mark and mix well by inverting flask many times.

16. Plastic flask is for trace analysis of ppb analytes that might be
adsorbed on glass.

17. (a) See Section 2.6   (b) transfer pipet

18. (a) Use forward mode described in text.   (b) Use reverse mode
described in text.

19. Trap prevents liquid from being sucked into vacuum system.
Watchglass keeps dust out of the sample.

20. phosphorus pentoxide

21. 9.979 9 mL

22. 0.2%, 0.499 0 M

23. 49.947 g in vacuum, 49.892 g in air

24. true mass = 50.506 g, mass in air = 50.484 g

25. (b) 54 days

26. 0.70%

Chapter 3

1. (a) 5   (b) 4   (c) 3

2. (a) 1.237   (b) 1.238   (c) 0.135   (d) 2.1   (e) 2.00

3. (a) 0.217   (b) 0.216   (c) 0.217

4. (b) 1.18 (three significant figures)   (c) 0.71 (two significant figures)

5. (a) 3.71   (b) 10.7   (c) 4.0×10^1 (d) 2.85×10^{-6}
(e) 12.625 1   (f) 6.0×10^{-4} (g) 242

6. (a) 175.324   (b) 140.093 6 or 140.094

7. (a) 12.3   (b) 75.5   (c) 5.520×10^3 (d) 3.04
(e) 3.04×10^{-10} (f) 11.9   (g) 4.600   (h) 4.9×10^{-7}

11. low, systematic

12. (a) 25.031 mL, systematic error; ±0.009 mL, random error
(b) 1.98 and 2.03 mL, systematic error; ±0.01 and ±0.02 mL, random
error   (c) random error   (d) random error

13. (a) Carmen   (b) Cynthia   (c) Chastity   (d) Cheryl

14. 3.124 (±0.005), 3.124 (±0.2%)

15. (a) 2.1 (±0.2 or ±11%)   (b) 0.151 (±0.009 or ±6%)
(c) 0.22_3 (±0.02_4 or ±11%)   (d) 0.097_1 (±0.002_2 or ±$2._2$%)

16. (a) 10.18 (±0.07 or ±0.7%)   (b) 174 (±3 or ±2%)
(c) 0.147 (±0.003 or ±2%)   (d) 7.86 (±0.01 or ±0.1%)
(e) 2 185.8 (±0.8 or ±0.04%)   (f) 1.464_3 (±0.007_8 or ±0.5_3%)
(g) 0.496_9 (±0.006_9 or ±1.3_9%)

18. (b) 0.450 7 (±0.000 5) M

19. 1.035 7 (±0.000 2) g

20. 16.2 ± 0.1 mg

21. 0.667 ± 0.001 M

22. $6.022\ 136\ 9\ (48) \times 10^{23}$

23. 255.184 ± 0.004

24. 0.1 ppm

Chapter 4

2. **(a)** 0.682 6 **(b)** 0.954 6 **(c)** 0.341 3 **(d)** 0.191 5 **(e)** 0.149 8

3. **(a)** 1.527 67 **(b)** 0.001 26 **(c)** 1.59×10^{-6} **(d)** $1.527_7 \pm 0.001_3$

4. **(a)** 0.044 6 **(b)** 0.417 3 **(c)** 0.404 0

5. **(a)** 0.5 **(b)** 0.8% **(c)** 8.7%

11. 90%: $0.14_8 \pm 0.02_8$; 99%: $0.14_8 \pm 0.05_6$

12. $\bar{x} \pm 0.000\ 10$ (1.527 83 to 1.528 03)

13. **(a)** dL = deciliter = 0.1 L **(b)** yes ($t_{calculated} = 2.12 < t_{table} = 2.262$)

14. Difference is <u>not</u> significant ($t_{calculated} = 0.99 < t_{table} = 2.57$).

16. Differences are <u>not</u> significant ($F_{calculated} = 2.43 < F_{table} = 9.28$ and $t_{calculated} = 1.55 < t_{table} = 2.447$).

17. Difference <u>is</u> significant ($F_{calculated} = 92.7 > F_{table} = 6.26$, so we use Equations 4-8a and 4-9a. $t_{calculated} = 11.3 > t_{table} = 2.57$ for 5 degrees of freedom).

18. yes (90%: $\bar{x} \pm 1.1_8\%$)

19. 1–2 difference <u>is</u> significant ($F_{calculated} = 5.3 > F_{table} \approx 2.2$, so we use Equations 4-8a and 4-9a. $t_{calculated} = 18.2 > t_{table} = 2.02$ for 40 degrees of freedom); 2–3 difference is <u>not</u> significant ($F_{calculated} = 1.3 < F_{table} \approx 2.2$, so we use Equations 4-8 and 4-9. $t_{calculated} = 1.39 < t_{table} \approx 2.02$ for 43 degrees of freedom)

20. Difference <u>is</u> significant at 95% and 99% levels ($t_{calculated} = 2.88$).

21. Difference <u>is</u> significant in both cases.

22. **(a)** Differences are <u>not</u> significant. **(b)** yes

23. Retain 216.

24. $m = -1.299\ (\pm 0.001) \times 10^4$ or $-1.298_7\ (\pm 0.001_3) \times 10^4$; $b = 3\ (\pm 3) \times 10^2$

25. $m = 0.6_4 \pm 0.1_2$; $b = 0.9_3 \pm 0.2_6$; $s_y = 0.27$

27. $m = -0.1379 \pm 0.0066$; $b = 0.195 \pm 0.163$; $s_y = 0.198$

30. 10.1 µg

31. **(a)** $2.0_0 \pm 0.3_8$ **(b)** 0.2_6

32. 10.1 ± 0.2 µg

33. **(a)** $m = 869 \pm 11$, $b = -22._1 \pm 8._9$ **(b)** 145.0 mV **(c)** $0.19_2\ (\pm 0.01_4)$ vol%

34. 21.9 µg

35. **(a)** Entire range is linear. **(b)** log(current) = 0.969 2 log(concentration) + 1.339 **(c)** 4.80 µg/mL

36. $15.2_2 \pm 0.8_6$ µg, $15._2 \pm 1._5$ µg

Chapter 5

11. statement c

14. yes: 7 consecutive measurements all above or all below the center line

15. $R^2 = 0.993\ 2$, y error bar $= \pm ts_y = \pm 162$

16. **(a)** 22.2 ng/mL: precision = 23.8%, accuracy = 6.6%
    88.2 ng/mL: precision = 13.9%, accuracy = -6.5%
    314 ng/mL: precision = 7.8%, accuracy = -3.6%
    (b) signal detection limit = $129._6$; detection limit = 4.8×10^{-8} M; quantitation limit = 1.6×10^{-7} M

17. **(a)** 4%, 128% **(b)** 1.4%

18. recovery = 96%; concentration detection limit = 0.064 µg/L

19. detection limit = 130 counts; minimum detectable concentration = 4.8×10^{-8} M

20. detection limits: 0.086, 0.102, 0.096, and 0.114 µg/mL; mean = 0.10 µg/mL

21. Testing 2 independent samples drawn at the same time from each athlete would reduce the rate of false positives to $0.01 \times 0.01 = 0.000\ 1$.

22. Lab C vs. Lab A: $F_{calculated} = 31._0 > F_{table} = 3.88$ (2 degrees of freedom for s_C and 12 degrees of freedom for s_A). $t_{calculated} = 2.4_1 < t_{table} = 4.303$ for 95% confidence and 2 degrees of freedom $\Rightarrow$ difference <u>is not</u> significant

Lab C vs. Lab B: $F_{calculated} = 1.9_4 < F_{table} = 4.74$ (2 degrees of freedom for s_C and 7 degrees of freedom for s_A). $s_{pooled} = 0.61_6$. $t_{calculated} = 2.4_7 > t_{table} = 2.262$ for 95% confidence and 9 degrees of freedom $\Rightarrow$ difference <u>is</u> significant

The conclusion that C is greater than B but C is not greater than A makes no sense. The problem is $s_C \gg s_A$ and number of replicates for C $\ll$ number of replicates for A. I suggest that C > A and C > B. I would ask for more replicates from C.

23. Addition of small volume keeps matrix nearly constant by not diluting the sample.

24. **(c)** 1.04 ppm

25. **(a)** 8.72 ± 0.43 ppb **(b)** 116 ppm **(c)** ± 6 ppm **(d)** ± 18 ppm

26. **(a)** tap water, 0.091 ng/mL; pond water, 22.2 ng/mL **(b)** This is a matrix effect. Something in pond water decreases the Eu(III) emission.

27. **(a)** 0.140 M **(b)** standard deviation = ± 0.005 M; 95% confidence = ± 0.015 M

28. **(a)** 8.2 ± 0.6 mg alliin/g garlic **(b)** 3.8 ± 0.3 mg allicin/g garlic

30. **(a)** 0.168_4 **(b)** 0.847 mM **(c)** 6.16 mM **(d)** 12.3 mM

31. 9.09 mM

32. response factor = slope of graph = 1.07_6; standard deviation = $0.06_7 = 6._2\%$

34. [A] = $0.83_8 \pm 0.02_7$ M, [B] = $0.44_0 \pm 0.02_7$ M, [C] = $0.25_6 \pm 0.02_7$ M

Chapter 6

4. **(a)** $K = 1/[Ag^+]^3\ [PO_4^{3-}]$ **(b)** $K = P_{CO_2}^6/P_{O_2}^{15/2}$

5. 1.2×10^{10}

6. 2.0×10^{-9}

7. **(a)** decrease **(b)** give off **(c)** negative

8. 5×10^{-11}

9. **(a)** right **(b)** right **(c)** neither **(d)** right **(e)** smaller

10. **(a)** 4.7×10^{-4} bar **(b)** 153°C

11. **(a)** 7.82 kJ/mol **(b)** A graph of ln K vs. $1/T$ will have a slope of $-\Delta H°/R$.

12. **(a)** right **(b)** $P_{H_2} = 1\ 366$ Pa, $P_{Br_2} = 3\ 306$ Pa, $P_{HBr} = 57.0$ Pa **(c)** neither **(d)** formed

13. 0.663 mbar

14. 5×10^{-8} M

15. 8.5 zM

16. 3.9×10^{-7} M

17. **(a)** 2.1×10^{-8} M **(b)** 8.4×10^{-4} M

18. BX_2 coprecipitates with AX_3

19. no, 0.001 4 M

20. no

21. $I^- < Br^- < Cl^- < CrO_4^{2-}$

23. **(a)** BF_3 **(b)** AsF_5

24. 0.096 M

25. $[Zn^{2+}] = 2._{93} \times 10^{-3}$ M, $[ZnOH^+] = 9 \times 10^{-6}$ M, $[Zn(OH)_2(aq)] = 6 \times 10^{-6}$ M, $[Zn(OH)_3^-] = 8 \times 10^{-9}$ M, $[Zn(OH)_4^{2-}] = 9 \times 10^{-14}$ M

26. 15%

27. 1.1×10^{-5} M

29. (a) an adduct (b) dative or coordinate covalent (c) conjugate
(d) $[H^+] > [OH^-]$, $[H^+] < [OH^-]$

33. (a) HI (b) H_2O

34. $2H_2SO_4 \rightleftharpoons HSO_4^- + H_3SO_4^+$

35. (a) (H_3O^+, H_2O); $(H_3\overset{+}{N}CH_2CH_2\overset{+}{N}H_3, H_3\overset{+}{N}CH_2CH_2NH_2)$
(b) $(C_6H_5CO_2H, C_6H_5CO_2^-)$; $(C_5H_5NH^+, C_5H_5N)$

36. (a) 2.00 (b) 12.54 (c) 1.52 (d) −0.48 (e) 12.00

37. (a) 6.998 (b) 6.132

38. 1.0×10^{-56}

39. 7.8

40. (a) endothermic (b) endothermic (c) exothermic

43. $Cl_3CCO_2H \rightleftharpoons Cl_3CCO_2^- + H^+$

⬡$-\overset{+}{N}H_3 \rightleftharpoons$ ⬡$-NH_2 + H^+$

$La^{3+} + H_2O \rightleftharpoons LaOH^{2+} + H^+$

44. ⬡$N + H_2O \rightleftharpoons$ ⬡$\overset{+}{N}H + OH^-$

$HOCH_2CH_2S^- + H_2O \rightleftharpoons HOCH_2CH_2SH + OH^-$

45. K_a: $HCO_3^- \rightleftharpoons H^+ + CO_3^{2-}$
K_b: $HCO_3^- + H_2O \rightleftharpoons H_2CO_3 + OH^-$

46. (a) $H_3\overset{+}{N}CH_2CH_2\overset{+}{N}H_3 \overset{K_{a1}}{\rightleftharpoons} H_2NCH_2CH_2\overset{+}{N}H_3 + H^+$
$H_2NCH_2CH_2\overset{+}{N}H_3 \overset{K_{a2}}{\rightleftharpoons} H_2NCH_2CH_2NH_2 + H^+$

(b) $^-O_2CCH_2CO_2^- + H_2O \overset{K_{b1}}{\rightleftharpoons} HO_2CCH_2CO_2^- + OH^-$
$HO_2CCH_2CO_2^- + H_2O \overset{K_{b2}}{\rightleftharpoons} HO_2CCH_2CO_2H + OH^-$

47. a, c

48. $CN^- + H_2O \rightleftharpoons HCN + OH^-$; $K_b = 1.6 \times 10^{-5}$

49. $H_2PO_4^- \overset{K_{a2}}{\rightleftharpoons} HPO_4^{2-} + H^+$
$HC_2O_4^- + H_2O \overset{K_{b2}}{\rightleftharpoons} H_2C_2O_4 + OH^-$

50. $K_{a1} = 7.04 \times 10^{-3}$, $K_{a2} = 6.25 \times 10^{-8}$, $K_{a3} = 4.3 \times 10^{-13}$

51. 2.9×10^{-6}

52. (a) 1.2×10^{-2} M (b) Solubility will be greater.

53. 0.22 g

Chapter 7

2. (a) true (b) true (c) true

3. (a) 0.008 7 M (b) 0.001 2 M

4. (a) 0.660 (b) 0.54 (c) 0.18 (d) 0.83

5. 0.88_7

6. (a) 0.42_2 (b) 0.43_2

7. 0.20_2

8. increase

9. 7.0×10^{-17} M

10. 6.6×10^{-7} M

11. $\gamma_{H^+} = 0.86$, pH = 2.07

12. 11.94, 12.00

13. 0.329

15. 0.63

18. $[H^+] + 2[Ca^{2+}] + [Ca(HCO_3)^+] + [Ca(OH)^+] + [K^+] = [OH^-] + [HCO_3^-] + 2[CO_3^{2-}] + [ClO_4^-]$

19. $[H^+] = [OH^-] + [HSO_4^-] + 2[SO_4^{2-}]$

20. $[H^+] = [OH^-] + [H_2AsO_4^-] + 2[HAsO_4^{2-}] + 3[AsO_4^{3-}]$

21. (a) charge: $2[Mg^{2+}] + [H^+] + [MgBr^+] + [MgOH^+] = [Br^-] + [OH^-]$; mass: $[MgBr^+] + [Br^-] = 2\{[Mg^{2+}] + [MgBr^+] + [MgOH^+]\}$
(b) $[Mg^{2+}] + [MgBr^+] + [MgOH^+] = 0.2$ M; $[MgBr^+] + [Br^-] = 0.4$ M

22. 2.3×10^6 N, 5.2×10^5 pounds, no

23. $[CH_3CO_2^-] + [CH_3CO_2H] = 0.1$ M

24. $[Y^{2-}] = [X_2Y_2^{2+}] + 2[X_2Y^{4+}]$

25. $3\{Fe^{3+}\} + [Fe(OH)^{2+}] + [Fe(OH)_2^+] + 2[Fe_2(OH)_2^{4+}] + [FeSO_4^+]\} = 2\{[FeSO_4^+] + [SO_4^{2-}] + [HSO_4^-]\}$

26. (c) $[H^+] = 4.19 \times 10^{-9}$ M, $[OH^-] = [HA] = 2.39 \times 10^{-6}$ M, $[A^-] = 1.00 \times 10^{-2}$ M

27. (b) $[Ca^{2+}] = 0.010\ 1$ M, $[CaOH^+] = 0.005\ 1$ M, $[OH^-] = 0.025\ 4$ M, $[H^+] = 4 \times 10^{-13}$ M; solubility = 1.1 g/L

28. (a) 0.004 8 M (b) $[Zn^{2+}] = 0.006\ 7$ M, 33% is ion paired, $\mu = 0.027$ M (c) $Zn^{2+} + H_2O \rightleftharpoons ZnOH^+ + H^+$ $K_a = \beta_1 K_w = 10^{-9.0}$
$SO_4^{2-} + H_2O \rightleftharpoons HSO_4^- + OH^-$ $K_b = K_w/K_{a2} = 10^{-12.01}$

29. $[Li^+] = [F^-] = 0.050$ M, $[LiF(aq)] = 0.002\ 9$ M

30. (a) 4.3×10^{-5} (b) 4.9×10^{-4} M = 20 mg/L (c) 0.023 bar

Chapter 8

2. (a) 3.00 (b) 12.00

3. 6.89, 0.61

4. (a) 0.809 (b) 0.791 (c) Activity coefficient depends slightly on counterion.

5. (a) ⬡$-CO_2H \rightleftharpoons$ ⬡$-CO_2^- + H^+$ K_a
(b) ⬡$-CO_2^- + H_2O \rightleftharpoons$ ⬡$-CO_2H + OH^-$ K_b
(c) ⬡$-NH_2 + H_2O \rightleftharpoons$ ⬡$-\overset{+}{N}H_3 + OH^-$ K_b
(d) ⬡$-\overset{+}{N}H_3 \rightleftharpoons$ ⬡$-NH_2 + H^+$ K_a

6. pH = 3.00, $\alpha = 0.995\%$

7. 5.50

8. 5.51, 3.1×10^{-6} M, 0.060 M

10. 99% dissociation when F = $(0.010\ 2)K_a$

11. 4.20

12. 5.79

13. (a) 3.03, 9.4% (b) 7.00, 99.9%

14. 5.64, 0.005 3%

15. 2.86, 14%

16. 99.6%, 96.5%

19. 11.00, 0.995%

20. 11.28, [B] = 0.058 M, $[BH^+] = 1.9 \times 10^{-3}$ M

21. 10.95

22. 0.007 6%, 0.024%, 0.57%

23. 3.6×10^{-9}

24. 4.1×10^{-5}

25. 0.999, 0.000 999

32. 4-aminobenzenesulfonic acid

33. 4.70

34. (a) 0.180 (b) 1.00 (c) 1.80

35. 1.5

36. (a) 14 (b) 1.4×10^{-7}

37. (a) NaOH (b) 1. Weigh out $(0.250\ L)(0.050\ 0\ M) = 0.012\ 5$ mol of HEPES and dissolve in ~200 mL. 2. Adjust the pH to 7.45 with NaOH. 3. Dilute to 250 mL.

38. 3.38 mL

39. (b) 7.18 (c) 7.00 (d) 6.86 mL

40. (a) 2.56 (b) 2.61 (c) 2.86

41. 16.2 mL

42. (a) pH = 5.06, [HA] = 0.001 99 M, [A$^-$] = 0.004 01 M

43. (a) approximate pH = 11.70, more accurate pH = 11.48

44. 6.86

Chapter 9

2. $H_3\overset{+}{N}\underset{}{\overset{R}{\underset{|}{C}}}CO_2^-$; pK values apply to —NH$_3^+$, —CO$_2$H, and, in some cases, R.

3. 4.37×10^{-4}, 8.93×10^{-13}

4. (a) pH = 2.51, [H$_2$A] = 0.096 9 M, [HA$^-$] = 3.11×10^{-3} M, [A^{2-}] = 1.00×10^{-8} M **(b)** 6.00, 1.00×10^{-3} M, 1.00×10^{-1} M, 1.00×10^{-3} M **(c)** 10.50, 1.00×10^{-10} M, 3.16×10^{-4} M, 9.97×10^{-2} M

5. (a) pH = 1.95, [H$_2$M] = 0.089 M, [HM$^-$] = 1.12×10^{-2} M, [M^{2-}] = 2.01×10^{-6} M **(b)** pH = 4.28, [H$_2$M] = 3.7×10^{-3} M, [HM$^-$] ≈ 0.100 M, [M^{2-}] = 3.8×10^{-3} M **(c)** pH = 9.35, [H$_2$M] = 7.04×10^{-12} M, [HM$^-$] = 2.23×10^{-5} M, [M^{2-}] = 0.100 M

6. pH = 11.60, [B] = 0.296 M, [BH$^+$] = 3.99×10^{-3} M, [BH$_2^{2+}$] = 2.15×10^{-9} M

7. pH = 3.69, [H$_2$A] = 2.9×10^{-6} M, [HA$^-$] = 7.9×10^{-4} M, [A^{2-}] = 2.1×10^{-4} M

8. 4.03

9. (a) pH = 6.002, [HA$^-$] = 0.009 8 M **(b)** pH = 4.50, [HA$^-$] = 0.006 1 M

11. [CO$_2(aq)$] = $10^{-4.9}$ M, pH = 5.67

12. (a) [CO$_3^{2-}$] = $K_{a2}K_{a1}K_HP_{CO_2}/[H^+]^2$ **(b)** 0°C: 6.6×10^{-5} mol kg^{-1}; 30°C: 1.8×10^{-4} mol kg^{-1} **(c)** 0°C: [Ca^{2+}][CO$_3^{2-}$] = 6.6×10^{-7} mol^2 kg^{-2} (aragonite dissolves, calcite does not); 30°C: [Ca^{2+}][CO$_3^{2-}$] = 1.8×10^{-6} mol^2 kg^{-2} (neither dissolves)

13. 2.96 g

14. 2.22 mL

15. Procedure: Dissolve 10.0 mmol (1.23 g) picolinic acid in ~75 mL H$_2$O in a beaker. Add NaOH (~5.63 mL) until the measured pH is 5.50. Transfer to a 100-mL volumetric flask and use small portions of H$_2$O to rinse the beaker into the flask. Dilute to 100.0 mL and mix well.

16. 26.5 g Na$_2$SO$_4$ + 1.31 g H$_2$SO$_4$

17. no

18.

Glutamic acid

Tyrosine

19. (a) 2.8×10^{-3} **(b)** 2.9×10^{-8}

20. (a) NaH$_2$PO$_4$ and Na$_2$HPO$_4$ would be simplest, but other combinations such as H$_3$PO$_4$ and Na$_3$PO$_4$ or H$_3$PO$_4$ and Na$_2$HPO$_4$ would work just fine. **(b)** 4.55 g Na$_2$HPO$_4$ + 2.15 g NaH$_2$PO$_4$ **(c)** One of several ways: Weigh out 0.050 0 mol Na$_2$HPO$_4$ and dissolve it in 900 mL of water. Add HCl while monitoring the pH with a pH electrode. When the pH is 7.45, stop adding HCl and dilute to exactly 1 L with H$_2$O.

21. pH = 5.64, [H$_2$L$^+$] = 0.010 0 M, [H$_3$L^{2+}] = 1.36×10^{-6} M, [HL] = 3.68×10^{-6} M, [L$^-$] = 2.40×10^{-11} M

22. 78.9 mL

23. (a) 5.88 **(b)** 5.59

24. (a) HA **(b)** A$^-$ **(c)** 1.0, 0.10

25. (a) 4.00 **(b)** 8.00 **(c)** H$_2$A **(d)** HA$^-$ **(e)** A^{2-}

26. (a) 9.00 **(b)** 9.00 **(c)** BH$^+$ **(d)** 1.0×10^3

27.

28. α_{HA} = 0.091, α_{A^-} = 0.909, [A$^-$]/[HA] = 10

29. 0.91

30. α_{H_2A} = 0.877, 0.049 6; α_{HA^-} = 0.123, 0.694; $\alpha_{A^{2-}}$ = 4.54×10^{-4}, 0.257

31. $\alpha_{H_2A^-}$ = 0.893, 0.500, 5.4×10^{-5}, 2.2×10^{-5}, 1.55×10^{-12}
α_{HA^-} = 0.107, 0.500, 0.651, 0.500, 1.86×10^{-4}
$\alpha_{A^{2-}}$ = 5.8×10^{-7}, 2.2×10^{-5}, 0.349, 0.500, 0.999 8

32. (b) 8.6×10^{-6}, 0.61, 0.39, 1.6×10^{-6}

33. 0.36

34. 96%

36. At pH 10: α_{H_3A} = 1.05×10^{-9}, $\alpha_{H_2A^-}$ = 0.040 9, $\alpha_{HA^{2-}}$ = 0.874, $\alpha_{A^{3-}}$ = 0.085 4

37. (b) [Cr(OH)$_3(aq)$] = $10^{-6.84}$ M **(c)** [Cr(OH)$_2^+$] = $10^{-4.44}$ M, [Cr(OH)$^{2+}$] = $10^{-2.04}$ M

39. The average charge is 0. There is no pH at which all molecules have zero charge.

40. isoelectric pH 5.59, isoionic pH 5.72

Chapter 10

2. 13.00, 12.95, 12.68, 11.96, 10.96, 7.00, 3.04, 1.75

6. 3.00, 4.05, 5.00, 5.95, 7.00, 8.98, 10.96, 12.25

7. $V_e/11$; $10V_e/11$; V_e = 0, pH = 2.80; $V_e/11$, pH = 3.60; $V_e/2$, pH = 4.60; $10V_e/11$, pH = 5.60; V_e, pH = 8.65; $1.2V_e$, pH = 11.96

8. 8.18

9. 5.4×10^7

10. 0.107 M

11. 9.72

14. 11.00, 9.95, 9.00, 8.05, 7.00, 5.02, 3.04, 1.75

15. $V_e/2$

16. 2.2×10^9

17. 10.92, 9.57, 9.35, 8.15, 5.53, 2.74

18. (a) 9.45 **(b)** 2.55 **(c)** 5.15

20. positive

21. isoionic

23. 11.49, 10.95, 10.00, 9.05, 8.00, 6.95, 6.00, 5.05, 3.54, 1.79

24. 2.51, 3.05, 4.00, 4.95, 6.00, 7.05, 8.00, 8.95, 10.46, 12.21

25. 11.36, 10.21, 9.73, 9.25, 7.53, 5.81, 5.33, 4.86, 3.41, 2.11, 1.85

26. 5.01

27. **(a)** 1.99

28. **(b)** 7.13

29. 2.72

30. **(a)** 9.54   **(b)** 7.9×10^{-10}

31. 6.28 g

32. $pK_2 = 9.84$

34. end point = 23.39 mL

35. end point = 10.727 mL

39. H_2SO_4, HCl, HNO_3, or $HClO_4$

40. yellow, green, blue

41. **(a)** red   **(b)** orange   **(c)** yellow

42. **(a)** red   **(b)** orange   **(c)** yellow   **(d)** red

43. no (end-point pH must be >7)

44. **(a)** 2.47

45. **(a)** violet   **(b)** blue   **(c)** yellow

46. **(a)** 5.62

47. 2.859 wt%

52. 0.079 34 mol/kg

53. 1.023_8 g, systematic error = 0.08%, calculated HCl molarity is low

54. 0.100 0 M

55. 0.31 g

56. **(a)** 20.254 wt%   **(b)** 17.985 g

57. **(a)** 204.221 ± 0.004 g/mol   **(b)** $1.000\,00 \pm 0.000\,03$

58. 15.1 wt%

61. **(a)** acetic acid   **(b)** pyridine

68. **(b)** $K = 0.279$, pH = 4.16

73. 0.139 M

74. 0.815

Chapter 11

2. **(a)** 2.7×10^{-10} **(b)** 0.57

3. **(a)** 2.5×10^7 **(b)** 4.5×10^{-5} M

4. 5.60 g

5. Neutral H_5DTPA is $DTPA(CO_2H)_2(CO_2^-)_3(NH^+)_3$. The predominant species at pH 14 is $DTPA^{5-}$. By analogy with EDTA, all carboxyl groups are probably ionized at pH 3–4, so the predominant species is H_3DTPA^{2-}, which is $DTPA(CO_2^-)_5(NH^+)_3$. At pH 14 and at pH 3, sulfate is in the form SO_4^{2-}. 10^{-3} M H^+ displaces Ba^{2+} from DTPA, but 10^{-14} M H^+ does not.

6. **(a)** 100.0 mL   **(b)** 0.016 7 M   **(c)** 0.041   **(d)** 4.1×10^{10}
   (e) 7.8×10^{-7} M   **(f)** 2.4×10^{-10} M

7. **(a)** 2.93   **(b)** 6.79   **(c)** 10.52

8. **(a)** 1.70   **(b)** 2.18   **(c)** 2.81   **(d)** 3.87   **(e)** 4.87   **(f)** 6.85
   (g) 8.82   **(h)** 10.51   **(i)** 10.82

9. ∞, 10.30, 9.52, 8.44, 7.43, 6.15, 4.88, 3.20, 2.93

10. 4.6×10^{-11} M

15. **(a)** 25   **(b)** 0.017

16. **(a)** 15.03   **(b)** 15.05   **(c)** 16.30   **(d)** 17.02   **(e)** 17.69

17. **(b)** $\alpha_{ML} = 0.28$, $\alpha_{ML_2} = 0.70$

18. **(d)** $[T] = 0.27_7$; $[Fe_aT] = 0.55_3$; $[Fe_bT] = 0.09_2$; $[Fe_2T] = 0.07_7$

20. **(b)** 1.34 mL, pNi = 7.00; 21.70 mL, pNi = 8.00; 26.23 mL, pNi = 17.00

24. 1. with metal ion indicators; 2. with a mercury electrode; 3. with an ion-selective electrode; 4. with a glass electrode

25. HIn^{2-}, wine-red, blue

26. Buffer (a): yellow $\to$ blue; other buffers: violet $\to$ blue, which is harder to see

27. analyte precipitates without EDTA, reacts slowly with EDTA, or blocks the indicator

28. analyte displaces a metal ion from a complex

30. hardness $\approx [Ca^{2+}] + [Mg^{2+}]$. Temporary hardness, due to $Ca(HCO_3)_2$, is lost by heating. Permanent hardness is derived from other salts such as $CaSO_4$ and is not affected by heat.

31. 10.0 mL, 10.0 mL

32. 0.020 0 M

33. 0.995 mg

34. 0.092 54 M

35. 21.45 mL

36. $[Ni^{2+}] = 0.012\,4$ M, $[Zn^{2+}] = 0.007\,18$ M

37. 0.024 30 M

38. 0.092 28 M

39. observed: 32.7 wt%; theoretical: 32.90 wt%

Chapter 12

1. $PbS(s) + H^+ \rightleftharpoons Pb^{2+} + HS^-$
   $PbCO_3(s) + H^+ \rightleftharpoons Pb^{2+} + HCO_3^-$

2. **(a)** pH = 9.98   **(b)** pH = 10.00   **(c)** pH = 9.45

3. pH = 9.95

4. predicted values: $pK_1' = 2.350$, $pK_2' = 9.562$

5. pH = 10.194 from spreadsheet and 10.197 by hand method

6. pH = 4.52

7. pH = 5.00

8. ionic strength = 0.025 M, pH = 4.94

9. **(a)** pH = 7.420   **(b)** pH = 7.403

10. pH = 4.44

11. **(d)** $[Fe^{3+}] = \dfrac{F_{SCN} - [SCN^-]}{\beta_1'[SCN^-] + 2\beta_2'[SCN^-]^2}$

    (e) $[H^+] = \dfrac{K_a'[Fe^{3+}]}{F_{Fe} - [Fe^{3+}] - \beta_1'[Fe^{3+}][SCN^-] - \beta_2'[Fe^{3+}][SCN^-]^2}$

    (f, g, h, i) $[SCN^-] = 2.03$ μM, $[Fe^{3+}] = 4.20$ mM, $[H^+] = 15.8$ mM, $[Fe(SCN)^{2+}] = 2.97$ μM, $[Fe(SCN)_2^+] = 0.106$ nM, $[FeOH^{2+}] = 0.802$ mM, ionic strength = 0.043 4 M, pH = 1.88

    $\dfrac{[Fe(SCN)^{2+}]}{\{[Fe^{3+}] + [FeOH^{2+}]\}[SCN^-]} = 293$

    (j) $[SCN^-] = 2.81$ μM, $[Fe^{3+}] = 4.45$ mM, $[H^+] = 15.5$ mM, $[Fe(SCN)^{2+}] = 2.19$ μM, $[Fe(SCN)_2^+] = 0.068$ nM, $[FeOH^{2+}] = 0.546$ mM, ionic strength = 0.243 9 M, pH = 1.94

    $\dfrac{[Fe(SCN)^{2+}]}{\{[Fe^{3+}] + [FeOH^{2+}]\}[SCN^-]} = 156$

12. **(a)** $[SO_4^{2-}] = 1.50$ mM, $[La^{3+}] = 0.57$ mM, $[H^+] = 1.14$ μM, $[La(SO_4)^+] = 1.36$ mM, $[La(SO_4)_2^-] = 67$ μM, $[LaOH^{2+}] = 1.13$ μM, ionic strength = 0.006 29 M, pH = 5.98   **(b)** ionic strength of strong electrolyte = 15.0 mM; actual ionic strength = 6.3 mM   **(c)** 28.5%
    (d) pK_a for HSO_4^- is 1.99 and we expect the solution to be near neutral pH.
    (e) no; $[La^{3+}][OH^-]^3\gamma_{La^{3+}}\gamma_{OH^-}^3 = 2.3 \times 10^{-28} < K_{sp}$ for $La(OH)_3 = 2 \times 10^{-21}$

13. pH = 7.06, ionic strength = 0.041 M; $[Ca^{2+}] = 0.010\,2$ M, $[CaOH^+] = 1.4 \times 10^{-8}$ M, $[SO_4^{2-}] = 0.010\,2$ M, $[HSO_4^-] = 5.0 \times 10^{-8}$ M

14. $[CN^-] = 1.51$ μM, $[H^+] = 1.29 \times 10^{-12}$ M, $[OH^-] = 0.012\,9$ M, $[Ag^+] = 0.241$ nM, $[AgOH] = 0.187$ nM, $[Ag(OH)(CN)^-] = 46.9$ μM, $[Ag(CN)_2^-] = 0.100$ M, $[Ag(CN)_3^{2-}] = 4.19$ μM, $[HCN] = 1.90$ nM, $[Na^+] = 0.013\,0$ M, $[K^+] = 0.100$ M, ionic strength = 0.113 M

15. fraction of Fe in each form: $[Fe^{2+}]$, 2.7%; $[FeG^+]$, 34.0%; $[FeG_2]$, 62.0%; $[FeG_3^-]$, 1.1%; $[FeOH^+]$, 0.2%
fraction of glycine in each form: $[G^-]$, 1.1%; $[HG]$, 18.2%; $[H_2G^+]$, 0.000 015%; $[FeG^+]$, 17.0%; $2[FeG_2]$, 62.0%; $3[FeG_3^-]$, 1.7%
HCl added = 18.1 mmol; ionic strength = 21.1 mM
chemistry: $FeG_2 \rightleftharpoons FeG^+ + G^-$ followed by $G^- + H^+ \rightleftharpoons HG$.
G^- released when FeG_2 dissolves requires HCl to lower the pH to 8.50.

16. **(b)** Fixing pK'_w at 13.797 causes $\bar{n}_H$(measured) to deviate systematically above $\bar{n}_H$(theoretical) at the end of the titration when $\bar{n}_H$ should approach 0.

17. **(a)** $\bar{n}_H$ (experimental) $= 3 + \dfrac{[OH^-] + [Cl^-]_{HCl} - [H^+] - [Na^+]}{F_{H_3A}}$

$\bar{n}_H$(theoretical) $= 3\alpha_{H_3A} + 2\alpha_{H_2A} + \alpha_{HA}$

(b) $pK'_w = 13.819$, $pK_1 = 8.33$, $pK_2 = 9.48$, $pK_3 = 10.19$

18. **(b)** pH = 4.61. In the absence of precipitation, at pH 4.61, the concentrations are $[Cu^{2+}] = 18.6$ mM, $[SO_4^{2-}] = 18.7$ mM, $[CuSO_4(aq)] = 6.3$ mM, $[HSO_4^-] = 20$ μM, $[CuOH^+] = 12$ μM, $[Cu_2(OH)^{3+}] = 28$ μM, and $[Cu_2(OH)_2^{2+}] = 5.8$ μM; the other species have lower concentrations. **(c)** The solubility of $Cu(OH)_{1.5}(SO_4)_{0.25}(s)$ is exceeded above pH ≈ 4.5. $CuO(s)$ solubility is exceeded above pH ≈ 5. $Cu(OH)_2(s)$ solubility is exceeded above pH ≈ 5.5. At pH 4.61, $Cu(OH)_{1.5}(SO_4)_{0.25}(s)$ will precipitate from 0.025 M $CuSO_4$.

19.

(b) $[T^{2-}] = \dfrac{F_{H_2T}}{\dfrac{[H^+]^2}{K_1K_2} + \dfrac{[H^+]}{K_2} + 1 + K_{NaT^-}[Na^+] + K_{NaHT}[Na^+]\dfrac{[H^+]}{K_2}}$

(c) $[HT^-] = \dfrac{F_{H_2T}}{\dfrac{[H^+]}{K_1} + 1 + \dfrac{K_2}{[H^+]} + K_{NaT^-}[Na^+]\dfrac{K_2}{[H^+]} + K_{NaHT}[Na^+]}$

$[H_2T] = \dfrac{F_{H_2T}}{1 + \dfrac{K_1}{[H^+]} + \dfrac{K_1K_2}{[H^+]^2} + K_{NaT^-}[Na^+]\dfrac{K_1K_2}{[H^+]^2} + K_{NaHT}[Na^+]\dfrac{K_1}{[H^+]}}$

(d) pH = 4.264, $[PyH^+] = 0.013\,4$, $[Na^+] = 0.018\,5$, $[K^+] = 0.010\,0$, $[OH^-] = 1.84 \times 10^{-10}$, $[HT^-] = 0.010\,0$, $[T^{2-}] = 0.007\,92$, $[Cl^-] = 0.015\,0$, $[NaT^-] = 0.001\,17$, $[H_2T] = 5.93 \times 10^{-4}$, $[Py] = 0.001\,56$, $[NaHT] = 2.97 \times 10^{-4}$ M

Chapter 13

2. **(a)** $6.241\,509\,48 \times 10^{18}$ **(b)** 96 485.338 3

3. **(a)** $71._5$ A **(b)** 4.35 A **(c)** 79 W

4. **(a)** 1.87×10^{16} e^-/s **(b)** 9.63×10^{-19} J/e^- **(c)** 5.60×10^{-5} mol **(d)** 447 V

5. **(a)** I_2 **(b)** $S_2O_3^{2-}$ **(c)** 861 C **(d)** 14.3 A

6. **(a)** NH_4^+ and Al, reducing agents; ClO_4^-, oxidizing agent **(b)** 9.576 kJ/g

8. **(a)** $Fe(s) \mid FeO(s) \mid KOH(aq) \mid Ag_2O(s) \mid Ag(s)$;
$FeO(s) + H_2O + 2e^- \rightleftharpoons Fe(s) + 2OH^-$; $Ag_2O(s) + H_2O + 2e^- \rightleftharpoons 2Ag(s) + 2OH^-$ **(b)** $Pb(s) \mid PbSO_4(s) \mid K_2SO_4(aq) \parallel H_2SO_4(aq) \mid PbSO_4(s) \mid PbO_2(s) \mid Pb(s)$; $PbSO_4(s) + 2e^- \rightleftharpoons Pb(s) + SO_4^{2-}$; $PbO_2(s) + 4H^+ + SO_4^{2-} + 2e^- \rightleftharpoons PbSO_4(s) + 2H_2O$

9. $Fe^{3+} + e^- \rightleftharpoons Fe^{2+}$; $Cr_2O_7^{2-} + 14H^+ + 6e^- \rightleftharpoons 2Cr^{3+} + 7H_2O$

10. **(a)** Electrons flow from Zn to C. **(b)** 1.32 kg

11. **(a)** anode: $C_6Li \rightleftharpoons C_6 + Li^+ + e^-$; cathode: $CoO_2 + Li^+ + e^- \rightleftharpoons LiCoO_2$ **(b)** 3.6 C **(d)** 0.51 **(e)** 0.52 W·h/g $LiCoO_2$

12. Cl_2 has the most positive $E°$.

13. **(a)** Fe(III) **(b)** Fe(II)

15. **(a)** $Cu^{2+} + Zn(s) \rightleftharpoons Cu(s) + Zn^{2+}$ **(b)** Zn^{2+}

16. **(b)** −0.356 V

17. **(a)** $Pt(s) \mid Br_2(l) \mid HBr(aq, 0.10\text{ M}) \parallel Al(NO_3)_3(aq, 0.010\text{ M}) \mid Al(s)$ **(b)** $E = -2.854$ V, e^- flow from Al to Pt **(c)** Br_2 **(d)** 1.31 kJ **(e)** 2.69×10^{-8} g/s

19. **(a)** 0.572 V **(b)** e^- flow from left to right **(c)** 0.568 V

20. 0.799 2 V

21. $HOBr + 2e^- + H^+ \rightleftharpoons Br^- + H_2O$; 1.341 V

22. $3X^+ \rightleftharpoons X^{3+} + 2X(s)$; $E_2° > E_1°$

23. 0.580 V, electrons flow from Ni to Cu

24. **(a)** 1.33 V **(b)** 1×10^{45}

25. **(a)** $K = 10^{47}$ **(b)** $K = 1.9 \times 10^{-6}$

26. **(b)** $K = 2 \times 10^{16}$ **(c)** -0.02_0 V **(d)** 10 kJ **(e)** 0.21

27. $K = 1.0 \times 10^{-9}$

28. 0.101 V

29. 34 g/L

30. 0.116 V

31. −1.664 V

32. $K = 3 \times 10^5$

33. **(a)** $Al_2O_3(s) + MgO(s) \rightleftharpoons MgAl_2O_4(s)$ **(b)** −29.51 kJ/mol **(c)** $\Delta H° = -23.60$ kJ/mol, $\Delta S° = 5.90$ J/(K · mol)

35. **(b)** 0.14_3 M

36. **(b)** A = −0.414 V, B = 0.059 16 V **(c)** Hg → Pt

37. 9.6×10^{-7}

38. 7.1×10^{14}

39. 0.76

40. 7.5×10^{-8}

42. **(c)** 0.317 V

43. −0.041 V

44. −0.268 V

45. −0.036 V

46. 7.2×10^{-4}

47. −0.447 V

48. **(a)** $[Ox] = 3.82 \times 10^{-5}$ M, $[Red] = 1.88 \times 10^{-5}$ M **(b)** $[S^-] = [Ox]$, $[S] = [Red]$ **(c)** −0.092 V

Chapter 14

1. **(b)** 0.044 V

2. **(a)** 0.326 V **(b)** 0.086 V **(c)** 0.019 V **(d)** −0.021 V **(e)** 0.021 V

3. 0.684 V

4. 0.627

5. 0.243 V

6. **(c)** 0.068 V

8. 0.481 V; −0.039 V

9. 3×10^{21}

10. **(a)** $Fe^{3+} + e^- \rightleftharpoons Fe^{2+}$ **(b)** 1×10^{11} **(c)** 6×10^{10}

11. $[CN^-] = 0.847$ mM; $[KOH] = 0.29_6$ M

12. Figure 13-4

14. left

15. **(a)** 42.4 s **(b)** 208 s

16. **(a)** $3._2 \times 10^{13}$ **(b)** 8% **(c)** 49.0, 8%

18. **(c)** 0.1 M HCl $\mid$ 1 mM KCl, 93.6 mV; 0.1 M HCl $\mid$ 4 M KCl, 4.7 mV

21. 10.67

22. potassium hydrogen tartrate and potassium hydrogen phthalate

24. +0.10 pH unit

25. **(a)** 274 mV **(b)** 285 mV

26. pH = 5.686; slope = -57.17_3 mV/pH unit; theoretical slope = -58.17 mV/pH unit; β = 0.983

27. **(b)** 0.465 **(c)** $Na_2HPO_4 = 0.026\,8$ m and $KH_2PO_4 = 0.019\,6$ m

29. Smaller $K_{A,X}^{Pot}$ is more selective for the intended ion.

33. **(a)** -0.407 V   **(b)** $1.5_5 \times 10^{-2}$ M   **(c)** $1.5_2 \times 10^{-2}$ M

34. $+0.029\ 6$ V

35. 0.211 mg/L

36. Group 1: K^+; Group 2: Sr^{2+} and Ba^{2+}; $[K^+] \approx 100[Li^+]$

37. 3.8×10^{-9} M

38. **(a)** $E = 51.10\ (\pm0.24) + 28.14\ (\pm0.08_5)\ \log[Ca^{2+}]\ (s_y = 0.2_7)$
(b) 0.951 **(c)** $2.43\ (\pm0.04) \times 10^{-3}$ M

39. -0.331 V

40. 3.0×10^{-5} M

41. **(a)** 0.36 ± 0.15 ppm   **(b)** There is too much added standard.

42. $\log K_{Na^+,Mg^{2+}}^{Pot} = -8.09, -8.15$; $\log K_{Na^+,K^+}^{Pot} = -4.87, -4.87$

43. Na^+ error $= 0.25\%$, Ca^{2+} error $= 2.5\%$

44. $E = 120.2 + 28.80\ \log([Ca^{2+}] + 6.0 \times 10^{-4}\ [Mg^{2+}])$

47. **(a)** 1.13×10^{-4} **(b)** 4.8×10^{4}

Chapter 15

1. **(d)** $0.490, 0.526, 0.626, 0.99, 1.36, 1.42, 1.46$ V

2. **(d)** $1.58, 1.50, 1.40, 0.733, 0.065, 0.005, -0.036$ V

3. **(d)** $-0.120, -0.102, -0.052, 0.21, 0.48, 0.53$ V

4. **(b)** $0.570, 0.307, 0.184$ V

5. **(d)** $-0.143, -0.102, -0.061, 0.096, 0.408, 0.450$

6. diphenylamine sulfonic acid: colorless $\rightarrow$ red-violet; diphenylbenzidine sulfonic acid: colorless $\rightarrow$ violet; tris(2,2′-bipyridine)iron: red $\rightarrow$ pale blue; ferroin: red $\rightarrow$ pale blue

7. no

13. **(a)** $MnO_4^- + 8H^+ + 5e^- \rightleftharpoons Mn^{2+} + 4H_2O$
(b) $MnO_4^- + 4H^+ + 3e^- \rightleftharpoons MnO_2(s) + 2H_2O$
(c) $MnO_4^- + e^- \rightleftharpoons MnO_4^{2-}$

14. $0.011\ 29$ M

15. $0.586\ 4$ M

16. **(a)** Scheme 1: $6H^+ + 2MnO_4^- + 5H_2O_2 \rightarrow 2Mn^{2+} + 5O_2 + 8H_2O$
Scheme 2: $6H^+ + 2MnO_4^- + 3H_2O_2 \rightarrow 2Mn^{2+} + 4O_2 + 6H_2O$
(b) Scheme 1: 25.43 mL; Scheme 2: 42.38 mL

17. 3.826 mM

18. 41.9 wt%

19. 78.67 wt%

20. oxidation number $= 3.761$; 217 μg/g

21. **(a)** $0.020\ 34$ M   **(b)** $0.125\ 7$ g   **(c)** $0.019\ 82$ M

24. iodometry

25. **(a)** 1.433 mmol   **(b)** $0.076\ 09$ M   **(c)** 12.8 wt%
(d) Do not add starch until right before the end point.

26. 11.43 wt%; just before the end point

27. $0.007\ 744$ M; just before the end point

28. **(a)** 7×10^{2} **(b)** 1.0 **(c)** 0.34 g/L

29. mol $NH_3 = 2$(initial mol H_2SO_4) $-$ mol thiosulfate

30. **(a)** no, no   **(b)** $I_3^- + SO_3^{2-} + H_2O \rightarrow 3I^- + SO_4^{2-} + 2H^+$
(c) $5.07\ 9 \times 10^{-3}$ M, 406.6 mg/L   **(d)** no: $t_{calculated} = 2.56 < t_{table} = 2.776$

31. 5.730 mg

32. **(a)** 0.125 **(b)** 6.875 ± 0.038

34. **(a)** $0.191\ 5$ mmol   **(b)** 2.80 **(c)** 0.20 **(d)** $0.141\ 3$, difference is experimental error

35. Bi oxidation state $= +3.200\ 0\ (\pm0.003\ 3)$
Cu oxidation state $= +2.200\ 1\ (\pm0.004\ 6)$
formula $= Bi_2Sr_2CaCu_2O_{8.400\ 1\ (\pm0.005\ 7)}$

Chapter 16

1. Difference is due to overpotential.

2. 2.68 h

3. $-1.228\ 8$ V

4. **(a)** -1.906 V   **(b)** 0.20 V   **(c)** -2.71 V   **(d)** -2.82 V

5. V_2

6. **(a)** 6.64×10^{3} J   **(b)** $0.012\ 4$ g/h

9. anode, 54.77 wt%

10. -0.619 V, negative

11. -0.744 V

12. 94%

15. **(a)** $5._2 \times 10^{-9}$ mol   **(b)** $0.000\ 2_6$ mL

16. **(a)** 5.32×10^{-5} mol   **(b)** 2.66×10^{-5} mol   **(c)** 5.32×10^{-3} M

17. 151 μg/mL

18. **(a)** current density $= 1.00 \times 10^{2}$ A/m^2, overpotential $= 0.85$ V
(b) -0.036 V   **(c)** 1.160 V   **(d)** -2.57 V

19. $96\ 486.6_7 \pm 0.2_8$ C/mol

20. **(a)** $H_2SO_3 <$ pH 1.86; pH $1.86 < HSO_3^- <$ pH 7.17; $SO_3^{2-} >$ pH 7.17
(b) cathode: $H_2O + e^- \rightarrow \frac{1}{2}H_2(g) + OH^-$; anode: $3I^- \rightarrow I_3^- + 2e^-$
(c) $I_3^- + HSO_3^- + H_2O \rightarrow 3I^- + SO_4^{2-} + 3H^+$;
$I_3^- + 2S_2O_3^{2-} \rightleftharpoons 3I^- + S_4O_6^{2-}$ **(d)** 3.64 mM

21. **(a)** $B = c$; $C = x$; $D = n$
$o + A = 2B \Rightarrow o + A = 2c \Rightarrow A = 2c - o$
$h + 2A = 3D + E \Rightarrow h + 2(2c - o) = 3n + E \Rightarrow$
$E = h + 4c - 2o - 3n$
$F = E - C = h + 4c - 2o - 3n - c = h - c/2 + o - 3n$
(b) $F/4$ **(c)** $2.22_3 \times 10^{-8}$ mol   **(d)** 52.7 mg O_2/L   **(e)** 2.26×10^{-4} M

23. **(e)** 321 μC

24. 15 μm, 7.8×10^{2} A/m^2

27. 0.12%

28. 0.096 mM

30. **(a)** $Cu^{2+} + 2e^- \rightarrow Cu(s)$ **(b)** $Cu(s) \rightarrow Cu^{2+} + 2e^-$ **(c)** 313 ppb

31. Estimated relative peak heights are $1, 1.5_6$, and 1.9_8. Fe(III) in seawater $= 1.0 \times 10^{2}$ pM

32. peak B: $RNHOH \rightarrow RNO + 2H^+ + 2e^-$; peak C:
$RNO + 2H^+ + 2e^- \rightarrow RNHOH$. There was no RNO present before the initial scan.

33. 7.8×10^{-10} m^2/s

Chapter 17

1. **(a)** double   **(b)** halve   **(c)** double

2. **(a)** 184 kJ/mol   **(b)** 299 kJ/mol

3. 5.33×10^{14} Hz, 1.78×10^{4} cm^{-1}, 3.53×10^{-19} J/photon, 213 kJ/mol

4. rotation, vibration, electronic excitation, electronic excitation and bond breaking

5. $\nu = 5.088\ 491\ 0$ and $5.083\ 335\ 8 \times 10^{14}$ Hz, $\lambda = 588.985\ 54$ and $589.582\ 86$ nm, $\tilde{\nu} = 1.697\ 834\ 5$ and $1.696\ 114\ 4 \times 10^{4}$ cm^{-1}

9.

Curve	Absorption peak (nm)	Predicted color (Table 17-1)	Observed color
A	760	green	green
B	700	green	blue green
C	600	blue	blue
D	530	violet	violet
E	500	red or purple red	red
F	410	green-yellow	yellow

11. 3.56×10^{4} M^{-1} cm^{-1}

12. violet-blue

13. 2.19×10^{-4} M

14. **(a)** 325 nm: $T = 0.90$, $A = 0.045$; 300 nm: $T = 0.061$, $A = 1.22$
(b) 2.0%   **(c)** $T_{winter} = 0.142$; $T_{summer} = 0.095$; 49%

15. neocuproine masks Cu(I)

16. **(a)** 6.97×10^{-5} M   **(b)** 6.97×10^{-4} M   **(c)** 1.02 mg

17. yes

18. **(a)** 7.87×10^4 M^{-1} cm^{-1} **(b)** 1.98×10^{-6} M

19. **(a)** 4.97×10^4 M^{-1} cm^{-1} **(b)** 4.69 µg

20. **(b)** 2.33×10^{-7} mol Fe(III)   **(c)** 5.83×10^{-5} M

21. theoretical equivalence point = 13.3 µL; observed end point = 12.2 µL = 1.83 Ga/transferrin. Ga does not appear to bind in the absence of oxalate.

22. **(a)** end point = 21.4 µL, 2.29 mmol Au(0)/g   **(b)** 1.24 mmol $C_{12}H_{25}S$/g   **(c)** 1.52 mmol Au(I)/g, mole ratio Au(I):$C_{12}H_{25}S = 1.23$

23. $\Delta E(S_1 - T_1) = 36$ kJ/mol

27. wavelength: absorption < fluorescence < phosphorescence

28. Excitation spectrum resembles absorption spectrum.

29. Fluorescence is proportional to concentration up to 5 µM (with 5%); yes.

30. $3.56 (\pm 0.07) \times 10^{-4}$ wt%; 95% confidence interval: $3.56 (\pm 0.22) \times 10^{-4}$ wt%

Chapter 18

1. $[X] = 8.03 \times 10^{-5}$ M, $[Y] = 2.62 \times 10^{-4}$ M

2. $[Cr_2O_7^{2-}] = 1.78 \times 10^{-4}$ M, $[MnO_4^-] = 8.36 \times 10^{-5}$ M

5. $[A] = 9.11 \times 10^{-3}$ M, $[B] = 4.68 \times 10^{-3}$ M

6. $[TB] = 1.22 \times 10^{-5}$ M, $[STB] = 9.30 \times 10^{-6}$ M, $[MTB] = 1.32 \times 10^{-5}$ M

7. [p-xylene] = 0.062 7 M, [m-xylene] = 0.079 5 M, [o-xylene] = 0.075 9 M, [ethylbenzene] = 0.076 1 M

8. $[In^-] = 0.794$ µM, $[HIn] = 0.436$ µM, $pK_a = 4.00$

9. **(f)** $[CO_2(aq)] = 3.0$ µM   **(g)** $\mu \approx 10^{-4}$ M, yes

10. $K = 4.0 \times 10^9$ M^{-1}, $S = 0.29$ to 0.84

11. **(a)** $K = (4.3 \pm 0.8) \times 10^4$ M^{-1} **(b)** fraction bound = 0.21 and 0.77

12. **(b)** $K = 0.464$, $\varepsilon = 1.074 \times 10^4$ M^{-1} cm^{-1}

13. **(b)** $K = 0.464$, $\varepsilon = 1.073 \times 10^4$ M^{-1} cm^{-1}

14. **(a)** 1:1   **(b)** K must not be very large   **(c)** to maintain constant ionic strength

15. (xylenol orange)$_2$Zr$_3$ has a xylenol orange mole fraction of 0.4

17. yes

22. $pK_a \approx 10.8$

23. **(b)** $N_{av} = 55.9$ **(c)** [M] = 0.227 mM; $\overline{Q} = 0.881$ molecules per micelle   **(d)** $P_0 = 0.414$; $P_1 = 0.365$; $P_2 = 0.161$

Chapter 19

3. D_2, silicon carbide globar

9. **(a)** 2.38×10^3 lines/cm   **(b)** 143

11. **(a)** 1.7×10^4 **(b)** 0.05 nm   **(c)** 5.9×10^4 **(d)** 0.000 43°, 0.013°

12. **(a)** $T = 0.036\,4$, $A = 1.439$ **(b)** 0.002 3%   **(c)** 0.000 022, 0.000 22

13. 0.124 2 mm

14. 77 K: 1.99 W/m^2; 298 K: 447 W/m^2

15. **(a)** $M_\lambda = 8.79 \times 10^9$ W/m^3 at 2.00 µm; $M_\lambda = 1.164 \times 10^9$ W/m^3 at 10.00 µm   **(b)** 1.8×10^2 W/m^2 **(c)** 2.3×10^1 W/m^2
(d) $M_{2.00\,µm}/M_{10.00\,mm} = 7.55$ at 1 000 K; $M_{2.00\,µm}/M_{10.00\,mm} = 3.17 \times 10^{-22}$ at 100 K

16. 2.51×10^{-6}

17. **(a)** 34°   **(b)** 0°

19. $\theta_{critical}$ (solvent/silica) = 76.7°, $\theta_{critical}$ (silica/air) = 43.2°. Total reflection is from silica/air interface.

22. **(a)** 80.7°   **(b)** 0.955

23. **(a)** 61.04°   **(b)** 51.06°

24. $n_{prism} > \sqrt{2}$

25. **(b)** 76°   **(c)** 0.020

26. **(a)** 0.964   **(b)** 343 nm, 5.83×10^{14} Hz

27. **(b)** blue

28. **(a)** ±2 cm   **(c)** 0.5 cm^{-1} **(d)** 2.5 µm

33. 7

35. [at cycles, predicted S/N] at 1 000, 60.0 (observed); at 300, 32.9; at 100, 19.0; at 1, 1.90

Chapter 20

12. Pb: 1.2 ± 0.2; Tl: 0.005 ± 0.001; Cd: 0.04 ± 0.01; Zn: 2.0 ± 0.3; Al: $7 (\pm 2) \times 10^1$ ng/cm^2

13. 589.3 nm

14. 0.025

15. Na: 0.003$_8$ nm; Hg: 0.000 5$_6$ nm

16. **(a)** 283.0 kJ/mol   **(b)** 3.67×10^{-6} **(c)** +8.4%   **(d)** 1.03×10^{-2}

17.
wavelength (nm):	591	328	154
N^*/N_0 at 2 600 K in flame:	2.6×10^{-4}	1.4×10^{-7}	1.8×10^{-16}
N^*/N_0 at 6 000 K in plasma:	5.2×10^{-2}	2.0×10^{-3}	1.2×10^{-7}

19. 20 GW/cm^2, 5 ng

21. 0.429 ± 0.012 wt%

22. **(a)** 7.49 µg/mL   **(b)** 25.6 µg/mL

23. 17.4 ± 0.3 µg/mL

24. **(a)** CsCl inhibits ionization of Sn.   **(b)** $m = 0.782 \pm 0.019$; $b = 0.86 \pm 1.56$; $R^2 = 0.997$ **(c)** Little interference at 189.927 nm, which is the better choice of wavelengths. At 235.485 nm, there is interference from Fe, Cu, Mn, Zn, Cr, and, perhaps, Mg.   **(d)** limit of detection = 9 µg/L; limit of quantitation = 31 µg/L   **(e)** 0.8 mg/kg

25. Ti/transferrin = 2.05

Chapter 21

3. 1 Da = $1.660\,54 \times 10^{-24}$ g, $83.5 (\pm 2.3)$ fg

4. 58.5$_1$ from the spectrum in the text

5. 1.5×10^4, yes

6. ~3 100

7. 2.0×10^6, 3.4×10^6

8. $C_6H_{11}^+$

9. $^{31}P^+ = 30.973\,21$, $^{15}N^{16}O^+ = 30.994\,47$, $^{14}N^{16}OH^+ = 31.005\,25$

10. 1: 1.946: 0.946 3

11. 1: 8.05: 16.20

12. **(a)** 4   **(b)** 6   **(c)** $1\frac{1}{2}$

13. **(a)** ⬡—Cl, C_6H_5Cl: $M^{+\cdot} = 112$

(b) Cl—⬡—Cl, $C_6H_4Cl_2$: $M^{+\cdot} = 146$

(c) ⬡—NH$_2$, C_6H_7N: $M^{+\cdot} = 93$

(d) $(CH_3)_2Hg$: $M^{+\cdot} = 228$ **(e)** CH_2Br_2: $M^{+\cdot} = 172$

(f) , 1,10-phenanthroline, $C_{12}H_8N_2$: $M^{+\cdot} = 180$

(g) Fe , ferrocene, $C_{10}H_{10}Fe$: $M^{+\cdot} = 186$

14. $M^{+\cdot} = 206$, $CH^{79}Br_2^{35}Cl$

16. **(a)** mass of $p^+ + e^-$ = mass of ^1H **(b)** mass of $p^+ + n + e^-$ = 2.016 489 963 Da; mass of ^2H = 2.014 10 Da   **(c)** 2.15×10^8 kJ/mol
(d) 1.31×10^3 kJ/mol   **(e)** 5×10^5

17. 0.342 7: 1: 0.972 8: 0.315 4

18. (mass, intensity): (84, 1) (85, 0.152) (86, 0.108) (87, 0.010 3) (88, 0.003 62) (89, 0.000 171) (90, 0.000 037)

20. 4.39×10^4 m/s; 45.6 μs; 2.20×10^4 spectra/s; 1.56×10^4 spectra/s

22. (a) 93 m   (b) 93 km

27. (a) negative ion mode, neutral solution   (b) 14.32

28. $n_A = 12$ and $n_I = 20$; mean molecular mass (disregarding peak G) is 15 126 Da

29. charge = 4; molecular mass = 7 848.48 Da

30. 37:3: $[MNH_4]^+ = C_{37}H_{72}ON$ $[MH]^+ = C_{37}H_{69}O$
    M+1:   41.2% predicted   40.8% predicted
            35.8% observed   23.0% observed
    M+2:   7.9% predicted   7.9% predicted
            7.0% observed   8.0% observed

    37:2: $[MNH_4]^+ = C_{37}H_{74}ON$ $[MH]^+ = C_{37}H_{71}O$
    M+1:   41.3% predicted   40.8% predicted
            40.8% observed   33.4% observed
    M+2:   7.9% predicted   7.9% predicted
            3.7% observed   8.4% observed

32. (d) 7.63_9 μmol V/g

Chapter 22

1. three

2. 3

3. (a) AlY^- is an anion, but AlL_3 is neutral   (b) hydrophobic cation

7. (a) 0.080 M   (b) 0.50

8. 0.088

9. (c) 4.5   (d) greater

11. (a) 0.16 M in benzene   (b) 2×10^{-6} M in benzene

12. 2 pH units

13. (a) 2.6×10^4 at pH 1 and 2.6×10^{10} at pH 4   (b) 3.8×10^{-4}

15. n: 1   2   3   4   5   6   10
    fraction extracted: 0.667 0.750 0.784 0.802 0.814 0.822 0.838
    $q_{limit} = 0.865$; 95% of q_{limit} is attained in 6 extractions

16. 1-C, 2-D, 3-A, 4-E, 5-B

19. (a) 17.4 cm/min   (b) 0.592 min   (c) 6.51 min

20. (a) 13.9 m/min, 3.00 mL/min   (b) $k' = 7.02$, fraction of time = 0.875 (c) 295

21. (a) 40 cm long × 4.25 cm diameter   (b) 5.5 mL/min (c) 1.11 cm/min for both

22. (a) 2.0   (b) 0.33   (c) 20

23. 19 cm/min

24. 0.6, 6

25. $k' = 3.59$, $K = 4.69$

26. 603, 0.854

27. (a) $1.2_6 \times 10^4$ (b) 40 μm   (c) 0.72

28. (a) 1   (b) 2   (c) 1   (d) neither   (e) B   (f) B   (g) 1.25

30. 0.1 mm

33. 33 mL/min

37. 2.65 mm

38. (a) 1.1×10^2 (b) 0.89 mm

39. (a) 4.1×10^3 (b) Eq. 22-28: $N = 7.72 \times 10^3$; Equation 22-29: $N = 7.75 \times 10^3$

40. resolution = 0.83

41. 10.4 mL

42. 110 s^2, 43 s^2, 26.9 s

43. 1.3×10^3

44. (a) $\gamma = 1.01$, $N = 640\ 000$; $\gamma = 1.05$, $N = 25\ 600$; $\gamma = 1.10$, $N = 6\ 400$

45. (a) $k = 11.25$, 11.45   (b) 1.018   (c) 1.017   (d) C_6HF_5: 60 800 plates, height = 0.493 mm; C_6H_6: 66 000 plates, height = 0.455 mm (e) C_6HF_5: 55 700 plates, C_6H_6: 48 800 plates   (f) 0.96   (g) 0.97

Chapter 23

8. (a) hexane < butanol < benzene < 2-pentanone < heptane < octane (b) hexane < heptane < butanol < benzene < 2-pentanone < octane (c) hexane < heptane < octane < benzene < 2-pentanone < butanol

9. (a) 3, 1, 2, 4, 5, 6   (b) 3, 4, 1, 2, 5, 6   (c) 3, 4, 5, 6, 2, 1

10. (a) 4.7 min, 1.3   (b) 1.8

11. 836

12. $a = 1.69 \times 10^3$ K, $b = -3.36$, 27.1 min

17. (b) Injections: 0.16 ng (narrow bore), 56 ng (wide bore)

18. Use narrower or longer column or different stationary phase.

19. (b) air: 9.3 cm/s, 0.036 cm; H_2: 17.6 cm/s, 0.051 (c) air: 8 300; H_2: 5 900   (d) air: 32 s; H_2: 17 s   (e) C_s becomes negligible

20. (a) 1.25_3 (c) 77.6 mM

21. 0.41 μM

22. 932

24. (d) nonsmoker: 78 ± 5 μg/L; nonsmoker whose parents smoke: 192 ± 6 μg/L

25. $[^{14}NO_2^-] = 1.8$ μM; $[^{14}NO_3^-] = 384$ μM

26. (a) $A = 0$ (b) $B = 2D_m$

(c) $C = C_s + C_m = \dfrac{2k}{3(k+1)^2}\dfrac{d^2}{D_s} + \dfrac{1+6k+11k^2}{24(k+1)^2}\dfrac{r^2}{D_m}$

(d) u_x (optimum) $= \sqrt{\dfrac{B}{C}}$; $H_{min} = 2\sqrt{B(C_s + C_m)}$

27. (a) 0.58, 1.9   (b) 0.058 mm, 0.19 mm   (c) 3.0×10^5 (d) 4.0

28. $D_m = 3.0 \times 10^{-5}$ m²/s, $D_s = 5.0 \times 10^{-10}$ m²/s

29. (b) limiting $m = KV_fC_o$ (c) 0.69%, 41%

30. (a) 6   (b) 350   (c) 350, 315, 280, 245, 210 = M^+, $(M-Cl)^+$, $(M-2Cl)^+$, $(M-3Cl)^+$, $(M-4Cl)^+$, $(M-5Cl)^+$ (d) For 5Cl, predicted relative abundances = 62.5 : 100 : 64.0 : 20.4 : 3.3 : 0.2. For 4Cl, abundances = 78.2 : 100 : 48.0 : 10.2 : 0.8. For 3Cl, abundances = 100: 95.9: 30.7: 3.3. For 2Cl, abundances = 100: 64.0: 10.2.

Chapter 24

1. (b) pressure gradient (= density gradient)

5. (a) L = 33, 17, 10, 5 cm

6. 0.14 min, 0.30 min

11. (a) 1 560 for L enantiomer and 1 310 for D enantiomer   (b) 1.25 (c) 1.35

12. (a) 18   (b) 10

13. (a) bonded reversed-phase chromatography (b) bonded normal-phase chromatography   (c) ion-exchange or ion chromatography   (d) molecular exclusion chromatography (e) ion-exchange chromatography   (f) molecular exclusion chromatography

14. 0.27 m²

15. (a) shorter   (b) amine

16. (a) 1.76 mm/s   (b) 3.51   (c) 192 000, 22.9 μm   (d) 0.59 (e) $\alpha = 1.006_6$, $\gamma = 1.005_2$ (f) $5.9_2 \times 10^5$, 13.6 m   (g) decrease flow or change solvent   (h) 0.91

17. (a) (R)-Gimatecan would be eluted at 6.96 min from (R,R)-stationary phase and 6.10 min from (S,S)-stationary phase.   (b) (R)-Gimatecan at 6.96 min is lost beneath tail of (S)-gimatecan   (c) $\alpha = 1.25_8$, $\gamma = 1.14_1$ (d) 2.91 (better than baseline separation)

18. 126 mm

20. **(a)** $t_m = 0.38$ min for column A and 0.26 min for column B
(d) $k' = 1.3$ for morphine 3-β-D-glucuronide and 3.3 for morphine   **(e)** 6.2

21. **(b)** 5.8 W

22. **(a)** m/z 304 is BH^+ (cocaine protonated at N)   **(b)** loss of $C_6H_5CO_2H$
(e) phenyl group

23. **(a)** m/z 234 is MH^+, m/z 84 is $C_5H_{10}N^+$ **(b)** $237 \to 84$

24. **(c)** 1.1 min for both

32. 27.8 min predicted, 20.2 min observed

33. **(a)** ~36 min   **(b)** 43.0 min using points from $\Phi = 0.35$ to 0.6

34. **(b)** between B and F: 30% methanol/8% tetrahydrofuran/62% buffer; between F and C: 10% methanol/24% tetrahydrofuran/66% buffer

35. D: 25% acetonitrile/30% methanol/45% buffer; E: 25% acetonitrile/20% tetrahydrofuran/55% buffer; F: 30% methanol/20% tetrahydrofuran/50% buffer; G: 16.7% acetonitrile/20% methanol/13.3% tetrahydrofuran/50% buffer

36. 38%

37. **(a)** lower   **(b)** higher

38. **(a)** $\Delta t/t_G = 0.32 > 0.25 \Rightarrow$ gradient elution   **(b)** 40 to 70% acetonitrile in 60 min

39. **(a)** Change solvent strength, temperature, or pH. Use a different solvent or a different kind of stationary phase.   **(b)** slower flow rate, different temperature, longer column, smaller particle size

40. **(a)** ~29 min   **(b)** $k^* = 12.9$, $F = 4.7$ mL/min, $m = 4.7$ mg, $t_G = 11.5$ min

Chapter 25

6. 38–75 μm; 200/400 mesh

8. **(a)** 30   **(b)** 3.3   **(c)** increase

9. cation charge = 0.002 02 M, anion charge = −0.001 59 M. Either some concentrations are inaccurate or some ionic material was not detected.

13. **(b)** 29 ng/mL

15. **(a)** 3.8×10^2 h   **(b)** 8.0 mA, 0.16 A

17. **(b)** 10 μm

18. **(a)** 40.2 mL   **(b)** 0.53

19. 0.16

20. **(a)** 2 000 Da   **(b)** 300 Da

21. **(a)** 5.7 mL   **(b)** 11.5 mL   **(c)** Solutes must be adsorbed.

22. 320 000

28. **(a)** 0.167 mm   **(b)** 0.016 s   **(c)** 0.000 40 s   **(d)** 0.064 s

31. **(a)** 1.15×10^4 Pa   **(b)** 1.17 m height (not possible), so use pressure of 11.5 kPa = 0.114 atm.

32. **(a)** 29.5 fmol   **(b)** 3.00×10^3 V

33. 9.2×10^4 plates, 4.1×10^3 plates (My measurements are about 1/3 lower than the values labeled in the figure from the original source.)

34. **(a)** maleate   **(b)** Fumarate is eluted first.   **(c)** Maleate is eluted first.

35. **(a)** pH 2: 920 s; pH 12: 150 s   **(b)** pH 2: never; pH 12: 180 s

36. **(a)** $t_{120\,kV}/t_{28\,kV} = 4.3$ (observed ratio = 3.9)   **(b)** $N_{120\,kV}/N_{28\,kV} = 4.3$ **(c)** $\sigma_{120\,kV}/\sigma_{28\,kV} = 0.48$ **(d)** Increasing voltage decreases migration time, giving bands less time to spread apart by diffusion.

38. $1.3_5 \times 10^4$ plates, 30 μm

39. 20.5 min

40. light chain = 17 300 Da, heavy chain = 23 500 Da

41. $z_0 = -3.28$

42. 2.0×10^5 plates

43. thiamine < (niacinamide + riboflavin) < niacin. Thiamine is most soluble.

44. Cyclobarbital and thiopental each separate into two peaks because each has a chiral carbon atom.

46. 5.55

48. **(d)** $N_{obs} = 1.45 \times 10^5$, $N_{theory} = 2.06 \times 10^5$ **(e)** 1.6

Chapter 26

10. 0.022 86 M

11. 1.94 wt%

12. 225.3 g/mol

13. 0.085 38 g

14. 50.79 wt%

15. 0.191 4 g calcium carbonate, 0.107 3 g calcium oxide

16. 104.1 ppm

17. 7.22 mL

18. 0.339 g

19. **(a)** 19.98%

20. **(a)** 5.5 mg/100 mL   **(b)** 5.834 mg, yes

21. 14.5 wt% K_2CO_3, 14.6 wt% NH_4Cl

22. 40.4 wt%

23. 22.65 wt%

24. **(a)** 40.05 wt%   **(b)** 39%

25. **(a)** 1.82   **(b)** $Y_2O_2(OH)Cl$ or $Y_2O(OH)_4$

26. **(a)** 0.854 976   **(b)** 0.856 3   **(d)** 0.856 ± 0.008, 0.86 ± 0.08

27. **(b)** 0.204 (± 0.004)

33. 11.69 mg CO_2, 2.051 mg H_2O

34. $C_8H_7NO_2SBrCl + 9\frac{1}{4} O_2 \to 8CO_2 + \frac{5}{2}H_2O + \frac{1}{2}N_2 + SO_2 + HBr + HCl$

35. $C_4H_9NO_2$

36. 10.5 wt%

37. $C_8H_{9.06 \pm 0.17}N_{0.997 \pm 0.010}$

38. 12.4 wt%

39. **(a)** 95% confidence: $10.16_0 \pm 1.93_6$ μmol Cl^- (Experiment 1), $10.77_0 \pm 2.29_3$ μmol Cl^- (Experiment 2)   **(b)** difference is *not* significant   **(c)** 24.2_{95} mg $BaSO_4$ **(d)** 4.35%

41. **(a)** 13.08   **(a)** 8.04   **(c)** 2.53

42. $[Ag^+] = 9.1 \times 10^{-9}$ M; $Q = [Ag^+][Cl^-] = 2.8 \times 10^{-10} > K_{sp}$ for AgCl

43. **(a)** 6.06   **(b)** 3.94   **(c)** 2.69

44. $[AgCl(aq)] = 370$ nM, $[AgBr(aq)] = 20$ nM, $[AgI(aq)] = 0.33$ nM

45. 56.28 wt%

46. $V_{e1} = 18.76$ mL, $V_{e2} = 37.52$ mL

47. **(a)** 14.45   **(b)** 13.80   **(c)** 8.07   **(d)** 4.87   **(e)** 2.61

48. **(a)** 19.00, 18.85, 18.65, 17.76, 14.17, 13.81, 7.83, 1.95   **(b)** no

49. $V_X = V_M^0(C_M^0 - [M^+] + [X^-])/(C_X^0 + [M^+] - [X^-])$

54. negative

56. 947 mg

Chapter 27

3. **(a)** 5%   **(b)** 2.6%

4. 1.0 g

5. 120/170 mesh

6. $10^4 \pm 0.99\%$

7. **(a)** 15.8   **(b)** 1.647   **(c)** 474 − 526

8. 95%: 8; 90%: 6

9. **(a)** 5.0 g   **(b)** 7

10. 0.34 ± 0.14 pg

11. **(a)** Na_2CO_3: 4.47 μg, 8.94×10^5 particles; K_2CO_3: 4.29 μg, 2.24×10^7 particles   **(b)** 2.33×10^4 **(c)** Na_2CO_3: 3.28%; K_2CO_3: 0.131%

12. Zn, Fe, Co, Al

13. prevents possible explosion

19. **(a)** 53

20. 64.90 wt%

Abbreviations

App. = Appendix   f = footnote   NR = Notes and References
b = box   i = illustration   p = problem
d = demonstration   m = marginal note   t = table